Second Edition

PRIMATES IN PERSPECTIVE

Edited by

Christina J. Campbell
California State University, Northridge

Agustín Fuentes
University of Notre Dame

Katherine C. MacKinnon
Saint Louis University

Simon K. Bearder
Oxford Brookes University

Rebecca M. Stumpf
The University of Illinois

New York Oxford
OXFORD UNIVERSITY PRESS
2011

Oxford University Press, Inc., publishes works that further Oxford University's
objective of excellence in research, scholarship, and education.

Oxford   New York
Auckland   Cape Town   Dar es Salaam   Hong Kong   Karachi
Kuala Lumpur   Madrid   Melbourne   Mexico City   Nairobi
New Delhi   Shanghai   Taipei   Toronto

With offices in
Argentina   Austria   Brazil   Chile   Czech Republic   France   Greece
Guatemala   Hungary   Italy   Japan   Poland   Portugal   Singapore
South Korea   Switzerland   Thailand   Turkey   Ukraine   Vietnam

Copyright © 2011 by Oxford University Press, Inc.

Published by Oxford University Press, Inc.
198 Madison Avenue, New York, New York 10016
http://www.oup.com

Oxford is a registered trademark of Oxford University Press

ISBN: 978-0-19-539043-8

Printing number: 9 8 7 6 5 4

Printed in the United States of America
on acid-free paper

CONTENTS

PART THREE   Methods

PART FOUR   Development and Reproduction

PREFACE

Welcome to the second edition of *Primates in Perspective*, offering a comprehensive overview of contemporary primate research. With significant improvements in methodology and increases in the number and types of primate studies conducted, primatology has continued to mature since the first edition of this book. We thought it was time to update the volume to reflect these changes.

OUR APPROACH

As in the first edition, we bring together here a broad range of primatologists who provide thorough reviews of their areas of expertise. We hope that the contributed chapters are accessible and useful to university students (undergraduate and graduate) and primate researchers alike, and that the chapters will stimulate new ideas and areas of research for the future. All chapters are written and edited in a manner that ensures the book is not only a valuable teaching asset but also a vital reference tool for professionals. The 1987 edited volume *Primate Societies* (University of Chicago Press) set the standard for a primate overview and reference text. An updated edition of that text was not published, nor had anything comparable appeared in the literature until the first edition of *Primates in Perspective*, an inclusive volume reflecting the taxonomic, methodological, and theoretical changes that occurred in primatology since the late 1980s.

We designed this edition so that it can be used by a wide range of people interested in the nonhuman primates. In particular, we feel the broad coverage of topical areas (e.g., taxonomy, methods, ecology, behavior) is well suited for general undergraduate primatology courses and graduate seminars. By providing extensive coverage of the primates via well-structured and inclusive overviews of taxonomy, methods, and theoretical issues, we remove the need for students and professors to deal with multiple books and/or photocopied readers. This text will also be of use as a reference source for professional primatologists, as it is an up-to-date, inclusive overview. The chapters will be useful for all primatologists looking for current and older literature pertaining to almost any area of interest within the discipline.

NEW TO THIS EDITION

- We have added new chapters on baboons, mandrills, and mangabeys; behavioral methods; ecological methods; juveniles and adolescents; kinship; and tool use.
- A 16-page insert features more than 30 full-color photos of primates.
- More than 50 new maps and figures are included

throughout.
- A glossary at the back of the text clearly defines new terms for students.
- Reading questions at the beginning of each chapter guide students to big-picture concepts and serve as discussion starters and assignment prompts.
- A "How to Use This Book" section offers tips on teaching from the text.
- An extensive, up-to-date bibliography is now included at the back of the text (rather than at chapter ends).

FEATURES

Throughout the volume we have ensured that there is a synthesis of modern data that will inform all readers of the state of affairs of nonhuman primate research. We have worked hard to ensure that this book is especially strong in behavioral ecology and theory. With specific taxonomic and behavioral chapters by leading researchers in the field, this volume provides unmatched detailed coverage of a multitude of topics. Because such incredible diversity is found among primates, the cursory taxonomic summaries found in most introductory textbooks can be of only limited value to students. Therefore, for students, these chapters provide a valuable and complete introduction to the primates, and an important foundation for the rest of the book. For professionals, this section provides current information and literature citations for a wide variety of taxa. No recently published primate volume available provides this type of overview.

We also include here varied theoretical positions on several topics in primatology (most notably in the reproduction, ecology, and social behavior and intelligence sections). We especially want student readers to be aware of the diversity of theoretical opinions that currently pervade our field. Dominant paradigms in modern primatology are well represented, as are less-entrenched positions that are nonetheless supported by existing data. Such a range of theoretical orientations is generally not found in primate texts, and we hope classroom discussion and critical evaluation of the topics will be facilitated by their inclusion here.

ORGANIZATION

The book is divided into seven main sections that reflect the major themes of primate research today; collectively they contain 47 chapters (three more than in the first edition) by 70 authors. We begin with the background section, which serves largely as a prologue to the study of primatology and to the evolutionary history of the primate order. We hope the chapters in this section will provide students with a comprehensive overview and introduction to the field (e.g., in the first week of an upper-division primatology course).

Next, and with that background in mind, we present the taxonomy chapters, each of which focuses on the major taxonomic groups within the order Primates. The individual chapters include an overview of natural history, life history parameters, and the like, and do not revolve heavily around any theoretical framework. We hope students will find these chapters to be a valuable and complete introduction to the diversity of the living nonhuman primates. For our professional audience, this section provides up-to-date information and literature on a wide variety of taxa. For all readers, this portion serves as an important foundation for the rest of the book.

In the third section we include four chapters on methods. In order to gain a full appreciation for the study of primates, students must have some understanding of how research is carried out in the field. In this edition we chose to include two separate chapters on field methods, one on studying behavior and the other on ecology. By doing this we believe we have been able to provide more detail about certain techniques than in the first edition. In addition, the final two chapters in this section will be of use to professional primatologists interested in learning more about recent advances in endocrinological and molecular technologies.

We then present chapters on development and reproduction. As theoretical topics, the many aspects of primate reproduction and development are vital components of a modern comprehensive-review book. A new chapter on juveniles and adolescents has been added to this section. The chapters included here introduce students to many aspects of primate reproduction, growth, life history, and skeletal and behavioral development, and the overviews and bibliographies provide a valuable reference source for professionals in a rapidly growing field of research.

The ecological topics of socioecology, diet, seed dispersal, predation, locomotion, and self-medication are found in our next section. All are important aspects of the interface between primate populations and the habitats they occupy. Students and established researchers alike will find in-depth explanations and important theoretical arguments within these chapters.

In the next section we present a selection of chapters that covers various aspects of primate social behavior and cognition. The vast majority of nonhuman primates live in stable social groups throughout their lives and exhibit complex behavioral relationships with other members of their group. Among mammals, primates are characterized by displaying an extended period of growth and development, during which virtually all social behaviors are learned. We have included a new chapter on kinship that we felt was sorely missing from the first edition. Thus, this section highlights the complex dynamics that characterize the social aspects of being a primate.

In the final section of the book we have three chapters that focus on the future of primates and primatology, how we interact with nonhuman primates, and issues related to conservation.

We feel this progression—from a general introduction and background, to taxonomic overviews, then to methodological issues, and finally to topics in ecology and behavior—gives the book a sound pedagogical structure and a user-friendly format in which to find information for classroom and/or professional use.

ACKNOWLEDGMENTS

When this venture was first discussed at the reception of the 71st Annual Meeting of the American Association of Physical Anthropologists in 2002, we knew it would end up involving an extremely large number of people. Our first round of acknowledgments and "thank yous" are graciously extended to all the authors, including those who came back for the second time and those new to this edition. The current volume would not have resulted in the astounding review of primate behavior and ecology without each and every one of them and the enormous effort they put into the excellent chapters that follow. In addition to their contributions, we greatly appreciate the authors' professional responses to our editorial suggestions and patience with the inevitable delays that accompany the production of such a large project. We would like to extend our deepest gratitude to our colleague and friend, Melissa Panger, who was an editor on the first edition of this volume. Her contributions are still omnipresent here and we thank her for her continued support.

Two people at Oxford University Press deserve special acknowledgment. Executive Editor Jan Beatty has been extremely supportive from the beginning, offering advice, encouragement, and the occasional prodding in order to keep the volume on schedule. Assistant Editor Lauren Mine is a miracle worker and her efforts in the actual publishing and organizational aspects of the project helped immensely, especially in keeping track of missing files, figures, and tables. This book would certainly not be so well organized without the hard work and dedication she contributed.

The following external reviewers provided invaluable comments and suggestions for this edition, and we extend our extreme appreciation for their time and interest: Michelle Bezanson, Santa Clara University; René Bobe, University of Georgia; Scott McGraw, The Ohio State University; Elizabeth Peters; Florida State University; Ulrich Reichard, Southern Illinois University; Melissa Remis, Purdue University; Linda Taylor, University of Miami; and Patricia Whitten, Emory University.

Stephanie Kalchik of the University of Notre Dame spent many hours combining and formatting references in order to create the master bibliography. We are all very grateful to her for making our lives easier.

We would be remiss if we did not respectfully acknowledge and extend appreciation to Barbara Smuts, Dorothy Cheney, Robert Seyfarth, Richard Wrangham and Thomas Struhsaker, the editors of *Primate Societies* (1987, University of Chicago Press). Their volume inspired us in part to undertake the original *Primates in Perspective*. We sincerely hope that *Primates in Perspective* will be as influential to the primatological community as theirs was. Various funding agencies have contributed to the work presented in this volume. These include (but are not limited to) the National Science Foundation, the National Institutes of Health, the LSB Leakey Foundation, National Geographic, the Wenner Gren Foundation, the Margot Marsh Biodiversity Fund, the American Primatological Society, Primate Conservation International, the Smithsonian Institution, and anthropology, psychology, zoology, and biology departments at many different universities. We thank all of these agencies (and any others that are not mentioned) for their continued support of this important area of research. Many of the contributing authors wanted to add their own acknowledgments, but due to space limitations, they were unable to be included. We would like to take this opportunity to thank all those individuals who helped the authors produce the chapters that follow.

In addition to those who have our united thanks, there are many who have contributed to this volume through their support and encouragement of each of us individually. C. J. C. acknowledges Phyllis Dolhinow's mentorship and guidance. In addition, she thanks members of the Department of Anthropology at California State University, Northridge for their support during this revision process. Additionally she would like to thank the CSUN Probationary Faculty Support Program for their support in completing this project. She also offers her undying gratitude to Tom and Summer Wake for their love and patience throughout this process.

A. F. thanks Devi Snively and Audrey, the wonder-dog, for inspiration and assistance. He also thanks Phyllis Dolhinow for her mentoring and guidance and all the students in his primate behavior classes at the University of California, Berkeley, Central Washington University, and the University of Notre Dame for keeping him on his toes and always wanting a better book.

K. C. M. thanks Phyllis Dolhinow, Linda Fedigan, and Bob Sussman for being exemplary mentors, true friends, and inspiring role models. She also thanks her fellow editors on this volume for patience and good humor during the entire editorial process. She acknowledges her anthropology and primatology students at Saint Louis University, who continually inspire, amuse, and stimulate discussion in the classroom. Finally, she expresses extreme gratitude to Matt Wyczalkowski, as well as Darby, Dexter, Caper, and Basil for holding down the fort, yet again.

S. K. B. thanks K. A. I. Nekaris, Kate Hill, Chris McDonaugh, and Iris Geens at Oxford Brookes University; the students and staff on the MSc in Primate Conservation; and research students at Brookes. He also acknowledges Bob Martin, Alan Dixson, Tom Butynski, and Gerald Doyle for their inspiration over the years. Last, but definitely not least, he expresses his deepest thanks to his wife, Catherine.

R. M. S. thanks the faculty, staff, and students of the Department of Anthropology at the University of Illinois for their inspiration, support, and assistance while she worked on this book. She is particularly grateful to Paul Garber and Steve Leigh for their helpful mentoring, humor, and friendship. Much appreciation also goes to Chrissy, Katie, Agustín, Simon and, of course, Melissa Panger, for the opportunity to work on this new edition. Jennifer Stumpf, with her talented photographic eye, provided invaluable help by winnowing down the pool of many terrific photos for the color insert and cover. With much gratitude, R. M. S. thanks her friends and family, particularly John and Annika Polk, for all of their support, patience, good humor, and love.

Finally, we would all like to thank the nonhuman primates that continue to intrigue and inspire us!

Christina J. Campbell
Agustín Fuentes
Katherine C. MacKinnon
Simon K. Bearder
Rebecca M. Stumpf

HOW TO USE THIS BOOK

One of the major obstacles to reading about and teaching primatology is the sheer volume of information out there. This short essay contains a few thoughts on how you might most effectively navigate the book and some suggestions for using it as a teaching tool.

Keep in mind that this book is simultaneously a textbook and a resource text. That is, it contains the most up-to-date and comprehensive primatological information available in book form, so it acts as a go-to source for a wide range of questions about primates. For students and instructors, the index, glossary, and table of contents are the gateways to the information throughout the book. For research purposes, the bibliography is a gold mine. Each of the chapters in Part Two (The Primates) contains a comprehensive overview of the taxa, but even more detail is available through the references. In general, when reading this text as a student, you want to focus on the big-picture concepts outlined by the questions at the start of each chapter rather than the tables and citations.

Both instructors and students can maximize the value of the book by deciding how it integrates with course goals. In most courses, you will probably not read every chapter, or you will spend more time on certain sections. For example, if a course focuses on ecology and behavior, you might want to select specific sections to spend the most time on. In this case, selected chapters from Part Two (to cover specific types of primates), Part Five (Ecology), and Part Six (Social Behavior and Cognition) could form the core of the assignments for lectures, with other chapters being assigned as part of course projects. Alternatively, if the course is a broader survey of the primates, you might start with Part One for a historical context and spend a good amount of time on each of the chapters in Part Two. You can then integrate all or parts of selected chapters from the other sections on methodology, ecology, behavior and reproduction as they best connect to the highlighted topics in the primate groups chapters. Part Three (Methods) can be used as a primer on current methods in preparing for the field or lab projects. For an advanced course, such as topics in primatology, you could create topical clusters by selecting chapters thematically. For example, a theme of sex and aggression might include Chapter 23 for an update on hormones and behavior; Chapters 27 through 30 to cover social development and sexual behavior; 32 and 35 for basic socioecology and predation; and 38 through 41 for kinship, aggression, cooperation, and reconciliation. A theme of ecological diversity and conservation might include Chapters 4, 5, 9, 11, 13, 15, 16, 17, and 20 to illustrate the primate diversity within and among major geographic areas; 32 through 37 for ecological basics and connections; and 45 and 46 for the conservation and human interface contexts. Numerous themes can be selected for either full-course length topics or sections within courses.

Finally, there are a number of ways to use the text in the classroom as a pedagogical tool. For example, using Part Two, you can have one or a few students in the class be responsible for each chapter. The students or groups would then create lectures on their taxonomic group, along with PowerPoint-based presentations and summary sheet handouts, and present to the rest of the class. Another possibility is to hand out reading guideline sheets at the start of each week for all assigned chapters. Use the chapter-opening questions to help frame three or four target questions about each chapter. Have the students write brief responses to your questions; these questions and responses can be used as discussion focal points during class time. Another option is to assign a literature-based research project on primate groups or specific topics in behavior, reproduction, or ecology, for example. Students can develop their bibliography by searching the chapters in the book for relevant information and following the reference trail to the original literature. This enables them to see the summary data covered in the book expanded out to much greater detail.

CONTRIBUTORS

Kate Arnold
School of Psychology
University of St. Andrews

Filippo Aureli
School of Biological and Earth
Sciences
Liverpool John Moores University

Thad Q. Bartlett
Department of Anthropology
University of Texas at San Antonio

Simon K. Bearder (General Editor)
Department of Anthropology
Oxford Brookes University

Carol M. Berman
Department of Anthropology
University at Buffalo, The State
University of New York

Irwin S. Bernstein
Department of Psychology
University of Georgia

Gregory E. Blomquist
Department of Anthropology
University of Missouri

Debra Bolter
Department of Anthropology
University of California, Santa Cruz

Christine A. Caldwell
School of Psychology
University of Exeter

Christina J. Campbell (General
Editor)
Department of Anthropology
California State University, Northridge

Colin A. Chapman
Department of Anthropology
and McGill School of Environment
McGill University

Margaret C. Crofoot
Smithsonian Tropical Research
Institute

Anthony Di Fiore
Department of Anthropology
New York University

Leslie J. Digby
Department of Biological
Anthropology and Anatomy
Duke University

Phyllis Dolhinow
Department of Anthropology
University of California, Berkeley

Nathaniel Dominy
Department of Anthropology
University of California, Santa Cruz

Peter J. Fashing
Department of Anthropology
California State University, Fullerton

Linda Marie Fedigan
Department of Anthropology
University of Calgary

Eduardo Fernandez-Duque
Zoological Society of San Diego
Center for Reproduction of
Endangered Species

Stephen F. Ferrari
Department of Genetics
Universidade Federal do Pará

Dorothy M. Fragaszy
Psychology Department
The University of Georgia

Orlaith N. Fraser
School of Biological and
Earth Sciences
Liverpool John Moores University

Agustín Fuentes (General Editor)
Department of Anthropology
University of Notre Dame

Pascal Gagneux
Conservation and Research for
Endangered Species (CRES)
San Diego Zoological Society and
Department of Cellular and Molecular
Medicine
University of California, San Diego

Paul A. Garber
Department of Anthropology
University of Illinois at Urbana-
Champaign

Lisa Gould
Department of Anthropology
University of Victoria

Harold Gouzoules
Department of Psychology
Emory University

Sarah Gouzoules
Department of Anthropology
Emory University

Sharon Gursky
Department of Anthropology
Texas A&M University

Walter Hartwig
Department of Basic Sciences
Touro University College of
Osteopathic Medicine

Michael A. Huffman
Primate Research Institute
Kyoto University

Tatyana Humle
PhD School of Anthropology and
Conservation
University of Kent

Lynne A. Isbell
Department of Anthropology
University of California, Davis

Katharine M. Jack
Department of Anthropology
Tulane University

Karin Enstam Jaffe
Department of Anthropology
Sonoma State University

Sonya M. Kahlenberg
Department of Biology
Bates College

R. Craig Kirkpatrick
TRAFFIC East Asia

Cheryl D. Knott
Department of Anthropology
Boston University

Joanna E. Lambert
Department of Anthropology
University of Wisconsin–Madison

Bill L. Lasley
Institute of Toxicology and
Environmental Health
University of California, Davis

Richard R. Lawler
Department of Anthropology
James Madison University

Steven R. Leigh
Department of Anthropology
The University of Illinois

Andrés Link
Department of Anthropology
New York University

Katherine C. MacKinnon
(General Editor)
Department of Sociology and
Criminal Justice, and Center for
International Studies
Saint Louis University

Joseph H. Manson
Department of Anthropology
University of California, Los Angeles

Lynne E. Miller
Department of Anthropology
Mira Costa College

K. Anna I. Nekaris
School of Social Sciences and Law
Oxford Brookes University

Marilyn A. Norconk
Department of Anthropology
Kent State University

Deborah J. Overdorff
Department of Anthropology
University of Texas–Austin

Joyce A. Parga
Department of Anthropology
University of Texas–Austin

Mary S. M. Pavelka
Department of Anthropology
University of Calgary

Erin P. Riley
Department of Anthropology
San Diego State University

Martha M. Robbins
Department of Primatology
Max Planck Institute for Evolutionary
Anthropology

Sabrina E. Russo
School of Biological Sciences
University of Nebraska-Lincoln

Wendy Saltzman
Department of Biology
University of California, Riverside

Michelle Sauther
Department of Anthropology
University of Colorado

Anne Savage
Disney's Animal Kingdom

Hogan M. Sherrow
Department of Sociology and
Anthropology
Ohio University

Karen B. Strier
Department of Anthropology
University of Wisconsin–Madison

Rebecca M. Stumpf
(General Editor)
Department of Anthropology and the
Program in Ecology, Evolution, and
Conservation
The University of Illinois

Robert W. Sussman
Primate Biology Program and
Department of Anthropology
Washington University

Larissa Swedell
Department of Anthropology
Queens College

Bernard Thierry
Equipe d'Ethologie des Primates
Centre d'Ecologie, Physiologie,
et Ethologie

Adrian Treves
Department of Forest Ecology and
Management
University of Wisconsin

Erin R. Vogel
Center for the Advanced Study of
Hominid Paleobiology
Department of Anthropology
The George Washington University

Catherine Wheaton
Disney's Animal Kingdom

Andrew Whiten
Scottish Primate Research
Group and Centre for Social
Learning and Cognitive
Evolution
University of St. Andrews

Linda D. Wolfe
Department of Anthropology
East Carolina University

Adrienne Zihlman
Department of Anthropology
University of California,
Santa Cruz

Introduction

Phyllis Dolhinow

Welcome to the field of primatology. It is now in the neighborhood of 50 years since I felt a special rush of excitement provoked by spotting a group of honest-to-god, real live monkeys running along rooftops next to a train station in India. I can recall it today as vividly as then. Happily, those infrangible feelings of intense engagement and fascination have not faltered; they continue to this day. I hazard a guess that you could consider this an addiction, and if you are very fortunate, you will acquire it.

Primatology has grown exponentially in the last 50 years, thanks to continued study of the nonhuman primates, prosimians, monkeys, and apes, in many different environments from rain forests to laboratories. Studies of different species, as well as groups within an individual species living in varied habitats under dissimilar conditions, reveal important variations in behavior among and within species. Even at the level of the individual primate, we are experiencing a new flow of information thanks to a fairly recent but steady infusion of techniques of investigation from other disciplines. Far-reaching vistas are opening into the hidden aspects of behavior: the neurological and physiological processes going on inside the animal out of range of human vision.

Kinship, the degree of relatedness among the members of groups, is an extremely important component for unraveling and analyzing social relations. Knowing how individuals are related biologically enables us to measure and evaluate the importance of family, lineages, friends, and strangers. For example, consider the question of paternity, knowing which male is the biological father of another animal. Fatherhood is a vital concern among humans, and nowadays it is relatively easy to resolve questions concerning human paternity. When investigators of nonhuman primates want to identify the father of a specific animal, the answer does not have to be puzzled out by observing many months of adult reproductive behavior. We can test data in the form of biological materials from the field using deoxyribonucleic acid (DNA) analysis.

Modern investigative techniques make it possible to uncover links in the relationship between biology and behavior that were previously invisible. We have windows into the internal mechanisms of male and female reproduction and are able to investigate critical components of the biochemistry of stress, loss, success, and failure. We can monitor physiological responses as an animal experiences changing levels of social power or control, as well as during the more mundane events of daily life. Most importantly, using the appropriate techniques allows biological investigations to occur in the field as well as in captivity. Today, we measure the unmeasurables of yesterday. Just imagine what will be available in the future.

Understandably, the newcomer to primatology might be intimidated by the masses of information already acquired and the technologies of investigation that are in place. Many of the methods and products of our research at the macroscopic as well as microscopic levels are carefully summarized in the following chapters. The wealth of information and techniques might induce a rather anxious state of epistemological aphasia, but do not fret: That should not last long, and it does offer a remarkable variety of niches for investigation.

Why are we so fascinated by our cousins, the nonhuman primates? Likely in part because when we watch them, every one of us is apt to recognize something of ourselves in what we see. We share and express many very similar, if not at times the same, emotions. We often readily grasp the significance of their actions and interactions with each other. This, in itself, is intriguing. We empathize when they care for their rambunctious youngsters, play, fight, threaten, struggle to survive danger, or compete for a few desirable things or places. Remember, the recognition of these similarities simultaneously presents us with major enticements to project ourselves onto them, to attribute to them our feelings, our attitudes, our values, and our concerns when they may not be accurate. Actions, events, or interactions that bother, please, or alarm us may not have the same effects on them.

Consider one of a multitude of possible conundrums that come about when we try to make sense of what we see happening or not happening—yes, not happening. In a very tense situation, one of the highest-ranking, normally hyper-controlling members of a group of langur monkeys remains nonreactive, impassive. He just sits calmly and does not even appear to be aware of the mayhem around him. One thing is certain, he will not tell us what is happening, why he is not reacting in any overt manner to skirmishes he normally would resolve with a low-level threat or glare. What do we do? Keep watching. We watch long enough to become familiar with what the nonresponse might mean in the context of his life and experiences. We will be that much closer to solving this puzzle.

Projecting ourselves into other animals is precisely what we must avoid if we are to maintain a scientific, objective perspective and avoid anthropomorphism. Should we become aware that we cannot be entirely objective, we at least must be conscious of our potential biases. If we are not, we will make some basic assumptions about the nature and meaning of their behavior that will come back to haunt us and potentially undermine the validity of our observations and analyses. Although the nonhuman primates are intriguing and engrossing, we must be aware and resist the temptation to push the limits of credulity to recreate them in our own image. It is too easy to invest them with human concerns that are created by our unique abilities to reason and communicate linguistically. They are like us in many ways, some primates more so than others, but they are not us. Being cautious in no way diminishes the immense complexity of their lives. Enjoy them as they are; marvel at and investigate their lives and their place in nature.

Studies of primate evolution and subsequent taxonomic designations provide a necessary understanding of the degrees of relationship among primate species. They offer another tool for the analysis of general patterns of behavior far beyond those observed for, say, a single species. Having learned the taxonomic relationships of the primates, we can adjust to occasional changes in the assignment of units of taxonomic nomenclature. These usually, but not always, occur at the genus or species level. We all become accustomed to these unpredictable changes that accompany our growing understanding of the evolutionary relationship among the primates.

For the majority of primates, life takes place in a social setting. Social groups, ranging greatly in size, composition, and permanency, are the setting for most, if not all, of a primate's life. Every member, young to old, male and female, serves as a part of this vital unit within which maturation and development occur, affecting the life cycle of every individual. For the most part, learning is a social, experiential activity. Recognizing and documenting variability in the context of rearing are critical to understanding the details of development and the outcomes of experience throughout the remainder of life. Variances in behavioral development drive evolution by forming the individual. To identify and evaluate the multitudes of effective developmental factors requires longitudinal study of individuals of many species. Observations of animals of different ages over shorter periods of time fill out our appreciation of primate variability. They also identify significant components of development for research.

Who would anticipate that an adult female langur monkey could, and most likely would, change her caregiving style, not once, but several times during her reproductive years? Only the study of her offspring as they matured could demonstrate the impact of a female's attentions or inattentions on the next generation. There are so many variables affecting individuals as well as groups, populations, and species throughout individual lifetimes that it takes many years of research to record the range of these differences. Imagine undertaking to chronicle the prodigious range of phenomena displayed by all primates. Any general statement will provoke a slew of contrary examples, making generalization an untidy and vulnerable enterprise. However, take comfort: The chapters which follow provide excellent and detailed examples and summaries of much of the variability and many of the complexities of primate life.

How does one go about researching primates? First comes the planning, and that means deciding what it is that we want to know, as described in the Methods section. It will be most productive if we formulate our questions as hypotheses. We already have a great armamentarium of theoretical paradigms, models, and methods of observation and data analysis that are prefigured on what we have learned from past studies. These will serve to guide future investigations, including ways of conceptualizing which tools will be relevant to answering questions. Hypothesis-driven research is invaluable, but there have been and remain other forms of research. We can call the latter "natural history," or what we will. It is the first stage in the study of a species or a new primate group. It provides basic needed information on the demographic, social, and ecological dimensions of life. This basic first-round work is less ideologically intoxicating but necessary, and it requires appallingly hard work, sometimes for long periods of time in unromantic conditions. When research continues in locations away from the initial observation site, it is possible to identify and measure variability in diverse aspects of life. This is well illustrated in the following chapters on specific groups of primate taxa.

Not all data are created equal. The *repertoire*, a set of words or labels used to record what is seen and how the continuous flow of action and interaction is divided into measurable units, reflects the researcher's evaluation of what is important. Labels are given to the pieces of what happens that we wish to record and later to analyze. It is the researcher who establishes the accuracy or validity of data categories. The meaning of words and labels reflects the assumptions that have been made, a priori, about behavior. It is said that words are mightier than swords. Words can do just as much damage if they are misleading.

To describe animals "cheating" or "deceiving" is actually an "as-if" observation. Animal A acts as if it intended its actions to deceive or cheat animal B. However, if deception or cheating is supported with detailed descriptions of many such interactions, the interpretation is strengthened. Providing detailed observations underpinning the terms leaves it to the reader to agree or disagree with the application of such descriptive labels. Interpretation is a necessary element in understanding the behavior of an animal that cannot talk; and when such interpretations are fully explained and illustrated, they provide a firm platform for comparison and analysis. Ideally, a majority of the elements of behavior patterns that characterize one species ought to be comparable to those of other species. If they are not, then the value of comparative studies will be minimal. Of course, many species may display some gestures or vocalizations not shared by others, but a good percent of all actions and interactions are common to all of them.

The words we choose to describe and thereby understand the nonhuman primates are of particular concern when we seek to understand the root causes of their actions. Reconstructing the history and causes of primate behavior and social systems is both worthwhile and exceedingly challenging. As suggested in later chapters of this book, it requires new perspectives and creative ways of conceptualizing behavior systems. Existing social networks or ecological systems of behavior have a host of varied biotic and social causes and functions. Untangling and identifying the elements of these multiplex behaviors demands careful delineation and explicit characterization of the levels or points of view of inquiry. The layered, interwoven nature of social life invites study.

As quoted in Chapter 32, "Functional labels should not substitute for a full analysis of the causes of behaviour." Using labels that refer to the presumed function of a behavior is likely to beget quagmires of misunderstanding unless these labels are appropriately applied. It is not sufficient to state that specific behavior Y is caused by X to do Z because "it makes sense," even if it does make apparent sense. To reason backward from effects, assumed outcomes, to probable cause creates seductive sequences; but the reasoning is going in the wrong direction. As stated above, it definitely is not sufficient to assert that an animal behaves "as if" anything. A case in point is the infrequent death of infants by aggression. Certainly, it occurs, in some locations far more often than in others. The data set remains sparse, especially when we recognize that the majority of deaths either were not soon after birth or were not witnessed and only assumed to have occurred. To incorporate these occurrences into a presumed evolved male reproductive strategy before sufficient evidence is acquired to prove or falsify the hypothesis begs the question. Alternative hypotheses are seldom considered, and the functional label of *infanticide* remains in constant use.

When it comes to analyses of living primates' behavior, two investigators may have drastically different ways of viewing and interpreting the same data set. This happens frequently and depends on a significant number of factors, such as training, experiences, and intellectual predispositions. More to the point, we call these *biases*, and we all have them. It is important to remember that even before data are analyzed there are many experiences and theoretical paradigms of the observer that are responsible for each researcher's initial recognition and collection of units of information that become the data of a study. Given our array of theoretical paraphernalia, what is the neophyte to do?

As in many disciplines, theory can easily outrun evidence when hypotheses are accepted before they have been investigated fully. Whereas theorizing and building models are fun and can be done from an armchair at home, they are the icing on the cake. Collecting evidence is also enjoyable, but it is labor-intensive and time-consuming. A hypothesis or, better yet, a set of alternative hypotheses generated by an initial study should be tested, ideally, on a body of data gathered from new groups that did not generate the hypotheses originally. With the constantly increasing amount of information from new and ongoing studies, it is possible to reanalyze correlations identified and accepted earlier. There are several excellent examples of such reanalysis in the chapters that follow, and they do not always uphold the earlier perceived truths.

Not all primatologists work in the field because not all research questions are best answered in the field. Consider briefly the appropriate role of investigations in captivity, which, of necessity, depends on the questions being asked. What do we want to see or to measure, and where are we most likely to achieve our objectives? Each research location, laboratory, colony, and field has advantages and disadvantages for investigating different questions. The behaviour systems characteristic of social animals are fractally complex, with any small section of the whole likely to turn out as complicated as the broader system was considered to be. We have ample evidence of the tremendous variability that exists within the geographic range for many species in the field. In terms of variability, but in far different ways, life in captivity can be characterized by a world of different conditions from a tiny, bare individual cage with minimal or no communication with a conspecific to large, outdoor/indoor enclosures with natural vegetation housing an intact social group. Because there can be atypical environments in the field as well as under captive conditions, it is critically important to choose the research location carefully.

Observation conditions in the field often interrupt continuous observation of individuals or groups. Forests did not evolve with the convenience of the observer in mind, nor does the wandering of the animals make it any easier to follow them. The result is we cannot and do not witness as much as we need to if we are to characterize the animals' behavior fully, and the devil all too often does reside in the undetected details.

With appropriate planning, good baseline information on the subject species, and a considerable amount of time, it is possible to witness animals without interruption from birth to death in captivity. Also, providing our own lives and work schedules permit, we can watch generations unfold. Of course, we can also watch generations in the field; and at some sites for a few species, we have long-term information. Even in field locations where teams of workers have produced decades of data, limited observation conditions and the mobility of the animals themselves seldom allow monitoring of all the tiny details we want to know. There is no doubt that visibility is far more problematic when following an arboreal species than one spending appreciable time on the ground. It is definitely easier on the neck to follow the latter.

No matter where primates live or how easy or difficult it is to study them, our aim is to generate a real feeling and appreciation for the organism. We have a head start because they are so easy to identify with and intriguing to watch. What is it like for a young capuchin monkey making its way in a complex social world? How does this differ from the life experiences and challenges facing an immature chimpanzee or langur? These and all the other nonhuman primates are engaging. Each, in its own way, offers both a challenge and an obligation to those who study them to express and make real the myriad dimensions of their lives.

We have devoted many years to discovering all we can about the lives of our subjects in the diverse habitats of nature. We sample the range of living conditions, measure ecological parameters, and record everything considered important in their existence. Ironically and indisputably, it is humanity that has placed the lives of these extraordinary animals in danger. The most important achievement of our research is that it enables us to do the most we can to slow a potentially catastrophic decline in biotic diversity and decrease in the quality of life for the survivors. Some of the important steps we need to take to put primate conservation into effect are discussed in this book.

The following chapters offer an excellent selection of what we know about the nonhuman primates, their evolution, present great diversity, and what must be done in the future to assure their continued survival and prosperity. A prodigious amount of work is summarized, emphasizing not only how many humans have devoted decades of work but also how many disciplines have contributed directly to our understanding of primate life. There is a challenge for everyone, regardless of training or interests. We can seek the most microscopic and embedded data or perhaps search for antipodal symmetries. We have just begun and, as they say, the best is yet to come.

PART ONE

Background

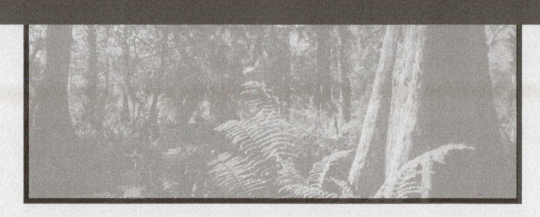

1

A Brief History of Primate Field Studies

Robert W. Sussman

1. When did the study of primates in their natural habitat begin?
2. Why were most studies of primates in their natural habitats conducted by anthropologists after the 1950s?
3. Is primatology a growing or diminishing field of science?

We might say that scientific interest in the natural behavior of primates began with an argument at the 1860 annual meeting of the British Association for the Advancement of Science. It was at this meeting that Thomas Henry Huxley defended Charles Darwin's *On the Origin of Species* from Bishop Wilberforce (Montagu 1959; Millar 1972; from an account written by Sir Charles Lyell, January 3, 1861). After a long-winded criticism of the theory of natural selection, the bishop turned to Huxley and begged to know, "Was it through his grandfather or his grandmother that he claimed descent from a monkey?" However, Huxley had been preparing for such an attack and whispered to a friend, "The Lord has delivered him into mine hands." He then answered, "If the question is put to me would I rather have a miserable ape for a grandfather or a man highly endowed by nature and possessing great means and influence for the mere purpose of introducing ridicule into a grave scientific discussion—I unhesitatingly affirm my preference for the ape."

Soon after, in 1863, Huxley's *Evidence as to Man's Place in Nature* was published. In this book, he put human and nonhuman primates into Darwin's story and compiled what could be thought of as the first "textbook" in biological anthropology. The book was organized into three sections: One was on comparative anatomy, illustrating the anatomical likeness of humans to other animals, and another included a summation of the human fossil remains known at the time; however, in the first section of the book, Huxley wrote the first systematic review of what was known about primates in their natural habitat. As might be expected, there was only anecdotal information available.

Before Darwin introduced the paradigm for modern evolutionary biology, primates were known to Western society only from adventurers' and colonialists' tales or from those brought back as pets or for menageries. Primates became first the subject of bestiaries and then around the 1400s, during the Renaissance, the subject of art and anatomical study (McDermott 1938, Janson 1952, Zuckerman 1998, Groves 2008). During the eighteenth and early nineteenth centuries, under the biblical paradigm of the fixity of species, early natural historians such as Linnaeus, Buffon, George Edwards, Geoffroy Saint-Hilaire, Cuvier, and John Gray attempted to put God's creatures into some order, naming and classifying them. These authors published extensive tomes, often with beautiful artistic and realistic renderings of various primates (see Groves 2008 for a history of this early period).

Immediately after Darwin, in the late nineteenth and early twentieth centuries, only a few scientists were interested in venturing into the forests to see what wild-living primates (and not those living in cages or already anatomical specimens) could tell us about human existence. The earliest of these adventurers were interested mainly in chimpanzees and gorillas. In the 1890s, Richard L. Garner (1848–1920), a zoologist and collector interested in the "speech" of monkeys and apes, went to Gabon, West Africa, to collect great apes and to observe them in the wild. Because it was commonly thought that gorillas were bold and violent, Garner (1896) armed himself and built a cage to sit in while waiting for the animals to come by. This he did much of the day and night for 112 successive days. A few individual apes approached and quickly wandered off. Needless to say, little was learned about gorillas or chimpanzees on this venture. Indeed, we have only recently begun to learn about the natural behavior of lowland gorillas (*Gorilla gorilla gorilla*). Garner (1892) also was the first to use a phonograph to study primate vocalizations. Around the same time, Sir Arthur Keith (1866–1955) observed the naturalistic behavior of Asian primates in Thailand, particularly the gibbon (*Hylobates lar*). Mainly an anatomist, he related his observations on behavior and ecology to his field-based anatomical dissections (Sheeran 1997).

The first American woman to observe and describe the behavior of great apes in the field was Mary Hastings

Bradley (188?–1976), a Chicago fiction writer and socialite. Her husband, Herbert Bradley, was a big-game hunter, who was asked to accompany Carl Akeley to the Virunga Volcanoes in East Africa to collect gorilla (*G. g. beringei*) specimens for the American Museum of Natural History. She wanted to see and write about Africa, and she and her 5-year-old daughter Alice became an integral part of the expedition. (Mary Hastings Bradley's biography, along with her daughter's, is a story in itself but, alas, for another time and place.) Akeley believed that the rumors of the gorilla's ferocity were greatly exaggerated and wanted not only to collect animals but also to observe and photograph them. He developed one of the first motion picture cameras for the field and was the first to film primates (gorillas) in their natural habitat (Bodry-Sanders 1998). Bradley (1922:131–132) was able to follow a group of gorillas one afternoon:

> I had seen six gorillas, one of them, at least, and probably two, the demon male, and five gorillas had certainly seen us. And we had not been attacked on sight. Not one had beat his breast or roared or tried to ambush us!...We had never heard of any of the native(s)...being raided by gorillas....Altogether Mr. Akeley's belief in the essential character of the gorilla was justified. He was simply the big monkey, the man ape, powerful beyond all words, dangerous when attacked, but not a bit the hellish demon or the malignant arch fiend!

Akeley returned to the volcanoes in 1926, this time to study the mountain gorilla and not to shoot it. However, he died at the beginning of the expedition and was buried in Albert National Park, Congo, which he had been instrumental in establishing the year before (Schaller 1964, Bodry-Sanders 1998).

In 1929, Robert Mearns Yerkes (1876–1956) and his wife Ada Yerkes (1874–1963) published *The Great Apes*, a compilation of the current knowledge about primates. However, as in Huxley's volume written over 60 years earlier, there was no systematic field research to report in this volume, and concerning knowledge of primates to that date, they wrote that we know "next to nothing, with certainty, concerning their instincts, habits, other individual modes of behavior, mental life, and social relations" (p. 582).

In the same year, Robert Yerkes was funded to begin a great ape breeding facility in Orange Park, Florida, as part of the Yale Laboratories of Primate Biology. (This later became the Yerkes National Primate Research Center of Emory University, Atlanta, Georgia.) He sponsored two expeditions to Africa to study the behavior of apes in their native habitat, to aid in his captive breeding program, and to collect some of these animals for his laboratories. In 1929, Harold C. Bingham set off to the Congo to conduct research on gorillas (*G. g. gorilla*), and a year later, Henry Nissen traveled to French Guinea, West Africa, to collect and study chimpanzees (*Pan troglodytes verus*). Yerkes, Bingham, and Nissen were psychologists, and their research on great apes was motivated mainly by an interest in the study of the evolution of intelligence. Yerkes had both scientific and political influence. He was a leader in the eugenics movement in the United States during the decades preceding World War II and a major proponent of the use of intelligence testing to restrict immigration into the United States and to support and rationalize racial prejudice. In 1923, he wrote a glowing foreword to Carl Brigham's racist tome *A Study of American Intelligence.*

Nissen's expedition was fairly successful. He was able to observe chimpanzees on 49 of 64 days in the field, dispelling a number of widely believed but fictitious ideas about them. His research was published as a monograph, the earliest on primate field behavior (Nissen 1931). Bingham had little more success in his study of gorillas than did Garner before him. Unlike Akeley and Bradley, he believed anecdotal accounts of the gorilla's ferocity and thought it necessary to track the animals with a large entourage of guides, gun bearers, and porters. Bingham carried a rifle, and his wife carried a pistol. After about 2 weeks in the field, his party inadvertently surprised a group of feeding gorillas and was threatened by a large male. Bingham shot the animal and returned home about a month later, learning very little about the natural behavior of gorillas (Bingham 1932).

At about the same time that Nissen and Bingham were attempting to study great apes in western and central Africa, Sir Solly Zuckerman (1904–1993), a medical doctor, anatomist, endocrinologist, and chief scientific advisor to the British government for many years, was studying a captive colony of baboons in Regent's Park Zoo, London. In 1930, he returned to his native South Africa to add some information on natural populations of these creatures. However, he spent only 9 days observing chacma baboons (*Papio ursinus*) in the field. His monograph *The Social Life of Monkeys and Apes,* published in 1932, focused on the importance of the social behavior of primates and emphasized the fact that higher primates lived in permanent social groups. Zuckerman theorized that sex was the major reason for group living in primates; his theory is no longer accepted, and the factors underlying social organization in primates are still being debated.

There was, in fact, one earlier field study on baboons. This was by Eugene Marais (1871–1936), a South African journalist, lawyer, poet, and natural historian. Sometime around 1907, shortly after the Boer War, Marais retreated to a remote area of the Transvaal, Waterberg. Here, he and a young colleague studied a troop of habituated chacma baboons (*P. ursinus*) for 3 years, essentially from his backyard. It is interesting to note that Marais was worried during his observations because he "had no libraries and no means of checking what work had already been accomplished in this field" (Marais 1968:63). Little did he know that there had been no previous studies of this sort. However, until his unfinished monograph *The Soul of the Ape* appeared in 1968, long after his death, the primate fieldwork of Marais remained relatively unknown (though Zuckerman made brief reference to his work).

Thus, although some field studies had been done during this early period, the individuals involved in the actual fieldwork were not interested in dedicating their lives to the study of primates in remote habitats and none contributed to field primatology in any permanent way. However, C. Raymond Carpenter (1905–1975) was an exception. After earning his PhD in psychology (on the sexual behavior of birds), Carpenter was granted a postdoctoral fellowship to work with Robert Yerkes at Yale. Yerkes and Frank M. Chapman, an ornithologist, convinced Carpenter to study monkeys at Chapman's field site, Barro Colorado, Panama. This began the first long-term involvement of a scientist in field primatology or, as Carpenter called it, "the naturalistic behavior of nonhuman primates" (Teleki 1981a). Carpenter spent a number of months studying howlers (*Alouatta palliata*) on Barro Colorado and making some observations on spider monkeys (*Ateles geoffroyi*) in western Panama between 1931 and 1935 (Carpenter 1934, 1935). He focused mainly on social organization, social interactions, and group censuses.

In 1937, Carpenter was invited to collect the behavioral data on a multidisciplinary expedition to study gibbons (*H. lar*) in Thailand. The "Asian expedition" was sponsored by Harvard, Columbia, and Johns Hopkins universities and included two physical anthropologists/anatomists who were to become very important to primatology, Adolph Schultz and Sherwood Washburn. Carpenter spent 4 months in the field studying white-handed gibbons, and this led to his monograph on this species, now considered a classic in the field (Carpenter 1940). In 1938, Carpenter was instrumental in exporting 450–500 rhesus monkeys (*Macaca mulatta*) from India and releasing them on Cayo Santiago, an island off the coast of Puerto Rico. This began the first long-term research on semicaptive primate populations. The colony is now part of the Caribbean Primate Research Center and is still an active research colony.

Because of World War II, field research essentially came to a halt between the late 1930s and early 1950s. In fact, before 1960, there was very little primate field research, though it is obvious that the natural behavior of primates was beginning to interest scientists. Between 1960 and 1965, a number of important scientific meetings and books appeared and the discipline of field primatology became solidly established. It has grown steadily since that time. Space will not allow me to detail the stages of this development and growth of the field. I thus simply will outline what I consider to be major events and influences.

Field research on primates during the 1950s stemmed mainly from four different areas. (*1*) Yellow fever had spread in many tropical areas during World War II, and nonhuman primates as well as humans were affected with the disease. Some studies of free-ranging primates were carried out in relation to the epidemiology of this disease, such as that done by the Virus Research Institute in Uganda (e.g.,

Haddow 1952). (*2*) Following Carpenter's research at the Smithsonian Institution field site of Barro Colorado, zoologist Charles Southwick and his professor Nicholas Collias recensused the howler monkeys (*A. palliata*) (Collias and Southwick 1952), and another zoologist, Stuart Altmann, studied the social behavior of the same species (Altmann 1959). Both Southwick and Altmann became major contributors to primate field studies during the 1960s and are still active in the field. (*3*) A few animal behaviorists and mammalogists teaching and residing in tropical countries did studies of primates as part of more general research on local fauna. For example, Niel Bolwig of Witwatersrand University studied baboons (*P. ursinus*) in South Africa and great ape (*Pan troglodytes schweinfurthii* and *G. g. beringei*) nests in East Africa (Bolwig 1959a,b), and Angus Booth of University College, Ghana, did sophisticated studies of synecology of West African primate communities in the early 1950s (e.g., Booth 1956). Booth surely would have had a major influence on primate field research if he had not died tragically in 1959 at the age of 30. (*4*) Under the leadership of Kinji Imanishi, in 1948, a number of Japanese animal behaviorists began to study the indigenous Japanese macaque (*Macaca fuscata*) (e.g., Imanishi 1953, Kawamura 1958). These were the first long-term studies of identified individuals of known kinship in natural primate populations and the beginning of what has turned out to be one of the most active centers of primate research. This group of researchers founded the first journal in primatology, *Primates*, which was published in Japanese in 1957 and in English in 1959.

Thus, studies of primates in their natural habitats began again in the 1950s, but field primatology did not really take off until the next decade. It appears that two developments in the 1950s were very important to the enormous growth of this subdiscipline in the 1960s. The first was the exchange of ideas and concepts between population biologists and anthropologists, especially concerning human evolution. This is exemplified by the Cold Spring Harbor Symposium of Quantitative Biology in June 1950. This meeting was attended by 129 of the most influential biologists and anthropologists in the world. The proceedings were published in 1951 and contained a number of papers calling for a more detailed look at primate behavior in interpreting nonhuman and human primate fossils. In 1953, a paper was published in *American Anthropologist*, coauthored by biologist George Bartholomew and anthropologist Joseph Birdsell (who had attended the 1950 symposium), entitled "Ecology and the Protohominids." This paper developed a method for reconstructing the behavior of early hominids by extrapolating from the behavior of extant primates and other mammals. The second development that stimulated primate field research was the discovery, in the early 1950s, that the Piltdown Man was a fake, accompanied by the concomitant acceptance of the australopithecines as true ancestors of humans. Piltdown Man was a skull discovered in 1912 and long thought to be the missing link between humans and apes, fitting the belief at that time that the earliest humans would have large brains and be more ape-like in the rest of their anatomy. In 1953, it was discovered to be the spurious amalgam of a fairly recent human skull and the jaw of a female orangutan (Millar 1972). The discovery of this hoax ended any idea that there was a major evolutionary "gap" between human and nonhuman primates and emphasized the continuity between ourselves and our ancestors (Sussman 2000).

These developments stimulated biologists, and especially biological anthropologists, to think about the behavior and ecology of early humans and to see living primates as potential windows into the study of the evolution of human behavior. For example, the two preeminent australopithecine hunters, Raymond Dart and Louis Leakey, enlisted Rosalie Osborn, a former secretary, and Jill Donisthorpe, a journalist, to conduct a study of gorillas (*G. g. beringei*) at the Virunga Volcanoes of Albert National Park in the mid-1950s (see Schaller 1964). This study did not yield a great deal of information, but it was followed up by University of Wisconsin zoologist George Schaller's study of the Virunga gorillas in 1959, the first detailed study of a great ape (Schaller 1963). Soon after this, in 1960, Leakey sponsored and Robert Hinde, of Cambridge University, supervised Jane Goodall's study of the chimpanzees (*Pan t. schweinfurthii*) at Gombe.

By the mid-1950s and early 1960s, many conferences and books focused on the relationship between primate behavior and human evolution. For example, in 1953 a symposium was presented at the annual meeting of the American Association for the Advancement of Science entitled "The Non-Human Primates and Human Evolution." The proceedings of this meeting (Gavan 1955) were dedicated to Earnest Hooton (1887–1954), an anthropologist who had been calling for primate field studies since his classic book *Man's Poor Relations* in 1942. He delivered a paper at this conference, shortly before his death, entitled "The Importance of Primate Studies in Anthropology." Other volumes to appear were *The Evolution of Man* (Tax 1960), *Social Life of Early Man* (Washburn 1961), *Ideas on Human Evolution* (Howells 1962), *Classification and Human Evolution* (Washburn 1963), and *African Ecology and Human Evolution* (Howell and Bourliere 1963).

Except for the volume edited by Howells, each of these books developed out of international conferences. One of the prime movers for this interchange was the anthropologist Sherwood Washburn (1911–2000), a functional anatomist who had been a student of Hooton. Washburn was also a member of the Asian expedition with Carpenter and, as a graduate student, had spent a semester at Oxford, where he worked in Zuckerman's lab. He also participated in the 1950 Cold Spring Harbor symposium. Papers by Washburn, stressing the need for primate field research, appeared in each of these volumes, including now classic papers by Washburn and his student Irven DeVore on baboon (*Papio anubis*) and early human ecology and social behavior.

These works stimulated students to study primates in their natural habitats; by the early 1960s a number of conferences on free-ranging primates were held and related books began to appear. The first two of these volumes to appear were based on international conferences held in 1962 in New York (Buettner-Janusch 1962) and London (Napier and Barnicot 1963). Other early collections were edited by Southwick (1963), DeVore (1965b), Jay (1968b), and Altmann (1967), the latter three based on meetings held between 1962 and 1965. The international journal *Folia Primatologica* began in 1963. The major figures in primatology at that time contributed to these books. Washburn was again a major catalyst for many of these meetings, and his influence on primate field biology cannot be overemphasized. In fact, the first eight dissertations in primatology after 1960 and 15 of the first 19 were stimulated by Washburn. By 1980, Washburn and his students had produced more than half of the active anthropological primatologists (Gilmore 1981), and his influence continues to the present.

In a recent academic genealogy, mainly of American field primatologists, Washburn's lineage is the largest, with 326 (60.6%) of the 538 field primatologists linked to his lineage (Kelley and Sussman 2007). Using this same database, we found that field primatology has been steadily growing since the mid-1950s. The number of field primatologists who graduated with doctoral degrees (mainly in the United States) from 1955 to 1964 was 13. This grew to 56 degrees granted from 1965 to 1974, 96 from 1975 to 1984, 118 from 1985 to 1994, and 215 from 1995 to 2005. Even more impressive is the fact that interest in field primatology has been steadily growing in many countries, especially in the tropical countries where primates occur naturally (e.g., Kinzey 1997c).

Research in field primatology during the late 1950s and early 1960s was mainly descriptive natural history, with few comparative and quantitative, or problem-oriented, studies. However, by the 1970s and 1980s, field primatology moved into a problem-oriented phase (see Sussman 1979, Smuts et al. 1987). Altmann (1965) and the late K. R. L. Hall (1962a,b; 1965a,b) were the first to collect quantitative data on free-ranging primate populations. Hall urged the use of quantitative methods, and at his suggestion, two of his colleagues in the Psychology Department at Bristol University, John Crook and Pelham Aldrich-Blake (1968), were the first to use the now common "scan sampling" method in a primate ecology field study. In 1974, Jeanne Altmann described various methods of collecting quantitative data on free-ranging primates. This is now one of the most referenced papers by field primatologists.

Problem-oriented studies focused mainly on determining relationships between behavior and morphology, on the one hand, and ecology and social structure, on the other, were an important component of field primatology during the 1970s and 1980s. More recently, many field primatologists have been formulating and testing theories developed out of classical sociobiology, such as those related to kin selection, reciprocal altruism, dominance and reproductive success, and the relationship of sexual selection to social organization and infanticide. In many cases, the theories and studies are elegant and elaborate, but the data are meager.

The philosopher of science F. S. C. Northrop (1965) suggests that any healthy scientific discipline goes through three stages during its development: the first stage involves analysis of the problem, the second is a descriptive natural history phase, and the third is the stage of postulationally prescribed theory, in which fundamental theories are tested. Although there is a movement in field primatology toward this final stage of enquiry, even today many primate species have not been studied in detail, and the most studied species normally are known only from a few localities (Sussman 2003a,b; Strier 2003a–c; see also Chapter 47).

The range of variation in the behavior and ecology of most free-ranging primates is still unknown, and basic natural history remains a necessary component of the subdiscipline. As stated by Northrop (1965:37–38):

> In fact, if one proceeds immediately to the deductively formulated type of scientific theory which is appropriate to the third stage of inquiry, before one has passed through the natural history type of science with its inductive Baconian method appropriate to the second stage, the result inevitably is immature, half-baked, dogmatic, and for the most part worthless theory.

With eugenics, social Darwinism, and the Piltdown Man being conventional wisdom during the first half of the twentieth century, these cautions still must be born in mind when developing theories in biology, the social sciences, and primate ecology and social behavior. Given the conservation status of most primate populations, let us hope that there is still enough time to learn more about these fascinating animals and to collect the data needed to properly test our theories.

More recent developments in primatology have been the focus of several articles (e.g., Haraway 1989; relevant chapters in Spencer 1997; Rodman 1999; Fedigan and Strum 2000; Strier 2003 a,b; and see Chapter 47). However, I will briefly give my own personal perspective of where I think primatology has been and might be going. My interests in primatology began in the mid-1960s and I went into the field to conduct my PhD dissertation research in Madagascar in August, 1969. These were extremely exciting times in a young field. It was a time when many pioneering studies had been done and they gave rise to more questions than they answered. At this time, collecting basic natural history data was essential. Few species had been studied, the lengths of most studies were generally quite short, and the methods were mainly descriptive, very much like qualitative ethnographies (see Strier 2003a). As stated above, the first detailed quantitative methods for collecting ecological and behavioral data in field primatology were just being developed (Crook and Aldrich-Blake 1968). I used these methods in my thesis

(Sussman 1972). To my mind, three of the most important developments that occurred in the 1960s and early 1970s were: more species began to be studied at more sites; relatively standardized quantitative methods began to be used; and a few long-term research sites were initiated. Of course, primatologists also began to realize that primates and their forests were vulnerable and began to become conservationists, an important trend that continues in our field.

Thus, as with all scientific fields, primatology began with its natural history phase. However, although natural history continued, continues today, and needs to continue into the future (see Chapter 47), with fairly standardized quantitative methods, comparative research emerged and problem-oriented studies began in earnest. So, besides obtaining natural history of more and more species, studies also began to focus on testing specific questions. Relationships between ecology and social structure, or ecological and morphological adaptations were explored. This led to comparative studies of the habitat preferences of closely (or not so closely) related species living sympatrically, or of different populations of the same species living in very different habitat types.

One aspect of this widening of focus is that the emphasis, in American primatology at least, on primates as models for human evolution often got lost. It was easy to understand the connection when one studied the great apes, after all, they were our closest relatives, or even baboons, since early humans must have lived in similar environments. However, why would any anthropologist be interested in ring-tailed lemurs? Well, in our natural history studies, we found that spider monkeys had social structures that mirrored those of chimpanzees, and ring-tailed lemurs (as well as baboons, vervet monkeys, and macaques) all lived in edge habitats and shared many characteristics that might have been shared by early hominins. After all, the earliest hominins were in most ways more like nonhuman primates than they were like modern humans.

Emerging in the 1970s with the popular books of Wilson (1975) and Dawkins (1976), sociobiology became accepted wisdom by much of primatology. This became an extension of the period of problem-oriented field studies and a catalyst for many research questions. However, by the 1980s, to my mind, many of the theories proposed in sociobiological models were being accepted as fact. In a Kuhnian (1962) sense, researchers became technicians within the field, attempting to gather evidence that helped support their theories rather than questioning the theories themselves. We began to answer ultimate evolutionary questions without understanding proximate mechanisms. For example, in many ways, it was thought that we could answer questions about social relationships in group-living primates without even knowing the specific contexts in which social interactions occurred.

There were a few of us remaining from the 1960s and early 1970s who were disappointed in this approach, and even a few young primatologists who were questioning some of the basic assumptions of sociobiology, but overall the tenets of sociobiology were becoming ensconced, both in the scientific literature of primatology and in the popular literature, much to the dismay of some anthropologists. In fact, this led to some primatologists questioning their role in anthropology and some anthropologists questioning the role of primatology in anthropology. This is not to say that sociobiology did not generate any interesting questions and theories, but only to stress that theories must continue to be tested and if they don't fit the data, they must be reformulated or abandoned.

I kept working on natural history and conservation, long-term demography of known populations, and problem-oriented research in which proximate questions often trumped evolutionary ones. I still believe that one must understand proximate mechanisms before we can begin to develop and answer evolutionary questions. We still have not studied most species in more than a few environments, nor do we understand the amount of variation that exists in species over space or time (see Strier 2003 a,b). Although many of our current theories are developed around the idea that competition for resources by females and for females by males drives social organization, we have little data for any species on the actual frequencies of intense versus mild aggression, or cooperation and how these relate to social or ecological context, kinship, familiarity, demography, etc., or how these values vary within and between species.

Some of the positive things that happened over the past 30 years of primate research are the expansion of the number of species studied—now we have some basic information on many species, though some have not been studied in more than one location and some have still not been studied at all. We need to continue to work diligently on sensible conservation efforts. We also need to continue natural history.

New technologies have been introduced that will enable us to answer many questions that previously were left to pure theory and speculation. These include population genetics, hormone studies, radiotracking and telemetry and marking individuals, use of GPS and satellite imagery, and veterinary field technology. Also, computers and new statistical techniques (e.g., randomization) have enabled us to explore questions not possible just a few years ago. How many of you remember using computer cards, and how many of you have no idea what I am talking about?

Two other very positive trends that have continued and increased are long-term field research sites with identified individuals (these, of course, started early in some areas but now there are many sites and data are increasingly becoming available from them), and interdisciplinary team research (see Chapter 47). I have had the opportunity to use many of these relatively new techniques and to participate in long-term and interdisciplinary research. These are exciting developments, and research using these techniques and approaches is yielding new types of data and bringing up exciting new questions, many of which are discussed in this volume. I am looking forward to new horizons in primatology, and a new appreciation in primatology of anthropology and in anthropology of primatology.

2

Primate Socioecology

Joyce A. Parga and Deborah J. Overdorff

Primatology is not something over there. It is not a logically conceived body of knowledge and techniques. Primatology is us, and we are deeply immersed in a particular European history. That history determined our education and set our social rewards; the greatest contribution to the study of the nonhuman primates might be to free us from some of the traditional limitations and points of view. Washburn 1973:180

1. What is primate socioecology?
2. How have attempts at modeling primate socioecology changed over time?
3. What are the various ecological and social factors that can affect primate behavior and social grouping patterns?

INTRODUCTION

Historically, *primate socioecology* has been the study of how environmental variables (usually food, its seasonality, distribution, and abundance) influence primate group size, composition, and social dynamics. Our understanding of primate socioecology largely comes from long-term work at well-established research sites on such anthropoid genera as *Pan, Macaca, Papio,* and other related catarrhines. As more studies on these species and others are conducted in a variety of habitats, researchers have recognized a wide range of behavioral and dietary variation, even within closely related taxa such as *Semnopithecus* spp. (Koenig and Borries 2001), *Papio* spp. (Barton et al. 1996), *Colobus* spp. (Chapman et al. 2002a,b), *Pan troglodytes* and *P. paniscus* (Boesch 2002; Mitani et al. 2002c; Muller and Mitani 2005; Stanford 1998a,c), *Pongo* spp. (Delgado and van Schaik 2000), and *Gorilla* spp. (Doran and McNeilage 1998; Harcourt and Stewart 2007a,b). Work on less well-studied groups, most notably platyrrhines and strepsirhines, has further underscored the potential for inter- and intraspecific social diversity (i.e., *Saimiri* spp., Boinski et al. 2002; *Cebus*, Panger et al. 2002b; *Callimico* and *Saguinus*, Porter 2004, Porter and Garber 2004; *Pithecia*, Norconk and Conklin-Brittain 2004; *Hapalemur*, Grassi 2006, Tan 1999; *Eulemur*, Overdorff and Johnson 2003; *Lemur* and *Propithecus*, Jolly et al. 2006, Richard 2003, Yamashita 2002).

Researchers now realize that not all primates live in large groups, maintain rigid dominance hierarchies, show male dominance over females, or exhibit female philopatry (Strier 1994b). Strepsirhines, for example (Fig. 2.1), are highly nonconvergent with haplorhines based on the prevalence of (*1*) small group sizes which converge on those documented for platyrrhines, (*2*) female dominance, and (*3*) nocturnal and cathemeral activity cycles (Curtis 2004; Curtis and Zaramody 1999; Overdorff and Erhart 2001; Pereira et al. 1990; Rasmussen 1999; Richard 1987; Tattersall 1987, 2006a). Consequently, identifying a common set of selection pressures that can explain such diversity has become more challenging.

Although primatology has come a long way theoretically since its inception in the 1940s and 1950s (Jolly 2000, Strum and Fedigan 2000), there are still several obstacles that primatologists must overcome. One inherent tendency for primatologists is to assume that nonhuman primates are "special"—set apart from other social mammals and birds due to their complex sociality, exceptional cognitive abilities, and close taxonomic affinity to humans. While no one would question that these are indeed uniquely primate traits, it has led to a general ignorance in our field of the value of a broader comparative perspective (see Harcourt 1998, Rowell 2000). Because primates are long-lived, reproduce slowly, and demonstrate extraordinary behavioral flexibility compared to similar-sized mammals, researchers must be tremendously patient, collect a much larger volume of data, and/or commit to data collection for a longer period of time when looking for emergent patterns. Primatologists also tend to place great value on studying their subjects under natural conditions; this devotion to naturalism poses its

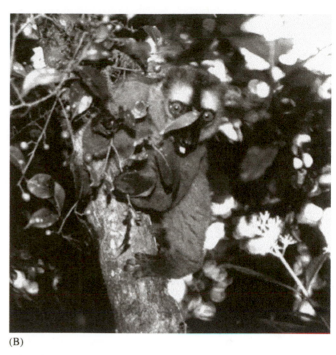

(A)   (B)

Figure 2.1 *A* Ring-tailed lemurs (*Lemur catta*) feeding on terrestrial foliage (photo by Joyce Parga). *B* A red-fronted lemur female (*Eulemur fulvus rufus*) feeding on *Ficus* sp. fruits (photo by Deborah Overdorff).

own set of problems experimentally. Despite years of study, small sample sizes for certain behaviors or age groups are common. This is exacerbated by the fact that wild subjects will die of unknown causes—or, worse, will mysteriously disappear—leaving the explanation for such events up to the researcher's discretion. Finally, the degree of historical or ongoing anthropogenic interference may also influence behavior in unpredicted ways, leading researchers to wonder if current conditions actually represent the original evolutionary context of the observed behavior (Janson 2000a).

As a result, primate socioecological studies have passed through various stages of increasing sophistication since the 1950s (see reviews by Dunbar 1988, Janson 2000a, Kappeler and van Schaik 2002, Strum and Fedigan 1999, Terborgh and Janson 1986). One constant throughout these stages has been the acceptance that the main selection pressures that influence primate behavior are predation and feeding ecology. Although these two variables go hand in hand, this chapter will primarily focus on the explanations offered to interpret primate behavior based on feeding ecology. We will describe briefly each of the historical stages, discuss the pros and cons of existing theoretical approaches, describe where primatology is heading, and discuss some newer approaches which, ironically, consider selection pressures other than feeding ecology (specifically, intersexual conflict).

IN THE BEGINNING

Primatological studies in their infancy (Carpenter 1942a, 1954; Zuckerman 1932) were largely descriptive and not

based on models or hypothesis testing. Subsequent studies attempted to synthesize broad categories of primate behavior and searched for correlations between primate sociality and elements of their ecology. Two such studies (Crook and Gartlan 1966, Eisenberg et al. 1972) provided the first major contributions that helped move primatology away from being a solely descriptive discipline by attempting to find common ground between different primate species.

Crook and Gartlan's (1966) approach was to develop specific dietary and ecological categories and then place primates into these categories based on activity, type of habitat used, and group size. Using this technique, they found broad consistencies within groups, such as the fact that solitary primates tended to be nocturnal and savanna groups tended to live in large social units. Eisenberg et al. (1972) expanded their analysis by including additional categories such as diet and reproductive units (mating system). Despite these additions, it was still an exercise in assigning primates to categories with little adaptive rationale. Nevertheless, each of these studies laid the groundwork for the idea that there existed an evolutionary grade among social systems as primates evolved from a more ancestral (nocturnal and solitary) social pattern to a more derived, socially complex pattern (diurnal and group living). This assumption still forms the basis for more recent socioecological models (van Schaik and Kappeler 1996).

Though ambitious, these first theoretical papers reflected work with only a small number of well-studied species and did not take into account phylogeny, which can explain the presence of many similar traits found in closely related taxa (Rendall and Di Fiore 1994, but see Korstjens et al.

2002). They also did not account for the possible causal mechanisms that might shape behavior, although Crook's later work (1972, Crook et al. 1976) made a more significant attempt to place the behaviors in an adaptive and comparative (nonprimate) framework.

At approximately the same time, a more inductive and comparative approach documented more specific correlations between ecological and behavioral variables within primates (Clutton-Brock 1974b, 1977, 1989; Clutton-Brock and Harvey 1980; Milton and May 1976). The implementation of more standardized behavioral data-collection methods (Altmann, J., 1974) and the increase in available data on a greater number of primate species also made such an approach more productive. The relationship between activity, diet, and ranging patterns was the main focus in these papers; and researchers relied on a more quantitative, statistical approach to their analyses. Other previously disregarded variables were included, such as daily ranging patterns, body size, and sex ratio. Some trends emerged during this time that are now taken for granted: nocturnal primates are smaller than their diurnal counterparts, frugivores range farther on a daily and long-term basis compared to folivores, smaller groups are more cryptic than larger groups, and sexual dimorphism increases with increasing group size where females outnumber males. Ongoing studies of the behavioral ecology of sympatric, congeneric species further confirmed that these have been consistent trends among primates (Crockett and Wilson 1980; Gautier-Hion 1978; Hladik 1977a; Kinzey and Gentry 1979; Milton 1981c; Mittermeier and Roosmalen 1981; Struhsaker and Oates 1975; Sussman 1972, 1974). One of the main assumptions in the study of sympatric congeners was that habitat and evolutionary forces remain broadly constant so that finer details of social behavior can emerge. However, as with other approaches, there were still drawbacks: (1) confounding variables were problematic (e.g., frugivory and monogamy often coexist), (2) which independent variable should be considered the most important (group size, resource size, social interactions, activity, etc.) was not well defined or agreed upon, and (3) the causal mechanisms behind the observed patterns were vaguely articulated.

An additional weakness during this comparative/correlation phase was that the potential costs and benefits of behaviors were not really considered. There were also no generalized models that could produce testable hypotheses to guide future study. Conversely, this was not the situation for other taxonomic groups being studied concurrently. Ornithologists and mammalogists were developing internally consistent, evolutionarily based models that were used to explain sociality in nonprimate mammals and birds, which were firmly grounded in a sociobiological approach (Alexander 1974, Orians 1969, Wittenberger 1980, Wittenberger and Tilson 1980). Although ecological variables were considered important (i.e., the degree of

territoriality and habitat quality), there was more focus on the role that sexual selection might play (especially male intrasexual competition for mates). More consideration was also given to kin selection and the role that inclusive fitness plays in determining social and mating systems. These papers ultimately suggested that grouping patterns among animals are a result of (1) specific selection pressures (ecological conditions, predation levels, etc.) and (2) the interplay between the costs and the benefits (in the form of reproductive success) of certain behaviors to males and females.

Primatologists were not completely ignorant of such considerations. In a paper not often cited, Denham (1971) detailed a model that broadly linked three variables into a predictive framework: energy acquisition, predation pressure (what he called "defensive response"), and reproductive strategies. He laid out several key ecological variables, such as food predictability, food density, and antipredator strategies—all are concepts that would play a central role in later, more high-profile models. Shortly thereafter, S. A. Altmann (1974) published a more specific test of the energetics of baboon movements in which he considered these same variables.

Wrangham's (1979b, 1980, 1982, 1987) work comprises one of the most pivotal and often-cited set of papers and demonstrates a significant turning point in primatological theory. In these papers, he developed the first ecological model for primates that provided for testable predictions and a clear direction for future research. Intergroup competition in this model is considered the most important factor in the evolution of primate grouping patterns. The roles of kin selection and inclusive fitness (which were essential elements in nonprimate models at the time) were also considered to be integral.

Generally, Wrangham proposed a strong link between the defendability of food resources, the degree of female bonding, and female dispersal patterns. Where food is defendable and of high quality, females are expected to cooperate in its defense with kin, resulting in female philopatry. Under these conditions, strong female bonds are expected to form and female dominance relationships will be well-defined. Where food is not defendable (e.g., where food is uniformly distributed and cannot be monopolized), females are not expected to cooperate in its defense, female philopatry is not expected, and female relationships should be loosely defined. Under both conditions, male presence is expected to be determined by the spatial distribution of females. Food was thought to be the primary limiting factor for female reproductive success, and the number of mates was considered to be the primary limiting factor for male reproductive success. It is worthwhile noting that the foundation for this basic premise was laid 30 years earlier in experimental work on sexual selection using *Drosophila* as a model (Bateman 1948, Tang-Martinez 2000). Wrangham's own work (1980), concurrent platyrrhine studies (Robinson 1988b; Terborgh 1983; Terborgh and Janson 1986), and more recent work back up the general premise of his model, although the picture has become more complex as more independent variables are considered (Dunbar 2000, Eberle and Kappeler 2002).

COMPETING COMPETITION MODELS

The next generation of modeling focused on the role predation avoidance should play (in conjunction with feeding ecology) and whether group size and social organization (how individuals interact within groups) function as independent variables. One such example is van Schaik and van Hooff's model (1983), which included predation pressure along with other ecological factors (e.g., food distribution and density). The combined influences of net food intake and predation risk will ultimately produce the optimal group size and social organization for a particular primate species. In a more rigorous test of this premise, van Schaik (1983) examined birthrates (which he used as a proxy for foraging efficiency) as a function of group size. By comparing the observed slopes to the expected slopes for two conditions (predation with intragroup feeding competition and intragroup feeding competition alone), he concluded that the lower limit to group size is set by predation pressure, while the upper limit is influenced by the degree of feeding competition. This paper demonstrated the need for direct, measurable variables to better test hypotheses.

Subsequent papers addressed this issue by including an explicit list of measurable variables (mostly related to feeding) that could be used to test predictions. Like their predecessors, Terborgh and Janson (1986) introduced a model for optimal group size that considered both predation and foraging efficiency but also introduced the idea that food resource size (or patch size) was an integral piece of the puzzle. Janson (1988a) subsequently produced a more complete list of variables related to net food intake that should be considered (e.g., encounter rate, amount of food consumed during each feeding bout, nutritional content of food, and time spent foraging).

Another pivotal moment occurred when primatologists finally integrated two concepts that had already been incorporated into group size models in other taxa: (1) the two main forms of competition and (2) that competition levels can differ within and between groups. Individuals can aggressively compete over food which is monopolizable (known as *contest* or *interference competition*), or they can indirectly compete by exploiting nonmonopolizable food on a "first come–first served" basis (known as *scramble competition*) (Nicholson 1954). Either of these forms of competition can occur within or between groups. Of note, avian models at this time emphasized that both the abundance and the divisibility of food resources were important variables (Elgar 1986, Mangel 1990), an idea featured prominently in the next round of seminal primate papers.

In 1988, a special issue of the journal *Behaviour* was dedicated to the question of which selection pressures most profoundly affect primate foraging and competitive regimes. These papers aptly demonstrated that it was possible to document differing levels of competition in a range of primate species (Chapman 1988a; Garber 1988a; Isabirye-Basuta 1988; Janson 1988a,b; Rowell 1988b; Symington 1988c; van Schaik and van Noordwijk 1988; White and Wrangham 1988; Whitten 1988). They also underscored the fact that researchers lacked a uniform methodological approach with which to investigate feeding competition from an evolutionary perspective. Accordingly, Janson and van Schaik (1988) emphasized the need for a more standardized methodology to measure the degree of competition for food within and between primate groups and provided a conceptual model that could more readily predict when different types of competition might occur. Using the *Behaviour* collection of articles as their starting point, they synthesized a comprehensive list of data that should be collected in future studies. This list included many previously published variables as well as a more sophisticated estimate for energy or gain rates while feeding. They also suggested that researchers should pay more attention to collecting data on food handling time, energy ingestion rate, and the amount of energy expended in different activity states. Although their intention was to challenge primatologists to develop and use more precise and quantitative methods in future competition studies, few studies (if any) have since incorporated *all* of the variables suggested. Of these variables, patch size continues to be commonly used, yet there remains little uniformity across studies as to how it is measured (Chapman 1990b; Chapman et al. 1995; Irwin 2007; Overdorff 1996b; Phillips 1995; Rodrigues de Moraes et al. 1998; Strier 1989; Wich et al. 2002; White 1989b).

Subsequent revisions of the competition model refined predictions regarding intergroup dynamics, most often relying on previously published data. Van Schaik (1989) revisited the concepts of between- and within-group competition and developed a more complex list of competitive regimes that included the possible strength of scramble and contest competition. Using these possible combinations, he made specific predictions about female intrasexual relationships based on food density, population density, and levels of between-group contest competition.

Isbell's (1991) variation on this approach also considered intra- and intergroup contest and scramble competition, but her model made more specific predictions about which ecological factors would influence competition. To test her model, she used data on day range and home range size for 20 primate species as proxies for food distribution and food abundance. After considering the variation in group size and ranging patterns, she concluded that food abundance shapes between-group relationships, while food distribution (clumped versus dispersed) shapes within-group relationships. One significant factor that sets her model apart from others is that outcomes are not dependent on levels of predation pressure. It is interesting to note that predation pressure becomes more or less a silent partner in many of the subsequent models: Either it is dismissed as an unimportant influence or it is recognized as important but not considered any further in deference to behavior (competition) and diet.

Sterck et al. (1997) offered one more refinement to the competition model by considering how competition levels vary along three discrete dimensions: (*1*) egalitarian versus despotic (see Vehrencamp 1983), (*2*) individualistic versus nepotistic (kin), and (*3*) the degree of social tolerance (as measured by the severity of aggression and amount of reconciliation observed). By including female dispersal patterns, they then created four categories in which primates can be grouped based on the degree of within- and between-group competition, which could be strong (high) or low (weak). These categories were dispersal-egalitarian, resident-egalitarian, resident-nepotistic (but also despotic), and resident-nepotistic-tolerant. Using an updated comparative mammalian database, the authors demonstrated that the predicted categories were supported by empirical data (e.g., Barton et al. 1996, Boinski et al. 2002).

There were, of course, some exceptions to predicted patterns, and two alternative explanations were provided to guide future work. One was that habitat fragmentation or saturation may lead to nonadaptive behaviors within populations (*behavioral disequilibrium*). This has subsequently been proposed as a possible explanation for the unique grouping patterns of Malagasy lemurs (van Schaik and Kappeler 1996, but see Kay and Kirk 2000, Kirk 2006). Another consideration is the role that sexual coercion (Smuts and Smuts 1993) or, more explicitly, infanticide plays in shaping female relationships. Sterck et al. (1997) hypothesized that infanticide avoidance may act as a strong selective pressure on female social behavior and may promote varying counterstrategies depending on the degree of perceived threat (see data from *Gorilla*, Watts 1989b, 1990a; *Presbytis*, Sterck 1999, Sterck and Steenbeek 1997).

FINE-TUNING

Although the "competition modeling" phase provided primatologists with a more unified perspective on primate social behavior, several major drawbacks remain (Isbell and Young 2002, Koenig 2002). One problem is the inconsistency in the primatological lexicon regarding social behavior categories. Despite early attempts at defining social systems, different researchers mean different things when they refer to "social structure," "social organization," and "mating system" (see Chapter 32). A second problem is that there is still a lack of consensus among primatologists with respect to which ecological variables should be considered, how they are measured, and which function as independent variables. Finally, there is still considerable focus on the proximate causes of behavior and not enough consideration of the ultimate, or evolutionary, causes.

The Importance of Words

Clear, operational definitions are a necessity when using the scientific method (Lehner 1996, Martin and Bateson 1993), yet primatologists over the years have been inconsistent in their use of definitions for varying aspects of social behavior and the ecological variables used within models. In earlier models, there was a tendency to portray social systems as a set of Russian dolls: Social organization (group size) determined social structure (social dynamics within groups), which in turn influenced mating system (van Schaik and van Hooff 1983). However, long-term work on populations with known individuals coupled with the recent ability to obtain paternity information through deoxyribonucleic acid (DNA) testing (see Chapter 24) demonstrates that this strictly linear approach is no longer appropriate. In many species, group size and composition vary seasonally; therefore, the subsequent mating system may not always be predicted based on group composition. Both monogamous and polygynous mating systems, for example, have been observed in "solitary" strepsirhines (Bearder 1999, Fietz 1999b, Mueller 1999). In other species, group composition and mating patterns can vary seasonally (Cords 2002b, Cords et al. 1986) or considerably between groups within populations, such as in *Propithecus* spp., where pair-bonded, unimale, and multimale groups are commonly observed (Pochron and Wright 2003, Pochron et al. 2004, D. J. Overdorff unpublished data).

Independent Variables

Authors also have used more subjective terms or categories when classifying social and feeding patterns. Primatologists seem to have a penchant for forcing variability into dichotomous categories. For example, food sources can be clumped or dispersed, social dominance can be strong or weak, and food patches can be small or large. Lack of consistent use of these terms between researchers and the less than quantitative way that these terms are used severely limit their application (Isbell and Young 2002). Additionally, there is a growing consensus that primatologists have relied on variables that are easy to measure (e.g., patch size) to the exclusion of others which may be harder to measure but are more important to our understanding of the ultimate mechanisms at work (Kappeler and van Schaik 2002, Koenig 2002). Energy exchange rates (as first proposed by Janson and van Schaik 1988) and actual reproductive success are key variables that primatologists need to measure more precisely. The "patchiness" of food not only is hard to determine but may not universally influence primate feeding patterns. Other feeding variables that could be more rigorously quantified, such as food depletion time or even the size of individual food items, may be more appropriate (Isbell and Young 2002; Overdorff 1996b; Pruetz and Isbell 2000; Vogel and Janson 2003).

ECOLOGY AND BEYOND

Janson (2000a) suggested that the business of primate socioecology as it has been conducted needs to be reassessed. By considering new paradigms and methodologies to think "outside of the box," some of the inconsistencies in primate social patterns might be better elucidated. To this end, the most recent wave of inquiry focuses on what have been traditionally considered rare occurrences or weak selection pressures, such as dispersal, predation, parasites, disease, and infanticide.

Dispersal

One of the basic assumptions behind the earlier competition models is that there is a strong association between female philopatry and female intergroup aggression. However, recent work with some species, especially colobines (Korstjens et al. 2002), indicates otherwise. Cooperation can occur in the absence of kinship. Also, female dispersal is more common in some species (especially platyrrhines) than previously thought. Consequently, Isbell (2004) proposed the "dispersal/foraging model" to explain variation seen in dispersal patterns not previously accommodated in other models. Instead of focusing on the inclusive benefits accrued by kin defending food sources, she explores how the locational (travel to unfamiliar home range) and social (harassment, see Pope 1992, Watts 1991c) costs of dispersal influence formation of kin groups. Foraging efficiency remains the main selection pressure, but success in intergroup competition is no longer the main benefit of living with kin. How well this new model holds up as a replacement for its predecessors depends on more ad hoc tests, which must include data on travel patterns (wandering versus goal-oriented), travel rates (slow versus fast), mother–infant tolerance, and dispersal patterns.

Predation

Although the topic of predation and its influence on primate social behavior is reviewed in more detail in this volume (see Chapter 35) and others (Gursky and Nekaris 2007, Miller 2002a), we will add that some of the more recent attempts to examine the influence of predation on primate social behavior are much more sophisticated in addressing previous gaps (Boinski and Chapman 1995, Isbell 1994). Improvements in understanding the dynamic between predators and their primate prey have included experimental studies of prey response to predator presence (i.e., playback calls, Gursky 2003b, Karpanty and Grella 2001) and studies that focus on predators. In Madagascar, it was believed that current predator pressure was minimal given the low diversity and small body size of potential extant predators. It had even been suggested that some of the unique behaviors exhibited by lemuriform primates, such as increasingly large group size in diurnal species and cathemeral

activity, are the result of the recent extinction of a giant eagle (*Aquila*, Csermely 1996, Goodman 1994). However, recent predator studies coupled with long-term studies on marked primate populations indicate that predation pressure is still very much an issue for lemuriforms (Britt et al. 2001b; Brockman 2003; Brockman et al. 2008; Colquhoun 2006; Dollar 1999; Goodman et al. 1993; Karpanty 2006; Karpanty and Goodman 1999; Wright 1998). The real future of understanding the interaction between predators and prey lies in studying both populations simultaneously.

Parasites and Disease

In the late 1970s, Freeland (1976) published a non-food-driven model to explain primate sociality that has been relatively overlooked until recently. His hypothesis was that primate sociality can be explained as an adaptation that minimizes the risk of acquiring new pathogens and limits the pathogenicity of diseases already harbored, thereby maximizing individuals' reproductive fitness. In Freeland's view, group size is limited by food availability and by infant mortality rates, which can increase as a result of the higher disease density brought about by the presence of too many group individuals. Group stability and fidelity to a home range are influenced by perceived disease risk. More stable groups with little turnover in membership are less likely to have new pathogens introduced. Site fidelity to a home range is also less likely to introduce members to new diseases. When primates do disperse to avoid inbreeding (and/or to counterbalance increased infant mortality due to higher disease density), the rate of dispersal and the acceptance of immigrants are influenced by perceived disease risk. Freeland also suggests that female mate choice and, more notably, promiscuity may play a strong role in regulating the numbers of males and their success upon entering a group. For example, males that survive the stress experienced during intense mating competition are likely to be "superior" (i.e., rapid, aggressive take-overs of unimale groups). In other cases, individuals attempting to enter a new group more subtly may be observed by resident females across a long transitional period, during which time their overall health can be assessed.

One of the drawbacks to this hypothesis is in devising ways to separate out the influence of food versus perceived pathogen risk; any one of the behavioral patterns described above could be parsimoniously explained by patterns of food abundance, density, and/or seasonality. Steps toward testing hypotheses regarding pathogen influences must include understanding the evolution of the primate immune system, identifying proxies that can be used to assess pathogen risk, and actual collection of parasites from feces, blood, and/or corpses (Stuart and Strier 1995). Much recent work has begun to tackle these issues (Nunn 2002, 2003a,b; Nunn and Altizer 2006; Nunn and Dokey 2006; Nunn and Heymann 2005; Nunn et al. 2000; Sauther et al. 2006; Snaith et al. 2008). The recent launching of a global database on mammalian parasites (Nunn and Altizer 2005) will go a long way toward helping us achieve a better understanding of the potential coevolutionary or symbiotic relationship between primates and parasites.

Infanticide and Intersexual Conflict

With an increasing number of long-term studies and with a wider range of species being studied, the number of observed infanticide cases by adult males and females has increased. Male infanticide has been observed or inferred in a number of species in all primate radiations (van Schaik and Janson 2000). Consequently, many primatologists have taken the stance that infanticide (or sexual coercion) exerts a strong selection pressure on behavior. This perspective has its supporters and its detractors and has become a divisive issue (Sommer 2000). Much of the controversy centers on whether or not one considers infanticide to be a product of selection.

Skeptics have taken aim at the infanticide database, which, in their view, is still considered weak evidence for sexual selection (Bartlett et al. 1993, Sussman et al. 1995). One of the major deficiencies is that there remains a small number of directly observed infanticides that fit the "sexual selection profile." Additionally, infanticide is often inferred when infants disappear, and inference should not be substituted for direct observation. The prevalence of observations in some species (such as *Presbytis*) and the rarity of instances in others (such as *Eulemur*) are also cited as evidence against considering behavioral responses to infanticide as a widespread primate adaptation. Also cited as problematic is the lack of substantial data documenting infanticidal counterstrategies. One final point considers the issue of how such behaviors are passed from one generation to the next (Sussman et al. 1995, Dolhinow 1999). For a behavior to be "selected for" and maintained in a population, it must be heritable and inherited by progeny. If infanticidal behavior were favored, such a scenario would lead to a form of runaway selection with increasingly aggressive males, which would eventually be a maladaptive strategy within a society. Instead, they suggest that infanticide is a by-product of generalized male aggression.

Responses to these criticisms emphasize that the database is, in fact, more robust than suggested (Borries and Koenig 2000, Crockett and Janson 2000, Palombit et al. 2000, Steenbeek 2000). Sommer (2000) points to a large number of incidences in *Presbytis*, *Pan*, and *Gorilla* that he considers support infanticide as an adaptive strategy (while other researchers do not), leading him to conclude that "What counts as unequivocal evidence is obviously largely in the eye of the beholder" (Sommer 2000:23). Therefore, it is up to the beholder to be explicit in his or her description of what happened and to which individual it happened and to be clear about why he or she interprets the behavior as infanticidal (or not). This is one of the benefits of the dialogue between those in favor of infanticide hypotheses and those opposed: It forces both sides to be more specific about actual observations and predictions.

Such dialogue also challenges primatologists to test predictions regarding infanticide in creative ways, such as identifying proxies for a relatively rare behavior. One example uses the ratio of lactation length to gestation length to estimate the vulnerability of infants to infanticide: When lactation is longer than gestation, infanticide should be more likely to occur (van Schaik 2000a,b). Using a broader mammalian database, van Schaik concluded that infanticide does appear to occur more often in species whose life history trajectory is such that a nonreproductive male would benefit by jump-starting a female's reproductive cycle via infant killing. Finally, actual documentation of changes in infant mortality and birthrate within populations as the result of infant killing as well as an individual infanticidal male's lifetime reproductive success compared to that of noninfanticidal males must be achieved to fully test the infanticide/sexual coercion hypothesis (Fedigan 2003). It is equally important to note that negative results that provide evidence against the presence of sexual coercion must also be equally weighted (e.g., *Ateles*, Campbell 2003).

Evidence for possible infanticidal counterstrategies is also mounting. Paternity confusion by females mating with multiple males (Hrdy 1979) has long been thought to be a viable counterstrategy to infanticide (see work by Borries et al. 1999b; Enstam et al. 2002; Heistermann et al. 2001b; van Schaik et al. 2000; Soltis 2002; Soltis et al. 2000a; Watts 2007). Additionally, Treves (2000a) speculates that there may be some self-protection strategies implemented by infants themselves, such as age deception, concealment, or simply stranger avoidance. Strong affiliative bonds between males and females in some species may be the result of protection from infanticide (Palombit 1996, 1999, 2000). Female dispersal (Sterck and Korstjens 2000, Isbell 2004), female alliances (Treves and Chapman 1996), and male–infant relationships (Paul et al. 2000) also may be shaped by infanticide risk.

One of the biggest drawbacks to the infanticide perspective is the risk of it becoming a "megaparadigm" (see Sommer 2000). Infanticide prevention currently seems to be the stock explanation for behavioral patterns when all others fail. Pair-bonding in gibbons (van Schaik and Dunbar 1990) and the presence of male–female dyads within several other primate taxa, such as *Lemur* and *Eulemur* spp., are such examples (Pereira and McGlynn 1997, van Schaik and Kappeler 1996).

Clearly, more data are needed to adequately test the infanticide-protection hypothesis to address the following: (*1*) What are the actual fitness costs and benefits to males performing infanticide across their lifetime? (*2*) How does male presence in pair-bonds actually influence infanticide risk? (*3*) How effective are counterstrategies to infanticide, and can these in fact be most parsimoniously explained by infanticide/sexual coercion hypotheses?

REMODELING

Primatologists continue to challenge the basic assumptions made by earlier models regarding the role that food or competition plays in shaping primate social systems, the nature of female relationships (Barrett and Henzi 2002), and the acceptance that there is not a typical or universal primate pattern (Strier 1994b). For example, Sussman et al. (2005) recently questioned the assumption that competition alone drives primate sociality. They proposed that cooperative and affiliative behaviors are equally likely to influence sociality given the higher proportion of time primates devote to these behaviors. Although some of these research avenues have generated heated debate (infanticide being one of the best examples), this is still the scientific process at its very best; paradigm shifts will not occur unless established paradigms are challenged forcefully (Kuhn 1962).

One of the continued obstacles primatologists face is determining ways in which to tease out the complicated effects of (*1*) easily documented (but not rigorously or systematically quantified) behaviors such as feeding, (*2*) rare and difficult-to-observe behaviors (due to observer affect) such as infanticide and predation, and (*3*) aspects of social living that are difficult to quantify (e.g., disease risk). The challenge to the current and future generations of primatologists is to devise creative and technologically savvy methods to tackle these issues. As with those who came before them, their efforts will help advance our knowledge of the socioecology of this eclectic and unique mammalian order, the primates.

3

Primate Evolution

Walter Hartwig

1. What is a likely time range for the origin of primates as a radiation of mammals? What data are used to determine this range?
2. If it is difficult to define all species of primates according to a common, shared, and derived anatomical complex, what does this difficulty itself convey about the position of primates relative to other mammal groups?
3. The different types of living primates did not all originate and diversify at the same time. What characterizes the types that appear to have diversified most recently compared to types that appear to have split apart very shortly after the order first radiated?

INTRODUCTION

Primates comprise an order of Mammalia. Although no single anatomical trait unites them, their "place" in the natural world is *ordered* in the taxonomic sense of the word. Primates are, with few exceptions, generalist foragers with relatively large brains and sophisticated social systems. They interface with ecology cognitively as much as biologically and are not the primary or usual target of predators. This "exemption" from the adaptive spiral of predators and their prey is just one of numerous interpretive themes that emerge from a study of primate evolution.

Because the fossil record is incomplete, large gaps of inference in primate evolution are spanned either by stretching deductive reasoning beyond its capacity or by accepting inductive scenarios that can never be tested. Scholars must attend the data themselves, approach all interpretations critically, and look beyond the constraints of the historical moment. Do not be afraid to challenge traditional interpretations, but also be aware that answers to major and very interesting questions (e.g., How did primates originate? and How did the earliest New World monkeys get to South America?) will always be beyond the reckoning of the fossil record.

Accounts of primate evolution should be constructed from multiple lines of evidence, but the end result must be able to include the fossil record. That is, the fossil record is an essential fabric of our understanding of primate evolution, however fragmentary or undersampled it may be. With this in mind, it is equally important to regard aspects of the fossil record for what they are, not what we hope they are. We tend to promote the current "earliest evidence" of a lineage as its ancestral population, only to revise when the next earliest fossil is found. Having studied the interpretive follies of their predecessors, students of the next generation have the opportunity to be more objective about rendering the fossil record into evolutionary accounts rooted in other databases, molecular in particular.

Perhaps the most important, and most obvious, context of the primate fossil record is how it relates to the living species. Primate fossils represent the tangible record of the entire radiation through time and space. Living primates, on the other hand, constitute a negligible amount of the total evolutionary diversity of the order. They are precious, indeed, and distinct among the thousands of extinct species by virtue of still being. So, while living primates must have a fossil record, there is no certainty that the fossils we find are closely related to living primates. This distinction is key to a meaningful understanding of both.

PRIMATE CLASSIFICATION

How we organize the living primates changes with time and is inherently subjective. We need classifications in order to communicate to each other about which groups of animals pertain to our studies, but we also need to recognize that no single classification scheme satisfies all biological perspectives of diversity, relatedness, and evolutionary history. The following classification draws from taxonomies that you are likely to see in primate databases and literature and is offered as a basic framework, not as the definitive reckoning of the order.

Table 3.1 Taxonomy of the Living Primates

ORDER	SUBORDER	INFRAORDER	SUPERFAMILY	FAMILY	SUBFAMILY	GENUS
Primates	Strepsirhini	Lorisiformes	Lorisoidea	Lorisidae	Lorisinae	*Loris*
						Nycticebus
					Perodicticinae	*Arctocebus*
						Perodicticus
						Pseudopotto
				Galagidae	Galaginae	*Galago*
						Euoticus
						Sciurocheirus
						Galagoides
						Otolemur
		Lemuriformes	Lemuroidea	Lemuridae		*Lemur*
						Eulemur
						Hapalemur
						Prolemur
						Varecia
				Indriidae		*Indri*
						Propithecus
						Avahi
				Megaladapidae		*Lepilemur*
				Cheirogaleidae		*Cheirogaleus*
						Microcebus
						Mirza
						Allocebus
						Phaner
				Daubentoniidae		*Daubentonia*
	Haplorhini	Tarsiiformes	Tarsiodea	Tarsiidae		*Tarsius*

Table 3.1 (*cont'd*)

ORDER	SUBORDER	INFRAORDER	SUPERFAMILY	FAMILY	SUBFAMILY	GENUS
		Platyrrhini	Ceboidea	Cebidae	Callitrichinae	*Callithrix*
						Cebuella
						Callimico
						Callibella
						Mico
						Leontopithecus
						Saguinus
					Cebinae	*Cebus*
						Saimiri
				Atelidae	Pitheciinae	*Pithecia*
						Chiropotes
						Cacajao
						Callicebus
					Aotinae	*Aotus*
					Atelinae	*Ateles*
						Brachyteles
						Lagothrix
		Catarrhini	Cercopithecoidea	Cercopithecidae	Colobinae	*Colobus*
						Piliocolobus
						Procolobus
						Semnopithecus
						Trachypithecus
						Presbytis
						Nasalis
						Simias
						Pygathrix
						Rhinopithecus
					Cercopithecinae	*Cercopithecus*
						Allenopithecus
						Miopithecus
						Erythrocebus
						Chlorocebus
						Macaca
						Mandrillus
						Cercocebus
						Lophocebus
						Papio
						Theropithecus
			Hominoidea	Hylobatidae		*Hylobates*
						Symphlangus
				Pongidae[1]		*Pongo*
						Gorilla
						Pan
				Hominidae[1]		*Homo*

[1] See text for explanation of why these two family designations are debated.

Differences of opinion exist at every level of this classification. Dividing primates into strepsirhine and haplorhine radiations emphasizes some of the anatomical similarities between tarsiers and the rest of the haplorhines (= anthropoids). The term "anthropoid" when used in the literature usually refers to monkeys, apes, and humans, whereas the term "prosimian" refers to lemurs, lorises, and tarsiers. Tarsiers resemble the other prosimians and anthropoids but in different ways. If classifications should reflect evolutionary history, Haplorhini and Strepsirhini may be the preferred division of the order. However, familiarity of referral in textbooks and the research literature is anchored to a traditional prosimian–anthropoid dichotomy, much like the dual existence of a metric system standard in science and the other one that is used much more commonly in all other settings (in the United States). The platyrrhine primates, or New World monkeys, are grouped into widely accepted subfamilies; but the placement of the subfamilies relative to one another at the family level is debated. For the Cercopithecoidea, the monkeys of Africa, Europe, and Asia, classification debates exist most dynamically at the genus level. Some primate radiations are subject to more extensive study and/or more biopolitical renderings of diversity. For example, a more realistic rendering of the gibbons (Hylobatidae) might divide them into three or four different genera given their spatial distribution and population isolation in peninsular Southeast Asia. The most obvious challenge to unite classification with phylogeny is found in how we refer to African apes and humans. To the extent that we are relatively confident that humans and chimpanzees are more closely related to each other than either is to gorillas, and likewise for the African apes relative to the orangutan, then a phylogenetic classification would need to tease apart the conservative Pongidae–Hominidae division. Pongidae would be limited to just genus *Pongo*, and Hominidae would have to include *Pan*, *Gorilla*, and *Homo*. Aggressive use of categories between family and genus would then sort out subfamilies Gorillinae from the Homininae (*Pan* and *Homo*) and then within the hominines would be the tribes Panini and Hominini. Because this phylogenetic classification radically alters the traditional use of the term "hominid," it has yet to replace fully the traditional classifications, in which "hominid" implies only living humans and their direct ancestors.

PRIMATE ORIGINS

How would we know the earliest primate if we held it in our hands? It would be a small mammal, given that all early mammals were small. It would have generalized gross anatomy, given that living primates today are the least specialized of all the mammalian radiations. Also, it would be 50–90 million years old, depending on which extrapolation we support. However, these are just attributes of what really interests us—the origin itself. We can never know the actual story, but historically theories have emphasized the physical characteristics that make primates different from other mammals: grasping thumbs and toes, stereoscopic vision. They have also assumed that only one aspect of early primate life, such as an arboreal habitat or foraging for insects by grabbing at them, could be the root of primate origins. We should think pluralistically as much as possible however.

Several recent publications (including a major new edited volume, Ravosa and Dagosto 2006) explore primate origins in terms of the fossil record, genetic data, and molecular clocks of evolution (Hamrick 2001; Rasmussen 2002a,b; Sargis 2002; Tavaré et al. 2002; Martin 2003; Soligo and Martin 2006). The fossil record poorly represents tropical latitudes (where primates probably originated), especially in the time ranges believed to encompass the major early eutherian mammal radiations. Rather, provocative concepts of when primates arose and to what they are most closely related draw upon analysis of genes and models of how long it takes genetic distances to "evolve."

The Molecular Data

With enough lines of genetic evidence, such as mitochondrial deoxyribonucleic acid (DNA), nuclear DNA, amino acid sequences, and chromosomes, and with hundreds of separate vertebrate species included, broad pictures of genetic relatedness emerge. With a certain leap of faith, the genetic "distance" between two major types of mammal can be calibrated to real time based on confidence that the fossil record for the origin of the two types of mammal can be trusted (Foote et al. 1999, Murphy et al. 2001, Steiper and Young 2008). This is sometimes referred to as the "molecular clock." Assuming that regions of the genetic code mutate at a constant rate, the "amount" of genetic difference between species can be measured as a unit of time; likewise, assuming that the fossil record accurately reflects some specific divergences between taxa, the combination of the genetic difference between those taxa and the presumed divergence date calibrates a "clock" for that analysis. While the relative genetic relatedness of orders of mammals is widely accepted, the accuracy of molecular clocks is widely disputed (Graur and Martin 2004).

Several different molecular studies suggest that primates may have originated 40 million years prior to the earliest known fossil primates (Arnason et al. 1998, Eizirik et al. 2001, Madsen et al. 2001, Murphy et al. 2001). The disconnect between the fossil record and the morphological record may be due to the lag between genotypic "roots" and phenotypic "branches," in the sense that genotypic markers of "primate-ness" may go much deeper in time than do recognizable primate morphological traits (Tavaré et al. 2002). Origin dates based on measurements of genetic similarity are likely to be the gold standard for the near future. The objective now is to narrow the ranges through replicating the analyses and expanding the samples.

Molecular studies also ally primates to other lineages, most frequently to tree shrews (Scandentia) and flying lemurs (Dermoptera) and less frequently to rodents, rabbits, and bats (Springer et al. 2003). The consistency with which these taxa fall close together in genetic comparisons suggests that they ultimately have a common ancestor.

The Comparative Data

The many theoretical constructs of primate origins have been based on how living primates may represent the current form of the ancestral adaptation or niche. A foundation of evolutionary biology that radiations are based on competitive exclusion requires that we identify what it is about primates, anatomically and/or behaviorally, that distinguishes them from other mammals. Circling around this definition will be species that converge on primate-ness and vice versa, thus presenting some amount of parallel evolution that we are willing to tolerate in our theory of primate origins.

Because most primate species depend on trees for survival, the first dominant theory of primate origins focused on adaptations for an arboreal habitat. The arboreal hypothesis argued that hallmark primate traits, such as stereoscopic vision and nails instead of claws, were basic adaptations to living in a three-dimensional and vertically stratified environment (see Rasmussen 2002a). This basic concept is logical but fails to explain the existence of other species which live in trees without these traits (e.g., squirrels, marsupials) or species with these traits which do not live in trees (e.g., stereoscopic vision in terrestrial carnivores).

Primate origin theories rooted in comparative anatomy were advanced by Cartmill (1974, 1992), who argued that a basic adaptation for visually preying upon small animals probably led the earliest primates to emerge from populations of other small insectivorous mammals. This would explain grasping hands and stereoscopic vision since no other mammals hunted in quite this same way. Lemelin (1999) and Hamrick (2001) studied proportional length of the extremities of several related nonprimate species and concluded that the unique proportions in primates conferred a selective advantage for exploiting the original fine branch/small food item–grasping niche. Sargis (2004) concluded that the grasping anatomy of some tree shrews reflects a model for the earliest primates, to which they are probably closely related.

A more inclusive ecological theory involving the contemporaneous rise of flowering plants, or angiosperms, was developed by Sussman (1991b), who argued that this primary food source revolution would have opened numerous niches to animals capable of reaching the flowering parts, with or without stereoscopic vision. Fruit bats may represent the evolution of a mammal that reached the target differently (through flight, not climbing) but evolved similar visual specializations. The visual predation/manual grasping theory and the angiosperm theory are good examples of arguments that tried to minimize the amount of parallel examples in evolution but could not eliminate them.

Rasmussen (1990, 2002a) saw the tropical marsupial *Caluromys* as convergent on prosimian primates in several biological and behavioral traits. Using it as an analogy for early primates, he argued for tightly linked coevolution among insects, flowering plants, and the mammals that preyed upon them. Given a proliferation of a small-branch milieu, the earliest primates may have radiated into a niche that incorporated aspects of both the visual predation/manual grasping argument and the angiosperm theory.

Thinking critically about primate origins is a good exercise in logical reasoning and the tenets of evolutionary biology. The nearest relatives of living primates must be considered in this process, for if the superordinal group known as Archonta (tree shrews, primates, flying lemurs) is legitimate, then many anatomical aspects of arboreality were in place at the time primates diverged from the other groups. The arguments are not resolvable, of course, but analyzing them helps to expose assumptions we make about evolutionary history and how much convergence or parallelism we are willing to tolerate.

The Paleontological Data

The search for early primates in the fossil record has focused on Tertiary sites in the Northern Hemisphere, given that they are more abundant, have yielded primate fossils historically, and fall within the time range once believed to be the source age of primate origins. Although no definitive fossils of modern eutherian mammal orders are older than the Cretaceous–Tertiary boundary (65 million years ago), future investigations likely will focus on the tropical latitudes and earlier time ranges, given the signals of the genetic data (Martin et al. 2007).

In general, the boundary of what to call a primate in the early fossil record has included a taxonomic group called Plesiadapiformes (see Szalay and Delson 1979, Fleagle 1999). These species radiated widely during the Paleocene of North America and thus were in the "right place at the right time" for early theories of primate origins. The species lacked definitive primate traits, such as a postorbital bar of bone, forward-facing orbits, and generalized dentition; but some aspects of their auditory canal anatomy and speculations about their grasping abilities kept them under consideration as primate ancestors. With the shift to studying molecules and to the tropics (Martin 1993) coupled with earlier and earlier evidence of definitive anthropoids in the fossil record, Plesiadapiformes were sidelined as formal members of the primate line (e.g., Covert 2002, Hartwig 2002). Recent discoveries (Bloch and Boyer 2002) have expanded our knowledge of this radiation, however, and lobby once again for a nearest-relative relationship to primates (Ravosa and Savakova 2004, Sargis 2004), as summarized by Kay (2003:840):

> ...signs now point to the possibility that some or all Plesiadapiformes may be more closely related to primates than to any other living order of mammals. That makes them stem primates, even though they lacked the shared-derived features that are hallmarks of crown primates—"euprimates."

Fossils recognizable as primates of modern aspect (euprimates) are found in diverse assemblages beginning in the Eocene epoch, dated to 45–55 million years ago. These are the adapids and omomyids of the Northern Hemisphere (Gunnell and Rose 2002, Rasmussen 2002a), the early primates from northern Africa (Rasmussen 2002b, Seiffert et al. 2003), and the early anthropoids *Eosimias* and *Phenacopithecus* from China (Beard 2002, Dagosto 2002, Beard and Wang 2004). The earliest representative of the euprimates may be *Teilhardina asiatica*, from deposits in China estimated to be 55 million years old (Ni et al. 2004). Elucidating the story of primate origins from the fossil record is really about collecting as much data (fossils) as possible in order to understand the deep past. A reasonable current working consensus about the search for early primate fossils can be found in the concluding statement of Tavaré et al. (2002:728):

> Direct reading of the known fossil record suggests that primates originated during the Palaeocene in the northern continents and subsequently migrated southwards. An alternative interpretation is that primates originated earlier in the poorly documented southern continents and expanded northwards when climatic conditions permitted.

Summary

A good general principle of consensus may be to return to what it means to be a primate. Because many primate species are omnivorous, primate origins may be predicated on a dietary approach more generalized than just angiosperm or insect predation. The earliest primates may have been as frugivorous as they were insectivorous, and a large part of their divergence from other mammals may have been due to foraging on fruits where no other mammal could go. Thus, the best explanation for primate origins may be one in which both habit (visual predation, manual grasping) and habitat (the slight outer branches of the forest canopy and undergrowth) drove the evolution of characteristic primate features.

Much of our understanding of primate origins reflects our sense of how they radiated afterward. Primates have diversified into a great variety of habits and habitats due to the potential of their persistent generality, both anatomically and behaviorally, compared to other mammal species that are more specialized into the predator–prey pyramid. Behavioral novelty without loss of versatility was a central theme in primate origins, even if we cannot pin the sequence of adaptations to a particular scheme of phylogeny or discovery in the distant fossil record.

PRIMATE ADAPTIVE RADIATIONS

Fossil prosimians, monkeys, and apes are the only direct records of primate ancestry, even if they are often fragmentary and difficult to interpret. Finding fossils cumulatively adds to our understanding of primate evolution. However, our desire to interpret new fossils into an existing sense of primate evolution often leads to fruitless debate since the old interpretive framework was based on a different sense of the fossil record. The intractable limitations of the fossil record mean that some localities that have produced numerous fossils exert a tremendous influence over our interpretation of primate evolution, even though we understand that no single locality can sample the complete diversity of species living at any one time.

The Paleogene (35–65 Million Years Ago)

The oldest known fossil primates date to the 40–50 million year time range. They were small-bodied, both nocturnal and diurnal, and in evidence in North America, Europe, North Africa, and Asia. A broad picture of two types of primate at that time can be argued from the similarities among the fossils. New discoveries continue to challenge this broad picture and further argue for Asia as a major area of early primate diversification (Table 3.2).

Historically speaking, some of the earliest discoveries of fossil primates were in exposures of the Eocene age (approximately 40 million years ago) in the northern latitudes (Europe and North America). These have come to be known as the Adapidae and Omomyidae (Gebo 2002, Gunnell and Rose 2002). The adapids bear a striking resemblance to extant Lemuridae, which are isolated on Madagascar. The omomyids are slightly smaller on average and bear a striking resemblance to the living nocturnal prosimians, especially *Tarsius*.

Despite being found only in northern latitudes, adapids and omomyids also have been linked to the origins of anthropoid primates (Fleagle and Kay 1994, Ross and Kay 2004). Although this radiation of extinct primates is historically significant, recent discoveries of ancient anthropoids have prompted new theories of the time, place, and ancestry of their origins.

In the last 20 years, numerous primate fossils dating to the middle or late part of the Eocene (40–45 million years ago) have been found in North Africa and Asia (see Covert 2002). The record from North Africa is sparse everywhere except at the productive El Fayum locality, where Elwyn Simons and colleagues have been collecting fossils for several decades. The earliest deposits there record fragmentary remains of species believed to be related to later Old World adapoids. From smaller localities in Morocco and Mongolia, dental remains resembling primitive primates may date to the 50 million year range, which would make them among the most ancient fossils attributed to primates. The most recent contender for earliest fossil primate is a species of *Teilhardina*, an Old World omomyid, found in

Table 3.2 A Geological Timeline of Key Fossils and Major Trends in Primate Evolution (Dates for Fossils Are Generalized)

GEOLOGICAL EPOCH (MILLIONS OF YEARS AGO, MYA)	NORTH AMERICA	EUROPE	ASIA	AFRICA	SOUTH AMERICA
Pleistocene (2.0–present)		Old World monkeys diminish, hominins arrive	Old World monkeys dominate tropical biomes of insular SE Asia; largest hominoid, *Gigantopithecus*, found in southern China and Vietnam	Old World monkeys and hominins proliferate on the continent, giant lemuroids radiate in Madagascar, hominoids limited to tropical belt	Poor fossil record except for large subfossil platyrrhines in Brazil and unusual species on Caribbean islands
Pliocene (5.0–2.0)		Old World monkeys proliferate, hominoids disappear	Poor fossil record, hominoid diversity probably declined	Old World monkeys greatly diversify, no obvious fossil pongids, earliest hominins (*Sahelanthropus, Orrorin*) at the Miocene–Pliocene boundary	No known primate fossils
Miocene (23.0–5.0)		*Dryopithecus* (10 mya), *Ouranopithecus* (9 mya) hominoids flourish in the middle and late Miocene, including the earliest known definite brachiator	*Sivapithecus* (9 mya) hominoids flourish in the middle and late Miocene, including a probable direct ancestor of living orangutans; no known Old World monkeys	*Proconsul* (20 mya), *Morotopithecus* (17 mya), *Victoriapithecus* (16 mya), *Kenyapithecus* (12 mya), and *Nacholapithecus* (15 mya), earliest Old World monkey; diverse radiation of large and small hominoids until late Miocene	*Homunculus* (16 mya), *Stirtonia* (13 mya), *Cebupithecia* (12 mya), and *Neosaimiri* (12 mya), diverse radiation in southern South America in early Miocene; numerous more modern-looking taxa from single middle Miocene site in Colombia
Oligocene (35.0–23.0)	Last surviving primate in North America	Remaining adapid and omomyid primates go extinct	Fragmentary anthropoid fossils in Myanmar	*Aegyptopithecus, Catopithecus*, and numerous other early anthropoids found in North Africa	*Branisella* (27 mya), Bolivia; earliest known fossil New World monkeys
Eocene (55.0–35.0)	*Notharctus* and *Omomys*, two different types of prosimian-like primate, adapid, and omomyid, flourish	*Adapis* and *Teilhardina*, two different types of prosimian-like primate, adapid, and omomyid, flourish	*Teilhardina* (55 mya), recently discovered earliest omomyid, in China; *Eosimias* (40 mya, China) and *Pondaungia* (35 mya, Burma) evidence early anthropoids	*Parapithecus* and *Algeripithecus* (36–40 mya), early anthropoids; *Saharagalago* (40 mya), earliest Lorisidae, all from northern Africa	
Paleocene (65.0–55.0)	Plesiadapiform fossils but no good candidates for earliest primate ancestor		South and SE Asia may be the most likely places to recover early primates	Localities are sparse, molecular data suggest that primates are not endemic African mammals	

deposits in China believed to be 55 million years old (Ni et al. 2004).

Because living prosimians are perceived to be more like primitive primates than are the monkeys and apes, the possible phylogenetic relationship of these early fossil primates to living lemurs, lorises, and tarsiers is often debated (see Fleagle 1999, Covert 2002). The lorisiform fossils (*Karanisia* and *Saharagalago*) recently discovered from El Fayum deposits believed to be 40 million years old (Seiffert et al. 2003) and the equally ancient tarsier-like fossils in China (Gunnell and Rose 2002) are among the earliest primate remains in Africa and Asia, and they suggest that the prosimian primates as we know them today had already diverged from one another 40 million years ago.

Origin and Evolution of Anthropoid Primates

Anthropoids differ from prosimians in having a relatively larger braincase, fewer teeth, and complete bony closure behind the eyes. Many aspects of their anatomy and behavior can be explained in terms of an emphasis on cognition, sociality, and learned behavior. The fossil record of the near

tropics in the 40–45 million year time range has grown dramatically in the last 10 years, but with it have come controversial early anthropoid fossils that, predictably, are raising as many questions as answers.

Provocative fossil evidence of small anthropoid primates from Asia is not new. *Amphipithecus* and *Pondaungia*, known from jaw fragments discovered decades ago, represent two such genera from probable Eocene deposits in Myanmar (see Beard 2002). A majority of researchers accept them as early anthropoids, but new fossils of these and other taxa from the same area continue to fuel the debate (Ciochon and Gunnell 2002, Takai et al. 2001, Shigehara et al. 2002, Marivaux et al. 2003, Ross and Kay 2004).

Over the last 10 years, a major survey effort in China has yielded fragmentary remains of what may be the oldest definitive anthropoid, *Eosimias* (Beard 2002, Beard and Wang 2004). This taxon is smaller than any known anthropoid (<200 g) and comes from deposits believed to be about 40 million years old. Jaw fragments and postcrania attributed to the genus show similarities to both adapids and later anthropoids. Given that it is difficult to define a shared and derived anatomical trait in early anthropoids, support for calling *Eosimias* an anthropoid depends largely on which aspects of its anatomy are emphasized. The same holds for all claims of early anthropoids. The most recent discoveries include several new genera, which, like many early anthropoid fossil species designations, are based only on isolated teeth (Beard and Wang 2004, Bajpai et al. 2008).

The Egyptian locality of El Fayum has dominated our understanding of early anthropoid evolution because of the numerous genera found there under exacting stratigraphic control for several decades (Simons and Rasmussen 1994). Historically, the fossils could be grouped into a primitive branch (the Parapithecidae, with three premolars) and an advanced branch (the Propliopithecidae, with two premolars), which reflected the roots of living New World monkeys and living catarrhines, respectively. More recent cranial and dental discoveries, however, such as *Arsinoea* and *Proteopithecus* from lower in the stratigraphy at 36 million years ago, have shown that earlier anthropoids do not fall so hopefully into the ancestries of living monkeys.

Perhaps the most tantalizing early anthropoid fossil recovered at El Fayum is *Catopithecus,* which demonstrates a combination of primitive (unfused mandible) and advanced (full postorbital closure, two premolars) features that upset traditional divisions among early anthropoids (fused mandibles), platyrrhines (retention of three premolars), and catarrhines (two premolars). In this case, much of the cranial anatomy of *Catopithecus* is known; and in general aspect, it resembles expectations of an early catarrhine (Rasmussen 2002b). Its mosaic anatomy and well-dated age of 36 million years push again for a critical sense of how old we think primate lineages really are and how we believe anatomy evolves. As more fossils are uncovered at El Fayum, the picture of early primate evolution changes accordingly.

The Neogene (2–35 Million Years Ago) and Quaternary (0–2 Million Years Ago)

South America

How primates got to South America is one of the great questions of primate evolution. Assuming that anthropoids originated in the Old World, no scenario involving plate tectonics, rafting, vicariance biogeography, or island hopping is well supported (Hartwig 1994). The implausibility of a Tertiary migration of monkeys to the New World indirectly supports a very early origin of anthropoids, at a time when the southern continents were much closer together. However, given that primates and rodents are the only major Old World lineages to appear in South America, at approximately the same time according to the known fossil record, perhaps a miraculous transport via floating mats of vegetation is the only explanation, however fantastic it seems today.

The fossil record of New World monkeys is sparse but growing steadily (Rosenberger 2002). Fossils are known from southern South America, Bolivia, Chile, Colombia, and Brazil. The original source deposits in Colombia (Takai et al. 2003), Bolivia (Takai et al. 2000), and Argentina (Tejedor 2002, 2003) continue to be surveyed productively. New frontiers of discovery include western Amazonia (Kay and Cozzuol 2006), the Caatinga of eastern Brazil (Cartelle and Hartwig 1996, Hartwig and Cartelle 1996), and the Caribbean islands (MacPhee and Horovitz 2002, 2004). The record is relatively immature in the sense that each new discovery tempts the naming of a new ancient species, but it is developed enough to portray a broad outline of how New World monkeys radiated into their present distributions.

The earliest South American monkeys known so far have been found in Bolivia and southern South America, date to approximately 26 million years ago, and do not resemble any living forms (Takai and Anaya 1996, Fleagle and Tejedor 2002). Numerous small and fragmentary remains from Argentina and Chile dated to 15–25 million years ago constitute several species with some basic affinities to pitheciines and suggestive affinities to squirrel monkeys (*Saimiri*) and owl monkeys (*Aotus*). However, in general, the anatomy that is known is too derived to be ancestral to existing lineages. Rather, it seems more likely that these species are ancient offshoots of an as yet undiscovered ancestral platyrrhine lineage.

By contrast, the fossil New World monkeys from the Miocene of Colombia bear distinct similarities to living species. Dating from 12–16 million years ago, the fauna from La Venta, Colombia, includes more than a dozen different platyrrhine taxa (Kay et al. 1997b). Potential ancestors of the lineages of *Alouatta* (*Stirtonia*), *Saimiri* (*Neosaimiri*), *Aotus* (*Aotus dindensis*, known mostly from dental and lower jaw remains), and *Pithecia* (*Cebupithecia*, known from a partial skeleton) have been nominated (Hartwig and Meldrum 2002).

The very productive Miocene locality of La Venta has been a boon to our awareness of extinct New World monkeys, but it also may distort our perspective of platyrrhine evolutionary history by virtue of being the only locality representing a major chunk of time and space. To the same extent that El Fayum stands for most of the early anthropoid evidence from Africa, our sense of how diverse New World monkeys were over the last 15 million years is largely a function of discovery within a single locality. Undeniably, the pattern of New World monkey evolution is conservative, but we must remember that vast areas of their current distribution are not fossiliferous and that large time gaps (the last 10 million years, for example) in their known fossil record persist.

The Quaternary fossil record holds a special interest because it documents biodiversity during the time of humans on the landscape (the last 2 million years). This time period also includes the well-documented global "ice ages," gradual but radical shifts in prevailing climatic conditions. Quaternary New World monkeys have been found on several Caribbean islands and in arid caves in eastern Brazil (MacPhee and Horovitz 2002). They are most interesting because they document anatomies that cannot be accommodated easily by existing phylogenies. Yet, they are the most "recent" examples of past biodiversity, so their uniqueness gives us reason to be cautious in using the known fossil record as a template of the past.

Two taxa from the Caribbean islands, *Xenothrix* and *Paralouatta,* continue to vex the experts because their cranial and dental anatomy can be interpreted as support for a synapomorphic or convergent relationship to living species. The debate about their phylogenetic affinity (Rosenberger 1977, 1992, 2002; Horovitz and MacPhee 1999; MacPhee and Horovitz 2002, 2004) is an excellent capsule of how primate comparative anatomy can be analyzed competently toward opposing conclusions.

Another member of the Quaternary platyrrhine class, *Protopithecus* from Toca da Boa Vista in Brazil, merits a deeper look for the sake of its nearly complete preservation (Hartwig and Cartelle 1996, MacPhee and Horovitz 2002). We often lament having to reconstruct a phylogeny from bits and pieces of individuals, but when an entire skeleton lies before us, we sometimes are equally frustrated. In this case, the largest known platyrrhine (20 kg, as estimated from the skeleton) presents a body very similar to the suspensory spider monkeys (*Ateles*), a cranium that is distinctly similar to *Alouatta*, and a face that is like neither. Does it belong on the atelin clade or the alouattin clade? Our temptation is always to link fossil forms as closely to living forms as possible, but even with this complete skeleton, it is not clear. The enduring lesson of *Protopithecus* may be that parallel evolution in closely related lineages is the rule, not the exception that we have tended to think it is.

Africa

From Oligocene deposits at El Fayum come large collections of definitive early catarrhine primates, such as *Aegyptopithecus* (see Fleagle 1999, Rasmussen 2002b). They are a good study in what a primitive Old World catarrhine might look like, before the time Cercopithecoidea and Hominoidea separated from one another. This glimpse is brief, however, and limited only to this location. After that, a major void in the fossil record leaves us guessing about the actual split of Cercopithecoidea and Hominoidea, both temporally and geographically.

Despite the fact that living cercopithecoids are relatively successful compared to other primate groups, their fossil record is limited to a few early specimens from the Miocene of Africa (Benefit and McCrossin 2002) and many specimens from the recent past (Pliocene and Pleistocene) in Africa, Europe, and Asia (Jablonski 2002). The fossil record of Cercopithecoidea also tracks a general trend in primate evolution in which hominoid taxonomic diversity dwindled over the last 10 million years while cercopithecoid diversity expanded, perhaps as a result of more temperate climates replacing more tropical ones over large areas of the Old World.

The fossil record of hominoid primates definitively reaches back to a landmark fossil from the early Miocene of the African Rift Valley, *Proconsul* (Harrison 2002). Like *Aegyptopithecus*, *Proconsul* "fits" expectations of an early member of a radiation, in this case the radiation of hominoids. It lacks the apparent dental specializations of the later cercopithecoids, and its skeletal configuration is primitive enough to have given rise to the unusual limb proportions of the living hominoids. The combination of its generalized anatomy and its early discovery in history makes it a strong reference taxon for the wide variety of other early through middle Miocene East African hominoid fossils.

Recent discoveries in Uganda that date to approximately the same time period as *Proconsul* are shedding new light on our sense of the basal hominoids. *Morotopithecus* (MacLatchy 2004) looks less like a base hominoid and more like a derived one, although this statement is qualified by the fact that the entire taxonomic radiation is based on preservation of primitive anatomical traits. The discovery of *Morotopithecus*, much like that of the early catarrhine *Catopithecus*, forces us to reconsider both when hominoids emerged and how much our sense of hominoid evolution has been influenced by thinking of *Proconsul* as its root.

Hominoids dominate the Miocene primate fossil record in East Africa, as exemplified by other "index" taxa such as *Kenyapithecus* (Ward and Duren 2002). Very well-preserved specimens and new taxa, such as *Nacholapithecus* (Ishida et al. 2004), ensure that any synthesis of early African hominoid evolution is subject to quick revision. However, even as the diversity of African hominoid fossils continues to grow in the flush period of 12–20 million years ago, the signs of general shrinking of forest cover, proliferation of semiterrestrial monkeys, and decline of hominoid diversity at the close of the Miocene (5–12 million years ago) are unmistakable.

Eurasia

It is somewhat artificial to divide the fossil records of Africa and the rest of the Old World. Hominoids and cercopithecoids radiated widely throughout Europe and Asia shortly after they appear to have emerged in Africa. Their fates there seem to have been similar in the sense that cercopithecoids eventually succeeded and hominoids are now relegated to the still dwindling tropical forest cover in peninsular and insular Southeast Asia. The Eurasian record does include one Miocene radiation all its own—the enigmatic pliopithecoids.

Fossils that are hard to interpret nonetheless make excellent study samples for evolutionary biology and systematics. They are hard to interpret, presumably because they conflict with what either the theoretical bases or the data on hand, or both, lead us to expect. One of the best examples of such a radiation in the primate fossil record is the group of hominoid-like pliopithecoids from the Miocene of Europe and mainland Asia (Begun 2002a).

Pliopithecoids are united to each other almost by default, i.e., they lack definitive traits that put them into the ancestry of the "mainstream" hominoids. They were catarrhines that may have radiated out of Africa before the root populations of later hominoids. They clearly found opportunities in the rich Miocene environments of Eurasia, but then suffered the same fate as the hominoids when those environments shriveled. At times, pliopithecoids have been considered primitive hominoids. However, as is so common with the fossil record, eventually their presumed descendents were found in equally old deposits and their evolutionary position returned to enigmatic.

Hominoids spread widely throughout the southern latitudes of Europe, Asia, and Southeast Asia beginning about 15 million years ago (Begun 2002b, Kelley 2002, Kelley et al. 2008). Eurasian hominoids such as *Dryopithecus*, *Ouranopithecus*, and *Sivapithecus* have been promoted as ancestral to every combination of pongids and hominids, suggesting at the very least migrations to and from Africa instead of just from it (Kordos and Begun 2002). After a zenith in the middle Miocene, however, hominoid diversity dwindled steadily until the fossil record virtually disappears at the end of the Miocene (Agusti et al. 2003). The living pongids and some Pleistocene fossils are the only evidence of the once predominant radiation.

Recent fossil discoveries of *Dryopithecus* and *Oreopithecus* have demonstrated that ape-like brachiation may date back to the earliest descendants of *Proconsul*, particularly in the form of a very long-armed skeleton of *Dryopithecus* from Europe (Begun 2002b). The locomotor repertoire of Miocene hominoids influences the probable scenarios for the ancestry of our own locomotion, which has long been a central debate in human evolution (Nakasukasa 2004).

While the Miocene hominoid fossil record is rich in some ways, it is poor in others. Hominoids clearly radiated throughout Asia and into Southeast Asia and neighboring islands, but the critical period between the end of hominoid diversity and the emergence of human-like primates, probably 5–8 million years ago, is poorly sampled.

PHYLOGENIES AND LINEAGE RECONSTRUCTIONS

This book is concerned with the behavior and natural history of living primates. Grounding the different species and families in an evolutionary background helps us to make sense of the similarities and differences across taxa. The relatedness of species groups back through time is their *phylogeny*, a "family tree" in loose terms. Advances in molecular biology have enabled us to resolve primate family trees in a way that comparative anatomy and the fossil record never could.

Evolution of Living Prosimians

Phylogeny

The prosimian primates include the lemuroids, lorisids, galagids, and *Tarsius*. *Tarsius* has a dry rhinarium, as do anthropoids. This trait, along with others, has led taxonomists to distinguish the Strepsirhini (lemuroids, lorisids, and galagids) from the Haplorhini (*Tarsius* and the anthropoids). Thus, phylogenetically, *Tarsius* may be more closely related to anthropoids than to the strepsirhine primates (Fig. 3.1). The prosimian association used here is based more on a composite sense of their behavioral ecology and niche position relative to anthropoids and other mammals.

The separation of anthropoids, *Tarsius*, and the strepsirhines probably followed shortly after the emergence of primates as a distinct form of early mammal. The fossil record now supports a separation of the lemuroid lineage from the lorisid and galagid lineage as early as 40 million years ago, and fossil tarsiers date from approximately the same time. Molecular estimates push the dates even earlier (Tavaré et al. 2002, Yoder and Yang 2004). With this as

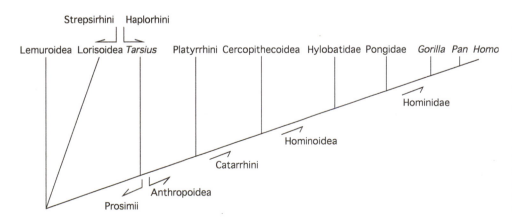

Figure 3.1 A macrophylogeny of living primates. This diagram applies higher taxonomic category terms to a standard cladogram design in order to express conventional interpretations of the relative time at which complementary categories arose or diverged from one another. Implicit in the diagram is the belief that major taxonomic groups of primates (e.g., anthropoids or catarrhines) experienced sequential dichotomous divisions (into platyrrhines and catarrhines as well as cercopithecoids and hominoids, respectively).

a benchmark, it appears that the three prosimian lineages have deeply exclusive evolutionary pathways.

Adaptive Radiations

Because they are found only on the island of Madagascar, the Lemuroids are a natural target of evolutionary investigation. They include examples of the smallest living primate, *Microcebus* (30–60 g), and analogs of woodpeckers, rabbits, pandas, raccoons, monkeys, and apes among the ecological profiles of the five major family groups: Cheirogaleidae, Lemuridae, Lepilemuridae, Indriidae, and Daubentoniidae (Fleagle and Reed 2004). In the recent past, they radiated into a spectacular variety of large-animal niches, until climatic change and human arrival drove them to extinction (Godfrey and Jungers 2002).

It is tempting to portray the Malagasy strepsirhines as "living fossils" or as the best examples of primitive primates, given their skeletal anatomy, reliance on olfaction, and relative brain size to body size compared to anthropoids. It is more likely that Lemuroidea became functionally isolated on a large island early in its evolutionary history and differentiated according to the currents of natural selection that have prevailed (see Goodman and Benstead 2003). In the absence of strong competition from other mammals, lemuroid populations have "succeeded" as much as could be predicted (and as much as have the mainland African prosimians or anthropoids in terms of taxonomic diversity). They have paralleled the evolutionary pathway of anthropoids in terms of occupying similar niches on Madagascar, but as a group they have remained below the anthropoid "grade" of neural complexity in doing so.

The strepsirhines of mainland Africa and Asia are the nocturnal Lorisidae and Galagidae. They are found in tropical forest settings, small in body size, and committed to arboreal habitats. Lorisids are found in Africa and Asia, but galagids are found only in Africa. Unlike the Malagasy strepsirhines, they have evolved in direct association with other primate and mammalian competitors. This may explain why they remain nocturnal, solitary, small-bodied, and limited to tropical forests (see Seiffert 2007 for a recent review).

Tarsiers (*Tarsius*) are limited geographically to Southeast Asia but play a pivotal role in our understanding of primate evolution and systematics (Wright et al. 2003). They are in many respects a typical prosimian primate—nocturnal, small-bodied, insectivorous, arboreal, and tropical (Gursky 2002b). However, genetically and anatomically, they share special similarities to anthropoids, such as a haplorhine (dry skin) rhinarium, often designated as "advances." For example, although they are nocturnal, their adaptation derives from a retinal fovea characteristic of diurnal primates rather than the reflective eye layer (tapetum lucidum) found in strepsirhine primates. Like most primates, *Tarsius* is highly endangered because it depends on tropical forest for its habitat.

Evolution of New World Monkeys

Phylogeny

Taxonomy and phylogeny weave together well in the Platyrrhini compared to other major primate groups. The phylogenetic integrity of the four major clades is consistent with how the animals have been classified based on morphology, molecules, and ecology. The monophyly of the Atelinae (*Alouatta, Ateles, Lagothrix, Brachyteles*), Pitheciinae (*Pithecia, Chiropotes, Cacajao*), Cebinae (*Cebus, Saimiri*), and Callitrichinae (*Callithrix, Cebuella, Saguinus, Leontopithecus, Callimico*) is supported by numerous genetic and morphological studies (Schneider et al. 2001). The "problem" taxa in phylogenetic reconstructions, both morphological and molecular, consistently have been *Callicebus* and *Aotus*. In this volume, *Callicebus* is included in the Pitheciinae and *Aotus* is placed in its own subfamily (Aotinae).

Adaptive Radiations

The tight fit of the genera within their subfamilies and the relatively deep genetic distance between subfamilies suggest that modern platyrrhines are the result of a rapid burst of evolutionary differentiation followed by sustained gradual change. As with the Malagasy strepsirhines, it seems likely that once the ancestral population of platyrrhines occupied the isolated landmass of South America, they radiated quickly and widely into distinct basic niches. From there, the lineages have evolved subspecialties within those superniches, preserving a close genetic relationship (see Norconk et al. 1996a).

All platyrrhines depend on trees. They have evolved to exploit different nutrient zones within the tropical forest as well as different vertical habitats within the same zones. Some genera, such as *Cebus* and *Alouatta*, have dispersed widely but for different reasons. *Alouatta* forages on low-quality foliage and so can range into more marginal forested habitats. *Cebus* is the most "cognitive" of the New World monkeys and may range widely because of its very flexible and fluid resource-exploitation strategies. Other genera, such as *Brachyteles*, found only in the narrow Atlantic Coastal Forest, are helpless victims of specific habitats that are diminishing around them. Platyrrhines are relatively small anthropoids, ranging in body size from the smallest of all anthropoids (*Cebuella*, the pygmy marmoset, at slightly over 100 g) to the 7–10 kg atelines.

The atelines are the large-bodied New World monkeys that occupy the uppermost canopies of the forest and exploit ripe fruits and leaves at the periphery. These conditions have led to their unique limb proportions for suspensory posture and brachiating locomotion—much like some of the hominoids. The pitheciines have radiated into a hard object–feeding niche that Rosenberger (1992) refers to as "sclerocarp foraging"—working through the tough outer coat of certain fruits and seeds to get to the nutritious pulp on the inside.

The callitrichines have radiated into a lower canopy habitat zone better suited to small body size, vertical clinging and leaping, and, in the case of *Callithrix*, specialized foraging on nectar and tree gum. The cebines are omnivorous, highly social, and significantly more encephalized than are other New World monkeys (Hartwig 1996).

Evolution of Old World Monkeys

Phylogeny

The two major clades of Cercopithecoidea, the Cercopithecinae and the Colobinae, are each monophyletic as determined both morphologically and molecularly (Hoelzer et al. 2004). In this case, the molecular data have revealed unexpected close relationships between species groups that do not share obvious morphological similarities. For example, the West African mangabeys were thought to be a monophyletic assemblage of two species groups, *Cercocebus* and *Lophocebus*. Molecular studies and revisited morphological studies confirm, however, that the *Cercocebus* group is more closely related to the West African drills and mandrills (*Mandrillus*), long thought to be near relatives of the baboons. Likewise, the *Lophocebus* group is more closely related to the baboons than it is to the other *Cercocebus* group (see Fleagle and McGraw 1999, 2002). The unfolding of these relationships is one of the best examples of why genetic reconstructions of phylogeny should be the bases upon which morphological studies construct and test hypotheses.

Adaptive Radiations

The cercopithecoids have exploited the more generalist habitat zones of both the tropical forest belts and the outlying savannas of Africa and Asia, as reflected in their classic quadrupedal skeletal profile that enables them to move effectively in both terrestrial and arboreal habitats. They are arguably the most successful primate radiation in terms of geographic distribution, numbers of species, and population sizes (see Whitehead and Jolly 2000).

At some point shortly after the divergence of catarrhines, the cercopithecoids diverged along two basic adaptive pathways. The cercopithecine Old World monkeys, exemplified by *Papio* and *Macaca*, stayed the course of the "generalized primate" in terms of anatomy and resource exploitation. They are the most widely dispersed of all primates. The colobines, by contrast, derived into a superniche based on more concentrated exploitation of local leafy food materials. As exemplified by *Colobus* and *Presbytis*, they are more committed to arboreal habitats, in general. They have speciated widely in the lush Southeast Asian and insular Asian tropical forests.

Evolution of Hominoids

The living pongids and hylobatids are instantly recognizable: chimpanzees, bonobos, gorillas, orangutans, and gibbons. However, these complex primates are just a fraction of the hominoid diversity that once stretched throughout the Old World tropics. This evolutionary context is key to understanding why they are so habitat-dependent and adaptively "overextended." In each case, the degree of specialization necessary to survive as a large-bodied, arboreal, highly cognitive, mostly frugivorous primate has cost them the anatomical variation and rate of reproductive turnover necessary to survive the loss of their immediate habitat.

Phylogeny

The basic assumption that hylobatids were the first branch of the hominoid family tree, *Pongo* the second, then some form of African pongid and human trichotomy has dominated depictions of hominoid phylogeny for several decades (Goodman et al. 2002). Molecular studies confirm that the Asian hominoids are less closely related to humans than are the African hominoids. Within the African clade, the preponderance of data suggest that *Homo* and *Pan* are more closely related to each other than either is to *Gorilla*. This implies that morphological complexes thought to be synapomorphic between the African great apes, such as the knuckle-walking complex, either evolved in parallel between them or were present in the hominin ancestor and then reversed completely (e.g., Begun 2004). Indeed, the important lesson of advances in molecular genetics is that anatomical traits probably converged, evolved in parallel, and reversed much more often than we ever suspected.

Adaptive Radiations

Within the bounds of the general ape superniche, the hylobatids (gibbons and siamangs) have derived ultraspecializations for a brachiating lifestyle in the dense tropical forests of Southeast Asia. They effectively have lost the grasping ability of their hands in order to swing aggressively from their elongated upper limbs. They are the most speciose of the hominoids, with two major clades, *Hylobates* and *Symphalangus*, derived along subtle differences in dietary emphasis on ripe fruit and leaves.

Pongo is the other Asian hominoid and is restricted to the islands of Borneo and Sumatra, undoubtedly due to the aggravating effects of human encroachment. *Pongo* is virtually omnivorous, climbs quadrumanously, and is perhaps the most vertically perambulatory hominoid (Knott 1999a). Its radiation would dominate the Asian tropical scene were it not for the faster-reproducing and less specialized cercopithecoids that radiated along with it.

The vestigial hominoids of Africa (excluding humans) are *Pan paniscus*, *P. troglodytes*, and *Gorilla*. As cognitive and complex as they are behaviorally, both *Pan* and *Gorilla* are on the rapid path to extinction. They are what remains after millennia of competition with African cercopithecoids, saddled the whole time with a slow reproductive rate and a morphological specialization for a life spent in or at least among the trees.

Gorilla probably represents the surviving hominoid that specialized into a mostly herbivorous, deep forest, reclusively arboreal niche (Taylor and Goldsmith 2003). *Pan* probably represents the surviving hominoid that stayed the course of moderate body size, multivory, and anatomical specialization to life in the trees. Out of this picture emerged humans, of course, probably as the surviving hominoid that exploited the patchy ecotones on the ever-expanding fringe of the forest, eventually specializing for a novel form of bipedal locomotion. As of now, the options were more productive for the human lineage and significantly more restrictive for *Pan* and *Gorilla*.

A SYNTHESIS

Through the fossil record, comparative anatomy, and molecular biology, we can reconstruct major aspects of how primates evolved and how living lineages relate to one another. The dates and places ascribed to major events in primate evolution continue to change as we collect more data and refine our analyses, but the broad patterns remain. Plurality rules the primate story. Primates exploit plural habitats, omnivorous diets, and fluid social group dynamics. A highly specialized primate species seems even more odd as we learn more about it, and the variation we have observed in other species expands by the same measure of further study.

In general, primates are a radiation of mammals in which growth of structures related to the brain has taken precedence over growth of structures related to the body. Comparatively speaking, other lineages of mammals have "spent" their genetic variation on more specialized biological adaptations—complex anatomies for preying upon their food and/or to avoid becoming prey. For primates the evolutionary path has stayed a general course in which neural development, rather than somatic, has characterized the emergence of new taxa (see Bush and Allman 2004). In the broadest possible context, the later a lineage appears, the more encephalized it and its descendents appear to be.

Primates have evolved to exploit rich but ephemeral food resources in well-protected tropical forest microhabitats, such as terminal branches and emergent canopies. From this core superniche, primates radiated throughout the tropical zones of Africa, Eurasia, and the Americas over the last 50 million years. Within each lineage, more specialized species groups radiated into narrower niches as determined by the balance of opportunity and competition. Madagascar offered more opportunity and less competition to lemuroid strepsirhines, for example, than did continental Africa to lorisid and galagid strepsirhines. The generalized species groups, such as *Macaca* and *Papio*, have survived major biome shifts in forest cover and thrived in the balance. Hominoids, dominant for millions of years in a stable forest milieu, enjoyed almost runaway selection for cognition, learning, and extended life histories—all luxuries of a certain independence of the predator–prey cycle. In the face of a constrained milieu and unable to undo their life history pattern, hominoids evolved as expected: Some lineages went extinct, some linger in the fragile remains of the original niche, and one abandoned its biological capacity completely in favor of its cognitive potential.

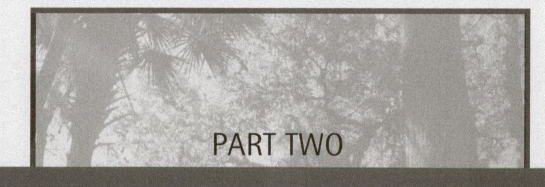

PART TWO

The Primates

4

The Lorisiform Primates of Asia and Mainland Africa

Diversity Shrouded in Darkness

K. Anna I. Nekaris and Simon K. Bearder

1. What methods have researchers employed to discover new galago, potto, and loris species?
2. What are the major behavioral features that distinguish galagos from pottos and lorises?
3. What aspects of the behavior and diet of lorisiforms are characteristic of their nocturnal activity patterns, and what aspects are more characteristic of diurnal primates?

INTRODUCTION

The primates known as galagos (or bushbabies), pottos (angwantibos and pottos), and lorises could easily vie for the position of "least known of all the primates." Despite the fact that the suborder Lorisiformes contains some of the most specialized primates, with a minimum of 34 species now recognized, some irresistible urge seems to possess the authors of textbooks to summarize what is known of this group in a hasty postscript to a chapter on their close cousins, the lemurs. One reason for this is that, unlike most lemurs, different taxa of lorisiforms can look very similar to one another (*cryptic* species), and for a long time they were misclassified as a few species and assumed to have little variation in genetics, behavior, and ecology. It is now known that superficial similarities are partly the result of extensive convergence due to the demands of nocturnal and arboreal niches and partly because members of each species recognize each other by more subtle visual, vocal, and olfactory signals. In this chapter, we show that the nocturnal strepsirhines of Asia and Africa are a diverse group of primates and represent an untapped resource for the aspiring field biologist. One-sentence synopses, steeped in the literature of the 1960s, branding this enigmatic group as no more than acrobatic leapers and slow creepers (e.g., MacDonald 2001) seem to have hindered interest in their study in the wild, even though early biologists recognized great variability within this group (e.g., Gray 1863). The lorisiforms display a multitude of social systems, life histories, and locomotor strategies, a diversity evident despite the fact that only a handful of species have been studied in detail.

The lorisiform primates are widely dispersed in Africa (excluding Madagascar), southern Asia, and Southeast Asia. The relatively few long-term field studies that have been published (Fig. 4.1) on the galagos, pottos, and lorises are summarized in Table 4.1. Detailed behavior and ecological data are available for only 18 species, fewer than half of those currently recognized. In some cases where researchers have set out to study behavior, their projects were confounded by the discovery of too many new species (e.g., Honess 1996, Ambrose 1999). Instead, these studies have led to extensive useful descriptions of the presence or absence of species across a large geographical range, with morphological data gathered from trapping regimes (Oates and Jewell 1967; Honess 1996; Ambrose 1999; Perkin 2000, 2001a,b, 2002; Perkin et al. 2002). Furthermore, despite advances in radiotracking, only nine species have been studied with this technology (Table 4.1) and only two studies have been able to take advantage of recent advances in molecular ecology (Pullen 2000, Pimley 2002). Clearly, an enormous avenue for research exists within this group.

Even what might appear to be the most fundamental questions regarding the evolutionary relationships among this group are far from resolved (Rasmussen and Nekaris 1998). For example, no consensus has yet been reached as to whether the pottos and lorises form a monophyletic clade to the exclusion of the galagos or if they form one of the most spectacular examples of parallel evolution among primates (Yoder et al. 2001b, Roos et al. 2004, Masters et al. 2005). Recent fossil discoveries have added new vigor to debates regarding the origins of the lorisiforms. Some authors contend that they may be among the most

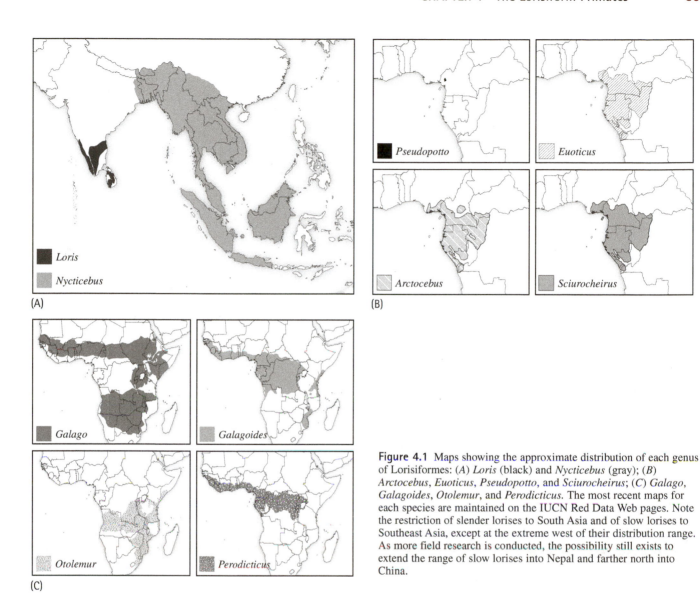

Figure 4.1 Maps showing the approximate distribution of each genus of Lorisiformes: (*A*) *Loris* (black) and *Nycticebus* (gray); (*B*) *Arctocebus*, *Euoticus*, *Pseudopotto*, and *Sciurocheirus*; (*C*) *Galago*, *Galagoides*, *Otolemur*, and *Perodicticus*. The most recent maps for each species are maintained on the IUCN Red Data Web pages. Note the restriction of slender lorises to South Asia and of slow lorises to Southeast Asia, except at the extreme west of their distribution range. As more field research is conducted, the possibility still exists to extend the range of slow lorises into Nepal and farther north into China.

ancient of the living primates, with origins extending back to the Eocene (Seiffert et al. 2003, Martin 2003, Seiffert 2007). Others propose an Asian origin for the Malagasy strepsirhines, with the deepest evolutionary relationships existing between the lemurs and lorises (Martin 2000, 2003; Tavare et al. 2002); yet another contrary view is that the lemurs are most closely related to the galagos (Charles-Dominique and Martin 1970, Roos et al. 2004). New molecular data have opened up questions about the genetic relationships between species. Mitochondrial deoxyribonucleic acid (DNA) analysis, for example, indicates that the galagos are not a single group of close relatives that have undergone recent speciation but can be divided into four deeply rooted clades which diverged over 30 million years ago (Bayes 1998). Further details of evolutionary relationships among galagos have been explored using comparisons of red blood cell enzymes (Masters et al. 1994, Masters and Brothers 2002) and

highly repeated DNA sequences (Crovella et al. 1994, DelPero et al. 2000), which led to a new appreciation of the age and extent of their divergence. Finally, the Asian lorises have long been regarded as essential to understanding questions regarding the evolution of primate characteristics due to their having the greatest degree of orbital convergence of all primates (Cartmill 1972, Ross 1996). Only recently has the ecological significance of their visual adaptations been tested in the field (Nekaris 2005, Bearder et al. 2006).

With such a potential for discovery within this group, we hope that this summary of what is now known about lorisiform behavior and ecology will stimulate a new era of research. Our chapter contains more gaps than it provides answers, but these indicate a new direction for research on the lorisiform primates in light of long-term field studies and surveys. Such research is urgently needed to help ensure their future protection.

Table 4.1 Taxonomy and Conservation Status

LATIN NAME	COMMON NAME	HABITAT	DISTRIBUTION	POPULATION DENSITY[1]	IUCN RED LIST STATUS (2008)
Galaginae					
Galagoides demidovii	Demidoff's dwarf	Understory/forest edge	Bioko, Cameroon, Gabon, Ivory Coast, Nigeria, Uganda	0.16/hr and 50–80/km²	Least concern
G. thomasi	Thomas's dwarf	Forest/mid- to high canopy	Bioko, Cameroon, Gabon, Ivory Coast, Nigeria, Uganda	0.46–2.0/h and 50–80/km²	Least concern
G. orinus	Taita Mountain dwarf	Submontane–montane forest/mid- to high canopy	Tanzania	0.4/h; 2.7–5.4/hr	Near threatened, trend decreasing
G. zanzibaricus (udzungwensis)	Zanzibar lesser	Secondary forest/mid- to high canopy	Tanzania	12.0/hr	Least concern
G. rondoensis	Rondo dwarf	Cloud coastal forest/understory	Tanzania	3–6/ha and 3–10/hr	Critically endangered, trend decreasing
G. sp. nov. 3	Ukinga or Rungwe dwarf	Montane forest	Tanzania		Critically endangered
G. cocos	Kenya coastal	Coastal forest/middle story	Kenya, Tanzania	170–180/km²	Least concern
G. granti	Mozambique lesser	Coastal forest/middle story	Tanzania		Least concern
G. nyasae	Malawi lesser	Woodland	Malawi		Data deficient
G. sp. nov. 1	Kalwe lesser	Forest/middle story	Malawi/SB		Not listed
G. sp. nov. 2	Mt. Thyolo lesser	Montane forest	Malawi		Not listed
Galago senegalensis	Senegal lesser	Miombo, acacia woodland to forest/all strata	Cameroon, Kenya, Tanzania, Uganda	0.03–0.67/hr	Least concern
G. gallarum	Somali lesser	Acacia woodland, thicket/all strata	Kenya	1.0/hr	Least concern
G. moholi	Southern lesser	Acacia woodland/all strata	Botswana, Malawi, Namibia, South Africa, Tanzania		Least concern
G. matschiei	Spectacled	Forest/all strata	Uganda		Least concern, trend decreasing
Euoticus elegantulus	Southern needle-clawed	Forest/mid- to high canopy	Cameroon, Gabon	15–20/km²	Least concern
E. pallidus	Northern needle-clawed	Forest/mid- to high canopy	Bioko, Cameroon	0.25/hr	Least concern
Sciurocheirus alleni	Allen's squirrel	Forest, forest edge/mid- to understory	Bioko, Cameroon	15/km²	Least concern
S. gabonensis	Gabon squirrel	Forest/ mid- to understory	Cameroon, Gabon	15–20/km²	Least concern
S. sp. nov.	Makande squirrel	Forest/ mid- to understory	Gabon		Not listed
Otolemur garnettii	Garnett's (small-eared) greater	Forest, farmland, plantation/mid- to high canopy	Kenya, Tanzania	>5/hr	Least concern
O. crassicaudatus	Thick-tailed greater	Woodland and forest edge/mid- to high canopy	Malawi, South Africa, Tanzania, Zimbabwe		Least concern
O. monteiri	Silver greater	Woodland/?	Kenya		Not listed
O. sp. nov.	Mwera (pygmy) greater	Woodland, farmland plantation/mid- to high canopy	Tanzania		Not listed
Perodicticinae					
Perodicticus potto potto	Western potto	Secondary colonizing or flooded primary forest	Guinea, Guinea Bissau, Nigeria	?	Least concern

Table 4.1 (cont'd)

LATIN NAME	COMMON NAME	HABITAT	DISTRIBUTION	POPULATION DENSITY[1]	IUCN RED LIST STATUS (2008)
P. p. edwardsi	Milne-Edwards or central potto	Swamp, lowland, mid-altitude montane rain forest	Nigeria, Zaire, Central African Republic	8–10/km², 4.7/km²	Least concern
P. p. juju	S. Nigerian potto	Forest edge	Guinea Coast of Nigeria	?	Least concern
P. p. faustus		Riverine forest	Congo Basin	?	Least concern
P. p. ibeanus	Bosman's or eastern potto	Semimoist deciduous forest	Zaire, Burundi, Rwanda	0.04–0.26/h, 1.8–17.7/km²	Least concern
P. p. stockleyi	Mt. Kenya potto	Secondary moist montane forest	Kenya	?	Data deficient
Arctocebus aureus	Golden angwantibo	Tree fall zones, forest edge understory	Gabon	2/km2	Least concern
A. calabarensis	Calabar angwantibo	Tree fall zones, forest edge understory	Cameroon, Gabon, Congo	0.7/km	Least concern
Lorisinae					
Loris lydekkerianus lydekkerianus	Mysore slender loris	Dry forest, acacia scrub jungle	South India	0.13–3.6/km or 28/km²	Least concern, trend decreasing
L. l. malabaricus	Malabar slender loris	Rain forests and coastal forests	South India	0.13–0.29/km	Least concern, trend decreasing
L. l. nordicus	Northern Ceylon gray slender loris	Low country dry zone, scrub forest, and grassland	Sri Lanka	0.33–5.3/km	Endangered, trend decreasing
L. l. grandis	Highland Ceylon slender loris	Montane forest mixed with patana grassland	Sri Lanka	0.11–3.3/km	Endangered, trend decreasing
L. tardigradus tardigradus	Western Ceylon red slender loris	Lowland rain forest, inter-monsoon forest	Sri Lanka	0.1–0.86/km	Endangered, trend decreasing
L. t. nycticeboides	Horton Plains slender loris	Montane rain and mist forests	Sri Lanka	0.08–0.16/km	Endangered, trend decreasing
Nycticebus bengalensis	Bengal slow loris	Bamboo forest mixed with hardwood trees; farmbush; mangrove swamps; tropical rain forest, semi- and wet evergreen	Bhutan, Cambodia, China, India, Laos, Myanmar, Thailand, Vietnam, possibly Nepal	0.1–0.13/km	Vulnerable, trend decreasing
N. coucang	Greater slow loris	Tropical rain forest with continuous canopy	Sumatra, West Malaysia, Thailand, Singapore	0.16–0.80/km	Endangered, trend decreasing
N. javanicus	Javan slow loris	Unknown	Indonesia (Java)	0.02–0.20/km	Endangered, trend decreasing
N. menagensis	Bornean slow loris	Unknown	Brunei, Malaysia (Borneo), Indonesia (Borneo, other small islands), Philippines	0.01–0.02/km	Vulnerable
N. pygmaeus	Pygmy slow loris	Bamboo forest mixed with hardwood trees, forest edge and dense scrub	China, Cambodia, Laos, Vietnam	0.05–0.22/km	Vulnerable, trend decreasing

[1] Because different survey methods were employed, some population densities are per kilometers squared, some are per kilometer, and some are a rate of animal encounters per hour.

TAXONOMY AND DISTRIBUTION

The Lorisiforms

Lorisiformes along with Lemuriformes of Madagascar comprise the infraorder Strepsirhini (Martin 1990). Although sometimes also classified as Prosimii along with the tarsiers, the strepsirhine primates are linked by a number of unique morphological traits, making them a monophyletic group to the exclusion of tarsiers, monkeys, and apes (Haplorhini). These traits include a moist nose, unfused mandibular symphysis and frontal bone, reduced upper incisors, a sloping talofibular facet (groove between the ankle and one of the lower limb bones), and a single grooming claw on the second digit of each hindfoot. Living strepsirhines are further united by the possession of a *toothcomb*—a forward-pointing dental structure comprised of the lower incisors and canines used for both dietary and grooming purposes (Fleagle 1999).

Nocturnal primate taxonomy in general has gone through intense revision in the last few years (e.g., Nietsch and Kopp 1998, Yoder et al. 2001a, Pastorini et al. 2003). While the revision of lemur and tarsier taxonomy and the subsequent reevaluation of their conservation status seem to have been accepted readily by the scientific community, the discovery of enormous taxonomic diversity among the Lorisiformes has met with the same skepticism as acknowledgment of the diverse behavior within this group. We emphasize this point because even recent textbooks have updated their lemur and tarsier taxonomy to the exclusion of the galagos, pottos, and lorises (e.g., Dunbar and Barrett 2000, Falk 2000, MacDonald 2001, Boyd and Silk 2003).

Members of the infraorder Lorisiformes are currently conservatively classified as one superfamily (Lorisoidea) and one family (Lorisidae), comprising three distinctive subfamilies (Galaginae, Perodicticinae, and Lorisinae) (Rasmussen and Nekaris 1998, Grubb et al. 2003) (Table 4.2). The galagos and pottos are restricted to Africa and range in size from 55 to 2,000 g. The lorises are found in Asia and range in size from 85 to 2,100 g (Table 4.3).

The Galagines

Before 1979, the accepted taxonomy of the Galaginae was monogeneric (*Galago*) and contained only six species (Petter and Petter-Rousseaux 1979). This classification is still reported in a number of prominent texts, despite several published revisions expanding the number of species to

Table 4.2 Long-Term Behavioral Studies of Lorisoid Primates; Data in Other Tables Are Drawn from These Studies Unless Otherwise Specified

TAXA	COUNTRY	STUDY SITE(S)	STUDY LENGTH	RADIO TRACKING?	REFERENCES
Galaginae					
Galagoides demidovii	Gabon	Makokou	42 months	Yes	Charles-Dominique 1972, 1977a
G. thomasi[1]	Gabon	Makokou	42 months	—	Charles-Dominique 1977a
G. cocos	Kenya	Gedi Ruins National Monument	20 months	Yes	Harcourt and Nash 1986a,b; Nash 1986a–c
G. cocos	Kenya	Diani	22 months	Yes	Harcourt 1984; Harcourt 1986a; Harcourt and Nash 1986a,b; Harcourt and Bearder 1989
G. rondoensis	Tanzania	Litipo, Rondo, Ziwani	During a 20-month survey	No	Honess 1996, Honess and Bearder 1996
G. zanzibaricus (*udzungwensis*)	Tanzania	Matundu	During a 20-month survey	No	Honess 1996, Honess and Bearder 1996
G. granti	Tanzania	Rondo, Mtopwa	During a 20-month survey	No	Honess 1996, Honess and Bearder 1996
G. orinus	Tanzania	Amani	During a 20-month survey	No	Honess 1996, Honess and Bearder 1996
Galago moholi	S. Africa	Mosdene	24 months, 12 months, 11 months	Yes	Bearder 1969, 1987; Harcourt 1980; Bearder and Martin 1980a,b; Charles-Dominique and Bearder 1979; Crompton 1980; Harcourt and Bearder 1989
Galago moholi	S. Africa	Nylsvley Nature Reserve	18 months	Yes	Pullen 2000, Pullen et al. 2000
Euoticus elegantulus	Gabon	Makokou	42 months	No	Charles-Dominique, 1977a

Table 4.2 (cont'd)

TAXA	COUNTRY	STUDY SITE(S)	STUDY LENGTH	RADIO TRACKING?	REFERENCES
Sciurocheirus gabonensis	Gabon	Makokou	42 months	Yes	Charles-Dominique, 1977a,b
S. alleni cameronensis	Cameroon	WWF Mt. Kupe Forest Reserve, Bakossiland	22 months	Yes	Pimley 2002; Pimley et al. 2005a,b
Otolemur garnettii	Kenya	Gedi Ruins National Monument	20 months	Yes	Nash 1986a–c, Nash and Harcourt 1986
O. garnettii	Kenya	Diani	22 months	Yes	Harcourt 1984, Nash and Harcourt 1986
O. crassicaudatus	S. Africa, Zimbabwe, Kwazula	Transvaal, Umtali, Mtunzini	15 months	Yes	Bearder 1974, Bearder and Doyle 1974b
O. c. umbrosus	S. Africa	Soutsanberg Range	16 months, 11 months	No	Clark 1978a,b, 1985; Crompton 1980, 1983, 1984
O. crassicaudatus	S. Africa	Louis Trichard	12 months	No	Harcourt 1980, 1986b
Perodicticinae					
Perodicticus potto edwardsi	Cameroon	WWF Mt. Kupe Forest Reserve, Bakossiland	22 months	Yes	Pimley 2002
P. p. edwardsi	Gabon	Makokou	42 months	Yes	Charles-Dominique 1974a,b, 1977a
Arctocebus aureus	Gabon	Makokou	42 months	No	Charles-Dominique 1977a
Lorisinae					
Loris lydekkerianus lydekkerianus	India	Ayyalur Interface Forestry Division	11–21 months	No	Nekaris 2001, 2002, 2003a,b; Nekaris and Rasmussen 2003; Rhadakrishna 2001; Rhadakrishna and Singh 2002, 2004
L. l. malabaricus (?)[1]	India	Kalaka-Mundanthurai Tiger Reserve	12 months	Yes	Kar Gupta 2007
L. l. nordicus	Sri Lanka	Several sites in northern Sri Lanka	5 months	No	Nekaris and Jayewardene 2003, Nekaris 2003b
L. tardigradus tardigradus	Sri Lanka	Masmullah Proposed Forest Reserve, Bangamukande Estate	8 months	No	Nekaris 2003b, Nekaris and Jayewardene 2003, Nekaris et al. 2005
L. t. tardigradus	Sri Lanka	Masmullah Proposed Forest Reserve	21 months	Yes	Bernede et al. 2008, Bernede 2009
Nycticebus coucang coucang	Malaysia	Manjung District, Perak	27.5 months	Yes	Wiens 1995, 2002; Wiens and Zitzmann 2003a
N. c. coucang	Malaysia	Pasoh Forest Reserve, Sungai Tekam Forestry Concession	16 months	Limited	Barrett 1984
N. javanicus	Indonesia	Sukakerti Ciamis and Kawungsari Tasikmalaya, West Java	6 months	No	Winarti 2008
N. menagensis	Malaysia	Danau Girang Field Center, Sabah	7 months	Limited	Munds et al. 2008; Nekaris et al. 2009
N. pygmaeus	Vietnam	Cuc Phuong National Park	24 months	Yes	Streicher 2004
N. pygmaeus	Cambodia	Siema Biodiversity Conservation Area, Mondulkiri Province	24 months	Yes	Starr et al. 2008, 2009; Nekaris et al. 2009
N. bengalensis	India	Trishna Wildlife Sanctuary, Tripura	7 months	No	Swapna 2008
N. bengalensis	India	Gibbon Wildlife Sanctuary, Assam	18 months	No	Das 2009
N. bengalensis	Thailand	Khao Ang Runai Wildlife Sanctuary	18 months	No	Pliwsungnoen 2007

[1] Subspecies in study not sufficiently identified.

Table 4.3 Physical Characteristics: Body Weights of Wild Caught Individuals, Unless Otherwise Noted

TAXA	ADULT MALE BODY WEIGHT (G)			ADULT FEMALE BODY WEIGHT (G)			BOTH SEXES WEIGHT (G)		
	AVERAGE	RANGE	N	AVERAGE	RANGE	N	AVERAGE	RANGE	N
Galaginae									
Galagoides demidovii	60	52–72	17	55	45–68	16	57	45–72	33
G. thomasi	82	74–88	6	75	59–85	6	78	59–88	12
G. orinus							89.6	74–98	3
G. zanzibaricus							149		23
G. rondoensis	69.2	60–73	7	66.5		3	69.1		7
G. cocos	150	130–183	35	137	118–155		38		
G. granti							134		5
Galago senegalensis	360[1]	+72	9	266[1]	+47	10	202[1]		1
G. moholi	186.1	+16.3	20	162.8	+16.3	20	200	177–250	
G. matschiei							196–225	210	
Euoticus elegantulus							300	270–360	39
E. pallidus								182–210	
Sciurocheirus alleni cameronensis	280.5	265–307	3	258	246–355	10	288	258–319	4
S. alleni (Bioko)	429		1	446	395–502	5	443	395–502	6
S. gabonensis							260	188–340	17
Otolemur garnettii	690–1060	846	14	805	604–985	11	842	604–1,060	25
O. crassicaudatus	1,510		9	1,258		8			
Perodicticinae									
Perodicticus potto potto							600		
P. p. edwardsi (Gabon)							1,100	850–1,600	33
P. p. edwardsi (Cameroon)	1,502	938–1,795	8	1,572	1,407–1,858	4	1,524	938–1,858	12
P. p. ibeanus	920		1	861	847–875	3			
Arctocebus aureus							210	150–270	30
A. calabarensis	318	315–320	2	298	270–325	9			
Lorisinae									
Loris lydekkerianus lydekkerianus	294.4	267–322	4	260	227–292	6	273	227–322	10
L. l. malabaricus	222.14	180–275	6	189	168–210	2			
L. l. nordicus		228–285	4		238–287	5			
L. l. grandis	204.1		1	238.1		1			
L. tardigradus tardigradus	162	153–172	2	118	103–148	3	137	103–172	5
L. t. nycticeboides	140		1	190		1	165	140–190	2
Nycticebus bengalensis	1,140		1	1,400		1		1,575–2,100	
N. coucang	737	+111	8	637	+61	11			
N. menagensis								265–800	6
N. pygmaeus	418 + 98	367–578	70	422 + 88	360–543	97			

[1] Captive.

11 (Olson 1979, Nash et al. 1989), 17 (Bearder et al. 1995), and 25 (Bearder et al. 2003). This extraordinary diversity of galagos makes them comparable to the guenon group (see Chapter 16) in having one of the widest distributions and greatest abundance of species found in Africa. For the galagos we adopt the most recent classification by Grubb et al. (2003; but see also Groves 2001a) and divide the taxa into five genera (Table 4.2). Bioacoustic studies, using Paterson's (1985) mate-recognition concept of species, have been at the forefront of this taxonomic revision (Masters 1988, 1991, 1998; Honess 1996; Bearder 1999; Ambrose 2003). This classification is also supported by behavioral studies (e.g., Harcourt and Bearder 1989), genetic research (Bayes 1998, Roos 2003), examination of hand, foot, and sexual organ morphology, and comparisons of hair scale structure (Dixson 1989, 1995, 1998; Anderson, M., 1998, 2000, 2001; Perkin 2007, Anderson, M. J., et al. 2000). In this chapter, we have carefully gone through previous studies, updating the taxonomy for each of them in the tables and text. This is extremely important as these animals are indeed distinct species but aspects of life history and morphology of even the best-studied taxa are often reported, even in the more recent literature, under the wrong name (e.g., *Galago moholi*, which was formerly known as *G. senegalensis*).

Galagos are distributed across the whole of Africa south of the Sahara, with the exception of southern regions of South Africa. They occupy a very wide range of habitats, from near-desert conditions in Somalia and northern Kenya through subtropical savannas, woodlands, and riverine and montane forests to dense tropical rain forests. Up to four species can occur in sympatry with each other, as well as with up to two pottos. One country, Tanzania, currently boasts 13 species of galagos within its borders and one species, *Otolemur garnettii*, has been found in association with any of 14 other galago species in different parts of its geographic range. With huge tracts of rain forest yet to be surveyed for these nocturnal primates and considering their secretive habits and relatively cryptic characteristics, it will not be astonishing if further research adds to the complexity of this emerging picture.

Perodicticines and Lorisines

Similar diversity is now being uncovered in the Perodicticinae in Africa and the Lorisinae in Asia, each of which was once thought to comprise two monospecific genera, one gracile and the other robust (Yoder et al. 2001a). Despite being less vocal than the galagos, vocalizations have again yielded important taxonomic information, as have differences in behavior, morphology, facial markings, and genetic data (Coultas 2002, Nekaris and Jayewardene 2003, Roos 2003, Nekaris and Jaffe 2007). It is almost without doubt that most of the currently recognized subspecies within the Perodicticinae will be elevated to species level. For this reason, we report data regarding the pottos at the subspecific level.

In Africa, the gracile forms are now recognized as two species, the golden angwantibo (*Arctocebus aureus*) and the Calabar angwantibo (*A. calabarensis*), both confined to the rain forests of central Africa. The taxonomy of the robust forms, the pottos, is being reevaluated but currently consists of one species with six subspecies. Following Kingdon (1997), these are *Perodicticus potto potto* in West Africa, *P. p. juju* in Nigeria, *P. p. edwardsi* in Cameroon and Gabon, *P. p. faustus* in the Congo Basin, and *P. p. ibeanus* in the eastern Democratic Republic of Congo, Uganda, and western Kenya. Schwartz (1996) recognizes another genus of potto, *Pseudopotto martini*, differing from *Perodicticus* in having relatively long upper first premolars, a reduced third molar, and a relatively longer tail. Other researchers doubt this taxon, suggesting the differences fall within the range of variation of *Perodicticus* (Sarmiento 1998, Smeenk et al. 2006); targeted searches in the wild have failed to yield any evidence of this putative genus (Pimley 2002).

Groves (2001a) recognizes two gracile lorisines in south Asia, although we follow the subspecific classification of Osman Hill (1953). The red slender loris, the smallest of the lorisines, resides only in the lowland (*Loris tardigradus tardigradus*), montane (*L. t. nycticeboides*), and highland (*L. t. grandis*) rain forests of Sri Lanka. One subspecies of the larger gray slender loris also is found on this island: the northern Ceylonese slender loris (*L. lydekkerianus nordicus*) (Nekaris and Jayewardene 2004). Southern India harbors an additional two slender loris taxa (Roonwal and Mohnot 1977, Schulze and Meier 1995a). The Malabar slender loris (*L. l. malabaricus*) is distributed in the wet southwest, including the Western Ghats, whereas the largest of all slender loris taxa, the Mysore slender loris (*L. l. lydekkerianus*), is distributed in the dry scrub forests of the southeast, including the Eastern Ghats.

The taxonomy of the robust Asian form, *Nycticebus*, is currently undergoing extensive revision, with some authorities recognizing three species based on morphology (e.g., Groves 1998, 2001a) and others recognizing five species based on genetic analyses (Roos 2003). Here, we follow the taxonomy suggested by Roos (2003). The Bengal or northern slow loris (*Nycticebus bengalensis*) has the largest geographic range, including Myanmar, Bhutan, Cambodia, southern China, northeast India, Laos, northern Thailand, and Vietnam. The greater slow loris (*N. coucang coucang*) is found in Indonesia, Malaysia, Singapore, and Thailand. The Javan slow loris (*N. javanicus*) occurs only in Java. The Bornean slow loris (*N. menagensis*) is found in the Philippines, Brunei, Indonesia, and Malaysia. Finally, the pygmy or lesser slow loris (*N. pygmaeus*) is found in Cambodia, China, Laos, and Vietnam. Most authorities do not recognize *N. intermedius* but class it together with *N. pygmaeus*. Streicher (2003, 2004) revealed that the characteristics that distinguished *N. intermedius* were in fact seasonal coat and body weight changes of *N. pygmaeus* (Fig. 4.2).

(A)

(B)

(C)

(D)

Figure 4.2 Photographs illustrating different genera within the Lorisidae: (*A*) *Loris tardigradus tardigradus* (K. A. I. Nekaris); (*B*) *Nycticebus coucang* (K. A. I. Nekaris); (*C*) *Galagoides rondoensis* (A. W. Perkin); (*D*) *Arctocebus calabarensis* (C. Wild);

(E)

(F)

(G)

(H)

Figure 4.2 (cont'd) (*E*) *Sciurocheirus* sp. (L. Ambrose); (*F*) *Perodicticus potto edwardsi* (E. R. Pimley); (*G*) *Otolemur crassicaudatus* (S. Bearder); (*H*) *Galago moholi* (S. Bearder).

PHYLOGENETIC RELATIONSHIPS

The evolutionary history of the lorisiform primates has been the subject of a comprehensive review (Rasmussen and Nekaris 1998) and will be only briefly recapped here. Until recently, most evidence for lorisiform origins pointed back to the Miocene of East Africa. Intense debate characterizes the subfamilial designation of the three best-known early Miocene forms, *Mioeuoticus*, *Progalago*, and *Komba*. Both cranial and postcranial features have allied these genera with either lorisines or galagines (Le Gros Clark 1956, Walker 1969, Gebo 1986, McCrossin 1992). Other authors have suggested that basal lorisiforms may have demonstrated a combination of lorisine cranial characteristics and galagine postcranial adaptations (Rasmussen and Nekaris

1998). Two newly discovered fossils may resolve this debate, making lorisiform origins even earlier than previously thought. *Karanisia* and *Saharagalago*, based on analysis of dental characteristics, are putative early lorises and galagos, respectively, from late Eocene sites at the Fayum Depression in Egypt (Seiffert et al. 2003). True, unrefuted lorisines (*Nycticeboides simpsoni*) and galagines (*Galago howelli* and *G. sadimensis*) occur in the fossil record of the late Miocene of Pakistan and early Plio-Pleistocene of Ethiopia and Kenya.

Some authors have attempted to resolve the phylogenetic relationships of the lorisiforms with molecular and morphological evidence (Bayes 1998, Masters and Brothers 2002, Roos et al. 2004). Although the standard practice is to consider the Lorisinae a monophyletic group to the exclusion of the Galaginae, most molecular and morphological studies cannot resolve the position of the African pottos (Rasmussen and Nekaris 1998). Behavioral evidence allies Asian slender lorises more closely with galagos than with pottos (Bearder et al. 2002, Pimley et al. 2005a). It is not implausible that the galagos, pottos, and lorises share a common ancestor and form three monophyletic groups (Yoder et al. 2001a). Rasmussen and Nekaris (1998) and Nekaris and Rasmussen (2003) suggest that the cause for the divergence of these groups may have been a deviation in foraging strategies, with the galagos specializing on evasive prey, resulting in an emphasis on hearing and leaping, and the pottos and lorises concentrating on toxic prey, with a subsequent reliance on olfaction and a reduced basal metabolic rate, coinciding with slow locomotion and life history. Further studies will surely elucidate these evolutionary relationships.

ECOLOGY AND BEHAVIOR

Diet

Only nine studies have focused in detail on the diet of lorisiforms, with one of these (Charles-Dominique 1977a) gaining most of its data from the analysis of stomach contents (Table 4.4). A number of brief studies provide us with preliminary knowledge of particular food preferences (e.g.,

Table 4.4 Diet and Activity Budgets of Wild Lorisiform Primates Based on Long-Term Studies

TAXON	DIET					ACTIVITY BUDGET				
	ANIMAL PREY	FRUIT	GUM	NECTAR	OTHER	REST	TRAVEL	FORAGE	SOCIAL	OTHER
Galaginae										
Galagoides demidovii/ thomasi[1]	70%	19%	10%						25%	
G. cocos	70%	30%	0%	0%						
G. moholi	52%	0%	48%	0%		4.5%	25%	63.9%	5.9–18%	0.6%
Euoticus elegantulus	20%	5%	75%	0%					24%	
Sciurocheirus gabonensis	25%	73%	0%	2%					14%	
S. alleni cameronensis	55%	55%	0%	0%					0.6%–30.5%	
Otolemur garnettii	50%	50%	0%	0%		9.4%	52.3%	21%	14.5%	2.8%
O. crassicaudatus	5%	33%	62%	0%					20%	
Perodicticinae										
Perodicticus potto edwardsi	40%	50%			10%				4%	
P. p. edwardsi	11%	67%	22%						0.2%–44%	
Arctocebus aureus	87%	13%							3%	
Lorisinae										
Loris lydekkerianus lydekkerianus	96%	1.2%	2.8%			36.4%	35.6%	26%	23%	2%
L. l. nordicus	95%	5%				20.8%	75.5% (includes forage)		49.5%	3.7%
L. tardigradus tardigradus	100%					19.1%	80% (includes forage)		43.7%	0.9%
Nycticebus coucang coucang	2.5%	22.5%	43.3%	31.7%		5.4%	70.6%	21%	3%	
N. c. coucang	29%	71%								
N. pygmaeus	33%	—	63%		4%					
N. bengalensis	3%	0.3%	87%	4%	5.7%					

[1] During his study, Charles-Dominique did not recognize *G. thomasi* as a distinct species, and thus all data were "lumped."

Happold and Happold 1992; Tan and Drake 2001; Ejidike and Okosodo 2007; Das 2009; Pliwsungnoen 2007; A. W. Perkin, personal communication; A. B. Rylands, R. A. Mittermeier, and B. R. Konstant, unpublished report). Both direct observations and stomach content analysis have their limitations, but what is clear is that dietary choice among the lorisiforms is varied, including gum-eating specialists, highly frugivorous taxa, and some that are among the most faunivorous of all the primates.

Galagos are extremely varied in their diet, but all species appear to consume at least some gum; the ability to consume and digest gum may be a fundamental adaptation of this group (Bearder and Martin 1980a; Harcourt 1980, 1984; Nash 1989; Nash and Whitten 1998). Apart from this, the smaller-bodied taxa (e.g., *Galagoides demidovii, G. thomasi, G. rondoensis*) rely more on insects, medium-sized taxa (e.g., *Galago moholi, Euoticus elegantulus*) add more exudates to their dietary repertoire, and the largest of the galagos (e.g., *Sciurocheirus* spp., *Otolemur* spp.) increase their intake of fruit. Nontoxic orthopterans and beetles comprise the invertebrate portion of galago diets (Bearder and Doyle 1974b, Harcourt and Nash 1986a), and the fruits eaten by galagines are in general sweet and soft (Charles-Dominique 1977a). Preliminary observations of a yet unnamed taxon in southeastern Tanzania indicate yet another feeding behavior, that of consuming floral nectar, suggesting an important role by this primate in pollination, not unlike that of the Malagasy *Eulemur mongoz* or the greater slow loris *N. coucang* (A. W. Perkin, personal communication; A. B. Rylands, R. A. Mittermeier, and B. R. Konstant, unpublished report).

Galagos have adapted to their varied diet through a variety of morphological and behavioral adaptations. All galagos are capable of localizing animal prey with the help of their particularly large and independently mobile ears and frequently use this sense to detect prey items that are out of sight. They also search for insects visually and find sources of gum using their keen sense of smell (Bearder 1969, Charles-Dominique 1977a, Hladik 1979, Pariente 1979). As with all strepsirhines, insects are grabbed in the hands in a stereotyped fashion involving control of the whole hand as the individual fingers cannot be moved independently (Martin 1990). The toothcomb plays an important role in scraping gum from trees, and gum can be cleaned from between the teeth using a serrated cartilaginous sublingual, a second type of tongue located underneath the main tongue that is notched at the tip like a saw. Gum is processed in an elongated cecum containing microorganisms capable of digesting the complex polymerized sugars. *Euoticus* spp., which eat mainly gum, have additional specializations in the form of enlarged canines and premolars for exposing sources of gum and *keeled* (pointed) nails, allowing the animals to cling to large tree trunks and reach exudates that would otherwise be inaccessible (Osman Hill 1953, Charles-Dominique 1977a, Ambrose 1999). Galagos living in seasonal environments in South Africa may rely almost completely on carbohydrate-rich gum in the cold winters and reduce their activity accordingly (Bearder and Martin 1980a, Mzilikazi et al. 2006). Squirrel galagos (*Sciurocheirus* spp.), which usually feed on fallen fruits, are reported to eat rapidly and even swallow fruits whole, allowing them to fill their stomachs within minutes and retreat to areas safer from potential predators (Charles-Dominique 1977a).

Only limited observations are available of potto feeding behavior (Jewell and Oates 1969, Charles-Dominique 1977a, Oates 1984, Pimley 2002). Generally, pottos (*Perodicticus* spp.) are mainly frugivorous but supplement their diet with a considerable amount of gums and animal prey, including ants, slow-moving arthropods, birds, and bats. Pottos are possessed with somewhat more powerful jaws than galagos and are able to consume fruits and stationary animal prey, in particular caterpillars and noxious beetles. Dietary conditioning is exhibited by all the African strepsirhines, whereby a young animal learns to eat by snatching food from its parent and examining novel food items with a curious, head-cocking movement (Bearder 1969, Charles-Dominique 1977a). This developmental behavior may be particularly important for the angwantibos (*Arctocebus* spp.), which process irritant prey in a specific manner (e.g., removing the hairs from caterpillars) before they can be consumed without discomfort.

In a detailed study of sympatric galagos (*Galagoides demidovii, Euoticus elegantulus, Sciurocheirus gabonensis*) and pottos (*P. p. edwardsi, A. aureus*) in Gabon, Charles-Dominique (1974a, 1977a) revealed classic dietary partitioning between nocturnal primate species that ensured they avoided competition. Species that spent most time in the forest canopy concentrated mainly on insects (*Galagoides*), gums (*Euoticus*), or fruits (*Perodicticus*). Species that preferred the undergrowth subsisted on caterpillars (*Arctocebus*) or fallen fruits (*Sciurocheirus*). Some years later, it was discovered that the dwarf galagos in Gabon were in fact two different species that live together throughout the tropical forests of central Africa (*Galagoides demidovii* and *G. thomasi*) (Wickings et al. 1998). Both these species prefer insects, but not surprisingly, one moves mainly in the canopy (*G. thomasi*) and the other is restricted to the undergrowth (*G. demidovii*), where it consumes fast-moving insects in contrast to the noxious forms eaten by angwantibos (*A. aureus*). Similar separations occur between sympatric species in other parts of Africa. For example, in the Rondo Forest of southeastern Tanzania, Garnett's galagos (*Otolemur garnettii*) forage in the canopy, Grant's galagos (*Galagoides granti*) use the middle story, and Rondo dwarf galagos (*G. rondoensis*) remain approximately 1 m above the ground and feed almost exclusively on insects and their larvae from the leaf litter (Honess 1996).

Detailed observations have been made on the diet of three slender loris taxa. *L. l. lydekkerianus* was the focus of a long-term study (Nekaris and Rasmussen 2003), whereas *L. l. nordicus* and *L. t. tardigradus* were the subjects of short-term studies (Petter and Hladik 1970, Nekaris 2002, Nekaris and Jayewardene 2003). These studies concur that slender lorises are among the most faunivorous of primates (very like tarsiers, see Chapter 6). They specialize on prey of small size classes and are highly tolerant of toxic prey such as ants and darkling beetles (Tenibrionidae). Prey items are consistently eaten head first, followed by the animal lapping at the innards. Those insects which emit irritant sprays are removed individually from the colony, taken several meters away, and consumed while the loris salivates excessively, closes its eyes tightly, and shakes its head, all combined to produce what can be aptly termed a "disgust face." Although gum comprised a portion of the diet of *L. l. lydekkerianus*, it was not seen to be consumed by other taxa. Consumption of plant material was minimal to nonexistent. Vertebrates, particularly geckos and lizards, were consumed by all three taxa but comprised a large portion of the diet of *L. t. tardigradus* (Nekaris and Rasmussen 2003, Nekaris and Jayewardene 2003).

According to Barrett (1984), the greater slow loris (*N. coucang coucang*) predominantly eats fruit, supplemented by insects. A more detailed study of this species in Malaysia was conducted via direct observation and fecal analysis (Wiens 2002, Wiens and Zitzmann 2003a). This population consumed mainly nectar, gum, and sap, with fruit and arthropods comprising only a small portion of the diet. Nectar from the flowers of the Bertram palm (*Eugeissona tristis*) comprised more observations than any other dietary item, with animals spending up to 30 min feeding from these nectaries (Wiens 2002, Wiens and Zitzmann 2003a). Preliminary results on pygmy lorises (*N. pygmaeus*) suggested that they too rely on nectar (particularly *Saraca dives*) and gum, visit the same sites often, and leave noticeable gouges in the tree trunks (Tan and Drake 2001, Streicher 2004). As for galagos, gum is probably an important component of the diet during cold Vietnamese winters (Streicher et al. 2009). Consumption of insects, including ants and moths, is relatively common; and processing of these prey items mirrors that of slender lorises (Streicher 2004). *N. pygmaeus* and *N. bengalensis*, sympatric in many parts of their range, are known from preliminary observations to share feeding sites; nothing is known about how they partition their niches (Duckworth 1994). A number of recent studies, still ongoing, provide strong evidence that *Nycticebus* is strongly specialized for extracting gum (Nekaris et al. 2009b).

ACTIVITY PATTERNS AND LOCOMOTION

Activity

Few activity budgets for the lorisiforms have been reported in the literature, but those that have are summarized in Table 4.4. What is clear is that all the lorisiforms are nocturnal in their activity patterns, with no diurnal or cathemeral species. Animals are not precluded from being active in

daylight, however. They may become active during the day in order to change position for thermoregulatory purposes, to eat during periods of intense food scarcity, and to avoid predators (Bearder et al. 2006). What is becoming clearer is that, at least in more open habitats, activity patterns change with the amount of light available. *G. moholi*, for example, increased its behavior and range of travel patterns during the light moon and during periods of twilight, while *L. l. lydekkerianus* maintained activity regardless of moon phase (Bearder et al. 2002). Galago species living in closed forest, on the other hand, do not appear to be influenced by changes in the level of moonlight (Nash 1986b). It is clear that further studies of nocturnal primates must take account of the importance of moonlight.

Locomotor Behavior

As the feature that is most often used to characterize this infraorder of strepsirhines, locomotion is possibly the best-studied aspect of their behavior (Table 4.5), forming the basis for entire field studies (Crompton 1980, 1983, 1984; Off and Gebo 2005) and for numerous captive studies (e.g., Dykyj 1980, Glassman and Wells 1984, Oxnard et al. 1990, Ishida et al. 1992, Demes et al. 1998). A complex suite of morphological traits linked to locomotion differentiates the galagos from the pottos and lorises (Charles-Dominique and Bearder 1979). All taxa of galagos have long tails and elongated tarsal bones and are characterized by intermembral indices <100, whereas pottos and lorises, to varying degrees, have reduced or lost their tails and have intermembral indices close to 100 (Martin 1990). As a result, galagos can cross gaps by hopping and leaping, while lorises and pottos do this by stretching. Similarly, galagos usually evade predators by swift locomotion, whereas lorises and pottos have developed a suite of morphological characteristics that provide camouflage, provide protection during attacks, and allow them to remain still for prolonged periods (Charles-Dominique 1977a, Bearder 1987, Nekaris 2001). For example, both pottos and lorises exhibit features that allow for prolonged grip with no fatigue, such as shortened second digits on the hands and feet, highly mobile ankles and wrists, and retia mirabilia of the proximal limb vessels (Rasmussen and Nekaris 1998). *Retia mirabilia* are arteries and veins in the arms and legs that subdivide extensively to form networks of intertwining vessels, which act as storage units, allowing blood to flow freely. Thus, the exchange of oxygen and waste materials in the muscles continues even though there is no bodily movement.

Perodicticus also possesses a scapular shield, a structure produced by a combination of raised apophyseal cervical spines, some of which protrude above the skin in the form of tubercles, which are covered by thick skin and bristles of sensory hair, that extend to a wider nuchal region. This structure is used to provide defense against predators and possibly other pottos (Charles-Dominique 1977a). Slow lorises have developed an even more elaborate defense

Table 4.5 Locomotion and Habitat Use

TAXON	CHARACTERISTIC LOCOMOTION	USE OF STRATA/SUPPORTS WHEN ACTIVE	SLEEPING SITES	SLEEPING ASSOCIATIONS	HABITAT TYPE OF STUDY SITE
Galaginae					
Galagoides demidovii	Fast-moving: mainly quadrupedal	0–5 m in dense secondary undergrowth, <1 cm fine branches and liane curtains, occur on roadside	Spherical leaf-nest or dense vegetation, few sites	2–10 females w/offspring, male often sleeps alone	Primary equatorial rain forest
G. cocos	Fast-moving: mainly hopping and quadrupedal running	Ground to canopy (0–13 m), prefers undergrowth	Tree hollows, few sites	Male sleeps with 1 or 2 females w/offspring	Lowland dry forest on coral rag
G. rondoensis	Vertical clinging and leaping from thin stems	Low-diameter perches (<3 m), small vertical supports (3 cm)	Flat and leafy nests in high trees (5 m)	At least 3 adults	Lowland evergreen and semi-evergreen forest
G. orinus	Quadrupedal running and walking	Canopy dwellers (20 m)	Nests of leaves and twigs set in lianes	At least 1–3 adults	Natural montane evergreen forest
G. granti	Agile jumping, quadrupedal climbing and walking	Mid-strata (5–7.5 m), 8.0 cm diameter vertical supports	Tree holes	4–5 individuals	Lowland and coastal forest
G. zanzibaricus (udzungwensis)	Quadrupedal walking and running	Upper strata (10 m), thick secondary growth and vine tangles, small horizontal perches	Tree hollows or secondary growth tangles	Male/female pair and offspring	Natural lowland evergreen forest
Galago moholi	Active leaping, bipedal hopping	Ground to upper canopy, prefer lower strata (0–4 m), use small vertical supports	Flat leaf nest, tree hollow, or branch fork in a thorn tree	1–8, males never together but with 2 or more females w/offspring	Acacia woodland savanna
G. senegalensis	Leaping and bipedal hopping	Ground to mid-canopy (1–4 m), can cross open ground	Tangled vegetation and tree holes at 1–2 m, human-made bee hives	At least 1–3 individuals	Savanna *Acacia* thorn scrub, *Cynometra* thicket, and open woodland
Euoticus elegantulus	Running, leaping, and climbing	Canopy (5–35 m), use large-caliber branches and vines	Branch fork in dense shelter of foliage	1–7 individuals	Primary equatorial rain forest
E. pallidus	Quadrupedal running and leaping	Prefers upper strata (4–12 m), use large horizontal supports	Branch fork in dense shelter of foliage	At least 1–4 individuals	Primary equatorial rainf orest
Sciurocheirus gabonensis	Active leaping	Undergrowth (1–2 m), prefers vertical supports	Tree holes	At least 2–3 individuals	Primary equatorial rain forest
S. type 3	Vertical clinging and leaping	Lower to mid-canopy (0–5 m), vertical substrates < 10 cm		1–4 individuals	Primary equatorial rain forest
S. alleni cameronensis	Vertical clinging and leaping	Ground to mid-canopy (0–5 m), 2–5 cm vertical supports	Tree holes and woody lianes at 1–4 m	At least 2–3 individuals	Secondary forest and farm bush
Otolemur garnettii	Quadrupedal running, leaping, and bipedal hopping	Mid-to upper canopy 50% of the time, horizontal substrates > 5 cm	Tangled vegetation, hollows rarely, many sites	1–4, male and female w/offspring	Coastal forest
O. crassicaudatus	Monkey-like quadrupedal walking and running, some leaping, no hopping	Low strata of canopy	Tangled vegetation or flat leaf nest, few sites	1–4, male and female w/offspring	Riverine forest
Perodicticinae					
Perodicticus potto edwardsi (Cameroon)	Slow climbing	Canopy at 6–10 m on 2–5 cm oblique branches	Leafy part of canopy at 10–30 m	1–3 individuals with male/female pairs	Farm bush, disturbed and secondary forest
P. p. edwardsi (Gabon)	Slow climbing	Canopy at 5–30 m, level branches and lianes of 1–15 cm diameter	Dense tangles or clumps w/branches or forks	1–3, male sleeps alone, female w/offspring	Farm bush, primary, and secondary forest
Arctoccbus calabarensis	Slow climbing	Small branches, twigs, climbers	Dense vegetation	1–2, adults sleep alone, female w/offspring	Forest edges and tree fall zone
A. aureus	Slow climbing	0–5 m in undergrowth, <5 cm branches and lianes, use ground often	Dense vegetation	1–2, adults sleep alone, female w/offspring	Forest edges and tree fall zone

Table 4.5 (*cont'd*)

TAXON	CHARACTERISTIC LOCOMOTION	USE OF STRATA/SUPPORTS WHEN ACTIVE	SLEEPING SITES	SLEEPING ASSOCIATIONS	HABITAT TYPE OF STUDY SITE
Lorisinae					
Loris lydekkerianus lydekkerianus	Quadrupedal climbing, walking, and cantilevering	Understory (<5 m), prefers oblique branches (1–5 cm), crosses open ground and roads	Vine tangles, dense branches, few sites	1–7, female w/offspring and 1 or more male	Acacia scrub forest
L. l. malabaricus	Quadrupedal climbing, walking, and cantilevering	0–10 m used equally, with average at 5 m, move on small branches	Use multiple sites, tangled branches	Male/female pair with offspring	Riverine, dry evergreen, dry deciduous, plantations, scrub forest
L. l. nordicus	Quadrupedal climbing, walking, and cantilevering	Dense understory (<5 m), oblique branches (1–5 cm), crosses open ground and roads	Vine tangles, dense branches, few sites	At least 1–6, female w/offspring and 1 or more male	Acacia scrub forest
L. tardigradus. tardigradus	Quadrupedal climbing and running	Understory to canopy (0–15 m), prefers horizontal and vertical branches, relies on vines	Vine tangles, dense branches, few sites	At least 1–4, female w/offspring and 1 male	Intermediate monsoon rain forest
Nycticebus coucang coucang	Quadrupedal climbing, cantilevering, rapid walking	Trunks, branches, and lianes <10 cm; mid- to upper canopy; will use understory in disturbed forest	Trees, palms, shrubs, lianes at 1.8–35 m; many sites	1–3, female w/offspring, sometimes 1 male	Primary forest, logged over forest, padang savanna
N. pygmaeus (Vietnam, Cambodia)	Quadrupedal climbing, cantilevering, and "race" walking	Branches, lianes, vines, bamboo; comes to understory if substrate is discontinuous	Dense scrub or bamboo, high thickets of branches	1–4, male/female pairs w/offspring	Forested limestone hills, plantation forest and scrub, deciduous forest, bamboo seasonal forests
N. bengalensis (Assam)	Quadrupedal climbing and walking, race walking, cantilevering	Ground to canopy (0–15 m); horizontal and vertical branches, trunk, and lianes	Tree hollows, more than one site	Not less than 2	Evergreen, semievergreen and wet evergreen forest

mechanism, that of being toxic. Before biting prey items or predators, slow lorises combine a secretion from brachial sebaceous glands with their saliva in order to produce a numbing poison that can send humans into anaphylactic shock (Alterman 1995, Fry and Fry 2003). It is rumored that they also use this solution to cover their parked infants, although this is yet to be verified by field data.

Although vertical clinging and leaping is considered the quintessential galago locomotor mode, it is used by most galagos only to negotiate gaps between trees. Only a few taxa, such as *Sciurocheirus* spp. and *Galagoides rondoensis*, use it as their stereotypic mode of locomotion (Charles-Dominique 1977a, Honess 1996, Perkin 2002, Pimley 2002). In fact, *Otolemur* spp. rarely use this mode of locomotion, although capable of leaping and bipedal hopping (Crompton 1983, Harcourt and Nash 1986a). These larger galagos are surprisingly monkey-like in their locomotion and regularly move quadrupedally through the trees on relatively broad and horizontal supports. Many of the smaller galagos (*Galagoides*) maneuver through the networks of tiny branches by quadrupedal running,

climbing, and agile jumping (Charles-Dominique 1972, Ambrose 1999). Several taxa can cross the ground by walking or running (e.g., *Otolemur crassicaudatus*) or bipedal hopping (e.g., *O. garnettii, G. moholi*, and *G. moholi*), whereas others are strictly arboreal (e.g., *Galagoides* spp. and *Euoticus* spp.).

The use of cantilevering (bridging or extending the body), rather than leaping, to move across arboreal gaps is the key feature that distinguishes perodicticine and lorisine locomotion from that of the galagines (Sellers 1996). Both pottos (*Arctocebus* spp., *Perodicticus* spp.) and lorises (*Loris* spp., *Nycticebus* spp.) use their long bodies and flexible limbs to stretch across gaps in the canopy, which requires a certain gauge of branch to sustain their weight during the crossing (Charles-Dominique 1974b, Nekaris 2001). Unlike galagos, which can charge through the trees changing directions, the "slowness" of loris and potto locomotion comes from testing branches and having to back up and move position in the canopy to find a suitable crossing point (Charles-Dominique 1977a, Nekaris 2001). However, this progression need not be slow-paced. Captive studies

Table 4.6 Range Size and Range Use Patterns for Wild Populations

TAXA	AVERAGE HOME RANGE (HA)	ADULT M (HA)	ADULT F (HA)	MM OVERLAP	FF OVERLAP	MF OVERLAP	METHOD	INFERRED SOCIAL ORGANIZATION
Perodicticinae								
Perodicticus potto edwardsi (Pimley 2002)	28.4	30.6	31.5	P: 29%	P: 25%	P: 47%	Kernel	Semidispersed unimale/unifemale
P. p. edwardsi (Charles-Dominique 1977a)		17.8	7.5	A	Limited	P	Mcp	Dispersed unimale/ multiple female
Lorisinae								
Loris lydekkerianus lydekkerianus	2.5	3.6	1.59	P: 20%	P: 57%	P: 14%	Mcp	Semidispersed multi-male
Loris l. nordicus				P	P	P		?
L. l. malabaricus		2.8–35.2	0.94–15.33	23–35%	11%–44%	38%–70%	Mcp, kernel	Paired, with 3 types of males
Loris tardigradus tardigradus				P	No data	P		?
Nycticebus coucang coucang (Wiens and Zitzmann 2003b)	2	0.8	2.1	No data	No data	P: 80.6%	Mcp, primary	Semidispersed unimale/unifemale
N. c. coucang (Wiens and Zitzmann 2003b)	6.4	7.35	4.8	No data	No data	P: 97.8%	Mcp, logged	Semidispersed unimale/unifemale
N. c. coucang (Wiens and Zitzmann 2003a)	18.1	22	10.4	No data	No data	P: 94.6%	Mcp, padang	Semidispersed unimale/unifemale
N. c. coucang (Barret 1984)			4.19	Rarely	P	P	Mcp	?
N. pygmaeus	3.1, 9–19 ha			P	P	P	Kernel/mcp	?
Galaginae								
Galagoides demidovii/ thomasi		0.5–2.7	0.6–1.4	P	P	P	Mcp	Dispersed multimale
G. cocos		2.2	1.8	P (slight)	P	P	Mcp	Spatial monogamy
G. moholi		9.5–22.9	4.4–11.7	P	P	P	Mcp	Dispersed multimale
Sciurocheirus gabonensis		30–50	8–16	A	P	P	Mcp	Dispersed harem
S. alleni cameronensis		2.84	1.97	A	P: 58%	P: 31%	Kernel	Dispersed multimale
Otolemur garnetti		17	12	P: different age classes	P: different age classes	P: extensive overlap	Mcp	Dispersed multimale

A, overlap absent; P, overlap present; C, ranges coincide; M, male; F, female.

have shown that the slender loris (*L. l. malabaricus*) is capable of a locomotor mode called the "race walk" (Demes et al. 1998). Wild *L. tardigradus* and *N. pygmaeus* regularly quadrupedally run and even negotiate gaps with mini-leaps, rearing up on their hindlegs and hurtling their bodies over gaps of several inches (Duckworth 1994, Nekaris and Stephens 2007). Even *P. p. edwardsi* has been described to have a mini-leap, when it simply cannot negotiate a gap with any amount of stretch (Charles-Dominique 1977a).

Locomotion has been implicated as a factor affecting the ranging behavior of the lorisiforms, with galagos able to cross a larger home range and return to dispersed sleeping sites with greater regularity than the pottos and lorises (Charles-Dominique 1977a,b; Oates 1984). Table 4.6 shows that, in fact, home ranges of similar-sized galagos and lorises are of comparable area. Despite initial suggestions that both pottos and lorises move as little as 10 m per

night, studies conducted with all-night follows (Nekaris 2003a, Bearder et al. 2006) and with radiotracking (Wiens and Zitzmann 2003b, Pimley et al. 2005b) have shown much more extensive ranging. For example, pottos (*P. p. edwardsi*) may move up to 6 km in one night, gray slender lorises (*L. lydekkerianus*) move several hundred meters, red slender lorises (*L. tardigradus*) travel up to 1 km per night, greater slow lorises (*N. c. coucang*) travel up to 400 m per hour, and pygmy lorises (*N. pygmaeus*) can travel up to 6 km (Starr, personal communication).

Habitat Use

As already noted when illustrating dietary partitioning, the lorisiforms also show a wide preference for use of both substrates and strata in the forest (e.g., Crompton 1983, Honess 1996, Ambrose 1999, Nekaris 2001, Pimley et al.

2005b, Nekaris et al. 2005). Substrate size selection is almost always related to the body weight of the animal, with smaller animals moving on smaller-gauged twigs, branches, and lianas and larger animals negotiating sturdier supports with greater girth. An exception is *Euoticus*, which makes more use of large vertical supports (Charles-Dominique 1977a). A number of species (e.g., *S. gabonensis*, *Galagoides demidovii*, *A. aureus*, *L. lydekkerianus*) thrive in the undergrowth and in tree fall zones, whereas others (*G. orinus*, *N. coucang*) prefer the canopy. This ecological division is what allows the African lorisiforms in particular to occur in sympatry in many places throughout their range (Charles-Dominique 1977a) and may influence the distribution of sympatric Asian lorises (Duckworth 1994).

SOCIAL ORGANIZATION

Because of the difficulty of nocturnal observation, especially of taxa living in dense tropical rain forest, direct observations of social behavior may be limited (Sterling et al. 2000). Observations of associations between conspecifics, especially at sleeping sites, contribute to our knowledge of the social lives of lorisiforms; but for the most part, indirect observation has been more fruitful. Therefore, studies of communication and patterns of home range overlap have provided the bulk of our knowledge of social behavior, supplemented by a few studies using radiotracking.

Olfactory Communication

One of the most understudied areas of lorisiform social behavior is that of olfactory or chemical communication aided by an acute sense of smell and Jacobson's organ in the roof of the mouth, which senses liquid chemicals transferred from the moist nose (Schilling 1979, Martin 1990). Nocturnal lorisiforms communicate both with a number of specialized scent glands as well as with urine, which has also been shown to play an important role in enhancing an animal's grip during locomotion (Welker 1973, Harcourt 1981). The visual systems of nocturnal lorisiforms are highly sensitive and are supplemented by olfactory communication (Bearder et al. 2006). The main advantage of olfactory communication via scent gland and urine marking in general is that it conveys information that is indirect and deferred in time, with the result that individuals do not have to come together in order to communicate. Although its prevalence has never been questioned, the difficulty of studying olfactory behavior has led to few systematic studies. Captive studies of pygmy lorises (Fisher et al. 2003a,b), Senegal galagos (Nash 1993), and thick-tailed galagos (Clark 1978a,b, 1982a,b) have shown the ability of nocturnal primates not only to differentiate the state of sexual receptivity of conspecifics using scent but also to recognize specific individuals of different age and sex classes. In fact, Clark (1985) suggested that the ability for fine olfactory differentiation contributed

to increased gregariousness among *O. crassicaudatus*. In the only systematic study of olfactory behavior in free-ranging lorisiforms, Charles-Dominique (1974b, 1977b) showed that, rather than using scent as trails, the sympatric taxa he studied scent-marked in specific areas, with clear signals serving for sexual attraction and avoidance. In slow lorises scent may have an important antipredator function (Nekaris et al. 2007a).

Vocal Communication

More easily studied than olfaction, vocalizations have been invaluable for understanding the social behavior of galagos and, to a lesser degree, of lorises (Bearder et al. 1995, 2002; Honess 1996; Zimmermann et al. 1988; Zimmermann 1990, 1995a; Anderson et al. 2000; Coultas 2002; Bearder 2007). Since animals can always remain silent, their calls invariably reflect circumstances where they benefit in some way and, therefore, provide a strong clue to important aspects of their ecology and social behavior. For example, calls are given when it is advantageous to attract and maintain contact with companions, increase distance between rivals, warn kin of the presence of dangers, and warn potential predators that they have been detected. In the case of galagos in particular, the safety provided by living in trees at night and the ability to escape rapidly if detected means that they can communicate effectively by sound even when they appear to be alone. They have a rich vocal repertoire of 8–25 structurally distinct calls, including sounds that are *discrete* (relatively invariable) and others that are *graded* (continuously changing from one form to another). Added to this, galagos are able to mix different calls into rapidly changing sequences that can sometimes last for over 30 min at a time. Calls are used during short-range social interactions, with some variation between animals of different age and sex; but each species also has some calls that are loud and used when mobbing predators, attracting partners, or repelling rivals. Fortunately for researchers, every species has one particular loud call that is common to both sexes and used to advertise their presence to companions and rivals. Since this call helps to bring mates together, it is invariably species-specific, remaining more or less constant across the entire geographical range of each species, thereby providing a convenient diagnostic tool for identifying new species (Courtenay and Bearder 1989; Masters 1991; Zimmermann 1995a,b; Anderson, M. J., et al. 2000; Wallace 2006).

The less agile pottos and lorises as a group are not so obviously vocal, but unlike galagos, some of their calls include sounds in the ultrasonic range that remain inaudible to humans without a bat detector (Zimmerman 1985, Schulze and Meier 1995b, Nekaris and Jayewardene 2004). Still, some species, such as three slender loris taxa (*L. t. tardigradus*, *L. l. lydekkerianus*, and *L. l. nordicus*), are known to call throughout the night. Although the calls of Mysore slender lorises have not been bioacoustically analyzed, they have several functions, including spacing, aggression, affiliation, and dawn assembly (Nekaris 2000, Bearder et al. 2002). At least six loud whistles with different functions have been identified for both *L. t. tardigradus* and *L. l. nordicus* (Coultas 2002, Bernede 2009). The latter taxon in captivity clearly uses one of these whistles for territorial spacing (Schulze and Meier 1995b), and one of these calls also has this function in the wild (Nekaris and Jayewardene 2003). Further studies of vocal repertoires within these species should prove to be rewarding.

Social Behavior

Nocturnal primates are typically described as "solitary," despite extensive efforts by individuals studying them to dispel the use of this term (Charles-Dominique 1978, Bearder 1987). Sterling (1993) recommended that three components be used to aid in emphasizing the diversity of nocturnal primate social organization. The first of these, the social system, relates to social behavior and relationships within a group. Many of the lorisiforms engage in considerable amounts of social behavior. Table 4.4 compares the percentage of the active period that lorisiforms were seen together or in close proximity. This percentage does not include time spent communicating by scent or vocal communication. Although these figures also include mothers with their dependent offspring (e.g., *A. aureus*), a number of authors have pointed out that many adult nocturnal primates spend time together outside the breeding season, foraging and feeding (e.g., lesser galagos, Bearder and Martin 1980b; thick-tailed galagos, Clark 1985, Rhadakrishna and Singh 2002; Mysore slender lorises, Nekaris 2003a). Some taxa spend up to 50% of their time in social proximity with adult conspecifics. Variability also exists in choice of companions. Among *Galago moholi*, for example, females were the most common social partners (Bearder and Doyle 1974b), whereas in *L. l. lydekkerianus*, females formed positive affiliations only with multiple adult males (Nekaris 2002, 2003a; Rhadakrishna and Singh 2004). When compared with diurnal primates, the figures for social interactions among nocturnal primates fall well within the range of diurnal monkeys and apes (see Chapter 39). This is excluding the fact that most nocturnal lorisiforms sleep in close proximity (e.g., *Perodicticus* and *Nycticebus*) or in gregarious groups (most galagos and slender lorises) (Table 4.5), where social cohesion behaviors such as grooming and huddling take place.

Ranging

Determination of home range overlap via radiotracking or extended observation in open environments further elucidates the varied social relationships of the lorisiforms and defines the second descriptor recommended by Sterling (1993), that of the spacing system. Building on pioneering work by Bearder (1987), Müller and Thalmann (2000) have constructed a framework by which home range overlap, or spacing system, can be used to illustrate the diversity among nocturnal mammal social organization. In this framework, grouping systems can be cohesive and gregarious, dispersed yet social, or solitary, meaning no social contacts are made outside the mating system (Müller and Thalmann 2000, Sterling et al. 2000). Adult sex composition mirrors that seen among diurnal primate social organizations, with single male and female units, single male and multiple female groups, single female with multiple male groups, and multiple male and female groups. Thirteen long-term studies have been conducted, which have determined the size and degree of overlap of the home ranges of lorisiform primates (Table 4.6). In the case of rain forest primates, where observation by any other means might prove impossible, radiotracking has become invaluable to infer social organization based on spacing patterns.

Table 4.6 summarizes the inferred social organizations of those lorisiforms studied to date; the social organization of these primates has also been the topic of two reviews (Bearder 1987, Müller and Thalmann 2000). Most galagos appear to exhibit a dispersed multimale system, whereby males have larger home ranges than females and females form matrilocal clusters of related females that may sleep together. These related females tend to be aggressive toward those from other groups, whereas males may be aggressive toward one another (Bearder and Doyle 1974a, Charles-Dominique 1974a, Bearder and Martin 1980b). These males may differ in age and status. For example, smaller resident males may be tolerated by the larger territorial males, others may be constantly on the move ("floaters"), and finally some males remain solitary during the process of dispersing from their natal groups (Charles-Dominique 1972, Bearder 1987). A one-male, multiple-female system may be present in *S. gabonensis*, where males are exclusively associated with small groups of females and have nothing but extremely aggressive contact with other males (Charles-Dominique 1974a, 1977a,b). Another exception is found in *Galagoides cocos*, which may form one-male/one-female or one-male/two- or three-female associations, although variability between study sites shows some convergence with the general multimale social organization (Harcourt and Nash 1986b).

Two different systems have been shown for *P. p. edwardsi*, the only potto for which home range data are available. Charles-Dominique (1977a) studied this species in a restricted forest environment where no matriarchies were present, with female home ranges isolated from one another. However, males may overlap their ranges with more than one female but tend to avoid one another, probably using scent. Not enough data are available from this study in order to classify the social organization (Müller and Thalmann 2000). A more recent study of the same subspecies of potto found that males and females shared their home ranges to the exclusion of other male/female pairs. These same pairs also slept together or very near one another on most occasions, suggesting a single-male/single-female spacing system (Pimley 2002).

Mysore slender lorises (*L. l. lydekkerianus*) exhibited limited range overlap between females, who were aggressive at territorial boundaries. Male ranges were much larger than those of females. One or more adult males shared sleeping sites with females; males were aggressive only to males from other sleeping groups. The spacing indicates a single-male/single-female and single-male/multiple-female system but is also combined with promiscuous mating, suggesting a multimale/multifemale social organization (Nekaris 2003a). Greater slow lorises (*N. c. coucang*) appear to exhibit a single-male/single-female social organization, with the most common groupings being an adult male and female pair and their dependent offspring. This assessment corresponds with low testes volume for this taxon (Wiens and Zitzmann 2003b). Nevertheless, a polygynous mating system may exist (Elliot and Elliot 1967).

The final aspect recommended by Sterling (1993) as necessary to understand nocturnal primate social complexity is knowledge of the mating system, that is, which animal actually mates and produces offspring with another. The study of molecular ecology for the understanding of lemur mating systems has recently taken off (e.g., Fietz et al. 2000; Radespiel et al. 2001, 2002). Due to difficulties in gaining permits, only two such studies are available for the lorisiforms. A recent elegant study (Pullen 2000, Pullen et al. 2000) showed that, despite their spatial advantages and despite fathering a majority of offspring in the study population, "alpha" lesser galagos, *Galago moholi*, were not always the fathers of infants. Furthermore, not all twins were fathered by the same individuals (Pullen et al. 2000). These results are in line with both the testicular and copulatory evidence for this species, which suggests polygynandry. Pimley's (2002) molecular data for *P. p. edwardsi* at Mt. Kupe showed that offspring of mothers were not fathered by the male with which they were spatially paired. These data were in contrast to testicular volume data, which implied monogamy, suggesting that the social system differed from the mating system (Pimley 2002).

REPRODUCTIVE STRATEGIES AND LIFE HISTORY

The life history strategies of the lorisiforms have been the focus of a number of captive studies (e.g., Manley 1966, 1967; Ehrlich and Musicant 1977; Doyle 1979; Izard and Rasmussen 1985; Rasmussen 1986; Rasmussen and Izard 1988; Ehrlich and Macbride 1989; Nash 1993; Weisenseel et al. 1998; Fitch-Snyder and Ehrlich 2003), yielding much of the information summarized in Table 4.7. A number of recent field studies, however, have supplemented the captive data, enhancing our knowledge of lorisiform life history parameters, mating behavior, mating systems, and infant care (Gursky and Nekaris 2003). Recent reviews have summarized in detail aspects of the development patterns of

nocturnal primates (Nash 1993) and the reproductive biology of the African lorisiforms (Bearder et al. 2003) and the slender lorises (Nekaris 2003b).

A number of reproductive parameters characterize the galagos, pottos, and lorises. All taxa, with few exceptions, give birth to either singletons or twins, with twin births being known from more than half the taxa studied at present. A number of taxa have two litters per year. Little is known about survivorship ratios of lorisiforms in the wild, but when it is mentioned, it is not uncommon for only one infant out of a potential four to reach sexual maturity. Infant parking is common among the lorisiforms. In general, the practice is for the mother to leave her infants on a branch or in a tree hole while she goes off to forage. The only variation seems to be whether the infant is parked throughout the night (most pottos and lorises) or carried with the mother for short distances and cached in multiple sites throughout the night (most galagos). Variation across taxa also exists in whether or not infants cling to the fur while carried or are transported in the mother's mouth. Contrasting rates of life history among the galagos, pottos, and lorises are considered by some authors to be related to other locomotor and ecological differences among the three subfamilies (see Rasmussen and Nekaris 1998 for a review). The pottos and lorises are noted for having among the longest life history of any primates of their body size, including long gestation lengths followed by low birth weights and long periods of lactation, in contrast to galagos, which fall more in line with other primates of their body size (Martin 1990).

Interestingly, another feature uniting the lorisiforms is the absence of a single observation of infanticide in the wild. Although adults may kill infants under captive conditions, this has been shown to be due to stress or poor management rather than infanticide as an evolutionary strategy (Nekaris 2003a). In fact, male slender lorises regularly play with infants outside their sleeping groups. Males dispersing to a new area also show this behavior, even though it is highly unlikely that they are the fathers of infants. High reproductive output among twin-bearing lorisiforms with much opportunity for males to sire offspring suggests that infanticide has not played an important role in this infraorder (Manley 1966, Nekaris 2002, Bearder et al. 2003).

Dixson (1995, 1998) has pointed out that a number of features of the genital morphology and the copulatory behavior of nocturnal lorisiforms may provide evidence that the spacing system does not necessarily coincide with the mating system. For example, larger testes size or increase of testes size during a breeding season should be linked with a multiple-male/multiple-female mating system (*polygynandry*). The elaborate penile morphology of most lorisiforms might also serve to enhance female receptivity or genital lock or to break up copulatory plugs left by other males and might also provide a clue to the mating systems of these primates.

Information on dispersal is limited for most taxa. Many galagos appear to be matrilineal in their social organization, with males dispersing at sexual maturity and females either

Table 4.7 Reproductive and Life History Parameters

TAXA	INFANTS/YEAR**	LITTER SIZE**	PARKING**	INFANT CARRIAGE**	GESTATION (DAYS)	WEANING (DAYS)	WEIGHT AT BIRTH (G)	AGE AT SEXUAL MATURITY (MONTHS)	DISPERSING SEX**	BREEDING SEASON?
Galaginae										
Galagoides demidovii	1–2	1	Yes	Mouth	111–114	40–50	5–10	8–10	F	—
G. thomasi	—	—	Yes	Mouth	111–114	—	5–12	—	—	—
G. cocos	2–4	1–2	Yes	Mouth	120	49	16.5	—	F	—
G. rondoensis	2	1?	Yes	Mouth	—	—	—	—	—	—
G. zanzibaricus (udzungwensis)	2	1	Yes	Mouth	—	—	—	—	—	—
G. granti	—	—	Yes	Mouth	—	—	—	—	—	—
G. orinus	—	—	Yes	Mouth	—	—	—	—	—	—
Sciurocheirus gabonensis	1–2	1	Yes	Mouth	133	—	24	8–10	—	—
S. alleni cameronensis	1–2	1	Yes	Mouth	—	—	—	—	F	Yes
Galago moholi	1–2	2	Yes	Mouth	120–126	—	11–12	8½	F	Jan–Feb/Oct–Nov
G. senegalensis	1–2	1	Yes	Mouth	141 + 2	70–98	19 + 2.6	12–18	F	Feb–Mar/Jun–Jul
G. matschiei	—	—	Yes	Mouth	—	—	—	—	—	—
Euoticus elegantulus	1	1	No	Fur	—	—	—	—	—	—
E. pallidus	1	1	No	Fur	—	—	—	10	—	—
Otolemur garnettii	1	1	Yes	Mouth/ fur	126–138	140	—	12–18	F	Possibly Oct/ Nov
O. crassicaudatus	2–3	1	Rare	Mouth/fur	136	70–134	—	18–24	F	Possibly Oct/ Nov
Perodicticinae										
Perodicticus potto edwardsi	1/(2)	1	Rare	Fur	197 (193–205)	120–180	52; 30–42	6	F	No; Aug–Jan high rate
Arctocebus aureus	1	1	Yes	Fur	131–136	100–130*	24–30*	9–10	—	No
A. calabarensis	1–2	1	Yes	Fur	130	115	35	—	—	?; common Jan–Apr
Lorisinae										
Loris lydekkerianus lydekkerianus	1–4	1–2*	Yes	Fur	164 (160–166)**	190			F or M	No**
L. l. malabaricus	1–4	1–2	Yes	Fur	166–169	120–150		11	F or M	No
L. l. nordicus	1–4	1–2*	Yes	Fur	157–172*	148–175*		12*	—	No**
L. tardigradus tardigradus	1–4	1–2*	Yes	Fur	167–175	185			—	No**
L. t. nycticeboides		2*	?		174*				—	—
Nycticebus coucang coucang	1	1	Yes	Fur	165–175*	85–180*	43.5–62.5	16–21	?	—
N. pygmaeus	1–4	1/2	Yes	Fur	185–197*	180	16.5–30	12–18*		

All data from captivity unless marked with a ** to indicate wild data or animals recently caught from the wild;—indicates no data are available.

sharing a range with their mother or moving into a neighboring range (Bearder 1987). In Mysore slender lorises (*L. l. lydekkerianus*) and greater slow lorises (*N. c. coucang*), both males and females have been seen to disperse (Wiens 2002, Nekaris 2003b, Rhadakrishna and Singh 2004).

CONSERVATION STATUS

A cursory examination of Table 4.1 is enough to emphasize that very little is known about the conservation status of most nocturnal lorisiforms. Despite a number of surveys conducted for galagos (e.g., Honess 1996; Honess and Bearder 1996; Weisenseel et al. 1998; Butynski et al. 1998; Ambrose 1999; Ambrose and Perkin 2000; Perkin 2001a,b, 2002), pottos (Oates and Jewell 1967), and lorises (e.g., Barrett 1981; Duckworth 1994; Nekaris 1997; Singh et al. 1999, 2000; Fitch-Snyder and Vu 2002; Nekaris and Jayewardene 2004; Nanda Kumar et al. 2004; Kumara et al. 2006; Nekaris et al. 2008; Rhadakrishna et al. 2006), a large proportion of species have been described as "data-deficient." Where systematic studies have been conducted, they have almost always resulted in worrying conservation rankings (e.g., Nekaris and Nijman 2007). For example, one Sri Lankan loris (*L. t. nycticeboides*) and an unnamed species of galago (*Galagoides* sp. nov. 3) from Tanzania are considered critically endangered and have been included on the recent list of the world's top 25 most endangered primates (Nekaris and Penera 2007, Honess et al. 2007). Although at this stage ranked as vulnerable, systematic surveys of pygmy lorises (*N. pygmaeus*) where virtually none have been seen suggest that they are more seriously threatened than the high availability in markets would suggest (Starr et al. in press).

The paucity of studies on these African and Asian primates may lead them to be ignored at a time when they are facing severe human-induced threats. The bushmeat trade in Africa and the pet and biomedical trades in Asia are having detrimental effects on lorisiform populations (Ratajszczak 1998, Schulze and Groves 2004, Parish and Nekaris in press). Habitat loss in both Africa and Asia as a result of human population pressures also poses a severe threat to these species, which often go unconsidered in habitat development and planning (Erdelen 1988, Butynski 1996/1997, Thorn et al. 2009). In Africa, human population growth rates are still increasing at 2.9% per annum (Butynski 1996/1997). In Africa and Asia, clearing of the land for agriculture and deforestation for logging are the chief causes of forest loss (Mill 1995). Nocturnal prosimians may be at the greatest risk as they are asleep during the times of mass forest clearance, whereas other primates have the chance to flee. Sleeping nocturnal primates may be more easily burned alive or chopped down with the trees, collected, and sent to animal markets (Ratajszczak 1998, Schulze and Groves 2004). The tendency for lorises to cling to trees as they are cut, rather than fleeing, makes them an easy target for removal for the pet trade. Thus, whereas other animals can escape capture, lorises can be completely drained from areas of deforestation (Fitch-Snyder and Vu 2002, Streicher 2004). Furthermore, logging and human disturbance have been shown to adversely affect lorisiform density (Weisenseel et al. 1993, Nekaris and Jayewardene 2004). It is inappropriate to assume that healthy diurnal primate populations signify a healthy nocturnal primate population as the substrate and sleeping site requirements of these two groups of primates differ and surveys have often shown an inverse relationship in the presence of strepsirhine and haplorhine primates (e.g., Singh et al. 1999, 2000; Perkin 2001a; Nekaris and Jayewardene 2004).

Luckily, an increasing number of sanctuaries and reintroduction programs are being developed for Asian lorises, where the trade for pets and medicines is especially dire (Sanfey 2003, Collins and Nekaris 2008). These sanctuaries operate in the face of stiff opposition from those who consider that priority should be given to "the more important" primates, such as gibbons and orangutans. Prosimians (strepsirhines and tarsiers) are in the unfortunate position of being relatively ignored by other conservation action groups because they are primates and ignored by primatologists because they are not anthropoids. With advances in the understanding of species-level biology and the uncovering of more and more species, there is a genuine chance that species can be lost or assigned the status of critically endangered before they are even named (Bearder 1999). Future studies of individual species, equivalent to those conducted for day-living primates, will ensure that this genetically diverse and interesting group is no longer excluded from conservation initiatives.

Species included in this chapter are shown in Color Plates 1 and 5 (*Nycticebus coucang* and *Galago matschiei*).

5

Lemuriformes

Lisa Gould, Michelle Sauther, and Alexandra Cameron

1. What are the hypotheses proposed to explain the evolution of unique behavioral characteristics in the lemurs?
2. Nocturnal and diurnal lemurs differ in many aspects of their morphology and behavior.
   a. What are some general morphological differences?
   b. What are some general behavioral differences?

INTRODUCTION

Behavioral and ecological research on the Malagasy primates began in the late 1950s, when Petter (1962) surveyed Madagascar's fauna and published preliminary information on several lemur species at a variety of sites around the island (Figs. 5.1, 5.2). In the 1960s and 1970s, a number of primate biologists undertook the first in-depth studies of *Lemur catta* and *Propithecus verreauxi* at the Berenty site in the far south of the island (Jolly 1966) and comparative studies of *P. verreauxi* in the northwest and south (Richard 1973, 1974), *L. catta* and *Eulemur fulvus rufus* in the southwest (Sussman 1972, 1974), and *Indri* in the eastern rain forest (Pollock 1975, 1977). Some of the nocturnal lemurs were also studied for the first time in the 1970s by Martin (1972a), who focused on *Microcebus*, while Charles-Dominique and Hladik (1971) documented early information on *Lepilemur*. On the nearby Comoro Islands, the only place outside of Madagascar where lemurs are found, Tattersall (1976, 1977b) conducted research on *Eulemur mongoz* on Moheli and Anjouan islands as well as on the one subspecies of brown lemur not found on Madagascar, *E. f. mayottensis*, the Mayotte brown lemur, on the island of the same name.

In the 1970s, the political situation in Madagascar precluded most lemur research (Jolly and Sussman 2006); but in the 1980s, many Malagasy, North American, European, and Japanese researchers began a new phase of field studies on most of the extant lemur species. Lemur research has grown remarkably in the past two and a half decades as topics such as geographic distribution; behavior and ecology of previously unknown species; correlations between climate, diet, and dominance patterns; long-term demographic and life history studies; hormonal correlates of behavior; health studies; and genetics have been and are being investigated.

In this chapter, we will first present information on the origins of the Malagasy primates, taxonomy and classification of the living lemurs, lemur morphology, and current explanations relating to the evolution of behavioral traits which distinguish lemurs from other primate groups. We will then give an overview of the ecology of the extant lemurs. A discussion of the taxonomy, morphology, and proposed ecology of the extinct subfossil lemurs is beyond the scope of this chapter; however, interested readers should consult Jungers et al. (2002, 2007), Godfrey et al. (1993, 1997, 2006), and Godfrey and Irwin (2007) for current information concerning these species.

ORIGIN OF THE LEMURS

The landmass that is now Madagascar split from mainland Africa approximately 165 million years ago (mya) and drifted southward to its present position, 400 km from mainland East Africa around 121 mya (Yoder et al. 2003). At this time, Madagascar was part of a larger landmass that included India. About 88 mya, the landmass split again, and India drifted northward, eventually colliding with Asia (Yoder et al. 2003). Madagascar has thus been isolated from any other landmass for 88 million years. How then did lemurs end up on Madagascar, when primates did not exist at the time of the landmass separation? Lemur ancestors arose in Africa either during the Eocene epoch (55–37 mya) or even earlier as Martin (2000) notes that deoxyribonucleic acid (DNA) sequencing suggests they may have begun to colonize Madagascar as early as 80 mya. The most accepted explanation as to how they got to Madagascar is via over-water dispersal, or "rafting," on large clumps of floating vegetation (Martin 1972b, 2000; Mittermeier et al. 1994; Yoder et al. 2003). It has been suggested that

Figure 5.1 National parks and reserves in Madagascar. Based on a map from Madagascar: *The Bradt Travel Guide*, 7th ed., 2002. Reprinted with kind permission of the editor, H. Bradt.

Figure 5.2 Map of vegetation zones in Madagascar. From Sussman (1999a), based on Humber's (1955) original.

the ancestral lemurs survived this long journey because they may have had the ability to go into a state of torpor for lengths of time, which would have offset the problem of food shortages (Martin 1972b, 2000; Kappeler 2000a). Tattersall (2004, 2006b) argues that some form of "island hopping" may have occurred by means of paleogeographic "stepping stones" in the form of seabed exposures in the Mozambique channel at different times during the Tertiary and that these small islands would have reduced the distances between points of land that mammals crossing over to Madagascar would have had to travel. Whether today's lemurs arose from just one or several separate waves of migration (Yoder et al. 2003; Karanth et al. 2005; Tattersall 2004, 2006b), once on Madagascar these ancestral species underwent a large adaptive radiation over many millions of years, resulting in the living lemurs of today and the extinct (subfossil) lemurs.

CLASSIFICATION OF THE LEMURS

There are five taxonomic families of living Malagasy primates: the Lemuridae, Indriidae, Cheirogaleidae, Lepilemuridae (sometimes classified as Megaladapidae), and Daubentoniidae. Within these families we find 15 genera of extant lemurs, which represent 25% of all extant primate genera in the world (Fleagle 1999) and possibly up to 97 or more species (Mittermeier et al. 2008a). There are also seven genera and at least 16 species of very recently extinct, or subfossil, lemurs (Karanth et al. 2005). Within

the past few years taxonomic revisions and the discovery of new species have greatly increased the projected number of lemurs from even a few years ago. Table 5.1 lists current extant lemur and extinct subfossil taxonomy.

Morphology

Like other strepsirhines, lemurs are morphologically characterized by a number of primitive features of the skull, including a postorbital bar rather than postorbital closure, a less derived mammalian nasal region, reduced upper incisors, and a toothcomb, made up of the lower incisors and canines (with the exception of the aye-aye). They also possess a moist rhinarium, or "wet nose," a primitive mammalian trait which aids in their very keen sense of smell (Fleagle 1999). Scent marking is used by all species in various forms and contexts: Scent glands are situated on heads and palms in male *Eulemur*, under wrist spurs in male *L. catta*, on chest glands in *Propithecus*, and in the anogenital area in all species, under the tail. Scent is used commonly to denote the presence of a group or individual and extensively during the mating season.

All lemurs possess a grooming claw on the second toe of each foot (Fleagle 1999).

Table 5.1 Taxonomy of the Malagasy Primates

Extant lemurs

Family Lemuridae

Lemur catta (ring-tailed lemur)

Eulemur fulvus (5 subspecies)

E. f. fulvus (common brown lemur), *E. f. rufus* (red-fronted brown lemur), *E. f. sanfordi* (Sanford's brown lemur), *E. f. albifrons* (white-fronted brown lemur), *E. f. mayottensis* (Mayotte brown lemur)

Eulemur collaris (collared brown lemur)

Eulemur cinereiceps (gray-headed lemur/white-collared brown lemur)

Eulemur macaco (2 subspecies)

E. m. macaco (black lemur), *E. m. flavifrons* (Sclater's black lemur)

Eulemur coronatus (crowned lemur)

Eulemur rubriventer (red-bellied lemur)

Eulemur mongoz (mongoose lemur)

Hapalemur griseus (3 subspecies)

H. g. griseus (eastern lesser bamboo lemur), *H. g. occidentalis* (western lesser bamboo lemur), *H. g. alaotrensis* (Lac Alaotra bamboo lemur)

Hapalemur meridionalis (southern lesser bamboo lemur)

Hapalemur gilberti (Gilbert's bamboo lemur)

Hapalemur aureus (golden bamboo lemur)

Prolemur simus (greater bamboo lemur)

Varecia variegata (2 subspecies)

V. v. variegata (black and white ruffed lemur), *V. v. rubra* (red ruffed lemur)

Family Indriidae

Propithecus verreauxi (4 subspecies)

P. v. verreauxi (Verreaux's sifaka), *P. v. deckeni* (Decken's sifaka), *P. v. coronatus* (crowned sifaka), *P. v. coquereli* (Coquerel's sifaka)

Propithecus diadema (4 subspecies)

P. d. diadema (diademed sifaka), *P. d. edwardsi* (Milne-Edwards sifaka), *P. d. candidus* (silky sifaka), *P. d. perreri* (Perrier's sifaka)

Propithecus tattersalli (golden-crowned sifaka)

Indri indri (indri)

Avahi laniger (eastern woolly lemur)

Avahi occidentalis (western woolly lemur)

Avahi meridionalis (southern woolly lemur)

Avahi unicolor (unicolor avahi)

Avahi cleesei (Bemaraha woolly lemur)

Avahi mooreorum (Moore's woolly lemur)

Avahi peyrierasi (Peyrieras' woolly lemur)

(cont'd)

Avahi ramanantsoavani (Ramanantsoavana's woolly lemur)

Avahi betsileo (Betsileo woolly lemur)

Family Megaladapidae

Lepilemur aeeclis (Aeecl's sportive lemur)

Lepilemur ahmansoni (Ahmanson's sportive lemur)

Lepilemur ankaranensis (Ankarana sportive lemur)

Lepilemur betsileo (Betsileo sportive lemur)

Lepilemur fleuretae (Fleurete's sportive lemur)

Lepilemur grewcocki (Grewcock's sportive lemur)

Lepilemur hubbardi (Hubbard's sportive lemur)

Lepilemur jamesi (James' sportive lemur)

Lepilemur manasamody (Manasamody sportive lemur)

Lepilemur milanoii (Daraina sportive lemur)

Lepilemur otto (Otto's sportive lemur)

Lepilemur petteri (Petter's sportive lemur)

Lepilemur randrianasoli (Randrianasoli's sportive lemur)

Lepilemur sahamalazensis (Sahamalaza's sportive lemur)

Lepilemur scottorum (Scott's sportive lemur)

Lepilemur seali (Seal's sportive lemur)

Lepilemur tymerlachsoni (Hawk's sportive lemur)

Lepilemur wrighti (Wright's sportive lemur)

Lepilemur mustelinus (weasel sportive lemur)

Lepilemur microdon (small-toothed sportive lemur)

Lepilemur leucopus (white-footed sportive lemur)

Lepilemur ruficaudatus (red-tailed sportive lemur)

Lepilemur edwardsi (Milne-Edwards sportive lemur)

Lepilemur dorsalis (gray-backed sportive lemur)

Lepilemur septentrionalis (northern sportive lemur)

Lepilemur hollandorum (Holland's sportive lemur)

Family Cheirogalidae

Microcebus arnholdi (Arnhold's mouse lemur)

Microcebus simmonsi (Simmons' mouse lemur)

Microcebus lehilahytsara (Goodman's mouse lemur)

Microcebus jollyae (Jolly's mouse lemur)

Microcebus margotmarshae (Margot Marsh's mouse lemur)

Microcebus macarthurii (MacArthur's mouse lemur)

Microcebus mittermeieri (Mittermeier's mouse lemur)

Microcebus mamiratra (Claire's mouse lemur)

Microcebus bongolavensis (Bongolava mouse lemur)

Microcebus danfossi (Danfoss' mouse lemur)

Microcebus lokobensis (Lokobe mouse lemur)

Microcebus murinus (gray mouse lemur)

Microcebus rufus (brown mouse lemur)

Table 5.1 (cont'd)

Microcebus myoxinus (pygmy mouse lemur)

Microcebus ravelobensis (golden-brown
mouse lemur)

Microcebus tavaratra (northern rufous mouse lemur)

Microcebus sambiranensis (Sambirano mouse lemur)

Microcebus berthae (Berthe's mouse lemur)

Microcebus griseorufus (gray-brown mouse lemur)

Allocebus trichotis (hairy-eared dwarf lemur)

Cheirogaleus major (greater dwarf lemur)

Cheirogaleus crossleyi (Crossley's greater dwarf lemur)

Cheirogaleus ravus (large iron-gray dwarf lemur)

Cheirogaleus minisculus (lesser iron-gray
dwarf lemur)

Cheirogaleus sibreei (Sibree's dwarf lemur)

Cheirogaleus medius (fat-tailed dwarf lemur)

Cheirogaleus adipicaudatus (southern dwarf lemur)

Phaner furcifer (4 subspecies)

   P. furcifer furcifer (eastern fork-marked lemur),
   P. furcifer pallescens (pale fork-marked lemur),
   P. furcifer parienti (Pariente's fork-marked lemur),
   P. furcifer electromontis (Amber Mountain
   fork-marked lemur)

Family Daubentoniidae

  Daubentonia madagascariensis (aye-aye)

Subfossil lemurs

  Subfossil Lemuridae

    Pachylemur insignis

    Pachylemur jullyi

  Subfossil Daubentoniidae

    Daubentonia robusta

  Subfossil Megaladapidae

    Megaladapis grandidieri

    Megaladapis madagascariensis

    Megaladapis edwardsi

  Family Paleopropithecidae (all subfossils)

    Mesopropithecus globiceps

    Mesopropithecus pithecoides

    Mesopropithecus dolichobrachion

    Babakotia radofilai

    Paleopropithecus ingens

    Paleopropithecus maximus

    Archaeoindris fontoynontii

Source: Adapted from Tattersall (1982), Mittermeier et al. (1994,
2008a,b), Groves (2000), Rasoloarison et al. (2000), Jungers et al. (2002),
Johnson, et al. (2008).

The dental formula is variable. In both the cheirogaleids and the lemurids, the dental formula is 2.1.3.3. Lepilemurids lack permanent upper incisors, so their upper dental formula is 0.1.3.3 and the lower, 2.1.3.3. The indriids have quite different dentition from the above lemurs. Their dental formula is the same as that found in the Old World monkeys, apes, and hominids, 2.1.2.3; and their toothcomb includes the incisors only, not the canine as in the other lemuroid taxa. The oddest dentition of all lemurs, and probably of all primates, is that of the aye-aye (*Daubentonia madagascariensis*), the only living member of the Daubentoniidae family. Its dental formula is 1.0.1.3/1.0.0.3 and the middle two incisors grow constantly, like those found in rodents and lagomorphs (rabbits and hares) (Fleagle 1999).

Mean weights of species captured in the wild can be found in Table 5.2.

Nocturnal Lemur Postcranial Morphology and Locomotion

Most nocturnal lemurs have relatively short, pointed snouts and large, moveable ears. The mouse lemurs (*Microcebus*) are branch runners. Their arms and legs are short relative to their trunk, and their tail is as long as their body. Dwarf lemurs (*Cheirogaleus*) have a shorter tail and legs that are longer than the arms. Sportive lemurs (*Lepilemur*) have an enlarged cecum, to help digest the cellulose in their mostly folivorous diet (Fleagle 1999). Sportive lemurs are not branch runners; rather, they travel by vertical leaping (Ganzhorn 1993). The woolly lemur (*Avahi*) is the only nocturnal indriid. It is much smaller than its close relatives, the diurnal sifakas (*Propithecus*) and indri (*Indri*); but its mode of locomotion is the same: vertical clinging and leaping. Using vertical clinging and leaping, the animal begins the leap with its back toward its destination tree, takes a large leap, twists in midair, and lands facing forward (Richard 1985a). The legs of both *Lepilemur* and *Avahi* are much longer than the arms, a necessary adaptation for this type of locomotion (Fleagle 1999). The aye-aye, an extremely unusual-looking primate, is covered in black, shaggy hair and has enormous bat-like ears, a large bushy tail, and an extremely elongated third digit on its hands, which it uses in extractive foraging for grubs, insects under bark, and egg yolks (Erickson 1991, 1994; Fleagle 1999).

Postcranial Morphology and Locomotion of the Diurnal Lemurs

Lemur, *Eulemur*, and *Varecia* are quadrupedal walkers and runners; but they also leap from branch to branch. *Varecia* also uses suspensory postures for feeding (Fleagle 1999). *L. catta*, the only species within the genus *Lemur*, spends about 30% of its time on the ground (Jolly 1966, Sussman 1974). Unlike other lemurs, the

Table 5.2 Mean Body Weights of Wild-Caught Lemur Species

SPECIES	MEAN WEIGHT (AND/OR WEIGHT RANGE)	SOURCE
Diurnal lemurs		
Lemur catta	2.2 kg	Sussman 1991a,b
Eulemur fulvus spp.	1.8–2.6 kg	Glander et al. 1992, Mittermeier et al. 1994, Freed 1996, Terranova and Coffman 1997, Vasey 2000
E. mongoz	1.5 kg	Terranova and Coffman 1997
E. macaco	2.4 kg	Mittermeier et al. 1994
E rubriventer	2.0 kg	Glander et al. 1992
E. coronatus	1.77 kg	Terranova and Coffman 1997
Hapalemur griseus griseus		
H. g. occidentalis	700–1,000 g	Mittermeier, et al. 1994, Tan 1999
H. g. alaotrensis	1.2 kg	Mutschler 2002
H. gilberti	967 g	Rabarivola et al. 2007
H. aureus	1.5–1.6 kg	Glander et al. 1992, Tan 1999
P. simus	2.4 kg	Meier et al. 1987, Tan 1999
Varecia variegata	3–4.5 kg	Tattersall 1982, Morland 1993, Terranova and Coffman 1997, Britt et al. 2001b, Vasey 2002
Propithecus verreauxi verreauxi	2.8 kg	Richard et al. 2002
P. v. deckeni	3.5–4.5 kg	Tattersall 1982
P. diadema edwardsi	5–6.5 kg.	Glander et al. 1992, Wright 1995
P. d. diadema	5–6 kg	Powzyk 1997
P. tattersalli	3.5 kg	Meyers and Wright 1993
Indri indri	6.5–6.9 kg	Powzyk 1997, Britt 2000
Nocturnal lemurs		
Cheirogaleus medius	Body weight changes seasonally range = 75–200 g	Hladik et al. 1980
C. major	Body weight changes seasonally mean = 400 g	Martin, R. D., 1984
C. crossleyi	Body weight changes seasonally mean = 400 g	Petter et al. 1977
Allocebus trichotis	75–80 g	Meier and Albignac 1991
Mirza coquereli	300 g	Tattersall 1982
Microcebus murinus	50–90 g, mean = 62.3	Martin 1973, Rasoloarison et al. 2000
M. rufus	50 g	Harcourt 1987
M. myoxinus	49 g	Rasoloarison et al. 2000
M. ravelobensis	71.7g	Rasoloarison et al. 2000
M. tavaratra	61.1g	Rasoloarison et al. 2000
M. sambiranensis	44.1 g	Rasoloarison et al. 2000
M. berthae	30.6 g	Rasoloarison et al. 2000
M. griseorufus	62.6 g	Rasoloarison et al. 2000
M. margotmarshae	41 g	Louis et al. 2008
M. arnholdi	49.7 g	Louis et al. 2008
M. simmonsi	64.8 g	Louis et al. 2006a

Table 5.2 (cont'd)

SPECIES	MEAN WEIGHT (AND/OR WEIGHT RANGE)	SOURCE
M. jollyae	60 g	Louis et al. 2006a
M. mittermeieri	44.1 g	Louis et al. 2006a
M. lehilahytsara	30–64 g	Roos and Kappeler 2005
M. sambiranensis	44 g	Rasoloarison et al. 2000
M. mamiratra	60.8 g	Andriantompohavana et al. 2006
Phaner furcifer	360–500 g	Petter et al. 1977
Avahi laniger	900–1,200 g and 600–700 g	Razanahoera-Rakotomalala 1981, Petter et al. 1977
A. occidentalis	700–900 g	Razanahoera-Rakotomalala 1981
A. betsileo	1,050 g	Andriantompohavana et al. 2007
A. ramanantsoavani	875–1,255 g	Zaramody et al. 2006
A. meridionalis	950–1,400 g	Zaramody et al. 2006
A. peyrierasi	850–1,125 g	Zaramody et al. 2006
A. mooreorum	920 g	Lei et al. 2008
A. cleesei	860 g	Lei et al. 2008
Daubentonia madagascariensis	3 kg	Tattersall 1982
Lepilemur mustelinus	1 kg	Jenkins 1987
L. dorsalis	500 g	Tattersall 1982
L. septentrionalis	700–800 g	Tattersall 1982
L. edwardsi	600–900 g	Tattersall 1982
L. leucopus	550 g	Petter et al. 1977, Petter and Petter-Rousseaux 1979
L. ruficaudatus	600–900 g	Petter et al. 1977, Petter and Petter-Rousseaux 1979
L. microdon	1 kg	Petter et al. 1977, Petter and Petter-Rousseaux 1979
L. hollandorum	990 g	Ramaromilanto et al. 2009
L. scottorum	880 g	Lei et al. 2008
L. wrighti	950 g	Louis et al. 2006b
L. tymerlachsoni	880 g	Louis et al. 2006b
L. seali	950 g	Louis et al. 2006b
L. sahamalazensis	700 g	Louis et al. 2006b
L. randrianasoli	920 g	Louis et al. 2006b
L. petteri	630 g	Louis et al. 2006b
L. milanoii	720 g	Louis et al. 2006b
L. jamesi	780 g	Louis et al. 2006b
L. hubbardi	990 g	Louis et al. 2006b
L. grewcocki	780 g	Louis et al. 2006b
L. fleuretae	980 g	Louis et al. 2006b
L. betsileo	1.15 kg	Louis et al. 2006b
L. ankaranensis	790 g	Louis et al. 2006b
L. ahmansoni	610 g	Louis et al. 2006b
L. aeeclis	860 g	Louis et al. 2006b

Because there is very little sexual dimorphism in terms of body weight, "mean weight" is presented here as the actual mean when data are combined for both males and females.

fleshy pads of its hands and feet extend upward to the wrist and beyond the heel.

The three *Hapalemur* species have shorter faces than the other lemurids. Their arms are short and legs are proportionally long (Jungers 1979), an adaptation to their primary mode of locomotion, vertical clinging and leaping, although these species also move quadrupedally along branches when feeding (Fleagle 1999).

The longest leg in proportion to arm length is found in the indriids and is an adaptation to the vertical clinging and leaping mode of locomotion: The indriids are extraordinary leapers. Some sifakas (*Propithecus* spp.) occasionally come to the ground, particularly Verreaux's sifaka, which lives in dry forests. Because their legs are so long, they must move along the ground by hopping bipedally. Indriids also possess very long fingers and toes, which aid them when adopting suspensory feeding postures.

EVOLUTION OF UNIQUE BEHAVIORAL TRAITS IN MALAGASY PRIMATES, INCLUDING FEMALE DOMINANCE

Not only do lemurs exhibit aspects of morphology that differ from anthropoids but some unique behavioral traits have evolved in this group of primates which are not found in other strepsirhines or the haplorhines. These traits include female dominance in the majority of species, targeted female–female aggression, lack of notable sexual dimorphism, strict seasonal breeding (in all but two species, the aye-aye and the Lac Alaotra bamboo lemur), high infant mortality, and *cathemerality* (exhibiting both day and night activity). Why do we find such a combination of characteristics in the lemurs? A number of explanations have been offered (Wright 1999, Sussman 1999a). These hypotheses are briefly presented.

The *energy conservation hypothesis* involves a synthesis of explanations presented by Jolly (1984), Richard (1987), Young et al. (1990), Wright (1993), Sauther (1993), and Pereira (1993a,b). This hypothesis proposes that the combination of extreme and largely unpredictable climatic seasonality in Madagascar and high pre- and postreproductive costs have resulted in the evolution of female dominance. More specifically, strong food resource seasonality and climatic factors have resulted in energetic stress with respect to reproductive females. In group-living lemurs, all females in a social group are pregnant and lactating at the same time, which leads to strong female–female feeding competition during both gestation and lactation periods. Furthermore, in both group- and non-group-living lemur species, females give birth to altricial, quickly growing infants for which they must lactate. Female lemurs may have responded to such reproductive stress through the evolution of female priority of access to food resources, which can help them offset energy demands experienced in this situation. In addition to

seasonal reproduction which is largely tied to availability of good weaning foods for growing infants, the relatively lengthy winter season in parts of Madagascar is proposed to have led to the evolution of seasonal energy storage in some species, strategies for temperature regulation, modulation of metabolic and growth rates and activity levels, and timing of aggressive behaviors.

Wright (1999) argues that not all of the unique behavioral traits found in lemurs are strictly related to the conservation of energy and suggests that some are more tailored to maximizing the extraction of scarce resources. She proposes that a low basal metabolic rate, small group size, torpor, sperm competition, and seasonal breeding are adaptations related to energy conservation and that others, such as female dominance, weaning synchrony, fibrous diets, territoriality, and cathemerality (found in some species), have evolved as strategies to maximize the use of scarce resources resulting from seasonal resource shortages. Thus, Wright (1999) suggests that it may be appropriate to consider the energy conservation hypothesis as more of an "energy frugality" hypothesis.

The *evolutionary disequilibrium hypothesis* (van Schaik and Kappeler 1996) suggests that recent extinctions, particularly of large diurnal predators such as raptors, have allowed many lemur species to switch from a nocturnal activity pattern to diurnality (and cathemerality). Because of such extinctions, adaptation to a diurnal activity pattern is suggested to have occurred rapidly and recently: between 1,000 and 500 years ago. Also, the social systems of today's diurnal lemurs (species living in relatively small mixed-sex groups) may be an outgrowth of an ancestral nocturnal, monogamous condition; and pair-living animals may have been sufficiently tolerant, once diurnal, to form larger groups. Van Schaik and Kappeler (1996) propose that cathemeral activity may be an ancestral and stable activity pattern among lemurs or, conversely, may have evolved relatively recently as an occasional habit of nocturnal animals. With respect to the evolution of female dominance, van Schaik and Kappeler suggest that in group-living lemurs today female dominance may be a relic of pair-living in ancestral times since female priority of access to resources seems to be the case in pair-living species that do not exhibit sexual dimorphism and where male polygyny, and male–male competition, does not occur. They argue that the expansion of female feeding priority to overall female dominance in *group*-living lemurs (emphasis ours) suggests that female needs in ancestral monogamous species were greater in lemurs than in other primate taxa.

OVERVIEW OF THE ECOLOGY, SOCIAL ORGANIZATION, AND SOME ASPECTS OF BEHAVIOR OF THE EXTANT LEMURS

Mean group size and home range size of extant lemur species (Fig. 5.3) can be found in Table 5.3, and mean weights of wild-caught animals are presented in Table 5.2.

Figure 5.3 *A*: *Lemur catta* (photo by L. Gould). *B*: *Eulemur coronatus* (photo by B. Z. Freed). *C*: *Hapalemur simus* (photo by D. Haring). *D*: *Propithecus verreauxi verreauxi* (photo by M. L. Sauther).

Figure 5.3 (*cont'd*) E: *Propithecus diadema edwardsi* (photo by S. Arrigo-Nelson). *F*: *Varecia variegata variegata* (photo by S. Arrigo-Nelson). *G*: *Lepilemur leucopus* (photo by L. Gould). *H*: *Microcebus ravelobensis* (photo by U. Radespiel).

Table 5.3 Group Size, Home Range Size, and Habitat of Several Lemur Species at Sites in Madagascar Where They Have Been Studied

SPECIES	MEAN GROUP SIZE (OR RANGE OF GROUP SIZE)	MEAN HOME-RANGE SIZE (+RANGE)	HABITAT AND WHERE STUDIED	STUDIED BY OR CITED BY
Diurnal lemurs				
Lemur catta	11.5 and 16 (at two sites), range 3–27	6–35 depending upon local habitat	South and southwestern riverine, xerophytic, spiny, and limestone forest Beza Mahafaly Special Reserve and Berenty Reserve	Budnitz and Dainis 1975; Sussman 1977, 1991a; Jolly et al. 2002; Koyama et al. 2002; Gould et al. 2003
Eulemur fulvus rufus (eastern)	6.8	100 ha	Southeastern submontane rain forest Ranomafana National Park	Overdorff 1993a, Overdorff et al. 2003
Eulemur f. rufus (western)	9.4	1–9 ha	Western dry deciduous forest Mangoky River, Kirindy forest	Sussman 1974, Kappeler and Erkert 2003
Eulemur fulvus fulvus	12	~7 ha (west), >20 (east)	Northwestern dry deciduous forest, eastern rain forest Ampijoroa (west), Andasibe (east)	Harrington 1975, Ganzhorn 1988, Mittermeier et al. 1994
E.. f. sanfordi	Range 5–9	5–9 ha	Northern dry deciduous forest Mt. d'Ambre	Freed 1996
E. f. albifrons	Range 7–11	13 ha	Northeastern rain forest Masoala Peninsula	Vasey 2000
E. collaris	3–7	–	Southeastern rain forest and littoral forest St. Luce	Mittermeier et al. 1994, Feistner and Schmid 1999, Donati and Borgognini-Tarli 2002a
E. cinereiceps	–	–	Restricted range, southeastern rain forest remnants	Tattersall 1982, Mittermeier et al. 1994
E. rubriventer				
	2–4	19 ha	Southeastern submontane rain forest Ranomafana National Park	Overdorff 1993b
E. coronatus	Range 5–9	6.5–15.5 ha	Northern dry deciduous forest Ankarana Reserve, Montagne d'Ambre National Park	Wilson et al. 1989, Freed 1996
E. macaco	10 (range 5–14)	3.5–7 ha	Northwestern dry forest Sambirano, Nosy Be	Colquhoun 1993, Andrews and Birkinshaw 1998
E. mongoz	Range 3–8	2.8 ha, 05–1.0	Northwestern dry forest on Comoran Islands of Anjouan and Mohéli, humid forest Ampijoroa Reserve	Tattersall 1977b, Harrington 1978, Curtis and Zaramody 1997
Hapalemur g. griseus	Range 2–9	6–10 ha, 14–20 ha	Eastern rain forest, southeastern submontane forest Andasibe National Park, Ranomafana National Park	Wright 1986, Tan 1999, Grassi 2001
H. g. occidentalis	Range 1–4	26 ha	Isolated forest regions Western Madagascar, Manongarivo Reserve	Petter and Peyrieras 1970b, Tattersall 1982, Raxworthy and Rakotondraparany 1988
H. g. alaotrensis	Range 3–9	0.6–8 ha	Reed beds Lac Alaotra, east-central Madagascar	Nievergelt et al. 1998, Mutschler 2002
H. aureus	Range 2–6	26–80 ha	Restricted range, found only at two sites in southeastern submontane rain forest Ranomafana and Andringitra National Parks	Wright et al. 1987, Meier and Rumpler 1987, Mittermeier et al. 1994, Tan 1999
H. gilberti	?	?	Eastern Madagascar Beanamalao	Rabarivola et al. 2007
H. meridionalis	?	?	Lowland and montane rain forest and littoral forest Between Fort Dauphin and Andohahela	Feistner and Schmid 1999; Mittermeier et al. 2008a,b

Table 5.3 (cont'd)

SPECIES	MEAN GROUP SIZE (OR RANGE OF GROUP SIZE)	MEAN HOME-RANGE SIZE (+RANGE)	HABITAT AND WHERE STUDIED	STUDIED BY OR CITED BY
P. simus	Range 4–12	62 ha	Southeastern submontane rain forest Ranomafana region, spotted in Andringitra National Park	Meier and Rumpler 1987, Wright et al. 1987, Tan 1999
Varecia variegata variegata	Range 2–6 and 8–16 (at Nosy Mangabe)	30–150 ha (depends upon site and habitat disturbance)	Lowland, mid-altitude, and higher-altitude rain forests in northern, eastern, and east-central Madagascar Nosy Mangabe, Ranomafana National Park, Manombo, Betampona	Morland 1991, Balko 1998, White 1989a, Ratsimbazafy 2002, Britt et al. 2001
V. v. rubra	2–6 (Rigamonti), 18–31 (including all core groups within a community)	25–58 ha	Northeastern rain forest Masoala Peninsula	Rigamonti 1993; Vasey 2002, 2006
Propithecus verreauxi	2–14, mean=6	3–10 ha	South, southwestern, western dry deciduous forest, spiny forest Beza Mahafaly Reserve, Kirindy Reserve	Richard et al. 2002
P. v. coquereli	Range 3–10	4–8 ha	Northwestern mixed-deciduous and evergreen forests, and brush and scrub forest Ankarafatsika Reserve	Petter 1962, Richard 1974, Albignac 1981b
P. v. coronatus.	2–8	1.2–1.5 ha	Northwestern Madagascar	Petter et al. 1977, Mittermeier et al. 1994, Curtis et al. 1998, Müller et al. 2002
P. v. deckeni	2–10	?	Western dry deciduous (fragments) Tsingy de Bemaraha Reserve	Mittermeier et al. 1994, Thalmann and Rakotoarison 1994, Curtis et al. 1998, Müller et al. 2002
P. diadema edwardsi	Range 2–9 mean = 5.3	400 ha	Southeastern submontane rain forest Ranomafana National Park	Wright 1995, Pochron and Wright 2003
P. d. diadema	Range 3–8 mean = 4.83	33–42 ha	Eastern rain forest Mantadia National Park	Powzyk 1997, Powzyk and Mowry 2003
P. d. perrieri	Range 2–6	30 ha	Northern dry forest Analamera Reserve, Ankarana Reserve	Meyers and Ratsirarson 1989, Hawkins et al. 1990
P. d. candidus	Range 3–7 mean = 4.3	?	Northeastern humid forest Marojejy Reserve	Safford et al. 1989 (unpub. report cited in Mittermeier et al. 1994). Patel, 2006
P. tattersalli	Range 3–10 mean = 5	9–12 ha	Northeastern dry deciduous forest and semievergreen forest patches Daraina region	Meyers 1993, Meyers and Wright 1993, Mittermeier et al. 1994, Vargas et al. 2002
Indri indri	2 + offspring	34–40 ha	Eastern rain forest Mantadia National Park	Powzyk 1997, Powzyk and Mowry 2003
Nocturnal lemurs				
Cheirogaleus medius	2 + offsping (monogamous)	4 ha	Dry deciduous forest South and southwestern Madagascar	Hladik et al. 1980; Muller 1998, 1999; Fietz 1999b
C. major	Solitary	?	Eastern lowland rain forest	Petter et al. 1977, Tattersall 1982
C. ravus	?	?	Eastern rain forest	Groves 2000
C. crossleyi	?	?	Northeastern Madagascar plateau	Groves 2000
C. adapicaudatus	?	?	Southern spiny forest	Groves 2000
C. minisculus	?	?	Central Madagascar Ambositra	Groves 2000
C. sibreei	?	?		
Mirza coquereli	Sleeps in groups up to 6 at some sites	~4 ha.	Western coastal forests	Tattersall 1982
M. zaza	Sleeps in groups of 2–8	?	Ambato and Pasandava	Kappeler et al. 2005
Allocebus trichotis	Sleeps in groups of 2–6	?	Northern tropical evergreen forest	Meier and Albinac 1991

Table 5.3 (cont'd)

SPECIES	MEAN GROUP SIZE (OR RANGE OF GROUP SIZE)	MEAN HOME-RANGE SIZE (+RANGE)	HABITAT AND WHERE STUDIED	STUDIED BY OR CITED BY
Microcebus murinus	Sleeps in groups of 1–15	0.22–3.2 ha (males) 0.24–1.8 ha (females)	Dry deciduous forest, spiny forest, littoral forest West and Southern Madagascar	Martin 1972a, 1973; Pages-Feuillade 1989; Radespiel et al. 1998, Ramanananjato and Ganzhorn 2001
M. rufus	Sleeps in groups of 1–4	?	Northwest and eastern rain forest	Tattersall 1982, Wright and Martin 1995, Atsalis 2000
M. myoxinus	?	?	Dry forests Southern and western Madagascar	Petter et al. 1971, Rasoloarison et al. 2000
M. ravelobensis	Sleep in groups	0.44–0.79 ha	Northwestern Madagascar	Petter 1962, Zimmerman et al. 1998
M. tavaratra	?	?	Northern dry deciduous forest Ankanrana	Rasoloarison et al. 2000
M. berthae	Solitary/dispersed, do not sleep in groups	Male home-ranges larger than female, size not known	Western dry deciduous forest Kirindy, Andranomena, Analabe	Schmidt and Kappeler 1994, Rasoloarison et al. 2000, *Schwab 2000*
M. griseorufus	?	?	Southestern dry deciduous forest, spiny forest Beza-Mahafaly	Rasoloarison et al. 2000, Rasoazanabary, under review
M. lehilahytsara	?	?	Eastern Madagascar Andasibe and Mantadia National Park	Kappeler et al. 2005
M. sambiranensis	?	?	Northwestern humid forest Manongarivo Special Reserve	Rasoloarison et al. 2000
M. arnholdi	?	?	Northern Madagascar Montagne d'Ambre National Park and Special Reserve	Louis et al. 2008
M. margotmarshae	?	?	Northwestern Madagascar Antafondro Classified Forest	Louis et al. 2008
M. mamiratra	?	?	Northwestern Madagascar Nosy Be	Andriantompohavana et al. 2006
M. bongolavensis	?	?	Western deciduous forest Bongolava and Ambodimahabibo	Olivieri et al. 2007
M. lokobensis	?	?	Northwestern Madagascar Lokobe Special Reserve and Nosy Be	Olivieri et al. 2007
M. danfossi	?	?	Western deciduous forest Ambarijeby	Olivieri et al. 2007
M. macarthurii	?	?	Northeastern Madagascar Anjiahely	Radespiel et al. 2008
M. jollyae	?	?	Eastern rain forest Mananjary and Kianjavato	Louis et al. 2006a
M. simmonsi	?	?	Eastern Madagascar Betampona Special Reserve and Zahamena National Park	Louis et al. 2006a
M. mittermeieri	?	?	Eastern Madagascar Anjanaharibe-Sud Reserve	Louis et al. 2006a
Phaner furcifer	Dispersed pairs (male/female pairs sleep together in nests)	3.8–4 ha	Humid, dry, and secondary forest Western and northern Madagascar	Petter and Petter-Rousseaux 1979, Schulke and Kappeler 2003
Avahi laniger	Male/female pairs and offspring	1–2 ha	Rain forest, coastal forest Eastern Madagascar	Albignac 1981a, Ganzhorn et al. 1985, Harcourt 1988

Table 5.3 (cont'd)

SPECIES	MEAN GROUP SIZE (OR RANGE OF GROUP SIZE)	MEAN HOME-RANGE SIZE (+RANGE)	HABITAT AND WHERE STUDIED	STUDIED BY OR CITED BY
A. occidentalis	Male/female pairs and offspring	?	Dry deciduous forest Northwestern Madagascar	Petter et al. 1977, Tattersall 1977c
A. unicolor	Male/female pairs and offspring	?	Dry deciduous forest Northwestern Madagascar, Sambirano region	Thalmann and Geissmann 2000
A. betsileo	?	?	Eastern Madagascar Bemosary Classified Forest	Andriantompohavana et al. 2007
A. ramanantsoavani	?	?	Southeastern humid forest Manombo Reserve	Zaramody et al. 2006
A. meridionalis	Male/female pairs and offspring	2–3.5 ha	Humid forests Andohahela Reserve and the area of Sainte Luce	Zaramody et al. 2006, Norscia and Borgognini-Tarli 2008
A. peyrierasi	?	?	Southeastern Madagascar Manara, Vatoalatsaka, Sangalampona, Mahasoarivo, and Ranomafana	Zaramody et al. 2006
A. mooreorum	?	?	Humid evergreen forest Masoala National Park	Lei et al. 2008
A. cleesei	?	?	Tsingy of Bemaraha National Park Western Madagascar	Thalmann and Geissmann 2005
Daubentonia madagascariensis	Solitary	Mean = 35.6 (females), 170.3 (males)	Eastern, western, and northern Madagascar; primary rain forest, deciduous forest, secondary growth, cultivation, and dry scrub forest Nosy Mangabe	Tattersall 1982, Sterling 1993, Sterling and Richard 1995
Lepilemur mustelinus	Solitary	1.5	Eastern rain forest	Ratsirarson and Rumpler 1988
L. dorsalis	Solitary	?	Humid forest, northwestern Madagascar Nosy Be	Petter et al. 1977, Tattersall 1982
L. septentrionalis	Solitary	1 ha	Northern Madagascar, dry deciduous forest and humid forest	Tattersall 1982, Ratsirarson et al. 1987, Hawkins et al. 1990
L. edwardsi	Dispersed pairs (male/female pairs sleep together in tree holes)	1 ha	Western Madagascar, dry deciduous forests	Petter and Petter-Rousseaux 1979; Albinac 1981a,b; Warren 1994
L. leucopus	Solitary/dispersed (male/female pairs or mother-daughter pairs sleep together)	0.18–0.3 ha	Southern Madagascar, spiny and gallery forests Beza Mahafaly Reserve	Petter and Petter-Rousseaux 1979; Sussman and Richard 1986
L. ruficaudatus	Solitary or male/female pairs	0.8 ha	Western Madagascar, dry forest	Petter and Petter-Rousseaux,1979
L. microdon	?	?	Eastern rain forest	Petter et al. 1977
L. aeeclis	?	?	Northwestern Madagascar, between the Betsiboka and Mahavavy du Sud rivers	Andriaholinirina et al. 2006
L. randrianasoli	?	?	Western deciduous forest Andramasay and Bemaraha	Andriaholinirina et al. 2006
L. sahamalazensis	?	?	Northwestern Madagascar Sahamalaza Peninsula	Andriaholinirina et al. 2006
L. scottorum	?	?	Humid evergreen forest Masoala National Park	Lei et al. 2008
L. hollandoru m	?	?	Lowland rain forest Mananara-Nord Biosphere Reserve	Ramaromilanto et al. 2009
L. ankaranensis	Solitary	1 ha	Dry deciduous and some evergreen forests Montagne D'Ambre, Ankarana, Analamera	Hawkins et al. 1990; Mittermeier et al. 2008a,b; Ratsirarson and Rumpler 1988
L. ahmansoni	?	?	Dry western forest Tsiombikibo forest	Louis et al. 2006b

Table 5.3 (cont'd)

SPECIES	MEAN GROUP SIZE (OR RANGE OF GROUP SIZE)	MEAN HOME-RANGE SIZE (+RANGE)	HABITAT AND WHERE STUDIED	STUDIED BY OR CITED BY
L. milanoii	?	?	Dry deciduous, gallery, and semievergreen forest Andrafiamena Classified Forest	Louis et al. 2006b
L. betsileo	?	?	Rain forest Fandriana region	Louis et al. 2006b
L. fleuretae	?	?	Rain forest Andohahela National Park	Louis et al. 2006b
L. grewcocki	?	?	Dry deciduous forest Anjiamangirana region	Louis et al. 2006b
L. jamesi	?	?	Eastern lowland rain forest Manombo Special Reserve	Louis et al. 2006b
L. petteri	?	?	Spiny and gallery forest Beza-Mahafaly area	Louis et al. 2006b
L. seali	?	?	Northeastern rain forest Anjanaharibe-Sud Special Reserve	Louis et al. 2006b
L. tymerlachsoni	?	?	Northwestern humid forest Lokobe Special Reserve, Nosy Be	Louis et al. 2006b
L. wrighti	?	?	Southeastern rain forest Kalambatritra Special Reserve	Louis et al. 2006b
L. hubbardi	?	?	Southwestern, dry transitional forest Zombitse National Park	Louis et al. 2006b
L. manasamody	?	?	Northwestern dry deciduous forest Ambongabe and Anjiamangirana	Craul et al. 2007
L. otto	?	?	Western deciduous forest Ambodimahabi	Craul et al. 2007

Lemuridae (Lemur, Eulemur, Hapalemur, Varecia)

Lemur catta (Ring-Tailed Lemur)

L. catta is one of the two lemur species that has been studied over the longest period of time, beginning with Jolly's (1966) pioneering work. Ring-tailed lemurs have been studied primarily at five sites and several different habitats in the south and southwest of Madagascar: Berenty, Beza Mahafaly, Antserananomby, Andringitra, and Cap Ste. Marie.

While this species is found primarily in riverine, *xerophytic* (drought-adapted), and scrub forests in south and southwestern Madagascar (Jolly 1966, Sussman 1977), it is also found in spiny forests and low-lying limestone forests; and one population has even been found at Andringitra National Park in the central southeast, living above the tree line at an elevation of 2,500 m (Goodman and Langrand 1996, Sussman et al. 2003a, Gould 2006).

L. catta has been defined as a very flexible "edge" species, able to withstand relatively extreme temperatures and to recover from serious droughts (Sussman 1977; Gould et al. 1999, 2003; Sauther et al. 1999a,b; Gould 2006). It is the most terrestrial species of lemur, spending an average of up to 30% of the time on the ground (Jolly 1966, Sussman 1974). The degree of terrestriality varies by season (Sauther 2002).

Social Organization. *L. catta* lives in multimale/multifemale groups (Jolly 1966, Sussman 1977). Group fission commonly occurs when groups reach a critical size, and they split along matrilines (Sussman 1991a, Koyama 1991, Hood and Jolly 1995, Jolly et al. 2002, Gould et al. 2003). This species is female-philopatric, and males disperse at 3–4 years of age (Sussman 1992).

Diet and Feeding Ecology. Ring-tailed lemurs have been described as "opportunistic omnivores" (Sauther et al. 1999a,b), feeding on fruit (particularly tamarind in gallery forest habitat), leaves and stems, flowers, some insects, and soil from both the ground and termite mounds (Jolly 1966; Sussman 1977; Sauther 1992; Sauther et al. 1999a,b). Food resources are extremely seasonal, and the regions where *L. catta* are found often experience severe droughts (Jolly 1966, Sussman 1977, Sauther 1992, Gould et al. 1999). As a result, ring-tailed lemurs are very flexible and can switch their primary food resources to follow ecological unpredictability. For example, *L. catta* groups will expand their home ranges into those of other groups when particular seasonal resources are unavailable (Budnitz 1978, Jolly et al. 1993, Sussman 1991a, Sauther and Sussman 1993).

Reproduction. Reproductive synchrony is marked in this species and strongly tied to the specific nature of seasonal food resources (Jolly 1966; Sauther 1992, 1998). Average gestation length is 141 days (Sauther 1991). Although females normally give birth to a single infant, twinning in the wild has occasionally been reported (Koyama et al. 2002, Jolly et al. 2002). Infants are born near the end of the dry season in September and October and weaned at 4–5 months, during the rainy season, when weaning foods are available (Jolly 1966, Sussman 1977, Gould 1990, Sauther 1992). Alloparental care is common (Gould 1992).

Sauther (1994) found that pregnant females in gallery forest habitat feed primarily on fruit and flowers but, when lactating, switch to low-cost, predictable, high-protein plant foods.

Infant mortality differs at the two different sites where it has been documented. Koyama et al. (2002) report 32%–37% at Berenty, where many groups are water-provisioned; but at Beza, where no provisioning occurs, a range of 52%–80% has been reported, depending on rainfall (Gould et al. 1999, 2003).

Eulemur

Within the *Eulemur* genus, we find seven species and several subspecies. *E. fulvus* (the brown lemur) is the most geographically widespread, and as a result, there are five subspecies (Table 5.1). All subspecies except *E. f. mayottensis*, the Mayotte brown lemur which lives on the island of Mayotte in the neighboring Comoros (Tattersall 1982, Tarnaud and Simmen 2002), inhabit continuous forest throughout Madagascar (Freed 1999).

Sexual Dichromatism. One interesting morphological feature of *Eulemur* is that most species and subspecies are sexually dichromatic, making it easy to distinguish males from females, even in relatively high canopy. The extent of dichromatism ranges from completely different pelage color in *E. macaco macaco* and *E. m. flavifrons* (males black, females russet brown) to different-colored ventrums, beards, heads, and face markings in *E. fulvus* spp., *E. collaris*, *E. cinereiceps*, *E. rubriventer*, and *E. mongoz*. *E. f. fulvus* is the only lemur in this group that exhibits no sexual dichromatism (Harrington 1975, Mittermeier et al. 1994).

Reproduction. Gestation in *Eulemur* in the wild is reported to be between 120 and 126 days, and infants in most species are born between September and November (Colquhoun 1993, Mittermeier et al. 1994, Sussman 1999a). Females

normally give birth to just one infant, which, like most diurnal lemurs, can cling immediately. Interbirth intervals are 1 year, though Overdorff et al. (1999) found that in an *E. f. rufus* population which had been studied for 7 years, the mean interbirth interval between surviving offspring was 2.1 years. In *E. mongoz*, which is often monogamous, adult males frequently carry the infant after it is 2 weeks old (Curtis and Zaramody 1997).

Cathemerality. Cathemerality has been observed in all *Eulemur* species. Most species exhibit year-round cathemeral activity but with some seasonal variation (Overdorff 1988; Overdorff and Rasmussen 1995; Freed 1996; Andrews and Birkenshaw 1998; Rasmussen 1999; Donati et al. 1999, 2001; Donati and Borgognini Tarli 2002b; Kappeler and Erkert 2003; Curtis 2006). For example, *E. mongoz* exhibits greater nocturnal activity during the cooler, dry seasons, which may correlate with thermoregulation during long, cool nights (Curtis et al. 1999). Low nocturnal light during the wet season compromises nighttime activity in *E. mongoz* (Rasmussen 1999), and Colquhoun (1998), Donati et al. (2001), Donati and Borgognini Tarli (2002b), and Kappeler and Erkert (2003) stress that nocturnal activity in *E. m. macaco*, *E. collaris*, and *E. f. rufus* is strongly dependent on phases of the moon and available light. Rasmussen (1999) also suggests that in seasonally dry forests cathemeral activity may function as an antipredator strategy during times when canopy cover is thin.

Rasmussen (1999) divides cathemerality into three types: seasonal differences in day and night activity, found in *E. mongoz*; seasonal shift from diurnal to 24 hr activity, found in *E. f. fulvus* (Rasmussen 1999) and *E. f. rufus* (Donati et al. 1999); and year-round 24 hr activity, found in all *Eulemur* species that have been studied in rain forest habitats (Andrews and Birkenshaw 1998, Freed 1996, Overdorff 1988, Overdorff and Rasmussen 1995) as well as in some dry forest habitats (Kappeler and Erkert 2003).

Van Schaik and Kappeler (1996) propose that cathemerality may have occurred in formerly nocturnal taxa, due to an "evolutionary disequilibrium" related to human activities causing the subsequent extinction of both the aforementioned large-bodied lemurs as well as large raptors. These authors suggest that the extinctions of large raptors allowed for greater diurnal activity in the relatively small-bodied *Eulemur* species. Colquhoun (1993) suggests that cathemeral activity may be an ancestral trait for the entire *Eulemur* genus. Kappeler and Erkert (2003) argue that cathemerality evolved from nocturnal ancestors, perhaps relatively recently, and may have occurred by adding some diurnal activity to a largely nocturnal baseline. Tattersall (1976), Curtis et al. (1999), and Curtis and Rasmussen (2002, 2006) argue that cathemerality may have arisen as an aid to thermoregulation: 24 hr activity could help prevent heat stress from high temperatures during the hot season and mitigate stress from cold temperatures in the cool season. Because cathemeral primates are primarily restricted to

Madagascar, Kappeler and Erkhart suggest that the unusual aspects of Madagascar's ecology, outlined in Richard and Dewar (1991) and Wright (1999), have allowed for such a transition in activity pattern.

Eulemur fulvus Group *(Brown Lemur, Five Subspecies: Common Brown Lemur, Red-Fronted Brown Lemur, White-Fronted Brown Lemur, Sanford's Lemur, Mayotte Brown Lemur), Collared Lemur, and Grey-Headed Lemur*

Of note is recent taxonomic work suggesting that *E. cinereiceps* and *E. albocollarus* are not distinct species and the two should be subsumed under the senior synonym *E. cinereiceps* (Johnson et al. 2008).

Social Organization. The social organization of all of these taxa is mixed-sex groups (Sussman 1974, 1999a; Harrington 1975; Overdorff 1992, 1993a, 1996a,b, 1998; Gerson 2001; Mittermeier et al. 1994; Vasey 2000) and fission–fusion groups in *E. f. mayottensis* (Tattersall 1977c) and *E. cinereiceps* (Johnson 2002). In *E. f. rufus* (red-fronted lemur), however, groups sometimes fission during periods of food scarcity (Overdorff 1998); and individuals of this subspecies also form strong affiliative dyadic relationships, primarily between males and females but also between other sex/age combinations (Overdorff 1998, Gerson 2001). Overdorff (1998) notes that dyads occurred more often in feeding contexts during the mating season and during periods of food scarcity and may be related to the unclear dominance hierarchies found in rufous brown lemurs, the distribution and density of food patches in the habitat, and vulnerability to predators.

Diet and Feeding Ecology. Diet in the wild has been noted in *E. f. fulvus* in a western dry forest (Harrington 1975); *E. f. rufus* in a southwestern dry forest (Sussman 1972, 1974); *E. f. rufus* at Ranomafana, a rain forest site (Overdorff 1992, 1993b, 1996a); *E. f. sanfordi* in a northern dry forest at Montagne d'Ambre (Freed 1996, 1999); *E. f. collaris* in a southeastern littoral (coastal) forest (Donati and Borgognini Tarli 2002a); *E. f. albifrons* in a northern montane rain forest (Vasey 2000); and three populations in three different areas of the southeastern rain forests (Johnson and Overdorff 2002). The diet of all subspecies is described as highly frugivorous, but leaves, buds, flowers, invertebrates, and nectar are also consumed. In most areas where *E. fulvus* occurs, there can be marked seasonal fluctuation in amount and type of food resources available. During times of fruit scarcity (usually the dry season), animals include more leaves, flowers, and figs in their diets; and *E. f. rufus* groups have been observed to move well out of their home ranges during these periods to seek alternative resources or fission into smaller groups (Overdorff 1993a,b, 1996a; Johnson and Overdorff 2002; Overdorff et al. 2003). *Eulemur* ingests a significant amount of tannins and alkaloids from unripe fruit and mature leaves in the diet, and Ganzhorn (1988) and Vasey (2000) suggest that tolerance of secondary compounds

combined with ecological flexibility may explain the wide geographic range of the group.

Sympatry between *E. fulvus* and other lemur taxa has been documented in a number of geographic areas. Sympatry and polyspecific associations between *E. f. sanfordi* and *E. coronatus* are discussed in the following section on crowned lemurs, and information on sympatry and niche separation in *E. f. rufus* and *E. rubriventer* is presented in the section on *E. rubriventer*.

E. coronatus *and* E. fulvus sanfordi: *Sympatry and Polyspecific Associations*

Sanford's brown lemurs (*E. f. sanfordi*) and crowned lemurs (*E. coronatus*) are sympatric throughout the same region at the northern tip of Madagascar (Wilson et al. 1989; Freed 1996, 1999, 2006). Freed studied these two species at Montagne d'Ambre National Park in northern Madagascar, while Wilson's group focused on the two species in the unusual habitat of Ankarana, which is composed of dry forest growing on and around limestone karst pinnacles (*tsingy*) as well as xerophytic scrub and semideciduous dry forest.

Both species inhabit dry forests that vary in elevation, climate, structure, and disturbance. Both live in small, multimale/multifemale groups (Freed 1996, Wilson et al. 1989); but group cohesion and spacing differ by species. In the dry season, crowned lemur groups are less cohesive than those of Sanford's lemurs, and crowned lemurs sometimes divide into small foraging subgroups during the day (Freed 1996, 2006).

Both species are highly frugivorous and occasionally feed on leaves and insects. Freed (1996, 2006) observed frequent polyspecific associations, the first report of such in sympatric lemurs. Polyspecific associations varied according to season and were most frequent during the wet season. Freed suggests that each species benefits from the other's familiarity with food resources in different forest levels but not in relation to enhanced predator protection since there were few predators in the area and actual predation on these lemurs was rare.

E. mongoz *(Mongoose Lemur)*

E. mongoz occurs in the subhumid, seasonal forests of northwestern Madagascar as well as on two of the Comoro Islands, Anjouan and Moheli, where they were likely introduced by humans (Tattersall 1982, Mittermeier et al. 1994, 2006).

Social Organization. The social organization of *E. mongoz* is variable as it has been observed in both pair-bonded (monogamous) family groups and larger mixed-sex groups (Harrington 1978, Tattersall 1977a, Curtis and Zaramody 1997). Offspring of both sexes disperse and establish their own social groups. Females leave the natal group at 27–30 months and males, at 31–42 months (Curtis and Zaramody 1997, 1998).

Diet and Feeding Ecology. The mongoose lemur diet can be categorized as highly nectivorous during the dry season and frugivorous/folivorous during the wet season (Sussman and Tattersall 1976, Curtis and Zaramody 1997).

Reproduction. Infants are born in October–November. At 3 weeks, infants begin to explore the environment. Adult males frequently carry the infant between weeks 2 and 5, and at 9 weeks infants begin to move and feed independently. Females give birth annually (Tattersall 1976, Curtis and Zaramody 1997).

E. macaco macaco *(Black Lemur)* and E. m. flavifrons *(Sclater's Black Lemur)*

E. macaco exhibits marked sexual dichromatism: Males are black with black ear tufts, and females are golden/reddish/rust brown with off-white ventrum and white ear tufts (Tattersall 1982, Mittermeier et al. 1994).

E. m. flavifrons and *E. m. macaco* × *E. m. flavifrons* hybrids are restricted to dry northwestern forests, just south of the geographic range of the black lemur. Sclater's black lemur differs from the black lemur in that it lacks tufted ears, but more strikingly, its eye color ranges from turquoise blue to gray, as opposed to the amber brown eyes of *E. m. macaco* (Koenders et al. 1985, Mittermeier et al. 1994). Hybrids exhibit either duller blue eyes and no beard or light brown eyes and a less prominent beard and ear tufts compared to the black lemur (Rabarivola et al. 1991).

Social Organization. Social organization is multimale/multifemale. At Ambato Massif in the northwest, Colquhoun (1993) found that larger groups often fissioned into smaller subgroups.

Diet and Feeding Ecology. Marked wet and dry seasons occur in this area, and seasonal variation was noted with respect to dietary patterns. Fruit was the dominant food item during the rainy season, supplemented by mushrooms and millipedes. During the dry season, flowers, nectar, seed pods, and some leaves were eaten (Colquhoun 1993). Andrews and Birkenshaw (1998) found differences in daytime and nighttime feeding, with more variation in fruit species and leaves consumed in the day and more nectar consumed at night.

Cathemerality. Colquhoun (1993) and Andrews and Birkenshaw (1998) noted year-round cathemeral activity in black lemurs, with nocturnal activity following phases of the moon. Cathemeral activity was seen more in the cooler, dry season and, as with *E. mongoz*, may be related to thermoregulation, allowing these lemurs to avoid cold stress by being physically active during cool nights in the dry season.

Reproduction. Females usually give birth annually to a single infant, in September or October, after a gestation period of 125–126 days (Colquhoun 1993, Mittermeier et al. 1994).

E. rubriventer *(Red-Bellied Lemur)*

One of the few pair-bonded lemurs, *E. rubriventer* has been closely studied at Ranomafana National Park by Overdorff (1992, 1993a,b, 1996a,b).

Social Organization and Group Size. The red-bellied lemur lives in monogamous pairs with offspring and maintains exclusive use of its home range, actively defending the boundaries.

Diet, Feeding Ecology, and Sympatry with E. f. rufus. *E.rubriventer* is a highly frugivorous primate and includes some leaves and nectar in the diet (Overdorff 1992). At the Ranomafana site, *E. rubriventer* and *E. f. rufus* are sympatric. Even though the composition of their diets is similar, Overdorff (1992, 1993b) notes that *E. f. rufus* ate more unripe fruits, mature leaves, and insects than did *E. rubriventer*; and she suggests that *E. f. rufus* may have a higher tolerance than *E. rubriventer* for secondary compounds, which may also help with niche separation.

Both of these species may serve as pollinators for some of the plant species that they use, but Overdorff (1992) notes that *E. rubriventer* may be a more efficient pollinator since it does not destroy the reproductive parts of the flower. Overdorff (1996a) suggests that subtle and consistent differences in diet throughout the seasons allow these two species to coexist.

Reproduction. Females give birth to one infant annually, in September or October. As in some pair-bonded anthropoid species, male red-bellied lemurs help with infant care, often holding or carrying the infant. Males have been noted to carry infants up to 100 days (Overdorff 1993a, Mittermeier et al. 1994).

Hapalemur *and* Prolemur *(Bamboo Lemurs)*

All species of *Hapalemur* and *Prolemur* are highly unusual because they specialize on bamboo, a dietary focus not found in other primates. There are five species of bamboo lemur and three subspecies (see Table 5.1). *H. aureus*, an extremely rare lemur so far found only in very small populations in two southeastern national parks (see Table 5.3), was discovered only in 1986 (Meier and Rumpler 1987, Wright et al. 1987); and both *H. meridionalis* and *H. gilberta* are newly described (Warter et al. 1987, Rabarivola et al. 2007); these species are not yet well-studied.

Social Organization. *H. griseus griseus* is reported to have flexible social organization. Grassi (2001) found monogamous pairs as well as polygynous and multimale/multifemale social groups at her field site. Alaotran gentle lemurs (*H. g. alaotrensis*) also live in varying kinds of group: monogamous pairs, groups with two breeding females, and some groups with three adult males, though there is only one breeding male per group (Mutschler et al. 2000, Mutschler 2002). Mutschler found that both sexes disperse: Females leave their natal group as subadults, and males make their first migration as adults.

Tan (1999) notes that *H. aureus* live in monogamous pairs, and the one group of *H. simus* which Tan studied was multimale/multifemale, with three adult males, two adult females, and offspring.

Diet and Feeding Ecology. As their common names suggest, all bamboo lemur species are bamboo specialists, and all species ingest the cyanide found in the giant bamboo without harm, a remarkable dietary adaptation (Glander et al. 1989, Tan 1999). More than 85% of *Hapalemur* diets are made up of bamboo and grasses (Tan 1999, Mutschler 2002). Tan found that the three species found at Ranomafana National Park are able to coexist sympatrically because each specializes on different parts of the bamboo plant. They also feed on several other plant species, fungus, and, at times, soil (Tan 1999, Grassi 2001). *H. griseus* and *H. aureus* both consume some fruit, and Grassi notes that both new and mature leaves were eaten by *H. griseus* at the higher-elevation Vato site in Ranomafana Park. She suggests that increased dietary diversity by female *H. g. griseus* during reproductive periods helps offset high metabolic needs.

Some preliminary research indicates that *H. meridionalis* focuses on seven species of terrestrial grass, which make up 75.9% of the diet (Eppley and Donati 2009). This is a lower-quality diet than observed in other species of this genus and may indicate enhanced digestive ability (Eppley and Donati 2009).

Reproduction. *H. griseus* and *H. aureus* have similar gestation lengths of 137–140 and 138 days, respectively, while the gestation period of the larger *H. simus* is somewhat longer, at 149 days (Tan 2001, Grassi 2001). *H. g. alaotrensis* (the Alaotran gentle lemur) does not have as strict and discrete a mating season as that found in most other lemur species. The mating season begins in September and ends in February (Mutschler 2002). Mutschler suggests that a year-round, consistent resource base is a key factor in the absence of strict breeding seasonality.

H. aureus mothers have been noted to nest their infants in thick foliage during the first 2 weeks of life (Tan 2001). Tan also found that *H. griseus* and *H. aureus* females both park and orally transport infants but that *H. simus* females carry their newborns.

Grassi (2001) reports high infant mortality (67%) in *H. g. griseus* at the Vato site, Ranomafana. Surviving infants were fully weaned by 5 months. There is a 1-year interbirth interval in *Hapalemur*.

Varecia *(Ruffed Lemur)*

The genus *Varecia* contains two subspecies: the black and white ruffed lemur (*V. variegata variegata*) and the red ruffed lemur (*V. v. rubra*), the latter having a very restricted range in the northern Masoala Peninsula. In all areas where these lemurs have been studied, populations have experienced occasional and sometimes devastating cyclones (Balko 1998; Ratsimbazafy 2001, 2002, 2006).

Social Organization. Both monogamy and multimale/multifemale mating systems have been reported (Morland 1991, White 1989a, Balko 1998, Britt 2000). Rigamonti's (1993) two study groups of red ruffed lemurs fissioned into subgroups of two or three animals. During the cool wet season, they lived in these small subgroups for several weeks at a time, and groups were cohesive in the transitional dry months. Vasey's (2006) study population lived in a dispersed social network, or community; the individual members interacted but were never observed all together at once, as a group would be. Some individuals within the community shared core areas and were designated as members of the respective "core group." Ratsimbazafy (2002) found that after a severe cyclone black and white ruffed lemurs at Manombo on the southeastern coast foraged singly rather than as a group as 95% of larger trees in the area stopped fruiting. As ruffed lemurs are both selective in their choice of food and devoted to food that is patchy both in space and through time (e.g., fruits, flowers), a fluid social structure may be an adaptation to resource procurement (Vasey 2006).

Diet and Feeding Ecology. *Varecia* is highly frugivorous, with fruit making up 75%–95% of the diet and the remainder being comprised of nectar, flowers, and some leaves (Morland 1991, Britt 2000, Vasey 2000). During times of low fruit availability, *Varecia* will consume large amounts of young leaves (Balko 1998). After the abovementioned cyclone at Manombo, ruffed lemurs at this site relied on fruit from nonendemic, invasive plant species, as well as fungus (Ratsimbazafy 2002). *Varecia* has been observed to come to the ground and ingest soil at particular times of the year (Morland 1991, White 1989a, Britt 2000). Such geophagy may serve to neutralize secondary compounds in the diet as well as provide a source of minerals (Ganzhorn 1988, Britt 2000). Britt suggests that even though *Varecia* are marked frugivores, their ability to use other food items may be an important adaptation for dealing with low or absent fruit productivity during times of environmental stress because these lemurs live in areas of Madagascar where cyclones are common and important food trees can be destroyed.

Ruffed lemurs are considered important seed dispersers and pollinators in the eastern Madagascar rain forests (Balko 1998, Britt 2000), and Britt (2000) stresses that, as such, it is of utmost importance to develop conservation strategies that will aid in the survival of ruffed lemurs.

Reproduction. Reproduction and infant care differ in *Varecia* from other diurnal lemurs. *Varecia* is the only diurnal prosimian in which females possess two sets of mammary glands and regularly exhibit multiple (two to four infants) births (Morland 1990, Mittermeier et al. 1994). Infants do not cling to the mother as do other diurnal lemur infants; rather, the mother transports infants one at a time by mouth and parks them in nests or in trees (Petter et al. 1977, Tattersall 1982). Morland (1990) noted frequent alloparental care consisting of guarding infants at nest sites and allonursing. Nests are constructed by pregnant females a few weeks prior to parturition 10–20 m aboveground (Morland 1990, Balko 1998). Furthermore, ruffed lemur infants develop more quickly than do other diurnal lemur offspring, and in captivity they have been noted to weigh up to 70% of adult body weight by 4 months of age (Pereira et al. 1987). In the wild, infants grow rapidly (Balko 1998) and appear to reach nearly adult size at about 6 months of age (Morland 1990). Gestation length is estimated at around 102 days in the wild, and infants are born in September and October (Morland 1990, Mittermeier et al. 1994, 2006). This is the shortest gestation period seen in the lemurids (Vasey 2006).

Ruffed lemurs commonly experience cyclones in their geographic range. In Ratsimbazafy's (2001, 2002, 2006) study, females ceased reproduction for 3 years after a cyclone destroyed most of their resource base. He suggests that plasticity in diet, small group size, solitary foraging, and reproductive cessation following a severe natural disaster are important reasons that *Varecia* groups can persist in such a highly disturbed habitat. Ratsimbazafy (2002) points out the link between environmental variability and female fertility in this species.

Indriidae (*Propithecus* and *Indri*)

Propithecus (*Sifakas*)

Propithecus species are extraordinary vertical clingers and leapers. With three species and eight subspecies (Table 5.1), sifakas are found in many habitats and many regions of Madagascar, although some, like *P. tattersalli*, are found only in very restricted ranges and several are very rare, with populations threatened by habitat destruction. Of note is that *P. tattersalli* has recently been reported to exist slightly outside of the original restricted area in the Daraina region of northeastern Madagascar, with an effective population size of these rare lemurs estimated at 2,520–3,960 individuals (Vargas et al. 2002). Powzyk (1997) studied sympatric *P. d. diadema* and *Indri* at Mantadia National Park in the eastern rain forest. She notes that in parts of their former distribution diademed sifaka populations have disappeared due to overhunting or habitat destruction.

The remainder of this section will focus on the four most-studied *Propithecus* species, *P. verreauxi verreauxi*, *P. tattersalli*, *P. diadema edwardsi*, and *P. d. diadema*, and will include some of the limited information reported for *P. candidus*.

Social Organization. There is some variety between species, though multimale/multifemale groups are the trend; *P. tattersalli*, *P. d. diadema*, *P. d. edwardsi*, and *P. candidus* all live in multifemale/multimale groups (Meyers and Wright 1993, Powzyk 1997, Patel 2006), although *P. candidus* has been observed living in pairs (Patel 2006) and *P. d. edwardsi* has been observed in unimale polygynous, polyandrous, and male/female pair groups (Pochron and Wright 2003,

Pochron et al. 2004) *P. verreauxi verreauxi* (Verreaux's sifaka) is found in small multifemale/multimale groups as well; however, this species also fissions into small foraging parties at times. The social organization of Verreaux's sifaka has been sometimes referred to as "neighborhoods" because of the fluidity of groups, the fact that males make temporary visits to adjacent groups, and the frequency of adult male intergroup transfer (Jolly 1966, Richard 1978, Richard et al. 1993). Regarding the variation reported for *P. d. edwardsi*, Pochron and Wright (2003) suggest that since females sometimes mate with males outside of their groups, such flexibility may reduce pressure for males to join groups with several females and result in the variability seen in group composition in this species. These authors suggest that Madagascar's harsh and unpredictable environment may have resulted in *P. d. edwardsi* living and foraging in small groups, which would reduce feeding competition yet help somewhat with predator protection (vs. living/foraging solitarily).

Diet and Feeding Ecology. Seasonal variability is common. Sifakas (and all indriids) are considered folivores; however, the inclusion of fruit and flowers in the diet is frequently reported, and in some species these foods make up the bulk of the diet when they are available. Verreaux's sifakas live in the dry west, south, and southwest of Madagascar and experience dramatic shifts in seasonal resource distribution between the wet and dry seasons: During the wet season, 60%–70% of the diet is made up of fruit and flowers, with young leaves accounting for 20%; during the dry season, mature leaves make up 70% of the diet, with fruit and flowers contributing only 20%. Bark makes up the remainder of the diet in both seasons (Richard 1978, Sussman 1999a). Similarly, Meyers and Wright (1993) note that the *P. tattersalli* diet consists of immature and mature leaves (22% and 17%, respectively), 37% unripe fruit and seeds, 9% fruit pulp, and 13% flowers. These items peaked in availability in the wet season, but seeds, available year-round, formed the staple food item. Bark is also sometimes eaten during the dry season (Mittermeier et al. 1994).

Some species show a preference for fruit, underscoring the variability of "folivorous" species: ripe fruit and seeds make up the majority (55%) of *P. d. edwardsi*'s diet, supplemented by vine leaves (15%), flowers (3%), and immature leaves (26%) (Meyers and Wright 1993). *P. d. diadema* are primarily frugivore–folivores (Irwin 2006a), with 42% of the diet consisting of immature leaves, supplemented with fruits and flowers (Powzyk 1997, Powzyk and Mowry 2003). Fruit supplementation is more pronounced in some habitats, possibly due to altitude-related variation in fruit species (Irwin 2006b). Fruit seems to be a preferred resource in this species (Irwin 2006a). Seasonal changes in diet vary in *P. d. diadema*, with some populations highly dependent on a single plant species during the dry season (Irwin 2006a). Feeding differences between *P. d. diadema* and sympatric *Indri* at Mantadia are presented in the section on *Indri*.

Reproduction. Gestation varies: 150–160 days in Verreaux's sifaka (Petter-Rousseaux 1964) and an average of 179 days in *P. d. edwardsi*, 1 month longer than in the smaller *P. verreauxi* (Wright 1995). Gestation length has yet to be described in *P. candidus*; however, it has been noted that the mean interbirth interval in one group was 1.67 years, shorter than the 2 years typically noted in other rain forest sifaka (Patel 2006). Females give birth to one infant, sometimes in consecutive years (Patel 2006). There may be a trend toward the mating season occurring in late January and infants being born in late July, as is the case in *P. tattersalli*.

In *P. tattersalli*, weaning occurs at 5–6 months and is timed to coincide with peak immature leaf availability. As in many other lemur species, late lactation/weaning occurs in the early wet season so that infants have access to abundant weaning foods (Meyers and Wright 1993). Weaning lasts beyond the point at which infants are otherwise independent: In *P. verreauxi*, infants are carried ventrally at first, then dorsally, and infants will continue to ride on the mother until 6–7 months (Jolly 1966). *P. d. edwardsi* infants are primarily independent by 7 months of age and fully weaned by 1 year.

Most *P. d. edwardsi* females begin reproducing at 4 years of age. However, Pochron et al. (2004) have found that only 24% of all females survive to the age of 4 years. Average interbirth interval is 1.5 years, and average infant mortality is 50%. In the *P. verreauxi* population at the Beza Mahafaly site in southwestern Madagascar, Richard et al. (2002) found that more than half of the females did not reproduce for the first time until they were 6 years old (adult males mate at 3–4 years of age). Such a delay in reproduction for such a small primate is considered by Richard et al. (2002) to be "bet hedging," i.e., a slowing down of female reproductive life history where first births are later than expected and females reproduce into old age, also later than expected. This unusual reproductive strategy is suggested to be an evolutionary response to the climatic unpredictability in this species' geographic region, such as extremely varied annual rainfall patterns and frequent droughts (Richard et al. 2002). Infant mortality is high, averaging 52% in the first year of life (Richard et al. 2002). High mortality may be related to a combination of starvation after weaning in particularly dry years, hypothermia in the cold and dry season, disease, and predation.

Indri indri

There is only one species of *Indri*, and the common name is also "indri." In addition to its large size (see Table 5.2), *Indri* can be distinguished from most other lemurs by its rudimentary tail (Pollock 1975). It is also known because of its loud, wailing morning calls, which can be heard up to 3 km away (Pollock 1975, Mittermeier et al. 1994, 2006).

Indri has been characterized as the largest extant lemur species; however, Glander and Powzyk (1998) and Powzyk (1997) found that both *Indri* and *P. d. diadema* were similar in body weight, and Powzyk suggests that both species be considered the largest two extant lemurs. In both species, females weigh slightly more than males.

Geographic Range/Habitat. *Indri* inhabit the eastern rain forests, from near, but not in, the Masoala Peninsula in the northeast to east central Madagascar (Petter et al. 1977, Tattersall 1982).

Social Organization. *Indri* is one of the few monogamous lemur species, living in pairs or small groups consisting of a pair and offspring (Pollock 1975, Powzyk 1997, Britt et al. 2001a). The mated pair use morning calls to announce both their location to other pairs and their mated status (Powzyk 1997).

Diet and Feeding Ecology. Like other indriids, *Indri*'s diet consists of leaves, flowers, fruit, bark, and seeds (Pollock 1975, Powzyk 1997, Britt et al. 2001a, Powzyk and Mowry 2003). Powzyk (1997) found that leaves made up 71% of the diet at Mantadia. Britt et al. (2001a) note that *Indri* at Betampona, farther north, fed on more mature leaves than those at Mantadia and that in the winter season they increased consumption of bark and fruit. Pollock (1975) notes that *Indri* regularly came to the ground to ingest earth; however, Powzyk and Mowry (2003) note that sympatric diademed sifaka at Mantadia engaged in geophagy twice as often as did *Indri*. Powzyk (1997) suggests that *Indri*'s specialization for plant fiber has allowed these two large-bodied diurnal lemur species to coexist in over 90% of their range.

Reproduction. A single infant is born in May, after a gestation period of 120–150 days (Pollock 1975, Mittermeier et al. 1994). Infants are carried on the ventrum until 4 months, then carried dorsally until 8 months. Infants are weaned between 8 and 12 months (Pollock 1975). Powzyk (1997) calculated *Indri* average birthrates as 0.33/year and infant mortality as 0.67 over a 3-year period.

Nocturnal Lemurs

During the past 10–15 years, the nocturnal lemurs of Madagascar have been the focus of a number of behavioral and phylogenetic studies that have greatly expanded what is known of their socioecology and biology. Exciting new data on their behavioral ecology indicate that nocturnal strepsirhines live in complex societies and exhibit high interspecific diversity in social and mating systems. Indeed, it has been suggested that nocturnal lemur social systems contain three types: gregarious, for animals living in cohesive groups; dispersed, for solitary foragers with social networks; and solitary, for completely solitary animals (Müller and Thalmann 2000).

There are three families that contain only nocturnal lemuriforms: Cheirogaleidae, Megaladapidae, and Daubentoniidae. The primarily diurnal Indriidae contains nine nocturnal species, all in the *Avahi* genus.

Cheirogaleidae

Members of the family Cheirogaleidae are small, quadrupedal lemurs that sleep in nests of leaves or in tree holes during the day. The genus *Cheirogaleus* has recently undergone a number of taxonomic changes, and 30 species are now recognized (Mittermeier et al. 2008a). The scientific and common names for these species are found in Table 5.1. *Allocebus trichotis*, the hairy-eared dwarf lemur, was originally thought extinct but was rediscovered in 1989 (Meier and Albignac 1991). The genus *Mirza* is comprised of two species, *Mirza coquereli* and *M. zaza*, Coquerel's giant mouse lemur and the northern giant mouse lemur, respectively. Fork-marked lemurs include *Phaner furcifer*, *Ph. pallescens*, *Ph. parienti*, and *Ph. electromontis*; the latter three were originally described as subspecies of *Ph. furcifer* (Groves and Tattersall 1991) but have since been recognized as discrete and elevated to full species (Groves 2001a).

Microcebus *(Mouse Lemurs)*

Recent phylogenetic analyses of mitochondrial DNA (mtDNA) sequence data and newly collected mouse lemur specimens have also resulted in designations of several new species of *Microcebus* (Schmid and Kappeler 1994; Zimmermann et al. 1998; Rasoloarison et al. 2000; Yoder et al. 2000; Pastorini et al. 2001; Kappeler et al. 2005; Roos and Kappeler 2005; Andriantompohavana et al. 2006; Louis et al. 2006a, 2008; Olivieri et al. 2007; Radespiel et al. 2008). There are now 19 species of mouse lemur recognized, and these are listed in Table 5.1.

Cheirogaleus medius *Group (*C. medius *and* C. adipicaudatus*) and* C. major *Group (*C. major *and* C. crossleyi*) (Dwarf Lemurs)*

Social Organization. Home ranges of *C. medius* and *C. adipicaudatus* may overlap, and same-sexed individuals are intolerant of one another (Hladik et al. 1980). *C. medius* is monogamous and lives in dispersed family groups (Müller 1998, 1999; Fietz 1999b). This species deals with seasonal variation in food resources by entering torpor, during which nesting size is variable, from one to as many as five individuals sharing a nest in a hollow tree trunk. Males, however, emerge from torpor sooner than females, and this may be a form of paternal investment as males patrol their home range and, by doing so, may maintain access to its resources for their family group (Müller 1999). Males dramatically lose weight during this time (Müller 1999). Little is known regarding the ranging or social behavior of the *C. major* group. They are essentially solitary and may nest together in groups of two (Petter et al. 1977).

Diet and Feeding Ecology. The *C. medius* group focuses on a variety of high-quality foods, including fruits, nectar, vertebrates, and insects (Hladik 1979, Hladik et al. 1980, Wright and Martin 1995). Of particular note is their ability to store substantial fat in their tails (during which the volume of the tail triples), which is used during the torpid state in seasons of low food abundance (Hladik et al. 1980, Wright and Martin 1995). Similarly, members of the *C. major* group feed on young leaves, fruit, nectar, pollen, and insects (Petter et al. 1977) and can tolerate a medium level of tannins in the diet (Ganzhorn 1988). They also enter a period of torpor during the dry season and store fat in the tail to accommodate this period.

Reproduction. Fat-tailed dwarf lemurs are seasonal breeders. The gestation period is 61–64 days (Petter 1978, Hladik et al. 1980), and a female normally produces twins, although this can vary from one to four infants (Foerg 1982). In captivity, these lemurs become sexually mature in their first year of life (Foerg 1982); however, in the wild, sexual maturity may not occur until 2 years of age (Müller 1999). *C. medius* exhibits male parental care, signaling high reproductive costs for both sexes; this is visible in the reduced fat stores and overall body condition of breeding pairs compared to nonreproducing individuals (Fietz and Dausmann 2006). In the *C. major* group, gestation length is 70 days, with two or three infants born in January (Petter-Rousseaux 1964, Petter et al. 1977). They are carried by the mother in her mouth as they are unable to cling at birth (Petter-Rousseaux 1964). Lactation lasts only 1.5 months, and infants develop quickly, being able to follow their mothers within a month and to eat fruit at about 25 days of age (Petter-Rousseaux 1964).

Allocebus trichotis *(Hairy-Eared Dwarf Lemur)*

Social Organization. As *Al. trichotis* has yet to be systematically studied in the wild, little information is available. It sleeps in tree holes in groups of two to six (Meier and Albignac 1991).

Diet and Feeding Ecology. Observations in captivity indicate that *Allocebus* feeds on fruit, honey, and locusts; and its long tongue is suggestive of nectar feeding (Meier and Albignac 1991). Seasonal body fat storage occurs but over the entire body, not just in the tail (Meier and Albignac 1991).

Reproduction. Little is known about its reproduction, but infants may be born in January or February (Meier and Albignac 1991).

Mirza coquereli *(Coquerel's Giant Mouse Lemur) and* M. zaza *(Northern Giant Mouse Lemur)*

Social Organization. In *M. coquereli* adult males are heavier than adult females, and this increases most dramatically prior to and during the mating season (Kappeler 1997). Female home ranges are 4 ha, remain stable over time, and overlap considerably, with little evidence of actively defended territories (Kappeler 1997). Male home ranges increase during the mating season and overlap with other home ranges only at that time (Kappeler 1997). Genetic data indicate that females are organized into matrilines, most females show philopatry, and dispersed multimale/multifemale is the social organization of this species (Kappeler et al. 2002). Little is known about *M. zaza*, although this species seems to be more social than *M. coquereli*, and it has been suggested that *M. zaza* occur at higher density than *M. coquereli* due to more frequent sociality (Markolf et al. 2008).

Diet and Feeding Ecology. Coquerel's dwarf lemur has an eclectic diet that includes fruit, flowers, buds, gums, insects and insect secretions, spiders, frogs, chameleons, and small birds (Pagès 1980, Andrianarivo 1981).

Reproduction. Mating occurs in October, followed by a 3-month gestation (Petter-Rousseaux 1980). Infants develop quickly and can leave their nests after 1 month (Pagès 1980). Females may become reproductive within their first year (Kappeler 1997).

Microcebus

Considerable advances in the study of the behavior and ecology of *Microcebus* have revealed great flexibility in this genus. Many of the new species have been described based on their difference at a genetic level because they are very similar in appearance and ecology to already recognized species. As well, many of the newer species have yet to be completely described, including *M. tavaratra*, *M. sambiranensis*, *M. griseorufus*, *M. arnholdi*, *M. simmonsi*, *M. lehilahytsara*, *M. jollyae*, *M. margotmarshae*, *M. macarthurii*, *M. mittermeieri*, *M. mamiratra*, *M. bongolavensis*, *M. danfossi*, and *M. lokobensis*. In order to be more representative of these species, we have conflated much of the information about individual species into a group description.

Social Organization. *M. murinus* (the gray mouse lemur), *M. rufus* (the brown mouse lemur), and *M. ravelobensis* (the golden-brown mouse lemur) have all been observed in multimale/multifemale groups; and this may be the system typical of mouse lemurs (Fietz 1999a; Radespiel 2000, 2006; Atsalis 2000; Radespiel et al. 2003a).

M. murinus appears to exhibit a multimale/multifemale system within a dispersed social network, foraging alone at night and sleeping in groups of up to 15 individuals (Martin 1972a, 1973; Radespiel et al. 1998; Fietz 1999a; Radespiel 2000).

Females will sleep with the same female partners, and these individuals often share home ranges; however, different female groups use nearly exclusive home ranges (Radespiel 2000). Males often sleep alone (Radespiel et al. 1998, Radespiel 2000), but during the mating season it is common to find mixed-sex groups in these nests, with a single male nesting with as many as seven females (Martin 1973). Home ranges overlap substantially (Barre et al. 1988; Fietz 1999a; Radespiel 2000), and male home ranges are larger than those of females (Table 5.3) (Pagès-Feuillade 1989). *M. ravelobensis* shares this system; conspecifics interact frequently, and sleeping groups can be made up of only females or both females and males and are maintained over time even though sleeping sites may change (Radespiel

et al. 2003a, Weidt et al. 2004). It is suggested that thermoregulation may explain such groupings and that sleeping groups are the basic social unit in mouse lemur society (Weidt et al. 2004).

Variation does occur: Current data indicate that *M. berthae* (Berthe's mouse lemur) is nongregarious and both forages and sleeps solitarily. It has been suggested that such a sleeping pattern may occur as a result of both competition for nest sites from other nocturnal sympatric animals, including lemurs, as well as an antipredator strategy in this smallest of the living primates (Schwab 2000). Male home ranges appear to be larger than female home ranges at least during the mating season (Schwab 2000). Indirect data (e.g., changes in testicle size, presence of sperm plugs) suggest this species has a multimale mating system that includes promiscuous mating and sperm competition (Schwab 2000).

Diet and Feeding Ecology. *M. murinus* stores fat in its tail and may enter torpor, but time spent in torpor varies by sex; also, while females become inactive during periods of low food availability, males are more active (Rasoazanabary 2004). This species is omnivorous, but fruit and invertebrates are a major component of the diet. Other foods include flowers, nectar, leaves, sap and gum, homopteran larvae secretions, and small invertebrates (Martin 1972a, 1973; Petter 1978; Hladik 1979; Barre et al. 1988; Corbin and Schmid 1995; Radespiel 2006). Insect prey is often caught on the ground (Martin 1972a, 1973). Similarly, *M. rufus* consumes fruits, insects, and flowers (Martin 1972a, Harcourt 1987) as well as, more rarely, young leaves (Ganzhorn 1988). *M. rufus* stores some fat in its tail and may enter torpor depending on the habitat (Atsalis 1998). The diet of *M. ravelobensis* is omnivorous and similar to that of *M. murinus* (Reimann 2002, Radespiel et al. 2003a). Individuals forage alone, remain active despite changes in environmental conditions, and do not appear to alter fat storage in their tails across different seasons and photoperiods (Randrianambinina et al. 2003). Daily torpor occurs in this species (Radespiel et al. 2003a). Omnivory has also been observed in *M. griseorufus*, which has been observed feeding on insects and gum, although not much else is known (Rasoazanabary 2004, Radespiel 2006). In-depth studies of *M. berthae* feeding ecology have yet to be carried out, but males and females forage separately (Schwab 2000).

Reproduction. In *M. murinus,* mating occurs in September, with a gestation of 59–62 days (Martin 1972a, Radespiel 2000). Normally, twins are born and are parked in tree holes and/or carried until 3 weeks of age. Infants develop quickly and exhibit adult behaviors by 2 months of age (Petter-Rousseaux 1964, 1980; Martin 1972a). In captivity, females first give birth at 18 months (Petter-Rousseaux 1964). No data are available on *M. rufus* reproduction in the wild. This species is also difficult to maintain in captivity, but data from a breeding colony of wild-caught *M. rufus* indicate an estrous cycle of 59 days, 2.5 cycles per season, seasonal reproduction with a seasonal change in testicular size, and a gestation length of 56.5 days, with litter size ranging from one to three offspring. Mating behavior varied among pairs, but copulation appeared to be limited to a single day per estrus (Wroegemann and Zimmermann 2001).

During the *M. berthae* mating season female cycles are not synchronized. Males are heavier than females during the mating period, but females are heavier than males during the nonreproductive season (Schwab 2000). Likewise, although there is a distinct mating season in *M. ravelobensis*, females' estrus does not appear to be strongly synchronized (Schmelting et al. 2000, Randrianambinina et al. 2003). Mating occurs in September, with a gestation of 59–62 days (Martin 1972a, Radespiel 2000). Normally, twins are born and are parked in tree holes and/or carried until 3 weeks of age. Infants develop quickly and exhibit adult behaviors by 2 months of age (Petter-Rousseaux 1964, 1980; Martin 1972a). In captivity, females first give birth at 18 months (Petter-Rousseaux 1964).

Phaner furcifer *(Fork-Marked Lemur, Four Subspecies: Eastern Fork-Marked Lemur, Pale Fork-Marked Lemur, Pariente's Fork-Marked Lemur, Amber Mountain Fork-Marked Lemur)*

Social Organization. The highly vocal *Ph. furcifer* (a mean of 30 loud calls an hour emitted by males has been counted in a radius of about 200 m) can be found in holes in baobab trees, old *Mirza coquerli* nests, and leaf nests (Petter et al. 1971, 1975; Schulke and Kappeler 2003). Meetings between neighboring family groups occur where home ranges intersect, during which females may interact affiliatively with females of other groups (Schulke and Kappeler 2003). Males often interact agonistically with neighboring males and females during such encounters (Schulke and Kappeler 2003). Male/female pairs can maintain vocal contact throughout the night and may nest together during the day (Charles-Dominique and Petter 1980, Schulke and Kappeler 2003). *Phaner*'s social organization may be described as "dispersed pairs" because although there is pair stability for as many as 3 years and their territories overlap nearly completely, actual interaction between male and female pairs is very low (Schulke and Kappeler 2003).

Diet and Feeding Ecology. The fork-marked lemurs' primary food is gum, particularly from *Terminalia* trees; but they also consume insects, sap, buds, flowers, and insect exudates (Charles-Dominique and Petter 1980). Gum feeding correlates with this species' highly specialized toothcomb, which is used to create holes to access tree gum and sap (Charles-Dominique and Petter 1980).

Reproduction. Mating is in June (Charles-Dominique and Petter 1980), with a single infant born in November or December that is first carried and then rides on the mother's back (Petter et al. 1971, 1975; Charles-Dominique and Petter 1980).

Avahi *Group (Woolly Lemurs)*

As in many of the other nocturnal genera, there has been a recent proliferation of newly described species and, for many, little is known. Newer species include *A. mooreorum* (Moore's woolly lemur), *A. peyrierasi* (Peyrieras' woolly lemur), *A. meridionalis* (southern woolly lemur), *A. ramanantsoavani* (Ramanantsoavana's woolly lemur), *A. betsileo* (Betsileo's woolly lemur), and *A. cleesei* (Bemaraha's woolly lemur), most of which are currently underrepresented in the literature.

Social Organization. Short-term studies indicate that woolly lemurs are monogamous. An adult male, female, and offspring make up the group; usually, it is encountered in pairs or trios, but as many as five individuals can be together (Pollock 1975, Petter and Charles-Dominique 1979, Albignac 1981a, Ganzhorn et al. 1985, Harcourt 1988, Thalmann 2001, Norscia and Borgognini-Tarli 2008). Individuals may forage alone and meet throughout the night to groom and rest together (Harcourt 1988, Razanahoera-Rakotomalala 1981), or males and females may move together continuously, staying in close proximity while feeding on the same plants (Thalmann 2006). Group members sleep together in dense foliage (Albignac 1981a). Vocalization may be minimal (Thalmann 2006).

Diet and Feeding Ecology. Although primarily active at night, woolly lemurs have also been observed feeding during the day (Ganzhorn et al. 1985). *Avahi* feeds primarily on leaves, an unexpected diet given its relatively small body size. Such folivory may explain its high level of resting during the evening (Albignac 1981a, Razanahoera-Rakotomalala 1981, Ganzhorn et al. 1985, Ganzhorn 1988, Harcourt 1988, Thalmann 2001). Males and females forage together and feed in the same trees (Thalmann 2001).

Reproduction. Woolly lemurs give birth to a single infant in August or September. Infants initially cling to the mother's ventrum and then later ride on her back (Martin 1972b, Petter et al. 1977, Ganzhorn et al. 1985, Harcourt 1988).

Daubentonia madagascariensis *(Aye-Aye)*

Originally believed to be extinct, the highly specialized aye-aye was rediscovered in 1957 (Petter and Petter-Rousseaux 1959). New taxonomic studies have recommended placing *Daubentonia* in its own infraorder, the Chiromyiformes, due its early separation from the ancestral form (Quinn and Wilson 2004, Warter et al. 2005). For the purpose of concision, this chapter has retained the more standard inclusion of *D. madagascariensis* with the Lemuriformes, a taxonomy that is still widely used (e.g. Mittermeier et al. 2008a).

Some recent work has investigated the retention of genes signaling dichromatic vision in this nocturnal species (Perry et al. 2007). This research has suggested that though the aye-aye has filled its current nocturnal niche for millions of years, dichromacy has remained beneficial during any cathemeral activity as a more effective means of detecting and avoiding predation after daytime disturbance at sleeping sites and possibly improving the efficiency of foraging by moonlight (Perry et al. 2007b).

Social Organization. The aye-aye builds its nest in the fork of trees and normally forages alone, but it can be found near other individuals (Petter et al. 1977; Iwano and Iwakawa 1988; Sterling 1992, 1993). Studies at Nosy Mangabe indicate that females have exclusive ranges and rarely interact with one another or do so aggressively (Sterling and Richard 1995). Males have large, overlapping ranges and interact both aggressively and affiliatively with one another (Sterling and Richard 1995). Male and female ranges overlap, and most interactions appear to be affiliative, with individuals communicating through vocalizations and scent marking (Sterling and Richard 1995). Recent observations of a captive population suggest that immature aye-ayes are significantly more dependant on observation-based learning of foraging behavior than some diurnal species and that social learning is of great importance in this species (Krakauer 2006).

Diet and Feeding Ecology. The aye-aye exhibits a number of specializations, including continuously growing rodent-like incisors and a long and thin third digit that allows it to forage for wood-boring larvae and to feed on hard seeds of the genus *Canarium* (Sterling et al. 1994). It also focuses on other high-quality foods that include fruit, especially coconuts; adult insects; fungus; and nectar (Petter et al. 1977, Iwano and Iwakawa 1988, Sterling 1993, Sterling et al. 1994). This species is able to inhabit a wide variety of habitats, from rain forest to cultivated areas (especially coconut groves) (Tattersall 1982).

Reproduction. Females in estrus give loud calls that attract males and will mate with some, but not all, attracted males (Sterling and Richard 1995). Aye-ayes are reported to give birth only every 2–3 years (Petter and Peyriéras 1970a,b; Petter et al. 1977). Births appear not to be seasonal, and infants may be weaned at 7 months (Petter and Peyriéras 1970a,b; Sterling 1993).

Lepilemur *(Sportive Lemurs)*

Many new lepilemur species have been recently proposed, bringing the number to 26 (Mittermeier et al. 2008a, Ramaromilanto et al. 2009). The majority of the newer species have yet to be extensively described.

Social Organization. *Lepilemur leucopus* (the white-footed sportive lemur) males and females may sleep in separate tree holes or bundled lianas and are sometimes found sleeping in pairs (Russell 1977, 1980). Females may share ranges with young offspring and perhaps even with adult daughters (Charles-Dominique and Hladik 1971). In *L. edwardsi* (the Milne-Edwards sportive lemur) as well, males and female commonly sleep together in tree holes or near one another in separate holes (Albignac 1981b, Petter et al. 1977, Warren 1994, Rasoloharijaona et al. 2003). Two to three individuals may forage together and regularly engage in grooming bouts (Warren 1994). Current studies indicate this species may be characterized by dispersed monogamy, with each pair defending its home range by branch shaking and vocal displays (Rasoloharijaona 1998, Zimmermann 1998, Rasoloharijaona et al. 2003). Fidelity of these pairs may last as long as 4 years (Altrichter 2001). *L. ruficaudatus* (the red-tailed sportive lemur) is pair-living (Hilgartner et al. 2008), and pairs maintain their small (≈1 ha) territories over several years (Zinner et al. 2003). Observations of *L. mustelinus* (the weasel sportive lemur) have thus far indicated a solitary form of social organization, in marked contrast to the pattern observed in *L. ruficaudatus* and *L. edwardsi* (Rasoloharijaona et al. 2008). This species has not been seen sleeping or foraging in groups or pairs, and it is possible that differences between rain and dry forests may be the reason (Rasoloharijaona et al. 2008).

Diet and Feeding Ecology. Lepilemurs are generally highly folivorous. *L. leucopus* focuses on low-quality leaves or flowers of the Didiereaceae species *Alluadia procera* and *Alluadia ascendens* (Charles-Dominique and Hladik 1971, Hladik and Charles-Dominique 1974). *L. edwardsi* as well focuses primarily on leaves, which are selected for their protein value and low alkaloid content (Ganzhorn 1988, 1993). Fruit, flowers, and fleshy seeds are also eaten but at much lower levels (Razanahoera-Rakotomalala 1981, Albignac 1981b, Ganzhorn 1988, Thalmann 2001). Lepilemurs studied in the wild often forage solitarily, as is the case in *L. edwardsi* (Thalmann 2001).

Reproduction. Mating in *L. leucopus* occurs in May–July, and gestation is 4.5 months, with singleton births in September–November (Petter et al. 1977). Individuals are sexually mature at 18 months (Richard 1984). Female *L. edwardsi* give birth at the end of September to a single infant (Rasoloharijaona et al. 2003), and infants are left in a tree hole or within dense foliage while the mother forages (Rasoloharijaona et al. 2003). *L. ruficaudatus* is a highly seasonal breeder, with mating restricted to 2 weeks in May–June (Hilgartner et al. 2008). Males have been observed guarding their pair-partner during the mating season, and females are noted to have conspicuous sexual swellings during the mating season (Hilgartner et al. 2008). Gestation takes place during the dry season and lasts an average of nearly 6 months, with females giving birth to single infants from November to mid-December (Hilgartner et al. 2008).

Subfossil Lemurs

There has been a great deal of work done to describe the subfossil lemurs, that is, those 16 or so lemurs that became extinct in the late Holocene due to anthropogenic effects (overhunting, habitat destruction, and aridification) (Burney 1997, Dewar 1997). This work is, unfortunately, beyond the scope of this chapter. (See Walker 1967; Gingerich and Martin 1981; Martin 1990; Godfrey et al. 1993, 1997, 2004b, 2006; Simons et al. 1995; Fleagle 1999; Jungers et al. 2002, 2007; Rafferty et al. 2002; Godfrey and Irwin 2007 for more information.)

While we will unfortunately never see these large and unusual extinct lemurs, research and conservation efforts are occurring in many areas of Madagascar today on extant species and new protected areas have been designated since 2002. Hopefully, these conservation programs, along with the continued work of both Malagasy and foreign primatologists, will inform us further as to the behavior and ecology of the lemurs and help protect these beautiful and fascinating primates through the twenty-first century and beyond.

Species included in this chapter are shown in Color Plates 2–4 (*Lemur catta, Varecia rubra*, and *Eulemur fulvus rufus*).

6

Tarsiiformes

Sharon Gursky

1. Why is the taxonomic classification of the tarsiers contested?
2. What do tarsiers eat, and why is their diet unusual for a primate?
3. Although nocturnal, tarsiers exhibit high degrees of social gregariousness. Describe social behavior in the tarsiers.

TAXONOMY

Tarsiers first became known to Western scientists through a description given by the missionary J. G. Camel, based on his observations of an individual from the island of Luzon in the Philippines (Hill 1955). It is important to note that the Philippine tarsiers are restricted to the southern Philippine islands and, consequently, this individual tarsier could *not* have been from Luzon! In his description, Camel described the tarsier as a small monkey with big, round eyes that rarely closed; skinny, hairless ears; and the length of the tail and hindfeet similar to the length the rest of the body. Petiver published Camel's description in 1705 and named the animal *Cercopithecus luzonis minimus* due to its primate-like features. Based on Petiver's account of Camel's discovery, Linnaeus (1758) coined the name *Saimiri syrichta*. In 1765, Buffon also described a juvenile tarsier from an unknown locale that he gave the name *Didelphis macrotarsus*, "*Didelphis*" because he believed that tarsiers were related to opossums (or perhaps jerboas, small rodents from the Dipodidae family) and "*macrotarsus*" because of the extremely long tarsal bones exhibited by this animal. Interestingly, in a later edition of his *Systema Naturae*, Linnaeus recognized Buffon's tarsier under the name *Didelphis macrotarsus*, also regarding it as an opossum! He did not apparently associate it with his own *Saimiri syrichta*, based on Camel's animal. Erxleben (1777) was the first scientist to link tarsiers with prosimian primates and gave them the name *Lemur tarsier*. Several years later, Storr (1780) recognized the numerous differences between lemurs and tarsiers and argued for their generic separation. He coined the genus name *Tarsius* for these unusual primates that possess exceptionally long tarsal bones.

Since 1780, when tarsiers were finally recognized as primates (Storr 1780), they have been the focus of numerous controversies. Most notably, there has been considerable debate over the classification of the tarsiers within the order Primates (Pocock 1918, Wolin and Massoupust 1970, Cave 1973, Luckett 1976, Cartmill and Kay 1978, Rosenberger and Szalay 1980, Aiello 1986). There are two major schools of thought. The first school of thought is based on the work of Pocock (1918), who believed that tarsiers are more similar to the simian primates than to the lemurs and lorises. To express this relationship, Pocock established two infraorders within the order Primates: the Haplorhini, which included the suborder Tarsoidea as well as the suborders Ceboidea, Cercopithecoidea, and Hominoidea, now all separated at the superfamily level, and the Strepsirhini, which included the lemurs and lorises. Pocock believed tarsiers were more similar to the anthropoid primates because of numerous shared derived characteristics. Among the characteristics shared by tarsiers and anthropoid primates are hemochorial placentation, dry rhinarium, absence of a tapetum lucidum, presence of a fovea centralis, lack of a toothcomb, reduced olfactory bulbs, a well-developed promontory artery, and lack of an attached upper lip (Pocock 1918, Wolin and Massoupust 1970, Cave 1973, Luckett 1976, Cartmill and Kay 1978, Rosenberger and Szalay 1980, Aiello 1986, MacPhee and Cartmill 1986, Fleagle 1999). The second school of classification is based on the idea that lemurs, lorises, and tarsiers are grouped together in the suborder Prosimii, with the simian primates all grouped together in the Anthropoidea (Simpson 1945, Cartmill and Kay 1978, Rosenberger and Szalay 1980, Fleagle 1999). This grouping pattern is based on numerous ancestral characters, such as the tarsier's nocturnal habits, its small body size, and its parenting strategy (infant parking and oral transport). This represents a gradistic classification scheme compared to the cladistic classification proposed by Pocock (1918). Anatomical characteristics that are shared by tarsiers and other prosimian primates include an unfused mandibular symphysis, a grooming claw, multiple pairs of mammae, and a bicornuate uterus (Pocock 1918, Wolin and Massoupust 1970, Cave 1973, Luckett 1976, Cartmill and Kay 1978, Rosenberger and Szalay 1980, Aiello 1986, MacPhee and Cartmill 1986, Fleagle 1999).

Molecular data have not been useful in elucidating the phylogenetic position of tarsiers. Specifically, a study of the apolipoprotein b gene (APOB) showed strong support for an association of tarsiers with anthropoids, thus supporting the Haplorhine–Strepsirhine dichotomy (Amrine-Madsen et al. 2003). On the other hand, the substitution rate of the cytochrome b gene has increased approximately twofold along lineages leading to simian primates in comparison to tarsiers, other primates, and nonprimate mammalian species, thus supporting the prosimian–anthropoid dichotomy (Andrews et al. 1998). Resulting from the large amounts of conflicting data and their interpretations, the question of tarsier phylogeny is tremendously complicated and unlikely to be resolved anytime soon.

There has been not only substantial disagreement concerning the taxonomic classification of the tarsier but also considerable disagreement regarding the number of species (Hill 1955; Niemitz 1984a; Musser and Dagosto 1987; Niemitz et al. 1991; Nietsch and Kopp 1998; Nietsch 1999; Groves 1998, 2001a,b; Shekelle 2003). At present, six species are formally recognized: *Tarsius syrichta* (Linnaeus 1758), the Philippine tarsier; *T. bancanus* (Color Plate 6; Horsfield 1821), the Bornean tarsier; *T. pumilus* (Miller and Hollister 1921, Musser and Dagosto 1987), the pygmy tarsier from central Sulawesi; *T. spectrum* (Pallas 1778), the spectral tarsier from Sulawesi; and *T. dianae* (Niemitz et al. 1991), another central Sulawesi lowland tarsier species; and most recently, *T. lariang* (Merker and Groves 2006).

According to Hill (1955), Cabrera suggested that Buffon's tarsier was actually an eastern tarsier (*T. spectrum*) and not a Philippine tarsier (*T. syrichta*). Despite the loss of this specimen but based on Buffon's written description, Groves (2001b) and Shekelle (2003) have reaffirmed this opinion. The nomenclatural implication of this error is that the name "*L. tarsier*" (Erxleben 1777) has seniority over "*L. spectrum*" (Pallas 1778). Thus, *T. spectrum* would be *T. tarsier*. I continue to use the name "*T. spectrum*" but recognize that the correct name might be "*T. tarsier.*"

The tarsier species all differ morphologically from one another in a variety of ways. They differ in their body weight (Table 6.1), absolute orbit size, absolute tooth size, proportion of tail that is covered by hair, and limb proportions. In terms of absolute orbit size and absolute tooth size, *T. bancanus* possesses the largest orbits and largest teeth of any of the species, followed by *T. syrichta*, *T. spectrum*, and then *T. pumilus* (Musser and Dagosto 1987). There is also a cline in the proportion of the tail that is covered by hair for each species. *T. spectrum* and the other Sulawesian tarsier species possess the most tail hair, followed by *T. bancanus* and *T. syrichta*, whose tail is often described as "naked" due to the sparse, fine hairs that cover it. Niemitz (1984a) also recorded numerous differences in limb proportions between the three major tarsier species (*T. spectrum*, *T. bancanus*, and *T. syrichta*). *T. bancanus* has the longest hindlimbs and hands, followed by *T. syrichta*, whose hindlimbs and hands are intermediate in length, followed by *T. spectrum*, whose hindlimbs and hands are shortest in length. Dagosto and colleagues (2001) also demonstrated a cline in the locomotor patterns of tarsiers. They found that *T. bancanus* is the most specialized, using vertical clinging postures the most frequently, followed by *T. syrichta* and then the Sulawesian tarsiers *T. spectrum* and then *T. dianae*.

According to Nietsch and Kopp (1998), two of the Sulawesian tarsiers (*T. spectrum* and *T. dianae*) can be distinguished from one another based on their vocalizations. Using playbacks of the vocal duets sung each morning, Nietsch and Kopp found that each species responded only to the vocalizations of their own species. During their study neither species responded to the vocalizations of the other Sulawesian tarsier species, while avidly responding to the vocalizations of their own species. Nietsch and Kopp also found that the tarsier population on Togian Island, adjacent to Sulawesi, did not respond to the playback vocalizations of *T. spectrum* or *T. dianae* and proposed that the Togian tarsier population may also represent a distinct species, *T. togianensis*. Specifically, Nietsch (1999) found that the degree of separation of the acoustic characteristics of the Togian tarsier population from the *T. spectrum* population was substantial enough to warrant their designation as a distinct species.

Although the vocalizations of *T. lariang* are quite distinct from those of *T. dianae*, Merker and Groves (2006) argue that the distinctiveness of *T. lariang* relative to the other Sulawesian tarsier species is primarily based on its

Table 6.1 Body Weights of Four Different Tarsier Species Separated by Sex

SPECIES	BODY WEIGHT (G)	SEX	*N*	REFERENCES
Tarsius bancanus	128	M	—	Niemitz 1984a
	117	F	—	Niemitz 1984a
	128	M	5	Crompton and Andau 1987
	118	F	2	Crompton and Andau 1987
T. dianae	110	F	1	Niemitz et al. 1991
	105	F	1	Muskita cited in Sussman 1999a
	104	M	1	Muskita cited in Sussman 1999a
	100	F	4	Shekelle 2003
	119	M	5	Shekelle 2003
	130	M	1	Gursky 1997
	110	F	1	Gursky 1997
T. spectrum	113	F	1	Niemitz et al. 1991
	110	M	1	Niemitz et al. 1991
	104	F	21	Shekelle 2003
	115	M	11	Shekelle 2003
	107	F	8	Gursky 1999
	126	M	5	Gursky 1999
T. syrichta	136	M	4	Neri-Arboleda 2001
	120	F	6	Neri-Arboleda 2001
	141	M	4	Dagosto et al. 2001
T. pumilus	58		1	Maryanto and Yani 2008

relatively long, narrow bullae. However, they also noted that *T. lariang* is the largest species of tarsier on mainland Sulawesi. This is evident not only in its total skull length but also in the M1 mesiodistal length as well as the length of the third finger.

Two additional Sulawesian tarsier species, *T. pelengensis* and *T. sangirensis*, have also been tentatively proposed and are awaiting additional studies utilizing larger sample sizes (Shekelle et al. 1997; Groves 1998, 2001a; Shekelle 2003). Some scientists naturally question whether there really are so many distinct tarsier species in Sulawesi or whether they are in fact subspecies. Based on preliminary genetic data (Shekelle 2003) as well as my own preliminary observations of behavior, vocalizations, and morphology, I believe that these populations, as well as several others in Sulawesi, will be shown to be distinct, representing new species. It is clear that the continued use of vocalizations is integral for the proper identification of species boundaries when dealing with cryptic primates such as the tarsier (Bearder 1999).

Another taxonomic argument relating to tarsiers concerns whether the Sulawesian tarsiers are distinct enough to warrant their own genus name. Groves (1998) argues in his analyses of the three historically recognized species (*T. bancanus*, *T. syrichta*, and *T. spectrum*) that *T. spectrum* is distinct enough from *T. bancanus* and *T. syrichta* to warrant its own genus name, *Rabienus*. At present, however, even Groves (2001b) continues to use the genus name *Tarsius* for the Sulawesian tarsiers (*T. spectrum*, *T. dianae*, and *T. pumilus*).

Interestingly, although there are only minor differences in morphology, there are substantial chromosomal differences between some tarsier species. *T. bancanus* and *T. syrichta* are each reported to have a karyotype of 80 chromosomes (Klinger 1963, Poorman et al. 1985, Dutrillaux and Rumpler 1988), whereas *T. dianae* is reported to have only 46 chromosomes (Niemitz et al. 1991). The karyotypes for both *T. spectrum* and *T. pumilus* are still unknown. Dutrillaux and Rumpler (1988) report that the chromosomes of *T. syrichta* and *T. bancanus* are distinct from over 100 other primate species and other mammals. It has been suggested therefore that the tarsiers do not have a primitive karyotype but have undergone a large number of chromosomal rearrangements (Dutrillaux and Rumpler 1988). A large number of fissions have certainly occurred since the diploid chromosome number of 80 is among the highest in mammals.

ECOLOGY

Despite the great interest in the phylogenetic relationships of tarsiers and their taxonomic nomenclature, only recently have primatologists begun to study their behavior, ecology, and conservation status. In part, this is a reflection of the difficulty of observing a small, nocturnal primate. In addition, tarsiers do not possess a *tapetum lucidum* (a reflective layer behind the eye), which helps observers locate most other nocturnal mammals in the dark. Over the last decade, several scientists have come to recognize that the tarsier's behavioral repertoire is relatively unusual and deserves equal attention. Behavioral and ecological studies of wild, free-ranging, and semicaptive tarsiers have concentrated on five species: *T. bancanus* (Fogden 1974; Niemitz 1972, 1974, 1979, 1984a; Crompton and Andau 1986, 1987), *T. spectrum* (MacKinnon and MacKinnon 1980b; Gursky 1995, 1997, 1998a,b, 1999, 2000a, 2002a,b, 2003c, 2005a,b, 2006, 2007a,b), *T. syrichta* (Dagosto and Gebo 1998, Dagosto et al. 2001, Neri-Arboleda 2001), *T. dianae* (Niemitz et al. 1991, Tremble et al. 1993, Gursky 1998b; Merker and Yustian 2008; Merker 2006; Merker et al. 2005; Merker and Muhlenberg 2000), (Table 6.2) and *T. pumilus* (Grow and Gursky 2009; N. Grow and S. Gursky unpublished data). These studies represent the primary behavioral and ecological literature for the remainder of this chapter, which compares the differences between the member species of the genus *Tarsius* with respect to their behavior. No field study has been conducted on *T. pumilus*.

ACTIVITY PATTERNS

Tarsiers are nocturnal and begin their activity around sunset. Although they are active throughout the night, they are known to have peaks of activity in the early evening and again just before sunrise. Niemitz (1984a) found that there was a minor peak in activity early and a major peak late in the evening in two *T. bancanus* individuals. Crompton and Andau (1987) observed more travel in the early evening in the same species. In *T. spectrum*, the most energetic periods occurred during the first half-hour of activity and then again during the last half-hour of activity (MacKinnon and MacKinnon 1980b, Gursky 1997). In *T. syrichta*, there is a big peak in activity early in the evening and then there are three minor peaks throughout the night (Neri-Arboleda 2001).

In *T. spectrum*, males spent approximately 54% of their night foraging, 30% traveling, 11% resting, and 5% socializing; females spent approximately 53% of their night foraging, 26% traveling, 15% resting, and 6% socializing (Gursky 1997). *T. dianae* spent 48% of the night foraging, 27% traveling, 21% resting, and 7% socializing (Tremble et al. 1993). *T. bancanus* spent approximately 62% of the night foraging, 27% traveling, and 11% resting (Crompton and Andau 1986). The similarity in time spent foraging and traveling is quite remarkable given that all these species live in very different habitats. The only observable difference in activity patterns across tarsier species is that *T. dianae* spent nearly twice as much time resting in comparison with the other species. No quantifiable data have been obtained for *T. syrichta* or *T. pumilus* on nightly activity patterns.

Table 6.2 Major Studies of Wild Tarsier Species and the Locations Where They Were Conducted

SPECIES	STUDY LOCATION	YEAR	REFERENCES
Tarsius bancanus	Semongok, Sarawak	1970–1985	Niemitz 1972,1974,1979,1984a
T. bancanus	Semongok, Sarawak	1970	Fogden 1974
T. bancanus	Sepilok, Sabah	1985	Crompton and Andau 1987a,b
T. spectrum	Tangkoko, Sulawesi	1978–1979	MacKinnon and MacKinnon 1980
T. spectrum	Tangkoko, Sulawesi	1994–2008	Gursky 1995,1997, 1998a,b, 1999, 2000a,b, 2002a,b, 2003a–c, 2005a,b, 2006, 2007a,b
T. dianae	Kamarora, Sulawesi	1992	Tremble et al. 1993
T. dianae	Kamarora, Sulawesi Posangke, Sulawesi	1997	Gursky 1997, 1998b
T. dianae	Kamarora, Sulawesi	1990	Niemitz et al. 1991
		2004–2008	Merker et al. 2005, Merker 2006, Merker and Yustian 2008
Tarsius syrichta	Mt. Pangasugan, Leyte	1995–2003	Dagosto and Gebo 1998, Dagosto et al. 2001
Tarsius syrichta	Corella, Bohol		Neri-Arbodela 2001

DIET

Tarsiers are also unusual among nocturnal prosimians in that they exclusively consume animal food (arthropods and vertebrates). At present, there are no observations in the literature of any tarsier species consuming fruit, leaves, or gum (MacKinnon and MacKinnon 1980b; Niemitz 1984a; Crompton and Andau 1987; Tremble et al. 1993; Gursky 2000b). Niemitz (1984a) has occasionally observed *T. bancanus* biting leaves but never chewing or digesting them. The same pattern has also been observed in *T. spectrum* (S. Gursky personal observation). Niemitz's observations (1979, 1984a) on semiwild *T. bancanus* indicated that their diet includes beetles (35%), ants (21%), locusts (16%), cicadas (10%), cockroaches (8%), and other miscellaneous vertebrates (11%: birds, smaller fruit bats, spiders, and poisonous snakes).

My own research on *T. spectrum* has shown that this species also prefers moths (Lepidoptera) as well as crickets and cockroaches (Orthoptera). These two arthropod orders comprised more than 50% of the study population's diet. I noted though that during the dry season, when these types of insect were less abundant, *T. spectrum* substantially increased their consumption of beetles (Coleoptera) and ants (Hymenoptera) (Gursky 2000b). The mean size of insects captured by *T. spectrum* was 1.9 cm and ranged 1–6 cm. The majority of the insects consumed either were located on or underneath a leaf (43%) or were flying insects (35%). A smaller percentage of the insects were located on branches (13%) or on the ground (9%).

Tremble et al.'s (1993) observation on the dietary items of wild *T. dianae* included moths, crickets, and a lizard. At present, nothing is known about the diet of wild *T. syrichta* or *T. pumilus*.

SLEEPING TREES

There is a large amount of variation among tarsier species with regard to sleeping-site preferences. This has important implications for determination of social groups and mating systems since tarsier groups are often identified by which animals sleep together. The majority of sleeping trees utilized by *T. spectrum* are *Ficus* spp., with *F. caulocarpa* being used much more frequently than all other species (Gursky 1997, MacKinnon and MacKinnon 1980b). *T. spectrum* primarily utilized one major sleeping site, although most groups had one alternate sleeping site as well. The mean circumference of these strangler fig sleeping trees was 287 cm, ranging 30–700 cm. The mean height of the sleeping trees was 20.17 m, ranging 6–38 m. MacKinnon and MacKinnon (1980a,b) also observed several *T. spectrum* groups utilizing vine tangles and grass platforms when groups were found in secondary forest and grassland areas.

In comparison, data from Crompton and Andau (1986) on *T. bancanus* indicate that the average height of sleeping trees was 4.0–5.5 m but occasionally as low as 2.5 m. Vine tangles and naturally formed platforms of creepers were the most common sleeping sites. Thus, the sleeping sites utilized by *T. bancanus* were much lower in the forest canopy and more open compared to those used by *T. spectrum*.

Niemitz et al. (1991), in their description of *T. dianae*, state that this species does not return to the same nest each night. This description agrees with MacKinnon and MacKinnon's (1980b) observation that tarsiers in central Sulawesi do not use the same sleeping site each night. Tremble et al. (1993), however, found that *T. dianae* did return to the same nest/tree each night. At two sites (Kamarora and Posangke), Gursky (1997, 1998b) also found that *T. dianae* returned to the same nest each morning. Nest sites for *T. dianae* include vine tangles, tree cavities, and fallen logs (Tremble et al. 1993). The type of sleeping site used might account for this variation in whether or not groups return to a sleeping site.

T. bancanus sleep singly unless it is a mother and her infant (Niemitz 1984a, Crompton and Andau 1987). Niemitz found that a population of *T. bancanus* sleeps clinging to vertical branches at 2 m or lower. In another population of *T. bancanus*, Crompton and Andau (1987) found that individuals sleep in vine tangles 4–5.5 m above the ground and on 50–90 degree supports. They also noted that the sleeping sites tend to cluster at the edge of a tarsier's home range, in areas of overlap with neighbors of the opposite sex. This contrasts with *T. spectrum*, whose sleeping sites tend to be located near the center of the group's home range.

Dagosto and Gebo (1998) report that *T. syrichta* in Leyte always sleep singly, never in groups, except for mother/infant pairs; and this result has been replicated by Neri-Arboleda (2001) with another population. *T. syrichta* in Leyte utilized three or four different sleeping sites. They were primarily *Arctocarpus*, *Pterocarpus*, and *Ficus*, low to the ground and surrounded by very dense vegetation. *T. syrichta* on the island of Bohol generally utilized one sleeping site but had several alternate sleeping sites. Males utilized seven or eight sleeping sites, while females tended to limit their sleeping sites to three or four different locations over the entire duration of the study. Five of the 14 tree species used as sleeping sites were *Ficus* spp. The average tree height of sleeping sites was 7.53 m. The sleeping sites of the Bohol *T. syrichta* were highly variable. They ranged from rock crevices near the ground (3%) to vine tangles (10%) and dense thickets of *Pandanus* palms (15%), tree trunks (45%), and fork branches of trees (27%). Pygmy tarsiers are also known to sleep in groups (N. Grow and S. Gursky, unpublished data). The sleeping tree of one group, a tree by the local name of "*luku*," had a diameter at breast height of 81 cm and an estimated height of 12 m. The pygmy tarsiers were observed to leave their sleeping tree each evening substantially before dusk and to return to the same sleeping tree each morning.

HOME RANGE

The early work of Fogden (1974) on *T. bancanus* suggested that the Bornean tarsier's home range is 2.5–3.0 ha (Table 6.3). This study was conducted over a 2-year period on 26 individuals. Niemitz (1979, 1984a) conducted his own observations on *T. bancanus* under semiwild conditions (within a 90 m² cage enclosure). He utilized urine and epigastric (abdominal) gland marking sites to determine the home range of the animal. His results indicated that the species' home range size varied 1–2 ha and was remarkably stable. Niemitz (1984a) found that the ranges were larger for males than females and that they overlapped extensively.

Crompton and Andau (1987) conducted the first study of tarsiers involving radiotelemetry in Sabah, Borneo. They collected 120 hr of observation on four individuals over the duration of their study. Males had a home range of 8.75–11.25 ha and females, 4.5–9.5 ha. Individuals traveled a mean distance of 1,800 m per night; males moved significantly farther (2,082 m) than females (1,448 m). Crompton and Andau (1987) report that males utilized 50%–75% of their total home range and females utilized 66%–100% of their range per night.

In a long-term radiotelemetry study of *T. spectrum*, Gursky (1998a) found that the home range size varied 1.6–4.1 ha, with males having larger ranges (mean 3.07 ha) than females (mean 2.32 ha). This is notably larger than the 1 ha range size previously estimated by MacKinnon and MacKinnon (1980b) based on occasional sightings. The mean nightly path length (NPL) for males was 791 m (n = 5), while that for females was 448 m (n = 8).

Dagosto and Gebo (1998) radiocollared four male *T. syrichta* on the island of Leyte during a 6-week study.

Male home range size was 0.6–2.0 ha. An 8-month study of *T. syrichta* was conducted by Neri-Arboleda (2001), who found that the home range was 6.86 ha for males (n = 4) and 2.76 ha for females (n = 6). Interestingly, Neri-Arboleda (2001) also had one home range of a subadult that averaged 13.4 ha, slightly larger than Crompton and Andau's (1987) 11.25 ha for male *T. bancanus*. Neri-Arboleda (2001) noted that the incremental plot of home range never reached an asymptote, suggesting that the home range would be even larger with additional data collection.

The mean NPL for *T. syrichta* females was 1,118 m, ranging 820–1,304 m (Neri-Arboleda 2001). The mean NPL for males was 1,636 m. While these figures are comparable with those of *T. bancanus*, they are substantially greater than those observed for *T. spectrum* (Gursky 1998a) or the Leyte population of *T. syrichta*. Dagosto and Gebo's (1998) preliminary study of tarsiers in Leyte suggests that the mean NPL was 301 m per night for two males.

The home range of Dian's tarsier averaged 1.1–1.8 ha (Merker 2006, Merker and Yustian 2008). There was substantial overlap between the male and female ranges and no statistical difference in range size usage between the sexes. However, the size of female home ranges varied significantly between habitat types, with the smallest ranges in slight- and intermediate-disturbance areas. The NPL varied 800–1,400 m.

According to Grow and Gursky (2009, unpublished data), female pygmy tarsiers have a home range encompassing at least 1.2 ha. While being radiotracked, females traveled high in the canopy, out of visual range. The male range exceeded the capability of the radiotracking equipment and therefore must have been substantially larger than 2 ha. It is apparent that tarsiers, in general, have relatively small home ranges. However, home range sizes are not all

Table 6.3 Reported Home Range Sizes, Nightly Path Length (NPL), Mean Group Size, and Dispersal Distance in Wild Tarsier Species

SPECIES	HOME RANGE (HA)	NPL (M)	MEAN SOCIAL GROUP SIZE	DISPERSAL DISTANCE (M)
Tarsius bancanus	Males and females = 2.5–3.0 Males and females = 1.0–2.0 Males and females = 4.5–11.25	Males = 2082 Females = 1,448	1	N/A
T. spectrum	Males and females = 1.6–4.1	Males = 791 Females = 448	3	Males = 641 Females = 255
T. dianae	Males and females = 0.5–0.8 Males and females = 1.1–1.8	800–1400	2	N/A
T. syrichta	Males and females = 0.6–2.0 Males = 6.86 Females = 2.76 Subadult = 13.4	Males = 1,636 Females = 1,118	1	N/A
T. pumilus	Females = 1.2	N/A	4	N/A

readily comparable because they were collected using different techniques (trap–retrap vs. scent-marked trees vs. radiotracking) as well as different sample sizes in studies of varying duration.

REPRODUCTIVE PARAMETERS

The gestation period for captive *T. bancanus* is 178 days (Izard et al. 1985), which is comparable to the mean gestation period of 191 days estimated (Gursky 1997) for wild *T. spectrum* (Table 6.4). There are no published estimates of gestation length in either wild or captive *T. syrichta*, *T. dianae*, or *T. pumilus*.

Litter size for tarsiers has been reported to be strictly limited to one offspring; that is, there have been no reports of any intraspecific variation in litter size (Catchpole and Fulton 1943, Ulmer 1963, Niemitz 1984a, , Crompton and Andau 1987, Haring and Wright 1989, Roberts and Kohn 1993, Roberts 1994, Gursky 1997). The lack of variation in litter size in *Tarsius* spp. is surprising for two reasons. First, it is surprising because tarsiers are morphologically capable of producing twins; that is, they possess a bicornuate uterus, enabling two offspring to be produced (Luckett 1976, Fleagle 1999). Second, all tarsiers have two or three pairs of mammary glands (Schultz 1948, Niemitz 1984a, S. Gursky, personal observation) such that they could nurse more than one infant without any difficulty. There is a well-known relationship between nipple number and litter size, where litter size is usually one-half of nipple number (Gilbert 1986, Lagreca et al. 1992, Sherman et al. 1999). Yet, tarsiers have not been observed producing litters of two or three offspring despite being morphologically capable. Instead, they exhibit a life history strategy that minimizes the number of offspring produced each year, while at the same time producing a single large infant (25% of adult weight at birth) over an exceptionally long gestation period, with rapid postnatal growth (Wright 1990). The presence of the bicornuate uterus and multiple mammary glands represents evidence that historically tarsiers produced multiple offspring and not a single large offspring as they do today.

BIRTH SEASONALITY

Although Neri-Arboleda (2001) did not have data for an entire year on *T. syrichta* (March–October), the data she does have are certainly suggestive of birth seasonality in this species. Of the 10 births that she observed over the 8-month study, all were between April and July: one in April, two in May, four in June, and three in July. Captive *T. syrichta* has not been reported to be a seasonal breeder (Catchpole and Fulton 1943, Haring and Wright 1989). Fogden (1974) reported that wild *T. bancanus* has a sharply defined breeding season, with mating in October through December and births in January through March. However, it was observed to be a nonseasonal breeder in semiwild conditions (Niemitz 1984a) and in captivity (Wright et al. 1986, Roberts and Kohn 1993). Wild *T. spectrum* also has been reported to be a seasonal breeder, with two distinct breeding seasons within a year (MacKinnon and MacKinnon 1980b, Gursky 1997): April–May and November–December.

Lactation length for wild tarsiers is known only for *T. spectrum* and *T. syrichta*. Gursky (1997) and Neri-Arboleda (2001) both report a lactation length of approximately 78 days. For *T. syrichta*, weaning was determined when infants no longer shared a sleeping site with their mother. In captive tarsiers, lactation length is reportedly much more variable, 49–120 days (Niemitz 1984a, Haring and Wright 1989, Roberts and Kohn 1993, Roberts 1994).

Table 6.4 Life History Variables for Captive and Wild Tarsier Species

SPECIES	MEAN BIRTH WEIGHT (G)	LITTER SIZE	GESTATION LENGTH (DAYS)	METHOD OF INFANT TRANSPORT	BIRTH SEASONALITY	LACTATION LENGTH (DAYS)	INTERBIRTH INTERVAL (MONTHS)
Captive							
Tarsius bancanus	25	1	178	Fur/oral	No	49–82	—
T. spectrum	—	—	—	—	—	—	—
T. dianae	—	—	—	—	—	—	—
T. syrichta	26	1	180	Oral	No	60	—
Wild							
T. bancanus	—	1	—	—	Yes	—	—
T. spectrum	24	1	191	Oral	Yes	78	13
T. dianae	—	1	—	Oral	—	—	—
T. syrichta	37	1	—	—	Yes	78	—

PARENTAL CARE

One of the most unusual reproductive behaviors of the tarsiers concerns their parenting strategy. Tarsier mothers do not continually transport their infants on their bodies (ventrally or dorsally). Historically, tarsiers were thought to be rather nonchalant parents, parking their infants through the entire night and coming back to get them at dawn (Charles-Dominique 1977a, MacKinnon and MacKinnon 1980b, Niemitz 1984a). However, a study of *T. spectrum* infants has demonstrated that *T. spectrum* are not negligent parents but optimize their behavior given numerous constraints. Specifically, Gursky (1997) found that although infants are parked throughout the territory, the mean length of each parking bout was a mere 27 min. Also, in a given night, the mean number of locations that the infant was parked was 11 per night and that the average distance between a mother and her parked infant was 4 m (standard deviation 3.9). In fact, a *T. spectrum* mother more often than not foraged in the same tree where her infant was parked. These results together indicate that *T. spectrum* are not inattentive parents. The mother caches the infant in one tree where she forages and then carries the infant (in her mouth) into another tree, parks it, and begins foraging again within the tree where she parked her infant. Thus, "cache and carrying" better describes the infant caretaking system of *T. spectrum* than does "infant parking."

SOCIAL ORGANIZATION

In *T. spectrum*, group size is highly variable, ranging from two individuals to as many as eight individuals per sleeping site. The mean observed group size is three, but composition varies from two adult individuals of the opposite sex to one adult male and two adult females and their numerous offspring. Approximately 15% of groups contain two adult females. Groups with two adult females and two young infants suggest that there may be several breeding females in a group. Nietsch and Niemitz's (1992) preliminary observations produced similar results but with one group containing three adult females. Whereas MacKinnon and MacKinnon (1980b) observed *T. spectrum* groups with multiple adult males, Gursky (1997) reports only groups with one adult male, although some groups also contained a subadult male. The variation in group size and composition observed in *T. spectrum* is inconsistent with the traditional interpretation of a strictly monogamous social structure for this species (MacKinnon and MacKinnon 1980b). Instead, "facultative polygyny" more aptly describes the social structure of this nocturnal primate. In a study of *T. dianae* in central Sulawesi, Gursky (1998b) censused nine groups containing a total of 22 individuals. The mean group size was two. Six groups contained two individuals, two groups contained three individuals, and one group contained four individuals.

In contrast to *T. spectrum*, *T. bancanus* and *T. syrichta* do not live in groups but have overlapping home ranges (Fogden 1974, Niemitz 1984a, Crompton and Andau 1987). Thus, they are traditionally considered to exhibit a dispersed social organization, where a male's range overlaps the ranges of several females. However, Fogden (1974) reports a male and female *T. bancanus* traveling together on eight occasions. In one situation, a male was seen within a few days with two different females. Niemitz (1984a) reports that *T. bancanus* possibly live in pairs in that they live synterritorially for long periods. As the abovementioned studies were conducted not only at different field sites but also with different groups, it is possible that this tarsier species exhibits both male/female pair bonds (monogamy) as well as a dispersed social system.

According to Neri-Arboleda (2001), the Bohol *T. syrichta* are also solitary. They foraged and slept solitarily except for females carrying their infants. She never observed an individual in close proximity (<10 m) with a conspecific during nocturnal activity. A single observation was made of a male and female having sleeping sites 5 m apart. The Leyte *T. syrichta* are also solitary (Dagosto et al. 2001). All of the tarsiers captured were alone at their sleep sites, and three of the four males were never within close proximity to another tarsier. However, Dagosto et al. (2001) did observe one male sharing a sleep site with another tarsier (sex unknown) on three out of eight nights of observation. At the sleep site, the two tarsiers were never closer than 1 m from each other.

Based on preliminary data, pygmy tarsiers live in groups. Grow and Gursky (2009, unpublished data) observed a group with at least four individuals. While the fourth individual's sex is unknown, the captured individuals included two males and one female, indicating a potential polyandrous group composition. However, based on the presence of a fourth group member and the group composition of other species of tarsier, it probably represents a multimale/multifemale group composition.

SOCIALITY

Recent studies of *T. spectrum* have demonstrated that this species exhibits substantial amounts of gregariousness, relative to its sister species as well as to other nocturnal prosimian primates (Gursky 2002a). For example, in *T. spectrum* groups, approximately 28% of their nightly forays outside the sleeping tree were spent at distances from one another of ≤10 m. Almost 40% of their night was spent at <20 m from one another, including 11% of their time spent in actual physical contact. These observations of encounters with other group members are not the result of chance given the home range size, NPL, and travel speed of this species (Gursky 2002b). Gursky compared the observed encounter frequency with that expected based on Waser's random gas model (Waser 1976, Jolly et al. 1993, Holenweg et al. 1996, Gursky 2002b). Waser's (1976) random gas model is based on the idea that if group members' movements are random and independent, then their territory can be considered a two-dimensional gas of tarsier individuals. Using this model, Gursky (2002b) found that *T. spectrum* individuals encountered one another more frequently than expected by chance alone for distances of <50 m. This is especially true at closer distances. *T. spectrum* also spent more time foraging when close to another group member (≤10 m) compared to when individuals were 50 m or 100 m apart. However, individuals foraging close to another adult group member had lower insect-capture rates compared to more isolated foraging individuals. The decrease in foraging efficiency experienced by *T. spectrum* when foraging close to other group members may explain the corresponding increase in time allocated to foraging.

This result naturally raises the following question: If *T. spectrum* individuals experience substantial intragroup competition over food resources when foraging in close proximity to other group members, then why do they continue to forage in gregarious situations? Group living is predicted to occur only when the benefits outweigh the costs incurred (Krebs and Davies 1984, Dunbar 1988, Kappeler and Ganzhorn 1993). If living in a group is costly to *T. spectrum*, then why do they not actively avoid one another while traveling throughout their territory? One possibility is that *T. spectrum* forages in proximity to other adult group members only when the benefits of being near another group member outweigh the costs of intraspecific food competition (lower insect-capture rates when foraging in proximity). A few situations under which it might benefit *T. spectrum* to be gregarious are (*1*) when predation pressure is high, (*2*) when prospects of meeting up with potentially infanticidal males are high, and (*3*) when females are

sexually receptive. Currently, only data addressing the first potential situation are available.

In an experimental study of the determinants of gregariousness for *T. spectrum*, Gursky (2002a) found evidence to support the importance of predation pressure. In response to artificially high predator pressure, *T. spectrum* individuals were observed to spend more time together, often meeting up to try and scare away the predator. Adult group members were significantly closer on nights when predation-pressure experiments were conducted (snake models, bird of prey models, or vocalizations of birds of prey were distributed or played back) than on nights when no experiments were conducted (Fig. 6.1). In particular, on 60% of the nights that rubber snakes were placed in a group's territory, *T. spectrum* were observed mobbing the rubber snake. Mobbing involved all group members traveling to the area of the individual that initially located the snake. Group members began vocalizing loudly and then lunging at and retreating from the snake. On three occasions, an adult male *T. spectrum* was observed literally biting the rubber snake (Gursky 2002a). In each instance, it was the group's adult male. In addition, the mean distance between group members was twice as far on nights when no rubber snakes were present in the territory as on experimental nights with rubber snakes (Gursky 2002a).

An interesting behavioral observation that lends support to the importance of predation pressure for *T. spectrum* sociality is this species' response to bright moonlight (Gursky 1999, 2003a,c). *T. spectrum* were observed to increase the amount of time spent foraging and to decrease the amount of time spent resting during full moons relative to other moon phases (Fig. 6.2). Similarly, they were also observed to increase the number of loud calls emitted, territorial

Figure 6.1 Mean distance (m) between adult group members on nights when rubber snakes, plastic birds of prey, or vocalizations of birds of prey were distributed/played throughout the focal group's territory. The mean distance between adult group members when no experiments were conducted is represented by the control.

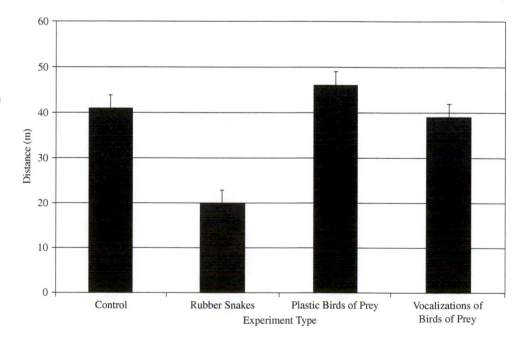

disputes, and intragroup encounters, while they decreased the number of scent marks compared to nights with a new moon (Gursky 2003a,c). These behavioral responses are quite surprising because nearly all other nocturnal mammals appear to avoid bright moonlight. Numerous studies of nocturnal mammals, including many nocturnal primates, have consistently shown that they restrict their foraging activity, restrict their movement, reduce their use of open space, reduce the duration of the activity period, or switch their activity to darker periods of the night in response to bright moonlight (Morrison 1978, Wolfe and Summerlin 1989). The only consistent exceptions to this pattern have been observed in three primates, the night monkey (*Aotus*), the southern lesser galago (*Galago moholi*), and the Mysore slender loris (*Loris tardigradus lydekkerianus*) (Trent et al. 1977, Erkert and Grober 1986, Nash 1986b, Bearder et al. 2002). Instead of reducing their activity during the full moon, all three species are reported to increase their activity levels during periods of moonlight compared to periods without moonlight.

The lunar phobia exhibited by most nocturnal mammals is believed to be a form of predator avoidance (Morrison 1978, Wolfe and Summerlin 1989, Daly et al. 1992, Kotler et al. 1993, Bearder et al. 2002). This implies that *T. spectrum*

(A)

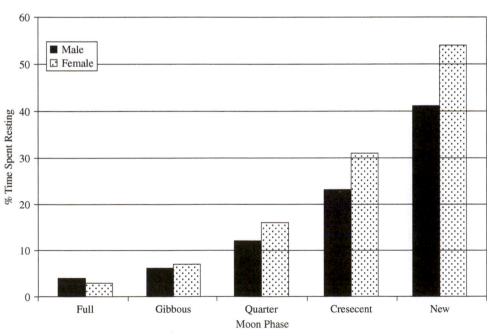

(B)

Figure 6.2 *A*: The mean percent time male (*black bars*) and female (*dotted bars*) spectral tarsiers allocated to foraging during various phases of the lunar cycle. *B*: The mean percent time male (*black bars*) and female (*dotted bars*) spectral tarsiers allocated to resting during various phases of the lunar cycle.

and a few other nocturnal prosimians increase their exposure to predators when they increase their activity during full moons. This raises two equally intriguing questions. First, are the benefits of foraging during the full moon so high that they outweigh the costs of increased predation pressure? Second, how do *T. spectrum* deal with the increasing predator pressure during full moons? Concerning the first question, a comparison of *foraging efficiency* (defined as the number of insects captured per unit time) during full moons relative to the other moon phases suggests that the benefits of foraging during the full moon are in fact tremendous. Gursky (2003a,c) observed that *T. spectrum* captured three times as many insects during full moons than during moonless nights (new moon). These numbers do not simply reflect a bias in observation conditions because this pattern was also observed in terms of the numbers of insects captured in insect traps during the different moon phases (Gursky 1997, 2000b, 2003a,c). Thus, for *T. spectrum*, the benefits of foraging during full moons are substantial.

To deal with the potential increase in predation during a full moon (when diurnal and nocturnal predators can take advantage of the additional light), *T. spectrum* modified other aspects of their behavioral repertoire. Most interestingly, they increased the frequency that group members traveled together. In particular, the frequency that any two group members were observed together increased substantially during full moons. It is believed that the more individuals there are in the group to scan for predators, the less time each individual will have to spend in vigilance (Krebs and Davies 1984, Dunbar 1988, Kappeler and Ganzhorn 1993). However, living in a group can be costly in terms of intragroup foraging competition (Janson 1988a). *T. spectrum* overcome this costly behavior by consuming insects that are abundant in moonlight.

In summary, *T. spectrum*, like three other nocturnal primates, do not exhibit lunar phobia. Instead, they increase their activity during full moons and decrease their activity during new moons. The increased activity during full moons may increase their exposure to predators. To counter this putative increased risk of predation, *T. spectrum* increase the frequency that they travel in groups. *T. spectrum* also increase consumption of insects that are abundant during full moons to counter the within-group competition for insects that may result from traveling in groups.

CONSERVATION

My study of the population density of *T. spectrum* indicates that they are relatively abundant (Table 6.5). I calculated population density as 156 individuals within a 100 ha study area, and the number of groups estimated within the study area was 56 (Gursky 1998b). This estimate is one-half of the lowest estimate calculated by Bearder (1987) (300–1,000 individuals/km² over 100 ha), based on density estimates determined by MacKinnon and MacKinnon (1980b) of 3–10 individuals/ha in north Sulawesi.

In comparison, studies of *T. bancanus* have also yielded varying estimates of population density. Crompton and Andau (1987) determined that there are 14–20 individual tarsiers/km². On the other hand, Niemitz (1984a) reported an estimate of 80 individuals/km². Given the significantly larger home range of individuals observed by Crompton and Andau, the reduced number of individuals in a given area can be better understood. Neri-Arboleda (2001) estimated the density of the Bohol *T. syrichta* as approximately 57 individuals/km². Gursky (1998b) also estimated the population density of Dian's tarsier at Kamarora as 129 individuals/km². No data on population density have been obtained for the pygmy tarsier.

These values are somewhat deceptive because tarsier population density is known to change substantially with both altitude and habitat type. For example, the density of

Table 6.5 Reported Population Density and Conservation Status of Wild Tarsier Species

SPECIES	IUCN CONSERVATION STATUS[1]	REVISED CONSERVATION STATUS[2]	POPULATION DENSITY (INDIV. / KM²)	REFERENCES
Tarsius bancanus	LR lc	LR cd	80, 15–20	Niemitz 1984a; Crompton and Andau 1987
T. spectrum	DD	VU	30–100, 156	MacKinnon and MacKinnon 1980b; Gursky 1998a
T. dianae	DD	VU	129	Tremble et al. 1993, Gursky 1998b
T. syrichta	LR cd	VU	57	Neri-Arboleda 2001
T. pumilus	DD	DD	—	

[1] World Conservation Union conservation status was taken from Appendix 1 in Cowlishaw and Dunbar (2000). LR lc, low risk, least concern; LR cd, low risk, conservation-dependent; DD, data-deficient.

[2] The revised conservation status is based on a workshop organized by Myron Shekelle and the Pusat Primata Schmutzer in Jakarta, in February 2002, attended by M. Shekelle, A. Nietsch, and S. Gursky.

both *T. spectrum* and *T. dianae* was significantly lower at higher altitudes than lower altitudes (Gursky 1998b). Their population density was also significantly higher in the secondary disturbed forest than in the primary habitat.

The conservation status of the six accepted tarsier species is presently categorized as either low risk (least concern), low risk (conservation-dependent), or data-deficient. Following a workshop organized by Myron Shekelle and the Pusat Primata Schmutzer in Jakarta, in February 2002, Shekelle, Gursky, and Nietsch (unpublished data) compiled and reanalyzed data on each species' distribution, abundance, and conservation status. Using one of the IUCN (International Union for the Conservation of Nature) criteria for calculating conservation status (estimates of the amount of land within each species' distribution), we determined that *T. bancanus* should be listed as low risk (conservation-dependent); *T. syrichta* as vulnerable; *T. spectrum* as vulnerable; *T. dianae* as vulnerable; and *T. pumilus* as data-deficient.

Neri-Arboleda (2001) conducted a population viability analysis (PVA) of *T. syrichta* using a metapopulation model, analysis of the likelihood of extinction (ALEX). Given the values of the life history and environmental parameters used for the analysis (Neri-Arboleda 2001) and in the absence of environmental catastrophe, the minimum habitat area that could attain a <5% probability of extinction within 100 years is 60 ha. Therefore, 60 ha is considered the minimum viable habitat area for *T. syrichta*. Each tarsier has a home range of approximately 2.5 ha, with little overlap; thus, the minimum viable habitat of 60 ha contains 24 female individuals. Neri-Arboleda (2001) notes that the model's output is very sensitive to changes in adult mortality. That is, slight increases in adult mortality made the population very unstable and caused substantial increases in the probability of extinction. Given numerous observations of predation attempts (Gursky 2003b) and the relatively high rate of predation in *T. spectrum*, it is probably fair to assume that adult female mortality is relatively high. This estimate of 60 ha takes into consideration that the forested area is fully occupied, which in the wild is most often not the case. This estimate is also based on life history parameters derived from one 8-month study on the island of Bohol in the Philippines.

DIRECTIONS FOR FUTURE RESEARCH

Directions for future research on tarsiers are numerous. There is the question of the phylogenetic relationship of tarsiers to other primates, both living and fossil. Numerous primate evolutionary biologists have been working on this problem (Beard 1998, Ross 1994, Beard and Wang 1990, Ginsburg and Mein 1986, Simons and Bown 1985). There is also the question for taxonomists concerning the number of tarsier species. The use of vocalizations for the identification of

distinct tarsier species is a method that needs to be systematically applied to all tarsier populations. It has already proved quite successful on galagos in Africa (Bearder 1999) and is beginning to be used within Sulawesi. Its use in Borneo, the Philippines, and southern Sumatra is needed. Using this method, I anticipate that several tarsier species will be identified in Borneo. While Crompton and Andau (1986, 1987) found that *T. bancanus* at Sepilok Forest Reserve regularly have calling concerts, Niemitz (1984a) reported that *T. bancanus* at Semongok, Sarawak, rarely vocalize. These vocalization differences may be the result of niche differences, thus representing different dialects. However, according to Bearder (1999), these observations suggest that these two populations are not members of the same species. Studies of the number of tarsier species as well as their distribution are critical if we are to develop any conservation schemes for the tarsiers. Thus, an avenue for future research would be an island-wide census of tarsiers on Borneo, Sumatra, Sulawesi, and several islands within the Philippines while controlling for altitude and forest type.

No detailed or long-term studies of the pygmy tarsier have been conducted. Recently, Grow and Gursky (unpublished data) found a living group of pygmy tarsiers in the highlands of central Sulawesi within Lore Lindu National Park. It is imperative that additional attempts be made to study this species in the wild, including studies of its upper and lower altitudinal limits, as well as its distribution on other mountaintops within central Sulawesi.

Another extremely important avenue for future research involves theoretically based studies. Most research on prosimians, including tarsiers, is generally descriptive and naturalistic. Only within the last 10 years have primatologists begun to ask theoretical questions about the nocturnal primates, trying to understand what factors pattern their behavior in space and time. One theoretical question that stands out concerns the variation in social systems among the different tarsier species. Is it due to ecology or perhaps phylogeny? Why are *T. spectrum* so much more social than the other tarsier species? Is it a reflection of their different habitat? *T. bancanus* and *T. syrichta* live in dipterocarp forests that fruit on 7-year cycles, whereas the Sulawesian tarsiers live in forests with very few dipterocarps and higher proportions of figs that fruit all year-round. As the tarsiers consume insects that are attracted to the fruit of these trees, this may account for some of the differences. Note that the Sulawesian tarsiers are east of Wallace's line, while the other tarsier species are west of Wallace's line. Crompton (1989) believes it may have to do with the different types of insect that each species consumes: *T. bancanus* eats large ground-dwelling insects, while *T. spectrum* consumes more flying and colonial insects. Research is needed to determine which, if any, of these explanations can account for the different social systems in these sister species.

7

Callitrichines

The Role of Competition in Cooperatively Breeding Species

Leslie J. Digby, Stephen F. Ferrari, and Wendy Saltzman

1. Marmosets and tamarins are said to have a high reproductive potential. What then limits the number of offspring that individuals actually have?
2. We often label species as having a specific mating system (e.g., monogamy, polygyny, or polyandry). Why is this system problematic in the marmosets and tamarins?
3. Callitrichines are classified as "cooperative breeders." In what ways is competition an integral aspect of this "cooperation"?

INTRODUCTION

The callitrichines are best known for their suite of reproductive, morphological, and behavioral characteristics that are unusual or unique among the primates. Social suppression of reproduction, postpartum ovulation, twinning, cooperative care of young, and flexible mating systems make this a useful group of animals for testing hypotheses about the evolution of reproductive strategies and social systems. In addition, these species are characterized by claw-like nails on all digits but the hallux, two molars instead of the three typical of platyrrhines (2.1.3.2 dental formula, except in *Callimico*), small body size (from 120 g in *Cebuella* to 650 g in *Leontopithecus*), and dramatic variation in coloration, ear tufts, and even "mustaches" (detailed physical descriptions in Eisenberg and Redford 1999). When reviewed by Goldizen (1987a), information on these species in the wild was limited to five long-term studies, most of which focused on the genus *Saguinus* (e.g., Dawson 1978, Neyman 1978, Terborgh and Goldizen 1985). The past 25 years have witnessed a surge of work on the subfamily Callitrichinae, including field studies of several species for which little or no data have been available (e.g., *Callimico goeldii*, Porter et al. 2009; *Callithrix kuhlii*, Raboy et al. 2008; *Leontopithecus* spp., Kleiman and Rylands 2002; *Saguinus niger*, Oliveira and Ferrari 2008). Nevertheless, intensive long-term ecological and behavioral studies are still restricted to just over a dozen species (e.g., Rylands 1993a, Kleiman and Rylands 2002), and the majority of taxa, including the 10 newly described species and subspecies (Rylands et al. 2000), are known primarily from general surveys or initial species descriptions.

Field studies of callitrichines have investigated topics as diverse as cognitive mapping (Bicca-Marques and Garber 2001), the influence of color vision polymorphism on foraging behavior (Surridge et al. 2005, Jacobs 2007), reproductive endocrinology (French et al. 2003, Araújo and de Sousa 2008), foraging biomechanics (Hourani et al. 2003, Vinyard et al. 2003), nutritional content of food (Smith, A. C., 2000, Porter et al. 2009), and genetics (Faulkes et al. 2003, Huck et al. 2007). Field data have been complemented by laboratory studies on energetics (Genoud et al. 1997, Power et al. 2003), nutrition and digestion (Power et al. 1997, Ullrey et al. 2000), phylogenetics (Seuánez et al. 2002, Araripe et al. 2008), neuroendocrine and behavioral control of reproduction (Abbott et al. 1998, Saltzman et al. 2009), and ovulation and pregnancy using ultrasonography (Jaquish et al. 1995, Oerke et al. 2002), as well as many biomedically oriented studies of reproduction, immune response, and disease (reviewed in Mansfield 2003).

Our goal here is to summarize the most recent information available on the behavior, ecology, and reproduction of callitrichine primates, with an emphasis on comparisons among genera. We then examine the extent to which competition plays a role in social interactions and reproductive success in these cooperatively breeding primates.

TAXONOMY AND DISTRIBUTION

Taxonomy

The systematics of the Platyrrhini has undergone extensive revision at all levels since the classic review of Hershkovitz (1977), and the callitrichines have been among the most controversial taxa. Because it combines morphological, genetic, and ecological perspectives, the recent revision by Rylands et al. (2000, Rylands and Mittermeier 2008, see also Groves 2001a) is perhaps most representative of the current consensus and is followed here (see Appendix 7.1). According to these authors, the subfamily Callitrichinae is a monophyletic group containing six genera (*Callimico* [Goeldi's monkey], *Callithrix* [Atlantic marmosets], *Cebuella* [pygmy marmoset], *Leontopithecus* [lion tamarins], *Mico* [Amazonian marmosets], and *Saguinus* [tamarins]), with a total of 61 species and subspecies. We also include the recently proposed genus name *Callibella* for the dwarf marmoset (van Roosmalen and van Roosmalen 2003).

Callimico goeldii has posed the major problem for callitrichine taxonomists. Its small body size and claw-like nails are characteristic of callitrichines, but its third molar and singleton births are typical of the larger-bodied platyrrhines. This led to the monospecific genus being placed at various times within the Callitrichidae (Hill 1957, Napier and Napier 1967), the Cebidae (Cabrera 1958, Simons 1972), or even its own family, the Callimiconidae (Hershkovitz 1977). The current, widely held consensus is that *Callimico* is a true, albeit atypical, callitrichine. This is strongly supported by a number of molecular studies (Schneider and Rosenberger 1996, Tagliaro et al. 1997, Canavez et al. 1999b, von Dornum and Ruvolo 1999) that clearly place *Callimico* as a sister group of the marmosets (*Callithrix*, *Mico*, *Callibella*, and *Cebuella*).

The marmosets have also been the subject of taxonomic revisions. Rosenberger (1981) proposed that *Cebuella* should be included within the genus *Callithrix*. This view has been supported by many genetic studies (e.g., Canavez et al. 1999b; Tagliaro et al. 2000) but never widely accepted. Schneider and Rosenberger (1996) and Rylands et al. (2000) not only excluded this proposition but instead reinstated the genus *Mico*, which is equivalent to Hershkovitz's (1977) Amazonian *Callithrix argentata* group. The inclusion of *Mico* avoids paraphyly and is consistent with differences in the dental morphology of *Callithrix* and *Mico* as well as with their allopatric distribution. Recent studies of Amazonian marmosets (Gonçalves et al. 2003; Almeida Noronha et al. 2008a,b) have adopted the new arrangement.

The newest genus to be added to the callitrichine subfamily is *Callibella* (van Roosmalen and van Roosmalen 2003). Although it was originally included in *Mico*, detailed genetic and morphological studies of the recently described black-crowned dwarf marmoset (intermediate in size between *Cebuella* and *Mico*) (van Roosmalen et al. 1998) support the creation of the new genus (Aguilar and Lacher 2003, van Roosmalen and van Roosmalen 2003).

Distribution

Current knowledge of the zoogeography of the callitrichine genera is little changed from that reviewed by Hershkovitz (1977). Perhaps the most significant alteration has been the extension of the southern limit of the range of *Leontopithecus*, following the discovery of *Leontopithecus caissara* in the southern Brazilian state of Paraná (Lorini and Persson 1990). *Leontopithecus* species are now found in

four distinct areas, corresponding to the geographic ranges of the four known species (Fig. 7.1).

Two genera, *Callimico* and *Cebuella*, have roughly equivalent geographic ranges in western Amazonia. Studies have confirmed that both genera range as far east as the left bank of the Madeira, where they are potentially parapatric with *Mico* (van Roosmalen and van Roosmalen 1997, Ferrari et al. 1999c).

Saguinus is sympatric with *Callimico* and *Cebuella* throughout their distributions but also ranges much farther north and east. There is now confirmation of sympatric zones between *Saguinus* and *Mico* on the upper Madeira River (Schneider et al. 1987) as well as the lower Toncantins–Xingu interfluviam (Ferrari and Lopes Ferrari 1990a).

Marmosets of the genus *Mico* are found in the southern Amazon Basin between the Madeira and the Tocantins rivers and as far south as northeastern Paraguay (Hershkovitz 1977, see also recent extensions by Almeida Noronha et al. 2007, 2008a,b). The newly described *Callibella* also inhabits the area west of the Rio Aripuanã and overlaps with at least one *Mico* species. *Callithrix* ranges farther south and

Figure 7.1 Distribution map of the Callitrichinae (based on Rylands et al. 1995, Ferrari 1993, Eisenberg and Redford 1999, van Roosmalen and van Roosmalen 2003, Infonatura 2004). Lighter shades indicate probable but unconfirmed presence in that area.

east than *Mico*. This includes sympatry with *Leontopithecus chrysomelas* and *L. rosalia* in some regions (Rylands 1989, Burity et al. 2007), though preferences for high or low elevation may limit range overlap.

ECOLOGY

Habitat

As might be expected from their wide geographic distribution, callitrichines inhabit a wide variety of neotropical habitats, with correspondingly variable patterns of occupation. There are some general ecological differences among the genera, with *Callithrix* and *Mico* tending to form larger groups, to occupy smaller home ranges, and consequently, to have higher population densities than *Saguinus* and *Callimico*. However, home range size and population density vary by up to two orders of magnitude not only within the Callitrichinae but also within some genera (Table 7.1). Consequently, it is difficult to identify genus-specific or even species-specific patterns, especially where data are based on observations of a single group or population.

Table 7.1 Habitat, Home Range, and Daily Path Length

SPECIES AND LOCATION	HABITAT[1]	HOME RANGE (HA)	DAILY PATH LENGTH (M/DAY)	HOME RANGE OVERLAP (%)	REFERENCES
Callimico goeldii					
Pando, Bolivia	SC	30–150	Approx. 2,000		Pook and Pook 1981, Porter 2001a
Virazon, Bolivia	RV	45–50			Christen 1999
Callithrix jacchus					
João Pessoa, Brazil	AF (2)	2–5	1,300		Maier et al. 1982, Alonso and Langguth 1989
Nisia Floresta, Brazil	AF (2)	0.7–5.2	912–1,243	46–86	Digby and Barreto 1996, Castro 2000
Dois Irmãos, Brazil	AF (2)	4.11		23–99	Mendes Pontes and Monteiro da Cruz 1995
C. penicillata					
Brasilia, Brazil	G	3.5			de Faria 1986
C. aurita					
Cunha, Brazil	AF (2)	35.3	958.8	15	Ferrari et al. 1996
Fazenda Lagoa, Brazil	SC	16.5	986		Martins 1998
C. flaviceps					
Caratinga, Brazil	AF (2)	33.86–35.5	883.8–1,222.5	80	Ferrari 1988, Ferrari et al. 1996, Corrêa et al. 2000, Guimarães 1998
C. kuhlii					
Lemos Maia, Brazil	WS	10	830–1,120	50	Rylands 1989
Una Biological Reserve		34–39, 58.3[3]	1498		Raboy et al. 2008
Cebuella pygmaea					
Peru	V	0.1–0.5	280–300	0	Soini 1982, 1988; Heymann and Soini 1999
Ecuador	TF	0.4–1.09			de la Torre et al. 2000
NE Ecuador	VTF	0.15–1.2			Yépez et al. 2005
Mico intermedius					
Mato Grosso, Brazil	ED	22.1	772–2,115	22	Rylands 1986a
M. argentatus					
Tapajós, Brazil	WS	4–24			Albernaz and Magnusson 1999
Caxiuanã, Brazil	AM	35			Veracini 2000
Saguinus fuscicollis					
Manu, Peru	AM (L)	30–100[2]	1,220		Terborgh 1983, Goldizen 1987a
Rio Blanco, Peru	AM (L)	40	1,849	23	Garber 1988a,b

Table 7.1 (cont'd)

SPECIES AND LOCATION	HABITAT[1]	HOME RANGE (HA)	DAILY PATH LENGTH (M/DAY)	HOME RANGE OVERLAP (%)	REFERENCES
Rio Urucu, Brazil	TF	149	1,150–2,700	76	Peres 2000a,b
Pando, Boliva		30			Porter 2007
S. f. weddelli					
Cachoeira Samuel, Brazil	TF	44?	1,312		Lopes and Ferrari 1994
S. imperator					
Manu, Peru	AM (L)	30–100[2]	1,420		Terborgh 1983, Goldizen 1987a
S. labiatus pando					
Bolivia		30			Porter 2007
S. mystax					
Rio Blanco, Peru	AM (L)	40[2]	1,946	23	Garber 1988a,b
Quebrada Blanco, Peru	AM	41–45	1,500–1,720		Heymann 2000, 2001
S. niger					
Caxiuanã, Brazil	AM	35			Veracini 2000
Fazenda Vitoria, Brazil	TF, SF	15.6		21.7	Oliveira and Ferrari 2008
S. midas midas					
French Guiana	AM (H)	31.1–42.5			Day and Elwood 1999
S. tripartitus					
Tiputini, Ecuador	AM	16–21	500–2,300		Kostrub 2003
Leontopithecus rosalia					
Poço das Antas, Brazil	AF (2)	21.3–73	955–2,405	61	Dietz et al. 1997; Peres 2000a,b
Fazenda União, Brazil	AF	65–229	1,873–1,745		Kierulff et al. 2002a,b
L. caissara					
Superagüi, Brazil	AF	125.5–300	1,082–3,398		Prado 1999a,b as cited in Kierulff et al. 2002a,b
L. chrysomelas					
Lemos Maia, Brazil	WS	36	1,410–2,175	7	Rylands 1989
Una, Brazil	AF	66–130.4	1,684–2,044	Approx. 10–14	Dietz et al. 1994b; Keirulff et al. 2002a,b Raboy and Dietz 2004
L. chrysopygus					
Morro do Diabo, Brazil	AF (1)	113–199	1,362–2,088		Valladares-Padua 1993, Valladares-Padua and Cullen 1994 as cited in Kierulff et al. 2002a,b
Caetetus, Brazil	SC	276.5–394	1,164–3,103		Passos 1997 as cited in Kierulff 2002a,b, Passos 1998

[1] AF, Atlantic Forest; AF (2), secondary Atlantic Forest; AF (1), interior Atlantic Forest; SC, sandy clay forest; RV, riverine forest; G, gallery/Cerrado forest; TF, terra firma forest; V, Varzea; ED, evergreen dryland forest; AM, Amazonian forest; AM (L), Amazonian lowland forest; AM (H), Amazonian highland forest; WS, white-sand forest; SF, secondary forest.

[2] Home range of mixed groups (two species) of tamarins.

[3] 34–39 based on kernal method, 58.3 ha based on minimum convex polygon method.

The key factor determining differences between marmosets and tamarins appears to be the marmosets' morphological specializations for the dietary exploitation of plant exudates (e.g., gums and saps) (Ferrari 1993). The ability to exploit exudates systematically as a substitute for fruit throughout the year allows marmosets to inhabit resource-poor or highly seasonal habitats in which tamarins may be unable to survive. Such habitats include not only the wooded ecosystems of the Cerrado, Caatinga, and Chaco biomes and Amazonian savannas but also forests that have suffered intense anthropogenic disturbance. Distributed throughout tropical Brazil south of the Amazon, marmosets—in particular *Callithrix jacchus* and *C. penicillata*—can thrive in such unlikely habitats as city parks, backyards, and even coconut plantations (Rylands and de Faria 1993, L. J. D. and S. F. F. personal observation). *Callibella* takes this ability to make use of disturbed habitats to an extreme and may be dependent on human occupation of a habitat (e.g., the presence of orchards and gardens) (van Roosmalen and van Roosmalen 2003).

Cebuella, in contrast to *Callithrix* and *Mico*, appears to be a habitat specialist. It may reach extremely high ecological densities in riparian forest but is usually absent from neighboring areas of terra firma forest (Soini 1988). This unusual distribution pattern appears to be related to *Cebuella*'s specialization for exudativory (groups may inhabit a single gum tree for long periods) and the avoidance of competition with sympatric callitrichines. In this context, the apparent specialization of *Cebuella* can be interpreted as an accentuated preference for a marginal habitat type, which may be relatively unsuitable for other callitrichines.

Ranging Patterns

With regard to both home range size and population density, the major division between *Callithrix* and *Mico* appears to lie not between the two genera but between patterns characteristic of *C. jacchus* and *C. penicillata* and the remaining species. Typically, *C. jacchus* and *C. penicillata* occupy home ranges of less than 10 ha, with correspondingly high population densities. Other marmosets (Table 7.1) tend to occupy much larger home ranges (up to 35 ha) with correspondingly lower population densities. However, the data available for *Mico argentatus* indicate that this species may be more similar to *C. jacchus* and *C. penicillata*, at least under equivalent ecological conditions (i.e., forest patches in savanna ecosystems and anthropogenic fragments) (Albernaz and Magnusson 1999, Corrêa et al. 2002, Gonçalves et al. 2003). With additional data on the ecology of these species, we may find that observed differences are more closely related to habitat characteristics than to taxon-specific variables.

These findings implicate habitat quality as a primary determinant of home range size and, consequently, population density. For marmosets, the key factors are likely to be the availability of exudate sources and arthropod abundance. Because many gum-producing plants (e.g., Leguminosae, Vochysiaceae) are abundant in the habitats favored by marmosets, arthropod abundance is probably the limiting factor. This may account for the relatively large home ranges recorded for *C. aurita* and *C. flaviceps*, which occur in comparatively seasonal ecosystems in which arthropods may be relatively scarce (Ferrari 1993, see also Raboy et al. 2008 for variation in less seasonal Atlantic Forest habitats).

Home range sizes of *Saguinus* and *Leontopithecus* are highly variable, with some species and populations showing small home ranges similar to those of the marmosets and others extending to over 100 ha (Table 7.1). The larger home ranges and correspondingly lower population densities are consistent with the tamarins' relatively high degree of frugivory (Tables 7.1 and 7.2). *Callimico* shows a similar pattern of large home range and relatively small group size (Tables 7.1 and 7.3), in spite of its ability to utilize fungus during periods of fruit scarcity (Porter 2001b, 2007).

Population densities of both *Callithrix* and *Leontopithecus* decrease farther south, where climate and, presumably, resource abundance are far more seasonal. Once again, differences are disproportionate relative to differences in body size, with home ranges reported for *L. caissara* and *L. chrysopygus* being the largest for any callitrichine (Passos 1998, Prado 1999b as cited in Kierulff et al. 2002b). It is interesting to note, however, that *L. rosalia* may be relatively abundant in anthropogenic forest patches, with home range sizes similar to those recorded for *C. aurita* and *C. flaviceps* (Table 7.1).

Territoriality

Callitrichines typically respond to intergroup exchanges with vocalizations, chases, and occasionally physical aggression (Garber 1988a, 1993a; Peres 1989, Lazaro-Perea 2001; Raboy et al. 2008; but see also van Roosmalen and van Roosmalen 2003). While such intergroup aggression can be interpreted as territoriality (e.g., Peres 1989, Lazaro-Perea 2001), the high degree of overlap between home ranges in some species (e.g., up to 80%–90% in *C. jacchus*; Mendes Pontes and Monteiro da Cruz 1995, Digby and Barreto 1996) (Table 7.1) may indicate that these behaviors are being used primarily for mate defense or defense of a specific set of currently fruiting trees or exudate sources (e.g., Garber 1988a, 1993a; Peres 2000b; Decanini and Macedo 2008a). Intergroup interactions may also allow individuals to monitor the composition and status of neighboring groups and to assess possible mating opportunities (Digby 1992, Ferrari and Diego 1992, Goldizen et al. 1996, Schaffner and French 1997, Lazaro-Perea 2001, Kostrub 2003, Decanini and Macedo 2008b).

Mixed-Species Troops

Mixed-species troops (two or more species associating in a nonrandom fashion, often coordinating activities; e.g., Pook and Pook 1982) play an important role in the ecology of several callitrichine species. Until recently all reported mixed-species troops included *Saguinus fuscicollis* interacting with either *S. mystax*, *S. imperator*, *S. labiatus*, or *Mico emiliae* (reviewed in Heymann and Buchanan-Smith 2000). At some sites, two of the *Saguinus* species also associate with *Callimico* (e.g., Porter 2001b, 2007; Rehg 2006). A new study by Raboy et al. (2008) has described *C. kuhlii* frequently traveling and sometimes feeding with *L. chrysomelas* within the Atlantic Forest of Brazil. The

Table 7.2 Diet

SPECIES AND LOCATION	% TIME FEEDING	REPRODUCTIVE PLANT PARTS[1]	% TIME FEEDING ON			REFERENCES
			EXUDATES	ANIMAL PREY	OTHER	
Callimico goeldii						
Pando, Bolivia	15	29	1	33	29 (fungus), 27 (unknown)	Porter 2001b, 2004
Callithrix jacchus						
Joao Pessoa, Brazil	27	18.1	76.4	5.4	12.9 (fungus)	Alonso and Langguth 1989
Nisia Floresta, Brazil	43[2]	23	68	9		Digby (unpublished data)
C. aurita						
Cunha, Brazil	17	40.5	30	29.5	2 (fungus)	Corrêa et al. 2000
Fazenda Lagoa, Brazil	6	11	50.5	38.5		Martins and Setz 2000
C. flaviceps						
Caratinga, Brazil (1)		14.4	65.7	19.9		Ferrari et al. 1996
Caratinga, Brazil (2)	11.8	2.0	83.2	14.7		Corrêa et al. 2000
C. kuhlii						
Lemos Maia, Brazil	23	58.2	28.3	13.5		Rylands 1989, Corrêa et al. 2000
Una Biological Reserve		14	5.7	7.4	0.9	Raboy et al. 2008
C. geoffroyii						
Espírito Santo, Brazil	21	15	68.6	15.4		Passamani 1998, Passamani and Rylands 2000
Cebuella pygmaea						
Rio Nanay, Peru	48[2]	Minor	67	33		Ramirez et al. 1977
NE Ecuator			20–34[4]	5–12[2]		Yépez et al. 2005
Mico intermedius						
Mato Grosso, Brazil		74.9	15.5	9.6		Rylands 1982 as cited in Corrêa et al. 2000
M. argentatus						
Caxiuanã, Brazil		36	59	5		Veracini 1997 as cited in Corrêa et al. 2000
Saguinus fuscicollis						
Quebrada Bl., Peru		64.2	30.3	5.8		Knogge and Heymann 2003
Rio Blanco, Peru	12.9	39.2	7.6	53.1		Garber 1988
Pando, Bolivia		63	12	26	6 (unknown)	Porter 2001b
S. f. weddelli						
Cachoeira Samuel, Brazil	9.8	69.6	15.8	7.3		Lopes and Ferrari 1994
S. labiatus						
Pando, Bolivia		73	8	11	8 (unknown)	Porter 2001b
S. mystax						
Quebrada Blanco Peru		77.6	19.8	2.7		Knogge and Heymann 2003
Rio Blanco, Peru	13.1	50.6	1.5	47.8		Garber 1988

Table 7.2 (cont'd)

SPECIES AND LOCATION	% TIME FEEDING	% TIME FEEDING ON				REFERENCES
		REPRODUCTIVE PLANT PARTS[1]	EXUDATES	ANIMAL PREY	OTHER	
S. niger						
Fazenda Vitória, Brazil		87.5	3.1	9.4		Oliveira and Ferrari 2000, 2008
Caxiuanã, Brazil	17.6	71.1	23.81	4.5		Veracini 2000
S. tripartitus						
Tiputini, Ecuador		62	12	21	5 (unknown)	Kostrub 2003
Leontopithecus rosalia						
Poço das Antas, Brazil	18.5	83.5	1.4	14.9		Dietz et al. 1997
União, Brazil	10.4	84.6	0	15.4		Kierulff 2000 as cited in Kierulff et al. 2002
L. caissara						
Superagüi, Brazil	29.4[2]	75.5	1.3	10.3	12.9 (fungus)	Prado 1999b as cited in Kierulff et al. 2002
L. chrysomelas						
Lemos Maia, Brazil	27	74–89,[3] 11 flowers	3–11[4]	13–15		Rylands 1989
L. chrysopygus						
Morro do Diabo, Brazil	6–10	78.5	7.8	13.5		Valladares-Padua 1993 as cited in Kierulff et al. 2002
Caetetus, Brazil	29.92	74.7	15.2	10.1		Passos 1999 as cited in Kierulff et al. 2002

[1] Includes fruit, seeds, flowers, and nectar.
[2] Includes foraging.
[3] Percent of plant records only; flowers made up 0%–20% of monthly diet (percent overall diet not reported).
[4] Percent of individual observations of feeding on exudates.

two species were even seen mobbing a potential predator together. Though expected to generate costs due to niche overlap, mixed-species troops appear to limit competition via differential use of forest strata and foraging techniques (Terborgh 1983, Heymann and Buchanan-Smith 2000). Associations may also increase when foraging on compatible food (e.g., increased association during periods when feeding on fruits; Porter 2007). During periods of fruit scarcity, Callimico focus on fungi, while tamarins focus on nectar and exudates, resulting in the species spending less time in proximity (Porter 2007). Potential benefits include increased protection from predators, increased foraging efficiency (including increased insect-capture rates), and resource defense (reviewed in Heymann and Buchanan-Smith 2000).

Foraging Behavior

Exudates

In the wild, callitrichines exploit a wide variety of food types, avoiding only nonreproductive plant parts such as leaves and bark (Garber 1993a,b; Rylands and de Faria 1993; Digby and Barreto 1998; Heymann and Buchanan-Smith 2000; Smith 2000; Porter 2001b; Kierulff et al. 2002b) (Table 7.2). The principal feature of callitrichine diets is their variety, and the only clear taxon-specific pattern is the marmosets' use of plant exudates (primarily gums, with some sap) as a dietary staple. All callitrichines eat some exudates, but Callithrix, Mico, and Cebuella are morphologically specialized for the systematic harvesting and digestion of gum and, thus, able to sustain high levels of exudativory throughout the year (Ferrari 1993).

The marmosets' specializations for exudativory include elongated, chisel-like lower incisors and a wide jaw gape that permit them to gouge through the bark of gum-producing plants, thus provoking exudate flow (Hershkovitz 1977; Vinyard et al. 2001, 2003; Taylor and Vinyard 2008). Gums contain complex polysaccharides, a potentially high-energy source but one which cannot be broken down enzymatically by other mammals (reviewed in Power and Oftedal 1996, Heymann and Smith 1999). Marmosets are able to digest gums more efficiently than other callitrichines because

Table 7.3 Group Size, Composition, and Mating Patterns

SPECIES	GROUP SIZE	ADULT MALES	ADULT FEMALES	MATING PATTERNS[1]	REFERENCES
Callimico goeldii	4–12	1–3	1–3	M/PG/PA	Reviewed in Porter 2001a
Callibella humilis	6–8			M?/PG	van Roosmalen and van Roosmalen 2003
Callithrix jacchus	3–16	2–7	2–6	M/PG/PGA	Barreto 1996, Digby and Barreto 1996, Lazaro-Perea et al. 2000, Faulkes et al. 2003
C. penicillata	3–16	3–5	2–3	M/PG?	Reviewed in de Faria 1986, Decanni and Macedo 2008b
C. aurita	4–11	1–2	2–3	M/PG	Ferrari et al. 1996, Martins and Setz 2000
C. flaviceps	5–20	3–5	1–6	M/PG	Ferrari and Diego 1992, Coutinho and Corrêa 1995, Guimarães 1998
C. geoffoyi	2–8	2	1		Passamani 1998, Chiarello and de Melo 2001, Price et al. 2002
C. kuhlii	2–16	1–2	1–2	M/PG	Rylands 1989, Raboy et al. 2008
Cebuella pygmaea	2–9	1–2	1–2	M/PG?[2]	Soini 1982, 1988; Heymann and Soini 1999; de la Torre et al. 2000; Yépez et al. 2005
Mico intermedius	9–15	2–3	2–5	M/PA	Rylands 1986b
M. argentatus	6–10	1–3	1–2		Albernaz and Magnusson 1999, Tavares and Ferrari 2002
Saguinus fuscicollis	2–10	1–4	1–2	M/PA/	Goldizen 1987a,b; 2003; Garber 1988a,b; Porter 2001a;
				PG/PGA	Heymann 2001
S. f. weddelli	4–11	1–2	1–2	M/PG	Garber and Leigh 2001, Buchanan–Smith et al. 2000, Lopes and Ferrari 1994
S. mystax	3–11	1–4	1–2	M/PG[3]	Garber 1988a,b, Heymann 2000; Smith et al. 2001
S. niger	5–7	3–4	1		Oliveira and Ferrari 2000, 2008
S. labiatus labiatus	2–13	2	1		Buchanan–Smith et al. 2000, Garber and Leigh 2001
S. oedipus	2–10			M/PG[3]	Savage et al. 1996a
S. tripartitus	2–9	1–4	1–2	PA	Heymann et al. 2002, Kostrub 2003
Leontopithecus rosalia	2–11	0–5	0–4	M/PA/PG	Baker et al. 1993, Dietz and Baker 1993
L. caissara	4–7				Prado 1999a
L. chrysomelas	3–10	1–3	1–3	M/PG	Dietz et al. 1994b, Baker et al. 2002, Raboy and Dietz 2004
L. chrysopygus	2–7	2	2		Passos 1994, reviewed in Kierulff et al. 2002

[1] Designations are based on reports from the field (direct observation of copulations and/or births unless otherwise noted). M, monogamy; PG, polygyny; PA, polyandry; PGA, polygynandry.

[2] PG based on short interbirth interval.

[3] PG based on presence of pregnant females.

their intestines have a comparatively enlarged and complex cecum, allowing for relatively slow gut-passage rates and microbial fermentation (Ferrari and Martins 1992, Ferrari et al. 1993, Power and Oftedal 1996). Once digested, plant gums provide not only carbohydrates but also minerals (particularly calcium) and proteins (Garber 1984a, Smith 2000). *C. jacchus* has also been found to have a high rate of passive paracellular nutrient absorption that may allow them to digest gums more efficiently (McWhorter and Karasov 2007).

Saguinus, *Leontopithecus*, and *Callimico* are opportunistic gummivores. Though exudates may provide as much as half of the diet during some periods (Porter 2001b), this is invariably a temporary phenomenon, and exudates are never a dietary staple (Porter et al. 2009). *Saguinus* has been observed parasitizing gouge holes made by *Cebuella* (Soini 1988), but most exudate sources are either rare (broken branches) or seasonal (e.g., insect bore holes, *Parkia pendula* fruit pods; Garber 1993a). In addition, the digestive tracts of *Saguinus* and *Leontopithecus* are not specialized for gum consumption, decreasing their digestive efficiency (Ferrari and Martins 1992, Power and Oftedal 1996). One compensatory strategy may be to concentrate gum feeding in the late afternoon, which allows longer retention time (overnight) and, presumably, better absorption of nutrients (Heymann and Smith 1999).

Fruits, Nectar, and Fungus

Fruit is the primary food item for many callitrichine species, especially among *Saguinus* and *Callimico* (e.g., Porter 2007, Oliveira and Ferrari 2008). Callitrichines typically exploit relatively small fruit patches, characteristic of the lower forest strata, and prefer disturbed and edge habitats (thus reducing competition with larger-bodied primates) (Ferrari 1993). Fruit can be a highly seasonal resource, especially in habitats of low diversity. Because *Saguinus* and *Leontopithecus* are unable to compensate systematically for fruit scarcity by gouging exudate sources, they are especially vulnerable to such seasonality. Small body size limits the potential for the exploitation of nonreproductive plant parts, as indicated by the lack of reports of folivory (Table 7.2). As a last resort under extreme seasonal fruit scarcity, *Saguinus* may turn to resources such as nectar (Terborgh and Goldizen 1985, Kostrub 1997). Although relatively nutritious, nectar is normally available in quantities too small to be harvested adequately by vertebrates as large as *Saguinus*.

Another potential resource is fungus, which may contribute significantly to the diet of some species during certain parts of the year (*Callimico*, Porter 2001b; *C. aurita*, Corrêa et al. 2000). It is interesting to note that, while fungi (including types of jelly and bamboo fungus) are a dietary staple of *Callimico* in some months, they are not exploited by sympatric tamarins (*S. fuscicollis* and *S. labiatus*), reinforcing the role of divergent diets in reducing competition among these species (Porter 2001b, 2007).

Availability of fruit tends to correlate with arthropod abundance, with both resources relatively scarce during the dry season. While seasonal fluctuations in resource abundance are faced by all primates, their effects may be relatively severe for callitrichines because of their small body size and high metabolic rate. Goldizen et al. (1988), for example, found that resource scarcity in the dry season resulted in potentially deleterious weight loss in *S. fuscicollis*. This is further illustrated in *Callimico*, where fruit fell from 53% of feeding time in the wet season to 14% during the dry season. This drop was compensated for with an increased consumption of "fallback" foods such as exudates and fungi (Porter et al. 2009). Thus, specialization for gum feeding among the marmoset genera may be the decisive factor permitting them to reproduce at 5- to 6-month intervals (Ferrari 1993, Ah-King and Tullberg 2000), in contrast with other callitrichines, which normally produce only one litter per year (but see Smith 2000).

Prey

Animal material, predominantly arthropods, is the third major component of callitrichine diets. Typical arthropod prey species are large-bodied and mobile and generally depend on camouflage as a predator-avoidance strategy. The characteristic callitrichine foraging behavior is the stealthy "scan-and-pounce" technique (Ferrari 1993, Rylands and de Faria 1993, Porter 2001b, Souto et al. 2007). In addition, callitrichines will often pursue disturbed prey that has fallen to the forest floor (Rylands and de Faria 1993). Callitrichines also take advantage of insects flushed into the open by other species, including army ants (de Melo and Zara 2007).

Extractive foraging is an important strategy for both *S. fuscicollis* and *Leontopithecus* species. In both cases, the animals spend a relatively large proportion of foraging time investigating concealed hiding places manually, rather than visually. For *S. fuscicollis*, the manipulative investigation of substrates, such as bark crevices, on vertical supports in the lower strata of the forest appears to be a key factor in niche separation with sympatric tamarin species (Terborgh 1983, Heymann and Buchanan-Smith 2000, Nadjafzadeh and Heymann 2008). Extractive foraging is also thought to contribute to more complex cognitive abilities (e.g., Gibson 1986, Day et al. 2003).

In addition to arthropods, vertebrates (e.g., bird nestlings, small lizards, and frogs) are included in callitrichine diets (Ferrari 1988, Digby and Barreto 1998, Smith 2000, Porter 2001b, Kierulff et al. 2002b, Nadjafzadeh and Heymann 2008) (Table 7.2). All large prey, but especially vertebrates, are highly valued food items; and their capture almost invariably provokes solicitation from younger group members, resulting in a variety of social interactions ranging from passive food transfer to agonistic behavior (Digby and Barreto 1998).

REPRODUCTION

Reproductive Potential

Callitrichines are characterized by a collection of traits that results in a high reproductive potential for some females while at the same time restricting breeding opportunities for others (one captive *C. jacchus* female produced 64 offspring, McGrew and Barnett 2008). All species studied, with the exception of *Callimico goeldii* and possibly *Callibella humilis*, typically ovulate multiple ova and produce litters of two or more infants. In captivity, 50%–80% of litters comprise dizygotic twins, with the remainder comprising singletons, triplets, or, less frequently, quadruplets (Ziegler et al. 1990a, Baker and Woods 1992, Tardif et al. 2003, De Vleeschouwer et al. 2003, Mattle et al. 2008). Only rarely, however, do more than two infants survive from a single litter (e.g., McGrew and Barnett 2008). Ovulation number correlates with maternal body mass and, thus, may vary in response to maternal nutritional status (e.g., *C. jacchus*, Tardif et al. 2003).

Callitrichines typically ovulate and may conceive within 2–4 weeks after giving birth (Heistermann and Hodges 1995, French et al. 2002, Tardif et al. 2003). Unlike other primates, ovulation in these species is not inhibited by lactation, although postpartum ovulation may be delayed slightly in females nursing more than one infant (Ziegler et al. 1990b,

Baker and Woods 1992) (Table 7.4). Combined with gestation lengths ranging from 125 days in *Leontopithecus* spp. to 184 days in *Saguinus oedipus* (Ziegler et al. 1987a, French et al. 2002) (Table 7.4), this pattern allows breeding females to produce litters at approximately 5- to 6-month intervals. Most species frequently produce two litters per year in captivity, and most genera (*Callithrix*, *Mico*, *Cebuella*, and *Callimico*) also do so in the wild (Stevenson and Rylands 1988, Soini 1993, Digby and Ferrari 1994, Porter 2001a). *Leontopithecus* and *Saguinus*, in contrast, typically breed only once per year in the wild, with most births clustered during the first half of the rainy season, corresponding with the period of maximal fruit availability (Snowdon and Soini 1988, Ferrari and Lopes Ferrari 1989) (Table 7.4).

Table 7.4 Reproductive Parameters (Based on Captive Animals, Unless Otherwise Noted, Average ± Standard Error)

SPECIES	GESTATION LENGTH (DAYS)	INTERBIRTH INTERVAL (DAYS)	AGE AT FEMALE MATURITY (MONTHS)[1]	CYCLE LENGTH (DAYS)	PEAK BIRTH PERIOD	POSTPARTUM OVULATION[2]	REFERENCES
Callimico goeldii	151–152		Approx. 13	23.9 ± 0.4	Sept–Nov (Bolivia)	22–23	Dettling 2002, Dettling and Pryce 1999, Pook and Pook 1981
Callithrix jacchus	143–144	162	Approx. 13	28.6 ± 1.0	Weakly bi-modal, most Oct–Feb	10–20	Abbott and Hearn 1978, Digby and Barreto 1993, Harlow et al. 1983, Lazaro-Perea et al. 2000, Tardif et al. 2003, Torii et al. 1987
C. kuhlii	143.1 ± 1.6	156.3 ± 2.9	Approx. 12–15	24.9 ± 0.6		13.6 ± 1.2	French et al. 1996, Smith et al. 1997
Cebuella pygmaea	141.9	212.7 ± 122.3	15–17	28.6 ± 4.1	May–Jun, Oct–Jan (NE Peru)	15.6 ± 4.1	Ziegler et al. 1990a, Carlson et al. 1997, Spurlock 2002
Saguinus fuscicollis	149.8 ± 2.4	185	13	25.7 ± 1.0	Nov–Feb (SE Peru)	17 ± 3.4	Epple and Katz 1980, Tardif et al. 1984, Goldizen et al. 1988, Heistermann and Hodges 1995, Kuederling and Heistermann 1997
S. oedipus	183.7 ± 1.1	240–267[3] 332.9 ± 53.6 (field)	15–17	23.6 ± 1.2	Mar–Jun (Colombia)	16.5 ± 1.6 to 30.8 ± 5.2[3]	French et al. 1983; Ziegler et al. 1987a,b, 1990b; Baker and Woods 1992; Savage et al. 1997
Leontopithecus rosalia	125.0	182–215[3] 311 ± 11.2 (field)	12–17	18.5 ± 0.3	Sept–Nov (SE Brazil)	Nonconceptive	French and Stribley 1985; French et al. 1989, 2002; Baker and Woods 1992; Dietz et al. 1994a; Monfort et al. 1996
L. chrysomelas	125.3 ± 3.0	160.6–257.8[3]	17	21.5 ± 2.5	Oct–Apr (E Brazil)	17.3 ± 3.5, nonconceptive	De Vleeschouwer et al. 2000a,b, 2003; Bach et al. 2001; French et al. 2002
L. chrysopygus		242		23.0 ± 2.0		Nonconceptive	Wormell and Price 2001, French et al. 2002

[1] Age at which females first undergo ovulatory cycles or reproductive hormone elevations, especially when housed in the absence of a dominant female.
[2] Days from parturition.
[3] Depending on lactation, litter size, and/or origin of dam (see text).

The reproductive potential of callitrichines is further enhanced by rapid maturation in both males and females (Tardif et al. 2008). Females are typically capable of ovulation and conception by 12–17 months of age (Table 7.4), while males produce sperm by 13–18 months and can sire infants by 15–25 months (Abbott and Hearn 1978, Epple and Katz 1980, Ginther et al. 2002). However, the onset of sexual maturity, especially in females, is often obscured by social suppression of reproductive function. Onset may be much later in free-ranging animals (e.g., 29–35 months in *C. jacchus*, Tardif et al. 2008; 2–4 years in *S. fuscicollis*, Goldizen et al. 1996). Ovarian cycles last from approximately 19 days in *L. rosalia* to 28 days in *C. jacchus* (Harlow et al. 1983, French and Stribley 1985) (Table 7.4). Like other platyrrhines, callitrichines do not menstruate and exhibit no conspicuous external signs of ovulation. Sexual behavior can occur throughout the ovarian cycle and pregnancy but is most common during the periovulatory period (Kendrick and Dixson 1983, Converse et al. 1995, Digby 1999, De Vleeschouwer et al. 2000a). Reproduction can continue into old age; however, ovulatory cycles may become irregular or cease and reproductive output may decline in the oldest females (Tardif and Ziegler 1992, Tardif et al. 2008). Thus, while callitrichines have the highest annual reproductive potential of any anthropoid primate, actual lifetime reproductive output is limited by their relatively short reproductive tenure, relatively high infant mortality rates, and suppression of reproduction in socially subordinate individuals (Tardif et al. 2003, 2008).

Mechanisms of Reproductive Suppression

One of the most striking features of callitrichine reproduction is the monopolization of breeding by a single, behaviorally dominant female in most social groups (reviewed by Saltzman et al. 2009). Although similar breeding patterns are found throughout the callitrichine subfamily (French 1997), the underlying mechanisms, particularly the relative contributions of physiological suppression and behavioral inhibition, appear to differ across genera. In captive *Leontopithecus* and *Callimico*, eldest daughters living with their natal families typically undergo ovulatory cycles indistinguishable from those of breeding females (Dettling and Pryce 1999, French et al. 2002, Mattle et al. 2008), suggesting that inhibition of sexual behavior is the primary cause of reproductive failure. Among captive *Saguinus*, in contrast, adult daughters routinely fail to ovulate while living with their natal families (Epple and Katz 1984, Ziegler et al. 1987b, Kuederling et al. 1995). Captive *Callithrix* females are intermediate, with up to 50% or more of eldest daughters ovulating while living with their natal families (Saltzman et al. 1997, Smith et al. 1997). Endocrine studies of free-living callitrichines, however, have not consistently supported findings from captivity. For example, data on wild *L. rosalia* indicate that periods of ovarian insufficiency occur in adult daughters living with their natal families (French et al.

2003), whereas both ovulatory cyclicity and pregnancy have been detected in wild *S. oedipus* and *C. jacchus* daughters (Savage et al. 1997, Sousa et al. 2005).

The physiological, sensory, and behavioral determinants of ovulation suppression have been investigated in several species. Circulating or excreted concentrations of cortisol, a stress-responsive hormone from the adrenal cortex, are similar in dominant and subordinate females or, in some cases, lower in subordinates (Saltzman et al. 1994, 1998; Ziegler et al. 1995; Smith and French 1997a), suggesting that ovulation suppression cannot be attributed to generalized stress (Abbott et al. 1997). Instead, anovulation appears to result from a specialized neuroendocrine mechanism activated by specific social cues (Abbott et al. 1997). In *C. jacchus*, anovulation in subordinate females is mediated by suppression of gonadotropin secretion from the anterior pituitary, which is associated with enhanced negative feedback and diminished positive feedback effects of estrogen. Hypothalamic secretion of gonadotropin-releasing hormone (GnRH) does not appear to be altered by social subordination, suggesting that pituitary responsiveness to GnRH may be dampened (Abbott et al. 1997). Olfactory cues from dominant females have been implicated in the initiation and maintenance of ovulation suppression in several species but may play a redundant role with other cues (Epple and Katz 1984; Savage et al. 1988; Barrett et al. 1990; Abbott et al. 1993, 1998).

Reproductive failure in subordinate females can occur in response to either intrasexual (i.e., rank-related suppression) or intersexual (i.e., inbreeding avoidance) influences. The specific roles of these two factors differ among species. Captive *S. oedipus* females require cohabitation with an unrelated male in order to commence ovulatory cyclicity, even after removal from the natal family (Widowski et al. 1990, 1992). Among captive *C. jacchus* females, in contrast, ovulation suppression is determined by intrasexual dominance relationships: Daughters living with their families frequently ovulate even in the absence of unrelated males but only if they are not behaviorally subordinate to another female (Saltzman et al. 2004, Alencar et al. 2006, Mattle et al. 2008). Nonetheless, *C. jacchus* daughters do not normally engage in sexual behavior unless they have access to an unrelated male (Saltzman 2003, Saltzman et al. 2004). Postconception control of reproduction has also been documented in *C. jacchus* where pregnant females have been observed killing the neonates of other breeding females (Sousa et al. 2005, Saltzman et al. 2008, Digby and Saltzman 2009).

Male callitrichines, like females, engage in little or no intersexual copulatory behavior while living with their natal families. This appears to reflect inbreeding avoidance rather than intrasexual, rank-related suppression and is not generally associated with suppression of testosterone or luteinizing hormone concentrations (Baker et al. 1999, Ginther et al. 2001, French et al. 2002). Recent research has found a significant increase in the ratio of testicular volume to body mass in breeding versus nonbreeding males, a pattern possibly related to resource defense or reproduction itself (Araujo and de Sousa 2008).

SOCIAL ORGANIZATION

Group Composition

Group size in callitrichines varies from two to 20 individuals, with *Mico* and *Callithrix* species tending to have larger groups than *Saguinus*, *Callimico*, or *Leontopithecus* (Table 7.3). *Cebuella* tends to have the smallest groups, with most containing a single breeding pair and young (Soini 1982, 1988; de la Torre et al. 2000) (Table 7.3). Solitary animals have been noted in several species, and male/female pairs appear to be rare and not always successful at raising young (Terborgh and Goldizen 1985, Goldizen 2003; see also Porter 2001a for *Callimico*). Differences in group size across species are likely tied to increased recruitment rates through biannual births in marmosets and to the diet and habitat preferences of the different genera (Koenig 1995, Heymann 2000, Goldizen 2003; see earlier section, "Ecology"). Intriguing preliminary data on *Callibella* note that, while group size usually falls within the typical callitrichine range (six to eight individuals), aggregations of up to 30 individuals sometimes form (van Roosmalen and van Roosmalen 2003).

Although some early reports indicated that group membership could be dynamic, with individuals frequently moving in and out of groups (e.g., Dawson 1978, Neyman 1978, Scanlon et al. 1988), more recent studies based on direct follows of animals indicate more stable, extended family groups (Ferrari and Digby 1996, Goldizen et al. 1996, Nievergelt et al. 2000, Baker et al. 2002; but see also Garber et al. 1993). Typical of family groups, both sexes may emigrate as individuals mature and as groups grow in size (e.g., Goldizen et al. 1996, Porter et al. 2001, Baker et al. 2002, Huck et al. 2007). In *L. rosalia*, some 60% of individuals born into a group will have dispersed by 3 years of age and 90% by 4 years of age (Baker et al. 2002, Tardif et al. 2008). Of those individuals that remain with their natal group, most become breeders by the time they are 4 years old (Baker et al. 2002). Similar patterns have been described for *S. fuscicollis* (Goldizen et al. 1996). Although both sexes may disperse, males have a higher probability than females of entering an established group (Ferrari and Diego 1992; Baker et al. 1993, 2002; Goldizen et al. 1996; but see Huck et al. 2007). Females may "float" as solitary individuals waiting for a breeding vacancy in an already established group, or they may form a new group with other recent emigrants (Lazaro-Perea et al. 2000, Baker et al. 2002). Breeding takeovers by individuals within their natal group are more likely to occur when unrelated mates are available (Ferrari and Diego 1992, Goldizen et al. 1996, Saltzman et al. 2004). The general pattern of extended family groups with occasional immigration (typically, but not always,

into a breeding position) has been supported by studies of genetic relatedness within *C. jacchus* groups (Nievergelt et al. 2000), but periods of instability can result in groups of mixed parentage (Faulkes et al. 2003).

Social Relationships

Groups are typically cohesive, and individuals often rest in physical contact with one another and allogroom (Ferrari 1988, Alonso and Langguth 1989, Digby 1995b, Heymann 1996, Lottker et al. 2007). Allogrooming has been described as asymmetrical in at least some wild groups, with females typically receiving more grooming than they perform and breeding individuals of both sexes being favored grooming partners (Goldizen 1989, Digby 1995b, Heymann 1996, Kostrub 2003, Lazaro-Perea et al. 2004, Löttker et al. 2007, Decanini and Macedo 2008b). Breeding females may be performing a "service" that entices nonbreeding females to remain in the group (Lazaro-Perea et al. 2004), but grooming during times of infant dependence turned out to be equal or more directed at the breeding females in *S. mystax* (Löttker et al. 2007). Breeding females preferentially groom one potential sexual partner over another during the periovulatory period in *S. mystax* (Löttker et al. 2007), but similar patterns were not found in other studies where reproductive status was not taken into account (Heymann 1996, Kostrub 2003).

Aggressive behavior in callitrichines is relatively rare (e.g., fewer than 0.1 acts/hr in *S. mystax*; Heymann 1996, Garber 1997) and mild, typically consisting of cuffs, "arch walks," piloerection, chases, avoidance, and submissive vocalizations; it occurs most often in feeding contexts (Goldizen 1989, Baker et al. 1993, Digby 1995b, Garber 1997, Kostrub 2003, Decanini and Macedo 2008b). Rates of aggression can increase following changes in group composition, for example, the loss of a breeding female (Lazaro-Perea et al. 2000). Experimental exposure to unfamiliar "intruders" or changes in membership of captive groups can also elicit physical attacks (reviewed in Anzenberger 1993, Caine 1993). The pattern of aggression suggests that it is used to limit access to mates and to control group membership (Anzenberger 1993, Lazaro-Perea et al. 2000, Baker et al. 2002). Where intersexual dominance has been described, either there is no clear pattern of male or female dominance (e.g., Digby 1995b, Kostrub 2003) or males are able to displace females at feeding sites (Baker and Dietz 1996). Intrasexual dominance can have profound implications for reproductive success in these species (Baker et al. 1993; Digby 1995a,b; Lazaro-Perea et al. 2000).

Mating Systems

Callitrichine mating systems have been described as monogamous, polyandrous, polygynous, and polygynandrous, with variation occurring both within and between groups and populations (reviewed in Garber 1997, Baker et al. 2002,

Goldizen 2003, Saltzman 2003) (Table 7.3) (note: we use these terms to describe patterns of copulation only). Such flexible mating strategies are linked to changes in both group composition and social relationships, and specific patterns appear to be more typical of some genera than others.

The presence of more than one reproductively active male (polyandry and polygynandry) has been described for a number of species but appears to be most prevalent in *Saguinus* and *Leontopithecus* (Table 7.3). In *L. rosalia*, "potentially polyandrous" groups (based on the presence of two or more potentially reproductive males) were noted in 46% of monthly censuses, but direct observations of the timing of copulatory behavior together with information on probable fertile periods indicated that most groups were "genetically" monogamous, including six of seven potentially polyandrous groups (Baker et al. 1993). Other species exhibit little or no competition over access to breeding females (e.g., Kostrub 2003, Schaffner and French 2004, Decanini and Macedo 2008b). Only a handful of polyandrous groups have been noted among *Mico* and *Callithrix* (*M. humeralifer*, Rylands 1986b; *C. jacchus*, two males copulating with two females in a newly formed group, Lazaro-Perea et al. 2000; see also Schaffner and French 2004 for captive *C. kuhlii*). The limited data available for *Cebuella* and *Callimico* suggest that groups typically contain a single breeding male (Soini 1988, Porter 2007). Possible reasons for differences across genera may involve differential costs of infant care: Species using larger home ranges and having smaller overall group sizes may benefit more from infant care shared by two "potential" fathers (Heymann 2000, Goldizen 2003).

Female reproductive strategies also vary across marmoset and tamarin species. Although breeding is typically restricted to a single female, the presence of multiple breeding females in some groups demonstrates that reproductive suppression is not absolute. It is notable that even when suppression is relaxed, breeding appears to be restricted to no more than two females, even in groups containing additional females of breeding age. Polygyny has been documented in multiple wild groups of *Callithrix*, *Callibella*, *Saguinus*, and *Leontopithecus* and inferred (based on a short interval between births) in *Cebuella* (Table 7.3). Additional groups have been documented to contain two pregnant/lactating females, although tenure in the group was unknown (i.e., a female may have immigrated into a group while pregnant) or one of the females failed to raise young (*S. mystax*, Garber et al. 1993; *S. oedipus*, Savage et al. 1996a; *Callimico goeldii*, Porter 2001a; plus additional groups of *S. fuscicollis*, Goldizen et al. 1996; and *L. rosalia*, Baker et al. 2002). Only in *Mico* have multiple breeding females not been reported. Though two breeding females may be present in a group, it is possible that mating within the group is still monogamous, with the second breeding female mating with males from outside the group (Arruda et al. 2005). The presence of two breeding females and their resulting young would presumably increase average costs of infant care in terms of time spent carrying and food sharing. Some species or populations may be better able to accommodate these costs because of larger group sizes, more steady food supply (e.g., exudates), and smaller home ranges that allow caretakers to reduce travel while maintaining contact with their group (Digby and Barreto 1996, Goldizen 2003).

A further complication in describing the mating system of callitrichines is the occurrence of extragroup copulations in some species. In *C. jacchus*, for example, both reproductive and nonreproductive males and females have been observed participating in extragroup copulations, often during or just after aggressive intergroup interactions (Digby 1999, Lazaro-Perea et al. 2000, Arruda et al. 2005). Thus, even some groups described as monogamous based on within-group copulation patterns may show more complex patterns of potential and actual paternity (Digby 1999, Arruda et al. 2005).

Infant Care

The high cost of infant care (due to twinning, high infant/maternal weight ratios, and frequent overlap of lactation and pregnancy in some species) is thought to be a key determinant of many aspects of callitrichine behavior, most notably the high level of cooperative care of young (reviewed in Tardif et al. 1993, 2002; Yamamoto 2005). Captive studies have provided detailed information on the proximate mechanisms and energetics of infant carrying (e.g., Tardif 1997, Nievergelt and Martin 1999, Mota et al. 2006) and have demonstrated several species differences in patterns of infant care, including latency to onset of alloparental behavior, degree of maternal involvement, and overall time spent carrying (Tardif et al. 1993, Yamamoto 2005).

In both *L. rosalia* and *Callimico*, mothers are typically the sole caretakers for up to the first 3 weeks of an infant's life (Schradin and Anzenberger 2001, Tardif et al. 2002; see also de Oliveira et al. 1999 for an exception in *L. chrysomelas*); but *Saguinus*, *Callithrix*, *Mico*, and *Cebuella* adult males and other group members may begin carrying the infant as early as the day of birth (reviewed in Tardif et al. 2002, Yamamoto 2005). The mother's social status can also influence the onset of allomaternal care, with subordinate breeding females avoiding potential helpers for the first 10 days postpartum (Digby 1995a). In some tamarin species, males may actually care for infants more than the mother (e.g., *S. oedipus*, Savage et al. 1996b; *S. fuscicollis*, Goldizen 1987a,b), whereas mothers act either as primary caretakers or as equal partners in *Callithrix*, *Mico*, and *Callimico* (*C. jacchus*, Digby 1995a, Yamamoto and Box 1997; *M. humeralifer*, Rylands 1986b; *Callimico*, Schradin and Anzenberger 2001, Porter 2007). The extent of maternal care can also be influenced by group size and composition, litter size, and maternal weight (reviewed in Tardif et al. 1993, Bales et al. 2002, Yamamoto et al. 2008). Both captive and field data confirm the general pattern that marmosets (e.g., *C. jacchus* and *M. argentata*) carry less frequently and for a shorter period of time than do tamarins (e.g., *S. fuscicollis* and *S. oedipus*), with *L. rosalia* being intermediate (Tardif et al. 1993, 2002; Savage et al. 1996b). Differences in carrying patterns are once again likely to be linked to daily path length, home range size, and ultimately differences in diet and habitat use (Tardif et al. 1993, Goldizen 2003, Yamamoto 2005).

Several hypotheses have been proposed to explain why group members other than the parents (alloparents) participate in infant care in the callitrichines: (*1*) enhancing direct fitness by gaining experience in caretaking behaviors, (*2*) increasing inclusive fitness, (*3*) maintaining group membership while waiting for a breeding spot to become vacant, and (*4*) using caretaking as a courtship strategy for breeding males (reviewed in Bales et al. 2000, Yamamoto 2005). While there is some evidence that caretaking experience and overall number of helpers increase infant survival, evidence for a connection between helping behavior and future reproductive opportunities (either inheritance of breeding position or as a courting strategy) is currently lacking (Tardif 1997, Bales et al. 2000).

CONSERVATION STATUS

The Callitrichinae include not only some of the world's most critically endangered species but also species such as *C. jacchus* that thrive under disturbed conditions and are unlikely to be threatened with extinction. A characteristic shared by all endangered callitrichine species is a relatively small geographic range combined with critical levels of anthropogenic habitat alteration. As a group, *Leontopithecus* faces the most serious threat, with one species listed as critically endangered (*L. caissara*) and three listed as endangered (*L. rosalia*, *L. chrysomelas*, and *L. chrysopygus*; Rylands and Chiarello 2003, IUCN Species Survival Commission 2008). *S. oedipus* was raised to critically endangered status in 2008 due to a severe decline (80%) in the population over the last two decades (IUCN Species Survival Commission 2008). Only four other species are allocated endangered status by the World Conservation Union (*S. bicolor, C. flaviceps*, and *S. oedipus, C. aurita*; IUCN Species Survival Commission 2008), but numerous species are classified as data-deficient, including several of the recently described species of *Mico*. While the more endangered species of callitrichines receive official protection and are the subjects of conservation-oriented research projects (e.g., the Golden Lion Tamarin Project), their survival still depends on careful metapopulation management (e.g., chapters in Kleiman and Rylands 2002).

COMPETITION IN COOPERATIVELY BREEDING SPECIES

Cooperatively breeding species are characterized, in part, by their unusual propensity to share in the rearing of offspring. In this system, parents, older siblings, members of the extended family, and in some cases unrelated individuals participate in resource and territory defense, infant carrying, food sharing, babysitting, and even allonursing (chapters in Solomon and French 1997). Such behaviors, along with low rates of physical aggression, are typical of the callitrichines (Garber 1997, Schaffner and Caine 2000). The cooperative nature of some aspects of callitrichine behavior, however, belies the fact that typically only two (or sometimes three) individuals receive the majority of the benefits from such a "cooperative" system. Unlike communal breeding (Price and Evans 1991), in which there is shared parentage of the young, most callitrichines invest time and energy into rearing the offspring of other group members. Though helpers may gain indirect benefits in the form of inclusive fitness or caretaking experience (reviewed in Tardif 1997, Yamamoto 2005), direct benefits accrue primarily to those individuals that are able to breed. This reproductive skew is expected to give rise to intense, if sometimes subtle, reproductive competition.

Control or manipulation of breeding opportunities can occur prior to conception as well as after birth. Competition can be manifest in acquisition and maintenance of social status, physiological suppression of ovulation, interference in the feeding and care of young, and, in extreme cases, infanticide. Next, we outline some of the means by which callitrichines compete for reproductive opportunities and examine potential causes for variation in these mechanisms across genera.

Social Status and Reproduction

Within groups, social status can play a profound role in determining the reproductive opportunities and reproductive success of both males and females. Breeding individuals (both males and females) are typically those that are socially dominant over all others within the group (Baker et al. 1993; Dietz and Baker 1993; Digby 1995a,b; Goldizen et al. 1996; Alencar et al. 2006).

The mechanisms by which females' social status results in differential reproductive success include both inhibition of sexual behavior and suppression of ovulation (Saltzman et al. 2009, see earlier section, "Mechanisms of Reproductive Suppression"). There has been some debate on whether subordinate females' failure to reproduce results from direct manipulation by the dominant female (e.g., dominant control model; Snowdon 1996) or whether it is more appropriately interpreted as a type of self-inhibition on the part of the subordinate female in an attempt to reduce wasted reproductive effort (e.g., conceiving or giving birth to infants that are unlikely to survive—self-restraint model; Snowdon 1996; see also Wasser and Barash 1983, Saltzman et al. 2009). In either scenario, socially dominant females are able to maintain reproductive sovereignty in most cases and, thus, also gain the benefits of higher reproductive success (Yamamoto 2005). Even in groups where subordinate females are able to conceive, the dominant female will typically produce more infants and have higher infant survival rates (Digby 1995a, Goldizen et al. 1996, Tardif et al. 2008). In *L. rosalia*, for example, dominant females had twice the reproductive success of subordinate breeding females (Dietz and Baker 1993); and in *C. jacchus*, dominant breeding females gave birth to twice as many infants and had nearly twice the overall survival rate (Digby 1995a, Sousa et al. 2005). Extrapair copulations may also result in subordinate pregnancies, but these infants have so far suffered a 100% mortality rate (Arruda et al. 2005).

For males in potentially polyandrous groups, dominance may determine the likelihood of paternity. As noted, dominance plays little or no role in access to females during the periovulatory period in some species (*S. mystax*, Garber et al. 1993; *S. tripartitus*, Kostrub 2003; *S. fuscicollis*, Goldizen 1989—but note that some consort behavior has been reported). In contrast, Baker et al. (1993) were able to demonstrate that the more dominant of the two sexually active males in *L. rosalia* groups was responsible for 94% of sexual behavior during periods when the female was most likely to conceive. In one of these groups, direct aggression was used by a dominant male to prevent the subordinate male's access to the breeding female. Mate guarding by dominant males has also been observed in *C. jacchus* (Digby 1999) and *Cebuella* (Soini 1987). Dominant males may be responding to extrapair copulations (both dominant and subordinate males have been observed to participate in extrapair copulations), but it is unclear if these copulations result in viable offspring (Digby 1999, Arruda et al. 2005).

In summary, both subordinate males and females may be subject to interference or inhibition of sexual behavior, and subordinate females may also undergo physiological suppression of reproduction. With both polyandrous and polygynous groups occurring in several genera, however, it is clear that these mechanisms are not always successful at maintaining reproductive sovereignty for the dominant male/female pair. When subordinates do breed, postpartum reproductive competition may play a role in determining which infants survive (Saltzman et al. 2008, 2009).

Infanticide

Perhaps the most extreme form of reproductive competition in callitrichines is the killing of infants by females other than the mother. Infanticide in primates is typically associated with the killing of infants by unrelated males that have recently joined or taken over a group (e.g., langurs, *Presbytis entellus*, Hrdy 1979; gorillas, *Gorilla gorilla beringei*, Watts 1989b). However, the sexual selection hypothesis (Hrdy 1979) put forth to explain these cases of infanticide is not applicable to most callitrichine incidents because the killing of dependent young will not necessarily bring females back into estrus more quickly (due to the limited influence

of lactation on ovulation in these species). Instead, competition for access to alloparents or other resources in polygynous groups could lead to enforced neglect and/or killing of infants (Digby 1995a, 2000; Digby and Saltzman 2009). Infanticide by females has now been reported in several free-ranging groups and populations of *C. jacchus* (Digby 1995a, Roda and Mendes Pontes 1998, Lazaro-Perea et al. 2000, Arruda et al. 2005, Bezerra et al. 2007), several captive groups of *C. jacchus* (reviewed by Saltzman 2003, Saltzman et al. 2008), and at least one captive group of *L. chrysomelas* (De Vleeschouwer et al. 2001). In five of the six cases where social status was known, the dominant female was observed or strongly suspected of killing the offspring born to the socially subordinate female (in the sixth case, the perpetrator became dominant following the infanticide) (Roda and Mendes Pontes 1998). In at least three cases, the perpetrator gave birth within days or weeks after the infanticide (Digby 1995a, Roda and Mendes Pontes 1998, Lazaro-Perea et al. 2000). In some cases, subordinate females that had lost their infants (either to infanticide or to unknown causes) subsequently carried and occasionally nursed infants born to the dominant female (Digby 1995a, Roda and Mendes Pontes 1998).

The harassment of subordinate females with young, the protective rearing strategies used by subordinate mothers (e.g., the extended period before alloparental care is tolerated), and the increased chance of infant loss when the births of dominant and subordinate females are closely spaced support the hypothesis that infanticide by females is a response to resource competition (Digby 1995a, Saltzman 2003, Saltzman et al. 2009). The intensity of competition between females is likely to vary across callitrichine genera due in part to such factors as typical group size (the smaller groups of *Saguinus* and *Cebuella* would result in less severe reproductive skew within the population), population density (higher densities will likely result in fewer reproductive vacancies within the population), and diet (with less seasonally influenced foods such as gum allowing for smaller home range sizes and, thus, lower infant care costs) (Digby and Saltzman 2009). Specifically, the marmosets' ability to maintain large groups in high-density areas may allow for a higher proportion of polygynous groups in these species. Once reproductive suppression is relaxed, however, infanticide by females may provide an alternative strategy by which dominant females may maintain their reproductive sovereignty (Digby 2000, Hager and Johnstone 2004, Saltzman et al. 2008, 2009).

Infanticide by females other than the mother is unusual among primate species but has been well documented in several other cooperatively breeding species (black-tailed prairie dogs, *Cynomys ludovicianus*, Hoogland 1985; meerkats, *Suricata suricatta*, Clutton-Brock et al. 1998). In addition to the reproductive state of the perpetrator (either in the late stages of pregnancy or lactating), the other striking pattern among these species is that the perpetrator is likely to be closely related to the victim (Digby 2000).

Given the potential loss of inclusive fitness in these cases, we can only assume that the benefits to the infanticidal female must be greater than the costs, suggesting particularly intense reproductive competition. This hypothesis remains to be tested.

CONCLUSION

Although social tolerance and cooperation are important aspects of callitrichine social organization, it is important to consider the reproductive skew that often results from this type of social system. Only a portion of the population directly benefits from the behavior of helpers, who may delay their own reproduction for many years or forfeit it altogether (e.g., Goldizen et al. 1996). As a result, intense competition over the limited number of reproductive positions may occur. This competition may be subtle (suppression of ovulation and inhibition of sexual behavior) or overt (infanticide), but it nonetheless has a profound impact on the reproductive success of these animals. Ultimately, our understanding of callitrichine social organization and reproductive strategies will need to balance the cooperative aspects of their social interactions with the inevitable reproductive competition inherent in all social systems.

Species included in this chapter are shown in Color Plates 7 and 8 (*Callimico goeldii* and *Saguinus tripartitus*).

Appendix 7.1 Species and Subspecies Names for the Callitrichines

SPECIES AND SUBSPECIES	COMMON NAME
Callimico goeldii	Goeldi's monkey
Callithrix aurita	Buffy-tufted-ear marmoset
Callithrix flaviceps	Buffy-headed marmoset
Callithrix geoffroyi	Geoffroy's tufted-ear marmoset
Callithrix jacchus	Common or white-tufted-ear marmoset
Callithrix kuhlii	Wied's black-tufted-ear marmoset
Callithrix penicillata	Black-tufted-ear or black-penciled marmoset
Cebuella pygmaea pygmaea	Pygmy marmoset
Cebuella pygmaea niveiventris	White-bellied pygmy marmoset
Callibella humilis	Black-crowned dwarf marmoset
Mico acariensis	Rio Acari marmoset
Mico argentatus	Silvery marmoset
Mico chrysoleucus	Golden-white tassel-ear marmoset
Mico emiliae	Snethlage's marmoset
Mico humeralifer	Black and white tassel-ear marmoset
Mico intermedius	Aripuanã marmoset
Mico leucippe	Golden-white bare-ear marmoset
Mico manicorensis	Manicoré marmoset
Mico marcai	Marca's marmoset
Mico mauesi	Maués marmoset
Mico melanurus	Black-tailed marmoset
Mico nigriceps	Black-headed marmoset
Mico saterei	Sateré marmoset
Saguinus bicolor	Pied bare-face tamarin
Saguinus fuscicollis avilapiresi	Ávila Pires' saddle-back tamarin
Saguinus fuscicollis cruzlimai	Cruz Lima's saddle-back tamarin
Saguinus fuscicollis fuscicollis	Spix's saddle-back tamarin
Saguinus fuscicollis fuscus	Lesson's saddle-back tamarin
Saguinus fuscicollis illigeri	Illiger's saddle-back tamarin

Appendix 7.1 (cont'd)

Saguinus fuscicollis lagonotus	Red-mantle saddle-back tamarin
Saguinus fuscicollis leucogenys	Andean saddle-back tamarin
Saguinus fuscicollis nigrifrons	Geoffroy's saddle-back tamarin
Saguinus fuscicollis primitivus	Saddle-back tamarin
Saguinus fuscicollis weddelli	Weddell's saddle-back tamarin
Saguinus geoffroyi	Geoffroy's tamarin
Saguinus graellsi	Graell's black-mantle tamarin
Saguinus imperator imperator	Black-chinned emperor tamarin
Saguinus imperator subgrisescens	Bearded emperor tamarin
Saguinus inustus	Mottled-face tamarin
Saguinus labiatus labiatus	Red-bellied tamarin
Saguinus labiatus rufiventer	Red-bellied tamarin
Saguinus labiatus thomasi	Thomas' mustached tamarin
Saguinus leucopus	Silvery-brown bare-face tamarin
Saguinus martinsi martinsi	Martin's bare-face tamarin
Saguinus martinsi ochraceus	Ochraceous bare-face tamarin
Saguinus melanocephalus Crandall	Crandall's saddle-back tamarin
Saguinus melanocephalus melanocephalus	White saddle-back tamarin
Saguinus midas	Golden-handed or Midas tamarin
Saguinus mystax mystax	Spix's mustached tamarin
Saguinus mystax pileatus	Red-cap mustached tamarin
Saguinus mystax Pluto	White-rump mustached tamarin
Saguinus niger	Black-handed tamarin
Saguinus nigricollis graellsi	Graell's black-mantle tamarin
Saguinus nigricollis hernandezi	Hernández-Camacho's black-mantle tamarin
Saguinus nigricollis nigricollis	Spix's black-mantle tamarin
Saguinus oedipus	Cotton-top tamarin
Saguinus tripartitus	Golden-mantle saddle-back tamarin
Leontopithecus caissara	Black-faced lion tamarin
Leontopithecus chrysomelas	Golden-headed lion tamarin
Leontopithecus chrysopygus	Black or golden-rumped lion tamarin
Leontopithecus rosalia	Golden lion tamarin

Source: Adapted from Rylands et al. (2000), van Roosmalen and van Roosmalen (2003), Rylands and Mittermeier (2008).

8

The Cebines
Toward an Explanation of Variable Social Structure
Katharine M. Jack

1. While the cebines are widely distributed throughout Central and South America, only members of the genus *Cebus* appear to be able to exploit variable habitat types (they are referred to as "habitat generalists"). What characteristics (e.g., behavioral, morphological, social) enable *Cebus* to occupy diverse habitat types, and what characteristics act to restrict *Saimiri* to mainly secondary lowland tropical rain forests?
2. What is parallel dispersal, and how does it affect the social relationships of the cebines?
3. As a genus, *Saimiri* are unique in that they display a full range of dispersal patterns observed in group-living primates. What are these patterns, and why do we see such diversity across this genus?

INTRODUCTION

Capuchins (genus *Cebus*) and squirrel monkeys (genus *Saimiri*), which together make up the subfamily Cebinae, are among the most widely recognized species of New World primates. Some may recognize capuchins as the hat-tipping, money-collecting partner of the organ grinder or may have seen them in movies or television ads spreading a deadly strain of the Ebola virus or as the "spokesperson" for any number of products, both of which have little to do with being a capuchin or a primate. Although perhaps not quite as recognizable as their capuchin cousins, squirrel monkeys are also well known because of their prominence in the illegal pet trade and their use in biomedical research; they are one of the most commonly used laboratory primates, second only to rhesus macaques (*Macaca mulatta*) (Kinzey 1997c). Although these two primate genera have much in common and can often be observed in the same forests in Central and South America, they comprise a very diverse subfamily, particularly in respect to their morphology and behavioral ecology.

TAXONOMY AND DISTRIBUTION

The taxonomic classification of the platyrrhines has seen a long history of debate, and the exact placement of the genera *Cebus* and *Saimiri* within the infraorder has been particularly problematic (see Janson and Boinski 1992 for

review). Ford and Davis (1992:434) sum the problem up succinctly, stating that "*Cebus* and *Saimiri* are in some ways enigmatic" and "[t]heir potential relationships to each other and to other New World monkeys remain unclear" (see also Tyler 1991). They argue that morphological and life history data indicate that *Cebus* may have separated from the other platyrrhines as many as 20 million years ago and, accordingly, they should be placed within their own subfamily. Groves (2001a) agrees, placing *Cebus* in the subfamily Cebinae and *Saimiri* in Chrysotrichinae (also Saimirinae; see Rosenberger's interpretation in Schneider and Rosenberger 1996), although he does group these two neotropical primates together in the family Cebidae, accompanied only by the small squirrel-like marmosets and tamarins (*Callithrix, Callimico, Leontopithecus,* and *Saguinus;* subfamily Hapalinae). Contrary to these classification schemes that place *Cebus* and *Saimiri* in different subfamilies, other taxonomists argue that current phylogenetic analyses of both morphological and molecular data support the placement of these two genera together within the Cebinae subfamily (Schneider et al. 1993; for recent reviews, see Rylands et al. 2000 and Schneider and Rosenberger 1996). The question of whether or not the marmosets and tamarins, and perhaps owl monkeys (*Aotus*), should accompany *Cebus* and *Saimiri* in the Cebinae subfamily is still highly debated; and here I follow Rylands et al. (2000) in considering *Cebus* and *Saimiri* to be the exclusive members of Cebinae and use the term "cebine" to refer solely to these two genera.

Four species of *Cebus* (*C. apella*, *C. albifrons*, *C. capucinus*, and *C. olivaceus*) and more than 30 subspecies are traditionally recognized (Ford and Hobbs 1996). A fifth species, *C. kaapori*, discovered in 1992 (Queiroz 1992, cf. Masterson 1995), is increasingly included in this list. Groves (2001a) and Rylands et al. (2000) have called for the recognition of *C. libidinosus*, *C. xanthosternos*, and *C. nigritus* as distinct species rather than subspecies of *C. apella*, as they have previously been classified (see Fragaszy et al. 2004b for review). The four better-known *Cebus* species are divided into two main groups according to the presence or absence of tufts on the top of the head. The tufted group contains only one species, *C. apella*, while the untufted group contains *C. albifrons*, *C. capucinus*, and *C. olivaceus* (Hershkovitz 1949 based on Elliot 1913).

The number of species included within the genus *Saimiri* has been widely investigated, and molecular studies conducted in the last decade have shed much light on the topic. Historically, the genus was divided into only two species based on geographic distribution: *S. oerstedii* in Central America and *S. sciureus* in South America. In some accounts, *S. oerstedii* is listed as an offshoot of *S. sciureus* that was thought to have been introduced to Central America by humans in pre-Columbian times (Hershkovitz 1969; cf. Cropp and Boinski 2000, Costello et al. 1993; see Boinski 1999 for review of *Saimiri* taxonomy); however, genetic analyses have shown this not to be the case, and it is now known to be a distinct species (Cropp and Boinski 2000). Taxonomists are now in general agreement on the division of *Saimiri* into five species, with as many as 12 subspecies: *S. oerstedii*, *S. sciureus*, *S. boliviensis*, *S. ustus*, and *S. vanzolinii* (Groves 2001a, Rylands et al. 2000). This division is based on a variety of molecular, morphological, and behavioral data (Boinski and Cropp 1999; see also Ayres 1985, Groves 2001a). It should be noted that the previous practice of classifying all South American squirrel monkeys together as *S. sciureus* does lead to some confusion when trying to decipher early reports of squirrel monkey taxonomy and behavior. In particular, most of our current information on *S. boliviensis* comes from Mitchell's (e.g., 1990) study at Manu, Peru, where the species was formerly referred to as *S. sciureus* (see Boinski 1999). The species designator "*sciureus*" is now reserved for squirrel monkey populations in the northeastern Amazon Basin (Colombia, Ecuador, and Venezuela) and the Guyana Shield (Boinski 1999, Boinski and Cropp 1999).

The geographic distribution of the cebines overlaps extensively; indeed, the two genera are often found in sympatry and commonly form mixed-species associations (see later section "Ecology"). The geographic range of *Cebus* is more extensive than that of *Saimiri* and, among New World monkeys, is second only to that of *Alouatta* (howler monkeys) (Sussman 2000). *Cebus* range throughout much of Central and South America, from Honduras in the north to Argentina in the south (Eisenberg 1989) (see

Fig. 8.1 for map of *Cebus* distribution). With the exception of *C. capucinus*, which ranges from Honduras through the northwestern coast of Ecuador (Rowe 1996), all *Cebus* species are indigenous to South America (see Table 8.1 for species distributions and common names). Despite the large geographic distribution of this genus, long-term field studies have been limited, with the majority of data coming from six main field sites: *C. capucinus* in Santa Rosa and Lomas Barbudal (Costa Rica); *C. olivaceus* in Hato Masaguaral (Venezuela); and *C. apella* in Iguazu (Argentina), Manu (Peru), and a study begun in 2000 at Raleighvallen (Suriname) (see Fragaszy et al. 2004b for a complete overview of field studies of *Cebus*). To date, only short-term studies of *C. albifrons* have been undertaken in Peru (Janson 1984, 1986), Colombia (Defler 1979a,b, 1982), Ecuador (Field 2008, Matthews 2008, Jack 2005), and Trinidad (Phillips and Abercrombi 2003). Despite this apparent paucity of studies focused on this genus in the wild, particularly in comparison to the long-term field studies of many Old World monkey and ape species, *Cebus* is one of the most extensively studied genera of New World primates, with studies of *C. capucinus* in Costa Rica well into their third decade.

Saimiri are found in two geographically separated pockets. A single species, *S. oerstedii*, is isolated in Central America in a relatively small area of lowland forests that extend from the central Pacific coastline of Costa Rica through the Pacific Coast of western Panama. The remaining

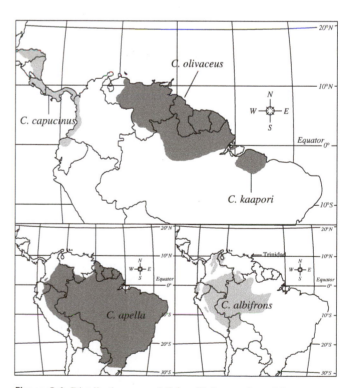

Figure 8.1 Distribution map of *Cebus* (B. Lenz, adapted from Fragaszy et al. 2004a).

species occur in the lowland forests of the Amazon Basin of South America from Guyana through Paraguay (Boinski et al. 2002) (Fig. 8.2 and Table 8.1). To date, long-term studies of *Saimiri* have been limited: *S. oerstedii* in Corcovado, Costa Rica (e.g., Boinski 1987c, 1988a); *S. boliviensis* in Manu, Peru (Mitchell 1990, 1994; Boinski 1991, 1994); and *S. sciureus* in Raleighvallen, Suriname (Boinski 1999), which is currently the only long-term investigation being undertaken on this genus (see Boinski et al. 2002 for a summary of shorter studies).

ECOLOGY

Morphological Adaptations to Habitat and Diet

Cebus are medium-sized primates, with females weighing 1.4–3.4 kg (mean = 2.3 kg) and males ranging 1.3–4.8 kg (mean = 3.0 kg) depending on the species under consideration (see Table 8.2). *Cebus* display moderate levels of sexual dimorphism, with males weighing 19.5%–27% more than females (mean = 24%, Table 8.2) and possessing canine teeth that are 16%–22% larger than female canines (Kay et al. 1988). *Saimiri* are considerably smaller than *Cebus*, with male weights ranging 620–1,200 g (mean = 884 g) and female weights ranging 600–880 g (mean = 700 g). However, the degree of sexual dimorphism in body size is comparable across the two genera (male *Saimiri* are on average 19.7% larger than females) (Table 8.2) (see also Janson 1984). Both *Cebus* and *Saimiri* show large brain

to body weight ratios, particularly for primates of their size, a trend that may be an adaptation to their relatively large home ranges, variable diets (Stephen et al. 1988; see Terborgh 1983 for home range comparisons), and/or complex

Figure 8.2 Distribution map of *Saimiri* (B. Lenz, adapted from Boinski 1999).

Table 8.1 Species, Common Names, and Geographic Distribution of the Most Commonly Recognized Cebines

SCIENTIFIC NAME	COMMON NAME	GEOGRAPHIC DISTRIBUTION
Cebus albifrons	White-fronted capuchin	Brazil, Colombia, Ecuador, Peru, Venezuela, northern Bolivia, and Trinidad
C. apella[1]	Tufted, black-capped, or brown capuchin	Northern and central South America from Colombia and Ecuador (west of the Andes) through coastal Brazil
C. capucinus	White-faced, white-headed, or white-throated capuchin	Honduras through northwestern Ecuador
C. olivaceus (also *C. nigrivittatus*)	Weeper or wedge-capped capuchin	Northeastern Brazil, Guyana, French Guiana, Suriname, and Venezuela
C. kaapori	Ka'apor capuchin	Brazil
Saimiri oerstedii	Red-backed or Central American squirrel monkey	Central Pacific Coast of Costa Rica through Pacific Coast of Panama
S. sciureus	Common or South American squirrel monkey	North–central South America (Brazil, Guyana, French Guiana, Suriname, Venezuela, Ecuador, and Colombia)
S. ustus	Golden-backed or bare-eared squirrel monkey	Northwestern Brazil
S. boliviensis	Bolivian or black-capped squirrel monkey	Brazil, Bolivia, Peru
S. vanzolinii	Black squirrel monkey, many classify with *S. ustus*	Northwestern Brazil

[1] *C. libidinosus, C. xanthosternos,* and *C. nigritus* are now considered by many taxonomists to be distinct species, rather than subspecies of *C. apella*. The above list follows more traditional taxonomy.

Sources: Cebus, Groves 2001a, Rowe 1996, Fragaszy et al. 2004b; *Saimiri*, Groves 2001a, Boinski 1999, Rowe 1996.

Table 8.2 Adult Weights and Sexual Dimorphism of the Cebines

GENUS	SPECIES	MALE BODY WEIGHT RANGE (MEAN) (G)	FEMALE BODY WEIGHT RANGE (MEAN) (G)	DEGREE OF SEXUAL DIMORPHISM
Cebus	albifrons	1,700–3,260 (2,480)	1,400–2,228 (1,814)	27%
	apella	1,350–4,800 (3,050)	1,758–3,400 (2,385)	22%
	capucinus	3,765–3,970 (3,668)	2,610–2,722 (2,666)	27%
	olivaceus	1,447–4,500 (2,974)	1,589–3,200 (2,395)	19.5%
Mean		3,043	2,315	24%
Saimiri	boliviensis	(992)	(751)	24.3%
	oerstedii	750–950 (829)	600–800 (695)	16%
	sciureus	(740)	(635)	14%
	ustus	620–1,200 (910)	710–880 (795)	13%
	vanzolinii	(950)	(650)	31.5%
Mean		884	705	19.7%

Sources: Ford and Davis 1992 (data based on wild individuals only). Boinski (1999); with the exception of those weights for female *S. oerstedii*, sexual dimorphism for *S. boliviensis*, male *S. oerstedii*, and *S. sciureus* has been calculated based on weights provided in Boinski (1999) rather than Ford and Davis (1992) because of earlier practices of lumping numerous species under *S. sciureus*.

foraging patterns (e.g., Fragaszy 1990, Parker and Gibson 1977, Terborgh 1983).

In addition to differences in absolute body size, *Cebus* and *Saimiri* display considerable morphological variability. For example, although both genera have skeletons adapted for quadrupedal locomotion, they differ markedly in their relative limb proportions. Unlike other platyrrhines, *Cebus* tend to have forelimbs and hindlimbs of fairly equal proportions, while *Saimiri* have comparatively longer hindlimbs (Fleagle 1999). This is likely an adaptation for the more frequent leaping that characterizes *Saimiri* locomotion, while the equal limb lengths of capuchins may be an adaptation for their more terrestrial locomotor patterns (Janson and Boinski 1992). Overall, *Cebus* postcranial anatomy is more similar to that of terrestrial Old World monkeys than it is to that of other platyrrhines, which is likely because capuchins, particularly the untufted species, spend more time on the ground than other platyrrhines. Capuchins also have a semiprehensile tail that, unlike their ateline cousins (see Chapter 11), is fully covered in fur and unable to sustain the full weight of an adult. Instead, the tail provides support and balance and is commonly used to anchor an individual to trees while foraging (see Fig. 8.3). Squirrel monkeys locomote quadrupedally along the surfaces of thin branches and lianas, often leap between them (Boinski et al. 2002), and rarely descend to the ground to forage. Although they are born with prehensile tails, the grasping ability is lost with age and the rather large tail of adults assists only in balance (Boinski 1989a) (Fig. 8.4).

The dentition of *Cebus* and *Saimiri* is also variable and appears to reflect dietary differences between the two genera. Although cebines are considered the most omnivorous

of the platyrrhines (Fleagle 1999) and both *Cebus* and *Saimiri* are traditionally categorized as frugivore–insectivores (Robinson and Janson 1987), *Saimiri* display a relatively heavier reliance on insects, while *Cebus* rely more on fruit (see Janson and Boinski 1992 for review). The cebines spend upward of 75% of their day engaged in active foraging (Robinson and Janson 1987), and *Saimiri* can spend up to 80% of this time foraging just for insects. The diets of both genera are extremely flexible, and individuals are able to alter their diet to become very specialized in response to changing environmental conditions. For example, during the peak of the dry season, *Saimiri* are able to become complete insectivores for up to a week at a time, while *Cebus*, which can survive in very marginal habitats, can rely almost completely on bromeliads (a plant in the family that includes pineapples) or become seed predators in times of food stress (see Sussman 2000 for review). *Cebus* are very adept at capturing small vertebrate prey, including birds (and eggs), lizards, squirrels (Fig. 8.5), and coatis, although such animal prey form only a small portion of their diet (~0.5%–2.5%) (Sussman 2000). The hunting behavior displayed by *Cebus* is opportunistic rather than cooperative in nature (see Perry and Rose 1994). Interestingly, at least among *C. capucinus*, the individual that captures the prey, regardless of age or sex class, is able to eat it with only minimal harassment from other group members (K. M. Jack personal observation). Although food sharing does occur in this genus, it is indirect in that bits of food are dropped or left behind for other group members to eat (Perry and Rose 1994).

Cebus are often referred to as "habitat generalists" (Chapman et al. 1989a) in that they occupy a great

Figure 8.3 An adult male *Cebus capucinus* using his semiprehensile tail to anchor him while foraging for insects (photo by Katharine M. Jack).

Figure 8.4 An adult female *Saimiri oerstedii* with 6-month-old infant. Note the lack of prehensile tails (photo by Katharine M. Jack).

diversity of habitats, including primary forests of all types (rain, cloud, dry, deciduous, etc.), highly disturbed and fragmented areas, as well as swamp and seasonally flooded forests (Freese and Oppenheimer 1981). Members of the genus *Cebus* are the most dextrous of the platyrrhine primates. They have shortened fingers, "pseudo-opposable thumbs," and the ability to move all digits independently of the others (Janson and Boinski 1992). This dexterity, accompanied by what seems to be an innate curiosity, is readily visible in their foraging habits. Capuchins are extractive foragers (Parker and Gibson 1977) and appear to specialize in consuming foods that "fight back"—foods that have hard protective coverings, toxins, and other predatory defenses or insects and vertebrates that bite and/ or sting (Fragaszy et al. 2004b). In order to access these foods, wild capuchins will pound hard husked or shelled objects onto a substrate, and they frequently rid food items of spines or poisonous hairs by rubbing them against branches (Panger et al. 2002b). In some cases, these food items are wrapped in leaves prior to rubbing, perhaps as a way of directly protecting the hands from the toxins they are attempting to remove (Panger et al. 2002b). Fragaszy et al. (2004a) recently described a wild group of *C. apella* in Brazil that are regularly observed to crack open palm nuts using rocks in a hammer and anvil–type fashion, further demonstrating the dexterity and ingenuity of these small New World primates. Capuchins have been observed using a variety of other tools in the wild, including a club to attack a venomous snake, leaves as containers, and probing tools to access imbedded foods; among the nonhuman primates, capuchins and the great apes are considered to be the most adept and varied tool users (see Panger et al. 2002b). Another characteristic that *Cebus* has evolved as a means of gaining access to protected foods (e.g., palm nuts) is their thick tooth enamel; in fact, when tooth size is

Figure 8.5 An adult male *Cebus capucinus* holds a young squirrel he has just killed (photo by Katharine M. Jack).

Figure 8.6 An adult male *Saimiri boliviensis* unrolling a dead leaf in search of imbedded invertebrate prey (photo by B. Lenz).

taken into account, *Cebus* show the thickest tooth enamel of any nonhuman primate (Kay 1981; see Janson and Boinski 1992 for review).

Despite their large geographic distribution, the habitat preferences of *Saimiri* show extreme ecological consistency (see Janson and Boinski 1992). *Saimiri* show a preference for tropical lowland rain forests (Boinski et al. 2002), with most activities concentrated in secondary forests and relatively little time in primary and late successional forests (Boinski 1987c; see Sussman 2000 for review). Although, in general, the cebines are more manually dextrous than other platyrrhines, *Saimiri* do not possess the manipulative abilities that are so characteristic of capuchins. Squirrel monkeys do not have independently mobile digits, nor do they possess opposable thumbs, a combination that prohibits them from generating a strong grip between the fingers and thumb (Janson and Boinski 1992). Although both *Saimiri* and *Cebus* rarely go after prey that is in motion, *Saimiri* are more prone to hunt potentially mobile prey that are visible on substrate surfaces (branches, leaves, tree trunks) than are *Cebus*. Squirrel monkeys also specialize in finding insects by unrolling both dead and living leaves to reveal invertebrate prey imbedded within them (Fig. 8.6) (Janson and Boinski 1992).

Predation: Susceptibility and Behavioral Adaptations

Both *Cebus* and *Saimiri* are at risk of being taken by similar types of predators, including raptors, felids, constricting and venomous snakes, coyotes, and tayras (an arboreal mammal that is part of the weasel family, *Mustelidae*) (Freese 1983, van Schaik and van Hooff 1983, Terborgh 1983, Chapman 1986, Boinski et al. 2002, Bianchi and Mendes 2007). However, given that an adult capuchin outweighs an adult squirrel monkey by as much as seven times, it is not surprising that *Saimiri* are much more susceptible to predation than are *Cebus* (Fedigan et al. 1996). Raptors are the main predators observed to kill *Saimiri*, and Boinski (1987b) reports that 50% of all infants born to *S. oerstedii* do not survive to 6 months of age due to confirmed or suspected predation by avian predators. Boinski et al. (2002) report similar predation rates on *S. boliviensis*, while *S. sciureus* of Suriname appear to be at lower risk despite an intact predator community. Perhaps as a response to their susceptibility to predation, *Saimiri* form groups that are usually between three and six times the size of *Cebus* troops (see later section "Social Organization"); and although both genera are known for being highly vocal primates that display a great repertoire of call types, *Saimiri* have more calls that serve as contact calls than do *Cebus*. While these vocalizations do not directly function in predator avoidance or detection, they are believed to function as a type of "security blanket," to make up for the visual separation that occurs while squirrel monkeys forage and enable group members to keep alerted to the presence and location of their group mates (see Fedigan et al. 1996 for review).

The South American populations of *Saimiri* have developed another way of combating their high risk of predation: In all areas where they occur sympatrically, *Saimiri* parasitize on the predator-detecting abilities of *Cebus* by joining them in their daily activities (Waser 1987). These mixed-species associations may last several days, and for the most part the two genera will interact peacefully. However, it is not uncommon for an adult male capuchin to chase away an entire group of squirrel monkeys, although the displaced squirrel monkeys usually quickly return to their capuchin comrades. Data from South American cebine populations, particularly those coming from Manu, Peru (e.g., Terborgh 1983, Terborgh and Janson 1986), have shown that it is *Saimiri* rather than *Cebus* that maintain these mixed-species associations. Terborgh (1983) reports that, in Manu, *Saimiri* actually seek out *Cebus* and that although *Saimiri* often appear to be leading the movements and direction of the mixed-species foraging excursions, they will usually backtrack if the capuchins lag behind or fail to follow them. The maintenance of these associations with capuchins is thought to benefit squirrel monkeys in terms of both increased foraging efficiency and predation avoidance. For example, when they are not in association with *Cebus*, squirrel monkeys travel twice as far in an hour. They also appear to benefit from the more detailed knowledge that *Cebus* have of fruiting tree locations during times of resource scarcity. In addition, *Saimiri* gain access to fruits dropped by *Cebus* that they would otherwise be unable to open (e.g., tough-husked palm nuts). Despite the obvious foraging advantages that these associations provide, the real advantage for *Saimiri* seems to lie in the increased predator-detection skills of *Cebus* (Terborgh 1983; see also Sussman 2000 for review). *Saimiri* appear to benefit greatly from the ever-vigilant capuchin males and the alarm calls that they emit; in fact, squirrel monkeys respond more readily to *Cebus* alarm calls than they do to the alarm calls of their own group members (Terborgh 1983). *Cebus*, on the other hand, do not seem to benefit much from these associations; they pay little attention to *Saimiri* alarm calls, and they travel up to 40% more when in the company of squirrel monkeys. It is no wonder, then, that capuchins do not wait for sidetracked squirrel monkeys to catch up when they lag behind.

Interestingly, in Central America, sympatrically occurring *Cebus* and *Saimiri* do not form mixed-species associations more frequently than predicated by chance. Given that *S. oerstedii* are under severe predation pressure, particularly from raptors (Boinski 1987b), would they not also benefit from the predator-detection skills of *C. capucinus*? Boinski (1989b) reports that in Corcovado National Park, Costa Rica, the two cebine genera spent only approximately 6% of their time in association with one another, which greatly contrasts with the 90% rate reported for *S. boliviensis* with *C. apella* and *C. albifrons* in Peru. Boinski (1989b) suggests

that the two Central American species do not form mixed troops simply because the presence of *C. capucinus* would impose very high foraging costs on *S. oerstedii*. Her study showed that *Saimiri* would have little to gain by associating with *Cebus* even in terms of increased predator detection and avoidance. Boinski concluded that, although *Cebus* in Costa Rica are susceptible to predation, their vigilance does not appear to function for detecting predators. Several studies focusing on male vigilance in *C. capucinus* have shown that while they are alert to potential predators, as evidenced by their elaborate alarms calls that differ according to predator types, the majority of their vigilance is directed at detecting extragroup males (Rose and Fedigan 1995) or monitoring the activities of their coresident males (Jack 2001). The foraging costs and limited benefits associated with detecting predators appear to have precluded the formation of mixed-species groups between the Costa Rican cebines, although this topic requires further detailed investigation.

REPRODUCTIVE PARAMETERS

Saimiri are highly seasonal breeders, with all matings being confined to an annual 2-month period and births being even more tightly synchronized (Table 8.3). For example, *S. oerstedii* are reported to have the most restricted birth season of any primate species, with group females giving birth within the same 1-week period (Kinzey 1997c, Boinski 1987b). *C. apella* in Argentina is the only capuchin population studied thus far that displays a true birth season, which in turn reflects mating seasonality (Di Bitetti and Janson 2000, 2001a). While the other *Cebus* species do show a yearly birth peak during which the majority of infants are born, they are not strict seasonal breeders as births occur throughout all months of the year (Freese and Oppenheimer 1981, Robinson and Janson 1987, Fedigan and Rose 1995). Like most New World monkeys, cebine birth peaks or seasons generally coincide with seasonal peaks in resource availability (Di Bitetti and Janson 2000). *Cebus* interbirth intervals range 18–26.4 months (mean = 22 months), while *Saimiri* interbirth intervals, with the exception of *S. boliviensis* where females give birth every other year, tend to be shorter and average 12 months (see Table 8.3).

Compared with other mammals, where life history variables strongly correlate with body size, primates show very long life histories, particularly in their tendency toward extreme delays in reproductive maturity (Promislow and Harvey 1990, Purvis et al. 2003). If we restrict the examination of life history variables and body size to the order Primates, the general pattern holds, with larger primates showing slower life histories than smaller ones (Eisenberg 1979). However, comparing published data available for New World species, *Cebus* do not fit this general pattern. Rather, their maturation rates and overall life history variables are more comparable with members of the subfamily

Table 8.3 Cebine Reproductive Parameters and Life History Variables

GENUS	SPECIES	BIRTH SEASON	INTERBIRTH INTERVAL (MONTHS)	FEMALE AT FIRST BIRTH (YEARS)	MALE ADULT SIZE, REPRODUCTIVE MATURITY (YEARS)	LIFE SPAN (CAPTIVE)
Cebus	albifrons	Peak[1]	18[3]	N/A	N/A	44[15]
	apella	Peak–Peru				
		Season–Argentina[1]	19.4[4]	7 (wild)[1]	7[11,12], 4.5[1] (captive)	45.1[15]
	capucinus	Peak[1]	26.4[5]	7 (wild)[1]	10 (wild)[13]	54.8[15]
	olivaceus	Peak[1]	26[6]	6 (wild)[1]	15 (wild)[13]	41[15]
Saimiri	boliviensis	Yes (2 months)[2]	~24[7]	2.5 (wild)[10]	6 (wild)[7]	N/A
	oerstedii	Yes (1 week)[2]	~12[8]	2.5 (wild)[10]	5, 2.5 (wild)[14]	N/A
	sciureus	Yes (1 week)[2]	~12[9]	N/A	N/A	21[16]

Sources:
[1] Fragaszy et al. 2004b.
[2] Boinski 1987b.
[3] Kappeler and Pereira 2003.
[4] Di Bitetti and Janson 2000.
[5] Fedigan and Rose 1995.
[6] Robinson and Janson 1987.
[7] Mitchell, C. L., 1994.
[8] Boinski 1999.
[9] Rowe 1996.
[10] Baldwin and Baldwin 1981.
[11] Patino et al. 1996.
[12] Nagle and Denari 1982.
[13] Jack and Fedigan 2004b.
[14] Boinski 1994.
[15] Hakeem et al. 1996.
[16] Harvey et al. 1987.

Atelinae (spider monkeys and woolly monkeys), which generally weigh about four times as much as the average capuchin (see data in Robinson and Janson 1987, Rowe 1996, Kinzey 1997c). Compare, for example, the largest of the New World monkeys, the muriqui (*Brachyteles* spp.), with *Saimiri* and *Cebus*. Female muriquis weigh about 9.5 kg (Rowe 1996) and are 7.5 years of age when they give birth to their first offspring (Strier 1997c). *Saimiri* females weigh on average 700 g and give birth to their first infant at 2.5 years, while *Cebus* females weigh 2.3 kg and are 7–8 years of age when they give birth for the first time (Tables 8.2 and 8.3). Even male *Cebus* show slow maturation rates, with most studies estimating sexual maturity to occur after 7 years of age; and in several species, males do not attain full adult body size until after 10 years of age (Jack and Fedigan 2004a,b). Interestingly, the small size and long life history patterns characteristic of *Cebus* continue to stand out even when we extend comparisons to include the Old World monkeys.

These slow maturation rates of *Cebus* may be related to their extremely long life spans, which average about 41 years in captivity (the record is nearly 55 years; Hakeem et al. 1996). In stark contrast, the smaller *Saimiri* are reported to live only 21 years, and most capuchin-sized (and larger) primates live approximately 30 years (see data presented in Rowe 1996). It is also possible that delayed maturation in *Cebus* is related to their unusually slow postnatal brain growth and motor skill development (see Hartwig 1996). *Cebus* display more postnatal brain growth and development than any of the other platyrrhines, while

Saimiri neonates are more precocial in terms of both brain growth and motor skill development (Hartwig 1995).

GROUP STRUCTURE

Both *Cebus* and *Saimiri* form mixed-sex groups, but the two genera show much diversity in group size and composition (see later section "Cebine Dispersal Patterns"). Although the mean size of *Cebus* groups does not vary greatly across the four better-known species (range = 16.4–21, mean = 18.8), there is considerable variation in adult sex ratios within groups. For example, *C. olivaceus* show the most skewed sex ratios, with approximately one adult male for every two adult females in a group, while *C. albifrons* show the most equitable sex ratios, with 1.08 adult males per female (Table 8.4). The size of *Saimiri* groups is much more variable across species, ranging from 15 to 75 individuals. *S. boliviensis* form the largest groups (mean = 54), while *S. sciureus* display the smallest mean group sizes, at 23 individuals per group. Several researchers have reported *Saimiri* group sizes of up to 300 individuals; however, in all instances, it appeared that these large groups were actually temporary unions of multiple groups (see Sussman 2000 for review). Sex ratios also vary across species, with *S. oerstedii* showing one male per 1.6 adult females and *S. boliviensis* showing a much more skewed ratio, with one male per 2.5 adult females (data are not available for *S. sciureus*) (see Table 8.4). Interestingly, *S. boliviensis* is also the only species of *Saimiri*, indeed of the platyrrhines, reported to form all-male groups; and even within mixed-sex groups, males

Table 8.4 Cebine Social Organization and Social Structure

SPECIES	MEAN GROUP SIZE	MEAN ADULT SEX RATIO (M:F)	DISPERSAL PATTERNS	DOMINANCE
Cebus albifrons[1]	19.8	1.08	Male	Both sexes display linear hierarchies, and males are individually dominant over females; males well-integrated into the group
C. apella[1]	18	0.85	Male	Both sexes display linear hierarchies, and males are individually dominant over females; alpha female may rank below alpha male; subordinate males are often peripheral group members
C. capucinus[1]	16.4	0.71	Male: parallel dispersal is common and lasts through multiple emigration events; female dispersal is rare, but it does occur	Both sexes display linear hierarchies, and males are individually dominant over females; alpha female may rank below alpha male; female coalitions can displace alpha male
C. olivaceus[1]	21	0.53	Male	Both sexes display linear hierarchies, and males are individually dominant over females; alpha female may rank below alpha male
Mean	18.8	0.79		
Saimiri oerstedii	35–65 (41)[2]	0.63 (1:1.6)[3]	Females disperse prior to first mating season/flexible male philopatry or dispersal[5]	Egalitarian; no female dominance hierarchy or coalitions; male hierarchies evident only during mating season[5]
S. boliviensis (Peru)	45–75 (54)[2]	0.40 (1:2.5)[4]	Male:[2] parallel dispersal is common and lasts through multiple emigration events[4]	Females dominant (matrilineal hierarchies and frequent coalitions); both sexes form stable linear dominance hierarchies;[5,6] males are peripheral group members
S. sciureus (Suriname)	15–50 (23)[2]	NA	Both sexes thought to disperse; female dispersal is flexible (may spend first mating season in natal group)[5]	Males dominant; both males and females form stable linear hierarchies[5]

Sources:
[1] All Cebus data are from Fragaszy et al. 2004b.
[2] Boinski et al. 2003.
[3] Boinski 1987a.
[4] Mitchell 1994.
[5] Boinski et al. 2002.
[6] Mitchell 1990.

of this species are generally peripheral and have little to no interaction with group females outside of the mating season (Boinski 1999).

MATING SYSTEMS

Similar to the other platyrrhines, the cebines do not show external cues that signal ovulation or pregnancy like those characteristic of many of the catarrhine primates, making it extremely difficult to discern reproductive from nonreproductive states (Dixson 1983a; see also Chapter 29). In *Cebus*, this lack of obvious fertility signaling, accompanied by the fact that members of the genus show limited reproductive seasonality and engage in frequent nonconceptive copulations (complete with copulatory displays), makes it nearly impossible for observers to discern fertile versus nonfertile matings (e.g., see Manson et al. 1997, Carnegie et al. 2006). Given that *Saimiri* are highly seasonal breeders, with all mating activity being confined to a 2-month period each

year, and that individual females are sexually active for only about a 2-day period throughout these short mating seasons (Boinski 1992) (Table 8.3), determining fertile versus nonfertile matings is not quite as challenging.

Mating behavior in *Cebus*, both conceptive and nonconceptive, is preceded, at least in some species, by elaborate courtship rituals. The most detailed data available to date come from studies of captive *C. apella*. In this species, a female begins the courtship ritual by gazing toward the male of her choice (usually the alpha male), often tilting her head from side to side and raising her eyebrows at him in an attempt to catch his gaze (see Fragaszy et al. 2004b). She slowly decreases the distance between herself and the male and, when she is close enough, reaches out, quickly touches him, and then runs away. Throughout this process, the female continuously rubs her chest and emits soft vocalizations. In these initial stages, her attempts to gain the male's attention seem futile; he appears utterly and completely disinterested. Such solicitation by a female can go on for hours, but the male eventually relents and begins to reciprocate her movements and vocalizations, adding a mutual gaze and a grin. At this point, the movements of the pair become mirrored in a coordinated dance display. One member of the pair advances, while the other retreats, all the while maintaining their locked gaze. Pirouettes of 180 degrees are added, with the dancer always coming back to face the partner. At this point in the courtship, the participants appear oblivious to what is going on around them. The pair eventually comes together and mating ensues. During copulation, the male emits a loud vocalization that is unique to the mating context; and following ejaculation, the pair remain together, continuing their dance display as if nothing has happened and beginning the process anew.

C. capucinus show a similarly elaborate pattern of courtship and mating, although the eyebrow raising and grinning characteristic of *C. apella* are exchanged for a "duck face," which involves a protrusion of the lips in the direction of the partner (Manson et al. 1997). During their dance displays, *C. capucinus* males and females coordinate their pirouettes and often face away from each other, only to look at their partner either over the shoulder or through the legs. A unique vocalization also accompanies mating in this species, and like the patterns described for *C. apella*, individuals involved in the mating and dance display do not make any attempts to hide their activities. Data on mating in *C. olivaceus* are completely absent from the published literature. The mating behavior and rituals of *C. albifrons*, although not well described, do not appear to be as elaborate as those of either *C. apella* or *C. capucinus*; but data for this species are very limited.

Unlike *Cebus*, *Saimiri* do not show any sort of preparatory behaviors that precede copulation; there are no special vocalizations, displays, or behaviors that indicate readiness or willingness to mate (Mendoza and Mason 1994). That said, *Saimiri* do display one of the most unique reproductive systems reported for nonhuman primates. Prior to and during the short annual mating season, male *S. boliviensis* and *S. oerstedii* undergo dramatic physiological and morphological changes; they gain up to 222 g, a shocking 22% of their body weight (Dumond and Hutchinson 1967) (see Table 8.3; there are no data available for other *Saimiri* species). Males begin accumulating "fat" prior to the start of the annual mating season, and this fattened state is maintained throughout the season, with males reaching their largest size during the same months that the majority of conceptions occur (Schiml et al. 1996). This "seasonal enhancement of male body size" (Boinski 1998:174) results not from the accumulation of fat, as has been reported for rhesus macaques and other mammalian species (see Berkovitch 1992), but from the deposition of water between the skin and muscles, particularly along the shoulders, back, and arms; and it has been likened to the water retention–induced swelling that women often experience before the onset of menstrual cycles (Boinski 1998). This swelling of males

does not occur because they are eating more or doing less. On the contrary, given the high level of reproductive synchrony observed among group females, males are extremely active during the mating season. Male "fattening" is also not related to a restricted seasonal ability to produce sperm. Mendoza et al. (1978) have shown that, although males do experience an increase in testes size and sperm production during the brief mating season, they are fertile year-round. Instead, the fattening appears to be in response to changing hormone levels within individual males, and males that become the fattest are those with the highest testosterone levels (Schiml et al. 1996).

During the nonmating season, there are no discernable size differences among coresident male *Saimiri*; however, with the onset of the mating season, some males become significantly larger than others and one male stands out as the largest. It is this male, the "fattest" male in a group, that is selectively preferred as a mating partner by group females. Indeed, Boinski (1992) reports that the largest male in her study group of *S. oerstedii* participated in over 70% of all observed copulations and the less swollen males were successful in gaining solicitations from females only after the females had been rejected by the largest male. For *S. boliviensis*, the dominant male becomes the most fattened and apparently suppresses the reproductive potential of less swollen subordinate males, although it remains to be determined by exactly what means this suppression occurs (DuMond and Hutchison 1967). Outside of the mating season, *S. boliviensis* males usually occupy peripheral positions within the group and rarely, if ever, interact with the females that dominate them. However, during the mating season, all males become more central within the group and interact regularly with females (DuMond and Hutchinson 1967). Within groups of *S. oerstedii*, there are no discernable dominance hierarchies either within or between the sexes, even during the mating season. Therefore, rather than dominance status, male enlargement in this species appears to be linked with the length of time a male has spent in a particular group, with long-term residents (>3 years) being the most enlarged and, therefore, the most popular during the mating season (Boinski 1992). At present, it is unclear whether all populations and species of *Saimiri* undergo these seasonal changes; data from additional field sites are needed to address these issues.

The proximate trigger for these morphological changes is still unknown; however, it is possible that males respond to subtle pheromonal or behavioral cues from group females. Male *S. oerstedii* frequently perform genital inspections of females, and this has been interpreted as a means of assessing female reproductive status (Boinski 1987a). Although these inspections occur throughout the year, during the 2 months prior to the commencement of the annual mating season, which is when males begin to fatten, there is a notable increase in the frequency of male coalitionary mobbing of females in order to perform such inspections. Boinski's (1987a) study showed that, at their peak, these aggressive inspections occur approximately twice per hour and may include up to 16 males. Boinski et al. (2002) even describe one female being wounded during such a mobbing. It is possible that these close inspections trigger the onset of seasonal swelling in males, although additional comparative data of intrasexual hormonal states are needed to test this suggestion. However, the fact that males are at their most swollen at the same time that group females achieve peak fertility (see Boinski 1992; Schiml et al. 1996, 1999) is a good indication that pheromones are at work.

Dominance appears to play a very strong role in the reproductive success of male *Cebus*. In all three species for which data on wild groups are available, a significant link between male dominance rank and reproductive success has been found, with alpha males siring the majority of infants born (*C. apella*, Escobar-Páramo 2000; *C. capucinus*, Jack and Fedigan 2003, 2006; *C. olivaceus*, Valderrama et al. 2000a). Given the rather divergent mating systems seen across the genus, the consistency of the positive correlation between male dominance rank and reproductive success is surprising. Although a distinctive male dominance hierarchy is discernable in *C. capucinus*, male/male intragroup competition for access to mates has not been documented. The alpha male does not monopolize matings, nor does he appear to be the exclusive target of female choice (Fedigan 1993); and all males within the group experience some degree of mating success (e.g., Rose 1998). However, paternity analysis in the Santa Rosa study groups has demonstrated that alpha males sire the majority of infants born into their groups (Jack and Fedigan 2003, 2006). Consequently, regardless of the apparently egalitarian mating system of this species, reproduction is not shared among group males (see Jack and Fedigan 2006). A recent study of *C. capucinus* by Carnegie et al. (2006) has shown that matings between females and alpha males occur during fertile periods, while those between females and subordinate males tend to occur during nonfertile times (when females are pregnant, lactating, or in the postovulatory phase). However, it is still unclear which sex is responsible for the timing of fertile matings, that is, whether females select to mate with alpha males during their fertile times or if a more subtle form of mating competition occurs and alpha males somehow exclude subordinates from mating with fertile females.

The two other well-studied species of *Cebus* (*C. apella* and *C. olivaceus*) show tense relationships among group males, and although both species reside in multimale groups, they are described as being functionally unimale, with the dominant male monopolizing matings and subordinate males being peripheral group members (e.g., *C. apella*, Janson 1985b, 1986; *C. olivaceus*, Robinson 1988a, O'Brien 1991). Reproductive success within groups of these two species largely reflects the observed social patterns, with alpha males monopolizing reproductive success (*C. apella*, Escobar-Páramo 2000; *C. olivaceus*, Valderrama et al. 2000a), although, at least for *C. apella*, beta males also sire

offspring, particularly with the daughters of alpha males (Escobar-Páramo 2000).

CEBINE DISPERSAL PATTERNS: TOWARD AN EXPLANATION OF VARIABLE SOCIAL STRUCTURE

The dispersal of one or both sexes from the birth group is a pattern common to all social mammals and birds, and many species show a bias toward the dispersal of one sex over the other (Greenwood 1980). The majority of Old World cercopithecines and prosimians are characterized by male-biased dispersal and female *philopatry* (the tendency to remain in the birth group for life), while New World monkeys and apes tend toward bisexual dispersal (e.g., *Alouatta* and *Gorilla*) or female-biased dispersal (e.g., *Ateles* and *Pan*) (see Pusey and Packer 1987 for review, Strier 1999). Dispersal is said to be among "the most important life history traits involved in both species persistence and evolution" (Clobert et al. 2001); however, despite its importance and the fact that it is one of the most studied phenomena, it remains among the most poorly understood issues in ecology and evolutionary biology (Clobert et al. 2001). Without a doubt, dispersal patterns have a profound influence on the type and nature of social relationships within groups. For example, in species characterized by male-biased dispersal, females within groups are related, most often show close bonds, and regularly form matrilineally based dominance hierarchies (e.g., *Macaca mulatta*, *M. fuscata*, and *Papio cynocephalus*). Within female philopatric groups, males are generally considered to be unrelated and their social relationships are usually characterized by very low levels of affiliation and cooperation. In general, the opposite pattern of social relationships is observed when males are philopatric (e.g., chimpanzees and muriquis, although bonobos are an exception to this pattern), while those species characterized by bisexual dispersal most often show much looser bonds among same-sexed individuals and closer bonds between the sexes (e.g., gorillas) (for reviews see Pusey and Packer 1987, van Hooff 2000).

Unlike most platyrrhine species, *Cebus* are characterized by female philopatry and male dispersal from the birth group; in this respect, they more closely resemble Old World monkeys. In general, *Cebus* are considered female-bonded (Wrangham 1980) or resident nepotistic (Sterck et al. 1997) in that (*1*) they exhibit female-biased philopatry and male dispersal, (*2*) affiliative bonds among females are generally stronger than they are between the sexes or among males, (*3*) females develop dominance hierarchies, and (*4*) females appear to be responsible for the direction of group movement (Fragaszy et al. 2004b; but see Phillips and Newlon 2000 on *C. albifrons trinitatis*). In addition, field studies are shedding light on the importance of kinship in the formation of female/female relationships and, although it may not be as decisive a factor as observed among the Old

World cercopithecines, this does appear to play a prominent role in at least some species of *Cebus* (*C. capucinus*, Rose 1998, Perry 1996; *C. olivaceus*, O'Brien and Robinson 1991). In terms of relationships between the sexes, males are generally individually dominant over adult females for all species; however, at least for the three better-studied species (*C. apella*, *C. capucinus*, and *C. olivaceus*), an alpha female will often rank directly below the alpha male and dominate individual subordinate males within the group (Fragaszy et al. 2004b). In addition, females will often form coalitions that enable them to displace even the alpha male from feeding trees and, at least for *C. capucinus*, females tend to direct more threats toward males than they receive from them (Perry 1997, Rose 1998). Overall, this results in somewhat egalitarian relationships between males and females; and across the genus, these relationships are best characterized as being affiliative in nature, with little physical aggression being exchanged between the sexes.

Throughout the genus, social relationships among group members are maintained through the exchange of frequent grooming, the maintenance of proximity, and the frequent formation of coalitions with preferred partners (Fragaszy et al. 2004b). Although social relationships among females and between the sexes across the better-studied *Cebus* species are remarkably consistent, male capuchins display extensive variation in their relationships with one another. Male/male relationships within the genus range from despotic (*C. apella*, Janson 1986; *C. capucinus*, Perry 1998b; *C. olivaceus*, Robinson 1988b) to highly affiliative and cooperative (*C. albifrons*, Janson 1986; *C. capucinus*, Jack 2003a), and, at least for *C. apella* and *C. capucinus*, which have been studied the most extensively and at multiple sites, this variability in male/male relationships exists both within and among species. For example, the first published accounts of male relationships in *C. apella* in Peru reported that alpha males directed high rates of aggression toward subordinate males, which remained on the periphery of the group, avoiding interactions with the alpha male and resident females (Janson 1985a 1986). However, additional studies on *C. apella* at multiple sites have yielded a more diverse view of male relationships within this species. In Colombia (Izawa 1980, 1994), Brazil (Lynch et al. 2002), and Suriname (Kauffman et al. 2004), males are reported to display low rates of aggression and frequently interact affiliatively, while in Argentina (Janson 1998a and personal communication), male *C. apella* exhibit greater levels of cooperation in resource defense, although relationships between dominants and subordinates are still described as agonistic.

Similarly, male relationships in *C. capucinus* have been reported as highly variable, and this variability occurs not only across study sites but also among groups within the same study population (e.g., Santa Rosa National Park in Costa Rica, where studies have been ongoing since 1983; see Jack 2003a). In terms of access to mates and resources, relationships among males are fairly egalitarian, with little

aggression being exchanged among group males. Although a distinct alpha male is discernable within groups, it is the extent to which he asserts his dominance that appears to be most variable. For example, some alpha males actively, and often aggressively, disrupt affiliative interactions among subordinate coresident males and spend the majority of their time affiliating with group females (Perry 1998a,b), while other alphas form close affiliative relationships with subordinates and, in some groups, spend more time interacting with them than with group females (Jack 2003a).

This diversity in male relationships across *Cebus* likely reflects the behavioral plasticity characteristic of the genus (see Fragaszy et al. 1990), and within species, male relationships are apt to change in response to ecological and/or social pressures (e.g., Janson 1998a, Di Bitetti and Janson 2001a). Ecological factors appear to play an important role in determining male/male relationships in *C. apella*, which has now been studied in five countries throughout South America (Argentina, Brazil, Colombia, Peru, and most recently Suriname). Janson (1986) suggests that it is the distribution of food resources and a male's ability to monopolize access to them that most profoundly influence primate mating systems, which in turn dictate the types of social relationships among group males. Janson (1986) argues that if a single male is able to dominate individual access to resources, it is likely that females will choose to mate with him to ensure resource access for themselves and their infants. Under these circumstances, the mating system essentially becomes unimale. Such a skewed mating system results in rather tense male/male relationships, with males tending to be less inclined to cooperate in group and resource defense when there is little reward for their assistance (i.e., little mating activity) (see van Hooff 2000). If, on the other hand, resources are distributed in such a way that they cannot be monopolized by a single individual (e.g., they occur in large patches), females need not choose to mate exclusively with the alpha male and a more egalitarian mating system may result. This mating system may lead to increased male cooperation in resource defense (Janson 1984), although it may also result in increased within-group mating competition and does not necessarily trigger affiliative relationships among group males.

Such differences in ecological factors, namely the defensibility of food resources, do not, however, explain the intergroup variability in relationships among coresident male *C. capucinus* that is reported to occur within the same study populations (see Jack 2003a; S. Perry, personal communication). It appears that within groups of *C. capucinus*, male familiarity, and perhaps kinship, best explains the observed variability (Jack 2003a). Although *C. capucinus*, like all *Cebus* species, is characterized by the emigration of males from their birth group, high rates of parallel dispersal can, in the absence of philopatry, promote the retention of kinship among group males (van Hooff 2000). Parallel dispersal can occur through the coordinated emigration of male siblings or the movement of males toward groups that contain familiar, previously dispersed males. This type of coordinated male dispersal has been reported for several primate species, although in most species it appears to be limited to particular life phases and to be most common among immature males (see Jack 2003b, Schoof et al. 2009). However, studies of *C. capucinus* in Santa Rosa National Park, Costa Rica, found that parallel dispersal occurs at extremely high levels (67%–80% of all emigrations depending on male age class), lasts through multiple dispersal events, and persists at high rates even among adult males (Jack and Fedigan 2004a,b). In addition, analysis of male/male interactions in this species has shown that familiar males do display more affiliative relationships (Jack 2003a), and Janson (personal communication) suspects that male familiarity and/or relatedness may well account for the more cooperative relationships he has observed among male *C. apella* in Argentina. *C. apella* groups in Argentina contain more males than those Janson first studied in Peru, which may enable parallel dispersal to occur more readily. Relatedness among group males in both *C. apella* in Argentina and *C. capucinus* in Costa Rica, and perhaps in *C. olivaceus* (although very little is known about the dispersal patterns of this species), may explain the high degree of cooperation among group males in the face of the unimale reproductive system each of these species exhibits (see earlier section "Mating Systems"). We eagerly await genetic analyses of male kinship within groups so that these predictions can be tested.

Despite the remarkable similarity that *Saimiri* species display in terms of morphology and general habitat preference, the genus shows an even greater diversity in social relationships than we see in *Cebus*. However, across the three species for which sufficient data are available (*S. oerstedii*, *S. boliviensis*, and *S. sciureus*), it is the social relationships among group females and between males and females that are extremely variable, while relationships among group males are surprisingly consistent (Table 8.4). Boinski, who has studied this genus extensively over the past two decades, claims that "squirrel monkeys arguably exhibit the most geographically variable social organization of any set of closely related primate populations" (1998:179). The variability that we see across the genus appears to directly reflect the extremely divergent dispersal patterns observed among the species (see Boinski et al. 2005a,b; Boinski 2005), which are likely determined by the nature and distribution of resources each species exploits. Interestingly, the formation and maintenance of social bonds throughout this genus are not based on grooming interactions, as is the case for many nonhuman primates. In fact, *Saimiri* are among the few primate species where social grooming is almost completely absent. Instead, social relationships are based on proximity patterns, tolerance around feeding sources, the frequent exchange of vocalizations, and the formation of alliances in some species (see Sussman 2000 for review; Boinski 1999).

S. oerstedii in Costa Rica are characterized by female emigration from the birth group (Boinski and Mitchell 1994) and a flexible pattern of male philopatry. While females typically disperse as juveniles before their first mating season, secondary dispersal is not uncommon and appears to be linked to the survivorship of offspring. In one long-term study, all females whose offspring died transferred to a new group before the start of the next mating season (see Boinski et al. 2005a). A natal male will generally remain in his birth group until he reaches sexual maturity (4–5 years), after which he will either take up one of the few reproductive positions within the group or disperse with members of his age cohort and attempt to take over breeding positions in another group (Boinski 1998). No matter which pattern is followed, it is thought that resident males of this species are related as a result of parallel dispersal. This dispersal pattern leads to strong affiliative bonds among group males, while female bonds are described as weak (Boinski 1999, Boinski and Mitchell 1994) and relationships between males and females are described as being extremely egalitarian, with neither males nor females being dominant over the other (Boinski 1987a, 1988a). With the exception of the 2-month period prior to the mating season when males are observed to form coalitions and frequently to mob females to perform genital inspections (see earlier section "Mating Systems"), aggression within and between the sexes very rarely occurs. Even intergroup interactions are described as neutral, although avoidance is generally practiced (Boinski, 1987a, 1988a).

In contrast, *S. boliviensis* in Peru display female philopatry; and accordingly, although perhaps more so than any other species of platyrrhine primate, females form tight-knit matrilineal relationships that cooperate in resource acquisition and defense (Mitchell, C. L., 1990, 1994; Mitchell et al. 1991). Intragroup resource competition is described as occurring at moderate levels within this species, and it occurs both among and between the sexes (Boinski 1999). Interestingly, despite the fact that this is the most sexually dimorphic of the *Saimiri* species, with males being 24% larger than females (Table 8.2), females form stable linear dominance hierarchies, dominate males, and aggressively force them to occupy peripheral positions within the group (Mitchell, C. L., 1990, 1994). Within these peripheral subgroups, males also form stable linear dominance hierarchies and, although social aggression is described as common (Boinski 1999), affiliative coalitions that are often maintained through multiple emigrations (i.e., parallel dispersal) may work to ensure kinship among group males (Mitchell, C. L., 1994). *S. boliviensis* males emigrate with other natal males of the same age and will typically join an all-male band or bachelor group before attempting to infiltrate a new group. More often than not, males enter a new group which contains familiar males from their natal cohort (Boinski et al. 2005a). *S. boliviensis* groups have home ranges that overlap extensively, and intergroup interactions are nonaggressive; in areas where fruit patch size allows, it is not uncommon to see groups foraging together (Mitchell 1990).

S. sciureus presents yet another configuration of dispersal patterns, social interactions, and dominance structures for the genus. Both male and female *S. sciureus* disperse from their birth group (Boinski et al. 2005a). Female natal emigration may occur before or after the first mating season and appears to be linked to relative access to food and overall food abundance (Boinski 2005). The dispersal of natal male *S. sciureus*, on the other hand, usually occurs well before sexual maturity. However, those males who delay natal dispersal are often forcibly evicted as they approach sexual maturity (Boinski et al. 2005a). *S. sciureus* display the highest level of intragroup aggression reported among the three *Saimiri* species. The majority of this aggression occurs within the context of resource acquisition, and wounding of females and immatures is common, something which is rarely seen in the other two species (Boinski et al. 2002). Overall, *S. sciureus* is best described as being male-dominant; and although both males and females form stable linear dominance hierarchies, males are reported to be more closely bonded in that they form more frequent cooperative coalitions with moderate to close affiliative relationships (Boinski 1999). However, the majority of males spend their adult life as solitary or peripheral individuals, and very few achieve secure resident status within a group (Boinski et al. 2005a). Unlike the other two species of *Saimiri*, which do not display tense interactions with neighboring groups, *S. sciureus* in Suriname are described as territorial. Their home ranges show minimal overlap with other groups, and intergroup interactions are described as highly agonistic (Boinski 1999).

In an attempt to explain the variable patterns of female relationships seen across this genus, Boinski et al. (2002) provide an extensive examination of the detailed ecological data available for the three species of *Saimiri*. Following Janson (1986) and his explanation of variable relationships among *Cebus* males, Boinski et al. (2002) suggest that the extreme variability in dispersal patterns and the corresponding social relationships among group members (at least among group females) that characterize *Saimiri* are largely attributed to ecological factors associated with the defensibility of food resources, namely fruit patch size and abundance. As mentioned, *Saimiri* are classified as insectivore–frugivores and insects form the bulk of their diets, with all species spending an average of 75%–80% of their time foraging for insects (Terborgh 1983, Boinski 1988a). The distribution of arthropods does not enable a single individual, or even groups of individuals, to effectively monopolize the food supply; therefore, direct competition (*contest competition*) for this resource is absent (Boinski 1988a, Mitchell 1990). However, fruit is also a key resource exploited by *Saimiri*; and even though the habitats of the three well-known species are quite uniform in terms of relative abundance and distribution of arthropods, these habitats do display extensive variability in fruit production.

A comparison of fruit patch size and abundance demonstrates that, overall, Peruvian forests have the highest levels of fruit abundance and the lowest fluctuations in seasonal availability of fruit, followed by Costa Rican forests, while Surinamese forests show the lowest abundance and greatest seasonal fluctuations (Boinski et al. 2002). Hence, it is not surprising that in Peruvian forests, where food patches are too large to be effectively monopolized by a single individual, females cooperate with kin to defend access to these resources and exclude males. Cooperation is necessary because males are larger and, therefore, it is beneficial for females to remain philopatric and cooperate with their kin to control access to resources (Boinski et al. 2002; see also Sterck et al. 1997). In Costa Rica, where fruit is distributed in such a way that it cannot be monopolized by either a group or a single individual (it can be eaten on the spot on a first-come, first-served basis), it is not advantageous for females to remain in their birth group, a condition which most often leads to male philopatry. In the Surinamese forests, fruit patches are small and in relatively low abundance, making an individual patch easily monopolized by single individuals and of great benefit for those that can gain access to them, namely dominants. This type of distribution leads to very strong contest competition; therefore, it makes sense for a female to disperse so that she is not directly competing with kin for access to necessary resources. According to Boinski et al. (2002), this type of situation will not promote stable coalitions among individuals because cheating (i.e., reneging on sharing resources with a coalition partner) would provide great benefits to dominant individuals.

Boinski and Cropp (1999) state that, across the three species, relationships among male squirrel monkeys are rather consistent—males show affiliative bonds—while female/female bonds vary from strong to weak and male/female relationships also vary from species to species. They provide a very good example of how even slight differences in ecological factors (e.g., quality, size, and defensibility of food patches) can have profound effects on primate social structure and how these factors influence the variable nature of female/female and female/male relationships across the genus. They do not, however, offer an explanation for the consistency of male/male relationships in the face of the vast ecological variability that can be seen across the genus. Familiarity and/or relatedness among group males likely explains the relative consistency in male/male relationships across the genus *Saimiri*. It seems likely that in all three species males within groups maintain some degree of relatedness to one another. In *S. oerstedii*, males either are philopatric or disperse together in cohorts, which also appears to be the case for male dispersers in *S. sciureus* and *S. boliviensis*. In *S. boliviensis*, where male emigration is the norm, relatedness among group males also appears to be maintained through parallel dispersal. C. L. Mitchell (1994) reports that males form dispersal alliances whose composition often remains constant over multiple emigration events, and this may act to ensure the relatedness of males in the face of dispersal. These data show that, regardless of ecological pressures and female dispersal patterns, males in the genus appear to have evolved a mechanism for retaining residence with their kin, regardless of dispersal patterns. A recent meta-analysis of parallel dispersal by male nonhuman primates found that this behavior may have evolved in species where it is important for males to retain coalition partners in the face of male dispersal (Schoof et al. 2009).

Additional comparative studies are needed to further our understanding of ecological pressures on social relationships, mating systems, and dispersal patterns. In addition, genetic data on the relatedness of group members (both males and females) and more detailed reports on dispersal patterns and the fates of dispersing individuals in other species are required in order to attain a more complete overview of the dynamics and interactions that these processes have on primate behavior.

Species included in this chapter are shown in Color Plates 9, 10, and 17 (*Cebus capucinus, Cebus apella*, and *Saimiri sciureus*).

9

Sakis, Uakaris, and Titi Monkeys
Behavioral Diversity in a Radiation of Primate Seed Predators

Marilyn A. Norconk

1. What is seed predation, and how do seeds provide both benefits and costs to sakis, bearded sakis, and uakaris?
2. How do pitheciines differ in their social group and ranging patterns using body size and group size as relevant variables?
3. How is male affiliation manifested in pitheciines and how does it compare with woolly spider monkeys, chimpanzees, and bonobos?"

INTRODUCTION

The pitheciines are a cohesive group of New World monkeys phylogenetically and in terms of diet but exhibit a range of variation in group size, use of space, and social dynamics. At one end of a continuum that ranges from small to large group sizes, titi monkeys (*Callicebus* spp.) form cohesive, pair-bonded groups that in many ways represent the "classic monogamous" pattern (Fuentes 1999b, van Schaik and Kappeler 2003) (Table 9.1). Pairs are generally territorial, adults are monomorphic in body size and color. They exhibit social and physiological mechanisms that promote and reinforce attachment between mates, and males are strongly paternalistic (Mason 1968, 1971; Fragaszy et al. 1982; Kinzey 1981; Menzel 1986; Mendoza and Mason 1986a,b; Mason and Mendoza 1998; Schradin et al. 2003; Bales et al. 2007).

Bearded sakis (*Chiropotes* spp.) and uakaris (*Cacajao* spp.) are at the other end of the group size continuum. They form large, more loosely structured groups that may fission into smaller feeding parties (Ayres 1986; Norconk and Kinzey 1994; Kinzey and Cunningham 1994; Defler 1999a, 2003a; Veiga 2006; Bowler et al., 2009). Groups travel through large home ranges and day ranges are as long as those of any platyrrhine (Ayres 1981, 1986; Norconk and Kinzey 1994; Aquino 1998; Boubli 1999; Defler 1999a; Peetz 2001; Barnett et al. 2002). Males do not take an active part in infant care. Bearded saki and uakari males are generally larger than females (Table 9.2) and exhibit sex-specific characteristics of robusticity such as enlarged temporal muscles (Figs 9.1 and 9.2). Both male and female bearded sakis exhibit well-developed beards upon sexual maturity and colorful, relatively large external genitalia (pink scrotum

Figure 9.1 *Cacajao calvus ucayalii* adult male (photo by Mark Bowler with permission, Lago Preto Conservation Concession, Peru).

Figure 9.2 *Chiropotes sagulatus* adult male (photo by M. Norconk, Paramaribo Zoo, Suriname).

and enlarged, pink labia majora) (van Roosmalen et al. 1981, Peetz 2001).

Members of the genus *Pithecia* are intermediate between titis and bearded sakis/uakaris. Groups are usually reported as being small (Table 9.1), and many, but not all, conform to the "two-adult group" suggested by Fuentes (1999b). White-faced sakis (*Pithecia pithecia*)

exhibit aggressive intergroup behavior both in the wild and in captivity (Homburg 1997, Savage et al. 1992, Shideler et al. 1994, Norconk et al. 2003). Mothers are primary caregivers (Brush and Norconk 1999), but adult male interest increases as infants mature; and adult males have been observed to play and share food with older infants (Buzzell and Brush 2000). Ryan (1995) reported that white-faced saki males carried infants, but Homburg (1997) did not find that to be the case. The expression of sexual dichromatism in sakis varies from subtle variation in the western (Amazonian) species to striking pelage differences in the eastern (Guiana Shield) *P. pithecia* ssp. (Hershkovitz 1987b, Gerald 2003) (Fig 9.3). Small group size, small body mass, and territorial behavior ally *Pithecia* with *Callicebus*, whereas diet and dental anatomy, relatively low level of male infant care, and bushy tails ally them with *Chiropotes* and *Cacajao*.

SYSTEMATICS AND GEOGRAPHIC DISTRIBUTION OF THE PITHECIINES

Callicebus (Titis)

In the process of preparing the second volume of *Living New World Monkeys (Platyrrhini)*, Hershkovitz wrote six taxonomic reviews of non-prehensile-tailed platyrrhines including the four genera reviewed here (Hershkovitz 1985, 1987a,b, 1990). Of the four genera of the Pitheciinae, *Callicebus* is the largest and the most complex group. Ten taxa distributed in two species were recognized in an earlier review (Hershkovitz 1963), but in 1990, after examining nearly

Table 9.1 Pitheciine Group Size, Group Composition, and Use of Space

SPECIES AND STUDY LOCATION AND LATITUDE, DURATION OF STUDY*	GROUP SIZE (N)	GROUP SIZE RANGE (N)	ADULT MALES	ADULT FEMALES	JUVENILES + SUBADULTS	INFANTS	HOME RANGE SIZE (HA)	DAY RANGE (KM)	POP. DENSITY (INDIVID./KM²) OR INDIVID. SIGHTING RATE (SR)/10 KM	SOURCE
Callicebus brunneus Peru (c. 11° S)–ST		2–5 (6)	1(–2?)	1	0–3		1.4			Ferrari et al. 2000, Lawrence 2003
Callicebus brunneus Peru (c. 11° 51' S)–LT	4.6 (n = 6)	2–5					6.8 (n = 6)	349–1108	25–30/km²	Wright 2009
Callicebus brunneus Bolivia (10° 35' S to 11° 31' S)–C	3	1–5 (39)								Buchanan-Smith et al. 2000
Callicebus brunneus Brazil (10°–12° S)–C	2.25 ± 0.97 (118)	max = 5								Ferrari et al. 2000
Callicebus caligatus Brazil (c.10 °S)–C	2.20 ± 0.84 (5)	max = 4								Ferrari et al. 2000
Callicebus cinerascens Brazil (c. 12 ° S)–	1.67 ± 0.33 (3)	max = 2								Ferrari et al. 2000
Callicebus cupreus Peru (5° 35' S)–	3.0–3.73	2–7 (18)							14.6/km²	Bennett et al. 2001
Callicebus donacophilus Brazil–(c. 13° S)–	2.0 ± 1.0	max = 3								Ferrari et al. 2000
Callicebus modestus Peru (~11° S)–C	3.0	1–6								Martinez and Wallace 2007
Callicebus moloch Brazil (c.12° S)–	2.47 ± 0.92 (15)	max = 4								Ferrari et al. 2000
Callicebus moloch Peru (11°52')–T	4.1 (2)	2–5 (2)	1	1	1–2	0–1	6–8	0.55 ± 0.1, 0.67 ± 0.2	20–26/km²	Wright 1984,1985,1986
Callicebus moloch Peru (11°52')–T	4.0	2–7					11.5 (6–18)			Bossuyt 2002
Callicebus olallae Peru (~11° S)–C	2.7	1–5								Martinez and Wallace 2007
Callicebus oenanthe Peru (6°01 S)–ST	4.5ᵃ	4–5	1	1	2	1				DeLuycker 2006
Callicebus oenanthe Peru (5° 58 S)–RC	3.6	1–8							113/km²	Aldrich et al 2008
Callicebus ornatus Colombia (c. 4° N)–T	3.2 (9)	2–4 (9)	1	1	0–1	0–1	4.4 (3.2–5.1)	0.57 (0.3–0.87)		Mason 1968
Callicebus ornatus Colombia (5° N)–LT							3.5–14			Defler 1994
Callicebus personatus melanochir Brazil (15° 18' S)–LT		2–6ᵃ	1	1	1–4		24	1.0		Müller 1996
Callicebus personatus personatus Brazil (19° S)–ST		6 (1)	1	1	3	1	4.7	0.69 ± 0.04 (0.52–0.80)		Kinzey 1981, Kinzey and Becker 1983
Callicebus personatus personatus Brazil (c. 20° S)–ST		3–5 (2)	1(–2?)	1(–2?)	1	1	10.7–12.3	1.0 ± 0.2 (0.8–1.3)		Price and Piedade 2001
Callicebus torquatus torquatus Peru (4° S)–LT		3–5ᵃ	1	1	1–2	1	29 (4–30)	0.8 ± 0.04 (0.5–1.4)	16/km²	Kinzey 1978, 1981
Callicebus torquatus Brazil (0.24N)–LT									1.6/km²	Boubli 2005
Callicebus torquatus Peru (2° 29S and 4° 15S)–C	2.9	2–5 (39)	1	1					2.5 to 2.8/km²	Aquino et al. 2008
Callicebus torquatus lugens Colombia (1° 5.55' S)–RC	4.8 (10)	3–5 (10)	1	1	0–1	1	14.2 (9–22)		6.08 (8 groups)/ km²	Defler 1983,2003a

Table 9.1 (cont'd)

SPECIES AND STUDY LOCATION AND LATITUDE, DURATION OF STUDY*	GROUP SIZE (N)	GROUP SIZE RANGE (N)	ADULT MALES	ADULT FEMALES	JUVENILES + SUBADULTS	INFANTS	HOME RANGE SIZE (HA)	DAY RANGE (KM)	POP. DENSITY (INDIVID./KM²) OR INDIVID. SIGHTING RATE (SR)/10 KM	SOURCE
Cacajao calvus calvus Brazil (3° 22' N)–LT		30–48 (3)	16	16	13	3	500–550	2.5–5.0	7–8/km²	Ayres 1986,989
Cacajao calvus Peru (4° 23' S, 4° 30' S)–ST	41.9 ± 16.7 (21)[b]	8–70								Aquino 1998, Aquino and Encarnación 1999
Cacajao calvus Peru (4° 27' S)–LT	43.5 ± 24.1	max = 150+	1/3	1/3	1/3[c]				14.3–116.8/ km²[c]	Bowler et al. 2009
Cacajao calvus Peru (5° 35' S)–C	15.8–33.1	2–55 (18)							7.44 and 25.78/km²	Bennett et al. 2001
Cacajao melanocephalus [hosomi] Brazil (00° 24' N)–LT		c. 70						4.4	14/km²	Boubli 1999, 2005
Cacajao melanocephus ouakary Colombia (1° 5.55' S)–RC	20–30	1–108+							4.15/km² overall, 12.0 in igapó habitat	Defler 2001
Cacajao melanocephus ouakary Brazil (1° 5.55' S)–LT		5–30 (wet season), 1–100+ (dry season)								Barnett et al. 2005
Chiropotes albinasus Brazil (10° 10' S)–LT		19–26 (4)	8	9	8		250–350	2.5–3.5	10–11/km²	Ayres 1981,1989
Chiropotes albinasus Brazil (9–13° S)–C	4.2 ± 3.1 (7)[d]								1.6–2.5 (SR)	Ferrari, et al. 1999a–c
Chiropotes chiropotes Venezuela (7° 21' N)–LT		15–22 (1)[a]	1–2	8–10	3–5	1–5	180	1.6 (0.5–2.7)		Norconk 1996, Peetz 2001
Chiropotes sagulatus Suriname (4°41' N) and Brazil (2° N)–LT		8–27+ (4)	8	9	5	2–3	200–250	2.5		van Roosmalen et al. 1981, Ayres 1981
Chiropotes sagulatus Brazil (2° 30' S)–LT							429 ± 129.4	2.99 ± 0.02		Boyle et al. 2009a,b
Chiropotes sagulatus Suriname (4°41' N)–ST	9 and 13 (2)							3.2 ±1.1		Norconk and Kinzey 1994
Chiropotes sagulatus Suriname (5° 01' N)–C	32.7 (3)	22–44 (3)							37.4 (SR)	Norconk et al. 2003
Chiropotes satanas Brazil (multiple sites, 1–5 ° S)–C									1.8–10.08, 0.3 (SR)	Ferrari and Lopes, 1996
Chiropotes satanas, Brazil (4° 30' S)–ST		30–34 (1)	12	9	16–20	0				Silva and Ferrari 2009
Pithecia aequatorialis Peru (2° S)–C		1–7 (4)								Heymann et al. 2002
Pithecia albicans Brazil (4° 51' S)–RC	4.6 ±1.5 (5)	3–7					172.4 (147–204)		4.1	Peres 1993a–c
Pithecia hirsuta Peru (c. 5° S)–RC	3.8	2–8	1–3	1–2			24.9 (9.7–42)		12.8/km²	Soini 1986

126 PART TWO The Primates

Table 9.1 (cont'd)

SPECIES AND STUDY LOCATION AND LATITUDE, DURATION OF STUDY*	GROUP SIZE (N)	GROUP SIZE RANGE (N)	ADULT MALES	ADULT FEMALES	JUVENILES + SUBADULTS	INFANTS	HOME RANGE SIZE (HA)	DAY RANGE (KM)	POP. DENSITY (INDIVID./KM²) OR INDIVID. SIGHTING RATE (SR)/10 KM	SOURCE
Pithecia irrorata Brazil (9°-13° S)–C	2.68 ±1.38								1.3 (SR)	Ferrari et al. 1999a–c
Pithecia irrorata Bolivia (10° 35'S to 11° 24'S)–C	3.5	2–5 (6)								Buchanan-Smith et al. 2000
Pithecia monachus Peru (5° 35' S)–C	3.75	2–5 (16)							9–17.2/km²	Bennett et al. 2001
Pithecia pithecia chrysocephala Brazil (2° 25'S)–LT	6 (1)	4–7a	1	1–3	0–2	0–2				Setz and Gaspar 1997, Gilbert and Setz 2001
Pithecia pithecia chrysocephala Brazil (c. 2° S)–ST	2.6 ± 0.5	2–3	0–2	0–2	0–1					Oliveira et al. 1985
Pithecia pithecia French Guiana (4° N)–LT	2.8 ± 1.0 (4)	1–4								Kessler 1998
Pithecia pithecia Guyana (various)–C	4.8 ± 2.4 (21)	2–12	2.0	1.8	1.0 (m) 1.3 (f)	1.0				Lehman et al. 2001
Pithecia pithecia Guyana (various)–C	3.3 ± 1.7 (10)	1–5								Muckenhirn et al. 1975
Pithecia pithecia Suriname (4° 41'N)–RC	2.7 ± 0.8	9	1–2	1	0–1					Mittermeier 1977
Pithecia pithecia Suriname (5° 01' N)–C	3.7 (10)	2–6	1–2	1–3	1		10.3		14.1 (SR)	Norconk et al. 2003
Pithecia pithecia pithecia French Guiana (5° 04' N)–ST	2.3 (35)	1–5						1.88	0.64/km²	Vié et al. 2001
Pithecia pithecia pithecia Venezuela (7° 21'N)–LT	9	5–9	3	2	2	2	15e			Homburg 1997
Pithecia pithecia pithecia Venezuela (7° 21'N)–ST	6							1.77 ± 0.27		Cunningham 2003, Cunningham and Janson 2006
Pithecia pithecia pithecia Venezuela (7° 21'N)–LT	6.9 ± 1.4 (1)a	5–9	1–4	2–3	0–2 (m), 0–2 (f)	0–2	12.8e	1.5		Norconk 2006

*C = census; RC = repeated census; ST = < 6 months of observation; LT = > 6 months of observation.

a Range = the change in group size of one group

b Lower number of range estimates used to calculate mean ± standard deviation. The count (100–120) of multiple groups was excluded.

c Estimated proportion of adult males, adult females, and juveniles and infants in the entire population at Lago Preto; population density in 1999 and 2007, respectively.

d An average of mean group sizes is reported (n = 7 locations).

e Island-bound groups.

1,200 specimens, Hershkovitz raised the number of taxa in the genus *Callicebus* to 25, which he distributed into 13 species (Hershkovitz 1990). van Roosmalen et al. (2002) reexamined the genus, named two new species, raised all subspecific taxa to species level, and divided the genus into five groups—for a total of 28 species. Wallace et al. (2006) added yet another species. On one level, *Callicebus* is a highly variable and wide-ranging genus with interspecific differences in body size, pelage color, cranial dimensions, and chromosomal number (van Roosmalen et al. 2002). At another level, the genus seems remarkably uniform, particularly in mating and parenting strategies. Until recently, pelage color was treated

Table 9.2 Social and Reproductive Characteristics of the Pitheciines

SPECIES	MALE BODY MASS IN GRAMS	FEMALE BODY MASS IN GRAMS	GROUP SEX RATIO (F/M)	AGE AT SEXUAL MATURITY (MONTHS)/ INTERBIRTH INTERVAL (MONTHS)	BIRTH PEAK (MONTHS) AND SEASON: WET OR DRY	PATERNAL CARE	TERRITORIAL INTERACTIONS WITH CONSPECIFIC GROUPS	LONG DISTANCE VOCALIZATIONS	SOURCE
Callicebus brunneus				c. 36 / 12	Copulations: Feb–June; pregnancy during dry season	Males carry, share food, and play with infants	7 encounters in 15 months of observation (females participated in 2:7)	90% calls before 0900 (calls on 15 mornings a month)	Wright 1985, 1986, 2009
Callicebus ornatus	1,178	1,163	1.0				Calling, chasing, piloerection, tail-lashing, chest-rubbing (5%–7% HR overlap); rate = 1.67/day	Regular dawn calls and during ITEs	Mason 1968, Hershkovitz 1990
Callicebus discolor	935	1,075		42/11.8	Winter: Dec–Mar (Davis, CA)				Hershkovitz 1990, Valeggia et al. 1999
Callicebus moloch	1016	877							Hershkovitz 1990
Callicebus ornatus	845	850	1.0		Nov–March	Males carry infants			Mason 1968, Hershkovitz 1990
Callicebus personatus	1,270	1,378			Oct (dry)		Vocal responses on 36 occasions to distant groups	Dawn calls given from sleeping trees	Hershkovitz 1990, Kinzey and Becker 1983
Callicebus personatus personatus							Rare: vocalizations only; no chasing, physical contact or threat displays	97% before 1000; not daily	Price and Piedade 2001
Callicebus torquatus torquatus	1,110	1,310		?/c. 36	Nov–March	Males carry infants	Vocalizations and movement away from intruders (playbacks), occasional active encounters (10% overlap of HR)	Solo male calls and duets, from sleeping trees	Kinzey 1981, Hershkovitz 1990, Kinzey and Robinson 1983, Easley and Kinzey 1986
Callicebus torquatus lugens					Jan (dry)		Not daily, duetting 10–20 m apart from stable position		Defler 1983
Cacajao calvus calvus	3,450	2,880	1.20		Oct–Nov (dry)				Ayres 1986
Cacajao calvus ucayalii			Variesᵃ	?/24	Aug–Novᵃ				Bowler et al. 2009
Chiropotes albinasus	3,175 ± 185	2,518 ± 192	1.26						Ayres 1981, 1989
Chiropotes albinasus	3,700								Ferrari 1995
Chiropotes chiropotes				c. 36/24+	Dec–April (late wet to dry)				Peetz 2001
Chiropotes sagulatus	2880	2660	1.08						Ayres 1981
Chiropotes sagulatus					Wet season				van Roosmalen et al. 1981
Chiropotes sagulatus					Late dry and short wet				T. Gregory, personal communication
Chiropotes satanas			0.75		March–May (late wet to dry)ᵇ				Silva and Ferrari 2009

Table 9.2 *(cont'd)*

SPECIES	MALE BODY MASS IN GRAMS	FEMALE BODY MASS IN GRAMS	GROUP SEX RATIO (F/M)	AGE AT SEXUAL MATURITY (MONTHS)/ INTERBIRTH INTERVAL (MONTHS)	BIRTH PEAK (MONTHS) AND SEASON: WET OR DRY	PATERNAL CARE	TERRITORIAL INTERACTIONS WITH CONSPECIFIC GROUPS	LONG DISTANCE VOCALIZATIONS	SOURCE
Pithecia albicans	3000								Peres 1993a–c
Pithecia irrorata	2,010 (2,920)	1,875 (1,980–2,160)							Hershkovitz 1987b, Ford 1994a
Pithecia hirsuta				?/24–36	Sep–Dec (late dry to early wet)		Rare		Soini 1986
Pithecia monachus	2,795 (2,500–3,100)	1,900 (1,300–2,000)							Hershkovitz 1987b, Ford 1994a
Pithecia pithecia, Venezuela	1,732 (1,380–1,866)	1,515 (1,347–1,875)	Reg cycles (birth)/23.2 (15–34)	Nov–April (dry)		Play starting c. 4 months of age	Calling (roaring), chasing, piloerection, neck-rubbing, urine-marking; rate = 9/100 hr	During travel and intergroup encounters	Hershkovitz 1987b, Ford 1994a, Norconk 2006
Pithecia pithecia, Suriname					Oct–Jan (dry to short wet)		Primarily male		C. Thompson personal communication
Pithecia pithecia, Captive						Play and share food with older infants	Aggressive toward same-sex intruders		Shideler et al. 1994, Savage et al. 1992

[a] Population-wide, authors suggest equal numbers of males and females. Inverse correlation between group size and number of males. Smaller groups have proportionately more males or males separate from main group and spend more time in smaller groups, some of which are all-male groups. Births tend to fall in a period of low fruit availability (see text).
[b] Inferred; authors witnessed copulations from October to December.

Figure 9.3 *Pithecia pithecia* adult male (photo by M. Norconk, Brownsberg Nature Park, Suriname).

as if it were a uniform character and little attention was given to within-species variation (but see DeLuycker 2007, Aquino et al. 2008). In a study of two endemic species of Bolivia, *Callicebus modestus* and *C. olallae* (often classified as subspecies of *C. donacophilus*)Martinez and Wallace (2007:48) comment that "similar morphological features...together with variable lighting conditions during field observations make accurate identification a challenge." Nevertheless, they

suggest that pelage coloration provides the best cue to species identity.

In addition to variation in pelage color, titis exhibit small differences in body size (Table 9.2) and chromosome number ($2n = 20$ in *Callicebus torquatus*, 46 in *C. cupreus*, 48 in *C. brunneus* and *C. moloch*, and 50 in *C. dubius donacophilus*) (Hershkovitz 1990:Table 12). Molecular and morphological studies place *Callicebus* at the base of the pitheciine radiation, diverging from the rest of the platyrrhines about 15 million years ago (Schneider et al., 1993, Schrago, 2007, also see Kay et al., 2008).

At a proximate level, Casado et al. (2007) provide insight into how difficult it might be to sort out these species using molecular techniques. Using cytochrome *b* DNA sequence data, they sampled eight individuals (four *C. lugens* from the left bank and four *C. lugens* from the right bank of the Rio Negro—a distance of 190 km separating the two populations. Casado et al. (2007) estimated genetic distance to be much higher across large rivers than among animals on the same bank. Although their estimated time of divergence (2.2 mya) seems far-fetched, van Roosmalen et al. (2002) contend that rivers provide substantial dispersal barriers for titis, which are unlikely to cross major rivers except at their headwaters.

Callicebus spp. occupy much of the Amazon and Orinoco river basins, ranging from the foothills of the Andes to the

Atlantic Forest of southeastern Brazil (see maps in van Roosmalen et al. 2002). The only large-scale discontinuous distribution among *Callicebus* species occurs between *C. personatus* and *C. moloch* groups in central Brazil (Ferrari et al. 2000). Titis are absent from the Guiana Shield, apparently confined to the right banks of Rios Negro and Branco (Hershkovitz 1985); and their distribution is also limited by local or regional habitat variation or flooding in both *várzea* and *igapó* forests (van Roosmalen et al. 2002). In support of the view that titis express specific preferences for habitat types despite their broad distribution, Kinzey and Gentry (1979) proposed that *C. torquatus* inhabited white-sand terraces preferentially. This was questioned by Defler (1994) in Colombia but recently supported by Aquino et al. (2008) working in Peru. Interestingly, van Roosmalen et al. (2002) proposed that *Callicebus* species colonize new areas by either replacing a closely related ecospecies or accommodating sympatry in two species with different ecological preferences (i.e., they found no evidence for hybrid zones).

Pithecia (Sakis)

Pithecia spp., like *Callicebus*, have a broad geographic distribution and occupy a range of habitats from tropical wet to tropical dry forests throughout the Amazon Basin and north into the Guianas and eastern Venezuela (approximately to 9E° N to 14E° S latitudes; see Table 9.1, Figure 9.4). Hershkovitz (1987b) revised the systematics of *Pithecia* and divided them into two groups based on geographic distribution: a Guianan group (two taxa) and an Amazonian group (six taxa). The distinction between these groups is partly based on pelage color. A recent study of cranial morphology advocates raising all members of the Amazonian (Monacha) group to species status and leaving the Guianan (Pithecia) group with two subspecies (Marroig and Cheverud 2004). L. Marsh (personal communication) is in the process of revising the taxonomy of *Pithecia*, with an expected increase in the number of species, particularly in the western Amazon Basin.

Guianan sakis (white-faced or pale-faced sakis), found north of the Amazon River and ranging into the Guianas, are strongly sexually dichromatic: Male body pelage is entirely black with a white or yellowish face; females are gray–brown with white or off-white facial markings above the brow and malar stripes on the side of the face. Amazonian males found south of the Amazon River (Brazil) and in the western Amazon Basin (Colombia, Ecuador, Peru) exhibit a range of variation in facial and extremity color but lack the black–white contrast exhibited by Guianan males (see Figure 9.3). Females of all *Pithecia* spp. are very similar in appearance (Hershkovitz 1987b).

According to Ford (1994a), sexual dimorphism in body mass ranges from 1.1 to 1.4 for three species of *Pithecia* sakis (see Table 9.2). *P. albicans* nearly bridges the body mass gap between *Pithecia* and larger sakis/uakaris: *Chiropotes* and

Cacajao. Peres' (1993b) research on *P. albicans* in central Amazonia suggested that relatively larger body size compared with other *Pithecia* spp., in addition to a preference for higher-canopy travel and larger home ranges, may reflect competitive release in the absence of the two larger-bodied genera (*Cacajao* and *Chiropotes*).

Chiropotes (Bearded Sakis)

Bearded sakis (*Chiropotes* spp.) are found primarily in upland, nonflooded habitats in eastern Amazonia, both north and south of the Amazon River (Hershkovitz 1985, Walker 1996, Auricchio 1995) (Figure 9.5). In the Guianas, they are absent from the region west of the Essequibo River in Guyana and their presence in Venezuela may have been due to their ability to follow the right (south) bank of the Orinoco River from Brazil into southern Venezuela (state of Amazonas) and then eastward into the state of Bolívar. Their present eastern boundary in Venezuela appears to be of the Caroní River (Norconk et al. 1996b). In Brazil, they range north and south of the Amazon River, east of Rio Madeira, and throughout eastern Amazonia (66° to 44° W latitude) (Ferrari and Lopes 1996 and distribution maps therein). The Rio Tocantins provides a boundary between two subspecies of *Chiropotes satanas*, *Ch. s. satanas* and *Ch. s. utahicki*.

It is unclear why bearded saki distribution is spotty in French Guiana. They are not found at the Nouragues Research Station in French Guiana (Bongers et al. 2001) despite geological similarities to sites in Suriname and Guyana and high species diversity of one of their most important plant food families, the Lecythidaceae (Mori 1989). de Granville's (1982) description of the forests of southern French Guiana as xeric and scrubby may constitute a barrier to the northern migration of bearded sakis from Brazil. If vegetation or riverine barriers do exist for *Chiropotes*, they have not limited *Pithecia* dispersal into either Guyana or French Guiana. Having a wider tolerance for seasonally dry habitats, *P. pithecia* is apparently broadly distributed across the entire region.

Hershkovitz (1985) reviewed the systematics of *Chiropotes*, dividing the genus into two species, *C. satanas* (with three subspecies) and *C. albinasus*, but the group was reevaluated by Bonvicino et al. (2003). Based on pelage coloration and karyotypic analysis, they suggested that the most westerly group could be a different species and recommended raising the present subspecies to species status. This revision increases the number of *Chiropotes* species to five (Silva and Figueiredo 2002, Bonvicino et al. 2003, Veiga 2006, Silva et al. 2009) (Figure 9.3) and perhaps more if the Guianan bearded sakis (from Venezuela and the Guianas) are included in future genetic analyses.

Cacajao (Uakaris)

Hershkovitz (1987a) summarized data from collection localities for uakaris in the Orinoco and western Amazon

■	*P. aequatorialis*	■	*P. pithecia pithecia*
■	*P. albicans*	░	*P. pithecia chrysocephala*
▨	*P. monachus*	▨	*P. irrorata*

Figure 9.4 Distribution of the six current taxa of *Pithecia* spp. across the Amazon Basin and throughout the Guiana Shield and northern Brazil (map adapted from Hershkovitz 1987b and the IUCN website, http://www.iucnredlist.org/, by E. Bailey). Note that *Pithecia* spp. overlap throughout much of their range with larger *Chiropotes* and *Cacajao* spp. (see Figure 9.5). Overlap with Callicebus spp. is limited to western Amazon Basin sakis (compare figures in van Roosmalen et al. 2002).

basins, ranging from southern Venezuela and western Brazil to eastern Colombia and Peru. He designated two species, black-headed uakaris (*Cacajao melanocephalus*) with two sub-species (*C. m. melanocephalus* and *C. mouakary*), and bald-headed uakaris (*C. calvus*) with four subspecies. *Chiropotes* and *Cacajao* are generally found to be allopatric, with *Cacajao* inhabiting the western Amazon Basin and *Chiropotes* inhabiting the eastern Amazon Basin (Hershkovitz 1985, 1987a; Auricchio 1995) (Figure 9.2). A permeable species boundary apparently exists on the eastern edge of Pico de Neblina National Park, Brazil (c. 65° W latitude), where Boubli (2002) found a few *Chiropotes* individuals in an area also occupied by *C. melanocephalus*.

Uakaris appear to range widely on a daily and seasonal basis but particularly inhabit areas that flood seasonally along white-water rivers (*várzea*) and black-water rivers (*igapó*) (Ayres 1986, Hershkovitz 1987a, Barnett and Brandon-Jones 1997, Boubli 1999). These habitats, tributaries of the Amazon, Orinoco, and Rio Negro, are flooded up to 9 months of the year to a depth of 6–20 m (Ferreira and Prance 1998) and support fewer primate species year-round than terra firma forests occupied by bearded sakis. Ayres et al. (1999) found that both uakaris and squirrel monkeys were endemic to the *igapó* habitats of the Mamirauá Reserve in central Amazonia but only uakaris traveled throughout the reserve, moving deeply into flooded forest.

Low concentrations of suspended nutrients and a low pH of black-water rivers result in relatively lower plant species diversity in *igapó*, although plant diversity is still much higher in flooded tropical forests than in temperate forests (Junk 1989). Plant strategies related to growth and reproduction differ in *várzea* and *igapó* forests. Parolin (2000, 2001)

found that habitats flooded by white-water rivers originating from the Andes (*várzea*) have rich soils due to a high load of suspended sediments. Selection for fast-growing stems allows young plants to reach sufficient height to survive seasonal floods. Rapid stem growth is replaced by high maternal investment in large seeds in the habitats drained by black-water streams (*igapó*) (Parolin 2000, 2001). It is not yet understood how these differences in plant growth rates and maternal investment might translate into habitat and dietary differences among populations of uakaris.

Várzea and *igapó* habitats may provide reliable resources for primate seed predators with few competitors, but long day ranges, large home ranges, and reports of variable population densities of uakaris (Table 9.1) suggest that seasonal variability in food abundance affect grouping patterns and local population densities (Bowler et al. 2009). Stevenson (2001) found that fruit production (as estimated by fruit traps) did not predict either pitheciine biomass or number of species found at 30 neotropical field sites (he excluded *Callicebus* in this analysis). Indeed, "the most striking result [of the study] was the association between the abundance of pitheciine species and *Eschweilera* trees" (Stevenson 2001:172). This observation accords well with research on the biogeography of the family Lecythidaceae. Mori (1989) suggested that many Lecythidaceae species had their origin in the ancient Guiana Shield forests and have recently migrated into the alluvial habitats of the Amazon Basin. *Eschweilera* and other Lecythidaceae genera are now abundant in *várzea* forests, with *E. turbinata* documented as the most abundant tree species by Ayres (1986). The geographic distribution of *Cacajao* and possibly *Chiropotes* may be more dependent on and limited by seed availability

Figure 9.5 Distribution of *Cacajao* (*Ca.*) and *Chiropotes* (*Ch.*) species (adapted from Hershkovitz 1985, 1987a; Silva and Figueiredo 2002; Barnett 2005; Veiga 2006; Figueiredo 2006; Boubli et al. 2008; and the IUCN Web site, http://www.iucnredlist.org/, by E. Bailey). A number of these nomens and distributions are in flux currently. The four subspecies of bald uakaris, *Cacajao calvus* (see text), are collapsed into a single distribution. Taxonomic designations for black uakaris are expressed variously, with some authors retaining subspecies designations for *Ca. melanocephalus melanocephalus* and *Ca. m. ouakary* (e.g., Barnett 2005, Barnett et al. 2005), others raising subspecies to species status (e.g., Figueiredo 2006), or designating new species due to observations of expanded geographic ranges and molecular genetics (e.g., Boubli et al. 2008). For bearded sakis, a number of changes have been made since Hershkovitz 1985: *Ch. sagulatus* = *Chiropotes satanas chiropotes* = northern bearded saki of eastern Brazil and the Guiana Shield; *Ch. satanas* = *C. satanas satanas* = black bearded saki; *Ch. chiropotes* = *Ch satanas chiropotes* bearded saki of western Brazil and Venezuela.

of specific plant families than that of either *Pithecia* or *Callicebus*.

PHYSICAL CHARACTERISTICS OF THE PITHECIINES

Pitheciines are small- to medium-sized platyrrhines, ranging in body size from about 850 to 3,500 g. Titis are the smallest-bodied of the pitheciines, ranging in size from approximately 800 to 1,300 g (see Table 9.2). The sexual dimorphism ratio in body mass ranges from 0.85–1.16. Adult females are heavier than males in four of six species for which body mass data exist, although Hershkovitz (1990:37) remarked that there were "no appreciable morphological differences between the sexes at comparable ages." In the other three genera, males are slightly larger than females (1.08–1.26 in *Chiropotes* and *Cacajao*, n = 3 species, and 1.07–1.47 in *Pithecia*, n = 3 species). Sexual dichromatism has been discussed for *Pithecia* spp. and is absent in *Callicebus* spp., *Chiropotes* spp., and *Cacajao calvus*; but there may be subtle color differences in the red faces of male and female bald-headed uakaris (Hershkovitz 1987b, Barnett and Brandon-Jones 1997) and differences in texture and golden-tipped hairs in the body pelage of male northern bearded sakis

compared with more olive-colored females (M. Norconk personal observation).

Similarities in dental anatomy unite the sakis/uakaris and distinguish them from titis (Rosenberger 1992, Kay et al. 2008). Titis lack the highly derived incisor/canine complex of the larger pitheciines but exhibit (with *Aotus*) tall incisors and enlarged incisor roots that suggest heavy use of anterior dentition during food acquisition or processing (Kinzey 1992).

Dental adaptations in the sakis/uakaris are strongly correlated with a high incidence of seed predation (see Table 9.3). Kinzey and Norconk (1990) and Norconk et al. (2009a) described them as "sclerocarpic" foragers, specialists in opening mechanically protected (i.e., brittle and/or thick-husked) fruit. The protected characteristics of fruit exocarp in their preferred plant families (Lecythidaceae, Chrysobalanaceae, Sapotaceae) contrast with the relatively soft seeds, particularly if they are young seeds. Mechanically protected fruits are breached using robust, widely flaring canines; but the canines of *P. pithecia* are also used very precisely to open small, multiloculed fruit, like many species of the Euphorbiaceae, to extract tiny seeds. Procumbent lower incisors are used to scrape adherent mesocarp from the inside of fruit husks and to plane or reduce the pericarp of tough or large fruit.

Table 9.3 Feeding Ecology of the Pitheciines

SPECIES AND COUNTRY OF RESEARCH	FEEDING GROUP SIZE*	% FEEDING					IMPORTANT PLANT FAMILIES (%)	SOURCE
		SEEDS	FLESHY FRUIT	FLOWERS	LEAVES	INSECTS		
Callicebus brunneus Peru	C				23–66 varies seasonally		Moraceae, Fabaceae, Annonaceae	Wright 1986
Callicebus personatus personatus Brazil	C	21.9	54.7		17.2		Myrtaceae, Sapotaceae, Moraceae (51.4%)	Müller 1996
Callicebus personatus personatus Brazil			81[a]	1	18		Sapotaceae (41%)	Kinzey and Becker, 1983
Callicebus torquatus torquatus Peru			71[b]		4	20	Moraceae, Guttiferae, Fabaceae, Euphorbiaceae, Convolvulaceae, Palmae (52%)	Kinzey, 1978; 1981
Callicebus torquatus lugens Colombia							Euphorbiaceae	Defler 1983, 2003b
Cacajao calvus Brazil	8.95 (1–50) F	66.9	18.4		3.3	5.2	Lecythidaceae, Moraceae, Hippocrateaceae, Sapotaceae, Annonaceae (62.4%)	Ayres 1986, 1989; Walker and Ayres, 1996
Cacajao calvus Peru		46.0	50.0				Sapotaceae, Fabaceae, Apocynaceae (38%)	Aquino 1998, Aquino and Encarnación 1999
Cacajao calvus ucayalii Peru	F						Important resource: Maurtia flexuosa palm fruits	Bowler et al. 2009
Cacajao melanocephalus [*hosomi*] Brazil	D	67.0	28.8	5.0	4.0	2.0	Euphorbiaceae (24%), Caesalpinoidea, Fabaceae, Lecythidaceae, Sapotaceae	Boubli 1999
Cacajao ouakary Colombia	F						Lecythidaceae	Defler 1999a,b, 2003a
Cacajao ouakary Brazil							Euphorbiaceae, Lecythidaceae, Fabaceae, Combretaceae	Barnett et al. 2005
Chiropotes albinasus Brazil	22.5 ± 3.5	35.9	53.9	3.0	7.2[a]		Palmae, Sapotaceae, Fabaceae, Caryocaraceae Moraceae (54%)	Ayres 1981, 1989

Table 9.3 (cont'd)

SPECIES AND COUNTRY OF RESEARCH	FEEDING GROUP SIZE*	% FEEDING					IMPORTANT PLANT FAMILIES (%)	SOURCE
		SEEDS	FLESHY FRUIT	FLOWERS	LEAVES	INSECTS		
Chiropotes chiropotes Venezuela[c]		50.7 (4–60)	0–52.0	0–8.1	0–1.5	1–21	Sapotaceae, Loranthaceae, Moraceae (76%)	Peetz, 2001
Chiropotes chiropotes Venezuela[c]	F	74.8	21.6	0.4	0.2	0.5	Sapotaceae, Loranthaceae, Moraceae, Meliaceae (74.5%)	Norconk 1996, Kinzey and Norconk 1993
Chiropotes sagulatus Brazil	2.88	63.3	9.3	11.4	16.1[a]		Moraceae, Fabaceae, Lecythidaceae, Sapotaceae (62%)	Ayres 1981
Chiropotes sagulatus Suriname	F	66.4	27.6	4.6				van Roosmalen et al. 1981
Chiropotes satanas	F	54.0/ 59.9[d]	25.0/ 13.7	12.3/ 17.4	2.9/3.6	4.6/3.7	Simaroubaceae, Fabaceae, Lecythidaceae, Sapotaceae, Arecaceae	Veiga 2006, Veiga and Ferrari 2006
Chiropotes utahickae		75.6	5.4	18.9	0.1			Santos 2002?
Chiropotes utahickae		36.2	43.3	16.6	3.1	0.7		Vieira 2005?
Pithecia albicans Brazil	F	46.2	28.6	6.5	9.5	0.4	Sapotaceae (21%), Fabaceae (20%)	Peres 1993a–c
Pithecia hirsuta Peru		38.0	55.0	3.0	4.0		Lecythidaceae, Fabaceae, Annonaceae	Soini 1986
Pithecia irrorata Peru		83.0	15.0	0.5	1.5		Moraceae, Sapotaceae, Fabaceae, Chrysobalanaceae	Palminteri et al. 2009
Pithecia pithecia chrysocephala Brazil		26–31		15.9	4.0–18.4			Setz 1993
Pithecia pithecia pithecia Venezuela[c]		53.3	31.0	2.0	10.4	3.7	Connaraceae, Erythroxylaceae, Rubiaceae, Chrysobalanaceae (53.7)	Homburg 1997
Pithecia pithecia pithecia (Venezuela)[c]	C	60.6	27.8	2.2	7.1	2.3	Connaraceae, Lecythidaceae, Loganiaceae, Fabaceae, Erythroxylaceae (57%)	Kinzey and Norconk 1993; Norconk 1996

*C = primarily cohesive; D = well dispersed, no obvious fissioning; F = fission into feeding subgroups

Percentage feeding on various resources was taken directly from sources; no attempt was made to total the "% feeding" resources to equal 100%.

[a] "Fruit" did not specify seed-eating.

[b] "Other" catagory included leaves, insects, bark.

[c] Studies at the same site, overlapped in time.

[d] Study compared island and mainland groups of bearded sakis (expressed as mainland/island dietary components).

Compared with anterior dentition, saki/uakari molars are flat, with undeveloped shearing crests (Kay et al. 1988, Rosenberger 1992). The adaptations of both anterior and posterior dentition, in addition to crenulated and crack-resistant enamel on molars (Kinzey 1992, Martin et al. 2003), form a highly specialized dental apparatus that allows these monkeys to prepare and ingest fruit or seeds that are often ignored by other platyrrhines (Norconk et al. 1998, 2009a).

Both bearded sakis and uakaris are above-branch quadrupeds (pronograde clamberers) and leapers, in the upper canopy levels of the forest (Walker 1996). Vertical clinging and leaping and extensive use of the understory in *P. pithecia* differ from habitat use and locomotion of other saki species that are primarily above-branch quadrupeds (e.g., *P. albicans*, Peres 1993b; *P. monachus*, Youlatos 1999; *P. irrorata*, S. Palminteri, personal comment). Davis and Walker-Pacheco (in press) documented a distinctive morphological pattern in the postcrania of *P. pithecia*, with these smallest of the sakis clustering (statistically) well away from *Chiropotes* spp., *P. monachus*, and *P. hirsuta*.

All sakis/uakaris have bushy tails—long and bushy in *Pithecia* and *Chiropotes*, short and bushy in *Cacajao*. Despite the reduced tail length in uakaris, tail wagging occurs in both bearded sakis and uakaris (Fontaine 1981, Fernandes 1993b, Defler 2003a). Tail wagging, whether below-branch or arched over the head (in bearded sakis) and accompanied by vocalizations, occurs in a variety of contexts from mild to severe agitation (e.g., predator sightings and in response to alarm calls, reunion of group members, and group reorganization after rest or feeding periods) (Fernandes 1993b, Walker and Ayres 1996, Peetz 2001, Defler 2003a). Tail wagging is absent in *Pithecia* and titi monkeys, but tail twining among group members is common in titis. Rather than providing a medium-distance visual cue, as it may in bearded sakis and uakaris, titi monkey tail twining provides a tactile cue "which, it seems reasonable to suppose, contribute[s] to the formation and maintenance of the bond between male and female" (Mason 1974:7).

Kinzey (1986:136) observed that scent marking "plays a major role in regulating social behavior in the marmosets and tamarins, and almost all species of platyrrhines utilize specialized skin scent glands and/or urine for chemical communication." To date, however, there is little information on the function of scent marking in pitheciines. White-faced sakis (*P. pithecia*) possess scent glands in the gular, sternal, and anogenital areas and at times combine scent marking with urine washing (Brumloop et al. 1994, Setz and Gaspar 1997, Gleason 1998). Adult males were scored more often than adult females in scent-marking activities by all of these observers. Gleason (1998) found that the frequency of scent marking peaked in an overlap zone between two groups, and most of his samples occurred while the sakis were traveling. Setz and Gaspar (1997) concluded that scent marking was related to sexual behavior, but there was only one group at their site, a forest fragment in central Amazonia.

In addition to scent marking branches, Gleason (personal communication) noted that adult male white-faced sakis huddled, rubbed their chests against each other, and possibly exchanged scent just prior to some intertroop encounters. Observations of group huddles occurred only in a year in which the study group had four adult males (and three females), an unusually large size for wild white-faced saki groups.

Bearded sakis do not scent mark branches, but they may exchange scent through body contact. Peetz (2001:146) observed "ritualized behavior patterns of hugging and lining up." The behaviors were not limited to adult males, but Veiga and Silva (2005) found that male bearded sakis were more likely to interact with each other socially (including hugging and lining up) than expected given group composition. They found that more than 90% of male/male interactions were affiliative. B. Bezerra (personal comment) and M. Bowler (personal comment) report that uakari adult males also engage in affiliative social interaction, although it is limited to spatial proximity, not body contact. Titi monkeys also "line up," but they engage in lengthy periods of body contact during resting periods, unlike bearded sakis whose contact periods are brief and active, taking on characteristics of a "reunion" (M. A. Norconk, personal observation).

FEEDING ECOLOGY AND DIET

Fleshy fruit comprises the largest component of titi monkey diets, particularly fruit of the Moraceae, Fabaceae, and Sapotaceae families. Leaves are the second highest food category, but leaf composition of the diet ranges 4%–66% depending on the season and titi monkey species (Table 9.3). The *moloch* and *cupreus* groups of titis appear to have a higher proportion of leaves in the diet than the *torquatus* and *personatus* groups (Table 9.3). Insect eating represented 20% of the diet of *C. torquatus* in a study by Kinzey (1978, 1981).

Only a few studies have examined *Callicebus* diets in a long-term, comprehensive manner (Wright 1986, Müller 1996); but most studies of titis have not found them to be significant seed predators (Table 9.3). Indeed, Carrillo-Bilbao et al. (2005) considered *C. discolor* to be a potentially important seed disperser. In contrast, seeds are often found to comprise a third of saki/uakari diets, sometimes more than two-thirds (see Table 9.3). Sakis and uakaris frequently ingest unripe, dull-colored (green or brown) fruit that has a woody or well-protected exocarp (Ayres 1986, 1989; Peres 1993b; Norconk et al. 1998; Boubli 1999; Barnett et al. 2005; Veiga 2006). Mature seeds and fruit pulp are also eaten, but they usually make up a smaller proportion of the diet.

Leaves, flowers or nectar, insects, bark, pith, termite nests, and wasp nests are secondary resources for sakis and uakaris but may be important seasonally. Ayres (1989),

Norconk (1996), Boubli (1999), and Veiga and Ferrari (2006) found that these secondary resources make up only about 10% of the annual diet; but Peetz (2001) found that insects made up over 20% of the diet of bearded sakis seasonally. B. Urbani (personal comment) suggests that secondary resources provide an intermittent nutritional boost to the white-faced saki diet. *Polistes* spp. wasp nests, taken opportunistically whether or not the wasps had deserted the nest, were higher in crude protein (10.9% dry matter, DM) than most other resources ingested by white-faced sakis. Grasshoppers (*Tropidacris* spp.) had the highest crude protein value (58.4% DM) and were also relatively high in lipids (8.0% DM). Nothing is known about the specific mineral requirements of wild sakis, but iron and manganese were found to be in significantly higher concentrations in termite nests than in fruit and leaves ingested by the sakis (B. Urbani, personal comment). Thus, incidental items are likely to provide important nutritional diversity to saki/uakari diets (also see Ayres and Nessimian 1982, Mittermeier et al. 1983, Frazão 1991, Barnett et al. 2005, Veiga 2006, and Veiga and Ferrari 2006 for information on insect eating in bearded sakis).

Venezuelan white-faced sakis ingest a diet rich in seeds, but they also ingest young leaves daily and insects and fleshy fruit seasonally (Norconk and Conklin-Brittain 2004). This combination of items provided an intake that was calculated, primarily from fruit, to be seasonally rich in lipids (11.4%–27.5% estimated DM basis), high in total dietary fiber (25.4%–40.8% DM), and seasonally low in both free simple sugars (4.0%–21.3% DM) and crude protein (4.0%–12.6% DM) (Norconk and Conklin-Brittain 2004).

Bearded sakis and uakaris may also have a high intake of and preference for lipid-rich seeds. Ayres (1986:191–192) found that lipid value was higher in large seeds than small seeds. Large seeds also had a significantly higher ratio of lipids + protein/condensed tannins + acid detergent fiber. This finding correlates well with the high proportion of large seeds found in *igapó* habitats, where plant maternal investment is high and compensates for poor nutrient availability in the soils of black-water river basins (Parolin 2001). Seed eating is equivalent to eating leaves with regard to dietary fiber intake so that (white-faced) sakis and uakaris may select fruit based on lipid value in spite of high fiber content (Ayres 1986; Norconk et al. 2002, 2009; Norconk and Conklin-Brittain 2004).

The major advantage of a diet rich in seeds may lie in the ability to reduce or shift the impact of seasonal fruit shortages relative to that experienced by other platyrrhine frugivores. For both bearded sakis and white-faced sakis inhabiting the tropical dry forests of Lago Guri, Venezuela, seeds represented more than 60% of the diet in the early and late dry seasons. Both species shifted to fleshy fruits in the late wet season, and their seed intake fell (Norconk 1996). The period of food shortage as measured by low food species diversity seems to occur at the end of the wet season and the beginning of the dry season in Lago Guri. At this time, many seeds are small, still early in their maturation, and fleshy fruits have declined in abundance. Ayres (1986:206) provides support for this observation by noting that *Cacajao calvus* "seem to accumulate extensive fat tissues, comparable to that reported in *Saimiri*" at a time when other frugivorous primates have difficulty finding fleshy fruits in the white-water flooded forests of Lake Teiú, Brazil.

A second advantage of seed eating is the duration of availability of at least some seed species. Boubli (1999) and Norconk (1996) noted that uakaris and sakis, respectively, can gain access to seeds in well-protected fruits that have slowly maturing seeds. Third, these pitheciines have the dental and gnathic strength to break open large, woody young pods of the Bignoniaceae, a family of primarily wind-dispersed seeds, fruiting in the dry season. Winged seeds are largely ignored by other primate frugivores and, as such, may be an important fallback food for sakis (Norconk and Conklin-Brittain 2004). Barnett et al. (2005:962) suggest that leaf eating "is highly seasonal and related to a dearth of fruits available in any forest type during the dry season" for golden-backed uakaris.

GROUP SIZE, USE OF SPACE, AND INTERGROUP RELATIONS

Titis and *Pithecia* spp. sakis form small social groups, have relatively small home ranges, and often exhibit behaviors that are associated with defense of territories (Tables 9.1 and 9.2). Group sizes, ranging from two to seven individuals in titis, conform to the expectation of pair-bonded primates (see Table 9.1); but *Pithecia* groups are more variable. A number of censuses have reported that *Pithecia* group sizes range from two to five individuals, consistent with a pair-bonded primate (e.g., Ferrari et al. 2000, Buchanan-Smith et al. 2000, Bennett et al. 2001). However, Lehman et al.'s (2001) extensive survey in Guyana documented a wide range of variation in group size. They observed 21 groups, of which one had 12 individuals. Only five groups had the expected one adult male:one adult female ratio typical of pair-bonded primates, but our recent observations on individually identified, free-ranging sakis in Suriname suggest that slow dispersal of young adults may account for the "extra" adults in the group (M. A. Norconk and C. L. Thompson, personal observations). In the absence of genetic data on paternity and long-term studies in intact (nonfragmented) habitats, it is best to interpret *Pithecia* social groups as small but flexible and responsive to variables such as population density, food distribution, and perhaps sympatry with the two larger-bodied pitheciines, *Chiropotes* and *Cacajao* (see Ferrari et al. 1999a).

Using playbacks, Robinson (1979) and Kinzey (1981, 1997a) defined two distinctive patterns of home range use and defense by *Callicebus* spp. *C. (moloch) ornatus* gave early-morning calls regularly, followed by intertroop encounters depending on the proximity of callers to their territorial boundary (Robinson 1979, Robinson et al. 1987). In contrast, a *C. torquatus* group moved away from playbacks of both adult males and male/female pairs using, what Kinzey and Robinson (1983) called, "proximity-dependent avoidance." *C. personatus* seems to be closer to the *C. torquatus* pattern of group dispersion (Ferrari et al. 2000), but other behavioral and spatial use differences may exist among these species suggesting that we have not yet exhausted all of the habitat-use strategies of species in this very widespread and diverse genus.

A few studies suggest that *Pithecia* spp. are also territorial (Norconk et al. 2003, Di Fiore et al. 2007, S. Palminteri personal communication), and Shideler et al. (1994) have reported aggressive interactions with same-sex intruders among captive sakis. Territorial activities that consist of behavior-specific vocalizations and chasing (Norconk 2006, Thompson and Norconk 2009) are strikingly similar to Mason's (1968) description of behaviors associated with territoriality in *C. ornatus* and Fernandez-Duque's (see Chapter 10) description of territoriality in *Aotus*.

Chiropotes and *Cacajao* live in relatively large groups and are as far-ranging as any platyrrhine (Table 9.1) (Boyle et al. 2009b). Daily travel distance appears to be driven by the density and dispersion of seed crops, and some authors have reported extensive group fissioning and wide variations in feeding group sizes seasonally (Ayres 1986, Defler 1999a, Barnett et al. 2005, Bowler et al. 2009; see Table 9.3). Ayres (1986) estimated white uakari home ranges to be 500–550 ha, with daily path lengths up to 5 km. Path lengths averaged 4.4 km for black-headed uakaris in Brazil (Boubli 1999), 2.5–4.0 km for bearded sakis in Suriname (Norconk and Kinzey 1994), and 2.99 km for bearded sakis in Brazil (Boyle 2008a).

The difficulty of tracking fast-moving, widely ranging bearded saki and uakari groups, as well as the paucity of studies on captive groups, has resulted in little information on how *Chiropotes* and *Cacajao* groups are organized socially. Ayres (1981) suggested that bearded saki groups may be based on an underlying structure of multiple male/female units. This hypothesis might provide information on how groups fission during feeding and travel, but logistical difficulties have thus far precluded the ability to address the question. To my knowledge, there have been no studies of wild bearded sakis or uakaris with individually identified animals.

Defler (2003b) documented seasonal variation in group size for black-headed uakaris in Colombia, but Boubli (1999) found them to be more cohesive in Brazil. Van Roosmalen et al. (1988) and Norconk and Kinzey (1994) found that bearded sakis often fissioned when multiple feeding trees were within about 100 m, but they traveled cohesively between feeding sites. Recent studies have consistently reported group fissioning in both bearded sakis and uakaris, although the mechanism does not appear to be as simple as Norconk and Kinzey (1994) suggested.

REPRODUCTION AND DEVELOPMENT

Pitheciine female reproductive physiology has been studied for the smaller members of the group: white-faced sakis (*P. pithecia*, Savage et al. 1992, Shideler et al. 1994) and titis (*C. cupreus*, Hoffman et al. 1995b; Valeggia et al. 1999). White-faced sakis have an ovarian cycle of 16–17 days, gestation of 150 days, and approximately 195 days of lactational amenorrhea. The interbirth interval in captivity was 15.1 months, and sexual maturity occurred at about 30 months. Interbirth intervals are longer in wild sakis (21.5 ± 9 months, n = 6), and first birth occurred at the age of 5 years (M. A. Norconk and S. E. Shideler unpublished data; Norconk 2006) (see Table 9.2).

The length of titi monkey ovarian cycles is very similar to that of white-faced sakis (*C. cupreus*, 17.2 ± 1.5 days), but gestation length and interbirth intervals were shorter (128.6 ± 4.4 days and 11.8 months, respectively) (Valeggia et al. 1999). Titi monkey females housed with an unfamiliar male had their first infants at the age of 3.7 years (range 2–6 years) (Valeggia et al. 1999).

Mason, Mendoza, and colleagues conducted a number of elegant studies on the behavior and physiology of filial attachment in *C. (moloch) cupreus* (e.g., Mason 1971, 1974; Mendoza and Mason, 1986a,b; Mason and Mendoza 1998; Hennessy et al. 1995; Hoffman et al. 1995a). They found that titi monkey infants exhibit preferences for fathers (Mason and Mendoza 1998) and that separation from the father, but not the mother, elicited a strong cortisol response (Hoffman 1998). Male caretaking peaked in the second month and continued until the sixth month (Fragaszy et al. 1982). More recently, Bales et al. (2007) found significant differences between bonded vs. nonbonded titi males in the activity of neural areas that are associated with pair-bonding in rodents. Interestingly, newly bonded males were intermediate, and more variable, in glucose-uptake values, suggesting individual variation in how male/female bonds are formed in these monogamous primates.

In a series of experiments, Hoffman (1998) studied physiological and behavioral parameters in the interaction between nearly mature titi monkeys and their parents. His experimental studies on captive *C. cupreus* "dispersal" accord well with data collected by Bossuyt (2002) on wild *C. moloch*. Hoffman's data suggested that sons were more reticent to leave the family group than daughters, but both daughters and sons showed strong behavioral attachment to parents until they were 3.5–4 years of age (Hoffman 1998:92). Bossuyt (2002), working in Cocha Cashu, Peru, found that both daughters and sons dispersed from their natal groups between 3 and 4 years of age, that daughters tended to disperse earlier than sons, and that, in light of high juvenile mortality, parents may benefit from retaining subadults in the group as individuals that could provide "critical aid" to siblings. Existing data on dispersal patterns in titis appear to be similar to preliminary data on white-faced sakis: Both sons and daughters appear to leave their natal group but do not do so until they are mature (at least 3 years of age). Like titis, white-faced sakis appear to retain affiliative relations within their natal group prior to dispersal (M. A. Norconk, personal observation).

Homburg (1997) documented developmental patterns for the first 5 months of life for wild white-faced sakis in Lago Guri, Venezuela. The mother was the only carrier for the first 8 weeks. In the third month, two other adult females carried the infant for 15–85 min once or twice a day (Homburg 1997:131). Distance between mother and infant increased in the fourth month, and solitary play was observed in the fifth month. Infants made brief contact with males, but no male infant carrying was observed (but see Ryan 1995). A wild white-faced saki infant was observed to spend about 50% of its time off the mother's back by the fifteenth week of life (Buzzell and Brush 2000). Captive infants showed an accelerated level of independence and were on the mother in only 3% of the samples at the end of the twelfth week of life. Carrying by other group members was seen during week 12 in the wild but was not seen in the captive sample (Buzzell and Brush 2000). Whereas white-faced saki infants became self-locomoting by their fourth month, both bearded saki and uakari infants were carried by the mother for at least 9 months. Larger body mass may account for the slower developmental pattern in bearded sakis and uakaris, but longer daily travel paths for these species may also influence the duration of maternal carrying.

The transition to the white face and black body pelage characteristic of adult male white-faced sakis (*P. pithecia*) appears to follow a prolonged but variable developmental path. Some young males in Lago Guri, Venezuela, exhibited the dark body pelage of adult males within a few months, and others maintained the orange–brown agouti coloration of females into their third year (M. A. Norconk, personal observation). Homburg (1997) estimated that body pelage changed at about 20 months of age but body hair was not entirely black until the third year. Development of the white facial mask appears to proceed at a variable rate with respect to changes in body pelage in Guianan sakis (M. A. Norconk, personal observation), but nothing is known about the development of pelage differences in Amazonian sakis.

Peetz (2001) provides the only information available on the development of wild bearded sakis, but Hick provides developmental data on captive sakis (cited in van Roosmalen et al. 1981). According to Peetz, bearded saki births peaked in the dry season (December to April); however, van Roosmalen et al. (1981) correlated the birth season with the beginning of the wet season in Suriname. Peetz (2001) estimated a minimum interbirth interval of 2 years, that infants were carried ventrally for the first 2 months of life and dorsally through the fifth month, and that they were still carried in the ninth month whenever mothers leapt between widely spaced tree crowns. Older juveniles traveled

independently by age 10–13 months (Peetz 2001) but continued to suckle into the second year. Testes descended at about 3 years of age.

Peetz (2001) did not document any courtship behavior pre- or postcopulation. After copulation, both males and females resumed precopulation activities (Peetz 2001). Adult male care of infants is not as extensive in bearded sakis; it was limited to grooming and playful interactions (Peetz 2001).

Uakari developmental data, reported by Fontaine (1981) on a semi-free-ranging group, are remarkably similar to Peetz's description for bearded sakis. Suckling continued until 22 months of age, and mothers persisted in carrying 12-month-old infants across wide gaps between tree crowns (Fontaine 1981). Phenotypic changes from young uakari to adult began at about 2 years of age: "a darkly pigmented glandular field develops in the sternal area, alopecia of the scalp develops, and the general body pelage of the dorsal torso and lateral limb surfaces fills out to form a mantle" (Fontaine 1981:457). Sexually dimorphic traits develop slowly in red uakaris, as they do in white-faced sakis. Infant red uakaris are born with only a trace of pink in the face, and color change begins gradually in the juvenile period (3–12 months). Over a period of 2 years, Fontaine (1981) observed young male uakaris to increase in body mass and musculature relative to females and to develop the characteristic paired muscle masses overlying the frontal and parietal areas of the skull. *P. pithecia* saki males complete the developmental process by about 3 years of age, when uakaris are just beginning it. Fontaine (1981) also noted that the developmental process in male red uakaris is remarkably plastic, as it appears to be in male white-faced sakis.

CONSERVATION STATUS

Given wide variation in habitat use, body size, geographic distribution, and ecological specializations, it is not surprising that pitheciines face a range of conservation problems. However, habitat fragmentation due to dam creation, ranching, agriculture, logging, and expansion of human populations appears to be the most widespread problem (Ferrari et al. 1999b). In addition to fragment size, long-term survival in fragments probably depends on availability of forest corridors, quality of matrix between terrestrial fragments (Boyle 2008b, Boyle et al. 2009b), and ability to move between fragments. On the positive side, studies of bearded sakis in the Biological Dynamics Forest Fragments Project (BDFFP) (Boyle 2008a,b), the Tucuruí Reservoir (Veiga 2006, Silva and Ferrari 2009), and the Celmar plantation complex (Port-Carvalho and Ferrari 2004) in Brazil and Lago Guri, Venezuela (Kinzey and Norconk 1993, Homburg 1997, Peetz 2001, Norconk 2006), found that the monkeys maintain strikingly diverse diets even with differences in activity and ranging patterns. Both Peetz

(2001) and Veiga (2006) found that bearded sakis fed on more than 100 species of plants in fragments ranging from 19 to over 1,000 ha. Veiga's study of *Chiropotes satanas* in both large and small fragments highlights the ability of sakis to shift feeding species preferences yet maintain a diet of more than 50% seeds. Feeding flexibility aside, stochastic forces operating on small group sizes and groups that receive no gene flow from immigrants eventually show population declines. For example, white-faced saki group size on the small island in Lago Guri peaked at nine and then declined to five adults 10 years later, at which time live births dropped to zero.

Unlike bearded sakis, specialized feeding and foraging strategies of uakaris may heighten their risk of extinction when forests are fragmented and migration routes are cut off. According to Barnett et al. (2005:961) black uakaris in Jaú National Park "make limited excursions far away from the igapó to exploit very specific food resources." Observations by Bowler et al. (2009) in Peru suggest that uakari birth rates are tied to *Mauritia flexuosa* palm fruit production. With the increasing number of field studies and censuses of pitheciines over the past 10 years, we are beginning to gain an appreciation for the complexity of their adaptations. The spotty distribution of white-nosed bearded sakis in southwestern Amazonia prompted Ferrari et al. (1999a) to suggest that complex interactions among several variables (e.g., soil conditions, phenology, presence or absence of key resources, and perhaps presence of a second pitheciine species) affect bearded saki survival and distribution. Hunting pressure appears to be most applicable to the larger-bodied species, and Bowler et al. (2009) describe what appears to be a stunning recovery of a uakari population in the Lago Preto Conservation Concession in Peru once hunting ceased.

Although titis should be able to handle habitat fragmentation better than larger sakis since their home ranges are much smaller, isolation of small populations seems to be their biggest conservation concern. Some species, like *Callicebus torquatus*, have specific habitat preferences that result in a mosaic of populated areas in Peru (Aquino et al. 2008) or, in the case of *C. olallae* in Bolivia (Felton et al. 2006) and *C. coimbrai* in Brazil (Jerusalinsky et al. 2006), where ranches have created open grasslands that are not easily crossed (Felton et al. 2006). Some newly named species are vulnerable because they inhabit very small areas (e.g., *C. aureipalatii*, Wallace et al. 2006) or are restricted geographically and in the path of intensive development (*C. oenanthe* in northern Peru, DeLuycker 2006, Aldrich et al. 2008). Most of these authors acknowledge that many species of *Callicebus* can tolerate degraded habitats but also state that isolation is a deepening problem. In order to address and potentially ease the pressure of persistent fragmentation coupled with development, many authors are calling for increased sizes of protected areas and active management of isolated groups at the metapopulation level.

CONCLUSIONS

I opened this chapter with the view that the pitheciines represent a closely related group of species arrayed along a continuum of smaller to larger body mass, with group compositions ranging from pair-bonded and territorial to multimale and relatively nomadic. *Pithecia*, *Chiropotes*, and *Cacajao* are clearly a very cohesive group from the perspective of feeding adaptations. Body size aside, they share the same dental adaptations. *Pithecia*, the smallest of the three, may have a broader diet that includes leaves on a regular basis as well as fewer habitat restrictions. Tropical dry and savanna habitats probably limit the distribution of bearded sakis and uakaris but do not seem to be a barrier to *Pithecia* dispersal, particularly for *P. pithecia*, the white-faced saki. This view is complicated by some diversity within the small radiation of *Pithecia* species, as suggested by Peres (1993b), Walker (1996), and Davis and Walker-Pacheco (in press). The smaller-bodied, vertical clinging and leaping white-faced sakis of the Guianas are well adapted to moving through low- to middle-canopy levels compared with the Amazonian sakis that appear to prefer upper forest levels.

There also appears to be a size-related continuum among titi monkeys, with the smaller ones (*C. moloch, donacophilus, cupreus*) inhabiting small, well-defended territories using voice and movement toward territorial boundaries and *C. torquatus* and *C. personatus* using voice to announce their presence in a larger home range. *Callicebus* contrasts with the other three pitheciines in aspects of feeding ecology, but *Pithecia* is closer to *Callicebus* in aspects of body size, group size, territorial behavior, and home range size.

Although data on aspects of the social behavior of wild pitheciines, which require well-habituated and identifiable individuals, continue to lag behind ecological data, there are a growing number of reports suggesting that males are tolerant of, or actively cooperate with, other intragroup males. Since there is also a considerable range in the number of males in pitheciine social groups (from generally one adult male in titis and sakis to many males in bearded sakis and uakaris) (see Tables 9.1 and 9.2), the expressions of male/male behavior are variable (Table 9.4). Bossuyt (2002) and Norconk (2006) suggested that even with a single breeding male in titi and white-faced saki groups, respectively, sons are retained in the group into subadulthood, although Wright (2009) found that all subadults in the eight titi groups she monitored dispersed as subadults. Retention of sons that are willing to invest in territorial behaviors (chasing and threatening nongroup males) may improve territory maintenance (Thompson and Norconk 2009, Wright 2009). In neither titi nor saki groups do adult females regularly participate in intertroop encounters. Intertroop encounters are rare in bearded sakis and uakaris, probably because home ranges in free-ranging conditions are very large and groups often fission into subgroups.

Table 9.4 Preliminary Assessment of Male/Male Behaviors in Pitheciines

MALE-MALE CHARACTERISTICS	TITIS	SAKIS	BEARDED SAKIS	UAKARIS
Adult male tolerance of young males	x	x	x	x
Male (son) cooperation in inter-troop encounters		x		
Cohesive groups	x	x		
Well-dispersed groups			x	x
Fission-fusion			x	(x)
Sex-specific male contact behaviors			x	
Male grooming partner preferences			x	x

Ayres (1981, 1986) believed that fissioning in bearded saki and uakari groups reflected variation in resource availability. But we still do not know what it is about resource availability that affects group fissioning or determines the size of subgroups: Is it the relative abundance of fruit crops, availability of fruit-producing crowns of a particular size, seasonal availability of preferred foods like seeds (e.g., *Mauritia flexuosa*) driving either group dispersion or cohesion, diversity of fruiting species, or absolute (regional) fruit abundance? Nor do we understand the factors that affect individual decisions about which subgroup to join. Bowler et al. (2009) describe fissioning that may reflect reproductive or social preferences (i.e., all-male groups that follow larger multimale/multifemale uakari groups), and Ayres (1981) suggested that bearded saki groups may fission into male/female units. One thing seems certain however: The suggestion that bearded saki groups fission and fuse locally while feeding in nearby trees (Norconk and Kinzey 1994) is too simplistic.

Preferences expressed by male bearded sakis and uakaris for other males as social partners are now supported by quantitative data (Peetz 2001, Veiga and Silva 2005). Veiga and Silva (2005) found that more than half of the social interactions recorded involved two or more bearded saki males and over 90% of those interactions were affiliative. Behaviors included a high incidence of male/male social resting, hugging, lining up, and grooming in bearded sakis; but Bowler (personal comment) suggests that, while uakari males are generally affiliative, they do not engage in social contact to the same degree as bearded sakis (Table 9.4). In light of the growing evidence for male/male affiliation in bearded sakis and uakaris, it is interesting to note that Hershkovitz (1993) found that males of these two genera bore large, hooked penile spines. It leads one to wonder whether the affiliation expressed by males in these large multimale groups masks a subtle form of competition. There are still many interesting discoveries to be made in feeding ecology, ranging behavior, intragroup relations, and sociosexual behavior in the pitheciines, particularly in large multimale groups of bearded sakis and uakaris—and we have not yet scratched the surface regarding female strategies.

Species included in this chapter are shown in Color Plates 14–16 (*Pithecia pithecia*, *Cacajao calvus*, and *Callicebus discolor*).

10

Aotinae
Social Monogamy in the Only Nocturnal Anthropoid
Eduardo Fernandez-Duque

1. What is cathemerality, and why does it exist in some *Aotus* populations and not others?
2. What is the typical social organization of owl monkeys, and how does this translate to patterns of parental care in the genus?
3. How is *Aotus* activity influenced by moon phase?

OVERVIEW

The two most salient features of owl monkeys are nocturnal habits and monogamous social organization. Owl monkeys concentrate their activities during the dark portion of the 24 hr cycle, with peaks of activity at dawn and dusk. Interestingly, our understanding of the evolution of nocturnality in the genus is further challenged by the existence of at least one owl monkey species that shows some remarkable temporal plasticity in its activity patterns. Differently from all owl monkey species in the tropics, which are strictly nocturnal, *Aotus azarai azarai* of the South American Gran Chaco is cathemeral, showing activity during the day as well as during the night (Fig. 10.1).

Owl monkeys are also one of the few socially monogamous primates in the world. They live in small groups that include only one pair of reproducing adults, one infant, one or two juveniles, and sometimes a subadult. Males show intense care of the infants. For a long time, the difficulties of studying a small arboreal and nocturnal primate limited our understanding of the evolution and maintenance of monogamy in this taxon. More recently, studies on the cathemeral owl monkeys of the Gran Chaco have begun to offer some insights into their social organization.

Figure 10.1 An owl monkey (*Aotus azarai azarai*) social group sunbasking during a cold morning in the Argentinean Chaco.

DISTRIBUTION AND TAXONOMY

Distribution

Owl monkeys range from Panama to northern Argentina and from the foothills of the Andes to the Atlantic Ocean (Aquino and Encarnación 1988, 1994; Aquino et al. 1992b; Barnett et al. 2002; Bennett et al. 2001; Castaño and Cardona 2005; Cornejo at al. 2008; de Sousa e Silva and Nunes 1995; Defler 2003a, 2004; Fernandes 1993a; Fernandez-Duque et al. 2008a; Ford 1994; García and Tarifa 1988; Hernández-Camacho and Cooper 1976; Hernández-Camacho and Defler 1985; Hershkovitz 1983; Kinzey 1997a; Peres 1993b; Villavicencio Galindo 2003; Wright 1981; Zunino et al. 1985). They inhabit a variety of forests of both primary and secondary growth, sometimes up to 3,200 m above sea level (Defler 2003a, Hernández-Camacho and Cooper 1976). At the southern end of their range, owl monkeys are distributed across the South American Gran Chaco of Argentina, Bolivia, and Paraguay, where they can be found in dry forests that receive only 500 mm of annual rainfall (Brooks 1996, Campos et al. 2004, Fernandez-Duque et al. 2002, Lowen et al. 1996, Neris et al. 2002, Stallings et al. 1989, Wright 1985) (Fig. 10.2).

	A. lemurinus
	A. griseimembra
	A. nancimaae
A. azarai azarai	A. vociferans
A. azarai boliviensis	A. trivirgatus
A. infulatus	A. zonalis
A. nigriceps	A. brumbacki
A. miconax	A. sp.?

Figure 10.2 The geographic distribution of owl monkey species and subspecies. The map has been redrawn based on Defler (2003a), Erkert (1999), Ford (1994b), and Rylands (unpublished data).

Taxonomy

Owl monkeys, also known as night monkeys, douroucoulis, *martejas*, *micos de noche*, and *mirikinás*, belong in the genus *Aotus*. The genus name is derived from combining the Latin words *a*, meaning "without," and *otis*, meaning "ear," making reference to the inconspicuous earlobes usually hidden by dense fur. Several taxonomic issues remain largely unsettled, including the classification of the genus at the family and subfamily levels and the number of recognized species and subspecies within it (Defler and Bueno 2007).

Regarding its suprageneric classification, the genus was for many years placed alternatively in the Atelidae or in the Cebidae family. Based on morphological data, owl monkeys were considered to be closely related to the pitheciines within the atelids (Rosenberger 1981, Schneider and Rosenberger 1996). A close affinity between owl monkeys and the atelines has also been suggested based on dental morphology (Tejedor 1998, 2001). On the other hand, analysis of molecular genetic data led researchers to place the genus within the cebids (Porter et al. 1997; Schneider and Rosenberger 1996; Schneider et al. 1993, 1996). More recently, *Aotus* was placed in its own separate family, Aotidae (Defler 2003a, Groves 2001a, Rylands et al. 2000).

The genus included only the species *A. trivirgatus* when first described. Following the discovery of various karyotypes (Brumback 1973, 1974; Brumback et al. 1971; Ma 1981; Ma et al. 1976a,b, 1977, 1978, 1985), Hershkovitz (1983) divided the genus into nine species organized in two groups based on their karyotypes, coloration of the neck, and susceptibility to *Plasmodium falciparum*, one of the pathogens of human malaria: The gray-necked group occurs to the north and the red-necked group to the south of the Amazon River. Later on, an independent evaluation of craniodental measures and color pelage and patterns led Ford (1994b) to accept at least five, and possibly seven, of the nine species identified by Hershkovitz.

However, because some of the owl monkey species in the gray-necked group are part of a sibling species complex, reliance on phenotypic traits for species recognition may be problematic (Defler 2003a, 2004; Defler and Bueno 2003; Defler et al. 2001). In fact, extensive and systematic research using both phenotypes and karyotypes (Defler and Bueno 2003, Defler et al. 2001, Giraldo et al. 1986, Torres et al. 1998) led researchers to recognize up to 11 species (Defler 2003a, Defler and Bueno 2003, Defler et al. 2001, Galbreath 1983, Rylands et al. 2000). There is enough information to recognize at least the following seven species in the gray-necked group: *A. lemurinus*, *A. zonalis*, *A. brumbacki*, *A. griseimembra*, *A. vociferans*, *A. trivirgatus*, and an undescribed eastern Colombian species. In turn, the red-necked group includes four species: *A. miconax*, *A. nancymaae* (the ending should be with "e," not "i"), *A. nigriceps*, and *A. azarai* (the ending should be with "i," not "e") (Mudry de Pargament et al. 1984, Mudry et al. 1990, Pieczarka et al. 1993) (Table 10.1).

The lumping until so recently of all owl monkey diversity as *A. trivirgatus* (e.g., Robinson et al. 1987) should be carefully considered when evaluating the published literature. Given the existing knowledge on significant differences among owl monkey species in body size and body mass (Smith and Jungers 1997), activity patterns (Erkert 2008, Fernandez-Duque 2003; Wright 1989, 1994a, 1996), and canine sexual dimorphism (Hershkovitz 1983), published data from *A. trivirgatus* should be checked against the geographic origin of the sample or the locality where the study was conducted before they are assumed to be from any particular species.

ECOLOGY

Population Density and Group Size

Few reliable estimates of population densities exist for most owl monkey species (Table 10.2). The most comprehensive population studies have been done on *A. vociferans* and *A. nancymaae* in Peru, where numerous researchers have

evaluated population densities over a long period of time and in a large number of different localities and habitats (Aquino and Encarnación 1986b, 1988, 1989, 1994; Aquino et al. 1990, 1992a,b, 1993; Moya et al. 1990).

Only a handful of studies have specifically evaluated the density of owl monkeys during the time when they are most active. On the other hand, several studies have reported densities estimated from sightings of nocturnal animals during the day while censusing diurnal primates. Unless the time of activity of the specific owl monkey species is considered, the estimates may be of little value (Aquino and Encarnación 1994, Weisenseel et al. 1993). For example, nocturnal censuses must consider moon phase since the probability of detection of owl monkeys is severely affected by available moonlight (Fernandez-Duque 2003, Fernandez-Duque and Erkert, 2006, Wallace et al. 2000). Given these serious limitations in some of the published data on owl monkey densities, there are no adequate data to evaluate the potential influence of habitat quality, predators, or competitors in determining population parameters across species. The value of discussing differences among estimates is highly questionable unless one has access to the raw sighting data obtained by the investigators (Peres 1999).

Table 10.1 Taxonomy, Body Mass, and Chromosome Numbers of Owl Monkey Species and Subspecies

SPECIES	CAPTIVE (C) OR WILD (W)	ADULT MALE BODY MASS (KG)			ADULT FEMALE BODY MASS (KG)			CHROMOSOME 2N[1]	SOURCE
		AVERAGE	RANGE	NUMBER OF INDIVIDUALS	AVERAGE	RANGE	NUMBER OF INDIVIDUALS		
Aotus azarai azarai	W	1,254 ± 118	990–1,580	40	1,246 ± 114	1,010–1,450	39	m 49, f 50	Fernandez-Duque 2004
A. a. boliviensis	W	1,180		4	1,230		8	m 49, f 50	Smith and Jungers 1997
A. a. boliviensis	C	1,091		7	1,141		7		Dixson 1983a,b
A. a. infulatus	W	1,190		1	1,240		1	m 49, f 50	Fernandes 1993a,b
A. brumbacki	W	875 (unsexed)		1	455		1	50	Hernández-Camacho and Defler 1985
A. griseimembra	C	1,009 ± 200		20	923 ± 63		16	52, 53, 54	Dixson 1983a,b
A. griseimembra	C	925 (median)	800–1,080	12					Dixson et al. 1980
A. lemurinus	W	920.7 ± 79.9	608–1,150	7	859 ± 87.6	578–1,050	6	58	Hernández-Camacho and Defler 1985
A. miconax								Unknown	
A. nancymaae	C				930 (median)[2]	1,190–760	16		Málaga et al. 1991
A. nancymaae	W	794		32	780		24	54	Aquino and Encarnación 1986b
A. nancymaae	C	946.5 ± 142.3	750–1,077	4	905.6 ± 123.6	706–1,055	6		S. Evans (unpublished)
A. nigriceps	W	875		1	1,040		2	m 51, f 52	Peres 1993a–c
A. trivirgatus	W	813		20	736		17	Unknown	Smith and Jungers 1997
A. trivirgatus	W	1,200		1	1,000		1		Fernandes 1993a,b
A. vociferans	C	708		20	698		20		Montoya et al. 1995
A. vociferans	W	697.5 ± 24	568–800	4				46, 47, 48	Hernández-Camacho and Defler 1985
A. zonalis	W	889		6	916		11	55, 56	Crile and Quiring 1940

[1] Chromosome numbers from Defler 2003, Erkert 1999, and Santos-Mello 1986.

[2] The median was visually estimated from the original data depicted in Figure 1 of Malaga et al. 1991.

Table 10.2 Population Density and Group Size of Owl Monkey Species

TAXON	STUDY SITE	GROUP DENSITY (G/KM²)	INDIVIDUAL DENSITY (IND/KM²)	AVERAGE GROUP SIZE	RANGE	NUMBER OF GROUPS SIGHTED	TYPE OF STUDY	TYPE OF HABITAT	SOURCES
A. a. azarai	Guaycolec, Formosa, Argentina	14[1]	32.3	—	—	—	c	Gallery forest	Zunino et al. 1985
A. a. azarai	Guaycolec, Formosa, Argentina	8.0	25.0	3.3	—	47	c	Gallery forest	Arditi and Placci 1990
A. a. azarai	Guaycolec, Formosa, Argentina	16.0	64.0	4.0	2–6	11	be	Gallery forest	Fernandez-Duque et al. 2001
A. a. azarai	Formosa Province, Argentina	5.5	12.8	2.3	—	12	c	Mix of forest types	Zunino et al. 1985
A. a. azarai	Formosa Province, Argentina		15.0	3.2	—	6	c	Forest patches	Brown and Zunino 1994
A. a. azarai	Formosa Province, Argentina	10.0	29.0	2.9	1–4	25	c	Mix of forest types	Rathbun and Gache 1980
A. a. azarai	Teniente Enciso, Paraguay	3.3	8.9	2.7	—	6	ce	High dry woodland	Stallins et al. 1989
A. a. azarai	Agua Dulce, Paraguay	4.7	14.4	3.1	—	21	ce	Low dry woodland	Stallins et al. 1989
A. a. azarai	Presidente Hayes, Paraguay	—	—	—	2–4	2	be	Forest patches	Wright 1985
A. a. azarai	Cerro Leon, Paraguay			3.0	3	1	c	Low canopy scrub thorny forest	Campos et al. 2004
A. a. boliviensis	Departamento Beni, Bolivia	68.9[1]	242.5[1]	3.5	—	21	be	Low gallery forest	García and Braza 1989
A. a. boliviensis	Departamento Beni, Bolivia	—	—	3.6	2–5	23	c	Low gallery forest	García and Tarifa 1988
A. brumbacki	Río Duda, Tinigua NP, Colombia	—	—	3	—	1	be		Solano 1995
A. griseimembra	Northern Colombia	—	150[2]	—	2–4	—	c		Heltne 1997
A. lemurinus	Cuenca media Río Cauca, Colombia			3.8	3–5	8	ce		Castaño and Cardona 2005
A. miconax	Departamento Amazonas, Peru			—	2–5	—	ce	Primary cloud forest and disturbed forest fragments	Cornejo et al. 2008
A. nancymaae	Isla Iquitos, Peru	10.0	—	—	—	—	c		Soini and Moya 1976
A. nancymaae	Río Tahuayo, Peru	7.5	29.0	3.4	—	42	c	Floodable low forest	Aquino and Encarnación 1986a

Table 10.2 (cont'd)

TAXON	STUDY SITE	GROUP DENSITY (G/KM²)	INDIVIDUAL DENSITY (IND/KM²)	AVERAGE GROUP SIZE	RANGE	NUMBER OF GROUPS SIGHTED	TYPE OF STUDY	TYPE OF HABITAT	SOURCES
A. nancymaae	Northeastern Peru	11.3	46.3	4.1	–	75	c	Lowland	Aquino and Encarnación 1988
A. nancymaae	Northeastern Peru	5.9	24.2	–	–	23	c	Highland	Aquino and Encarnación 1988
A. nancymaae	Ríos Marañón-Amazonas, Peru	–	–	4	2–6	142	c		Aquino et al. 1990
A. nigriceps	Cocha Cashu, Manú NP, Peru	10³	36–40	4.1	2–5	9	be	Lowland	Wright 1985
A. trivirgatus[4]	Departmento Bolivar, Colombia	0.5	1.5	2.5	–	8	c		Green 1978
A. trivirgatus[5]	Pto. Bermúdez, Dept. Pasco, Peru	–	–	3.3	2–4	3	be		Wright 1978
A. vociferans	Tiputini Biodiversity Station, Yasuni NP, Ecuador				3–5	4	be	Tropical rain forest	Fernandez-Duque et al. 2008a
A. vociferans	Ríos Nanay and Napo, Peru	–	–	3.3	2–5	82	c		Aquino et al. 1990
A. vociferans	Northeastern Peru	10.0	33.0	3.3	–	22	c	Lowland	Aquino and Encarnación 1988
A. vociferans	Northeastern Peru	2.4	7.9	–	–	11	c	Highland	Heltne 1977

[1] These data are based on censusing an island of forest of 0.33 ha. Thus, the estimates of density are based on repeated sampling of a very small area corresponding to the territory of one group.

[2] These data were collected in a forest remnant that may have served as a refuge, thus explaining the very unusual high density of 150 ind/km².

[3] The number of groups was not reported, so it was estimated dividing the density of individuals by the reported average group size.

[4] Only eight sightings while censusing during daylight. The source refers to A. trivirgatus, but it should be A. griseimembra.

[5] The source refers to A. trivirgatus, but it should be A. nigriceps given its locality.

c, census; be, behavior and ecology study; ce, census and ecology study.

Furthermore, the estimates presented in Table 10.2 span 30 years and were collected by a wide range of researchers using different methods. Before differences among densities are attributed to ecological factors, the possible effects of different methodologies and temporal fluctuations in population values must be at least considered.

All reports agree that owl monkey group size ranges between two and six individuals (Aquino and Encarnación 1986b; Brooks 1996; Brown and Zunino 1994; Castaño and Cardona 2005; Cornejo et al. 2008; Fernandes 1993a; Fernandez-Duque and Bravo 1997; Fernandez-Duque et al. 2001, 2008a; García and Braza 1989; García and Tarifa 1988; Peres 1993b; Rathbun and Gache 1980; Stallings et al. 1989; Wright 1994a). In every locality, owl monkeys were found in small groups generally composed of an adult

heterosexual pair, one infant, and one or two individuals of smaller size (Table 10.2). In an A. a. azarai population in Formosa, Argentina, many groups also included 3- and 4-year-old subadults (Fernandez-Duque and Huntington 2002, Fernandez-Duque 2009).

At least in one population of A. a. azarai, a significant number of adults (25%–30%) do not belong to a social group but range solitarily (Fernandez-Duque 2004, Fernandez-Duque and Rotundo 2003). These individuals are either young adults that have recently emigrated from their natal groups or relatively old adults that have been evicted from their groups by incoming adults. A recent survey of Aotus spp. in northern Colombia also found a very significant number of solitary individuals (Villavicencio Galindo 2003). It seems likely that the less conspicuous solitary individuals will be detected in other populations and species as more long-term studies of identified individuals are conducted.

Ranging and Territoriality

Owl monkeys are territorial, each group occupying a range that overlaps only slightly with the area used by neighboring groups (Table 10.3).

Groups regularly encounter other groups at range boundaries (García and Braza 1987; Garcia Yuste 1989; Robinson et al. 1987; Schwindt et al. 2004; Solano 1995; Wright 1978, 1985, 1994a), vocalizing and chasing each other (Wright 1978, 1985). Confrontations last from a few minutes to almost half an hour (Aquino and Encarnación 1986a, Wright 1985). Vocal confrontations include resonant whooping by both groups (Moynihan 1964, Wright 1981). In the strictly nocturnal species, encounters are more likely when the moon is full or directly overhead, whereas the cathemeral

Table10.3 Home Range Size and Ranging Patterns

TAXON	STUDY SIZE	HOME RANZE SIZE (HA)	NUMBER OF GROUPS	AVERAGE DAY RANGE (M)	AVERAGE NIGHT RANGE (M)	SAMPLING EFFORT	SOURCES
A. a. azarai	Presidente Hayes, Paraguay	5	1	199 (100–400)	420 (120–600)	10 day and 10 night follows during 2 weeks in July–August (n = 1 group)	Wright 1989, 1994a,b
A. a. azarai	Guaycolec, Formosa, Argentina	4–10	15			80 day and 40 night follows during 1 year (n = 5 groups)	Fernandez-Duque unpublished
A. a. azarai	Guaycolec, Formosa, Argentina	12	—	—	—	12 months, 262 hr	Arditi and Placci 1990
A. a. boliviensis	Departamento Beni, Bolivia	0.33[1]	1	—	337 (153–440)	10 nights during 5 months, only during full-moon nights (n = 1 group)	García and Braza 1987
A. brumbacki	Río Duda, Tinigua NP, Colombia	17.5	1	—	837.3[2]	53 night follows during 5 months (n = 1 group)	Solano 1995
A. nigriceps	Cocha Cashu, Manú NP, Peru	7–14	3	—	708 (340–1,025)	60 night follows during 1 year (n = 1 group)	Wright 1989, 1994a,b
A. trivirgatus[3]	Pto. Bermúdez, Dept. Pasco, Peru	3.1	1	—	252 (60–450)	34 night follows during 9 weeks (n = 1 group)	Wright 1978
A. vociferans	Tiputini Biodiversity Station, Yasuni	6.3	1		645 (150–1,358)	24 night follows during 10 months (n = 1 group)	Fernandez-Duque et al. 2008a

[1] The group occupied an island forest of 0.33 ha.

[2] The source refers to "average daily ranging." It is assumed it referred to night ranging.

[3] The source refers to A. trivirgatus, but it should be A. nigriceps given its locality.

A. a. azarai has intergroup encounters during the night as well as during daylight hours (E. Fernandez-Duque personal observation).

The function of territories and the behavioral mechanisms that maintain them remain unknown. It has been suggested that the animals defend feeding sources (Wright 1985), and it is also possible that at least some of the reported encounters are related to social or reproductive interactions with other groups or floaters. For example, in a population of *A. a. azarai*, some of the most aggressive interactions involve the resident group and a dispersing animal, possibly in search of reproductive opportunities (Fernandez-Duque 2003, 2004, 2009; Fernandez-Duque and Huntington 2002).

Ranging is strongly influenced by available moonlight in all examined species. The distance traveled was significantly longer during full-moon nights than during new-moon nights in *A. a. boliviensis* in Bolivia (García and Braza 1987),

A. a. azarai in Argentina (Fernandez-Duque 2003) and Paraguay (Wright 1985, 1989), *A. nigriceps* in Peru (Wright 1978, 1981, 1985), and *A. vociferans* in Ecuador (Fernandez-Duque et al. 2008a). In *A. a. azarai*, ranging distance was also influenced by ambient temperature. When the moon was full, animals traveled more during warm nights than during cold ones, but there were no effects of temperature during the new moon (Fernandez-Duque 2003).

Diet

There are still no satisfactory quantitative estimates of diet composition and foraging for any of the strictly nocturnal species (Table 10.4). Wright's work (1986, 1994a) on *A. nigriceps* in Manú National Park, Peru, is the only thorough attempt at quantifying diet in one of the nocturnal species; but the problems of obtaining quantitative estimates in a nocturnal primate were numerous. Cathemeral

Table10.4 Diet Composition

SPECIES AND SUBSPECIES	LOCATION	FRUIT	NECTAR AND FLOWERS	LEAVES	FUNGUS	INSECTS	LENGTH OF STUDY (MONTHS)	SOURCES
A. a. azarai	Presidente Hayes, Paraguay	16	33	40		1	2	Wright 1985
A. a. azarai	Guaycolec, Formosa, Argentina	45	14	41		1	12	Arditi and Placci 1990
A. a. azarai	Guaycolec, Formosa, Argentina	66	1	15	2	1	2	Giménez. and Fernandez-Duque 2003, Giménez 2004
A. brumbacki	Río Duda, Tinigua NP, Colombia	59	13			28	6	Solano 1995[5]
A. lemurinus	Quimbaya, Depto Quindio, Colombia		1				1	Marín-Gómez 2008
A. miconax	Departamento Amazonas, Peru	1	1				Ad libitum obs.	Cornejo et al. 2008
A. nigriceps	Cocha Cashu, Manú NP, Peru	60	3	4		1	15	Wright 1985
A. vociferans	Puerto Huamán y Mishana, Peru	83	17			—	9	Puertas et al. 1992
A. zonalis		65		30		5		Hladik et al. 1971[5]

[1] No quantitative estimate.

[2] On several occasions they have been observed ingesting an unidentified fungus.

[3] Nectar was mainly ingested in July and August; the average of those 2 months was 67%.

[4] No quantitative data available; the author considered it not extensive since there were no leaf veins or leaf refuse in 64 fecal samples examined.

[5] Stomach content of "several" individuals.

owl monkeys (*A. a. azarai*) have provided opportunities to examine the diet in some detail during daylight hours (Arditi 1992, Giménez and Fernandez-Duque 2003, Giménez 2004, Wright 1985), but determining their foraging habits during the night remains a challenge.

Owl monkeys are primarily frugivorous. Fruits are the most consumed item in *A. nigriceps* (Wright 1985, 1986, 1994a), *A. a. azarai* (Arditi 1992, Fernandez-Duque et al. 2002, Giménez and Fernandez-Duque 2003, Wright 1985), and *A. vociferans* (Puertas et al. 1992). *Ficus* spp. fruits are a highly preferred item in all examined species.

Leaf and insect eating is virtually impossible to quantify during the night (Arditi 1992, Wright 1985). The absence of leaf veins or leaf refuse in 36 fecal samples and the fact that she observed *A. nigriceps* eating a vine leaf only occasionally led Wright (1985) to conclude that leaf eating may not be important in *A. nigriceps*. Still, it remains possible that leaf eating in strictly nocturnal owl monkeys occurs but remains undetected.

On the other hand, leaf consumption has been regularly observed in the cathemeral *A. a. azarai*. A study of the *A. a. azarai* diet in Paraguay (Ganzhorn and Wright 1994, Wright 1985) and two studies in Argentina (Arditi 1992, Giménez and Fernandez-Duque 2003) found a significant consumption of leaves (Table 10.4). Arditi and Placci (1990) reported that the diet of diurnal and crepuscular owl monkeys in Argentina was more than 40% leaves. Giménez (2004) found that the consumption of leaves reached 30% for a group inhabiting a patch of thorn forest, whereas it was lower (15%) for a group inhabiting a type of forest where fruits are more abundant.

Although insects are undoubtedly part of the diet, quantifying their representation in the diet has proven so far impossible. Most authors have observed owl monkeys eating insects (Arditi 1992, Giménez and Fernandez-Duque 2003, Moynihan 1964, Puertas et al. 1992, Wright 1985), but none has been able to obtain quantitative estimates of its prevalence. Wright (1985:60) observed them eating lepidopterans (moths 2–8 cm), coleopterans (beetles, 2–4 cm), and spiders; but she "could not collect quantitative data on insect-foraging." Fernandez-Duque (unpublished data) observed *A. a. azarai* eating cicadas a few times during the daylight hours, and it is frequently reported that pet owl monkeys keep houses clean of spiders (Fernandez-Duque, unpublished data).

Flowers may be an important food item for *A. a. azarai* and *A. lemurinus* during certain times of the year (Ganzhorn and Wright 1994, Giménez and Fernandez-Duque 2003, Marín-Gómez 2008, Wright 1985). In August and September, when fruit availability is lower (Fernandez-Duque et al. 2002), *A. a. azarai* routinely feeds on trumpeter tree flowers (*Tabebuia ipe*).

REPRODUCTIVE PARAMETERS

The reproductive biology of some species has been well documented in captive animals (Table 10.5). In captivity,

A. lemurinus males enter puberty at a surprisingly early age (Dixson 1994, 1998; Dixson et al. 1980), with testosterone first increasing when individuals were between 7 and 11 months of age (*n* = 6 males). The timing of puberty does not seem to be affected by the social context since there are no differences in levels of testosterone or glandular growth between males caged together with the family or caged alone.

Full sexual maturation of *A. lemurinus* males is achieved at approximately 2 years of age, as indicated by measures of body mass, growth of the subcaudal scent gland, and circulating reproductive hormones (Dixson 1982, 1983b, 1994; Dixson and Gardner 1981; Dixson et al. 1980; Hunter et al. 1979). The timing of first reproduction in captive *A. vociferans* and *A. nancymaae* females occurred between 3 and 4 years of age.

Sexually mature captive males of *A. lemurinus* show partially arrested spermatogenesis with a very low sperm count and small testis size (Dixson 1986, Dixson and Gardner 1981, Dixson and George 1982, Dixson et al. 1979). The median testis volume was 514 mm^3 (range 378–673, *n* = 12 males). The average testis volume of 14 free-ranging adult *A. a. azarai* was 190 mm^3 (range 86–308) (Fernandez-Duque, unpublished data). Although estimates in the wild were obtained by applying the same methodology used with the captive animals, it seems reasonable that the differences in volume between species may be research artifacts.

The knowledge of reproductive parameters in wild individuals is much more limited. A physical examination of 115 individuals of *A. a. azarai* showed that they did not reach adult body mass or exhibit a fully developed subcaudal gland until they were approximately 4 years of age (Fernandez-Duque 2004, Juárez et al. 2003). In this population, age at first reproduction was at least 5 years, as indicated by age at dispersal, the average delay from pairing to parturition, and the observation of one female of known date of birth who reproduced for the first time when she was 58 months old (Table 10.5).

Mating is relatively infrequent in owl monkeys, although it is extremely likely that unobserved mating behavior takes place during the night. Over the course of 3 years and more than 2,000 hr of observations, mating behavior was recorded only on eight occasions (*n* = 5 groups) (Fernandez-Duque et al. 2002). Further evidence for the low frequency of mating comes from studies in captivity. Dixson (1994) recorded it on only 19 occasions during 278 hr of observation. Mating in captive *A. nancymaae* was more frequent among newly formed pairs than established ones (Wolovich and Evans 2007). Mating has been observed during pregnancy in free-ranging *A. a. azarai* and in captive *A. lemurinus* and *A. nancymaae* (Dixson 1994, Hunter et al. 1979, Wolovich and Evans 2007).

Owl monkey females' reproductive cycle lasts 13–25 days and gestation length 117–119 days depending on the species (Dixson et al. 1980, Wolovich et al. 2008a) (Table 10.5), and females usually produce one infant per

Table 10.5 Life History and Reproductive Parameters

SPECIES	CAPTIVE (C) OR WILD (W)	AVERAGE INTERBIRTH INTERVAL (IBI, DAYS, RANGE), NUMBER OF INTERVALS	CYCLE LENGTH	GESTATION LENGTH	BIRTH SEASONALITY	FEMALE AGE AT FIRST REPRODUCTION	SOURCES
A. griseimembra	C	253 (166–419), 36 IBIs	16 (range 13–19, $n = 3$)		No		Dixson et al. 1980, Bonney et al. 1979, Dixson 1983a,b
A. griseimembra	C	271 (median = 258 days), 48 IBIs		133	No		Hunter et al. 1979
A. trivirgatus	C			126, 138, and 148–159			Meritt personal communication as cited in Hunter et al. 1979: Hall and Hodgen 1979 as cited in Dixson 1983a,b Elliot et al. 1976 as cited in Dixson 1983a,b
A. vociferans	C	12.8 ± 6.5 ($n = 110$ IBIs w), 10.2 ± 3.2 ($n = 19$ IBIs c)		146, 132, and 151 ($n = 3$)	58.8% (87/148) December–March, 75.7% (112/148) December–May	48 ± 12 ($n = 15$)	Montoya et al. 1995
A. nancymaae	C	12.72 ± 5.72 ($n = 75$ IBIs)			52.3% (66/126) October–January	40.56 ± 7.82 ($n = 9$)	Gozalo and Montoya, 1990
A. nancymaae	C		18.5 ± 1.3 ($n = 3$)	117 ± 8 ($n = 4$)	May–October		Wolovich et al. 2008a
A. azarai	C		25.0 ($n = 1$)	121 ($n = 1$)			Wolovich et al. 2008a
A. nigriceps	W				100% (9/9) August–February		Wright, 1985
A. nancymaae	C			122–141	No		Málaga et al. 1997
A. nancymaae	W				December–March		Aquino et al. 1990
A. vociferans	W				November–January		Aquino et al. 1990
A. a. azarai	W	Median 370 (345–426), 13 IBI			88% (24/27) October–November	60 ($n = 1$)	Fernandez-Duque et al. (2002)

year. In captivity, twinning occurred in one of 169 births (Gozalo and Montoya 1990) and in one of 287 births (Málaga et al. 1997) in *A. nancymaae* and in three of 365 births in a colony of unknown karyotypes (Gibson, S. V., et al. 2008). In the wild, it has been reported only once in *A. vociferans* (Aquino et al. 1990). Two newborns, weighing 125 and 150 g, were removed from the male's back when the group was captured.

Survival tends to be higher among infants than juveniles, both in captivity and in the wild. Infant survival in captivity was 93.8% during the first week of life (Gozalo and Montoya 1990, Málaga et al. 1997) and 96% (26 of

1. Slow loris (*Nycticebus coucang*).
   (Photograph by K. A. I. Nekaris.)

2. Ring-tailed lemurs (*Lemur catta*).
   (Photograph by Lisa Gould.)

3. Red-ruffed lemur (*Varecia rubra*).
   (Photograph by Pete Meenen.)

4. Red-fronted brown lemur (*Eulemur fulvus rufus*).
   (Photograph by Lisa Gould.)

5. Spectacled galago (*Galago matschiei*).
   (Photograph by Simon Camp.)

6. Bornean tarsier (*Tarsius bancanus)* mother and infant.
   (Photograph by Rachel Munds.)

7. Goeldi's monkey (*Callimico goeldii*). (Photograph by Leila Porter.)

8. Golden-mantled tamarin (*Saguinus tripartitus*). (Photograph by Dylan Schwindt.)

9. White-faced capuchin monkeys (*Cebus capucinus*). (Photograph by Matthew A. Wyczalkowski.)

10. Brown capuchin monkey (*Cebus apella*). (Photograph by Erin Ehmke.

(Counter-clockwise from top)

11. Northern muriqui (*Brachyteles hypoxanthus*).
    (Photograph by Mariane Kaizer.)

12. White-bellied spider monkey (*Ateles belzebuth*).
    (Photograph by Dylan Schwindt.)

13. Mantled howler monkey (*Alouatta palliata*).
    (Photograph by Matthew A. Wyczalkowski.)

14. White-faced saki monkey (*Pithecia pithecia*).
(Photograph by Marilyn Norconk.)

15. Red uakari monkey (*Cacajao calvus*).
(Photograph by Mark Bowler.)

16. Red titi monkey (*Callicebus discolor*).
    (Photograph by Dylan Schwindt.)

17. Squirrel monkey (*Saimiri sciureus*).
    (Photograph by Marilyn Norconk.)

(Counter-clockwise from top)

18. Black and white colobus (*Colobus guereza kikuyensis*).
(Photograph by Yvonne de Jong/Tom Butynski.)

19. Yunnan snub-nosed monkey (*Rhinopithecus bieti*).
(Photograph by Cyril C. Grueter.)

20. Francoís langur (*Trachypithecus francoisi*) mother and infant.
(Photograph by Pete Meenen.)

21. Barbary macaques (*Macaca sylvanus*).
(Photograph by Agustín Fuentes.)

22. Sulawesi macaque (*Macaca nigra*) mother and infant.
(Photograph by Pete Meenen.)

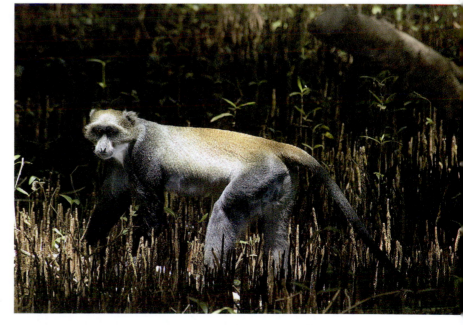

(Clockwise from top left)

23. Patas monkey (*Erythrocebus patas*).
    (Photograph by Karin Enstam Jaffe.)

24. Vervet monkey (*Chlorocebus pygerythrus hilgerti*).
    (Photograph by Yvonne de Jong/Tom Butynski.)

25. Blue monkey (*Cercopithecus mitis albogularis*).
    (Photograph by Yvonne de Jong/Tom Butynski.)

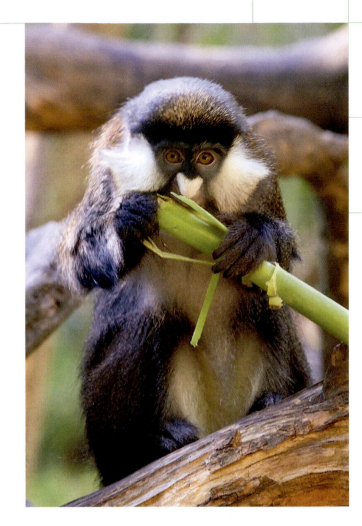

26. Schmidt's guenon (*Cercopithecus ascanius schmidti*).
(Photograph by Pete Meenen.)

27. Hamadryas baboons (*Papio hamadryas*).
(Photograph by Helga Peters.)

28. Olive baboon (*Papio anubis*).
(Photograph by Yvonne de Jong/Tom Butynski.)

29. Mandrill (*Mandrillus sphinx*).
(Photograph by Joanna Setchell.)

30. White-cheeked gibbon (*Nomascus leucogenys*). (Photograph by Pete Meenen.)

31. Siamang (*Symphalangus syndactylus*). (Photograph by Pete Meenen.)

32. Eastern gorillas (*Gorilla beringei*).
(Photograph by Martha Robbins.)

33. Orangutan (*Pongo pygmaeus*).
(Photograph by Karen Caster.)

34. Chimpanzee (*Pan troglodytes verus*) mother and offspring.
    (Photograph by Tatyana Humle.)

35. Bonobo (*Pan paniscus*).
    (Photograph by Nicholas Malone.)

27 infants) during the first 6 months of life in *A. a. azarai* in the wild (Fernandez-Duque et al. 2002). On the other hand, survival was lower for 1-year-old captive juveniles of *A. vociferans* (76.9%) (Montoya et al. 1995) and *A. nancymaae* (85.8%) (Gozalo and Montoya 1990). Further evidence for lower survival among juveniles comes from a wild population of *A. nancymaae*, where juveniles were the least numerous age category (6%–7%) in 14 captured and 23 partially captured groups (Aquino and Encarnación 1989).

There is a strong environmental influence on the timing of births in free-ranging individuals. All nine births recorded in four groups of *A. nigriceps* in Peru occurred between August and February (Wright 1985). Births were estimated to occur between December and March in *A. nancymaae* based on the presence of dependent and independent offspring in 75 captured groups in northeastern Peru (Aquino et al. 1990). In the Argentinean Chaco, most births (88%, n = 24) took place during an 8-week period between late September and late November (Fernandez-Duque et al. 2002).

In captivity, the tropical species *A. lemurinus* and *A. nancymaae* bred throughout the year when photoperiod was kept constant (Dixson 1994, Málaga et al. 1997) but *A. nancymaae* and *A. vociferans* adjusted their reproduction accordingly when housed under conditions of natural photoperiod (Gozalo and Montoya 1990, Montoya et al. 1995). Finally, when owl monkeys that had been living in indoor facilities were housed in an outdoor facility at 25° of latitude in the Northern Hemisphere, they became increasingly more seasonal the longer they lived outdoors (Holbrook et al. 2004). Very interestingly, most births were confined to the March–June period, a 6-month difference from the birth season in the Argentinean Chaco *A. a. azarai* population, which is also located at 25° of latitude but in the Southern Hemisphere.

SOCIAL ORGANIZATION

Serial Monogamy

Owl monkeys are one of the few socially monogamous primates in the world (Fuentes 1999b; Kappeler and Van Schaik 2002; Kinzey 1997a,b; Moynihan 1964, 1976; Robinson et al. 1987; Wright 1981). They live in small groups (two to six individuals), which include an adult heterosexual pair, one infant, one or two juveniles, and sometimes a subadult (Table 10.2). There is never more than one reproducing female in the group or more than one adult male caring for the offspring. Still, data from a long-term study of wild *A. a. azarai* suggest that the social system of owl monkeys may be more dynamic than commonly assumed, in good agreement with cumulating evidence from other socially monogamous primates (Brockelman et al. 1998, Palombit 1994a, Reichard 1995). In the aforementioned *A. a. azarai* population, turnover of resident adults is very frequent: 14 of 15 pairs had at least one of the mates replaced during a 3-year period. Males and females were similarly likely to be replaced. In almost all cases, mate replacement occurred as a consequence of an intruding adult expelling the same-sex resident individual and not because of desertion of one of the pairmates (Fernandez-Duque 2004, 2009).

Biparental Care

The intensive involvement of the male in infant care is one of the most fascinating and unique aspects of the social organization of the genus (Fernandez-Duque et al. 2009). It is comparable only to the pattern described in *Callicebus* spp., where there is intensive paternal care but no involvement of the siblings in the care of infants (Hoffman et al. 1995a, Mendoza and Mason 1986a, Wright 1984). Sibling care has been recorded infrequently in captive groups of *A. lemurinus* and *A. a. boliviensis* (Dixson and Fleming 1981, Jantschke et al. 1996, Wright 1984) and once in free-ranging *A. a. azarai* following the eviction of the putative father of the infant (Fernandez-Duque et al. 2008b).

There have been several detailed studies of parental behavior and infant development in captive groups (Dixson and Fleming 1981, Jantschke et al. 1996, Málaga et al. 1997, Meritt 1980, Robinson et al. 1987, Wright 1984). Although studies of free-ranging populations have been limited by the nocturnal habits of owl monkeys (Wright 1984, 1985), studies of the cathemeral *A. a. azarai* have confirmed the findings in captivity (Juárez et al. 2003; Rotundo et al. 2002, 2005; Schwindt et al. 2004; Wright 1981, 1984, 1986, 1994a).

Newborns are carried in a distinctive ventrolateral position, in the flexure of the mother's thigh during the first 2 postnatal weeks. In captivity, the mother is the main carrier of the single offspring only during the first week in *A. lemurinus* (Dixson and Fleming 1981) and during the first 2 weeks of life in *A. a. boliviensis* (Jantschke et al. 1996); thereafter, the male takes over the role. After the first week of life, free-ranging *A. a. azarai* males carry the infant 84% of the time (Juárez et al. 2003, Rotundo et al. 2002). Rather than providing comfort and extended periods of physical contact with their infants, it appears that female owl monkeys have evolved mechanisms to reject the developing offspring once suckling is complete and to encourage it to transfer to the adult male (Dixson and Fleming 1981, Wright 1984). For example, 40 of 146 suckling bouts were followed by maternal rejection in captive *A. lemurinus* (Dixson 1994).

The male not only carries the infant most of the time but also plays, grooms, and shares food with the infant. Food sharing was prevalent among captive *A. nancymaae* and *A. lemurinus*, as well as in free-ranging *A. azari* (Feged et al. 2002; Wolovich et al. 2006, 2008a; Wright 1984). Both in captivity and in the wild, adults shared food with each other and with the young. The closer relationship of the male with the infant continues as the infant approaches maturity (Dixson 1983b). Then, juveniles maintain more frequent social proximity with the male than the female (Juárez et al. 2003) and may even disperse from their natal group following eviction of their putative father by an incoming male (Fernandez-Duque 2009). The strong male/infant attachment is also manifested in the preference of the infant to stay in the group with the male following the eviction of the mother by an incoming female (Fernandez-Duque 2004), instead of dispersing with the mother. These observations suggest that, in owl monkeys, the infant may be primarily attached to the putative father, similar to what has been described in titi monkeys (Fernandez-Duque et al. 2009; Hoffman et al. 1995a,b; Mendoza and Mason 1986a,b).

The unquestionable intensive male care notwithstanding, the evolution and maintenance of paternal care in owl monkeys remain largely unexplained. It has been hypothesized that paternal care may be adaptive because it increases offspring survival or offers increased foraging opportunities for the lactating female (Fernandez-Duque et al. 2009; Tardif 1994; Wright 1984, 1986). Data from free-ranging *A. a. azarai* do not support the "increased foraging" hypothesis since the adult carrying the infant traveled most frequently in the middle of the group, sometimes first but rarely last as predicted by the hypothesis (Rotundo et al. 2005). At present, there is no evidence suggesting that male care reduces the risk of infanticide in free-ranging *A. a. azarai*. A third possibility is that male care could function as mating effort. This possibility is somewhat supported by the observation of an intruding *A. a. azarai* male that, following eviction of the putative father, took care of the infant in a manner that could not be distinguished behaviorally from the care provided by the putative father (Fernandez-Duque et al. 2008b).

Dispersal

Based on the species-specific social group size, it has usually been assumed that natal dispersal might occur when individuals are 2 or 3 years of age and have reached adult size. Additional inferences on age at dispersal were made from changes in group composition (Fernandez-Duque and Huntington 2002, Huntington and Fernandez-Duque 2001, Wright 1985). More recently, the first data on dispersal from identified individuals of known sex and age have been obtained from free-ranging, radiocollared *A. a. azarai* and have shown that the timing of dispersal is flexible and relies on local ecological, social, and demographic variables (Fernandez-Duque 2009).

All *A. a. azarai* males and females left their natal group, and they did so when they were between 2.2 and 4.9 years of age (Fernandez-Duque 2004, 2009; Juárez et al. 2003). Animals disperse from their natal groups before reaching full sexual maturation, around the time of sexual maturation, and well after it but never before they are 26 months old. Dispersal of relatively young animals seems to be associated with some major changes in group composition, like eviction of the resident male or female. Although most animals disperse when they are approximately 3 years of age, some stay until they are 4 or even 5 years old (Fernandez-Duque and Huntington 2002, Juárez et al. 2003). Aquino et al. (1990) also found more than two adults in captured groups, suggesting that nonreproductive adult offspring may be tolerated for a while. Thus, the available evidence suggests that age at dispersal may be flexible and dependent on the social context in which the predispersing individual finds itself.

There are no data on free-ranging populations revealing whether the observed peripheralization of subadults that precedes dispersal is triggered by aggression from within the group. In captivity, maturing offspring are tolerated by their parents and no increase in agonistic behavior occurs between pubertal monkeys and their parents (Dixson 1983b). Two- or 3-year-old *A. a. azarai* individuals sometimes lag behind the rest of the group or sleep in different trees, suggesting that some peripheralization may start taking place around that time. Some animals leave the group for a few days and then return. The lack of unoccupied areas of the forest where dispersing animals could establish their own territory may lead offspring to postpone dispersal in this population of *A. a. azarai*.

Most dispersal events in *A. a. azarai* occur immediately before or during the birth season and very few of them during the mating season. The concentration of these events around the birth season raises the possibility that births within the group may trigger the process of dispersal. On the other hand, it is also possible that dispersal is timed to take place in anticipation of the May–June mating season. Finally, given that dispersed individuals need to range solitarily for a few weeks to many months before moving into a new group, it is possible that dispersal takes place during the spring and summer months when temperatures are not as harsh and food resources are more abundant (Fernandez-Duque et al. 2002, Fernandez-Duque 2009).

Mate Choice and Pair Formation

A. a. azarai individuals appear to disperse from their natal groups to find reproductive opportunities. There is no indication that they may find reproductive opportunities within their natal territory, as has been described for gibbons (Brockelman et al. 1998, Palombit 1994a). Following dispersal, young adults range alone for various periods of time, in apparent search of a reproductive opportunity within a social group. Both male and female floaters replace resident adults through a process that may take a few days and involves aggressive interactions (Fernandez-Duque 2004, 2009). Frequently, mating takes place as soon as the new adult is accepted into the group. The observation of mating following pair formation agrees well with results from pair-testing experiments in captivity. *A. lemurinus* and *A. nancymaae* individuals mated more frequently in the presence of a new partner than when paired with their regular mate (Dixson 1994, Wolovich and Evans 2007).

New pairs of free-ranging *A. a. azarai* take at least 1 year until the female produces offspring, even if the pair is formed during the mating season. In captivity, the latency to reproduce could be taken as an indirect indicator of successful pair-bonding. Montoya et al. (1995) found that, on average, a captive reproducing *A. vociferans* female took 26 months to reproduce after pairing. Another indirect indication of an improved pair-bond with time could be the shortening of interbirth intervals in multiparous pairs. *A. nancymaae* captive females took, on average, 11 months between the first and second births but only 8.8 months following the third one (Málaga et al. 1997).

Intrasexual Competition and Aggression

Social groups sometimes interact aggressively with each other at territory boundaries. Interactions between members of social groups and floaters can also be aggressive. Both sexes participate in these interactions (Fernandez-Duque 2004, Fernandez-Duque and Rotundo 2003). Most resident adults in 15 social groups of *A. a. azarai* had broken or missing canines (78% of males and 56% of females), and approximately one-third had damaged earlobes (28% of males and 32% of females). Given the lack of observed intragroup aggression, the wounds are most likely the consequence of the severe fights that take place during encounters with other groups or with floaters trying to take over (Fernandez-Duque 2004). Fights have also been reported in *A. nigriceps* (Wright 1985) and *A. nancymaae* (Aquino and Encarnación 1989). In the latter case, when the authors captured the observed group, they confirmed that the adult male had wounds in the forelimb and earlobe due to bites.

The prevalence of aggressive interactions is supported by data from experiments in captivity. Most pairs (seven of eight) of confronting *A. lemurinus* males fought vigorously during testing so that four tests had to be terminated to avoid serious injuries to the animals (Hunter 1981, Hunter and Dixson 1983). Similarly, contact aggression occurred in five of eight pairs of confronting females. In 71% of male pairs and 40% of female pairs, the resident monkey became the aggressor.

Communication

Vocal, olfactory, and visual forms of communication are undoubtedly important for owl monkeys (MacDonald et al. 2008; Robinson et al. 1987; Wright 1985, 1989). Their vocal repertoire can be divided into eight different categories (Moynihan 1964). The two most salient calls are resonant whoops and hoots. Resonant whoops are usually produced by both sexes during intergroup encounters and occur together with visual displays like swaying and arching (Moynihan 1964, Wright 1985). Hoots are low-frequency

calls given by one individual in the social group or by a solitary individual that convey information over long distances. Playback of these calls elicits responses from animals in the area, both groups and solitaries (Depeine et al. 2008).

Because they are nocturnal, owl monkeys rely heavily on olfactory cues for communication, using both urine and cutaneous secretions in their scent-marking behaviors to relay information between mates and other conspecifics (Hill et al. 1959, Wolovich and Evans 2007). Captive male owl monkeys have been observed rubbing their glands against their mate's fur as well as drinking their urine (Wolovich and Evans 2007). The secretions from their sternal and subcaudal glands most likely play a role in mediating social interactions, as suggested by differences between sex, age, and family in the chemical constitution of the secretions (MacDonald et al. 2008). Olfaction also plays a prominent role in sexual recognition and aggression (Dixson 1994, Hunter 1981, Hunter and Dixson 1983, Hunter et al. 1984). During the confrontations of *A. lemurinus*, contact aggression between same-sex individuals was always preceded by some form of olfactory communication. Blocking olfactory input led to a reduction in intermale aggression (Hunter and Dixson 1983). In captivity, owl monkeys have been observed self-anointing with plants and millipedes (Zito et al. 2003).

Temporal Plasticity in Activity Patterns

Owl monkeys concentrate most of their activity during the dark portion of the 24 hr cycle, as indicated by observational studies of free-ranging *A. nigriceps* (Wright 1978, 1985, 1989, 1994a,b, 1996), *A. a. boliviensis* (García and Braza 1987, 1993; Garcia Yuste 1989), *A. a. azarai* (Arditi 1992; Fernandez-Duque 2003; Fernandez-Duque and Erkert 2006; Rotundo et al. 2000; Wright 1994a,b, 1996), *A. vociferans* (Fernandez-Duque et al. 2008a), and *A. miconax* (Cornejo et al. 2008) (Table 10.6). The nocturnal activity of these owl monkey species is strongly influenced by available moonlight. Activity is maximal during full-moon nights and minimal when there is no moonlight. *A. nigriceps* spent approximately half of the night active, but maximal levels of activity were recorded when the full moon was near the meridian (Wright 1985, 1989). *A. nancymaae, A. a. boliviensis, and A. vociferans* also showed intensive nocturnal activity during nights of full moon (Aquino and Encarnación 1986a; García and Braza 1987; Wright 1978, 1989; Fernandez-Duque et al. 2008a). Captive *A. lemurinus* in Colombia showed marked lunar periodic variations in activity patterns under natural lighting conditions (Erkert 1974, 1976).

The strictly nocturnal owl monkeys of Colombia (*A. griseimembra*) were the focus of a series of laboratory experiments analyzing circadian rhythms of locomotor activity as well as their entrainment and masking by light (Erkert 1976, 1991, 2008; Erkert and Grober 1986; Erkert and Thiemann-Jager 1983; Rappold and Erkert 1994; Rauth-Widmann et al. 1991). These studies replicated the patterns

Table 10.6 Activity Patterns of Free-Ranging Owl Monkeys (Percentage of Sampling Points when Group Was Active)

SPECIES NAME	MOON PHASE OR SEASON	1 hr	2 hr	3 hr	4 hr	5 hr	6 hr	7 hr	8 hr	9 hr	10 hr	11 hr	12 hr	13 hr	14 hr	15 hr	16 hr	17 hr	18 hr	19 hr	20 hr	21 hr	22 hr	23 hr	24 hr	SOURCE
Aotus azarai azarai	Full moon	100	100	50	40	50	100	0	0	0	0	0	0	0	100	0	0	75	100	75	100	80	0	60	90	Wright 1985, 1989
A. a. azarai	New moon	0	0	40	40	10	50	40	0	30	40	0	60	0	0	0	0	80	100	90	0	50	0	30	10	Wright 1985, 1989
A. a. azarai	Annual mean	50	50	45	40	30	75	20	0	15	20	0	30	0	50	0	0	78	100	83	50	65	0	45	50	Wright 1985, 1989
A. a. azarai	Summer	–	–	–	–	–	92	65	30	50	20	5	0	0	5	8	0	10	10	55	10	5	–	–	–	Arditi 1992
A. a. azarai	Fall	–	–	–	–	–	–	90	65	40	15	15	5	5	0	3	15	45	92	100	–	–	–	–	–	Arditi 1992
A. a. azarai	Winter	–	–	–	–	–	–	100	88	30	32	28	38	25	18	20	35	70	95	100	–	–	–	–	–	Arditi 1992
A. a. azarai	Spring	–	–	–	–	–	95	55	25	60	50	5	0	0	0	0	0	10	80	95	–	–	–	–	–	Arditi 1992
A. a. azarai	Annual mean	–	–	–	–	–	94	78	52	45	29	13	11	7.5	5.8	7.8	13	34	69	88	10	5	–	–	–	Arditi 1992
A. a. azarai	Annual mean	45	48	37	43	51	70	70	47	29	24	14	12	16	11	12	18	46	73	87	68	45	42	45	34	Fernandez-Duque 2003
A. a. boliviensis	Full moon	60	38	35	25	38	40	0	0	–	–	–	–	–	–	0	–	–	12	85	92	50	50	65	45	Garcia and Braza 1987
A. nigriceps	Full moon	80	100	100	100	70	0	0	0	0	0	0	0	0	0	0	0	0	60	60	90	90	75	0	25	Wright 1985, 1989
A. nigriceps	New moon	90	0	0	75	75	0	0	0	0	0	0	0	0	0	0	0	0	60	100	100	70	80	80	70	Wright 1985, 1989
A. nigriceps	Mean	85	50	50	88	73	0	0	0	0	0	0	0	0	0	0	0	0	60	80	95	80	78	40	48	Wright 1985, 1989

TIME OF DAY

152

observed under natural lighting conditions in the controlled setting of the laboratory (Erkert and Grober 1986, Erkert and Thiemann-Jager 1983). When new-moon conditions are simulated by means of continuous low light intensity, there is suppression of locomotor activity, indicating that low light intensity or a brightness change during the dark phase causes particularly strong masking of the light/dark-entrained circadian activity rhythm. Under nonmasking lighting conditions, *A. lemurinus* females show an increase in locomotor activity every 2 weeks, which corresponds approximately to the ovarian cycle length (Rauth-Widmann et al. 1996).

Despite a preference for being active during the night, at least one owl monkey species is also active during daylight. *A. a. azarai* of the Argentinean and Paraguayan Chaco shows peaks of activity during the day as well as during the night (Arditi 1992, Erkert 2004, Fernandez-Duque 2003, Fernandez-Duque and Bravo 1997, Fernandez-Duque and Erkert 2006, Rotundo et al. 2000, Wright 1985). The results from observational studies were confirmed by fitting owl monkeys with accelerometer collars that allow uninterrupted activity recordings over a span of 6 months (Erkert 2004, 2008; Fernandez-Duque and Erkert 2006). Owl monkeys showed a clearly bimodal pattern of motor activity, with two main peaks of activity from 1800 to 2100 hr and from 0500 to 0800 hr (Fig. 10.3, bottom panel). Significant amounts of activity occurred regularly only during the moonlit parts of the night (Fig 10.3, top panel).

An understanding of the mechanisms regulating cathemerality in this species is emerging (Erkert 2004, Fernandez-Duque 2003, Fernandez-Duque and Erkert 2006). The monkeys show marked lunar periodic and seasonal modulations of their activity pattern. At full moon, they are active throughout the night and show reduced activity during the day. With a new moon, activity decreases during the dark portion of the night, peaks during dawn and dusk, and extends over the bright morning hours. Waxing and waning moons induce a significant increase in activity during the first and second halves of the night, respectively. During the cold winter months, the monkeys display twice as much activity throughout the warmer bright part of the day than during the other months. These findings suggest that *A. a. azarai* is mainly a dark-active species but still able to shift a considerable portion of activity into the bright part of the day if unfavorable lighting and/or temperature conditions prevail during the night.

On the other hand, understanding the evolution of cathemerality remains a challenge (Ankel-Simons and Rasmussen 2008). There have been several hypotheses proposed and evaluated to explain cathemerality in primates (Curtis and Rasmussen 2002; Curtis et al. 1999; Donati et al. 2001; Ganzhorn and Wright 1994; Morland 1993; Overdorff 1996a; Overdorff and Rasmussen 1995; Tattersall 1987; van Schaik and Kappeler 1993, 1996; Warren and Crompton 1997; Wright 1989, 1999). One of them suggests

F: 1040903.011 C: 11 D: 23-11-03 12:05 - 21-02-04 12:00 R: 3 - 100 (ACTI)

Figure 10.3 *Top*: Double plot of original activity recordings in a male *Aotus azarai azarai* over a period of 3 months from November 2003 to February 2004 (*top to bottom*). Line spaces between the histograms correspond to 100 counts per 5 min interval. Please note the pronounced lunar periodic course of nighttime activity and that around the new moon (indicated by *black circles* on the *right*) only little motor activity is produced throughout the night, while much more activity occurs throughout the morning hours than at the other phases of the moon. *Bottom*: Double plotted bimodal activity pattern of this owl monkey as averaged over the whole recording period shown above. During the nighttime (2100–0600), the average level of activity is considerably higher than during the bright hours of the day. *Abscissa*: Argentinean official time, which is about 1 hr advanced in relation to the local time at the study site. *Ordinate*: Relative activity, in percent of the mean 24 hr total as averaged over the whole recording period.

that cathemerality may result from unusually harsh climatic conditions, whereas a second one poses that cathemerality may be the consequence of a pronounced seasonality in resource availability (Engqvist and Richard 1991, Ganzhorn and Wright 1994, Overdorff and Rasmussen 1995).

If cathemerality in owl monkeys is, at least partially, a response to food availability and digestive constraints, diurnal activity should increase during months of less fruit and insect availability. Owl monkeys do not show changes in activity patterns that can be interpreted this way. The total amount of diurnal activity in *A. a. azarai* in the Argentinean Chaco (Fernandez-Duque 2003, Fernandez-Duque and Erkert 2006, E. Fernandez-Duque and H. G. Erkert unpublished data) remained fairly constant throughout the year despite seasonal changes in food availability, temperature, and rainfall.

Instead, owl monkeys adjust their periods of activity to changes in ambient temperature, although the effects of temperature are contingent on moon phase. Owl monkeys apparently have a very narrow thermoneutral zone that may range 28°–30°C and a relatively low resting metabolic rate (Le Maho et al. 1981, Morrison and Simoes 1962). In captivity, *A. lemurinus* was most active when ambient temperature was 20°C and least active when it was 30°C (Erkert 1991). Still, the existing characterization of the thermoneutral zone and the relatively low basal metabolic rate should be interpreted with special attention to the geographic origins of the specimens studied in view of the various activity patterns observed in the genus.

It is also reasonable to hypothesize that the risk of predation may influence the activity patterns of owl monkeys. Wright (1989) suggested that a release from diurnal predation pressure may have been a selective force favoring cathemerality in *A. a. azarai*. Unfortunately, no adequate data exist on predation risk, predator activity patterns, or predation events. There is some anecdotal evidence that runs counter to the hypothesis that the absence of large diurnal raptors may have favored cathemerality in owl monkeys. At El Beni in Bolivia, where harpy eagles are present, owl monkeys (*A. a. boliviensis*) showed diurnal activity when the climate was unusually cold (Mann 1956).

CONCLUSIONS

Our understanding of owl monkeys' behavior, ecology, and evolution remains severely limited. Although a picture is emerging about the social organization, behavior, and ecology of the southernmost taxon (*A. a. azarai*), several intriguing aspects of this subspecies will need to be examined in other species before any broad generalizations can be made for the genus as a whole.

Although owl monkeys are undoubtedly socially monogamous, the unexpected fast rate of adult replacement in the *A. a. azarai* population suggests that serial monogamy may be the norm. Thus, the long-held assumption of stable, lasting pair-bonds in monogamous primates will need to be once more revised. As a consequence, it may also become necessary to reevaluate many of our assumptions about the evolutionary forces leading to monogamy. For example, if it is confirmed that putative and nonputative males provide similar care to infants, the function of the intensive care provided by owl monkey males will have to be reevaluated. Logically, a thorough evaluation of the costs and benefits of intensive male care will not be complete until there is an understanding of the genetic structure of the different social groups.

Ongoing genetic studies of 20 social groups in one *A. a. azarai* population (Babb et al. 2008, Lau et al. 2004, Lau 2002, Sharma et al. 2003) will hopefully shed light on the genetic aspects of serial monogamy. Preliminary analyses have indicated a very low level of genetic diversity in the population, with a relatively high degree of relatedness among individuals in different groups. Is it possible that the potential costs of serial monogamy are attenuated through kin selection effects resulting from neighboring groups being formed by closely related individuals?

The function of territoriality in owl monkeys will also need careful examination. To successfully identify some of the relevant factors driving and maintaining territoriality, it will be necessary to develop a semiexperimental approach to examine some of the unresolved issues. For example, playback experiments to simulate intruders or food-provisioning experiments to manipulate available food resources will need to be considered and implemented.

Finally, advancing our understanding of the evolutionary forces favoring monogamy in owl monkeys will require a comparative approach that considers some of the other more tropical, strictly nocturnal owl monkey species as well as some of the other socially monogamous primates. Recent efforts to begin a long-term research program on the behavior and ecology of *A. vociferans*, *Pithecia monachus*, and *Callicebus discolor* in Yasuní National Park, Ecuador (Di Fiore and Rodman 2001; Schwindt et al. 2004; Fernandez-Duque et al. 2008a, 2009), are one step toward accomplishing a more general and broad description of owl monkey ecology and behavior.

11

The Atelines
Behavioral and Socioecological Diversity in a New World Monkey Radiation

Anthony Di Fiore, Andrés Link, and Christina J. Campbell

1. What are fission–fusion societies, and how do these relate to competition for food?
2. What are the two predominant ecological strategies employed by the ateline primates?
3. How do intrasexual (i.e., within the same sex) relationships and mating patterns differ among the ateline genera?

INTRODUCTION

Primates of the subfamily Atelinae are the largest monkeys in the New World and include the howler monkeys (genus *Alouatta*), spider monkeys (genus *Ateles*), woolly monkeys (genus *Lagothrix*), and muriquis (genus *Brachyteles*); additionally, according to some authors, the yellow-tailed woolly monkey is sufficiently distinct from other woolly monkeys (and other atelines) to warrant its elevation to a separate genus, *Oreonax* (e.g., Groves 2001a). Atelines are a *monophyletic* group (descended from a common ancestor), constituting one of the three major platyrrhine radiations recognized in modern molecular systematic studies (e.g., Schneider 2000, Schneider et al. 2001). Morphologically, atelines are characterized by having a muscular prehensile tail, which, in all genera, is commonly used to support the full weight of the body during feeding and as an additional support during locomotion (Fig. 11.1).

In two genera (*Ateles*, *Brachyteles*), the use of rapid suspensory, semibrachiating locomotion is particularly common. Both genera show dramatic modifications of the hand (e.g., a reduced thumb and elongation of the remaining digits) and shoulder (e.g., an elongated and dorsally positioned scapula), as well as elongation of the tail and limbs relative to trunk length, as adaptations to this distinctive locomotor pattern (Hill 1962, Erickson 1963). Adult body size in atelines ranges from just over 3 kg to an estimated 15 kg (Table 11.1). In howler monkeys and woolly monkeys males are considerably larger than females, while in spider monkeys and muriquis there is little sexual dimorphism in body size (Table 11.1).

Figure 11.1 An ateline primate (*Ateles belzebuth belzebuth*) hanging from its prehensile tail (D. Schwindt).

Table 11.1 Currently Recognized Genera and Species of Ateline Primates Based on Recent Molecular and Morphological Studies[1]

SPECIES	COMMON NAME[2]	ADULT MALE BODY WEIGHT (G)[3]			ADULT FEMALE BODY WEIGHT (G)[3]			BODY SIZE DIMORPHISM[4]	BODY WEIGHT SOURCES	IUCN RED LIST STATUS 2008[5]
		AVERAGE	RANGE	NUMBER OF INDIVIDUALS	AVERAGE	RANGE	NUMBER OF INDIVIDUALS			
Alouatta belzebul	Red-handed howler monkey	7,270	6,540–8,000	27	5,525	4,850–6,200	26	1.32	Ford and Davis 1992	VU
Al. caraya	Brown (black) howler monkey	6,800	4,000–9,600	≥19	4,605	3,800–5,410	13	1.48	Ford and Davis 1992	LC
		6,420		58	4,330		117	1.48	Rumiz 1990	
Al. guariba[6]	Red-and-black (brown) howler monkey	6,175	5,200–7,150	4	4,550	4,100–5,000	3	1.36	Ford and Davis 1992	LC
		6,730		4	4,350		5	1.55	Smith and Jungers 1997	
Al. macconelli	Guyanan red howler monkey	7,585	7,170–8,000	3	5,000	5,000	2	1.52	Ford and Davis 1992	LC
Al. nigerrima	Amazon black howler monkey	—			—			—		LC
Al. palliata[7]	Mantled howler monkey	7,150	4,500–9,800	≥56	5,350	3,100–7,600	≥67	1.34	Ford and Davis 1992	LC
Al. pigra	Guatemalan black howler monkey	11,352	11,113–11,590	2	6,434	6,290–6,577	4	1.76	Ford and Davis 1992	EN
Al. sara	Bolivian red howler monkey	—			—			—		LC
Al. seniculus	Venezuelan red howler monkey	7,200	5,400–9,000	61	5,600	4,200–7,000	61	1.29	Ford and Davis 1992	LC
		5,617		31	4,034		29	1.39	Braza et al. 1983	
		7,540	5,000–12,500	8	6,298	4,000–10,000	9	1.20	Hernandez–Camacho and Defler 1985	
		6,306		64	4,670		46	1.35	Rodriguez and Boher 1988	

Ateles belzebuth[8]	White-fronted (white-bellied) spider monkey	8,260	–	10	7,880	–	16	1.05	Smith and Jungers 1997	EN
		8,532	7,264–9,800	12	8,112	5,824–10,400	15	1.05	Ford and Davis 1992	
		8,160	7,600–8,600	5	8,275	7,800–8,750	2	0.99	Karesh et al. 1998	
At. geoffroyi[9]	Geoffroy's (black-handed) spider monkey	8,210	7,420–9,000	56	7,700	6,000–9,400	≥101	1.07	Ford and Davis 1992	EN, CR
At. hybridus	Brown or variegated spider monkey	8,250	7,875–8,625	2	9,151	7,500–10,500	7	0.90	Hernandez-Camacho and Defler 1985[10]	CR
At. paniscus	Red-faced (red-faced black) spider monkey	9,110		20	8,440		42	1.08	Smith and Jungers 1997	VU
		7,335	5,470–9,200	8	8,750	6,500–11,000	12	0.84	Ford and Davis 1992	
Brachyteles arachnoides	Southern muriqui or woolly spider monkey	10,200	10,200	1	8,500	8,500	1	1.20	Lemos de Sá and Glander 1993	EN
B. hypoxanthus	Northern muriqui or woolly spider monkey	9,416	9,250–9,600	3	8,333	6,900–9,300	3	1.13	Lemos de Sá and Glander 1993	CR
		13,800	13,800	1	13,500	~12,00–15,000[11]	–	–	Ruschi 1964	
Lagothrix cana	Gray (Geoffroy's) woolly monkey	9,493	8,930–10,200	3	7,650	7,650	1	1.24	Peres 1994a,b	EN
L. lagotricha	Brown (Humboldt's) woolly monkey	9,000	8,000–10,000	3	5,750	5,000–6,500	6	1.57	Ford and Davis 1992	VU
L. lugens	Colombian woolly monkey	–			6,000	6,000	1	–	Ford and Davis 1992	CR

Table 11.1 (cont'd)

SPECIES	COMMON NAME[2]	ADULT MALE BODY WEIGHT (G)[3]			ADULT FEMALE BODY WEIGHT (G)[3]			BODY SIZE DIMORPHISM[4]	BODY WEIGHT SOURCES	IUCN RED LIST STATUS 2008[5]
		AVERAGE	RANGE	NUMBER AT INDIVIDUALS	AVERAGE	RANGE	NUMBER AT INDIVIDUALS			
L. poeppigii	Silvery (Poeppig's) woolly monkey	7,100		6	4,533		9	1.57	Lu 1999	VU
Oreonax flavicauda	Yellow-tailed or Hendee's woolly monkey				~10000[12]			–	Leo Luna 1984	CR

[1] Taxonomy follows that of Rylands et al. 2000 and Groves 2001a with recent modifications based on molecular data for *Alouatta* Cortés-Ortiz et al. 2003 and *Ateles* Collins and Dubach 2000b.

[2] First name noted is that listed in Groves 2001a. Additional common names, listed in parentheses, come from Rylands et al. (2000).

[3] Body weight averages and ranges are based only upon free-ranging animals or those presumed in the cited source to be free-ranging. Where Ford and Davis's 1992 compilation of platyrrhine body weights is cited as the data source, we recalculated the appropriate range of weights based on the taxonomy adopted here and calculated average weights as the mid-value of the range reported. The mid-value of Ruschi's 1964 estimated size for female northern muriquis is also used as an estimate of average female body weight for that species. For all other sources, average weights represent the arithmetic mean of individual body weights either reported in the cited source or calculated from the original data. Where both Ford and Davis 1992 and Smith and Jungers 1997 have compiled data on a taxon from some of the same original sources, care was taken to include, those data only once in the summary presented here.

[4] Dimorphism calculated as the ratio of average male to average female body weight.

[5] Data compiled from IUCN online resources at www.iucnredlist.org according to 2001 criteria. Listings apply either to entire species or to one or more recognized subspecies; thus, more than one listing is possible for a taxon. For some species, conservation status has been assessed only for a subset of currently recognized subspecies. Abbreviations used are as follows: CR, critically endangered; EN, endangered; VU, vulnerable; LC, least concern

[6] *Alouatta guariba* was formerly called *Al. fusca* and appears in many earlier publications as such.

[7] *Alouatta palliata* includes *Al. coibensis* recognized by Rylands et al. 2000, Groves 2001a, and other authors.

[8] *Ateles belzebuth* includes *At. chamek* and *At. marginatus* recognized by Rylands et al. 2000, Groves 2001a, and other authors.

[9] *Ateles geoffroyi* includes *At. fusciceps* recognized by Groves (2001a) and other authors.

[10] Source refers to taxon as *Ateles paniscus brunneus*.

[11] Estimated range for the species. This estimate and reference are included here only for completeness of the data set.

[12] The only body weight available for *Oreonax flavicauda* is an estimate for which the sex is not specified. This estimate and reference are included here only for completeness of the data set.

DISTRIBUTION, TAXONOMY, AND PHYLOGENY

Alouatta is the most widely distributed ateline genus, with a geographic range extending from the dry, deciduous forests of the northern Argentine Chaco to Central America as far north as eastern Mexico. *Ateles*, too, is distributed widely in moist tropical and deciduous forests throughout South and Central America, from northern Bolivia to the coastal regions of southern Mexico and the Yucatan Peninsula. *Lagothrix* shows a broad distribution in rain forest habitats throughout the western Amazon and upper Orinoco basins, and some isolated populations occur west of the eastern cordillera of the Andes in Colombia, while *Oreonax* is restricted to only small areas of moist, evergreen forest on the eastern Andean slopes in northern Peru. The geographic range of *Brachyteles* is limited to several small remnants of Brazilian Atlantic Forest (Fig. 11.2).

Most modern taxonomic treatments divide the atelines into two groups, one comprising the various species of *Alouatta* and one containing the remaining genera, which had a common ancestor roughly 15–16 million years ago (Goodman, M., et al. 1998, Schneider 2000) (Fig. 11.3). These groups are typically separated at the level of the tribe (Alouattini and Atelini, Schneider et al. 1993, 1996; Schneider and Rosenberger 1996; Goodman, M., et al. 1998; Rylands et al. 2000), although Groves (2001a) assigns the howler monkeys to their own subfamily. Here, we adopt the former and more commonly used tribe-level division, and we thus use the term "atelin" to refer to the non-alouattin genera *Brachyteles*, *Ateles*, *Lagothrix*, and *Oreonax*.

Both morphological and molecular phylogenetic studies consistently place *Alouatta* as the basal genus within the ateline clade, but the relationships among the remaining genera are more controversial. Morphological studies have tended to place *Brachyteles* and *Ateles* or *Brachyteles* and *Lagothrix* as sister taxa (e.g., Ford 1986); nonetheless, Kay (1990) has noted a possible sister relationship between *Brachyteles* and *Alouatta* on the basis of dental characteristics, and cranial features suggest a closer relationship between *Lagothrix* and *Ateles* (Groves 2001a). Rosenberger and Strier (1989) argue that a sister grouping of *Brachyteles* and *Ateles* is most likely on the basis of shared morphological and behavioral features associated with their extensive use of suspensory locomotor behavior and feeding postures. In contrast, several recent molecular studies (Schneider et al. 1993, 1996; von Dornum and Ruvolo 1999; Canavez et al. 1999a; Meireles et al. 1999; de Lima et al. 2007) have linked *Brachyteles* and *Lagothrix* as sister taxa, although Collins (2003) argues that the molecular evidence is inconclusive and favors instead an unresolved trichotomy among the atelins. No molecular studies to date have examined *Oreonax*, but on the basis of limited morphological data, Groves (2001a) suggests it may be linked to *Ateles* rather than occupying a basal position within the woolly monkey clade. However, the position has recently been challenged (Rosenberger and Matthews 2008, Matthews and Rosenberger 2008) in favor of rejecting

(A)

(B)

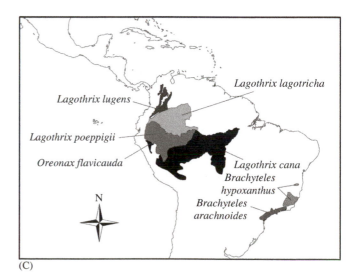
(C)

Figure 11.2 The geographic distributions of extant atelines. *A: Alouatta. B: Ateles. C: Brachyteles, Lagothrix,* and *Oreonax.* Maps are redrawn based on Fooden (1963), Defler and Defler (1996), Kinzey (1997b), Collins and Dubach (2000a), and Cortés-Ortiz et al. (2003).

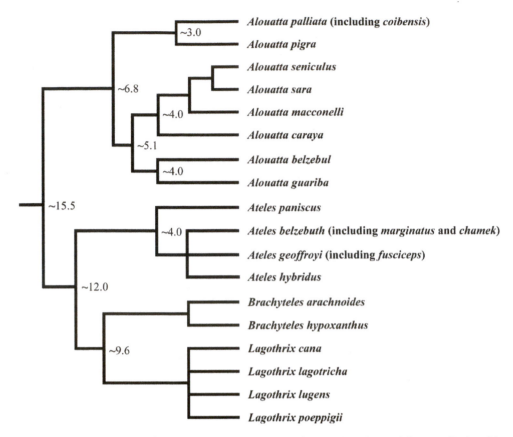

Figure 11.3 Phylogenetic relationships among extant ateline species (excluding *Alouatta nigerrima* and *Oreonax flavicauda*) and selected approximate crown group ages in millions of years. Branching order and dates are derived from Meireles et al. (1999), Collins and Dubach (2000a,b), Collins (2001), and Cortés-Ortiz et al. (2003). Where multiple estimates of the age of a crown group were available from different sources, the midpoint of the range of estimated dates is presented. Note that branch lengths are not drawn to scale.

Oreonax as a distinct atelin genus. Several of the atelines have been the subject of recent molecular analyses concerned with elucidating species-level phylogenies and the biogeographic histories of the genera (*Alouatta,* Meireles et al. 1999, Cortés-Ortiz et al. 2003; *Ateles,* Collins and Dubach 2000a,b; *Brachyteles,* Lemos de Sá et al. 1990, Pope 1998b) (Fig. 11.3).

GROUP COMPOSITION AND SOCIAL STRUCTURE

One of the most remarkable characteristics of atelines is the diversity of grouping patterns and social systems represented in the clade (Table 11.2). Howler monkeys live in cohesive social groups usually containing one or a small number of adult males and several adult females, although occasionally groups with just one reproductive-age individual of each sex have been reported. In most species of howler monkeys (e.g., *Alouatta seniculus, A. caraya, A. guariba, A. pigra*), group size is limited to fewer than 10–15 animals and there is commonly only one adult male per group and seldom more than three. In *A. palliata,* however, groups are much larger and typically contain three or more adult males and up to nine or more adult females; groups of this species containing over 40 members have also been reported (Fedigan et al. 1985;

Chapman 1988b; see also review in Neville et al. 1988). In this genus, individuals of both sexes typically disperse from their natal social groups prior to first reproduction, although some males and females may inherit breeding positions in their natal groups (Rudran 1979; Glander 1980, 1992; Crockett 1984; Clarke and Glander 1984; Pope 1992, 2000b; Rumiz 1990; Crockett and Pope 1993; Calegaro-Marques and Bicca-Marques 1996; Brockett et al. 2000a). In general, females appear more likely to disperse than males, and they move farther distances than males do upon dispersal (Gaulin and Gaulin 1982; Crockett 1984; Glander 1992; Pope 1989, 1992; Crockett and Pope 1993). In this respect, the dispersal pattern can be characterized as "female-biased." Males that disperse typically try to take over existing social groups by evicting the resident dominant male, sometimes forming coalitions with one another to do so (Rudran 1979, Clarke 1983, Pope 1990, Agoramoorthy and Rudran 1993). Among red howler monkeys (*A. seniculus*), dispersing females have often been ousted from their natal groups by aggression from older females and are seldom able to integrate themselves into established social groups. Instead, they must form new social groups with other dispersers and successfully defend a home range against other groups before they begin breeding (Crockett 1984; Crockett and Pope 1988, 1993; Pope 2000b).

Table 11.2 Population Characteristics and Group Composition in Atelines Based on Both Population Censuses and Long-term Behavioral or Ecological Studies[1]

GENUS AND SPECIES	STUDY SITE	POPULATION DENSITY (IND/KM²)	AVERAGE GROUP SIZE (AND RANGE)	GROUP STRUCTURE[2]						NUMBER OF GROUPS	% UNIMALE GROUPS	ADULT SEX RATIO WITHIN GROUPS (FEMALES/MALES)	DATES OF STUDY	SOURCES
				AM	AF	SM	SF	JUV	INF					
Alouatta belzebul	Sapé, Paraiba, Brazil	—	7.4 (5–14)	1.2 (1–2)	—	—	—	—	—	5	80%	—	1985–1986	Bonvicino 1989
Al. belzebul	Paranaíta, Mato Grosso, Brazil	—	7	1	2	0–1	1	2	0–3	1	—	1.5	1999–2000	Pinto and Setz 2004
Al. caraya	Rio Riachuelo, Corrientes, northeastern Argentina	81	6.4 (3–10)	1.4 (1–2)	2.0 (1–3)	0.5 (0–3)		1.9 (0–4)	0.6 (0–2)	11	64%	1.4	1982	Rumiz 1990
Al. caraya	Rio Riachuelo, Corrientes, northeastern Argentina	102	8.4 (5–13)	1.4 (1–3)	2.3 (1–3)	0.9 (0–2)		2.4 (0–5)	1.4 (0–3)	11	73%	1.6	1984	Rumiz 1990
Al. caraya	Rio Riachuelo, Corrientes, northeastern Argentina	—	7.1 (2–12)	1.7 (0–4)	2.4 (1–5)	0.3 (0–2)	0.3 (0–2)	1.6 (0–4)	0.8 (0–3)	24	54%	1.4	1997	Agoramoorthy and Lohmann 1999
Al. caraya	Isla Guascára, Corrientes, northeastern Argentina	280	10.2 (5–15)	2.7 (1–4)	3.8 (2–6)		2.5 (0–5)		1.2 (0–3)	11	27%	1.4	1981	Rumiz 1990
Al. caraya	Ríos Paraná and Paraguay, northeastern Argentina	—	18.5 (16–21)	2 (2–2)	5.5 (5–6)	1 (1–1)	0.5 (0–1)	6 (5–7)	3.5 (3–4)	2	0%	2.8	1998–2000	Bravo and Sallenave 2003
Al. guariba	Cantareira Reserve, São Paulo, Brazil	48–113	5.8 (2–11)	1.8 (1–3)	2.4 (1–4)	—	—	1.2 (0–3)	0.4 (0–1)	25	36%	1.3	1979	da Silva 1981
Al. guariba	Estação Biológica Caratinga, Minas Gerais, Brazil	117	6.8	1.2 (1–2)	2.3 (1–3)	0.4 (0–1)	—	2.0 (0–4)	0.9 (0–2)	19	84%	1.9	1983–1984	Mendes 1989
Al. guariba	Estação Biológica Caratinga, Minas Gerais, Brazil	—	7.0 (4–11)	1.1 (1–2)	1.9 (1–3)	0.5 (0–1)	—	3.4 (2–6)		10	90%	1.7	2000	Strier et al. 2001

161

Table 11.2 (cont'd)

GENUS AND SPECIES	STUDY SITE	POPULATION DENSITY (IND/KM2)	AVERAGE GROUP SIZE (AND RANGE)	GROUP STRUCTURE[2]						NUMBER OF GROUPS	% UNIMALE GROUPS	ADULT SEX RATIO WITHIN GROUPS (FEMALES/MALES)	DATES OF STUDY	SOURCES
				AM	AF	SM	SF	JUV	INF					
Al. palliata	Santa Rosa National Park, Guanacaste, Costa Rica[3]	4.9	13.6	22%	44%	–	–	20%	14%	25	–	2.0	1984	Fedigan et al. 1985, Fedigan 1986
Al. palliata	Santa Rosa National Park, Guanacaste, Costa Rica	7.9	16.3	24%	41%	–	–	16%	19%	34	–	1.7	1992	Fedigan et al. 1998
Al. palliata	Cabo Blanco, Costa Rica	–	14.9 (8–19)	2.5 (1–4)	7.8 (4–12)	–	–	2.4 (0–7)	2.3 (1–4)	8	12.5%	3.1	1987–1988	Lippold 1988, 1989
Al. palliata	Various Locations, Guanacaste, Costa Rica[4]	–	21.8 (13–35)	3.1 (2–6)	10.2 (6–15)	–	–	5.8 (3–12)	2.7 (1–6)	11	0.0%	3.3	Not specified	Jones 1996
Al. palliata	Hacienda La Pacifica, Guanacaste, Costa Rica	–	15.5 (4–29)	2.6	7.6	–	–		4.3	15	–	2.9	1974–1976	Clarke et al. 1986
Al. palliata	Hacienda La Pacifica, Guanacaste, Costa Rica	–	15.7 (4–31)	2.1	8.9	–	–		4.0	16	6%	4.2	1984	Clarke et al. 1986
Al. palliata	Hacienda La Pacifica, Guanacaste, Costa Rica	26	12.6 (2–45)	2.5	6.8	–	–		3.1	27	26%	2.7	1991	Clarke and Zucker 1994[5]
Al. palliata	Hacienda La Pacifica, Guanacaste, Costa Rica	30	10.2 (4–28)	1.6	4.8	–	–	1.5	2.6	34	44%	3.0	1998	Clarke et al. 2002
Al. palliata	Finca Taboga, Guanacaste, Costa Rica	–	11.5 (2–39)	21%	48%	–	–	21%	10%	7–22	12%–50%	2.3	1966–1971	Heltne et al. 1976[6]

Al. palliata	La Selva Biological Reserve, Heredia, Costa Rica	7–15	11 (6–15)	3.3 (1–5)	4.0 (3–6)	2.3 (0–4)	—	—	1.3 (0–2)	7	14%	1.2	1990	Stoner 1994
Al. palliata	Inland Lowland Forest, Chiriqui, Panama	—	18.9 (7–28)	3.9 (2–6)	8 (3–13)	0	—	3.8 (2–6)	3.3 (0–6)	8	0%	2.1	1970–1971	Baldwin and Baldwin 1976
Al. palliata	Barro Colorado Island, Panama[7]	—	21.3 ± 4.5 SD (16–28)	3.9 ± 1.1 SD	8.4 ± 2.7 SD	—	—	2.6 ± 1.3 SD	7.0 ± 1.3 SD	13	—	2.2	1980	Milton 1982[8]
Al. palliata	Barro Colorado Island, Panama	—	19.4 ± 6.3 SD	3.1 ± 1.2 SD	8.6 ± 3.1 SD	—	—	2.5 ± 1.5 SD	5.3 ± 2.4	—	—	2.8	1977–1993	Milton 1996[8]
Al. palliata	Los Tuxtlas, Veracruz, Mexico	23.3	9.1 (5–16)	3.0 (1–5)	4.1 (2–6)	—	—	0.9 (0–3)	1.2 (0–3)	17	6%	1.4	1978–1981	Estrada 1982, 1984
Al. pigra	Palenque, Chiapas, Mexico	23	7.0 (2–12)	2.0 (1–4)	1.9 (1–4)	—	—	1.8 (0–4)	1.0 (0–3)	20	35%	1.0	2000	Estrada et al. 2002a
Al. pigra	Calakmul, Campeche, Mexico	15.2	7.5 (4–9)	2.5 (1–3)	2.2 (1–4)	—	—	1.8 (0–3)	1.0 (0–2)	8	25%	0.9	2001–2002	Estrada et al. 2004
Al. pigra	Yaxchilán, Chiapas, Mexico	12.8	6.6 (4–10)	2.8 (1–5)	2.0 (1–3)	—	—	0.9 (0–2)	1.3 (0–2)	8	25%	0.7	2001–2002	Estrada et al. 2002b, 2004
Al. pigra	Muchukux Forest, Quintana Roo, Mexico	16.5	3.2 (1–8)	—	—	—	—	—	—	25	—	—	1995	Gonzalez-Kirchner 1998
Al. pigra	Tikal National Park, Guatemala	5	6.3	—	—	—	—	—	—	4	—	—	1973	Coehlo et al. 1976a,b
Al. pigra	Tikal National Park, Guatemala	17.8	8.7 (6–12)	2.2 (1–3)	2.9 (1–4)	—	—	1.8 (0–3)	1.8 (1–3)	10	10%	1.3	2002	Estrada et al. 2004
Al. pigra	Bermuda Landing, Gulf Coast, Belize	—	6.8 (4–10)	1.7 (1–3)	2.1 (1–3)	0.3 (0–1)	0.2 (0–1)	1.4 (0–3)	0.4 (0–1)	9	44%	1.3	1981	Horwich and Gebhard 1983
Al. pigra	Bermuda Landing, Gulf Coast, Belize	8.1	4.4 (2–7)	1.1 (1–2)	1.2 (1–2)	—	—	1.3 (0–4)	0.7 (0–2)	13	92%	1.1	1978–1979	Bolin 1981
Al. pigra	Community Baboon Sanctuary, Gulf Coast, Belize	31.9–178.2	6.0 (5.2–6.9)	1.6 (1.3–1.8)	2.1 (1.8–2.4)	—	—	—	—	10–36	—	1.3	1985–2001	Horwich et al. 2000[9]
Al. pigra	Community Baboon Sanctuary, Gulf Coast, Belize	47–257	5.8 (3–16)	1.5 (1–3)	2.0 (1–4)	—	—	2.2 (0–10)	—	38	58%	1.3	1994–1997	Ostro et al. 2001[10]

Table 11.2 (cont'd)

GENUS AND SPECIES	STUDY SITE	POPULATION DENSITY (IND/KM²)	AVERAGE GROUP SIZE (AND RANGE)	GROUP STRUCTURE[2]						NUMBER OF GROUPS	% UNIMALE GROUPS	ADULT SEX RATIO WITHIN GROUPS (FEMALES/MALES)	DATES OF STUDY	SOURCES
				AM	AF	SM	SF	JUV	INF					
Al. pigra	Monkey River, Gulf Coast, Belize	102	6.4 (2–9)	–	–	–	–	–	–	8	–	–	1999–2001	Pavelka et al. 2003
Al. seniculus	Bush Bush Forest, Trinidad	–	8.5 (8, 9)	3.5 (4, 3)	2.5 (2, 3)	–	–	1 (1, 1)	1.5 (1, 2)	2	0%	0.7	1968	Neville 1972a
Al. seniculus	Nourague Station, French Guiana	–	6–8	1	1–2	0–2	–	2–3	1	1	–	1.2	1988–1990	Julliot 1996a
Al. seniculus	Rios Tuparro and Tomo, Vichada, Eastern Colombia	23–27	6.3 (3–9)	1.9 (1–3)	2.4 (1–4)	0.1 (0–1)	–	1.0 (0–2)	1.1 (0–2)	10	30%	1.3	1977–1978	Defler 1981[11]
Al. seniculus	Río Peneya, Meta, Colombia	–	5.5 (3–11)	1.2 (1–3)	1.6 (1–4)	0.1 (0–1)	0.5 (0–2)	1.3 (0–3)	0.3 (0–1)	29	87%	1.3	1971–1974	Izawa 1976
		G: 36	7.7 (4–12)	1.2	2.5	–	–	–	–	25	–	2.1		
Al. seniculus	La Macarena National Park, Meta, Colombia	–	7.5 (2–13)	1.5 (1–2)	2.5 (1–5)	0.6 (0–2)	0.6 (0–2)	1.4 (0–4)	0.9 (0–2)	8	50%	1.7	1986	Izawa and Nishimura 1988
Al. seniculus	Finca Merenberg, Huila, Colombia	–	9 (9, 9)	2 (2, 2)	2.5 (3, 2)	0 (0, 0)	0 (0, 0)	3 (3, 3)	1.5 (1, 2)	2	0	1.3	1975	Gaulin and Gaulin 1982
Al. seniculus	Hato El Frío, Apure State, Venezuela	25	7.6 (5–11)	1.8 (1–3)	3.0 (2–4)	–	–	1.4 (0–3)	1.4 (1–2)	5	40%	1.7	1975–1976	Braza et al. 1981[12]
Al. seniculus	Hato Masaguaral, Guárico State, Venezuela	M: 112	10.5 (6–16)	1.6	2.9	–	–	–	–	27	–	1.8	1978–1981	Crockett 1984[13]
Al. seniculus	Hato Masaguaral, Guárico State, Venezuela	58–223	UM: 7.9 (5.9–9.6); MM: 9.1 (8.5–10.5)	UM: 1; MM: 2.2	UM: 2.5; MM: 2.7	UM: 1.3; MM: 1.2	UM: 1.0; MM: 1.1	UM: 7.1; MM: 7.7	UM: 2.5; MM: 1.5	15–36	36%–93%	UM: 2.5; MM: 1.2	1976–1999	Rudran and Fernandez-Duque 2003[14]

Taxon	Location											Reference
Ateles belzebuth belzebuth	La Macarena National Park, Meta, Colombia	11.6–15.4	23.5 (20, 27)	4 (3, 5)	11.5 (11, 12)	4 (3, 5)		4 (3, 5)	2	2.9	1968	Klein 1972, Klein and Klein 1976[15]
At. belzebuth belzebuth	La Macarena National Park, Meta, Colombia	–	16	4	5	1	3	3	1	1.3	mid 1980s	Izawa and Nishimura 1988, Ahumada 1989
At. belzebuth belzebuth	La Macarena National Park, Meta, Colombia	–	30	5	10	2	4	9	1	2.0	2001–2002	Shimooka 2005
At. belzebuth belzebuth	Maracá Ecological Station, Roraima, Brazil	–	19–23	6	8	5–9			1	1.3	1987–1989	Nunes 1995
At. belzebuth belzebuth	Yasuni National Park, Ecuador	11.5	~28	3	10–11	~15			1	3.5	1995–1996	Dew 2001
At. belzebuth chamek	Cocha Cashu, Manu National Park, Peru	25–31	38.5 (37, 40)	5 (5, 5)	15.5 (15, 16)	4 (4, 4)	14 (13, 15)		2	3.1	1983–1986	Symington 1987a, 1988b
At. belzebuth chamek	Noel Kempff Mercado National Park, Bolivia	–	55	15	15	5	12	3	1	1.0	1996–1997	Wallace 2007
At. belzebuth chamek	Guarayos Forest Reserve, Bolivia	–	55	8+	10+	–	–	–	1	–	2003–2004	Felton et al. 2008
At. geoffroyi	Santa Rosa National Park, Guanacaste, Costa Rica	–	42	4	17–18	3–4	5–8	5–8	1	4.4	1983–1988	Chapman 1990a,b
At. geoffroyi	Barro Colorado Island, Panama	2.1–2.3	21–24	4–5	7–10	0–3	5	0–2	1	1.9	1991–1998	Campbell 2000b
At. geoffroyi	Otoch Ma'ax Yetel Kooh Reserve, Quintana Roo, Mexico	89.5, 6.3	28.5 (16, 41)	4.5 (3, 6)	10 (5, 15)	–	7 (4, 10)	7 (4, 10)	2	2.2	1997–1998	Ramos-Fernández and Ayala-Orozco 2003[16]
At. geoffroyi	Otoch Ma'ax Yetel Kooh Reserve, Quintana Roo, Mexico	–	34.5 (19–22, 41–56)	6 (4, 8)	~12 (7–8, 15–17)	~18 (8–11, 18–33)			2	2.0	2003–2004	Slater et al. 2008, 2009

Table 11.2 (cont'd)

GENUS AND SPECIES	STUDY SITE	POPULATION DENSITY (IND/KM²)	AVERAGE GROUP SIZE (AND RANGE)	GROUP STRUCTURE[2]						NUMBER OF GROUPS	% UNIMALE GROUPS	ADULT SEX RATIO WITHIN GROUPS (FEMALES/ MALES)	DATES OF STUDY	SOURCES
				AM	AF	SM	SF	JUV	INF					
At. geoffroyi	Calakmul, Campeche, Mexico	17.2	–	22%	45%	–	–	21%	12%	–		2.1	2001–2002	Estrada et al. 2004
At. geoffroyi	Yaxchilán, Chiapas, Mexico	17.0	–	35%	29%	–	–	18%	18%	–		0.8	2001–2002	Estrada et al. 2004
At. geoffroyi	Muchukux Forest, Quintana Roo, Mexico	27.1	–	–	–	–	–	–	–	–		2.6	1995, 1997	Gonzalez-Kirchner 1998
At. geoffroyi	Najil Tucha Forest, Quintana Roo, Mexico	14.5	–	–	–	–	–	–	–	–				
At. geoffroyi	Tikal National Park, Guatemala	28	–	18%	32%	–	–	37%	12%	–		1.8	1975–1976	Cant 1977
At. geoffroyi	Tikal National Park, Guatemala	45	–	15%	33%	–	–	30%	22%	–		2.2	1973	Coehlo et al. 1976a,b
At. geoffroyi	Tikal National Park, Guatemala	56.4	–	27%	43%	–	–	17%	13%	–		1.6	2002	Estrada et al. 2004
At. paniscus	Raleighvallen-Voltzberg Nature Reserve, Surinam	8.2	18	3	8	0	1	4	2	1		2.7	1976–1977	van Roosmalen 1985
Brachyteles arachnoides	Serra de Paranapiacaba, São Paulo, Brazil	2.3	7.0	–	–	–	–	–	–	8		–	1998	González-Solis et al. 2001
B. arachnoides	Fazenda Barreiro Rico, São Paulo, Brazil	–	~13	4	3 or 4	–	–	2	3	1		0.9	1979	Torres de Assumpção 1983
B. arachnoides	Parque Estadual de Carlos Botelho, São Paulo, Brazil	2.0–3.3	11.3	–	–	–	–	–	–	–		–	1985–1986	Paccagnella 1991

Species	Locality											Study period	Source
B. arachnoides	Parque Estadual de Carlos Botelho, São Paulo, Brazil	—	33-39	9	12		12-18			1	1.3	1995	Talebi et al. 2005
B. hypoxanthus	Estação Biológica Caratinga, Minas Gerais, Brazil	—	22-27	6	8	0-2	2-7		1-6	1	1.3	1982-1984	Strier 1991a
B. hypoxanthus	Estação Biológica Caratinga, Minas Gerais, Brazil	—	32-33	7-8	9	0	9-13		1-6	1	1.2	1986-1987	Strier 1991a
B. hypoxanthus	Estação Biológica Caratinga, Minas Gerais, Brazil	—	31-43	6-8	10-12	0-2	8-15		2-7	1	1.6	1988-1990	Strier 1991a
B. hypoxanthus	Estação Biológica Caratinga, Minas Gerais, Brazil	—	52	7	17		28			1	2.4	1994	Strier 1996b
B. hypoxanthus	Estação Biológica Caratinga, Minas Gerais, Brazil	—	~64-~73	~13-~20	~19-~28	~11-~17			~5-~14	2	1.4	August 1999	Strier et al. 2002
B. hypoxanthus	Estação Biológica Caratinga, Minas Gerais, Brazil	—	~70-~80	~19-~21	~20-~24	~14-~18			~14	2	—	September 2002	Strier et al. 2002[17]
Lagothrix cana	Rio Urucu, Amazonas, Brazil	—	44-49	7	12-14	2	15-18		5-8	1	1.9	1988-1989	Peres 1996
L. lagotricha	Estación Biológica Caparú, Colombia	—	24	4	11	0	3		6	1	2.8	1984-1987	Defler 1996
L. lagotricha	Río Peneya, Meta, Colombia	—	42-43	11	15	3	5-6		6	1	1.4	1971-1972	Izawa 1976
L. lagotricha	Río Peneya, Meta, Colombia	—	29 (13, 45)	5.5 (4, 7)	7.5 (3, 12)	2 (1, 3)	3 (3, 3)	7 (2, 12)	4 (0, 8)	2	1.4	1975-1976	Nishimura 1990a

Table 11.2 (cont'd)

| | | | | | | | | | | | | ADULT SEX RATIO WITHIN GROUPS | | |
| | | | **GROUP STRUCTURE[2]** | | | | | | | | | | | | |
GENUS AND SPECIES	STUDY SITE	POPULATION DENSITY (IND/KM²)	AVERAGE GROUP SIZE (AND RANGE)	AM	AF	SM	SF	JUV	INF	NUMBER OF GROUPS	% UNIMALE GROUPS	(FEMALES/MALES)	DATES OF STUDY	SOURCES
L. lugens	Tingua National Park, Colombia	–	~21 (~12–~33)	5 (4–8)	6.5 (4–10)	1.0 (0–2)	1.3 (0–4)	4.1 (2–6)	2.0 (1–4)	5		1.3	1988–1991	Stevenson et al. 1994, Izawa and Nishimura 1988[18]
L. poeppigii	Yasuni National Park, Ecuador	31+	23 (22, 24)	3 (4, 2)	9.5 (8, 11)	5 (5, 5)	–	4.5 (4, 5)	1.0 (1, 1)	2		3.2	1995–1996	Di Fiore 1997[19]
Oreonax flavicauda	Amazonas State, northern Peru	–	4–14	1–3	–	–	–	–	–	11		–	1978–1980	Leo Luna 1980, 1982
O. flavicauda	Amazonas State, northern Peru	–	6 +	1	5	–	>1	–	–	1		–	1976	Graves and O'Neill 1980
O. flavicauda	San Martin State, northern Peru	–	10–30	–	–	–	–	–	–	3		–	2004	DeLuycker 2007
O. flavicauda	Amazonas and San Martin States, northern Peru	–	8, 9	–	–	–	–	–	–	2		–	2007	Shanee et al. 2007

[1] Crockett and Eisenberg 1987, Neville et al. 1988, and Chapman and Balcomb 1998 provide excellent overviews of population characteristics for *Alouatta* based upon data published up to 1985–1987, including longitudinal data for several well-studied populations. Instead of repeating all of that information, we concentrate here on data published since 1985 and those not included in previous reviews and direct the reader to earlier surveys. For less well-studied populations of *Alouatta*, we include data from pre-1985 publications that may also have been cited in earlier reviews in order to provide a thorough overview of the variation in population characteristics. Far fewer population surveys have been published for other atelines; thus we include all available data on group size and composition, including those gleaned from studies of only a small number of groups.

[2] AM, adult male; AF, adult female; SM, subadult male; SF, subadult female; JUV, juvenile; INF, infant. We report data as presented in the original sources and make no attempt to standardize the definitions of various age sex classes across studies. The symbol "—" is used to indicate that these data were not available in the cited source. Wherever possible, the average and range (in parentheses) of group sizes and number of individuals per class are given, though some studies report only the percentage of the population belonging to each age sex class. Where two groups from the same population were studied, the average and actual number of individuals per group (in parentheses, separated by commas) are noted, rather than the range. In general, few howler monkey population surveys have explicitly recognized "subadults" as a distinct age class, while many studies of other atelines have. Sex ratio calculation based on average number of adult females and males per group, as given in the table.

[3] Data from an earlier population survey at this site are presented in Freese 1976.

[4] Data from an earlier population survey at this site are presented in Heltne et al. 1976.

[5] Another three groups in this population had no adult males and do not contribute to the summary values included in the table. Results are presented for the more extensive "repeat" rather than "initial" survey at this site.

[6] Values noted in table represent the mean, across years, of average group size and proportion of individuals in each age sex class reported within years. Ranges span the minimum to maximum values recorded across years.

168

7 Data from earlier population surveys at this site are presented in Carpenter (1934, 1965), Collias and Southwick (1952), Mittermeier (1973), and Smith (1977) and are summarized in Milton (1982).

8 Source reports average number of individuals in each age sex class (± SD) per group during the census period but does not provide the range.

9 Values in table represent averages from across the 14-year census period. Ranges given in the group size and age sex composition columns represent the minimum to maximum average yearly values recorded across 11 census years within the census period. The total range of observed group sizes across the census period was 2–16 individuals. The total range in number of adult males per group and adult females per group was 1–4 individuals.

10 Summary of surveys conducted at five local sites within the same general region.

11 Summary of surveys conducted at two local sites within the same general region.

12 Estimate of population density comes from a larger survey of 159 groups with an average group size of 6.3 individuals and range of sizes from 3–13.

13 Data from earlier population surveys at this site are presented in Neville (1972a, 1976) and Rudran (1979). Group size and composition data are separated for two habitat types: "open shrub woodland" or "mata" (M) and "gallery forest" (G).

14 In the original source, group size and composition data are separated for unimale (UM) and multimale (MM) groups and are thus presented separately here. Values in table represent averages from across the ~30-year census period. Ranges given in the group size column represent the minimum to maximum average yearly group size recorded across years of the census. The total range of observed group sizes across the census period was 2–18 individuals.

15 Population density calculation for this population is taken from van Roosmalen (1985).

16 Population densities reported for two types of forest: "old" and "successional" forest.

17 Population in 2002 also included about eight additional individuals not assigned to a particular age sex class.

18 Two of these groups were censused in each of several years. For these groups, averages were taken for each age sex class for the two most complete censuses (Stevenson et al. 1994), and these were then averaged with censuses of other groups to find the mean number of individuals in each age-sex class. Ranges reported cover minimum to maximum numbers of individuals recorded in each age sex class across all groups.

19 Counts taken at end of study. Some individuals assigned to adult female age sex class perhaps are better classified as subadults.

The situation is somewhat different in mantled howler monkeys (*A. palliata*), where female dispersers sometimes succeed in joining established groups (Jones 1980b, Glander 1992).

Spider monkeys (*Ateles* spp.) live in multimale/multifemale societies, which are characterized as having fission–fusion grouping patterns in which the members of a large and stable social group associate on a daily basis in small, flexible parties that change size and composition frequently (Klein 1972, Cant 1977, van Roosmalen 1985, McFarland 1986, Ahumada 1989, Symington 1990, Chapman 1990a, Aureli and Schaffner 2008). Fission–fusion sociality is thought to represent an adaptation to individually optimize the difference between the benefits and costs of group living (Lehmann and Boesch 2004, Aureli et al. 2008). Costs are presumed to be associated with competition over food items when high-quality resources are scarce and presented in patches too small to support all of the individuals in a feeding party (Klein and Klein 1977; Wrangham 1980, 1987; Symington 1990), while benefits may involve the ability to adaptively respond to situations of high predation pressure (Link et al. 2006; A. Link and A. Di Fiore, unpublished data). Alternatively, such systems may help mitigate the increased daily travel costs that large parties would incur under these same conditions of resource abundance and distribution (Chapman et al. 1995, Asensio et al. 2009). Supporting these hypotheses is the fact that the size of foraging and traveling parties is positively correlated with the habitat-wide abundance of preferred, high-value food items, such as ripe fruits (Symington 1987a, 1988b; Chapman et al. 1995; Shimooka 2003). When intragroup feeding competition is reduced due to an extensive home range and lack of neighboring communities, spider monkeys appear to forage as a more cohesive unit (Campbell 2002). Party size and composition in *Ateles* are also influenced by female reproductive status (Symington 1987a, Chapman 1990a, Shimooka 2003). Dispersal in *Ateles* is largely or solely by females, while males mature and begin breeding in their natal communities (Symington 1987b, 1988a, 1990; Ahumada 1989; Shimooka et al. 2008; Di Fiore 2009a; Di Fiore et al. 2009). The fission–fusion social system of spider monkeys has sometimes been referred to as "sex-segregated" in that adult females and their dependent offspring often forage and travel independently of the group's adult males, which tend to range farther and more widely than females (Fedigan and Baxter 1984, Symington 1988a, Ahumada 1989, Chapman 1990a).

Woolly monkey (*Lagothrix* spp.) group size is large (18–45 individuals), and groups contain multiple reproductive-age animals of both sexes (Ramirez 1980, 1988; Nishimura 1990a; Peres 1994a; Stevenson et al. 1994; Defler 1995, 1996; Di Fiore 1997). Little is known of the natural history of yellow-tailed woolly monkeys (*Oreonax*). Early published reports suggest that group size may be much smaller than in other woolly monkeys (Leo Luna 1980, 1982), but recent field studies report group sizes of up to 30 animals (DeLuycker 2007; A. Di Fiore and C. Schmitt, unpublished data). Some early reports suggested that woolly monkeys might have fission–fusion grouping patterns like those of spider monkeys (Kavanaugh and Dresdale 1975). This suggestion, however, may have been prompted by the fact that observers often can see only a few individual woolly monkeys at any given time during an encounter due to the extremely diffuse spatial pattern that groups assume while foraging. Woolly monkey group members are often spread out over hundreds of meters during the course of their daily activities (Peres 1996, Di Fiore 1997 and unpublished data); nonetheless, groups appear to maintain social cohesion through the frequent use of contact vocalizations. Social groups of woolly monkeys occasionally do split into "dispersed subunits" (Defler 1996) or independently traveling subgroups that remain apart for periods of several hours to several days (Ramirez 1980, 1988; Di Fiore 1997 and unpublished data). In addition, individual animals sometimes visit other social groups for variable periods of time (Nishimura 1990a, 2003). In this sense, woolly monkey grouping patterns and social organization can be considered relatively flexible (Di Fiore and Strier 2004). Nonetheless, members of the same group forage and travel as a socially cohesive unit the vast majority of the time; although group subdivision does occasionally occur, it happens with nowhere near the frequency nor the associated small party sizes seen in the fission–fusion grouping patterns of spider monkeys. Observed cases of disappearances and transfer among woolly monkey groups suggest that dispersal is predominantly by females (Nishimura 1990a, 2003; Stevenson et al. 1994; Stevenson 2002; Di Fiore et al. 2009), and genetic studies confirm that the level of female transfer is substantial (Di Fiore 2002, Di Fiore and Fleischer 2005; Di Fiore et al. 2009). Nonetheless, solitary males, including adults, have been seen in at least one species (*Lagothrix poeppigii*, Di Fiore 2002; Di Fiore et al. 2009), suggesting some degree of male transfer as well.

Grouping patterns among muriquis (*Brachyteles* spp.) are quite flexible. Among southern muriquis, at least two populations have been characterized as living in the same kinds of fission–fusion society as spider monkeys (Torres de Assumpção 1983, Milton 1984a, de Moraes et al. 1998, Coles and Talebi 2004). Northern muriquis, by contrast, were initially reported to live in large, cohesive, multimale/multifemale groups (Strier 1986a, 1992c). Later research on the same population, however, showed that these groups began fissioning into smaller foraging units as group size increased (Strier et al. 1993). Nonetheless, these subgroups tend to be much larger, on average, than the foraging parties of spider monkeys. As in spider monkeys, females are the dispersing sex, while males mature and begin breeding in their natal groups (Strier 1987b, 1990, 1991a).

ECOLOGY AND BEHAVIOR

Diet and Ecological Strategies

Broadly speaking, all atelines have diverse and seasonally variable diets, and even the most folivorous atelines (*Alouatta*, *Brachyteles*) eat considerable amounts of fruit and other plant parts (Table 11.3). Immature and, to a lesser extent, mature leaves are the predominant component in the diets of both *Brachyteles* and *Alouatta*, and both genera show morphological adaptations for folivory in the teeth (e.g., high, shearing crests on the molars) (Zingeser 1973, Kay 1990). While howler monkeys lack extreme modifications in gut morphology (Milton 1998), they do show very long retention times for digesta in the alimentary canal (Milton 1984b, Crissey et al. 1990). This allows the gut bacteria to more thoroughly break down the structural carbohydrates common in leaves (Milton 1998, Lambert 1998). *Brachyteles* gut-passage rates are much faster than those of howler monkeys and more in line with those of other atelins (Milton 1984a,b), suggesting an important difference between the folivorous dietary strategies of howler monkeys and muriquis. Whereas howler monkeys rely on slow and efficient digestion to extract energy locked up in leaf structural carbohydrates, muriquis appear to rely on faster intake, processing, and elimination of a greater volume of less thoroughly digested material (Milton 1984a). Not surprisingly, howler monkeys are highly selective feeders and focus their feeding on immature, more easily digested leaves with a high ratio of protein to fiber (Milton 1979, 1980) and on plant species and leaf parts containing lower concentrations of certain plant secondary compounds (Glander 1978, 1982).

The remaining atelines are predominantly frugivorous. Spider monkeys are considered "ripe fruit specialists" (Cant 1977, 1990; Klein and Klein 1977; van Roosmalen and Klein 1988; Kinzey 1997c; Dew 2001; Di Fiore et al. 2008), and at all sites where they have been studied long-term, more than 70% of the annual diet consists of fruits, predominantly ripe ones (Table 11.3) (Russo et al. 2005, Di Fiore et al. 2008). Nonetheless, in some months at some sites (e.g., Chapman 1987, 1988b; Chapman et al. 1995), as well as in areas with severe habitat fragmentation (A. Link, unpublished data), immature leaves and other plant parts comprise the majority of the diet. At several sites, spider monkeys are also reported to regularly consume decaying wood and to visit terrestrial mineral licks (*salados* or *saladeros*) to drink water or consume soils (Izawa 1993, Suarez 2003, Campbell et al. 2005, Ferrari et al. 2008, A. Link and A. Di Fiore, unpublished data). This behavior has also been reported in howler monkeys (Izawa 1993), which sometimes co-feed at mineral licks with spider monkeys (A. Link and A. Di Fiore, unpublished data). The diet of *Lagothrix* also consists predominantly of ripe fruit, and throughout their geographic range this genus focuses almost as heavily on ripe fruits as *Ateles* (Soini 1986, 1990; Stevenson 1992; Stevenson et al. 1994; Peres 1994a;

Defler and Defler 1996; Di Fiore 1997, 2004; Dew 2001). However, in some populations of *Lagothrix*, foraging for animal prey, particularly insects, is also clearly an important component of the ecological strategy (Stevenson 1992, Stevenson et al. 1994, Di Fiore and Rodman 2001, Di Fiore 2004). Where *Ateles* and *Lagothrix* occur sympatrically, there is considerable overlap in the set of plant species contributing to the fruit diet of the two genera. Partitioning of the ripe fruit frugivore niche appears to take place, at least in part, along a dimension of fruit phytochemical composition, with *Ateles* focusing more on lipid-rich, or "fatty," fruits and *Lagothrix*, on fruits containing high concentrations of easily digestible sugars (van Roosmalen 1985; Castellanos and Chanin 1996; Di Fiore 1997, 2004; Dew 2001).

Strier (1992a) has suggested that the two groups of atelines exemplify contrasting ecological strategies, that is, that *Alouatta* follows a strategy of energy minimization, moving little and resting for long periods of time each day, thereby allowing them to subsist predominantly on a diet of hard-to-digest leaves, while *Lagothrix*, *Oreonax*, and *Brachyteles* typically adopt a strategy of traveling widely and rapidly between patches of higher-quality, more easily digested resources (e.g., fruits, flowers) to maximize energy intake. Under this scenario, the more folivorous diet of *Brachyteles* relative to the other atelins, its larger body size, and its convergence with *Alouatta* in craniodental features are interpreted as convergent adaptations that permit intensive leaf eating during periods of scarcity of higher-quality foods (Strier 1991b, 1992a). The fact that populations of muriquis living in primary forest appear more frugivorous than those living in smaller and more disturbed forest fragments supports this interpretation (Talebi 2005). Given the association seen in other primate radiations between body size and degree of folivory, it is interesting that howler monkeys, the smallest of the atelines, are also the most folivorous.

Activity Patterns and Ranging Behavior

Consistent with their ecological strategy, howler monkeys spend a large portion of their day (66%–80%) resting—more, on average, than any of the other atelines (Table 11.4). *Brachyteles* and *Ateles*, too, are typically inactive for more than half of the day, while *Lagothrix* are the most active atelines, spending upward of 60% of their daily activity either moving or engaged in subsistence activity (Stevenson et al. 1994, Defler 1996, Di Fiore and Rodman 2001). The greater activity of *Lagothrix* compared to the other atelines has several likely explanations. First, the large size of their social groups, combined with their high degree of frugivory, means that, all else being equal, *Lagothrix* groups need to visit more food patches per day to meet the nutritional requirements of group members than do those of other atelines. Moreover, woolly monkeys tend to use slower and more deliberate quadrupedal locomotion and less suspensory locomotion than do spider monkeys (Defler 1999b,

Table 11.3 Diets of Wld Aeline Primates Based on Studies Lasting Six Months or More[1]

SPECIES	LOCATION	PERCENT FRUIT[2]		PERCENT LEAVES[3]		PERCENT PREY[4]		PERCENT FLOWERS[5]		PERCENT OTHER[6]		STUDY LENGTH[7]	SOURCES
		MEAN	RANGE	MEAN	RANGE	MEAN	RANGE	MEAN	RANGE	MEAN	RANGE		
Alouatta belzebul	Brazil	55.6 (55.0)	25–80	24.8	11–54	—	—	5.7	0–13	14.0	4–26	10	Pinto 2002a,b, Pinto and Setz 2004
Al. belzebul	Brazil	59.0	43–92	13.3	8–15	0	0	27.6	0–41	0	0	13	Bonvicino 1989[8]
Al. caraya	Brazil	28.5	~6–~63	67.3	~37–~86	0	0	2.7	~0–~13	1.6	~0–~7	12	Bicca-Marques and Calegaro-Marques 1994[9]
Al. caraya	Argentina	19 (18)	~0–~60 (~0–~60)	68	~36–~95	0		12	0–44	1	~0–~4	17	Bravo and Sallenave 2003[10]
Al. caraya	Argentina	24	7–27	76	72–91	0		0	0	0	0	6	Zunino and Rumiz 1986[8]
Al. guariba	Brazil	15.6	1–30	76.0	64–88	0		8.4	6–11	0	0	7	Mendes 1989[8]
Al. guariba	Brazil	5.2	~0–~12	73.0	~56–~92	0		11.7	~0–~27	10.1	~3–~20	12	Chiarello 1994
Al. guariba	Brazil	8.0	–	80.7	–	0	–	7.5	–	3.7	–		Martins 2008
Al. palliata	Mexico	40.6 (34.8)	3–87	54.4	17–87	0		0.7	0–3	4.3	0–11	12	Estrada et al. 1999
Al. palliata	Mexico	49.9 (41.4)	~0–~80 (~0–~66)	49.3	~20–~100	0		<1	–	0	0	12	Estrada 1984
Al. palliata	Panama	42.1	~10–~66	48.2	~26–~84	0		9.6	~0–~24	–	–	10	Milton 1980
Al. palliata	Costa Rica	12.5	~9–~16	69.3	~64–~72	0		18.2	~17–~21	0	0	12	Glander 1978[8]
Al. palliata	Costa Rica	23.0	–	68.5	–	0	–	8.5	–	0	–	15	Stoner 1996a[11]
Al. palliata	Costa Rica	28.5	0–55	49.0	0–95	0		22.5	–	0	0	24	Chapman 1987, 1988b
Al. palliata	Nicaragua	34.8 (26.6)	16–48	55.8	~32–~82	0		7.9	0–28	1.6	~0–~8	13	Williams-Guillén 2003
Al. pigra	Belize	40.8	11–65	45.1	~32–~66	0		10.6	–	3.4	–	14	Silver et al. 1998
Al. seniculus	French Guiana	25.5 (21.5)	~0–~74 (0–51)	57.0	~27–~79	–		12.6	0–39	~5	~0–~17	19	Julliot and Sabatier 1993
Al. seniculus	Colombia	52.3	~17–~76	35.3	~25–~69	0		1.1	~0–~11	11.3	~0–~22	10	Palacios and Rodríguez 2001
Al. seniculus	Colombia	42.3 (28.4)	–	52.1	–	0	–	5.4	–	<1	–	10	Gaulin and Gaulin 1982

172

Species	Country										N	Reference
Ateles belzebuth belzebuth	Colombia	83	78–100	7	0–22	0	<1	–	10	0–18	10	Klein and Klein 1977
At. belzebuth belzebuth	Colombia	73.0	41–96	13	3–25	1.4	12	0–9	1.2	0–4	12	A. Link unpublished data
At. belzebuth belzebuth	Brazil	91.7 (88.5)	~74–~100	8.3	~0–26	0	0	0	0	0	12	Nunes 1998
At. belzebuth belzebuth	Ecuador	87	64–100	9	0–23	<1	1	~0–~6	~3	~0–~17	12	Dew 2001
At. belzebuth belzebuth	Ecuador	78.8	52–92	7.7	0–22	0	3.2	0–25	10.3	0–24	17	Suarez 2003, 2006[12]
At. belzebuth chamek	Peru	74.7	54–99	15.5	<1–38	–	4.5	~0–~22	4.5	~0–~21	12	Symington 1987a, 1988b[13]
At. belzebuth chamek	Bolivia	85.8 (78.4)	63–99 (37–98)	10.7	1–37	<1	2.9	–	0.6	0–3	11	Wallace 2005
At. belzebuth chamek	Bolivia	82.0 (59.1)	~44–~98	13.0	~0–29	<1	4.7	–	0.2	~0–~4	8	Felton et al. 2008
At. geoffroyi	Guatemala	54.9	31–84	15.1	1–34	1.9	6.3	0–17	21.8	1–66	9	Cant 1977[8]
At. geoffroyi	Costa Rica	71.4	14–100	12.5	0–86	2.1	14	0–30	0	0	38	Chapman et al. 1995[14]
At. geoffroyi	Panama	82.2	69–91	17.2	6–32	0.6	1.0	0–2	0	0	14	Campbell 2000b
At. paniscus	Surinam	79.8	54–92	7.9	1–23	<1	6.4	–	5.6	1–17	26	van Roosmalen 1985
Brachyteles arachnoides	Brazil	20.8	4–59	67.1	41–93	0	12.2	0	0	0	10	Milton 1984a[14]
B. arachnoides	Brazil	12.1	–	55.3	–	0	16.1	–	16.5	–	12	Martins 2008
B. arachnoides	Brazil	59.2	57–69	33.2	28–34	0	4.1	0	3.6	2–4	22	de Carvalho et al. 2004[8]
B. arachnoides	Brazil	73.2	~40–~85	21.7	~14–~41	0	1.7	0	3.4	~0–~9	12	Talebi et al. 2005
B. hypoxanthus	Brazil	26.8 (14.8)	11–66	50.9	28–78	0	10.9	0	11.4	2–33	14	Strier 1991b
B. hypoxanthus	Brazil	19	–	51	–	–	28	–	–	–	7	Lemos de Sá 1988[14,15]
Lagothrix cana	Brazil	67.4 (66.6)	~6–~96 (2–95)	16.2	~2–~47	<1	3.1	–	13.4	~2–~59	11	Peres 1994a[16]
L. lagotricha	Colombia	78.9	–	11.4	–	4.9	0.1	–	4.7	–	12	Defler and Defler 1996
L. lugens	Colombia	53	–	13	–	25	2	–	7	–	36	Stevenson 2002, 2006[17]

Table 11.3 (cont'd)

SPECIES	LOCATION	PERCENT FRUIT[2]		PERCENT LEAVES[3]		PERCENT PREY[4]		PERCENT FLOWERS[5]		PERCENT OTHER[6]		STUDY LENGTH[7]	SOURCES
		MEAN	RANGE	MEAN	RANGE	MEAN	RANGE	MEAN	RANGE	MEAN	RANGE		
L lugens	Colombia	60	~37–~78	~16	~5–~54	23	~4–~36	~1	–	<1	–	13	Stevenson et al. 1994[18,19]
L lugens	Colombia	78	72–84	9	5–18	12	9–22	–	–	<1	–	6	Stevenson 1992[19,20]
L poeppigii	Peru	77	69–79	7	4–20	0	–	2	1–5	14	5–24	72	Soini 1990[21]
L poeppigii	Ecuador	75.5 (71.3)	64–89 (61–86)	9.8	5–17	9.3	4–16	3.5	0–11	1.9	0–12	12	Di Fiore 1997, 2004
L poeppigii	Ecuador	73	47–88	10	0–20	6	1–10	5	~0–~19	~6	~0–~17	12	Dew 2001

[1] Means and ranges are taken from or calculated based upon data in original sources. Approximate means and ranges are indicated by ~ and are based on extrapolation of values from figures presented in original sources when raw data are not given. Values that are not available or could not be calculated or estimated from the original source are indicated by –.

[2] Wherever possible, the overall percentage and range of the total diet comprising "ripe" or "mature" fruit are separated out and noted in parentheses.

[3] Includes leaves and other plant vegetative parts (shoots, leaf buds, petioles, meristem of epiphytes).

[4] Includes both invertebrate and vertebrate prey, although the latter is negligible.

[5] Includes nectar where distinguished in original source.

[6] Includes seeds, seed pod exudates, bark, fungi, aquatic plants, moss, termitaria soil, ground soil, and undetermined food items.

[7] The number of months during which systematic dietary data were collected, or the overall length of the study in months not including those periods when data collection was suspended, if that information was available from the original source.

[8] The listed range of values for each food category is based on averages for different seasons (e.g., "dry" and "wet") noted in original source. The actual range across the years should be somewhat greater.

[9] In the original source, "seeds" were included within the category "fruit." Based on reported values for "seed" consumption averaged across four seasonal samples, we recalculated dietary proportions including "seeds" in "other."

[10] Original study does not specify whether "buds" consumed were from leaves or flowers. They are included here in "leaves."

[11] Means reported represent the average of means for two study groups.

[12] Based on 10 follows of up to 2 weeks spread among three focal individuals.

[13] Means reported represent the average of means for two study groups, while ranges cover the minimum to maximum percentages noted in both groups. Some animal prey (caterpillers) were consumed by members of one study group in 2 months of the year but are included in "other" as the original source does not allow these to be separated out into "prey."

[14] Some "seeds" are included in the "fruit" category but cannot be separated out given how the data are presented in the cited source.

[15] Original source not available, but the values included in table were taken from multiple secondary sources citing this original.

[16] In the original source, the author includes "seeds" with "fruit." Here, percentages have been recalculated including "seeds" as "other" for comparability across studies.

[17] Values included in table are from Stevenson 2002, 2006, which incorporate most of the data also presented in Stevenson et al. 1994 but explicitly exclude "seeds" and "unripe fruit" from the category "fruit." These are included in "other."

[18] Authors of original source included leaves, flowers, and other plant vegetative parts all in same dietary category. Here, we have pulled out the approximate percentage of the diet comprising flowers based on information provided in the original source.

[19] Some "seeds" are included in the "fruit" category but cannot be separated out accurately given how the data are presented in the cited source. For Stevenson et al. 1994 this is presumed to be about 5%.

[20] Author of original source included leaves, flowers, and other plant vegetative parts all in same dietary category. Here, all of these items are assigned to the category "leaves" as leaves consitituted the vast majority of the items consumed, and no information was available in the original source that allowed us to estimate the contribution of "flowers" to this total.

[21] Diet presented is based on periodic observations of feeding accumulated over a period of 7 years.

174

Table 11.4 Percent of Daily Time Budget Spent in Various Activities for Wild Ateline Primates Based on Studies Lasting 6 Months or More[1]

SPECIES	ACTIVITY CATEGORY					SOURCES
	EAT + FORAGE	MOVE	REST	SOCIAL + OTHER	STUDY LENGTH[2]	
Alouatta belzebul	20.0	18.2	58.7	3.1[3]	10	Pinto 2002a,b
Al. belzebul	7.5	19.3	55.9	16.9	13	Bonvicino 1989[4]
Al. caraya	15.9	17.6	61.6	4.9	12	Bicca-Marques 1993
Al. guariba	17.3	11.0	71.8	0	13	Mendes 1989
Al. guariba	19.0	18.8	57.6	4.6[3]	12	de Marques 1995, 1996
Al. guariba	18.7	13.2	63.7	4.4	12	Chiarello 1993a[5]
Al. palliata	16.2	10.2	65.5	8.1[3]	10	Milton 1980
Al. palliata	17.3	2.2	79.7	0.9	12	Estrada et al. 1999
Al. palliata	18.1	7.8	72.5	1.6[3]	49[6]	Teaford and Glander 1996
Al. palliata	~24.5	~14	~56	~5.5	15	Stoner 1996a[7]
Al. palliata	13.6	27.4	57.0	2.1	13	Williams-Guillén 2003
Al. pigra	24.4	9.8	61.9	3.8	14	Silver et al. 1998
Al. seniculus	12.7	6.2	78.5	2.5	10	Gaulin and Gaulin 1982
Ateles belzebuth belzebuth	22.2	9.7	61	7.1	10	Klein and Klein 1977
At. belzebuth belzebuth	18	36	45	1[3]	12	Nunes 1995
At. belzebuth belzebuth	16.7	24.9	58.2	0.1	17	Suarez 2003, 2006[8]
At. belzebuth chamek	29	26	45	<1	12	Symington 1988a
At. belzebuth chamek	18.9	29.7	45.5	5.9	11	Wallace 2001
At. belzebuth	50.5	18.1	23.7	7.7	~16	Castellanos 1995 cited in Wallace 2001
At. geoffroyi	33.5	32.6	24.1	9.8[3]	36	Chapman et al. 1989a
Brachyteles arachnoides	27.8	9.7	61.0	1.6	10	Milton 1984a
B. hypoxanthus	18.8	29.4	49.3	2.5	14	Strier 1987a
Lagothrix lagotricha	25.8	38.8	29.9	5.5	12	Defler 1995
L. lugens	36	26	35	3	36	Stevenson 2002, 2006[9]
L. lugens	36	24	36	4	13	Stevenson et al. 1994
L. poeppigii	36.2	34.5	23.2	6.1	12	Di Fiore and Rodman 2001

[1] Values are taken from or calculated based upon data in original sources. Approximate values are indicated by ~ and are based on extrapolation of from figures presented in original sources when raw data are not given.

[2] The number of months during which systematic behavioral data were collected or the overall length of study in months not including those periods when data collection was suspended, if that information was available from the original source.

[3] Value calculated by subtracting the sum of the other category percentages, reported in the original source, from 100%.

[4] Budget calculated by averaging "dry" and "wet" season budgets for each of two groups sampled for very different periods of time and then taking the mean of the group budget, weighted by sampling time.

[5] Time budget calculated here as the mean of four seasonal budgets presented in the original source. The total yearly budget presented therein, while similar to that reported here, did not sum to sufficiently close to 100%.

[6] Activity budget is based on four groups sampled periodically over this time period.

[7] Values reported are the average of means for two study groups, extrapolated from a figure presented in the original source.

[8] Based on 10 follows of up to 2 weeks spread among three focal individuals.

[9] Incorporates data from Stevenson et al. 1994.

Cant et al. 2001) and presumably muriquis, and they may take longer, less direct routes to traverse tree crowns (Cant et al. 2003). As a result, they may commonly spend more time and/or energy covering the same distance as these other atelins.

As in other features of their natural history, there is considerable variation within and between genera of atelines in home range size and ranging behavior (Table 11.5), and generalizations about the range-use patterns characteristic of any one taxon are thus difficult to make. In howler monkeys, for example, variation in home range size is extensive,

and the variation within species is greater than that between species (Crockett and Eisenberg 1987). Similarly, the degree of range overlap between groups of howler monkeys at the same study site varies from slight to complete. A cogent explanation for the variation in range size and range overlap among howler monkeys remains elusive. Although howler monkey groups do not appear to defend exclusive territories (Milton 1980), groups are site-faithful, occupying roughly the same range from year to year (Crockett 1996, Fedigan and Jack 2001, Clarke et al. 2002, Pavelka et al. 2003). Interactions between adjacent groups in the best-studied

Table 11.5 Range Size and Range Use Patterns for Wild Ateline Primates Based on Studies Lasting 6 Months or More

TAXON	GROUP SIZE	GROUP OR COMMUNITY RANGE SIZE (HA)	SEX DIFFERENCES IN RANGING[1]	SPACE USED (HA/INDIVIDUAL)[2]	RANGE OVERLAP	AVERAGE DAY RANGE (M)	MAX. DAY RANGE (M)	SOURCES
Alouatta belzebul	7–9	63.2	—	7.9	—	761	—	Pinto 2002a,b
Al. belzebul	6–8	9.5	—	1.4	11%	—	—	Bonvicino 1989
Al. belzebul	6–8	4.7+	—	0.7	21%	—	—	Bonvicino 1989
Al. caraya	15–17	2	—	0.1	—	454	893	Bicca-Marques 1994
Al. caraya	16	1.7	—	0.1	—	513	—	Bravo and Sallenave 2003
Al. caraya	21	2.2	—	0.1	—	602	1,515	Bravo and Sallenave 2003
Al. caraya	5.5	5.5	—	1.0	<20%	345	—	Zunino and Rumiz 1986[3]
Al. guariba	6	4.1	—	0.7	33%	467	808	Chiarello 1993b
Al. guariba	7	7.9	—	1.1	30%	364	540	Mendes 1989
Al. pallitata	40	108	—	2.7	—	—	—	Chapman 1988a,b
Al. seniculus	8	25	—	3.1	—	—	—	Izawa and Nishimura 1988
Al. seniculus	6–8	45	—	~6.4	—	—	1,620	Julliot 1996
Al. seniculus	9	~22	—	~2.4	—	706	—	Gaulin and Gaulin 1982[4]
Al. seniculus	—	25	—	—	—	542	655	Crockett and Eisenberg 1987
Al. seniculus	9.5	5.1	—	0.5	63%	375	610	Sekulic 1982a[5]
Al. seniculus	11.5	7.4	—	0.6	64%	355	790	Sekulic 1982a[5]
Al. seniculus	9.5	3.9	—	0.4	56%	340	520	Sekulic 1982a[5]
Al. seniculus	10.5	5.8	—	0.6	28%	445	840	Sekulic 1982a[5]
Al. seniculus	7–8	182.0	—	24.3	—	1,150	2,200	Palacios and Rodriguez 2001
Al. palliata	13	9.9	—	0.8	—	596	1,261	Glander 1978
Al. palliata	18	31.7	—	1.8	Extensive	392	792	Milton 1980
Al. palliata	18	31.1	—	1.7	100%	488		Milton 1980
Al. palliata	14	60	—	4.3	—	123	503	Estrada 1984
Al. palliata	26	21.9	—	0.8	Minimal	752	—	Williams-Guillén 2003
Al. palliata	20	21.3	—	1.1	Minimal	572	—	Williams-Guillén 2003
Average		28		2.7		526	968	
Ateles belzebuth belzebuth	12–16	300	—	21.4	—	3,311	6,039	Suarez 2003, 2006[6]
At. b. belzebuth	15–16	80–90	—	5.5	—	—	—	Ahumada 1989
At. b. belzebuth	30	163	M spend more time traveling, travel farther, and travel faster than F	5.4	<10%	—	—	Shimooka 2005
At. b.belzebuth	19–23	316	M: 120, F: 88	15.0	—	1,750	~2,600	Nunes 1995
At. b.belzebuth	15–18	259–388	—	19.6	20%–30%	—	4,000+	Klein 1972, Klein and Klein 1976
At. b.chamek	37	231	M "core": 57, F "core": 28	6.2	10%	1,977	4,070	Symington 1988a
At. b.chamek	40	153		3.8	16%			Symington 1988a
At. b.chamek	55	234	—	4.3	—	2,338	5,690	Wallace 2006, 2007

Table 11.5 (cont'd)

TAXON	GROUP SIZE	GROUP OR COMMUNITY RANGE SIZE (HA)	SEX DIFFERENCES IN RANGING[1]	SPACE USED (HA/INDIVIDUAL)[2]	RANGE OVERLAP	AVERAGE DAY RANGE (M)	MAX. DAY RANGE (M)	SOURCES
At. b.chamek	55	360	—	6.5	—	—	—	Symington 1988a
At. geoffroyi	20–24	963	—	43.8	—	2,055	4,500	Campbell 2000b
At. geoffroyi	42	170	M: 81, F: 55	4.0	—	1,297	—	Chapman et al. 1989a, Chapman 1990a
At. geoffroyi	—	—	M: 93, F: 50	—	—	—	—	Fedigan et al. 1988[7]
At. geoffroyi	16	95	M "core": 15, F "core": 8	5.9	Minimal	2,302	3,872	Ramos-Fernández and Ayala-Orozco 2003
At. geoffroyi	41	166	M "core": 21, F "core": 12	4.0	Minimal			Ramos-Fernández and Ayala-Orozco 2003
At. paniscus	12	—	—	—	Minimal	2,300	—	Norconk and Kinzey 1994
At. paniscus	18	255	M > F	14.2	0%	—	5,000	van Roosmalen 1985
Average		278		13.9		2,142	4,297	
Brachyteles arachnoides	7–10	70+	M > F	~8.2+	—	630	~1,350	Milton 1984a[8]
B. hypoxanthus	26	168	—	6.5	46%	1,283	3,403	Strier 1987b
B. hypoxanthus	34	184	—	5.4	—	—	—	Strier 1991a
B. hypoxanthus	62–63	309	—	~4.9	—	1,313	2,835	Dias and Strier 2003
B. hypoxanthus	15–18	40	—	~2.4	—	—	—	Lemos de Sá 1988[9]
Average		154		5.5		1,075	2,529	
Lagothrix cana	44–49	1,021+	—	~22.0+	—	—	—	Peres 1996
L. lagotricha	13	350	—	26.9	65%–100%	—	—	Nishimura 1990a
L. lagotricha	45	450	—	10.0	50%–100%	—	—	Nishimura 1990a
L. lagotricha	24	760	—	31.7	90%–100%	2,880	3,582	Defler 1989, 1996[10]
L. lugens	17–19	169	—	~9.4	100%	1,633	1,853	Stevenson et al. 1994[11]
L. lugens	14	—	—	—	—	2,500	3,450	Stevenson and Castellanos 2000[12]
L. lugens	17–33	—	—	—	—	1,750	2,200	Stevenson and Castellanos 2000[12]
L. lugens	17–21	—	—	—	—	1,900	2,250	Stevenson and Castellanos 2000[12]
L. lugens	27–32	—	—	—	—	2,450	3,200	Stevenson and Castellanos 2000[12]
L. poeppigii	24–25	124	—	~5.1	45%	1,792	2,738	Di Fiore 2003b[13]
L. poeppigii	23	108	—	4.7	47%	1,878	2,859	Di Fiore 2003b[13]
L. poeppigii	17–23	350	—	~17.5	—	540	950	Soini 1986[14]
L. poeppigii	14	400+	—	28.6+	—	—	—	Ramirez 1980
L. poeppigii	10	250+	—	25+	—	—	—	Ramirez 1980
Average		398		18.1		1,925	2,565	

[1] Differences in the size, in hectares, of the home range used by males (M) versus females (F) are noted for populations in which males and females often range separately. For some populations, data are not available on actual range sizes but the source indicates the relative size of the ranges of males versus females.

[2] Calculated by dividing the group or community range size by group size. If a range of values is given for group size or home range size, the midpoint of the range is used in the calculation.

[3] Based on average values reported for two groups studied for 6 months in a forest fragment.

[4] Day range estimated by dividing the total horizontal distance traversed throughout the entire study by total observation time, normalized by hour of the day.

[5] Similar home range sizes were reported by Neville 1972a for a larger number of groups studied for a shorter period of time in the same population. Average day ranges values reported in Sekulic 1982a are medians rather than means.

[6] Based on 10 follows of up to 2 weeks spread among three focal individuals.

[7] Based on radiotelemetric data from two males and five females. Same population as studied by Chapman et al. 1989a and Chapman 1990a.

[8] Mean day range calculated as the grand mean of the average day ranges recorded during 12 ~1-week sampling periods across 10 months. Maximum day range estimated as the largest 1-week average (814 m) plus two standard deviations for that period.

[9] Original source not available, but the values included in table were taken from multiple secondary sources citing this original.

[10] Table entries reflect the grand mean and maximum of average monthly day ranges. Without first averaging within months, the overall mean day range for this population was 2,670 m and the maximum distance traveled in a day was 5,130 m.

[11] The maximum day range value reflects the largest of four trimester means.

[12] Mean and maximum day range values estimated from Figure 3 of the original source. One of the groups is the same as that followed by Stevenson et al. 1994.

[13] Day ranges calculated for period from 0700 to 1700 only.

[14] Day ranges calculated as the straight line displacement between sleeping sites on successive days.

species (*Alouatta palliata*, *A. seniculus*, and *A. guariba*) are typically agonistic and often involve loud, coordinated vocal displays that may either influence or reflect the eventual outcome of the encounter (Baldwin and Baldwin 1972b; Sekulic 1982a,b; Chiarello 1995b). Day ranges in howler monkeys typically average 700 m or less (Crockett and Eisenberg 1987) and are much shorter than those of other atelines (Table 11.5).

Among different species of *Lagothrix*, too, there is considerable variation in range size (Stevenson et al. 1994, Defler 1996, Peres 1996, Di Fiore 2003b) (Table 11.5). This is probably due, in part, to the fact that a larger supplying area is needed to support larger group biomass (Milton and May 1976, Dunbar 1988, Isbell 1991, Janson and Goldsmith 1995). No single species of woolly monkey has yet been studied well enough to address the extent of range-size variation within species. Despite variation among the three species of woolly monkeys for which data are available, average day range length is clearly much longer for *Lagothrix* than for *Alouatta* or *Brachyteles* and is comparable to, or greater than, that seen for *Ateles* (Table 11.5). Di Fiore (2003b) has suggested that the day ranges of western Amazonian *L. poeppigii* may be associated with the high level of insect prey foraging noted in this species or with the monitoring of potential feeding trees. As with home range size and day range length, the degree of range overlap between adjacent groups varies within *Lagothrix* but can be extensive. Although groups do not defend exclusive areas from other groups, interactions between groups occasionally occur in areas of range overlap. Sometimes these encounters are peaceful, with groups associating passively with one another for up to several hours (Nishimura 1990a, Di Fiore 1997, A. Di Fiore and C. Schmitt, unpublished data). At other times, these encounters can be quite hostile, accompanied by chases and loud vocal behavior (Di Fiore 1997). Males seem to be the more active participants, and aggressive intergroup encounters are one of the few instances where adult males from the same social group show either affiliative or cooperative behavior toward one another (Di Fiore personal observation).

Among spider monkeys, females from the same community often range within separate but overlapping areas that comprise but a fraction of the entire community range (McFarland 1986, Symington 1988a, Fedigan et al. 1988, Chapman 1990a, Shimooka 2005). Symington (1988a), for example, found the average area used by female spider monkeys in two adjacent communities at Cocha Cashu, Peru to be only 34% and 75% of their respective community home ranges; and the "core" areas used by these females were considerably smaller, averaging just 21% and 33% of their community ranges, respectively. On Barro Colorado Island, Panama, females do not appear to use core areas but, rather, travel in larger, more cohesive groups throughout the community's range; this is probably a result of the fact that they have no conspecific competitors on the island (Campbell 2000b, 2002). Male spider monkeys typically spend more

time moving, range farther each day, and use a larger overall area than do most females (Fedigan et al. 1988, Symington 1988a, Chapman 1990a, Nunes 1995, Ramos-Fernández and Ayala-Orozco 2003) (Table 11.5). Moreover, dominant individuals tend to have larger individual ranges than subordinate individuals in at least one species of spider monkey, *Ateles geoffroyi* (Chapman 1990a). Although average day range length in spider monkeys is broadly comparable to that of woolly monkeys, which are also highly frugivorous, *Ateles* sometimes range much farther in a given day than the longest day range lengths reported for *Lagothrix*. The overlap in home range between adjacent communities of spider monkeys is far less than that seen in woolly monkeys, and intercommunity interactions in spider monkeys are typically aggressive. Males are the main participants in intergroup encounters, and some of the ranging behavior of males appears to be associated specifically with territorial defense as males cooperatively participate in boundary patrols (Shimooka 2005, Wallace 2007, Link et al. 2009) and intrusions into neighboring groups' territories (Aureli et al. 2006, A. Link and A. Di Fiore, unpublished data). Males also spend more time in border areas than do females (Fedigan and Baxter 1984, Chapman 1990a).

For one southern muriqui population studied for 10 months, Milton (1984a) found range-use patterns very comparable to those that characterize *Ateles*, with small groups of females occupying predictable core areas and "itinerant" males that ranged over a much larger area, sometimes joining females and other times traveling independently or in parties with other males. In early studies of a population of northern muriquis, however, females were not found to maintain individual core areas (Strier et al. 1993). Moreover, in this species, large social groups with multiple adult males and females remained socially cohesive and ranged as a single unit at lower population densities (Strier 1989). As group size has increased in this population, groups have begun to split into several independent foraging subunits, which are nonetheless considerably larger than those seen in spider monkeys (Dias and Strier 2003). *Brachyteles* day ranges tend to be longer, on average, than those of *Alouatta* but are considerably shorter than those of the more frugivorous *Lagothrix* and *Ateles* (Strier 1987c). For one group of northern muriquis monitored over several years, home range size increased as the community grew in size (Strier 1987c, Strier et al. 1993, Dias and Strier 2003). Interestingly, however, even though recently observed subgroups in this population were, on average, larger than the original community size, average day range length did not change appreciably, suggesting that factors other than intragroup feeding competition may also influence subgroup size (Dias and Strier 2003). Dias and Strier (2003) suggest that the shift to a more fluid subgrouping pattern in this growing community while day range length has remained relatively stable may actually reflect the existence of an upper limit to the number of individuals that can coordinate their movements rather than a response to increased intragroup feeding competition.

Intergroup interactions among muriquis are similar to those among spider monkeys and woolly monkeys: Males are the principal participants in encounters between groups and cooperate with one another against male intruders into their group's range (Strier 1994a, Strier et al. 2002). Depending on the local population density, home range overlap between adjacent groups can be slight or extensive (de Moraes et al. 1998).

Kin Structure and Social Behavior

All atelines are characterized by a high incidence of female dispersal and, in the atelin genera at least, by a greater degree of male than female philopatry (Strier 1990, 1994a,b; Di Fiore 2009a). Under conditions of female-biased dispersal, males within social groups and within local populations are expected to be, on average, more closely related to one another than are females (Morin et al. 1994; Di Fiore 2003a, 2009a). This, theoretically, should have important implications for the patterning of social behavior within ateline groups. Indeed, despite considerable variation among ateline genera in the expression of affiliative and agonistic social relationships, these relationships are, broadly speaking, consistent with expectations based on kin structure patterns. However, the presumption of greater average male/male versus female/female relatedness in atelines has been tested genetically only for *Lagothrix* and *Ateles* and shown to be true for several (but not all) social groups (Di Fiore and Fleischer 2005, Di Fiore 2009a, Di Fiore et al. 2009). A comparative overview of some aspects of social relationships in atelines is presented in Table 11.6.

Among howler monkeys, animals of both sexes disperse (Clarke and Glander 1984; Crockett and Pope 1993; Rumiz 1990; Glander 1980, 1992); thus, many individuals are likely to live a good portion of their adult lives in groups without close kin. In most well-studied populations, both males and females are reported to form dominance hierarchies, although in contrast to the pattern commonly seen among many Old World primates, rank is inversely related to age among males in *Alouatta palliata* and among females in both *A. palliata* and *A. seniculus* (Glander 1980; Jones 1980a, 1981; Crockett 1984; Zucker and Clarke 1998). Moreover, for several populations where the relationship has been studied, both mating and reproductive success appear to be related to dominance rank among males (Clarke 1983, Jones 1985, Pope 1990), and among females dominant individuals commonly direct aggression toward subordinates and their offspring (Crockett 1984, Clarke et al. 1998). However, in one large group of *A. palliata*, Wang and Milton (2003) could not discern dominance relationships among the adult males, apart from the existence of an alpha male.

In general, intrasexual social relationships in howler monkeys are weak and affiliation is rare (Zucker and Clarke 1998, Wang and Milton 2003). Grooming, when it occurs, is frequently between the sexes, most often from females to adult males; and proximity partners among adults tend to be opposite-sex individuals (Neville 1972b, Chiarello 1995a, Zucker and Clarke 1998, Wang and Milton 2003). Kin networks are generally not as developed in howler monkeys as in primate species where dispersal is predominantly by members of one sex. Despite these observations, howler monkeys provide some of the clearest examples yet documented in primates of the actual fitness benefits that may accrue as a result of kin-directed nepotism (Pope 1998a, 2000b). For example, in howler monkeys, males often form coalitions to take over groups of females from other males in order to begin breeding (Pope 1990, Horwich et al. 2000). Pope (1990) has demonstrated that in *A. seniculus* coalitions comprising male kin last, on average, 3.6 times longer than coalitions made up of unrelated males. Similarly, females in new groups are generally unrelated individuals that join together with one or more emigrating males and begin defending a range. As groups become more established, however, founding females try to recruit their own daughters as breeders and to force the emigration of other females' daughters (Crockett 1984; Pope 2000b). As a result, the degree of relatedness among females within groups of red howler monkeys tends to increase over time, as does females' net reproductive output, indicating a clear fitness benefit for kin-directed recruitment (Pope 2000b).

Among spider monkeys and muriquis, it appears that females disperse while males are philopatric. Not surprisingly, in these two genera, social interactions among males are more affiliative and cooperative than those among females or between the sexes. In spider monkeys, males are each other's most common associates and interact affiliatively with each other much more often than do females (Slater et al. 2009, Link et al. 2009). For example, Symington (1990) found that "association indices" for male/male dyads were much higher, on average, than indices for either male/female or female/female dyads. Similarly, for grooming bouts not occurring between mothers and offspring, Symington (1987a, 1990) noted that the most common grooming dyads involved pairs of males rather than pairs of females or mixed-sex pairs. Fedigan and Baxter (1984) and Slater et al. (2009) also found that males engaged in more affiliative behaviors than did females and directed these more toward other males than toward females. For example, males "greet" each other significantly more than females with an embrace in which pectoral scent glands are often inspected (Eisenberg and Kuehn 1966, Fedigan and Baxter 1984). Male *Ateles* also cooperate with one another to patrol and defend a large community range (Fedigan and Baxter 1984; Symington 1987a, 1988a; Chapman 1990a; A. Link and A. Di Fiore, unpublished data).

Grooming (auto- and allo-) is absent from the *Brachyteles* behavioral repertoire (Strier 1994a); however, the same general pattern seen among male spider monkeys is apparent for other affiliative behaviors, at least among the well-studied northern muriqui. In this species, for instance, males are the most common proximity and

Table 11.6 Comparative Social and Reproductive Behavior of Ateline Primates

	ALOUATTA	ATELES	LAGOTHRIX	BRACHYTELES
Female dispersal	Yes	Yes	Yes	Yes
Male dispersal	Yes	No	Some likely	No
Intersexual dominance relationships	Males > females	Males > females	Males > females	Codominant
Female competition over group membership				
Female response to immigrants	Intense hostility	Harass immigrants	Indifferent	Harass immigrants
Females force emigration of maturing females	Usually	No?	No?	No
Female dominance relationships within groups	Hierarchical	Hierarchical	Hierarchical?	Egalitarian
Male competition over group membership				
Male reponse to immigrants	Intense Hostility	—	Tolerant to hostile	—
Male reaction to males in other groups	Hostile	Hostile	Tolerant to hostile	Hostile
Male dominance relationships within groups	Hierarchical and competitive	Hierarchical and tolerant?	Hierarchical and tolerant?	Egalitarian and tolerant
Affiliative bonds among adult females	Weak	Weak	Weak	Weak?
Affiliative bonds among adult males	Weak	Strong	Weak	Strong
Intersexual affiliation	Strong	Weak	Strong?	Weak
Strongest grooming and proximity relationships	Male/Female	Male/Male	Male/female and adult male/subadult male	Male/Male
Contexts of male-male cooperation	Related or unrelated males may form coalitions with each other to take over a group and oust resident male	Males of a community may jointly patrol community range boundaries, aggress against intruders, and cooperate in intergroup encounters	Males from the same social group may cooperate during intergroup encounters, at other times, are tolerant but not affiliative with each other	Males of a community jointly aggress against intruders, cooperate in intergroup encounters, spend considerable time in proximity with one another, are overtly affiliative with each other
Females solicit copulations from males?	Yes	Yes	Yes	Yes
Females copulate outside of periovulatory period?	Yes	Yes	Yes	Yes
Females mate with multiple males?	Occasionally	Yes	Yes, both within and across receptive periods, sometimes on the same day and in quick succession	Yes, both within and across receptive periods, sometimes on the same day
Social context of copulations	Dominant male may form consortship with receptive female and keep other males from mating	Often take place during consortships or in seclusion, away from other group members	Take place in exposed positions, in view of other group members, males tolerant of mating by other males	Take place in exposed positions, in view of other group members, males tolerant of mating by other males
Differential male mating success?	Yes, related to dominance rank or tenure as alpha male	Unknown	Some, may be related to male age, dominance, or size	Yes, based on female preferences
Differential male breeding success (based on genetic assessment of paternity)?	Dominant male sires all offspring in *A. seniculus*, unknown for other species	More than one male sires offspring born within groups, degree of reproductive skew among males varies	Unknown	Unknown
Importance of female choice in mating system	Weak	Strong	Strong	Strong
Direct female competition over matings?	Yes	No	Yes	No

embrace partners of other males (Strier 1992b, 1994a, 1997a), and they cooperate to defend group females from males of neighboring groups (Strier 1992c, 1994a). Female social relationships in muriquis and spider monkeys are less well described. In both taxa, females do interact affiliatively with one another, although, at least in *Ateles*, the frequency with which affinitive behavior is seen is far less among females than among males (Fedigan and Baxter 1984). Northern muriqui females spend more time in proximity to other females than they do to males, and common proximity partners tend to be common embracing partners as well (Strier 1990, 1992b); however, the relative strength of female/female versus male/male bonds has not been addressed. Among southern muriquis, the pattern appears to be somewhat different, with intersexual relationships apparently more important than same-sex ones in at least one population (Coles and Talebi 2004). Although low in both taxa, the rate of intragroup aggression is much higher in *Ateles* than *Brachyteles*, in which very little aggression has ever been seen (Fedigan and Baxter 1984, Strier 1992c). Aggression in spider monkeys is typically directed from males toward females (Fedigan and Baxter 1984, Campbell 2003, Wallace 2007, Link et al. 2009). Although most aggression is directed toward potentially cycling females, most of these events do not lead to sexual behavior or copulation (Campbell 2003), thus suggesting that male-to-female aggression may act as a cooperative "social control" from males to their own group females (Link et al. 2009). Unlike in *Ateles*, the sexes are codominant in *Brachyteles* and the relationships between males and females appear to be more egalitarian (Strier 1994a).

Grooming and other affiliative behaviors are rarely observed in studies of *Lagothrix* social behavior. When grooming does occur, adult males are the most common recipients, being groomed more often by subadult males than by females; adult males rarely, if ever, groom one another, and grooming among adult females is also uncommon (Nishimura 1990a, 1994; Di Fiore 1997; Di Fiore and Fleischer 2005). Male bonds are not nearly so well developed in woolly monkeys as in the other atelins, and adult males from the same social group rarely interact with one another except during hostile intergroup encounters. Nonetheless, while not overtly affiliative, adult male woolly monkeys are extremely tolerant of one another, even in the context of mating (see "Reproductive Strategies and Life History"). In general, the most developed affiliative social interactions among nonjuvenile woolly monkeys appear to be those between subadult and adult males (but not among adults) and between males and females (Di Fiore and Fleischer 2005); interestingly, relationships among females seem to be even less developed than in other atelins. Nishimura (1990a, 1994) suggests that male woolly monkeys can be ranked into a dominance hierarchy that correlates with age and body size; while likely to be true, interactions among males are sometimes too infrequent to allow determination of a hierarchy (Di Fiore 1997 and unpublished data).

REPRODUCTIVE STRATEGIES AND LIFE HISTORY

Among the atelines, mating patterns are varied and the two ateline tribes demonstrate behaviorally divergent reproductive strategies. In howler monkeys, reproduction appears to be strongly skewed toward socially dominant males. In the best-studied case of red howler monkeys (*A. seniculus*) living in the llanos region of central Venezuela, the single resident male in one-male groups, and the dominant male in multimale groups, is responsible for most of the observed mating and sires all of the infants born in the group during his tenure (Pope 1990). Additionally, aggression among red howler monkey males for breeding positions is intense, and infanticide has been reported when a new male takes over a group (Crockett and Sekulic 1984, Agoramoorthy and Rudran 1995), with 44% of infant mortality presumably due to infanticide in one well-studied population (Crockett and Rudran 1987b). Infanticide has also been reported in three other howler monkey species living in groups with one or a small number of adult males (e.g., *A. caraya*, Zunino et al. 1986, Rumiz 1990, Calegaro-Marques and Bicca-Marques 1996; *A. guariba*, Galetti et al. 1994; *A. pigra*, Brockett et al. 1999, Knopff et al. 2004). Consistent with the sexual selection hypothesis, which interprets infanticide as an adaptive male reproductive strategy (Hrdy 1977a, 1979), red howler monkey females are able to breed more quickly following the loss of an infant, and they mate with infanticidal males following group takeovers (Clarke 1983, Agoramoorthy and Rudran 1995, Crockett and Sekulic 1984, Crockett and Rudran 1987b).

Overt male/male competition is less extensive in mantled howler monkeys (*A. palliata*) where larger, multimale groups are common. In some populations of mantled howlers, males can be ranked into dominance hierarchies, male mating success is associated with rank, and group takeovers and subsequent infanticidal attacks by new males have been observed (Jones 1980a, 1985; Clarke 1983; Clarke and Glander 1984; Glander 1980, 1992; Clarke et al. 1994). However, in one large group of mantled howler monkeys on Barro Colorado Island, Panama, all six adult males, regardless of rank, mated with receptive females with no overt competition and the highest-ranking male did not copulate the most (Wang and Milton 2003). Taken together, these results suggest a general importance of direct reproductive competition among males across howler species, the intensity of which appears to decrease with increasing male group size. Among female howler monkeys, as among males, direct competition over group membership and/or breeding positions is fierce. In *A. seniculus*, maturing females are often forced from their natal groups by resident females and rarely are able to immigrate into existing groups, instead securing breeding opportunities by establishing new groups with other emigrants (Crockett 1984, Crockett and Sekulic 1984, Crockett and Pope 1993, Pope 2000b). Female *A. palliata*, by contrast, are able to immigrate successfully into established groups, where they either fight their way to the top of the female dominance hierarchy or, if unsuccessful, disperse again and try to enter a different group (Jones 1980b, Glander 1992).

Within the atelins, observed mating is not nearly so skewed toward specific males as it appears to be in *Alouatta*, nor is direct intrasexual competition over mating opportunities nearly so overt. Nonetheless, mating behavior and the intensity of male/male competition are qualitatively different in the different atelin genera. In all three genera, females have been observed to mate promiscuously with multiple males during a given fertile period, sometimes even on the same day (*Ateles*, Campbell 2000b, A. Di Fiore, A. Link, and S. Spehar, unpublished data; *Brachyteles*, Aguirre 1971, Milton 1985a,b, Strier 1986b; *Lagothrix*, Nishimura 1988, 1990b, Di Fiore 1997 and unpublished data). Moreover, in *Brachyteles* and *Lagothrix*, males are extremely tolerant of mating by other males (Strier 1993, 1994a; Nishimura 1994; Di Fiore 1997; Di Fiore and Fleischer 2005; Strier et al. 2000; Possamai et al. 2007); and in muriquis, multiple males have even been observed mating with the same female in quick succession (Aguirre 1971, Milton 1985a,b). Among muriquis, certain males have been recorded mating more often than others, but relative mating success does not appear to be related to male dominance rank in any way; indeed, no dominance hierarchy is discernable among muriqui males, among which very few agonistic interactions have ever been seen (Strier 1990). Rather, this differential mating pattern has been attributed to female preferences for particular males or classes of males (Strier 1994a, 1997b). Although various males of all ages mate with females among woolly monkeys (Nishimura 1990b, Di Fiore 1997), Nishimura (1990b, 1994) has suggested that mating success may be skewed toward older, perhaps more socially dominant, males. Observations in captivity (Williams 1967) also suggest that socially dominant males may monopolize copulations during a female's periovulatory period, implying a somewhat more dominance-based but promiscuous mating system for *Lagothrix* than for *Brachyteles* (see also Strier 1996a). Very little is known of mating patterns in *Ateles* as mating tends to take place in seclusion, away from other group members, especially other group males (van Roosmalen 1985; Campbell 2000b, 2006a; Campbell and Gibson 2008; A. Link and A. Di Fiore unpublished data). However, females have also been reported to mate with multiple males, sometimes on the same day (van Roosmalen 1985, Robinson and Janson 1987, Campbell 2000b, Campbell and Gibson 2008), and perhaps to form consortships with specific, preferred males (Symington 1987a; Campbell 2000b; Campbell and Gibson 2008; Gibson, K. N., et al. 2008). Recent molecular genetic work (Di Fiore et al. 2008 and unpublished data) has shown that multiple males sire offspring in groups of white-bellied spider monkeys (*Ateles belzebuth*).

Unlike the situation in most populations of howler monkeys, overt male/male competition appears to be of little importance in atelin societies. In fact, among spider monkeys and muriquis, social relationships among males are overtly affiliative and the rate of aggressive interactions among males is extremely low. No cases of aggression have been reported in the context of copulation for either species (Fedigan and Baxter 1984; Strier 1986a, 1990, 1992b; Strier et al. 2000), although for spider monkeys this may be due, in part, to the secluded nature of copulation. It is worth noting, however, that several cases of serious, even lethal aggression among males, particularly by older toward younger males, have recently been reported in spider monkeys (Campbell 2006b; Valero et al. 2006; Gibson, et al. 2008), which highlights the potential for intense within-group intrasexual competition. Among woolly monkeys, too, adult males are very tolerant of one another, especially in the context of mating. Males will often mate within sight of other males without receiving overt aggression from them (Di Fiore 1997, Di Fiore and Fleischer 2005), although Nishimura (1990b) has observed copulation by a subadult male being interrupted by an adult male, and Stevenson (1997) has reported that the only case of "strong" aggression observed between adult males of the same group during a year-long study occurred within half an hour of a copulation involving one of the males.

Instead of competing overtly with groupmates over reproductive opportunities, male atelins (and perhaps male howler monkeys in the largest groups of *A. palliata*) appear to compete more subtly. Atelins in general are characterized by having very long intromission times relative to howler monkeys (Table 11.7), suggesting that mate guarding may be an important component in the male reproductive strategy. Similarly, at least in muriquis, males have large testes relative to their body size, an indication that competition among males over fertilizations may be relegated largely to competition among sperm (Milton 1985b; Strier 1992b,c; Strier et al. 2002; see also Chapter 29). Also, female choice appears to be an extremely important component of the mating systems of all atelin primates (Strier 1990, Di Fiore 1997 and unpublished data). In spider monkeys, woolly monkeys, and muriquis, females commonly (and sometimes insistently) solicit copulation from males and may advertise their receptivity broadly to males through other behavioral signals (e.g., by manually stimulating male genitals, *Ateles geoffroyi*, Campbell 2000b) as well as olfactory cues (e.g., Williams 1967; Klein and Klein 1971; Milton 1987a; Ramirez 1988; Di Fiore 1997; Campbell 2000b). Based on tracking ovarian cycles using fecal steroid assays, Strier and Ziegler (1997) and Campbell (2004) found that female northern muriquis (*Brachyteles hypoxanthus*, previously named *B. arachnoides hypoxanthus*) and black-handed spider monkeys (*A. geoffroyi*) often mate outside of their periovulatory periods. In captivity, female *Lagothrix* also have been reported to mate outside of their 3- to 4-day periovulatory period (Williams 1967, 1974), but this has not been confirmed in the wild. Interestingly, female woolly monkeys have been observed to compete directly over reproductive opportunities. It is not uncommon

Table 11.7 Reproductive and Life History Parameters for Ateline Primates[1]

PARAMETER	ALOUATTA	SPECIES	ATELES	SPECIES	LAGOTHRIX	BRACHYTELES
				TAXON		
Female cycle length (days)	16.3 ± 3.5	*palliata* (La Pacifica)	~25	*paniscus* (Raleighvallen)	~18 d[2]	21.0 ± 5.4
			22.7	*geoffroyi* (BCI)		
Copulatory or receptive period (days)	2–4	*palliata* (La Pacifica)	8–10	*paniscus* (Raleighvallen)	3.1 ± 2.4	2.1 ± 1.2
Number of cycles or months to conception	8–15 months	*palliata* (La Pacifica)	At least 3–8 cycles	*paniscus* (Raleighvallen)	~12[3] cycles	2–10 cycles
			3–6 cycles	*geoffroyi* (BCI)		
Average intromission time	< 50 sec	*palliata* (La Pacifica)	10 min	*paniscus* (Raleighvallen)	6.6 min (max >20 min)	6.5 min (max 18 min)[4]
	41 ± 11 sec (max 70 sec)	*caraya* (Brazil)	19 min	*geoffroyi* (BCI)		4.1 min (max 8.3 min)[5]
			14.7 min (max 23 min)	*belzebuth* (Manu)		
Gestation length (days)	186 ± 6	*palliata* (La Pacifica)	226–232	*geoffroyi* (captive, BCI)	~210–225	216.4 ± 1.5
	184–194	*seniculus* (Hato Masaguaral)				
	152–195	*caraya* (Brazil)				
Sex ratio at birth (M:F)	1:0.88	*caraya* (NE Argentina)	1:2.67	*belzebuth* (Manu)	1:1.18	1:1.76
	1:1.67	*palliata* (La Pacifica)	1:1.00	*geoffroyi* (Santa Rosa)		
	1:0.94	*seniculus* (Hato Masaguaral)				
Birth seasonality	Dry season	*caraya* (NE Argentina, mainland gallery forest)	Late wet/early dry season	*belzebuth*		
(La Macarena)	Late wet/early dry season	Dry season				
	No seasonality	*caraya* (NE Argentina, island flooded forest)	Late wet/early dry season	*geoffroyi* (BCI)		
	No seasonality	*guariba* (EBC)	Dry season	*geoffroyi* (Santa Rosa)		
	No seasonality	*palliata* (La Pacifica)	Late dry/early wet season	*paniscus* (Raleighvallen)		
	No seasonality	*palliata* (BCI)	Little seasonality, but more births in early wet season	*belzebuth (Manu)*		
	Dry season	*palliata* (Santa Rosa)				
	Early dry to early wet season	*pigra* (Belize)				
	Little seasonality, but more births in dry season	*seniculus* (Hato Masaguaral)				
Age at weaning (years)	12	*palliata* (La Pacifica)	26+	*geoffroyi* (BCI)	—	18–24
	11–14	*seniculus* (Hato Masaguaral)	24+	*geoffroyi* (Santa Rosa)		
			~ 36	*paniscus* (Raleighvallen)		

Table 11.7 (cont'd)

PARAMETER	ALOUATTA	SPECIES	ATELES	SPECIES	LAGOTHRIX	BRACHYTELES
				TAXON		
Sex and age at emigration (years)	M: 1–2, F: 2–3	*palliata* (La Pacifica)	F: 4–5	*belzebuth* (Manu)	F: 6.0 ± 0.4	F: 6.1 ± 0.6
	M: 4–6, F: 2–4	*seniculus* (Hato Masaguaral)				
Age at menarche (years)	~ 2.8	*palliata* (La Pacifica)	Earliest at ~ 6.5	*geoffroyi* (BCI)	–	~ 6
Female age at first reproduction (years)	~ 3.6	*palliata* (La Pacifica)	Earliest at ~ 7.1	*geoffroyi* (BCI)	9.0 ± 0.7	9.0 ± 1.8, earliest at 7.25 for philopatric female
	5.2 (4–7)[6]	*seniculus* (Hato Masaguaral)			9.0 (captive, wild-born)	
Male age at sexual maturity (years)	3.5	*palliata* (La Pacifica)	Earliest behavioral correlates of sexual maturity at 4–5	*geoffroyi* (BCI)	First copulation without ejaculation at 3.5 y, first "true" copulation at 5.5–6	Earliest observed mating at 5.5
	7 (6–8)[6]	*seniculus* (Hato Masaguaral)				
Interval from female immigration to first copulation (months)	3–4	*palliata* (La Pacifica)	–	–	Typically week, else female does not remain in the group	11.2 ± 2.2
Interval from immigration to first birth (months)	19.7	*palliata* (Santa Rosa)	–	–	22.8 ± 7.2	33.8 ± 7.3
Interval from parturition to next conception (months)	12.8	*palliata* (La Pacifica)	27.1	*geoffroyi* (Santa Rosa)	23.4	–
	13.7	*palliata* (Santa Rosa)				
Interbirth interval (months)	15.9	*caraya* (NE Argentina)	34.5 ± 5.8	*belzebuth* (Manu)	36.7 ± 4.7	36.4 ± 4.3
	21.2	*guariba* (EBC)	31.9 ± 3.0	*geoffroyi* (BCI)		
	22.5	*palliata* (La Pacifica)	34.7	*geoffroyi* (Santa Rosa)		
	19.9	*palliata* (Santa Rosa)	46–50	*paniscus* (Raleighvallen)		
	17.0	*seniculus* (Hato Masaguaral)				
Infant survivorship to 12 months	74%+	*guariba* (EBC)	67%	*belzebuth* (Manu)	93.7% (captive)	Close to 100%
	M: 53%, F: 100%	*palliata* (La Pacifica)				
	70%	*palliata* (Santa Rosa)				
	80%	*seniculus* (Hato Masaguaral)				
Females cycle synchronously?	–	–	Yes	*paniscus* (Raleighvallen)	No	No
% of Matings with extragroup males	None reported		None reported		None reported	13%
Average male life span (years)	16.6	*palliata* (BCI)	–	–	–	–
Average female life span (years)	15.5	*palliata* (BCI)	–	–	F: 13 (captive)	–

Table 11.7 *(cont'd)*

	TAXON					
PARAMETER	ALOUATTA	SPECIES	ATELES	SPECIES	LAGOTHRIX	BRACHYTELES
Maximum life span (years)	M: > 20.5, F: > 19	*palliata* (BCI)	M: > 24, F: >22	*geoffroyi* (BCI)	F: > 30 (captive)	–
	F: > 16	*palliata* (La Pacifica)				
	4.2–6.2	*seniculus* (Hato Masaguaral)				
Intrinsic rate of natural increase	0.18	*palliata*	0.11	*geoffroyi*	0.16[7]	–
	0.18	*seniculus*	0.09	*paniscus*		

[1] Data in table compiled from the sources listed below. In some cases, we have converted from time in months to time in years or vice versa. Parameters for which data are not available are indicated by "–". Where data on both wild and captive animals are available for a taxon, only those from wild animals are reported. If only captive data are available, then these are reported in lieu of "–".

[2] Calculated as the sum of the average copulation period plus the average length of the interval between periods.

[3] Calculated as the number of cycles in 7.2 m, the average number of months from resumption of sexual activity postpartum to next conception.

[4] *Brachyteles hypoxanthus* at the Estação Biológica Caratinga, Minas Gerais, Brazil.

[5] *Brachyteles arachnoides* at Fazenda Barreiro Rico, São Paulo, Brazil.

[6] Average value reported is median rather than mean.

[7] Species of *Lagothrix* for which this value was derived was not specified.

Sources: *Alouatta*, Neville 1972a, Glander 1980, Froelich et al. 1981, Crockett and Sekulic 1982, Milton 1982, Crockett 1984, Clarke and Glander 1984, Jones 1985, Crockett and Rudran 1987a,b, Pope 1990, Rumiz 1990, Ross 1991, Calegaro-Marques and Bicca-Marques 1993, Crockett and Pope 1993, Fedigan and Rose 1995, Zucker et al. 1997, Fedigan et al. 1998, Brockett et al. 2000b, Strier et al. 2001; Kowaleski and Zunino 2004; *Ateles*: Eisenberg 1973, Milton 1981a–c, van Roosmalen 1985, Symington 1987a,b, 1988a, Chapman et al. 1989b, Ross 1991, Campbell 2000a,b, Izawa 2000 cited in Nishimura 2003; *Lagothrix*: Ross 1988, Nishimura 1988, 1990b, 2003, Nishimura et al. 1992, Stevenson 1997, Di Fiore 1997, Mooney and Lee 1999, Di Fiore and Fleischer 2005; *Brachyteles*: Milton 1985a, Strier 1986b, 1991a, 1992b, 1996b, 1997b, Strier and Ziegler 1997, 2000, Strier et al. 2003, Martins and Strier 2004, Possamai et al. 2007.

for female woolly monkeys to harass copulating pairs, to the point where mating is interrupted prior to ejaculation by the male, suggesting that direct female mate competition, in addition to female choice, may be a significant part of the mating system of *Lagothrix* (Di Fiore 1997, Di Fiore and Fleischer 2005).

Life history parameters are rather poorly studied for ateline primates, even in captivity. Nonetheless, it is clear that spider monkeys, woolly monkeys, and muriquis have "slow" life histories compared to other platyrrhines and other nonhominoid primates generally, with a late age at first reproduction and a long interbirth interval for their body size (Fig. 11.4). Howler monkeys, in contrast, begin reproducing somewhat earlier and have shorter interbirth intervals than might be predicted on the basis of body size; and these features of life history may help explain their widespread distribution and persistence in regenerating secondary habitats. While female howler monkeys may begin breeding even before they reach 4 years of age and thereafter produce offspring at 1.5- to 2-year intervals, female atelins begin breeding much later and typically have interbirth intervals of 3 years or more; woolly monkey and muriqui females, for example, do not typically begin breeding until about age 9 (Table 11.7). For all of the atelins, too, realized female fertility over the reproductive life span is low.

CONSERVATION

Throughout their geographic ranges, atelines are the New World monkeys most susceptible to the negative impacts of anthropogenic activities. Hunting and habitat loss and fragmentation are the most important factors in the current decline of wild ateline populations. Atelines are the preferred and most common mammals in the diets of many indigenous groups in South America and are among the first primates experiencing population declines to go locally extinct in the face of human hunting pressure (Peres 1990b, 1991b; Mena et al. 2000; Peres and Lake 2003; Peres and Palacios 2007). In fact, throughout the neotropics, the number of ateline primates killed by subsistence hunting alone is staggering. For example, in lowland Ecuador, Yost and Kelley (1983) recorded 562 woolly monkeys and 146 howler monkeys being killed in a 275-day period by hunters in three villages of indigenous Huaorani. Similarly, Peres (1991b) noted that during an 18-month period in the mid-1980s a single family of rubber tappers in rural Brazil killed about 200 woolly monkeys, 100 spider monkeys, and 80 howlers; and Mena et al. (2000) reported that 395 woolly monkeys, 85 howler monkeys, and 10 spider monkeys were killed during an 11-month period in yet another Huaorani community in Ecuador in the mid-1990s. On a regional scale, Peres (2000a) has estimated that 1.1–2.6 million individual ateline

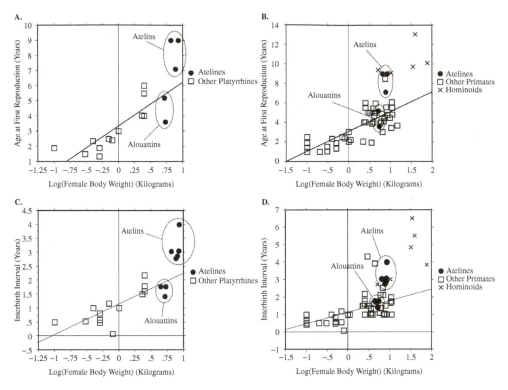

Figure 11.4 Ateline life history in a comparative context. Relationship between female age at first reproduction and female body weight (log-transformed) for (*A*) atelines relative to other platyrrhines and (*B*) atelines and hominoids relative to other primates. Relationship between interbirth interval and female body weight (log-transformed) for (*C*) atelines relative to other platyrrhines and (*D*) atelines and hominoids relative to other primates. In all cases, the atelines (spider monkeys, woolly monkeys, and muriquis) fall substantially above the regression line, indicating they are characterized by having long prenatal and postnatal developmental periods compared to other primates of comparable size. Similar life history features characterize the hominoids. Data on female body weight, age at first birth, and interbirth interval come from Ross and Jones (1999) and Nunn and Barton (2001), except that data from Table 11.1 are used in lieu of the ateline data presented in those sources.

primates (corresponding to a biomass of 6,812–16,710 tons) are harvested each year by the low-income rural population of the Brazilian Amazon.

Such a large harvest of ateline primates is clearly not sustainable and can dramatically influence primate community structure in local areas. In a study, in part, of the density of ateline primates at 25 western Amazonian sites subject to varying degrees of hunting pressure, Peres (2000a) found that population densities were nearly 10 times lower at sites subject to moderate or heavy hunting pressure compared to sites where little or no hunting occurred. Combined with their susceptibility to forest fragmentation, habitat disturbance, and hunting, the slow reproductive rates of atelines can make it difficult for populations to recover even after hunting in a local area has been curtailed. Tellingly, four of the world's 25 most endangered primates identified in the last decade by Conservation International are atelines: the yellow-tailed woolly monkey, *Oreonax (Lagothrix) flavicauda*; the northern muriqui, *Brachyteles hypoxanthus*; the variegated spider monkey, *Ateles hybridus*; and the brown-headed spider monkey, *Ateles fusciceps* (Mittermeier et al. 2000a, 2002, 2006, 2007; Konstant et al. 2002). Moreover, the majority of the ateline species listed in Table 11.1 for which data are available are considered vulnerable,

endangered, or critically endangered by the International Union for the Conservation of Nature (IUCN 2004, 2008), and all are included in either Appendix I or II of the Convention on International Trade in Endangered Species of Wild Fauna and Flora (IUCN 2008).

SOCIOECOLOGICAL CONVERGENCE: ATELINES AND THE AFRICAN GREAT APES

One of the interesting aspects about ateline primates is their convergence with some of the African great apes (subfamily Homininae) in various aspects of their morphology, ecology, social organization, life history, and behavior (Table 11.8). For example, a high level of female dispersal and at least some degree of male philopatry characterize all atelines; and these features likewise characterize all hominines (see Chapters 19 and 20). Similarly, both atelines and African great apes show marked morphological adaptations for more upright, suspensory locomotion and feeding postures, including a shortened lumbar portion of the vertebral column, more dorsally placed scapulae, and shoulder joints capable of full rotation. Finally, at least some members of both clades take to extremes the primate tendency of having "slow" life histories relative to other mammals, with long

Table 11.8 Socioecological Convergence of Atelines and African Apes

	ALOUATTA AND *GORILLA GORILLA*	*ATELES* AND *PAN TROGLODYTES*	*BRACHYTELES* AND *PAN PANISCUS*
Ecological strategy	Largely folivorous	Frugivorous, focus on ripe fruits	Both frugivorous and folivorous, but both taxa rely heavily on plant vegetative parts at times
Dispersal pattern	Dispersal by both females and males	Dispersal by females	Dispersal by females
Social organization	Unimale, age-graded male, and multimale social groups	Fission–fusion communities	Both cohesive social groups and fission–fusion communities in muriquis, only the latter reported for bonobos
Male/male relations	Intense male competition over group membership and thus breeding opportunities, age-related male dominance hierarchies	Strong male/male affiliation in both taxa, male dominance hierarchies present among chimpanzees but unknown among spider monkeys	Tolerant to affiliative relations among within-group males, no overt dominance hierarchies among males
Infanticide	Reported and common	Reported in chimpanzees, one case reported in spider monkeys along with several other instances of lethal intragroup aggression	Not reported for either taxon
Male intergroup relations	Resident males aggressive toward males in other groups and solitaries	Intergroup relationships are very hostile and males cooperate in patrolling and defense of community range against other groups of males	Male muriquis cooperate against extragroup males, unknown for bonobos
Male/female dominance	Males dominant to females	Males dominant to females	Males and females are codominant
Ranging patterns	Groups are cohesive, sexes range together	Average subgroup size is small, individual females typically utilize different core areas within community range, male ranges larger and up to size of community range, males and females often range separately	Subgroup size is large, individual females do not have own core areas, societies not sex-segregated
Mating patterns	Male mating success is largely dominance-based	Consortships reported in both species, some promiscuous and dominance-based mating in chimpanzees	Promiscuous matings and tolerant male/male relationships

periods of juvenile development, late age at first reproduction, long interbirth intervals, and low realized lifetime fertility.

Several specific examples further illustrate the marked socioecological convergences between atelines and distantly related hominines. The first of these highlights the similarities in foraging ecology, social structure, and mating system between the various howler monkey species (*Alouatta*) and gorillas (*Gorilla gorilla*); the second of these considers the convergence of spider monkeys (*Ateles*) and muriquis (*Brachyteles*) with common chimpanzees (*Pan troglodytes*) and bonobos (*Pan paniscus*), respectively, in light of contemporary socioecological theory.

Howler monkeys (*Alouatta* spp.) and gorillas (*G. gorilla*) are more folivorous than their closest relatives and, depending on the species, subspecies, or population being considered, may live in groups containing one or multiple reproductive-age males. Red, brown, and red-and-black howler monkeys (*A. seniculus*, *A. caraya*, and *A. guariba*, respectively), like mountain gorillas (*G. g. beringei*), typically live in groups containing one or a few adult males; and group membership is essential for male reproduction. Competition among males for breeding positions is strong, and infanticide has been reported when a new male enters a group of females (*Alouatta*, Crockett and Sekulic 1984, Zunino et al. 1986, Crockett and Rudran 1987b, Rumiz 1990, Galetti et al. 1994, Agoramoorthy and Rudran 1995, Calegaro-Marques and Bicca-Marques 1996; *Gorilla*, Fossey 1984, Watts 1989b). Within groups of mantled howlers (*A. palliata*) and western lowland gorillas (*G. g. gorilla*) (and in some mountain gorilla groups), there may be multiple breeding males and male reproductive success is likely to be correlated with male competitive ability. Additionally, in both *Alouatta* and *Gorilla*, intrasexual relationships tend to be weak and adult males appear to be the primary focus of social attention from other group members.

Spider monkeys (*Ateles* spp.) and chimpanzees (*Pan troglodytes*) are ripe fruit specialists that typically live in fission–fusion social systems characterized by the formation of subgroups whose composition is relatively fluid (*Ateles*, Klein 1972, Cant 1977, van Roosmalen 1985, McFarland 1986, Symington 1990, Chapman 1990a; *Pan*, Nishida 1968, Wrangham 1977, Wrangham and Smuts 1980, Goodall 1986, Aureli and Schaffner 2008, Aureli et al. 2008). For both taxa, subgroup and feeding party size closely track habitat-wide measures of the availability of ripe fruits (Symington 1987a, 1988c; Chapman et al. 1995), suggesting that fission–fusion sociality may represent an adaptation for reducing the level of intragroup feeding competition experienced during times of scarcity. Moreover, in both taxa, the sexes often range separately (Wrangham and Smuts 1980, Symington 1988a, Fedigan et al. 1988, Chapman 1990a, Nunes 1995); and males in each taxon are affiliative and cooperative with one another (Nishida 1979, Fedigan and Baxter 1984, Goodall 1986, Watts 1998b, Slater et al. 2008, Link et al. 2009), commonly direct high levels of aggression toward females (Goodall 1986, Campbell 2003, Slater et al. 2009, Link et al. 2009), and join with other males to patrol community boundaries and defend these areas against males from other groups (Bygott 1979, Watts and Mitani 2001, Shimooka 2005, Watts et al. 2006, A. Link and A. Di Fiore, unpublished data). Male behavior in intergroup encounters and male strategies to defend their territories and initiate coalitionary attacks on males in other groups (A. Link and A. Di Fiore, unpublished data) greatly resemble the behavior of common chimpanzee males (Wrangham 1999, Watts and Mitani 2001, Wilson et al. 2004, Watts et al. 2006). Additionally, mating strategies of both chimpanzees and spider monkeys include the formation of consortships (Tutin 1979; van Roosmalen 1985; Symington 1987a; Goodall 1986; Campbell 2000b, 2006a; Campbell and Gibson 2008; Gibson, K. N., et al. 2008), although the overt promiscuity seen occasionally in chimpanzees is not reported for *Ateles*. Both within and across periovulatory periods, females of both species mate with multiple males; however, copulation usually does not take place in the presence of other adult males in spider monkeys (Campbell 2000b, 2006a; Campbell and Gibson 2008; Gibson, K. N., et al. 2008). Chimpanzee males may become very possessive of copulatory partners close to the presumed time of ovulation (Tutin 1979). The convergence between the social behavior of spider monkeys and chimpanzees seems to be related to a pattern of high competition between groups (for males) that has led to the evolution of strong bonds between males within social groups, while differences between these two species in male behavior seem to be associated with the patterns of intragroup competition among males (Link et al. 2009).

In contrast, muriquis (*Brachyteles* spp.) and bonobos (*Pan paniscus*) feed more on leafy material found in larger patches. Probably as a result of the more uniform distribution of their food, the social system of these two primate taxa differs subtly from that of *Ateles* and *P. troglodytes*. While both bonobos and muriquis may be found in fission–fusion societies, at least in some populations (bonobos, Kuroda 1979, Kano 1982, Badrian and Badrian 1984, Nishida and Hiraiwa-Hasegawa 1987, White 1986, 1988; muriquis, Milton 1984a, Strier et al. 1993), subgroup size tends to be much larger than that seen in comparably sized communities of spider monkeys or chimpanzees (Dias and Strier 2003). Social life in these two species is likewise dramatically different from that of chimpanzees and spider monkeys. Neither muriqui nor bonobo society is considered sex-segregated to the extent seen in spider monkeys and chimpanzees: Females do not typically maintain individual core areas within the community range, and foraging parties are far more likely to contain individuals of both sexes. Additionally, relationships between males and females are more egalitarian, and aggressive behaviors, whether within or between the sexes, are observed only rarely (bonobos, Furuichi 1997; muriquis, Strier 1992c). Finally, sexual behavior in both taxa is often overtly promiscuous and often takes place in the presence of other adult males (Milton 1985a,b; Furuichi 1992).

Convergences such as these are clearly interesting as they imply that similar ecological pressures have resulted in the evolution of similar social systems and similar mating patterns in widely divergent taxa. Nonetheless, it is important to stress that the atelines should not be thought of as simply New World equivalents of the African great apes (nor, for that matter, should the African apes be thought of as merely "Old World atelines"). Rather, the atelines represent a unique radiation of primates that fill important roles in the ecology of contemporary neotropical forests and that have social lives we are now beginning to understand.

Species included in this chapter are shown in Color Plates 11–13 (*Brachyteles hypoxanthus*, *Ateles belzebuth*, and *Alouatta palliata*).

12

The Asian Colobines
Diversity Among Leaf-Eating Monkeys
R. Craig Kirkpatrick

1. What differentiates the Asian colobines from other Asian primates? How might this affect their ecological range?
2. What behavioral and ecological patterns unite the Asian colobine genera, and in which ways do they differ from one another?
3. How does the one-male unit basal system allow for both stability and change in Asian colobine social organization?

INTRODUCTION

Colobines are known as "leaf-eating monkeys." This is because they have physiological adaptations—most notably, a multichambered stomach with fermenting bacteria—that help them digest leaves. Yet, this generalization masks the remarkable diversity within colobines.

Colobines are found throughout Asia, in a broad range of habitats. These monkeys also have a broad range of diets and social organizations. The diets of Asian colobines are often seasonal, and the monkeys typically prefer young leaves, seeds, and fruits when these foods are available. Relative to cercopithecines, groups of Asian colobines are usually small, with just several females, their offspring, and one male. As might be expected, these groups have relatively small home ranges and short daily travel distances.

Asian colobines are important to our understanding of primates in general. Although Asian colobines have modal patterns in diet and in social organization, there are variations both within and among species. This makes Asian colobines useful in the study of one of primatology's enduring themes: relationships among ecological variables, feeding strategies, and social organization. Asian colobines play a key role in primatology's quest to better understand food choice, such as trade-offs between food availability and food quality. Additionally they are central to investigations of social behavior, such as male/male competition and alloparental care.

TAXONOMY AND DISTRIBUTION

The taxonomy of the Asian colobines has been, and continues to be, a topic of exciting debate. All taxonomies are hypotheses, revised as new information becomes available (Groves 2001a). An example of this comes from the "odd-nosed colobines." Discussion of whether these species should be lumped in the two genera *Nasalis* (proboscis monkeys) and *Pygathrix* (doucs) or split into four with the addition of *Rhinopithecus* (snub-nosed monkeys) and *Simias* (pig-tailed langurs) took place over 25 years (see Groves 1970; Jablonski 1998a). Another example comes from discussion over the past 15 years of whether species historically grouped under the genus *Presbytis* are better represented as two genera, *Presbytis* (surilis) and *Trachypithecus* (lutongs).

In the absence of a unified framework, researchers have used idiosyncratic and varying models, leading to competing taxonomies and a literature that at times resembles the Tower of Babel. Groves (2001a) provides comprehensive structure, guiding a path through recent debates (Table 12.1). Groves's taxonomy is in overall accord with comprehensive solutions proposed by others (e.g., Brandon-Jones et al. 2004) and is used in this chapter.

Asian colobines are divided into seven genera: *Nasalis*, *Simias*, *Pygathrix*, *Rhinopithecus*, *Presbytis*, *Trachypithecus*, and *Semnopithecus*. The genus *Semnopithecus* (gray langurs) has seven species, previously described as one, *Presbytis entellus*. Gray langurs are distributed widely in southern Asia, from the temperate Himalayas to tropical Sri Lanka; growing analysis suggests that different populations of the former *P. entellus* are better represented by separate species. There is also a strong case for separating *Presbytis* and *Trachypithecus* on the basis of both dental and genital morphology. The remaining genera—*Nasalis*, *Simias*, *Pygathrix*, and *Rhinopithecus*—are grouped as the "odd-nosed colobines," although questions remain as to the phyletic relations within this clade.

Table 12.1 Taxonomy of Asian Colobines

LATIN NAME	COMMON NAME	NOTES
Semnopithecus	**Gray langur or Hanuman langur**	
S. entellus	Northern plains gray langur	Formerly *Presbytis entellus*
S. schistaceus	Nepal gray langur	Formerly in *S. entellus*
S. ajax	Kashmir gray langur	Formerly in *S. entellus*
S. hector	Tarai gray langur	Formerly in *S. entellus*
S. hypoleucos	Black-footed gray langur	Formerly in *S. entellus*
S. dussumieri	Southern plains gray langur	Formerly in *S. entellus*
S. priam	Tufted gray langur	Formerly in *S. entellus*
Trachypithecus	**Lutong**	
Vetulus group		
T. vetulus	Purple-faced langur	Formerly *Presbytis senex*
T. johnii	Nilgiri langur	Formerly *Presbytis johnii*
Cristatus group		
T. auratus	Javan lutong	Formerly in *T. cristatus*
T. cristatus	Silver leaf monkey	Formerly *Presbytis cristatus*
T. germaini	Indochinese lutong	Formerly in *T. cristatus*
T. barbei	Tenasserium lutong	Formerly in *T. cristatus*
Obscurus group		
T. obscurus	Dusky, or spectacled, leaf monkey	Formerly *Presbytis obscurus*
T. phayrei	Phayre's leaf monkey	Formerly in *T. obscurus*
Pileatus group		
T. pileatus	Capped langur	Formerly *Presbytis pileatus*
T. shortridgei	Shortridge's langur	Formerly in *T. pileatus*
T. geei	Gee's golden langur	Formerly *Presbytis geei*
Francoisi group		
T. francoisi	Francois's langur	Formerly *Presbytis francoisi*
T. hatinhesis	Hatinh langur	Formerly in *T. francoisi*
T. poliocephalus	White-headed langur	Formerly in *T. francoisi*
T. laotum	Laotian langur	Formerly in *T. francoisi*
T. delacouri	Delacour's langur	Formerly in *T. francoisi*
T. ebenus	Indochinese black langur	Formerly in *T. auratus*
Presbytis	**Surili**	
P. melalophos	Sumatran surili	
P. femoralis	Banded surili	Formerly in *P. melalophos*
P. chrysomelas	Sarawak surili	Formerly in *P. femoralis*
P. siamensis	White-thighed surili	Formerly in *P. femoralis*
P. frontata	White-fronted langur	
P. comata	Javan surili	Formerly *P. aygula*
P. thomasi	Thomas's langur	Formerly in *P. aygula*
P. hosei	Hose's langur	Formerly in *P. aygula*
P. rubicunda	Maroon leaf monkey	
P. potenziani	Mentawai langur	
P. natunae	Natuna Islands surili	Formerly in *P. femoralis*

Table 12.1 (cont'd)

LATIN NAME	COMMON NAME	NOTES
Pygathrix	Douc	
Py. nemaeus	Red-shanked douc	
Py. nigripes	Black-shanked douc	Formerly in Py. nemaeus
Py. cinerea	Gray-shanked douc	Formerly in Py. nemaeus
Rhinopithecus	Snub-nosed monkey	
R. roxellana	Golden snub-nosed monkey	Formerly Py. roxellanae
R. bieti	Black snub-nosed monkey	Formerly Py. bieti
R. brelichi	Gray snub-nosed monkey	Formerly Py. brelichi
R. avunculus	Tonkin snub-nosed langur	Formerly Py. avunculus
Nasalis		
N. larvatus	Proboscis monkey	
Simias		
S. concolor	Pig-tailed langur	Formerly Nasalis concolor

Source: Groves 2001a.

Semnopithecus is found from Pakistan through India into Nepal, Bhutan, and Sri Lanka. *Presbytis* and *Trachypithecus* are broadly sympatric in Indonesia and Malaysia, with *Trachypithecus* extending farther north into Indochina, Burma, and Bangladesh. The odd-nosed monkeys are dispersed in distribution, with *Rhinopithecus* in southern China and northern Vietnam; *Pygathrix* in Vietnam, Laos, and Cambodia; *Nasalis* on the island of Borneo; and *Simias* only on the Mentawai Islands of Indonesia.

Some species of Asian colobines are widely distributed. *Semnopithecus dussumieri* (southern plains gray langur) is found throughout northern India, *Trachypithecus cristatus* (silver leaf monkey) across Indonesia, and *Presbytis rubicunda* (maroon leaf monkey) throughout Borneo. More often, however, species have restricted ranges, such as *Trachypithecus poliocephalus* (white-headed langur) in southern China, *Trachypithecus barbei* (Tenasserium lutong) in the northern peninsular areas of Burma and Thailand, and *Pygathrix cinerea* (gray-shanked douc) in central Vietnam. Although deforestation in Asia has severely constricted habitat, the limited ranges are not simply the result of human action. The restricted ranges of Asian colobines and the discontinuous distributions of those species with larger ranges may be an artifact of habitat refugia resulting from climate fluctuations during the Pleistocene (Brandon-Jones 1996).

ECOLOGY

Habitats and Range Use

In keeping with their distribution throughout Asia, the colobines are found in many types of habitat (Table 12.2).

Most species live in tropical or subtropical forests, expected habitat for monkeys. Less conventional habitats are home to some species, too. These include the temperate Himalayan forests of Nepal (*Semnopithecus schistaceus* [Nepal gray langur]) and southwest China (*Rhinopithecus bieti* [black snub-nosed monkey]), the peat swamps of Borneo (*Nasalis larvatus* [proboscis monkey]), and the dry scrub deserts of western India (*S. dussumieri*). Perhaps most striking, members of the Franciosi group of *Trachypithecus* live in forests on vertical cliffs of karst limestone in Vietnam and China, regularly using large caves as sleeping sites (e.g., *T. poliocephalus*, Huang et al. 2003).

The population densities of Asian colobines vary by an order of magnitude (Table 12.2). The Chinese species of *Rhinopithecus*, for example, have relatively low densities of about 10 individuals/km² (Kirkpatrick 1998), perhaps because they live in temperate forests of low productivity. At the other extreme, *Trachypithecus johnii* (Nilgiri langur) reach densities of over 70 individuals/km² (Oates et al. 1980) and *Presbytis siamensis* (white-thighed surili) reach densities of over 100 individuals/km² (Davies 1994). Population densities vary not only among species but also among different populations of the same species. Densities of *S. dussumieri* span from under 5/km² (Orcha, India; Dolhinow 1972a) to over 45/km² (Kanha, India; Newton 1992); the habitats of Orcha and Kanha are both moist deciduous forest. Similarly, in both *N. larvatus* and *Trachypithecus geei* (Gee's golden langur), densities vary substantially between populations, from about 10 to over 60 individuals/km², without notable differences in habitat (see Table 12.2).

The range sizes used by groups of Asian colobines vary widely (Table 12.2). Most Asian colobines have range sizes below 1 km². Some species, however, such as *S. schistaceus*

Table 12.2 Range Use and Diet in Asian Colobines[1]

SPECIES	SITE	HABITAT	RANGE SIZE (HA)	OVERLAP (%)	DAY RANGE (M)	POPULATION DENSITY (INDIVIDUALS/KM²)	L (%)	YL (%)	ML (%)	S/F (%)	FL (%)	B/B (%)	L/F (%)	O/U (%)	REFERENCE
Semnopithecus															
Semnopithecus dussumieri	Orcha	Moist deciduous forest	~375	Extensive		3–6									Dolhinow 1972a,b
	Kaukori	Dry scrub forest	~775			3									Dolhinow 1972a,b
	Kanha	Moist deciduous forest	74		1,083	46	39	4	35	24	9	11	—	3	Newton 1992, 1994
	Gir	Dry deciduous forest	20	0		121									Starin 1978
	Dhawar (open)	Tropical broadleaf forest	149			11	58			29	7	2	—	4	Sugiyama 1964
	Dhawar (closed)	Tropical broadleaf forest	19			76									Sugiyama 1964
S. schistaceus	Junbesi	Conifer/ broadleaf forest	~1,275			~2									Curtin 1975
	Ramnagar	Semievergreen forest				26									Borries et al. 2001
S. priam	Polonnaruwa						48	27	21	45	7	—	—	—	Hladik 1977
Trachypithecus															
Vetulus group															
Trachypithecus vetulus	Polonnaruwa	Dry deciduous forest	2				60	20	40	28	12	—	—	—	Hladik 1977
T. johnii	Nilgiri district	Moist forest	65–250	Considerable											Poirier 1970
	Kakachi	Evergreen forest	24	10		71	52	25	27	25	9	7	—	6	Oates et al. 1980
Cristatus group															
T. auratus	Pangandaran	Secondary forest					56	46	<10	32	14	—	—	8	Kool 1992, 1993
T. cristatus	Rantau Panjang	Mangrove forest			<200										Furuya 1961
Obscurus group															
T. obscurus	Kuala Lompat	Dipterocarp forest	17–33	3	560	87	58	36	22	35	7	—	—	—	Curtin 1976, 1980
Pileatus group															
T. pileatus	Madhupur	Tropical moist forest	22	84	325	52	53	11	42	34	7	4	—	1	Stanford 1991b,c
	Manas	Tropical moist forest	64	~60											Mukherjee 1978
T. geei	Ripu	Tropical moist forest				46									Srivastava et al. 2001
	Chirrang	Tropical moist forest				64									Srivastava et al. 2001
	Manas 01	Tropical moist forest				20									Srivastava et al. 2001
	Manas 02	Tropical moist forest				8									Srivastava et al. 2001
Presbytis															
Presbytis femoralis	Perawang	Rubber plantation	22	25	935	42	29	26	3	58	<1	8	—	5	Megantara 1989
P. siamensis	Kuala Lompat	Dipterocarp forest	27	~80	750		36	28	8	49	12	—	—	3	Davies et al. 1988, Bennett 1983

Table 12.2 (cont'd)

SPECIES	SITE	HABITAT	RANGE SIZE (HA)	OVERLAP (%)	DAY RANGE (M)	POPULATION DENSITY (INDIVIDUALS/ KM²)	L (%)	YL (%)	ML (%)	S/F (%)	FL (%)	B/B (%)	L/F (%)	O/U (%)	REFERENCE
	Kuala Lompat	Dipterocarp forest	21	33	750		35	24	11	56	6	—	—	2	Curtin 1980
P. comata	Kutai	Dipterocarp forest				20	65			27	—	—	—	8	Rodman 1978
	Kamojang	Montane forest	38	10	500	11	65	59	6	14	7	1	4	7	Ruhiyat 1983
P. thomasi	Ketambe	Primary rain forest													Steenbeek et al. 2000
	Bohorok	Rubber plantation	14		683	21	36			53	8	—	—	3	Gurmaya 1986
P. hosei	Lipad	Dipterocarp forest	35		740	~26	78	45	5	19	3	—	—		Mitchell 1994
P. rubicunda	Sepilok	Dipterocarp forest	67	10	890		37	36	1	49	11	—	—	2	Davies 1984, Davies et al. 1988
	Tanjung Puting	Dipterocarp forest and peat swamp	33–99				36			52	12	—	—	—	Supriatna et al. 1986
P. potenziani	Betumonga	Dipterocarp forest	34	~40	540	11	55			32	—	—	—	13	Fuentes 1994
	Sarabua	Dipterocarp forest	13			14									Watanabe 1981
Pygathrix															
Pygathrix nemaeus	Son Tra	Tropical moist forest													Lippold 1977
Rhinopithecus															
Rhinopithecus roxellana	Shennongjia	Conifer/ broadleaf forest	2,600	100		8									Ren et al. 1998
R. bieti	Wuyapiya	Conifer forest	2,500	100	1,310	7	>6						86	8	Kirkpatrick et al. 1998, Kirkpatrick 1996
R. brelichi	Fanjingshan	Temperate broadleaf forest	3,500 ~1,275	100	1,290	11	71			15	7		0.2	6	Bleisch et al. 1993, Bleisch and Xie 1998
R. avunculus	Ta Ke/Nam Trang	Tropical broadleaf forest	>1,000			<8	38			62					Boonratana and Le 1998
Nasalis															
Nasalis larvatus	Kinabatangan-Sukai	Peat swamp	220	~100	910	34	74	73	<1	11	8	—	—	8	Boonratana 1994
	Tanjung Puting	Peat swamp	130	96		63	52	42	10	40	3	—	—	5	Yeager 1989, 1990
	Samunsam	Mangrove/ heath forest	770	87	485	6	41	38	3	58					Bennett and Sebastian 1988
Simias															
Simias concolor	Pagai Islands	Swamp and lowland forest	7–20			21									Tenaza and Fuentes 1995
	Sarabua	Dipterocarp forest	13	~5		8									Watanabe 1981
	Grukna	Dipterocarp forest	3.5			220									Watanabe 1981

L, leaf; YL, young leaf (a subset of leaf); ML, mature leaf (a subset of leaf); S/F, seed/fruit; FL, flower; B/B, buds and bark; L/F, lichens and fungus; O/U, other or unknown.

in the Himalayas and *R. bieti* in southwest China, have ranges that are substantially larger, over 10 km² (Bishop 1979, Kirkpatrick 1998). In general, species with lower densities have larger range sizes (Yeager and Kirkpatrick 1998), suggesting that resource availability may regulate the number of individuals in an area.

Colobines typically have daily travel distances of under 1 km (Table 12.2). Even the Asian colobines with large home ranges do not have a correspondingly large increase in the length of their daily travel distances. Daily travel distances appear to be related to food resources. *P. rubicunda*, *P. siamensis*, and *S. dussumieri* all travel farther in months when diet is based on fruits and flowers rather than leaves, for example (Davies 1984, Bennett 1986, Newton 1992). Similarly, groups of *Trachypithecus pileatus* (capped langur) move their farthest distances when feeding on fruits and their shortest distances when feeding on mature leaves (Stanford 1991a).

Diet

Colobines are widely known for their physiological adaptations to eat leaves. These adaptations include sharp molars to chew leaves, enlarged salivary glands that help degrade them, and a multichambered stomach with symbiotic microbes that break down leaf fibers (Kay and Davies 1994). Yet, the generalization "leaf-eaters" does injustice to the diversity of diets found among the Asian colobines. The feeding adaptations that aid the digestion of leaves also aid the digestion of other food types. The capacity for diet variability is reflected in the ability of Asian colobines to live in diverse habitats and to endure seasonality in those habitats (Kirkpatrick 1999).

Certainly, many Asian colobines subsist primarily on leaves (Table 12.2). Most of these leaves are young, easier to digest than mature leaves, and often quite seasonal. Asian colobines also typically get a quarter of their diet from seeds and fruits. This reaches an extreme in *P. rubicunda*, whose diet is over 80% seeds in some months (Davies 1991). The most striking examples of diet diversity in Asian colobines come from *Rhinopithecus roxellana* (golden snub-nosed monkey) and *R. bieti*. *R. roxellana* at Baihe, China, switch from leaves in summer to lichens and the green bark and buds of dicot plants in winter (Kirkpatrick et al. 1999); at Wuyapiya, China, *R. bieti* feeds almost exclusively on lichens (Kirkpatrick 1996). Animal matter is rarely a major part of the diet for Asian colobines, although caterpillars and other insects can provide up to a quarter of the diet of *S. dussumieri* at Kanha, India, during the monsoons, when insects are plentiful (Newton 1992).

The diets of Asian colobines are often marked by seasonality. *T. pileatus* at Madhupur, Bangladesh, eats over 80% mature leaves from November to March but switches to over 50% young leaves or fruit when these foods are available (Stanford 1991b). *N. larvatus* at Kinabatangan, Malaysian Borneo, typically feeds on young leaves, switching to fruits

when seasonally available (Boonratana 1994). The fact that switching between mature leaves, young leaves, and seeds or fruits is associated with availability (e.g., *T. johnii*, Oates et al. 1980; *N. larvatus*, Yeager 1989; *T. pileatus*, Stanford 1991b; *S. dussumieri*, Newton 1992) shows that there is a preference for these foods.

Although leaves may be ubiquitous, they are often defended by toxins (e.g., phenols) and digestion-inhibiting materials (e.g., fiber). Asian colobines are therefore selective in the plant parts they choose and in the plant species from which they choose. This can be quantified with "selectivity ratios," comparing the representation of a species in the habitat to the representation of that species in the diet. Such selectivity ratios show that Asian colobines are picky eaters. This is, of course, variable between species and sites. *P. rubicunda* at Sepilok in Malaysian Borneo, for example, derives only 7% of its diet from the five most common tree families (83% of basal area), while *P. siamensis* at Kuala Lompat in Peninsular Malaysia derives 46% of its diet from the five most common tree families (49% of basal area) (Davies et al. 1988).

Sharp molars, enlarged salivary glands, and multichambered stomachs are common to all colobines. Species differences in diet may come from physiological specializations within these general colobine adaptations (Chivers and Hladik 1980, Caton 1999). *Trachypithecus* may be better adapted to leaf eating than is *Presbytis*, for example (Bennett and Davies 1994, Yeager and Kool 2000). The molars of *Trachypithecus* are better at shearing than the molars of *Presbytis*, and the forestomachs of *Trachypithecus* are larger than the forestomachs of *Presbytis* (Chivers and Hladik 1980, Davies 1991, Chivers 1994). Both *Trachypithecus obscurus* (dusky leaf monkey) and *T. pileatus* eat more mature leaves than any known *Presbytis*; *T. obscurus* also eat more leaves than *P. siamensis* at Kuala Lompat, where they are sympatric (Curtin 1980). If indeed *Trachypithecus* is better adapted to leaf eating than *Presbytis*, this may underlie niche separation in sympatric Asian colobines and explain, in part, the broader distribution of *Trachypithecus* relative to *Presbytis*. Although this hypothesis is tantalizing, there is some evidence that contradicts it. *Trachypithecus auratus* (Javan lutong) does not eat more leaves than the broadly sympatric *Presbytis comata* (Javan surili), for example (Rodman 1978, Ruhiyat 1983, Kool 1993). Kool's study of *T. auratus* covered only 8 months, however, possibly underrepresenting the leaf portion of the annual diet.

Predation

While the feeding behavior of Asian colobines is well documented, information on predation rates remains elusive. Cases of predation have only rarely been observed. Examples include a female *N. larvatus* preyed upon by a crocodilian during a river crossing at Tanjung Puting in Indonesian Borneo (Yeager 1991a), a juvenile *T. pileatus* preyed upon by jackals at Madhupur in Bangladesh (Stanford 1991a), and a juvenile *R. roxellana* preyed upon by a goshawk at Qinling in China (Zhang et al. 1999).

Predation pressure is best measured by the importance of a species to a predator's diet. This combined with the density of the predator population gives an indication of the frequency with which individuals are at risk from predation events. Using frequency in feces as an estimate of proportion in diet, *S. dussumieri* provided 6% and 27% of the diet of leopards and tigers, respectively, at Kanha in India (Dolhinow 1972a). *R. roxellana* provided 2% of the diet for leopards at Wolong in China from 1981 to 1983 and 0% from 1984 to 1987 (Johnson et al. 1993).

At least relative to African colobines (see Chapter 13), predation rates for Asian colobines generally seem low. Predation is a significant risk for Asian colobines, nonetheless, as evidenced by the evolution of an array of antipredator behaviors. Leopards evoke alarm calls from *S. dussumieri* (Starin 1978), and birds of prey evoke alarm calls in *R. bieti* (Cui 2003). *N. larvatus* sit and scan rivers before crossing, possibly to detect predators (Yeager 1991a). *Presbytis thomasi* (Thomas's langur) have more neighbors when feeding on the ground, presumably an antipredator strategy (Sterck 2002).

REPRODUCTIVE PARAMETERS AND SOCIAL BEHAVIOR

Social and Sexual Behavior

As measured by grooming, Asian colobines are an unsocial lot. They typically spend no more than 2% of their time in social grooming (Table 12.3). Some species, such as *T. johnii* and *P. siamensis*, have an almost complete absence of social grooming (Poirier 1970, Bennett 1983). *S. dussumieri* and *R. bieti* are relatively social, however, with grooming occupying about 6% of their time (Newton 1994, Kirkpatrick et al. 1998).

Social grooming in Asian colobines is primarily a female affair. Males are generally peripheral to social interaction, whether measured by lack of involvement in grooming (e.g., *Presbytis femoralis* [banded surili], Megantara 1989) or in greater distances between males and the other members of a group (e.g., *P. rubicunda*, Davies 1984). Since males apparently do not regulate their relationships with females through social grooming, it is perhaps surprising that male *herding* (i.e., aggressively maintaining proximity of females, particularly in the presence of extragroup males) is reported only for *T. pileatus* (Stanford 1991c).

Almost universally, Asian colobine sex is initiated by females. Females use species-typical behavior patterns, such as the "head shake" in *S. dussumieri* and the "crouch" in *R. roxellana*, to signal sexual proceptivity and encourage male mounting. With the head shake solicitation, the female directs her hindquarters toward the male while "frantically shuddering" her head (Hrdy 1977b:49). The crouch is similar, although there is no shaking of the head and the female's forearms are bent to lower the upper body (Ren et al. 1995). Copulations are almost always preceded by female solicitation, although not all solicitations result in copulation (e.g., *S. schistaceus,* Borries et al. 2001). *Simias concolor* (pig-tailed langur) is the only Asian colobine in which sexual swellings indicate receptivity (Tenaza 1989).

Gestation in Asian colobines typically lasts 6–7 months (e.g., *Trachypithecus vetulus* [purple-faced langur], Rudran 1973; *R. roxellana*, Qi 1986; *S. dussumieri*, Sommer et al. 1992; *S. schistaceus*, Borries et al. 2001). Births—and, presumably, copulations—are often concentrated in "seasons" lasting 2–6 months (Table 12.3). It is unclear what regulates birth seasonality, although it is possibly based on nutrition (e.g., *S. schistaceus*, Ziegler et al. 2000b). Seasonality in copulations and births is typically retained by captives, however (e.g., *R. roxellana*, Zhang et al. 2000), so regulatory mechanisms are probably not directly tied to nutritional intake. In general, interbirth intervals are around 2 years, sometimes less (Table 12.3). Interbirth intervals often vary between different populations of the same species. *T. vetulus* in Sri Lanka, for example, has an interbirth interval of 2 years at Polonarruwa but only 1.3 years at Horton Plains (Rudran 1973).

Infants and Adults

Asian colobines often wean their infants within the first year (e.g., *T. johnii*, Poirier 1970; *T. vetulus*, Rudran 1973; *P. femoralis*, Megantara 1989). Some studies have reported suckling beyond 1 year, however (e.g., *P. comata*, Ruhiyat 1983; *S. dussumieri*, Dolhinow 1972a; Sommer et al. 1992). Weaning may be based on nutritional constraints, with monkeys at poorer sites weaning later. The longest reported lactation period is for *S. schistaceus* at Ramnagar, Nepal, a relatively poor site with long interbirth intervals. Females at Ramnagar typically wean their infants only after again becoming pregnant, an average of 25 months (Borries et al. 2001).

Nonmaternal caretakers, or *alloparents*, are found in virtually all Asian colobines for which we have good records (Table 12.3). The alloparent may hold, groom, or carry the infant. Alloparents are usually adult females, at times juveniles (mainly female, sometimes male). Alloparenting can occur soon after birth and provide a substantial portion of parental care. In the first 4 weeks of life, *P. femoralis* infants are held about a third of the time by individuals other than the mother (Megantara 1989). In *R. bieti*, juveniles provide about a quarter of infant grooming and 15% of infant carriage during long-distance travel (Kirkpatrick et al. 1998).

Adult male Asian colobines are typically "passively tolerant" of infants (Whitten 1987). Males are alloparents only on the rarest occasions. More common are reports of adult males being "protective" of infants, giving alarm calls

Table 12.3 Social Behavior in Asian Colobines

SPECIES	SITE	BIRTH SEASON	WEANING AGE (MONTHS)	INTERBIRTH INTERVAL (YEARS)	ALLOPARENTAL CARE?	MALE-INFANT RELATIONS	SOCIAL TIME (%)	REFERENCE
Semnopithecus								
Semnopithecus	Orcha	Weak	11–15	~2	Yes			Dolhinow 1972a,b
dussumieri	Kaukori	Apr–May			Yes			Dolhinow 1972a,b
	Kanha						6	Newton 1994
	Jodhpur	Mar (peak)	13	1.4			6.5	Sommer and Rajpurohit 1989, Sommer et al. 1992, Borries et al. 1994
	Gir						<2	Starin 1978
	Dhawar (open)					Violent during male replacements		Sugiyama 1964
S. schistaceus	Melemchi	Feb–Apr						Bishop 1979
S. schistaceus	Ramnagar	Jan–Jun (47% Mar)	25	2.4	To 5 months	At times protective, at times violent		Borries et al. 1999a, 2001
Trachypithecus								
Vetulus group								
Trachypithecus	Polonnaruwa	May–Aug (40% Jun)	7–8	2				Rudran 1973
vetulus	Horton Plains			1.3				Rudran 1973
T. johnii	Nilgiri district	May–Jun	10–11	~1.75	Mostly at 3–7 weeks	Passively tolerant	0.1	Poirier 1970
Pileatus group								
T. pileatus	Madhupur	Dec–May		2	Yes	Passively tolerant	0.4	Stanford 1991a,c
Presbytis								
Presbytis femoralis	Perawang	No	9–10		To 3 months		1.5	Megantara 1989
P. siamensis	Kuala Lompat				Rare		Rare	Bennett 1983
P. comata	Kamojang		>12					Ruhiyat 1983
P. thomasi	Bohorok	No						Gurmaya 1986
P. rubicunda	Sepilok				Yes		Rare	Davies 1984
P. potenziani	Betumonga				No		<2	Fuentes 1994
Rhinopithecus								
Rhinopithecus bieti	Wuyapiya	Mar/Apr		~3	Yes	Passively tolerant	6.1%	Kirkpatrick et al. 1998
Nasalis								
Nasalis larvatus	Kinabatangan/ Sukai	Weak			Yes		2.2	Boonratana 1994
	Tanjung Puting						Rare	Yeager 1990
	Samunsam				Yes		Rare	Rajanathan and Bennett 1990

or standing between the infant and harm. Male protection is context-specific, however, and may depend on relatedness. While certain males (i.e., probable fathers) may protect infants at times, other males (i.e., nonfathers) may be extremely aggressive toward them. At Ramnagar in Nepal, for example, *S. schistaceus* males in multimale groups are principal defenders of infants against attack but only if they are the genetic father or were resident when the infant was conceived (Borries et al. 1999a).

The question of infanticide in Asian colobines has generated heated debate (e.g., Sussman et al. 1995, Hrdy et al. 1995). The social pathology hypothesis suggests that infanticide results from crowded living conditions. The sexual selection hypothesis ties primates into the larger body of literature on infanticide in mammals, suggesting that males will kill unrelated infants if this increases their reproductive success. An infanticidal male may benefit his reproductive success if (*1*) the female losing the infant becomes sexually receptive sooner than if she had not lost her infant and (*2*) the infanticidal male gains sexual access to her.

As with other types of predation, infanticide is rarely observed. In Asian colobines, it is reported primarily in some populations of *Semnopithecus* (e.g., *S. dussumieri* of Dharwar and of Abu, both in India, Sugiyama 1964, Hrdy 1977b; *S. schistaceus* of Ramnagar in Nepal, Borries et al. 1999a). Infanticide is estimated to cause 20%–30% of infant mortality at Ramnagar (*S. schistaceus*) and at Kanha and Jodhpur (both sites of *S. dussumieri* in India) (Borries 1997). Data from both *S. schistaceus* and *S. dussumieri* tend to support the sexual selection hypothesis. In both species, a mother losing an infant to infanticide has a shorter interbirth interval than a mother not losing an infant (Sommer et al. 1992, Borries 1997). Further, genetic analysis of *S. schistaceus* at Ramnagar clearly shows that a male will not attack his own infant and that, when a female loses her infant, the male that attacked the infant has a higher chance of siring the female's next offspring (Borries et al. 1999b).

Whatever the pressures leading to infanticide, they are variable among sites; many studies of *Semnopithecus* have not reported infanticide. Also, for all the discussion of infanticide in Asian colobines, it is rarely reported for species other than *Semnopithecus*, with the exception of *P. thomasi* (Sterck 1995). Single cases of infanticide have been observed in the wild for *N. larvatus* and *R. bieti* (Agoramoorthy and Hsu 2005, Xiang and Grueter 2007). Infanticide has also been inferred both from the disappearance of infants after new males become resident in groups (e.g., *T. cristatus*, Wolf and Fleagle 1977) and from the synchrony of births within groups and not among them (e.g., *T. vetulus*, Rudran 1973). While there are valid theoretical arguments as to why infanticide can be an adaptive strategy for some males, the data do not yet allow a detailed evaluation of its importance to social organization in Asian colobines as a whole.

SOCIAL ORGANIZATION

The modal pattern of mixed-sex groups in Asian colobines is the one-male unit (OMU). In such groups, one male associates with multiple females and their offspring. In many Asian colobines, over 50% of the males in mixed-sex groups are in OMUs, suggesting that this is the primary strategy (Table 12.4). There is much variation around this central theme, however. In some species, mixed-sex groups are mainly multimale, with two or, rarely, three or more males. In some species, OMUs have overlapping ranges and a second level of social organization, the "band." In such species, OMUs have relatively nonaggressive relationships with a subset of OMUs in the surrounding area, at times traveling together. *T. pileatus* males in OMUs at Madhupur, Bangladesh, for example, respond less aggressively to familiar OMUs than to unfamiliar OMUs (Stanford 1991c). The band form of social organization reaches an extreme in *Rhinopithecus*, with OMUs having complete range overlap and traveling together as a single entity of 200 or more individuals in ranges of over 20 km² (Kirkpatrick 1998).

As in most Old World monkeys, male Asian colobines are the "immigrating sex." Male departures typically occur before physical maturity (e.g., *N. larvatus*, Boonratana 1994; *P. thomasi*, Steenbeek et al. 2000). Adult, subadult, and juvenile males that are not part of mixed-sex groups may be solitary or found in all-male units (AMUs). Juvenile females are present in AMUs at times (e.g., *N. larvatus*, *T. auratus*, *T. johnii*, *P. thomasi*, Yeager and Kool 2000, Steenbeek et al. 2000), suggesting that "AMU" is an imprecise term that masks more complex demographic and life history processes. Although juvenile females may be present, heterosexual sex has not been reported for AMUs.

OMUs vary widely in their size and in the number of females within them (Table 12.4). The OMUs of *T. pileatus* at Madhupur, Bangladesh, are typical of Asian colobines as a whole, with up to 13 individuals and up to five adult females (Stanford 1991a). In contrast, the OMUs of *S. dussumieri* at Kanha, India, are relatively large, with up to 34 individuals and up to 15 adult females (Newton 1987). At the other extreme, *S. concolor* and *Presbytis potenziani* (Mentawai langur) typically, though not exclusively, have OMUs with only one female, giving the appearance of monogamous pairs (Watanabe 1981, Tenaza and Fuentes 1995).

Multimale groups of Asian colobines generally have neither more nor fewer females than do OMUs (Sterck and van Hooff 2000). This suggests that the multimale group is not a strategy to maintain access to more females. In many cases, multimale groups may be relatively transitory, signaling the end of a resident male's tenure, to be replaced by either a maturing son or, more commonly, a male that has recently entered a group that was an OMU prior to the entry of the second male. The dynamic, unstable nature of multimale groups in most species of Asian colobines is brought about by the reproductive interests of individual males at different

Table 12.4 Social Organization in Asian Colobines

SPECIES	SITE	AVERAGE GROUP SIZE (MSG)	% MALES IN MSG IN OMUS	AM/AF RATIO (MSG)	ADULT/ IMMATURE RATIO (MSG)	I/AF RATIO	INTERGROUP RELATIONS	LENGTH OF MALE TENURE	REFERENCE
Semnopithecus									
Semnopithecus dussumieri	Orcha	19	0	1:1.6	1:0.9	1:1.4	Avoidance, nonaggressive		Dolhinow 1972a,b
	Kaukori	54	0	1:3.2	1:1.2	1:1.4			Dolhinow 1972a,b
	Kanha (meadow)	22	87	1:7.9	1:1.1	1:1.7		8.7 years	Newton 1994 (data for 1982)
	Gir	28	33	1:6.0	1:0.6	1:2.0			Starin 1978 (data for April)
	Jodhpur	38	100	1:4.9				2.2 years	Sommer and Rajpurohit 1989, Sommer et al. 1992
	Abu							28–31 months	Hrdy 1977b
	Dhawar	15	42	1:3.1	1:1.1			27–46 months	Sugiyama 1964, Newton 1994
	Raipur	29	23	1:5.7	1:0.6	1:2.6			Sugiyama 1964
S. schistaceus	Junbesi			1:1.6			Rare (once in 16-month study with 243 days of contact)		Curtin 1975
S. schistaceus	Ramnagar	18	38	1:2.6					Borries 1997, et al. 2001 Borries
Trachypithecus									
Vetulus group									
Trachypithecus vetulus	Horton Plains		100	1:3.4	1:1.0	1:1.4			Rudran 1973
T. johnii	Nilgiri district	9	35	1:1.2	1:1.0	1:1.4	Primarily vocal		Poirier 1970
	Kakachi	17							Oates et al. 1980
Cristatus group									
T. auratus	Pangandaran	14	100	1:6.5	1:0.7	No infants			Kool 1993
T. cristatus	Rantau Panjang	35	0	1:4.6	1:1.4	1:2.6			Furuya 1961
Obscurus group									
T. obscurus	Kuala Lompat	17	0	1:2.4	1:1.0	1:1.7			Curtin 1980
Pileatus group									
T. pileatus	Madhupur	8	100	1:3.6	1:0.8	1:1.4	~1 encounter per day, OMUs with variable aggression based on familiarity, no male loud call	At least to 26 months	Stanford 1991c
	Manas	10	71	1:3.3	1:0.9	1:1.2			Mukherjee 1978
T. geei	Ripu	7		1:2.2	1:0.6	1:2.4			Srivastava et al. 2001
	Chirrang	9		1:1.5	1:0.7	1:2.3			Srivastava et al. 2001

Table 12.4 (cont'd)

SPECIES	SITE	AVERAGE GROUP SIZE (MSG)	% MALES IN MSG IN OMUS	AM/AF RATIO (MSG)	ADULT/ IMMATURE RATIO (MSG)	I/AF RATIO	INTERGROUP RELATIONS	LENGTH OF MALE TENURE	REFERENCE
	Manas 01	11		1:2.2	1:0.8	1:1.9			Srivastava et al. 2001
	Manas 02	7		1:3.8	1:0.6	1:4.6			Srivastava et al. 2001
	Jam Duar	10	71	1:3.4	1:1.0	1:1.7	Common, nonaggressive		Mukherjee 1978, Mukherjee and Saha 1974
Presbytis									
Presbytis femoralis	Perawang	11	71	1:4.8	1:0.8		Males aggressive		Megantara 1989
P. siamensis	Kuala Lompat	15	100	1:7.8	1:0.7	1:2.9	~1 encounter per day		Bennett 1983
	Kuala Lompat	16	11	1:2.6	1:1.2	1:3.3	Mostly mutual avoidance, mediated vocally, relations typically aggressive		Curtin 1980
	Kuala Lompat	14	38	1:3.1	1:1.2	1:1.7			Curtin 1976
P. comata	Kamojang	8	100	1:3.0	1:1.0	1:2.4	Males aggressive, ~1 encounter every 30 hr of observation		Ruhiyat 1983
P. thomasi	Ketambe						Both neutral and with male chases	4.9 years (0.3–7.7)	Steenbeek et al. 2000
	Bohorok	8	70	1:3.0	1:0.7	1:2.3	~1 encounter every 130 hr of observation, typically aggressive		Gurmaya 1986
P. rubicunda	Sepilok	6	100	1:2.0	1:1.0	1:2.0	~1 encounter every 20 days		Davies 1984, 1991
	Tanjung Puting		100	1:2.6	1:0.7	1:1.5			Supriatna et al. 1986
P. potenziani	Betumonga		100	1:1.0	1:0.8	1:3.3	Mostly mutual avoidance, mediated vocally, relations typically aggressive		Fuentes 1994
	Sarabua/ Grukna	3	100	1:1.0	1:0.6	1:2.3			Watanabe 1981
Pygathrix									
Pygathrix nemaeus	Son Tra	9	50	1: 2.7	1: 0.9	1:1.6			Lippold 1977
Rhinopithecus									
Rhinopithecus roxellana	Shennongjia	12 (B: 340)	"Most"				OMUs travel together as band		Su et al. 1998
R. bieti	Wuyapiya	<15 (B: 175)	100	1:3.1	1:1.0	1:2.3	OMUs travel together as band		Kirkpatrick et al. 1998
R. brelichi	Fanjingshan	6 (B: 225)	100	1:2.2	1:0.9	1:2.4	OMUs travel together as band, OMUs "rarely aggressive" to one another		Bleisch et al. 1993, Bleisch and Xie 1998

Table 12.4 (cont'd)

SPECIES	SITE	AVERAGE GROUP SIZE (MSG)	% MALES IN MSG IN OMUS	AM/AF RATIO (MSG)	ADULT/ IMMATURE RATIO (MSG)	I/AF RATIO	INTERGROUP RELATIONS	LENGTH OF MALE TENURE	REFERENCE
R. avunculus	Ta Ke/Nam Trang	15 (B: 65)					OMUs occasionally travel together as band		Boonratana and Le 1998
Nasalis									
Nasalis larvatus	Kinabatangan/ Sukai	17	100	1:7.5	1:1.1	1:2.3	Associations among some OMUs		Boonratana 1994
	Tanjung Puting	12	100	1:3.6	1:0.8	1:1.3	Associations among some OMUs, commonly seen at riverbanks, typically neutral encounters with low-intensity displays by males		Yeager 1990
	Samunsam	9	100	1:3.8	1:0.9	1:1.9	Associations among some OMUs, commonly seen at riverbanks, male displays when OMUs approach one another		Bennett and Sebastian 1988
Simias									
Simias concolor	Pagai Islands	4	95	1:1.8	1:0.4	1:3.2	Apparently rare		Tenaza and Fuentes 1995
	Sarabua/Grukna	5 (2–20)	100	1:1.7	1:0.8	1:3.2	Rare, both neutral and aggressive		Watanabe 1981

MSG, mixed-sex group; OMU, one-male group; B, band (i.e., an association of OMUs); AM, adult male; AF, adult female; I, infant.

times in their lives and is reviewed more fully later (see "Population Consequences of Individual Strategies").

The home ranges of Asian colobines often overlap, although the degree of overlap varies both across species and across populations of the same species living in different places. In places where home ranges do not overlap much, physical encounters between mixed-sex groups are rare. When range overlap is greater, encounters are more frequent. Usually, encounters are mediated by loud calls given by males. In some species, mixed-sex groups are aggressive when they meet one another, with chases by the males of differing groups. In other species, mixed-sexed groups are tolerant of one another, even traveling together (N. larvatus, Yeager 1991b; T. pileatus, Stanford 1991c; T. geei, Mukherjee and Saha 1974). Although OMUs in these species may travel together, the OMUs apparently maintain integrity and are still at times aggressive to one another (e.g., Rhinopithecus brelichi [gray snub-nosed langur], Bleisch et al. 1993).

This striking combination—spatial overlap of groups with maintenance of OMU boundaries—has led to the hypothesis that intergroup encounters are based on mate

defense by males and not, as widely argued, on food defense (van Schaik et al. 1992). Groups do not fight over food; rather, males defend their access to females. Two examples of this are P. siamensis and T. pileatus. In P. siamensis, two OMUs will at times come together in a single feeding tree, apparently not using contest competition for access to food, and intergroup aggression is a male affair (Bennett 1983). In T. pileatus, OMUs do not defend food resources and only males, not females, are involved in aggressive intergroup encounters (Stanford 1991a).

POPULATION CONSEQUENCES OF INDIVIDUAL STRATEGIES

In making decisions about whether to join or leave a group—or to contest the entry of or exit by another in one's group—different individuals have different interests. Male and female interests, for example, can overlap but they are not identical. The interests of an individual will also change over time—as a young adult becomes fully mature, for example.

Population structure emerges from the choices made by individuals within that population. This is seen most clearly, perhaps, in changes of group composition over time, including the establishment of new groups. As with other primates, groups of Asian colobines are dynamic. In addition to births and deaths, group change is driven by reproductive decisions: by male/male competition for access to females, often seen in aggressive contests, and by female choice, with females leaving their current group to join another group or to join a lone male, forming a new group.

Male/Male Competition

If an adult male is to have offspring, he needs residence in an OMU or high rank in a multimale group. For *S. dussumieri* at Jodhpur, India, the 25% of males that are never adult residents of OMUs—and have a longer life span because of this—share only 5% of conceptions (Sommer and Rajpurohit 1989). In multimale groups of *S. schistaceus* at Ramnagar, Nepal, male dominance rank correlates with number of matings and, presumably, offspring (Borries 1997).

An adult male generally should tolerate other males in his group only if the potential fitness benefits are greater than the costs. Theoretically, benefits can come from improved predator protection, enhanced intergroup competition for food, or support against infanticidal assaults by outside males. There is little reason to believe that an additional male provides more or better predator protection than an additional female. This is particularly true in Asian colobines without sexual dimorphism (e.g., *Presbytis* spp.). Also, groups of Asian colobines do not typically compete aggressively for food. More likely, multimale groups result when more than one male has a fitness interest in the group, either because one male cannot monopolize the females in a group or because a male has reached maturity within his natal group. Although some multimale groups are stable over time, many are not, with the males in these groups engaging in either sudden contests or gradual transitions that result in OMUs.

Stable multimale groups are best known for *Semnopithecus*. Groups of *S. schistaceus* are relatively large, possibly because they live in low-productivity environments (Bishop 1979). A single male cannot monopolize the females in these groups. Multimale groups of *S. schistaceus* at Ramnagar have up to five adult males, with a hierarchy of sexual access to females (Borries et al. 1999a). Low rank may cause some males to transfer; new resident males in groups at Ramnagar are former residents of adjacent groups about a third of the time (Borries et al. 1999b).

Although the multimale groups of some Asian colobines are relatively stable, most are groups in transition. When there are two or more males in a group, their relations can become quite aggressive, to the point of group split or one of the males being expelled. This transition can happen slowly, as with most age-graded multimale groups, or quickly, as when extragroup males aggressively join OMUs.

When the dominant male of an OMU is tolerant of his sons, allowing them to stay after maturation, the OMU becomes an age-graded, multimale group. This occurs about half the time in *P. thomasi* at Ketambe, Indonesia (Steenbeek et al. 2000). These young adult males are aggressive with extragroup males, from both mixed-sex groups and all-male bands, keeping them away from females and infants. Age-grade groups in *P. thomasi* merely delay immigration; young adult males rarely replace their fathers as the sexually active male of the group (Steenbeek et al. 2000, cf. Gurmaya 1986). This is variable among species, however. In *T. johnii*, age-graded groups may split, with both the older male and younger male resident in the resulting OMUs (Hohmann 1989).

The transitory multimale groups of Asian colobines are not always age-graded. In *T. johnii* and *P. femoralis*, extragroup males at times join OMUs with surprisingly little aggression (Poirier 1970, Hohmann 1989, Megantara 1989). Over a period of months, however, these groups split due to gradually increasing aggression between the males. Alternately, the transition from multimale group to OMU or group split can happen quite quickly. Violence may start immediately after an extragroup male joins a group, continue for days or weeks, and lead to replacement of the OMU's resident male. The multimale phase in some populations of *S. dussumieri*, for example, is generally short-lived and seems to occur only in the context of aggressive contests between males (e.g., at Dharwar, Sugiyama 1964; at Abu, Hrdy 1977b; at Kanha, Newton 1987—all three sites in India). At Dharwar, male invasions from AMUs will split a group; aggression between the resident male and invading males can result in two OMUs, with females divided between them (Sugiyama 1964). Although age-graded multimale groups are common in *P. thomasi*, violent change in resident males of OMUs occurs as well (Steenbeek et al. 2000).

Female Choice

Females are active participants in decisions about group size and structure. In mammals, variance in reproductive success is generally lower for females than for males, leading to female philopatry and male immigration. In primates, females generally are thought to resist movement between groups because foods and predators are unknown in the range of the new group; there is also less chance to find related individuals with which to form alliances for contest competition (Wrangham 1980). Yet, colobine females may have lower resistance to changing groups because foods are relatively ubiquitous and predation pressure is low. Scramble competition, rather than contest competition, is more relevant. Particularly in species or populations in which groups have overlapping ranges, female resistance to movement between groups may be negligible. Although males are consistently the immigrating sex in colobines (i.e., the sex that almost invariably changes group residency prior to becoming sexually active), females may also transfer (i.e., change group residency either before or after becoming sexually active).

Female transfer is common in Asian colobines (Yeager and Kool 2000, Sterck and van Hooff 2000), and female mate choice is important in determining group size (e.g., *T. pileatus*, Stanford 1991a). Females interact with lone males following established OMUs, at times leaving their current group to join a lone male or transferring to a small, newly formed group (*T. pileatus*, Stanford 1991c; *P. thomasi*, Steenbeek et al. 2000). In *T. johnii*, *P. siamensis*, and *N. larvatus*, even females with infants will transfer (Poirier 1970, Bennett 1983, Bennett and Sebastian 1988). By inference, this suggests that the risk of infanticide is low.

Female choice is important during division of multimale groups as well. In *T. johnii* and *P. femoralis*, division of multimale groups is presaged by consistent membership in subgroups—seen during feeding, for example (Hohmann 1989, Megantara 1989). Females make choices about which male to join and, therefore, are agents in deciding whether the resident male will be replaced or whether the group will split (e.g., *S. dussumieri*, Sugiyama 1964; *T. johnii*, Hohmann 1989).

What do females need from males? A group's male defends females and their infants from the harassment of males from other mixed-sex groups and from AMUs. In some cases, Asian colobine females may also compete for sperm (e.g., *S. dussumieri*, Sommer et al. 1992). In larger groups, one male may not provide enough sperm to impregnate all receptive females. If so, species with tightly constrained breeding seasons should have the smallest groups; females choose smaller groups to ensure access to sperm during the brief window for impregnation.

The male characters subject to female choice remain unclear, as are the mechanisms by which Asian colobine males maintain the allegiance, or forbearance, of females. Male Asian colobines are unsocial, are minor participants in grooming networks, and do not "herd" females. When adult females transfer between existing groups, they may be evaluating female companions as much as male characteristics. Given the general consistency in size of groups of different species of Asian colobines, it seems likely that females balance mate choice with relative group size.

Stability and Change in Social Organization

Population structure results from the interactions of strategies of different individuals. Further, different processes are at work in the same group at different times. Since group changes are based on individual decisions, it is unsurprising that the speed and extent of group change vary over time and across both populations and species. In both *N. larvatus* and *P. thomasi*, for example, some groups remain stable, while others change radically from year to year (Rajanathan and Bennett 1990, Yeager 1990, Sterck 1995). Groups of *T. johnii* also change substantially over varying periods of time, following unpredictable and idiosyncratic patterns (Poirier 1970, Hohmann 1989).

Notwithstanding social change within groups, social organization at the level of the population can be relatively constant. In *N. larvatus*, for example, multitiered social organization—OMUs associating in bands—is consistent among the three sites with long-term studies (Bennett and Sebastian 1988, Yeager 1991b, Boonratana 1994). In some species, however, there is variation in social organization among sites. *S. dussumieri* is the best-studied Asian colobine and is found only in OMUs at Jodhpur, only in multimale groups at Orcha, and in a variable mix of the two at Kanha, at Gir, and at other sites (Sommer and Rajpurohit 1989, Dolhinow 1972a, Newton 1987, Starin 1978; see Newton and Dunbar 1994 for a comprehensive review).

Grooming relationships and the lack of male herding behavior suggest that female decisions maintain internal group structure, including the number of females in a group. Intergroup relations are the province of both males and females, with male/male aggression creating distance between groups and female transfer possibly bringing groups closer together. *Rhinopithecus* provides an instructive example. Given the size of bands of *Rhinopithecus*, in which over 200 individuals travel together, it is perhaps surprising that these are not multimale groups. Weak competition for food among females appears central to the large band sizes (Kirkpatrick 1998). Females maintain subgroups within the band, however, as reflected by grooming cliques. Males attach to and defend these subgroups as a means of assuring access to female mates. The use of OMUs—a character fundamental in colobines—to structure bands may also be based, in part, on phylogenetic constraints whereby *Rhinopithecus* males are relatively ineffective at forming the cooperative relationships necessary to jointly exclude additional males from a large group of females.

As we build a greater understanding of Asian colobines, it is worthy and justifiable to look for attributes at the level of genus and species. We must be wary of an "essentialist" approach that dismisses variation between studies, however. The most widely studied Asian colobine, *S. dussumieri*, shows that the more information we have, the more diversity we see. Similarly, the studies that occur over the longest time spans, such as those for *P. thomasi* at Ketambe, Indonesia, show that individual behavior is flexible and opportunistic, leading to much range around species means. This diversity on common themes is the central lesson of our studies of Asian colobines.

Species included in this chapter are shown in Color Plates 19 and 20 (*Rhinopithecus bieti* and *Trachypithecus francoisi*).

13

African Colobine Monkeys
Their Behavior, Ecology, and Conservation
Peter J. Fashing

1. What anthropogenic factors are responsible for the disappearance of many African colobines from the wild?
2. In what ways do nutritional factors influence African colobine food choice and biomass?
3. In what ways are the results of recent studies of male and female strategies during intergroup encounters in African colobines inconsistent with the predictions of traditional models of primate socioecology?

INTRODUCTION

The Colobinae, a subfamily of monkeys found primarily in the rain forests of Africa and Asia, include many of the most intriguing and endangered taxa in the primate order (Struhsaker 1987, Davies and Oates 1994). Perhaps best known for the multichambered stomachs that enable them to digest cellulose and detoxify secondary compounds in leaves and other food items (Chivers 1994; Kay and Davies 1994), colobines have played key roles in advancing our understanding of several fundamental issues in biology (Fashing 2007). In particular, colobines provided some of the earliest and most persuasive evidence that infanticide by males is an adaptive behavior (Hrdy 1974, Borries et al. 1999) and yielded important insights into the ecological determinants of folivorous primate biomass (McKey 1978a, 1979, Oates et al. 1990). Furthermore, colobine researchers raised some of the earliest and most strident alarms regarding the dire conservation status of many of the world's primates and their habitats (Struhsaker 1972, 1997, Oates 1999, Oates et al. 2000).

The colobines of Africa differ from their Asian counterparts (see Chapter 12) in several ways. First, due to Africa's more contiguous geography and the lower overall ecological diversity of its forests, less taxonomic divergence has taken place among African colobines than among those in Asia (Oates 1994). Second, while all African colobines are arboreal, there is one Asian colobine genus, *Semnopithecus* or the Hanuman langur, that is predominantly terrestrial. Lastly, there are several minor morphological differences between African and Asian colobines, including a more pronounced reduction of the thumb in African taxa (Szalay and Delson 1979, Oates et al. 1994).

TAXONOMY AND DISTRIBUTION

African colobine taxonomy has long been in a state of flux (Lydekker 1905, Schwartz 1929, Rahm 1970, Szalay and Delson 1979, Struhsaker 1981b, Oates and Trocco 1983, Groves 2001a, Grubb et al. 2003, Groves 2007, Nowak et al. 2008, Ting 2008). Conservative taxonomists split African colobines into only two genera, *Colobus* (black and white colobus) and *Procolobus* (red and olive colobus), with five species of *Colobus* and two of *Procolobus* (Oates et al. 1994). A recent taxonomy and the one adopted for this chapter (Groves 2007), however, splits the African colobines into three genera, *Colobus* (black and white), *Procolobus* (olive), and *Piliocolobus* (red). It argues that there are five species of *Colobus*, one of *Procolobus*, and 16 of *Piliocolobus* (Table 13.1). Based on means of adult male and female weights, *Colobus* spp. are the heaviest at an average of 8.9 kg, *Piliocolobus* spp. are intermediate at 7.6 kg, and *Procolobus verus* is the lightest at 4.5 kg (Oates et al. 1994).

Most species can be identified by differences in their pelage. Four of the five species of *Colobus* feature differing black and white pelage patterns, while the fifth is entirely black (*C. satanas*). *Piliocolobus* exhibits greater color variation than the other genera, with pelages of the different forms incorporating various combinations of red, orange, brown, black, gray, and white. The single species of *Procolobus*, *Pr. verus*, has a dull olive-brown pelage (Groves 2007).

The African colobines inhabit many of the forested regions of equatorial and West Africa (Table 13.1). *Colobus* occur from Ethiopia to Guinea and are allopatric, with the exception of one area of sympatry between *C. angolensis*

Table 13.1 The African Colobines and Their Distributions, Habitats, Major Study Sites, and Active Long-Term Researchers

LATIN NAME	COMMON NAME	GENERAL DISTRIBUTION	HABITATS	MAJOR STUDY SITES	ACTIVE LONG-TERM RESEARCHERS
C. angolensis	Angolan colobus	East and Central Africa	Lowland rain, coastal, gallery, and montane forests	Ituri (DRC), Kwale (Kenya), Nyungwe (Rwanda), Salonga (DRC)	None
C. guereza	Guereza	East and Central Africa	Lowland rain, gallery, and montane forests	Budongo (Uganda), Ituri (DRC), Kakamega (Kenya), Kibale (Uganda)	C. Chapman, P. Fashing, T. Harris
C. polykomos	King colobus	West Africa	Lowland rain and gallery forests	Taï (Ivory Coast), Tiwai (Sierra Leone)	None
C. satanas	Black colobus	Central Africa	Lowland rain and montane forests	Douala-Edea (Cameroon), Fôret des Abeilles (Gabon), Lopé (Gabon)	None
C. vellerosus	Ursine colobus	West Africa	Lowland rain, dry, and gallery forests	Bia (Ghana), Boabeng-Fiema (Ghana)	P. Sicotte
Pi. badius	Western red colobus	West Africa	Lowland rain and gallery forests	Abuko (Gambia), Fathala (Senegal), Taï (Ivory Coast), Tiwai (Sierra Leone)	None
Pi. bouvieri	Bouvier's red colobus	DRC	Lowland rain forests	None	None
Pi. epieni	Delta red colobus	Nigeria	Lowland rain forests	Gbanraun (Nigeria)	None
Pi. foai	Central African red colobus	DRC and Zambia	Lowland rain and montane forests	None	None
Pi. gordonorum	Udzungwa red colobus	Tanzania	Lowland rain and montane forests	Udzungwa Mountains (Tanzania)	T. Struhsaker
Pi. kirkii	Zanzibar red colobus	Zanzibar	Dry coastal, coral rag, and mangrove forests	Jozani (Zanzibar), Kiwengwa (Zanzibar), Uzi (Zanzibar)	K. Siex, K. Nowak
Pi. langi	None	DRC	Lowland rain forests	None	None
Pi. oustaleti	None	DRC	Lowland rain forests	None	None
Pi. parmentieri	None	DRC	Lowland rain forests	None	None
Pi. pennantii	Pennant's red colobus	Bioko, Equatorial Guinea	Lowland rain and montane forests	None	None
Pi. preussi	Preuss's red colobus	Cameroon and Nigeria	Lowland rain forests	None	None
Pi. semlikiensis	None	DRC	Lowland rain forests	None	None
Pi. tephrosceles	Ugandan red colobus	Tanzania and Uganda	Lowland rain forests	Gombe (Tanzania), Kibale (Uganda)	C. Chapman, S. Teelen
Pi. rufomitratus	Tana river red colobus	Kenya	Gallery forests	Tana River (Kenya)	D. Mbora
Pi. tholloni	Thollon's red colobus	DRC	Lowland rain forests	Salonga (DRC)	None
Pi. waldronae	Miss Waldron's red colobus	Ghana and Ivory Coast	Lowland rain forests	None (extinct or near extinct)	None
Pr. verus	Olive colobus	West Africa	Lowland rain forests	Taï (Ivory Coast), Tiwai (Sierra Leone)	None

Taxonomy based on Groves (2007) (*C.* = *Colobus*; *Pi.* = *Piliocolobus*; *Pr.* = *Procolobus*). Distribution given as general region unless the species occupies two countries or less in which case the individual countries are given. Note that most of these species are not currently being studied and in many cases have never been studied. DRC, Democratic Republic of Congo (all tables).

and *C. guereza* in the Ituri Forest, Democratic Republic of the Congo (DRC) (Oates 1994, Bocian 1997). *Piliocolobus* are entirely allopatric and occur from Zanzibar to Senegal. *Procolobus* are limited to West Africa, occurring from Nigeria to Sierra Leone (Oates et al. 1994). At some rain forest locations in West Africa, all three African colobine genera are sympatric (Davies et al. 1999, Korstjens 2001).

ECOLOGY

Habitat Requirements

African colobines live in a variety of forested habitats, though some taxa exhibit greater habitat flexibility than others (Table 13.1). Olive colobus are restricted to lowland rain forests and seem to gravitate toward disturbed areas, where they can feed and move stealthily through dense vegetation in the lower and middle canopy (Booth 1957; Oates 1988a, 1994; Korstjens 2001). Most forms of red colobus are also limited to lowland rain forests (Oates 1994), though some are capable of occupying montane forests (*Piliocolobus pennantii*, Gonzalez-Kirchner 1997), dry coastal and mangrove forests (*Pi. kirkii*, Siex 2003; Nowak 2008), or gallery forests (*Pi. badius*, Gatinot 1976; *Pi. rufomitratus*, Mbora and Meikle 2004a). Unlike olive colobus, red colobus tend to spend most of their time in the upper and middle forest canopy (Struhsaker and Oates 1975, Oates 1994), though some species are known to occasionally descend to the ground to cross forest gaps or feed on aquatic plants, soil, or charcoal (Gatinot 1976, Galat-Luong and Galat 1979, 2005, Starin 1991, Struhsaker et al. 1997). Red colobus also tend to be more averse to disturbed areas than olive colobus and may require many decades to recover from population declines caused by human disturbance (Chapman et al. 2000, 2005; Mbora and Meikle 2004b) (Fig. 13.1).

Black and white colobus (Color Plate 18) generally occupy a wider variety of habitats than the other African colobines, with *C. angolensis* and *C. guereza* exhibiting particular habitat lability (Table 13.1). *C. guereza* is unusual among the African colobines in that it seems to thrive most in fragmented and secondary forests (Oates 1977a, 1994). *C. guereza* densities are typically low in primary rain forest blocks (Bocian 1997, Poulsen et al. 2001b), intermediate in disturbed rain forest blocks (Oates 1974, Fashing and Cords 2000), and high in forest fragments (Schenkel and Schenkel-Hulliger 1967, Leskes and Acheson 1971, Dunbar 1987, Krüger et al. 1998). In fact, *C. guereza* is one of the few species of primates known to reach higher densities in logged than unlogged forest (Skorupa 1986, Plumptre and Reynolds 1994, Chapman et al. 2000). Though they spend considerable amounts of time in the canopy, black and white colobus tend to occupy a lower strata than red colobus at sites where they are sympatric (Struhsaker and Oates 1975, Korstjens 2001) and are the African colobines that most often descend to ground level to cross forest gaps or to feed on soil, swamp plants, climbers, or terrestrial herbaceous vegetation (Booth

Figure 13.1 *Piliocolobus kirkii* inhabit the dry coastal, coral rag, and mangrove forests of Zanzibar. With fewer than 2,000 individuals remaining, they are classified by IUCN as 'Endangered' (Siex and Struhsaker 1999a, Struhsaker and Siex 2008, photo by Kirstin Siex).

1956; Dunbar and Dunbar 1974c; McKey 1978a,b; Oates 1978; Harrison and Hladik 1986; Fashing 2001b; Fimbel et al. 2001; Anderson 2007c; Fashing et al. 2007a,b).

Predators

The natural predators of African colobines are crowned-hawk eagles (*Stephanoaetus coronatus*), chimpanzees (*Pan troglodytes*), leopards (*Panthera pardus*), lions (*Panthera leo*), and hyenas (*Crocuta crocuta*) (Oates 1996, Schultz et al. 2004, Galat-Luong and Galat 2005, Bodendorfer et al. 2006). In *C. guereza*, for example, the primary predator appears to be crowned-hawk eagles. Studies in nearby forest compartments at Kanyawara (Kibale Forest, Uganda) found that guerezas accounted for 39% and 10%, respectively, of the prey of crowned-hawk eagles (Skorupa 1987, Struhsaker and Leakey 1990). In comparison, the rate of chimpanzee predation on *C. guereza* at another Kibale site (Ngogo) proved to be quite low, with guerezas accounting for only 4% of the chimpanzee's mammalian prey items (Watts and Mitani 2002). Leopards prey on black and white colobus at several sites. *C. angolensis* and *C. guereza* combined accounted for 1% of prey items in leopard scats at Ituri Forest (Hart et al. 1996), *C. polykomos* remains were found in 2%–8% of leopard scats at Taï Forest, Ivory Coast (Hoppe-Dominik 1984, Zuberbühler and Jenny 2002), and *C. vellerosus* accounted for up to 10% of the prey items found in leopard scats at Comoé National Park, Ivory Coast (Bodendorfer et al. 2006). *C. vellerosus* are also eaten by lions at Comoé accounting for 5% of the prey items in lion scats at this site (Bodendorfer et al. 2006).

Unlike black and white colobus, red colobus are often preyed upon heavily by chimpanzees (Fig. 13.2). At the four sites where chimpanzee hunting behavior has been studied intensively (Taï, Mahale [Tanzania], Gombe [Tanzania], Kibale), red colobus account for 53%–91% of the chimpanzee's mammalian prey items (Boesch and Boesch 1989, Uehara 1997, Stanford 1998a, Watts and Mitani 2002, Pobiner et al. 2007). Stanford (1998a, 2002) provided compelling evidence that chimpanzee hunting patterns exert a major influence on red colobus population size, demography, and behavior at Gombe. He found that mean annual mortality due to chimpanzee predation was 18% for the Gombe red colobus, with immatures disproportionately represented among the individuals killed. Indeed, recent analyses of long-term datasets from Gombe (and Kibale) suggest that, like humans, chimpanzees are prone to hunting red colobus at unsustainable rates, thereby raising the specter of one endangered species hunting another to local extirpation (Teelen 2007, 2008, Fourrier et al. 2008, Struhsaker 2008).

Red colobus behavioral reactions to chimpanzee hunts differ considerably across sites. At Gombe, once chimpanzees are detected by red colobus, the latter often engage in long bouts of alarm calling and males band together to counterattack chimpanzee hunters that approach their group (Stanford 1998a). The more red colobus males that counterattack, the less likely a hunt is to be successful (Stanford 1998a). At Taï, where the canopy is more closed than at Gombe, male red colobus do not counterattack but rather retreat quietly along with other group members when chimpanzees are detected (Bshary and Nöe 1997a). Another strategy employed by Taï red colobus to reduce risk of predation by chimpanzees is to form polyspecific associations with a guenon species (*Cercopithecus diana*) that is particularly vigilant and utters alarm calls when chimpanzees are detected (Bshary and Nöe 1997a,b).

Though chimpanzees are typically their main predator, red colobus are also hunted by crowned-hawk eagles. In the aforementioned studies of eagles in two forest compartments at Kibale, red colobus accounted for 18% and 20% of eagle prey items, respectively (Skorupa 1987, Struhsaker and Leakey 1990). Red colobus are also eaten by leopards, and their remains were found in 4%–11% of leopard scats at Ituri and Taï (Hoppe-Dominik 1984, Hart et al. 1996, Zuberbühler and Jenny 2002). Hyenas prey on red colobus

Figure 13.2 Western red colobus (*Piliocolobus badius*) are the primary targets of both human and chimpanzee hunters in Taï Forest, Ivory Coast (photo courtesy of www.florianmoellers.com).

at Fathala, Senegal where nails from red colobus were found in several hyena regurgitations (Galat-Luong and Galat 2005). Little is known about the predators of the secretive olive colobus. However, it has been speculated that the small group sizes, tendency to form polyspecific associations with *Cercopithecus diana*, and cryptic habits of this species are responses to the threat of chimpanzee predation (Oates and Whitesides 1990, Korstjens 2001).

DISEASE

Colobine populations across Africa have been reported to support a number of disease-causing agents, including simian immunodeficiency virus (SIV), poxviruses, and gastrointestinal (GI) parasites. In a recent study at Taï, for example, over a quarter of *Piliocolobus badius* individuals tested positive for SIV (Locatelli et al. 2008). In addition, scientists recently identified a novel poxvirus in *Pi. tephrosceles* at Kibale, and speculated that the same or a similar virus may have been responsible for an undiagnosed infectious disease outbreak that killed nearly 10% of the red colobus population there in the 1970s (Goldberg et al. 2008). In recent years, researchers have begun to employ non-invasive methods to document the prevalence, and identify the causes and consequences, of parasitism in individuals, groups, and populations of colobines. These studies have specifically focused on GI parasite infections which provide relatively simple and inexpensive indicators of population health (Gillespie 2006). Comparison of data on GI parasite fauna from preliminary surveys of *Pi. rufomitratus* at Tana River, Kenya (Mbora and Munene 2006) and *C. angolensis* at Kwale, Kenya (Okanga et al. 2006) with those from more intensive investigations of *C. guereza* and *Pi. tephrosceles* at Kibale (Gillespie et al. 2005) reveal that similar taxa of GI parasites are found across these different colobine species. Where red colobus co-occur with black and white colobus, the former appear susceptible to a greater richness of GI parasite taxa than the latter (Gillespie et al. 2005, Salzer et al. 2007).

At Kibale, GI parasite infections are often associated with habitat type and level of anthropogenic disturbance, though this pattern varies somewhat across species (Gillespie and Chapman 2008) and even studies (Gillespie et al. 2005, Gillespie and Chapman 2006). Studies in nine Kibale forest fragments that had suffered varying degrees of degradation revealed that *Pi. tephrosceles* populations living in more degraded fragments had higher prevalences of nematodes, a class of potentially pathogenic GI parasites, than those living in less degraded fragments (Gillespie and Chapman 2006). Furthermore, helminths, another class of GI parasites, were more frequent among *Pi. tephrosceles* in fragmented forests than in unfragmented forests, though forest fragmentation had no detectable effect on helminth prevalence in a second colobine species (*C. guereza*) at the same site (Gillespie and Chapman 2008). Differences

in parasitic infections have also been documented between colobine groups living in the interior and those living on the edge of the forest. Specifically, individuals of both *Pi. tephrosceles* and *C. guereza* at Kibale were more likely to be infected with multiple parasite taxa if they lived on the edge of the forest than if they lived in the interior, a pattern that may be attributable to direct transmission of parasites from humans living on the forest edge (Chapman et al. 2006b). Surprisingly, in light of the results described above, one additional study of *Pi. tephrosceles* and *C. guereza* at Kibale found no differences in either parasite infection prevalence or parasite richness between populations living in logged versus unlogged forests (Gillespie et al. 2005). To achieve a fuller understanding of the causes and consequences of disease in African colobines, data are needed from additional species and sites.

CONSERVATION STATUS AND THREATS

While predation and disease have considerable effects on African colobine populations, even greater threats to their survival come directly from anthropogenic sources, including habitat destruction, hunting, and global climate change (Chapman et al. 2006a). Colobines are disproportionately represented among the world's 25 most endangered primates (Fashing 2007), with six Asian forms and three African forms (*Piliocolobus waldronae*, *Pi. pennantii*, and *Pi. rufomitratus*) falling into this category (Mittermeier et al. 2007). Several additional *Piliocolobus* spp. must also be regarded as at least 'threatened with extinction', including the critically endangered *Pi. kirkii* (Struhsaker 2005; Fig. 13.1). Even the most widespread and adaptable species of African colobine, *Colobus guereza*, contains several subspecies that are endangered or of uncertain status (Kingdon et al. 2008, Fashing and Oates in press).

A recent model found that colobine extinction risk is positively related to female body mass and negatively related to mature leaf consumption, number of habitat types occupied, and maximum latitude occupied (Kamilar and Paciulli 2008). Thus, those African colobines at greatest risk of extinction tend to be large-bodied, unable to subsist heavily on mature leaves during lean times, have very narrow habitat requirements, and live near the equator. Habitat destruction, whether for logging or conversion to agriculture, poses a grave threat to most African colobines (Oates 1996, Anderson et al. 2007a). Logging has been shown to have especially adverse effects on red colobus populations across Africa (Skorupa 1986, Davies 1994). For example, several decades after logging ended at Kibale, *Pi. tephrosceles* population densities remained substantially lower in formerly logged areas than in unlogged areas (Chapman et al. 2005). Furthermore, forest clearance by local people has led to a drastic reduction in *Pi. rufomitratus* density over the past several decades at Tana River, Kenya (Decker 1994, Wieczkowski and Mbora 1999/2000, Mbora and Meikle 2004b).

Because human disturbance has left primate populations across Africa isolated in small patches of forest, much recent research has focused on the value of forest fragments for African colobine conservation. One of the principal conclusions from this research is that for fragments to be of conservation value, corridors must exist that allow monkeys to disperse through the matrix between fragments (Baranga 2004, Mbora and Meikle 2004b, Wong and Sicotte 2006). An especially intensive study by Anderson et al. (2007b) found that matrix habitat containing wooded shrubland and perennial plantations provided not only means of travel between forest fragments but also important food sources for *C. angolensis* in Kwale District, Kenya. This finding led Anderson and her co-authors (2007b,c) to recommend that the 55 *C. angolensis* populations in Kwale be managed within a metapopulation framework rather than through a more traditional fragment-by-fragment approach (Anderson et al. 2007b,c). Chapman et al. (2007) offer a sobering view of the conservation prospects for colobines in fragments. Having witnessed local people clear 4 of their 20 study fragments near Kibale over an 8-year period, Chapman et al. (2007) contend that colobines in fragments face a bleak future, especially those outside protected areas.

Hunting of colobines is especially prevalent in Central and West Africa, where the wild meat trade is rampant and primates serve as a protein source for many local people and migrant workers (Chapman et al. 2006a). Davies (1994) noted that in Sierra Leone commercial hunting led to major declines in *Pi. badius* populations in accessible forests. Refisch and Koné (2005) recently quantified the intensity of market hunting for several primate species including *Pi. badius* and *C. polykomos* in Taï and adjacent forests. They found that these colobines were the two most intensively hunted monkeys in the area, and that *C. polykomos* was being hunted at an unsustainable rate while *Pi. badius* was being hunted at a rate that approached the threshold of sustainability. This intensive level of colobine hunting is especially alarming given that it is taking place in one of the most intensively protected rain forests in all of West Africa (Refisch and Koné 2005). Sadly, another West African red colobus species, *Pi. waldronae*, already appears to have been hunted to the very brink of extinction (Oates et al. 2000a, McGraw and Oates 2002, McGraw 2005).

Global climate change poses yet another more long-term threat to African colobines and other primates. Indeed, Africa is thought to be the continent expected to experience the most adverse effects of global warming including extreme shifts in rainfall and other climatic variables (Hannah et al. 2002, Lovett et al. 2005). Colobines adapted to life at high elevations like the unique *C. angolensis* population that forms groups of >300 individuals in Nyungwe Forest, Rwanda (Fashing et al. 2007b) are anticipated to be most adversely affected (Chapman et al. 2006a). As the climate warms, species adapted to montane areas will have nowhere to migrate to once their habitat becomes unsuitable and are predicted to be the first to go extinct (Chapman et al. 2006a).

Despite the varied threats, hope for African colobine conservation does, however, exist. Several African colobines act as 'flagships' for their ecosystems (e.g. *C. angolensis* at Kwale; *C. angolensis* at Nyungwe; *C. vellerosus* at Boabeng-Fiema, Ghana; *Pi. kirkii* at Jozani, Zanzibar) attracting ecotourists and their conservation money to these areas (Saj et al. 2006, Anderson et al. 2007b). Furthermore, though their effectiveness has rarely been quantified, community conservation programs involving people living near African forests would seem to have the potential to influence local attitudes about conservation of colobines and their ecosystems (Kasanene and Ross 2008; though see Struhsaker 2005). Lastly, a growing number of African wildlife professionals and scientists are obtaining the education and training necessary to become more informed stewards and effective managers of their local or national heritage (Trevelyan and Nuttman 2008).

Diet

In all African colobine species studied, leaves make up a substantial portion (26%–92%) of the diet (Table 13.2). Though their diet has been investigated at only two sites, olive colobus may be the most consistently folivorous (74%–85%) of the African colobine genera. At both sites, they consumed far less fruit than sympatric *C. polykomos* and *Piliocolobus badius*. Most forms of red colobus are also primarily folivorous, with an apparent preference for young leaves (e.g., Struhsaker 1975, Marsh 1981a, Werre 2000). *Pi. tephrosceles* appears to be especially folivorous and at one site, Kibale, has exhibited an increasing reliance on leaves over the past several decades. *Pi. badius* populations in West Africa are unusual in that they tend to consume fruit or seeds nearly as often as leaves. Many red colobus populations differ from other African colobines in that flowers comprise ≥10% of their diet.

Black and white colobus appear to have the most variable diets of the African colobines (Table 13.2). While most are primarily folivorous, numerous populations, especially those of *C. satanas*, feed heavily on seeds when they are available. This granivory often occurs at central and West African sites where leguminous seed-producing tree species are abundant and is not a response to scarcity of high-quality foliage (Oates et al. 1990). Studies in Nyungwe Forest, Rwanda, show that *C. angolensis* diets vary widely over time even among groups occupying the same general area and, in some instances, include large quantities of lichen, an unusual food for primates. Even *C. guereza*, which early studies suggested was among the most folivorous colobines, appears to be quite flexible dietarily. Guerezas at three different sites across their range include large quantities of whole fruits and/or seeds in their annual diets (Budongo, Uganda, Plumptre 2006; Dja, Cameroon, Poulsen et al. 2002; Kakamega, Kenya, Fashing 2001b), with those in Kakamega Forest, Kenya, clearly selecting whole fruits over young leaves when both are available (Fashing 2001b).

Table 13.2 Diets of African Colobine Monkeys (9–Month Study Minimum)

SPECIES	STUDY SITE	% COMPOSITION OF DIET BY FOOD PART								NUMBER OF SPECIES IN DIET	REFERENCES
		YL	ML	UL	TL	FL	FR	(SD)	OT		
C. angolensis	Ituri, DRC	26	2	22	50	7	28	(22)	15	37	Bocian 1997
C. angolensis	Nyungwe, Rwanda[2] (1987–88)	30	7	1	38	1	23	(20)	37[3]	45+	Vedder and Fashing 2002
C. angolensis	Nyungwe, Rwanda (1993)	25	40	7	72	5	17	—	6	59+	Fimbel et al. 2001
C. guereza	Kibale (Kanyawara), Uganda (1971–72)	65	13	3	81	2	15	—	2	43	Oates 1977a, 1994
C. guereza	Kibale (Kanyawara), Uganda (1998–99)	81	5	1	86	1	5	—	8	—	Wasserman and Chapman 2003
C. guereza	Kibale (fragment), Uganda	65	14	1	80	6	12	—	2	—	Wasserman and Chapman 2003
C. guereza	Kibale (logged), Uganda	78	5	3	87	3	10	—	0	—	Wasserman and Chapman 2003
C. guereza	Ituri, D.R. Congo	30	4	24	58	3	25	(22)	15	31	Bocian 1997
C. guereza[4]	Kakamega, Kenya	24	7	23[5]	54	1	39	(1)	8	37+	Fashing 2001b
C. guereza	Budongo (unlogged), Uganda	37	25	0	62	9	26	(12)	4	48	Plumptre 2006
C. guereza	Budongo (logged), Uganda	33	20	0	53	8	36	(9)	3	47	Plumptre 2006
C. polykomos	Tiwai, Sierra Leone	30	26	2	58	3	35	(32)	3	46+	Dasilva 1989
C. polykomos	Taï, Ivory Coast	28	20	0	49	3	48	—	1	—	Korstjens et al. 2007
C. satanas	Douala-Edea, Cameroon	21	18	0	39	3	53	(53[6])	5	84+	McKey et al. 1981
C. satanas	Lopé, Gabon	23	3	0	26	5	64	(60)	4	65	Harrison in Oates 1994
C. satanas	Fôret des Abeilles, Gabon	35	3	0	38	12	50	(41)	0	109	Gautier-Hion et al. 1997, Fleury and Gautier-Hion 1999
C. vellerosus	Boabeng-Fiema, Ghana	40	34	5	79	6	15	(4)	0	56	Saj and Sicotte 2007
Pi. badius	Fathala, Senegal (1973–74)	42	5	0	47	9	36	(19)	9	39	Gatinot 1978, Oates 1994
Pi. badius	Fathala, Senegal (1994–96)	70	6	0	76	1	19	—	4	17	Diouck in Galat-Luong and Galat 2005
Pi. badius	Abuko, Gambia	35	12	0	47	9	42	(3)	3	89+	Starin 1991
Pi. badius	Taï, Ivory Coast (1992–93)	24	7	0	31	30	37	—	2	153	Wachter et al. 1997
Pi. badius	Taï, Ivory Coast (1996–98)	46	4	0	50	20	29	—	1	—	Korstjens et al. 2009
Pi. badius	Tiwai, Sierra Leone	32	20	0	52	16	31	(25)	1	51	Davies et al. 1999
Pi. epieni	Gbanraun, Nigeria	56	10	0	66	9	16	(12)	9	19+	Werre 2000
Pi. kirkii[4,7]	Jozani (forest), Zanzibar (1980–81)	50	7	7	63	8	32	—	0	63	Mturi 1993
Pi. kirkii[8]	Jozani (forest), Zanzibar (1999)	51	9	4	64	5	26	(2)	6	21	Siex 2003

Table 13.2 (cont'd)

SPECIES	STUDY SITE	% COMPOSITION OF DIET BY FOOD PART								NUMBER OF SPECIES IN DIET	REFERENCES
		YL	ML	UL	TL	FL	FR	(SD)	OT		
Pi. kirkii[9]	Jozani (*shamba*), Zanzibar (1999)	55	7	6	67	6	5	(0)	22[10]	27	Siex 2003
Pi. preussi[11]	Korup, Cameroon	89	0	0	89	10	1	—	0	17	Usongo and Amubode 2001
Pi. rufomitratus	Tana River (Mchelelo), Kenya (1973–74)	52	11	1	65	6	25	(1)	4	22	Marsh 1981b
Pi. rufomitratus[12]	Tana River (Mchelelo), Kenya (1986–88)	61	2	0	63	13	22	—	2	28	Decker 1994
Pi. rufomitratus[13]	Tana River (Baomo S.), Kenya	46	1	0	47	27	26	—	1	26	Decker 1994
Pi. tephrosceles[14]	Gombe, Tanzania	35	44	0	79	7	11	—	3	58+	Clutton-Brock 1975a
Pi. tephrosceles	Kibale (Kanyawara), Uganda (1972–75)	42	24	8	73	16	6	(1)	5	57	Struhsaker 1978a,b
Pi. tephrosceles	Kibale (Kanyawara), Uganda (1998–99)	70	10	7	87	2	8	—	3	—	Wasserman and Chapman 2003
Pi. tephrosceles	Kibale (Kanyawara), Uganda (2005–06)	77	3	12	92	1	5	(2)	0	95	Snaith and Chapman 2008
Pi. tephrosceles	Kibale (fragment), Uganda	60	22	2	84	2	7	—	7	—	Wasserman and Chapman 2003
Pi. tephrosceles	Kibale (logged), Uganda	79	7	6	92	2	6	—	1	—	Wasserman and Chapman 2003
Pi. tholloni	Salonga, DRC	54	6	0	61	1	38	(31)	0	84	Maisels et al. 1994
Pr. verus[15]	Tiwai, Sierra Leone	59	11	4	74	7	19	(14)	0	50+	Oates 1988a
Pr. verus	Taï, Ivory Coast	83	1	1	85	4	8	—	3	—	Korstjens et al. 2009

[1] YL, young leaves; ML, mature leaves; UL, unidentified leaves; TL, total leaves; FL, flowers; FR, fruit including seeds; (SD), seeds alone where reported; OT, other.

[2] Different group studied several years earlier than the Fimbel et al. 2001 study.

[3] Includes the 32% of the annual diet composed of lichen.

[4] Mean diet for two groups.

[5] Believed to be mostly mature leaves.

[6] Includes primarily seeds but some records of seeds and pericarp.

[7] Totals exceed 100%.

[8] Mean for 3 groups.

[9] Mean for 4 groups.

[10] Includes the 15% of the annual diet composed of herbs.

[11] Based on 17-month study.

[12] Based on 24-month study.

[13] Based on 19-month study.

[14] Based on 9 months of data between August 1969 and June 1970.

[15] Based on 11 months of data between June 1983 and October 1985.

While young leaves, seeds, or whole fruits generally appear to be their preferred foods, most red as well as black and white colobus are capable of switching their focus to lower-quality mature leaves in times of preferred food scarcity (Struhsaker 1975, Oates 1977a, Marsh 1981b, McKey et al. 1981, Dasilva 1994, Fashing 2004, Preece 2006). However, even at times of preferred food scarcity, olive colobus appear limited in their ability to exploit mature leaves (Oates 1988a). This limitation may result from the small body size of olive colobus, which limits their ability to efficiently engage in the forestomach fermentation necessary to extract nutrients from mature leaves (Oates 1988a, Kay and Davies 1994).

When feeding on leaves, African colobines select those that are high in protein (*C. angolensis*, Maisels et al. 1994; *C. guereza*, Bocian 1997; *Pi. preussi*, Usongo and Amubode 2001; *Pi. tholloni*, Maisels et al. 1994), low in fiber (*C. angolensis*, Bocian 1997; *Pr. verus*, Oates 1988a), or both (*C. guereza*, Chapman et al. 2004; Fashing et al. 2007a; *C. satanas*, McKey et al. 1981; *Pi. rufomitratus*, Mowry et al. 1996). Most minerals do not appear to exert a strong influence on African colobine leaf choice, though a notable exception is provided by zinc, a mineral for which *C. guereza* exhibit striking preferences at both Kibale and Kakamega (Rode et al. 2003, Fashing et al. 2007a). Several colobines exhibit some avoidance of leaves high in secondary compounds (*C. guereza*, Oates et al. 1977; Fashing et al. 2007a; *C. polykomos*, Dasilva 1994; *Pr. verus*, Oates 1988a), whereas others appear undeterred by these compounds (*C. guereza*, Bocian 1997; *C. angolensis* and *Pi. tholloni*, Maisels et al. 1994; *Pi. rufomitratus*, Mowry et al. 1996; *Pi. tephrosceles*, Chapman and Chapman 2002a). The consumption of secondary compounds by colobines is probably facilitated by the fermentative abilities of their forestomachs, which may be capable of neutralizing the effects of some plant toxins before they are absorbed into the stomach (Kay and Davies 1994).

Ranging

Colobines typically travel shorter distances each day and occupy smaller home ranges than sympatric cercopithecines and apes (Clutton-Brock and Harvey 1977b, Chapman and Chapman 2000b). It is unusual for an African colobine group to have a home range greater than 100 ha or to travel over 1,000 m/day (Table 13.3). Even within their generally small ranges, many colobines spend most of their time in a "core area" that covers only a small portion of their entire range. These limited ranging patterns likely result from the relative abundance and even distribution of colobine food sources as well as the energy conservation strategies of some species (Dasilva 1992, 1993; Kay and Davies 1994).

As the most selective feeders among the African colobines, it is not surprising that olive colobus have the longest daily path lengths (mean = 1,212 m, *n* = 1 site). The generally larger group sizes of red colobus probably explain why they travel longer distances per day (mean = 920 m, *n* = 5) than most black and white colobus (mean = 586 m, *n* = 12) despite their relatively similar diets. Groups of *C. satanas* in Fôret des Abeilles, Gabon, and of *C. angolensis* in Nyungwe Forest, Rwanda, are unusual in that they adopt seminomadic lifestyles involving the occupation of very large ranges that continue to expand in size even after years of study. In Fôret des Abeilles, *C. satanas*'s seminomadism may be explained by the irregular fruiting patterns of the Caesalpinaceae that dominate this site and provide an important component of

their diet (Fleury and Gautier-Hion 1999). In Nyungwe, however, the extraordinarily large groups (>300 individuals) formed by *C. angolensis*, and the feeding competition that results, probably explains why they must range so widely (Fashing et al. 2007b) even though food is of unusually high quality at this site (Fimbel et al. 2001).

Several colobine species exhibit a tendency for home range size to become compressed as population density increases (Dunbar 1987, Newton and Dunbar 1994). For example, in a comparison of nine East African *C. guereza* populations, Fashing (2001a) found a significant negative correlation between population density and mean home range size. Moreover, the mean range size of *Pi. kirkii* groups in Jozani Forest decreased from 60 ha to 25 ha over a recent 19-year period during which population density increased from 100 individuals/km² to 176 individuals/km² (Mturi 1993, Siex 2003). Consistent with this pattern is the fact that range size averaged only 13 ha in nearby *shambas* (farms) at Jozani, where *Pi. kirkii* density reached an astounding 784 individuals/km² in 1999 (Siex 2003).

Like many other primates, most African colobines exhibit seasonal variation in ranging patterns, corresponding with changes in the consumption of major dietary items and/or food availability (Clutton-Brock 1975b, Marsh 1981c, McKey and Waterman 1982, Dasilva 1989, M. Harrison personal communication cited in Oates 1994, Bocian 1997, Fleury and Gautier-Hion 1999). However, for several populations, ranging behavior is not clearly linked to these primary food-related variables (Struhsaker 1974, Oates 1977a, Bocian 1997, Fashing 2001a). In *C. guereza*, for example, temporal variability in ranging patterns appears to be more related to the distribution of infrequently eaten food items that are spatially rare yet nutritionally important, like sodium-rich swamp plants or eucalyptus bark (Oates 1978; Fashing 2001a, 2007a; Harris and Chapman 2007). Olive colobus and some red colobus (e.g., *Pi. badius* at Taï) are also unusual in that their ranging patterns appear to be dictated primarily by the movements of the guenon (*Cercopithecus diana*) groups with which they form near-permanent polyspecific associations as an apparent antipredation strategy (Oates and Whitesides 1990, Holenweg et al. 1996, Bshary and Nöe 1997a, Korstjens 2001).

Activity Patterns

Colobines are notorious for their lethargic lifestyles. Among the African colobine genera, activity level appears to be inversely related to body size: Mean time spent resting is 54% for nine *Colobus* populations, 47% for eight *Piliocolobus* populations, and 40% for one *Procolobus* population (Table 13.4). This pattern may result from the greater intake of mature leaves by *Colobus* (mean = 14%, *n* = 15 sites; Table 13.2) than by the other two genera (*Piliocolobus*, mean = 11%, *n* = 15; *Procolobus*, mean = 6%, *n* = 2) and the concomitant greater reliance by *Colobus* on foregut fermentation, which requires large amounts of resting time to facilitate digestion. Though some have suggested

Table 13.3 Ranging Behavior of African Colobine Monkeys (6-Month Study Minimum) (Home Range Sizes Based on Number of 50 x 50 m Grid Cells Entered Unless Otherwise Noted)

SPECIES	STUDY SITE	DENSITY (KM²)	HOME RANGE SIZE (HA)	RANGE OVERLAP (%)	DAILY PATH LENGTH (M) MEAN	RANGE	NUMBER OF GROUPS	CORE AREA	REFERENCES
C. angolensis	Ituri, DRC	3	371[1]	100	983	312–1914	1	No	Bocian 1997
C. angolensis	Nyungwe, Rwanda	–	2440[2]	–	–	–	1	Yes	Fashing et al. 2007
C. guereza	Kibale, Uganda (1971–72)	100	28	74	535	288–1004	1	Yes	Oates 1977a,c
C. guereza	Kibale, Uganda (1998–99)	–	10[3]	35	–	–	2	–	Chapman and Pavelka 2005
C. guereza	Ituri, DRC	17	100[1]	≥22	609	268–1112	1	Yes	Bocian 1997
C. guereza	Kakamega, Kenya	150	18[3]	67	588	166–1360	5	Yes	Fashing 2001a,c
C. guereza	Entebbe, Uganda	63[4]	7.5	+	386	62–1036	3	–	Grimes 2000
C. guereza	Budongo, Uganda (1964–65)	–	14	Minimal	–	–	5	–	Marler 1969
C. guereza	Budongo, Uganda (1967–73)	49	14[5]	Minimal	–	–	25	–	Suzuki 1979
C. guereza	Budongo, Uganda (1994–95)	20–50	10–33	–	–	–	6	–	Plumptre unpublished data
C. polykomos	Tiwai, Sierra Leone	67	24	≥20[6]	832	350–1410	1	Yes	Dasilva 1989, Oates 1994
C. polykomos	Taï, Ivory Coast	47	77[7]	≥21[8]	617	241–1341	2	Yes	Korstjens 2001
C. satanas	Douala-Edea, Cameroon	38	60	+	459	<100->800	1	Yes	McKey 1978a,b, McKey et al. 1982
C. satanas	Lopé, Gabon	11[9]	184	–	510	40–1100	1	–	Harrison (1986); Harrison in Oates 1994, White 1994
C. satanas	Fôret des Abeilles, Gabon	8	573[10]	≥65	852	20–1983	1	No[11]	Fleury and Gautier-Hion 1999
C. vellerosus	Bia, Ghana	–	48	–	307	75–752	1	–	Olson 1986, Oates 1994
C. vellerosus	Boabeng-Fiema, Ghana	119	11	–	359	101–712	5	–	Wong and Sicotte 2006, Teichroeb and Sicotte 2009
Pi. badius	Fathala, Senegal	–	20	–	–	–	1	–	Gatinot 1977
Pi. badius	Abuko, Gambia	124	34	60	–	–	1	–	Starin 1991
Pi. badius	Tiwai, Sierra Leone	66	53	10–20	–	–	1	–	Davies et al. 1999
Pi. badius	Taï, Ivory Coast (1992–93)	–	64[10]	–	967	–	1	–	Holenweg et al. 1996
Pi. badius	Taï, Ivory Coast (1992–93)	–	65[10]	+	–	–	2	–	Höner et al. 1997

Table 13.3 (cont'd)

SPECIES	STUDY SITE	DENSITY (KM²)	HOME RANGE SIZE (HA)	RANGE OVERLAP (%)	DAILY PATH LENGTH (M) MEAN	RANGE	NUMBER OF GROUPS	CORE AREA	REFERENCES
Pi. badius	Taï, Ivory Coast (1996–98)	158	58[7]	>6[8]	872	300–1532	2	—	Korstjens 2001
Pi. epieni	Gbanraun, Nigeria	120	73	>42	1040	450–1900	1	—	Werre 2000
Pi. kirkii	Jozani (forest), Zanzibar (1980–81)	100	60[10]	70	1044	—	2	—	Mturi 1993, Mturi in Siex 2003
Pi. kirkii	Jozani (forest), Zanzibar (1999)	176	25	31–65	565	50–1270	3	Yes	Siex 2003
Pi. kirkii	Jozani (*shamba*), Zanzibar (1999)	784	13	70–95	310	0–690	4	Yes	Siex 2003
Pi. kirkii	Uzi, Zanzibar	—	5	7	—	—	3	Yes	Nowak 2008
Pi. rufomitratus	Tana River (Mchelelo), Kenya (1973–74)	253	9[12]	36	603	≤200–<1100	1	Yes	Marsh 1979a,b, 1981c, Marsh in Decker 1994
Pi. rufomitratus	Tana River (Mchelelo), Kenya (1986–88)	56	12	—[13]	532	210–870	1	—	Decker 1994
Pi. rufomitratus	Tana River (Baomo S.), Kenya	33	13	—[13]	461	180–780	1	—	Decker 1994
Pi. tephrosceles	Gombe, Tanzania	—	114[14]	—	—	—	1	—	Clutton-Brock 1975b
Pi. tephrosceles	Kibale (Kanyawara), Uganda (1970–71)	≥260	65	99	649	223–1185	1	Yes	Struhsaker 1975
Pr. verus	Tiwai, Sierra Leone	11	28	—	—	—	1	—	Oates and Whitesides 1990
Pr. verus	Taï, Ivory Coast	14	56[7]	>14[8]	1212	482–2105	2	—	Korstjens 2001

[1] Based on number of 1.56 ha cells entered.
[2] Based on the 95% utilization distribution derived from fixed kernel analysis.
[3] Mean for 2 groups.
[4] Calculated from number of individuals in the 35 ha study area.
[5] Based on mean home range size of 25 groups, though the number of hours of observation for each group was not reported.
[6] Calculated based on map of intergroup encounter locations from Dasilva 1989:305.
[7] Based on number of 1 ha cells entered per year.
[8] Based on 2 years of data; measures only overlap with one of the multiple groups that overlap study groups' home ranges.
[9] Based on mean for 5 sites at Lopé.
[10] Based on number of 1 ha cells entered.
[11] Certain areas were intensively used, but were not contiguous.
[12] Based on 12 months.
[13] No other red colobus groups were adjacent to Decker's study groups.
[14] Based on number of 100 x 100 yard grid cells.

that colobines have abnormally low metabolic rates (McNab 1978, Müller et al. 1983), Dasilva (1992, 1993) argues that most available evidence contradicts this contention. Instead, she provides evidence that the low activity levels of one species, *C. polykomos*, are part of a behavioral, rather than a physiological, strategy of conserving energy. *C. polykomos*'s strategy of behavioral thermoregulation involves actions like reducing energy expenditure at times of preferred food scarcity and adopting hunched positions during rainstorms and when the temperature is low. *Procolobus* do not appear to have this option due to both their small body size, which requires them to maintain a higher-quality diet, and their need to move a great deal to keep up with the guenons with whom they typically travel (Oates 1994).

Table 13.4 Activity Budgets of African Colobines from Studies of ≥4 Months (Based on Scan Sampling of the First Behavior Performed for ≥5 Sec Continuously Unless Otherwise Noted.

SPECIES	STUDY SITE	REST	FEED	MOVE	SOCIAL	OTHER	REFERENCES
C. angolensis[1]	Ituri, DRC	43	27	24	5	1	Bocian 1997
C. angolensis[2,3]	Nyungwe, Rwanda	32	42	20	5	1	Fashing et al. 2007b
C. guereza	Kibale, Uganda	57	20	5	11	7	Oates 1977a
C. guereza[1]	Ituri, DRC	52	19	22	5	2	Bocian 1997
C. guereza[1,4]	Kakamega, Kenya	63	26	2	7	2	Fashing 2001a
C. guereza[5,2]	Entebbe, Uganda	58	20	12	10	1	Grimes 2000
C. polykomos[6]	Tiwai, Sierra Leone	61	28	9	1	1	Dasilva 1989, 1992
C. satanas	Douala-Edea, Cameroon	60[7]	23	4	14[8]	0	McKey and Waterman 1982
C. vellerosus[6,9]	Boabeng-Fiema, Ghana	59	24	15	3	0	Teichroeb et al. 2003
Pi. badius	Abuko, Gambia	52	21	13	13	1	Starin 1991
Pi. badius	Tiwai, Sierra Leone	55	37	5	—	3	Davies in Oates 1994, Davies et al. 1999
Pi. epieni[10]	Gbanraun, Nigeria	33	37	25	6	0	Werre 2000
Pi. kirkii	Jozani (forest), Zanzibar	47	29	12	7	5	Siex 2003
Pi. kirkii	Jozani (shamba), Zanzibar	44	29	6	15	7	Siex 2003
Pi. rufomitratus	Tana River (Mchelelo), Kenya (1973–74)	55	30	7	8	0	Marsh 1978a,b
Pi. rufomitratus[10]	Tana River (Mchelelo), Kenya (1986–88)	48	29	22	2	0	Decker 1994
Pi. rufomitratus[10]	Tana River (Baomo S.), Kenya	50	23	24	3	0	Decker 1994
Pi. tephrosceles[10]	Gombe, Tanzania	54	25	8	9	5	Clutton-Brock 1974a,b
Pi. tephrosceles	Kibale (Kanyawara), Uganda	38	45	9	8	0	Struhsaker 1975
Pr. verus	Tiwai, Sierra Leone	40	27	25	—	8	Oates 1994, Davies et al. 1999

[1] Recorded first behavior performed for >3 sec continuously.

[2] Conducted instantaenous scan samples.

[3] Based on data collected primarily between 10:00 a.m. and 3:00 p.m.

[4] Mean of activity budgets of 2 groups.

[5] Mean of activity budgets for 3 groups studied for 6 months each. Values recalculated after discarding 18.6% of 5,790 total scans which were categorized as "out of sight".

[6] Recorded first behavior performed 5 sec after spotting an individual.

[7] Total of time spent sitting, lying, and clinging.

[8] Authors included time spent self-cleaning and grooming as one category, so time spent social (playing and grooming) is overestimated here.

[9] Mean of activity budgets of 3 groups studied for 4 months each.

[10] Scanning method not described.

Competition over Food Within Groups

Colobines are gregarious primates whose groups vary widely in size (see Social Organization). While group-living has several probable benefits (Wrangham 1980, van Schaik 1983, Brereton 1995), it is not without its costs including the potential for increased disease transmission and feeding competition among group mates. Feeding competition can take two forms, scramble or contest competition. In the former, individuals compete indirectly by depleting resources before others can encounter them, while in the latter, individuals compete directly through agonistic interactions over food (van Schaik 1989). Colobines were long thought to be less prone to feeding competition than other primates given the perceived relative ubiquity and even distribution of the leaves that often dominate colobine diets. Early tests for scramble competition in several African colobine species seemed to confirm this impression, finding that daily path length did not increase with increasing group size (*Pi. tephrosceles*, Struhsaker and Leland 1987; *C. guereza*, Fashing 2001a; *Piliocolobus kirkii*, Siex 2003). However, recent studies capitalizing on refinements in the techniques used to test for scramble competition have begun to paint a different picture (Snaith and Chapman 2007). For example, a study focusing on the same *P. tephrosceles* population previously studied by Struhsaker and Leland (1987) found that after controlling for differences in habitat quality between study groups, highly significant correlations emerge between group size and both daily path length and home range size, suggesting scramble competition is indeed occurring within groups in this population (Snaith and Chapman 2008). Another recent study using similar research techniques produced comparable evidence for within-group scramble competition among *C. vellerosus* at Boabeng-Fiema, Ghana (Teichroeb and Sicotte 2009).

Studies specifically testing for contest competition within African colobine groups have seldom been conducted, probably due to the rarity of intragroup aggression in most populations (Struhsaker and Leland 1979, Dunbar 1987, Fashing 2001a, 2007b, Saj et al. 2007). A study of sympatric *C. polykomos* and *P. badius* at Taï provides a notable exception (Korstjens et al. 2002). In this study, *C. polykomos* were more frugivorous and consumed foods that required greater handling time and longer food patch residence than *Pi. badius*. Rates of intragroup aggression were also over three times higher in *C. polykomos*, suggesting that intragroup contest competition is more intense for this species than for *Pi. badius* at Taï (Korstjens et al. 2002). Because the *C. polykomos* at Taï are unusually frugivorous for an African colobine (Table 13.2) and often consume seeds that may incite competition owing to their long handling time, they are probably one of the few African colobine populations experiencing such intense contest competition. Competition between groups will be discussed in detail in the final section of this chapter.

Biomass and Its Determinants

The biomass of African colobines varies widely among sites (Fashing and Cords 2000). At Kibale's Kanyawara study site, for example, two colobine species (*Pi. badius* and *C. guereza*) combine to reach a biomass of 2,386 kg/km², or 82% of the total primate biomass (Struhsaker 1975, 1997). In contrast, at Lomako, DRC, the only colobine species (*C. angolensis*) occurs at a biomass of 57 kg/km², or 6% of the total primate biomass (McGraw 1994). While it has been suggested that a variety of biotic and abiotic factors probably contribute to this intersite variation in colobine

biomass, the protein to fiber ratio (an indicator of nutritional quality) in mature foliage is by far the strongest predictor of colobine biomass across sites (Oates et al. 1990). The predictive powers of this ratio have proven to be remarkably robust and have held not only between forests (Oates et al. 1990, Davies 1994, Fashing 2007a) but also among sites within the same forest (Chapman et al. 2002a, 2004; Wasserman and Chapman 2003). The link between mature leaf protein:fiber ratio and colobine biomass is founded on the assumption that since mature leaves are the fallback foods of many colobines (see Diet), their quality plays a critical role in determining the density of the colobine populations that can be supported through lean periods (McKey 1978a,b; Davies 1994). Intriguingly, intersite variation in biomass of folivorous lemurs in Madagascar can also be predicted from mature leaf protein:fiber ratio, suggesting that a pattern most clearly established among the colobines is probably of general applicability to arboreal folivorous primates (Ganzhorn 1992).

REPRODUCTIVE PARAMETERS

The reproductive parameters of African colobines have not been studied extensively (Table 13.5). Though rare in the Asian colobines, periodic swellings of the sex skin occur in all *Piliocolobus* and *Procolobus* spp. as well as in *C. angolensis* and *C. satanas* (Bocian 1997, Dixson 1998). Among *Pi. badius* in Abuko Forest, The Gambia, males direct most of their sexual interest at inflating and maximally tumescent females (Starin 1991). It is generally assumed that ovulation occurs at some point while a female is swollen, though hormonal studies are still needed to test this idea (Oates 1994).

Little is known about the copulatory behavior of most African colobines. Both sexes solicit copulations in *C. guereza*, *C. vellerosus*, *Pi. badius*, *Pi. tephrosceles*, and *Pr. verus*, while only males have been observed soliciting copulations in *C. angolensis* and *C. polykomos* (Table 13.5). All three species for which mounting behavior has been described are multiple mounters. Female *C. guereza* at Kibale copulate primarily during the portion of their ovarian cycle from 5 days before to 3 days after their (hormonally) estimated date of ovulation (Harris and Monfort 2006). Among African colobine genera, only female red colobus give copulation calls (Struhsaker 1975, Korstjens 2001). These calls are audible up to 400 m away in *Pi. badius* at Taï and may function to attract other males (Korstjens 2001). Copulatory harassment by immatures has been described for *C. guereza*, *Pi. badius*, and *Pi. tephrosceles* (Oates 1977c; Struhsaker and Leland 1979, Starin 1991, Harris and Monfort 2003). In *Pi. badius* and *Pi. tephrosceles*, copulating couples are also harassed by adult males (Struhsaker and Leland 1979, Starin 1991, Firos 2001). Consort pairs, in which a male and a female leave their group for 1 day or longer and mate exclusively with one another, have been reported only for *Pi. badius*

Table 13.5 Reproductive Parameters of the African Colobines

SPECIES	STUDY SITE	BIRTH SEASON	INTRAGROUP SYNCHRONY	SOLICITING SEX	COPULATORY HARRASSMENT	SINGLE OR MULTIPLE MOUNTS	AGE AT SEXUAL MATURITY (MONTHS) MALE	AGE AT SEXUAL MATURITY (MONTHS) FEMALE	GESTATION LENGTH (DAYS)	INTERBIRTH INTERVAL (MONTHS)	SEX SKIN SWELLING	REFERENCES
C. angolensis	Ituri, DRC	–	–	M	–	–	–	–	–	–	Yes	Bocian 1997
C. angolensis	Nyungwe, Rwanda	None?[1]	–	–	–	–	–	–	–	–	No	Fimbel et al. 2001
C. guereza	Kibale, Uganda	None	Yes	M,F	Yes	Multiple	est. 72	est. 48	158[2]	21.5[11]	No	Oates 1977c, Struhsaker and Leland 1979a,b, 1987, Harris and Monfort 2003, 2006
C. guereza	Tigoni, Kenya (captivity)	None?[3]	–	–	–	–	–	–	est. 170[4]	15.6[5]	No	Rowell and Richards 1979
C. guereza	Ituri, DRC	–	–	M,F	–	–	–	–	–	–	No	Bocian 1997
C. guereza	Kakamega, Kenya	None	Yes	M,F	Yes	Multiple	–	–	–	17[6]	No	Fashing 2002, unpublished data
C. polykomos	Tiwai, Sierra Leone	Strict	–	M	No	–	–	–	–	24	No	Dasilva 1989
C. polykomos	Taï, Ivory Coast	None	–	–	–	–	–	–	–	–	No	Korstjens 2001
C. vellerosus	Boabeng-Fiema, Ghana	None	Yes	M,F	–	–	–	–	–	–	No	Saj and Sicotte 2005, Teichroeb and Sicotte 2008b
Pi. badius	Abuko, Gambia	Peak	Yes	M,F	Yes	Multiple	~28[7]	34[12]	~174	29.4[8]	Yes	Starin 1988, 1991
Pi. epieni	Gbanraum, Nigeria	Peak	–	–	–	–	–	–	–	–	Yes	Werre 2000
Pi. kirkii	Jozani (forest), Zanzibar	Peak	–	–	–	–	–	–	–	26–55[6]	Yes	Siex and Struhsaker 1999a,b, Siex 2003
Pi. kirkii	Jozani (shamba), Zanzibar	Peak	–	–	–	–	–	–	–	32–35[6]	Yes	Siex and Struhsaker 1999a,b, Siex 2003
Pi. rufomitratus	Tana River, Kenya (Mchelelo)	Peak[9]	–	–	–	–	–	–	–	–	Yes	Marsh 1979a
Pi. tephrosceles	Kibale, Uganda	Peak	No	M,F	Yes	Multiple	>35–58	~38–46	–	25.5±5.1[10]	Yes	Struhsaker 1975, Struhsaker and Leland 1979a,b, 1987
Pr. verus	Tiwai, Sierra Leone	Strict	–	F	–	–	–	–	~180	–	Yes	Oates 1994
Pr. verus	Taï, Ivory Coast	Strict	–	M,F	–	–	–	–	–	20[6]	Yes	Korstjens 2001, pers. comm.

[1] Fimbel et al. 2001 mention that white infants were seen throughout the year but do not address whether or not peaks occurred.

[2] Median value for 4 females based on hormonal monitoring.

[3] Rowell and Richards 1979 argue that breeding is aseasonal in *C. guereza*, though based on Table 2, 12 of 21 births during the study occurred between June–September.

[4] No explanation given for how estimate was derived.

[5] Calculated based on 14 *C. guereza* births included in Figure 2 (but excluding two stillbirths).

[6] Estimated based on total adult female months divided by total number of infants born during the study period.

[7] $n = 3$.

[8] $n = 4$.

[9] Based on my interpretation of data presented in Marsh 1979a: Figure 7.

[10] $n > 20$.

[11] Median for 5 females for whom interbirth interval was directly observed.

[12] $1 < n < 5$

and *Pi. rufomitratus* (Starin 1991, Mbora and McGrew 2002).

Gestation lengths have been estimated at 5.5–6 months for African colobines (Table 13.5). Olive colobus are the only African colobine known to exhibit strict birth seasonality at multiple sites. In red colobus, the typical pattern is for births to occur throughout the year but to peak during certain periods. *C. polykomos* on Tiwai Island, Sierra Leone, are unusual among black and white colobus populations in that they have a strict birth season, whereas another *C. polykomos* population at Taï and two populations of *C. guereza* feature births throughout the year with no obvious peaks. There is, however, evidence of intragroup birth synchrony for several black and white colobus populations and one red colobus population. Few African colobine studies have been of sufficient duration to record interbirth intervals for a large number of individually recognized females. An exception is provided by Struhsaker and Leland (1987), who found that, based on over 20 births, *Pi. tephrosceles* interbirth intervals averaged 25.5 ± 5.1 months at Kibale. Two 1-year studies of *Pi. kirkii* on Zanzibar conducted 7 years apart suggest that interbirth intervals can change dramatically within a population over time (Siex and Struhsaker 1999b, Siex 2003). Estimated interbirth interval in Siex's (2003) *shamba*-dwelling *Pi. kirkii* population decreased from 55 months in 1992–1993 to 26 months in 1999. Dunbar (1987) inferred that interbirth intervals are significantly longer in multimale groups than in one-male groups of *C. guereza*, perhaps due to female reproductive suppression resulting from the stress created by competition among males in multimale groups.

Infants of all *Colobus* spp. except *C. satanas* are born with entirely white coats, which contrast strikingly with those of adults (Oates 1994). Red colobus newborns also differ from adults in coat color, though less strikingly. In olive colobus, the coats of newborns are darker than those of adults. During the first few months of life, when black and white colobus infants remain flamboyantly colored, they are very attractive to, and are often held by, females other than their mothers (Horwich and Manski 1975, Oates 1977c, Dasilva 1989, Korstjens 2001, Brent et al. 2008). Similar "allomothering" behavior is rarely tolerated in red colobus (though *Pi. kirkii* provides an interesting exception [Siex 2003]) and has never been observed in olive colobus (Oates 1994). African colobine males typically exhibit little interest in infants, though preliminary observations suggest that *C. angolensis* may be different in this respect. Adult male *C. angolensis* were observed carrying infants on several occasions during 5 weeks of intermittent observations at Nyungwe, and a captive juvenile male of the same species was observed to persistently carry an infant during its first 3 months of life (Fashing unpublished data). Olive colobus are unique among anthropoid primates in that mothers carry young infants in their mouths when traveling (Booth 1957,

Oates 1994). The reasons for this behavior are unclear, though Booth (1957) has speculated that the short fur and severely reduced pollex typical of the species make it too difficult for infants to maintain a grip on their mothers, especially since the species spends much of its time moving through dense vegetation.

SOCIAL ORGANIZATION

The social organizations of African colobines are extremely diverse (Table 13.6). Small groups averaging fewer than a dozen individuals, including one or two adult males, several adult females, and their dependent offspring, are typical of some taxa, such as olive colobus and *C. guereza*. Group sizes in several other black and white colobus, including *C. polykomos*, *C. satanas*, and *C. vellerosus*, tend to be larger, averaging ~15 individuals, including multiple males, at most sites. *C. angolensis* has a remarkably labile social organization, with group sizes ranging from fewer than 10 individuals at several sites to more than 300 individuals at Nyungwe. Most red colobus species live in large multimale groups averaging 25–50 individuals.

Habitat type, logging history, predation risk, dietary choices, and feeding competition all appear to influence African colobine group sizes. For example, groups of red colobus in marginal, strongly seasonal habitats, like *Pi. rufomitratus* in the gallery forests at Tana River and *Pi. foai* in the seasonally flooded forests at Badane, Central African Republic, are often only one-fourth to one-half the size of red colobus groups in rain forest habitats (Table 13.6). Similarly, among *C. guereza* populations across East Africa, mean group size tends to be lower in gallery or scrub forest than in rain forest (Dunbar 1987). Evidence that logging history influences group size comes from Kibale, where *C. guereza* groups in lightly logged forest are 25% smaller than those in unlogged forest (Struhsaker 1997). A similar, but even more extreme, pattern was reported for *P. gordonorum* which lives in groups that are 48% smaller in disturbed forest than in undisturbed forest in the Udzungwa Mountains, Tanzania (Marshall et al. 2005).

The influence of predation risk on group size can be inferred by comparing colobine group sizes at locations with and without certain predators. Struhsaker (2000) has shown that on Bioko, where crowned-hawk eagles are absent, groups of *C. satanas* and *Pi. pennantii* are distinctly smaller than in forests on the African mainland, where crowned-hawk eagles occur. Furthermore, Stanford (1998a) has noted that red colobus groups tend to be smaller in forests where chimpanzees are absent.

Dietary choices and their interaction with digestive physiology also appear to influence colobine group size. Korstjens and Dunbar (2007) used data from a variety of ecological and climatic variables to create a systems model which predicted that *Piliocolobus* spp. should have a median group size of 53 and *Colobus* spp. a median group

Table 13.6 Social Organization of the African Colobines (For Sites Where Multiple Studies Have Been Conducted, Only the Most Recent Data are Presented)

| | | | GROUP SIZE | | | GROUP COMPOSITION | | | | | | |
| | | | | | | ADULT MALES | | ADULT FEMALES | | IMMATURES | | |
SPECIES	STUDY SITE	HABITAT	MEAN	RANGE	N	MEAN	RANGE	MEAN	RANGE	MEAN	RANGE	REFERENCES
C. angolensis	Salonga, DRC	Lowland moist primary forest	–	3–7	5	–	–	–	–	–	–	Maisels et al. 1994
C. angolensis	N.E. Tanzania	–	4.9	2–9	10[1]	1.3	1–2	1.6	1–3	1.6	1–3	Groves 1973
C. angolensis	Diani, Kenya	Fragmented lowland dry forest	6.5	3–11	25	–	–	–	–	–	–	Kanga 2001, Kanga and Heidi 1999/2000
C. angolensis	Ituri, DRC	Lowland moist primary forest	13.9	6–20	8[2]	–	2–≥5	–	6–≥8	–	≥6	Bocian 1997
C. angolensis	Nyungwe, Rwanda	Montane moist mixed forest	>300	–	2	Many	–	Many	–	Many	–	Fimbel et al. 2001, Fashing (personal observation)
C. guereza	Arusha, Tanzania	–	5.4	4–7	5[3]	1.5	1–2	2.0	1–3	1.8	1–2	Groves 1973
C. guereza	Kibale (fragments), Uganda	Fragmented lowland moist forest	6.2	3–11	29[4]	1.2	1–2	1.8	1–4	3.0	1–8	Onderdonk and Chapman 2000
C. guereza	Budongo, Uganda	Moist medium-altitude mixed forest	6.9	2–13	25	1.1	1–2	3.6	1–6	2.2	0–6	Suzuki 1979
C. guereza	Bole, Ethiopia	Lowland gallery forest	7.0	3–10	12	1.6	1–4	2.1	1–3	3.3	0–6	Dunbar 1987
C. guereza	Kibale (forest), Uganda	Moist medium-altitude secondary forest	7.0	6–9	6	1.2	1–2	2.2	1–3	3.7	3–5	Harris 2006
C. guereza	Entebbe, Uganda	Botanical garden/ natural forest	7.3[5]	3–12	3	1.7	1–2	2.3	1–4	3.3	1–6	Grimes 2000
C. guereza	Lake Shalla, Ethiopia	Lowland gallery forest	7.8	6–10	6	1.0	1	2.0	1–3	4.8	3–6	Dunbar 1987
C. guereza	Ituri, D.R. Congo	Lowland moist mixed forest	7.8	5–11	5[2]	1.5	1–2	2.5	2–3	2.5	2–3	Bocian 1997
C. guereza	Kyambura Gorge, Uganda	Lowland gallery forest	8.0	3–13	24	1.1	1–2	4.7	2–7	2.2	0–5	Krüger et al. 1998
C. guereza	Limuru, Kenya	Fragmented highland forest	9.8	6–15	4[6]	3.1	2–5	3.3	2–5	3.4	2–5	Schenkel and Schenkel-Hulliger 1967
C. guereza	Kibale (forest), Uganda[12]	Moist medium-altitude secondary forest	11.4	9–15	7	1.4	1–4	3.4	3–4	6.5	3–8	Oates 1974, 1977a,c, 1994, Struhsaker 1997

Table 13.6 (cont'd)

| SPECIES | STUDY SITE | HABITAT | GROUP SIZE | | | GROUP COMPOSITION | | | | | | REFERENCES |
| | | | | | | ADULT MALES | | ADULT FEMALES | | IMMATURES | | |
			MEAN	RANGE	N	MEAN	RANGE	MEAN	RANGE	MEAN	RANGE	
C. guereza	Kakamega, Kenya	Moist medium-altitude secondary forest	12.8	5–23	5	2.4	1–6	3.8	3–5	6.8	1–13	Fashing 2001a, Fashing and Cords 2000
C. guereza	Lake Naivasha, Kenya	Medium-altitude gallery forest	19	19	1	2	2	3	3	14	14	Rose 1978
C. polykomos	Tiwai, Sierra Leone	Lowland moist secondary forest	11	11	1	3	3	4	4	4	4	Dasilva 1989, 1994, Whitesides et al. 1988
C. polykomos	Taï, Ivory Coast	Lowland moist primary forest	16.2	14–19	10	1.2	1–2	4.7	4–6	10.3	9–12	Korstjens 2001
C. satanas	Lopé, Gabon	Lowland moist mixed forest	9	9	1	2	2	5	5	2	2	Harrison and Hladik 1986, Harrison in Oates 1994, White 1994
C. satanas	Douala-Edea, Cameroon	Lowland moist primary coastal forest	15	11–16	1	2.3	1–3	6.0	6	7.3	6–8	McKey 1978a,b, McKey and Waterman 1982
C. satanas	Foret des Abeilles, Gabon	Lowland moist primary forest[7]	17	7–25	3	–	>1–6	–	–	–	–	Fleury and Gautier-Hion 1999, Brugière et al. 2002
C. vellerosus	Boabeng-Fiema, Ghana	Dry semi-deciduous secondary forest	14.3	4–33	14	–	1–4	–	>2–9	–	1–8	Saj et al. 2005a,b
C. vellerosus	Bia, Ghana	Lowland moist rain forest	16	16	2	3.0	2–4	6.5	6–7	6.5	–	Olson in Oates 1994
Pi. badius	Fathala, Senegal	Gallery forest	25.2	14–62	6	5.5	4–13	10.7	5–27	9.0	5–22	Gatinot 1977
Pi. badius	Abuko, Gambia	Tree and shrub savannah	25.9	18–32	2	2.1	2–3	10.8	8–14	12.6	10–16	Starin 1991, personal communication
Pi. badius	Tiwai, Sierra Leone	Lowland moist secondary forest	33	33	1	7	–	13	–	13[8]	–	Whitesides et al. 1988, Davies in Oates 1994, Davies et al. 1999
Pi. badius	Taï, Ivory Coast	Lowland moist primary forest	52.3	41–64	4	10.5	6–15	18.3	14–22	23.5	20–27	Korstjens 2001

Table 13.6 (cont'd)

| SPECIES | STUDY SITE | HABITAT | GROUP SIZE | | | GROUP COMPOSITION | | | | | | REFERENCES |
| | | | | | | ADULT MALES | | ADULT FEMALES | | IMMATURES | | |
			MEAN	RANGE	N	MEAN	RANGE	MEAN	RANGE	MEAN	RANGE	
Pi. epieni	Gbanraun, Nigeria	Lowland disturbed marsh forest	~46	~15–~80	4[9]	7	7	26	26	27	27	Werre 2000
Pi. foai	Badane, C.A.R.[13]	Seasonally flooded forest	11.7	3–18	3	–	–	–	–	–	–	Galat-Luong and Galat 1979
Pi. gordonorum	Udzungwa (NDUFR)[14], Tanzania	Disturbed montane moist forest	11.8	5–24	6	–	–	–	–	–	–	Marshall et al. 2005
Pi. gordonorum	Udzungwa (Ndundulu), Tanzania	Montane moist mixed forest	22.5	12–34	8	–	–	–	–	–	–	Marshall et al. 2005
Pi. gordonorum	Magombera, Tanzania	Disturbed ground-water forest	24.8	21–33	4	2.8	1–5	9.3	7–11	12.8[8]	–	Struhsaker and Leland 1980, Struhsaker in Oates 1994
Pi. kirkii	Jozani (forest), Zanzibar	Secondary ground-water forest	31.1	23–36	3	3.9	–	12.0	–	13.7	–	Siex 2003
Pi. kirkii	Jozani (*shamba*), Zanzibar	Cultivated areas and regenerating forest	37.5	20–65	4	4.7	–	14.9	–	17.7	–	Siex 2003
Pi. preussi	Korup, Cameroon	Lowland moist primary forest	>47.0	>24–≥80	7	–	–	–	–	–	–	Struhsaker 1975
Pi. rufomitratus	Tana River, Kenya	Lowland disturbed gallery forest	11.2	4–24	9	1.1	1–2	5.6	2–11	4.5	1–11	Decker 1994
Pi. tephrosceles	Kibale, Uganda	Moist medium-altitude mixed forest	≥44	9–80	20[10]	–	≥2–≥7	–	≥3–≥16	–	?–≥21	Struhsaker 1975, 2000
Pi. tephrosceles	Gombe, Tanzania	Disturbed mixed forest	82	–	1	11	–	24	–	47[8]	–	Clutton-Brock 1975a
Pi. tholloni	Salonga, DRC	Lowland moist primary forest	–	>60	1	–	–	–	–	–	–	Maisels et al. 1994
Pr. verus	Taï, Ivory Coast	Lowland moist primary forest	7.1	2–12	10[11]	1.5	1–3	3.0	1–6	2.6	1–4	Korstjens 2001, Korstjens and Schippers 2003
Pr. verus	Tiwai, Sierra Leone	Lowland moist secondary forest	8.5	3–11	1	2.0	1–2	3.5	1–5	3.0	1–4	Oates 1988a,b, Whitesides et al. 1988, Davies et al. 1999

[1] Group composition based on the 7 groups for which composition data were complete.

[2] Group composition data based on only 2 groups since age/sex class of some individuals in the other groups could not be determined.

[3] Group composition based on the 4 groups for which composition data were complete.

[4] Group composition based on only 26 groups.

[5] Group size and composition data based on values at start of study.

[6] Excludes 2 groups for which widely variable sizes and no information on group composition were reported.

[7] Authors describe forest as primary despite selective logging occurring there 3 years before their study.

[8] Includes both immatures and individuals of indeterminate age/sex class.

[9] Group composition data based on only 1 group.

[10] Group composition based on only 14 groups.

[11] Group composition data based on only 6 groups since age/sex class of some individuals in the other groups could not be determined.

[12] Different, taller-canopy section of same study site as Harris 2006.

[13] C.A.R., Central African Republic.

[14] NDUFR, New Dabaga Ulangambi Forest Reserve.

size of 19. While slightly high, these predicted values approximate the actual degree of disparity in group sizes observed between *Piliocolobus* and *Colobus* in nature. Korstjens and Dunbar (2007) concluded that the primary factor constraining *Colobus* to life in smaller groups is the lower digestibility of their food sources (i.e. their greater reliance on mature leaves as fallback foods), a choice which requires them to devote most of their time to resting. Lastly, there is considerable evidence (see Ecology) that feeding competition may set the upper limit of group size in several African colobines (e.g., *C. vellerosus*: Teichroeb and Sicotte 2009; *Pi. tephrosceles*: Snaith and Chapman 2008).

The composition of African colobine groups is also influenced by several factors. As in many other primates (Mitani et al. 1996), the number of adult females appears to have a strong influence on the number of adult males within some African colobine populations (e.g., *Pr. verus*, Korstjens 2001). Furthermore, adult sex ratios in red colobus populations appear to be influenced by the presence or absence of chimpanzees. At sites where red colobus co-occur with chimpanzees, ratios of adult red colobus males to females tend to be much higher than at sites where chimpanzees are absent (Stanford 1998a). Crowned-hawk eagles do not, however, appear to have a similar influence on red colobus sex ratios (Struhsaker 2000).

Mating systems have been identified for only a handful of African colobines and are related largely to group composition. One-male groups are typified by polygynous mating, in which the group male generally monopolizes copulations with females in his group (e.g., *C. guereza*; Bocian 1997, Fashing 2001c). Multimale groups are characterized by promiscuous mating (*C. angolensis*, Bocian 1997; *C. polykomos*, Dasilva 1989; *Pi. badius*, Starin 1991; *Pi. tephrosceles*, Struhsaker 1975; *Pr. verus*, Korstjens and Noë 2004), though often the dominant male does most of the copulating with the females in his group (Struhsaker 1975, Starin 1991, Korstjens and Noë 2004). Occasional extragroup copulations have been reported (*C. guereza*, Fashing 2001c; Taï *Pr. verus*, Korstjens and Noë 2004) but are not known to be common in any species of African colobine.

Olive colobus are unusual in that group membership appears to be remarkably fluid (Oates 1994, Korstjens and Schippers 2003) (Fig. 13.3). Juveniles and adults of both sexes are known to frequently disperse to new groups at Taï, with females being the sex that disperses most often (Korstjens and Schippers 2003). Costs of dispersal appear to be low for olive colobus because even while between groups, individuals typically associate with a guenon (*Cercopithecus diana*) group that provides protection against predation. Adult female olive colobus appear to disperse as a means of reducing feeding competition, whereas adult males seem to disperse as a means of improving their mating opportunities (Korstjens and Schippers 2003). Dispersal is believed to be far more common among females than males in most red colobus taxa (e.g., *Pi. badius*, Starin 1994; *Pi. rufomitratus*, Marsh 1979a,b; Decker 1994; *Pi. tephrosceles*, Struhsaker and Leland 1987). At Kibale, for example, several females immigrated into Struhsaker and Leland's (1987) *Pi. tephrosceles* study groups each year, whereas male immigration occurred roughly only once every 9 years. Female *Pi. badius* at Abuko also transfer more often than males, though most juveniles of both sexes typically disperse from their natal groups in this population (Starin 1994). Studies of *Pi. rufomitratus* at Tana River show that rates of female transfer can vary widely over time; female migrations were 10 times more common in 1973–1975 (11.6/year, Marsh 1979b) than in 1986–1988 (1.1/year, Decker 1994). Patterns of dispersal in the *Pi. kirkii* population inhabiting *shambas* at Jozani are unusual for red colobus in that males transfer more often than females (Siex 2003). The unusual degree of female philopatry in the *shambas* may have arisen in response to intense intergroup feeding competition resulting from the extraordinarily high population density at this site (Siex 2003).

Transfer patterns among black and white colobus remained poorly known, until recently with most long-term studies having reported only a few observed or inferred instances of dispersal. Recent observational and genetic work by Harris et al. (in press) on *C. guereza* at Kibale offers unprecedented insight into the dispersal patterns of this species. Harris

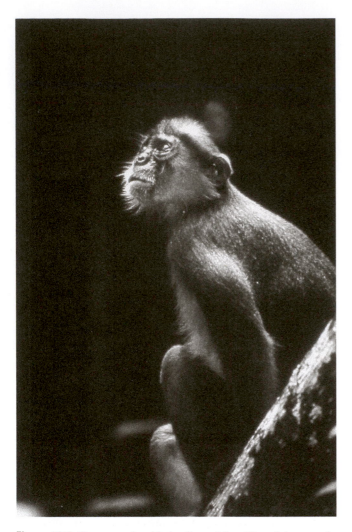

Figure 13.3 Group membership in olive colobus (*Procolobus verus*) is unusually fluid, with juveniles and adults of both sexes regularly dispersing to new groups (photo by Melanie Krebs).

1989, M. Harrison personal communication cited in Oates 1994, Korstjens et al. 2005).

SOCIAL BEHAVIOR

African colobines vary widely among species and among populations of the same species in the amount of time they devote to social activities (Table 13.4). Below, I discuss two of the better-studied topics in African colobine social behavior: grooming and aggression.

Grooming

Grooming is the predominant social behavior in most African colobines (Oates 1977c, Struhsaker and Leland 1979, Dasilva 1989, Starin 1991, Grimes 2000, Werre 2000, Fashing 2001a, Siex 2003). In black and white colobus, adult females are the primary groomers while adult males rarely groom others (*C. guereza*, Leskes and Acheson 1971, Oates 1977c, Harris 2005; *C. polykomos*, Dasilva 1989, Korstjens 2001; *C. vellerosus*, Saj et al. 2007). The primary recipients of grooming are typically adult females and/or adult males, though the exact pattern varies across black and white colobus study sites (Oates 1977c, Dasilva 1989, Grimes 2000; Saj et al. 2007).

Red colobus grooming patterns exhibit considerable interpopulational variability and are often strikingly different from those of black and white colobus. In *Pi. epieni* at Gbanraun, grooming occurred far more often among adult males than among adult females (Werre 2000). Similarly, at Kibale, adult male *Pi. tephrosceles* groomed each other more than expected by chance, while adult females primarily directed their grooming at adult males rather than each other (Struhsaker and Leland 1979). Struhsaker and Leland (1979) argue that the male-centered grooming patterns of *Pi. tephrosceles* are probably an outcome of the fact that males are philopatric and females disperse in this species. Among *Pi. badius* at Abuko, where both sexes usually disperse, adult males rarely groomed each other and females mostly groomed their offspring and adult males (Starin 1991). Similarly, among *Pi. badius* at Taï, where dispersal patterns are unknown, adult males were no more likely to groom other males than they were to groom females, though females were still more likely to groom males than other females (Korstjens 2001). At Jozani, where dispersal is male-biased, *Pi. kirkii* females still exhibit the typical red colobus pattern of directing much of their grooming at adult males, while adult males rarely groom each other as at Abuko and Taï (Siex 2003). Grooming patterns have been described for only one population of olive colobus. Oates (1994) reported that adult males of this species at Tiwai primarily groomed adult females, while adult females were more equitable in whom they groomed, albeit with some bias toward adult males. Oates (1994) also noted that male/female grooming bouts frequently occurred before or, particularly, after copulation.

et al. (2009) witnessed several male transfer events between groups, but never a female transfer event. Genetic data corroborated the observational evidence that dispersal is male-biased in this population, though also revealed evidence consistent with some degree of female dispersal as well. More limited observational data on *C. guereza* at Kakamega are consistent with the results from Kibale, with males being the primary dispersers and females appearing to disperse at times as well (Fashing and Oates in press). *C. vellerosus* at Boabeng-Fiema also exhibit a pattern of male-biased dispersal, though female dispersal is not uncommon either (Teichroeb et al. in press). Patterns of female dispersal in *C. vellerosus* suggest that they are attracted to groups with 'high quality' males, and that dispersal may help females to both reduce their vulnerability to infanticide and improve their access to food (Teichroeb et al. in press). Comparatively little is known about the dispersal patterns of *C. polykomos* and *C. satanas*, though members of both sexes appear to transfer at least occasionally in these species as well (Dasilva

Aggression

Aggression typically occurs at relatively low rates within African colobine groups (Marler 1972, Oates 1977c, Dunbar 1987, Dasilva 1989, Werre 2000, Fashing 2001a, Harris 2005, Saj and Sicotte 2005). As in most primates, males tend to be the more aggressive sex (Oates 1977c, Dasilva 1989, Werre 2000, Siex 2003). In an Ethiopian population of *C. guereza*, for example, Dunbar and Dunbar (1976) found that rates of aggression were higher in multimale groups than in one-male groups, due primarily to aggression among males. Most aggression in *Pi. tephrosceles* groups at Kibale is directed by adult males at other adult males, often in the form of harassment during copulation (Struhsaker and Leland 1979, 1987). Most aggression by adult male *Pi. badius* at Abuko and Taï occurs in the context of sex or feeding (Starin 1991, Korstjens 2001). In most African colobine groups where multiple males coexist, one male is clearly dominant to all others (*C. guereza*, Dunbar and Dunbar 1976; Bocian 1997; *Pi. badius*, Starin 1991), and in some cases, a linear dominance hierarchy is apparent (*C. polykomos*, Dasilva 1989; *C. vellerosus*, Teichroeb and Sicotte 2008b; *Pi. tephrosceles*, Struhsaker and Leland 1979). In *C. vellerosus*, Teichroeb and Sicotte (2008b) found no association between a male's rank and his fecal testosterone levels. Instead, they found that the amount of male/male aggression individual males experienced one day predicted their testosterone levels the next; as aggression experienced increased, so did testosterone levels (Teichroeb and Sicotte 2008b).

Patterns of adult female aggression have been described for several populations of African colobines. Among *Pi. badius* at Abuko, most aggression by females occurs in the contexts of protecting their offspring and weaning (Starin 1991). Struhsaker and Leland (1979) contend that aggression by female *Pi. tephrosceles* at Kibale may also be related to restricting access to their infants and possibly to the weaning process. Female aggression occurs primarily in the feeding context in both *C. polykomos* and *Pi. badius* at Taï (Korstjens et al. 2002). Strong linear dominance hierarchies exist among female *C. polykomos* at Taï and *Pi. badius* at Abuko (Starin 1991, Korstjens et al. 2002). Only a weakly developed dominance hierarchy has been detected for female *Pi. tephrosceles* at Kibale (Struhsaker 1975), and no dominance hierarchy appears to exist among female *Pi. badius* at Taï (Korstjens et al. 2002). The benefits of high rank in the populations that feature female hierarchies are as yet unclear. Starin (1991) found no correlation between female rank and age at first parturition, birthrate, or total number of surviving offspring. No information is yet available on aggression and dominance in olive colobus.

INFANTICIDE

Among the primates, infanticide by males has long been known to commonly occur among Asian colobines,

especially those belonging to the genus *Semnopithecus* (Sugiyama 1965, Hrdy 1974, van Schaik 2000). In addition, the strongest evidence to date for the hypothesis that infanticide by males is an adaptive, sexually-selected strategy comes from *Semnopithecus* (Borries et al. 1999). It is therefore rather surprising that, until recently, infanticide had only very rarely been reported among the African colobines (Struhsaker and Leland 1985). Even with the number of observed or inferred African colobine infanticides having risen rapidly over the past decade, there are still only two African forests (Kibale and Boabeng-Fiema) in which incidents of infanticide among colobines have been documented (Table 13.7). Six confirmed cases of infanticide or attempted infanticide involving colobines have occurred at Kibale. These incidents include four successful infanticide attempts in *C. guereza* (Onderdonk 2000, Harris and Monfort 2003, Chapman and Pavelka 2005) and one successful and one unsuccessful attempt in *Pi. tephrosceles* (Struhsaker and Leland 1985). In all cases at Kibale in which the identity of the perpetrator could be determined during attacks on infants, the infanticidal individual was an adult or subadult male (Struhsaker and Leland 1985, Onderdonk 2000, Harris and Monfort 2003, Chapman and Pavelka 2005). However, in only one instance (*Pi. tephrosceles*) was the infanticidal male observed to copulate with the mother whose infant he killed subsequent to the attack (Struhsaker and Leland 1985), as would be expected by the sexual selection hypothesis (van Schaik 2000).

Recent evidence from 11 observed or inferred infanticide attempts (10 successful) among *C. vellerosus* at Boabeng-Fiema is more consistent with the sexual selection hypothesis (Saj and Sicotte 2005, Sicotte et al. 2007, Teichroeb and Sicotte 2008b). First, in this population, infanticidal males were always immigrants to the group and thus unlikely to be related to the infants they killed. Second, in at least 5 instances, infanticidal males subsequently mated with the female whose infant they had killed once she resumed cycling. Lastly, interbirth intervals averaged only 10 months for females (n=5) who lost their infants to infanticide, or about half the length of a typical black and white colobus interbirth interval (Teichroeb and Sicotte 2008b). Thus, while the genetic data necessary to conclusively test the sexual selection hypothesis for infanticide (e.g., Borries et al. 1999) in *C. vellerosus* at Boabeng-Fiema are not available, the behavioral data on this population are generally consistent with this hypothesis.

Outside of Boabeng-Fiema and Kibale, no other instances of infanticide have been reported among African colobines, even at other long-term study sites like Kakamega or Taï. Even at Kibale, Chapman and Pavelka (2005) noted that *Pi. tephrosceles* has been studied for over 6,000 hours in recent years without any additional observations of infanticide in this population. Given the relative rarity of observed infanticides in red colobus at Kibale and elsewhere, it seems reasonable to cautiously suggest that infanticide may not be as common a behavior in red colobus as it is in some black and

Table 13.7 Long-Term (≥6 Months) Studies of Intergroup Encounters (IEs) in African Colobines[1]

SPECIES	STUDY SITE	STUDY LENGTH (MONTHS.)	HOME RANGE SIZE (HA)	RANGE OVERLAP (%)	DAILY PATH LENGTH (M) MEAN	NUMBER OF GROUPS	CORE AREA	D INDEX	TERRITORIAL	NUMBER OF IES	IES/HR	% IES AGGRESSIVE MALE(S)	% IES AGGRESSIVE FEMALE(S)	% IES AGGRESSIVE ANYONE	DISPERSING SEX[1]	INFANTICIDE REPORTED	REFERENCES
C. angolensis	Ituri, DRC	11	371[2]	100	983	1	No	0.45	No	≥25	>0.03	+	*	+	M?,F?	No	Bocian 1997, Oates et al. 2000a,b
C. guereza	Budongo, Uganda (1964–65)	8	14	Minimal		5			Maybe					+		No	Marler 1969
C. guereza	Kibale, Uganda (1971–72)	12	28	74	535	1	Yes	0.90	No	65	0.09	+	*	62	M, F	Yes[3]	Oates 1977a,c
C. guereza	Kibale, Uganda (2002–03)	16[22]	14[23]	83	588	6	Yes	0.54	No	220	0.05	~75[24]	~20[24]	~79[24]	M, F	Yes	Harris 2005, 2006a, Harris et al. 2009
C. guereza	Ituri, DRC	11	100[2]	≥22	609	1	Yes		No	2	0.003	+			M?,F?	No	Bocian 1997
C. guereza	Kakamega, Kenya	12	18[4]	67	588	5	Yes	1.23	No	136	0.07	70	18	70	M,F?	No	Fashing 2001a,c, Fashing and Oates in press
C. guereza	Entebbe, Uganda	6	7.5	+	386	3	Yes	1.25		12[5]		+	–	+	M	No	Grimes 2000
C. polykomos	Tiwai, Sierra Leone	12	24	≥20[6]	832	1	Yes	1.50	Maybe	9	0.010	67	0	67	M,F	No	Dasilva 1989, Oates 1994
C. polykomos	Taï, Ivory Coast	78	77[7]	≥47[8]	617	2	Yes	0.62	No	83	0.012	69[9]	52[9]	75	M,F	No	Korstjens 2001, Kostjens et al. 2005
C. satanas	Douala-Edea, Cameroon	11	60	+	459	1	Yes	0.53		4	0.006	+	–	+		No	McKey 1978a,b, McKey et al. 1982
C. satanas	Fôret des Abeilles, Gabon	18	573[10]	≥65	852	1	No[11]	0.32	No	19[12]	0.02	+	–	+		No	Fleury and Gautier-Hion 1999
C. vellerosus	Boabeng-Fiema, Ghana	6	9	≥33	300	1		2.61	No	47	0.12	83	9	83	M,F	Yes	Sicotte and MacIntosh 2004, Teichroeb et al. 2009
Pi. badius	Abuko, Gambia	51	34	60		1			No	≥64[13]		+	+	+	F,M	Suspected	Starin 1991, 1994
Pi. badius	Taï, Ivory Coast (1996–98)	35	58[7]	>6[8]	872	2		1.01	Maybe			+	*	+	F?	No	Korstjens 2001, personal communication
Pi. epieni	Gbanraun, Nigeria	12	73	>42	1040	1		1.08	No	6	0.008	17[15]	17	17		No	Werre 2000
Pi. kirkii	Jozani (forest), Zanzibar (1999)	12	25	31–65	565	3	Yes	1.00	No			+	–	+		No	Siex 2003
Pi. kirkii	Jozani (shamba), Zanzibar (1999)	12	13	70–95	310	4	Yes	0.76	No			+[14]	+	+	M,F	No	Siex 2003
Pi. rufomitratus	Tana River (Mchelelo), Kenya (1986–88)	15	9[16]	36	603	1	Yes	1.78	No	8[17]	Rare	+?[18]		50	F,M	Suspected	Marsh 1979a, 1981b
Pi. tephrosceles	Kibale (Kanyawara), Uganda (1970–71)	12	65	99	649	1	Yes	0.71	No	≥64[19]		≤70	≤2	≤70	F,M	Yes	Struhsaker 1975, Struhsaker and Leland 1985, 1987

Pr. verus	Tiwai, Sierra Leone	8[20]	28			1		No		−	−			F,M	No	Oates and Whitesides 1990
Pr. verus	Taï, Ivory Coast	21	56[7]	>14[8]	1,212	1.43	2	Maybe	10[21]	+	+	−	+	F,M	No	Korstjens 2001; personal communication

[1] Bold lettering is reserved for cases in which one sex clearly disperses more often than the other.

[2] Based on number of 1.56 ha cells entered.

[3] Infanticide suspected by Oates 1977c, but definitive reports of infanticide for this population come from Onderdonk 2000 and Harris and Monfort 2003.

[4] Mean for 2 groups.

[5] Encounters appear to have been defined by visual contact between groups rather than by groups being within 50 m of each other.

[6] Calculated based on map of IE locations from Dasilva 1989:305.

[7] Based on number of 1 ha cells entered per year.

[8] Based on 2 years of data; measures only overlap with two of the five groups that overlap study groups' home ranges.

[9] Calculated from data in Table I in Korstjens et al. 2005.

[10] Based on number of 1 ha cells entered.

[11] Certain areas were intensively used but were not contiguous.

[12] Encounters defined as when two groups were within 200 m of each other.

[13] Starin 1994 gives only the number of aggressive encounters (n = 64).

[14] Rate of adult male aggressive acts was over three times higher than the rate for adult females.

[15] In the 1 of 6 encounters involving aggression, males were more persistently aggressive than females, though both sexes chased the other group.

[16] Based on 12 months.

[17] Does not include 15 encounters with solitaries and small parties since a larger cut-off distance (<300 m) appears to have been used to define these encounters relative to those with bisexual groups (<50 m).

[18] Does not say which sex(es) participated in aggression during IEs between bisexual groups but mentions male aggression (chasing) during 4 of 15 interactions with solitaries and small parties.

[19] Struhsaker 1975 notes that encounters (aggressive or tolerant) occurred on 49 of 83 study days and that on 30 of these days 45 aggressive encounters occurred.

[20] Based on 33 days during 8 months over a 29-month period.

[21] Encounters defined by loud ("laughing") call bouts between olive colobus males of different groups rather than distances between groups.

[22] Studied 6 groups for 3 months each.

[23] Mean for 6 groups.

[24] Estimated from data in Harris 2005: Chapter 7.

white colobus populations. Only further long-term research on a variety of colobine taxa at multiple sites will produce a more definitive understanding of infanticide in African colobines.

Intergroup Relationships

Interactions between individuals in different groups are less common and more difficult to observe than those occurring among members of the same group. As a result, far more is known about relationships within than between primate groups (Cheney 1987). Our limited knowledge of intergroup relationships is unfortunate considering that most models of primate socioecology feature predictions about intergroup interactions that have not been well tested (e.g., Wrangham 1980, van Schaik 1989, Isbell 1991, Sterck et al. 1997). Nevertheless, intergroup relationships have been surprisingly well studied for African colobines relative to other forest primates. The relatively small home range sizes and often predictable ranging patterns, as well as the small group sizes and spatial cohesion, of many African colobines have facilitated this unusually intensive investigation into intergroup relationships. Patterns of intergroup relations in African colobines have been found to vary at several levels: among different species (Struhsaker and Oates 1975), among different populations of the same species (Dasilva 1989, Korstjens 2001), and even among different subpopulations of the same species in a single forest (Siex 2003). Here, I review the patterns of intergroup relations among the African colobines and evaluate how consistent they are with the predictions of recent models of primate socioecology.

Loud Calls

Though encounters between colobine groups are usually defined as occasions when two or more groups are within 50 m of each other (e.g., Oates 1977c, Stanford 1991a), the vocal repertoire of adult male African colobines allows groups to remain in contact over much larger distances. Male black and white colobus utter low-frequency, high-amplitude loud calls, referred to as "roars," which in *C. guereza* have been estimated to be audible up to 1.6 km from the source (Marler 1969, Oates and Trocco 1983). Though black and white colobus roars appear to play roles in male/male competition and predator intimidation, there is little doubt they also serve as intergroup spacing mechanisms, especially in *C. guereza* (Marler 1969, 1972; Oates 1994). Recent evidence suggests that the formant dispersion (i.e., average spacing between frequency bands) of a typical *C. guereza* roar is substantially lower than expected for an animal its size (Harris et al. 2006). This result suggests that through their roars, guerezas may be attempting to send exaggerated signals of body size to potential predators. However, comparisons of roars produced by captive male guerezas of known body mass found that the formant dispersion in roars provides an honest advertisement of body mass relative to conspecifics (Harris et al. 2006).

Lacking a similar low-frequency, high-amplitude loud call, red colobus males utter several other calls (e.g., *bark*, *chist*, *nyow*) that carry over distances of >300 m and facilitate intergroup communication (Struhsaker 1975, Oates 1994). The loudest call uttered by olive colobus males appears to be their "laughing" call (Oates 1994), which carries ~150 m (M. Korstjens unpublished data cited in Wich and Nunn 2002). With their travel patterns largely dictated by the movements of the diana monkey group with which they associate, olive colobus are probably less likely to use their loud calls to mediate intergroup spacing than the other African colobine taxa. In fact, laughing calls have primarily been mentioned as occurring when olive colobus groups are already in proximity (e.g., during intergroup encounters involving neighboring olive colobus–diana monkey associations [Oates and Whitesides 1990]).

Territoriality and Other Spacing Systems

The word "territory" is often used (and misused) when discussing intergroup relationships (Maher and Lott 1995). It has been employed to describe a variety of patterns of range use and defense, the most common of which are (*1*) an area defended by a group, (*2*) an area exclusively used by a group, or (*3*) an area in which a group exhibits site-specific dominance. Since use of a common terminology is critical for making comparisons of spacing systems across studies (Maher and Lott 1995), I conceptually and operationally define as *territorial* any populations of African colobines in which groups consistently defend large parts of their home ranges against conspecific intruders and maintain ≥80% of their ranges as areas of exclusive use. This definition seems appropriate since it implies that a group actively defends its range and has considerable success keeping out intruders, two features that are at the heart of the concept of territoriality (Maher and Lott 1995).

Mitani and Rodman (1979) developed a simple yet powerful model that predicts which primate populations are capable of being territorial. This model treats home ranges as circular and assumes that only those groups that travel distances equivalent to at least the length of the diameter of their home range on a typical day can be territorial. The formula for the model is $D = d / (4a/\pi)^{0.5}$", where D represents defendability, a represents home range area, and d represents mean daily path length. Mitani and Rodman found that only primate groups with $D \geq 1$ maintain a territory, though some groups with $D \geq 1$ are not territorial.

Although according to the defendability index about half the African colobine populations studied to date are theoretically capable of territoriality, very few, if any, are actually territorial (Table 13.7). Based on the 21 long-term studies summarized in Table 13.7, only four populations,

C. guereza at Budongo, *C. polykomos* at Tiwai, *Pr. verus* at Taï, and *Pi. badius* at Taï, qualify as possibly territorial, though there is some uncertainty as to the amount of home range overlap that occurs in these populations. It is possible that the energetic constraints placed on colobines by their specialized digestive systems make it difficult for them to be territorial even when they are theoretically capable of doing so. Because many colobines seem to require long resting periods after meals to allow time for bacterial fermentation, they may simply not be energetically capable of evicting intruders from their home range at all times.

Colobine populations in which groups at least sometimes behave aggressively toward one another but are not territorial can be categorized as exhibiting (*1*) core area defense, (*2*) intergroup dominance, or (*3*) "supertroop" formation. *Core area defense* is less intensive than territoriality in that it involves defending only frequently used portions of the home range. It occurs in *C. guereza* populations at Kakamega, Entebbe, and in some instances at Kibale, and is the most common spacing system for this species (Oates 1977a,c; Fashing 1999; Grimes 2000, Harris 2006a). *C. polykomos* at Taï also appear to feature core area defense (Korstjens 2001).

Intergroup dominance, in which the outcome of intergroup encounters is determined by a linear dominance hierarchy among groups, has been described for two red colobus populations, *Pi. tephrosceles* at Kibale and *Pi. badius* at Abuko (Struhsaker and Leland 1979, Starin 1991, Snaith and Chapman 2008). The dominance hierarchy among *Pi. tephrosceles* groups at Kibale is based on the number of adult males in each group, their fighting abilities, and their inclination to involve themselves in encounters (Struhsaker and Leland 1979). In contrast, among *Pi. badius* at Abuko, number of adult males does not influence intergroup dominance. Instead, the physical condition and identities of adult males, as well as social changes resulting from incidences of female dispersal, all appear to contribute to the patterns of intergroup dominance in this population (Starin 1991). Intergroup dominance also appears to play variable roles in determining the outcome of encounters among *C. guereza* groups at Kibale (Oates 1977c). In Oates's (1977c) original study, core area defense usually determined the outcome of encounters, though groups seemed to have differentiated dominance relationships that sometimes influenced the outcome of encounters instead. A recent study by Harris (2005, 2006a) in a nearby forest compartment, however, found that intergroup dominance was entirely responsible for the outcomes of encounters, though all groups at least attempted to defend their core areas. Interestingly, intergroup dominance rank in this study appeared to be associated with group male stamina during early morning loud call choruses; the longer a male roared on average during choruses, the higher ranking his group (Harris 2006b).

Supertroops (or "associations") occur when two or more groups rest, feed, or travel while intermingled for periods ranging from more than 1 hr to more than 1 day. Among the African colobines, these peculiar associations appear to be unique to *C. angolensis* and have been reported for populations at Diani Beach, Kenya; Sango Bay, Uganda; and Ituri Forest, DRC (Moreno-Black and Bent 1982, Oates 1994, Bocian 1997). Despite the intermingling of group members that occurs when supertroops form, there have been no reports of affiliative behaviors between members of different groups. In contrast, intergroup aggression escalating to the level of grappling and even biting among males has been reported within supertroops at Ituri (Bocian 1997). The largest supertroop yet reported was observed at Sango Bay and included at least 51 individuals (Oates 1994). The fascinating >300-member groups of *C. angolensis* at Nyungwe appear to be permanent associations rather than supertroops (Fimbel et al. 2001, Fashing et al. 2007), and nothing has been written about intergroup relationships in this population.

Male and Female Strategies During Intergroup Encounters

As noted before, encounters between African colobine groups range in intensity from peaceful comminglings to chases resulting in physical attacks and even infanticide (Bocian 1997, Fashing 2001c, Harris and Monfort 2003). When encounters are aggressive (i.e., when individuals in one group threaten, chase, and/or physically attack members of another group), not all individuals participate equally in the aggression. A comparison across African colobine populations reveals that adult males are nearly always the primary aggressors during intergroup encounters (Table 13.7). Among the 19 populations for which adult participation in intergroup encounters has been reported, males exhibit intergroup aggression in 18 and females in only 12. Furthermore, in the seven populations for which data on the percentage of encounters in which males and females behaved aggressively are available, males (64%) were aggressive during a far greater percentage of encounters than females (17%). In fact, regular female participation in intergroup aggression appears to occur in only three of the populations studied to date (*C. polykomos*, Korstjens et al. 2005; *Pi. badius*, Starin 1991; *shamba*-dwelling *Pi. kirkii*, Siex 2003). More often, females are described as only occasional aggressive participants, while males are regarded as regular aggressive participants, in encounters.

Predictions about the behavior of adults of each sex during intergroup encounters play important roles in models of primate socioecology (Wrangham 1980, van Schaik 1989, Isbell 1991, Sterck et al. 1997). Though the specifics of the various models differ, all agree that if food is limited and distributed in such a way that it can be defended against other groups, females will spend their lives in their natal groups and form coalitions with female relatives to defend food sources against other groups. The rationale here is that as the sex that makes the greatest energetic investment in offspring, females are more dependent on food and high-quality nutrition for enhancing their lifetime reproductive success than males (Trivers 1972). Conversely, as the sex whose reproductive success is limited more by access to mates than by access to food (Trivers 1972), males are expected to focus primarily on directly defending access to females during intergroup encounters.

Studies of intergroup encounters in primates have often used Trivers's (1972) differential investment theory as a means of inferring the behavioral strategies adopted by males and females during encounters (Cheney 1987, van Schaik et al. 1992). Female participation in intergroup aggression has typically been assumed to reflect food defense, whereas male participation has generally been assumed to reflect mate defense. Only recently have researchers begun to empirically test these assumptions by examining patterns of male and female participation in intergroup encounters with regard to the ecological and social contexts in which encounters occur (for reviews of this research across the primates see Fashing 2001c and Kitchen and Beehner 2007). Several such intensive studies have now been conducted on African colobines, and the results offer evidence that the strategies adopted during encounters may be more complex than predicted by models of primate socioecology.

Recent studies of *C. guereza* at Kakamega and Kibale, and of *C. vellerosus* at Boabeng-Fiema, have all produced evidence consistent with the surprising conclusion that males, not females, may be the primary defenders of food resources during intergroup encounters in these taxa (Fashing 1999, 2001c; Sicotte and MacIntosh 2004; Harris 2005, 2006a) (Fig. 13.4). In all three populations, males are far more frequent aggressive participants in encounters than females (Table 13.7). Furthermore, in *C. guereza* at both Kakamega and Kibale, levels of male, but not female, aggression during

encounters were positively related to the amount of time group members spent feeding that month in encounter grid cells (Fashing 2001c, Harris 2005). Among *C. vellerosus* at Boabeng-Fiema, intergroup encounters were significantly more likely to occur in the feeding context than predicted by chance alone and it was common for males to direct aggression toward female and male members of opposing groups during encounters (Sicotte and MacIntosh 2004). Curiously, patterns of intergroup encounters among *C. polykomos* at Taï appear to differ considerably from those in other black and white colobus populations (Korstjens et al. 2005). At Taï, female *C. polykomos* participated in intergroup aggression nearly as often as their male groupmates. Furthermore, females, but not males, were more likely to become involved in intergroup aggression during months when the nutrient-rich seeds of a favored, clumped tree species were available.

While differences in methodology across the studies make them difficult to compare (Harris 2007), taken collectively, these results suggest that in most *Colobus* populations, males play a larger role in food defense than females. Fashing (2001c) proposed that male resource defense in *C. guereza* may be related to an indirect form of mate defense in which males that consistently successfully defend food sources during intergroup encounters are preferred as mates by females. One of the predictions of this male resource defense model is that females should be capable of transferring away from groups with poor resource-defending males and to groups with strong resource-defending males. Intriguingly, a recent study of dispersal patterns among females in *C. vellerosus* at Boabeng-Fiema found that females tended to transfer away from groups containing males who fared poorly in intergroup encounters and

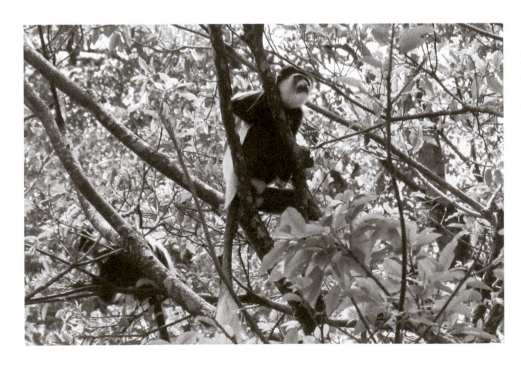

Figure 13.4 Females, like the ones pictured here, rarely behave aggressively during encounters between guereza (*Colobus guereza*) groups. Instead, males appear to play the primary role in intergroup food defense in this species (photo by Peter Fashing).

to groups with males who succeeded in intergroup conflict (Teichroeb et al. in press). If indirect mate defense via resource defense is indeed a regular strategy of black and white colobus males, they appear to adopt it in addition to the strategy of direct mate defense predicted by traditional socioecological models. For each of the studies discussed above, in addition to the evidence suggesting that males play an important role in food defense, there was also (not mutually incompatible) evidence that males were involved in directly defending mates (e.g., chasing away bachelor males) during at least some encounters (Fashing 2001c, Sicotte and MacIntosh 2004, Harris 2005).

Still, based on the available evidence, primate socioecological models do a poor job of explaining patterns of between-group competition in the several African colobines for which intergroup relationships have been intensively investigated. In particular, the assumption that females will take the lead in intergroup conflict over food when food is limiting and defensible comes into question. It appears that, at least among black and white colobus, males are primarily responsible for resource defense during intergroup encounters, perhaps as a form of indirect mate defense. Furthermore, it seems unlikely that the male resource defense strategy is exclusive to colobines (Fashing 2001c). Indeed, there is evidence that males defend food during intergroup encounters in other primate taxa as varied as *Hapalemur griseus*, *Cebus olivaceus*, *Chlorocebus aethiops*, *Macaca radiata*, and *Pan troglodytes* (Harrison 1983b, Robinson 1988b, Nievergelt et al. 1998, Mutschler et al. 2000, Cooper et al. 2004a, Williams et al. 2004). With the growing evidence that males play a prominent role in between-group feeding competition in many primate taxa, it may ultimately prove necessary to reconsider existing socioecological models whose foundations rest on the importance of female bonds and female intergroup food defense.

14

The Macaques
A Double-Layered Social Organization
Bernard Thierry

1. How do the macaque species differ in ecological adaptation, social relationships, and mating patterns?
2. Are differences in social relationships between species better explained by phylogeny or by ecology in macaques?
3. Mating patterns vary with reproductive seasonality in macaques. Does this mean that they are largely independent from social relationships?

INTRODUCTION

The genus *Macaca* is one of the most successful primate radiations. Its representatives are present in both Asia and Africa, and it has the widest geographical range of primates after *Homo* (Fooden 1982) (Fig. 14.1). Whereas many macaques live at the heart of the tropics, others inhabit northern regions where snowfall occurs during winter. In spite of their spread, macaques remain interfertile (Bernstein and Gordon 1980). All possess cheek pouches, allowing them to harvest food and minimize feeding time.

Their social organization is characterized by both a profound unity and a great diversity, which may be best described as variations on the same theme (Thierry et al. 2004). On the one hand, macaques share the same basic patterns of organization. They form multimale/multifemale groups, neighboring groups have overlapping home ranges, females are philopatric—consequently forming kin-bonded subgroups within their natal group, and most males disperse and periodically transfer from one group to another. On the other hand, macaques vary widely in the degree to which nepotism governs their societies (Matsumura 2001). Though

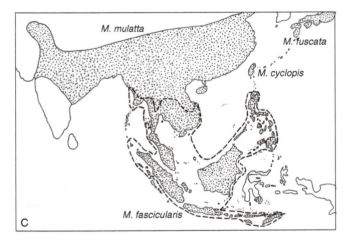

Figure 14.1 Present geographical distribution of macaques (from Fooden 1982). *A: Silenus–sylvanus* lineage. *B: Sinica–arctoides* lineage. *C: Fascicularis* lineage. *Arrow* in *B* indicates the narrow range of *Macaca munzala* (in black) within the western distribution of Assamese macaques.

there were early hints that some macaques might be "nicer" than others (Rosenblum et al. 1964), two decades passed before interspecific contrasts in the conciliatory dispositions of macaques were directly addressed (Thierry 1986, 2000; de Waal and Luttrell 1989). A number of contrasts in the social relationships of macaques are now documented, making the genus an especially interesting taxon

to compare homologous social features and investigate their determinism. In what follows, I review the evolution, ecology, life history, reproduction, and social behaviors of macaques, with the aim of unraveling the two layers that structure their social organization: social styles and female defensibility by males.

EVOLUTION AND BIOGEOGRAPHY

Presently, 22 species of macaques are recognized (Table 14.1) (Fooden 1976, Groves 2001a, Ziegler et al. 2007, Mishra and Sinha 2008), representing a monophyletic group (Delson 1980, Morales and Melnick 1998). The fossil record indicates that they diverged from other papionins about 7 million years ago in northern Africa. Macaques invaded Eurasia about 5.5 million years ago, probably via the Near East. They then branched into several phyletic lineages that have been identified from morphological (Fooden 1976, Delson 1980) and molecular (Hoelzer and Melnick 1996; Ziegler et al. 2007) evidence.

Three main phyletic groups of extant macaques are distinguished, which correspond to three dispersal waves in Asia (Fooden 1976) (Table 14.1). The *silenus* lineage has the most disjunct geographical distribution (Fig. 14.1). Its 11 species are situated far away from each other, and most are represented by relict distributions, indicating an early dispersal. Only *M. nemestrina* has a large range. *M. silenus* is found in the evergreen forests of south India. The other species of the lineage inhabit the evergreen forests of the Sulawesi and Mentawai Islands. The *sinica* lineage has a moderately fragmented distribution in south Asia; it is thought to be the second to have dispersed (Fig. 14.1). *M. radiata*, *M. assamensis*, *M. munzala*, and *M. thibetana* live in tropical and subtropical continental areas; only *M. sinica* is present on an island. The *fascicularis* lineage has the most broadly continuous distribution; it is likely the third to have dispersed (Fig. 14.1). *M. fascicularis* inhabits equatorial and tropical regions. The other species occur in subtropical and temperate Asia; while *M. fuscata* and *M. cyclopis* are confined to islands, *M. mulatta* occurs from Pakistan to China. Species from different lineages may be found sympatrically in many areas— for example, *M. silenus* and *M. radiata* in south India and *M. mulatta*, *M. assamensis*, *M. nemestrina*, and *M. arctoides* in the evergreen forests of the Brahmaputra Valley in Assam.

The taxonomic position of two further species is still debated. *M. sylvanus* lives in the montane forests of North Africa; it is likely the most ancient taxon of the genus and, in fact, its last African representative (Fig. 14.1). It is alternatively classified as either being the only member of its own species group or belonging to the *silenus–sylvanus* lineage. *M. arctoides* inhabits south Asian broadleaf evergreen forests (Fig. 14.1); it is either ascribed to its own species group

Table 14.1 Species and Phyletic Lineages of the Genus *Macaca*

SPECIES

Silenus–sylvanus lineage

　M. sylvanus (Barbary macaque)

　M. silenus (lion-tailed macaque)

　M. nigra (crested macaque)

　M. nigrescens (Gorontalo macaque)

　M. hecki (Heck's macaque)

　M. tonkeana (Tonkean macaque)

　M. maurus (moor macaque)

　M. ochreata (booted macaque)

　M. brunnescens (Muna-Butung macaque)

　M. pagensis (Mentawai macaque)

　M. siberu (Siberut macaque)

　M. nemestrina (pig-tailed macaque)[1]

Sinica–arctoides lineage

　M. sinica (toque macaque)

　M. radiata (bonnet macaque)

　M. assamensis (Assamese macaque)

　M. munzala (Arunachal macaque)

　M. thibetana (Tibetan macaque)

　M. arctoides (stump-tailed macaque)

Fascicularis lineage

　M. fascicularis (long-tailed macaque)

　M. mulatta (rhesus macaque)

　M. fuscata (Japanese macaque)

　M. cyclopis (Taiwanese macaque)

Sources: Fooden 1976, 1982, Delson 1980, Groves 2001a, Ziegler et al. 2007, Mishra and Sinha 2008.
[1] Pig-tailed macaques may be split into *M. nemestrina* and *M. leonina* (Groves 2001a).

or included in the *sinica–arctoides* lineage (see Fooden 1976, Delson 1980, Deinard and Smith 2001).

The replacement of older lineages by younger ones was for some time explained by the occurrence of interspecific competition (Fooden 1976, Delson 1980). The most recent colonizers were supposed to have been better adapted to the new environmental conditions. Only geographical barriers would have prevented a later lineage from invading some refuges where remnant populations of an earlier lineage still survived. It has been pointed out, however, that the climatic changes of the Pleistocene significantly affected the distribution of primates (Eudey 1980, Brandon-Jones 1996). The aridity associated with glaciations eliminated species from their distribution ranges and split up areas where relict populations were able to survive. A reappraisal

of the circumstances of the deployment of macaques in Southeast Asia indicates that the disappearance of the first lineage from large areas was caused by glacially induced deforestation (Abegg and Thierry 2002a). Several populations took refuge in relict moist habitats. When forests recovered during wetter and warmer periods, some species could not disperse due to geographical obstacles and oceanic gaps, whereas others were lucky enough to find migration routes and colonize a wealth of habitats devoid of primates. Historical contingencies likely played a major role in the deployment and present distribution of macaque species.

With the lowering of sea levels in recent geological time, certain species have become marginally sympatric and low rates of hybridization may occur at their boundary. A narrow contact zone extends 2,000 km across the Indochinese peninsula between the ranges of *M. mulatta* and *M. fascicularis* (Fooden 1997). Hybrids displaying an intermediate morphology are found on this border, but their number remains limited. Such a sharp transition between species may result from the removal of a biogeographical barrier that was present during a Pleistocene glacial period. Genetic evidence indicates that gene flow goes from *M. mulatta* into *M. fascicularis* (Tosi et al. 2002). Another case of secondary intergradation is provided by the seven parapatric species inhabiting Sulawesi. This island was subdivided into an archipelago during the Pleistocene, inducing allopatric speciation (Fooden 1969). A recent uplifting of the lands has reunited the sister species. As a consequence, contemporary hybridization is observed in all the contact zones between Sulawesi species (Camperio Ciani et al. 1989). Gene flow appears to be mostly limited to narrow contact zones, the morphology of parental species changing abruptly on either side of these zones (Bynum et al. 1997, Evans et al. 2001).

HABITAT AND ECOLOGY

Macaques are semiterrestrial primates that inhabit a wide range of habitats, including evergreen, deciduous, and coniferous forests; grasslands; swamps; semideserts; and areas settled by humans. Ecological preferences do not relate directly to phylogeny (Fooden 1982, Richard et al. 1989). Species such as *M. mulatta*, *M. fascicularis*, and *M. nemestrina* have extensive distributions, which include diverse types of habitat. Other species have a more limited geographical distribution, yet they may live in diverse habitats. *M. maurus*, for instance, is found both in dipterocarp and tropical seasonal forests of south Sulawesi from sea level to near 1,000 m elevation (Supriatna et al. 1992). Alterations of the environment induced by human activities have further increased the diversity of ecological niches. *M. mulatta*, *M. fascicularis*, *M. radiata*, and *M. sinica* may live commensally with humans; and they are able to cope with degraded habitats and even urbanized areas (Richard et al. 1989). Other species have a narrower ecological range. Fooden (1976) stresses that some are restricted to primary broadleaf evergreen forests (e.g., *M. silenus, M. nemestrina, M. arctoides*). When macaque species are sympatric, they often exploit different ecological niches or habitats (Fooden 1982). In Borneo, Sumatra, and Malaysia, *M. nemestrina* prefers primary montane forests whereas *M. fascicularis* inhabits riverine and secondary forests at lower altitudes. Both species additionally use different forest strata; *M. nemestrina* travel more often on the ground between more dispersed food resources (Rodman 1979b, Crockett and Wilson 1980).

Although macaques are usually considered to be mainly frugivorous animals, the actual picture is more complex (Ménard 2004). Most species show a highly flexible diet with marked annual and interannual variations. Fruits represent a major proportion of the feeding time in tropical species such as *M. nemestrina, M. fascicularis,* and *M. cyclopis.* These appear to be the most frugivorous among known species (Table 14.2). During periods of fruit shortage, however, they switch to a diet that is mainly leaves, and *M. cyclopis* may even become completely folivorous (Su and Lee 2001). *M. silenus, M. nigra, M. radiata,* and *M. assamensis* are also mainly frugivorous; but insects represent a significant part of their diet (O'Brien and Kinnaird 1997, Ménard 2004). In *M. fuscata,* populations from warm-temperate areas such as Yakushima are more frugivorous than those inhabiting cool-temperate areas such as Kinkazan, which are mainly granivorous (Table 14.2). Food availability is highly variable from month to month and from one season to another; in northern regions, the diet reduces to bark and buds during winter. Fruit production is limited in temperate regions, and the animals must rely on other major food sources. *M. thibetana* is primarily a leaf eater (Zhao 1996). *M. sylvanus* appears to be either mainly granivorous or folivorous depending on the habitat; the proportion of herbaceous plants in the diet is consistently elevated (Ménard and Vallet 1996) (Table 14.2). Lastly, *M. mulatta* appears to be mainly folivorous in the deciduous forests of the Himalayan foothills (Table 14.2), but it may be largely frugivorous on occasion; in many parts of its range, it relies heavily on agricultural crops and food from people (Richard et al. 1989, Fooden 2000). Like *M. fuscata* and *M. sylvanus, M. mulatta* should be considered a nonspecialized feeder (Ménard 2004).

The daily feeding and foraging time usually ranges between 2 and 6 hr (Ménard 2004). Comparisons between seasons and species indicate that feeding and foraging time increases when the proportion of leaves rises and that more frugivorous diets are associated with more time spent moving. Animals travel through greater areas when food is clumped (e.g., Agetsuma 1995b). The average daily path length ranges from 100 m to 3 km according to the

Table 14.2 Diet of Nonprovisioned Macaques: Percentage of Feeding Time

SPECIES (REFERENCE AND STUDY SITE)	FRUITS	SEEDS	LEAVES AND BUDS	ROOTS, FLOWERS, AND HERBS	INVERTEBRATES	OTHERS
Macaca sylvanus (Ménard and Vallet 1996: Akfadou, Algeria)	0.8	32.2	8.8	28.9	10.5	18.8
M. sylvanus (Ménard and Vallet 1996: Djurdjura, Algeria)	4.3	26.7	13	44.4	5.6	6.0
M. nemestrina (Caldecott 1986a: Pasoh, Malaysia)	74.2	0	11.1	1.1	12.2	1.4
M. fascicularis (Yeager 1996: Tanjung Puting, Kalimantan)	66.7	0	17.2	8.9	4.1	3.1
M. fascicularis (Wheatley 1980: Kutai, Kalimantan)	87.0	0	1.6	5.4	4.0	2.0
M. mulatta (Goldstein and Richard 1989: Murree Hills, Pakistan)	8.5	0	84.4	5.9	0	1.1
M. fuscata (Hill 1997: Yakushima, Japan)	30.2	13.2	35.0	5.5	10.3	5.8
M. fuscata (Agetsuma 1995a,b: Yakushima, Japan)	28.6	28.2	22.0	4.9	8.9	7.4
M. fuscata (Agetsuma and Nakagawa 1998: Kinkazan, Japan)	10	44	14	18	2	12
M. cyclopis (Su and Lee 2001: Jentse, Taiwan)	53.8	0	29.1	7.2	9.8	0

Source: Ménard 2004.

dispersion of resources. Foraging efficiency is also affected by group size; larger groups travel more than others as a result of feeding competition (van Schaik et al. 1983a, O'Brien and Kinnaird 1997). Home range size usually fluctuates between some dozen hectares and some square kilometers, but minima and maxima may be as extreme as a few hectares and 22 km^2 in a single species such as *M. mulatta* (Caldecott 1986a; Fooden 1986, 1995, 2000; Kurup and Kumar 1993; Fa and Lindburg 1996; O'Brien and Kinnaird 1997; Yamagiwa and Hill 1998). The degree of arboreality represents a further source of diversity. Many macaques spend more time on the ground than in the trees during the day. Species such as *M. arctoides* and *M. thibetana* are mostly terrestrial, foraging on the forest floor with their heavy bodies and short tails. However, other species are primarily arboreal. *M. fascicularis* and *M. silenus* are gracile animals whose long tails serve as rudders during leaping; they may spend more than 90% of their time in the trees.

The availability of food resources is a main factor determining the size of populations. For example, a population of *M. sinica* declined by 15% during an 11-month drought (Dittus 2004). When animals are artificially fed, groups grow to unusual sizes. Provisioned groups of *M. fuscata* may reach several hundred individuals (Asquith 1989). In undisturbed habitats, groups rarely exceed 100. Wide intraspecific variations are reported for group size. The mean size is quite comparable across macaque species, most groups ranging between 15 and 50 individuals. Maximal group size before fission, however, may differ between species. Social groups usually do not go above 30–40 individuals in *M. radiata*, *M. sinica*, and *M. silenus*, whereas groups of 70–90 are not rare in some habitats for *M. mulatta*, *M. fuscata*, and *M. sylvanus* (Fooden 1986, 1995, 2000; Yamagiwa and Hill 1998; Ménard 2004).

In species such as *M. nemestrina* and *M. fascicularis*, groups may temporarily split into smaller foraging units that gather at night (Caldecott 1986a, van Schaik and van Noordwijk 1986). Smaller parties may allow individuals to maximize feeding efficiency and cope with the diverse nutritional requirements of individuals. The facultative partitioning documented in *M. sylvanus* and *M. fuscata* supports this argument. The latter two species are characterized by cohesive groups, but instances of temporary partitioning have been reported in conditions of increased food competition because of either overcrowding (Fukuda 1989) or limited food resources (Ménard and Vallet 1996).

It should be added that knowledge about the impact of disease and predation on the size of populations and groups is limited. No information exists about how acute and chronic diseases affect adults and immatures. It is known that moderate intestinal parasitism may not induce heightened mortality, but there are no quantified data about its toll on food intake (Dittus 2004). Cases of predation by leopards, tigers, eagles, crocodiles, and pythons have been reported (Fooden 1986, 1995, 2000). Groups of *M. fascicularis* appear smaller when predators are absent (van Noordwijk and van Schaik 1985). Nevertheless, it is not possible to provide accurate estimations of predation rates in the wild because a majority of studies have been carried out in habitats where large predators had been eliminated (Dittus 2004). Actually, feral dogs and humans represent the primary threat for most macaque populations.

LIFE HISTORY AND REPRODUCTIVE BIOLOGY

The variation in adult body weight is quite large in the genus (Table 14.3). Patterns of growth and reproduction are heavily dependent on food intake. In general, food-enhanced females grow faster, mature earlier, survive longer, and produce more offspring (Lyles and Dobson 1988, Asquith 1989). Reproduction starts prior to full body growth, with females reaching sexual maturity between 2 and 5 years of age depending on species and nutritional condition (Bercovitch and Harvey 2004). The first birth usually occurs between 4 and 6 years (Table 14.3), but it may be delayed beyond 10 years of age when food is scarce. Females nurse infants until 6–12 months of age. Mean interbirth interval varies from 1 to 2 years. Gestation length averages 24 weeks, with 1 further week for heavier species and 1 less week for lighter ones (Table 14.3). Female reproduction ceases around the age of 20–25 years. Though maximum life span ranges 25–40 years, in free-ranging conditions few females live long enough to become postreproductive (Bercovitch and Harvey 2004; see also Chapter 31).

In some macaques, females display cyclical changes in coloration and swelling of the sexual skin in the anogenital region and the rump (see Chapter 29). The genus is notable for its diversity regarding these changes (Table 14.3). In all species of the *silenus–sylvanus* lineage, adult females display an enlarged anogenital swelling, which develops gradually during the follicular stage of the ovarian cycle. This characteristic has been lost to various degrees in the other groups. In the *fascicularis* lineage, the perineum reddens during estrus but only adolescent females display some swelling of the sexual skin. In most species of the *sinica–arctoides* lineage, no changes of the female appearance are visible during the cycle (Dixson 1998, Bercovitch and Harvey 2004). The occurrence of sexual swellings in a species is generally associated with the formation of multimale groups, its arousing effect on males potentially raising competition among them. The absence of sexual swellings in some lineages remains unexplained, however.

An extension of the follicular period might favor the development of signals advertising the female reproductive state (Aujard et al. 1998). Females show a long-lasting ovarian cycle in species displaying voluminous swellings (Table 14.3). Macaque females may display a second conspicuous sexual signal, the estrous call. This staccato vocalization is triggered by copulation in some macaque species; it additionally occurs outside the mating context in others, and it

Table 14.3 Life History Traits of Macaques

SPECIES	MALE BODY WEIGHT (KG)[1]	FEMALE BODY WEIGHT (KG)[1]	SEXUAL DIMORPHISM[2]	SOCIONOMIC SEX RATIO (F/M)[3]	FIRST BIRTH (YEARS)[4]	GESTATION LENGTH (DAYS)[4]	OVARIAN CYCLE LENGTH (DAYS)[4]	SEX SKIN SWELLING[4]	ESTROUS CALL[5]	COPULATION PATTERNS[6]	BIRTH SEASONALITY[7]
Macaca sylvanus	16.0	11.0	0.37	1.1	5	165		Large	C, NC	SM, MM[8]	Discrete
M. silenus	8.9	6.1	0.38	9.9	3.9	172	32	Large	C, NC	MM	Year-round
M. nigra	9.9	5.5	0.59	3.5		176	36	Large	C	MM	Year-round
M. nigrescens								Large	None		
M. hecki								Large			
M. tonkeana	14.9	9.0	0.50	1.1[9]	5	176	37.4	Large	C, NC	MM	Year-round
M. maurus	9.7	6.1	0.46	4.0	6–7	176	36.2	Large	None	MM	Year-round, birth peak
M. ochreata								Large			
M. brunnescens								Large			
M. nemestrina	11.2	6.5	0.54	6.3	3.9	171	38	Large	C, NC	MM	Year-round, birth peak
M. pagensis								Large			
M. sinica	5.7	3.2	0.58	2.4	5	168		None	None	SM	Birth peak
M. radiata	6.7	3.9	0.54	1.7	4	168	27.8	None	None	SM	Discrete
M. assamensis	11.3	6.9	0.49	2.4			32.0	Adolescence[10]		SM	Seasonal
M. thibetana	18.3	12.8	0.36	1.9	5.5		26.4	None	None	MM	Discrete
M. arctoides	12.2	8.4	0.37	5.7	4.9[11]	175	29.5	None	None	SM	Year-round
M. fascicularis	5.4	3.6	0.41	4.8	5.2	163	33.5	Adolescence[10]	C	SM, MM	Birth peak
M. mulatta	11.0	8.8	0.22	2.9	4	167	28.3	Adolescence[10]	None	MM	Discrete
M. fuscata	11.0	8.0	0.32	1.3	5	173	26.5	Adolescence[10]	NC	MM	Discrete
M. cyclopis	6.0	4.9	0.20	1.7	4–6	163	29.4	Adolescence[10]		MM	Discrete

Sources:

[1] Smith and Jungers 1997.

[2] Weight dimorphism is calculated as the natural logarithm of the male body weight minus that of the female body weight.

[3] Singh and Sinha 2004.

[4] Bercovitch and Harvey 2004.

[5] Oi 1996, Aujard et al. 1998, Thierry et al. 2000a. C, copulatory call; NC, noncopulatory call.

[6] Soltis 2004. SM, single mount; MM, multiple mount.

[7] Bercovitch and Harvey 2004: birth peak means a period of the year in which a high proportion of births, but not all births, are concentrated, contrary to discrete periods, in which all births are confined.

[8] Ménard personal communication.

[9] Pombo et al. 2004, Riley 2007.

[10] Swelling occurs in adolescent females only.

[11] Fooden 1990.

is absent in a third category of species (Table 14.3). The rise of female sexual displays is tracked to an increase of estrogen levels. Several functions have been proposed for their occurrence: signaling fertility, indicating female quality, promoting intermale competition, and facilitating paternity confidence or paternity confusion (Dixson 1998, Soltis 2004). In several species, the climax of sexual displays corresponds with the periovulatory period of adult females; this provides males with reliable cues that allow them to concentrate their mating effort within the female conception time (Aujard et al. 1998, Bercovitch and Harvey 2004, O'Neill et al. 2004, Engelhardt et al. 2005, Heistermann et al. 2008). In theory, females should be selective in their choice of mating partners. Until now, however, there is little evidence that sexual selection has favored choosiness in female macaques (Small 1990, Bercovitch 1997, Nikitopoulos et al. 2005, Engelhardt et al. 2006, Brauch et al. 2008).

Puberty starts in males at around 3–4 years of age, at the time of testicular enlargement; rates of agonistic interactions with other males increase, and in the next 2 years adolescents undergo a rise in weight and testosterone levels associated with an elevation in dominance rank (e.g., Nieuwenhuijsen et al. 1988, Bercovitch 1993). Like females, growth and sexual maturation in males are both linked to feeding and social conditions. In *M. mulatta*, a high dominance status provided by maternal protection accelerates maturation (Bercovitch 1993). In *M. fuscata*, the age at testicular descent may be delayed by 1 or 2 years when food supply is limited (Nigi et al. 1989). Age at reproduction is significantly delayed relative to sexual maturation in males. Most have to wait until 7–11 years of age to reach full body size and compete with other adult males for access to reproductive females.

Macaque species differ by their patterns of copulation. In some, males need only one mount to ejaculate, whereas others show multimount ejaculatory patterns (Table 14.3). *M. arctoides* is featured by a further peculiarity: After ejaculation, the male sits and brings the female onto his lap while the intromission continues over the next 1–2 min (Fooden 1990). The significance of such variation remains ill-understood (see Soltis 2004).

Other differences among macaques relate to the periodicity of reproduction. In species from temperate regions, breeding is annual (Table 14.3). A discrete mating season centered on fall produces a discrete birth season around spring. In *M. mulatta* and *M. fuscata*, testis volume, testosterone levels, and spermatogenesis increase at the onset of the mating season and the face of males and their scrotal area redden (Nigi et al. 1989, Lindburg 1983, Bercovitch and Harvey 2004); also, females may pay more attention to red-colored males in *M. mulatta* (Waitt et al. 2003). Females are seasonally polyestrous, several of them coming simultaneously into estrus. No male can monopolize all of them, and mate guarding is limited. Associations between males and females typically last some hours, but this may reduce to momentary encounters or, on the contrary, extend to several days. Both males and females mate with multiple partners. Dominant males can interfere in the mounts of subordinates and chase them away from females. Nevertheless, males do not fight for access to females. The latter may express mate choice, favoring some males and rejecting others; and sneaky copulations are not rare (Lindburg 1983, Huffman 1991, Manson 1992, Bercovitch 1997, Soltis et al. 2001, Hayakawa 2007). A similar diversity in sexual relationships is documented in *M. thibetana* and *M. radiata*, albeit with a somewhat higher degree of tolerance among males (Glick 1980, Zhao 1996). In *M. sylvanus*, the lowest-ranking males have to attract females in remote places to mate, but medium-ranking males may copulate in full view of the highest-ranking males; associations with females are brief, the latter commonly shifting from one partner to another every quarter- or half-hour (Taub 1980a). Priority of access and furtive mating are not the only tactics open to males. Males can compete by endurance rivalry. High energetic expenditures over extended periods of time lead some males to forego mating opportunities. In *M. mulatta*, males increase their fat reserves before the mating season; they may experience a loss of 10% of their body weight during the mating season (Bercovitch 1997; see also Cooper et al. 2004b). There is also evidence that paternity can be influenced at the postcopulatory level through sperm competition (Soltis 2004, Brauch et al. 2008).

In species living in the tropics, reproduction occurs year-round (Table 14.3). When there is no more than one sexually receptive female at the time, males and females form long-lasting consortships in which one male typically follows and mates with the female during days or weeks, excluding other males from reproduction. In such a case, the close control exerted by the male leaves little room for female mate choice, as observed in a group of *M. tonkeana* (Aujard et al. 1998, Thierry and Gachot-Neveu 2004). In this species, male dominance rank functions as a queue for mating opportunities; top-ranking males have priority of access to fertile females and exclude rivals from reproduction. When the number of males is especially high, however, some of them form coalitions against top-ranking males and the queuing system breaks down (Thierry and Gachot-Neveu 2004).

The *operational sex ratio*, that is, the ratio of adult males to fertile females, appears to be a determinant factor in the choice of mating tactics. In several species, mounts are more frequent during a particular period of the year, and limited peaks of births are reported (Table 14.3); there can be one or several fertile females depending on time, female defensibility, and the extent of consortship, exclusivity being consequently variable. In *M. fascicularis* and *M. nemestrina*, the duration of associations between males and females may be brief or long; consorting females may copulate with further partners if the attention of the male is diverted by another female or if he is displaced by a higher-ranking male (van Noordwijk 1985; Caldecott 1986b; Fooden 1995; Oi 1996; Engelhardt et al. 2004, 2006). In captive *M. arctoides*, subordinate males can mount fertile females only when out of sight of the higher-ranking male, but neither consortships nor reproductive seasonality has been reported (Nieuwenhuijsen et al. 1986).

DEMOGRAPHIC STRUCTURE AND GENE FLOW

Macaques live in bisexual social groups where the socionomic sex ratio is in favor of females (Table 14.3). The ratio is elevated in *M. nemestrina* and *M. silenus*, for example, whereas it is quite balanced in *M. sylvanus*. Although sexual dimorphism is commonly held to arise from mating competition, there is no direct relation between sex ratio and sexual dimorphism (see Plavcan 2001). The level of competition may, however, differ between species according to the extent of monopoly of males over females (Oi 1996, Thierry et al. 1996, Paul 2004, Thierry 2007). In general, the number of offspring sired by males weakly correlates with their dominance ranks in seasonally breeding macaques. This should not come as a surprise given the diversity of mating tactics that rivals may use to break down the dominants' priority of access to females. When the number of females cycling simultaneously is limited, alternative mating tactics are less effective than overt competition and higher-ranking males gain more paternity of offspring (Soltis et al. 2001). The variance in the reproductive success of males should be especially strong when mating occurs year-round. In three groups of tropics-living species, a positive correlation was indeed found between dominance and paternity (wild *M. fascicularis*, de Ruiter et al. 1994; captive *M. arctoides*, Bauers and Hearn 1994; captive *M. tonkeana*, Gachot-Neveu et al. 2004). If future studies confirm that priority of access best applies to year-round breeders (see Ostner et al. 2008), the genetic structure of macaque groups would then be contrasted. In year-round breeders, each age cohort would represent a separate paternal sibship, contrary to seasonal breeders, in which offspring may have several fathers at each generation.

Competition and dispersal make the variance in reproductive success greater in males than in females. The latter usually remain and reproduce in their natal group throughout life, notwithstanding some exceptions (Sinha et al. 2005). The lifetime reproductive success of females primarily depends on their longevity, and social rank has no significant impact on it (in provisioned groups at least) (Fedigan et al. 1986, Bercovitch and Berard 1993, Paul and Kuester 1996a; see also van Noordwijk and van Schaik 1999 for limited effects of dominance in wild *M. fascicularis*). By contrast, most males break their social ties and emigrate. Young males commonly disperse at sexual maturation; they often leave in the company of same-aged peers. In *M. fuscata*, *M. mulatta*, and *M. cyclopis*, a majority of males have left their natal group by 5 years of age and may join small all-male bands or be solitary for months or even years (Drickamer and Vessey 1973, Sugiyama 1976, Hsu and Lin 2001). In other species, first dispersal occurs at least 1 or 2 years later, all-male bands are rarely reported, and males may directly transfer into another group within 1 day or spend months solitary or semisolitary (*M. fascicularis*, van Noordwijk and van Schaik 1985; *M. radiata*, Simonds 1992; *M. sinica*, Dittus 1975; *M. sylvanus*, Ménard and Vallet 1996; *M. thibetana*, Zhao 1996; *M. maurus*, Okamoto et al. 2000). Dispersal is not limited to younger individuals; males continue to shift groups during their whole life. This secondary dispersal may occur every 2–5 years. Though a drop in rank or a lack of mating opportunities in the home group sometimes prompts a male to leave, even high-ranking males regularly transfer between groups (Sugiyama 1976, Zhao 1996, Berard 1999).

Males often immigrate into groups adjacent to their original group, and it is not rare that they find ex-groupmates in the new group (Sugiyama 1976; Melnick et al. 1984b, van Noordwijk and van Schaik 1985, Mehlman 1986, Zhao 1996). Two tactics are open to newcomers: They may either settle at the bottom of the male hierarchy and increase rank with tenure or challenge the highest-ranking males and directly attain the top of the hierarchy (van Noordwijk and van Schaik 2004, Thierry 2007). The first tactic is unobtrusive; it is used by males of any age. The second one is risky, and it is mainly reported among full-grown males at their prime (van Noordwijk and van Schaik 1985, 2001; Sprague 1992, 1998). Tactical choices are commanded by the strength of individuals but also by the demographic context, a main parameter of which may be the operational sex ratio. Males usually enter as subordinates into groups of *M. mulatta* and *M. fuscata*; this might be typical of seasonal breeders (see Drickamer and Vessey 1973; Sprague 1992, 1998; Berard 1999). By contrast, aggressively outranking the top-ranking males is the only way to reproduce when priority of access to females represents the prevailing mating tactic. Challenge rates appear higher in smaller groups of *M. fuscata*, in which the dominant males could control females more effectively (see Sprague 1992). The relatively high rate of outranks in *M. fascicularis* may be explained by the fact that the top-ranking males sire the majority of offspring in this species (de Ruiter et al. 1994, Engelhardt et al. 2006). Take-overs also seem to be the rule in groups of *M. silenus*, which usually contain no more than one adult male (Kumar et al. 2001). *M. tonkeana* provides one further case. In a captive group of this species, the highest-ranking male most of the time had sexual monopoly over fertile females. To obtain sexual access to females, other males had no choice but to challenge him directly (Thierry and Gachot-Neveu 2004). If the occurrence of aggressive outranking among male rivals depends on the extent of female defensibility, it should be more frequent in nonseasonal breeders than in seasonal ones (Thierry 2007).

Solitary life and immigration increase the likelihood of succumbing to predation, starvation, and injuries. Such costs should then be offset by reproductive benefits (Dittus 2004, Bercovitch and Harvey 2004). It is likely that dispersal increases mating opportunities. As a general finding, females prefer to mate with novel males, which contributes to maintaining genetic diversity (Huffman 1991, Bercovitch 1997, Sprague 1998, Berard 1999, Jack 2003b, Singh et al. 2006; but see Manson 1995). Regardless of the dominance rank at which they join a group, newcomers display elevated rates of copulation. Correlatively, the reproductive activity of males decreases with increasing tenure length. This may be an effect of a lower sexual attractiveness of familiar males when female mate choice matters. It may be also the outcome of an age-related decline in fighting ability whenever overt competition is the main determinant of reproductive success.

A significant amount of gene flow may additionally occur through extragroup copulations. In all species documented so far, mating occurs between females and male visitors (Dittus 1975, Mehlman 1986, Simonds 1992, Sprague 1992, Oi 1996, Engelhardt et al. 2006, Singh et al. 2006, Hayakawa 2007). The visitors may be either bachelors or males that move to other groups for some hours or days, then return to their original group. They are mainly attracted by sexually active females. Among seasonal breeders, the mobility of males peaks at the mating season. In *M. fuscata*, it sometimes happens that nonresident males invade groups during this season and become involved in as much as one-third of the resident females' copulations (Sprague 1992). Nonresident males actually sire a significant number of offspring (Soltis et al. 2001).

Whereas male transfer tends to homogenize the genetic structure of populations, the differential reproduction of males is liable to increase within-group relatedness (de Ruiter and Geffen 1998). Another phenomenon liable to augment intergroup genetic differentiation is group fission. This is a rare but important event, usually preceded by an increase in group size. The division of a large group into smaller groups follows genealogical lines, maternal relatives staying in the same groups. Such nonrandom splits create daughter groups, which are genetically distinct. By contrast, patrilineal relatives are randomly distributed into the new groups (Gachot-Neveu and Ménard 2004). In many cases, females promote fission; lower-ranking kin subgroups join in one group, whereas higher-ranking kin subgroups form another group. In *M. maurus*, however, the dominance status of females does not influence group membership (Okamoto and Matsumura 2001).

The past history of groups and the regular transfer of individuals make their relationships complex. Individuals meeting after a transfer may be strangers or ex-groupmates. It is understandable that encounters between groups have quite variable outcomes among macaques, going from peaceful to neutral or agonistic, as in many other primates. When conflicts occur, subadult and young adult males represent the main category of individuals involved (Cooper et al. 2004b). Larger groups usually supplant smaller ones, which may have significant effects regarding access to the food supply (Dittus 2004). In *M. nigra*, the rate of intergroup conflicts was found to increase when fruit availability decreased (Kinnaird and O'Brien 2000).

SOCIAL BEHAVIORS AND RELATIONSHIPS

Dominance and kinship relations structure macaque societies. Decades of study have documented this point in *M. mulatta* and *M. fuscata* (Kawamura 1958, Kawai 1958, Missakian 1972, Sade 1972, Kurland 1977, Chapais 1988b, Datta 1992). They form strong hierarchies in which higher-ranking individuals may be recognized by their steady walk and held-up tail carriage. By contrast, lower-ranking individuals are prompt to flee and avoid confrontations and use a particular facial expression to submit, the silent bared-teeth display; by retracting the lips and exposing the teeth, they formally acknowledge their lower dominance status relative to higher-ranking individuals (de Waal and Luttrell 1985). Dominance asymmetry determines who may interact

with which others. It affects how an individual chooses partners for proximity, affiliation, or play and whether the distribution of choices is skewed in favor of higher-ranking individuals. Dominance status partly rests upon individual power, that is, on factors such as physical strength, personality, and experience. Differences in dominance status also depend on variations in social power, that is, the strength of groupmates that an individual can summon to support him or her (Kawai 1958; Chapais 1988b).

Third-party interventions in conflicts have a decisive role in social relationships. Since individuals are all the more prone to help mates in conflicts as they are nearer relatives, this results in the formation of alliance networks based on kinship. These networks constitute subgroups of kin-related individuals called *matrilines* that descend from one female and represent the core of macaque social groups. In this female-bonded system, mothers, daughters, and sisters maintain strong ties during their lifetime; they are often found together, frequently groom each other, and offer mutual support in conflicts. Strict rules of inheritance determine the acquisition of dominance rank within matrilines (Kawamura 1958, Datta 1992). By adulthood, females achieve a dominance rank just below their mothers and rarely outrank them. Because females choose to support their youngest relatives in contests, dominance ranks are ordered inversely to age within matrilines, and younger daughters dominate their elder sisters. Matrilines are themselves ordered along a linear hierarchy. Chapais (1988b) has experimentally shown that the ranks of matrilines depend on the number of kin able to support each other. It should be added that, to apply, rank inheritance needs the occurrence of strong matrilines. In small groups and/or when the demographic rates do not allow females to have many adult relatives, rules of rank inheritance become less reliable (Datta 1992, Hill and Okayasu 1995).

The picture drawn from *M. mulatta* and *M. fuscata* does not strictly apply to other macaque species. In several species, social relationships are characterized by higher levels of tolerance between dominants and subordinates, and the record provides regular exceptions to the previous rules of rank inheritance (Angst 1975, Silk et al. 1981a, Nieuwenhuijsen et al. 1988). In some species, the exceptions are so numerous that the rules are no longer valid. In *M. sylvanus*, adult daughters often outrank postreproductive mothers and females are usually subordinate to their older sisters (Paul and Kuester 1987; Prud'homme and Chapais 1993b). Whereas the society remains based on mutual support between kin-related partners, matrilines are more open and the gradient of dominance is weaker in species such as *M. sylvanus*, *M. arctoides*, *M. radiata*, *M. maurus*, and *M. tonkeana*; this favors social tolerance and the development of alliances even between nonrelatives (e.g., de Waal and Luttrell 1989, Thierry et al. 1994, Chapais 1995). In *M. sylvanus* and *M. silenus*, the silent, bared-teeth display may have a positive meaning or express subordination according to the context (Preuschoft 1995). In Sulawesi macaques (e.g., *M. nigra*, *M. tonkeana*, *M. maurus*), the same display does not have a communicative function about dominance status; it signals the sender's peaceful intentions and is used like a smile to initiate affiliative interactions (Thierry et al. 1989, Petit and Thierry 1992).

The degree of kin bias in social relationships is a main variable of macaque societies. According to Chapais and colleagues (1997), there is a relatedness threshold for kin preference. In *M. fuscata*, maternal relatives are favored until a degree of relatedness equal to 0.25, that is, those corresponding to half-sibling and grandmother/granddaughter bonds. This number also applies to the avoidance of sexual interactions between relatives. The authors propose that the individuals' ability to recognize maternal kin only would not allow them to go beyond the 0.25 threshold. In *M. sylvanus*, however, mating inhibition occurs in niece/uncle and aunt/nephew dyads (maternal relatedness = 0.125) and even between cousins (maternal relatedness = 0.063) (Kuester et al. 1994). This discrepancy may reflect a different kin bias in the relationships of *M. fuscata* and *M. sylvanus*, the former being more nepotistic than the latter. It is worth adding that the social networks of mothers and offspring show a significant degree of similarity in *M. mulatta* and *M. fuscata* (de Waal 1996, Berman and Kapsalis 1999, Schino et al. 2004). On the one hand, this points to the possible occurrence of an intergenerational transmission of social preferences. On the other hand, relatives spend much time in proximity to each other in these species, and their association may account as well for the similarity of mothers and offspring in partner preferences. In fact, no shared partner preferences were found in less nepotistic species for which grooming associations were studied (Schino et al. 2004).

Adult social relationships influence the socialization of immatures. In macaque species typified by strong kin bias and strict hierarchies, mothers appear quite protective; they frequently retrieve their infants and restrict their interactions mostly to relatives. In species displaying more balanced relationships, the mother is quite permissive, alloparental care is common, and even nonkin females may handle and carry infants from an early age (Thierry 2004). Play fighting is also more competitive in a hierarchical species such as *M. fuscata* compared with more tolerant species such as *M. tonkeana* and *M. nigra* (Petit et al. 2008, Reinhart, C.J., Thierry, B, Gauthier, C.A., Pellis, V.C., and Pellis, S.M., unpublished data). Adult males rarely take care of infants in macaques, but there are some noteworthy exceptions. In *M. sylvanus* and *M. thibetana* in particular, males display a strong commitment toward them, providing active support and frequently handling infants. They use infants to regulate their relations: Typically, an adult male picks and carries an infant so as to approach another male and engage in an affiliative interaction (Deag 1980, Ogawa 1995). No correlation was found between male caretaking and paternity in *M. sylvanus*. The behavior of males cannot be held as paternal investment; neither is it a successful strategy, which

would increase the chances of mating with the mother (Paul et al. 1996, Ménard et al. 2001). The interest of males for infants may be a side effect of their exploitation in males' interactions (Paul et al. 1996, Thierry 2000).

Affinitive associations between adult males and females may develop outside the mating season. Such "friendships" are characterized by frequent proximity and grooming between partners. It is reported in *M. fuscata* (Takahata 1982a), *M. mulatta* (Chapais 1986; Manson 1994b), and *M. sylvanus* (Paul et al. 1996, Ménard et al. 2001). There is little correspondence between male/female associations and paternity, however. This finding is consistent with the fact that males do not care for mothers' infants as part of their mating effort and that individuals preferentially choose sex partners among less familiar conspecifics. Support in conflicts may be the main benefit drawn by males and females.

Levels of social tolerance are relatively low between adult males in *M. mulatta* and *M. fuscata*, and coalitions almost never occur. In other species, males reduce social tension through affiliative contacts; they establish tolerant social relationships, and they can form coalitions (Thierry 2000, Berman et al. 2006, Cooper and Bernstein 2008). Support in conflict between males is uncommon in *M. fascicularis* (van Noordwijk and van Schaik 2001), whereas it frequently occurs in other species (*M. radiata*, Silk 1992a,b; *M. sylvanus*, Widdig et al. 2000; *M. thibetana*, Berman et al. 2007; *M. tonkeana*, B. Thierry unpublished data). Most coalitions aim to reinforce current hierarchy, and fair levels of reciprocity in support have been reported. Forming coalitions to take over consortships with estrous females occurs in Tonkean macaques (Thierry and Gachot-Neveu 2004) but is rarely observed in other species (Taub 1980a, Berman et al. 2007).

SOCIAL STYLES AND COVARIATION

The comparative study of agonistic interactions in captive groups of macaques has yielded marked contrasts between species (Thierry 2000, Thierry et al. 2008). In *M. mulatta* and *M. fuscata*, most conflicts are unidirectional, high-intensity aggression is common, and reconciliations are not frequent, the conciliatory tendency (see Chapter 41) among unrelated individuals scoring 4%–12%. These results depart from those found in Sulawesi macaques (*M. nigra*, *M. tonkeana*, and *M. maurus*), in which a majority of conflicts are bidirectional; that is, most aggressive acts induce protests or counterattack. Aggression is generally of low intensity, and measuring rates of reconciliation yields especially high values: Conciliatory tendencies score around 50% among unrelated partners. Other macaques are intermediate between the previous species. *M. fascicularis* and *M. nemestrina* are more similar to *M. mulatta* and *M. fuscata*, whereas *M. arctoides*, *M. sylvanus*, and *M. silenus* tend toward the Sulawesi macaques. Behavioral constraints may account for the covariation of aggression patterns and reconciliation rates at a functional level. On one side, asymmetric conflicts and increased risk of injury may inhibit the occurrence of affiliative contacts between opponents, while approximate symmetry and uncertainty about outcomes create room for negotiation (Silk 1997). On the other side, conciliatory behaviors may reduce the probability of conflict escalation by facilitating information exchange between adversaries (de Waal 1986c, Aureli et al. 2002, Thierry et al. 2008).

Macaque species may be set along a four-grade scale based on their patterns of aggression and reconciliation (Table 14.4) (Thierry 2000). Species from grades 3 and 4 display high rates of affiliative contacts such as clasps and embraces; they are also characterized by the development of behaviors that reduce social tension, such as the use of infants in *M. sylvanus*, ritualized biting in *M. arctoides*, and peaceful interventions by third parties in conflicts among *M. tonkeana* and *M. nigra*. Such behavior patterns are absent in species from grade 1 (*M. mulatta* and *M. fuscata*) and occur at relatively low rates in species from grade 2 (*M. fascicularis* and *M. nemestrina*).

The coupling between social characters may stem from a link between the various traits of the individual, either at the genetic level (e.g., gene pleiotropy and linkage disequilibrium) or the phenotypic level (e.g., trade-offs and structural constraints). It may also arise through the mutual influence of individuals at the social level. This induces constraints which favor some behaviors while forbidding others, as specified for agonistic interactions. In view of the numerous constraints liable to link the characters of a social organization, the covariation hypothesis states that any significant variation in a single character induces a set of correlated changes (Thierry 2004). The contrasts observed in aggression and conciliation patterns extend to a consistent suite of interspecific variations. Many of the differences previously reported among macaques at the level of dominance asymmetry, the degree of kin bias, the modes of socialization,

and the patterns of male dispersal make sense from such a viewpoint (Table 14.4).

At the level of the individual, interspecific variations in temperament have the potential to explain many differences in social behaviors. Consistent differences have been found between species in response to stress and novelty when measured by arousal, exploration behaviors, corticosteroid levels, and heart rates. *M. mulatta* and *M. fascicularis* are less explorative with regard to their physical environment when compared with *M. silenus* or *M. tonkeana* (Thierry et al. 1994, Clarke and Boinski 1995). As a general rule, species from grades 3 and 4 are less easily aroused than are species from grades 1 and 2 (de Waal 1989c, Clarke and Boinski 1995). It is likely that macaque species differ in a number of psychobiological variables. For instance, lower serotoninergic activity is associated with higher aggression intensity in *M. mulatta* compared with *M. nemestrina* (Westergaard et al. 1999).

At the social level, the linkage between the degree of nepotism and the gradient of dominance may be accounted for by the frequent occurrence of coalitions within groups of macaques (Thierry 2004). When most alliances involve relatives, the dominance status of individuals primarily depends on the power of the kin subgroup to which they belong. This increases rank differences between nonrelatives and further develops kin alliances, generating group structures based on strong hierarchies. Conversely, when kin bias is less pronounced, coalitions involving nonrelatives are more common, dominance appears to be more a question of individual attributes, and the individual retains some degree of freedom with regard to power networks. Dominance relationships remain balanced among group members, and close ties exist even between nonrelatives. Comparisons between species support the view that dominance and nepotism among females are connected by positive feedback (Aureli et al. 1997, Demaria and Thierry 2001, Thierry et al. 2008).

Table 14.4 Tentative Scaling of Macaque Social Styles

GRADE 1	GRADE 2	GRADE 3	GRADE 4
M. mulatta	*M. fascicularis*	*M. arctoides*	*M. tonkeana*
M. fuscata	*M. nemestrina*	*M. sylvanus*	*M. maurus*
(*M. cyclopis*)	(*M. thibetana*)[1]	*M. silenus*	*M. nigra*
	(*M. assamensis*)[2]	*M. radiata*	*M. brunnescens*[3]
		(*M. sinica*)	(*M. ochreata*)
			(*M. hecki*)
			(*M. nigrescens*)
			(*M. siberu*)[4]

Source: Thierry 2000.

Species are ordered mainly based on conciliatory tendency and social tolerance, which increase from the left (grade 1) to the right (grade 4), and based on asymmetry of conflicts, dominance gradient, and kin bias, which decrease from the left to the right. The least-known species are indicated in parentheses; their location on grade is predicted from a few behavioral characters or studies, e.g., [1] Berman et al. (2004, 2006), [2] Cooper and Bernstein 2002, 2008; [3] Slater 2002, [4] Abegg and Thierry 2002b.

A number of characters thus appear interrelated: individual temperament, rate of conciliation behaviors, level of aggression asymmetry, gradient of dominance, degree of nepotism, degree of mother permissiveness, amount of female alloparental care, form of social play, patterns of female rank inheritance, and male dispersal and coalitions. Each species is featured by a given set of characters that may label its dominance style (de Waal and Luttrell 1989) or more generally its social style (Thierry 2004). Direct comparisons have been made only for a limited sample of species, groups, and characters; and intraspecific variability calls for caution (Thierry 2000, Ménard 2004). Rating species along a discrete and bipolar scale is an idealization. Although each species is assigned to one grade, a more accurate picture would represent the various study populations of each species using a cluster of points centered on one modal location and would allow for overlaps with other clusters centered on neighboring locations.

PHYLOGENETIC INERTIA AND THE SOCIOECOLOGICAL MODEL

The finding that macaque social styles represent covariant sets of characters indicates that they belong to a single family of forms. By limiting the changes possible to social organizations, interconnections between characters act as constraints that channel evolutionary processes and allow only a subset of organizations to arise. Thus, it should not come as a surprise that the three phyletic lineages of macaques have a different distribution on the four-grade scale (compare Tables 14.1 and 14.4). Variations in the social styles of macaques significantly correlate with their phylogeny (Matsumura 1999, Thierry et al. 2000b). By tracing each of the characters on the phylogenetic tree of macaques, it is possible to recognize their most ancient states and reconstruct their typical ancestral organization. The resulting set of characters closely matches grade 3 on the scale, which may be tentatively considered the ancestral state (Thierry et al. 2000b). The fact that *M. sylvanus* and *M. silenus* are grade 3 reinforces the previous finding since both species come closer to the root of the phylogenetic trees established from morphological and molecular data. The other species of the first lineage have diverged either on grade 4 (Sulawesi macaques) or on grade 2 (*M. nemestrina*). Members of the second lineage have remained on grade 3, except *M. assamensis* and *M. thibetana*, which would have diverged on grade 2. The third lineage evolved toward grades 1 and 2. The location of every lineage is mostly restricted to two grades in the scale. The good match between phylogeny and the four-grade scale of macaques points to the occurrence of phylogenetic inertia in their social patterns, indicating that the core of the species-specific systems of interconnections underwent limited changes during several hundred thousands or even millions of years in several taxa.

The powerful action of internal attractors pointed out by phylogenetic inertia does not mean that external factors play no role in the determinism of social relationships. Attempts have been made to correlate the social styles of macaques with the main ecological features of their habitats, predation risk, and food distribution (de Waal and Luttrell 1989, van Schaik 1989, Sterck et al. 1997; see also Caldecott 1986b). According to a basic ecological argument, animals live in groups to protect themselves from predators, whereas group living in turn induces competition between individuals and groups (see Chapters 2, 32, and 39). In mainly frugivorous species such as macaques, females should constitute kin-bonded coalitions to face overt competition that arises for resources within and between groups. If predation risks are high, the costs of leaving the group are elevated for subordinates. As a consequence, dominants take the lion's share and the relationships between unrelated group members are despotic. On the contrary, when predation risks are low, subordinates are not forced to remain in the group. Dominants in this case benefit from subordinates' cooperation against external threats and so must accept a relatively equal exploitation of resources. This condition would produce rather tolerant relationships even among nonkin. Available data on the ecology of macaques do not fit the expectations of the model, however. No correspondence has been found between social styles and between-group competition, food patchiness, or the diet of macaques (Matsumura 1999, Ménard 2004) (see Table 14.2). The ecological tolerance of *M. mulatta* would single-handedly challenge any ecological theory of macaque social organization.

Several reasons may be proposed to explain the failure of the socioecological model. Some relevant ecological factors may still go unidentified. It could also be that social styles have been selected under past ecological conditions and that phylogenetic inertia has maintained them in spite of subsequent ecological changes (Sterck et al. 1997, Ménard 2004). This argument of anachronism is not satisfactory, however. By forbidding any possibility of falsification, it makes the hypothesis of ecological determinism a dogma. Socioecological models commonly envision social organizations as sets of coadapted characters that result from trade-offs between the selective advantages of each of them. Future models should integrate the structural coupling of characters liable to produce phylogenetic inertia. Evolutionary changes depend on a balance between the action of selective processes and the degree of entrenchment of characters. The important role of historical contingencies in the deployment of macaque species may help explain the lack of correlation between ecology and social style (Abegg and Thierry 2002a, Thierry 2008).

CONCLUSION

The study of macaque societies reveals a double-layered organization. On one side, individual behaviors are linked through multiple influences that produce sets of covarying characters. On the other side, patterns of competition and reproduction heavily depend on the more or less seasonal nature of the climate. By influencing the extent of female defensibility by males, the latter is the ecological variable that most directly affects macaque social organization. Both layers are disconnected. A first consequence is that knowing the typical social relationships of a species is not sufficient to predict the outcome of intermale competition. The dominance status has more influence on the reproductive success of males in nonseasonal breeders with limited dominance asymmetry than in species in which asymmetry is marked but where females' fertility is synchronous. A second consequence is that layers may vary at different

speeds. Macaque populations have experienced more than once a change between warm and temperate climate during recent geological time (Eudey 1980, Richard et al. 1989). Switching between seasonal and aseasonal reproduction could represent an easy modular change regarding physiological mechanisms. On the contrary, modifying the social style, and by the way the whole system of social relationships and the individual characters on which they rest, would be more difficult to achieve at the evolutionary level.

Species included in this chapter are shown in Color Plates 21 and 22 (*Macaca sylvanus* and *Macaca nigra*).

15

African Papionins
Diversity of Social Organization and Ecological Flexibility
Larissa Swedell

1. What is meant by ecological flexibility, which African papionins can be best characterized in this way, and what are the consequences of such flexibility with regard to human-wildlife conflict and conservation?
2. How are differences in male and female reproductive strategies across the African papionins related to differences across species in behavioral elements such as coalitions, coercion, and infanticide?
3. What are the similarities in social structure across the African papionins? What are the differences in underlying bonding patterns across this same group, and how are those patterns related (or not) to the observable social structure?

INTRODUCTION

The African papionins are a diverse group of large-bodied monkeys inhabiting a wide variety of habitats across sub-Saharan Africa and displaying a wide array of social systems and ecological adaptations. All are sexually dimorphic, with females about 50%–70% of male size in all except drills and mandrills, in which females are 40% and 30% of male size, respectively (Table 15.1). African papionins are among the most visually striking monkeys, with male geladas, hamadryas baboons, and mandrills as prominent examples. Mandrills are adorned with various shades of red, blue, pink, and purple coloration on the face, perineum, and

genitalia, as well as sternal scent glands, larger in males, demarcated by long, bristly hair (Wickings and Dixson 1992b; Setchell and Dixson 2001a). In drills, these coloration patterns occur on the perineum but not the face (Grubb 1973). The striking coloration of male mandrills as well as the showy manes and variable skin coloration of hamadryas baboons and geladas have likely evolved via sexual selection (Darwin 1871; Jolly 1963, 2007; Jouventin 1975b; Setchell and Dixson 2001b; Setchell 2005).

Habitats of African papionins range from the deserts of Saudi Arabia and Namibia to the tropical rain forests of central Africa. *Lophocebus* mangabeys and kipunjis are mainly tree-dwelling and use a combination of

Table 15.1 Body Size and Sexual Dimorphism

TAXON AND STUDY SITE	MEAN MALE BODY WEIGHT (KG)	RANGE MALE BODY WEIGHT	MALE NUMBER OF SPECIMENS	MEAN FEMALE BODY WEIGHT (KG)	RANGE FEMALE BODY WEIGHT	FEMALE NUMBER OF SPECIMENS	SEXUAL DIMORPHISM (MALE/FEMALE)	SEXUAL DIMORPHISM (FEMALE/MALE)	REFERENCES
Cercocebus									
C. agilis, agile mangabey									
Makokou, Gabon	8.8	4.8–10.0	7	5.4	4.3–5.2 4.3–6.2	5	1.63	0.61	Gautier-Hion et al 1999
N'Gotto Forest, CAR	9.5	9.0–10.0	2	5.7	N/A	2	1.67	0.60	Colyn 1994
Captive (France)	9.7	N/A	1	6.3		2	1.54	0.65	Gautier-Hion & Gautier 1976
C. atys, sooty mangabey									
Taï Forest, Côte d'Ivoire	10.6	9.5–11.4	4	6.2	5.6–7.0		1.71	0.58	McGraw in press
C. chrysogaster, golden-bellied mangabey									
No data									
C. galeritus, Tana River mangabey									
Tana River, Kenya				3.7		1			Wieczkowski & Butynski in press
C. sanjei, Sanje River mangabey									
No data									
C. torquatus, red-capped (or white-collared) mangabey									
Various sites	10.6	8.8–12.5	4	5.8		1	1.83	0.55	Ehardt in press
C. lunulatus, white-naped mangabey									
Captive	9.9	7.1–11.8	7	5.3	3.9–6.3	8	1.87	0.54	Butynski in press
Lophocebus									
L. albigena, gray-cheeked mangabey									
Kibale, Uganda	10.5			7.0			1.50	0.67	Struhsaker and Leland 1979a,b
Kibale, Uganda	8.7	7.5–11.0	31						Olupot 2000
Makokou, Gabon	9.0		5	6.4		6	1.40	0.71	Gautier-Hion and Gautier 1976
N'Gotto Forest, CAR	7.3	6.1–8.35	4	4.9	4.7–5.25	4	1.49	0.67	Colyn 1994
Kisangani, DRC	7.7	5.75–8.65	4	6.5	5.05–7.85	6	1.18	0.85	Colyn 1994
Captive (France)	9.3		2	7.0		4	1.33	0.75	Gautier-Hion and Gautier 1976
L. aterrimus, black (or black-crested) mangabey									
Kisangani, DRC	7.9		1	5.6	4.5–6.7	4	1.40	0.71	Colyn 1994
Mandrillus									
M. leucophaeus, drill									
Bioko Island, Equatorial Guinea	19.8	14.0–27.0	27	8.7	6.5–12.0	36	2.28	0.44	T. Butynski, Y. de Jong, and G. Hearn, pers. comm.

Table 15.1 (cont'd)

TAXON AND STUDY SITE	MEAN MALE BODY WEIGHT (KG)	RANGE MALE BODY WEIGHT	MALE NUMBER OF SPECIMENS	MEAN FEMALE BODY WEIGHT (KG)	RANGE FEMALE BODY WEIGHT	FEMALE NUMBER OF SPECIMENS	SEXUAL DIMORPHISM (MALE/ FEMALE)	SEXUAL DIMORPHISM (FEMALE/ MALE)	REFERENCES
M. sphinx, mandrill									
CIRMF, Franceville, Gabon		30–35			10–15				Wickings and Dixson 1992a,b
CIRMF, Franceville, Gabon	30.9	27–35	10						Setchell and Dixson 2001a
CIRMF, Franceville, Gabon	32.2		9	9.0		21	3.59	0.28	Setchell et al., 2001
Lopé Reserve, Gabon	29.1		12	9.0		7	3.25	0.31	Abernethy and White in press
Papio									
P. anubis, olive (or anubis) baboon									
Awash, Ethiopia	21.2	16.2–26.4	39	11.7	9.3–14.1	35	1.81	0.55	Phillips-Conroy and Jolly 1981
Gilgil, Kenya	24.3	21.2–29	11	12.8	9.8–15.2	30	1.90	0.53	Smuts 1985
Gilgil, Kenya	22.4		56	13.7		64	1.64	0.61	Bercovitch 1987a (females), 1989 (males)
Mara, Kenya	27.1	21.8–32.0	54	14.0	10.9–18.0	23	1.94	0.52	Popp 1983
P. cynocephalus, yellow baboon									
Amboseli, Kenya	25.8		20	11.9		18	2.17	0.46	Altmann et al. 1993
Tana River, Kenya				9.0	9.5–11.0	8			Bentley-Condit and Smith 1997
Tana River, Kenya	21.9		5	10.4		12	2.11	0.47	Bentley-Condit, unpubl. data
Captive (USA)	26.7		43	15.9		40	1.68	0.60	Glassman et al. 1984
P. hamadryas, hamadryas baboon									
Erer Gota, Ethiopia	19.5		3	11.5		14	1.70	0.59	Sigg, pers. comm. to Popp 1983
Harar, Ethiopia	21.0	18.3–24.0	7	11.4	9.5–13.5	13	1.84	0.54	Popp 1983
Awash, Ethiopia	16.9	13.2–24.0	41	9.9	7.3–13.2	39	1.71	0.59	Phillips-Conroy and Jolly 1981
P. papio, Guinea baboon									
Captive (France)	26.0	25–27	2	14.0	7–21	21	1.86	0.54	Galat-Luong and Galat in press

Table 15.1 (cont'd)

TAXON AND STUDY SITE	MEAN MALE BODY WEIGHT (KG)	RANGE MALE BODY WEIGHT	MALE NUMBER OF SPECIMENS	MEAN FEMALE BODY WEIGHT (KG)	RANGE FEMALE BODY WEIGHT	FEMALE NUMBER OF SPECIMENS	SEXUAL DIMORPHISM (MALE/ FEMALE)	SEXUAL DIMORPHISM (FEMALE/ MALE)	REFERENCES
P. ursinus, chacma baboon									
Okavango (Moremi), Botswana	28.8	25.8–35.0	19	15.0	12.2–18.3	26	1.92	0.52	Bulger and Hamilton 1987
Okavango (Moremi), Botswana	26.7	22.6–31.2	9						Kitchen et al. 2003
Honnett, Northern Transvaal, South Africa	26.3			14.5			1.81	0.55	Stoltz and Saayman 1970
Suikerbosrand, Southern Transvaal, South Africa	27.3			15.4			1.77	0.56	Anderson 1981b
de Hoop, South Africa	29.7	26.8–34.1	18	16.1	13.2–20.5	12	1.84	0.54	Barrett and Henzi 1997
Giant's Castle, Drakensberg, South Africa	23.0	20.5–26.4	6	15.9		1	1.45	0.69	Barrett and Henzi 1997
Rungwecebus									
R. kipunji, kipunji (formerly highland mangabey)									
No data									
Theropithecus									
T. gelada, gelada									
Gich Plateau, Ethiopia	26.1		1	14.0	13.1–14.8	2	1.87	0.53	Ohsawa 1979b

DRC, Democratic Republic of Congo; CAR, Central African Republic (all tables).

arboreal quadrupedalism, climbing, and leaping (Horn 1987a, McGraw 1994, Gebo and Chapman 2000); but all other African papionins are terrestrial or semiterrestrial quadrupeds that forage both arboreally and terrestrially.

Baboons of the genus *Papio* are particularly well-studied, with several long-term field projects contributing greatly to our understanding of social behavior, ecology, and evolution. The broad geographic distribution, wide array of habitats, and behavioral diversity of baboons make them particularly useful for comparative studies of socioecology. Baboons have also been used as models of human ecological and social evolution as they are large-bodied, terrestrial, behaviorally flexible omnivores that exploit habitats broadly similar to those of early hominins (Washburn and DeVore 1961; DeVore and Washburn 1963; Jolly 1970c, 2001; Rose 1976; Strum and Mitchell 1987; Plummer 2004; Elton 2006; Codron et al. 2008).

TAXONOMY AND DISTRIBUTION

Systematics and Taxonomy

Together with *Macaca*, the African papionins comprise the tribe Papionini. Recent analyses identify two subclades: one comprising *Papio*, *Theropithecus*, and *Lophocebus* and one comprising *Mandrillus* and *Cercocebus*. *Lophocebus*, the arboreal mangabeys, comprise at least two species and *Cercocebus*, the semiterrestrial mangabeys, comprise seven species or subspecies. The genus *Mandrillus* includes drills and Mandrills. (Table 15.2).

Papio baboons occur in five main forms and at least six variants of these forms and are traditionally classified as the five species listed in Table 15.2 (see references therein). The alternate classification of two species, *P. hamadryas* and four subspecies of *P. cynocephalus* (e.g., Kingdon 1974, Smuts et al. 1987), is not supported by the pattern of morphological

Table 15.2 Taxonomy, Distribution, and Habitats of African Papionins

SPECIES	COMMON NAME	SPECIES GROUP (FOR ALTERNATE CLASSIFICATION AS SUBSPECIES)	SUBSPECIES OR VARIANTS	SUBSPECIES/ VARIANT DISTRIBUTION	SPECIES-WIDE DISTRIBUTION	HABITAT	REFERENCES
Semi-terrestrial mangabeys						Lowland rain forests, coastal and gallery forests, floodplain forests, mangrove forests, and swamp forests; Sanje River mangabey in more mosaic habitats, including miombo woodlands and mixed evergreen and semideciduous forests	Groves 1978, 2001a; Harding 1984; Disotell 1994, 2000, 2008; Fleagle and McGraw 1999; Gautier-Hion et al. 1999; Grubb et al. 2003; Butynski in press; Ehardt in press; Ehardt and Butynski in press a,b; Shah in press; Wieczkowski and Butynski in press
Cercocebus agilis	Agile mangabey	galeritus			From SE Cameroon and NE Gabon through northern Congo (Brazzaville) to southern CAR and northern DRC		
C. chrysogaster	Golden-bellied mangabey	galeritus			Central DRC		
C. galeritus	Tana River mangabey	galeritus			Lower Tana River in Kenya		
C. sanjei	Sanje River mangabey	galeritus			Udzungwa Mountains in Tanzania		
C. atys	Sooty mangabey	atys			Guinea, Sierra Leone, Liberia, and Côte d'Ivoire		
C. lunulatus	White-naped mangabey	atys			Eastern Côte d'Ivoire and Ghana		
C. torquatus	Red-capped or white-collared mangabey	torquatus			Coastal Nigeria, Cameroon, Equatorial Guinea, and Gabon		
Arboreal mangabeys					SE Nigeria east of the Cross River through Cameroon, Equatorial Guinea and Gabon, northern Congo (Brazzaville), southern CAR, NE DRC north of the Congo River and east to Uganda	Low- and medium-altitude rain forests, gallery forests, and some swamp forests	Groves 1978, 2001a; Horn 1987b; Disotell 1994, 2000, 2008; Fleagle and McGraw 1999; Grubb et al. 2003; Olupot and Waser in press
Lophocebus albigena	Gray-cheeked mangabey						
L. aterrimus	Black or black-crested mangabey		L. a. aterrimus (northern black mangabey)	Western and central DRC	Western and central DRC south of the Congo River and NE Angola		
			L. a. opdenboschi (southern black mangabey)	SW DRC and NE Angola			
Drills and mandrills							Sabater Pi 1972; Grubb 1973; Harrison 1988; Rogers et al. 1996; Gadsby and Jenkins 1997-1998; Grubb et al. 2003; Wild et al. 2005
Mandrillus sphinx	Mandrill				Gabon, Equatorial Guinea, southern Congo, and southern Cameroon south of Sanaga River	Mainly primary and secondary dense rain forests but also gallery forests and forest-savanna mosaic, especially during dry season	

Table 15.2 (cont'd)

SPECIES	COMMON NAME	SPECIES GROUP (FOR ALTERNATE CLASSIFICATION AS SUBSPECIES)	SUBSPECIES OR VARIANTS	SUBSPECIES/ VARIANT DISTRIBUTION	SPECIES–WIDE DISTRIBUTION	HABITAT	REFERENCES
M. leucophaeus	Drill		*M. l. leucophaeus* (mainland drill)	South of Cross River in Nigeria and north of Sanaga River in Cameroon	South of Cross River in Nigeria, north of the Sanaga River in Cameroon, and Bioko Island, Equatorial Guinea	Primarily lowland, coastal, and riverine primary rain forests, though also semi-montane and submontane forests up to 2,000 m	See above
			M. l. poensis (Bioko Island drill)	Bioko Island, Equatorial Guinea			
Baboons							Crook and Aldrich–Blake 1968; Kummer 1968a; Altmann and Altmann 1970; Davidge 1978b; Wolfheim 1983; Harding 1984; Kummer et al., 1981; Groves 1993, 2001a; Jolly 1993, 2003, 2007; Disotell 2000; Disotell 2000; Zinner et al. 2001a; Grubb et al. 2003; Galat-Luong et al., 2006; Burrell 2009
Papio anubis	Olive or anubis baboon	*hamadryas*			From Rift Valley in East Africa westward across central Africa to West Africa	Wide range of ecological zones, from savanna and savanna–woodland to tropical dry forests, rain forests, and high-altitude bamboo forests	
P. cynocephalus	Yellow baboon	*hamadryas*	*P. cynocephalus ibeanus* (Ibean or northern yellow baboon)	Somalia, Kenya and northern Tanzania	From SE Somalia and Kenya and SE Tanzania east of the Rift Valley southward across south central Africa (southern DRC, northern Angola and Zambia, Malawi, and northern Mozambique)	Savanna, savanna–woodland, and miombo woodlands	
			P. cynocephalus cynocephalus ('typical' yellow baboon)	Tanzania, eastern Zambia, Malawi, and Mozambique			
			P. cynocephalus kindae ("kinda" or small yellow baboon)	Southern DRC, Angola, and northern Zambia			
P. hamadryas	Hamadryas or sacred baboon	*hamadryas*			Eastern Ethiopia, northern Somalia, Djibouti, Eritrea, and SW Yemen and Saudi Arabia	Mainly lowland, semi-desert regions, though also in montane forests and woodlands in Eritrea and montane regions up to 3000 m above sea level in Ethiopia	
P. papio	Guinea or red baboon	*hamadryas*			SE Mauritania, Senegal, Gambia, Guinea-Bissau, western Guinea, eastern Mali, and northern Sierra Leone	Shrubby savanna, mosaic woodlands, secondary high forest, and mangrove forests	
P. ursinus	Chacma baboon	*hamadryas*	*P. ursinus griseipes* (gray-footed chacma baboon)	Southern Zambia and Mozambique, Zimbabwe, SE Angola, and	Southern Zambia, southern Mozambique, Zimbabwe, Botswana,	Widest range of habitats among African papionins: savanna grasslands and woodlands;	

Table 15.2 (*cont'd*)

SPECIES	COMMON NAME	SPECIES GROUP (FOR ALTERNATE CLASSIFICATION AS SUBSPECIES)	SUBSPECIES OR VARIANTS	SUBSPECIES/ VARIANT DISTRIBUTION	SPECIES-WIDE DISTRIBUTION	HABITAT	REFERENCES
				northern Botswana	southern Angola, Namibia, South Africa, Swaziland, and Lesotho	semiarid to arid deserts in Kuiseb River canyon in Namib Desert; endemic fynbos in South Africa; montane grasslands and temperate forests of Drakensberg	
			P. ursinus ursinus ("typical" or Cape chacma baboon)	SE Botswana, southern Mozambique, South Africa, Swaziland, and Lesotho			
			P. ursinus ruacana (Ruacana or southern chacma baboon)	SW Angola and Namibia			
Kipunji *Rungwecebus kipunji*	Kipunji (formerly highland mangabey)				Mount Rungwe-Kitulo Forests in southern highlands and Ndundulu in Udzungwa Mountains in southern Tanzania	Submontane, montane, upper montane and bamboo forests	Jones et al., 2005; Davenport et al. 2006, 2008; De Luca et al., in press; Olson et al. 2008
Gelada *Theropithecus gelada*	Gelada, gelada baboon, or gelada monkey		*T. g. gelada*	West of Rift Valley	Endemic to Ethiopia; Ethiopian highland plateaux; isolated population further south along banks of Wabi-Shebeli river in Arsi region	Montane grasslands ranging in altitude from 1,500 to 4,500m, characterized by open grassland and patches of woodland	Yalden et al. 1977; Mori and Belay 1990; Iwamoto 1993; Groves 2001a
			T. g. obscurus	East of Rift Valley			

CAR, Central African Republic; DRC, Democratic Republic of Congo.

and genetic variation among populations. Recent approaches have classified all baboons as a single species, *P. hamadryas*, with five or more phenotypically and geographically distinct subspecies (e.g., Williams-Blangero et al. 1990, Groves 1993, Jolly 1993, Disotell 2000, Frost et al. 2003). Taxonomy aside, baboon populations are not reproductively isolated, forming hybrid zones at most species/subspecies borders for which data are available (Phillips-Conroy and Jolly 1986; Samuels and Altmann 1986; Jolly 1993, 2003; Alberts and Altmann 2001; Burrell 2009).

Lophocebus kipunji, the highland mangabey, was discovered in 2005 (Jones et al. 2005) and subsequently assigned a new genus, *Rungwecebus* (and a new name, the "kipunji"), based on stronger affinities with *Papio* than with *Lophocebus* (Davenport et al. 2006, 2008; Olson et al.

2008; Hoffman et al. 2009). Zinner et al. (2009) found the kipunji's mitochondrial DNA to cluster within *Papio* and suggested prior introgression with neighboring baboon populations. It has recently been proposed that the kipunji may be a product of past hybridization between "kinda" yellow baboons (see Table 15.2) and an isolated population of *Lophocebus* mangabeys (Burrell et al. 2009), but more recent DNA sequence data from multiple sites suggests that local introgression, not hybrid origin, explains mitochondrial DNA variation in this taxon (Roberts et al 2009).

Geographic Distribution

Papio baboons are distributed clinally throughout sub-Saharan Africa, from the Horn of Africa and southwestern

Arabian Peninsula to Senegal in West Africa and southward to the Cape of Good Hope (Table 15.2 and references therein). Mangabeys are distributed across equatorial Africa, with *Lophocebus* mainly in forested central Africa and *Cercocebus* patchily distributed from West to East Africa. Drills and mandrills are allopatric in a small portion of forested central Africa, with the Sanaga River dividing their distribution. Geladas are endemic to Ethiopia, while the kipunji is found in only two isolated areas in southern Tanzania. The kipunji (*Rungwecebus kipunji*), the Tana River mangabey (*Cercocebus galeritus*), and the Sanje River mangabey (*Cercocebus sanjei*) have been given critically endangered status in the International Union for the Conservation of Nature (IUCN) Red List of Threatened Species (www.iucn.org). Recent population surveys indicate that there are less than 1,500 members of each of these species remaining (Ehardt et al. 2005; Davenport et al. 2008).

ECOLOGY

Habitat

Papio baboons occupy the broadest array of habitats among African papionins (Table 15.2). Olive, yellow, and chacma baboons are often referred to as "savanna baboons" because most (though not all) populations are found in habitats broadly characterized as savanna-woodlands; other than this ecological association, the term "savanna baboon" has little biological meaning. In a similar vein, populations of chacma baboons inhabiting the Drakensberg Mountains of South Africa are called "mountain baboons" and both

hamadryas and chacma baboons (in Ethiopia and Namibia, respectively) have been termed "desert baboons."

Mangabeys are generally found in more densely forested habitats across equatorial Africa, with *Cercocebus* mangabeys inhabiting mainly riverine and swamp habitats (Table 15.2). Other African papionins are more restricted ecologically, with geladas found only in the montane grasslands of the Ethiopian highland plateaux and drills and mandrills found mainly in primary rain forests in central Africa. The kipunji is limited to montane and bamboo forests in southern Tanzania, which are becoming increasingly fragmented, posing a significant threat to the survival of this critically endangered species.

Diet and Feeding Ecology

Baboons are omnivorous and opportunistic in their feeding, incorporating into their diet a combination of fruit, flowers, seeds, pods, leaves, gum, and underground storage organs from a wide array of plant taxa as well as almost any small animal that they can catch (Hall 1962a; DeVore and Washburn 1963; Rowell 1966; Altmann and Altmann 1970; Harding 1975; Davidge 1978b; Hamilton and Busse 1982; Norton et al. 1987; Whiten et al. 1987, 1991; Altmann 1998) (Table 15.3). Baboons are also adaptable foragers, sometimes dramatically shifting their diets in response to changes in available resources (e.g., Hamilton et al. 1978) and readily feeding on human-produced garbage. Many baboons exploit foods that are difficult to harvest and process, including palm nuts and other fibrous fruits (Hamilton et al. 1978, Swedell et al. 2008) as well as underground plant parts such as corms, bulbs, rhizomes, and tubers that require strength and dexterity to extract (Hall 1962a; Altmann and

Table 15.3 Diet[1]

TAXON AND STUDY SITE	% DIET FRUIT	% DIET NUTS + SEEDS + PODS	% DIET FRUIT + SEEDS	% DIET LEAVES	% DIET FLOWERS	% DIET GRASSES AND HERBS[2]	% DIET UNDERGROUND PLANT PARTS	% DIET FUNGI AND LICHENS	% DIET SAP/ GUM	% DIET HERB, PITHS, STEMS, AND SHOOTS	% DIET BARK, PITH, CAMBIUM	% DIET ANIMAL FOOD	% DIET OTHER (OR UNIDENT.)	NUMBER OF PLANT SPECIES IN DIET	REFERENCES
Cercocebus															
C. agilis, agile mangabey															
Mondika, Ndoki Forest, CAR/Congo	76						2	2		16		5			Shah 2003
Mondika, Ndoki Forest, CAR/Congo	64–76		76												Shah in Press
C. atys, sooty mangabey															
Taï Forest, Côte d'Ivoire			68									26			Bergmüller et al. unpubl. data, cited in Range and Noë 2002
Taï Forest, Côte d'Ivoire	68			2	2			1				26			Bergmüller, cited in McGraw 1998a–c

Table 15.3 (cont'd)

TAXON AND STUDY SITE	% DIET FRUIT	% DIET NUTS + SEEDS + PODS	% DIET FRUIT + SEEDS	% DIET LEAVES	% DIET FLOWERS	% DIET GRASSES AND HERBS[2]	% DIET UNDERGROUND PLANT PARTS	% DIET FUNGI AND LICHENS	% DIET SAP/ GUM	% DIET HERB, PITHS, STEMS, AND SHOOTS	% DIET BARK, PITH, CAMBIUM	% DIET ANIMAL FOOD	% DIET OTHER (OR UNIDENT.)	NUMBER OF PLANT SPECIES IN DIET	REFERENCES
C. chrysogaster, golden-bellied mangabey															
No data															
C. galeritus, Tana River mangabey															
Tana River, Kenya			76	3	2					12		2	5		Homewood 1978
Tana River, Kenya	34	42		2	3				1	4	2	11		58	Kinnaird 1990a
Tana River, Kenya (summary of several studies)	44	32												96	Wieczkowski and Butynski in press
C. sanjei, Sanje River mangabey															
Udzungwa Mountains, Tanzania	57	19		3	2			4		4		6			Ehardt and Butynski in press b
C. torquatus, red-capped (or white-collared) mangabey															
Campo, Cameroon	60	20													Mitani 1989
C. lunulatus, white-naped mangabey															
No Data															
Lophocebus															
L. albigena, gray-cheeked mangabey															
Lopé, Gabon	30	36		4	2							28			Ham 1994
Mondika, Ndoki Forest, CAR/ Congo			76	7	1						3	13			Shah 2003
Dja Reserve, Cameroon	33	29		3–19	3–15						2–3			132	Poulsen et al. 2001
Kibale (Ngogo), Uganda	58				10							25		29	Freeland 1979
Kibale (Kanyawara and Dura), Uganda	59 (35–78)			2–10	1–8						0–9	12–44		65	Waser 1975
Kibale (Kanyawara), Uganda	59			5	5						4			51	Olupot et al. 1998
L. aterrimus, black (or black-crested) mangabey															
Lake Tumba, DRC	57	31		3	4							2		31	Horn 1987a
Mandrillus															
M. leucophaeus, drill															
Bioko Island, Equatorial Guinea	58			16% leaves, stems, and other nonfruit plant parts								26			Gonzalez-Kirchner and Sainz de la Maza 1996
M. sphinx, mandrill															
Campo, Cameroon	84											8		113	Hoshino 1985 (from fecal samples)

Table 15.3 (cont'd)

TAXON AND STUDY SITE	% DIET FRUIT	% DIET NUTS + SEEDS + PODS	% DIET FRUIT + SEEDS	% DIET LEAVES	% DIET FLOWERS	% DIET GRASSES AND HERBS[2]	% DIET UNDERGROUND PLANT PARTS	% DIET FUNGI AND LICHENS	% DIET SAP/GUM	% DIET HERB, PITHS, STEMS, AND SHOOTS	% DIET BARK, PITH, CAMBIUM	% DIET ANIMAL FOOD	% DIET OTHER (OR UNIDENT.)	NUMBER OF PLANT SPECIES IN DIET	REFERENCES
Gabon	88			5				2.5				5		63	Lahm 1986 (from stomach contents)
Gabon	47	34		6	1							5			Tutin et al. 1997
CIRMF, Franceville, Gabon		7.5					7.5					8	37		Norris 1988
Lopé Reserve, Gabon														47	Rogers et al. 1996
Papio															
P. anubis, olive (or anubis) baboon															
Bole Valley, Ethiopia			55	33	7		2				1	3			Dunbar and Dunbar 1974a
Laikipia, Kenya	13.6	10.7		7.2	29.5	19.1	16.5	0.1		5.3					Barton et al. 1993
Gombe, Tanzania														75	Ransom 1981
Budongo Forest, Uganda	34	13	47	17			1			7	2		17 (garbage)	51	Okecha and Newton-Fisher 2006
Comoé, Côte d'Ivoire	29	18		5	6	29			1			1	10	84	Kunz and Linsenmair 2008
P. cynocephalus, yellow baboon															
Amboseli, Kenya	10	18		3	5	12	31	15				1		30+	Post 1982
Amboseli, Kenya	28	18		1	3	41		8				1	1		Stacey 1986
Mikumi, Tanzania	14	29		14	20		6	3	6		2		7	185	Norton et al. 1987
Tana River, Kenya	34	17				30	12 (corms)	1				1		67	Bentley-Condit 2009, unpubl. data
P. hamadryas, hamadryas baboon															
Filoha, Ethiopia	49 (includes palm nuts)													44	Swedell et al. 2008
Filoha, Ethiopia	23.1 (includes palm nuts)	36		21.2	12	4.1	1.7					0.6		41	Schreier 2009
P. papio, Guinea baboon															
Niokolo-Koba, Senegal	60	17												58+	Sharman 1981
Niokolo-Koba, Senegal														17	Galat-Luong and Galat in press

Table 15.3 (cont'd)

TAXON AND STUDY SITE	% DIET FRUIT	% DIET NUTS + SEEDS + PODS	% DIET FRUIT + SEEDS	% DIET LEAVES	% DIET FLOWERS	% DIET GRASSES AND HERBS[2]	% DIET UNDERGROUND PLANT PARTS	% DIET FUNGI AND LICHENS	% DIET SAP/ GUM	% DIET HERB, PITHS, STEMS, AND SHOOTS	% DIET BARK, PITH, CAMBIUM	% DIET ANIMAL FOOD	% DIET OTHER (OR UNIDENT.)	NUMBER OF PLANT SPECIES IN DIET	REFERENCES
P. ursinus, chacma baboon															
Mkuzi, South Africa	90			6	3		1					1			Gaynor 1994
de Hoop, South Africa	35			13	5		42								Hill 1999
Cape Peninsula, South Africa														94	Hall 1962a
Cape Peninsula, South Africa	5	40		3	14	9	25					1	3	72	Davidge 1978b
Giant's Castle, Drakensberg, South Africa	3			26	14		53					4			Whiten et al. 1987, 1991
Giant's Castle, Drakensberg, South Africa							65								Whiten et al. 1987, Henzi et al. 1992 (high-altitude troop)
Giant's Castle, Drakensberg, South Africa							42								Whiten et al. 1987, Henzi et al. 1992 (low-altitude troop)
Rungwecebus															
R. kipunji, kipunji (formerly highland mangabey)															
Southern Highlands, Tanzania	27			34	9			1		16	11	2		122	Davenport et al 2010
Theropithecus															
T. gelada, gelada[2]															
Bole Valley, Ethiopia	2					98									Dunbar and Dunbar 1974a
Sankaber, Simien Mountains, Ethiopia						90–95									Dunbar 1977
Sankaber, Simien Mountains, Ethiopia						92									Wrangham 1976
Sankaber, Simien Mountains, Ethiopia	3	2			1										Hunter 2001
Gich Plateau, Ethiopia						100								35	Iwamoto 1979

[1] Unless otherwise noted, figures are percentages of time or percentages of feeding records spent eating each type of food.

[2] For geladas, % diet grasses and herbs includes *all* parts of grasses and herbs: blades, flowers, seeds, roots, and rhizomes.

Altmann 1970; Hamilton et al. 1978; Rasmussen, 1978; Post 1982; Whiten et al. 1987, 1991). In the Cape Peninsula of South Africa, baboons feed on shellfish, including mussels, limpets, and crabs (Hall 1962a, Davidge 1978b).

Despite such dietary diversity and flexibility, baboons are also highly selective in their dietary choices, with nutrient composition playing a large role in food selection. Baboons choose foods that are high in protein and lipids and low in fiber, alkaloids, and phenolics—that is, foods with high protein-to-fiber ratios and low toxicity (Hamilton et al. 1978; Altmann et al. 1987; Norton et al. 1987; Whiten et al. 1987, 1990, 1991; Johnson 1990; Barton et al. 1993; Barton and Whiten 1994; Altmann 1998).

Drill and mandrill diets appear to be similarly diverse, dominated by fruit but also including seeds, leaves, stems, bark, and roots from a wide array of plant species supplemented with fungi and both invertebrate and vertebrate animal matter (Hoshino 1985; Lahm 1985, 1986; Harrison 1988; Gonzalez-Kirchner and Sainz de la Maza 1996; Rogers et al. 1996; Abernethy et al. 2002; White 2007) (Table 15.3). Both drills and mandrills forage mainly in the lower strata of the forest, spending much of their time turning over leaf litter in search of invertebrate prey (Sabater Pi 1972; Jouventin 1975a; Norris 1988). The kipunji is also omnivorous, consuming ripe and unripe fruits, young and mature leaves, seeds, shoots, flowers, bark, moss, lichens, and invertebrates (Jones et al. 2005, Davenport et al. 2006, Davenport and Butynski in press).

The general pattern among *Lophocebus* mangabeys is omnivory, with a strong bias toward fruit and seeds and a secondary focus on invertebrate prey found inside dead trees, in leaf litter, and in tangles of lianas and masses of epiphytes (Waser 1975b, 1984; Struhsaker and Leland 1979; Poulsen et al. 2001a; Olupot and Waser in press) (Table 15.3). Gray-cheeked mangabeys in central Africa consume less fruit pulp and more leguminous seeds than their East African counterparts and focus more on seeds as well as fruits that are hard to crack (Chalmers 1968a; Poulsen et al. 2001; Brugière et al. 2002; Shah 2003). Seed eating in this species may be facilitated by a relatively thick dental enamel: When fruit is less available, mangabeys fall back on hard foods such as bark and seeds in addition to the leaves used as fallback foods by sympatric guenons (Lambert et al. 2004, Lambert 2007; Shah 2003). Black mangabeys (*L. aterrimus*) in the Salonga National Park, Democratic Republic of Congo, focus heavily on the nectar of *Daniellia pynaertii* flowers, with nectar averaging 20% of feeding records, and are probably important pollinators for this species (Gautier-Hion and Maisels 1994).

Cercocebus mangabeys are variable and flexible in their diets, focusing mainly on fruit and seeds (including unripe, ripe, and overripe fruits) but also on leaves, stems, shoots, flowers, fungi, roots, tubers, and invertebrate and vertebrate prey (Homewood 1976; 1978; Harding 1984; Mitani 1989; Ehardt in press b; Ehardt and Butynski in press b; Shah in press; Wieczkowski and Butynski in press). *Cercocebus* mangabeys in general appear able to crack hard foods that other sympatric monkeys cannot (Ehardt and Butynski in

press, Shah in press). Along with mandrills, *Cercocebus* mangabeys have large premolars compared to other African papionins, possibly an adaptation to feeding on hard foods on the forest floor such as nuts, seeds, and dry fruits, seeds, and pods (Fleagle and McGraw 2002).

In contrast to this general pattern of dietary breadth and flexibility, the gelada is highly granivorous, focusing primarily on the blades, rhizomes, flowers, and seeds of grasses (Dunbar and Dunbar 1974a; Dunbar 1977; Iwamoto 1979, 1993) (Table 15.3). Geladas forage while seated in a characteristic shuffling motion, using their hindlimbs for locomotion while foraging with their forelimbs (Dunbar and Dunbar 1974a). Geladas are particularly well-adapted to a diet of grasses in that their large thumbs and short second digits provide them with the highest opposability index among primates, giving them greater dexterity and allowing them to pick grasses (which they do using a precision grip with the thumb and second digit) at a higher rate than other sympatric primates (Jolly 1970c, Dunbar and Dunbar 1974a). Geladas also show digestive specializations for grazing (Mau et al 2009).

Socioecology and Community Ecology

Several comprehensive reviews discuss the effects of ecological variables on various aspects of behavior in baboons (Dunbar 1992b, 1993a; Barton et al. 1996; Bronikowski and Altmann 1996; Barton 2000; Hill and Dunbar 1998, 2002; Henzi and Barrett 2003, 2005; Alberts and Altmann 2006). Most obviously, group sizes vary in predictable ways with both predation risk (e.g., Hill and Lee 1998) and habitat quality, with larger groups of 'savanna' baboons, for example, found in richer habitats such as the fertile swamps of the Okavango Delta and the smallest in the arid Namib Desert and the highly seasonal Drakensberg Mountains (Table 15.4). Where food abundance and distribution vary over space and time, temporary subgrouping can occur during foraging (e.g., Aldrich-Blake et al. 1971; Homewood 1976; Anderson 1981b, 1982; Horn 1987a; White 2007).

Habitat Use

Home range size and travel patterns in African papionins are influenced by the distribution of food resources, water sources (in drier habitats), and suitable sleeping sites, with ranging patterns and daily path lengths largely a function of habitat quality and group size. In baboons, home range size and daily path lengths increase with decreased resource quality (Barton et al. 1992); and in geladas, home range size and daily path lengths increase with group size (Iwamoto and Dunbar 1983). Daily path lengths also increase with group size in gray-cheeked mangabeys (Olupot et al. 1994), in which ranging patterns are also influenced by *site attachment* (the return to preferred locations), avoidance of nearby social groups, and fruit availability (Waser 1976, Olupot et al. 1997). Sleeping sites of African papionins

Table 15.4 Group Size, Group Composition, and Sex Ratio

TAXON AND STUDY SITE	MEAN GROUP SIZE[1]	NUMBER OF GROUPS	RANGE GROUP SIZE	MEAN NUMBER OF ADULT MALES	MEAN NUMBER OF ADULT FEMALES	MEAN NUMBER OF IMMATURES (INCLUDING SUBADULTS)	ADULT SEX RATIO (ADULT FEMALES/ ADULT MALES)	REFERENCES
Cercocebus								
C. agilis, agile mangabey								
Northeast Gabon	10.5	3	7–18	1	2.2		2.2	Quris 1975, 1976
Mondika, Ndoki Forest, CAR/Congo	21.5	1	21–22	2	7–8	11–12	3.8	Shah 2003
Mondika, Ndoki Forest, CAR/Congo			8–230					Shah in press
C. atys, sooty mangabey								
Taï Forest, Cote d'Ivoire	89	2		9	25	55	2.8	Range and Noë 2002, Range 2006
Taï Forest, Cote d'Ivoire			75–125					McGraw in press
C. chrysogaster, golden-bellied mangabey								
Various sites, DRC			15–100 (estimates)					Ehardt and Butynski in press a
C. galeritus, Tana River mangabey								
Tana River, Kenya	26.5	4	17–36	3	10 (includes subadults)	12	3.3	Homewood 1976
Tana River, Kenya	38	1		2	8	28	4.0	Gust 1994
Tana River, Kenya	21	3	12–28					Kinnaird in press a
Tana River, Kenya (summary of several studies)	27	9	6–62	2.2	7	17.3	3.2	Wieczkowski and Butynski in press
C. sanjei, Sanje River mangabey								
Udzungwa Mountains, Tanzania			1–58	5	23	19	4.6	Ehardt and Butynski in press b
C. torquatus, red-capped (or white-collared) mangabey								
Rio Muni, Equatorial Guinea			14–23	Several				Jones and Sabater Pi 1968
Campo, Cameroon	25	1					1.75	Mitani 1989
Various sites			14–60					Ehardt in press
C. lunulatus, white-naped mangabey								
Various sites			3–58					Butynski in press
Lophocebus								
Lophocebus albigena - gray-cheeked mangabey								
Rio Muni, Equatorial Guinea	10	1		1				Jones and Sabater Pi 1968
Northeast Gabon	17	1		2				Quris 1976

Table 15.4 (*cont'd*)

TAXON AND STUDY SITE	MEAN GROUP SIZE[1]	NUMBER OF GROUPS	RANGE GROUP SIZE	MEAN NUMBER OF ADULT MALES	MEAN NUMBER OF ADULT FEMALES	MEAN NUMBER OF IMMATURES (INCLUDING SUBADULTS)	ADULT SEX RATIO (ADULT FEMALES/ ADULT MALES)	REFERENCES
Lopé, Gabon	20.5	1	18–23	1–3	9–10		4.8	Ham 1994
Dja Reserve, Cameroon	19	1	16–22	2	10	10	5	Poulsen et al. 2001a,b
Mondika, Ndoki Forest, CAR/Congo	13	1	12–14	2–4	4		1.3	Shah 2003
Bujuko, Uganda	21	2	16–25	5	8	7.5	1.6	Chalmers 1968a
Kibale, Uganda	15							Struhsaker and Leland 1979a,b
Kibale, Uganda	14	6	6–28					Waser 1977
Kibale, Uganda	15	1						
Kibale, Uganda	16	5	9–22					Arlet et al. 2009
Kibale, Uganda	16	6	5–30	3	6	7	2	Olupot and Waser in press
L. aterrimus, black (or black-crested) mangabey								
Lake Tumba, Zaire	16.5	4	14–19	3	5	11	1.7	Horn 1987a
Lomako Forest	11							McGraw 1994
Mandrillus								
M. leucophaeus, drill								
Cameroon	23	12	9–55					Struhsaker 1969
Cameroon	63.5	12	14–179	2.5				Gartlan 1970
Bakossiland, Cameroon	93.1	105	5–400					Wild et al. 2005
Korup, Cameroon	52.3	1–4	25–77					Astaras et al. 2008
Bioko Island, Equatorial Guinea			2–20					Gonzalez-Kirchner and Sainz de la Maza 1996
M. sphinx, mandrill								
Campo, Cameroon	52	4	15–95				13.9	Hoshino et al. 1984
Gabon			15–250					Harrison 1988
Lopé Reserve, Gabon	620	20	338–845	7	182	410	26	Abernethy et al. 2002
Lopé Reserve, Gabon	702–845	1						White 2007
Papio								
P. anubis, olive (or anubis) baboon								
Metahara, Ethiopia	44	6	14–87	8	16	24	2.0	Aldrich-Blake et al. 1971

Table 15.4 (*cont'd*)

TAXON AND STUDY SITE	MEAN GROUP SIZE[1]	NUMBER OF GROUPS	RANGE GROUP SIZE	MEAN NUMBER OF ADULT MALES	MEAN NUMBER OF ADULT FEMALES	MEAN NUMBER OF IMMATURES (INCLUDING SUBADULTS)	ADULT SEX RATIO (ADULT FEMALES/ ADULT MALES)	REFERENCES
Awash, Ethiopia	56	5	16–81	17 (includes subadults)	26 (includes subadults)		1.5	Nagel 1973
Bole Valley, Ethiopia	20	7	15–24					Dunbar and Dunbar 1974a
Athi River, Kenya	31	3	5–60	4	38		9.5	Maxim and Buettner-Janusch 1963
Nairobi Park, Kenya	39	9	12–87					DeVore and Washburn 1963
Nairobi Park, Kenya	41	9	12–87					DeVore and Hall 1965
Gilgil, Kenya	49	7	35–121	4	18	27	4.5	Harding 1975, 1976
Gilgil, Kenya	115	1		14	34	67	2.4	Smuts 1985
Laikipia, Kenya	103	1			26			Barton et al. 1993
Gombe, Tanzania	62	2	49–71	12	20.5	30	1.7	Ransom 1981
Queen Elizabeth Park, Uganda	45	2	32–58	9.5	10.5	23.5	1.1	Rowell 1966
Gasha Gumti, Nigeria	21	2	14–28	5			1.1	Warren 2003, Higham et al. 2009a,b
Comoé, Côte d'Ivoire	15	10	7–40	2.3	3.9		1.7	Kunz and Linsenmair 2008
P. cynocephalus, yellow baboon								
Amboseli, Kenya	80		12–185					DeVore and Hall 1965
Amboseli, Kenya	51	51	18–198	12	16	27	1.3	Altmann and Altmann 1970
Amboseli, Kenya	35.5	1	35–36	8	11.5	16	1.4	Hausfater 1975a,b
Amboseli, Kenya	31	8						Altmann et al. 1985
Amboseli, Kenya	55	3	38–78					Bronikowski and Altmann 1996
Tana River, Kenya	79	1	74–83	5	22	52	4.4	Bentley-Condit and Smith 1997
P. hamadryas, hamadryas baboon[1]								
Erer Gota, Ethiopia	64	4	20–90				1.2	Kummer 1968a
Erer Gota, Ethiopia	78	3	62–90	9	26	43	.2.9	Abegglen 1984

Table 15.4 (*cont'd*)

TAXON AND STUDY SITE	MEAN GROUP SIZE[1]	NUMBER OF GROUPS	RANGE GROUP SIZE	MEAN NUMBER OF ADULT MALES	MEAN NUMBER OF ADULT FEMALES	MEAN NUMBER OF IMMATURES (INCLUDING SUBADULTS)	ADULT SEX RATIO (ADULT FEMALES/ ADULT MALES)	REFERENCES
Erer Gota, Ethiopia	80.5	2	66–95					Sigg and Stolba 1981; Sigg et al. 1982
Filoha, Ethiopia	146	5	50–220	30	46	99	1.5	Swedell 2002b, 2006
Awash Station, Ethiopia	54	2	51–57	27 (includes subadults)	28.5 (includes subadults)		1.1	Nagel 1973
Saudi Arabia (various sites)			13–70					Kummer et al. 1981
Taif, Saudi Arabia	38	48	9–102				2.2	Biquand et al. 1992b, Boug et al. 1994a
Northern Yemen			22–89					Al-Safadi 1994
Eritrea	139	6	(SD=111)	20	58 (includes subadults)	60	2.8	Zinner et al. 2001b
P. papio - Guinea baboon[1]								
Niokolo-Koba, Senegal	62	111	22–249					Galat-Luong et al. 2006
P. ursinus, chacma baboon								
Okavango (Moremi), Botswana	79	7	7–128					Hamilton et al. 1976
Okavango (Moremi), Botswana	58	1	46–71					Bulger and Hamilton 1987
Okavango (Moremi), Botswana	75	1		3–12	19–26			Cheney et al. 2004, 2006
Okavango (Moremi), Botswana	32	12	5–70					Gwenzi et al. 2007
Kuiseb River Canyon, Namibia	26	3	12–35					Hamilton et al. 1976
Tsaobis Leopard Park, Namibia	34.3	4	22–55	3.3	11.8	19.3	3.6	Cowlishaw 1997a,b
Honnett, Northern Transvaal, South Africa	46	6	24–77	13.3	20.5	12.2	1.5	Stoltz and Saayman 1970
Honnett, Northern Transvaal, South Africa	77	1		3	31	43	10.3	Saayman 1970
Suikerbosrand, Southern Transvaal, South Africa	78	4	68–88	12.3	28.3	37.5	2.3	Anderson 1981a,b
Mkuzi, South Africa	76			8	17	51	2.1	Ron et al. 1996

Table 15.4 (cont'd)

TAXON AND STUDY SITE	MEAN GROUP SIZE[1]	NUMBER OF GROUPS	RANGE GROUP SIZE	MEAN NUMBER OF ADULT MALES	MEAN NUMBER OF ADULT FEMALES	MEAN NUMBER OF IMMATURES (INCLUDING SUBADULTS)	ADULT SEX RATIO (ADULT FEMALES/ ADULT MALES)	REFERENCES
Mkuzi, South Africa	55	3		8	17		2.1	Henzi et al. 1999
Mountain Zebra Nat Park	24	1		2	8	14		Seyfarth 1976
de Hoop, South Africa	34	6		4	19		4.8	Henzi et al. 1999
de Hoop, South Africa	38	4	23–50	5	11		2.2	Weingrill et al. 2003
Cape Peninsula, South Africa	32	4	20–53	3	13	16	4.3	Hall 1962a
Cape Peninsula, South Africa	85	1		12	28	45	2.3	Davidge 1978a,b
Drakensberg survey	20.5	27	4–40					Henzi et al. 1990
Drakensberg survey	22.5	61					2.1	Henzi and Lycett 1995
Cathedral Peak, Drakensberg	29–36	1	29–36	5.6	12		2.1	Weingrill 2000
Rungwecebus								
R. kipunji, kipunji (formerly highland mangabey)								
Rungwe Kitulo Forest, Southern Highlands, Tanzania	31	34	25–39					Davenport et al. 2008
Ndundulu Forest, Udzungwa Mountains, Tanzania	19	4	15–25					Davenport et al. 2008
Theropithecus								
T. gelada, gelada[1]								
Bole Valley, Ethiopia	60	1						Dunbar and Dunbar 1974a
Sankaber, Simien Mountains, Ethiopia	171	2	79–262	16	59	97	3.7	Dunbar and Dunbar 1975
Sankaber, Simien Mountains, Ethiopia	271	1						Wrangham 1976
Sankaber, Simien Mountains, Ethiopia	168	1						Hunter 2001
Gich Plateau, Ethiopia	111	2						Iwamoto 1979

[1] For hamadryas baboons and geladas, group size and composition refers to the *band* level of social structure; for Guinea baboons, it refers to the third level grouping, analogous to bands (see text).

are most commonly trees, although geladas sleep on cliffs and baboons use both cliffs and trees, as well as caves or rocky overhangs (Hall 1962a, Bert et al. 1967, Altmann and Altmann 1970, Hamilton 1982, Anderson and McGrew 1984, Wahungu 2001, Barrett et al. 2004; Schreier and Swedell 2008).

Activity Budgets

Among baboons, populations in lower-quality habitats spend more time feeding and foraging than those in higher-quality habitats, and within-population variation shows the same pattern (Hall 1962a, Whiten et al. 1987, Norton et al. 1987, Bronikowski and Altmann 1996, Barrett et al. 2006, Swedell et al. 2008). Hill et al (2003) demonstrated that activity budgets in baboons can be at least in part predicted by the mean annual temperature. Amboseli yellow baboons spend more time resting in years with greater daily maximum temperatures (Bronikowski and Altmann 1996), and desert chacmas are far less active during intergroup encounters after several days of water deprivation (Brain 1991). Baboons living in hot climates adjust their activity levels and postures to changes in temperature and solar radiation in ways that suggest active thermoregulation (Stelzner and Hausfater 1986, Stelzner 1988, Hill 2006). Baboons in the Namib Desert "sand bathe," apparently to lower body temperature (Brain and Mitchell 1999).

Seasonality

Diets vary seasonally in many African papionins, with fallback foods replacing preferred foods during the dry season or winter months (Kummer 1968a, Altmann and Altmann 1970, Norton et al. 1987, Whiten et al. 1987, Hill and Dunbar 2002, White 2007, van Doorn et al. 2009). In some populations, few species are consumed during the dry season compared to the wet season, reflecting the reduced food supply (Post 1982, Norton et al. 1987, Swedell et al. 2008). Daily path lengths of Laikipia olive baboons increase after reduced rainfall and with lower plant biomass (Barton et al. 1992), and Tana River mangabeys increase daily path lengths, feeding time, and group spread during the dry season (Homewood 1978, Waser 1984). In the Drakensberg of South Africa, where the dry season corresponds with winter, one troop of chacma baboons migrated to higher altitudes during the summer (and wetter) months and to lower altitudes during the winter (and drier) months (Whiten et al. 1987). In extreme environments such as the Drakensberg, thermoregulatory and nutritional stress during the cold, dry winters appears to force baboons to lower elevations on a seasonal basis (Whiten et al. 1987, Henzi et al. 1992) and increase levels of stress hormones (Weingrill et al 2004). At Amboseli, thermoregulatory (heat) and nutritional stress during the dry season is similarly associated with higher levels of stress hormones among females (Gesquiere et al. 2008).

Altitude

Social structure and activity budgets vary with reduced food availability and climatic extremes at high altitudes such as the Drakensberg in South Africa and the Simien Mountains in Ethiopia. Whiten et al. (1987) and Byrne et al. (1987) found an inverse relationship between altitude and group size in chacma baboons in the Drakensberg. Although Henzi et al. (1990) found no such relationship in this population, it nevertheless appears that at higher altitudes under conditions of nutritional stress, baboons are found in smaller group sizes and/or lower population densities (Byrne et al. 1993). In geladas, time spent feeding increases with altitude, likely resulting from both thermoregulatory costs and decreased habitat quality at higher altitudes (Iwamoto and Dunbar 1983).

Territoriality

African papionins are generally not territorial, and home ranges of neighboring troops usually overlap, sometimes extensively. Chacma baboons, however, have been reported to defend territorial boundaries in a desert habitat in Namibia and in the Okavango Delta in Botswana, where population densities were relatively high (Hamilton et al. 1976). Mangabeys and kipunjis are not overtly territorial; however, gray-cheeked mangabeys engage in aggressive intergroup encounters and mangabeys give loud *whoop–gobble* vocalizations and kipunjis 'honk-barks' that, together with a tendency to remain in familiar areas, may function as intergroup spacing mechanisms (Waser 1975a, 1976, 1982b; Ehardt and Butynski 2006, in press b; Davenport and Butynski in press; De Luca et al in press).

Polyspecific Associations

Most mangabeys and kipunjis form polyspecific associations with other primate taxa, primarily guenons (*Cercopithecus* spp.) but also colobus monkeys (*Procolobus* and *Colobus*). Most commonly, mangabeys are found in association with crowned guenons (*Cercopithecus pogonias*) where they are sympatric (Table 15.5).

Predators

Observed predators of African papionins include the crowned hawk-eagle (*Stephanoaetus coronatus*), central African rock python (*Python sebae*), leopard (*Panthera pardus*), lion (*Panthera leo*), spotted hyaena (*Crocuta crocuta*), and humans (Altmann and Altmann 1970; Horn 1987a, Fischer et al. 1999–2000, Davenport et al. 2006, Paterson 2006, Davenport and Butynski in press, Ehardt in press, Wieczkowski and Butynski in press).

SOCIAL ORGANIZATION

Variation in Social Structure: One-Male, Multimale, and Multilevel

Most baboons and mangabeys live in large social groups, or "troops," of multiple adult males and females, with group sizes varying widely across populations and habitats (Table 15.4). While groups generally function as cohesive social units, they often break up into smaller parties during foraging.

In baboons, the ratio of adult males to adult females varies across populations (Table 15.4). One-male groups, for example, occur in olive (DeVore and Hall 1965a, Nash 1976, Kunz and Linsenmair 2008), yellow (Stein and Stacy 1981), and chacma (Hall 1962a; Whiten et al. 1987; Byrne et al. 1987, 1989; Henzi et al. 1990) baboons. In particular, one-male groups are common among mountain chacmas in the Drakensberg of South Africa, likely a consequence of the high altitude and marked seasonality of this region (Anderson 1981b, 1989; Byrne et al. 1987, 1989; Whiten et al. 1987; Henzi et al. 1990). Rather than being a normative

Table 15.5 Ranging Behavior, Population Density, Polyspecific Associations and Activity Patterns

TAXON AND STUDY SITE	HOME RANGE SIZE (KM²)	NUMBER OF GROUPS	MEAN OR MEDIAN DAILY PATH LENGTH (KM)	RANGE DAILY PATH LENGTH	NUMBER OF PATHS	POPULATION DENSITY (INDIVIDUALS/ KM²)	POLYSPECIFIC ASSSOCIATIONS	% TIME FORAGING/ FEEDING	% TIME MOVING/ TRAVELING	% TIME RESTING	% TIME SOCIAL	REFERENCES
Cercocebus												
C. agilis, agile mangabey												
Northeast Gabon	1.98	1	1.2	0.9–1.7			*Cercopithecus pogonias* and *C. nictitans*					Quris 1975
Mondika, Ndoki Forest, CAR/ Congo	3	1	1.2	0.4–2.0	54	6.9		41	31	13	10	Shah 2003
Ituri Forest, DRC						2–38						Shah in press
C. atys, sooty mangabey												
NW Sierra Leone							*Cercopithecus campbelli*					Harding 1984
Taï Forest, Côte d'Ivoire								74				Range and Noë 2002
Taï Forest, Côte d'Ivoire	4.92		2.2		58	11.9		24.5	10	18.5	8	McGraw 1998a–c, in press
Taï Forest, Côte d'Ivoire							*Cercopithecus diana* and *Piliocolobus badius*					McGraw and Bshary 2002
C. chrysogaster, golden-bellied mangabey												
No data												
C. galeritus Tana River mangabey												
Tana River, Kenya	0.53	2	1.2	0.6–1.9	59			45.5	21	19	11	Homewood 1976
Tana River, Kenya	0.36							60	18	8	14	Kinnaird 1990a
Tana River, Kenya	0.47											Wieczkowski 2005
Tana River, Kenya (summary of several studies)	0.47	10	1.5	1.0–2.6		<.07	none					Wieczkowski and Butynski in press
C. sanjei, Sanje River mangabey												
Udzungwa Mountains, Tanzania	2	1	1.35	0.5–1.65			*Cercopithecus mitis moloneyi, Procolobus gordonorum,* and *Colobus angolensis palliatus*					Ehardt and Butynski in press b
C. torquatus, red-capped (or white-collared) mangabey												
Rio Muni, Equatorial Guinea							Most commonly *Cercopithecus pogonias*					Jones and Sabater Pi 1968
Campo, Cameroon	2.47											Mitani 1989

Table 15.5 (cont'd)

TAXON AND STUDY SITE	HOME RANGE SIZE (KM²)	NUMBER OF GROUPS	MEAN OR MEDIAN DAILY PATH LENGTH (KM)	RANGE DAILY PATH LENGTH	NUMBER OF PATHS	POPULATION DENSITY (INDIVIDUALS/ KM²)	POLYSPECIFIC ASSSOCIATIONS	% TIME FORAGING/ FEEDING	% TIME MOVING/ TRAVELING	% TIME RESTING	% TIME SOCIAL	REFERENCES
Various sites (summary)							Cercopithecus pogonias, C. cephus, C. nictitans, C. neglectus, C. erythrogaster, and Mandrillus sphinx					Ehardt in press
C. lunulatus, white-naped mangabey												
Various sites (summary)	2					5	Cercopithecus lowei, C. petaurista, C. roloway, Papio anubis, Colobus vellerosus, and Chlorocebus sabaeus					Butynski in press
Lophocebus												
L. albigena, gray-cheeked mangabey												
Rio Muni, Equatorial Guinea	1						Various guenons; most commonly C. pogonias					Jones and Sabater Pi 1968
Dja Reserve, Cameroon								52	27	11	9	Poulsen et al. 2001a,b
Lopé, Gabon	2.25	1					most commonly Cercopithecus pogonias					Ham 1994
Mondika, Ndoki Forest, CAR/ Congo	1.67	1	1.14		46		most commonly Cercocebus agilis, Cercopithecus pogonias	49	28	14	6	Shah 2003
Bujuko, Uganda	0.13					38.6	None					Chalmers 1968a
Kibale (Ngogo), Uganda		1	1.3					47	27			Freeland 1979
Kibale (Kanyawara and Dura), Uganda	4.1		1.3	0.8–1.5		8.5	commonly Cercopithecus ascanius; also C. mitis, Procolobus badius, and Colobus guereza	42	22			Waser 1975, 1976, 1980
Kibale (Kanyawara), Uganda	1.55	7	1.0			10		43	24			Olupot et al. 1994 (averages across two study periods: 1971–74 and 1991)
Kibale (Kanyawara), Uganda								34	29	35	5	Olupot and Waser 2001a (data on males only)

Table 15.5 (cont'd)

TAXON AND STUDY SITE	HOME RANGE SIZE (KM²)	NUMBER OF GROUPS	MEAN OR MEDIAN DAILY PATH LENGTH (KM)	RANGE DAILY PATH LENGTH	NUMBER OF PATHS	POPULATION DENSITY (INDIVIDUALS/ KM²)	POLYSPECIFIC ASSSOCIATIONS	% TIME FORAGING/ FEEDING	% TIME MOVING/ TRAVELING	% TIME RESTING	% TIME SOCIAL	REFERENCES
L. aterrimus, black (or black-crested) mangabey												
Lake Tumba, Zaire	0.6	2				69	Cercopithecus wolfi and C. ascanius	36	1.5	38	24	Horn 1987a
Lomako Forest						73	Cercopithecus wolfi and C. ascanius					McGraw 1994
Mandrillus												
M. leucophaeus, drill												
Korup, Cameroon						1.7						Astaras et al. 2008
Bioko Island, Equatorial Guinea							3 occasions only (Colobus satanas, Cercopithecus pogonias, C. erythrotis)					Gonzalez-Kirchner and Sainz de la Maza 1996
M. sphinx, mandrill												
CIRMF, Franceville, Gabon								64				Norris 1988
Lopé Reserve, Gabon			4.2									Abernethy and White in press
Lopé Reserve, Gabon	181	1	2.9	0.1–10.4	101							White 2007
Papio												
P. anubis, olive (or anubis) baboon												
Metahara, Ethiopia	4.3	1	5.8		12							Aldrich-Blake et al 1971
Awash, Ethiopia			5.3		7	5.6		32	25	31	12	Nagel 1973
Nairobi Park, Kenya	25.8	7										DeVore and Washburn 1963
Nairobi Park, Kenya	23.3	8				3.9						DeVore and Hall 1965
Gilgil, Kenya	19.7	1	4.6	2.2–7.8		10.3						Harding 1976
Gilgil, Kenya	31											Smuts 1985
Laikipia, Kenya	43.75	1	5.6									Barton et al. 1992
Gombe, Tanzania*	3.9–5.2	1				15–23						Ransom 1981*
Queen Elizabeth Park, Uganda	4.6	2				10.8						Rowell 1966
Comoé, Côte d'Ivoire	10.4	2				1.2		42	26	19	14	Kunz and Linsenmair 2008
P. cynocephalus, yellow baboon												
Amboseli, Kenya						10+						DeVore and Hall 1965
Amboseli, Kenya	24.1	1	5.5	4.6–6.0	12							Altmann and Altmann 1970
Amboseli, Kenya (1963)*						73						Altmann et al. 1985*
Amboseli, Kenya (1983)*						2.2						Altmann et al. 1985*

Table 15.5 (cont'd)

TAXON AND STUDY SITE	HOME RANGE		DAILY PATH LENGTH			POPULATION DENSITY (INDIVIDUALS/ KM²)	POLYSPECIFIC ASSSOCIATIONS	ACTIVITY PATTERNS				REFERENCES
	HOME RANGE SIZE (KM²)	NUMBER OF GROUPS	MEAN OR MEDIAN DAILY PATH LENGTH (KM)	RANGE DAILY PATH LENGTH	NUMBER OF PATHS			% TIME FORAGING/ FEEDING	% TIME MOVING/ TRAVELING	% TIME RESTING	% TIME SOCIAL	
Amboseli, Kenya								46	24	22	7	Post 1981
Amboseli, Kenya	16	3						35	24	32	9.3	Stacey 1986
Amboseli, Kenya			5	3–7				46.5	26	8.9	18.6	Bronikowski and Altmann 1996
Mikumi, Tanzania									26			Norton et al. 1987
Tana River, Kenya			5.3									Wahungu 2001
P. hamadryas, hamadryas baboon												
Erer Gota, Ethiopia			13.2	4.1–19.2	9	1.8						Kummer 1968a
Erer Gota, Ethiopia	28	1	9.5		70							Sigg and Stolba 1981
Filoha, Ethiopia	30	1	7.5	3.2–11.2	30							Swedell 2002b, 2006
Filoha, Ethiopia	38.6	1	8.3	4.6–14.2	105			29	27.6	23.5	19.9	Schreier 2009
Awash, Ethiopia*			6.5		7	3.4		31	25	26	17	Nagel 1973*
Taif, Saudi Arabia	9.3	1		1.0–14.0								Boug et al. 1994a
Furrus and Durfo, Eritrea						10.2 and 23.9						Zinner et al. 2001a
P. papio, Guinea baboon												
Niokolo-Koba, Senegal	32.5	2	8	4–13	49					21	19	Sharman 1981
Niokolo-Koba, Senegal								21–39	22–58	19–26	2–13	Galat-Luong et al. 2006
P. ursinus, chacma baboon												
Okavango (Moremi), Botswana	4.5	4				24						Hamilton et al. 1976
Okavango (Moremi), Botswana	10											Bulger and Hamilton 1987
Okavango (Moremi), Botswana	5	1										Cheney et al. 2004, 2006
Chimanimani highlands, Zimbabwe*	0.78	12										Gwenzi et al. 2007*
Kuiseb River Canyon, Namibia	6.7	2				5.3						Hamilton et al. 1976
Honnett, Northern Transvaal, South Africa	17	3	8.45	3.5–7.5	54							Stoltz and Saayman 1970
Suikerbosrand, Southern Transvaal, South Africa	24.6	4	3.6	2.3–4.6								Anderson 1981b, 1982

Table 15.5 (*cont'd*)

TAXON AND STUDY SITE	HOME RANGE		DAILY PATH LENGTH			POPULATION DENSITY (INDIVIDUALS/ KM²)	POLYSPECIFIC ASSSOCIATIONS	ACTIVITY PATTERNS				REFERENCES
	HOME RANGE SIZE (KM²)	NUMBER OF GROUPS	MEAN OR MEDIAN DAILY PATH LENGTH (KM)	RANGE DAILY PATH LENGTH	NUMBER OF PATHS			% TIME FORAGING/ FEEDING	% TIME MOVING/ TRAVELING	% TIME RESTING	% TIME SOCIAL	
Mkuzi, South Africa	24.4		4.9		70			36	31	21	12	Gaynor 1994
de Hoop, South Africa								43	27	13	15	Hill et al. 2003
Cape Peninsula, South Africa	19.2	3	4.5	1.6–8.0	33			80				Hall 1962a; DeVore and Hall 1965
Cape Peninsula, South Africa	37	1	7.9	3.0–13.8	51			34	32	24	10	Davidge 1978a,b
Giant's Castle, Drakensberg, South Africa	18.9	1	4.3	3.0–6.0	101	0.95						Whiten et al. 1987; Henzi et al. 1992 (high-altitude troop)
Giant's Castle, Drakensberg, South Africa	10	1	3.8	2.7–5.0	29	1.9						Whiten et al. 1987; Henzi et al. 1992 (low-altitude troop)
Giant's Castle, Drakensberg, South Africa								57	18	17	8	Whiten pers. comm. to Dunbar 1992a,b
Drakensberg (various sites)						2.7						Henzi and Lycett 1995
Rungwecebus												
R. kipunji, kipunji (formerly highland mangabey)												
Livingstone (Kitulo) Forest, Southern Highlands, Tanzania	0.54	3					*Colobus angolensis, Cercopithecus mitis moloneyi*					Davenport et al. 2006
Rungwe Kitulo Forest, Southern Highlands, Tanzania			1.3			83.9	*Colobus angolensis, Cercopithecus mitis moloneyi*					Davenport et al. 2008
Rungwe Kitulo Forest, Southern Highlands, Tanzania	3.1	4	1.3									De Luca et al 2009
Ndundulu Forest, Udzungwa Mountains, Tanzania						14.2	*Cercopithecus mitis moloneyi, Procolobus gordonorum, & Colobus angolensis palliatus*					Davenport et al. 2008; Davenport and Butynski in press
Theropithecus												
T. gelada, gelada												
Sankaber, Simien Mountains, Ethiopia	2.9	3	2.2			77.6		45	20	14	21	Dunbar and Dunbar 1975

Table 15.5 (*cont'd*)

TAXON AND STUDY SITE	HOME RANGE SIZE (KM²)	NUMBER OF GROUPS	MEAN OR MEDIAN DAILY PATH LENGTH (KM)	RANGE DAILY PATH LENGTH	NUMBER OF PATHS	POPULATION DENSITY (INDIVIDUALS/ KM²)	POLYSPECIFIC ASSSOCIATIONS	% TIME FORAGING/ FEEDING	% TIME MOVING/ TRAVELING	% TIME RESTING	% TIME SOCIAL	REFERENCES
Sankaber, Simien Mountains, Ethiopia								55	7	18	12	Wrangham 1976
Sankaber, Simien Mountains, Ethiopia	9.3	1	2.1									Hunter 2001
Gich Plateau, Ethiopia			1.0					80	9	2	8	Kawai and Iwamoto 1979
Gich Plateau, Ethiopia	1.0		1.0			63						Iwamoto 1979

*includes semi-commensal troops feeding on human-derived foods

social structure, one-male groups in this case appear to result from reduced population densities due to low habitat quality combined with reduced predator pressure (Hamilton 1985; Byrne et al. 1987, 1993; Hamilton and Bulger 1987, 1992; Whiten et al. 1987; Anderson 1989; Henzi et al. 1990, 1998; Henzi and Barrett 2003). The variable distribution of females on a landscape combined with the variable ability of one male to exclude other males from a group of females may result in one-male or multimale groups depending on population density and group size (Hamilton and Bulger 1992, 1993). Altmann (1990) pointed out that, in cercopithecines as a whole and baboons in particular, groups of five or fewer females virtually always have only one male, whereas groups of more than 10 females include multiple males (Altmann 1990, 2000; see below).

Hamadryas baboons and geladas are characterized by a multitiered social system consisting of three to four levels of social structure (Dunbar and Dunbar 1975; Kummer 1968a,b, 1984, 1990; Dunbar 1984b, 1993b). In both taxa, the smallest and most stable social unit is the *one-male unit* (OMU), comprising a "leader male," several females, dependent offspring (Figure 15.1), and sometimes one or more "follower males" that socialize, but do not typically copulate, with the females (Kummer 1968a,b; Mori 1979b,d). In hamadryas, OMUs comprise one to nine females, averaging two to three (Kummer 1968a, Nagel 1973, Swedell 2006); and in geladas, they comprise one to 12 females, averaging 3.5–4 (Dunbar and Dunbar 1975, Ohsawa 1979a). In hamadryas, several OMUs and additional "solitary" (unaffiliated) males may together form a *clan*, a spatial and social association among males, possibly kin-based (Abegglen 1984, Colmenares 1992, Schreier and Swedell 2009); similarly, two closely associated gelada OMUs, possibly the result of fission, may form a *team* (Kawai et al. 1983). In both taxa, the social unit analogous to troops of other baboons (see Table 15.4) is the *band*, which shares a common home range, travel pattern, and sleeping site (Kummer 1968a, Kawai et al. 1983). A hamadryas band consists of multiple OMUs and a number of solitary males; in geladas, these extra males form relatively cohesive all-male groups. Gelada bands are less consistent in their composition than those of hamadryas and are best thought of as a set of OMUs sharing a foraging area rather than a closed social unit (Mori 1979c,e; Ohsawa 1979a; Kawai et al. 1983). The largest grouping in both taxa, an aggregation of bands at a sleeping site or foraging area, can number up to 800 individuals and is called a *troop* in hamadryas and a *herd* in geladas (Kummer 1968a, Kawai et al. 1983).

Although less is known about them, Guinea baboons may also be characterized by a multilevel social system. The largest social unit is the aggregation at a sleeping site, which consists of several hundred baboons (Sharman 1981, Anderson and McGrew 1984). These aggregations subdivide into clans and OMUs that appear similar, and possibly homologous, to those of hamadryas (Dunbar and Nathan 1972; Sharman 1981; Boese 1973, 1975; Galat-Luong et al. 2006). Current evidence suggests an intermediate social organization between the unstructured multimale groups of most *Papio* baboons and the more rigid multilevel societies of hamadryas and geladas, with greater flexibility in female behavior than in hamadryas baboons (Jolly and Phillips-Conroy 2006, Galat-Luong et al. 2006).

In mangabeys, some species have been observed only in multimale groups (*L. aterrimus*, Horn 1987a, McGraw 1994;

Figure 15.1 A hamadryas baboon (*Papio hamadryas*) one-male unit at the Filoha field site in central Ethiopia (photo by Mathew Pines).

C. atys, Range 2006) and others have been observed in both multimale and one-male groups, though most often multimale (*C. agilis*, Quris 1975, 1976, Shah in press; *C. galeritus*, Wieczkowski and Butynski in press; *L. albigena*, Oluput and Waser in press:Table 15.4). In the gray-cheeked mangabey, the number of males in a group appears to be determined both by overall group size, as only the smallest groups contain only one male (Olupot and Waser in press; Alert et al 2008), and by the number of estrous females (Olupot and Waser 2001b). Although kipunjis are only now being studied in detail, recent surveys show that they live in multimale groups as well (Davenport et al 2008; Table 15.4).

The social system of mandrills and drills has historically been unclear with contradictory observations, as both live in dense rain forests and are difficult to habituate and observe (Jouventin 1975a, Hoshino et al. 1984, Harrison 1988). As a result, most field studies have yielded little reliable information and nearly all concrete data on mandrills (e.g., data on reproduction and social behavior reported elsewhere in this chapter) derive from the Centre International de Recherches Médicales de Franceville (CIRMF), a semi-free-ranging colony of mandrills in Franceville, Gabon, where stable groups of multiple adult males and females living in large forested enclosures have been under observation since 1982 (Norris 1988; Wickings and Dixson 1992a,b; Setchell et al. 2001). Early interpretations of wild mandrills suggested a multilevel social structure similar to that of hamadryas and geladas (Jouventin 1975a, Hoshino et al. 1984, Stammbach 1987, Harrison 1988), though most of these studies were based on very limited observations. What we know from field studies is that both mandrills and drills are unusual among papionin monkeys in that (1) they aggregate into extremely large groups, called "hordes," "supergroups," "*grandes hordes*," or "*grandes*

hordes bruyantes" and (2) adult males are often solitary (Gartlan 1970, Sabater Pi 1972, Jouventin 1975a, Harrison 1988, Rogers et al. 1996, Gadsby and Jenkins 1997–1998, Abernethy et al. 2002, Wild et al. 2005, White 2007, Astaras et al. 2008) (Table 15.4).

Most recently, observations from Lopé National Park in Gabon by Abernethy et al. (2002) and White (2007) have shown that mandrill hordes are stable, cohesive, and exceptionally large, averaging 620 individuals. These observations have also revealed that hordes do split up into smaller groups on a temporary basis and that adult males reside in mandrill hordes only during the mating season and are otherwise solitary (Abernethy et al. 2002, White 2007).

The social structure of drills was originally suggested by Gartlan (1970) to be multilevel, as he observed OMUs of fewer than 20 individuals aggregating into larger groupings of up to 179 individuals. Since then, additional data have supported a two-level social structure in drills, with supergroups fissioning into smaller subgroups (Gadsby and Jenkins 1997–1998, Astaras et al. 2008). Whether these subgroups typically consist of only one or multiple adult males cannot be determined with current data (Wild et al. 2005, Astaras et al. 2008). Regardless, the near universality of observations of solitary adult males suggests that in drills, as in mandrills, the sex ratio within social groups is strongly female-biased and most adult males are solitary for most of the year.

Dispersal and Philopatry

In most African papionins for which data are available, females typically remain in their natal groups throughout their lives (Hausfater 1975a, Smuts 1985, Dunbar 1993b, Hamilton and Bulger 1993, Setchell 1999, Olupot and Waser

2001b, Range 2006). Female migration has been observed in some baboon populations but only rarely (Anderson 1981a, Rasmussen 1981, Byrne et al. 1987). Males typically disperse from their natal groups and may disperse several times throughout their lifetime (Hausfater 1975a; Packer 1979a; Dunbar 1984b; Smuts 1985; Manzolillo 1986; Alberts and Altmann 1995a; Sapolsky 1996; Charpentier et al. 2004; Olupot and Waser 2001b, 2005; Janmaat et al 2009).

The one known exception to this general pattern occurs in hamadryas baboons, in which females are transferred, usually coercively, among OMUs by males (Swedell and Schreier 2009) and males sometimes disperse as well (Sigg et al. 1982; Phillips-Conroy et al. 1991, 1992; Phillips-Conroy and Jolly 2004). Transfer of females generally occurs within clans or bands rather than between them (Sigg et al. 1982, Swedell 2005, Schreier and Swedell 2009) and is thus not analogous to the male transfer seen in other African papionins (Swedell 2005, 2006). A combination of behavioral and genetic evidence suggests limited bisexual dispersal with primarily female-mediated gene flow (Sigg et al. 1982; Woolley-Barker 1999; Hapke et al. 2001; Hammond et al. 2006; Swedell 2005, 2006).

Social Behavior Within Groups

Patterns of Bonding

The apparent similarity in social structure between hamadryas, geladas, and Guinea baboons, on the one hand (OMUs within larger groups), and most *Papio* baboons and mangabeys, on the other (large, cohesive, stable social groups), tells us nothing about—and in some cases actually obscures—the underlying patterns of bonding characterizing each species. For example, in the hamadryas social system, cross-sex bonds underlie the cohesion of OMUs, whereas it is bonds among females that underlie OMU cohesion in geladas (Mori 1979e, Dunbar 1984b).

In mangabeys, geladas, mandrills, and most baboons, dyadic bonds among philopatric, related females form the core of a social group (Seyfarth 1976; Dunbar 1978, 1993b; Altmann 1980; Strum 1987; Barton et al. 1996; Setchell 1999; Silk et al. 1999, 2003a,b; Henzi et al. 2000; Range and Noë 2002; Range 2006). Females maintain relatively strong, differentiated, affiliative relationships, mainly expressed via allogrooming, and interact preferentially with kin (Chalmers 1968b; Boese 1975; Seyfarth 1977; Smuts 1985; Barrett et al. 1999, 2000, 2002; Bentley-Condit and Smith 1999, Silk et al. 1999; Henzi et al. 2000; Leinfelder et al. 2001; Range and Noë 2002; Smith, K., et al. 2003; Chancellor and Isbell 2009). These taxa are characterized as *female-bonded*; that is, females remain in their natal groups and maintain differentiated social relationships with other females. While less is known about individual social interactions in Guinea baboons and drills, current evidence supports a female-bonded social organization for these species as well (Boese 1973, Wild et al. 2005).

The one exception to this pattern once again lies with hamadryas baboons, in which female bonding is reduced

due to coercive transfer of females among social units by males. Leader males aggressively herd females between OMUs, breaking up female kin groups in the process. Related females, however, do often end up in the same OMU (Sigg et al. 1982, Chalyan et al. 1994), and the highly variable level of social interaction among female dyads likely reflects variation in kinship among them (Swedell 2002a, 2006).

Bonds among males vary widely among African papionins. At one end of the spectrum, drill and mandrill males appear relatively intolerant of one another, as they spend most of their time completely solitary and immigrate into social groups only temporarily during the mating season (Abernethy et al. 2002, White 2007). In baboons and mangabeys, adult males exhibit greater tolerance, as multiple males remain together in social groups year-round and maintain dominance relationships that mediate competitive interactions. Both mangabey and baboon males are less social than females, however, especially with members of their own sex (Chalmers 1968b; Aldrich-Blake et al. 1971; Saayman 1971; Davidge 1978a; Struhsaker and Leland 1979; Smuts 1985); and interactions among males in these species are more likely to be agonistic than affiliative (Chalmers 1968b, Hausfater 1975a). In hamadryas and geladas, by contrast, males groom more with other males, maintain differentiated affiliative relationships, and engage in ritualized greetings that, for leader males, replace grooming relationships (as males are less tolerant of other males once they acquire females) (Abegglen 1984; Mori 1979a,b; Colmenares 1990, 1991). Guinea baboon males appear to be more similar to hamadryas and geladas in this respect, although additional field data are needed to confirm this (Boese 1973, 1975; Sharman 1981; Whitham and Maestripieri 2003).

Affiliative relationships between the sexes are usually limited to sexual consortships, but strong cross-sex bonds outside of this context occur as well, particularly in one-male groups (Byrne et al. 1989, Barton et al. 1996). Compared to lowland chacma baboons, for example, mountain chacmas are more commonly found in one-male groups, cross-sex bonds are stronger, female-female bonds are weaker, and herding behavior by males occurs more frequently (Anderson 1981a, 1990; Byrne et al. 1987, 1989; Whiten et al. 1987; Henzi et al. 1990, 1999; Hamilton and Bulger 1992). This suggests a relationship between the one-male group structure and (*1*) a male's motivation to invest in and defend his females and infants, probably due to increased paternity certainty, and (*2*) a female's motivation to invest in relationships with the resident male, possibly due to the increased importance of protection against predation and infanticide in a one-male group.

Cross-sex affiliative relationships occur in multimale groups as well. In mangabeys as well as olive, yellow, and chacma baboons, some male-female dyads maintain grooming relationships, or "friendships," outside of estrus (Chalmers 1968b; Chalmers and Rowell 1971; Seyfarth 1978b; Altmann 1980; Smuts 1983a, 1985; Byrne et al.

1989; Henzi et al. 2000). These relationships likely benefit a female via protection for herself and her offspring (Smuts 1983b, 1985; Nguyen et al 2009). The effort devoted by males to maintaining these relationships has been alternately interpreted as mate investment or parental investment, with the relative importance of each factor possibly reflecting differing infanticide risk—and, hence, cross-sex bonding to counteract this risk—among baboon populations (Smuts 1983b, 1985; Anderson 1992; Palombit et al. 1997; Weingrill 2000; Henzi and Barrett 2003).

Cross-sex bonds persist year-round in hamadryas baboons. Hamadryas leader males use physical coercion, or the threat thereof, to maintain proximity to females, which are essentially conditioned to remain within the spatial and social sphere of their own OMU (Swedell and Schreier 2009). In Guinea baboons, the degree of male aggression toward females appears to be lower, but the basis for the OMU social structure appears to be similar to that of hamadryas in that it is centered on the bonds between the leader male and each of his females (Boese 1975, Maestripieri et al. 2007).

Dominance and Its Correlates

In most African papionins for which detailed data on social behavior are available, dominance hierarchies characterize members of each sex. These are most obvious among baboon males, which aggressively compete over positions in the dominance hierarchy, engage in frequent contest competition over females, and use vocalizations to signal competitive ability (Hall 1962b; DeVore and Washburn 1963; Hall and DeVore 1965; Hausfater 1975a,b; Bernstein 1976; Kitchen et al. 2003; Fischer et al. 2004). A similar suite of behaviors characterizes male mandrills at the CIRMF (Setchell et al. 2005a,b, 2006). In gray-checked mangabeys, however, aggression among males is less common and males do not maintain obvious dominance relationships (Waser 1976, Olupot and Waser 2005). Similarly, hamadryas baboons (and possibly Guinea baboons) can be best characterized as having two social strata for males: those with females and those without (Boese 1973; Kummer 1968a, 1973; Abegglen 1984). Leader males of OMUs are dominant to follower and solitary males in the sense that the former have priority of access (in this case, near exclusive access) to females, but this relationship is maintained via ritualized "notifications" rather than traditional dominant–submissive interactions.

In mangabeys, geladas, mandrills, and most baboons, females maintain dyadic dominance relationships within linear, stable dominance hierarchies (Chalmers and Rowell 1971; Seyfarth 1976; Dunbar and Dunbar 1977, Mori 1979b; Altmann 1980; Dunbar 1980b; Hausfater et al. 1982; Smuts 1985; Byrne et al. 1989; Bentley-Condit and Smith 1999; Setchell 1999; Range 2006; Chancellor and Isbell 2009) that mediate competitive interactions over food resources (Barton 1993, Barton et al. 1996,

Range and Noë 2002, K.L. Wood personal communication). In geladas, dominance relationships appear to mediate female reproductive, rather than feeding, competition (Dunbar 1980b, 1989).

In baboons, drills, and mandrills, female dominance rank is inherited matrilineally. Females generally rank just below their mothers, and the process by which this rank is attained involves substantial maternal support (Cheney 1977; Altmann 1980; Walters 1980; Pereira 1986, 1988a, 1989; Setchell 1999). Higher rates of grooming among adjacently ranked females suggest that dominance is matrilineally based in sooty mangabeys as well (Range and Noë 2002, Range 2006). In populations for which long-term data are available, female dominance hierarchies remain stable for many generations (Hausfater et al. 1982, Samuels et al. 1987, Silk et al. 1999).

In hamadryas baboons, by contrast, dominance relationships among females are either absent or extremely subtle in the wild. Hamadryas females do, however, develop and maintain dominance relationships in captivity (e.g., Vervaecke et al. 1992, Leinfelder et al. 2001), as do Guinea baboons (Boese 1973).

Dominance carries fitness benefits for both sexes. Among females, lower-ranking individuals suffer greater mortality via predation (baboons, Ron et al. 1996); lower food intake via lower bite rates, shorter feeding bouts, or less-suitable feeding sites (baboons, Post et al. 1980, Barton 1990, 1993, Barton and Whiten 1993; sooty mangabeys, Range and Noë 2002); reduced fertility via lower nutritional status and/or body weight (baboons, Strum and Western 1982; Bercovitch 1985, 1987a; Barton 1990; Bercovitch and Strum 1993); and consequent reduced fitness (Altmann 1991, 1998). Higher-ranking females begin reproducing at an earlier age, reproduce at shorter intervals, and/or produce offspring that grow faster (and, in baboons, are weaned at earlier ages) compared to lower-ranking females (baboons, Altmann et al. 1988, Smuts and Nicolson 1989, Packer et al. 1995, Altmann and Alberts 2003b, 2005, Johnson 2003, 2006, Cheney et al. 2004, 2006; mandrills, Setchell et al. 2001, 2002, Setchell and Wickings 2004). In geladas, higher-ranking females produce more offspring, likely due to aggression-induced reproductive suppression.

Among males, higher-ranking baboons and mangabeys are more successful at establishing and maintaining consortships (Altmann et al. 1996, Hausfater 1975a) and usually achieve higher mating and/or reproductive success (Packer 1979b, Altmann et al. 1996, Alberts et al. 2003, Altmann and Alberts 2003b, Arlet et al. 2007). Lower-ranking males, however, can form coalitions and overturn consortships. Higher rank is also associated with lower stress hormone levels and more adaptive endocrinological profiles during periods of social stability (Sapolsky 1982, 1993; Bergman et al. 2005), though social instability negates these rank-related advantages (Sapolsky 1992, Bergman et al. 2005) and the reverse relationship characterizes gray-cheeked mangabeys (Arlet et al. 2009).

Coalitions

Gelada males cooperate to overthrow leaders of OMUs, and hamadryas leader and follower males may cooperate to defend an OMU from potential take-overs. However, neither geladas nor hamadryas form coalitions in the same way that has been described for East African olive and yellow baboons (Packer 1977; Ransom 1981; Smuts 1985; Noë 1986, 1992; Bercovitch 1988; Noë and Sluijter 1990, 1995). Guinea baboon males have not been reported to engage in coalitionary behavior, and coalitions are notably absent in chacma baboons (Henzi et al. 1999, Henzi and Barrett 2003). Male coalitionary behavior has not been described for mangabeys, drills, or mandrills.

Among females, coalitions have been observed in olive and yellow baboons, geladas, drills, mandrills, sooty mangabeys, and gray-cheeked mangabeys, but not hamadryas, chacma, or Guinea baboons (Dunbar 1980b,

1984b; Smuts 1985; Wasser and Starling 1986; Barton et al. 1996; Henzi et al. 1997a; Range and Noë 2002; Silk et al. 2004; Range 2006; K.L. Wood, personal communication). In female baboons, coalitions occur most often among members of the same matriline and kin support one another more than nonkin (Smuts 1985, Silk 2002b).

REPRODUCTION AND LIFE HISTORY

Sexual Behavior

All African papionins except *Mandrillus* breed year-round, though some populations show seasonal peaks in mating and births (Table 15.6). Copulations are concentrated during a discrete period of estrus (see Beach 1976), occurring cyclically and broadly corresponding to the periovulatory period (Saayman 1975). Copulations may be initiated by either sex; females often initiate copulations via sexual "presents," in which they orient the perineum toward the male and pause

Table 15.6 Reproductive Parameters

TAXON AND STUDY SITE	MEAN OR MEDIAN AGE AT MENARCHE (YEARS)	N	MEAN AGE AT FIRST BIRTH (YEARS)	N	MEAN OR MEDIAN POST PARTUM AMENORRHEA AFTER SURVIVING INFANT (MONTHS)	N	MEAN INTERBIRTH INTERVAL AFTER SURVIVING INFANT (MONTHS)	N	MEAN LENGTH OF OVARIAN CYCLE (DAYS)	N	MEAN OR MEDIAN LENGTH OF GESTATION (DAYS)	N	SEASONALITY OF REPRODUCTION	REFERENCES
Cercocebus														
C. agilis, agile mangabey														
Mondika, Ndoki Forest, CAR/ Congo							21				180		Most births Dec–Feb and June–Aug	Shah in press
Captive (France)									29.8 (22–39)	6				Gautier-Hion and Gautier 1976
C. atys, sooty mangabey														
Captive (Yerkes, U.S.)	2.7 (2.5–3.3)	6			4.6		12.8		34.5		167		Peak July–Dec	Hadidian and Bernstein 1979
Captive (Yerkes, U.S.)	3.03 (SD 1.1 mo)	14	4.7 (SD 1.8 mo)	28	7.1 (SD 0.3)	110	18 (SD 0.7)				167		Peak March–August	Gust et al. 1990
C. chrysogaster, golden-bellied mangabey														
Captive (U.S.)	2.6	2	5.4		8.6 (5.5–10)		19.9 (16.6–24.8)		30.7		178			Walker et al. 2004
C. galeritus, Tana River mangabey														
Tana River, Kenya											180 (SE 4.49)	7		Kinnaird 1990b
Tana River, Kenya							18–24						Peak Aug–April	Wieczkowski and Butynski in press

Table 15.6 (*cont'd*)

TAXON AND STUDY SITE	MEAN OR MEDIAN AGE AT MENARCHE (YEARS)	N	MEAN AGE AT FIRST BIRTH (YEARS)	N	MEAN OR MEDIAN POST PARTUM AMENORRHEA AFTER SURVIVING INFANT (MONTHS)	N	MEAN INTERBIRTH INTERVAL AFTER SURVIVING INFANT (MONTHS)	N	MEAN LENGTH OF OVARIAN CYCLE (DAYS)	N	MEAN OR MEDIAN LENGTH OF GESTATION (DAYS)	N	SEASONALITY OF REPRODUCTION	REFERENCES
C. sanjei, Sanje River mangabey														
no data														
C. torquatus, red-capped (or white-collared) mangabey														
Captive (France)									32.5 (26–35)	8	168 (160–182)	3	peak in autumn (Oct–Nov)	Gautier-Hion and Gautier 1976
Captive (U.S)	2.2–3.0						13–16		30–33		170–175			Ehardt in press
C. lunulatus, white-naped mangabey														
Captive (U.S.)									19		152–180			Butynski in press
Lophocebus														
L. albigena, gray-cheeked mangabey														
Lopé, Gabon							36		31 (SD=1.5)	12	183	3		Ham 1994
Kibale, Uganda							33 (extrapolated)		47 (8–184)	26	186	3		Wallis 1983
Captive(France)	3.6 (3.15–4.21)	8			7.4	8	28	13	31 (11–72)	335	175	25	Peak in autumn	Deputte 1991, 1992
Captive (France)									29.8 (25–42)	44	177 (170–181)	3	Peak in autumn (Oct–Nov)	Gautier-Hion and Gautier 1976
L. aterrimus, black (or black-crested) mangabey														
Captive (U.S.)									26 (21–40)	26				Calle et al. 1990
Mandrillus														
M. leucophaeus, drill														
Nigeria	35.5		4.5				39.4				176		Most births May–Sept	Wood and Gadsby unpub. data
Captive (Yerkes)									35		173			Hadidian and Bernstein 1979
M. sphinx, mandrill														
CIRMF, Franceville, Gabon	3.6 2.75–4.5)	10	4.4 (3.25–5.5)	10			12.4 (11–15)	15			167 (152–176)	6	25 of 26 births Jan–May	Wickings and Dixson 1992a,b
CIRMF, Franceville, Gabon			4.6 (3.3–6.1)				15.5						birth peak Jan–March	Setchell et al. 2001

Table 15.6 (cont'd)

TAXON AND STUDY SITE	MEAN OR MEDIAN AGE AT MENARCHE (YEARS)	N	MEAN AGE AT FIRST BIRTH (YEARS)	N	MEAN OR MEDIAN POST PARTUM AMENORRHEA AFTER SURVIVING INFANT (MONTHS)	N	MEAN INTERBIRTH INTERVAL AFTER SURVIVING INFANT (MONTHS)	N	MEAN LENGTH OF OVARIAN CYCLE (DAYS)	N	MEAN OR MEDIAN LENGTH OF GESTATION (DAYS)	N	SEASONALITY OF REPRODUCTION	REFERENCES
CIRMF, Franceville, Gabon											175 (168–183)	61	mating peak July-Sept; birth peak Jan-March	Setchell et al. 2002
CIRMF, Franceville, Gabon	3.7 (3.2–4.6)	31			7.9	92							Mating peak July-Sept	Setchell and Wickings 2004
Lopé Reserve, Gabon													Most sexual swellings June-Nov	Abernethy et al. 2002
Captive (Yerkes)									33.5		175			Hadidian and Bernstein 1979
Papio														
P. anubis, olive (or anubis) baboon														
Athi River, Kenya									34					Maxim and Buettner-Janusch 1963
Gilgil, Kenya			6.7				22.9							Strum and Western 1982
Gilgil, Kenya					13.4		25		42	33	180			Smuts and Nicolson 1989
Gilgil, Kenya	4.4								40					Scott 1984
Gilgil, Kenya									37.5 (SD 3.6)	79			None	Bercovitch 1991
Gilgil, Kenya	4.79		6.92										None	Bercovitch and Strum 1993
Gombe, Tanzania	4.96 (4.5–5.6)												None	Packer 1979b
Gombe, Tanzania	3.5								39 (28–56)	28	182		None	Ransom 1981
Gombe, Tanzania							25							Collins et al. 1984
Gashaka-Gumti, Nigeria					11.3 (7–17)	7	25 (23–27)	2	42.3 (39–46)	3	185.3 (180–188)	4		Higham et al. 2009a,b
Captive									31.5					Shaikh et al. 1982
P. cynocephalus, yellow baboon														
Amboseli, Kenya					12		15						Birth peak June-July	Altmann 1980
Amboseli, Kenya	4.8 (4.0–5.75)	10	5.9	14	12–12.5 (6–16)	20	21	9			175	23	Slight	Altmann et al. 1977, 1981

Table 15.6 (cont'd)

TAXON AND STUDY SITE	MEAN OR MEDIAN AGE AT MENARCHE (YEARS)	N	MEAN AGE AT FIRST BIRTH (YEARS)	N	MEAN OR MEDIAN POST PARTUM AMENORRHEA AFTER SURVIVING INFANT (MONTHS)	N	MEAN INTERBIRTH INTERVAL AFTER SURVIVING INFANT (MONTHS)	N	MEAN LENGTH OF OVARIAN CYCLE (DAYS)	N	MEAN OR MEDIAN LENGTH OF GESTATION (DAYS)	N	SEASONALITY OF REPRODUCTION	REFERENCES
Amboseli, Kenya									32.5 (26–52)	34				Hausfater 1975a
Amboseli, Kenya	4.5	174	5.8	97										
Mikumi, Tanzania	4.9	29	6.5	27									Birth peak June-Aug (early dry season)	Rhine et al. 1988, 2000
Tana River, Kenya	6.0 (4.4–7.5)	6			14.6		28		38.9	120	181.5	28	Birth peaks Jan-Feb and Sept-Oct	Bentley-Condit and Smith 1997
P. hamadryas, hamadryas baboon														
Erer Gota, Ethiopia	4.3	13	6.1 (5.5–7)	8	14 (10–16)	13	22 (18–28)	12						Sigg et al. 1982
Filoha, Ethiopia					8.75 (8.7–8.8)	2	18.8 (15.5–21.5)	3	39	17			None	Swedell 2006
Captive (Gumista, Georgia)	3.4		4.9				14		37				None	Asanov 1972; Caljan et al. 1987
Captive (Deutsches Primatenzentrum)	3	4	4.4	1			13.8	18			182	52		Kaumanns et al. 1989
Capative (Deutsches Primatenzentrum)									42 (33–49)	9				Zinner, pers. comm.
P. papio, Guinea baboon														
Captive (France)	3.8	8			5.3		13.5		29.8	46	182	10		Gauthier 1999
P. ursinus, chacma baboon														
Okavango (Moremi), Botswana							21.7							Bulger and Hamilton 1987
Okavango (Moremi), Botswana			6.75	28									most births July-Dec (dry season)	Cheney et al. 2004, 2006
De Hoop, South Africa							21		39.7	43	190	10	none	Weingrill et al. 2004; Barrett et al. 2006
Cape Peninsula, South Africa							27.7						most births July-Nov (end of wet months)	Davidge 1978b
Drakensberg, South Africa							38						birth peak Sept-Jan (wet months)	Lycett et al. 1998, 1999
Drakensberg, South Africa							38.5						most births Sept-Jan (wet months)	Barrett et al. 2006

Rungwecebus

R. kipunji, kipunji (formerly highland mangabey)

No data

Table 15.6 (cont'd)

TAXON AND STUDY SITE	MEAN OR MEDIAN AGE AT MENARCHE (YEARS)	N	MEAN AGE AT FIRST BIRTH (YEARS)	N	MEAN OR MEDIAN POST PARTUM AMENORRHEA AFTER SURVIVING INFANT (MONTHS)	N	MEAN INTERBIRTH INTERVAL AFTER SURVIVING INFANT (MONTHS)	N	MEAN LENGTH OF OVARIAN CYCLE (DAYS)	N	MEAN OR MEDIAN LENGTH OF GESTATION (DAYS)	N	SEASONALITY OF REPRODUCTION	REFERENCES
Theropithecus														
T. gelada, gelada														
Sankaber, Simien Mountains, Ethiopia					12–18		18–24		29		159			Dunbar and Dunbar 1974b
Sankaber, Simien Mountains, Ethiopia													peaks in June and January	Dunbar and Dunbar 1975
Gich Plateau, Ethiopia									usually 20–49				mating peak Jan–Feb	Mori 1979d
Captive (Yerkes, U.S.)							17.3						none	Hadidian and Bernstein 1979
Captive (NYZS, U.S.)			16				37				182			McCann 1995

with the tail lifted (Dunbar 1978, Mori 1979d, Ransom 1981, Arlet et al. 2007).

Sexual Swellings

In baboons, mangabeys, drills, mandrills, and kipunjis, females exhibit exaggerated sexual swellings, associated with increased sexual activity, corresponding to the periovulatory period (Chalmers 1968b; Hendrickx and Kraemer 1969; Chalmers and Rowell 1971, Saayman 1970, 1972, 1973; Wildt et al. 1977; Wallis 1983; Horn 1987a; Whitten and Russell 1996; Gauthier 1999; Arlet et al. 2007; Davenport and Butynski in press). In captive chacma baboons, artificially exaggerated swellings increased male sexual arousal, showing that swellings function proximately to attract males (Bielert and Anderson 1985, Girolami and Bielert 1987; see Chapter 29 for more on sexual swellings). In geladas, the signal of sexual receptivity takes the form of patches of bare skin on the chest encircled by vesicles. Also found on the perineum, these vesicles fill with water and swell during the periovulatory period (Dunbar and Dunbar 1974b, Dunbar 1978; McCann 1995).

Consortships

In multimale groups of baboons and mangabeys, males and females form sexual consortships, characterized by close spatial association, increased frequency and length of grooming bouts, and frequent copulation (Seyfarth 1978a; Ransom 1981; Rasmussen, K. L. R., 1983, 1985, 1986;

Bercovitch 1983, 1991, 1995). Consortship duration is largely related to the ability of a male to exclude other males from an estrous female, which may in turn be related to factors such as number of females simultaneously in estrus and alternative male strategies such as coalitions. Lengthy consortships are most common in chacma baboons, in which alpha males can monopolize estrous females for several days. In gray-cheeked mangabeys, females typically solicit copulations from multiple partners and males are not able to monopolize females during consortships as in baboons (Arlet et al 2007, 2008; P. Waser, pers. comm.)

Copulation Calls

Many African papionins give copulation calls, vocalizations by females during and/or after copulation (Hall 1962b, Saayman 1970, Boese 1973, Hamilton and Arrowood 1978, Collins 1981, Ransom 1981, Bercovitch 1985, O'Connell and Cowlishaw 1994, Semple 2001, Semple et al. 2002, Maestripieri et al. 2005, Swedell and Saunders 2006, Arlet et al. 2007, Ehardt and Butynski in Ehardt and Butynski in press b). In baboons, these calls have been described as "intermittent roars" (Bolwig 1959a), "staccato grunts" (Saayman 1970), and "gurgling growls" (Hall 1962b). Females give copulation calls while defecating as well (Hall 1962b, Boese 1973, Bercovitch 1985), suggesting that they may be an involuntary reaction to compression of the vaginal wall. If copulation calls are costly signals, however, then they must also serve a communicative function, one of which may be the incitation of competition among males (Hamilton and Arrowood 1978, Dunbar 1988, Dixson 1998, O'Connell and Cowlishaw 1994).

Life History and Reproduction

Males

Males reach adult size between 7 and 11 years of age, with an adolescent growth spurt just prior to attainment of adult body size (Altmann et al. 1981, Sigg et al. 1982, Deputte 1992, Setchell et al. 2001). The extreme sexual dimorphism of mandrills results from an exceptionally prolonged growth spurt in males (Setchell et al. 2001).

In baboons, male testes enlargement typically occurs between 4 and 6.5 years of age (Alberts and Altmann 1995b; Jolly and Phillips-Conroy 2003, 2006). Hamadryas testes enlargement begins at least a year earlier than in olive and yellow baboons; this may relate to a life history strategy in which subadult males copulate surreptitiously prior to attaining leader male status, an option for hamadryas because males breed in their natal groups (Jolly and Phillips-Conroy 2003, 2006). Supporting this notion is the fact that, by age 4, hamadryas males have sperm that are indistinguishable from those of an adult and are capable of siring offspring (Zinner et al. 2006).

Among semi-free-ranging mandrills in Gabon, the development of the striking coloration on the face, perineum, and genitalia is associated with both testosterone and male dominance rank (Setchell and Dixson 2001a, Setchell et al. 2006, 2008). With a rise in rank to alpha status, males develop larger testes, higher testosterone concentrations, brighter coloration, and fully functional sternal scent glands that produce glandular exudate (Setchell and Dixson 2001a,b). Upon a loss in alpha status, males undergo a decrease in both testicular volume and body mass as well as a reduction in red facial and perineal coloration and gland functioning (Setchell and Dixson 2001b). Genital and perineal coloration in drills is likely also tied to testosterone and dominance as it can be induced experimentally with testosterone injections (Zuckerman and Parkes 1939).

Females

African papionin females typically reach menarche between 4 and 5 years of age in the wild and at least a year earlier in captivity and give birth to their first surviving infant between 5 and 7 years of age after a period of adolescent subfertility (Altmann et al. 1977, 1981; Hadadian and Bernstein 1979; Scott 1984; Smuts and Nicolson 1989; Gust et al. 1990; Gauthier 1999) (Table 15.6). Females reach adult body size between 5 and 6 years of age (Sigg et al. 1982; Wickings and Dixson 1992a).

Ovarian cycles are typically 30–40 days in length (Table 15.6), with sexual swellings lasting anywhere from 1–5 days in geladas to 19 days in olive baboons (Hendrickx and Kraemer 1969, Mori 1979d, Bercovitch 1991). Gestation lengths vary from 165 to 185 days. In baboons and geladas,

the paracallosal skin, lateral to the ischial callosities, reddens during pregnancy (Altmann 1970, Dunbar and Dunbar 1974b, Gauthier 1999). In drills, mandrills, and some mangabeys, post-conception sexual swellings occur, usually during the first 1-2 months of pregnancy (Hadadian and Bernstein 1979; Wallis 1983; Kinnaird 1990b; Setchell and Wickings 2004; Ehardt in press). Interbirth intervals (after infant survival) range from 12 to 40 months (Table 15.6).

Virtually all female reproductive parameters respond to changes in local environmental conditions. Age at menarche, interbirth intervals, infant growth rates, and infant survival all vary with food supply (Bercovitch and Strum 1993; Lycett et al. 1998; Gauthier 1999; Altmann and Alberts 2003b; Barrett et al. 2006; Higham et al. 2009b); interbirth intervals vary with mean annual temperature (Hill et al. 2000); and extreme heat or drought reduces female fertility (Beehner et al. 2006).

Mating Systems and Reproductive Strategies

Mating systems in African papionins take essentially three main forms. Most mangabeys and olive, chacma, and yellow baboons are polygamous within multimale groups. Hamadryas baboons, geladas, and possibly Guinea baboons are generally polygynous within OMUs. Drills and mandrills exhibit a seasonal polygamy in which adult males enter groups and copulate only during the mating season.

Male Reproductive Strategies

Aggressive Competition in Multimale Groups

In multimale groups, aggression among males increases in frequency and intensity during the mating season in general and/or specifically when estrous females are present (Hall 1962b; Hall and DeVore 1965; Chalmers 1968b; Hausfater 1975b; Packer 1979b; Berenstain and Wade 1983; Bulger 1993; Olupot and Waser 2001b; Arlet et al. 2008, 2009; K.L. Wood, personal communication). This competition is usually mediated by a linear dominance hierarchy among males in which rank often, but not always, correlates with priority of access to females (DeVore 1965a; Hausfater 1975a; Packer 1979b; Berenstain and Wade 1983; Bercovitch 1986; Cowlishaw and Dunbar 1991; Bulger 1993; Arlet et al. 2007, 2008) and reproductive success (Dixson et al. 1993, Wickings et al. 1993, Altmann et al. 1996). Males may maintain sexual access to females via consortships in which males mate guard females for several hours to several days (Hausfater 1975a, Bercovitch 1987b, Alberts et al. 1996, Arlet et al. 2008). Efforts by high-ranking males to monopolize matings are constrained, however, by female mate preferences, surreptitious copulations, and male coalitions that overturn consortships (Packer 1977; Bercovitch 1986, 1988; Cowlishaw and Dunbar 1991;

Bulger 1993). In olive and yellow baboon males, for example, consortship duration is relatively short, largely due to coalitions by lower-ranking males (Hall and DeVore 1965, Smuts 1985, Bercovitch 1988, Noë and Sluijter 1990). Chacma baboon males, by contrast, do not form coalitions and alpha males are able to monopolize estrous females for most of peak estrus, resulting in a higher reproductive skew in chacmas compared to olive and yellow baboons (Bulger 1993, Weingrill et al. 2000). A similar degree of reproductive skew probably characterizes mandrills, as alpha males in semi-free-ranging conditions succeed in monopolizing estrous females (Setchell et al. 2005a,b). As noted, males may also forge friendships with females as an alternate means to increase mating success, effectively "opting out" of the power struggle (Smuts 1983a, 1985).

Sperm Competition

Males in multimale groups may also compete via sperm competition (Harcourt et al. 1981a; Bercovitch 1989; see Chapter 29). That sperm competition occurs in baboons is supported by a relationship between testis size and mating system across species. Jolly and Phillips-Conroy (2003, 2006), for example, found that testis size is larger in olive and yellow baboons, in which females mate with multiple males during a single estrous period, than in hamadryas baboons, in which females typically mate with only one male (Swedell and Saunders 2006), and that testis size in Guinea baboons is intermediate between the two.

Exclusion Strategies

In hamadryas baboons and geladas, and possibly in Guinea baboons, males adopt an exclusion strategy whereby they defend a small group of females from other males at all times. In this way, a male maintains (nearly) exclusive reproductive access to a set of females, with reproductive success largely determined by the length of his tenure as leader male. In hamadryas, males "respect" the "possession" of females in that they are largely inhibited from challenging a leader male for his females or openly attempting to mate with females that are part of an OMU (Kummer et al. 1974, Sigg and Falett 1985).

In both hamadryas and geladas, males may acquire females via aggressive take-overs, via assuming a follower role and waiting to inherit females, or via "adopting" a young prereproductive female and forming an "initial unit" with her that will eventually develop into a reproductive unit (Kummer 1968a; Mori 1979f; Sigg et al. 1982; Dunbar 1984b, 1993b). Despite the apparent reproductive monopoly in both species, however, the actual degree of reproductive skew is not yet known. Females do copulate with males other than their leader (hamadryas, Swedell and Saunders 2006), at least during attempted take-overs (geladas, J. C. Beehner and T. J. Bergman, personal communication), and hamadryas subadult males as young as 4 years are likely fertile (Jolly and Phillips-Conroy 2003, 2006; Zinner et al. 2006).

While one-male groups are relatively common in chacma baboons, they do not appear to be an exclusion strategy but simply the outcome of males mapping themselves onto females found at a lower density due to low habitat quality and reduced predator pressure. When this occurs, patterns of bonding often change as well, with stronger cross-sex social bonds than in multimale groups, likely a result of increased paternity certainty and reproductive investment by males (Byrne et al. 1989; Anderson 1992; Barton et al. 1996; Henzi et al. 1997b, 2000).

Male mandrills rely heavily on olfactory, vocal, and visual displays to acquire mates. For example, males scent-mark with their sternal glands, grunt continuously in the presence of estrous females, and display colorful visual signals on their faces, perineum, and genitalia. These signals have likely replaced the aggressive behavioral competition of other papionin males and have likely evolved via sexual selection (Setchell and Dixson 2001a,b; Setchell et al. 2008). Whether dominance relationships and rank-related differences in physiology and behavior characterize wild mandrills, however, is not known (K. A. Abernethy, personal communication).

Herding and Coercion

In baboons and some mangabeys, males are often aggressive toward females that consort or attempt to copulate with other (usually lower-ranking or extragroup) males (Kummer 1968a, Berenstain and Wade 1983, Arlet et al 2008; Kitchen et al. 2009). Herding behavior, in which males use visual threats, chasing, and aggression to control female movements, is most prevalent and stereotyped in hamadryas, in which leader males routinely herd females to maintain cohesion of OMUs. Herding also occurs in chacma and olive baboons, especially during intergroup encounters (Saayman 1971, Packer 1979a, Byrne et al. 1987, Henzi et al. 1998, Kitchen et al. 2004), at the initiation of troop movement (Stoltz and Saayman 1970, Buskirk et al. 1974), and during consortships (Bercovitch 1995). Male herding of females has also been reported for Guinea baboons (Boese 1975) and drills (Gartlan 1970). In chacmas, herding occurs more in one-male than multimale groups (Byrne et al. 1987, Hamilton and Bulger 1992, Barton et al. 1996).

Female Reproductive Strategies

African papionin females commonly mate with multiple males during a given estrous period (Hall 1962b; Hall and DeVore 1965; Saayman 1970; Hausfater 1975a; Seyfarth 1978a; Smuts 1985; Bercovitch 1987b, 1995; Noë and Sluijter 1990; Arlet et al. 2007). Although not generally selective, they sometimes exhibit preferences for unfamiliar or immigrant males (Bercovitch 1991, 1992; Arlet et al 2007). Females exert choice in mating (Halliday 1983, Small 1989) and may initiate or refuse copulations (Hall 1962b; Hausfater 1975a; Seyfarth 1978a; Scott 1984; Smuts 1985; Bercovitch 1991; Arlet et al. 2007). Females (and males) may also circumvent mating strategies of other individuals by engaging in surreptitious copulations (Arlet et al. 2007, 2008).

African papionin female reproduction is also mediated by competition with other females (Seyfarth 1976, Wasser 1983, Barton 1993). In yellow baboons, for example, aggressive coalitions among females increase in frequency with the number of simultaneously estrous females (Wasser 1983, Wasser and Starling 1986). In olive baboons, follicular phases are longer than usual in females who suffer severe aggression from other females before, during, or just after menstruation (Rowell 1970a). In wild geladas, Dunbar and Dunbar (1977, Dunbar 1980b) suggested that aggression by high-ranking females delays conception in low-ranking females. This has since been confirmed in captive geladas by McCann (1995), who found that low-ranking females receive more aggression, especially during the follicular phase, and have longer cycles than high-ranking females.

African papionin females may also compete for access to mates. In geladas, the alpha female of each OMU usually has a closer relationship with the male than do the other females, and she often threatens females that approach or copulate with him (Mori 1979b,d,e). Low-ranking gelada females not only receive more aggression than do high-ranking females but copulate less frequently as well (McCann 1995). Patterns of aggression among females suggestive of competition for access to males have also been observed in chacma (Hall 1962b; Seyfarth 1976) and olive (Rowell 1967) baboons.

Finally, Dunbar and Sharman (1983) found a negative correlation between adult sex ratio and birthrate in olive, yellow, and chacma baboons: The more females in a group relative to males, the lower the birthrate. Similarly, Zinner et al. (1994) found that the degree of estrus synchrony within captive hamadryas OMUs is lower during conceptive compared to nonconceptive cycles, suggesting that sperm in hamadryas OMUs is in limited supply or that females vary in their access to the leader male during estrus (Small 1988, Zinner et al. 1994).

Infanticide by Males and Female Counterstrategies

Infanticide and attempted infanticide by males—either directly observed or strongly inferred—has been reported for a number of populations of baboons (Palombit 2003; Swedell and Saunders 2006) as well as geladas (Beehner and Bergman 2008, and has been suspected in mangabeys as well (Arlet et al 2007). While baboon populations vary widely in the occurrence of infanticide, its prevalence in the genus *Papio* as a whole suggests that it is a behavioral predisposition shared by all baboon males (Palombit 2003). For example, in savanna baboon populations such as that at Moremi (e.g., Palombit et al. 1997, 2000), where adult males are commonly in contact with infants they likely did not sire, the risk of infanticide is chronic and it is the primary cause of infant mortality (Cheney et al. 2006). Females are

in fact attuned to this risk as stress hormone levels rise in lactating females following both male immigration (Engh et al. 2006) and rises in rank by immigrant males (Beehner et al. 2005). In hamadryas baboons and geladas, by contrast, the actual rate and chronic risk of infanticide are low but the acute risk after take-overs is high. In fact, the normally high rate of infant survival in hamadryas compared to other baboons (Sigg et al. 1982, Swedell, 2006) coupled with the observed and inferred infant mortality after take-overs in hamadryas and geladas (Swedell 2000, Swedell and Tesfaye 2003) suggest that infanticide may be the primary cause of infant mortality for both geladas and hamadryas baboons as well (Swedell and Saunders 2006, Beehner and Bergman 2008).

In female baboons, mangabeys, and geladas, adaptive responses to male infanticide may include minimizing losses via an earlier return to reproductive condition following immigration or take-overs (Pereira 1983; Dunbar 1984a; Colmenares and Gomendio 1988; Alberts et al. 1992; Swedell, 2000, 2006; Beehner and Bergman 2008), the manipulation of paternity assessment through "pseudo-estrus" (Kinnaird 1990b, Zinner and Deschner 2000, Swedell 2006) or mating with multiple males (Hausfater 1975a, Smuts 1985, Bercovitch 1987b, Swedell 2006), and social bonding with a protective male for infant defense (Smuts 1985, Palombit et al. 1997, Weingrill 2000, Swedell 2006; Arlet et al 2007). Within the paternity confusion–concentration dichotomy proposed by van Schaik et al. (1999b), most baboons use a predominant strategy of paternity *confusion*, characterized by behavioral elements such as presentations to multiple males, copulation with multiple males, postcopulatory darts, and copulation calls, all of which elicit contest and/or sperm competition and promote mating with multiple partners (Hall and DeVore 1965; Hausfater 1975a; Cox and LeBoeuf 1977; Bercovitch 1991, 1995; Oda and Masataka 1992; O'Connell and Cowlishaw 1994, 1995; Dixson 1998; Soltis 2002; Swedell and Saunders 2006). Even in chacma baboons, in which alpha males monopolize females during peak estrus, females solicit and copulate with other males during early and late estrus (Saayman 1970; Bulger 1993; Weingrill et al. 2000, 2003).

In hamadryas baboons and geladas, by contrast, females use a predominant strategy of paternity *concentration*, with both males and females focusing their reproductive effort more narrowly than in other African papionins, a pattern that likely characterizes Guinea baboons as well (Boese 1975, Swedell and Saunders 2006). Among *Papio* baboons, such a system whereby males and females engage in mutualistic strategies of close association, protection, and (relatively) exclusive mating has evolved to an extreme only in hamadryas, but it appears to exist to varying degrees in other baboon species (Boese 1973; Smuts 1985; Anderson 1989, 1990; Palombit et al. 1997). For example, strong social bonds or "friendships" characterize some baboon females and specific adult males; these relationships may benefit females via protection from infanticide or aggression by males (Smuts 1985, Palombit et al. 1997, Weingrill 2000). In some cases, males that pursue a friendship strategy have greater access to those females as mates. In chacma baboons in particular, females may be in effect "concentrating" paternity by focusing their periovulatory mating on the alpha male (or the resident male in a one-male group) while simultaneously "confusing" paternity by mating with other males at other times.

BABOON COMMENSALISM

Baboons, like macaques, are remarkable for their ecological and behavioral flexibility (Alberts and Altmann 2006, Jolly in press). As a result, human population expansion does not adversely affect baboon populations in the same way as it does less behaviorally flexible, more habitat-limited primates such as mangabeys or geladas (Iwamoto and Dunbar 1983, Wieczkowski 2004; Mau et al 2009). Baboon omnivory and opportunism in feeding, behavioral plasticity, and versatile substrate use all predispose them to use changing environments in new ways (Strum 1987, Else 1991, Chism 2005). One of the consequences of baboon flexibility, therefore, is a level of contact between humans and baboons that does not occur in many other primates (Fig. 15.2).

The most widely cited form of commensalism in baboons (and other primates) is crop raiding, which occurs virtually everywhere that baboons are found adjacent to agricultural areas (Maples et al. 1976; Forthman Quick 1986; Strum 1986, 1987; Else 1991; Biquand et al. 1992a, 1994; Naughton-Treves 1998; Hill 2000, 2005; Zinner et al. 2001b; Higham et al. 2009b). Baboons also raid garbage dumps (Altmann and Muruthi 1988; Biquand et al. 1992a, 1994;

Zinner et al. 2001b; Okecha and Newton-Fisher 2006) and even resort to the direct theft of food from homes, game lodges, and picnic spots in national parks (Else 1991, Kansky and Gaynor 2000). In some cases, humans deliberately provision baboons (Biquand et al. 1992a, 1994; Boug et al. 1994b; Kamal et al. 1997), thereby reinforcing a mental association between humans and food and attracting baboons to human-frequented areas.

Commensalism increases feeding efficiency, which influences activity budgets. Olive and yellow baboons that feed on human-derived foods spend far less time foraging and feeding and more time resting than neighboring wild-foraging troops (Forthman Quick 1986, Altmann and Muruthi 1988, Muruthi et al. 1991; Bronikowski and Altmann 1996). In addition to activity budget changes, a concentration of high-quality food in one area leads to reduced home range sizes and travel distances (Forthman Quick and Demment 1988, Altmann and Muruthi 1988).

The availability of human food waste can also have cascading effects on reproductive parameters, life history variables, health, mortality, and social organization (Bishop et al. 1981). At Amboseli, baboons with regular access to food from a garbage dump experience higher infant growth rates, earlier maturation, higher infant survival, and reduced interbirth intervals compared to wild-foraging troops (Altmann and Alberts 2003b); and crop-raiding olive baboons in Nigeria have shorter interbirth intervals than wild-foraging baboons (Higham et al. 2009b). Moreover, baboons at Amboseli that subsist largely on food from garbage dumps are 50% heavier with 21% more body fat (Altmann et al. 1993) and higher concentrations of serum insulin and cholesterol compared to wild-foraging troops, suggesting a greater risk of cardiovascular disease (Kemnitz et al. 2002).

Figure 15.2 A chacma baboon (*Papio ursinus*) waiting for a raiding opportunity in the Cape Peninsula of South Africa (photo by Larissa Swedell).

The hunting of nonhuman primates by humans for food, most common in west and central Africa, is a direct source of human-mediated mortality for drills, mangabeys, and baboons (Gonzalez-Kirchner and Sainz de la Maza 1996, Paterson 2006, Fischer et al. 1999–2000). Even when hunting does not occur, wherever humans and baboons live in close proximity and there are no inhibitory cultural factors (e.g., Bishop et al. 1981, Wolfe and Fuentes 2007), people generally regard baboons as pests and often kill them either to eliminate raiding behavior or for sport. Human-induced mortality is usually biased toward males as the former are typically more aggressive as well as larger and more conspicuous, thus more easily identified and shot.

One of the best examples of baboon commensalism can be found in the Cape Peninsula of South Africa, where an expanding human population and heavy tourism have dramatically impacted the ecology and behavior of the baboon population (Kansky and Gaynor 2000). Virtually every troop in the Cape Peninsula has some contact with humans and human-produced food, whether it be along the highways, in urban areas, or in Table Mountain National Park. Commensalism in the Cape Peninsula has led to extensive research on baboon–human conflict, the establishment of

a Baboon Management Team, and the employment of "baboon monitors" to herd baboons away from towns and other human-frequented areas (Kansky and Gaynor 2000, van Doorn 2009).

As human populations continue to expand, nonhuman primate commensalism will become increasingly common. Baboons are not the only primates to which these issues apply: The types of human–wildlife conflict described here, as well as the associated behavioral changes, are also found in other omnivorous and behaviorally flexible primates, such as vervet monkeys (Brennan et al. 1985, Lee et al. 1986) and macaques (Fa 1991, Schlotterhausen 2000). One role of conservation efforts, in addition to habitat preservation and education, should thus be to continue to devise and refine management solutions to conflict between humans and nonhuman primates, especially such highly flexible primates as baboons (Strum 1986, 1987, 1994, 2005; Biquand et al. 1994; Hill 2000; Forthman et al. 2005; Webber et al. 2007; van Doorn 2009).

Species included in this chapter are shown in Color Plates 27–29 (*Papio hamadryas*, *Papio anubis*, and *Mandrillus sphinx*).

16

The Guenons

Polyspecific Associations in Socioecological Perspective

Karin E. Jaffe and Lynne A. Isbell

---

1. What genus is normally associated with the term "guenon," and what are the other genera the authors have included in their broader definition of this term? Why have the authors included these other genera within their discussion of the guenons?
2. Describe the pattern of social organization among the guenons. What is the most common type of social system for guenons? Which species differ from this "norm," and what type of social system do they possess?
3. What is a polyspecific association? What are the benefits of polyspecific associations, and how might we determine if the guenons are benefiting from them?

INTRODUCTION

This chapter deals with the taxonomy, distribution, ecology, reproduction, and social organization of the guenons and their allies. Strictly speaking, the term "guenon" includes catarrhine

monkeys of the genus *Cercopithecus* (Butynski 2002); but in this chapter, we will expand our discussion to include three other closely related genera of the tribe Cercopithecini—*Allenopithecus*, *Erythrocebus*, and *Miopithecus* (Grubb et al. 2003) (Table 16.1)—and refer to all members of the tribe

Table 16.1 Comparison of Two Most Recent Guenon Taxonomies

SCIENTIFIC NAME	COMMON NAME	GRUBB ET AL. 2003	GROVES 2005	DISCREPANCIES[2]
Allenopithecus nigroviridis	Allen's swamp monkey	x[1]	x	
Miopithecus talapoin	Angolan talapoin	x	x	
Miopithecus ogouensis	Gabon talapoin	x	x	
Erythrocebus patas	Patas monkey	x	x	
C. aethiops superspecies		X		Groves 2005: *Chlorocebus* (six species)
C. aethiops	Vervet/grivet/green monkey	x	x	Ch. aethiops (Grivet)
			x	*Ch. cynosuros* (Malbrouck)
			x	*Ch. djamdjamensis* (Bale Mt. vervet)
			x	*Ch. pygerythrus* (Vervet)
			x	*Ch. sabaeus* (Green monkey)
			x	*Ch. tantalus* (Tantalus monkey)
C. cephus superspecies		X[2]	X	
C. ascanius	Red-tailed monkey	x	x	
C. cephus	Mustached guenon	x	x	
C. erythrogaster	White-throated guenon	x	x	
C. erythrotis	Red-eared guenon	x	x	
C. petaurista	Lesser spot-nosed guenon	x	x	
C. sclateri	Sclater's guenon	x	x	
C. diana superspecies		X	X	
C. diana	Diana monkey	x	x	
C. roloway	Roloway monkey	x	x	
C. dryas superspecies		X	X	
C. dryas	Dryas guenon	x	x	
C. hamlyni superspecies		X	X	
C. hamlyni	Owl-faced guenon/ Hamlyn's monkey	x	x	
C. lhoesti superspecies		X	X	Grubb et al. 2003: *C. pruessi* superspecies
C. lhoesti	L'Hoest's monkey	x	x	
C. preussi	Preuss's monkey	x	x	
C. solatus	Sun-tailed guenon	x	x	
C. mitis superspecies		X	X	Grubb et al. 2003: *C. nictitans* superspecies
C. albogularis	Sykes monkey		x	Grubb et al. 2003: *C. mitis* subspecies
C. doggetti	Silver monkey		x	Grubb et al. 2003: *C. mitis* subspecies
C. kandti	Golden monkey		x	Grubb et al. 2003: *C. mitis* subspecies
C. mitis	Blue monkey	x	x	
C. nictitans	Putty-nosed/greater spot-nosed guenon	x	x	
C. mona superspecies		X	X	
C. campbelli	Campbell's guenon/ Campbell's Mona monkey	x	x	
C. denti	Dent's guenon/Dent's Mona monkey		x	Grubb et al. 2003: *C. pogonias* subspecies

Table 16.1 (cont'd)

SCIENTIFIC NAME	COMMON NAME	GRUBB ET AL. 2003	GROVES 2005	DISCREPANCIES[2]
C. lowei	Lowe's guenon/Lowe's Mona monkey		x	Grubb et al. 2003: *C. campbelli* subspecies
C. mona	Mona monkey	x	x	
C. pogonias	Crowned guenon/Crested Mona monkey	x	x	
C. wolfi	Wolf's guenon/Wolf's Mona monkey		x	Grubb et al. 2003: *C. pogonias* subspecies
C. neglectus superspecies		X	X	
C. neglectus	De Brazza's monkey	x	x	
Total *Cercopithecus* species		20	25	
Total guenon species		24	35	

[1] An x indicates that taxonomy includes the species (or subspecies) listed. An X indicates the taxonomy includes the superspecies group listed.

[2] Indicates the differences between the two taxonomies and where the discrepancy lies. For example, Grubb et al. (2003) lists one species (*C. aethiops*) in the *C. aethiops* superspecies group, with six subspecies. Groves (2005) lists six species of *Chlorocebus*.

Cercopithecini as "guenons" (see also Butynski 2002). Specific genera and species will be discussed as warranted.

Guenons are small to medium-sized monkeys (average 3.6 kg for females, 5.9 kg for males; Haltenorth and Diller 1988) distributed throughout sub-Saharan Africa. The members of the genus *Cercopithecus* are the most colorful of all primates (Rowell 1988a), particularly in the face; and the variation in coloration may be used in intraspecific communication (see Vine 1970, Kingdon 1987) for recognition of individuals, species, and appropriate mates (e.g., sexual selection, see Ryan and Rand 1993). *Allenopithecus* and *Miopithecus* are drab, with coat colors ranging from greenish gray in *Allenopithecus* to varying shades of olive-yellow in *Miopithecus* (Haltenorth and Diller 1988). *Erythrocebus* is unique among guenons with its reddish brown to orange coat color (Haltenorth and Diller 1988).

TAXONOMY AND DISTRIBUTION

Considerable debate still exists regarding the taxonomy of the tribe Cercopithecini (see Butynski 2002), but one thing is clear: The guenons are the largest group of African primates. Recent taxonomic work lists as few as 24 (Grubb et al. 2003) and as many as 35 (Groves 2005) species of guenons (Table 16.1). Within the guenons, the genus *Cercopithecus* is divided into several superspecies groups. According to Groves (2005), there are eight superspecies groups: *C. cephus*, *C. diana*, *C. dryas*, *C. hamlyni*, *C. lhoesti*, *C. mitis*, *C. mona*, and *C. neglectus*. Grubb et al. (2003) give a similar, but slightly different, organization of the superspecies groups, dividing the genus into *C. aethiops*, *C. cephus*, *C. diana*, *C. dryas*, *C. hamlyni*, *C. preussi* (*C. lhoesti* group in Groves [2005] taxonomy), *C. nictitans* (*C. mitis* group in Groves [2005] taxonomy), *C. mona*, and *C. neglectus* (see Table 16.1). One should note that there are

nine superspecies groups in Grubb et al.'s (2003) taxonomy of *Cercopithecus* because they retain *Cercopithecus* as the genus name for vervet/grivet/green monkeys (*C. aethiops*). Groves (2005) elevates *C. aethiops* to its own genus, *Chlorocebus*, due to certain morphological differences (e.g., shape of the auditory tube and eye orbits, shape of the second upper incisor) between this species group and the genus *Cercopithecus* (see Groves 1989).

Likewise, there is dispute between these two studies regarding the division within the vervet/grivet/green monkey group. Grubb et al. (2003), following Napier (1981), consider the *C. aethiops* group to be one highly polytypic species, partly because of uncertainty about the boundaries between the species and partly because Struhsaker (1970) was unable to distinguish between populations of *C. aethiops* and *C. tantalus* based on their vocalizations. According to Grubb et al. (2003), this group consists of six subspecies of *C. aethiops*, all of which Groves (2005) has elevated to a species within the genus *Chlorocebus* (Table 16.1). Other divisions of the vervet/grivet/green monkey group also exist. Kingdon et al. (in press) retain the *Cercopithecus* genus name for the vervet/grivet/green monkey group but divide the group into five species rather than subspecies.

There is less dispute over the taxonomy of *Allenopithecus*, *Erythrocebus*, and *Miopithecus*. Both *Allenopithecus* and *Erythrocebus* are monotypic, but *Erythrocebus* may contain from zero (Groves 2001a, 2005; Grubb et al. 2003) to four (Kingdon 1997) subspecies. Isbell (in press) describes three subspecies of *Erythrocebus* (*E. patas patas* in West Africa, *E. p. baumstarki* in Tanzania, and *E. p. pyrrhonotus* in the rest of East Africa) based on differences in facial coloration. *Miopithecus*, once believed to be monotypic, is now believed to contain two species, *M. talapoin* in the northern part of its range and *M. ogouensis* in the south (Kingdon 1997, Groves 2001a, Grubb et al. 2003). In this chapter, we follow the taxonomy of Grubb et al. (2003) and use the genus name *Cercopithecus* rather than *Chlorocebus* when referring to the vervet/grivet/green monkey superspecies group (Table 16.1).

All guenons are endemic to sub-Saharan Africa (Table 16.2). Within the genus *Cercopithecus*, the *C. aethiops* superspecies group is by far the most widespread, being found throughout sub-Saharan Africa, except in the Namib Desert in southern Africa and the dense rain forests of the West African nations of Nigeria, Cameroon, Gabon, Congo, and the Democratic Republic of Congo. The rest of the *Cercopithecus* species are found primarily in central and western Africa. The distributions of a few species, including *C. lhoesti*, *C. mitis*, and *C. neglectus*, extend as far east as Rwanda, Uganda, and Kenya. *Allenopithecus nigroviridis* is found in central Africa, from Angola in the southwest to the Democratic Republic of Congo to the northeast of its distribution. *E. patas* is found in a relatively narrow band just south of the Sahara Desert, from Senegal and Guinea in West Africa to Ethiopia, Kenya, and Tanzania in East Africa. *Miopithecus* is found in West Africa, with *M. talapoin* distributed in Cameroon and Gabon in the north and *M. ogouensis* in Angola and the Democratic Republic of Congo in the south.

ECOLOGY

Habitat

All members of the genus *Cercopithecus*, except the vervet/grivet group, occupy varying forest habitats, from primary, secondary, and gallery rain forest (e.g., *C. ascanius* and *C. pogonias*) to bamboo forest (e.g., *C. hamlyni* and *C. neglectus*) to flooded and swamp forest (e.g., *C. ascanius*, *C. neglectus*, and

Table 16.2 Guenon Distributions

SPECIES	DISTRIBUTION
C. aethiops	Widespread through sub-Saharan Africa, except Namib Desert in S. Africa, and the dense rain forests of West Africa
C. ascanius	Central Africa: Angola in the SW to Central African Republic in the north to Uganda and Zambia in east
C. campbelli	Coastal NW Africa: Gambia to Ghana
C. cephus	West Africa: Cameroon in the north to Angola in south
C. diana	Coastal NW Africa: Sierra Leone to Ghana
C. dryas	Central Africa: Northern Democratic Republic of Congo
C. erythrogaster	West Africa: Benin and Nigeria
C. erythrotis	West Africa: Nigera, Cameroon, Equatorial Guinea
C. hamlyni	Central Africa: Democratic Republic of Congo and Rwanda
C. lhoesti	Central/East Africa: Democratic Republic of Congo, Rwanda, Uganda, Burundi
C. mitis	Widespread through central and eastern Africa, Ethiopia in north, NE coast South Africa in south and Angola in west
C. mona	Coastal West Africa: Ghana to Cameroon
C. neglectus	Central Africa: Cameroon in NW to Uganda and Kenya in NE to Angola in south; small population in Sudan
C. nictitans	West Africa: Liberia to Ivory Coast and Nigeria to Angola
C. petaurista	Coastal West Africa: Gambia to Benin
C. pogonias	West Africa: Nigeria to Gabon to Central African Republic
C. preussi	West Africa: Cameroon and Equatorial Guinea
C. sclateri	West Africa: southern Nigeria
C. solatus	West Africa: Gabon
C. wolfi	Central Africa: Democratic Republic of Congo to Uganda
A. nigroviridis	Central Africa: Democratic Republic of Congo to Angola
M. talapoin	West Africa: Angola and Democratic Republic of Congo in south and Cameroon and Gabon in north
E. patas	Band just below Sahara: Senegal and Guinea in west to Ethiopia, Kenya, and Tanzania in east

Source: Wolfheim (1983).

C. wolfi) (Table 16.3). Members of the *C. aethiops* superspecies group prefer closed savanna woodland habitat to primary and secondary rain forest, in contrast to other members of their genus, and sleep in trees along rivers or other water courses but also forage away from rivers in closed woodland during the day. Like most members of the *Cercopithecus* genus, *A. nigroviridis* and *M. talapoin* occupy forest habitats. *Miopithecus* are found in a variety of forest types, from primary and secondary forest to swamp forest. *A. nigroviridis* in particular prefer primary lowland swamp forest, which is a likely contributor to the relative lack of information about this species in the wild. *Erythrocebus patas* differ from the forest guenons, *Allenopithecus* and *Miopithecus*, but are similar to

the *C. aethiops* superspecies group in their preference for more open habitats. *E. patas* occupy nonriverine grassland and open woodland in both western (Cameroon, Nakagawa 1999) and eastern (Kenya, Chism and Rowell 1988, Enstam and Isbell 2002) Africa. *E. patas* and *C. aethiops* avoid dense rain forest habitat (although *C. aethiops* in South Africa can be found in forest habitats [e.g., van der Zee and Skinner 1977]), unlike all other members of the Cercopithecini tribe.

Diet

The guenons are members of the subfamily Cercopithecinae (the cheek-pouch monkeys), which also includes the tribe

Table 16.3 Guenon Habitats

SPECIES	HABITAT	REFERENCE
C. aethiops	Closed riparian savana woodland	Struhsaker 1967a, Nakagawa 1999, Enstam and Isbell 2002
C. ascanius	Moist deciduous forest / Swamp forest	Cords 1986 / Zeeve 1991
C. campbelli	Primary, secondary, swamp forest / Tropical evergreen seasonal lowland forest	Gautier-Hion et al. 1981, Zuberbühler 2001, Buzzard 2006
C. cephus	Primary tropical rain forest	Fleury and Gautier-Hion 1997
C. diana	Tropical, moist forest / Secondary forest / Tropical evergreen seasonal lowland forest	Wolters and Zuberbühler 2003 / Whitesides 1989, Oates and Whitesides 1990 / Buzzard 2006
C. dryas	Tropical rain forest	Colyn et al. 1991
C. erythrgaster	Lowland rain forest	Oates 1985
C. erythrotis	Lowland, tropical forest	Struhsaker 2000
C. hamlyni	Bamboo forest, lowland forest	Wolfheim 1983
C. lhoesti	Tropical montane forest, mature lowland rain forest	Struhsaker 1969, Kaplin 2001, Kaplin and Moermond 2000
C. mitis	Moist, semideciduous forest, evergreen rain forest / Tropical montane forest	Cords 1986, Butynski 1990 / Kaplin 2001
C. mona	Mangrove swamp forest	Struhsaker 1969
C. neglectus	Swampy, flooded areas of riverine forest / Montane forest	Wahome et al. 1993 / Zeeve 1991
C. nictitans	Primary tropical rainforest	Fleury and Gautier-Hion 1997
C. petaurista	Primary rain forest / Tropical evergreen seasonal lowland forest	Zuberbühler et al. 1999 / Buzzard 2006
C. pogonias	Primary tropical rain forest	Fleury and Gautier-Hion 1997
C. preussi	Primary and secondary forest	Fleury and Gautier-Hion 1997
C. sclateri	Moist, tropical, swamp, riverine forest	Oates et al. 1990
C. solatus	Primary tropical rain forest	Fleury and Gautier-Hion 1997
C. wolfi	Primary, secondary lowland rainforest, swamp forest	Zeeve 1991
A. nigroviridis	Primary lowland riverine swamp forest	Gautier 1985, Zeeve 1991
M. talapoin	Inundated primary riverine forest	Gautier-Hion 1971, 1973
E. patas	Open woodland, wooded savanna	Hall 1965a, Nakagawa 1999, Chism and Rowell 1988, Enstam and Isbell 2002

Papionini (*Papio* [baboons], *Cercocebus* [mangabeys], *Lophocebus* [mangabeys], and *Macaca* [macaques]) (Grubb et al. 2003). Like the other members of the Cercopithecinae, the guenons can be distinguished from the Colobinae (the leaf-eating monkeys) by morphological adaptations to a frugivorous diet (e.g., low and rounded molar cusps, simple stomachs, pouches in their cheeks for storing food) (Fleagle 1999). For the forest guenons (including *A. nigroviridis* and *Miopithecus*), the majority of the diet is composed of fruit (range: 24.5%–91%). But the guenons do display diverse dietary preferences, consuming leaves, flowers, vertebrates, and invertebrates to varying degrees (Table 16.4), even within species. Dietary data for *C. mitis*, for example, are given for seven different sites; and the amount of fruit in the blue monkeys' diet at these sites is highly variable, accounting for the entire range of fruit composition in forest guenon diets (24.5%–91%). Similarly, the composition of fruit in the diets of *C. ascanius* and *C. cephus* across several sites ranges from 29.5%–57% and 49%–78%, respectively. Even at the same site, dietary composition varies among age/sex classes. At Taï National Park, adult male *C. campbelli* and *C. diana* eat more fruit but less foliage and fewer invertebrates than adult females and juveniles (Buzzard 2006).

Despite being highly frugivorous, forest guenons also feed on leaves; and in some populations of some species, leaves can make up as much as a third of the diet (e.g., *C. preussi* 41%, Beeson et al. 1996; *C. lhoesti* 35%, Kaplin and Moermond 2000; *C. petaurista* 40%, Buzzard 2006). In only one species, however, do leaves compose over 50% of the diet (i.e., *C. ascanius* in Budongo, up to 74%, Sheppard 2000). Mature leaves are more difficult to digest than immature leaves because of their higher levels of cellulose and potentially higher levels of toxic secondary compounds (Richard 1985a, Strier 2003a), and since guenons do not possess adaptations for digesting leaves, when leaves are eaten, young, immature leaves are preferred (see also Butynski 1988). For example, in a study of *C. mitis* and *C. ascanius* in Kakamega Forest, Kenya, Cords (1987a) found that leaves made up almost 23% and 10% of the plant diet, respectively. However, the majority of the feeding scores for both species were on young leaves (*C. mitis* 20% young, 2.7% mature; *C. ascanius* 9.1% young, 0.5% mature). Similar results have also been reported for two guenon species in Gabon (*C. nictitans* 10.2% young, 0.1% mature; *C. pogonias* 12.5% young, 0.1% mature; Brugière et al. 2002). Unusually for guenons, seeds comprise the majority of the diet of *C. nictitans* (50.2%) and *C. pogonias* (49.8%) in Makande, Gabon (Brugière et al. 2002). Although seeds can be difficult for guenons to digest because they can contain toxic compounds (Janson and Chapman 1999), irregular and recurrent food shortages due to a lack of fruit-bearing trees at Makande means *C. nictitans* and *C. pogonias* must sometimes eat less desirable foods (i.e., seeds) (Brugière et al. 2002).

More than half of the forest guenons (11 of 21, including *Allenopithecus* and *Miopithecus*) have been observed feeding on arthropods or other invertebrates (Table 16.4; see also Butynski 1982). In some cases, insects make up a large proportion of the diet (e.g., *C. mitis* at Kibale 35.9%–37.74%, *E. patas* at Segera Ranch, Kenya, 35% [Isbell 1998]; *C. campbelli* at Taï National Park 33% [Buzzard 2006]). The diet of *Miopithecus* consists primarily of fruit and seeds (43%), but a third comes from arthropods (Gautier-Hion 1988a). The high proportion of arthropods in the *Miopithecus* diet is not surprising considering their small body size.

Few studies provide detailed information on the arthropod foraging and capture methods of guenons. However, Cords (1986) found that *C. mitis* and *C. ascanius* at Kakamega Forest, Kenya, capture the majority of their arthropod prey from the surface of mature leaves (61.9% of captures for *C. mitis*, 60.1% of captures for *C. ascanius*). The next most common substrate for insect capture in both species was tree trunks or branches (10.1% of captures for *C. mitis*, 13.2% of captures for *C. ascanius*). Cords (1986) documented 11 capture techniques, all of which were used by both species, to varying degrees. These techniques vary in (*1*) the speed of capture, (*2*) whether the prey item is ingested with or without handling, and (*3*) whether the monkey holds or braces the substrate containing the prey item. Both species used an intermediate speed to capture prey, although blue monkeys preferred to hold the substrate containing the prey item and ingest the insect directly from the substrate without handling it (42.4% of captures). *C. ascanius* also used this method of capture (29.1% of captures) but used their hands to "pick" insect prey from the substrate almost as frequently (28.1% of captures).

Unlike *C. mitis* and *C. ascanius*, which actively hunt for their arthropod prey, the primary arthropod prey of *E. patas* in Kenya is colonially living ants of the genus *Crematogaster*, which *E. patas* obtain by biting open swollen thorns of *Acacia drepanolobium* trees (Isbell 1998). Holding on to the swollen thorn or branch with their hands, *E. patas* extract adult and larval ants with their tongues (K. E. Jaffe personal observation). Since the ants defend their nests and the tree by biting (Young et al. 1997), *E. patas* usually open only one or two swollen thorns per tree (97% of feeding observations) before moving on to the next one (Isbell 1998). The contribution of ants to the caloric intake of adult female *E. patas* is substantial, accounting for 20%–30% of their daily energy requirement (Isbell and Young 2007). *E. patas* will occasionally eat other invertebrates, especially grasshoppers, which they usually capture in the grass, sometimes after chasing them (K. E. Jaffe personal observation).

Gum of *A. drepanolobium* also comprises a high proportion (37%) of the *E. patas* diet in Kenya (Isbell 1998). For both *E. patas* and *C. aethiops*, fruit makes up a very small percentage of the diet (10% or less in both species: patas, Nakagawa 1989, Isbell 1998; vervets, Whitten 1983, Isbell et al. 1998), in stark contrast to most of the forest guenons (see Table 16.4). *Acacia* trees are also important food sources for *C. aethiops*. Leaves, seeds, fruits, flowers, gum, and bark of *A. xanthophloea* and *A. tortilis* account

Table 16.4 Guenon Diets

SPECIES	STUDY SITE	LEAVES	SEEDS	FRUIT	FLOWERS	INSECTS	GUM	OTHER	REFERENCE
C. aethiops	Amboseli National Park, Kenya (n = 3)[1]	26.6%	2.6%	11.1%	14.3%	7.7%	30.0%	0.2% bark	Wrangham and Waterman 1981
	Samburu-Isiolo Reserve, Kenya (n = 1)	0.0%	19.6%	5.8%	44.7%	NA	NA	8.3% grass 1.3% swollen thorns	Whitten 1983
	Segera Ranch, Kenya (n = 1)	2.0%	8.0%	10.0%	8.0%	23.0%	37.0%	2% swollen thorns 8% grass	Isbell et al. 1998
C. ascanius	Kakamega, Kenya (n = 1)	7.2%	0.4%	61.3%	2.0%	25.1%	2.8%	0.7% stems, shoots 0.5% unidentified	Cords 1986
	Kibale, Uganda (n = 5)	21.2% (range 13%–35%)	0.2% (range 0%–1%)	48.2% (range 36%–60%)	8.4% (range 3%–15%)	21.4% (range 15%–31%)	NA		Struhsaker 1978b, Chapman et al. 2002a–c
	Kibale, Uganda (n = 2)	9.6%[3]	12.2%[3]	22.6%[3]	7.5%[3]	NA	NA	NA	Lambert 2002a
	Budongo, Uganda (n = 2)	58.0% (range 42%–74%)	0.0%	29.5% (range 13%–46%)	7.0% (range 5%–9%)	NA	NA		Sheppard 2000
	Lomako, Democratic Republic of Congo (n = NA)	32%	5.7%	39.6%	0.0%	22.6%	0.0%		Zeeve 1991
C. campbelli	Not specified (n = NA)	1.0%	Combined with fruit category	78.0%		15.0%	NA		Gautier-Hion 1988a
	Taï, Ivory Coast (n = 2)	8.0%	0.0%	46.0%	1.0%	33.0%	0.0%	Frogs included with insects 2% fungi 9% unidentified cheek pouch 1% mushroom, bark, moss	Buzzard 2006
C. cephus	Makokou, Gabon (n = 1)	6.0%	0.0%	78.0%	1.0%	13.0%	NA		Gautier-Hion 1980
	Dja Reserve, Cameroon (n = NA)	5.2%	3.3%	80.4%	2.6%	5.2%	NA	3.3% other	Poulsen et al. 2002
	Lopé, Gabon (n = 1)	11.0%	7.0%	67.0%	6.0%	9.0%	NA		Tutin et al. 1997
C. diana	Bia National Park, Ghana (n = 1)	8.0%	24.3%	36.1%	5.9%	25.2%	NA	0.6% other	Curtin 2002
	Taï, Ivory Coast (n = 2)	16.0%	0.0%	59.0%	3.0%	16.0%	0.0%	Frogs included with insects 6% unidentified cheek pouch 0.2% mushroom, bark, moss	Buzzard 2006
C. dryas	No data[2]								
C. erythrogaster	No data								
C. erythrotis	No data								
C. hamlyni	No data								
C. lhoesti	Nyungwe Forest, Rwanda (n = 1)	35.2%	17.7%	24.5%	4.0%	8.8%	NA	9.8% unspecified	Kaplin and Moermond 2000, Kaplin 2001
C. mitis	Kakamega, Kenya (n = 1)	18.9%	2.5%	54.6%	3.7%	16.8%	1.9%	1.1% stems, shoots 0.5% unidentified	Cords 1986

Table 16.4 (cont'd)

SPECIES	STUDY SITE	LEAVES	SEEDS	FRUIT	FLOWERS	INSECTS	GUM	OTHER	REFERENCE
	Kibale, Uganda (n = 2)	0.0%	20.5%[3]	43.5%[3]	NA	NA	NA	NA	Lambert 2002a
	Nyungwe Forest, Rwanda (n = 1)	6.2%	9.3%	47.4%	6.2%	24.9%	NA	6.2% unspecified	Kaplin and Moermond 2000, Kaplin 2001
	Cape Vidal, South Africa (n = 1)	26.0%	0.0%	57.0%	13.0%	6.0%	NA		Lawes 1991
	Zomba Plateau, Malawi (n = 1)	32.6%	Combined with fruit category	53.5%	10.2%	1.0%	NA	2.9% other plant material	Beeson et al. 1996
	Ngoye Forest, South Africa (n = 1)	3.0%	0.0%	91.0%	2.0%	0.0%	NA		Lawes et al. 1990
	Kanyawara, Kibale (high biomass) (n = 4)	33.0% (range 22.4%–35.4%)	NA	27.7% (range 15%–28.9%)	6.9% (range 2.9%–7.8%)	37.7% (range 35.1%–45.4%)	NA	0.6% unspecified (range 0.4%–1.7%)	Butynski 1990
	Ngogo, Kibale (low biomass) (n = 1)	22.8%	NA	30.1%	9.8%	35.9%	NA	1.3% unspecified	Butynski 1990
C. mona	No data								
C. neglectus	Mpassa, Gabon (n = 1)	9.1%	Combined with fruit category	74.4%	3.0%	4.7%	NA	3.7% soil	Gautier-Hion and Gautier 1978
	Lomako, Democratic Republic of Congo (n = NA)	0.0%	0.0%	33.3%	0.0%	0.0%	66.7%		Zeeve 1991
C. nictitans	Makokou, Gabon (n = 1)	17.0%	0.0%	71.0%	1.0%	9.0%	NA		Gautier-Hion 1980
	Lopé (continuous), Gabon (n = 1)	16.0%	11.0%	59.0%	10.0%	3.0%	NA		Tutin et al. 1997
	Lopé (fragment), Gabon (n = 1)	17.0%	4.0%	44.0%	9.0%	24.0%	NA		Tutin et al. 1997
	Dja Reserve, Cameroon (n = NA)	13.1%	3.7%	69.3%	7.2%	2.5%	NA	4.3% other	Poulsen et al. 2002
	Makande, Gabon (n = 1)	10.3%	50.2%	35.5%	4.1%	NA			Brugière et. al. 2002
	Taï, Ivory Coast (n = NA)	4.2%	NA	58.9%	2.5%	31.3%	NA	3.1%	McGraw and Zuhberbühler 2007
C. petaurista	Taï, Ivory Coast (n = NA)	6.0%	Combined with fruit category	77.0%	NA	7.0%	NA		Gautier-Hion 1988a
	Taï, Ivory Coast (n = 2)	40.0%	0.0%	34.0%	6.0%	12.0%	0.0%	Frogs included with insects 1% fungi 6% unidentified cheek pouch 0.1% mushroom, bark, moss	Buzzard 2006
C. pogonias	Makokou, Gabon (n = 1)	1.0%	3.0%	80.0%	1.0%	16.0%	NA		Gautier-Hion 1980
	Lopé Reserve, Gabon (n = 1)	7.0%	9.0%	69.0%	9.0%	7.0%	NA		Tutin et al. 1997
	Dja Reserve, Cameroon (n = NA)	6.0%	3.3%	84.0%	2.5%	4.0%	NA	0.2% other	Poulsen et al. 2002
	Makande, Gabon (n = 1)	12.6%	49.8%	26.9%	4.7%	NA	NA		Brugière et. al. 2002
C. preussi	Kilum, Cameroon (n = 1)	41.4%	Combined with fruit category	51.9%	1.2%	1.0%	NA	4.6% other plant material	Beeson et al. 1996
C. sclateri	No data								

Table 16.4 (cont'd)

SPECIES	STUDY SITE	LEAVES	SEEDS	FRUIT	FLOWERS	INSECTS	GUM	OTHER	REFERENCE
C. solatus	No data								
C. wolfi	Salonga, Democratic Republic of Congo (n = 1)	30.0%	27.0%	32.0%	11.0%	NA	NA		Gautier-Hion et al. 1993
	Lomako, Democratic Republic of Congo (n = NA)	26.7%	0.0%	46.7%	6.7%	20.0%	0.0%		Zeeve 1991
A. nigrovirdis	Zaire (n = NA)	<5.0%	Combined with fruit category	81.0%	NA	18.0%	NA		Gautier-Hion 1988a
	Lomako, Democratic Republic of Congo (n = NA)	0.0%	0.0%	40.0%	20.0%	0.0%	6.7%	13.3% other plant material	Zeeve 1991
M. talapoin	not specified (n = NA)	<5.0%	Combined with fruit category	43.0%		35.0%	NA		Gautier-Hion 1988a
E. patas	Segera Ranch, Kenya (n = 1)	2.0%	10.0%	Included in seeds category	7.0%	35.0%	37.0%	1% vertebrates	Isbell 1998
	Kala Maloue, Cameroon (n = 1)	6.0%	0.0%	0.0%	65.0%	12.0%	7.0%		Nakagawa 1989

Data for each food type are provided as percent of the total diet. Note that different authors may have used different methods, so the data may not be comparable. NA indicates authors did not provide data for that food type.

[1] "n" refers to the number of study groups at the given study site for which data were provided. If data were provided for multiple groups, the percentages indicate the average percent of the diet across all study groups, and the range is given. If the author(s) did not indicate the number of study groups on which the data are based, "n = NA" follows the study site.

[2] "No data" indicates that dietary data are not available for that species.

[3] For C. mitis and C. ascanius at Kibale, data by Lambert (2002a) are provided only as percent of plant diet.

for as much as 75% of the diet of *C. aethiops* in Amboseli National Park, Kenya; and *Acacia* gum alone accounts for 30% (Wrangham and Waterman 1981). Swollen thorns, seeds, flowers, leaves, and gum from *A. xanthophloea* and *A. drepanolobium* trees account for 57.2% of the diet of *C. aethiops* at Segera Ranch, Kenya (Pruetz and Isbell 2000). *C. aethiops* at the Samburu-Isiolo Reserve in Kenya also rely primarily on *Acacia* trees, but the majority of their diet (44%) comes from *Acacia* flowers (Whitten 1983).

Ranging

Previous research has shown a link between diet and ranging behavior, particularly home range size and daily travel distance. Frugivores tend to have larger home ranges and longer day ranges than folivores with the same group biomass (Clutton-Brock and Harvey 1977b), perhaps because their preferred foods are more patchily distributed (Richard 1985b). However, Kaplin (2001) found the opposite to be true of *C. lhoesti* and *C. mitis*. In her study, *C. lhoesti* ate greater amounts of leaves than *C. mitis*, which incorporated more fruit and insects into their diet. However, the more folivorous *C. lhoesti* had larger home ranges and longer day ranges than *C. mitis*. Kaplin (2001) suggests that *C. lhoesti* travel farther because the terrestrial herbs on which they feed are not as densely distributed as other types of foliage. This suggests that resource abundance and distribution may

be more central than food type in determining home range size and day range length (see also Isbell 1991).

Ranging behavior varies among the guenons. Home range size varies from as little as 4 ha in *C. neglectus* (Wahome et al. 1993) to as much as 4,000 ha in *E. patas* (Chism and Rowell 1988, L. A. Isbell unpublished data), which have the largest home ranges for their group biomass of any primate (Clutton-Brock and Harvey 1977b, Isbell et al. 1998) (Table 16.5). Daily travel distance follows a similar pattern, being smallest in *C. neglectus* (530 m, Gautier-Hion and Gautier 1978) and largest in *E. patas* (2,250–4,025 m, Chism and Rowell 1988) (Table 16.5).

Data on daily ranging behavior other than daily travel distance are sparse. Cords (1987a) found that *C. mitis* and *C. ascanius* tend not to cross over their path during their daily travels. *C. mitis* tend to move in rather circular courses, covering about one-third of their home range per day. *C. ascanius* also traveled in a circular motion but in a more linear fashion (Cords 1987a). *C. ascanius* covered less of their home range (about one-fourth) than *C. mitis* per day.

E. patas have a more meandering, circuitous ranging pattern and often recross portions of their path during a single day of ranging (Isbell et al. 1999). *E. patas* ranging behavior also appears to be unique among guenons in that they seem to have a number of "sub-home ranges" in which they forage for various lengths of time (K. E. Jaffe unpublished data). Compared to *E. patas*, *C. aethiops* tend to follow a straighter

Table 16.5 Guenon Home Range Sizes and Day Range Lengths

SPECIES	STUDY SITE	AVERAGE GROUP SIZE	HOME RANGE SIZE (HA)	DAY RANGE LENGTH (M)	RANGE DEFENSE?	INTERGROUP AGGRESSION?	SOURCE
C. aethiops	Segera Ranch, Kenya	8	10	1,025[1]	Yes	Yes	Isbell unpub. data, Isbell et al. 1999
	Segera Ranch, Kenya	27	40	1,632[1]	Yes	Yes	Isbell unpub. data, Isbell et al. 1999
	Amboseli, Kenya	7.3	34	NA	Yes	Yes	Struhsaker 1967b
	Amboseli, Kenya	16.4	96	NA	Yes	Yes	Struhsaker 1967b
	Amboseli, Kenya	17.2	19	NA	Yes	Yes	Struhsaker 1967b
	Amboseli, Kenya	49.9	34	NA	Yes	Yes	Struhsaker 1967b
C. ascanius	Kibale, Uganda	15	28	1,595	Yes	Yes	Struhsaker and Leland 1988
	Kibale, Uganda	45–50	68	1,198	Yes	Yes	Struhsaker and Leland 1988
	Kakamega, Kenya	25	60	NA	Yes	Yes	Cords 1987a
	Kibale, Uganda	19–29	NA	1,170	NA	NA	Jones and Bush 1988
	Ngogo (Kibale), Uganda	32	NA	NA	NA	NA	Mitani et al. 2001
C. campbelli	Taï Forest, Ivory Coast	9	50	NA	NA	NA	Shultz and Noe 2002, Shultz et al. 2004
	Taï Forest, Ivory Coast	9	67	NA	Yes	Yes	Buzzard and Eckardt 2007
	Taï Forest, Ivory Coast	10	52	NA	Yes	Yes	Buzzard and Eckardt 2007
C. cephus	Makokou, Gabon	4	18	800	NA	NA	Gautier-Hion and Gautier 1974
	Makokou, Gabon	8	45	NA	NA	NA	Gautier-Hion and Gautier 1974
C. diana	Sierra Leone	20	41	1,019	NA	NA	Whitesides 1989
	Sierra Leone	27	29	1,513	NA	NA	Whitesides 1989
	Taï Forest, Ivory Coast	28	62	NA	NA	NA	Shultz and Noe 2002, Shultz et al. 2004
	Taï Forest, Ivory Coast	21	59	NA	Yes	Yes	Buzzard and Eckardt 2007
	Taï Forest, Ivory Coast	26	59	NA	Yes	Yes	Buzzard and Eckardt 2007
	Taï Forest, Ivory Coast	20–25	NA	NA	NA	NA	McGraw et al. 2002
C. dryas	No data						
C. erythrogaster	No data						
C. erythrotis	No data						
C. hamlyni	No data						
C. lhoesti	Nyungwe Forest, Rwanda	29	50[2]	2,092	NA	NA	Kaplin 2001
	Ngogo (Kibale), Uganda	18	NA	NA	NA	NA	Mitani et al. 2001
C. mitis	Ngogo (Kibale), Uganda	18	335	1,406	No	No	Butynski 1990
	Kanyawara (Kibale), Uganda	17	44	1,300	Yes	Yes	Butynski 1990

Table 16.5 (cont'd)

SPECIES	STUDY SITE	AVERAGE GROUP SIZE	HOME RANGE SIZE (HA)	DAY RANGE LENGTH (M)	RANGE DEFENSE?	INTERGROUP AGGRESSION?	SOURCE
	Kakamega, Kenya	24–43	38	NA	Yes	Yes	Cords 1987a, 2002c
	Nyungwe Forest, Rwanda	27	26[2]	1,307	NE[3]	NE	Kaplin 2001
C. mona	Cameroon, West Africa	8.8	NA	NA	NA	NA	Struhsaker 1969
	Cameroon, West Africa	10.8	NA	NA	NA	NA	Struhsaker 1969
C. neglectus	Kisere Forest, Kenya	11	>5.0	NA	No	NA	Wahome et al. 1993
	Kisere Forest, Kenya	13	4	NA	No	NA	Wahome et al. 1993
	Makokou, Gabon	3–4	4–10	530	No	No	Gautier-Hion and Gautier 1978
C. nictitans	Makokou, Gabon	38[4]	148	1,825	NA	NA	Gautier-Hion et al. 1983
	Makokou, Gabon	53[5]	119	1,980	NA	NA	Gautier-Hion et al. 1983
	Taï Forest, Ivory Coast	12	93	NA	NA	None observed	Buzzard and Eckardt 2007
C. petaurista	Taï Forest, Ivory Coast	9	100	NA	NA	NA	Shultz and Noe 2002, Shultz et al. 2004
	Taï Forest, Ivory Coast	16	74	NA	Yes	Uncommon	Buzzard and Eckardt 2007
	Taï Forest, Ivory Coast	7	64	NA	Yes	Uncommon	Buzzard and Eckardt 2007
C. pogonias	Makokou, Gabon	38[4]	148	1,825	NA	NA	Gautier-Hion et al. 1983
	Makokou, Gabon	53[5]	119	1,980	NA	NA	Gautier-Hion et al. 1983
C. sclateri	No data						
C. solatus	No data						
C. wolfi	No data						
A. nigroviridis	No data						
M. talapoin	Makokou, Gabon	64	140	2,323	NA	NA	Gautier-Hion 1971
E. patas	Mutara Ranch, Kenya	16	2,340	3,830	No	Yes	Chism and Rowell 1988
	Mutara Ranch, Kenya	47	3,200	4,220	No	Yes	Chism and Rowell 1988
	Segera Ranch, Kenya	45	4,000	3,188[1]	No	Yes	Isbell et al. 1999, 2009
	Murchison Falls, Uganda	31	5,200	2,250	No	Yes	Hall 1965a
	Kala Maloue, Cameroon	11.00	266–440	NA	NA	NA	Nakagawa 2008

[1] Day range length data based on Isbell et al. (1999) were calculated by multiplying "straight-line distance of groups" (per 30 min) for each group by 22 to obtain an estimate of day range length for an 11 h day.

[2] Home range size is an average of two methods, the minimum convex polygon method and grid-cell analysis.

[3] NE, no encounters, indicating that the researcher did not observe encounters between conspecific groups.

[4] Data on group size, home range size, and day range length for *C. nictitans* and *C. pogonias* are given for a mixed-species group containing 20 *C. nictitans* individuals and 18 *C. pogonias* individuals. No data on *C. nictitans* or *C. pogonias* in a monospecific group were provided.

[5] Data on group size, home range size, and day range length for *C. nictitans* and *C. pogonias* were given for a mixed-species group containing 20 *C. nictitans* individuals, 18 *C. pogonias* individuals, and 15 *C. cephus* individuals. No data on *C. nictitans* or *C. pogonias* in a monospecific group were provided.

daily path (Isbell et al. 1999, although *C. aethiops* will alter their ranging patterns to take advantage of reduced predator presence and/or enhanced feeding opportunities in recently burned areas [Jaffe and Isbell 2009]).

Predators

Guenons are susceptible to predation from a large number of mammalian, reptilian, and aerial predators. Confirmed and potential predators of various guenon species are listed in Table 16.6. Although data on predators of guenons are sparse, crowned eagles (*Stephanoaetus coronatus*) appear to be the most pervasive predator of forest guenons (see Struhsaker and Leakey 1990, Shultz et al. 2004). Crowned eagles have been documented as predators of nine species

(Table 16.6). Confirmed mammalian predators include chimpanzees (*Pan troglodytes*), leopards (*Panthera pardus*), and humans (*Homo sapiens*). Potential predators of forest guenons include African rock pythons (*Python sebae*) and golden cats (*Profelis aurata*). Given their small size, *Miopithecus* is presumably susceptible to a fairly large number of predators, and potential predators include *S. coronatus*, *Panthera pardus*, *Profelis aurata*, domestic dogs (*Canis familiaris*), *Python sebae* (Wahome et al. 1993), and Nile monitors (*Varanus niloticus*) (Gautier-Hion et al. 1983).

Guenons living in more open habitats tend to be preyed upon by a different set of predators. Confirmed predators of *E. patas* include black-backed jackals (*Canis mesomelas*) and *Canis familiaris* (Chism and Rowell

Table 16.6 Guenon Predators

SPECIES	BODY WEIGHT (KG)	SOURCE	CONFIRMED[1]	SOURCE	POTENTIAL[2]	SOURCE
C. aethiops	m: 5.6 f: 2.79	Haltnorth and Diller 1988 Gevaerts 1992	Leopard, martial eagle, python, yellow baboon	Seyfarth et al. 1980 Haufstater 1975a,b	Lion, spotted hyena, African wild cat, serval, black-backed jackal, cheetah, caracal	Isbell and Enstam 2002
C. ascanius	m: 3.7, f: 2.8 m: 4.4	Gevaerts 1992, Colyn 1994 Jones and Bush 1988	Crowned eagle Chimpanzee	Cords 1987a 1990a, 2002a, Struhsaker and Leakey 1990, Wrangham and Riss 1990, Boesch and Boesch 1989, Mitani et al. 2001		
C. campbelli	m: 4.5, f: 2.7	Oates et al. 1990, McGraw and Zuberbühler 2007	Leopard, crowned eagle	Hoppe-Dominik 1984, Zuberbühler 2001, Shultz et al. 2004, McGraw et al. 2006		
C. cephus	m: 4, f: 2.9	Gautier-Hion and Gautier 1976	Crowned eagle, human	Gautier-Hion et al. 1983	Python, golden cat, leopard	Gautier-Hion et al. 1983
C. diana	m: 5.2, f: 3.9	Oates et al. 1990, McGraw and Zuberbühler 2007	Chimpanzee, human, leopard, crowned eagle	Zuberbühler et al. 1999, Boesch and Boesch 1989, Shultz et al. 2004, McGraw et al. 2006, Zuberbühler et al. 1997		
C. dryas	m: 3, f: 2.2	Colyn et al. 1991				
C. erythrogaster	m: 4.1, f: 2.4	Oates 1985				
C. erythrotis	m: 4.0-4.5	Napier 1981				
C. hamlyni	m: 5.5, f: 3.3-3.68	Gevaerts 1992, Colyn 1994				

Table 16.6 (*cont'd*)

SPECIES	BODY WEIGHT (KG)	SOURCE	CONFIRMED[1]	SOURCE	POTENTIAL[2]	SOURCE
C. lhoesti	m: 6, f: 3.5	Gevaerts 1992; Colyn 1994	Crowned eagle	Struhsaker and Leakey 1990, Mitani et al. 2001	Leopard, golden cat, python	Haltenorth and Diller 1980
C. mitis	m: 5.9-8.9	Napier 1981	Crowned eagle	Cords 1987a, 1990a, 2002a;		
	f: 3.83	Gevaerts 1992	Chimpanzee, human	Struhsaker and Leakey 1990; Wrangham and Riss 1990; Mitani et al. 2001; Struhsaker 2000		
C. mona	m: 2.4-5.3	Napier 1981				
C. neglectus	m: 7.0, f: 3.5 m: 7.0, f: 3.96	Gautier-Hion and Gautier 1976, Gevaerts 1992, Colyn 1994 Leutenegger and Lubach 1987			Leopard, golden cat, crowned eagle, python	Haltenorth and Diller 1980
C. nictitans	m: 6.6, f: 3.65-4.2 m: 6.7, f: 4.2	Gautier-Hion and Gautier 1976, Gevaerts 1992 McGraw and Zuberbühler 2007	Crowned eagle, human	Gautier-Hion et al. 1983	Python, golden cat, leopard	Gautier-Hion et al. 1983
C. petaurista	m: 4.4, f: 2.9	Oates et al. 1990, McGraw and Zuberbühler 2008	Crowned eagle, leopard, chimpanzee	McGraw et al. 2006, Shultz et al. 2004		
C. pogonias	m: 4.5, f: 3.0	Gautier-Hion and Gautier 1976	Crowned eagle, human	Gautier-Hion et al. 1983	Python, golden cat, leopard	Gautier-Hion et al. 1983
C. preussi	N/A					
C. sclateri	N/A					
C. solatus	NA					
C. wolfi	m: 3.8, f: 2.76-2.9	Gevaerts 1992, Colyn 1994	Crowned eagle	Zeeve 1991	Leopard	Zeeve 1991
A. nigroviridis	m: 5.9, f: 3.58-3.7	Gautier-Hion 1988a Gevaerts 1992				
M. talapoin	m: 1.4 f: 1.1	Gautier-Hion and Gautier 1976 Gautier-Hion 1971			Leopard, golden cat, genet, Nile monitor, crowned eagle	Haltenorth and Diller 1988
E. patas	m: 10.25, f: 5.75	Haltnorth and Diller 1988	Black-backed jackal, domestic dog	Chism and Rowell 1988 L.A. Isbell unpub. data	Leopard, serval, caracal, African wild cat, lion, spotted hyena, martial eagle, chimpanzee (in West Africa), wild dog, baboon	Chism and Rowell 1988, Isbell and Enstam 2002, Isbell unpub. data, Chism et al. 1983

1 "Confirmed" predators include species that have been observed preying on a specific guenon species, whether the attack was successful or not. Confirmed predators also include species that have left remains of monkeys in their nests or dung.

2 "Potential" predators are species that researchers have listed as possible predators but have not been observed attacking or attempting to attack the guenon species in question. In general, potential predators are those that co-occur with guenons and are known to take prey of equal or greater size than a guenon, even if they have not been observed preying on guenons.

1988, Enstam and Isbell 2002). *Panthera pardus* have not been observed to attack *E. patas* but are strongly suspected in overnight disappearances (Chism et al. 1983, L. A. Isbell unpublished data). Lions (*Panthera leo*), spotted hyenas (*Crocuta crocuta*), wild dogs (*Lycaon pictus*), martial eagles (*Polemaetus bellicosus*), olive baboons (*Papio anubis*), and in West Africa *Pan troglodytes* are all potential predators (Enstam and Isbell 2002). Known predators of the *C. aethiops* superspecies group include *Panthera pardus, S. coronatus, Polemaetus bellicosus, Python sebae,* and yellow baboons (*Papio cynocephalus*) (Struhsaker 1967a, Altmann and Altmann 1970, Hausfater 1975a, Seyfarth et al. 1980, Baldellou and Henzi 1992), whereas *Panthera leo, Crocuta crocuta, Canis mesomelas,* and cheetahs (*Acinonyx jubatus*) are potential predators (Struhsaker 1967a, Seyfarth et al. 1980).

REPRODUCTIVE PARAMETERS

Age at Sexual Maturity

Within the genus *Cercopithecus*, age at sexual maturity averages 64.4 months for males (range 48 months in *C. campbelli* [Hunkeler et al. 1972] to 72 months in *C. ascanius* [M. Cords unpublished data in Cords 1987b]) and 47.6 months for females (range 42 months in *C. neglectus* [Gautier-Hion and Gautier 1976] to 54 months in *C. ascanius* [T. T. Struhsaker, unpublished data in Cords 1987b]) (Table 16.7). *E. patas* males and females mature faster than the forest guenons, with males maturing at 48 months and females at 30 months (Hall 1965a, Chism et al. 1984, Isbell et al. 2009). Female *M. talapoin* reach maturity at 48 months, which is comparable to females in most *Cercopithecus* species. Data on age at sexual maturity are unavailable for *Allenopithecus*.

Age at First Birth

Age at first birth for female *Cercopithecus* species averages 56.1 months (range 36 months in *C. campbelli* [Bourlière et al. 1970] and *E. patas* [Chism et al. 1984, Isbell et al. 2009] to 61.2 months in *C. aethiops* [Isbell et al. 2009]). Habitat quality is known to affect age at first birth in *C. aethiops*. At Amboseli National Park in Kenya, Cheney et al. (1988) found that females in a group without access to permanent water sources and associated flora gave birth for the first time later than females in two groups with access to permanent water and associated flora. Similarly, after the heavy rains of an El Niño weather event, a female in one study group in Laikipia, Kenya gave birth for the first time at 3 years of age (Isbell et al. 2009), apparently because of plentiful resources. This is in contrast to the late ages (5.8 and 6.1 years) at first birth of the two females that matured after the El Niño rains. *Miopithecus* females fall within the range of ages for *Cercopithecus* females, with an average age at first birth of 53.5 months (see Table 16.7). Data on age at first birth are not available for *Allenopithecus*.

Gestation Length

There are few data on gestation length for the guenons, but for the six species for which these data are available, average gestation length is 163.2 days (range 140 days in *C. mitis* [Rowell 1970b] to 180 days in *C. campbelli* [Bourlière et al. 1970]). Gestation lengths for *Erythrocebus* and *Miopithecus* (167 days [Sly et al. 1983] and 165 days [Gautier-Hion and Gautier 1976], respectively) are well within the range of guenons (Table 16.7). Data on gestation length are not available for *Allenopithecus*.

Interbirth Interval

The average interbirth interval for *Cercopithecus* species is 25 months (range 12 months in *C. campbelli* [Hunkeler et al. 1972] to 52 months in *C. ascanius* [Cords and Rowell 1987]) (Table 16.7). Interbirth intervals in vervet monkeys can also be affected by availability of water and food. In Amboseli, Cheney et al. (1988) found that a *C. aethiops* group that did not have access to permanent water and associated flora had an interbirth interval of almost 2 years, significantly longer than that of the females in two groups with access to water year-round (average 16.3 and 13.8 months, respectively). In Laikipia, Kenya, interbirth intervals of *C. aethiops* averaged 13.3 months (Isbell et al. 2009). *E. patas* at Mutara and Segera Ranches, also in Laikipia, had shorter mean interbirth intervals of 11.8 and 12.9 months, respectively (Chism et al. 1984, Isbell et al. 2009). There are no data on interbirth intervals for *A. nigroviridis*.

Mating/Birth Seasonality

The vast majority of guenons have distinct mating and birth seasons (Table 16.7). *C. ascanius* is the only guenon species known to give birth year-round (Butynski 1988). Even so, only 3% of births occur outside of the birth peak from April to November. Variation in birth seasons (and therefore mating seasons) tends to coincide with variation in rainfall and resource availability (Butynski 1988). That rainfall and subsequently resource availability are important correlates of guenon reproduction is highlighted by the response of vervet females after a particularly heavy rainy season, during an El Niño weather event. Normally, the birth season of *C. aethiops* on Segera Ranch, Kenya, falls between January and March (L. A. Isbell unpublished data); but two female vervets in one study group unusually gave birth in July, during the normal mating season (Isbell et al. 2009).

SOCIAL ORGANIZATION

Social and Mating Systems

Most guenons live in single-male/multifemale groups (Rowell 1988a, Cords 2002b). The exceptions are *C. aethiops* (Struhsaker 1967b), *M. talapoin* (Gautier-Hion 1971,

Table 16.7 Guenon Reproductive Parameters

SPECIES	SITE	SEXUAL MATURITY (MONTHS)	SOURCE	1ST BIRTH (MONTHS)	SOURCE	GESTATION LENGTH (DAYS)	SOURCE	INTERBIRTH INTERVAL (MONTHS)	SOURCE	MATING/ BIRTH SEASON	SOURCE
C. aethiops	Amboseli, Kenya	m: 60, f: 48	Cheney et al. 1988	60.8	Cheney et al. 1988	NA[1]		17.1	Cheney et al. 1988	MS: Apr–Oct	Andelman 1987
										BS: Oct–Dec	Cheney et al. 1988
	Samburu-Isiolo Reserve, Kenya	NA		NA		NA		NA		MS: Mar–Jun	Whitten 1984
	Segera Ranch, Kenya	NA		61.2	Isbell et al. 2009	NA		13.3	Isbell et al. 2009	BS: Jan–Mar	Isbell unpub. data
	Austin, Texas	NA		NA		163[2]	Bramblett et al. 1975	NA		BS: Oct–Jan	Whitten 1984
C. ascanius	Kakamega, Kenya	m: 72	Cords unpub. data in Cords 1987b	NA		NA		>24	Cords unpub. data in Cords 1987b	Birth peak: Dec–Mar,	Cords and Rowell 1987
								52	Cords and Rowell 1987	MS: Jun–Oct	Cords 2002b
	Democratic Republic of Congo	NA		NA		NA		NA		Birth peak: Apr–Nov	Gevaerts 1992 Struhsaker
	Kibale, Uganda	f: 48–60	Struhsaker unpub. data in Cords 1987b	NA		NA		17.8 (range 11.6–24)	Struhsaker unpub. data in Cords 1987b	Birth peak: Nov–Feb	unpub. data in Cords 1987b
C. campbelli	Ivory Coast	m: 48, f: 36	Hunkeler et al. 1972	36–48	Bourlière et al. 1970	180	Bourlière et al. 1970	12	Hunkeler et al. 1972	MS: Jun–Sept, BS: Nov–Jan	Bourlière et al. 1970
C. cephus	Paimpont, France	NA		60[2]	Cords 1988a	NA		27.4[1]	Cords 1988a	NA	
C. diana	Bia National Park, Ghana	NA		NA		NA		NA		MS: Feb–Jul, Mating peak: Jul	S. Curtin pers. com
										BS: Nov–Dec, Mar	S. Curtin pers. com
C. dryas	No data[3]										
C. erythrgaster	No data										
C. erythrotis	No data										
C. hamlyni	Zaire	NA		NA		NA		NA		BS: Jun–Nov	Gevaerts 1992
C. lhoesti	Zaire	NA		NA		NA		NA		BS: Apr–Dec	Gevaerts 1992
C. mitis	Kakamega, Kenya	m: 72, f: 48–60	Cords unpub data in Cords 1987b	NA		NA		47	Cords and Rowell 1987	Birth peak: Dec–Mar	Cords and Rowell 1987
C. mitis	Makerere University, Uganda	NA		NA		140[2]	Rowell 1970b	NA		NA	
C. mona		NA		NA		NA		NA		NA	
C. neglectus	Makokou, Gabon	m: 60–72, f: 48	Gautier-Hion and Gautier 1976, 1978	54[5]	Gautier-Hion and Gautier 1976	182	Haltenorth and Diller 1988	NA		NA	
	Kisere Reserve, Kenya	NA		NA		NA		>24	Wahome et al. 1993	Birth peak: Dec–Mar	Wahome et al. 1993
	Nairobi, Kenya	NA		NA		170[2,4]	Rowell and Richards 1979	NA		NA	
C. nictitans	Makokou, Gabon	m: 60–72, f: 48	Gautier-Hion and Gautier 1976	NA 60[2]	Cords 1988a	NA		24	Butynski 1988	Birth peak: Dec–May NA	Butynski 1988
	Paimpont, France	NA				NA		NA			
C. petaurista	No data										
C. pogonias	Makokou, Gabon	m: 60–72, f: 48	Gautier-Hion and Gautier 1976	NA		NA		NA		BS: Dec–Apr	Napier and Napier 1985, Cords 1988a

Table 16.7 (*cont'd*)

SPECIES	SITE	SEXUAL MATURITY (MONTHS)	SOURCE	1ST BIRTH (MONTHS)	SOURCE	GESTATION LENGTH (DAYS)	SOURCE	INTERBIRTH INTERVAL (MONTHS)	SOURCE	MATING/BIRTH SEASON	SOURCE
	Paimpont, France	NA		60[2]	Cords 1988a	NA		NA		NA	
C. preussi	No data										
C. sclateri	No data										
C. solatus	No data										
C. wolfi	Democratic Republic of Congo	NA		NA		NA		NA		BS: Jun–Dec	Gevaerts 1992
A. nigroviridis	No data										
M. talapoin	Makokou, Gabon	f: 48	Gautier-Hion and Gautier 1976	53.5[5]	Gautier-Hion and Gautier 1976	167	Gautier-Hion and Gautier 1976	NA		BS: Nov–Apr	Gautier-Hion 1973
	Mbalmayo Reserve, Cameroon	m: 48, f: 48	Rowell and Dixson 1975	NA		NA		NA		BS: Jun–Aug MS: Jan–Mar	Rowell and Dixson 1975 Rowell and Dixson 1975
E. patas	Mutara Ranch, Kenya	f: 30	Chism et al. 1984	36	Chism et al. 1984	NA		11.8	Chism et al. 1984	MS: Jun–Aug, BS: Dec–Feb	Chism et al. 1983, 1984; Chism and Rowell 1986
	Segera Ranch, Kenya	NA		36	Isbell et al. 2009	NA		12.9	Isbell et al. 2009	MS: Jul–Sep, BS: Jan–Mar	Enstam et al. 2002; Isbell unpub. data
	Meloy Laboratories, Maryland	NA		NA		167[2]	Sly et al. 1983	NA		NA	
	Murchison Falls, Uganda	m: 36[6]	Hall 1965a	NA		NA		NA		BS: Dec–Feb	Hall 1965a
	Waza Reserve, Cameroon	NA		NA		NA		NA		BS: Nov–Jan	Struhsaker and Gartlan 1970
	Kala Maloue, Cameroon	NA		36.5	Nakagawa et al. 2003	NA		14.4	Nakagawa et al. 2003	BS: Dec–Feb	Nakagawa et al. 2003

[1] NA indicates data for that particular reproductive parameter are not available for that species.
[2] Data come from captive animals.
[3] "No data" indicates that there are no data on any reproductive parameters for that species.
[4] Estimated by the authors of the study.
[5] Age at first birth for de Brazza's and talapoin monkeys was estimated by the authors by combining age at sexual maturity and gestation length data.
[6] For patas males, age at dispersal is used as an estimate of age at sexual maturity.

1973; Rowell 1971; Rowell and Dixson 1975), and *A. nigroviridis* (Gautier 1985; see also Zeeve 1991), which live in multimale/multifemale groups. The existence of apparently single-male/single-female groups of *C. neglectus* (Gautier-Hion and Gautier 1978) is odd given their extreme sexual dimorphism. It is possible that single-male/single-female groups of *C. neglectus* largely occur in declining populations, and other researchers have reported *C. neglectus* living in single-male/multifemale groups (Wahome et al. 1993). In all species for which there are data, males disperse both socially and locationally at sexual maturity, whereas females for the most part are socially philopatric, living their entire lives in their natal groups. Females rarely leave their natal home range or group. When they do disperse, it is usually under unusual conditions. For example, female vervet monkeys in Amboseli National Park, Kenya, transferred

with juveniles into neighboring groups when the population declined substantially (Hauser et al. 1986, Isbell et al. 1991) and in Laikipia, Kenya, after a brief but intense period of predation (Jaffe and Isbell in press).

The single-male/multifemale social system can lead to extreme competition among males for access to females. In some guenons, including *C. ascanius*, *C. mitis*, and *E. patas*, heterosexual groups at times experience influxes of multiple extragroup males during the breeding season (Chism and Rowell 1986; Harding and Olson 1986; Cords 1988a, 2002b; Jones and Bush 1988; Ohsawa et al. 1993; Kaplin et al. 1998; Carlson and Isbell 2001; Enstam et al. 2002). In some cases, an influx of multiple males during the breeding season results in the resident male being replaced (e.g., Enstam et al. 2002), while in other cases, the resident male may not be expelled but may remain in the group after the influx (e.g., Tsingalia and Rowell 1984). Since multimale influxes occur only during *some* breeding seasons, Cords (2002b) was able to divide her study of Kakamega *C. mitis* into "influx" and "noninflux" periods. Multiple males were present in 50%–94% of days during the six influx breeding seasons but in only 4%–28% of days during noninflux breeding seasons (Cords 2002b). Breeding seasons among *E. patas* have also been divided into influx and noninflux periods, based on the number of males observed with the group during the breeding season (Harding and Olson 1986, Carlson and Isbell 2001, Enstam et al. 2002).

Among *C. mitis*, the most likely explanation for multimale influxes appears to be a single male's inability to defend multiple cycling females (Cords 2002b). The presence of two or more females cycling at the same time is highly correlated with the presence of more than one male in a group, apparently because it is too difficult for a single male to maintain exclusive access to multiple receptive females (Cords 2002b). This explanation is supported by data from other species that sometimes experience multimale influxes during the breeding season. Studies of *C. mitis labiatus* (Henzi and Lawes 1987, 1988), *C. ascanius* (Struhsaker 1988), and *E. patas* (Chism and Rowell 1988, Harding and Olson 1986) have shown a positive correlation between the number of males in a heterosexual group during the breeding season and the number of receptive females. The ability of a male to defend reproductively cycling females from extragroup males may be made even more difficult if the ratio of extragroup males to groups of females is high (Cords 2000a). Surveys of the home ranges of *C. mitis* groups routinely turned up more extragroup males during influx breeding seasons than noninflux seasons (Cords 2002b). Similarly, during a study of *E. patas* that spanned three breeding seasons, Carlson and Isbell (2001) found that the one breeding season during which multimale influxes were observed also had the highest number of sightings of extragroup males per month.

Although the number of receptive females or the density of extragroup males appears to influence multimale influxes in the species noted, it does not explain the year-round existence of multiple males in *C. aethiops* (Struhsaker 1967b, Isbell et al. 2002), *Miopithecus* (Gautier-Hion 1971, 1973; Rowell 1971; Rowell and Dixson 1975), and *Allenopithecus* (Gautier 1985; see also Zeeve 1991) groups. A proposed explanation for the formation of multimale groups in *C. aethiops* is the limited dispersal hypothesis, which states that *C. aethiops* males have limited dispersal abilities because of the configuration of the habitat and the costs of dispersal (Isbell et al. 2002). Unlike forest guenons, where groups can live adjacent to multiple other groups, vervets' restricted use of habitats along water courses in otherwise unsuitable savanna habitats means that *C. aethiops* groups often line up, one after another, along rivers, rather than being scattered throughout the savanna habitat (a notable exception is Amboseli, where vervet groups occupied more swampy habitat; Cheney and Seyfarth 1983). The result is that *C. aethiops* males have

limited options when they disperse from their natal groups (i.e., only two directions) and usually disperse into adjacent groups in which other males (including male relatives) already reside, thus perhaps increasing the inclusive fitness benefits of sharing residence with other males (Isbell et al. 2002). Further research into the dispersal patterns of *Miopithecus* and *Allenopithecus* needs to be conducted to determine whether similar limitations are also responsible for multiple males in these species.

Miopithecus are even more unusual guenons than their multimale/multifemale social system suggests, however. Male and female *M. talapoin* interact only during the breeding season. Outside of the breeding season, *M. talapoin* society is highly segregated, with males interacting and traveling with other males and females only with females (Rowell 1971, Rowell and Dixson 1975). Given the high level of interaction between males, it would be interesting to examine the behavioral genetics of *M. talapoin* to determine the degree of relatedness among males in the same group. Few field studies of *A. nigroviridis* have been conducted, but those that have been indicate that this species lives in multimale/multifemale groups (e.g., Gautier 1985).

Intragroup and Intergroup Relations

Given the fact that most guenon species live in single-male/multifemale social groups, it is not surprising that the majority of adult social interactions occur between females (Rowell 1988a) and between females and the resident male primarily during the mating season (e.g., Chism et al. 1984). In such species, the resident male is socially peripheral to the rest of the group (e.g., Hall 1965a, Struhsaker and Leland 1979). Intragroup interactions between females are generally amicable, as expected of female-philopatric species, consisting primarily of grooming (Chism and Rogers 2002, Buzzard and Eckardt 2007) and, in some species, allomaternal care of infants (Chism 1978, Struhsaker and Leland 1979, Zucker and Kaplan 1981). Most guenon species differ from the closely related *Papio* and *Macaca* in lacking strong, linear dominance hierarchies (*C. ascanius*, Cords 1987a, Buzzard and Eckardt 2007; *C. campbelli*, Cords 1987a; *C. mitis*, Cords 2000b). Isbell and Pruetz (1998) reported that *E. patas* in Laikipia, Kenya, also lacked some strong, linear dominance hierarchies, perhaps because they feed largely on small, dispersed food resources, including *Acacia* ants (see also Isbell 1998). In contrast, *E. patas* feeding on clumped fruit trees in Kala Maloue, Cameroon, do display a linear dominance hierarchy (Nakagawa 2008). Both male and female *C. aethiops* exhibit linear dominance hierarchies (Struhsaker 1967b, Cheney et al. 1988).

Conspecific groups of guenons are generally intolerant of one another, with both males and females behaving aggressively in intergroup encounters (see Cheney 1987 for a review), although among *C. mitis* and *C. diana*, males rarely participate in aggressive encounters between groups

(Cords 2002c, Buzzard and Eckardt 2007). Several species of guenons are known to engage in territorial range defense, including *C. aethiops* (Struhsaker 1967b; Cheney 1981, 1992), *C. diana* (Hill 1994, Buzzard and Eckardt 2007), *C. mitis* (Struhsaker and Leland 1979; Cords 1987a, 2002c; Butynski 1990), *C. campbelli* (Buzzard and Eckardt 2007), *C. nictitans* (Buzzard and Eckardt 2007), and *C. ascanius* (Struhsaker and Leland 1979, Cords 1987a). In some guenons, intergroup encounters are marked primarily by territorial calling. Among *C. diana*, for example, range boundaries are advertised and defended primarily through intergroup calling and secondarily through aggressive interactions (Hill 1994, Buzzard and Eckardt 2007). In this species, females play an important role in range defense; intergroup calls by females apparently stimulate the group's male to give his intergroup vocalization or loud call (Hill 1994). In other species, such as *C. aethiops*, range defense includes both intergroup vocalizations and active defense at territorial boundaries, in which both males and females participate (Cheney 1981). Territorial encounters are similar in *C. ascanius*. After a large *C. ascanius* group (50 individuals) fissioned into two smaller groups of similar size, intergroup territorial encounters increased significantly as the two daughter groups struggled to establish a new territorial border between them (Windfelder and Lwanga 2002). Adult females were the primary participants during the encounters, which included vocalizations, chases, and physical aggression (Windfelder and Lwanga 2002).

Although other guenons, including *E. patas* and *C. neglectus*, do not defend territorial boundaries (Rowell 1988a, Chism 1999), they may still engage in aggressive intergroup encounters. Among *E. patas*, whose home ranges are apparently too large to actively defend, adult females and juveniles chase the members of the other group over long distances (Chism 1999). Although resident male *E. patas* are invariably aggressive toward extragroup males (Isbell et al. 2009), they are seldom involved in intergroup encounters, except during the breeding season (Struhsaker and Gartlan 1970, Chism 1999). *C. nictitans* in the Taï Forest do not have overlapping home ranges and intergroup interactions were not observed (Buzzard and Eckhardt 2007).

Although female guenons are philopatric and generally intolerant of conspecifics in other groups, occasionally females do transfer between groups. Sometimes such transfers result in apparently peaceful coexistence between established and new residents of the groups (Hauser et al. 1986, Isbell et al. 1991). However, transfers by females are not always without incident. At Segera Ranch, in Laikipia, Kenya, an adult female *C. aethiops* and her dependent offspring recovered after they were attacked by three other females and dropped 12 m from a fever tree a few days after her group fused with the aggressors' group (K.E. Jaffe and L.A. Isbell, in press). In the Cape Vidal Nature Reserve in South Africa, a female *C. mitis*, apparently attempting to transfer into another group, was attacked by the females

of the new group and a few hours later died of her injuries (Payne et al. 2003).

SOCIOECOLOGY OF POLYSPECIFIC ASSOCIATIONS

Polyspecific associations, or mixed-species groups, are associations between two or more different species that involve behavioral changes by at least one of the participating species (Strier 2003a). For example, associations between brown capuchins (*Cebus apella*) and squirrel monkeys (*Saimiri sciureus*) are maintained because *S. sciureus* change their ranging behavior and actively follow the *C. apella* group (Terborgh 1983). Such associations are well known in both New World and Old World monkeys. In some cases, associations between different species are not true associations but occur by chance as a result of high primate biomass (e.g., *Lophocebus albigena* with *C. mitis*, *Procolobus badius*, and *Colobus guereza*, Waser 1982a; *Procolobus verus* with *Cercocebus atys*, *Colobus polykomos*, and *Procolobus badius*, Oates et al. 1990). In other cases, they are true associations, occurring at frequencies greater than expected (e.g., Goeldi's monkeys [*Callimico goeldii*] with saddleback tamarins [*Saguinus fuscicollis*] and red-bellied tamarins [*S. labiatus*], Porter 2001c, Porter and Garber 2007; *S. fuscicollis* and mustached tamarins [*S. mystax*], Smith et al. 2005; *L. albigena* with *C. ascanius*, Waser 1982a; *C. ascanius* with *C. mitis*, Cords 1990a; *Procolobus verus* with *C. campbelli*, *C. diana*, and *C. petaurista*, Whitesides 1989, Oates et al. 1990), with one or more species gaining real benefits. In some cases, these associations may be nearly permanent (e.g., *Saguinus mystax* with *S. fuscicollis*, Peres 1992, Smith et al. 2005; emperor tamarins [*S. imperator*] with *S. fuscicollis*, Terborgh 1983).

Within the Cercopithecini, the tendency to form polyspecific associations is more common in the forest guenons, though still variable. *C. neglectus* do not form associations with other species (Rowell 1988a, McGraw 1994, Mugambi et al. 1997, Zeeve 1991), and *C. lhoesti* spends little time in association with other species (Gautier-Hion 1988b). In East Africa, *E. patas* and *C. aethiops* have not been observed in polyspecific associations (K. E. Jaffe personal observation), although sometimes individual vervets or patas join groups of different species (L. A. Isbell personal observation) or vice versa (Starin 1993). In Senegal, however, Galat-Luong and Galat (2005) observed Temminck's red colobus (*P. colobus temmincki*) in association with *E. patas* and *C. sabaeus* (a member of the *C. aethiops* superspecies group, Table 16.1) in 3.7% and 31.9% of *P. colobus temmincki* observations, respectively. Furthermore, they observed all three species together on one occasion (0.5% of observations, Galat-Luong and Galat 2005). In contrast, *C. pogonias* and *C. nictitans* spend virtually all of their time in association with each other or other guenon species (Gautier-Hion and Gautier 1974, Gautier-Hion et al. 1983).

Guenons also vary in the species with which they choose to associate (Table 16.8). *C. pogonias* (Gautier-Hion 1988b) and *A. nigroviridis* (Zeeve 1991) have been

Table 16.8 Guenon Polyspecific Associations

SPECIES	ASSOCIATES WITH	SOURCES
C. aethiops	Temminck's red colobus (Procolobus badius temmincki)	Galat-Luong and Galat 2005: C. sabaeus
C. ascanius	Red colobus (Procolobus badius)	Struhsaker 1981a; Chapman and Chapman 2000c; Teelen 2007
	Gray-cheeked mangabeys (Lophocebus albigena)	Struhsaker 1981a; Chapman and Chapman 2000c
	Blue monkeys	Struhsaker 1981a (Uganda), Cords 1990a (Kenya); Chapman and Chapman 2000c
	Black and white colobus (Colobus guereza)	Cords 1990a (Kenya), Chapman and Chapman 2000c
	Allen's swamp monkey	Zeeve 1991
	Wolf's guenon	McGraw 1994
	Angolan colobus (Colobus angolensis)	McGraw 1994
C. campbelli	Lesser spot nosed guenons	Gautier-Hion 1988b
	Olive colobus (Procolobus verus)	Oates and Whitesides 1990
	Diana monkeys	Gautier-Hion 1988b
C. cephus	Putty-nosed guenons	Gautier-Hion 1988b
	Crowed guenons	Gautier-Hion 1988b
	White-collared mangabeys (Cercocebus torquatus)	Mitani 1991
	Gray-cheeked mangabeys	Mitani 1991
C. diana	Campbell's guenons	Gautier-Hion 1988b, Wolters and Zuberhbühler 2003, McGraw et al. 2002
	Lesser spot nosed guenons	Gautier-Hion 1988b
	Olive colobus	Oates and Whitesides 1990, McGraw et al. 2002
	Red colobus	Oates and Whitesides 1990, Bshary and Noë 1997, Noë and Bshary 1997, McGraw et al. 2002
	Putty-nosed guenons	Eckardt and Zuberbühler 2004
	King colobus	McGraw et al. 2002
C. dryas	No data[1]	
C. erythrogaster	Mona monkeys	Oates 1985
	White-collared mangabeys	Oates 1985
C. erythrotis	Gray-cheeked mangabeys	Waser 1980
C. hamlyni	No data	
C. lhoesti	Rarely associates with other guenons	Gautier-Hion 1988b
C. mitis	Red-tail guenons	Struhsaker 1981a (Uganda); Cords 1990a (Kenya); Chapman and Chapman 2000c
	Red colobus	Struhsaker 1981a; Chapman and Chapman 2000c
	Black and white colobus	Cords 1990a
	Gray-cheeked mangabeys	Struhsaker 1981a, Chapman and Chapman 2000c
	Angolan colobus	Zeeve 1991
C. mona	White-throated guenons	Oates 1985
	White-collared mangabeys	Harding 1984
C. neglectus	Does not associate with other primates[2]	Rowell 1988a, Zeeve 1991, Wahome et al. 1993, McGraw 1994
C. nictitans	Crowned guenons	Gautier-Hion et al. 1983
	Gray-cheeked mangabeys	Mitani 1991
	Diana monkeys	Eckardt and Zuberbühler 2004
	Mustached guenons	Gautier-Hion 1988b
C. petaurista	Campbell's guenon	Gautier-Hion 1988b
	Diana monkeys	Gautier-Hion 1988b
	Olive colobus	Oates and Whitesides 1990

Table 16.8 (cont'd)

SPECIES	ASSOCIATES WITH	SOURCES
C. pogonias	Mustached guenons	Gautier-Hion 1988b
	Putty-nosed guenons	Gautier-Hion 1988b
C. preussi	No data	
C. sclateri	No data	
C. solatus	No data	
C. wolfi	Black mangabeys (Lophocebus aterrimus)	Gevaerts 1992, Zeeve 1991
	Red-tailed guenons	McGraw 1994, Zeeve 1991
	Angolan colobus	McGraw 1994, Zeeve 1991
	Allen's swamp monkey	Zeeve 1991
A. nigroviridis	Red-tailed guenons	Gautier 1985, Zeeve 1991
	Wolf's guenon	Zeeve 1991
M. talapoin	No data	
E. patas	Temminck's red colobus	Galat-Luong and Galat 2005

[1] "Does not associate" indicates that groups of a specific species have not been observed forming polyspecific associations with other groups. Individuals of these species may associate with allospecifics (see text).

[2] "No data" indicates data on polyspecific association formation are not available for that species.

Table 16.9 Proposed Benefits of Polyspecific Associations

FORAGING EFFICIENCY ADVANTAGES	ANTIPREDATOR ADVANTAGES
Reduced scramble competition	Increased predator detection
Increased detection of resources	Increased predator confusion
Access to otherwise unavailable resources	Decreased chance of "being the victim"
Increased prey-capture rate	Increased ability to defend against predators
Increased competitive ability	

observed associating only with other guenons, although a single male *C. pogonias* was found with *Colobus satanas* groups in the Fôret des Abeilles in central Gabon (Fleury and Gautier-Hion 1997). *C. ascanius* (Struhsaker 1981a, Cords 1990a; Zeeve 1991), *C. mitis* (Struhsaker 1981a, Cords 1990a), and *C. wolfi* (Zeeve 1991, Teelen 2007) form associations with both other guenons and nonguenon species. *C. erythrotis* is reported to associate only with a nonguenon species, *L. albigena* (Waser 1980), despite the fact that it is sympatric with *C. preussi* and *C. pogonias* in parts of its range (Oates 1988b). Finally, although rare, some guenon species have been observed associating with nonprimate species (e.g., *C. ascanius* with red forest duikers [*Cephalophus natalensis*] in Kibale Forest, Uganda). In this case, the association appears to be maintained by *Ceph. natalensis*, which forage on food dropped by *C. ascanius* (Struhsaker 1981a).

In some cases, association between two species, or simply sympatry, can lead to interspecific matings and even fertile hybrids (*C. mitis* and *C. ascanius*, Aldrich-Blake 1968, Struhsaker et al. 1988). However, in the vast majority of cases, polyspecific associations do not result in viable offspring or even interspecific matings, so a primary question is what (nonreproductive) benefits do individuals of different species acquire by forming long-term, stable polyspecific associations?

Explanations for such associations center on the benefits that individuals acquire from increased group size without suffering the costs associated with traveling in larger groups. The benefits of polyspecific associations can be divided into two broad categories which mirror primary benefits that have been forwarded to explain the evolution of group living in primates: increased access to foods and enhanced predator detection and avoidance (Table 16.9). The specific benefits will likely differ for each polyspecific association, depending on the species involved, the habitat type, and potential predators, and may involve any combination of benefits listed. Associations between guenon species and between guenons and other catarrhines can involve both categories of benefits.

Increased competition for resources is thought by most to be a primary cost of increased group size (Alexander 1974, van Schaik 1983, Terborgh and Janson 1986, Isbell 1991; but see Isbell 2004). Such a cost may be alleviated to some extent if group size is increased by associating with allospecifics that eat different foods (Oates and Whitesides 1990, Wolters and Zuberbühler 2003). Even if species have similar diets, however, they may still obtain foraging advantages by associating. For example, in a community of *Cercopithecus* species in Makokou, Gabon, polyspecific associations occur between three species: *C. cephus*, *C. nictitans*, and *C. pogonias* (Gautier-Hion et al. 1983). Despite having a high degree of dietary overlap (97%), monkeys in polyspecific associations increased foraging efficiency by reducing their chances of visiting a food source that had already been depleted (Gautier-Hion et al. 1983; see also Isbell 2004 for a similar argument for group living in general).

Studies of associations between *C. mitis* and *C. ascanius* suggest also that one species may use another to locate food sources. In Kakamega Forest, Kenya, *C. ascanius*, which travel farther per day than *C. mitis*, apparently use "local" *C. mitis* groups as guides to food sources and by doing so may avoid feeding at sites that have been recently visited by *C. mitis* (Cords 1987a). A study of polyspecific associations in *C. mitis* and *C. ascanius* in Kibale Forest, Uganda, on the other hand, suggests that *C. mitis* use *C. ascanius* as guides to food sources. In Kibale, unlike Kakamega, *C. ascanius* have smaller home ranges than *C. mitis* and may have better knowledge of food source locations (Struhsaker 1981a; see also Cords 1990a).

The other category of benefits obtained from polyspecific associations is increased predator avoidance. Like foraging benefits, anti-predator-related benefits of polyspecific associations come in many forms. Studies of birds have shown that increased group size reduces the amount of time any individual needs to spend scanning because there are more eyes with the potential to detect a predator (e.g., Powell 1974, Kenward 1978). Guenons may gain a similar benefit. In their study of associations between *C. campbelli* and *C. diana*, Wolters and Zuberbühler (2003) found that both species spent less time being vigilant when in association. The available time was apparently used (at least by *C. diana*) to search for food as their time spent foraging increased when in association. Wolters and Zuberbühler (2003) interpret this as a benefit to *C. diana* of associating, but the possibility exists that increased time spent foraging could actually represent a cost. Given that *C. campbelli* are known to be largely responsible for the formation and maintenance of associations with *C. diana* (Bshary and Noë 1997b), *C. diana*'s increased foraging may be a means of off-setting the cost of increased competition for resources they incur by associating with *C. campbelli*.

The increased chance of detecting a predator when in association also seems to lead to a change in *C. campbelli*'s behavior. When in association with *C. diana*, *C. campbelli* apparently experience a reduced risk of predation and (*1*) spend more time in exposed (i.e., more dangerous) areas of the habitat, (*2*) vocalize more often, and (*3*) display greater interindividual distances, apparently indicating reduced perceived risk of predation when in association. While all three of these results are expected if perceived risk of predation is reduced, the last result, at least, may also be due to *C. diana* being interspersed among *C. campbelli*, forcing the *C. campbelli* individuals to spread out. In addition, both species apparently benefit from associating by

using a broader ecological niche (i.e., both species increase their use of intermediate strata; Wolters and Zuberbühler 2003). Increased use of intermediate strata may be an antipredator benefit of polyspecific associations because of the seeming protection from both terrestrial and arboreal predators.

As with the study of *C. diana*–*C. campbelli* associations, other researchers have also noted increased used of vertical space as an antipredator benefit because, when in association, individuals are able to use strata that are otherwise unavailable to them. For example, polyspecific associations between *C. cephus*, *C. nictitans*, and *C. pogonias* in Gabon indicate that in association with the other two species *C. cephus* expand their vertical use of space and are less dependent on low, dense vegetation, perhaps because they are less susceptible to predation from aerial predators (Gautier-Hion et al. 1983). Similarly, *C. diana*, which are primarily arboreal, descend to low forest levels and the forest floor significantly more often when in association with the highly terrestrial *Cercocebus atys*, apparently because the latter are better at detecting terrestrial predators (McGraw and Bshary 2002). In another study of *C. diana*, when in association with *Procolobus badius*, researchers found that both species changed their habitat use, with *C. diana* spending more time in higher strata and *Procolobus badius*, in lower strata (Bshary and Noë 1997b). According to the authors, this change in strata use indicates that *C. diana* experience a reduced risk of predation by eagles when in association (Bshary and Noë 1997b). Since *Procolobus badius* and *C. diana* share responsibility for the formation and maintenance of these associations (Bshary and Noë 1997b), it is likely that both species gain some advantage from associating. It is difficult to determine, however, whether individuals in polyspecific associations are forced to change their stratum use because of the association or are freed up to do so.

Increased use of otherwise inaccessible strata is apparently a common result of polyspecific associations, but these examples highlight the difficulty of teasing apart foraging benefits from antipredator benefits. The use of an expanded ecological niche during polyspecific associations may be the result of antipredator benefits, but such habitat use clearly has foraging benefits as species would likely encounter additional foods when they forage in strata they do not usually occupy.

The difficulty of separating predation from feeding as the ultimate cause of polyspecific associations or behaviors while in association can be seen with the association between *C. diana* and *C. campbelli*. In addition to other behavioral changes, Wolters and Zuberbühler (2003) note that both species increased their day range lengths when in association and cite this as an antipredator benefit because moving is a dangerous, attention-grabbing behavior and primates should display adaptations to reducing predation risk while traveling (see also Boinski et al. 2000b). On the other hand, increased daily travel distance could be a true cost of polyspecific associations, just as it appears to be for many single-species groups (Isbell 1991, Janson and Goldsmith 1995). Dietary overlap is high in these two species (see Gautier-Hion 1988b), so a larger mixed-species group of *C. diana* and *C. campbelli* may need to travel farther to obtain enough food for all group members than smaller, single-species groups. If this is the case, increased day range lengths and increased (scramble) competition for resources are apparently costs that both species are willing to incur given the other benefits they obtain from associating.

In other cases, it is easier to determine the ultimate cause of polyspecific associations. Despite a high degree of dietary overlap, and thus resource competition, between *C. diana* and *C. nictitans* in the Taï Forest, Ivory Coast, *C. diana* tolerates the presence of *C. nictitans* because *C. nictitans* provides reliable notification of the presence of crowned eagles (Eckardt and Zuberbühler 2004). Adding to their antipredator benefit for *C. diana*, *C. nictitans* males attack crowned eagles, and when in association, *C. diana* and *C. nictitans* males coordinate their attacks.

Perhaps the most obvious evidence that polyspecific associations provide antipredator benefits is supplied by alarm-calling behavior. In the associations of *C. cephus*, *C. nictitans*, and *C. pogonias* in Makokou, Gabon, the species had different predator-detection roles, depending on the strata they typically occupied (Gautier-Hion et al. 1983). When the potential predator was terrestrial, *C. cephus*, which tend to be closer to the ground, gave the first alarm call 75% of the time. In contrast, in the case of aerial predators, *C. cephus*, which were higher in the canopy, were the first to alarm call 88% of the time and the first caller was invariably the male. In a study of associations between single male *C. pogonias* and black colobus (*Colobus satanas*) groups, *C. pogonias* were better at detecting aerial predators and *Colobus santanas* were better at detecting *Pan troglodytes*, despite the fact that both were found at the same height in the canopy (Fleury and Gautier-Hion 1997).

Similarly, in *Procolobus badius*–*C. diana* associations, both species benefit from the other's alarm calls at crowned eagles by being alerted to predator presence, and *C. diana* may further reduce their susceptibility to terrestrial predators simply by associating with *Procolobus badius* (Bshary and Noë 1997b), even though this seems counterintuitive (i.e., *C. diana* would seem to be in harm's way more often by associating with a preferred prey species). Although *Pan troglodytes* eat *C. diana*, they prefer *Procolobus badius* as prey (Bshary and Noë 1997b, Stanford 1998a). Given this inclination, *C. diana* may decrease their chances of being preyed upon by associating with *Procolobus badius* because, given the choice, *Pan troglodytes* prefer to hunt and eat *Procolobus badius*, at least in part, it seems, because they are easier to catch (Boesch 1994a).

CONSERVATION

The conservation status of the guenons is mixed. Most species are listed as being of "least concern" by the World Conservation Union (IUCN), indicating that they are not endangered or threatened (IUCN 2009; Table 16.10). A few species are in jeopardy, however. *C. diana*, *C. erythrogaster*, *C. erythrotis*, *C. hamlyni*, *C. lhoesti*, *C. sclateri*, and *C. solatus* are all listed as "vulnerable" and face a high risk of extinction in the wild. *C. preussi* is "endangered," facing a very high risk of extinction, and *C. dryas* is "critically endangered," facing an extremely high risk of extinction (IUCN 2009). Only two species, *C. aethiops* and *M. ogouensis*, have stable populations in the wild. All other guenons, even some of "least concern," have either unknown or decreasing population sizes (IUCN 2009); thus, some of these species will undoubtedly change categories if the threats to their populations are not reduced or eliminated. In most cases, habitat loss or degradation and human hunting are the main threats to guenons (Table 16.10). Two species (*C. neglectus* and *E. patas*) are also killed by humans when they are caught raiding crops (Kingdon et al. 2008b, Struhsaker et al. 2008). With regard to *C. sclateri*, habitat fragmentation (with accompanying population isolation and reduced gene flow) is a major concern (Oates et al. 2008j).

SUMMARY

The guenons are a diverse group of African monkeys in a number of ways, including pelage coloration (ranging from the drab and cryptic *A. nigroviridis* to the brightly colored and conspicuous *C. diana*), diet (ranging from the highly frugivorous *C. cephus* to the highly gummivores and insectivorous *E. patas*), habitat preferences (dense rain forests of the forest guenons to open woodland occupied by *C. aethiops* and *E. patas*), and social structure (single-male/multifemale in most forest guenons to multimale/multifemale in *C. aethiops* and *Miopithecus* to possible single-male/single-female in some *C. neglectus* populations). In addition, guenons display diversity in their willingness to form associations with other guenon (and nonguenon) species. Polyspecific associations are relatively widespread among the guenons, and benefits resulting from such associations fall into two broad categories: resource acquisition and predator avoidance.

In this chapter, we have provided an overview of the taxonomy, ecology, and behavior of the guenons. This overview has shed light on two areas in particular that require further attention. First, while many guenon species are relatively well known, the behavior and ecology of a number of others largely remain a mystery. We hope this review encourages further research on all species of guenons, but we especially hope some intrepid explorers will take the lead in studying *A. nigroviridis*, *C. erythrogaster*, *C. erythrotis*, *C. sclateri*,

Table 16.10 Conservation of Guenon Species, Including Population Trends (if Known) and Current Threats

SPECIES	IUCN STATUS[1]	POPULATION TREND	MAIN THREATS[2]	REFERENCES
C. ascanius	Least Concern	Unknown	Habitat loss, hunting	Oates et al. 2008a, IUCN 2009
C. campbelli	Least Concern	Unknown	Habitat loss (deforestation), hunting	Oates et al. 2008b, IUCN 2009
C. cephus	Least Concern	Unknown	Habitat loss (deforestation), possibly hunting	Oates et al. 2008c, IUCN 2009
C. diana	Vulnerable	Decreasing	Hunting, habitat loss	Oates et al. 2008d, IUCN 2009
C. dryas	Critically Endangered	Unknown	Hunting, habitat loss (agriculture)	Hart et al. 2008a, IUCN 2009
C. erythrogaster	Vulnerable	Decreasing	Habitat loss (deforestation, agriculture), hunting	Oates et al. 2008e, IUCN 2009
C. erythrotis	Vulnerable	Decreasing	Habitat loss (deforestation, agriculture), hunting	Oates et al. 2008f, IUCN 2009
C. hamlyni	Vulnerable	Decreasing	Habitat loss (agriculture), hunting	Hart and Butynski, 2008b, IUCN 2009
C. lhoesti	Vulnerable	Decreasing	Habitat loss (agriculture), hunting	Hart et al. 2008b, IUCN 2009
C. mitis	Least Concern	Decreasing	Habitat loss, hunting	Kingdon et al. 2008a, IUCN 2009
C. mona	Least Concern	Unknown	Habitat loss, hunting	Oates et al. 2008g, IUCN 2009
C. neglectus	Least Concern	Unknown	Habitat loss (deforestation, agriculture), hunting, pest control	Struhsaker et al. 2008, IUCN 2009
C. nictitans	Least Concern	Decreasing	Habitat loss (deforestation, agriculture), hunting	Oates and Groves 2008a, IUCN 2009
C. petaurista	Least Concern	Unknown	Habitat loss (deforestation, human settlement), hunting	Oates et al. 2008h, IUCN 2009
C. pogonias	Least Concern	Unknown	Habitat loss (deforestation, agriculture, human settlement), hunting	Oates et al. 2008i, IUCN 2009
C. preussi	Endangered	Decreasing	Habitat destruction (cultivation, fire), hunting	Oates 2008, IUCN 2009
C. sclateri	Vulnerable	Decreasing	Habitat destruction (logging, agriculture, oil exploration) and habitat fragmentation, hunting	Oates et al. 2008j, IUCN 2009
C. solatus	Vulnerable	Unknown	Hunting, habitat loss (logging)	Oates and Bearder 2008, IUCN 2009
C. aethiops[3]	Least Concern	Stable	No major threats	Kingdon and Butynski 2008, IUCN 2009
A. nigroviridis	Least Concern	Unknown	Hunting	Oates and Groves 2008b, IUCN 2009
E. patas	Least Concern	Decreasing	Hunting, pest control, habitat loss (overgrazing, agriculture)	Kingdon et al. 2008b; IUCN 2009
M. ogouensis	Least Concern	Stable	No major threats	Oates and Groves 2008c; IUCN 2009
M. talapoin	Least Concern	Unknown	No major threats	Oates and Groves 2008d; IUCN 2009

[1] IUCN status definitions: *Critically Endangered*, species is considered to be facing an extremely high risk of extinction in the wild; *Endangered*, species is considered to be facing a very high risk of extinction in the wild; *Vulnerable*, species is considered to be facing a high risk of extinction in the wild; *Least Concern*, species does not qualify for Critically Endangered, Endangered, Vulnerable, or Near Threatened—sapecies in this category are typically widespread and abundant.

[2] IUCN main threat definitions: hunting refers to *hunting* by humans, typically for meat; when specified, the reason behind *habitat loss* or *habitat degradation* is given in parentheses; *pest control* refers to hunting by humans because the species is viewed as an agricultural pest; *no major threats* indicates that there are no major conservation threats to the species.

[3] *Cercopithecus aethiops* is listed as *Chlorocebus aethiops* and includes several species from the *Cercopithecus aethiops* superspecies group (see Table 16.1). All but one of these species are listed as being of least concern, stable, and no major threats. The exception is *C. djamdjamensis*, which is listed as vulnerable with a decreasing population due to habitat loss and degradation (Butynski et al. 2008).

C. solatus, *C. dryas*, and *C. hamlyni*, for which little data currently exist.

Second, although much work has been conducted on polyspecific associations in guenons in the past 25 years, teasing apart the costs and benefits of such associations continues to be a difficult task and will require further attention. The confounding nature of many of the costs and benefits listed in Table 16.9 means that researchers must be careful not to assume that changes in behavior (e.g., feeding behavior, habitat use) are benefits. In order to determine whether such behavioral changes are costs or benefits, it is important to take a number of issues into account, including which species is responsible for forming and maintaining the association. For example,

increased time spent feeding in association may indeed be a benefit if the species that increases feeding is responsible for the association. However, if the species that increases feeding in association is not responsible for the association, this same change in behavior more likely represents a cost in terms of resource competition. In light of this, we encourage future research into the costs and benefits of polyspecific associations, with specific attention paid to the formation and maintenance of these associations.

Species included in this chapter are shown in Color Plates 23–26 (*Erythrocebus patas*, *Chlorocebus pygerythrus hilgerti*, *Cercopithecus mitis albogularis*, and *Cercopithecus ascanius schmidti*).

17

The Hylobatidae
Small Apes of Asia
Thad Q. Bartlett

1. What traits differentiate gibbons from the great apes?
2. What behaviors are characteristic of pair-bonding in gibbons?
3. What are potential explanations for the evolution of pair-bonds in gibbons?

INTRODUCTION

Members of the family Hylobatidae, known commonly as gibbons, include 12 species of small arboreal ape that inhabit the rain forests of eastern and southeastern Asia. Owing to shared locomotor morphology, gibbons, along with the great apes and humans, make up the superfamily Hominoidea (Simpson 1945, Groves 1972). Yet, gibbons are far smaller than other hominoids, a trait that has given rise to the label "lesser apes" (Preuschoft et al. 1984). Depending on the species, gibbons weigh anywhere from 5 to 15 kg. A related characteristic is that gibbons are almost exclusively arboreal. They move quickly through the trees via a mixture of brachiation, bipedal running, and leaping (Fleagle 1976a). On the rare occasions that gibbons do come to the forest floor, they walk bipedally, which contrasts with most great apes that routinely come to the ground to travel and, when they do, walk quadruped-ally on either fists (*Pongo* spp.) or knuckles (*Gorilla* spp., *Pan* spp.).

While hylobatids are easily distinguished from other hominoids by their small body size, the label "lesser apes" is perhaps just as much an indication of their perceived relevance to understanding human evolution (Yerkes and Yerkes 1929). For example, gibbons do not display the cognitive abilities typically exhibited by other ape species (Povinelli and Cant 1995). However, while the relevance of gibbons as a model for human evolution may be debated, their unique suite of morphological and behavioral characters makes them of particular interest in the study of primate behavioral ecology.

In addition to their small size, gibbons can be distinguished morphologically from the great apes based on three sets of traits (Groves 1972, 1984). First, gibbons possess a number of ancestral characteristics shared with cercopithecoid monkeys but absent in most great apes. Among such traits are the *ischial tuberosities*, bony structures on the lower portion of the pelvis that serve as areas of attachment for callus-like sitting pads. This trait is probably linked to the fact that gibbons, like many monkeys, sleep while sitting on bare branches, rather than in nests like other apes. Also, gibbons are distinguished by unique derived traits related to brachiation. While all apes share an *orthograde* (or upright) posture, the pendulum-like swing of gibbon locomotion is unique among the hominoids. Related morphological traits include extremely long forearms, a highly mobile shoulder joint, and long hook-like hands with a deep cleft between the first and second digits. Finally, gibbons are unique among apes in the morphological similarity of males and females. Gibbon males and females are the same size, and both sexes have long saber-like canines with sectorial premolars to hone the upper canines (Groves 1972, 1984).

Behaviorally, gibbons can be distinguished from other primates based on a complex of characters that include frugivory, territoriality, stable pair-bonds, and regular vocal displays or songs. Much of the scientific literature on gibbons has focused on the selective pressures that gave rise to this unique suite of traits (Preuschoft et al. 1984, Leighton 1987), and this subject remains an area of active inquiry (Fuentes 2000, Reichard 2003a, Brockelman 2009).

TAXONOMY

Relative to most other primate families, the Hylobatidae are fairly homogenous as a group in respect to both anatomy and behavior (Chivers 1984). As Brockelman (2009) has observed, "the study of gibbons is in many respects a study in adaptive constraints rather than of adaptive radiation." Yet, in terms of gibbon systematics, four different taxa are readily recognized based on chromosome number (Prouty et al. 1983a,b). Given the overall ecological

similarity among gibbons, these four groups have typically been distinguished at the level of the subgenus (Marshall and Sugardjito 1986, Geissmann 1995, Chivers 2001, Groves 2001a). However, recent molecular data suggest that the temporal split between gibbon subgenera is as deep as that between chimpanzees and humans (Hayashi et al. 1995, Hall et al. 1998, Roos and Geissmann 2001). As a result, there is an emerging consensus that the recognized subgenera should be promoted to the genus level (Roos and Geissmann 2001, Geissmann 2002b, Groves 2001a, Brandon-Jones et al. 2004, Mootnick and Groves 2005). Nevertheless, investigators continue to differ on how many gibbon species should be recognized. Marshall and Sugardjito (1986) identify nine, Groves (2001a) 14, and Geissmann (1995, 2002b) 12. Further complicating matters is the observation by Groves (2001a, see also Brandon-Jones et al. 2004) that the type specimen for *Bunopithecus sericus* is not congeneric with other members of the genus. Consequently, the scientific name for the hoolock gibbon has recently been changed to *Hoolock hoolock* (Mootnick and Groves 2005). At the species level, the greatest area of disagreement over the taxonomy of the living gibbons concerns the crested group (*Nomascus* spp.). The taxonomy of this genus has been studied most extensively by Geissmann (1995, 2002b; Takacs et al. 2005), and therefore his taxonomy—itself under revision (see Geissmann 2009)—is provisionally accepted here (Table 17.1).

Most gibbon species are very similar in skeletal anatomy and more readily differentiated based on pelage and characteristics of their vocal repertoire (Fooden 1971, Chivers 2001). Hoolock, pileated (*Hylobates pileatus*), and crested gibbons are all sexually dichromatic. The predominant

Table 17.1 Species of the Family Hylobatidae[1]

GENUS	SPECIES	COMMON NAME(S)	WEIGHT[2] (KG)	COAT COLORATION	DUET
Hoolock (diploid number 38)	*Ho. hoolock*	Hoolock or white-browed gibbon	6.1–6.9	Males black, females buff	Yes
Hylobates (diploid number 44)	*H. agilis*	Dark-handed or agile gibbon	5.5–6.4	Varies, males have white cheeks	Yes
	H. klossii	Kloss's gibbon or biloh	5.8	Both sexes black	No
	H. lar	White-handed or lar gibbon	4.4–7.6	Both sexes black or buff	Yes
	H. moloch	Silvery or Javan gibbon	5.7	Both sexes silver–gray	No
	H. muelleri	Mueller's or Bornean gibbon	5.0–6.4	Both sexes brown to gray	Yes
	H. pileatus	Pileated or capped gibbon	6.3–10.4	Males black, females buff	Yes
Nomascus (diploid number 52)	*N. concolor*	Western black-crested gibbon	4.5–9.0	Males black, females buff	Yes
	N. cf. nasutus	Eastern black-crested gibbon	–	Males black, females buff	Yes
	N. gabriellae	Yellow-cheeked crested gibbon	5.8	Males black, females buff	Yes
	N. leucogenys	White-cheeked crested gibbon	5.6–5.8	Males black, females buff	Yes
Symphalangus (diploid number 50)	*S. syndactylus*	Siamang	10.0–14.7	Both sexes black	Yes

[1] Taxonomy disputed. See text.
[2] See references in Rowe (1996).

color of males is black, while females are mostly buff-colored. In these species, the sexes also differ in the markings of the crown, face, and chest. Sexually dichromatic species are the most northerly distributed, while monochromatic species are restricted to more southerly regions (Fooden 1971). Chivers (2001) speculates that increased visibility in more seasonal forests may increase selection for visible indicators of sex. Lar gibbons (*Hylobates lar*), which have the broadest north–south distribution of all gibbon species, may represent an evolutionarily intermediate condition. This species is asexually dichromatic; that is, both sexes can be either black or buff. Extensive demographic data on lar gibbons indicate that coat color is a simple Mendelian trait, with black dominant to buff (Fooden 1971; Brockelman 2004). Coat color changes more rapidly over the course of evolution than other traits and therefore, taken alone, may be less useful than vocal or genetic traits in reconstructing the phylogenetic relationships between gibbon species (Geissmann 2002b).

Songs

Gibbons exhibit a broad variety of loud calls, or songs, that are structurally complex and can be heard over long distances. Current speculation is that gibbon calls serve both to cement the bond between pairmates and to advertise the location of the gibbon territory (Wickler 1980, Raemaekers et al. 1984). Studies of hybrid gibbons demonstrate that there is a significant genetic component to gibbon songs (Brockelman and Schilling 1984; Geissmann 1984, 1993; Tenaza 1985), and phylogenetic analyses of song characteristics have delineated species by genus (Geissmann 2002b). The kinds of song produced by gibbons can be divided minimally into solos and duets. Among *Hylobates* spp., mated males sing solos frequently, though not exclusively, at or before dawn (Geissmann 2002a). In the remaining genera, solos are absent among mated males. Mated females are reported to sing solos in two species, Kloss (*H. klossii*) and silvery gibbons (*H. moloch*). Notably, these two species lack the highly coordinated duets seen in other taxa (Table 17.1).

Duets differ broadly between species but share many elements of their structure, which typically includes an introductory sequence, a female great call, and the male's reply or coda (Marshall and Marshall 1976, Raemaekers et al. 1984). Male and female contributions are distinctive but often blend together or overlap so that different parts of the song are not apparent to the novice listener. Duets are sung mostly in the morning and may last up to 30 min. Immatures do not customarily sing during duets but do occasionally contribute during the vocal interlude to produce a group chorus. It is not uncommon to hear maturing females, and less often males, sing great calls with their mothers. Groups also sing when disturbed by a predator or the alarm calls of another species. As Raemaekers et al. (1984) observe, calls with apparently different functions blend into one another

such that it is difficult to mark objectively the boundary between different classes of vocalization (but see Clarke et al. 2006). Calling rates vary seasonally, with the highest rates during periods of resource abundance (Chivers 1974, Bartlett 1999a, 2009b).

Biogeography

Gibbons are confined to areas of eastern and southeastern Asia (Fig. 17.1). Most gibbon species are distributed allopatrically, with the main exception being siamangs (*Symphalangus syndactylus*), which are sympatric with two smaller species, lar gibbons in northern Sumatra and on the Malay Peninsula and agile gibbons (*H. agilis*) south of Lake Toba on Sumatra (Gittins and Raemaekers 1980). Brockelman and Gittins (1984) describe the known contact zones between species of the genus *Hylobates* at three separate sites, including a small area of sympatry between lar and pileated gibbons in Khao Yai National Park, Thailand, and a second between lar and agile gibbons along the Thai–Malay border (Gittins 1978). A third contact area occurs between Mueller's (*H. muelleri*) and agile gibbons in the Barito watershed, central Kalimantan, Borneo, Indonesia. This population, first documented by Marshall and Sugardjito (1986), has been studied intensively by others

Figure 17.1 Approximate historical distribution of four gibbon genera including major study sites. See text for details (distribution follows Chivers 1977, 1984; Geissmann 2004).

(Mather 1992, Bricknell 1999, McConkey 2000, Chivers 2001). An extensive survey by Mather (1992) showed that the hybrid zone extends over at least 3,500 km². Because no pure-species individuals were documented in the Barito area, it has been argued that the species rank of these two taxa should be reconsidered (see Chivers 2001). Finally, there has long been speculation about the occurrence of a hybrid zone between *N. leucogenys* and *N. gabriellae* in the triborder area between Cambodia, Laos, and Vietnam; but available data are not sufficient to draw firm conclusions (Groves 2001a, Konrad and Geissmann 2006, Mootnick 2006).

Field Studies

With the exception of some members of the crested group, most gibbon species have now been the subject of studies of at least several months. Studies amounting to several years have been completed on co-occurring lar gibbons and siamangs at Kuala Lompat, Malaysia (Chivers 1980, Raemaekers 1980) and the Ketambe Research Station, Sumatra, Indonesia (West 1981, Palombit 1992). Several other long-term projects are ongoing, including studies on lar gibbons in Khao Yai National Park, Thailand (Reichard and Sommer 1997; Brockelman et al. 1998; Bartlett 1999a, 2009b; Clarke et al. 2006; Barelli et al. 2007; Savini et al. 2008); agile gibbons in the Gunung Palung Nature Reserve, West Kalimantan, Borneo, Indonesia (Mitani 1990, Marshall and Leighton 2006) and at the Natural Laboratory of Peat-Swamp Forest (NLPSF), central Kalimantan, Indonesia (Cheyne 2007, Cheyne et al. 2008); hybrid gibbons (*H. muelleri* × *H. agilis*) at the Barito Ulu Research Area, central Kalimantan, Borneo, Indonesia (Mather 1992, Bricknell 1999, McConkey and Chivers 2007); black crested gibbons (*Nomascus concolor*) at Mt. Wuliang, Yunnan Province, China (Lan and Sheeran 1995, Fan et al. 2008); and siamangs at the Way Canguk Research Station, Sumatra, Indonesia (O'Brien et al. 2003, Lappan 2007). Much of this research has been reviewed previously (Chivers 1977, 2001; Leighton 1987; Bartlett 1999b; Geissmann 2004). The present chapter will emphasize developments in our understanding of the behavioral ecology of gibbons during the last two decades.

ECOLOGY

Because gibbons live in small social groups, are mostly active high in the canopy, and have the ability to move very quickly through the forest, detailed studies of their social behavior have been slow to materialize. Most studies have focused on aspects of ecology, and comparative data are now available for all species except for some members of the crested group (Table 17.2). As a rule, gibbons occupy small, well-defended territories that they travel via regular pathways in search of sources of ripe fruit (Chivers 1984,

2001; Leighton 1987). The limited data available on the ecology of crested gibbons suggest that these species may diverge from this pattern; consequently, crested gibbons are afforded separate attention later in this chapter.

The Gibbon Feeding Niche

All long-term studies of gibbon feeding behavior have found that they rely heavily on ripe fruit pulp (Chivers 1984, 2001). On average, fruit comprises 59% of the annual diet (Table 17.3). Most comparative studies have revealed that gibbons feed more heavily on fruit than other frugivorous primates, including macaques (*Macaca* spp., Feeroz et al. 1994, Ungar 1995) and orangutans (*Pongo pygmaeus*, Ungar 1995). However, quantitative differences are only part of the story and, in fact, do not always apply. For example, MacKinnon and MacKinnon (1980a) found that long-tailed macaques included as much fruit in the diet as lar gibbons and were much more frugivorous than sympatric siamangs. Where gibbons and sympatric species differ most is in the supplementary foods: Macaques have a more diverse diet overall, and orangutans typically make greater use of hard or unripe fruit and seeds (MacKinnon 1977, Feeroz et al. 1994, Ungar 1995). Gibbons may also avoid competition with sympatric primates by feeding higher in the canopy (MacKinnon and MacKinnon 1980a, Ungar 1996) and by making greater use of the small terminal branches of the crown margin (Cant 1992). The efficiency with which gibbons exploit the terminal branch niche has long been recognized as an essential component of their adaptive complex (Ellefson 1974).

The remainder of the gibbon diet comes mostly from young leaves, flowers, and insects. The contribution of these additional foods varies considerably between species and study sites. Rates of insectivory may account for up to 26% of feeding time but more typically less than 10%. To date, there has been no attempt to compare insect feeding to insect abundance, but at Khao Yai, Thailand, insect feeding varied seasonally (Bartlett 1999a, 2009b). It is likely that, as with other foods, the amount of insects in the diet varies in relation to availability. The significance of insectivory to gibbon socioecology remains an important area for future research.

Leaves are the second most common food item for most species, accounting for 30% of the diet on average. Siamangs tend to feed more heavily on leaves than sympatric lar gibbons (MacKinnon and MacKinnon 1980a, Palombit 1997), but very high rates of folivory have been recorded for *Hylobates* spp. where they occur in the absence of larger siamangs (*H. agilis*, 39%; *H. moloch*, 38%). In general, there is a negative correlation between the contribution of leaves and insects in the diet, suggesting that these food items are exploited as alternative sources of protein. Flowers generally contribute least to the gibbon diet; however, this is likely due to the short period of availability. Selection of flowers during periods of abundance is generally high

Table 17.2 A Summary of Gibbon Feeding and Ranging Behavior

SPECIES	STUDY SITE	DIET (%)				RANGE USE			SOURCE
		FRUIT (FIG1)	LEAVES	FLOWERS	INSECTS	DAY RANGE (KM)	HOME RANGE (HA)	TERRITORY (HA, %2)	
Ho. hoolock	Lawachara, Bangladesh	89 (38)	6	5	0	1.2	35	32 (91%)	Islam and Feeroz 1992
		77 (44)	12	4	5	1.7	63	51 (81%)	Ahsan 2001, Chivers 2001
	Chunati, India	71 (30)	13	14	1	0.9	26	—	Ahsan 2001, Chivers 2001
H. agilis	Sungai Dal, W. Malaysia	57 (17)	39	3	1	1.3	29	22 (76%)	Gittins 1982
H. klossii	Paitan, Siberut, Indonesia	72 (23)	2	0	25	1.5	32	21 (66%)	Whitten 1984
H. lar	Tanjon Triang, W. Malaysia	67 (–)	33	0	0	—	—	—	Ellefson 1974
	Kuala Lompat, W. Malaysia	50 (22)	29	7	13	1.5	57	—	Raemaekers 1979
		64 (27)	31	1	5	1.8	53	28 (53%)	MacKinnon and MacKinnon 1980a,b
	Ketambe, Sumatra, Indonesia	71 (45)	4	1	24	—	—	—	Palombit 1997
	Mo Singto, Khao Yai, Thailand	66 (19)	24	1	9	1.2	23	19 (83%)	Bartlett 1999a,b
	Klong Sai, Khao Yai, Thailand	66 (20)	27	2	7	1.3	43	40 (93%)	Suwanvecho 2003
H. moloch	Ujong Kulan, Java, Indonesia	61 (–)	38	1	0	1.4	17	16 (94%)	Kappeler 1984
H. muelleri	Kutai, Kalimantan, Indonesia	62 (24)	32	4	2	0.9	44	39 (89%)	Leighton 1987
H. muelleri x agilis	Barito Ulu, Kalimantan, Indonesia	62 (17)	24	13	1	—	53	42 (79%)	McConkey 2000, McConkey et al. 2002
H. pileatus	Khao Soi Dao, Thailand	71 (26)	13	0	15	—	—	—	Srikosamatara 1984
	Klong Sai, Khao Yai, Thailand	64 (21)	27	2	7	1.3	38	37 (96%)	Suwanvecho 2003
N. leucogenys	Meng La, Yunnan, China	39 (–)	53	5	4	—	—	—	Hu et al. 1989, 1990
N. concolor	Mt. Wuliang, Yunnan, China	44 (19)	47	9	0	1.4	129	—	Fan et al. 2008, 2009
	Ailao, Mt., Yunnan, China	24 (0)	54	6	14	1.3	87	—	Chen 1995
S. syndactylus	Ulu Sempan, W. Malaysia	47 (41)	50	2	1	0.8	15	13 (87%)	Chivers 1974
	Kuala Lompat, W. Malaysia	32 (24)	58	9	2	0.9	35	26 (76%)	Chivers 1974
		36 (22)	43	6	15	0.7	47	—	Raemaekers 1979
		45 (31)	44	4	8	0.6	28	18 (65%)	MacKinnon and MacKinnon 1980a,b
	Ketambe, Sumatra, Indonesia	61 (43)	17	1	21	—	—	—	Palombit 1997

[1] Percent of total diet.

[2] Percent of home range.

(Bartlett 1999a, 2009b; Bricknell 1999; McConkey et al. 2003).

Attempts to further define the gibbon feeding niche have focused principally on three areas: fruit size, patch size, and the contribution of figs (*Ficus* spp.) to the diet. Ungar (1995) studied fruit choice by four sympatric primates at Ketambe, Sumatra. He argues that gibbons demonstrate a preference for small fruit (<7.92 cm^3). Though gibbons fed equally on fruits of all sizes, they selected small fruit more often than all other sympatric primate species except smaller-bodied macaques (*Macaca fascicularis*). Ungar (1995) speculates that the gibbon preference for small fruit may be related to suspensory foraging and "a need to limit hand use during ingestion." However, research on fruit selection by hybrid gibbons (*H. muelleri* × *H. agilis*) at Barito Ulu, Borneo, failed to confirm Ungar's findings. According to McConkey et al. (2002), gibbons at Barito Ulu selected medium-sized fruit overall and favored large fruit during periods when fruit abundance was low. Methodological differences between the two studies make it difficult to draw firm conclusions; McConkey et al.'s estimate of fruit size was based on weight rather than volume. In addition, the two study sites differ in terms of the presence of primate competitors. At Ketambe, sympatric primates (e.g., *P. pygmaeus*,

Table 17.3 Average Diet and Ranging Parameters in Gibbons

SPECIES	DIET (%)				RANGE USE		
	FRUIT (FIGS[1])	LEAVES	FLOWERS	INSECTS	DAY RANGE (KM)	HOME RANGE (HA)	TERRITORY (HA, %[2])
Ho. hoolock	77 (36)	11	9	2	1.2	38	42 (86)
H. agilis	58 (17)	39	3	1	1.3	29	22 (76)
H. klossii	72 (26)	2	0	25	1.5	32	21 (66)
H. lar	66 (27)	24	1	9	1.4	40	29 (76)
H. moloch	61 (–)	38	1	0	1.4	17	16 (94)
H. muelleri	62 (24)	32	4	2	0.9	44	39 (89)
H. muelleri × agilis	62 (17)	24	13	1	–	45	–
H. pileatus	68 (24)	20	1	11	1.3	38	37 (96)
N. leucogenys	39 (–)	53	5	4	–	–	–
N. concolor	34 (10)	51	8	7	1.3	108	–
S. syndactylus	49 (37)	38	3	10	0.8	26	18 (79)
Average	**59 (24)**	**30**	**4**	**7**	**1.3**	**42**	**28 (83)**

[1] Percent of total diet.
[2] Percent of home range.

M. fascicularis) shared many of the same food sources, while at Barito Ulu, primate competitors were largely absent from the study area.

Elsewhere it has been argued that the key aspect of the gibbon niche is a preference for food distributed in small patches. According to Gittins and Raemaekers (1980:98), "the gibbons' feeding niche appears to be the succulent fruit occurring in smaller scarcer sources"; yet, despite the persistence of this view (e.g., Leighton 1987, Chivers 2001), the size and dispersion of gibbon feeding sources are rarely quantified. In fact, available data do not support the characterization of gibbons as small-patch specialists. At Khao Yai, Thailand, for example, lar gibbons routinely fed on the largest tree species in the forest (Bartlett 1999a, 2009b). Furthermore, the characterization of gibbons as small-patch specialists stands in stark contrast to the fact that they feed heavily on large fig trees throughout most of their range (Gittins and Raemaekers 1980).

Fig Exploiters?

On average, fig fruit makes up a quarter of the gibbon diet, and gibbons feed more heavily on figs than co-occurring macaques or langurs (Kuala Lompat, MacKinnon and MacKinnon 1980a; Ketambe, Ungar 1995; Lawachara, Feeroz et al. 1994). Accordingly, MacKinnon and MacKinnnon (1980a) conclude that gibbons are best described as "fig exploiters" (see also Chivers and Raemaekers 1986, Palombit 1997). It is not clear, however, if fig consumption alone distinguishes gibbons and larger orangutans, which also rely heavily on figs (MacKinnon

1977, Leighton 1993). Both gibbons (*H. muelleri*, Leighton and Leighton 1983; *H. lar*, Bartlett 1999a, 2009b) and orangutans (*P. pygmaeus*, Leighton and Leighton 1983, Leighton 1993) turn to figs most when preferred foods are unavailable. On the other hand, seasonal fruit shortages may impact gibbons more than orangutans, which are able to exploit foods such as wood pith and bark, for which gibbons lack the necessary anatomical traits to exploit. In any case, high rates of fig feeding are not universal for gibbons. At many sites, rates of fig feeding are less than half that reported for Ketambe (Table 17.2). It is unlikely, therefore, that fig feeding per se defines the gibbon feeding niche.

Range Use and Territoriality

The average home range size for all gibbon species is 42 ha, but estimates differ considerably both within and between species (Table 17.2). In lar gibbons, for example, average home range size at Kuala Lompat, Malaysia (55 ha), is over twice that reported for Mo Singto in Khao Yai, Thailand (23 ha). This range is comparable to that exhibited by the Hylobatidae as a whole. In general, it appears that ranges are larger at sites where at least two hominoid species are present (e.g., *H. lar* and *S. syndactylus* in Kuala Lompat, *H. muelleri* and *P. pygmaeus* in Kutai). However, ranges for hybrid gibbons at Barito Ulu, Borneo, are relatively large despite the fact that orangutans are only rarely sighted. According to McConkey (2000), the forest at Barito Ulu is less productive than other sites. Mather (1992) demonstrated that gibbon biomass is related to the local abundance of gibbon food trees, especially fig trees (Chivers 2001, see also Marshall and Leighton 2006). Figs contribute relatively little to the gibbon diet at Barito Ulu (17%), which may account for the large home range size at that site.

Territory can be distinguished from home range as the portion of the range actively defended against incursion by conspecifics (Bourliere 1964). On average, gibbon territory size is 82% of home range size (range 53%–96%). Day range is typically long relative to home range size, with the result that gibbons routinely enter all parts of their range. This is likely a prerequisite for territoriality in primates (Mitani and Rodman 1979).

Patterns of range use by gibbons appear to be influenced most by the location of food sources (Chivers 1977, Whitten 1982a), but factors such as patrolling of territorial borders (MacKinnon 1977) or the distribution of sleeping trees (Tenaza 1975) may also be important determinants of gibbon travel routes. Patterns of range use do not differ considerably by season (Whitten 1982b), but several investigators have shown that distance traveled per day does change over the course of the year. At Kuala Lompat, Malaysia, both lar gibbons and siamangs reduced travel during months when the abundance of preferred food was low (Chivers 1974, Raemaekers 1980). Similarly, at Khao Yai, Thailand, lar gibbons reduced travel during the cool, dry season when fruit abundance was lowest; but day length was also shortest during the cool, dry season and may have influenced daily path length (Bartlett 1999a, 2009b). Gittins (1982) concludes that day range length for agile gibbons was shortest during the wet season and correlated with activity period.

During directed travel, gibbons move in single file along shared pathways. Studies of siamang (Chivers 1974), Kloss (Tenaza 1975), lar (Reichard and Sommer 1997), and hoolock (Ahsan 2001) gibbons determined that females lead the majority of group travel bouts. Given the greater nutritional stress on females due to pregnancy, lactation, and infant carrying, female primates tend to devote more time to feeding behavior relative to males (Lee 1996, Strier 2003a), which may account for female group leadership in gibbons. Alternatively, Tenaza (1975) has suggested that Kloss gibbon males travel behind the group as a form of predator defense. It is possible that both factors influence group leadership in gibbons; however, female group leadership is not universal. Gittins (1979) determined that the male leads the majority of travel bouts in the group of agile gibbons he studied. This raises the possibility that other factors, such as detailed knowledge of resource location, may influence group leadership in gibbons (Brockelman 2009).

Activity Pattern

Unlike most primates, gibbons lack a prolonged midday rest period and retire 2–3 hr before sunset (Gittins and Raemaekers 1980). Optimal foraging theory predicts that feeding and travel time should be balanced with other behaviors as a means to optimize energy intake (MacArthur and Pianka 1966, Schoener 1971, Pyke et al. 1977). Most often, this approach views animals as energy maximizers. That is, they use their time in a way that insures maximal energy intake in the available time, where "available time" is assumed to be dawn to dusk. Gibbons, on the other hand, retire early even when resources are abundant. In seasonal environments, activity is further reduced during periods of food scarcity (Chivers 1974, Bartlett 1999a, 2009b). Thus, gibbons may best be characterized as time minimizers (see Raemaekers 1979, 1980). That is, they are active only for as long as it takes to fulfill their minimal caloric needs. If this were strictly true, however, the overall activity period should be most limited during periods of fruit abundance. This is not the case, but gibbons do have more time for nonsubsistence activities such as grooming and playing during periods when ripe fruit is plentiful (Bartlett 1999a, 2009b). It is not known if gibbons store calories in the form of fat during periods of fruit abundance as, for example, orangutans do (Knott 1998a); but this might account, in part, for longer activity periods when abundance is high. Alternatively, a short activity period relative to sympatric primates may be necessitated by territoriality. In the small territories occupied by gibbons, increased foraging time would translate into much more intensive use of the available resources. Gibbons appear to ensure continued access to highly favored resources, namely ripe fruit, by exploiting their ranges less intensively. Access is nevertheless ensured as a result of territorial defense. Accordingly, comparisons across species suggest that range defense is less intense when high-quality resources are less abundant. At Kuala Lompat, for example, higher rates of folivory by siamangs relative to lar gibbons have been correlated with reduced rates of travel, longer feeding bouts, smaller home range size, and fewer territorial encounters (Chivers 1974, MacKinnon 1977, Raemaekers 1979, Gittins and Raemaekers 1980).

Gibbons and the Forest

The role of gibbons as seed dispersers (see Chapter 34) has been studied only at a few sites, but the evidence indicates that hylobatids are important dispersal agents in Asian forests (Whitington and Treesucon 1991, Corlett 1998). McConkey's (2000) extensive study of the seed shadow generated by hybrid gibbons at Barito Ulu, Borneo, determined that gibbons dispersed up to 81% of the species they consumed, while the seeds of 12% of species were destroyed. Furthermore, germination experiments have shown that gibbons improve the chance of germination of at least some species in their diet (Whitington and Treesucon 1991, McConkey 2000, Ahsan 2001). According to McConkey (2000), gibbons may be important fig dispersers as many fig seeds will not germinate without prior passage through the gut. Given the loss of primates in many Asian forests, the role of gibbons in maintaining biodiversity is an area of pressing interest.

Predator Pressure

As with all primates, predator pressure on gibbons is difficult to assess. It has been argued that small, two-adult groups are more likely to occur when predator pressure is relaxed (van Schaik and van Hooff 1983). As a result, the vulnerability of gibbons to predators is a topic of considerable interest. According to van Schaik and Dunbar (1990:40–41), the fact that they are extremely fast and occupy the upper canopy renders gibbons "virtually immune from cursorial and arboreal predators" (see also van Schaik et al. 1983b). Yet, it is unlikely that any arboreal animal's key defensive strategy would be to outrun its pursuers. Snakes, to which gibbons respond with vigorous displays, are stealth hunters; and raptors, too, rely on surprise. It is true that gibbons spend considerable time high in the canopy, but they regularly come close enough to the ground to be at risk. Jiang et al. (1994) observed a pair of adult crested gibbons being chased by a clouded leopard at Mt. Wuliang, and Rabinowitz (1989) found the hair of a lar gibbon in leopard feces. In addition, Uhde and Sommer (2002) report a case where a group of lar gibbons mobbed a tiger. The reticence of gibbons to come to the ground, even to recover a fallen infant (Bartlett 2009b), further indicates the threat posed by terrestrial predators. It is also true that gibbons are relatively large compared to other arboreal primates and may experience reduced predator pressure as a result. However, a small gibbon could certainly be taken by a large snake, and even adults are vulnerable. For example, Schneider (1906) recovered a full-grown siamang from the belly of a python (see Uhde and Sommer 2002).

One possible behavioral response to predator pressure is increased vigilance by adult males. In fact, male gibbons do feed less than females (Bartlett 1999a, 2009b), which may mean they are better able to detect predators (see Dunbar and Dunbar 1980). However, Uhde and Sommer (2002) were unable to connect reduced activity by male gibbons to increased vigilance (as defined by visual scanning of the environment). In contrast to Tenaza (1975), they concluded that the fact that male gibbons most often travel behind females argues against the view that male gibbons play a major role in predation detection and defense because the first animal in progression is the most likely to encounter a predator.

Ecology and Behavior of Crested Gibbons

To date, comparatively little is known about the feeding and ranging behavior of crested gibbons. These species are critically endangered, and most published data come from the far northern extreme of their distribution. Nevertheless, available data hint at significant differences between crested

gibbons and other gibbon species (Sheeran et al. 1998; Fan et al. 2008, 2009). Studies at three sites in Yunnan, China, have revealed comparatively high rates of folivory (Meng La, 53%; Mt. Wuliang, 47%; Ailao, 54%). Furthermore, Chen (1995) noted the complete absence of figs in the home range of the group he studied, and fruit of any kind was available only during 4 months of the year, accounting for just 24% of the annual diet. Though the overall rate of frugivory at Mt. Wuliang is much higher, Fan et al. (2009) report that during fruit-poor months leaves and buds accounted for as much as 83% of the diet. Interestingly, Fan et al. also observed gibbons preying on a variety of animals, including insects, lizards, birds, and flying squirrels (see Fan and Jiang 2009).

Differences in diet appear to correspond to differences in range use. Home range size is larger for crested gibbons than any other species (Chen 1995, Sheeran et al. 1998, Fan and Jiang 2008). In addition, Sheeran et al. (1998) recorded increased travel during the cold, dry season by black crested gibbons at Mt. Wuliang. They reasoned that gibbons increase day range to find more fruit. This contrasts with the apparent energy-minimizing strategy of other gibbon species. In the context of marked drops in resource abundance during which minimal caloric needs are otherwise unmet, increased search time may be the only viable foraging strategy (but see Fan and Jiang 2008).

REPRODUCTIVE PARAMETERS

Gibbons live in pair-bonded social groups that typically comprise two adults and one to three offspring. Mean group size is approximately four individuals (MacKinnon 1977, Gittins 1980, Tilson 1981, Mitani 1990, Fuentes 2000) and rarely exceeds six even in populations where the occurrence of extra males is common (e.g., Khao Yai, Reichard and Barelli 2008; Way Canguk, O'Brien et al. 2003). However, group size may be somewhat larger among multifemale black crested gibbon groups. Fan et al. (2006) observed five groups at Mt. Wuliang over 2 years. Each group included one adult male and two adult females, and both females in three of the groups gave birth and/or carried infants over the course of the study. As a result, mean group size for these five groups was just over six individuals. Whether the occurrence of extra females is common in other members of the genus is not known (but see Zhou et al. 2008).

Female gibbons give birth to a single young after a gestation of approximately 7 months (Napier and Napier 1985, Geissmann 1991) and have an interbirth interval of 2–4 years (*H. lar*, 3.4 years, Reichard and Barelli 2008; *H. agilis*, 3.2 years, Mitani 1990; *H. klossii*, 3.3 years, Tilson 1981; *S. syndactylus*, 2.8 years, O'Brien et al. 2003, Lappan 2008), though there is considerable variation between individuals (Raemaekers and Chivers 1980, Ahsan 1994, Palombit 1995). Data from lar gibbons indicate that reproduction may be somewhat seasonal, with conception most often occurring during peaks in resource abundance (Chivers and Raemaekers 1980, Savini et al. 2008).

Infants are carried ventrally by females until they are weaned sometime during their second year. In siamangs, adult males as well as older juveniles are known to carry infants (Chivers 1974, Dielentheis et al. 1991). Data from both captive (Alberts 1987) and wild (Lappan 2008) populations suggest that infant siamangs seek out male contact beginning in their second year of life. Males in all species may contribute parental care in the form of play, grooming, protection, and territorial defense (Rutberg 1983, Brockelman 2009).

Dispersal

Typically, five age grades are recognized in gibbons (Brockelman et al. 1998): infant (0–2 years), juvenile (2–5 years), adolescent (5–8 years), subadult (8+ years), and adult (mated with territory). Nevertheless, estimates of the age at sexual maturity for gibbons vary considerably between captive and wild populations. For example, Geissmann's (1991) literature review of captive gibbons showed that females generally give birth to their first offspring between 6 and 8 years of age, while Reichard and Barelli's (2008) long-term study of wild lar gibbons at Khao Yai revealed a mean age at first reproduction of 11 years. The greater mean age in the wild may be related to social factors. Barelli et al. (2007) monitored reproductive hormone profiles in two subadult females (ca. 8 years old) and found that, while they exhibited sexual swellings, both subadults had irregular ovarian cycles and were not reproductively viable. However, once they dispersed and acquired a territory, both females conceived relatively quickly (within 13 months).

Whether wild males achieve sexual viability prior to dispersal is not presently known, but it appears that dispersal is typically delayed well beyond the age at which sexual viability is reached in captivity (ca. 5 years, Geissmann 1991). For example, Brockelman et al. (1998) observed five dispersals by lar males over a period of 18 years. They estimated that the males reached mature body size by 8 years of age but delayed dispersal until 10 years on average.

Contrary to previous studies (e.g., Tilson 1981), Brockelman et al. (1998) did not find evidence that subadults are forced out via intrasexual aggression upon reaching sexual maturity. They speculate that extra subadults provide a beneficial service to their parents by acting as social partners for younger infants and contributing to territorial defense. Even when they do disperse, subadults may remain close by. Demographic data for Kloss (Tilson 1981) and lar (Brockelman et al. 1998) gibbons suggest that most subadults transfer into neighboring territories, and genetic data collected from siamangs at Way Canguk, Indonesia, are consistent with short dispersal distances (one or two territories) for emigrating males (Lappan 2007). At Khao Yai, three lar males transferred into a neighboring group together. Subsequently, the two younger animals transferred again (Brockelman et al. 1998). Codispersal has also been observed in lar gibbons at Ketambe, Indonesia (Palombit 1994a). To date, the majority of data on dispersal reflect emigration by subadult males. The sex difference in dispersal distance remains an important area for future research.

Pair Formulation

Early considerations of pair formation in gibbons focused on new pairs formed via coincident emigration by maturing subadults (Raemaekers and Chivers 1980, Tilson 1981, Palombit 1994a, Brockelman et al. 1998, Ahsan 2001), but subsequent work has revealed that gibbon pair-bonds are less stable than originally thought and that new bonds may form by a number of alternative routes, including replacement of one pairmate by an outside usurper (Brockelman et al. 1998, Ahsan 2001), filling of a vacancy left open by a dead or departed pairmate (Palombit 1994a), or replacement of one pairmate by a maturing offspring (Raemaekers and Chivers 1980, Tilson 1981, Raemaekers and Raemaekers 1984, Palombit 1994a, Ahsan 2001). In fact, it now appears that bonds formed by two neophytes are relatively rare. Long-term research on the population dynamics of the community of lar gibbons at Khao Yai, Thailand, determined that just one of seven documented pair formations was the result of two dispersing subadults jointly entering an unoccupied territory (Brockelman et al. 1998). At Ketambe, Indonesia, just one of six new pair-bonds was established in such a manner (Palombit 1994a). At Khao Yai, bonds were formed most often via take-overs during which an established mate was usurped by a younger individual (take-overs were clearly observed in two cases and strongly suspected in three others), while at Ketambe, new bonds resulted most often when an adult or subadult filled a vacancy after the previous pair-bond was terminated due to death, disappearance, or desertion.

Though apparently not uncommon, the extent to which mate desertion is a successful reproductive strategy will require further long-term data. In two of the cases reported by Palombit (1994a), the deserting adult returned to its mate; and in two others, the males that left their territories failed to acquire a new mate during the period of the study. On the whole, existing data suggest that abandoning a mate entails considerable risk: Lone males may experience considerable delay in attracting a suitable mate (Aldrich-Blake and Chivers 1973, Tilson 1981), lone females may reject the approaches of lone males (Tilson 1981), and even successful pairings may experience considerable delay before first birth (Raemaekers and Chivers 1980).

Pair-Bonds and Sexual Monogamy

As illustrated, one means by which a paired gibbon may increase its reproductive access is to abandon its current mate for real or potential opportunities to mate with another adult. However, while the potential for gibbons to practice serial monogamy has long been recognized (Kloss 1908), recent observations of lar gibbons and siamangs demonstrate that an alternative sexual strategy employed by gibbons is to mate opportunistically with already paired neighbors during intergroup encounters. These findings have helped to erode the presumed connection between pair-bonds and sexual fidelity in gibbons.

Over a period of 2.5 years, Palombit (1994b) documented five instances of extrapair copulation (EPC) involving a single pair-bonded siamang female and three males of a neighboring group. Similarly, Reichard (1995, 2003b) determined that EPCs accounted for 9% of the copulations observed in a population of lar gibbons over an 11-year period at Khao Yai, Thailand. As of yet, the genetic consequences of EPCs have not been reported. Palombit (1994b) notes that EPCs observed at Ketambe did not result in the birth of an infant. However, in the case of lar gibbons, Reichard (1995) concludes that some of the EPCs took place over the period during which the female was believed to have conceived, though she also copulated with her own mate during this period.

Ongoing research at Khao Yai has corroborated the pattern of EPCs previously described. Indeed, Reichard and Barelli (2008:831) report that "All sexually active long-term resident females and half of the short-term resident females were sexually polyandrous (N=11), i.e., they maintained simultaneous sexual relationships with ≥2 males." Significantly, these values include (in addition to EPCs) females living in socially polyandrous groups, which accounts for 21% of all social groups at the site. High rates of social and sexual polyandry have also been reported for siamangs at Way Canguk. However, in contrast to the reports for Khao Yai gibbons, EPCs have not been recorded (Lappan 2008). Socially polygynous gibbon groups have also been documented (Brockelman and Srikosamatara 1984, Sommer and Reichard 2000, Ahsan 2001), but with the exception of crested gibbons (e.g., Haimoff et al. 1987, Jiang et al. 1999, Fan et al. 2006) such groupings are comparatively rare. While demographic data indicate that the two-adult group is the most common pattern in gibbons, frequent exceptions undermine the description of gibbons as monogamous and have stimulated extensive discussion about gibbon sociality and social organization (Palombit 1994a, Brockelman et al. 1998, Fuentes 2000, Sommer and Reichard 2000).

SOCIALITY AND SOCIAL ORGANIZATION

Intragroup Social Behavior

In gibbons, heterosexual pair-bonds are signaled by a host of affiliative interactions, including joint defense of territory, duetting, social proximity, and affiliative social interactions (Palombit 1994a, Geissmann and Orgeldinger

2000, but see Reichard and Sommer 1997). In general, data on affiliative social behavior (e.g., grooming and play) have been slow to materialize. According to Leighton (1987), the paucity of data on gibbon sociality is largely due to a lack of social contact in gibbons. Gittins and Raemaekers (1980), for example, concluded, based on a 2-year study, that social behavior was almost wholly lacking in agile gibbons. Yet, low rates of social activity are not universal. At Khao Yai, for example, rates of affiliative social behavior have been found to be comparable to levels reported for primates generally (see Chapter 39). Bartlett (2003) found that social activity accounted for 11% of the activity budget in lar gibbons and that, during periods of fruit abundance, play and grooming accounted for as much as 20% of the activity time. Bouts of mutual grooming between adults tend to parallel juvenile play bouts, though adults, especially adult males, do engage in various forms of play with juveniles (Ellefson 1974, Bartlett 2003).

The available data suggest that the effort contributed to grooming relationships differs between males and females. In lar and hoolock gibbons, males groom females more often than the reverse (Palombit 1996, Ahsan 2001). Among siamangs, however, the time devoted to the grooming relationship is more equal (Gittins and Raemaekers 1980, Palombit 1996). Palombit (1996) argues that differences in grooming effort indicate that lar gibbon males contribute more to the maintenance of the pair-bond than females do, whereas in siamangs the effort is more reciprocal. He suggests that higher rates of intragroup feeding competition may indicate a higher cost for bonding in smaller-bodied lar gibbons. Palombit allows that there is a certain amount of intraspecific variability (see Chivers 1974, Fischer and Geissmann 1990); nevertheless, he concludes that past research has overestimated the similarity in the social system of different gibbon species.

Intergroup Encounters

Gibbons actively defend a fixed range or territory via a host of behaviors that include duets and male and female solos, in addition to protracted chases between adults, especially males (Ellefson 1974, Gittins 1984). Disputes are initiated when two groups come into visual contact or when the male of one group approaches the calls of rival gibbons near the area of the territorial boundary. Adult males take the lead role in virtually all disputes; however, subadult males also participate in chases and calls. Though females are not regular participants in disputes, it is not uncommon for neighbor males to approach females over the course of an encounter (Reichard and Sommer 1997, Bartlett 2003). Such incursions are generally repelled by the resident male or by the female herself, though observations of EPCs in lar gibbons and siamangs demonstrate that some incursions are tolerated, if not welcomed, by females. During much of the dispute, rival males sit or hang within view of one another. Such standoffs, which may or may not be accompanied by conflict vocalizations, are punctuated by rapid chases back and forth between the two territories. Encounters tend to end rather suddenly, with the males quickly moving to rejoin their respective groups. In rare cases, encounters may be lethal to one of the participants (Palombit 1993, Brockelman et al. 1998, Reichard 2003b).

Comparing data from eight species, Gittins (1984) concludes that gibbon groups sing on average once a day and engage in territorial disputes on average once every 5 days. However, rates of intergroup encounters are subject to numerous complicating factors. For example, the main study group observed by Chivers (1974) had no immediate neighbors and therefore engaged in no territorial disputes. All disputes recorded by Kappeler (1984) were terminated once his presence was revealed due to the fearfulness of the unhabituated neighboring groups. Thus, while territorial activity is one of the most easily recognized gibbon traits, it remains poorly understood (Gittins 1984). Studies have rarely involved simultaneous observations of multiple groups, and as a result, it is often difficult to evaluate the cause of a given encounter or its outcome.

Playback Experiments

One methodology that has been employed to better elucidate the function of territorial encounters is the use of playback experiments. Mitani (1984, 1985a) used playbacks of male and female calls to determine the response of mated Mueller's gibbons to simulated intrusions from lone gibbons and neighbors. Responses differed depending on the location of the playback and the sex of the singer. Female solos or duets played at least 100 m inside the group's territory elicited approach, typically led by females, and countersinging from the resident group. Songs played from the territorial boundary elicited duets but were less likely to elicit approaches. In contrast, residents did not sing in response to male solos played inside the group's territory. Instead, males led silent approaches to the position of the playback or approached alone. These findings corroborate early suggestions that territoriality among gibbons is maintained by intrasexual intolerance (Ellefson 1974).

Additional playback experiments conducted on lar (Raemaekers and Raemaekers 1985) and agile (Mitani 1987) gibbons confirm the location-dependent response of mated gibbons to neighbor songs. Gibbons respond most quickly to territorial intruders. In contrast to Mueller's gibbons, however, both lar and agile gibbon males lead the majority of approaches, including those elicited by female solos or duets. Mitani (1987) notes that in several cases in which male Mueller's gibbons led approaches, the female in question carried a young infant. It is likely that the role of female gibbons in territorial defense is mediated in part by their reproductive condition. Females without small infants are more likely to engage in territorial defense (Mitani 1987, Bartlett 1999a, 2009b).

Agonistic encounters are by their nature conspicuous, but extensive data on lar gibbons at Khao Yai indicate that not all group encounters are agonistic (Reichard and Sommer 1997). According to Bartlett (2003), 17% of encounters occurring over a 12-month period involved affiliative activity between neighboring groups. The majority of affiliative activity included chase play between juveniles, but during two encounters a neighbor male groomed the adolescent and juvenile of the neighboring group. Affiliative encounters between neighboring adults were not observed; nevertheless, neighboring adults do exhibit unappreciated levels of tolerance, which is likely a product of the genetic relatedness of the animals in adjacent groups (Brockelman et al. 1998).

Social Monogamy

While it has long been customary to refer to gibbons as a monogamous species, authors have not always taken care to explain what precisely a monogamous social system entails. Often masked by an ostensibly simple label are hidden assumptions about the sexual and genetic relationships of group members. While warnings about the imprecision of the term "monogamy" are not new (Wickler and Seibt 1983), data accumulated over the last two decades have forced a wholesale reevaluation of hylobatid social organization. The most important developments can be summarized as follows. First, observations of EPCs reveal that pair-bonding neither implies nor depends upon sexual fidelity. Second, long-term studies show that adult membership changes regularly as a result of death, mate desertion, and displacement. Third, when adults or subadults enter new groups, they are occasionally accompanied by one or more immatures, including siblings. Fourth, short dispersal distances mean that the members of neighboring groups may be close relatives. Fifth, in some populations socially polygamous groups are both relatively common and enduring. Emerging from these findings is the recognition that during the course of their lives most gibbons will inhabit a nonnuclear family group at some point. In at least one case, this has been confirmed using deoxyribonucleic acid (DNA) analysis (Okra and Takenaka 2001). In addition, it is likely that neighboring family groups frequently consist, in part, of unrelated individuals that have lived in the same social group during a portion of their lives (Reichard and Sommer 1997, Brockelman et al. 1998). Apart from genetic relationships, an individual's knowledge of the other animals within a community is likely to extend beyond those individuals it has seen. Group choruses, duets, and solo vocalizations provide opportunities for gibbons to learn the composition of nonadjacent groups. While it remains to be demonstrated, it is likely that gibbons know individuals they have never seen based on their vocal signature alone (Dallmann and Geissmann 2001).

These findings have stimulated many authors to reject the label "monogamy" in favor of specific models more reflective of the complex realities of gibbon social systems (Fuentes 1998, 2000; Sommer and Reichard 2000;

Reichard 2003b). One of the earliest attempts to accommodate groups with complex origins was that of Quiatt (1985, 1987) and Quiatt and Kelso (1987), who proposed the term "household," which they defined as a group that travels and exploits territorial resources as a single unit. As described by Quiatt and Kelso (1987:429), the advantage of the term "household" is that membership is less likely to imply either a mating relationship or genes in common with some other members. This analogy can be extended to the level of the neighborhood to describe a nested hierarchy in which several households occupy a shared social and physical landscape (Bartlett 2003). Conceptually, the household model is similar to population-level approaches advocated by Fuentes (2000) and Reichard (2003b). Despite different emphases, most investigators converge on two main points. First, gibbon groups should no longer be portrayed as socioreproductive isolates. Neighboring groups frequently consist of social and sexual partners as well as kin. Second, future discussions of gibbon social organization must separate mating systems from social systems because the conditions that favor one do not necessarily favor the other.

Frugivory, Territoriality, and the Evolution of Gibbon Pair-Bonds

However labeled, the social organization of gibbons is rare, if not unique, among primates; and it has been the subject of numerous attempts to account for its evolution (Table 17.4). The most widely cited explanation for pair-bonding maintains that females represent an "overdispersed resource" (van Schaik and Dunbar 1990) and that males are forced to accept a single mate due to mutual female intolerance (see Emlen and Oring 1977, Wrangham 1979b). As detailed by Leighton (1987:144), female antagonism is the product of

resource competition due to a feeding preference for ripe fruits that occur in small patches:

> The critical feature of these "small" patches is that they are large enough for one female and her offspring to use regularly, but small enough that if two females and their offspring visited the same patches, the dominant could not be assured a sufficient meal or high feeding rates.

As a result, females distribute themselves in exclusive ranges. Males accept a single mate only because they are unable to successfully defend the ranges occupied by two or more females. While playback experiments offer some support for this model, other investigators have questioned the assumption that gibbon males are unable to simultaneously defend multiple female ranges (van Schaik and Dunbar 1990, Reichard 2003b). van Schaik and Dunbar (1990) argue, for example, that gibbons travel far enough each day to defend the home ranges of as many as seven females, on average (but see Bartlett 2009a). They conclude, therefore, that some other factor must compel male/female spatial cohesion. Based on a comparison of four alternative hypotheses, van Schaik and Dunbar assert that the most likely explanation for pair-bonding in primates, including gibbons, is the near universal threat of infanticide posed by extragroup males. While a specific test of the infanticide-protection hypothesis is confounded by the almost complete lack of evidence for infanticide in gibbons, van Schaik and Dunbar argue that the wariness of females during intergroup encounters is clear evidence that infants are at risk.

The case for primate pair-bonding as a form of infanticide defense has been repeated many times (e.g., Sommer and Reichard 2000, van Schaik and Kappeler 2003), but as a general explanation for pair-bonding in gibbons, infanticide defense faces a number of significant hurdles (Palombit 1999, Brotherton and Komers 2003, Brockelman 2009).

Table 17.4 Support for Alternative Hypotheses for Pair Bonding in Gibbons

HYPOTHESIS	SELECTED REFERENCE(S)	SUPPORT FOR HYPOTHESIS			
		WITTENBERGER AND TILSON 1980	VAN SCHAIK AND DUNBAR 1990	FUENTES 2000	REICHARD 2003a,b
Male paternal care	Kleiman 1977	−	−	+/−	n.c.
Females as overdispersed resource	Emlen and Oring 1977, Rutberg 1983	+	−	−	−
Mate guarding	Wittenberger and Tilson 1980, Palombit 1999	−	n.c.	+	−
Predation, protection	Tenaza 1975, Dunbar and Dunbar 1980	−	+/−	−	+/−
Exclusive resource defense	Rutberg 1983, Brockelman 2009	n.c.	−	+/−	−
Infanticide defense	Wrangham 1979, van Schaik and Dunbar 1990	n.c.	+	−	+/−

+, hypothesis was supported; +/−, evaluation of the hypothesis was equivocal; −, hypothesis was rejected; n.c., hypothesis was not considered.

First, adult gibbons are monomorphic, and females may therefore be capable of defending infants themselves or making the cost of infanticidal attacks prohibitive. Second, unmated males, those who are expected to pose the greatest risk of infanticide, are rare in most gibbon populations. Third, males and females maintain spatial proximity even when no infants are present. Fourth, infanticide protection does not necessitate that gibbons live in two-adult groups. Furthermore, pair living in gibbons is uniformly associated with territoriality and, as Brockelman (2009) has observed, territorial defense and infanticide protection have conflicting aims—territorial defense requires males to leave females and their infants during intergroup encounters, precisely when they are most vulnerable to attack by neighbor males.

As an alternative to the infanticide-protection hypothesis, Palombit (1996, 1999) argues that pair-bonding in gibbons is a form of male mate guarding. Given the potential for EPCs, it is argued that gibbon males may improve their fitness by limiting access to their female partner. The benefits of mate guarding are realized in two ways. In the short term, mate guarding allows a male to increase paternal certainty. In the long term, it limits a female's ability to evaluate potential sexual partners, thus ensuring the resident male future breeding opportunities. According to Palombit, this may account for the persistence of male/female bonds outside of estrus. For their part, females accept male companionship, not because of any tangible benefits but because the cost of rejecting it is too high (Gowaty 1996). However, the mate-defense hypothesis is vulnerable to some of the same criticisms detailed here. In particular, field observations indicate that EPCs typically occur during territorial encounters (Palombit 1994b, Reichard 1995, Brockelman 2009); thus, males invite EPCs whenever they lead group travel toward territorial boundaries.

Pair Living and Resource Defense

The failure of either infanticide defense or mate defense to adequately account for gibbon behavior during intergroup encounters suggests that multiple selection pressures may act together to shape gibbon social organization (Fuentes 2000, 2002b; Brockelman 2009). Almost forgotten in the historical shift toward models derived from sexual selection theory (Janson 2000a, Paul 2002) is the importance of ecological pressures in limiting group size in primates (Chapman and Chapman 2000b). Only recently have investigators refocused attention on the relationship between foraging efficiency and pair living in primates (Kappeler and

van Schaik 2002, Isbell 2004, Brockelman 2009, Bartlett 2009b). Irrespective of mating strategies, it can be argued that the combination of small foraging parties (i.e., two-adult groups) and territoriality entails benefits to both males and females that would be lost if adults foraged alone. For example, gibbons may increase foraging efficiency by sharing knowledge about well-defended resources (Brockelman 2009) or by ensuring that resources are not unknowingly depleted by adults foraging apart (Terborgh 1983). This latter proposal finds support in the finding that adult female gibbons lead group travel and have priority of access to feeding trees (Reichard 2003b, Bartlett 2009b). A point further bolstered by Barelli et al.'s (2008a) discovery that the female bias in group leadership among lar gibbons is greatest when resource abundance is low.

Additional support for models based on foraging efficiency may be found in the observation that gibbons alter foraging in response to resource density. Drawing on the long-term demographic record for lar gibbons at Khao Yai, Savini et al. (2008) compared range use and reproductive performance in seven social groups and determined that there is a significant positive correlation between home range size and juvenile mortality. The authors conclude that increased mortality was a direct consequence of increased travel, which was required in order "to visit more dispersed resources to satisfy their nutritional needs" (p. 7). But even if these findings help to forge a link between foraging strategies and reproductive success in gibbons, it remains the case that essential data on the relationship between resource defense and reproductive success are almost completely absent from the study of gibbon social organization (van Schaik and Dunbar 1990). It is remarkable that one of the most conspicuous behaviors seen in gibbons—active territorial defense—should be so little studied and so poorly understood.

Though the last two decades have yielded significant new findings that have forwarded our understanding of the behavior and ecology of gibbons in important, and at times unanticipated, directions, there is still much to be learned about the mechanisms underlying even well-documented gibbon behaviors. Furthermore, it remains the case that the majority of information on this primate family comes from only a few study sites and even fewer species. One may hope that research over the next two decades will draw as often from less studied species as it does from those already well represented here.

Species included in this chapter are shown in Color Plates 30 and 31 (*Nomascus leucogenys* and *Symphalangus syndactylus*).

18

Orangutans
Understanding Forced Copulations
Cheryl D. Knott and Sonya M. Kahlenberg

1. Why are orangutans largely solitary in their foraging patterns?
2. How do Bornean and Sumatran orangutans differ behaviorally, and what might explain these differences?
3. What are physical and behavioral similarities and differences between flanged and unflanged males?

INTRODUCTION

Orangutans (genus *Pongo*) represent the extreme of many biological parameters. They are the largest primarily arboreal mammal, have the longest interbirth interval, are the most solitary of the diurnal primates, and are the most sexually dimorphic of the great apes. They range over large areas and have diffuse communities. Additionally, adult male orangutans come in two morphologically distinct types, an unusual phenomenon known as *bimaturism*. This intriguing suite of features has proven difficult for researchers to fully understand. The semi-solitary lifestyle of orangutans, combined with their slow life histories and large ranges, means data accumulate slowly. Also, rapid habitat destruction in Southeast Asia has left only a handful of field sites currently in operation (Figure 18.1 and Table 18.1). Despite these difficulties, long-term behavioral studies and recent genetic and hormonal data now allow us to explore some of the most theoretically interesting questions about this endangered great ape. Here, we summarize our current state of knowledge about wild orangutans and discuss advances in our understanding of one distinctive orangutan behavior—forced copulations.

DISTRIBUTION, TAXONOMY, AND MORPHOLOGY

Present Distribution

Though once found throughout Asia, orangutans now occur only on the islands of Sumatra (Indonesia) and Borneo (specifically the Malaysian states of Sarawak and Sabah and the Indonesian provinces of Kalimantan). Both ecological and anthropogenic factors seem responsible for their range collapse (Delgado and van Schaik 2000). Tropical and subtropical zones shifted during the Pleistocene (Jablonski 1998b, Jablonski et al. 2000), and mean annual temperature rose (Harrison 1999, Medway 1977), leading to habitat constriction. Among other evidence of human influence (Delgado and van Schaik 2000), orangutan remains found in caves in northern Borneo show they were preyed upon by humans (Hooijer 1960, Medway 1977).

Orangutans live in true wet rain forests, with reported average annual rainfall between 2,000 mm (Galdikas 1988) and 4,500 mm (Lawrence and Leighton 1996). The primary habitat types they occupy are peat and freshwater swamps and lowland forests. Orangutans have been found living up to an altitude of 1,200 m; however, in Borneo (but not Sumatra) orangutan density declines with increasing elevation (Husson et al. 2009).

Taxonomic Status

Most researchers classify Bornean (*P. pygmaeus*; Color Plate 33) and Sumatran (*P. abelii*) orangutans as separate species based on molecular and morphological analyses that show interisland differentiation that exceeds or is equal to differentiation between the two species of chimpanzee (e.g., Goossens et al. 2009, Groves 2001a, Guy et al. 2003, Ruvolo et al. 1994, Xu and Arnason 1996). A recent genetic meta-analysis estimates a divergence time between the two species of 2.7–5 million years ago (Steiper 2006). Bornean orangutans show further genetic (Warren et al. 2000, 2001; Zhi et al. 1996) and morphological (Groves 1986, Groves et al. 1992, Uchida 1998) variation that corresponds with populations separated geographically by major rivers. Groves (2001a) thus classifies Bornean orangutans into three subspecies: *P. p. pygmaeus* in Sarawak and northwestern Kalimantan, *P. p. wurmbii* in southwestern and

A

central Kalimantan, and *P. p. morio* in Sarawak and East Kalimantan (Figure 18.1). However, orangutan taxonomy remains controversial. Warren et al. (2001) found genetic evidence for four subpopulations on Borneo that diverged 860,000 years ago, with the *P. p. morio* subspecies being further subdivided into Sabah and East Kalimantan populations. Also, some researchers favor only a subspecies classification between Bornean and Sumatran orangutans (*P. p. pygmaeus* and *P. p. abelii*) (Courtenay et al. 1988, Muir et al. 2000).

Morphology

Often referred to as the "red ape," orangutans have thick reddish-orange hair. They negotiate the canopy through quadrumanual clambering, grasping supports with both their hands and feet (Knott 2004), and by adopting a wide array of body postures (Thorpe and Crompton 2005, 2006). There are no gross differences in positional behavior between orangutan species, but there are differences in the proportion of time spent in various body postures, likely due to habitat differences (Thorpe and Crompton 2005, 2009). Bornean flanged males are reportedly more

terrestrial than their counterparts in Sumatra, but this does not hold for all Bornean populations (Knott 2004). Morphological adaptations for arboreality include arms, hands, and feet that are longer than those of humans and the other great apes (Fleagle 1999) as well as a shallow hip joint that allows leg extension of over 90 degrees (MacLatchy 1996). Compared to chimpanzees, orangutans have thicker molar enamel and other dental adaptations associated with their routine ingestion of mechanically hard and tough foods, such as seeds and bark (Vogel et al. 2008).

Orangutans are unusual because males appear to have indeterminate growth, meaning that they gain weight throughout their lives (Leigh and Shea 1995). This growth pattern has been described for only a few other mammals and is thought to be associated with intense male/male competition. Orangutans are also exceptional for having two morphologically distinct types of adult male (Galdikas 1985a,b; Graham and Nadler 1990; Kingsley 1982, 1988; MacKinnon 1979; Maggioncalda et al. 1999, 2000, 2002; Mitani 1985a, b; Schürmann and van Hooff 1986; te Boekhorst et al. 1990). One type is large, weighing over 80 kg in the wild (Markham and Groves 1990), and

Figure 18.1 Maps of (*A*) Borneo and (*B*) Sumatra, showing species and subspecies distinctions, study sites, densities, and distributions (modified from Caldecott and Miles 2005 and Singleton et al. 2004 by L. Kee and K. Hendrickson).

Table 18.1 Location of Orangutan Field Sites as Referred to in Text

TAXON	FIELD SITE	ISLAND	PROVINCE/COUNTRY
P. pygmaeus morio	Kutai National Park	Borneo	East Kalimantan, Indonesia
	Kinabatangan Wildlife Sanctuary	Borneo	Sabah, Malaysia
P. pygmaeus wurmbii	Gunung Palung National Park	Borneo	West Kalimantan, Indonesia
	Tanjung Puting National Park	Borneo	Central Kalimantan, Indonesia
	Sabangau Ecosystem	Borneo	Central Kalimantan, Indonesia
	Tuanan Field Station	Borneo	Central Kalimantan, Indonesia
P. abelii	Suaq Balimbing Research Station	Sumatra	Gunung Leuser, Aceh, north Sumatra, Indonesia
	Ketambe Research Station	Sumatra	Gunung Leuser, Aceh, north Sumatra, Indonesia

possesses secondary sexual characteristics, including cheek pads called "flanges," a pendulous throat pouch, long hair, and a musky odor (reviewed in Crofoot and Knott in press) (Figure 18.2a). These "flanged" males also produce loud vocalizations, known as "long calls," that are audible to humans up to 800 m away (Mitani 1985b).

The second type of male is half the size of flanged males and does not have secondary sexual characteristics or produce long calls (Figure 18.2b). These "unflanged" males were originally called "subadults" (e.g., Galdikas 1985a,b), but this term is misleading since they are fully capable of siring offspring (Goossens et al. 2009; Kingsley

1982, 1988; Utami et al. 2002). Endocrinological data from unflanged males in zoos show that their gonadotropin and testicular steroid levels are sufficient for spermatogenesis but inadequate for triggering the development of secondary sexual characteristics (Kingsley 1982, Maggioncalda et al. 1999).

Wild orangutan females weigh on average 39 kg, which is 45% of flanged male body size (Markham and Groves 1990). This makes orangutans one of the most sexually dimorphic species on record. The evolution of extreme sexual dimorphism has been attributed to male/male competition (Rodman and Mitani 1987), female choice (Fox 1998, Utami et al. 2002), and sexual coercion (Smuts and Smuts 1993).

Phenotypic differences between the two orangutan species have been proposed, but no systematic study comparing soft tissue and appearance has been completed. Cranial morphology is highly variable, and more skeletal variation is sometimes found within the Bornean species than between *P. pygmaeus* and *P. abelii* (Courtenay et al. 1988). Among these differences are the shape and profile of the face, the degree of prognathism, the shape of the braincase, and several features of the teeth (Groves 1986, Jacobshagen 1979, Rorhrer-Ertl 1988, van Bemmel 1968). There are differences in jaw structure between the two species, indicating that Bornean orangutans may rely more heavily on hard-to-process foods (Taylor 2006, 2009).

ECOLOGY

Food Availability

The Southeast Asian rain forest is characterized by dramatic fluctuations in the availability of fruit, the preferred food of orangutans. Bornean and Sumatran forests are dominated by trees in the Dipterocarpaceae family that periodically experience a mast fruiting in which up to 88% of the trees of this and other plant families flower and fruit in synchrony (Appanah 1985, Ashton 1988, Cannon et al. 2007, Medway 1972, van Schaik 1986). Mast fruiting occurs every 2–10 years (Ashton 1988), appears to be driven by climatic events associated with the El Niño weather pattern (Curran et al. 1999), and is often followed by periods of extremely low fruit availability (Knott 1998a). Peat swamps are less subject to mast fruiting because of lower dipterocarp density, but fruit availability is still variable (Galdikas 1988). Across study sites, orangutan population density is positively correlated with fruit availability, particularly during periods of fruit scarcity, suggesting that "crunch periods" are important

(A)

(B)

Figure 18.2 Photos of (*A*) flanged and (*B*) unflanged males from Gunung Palung National Park (photos by Tim Laman).

for establishing a population's carrying capacity (Marshall et al. 2009a).

Sumatran forests are more productive than Bornean forests since the former have more trees in fruit at any given time, experience more frequent periods of high fruit availability, and have fewer periods of low fruit availability than the latter (Marshall et al. 2009a). Sumatran forests also have higher densities of important orangutan food trees and lower densities of dipterocarp trees than do Bornean forests (Marshall et al. 2009a). This difference in habitat quality likely explains why, in similar habitat types, orangutan densities are higher in Sumatra than in Borneo (Husson et al. 2009). The richer volcanic soil in Sumatran forests seems to underlie inter-island differences in productivity (Delgado and van Schaik 2000).

Diet and Nutrient Consumption

Orangutans feed on twice as many plant genera as do other great apes, and Bornean orangutans feed on more plant species and parts than Sumatran orangutans (Russon et al. 2009b). Sumatran orangutans show less variance in the time spent eating fruit, bark, and leaves and are generally able to maintain a higher level of frugivory in their diet (Morrogh-Bernard et al. 2009). On Borneo, there are periods when orangutan fruit intake reaches 100%, exceeding the highest rates on Sumatra. During low fruit periods on Borneo orangutans eat predominantly leaves and bark (Knott 2005a). Fruit consumption can drop to 10% on Borneo, but it has never been reported to reach below 57% on Sumatra (Morrogh Bernard et al. 2009). Orangutans also incorporate other foods into their diet when fruit is low, such as pithy vegetation and insects, primarily termites (Knott 2005a).

Nutrient intake may vary tremendously. Fruits produced during masts are not only more abundant but also higher in calories, carbohydrates, and lipids (Knott 1999b). When orangutans are subsisting off leaves and bark, fiber comprises a large portion of their diet. Vertebrate animal matter is rarely eaten, but orangutans in Sumatra have been seen opportunistically hunting slow lorises (Utami and van Hooff 1997) and eating a gibbon (Sugardjito and Nurhada 1981). Only one meat-eating observation (a tree rat) has been reported from Borneo (Knott 1998b). Daily caloric consumption has been calculated for Gunung Palung, Borneo, showing dramatic fluctuations associated with fruit availability, with orangutans consuming two to five times more kilocalories during fruit-rich periods compared to fruit-poor periods (Knott 1998a).

Physiologically, orangutans are particularly adapted to fat storage. During periods of high fruit availability, they gain weight, as measured by their intake of significantly more calories than they expend (Knott 1998a). This adaptation appears to have evolved as a way to sustain them through extended periods of low fruit availability when they have been shown, through the production of ketones (Knott 1998a) and the presence of C-peptide in urine

(Emery Thompson and Knott 2008), to metabolize their fat reserves. Orangutans on Sumatra may not rely as heavily on fat deposits because they are able to maintain a relatively stable dietary intake (Wich et al. 2006).

Activity Patterns

Orangutans cope with low food availability by being primarily solitary foragers. Across 10 study sites, age/sex class, fruit availability, habitat quality, logging impact, masting versus nonmasting forests, island of origin, and sociality were found to be important predictors of diet and activity patterns (Morrogh-Bernard et al. 2009). Sites where orangutans fed for less than 50% of their active period, on average, were masting, mixed-dipterocarp forests, whereas sites where they fed for greater than 50% of the time were nonmasting, peat swamp forests. Ketambe, in Sumatra, was an exception, a mixed-dipterocarp forest where they fed for greater than 50% of the day. This has been attributed to the site's very high strangler fig density (Wich et al. 2006, Morrogh-Bernard et al. 2009). Orangutans from peat swamp forests generally spent more time feeding and traveling, with Ketambe again being an exception. Orangutans also modify time spent feeding (Galdikas 1988, MacKinnon 1974, Mitani 1989, Rodman 1977), traveling (Galdikas 1988, Knott 1999b, Mitani 1989), and resting (Mitani 1989), as well as their active day length (Morrogh-Bernard et al. 2009), during periods of increased sociality. Across sites, there is a trend for non-sexually active females to feed the most and rest the least, for flanged males to rest the most and feed and travel the least, and for unflanged males to travel the most (Morrogh-Bernard 2009, van Schaik et al. 2009c).

Variability in the food supply seems to explain many of the differences in orangutan foraging strategies and activity patterns (Morrogh-Bernard et al. 2009). At Gunung Palung, a site with highly variable fruit production, orangutans spent more time awake, feeding, traveling, socializing, and mating when fruit availability was high (Knott 1999b). When fruit was low, orangutans conserved energetic resources by spending less time awake and traveling shorter distances per day (Knott 1998c, 1999b). Morrogh-Bernard et al. (2009) call this the "sit and wait strategy" and found it is typical of mixed-dipterocarp forests where mast fruiting results in high variability in fruit production. In peat swamp forests, fruit availability is less variable, leading to a continual "search and find" strategy (Morrogh-Bernard et al. 2009).

LIFE HISTORY

Juvenile Development

Wild orangutans experience very low infant mortality, ranging 7%–17% in Sumatra (van Noordwijk and van Schaik 2005, Wich et al. 2004b); but rates are much higher in captive (Anderson et al. 2008, Cocks 2007) and rehabilitant populations (Kuze et al. 2008). Orangutans spend their first few years in close contact with their mothers, especially during travel when they cling to their mother for the first two years (van Adrichem et al. 2006, van Noordwijk et al. 2009, van Noordwijk and van Schaik 2005). Locomotor competence emerges at around age 3, although some clinging may still be observed up to age 6 and mothers will occasionally use their bodies to bridge tree gaps for older juveniles (van Adrichem et al. 2006, van Noordwijk and van Schaik 2005). Other developmental and social skills gradually develop. Orangutans start building practice nests by age 1 (Prasetyo et al. 2009), but they do not sleep in their own nests until they are 6–8 years old (van Noordwijk and van Schaik 2005). In one study, solitary play was especially prevalent in infants under age 2, and social play began after age 6, peaking at age 8 (van Adrichem et al. 2006).

Infants begin incorporating solid food into their diet at about age 1 or earlier (van Noordwijk and van Schaik 2005). Nursing has been observed up to age 8, although nursing rates appear to be quite low (van Adrichem et al. 2006, van Noordwijk and van Schaik 2005). Juveniles develop feeding independence between ages 3 and 6 (van Adrichem et al. 2006, van Noordwijk and van Schaik 2005). Jaeggi et al. (2008) found that Bornean juvenile orangutans occasionally solicited food from their mothers, with high (>88%) rates of success. Juveniles were more likely to request difficult-to-obtain items, and food-transfer requests decreased with increasing ecological competence (Jaeggi et al. 2008). Overall, immatures seem to be ecologically dependent until ages 8–10, at which time their diet becomes similar to that of their mothers (van Adrichem et al. 2006, van Noordwijk et al. 2009). By age 11, orangutans in Sumatra spend 65% of their time away from their mothers (van Adrichem et al. 2006).

The timing of juvenile development shows important individual and population-level differences (van Adrichem et al. 2006, van Noordwijk and van Schaik 2005). Orangutans in two Bornean peat swamp populations appear to wean their infants earlier than do populations in comparable habitat types in Ketambe (van Noordwijk et al. 2009). However, the general pattern that emerges is one of infant dependence on the mother from 0 to 3 years, a juvenile period between 4 and 8 years marked by increasing competence in foraging and locomotory skills, followed by increasing independence at 8–10 years. Interestingly, orangutans show characteristic changes in the color and distribution of their head and facial hair as well as changes in skin coloration that correspond with these behavioral stages (Kuze et al. 2005).

Male Development

For males, the onset of sexual maturity occurs around age 14 in the wild (Wich et al. 2004b), although the timing of full development of secondary sexual characteristics is highly variable. Some develop into flanged males immediately following adolescence, while others experience maturational delay that can last for a few or many years. At the extreme,

wild adult males have been known to remain unflanged for more than 20 years (Utami Atmoko and van Hooff 2004). The variable timing of male development contrasts sharply with the more predictable development of the other great ape males. Maturational delay may be a temporary state, but the possibility that some males never become flanged cannot be rejected (Crofoot and Knott in press).

Unflanged males can develop secondary sexual characteristics in less than a year (Kingsley 1988, Utami and Mitra Setia 1995), caused by a surge in testicular steroids and growth hormones (Maggioncalda et al. 1999, 2000). However, the underlying mechanism that triggers or, alternatively, suppresses development of unflanged males remains elusive. It may be that the presence of flanged males inhibits unflanged male development (Graham and Nadler 1990; Kingsley 1982, 1988; Maggioncalda et al. 1999; Utami Atmoko 2000). However, unflanged males in captivity have fully matured while a flanged male was present (Kingsley 1982, Maggioncalda et al. 1999). Maggioncalda and colleagues (2002) rejected the possibility that "arrested development" in unflanged males is due to chronic stress imposed upon them by flanged males since cortisol, a hormone associated with metabolic and psychosocial stress, is lower in captive unflanged than flanged males. They suggest selection has maintained arrested development as an alternative male reproductive strategy (Maggioncalda et al. 1999). Under this model, males that "play it safe," by postponing the development of secondary sexual features at puberty, avoid the high costs associated with being a flanged male (e.g., male/male aggression, immunosuppression caused by high androgen levels) but are less reproductively successful due to female mate choice for flanged males and reduced competitive ability (Maggioncalda et al. 1999). However, males that experience developmental arrest may achieve approximately the same lifetime fitness as males that "take a chance" (i.e., immediately undergo secondary maturation at puberty) because they end up living longer (Maggioncalda et al. 1999).

Female Development and Ovarian Function

Wild female orangutans reach sexual maturity at approximately 10–11 years (Knott 2001). Like most other primates, they experience a period of adolescent subfecundity, which lasts 1–5 years (Knott 2001). Age at first birth has been reported as between 12 and 16 years in the wild. (Knott 2001, Wich et al. 2004b). As in humans, the menstrual cycle averages 28 days (Knott 2005b, Markham 1990, Nadler 1988). Gestation in orangutans is approximately 8 months (Markham 1990). Pregnancy is recognizable to human observers by a slight but conspicuous swelling and lightening in color of the perineal tissues (Fox 1998, Galdikas 1981b, Schürmann 1981). Orangutans exhibit no sexual swellings or other outward indicators of ovulation (Graham-Jones and Hill 1962, Schultz 1938a). Thus, ovulation appears to be effectively concealed from males (Nadler 1982, 1988; Fox 1998).

Orangutans give birth only once every 7–9 years, on average (Galdikas and Wood 1990, Knott 2001, van Noordwijk and van Schaik 2005, Wich et al. 2004b). Why should orangutans give birth so rarely? Due to the low fruit availability in Southeast Asian rain forests, orangutans may spend significant periods in negative energy balance, when their ovarian function is suppressed (Knott 1999b; Knott et al. 2009). This, in combination with postpartum amenorrhea, may lengthen the time between conceptions (Knott 1999a,b). Additionally, many tree species have long intervals between fruiting, and juveniles may need extended periods to learn the location and processing techniques of difficult foods (van Adrichem et al. 2006, van Noordwijk and van Schaik 2005). Extended time may also be needed to develop social skills before becoming largely solitary (van Noordwijk and van Schaik 2005). Orangutan females in Sumatra have longer interbirth intervals than their Bornean counterparts (Wich et al. 2004b), which is surprising given the greater habitat productivity of Sumatra and, thus, expected higher energetic status of Sumatran females (Knott et al. 2009). Wich et al. (2004b) suggested that this difference may reflect selection for a slower life history in Sumatran orangutans; however, there are no other life history differences between the two species in either the wild or captivity (Anderson et al. 2008, Cocks 2007, Knott et al. 2009, Wich et al. 2009) as would be expected if this was a species difference. Current studies are investigating this paradox.

SOCIAL ORGANIZATION

Grouping Patterns and Philopatry

Although orangutans are predominantly solitary, three types of groupings can be distinguished (Utami et al. 1997): *travel bands* (individuals co-feed and travel together during periods of high food availability), *temporary feeding aggregations* (individuals feed together during times of food scarcity but travel independently), and *consortships* (a receptive female travels in a coordinated fashion with a flanged or unflanged male for several hours, days, or weeks). Additionally, a mother and her dependent offspring may travel with an older daughter and her offspring. At some sites, such as Suaq, orangutans have been observed to be particularly gregarious (van Schaik 1999).

Feeding competition is intense in orangutans and likely explains their largely solitary lifestyle. Orangutans feed in trees that are significantly smaller than those utilized by chimpanzees and bonobos (Knott 1999a,b) and experience periods of much sparser food availability. Orangutan feeding competition mainly takes the form of *scramble competition*, where individuals compete indirectly for patchily distributed food sources. Contest competition also occurs in large fig trees (Sugardjito et al. 1987, Utami et al. 1997). Predation probably plays an insignificant role in regulating grouping patterns; the only reports are of clouded leopards on Sumatra killing ex-captive juveniles traveling without a mother's protection (Rijksen 1978).

In contrast to the other great apes, orangutans appear to be female-philopatric (Galdikas 1988, Knott et al. 2008, Rijksen 1978, Rodman 1973, Singleton and van Schaik 2002, van Schaik and van Hooff 1996). Yet, genetic data show similar relatedness in both sexes (Goossens et al. 2006b, Utami et al. 2002). However, complicating factors exist at both of the sites where genetic data have been reported so far. At Ketambe, individuals were largely unrelated (Utami et al. 2002), but for females this is likely due to the introduction of two rehabilitant females in the late 1970s (Mitra Setia et al. 2009). In Kinabatangan, individuals showed high levels of relatedness, but anthropogenic factors may have affected dispersal options (Goossens et al. 2006b). Dispersal patterns should become clearer as genetic data accumulate.

Social Relationships

Orangutan social interactions are more limited than in many other primates, with some individuals spending days or weeks without contact with conspecifics. Mean party size ranges 1.5–2.0 in Sumatra and 1.05–1.3 in Borneo (van Schaik 1999). Mothers with unweaned offspring are less gregarious than females with older or no offspring (Mitra Setia et al. 2009, van Noordwijk et al. 2009). Orangutans also rarely groom themselves or others. Despite their semi-solitary nature, behavioral and experimental evidence suggests that individualized relationships exist (Delgado and van Schaik 2000, van Schaik and van Hooff 1996, Mitra Setia et al. 2009).

Female Ranging and Relationships

Adult female home ranges overlap considerably (Galdikas 1988; Singleton and van Schaik 2002; van Schaik and van Hooff 1996), for example, 67.8% at Gunung Palung (Knott et al. 2008); and clusters of related females may interact frequently (Knott et al. 2008, Singleton et al. 2009, Singleton and van Schaik 2001). Female ranges vary from less than 200 ha to greater than 850 ha and are smallest in eastern Borneo, where more bark and leaves are consumed, compared to Sumatra, where orangutans are more frugivorous (Singleton et al. 2009). Populations living in more heterogeneous habitats also have larger home ranges (Singleton et al. 2009). Females' day ranges vary from 150 to 1,000 m (Singleton et al. 2009).

Knott et al. (2008) report that despite significant home range overlap, some females at Gunung Palung rarely encounter each other because females maintain distinct core areas with minimal overlap (13.4%) within their ranges. Most females encountered other females less than expected by chance, except for mothers and daughters, which had higher than expected encounter rates. One dyad with encounters at the predicted level also had the highest level of agonism. Though aggression is rare, resident females were more likely to win fights within their own core areas (Knott et al. 2008). Mothers may also intercede in altercations between their daughters and other females (Knott et al. 2008, Singleton et al. 2009). Evidence of female dominance relationships exists (Utami et al. 1997, Knott et al. 2008). It is possible that orangutan females use passive range exclusion to reduce the effect of scramble competition and lower the frequency of contest competition (Knott et al. 2008).

Male Ranging and Relationships

Male orangutans were once thought to defend exclusive home ranges (Rodman 1973), but it is now known that flanged male ranges overlap considerably (Knott 1998c, Singleton and van Schaik 2001). Much less is known about male range size because males travel outside the boundaries of orangutan study sites. Singleton and van Schaik (2001) estimated that ranges for flanged and unflanged males are at least 2,500 ha. Flanged male ranges are three to five times the size of female ranges (Utami Atmoko et al. 2009a) and, thus, overlap the ranges of multiple females (Galdikas 1979, 1985a; Mitani 1985a; Utami and Mitra Setia 1995). Some flanged males remain in a fixed area (residents), while others range over greater regions (nonresidents). However, nonresidents may not be true transients but just rarely encountered males with very large ranges (Mitra Setia et al. 2009). The home range of the dominant flanged male may be smaller than those of other flanged males, which enables the dominant male to closely monitor females within his range (Mitra Setia et al. 2009, Utami Atmoko et al. 2009a). Patterns of male residency are not permanent, however, as males are known to leave an area voluntarily (Galdikas 1979) or can be forced out by nonresident or resident challengers (Utami and Mitra Setia 1995).

Male long calls appear to act as an important mechanism by which relationships are communicated within the dispersed society (Delgado and van Schaik 2000). One flanged male can sometimes be recognized as dominant over all other males (Mitra Setia et al. 2009), although the length of the dominant male's tenure varies. When there is a dominant flanged male, he gives long calls at the highest rate (Delgado 2006, Galdikas 1983, Mitani 1985b, Mitra Setia and van Schaik 2007, Utami and Mitra Setia 1995). Individual long calls show clear acoustic differences (Delgado 2007) that can be identified at least until 300 m (Lameira and Wich 2008). Flanged males are totally intolerant of other flanged males and subordinates may monitor the dominant male's long calls to avoid altercations (Mitani 1985b). Dominant males are most likely to approach the long calls of subordinate flanged males (Mitra Setia and van Schaik 2007). Males are less likely to call when in the presence of a female (Mitani 1985a), presumably to avoid attracting the attention of more dominant males (Mitra Setia and van Schaik 2007). Recent data from Sumatra confirm that the primary function of long calls is to attract females

(Mitra Setia and van Schaik 2007), as has been proposed by earlier researchers (Delgado 2006, Fox 2002, Galdikas 1983, Utami Atmoko 2000, Utami and Mitra Setia 1995).

Flanged males are invariably dominant to unflanged males, as evidenced by observations of unflanged males fleeing from flanged males (Galdikas 1979, 1985a,b; Mitani 1985a; Utami Atmoko 2000) and being supplanted in fruit trees (Utami et al. 1997). However, flanged males are markedly more tolerant of unflanged than flanged males and often allow them to remain in proximity undisturbed (Galdikas 1985a). This tolerance is not due to the males being closely related (Utami et al. 2002), and it wanes when a potentially reproductive female is present (Galdikas 1985a,b; Utami Atmoko 2000). Unflanged males may try to remain near a flanged male when he is consorting with a female, and aggressive encounters between these males occur in this context more often than when females are absent (Utami Atmoko et al. 2009a). Unflanged males often travel together (Utami Atmoko et al. 2009b), usually without aggression (Galdikas 1985b, Mitani 1985a), although aggression is more common in the presence of females (Utami Atmoko et al. 2009a). Dominance relationships among unflanged males have not been fully described, but observations of approach–avoidance interactions, aggression, and mating interruptions involving these males suggest that dominance relationships exist (Galdikas 1979, 1985b; Utami Atmoko 2000). The ratio of flanged to unflanged males varies between populations (Utami Atmoko et al. 2009b) and is highly variable within a population, probably related to food availability and the presence of receptive females. Formerly dominant flanged males are called "past prime" because they show greatly diminished cheek flanges, give fewer long calls, have lower testosterone levels, mate infrequently, and are subordinate to "prime" flanged males (Knott 2009).

Male–Female Relationships

It has long been assumed that there were no long-term relationships between male and female orangutans, but recent reports suggest that, at least in Sumatra, females may maintain loose associations with the dominant flanged male in the area. Females seek out this male for protection from male harassment and possibly from infanticide. At Suaq, females that associated with a flanged male received less harassment from unflanged males seeking copulations (Fox 2002). Additionally, some of these associations did not involve mating with the dominant male but were female-initiated and disbanded after the unflanged male left the area (Fox 2002). Mitra Setia and van Schaik (2007) argue that males play a protective role since Ketambe females of all reproductive states responded to male long calls by traveling toward those males. Additionally, although rare, food transfer sometimes occurs between adult orangutans, particularly between sexually active females and unflanged males (van Noordwijk and van Schaik 2009).

Mating Behavior

Orangutan copulations normally occur face-to-face, are extremely long in duration (average 8–11 min), and often involve extensive precopulatory behavior and forced copulations (Fox 1998, Galdikas 1981b, Knott 2009, MacKinnon 1979, Mitani 1985a, Stumpf et al. 2008, Utami and van Hooff 2004). Copulatory frequency is fairly low, occurring at a rate of 0.002–0.14 copulations per observation hour (Stumpf et al. 2008). Young females are often particularly proceptive (Galdikas 1979, 1981b; Schürmann 1982; Schürmann and van Hooff 1986). Females sometimes play an active role during mating by aiding with intromission, manually or orally stimulating male genitalia, and performing pelvic thrusts (Fox 1998; Galdikas 1979, 1981b; Schürmann 1981, 1982; Utami Atmoko 2000).

Sexually receptive females express mating preferences for the resident dominant flanged male (Fox 1998, Utami Atmoko 2000, Knott 2009, Knott et al. 2010). Flanged males predominantly use a "sit-call-and-wait" strategy, whereby they advertise their presence with long calls and wait for females to approach and establish a consortship (Galdikas 1985a, Utami and van Hooff 2004). At Gunung Palung, females encounter prime flanged males at a higher rate around the time of ovulation, and this is likely a female-driven outcome (Knott et al. 2010).

While on consort, flanged males closely guard females and react aggressively to the calls or presence of other males, including unflanged males that often follow consorting pairs (Galdikas 1981b). If two flanged males meet in the presence of a receptive female, escalated contests often ensue, and the dominant male will take over a subordinate male's consortship (Galdikas 1985a). Wounds suffered during male/male fights can be serious (Galdikas 1985a) and even fatal (Knott 1998b). Participating in consortships is costly for flanged males because they must adjust their normal activity budgets to accommodate the longer day ranges of females (Mitani 1989, Utami Atmoko and van Hooff 2004). Flanged males are therefore somewhat selective and sometimes reject or react passively to nulliparous females but readily accept parous females as consort partners (Galdikas 1981b, 1985a, 1995b; Schürmann 1981, 1982). Such selectivity is likely adaptive since young females have a higher probability of being subfecund.

Unflanged males usually adopt a lower-profile mating strategy in which they do not emit long calls or form consortships but instead actively search for mating opportunities (Galdikas 1985a,b). Females are generally not proceptive toward these males when they are ovulating (Knott et al. 2010), but unflanged males try to copulate despite female resistance. Outside ovulation, females may show little resistance and even proceptive behavior toward these males (Knott et al. 2010). Female orangutans show strong mating preferences as evidenced by both forceful resistance and extreme proceptivity. Functional explanations for this behavior are explored in the are explored below in the "Discussion" section.

TOOL USE AND CULTURE

Tool use for extractive foraging is relatively rare in wild orangutans (van Schaik et al. 1999a). The site with the most habitual tool use is Suaq, where orangutans strip sticks and use them to harvest insects and *Neesia* seeds (Fox et al. 1999; van Schaik and Knott 2001; van Schaik et al. 1996a). Tool use for *Neesia* seed extraction seems to be a cultural difference as orangutans at Gunung Palung regularly eat this fruit but do not use tools to access seeds (van Schaik and Knott 2001). Orangutans may invent tool use at a fairly constant rate, but the high density of individuals at Suaq may permit increased transmission and diffusion throughout the population (Fox et al. 2004, van Schaik 2002, Whiten and van Schaik 2007). The strong association of mean female party size with tool-use specialization supports this hypothesis (van Schaik et al. 2003b).

Orangutans show many population-level innovations. These are novel learned behaviors that are not simply a result of social learning or environmental induction (Ramsey et al. 2007, Russon et al. 2009a, van Schaik et al. 2006) and include other types of tool use. van Schaik et al. (2009a) recently identified 54 innovations across eight orangutan sites, with 26 reaching cultural significance. Some cultural innovations involve leaves used in agonistic displays (Galdikas 1982a, MacKinnon 1974, Rijksen 1978), for self-cleaning (MacKinnon 1974, Rijksen 1978), as protection in food acquisition (Rijksen 1978), and as drinking vessels (Knott 1999a). Sticks have been used for scratching (Galdikas 1982a) and dead wood for opening up durian fruits (Rijksen 1978). Other behaviors, such as the use of "leaf umbrellas" during rainstorms, are considered a cultural universal (van Schaik et al. 2003a). Orangutans also show flexibility in their sounds and vocalizations. Thirty-two calls have been identified, many of which may be cultural innovations (Hardus et al. 2009). Intriguingly, ex-captive rehabilitant orangutans generate more innovative behaviors than their wild counterparts (Russon et al. 2007), many involving groundwater, which wild orangutans typically avoid (Russon et al. 2009a). This may be due to a lack of maternal guidance that allows these young individuals to invent their own solutions (Russon et al. 2009a). Differences between wild populations in the size of the cultural repertoire are predicted by the opportunity for oblique and horizontal transmission during development (van Schaik et al. 2003b), which may also be applicable to the rehabilitant case.

CONSERVATION

The IUCN Red List (IUCN Species Survival Commission 2008) classifies Bornean and Sumatran orangutans as endangered and critically endangered, respectively. In 2004, there were approximately 54,000 orangutans remaining on

Borneo and 6,500 on Sumatra (Wich et al. 2008). The minimum viable population size is 250 orangutans (Singleton et al. 2004, Marshall et al. 2009b), and presently only six populations on Sumatra and 32 on Borneo meet this criterion (Wich et al. 2008). Genetic analyses indicate that orangutan populations in Sabah have declined by 95% in the last few centuries, with a massive collapse in the last few decades coincident with anthropogenic forest exploitation (Goossens et al. 2006a). Sharp population declines have also been reported for other parts of Borneo and Sumatra and have been linked to recent human activities (Rijksen and Meijaard 1999, Robertson Yarrow and van Schaik 2001, van Schaik et al. 2001).

The main threat to orangutans is habitat loss and fragmentation. Between 1992 and 2002, 2.8% of orangutan habitat in Kalimantan was destroyed annually (Wich et al. 2008). Between 1985 and 2001, 1%–1.5% of orangutan habitat in Sumatra was lost per year (Singleton et al. 2004). Forests have been cleared for logging, mining, and agriculture. Habitat loss has accelerated recently due to the rapid expansion of oil palm plantations. Palm oil is in high demand globally for its use in cooking, in cosmetics, and as a biofuel. Indonesia and Malaysia produce the vast majority of the world's palm oil. On Borneo, the total area planted with oil palm increased from 2,000 km² in 1984 to 27,000 km² in 2003, and much of this converted area was prime orangutan habitat (Ancrenaz et al. 2008). Orangutan habitat has also been destroyed by fire, linked to droughts caused by El Niño climatic events. As a result of the 1997–1998 fires, the number of Bornean orangutans was reduced by about 33% in a single year (Ancrenaz et al. 2008).

About 75% of orangutans occur outside national parks (Meijaard and Wich 2007) and often in forests that are legally exploited for timber or agriculture. Logging is generally associated with a decline in orangutan density (Felton et al. 2003, Husson et al. 2009, Johnson et al. 2005, Morrogh-Bernard et al. 2003, Rao and van Schaik 1997, van Schaik et al. 2001), likely because it reduces food availability and introduces gaps into the canopy, making travel more difficult (Rao and van Schaik 1997). However, the extent of orangutan population disturbance depends on the intensity of logging, with lightly logged forests supporting more orangutans than heavily logged forests (Ancrenaz et al. 2005, Husson et al. 2009). Bornean orangutans display greater dietary flexibility, which might explain why *P. pygmaeus* populations are less disturbed by logging than *P. abelii* populations (Husson et al. 2009). Orangutan densities can recover to prelogging levels if forests containing adequate orangutan foods are allowed to regenerate (Knop et al. 2004). Habitat loss is a major threat to orangutans even within designated protected areas since most national parks have been seriously degraded as a result of inadequate law enforcement (Jepson et al. 2001, Nellemann et al. 2007, Robertson Yarrow and van Schaik 2001). Between 1985 and 2001, over 56% of Kalimantan's lowland rain forests within protected areas were destroyed (Curran et al. 2004).

Although prohibited by law, hunting is another major factor contributing to orangutan population decline (Marshall et al. 2006, 2009b). Orangutans are killed for food, for use in traditional medicines, as retribution for crop raiding, and to obtain infants for the illegal pet trade (Rijksen and Meijaard 1999). In 2003–2004, 200–500 orangutans from Kalimantan were lost to the pet trade (Nijman 2005). Orangutan mothers are typically killed in the capture process. Logging often exacerbates hunting pressures because it creates easier access to forests via construction of new roads. Even if habitat is not lost, orangutan populations cannot survive an annual removal rate of 1%–2%, due to their very long inter-birth intervals (Marshall et al. 2009b). This rate is likely exceeded in many populations.

Although the situation facing orangutans is dire, progress in orangutan conservation has been made. Forest loss has slowed in some orangutan strongholds (Meijaard and Wich 2007, Wich et al. 2008), and illegal logging within national parks has also dramatically declined in several areas (Wich et al. 2008). These changes can largely be attributed to increased political and financial support and efforts by local and international nongovernmental organizations. Though encouraging, such improvements must be considered fragile as threats to orangutans abound and conservation progress can be quickly reversed.

DISCUSSION: UNDERSTANDING FORCED COPULATIONS

One of the most striking features of orangutan biology is the regular occurrence of forced copulations. Male aggression in the mating context is not unique to orangutans (van Schaik et al. 2004, Muller and Wrangham 2009), but forced copulations are remarkably rare in primates. Although forced copulations have been thought to primarily be a mating strategy of unflanged males (Rijksen 1978; Galdikas 1979, 1981b; Schürmann 1982; Fox 2002), observations across study sites show that both flanged and unflanged males engage in this behavior (Figure 18.3). Orangutan mating involves both resistant and cooperative female behavior (Table 18.2), and the nature of female behavior can switch within a given mating interaction as well as between copulations on the same day (Rijksen 1978; Galdikas 1981b; Mitani 1985a; Fox 2002; Knott 2009). These observations are puzzling. Given that females seem to prefer flanged males, why do they sometimes resist these males and, alternatively, why do they sometimes mate cooperatively with less preferred unflanged males? Recent evidence demonstrates that female orangutans adopt multiple strategies, depending on their own reproductive state, the identity of their mating partner, and other considerations (Stumpf et al. 2008, Knott 2009).

A Flanged Males

B Unflanged Males

Figure 18.3 Proportion of forced and cooperative matings at major orangutan study sites for (*A*) flanged and (*B*) unflanged males. White bars indicate percentage of matings that were forced (i.e., females resisted), and black bars indicate percentage of matings that were cooperative. Sample sizes are for copulations of known outcome and are as follows (unflanged/flanged): Gunung Palung, *n* = 24/19 (Knott et al. 2009a); Suaq Balimbing, *n* = 90 with known outcomes/66 (Fox 2002); Ketambe, *n* = 38/50, 94/70, averaged between two studies (Schürmann and van Hooff 1986, Utami Atmoko 2000); Tanjung Puting, *n* = 22/30 (Galdikas 1985a,b); Kutai, *n* = 151/28 (Mitani 1985a).

Female Proceptivity and Cooperative Mating

The Good Genes Hypothesis

The preference that orangutan females have for flanged males has been interpreted as a female's effort to obtain "good genes" (Galdikas 1981b; Mitani 1985b). Female orangutans give birth to only a few offspring over their lifetime, invest heavily in each one, and have very high infant and juvenile survivorship. Such heavy investment is predicted to lead females to choose males with the best genetic makeup to father their young. Since it has been assumed that dominant flanged males sire the majority of offspring, selection for good genes has been a plausible explanation for why females resist subordinate males and mate cooperatively with the dominant flanged male. Recent genetic data (Goossens et al. 2006b) confirm greater reproductive success for flanged males. Because not all males can become flanged and the flanged stage is energetically

Table 18.2 Possible Explanations for Orangutan Female Cooperation and Resistance During Mating

FEMALE COOPERATION	FEMALE RESISTANCE
Paternity concentration in dominant male	Paternity avoidance of subordinate or unpreferred/unflanged male
Paternity confusion with potentially infanticidal males	Avoidance of unnecessary contact to limit disease transfer
Avoidance of sexual coercion/harassment if no risk of conception	Elicitation of support from dominant male
	Prevention of feeding/energetic costs
	Prevention of ejaculation
	Mate assessment

costly, this may serve as an honest signal of male quality (Knott et al. 2010).

Paternity Confusion to Prevent Infanticide

There is substantial evidence, across mammalian taxa, that males sometimes use infanticide as a reproductive strategy (see reviews by Hiraiwa-Hasegawa 1988, Hrdy and Hausfater 1984a, van Schaik 2000b; but see Bartlett et al. 1993, Sussman et al. 1995). By killing infants they have not sired, males can create a new reproductive opportunity by interrupting a mother's lactational amenorrhea, causing her to return to a fecund state sooner than would occur otherwise (Hrdy 1979). Given their slow life histories and high lactation-to-gestation ratio, female orangutans, along with the other great apes, should be among the primates most vulnerable to infanticide (van Schaik 2000b).

One way for females to decrease the risk of infanticide is to use sexual behavior (Hrdy 1979). Because males do not recognize their offspring, females are expected to mate with potentially infanticidal males when possible, including when conception is unlikely, to increase these males' assessments of likely paternity. If a newly dominant male is reasonably uncertain about whether he sired an infant, it benefits him, in terms of fitness, to refrain from killing the infant (van Schaik et al. 2004). Thus, if female orangutans can recognize that a dominant male may be overthrown before the birth of their next infant, the infanticide-avoidance hypothesis predicts that they should willingly copulate with up-and-coming unflanged males or flanged males that are likely to be successful in challenging the dominant male. Observations from Ketambe in Sumatra are consistent with these predictions. When a single dominant male could not be recognized, females, including those that were lactating and pregnant, readily copulated and formed consortships with unflanged and subordinate flanged males (Utami Atmoko 2000). Paternity data also indicate that female orangutans may be able to recognize males rising in rank because one unflanged male at Ketambe, which later became dominant, fathered 50% of offspring sired by unflanged males (Utami et al. 2002).

Recent data from Gunung Palung (Knott et al. 2010, Knott 2009) provide new support for the infanticide-avoidance hypothesis. The lower the likelihood of females getting pregnant, the more proceptive and cooperative they were to unflanged and past-prime flanged males during mating. Females mated most frequently with unflanged males when their conception risk was low, and these copulations were often consensual. At the beginning of pregnancy, before females show the characteristic labial swelling, they were extremely proceptive. Females facilitated penetration and engaged in oral contact with male genitalia, particularly when mating with prime flanged males. Other evidence of perceived infanticide risk comes from play-back experiments, which demonstrate that while females and infants ignore the long calls of resident flanged males, they show strong negative reactions to the vocalizations of nonresident males (Delgado 2003). Further, females with infants move toward flanged male long calls, which Mitra Setia and van Schaik (2007) interpret as an anti-infanticide strategy.

The most serious problem with the infanticide-avoidance hypothesis is that, despite nearly four decades of study, successful or attempted infanticides have never been reported in wild orangutans (Beaudrot et al. 2009), although there is one case of infanticide reported from captivity (Mallinson 1984). The absence of infanticide is surprising since it has been recorded in other great apes (chimpanzees, reviewed by Stumpf this volume; gorillas, Watts 1989b, Robbins this volume), with the exception of bonobos. However, because the average party size is much smaller in orangutans, the number of "ape-hours" observing chimpanzees and gorillas is much higher than in orangutans, and slow reproduction and low interaction rates may present limited opportunities for infanticide to occur and be observed (Stumpf et al. 2008, Knott et al. 2009). In a recent consideration of this hypothesis, Beaudrot et al. (2009) argue that the conditions for infanticide are so rarely met that we should not expect it to occur. However, the mounting evidence that females are engaging in infanticide-avoidance strategies suggests that the high potential cost of an infant loss may result in counterstrategies to infanticide, even when the probability of such an event is low.

Avoidance of Harassment

Females may mate cooperatively as a way to avoid harassment. Such a strategy would lead to the prediction that as male density, and thus potential harassment, increases, females should show less resistance. This behavior is seen in some species and has been termed "convenience polyandry" (Low 2005). However, Knott (2009) found that populations that had a higher male-to-female ratio had more, not less, female resistance to both flanged and unflanged males. Instead of acquiescing, females respond to male harassment

by either resisting during mating or increasing their proximity to the flanged male. Fox (2002) shows that females at Suaq received lower harassment if they remained close to or consorted with the dominant flanged male, and Mitra Setia and van Schaik (2007) suggest that females at Ketambe may remain within vocal contact with flanged males to reduce harassment.

Female Mating Resistance

Avoidance of Undesirable Paternity

Given that females prefer flanged males and their potentially superior genes, why do they sometimes resist mating with these males? Females may not prefer just any flanged male. When there is a clearly dominant resident flanged male, females show the highest preference for that male, rejecting other flanged males (Mitani 1985a; Utami Atmoko et al. 2009a). In addition, females may reject some flanged males because they are past-prime (Knott 2009). Females also resist mating with some, but not all, unflanged males. The category of unflanged males encompasses multiple developmental stages. Without knowing individual life histories, it is difficult to distinguish unflanged males that are adolescents from those that are older but have remained in an unflanged state and those that are in the process of becoming flanged (Knott 2009). The distinction may be very important to females in their mate assessment. Fox (1998) found that females behaved very differently toward small, medium, and large unflanged males. Female behavior is consistent with the possibility that they are aware of changing male status before researchers are since they seem to anticipate these changes. At Ketambe, Utami Atmoko (2000) found that an unflanged male with whom females mated cooperatively developed cheek pads and displaced the dominant flanged male six months later. Additionally, females started preferentially approaching the long calls of the new dominant male in the period *before* he became dominant (Mitra Setia and van Schaik 2007).

Disease Avoidance

Orangutan females seem to be surprisingly mating-averse. A female who mates cooperatively with a dominant flanged male may resist the same male later in the day. Knott et al. (2009) found that females were more resistant toward prime males outside of ovulation. Females that are pregnant or lactating also sometimes resist copulations, despite having no risk of conception. What accounts for this? One possibility is disease avoidance (Fox 2002, Knott and Kahlenberg 2007, Knott 2009). Orangutans at most sites have little physical contact with other members of their species. Females conceive only once every 5–8 years and typically have brief periods of sexual activity during those conceptive periods. Lengthy matings in orangutans and contacts with multiple males during these periods would increase the opportunity for transmission of both macro- and microparasites and disease vectors (Knott 2009). Orangutans have fewer white blood cells than do chimpanzees (Nunn et al. 2000), and thus, their ability to mount an immune response may be lower than that of other primates, perhaps leading them to avoid unnecessary contact.

Eliciting Support from the Dominant Male

van Schaik (2004) argues that females may resist matings with nonpreferred males to elicit the intervention of the dominant male and to convey to the dominant her unwillingness to mate with other males. Data from Tuanan also show that when loud vocalizations were made during a male/female interaction, they were more likely to be terminated by the intervention of another male (van Noordwijk and van Schaik 2009). In many circumstances, though, the dominant male would be too far away to be aware of this interaction. It would also seem to be in the female's interest to prevent the dominant male from being aware of matings with other males in order to increase confidence in his own paternity (Knott 2009).

Avoidance of Energetic Costs

Orangutans may resist some matings in order to avoid the loss of feeding time. Fox (1998) describes a female that showed no resistance to mating with an unflanged male and continued to feed on fruit throughout the mating. However, after 3 min the male grabbed her feeding hand to reposition her, thus preventing her from eating. This was followed by 12 min of struggle between the female and the male. Fox (1998) attributes this female's objection to her need to continue feeding. However, Fox (1998) tested the energetic cost hypothesis but did not find decreased foraging efficiency on days when females had forced consortships. It is also unclear whether female mating resistance would actually lower energetic costs as females are normally not able to prevent an unwanted mating and resistance expends more energy than cooperation. Thus, energetic constraints may explain why females would want to avoid unwanted matings and consortships but does not explain actual resistance during mating.

Decreasing Mating Duration

Resistance is rarely effective. For example, at Kutai, Mitani (1985a) observed females struggling free in less than 10% of forced mating attempts. What benefit might females be gaining from trying to resist copulation, even if resistance is unsuccessful? At Gunung Palung, resisted copulations were shorter in duration than unresisted copulations (Knott et al. 2010). Thus, resistance may serve two possible functions. First, it would reduce the total time of physical contact between the male and female, which may be beneficial for disease avoidance. Second, it may be that orangutans copulate for 8–11 min because it is necessary for ejaculation; thus, it is possible that, by resisting, females can shorten the mating and thus decrease the likelihood of ejaculation (Knott 2009).

Mate Assessment

In some species, females resist mating in order to gain information about the health and vigor of prospective mates (Smuts and Smuts 1993). This possibility may be applicable to orangutans because the sexes have little contact with each other outside of mating. If mate assessment is why orangutan females resist, then stronger, more aggressive males are expected to achieve the highest reproductive success and be chosen by females for their good genes. Fox (1998) predicted that if female resistance was due to mate assessment, females would decrease their resistance to males that had mated aggressively with them in the past, but she did not find support for this hypothesis. However, it may be that mate assessment explains why females often retreat from an advancing male but then become proceptive once contact is made.

Low Costs of Resistance

The cost of resistance must be considered in a female's decision to resist a mating. It is puzzling that females resist copulations given that resistance has been thought to be costly. However, Knott's (2009) recent review of the published data shows, perhaps surprisingly, that there are no reports of females receiving visible wounds from resisted matings. Although forced matings may hurt the female, there is no evidence that they regularly inflict injuries that compromise health or mobility (Knott 2009). Knott (2009) argues that coercion in the mating context in orangutans is direct; that is, males force females to copulate when they resist. They do not appear to use indirect coercion, as has been seen in chimpanzees and humans, where aggression toward females occurs outside of mating to coerce females into mating at a later date (Muller and Wrangham 2009). Thus, females may suffer little physical cost from resisting male mating attempts.

CONCLUSION

Almost 40 years of research on wild orangutans has revealed a range of extraordinary features of their behavior and biology. However, many questions remain unresolved, such as uncovering the mechanism that triggers male development, understanding ranging patterns, and completing genetic and hormonal analyses at multiple sites to understand male and female mating strategies. Many questions, such as juvenile developmental patterns, are just now being addressed and, through the power of long-term data, rare but important behaviors are being revealed, such as female/female aggression and food sharing between adults. Comparative studies have uncovered differences between populations, species, and subspecies (Wich et al. 2009, van Schaik et al. 2009a,b,c). One of the current challenges is to disentangle the sources of this variation. What differences can be attributed to variations in local ecology and, thus, serve as a reflection of phenotypic plasticity? Alternatively, what differences have a genetic foundation reflecting local adaptation? Finally, which behaviors reflects cultural differences between

populations? All of these questions deserve our attention; however, the issue that must move to the forefront of all agendas is conservation—habitat protection, in particular—for without increased efforts to save orangutans, we will lose the opportunity to fully understand and appreciate this fascinating great ape.

19

Gorillas
Diversity in Ecology and Behavior
Martha M. Robbins

1. What is the nature of intergroup interactions among gorillas?
2. How do female/female relationships compare to male/female relationships in gorillas?
3. How does male and female dispersal differ among gorillas?

INTRODUCTION

Gorillas are found in 10 central African countries, in a broad diversity of habitats ranging from coastal lowland forests to high-altitude, Afromontane rain forests (Fig. 19.1). The distribution of gorillas makes them a particularly appealing species to study, both in terms of understanding how they have adapted to such a variety of habitats and by providing an excellent opportunity to test many hypotheses of primate behavioral ecology that assume variation in ecology will lead to variation in behavior and demography (e.g., Sterck et al. 1997; *Papio*, Barton et al. 1996; *Presbytis entellus*, Koenig et al. 1998; *Saimiri*, Boinski et al. 2002). However, perhaps one of the biggest ironies of primatology is that the majority of information on gorillas has come from a very small population studied for over 35 years living at a unique ecological extreme (high altitude; Karisoke Research Center, Rwanda). Fortunately, in recent years, several studies of other gorilla populations have begun to provide us with information that is more representative of the genus as a whole, especially on feeding ecology; but because of the difficulty of habituating lowland gorillas, only a limited number of studies involving direct observations have been conducted and much remains unknown. One obvious conclusion, though, is that due to the large differences in ecology between western and eastern gorillas, we can no longer assume that all knowledge of the

behavior and demography of mountain gorillas applies to all gorilla populations.

TAXONOMY AND DISTRIBUTION

The taxonomic classification of gorillas has changed many times over the past century (Groves 2003). For much of the past few decades, gorillas were considered only one species with three subspecies (western lowland gorillas, eastern lowland gorillas, and mountain gorillas). A recent taxonomic reclassification now groups gorillas as two species and four subspecies (Groves 2001a) (Fig. 19.1, Table 19.1). Western gorillas include *Gorilla gorilla gorilla* (western lowland gorilla), found in Equatorial Guinea, Gabon, Angola, Cameroon, Central African Republic, and Democratic Republic of Congo, and *G. g. diehli* (Cross River gorilla), found in a handful of small populations in Nigeria and Cameroon. Eastern gorillas (Color Plate 32) include *G. beringei graueri* (eastern lowland or Grauer's gorilla), found in the east Democratic Republic of Congo, and *G. b. beringei* (mountain gorilla), found in two small populations at the Virunga Volcanoes of Rwanda, Uganda, and Democratic Republic of Congo and in the Bwindi Impenetrable National Park of Uganda (Figs. 19.2 and 19.3). Genetic studies suggest that the initial split between eastern

Figure 19.1 Map of distribution of gorillas in Africa. The locations of major field sites discussed in the text are indicated.

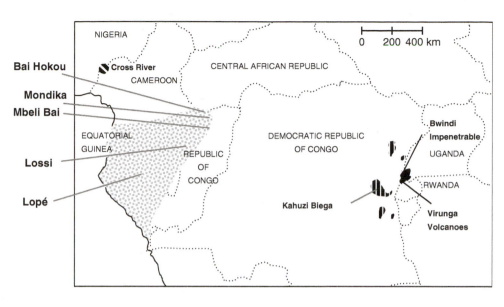

Bai Hokou

Mondika

Mbeli Bai

Lossi

Lopé

:::::: Western lowland gorilla (*Gorilla gorilla gorilla*)

NN Cross River gorilla (*G. g. diehli*)

llll Grauer's gorilla (*G. beringei graueri*)

▬▬ Mountain gorilla (*G. beringei beringei*)

and western gorillas occurred between 0.9 and 1.6 million years ago (Thalmann et al. 2007).

FEEDING ECOLOGY

Gorillas are largely vegetarian, with the only nonvegetative foods in their diet being ants and termites (Williamson et al. 1990, Deblauwe and Janssens 2008 Yamagiwa et al. 1991, Ganas and Robbins 2004). Their large body size, enlarged hindgut, and long hindgut fermentation time enable them to meet their nutritional requirements with a diet high in bulky foods and structural carbohydrates (Watts 1996). Mountain gorillas spend less time feeding than western gorillas (approximately 55% vs. 67% of daylight hours feeding) (Watts 1988, Masi et al. 2009). Gorillas eat nonreproductive plant parts (leaves, stems, pith, and bark) as well as fruit. A comparison of the diet of gorillas across Africa is best done by considering an elevation continuum from the lowland forests of central Africa to the Afromontane forests of the Virunga Volcanoes (Table 19.2). In general, the degree of frugivory decreases as altitude increases because of reduced fruit availability (Doran and McNeilage 2001, Goldsmith 2003, Ganas et al. 2004). Detailed studies on mountain gorillas at Karisoke Research Center initially labeled them as folivore–herbivores that included only a negligible amount of fruit in their diet (Fossey and Harcourt 1977; Watts 1984; Vedder 1984; McNeilage 1995, 2001). However, these studies were conducted in the only gorilla habitat that contains almost no fruit (>2,500 m in elevation). Moving only 30 km away, the mountain gorillas of Bwindi Impenetrable National Park, whose range is between 1,400 and 2,500 m in altitude, eat fruit on approximately 60%–80% of days (Robbins and McNeilage 2003, Ganas et al. 2004). Several studies of both Grauer's gorillas and western gorillas in lowland habitats reveal an almost daily consumption of fruit (Nishihara 1995; Remis 1997; Doran et al. 2002b; Rogers et al. 2004; Yamagiwa et al. 1994, 1996, 2003a). Cross River gorillas consume fruit slightly less frequently than other lowland gorillas (approximately 90% of days), probably because they are faced with the ecological constraint of a very long dry season due to the northern location (5 months/year with <100 mm of rainfall compared to 3 months at other sites) (Oates et al. 2003). It is now clear that all gorilla subspecies should be considered folivore–frugivores. However, these studies have measured fruit consumption by fecal analysis and from trail signs of feeding remains, which do not tell us the proportion and quantity of the gorilla diet that is fruit versus herbaceous vegetation. The few studies of direct observations found that Bwindi gorillas may spend about 11% of their foraging time eating fruit (but as much as 60% of time in any month) (Robbins 2008), whereas western gorillas may consume fruit during 35% of foraging time, with values reaching 70% during a 4-month fruiting season (Masi 2008, Masi et al. 2009).

While much emphasis has been placed on the degree of frugivory by gorillas, all populations do incorporate a significant amount of fibrous food (nonfruit foods including herbaceous vegetation) in their diet (Watts 1984, Ganas

Table 19.1 Subspecies of Gorillas, Countries Which They Inhabit, Estimated Area of Habitat, and Estimated Population Size

SPECIES/SUBSPECIES	COUNTRIES FOUND	AREA OF HABITAT (KM²)	ESTIMATED POPULATION SIZE
Eastern gorilla (*Gorilla beringei*)			
Mountain gorilla[1] (*G. b. beringei*)	Rwanda, Uganda, Democratic Republic of Congo	700	700
Grauer's gorilla[2] (*G. b. graueri*)	Democratic Republic of Congo	21,000	5,000–28,000
Western gorilla (*G. gorilla*)			
Cross River gorilla[3] (*G. g. diehli*)	Nigeria, Cameroon	140–200	200–250
Western lowland gorilla[4] (*G. g. gorilla*)	Gabon, Cameroon, Equatorial Guinea, Central African Republic, Angola, Democratic Republic of Congo	445,000	<200,000?

Sources:

[1] Guschanski et al., 2009, Gray et al., 2010

[2] Hall et al. 1998, Plumptre et al. 2003.

[3] Oates et al. 2003.

[4] Butynski 2001, Plumptre et al. 2003, Walsh et al. 2003.

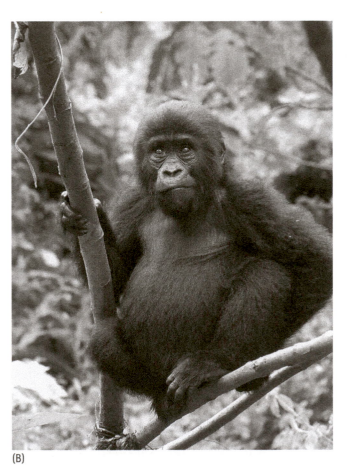

Figure 19.2 Infant gorillas from different populations. *A*: Mountain gorilla from Virunga Volcanoes. *B*: Mountain gorilla from Bwindi Impenetrable National Park, Uganda.

(C)   (D)

Figure 19.2 (*cont'd*) *C*: Juvenile western gorilla from Central African Republic. *D*: Infant western gorilla from Central African Republic.

Figure 19.3 Silverback mountain gorilla from Bwindi Impenetrable National Park, Uganda, descending a tree after foraging for fruit.

et al. 2004, Doran et al. 2002b, Oates et al. 2003, Yamagiwa et al. 2003a, Rogers et al. 2004, Masi 2008). In particular, during times of fruit scarcity, gorillas rely on fibrous foods as fallback foods. Cross River gorillas likely experience a more drastic shift in diet from fruit to fibrous foods because of the prolonged dry season compared with other lowland gorillas (Oates et al. 2003). There is almost no temporal variability in food availability for Virunga mountain gorillas, which almost entirely lack fruit in their diet (Watts 1984, 1991a, 1998a; Vedder 1984; McNeilage 1995, 2001). The only seasonal food for Virunga mountain gorillas is bamboo shoots (Vedder 1984). While bamboo is considered to be an important component of mountain gorilla diet because it is high in protein, not all mountain gorillas consume it due to lack of availability (Watts 1984; McNeilage 1995, 2001; Ganas et al. 2004). The number of herb species consumed by gorillas also decreases with increasing altitude, probably as a result of decreased plant diversity as altitude increases.

Gorillas are selective in their feeding behavior (Watts 1984, 1991a; Vedder 1984; Yamagiwa et al. 1996, 2003a; Ganas et al. 2004; Rogers et al. 2004). Their diet is primarily influenced by availability of food resources, but it is clear that gorillas seek out foods with particular nutritional compositions and often forage on rare species. Gorillas select leaves and herbaceous material that are high in protein and fruits that are high in soluble sugars (Waterman et al. 1983; Calvert 1985; Rogers et al. 1990; Watts 1991a; Remis et al. 2001; Remis 2003; Ganas et al. 2008, 2009b). Tests with captive gorillas have shown them to be tolerant of relatively high levels of tannins if accompanied by high levels of sugar (Remis and Kerr 2002). Certain foods that are found only in swamps offer high levels of particular minerals (e.g., sodium and potassium), which may influence the use of these areas by gorillas and their ranging patterns

Table 19.2 Dietary and Ranging Patterns of Gorillas

SUBSPECIES/STUDY LOCATION	ALTITUDE (M)	NUMBER OF FOOD SPECIES EATEN	NUMBER OF FRUIT SPECIES EATEN	DEGREE OF FRUGIVORY[1]	DAY JOURNEY LENGTH (M)	HOME RANGE SIZE (KM²)
Mountain gorilla (*Gorilla beringei beringei*)						
Virungas–Karisoke[2,3]	2,680–3,710	36	1	<1%*	570	3–15
Virungas–tourist group[3]	2,500–2,800	42	2	<1%**	756	5
Bwindi–Ruhija[4]	2,100–2,500	112	30	16–66%	1,034	21–40
Bwindi–Buhoma[5]	1,450–1,800	140	36	68–89%	547–978	16–22
Grauer's gorilla (*G. b. graueri*)						
Kahuzi-Biega–Kahuzi[6]	1,800–2,600	79	24	96%	850	23–31
Kahuzi-Biega–Itebero[6]	600–1,300	142	67	89%	–	23–31
Cross River gorilla (*G. gorilla diehli*)						
Cross River–Afi[7]	400–1,300	166	83	90%	–	~32
Western lowland gorilla (*G. g. gorilla*)						
Lopé, Gabon[8]	100–700	134	95	98%	1,105	22
Mondika, Central African Republic[9]	<400	100	70	99%	2,014	16
Bai Hokou, Central African Republic[10]	460	138	77	99%	1,580	11

[1] Degree of frugivory, percent of fecal samples that contained seeds as indication of daily fruit consumption, except for Virungas, where it was based on percent time of observation (*) or of feeding remains on trails that contained fruit (**).
[2] Vedder 1984, Watts 1984, 1991a, 1998b.
[3] McNeilage 1995, 2001.
[4] Robbins and McNeilage 2003, Ganas et al. 2004, Nkurunungi 2004, Ganas and Robbins 2005.
[5] Ganas et al. 2004, Ganas and Robbins 2005.
[6] Yamagiwa et al. 1992, 1994, 1996, 2003a.
[7] Oates et al. 2003.
[8] Rogers et al. 1990, Williamson et al. 1990, Tutin 1996.
[9] Doran et al. 2002b, 2004.
[10] Remis 1997, Cipolletta 2003, 2004.

(Magliocca and Gautier-Hion 2002, Doran-Sheehy et al. 2004).

A species' ranging patterns are strongly influenced by its diet and the availability of food resources. Karisoke mountain gorillas forage efficiently by preferentially feeding in areas containing a high biomass of food, especially those foods that are high in protein (Vedder 1984, Watts 1991a, J. Ganas and M. Robbins unpublished data). Additionally, it is hypothesized that a frugivorous diet should affect travel distances and home range size of primates because fruit resources are typically more dispersed in the environment than herbaceous vegetation (Clutton-Brock and Harvey 1977b, Janson and Goldsmith 1995). Because gorillas exhibit such variability in frugivory, they are an interesting species in which to examine the relationship between diet and ranging patterns. Gorillas are not territorial and have overlapping home ranges (Watts 1998b, Tutin 1996, Doran-Sheehy et al. 2004, Ganas and Robbins 2005). Current evidence suggests that gorillas are probably not opportunistic in their frugivory patterns but, rather, fruit "pursuers" that incur costs to forage on more dispersed resources (Rogers et al. 2004, Doran-Sheehy et al. 2004, Ganas and Robbins 2005). Daily travel distance is positively correlated to fruit

consumption at several sites (Goldsmith 1999, Yamagiwa et al. 2003a, Doran-Sheehy et al. 2004, Ganas and Robbins 2005). Across study sites there is a positive relationship between the degree of frugivory (and decreased density of herbaceous vegetation) and daily travel distance, with Karisoke mountain gorillas traveling only approximately 0.5 km/day, Grauer's and western gorillas traveling 1–2 km/day, and Bwindi mountain gorillas being intermediate (Table 19.2). We can predict that gorillas forage efficiently so as to maximize energy gain for distance traveled. Masi (2008) has shown that net energy gain is relatively constant across time, such that the additional travel costs incurred during the frugivory season are compensated for with the increased caloric intake and reduced time spent feeding.

The relationship between frugivory and home range size is less clear (Table 19.2). Karisoke mountain gorillas include only a negligible amount of fruit in their diet, their habitat contains a very high herb density, and they have the smallest home ranges of any population studied. Otherwise, comparable values have been observed among the size of home ranges of all other populations examined.

According to the ecological constraints model, one cost of group living is that as group size increases, the amount of food needed collectively by the group also increases and, therefore, the day journey length and home range size should increase accordingly (Clutton-Brock and Harvey 1977b, Chapman and Chapman 2000b). It has been possible to test this hypothesis with gorillas only at the two mountain gorilla sites because of the lack of more than one group of habituated gorillas at other sites. At both Karisoke and Bwindi, a positive relationship was found between group size, daily travel distance, and home range size (Fossey and Harcourt 1977; McNeilage 1995; Watts 1991a, 1998b; Ganas and Robbins 2005). However, solitary males travel farther per day and have larger home ranges than necessary to sustain a single individual because their movement patterns are also influenced by mate-acquisition strategies (Watts 1994a).

The population density of gorillas is also expected to be correlated with food availability. Studies of western gorillas suggest that their density is highest where particular herbaceous, staple foods are found in high abundance (Rogers et al. 2004, Bermejo 1999). The relationship between the density of seasonal and fallback fruits and gorilla density is not yet known. Additionally, the density of gorillas is likely to be affected by interspecific competition for food resources, particularly with chimpanzees, depending on the degree of niche separation for fruit and herb species (Tutin et al. 1991, Yamagiwa et al. 1996, Stanford and Nkurunungi 2003).

REPRODUCTIVE, DEMOGRAPHIC, AND LIFE HISTORY PARAMETERS

As expected, given their large brain and body size, gorillas have a long maturation time and are long-lived, have long interbirth intervals, and reproduce relatively few times in their lives; but they are faster or shorter in most life history parameters than other ape species (Bentley 1999). Due to the lack of long-term studies of habituated western gorillas, nearly all the information available on reproductive and life history parameters comes from mountain gorillas (but see Yamagiwa and Kahekwa 2001 and Yamagiwa et al. 2003b for Grauer's gorillas). Age/sex classifications typically used for gorillas include infants (0–3.5 years), juveniles (3.5–6 years), subadults (6–8 years), adult females (>8 years), blackback males (8–12 years), and silverback males (>12 years). However, maturation appears to be a slower process for western gorillas, with time of weaning, age at first birth, and becoming a silverback all occurring 1–2 years later (Nowell and Fletcher 2007, Breuer et al. 2009).

Females become sexually active at approximately age 6 and go through a period of adolescent sterility for usually at least 2 years (Watts 1990a, 1991b). Menstrual cycles are approximately 28 days long, and the length of the proceptive period is typically only 1–2 days (Watts 1990a). Based on hormonal analysis, mating behavior predominantly correlates to the time of ovulation (Czekala and Sicotte 2000). Nulliparous females may have small sexual swellings, but parous females show no swellings or other external signs of ovulation (Czekala and Sicotte 2000). Parous females typically conceive after four to six menstrual cycles (Watts 1990a, 1991b). Females show no obvious signs of being pregnant and usually mate while pregnant (Watts 1990a, 1991b). Gestation length is 8.5 months. The average age at first parturition is approximately 10 years (range 8–13). Typically, only one infant is born at a time, although twinning has been observed rarely (Watts 1988). Females go through lactational amenorrhea for approximately 3 years so that the typical interbirth interval is 4 years (Stewart 1988, Robbins et al. 2006, Watts 1991b; but this appears to be slightly longer in other populations [Table 19.3]). If an infant dies before weaning, the female typically produces another offspring 1 year after the infant death (Watts 1991b). Fertility is lower for younger females, improves as they mature, and then declines as they reach older age; but females do not appear to have an extended postreproductive life span (Robbins et al., 2006). Female mountain gorillas that survive to adulthood (60%) have an average reproductive life span of 24 years and produce an average of 4.6 offspring that survive past infancy (Gerald 1995, Robbins et al. 2006).

Differences in ecological conditions are expected to lead to differences in reproductive parameters. In particular, there should be a negative relationship between ecological conditions and both age at first parturition and interbirth intervals (Doran and McNeilage 2001). A comparison among western, Grauer's, and Virunga mountain gorillas did not find a difference in interbirth interval or birthrate (Robbins et al. 2004) (Table 19.3). However, a recent study showed that the interbirth interval was longer and the growth rate significantly lower in Bwindi than for the Virunga mountain gorillas (Robbins, M.M. et al. 2009)

Males experience a longer period until maturity than females and do not go through such well-defined physiological events to mark certain points of maturation. Therefore, it is difficult to determine when males reach sexual maturity. Male gorillas begin the development of secondary sexual characteristics at approximately age 10 years, including a large increase in body size, the silvering of hair on the back and legs, and the formation of a peaked sagittal crest. They are typically considered to be silverbacks by age 12–13 years, but they may not obtain full adult size until 15 years of age (Watts 1991b, Watts and Pusey 1993); Western gorillas may not be considered mature silverbacks until 18 years of age (Breuer et al. 2009).

Whereas female reproductive success is constrained by the costs of gestation and lactation, male reproductive success is limited by access to females. Male gorillas exhibit sociosexual behavior as juveniles and have been observed to start copulating at the age of 8–9 years (Watts 1990a, M. M. Robbins personal observation), long before attaining final

Table 19.3 Reproductive Parameters and Summary of Dispersal Patterns and Transitions of Group Structure Observed in Western Gorillas, Grauer's Gorillas, and Mountain Gorillas

	WESTERN GORILLA	GRAUER'S GORILLA[1]	MOUNTAIN GORILLA
Birth rate (births/adult female/year)	0.19 (Lossi)	Not calculated	0.257[2]
	0.198 (Mbeli)		
Interbirth interval (surviving births only)	4–6 years (n = 6)	4.6 years (n = 9, range 3.4–6.6)	4 years (n = 88)[2]
Infant mortality (to age 3 years)	22.0% (Lossi)	26.1%	27%[2]
	65% (Mbeli)		
Infanticide	Suspected cases	Observed (n = 3)	Observed (n = 6)[7]
Adult male emigration	Common	Common	<50%[4,5]
Female natal/primary transfer	Common	Common	Common[3]
Female secondary transfer	Common	Common	Common[3]
Multimale groups	Rare	Rare	Common[4,6]
Nonreproductive groups	Observed	Not observed	Observed[4,8]
Group formation by solitary male	Observed (n = 2)	Observed (n = 6)	Observed (n = 2)[4]
Group formation by group fissioning	Not observed	Observed (n = 2)	Observed (n = 5)[6]
Group disintegration	Observed (n = 6)	Observed (n = 3, due to poaching)	Observed (n = 5)[6]
Group take-over by outsider male	Not observed	Not observed	Not observed

Sources: Modified from Robbins et al. 2004.
[1] Yamagiwa and Kahekwa 2001.
[2] Robbins et al., 2007.
[3] Sicotte 2001; Robbins A.M. et al., 2009a,b.
[4] Robbins 1995, 2001.
[5] Watts 2000a.
[6] Kalpers et al. 2003.
[7] Watts 1989b.
[8] Yamagiwa 1987a,b.

body size and secondary sexual characteristics. Captive western gorillas are known to sire offspring by 7 years of age (Beck 1982, Kingsley 1988). In one-male groups, the silverback is assumed to sire all offspring (western gorillas, Bradley et al. 2004) because extragroup copulations have been observed only very rarely (Sicotte 2001). In multimale mountain gorilla groups, dominant males mate more than subordinate silverbacks but subordinates do mate, including at the likely time of conception (Robbins 1999, Watts 1990a). Paternity-determination studies at both Karisoke and Bwindi show that subordinate males sire offspring, even while still blackbacks (Bradley et al. 2005, Nsubuga et al. 2008). Variance in male reproductive success is high because some adult males never reproduce and others may be dominant males of large groups for many years (Robbins 1995, Robbins and Robbins 2005). Male lifetime reproductive success is correlated with the dominance tenure of silverbacks and the number of adult females in the group (Robbins 1995, Watts 2000a, Robbins and Robbins 2005). Innovative studies using photogrammetric methods in the field have shown a positive relationship between the size of the sagittal crest of males and their reproductive success,

providing evidence for the hypothesis that crests are a sexually selected trait (Caillaud et al. 2008).

Understanding patterns of mortality is one of the greatest challenges for researchers of long-lived species. As for most large-bodied mammals, mortality rates of gorillas are highest for infants and older adults (Gerald 1995). Infant mortality (to age 3 years) of mountain gorillas is 27% (Robbins et al. 2006), and preliminary data on Grauer's and western gorillas do not appear to differ significantly (Robbins et al. 2004). Approximately 60% of mountain gorillas survive to age 8 years (Gerald 1995). Mortality rates are very low through adulthood. Despite the expectation that adult mortality would be higher for males than females because of the high level of male/male competition, no sex differences in mortality have yet been observed; but this may be because the majority of available data are for gorillas under the age of 12 years, when competition is low (Gerald 1995).

The impact of predation and disease on mortality rates may also vary among gorilla populations. For example, respiratory diseases that are often fatal for mountain gorillas (Watts 1998a, Mudakikwa et al. 2001, Cranfield 2008) have not been reported for western gorillas. Ebola has had devastating effects on certain western gorilla populations (Walsh et al. 2003, Huijbergts et al. 2003, Leroy et al. 2004, Caillaud et al. 2006). Attacks by leopards on mountain gorillas were recorded in the 1950s and 1960s, but none has been noted in recent decades (Schaller 1963, Sholley 1991). Predation by leopards currently is likely to have a greater impact on western gorillas (Fay et al. 1995, Robbins et al. 2004) than on mountain gorillas, which probably have no predators other than humans.

GROUP STRUCTURE AND DISPERSAL PATTERNS

Gorillas live primarily in stable, cohesive social units, although adult males may be solitary. Groups typically contain several adult females, their immature offspring, and always at least one silverback. Groups may be one-male, multimale, or all-male (nonreproductive, containing no adult females) (Stewart and Harcourt 1987; Robbins 1995, 2001; Robbins et al. 2004; Yamagiwa 1987a; Yamagiwa et al. 2003b; Gatti et al. 2004; Levrero et al. 2006).

The variability in the social system of mountain gorillas is due to the following transitions (Robbins 1995, 2001; Yamagiwa 1987a; Yamagiwa and Kahekwa 2001) (Table 19.3). New social groups form when females transfer to lone silverbacks. Such groups remain one-male until male offspring mature into silverbacks, and the group is then multimale. Multimale groups return to a one-male structure if males emigrate, the original adult male dies, or the group fissions. When the silverback of a one-male group dies, the group disintegrates. If a breeding group loses all of its adult females, it becomes an all-male, or nonreproductive, group. All-male groups may also form through a merger of immature males (evicted from a heterosexual group taken over by a new silverback following the death of the previous silverback) and a solitary silverback (Robbins 1995, Yamagiwa 1987a, Gatti et al. 2003, Levrero et al., 2006). All-male groups can become heterosexual if females transfer into them. If a dominant male loses all of his group members, he becomes a lone silverback. Group fissions have been reported only with eastern gorillas (Robbins 2001, Kalpers et al. 2003, Yamagiwa and Kahekwa 2001). Takeovers by outsider males of established groups containing a mature silverback have not been reported for any gorilla population.

The average group size of both western and eastern gorillas is 8–10 individuals (range 2–60+) (Tutin 1996, Parnell 2002a, Kalpers et al. 2003, Yamagiwa et al. 2003b, Gatti et al. 2004). However, groups with more than 20 individuals have been found in several eastern gorilla populations but at only one location of western gorillas, Odzala National Park in the Democratic Republic of Congo (Bermejo 1999, Magliocca et al. 1999, Yamagiwa and Kahekwa 2001, Kalpers et al. 2003). Ecological constraints on western gorillas may prevent groups from surpassing a particular size. The percentage of groups in the Virunga population

containing over 20 gorillas has been increasing in the past three decades (3.5% in 1978–1979, 9% in 1989, and 16% in 2003) (Gray et al. 2009), which may be a reflection of increased protection and reduced poaching (Weber and Vedder 1983, Sholley 1991, Kalpers et al. 2003).

It is hypothesized that females form permanent associations with males as a means to avoid infanticide by extragroup males (Watts 1989b, 2000a; van Schaik and Kappeler 1997; van Schaik 2000a) and to benefit from protection against predators. Males benefit from the long-term associations with females. Males can use the risk of infanticide as an effective strategy to retain mates because of high male/male competition for mates and the long interbirth interval. Most infanticides in mountain gorillas have occurred when the only silverback of a group died and the new male (from outside the group) that took over killed the unweaned offspring (Watts 1989b). In Grauer's gorillas, a silverback was observed to kill three young infants of recent immigrant females, but females with unweaned offspring were also observed without a silverback for extended periods of time and did not suffer from infanticide (Yamagiwa and Kahekwa 2001, 2004). Infanticide has been inferred in western gorillas twice when the infants disappeared following their mothers' transfer after the death of a silverback; however, infants survived in two other cases in similar circumstances (Stokes et al. 2003). Furthermore, Harcourt and Greenberg (2001) used a mathematical model to show that females traveling unaccompanied by a silverback would be at three times greater risk of infanticide than those in social groups. Infanticide does not need to occur in all possible cases in order for it to be an evolutionarily stable strategy (van Schaik 2000a).

Gorillas are one of only a few primate species in which both males and females may be philopatric or disperse (Harcourt et al. 1976; Tutin 1996; Harcourt and Stewart 2007a,b). Females transfer directly to a solitary male or to another group during interunit encounters (Harcourt et al. 1976, Harcourt 1978). Female primary (natal) and secondary dispersals have been observed in mountain, Grauer's, and western lowland gorillas (Harcourt et al. 1976; Sicotte 2001; Yamagiwa and Kahekwa 2001; Stokes et al. 2003; Robbins et al. 2009a,b). Female transfer is considered to be an important method of female mate choice (Sicotte 2001). Female primary transfer may occur as early as age 5.5 (Sicotte 2001, Watts 1991b). Females may transfer several times in their lives, but because of the risk of infanticide by extragroup males, the window of opportunity for females to disperse is quite narrow, limited to the period of 3–6 months per 4-year interbirth interval when the female is neither pregnant nor lactating (Watts 1989b, Sicotte 2001; but see Sicotte 2000 and Yamagiwa and Kahekwa 2001 for cases of transfer with unweaned infants). Transfer possibilities are further dependent on the occurrence of intergroup encounters, which may be influenced by male strategies to seek out or avoid other groups as well as the use of clumped resources such as *bais* or fruit tree patches. The average rate of intergroup encounter per group varies among groups and populations (0.58–4.0 per month), but in some cases many months may pass without any intergroup encounters, which would further constrain the possibility for female transfer (Sicotte 2001, Robbins and Sawyer 2007).

Currently, more is known about what group conditions influence female transfer decisions than which male traits females prefer (e.g., size, sexually selected characteristics). Female western gorillas appear to prefer smaller groups, presumably because of reduced feeding competition (Tutin 1996, Stokes et al. 2003). In contrast, female mountain gorillas, which face low feeding competition, show no group size preference, yet they may prefer multimale groups over one-male groups and solitary males (Watts 2000a, Robbins et al. 2009a). This is likely due to females in multimale groups deriving greater protection against infanticide than those in one-male groups because in the event of one silverback death in a multimale group, the other silverback can assume leadership and prevent group disintegration and take-over by an external male (Robbins 1995, Watts 2000a). Females may also transfer to avoid inbreeding or to increase the genetic variability among offspring (Sicotte 2001). An analysis of the Bwindi mountain gorilla population found that there was genetic structure among the females that could be explained not only by distance (e.g., females primarily dispersing to neighboring groups) but also by the plant community composition, suggesting that dietary preference may influence female dispersal decisions (Guschanski et al., 2008).

Subordinate silverbacks emigrate to become solitary males at approximately 12–18 years of age (Robbins 1995, Stokes et al. 2003, Yamagiwa and Kahekwa 2001). It is unclear whether males voluntarily emigrate or if they are forcibly evicted, but dispersal decisions are presumably influenced by male/male competition, likelihood of mate acquisition within the natal group versus in neighboring areas, and inbreeding avoidance. Interestingly, in western gorillas, there are reports of immature males voluntarily immigrating into heterosexual groups (Gatti et al. 2004, Robbins et al. 2004). The proportion of males that emigrate is higher in western gorillas than in mountain gorillas, and correspondingly, multimale groups are rare among western gorillas (Tutin 1996, Magliocca et al. 1999, Parnell 2002a, Gatti et al. 2004, Robbins et al. 2004) (Table 19.3) and Grauer's gorillas (Yamagiwa et al. 1993, Yamagiwa and Kahekwa 2001) but common in mountain gorillas (8%–50% of groups) (McNeilage et al. 2001, Gray et al. 2009). These comparisons suggest that male western gorillas may typically attain higher reproductive success by dispersing and trying to form groups than by queuing for dominant positions within multimale groups, as is the case in mountain gorillas (Robbins 1995, Watts 2000a, Robbins and Robbins 2005). This is likely to be influenced by how females distribute themselves in groups. Both male eastern and western gorillas have low success rates in acquiring females after emigration; males have been observed as solitary for 5 years or more (Robbins 1995, Watts 2000a, Robbins et al. 2004). The rate of group disintegration appears to be higher in western gorillas than mountain gorillas, and group fissions have not been observed in western gorillas; these findings are to be expected given the rare occurrence of multimale groups in western gorillas (Robbins et al. 2004) (Table 19.3).

INTERGROUP INTERACTIONS

In most cases, intergroup interactions do not appear to be related to defending a group's range or to food resources (but see Bermejo 2004). They are an important time for female choice and male/male competition because these interactions are the only time that female transfers occur (Watts 1989b, Sicotte 2001, Robbins 2003, Robbins and Sawyer 2007). However, female choice may be limited by males that may herd females away from a neighboring group to prevent transfers (Sicotte 1993). Solitary males have been observed to pursue groups for several days at a time in both mountain and western gorillas, presumably in an attempt to acquire mates (Watts 1994a, Tutin 1996, Bermejo 2004). The frequency of intergroup encounters is likely to vary depending on the density of groups and of solitary males within a particular area.

During intergroup interactions, silverbacks typically perform chest-beating displays at each other, which sometimes elevates to contact aggression. However, interactions may also be peaceful, with group members intermingling. Sicotte (1993) found that the intensity of intergroup encounters was positively related to the number of potential migrant females. Caillaud et al. (2008) showed that the number of females present positively influenced the number of agonistic displays given by males but not the intensity. Established groups may be less tolerant of solitary males because they may be more likely to take risks to acquire females (Robbins and Sawyer 2007). Interestingly, studies of western gorillas have reported that intergroup encounters are often peaceful and involve aggression less frequently than has been observed in mountain gorillas (Parnell 2002b, Doran-Sheehy et al. 2004, Bermejo 2004). Bradley et al. (2004) hypothesized that the lower frequency of aggression during encounters observed in western gorillas may be due to males interacting differently with relatives and nonrelatives occupying neighboring home ranges. However, intergroup male/male competition occurs in western gorillas, based on observations of wounds on silverbacks that were likely to have been caused by other silverbacks (lacerations and bite marks on the head, neck, and back) (Parnell 2002b). Silverbacks have died as a result of serious injuries incurred during encounters in both western and mountain gorillas (Tutin 1996, Robbins 2003). Additional studies at all locations are needed to understand how the level of aggressiveness during intergroup encounters is related to the following variables: relatedness between males, number of silverbacks and potential migrant females in each group, familiarity between groups (e.g., frequency of encounters), competition for food resources, and presence of researchers around unhabituated gorillas, which may inhibit normal behavior.

SOCIAL RELATIONSHIPS

Many current socioecological models predict that the types of social relationship exhibited by primates are most influenced by three variables: distribution of food resources, predation risk, and risk of infanticide (van Schaik 1989, Sterck et al. 1997, Koenig 2002, Isbell and Young 2002). The extensive research on the mountain gorillas of Karisoke Research Center, Rwanda, has contributed significantly to the development of these models (Watts 2003, Harcourt and Stewart 2007a,b). Given the known variation in ecological conditions and dietary patterns of gorillas, the opportunity exists to use them as a test of these models. Unfortunately, data on social relationships in other populations of gorillas are minimal at this time; therefore, the focus here is on what is known from Karisoke, and comparisons with other populations are made when possible.

Female/Female Social Relationships

Overall, at the Virunga Volcanoes, female mountain gorillas have very weak social relationships with one another (Harcourt 1979a; Stewart and Harcourt 1987; Watts 1991c, 1994a,b, 2001, 2003), which are markedly different from those observed among many other primate species. They are classified as being "non-female-bonded" (Wrangham 1979b, van Schaik 1989) and "dispersal egalitarian" (Sterck et al. 1997). Food resources are abundant and evenly distributed, which leads to very low levels of contest competition (Watts 1984, 1985a, 1994b). Therefore, there are few benefits of associating with other female kin, and the social and ecological costs of transferring to new habitats are low (Watts 1990b). This allows for female transfer. As a result, females are typically in social groups containing unrelated individuals, although mother/daughter and sister pairs are not uncommon because females do not always transfer and related females may transfer into the same group (Watts 1996, 2001; see also Bradley et al. 2007 for western gorillas). Levels of aggressive and affiliative interactions among female/female dyads are highly variable and may change over time (Stewart and Harcourt 1987, Watts 2001). The long-term studies and variable group compositions at Karisoke have enabled researchers to conduct detailed studies on many factors influencing female social relationships.

Despite low levels of feeding competition, aggression among females most commonly occurs while feeding; but it also occurs during resting bouts and traveling, when females may be intolerant of close proximity to one another (Watts 1994b). The majority of aggressive interactions among females do not have a clear winner; they are undecided because the recipient ignores the action or retaliates. Submissive behavior (cowering, backing away, or an appeasement vocalization, "grumbling") is rarely exhibited among females (Watts 1994b). Dominance relationships are determined according to patterns of displacement and avoidance (approach–retreat interactions) and are not based on any formal submissive behavior, as is used in many other primate species. While females do not displace each other often, resulting in dominance relationships that are weaker than what is observed in many other species (Harcourt 1979a; Harcourt and Stewart 1987, 1989; Watts 1985a, 1994b), these relationships can be consistent and stable for more than 15 years (Robbins et al. 2005). Dominance rank is also positively correlated with some measures of female reproductive success (Sterck et al. 1997, Robbins et al. 2007). Additionally, aggression within individual dyads is frequently bidirectional (both females of a dyad initiate aggression against each other) so that females cannot be placed into linear hierarchies using all agonistic behavior (Watts 1994b). These relationships, described as being egalitarian (Sterck et al. 1997; Watts 2001, 2003), may be explained by two factors which show that there is little benefit of targeting aggression toward particular individuals (Watts 2001). First, there are low gains from agonistic behavior in an environment with highly abundant food resources (Watts 1985a, 1994b). Second, males frequently intervene in aggressive contests between females and do not clearly support either female (*control intervention*) (Harcourt and Stewart 1987, 1989; Watts 1994b, 1997). The fact that relationships among females are of little value is further emphasized by the very low rates of reconciliation following conflicts (Watts 1995a).

Even in an environment of high food availability, scramble competition is expected to increase as group size increases. As the number of females per group increased, the rates of aggression among females increased but not significantly (Watts 1985a, 2001). While females that immigrate into average-sized groups do not receive high levels of aggression from resident females, five females that immigrated into a large group within a few months of each other received high levels of harassment from the resident females (Watts 1991c, 1994c). However, this may have occurred in this case because several related resident females often formed coalitions against the immigrants or because the dominant male may not have been able to effectively intervene in aggression among so many females. Aggression among females in contexts other than feeding, for example, during rest sessions and traveling, indicates that they may be competing for male services, in particular protection against predators and extragroup males (Watts 1994b, 2001). A male will be limited in his ability to maintain close proximity to all females as group size exceeds a certain number.

Affiliative interactions, particularly grooming and resting in close proximity, occur at higher rates among maternally related individuals than unrelated individuals, with intermediate values for putative paternally related females (Watts 1994c). Grooming is typically reciprocal among dyads, but this is due to the higher occurrence of grooming among relatives. In general, female relatives (mother/daughter and sister dyads) most commonly have "good" relationships with each other (higher than median values of affiliative interactions and lower than median values of aggressive interactions for both partners) and unrelated females typically have "bad" relationships with each other (lower than median values of affiliative interactions and higher than median values of aggressive interactions), though sometimes they do have good relationships (Watts 2001).

Because fruit is usually distributed in a clumped fashion in time and space, the increase in frugivory experienced by all other gorilla populations besides those at the Virunga Volcanoes is predicted to lead to higher levels of within-group contest competition and differentiated female/female social relationships that include linear agonistic dominance hierarchies (Doran and McNeilage 1998, 2001). The lack of habituated gorillas at most study sites currently limits tests of this hypothesis to two locations. Female mountain gorillas in Bwindi Impenetrable National Park, Uganda, exhibited higher rates of aggression when foraging on fruit versus herbaceous vegetation; but overall, they exhibited weak dominance relationships, a high degree of bidirectionality in aggression, and low rates of affiliative behavior, as has been observed in Karisoke mountain gorillas (Robbins 2008). Bwindi also has abundant quantities of evenly distributed herbaceous vegetation (Ganas et al. 2004, Nkurunungi 2004), and the gorillas spend only approximately 10% of their foraging time eating fruit, which may limit the effect that frugivory has on their social relationships (Robbins 2008). Observations of western gorillas at Mbeli Bai, Democratic Republic of Congo, also failed to show a noticeable difference in social relationships among females from those observed at Karisoke (Stokes 2004). However, this study provides only a limited view of their social behavior because the gorillas were observed only in a bai (swampy clearing) feeding on abundant, evenly distributed aquatic vegetation and the observations do not include any frugivorous behavior and are not representative of the majority of western gorilla daily activity (a group typically spends only 1% of its time in a bai) (Magliocca et al. 1999, Parnell 2002b). Further detailed studies of habituated western gorillas are necessary to understand the influence that frugivory has on their social relationships.

Male/Female Social Relationships

Social relationships among males and females are the foundation of gorilla groups (Harcourt 1979b; Fossey 1983; Stewart and Harcourt 1987; Watts 1992, 2003; Harcourt and Stewart 2007a,b). These relationships reflect the costs and benefits of the reproductive strategies of both males and females. Females benefit from the protection against predators and infanticide by extragroup males, and males benefit by permanent association with females as a long-term mating strategy. Male/female social relationships vary depending on the reproductive status of females, the residency status of females, the number of males in the group, and kinship (Harcourt 1979a; Watts 1992, 2003). Particular male/female dyads may coinhabit a social group for 10 or more years. Male/female relationships may influence female transfer patterns (Watts 1992, 1996, 2003) and, if a multimale group fissions, with which male each female remains (Robbins 2001, Watts 2003).

The high level of affiliation between males and females is emphasized by proximity patterns. Females may spend up to 20% of their feeding time and over 50% of resting time within 5 m of the dominant silverback (Watts 1992). Females increase the amount of time they spend in close proximity to the silverback upon the birth of a new infant (Harcourt 1979b). Females are more responsible than males for maintaining proximity to the silverback in one-male groups (Harcourt 1979b), but in multimale groups males are more responsible for maintaining proximity to females when they are proceptive (Sicotte 1994). Grooming among adult gorillas is most common between males and females. However, there is high variability among groups in whether the silverback does more grooming of adult females or vice versa (Harcourt 1979b, Watts 1992).

Silverbacks are clearly dominant over adult females based on approach–retreat interactions, directionality of aggressive behavior, and submissive behavior (Harcourt 1979b; Watts 1992, 1994b). The majority of aggression by males toward females consists of mild vocalizations and displays (e.g., chest-beating displays) and only rarely results in physical injuries (Watts 1992; Robbins 2008, 2009). Females frequently "grumble" after receiving aggression from males, especially following displays; and this is the only formal submissive behavior observed in gorillas (Watts 1994b). Females frequently reconcile with males following aggressive encounters (Watts 1995a).

Given that males are clearly larger in size and dominant over females, why do they need to frequently exhibit aggression toward females? This aggression is not predominantly due to feeding competition, but it likely serves as a means of reasserting their protective abilities as well as courtship aggression or sexual coercion (Robbins 2009). In fact, only a minority of agonistic behavior that silverbacks direct toward females occurs in the context of feeding, and more than 60% may be classified as courtship aggression (displays and interventions in sexual interactions) (Watts 1992). Additionally, the levels of aggression directed toward individual females vary depending on their reproductive status, further emphasizing the role of courtship aggression. Recent immigrant females receive higher levels of aggression from males than do resident females in both one-male and multimale groups (Harcourt 1979b, Sicotte 2000, Watts

1992). While in one-male groups, the silverback does not increase the rate of aggression toward females when they are sexually active (Harcourt 1979b); but in multimale groups, both dominant and subordinate males do (Robbins 2003, 2009).

Females in multimale groups will commonly mate with more than one male, which may be a result of sexual coercion by the males and/or female choice and may serve to confuse paternity (Robbins 1999, 2003, 2009). Females may improve infanticide protection by blurring paternity (e.g., Pradhan and van Schaik 2008, Sicotte 2001). Subordinate males may use affiliative behavior such as grooming to influence female choice (Watts 1992). Dominant males are more likely than subordinate males to mate with cycling and pregnant females, whereas subordinate males are more likely to mate with subadult females (Robbins 1999). Mate guarding may occur in multimale groups but not in one-male groups (Harcourt et al. 1980, Sicotte 1994, Watts 1996).

Observations of male/female relationships in a multimale group in Bwindi Impenetrable National Park largely corroborate those at Karisoke (Robbins 2008, 2009), which is unsurprising given that reproductive strategies, not ecological conditions, are likely to have a larger influence on these social relationships. At Mbeli Bai, males exhibited higher rates of aggression toward females than was observed among females, but most of it consisted of mild aggression and could be considered courtship aggression (Stokes 2004). Again, until more detailed studies of western gorillas are conducted, it is not possible to draw conclusions about the differences in male/female social relationships between the species.

Male/Male Social Relationships

In many primate species, male/male social relationships are very weak and characterized by high levels of competition for mates (van Hooff and van Schaik 1994, Kappeler 2000b). Gorillas are not an exception to this pattern. In multimale groups, males may be related as father/son or half-brothers (maternal or paternal), but in some cases they may be unrelated (Robbins 1995, Bradley et al. 2005, Nsubuga et al. 2008). The coexistence of unrelated males may be caused by immigration of immature or blackback males into social groups (rarely) or lower levels of reproductive skew in multimale groups that lead to cohorts of young being sired by more than one male.

In the majority of social groups that are one-male, the only adult male present has no other males to form social relationships with except for maturing males. Silverback/blackback relationships are weak, based on low rates of close proximity and affiliation and unidirectional aggression from the silverback to the blackback. Silverbacks are always dominant over blackbacks, which spend considerable amounts of time on the periphery of the group, perhaps to

avoid conflict with the silverback or adult females. More affiliative relationships may exist between males that are related or more familiar with each other, which may influence dispersal decisions of young adult males (Harcourt 1979c, Harcourt and Stewart 1981). Unfortunately, male behavior shortly before emigration has not been systematically studied to determine if dispersal is related to increased rates of aggression and intolerance by the dominant male. If dominant males want to evict subordinate individuals, it would seem logical that they do it before the younger males become formidable competitors.

Heterosexual social groups that contain at least two fully mature silverbacks may exist for 10 or more years in mountain gorillas (Robbins 1995, 2001). Among males, dominance hierarchies are obvious, but rank does not always positively correlate with age (Harcourt 1979c, Sicotte 1994, Watts 1992, Robbins 1996). Silverbacks spend little time in close proximity to each other (<5 m), and they rarely affiliate with each other (Sicotte 1994, Robbins 1996). Rates of aggressive behavior between males depend on a variety of factors, including the number and reproductive status of females, the age of males, and the stability of their relationships. Aggression between males may increase when females are sexually active but not always (Sicotte 1994, Robbins 2003). In some cases, dominant males increase aggression toward subordinates; in other cases, subordinate males direct more aggression toward dominant males, or there may be an increase in either direction. Both dominant and subordinate males may harass each other during mating, but dominant males are more successful at stopping copulations than subordinate males (Robbins 1999). This variability may be due to the stability of the dominance relationship among the males and/or the reproductive status of the female (subadult females, cycling adult females, pregnant females). Reconciliation between adult males has not been observed (Watts 1995a). In general, when males have stable dominance relationships, they appear to coexist predominantly through avoidance or tolerance rather than frequent, high levels of aggression or by forming affiliative relationships. One benefit of a multimale structure is that males will cooperate during intergroup encounters against extragroup males (Sicotte 1993; Robbins 2001; Watts 1989b, 2000a, but see Robbins and Sawyer 2007).

Silverbacks may be dominant for 10 or more years. Usurpation of the alpha male by younger silverbacks in multimale groups has been observed three times in three decades at Karisoke (Watts 1990a, 1992; Sicotte 1993; Robbins 1996); however, the transitional process was not well documented, so the strategies that younger males use to attain alpha status or how dominant males resist are not yet understood. Observations in Bwindi suggest that usurpation may involve a lengthy process (several years) of agonistic encounters, rather than a quick turnover (M. M. Robbins personal observation). Deposed males are usually not evicted and typically retain positive relationships with adult females. In Bwindi the deposed male was effectively evicted following an injury that prevented him from traveling with the group. If deposed males sire offspring and reduce the direct fitness of the dominant male, the question remains of why these males are tolerated in the group. Eviction may be costly to the dominant male, and/or the benefits of retaining a multimale group status (e.g., more attractive to females, reduced risk of infanticide, and additional defense during intergroup encounters) may outweigh the costs.

Relationships among males in all-male groups are markedly different from those in heterosexual groups, presumably because the absence of females reduces competitive interactions (Yamagiwa 1987b, Robbins 1996). Dominance relationships are still apparent among the males, but they may be less clear within age classes, especially with blackbacks. While rates of aggression may be higher than those observed in heterosexual groups, most of the aggression consists of vocalizations and displays, and lower incidences of wounding caused by serious fights are observed (Robbins 1996). Males exhibit higher rates of grooming and playing and spend more time in close proximity than in heterosexual groups. Sociosexual behavior occurs but at low rates. Despite the lack of reproductive opportunities, all-male groups may be a better alternative for males than being solitary because they provide a social setting in which males can gain experience in aggressive and affiliative interactions (Robbins 1996, Levrero et al. 2006).

Development and Social Relationships of Immature Gorillas

Gorillas have an extended period of immaturity, during which time they develop their social skills. Infants are completely dependent on their mothers at birth and are unlikely to survive the death of the mother until they reach the age of 2–4 years because of the prolonged lactation period as well as their dependence for locomotion. The first 6 months of an infant's life are spent in constant contact with the mother, and then slowly the infant begins to spend increasing amounts of time farther away so that by the age of 30 months it spends only approximately 50% of its time within 5 m of the mother (Fletcher 2001, Nowell and Fletcher 2007). Mothers may continue to serve as an important social partner for offspring through immaturity, and relationships with other adult females are usually weak because female dispersal limits the number of female kin within a social group (Fossey 1979, Fletcher 1994).

Play likely serves as one of the most important behaviors for the development of social skills in gorillas. Play begins typically when infants are about 9 months of age, peaks in frequency during juvenility, and then decreases during adolescence (Fletcher 1994). Sex differences in play exist, with males playing more than females, especially during adolescence (Watts and Pusey 1993, Fletcher 1994).

While silverbacks' primary contribution to immature gorillas is to provide protection against infanticide from outsider males (Watts 1989b), there are other positive benefits to silverback/immature associations. As the time spent near the mother decreases in late infancy, the time spent in close proximity to the silverback increases. Immature gorillas are attracted to the silverback as the focal point of the group and spend considerable time in close proximity to him even when their mothers are not nearby (Watts and Pusey 1993, Fletcher 1994, Stewart 2001). Juveniles groom silverbacks more than vice versa in most cases (Stewart 2001). Silverbacks frequently intervene during aggressive conflicts involving immatures and usually support the younger opponent, which serves to protect immatures from high levels of aggression from older individuals (Harcourt and Stewart 1987, Watts 1997, Stewart 2001). Silverbacks may increase affiliative behavior toward young that have lost their mothers through death or dispersal, including grooming and co-nesting (Watts and Pusey 1993, Stewart 2001). Presumably, silverbacks benefit from their relationships with immatures as a form of paternal investment. Genetic studies in multimale groups will enable us to test if males have differential relationships with young that are their offspring versus those that are not.

CONSERVATION

All subspecies of gorillas are considered to be endangered (IUCN Species Survival Commission 2000). However, the population size and area inhabited vary between subspecies, country, and protected area (Table 19.1). The most critically endangered subspecies are the Cross River gorilla, with only approximately 200–250 individuals remaining in nine small, isolated pockets of forest (Oates et al. 2003), and mountain gorillas, with approximately 700 remaining in two isolated populations (Virunga Volcanoes and Bwindi Impenetrable Forest) (Gray et al. 2009, Guschanski et al. 2009). However, because of the small populations of Cross River and mountain gorillas, the loss of even a few gorillas represents a significant decline in the population. The number of Grauer's gorillas is believed to have declined dramatically in recent years, with current estimates ranging widely between 5,000 and 28,000 gorillas remaining (Mehlman 2008). The decline has largely been due to war and political instability in the area, which has also prevented an accurate assessment of population numbers. A survey was conducted in the highland sector of Kahuzi-Biega National Park in 2000 that indicated the population had declined by half since the mid-1990s (from 250 to 130 gorillas) (Plumptre et al. 2003). The most numerous subspecies is the western gorilla, but population estimates vary hugely (Plumptre et al. 2003, Walsh et al. 2003) so that the most conservative estimate to give may be fewer than 200,000. However, populations of western gorillas have been experiencing a dramatic decline, perhaps by as much as half between 1983 and 2000, and could be faced with extinction within the next decade (Walsh et al. 2003).

The current population estimates of western gorillas and Grauer's gorillas should be viewed with caution because the survey methods used have many biases and lack precision, which result in huge error estimations that render the final numbers not particularly reliable (Devos et al. 2008; Morgan et al. 2006; Kuehl et al. 2007, 2008). Accurate population estimates are necessary in order to monitor the changes in populations and assess the effectiveness of conservation strategies. Combining techniques such as genetic analysis or camera traps with the more traditional survey methods will improve the accuracy of population estimates (Kuehl et al. 2008, Guschanski et al. 2009).

The major threats to gorillas are habitat destruction, poaching, disease, and political instability; but there is variation in the degree to which each of these threats impacts each gorilla population (Plumptre and Williamson 2001, Plumptre et al. 2003, Walsh et al. 2003, Caldecott and Miles 2005). While conservation issues are largely similar across gorilla habitats, there is variation in the degree to which each of these threats impacts each gorilla population. Habitat destruction and degradation is largely caused by logging and currently has the greatest impact on western gorillas. For both Cross River and mountain gorillas the remaining habitats are already tiny and are surrounded by some of the highest human population densities found in agricultural areas (e.g., 300–600 people/km² around the Virunga Volcanoes).

Gorillas may be poached intentionally for bushmeat, as trophies, or for the pet trade or unintentionally caught in snares set for other animals such as antelope. The bushmeat crisis is greatest for western and Grauer's gorillas because it is usually exacerbated by the presence of logging companies and most people that live in their habitats have no taboos against eating great apes. While direct poaching of mountain gorillas for the pet trade was a serious problem in the 1970s, it declined dramatically through the 1980s and 1990s; but unfortunately there were several incidents again in the early 2000s. Getting caught in antelope snares is still a problem for mountain gorillas, especially at the Virunga Volcanoes.

The risk of disease to apes intensifies as there is increasing contact between humans, livestock, and wildlife. This may be caused by an increase in the number of humans inhabiting areas that neighbor ape habitat and by the presence of researchers and tourists who come into close contact with the gorillas (Butynski and Kalina 1998, Cranfield 2008). Ebola has caused a rapid decline in certain western gorilla populations (Walsh et al. 2003, Huijbregts et al. 2003, Leroy et al. 2004, Caillaud et al. 2006, Bermejo et al. 2006). Respiratory disease outbreaks that may have been caused by humans have occurred in mountain gorillas (Mudakikwa et al. 2001, Cranfield 2008).

Lastly, war and political instability threaten gorillas because conservation activities are increasingly difficult to conduct under such circumstances. Local people may be forced to rely on protected areas for resources such as food, water, firewood, and/or shelter; and military groups may use protected areas as hiding places. Amazingly, the mountain gorilla population of the Virunga Volcanoes actually increased in size during the 1990s despite intense instability and illegal activity in the protected areas (Kalpers et al. 2003, Gray et al. 2009). Unfortunately, the situation for Grauer's gorillas is likely to be bleak, given the continuing political instability in eastern Democratic Republic of Congo and the associated impact on gorilla habitat.

What can be done to ensure that gorillas do not go extinct? Currently, there are many conservation activities that focus on wild gorillas and their habitat, but clearly efforts must be expanded given the evidence of declining populations. Most importantly, law enforcement is necessary to reduce the bushmeat trade and ensure the protection of gorilla habitats. While research alone will certainly not prevent the extinction of gorillas, it can provide information on their ecology, habitat requirements, demographic patterns, and social behavior and can be applied to creating informed conservation management decisions. Given the widespread interest in charismatic gorillas, research results can be used in educational programs and may also provide information for the media, which in turn may generate the funding necessary for the protection of gorillas. Reducing the threat of disease will require a better understanding of host/reservoir dynamics, preventative measures programs, and perhaps even vaccination of wild gorillas. Integrated conservation and development programs may improve people's attitudes toward protected areas and can be a useful conservation strategy if done concurrently with law enforcement and other measures. Effective conservation of gorillas and their habitat will ultimately depend on the will and efforts of both local and the international communities.

20

Chimpanzees and Bonobos
Inter- and Intraspecies Diversity
Rebecca M. Stumpf

1. How do chimpanzees and bonobos differ behaviorally and morphologically? How are they similar?
2. What is the nature of the male/female sexual and social relationships in chimpanzees and in bonobos?
3. What might explain the observed behavioral differences within and between the two species?

INTRODUCTION

Field research on wild chimpanzees (*Pan troglodytes*) and bonobos (*Pan paniscus*) has been conducted since the 1960s and 1970s. Until recently, 1980, virtually everything that was known about wild chimpanzee natural history and behavior came from a handful of long-term field sites, largely in East Africa. Far less was known about wild bonobos. As more and more chimpanzee and, to a lesser extent, bonobo communities are studied in captivity and across equatorial Africa (Table 20.1, Figure 20.1), a far greater understanding is emerging of the substantial diversity both within and between the two species that comprise the genus *Pan*. Such variation within *Pan* has led to an ongoing debate surrounding the distinctiveness of chimpanzees and bonobos, as well as the remarkable behavioral variation among chimpanzee subspecies. Researchers are continuing their efforts to determine which combination of ecological, cultural, genetic, and/or demographic factors are the primary bases for these differences.

The principal goal of this chapter is to convey what is presently known of the taxonomy, morphology, ecology, and social and sexual behaviors of wild chimpanzees and bonobos and to examine the substantial range of variation both within and between the two species. This chapter is primarily concerned with studies of wild chimpanzees and bonobos. Since tool use and cognition, communication, and culture are reviewed extensively elsewhere in this volume (see Chapters 42, 43, and 44), these areas are not emphasized here.

CHIMPANZEE AND BONOBO RESEARCH: A HISTORICAL PERSPECTIVE

Research began on wild chimpanzees in the 1960s when Jane Goodall initiated a field study at Gombe Stream Reserve, Tanzania. During the same year, a Japanese team led by Toshisada Nishida began field research on chimpanzees in the Mahale Mountains National Park, at a site 200 km from Gombe (see Figure 20.1), both near the far eastern range of chimpanzee geographic distribution. This initial glimpse into chimpanzee daily life, diet, and intra- and intergroup social interactions laid the foundation for many subsequent studies of wild chimpanzee behavior.

In the late 1970s, Christophe Boesch (in the Taï National Park, Ivory Coast) and Yukimaru Sugiyama (in Bossou, Guinea) began field research on chimpanzees in West Africa. Although many aspects of behavior appeared to be relatively similar across all chimpanzee sites, substantial differences in behavior emerged between the western and eastern populations (Sugiyama and Koman 1979, Boesch and Boesch 1990). During the last two decades, studies in East Africa, such as those in Budongo and Kibale (containing the Kanyawara and Ngogo sites), and West Africa (e.g., Mt. Assirik and Fongoli in Senegal; see Figure 20.1) have added further insight into the complexity and diversity of chimpanzee grouping patterns, behavior, and feeding ecology, even within the same subspecies. At present, little is known about the central (*P. t. troglodytes*) and Nigerian-Cameroon (*P. t. ellioti*) subspecies, but emerging data on these two subspecies provide even more evidence of behavioral variation within chimpanzees (e.g. Fowler & Sommer, 2007; Sanz et al., 2009).

Bonobos have been studied to a far lesser extent than chimpanzees. This is largely due to their limited distribution and remote location, along with a civil war in Congo which prohibited continuous research throughout most of the 1990s. Current knowledge of bonobo behavior and ecology comes largely from observations in Lomako, Wamba, and Lui Kotal, Congo (see Figure 20.1), as well as from

Table 20.1 Some Chimpanzee and Bonobo Field Projects

SPECIES	FIELD SITE	COUNTRY	PRINCIPAL RESEARCHER(S)	PROJECT DATES	REFERENCES
Chimpanzee (*Pan troglodytes*)					
P. t. schweinfurthii	Gombe National Park[1]	Tanzania	J. Goodall, A. Pusey	1960–present	Goodall 1986; Pusey et al. 1997, 2008
P. t. schweinfurthii	Mahale Mountains National Park[1]	Tanzania	T. Nishida	1965–present	Nishida 1990a,b; Nishida et al. 2009
P. t. schweinfurthii	Ugalla	Tanzania	J. Moore	1985+, intermittent	Moore 1996
P. t. schweinfurthii	Kanyawara, Kibale National Park[1]	Uganda	R. Wrangham M. Muller	1987–present	Wrangham et al. 1996, Muller et al 2007
P. t. schweinfurthii	Ngogo, Kibale National Park[1]	Uganda	J. Mitani, D. Watts	1994–present	Mitani and Watts 2005, Mitani 2009, Watts et al., 2006, Watts 2007
P. t. schweinfurthii	Budongo Forest Reserve[1]	Uganda	V. Reynolds	1990–present	Reynolds 1992; Newton-Fisher 2002a,b; Wallis 2002
P. t. schweinfurthii	Semliki-Toro[1]	Uganda	K. Hunt	1996–present	Hunt and McGrew 2002; Hunt et al. 1999
P. t. schweinfurthii	Kalinzu Forest	Uganda	C. Hashimoto, T. Furuichi, Y. Tashiro	1995–present	Furuichi et al. 2001
P. t. schweinfurthii	Bwindi Impenetrable Forest	Uganda	C. Stanford	1996–present	Stanford and Nkurunungi 2003
P. t. schweinfurthii	Ishasha River	Uganda, Democratic Republic of Congo	J. Sept	1989+, intermittent	Sept 1998
P. t. schweinfurthii	Kahuzi-Biega National Park	Democratic Republic of Congo	J. Yamagiwa, A. Basabose	1991–present	Basabose 2002
P. t. troglodytes	Nouabalé-Ndoki National Park	Congo (Brazzaville)	S. Koruda	1989–present, intermittent	Kuroda 1992; Kuroda et al. 1996
P. t. troglodytes	Lopé Reserve	Gabon	C. Tutin	1984–present	Tutin and Fernandez 1993
P. t. troglodytes	Goualougo Triangle[1]	Democratic Republic of Congo	C. Sanz, D. Morgan	1999–present	Morgan and Sanz 2003; Sanz and Morgan 2007
P. t. troglodytes	Ngotto Forest	Central African Republic	R. and D. Fouts	2000–present	Hicks et al. 2005
P. t. troglodytes	Ntonga, Dja Reserve	Cameroon	van Elsacker	2002–present	Dupain et al 2004
P. t. troglodytes	Loanga National Park	Gabon	C. Boesch	2002–present	Boesch et al. 2007
P. t. troglodytes	Ebo Forest	Cameroon	B. Morgan	2002–present	Morgan and Abwe 2006
P. t. ellioti	Gashaka-Gumti National Park	Nigeria	V. Sommer	2002–present	Sommer et al. 2004; Fowler and Sommer 2007
P. t. verus	Taï National Park[1]	Ivory Coast	C. Boesch	1979–present	Boesch and Boesch-Achermann 2000; Boesch 1994; Boesch et al. 1994
P. t. verus	Bossou National Park, Mt. Nimba	Guinea, Ivory Coast	Y. Sugiyama, T. Matsuzawa	1976–present	Sugiyama 1989, 2004, Koops et al. 2007
P. t. verus	Mt. Assirik, Fongoli	Senegal	W. McGrew, J. Pruetz	1976–1979, 2001–present	McGrew et al. 1988; Pruetz et al. 2002, 2008
Bonobo (*Pan paniscus*)					
P. paniscus	Wamba Forest[1]	Democratic Republic of Congo	T. Kano	1973–present, intermittent	Kano 1992

Table 20.1 *(cont'd)*

SPECIES	FIELD SITE	COUNTRY	PRINCIPAL RESEARCHER(S)	PROJECT DATES	REFERENCES
P. paniscus	Lomako Forest[1]	Democratic Republic of Congo	A. and N. Badrian, R. Susman, G. Hohmann F. White	1974–present, intermittent	Susman 1984; Doran 1993; Hohmann and Fruth 1993, 2000, 2002; White 1988
P. paniscus	Lukuru Forest	Democratic Republic of Congo	J. Thompson	1992–present	Thompson 2002, 2003
P. paniscus	Salonga National Park	Democratic Republic of Congo	G. Hohmann, B. Fruth, J. Eriksson	2002–present	Hohmann and Fruth 2002, 2007

[1] Denotes a long-term field site (more than 10 years of continuous, habituated chimpanzee or bonobo research).

Figure 20.1 Map of equatorial Africa showing distribution of chimpanzee subspecies and bonobos. Predominant study sites follow taxonomic names. Habitats are highly fragmented in many of the chimpanzee ranges, particularly for the Nigerian-Cameroon and western chimpanzees. It is unclear whether chimpanzees west of the Niger River belong to *Pan troglodyte ellioti, P. t. verus,* or a new subspecies (Gonder et al. 2006). (Figure modified after de Waal and Lanting 1997.)

captive populations. Initial data from both wild and captive bonobos suggested that their behavior differed substantially from that of chimpanzees, particularly in the patterns of dominance, aggression, sexual behavior, and intercommunity interactions (Kuroda 1979, 1980; de Waal 1989a, 1998; Kano 1992; Thompson-Handler et al. 1984; White 1996a,b; Wrangham and Peterson 1996; Nishida 1997).

In the last 15 years, the behavioral disparities between chimpanzees and bonobos have been challenged (Boesch 1996a, Fruth 1998, McGrew 1998a, Stanford 1998c). Both species show considerable behavioral overlap (Stanford 1998c, Hohmann and Fruth 2002). Some behaviors observed in chimpanzees may never have been seen in bonobos because bonobos have been less studied than chimpanzees (Fruth 1998, McGrew 1998a). Some chimpanzee behaviors, like lethal aggression, were seen only after many years of close observation (Goodall 1986). Moreover, behavioral variation across chimpanzee field sites is considerable

(Wrangham et al. 1994b, Boesch 2002). Differences in bonobo behavior between research sites also have been reported (White 1992, 1996b; Hohmann and Fruth 2003a), and more differences are likely to be observed as bonobo research progresses in more diverse habitats. Bonobo studies have begun in more open and drier habitats (Thompson 2002), and such ecological variation may substantially influence bonobo behavioral variation. Thus, the extent of behavioral diversity within chimpanzees or within bonobos may rival the behavioral diversity between the two species. What is presently known of the nature of the similarities and differences within and between these sister taxa will be discussed in more detail.

CHANGING TAXONOMIC PERSPECTIVES

Debate has long surrounded the taxonomic and evolutionary relationships among all the great apes. Historically, chimpanzees and bonobos were placed with gorillas and orangutans in the family Pongidae, distinct from the family Hominidae, which included only humans and our fossil ancestors. However, recent genetic analyses indicate that chimpanzees and bonobos are more closely related to humans than either are to gorillas or orangutans (Sibley and Ahlquist 1984, Sarich et al. 1989, Caccone and Powell 1989, Horai et al. 1995, Takahata 1995, Ruvolo 1997b, Chen and Li 2001). The recent taxonomy proposed by Groves (2001a) (Figure 20.2) subsumes humans, the great apes, and their ancestors all within the family Hominidae. The subfamily Ponginae contains only the orangutans and their ancestors. The African apes along with humans and our fossil ancestors are grouped within the subfamily Homininae. While Groves (2001a) does not address subfamilial distinctions within the Homininae, molecular analyses suggest that chimpanzees, bonobos, and humans are united within the tribe Hominini. Chimpanzees and bonobos should be further united within the subtribe Panina.

Bonobos were first formally recognized as different from chimpanzees in 1929, based on skull morphology. Initially considered a chimpanzee subspecies (Schwartz 1929), bonobos were later classified as a separate species (Coolidge 1933) (Figure 20.2). This classification is supported by additional morphological, histological, and genetic data (Zihlman and Cramer 1978; Zihlman 1984; Sussman 1984; Moor-Jankowski et al. 1972, 1975; Ruvolo et al. 1994; Burrows and Ryder 1997; Jensen-Seaman and Li 2003).

Chimpanzees are currently divided into four subspecies. These include the East African (*P. t. schweinfurthii*), central African (*P. t. troglodytes*), West African (*P. t. verus*), and the recently recognized fourth subspecies from Nigeria and Cameroon (*P. t. ellioti*, formerly *vellerosus* (Oates et al. 2009), based on molecular data (Gonder et al. 1997) and supported by morphological data (Pilbrow 2006).

Several molecular studies have estimated that chimpanzees and bonobos diverged between 800,000 years ago and 2.7 million years ago (mya) (mitochondrial deoxyribonucleic acid [DNA], Morin 1994, Gagneux et al. 1999b; X chromosome, Kaessmann et al. 1999; Y chromosome, Stone et al. 2002; random nuclear genome, Yu et al. 2003, Fischer et al., 2004). The split between western (*P. t. verus*

and *P. t. ellioti*) and central/eastern (*P. t. troglodytes* and *P. t. schweinfurthii*) chimpanzees is estimated between 500,000 years ago and 1.6 mya (Morin 1994, Gagneux et al. 1999b). Chimpanzee mitochondrial and nuclear gene differentiation is greatest between western and eastern/central chimpanzees (Morin 1994, Fischer et al. 2004, Gonder et al. 2006), which confers with morphological findings (Guy et al. 2003, Taylor and Groves 2003). This led Morin (1994) and others (Gonder et al. 2006) to recommend bifurcation at the species or subspecies level, but this remains controversial (Fischer et al. 2006). The genetic variation within chimpanzee subspecies is greatest within *P. t. troglodytes* (Kaessmann et al. 1999, Gonder 2000, Fischer et al. 2004) and least within *P. t. verus* (Fischer et al. 2006), though *P. t. ellioti* were not included in the analyses. Bonobos have substantially less genetic variation than chimpanzees (Reinartz et al. 2000).

MORPHOLOGICAL VARIATION

Although males of both *Pan* species are larger and have longer canines than females, bonobos show lower levels of craniodental dimorphism than chimpanzees (White 1996b). Yet body mass dimorphism can be greater than or equal to that of chimpanzees, with male bonobos weighing about 15% more than females while male chimpanzees weigh between 5% and 14% more than females (Table 20.2). Of all of the chimpanzee subspecies, *P. t. troglodytes* is the largest and most sexually dimorphic in body mass, while *P. t. schweinfurthii* is the smallest.

Head plus body length ranges 635–940 mm for chimpanzees and 700–830 mm for bonobos. Although bonobos have a more slender build than chimpanzees, the body mass of the two species is comparable (Table 20.2), which may be explained by the bonobos' longer thigh bones, heavier thigh muscles (Zihlman 1984, 1996), and longer feet (Fleagle 1999). Due to their relatively longer legs, the bonobo intermembral index—[(humerus + radius)/(femur + tibia)] * 100—of 102.2 is lower than that of chimpanzees at 108.7 (Coolidge 1933, Zihlman and Cramer 1978). The crania and mandible of bonobos are clearly distinct

Figure 20.2 Recent taxonomy subsumes humans, the great apes, and their ancestors all within the family Hominidae. The subfamily Ponginae contains only the orangutans and their ancestors. African apes, humans, and our fossil ancestors are grouped within the subfamily Homininae. Chimpanzees, bonobos, and humans are united within the tribe Hominini. Chimpanzees and bonobos should be further united within the subtribe Panina (after Groves 2001a, Mann and Weiss 1996).

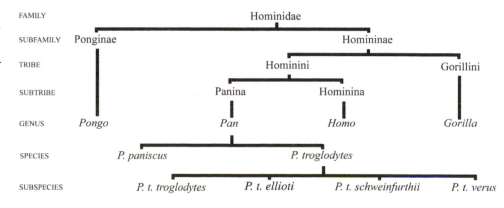

Table 20.2 Average Adult Body Mass (kg) for Bonobos and Chimpanzee Subspecies

	MALE	FEMALE
Pan paniscus	45	33.2
P. troglodytes schweinfurthii	42.7 (39.0–42.0*)	33.7 (31.3–35.2*)
P. t. troglodytes	59.7	45.8
P. t. verus	46.3	41.6
P. t. ellioti	?	?

Source: Smith and Jungers 1997 (*wild data from Uehara and Nishida 1987, Pusey et al. 2005).

Figure 20.3 A young male eastern chimpanzee (*Pan troglodytes schweinfurthii*) from Kanyawara, Kibale National Park, Uganda (photo by R. Stumpf).

from those of chimpanzees, even when data are adjusted for differences in size (Cramer 1977, Shea 1984, Shea and Coolidge 1988, Shea et al. 1993, Taylor and Groves 2003, Guy et al. 2003). Bonobos have a more anteriorly placed vulva (de Waal and Lanting 1997). Bonobo vocalizations are also notably higher in pitch than those of chimpanzees (de Waal 1988).

Facial patterns vary across the chimpanzee subspecies. Most infants of all chimpanzee subspecies are born with light pink faces. *P. t. verus* generally has a darker mask around the eyes and lighter skin tone in the rest of the face, and this pattern often remains into adulthood. The face of *P. t. troglodytes* generally darkens completely to deep black when adult. The pink face of *P. t. schweinfurthii* is often freckled and darkens with age (Figure 20.3). In contrast to most chimpanzees, bonobo faces are black from birth (Figure 20.4). The lips of bonobos are pink, and the hair on the head is parted in the middle. While both chimpanzees and bonobos are born with a white tail tuft, only bonobos maintain it into adulthood.

ECOLOGY

Habitat and Distribution

Chimpanzees occupy very diverse habitats across equatorial Africa, including dense lowland tropical rain forest as well as savanna woodland, and are found at altitudes of up to 3,000 m (Suzuki 1969, Kano 1971, McGrew et al. 1979) (Table 20.3, Figure 20.1). They are distributed as far northwest as Senegal, south to the Congo River, across northern Congo, and east of Lake Tanganyika, Tanzania (see Figure 20.1). *P. t. schweinfurthii* is thought to have moved into its seasonally dry habitat relatively recently (Goldberg and Ruvolo 1997b).

Bonobos live in lowland rain forest, swamp forest, dry forest, and grassland habitats. Bonobo distribution is more limited than that of chimpanzees, being bordered to the north and west by the Congo River, to the east by the Lomami River, and to the south by the Kasai and Sankuru

Figure 20.4 Young male bonobo (*Pan paniscus*) (photo by Henrike Wolf).

Rivers, completely within the Democratic Republic of Congo (Groves 2001a). Chimpanzee and bonobo ranges are not known to overlap. The formation of the Congo River separating the two species is estimated geologically at 1.5 mya (Beadle 1981), which is also the time of large climatic, vegetation, and habitat changes that may have initiated speciation in other hominins (Vrba 1995).

Diet

Chimpanzees and bonobos are highly frugivorous (Wrangham et al. 1998, Conklin-Brittain et al. 1998) (Table 20.4). Both species obtain substantial amounts of protein from terrestrial herbaceous vegetation (THV), although

Table 20.3 Pan Study Sites

	PAN TROGLODYTES VERUS			P. T. SCHWEINFURTHII							P. T. ELLIOTI	P. PANISCUS		
	TAÏ	BOSSOU[2]	MT. ASSIRIK[3]	GOMBE[4]	MAHALE[5]	KANYAWARA[6]	NGOGO[7]	SEMLIKI[8]	BUDONGO[9]	KAHUZI-BIEGA[10]	GASHAKA-GUMTI[11]	WAMBA[12]	LOMAKO[13]	LUKURU[14]
Study site location	Ivory Coast	Guinea	Senegal	Tanzania	Tanzania	Uganda	Uganda	Uganda	Uganda	D. R. Congo	Nigeria	D. R. Congo	D. R. Congo	D. R. Congo
Environment	Tropical rain forest	Tropical forest	Savanna woodland	Riverine forest and woodland	Riverine forest and woodland	Moist evergreen forest	Moist evergreen forest	Savanna with riverine forest	Semideciduous moist tropical rain forest	Montane and bamboo forest	Woodland, lowland, gallery forest	Dry, swamp, and secondary forest, cultivated fields	Primary evergreen forest, swamp forest	Evergreen forest and dry grassland
Annual mean rainfall (mm)	1,829	2,230	954	1,775	1,836	1,671	1,800	1,452	1,842	1,619	1,826	2,000	1,844	1,775
Elevation (m)	202	550	100–311	1,137	1,040	1,500	1,400	1,200	1,100	2,200	300–1,000	400	390	564
Dry season months (rainfall <60 mm)	3	3	7	4	4	4	4	4	3	3	4–5	0	0.5	
Provisioned?	No	Oil palm	No	Until 2000	Until 1987	No	No	No	No	No	No	Yes	No	No
Community size	29–82	16–22	>15	38–60	45–101	44	>140	>29	32–56	22	>35	28	33	?
Average party size	8.3	4	4	4.5	6.1	5.1	10.3	4.8	5.66	4.4	3.7	15	7.1	6.4
Home range size (km²)	13–26	15–20	50	4–24	7–14	16	35	38.3	7	10	26	22	14.7	?
Average day range (m)	2,400	1,000	–	3,450	4,825			?		400	?	2,400	3,035	?
Number of plant species consumed		>200	43	103	198	102		45	83	114	?	>110	>110	?

1 Boesch 1997, Boesch and Boesch-Achermann 2000, Herbinger et al. 2001, Lehmann and Boesch 2003, 2004.
2 Sugiyama 1989, 1994, 2004, Sugiyama and Koman 1992.
3 McGrew et al. 1996, 2004, Tutin et al. 1983.
4 Wrangham 1979a,b, Goodall 1986, Wallis 1997.
5 Nishida et al. 2003, Hasegawa and Hiraiwa-Hasegawa 1983.
6 Chapman and Wrangham 1993, Wrangham et al. 1996.
7 Watts 1998, Watts and Mitani 2001, Mitani et al. 2002a, Mitani (unpublished data cited in Nishida et al. 2003).
8 Hunt and McGrew 2002.
9 Newton-Fisher 1999a, 2003.
10 Basabose 2002, Basabose and Yamagiwa 2002.
11 Sommer et al. 2004.
12 Kano 1992, Badrian and Malenky 1984.
13 Thompson-Handler 1990.
14 Myers Thompson 2002.

eastern chimpanzees appear to rely on it more as a fallback food in times of food scarcity (Wrangham et al. 1994b, Newton-Fisher 1999a). The range of plant species eaten by chimpanzees varies widely across sites and depends on both local and seasonal availability (Table 20.4). During times of preferred-food scarcity, chimpanzees increase the proportion of leaves and other lower-quality foods in their diet (Gombe, Wrangham 1977; Taï, Doran 1997a) and/ or eat a wider range of foods (Gombe, Wallis 1995; Taï, Doran 1997a). At very arid sites, such as Semliki, chimpanzees regularly consume a limited range of foods (about 45 items), while at Mahale, more than 300 food items are eaten (Hunt and McGrew 2002). Animal protein (e.g., vertebrates, insects, eggs) constitutes approximately 8%–10% of the chimpanzee diet (Goodall 1986), with considerable intersite variation. Approximately 1%–3% of the chimpanzee diet consists of mammalian flesh (McGrew 1992). Where chimpanzees and gorillas overlap, their diets are seasonally similar but diverge during food scarcity (Tutin and Fernandez 1993, Basabose 2002, Stanford and Nkurunungi 2003).

Bonobos also eat a wide variety of foods (over 110 species) (Badrian and Malenky 1984, Kano 1992), including large ripe *Anonidium* (10 kg) and *Treculia* (30 kg) fruits. Bonobos rely heavily on THV and seem to prefer plant foods high in carbohydrates and low in fiber (Conklin-Brittain et al. 2001). Less than 1% of the bonobo diet is derived from animal protein (Kano 1992).

Hunting has been documented at all long-term chimpanzee research sites (Goodall 1986, Boesch and Boesch 1989,

Nishida et al. 1979), although there are considerable intersite differences in prey type, habitat characteristics, and hunting methods (Boesch et al. 2002, Pruetz and Bertolani 2007). At Mahale, chimpanzees hunt at least 17 vertebrate species, whereas only about five species are hunted at Taï (Boesch and Boesch 1989). Red colobus monkeys are the favored prey at all study sites where they are present (Nishida and Uehara 1983, Boesch 1994b, Stanford et al. 1994), though Gombe chimpanzees prefer immature monkeys while Taï chimpanzees prefer adults (Stanford et al. 1994, Boesch and Boesch-Achermann 2000).

Over 50% of all hunts are successful across all chimpanzee study sites. However, hunting success appears to depend on the level of forest cover. Hunting success is lower where the canopy is more continuous than in more interrupted forest cover. Presumably, this is because the continuous canopy allows more escape routes for arboreal prey species (Watts and Mitani 2002). In the east, where forest canopies are more interrupted, individuals hunt more opportunistically, whereas in the taller, more continuous forest canopy of Taï, Ivory Coast, more cooperative hunting may be necessary to successfully trap the prey (Boesch and Boesch 1989, Boesch 1994b, Boesch and Boesch-Achermann 2000).

Chimpanzee males hunt more often than females (Stanford et al. 1994). The frequency of chimpanzee hunting varies throughout the year and depends upon food availability, group composition, and the presence of estrous females. The relative importance of each of these factors differs across sites. For example, at Mahale, hunting is positively associated with the presence of larger groups, estrous females, and higher fruit availability (Mitani and Watts 2001, Watts and Mitani 2002). However, estrous female presence does not influence (at Ngogo) or negatively influences (at Gombe) the likelihood of male hunting (Mitani and Watts 2001, Gilby et al. 2006). Rather, Mitani and Watts (2001) suggest that hunting coincides with the presence of larger parties containing more males, which facilitates hunting success (but see Gilby et al. 2006). That large parties are able to form during these times suggests that feeding competition does not substantially limit grouping and, thus, that hunting is not purely a subsistence activity.

Compared to chimpanzees, bonobos appear to hunt rarely (Kano 1992, Ingmanson 1996), but intraspecies variation is likely (Surbeck et al. 2009). Duikers (small ungulates) are the more common prey (Fruth and Hohmann 2002, Hohman and Fruth 2007), even though colobus monkeys are present in their habitat (Ihobe 1990) and sometimes hunted (Surbeck and Hohmann 2008). Though low, the hunting frequency of bonobos falls within the range of chimpanzees (Fruth and Hohmann 2002, Hohmann and Fruth 2007) because chimpanzees at some sites hunt only rarely (e.g., Mt. Assirik (McGrew et al. 1988), Budongo, (Newton-Fisher 1999a). Ecological differences do not adequately account for this variation in hunting propensity among chimpanzees and bonobos: Bonobos hunt infrequently despite the apparent

Table 20.4 Chimpanzee and Bonobo Dietary Composition

	CHIMPANZEES[1]		BONOBOS[2]	
	(%) OF DIET	RANGE (%)	(%) OF DIET	RANGE (%)
Fruit	64	19–99	55	0–100
Leaves	16	0–56	14	0–28
THV[3]	7	0–27	25	0–100
Bark and misc.	4	0–41	2	0–11
Prey	4	0–28	2	0–3
Flowers	2	0–14	2	0–7
Seeds	3	0–30	–	–

Sources: All dietary data compiled by Conklin-Brittain et al. 2006.
[1] Chimpanzee data from Conklin-Brittain et al. 1998; Galdikas and Teleki 1981; Ghigliari 1984; van Lawick-Goodall 1968; Hladik 1973, 1977a,b; Newton-Fisher 1999a; Isabirye-Basuta 1989; Kuroda 1992; Kuroda et al. 1996; Matsumoto-Oda and Hayashi 1999; Mc-Grew et al. (1981), Peters and O'Brien 1981; Rodman 1984; Sabater-Pi 1979; Sugiyama and Koman 1987; Tutin and Fernandez 1993; Tutin et al. 1984, 1991, 1997; Wrangham 1977, 1996; and Yamagiwa et al. (1992).
[2] Bonobo data from Badrian and Malenky 1984; Badrian et al. 1981; Hashimoto et al. 1998; Kano 1992; Kano and Mulavwa 1984, and Uehara 1990.
[3] THV, terrestrial herbaceous vegetation.

similarity of their moist tropical forest habitats to those of chimpanzees with high hunting propensities.

SOCIAL ORGANIZATION AND GROUPING PATTERNS

Primate social organization and grouping patterns are strongly influenced by ecological factors. Chimpanzees in most of the eastern sites experience greater seasonality of rainfall and a greater number of dry season months compared to the west, whereas bonobos have few, if any, dry months per year (Doran et al. 2002a) (Table 20.3). This seasonal variation results in substantial spatial and temporal variation in food availability, as well as a greater period of food scarcity (Goodall 1986, Nishida 1990a, Wrangham et al. 1996, Doran et al. 2002a, Anderson et al. 2002). This section will discuss how these habitat differences and seasonal variation in food availability in turn appear to affect female grouping patterns and social behaviors across *Pan*.

Community Structure

Chimpanzees and bonobos live in multimale/multifemale communities, which vary greatly in size (Table 20.3). Chimpanzee communities range from a handful of individuals to as many as 150 (Watts 2002a). Bonobo communities show similar variation in size, ranging from 25 to 75 or perhaps 120 individuals (Furuichi 1989). In both species, males are the philopatric sex, remaining permanently in their natal group (but see Sugiyama 1999). Females generally transfer into a new community after reaching sexual maturity, presumably to avoid inbreeding (Nishida and Kawanaka 1972, Pusey 1979, Stumpf et al. 2009). Upon transfer to a new group, immigrant female chimpanzees establish initial relationships with males. Resident chimpanzee females are at

first largely intolerant of immigrant females and sometimes attack them (Pusey 1980; Pusey et al. 2008; Kahlenberg et al. 2008a,b; Nishida 1989). In contrast, young immigrant female bonobos primarily establish strong bonds with older resident females in order to integrate themselves into the community (Furuichi 1989, Idani 1991).

Although females of both species generally disperse during adolescence, the likelihood of female transfer varies between sites (Table 20.5). At most chimpanzee sites (e.g., Mahale, Ngogo, and Taï), the majority of females do transfer (Nishida et al. 2003, Boesch and Boesch-Achermann 2000). However, only about half of all Gombe females transfer (Pusey et al. 1997), while none is confirmed to have transferred at Bossou (Sugiyama 1999). Intersite differences may be influenced by the extent of community isolation as well as by the potential benefits of remaining in the natal group, such as inheriting the mother's core area and avoiding having to settle in the periphery of a new territory, as most Gombe immigrants do (Williams et al. 2002b). Male transfer has been observed twice at Lomako and potentially in Bossou, Guinea (Sugiyama 1999), suggesting that patterns of male philopatry may be more flexible for bonobos (Hohmann and Fruth 2002) and some chimpanzee populations.

Ranging Patterns, Day Ranges, and Nesting

Chimpanzee home ranges are relatively large in comparison to those of other apes, but there is considerable variation across sites (Table 20.3). Home ranges vary from the smallest (7–10 km²) in Gombe and Budongo (Goodall 1986, Newton-Fisher 2003) to 16–30 km² in primary forested habitat of Kanyawara and Taï (Herbinger et al. 2001, Wilson 2001) to the largest estimates of over 50 km² in more arid habitats such as Mt. Assirik and Fongoli (both in Senegal) as well as Semliki, Uganda (Baldwin et al. 1982; Hunt and

Table 20.5 Grouping Patterns of Chimpanzees and Bonobos

	P. T. VERUS		P. T. SCHWEINFURTHII		P. PANISCUS	
	TAÏ[1]	BOSSOU[2]	GOMBE[3]	MAHALE[4]	WAMBA[5]	LOMAKO[6]
Adult sex ratio (M/F)	1:4	1:2.3	1:1.35	1:3.6	1:3	1:1.2
Mixed parties[7] (%)	52	42	32	52	74	75.6
All-male parties[7] (%)	20	0	10	11	2	4.1
All-female parties[7] (%)	18	49	24	13	5	20.2
Solitary[7] (%)	18	3.5	18	21	6	4
Time spent alone by females[7] (%)	4	10	65	31.8	2	3

Sources:

[1] Boesch 1997; Boesch and Boesch-Achermann 2000; Herbinger et al. 2001; Lehmann and Boesch 2003, 2004.
[2] Sugiyama 1994, 2004.
[3] Wrangham 1979a,b; Goodall 1986; Wallis 1997.
[4] Nishida 1990a,b; Nishida et al. 2003; Takahata et al. 1996.
[5] Furuichi et al. 1998.
[6] Hohmann and Fruth 2002.
[7] Data compiled by Doran et al. 2002a.

McGrew 2002; Pruetz 2004; Hunt 2004). Chimpanzee home range sizes also vary considerably within sites (e.g., Herbinger et al. 2001). This variation may be influenced by a combination of factors such as chimpanzee density and the number of males in the community (Lehmann and Boesch 2003, but see Williams 1999).

Male and female chimpanzees differ substantially in their ranging patterns and day range lengths (Wrangham and Smuts 1980). Male chimpanzees have longer day ranges than anestrous females, whereas estrous females often travel in larger groups with males, so their day ranges are comparable (Goodall 1986). These differences also vary across sites (Table 20.3). West African chimpanzee females associate frequently with both males and females and utilize the entirety of their community's range (Sugiyama 1988; Boesch 1996a; Boesch and Boesch-Achermann 2000; Lehmann and Boesch 2004, 2005). In contrast, eastern female chimpanzees (Mahale, Gombe, Kibale) are less social than males, spending the majority of their time when not in estrus either alone with dependent offspring or with other mothers and young (Goodall 1986, Pusey 1980, Wrangham and Smuts 1980, Nishida et al. 1985, Pepper et al. 1999). Eastern (but not western) chimpanzee females occupy small, overlapping core areas within the home range and, unlike the males, do not utilize the entire community's territory (Wrangham 1979a, Chapman and Wrangham 1993, Wilson 2001, Williams et al. 2002a, Lehmann and Boesch 2004). Consequently, eastern female chimpanzee association rates are lower than at other chimpanzee sites (Williams et al. 2002a). Each night, both chimpanzees and bonobos sleep in fresh, individual arboreal nests interwoven from tree branches. Where trees are scarce, nests are sometimes reused (Fruth and Hohmann 1996). At times, chimpanzees at Mt. Nimba (*P. t. verus*) nest on the ground (Koops et al. 2007).

Grouping Patterns and Variation in Party Size

Ecological factors, such as food abundance and distribution, the presence of estrous females, and predation pressure, are thought to influence primate grouping patterns (Wrangham 1980, van Schaik 1983, Symington 1988c, Strier 1989, Chapman et al. 1995, Newton-Fisher et al. 2000). Heavy dependence on fruit presents a challenge for both chimpanzees and bonobos. Both species exhibit a flexible, fission–fusion social system whereby individuals form smaller subgroups (parties) that change in size and composition throughout the day (Nishida 1968, Goodall 1986). This fission–fusion strategy may limit intragroup feeding competition and improve foraging efficiency. Smaller parties form during times of food scarcity and increase in times of food abundance (White 1996a,b; Doran 1997a; Wrangham 2002; Mitani et al. 2002c; Chapman et al. 1995; Anderson et al. 2002).

Party size and composition are also influenced by the number of estrous females. Chimpanzee party sizes increase and numbers of both males and anestrous females increase with the presence of estrous females (Goodall 1986; Wrangham et al. 1992; Matsumoto-Oda et al. 1998; Wrangham 2000; Anderson et al. 2002; Mitani et al. 2002c; Newton-Fisher 2002a,b). Bonobo party sizes seem to be independent of the number of estrous females, although the proportion of males in the party increases (Hohmann and Fruth 2002).

Predation risk is also thought to affect primate group size, with larger groups providing increased vigilance and decreased individual risk of predation (van Schaik 1983). The relatively large body size of chimpanzees and bonobos may reduce predation risk, permitting the formation of smaller parties during diurnal foraging (Terborgh and Janson 1986). At night, however, both chimpanzee and bonobo party sizes increase (Fruth and Hohmann 1993, Anderson et al. 2002), perhaps as a response to the real threat of nocturnal predation (e.g., leopards, Boesch 1991).

Bonobo parties are generally larger and less variable than those of chimpanzees (Furuichi 1989, Hohmann and Fruth 2002) (Table 20.3), and unlike most chimpanzees, bonobos are very rarely alone (Hohmann and Fruth 2002, Goodall 1986; but see Lehmann and Boesch 2004) (Table 20.5). However, food availability is also an important factor influencing the size of bonobo parties. Though less well understood, their less seasonal habitat, more abundant and larger fruiting trees, and more even distribution of THV may result in lower within-group feeding competition among bonobos and permit larger and more cohesive groups than among most chimpanzees (Chapman et al. 1994a, Malenky and Wrangham 1994).

Generally, chimpanzees in West Africa form larger parties than East African chimpanzees (Boesch and Boesch-Achermann 2000, Wrangham 2000) and, like bonobos, are rarely alone. This variation in party size may be affected by ecological differences across sites. Large party sizes may be permitted where food availability and diet diversity are higher, reliance on THV is greater, predation risk is higher, or within-group competition is lower (Boesch 1991, Wrangham 2000, Williams et al. 2002a, Mitani et al. 2002c). Some intersite differences in bonobo grouping patterns have also been reported (White 1992, 1996b). Fissioning is less common in Wamba compared to Lomako (Kano 1992), and party sizes are generally smaller at Lomako (White 1992, 1996b; Hohmann and Fruth 2002).

Party composition also varies considerably between chimpanzees and bonobos (Table 20.5). Bonobo parties most frequently comprise both sexes, usually containing more females than males (Kano 1982, White 1988, Fruth 1995). All-female parties are more common among bonobos than chimpanzees (Hohmann and Fruth 2002). When food is scarce, male bonobos are more likely to disperse, whereas females remain together (White 1988). Intersite differences in bonobo party composition are also apparent: At Wamba, parties contain equal proportions of males and females (Kano 1992), whereas the proportion of females in a party is higher for Lomako bonobos (Fruth 1995, Hohmann et al. 1999). Eastern chimpanzee parties contain mostly males, whereas western chimpanzee parties more closely resemble bonobo parties in their more mixed composition (Sugiyama and Koman 1979; Boesch 1991, 1996a). Very little is known about grouping patterns of the central and Nigerian-Cameroon chimpanzee subspecies.

Intracommunity Social Relationships

Male dominance and strong male/male bonds characterize chimpanzees across all sites (Goodall 1986, Wrangham 1986, Nishida and Hiraiwa-Hasegawa 1987, Nishida and Hosaka 1996, Boesch and Boesch-Achermann 2000, Watts 2000b, Arnold and Whiten 2003). In most chimpanzee communities, males associate more with each other than in mixed parties. Male chimpanzees also groom one another more often than they groom females and than females groom other females (Takahata 1990; Watts 2000b,c; Arnold and Whiten 2003). Males groom those with whom they associate (Mitani et al. 2000), reciprocally (Watts 2000b,c, 2002a; Arnold and Whiten 2003), and up the hierarchy, with higher-ranking males being groomed more than lower-ranking males (Newton-Fisher 2002a, but see Takahata 1990). Male chimpanzees join in coalitionary mate guarding (Watts 1998b), hunting (Boesch and Boesch-Achermann 2000), and intergroup aggression (Wrangham 1999, Muller 2002). Moreover, prey capture and possession is male-biased, and males share food predominantly with other males (and some high-ranking or estrous females) (Boesch 1994b, Hockings et al. 2007, Gomes and Boesch 2009).

Since males are philopatric, kin selection has been proposed to explain the close relationships among male chimpanzees. However, chimpanzee male/male alliances can change easily, which does not fit well with the predictions of kin selection (Nishida 1983). Since chimpanzee interbirth intervals are long and preadult mortality is high (Goodall 1986, Wallis 1997, Hill et al. 2001), it is likely that brothers do not reach adulthood together. Thus, male chimpanzees rely on close age and rank cohorts (rather than kin) when forming alliances (Goldberg and Wrangham 1997, Watts 2000b, Mitani et al. 2002c), although additional studies have found that both maternal kin and nonkin alliances are important for males (Langergraber et al. 2007a, Mitani 2009).

Male chimpanzee hierarchies are linear, and dominance is valuable for males since dominant males generally sire more offspring (Constable et al. 2001, Vigilant et al. 2001, Boesch et al. 2006). Male coalitions in chimpanzees are very common and important in rank ascension and maintenance (de Waal 1982, Nishida 1983). Male chimpanzees are aggressive more frequently than females (Muller 2002). Though females are often victims of male aggression (Muller 2002, Stumpf and Boesch 2005), most chimpanzee aggression is between males and is related to male/male competition and rank (Goodall 1986; Nishida and Hosaka 1996;

Muller 2002; Wittig and Boesch 2003a). The fission–fusion social system and the fluctuating nature of male coalitions may impede a male's ability to monitor other males' social relationships, and frequent aggression may be necessary to maintain or reestablish rank (Muller and Wrangham 2001, 2004a). Lethal within-group aggression has been reported at Budongo (Fawcett and Muhumuza 2000), Mahale (Kitopeni et al. 1995), and Ngogo (Watts 2004) and may function to reduce mating competition (Watts 2004).

In contrast to males, female chimpanzee relationships are traditionally classified as egalitarian and weak (Wrangham 1980, Sterck et al. 1997). This is usually attributed to female emigration from their natal groups, decreased female sociality and female/female affiliative behaviors at most sites, and less research focused on female relationships. However, this view of female/female relationships has recently come into question. Relationships between female primates may be influenced by the level and type of feeding competition (see Chapter 2; Wrangham 1979b, van Schaik 1989). For example, the strength of female bonds and hierarchies is predicted to vary directly with the level of contest competition. When females are able to group more, as in the case of western chimpanzees and bonobos, the increased likelihood of contest competition leads to alliances among unrelated females (Williams et al. 2002a,b; Wittig and Boesch 2003a). Western chimpanzee females, in particular, do have higher than expected grooming frequencies and do form alliances and strong bonds (Sugiyama 1988, Boesch 1996a, Boesch and Boesch-Achermann 2000, Lehmann and Boesch 2008). Interestingly, it has been suggested that Taï females have a linear dominance hierarchy, implying that while food availability permits larger subgrouping in western chimpanzees, intragroup contest competition may still be considerable (Wittig and Boesch 2003a). Recent data from Gombe indicate that hierarchies are important for females since their reproductive success, longevity, infant survivorship, body weight, and interbirth intervals are strongly affected by female rank (Pusey et al. 1997, 2005; Williams et al. 2002a,b; Murray et al. 2006, 2007a). Dominant females travel with other high-ranking females, support each other in food competition, and occupy prime feeding sites, while low-ranked females spend little time with higher-ranked females (Goodall 1986; Boesch and Boesch-Achermann 2000; Williams et al. 2002a,b; Kahlenberg et al. 2008a,b) and have a poorer diet (Murray et al. 2006). When infanticide by females occurs, lower-ranking and immigrant females are the targets (Goodall 1977; Pusey et al. 1997, 2008; Townsend et al., 2007). Most female/female aggression among chimpanzees occurs over resources (e.g., food) and defense of offspring (Williams et al. 2002a,b; Wittig and Boesch 2003a,b; Kahlenberg et al. 2008a,b). Presumably, the positive correlation between female reproductive success and rank is due to the increased access to better feeding sites (Pusey et al. 1997; Williams 1999; Williams et al. 2002a,b). These findings suggest that intragroup contest competition over resources is not inconsequential and that female relationships among chimpanzees are not as egalitarian or as weak as originally considered (Williams et al. 2002a,b; Wittig and Boesch 2003a,b; Kahlenberg et al. 2008a).

Relationships within bonobo communities are markedly different from those of chimpanzees. Whereas male/male relations and male hierarchy are most evident among chimpanzees, female bonobos are characterized as dominant (Parish 1996, Kano 1998), codominant (Furuichi 1989, 1997), or dominant at feeding sites (White 1996b) to males. Despite male philopatry, female/female relations are paramount among bonobos (Kano 1992, Furuichi 1992, Hohmann and Fruth 1996, Parish 1996). The lower seasonality, presumed greater food availability, and fewer costs of grouping appear to increase the potential for female bonding. Female bonobos forage together without much competition (White and Wrangham 1988). Females can easily take food from males, more often possess shareable foods (Kuroda 1980, Fruth and Hohmann 2002), and regularly band together to chase away harassing males (Idani 1991, Kano 1992, Furuichi 1992, Hohmann and Fruth 1996). In contrast to chimpanzees, male bonobos are least likely to share food with other males (Kuroda 1980).

Bonobo grooming bouts are most common between the sexes, followed by female/female grooming, then lastly by grooming among males (Badrian and Badrian 1984, Thompson-Handler 1990, Kano 1992). The duration of these bouts suggests a different pattern of affiliation, with male/male grooming bouts lasting longer than either mixed-sex or female/female bouts (Wamba, Furuichi and Ihobe 1994; Lomako, Hohmann et al. 1999, Kano 1992). Although chimpanzee and bonobo grooming patterns appear to differ, grooming rates may be influenced by the bonobos' greater female-biased sex ratio, as well as larger and mixed-composition parties. Thus, grooming patterns for bonobos show some overlap with those for chimpanzees (Hohmann and Fruth 2002).

Although alpha-male bonobos have higher reproductive success than subordinates (Gerloff et al. 1999), male hierarchies appear to be less important for bonobos than for chimpanzees. Male bonobos do not assist one another during conflicts and, apart from the alpha position, male bonobo ranks are not clearly defined (de Waal and Lanting 1997).

Wild and captive bonobos are characterized by less aggression and higher conciliation rates than are chimpanzees (de Waal 1998, Hohmann and Fruth 2002). When aggression occurs, most fights are either between individual males or directed by females toward males (Parish 1994, de Waal 1998), while males rarely attack females (Wrangham 1993; Parish 1994, 1996; Vervaecke et al. 1999). Male/male aggression increases on mating days (Hohmann and Fruth 2003b). Very few fights are female/female (3.4%, Kano 1992) except in the context of mating (Hohmann and Fruth 2003b, Vervaecke et al. 2003).

In both chimpanzees and bonobos, enduring bonds exist between mothers and their offspring, particularly their sons (Goodall 1986, Boesch 1997). However, the period of mother/son dependence is longer for bonobos than for chimpanzees. Both bonobo and chimpanzee mothers appear to play a role in influencing a son's dominance rank (Furuichi 1989, Goodall 1986, Boesch and Boesch-Achermann 2000; but see Williams et al. 2002a,b).

Due to the variation in chimpanzee and bonobo grouping patterns across sites, three social system models have been applied to Pan. Because of male philopatry and the patterns of association, grooming, meat sharing, and boundary patrols at some sites, as well as the relatively weak egalitarian relationships among females (Nishida 1968; Wrangham and Smuts 1980; Boesch 1996a; Nishida and Hosaka 1996; Watts 2000b,c; Boesch and Boesch Achermann 2000, Watts and Mitani 2001), chimpanzee social organization historically has been considered male-bonded (Nishida 1968, Wrangham 1979b, Wrangham and Smuts 1980, Kawanaka 1984). However, because Taï chimpanzee females and males are equally social and maintain high range overlap between the sexes (Lehmann and Boesch 2005, 2008, 2009), Boesch (1996a) labeled the social system of P. t. verus "bisexually bonded." Bonobos are considered female-bonded because of the female-focused patterns of association and social interactions (Parish 1998).

Intercommunity Interactions

Chimpanzees are very territorial, and intercommunity interactions are usually hostile and sometimes deadly (Goodall 1986, Nishida 1979, Wrangham 1999, Watts and Mitani 2000, Muller 2002, Wilson and Wrangham 2003, Herbinger 2004, Watts et al. 2006). Chimpanzees patrol their boundaries about once a month (Herbinger et al. 2001, Anderson et al. 2002) and encounter other communities in 26%–40% of patrols (Boesch and Boesch-Achermann 2000, Wilson and Wrangham 2003). At other times, chimpanzees generally avoid the periphery of their territory (Herbinger et al. 2001). Such patrolling and avoidance of the periphery may reflect the risk of aggression from other communities (Nishida et al. 1985, Watts and Mitani 2000). The prevalence of intergroup encounters varies across chimpanzee sites: 9.5% of follows at Gombe (Goodall 1986), 11.8% at Taï (Boesch and Boesch-Achermann 2000, Herbinger 2004), and far less frequently at Kanyawara (2.8% cited in Wilson and Wrangham 2003). The composition of patrols also varies across sites. All-male patrols are the norm for eastern chimpanzees (Goodall 1986, Watts and Mitani 2001, Mitani and Watts 2005), while mixed-sex patrols are more common among Taï chimpanzees, though females rarely participate in physical attacks (Boesch and Boesch-Achermann 2000). In addition, intercommunity rates of aggression vary substantially (Wilson and Wrangham 2003). Lethal aggression against adults, while observed multiple times in the east, is rare in western chimpanzees (Boesch et al. 2008). Territoriality and lethal aggression in chimpanzees may improve access to food and females and thus increase both male and female reproductive success (Nishida et al. 1985, Manson and Wrangham 1991, Williams 1999, Goodall 1986, Wrangham 1999, Boesch and Boesch-Achermann 2000, Pusey 2001, Williams et al. 2004).

Chimpanzee strategies during intercommunity encounters depend upon the perceived balance of power between themselves and their opponents (Manson and Wrangham 1991, Wilson and Wrangham 2003). During intercommunity encounters, males will approach, attack, or retreat depending on the size of their patrol and the size and composition of the foreign group (Wilson et al. 2001). Chimpanzees will usually approach neighboring groups even when they are outnumbered, implying that the benefits of resource defense are substantial (Herbinger 2004). In the case of an extreme imbalance of power, such as finding a strange male traveling alone, lethal aggression may occur (Manson and Wrangham 1991, Wrangham 1999, Boesch et al. 2007). When nongroup females are encountered, males may tolerate, groom, and mate with them or, depending on the female's reproductive status, may attack her or her infant (Goodall 1986, Boesch et al. 2008).

Bonobos, like chimpanzees, defend their territories (Kano 1992, Hohmann et al. 1999). However, compared to chimpanzees, bonobos are less territorial, intergroup interactions are less aggressive (Kano 1992, Kano and Mulavwa 1984, Ingmanson 1998), and lethal aggression has never been observed (Stanford 1998c, Wrangham 1999). Bonobo intergroup interactions can last for several hours and include varied behaviors such as grooming, genitogenital (g–g) rubbing, mating, as well as some aggression (Idani 1990, Kano 1992, White 1996b). Bonobos have greater home range overlap than do chimpanzees, and male bonobos have not been observed to patrol borders (Hohmann and Fruth 2002). In addition, while most chimpanzee interactions occur on the periphery, encounters with other bonobo groups can occur in the center of the territory (Hohmann 2001). Differences between chimpanzees and bonobos in territoriality and intergroup aggression may result from lower feeding and mating competition among bonobos compared to chimpanzees.

Infanticide

Infanticide and attacks on juveniles have been observed in the majority of long-term chimpanzee sites, both within (Goodall 1977, 1986; Arcadi and Wrangham 1999; Takahata 1985; Hiraiwa-Hasegawa 1987; Hamai et al. 1992; Kawanaka 1981; Murray et al. 2007b, Pusey et al. 2008) and between (Goodall 1986; Watts and Mitani 2000, 2001; Nishida et al. 1979; Newton-Fisher 1999b; Kutsukake and Matsusaka 2002; Wilson et al. 2004; Sherrow and Amsler 2007) communities. These attacks can have a substantial effect on infant mortality. For example, at Mahale, conspecifics were responsible for 16% of community deaths, and the majority of these were infants killed by mature males (Hamai et al. 1992, Nishida et al. 2003).

Male between-group infanticide may function to eliminate future competitors and drive neighboring females from territorial borders (Muller and Mitani 2005). Infanticide may also be a strategy to obtain higher reproductive success over same-sex conspecifics (Hrdy 1979; but see Sussman et al. 1995). For males, killing dependent infants sired by other males results in a shortening of the female interbirth interval. Females resume cycling sooner and may permit the infanticidal male a possibility to sire the female's next infant.

Although males are responsible for the majority of infanticides observed, female chimpanzees also kill infants (Goodall 1986, Townsend et al. 2007, Pusey et al. 2008). Infanticide by females may be more common among eastern chimpanzees and may function to diminish within-group competition (Pusey et al. 2008). Differences between sites may be explained by lower intergroup competition and, for western chimpanzees, increased female/male sociality, which is thought to function as an anti-infanticide strategy (van Schaik 1996). Infanticide has never been observed among bonobos. This lack of evidence may be due to less observation time for bonobos or to their less aggressive nature, less mating and feeding competition, and more effective female counterstrategies.

REPRODUCTIVE PARAMETERS

Differences in Reproductive Parameters Within *Pan*

On average, chimpanzee females give birth only once every 5–6 years (Tutin and McGinnis 1981, Nishida et al. 1990, Boesch and Boesch-Achermann 2000). Interbirth intervals show considerable variation among sites, but are generally shortest among Bossou chimpanzees, followed by Taï, Mahale, and then Kanyawara (Wrangham et al. 1996) (Table 20.6). Bonobo interbirth intervals are, on average, slightly shorter than those of chimpanzees (Kano 1992), but they also vary substantially and do not differ significantly from those of Bossou chimpanzees (Fruth and Hohmann 2002, Hohmann and Fruth 2002). Resumption of cycling after parturition also varies considerably within *Pan*. Bonobos and female Taï chimpanzees often resume sexual swelling 1 year after parturition (Kano 1992, Fruth and Hohmann 2002, Boesch and Boesch-Achermann 2000). However, resumption of swelling among eastern chimpanzees is considerably later (see Table 20.6). Although bonobos have longer swelling phases than chimpanzees, there is also wide variation within chimpanzees, with longer swellings observed among western than eastern chimpanzee females (Wrangham 2002). Both chimpanzees and bonobos continue to have swellings and sexually active phases after conception. However, the period of postconception (pregnancy) swellings is longer for bonobos than for chimpanzees (Kano 1992, Furuichi and Hashimoto 2002).

Table 20.6 Reproductive Parameters

	PAN TROGLODYTES VERUS		P. T. SCHWEINFURTHII		P. PANISCUS
	TAÏ[1]	BOSSOU[2]	GOMBE[3]	MAHALE[4]	WAMBA[5]
Onset of menses[6] (age in years)	10	8.5	10.8	10.7	~10
Adolescent sterility (between first max. swelling and birth, years)	2.6	2.4	2.4	2.5	~4
Age at emigration/immigration (years)	11	>8?[a,b]	9–13	11 (median), 11.27 (mean)	8–13
Emigration rate (%)	~90	0–70[b]	50	90	100
Time between immigration and first birth (years)	2.7	n/a		2.7 (median), 3.3 (mean)	
Age at first birth (years)	13.7	13 (median), 10.9 (mean)[c]	13.3	13 (median), 15.6 (mean)	13–15
Estrous cycle	35		34–36	31.5	41.5 (range 33–49)
Days of maximal swelling	12	12	13	12.5	12.9–14.6
Maximal swelling % of whole cycle	34.3		36.1	39.7	40.7 (range 35–48)
Maximal swelling % of interbirth interval	4.9		4.2	6.4	27
Estrous sex ratio (M/max. tumescent F)	1.6		12.3	4.2	2.8
Female copulation rate per hour (within the cycle)	0.43		0.52	0.79	0.11–0.37
Number of pregnancy cycles	1.5		2	2.6	6.5 (months)
Gestation length (months)	8		7.6	7.6	7.6
Infant survival M (%)	60	80	66	36	81.8[7]
Infant survival F (%)	60	80	76	49	81.8[7]
Resumption of swelling after parturition (months)	24.5	36	46.8	55	12
Cycle resumption to conception (months)	5.1		3.6–4.75	8.9	38
Interbirth interval (months)	69.1	61.2	61.8	69.1	57.6
Annual birth rate per adult F	0.2	0.194		0.2	0.18
Age at last birth (years, max. observed)	~43	~41	37	39	~39
Lifetime reproductive success[8]	1.46	4.25	3.32	2.2	

Sources:

[1] Boesch 1997; Boesch and Boesch-Achermann 2000; Deschner et al. 2003; R. Stumpf 2005; R. Stumpf, unpublished data.

[2] Sugiyama 1994, 1999, 2004. [a]no females are known to have immigrated into Bossou, [b]70% of females disappeared by adolescence but emigration not confirmed, [c]includes females that remained in Bossou.

[3] Tutin 1979; Wrangham 1979a,b; Tutin and McGinnis 1981; Goodall 1983, 1986; Pusey 1990; Pusey et al. 1997; Wallis 1997.

[4] Nishida et al. 1990, 2003; Hasegawa and Hiraiwa-Hasegawa 1983; Hiraiwa-Hasegawa et al. 1984; Furuichi and Hashimoto 2002; Takahata et al. 1996.

[5] Kano 1989, 1992, 1996; Takahata et al. 1996; Furuichi 1987, 1989, 1992; Furuichi et al. 1998; Furuichi and Hashimoto 2002; Kuroda 1989.

[6] First full-sized swelling.

[7] Data for both sexes combined.

[8] Lifetime reproductive success was calculated as [(40 − (age of primiparity/interbirth interval)] + 1 × survival to primiparity (see Sugiyama 2004).

Ecological constraints influence female fecundity and reproductive parameters. The onset of postpartum cycling is seasonal in chimpanzees (Wallis 1992b, 1995; Nishida 1990; Wallis and Reynolds 1999; Anderson et al. 2006), coinciding most frequently with the end of the dry season. There is also a strong seasonal influence on chimpanzee sexual activity and conception/birth cycles (Wallis 1995, 2002; Wallis and Reynolds 1999; Boesch and Boesch-

Achermann 2000; Anderson et al. 2002; Emery Thompson and Wrangham 2008). Since party sizes in many chimpanzee communities increase with the presence of estrous females (Anderson et al. 2002, Mitani and Watts 2001), these females would be subjected to higher feeding competition. Females appear to avoid this by cycling in times of food abundance (Anderson et al. 2002, 2006). For chimpanzees, increased feeding competition and decreased food availability may constrain the optimal period for resuming cycling and producing new offspring more so than for bonobos (Furuichi and Hashimoto 2002, Fruth and Hohmann 2002). Thus, the shorter interbirth interval and period of lactational amenorrhea among bonobos and some chimpanzees (Taï and Bossou) may be related to the lower costs of grouping and loosened constraints for females in denser forest habitats with more reliable food availability (Wrangham 2002).

Swelling and Ovulation

Female chimpanzees have, on average, a 35-day menstrual cycle. Their sexually active phase (estrus) lasts about 10–15 days and is marked by a large, pink perineal swelling. Ovulation occurs near the end of maximal swelling (Graham 1981, Tutin and McGinnis 1981, Stumpf and Boesch 2005). Female chimpanzees are most sexually receptive during the period of maximal tumescence (Goodall 1986, Wallis 1997, Stumpf and Boesch 2005, Watts 2007), yet much of chimpanzee (and bonobo) sexual behavior is nonreproductive since sperm longevity is short and ovulation and conception occur only within a limited time window (Gomendio and Roldan 1993). In chimpanzees (and possibly bonobos), nonconceptive, multimale mating appears to be a female strategy to avoid infanticide (Stumpf and Boesch 2005) and strengthen social bonds.

The sexual cycle of female bonobos (averaging 40 days) is somewhat longer than that of chimpanzees, and sexual swelling occurs through most of the cycle (Table 20.6) (Kano 1996, but see Stanford 1998c, Hohmann and Fruth 2000). Because party sizes generally increase with increasing numbers of estrous females, the more relaxed constraints of feeding competition and lower costs of grouping compared to chimpanzees may permit a more lengthy maximal swelling in bonobos (e.g., Stanford 1998c, see also Wrangham 2002). Swellings in chimpanzees reliably advertise the probability of ovulation (Emery and Whitten 2003, Deschner et al. 2003). Female bonobo ovulation is less reliably tied to the swelling (thus more concealed) and often occurs after detumescence (Reichert et al. 2002).

After menarche, adolescent females in both *Pan* species go through a period of subfecundity where conception is delayed. Although they have frequent swellings, they are generally not yet fertile. During this intermediate period, when females are still in their natal group, adolescent subfecundity may protect these females from inbreeding (Pusey 1980).

Sexual Behavior and Copulation Rates

Some aspects of sexual behavior among chimpanzees and bonobos differ substantially. For example, among bonobos, almost all mating attempts are initiated by males, whereas females initiate one-quarter to one-third of all mating attempts among chimpanzees (Takahata et al. 1996, Nishida 1997, Stumpf and Boesch 2006). Dorsoventral copulation is the most common pattern for both species. However, ventroventral mating is also relatively common in bonobos (Kano 1980, Thompson-Handler et al. 1984) (Table 20.7). Sex is often nonreproductive in both *Pan* species, but, in contrast to chimpanzees, sex in bonobos frequently functions to appease and reduce aggression and tension (de Waal 1987).

Another aspect of sexuality among bonobos that is quite distinct from chimpanzees is the preponderance of same-sex, particularly female/female, sexual behavior. This behavior, g–g rubbing, occurs when females rapidly rub their clitorises together in a ventroventral position. G–g rubbing occurs regularly (Kano 1980, Thompson-Handler et al. 1984, de Waal 1987, Parish 1994, 1996) but is very rare among both wild and captive chimpanzees (R. Stumpf unpublished data, Anestis 2004). G–g rubbing may function to regulate tension (White 1988, Hohmann and Fruth 2000), as well to enhance and reconcile female/female relationships (Hohmann and Fruth 2000) and access to preferred food patches (Parish 1994). Rates of g–g rubbing are particularly high when access to food is limited and contested (Hohmann and Fruth 2000). However, White and Lanjouw (1992) found that g–g rubbing occurred more when food was abundant and decreased when food was rare. This suggests that g–g rubbing may depend on the quality of food, type of feeding competition, food distribution and accessibility, and extent of monopolization of the resource (Hohmann and Fruth 2000).

Bonobos are often considered to be more sexually active than chimpanzees. This determination depends upon the measures being considered. The majority of copulations for both species occur during the period of maximal tumescence, though this varies between them (96% for chimpanzees, Goodall 1986; 67%–82% for bonobos, Furuichi 1987, Takahata et al. 1996). Female chimpanzees have higher copulation rates during maximal tumescence than bonobos (Takahata et al. 1996) (Table 20.6). In contrast, female copulation rates between successive births (across the interbirth interval) are higher for bonobos than for chimpanzees (Wamba, 0.14/hr [Kano 1992]; Mahale, 0.04/hr [Takahata et al. 1996]). This latter difference can be explained by the fact that bonobos have longer tumescent periods (Kano 1996) and, on average, resume cycling earlier than chimpanzees (Takahata et al. 1996, Furuichi and Hashimoto 2002). Male copulation rates do not differ between chimpanzees and bonobos and range between 0.1 and 0.2/hr for both taxa (Furuichi and Hashimoto 2002).

Male and Female Sexual Strategies

There are at least four types of mating described for chimpanzees (Tutin 1979). *Opportunistic mating* occurs when females mate with many males in succession, with little outright aggression among the males. *Consortship* occurs when

Table 20.7 Behavioral Similarities and Distinctions Between Chimpanzees and Bonobos: Whether All of These Interspecies Distinctions Are Supported Over Time as More Information Is Gained Across All *Pan* Study Sites Remains to Be Seen

	CHIMPANZEES	BONOBOS
Activity pattern	Diurnal	Diurnal
Social organization	Fission–fusion	Fission–fusion
Community structure	Multimale/multifemale	Multimale/multifemale
Dispersing sex	Adolescent females	Adolescent females
Locomotion	Knuckle walking	Knuckle walking
Substrate use	Arboreal/semiterrestrial	Arboreal/semiterrestrial
Sleep site	Arboreal nests	Arboreal nests
Diet	Primarily frugivorous	primarily frugivorous
Sympatric gorillas?	Variable, only in some areas	No
Dominant sex	Male	Female or codominant
Intercommunity interactions	Usually antagonistic	Usually tolerant
Male/male association	Very strong	Moderate
Male/female association	Weak in eastern, strong in western	Strong
Female/female association	Very weak in eastern, strong in western	Strong
Resolve tension with	Aggression	Sex
Male/female aggression	Common	Rare
Female/male aggression	Rare	Common
Pant-grunt submissive vocalization	Common	Never observed
Grooming	Male/male most, female/female least	Male/female most, male/male least
Infanticide	Rare but present at all sites	Never observed
Tool use	Habitual	Infrequent
Consorts	Some but not all sites	Rare
Boundary patrolling	Habitual	Never observed
Hunting	Occurs at most sites	Rarely observed
Preferred prey	Monkeys	Duikers
Food sharing	Yes (mainly meat)	Yes (mainly fruit)
Vocalizations	Lower pitch	Higher pitch
Sexual behavior during nonswelling	Very rare	Common (33%)
Copulation posture	Dorsoventral	Dorsoventral (70%–89%), ventroventral (11%–30%)
Female/female sexual behavior	Very rare	Genitogenital rubbing common

a male and female leave the group and travel together for a period of days or weeks. *Possessive mating* occurs when a male guards a female and prevents mating by other males. A type of possessive mating, *coalitionary mate guarding*, has been described by Watts (1998b) and occurs when two or three males unite to guard a female from other males. Finally, *extragroup mating* (with individuals from outside the community) occurs rarely and results in a small proportion (<10%) of all paternities (Constable et al. 2001, Vigilant et al. 2001). Between sites, the propensity for consortships

differs substantially. Bonobo mating patterns are similar to chimpanzees', although possessive mating and consortships are rare (Kano 1992, Takahata et al. 1996). Among bonobos, extragroup mating is both common and tolerated.

At times, no relationship has been found between male dominance rank and mating success (Tutin 1979, Hasegawa and Hiraiwa-Hasegawa 1990, Nishida 1997, Boesch and Boesch-Achermann 2000, Newton-Fisher 2004). However, mating success for high-ranking males is improved by a stable hierarchy, fewer males, and the likelihood of conception

(Goodall 1986, Nishida 1997, Kano 1996, Gerloff et al. 1999, Mitani et al. 2002c, Boesch et al., 2006). Male interest intensifies around ovulation (Wallis 1992b), and primarily high-ranked or coalitionary males respond by increasing mate guarding (Tutin and McGinnis 1981, Watts 1998b). High-ranking males have greater mating success during the females' conceptive period (Tutin and McGinnis 1981, Nishida 1997, Muller and Wrangham 2001, Stumpf 2004, Deschner et al. 2004) and higher paternity success (Constable et al. 2001, Vigilant et al. 2001), which is similar for bonobos (Gerloff et al. 1999). Although dominant males benefit from increased reproductive success, dominance comes at a cost as high rank in chimpanzees is positively correlated with high cortisol levels (Muller and Wrangham 2004b).

Chimpanzee females mate with multiple males, yet females exhibit preferences and alter the frequency with which they mate with particular males (Stumpf and Boesch 2006). Female chimpanzees may influence paternity by following a mixed mating strategy of being selective when conception is likely and more promiscuous when conception is unlikely (Stumpf and Boesch 2005, Pieta 2008, see also Nunn 1999).

Female preferences during periovulatory phases correlate positively with male mating success in chimpanzees at Taï (Stumpf and Boesch 2006). This appears to result in part from females' resistance to solicitations of nonpreferred males. Female choice may be less effective in eastern chimpanzees due to the increased number of males, greater male/male competition, more solitary nature of eastern females, and more intense sexual coercion (Wrangham and Smuts 1980, Goodall 1986, Matsumoto-Oda and Oda 1998, Muller et al. 2007). For bonobos, increased female cohesiveness and female dominance may afford females more opportunity to implement their mating preferences (Hohmann and Fruth 2002).

CONSERVATION STATUS

Both chimpanzees and bonobos are characterized by long interbirth intervals, slow life histories, substantial maternal care, and relatively high infant mortality (Harvey et al. 1987) (Table 20.8). They are slow to reproduce and, therefore, especially vulnerable. Bonobos and chimpanzees are presently classified as endangered by the World Conservation Union and threatened with extinction by the Convention on International Trade in Endangered Species. The total wild chimpanzee population was recently estimated to be 173,000–300,000 (Butynski 2003), though populations have declined by as much as 66% in the last three decades due to increasing human populations, habitat destruction, and infectious diseases such as Ebola (Walsh et al. 2003, Butynski 2003, Oates 2006). Chimpanzees are also hunted for meat and medicinal and animistic religious purposes

Table 20.8 Life History Parameters (*Pan*)

Female growth complete (stature, age in years)[1]	11–12
Male growth complete (stature, age in years)[2]	13–17
Growth cessation (adult mass reached, age in years)[3]	21 F, 24 M
Age at locomotor independence (years)[4]	>5
Lifetime offspring produced[5]	3.9
Lifetime offspring survive weaning[6]	1.4–2
Longevity (years)[7]	40–55 estimate
Age at weaning[8] (years)	5
Birth weight (g)[9]	1,800,* 1,381**

Sources:

[1,2] Leigh and Shea 1995; Pusey 1990, 2004; Pusey et al. 2005; Goodall 1986.

[3] Pusey et al. 2005; Kuroda 1989.

[4] Doran 1997b.

[5] Hill et al. 2001; Nishida et al. 2003.

[6] Nishida et al. 2003.

[7] Hill et al. 2001, cited in de Waal and Lanting 1997.

[8] Pusey 1983; Hiraiwa-Hasegawa 1989; Kuroda 1980.

[9] Graham 1970 (*captive *P. troglodytes*), Thompson-Handler 1990 (**wild *P. paniscus*).

(Nishida and van Hooff 2002). In West Africa, chimpanzee populations have declined by 50% in the last two decades (Walsh et al. 2003) and are now extinct in Benin and Togo. Only 38,000 West African chimpanzees are estimated to remain (Butynski 2003). Eastern and central chimpanzees are estimated at 98,000 and 93,000, respectively (Butynski 2003). The estimated number of *P. t. ellioti* remaining is about 6,000 (Butynski 2003).

In 1998, the bonobo population was estimated at 10,000–20,000 (Bowen-Jones 1998). Habitat destruction and the bushmeat trade are the biggest threats to bonobos (Walsh et al. 2003). Although many local taboos prohibit eating bonobos, hunting bonobos increased markedly after the Congolese civil war. Salonga National Park, established in 1970, and the newly established Faunal Reserves of Lomako-Yokokala and Tumba-Lediima are currently the only legally protected reserves for bonobos. Based on the recent rate of African ape decline, Walsh et al. (2003) recommend changing the status of chimpanzees and bonobos from endangered to critically endangered. More data are needed on their distribution, abundance, ecological requirements, threats to survival, and areas of high potential for conservation.

CONCLUSION

Presently, there are over 40 chimpanzee and bonobo research sites, and more are being established. Moreover, within the last 15 years, the inclusion of hormonal and genetic methodologies together with more extensive behavioral studies has further expanded our understanding of relationships and behavioral strategies within *Pan*.

Until recently, chimpanzees and bonobos were thought to be quite behaviorally distinct from one another. However, it is becoming increasingly evident that there is substantial behavioral overlap between the two species and that the interspecies differences are not as marked as were originally thought (Stanford 1998c, Boesch 2002, Hohmann and Fruth 2002, Matsumoto-Oda 2002, Williams et al. 2002a) (Table 20.7). Many behavioral patterns once thought to distinguish chimpanzees and bonobos, such as their propensity for hunting, consorts, and tool use, as well as party sizes and composition vary substantially across *Pan* sites and thus may not support the distinction between these two taxa. These considerable behavioral variations observed within *Pan* have been attributed to ecological, cultural, genetic, and demographic factors (Mitani 2006, Laland and Janik 2006, Lycett et al. 2007, Watts 2007). Because of their close phylogenetic relationship, the two species are likely to have similar behavioral responses to similar environmental pressures (Boesch 2002). Additionally, the perceived behavioral distinctions may be the product of differing levels of research efforts (Stanford 1998c). Perhaps with equivalent observation and more comparative data, other aspects of chimpanzee and bonobo behavior may prove to be more similar than is currently thought.

In spite of the substantial behavioral overlap, it is apparent that many key differences do exist between these two species (de Waal 1998, Kano 1998, Parish 1998, Doran et al. 2002a) (Table 20.7). These nontrivial differences include dominance relations between the sexes, levels and patterns of aggression, the nature of intercommunity interactions, and female sociosexual behaviors. These differences may be influenced by variation in ecological factors such as rainfall, food availability, and seasonality (Wrangham 1986, Wrangham et al. 1996, White and Wrangham 1988, Malenky 1990, Malenky and Wrangham 1994, Chapman et al. 1994a, Doran et al. 2002a). Boinski's (1999) work on *Saimiri* has demonstrated that even subtle ecological differences can have substantial effects on social organization. In addition, the divergent evolutionary pathways of the two species may be a factor in their differing responses to comparable ecological pressures. Nevertheless, understanding the specific effects of ecological differences on chimpanzee and bonobo behavior requires more detailed research.

Further attempts to quantify food availability, defensibility, quality, and size of food patches using comparative methods across sites are in progress; and results may lead to a clearer explication of interspecific differences. With time, we hope to better understand whether the perceived behavioral distinctions between chimpanzees and bonobos will hold or whether the behavioral diversity within *Pan* is simply a continuum.

Species included in this chapter are shown in Color Plates 34 and 35 (*Pan troglodytes verus* and *Pan paniscus*).

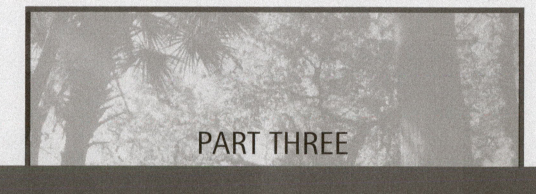

PART THREE

Methods

21

Behavioral Data Collection in Primate Field Studies

Christina J. Campbell, Margaret Crofoot,
Katherine C. MacKinnon, and Rebecca M. Stumpf

1. What is habituation, how is it carried out, and why is it necessary?
2. What are the main sampling methods employed in studies of primate behavior?
3. In what ways is technology aiding the study of primate behavior?

INTRODUCTION

Since the early days of primate field studies (see Chapter 1), the methods employed for the collection of data have changed dramatically. Prior to the pioneering paper by Jeanne Altmann (1974), there was no systematic way in which field researchers sampled behavioral information, thus making comparisons across studies problematic. Altmann's seminal paper provided researchers with several sampling techniques that would minimize observer bias and standardize data collection, thus facilitating cross-study comparisons (details on these different techniques are discussed in this chapter).

Historically, the field primatologist's most vital pieces of equipment were pen, paper, and a pair of binoculars (Figure 21.1). Today, although these items are still of vital importance, there are many other technologies that have been added to the equipment list. Those studying nocturnal species may use night-vision technology in order to actually see their study animals, electronic data recorders may replace pen and paper, and radiotelemetry and global positioning system (GPS) technology have not only made the life of a field primatologist easier but in some cases allowed the collection of data that were previously unattainable. The exact ways in which researchers collect their data and the technology they use (if any) will depend on a number of factors, including (but not limited to) their study species, their study question, their field site (e.g., the availability of electricity), and their research budget.

The first stage in any primatological field project is deciding what the research question and species are going to be. Some primatologists tend to be more question-oriented, studying their topic of interest in any species for which it is applicable. Others are more species-oriented, interested in anything and everything about their species of choice.

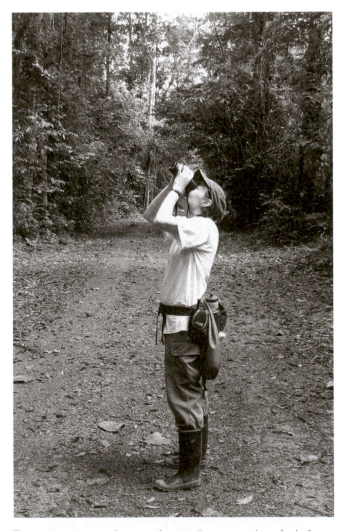

Figure 21.1 A researcher uses the most important primatological tool—the binoculars—to observe arboreal primates in Costa Rica (photo by K. C. MacKinnon).

Neither is better than the other; however, for both it is important that the question and the species are appropriate for each other. Once the question and species are decided, it needs to be determined how the data are going to be collected in a way that permits the question(s) to be answered. What follows are some of the methods and technologies that are available to help in primatological behavioral (primarily field-based) research (see also Paterson 2001, Setchell and Curtis 2003).

THE EARLY STAGES OF A RESEARCH STUDY

Habituation

Wild primates are often wary of humans and flee when approached, thus making naturalistic observation of primate behavior difficult. Compounding this problem, primates often show altered and affected patterns of behavior when they are disturbed by human observers (Williamson and Feistner, 2003). Over time, primates can be habituated to human presence through repeated neutral interactions (see Thorpe 1963) with human observers, thus permitting observation of the animal's natural behavior relatively uninfluenced by the observer's presence. Primates accustomed to human observers present several specific advantages. First, observations can occur at close range, thereby permitting recognition and identification of individuals and observation of rare, cryptic, or furtive behaviors (e.g., facial expressions, quiet vocalizations, complex food-processing techniques, subtle social interactions) (Williamson and Feistner, 2003). Moreover, habituated individuals or groups of primates can be followed throughout their range by observers, thus permitting more accurate information about primate day and home ranges, core areas, and habitat use. By habituating all individuals in the group, data are not skewed by an overemphasis on one sex or by central or dominant individuals (Williamson and Feistner 2003, Bertolani and Boesch 2008).

While the scientific benefits of habituation are clear, habituation should not be undertaken lightly. Habituation entails many risks and ethical dilemmas, which should be carefully considered prior to the initiation of contact with the study population (Martin and Bateson 2009). For example, because habituated primates are more accustomed to human presence, they are easier prey for poachers. In addition, closer and more frequent contact with humans provides increased opportunity for disease transmission (Butynski 2001; Fuentes 2006b, Jones-Engel et al. 2008), and both disease and poaching could significantly affect the long-term viability of the study population—even in protected populations. Moreover, primate day ranges, resting, and foraging efforts can be affected during the process of habituation (Cipolletta 2003, Blom et al. 2004). Thus, consideration of these risks as well as the development and implementation of appropriate protection measures for both observers and subjects are essential (see Williamson and Feistner, 2003).

Once a decision has been made to habituate, the effort and time commitment required for habituation vary substantially, from a few days or weeks to several years. This variation depends on several factors, including characteristics of the species (arboreal vs. terrestrial, frugivorous vs. folivorous, stable vs. fissioning groups), characteristics of the particular group (habitat type, group size, cohesiveness), prior exposure to negative human stimuli (e.g., hunting or logging), individual characteristics (size, rank, age, sex, reproductive status, and past experiences), and characteristics of the human habituator (methods, manner) (Tutin and Fernandez 1991, Cipolletta 2003, Doran-Sheehy et al. 2007, Bertolani and Boesch 2008).

Methods of habituation vary depending on the specific characteristics of the target population but generally follow several steps. Prior to even commencing, one can learn much about the target species' natural history, behavioral and vocal repertoire, locomotion, polyspecific associations, feeding ecology, and habitat by visiting conspecifics in captivity (Williamson and Feistner 2003, Martin and Bateson 2009). This preliminary step can also facilitate one's ability to distinguish individuals of that species based on distinct features of the face and body.

To commence habituation, little equipment is needed. Binoculars, a notebook, and a compass are essential. Because habituation requires repeated contact with the same individuals, details such as location, group size, age, sex, and distinct markings of each individual can be recorded at each sighting. Preferred tree species can be identified, located, and visited regularly to enhance the opportunity for observation. Additional useful equipment may include a GPS for orienting oneself and inputting ecological information and trip cameras to assist with determining habitat use and for obtaining data while the habituation is under way (see Sanz et al. 2004).

When in contact with the target group, the observer's comportment matters (Blom et al. 2004, Bertolani and Boesch 2008, Ando et al. 2008). The primates should ideally have a clear view of, and adequate distance from, the observer. By calmly and quietly observing and avoiding rapid movements and loud noises, primates will eventually relax in the observer's presence. Cutting of foliage for trails or passage should be kept to a minimum to reduce disturbance, though some cutting may be necessary in dense undergrowth. While provisioning may diminish the length of time required for habituation, it has several significant drawbacks. For example, provisioning can lead to disease transmission, nutritional dependence, and behavioral problems related to overhabituation, including crop raiding and dangerous interactions between the target primates and humans (Williamson and Feistner 2003).

Individual Identification

For many behavioral studies, identification of individuals beyond the standard age/sex class level (e.g., adult male, juvenile female) is required. Individual identification is significantly easier in some species (e.g., chimpanzees, spider monkeys, capuchin monkeys) than in others (e.g., howler monkeys, owl monkeys). Whatever the study species, a researcher needs to be attuned to features that vary among individuals (Martin and Bateson 2009). For example, in some species (e.g., chimpanzees and spider monkeys) facial coloration and patterning can be important. In others (e.g., capuchins) fur color and/or scar patterns may be more important. In some species, genitalia may be distinct enough to aid in the identification of individuals (e.g., the hypertrophied clitoris of female spider monkeys). Unfortunately, in many species (e.g., owl monkeys, *Aotus* spp.) individual identification can be so difficult that marking individuals with colored ear tags, collars, anklets, or freeze-dried or shaved marks on the tail may be necessary (e.g., Fernandez-Duque and Erkert 2006). However, such markings should be used only when absolutely necessary as marking in this way requires capturing the animal, which carries with it many potential dangers (see later section "Ranging Data").

The Ethogram

Simply put, an *ethogram* is a behavioral repertoire of the species being studied, and constructing such a repertoire is the first step to collecting data in a systematic way. Developing a useful ethogram entails watching the study animals before systematic data collection begins and taking very careful notes on what the animals do in the specific environment where they will be studied. The animals should be sufficiently habituated so that a broad range of "natural" behaviors can be observed. For researchers to effectively compare their data with those of other scientists, there must be some agreement between them in the behaviors they are observing. Thus, many categories are somewhat standardized in description (e.g., "groom," "display," "rest"). At the same time, many behaviors will be unique to the study species in question (e.g., "double-threat" in white-faced capuchins, "head toss" in langurs). Ethograms are drawn from personal observations of the species but can also be supplemented by previously published reports (if properly cited) and input from other researchers who also study the species. When doing preliminary observations to construct the ethogram, all that is needed is a notebook, pen, and binoculars. For each behavioral category, the observer must come up with a specific, descriptive definition that does not include the name itself (e.g., a definition for "rest" cannot include the word "rest"); behavioral definitions must be mutually exclusive, with no overlap. Behavioral categories are usually coded by abbreviation ("gr" = groom) for easier entry either into a handheld data-taking device or when writing via pen and paper.

Generally, behaviors fall into two types of categories: *states*, which are behaviors with a measurable duration (also called "durational behaviors"), and *events*, which are instantaneous behaviors that occur without measurable duration (also called "frequency behaviors"). These are different kinds of data and must be treated as such (Martin and Bateson 2009). *Bouts* are most often related durational behaviors that occur in sequence (e.g., interchange of grooming between two partners).

Individual behaviors can include anything that is done outside a social context (e.g., travel, sit, rest, eat, drink). Participant behaviors require at least one other animal and include affiliative behaviors (e.g., groom, rest-in-contact), agonistic behaviors (e.g., chase, face-threat), play behaviors (e.g. chase-play, wrestle-play), and sexual behaviors (e.g., sexual mount). Certain behaviors may occur in both solitary and social contexts and should be differentiated accordingly. For example, if an animal is eating in proximity with another individual, it can be classified as eating in a social context ("eat-prox") versus eating without any others nearby ("eat").

A SURVEY OF SAMPLING METHODS

Focal Animal Sampling

Focal animal sampling is the standard data-collection technique for most primate behavior studies, especially those involving social behavior (see Altmann 1974, Paterson 2001, Martin and Bateson 2009). Unfortunately, it is also one of the more challenging techniques to use. The method requires uninterrupted, concentrated attention on a single individual subject for a specific and consistent "sample time period" (e.g., 10 min, 30 min, 60 min, all day). The researcher records, as best she or he can, everything that the focal animal does and everything that is *directed to* it, noting down the times at which behaviors change. Once an activity is noted, it need not be repeatedly marked down on the data sheet until it ends, changes to another activity, or changes from a passive to an active form or vice versa (e.g., "groom" to "receive grooming"). Additionally, qualifying whether the focal animal was the actor or receiver in another column on the data sheet can designate the actor/receiver status.

This technique allows for the collection of both event and state behaviors and remains the most comprehensive way to collect data on behavioral sequences. Thus, the observer can collect a maximal amount of information on the behavior of each subject animal. The only drawback of this method is that it is both time- and labor-intensive and requires a high level of continuous concentration. Regular breaks are crucial not only to ensure that the observer's observation and recording skills stay sharp but also to ensure statistical validity of focal samples (as discrete individual units, unaffected by the preceding sampling bout).

Two additional categories that should be included for many data-recording techniques (including focal animal sampling) are "out of sight" and "other/unknown behavior." Primates move around quite a bit, and some individuals/species are very active throughout the day. When making structured behavioral observations, one of the most frustrating occurrences is when a target invariably goes up into a thick tangle of vines or retreats to its sleeping quarters behind the scenes at the zoo, right in the middle of a sample. If the animal reappears in a few seconds (usually up to 1 min), it is generally agreed that the sample bout can be completed. However, if the focal animal is out of sight of the observer for longer, the sample must be discarded. Also, if the target engages in a behavior that is not included in the ethogram, it is useful to have a catch-all category, like "other" or "unknown behavior," since the goal of focal animal sampling is to have a continuous sequence of events for a set time—there should be no gaps within a sample bout. Often, researchers include a qualifying column on data sheets or program spreadsheets for brief descriptions of unknown behaviors.

Instantaneous Scan Sampling

Scan sampling is an effective way to measure the rate per hour or minute and the percent of scan samples in which animals engage in specific activities as it results in a series of data snapshots taken at standardized intervals. Paterson (2001) provides a useful analogy, which likens scan sampling to the "repeated sweeps of an air traffic controller's radar screen" (p. 135). At each interval of time (e.g., every 30 sec for 10 min), the observer attempts to record what each animal in the group is doing (i.e., eating, resting, social, traveling, other). These scans are generally used for describing group and subgroup (e.g., age/sex class) behavior rather than individual behavior (Martin and Bateson 2009) and can be useful for general questions like discerning preferential use of various areas of a cage, tree, or habitat. However, this is typically not a good method of data collection if the research questions concern interindividual social interactions, simply because many behaviors are missed.

For some difficult-to-follow species, researchers often use a compromise between instantaneous scan and focal animal sampling. In this combined method a particular animal (the focal animal) is followed and its behavior is recorded at set intervals rather than continuously. Thus, for species where continuous, uninterrupted observations of the focal animal may prove too difficult, this technique provides a useful alternative.

All-Occurrences Sampling

With this technique, an observer notes each time a specific behavior is performed and records as much information about the participants and directionality of the behavior as possible. However, for this sampling method to be effective, observation conditions must be excellent and the behaviors need to be very conspicuous and relatively slow enough such that "all occurrences" of them can be adequately recorded (e.g., male attacks on female spider monkeys) (Campbell 2003, Slater et al. 2008). The duration is not critical, but some time control is necessary (to convert raw frequencies of data into rates of behavior).

Ad Libitum Records

Anytime a focal animal engages in behaviors that are likely rare but of great importance (e.g., copulation, intergroup encounter, fight, response to potential predator) as much information as possible should be recorded. Ideally, the information should include the animal's ID, age, sex, time and date of event, age and sex (or specific ID) of other participants, and anything else the observer thinks is relevant (Martin and Bateson 2009).

DATA COLLECTION

Interobserver Reliability

Primatological field projects often include more than one individual collecting data. Although there are many benefits associated with having multiple observers (e.g., safety, the ability to follow multiple animals and/or groups simultaneously, the ability to follow animals 7 days per week), there are also some potential problems that should be addressed early in the study. Before any project involving multiple observers can proceed, it must be determined that each researcher observes and collects data accurately and that all observers' observational accounts accord with one another. Clearly, when two observers are observing different animals (or even the same animal at different times) with the end goal of comparing the same behaviors among different individuals, it is imperative that the behaviors they record concur with one another.

Testing for agreement between two or more observers is a fairly easy procedure (Martin and Bateson 2009). The agreement in data record between any set of two observers can be tested by having them watch the same individual(s) over a prescribed set of time. Their records are then tested for reliabiltiy using the following equation:

$$R = \frac{s}{t/2}$$

where R is the coefficient of reliability, s is the number of behavior records scored by two observers that are the same, and t is the total number of behavior records scored by the two observers. The higher the value of R, the more similar will be the records of the two observers for the same events. Although there is no standard value that is accepted or rejected, any R value below 0.90 should be viewed as insufficient and the observers should be given more instruction and practice time until sufficient interobserver reliability is met.

Creating a research design that includes clear and concise definitions of behaviors is an obvious starting point for avoiding confusion among multiple observers. Additionally, allowing researchers time to "learn" not only the behaviors but also the ethogram for the study species will promote increased reliability and agreement in the data they record.

Additionally, multiple researchers can assure coordinated data recording by communicating effectively while in the field. Typically, it is not recommended that researchers yell to each other or use any other overly noisy form of communication (e.g., loud whistles, unless in an emergency) as this can impact the behavior of the study population. Basic walkie-talkies can allow researchers to communicate; however, they are usually not effective in a dense forested situation or over large distances (more than 100–200 m, even less in the forest). More advanced radios can be used and are often more effective over greater distances (Figure 21.2). However, radio frequencies are often controlled in many regions of the world, and in many cases one must register the frequency being used in order not to compete with frequencies used by other agencies (e.g., military, emergency services).

Figure 21.2 A researcher uses a radio to communicate with fellow researchers in the field about the animals they are studying (photo by K. C. MacKinnon).

Electronic Data Recorders

One of the most important considerations when planning behavioral research on primates is how the data will be physically collected, managed, and analyzed. Because primates are often studied outside of the office or laboratory, reliance on a portable, reliable, and durable system for data collection is essential. Perhaps the simplest and arguably one of the safest ways to collect data in the field is with a waterproof notebook and pen. However, the past decade or so has seen the emergence of several commercial software systems specifically designed to assist with collecting, analyzing, and presenting field behavioral data that can vastly modify and improve the process.

For example, one of the most frequently used software programs for collecting animal behavioral data is Observer from Noldus (http://www.noldus.com). Less expensive (or free) programs include Jwatcher (http://www.jwatcher.ucla.edu) and Cybertracker (http://www.cybertracker.org), or one can use all-purpose software applications such as Pendragon Forms (http://www.pendragon-software.com) or program and customize one's own data-collection system. The software programs can be loaded onto powerful, compact personal digital assistants (PDAs); palm computers such as the HP iPac (http://welcome.hp.com), Palm Pilot (www.palm.com), Psion Workabout (www.psionteklogix.com), or Trimble Recon (www.trimble.com); or even certain cell phones. Data are temporarily stored in the handheld computer in a standard format (i.e., comma delimited values), which can then be exported to desktops or laptops for analysis with various database or spreadsheet functions or geographic information systems (GIS) applications. Some of the diverse types of data which can be collected using these programs include association, time, duration, quantity and quality of foods eaten, aggressive and sexual events, postures, and movements. With electronic data recorders, digital images and video can also be automatically captured, associated with, and linked to the data. Many software programs can also be used to analyze sequential data, calculate time budgets, and code and analyze video footage. Some programs, such as Cybertracker, link geo-referenced data with real-time GPS data streamed into the same PDA to integrate spatial data and create detailed maps and indices of abundance.

There are many advantages of electronic data collection. Perhaps the greatest advantage is the efficiency of data entry. Transferring data and field notes from a notebook into an electronic database by hand is monotonous, extraordinarily time-consuming, and prone to errors in transcription; however, with a PDA, data can be transferred to a database within seconds. Electronic data can easily be backed up in several locations and can even be sent by e-mail for remote storage, eliminating the risk of theft, loss, or malfunction of the data-storage devices in the tropics. Digital data collection also saves paper. For research teams, additional advantages are apparent as standardized data-collection forms can increase the potential for more uniform data collection among multiple researchers and the quality and interobserver reliability can be quickly assessed.

Disadvantages of electronic data collection include the potential for data to be lost through computer, software, and human errors; power loss; and theft. Data files need to be backed up frequently, preferably after each data-collection session. Moreover, electronic equipment is susceptible to rain, heat, and moisture (all par for the course in most primate habitats) and can break easily without proper protection. Another disadvantage is the cost. For example, Noldus Observer software alone costs thousands of dollars (although free or inexpensive software can now be located on the Internet). Initially, considerable time is needed to format the ethogram or customize the coding system and data-collection process. PDAs may also exhibit slow data-processing speeds depending on the number of files in use and memory capacity. PDAs also require frequent battery changes, which adds to their expense and environmental impact, though this can be mitigated through the use of rechargeable batteries. Lastly, primatologists working nocturnally may not find the backlighting setting functional as battery drain occurs rapidly.

Factors for consideration when choosing a particular system include assessing the main form of use (e.g., video coding, focal behavioral follows, sequential data analyses, ecological or ranging data, census measures) as some have more specific functions or better interface capabilities for certain uses than others. Additional considerations include cost and durability; a comfortable and clear user interface and ease of use, paying particular attention to the ergometrics and keypad design to facilitate straightforward data collection and elimination of errors during keying; and the simplicity of transferring data to a laptop or desktop computer. Additional memory cards can be purchased to back up the handheld memory and prevent data loss, and it is advisable to have a spare PDA or two if one is relying heavily on these for data collection. Resealable plastic storage bags are indispensable for protection against rain, or better yet, invest in a shatterproof, crushproof, and waterproof container or slicker for the PDA (e.g., the Otterbox 1900 [http://www.otterbox.com] or the Aquapac [http://www.aquapac.net]).

After years of field use and feedback from field researchers to software designers, combined with increasing versatility of portable data-collection devices, electronic data-collection systems are getting cheaper, more reliable, and more adaptable. This is clearly a significant advantage for field researchers because these devices should result in more accurate, efficient, and economical data collecting and processing. Regardless of which system is chosen, always bring pen and paper as a backup in case the data recorder fails.

Auditory Data

Another area of behavioral research aided by advances in technology is the study of the form and function of vocalizations. There are two major ways of investigating the behavioral significance of certain vocalizations. The first is to record vocalizations in addition to noting the behavioral context under which they occur. When recorded, each vocalization can be "mapped" out on a sonogram to help distinguish different kinds of vocalizations that cannot be differentiated by the human ear. When they are indicated as being different on a sonogram, they can be compared to the behavioral contexts in which they occurred and a basic correlation can be formed (e.g., Rasoloharijaona et al. 2006, Gros-Luis et al. 2008). Once a hypothesis regarding the function of a certain vocalization has been formulated, the next level is to test this through play-back experiments. In a *play-back experiment*, vocalizations recorded in the field under known behavioral situations are played back to individuals in the group and the responses to the call are recorded. If the behavioral reaction is consistent with the initial hypothesis of the function of the call, then this hypothesis is strengthened. If not, one reevaluates!

Play-back experiments can also used to be test other aspects of primate behavioral ecology. For example, Goerlitz and Siemers (2007) played insect rustling noise to gray mouse lemurs (*Microcebus murinus*) in order to determine the auditory cues that the lemurs use while foraging for insects. Their results showed that the mass–amplitude correlation of prey-generated rustling sound is used by the lemurs to evaluate the average mass of insect prey and to guide foraging-based decisions.

Gil-da-Costa et al. (2003) used play-backs of harpy eagle (*Harpia harpyja*) calls to investigate the behavioral response of howler monkeys that had experience with this predator versus those who had not. They showed that upon introduction of two harpy eagles to a previously naive population, howler monkeys (*Alouatta palliata*) rapidly developed antipredatory responses specific to the call of this avian predator; that is, it was not a generalized reaction to avian vocalizations. These monkeys responded with significantly more vigilance and took much longer to return to baseline behaviors after the play-back than a control group that had never experienced harpy eagle predation.

Activity and Ranging Data

How individual primates spend their waking hours can reflect many things about their diet, ecology, and evolution. Activity/time budgets can be compared between all levels (i.e., between individuals, different age/sex classes, different groups, different populations, different species). For diurnal species, gathering activity data is a fairly simple procedure. Using either focal animal or instantaneous scan sampling, a traditional set of behaviors are recorded—for example, travel, rest, feeding, and other (usually social) behaviors— and the percentage of the total observation time that each was observed is calculated.

Uninterrupted remote recording of individual activity can also be achieved with accelerometer collars that store activity data until they are downloaded upon removal of the collar. With proper calibration, three-dimensional accelerometer data may be able to distinguish between behavioral categories (Sellers and Crompton 2004, Watanabe et al. 2005). Data can be stored for as long as 6 months before download is necessary (e.g., Kappeler and Erkert 2003, Erkert and Kappeler 2004, Fernandez-Duque and Erkert 2006). Such technology is especially useful for nocturnal species where changes in activity levels may not be easily observed at night (e.g., Fernandez-Duque and Erkert 2006).

Ranging Data

An important part of any behavioral study is to determine the daily path lengths and home ranges of the study animals. Traditionally, this has been done by recording the location of a focal animal, subgroup, or group at a fixed time interval in relation to locations on a trail system (a grid system is preferred). Recently, more precise locations can be determined using a handheld GPS. The effectiveness of GPS has in the past been negatively impacted by forest cover; however, recent models appear to function well even in dense canopy cover. Additionally, radiotracking (manual or automated) and GPS tracking via a GPS tag included in an animal's collar can provide precise ranging data.

Manual Radiotelemetry

Traditional, manual radiotracking on animals involves radio transmitters attached to or implanted in study animals that emit radio signals at a predefined frequency. The researcher receives signals from these transmitters via antennas and

receivers (e.g., Campbell and Sussman 1994, Fedigan et al. 1988, Bearder and Martin 1980b, Li et al. 1999, Honess and MacDonald 2003 for review). Such technology is especially useful for the study of nocturnal species, where changes in activity levels may not be easily observed at night (e.g., Fernandez-Duque and Erkert 2006) (Figure 21.3). Also, during the day, sleeping nocturnal species can be located easily and, thus, sleeping patterns can be determined using this technology (e.g., Kappeler 1997, Braune et al. 2005, Rasoloharijaona et al. 2006, 2008).

Since the early days of wildlife radiotelemetry, researchers have envisioned using the same technologies which allowed them to manually track study animals by listening to the beeps of a radio-tag to create a remote tracking system (Cochran and Lord 1963, Cochran et al. 1965). A number of such systems have been established (Cochran and Lord 1963, Johnson et al. 2000, Briner et al. 2003), but these attempts were hampered by, among other factors, insufficient computing power to process the large data sets that were produced. Modern database tools make these problems tractable and have led to the blossoming of a number of ecological sensor networks (Porter et al. 2005) including several automated radiotracking systems (e.g., Crofoot et al. 2008, Kenward 2001, Honess and MacDonald 2003).

The benefits of a remote system are numerous. Many behaviors that have a major impact on primate survival and reproduction are difficult to study because they are infrequent, are cryptic, or occur over large spatial or temporal scales. For example, behaviors that take place only once in an animal's lifetime, like dispersing from a natal home range or being predated, are nonetheless

Figure 21.3 An adult male pygmy loris (*Nycticebus pygmaeus*) wearing a radio collar at Seima Biodiversity Conservation Area, Cambodia (photo by K. A. I. Nekaris).

critical determinants of fitness (Cheney and Wrangham 1987, Pusey and Packer 1987, Miller and Treves 2007). In wide-ranging species, observing dispersal and predation events, extragroup copulations, or intergroup competition may be impossible because monitoring sufficiently large areas is not logistically feasible. Additionally, the very presence of human observers may influence the rate or bias the outcome of a number of key behaviors including, most notably, attacks by predators and intergroup competition (Isbell and Young 1993a, Zinner et al. 2001a). These logistical difficulties have hampered research on important but difficult-to-observe behaviors, and the lack of data about predation, dispersal, and intergroup competition in most primate species hampers the development and testing of ecological theories concerning these behaviors (Isbell 1994; Boinski and Chapman 1995; Boinski et al. 2000b; Miller and Treves 2007; Overdorff and Parga 2007).

Tracking systems that can remotely, continuously, and simultaneously monitor the movements and activity of large numbers of study subjects ameliorate this problem by facilitating research on topics that have been difficult to study using traditional methods of field observation and data collection (and even manual radiotracking). The number of people that can be devoted to following study subjects inherently limits studies of primate behavior that rely on human observers. This often means that, in studies of rare behaviors, samples sizes are small, limiting the number of parameters researchers can investigate and the strength of the conclusions they are able to draw. Automated tracking systems increase the number of study subjects a researcher can monitor and thus, in cases where the behaviors of interest can be detected remotely, greatly increase sample sizes. They also allow researchers to monitor animal movements over larger areas than is possible when following study subjects on foot and to "observe" study subjects without the risk that the presence of a human biased the individual's behavior.

Currently, two main types of technology exist for remotely tracking animal movements: automated radiotelemetry and satellite-based (GPS, ARGOS) tracking systems.

Automated Radiotelemetry

Automated radiotelemetry relies on the same basic technological components that are used in traditional, manual tracking. The primary difference between these two methods of tracking is that, rather than relying on a human to listen on a receiver to the beeps emitted by a radio transmitter, automated systems instead employ automated receivers which monitor a preprogrammed set of radio frequencies. By recording either the relative signal strength or the time of arrival of radio signals across an array of antennas, these receivers collect data that can be converted into directional bearings (Rodgers 2001). The locations of radio-tagged animals are then determined via triangulation, using the bearing estimates from several receivers positioned at different sites in the study area.

While in theory straightforward, tracking primates via automated radiotelemetry presents several challenges. First, some primate species can be wide-ranging, and thus, automated receivers must be distributed over relatively large spatial scales. The range of automated receivers is greater in open environments than in forests, where the leaf cover degrades radio signals and decreases reception range. Mounting antenna arrays atop radio towers significantly increases the reception range of the automated receivers in forest habitats, but such towers are expensive to erect and require regular upkeep. Second, powering automated receivers can be difficult at remote field sites. Car batteries recharged by solar panels have proved to be an effective solution to this problem. Finally, maintaining the electronic equipment, including the computers, required for automated radiotelemetry is difficult in the humid conditions of many primate habitats.

Although automated radiotracking requires extensive infrastructure which is expensive and time-consuming to maintain, it has several advantages over satellite-based tracking systems that make it better suited for certain types of research. Radio-tags are smaller, longer-lived, and less expensive than GPS tags and, thus, can be used to study a wider range of species over longer time scales (see Kenwood 2001). Dense canopy degrades the VHF signals used in radiotracking less than the UHF signals used by satellite tracking systems (Rodgers 2001); thus, it may be possible to implement successful automated radiotelemetry systems in habitats where satellite systems cannot reliably acquire location estimates. Finally, real-time collection of automated radiotracking data is possible with the addition of a wireless network that transmits data from receivers in the field to a central computer bank. Real-time data allow researchers to respond promptly to important occurrences, such as mortality events (Lambert et al. 2009).

Given these costs and benefits, automated radiotelemetry is probably best suited to well-established, accessible field sites for use in studying animals too small to be fit with GPS collars or in long-term studies where the battery capacity of GPS tags cannot provide a sufficient life span. Only one automated radiotracking system is currently being used to study primate ranging behavior (Wrangham et al. 2007, Crofoot et al. 2008).

GPS Tracking

Satellite-based tracking has revolutionized the study of animal movement ecology by allowing automated collection of location data over global scales (Wikelski et al. 2007). Most satellite-based animal tracking today relies on the Navstar GPS, an array of 24 geosynchronous satellites owned and maintained by the U.S. Department of Defense. GPS tags worn by study animals (Figure 21.4), rather than emitting signals like radio-tags, instead act as receivers for the signals broadcast by these 24 satellites. Locations are calculated based on the time lag between the radio signals being sent by the satellites and received by the tag and information about the orbital path of each satellite (Rodgers 2001, Hughes 2003). Location data are typically stored on the collars and either downloaded remotely, via radio or satellite link, or manually when the collar is recovered (Rodgers 2001).

Until recently, the weight and cost of satellite tags and their poor performance in forested environments have limited the utility of these systems for tracking most primate species (Phillips et al. 1998, Sprague 2004, Sprague et al. 2004, Markham and Altmann 2008). Although the U.S. Department of Defense discontinued its policy of degrading the accuracy of civilian GPS receivers in 2000, decreasing the error of GPS location estimates to less than ±10 m (Rodgers 2001), GPS reception under forest canopy remained poor (Rempel et al. 1995, Phillips et al. 1998, Jiang et al. 2008). Sprague and colleagues (2004) reported a 20% fix rate for a GPS collar worn by a female Japanese macaque (*Macaca fuscata*) and noted that all but one instance when the GPS tag failed to obtain a location estimate occurred in forested areas. However, new high-sensitivity GPS microcontrollers have improved the rate at which the newest satellite tags acquire location estimates, making them appropriate for use in forest habitats. A recent study of Yunnan snub-nosed monkeys reported a fix rate of 82% in a temperate, coniferous forest (Ren et al. 2008), while a collar deployed on a white-faced capuchin monkey living in a lowland, tropical forest in Panama achieved an 83% fix rate (Crofoot unpublished data). Reductions in the energy requirements of these new GPS receivers have also allowed the development of significantly lighter tags (<10 g) (Brown and Brown 2005, Tobler 2009), which are small enough to use on a wide range of primate species. The life span of GPS collars is highly dependent on sampling frequency,

but many of these smaller collars have life spans ranging from several months up to 2 years (Brown and Brown 2005, Crofoot unpublished data). Thus, while GPS tracking has previously been limited to medium- to large-bodied primates living in relatively open habitats (e.g., baboons, Markham and Altmann 2008; Japanese macaques, Sprague et al. 2004; Yunnan snub-nosed macaques, Ren et al. 2008), advances in technology have now made it possible to use GPS collars to track even small- to medium-sized, forest-living species (see Hughes 2003 for review).

The main advantage of satellite tracking is its relative simplicity—all that is required are collars and a mechanism for extracting data from the collars. The major drawback remains the more limited battery life compared to radio transmitters. Automated radiotelemetry, on the other hand, requires extensive infrastructure, which is expensive and time-consuming to maintain. However, it has certain advantages over satellite-based systems, including the ability to track very small animals for long periods of time in heavily forested environments and to provide real-time data on the location of study animals.

A Special Consideration: Studying Nocturnal Primates

Studying the behavior of nocturnal primates suffers from the obvious problem of collecting data from animals that are active at night and thus difficult to see. Even the best flashlights do not provide the kind of light available during the day. Even if they did they would be impractical to use while studying most of the strepsirhine primates as their reflective eyes would render them practically blind in such a bright light and clearly this would lead to drastically altered

Figure 21.4 A wild capuchin monkey (*Cebus capucinus*) wearing a GPS tag collar (photo by M. Crofoot).

behavior. Thus, researchers of these primates have traditionally relied on dimmed flashlights/headlamps (e.g., Kappeler 1997, Rasoloharijaona et al. 2008) or covering a flashlight with a red filter when close to the study animals (e.g., Pimley et al. 2005a). Night-vision technology has become more affordable and widely available and is being used in field studies (e.g., Fernandez-Duque 2003). However, night-vision goggles/binoculars still do not allow a researcher to see the kind of detail necessary to record certain behavioral data (E. Fernandez-Duque personal communication).

Related to this topic is the difference between the strepsirhine and haplorhine nocturnal primates in their reaction to light availability. Because the haplorhine nocturnal primates (*Tarsius* spp. and *Aotus* spp.) lack a tapetum lucidum (a reflective layer in the back of the eye that concentrates light, making night vision more effective), they are not negatively impacted by increased moonlight. Indeed, their activity actually increases on moonlit nights (Fernandez-Duque and Erkert 2006; see also Chapters 6 and 10). In contrast, the majority of lorisiform and nocturnal lemuriform species suffer from what is known as "lunar phobia"—they are less active on moonlit nights (probably as a result of predator avoidance). Thus, scheduling one's field research around the phases of the moon can be an important aspect of nocturnal fieldwork. While yearly moon phase maps are available, computer programs such as MoonDock (for Macs only, copyright Gary C. Morton) or Lunar Phase Pro (for PCs only, copyright Night Sky Observer) can be loaded onto one's laptop computer, allowing daily access to moon phases on any day in any region of the world.

CONCLUSIONS

Primate behavioral field research has always been an exciting field of study. Today, it is even more so because advances in technology have facilitated the collection of data never before possible. Advances in behavioral data collection combined with rapidly advancing methods for endocrinological and molecular data analyses now allow primatologists to investigate questions not previously open to field research. However, even with all of this wonderful technology, the reality it that nothing can replace the careful and direct observation of behavior. Only with hours (hundreds, if not thousands) will researchers become familiar enough with their subjects to be attuned to the subjects' important, but subtle, social interactions.

22

Measuring Ecological Variables for Primate Field Studies

Erin R. Vogel and Nathaniel J. Dominy

1. Why is it important to measure abiotic variables when observing primates in the wild?
2. What are the different methods that can be used to measure resource abundance, and what are the advantages and disadvantages of each method?
3. What are the most important factors to consider when setting up phenology plots?

INTRODUCTION

A central goal of primatology is to understand why such immense variation exists in all aspects of primate behavior including reproduction (Chapters 29, 30), feeding ecology (Chapters 33, 34), predator avoidance (Chapter 35), locomotion (Chapter 36), social interactions (Chapters 38–42), and social learning and intelligence (Chapters 43, 44). Yet the factors that drive such variation are challenging to study. Some behaviors may be explained by historical contingencies or genetic drift, but most appear to improve the survival and reproductive fitness of individuals. As a result, there is a long-standing tradition among fieldworkers to contextualize behavioral observations—that is, to describe and quantify the ecological characteristics of a habitat that can potentially account for the evolution of, or variation in, a behavior in a particular population or species. Yet habitats can be complex, especially in the tropics, where most primate species live. Notwithstanding the fact that habitats are dynamic and changing through time, it can be difficult to characterize or compare them.

Descriptions of tropical habitats are based historically on climatic, elevational, and phenological characteristics (e.g., montane forest, dry/wet/wet seasonal, moist forest, deciduous or evergreen forest); soil type (e.g. peat/swamp forest); and structure (open or closed canopy). In general, tropical forests have a predictable seasonal pattern, with peaks in rainfall following drier periods of high solar radiation (Whitmore 1990, Richards 1996). Annual variation in rainfall, in combination with soil quality, can have dramatic effects on the availability and quality of food for all primates, whether they depend on fruit, leaves, flowers, insects, or some combination of food types. Likewise, these abiotic and biotic variables can also influence the internal structural complexity of a habitat, which in turn could exert a selective pressure on primate physiology, morphology, and behavior. Although temperature is less likely to affect annual variation in food availability relative to precipitation and soil type (van Schaik and Pfannes 2005, Paoli and Curran 2007, Cannon et al. 2007, but see Ting et al. 2008), it can help explain many aspects of primate behavior, including variation in activity budgets (Dunbar 1998, Erkert and Kappeler 2004, Fernandez-Duque and Erkert 2006, Campos and Fedigan 2009), sleeping behavior (Randrianambinina et al. 2003, Schmid 2001, Schulke and Ostner 2007), ranging behavior (Janmaat et al. 2006, Li et al. 2008), reproductive behavior (Fernandez-Duque et al. 2002, Kowalewski and Zunino 2004, Anderson et al. 2006), and social behavior (Nunn and Deaner 2004, Hill 2006). Thus, variation in abiotic factors leads to variation in biotic factors (e.g., food availability, food distribution, forest structure), which affect almost all aspects of primate behavior.

Real-time monitoring of environmental factors and describing the physical characteristics of tropical forests are critical to understanding both intra- and interannual variation in primate behavior, within and between populations. The goal of this chapter is to provide a range of methods to measure the abiotic and biotic factors that may influence primate behavior in the wild. We will discuss methods ranging from simple, low-cost procedures to more advanced, recent technologies.

MEASURING ABIOTIC VARIABLES THAT AFFECT PRIMATE BEHAVIOR

Researchers should begin by constructing a standard weather station at their field site, following the specifications of the National Weather Bureau. The location of the weather station should be representative of the forest that the study species inhabit. Researchers typically construct the weather station (also known as a "Stevenson screen") along a trail that is close to base camp yet within the forest, to make the monitoring of abiotic variables more convenient. Instrumentation to measure precipitation should be kept separate from the main weather station if the study species inhabits a forest environment.

The role of the instrument shelter is to protect the thermometers and/or other instrumentation from direct sunlight, as well as from precipitation and flying debris. It is recommended to construct an instrument shelter that is made of wood, covered on top, and with well-ventilated materials (e.g., louvered panels or aluminum screen) on all sides to help protect equipment. If louvered panels are used, there should also be a gap between the slats of approximately 1 cm and the slat angle should be roughly 50–60 degrees from horizontal. Good ventilation is critical for accurate temperature and relative humidity readings; thus, if necessary, the front can be left open. However, if animals are observed tampering with the equipment, the front should be covered with ventilated material and locked. The entire shelter should be painted white, to maximize the reflection of sunlight. The instrument shelter should be mounted so that the thermometer inside is 1.5 m above the ground, to prevent ground heat from affecting the temperature readings.

Temperature and Relative Humidity

The maximum and minimum temperatures should be measured over a 24 hr period at the same time each day. Most thermometer models display both current temperature and maximum/minimum values in both Fahrenheit and Celsius scales and should be read to one decimal place on the Celsius scale. The thermometer should be mounted within the weather station shelter and aligned next to a certified calibration thermometer. Digital thermometers can also be used, but the accuracy of such thermometers varies with the measurable range and price. In addition, like all electrical equipment in tropical climates, it is our experience that models that are not water-resistant tend to cease working after about a year. If any battery-operated temperature devise is used, we recommend always purchasing a low-cost maximum/minimum thermometer as a backup method.

Recently, the development of compact and durable data-logging temperature devices has made daily temperature collection quite easy (Strangeways 2000). These devices have also been used for measuring body temperature (Dausmann 2005; Schmid 1998, 2000, 2001; Schmid and Speakman 2000), nest temperature (Florez 2007), and sleep-site temperature. Both HOBO® (Onset, Bourne, MA) (e.g., v2 submersible temperature logger) and iButtons® (Maxim Integrated Products, Sunnyvale, CA) make compact, weather-resistant temperature data loggers about the size of a large watch battery (Figure 22.1). They measure temperature with automated sensors and record the data in a protected memory section. Operating ranges typically span −40°C to 85°C, and they can record

Figure 22.1 Picture of iButtons® temperature and humidity data logger (photograph taken by Erin Vogel).

7,000–42,000 readings depending on the model, which are taken at equidistant intervals at a user-defined rate. This rate typically will determine battery life. In addition, some models (iButtons) have a programmable alarm that indicates the maximum and minimum temperatures. All models require additional software and data-transfer devices, which can be downloaded and/or purchased separately. Data can be transferred to a computer every 2 weeks or monthly, depending on the storage capacity. Like all equipment, these devices should be housed within the weather shelter to maximize battery lifetime and accuracy of readings.

Relative humidity is defined as the ratio of the partial pressure of water vapor in the mixture to the saturated vapor pressure of water at a prescribed temperature. Thus, it is highly dependent on temperature. If a measure of relative humidity is desired, the simplest measurement can be taken using a hygrometer, which is a set of wet and dry bulb thermometers mounted in a molded plastic case with a wick and water cistern. Wet and dry bulb observations are then compared to a chart to determine percent relative humidity. A whirling hygrometer can also be used, which maximizes evaporation and the accuracy of readings. There are some models of data loggers that also record relative humidity in addition to temperature. MAXIM makes an iButtons model that records relative humidity at the same time interval that it records temperature. We have used this model at the Tuanan Field Station in central Kalimantan, Indonesia, since 2005. Other companies make similar models that can be found in forestry supplier catalogs, although they may not be weather-proof.

Precipitation

The amount of precipitation can have dramatic effects on the availability of food for primates living in tropical climates. For example, periods of drought have been associated with low fruit availability, which can affect ranging behavior. Alternatively, bouts of heavy rain can reduce potential travel time in some species, while others continue with their daily activities unaffected. The amount of precipitation is perhaps one of the simplest and least costly abiotic variables to record. As with all abiotic variables, it is important to use methods that are compatible with conventional meteorological methods used in official observing sites. For precipitation, this includes making observations at least once at the same time each day (typically at the end of each day), setting up the rain gauge approximately 30 cm above ground level to avoid water splashing in when too low and the effects of wind when placed too high, and making sure it is level. We recommend hanging the rain gauge on a post in an area where there will be no runoff of water from buildings or trees. The general rule is that the rain gauge should be placed away from any potential obstructing object such that the distance between the gauge and the object is two times greater than the height of the object.

The amount of precipitation can be measured with a simple standard rain gauge or more costly data-logging rain gauges (Table 22.1). A low-cost rain gauge is usually made of plastic or aluminum and consists of a funnel that runs into a cylindrical inner measuring tube that typically measures up to 1.25–2.54 cm and should be graduated to 0.1 mm. The level of rainfall should be measured from the bottom of the meniscus of the inner tube. This inner tube rests inside of an overflow container that has a capacity of 17–28 cm, depending on the model. Thus, overflow can be poured into the inner measuring tube to get total daily precipitation. During periods of heavy rainfall, the rain gauge may have to be monitored every 12 hr. It is important to clean the rain gauge approximately once a month with water and a bottle brush, to eliminate any debris that has fallen into the collection container.

More expensive battery-powered data-logging rain gauges are also available. The advantage of these devices is that they can be programmed to record the amount of rain at specific times, and some models can store data and do not have to be monitored daily. Thus, if the rain gauge is placed farther away from the main study area, researchers can monitor the gauge less often. Other models have a wireless receiver that transmits the total amount of rainfall, but the data may still need to be recorded manually every 24 hr. The most advanced model involves a tipping bucket system that measures each 0.2 mm of rainfall and stores the data with a date and time stamp as each tip occurs. The advantage of this method is that the intensity of rainfall can also be deduced. Data can be easily offloaded into a PC computer and then into an Excel file. While this model is convenient, many digital data loggers tend to fail in tropical environments due to high humidity; thus, a standard backup source should also be made available.

Table 22.1 Equipment Needed to Measure Abiotic and Biotic Variables

VARIABLE TO MEASURE	DESCRIPTION OF EQUIPMENT NEEDED	EASE OF USE	COMMENTS
Precipitation	Plastic or aluminum rain gauge	Extremely easy to use	Pros: Accurate measurements and easy to use Cons: Must check every 12-24 hr at the same time each day depending on amount of precipitation; must mount in an area that is open; cannot measure intensity of rainfall
	Digital and data-logging rain gauge	Medium: software can be difficult to program correctly	Pros: Do not have to check rain gauge every 24 hr at the same time of day as data are collected continuously in most models (some digital models have to be monitored remotely every 24 hr); can measure intensity of rainfall Cons: Should use a backup manual gauge in case of equipment failure; although these claim to be weather-proof, some data loggers cannot withstand the constantly high relative humidity in some field sites
Temperature/Relative Humidity	Standard manual maximum/ minimum thermometer	Extremely easy to use	Pros: Reliable, accurate data collection; inexpensive option Cons: Can get air bubbles in transit, which can affect readings. Must check at the same time each day, and if reading is not made, data are lost at least 2 days
	Temperature and/or relative humidity data logger	Medium: software can be difficult to program	Pros: USB interface; do not have to check every 24 hr at the same time of day as data are collected continuously in most models (some digital models have to be monitored remotely every 24 hr) Cons: Battery life can be 5-10 years, and for some models batteries are not replaceable; should use a backup manual device in case of equipment failure; although these claim to be weather-proof, some data loggers can not withstand the constantly high relative humidity in some field sites; some models are difficult to program
	Humidity hygrometer/ psychrometer	Extremely easy to use	Pros: No reliance on external power source and, thus, should not malfunction due to weather conditions Cons: Have to check daily
	Temperature/relative humidity data logger	Medium: software can be difficult to program	Pros: Do not have to check every 24 hr at the same time of day as data are collected continuously in most models (some digital models have to be monitored remotely every 24 hr); once software is programmed, device is very easy to use Cons: Should still use a low-cost backup device or check the devise every few days in case of malfunction; devises are costly, and software can be difficult to use
Canopy Cover	Digital camera with fish-eye lens attachment; tripod		Pros: Affordable, easy to use, tripod and camera can easily be set up anywhere in the forest Cons: For most algorithms used to translate pixel count into leaf area, leaf plane, must be parallel to camera plane to satisfy assumptions of the models used; thus, count method of translation can be subjective and not as accurate as other methods
	Ground-based LiDAR system (includes range finder, laptop computer, stable platform system, software)		Pros: High accuracy of leaf area and direction Cons: Need to be able to walk straight lines at a constant speed; expensive
Light	Hand held light meter		Pros: Light weight, easy to carry around the forest Cons: Not sensitive to low-light conditions and, thus, not adequate for nocturnal studies
	Spectroradiometers		Pros: Portable, can measure the radiance spectra of difficult-to-reach surfaces and can instantly convert radiance spectra to human chromaticity units Cons: Can be very expensive
	Spectrometer		Pros: Portable and affordable Cons: Contact with a stimulus object is necessary to measure its luminance or spectral reflectance
Tree abundance and phenological plots	100-meter tape measurer, DBH tape measurer, tree height clinometer range finder, aluminum nails (1 lb box), aluminum tags (100 tags)	Extremely easy to use	

Light

The intensity and spectral characteristics of visible light can affect the onset and offset of activity and the visual perception of predators, mates, or food (Pariente 1979, 1980; Endler 1997; Gursky 2003c; Osorio et al. 2004). At the simplest level, a handheld light meter can be used to measure the amount of diffuse light in the environment, that is, the illuminance, measured in lux units (Dominy and Lucas 2004). More sophisticated spectroradiometers (e.g., the PR-655 SpectraScan; Photo Research, Chatsworth, CA) measure not only the amount of light entering the eye, i.e., the luminance (measured in candelas m^{-2}), but also the spectral characteristics of the incoming light (the radiance spectrum). Such instruments typically cost more than US$10,000, but they have the advantage of portability, the ability to measure the radiance spectra of difficult-to-reach surfaces, and the instant conversion of radiance spectra to human chromaticity units (e.g., International Commission on Illumination [or Commission internationale de l'éclairage, CIE] coordinates). Yet, these data may not be relevant to some primates. Many primates lack the human dimension of trichromatic color vision.

To model how primates with different kinds of color vision might perceive chromatic and achromatic stimuli, it is useful to estimate the quantum catches of individual pigment types (Sumner and Mollon 2000; Dominy and Lucas 2001, 2004; Osorio et al. 2004). A modular system based on a spectrometer (e.g., Jaz or USB2000; Ocean Optics, Dunedin, FL) outfitted with peripheral accessories is useful for measuring the spectral reflectance of a stimulus, such as a fruit or insect, and the spectral irradiance of down-welling light (e.g., sunlight, twilight) under different conditions (e.g., extent of cloud or canopy cover). The advantage of the Ocean Optics system is that it is portable and more affordable than a spectroradiometer. As a result, it has become relatively popular among field biologists studying the visual ecology of primates (Lucas et al. 2003, Dominy et al. 2006, Osorio et al. 2004, Stoner et al. 2005, Hiramatsu et al. 2008). A disadvantage of the system is that contact with a stimulus object is necessary to measure its luminance or spectral reflectance—stimuli such as fruit in the canopy or the skin color of a wild animal may be inaccessible.

MEASURING BIOTIC VARIABLES THAT AFFECT PRIMATE BEHAVIOR

This volume highlights the great behavioral variation that exists in the Primate order. Given this variation, it is extremely important to consider both the species under investigation and the research question when deciding upon the precise methods and scale that should be used when measuring biotic variables. In this section, we describe methods for quantifying forest structure, plant abundance and distribution, and food availability for primate field studies.

Canopy Continuity and Forest Structure

The continuity and vertical structure of a habitat can affect the foraging ecology and locomotor behavior of primates (Gautier-Hion et al. 1981, Emmons and Gentry 1983, Grand 1984, Cannon and Leighton 1994). Typically, the continuity of a canopy is estimated indirectly by measuring the amount of photosynthetically active radiation (PAR) transmitted to the ground or the *leaf area index* (LAI), which is the ratio of standing vegetation divided by the surface area of the earth beneath. Measures of PAR (e.g., with the LI-191 quantum sensor; LI-COR Biosciences, Lincoln, NE) and LAI tend to be correlated with each other. Recently, numerous methods have emerged for measuring LAI and the internal structure of a habitat, ranging from satellite-based remote sensing techniques at the scale of continents (Weishempel et al. 2001) to the use of sophisticated laser scanners and range finders at smaller regional scales (e.g., LiDAR) (Lefsky et al. 2000, Dial et al. 2004, Parker et al. 2004). For the field biologist, digital hemispheric photography (DHP) is perhaps the most affordable means of measuring LAI (Figure 22.2). For example, the Nikon (Tokyo, Japan) Coolpix series of digital cameras can be outfitted with an FC-E9 fisheye lens. This camera–lens combination has proved very popular among plant ecologists, partially because free software exists for analyzing the images (e.g., Gap Light Analyzer, available at http://www.rem.sfu.ca/forestry/downloads/gap_light_analyzer.htm).

Abundance of Plant Resources

Data on the abundance and distribution of plant resources can be very important when interpreting primate behaviors, including ranging behavior, social interactions (e.g., food competition), and both intra- and interspecific interactions. Given the great variation in home range size, group size and spread, habitat types, and body mass, it is not surprising that the different methods used to quantify resource abundance can lead to very different results for different species. Thus, it is important to keep in mind that different research questions and/or study species will require variation in the methods and scales used when quantifying the relative abundance, spatial distribution, and/or density of plant resources. Before setting up a sampling protocol, one should think carefully about the types of questions to be answered by the project. In this section, we will discuss several different methods that can be used to measure plant abundance and when they are most appropriate. Different statistical methods used to estimate abundance are not covered in this section as they are highly variable depending on the research question but are discussed in detail in Krebs (1999).

Plot-Based Methods

When sampling the abundance of plant resources, it is important to consider three main factors. First, what shape

Figure 22.2 Example of digital hemispheric photography for measuring leaf area index. Photograph was taken with the Nikon Coolpix series of digital cameras outfitted with an FC-E9 fisheye lens (photograph taken by Nathaniel Dominy).

is most appropriate? Second, what is the relevant size of the plot that should be used? Third, what forms of vegetation need to be measured (e.g., if the study species feeds from lianas, the method should include the sampling of herbaceous vegetation)? The answers to these questions will vary depending on the study species and research question(s). If a researcher is simply interested in the relative abundance of feeding tree species within the study area, then as long as the sampling area is representative of the study area used by the species, the size of the plots is not as important. However, if the study focuses on food competition among group members, then a plot size more relevant to the spatial spread of the study species may be more useful (Koenig et al. 1998). In general, longer and narrower plots with a 2:1 ratio tend to be better than square or circular ones of the same area as they will cover more patches, especially in heterogeneous habitats.

Primate researchers commonly use rectangular plots that cover an area of 100 m² and record data from trees with a diameter at breast height (DBH) greater than 10 cm. However, if the study species uses additional plant resources, additional, smaller plots should be used as well (e.g., 10 m² for shrubs, bushes, and small trees [2–4 m tall] and smaller plots of 1–2 m² for measuring herbaceous vegetation). Tree circumference or DBH is typically measured at 1.37 m above the forest floor on the uphill side of the tree. A DBH tape measurer can be purchased from a forestry supplier to facilitate data collection, although the DBH can easily be calculated as the circumference at breast height divided by π (3.14159). The number of plots used will depend on the heterogeneity of the habitat.

In plot methods, all individual trees, shrubs, or woody vegetation (depending on what is being sampled) within the plot are marked with aluminum tags or flagging tape, measured, and identified. We recommend hiring a local botanist once the plots are initially set up, to assist with species identification. The general rule is that if the center of the tree lies within the plot, it should be included within the sample. Typically, researchers tend to focus on trees with a DBH greater than 10 cm, to reduce sampling effort. However, if one is also interested in quantifying smaller trees, herbaceous vegetation, and/or vines (lianes), smaller plots or a nested plot design, known as a "modified Whittaker plot," can also be used (Stohlgren et al. 1998, Barnett and Stohlgren 2003). This method has the advantage that it can greatly increase the area sampled while decreasing the problems of spatial autocorrelation common with nested plot methods (Stohlgren et al. 1998). Briefly, the plot contains ten 1 m² nonoverlapping subplots, two 10 m² (2 × 5 m) nonoverlapping subplots, and one 100 m² (5 × 20 m) subplot in the center, all nested within a 1,000 m² (20 × 50 m) main plot (see Fig. 1a in Barnett and Stohlgren 2003). Using this method, one can use the smaller plots to sample all vegetation and the larger plots to record large trees. Because this method is labor-intensive, researchers can modify the size of the trees sampled in each of the different-sized plots or the size of the plots such that the area covered is relevant to their study species. Plot methods are advantageous in that they can be used to measure many different types of vegetation and they can easily be broken down into subplots of relevant sizes for determining spatial distribution. However, setting up large plots can be very labor-intensive, and different plot shapes can result in different plant-density estimates. Thus, if plots are set up to compare field locations or sites, the plot dimensions should be the same.

An alternative plot method that has been used by primatologists for both abundance and phenology data is the belt-transect method (Whitten 1983, Wieczkowski 2004, Vogel et al. 2008). Most primatologists who have used this method use 0.1-hectare plots (plot 10 × 100 m), which are more manageable and can easily be used to sample across heterogeneous habitats. As with all vegetative sampling methods, belt transects should also be set up randomly and can be set up along preexisting trails. All trees within the plot should be tagged and identified, and the DBH should be measured. In addition, the same trees can be used for phenology data. If subplots are to be used for smaller vegetation, they should be spaced outside of the plot to avoid autocorrelation issues.

Plot-less Methods

In plot-less methods, plant communities are not sampled with delimited plots but with sampling points, one-dimensional transect lines, or certain distances within the stand (Knapp 1984). The two most common types of plot-less methods for sampling vegetation are the point-quarter sampling method and the nearest-individual method. For both methods, a number of sample points are located at random within the study area. This can be done by placing a grid over the study area and using a random-number generator to determine the location of points. With both methods, the density of all trees or of individual tree species can be estimated. The number of sample points necessary for an adequate assessment is partly dependent on the species cover within a survey area, but a minimum of 50 points is recommended for medium-dense vegetation (Knapp 1984).

For the point-quarter method, two perpendicular straight lines that cross at a predetermined, randomly placed point are measured. By doing this, four quadrants are created, and in each of these the distance to the nearest tree from the center point is measured. The orientation of the lines should be determined in advance, and most researchers use compass points to lay down the straight lines. The distances of these four trees are measured, the species are identified, and the DBH are measured. These distances are then averaged, and the mean of these averages over all samples is equal to D1. Density is then equal to $A/(D1)^2$, where A is equal to the unit area. For example, suppose the mean distance of trees with DBH ≥10 cm to the sampling point is 3.72 m. If the study area is equal to 1 hectare, the density is equal to 723 trees per hectare. For the nearest-individual method, the nearest tree to the sample point is located and the distance between it and the sample point is measured. The tree also should be identified and the DBH should be measured. The mean of the distances over all of the samples is D2, and the density is equal to $A/(2D2)^2$. While each sample will take longer for the point-quarter method compared to the

nearest-individual method, the nearest-individual method tends to give a more variable estimate, and thus, more sampling points are required. These methods are probably best for surveys of overall tree density or if certain species are of interest (Ganzhorn 2002).

The advantages of plot-less sampling are that these methods can be much faster at estimating plant density than the quadrate method because they avoid the labor-intensive work of setting up plots and the researcher is able to eliminate the subjective error associated with the sample plot boundaries. The disadvantages of these methods are that rare species may not be sampled and that it is almost impossible to set up a truly random sample, which can complicate statistical analyses.

Vogel and collaborators (2004, 2006, 2007) used a "focal-tree" method modified from Strier (1988) to examine alternative resource availability in relationship to the frequency of food-related aggression. In this modified plot-less method, data are focused on the neighborhood surrounding the feeding tree. Within a short time span (usually less than 48 hr) after the behavioral observations (in this case, aggression), a researcher revisits the focal tree to map and describe, in detail, the ecological neighborhood of the tree (Vogel 2004; Janson and Vogel 2006a, Vogel and Janson 2007) (Figure 22.3). The area of the ecological neighborhood will vary depending on the plot size that is relevant to the species under investigation. For example, Vogel and Janson (2007) used a plot size of 0.5 hectares, or a circle with a diameter of 80 m. This distance was used

Figure 22.3 Focal tree method as described in Vogel and Janson 2007. The larger, center tree in the plot is the focal tree. The distance (d_i) to all alternative resources within average group spread (in this case 80 m) is recorded, along with diameter at breast height, crown volume, and fruit abundance.

because 80 m was determined to be the average group spread of the capuchin monkey groups while traveling. Thus, the researchers assumed that individuals would be more hesitant to spread out to alternative resources beyond this area. Using this method, the researcher has a quantitative assessment of the availability of food resources within the vicinity of the contested feeding tree (e.g., alternative resource availability). These data can also be used to determine the general abundance of each tree species; however, this method is also subject to the problems of nonrandom sampling and autocorrelation and, thus, should be used in combination with a plot method to obtain general abundance data of plant species.

Distribution: Patchiness

The concept of resource patchiness is primarily of concern when attempting to understand variation in primate social behavior in relation to plant resources, specifically patterns of food competition. The study of primate socioecology assumes that most behavior is influenced by an animal's surrounding environment and ecological conditions. The distribution and abundance of food resources are two of the many variables proposed to explain patterns of agonism in primates, which are used to determine a species' competitive regime (Wrangham 1980, van Schaik 1989, Sterk et al. 1997, Koenig 2002, Isbell and Young 2002; see also Chapter 32). However, measuring "patchiness" can be difficult. In this section, we discuss measures of resource patchiness in relation to both inter- and intraspecific food competition.

Traditionally, a botanical view has been used to measure the distribution of food sources used by primates. Individual-based methods (plot-less methods) to determine patchiness (used by Cottam and Curtis 1956, Shopland 1987, Pruetz and Isbell 2000) almost invariably focus on spatial dispersion at the scale of mean interindividual distances of the plant. Although these methods are the most sensitive at discerning dispersion relevant to the plant species, they may be irrelevant to how primates perceive the available resources at the time of the feeding bout. The more commonly used plot-based methods offer more flexibility because the plots can be of different scales (Robinson 1986, Janson 1987, Symington 1988c, Kinnaird 1992, Nunes 1995, Chapman et al. 1995, Doran et al. 2002b, Di Fiore 2003b). However, even these methods may be irrelevant for studying intraspecific food competition unless the chosen plot size is relevant to how primates perceive the availability of alternative resources (Koenig et al. 1998). Another difficulty with both of these methods is that they do not take into account local food availability at the time of an aggressive interaction. Most phenological plots are measured monthly or bimonthly, whereas tropical trees often have a ripe fruit turnover rate that is faster than this sampling period, at least for fruit consumed by the larger-bodied frugivores (Terborgh 1983). In fact, Milton (1981a) found that in

Panama ripe fruits were available on individual trees for on average about 22 days and in some cases certain plant foods used by primates were optimally edible for no more than 72 hr per year.

Few researchers have evaluated resource distribution by taking a primate-based view of resource "patchiness." Koenig et al. (1998) and Janson (1987) addressed this issue in part by measuring resource distribution within an area equal to the average group spread. Thus, they took into consideration a scale that was relevant to the study species. However, by using a monthly or bimonthly monitored plot-based system, their estimates of alternative resource availability and dispersion were at times spatially and temporally independent of the feeding tree where an aggressive interaction occurred. Isbell et al. (1998) addressed this issue by measuring distance traveled between alternative resources, feeding bout length, and fruit-ingestion rates directly from the focal animals. From these data, they inferred food quality and alternative resource availability. Isbell et al. (1998) reasoned that a shorter distance between feeding sites should reflect denser food sources and longer feeding bouts at a feeding site should reflect higher fruit abundance or longer handling times. Although this reasoning is justified for a habitat with few food species where the animals may potentially feed in all available resources, in a site where there are several synchronously fruiting species primates may skip a potential food source because a higher-quality feeding tree may be in close proximity. In this case, a *potential* alternative resource is ignored and not recorded (Garber 1988b, Janson 1998b). In addition, individuals in social foraging contexts may often have to skip over resources that have been exhausted by other group members, leading to "leap-frogging" behavior (e.g., Janson 1990a,b; cf. Murton et al. 1971). In this case, the observed distances between used resources becomes a function of the competitive regime itself, leading to a real possibility of circular analysis. However, if leap-frogging behavior is not occurring, then the behavior of the animal may be used as an excellent guide to resource availability and distribution (Isbell and Enstam 2002, Isbell and Young 2002).

To deal with the problem of temporal mismatch between behavioral and ecological observations, the focal-tree method can be used. Solving the problem of temporal mismatch, however, still does not deal with the issue that common indices of patchiness weight all trees equally. For this problem, a new index is needed.

Here, we describe a primate-based index of alternative food availability that is based on Lloyd's (1967) index of patchiness. While a few primate studies have used Morisita's index as an index of patchiness (Robinson 1986, Whitten 1983), it cannot be modified to incorporate resource quality and it changes when some of the trees are removed at random, even if the same spatial distribution remains (see Pielou 1969). Lloyd (1967) proposed two indices: crowding and patchiness (referred to hereafter as "Lloyd's index").

Crowding is the sum of the mean density of individual trees in a specified area and is experienced by each individual. Crowding changes when some of the population members are removed at random; thus, it depends on the total number present. Lloyd's index of patchiness is the ratio of mean crowding to mean density. By dividing by density, Lloyd's index becomes effectively independent of density and, thus, is a property of the spatial pattern alone (Pielou 1969). Thus, Lloyd's index of patchiness should remain relatively stable (excepting sampling variation) for ripe fruit trees of a given species even if the percentage of trees bearing fruit differs notably from year to year.

Lloyd's index, like all other plot-based methods, is sensitive to the plot size chosen. Lloyd (1967) emphasizes that the most appropriate size to make the outer boundary of the sample area would have to be determined empirically for each species and must be relevant to the species under investigation. For example, imagine sympatric groups of spider monkeys (*Ateles* spp.) and pygmy marmosets (*Cebuella* spp.). What may appear to be a more uniform distribution of resources to pygmy marmosets may appear clumped to a large group of spider monkeys (Figure 22.4). Thus, when determining the sample area for measuring spatial spread within plots, it is important to take into account how far the group will actually spread out to use the resources (Vogel and Janson 2007).

A major remaining problem with all plot-based methods for measuring spatial dispersion is that each individual tree is weighted equally. However, for an analysis of primate foraging benefits, a small food tree should not be considered equally valuable as a large one. Lloyd's extended index (LEI) incorporates food patch qualities into Lloyd's patchiness index (Vogel and Janson 2003, Vogel 2004). The LEI is as follows:

$$LEI = \frac{\sum_{j}\sum_{i}\left(s_{ij}\sum_{k \neq i}s_{kj}\right)}{\sum_{j}\sum_{i}s_{ij}} * \frac{n}{\sum_{j}\sum_{i}s_{ij}}$$

where s_{ij} is the size of food source i in patch j, k is an index of all food sources other than the focal tree within a plot, and n is the number of quadrates or plots. The first term in the equation is the equivalent to Lloyd's index of crowding, whereas the second term is the reciprocal of mean density, both expressed in terms of size or productivity rather than individuals. If s_{ij} is defined to be 1.0 for all individuals, then the LEI yields precisely the same result as Lloyd's index of patchiness.

Lloyd's index, whether extended or not, has a scale from zero (perfectly uniform dispersion) to potentially near infinity (clumped). The regular Lloyd's index reaches a value of 1.0 when resources are randomly dispersed. For the LEI, it is not clear what "random" means, as individual units of "size" (biomass, fruit numbers, DBH, or other scaling) are

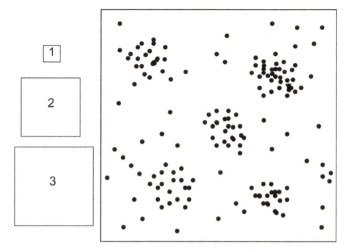

Figure 22.4 A hypothetical clumped population of fruiting trees with regularly distributed clumps. Depending on the quadrate size used to sample the population, different indices of dispersion would result. For example, quadrate size 1 would result in a random pattern, quadrate size 2 in a more clumped pattern, and quadrate size 3 in a more uniform pattern (redrawn from Elliot 1977).

not independent of each other. However, the interpretation of LEI remains the same as the regular index: It is the multiple of neighboring units per average resource individual per plot relative to that expected if dispersion were random. Thus, an LEI of 2.5 would indicate that the summed productivity or size of neighboring individuals per focal plant per plot is 2.5 times that expected if productivity or size were randomly divided across all plots. Vogel (2004) showed that for a given species LEI was sensitive to changes in plot size and that the frequency of aggression in the feeding tree was significantly related to LEI only when calculated using plot sizes smaller than average group spread.

To use this equation, we must have a measure of productivity that is relevant to the animals themselves. Fruit mass, fruit abundance counts, nutritional quality per unit fruit mass, DBH, and crown volume have all been used as measures of tree productivity (Chapman et al. 1992; Janson 1985a,b; Robinson 1986; Stevenson 1998). We propose using fruit abundance or fruit mass as a measure that best estimates tree productivity as it has been demonstrated to predict many aspects of behavior in primates (Janson 1988a,b; Koenig et al. 1998; Vogel 2005; Vogel and Janson 2007). In the following section, we describe several methods that have been used to quantify fruit availability.

Phenological Patterns

Most tropical forests display seasonal variation in the presence of new leaves, fruits, and flowers. This variation has been tightly linked to abiotic factors including light (Janzen 1972, Raich et al. 1991, van Schaik 1986, van Schaik et al. 1993, Wright and van Schaik 1994), rainfall (Frankie et al. 1974, Foster 1982, Reich and Borchert 1984, Wich and van Schaik 2000), and temperature (Ashton et al. 1988, Rathcke and Lacey 1985, Chapman et al. 2005, Miller-Rushing and Primack 2008, Post et al. 2008). Phenological data collected during primate field studies are typically combined with data collected on the abundance of plant species to quantify inter- and intra-annual variation in forest productivity and food availability. We can then examine the potential interactions between resource availability and several aspects of primate behavior including ranging (MacKinnon 1974, Raemaekers 1980, Vedder 1984, Twinomugisha and Chapman 2007, Fan and Jiang 2008), food competition (van Noordwijk and van Schaik 1987; Janson 1985a,b, 1988a,b; Isbell 1991; Vogel and Janson 2007; Vogel et al. 2007), mating and reproduction (van Schaik and van Noordwiijk 1985, Di Bitetti and Janson 2000, Janson and Verdolin 2005), and general social behavior (van Schaik and Pfannes 2005). Thus, quantifying the availability of food resources is a basic component to most research questions involving primate behavior in the wild.

The materials needed to construct phenological plots are quite basic and relatively inexpensive: a 100 m tape measurer, numbered aluminum tags and nails (which can be purchased from any forestry supplier), and a field notebook to record the species, DBH, and number of each tree. Additional information that may be of use includes tree height, which can be measured with a clinometer or laser range finder, and crown volume (Janson 1988a,b; Chapman et al. 1992). In addition, high-precision and relatively low-cost models are now available for handheld global positioning systems (GPS), and we recommend recording the location of each tree within a plot with such a devise. These data are easy to collect and could provide valuable information for future studies (Hughes 2003).

While the size and shape of phenological transects may vary, the most important aspect to consider is similar to habitat description plots: they must be representative of all forest types that are inhabited by the study species within the study area. Thus, several phenological transects may be needed to adequately represent the home range(s) of the study animal(s). Trees monitored for phenology can be part of larger plots used for quantifying tree species abundance and distribution as well. This facilitates not only data collection but also synchronization of data used for phenological patterns and vegetation descriptions. It is important to set up phenological plots using a random design across all habitat types and, if possible, to place two replicate transects for each habitat type (Cannon and Leighton 1994). Typically, each transect should be 100–200 m long and 10 m wide and can be set up along narrow, preexisting trails to facilitate the monitoring of trees. The advantage to using plot-like transects is that they can also be used for habitat descriptions, although sampling effort may need to be increased for habitat description plots to obtain sufficient data for all species. Most field studies label all trees of greater than 10 cm DBH. Trees and woody lianas should be marked using numbered, aluminum tags and nails as they will not rust and will cause

minimal damage to the trees. If the primate species under investigation also consume food items from lianas, they should also be labeled and monitored from the same location for each observation month.

Phenological plots should be monitored at least once per month and on the same day of each month, although given the rapid fruit turnover in tropical forests, twice a month is preferred. However, the amount of time it takes to monitor plots can be high; thus, the majority of researchers monitor plots once a month. The abundance of plant parts is usually recorded on relative scales to facilitate data collection and minimize interobserver reliability estimates. For example, to estimate the percentage of the canopy with young leaves, one may use a scale similar to 0, <5% = 2.5, 6% – 25% = 15, 26% – 50% = 37.5, 51% – 75% = 62.5, 76% – 100% = 87.5. It is standard to record the percentage of new leaves, the number of flowers, and the abundance of both ripe and unripe fruits. Fruit abundance can be estimated by counting all of the fruits in small trees or for larger trees with many fruits by counting the fruits within a given area of the tree crown using binoculars and then extrapolating the count to the total crown volume (Chapman et al. 1992). Chapman et al. (1992) discuss several ways to estimate fruit abundance and conclude that if there are several researchers collecting data, DBH is the preferred method due to low interobserver reliability in measuring the crown volume that is required with the alternative methods. However, if one person is collecting data, fruiting crown volume and visual counts should provide the data that best reflect actual food availability. In addition, different methods may be appropriate for trees with different quantities and size of fruits; thus, a researcher should choose a method based on the amount of effort available and the precision required by the study. If multiple observers are recording data, counts of items are also commonly converted to a logarithmic scale, resulting in approximate absolute value, which may minimize interobserver reliability issues (Chapman et al. 1994b, Janson and Chapman 1999). Chapman et al. (1994a,b) provide a thorough review of the pros and cons of using different methods to quantify fruit abundance.

More recently, advances in satellite-derived remote sensing have enabled biologists to examine large-scale phenological patterns in a variety of habitat types (Bohlman et al. 1998; Joshi et al. 2001; Reed, B. C., et al. 1994, 2003; Verstraete et al. 2008). These methods offer the advantage over traditional observation methods that quantitative data can be acquired remotely and can cover much larger areas, dramatically reducing sampling effort and time. This would be especially important if the study individuals or groups have large home ranges that span several different habitat types. However, these methods are still quite limited in the resolution of phenological events. Thus, we recommend that they always be used in combination with the ground, observational methods. At present, these methods are primarily limited to identifying changes in foliage coverage and the flushing of young leaves (see Bohlman et al. 1998), yet future advances in the resolution of images may enable researchers to detect variation in fruiting phases as well. In order to validate the models currently used by remote sensing scientists, we must continue to collect reliable and detailed ground data at our study sites.

Researchers commonly set up fruit and leaf traps to obtain an additional measure of litter fall and fruit availability (Hladik 1980; Terborgh 1983; Stevenson 1998; Chapman et al. 1992, 1994b). Traps can be set up randomly near phenological plots and should be checked at least twice a month. Fruit and litter traps consist of a plastic bag or a fine mesh material suspended above the ground by metal poles. Chapman et al. (1992) found that fruit trap data were not correlated to fruit availability data collected from trees in phenological plots. Thus, we recommend that if this method is used to monitor fruit abundance, it is used only in combination with fruit counts from trees in phenological plots. Litter traps are quite useful if the researcher is interested in site productivity, soil characteristics, and leaf turnover rates.

CONCLUSIONS

The aim of this chapter was to describe the basic methods and equipment necessary to monitor biotic and abiotic variables at a field study site. These data are valuable for interpreting the broad variation found in many aspects of primate behavior. We end this chapter with a reminder that while the initial setup and construction of vegetative and phenological plots is time-consuming and labor-intensive, once these plots are constructed they can provide years of valuable data. Plots should be monitored at least once a month, and new trees should be added when they reach the appropriate size classes for recording. Technologies for quantifying abiotic and biotic variables are constantly being upgraded and developed; thus, we recommend conducting research and testing out equipment before going to the field. Finally, we encourage field biologists to modify the methods we have provided as necessary to fit their study species and to always consider the study species' natural history when choosing the appropriate scale and sampling protocol.

23

Advances in the Understanding of Primate Reproductive Endocrinology

Catharine J. Wheaton, Anne Savage, and Bill L. Lasley

---

1. Why is hormone monitoring necessary to reveal ovarian function in many primate species?
2. What are two advantages of urinary hormone monitoring compared to relying on blood collection?
3. What is the major limitation of fecal hormone monitoring?

INTRODUCTION

The field of primate reproductive endocrinology has grown exponentially from initial studies that focused on collecting samples using chemical immobilization and/or physical restraint to study reproductive functioning, to a new era where noninvasive means are used to collect samples. Today, technology plays an integral role in how we manage animals in captivity and allows us to study various aspects of reproduction in free-ranging animals. This technology

has provided us with information that is essential for (*1*) understanding the basic biology of reproduction in captive and wild primates, (*2*) developing husbandry strategies that maximize successful production of offspring, (*3*) diagnosing departures from normal reproduction, and (*4*) developing assisted reproductive technologies when necessary. The development of reliable methods for monitoring reproductive events such as ovulation and pregnancy has allowed for increased reproductive performance but also has provided the basis for studies designed to accelerate reproduction using assisted reproductive technologies. This chapter will review the history of measuring hormones in primates and discuss the current technology available to assist in monitoring reproductive function in captive and wild primates.

HORMONE MEASUREMENT

Two assumptions of comparative reproductive physiology have been used to develop general strategies that allow us to monitor hormone levels in all mammals (Lasley and Kirkpatrick 1991, Lasley and Shideler 1993). The first assumption is that gonadal steroid hormones have the same molecular structure throughout the animal kingdom. While secretion and metabolism of steroids may vary among species, it is universally accepted that testosterone, estradiol, and progesterone are the primary bioactive sex steroids and that they reflect gonadal function in all mammals. The second assumption is that all steroid hormones are excreted rapidly through the kidneys or gut and then cleared from the body after undergoing only minor physical change. The resulting metabolites are concentrated in the urine, saliva, or feces and are relatively stable for a prolonged period of time (Lasley and Kirkpatrick 1991, Lasley and Shideler 1993). Although there are commonalities in the reproductive endocrinology of mammals, there are significant differences in both the pattern and levels of hormones produced, making extrapolation of findings from one species to another challenging and potentially misleading. Hormones are present and can be measured in a variety of biological fluids; however, the choice of which to use depends on various factors, such as the type of information desired, the assay techniques to be used, species differences in steroid metabolism and route of excretion, and the ease of sample collection (Asa 1996, Hodges 1996).

HORMONE METABOLISM AND EXCRETION

Circulating steroid hormones are found free or noncovalently bound to steroid binding proteins or in modified form as conjugated metabolites of the "parent" steroid. Many parent steroid hormones are rapidly metabolized by the liver or kidney to inactive forms. In urine and feces, most of the steroid exists in these modified forms. The principal conjugated forms in urine are sulfates or glucuronides, with only a small percent found unconjugated. The reverse is generally true in feces and saliva, where the predominant form is unconjugated. The conjugated steroids in circulation are no longer bioactive and represent a water-soluble form that is easily eliminated through the kidney. However, some steroid conjugates can be unconjugated and then converted to a form more biologically active than the parent steroid. Such hormones act as an intermediate step in the steroid metabolic pathway (Siiteri and MacDonald 1963) and act as a precursor or represent the end point of the pathway for excretion. Since the steroid metabolic pathway is governed

by a complex series of enzyme-mediated steps, the metabolites of the same parent steroid can be markedly different in different species (Bahr et al. 2000, Wright et al. 1981).

In the 1950s and 1960s, sophisticated techniques such as radiolabeled steroid clearance studies were used to identify and compare pathways of conversion and catabolism of circulating steroids in humans and nonhuman primates. In these experiments, radiolabeled steroid precursors (Bloch 1945) or the steroids themselves (Goldzieher and Axelrod 1969, Jirku and Layne 1965) were injected into the circulation. Serum, urine, and fecal samples were then collected at regular intervals; and the general steroid metabolic pathway, clearance rate, percent recovery, and relative specific activity of natural or synthetic (Musto et al. 1977) steroid metabolites could be characterized. Individual pathways were determined and compiled to develop our current understanding of steroid metabolism and excretion profiles for primates (Albrecht and Pepe 1998, Stouffer and Hearn 1998, Wright et al. 1981).

However, when a new species is studied, serum, fecal, and urine samples should be collected and assessed for variation of metabolism and conjugation pathways. In a laboratory setting, researchers are often able to collect adequate samples to permit validation of analytic methods to be employed and later choose to collect the biological sample that best fits their needs, project goals, and species. Often, the reproductive endocrinology of a species is completely characterized in captivity before it is applied to the field. This can be critical for successful reproductive monitoring in wildlife species. When only noninvasive monitoring is available, it is important that the choice of sample (*1*) contains a hormone or metabolite that can be quantitatively measured and accurately reflects the steroidal patterns measured in the blood, (*2*) indicates changing physiological condition (e.g., ovarian cycles or pregnancy) or reproductive behaviors, and (*3*) can be reliably and regularly collected in a method that is free of contamination from other biological samples, allows for identification of the individual, and can be collected frequently enough for reliable reproductive assessment.

SAMPLE PREPARATION

Prior to quantitative analysis, serum, urine, and fecal samples usually require preparation. Without adequate preparation, interfering factors or matrix effects will prevent accurate detection and quantification. Since hormone metabolites are concentrated in urine and feces, the universal preparation is dilution. However, if the targeted analyte is not sufficiently more concentrated than the interfering compounds, this approach is insufficient. Additional steps that can be taken include different forms of extraction with or without chromatography to separate the steroids or metabolites from the solids or interfering substances. The most common and simplest method is stepwise solvent extraction. In this technique, free steroids and their metabolites are partitioned according to their solubility in a series of solvents with varied polarity or fractioned between immiscible solvents (Mather 1942). The classic example of this technique is the use of ethyl ether or other nonpolar organic solvents to extract free steroids from serum or steroids released from urine and fecal samples after hydrolysis or solvolysis (Kellie 1966). Treatment of samples with sequential enzyme hydrolysis (ß-glucuronidase) (Graef et al. 1977, Stitch et al. 1956) and high-temperature or acid solvolysis to partition the primary sulfate and glucuronide conjugates can be used successfully to measure the proportions of released steroid (Eastman et al. 1984, Pryce et al. 1995, Shackleton 1986). Results from these studies will often give good comparative data but cannot be considered a complete quantitative estimate of the steroid in the sample. Enzyme hydrolysis with ß-glucuronidase from *Helix pomatia* or *Escherichia coli* will not completely hydrolyze all sulfates, not affecting some and underestimating others. Hot acid solvolysis may complete the deconjugation of some sulfates but may not fully cleave glucuronide conjugates. Together they may alter or degrade some of the steroid metabolites, resulting in procedural losses with each sample treatment step (Bradlow 1970, Shackleton 1986). These losses will be specific to each steroid metabolite and must be thoroughly tested and validated in every new species. As an alternative, steroid metabolites can be partitioned and purified from aqueous samples using solid phase extraction cartridges (Beehner and Whitten 2004, Ciotti et al. 1989, Heikkinen et al. 1981, Shackleton and Whitney 1980) prior to immunoassay or metabolite analysis by chromatography and mass spectrometry (Gomes et al. 2009, Shackleton 1986, Smith 2003).

ENDOCRINE ASSAYS

Endocrine assays are now applied in a wide range of disciplines to study the broadest possible range of species included in every conceivable physical setting (Lasley 1985, Lasley and Kirkpatrick 1991, Lasley and Shideler 1993). Blood, urine, saliva, and feces represent the range of possible biological samples that are collected in laboratories, zoos, forests, and savannas. Some of the earliest assay formats remain in use today, and new methods are constantly being developed. There are few known endocrine analytes that have defied assay development, and assay methods are available to virtually anyone who requires their use.

IMMUNOASSAYS

While few reports predate the advent of immunoassays, most of the endocrine data that are still in use today were reported after 1965, when the first truly practical steroid hormone assays were developed. The first of these were the protein binding assays. These assays were developed to measure only a few steroid hormones. The methods associated with this particular technology generally used corticobinding protein from the dog to measure either circulating cortisol or progesterone. Tissue extracts were also used for other analytes, such as estradiol. While the method was completely replaced by radioimmunoassay within five years, the basic concept remains valid today. Prior to this time, physical–chemical methods, such as thin-layer or paper chromatography followed by fluorescence or gas–liquid chromatography with flame or ion detection, were required. The advantages of protein binding technology included increased sensitivity and reduced sample size preparation. While this technology was limited to progesterone and estradiol as the only reproductive hormones measured, it was sufficient to make it an important contribution since little beyond ovulation and the luteal phase of ovarian cycles was well defined at this time.

PROTEIN HORMONE ASSAYS

The principle of protein binding assays was simple: Add insufficient binding protein to permit all of the analyte to bind and add a similar amount of radiolabeled analyte to compete with the endogenous analyte. The resulting competition would permit only a fraction of the label to be bound, and once this bound portion was separated from the original mixture, the amount of label bound would permit calculation of the amount of endogenous analyte (e.g., luteinizing hormone [LH], follicle-stimulating hormone [FSH]) that had competed. With two small exceptions, this basic principle still prevails in small analyte detection today. The weaknesses of protein binding assays included (1) the limited number of types of specific protein binding globulin that exist, (2) the difficulty of removing all endogenous analyte in order to obtain protein binding sites for the assays, and (3) separating the "bound" from the "free" components without changing the dynamic equilibrium that had been established. This method was made obsolete when Berson et al. (1956) revealed that specific antibodies could be produced to replace the binding globulins that had more specificity and increased binding avidity.

ANTIBODY ENGINEERING

Once a technique is shown to be useful in one field of study, it is often picked up and utilized in another. Unlike proteins, steroids are small molecules and must be conjugated to a larger protein that is foreign to the host animal (e.g., bovine serum albumin or other molecule) to induce the immune response (Beiser and Erlanger 1967, Erlanger 1980). The steroid-immunogen is then mixed with Freund's adjuvant (Freund 1947) and injected into a host species such as rabbit or mouse. After a series of booster shots and tests for titer, serum is collected and the antibodies are salted out, separated, and purified for assay (Vaitukaitis et al. 1971).

The key to developing specificity of the antisera developed has been determined to lie in the linkage between the steroid and the immunogenic molecule. Steroids do not have a functional group available for the linkage, so steroids modified with a carboxylic acid group are used. The position of the carboxylic acid on the steroid largely determines the nature and specificity of the antibodies produced. Better results come from linking to a position remote from the functional group on the steroid that gives rise to its specific steroidal properties (e.g., carboxymethyl oxime at the six- vs. three-carbon position for estradiol) or on another ring (e.g., progesterone-11"-hemisuccinate, C-ring) on the steroid (Kellie et al. 1975, Niswender 1973, van Weemen and Schuurs 1975). Antibodies specific to steroid glucuronides can readily be made by directly linking the steroid metabolite to the immunogen (Kellie et al. 1972). There will be a variable response in antisera production in individual host animals in terms of titer, number, and specificity of antibodies produced. When a strong, specific immune response is produced to the target analyte of choice, a monoclonal antibody can be produced by fusing a spleen B lymphocyte cell from the host with myeloma cells and then culturing and purifying the antibodies (e.g., LH; Matteri et al. 1987, reviewed in Kohler 1993).

Over the years, antibodies were produced and radio- and enzyme immunoassays were developed to measure every known human circulating reproductive hormone (Bersen and Yalow 2006, Munro and Lasley 1988, Pratt 1978). With immunoassays, peptide and protein hormone assays could be developed as easily as steroid hormone assays, once a pure form of the targeted analyte was made available for antibody production. Beginning in the late 1960s, radioimmunoassays were applied with varied success to a wide range of nonhuman primate species.

STEROID HORMONE ASSAYS

Assays for steroid hormones have always been highly adaptable because steroid hormones (estrogen, progesterone, testosterone, etc.) are physically identical in all mammalian species. There are species-specific differences, in the concentrations and potential interfering analytes and matrices. However in general, steroid hormone immunoassays have been used to measure circulating levels of sex steroids in all well-studied primate species. With the advent of radioimmunoassay, both competitive radiolabels and specific antisera became available for all clinically relevant steroid hormones. Unfortunately, this was not true for the peptide, protein, and glycoprotein hormones. Because peptides and proteins are derived from the genome and alterations of the genome are the basis of speciation, the physical structure of larger peptides and all protein hormones varies between species. While the physical difference in molecular structure between the same hormone in different species can be modest, if that difference in structure between species

changes the immunological properties of the analogous hormone, then assay reagents will not be adaptable between those species. This has been, and remains, the major problem in the application of existing hormone assays to a wide range of nonhuman primate species. This problem led to the creation of bioassays that permit the measurement of peptide/protein hormones in a wide range of species.

BIOASSAYS

The principle underlying bioassays is that while the overall physical structure of the hormone molecule changes, or "evolves," with speciation, the portion of that molecule that actually participates in sending the biological signal remains conserved. In other words, the portion of the molecule that endows it with biological activity is highly conserved and will be capable of exerting that biological activity in target cells of many species. Making use of this principle, target cells from laboratory animals were used to respond to the hormones of other species. The initial assays were in vivo assays and were used for both protein and steroid hormones. This use of live animals for assays was labor-intensive and short-lived. In a short time, steroid hormone assays were replaced by immunoassays and the protein hormones were directed to in vitro bioassays in which the target tissues were incubated with the sample to obtain results. The best example of such an assay is the use of minced mouse testes to respond to the serum of a wide range of species to produce testosterone as a function of the amount of LH present in the serum. The receptors that peptide and protein hormones bind to and transduce their signals, have been cloned and transfected into otherwise inert cells. When cotransfected with reporter genes that produce light or other signals, these transfected immortalized cell lines can be used to obtain the same result that was obtained using whole animals or fresh tissues from animals (Christin-Maitre et al. 2000). While this technology is still growing in popularity, it is likely to be replaced by immunoassays as a result of new molecular technologies.

THE NEED TO MEASURE HORMONES

The occurrence and timing of reproductive events such as ovulation, conception, early pregnancy, and losses, which are largely concealed, can be accurately detected and documented only through hormone analysis (Ardito 1976, Fite et al. 2005b). Early studies of female reproductive function required repeated blood draws from restrained animals to obtain sufficient data (Eaton and Resko 1974) to answer many of these questions. By their nature and by endocrine definition, the results of assays that measure circulating hormones from blood provide the "gold standard" for all other hormone assay results. While a single blood sample provided useful information, only serial blood samples provided information on hormonal profiles. Repeated capture and restraint to obtain blood samples was often stressful and led to potential behavioral abnormalities that were thought to influence reproductive events (Smith and French 1997a,b). Moreover, when studying small primates, repeated blood sampling can lead to significantly lower hematocrit values (French et al. 1992). Thus, techniques were developed to noninvasively monitor reproductive function in captive and wild primates (Czekala et al. 1983; Hodges et al. 1986; Mitchell et al. 1982; Savage et al. 1995, 1997; Shideler et al. 1983a,b, 1994; Wasser et al. 1988; Ziegler et al. 1987a,b, 1997), which have revolutionized our ability to understand the complex reproductive biology of primates. Using urine and fecal samples to measure hormones is now commonplace, and the changes in the concentration of specific analytes excreted in urine and feces are thought to accurately reflect the physiological processes and time course of events associated with those analytes.

URINARY ASSAYS

During the early 1970s, the success of immunoassays for measuring circulating hormone levels made it clear that daily measurement of reproductive hormones was required if the reproductive biology of any species was to be completely understood. By 1975, zoological parks and gardens aggressively took steps to study and understand the reproductive biology of endangered species and developed first-rate captive breeding programs (Hodges et al. 1979, Lasley 1985, Lasley and Kirkpatrick 1991, Lasley and Shideler 1993, Lasley et al. 1981a). The zoo community sponsored the effort to reevaluate their current management strategies and asked whether the new technologies developed for investigating reproductive physiology could also improve the ability to obtain useful information from urine-based assays. This new perspective in the field of reproductive biology soon became a major factor in the development of urinary hormone assays.

The majority of steroids in urine are present in the conjugated form. Initial analyses of steroids in urine involved a very labor-intensive process of hydrolysis and solvent extraction prior to assay. With the advent of nonextraction assays, it became possible to directly measure steroid conjugates (see Hodges 1996 for a complete review). Direct assays for steroid conjugates have the distinct advantage of producing a more informative hormonal profile than is possible with extraction methods (Hodges and Eastman 1984, Lasley et al. 1981b, Shideler et al. 1983b). Less complicated assays based on the use of enzymatic procedures are available as an alternative to assays using radioisotopes and are commonly used in most endocrine service laboratories (Munro and Stabenfeldt 1984, Munro and Lasley 1988). The advantages of these enzyme immunoassays (EIAs) include low cost, ease of performance, and no use of radioactive material. These assay systems require minimal instrumentation and are more likely to be used in the field. The

application of EIAs to reproductive monitoring in primates has been carried out by many researchers (e.g., Czekala et al. 1988, Heistermann et al. 1995b, Hodges 1985, Hodges and Green 1989, Lasley 1985, Lasley and Kirkpatrick 1991, Lasley et al. 1980a, Shideler et al. 1990).

Gonadotropic hormones, such as pituitary LH and FSH, are also excreted in urine. However, they have proven to be more challenging to measure by immunoassay methods since the species specificity in the immunological activities of these hormones makes multispecies application of assays difficult (Hodges 1996, Matteri et al. 1987). In contrast, biological activity is less species-specific, and in vitro bioassays have been established for urinary LH (Harlow et al. 1984; Hodges et al. 1979; Ziegler et al. 1993a,b) and FSH (see Dahl and Hseuh 1987 for a complete review, Hodges 1996).

ADVANTAGES OF URINE SAMPLES

Urine has favorable qualities that make it ideal as a sample for monitoring reproduction. It is constantly produced by the accumulation of metabolic products. This insures that the time course of metabolism is preserved in the components of urine, and it is a concentrate of metabolic by-products. In general, urine can be collected without capture or restraint and in relatively large quantities over indefinite periods of time when voluntary urination is trained (Smith et al. 2004). Because urine is largely water with concentrated analytes in solution, assay formats can use urine directly or, following a simple dilution, in assay buffer. Because the volume of urine is determined by water intake and need, changes in intake will be reflected in urine volume. The potential problem presented by variation in urine volume is usually addressed by indexing the concentration of the targeted analyte to some measure of the urine volume. The specific gravity of urine (Anestis et al. 2009) or urinary creatinine concentration is used as a divisor and produces a quotient that is a normalized concentration of the analyte (Taussky 1954). When daily changes in the concentration of the circulating parent analyte are compared to daily changes in the indexed urinary analyte, the patterns are parallel, thus verifying the validity of this method (Munro et al. 1991, 1996, 1997). Despite certain limitations in using creatinine measurements, good correlations of circulating hormone and hormone/creatinine index have been demonstrated for several species (Hodges and Eastman 1984, Shideler and Lasley 1982).

COLLECTING AND STORING URINE SAMPLES

The species and individual subjects usually determine the timing and method of collecting urine samples (Hodges 1985, Lasley 1985). For most studies conducted in captivity, it is relatively easy to condition most primates to urine collection. Primates can be trained to enter individual cages that allow urine to be collected (Anzenberger and Gossweiler 1993, Pryce et al. 1995) and to accept a caretaker collecting a urine sample upon waking (Pryce et al. 1995, Ziegler et al. 1993a), or urine samples may be collected from a clean floor of a cage when the animal has been moved (Bahr et al. 2001, Pryce et al. 1995).

While collection of urine in the field is not commonplace with most small-bodied primates, it has been used successfully with great apes. Urine is collected either upon waking or *ad libitum* by placing a collecting tube in the stream of urine upon urination (mountain gorillas, Robbins and Czekala 1997) or by pipetting urine found on nearby vegetation or soil (orangutans and chimpanzees, Knott 1997a, bonobos, Marshall and Hohmann 2005).

Most species have predictable voiding habits, and these can be used to insure the collection of clean samples. Frequently, portions of early-morning voids are collected because they represent an integrated accumulation of analytes over several hours of minimal physical activity (Czekala et al. 1983; Savage et al. 1995; Shideler et al. 1983a,b; Ziegler et al. 1987a,b). This collection schedule obviates concerns relating to wide changes in fluid intake and minimizes the need for adjustments for differences in analyte concentration due to changes in urine volume production. If intact LH or FSH is to be measured, then preservative (glycerol) needs to be added immediately to prevent dissociation of the subunits. Usually, the entire amount of urine collected is not needed and an aliquot is centrifuged to remove cell debris and other particulate matter prior to storage at a temperature that will insure the sample is frozen solid.

Urine samples can be stored indefinitely if kept frozen and rethawed multiple times without loss of the ability to accurately measure steroid conjugates. Multiple thaws are not recommended if the intact gonadotropins are to be measured. However, recent advances have also shown that small amounts of urine (200 ml) can be saved on absorbent filter paper and stored in airtight plastic containers with silica gel in the field (Marshall and Hohmann 2005, Shideler et al. 1995). The resulting profiles generated by urinary hormone analysis generally parallel those of circulating hormones.

URINARY STEROIDS

The majority of steroids in urine are present in the conjugated form. Early analyses of steroids in urine involved the laborious process of hydrolysis and solvent extraction prior to assay; however, the introduction of nonextraction assays allowing direct measurement of steroid conjugates has now greatly simplified procedures. Direct assays for steroid conjugates have the additional advantage of often generating a more informative hormone profile than previously possible with extraction methods (Hodges and Eastman 1984, Lasley 1985, Shideler et al. 1983b). Even simpler assays based on the use of enzymatic procedures are now

available as an alternative to conventional assays using radioisotopes. The advantages of these EIAs, including low cost, ease of performance, convenience, as well as avoidance of the problems associated with the use and disposal of radioactive materials, make them particularly suitable for institutions with limited laboratory facilities. Furthermore, the end point of these tests is a color change that is simple to quantify. These tests require minimal instrumentation and are therefore amenable to use under field conditions. EIAs have been applied to reproductive monitoring in primates by several researchers (e.g., Czekala et al. 1986, 1988; Hodges 1985; Hodges and Green 1989; Shideler et al. 1993a,b).

FECAL ASSAYS

In many captive situations, urine samples can be collected regularly for most primate species; however, in the field, it may be more practical to collect fecal samples (de Sousa et al. 2002, Fujita et al. 2001, Savage et al. 1997, Strier and Ziegler 1997, Wasser et al. 1991, Ziegler et al. 1997). In addition, there are some species (*Nycticebus coucang*) that excrete an insignificant amount of hormones in the urine, with primary hormonal excretion found in the feces (Perez et al. 1988). In such cases, fecal samples can be collected and analyzed in much the same fashion as urine. The primary differences in fecal analysis are the need to remove solid materials, the limitation of measurements to the metabolites of steroid hormones, and usually the need to measure unconjugated rather than conjugated forms of the steroid metabolites. As with urinary measurements, there is a need to index or adjust the amount of analyte to compensate for the total volume of feces produced. In most cases, this is accomplished by simply adjusting the mass of analyte to the mass of the fecal sample analyzed. While this is clearly less accurate than the adjustment of urine by creatinine, it is currently the only practical method available and has been successfully applied to several primate species.

COLLECTION AND STORAGE OF FECAL SAMPLES

Fecal samples can be collected systematically or randomly, stored in a secure container, and frozen solid until processed (Savage et al. 1997). Although the best method is to collect and freeze immediately, that option may not be possible in the field or in situations where access to refrigeration is limited or fecal samples have been stored in 95% ethanol and 0.2% sodium azide (Wasser 1996, Wasser et al. 1988), dried in gas ovens (Stavisky et al. 1995), or collected and stored on filter paper (Shideler et al. 1995). Due to the presence of bacterial enzymes native to the samples, metabolism of the steroids may occur, and all sample treatments and storage methods must be carefully evaluated and validated for each species (Palme 2005, Strier and Ziegler 2005a, Whitten et al. 1998, Ziegler and Wittwer 2005).

SALIVA SAMPLES

Collecting saliva for hormone analysis has been successful for studies investigating stress and cortisol levels (Tiefenbacher et al. 2003, Cross et al. 2004) and has also been used to assess reproductive function in selected species (DiGiano et al. 1992). Saliva is typically collected by enticing animals to lick a screen-covered gauze or bite on a dental rope (Behringer et al. 2009, Lutz et al. 2000). The rope is then centrifuged and the saliva removed and stored for analysis (Tiefenbacher et al. 2003). Others have trained animals, such as gorillas, to chew on dental rope and "trade" the dental rope for a food reward (Bettinger et al. 1998).

STUDYING PRIMATE REPRODUCTION IN CAPTIVITY AND THE FIELD

With more than 300 primate species and more than 600 primate taxa, it is surprising how little we know about the basic reproductive biology of primates (see Table 23.1). Less than 20% of all the primate species have had their basic reproductive biology characterized, either wild or

Table 23.1 Overview of Primate Reproductive Profiles

PRIMATES	REFERENCES	MEDIA/HORMONE	PURPOSE
Prosimians			
Propithecus verrauxi	Brockman et al. 1998	Fecal testosterone	Hormones and behavior in the field
	Brockman et al. 1995, 1996	Fecal estradiol and progesterone	Hormones and behavior in the field
Microcebus murinus	Perret 1986	Plasma progesterone	Hormones and behavior in captivity
Eulemur fulvus rufus	Ostner et al. 2008	Fecal testosterone and cortisol	Endocrine knowledge, hormones and behavior in the field
	Ostner and Heistermann 2003	Fecal estrogens and progesterone	Endocrine knowledge, hormones and behavior in captivity and in the field
E. macaco	Lasley et al. 1978	Serum progesterone, estrogens	Endocrine knowledge in captivity
Lemur catta	Cavigelli and Pereira 2000	Fecal testosterone	Hormones and behavior in semi-free-ranging animals
	Norman et al. 1978	Serum luteinizing hormone	Endocrine knowledge in captivity
	Lasley et al. 1978	Serum progesterone, estrogens	Endocrine knowledge in captivity
Varecia variegatus	Lasley et al. 1978	Serum progesterone, estrogens	Endocrine knowledge in captivity
	Shideler et al. 1983a,b	Urinary estrogens	Endocrine knowledge in captivity
Nycticebus pygmaeus	Fitch-Snyder and Jurke 2003	Fecal testosterone	Hormones and behavior in captivity
Galago crassicaudatus	Eaton et al. 1973	Plasma progesterone, estradiol	Hormones and behavior in captivity
	Van Horn and Eaton 1979	Plasma progesterone, estradiol	Hormones and behavior in captivity
G. moholi	Lipschitz 1996	Serum luteinizing hormone, estrogens, progesterone, testosterone	Endocrine knowledge in captivity
New World primates			
Callimico goeldii	Jurke et al. 1994	Urinary pregnanediol-3-α-glucuronide	Endocrine knowledge in captivity
	Pryce et al. 1993, 1994	Fecal estrogen, urinary and plasma progesterone, estrogen	Endocrine knowledge in captivity
	Ziegler et al. 1990b	Urinary luteinizing hormone, estrone conjugates	Endocrine knowledge in captivity
Callithrix jacchus	Power et al. 2006)	Plasma corticotropin-releasing hormone	Endocrine knowledge in captivity
	Tardif et al. 2005	Urinary estradiol, cortisol, chorionic gonadotropin	Endocrine knowledge, hormones and behavior in the field
	Sousa et al. 2005	Fecal progesterone and cortisol	Endocrine knowledge in captivity
	Rosenbusch et al. 1997		Endocrine knowledge in captivity
	Eastman et al. 1984	Plasma and urinary follicle-stimulating hormone	Endocrine knowledge in captivity
	Möhle et al. 2002	Serum estrone-3-sulfate, urinary estrone, estradiol, estrone-3-sulfate, pregnanediol-3-glucuronide	Endocrine knowledge in captivity
	Harding et al. 1982	Serum progesterone	Endocrine knowledge in captivity

Table 23.1 (cont'd)

PRIMATES	REFERENCES	MEDIA/HORMONE	PURPOSE
	Abbott et al. 1988	Serum luteinizing hormone	Hormones and behavior in captivity
	Heistermann et al. 1993	Fecal estradiol, progesterone, pregnanediol, plasma progesterone	Endocrine knowledge in captivity
	Hodges and Eastman 1984	Urinary estrogen	Endocrine knowledge in captivity
	Ziegler et al. 1996	Serum and fecal progesterone, pregnanediol, estradiol, estrone; serum luteinizing hormone	Endocrine knowledge in captivity
	Boes et al. 1993	Urinary estradiol	Endocrine knowledge in captivity
C. kuhlii	Fite et al. 2005a,b	Urinary estradiol, pregnanediol-3-glucuronide, testosterone, cortisol	Endocrine knowledge, hormones and behavior in captivity
	Puffer et al. 2004	Urinary estradiol, pregnanediol-3-glucuronide, testosterone, cortisol	Hormones and behavior in captivity
	Fite and French 2000	Urinary estradiol, pregnanediol-3-glucuronide, cortisol	Endocrine knowledge, hormones and behavior in captivity
	Smith and French 1997a	Urinary pregnanediol 3-α-glucuronide, luteinizing hormone/chorionic gonadotropin	Endocrine knowledge in captivity
	Nunes et al. 2002	Urinary testosterone, estradiol	Endocrine knowledge in captivity
C. pygmaea	Carlson et al. 1997	Urinary luteinizing hormone, pregnanediol-3-glucuronide	Endocrine knowledge in captivity
	Ziegler et al. 1990a	Urinary estrone, luteinizing hormone	Endocrine knowledge in captivity
Leontopithecus chrysomelas	Chaoui and Hasler-Gallusser 1997	Fecal progesterone	Hormones and behavior in captivity
L. rosalia	French et al. 2003	Fecal estrone conjugates, pregnanediol-3-glucuronide	Endocrine knowledge in the field
	French et al. 1992	Urinary and plasma luteinizing hormone, estrogens	Endocrine knowledge in captivity
Saguinus bicolor	Heistermann et al. 1987	Urinary estrogen, progesterone	Endocrine knowledge in captivity
S. fuscicollis	Heistermann et al. 1995b	Urinary estrone conjugates, pregnanediol glucuronide; fecal estradiol, progesterone, pregnanediol, plasma progesterone	Endocrine knowledge in captivity
	Tardif and Ziegler 1992	Plasma progesterone, urinary estrone conjugates, luteinizing hormone	Endocrine knowledge in captivity
	Rosenbusch et al. 1997	Plasma and urinary follicle-stimulating hormone	Endocrine knowledge in captivity
	Heistermann et al. 1995b	Urinary estrone conjugates, pregnanediol glucuronide, fecal estradiol, progesterone, pregnanediol, plasma progesterone	Endocrine knowledge in captivity
	Epple and Katz 1983	Urinary progesterone, testosterone, estradiol	Endocrine knowledge in captivity
	Lerchl et al. 1991	Urinary testosterone	Endocrine knowledge in captivity
S. labiatus	Kuederling et al. 1995	Urinary estrogens	Endocrine knowledge in captivity
S. oedipus	Rosenbusch et al. 1997	Plasma and urinary follicle-stimulating hormone	Endocrine knowledge in captivity
	Tardif and Ziegler 1992	Plasma progesterone, urinary estrone conjugates, luteinizing hormone	Endocrine knowledge in captivity
	Savage et al. 2004	Fecal estrone conjugates, pregnandiol-3-glucuronide, levonorgestrel metabolites	Endocrine and contraception knowledge in captivity
	Savage et al. 1997	Fecal estrone conjugates, pregnanediol-3-glucuronide	Endocrine knowledge in the field
	Almond et al. 2008, Ziegler et al. 1987a,b, 1990c, 1993a,b, 1996, 1997, 2000c	Fecal and serum progesterone, estrone, estradiol, luteinizing hormone/chorionic gonadotropin, urinary estrone conjugates, luteinizing hormone, estrone, estradiol prolactin, cortisol, corticosterone	Endocrine knowledge, hormones and behavior in captivity

Table 23.1 (cont'd)

PRIMATES	REFERENCES	MEDIA/HORMONE	PURPOSE
Alouatta palliata	Clarke et al. 2007	Fecal estradiol, testosterone	Endocrine knowledge in the field
A. pigra	Van Belle et al. 2009	Fecal estradiol, progesterone	Endocrine knowledge, hormones and behavior in the field
A. seniculus	Herrick et al. 2000	Urinary progesterone	Endocrine knowledge in the field
Aotus trivirgatus	Dixson 1994	Urinary and serum testosterone, estrogens, progesterone	Endocrine knowledge in captivity
	McConnell et al. 1990	Serum progesterone	Endocrine knowledge in captivity
	Setchell and Bonney 1981	Urinary estrogens, progesterone, cortisol	Endocrine knowledge in captivity
	Bonney et al. 1980	Plasma estradiol-17β, estrone, progesterone, testosterone	Endocrine knowledge in captivity
	Dixson and Gardner 1981	Plasma testosterone	Endocrine knowledge in captivity
Ateles geoffroyi	Campbell 2003, 2004	Fecal estrone conjugates, pregnanediol-3-glucuronide	Hormones and behavior in the field
	Campbell et al. 2001	Urinary and fecal estrone conjugates, pregnanediol-3-glucuronide	Endocrine knowledge in captivity
	Hernández-López et al. 1998	Serum estrogens, progesterone	Endocrine knowledge in captivity
Brachyteles arachnoids	Strier and Ziegler 1994, 1997, 2005	Fecal estrone, estradiol, progesterone, androgens	Hormones and behavior in the field
	Strier et al. 2003	Fecal estradiol, progesterone, testosterone, cortisol	Hormones and behavior in the field
	Strier and Ziegler 1997	Fecal estradiol, progesterone	Hormones and behavior in the field
	Ziegler et al. 1997	Fecal estrogens, progesterone	Endocrine knowledge in the field and in captivity
Callicebus moloch	Valeggia et al. 1999	Urinary estrone conjugates, pregnanediol-3-glucuronide	Endocrine knowledge in captivity
Cebus apella	Oritz et al. 2005, 2007	Serum estradiol, progesterone	Endocrine knowledge in captivity
	Lynch et al. 2002	Fecal testosterone, cortisol	Endocrine knowledge in the field
	DiGiano et al. 1992	Salivary and plasma progesterone	Endocrine knowledge in captivity
	Nagle et al. 1989	Serum progesterone	Endocrine knowledge in captivity
Saimiri boliviensis	Yeoman et al. 2000	Serum follicle-stimulating, luteinizing hormones, estradiol, progesterone	Endocrine knowledge in captivity
S. sciureus	Moorman et al. 2002	Urinary and fecal estradiol and progesterone metabolites	Endocrine knowledge in captivity
	Diamond et al. 1987	Serum estradiol, progesterone, chorionic gonadotropin, prolactin	Endocrine knowledge in captivity
	Chen et al. 1981	Plasma testosterone	Endocrine knowledge in captivity
Pithecia pithecia	Savage et al. 1995	Urinary and fecal estrone conjugates, pregnanediol-3-glucuronide	Endocrine knowledge in captivity
	Shideler et al. 1994	Fecal estrone conjugates, pregnanediol-3-glucuronide	Endocrine knowledge in captivity
Old World primates			
Cercocebus atys	Whitten and Russell 1996	Fecal estrogen, progesterone	Endocrine knowledge in captivity
	Mann et al. 1982	Serum luteinizing hormone, follicle-stimulating hormone	Endocrine knowledge in captivity
	Aidara et al. 1981	Serum estrogen, progesterone, prolactin, luteinizing hormone, follicle-stimulating hormone	Endocrine knowledge in captivity
	Stabenfeldt and Hendrickx 1973	Serum progesterone	Endocrine knowledge in captivity
Cercopithecus mitis	Pazol 2003	Fecal estradiol, pregnanediol-3-glucuronide	Endocrine knowledge, hormones and behavior in the field

Table 23.1 (cont'd)

PRIMATES	REFERENCES	MEDIA/HORMONE	PURPOSE
Chlorocebus aethiops	Molskness et al. 2007	Serum estradiol, progesterone, luteinizing hormone, follicle-stimulating hormone	Endocrine knowledge in captivity
	Seier et al. 2000	Serum estradiol, progesterone	Endocrine knowledge in captivity
	Setchell et al. 1980	Urinary estrone, estradiol and estriol conjugates, pregnanediol-3-glucuronideEndocrine knowledge in captivity	
	Eley et al. 1989b	Serum estradiol, progesterone, luteinizing hormone	Endocrine knowledge in captivity
	Andelman et al. 1985	Urinary pregnanediol-3-α-glucuronide	Hormones and behavior in the field
Colobus guereza	Harris and Monfort 2003	Urinary estrone conjugates, pregnanediol-3-α-glucuronide	Hormones and behavior in the field
Erythrocebus patas	Winterer et al. 1985	Serum estrogen, progesterone	Endocrine knowledge in captivity
Macaca arctoides	Wilks 1977	Serum estrogen, progesterone, luteinizing hormone, follicle-stimulating hormone	Endocrine knowledge in captivity
M. assamensis	Wehrenberg et al. 1980	Serum estrogen, progesterone, luteinizing hormone, follicle-stimulating hormone	Endocrine knowledge in captivity
M. fascicularis	Lohstroh et al. 2007	Urinary estrone conjugates, progesterone (hygeia pregnanediol-3-glucuronide), chorionic gonadotropin	Endocrine knowledge in captivity
	Shimizu et al. 2001	Urinary monkey chorionic gonadotropin	Endocrine knowledge in captivity
	Todd et al. 1999	Urinary follicle-stimulating hormone	Endocrine knowledge in captivity
	Munro et al. 1997	Serum and urinary chorionic gonadotropin, urinary estrone conjugate	Endocrine knowledge in captivity
	Munro et al. 1996	Serum and urinary levonorgestrel metabolites	Endocrine and contraception knowledge in captivity
	van Schaik et al. 1991	Urinary cortisol, (immunoreactive) prolactin, testosterone	Endocrine knowledge in the field
	Mehta et al. 1986	Serum progesterone, urinary pregnanediol	Endocrine knowledge in captivity
	Möhle et al. 2002	Urinary and fecal testosterone	Endocrine knowledge in captivity
	Shideler et al. 1993a,b	Fecal estrone conjugates and pregnanediol-3-glucuronide; urinary progesterone, estradiol, estradiol, estrone conjugates, estrone conjugates and pregnanediol-3-glucuronide	Endocrine knowledge in captivity
	Michael et al. 1987	Serum testosterone	Endocrine knowledge in captivity
M. fuscata	Fujita et al. 2001, 2004	Urinary and fecal estrone conjugates, pregnanediol-3-glucuronide, estrone conjugates, plasma estradiol-17β, progesterone, and luteinizing hormone	Endocrine knowledge, hormones and behavior in captivity and in the field
	Yoshida et al. 2001	Fecal estradiol and progesterone	Endocrine knowledge in the field
	Matsubayashi and Enomoto 1983	Plasma testosterone	Endocrine knowledge in captivity
	Nigi and Torii 1983	Plasma estradiol, progesterone	Endocrine knowledge in captivity
M. mulatta	Shideler et al. 1990	Serum estradiol, progesterone, urinary estrone conjugates, pregnanediol-3- glucuronide	Endocrine knowledge in captivity
	Monfort et al. 1987	Serum estradiol, luteinizing hormone, progesterone; urinary estrone conjugates	Endocrine knowledge in captivity
	Plant 1981	Serum prolactin, testosterone, cortisol	Endocrine knowledge in captivity
M. nemestrina	Chandrashekar et al. 1987	Serum estrone, estradiol, follicle-stimulating hormone	Endocrine knowledge in captivity
	Risler et al. 1987	Serum estrogens, progesterone	Endocrine knowledge in captivity
	Eaton and Resko 1974	Serum estrogen, progesterone	Endocrine knowledge in captivity
M. nigra	Thomas et al. 1992	Serum estradiol, progesterone, follicle- stimulating hormone, luteinizing hormone	Endocrine knowledge in captivity

Table 23.1 (cont'd)

PRIMATES	REFERENCES	MEDIA/HORMONE	PURPOSE
M. radiata	Jagannadha Rao et al. 1984	Serum chorionic gonadotropin, estradiol-17β, progesterone	Endocrine knowledge in captivity
	Kholkute et al. 1981	Serum testosterone	Endocrine knowledge in captivity
M. silenus	Heistermann et al. 2001a	Fecal estradiol, 5α-pregnane-3α-ol-20-one	Endocrine knowledge in captivity
	Shideler et al. 1983a, 1985	Urinary estrone conjugatesm, pregnanediol-3-glucuronide	Endocrine knowledge in captivity
M. sylvanus	Heistermann et al. 2008, Möhle et al. 2005	Fecal total estrogens, progesterone metabolites	Endocrine knowledge, hormones and behavior in the field
M. tonkeana	Thierry et al. 1996	Urinary estrone conjugates, pregnanediol glucuronide	Endocrine knowledge in captivity
Mandrillus sphinx	Phillips and Wheaton 2008	Urinary estrone conjugates, pregnanediol glucuronide	Endocrine knowledge, hormones and behavior in captivity
	Setchell et al. 2008	Fecal cortisol	Endocrine knowledge in the field
Miopithecus talapoin	Dixson and Herbert 1977	Serum estrogen, testosterone	Endocrine knowledge in captivity
Papio hamadryas	Pearson et al. 2008	Salivary cortisol	Endocrine knowledge in captivity
	Khan et al. 1985	Serum luteinizing hormone	Endocrine knowledge in captivity
	Cekan et al. 1977	Serum estrogen, progesterone, testosterone	Endocrine knowledge in captivity
P. h. anubis	Higham et al. 2009a,b	Fecal estrogen conjugates and pregnanediol-3α-glucuronide	Endocrine knowledge in the field
	Beehner and Whitten 2004	Fecal estradiol, progesterone, testosterone, glucocorticoid metabolites	Endocrine knowledge in the field
	Fortman et al. 1993	Serum estradiol, progesterone, chorionic gonadotropin	Endocrine knowledge in captivity
	Hodges et al. 1986	Urinary estrone conjugates, pregnanediol-3α-glucuronide	Endocrine knowledge in captivity
P. h. cynocephalus	Beehner et al. 2006	Fecal estrogen, progestin, glucocorticoid metabolites	Endocrine knowledge in the field
	Gesquiere et al. 2007	Fecal estrogen, progestin, testosterone, glucocorticoids	Endocrine knowledge in the field
	Altmann et al. 2004	Fecal estrogen metabolites, progestins, glucocorticoids and testosterone metabolites	Endocrine knowledge in the field
	Wasser 1996	Fecal estrogens and progestins	Endocrine knowledge in the field
	Stavisky et al. 1995	Fecal estradiol and progesterone	Endocrine knowledge in the field
Theropithecus gelada	Klecha and McCann 2001	Urinary progesterone	Endocrine knowledge in captivity
Rhinopithecus bieti	He et al. 2001	Urinary estrogen, progesterone; bioactive follicle-stimulating hormone, luteinizing hormone	Endocrine knowledge in captivity
Semnopithecus entellus	Ziegler et al. 2000a	Fecal pregnanediol glucuronide	Endocrine knowledge in the field
	Heistermann et al. 1995a	Urinary and fecal estrone conjugates, pregnanediol glucuronide, estradiol, progesterone, pregnanediol, 20α-hydroxsprogesterone	Endocrine knowledge in captivity
	Lohiya et al. 1988, 1998	Serum estradiol, progesterone, testosterone, cortisol	Endocrine knowledge in captivity
Great apes			
Gorilla gorilla beringei	Robbins and Czekala 1997	Urinary testosterone and cortisol	Endocrine knowledge in the field
G. g. gorilla	Margulis and Atsalis 2007, Atsalis et al. (2004)	Fecal progesterone and estradiol	Hormones and behavior in captivity
	Kuhar et al. 2005	Salivary cortisol	Hormones and behavior in captivity
	Czekala al. 1983, 1987, 1991	Urinary estrone conjugates, estradiol, pregnanediol-3α-glucuronide	Endocrine knowledge in captivity
	Bahr et al. 2001	Urinary estrone conjugates, pregnanediol-3α-glucuronide	Endocrine knowledge in captivity

Table 23.1 (cont'd)

PRIMATES	REFERENCES	MEDIA/HORMONE	PURPOSE
	Miyamoto et al. 2001	Fecal estrogens and progesterone	Endocrine knowledge in captivity
	Bellem et al. 1995	Urinary estrogens and progesterone	Endocrine knowledge in captivity
	Czekala et al. 1983, 1987	Urinary estrone conjugates, estradiol	Endocrine knowledge in captivity
Pan paniscus	Behringer et al. 2009	Salivary cortisol	Endocrine knowledge in captivity
	Sannen et al. 2003	Urinary testosterone conjugates	Endocrine knowledge in captivity
	Jurke et al. 2000, 2001	Urinary and fecal estrogens, progesterone, cortisol	Hormones and behavior in captivity
	Heistermann et al. 1996	Urinary estrone conjugates, pregnanediol glucuronide, estriol; fecal progesterone	Endocrine knowledge in captivity
P. troglodytes	Muller and Lipson, 2003	Urinary testosterone metabolites	Endocrine knowledge in the field
	Emery and Whitten 2003	Fecal estradiol, progesterone	Endocrine knowledge in captivity
	Muller and Wrangham 2003	Urinary testosterone, cortisol metabolites	Endocrine knowledge, hormones and behavior in the field
	Dahl et al. 1991	Urinary estrone conjugates, pregnanediol-3-glucuronide, luteinizing hormone	Endocrine knowledge in captivity
	Thompson et al. 2008, Thompson 2005	Urinary estrone conjugates, pregnanediol-3α-glucuronide and testosterone conjugates, fecal estradiol, progesterone	Endocrine knowledge, hormones and behavior in the field
	Möhle et al. 2002	Urinary and fecal testosterone	Endocrine knowledge in captivity
	Nadler et al. 1985	Serum estrogens, progesterone, luteinizing hormone, follicle-stimulating hormone, testosterone	Endocrine knowledge in captivity
Pongo pygmaeus	Knott 2005a,b	Urinary estrone conjugates	Endocrine knowledge in captivity and in the field
	Czekala et al. 1983	Urinary estrogens	Endocrine knowledge in captivity
	Bonney and Kingsley 1982	Urinary estrogens, progesterone	Endocrine knowledge in captivity

in captivity, and even fewer have studies that compare the reproductive biology of wild and captive animals. For example, Ziegler and colleagues (1987a,b, 1990c, 1997, 2000b,c) have spent many years examining the reproductive biology and behavior of cotton-top tamarins (*Saguinus oedipus*) in captivity and have provided the most complete information on any callitrichid studied in captivity. Savage et al. (1997) compared data from wild caught females to the data published for captive cotton-top tamarins and found some interesting differences. Wild female cotton-top tamarins are seasonal breeders, giving birth every 48 weeks, compared to captive females, which give birth every 28 weeks. Wild cotton-top tamarins have a postpartum suppression of fertility that averages approximately 144 days, while captive females typically conceive within 1 month of parturition. The postpartum period of ovarian inactivity in wild cotton-top tamarin females is influenced by environmental conditions (e.g., drought), and similar to captive females, it is possible to experience a 19-day postpartum ovulation under specific environmental conditions. Interestingly, these studies allow us greater insight into the factors that influence reproduction both in captivity and in the wild and demonstrate how both social and

environmental factors may cause females to regulate their own fertility. Wild female tamarins are observed to terminate their pregnancies when the environmental conditions (e.g., drought) are not conducive to a successful outcome. This type of comparative information is critically important when one is trying to develop conservation plans for a species. Understanding the reproductive capacity of a species in addition to a variety of other important factors provides the information necessary to develop models for long-term population viability.

HUSBANDRY

In many captive facilities in the United States and abroad, the hope is to develop self-sustaining captive breeding colonies that are genetically, demographically, and behaviorally viable. Attempts to breed animals without a clear understanding of the reproductive biology is often challenging since without hormonal data it is often difficult to document puberty, ovarian cycles, etc. in species that do not exhibit clear physical or behavioral signs of estrus or pregnancy. Early studies in white-faced sakis (*Pithecia pithecia*) estimated that sexual maturation, based on first conception, occurred within 26–32 months (Shoemaker 1982) after pairing with a male. However, it was not until a comprehensive study was conducted on the reproductive biology of the species (Savage et al. 1995, Shideler et al. 1994) that it was demonstrated that puberty occurred at approximately 32 months of age and that it did not appear that puberty could be accelerated in this species by pairing a prepubescent female with an adult male.

The ability to diagnose pregnancy in both captive and field conditions is essential. In captive colonies, managers are tasked with diagnosing pregnancy so that they can make the appropriate husbandry decisions from an animal welfare perspective. While pregnancy confirmation through a collection of serial samples is commonplace for many species, the ease and rapidity with which one can diagnose pregnancy were suggested as being of high priority for technological development. Shimizu et al. (2001) developed a noninstrumented enzyme-linked immunosorbant assay (NELISA) for the measurement of urinary monkey chorionic gonadotropin (mCG) for the detection of early pregnancy in macaque monkeys (*Macaca mulatta* and *M. fascicularis*) for use in both the laboratory and the field. The advantage of this technique is that the color change can be detected without the use of a spectrophotometer. The mCG in pregnant monkey samples was indicated by a dark green color. In contrast, nonpregnant monkey urine samples exhibited no color change. These findings suggest that the simple, economical, and reliable urinary mCG NELISA may be useful for diagnosing early pregnancy in these and related species.

ASSISTED REPRODUCTIVE TECHNOLOGIES

While assisted reproductive technology (ART) has revolutionized the breeding of domestic animals and the treatment of infertile human couples (Trouson 1993), the use of ART to assist with the propagation of threatened and endangered nonhuman primate species has been limited (Bavister and Boatman 1993, Reddy et al. 1992), with most research directed primarily as a model for understanding and developing ART in humans (Nyachieo et al. 2009, Hewitson 2004, Graham 1981).

While infertility is commonplace with humans, documented infertility in captive and wild primates is rare. In general, there has not been the need to develop ART for most primate species and, as a consequence, this field has progressed slowly. Documented individual infertility in free-ranging primates is either rare or seldom a breeding problem as the relatively complex interrelated social and sexual behaviors seem to override physiological deficiencies. Nonetheless, there is a significant amount of information on the production of embryos/infants with the great apes (Gould and Dahl 1993, Joslin et al. 1995, Pope et al. 1997), Old World primates (Cranfield et al. 1993, Curnow et al. 2002, Fazleabas et al. 1993, Ogonuki et al. 2003, Torii et al. 2000), and New World primates (Dukelow 1993, Marshall et al. 1999, Morrell et al. 1998).

SUMMARY AND CONCLUSIONS

Assays for monitoring reproductive events in primates continue to evolve as new technological developments make them less expensive, less labor-intensive, and more efficient to provide results. New sophisticated automated platforms will revolutionize the standard assay lab into one that will have the ability to process hundreds of samples in each run with national and international standards of quality control. In the future, participating scientists will be able to compare results without concern for assay differences, and those results will be obtained within weeks, if not days.

With the advent of new and less expensive technology, it is our hope that more scientists will develop projects to assess reproductive functioning in wild and captive primates that have not yet been evaluated. Many of these species are highly endangered and in areas that require protection or assistance with developing self-sustaining captive breeding programs. As part of any long-term conservation program, it is critical that we have an understanding of the reproductive biology of the species to insure its protection for the future.

24

Molecular Primatology

Anthony Di Fiore, Richard R. Lawler, and Pascal Gagneux

1. How can molecular data help us understand behavioral topics such as reproduction, dispersal, and social structure?
2. How may the increase in genomic studies influence our understanding of primate evolution and phylogeny?
3. At what levels can variation in molecular information be detected and analyzed?

THE UTILITY AND PROMISE OF MOLECULAR PRIMATOLOGY

The fundamental aim of molecular primatology is to describe and interpret the patterns of molecular variation found within and between primate taxa. There are a number of areas in contemporary primatology where molecular studies are particularly useful.

First, molecular studies provide data that can be used for inferring phylogenetic relationships among groups of organisms. One advantage of molecular data is that the alternative states of a molecular character are typically far less ambiguous than for many morphological traits (e.g., a nucleotide base that is "adenine" versus a molar crest that is "high"). Another advantage is that, at least for some loci, reasonably good estimates of the rate of change (mutation rates) for deoxyribonucleic acid (DNA) and amino acid sequences are available. If these rates of change are reasonably consistent over evolutionary time (i.e., approximate a so-called molecular clock), researchers can infer the timing of lineage-splitting events within a phylogeny. Over the last decade, molecular phylogenies for extant primates have largely replaced morphology-based ones.

Second, molecular studies provide a direct way to investigate numerous features of primate social systems not easily observed in the field. For example, using various kinds of molecular markers, researchers can conduct parentage assessments, document kin relationships among individuals, examine population substructuring, assess levels of inbreeding within and gene flow between populations, and evaluate sex biases in dispersal behavior, allowing inferences to be made about mating systems, social organization, and dispersal patterns that are typically impossible to make based on observational studies alone (Di Fiore 2003a, de Ruiter 2004).

Third, molecular techniques can contribute significantly to applied primate conservation efforts. For example, researchers might use genotyping of noninvasively collected samples (e.g., feces, hair from nests) to perform molecular mark-recapture studies (Pearse et al. 2001) for determining population density and sex composition or, if species-specific markers are available, to assay the presence or absence of a particular species in an area (Waits and Paetkau 2005). Similarly, with genotype data, captive colony managers and zoo personnel can make more informed decisions about which animals to breed in order to minimize inbreeding and preserve the genetic variation present in captive populations. Routine health monitoring and veterinary care for captive primates benefit from molecular studies as well. A variety of polymerase chain reaction (PCR)–based tests are already available that allow animals to be quickly screened for various hereditary disorders as well as for infection by numerous pathogens.

Fourth, molecular studies of primates are increasingly relevant to questions of human health and disease as well as to the tracking and management of disease outbreaks in wild animal populations. Primates are natural reservoirs and zoonotic sources for a variety of current and emerging human diseases (e.g., Ebola, Marburg, simian or human immunodeficiency virus, herpes simiae virus) (Wolfe et al. 1998, Daszak et al. 2000, Hahn et al. 2000). Molecular studies of both pathogen and host populations can lead to better understanding of the source and mode of transmission of these diseases, as well as the factors responsible for differential susceptibility to various pathogens.

Finally, the current era of *genomics* (the study of the complete DNA sequence of the genome of an organism) holds promise for evolutionary biologists to be able to undertake far more complete and sophisticated comparisons of the organization, function, and evolution of genomes across taxa. Complete or nearly complete genomes have been sequenced for a wide variety of organisms, ranging from microbes to invertebrates to mammals. Among mammals alone, close to three dozen genomes have been completed or are in draft assembly with >1X coverage (Table 24.1) and whole genome

Table 24.1 Overview of Mammalian and Primate Genomic Projects

ORDER	SCIENTIFIC NAME	COMMON NAME	NCBI TAXON ID	SEQUENCING TARGET	NCBI OVERVIEW ID	NCBI GPID(S)	GOLD ID(S)	STATUS	COVERAGE/TARGET COVERAGE	NCBI REFERENCE SEQUENCE	INITIAL RELEASE DATE	KEY SEQUENCING INSTITUTIONS/CENTERS	KEY FUNDING AGENCIES	REFERENCE
Afrosoricida	*Echinops telfairi*	Small Madagascar hedgehog	9371	Genome	12589	12590	Gi00745	Draft	2X	AAIY00000000	08/16/05	MIT-BI	NIH-NHGRI	
Artiodactyla	*Bos taurus*	Domestic Cow	9913	Genome	10708	12555	Gi00086	Quality draft	7.1X	AAFC00000000	10/01/04	BCM-HGSC, USDA, US Meat Animal Research Center	USDA, NIH-NHGRI	Bovine Genome Sequencing and Analysis Consortium 2009
Artiodactyla	*Lama pacos*	Alpaca	30538	Genome	32137	30567	Gi03080	Draft	2.51X		09/24/08	MIT-BI, WU-GSC	NIH-NHGRI	
Artiodactyla	*Sus scrofa*	Domestic pig	9823	Genome	10718	13421	Gi00635	In progress				Porcine Genome Sequencing Center, WTSI		
Carnivora	*Canis lupus familiaris*	Domestic dog	9615	Genome	10726	13179	Gc00597	Quality draft	7.6X	AAEX00000000, NC_006583-621	07/10/04	MIT-BI, Agencourt Bioscience, CNRS, CHORI, Fred Hutchinson Cancer Research Center, NC State U	NIH-NHGRI	Lindblad-Toh et al. 2005
Carnivora	*Canis lupus familiaris*	Domestic dog	9615	Genome	10726	10628	Gi00131	Draft	1.5X	AACN010000000	09/27/03	JCV Institute	NIH-NHGRI	Kirkness et al. 2003
Carnivora	*Felis catus*	Domestic cat	9685	Genome	10703	16726, 32759	Gi00748	Draft	1.87X	AANG00000000	04/27/06	MIT-BI, Agencourt Bioscience	NIH-NHGRI	Pontius et al. 2007
Cetacea	*Tursiops truncatus*	Bottlenose dolphin	9739	Genome	20365	20367	Gi02181	Draft	2.59X		09/23/08	BCM-HGSC, MIT-BI	NIH-NHGRI	
Chiroptera	*Myotis lucifugus*	Little brown bat	59463	Genome	16953	16951	Gi01615	Draft	1.7X	AAPE01000000	06/01/06	MIT-BI	NIH-NHGRI	
Chiroptera	*Pteropus vampyrus*	Large fruit bat	132908	Genome + selected regions	20277	20325, 20327		Draft	2.63X		09/12/08	BCM-HGSC, MIT-BI, NISC-NIH	NIH-NHGRI	
Didelphimorphia	*Monodelphis domestica*	Short-tailed opossum	13616	Genome	12560	12561	Gc00602	Quality draft	7.33X	AAFR00000000	11/08/04	MIT-BI, WU-GSC	NIH-NHGRI	Mikkelsen et al. 2007
Diprotodontia	*Macropus eugenii*	Tammar wallaby	9315	Genome	12586	12587	Gi00341	Draft	2X	ABQO0000000000	11/26/08	Australian National U, AGRF, Macquarie U BCM-HGSC, U Melbourne	NIH-NHGRI	
Hyracoidea	*Procavia capensis*	Cape rock hyrax	9813	Genome	13973	13972	Gi01612	Draft	2.19X		09/19/08	BCM-HGSC, MIT-BI	NIH-NHGRI	
Insectivora	*Erinaceus europaeus*	European hedgehog	9365	Genome	12575	12576	Gi00743	Draft	1.86X	AANN00000000	02/17/06	MIT-BI	NIH-NHGRI	
Insectivora	*Sorex araneus*	European shrew	42254	Genome + selected regions	12573	12574, 13689	Gi02450	Draft	1.9X	AALT00000000	11/07/05	NISC-NIH, BCCA-GSC, MIT-BI	NIH-NHGRI	
Lagomorpha	*Ochotona princeps*	American pika	9978	Genome	20373	19235	Gi01816	Draft	1.93X	AAYZ00000000	08/03/07	MIT-BI	NIH-NHGRI	
Lagomorpha	*Oryctolagus cuniculus*	European rabbit	9986	Genome + selected regions	12818	12819, 13686	Gi00747	Draft	2X	AAGW00000000	05/13/05	NISC-NIH, BCCA-GSC, MIT-BI	NIH-NHGRI	

Table 24.1 (cont'd)

ORDER	SCIENTIFIC NAME	COMMON NAME	NCBI TAXON ID	SEQUENCING TARGET	NCBI OVERVIEW ID	NCBI GPID(S)	GOLD ID(S)	STATUS	COVERAGE/TARGET COVERAGE	NCBI REFERENCE SEQUENCE	INITIAL RELEASE DATE	KEY SEQUENCING INSTITUTIONS/CENTERS	KEY FUNDING AGENCIES	REFERENCE
Monotremata	Ornithorhynchus anatinus	Duck-billed platypus	9258	Genome + selected regions	10803	12885, 13682	Gc00781	Quality draft	6X	AAPN01000000	04/10/06	WU-GSC, NISC-NIH	NIH-NHGRI	Warren et al. 2008
Perissodactyla	Equus caballus	Domestic horse	9796	Genome	11760	18661	Gi00232	Draft	6.79X	AAWR00000000	01/12/07	MIT-BI, U Veterinary Medicine Hannover	USDA	
Primates	Aotus nancymaae	Owl monkey	37293	Selected regions			Gi01809	In progress				NISC-NIH	NIH-NHGRI	
Primates	Ateles geoffroyi	Black-handed spider monkey	9509	Selected regions			Gi01807	In progress				NISC-NIH	NIH-NHGRI	
Primates	Callicebus moloch	Dusky titi monkey	9523	Selected regions			Gi01832	In progress				NISC-NIH, JCV Institute	NIH-NHGRI	
Primates	Callithrix jacchus	Common marmoset	9483	Genome	13629	13630, 13690, 20401	Gi00992	In progress				BCM-HGSC, WU-GSC, NISC-NIH	NIH-NHGRI	
Primates	Cercopithecus (Chlorocebus) aethiops	Vervet	9534	Selected regions			Gi01810	In progress				NISC-NIH	NIH-NHGRI	
Primates	Colobus guereza	Guereza	33548	Selected regions			Gi01808	In progress				NISC-NIH	NIH-NHGRI	
Primates	Eulemur macaco macaco	Black lemur	30603	Selected regions			Gi01830	In progress				NISC-NIH	NIH-NHGRI	
Primates	Gorilla gorilla gorilla	Western lowland gorilla	9593	Genome	18247	18247	Gi02158	Draft		AAGL00000000	11/09/06	Fritz Lipmann Institute (IMB)		
Primates	Gorilla gorilla gorilla	Western lowland gorilla	9595	Genome	31265	31265	Gi03640	Draft	2X	CABD00000000	11/03/08	WTSI		
Primates	Gorilla gorilla gorilla	Western lowland gorilla	9593	Selected regions			Gi01828	In progress				NISC-NIH	NIH-NHGRI	
Primates	Homo sapiens	Human	9606	Genome	9558	9558, 13178	Gc00044	Quality draft	10X		12/11/99	International Human Genome Sequencing Consortium		Lander et al. 2001
Primates	Homo sapiens	Human	9606	Genome	9558	1431		Draft			02/22/01	Celera Genomics		Venter et al. 2001
Primates	Homo sapiens	Human	9606	Genome	9558	19621	Gc00655	Quality draft	7.5X	ABBA01000000	05/22/07	JCV Institute		Levy et al. 2007
Primates	Homo sapiens	Human	9606	Genome	9558	28335	Gc00892	Quality draft	7.4X	ABKV00000000	04/17/08	BCM-HGSC, 454 Life Sciences		Wheeler et al. 2008
Primates	Homo sapiens	Human	9606	Genome	9558	29429	Gc00893	Quality draft	38X	SRA000271		Illumina Inc.		Bentley et al. 2008
Primates	Lemur catta	Ring-tailed lemur	9447	Genome	12896	12897	Gi01146	In progress				MIT-BI	NIH-NHGRI	
Primates	Macaca fascicularis	Long-tailed macaque	9541	Genome	20403	20409	Gi03083	In progress				WU-GSC	NIH-NHGRI	
Primates	Macaca fuscata	Japanese macaque	9543	Genome			Gi01822	In progress				RIKEN	MEXT	

Primates	Macaca mulatta	Rhesus macaque	9544	Genome	12536	12537	Gc00600	Quality draft	6X	NC_007858-78	02/08/06	BCM-HGSC, JCV Institute, WU-GSC	NCRR, NIH-NHGRI	Rhesus Macaque Genome Sequencing and Analysis Consortium 2007
Primates	Microcebus murinus	Gray mouse lemur	30608	Genome	20413	19967	Gi01831	Draft	1.93X	ABDC00000000	07/20/07	NISC-NIH, MIT-BI	NIH-NHGRI	
Primates	Nomascus leucogenys	Northern white-cheeked gibbon	61853	Genome	13974	13975, 20423	Gi01486	In progress				BCM-HGSC, WU-GSC	NIH-NHGRI	
Primates	Otolemur garnettii	Small-eared galago	30611	Genome	13680	16955	Gi01827	Draft	1.5X	AAQR01000000	06/02/06	MIT-BI	NIH-NHGRI	
Primates	Otolemur garnettii	Small-eared galago	30611	Selected regions	13680	13679	Gi01147	In progress				NISC-NIH	NIH-NHGRI	
Primates	Pan paniscus	Bonobo	9597	Genome			Gi02966	In progress				Max Planck Institute		
Primates	Pan troglodytes	Common chimpanzee	9598	Genome	12467	12552, 13184	Gc00604	Quality draft	6X	NC_006468-92	12/05/03	BCM-HGSC, WU-GSC, MIT-BI	NIH-NHGRI	Chimpanzee Sequencing and Analysis Consortium 2005
Primates	Pan troglodytes	Common chimpanzee	9598	Selected regions	12467	13698	Gi00425	In progress				Celera Genomics		
Primates	Papio anubis	Olive baboon	9555	Genome	12965	13688		In progress				NISC-NIH		
Primates	Papio anubis	Olive baboon	9555	Genome	12965	12966	Gi00429	In progress				U Oklahoma		
Primates	Papio hamadryas	Hamadryas baboon	9557	Genome	20427	20425	Gi01378	In progress				BCM-HGSC	NIH-NHGRI	
Primates	Pongo abelii	Sumatran orangutan	9601	Genome	12830	20869	Gi00749	Draft	6X	ABGA00000000	11/14/07	BCM-HGSC, WU-GSC	NIH-NHGRI	
Primates	Pongo pygmaeus	Bornean orangutan	9602	Selected regions			Gi01825	In progress				NISC-NIH, MIT-BI	NIH-NHGRI	
Primates	Pongo pygmaeus	Bornean orangutan	9602	Selected regions			Gi02166	In progress				BCM-HGSC	NIH-NHGRI	
Primates	Saimiri boliviensis boliviensis	Bolivian squirrel monkey	39432	Selected regions			Gi01826	In progress				NISC-NIH	NIH-NHGRI	
Primates	Tarsius syrichta	Philippian tarsier	9478	Genome	20335	20339	Gi02182	Draft	1.82X		09/17/08	NISC-NIH, WU-GSC, MIT-BI	NIH-NHGRI	
Proboscidea	Loxodonta africana	African elephant	9785	Genome + selected regions	12568	12569, 13683	Gi02449	Draft	2X	AAGU01000000	05/18/05	MIT-BI, NISC-NIH	NIH-NHGRI	
Rodentia	Cavia porcellus	Guinea pig	10141	Genome	12582	12583	Gi00744	Draft	6.87X	AAKN00000000	11/04/05	MIT-BI	NIH-NHGRI	
Rodentia	Dipodomys ordii	Ord's kangaroo rat	10020	Genome	20377	20385	Gi02183	Draft	1.85X		09/18/08	BCM-HGSC, MIT-BI	NIH-NHGRI	

Table 24.1 (cont'd)

ORDER	SCIENTIFIC NAME	COMMON NAME	NCBI TAXON ID	SEQUENCING TARGET	NCBI OVERVIEW ID	NCBI GPID(S)	GOLD ID(S)	STATUS	COVERAGE/TARGET COVERAGE	NCBI REFERENCE SEQUENCE	INITIAL RELEASE DATE	KEY SEQUENCING INSTITUTIONS/CENTERS	KEY FUNDING AGENCIES	REFERENCE
Rodentia	*Mus musculus*	House mouse	10090	Genome	9559	13183, 20689	Gc00112	Quality draft	5X	NC_000067-87	05/15/02	WTSI, MIT-BI, WU-GSC, NISC-NIH, BCM-HGSC, MRC Laboratory of Molecular Biology, Harvard Partners Genome Center, Cold Spring Harbor Lab, GTC, DOE Joint Genome Inst, U Oklahoma	NIH	Mouse Genome Sequencing Consortium 2002
Rodentia	*Mus musculus*	House mouse	10090	Genome	9559	11785	Gi02163	Draft		AAHY00000000	07/07/05	Celera Genomics		
Rodentia	*Rattus norvegicus*	Norway rat	10116	Genome	10621	10629, 13999	Gc00181	Quality draft	7X	NC_005100-20	04/02/04	GTC, U Utah, BCM-HGSC, Medical College of Wisconsin, JCV Institute, Celera Genomics, CHORI, BCCA-GSC	NHGRI, NIH, NHLBI	Gibbs et al. 2004
Rodentia	*Spermophilus tridecemlineatus*	Thirteen-lined ground squirrel	43179	Genome	13936	13937	Gi01145	Draft	1.9X	AAQO01000000	06/06/06	MIT-BI	NIH-NHGRI	
Scandentia	*Tupaia belangeri*	Northern tree shrew	37347	Genome	13938	13971, 20273	Gi01833	Draft	2X	AAPY01000000	06/08/06	MIT-BI, WU-GSC	NIH-NHGRI	
Xenarthra	*Choloepus hoffmanni*	Two-toed sloth	9358	Genome	32151	30809	Gi01484	Draft	2.05X	ABVD01000000	09/15/08	MIT-BI, WU-GSC	NIH-NHGRI	
Xenarthra	*Dasypus novemcinctus*	Nine-banded armadillo	9361	Genome + selected region's	12593	12594, 13687	Gi00746	Draft	2X	AAGV09000000	05/14/05	NISC-NIH, MIT-BI	NIH-NHGRI	

Information for this table was compiled from the following repositories of genomic data sources and metadata and is current as of May 2009: National Center for Biotechnology Information (NCBI, http://www.ncbi.nlm.nih.gov/); GOLD, Genomes Online (GO, http://www.genomesonline.org/); International Sequencing Consortium (ISC, http://www.intlgenome.org/); European Bioinformatics Institute (EMBL-EBI, http://www.ebi.ac.uk/); Ensembl Genome Browser (ENSEMBL, http://www.ensembl.org/).

AGRF, Australian Genome Research Facility; BCCA-GSC, British Columbia Cancer Agency Genome Sciences Centre; BCM-HGSC, Baylor College of Medicine—Human Genome Sequencing Center; CHORI, Children's Hospital Oakland Research Institute; CNRS, Centre National de la Recherche Scientifique; GTC, Genome Therapeutics Corporation; JCV Institute, J. Craig Venter Institute; MIT-BI, Massachusetts Institute of Technology-Broad Institute; NISC-NIH, National Intramural Sequencing Center—National Institutes of Health; WTSI, Wellcome Trust Sanger Institute; WU-GSC, Washington University—Genome Sequencing Center.

sequencing projects for more than a dozen additional taxa are nearing completion. For primates, whole-genome data are or soon will be available for taxa spanning all of the major radiations within the order, as well as several key outgroups (Table 24.1) (B. Bradley and R. Lawler , unpublished data). Comparative genomic studies are providing insights into the history of genes associated with important diseases in humans and are shedding light on a variety of candidate genes that have been implicated in the development of many aspects of biology and behavior that appear to distinguish humans from the great apes. For example, despite sharing close to 99% of their genomes in coding regions, humans and chimpanzees differ notably in many aspects of life history, physiology, behavior, and, fundamentally, their capacity for symbolic and acoustic communication. Comparative genomic studies have revealed some of the important molecular differences between modern humans and chimpanzees (Gagneux and Varki 2001, Chimpanzee Sequencing and Analysis Consortium 2005, Khaitovich et al. 2005, Chen et al. 2006).

BREADTH OF THE FIELD

Current use of the term "molecular primatology" typically equates molecular studies with studies of DNA, the fundamental molecule of genetic heredity. Indeed, in the last 15 years researchers have increasingly employed genetic techniques (particularly microsatellite genotyping and direct DNA sequencing) to investigate mating systems and social structure (Altmann et al. 1996; Keane et al. 1997; Gagneux et al. 1997b, 1999a; Gerloff et al. 1999; Nievergelt et al. 2000; Launhardt et al. 2001; Constable et al. 2001; Vigilant et al. 2001; Utami et al. 2002; Radespiel et al. 2002; Kappeler et al. 2002; Wimmer et al. 2002; Muniz et al. 2006; Goossens et al. 2006b), dispersal patterns (Morin et al. 1994, Melnick and Hoelzer 1996, Lawler et al. 2003, Di Fiore and Fleischer 2005, Hammond et al. 2006, Douadi et al. 2007, Langergraber et al. 2007b, Di Fiore et al. 2009, Harris et al. 2009), and within-group social behavior, kin recognition, and relatedness (Goldberg and Wrangham 1997, Alberts 1999, Hohmann et al. 1999, Mitani et al. 2000, Vigilant et al. 2001, Widdig et al. 2001, Buchan et al. 2003, Smith, K., et al. 2003, Di Fiore and Fleischer 2005, Lukas et al. 2005, Langergraber et al. 2007a, Di Fiore 2009b). Genetic data are also contributing increasingly to attempts to infer phylogenetic relationships among all major radiations of primates (strepsirhines, Yoder et al. 1996, Yoder 1997, Roos et al. 2004, Yoder and Yang 2004; New World monkeys, von Dornum and Ruvolo 1999, Schneider et al. 2001, Ray et al. 2005, Hodgson et al. 2009; Old World monkeys, Stewart and Disotell 1998, Disotell 2000, Raaum et al. 2005; Sterner et al. 2006; Ting et al. 2008; hominoids, Arnason et al. 1996, Ruvolo 1997b, Satta et al. 2000), as well as to evaluate the demographic (Gagneux et al. 1999a, Storz et al. 2002a,b) and biogeographic (Rosenblum et al. 1997a,b; Goldberg

and Ruvolo 1997a; Evans et al. 1999; Collins and Dubach 2000a,b; Hapke et al. 2001; Gagneux et al. 2001b; Tosi et al. 2000, 2003; Cortés-Ortiz et al. 2003; Wildman et al. 2004; Clifford et al. 2004; Bergl and Vigilant 2007; Ting et al. 2008) histories of various primate groups.

The true scope of molecular primatology is much broader than simply the study of DNA sequence-level genetic variation, and the discipline has deep historical roots (Table 24.2). First, there are many other types of biomolecules besides DNA that investigators might use to explore patterns of natural variation within and between primate species, from various types of ribonucleic acid (RNA) to regulatory enzymes to cell membrane proteins to "classic" serum protein markers. Moreover, there are numerous ways in which many of these molecules can be modified following transcription and translation through post-translational modifications (e.g., addition of sugars, phosphates, sulfates, fatty acids). The study of variation among and within taxa in all of these molecules also falls under the rubric of "molecular primatology" and can contribute to our understanding of the evolutionary relationships among primate taxa, the selective pressures which have acted on populations within taxa, and the differences in physiology and behavior observed between taxa. Additionally, advances in the study of chromosomes (e.g., cytogenetic techniques such as chromosome painting through fluorescence in situ hybridization [FISH]) and new bioinformatic and computational methods for mining and comparing large sets of sequence data are providing greater insights into variation among primate taxa in genome structure. Among such findings are the differences among taxa in the type, frequency, and distribution within the genome of various kinds of repetitive DNA, transposons (i.e., mobile pieces of DNA, or "jumping genes"), and endogenous retroviruses (Gagneux and Varki 2001, Trask 2002). In addition, we have come to realize the significance of variation in both coding and noncoding (e.g., promoter and enhancer binding sites, pre-messenger RNA [mRNA] splicing signals present in introns and at intron/exon boundaries) regions of the genome for producing phenotypic variation, either through their regulation of transcription or through their effects on pre-mRNA splicing (Hastings and Krainer 2001, Modrek and Lee 2002, Cartegni et al. 2002). As these loci play a central role in the link between genotype and phenotype and likely explain much of the evolutionarily significant phenotypic variation seen among closely related primate taxa, they should be important targets for future research. Finally, the growing availability of complete primate genomes (Table 24.1) has allowed for more refined comparative studies of genome architecture, which in turn has highlighted the significance of duplications, chromosomal rearrangements, and variation in gene copy number as important additional components of genetic differentiation between taxa (Cheng et al. 2005, Newman et al. 2005, Bailey and Eichler 2006).

In the remainder of this chapter, our aim is to provide a brief but broad summary of modern molecular primatology.

Table 24.2 Brief Overview of Significant Developments in Molecular Primatology

DEVELOPMENT	REFERENCE(S)
Discovery of human blood groups	Landsteiner 1900
Early immunological experiments on blood serum cross-reactivity	Nuttal 1904
First correct assessment of the human chromosome number ($2n =46$)	Tjio and Levan 1956
Immunological cross-reactivity experiments to elucidate the phylogeny of hominoid primates; results suggested a much more recent common ancestry for humans and great apes than previously thought	Sarich and Wilson 1967
Dramatic improvements in cytogenetic techniques for staining metaphase chromosomes	Caspersson et al. 1968
Phylogenetic studies of many of the major primate radiations using immunological methods	Goodman and Moore 1971
	Goodman et al. 1974
	Dene et al. 1976a,b, 1980
	Baba et al. 1979
	Sarich and Cronin 1976
Early applications of amino acid sequence data to primate phylogenetic studies	Goodman et al. 1974
	Goodman 1976
	Hewett-Emmett et al. 1976
	Romero-Herrera et al. 1976
	Tashian et al. 1976
Allozyme methods applied to studying the population structure, dispersal pattern, and mating system of a variety of wild primates (e.g., baboons, lion tamarins, howler monkeys, rhesus macaques, long-tailed macaques)	Jolly and Brett 1973
	Brett et al. 1976
	Melnick et al. 1984a,b
	Forman et al. 1987
	Pope 1990, 1992
	de Ruiter et al. 1992, 1994
Development of chain-terminator method for determining the sequence of DNA	Sanger et al. 1977
Cytogenetic studies of hominoid primates suggest a sister-taxon relationship for humans and chimpanzees to the exclusion of gorillas	Yunis and Prakash 1982
DNA–DNA hybridization studies of the phylogenetic relationships among African great apes provide additional support for a sister-taxon relationship between humans and chimpanzees	Sibley and Alquist 1984, 1987
Development of the polymerase chain reaction, which allows the in vitro cloning of specific fragments of the genome from a complex DNA sample	Saiki et al. 1985, 1988
	Mullis and Faloona 1987
Evolutionary relationships among hominoids investigated using 2D allozyme electrophoresis; results provide additional support for a chimpanzee-human clade within the hominoids	Goldman et al. 1987
Restriction fragment length polymorphism studies of mitochondrial DNA variation suggest a recent African origin for all modern humans	Cann et al. 1987
	Vigilant et al. 1991
Publication of draft sequence of complete *Homo sapiens* genome	Venter et al. 2001
	Lander et al. 2001
Publication of the initial sequence of the chimpanzee genome	Chimpanzee Sequencing and Analysis Consortium 2005
Publication of the macaque genome	Rhesus Macaque Genome Sequencing and Analysis Consortium 2007
Initial exploration of "personal genomics," where genomes of different individuals within a population are sequenced and compared for medical and evolutionary insights; personal genomics, at some point in the future, will likely be applied to nonhuman primate populations	McGuire et al. 2007

We begin by summarizing some of the major laboratory methods that are currently used to analyze molecular variation, dividing our discussion into four sections: variation at the level of the genome, variation at the level of the gene or DNA sequence, variation in gene structure and expression, and variation in other types of non-DNA molecules. We then briefly detail the kinds of samples primatologists should collect in the field in order to be able to carry out these types of molecular studies and provide some methodological suggestions for sample collection and storage. Finally, we point to a number of resources where students and researchers new to this field can find more information on laboratory and analytical methods.

MODERN TECHNIQUES FOR ASSAYING MOLECULAR VARIATION

Depending on the question being addressed, researchers will be interested in different types and levels of molecular variation. For example, comparative karyotypic or cytogenetic data are useful for broad phylogenetic studies as well as for comparing genome organization across taxa. Regarding the latter, bioinformatic methods (e.g., whole-genome shotgun detection and whole-genome assembly comparison) are increasingly being used to examine aspects of genome organization—namely, the position and extent of large, duplicated regions within taxa (Bailey et al. 2001, 2002; Cheng et al. 2005). Systematists can utilize DNA sequence data from entire genomes or particular genomic regions in order to evaluate phylogenetic relationships among taxa at both shallow and deep levels of the Linnean hierarchy (e.g., Janecka et al. 2007, Horvath et al. 2008). Of likely relevance to primate behavioral ecologists, DNA sequence data—particularly when drawn from multiple genetic loci—are useful for estimating how different evolutionary forces (i.e., mutation, migration, drift, selection) and demographic processes (e.g., population bottlenecks, population growth) act, and have acted, on primate populations. In this case, the power to reconstruct evolutionary and demographic processes from genetic sequence data is facilitated by two basic properties of population genetic theory. First, different loci and/or nucleotides within a locus can be differentially shaped by various evolutionary processes; for example, mutation, selection, and recombination act only on specific loci or genetic regions, while other regions are unaffected. Second, there are explicit (though assumption-laden) models of how genetic sequence data should be patterned under a variety of evolutionary and demographic conditions, which can be used to infer past population demographic and evolutionary history, for example, by comparing observed patterns of DNA sequence polymorphism to expectations under the neutral or nearly neutral models of molecular evolution (Hudson et al. 1987, Tajima 1989, McDonald and Kreitman 1991, Fu and Li 1993, Fay and Wu 2000, Nielsen and Beaumont 2009). For example, haplotype data (generated either through DNA sequencing or single-nucleotide polymorphism [SNP] typing) also can be used to evaluate selection on particular alleles by examining the relationship between allele frequency at a locus and the extent of linkage disequilibrium (LD) surrounding the locus. LD refers to the phenomenon whereby a stretch of DNA containing the locus of interest shows evidence of a relative lack of expected recombination, even at large distances from the locus (i.e., the length of nonrecombined haplotypes exceeds expectations). Alleles under positive selection should show greater than expected regions of LD given their frequency, a phenomenon referred to as "extended haplotype homozygosity" (Sabeti et al. 2002, Nielsen 2005, Zhang et al. 2006, Hancock and Di Rienzo 2008). The software DnaSP (Rozas et al. 2004) implements a number of these methods of testing for evidence of selection and for inferring population demographic history from DNA sequence data.

In an increasing number of cases, specific loci or genetic regions have been linked to phenotypic attributes. Depending on the method of discovery and type of variation, such genetic loci/regions are either dubbed "candidate genes" or quantitative trait loci (QTLs). Candidate genes often contain a functional polymorphism in which a particular change in one or more amino acids influences a particular phenotype, such as color vision (Fitzpatrick et al. 2005). QTLs are genomic regions that are statistically associated with variation in a particular trait. QTLs are often found by tracking the correspondence between a particular trait and a set of particular genetic markers, such as microsatellite loci or SNPs, across a pedigree (Slate 2005). Finally, some genetic loci and/or genetic regions are highly variable due to a high mutation rate. These loci are particularly useful for detailed analyses of identity, parentage, and relatedness within populations of a particular species. Examples of such genetic loci include microsatellites and minisatellites.

Variation at the Level of the Genome

Studying the organization and comparative composition of whole genomes is the purview of the fields of cytogenetics and comparative genomics. Classical cytogenetics uses precipitated metaphase chromosomes, which are harvested from actively dividing cells and then dropped on a glass slide, causing the chromosomes to form a spread or splatter. The chromosomes can then be treated with enzymes (e.g., trypsin) to partially digest the histone proteins around which DNA is coiled and then stained with various dyes to create banding patterns that reflect the different base pair composition of particular areas (e.g., GC-rich bands, which contain the majority of active genes). The species-specific number and appearance (morphology) of such chromosome spreads is called a *karyotype*.

Studies of the primate genome classically done by cytogenetic analysis are rapidly evolving to merge with molecular genetic techniques such as fluorescent nucleotide probe hybridization on whole chromosomes (FISH) (Ried et al.

1993; Wienberg and Stanyon 1995, 1997, 1998). FISH studies use PCR to produce chromosome-specific probes from flow cytometry–sorted chromosomes of a particular species. Fluorescent dyes incorporated in the amplified fragments can then be used to "paint" the target chromosomes of other species, where the fluorescent probes will hybridize to complementary sequences found on any of the chromosomes of the species under investigation. This technique allows the detection of major rearrangements in chromosomes between taxa by highlighting where genetic material with similar sequence has been maintained through evolution (Figure 24.1). The use of reciprocal chromosome painting via FISH has produced insights into chromosome evolution within a number of major primate groups, including strepsirhines (Müller et al. 1997, Stanyon et al. 2002), platyrrhines (Stanyon et al. 2000, 2001; Neusser et al. 2001), colobines (Bigoni et al. 2003, 2004), and hylobatids (Arnold et al. 1996; Koehler et al. 1995a,b; Müller et al. 2003).

At a more detailed level, comparative genomic studies have revealed a surprising degree of variation among taxa in genome organization due to dynamic processes of inter- and intrachromosomal segmental duplications. These duplications, which range in size from a few thousand to a few million base pairs, are widespread within the genome, suggesting that the genome is much more plastic than previously thought (Samonte and Eichler 2002, van Geel et al. 2002). While segmental duplications are more common in gene-poor regions of the genome, some duplications also involve regions containing functional genes, which may have surprisingly high rates of evolutionary change (e.g., the *morpheus* gene family, Johnson et al. 2001). Some recent work has focused on the relationship between chromosomal rearrangements and the quick pace of sequence evolution in genes located near rearrangement sites (Navarro and Barton 2003, Marquès-Bonet et al. 2004), but the jury is still out on the importance of such effects for gene expression evolution and its potential role in the speciation process (Zhang et al. 2004). More recently, Marques-Bonèt et al. (2009) found evidence for an increased rate of segmental duplication along the ancestral lineage leading to modern African great apes relative to other primates (orangutans, macaques), as well as additional shared and lineage-specific duplications within the African ape clade. Gene copy number variation (CNV) associated with these segmental duplications is suspected to be a significant factor influencing phenotypic differences among primate taxa (Bailey and Eichler 2006), and several recent studies have discussed the significance of CNV for patterns of gene expression, susceptibility to disease, developmental disorders, and adaptive molecular evolution in humans and chimpanzees (Perry et al. 2006, 2008; Wong et al. 2007; Varki et al. 2008).

Variation at the Level of the Gene or DNA Sequence

Modern techniques for studying DNA sequence-level variation take advantage of a relatively small set of fundamental

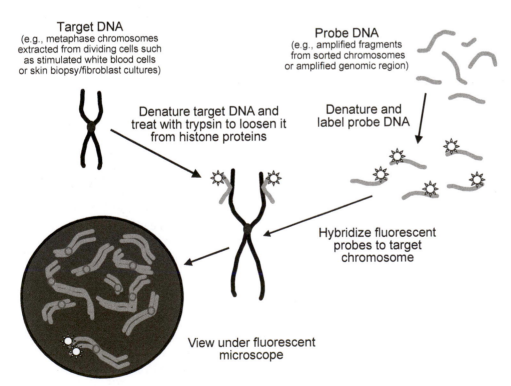

Target DNA
(e.g., metaphase chromosomes extracted from dividing cells such as stimulated white blood cells or skin biopsy/fibroblast cultures)

Probe DNA
(e.g., amplified fragments from sorted chromosomes or amplified genomic region)

Denature target DNA and treat with trypsin to loosen it from histone proteins

Denature and label probe DNA

Hybridize fluorescent probes to target chromosome

View under fluorescent microscope

Figure 24.1 Fluorescence in situ hybridization (FISH). This technique allows the localization of DNA sequences on chromosomes with the help of fluorescent dyes attached to short DNA probes. It can be used to study chromosome rearrangements (as in reciprocal chromosome painting) or to detect the precise chromosomal location of a known stretch of DNA, such as a fragment of genomic DNA containing a gene.

concepts of molecular biology: the cutting of DNA with restriction enzymes, the joining of DNA fragments with DNA ligase, the separation (denaturing) and joining (annealing) of complementary strands of DNA by heating and cooling, the replication of DNA using DNA polymerases, and the separation of fragments of DNA by size through gel electrophoresis. Since there are many excellent reviews of the basic molecular biological concepts applied in modern genetic analysis, we will not undertake such a task and instead direct the reader to those reviews (Hoelzel 1998, Hillis et al. 1996, Ferraris and Palumbi 1996, Baker 2000). We focus here on only a small set of techniques that are currently used to generate marker data (e.g., genotypes) and DNA sequences.

The workhorse tool of virtually all modern genetic studies is PCR (Saiki et al. 1985, 1988; Mullis and Faloona 1987). PCR is essentially a tool for in vitro cloning, allowing the many million–fold amplification of a specific target fragment of the genome from the midst of a complex DNA sample. PCR takes advantage of the discovery of various thermostable polymerases whose activity is not degraded by the high temperatures required to denature double-stranded DNA prior to replication. In PCR, template DNA and DNA polymerase are mixed with free nucleotides (deoxynucleotide triphosphates, dNTPs) and *primers*, short oligonucleotides of DNA that are complementary to the sequence flanking the region to be amplified, thus providing the double-stranded template necessary for the polymerase to function properly (Figure 24.2).

Figure 24.2 Polymerase chain reaction (PCR). The most common element in almost all modern genetic studies of primates, PCR takes advantage of a heat-stable, DNA-replicating enzyme isolated from a hot-spring bacterium (*Thermophilus aquaticus*) from Yellowstone National Park to generate large numbers of identical copies of a given target DNA sequence.

Microsatellite Genotyping

Microsatellites, also known as simple sequence repeats (SSRs) or short tandem repeats (STRs), are one class of molecular markers known more generally as variable number of tandem repeat (VNTR) loci. These loci consist of a

Figure 24.3 Genome and genetic marker vocabulary. *A*: Schematic overview of genome and gene structure. *B*: Various classes of molecular genetic markers. Each line represents a region of homologous DNA on one member of each autosomal chromosome pair. From top to bottom: (*1*) Microsatellite marker locus involving the repeat of a dinucleotide (CA) motif. This individual is a (CA)3/(CA)5 heterozygote. (*2*) Biallelic single-nucleotide polymorphism, or SNP, site. This individual is an A/G heterozygote. (*3*) Functional polymorphism. This individual carries one "normal" and one mutant copy of a protein-coding region of DNA. An example might be an individual heterozygous for the normal (*Hb*A) and sickle (*Hb*S) alleles coding for the β chain of a hemoglobin molecule. (*4*) Alternative haplotypes at a locus. The term "haplotype" refers to a set of markers that are "linked" or inherited together most of the time because they are located close to one another on the same chromosome and, therefore, less likely to undergo recombination during the process of gamete formation.

short 1–6 bp motif repeated in series a variable number of times (Figure 24.3).

Microsatellite regions tend to evolve far more rapidly than other portions of the genome, presumably due to slippage of DNA polymerase as DNA is replicated during the process of gametogenesis (Levinson and Gutman 1987, Kruglyak et al. 1998, Toth et al. 2000). The various alleles present in a population at a given locus can be assayed through PCR amplification of the locus followed by electrophoretic separation of the *amplicons* (the millions of copies of the same DNA fragment produced by PCR). Because alleles will differ in length due to the presence of different numbers of repeats of the microsatellite motif, these fragments will migrate to different positions in a gel during electrophoresis and individual genotypes can be scored directly (Figure 24.4A).

PCR-based microsatellite genotyping has been increasingly used to provide insight into various aspects of the behavioral biology of nonhuman primates, including mating systems, dispersal patterns, and population genetic structure (Di Fiore 2003a, de Ruiter 2004). However, one limitation is that primers need to be available for amplifying variable microsatellite loci. Many such primer pairs are known for a variety of hominoid and cercopithecoid taxa since primers used in human genotyping and linkage studies often amplify homologous regions in other catarrhines. Researchers working on more distantly related primates, however, often need to invest considerable time and money isolating novel microsatellites either from taxon-specific *genome libraries* (sets of short fragments of DNA covering the entire genome that are inserted into biological vectors and cloned; e.g., Strassman et al. 1996, Hamilton et al. 1999, Paetkau 1999) or from subsets of genome fragments generated through a variety of PCR procedures (e.g., Fisher et al. 1996, Lunt et al. 1999, Zane et al. 2002). A primate database of microsatellite loci has been compiled by Di Fiore (2003a, www.nyu.edu/projects/difiore/yearbook/appendix.html), and a similar database is currently under development by

Figure 24.4 *Right*: Molecular genetics techniques routinely used in primatological studies. *A*: Microsatellite genotyping by electrophoretic separation of amplified DNA fragments. Fragments are separated by size in an electrophoretic medium such as a polyacrylamide gel. DNA is negatively charged and will migrate toward the positive pole in a gel. Microsatellite alleles are codominant, allowing genotypes to be scored directly. Offspring must inherit one allele from each parent; thus, offspring 2 in this example appears to have been fathered by an unsampled individual. *B*: Direct DNA sequencing by the dideoxy method, which causes DNA replication of a template to be terminated when a dideoxynucleotide base analog is incorporated into a growing DNA strand (Sanger et al. 1977). Electrophoresis separates a mixture of partially copied DNA sequence fragments differing in size by 1 bp in a medium sensitive enough to resolve these differences. In modern automated sequencing, the four possible dideoxynucleotide triphosphates (ddNTPs; A, T, G, and C) are each tagged with a different fluorescent molecule, allowing the type of nucleotide at the terminal position of a DNA fragment to be resolved by laser detection. From shortest to longest fragments, the order of the passing fragments gives the complementary sequence of the template DNA.

INPRIMAT (www.inprimat.org), a European consortium to promote primate molecular biology research. With advances in direct DNA sequencing, it is now possible to identify microsatellite loci from sequences in silico (Raveedran et al. 2008) using one of many computer programs that scan whole genomes (e.g., MSAT Commander, Faircloth 2008; SciRoKo, Kofler et al. 2007).

Single Nucleotide Polymorphism Genotyping

Faster methods of detecting genetic variation have been developed that do not require separation of PCR products via electrophoresis. Emphasis has been placed recently on the search for SNPs (or "snips") in the human genome for use as molecular markers in large-scale linkage studies. SNPs, as their name implies, are single-nucleotide positions in the genome that differ between different chromosomes in a population (Figure 24.3). Most SNPs are biallelic, existing in only two alternative forms. In humans, SNPs occur approximately once every 1,000–2,000 bp in the genome, and over 1.4 million SNPs have already been putatively identified in the roughly 2.85 billion bp comprising the completed sequence of the euchromatic portion of the human haploid genome (International Human Genome Sequencing Consortium 2001, 2004; International SNP Map Working Group 2001).

SNPs have been used extensively in human studies to examine population genetic structure and migration patterns. As yet, SNPs have not been used widely in studies on nonhuman primates, primarily due to the lack of known markers; however, as improved protocols for isolating SNPs are developed (e.g., Aitken et al. 2004), this will likely change.

Because of their utility in linkage studies in humans, a number of high-throughput, automated methods have been derived to directly assay variation at multiple SNP loci simultaneously, including the use of allele-specific extension on DNA microarrays (Pastinen et al. 2000), hybridization of different alleles to complementary oligonucleotides bound on microarrays (Hacia et al. 1999), and the detection of alternative alleles created via PCR using mass spectrometry (Haff and Smirnov 1997, Bray et al. 2001). Gut (2001) and Syvänen (2001) provide reviews of these and other automated SNP typing methods.

Direct DNA Sequencing

The most detailed form of genetic variation that molecular primatologists can assay is variation in an actual DNA sequence between individuals or among taxa. Modern phylogenetic studies commonly use direct DNA sequencing to generate DNA sequence data for a particular site (locus) of interest and then compare the sequence composition, the distribution of variable sites, and the relative genetic distance between individuals or groups of sequences. Modern DNA sequencing relies on the controlled replication of a DNA template of interest using PCR. Template DNA is mixed with polymerase and other reaction components,

along with free nucleotides (A, T, G, and C) in the form of dNTPs. Some small fraction of these free nucleotides are ddNTPs, which, when randomly incorporated into a growing DNA strand, prevent further extension as they lack the 3′-hydroxyl group needed to form a phosphodiester bond with a subsequent nucleotide (Sanger et al. 1977). The result of the sequencing reaction is thus a mixed population of copied DNA fragments differing in length by a single base pair, from one base to the full complementary length of the template. These fragments are then separated by electrophoresis, allowing the sequence of DNA bases to be determined based on which ddNTP is incorporated at the end of each fragment (Figure 24.4B). Studies of sequence variation are useful at multiple levels (e.g., for inferring phylogenetic relationships among higher-order taxa, such as species or genera, as well as for inferring intraspecific phylogenies that can be used in phylogeographic studies). This type of sequencing technique is known as "Sanger sequencing" or "chain-terminator sequencing" (or "dye-terminator sequencing" if ddNTPs are labeled with a fluorescent biomolecule for automated sequence detection).

Since the late 1990s, several new DNA sequencing technologies have emerged (reviewed by Mardis 2008a,b), dubbed "next-generation sequencing" (NGS) methods for the speed with which they can produce DNA sequence data. In the course of a weekend, a day, or less, an entire genome can be sequenced. Several companies (e.g., Roche/454, Solexa, Applied Biosystems) have produced "next-generation" sequencers that are in commercial and research use, and while the specific technologies used to actually "read" DNA differ from company to company, all next-generation sequencers have the following properties that distinguish them from previous methods: (1) massively improved speed, the power to produce sequence data very rapidly because NGS methods process and sequence millions of DNA fragments simultaneously, where previous commercial technologies for DNA sequencing based on the chain-termination method could process only a couple of hundred sequences at a time (typically 96–384); (2) shorter "read lengths," they produce DNA sequence data from short fragments (often 35–250 bp) rather than the longer 600–900 bp reads used in chain-terminator sequencing; (3) "electrophoresis-free" sequencing, they do not rely on the electrophoretic separation of different-sized DNA fragments in either a gel or polymer-filled capillary, unlike chain-terminator sequencing (or most genotyping methods); (4) "cloning-free" sequencing, cycle-terminator sequencing relies on an intermediary step of either cloning the template DNA in a natural vector such as *Escherichia coli* or PCR amplification of the target locus prior to performing the final chain-termination PCR reaction that incorporates ddNTPs, whereas NGS methods sequence directly from template DNA initially broken up into small fragments. The basic methods of NGS sequencing (exemplified by Roche/454 Life Sciences ("pyrosequencing") are shown in Figure 24.5.

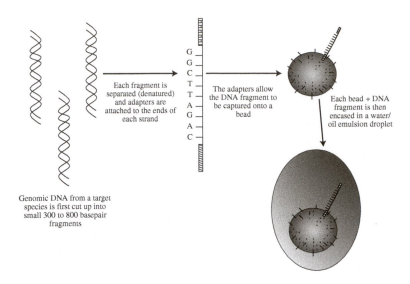

Genomic DNA from a target species is first cut up into small 300 to 800 basepair fragments

Each fragment is separated (denatured) and adapters are attached to the ends of each strand

The adapters allow the DNA fragment to be captured onto a bead

Each bead + DNA fragment is then encased in a water/ oil emulsion droplet

Figure 24.5 The general process of next-generation sequencing (NGS) methods, using Roche/454 Life Science's "pyrosequencing" technology as an example. Many of the NGS methods first cut up genomic DNA into small fragments and then amplify these fragments in parallel using a particular proprietary technology. The amplified fragments are then assembled to produce the genetic sequence of the species under study.

Within each droplet, "emulsion PCR" is performed to make many copies of each fragment

The droplet is removed and each bead and associated cloned DNA fragments is the loaded onto a microtitre plate containing thousands of wells

Ideally, one bead gets loaded into one well

Microtitre plate containing thousands of wells

one well

Within each well, each nucleotide fragment undergoes sequencing in order to create a duplex (double-stranded molecule)

This is done by sequentially "washing" the fragment with nuceleotide bases

Every time a nucleotide base is incorporated into the duplex, it emits a photochemical signal

polymerase

This signal is captured on a computer in order to generate the complementary sequence of the DNA fragment contained within the well

Peak height of the fluorescent signal is proportional to the number of identical sequential nucleotides

This process occurs in parallel across all wells, thus generating sequences for the many thousands of different DNA fragments on the microtitre plate

The fragments can be assembled using bioinformatic methods to produce the complete genomic DNA sequence of the target species

There are still some challenges that remain to be surmounted with respect to this new NGS technology. First, because the read lengths of each DNA fragment are short, it can be problematic to use NGS methods to assemble a genome de novo. Chain-terminator sequencing methods produce long stretches of DNA sequence, and these long stretches can be more easily overlapped (i.e., the end of one sequence "read" can be aligned with the beginning of another sequence "read") in order to produce the entire sequence that characterizes the genome under study. With very short DNA sequence fragments, it becomes tricky to assemble and align these fragments in order to produce a longer sequence. Thus, new computational methods are required for sequence alignment when the sequences themselves constitute very short runs of sequence information (Pop and Salzberg 2008). Further, because NGS methods are less accurate than chain-terminator sequencing, they require much oversampling or redundant sequencing in order to control for sequencing error. Usually, because of the short read lengths and the relatively more error-prone sequences, NGS data are often compared to sequence data produced by more traditional methods. In addition, some NGS methods have trouble reading long stretches of nucleotides of the same type (e.g., CGTTTTTTTTTA). Here, errors can occur with respect to how many nucleotides ("T" in this case) are registered by the instrument. This is problematic if one considers that a stretch of repetitive DNA can be longer than the read length produced by the instrument. How does one meaningfully assemble multiple stretches of DNA, each of which consists only of the same nucleotide (e.g., five stretches of the base "T" repeated 20–40 times)? The analysis of sequence data produced by NGS methods, at present, constitutes a major challenge for bioinformatics (Pop and Salzberg 2007).

Alternative Methods for Detecting Sequence Variation

Several alternative methods have been developed as shortcuts for rapidly evaluating whether sequence variation exists within a population at a given locus without having to directly sequence every individual sample. These include a variety of techniques—single-strand confirmation polymorphism analysis (SSCP), denaturing gel gradient electrophoresis analysis (DGGE), and conformation-sensitive gel electrophoresis analysis (CSGE)—that take advantage of the fact that fragments of DNA differing by as little as a single base pair will migrate differently through a variety of electrophoretic media.

Alternative methods for fast typing of well-characterized sequence variants, such as SNPs, include "real-time" or "quantitative" PCR and flow cytometry. Real-time PCR is a modified form of PCR that typically incorporates a third oligonucleotide that is complementary to a short segment of the target. This third oligonucleotide, or "probe," carries both a fluorescent dye molecule and a "quenching" molecule, which prevents the dye from emitting light as long as the probe is intact. During PCR amplification, the DNA polymerase, which also has exonuclease activity, digests the central probe as it works its way down the template synthesizing a new complementary strand (Figure 24.6). Digestion of the probe separates the fluorescent dye from the quencher, causing it to emit light when stimulated by a laser. The intensity of the fluorescent emission is proportional to the amount of template DNA present at each cycle; thus, the initial quantity of template in an experimental sample can be evaluated by comparing its amplification profile to that of a standard template of known DNA quantity. The coupling of allele-specific amplification (e.g., for an SNP locus) with real-time PCR detection allows for fast and efficient screening of a population for sequence variation.

Flow cytometry is a general procedure that can be used for the analysis of a variety of particulate samples, from fragments of DNA to organelles to bacteria to whole eukaryotic cells (Figure 24.7). The technique relies on samples being kept as a suspension in a "sheath" fluid, the flow of which is controlled such that the particles to be studied flow through a chamber one-by-one at a rate of several hundred per second. Laser light illuminates each passing particle, and a collection of light detectors (photomultipliers) measures the emission on the other side of the chamber.

Process is repeated for 25 to 50 cycles

TEMPLATE DOUBLE-STRANDED DNA
Region to be amplified bounded by known primer sequence

DENATURE DOUBLE-STRANDED DNA (~94C)

ANNEAL PRIMERS AND PROBE

EXTENSION (~72C)
Taq DNA polymerase digests probe and frees fluorescent molecule from quencher as it assembles new second strand, llowing it to be excited by a laser

Figure 24.6 Real-time or quantitative polymerase chain reaction (Q-PCR). In addition to flanking PCR primers, a third oligonucleotide is used that is complementary to a portion in the middle of the target region to be amplified and contains both a fluorescent and a "quencher" molecule. As replication proceeds, the DNA polymerase digests this third primer and releases the quencher molecule, allowing the fluorescent molecule to emit detectable light when stimulated by a laser. The intensity of this light can be measured to provide an accurate estimate of the number of amplified copies of a DNA sequence present after each cycle of replication.

Figure 24.7 Flow cytometry (literally "cell measuring") is a technique used to measure various aspects of individual particles or cells at a rate of several hundred per second. Sample particles in suspension are injected into a sheath fluid (usually an isotonic buffer or medium). By adjusting the pressure on the sheath and sample channels, the sample fluid can be directed to flow through the center of the sheath fluid column without mixing and contain a focused stream of separated molecules. These molecules flow one-by-one past a source of illumination (e.g., a laser). The amount of light scattered in the forward and side directions can be detected and used to characterize the size and granularity of the sample particles. Additionally, particles of interest can be assayed and quantified by measuring the color and amount of light emanating from laser-stimulated fluorescent probes used to stain those particles. Flow cytometry can be applied to many different tasks, from identifying and sorting different types of cell (on the basis of their forward and side scatter illumination profiles) to evaluating what stage of the cell cycle particular cells are in (based on fluorescent quantification of the amount of DNA present in the cell) to sorting chromosomes (based on fluorescent quantification of two different dyes, one that binds preferentially to GC-rich areas of chromosomes and one that binds to AT-rich areas) to multiplex genotyping of a mixture of polymerase chain reaction–amplified DNA fragments (based on the incorporation of allele-specific fluorescent molecules and their binding to similarly tagged microspheres).

This procedure allows the rapid determination of particle size and granularity and, based on the emission profile, can be used to analyze samples by detecting molecules present on the particles and using fluorescence intensity to quantify their abundance. Methods for SNP genotyping by flow cytometry make use of microspheres that are tagged with a distinct fluorescent color and that have short, allele-specific oligonucleotide capture probes attached to them. Populations of microspheres bearing probes for the alternative alleles at an SNP site are then hybridized to sample DNA that is subject to allele-specific PCR, which also incorporates a different fluorescent marker. The flow cytometer is then used to determine the presence of each alternative allele, based on the combined microsphere and allelic fluorescent signals recorded. Taking advantage of the fact that microspheres with 100 distinct fluorescent tags are commercially available (Luminex, Austin, TX), Ye et al.

(2001) have demonstrated the potential of the flow-cytometric method for multiplex genotyping of DNA samples at up to 15 SNP loci simultaneously.

Variation in Gene Structure and Expression

Studying the structure and expression of genes requires the copying of large stretches of chromosomal DNA that contain the full coding sequence as well as the regulatory regions and noncoding exons and introns of a gene. This is because regulation of gene expression is governed by the binding of various transcription factors (mostly proteins) to well-defined binding regions upstream from the gene itself. Variants of basic PCR procedures, such as "long-distance" or "long-range" PCR, make such amplification possible. Amplified stretches of that genomic DNA can then be probed for the presence of known sequence elements (e.g., regulatory sequences). This is done by separating the fragments on agarose gels, transferring the fragments to a membrane by Southern blotting, and screening the membrane with labeled probes for the DNA regions of interest. Such information is crucial for understanding how genes are structured.

Furthermore, many genes can be "read" in more than one way by the process of *alternative splicing* (the joining together of different exonic portions of a primary mRNA transcript in different ways). The DNA sequence at splice sites can provide information on the total number of potential transcripts (mRNAs) generated by alternative ways of transcribing the same gene. Different patterns of mutation in the exons and introns, for example, can be used to infer selection affecting a particular gene or even a particular functional region of the protein (Sonnenburg et al. 2004).

In order to more directly study the nature of expressed genes and the process of alternative splicing of genetic messages to generate more than one product from the same gene, mRNA can be extracted from a tissue of interest, separated by electrophoresis on agarose gels, and then transferred to a membrane. There, it can be probed with single-stranded DNA oligonucleotides labeled with either radioactive or fluorescent biomolecules to determine what genes (and combinations of genes) contribute and in what quantity they contribute to the mRNA transcripts expressed in particular tissues. Single-stranded mRNA, which is far less stable than DNA, can also be copied back into DNA via reverse-transcription PCR (RT-PCR), which allows it to be studied more easily. Real-time RT-PCR can also be used to quantify the genetic message by providing an estimate of the amount of starting mRNA template isolated from a particular tissue (Cáceres et al. 2003). The term "transcriptomics" is used to describe such studies of the complete collection of genetic messages, or "transcripts," present in a given tissue at a given time.

Increasingly, transcriptomic studies are being carried out using microarray technology to measure the presence and relative quantity of different mRNAs that are expressed in

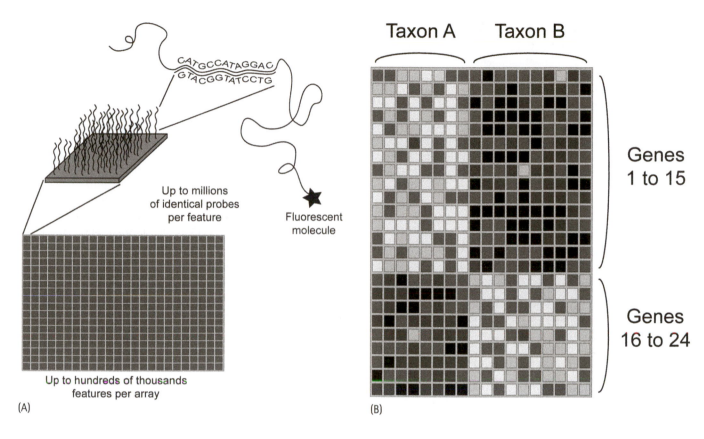

(A)

Up to millions of identical probes per feature

Fluorescent molecule

Up to hundreds of thousands features per array

(B)

Taxon A   Taxon B

Genes 1 to 15

Genes 16 to 24

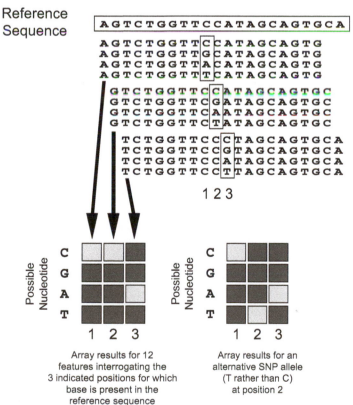

Reference Sequence

AGTCTGGTTCCATAGCAGTGCA

AGTCTGGTTCCATAGCAGTG
AGTCTGGTTGCATAGCAGTG
AGTCTGGTTACATAGCAGTG
AGTCTGGTTTCATAGCAGTG

GTCTGGTTCCATAGCAGTGC
GTCTGGTTCGATAGCAGTGC
GTCTGGTTCAATAGCAGTGC
GTCTGGTTCTATAGCAGTGC

TCTGGTTCCCTAGCAGTGCA
TCTGGTTCCGTAGCAGTGCA
TCTGGTTCCATAGCAGTGCA
TCTGGTTCCTTAGCAGTGCA

1 2 3

Possible Nucleotide

C
G
A
T

1 2 3

Array results for 12 features interrogating the 3 indicated positions for which base is present in the reference sequence

Possible Nucleotide

C
G
A
T

1 2 3

Array results for an alternative SNP allele (T rather than C) at position 2

(C)

Figure 24.8 DNA microarrays and gene chips. Microarrays are contraptions allowing miniature hybridization experiments of sample DNA against an array of up to thousands of known DNA probes fixed in a grid. *A*: Single-stranded DNA oligonucleotides are either blotted onto a small, flat substrate (e.g., a glass slide) in well-defined positions (a "DNA microarray") or synthesized directly on a silica-based chip using photolithography (a "gene chip"). Multiple copies of the same oligonucleotide probe bound to the same region of the microarray or chip constitute a "feature." A single photolithographically produced chip may contain up to several hundred thousand features. The microarray or chip is washed with a denatured nucleic acid sample of interest that has been tagged with a radioactive or fluorescent label. Sample DNA or RNA will hybridize only where a complementary sequence is found and can be detected based on the corresponding luminescent signal. *B*: As discussed in the text, microarrays and chips can be used in comparative studies of gene expression. As a hypothetical example, a cDNA-based microarray containing probes for 24 genes (rows) is used to interrogate fluorescently labeled mRNA extracted from tissue samples of 18 individuals from two different taxa (columns 1–8, taxon A; columns 9–18, taxon B). Lighter coloration for a feature indicates that more of the labeled mRNA sample bound to the probe. In this case, genes 1–15 show greater expression in taxon A, while genes 16–24 show greater expression in taxon B. *C*: Microarrays and chips can also be used for single-nucleotide polymorphism (SNP) genotyping of polymerase chain reaction–amplified samples. In this case, features consist of staggered sets of oligonucleotides that span a known reference sequence containing the SNP site. For each of these sets, there are four features on the array, one corresponding to each potential nucleotide present at a "query" position located in the middle of the oligonucleotide. Amplified samples will hybridize only to the feature containing the appropriate complementary base at each query position, which allows for rapid SNP genotyping. A similar procedure can be used for "resequencing" an entire DNA amplicon.

specific tissues or cell types. This involves the hybridization of reverse-transcribed mRNA to arrays of unique short oligonucleotide or complementary DNA (cDNA) probes specific to genes of interest. These arrays may be commercially available "DNA chips" containing thousands of probes or homemade dot plots of a much smaller number of probes bound to a membrane or slide (Figure 24.8). Karaman et al. (2003) used this procedure to evaluate the relative expression of a large number of genes in gorillas, bonobos, and humans by interrogating Affymetrix (Santa Clara, CA) U95Av2 DNA chips (which contain short probes for over 10,000 human gene transcripts) with mRNA isolated from multiple fibroblast cell lines (cultures of skin cells initiated from biopsy samples) of each of these three taxa. Their analysis identified a number of candidate genes (many involved in metabolic pathways, neural signal transduction, and cell and tissue growth) that were differentially expressed among these species, providing some directions for future research on the underlying causes of phenotypic differentiation within hominoids.

In independent studies using similar DNA chips, Enard et al. (2002) and Cáceres et al. (2003) compared the expression of the same set of human genes in the brains of humans, chimpanzees, orangutans, and macaques and compared these results to those for other body tissues. In both studies, nearly 170 genes were expressed differentially in the brains of chimpanzees and humans, with humans showing higher expression levels (upregulation) for the vast majority of these genes, many of which appear to be related to cell growth and maintenance as well as to neural function. These differences in expression levels, however, did not extend to heart or liver tissue, where only just over half of the genes with different expression levels were upregulated in humans (Cáceres et al. 2003). If real, these differences in expression levels suggest a possible neurological basis for some of the behavioral and cognitive adaptations thought to be associated with the divergence of humans from other African apes. Interestingly, the locations of these upregulated genes appear to be associated with areas of chromosomal rearrangement and segmental duplication between humans and chimpanzees (Khaitovich et al. 2004, Marquès-Bonet et al. 2004). A similar study by another group demonstrated that genes expressed in cortical tissue are upregulated in all of the African great apes relative to macaques, though in humans to a greater extent than in either chimpanzees or gorillas (Uddin et al. 2004). Preuss et al. (2004) provide a comprehensive review of these and other microarray studies of gene expression differences among primates. Data from these latter studies have been reanalyzed using network analysis to discern fascinating gene-coexpression networks in the brains of humans and chimpanzees (Oldham et al. 2006).

One of the important limitations of the microarray approach is that, to date, it is still impossible to detect subtle changes (e.g., less than twofold differences) in gene-expression levels (Forster et al. 2003). A further limitation is that these studies are heavily biased toward "prominent" messages (i.e., those genes that are expressed at very high levels, producing high copy numbers of the same mRNA transcript). Other genes with relatively low and much more narrowly localized expression may nonetheless have important phenotypic consequences (e.g., glycosyl transferases and other protein-modifying enzymes).

Recently, some tantalizing RNA genes have been described, some of them among the most rapidly evolving pieces of the human genome (Pollard et al. 2006). This is contributing an increased realization of the large number of expressed RNAs, which are never translated into proteins but rather act as regulators of gene expression (e.g., micro-RNAs). Finally, gene-expression studies will ultimately have to be combined with the emerging field of *epigenetics*, which considers sequence changes outside canonical gene loci, where transcription is regulated by the influence that such sequence changes exert on the binding of transcription factors and their complex assembly in the initiation of transcription (Peedicayil 2006, Yeo et al. 2009).

Studying Primate Pathogens, Parasites, and Symbionts

The same technical advances that have improved our ability to characterize primate genome variation are also used to study the diverse collection of microbes (bacteria, protozoans, and viruses) that exist on and in different primate species and/or populations. Piecing together the history of cross-species infections giving rise to the human immunodeficiency virus 1 (HIV1) pandemic has relied on molecular characterization of both the primate lentiviruses (simian immunodeficiency viruses [SIVs]) and the over 40 species of African nonhuman primate hosts to these viruses (Hahn et al. 2000, Apetrei et al. 2004). Recent studies using molecular techniques have revealed that non-African strains of pandemic human respiratory viruses—likely transmitted by field primatologists and/or tourists—can kill wild chimpanzees (Köndgen et al. 2008).

Finally, interest is growing in the comparative study of the *microbiome* (the complete set of microbial symbionts and parasites living in and on an animal) of humans and other primates. For example, molecular genetic techniques, in combination with classical methods from microbiology such as serology, can be used to identify the types and strains of bacteria inhabiting the bodies of healthy and sick primates, as well as potential adaptations of the host primate genome to their presence (e.g., Stewart et al. 1987, Zhang et al. 2002). This burgeoning research area, which has close ties to the broader fields of *metagenomics* or *environmental genomics* (i.e., study of the collection of nucleic acid sequences that characterize a particular ecological community), promises to uncover a vast and brand new realm of primate biodiversity, one with clear significance for understanding aspects of primate evolution relevant to health and diseases. For example, with the completion of the human microbiome project (Turbaugh et al. 2007), data gathered from nonhuman primate microbiomes offer a promising comparative database from which to assess the reciprocal adaptations of microbial communities and primate populations across the primate order (e.g., McKenna et al. 2008).

Assaying Variation in Non–DNA Molecules

In addition to nucleic acids (DNA and RNA), the cells of living organisms contain several other classes of important biomolecules, including proteins, lipids, and sugars, which exist in various combinations (Figure 24.9). Protein comparisons have become much more sophisticated since the days of early allozyme studies, contributing to the rapidly expanding field of *proteomics* (the study of the complete set of proteins found in a particular type of cell or tissue). Soluble and extracted proteins can be separated in two dimensions via isoelectric focusing and sodium dodecyl sulfate polyacrylamide gel electrophoresis (SDS-PAGE) and then transferred to a membrane and probed with various labeled antibodies to assay the presence of particular proteins as well as their electrophoretic signatures. An example of this approach is the comparison of soluble proteins in the blood plasma of great apes and humans (Gagneux et al. 2001a). Sugar molecules on glycoproteins can be assayed similarly, by probing with carbohydrate-specific lectins and certain antibodies that bind to these molecules (Boenisch 2001).

Variation in soluble or extracted proteins can also be screened immunologically using the enzyme-linked

immunosorbent assay (ELISA), where the presence of a protein or protein variant in a sample is detected via its reactivity with antibodies that are labeled with a *chromogenic* (color-producing) enzyme. In the line immunoassay (LIA), probes consisting of known proteins are blotted on a membrane, over which a sample is then washed. This technique has been used to test for the presence of antibodies against SIV in a large number of wild primates in Africa (Peeters et al. 2002). Animals that tested positive for antiviral antibodies in their serum or urine were then further investigated with PCR or RT-PCR to confirm the presence of the virus and to characterize its nucleotide sequence (Santiago et al. 2003).

Flow cytometry, a form of "liquid ELISA," allows the study of molecules on and inside living or fixed cells by taking advantage of fluorescently labeled probes that can be bound to the molecule of interest. Flow cytometry has recently been used to document differences among several primate species in the distribution and abundance of immune molecules present on white blood cells and in the ability of these cells to bind certain well-defined ligands (Brinkman-van der Linden et al. 2000). Analyses of primate sperm are also now conducted with the help of flow cytometry as this powerful method allows for the quantitative analysis of large numbers of cells in a very short time (~10,000 per minute) (Shankar et al. 2004). Flow-cytometric methods in combination with fluorescent dyes can also be used to study metabolic activity such as mitochondrial function within live cells of different primate species. For example, Anderson et al. (2006) recently used this method to investigate the effects of sperm competition on hominoid sperm function.

Specific molecules in and on cells can also be detected and visualized using immunohistochemical methods. These methods rely on enzymatic digestion, and they are sensitive enough to detect extremely low amounts of target molecules. This sensitivity is critical because certain components of a tissue may be present at very low concentration (and thus relatively underrepresented in a tissue extract) and yet be of the utmost biological importance (e.g., receptors found only on the apical side of a single-cell epithelial cell layer such as hormone receptors, olfactory receptors, endogenous lectins, or glycoconjugates) (Gagneux et al. 2003). Additionally, proteins with known genetic sequence can be cloned in vitro (i.e., by copying and pasting their genes in mRNA or cDNA to form plasmids, which are then introduced into a bacterial vector and replicated) and then expressed in an expression system (e.g., Chinese hamster ovary cells) for comparative functional and/or structural studies.

Fibroblast lines established from skin biopsy samples form precious repositories of live cells, which can be stored indefinitely and regrown as needed as a source of DNA for genetic studies or for use in comparative cytogenetic, transcriptomic, and proteomic studies as well as studies of gene expression in various tissues. With support from the National Science Foundation, numerous institutions,

Figure 24.9 Molecules in context. Living organisms are made up of four classes of molecules—nucleic acids (DNA and RNA), proteins, lipids, and carbohydrates/sugars—which exist in various combinations. The assembly, distribution, and abundance of the final molecules are affected by many factors, ranging from genetic control to cellular dialog within the organism to interactions with the biotic and abiotic environment faced during development. DNA (*light gray shading*) is found in the nucleus (and mitochondria, not shown), lipids (*dark gray shading*) are prevalent in the nuclear and cellular membranes, and carbohydrate-containing glycans (*black threads*) are attached to proteins expressed on the surfaces of cells. (Figure kindly provided by A. Varki.)

such as the Frozen Zoo of the Zoological Society of San Diego, the Coriell Institute, Princeton University, and the International Species Information System are currently collaborating to establish the Integrated Primate Biomaterials and Information Resource (www.ipbir.org), containing a large collection of such cell lines from animals with known provenience and, often, known life histories. The European INPRIMAT consortium is also establishing a similar biomaterials resource bank. Additionally, for hominoid primates, immortal cell lines created from white blood cells form a crucial resource for comparative studies of gene function, particularly for studies of the many important genes of the immune system, such as *MHC* and *KIR* (Adams and Parham 2001, Rajalingam et al. 2001).

FIELD METHODS: SAMPLE TYPES AND TECHNIQUES

In order to take advantage of any of these molecular techniques, it is crucial to collect appropriate samples. The collection, exportation, and importation of nonhuman primate samples are regulated by a variety of national and international laws and conventions. For both practical and ethical reasons, it is imperative that researchers comply fully with these regulations. For the United States, permission to import various nonhuman primate samples needs to be obtained from one or more of the following agencies: the Centers for Disease Control and Prevention of the U.S. Public Health Service, the U.S. Fish and Wildlife Service, and the U.S. Department of Agriculture. Also, as many nonhuman primates appear in Appendices I and II of the Convention on International Trade in Endangered Species of Wild Flora and Fauna (CITES), the U.S. Fish and Wildlife Service requires that CITES export permits be obtained from the country of origin for most primate samples. Furthermore, nonhuman primates, by virtue of their close taxonomic proximity to humans, can carry a large number of pathogens of great concern (e.g., herpes simiae and polioviruses in Old World monkeys; SIV and other retroviruses such as human T-cell leukemia viruses [HTLVs] and foamy virus, Ebola, and anthrax in great apes; leprosy in several catarrhines; malarial parasites in a variety of Old and New World anthropoids). Before embarking on a molecular study utilizing primate samples, researchers must be trained in safe laboratory practices, follow special primate protocols that are now required at many research institutions, and be vaccinated against a number of infectious agents likely to be encountered in primate tissues or secretions.

Sampling for Genetic Analyses

A variety of sample types have been used as DNA sources, including blood, tissue, hair, feces, urine, semen, buccal swabs, and masticated food wadges. Blood and tissue are the best sources for large quantities of high-quality genomic DNA, but they are the most difficult samples to acquire.

Blood collection requires that animals be immobilized, often under anesthesia, which may be difficult to do in field conditions. Additionally, collectors must have received professional veterinary training in how to draw blood without injuring the animal. Tissue sampling (e.g., collection of ear punches) can also be done on anesthetized animals and requires less training and practice to perform safely. Nonetheless, any such invasive procedures carry risks to the health of both animal and researcher. Such procedures will also be subject to Institutional Animal Care and Use Committee approval, which may require the presence of professional animal-care technicians during sampling. The logistical complexities associated with anesthetizing and collecting blood or tissue from wild animals may preclude the use of these kinds of sample if large numbers of samples are required. Despite these problems, the benefit to temporarily capturing and releasing individual animals is that both phenotypic data and tissue/blood samples can be collected from each animal. As such, it is possible to combine genetic data and phenotypic data in order to study questions regarding the action of selection (Lawler et al. 2005). Techniques for obtaining phenotypic information without manipulation have also been developed (Caillaud et al. 2008, Rothman et al. 2008).

Biopsy darts propelled from a compressed air- or CO_2-powered pistol or rifle are a minimally invasive alternative for retrieving small tissue samples that are suitable sources of large amounts of high-quality genomic DNA.

Some of the earliest field studies of primates to employ molecular techniques used noninvasively collected hair samples as a source of DNA (e.g., Morin et al. 1994; Gagneux et al. 1997b, 1999a; Goldberg and Wrangham 1997; Mitani et al. 2000). Unfortunately, it is now clear that shed hairs are generally a very poor source of nuclear DNA; follicle-containing plucked hairs are more suitable. A variety of creative techniques have been employed to collect fresh hair samples from wild nonhuman primates, including shooting animals with flat-tipped darts that have sticky tape mounted on the front, presenting animals with food bait wrapped in tape, and coating the edges of the entrance to a food corral with tape to capture an animal's hair as it enters (Valderrama et al. 1999).

Fecal samples have also been used as a source of DNA for numerous genetic studies of wild primates. Feces typically contains small numbers of cells sloughed from the epithelial lining of an animal's lower digestive tract as a bolus of fecal material passes through. An extraction of DNA from feces will thus contain a complex mix of DNA from a variety of sources, including the animal itself, as well as gut microbes and plants or insects that the animal consumed. PCR allows the preferential amplification of a specific target sequence present at low copy number (i.e., a piece of the genome of the animal) from amidst this complex sample. The low template copy number in hair and fecal samples poses a potential problem for determining reliable genotypes on the basis of fecal DNA due to allelic dropout

(Taberlet et al. 1996, 1999; Gagneux et al. 1997a). However, real-time PCR can be used to prescreen fecal DNA extracts, as well as extracts from any other noninvasive samples, to identify those containing sufficient template for subsequent analysis (Morin et al. 2001). Over the last several years, improved methods for storing (Roeder et al. 2004, Nsubuga et al. 2004) and extracting DNA from fecal samples (e.g., commercially available DNA extraction kits, such as the DNA Stool Mini Kit™ from Qiagen, Chatsworth, CA) have been developed. As a result, fecal samples are now routinely used as a source of DNA for molecular analysis in a number of field studies (e.g., Vigilant et al. 2001, Bradley et al. 2004, Di Fiore et al. 2009). Coupled with preamplification methods (Piggott et al. 2004) and new non-PCR-based methods for whole-genome amplification from very limited copy number templates (e.g., multiple displacement amplification [MDA]) (Dean et al. 2002), fecal samples will probably become even more widely used in future studies (see Rönn et al. 2006 for an encouraging assessment of the value of MDA as a method of whole-genome amplification).

A study of captive Japanese macaques by Hayakawa and Takenaka (1999) demonstrated that urine could be used as a source of DNA for microsatellite genotyping and that it yielded concordant genotypes to fecal samples from the same individuals (see also Valière and Taberlet 2000). More recently, Domingo-Roura et al. (2004) demonstrated the efficacy of amplifying mitochondrial DNA and microsatellite loci from DNA extracted from the semen of wild Japanese macaques. DNA extracted from buccal cells collected either via cheek swabs or from chewed food wadges (Takasaki and Takenaka 1991, Sugiyama et al. 1993, Vigilant et al. 2001) has also been used on occasion for microsatellite genotype and mitochondrial DNA sequencing. Woodruff (2004) and Goossens et al. (2003) provide excellent overviews of the use of noninvasive techniques for sampling primates for genetic studies.

Sampling for Other Kinds of Molecular Study

Samples for immunohistochemical studies are most valuable when they are rapidly frozen in an appropriate medium (e.g., embedded in optimal cutting temperature [OCT] compound and flash-frozen in a dry ice and 2-methylbutane/isopentane slush). Such treatment will maintain the cellular architecture of the tissue, and subsequent frozen sectioning on a cryostat allows for the detection of proteins and nucleic acids in situ. Immediate fixation in buffered formalin will result in the loss of a large number of protein epitopes as these are denatured by the preservative but will maintain much of the cellular architecture, which can be studied by embedding the fixed samples in paraffin for sectioning at a later point in time. Samples taken for other kinds of biochemical analysis (e.g., proteomics, lipid composition) should ideally be kept and transported frozen. Samples of glandular secretions for chromatographic analysis can be collected with a sterile dry swab and then stored in an airtight glass vial for long periods of time in the field.

Many kinds of sample typically collected for nucleic acid extraction also provide abundant opportunities for the study of other molecules. For example, because certain hormones (e.g., prolactin, testosterone, cortisol, progesterone, estradiol) and metabolic by-products (e.g., ketones from metabolism of fat, C-peptide from the body's production of insulin) are present in feces and urine, these samples can be used for endocrinological studies (see Chapter 23). It is important to keep aliquots of the various fractions created during DNA extraction as a potential source for some of these other biomolecules. Unlike larger and more fragile molecules, mono- and oligosaccharides are generally robust and subject to slower decay, unless there is bacterial or fungal contamination of the samples. As such contamination almost always renders the samples useless, an aseptic technique should be followed in the field for preserving these molecules. Skin biopsies, if taken with proper caution (aseptic techniques) and kept in cell culture medium, can survive at room temperature for a week to 10 days and still yield viable fibroblast cultures. Alternatively, if liquid nitrogen is available, skin biopsy samples can be minced and frozen in dimethyl sulfoxide in the field to prevent crystal formation and resulting cell death and then shipped in dry ice (Houck et al. 1995). With the advent of induced pluripotent stem cell technology, fibroblasts or other cell types collected from primates in the field will allow researchers to derive various cell types in vitro and to study the molecular changes associated with such differentiation across primate taxa (Yamanaka 2009).

Sample Storage

Storing samples while in the field can present challenges as electricity for running refrigerators or freezers for preserving samples is typically not available. However, many sample types can be kept for long periods of time at room temperature without substantial degradation of DNA and other molecules of interest if stored under the proper conditions. For example, all samples potentially containing DNA and RNA can be kept at room temperature for several weeks if stored in a sufficient volume of RNA*later*® (Ambion, Austin, TX), which is a nonflammable, nontoxic buffer that will maintain RNA and DNA for later isolation and amplification. Table 24.3 summarizes a set of appropriate field storage conditions (most at room temperature) for a range of sample types and includes recipes for a number of several useful field buffers. Whenever possible, these samples should be transferred to appropriate long-term storage temperatures (e.g., 4°C–20°C) as soon as possible.

A BRIEF NOTE ON LABORATORY AND ANALYTICAL METHODS

Because a number of excellent publications present detailed protocols for many of the laboratory methods discussed here—and because very few of these protocols are specific to primates—we saw little need to repeat that information.

Table 24.3 Suggested Field Storage Conditions for a Variety of Sample Types

SAMPLE TYPE	USED FOR	STORAGE MEDIUM	RECIPE	HOW TO USE AND REFERENCE
Blood	DNA	Blood field buffer	0.2 M NaCl, 0.1 M EDTA, 2% SDS, store at RT	Mix 1 volume whole blood with 5 volumes of buffer and store at RT.
	DNA	STE buffer	0.1 M NaCl, 10 mM Tris, 1 mM EDTA (pH 8.0), adjust pH with HCl and store at RT	Mix whole blood with equal volume of buffer and store at RT (Goossens et al. 2003).
	DNA	Citrate blood buffer	0.48 g citric acid, 1.32 g sodium citrate, 1.47 g glucose, add ddH2O to 100 ml, and store at RT	Mix 1 volume whole blood with 6 volumes of buffer and store at RT.
	DNA	Filter paper cards (e.g., Whatman FTA® cards)		Place drops of sample on card, allow to air dry, and store in air-tight container at RT with desiccant. FTA cards are impregnated with chemicals that lyse cells, denature proteins, and protect DNA from damage. Suitable for long-term storage.
	DNA	Queen's lysis buffer	10 mM NaCl, 10 mM Tris, 10 mM EDTA 1% n-lauroylacrosine (pH 8.0), adjust pH with HCl, and store at RT	Mix whole blood with 20 volumes of buffer (e.g., 50 ul in 1 ml) (Seutin et al. 1991).
	White blood cells, DNA	ACK lysis buffer, followed by storage in field buffer or PBS	150 mM NH4Cl, 1 mM KHCO3, 0.1 mM EDTA (pH 7.2–7.4), adjust pH with HCl, sterilize with 0.22 mm filter, and store at RT	Spin down fresh blood anticoagulated in EDTA or citrate. Remove supernatant and estimate remaining volume of cell pellet. Mix 1 volume of pellet with 5 volumes of ACK lysis buffer, vortex, spin, and remove supernatant. Repeat until no more red blood cells are visible in the remaining pellet of concentrated white blood cells. Resuspend the pellet in 5 volumes of field buffer at RT or frozen in PBS.
Blood serum	Proteins (albumin, globulins, hormones)			Collect whole blood into special serum separator or blood collection tubes. Allow blood to coagulate at RT for 2–4 hr, then carefully pipet off liquid phase (serum) and store in liquid nitrogen or at –80 C.
Tissue	DNA	NaCl-saturated DMSO	20% (wt/vol) DMSO, 0.25,M EDTA, NaCl to saturation (pH 7.5), adjust pH with HCl, and store at RT	Score tissue with razor to expose more of surface to solution. Immerse sample and store at RT (Seutin et al. 1991).
	DNA	90–100% ethanol		Score tissue with razor to expose more of surface to solution. Immerse sample and store at RT. Note: DNA yield will be lower and DNA more degraded (fewer large fragments) using this storage medium.
	DNA	Queen's lysis buffer	See above	Store small tissue biopsies in 1 ml of buffer (Bruford et al. 1998).
	DNA, RNA, proteins	RNA*later*®		Follow manufacturer's protocols for sample preparation and storage.
	Histology	OCT (optimal cutting temperature) compound	Available from several commercial suppliers (e.g., Pelco)	Embed sample in OCT compound, flash-freeze in dry ice and 2-methyl butanol slush by floating it in small plastic sample holder/boat, and store in liquid nitrogen or at –80° C.
Hair	DNA	Paper or glassine envelopes.		Collect sample using flame-sterilized forceps, being careful to include root. Store envelopes in air-tight container desiccated with silica gel beads.
Feces	DNA	RNAlater® buffer	Available from several commercial suppliers (e.g., Ambion)	Add 1 ml of sample to 5 ml of sample. Buffer should cover sample.
	DNA	70–100% ethanol		Add 1 ml of sample to 5 ml of sample. Buffer should cover sample.
	DNA	Desiccating silica gel beads	Available from several commercial suppliers (e.g., Sigma Chemical)	Place sample in air-tight plastic bag or collection tube and add sufficient silica gel beads to thoroughly desiccate sample (6–8 times weight of sample).

Table 24.3 (*cont'd*)

SAMPLE TYPE	USED FOR	STORAGE MEDIUM	RECIPE	HOW TO USE AND REFERENCE
	DNA	90% ethanol and desiccating silica gel beads		Place 200–250 mg of feces in 4 ml ethanol for 1 day, then decant ethanol and place sample on tissue in a collection tube containing 25–30 ml of silica gel beads and store at RT (Roeder et al. 2004).
	DNA	DETs solution	20% (wt/vol) DMSO, 0.25 M EDTA, 100 mM Tris, NaCl to saturation (pH 7.5)	Add ~2 g wet weight sample to 1 ml of DETs solution (Frantzen et al. 1998).
Buccal swabs	DNA	Resealable plastic bag containing desiccating silica beads		Using a sterile foam-tipped swab or nylon cytology brush, rub up and down along the inner cheek of the subject for ~15 sec. Allow to air-dry for 10–15 min, then remove the tip and store in air-tight environment with desiccating silica beads.
		Filter paper cards (e.g., Whatman FTA® cards)		Collect sample as above and then express directly onto card. Allow card to air-dry and store in air-tight container at RT with desiccant.
		Cell lysis buffer (or Queen's lysis buffer or STE buffer)	0.1 M NaCl, 0.1 M Tris, 10 mM EDTA, 1% SDS, store at RT	Collect sample as above, then swish the tip around in the buffer thoroughly, express excess fluid against side of collection tube, and store at RT.
Skin biopsy	Fibroblast cell culture	Sterile growth medium	Alpha MEM (GIBCO BRL) supplemented with 10% fetal calf serum, 1% penicillin/ streptomycin, and 200 mM L-glutamine	Aseptic technique for skin biopsy collection from a properly anesthesized animal: 1. Clean area on the skin with alcohol swabs 2. Shave area clear of any hair and repeat cleaning with second sterile alcohol pad 3. Using sterile forceps pinch and lift the skin 4. Using a new scalpel blade, cut a 5 X 5 mm piece of skin 5. Place the cut sample directly into a flask containing sterile medium 6. Disinfect cut and suture if necessary
		DMSO for freezing on site of minced tissue	Alpha MEM (GIBCO BRL) supplemented with 10% fetal calf serum, 1% penicillin/ streptomycin, and 200 mM L-glutamine 10% (wt/vol) DMSO	

RT, room temperature; EDTA, ethylene diamide tetra acetate (a chelating agent); SDS, sodium dodecyl sulfate (a detergent for lysing cells); PBS, phosphate buffered saline; DMSO, dimethyl sulfoxide.

Instead, we have compiled a short list of useful references where readers can go for more detail on how to implement some of the general methods discussed here (Table 24.4). In addition to these, Humana Press's *Methods in Molecular Biology* series and a variety of titles from Cold Spring Harbor Laboratory Press provide even more detail on using particular methods (e.g., RT-PCR) for specific applications. Table 24.4 also highlights a set of excellent sources providing overviews of important methods for analyzing molecular data generated using these techniques.

In addition, there now exists a wealth of software packages useful for analyzing various kinds of molecular data, many of which are publicly available for downloading from the Internet (see review in Excoffier and Heckel 2006).

These include packages for multiple sequence alignment, phylogeny inference (using a variety of algorithms, including distance, parsimony, and likelihood-based methods), basic population genetic analyses, evaluation of sex-biased dispersal, paternity assignment, inference of past demographic history, and tests of selection (Table 24.5). With respect to the numerous programs listed in Table 24.5, it is important to note that several "meta-programs" are also available for the conversion of input files from one program's format to another (Manoukis 2007, Coombs et al. 2008). Additionally, some journals (e.g., *Molecular Ecology Resources, Bioinformatics*) routinely publish notes on new software suitable for particular types of analysis.

Table 24.4 Texts Providing Laboratory Protocols and Overviews of Theoretical and Analytical Methods Useful to Molecular Primatologists

TITLE	REFERENCE	DESCRIPTION
An Introduction to Forensic Genetics	Goodwin et al. 2007	A recent treatment of the analysis of genetic data for paternity, relatedness, and individual identification from genetic, statistical and legal perspectives. In particular, it emphasizes the analysis of data from microsatellite and SNP loci.
Handbook of Statistical Genetics (3rd ed.)	Balding et al. 2007	A massive resource pertaining to the theory of genetic and genomic analysis from mathematical and statistical perspectives.
Fundamentals of Molecular Evolution	Graur and Li 2000	Provides a clearly written overview of many aspects of molecular evolution, including the structure of genes and genomes, genome evolution, population genetics, and methods of phylogeny reconstruction.
Phylogenetic Trees Made Easy (3rd ed.)	Hall 2007	Provides an easy introduction to the construction of phylogenetic trees from genetic sequence data. Emphasizes different analytical methods for constructing phylogenetic trees (parsimony, neighbor-joining, Bayesian methods) as well as more specialized topics such as ancestral sequence reconstruction and adaptive evolution.
Molecular Cloning (3rd ed.)	Sambrook and Russell 2001	Serves as a central source of useful, basic protocols for manipulating and studying DNA in vitro, including DNA extraction, PCR, electrophoresis, visualization, and vector cloning. Also a comprehensive source of basic buffer and reagent "recipes."
Molecular Genetic Analysis of Populations (2nd ed.)	Hoelzel 1998	Provides a short but comprehsive set of laboratory protocols for finding and assaying variation in allozymes and DNA.
Molecular Markers, Natural History, and Evolution (2nd ed.)	Avise 2004	Offers a comprehensive theoretical and practical overview of the various kinds of molecular markers evolutionary biologists might be interested in and of the utility of these markers for investigating many aspects of a taxon's biology and history.
Molecular Methods in Ecology	Baker 2000	Provides a general overview of both basic and newly developed molecular biology techniques that can be useful in behavioral and ecological research that is accessible to nonspecialists.
Molecular Ecology	Freeland 2005	One of several recent, introductory books on the growing field of molecular ecology. Provides an excellent and up-to-date overview of the field and is accessible to nonspecialists.
Molecular Systematics (2nd ed.)	Hillis et al. 1996	Provides a large set of protocols useful for assaying variation in a variety of molecule types, including allozymes, entire chromosomes, and DNA fragments. Also includes thorough overviews of many of the analytical techniques used for inferring phylogenies using molecular data.
A Primer of Genome Science (3rd ed.)	Gibson and Muse 2009	Provides an overview of traditional population genetic methods and how these methods integrate with genomics. Contains sections on genome annotation, variation, and organization, as well as sections pertaining to gene expression, proteomics, and integrated approaches to genome analysis.
Bioinformatics for Geneticists (2nd ed.)	Barnes 2007	A comprehensive resource for the analysis of genomic data from a variety of statistical and analytical perspectives.

CONCLUSIONS

Molecular primatology has already contributed much to our understanding of primate evolution. For the future, the promise of combining molecular genetic data with information gleaned from other biomolecules, such as proteins, lipids, and carbohydrates and their various combinations, is likely to reveal additional important insights into primate adaptations. The acute conservation crisis faced by most nonhuman primate species makes it imperative that we use all the tools at our disposal to further the understanding of primate biology and support conservation management decisions with key scientific knowledge. Both molecular and whole-organism primatologists have much to gain from the many possibilities for fertile collaboration.

We conclude this chapter by making one caveat. While the various techniques described here clearly constitute an impressive set of tools for examining and interpreting

Table 24.5 Software Packages Useful for the Analysis of Molecular Data

PROGRAM	DESCRIPTION	REFERENCE	PLATFORM AND SOURCE
Sequence alignment			
CLUSTAL	Program for multiple alignment of nucleotide and protein sequence data using a progressive sequence alignment algorithm, can also be used for inferring the phylogenetic relationships among sequences using the neighbor-joining method	Higgins et al. 1996	Web: www.ebi.ac.uk/Tools/clustalw2/index.html; Windows, Mac, Linux: www.clustal.org/
MALIGN	Program for multiple alignment of nucleotide sequence data that builds phylogenetic trees as it constructs an alignment and uses parsimony criteria to assess the best alignment	Wheeler and Gladstein 1994	Windows, Linux, Sun: research.amnh.org/scicomp/projects/malign.php
TreeAlign	Program for multiple alignment of nucleotide and protein sequence data that builds phylogenetic trees as it constructs an alignment and uses approximate parsimony criteria to assess the best alignment	Hein 1994	Web: mobyle.pasteur.fr/cgi-bin/portal.py?form=treealign
Sequencher	Commercially available program for multiple sequence alignment and editing and for archiving of sequence data		Windows, Mac: www.genecodes.com/
Bioinformatics software suites			
EMBOSS	European Molecular Biology Open Software Suite; a comprehensive, integrative set of open source programs for bioinformatics, including search and retrieval of sequence data from online databases, pattern analysis, trace file viewing, and primer design	Rice et al. 2000	Windows, Mac, Linux, Web: emboss.sourceforge.net/index.html
Genious	Commercial software for molecular data management and analysis; among the features included are multiple sequence alignment, contig assembly, search of online genetic databases (e.g., NCBI), in silico cloning, phylogenetic analysis, DNA and protein sequence visualization, and primer design		Mac: www.geneious.com/
VectorNTI	Commercial software for molecular data management and analysis; among the features included are multiple sequence alignment, contig assembly, search of online genetic databases (e.g., NCBI), in silico cloning experiments, restriction site profiling, and primer design		Windows, Mac: www.invitrogen.com/site/us/en/home/LINNEA-Online-Guides/LINNEA-Communities/Vector-NTI-Community/Vector-NTI.html
DNASIS	Commerical software for multiple alignment of nucleotide and protein sequence data using a progressive sequence alignment algorithm, also used for archiving of sequence data and for various kinds of sequence analysis (e.g., protein structure)		Windows, Mac: www.helixxtec.com/Hitachi/dnasis.htm
CLC Workbench	Commercial software for molecular data management and analysis; several different versions of the Workbench software are available with applications to genomic, DNA, RNA, or protein-oriented research; among the features included are multiple sequence alignment, search of online genetic databases (e.g., NCBI), and primer design		Windows, Mac, Linux: www.clcbio.com/
eBioTools, eBioX	Comprehensive set of bioinformatics tools plus a Graphical User Interface for Mac OSX users; provides integrated access to several of the other softwares included in this table (Clustal, PHYLIP, EMBOSS, MrBayes) as well as many additional ones not reviewed here	Lagercranz 2008	Mac: www.ebioinformatics.org/ebiox/
Phylogenetic inference			
PAUP*	Very widely used and general-purpose program for phylogenetic analysis; calculates numerous measures of genetic distance between taxa according to a user-specified model of evolution; performs phylogeny estimation using distance, parsimony, and likelihood methods	Swofford 2002	Windows, Mac, Unix, DOS: paup.csit.fsu.edu/
MEGA4	General-purpose program for DNA and protein sequence analysis, calculates various measures of genetic distance between sequences from molecular data according to a user-specified model of evolution, performs phylogeny estimation using distance and parsimony methods	Tamura et al. 2007, Kumar et al. 2008	Windows: www.megasoftware.net/

Table 24.5 (*cont'd*)

PROGRAM	DESCRIPTION	REFERENCE	PLATFORM AND SOURCE
PHYLIP	Large set of programs for phylogeny estimation that can use sequence data, allele frequency data, or character data; performs phylogeny estimation using distance, parsimony, and likelihood methods	Felsenstein 2004a	Windows, Mac, Linux: evolution. genetics.washington.edu/phylip. html
MrBayes	Program to estimate phylogenetic trees based on Bayesian inference	Ronquist and Huelsenbeck 2003	Windows, Mac, Unix: mrbayes. csit.fsu.edu/
PAML	Set of programs for phylogeny estimation using maximum likelihood methods and implementing many different models of sequence evolution	Yang 1997	Windows, Mac, Linux: abacus. gene.ucl.ac.uk/software/paml. html
TREE-PUZZLE	Program to estimate phylogenetic trees from DNA and protein sequence data using a maximum likelihood quartet-puzzling method	Schmidt et al. 2002	Windows, Mac, Linux: www. tree-puzzle.de/
POY	Program for estimating the topology of phylogenetic trees from molecular or morphological data using a "direct optimization" algorithm that estimates sequence alignments and phylogentic relationships simultaneously, unlike most other methods for phylogenetic inference that start with a multiple sequence alignment	Varón et al. 2009	Windows, Mac, Linux: research. amnh.org/scicomp/projects/poy. php
Population genetics and analysis of population structure			
GenAlEx	Very versatile add-in for Microsoft Excel, useful for conducting a wide variety of population genetic analyses; calculates allele frequencies as well as observed and expected genotype frequencies and evaluates whether genotype data deviate significantly from Hardy-Weinberg equilibrium expectation; also can be used to perform population assignment tests, test for sex-biased dispersal, generate simulated genotype data, conduct analysis of molecular variance (AMOVA), and run Mantel matrix correlations and other basic statistics; the program also allows for easy export of data to formats used by other population genetics software	Peakall and Smouse 2006	Windows, Mac: www.anu.edu. au/BoZo/GenAlEx/
GENEPOP	General-purpose software package for population genetic analysis; calculates allele frequencies as well as observed and expected genotype frequencies and evaluates whether genotype data deviate significantly from Hardy-Weinberg equilibrium expectation; also calculates various estimators of population differentiation and gene flow among populations and can be used to test for linkage among loci; web version includes useful routines for converting among file formats used by different population genetics programs	Raymond and Roussett 1995	Web: genepop.curtin.edu.au/
FSTAT	Versatile general-purpose program for population genetic analysis; calculates allele frequencies as well as observed and expected genotype frequencies and goodness of fit to Hardy-Weinberg equilibrium expectations; also calculates various estimators of population subdivision and inbreeding as well as estimators of genetic distance and relatedness between populations using allelic data; can also be used to test for linkage among loci and to evaluate potential sex biases in dispersal	Goudet 1995, 2001	Windows: www.unil.ch/izea/ softwares/fstat.html
Genetix	General-purpose software package for population genetic analysis; provides many of the same analyses as GENEPOP and FSTAT, as well as the ability to perform Mantel tests; also includes utilities for converting among common molecular data file formats; interface is in French	Belkhir et al. 2005	Windows: www.genetix.univ-montp2.fr/genetix/genetix.htm
RSTCALC	Program for analysis of population structure and gene flow specifically designed for use with microsatellite marker data	Goodman 1997	DOS: www.biology.ed.ac.uk/ research/institutes/evolution/ software/rst/rst.html

Table 24.5 (*cont'd*)

PROGRAM	DESCRIPTION	REFERENCE	PLATFORM AND SOURCE
ARLEQUIN	Versatile multiplatform program for population genetic analysis; calculates allele frequencies as well as observed and expected genotype frequencies, various estimators of population subdivision and gene flow, and various estimators of genetic distance between populations using allelic, haplotype, or sequence data; conducts tests for goodness of fit to Hardy-Weinberg expectations and can also be used to perform assignment tests and to test for isolation by distance	Schneider et al. 2000	Windows, Mac, Linux (via Java): lgb.unige.ch/arlequin/
Structure	Program for analysis of population structure using multilocus genotype data from various kinds of genetic markers, can evaluate whether a population is subdivided and can be used to assign individuals to subpopulations or to investigate hybridization and admixture	Pritchard et al. 2000	Windows: pritch.bsd.uchicago.edu/
MS-VAR	A Bayesian program for estimating numerous demographic parameters (e.g., population growth rate, time of population expansion) from microsatellite data	Beaumont 1999	Windows: www.rubic.rdg.ac.uk/~mab/software.html
GeneClass2	Program to detect migrants into a population and to assign individuals to a population based on multilocus genotype data, also useful for converting among several file formats used by different population genetics programs	Piry et al. 2004	Windows, Linux: www.montpellier.inra.fr/URLB/index.html
SIMCOAL2	A versatile coalescent-based program that estimates a wide variety of population genetic statistics using DNA sequence or microsatellite data		Windows, Linux: cmpg.unibe.ch/software/simcoal2/
Bottleneck	A program that uses allele frequency data to evaluate whether a population has experienced a recent reduction in effective population size (i.e., a genetic "bottleneck")		Windows: www1.montpellier.inra.fr/URLB/bottleneck/bottleneck.html
GenoDive	User-friendly program for manipulation of genotype data, distance matrices, ecological data, and spatial data for population genetic analysis; can be used to estimate pairwise relatedness among individuals, genetic distances among groups of individuals, and to run AMOVA and Mantel tests; supports input and output of a wide range of data formats	Meirmans and Van Tienderen 2004	Mac: www.bentleydrummer.nl/software/software/Home.html

Identity and parentage analysis

PROGRAM	DESCRIPTION	REFERENCE	PLATFORM AND SOURCE
RELATEDNESS	Estimates pairwise relatedness between individuals or average pairwise relatedness between groups of individuals using a regression-based estimator of relatedness; software is no longer being updated	Queller and Goodnight 1989	Mac: www.gsoftnet.us/GSoft.html
KINSHIP	Evaluates user-specified hypotheses about pedigree relationships using likelihood methods and can be used for parentage assignment; software is no longer being updated	Goodnight and Queller 1999	Mac: www.gsoftnet.us/GSoft.html
CERVUS	Conducts parentage assignments using likelihood-based methods that allow for incomplete sampling of potential parents and for some error in genotyping, also can be used to test for the existence of null alleles at a locus and for evaluating whether genotype data deviate significantly from Hardy-Weinberg expectations	Marshall et al. 1998, Kalinowski et al. 2007	Windows: www.fieldgenetics.com/pages/home.jsp
Gimlet	Software for constructing consensus genotypes for individuals from repeated genotyping, for estimating genotying error rates, and for estimating the kinship relationship among pairs of individuals from genotypic and demographic data	Valière 2002	Windows: pbil.univ-lyon1.fr/software/Gimlet/gimlet%20frame1.html
FaMoz	Conducts parentage assignments using likelihood-based methods; can be used with a variety of marker types (e.g., either dominant or codominant markers, or a mixture of types)	Gerber et al. 2003	Windows, Linux: www.pierroton.inra.fr/genetics/labo/Software/Famoz/index.html
P.A.E.	Program for determining paternity when the mother is known; uses microsatellite data, and can be run in Microsoft Excel	Rocheta et al. 2007	Windows, Mac: www.math.ist.utl.pt/~fmd/pa/pa.zip

Table 24.5 (*cont'd*)

PROGRAM	DESCRIPTION	REFERENCE	PLATFORM AND SOURCE
ML-RELATE	Program for determining relatedness and parentage using codominant molecular markers, based on a likelihood algorithm	Kalinowski et al. 2006	Windows: www.montana.edu/ kalinowski/ML-Relate/ ML-Relate_Home.htm
GenAlEx, GenoDive	See above; in addition to being very useful for population genetics applications generally, these softwares can also be used to derive estimates of relatedness using several different algorithms		

the molecular variation seen within and among different primate taxa, implementation of any of these techniques often requires substantial investment in time, energy, and resources to become proficient with both the theoretical and practical aspects of a particular kind of analysis. Thus, we caution fieldworkers to beware of succumbing to the idea that it is easy to simply "do genetics on the side." To be sure, molecular data can provide important insights into many aspects of the biology of wild primates that may be difficult to realize through field studies alone, but the rationale for adding a molecular component to any field study must be clearly scientifically motivated and worth the investment.

This is especially true when the collection of samples for molecular analyses requires anything but completely non-invasive procedures. Thus, as a point of caution, we urge students starting out in the field to not underestimate the workload involved in the molecular part of field projects, and we suggest that students undertake their laboratory training before taking off to the field to start a project. Likewise, we caution molecular biologists to not underestimate the crucial importance of reliable behavioral and ecological data from the field. Some of the most provocative molecular findings may be meaningless without their context in natural history.

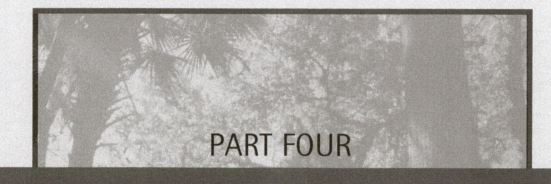

PART FOUR

Development and Reproduction

25

Life History

Steven R. Leigh and Gregory E. Blomquist

1. What is the intrinsic rate of increase (r) and how is it estimated?
2. What is the concept of a "trade-off," and why is it important in life history theory?
3. In what way(s) does brain size correlate with life history traits?

INTRODUCTION

And then increasingly, it was screaming at me, "These are the most interesting individuals; this has the greatest evolutionary impact; this is where the ecological pressures are." Jeanne Altmann, interview with D. Haraway on baboon mothers and infants (Haraway 1989:312)

The events, transitions, and phases that occur during the life course are fundamental to evolution (Raff 1996). Primates are especially interesting in this regard because they change substantially during life. A special class of theory, termed "life history theory," explains the evolution of life course changes by analyzing demography, genetics, behavior, and morphology developmentally and mathematically (Lande 1982; Roff 1992, 2002; Stearns 1992; Charnov 1993). Most broadly, life history theory includes birth and death rates along with "the entire sequence of changes through which an organism passes in its development from conception to death" (Lande 1982:608).

THEORETICAL CONTEXT OF PRIMATE LIFE HISTORY

Life history studies have a long tradition in both primatology and biological anthropology, with three different focuses. The first theoretical focus owes much to Jeanne Altmann's interest in baboon infants and their mothers, with projects centering on long-term field studies such as the Amboseli Baboon Project (Altmann, J., 1980; Altmann, S. A., 1998; Altmann and Alberts 2002; Silk et al. 2003a), the Cayo Santiago rhesus macaque colony (Sade et al. 1985), and Jane Goodall's (1986) chimpanzee research at Gombe. These studies couple information on vital population parameters, particularly (birth and death rates), with fine-grained behavioral observations on the day-to-day lives of study subjects. A second research area explores human life histories, concentrating on how an apparently distinctive set of human demographic and developmental characteristics might have evolved (Bogin 1999, Hawkes et al. 2002, Hill and Hurtado

1996, Kaplan et al. 2000, Leigh 2001). Studies of human life histories also utilize both demographic and behavioral data, but many studies also consider patterns of somatic growth in relation to demography, behavior, and culture. In this chapter, we emphasize a third research tradition that highlights interspecific comparisons of *ontogenetic* (growth-related) and *allometric* (size-related) variations in life history traits. These studies emphasize key events in life histories, such as birth, age at maturation, birthrates, and longevity, to understand how evolutionary forces shape the life courses of primates (Harvey et al. 1987; see also Godfrey et al. 2002, 2004a; Harvey and Clutton-Brock 1985; Martin, R. D., 1983, 1996; Martin and MacLarnon 1990; Ross 1988, 1991, 1992b, 2002, 2004; Ross and Jones 1999; Shea 1987, 1990). This perspective departs somewhat from theoretical priorities on genetic and demographic data in traditional life history theory, which emphasize measures of heritability and rates of gene substitution in populations (Stearns 1992). However, these approaches are compatible, and their interrelations have received some theoretical treatment (e.g., Charnov 1993). We stress theoretical developments and current ideas that emerge from comparative analyses but briefly incorporate advances from other research areas and explore ways in which genetic and demographic studies articulate with interspecific analyses.

HISTORICAL CONSIDERATIONS

Life history studies have a relatively long history in primatology. Several theories about the primate life course have been proposed, and researchers in primate biology continue to use a diversity of ideas, even some that are now relatively old. We discuss concepts about the primate life course and its evolution in a historical context because we see clear movement in theoretical ideas away from traditional viewpoints to a quantitative and developmental approach. Older theories are problematic, despite relatively frequent use in the literature.

Research conducted in the late 1970s through the 1980s broke with previous studies of primate life histories (Martin, R. D., 1983; Harvey and Clutton-Brock 1985; Harvey et al. 1987). Prior to this period, such studies focused most on comparative analyses of human growth and development (Gavan and Swindler 1966, Schultz 1969, Watts and Gavan 1982). Adolph Schultz made major contributions to this tradition, encapsulated by his often-repeated line drawing that shows "progressively" extended life stages in primates (1969:Fig. 57). The figure shows unidimensional, evenly proportioned, and progressive increases in each life stage during primate evolution, conveying the erroneous impression that primate life histories follow an orderly, orthogenetic, and unilinear advance. Schultz's view was consistent with bioanthropology's focus on evolutionary trends prior to the 1960s. However, cladistic thinking showed flaws with this and other "trends" by revealing that several "trends" were, in fact, independently evolved similarities (*homoplasies*) (see Leigh 2001).

Unfortunately, advances in biogeography, particularly the concepts of r and K selection, seemed to reinforce Schultz's general idea (MacArthur and Wilson 1967, Pianka 1970). These concepts were used widely in primatology, particularly in studies of human evolution. For example, the late Stephen Jay Gould (1977) claimed that K selection produced a slowing of developmental change during the course of human evolution. Similarly, Owen Lovejoy's (1981) influential model linked early hominin monogamy with life history, by invoking r–K selection under Schultz's orthogenetic scheme. Regrettably, these theoretical turns came at the expense of more powerful genetic and demographic concepts, advanced mainly by George C. Williams, who emphasized the unequal force of selection throughout the life span (1957, 1966a; see also Lack 1954; Medawar 1946, 1952). Differences in the effects of selection could explain why life stages differed, especially how genes deleterious late in life, including those producing senescence, could become fixed in populations. Senescence and diminishing reproductive output might occur because genes with tiny advantages during phases of relatively high reproduction can be selectively favored during these phases and, thus, increase in frequency. However, such genes may have detrimental effects later in life, with little consequence from selection because of low reproductive output. This basic insight gave life history theory notions of genetic trade-offs (antagonistic pleiotropy), juvenile phases, and risks of juvenility. Trade-offs are fundamental to life history theory, particularly when formalized by the concepts of reproductive value (Fisher 1930) and residual reproductive value (Williams 1966b). A *trade-off* means that energy invested in one area impacts, and usually limits, expenses in other areas. For example, an investment in a current offspring, like delaying weaning, may limit investment in future offspring.

The ideas of Williams, Fisher, and Medawar seem to have gone unrecognized in early primate life history studies, probably owing both to Schultz's priorities and to the general appeal of r and K selection. Primates are long-lived relative

to other mammals (Austad and Fischer 1992, Flower 1931, Prothero and Jürgens 1987, Shattuck and Williams, 2010), a finding superficially consistent with r–K ideas. In addition, the suggestion that the brain served as a pace-setter of life histories was consistent with r and K selection. In this model, larger brains were associated with greater longevity by increasing "precision of physiological regulation" (Sacher 1959:129, Sacher and Staffeldt 1974; see also Allman and Hasenstaub 1999, Deaner et al. 2002). This suggestion had obvious implications for an order classically defined in part by a trend toward large relative brain size (Le Gros Clark 1959; Martin, R. D., 1983; Shea 1987), but it relied upon an ageing paradigm uninformed by Williams's ideas.

Influential analyses in the 1980s marked a change in primate life history perspectives, although these analyses worked within a general framework that derived mainly from the traditions of Schultz and biogeography (Harvey et al. 1987). These studies concentrated on questions relating primate life history to brain and body size, but they also presented data enabling researchers to test novel hypotheses and raised awareness of the statistical complications posed by phylogeny. Harvey et al. assembled large databases representing estimates of "life history variables" for many primate species. Variables included markers of time points or intervals (gestation length, weaning age, age at maturation, age at first breeding, interbirth interval, and maximum recorded life span) and morphological variables (adult brain size, neonatal brain size, adult body size, and neonatal body size). They interpreted significant statistical variation among species as an outcome of evolutionary changes in body size, given high correlations between mass and other variables. Analyses of these data suggested considerable complexity in the relations of brain size to life history. Specifically, brain size and maturation age seemed to be linked, but further analyses indicated that this correlation reflected prenatal factors to a greater degree than postnatal factors. Essentially, later-maturing species have larger-brained infants, but the majority of brain growth seems to occur prenatally in larger-brained species. The mechanistic bases and implications of these correlations for life history have been difficult to comprehend fully and unambiguously.

POPULATION DYNAMICS AND PRIMATE LIFE HISTORIES

Despite Harvey et al.'s (1987) major advances, several areas of theoretical concern went unaddressed. Perhaps most importantly, while the article drew inspiration from population biology, it neglected questions regarding demography, quantitative genetics, and developmental biology. Fortunately, an influential study by Caroline Ross (1988) made a major step in this direction by merging comparative approaches with ideas about population dynamics from L. C. Cole (1954). Specifically, Ross (following Hennemann 1983, 1984) solved for r, the intrinsic rate of natural population increase (Table 25.1), by measuring key variables (age

Table 25.1 Important Life History Equations

Equation 1: Cole's equation, used by Ross (1988) for primates:

$$1 = e^{-r} + be^{-r\alpha} + be^{-r(n+\alpha)} \tag{1}$$

In this equation, e is the base of the natural logarithm, α is age at first reproduction, b equals the birth rate of female offspring, n is estimated by calculating $\omega - (\alpha + 1)$ (where ω = age at last reproduction), and the value of r is estimated through iteration (Cole 1954:eq. 21).

Equation 2: Euler-Lotka or characteristic equation:

$$\int_0^\infty l(x)m(x)e^{-rx}dx = 1 \tag{2}$$

This equation summarizes population rates of age-specific survivorship, $l(x)$ and production of female offspring, $m(x)$, for age x and measures the intrinsic rate of population increase (r).

Equation 3: Algebraic expression of the Euler-Lotka equation:

$$\sum_{x=0}^{\omega} l(x)m(x)e^{-rx} = 1 \tag{3}$$

at maturation, age at death, and birthrate) from published studies of many primate species. She suggested that values for each species provided a measure of maximum possible reproductive output (r_{max}). Ross discovered that species with low body mass tended to have high r_{max} values while large-bodied species had low values (Pearson product–moment correlation $r = 0.869$). The details of this correlation implied to her that the general idea behind r and K selection may hold for primates. Specifically, given the pattern of high r_{max} residuals from species designated as occupying "unpredictable" habitats, she suggested that such species tended to be r-selected. Covariation between r_{max} values and environments thus seemed to fit basic predictions of r and K selection.

Despite these important findings, some results were tenuous. Our objective in pointing this out is not to diminish Ross's important research. Rather, we wish to recognize the limitations of r and K selection for primates and follow Ross in this regard (Ross and Jones 1999; see also Stearns 1992). First, the measure of habitat predictability employed was imprecise, being derived from a tertiary literature source (see Ross 1992b for improved habitat estimates). Second, the association between higher relative r_{max} values and habitat predictability was modest at best. In fact, a species occupying a predictable habitat seemed to present the highest positive residual r_{max} value. Third, as with most studies at this time, phylogenetic adjustment methods could only roughly counter the effects of phylogeny. More recent analyses that include phylogenetic adjustments show "no significant links" between habitat and measures of reproductive rate (Ross and Jones 1999:94).

Ross's original analysis, despite these difficulties, was extremely important because it incorporated Cole's research (1954) and that of demographically oriented theoreticians (Blomquist 2009) into considerations of primate life histories. Unfortunately, it seems simultaneously to have reified r and K selection among primatologists. The role of r–K selection is sometimes abstracted by the catchy phrase that primates experience "slow" life histories, with some "living in the slow" lane, and variation among species described

along a "fast–slow continuum" (Ross 1992b:383, Kelley 2004, Promislow and Harvey 1990). In any case, the inclusion of ideas about population dynamics and life histories significantly advanced the field, sustaining many further theoretical developments and forcing a reconsideration of the brain's role in primate life history.

URGENT LIFE HISTORY THEORY

Ross's studies were complemented by two influential life history theory texts published in 1992 (Roff 1992, Stearns 1992). Charnov's theoretical monograph (1993) and two major edited volumes devoted to primate life histories and juvenility (DeRousseau 1990, Pereira and Fairbanks 1993a) supplemented these works, as did quantitative analyses at the population level (Sade 1990, Stucki et al. 1991). Moreover, phylogenetic adjustment techniques advanced tremendously (Felsenstein 1985, Garland and Adolph 1994).

Volumes by Roff (1992, 2002) and Stearns (1992) on contemporary life history theory firmly phase with ideas and methods of demography and quantitative genetics. This foundation derives directly from R. A. Fisher's (1930) interest in the demographic dimensions of the maintenance of genetic variation in populations. As Stearns notes, the main idea is that demographic methods facilitate a concern with "marginal effects of gene substitutions, not with numbers of organisms in a population" (1992:21). Stearns's six-point "rejection of r and K selection" is an exceptionally valuable contribution (1992:206–207). In showing the limitations of this idea beyond biogeographic studies, Stearns advocates a life history paradigm shift from the idea (latent in r–K selection theory) of direct habitat effects on life history (habitat → life history) to that of habitat → mortality regime → life history (1992:208). In a related advance, DeRousseau argued that a life history perspective differs considerably from traditional approaches to evolution, relaxing many of the simplifying assumptions of the evolutionary synthesis by introducing sex and, in particular, age structure to populations (1990; see also Shea 1990).

These advances improve our understanding of primate life history variation. We review key theories and briefly consider the problem of trade-offs and the problem of how to measure trade-offs. While not a theory per se, the problem of trade-offs requires more extensive theoretical and empirical investigation. Trade-offs may be especially important for long-lived organisms such as primates and have considerable relevance for understanding the evolution of human life histories (Hawkes et al. 2002). We use this chapter to point to new ways of exploring primate life histories.

Charnov and Life History Invariants

Models formulated in the early 1990s are still debated, as are long-standing ideas about learning and the brain as a direct pace-setter of life history (Allman and Hasenstaub 1999, Leigh 2004, Ross and Jones 1999, Sacher 1959, Sacher and Staffeldt 1974). These models have generally undergone initial rounds of empirical testing with primate species (Ross and Jones 1999), but critical evaluation of alternative life history models shows that much research remains to be done. A major theoretical development was the integration of optimality theory and comparative data in Charnov's life history invariants model (Charnov 1991, 1993; Charnov and Berrigan 1993; Berrigan et al. 1993). Charnov argued that primate life histories are impacted by a single trade-off between delaying reproduction to increase body size (an investment offset by higher fecundity) and the risk of mortality during the waiting period prior to reproduction. Primates could grow for a long time, to enjoy larger size and higher fecundity; but the risk of dying limits how much they can delay reproduction. Mathematically, the model proposes that $dM/dt = AM^{0.75}$, where M is the mass of either a growing (female) individual or an adult, dM/dt is the growth rate or annual litter mass, and A describes the height of the production function. The *production function* describes the fraction of total energy put into either growth or reproduction. Growth, by definition, ceases at maturity, diverting energy from adding body mass (growth) into producing offspring. Externally imposed adult mortality rates set the optimal age of maturation, setting an optimal body size. Optimal maturation age is also conditioned by the reproductive benefits of larger size, which translates to higher levels of energy available for production. Finally, juvenile mortality is assumed to be density-dependent and to keep populations stable ($R_o = 1$) (see Mylius and Diekmann 1995 for problems of this assumption).

Empirical analyses of Charnov's model indicate exceptionally low values of A for primates ($A = 0.42$), with $A = 1$ for other mammalian taxa (Charnov 1993). So the amount of energy a primate can allocate to either growth or reproduction is small. This forces primate females to spend much time growing until they reach sizes large enough to produce offspring efficiently. Consequently, they grow for longer time frames and produce fewer offspring than other mammals. Comparatively low mortality should mean that primates have the opportunity to live longer, achieving sufficient fitness to compensate for lost reproduction during long preadult periods.

A powerful element of Charnov's (1993) model lies in its potential to fuse growth or ontogenetic phenomena to population dynamics. Charnov demonstrates the compatibility between these fields by deriving the value of A through several different means. For example, he shows that the value of A can be calculated directly from data that describe ontogeny, including size at weaning (δ), body mass at maturation, and age at maturation (Charnov 1993:eq. 5.3). However, A also emerges from population dynamics, such as his value of A_3, obtained by estimating the height of a regression line of Ross's (1992b) r_{max} against body size (Charnov 1993:Fig. 6.4, Charnov and Berrigan 1993). Articulations between ontogenetic and population dynamic perspectives are significant because they tie ontogenetic patterns, particularly aspects of growth and developmental energetics, directly to population parameters. Consequently, the impact of life history variation on individual fitness can be assessed with respect to population dynamics. Finally, Charnov and Berrigan (1993) bluntly question life history models that involve the brain (cf. Deaner et al. 2002, Ross 2002), contrasting with older models (Sacher 1959, Sacher and Staffeldt 1974; see also Allman and Hasenstaub 1999, Kelley 2004).

Empirical tests of Charnov's model among mammals (Purvis and Harvey 1995) and within primates (Ross and Jones 1999) support some of its predictions but reveal a number of deficiencies, at least given available comparative data sets. One particular prediction of Charnov's model lacking empirical support is the invariance of mass at independence (weaning, δ) with adult mass (but see Hawkes et al. 2002). In addition, many of the simplifying assumptions of the original Charnov model have been criticized and reevaluated. The production constant A and the constant exponent of about 0.75 permit evolutionary change in body size only through differences in duration of growth and do not allow for change in growth rates. However, primate growth rates vary substantially: Some species reach different adult sizes in the same amount of time (Figure 25.1A,B), but others reach the same size in different amounts of time (Figure 25.1C) (Pereira and Leigh 2003). Charnov (2001) noted how changes in the exponent affect the scaling of body size and annual fecundity. Jones and MacLarnon (2001), applying Charnov's logic to analyses of bat life histories, found a higher exponent (~1), which still seemed to follow a power function relation with fecundity. Finally, a loose interpretation of Charnov's (1993) model could be construed as a restatement of r–K selection, possibly promoted by the subtitle of Charnov and Berrigan's (1993) article, "Life in the Slow Lane."

Juvenile Risks

The juvenile risk aversion model, proposed by Janson and van Schaik (1993), answers some questions unaddressed by Charnov's model, particularly growth rate differences among species. Their model also accommodates Williams's (1966a) notions about life stage duration and risk. Janson and van Schaik attribute the long period of primate juvenility to a low growth rate that serves as metabolic risk adaptation during ontogeny. Juvenile primates face a simultaneous trade-off between predation risks and the metabolic costs of feeding competition from conspecifics: They can forage at the center of groups, encountering competition from adult group members, or they can forage at the periphery, increasing their susceptibility to predators. Predators (and infanticide) should select against a low

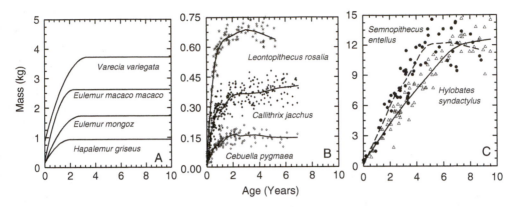

Figure 25.1 Comparative body mass growth curves showing how body mass increases with age for strepsirrhines (A), platyrrhines (B), and catarrhines (C). Primates can reach the same size through differences in rates of growth, either with no difference in total growth time (A, B) or with a difference in growth duration (C). Data on females are shown for *Semnopithecus entellus*, but sexes are combined in these plots, given the absence of dimorphism in these taxa. In A, lines represent piecewise regressions (Leigh and Terranova 1998), but other lines represent lowess regression lines (Leigh 1992).

growth rate, favoring juveniles that grow out of harm's way (a constant growth rate is assumed for each species). On the other hand, growing rapidly raises metabolic risks because growth is metabolically costly as a result of protein turnover (Tanner 1978). Thus, feeding competition should favor a slow growth rate, especially if the metabolic expenses of growth mean that juveniles face disproportionately high costs of intragroup feeding competition. Juveniles face extra burdens if they forage less efficiently than adults. In this model, trade-offs involving predation pressure and possibly infanticide, contrasted with the metabolic risks of feeding competition, select on growth rate. For primates, low growth rates seem to predominate, resulting in deferred maturation by forcing extended juvenile phases.

This model provides a compelling way to account for growth rate variation among primate species and follows from Williams's recognition that mortality risk influences the pace of development, affecting either rates of morphological change, duration of high-risk developmental phases, or both. However, the model does not explain growth spurts or growth rate variability within a species (Leigh 1996). Analyses of restricted data sets show mixed support for the model, and field-based evaluations have been slow in coming. For example, growth patterns in leaf-eating anthropoids are consistent with the model (Leigh 1994), and variation among ape species aligns with its predictions (Leigh and Shea 1996). In contrast, neither small-bodied New World monkey growth (Garber and Leigh 1997) nor Malagasy lemur growth (Godfrey et al. 2004a) seems to fit the model. A recent field study of *Saimiri sciureus* (squirrel monkeys) shows mixed support for components of the model. Juvenile squirrel monkeys experience negligible intraspecific feeding competition, and few differences in foraging proficiency occur across age classes. In squirrel monkeys, predation pressure seems to influence group dynamics, so elements of the model might fit (Stone 2004). Interspecific comparative analyses also show modest support for this model (Ross and Jones 1999).

Body Mass Optimization

Kozlowski and Wiener (1997) modified Charnov's model, interpreting it as a special case of their own. Kozlowski and Weiner's computer simulations indicate that adult body mass is optimized within a species in response to ecological parameters that affect the growth function. These ecological parameters can generate interspecific correlations between life history and morphological and physiological variables of body mass. These interspecific allometries are, however, by-products of selection that act only within species to optimize body size. The model also predicts some patterns of growth rate scaling and species body mass distributions (Kozlowski and Galwelczyk 2002, Kindleman et al. 1999). Most importantly, their research generally supports the idea, following Charnov, that age and size at maturity are strongly influenced by optimal resource allocations to either growth or reproduction. We expect difficulties in formulating critical tests between the Kozlowski-Wiener and Charnov models and note that applications to primates have yet to be undertaken.

Charnov and Sigmoidal Growth

Charnov's (2001) model of mammalian life history evolution incorporates a sigmoidal growth law developed by West et al. (2001; but see Ricklefs 2003). In sigmoidal growth, rates of change follow a curvilinear pattern, much like an italicized letter *S*. Along with this somewhat more realistic assumption about growth, the model does not reflect reproductive output as merely diverted self-growth but as some proportion of it. Charnov proposes several new invariant relationships based on adult size and mortality, cellular maintenance costs, offspring production, and growth rate and age at adulthood. This model has not been subjected to testing, and many of its predictions may be untestable until both accurate life tables and growth data are available on a large number of species.

Life History Trade-Offs

The idea of a trade-off is fundamental to life history thinking. This concept requires discussion because trade-offs are evident in these models and complicate attempts to test them. The basic problem is that fitness components (variables directly related to fitness like maturation age, interbirth interval, and litter size) cannot all be optimized at the same time, which may lead to negative correlations among them so that an increase in one variable decreases another. In other words, organisms face "decisions" of how, when, and in what proportions energy should be allocated to various fitness components and the physiological processes that impact these components. This produces classic patterns like the trade-off of current versus future reproduction or current reproduction versus survival (Fisher 1930, Williams 1966a). Furthermore, trade-offs may not act directly on reproduction and survival but can be mediated through other variables, such as body size (Roff 2002).

Perhaps the most significant problem in explaining primate life histories, especially human life histories, is that researchers have had difficulty in actually finding predicted trade-offs (Hill and Hurtado 1996). For example, humans, and possibly other primates, do not show expected trade-offs between current and future reproduction. Remarkably, the cost of current reproduction does not seem to reduce resources for future reproduction. New models are beginning to provide an answer to this paradox, by suggesting that variation in resource allocation and acquisition must be analyzed. Specifically, two conditions may obscure trade-offs. First, resource acquisition may be high enough to hide trade-offs. Second, and more subtly, trade-offs may not be evident when variation in resource acquisition exceeds variation in allocation (van Noordwijk and de Jong 1986). *Variation in allocation* refers to differences in how organisms devote energy to either survival (e.g., body maintenance, size, condition) or reproduction. If there are huge differences in populations in terms of energy acquired but few differences in how it is allocated, then trade-offs may not be apparent at the population level.

Quantitative genetic models help resolve this, showing that trade-offs among fitness components may be hidden under two conditions. First, more genes may contribute to variation in acquisition than to resource allocation. Second, there could be greater variation at higher levels in an allocation hierarchy (Houle 1991, de Jong and van Noordwijk 1992, de Jong 1993, de Laguerie et al. 1993, Worley et al. 2003). For example, an organism may allocate a certain fraction of resources to reproduction, with the remainder going to somatic maintenance, but the fraction allotted to reproduction may then be subdivided among traits that are traded off against one another, such as offspring size and number. Establishing whether or not primates face such trade-offs is a crucial area of future research, requiring ecological, demographic, and quantitative genetic information.

PATTERNS OF LIFE HISTORY VARIATION

This brief theoretical review establishes a context in which we can present summary analyses of life history correlations. Our objective is not to test various models. Instead, these correlations offer empirical experience with the relevant data and show the limitations of data used to analyze life histories. These deficiencies reveal a need for new kinds of data for testing and extending life history theories. They also raise questions about precisely how life history analyses should be conducted and interpreted.

Space limitations preclude detailed descriptions of data and methods. However, we use a high-quality data source (Kappeler and Pereira 2003:Appendix). This compendium includes carefully vetted sources for both adult and neonatal body mass data (Smith and Jungers 1997, Smith and Leigh 1998, respectively). We conduct analyses of both raw and phylogenetically adjusted estimates, with adjustments undertaken with independent contrasts computed through the PDAP module of Mesquite phylogenetic software (Maddison and Maddison 2004). The phylogenies utilized are modified from Smith and Cheverud's (2002) rendition of primate phylogeny. Statistically, species values are not independent data points, but standard statistical techniques can be used if we can measure the degree to which data points are independent. In essence, phylogenetic adjustments account for the fact that we expect closely related species to share greater similarity than distantly related species. Controlling for phylogenies is essential in order to account for differences in the degree to which species are related. Throughout, we use Pearson product–moment correlations to measure associations on both kinds of data.

Correlation estimates are consistent with numerous previous studies of primate life history variation (Table 25.2). However, phylogenetically adjusted correlations document surprisingly weak associations among certain variables traditionally regarded as fundamentally important (Table 25.2). For example, the correlation between age at reproductive maturation and body size, measured in Harvey et al.'s (1987) study at $r = 0.92$, diminishes to $r = 0.059$ given our phylogenetic adjustment. Ross and Jones (1999) report a higher adjusted value ($r = 0.42$), estimated after a special statistical procedure (trimming contrasts among terminal nodes, which may be problematic).

RETHINKING PRIMATE LIFE HISTORIES

Our review points to problems with theories, models, and empirical analyses that require new approaches to life history problems. Both the ontogenetic and population dynamic sides of the field require critical theoretical examination, but we prioritize ontogenetic studies. Recent theoretical advances in developmental biology (Raff 1996) have yet to be considered by life historians but provide important insights into life history problems. More specifically,

Table 25.2A Correlations Estimated by the Current Study with Values Published by Kappeler and Pereira (2002:Appendix)

	ADULT BODY MASS (G)	ADULT BRAIN MASS (G)	AGE AT FIRST REPRODUCTION (YEARS)	GESTATION LENGTH (DAYS)	INTERBIRTH INTERVAL (MONTHS)	LITTER SIZE	NEONATAL MASS (G)	WEANING AGE (DAYS)	NEONATAL BRAIN MASS (G)
Adult body mass (g)	—	0.961	0.881	0.683	0.821	−0.579	0.960	0.860	0.973
Adult brain mass (g)	0.846	—	0.898	0.707	0.745	−0.465	0.968	0.865	0.989
Age at first reproduction (years)	0.059	0.202	—	0.727	0.829	−0.571	0.914	0.861	0.948
Gestation length (days)	0.484	0.429	0.074	—	0.604	−0.537	0.735	0.733	0.760
Interbirth interval (months)	0.389	0.342	0.345	0.394	—	−0.400	0.779	0.813	0.837
Litter size	−0.149	−0.191	−0.023	−0.198	0.040	—	−0.647	−0.560	−0.633
Neonatal mass (g)	0.798	0.867	0.255	0.386	0.277	−0.280	—	0.879	0.987
Weaning age (days)	0.366	0.207	0.055	0.328	0.436	−0.118	0.404	—	0.872
Neonatal brain mass (g)	0.891	0.972	0.714	0.454	0.402	−0.488	0.853	0.574	—

Estimates above the diagonal represent unadjusted or "species" values, and estimates below the diagonal are adjusted using independent contrasts; raw data concern values for females unless otherwise noted.

Table 25.2B Correlations Published by Harvey et al. (1987) (Based on Subfamily Data)

	MALE ADULT BODY MASS (G)	GESTATION LENGTH (DAYS)	INDIVIDUAL NEONATAL MASS (G)	NUMBER OF OFFSPRING PER LITTER	WEANING AGE (DAYS)	AGE AT FIRST BREEDING (YEARS)	AGE AT SEXUAL MATURITY (YEARS)	MAXIMUM RECORDED LIFE SPAN (YEARS)	INTERBIRTH INTERVAL (MONTHS)	AGE AT SEXUAL MATURITY (MALE, YEARS)	NEONATAL BRAIN WEIGHT (G)	ADULT BRAIN WEIGHT (G)
Adult body mass (g)	0.996	0.74	0.97	−0.52	0.91	0.92	0.89	0.78	0.86	0.89	0.95	0.96
Male adult body mass (g)		0.73	0.97	−0.51	0.92	0.92	0.89	0.78	0.85	0.91	0.95	0.96
Gestation length (days)			0.82	−0.61	0.84	0.81	0.81	0.62	0.63	0.84	0.84	0.8
Individual neonatal mass (g)				−0.56	0.94	0.95	0.94	0.8	0.87	0.95	0.99	0.98
Number of offspring per litter					−0.56	−0.49	−0.47	−0.3	−0.41	−0.44	−0.51	−0.5
Weaning age (days)						0.9	0.92	0.7	0.89	0.93	0.89	0.91
Age at first breeding (years)							0.97	0.87	0.88	0.94	0.93	0.96
Age at sexual maturity (years)								0.83	0.85	0.96	0.95	0.94
Maximum recorded life span (years)									0.72	0.78	0.82	0.85
Interbirth interval (months)										0.83	0.86	0.86
Age at sexual maturity (male, years)											0.97	0.96
Neonatal brain weight (g)												0.99

Table 25.2C Correlations Published by Ross and Jones (1999) Comparing Body Mass to Other Variables

	BODY MASS	
	SPECIES VALUES	INDEPENDENT CONTRASTS
Length of Juvenile Period (years)	0.71	0.60
Birth rate	0.81	0.60
R_{max}	0.89	0.65
Age at first reproduction (years)	0.87	0.42
Average instantaneous adult mortality rate	0.22	0.32
Prereproductive mortality rate	0.41	0.20
Average infant mortality rate	0.37	0.46
Survival to reproductive age	0.16	0.35

developmental biology sees animal development as modular, with dissociation among developing parts playing a major role (Raff 1996, Raff and Raff 2000). *Modularity* means that morphological structures or organ systems may vary in the degree to which they are interrelated during development. Therefore, different levels of correlation, integration, and interaction among tissues should be expected during ontogeny. This means that different organs or organ systems may grow over separate age spans and at very different rates (i.e., they may be dissociated during ontogeny).

Dissociation of morphological structures causes the emergence of particular patterns or modes of ontogeny during development. These modes are mediated by adjustments in the growth of metabolically expensive tissues (e.g., the brain) so that there may be many ways to be a juvenile. The "fast versus slow" continuum that has dominated discussion of primate life histories does not account for these different modes and, thus, appears to be inaccurate. The emerging view of primate life history is much more complicated than is revealed by the "fast versus slow" perspective so that understanding life histories requires high-quality developmental data.

In this view, dissociation of developing structures is especially important for primates because of their extended life spans. Primate longevity may be a benefit conferred by life in the trees (or at least, typically, not on the ground [Austad and Fischer 1992 , Shattuck and Williams, 2010]). Bats and birds also have relatively long life spans (Williams 1957), but developing anatomy for flight may actually favor reduced juvenile periods. In terrestrial taxa, selection probably favors a short developmental period coupled with a very rapid shift of energetic resources from growth to reproduction, especially if juvenile mortality is high (Williams 1966a,b). We further expect few opportunities for dissociation and the emergence of life

history modes in these taxa, so these species should meet the assumptions of Charnov's model well. In primates, life history modes minimize risks from energetic competition between still-growing structures and offspring production, offering flexibility as to when to grow different structures and when to complete the juvenile period. Reduced primate mortality provides opportunities for the evolution of diverse patterns of development, depending on energetic risks and mortality profiles at different stages. Therefore, instead of a "fast–slow continuum," "modes" and phases characterize primate life histories (see Pereira and Leigh 2003, Leigh and Bernstein 2006). We define a *life history mode* as a distinctive pattern or arrangement of ontogeny with respect to the rate and scheduling of growth for various organs, organ systems, or modules.

Unfortunately, the theoretical infrastructure devoted to modularity in life history remains underdeveloped, as lamented by Stearns over a decade ago (1992, see also Watkinson and White 1985). Despite these difficulties, it is important to note that previous primate life history theory tends to assume, either implicitly or explicitly, that primate life histories are tightly integrated, with low dissociation and minimal modularity (e.g., Kelley 2004). High interspecific correlations reported by previous studies but shown to be problematic by the current study are at the heart of this interpretation. In addition, correlation studies virtually always lack true ontogenetic dimensions, blinding them to modularity. Correlation studies assume that all structures cease growth synchronously at the age of reproductive maturation. In contrast, ontogenetic studies accommodate variation in pathways through the juvenile period that arise from modularity. Moreover, an ontogenetic view incorporates ideas about metabolic risks and extrinsic mortality (Janson and van Schaik 1993, Williams 1966a).

We expect that alternative life history modes among primates have evolved in response to combinations of metabolic risks and differences in mortality at various phases of life histories. This idea relies on classic trade-off theory, anticipating that organisms face "decisions" as to how and when to allocate effort to growth and reproduction. We suggest that the situation for primates is more complicated than this, involving hierarchies of trade-offs. Specifically, a relatively lengthy period of primate ontogeny means that primates (including humans) have options (unavailable to other species) regarding energy allocation among modules. Three factors in primate energy allocation may be important. First, energy allocation may follow temporal patterns, involving a simple energetic shift from growth to reproduction, as predicted by classic theory. Second, dissociation may enable trade-offs among modules, with expensive tissues growing during ages of relatively low risk. Third, variation in total body mass growth rates may mediate energy allocation. Obviously, different combinations of these factors may play roles, but the important point is that the idea of a life history mode captures these different ways of arranging ontogeny and, thus, of "assembling" adults.

THE BRAIN AND MODULAR LIFE HISTORY

A cursory empirical example focusing on the brain illustrates advantages of the life history mode concept (see Leigh 2004, 2006; Leigh and Bernstein 2006; Pereira and Leigh 2003 for additional examples). Full documentation of modular ontogenies mandates analyses of many species and several organ systems, or structures (modules). However, consideration of body and brain size growth can provide preliminary insights into this idea. First, body mass growth rates vary substantially among species to produce comparably sized organisms (Figure 25.1). Growth rates also fluctuate (Leigh 1996) so that, when measured by the progression of size, life histories can consist of both "fast" and "slow" phases. Different combinations of these phases imply alternative pathways through the juvenile period. Second, turning to the brain (and mindful of editorial space limitations), we assume the brain is an energetically costly module that is responsive to selection partly on metabolic performance (Aiello and Wheeler 1995).

Brain modularity can be illustrated by summarizing analyses of brain growth in relation to key life history variables. Our example tests the hypothesis that the age at brain growth cessation, adult brain size, and age at reproductive maturation are tightly intercorrelated. Rejection of this hypothesis implies that modularity plays a role in primate life history evolution, while failure to reject it supports models that see the brain as a pace-setter of life histories (see also studies by Deaner et al. 2002; Ross 2002, 2004; Ross and Jones 1999; Sacher 1959; Sacher and Staffeldt 1974). Moreover, this test has significant implications for refining ideas proposed by R. D. Martin (1983, 1996; Harvey et al. 1987) that reveal ties between gestation, maternal reproductive or energetic effort, brain size, and life history.

Ontogenetic data for brain size can be gathered for only a handful of species. We follow procedures outlined previously (Leigh 2004), which involve estimating age at brain growth cessation, then comparing these estimates against adult brain size and literature-reported estimates of age at reproductive maturation. We present both raw and phylogenetically adjusted correlations. The latter correlations take into account differences in degrees of relatedness among species.

Evaluations of brain growth curves reveal variation independent of reproductive maturation age. An especially obvious case concerns comparisons between squirrel monkeys (*Saimiri sciureus*) and tamarins (*Saguinus fuscicollis*). Brain growth curves show that the former grow brains at higher rates than the latter but over a much shorter age interval (Figure 25.2). Moreover, squirrel monkey brains

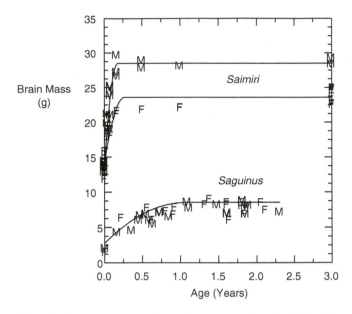

Figure 25.2 New World monkey brain growth trajectories. Squirrel monkeys (*Saimiri sciureus*) cease brain growth earlier than saddle-back tamarins (*Saguinus fuscicollis*). Lines represent best-fit piecewise regressions. M, males; F, females. Sexes are combined in the tamarin sample. See Leigh (2004) for details on data utilized.

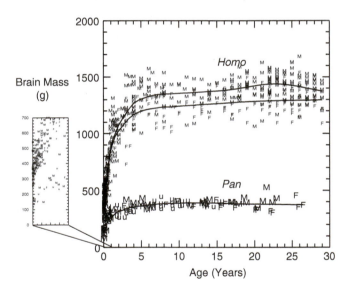

Figure 25.3 Brain mass growth data for humans (*Homo sapiens*) and chimpanzees (*Pan troglodytes*). Lines represent best-fit locally weighted scatterplot smoothing (LOESS) regressions through data. Symbols represent females (F), males (M), or unrecorded (u). Inset shows brain mass growth for each species during the first postnatal year. See Leigh (2004) for details on data utilized.

and bodies are larger as adults than those of tamarins, with squirrel monkeys reaching reproductive maturation later than tamarins (Garber and Leigh 1997). Chimpanzees and humans provide another clear example, showing major differences in brain size produced by different growth rates (Figure 25.3). Major distinctions in total time of brain growth are difficult to discern between these species,

Figure 25.4 Regression analyses of brain attributes and age at reproductive maturation. *A*: Age at reproduction is highly correlated with adult brain size. *B*: Age at reproduction is also correlated with age at brain growth cessation, but the relationship is not as strong. Plot abbreviations are as follows: T, *Saguinus fuscicollis*; S, *Saimiri sciureus*; R, *Macaca mulatta*; M, *Lophocebus/Cercocebus* (mangabeys); B, *Papio*; P, *Pan troglodytes*; H, *Homo sapiens*. All values are measured on females.

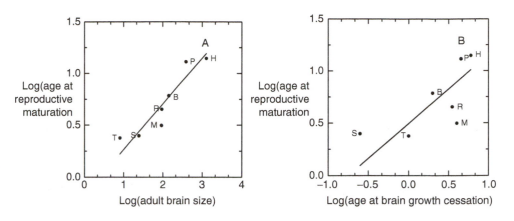

despite differences in brain size, age at reproductive maturation, age at body mass growth cessation, and adult body size. Both comparisons indicate some level of independence of brain and somatic growth.

Statistical analyses for a larger sample show that adult brain size and age at reproductive maturation are strongly correlated (Figure 25.4A, $r = 0.93$) and that age at brain growth cessation and age at reproductive maturation are correlated (Figure 25.4B, $r = 0.64$). Phylogenetic adjustment does not greatly alter the first correlation (brain size and reproductive maturation age, $r_{adj} = 0.70$), but the latter correlation evaporates ($r_{adj} = 0.12$).

Our small sample size presents obvious limitations. Nevertheless, these results refute models that designate the brain as a direct pace-setter of life histories. The time it takes to grow a brain and the length of the juvenile period are unrelated. However, these results point toward models that see roles for energetics and risk aversion in driving life history variation. Specifically, adult brain size and reproductive maturation age are correlated, no matter how we measure the association. This suggests indirect effects of brain ontogeny on both body size and reproductive maturation age. Thus, larger, faster-growing brains seem to require larger, later-maturing mothers (Leigh 2004). Species in which mothers invest relatively little in prenatal brain growth seem to produce offspring that grow brains slowly but through much of the postnatal period. In these cases, the costs of brain growth are borne by the offspring itself; and in the specific case of tamarins, males and other group members subsidize offspring (Garber and Leigh 1997).

These results suggest a significant role for the brain in primate life history. Species that produce large-brained offspring during the prenatal period invest heavily and "single-handedly" in brain size, a situation ameliorated by late maturation and large maternal size (see also Leigh 2004, Leigh and Bernstein 2006), while other species mature early in part because of limited investment in prenatal brain growth. These inferences rely on R. D. Martin's (1983, 1996) ideas tying brain size to life histories through maternal energetics (see Leigh 2004). Perhaps most importantly,

these results suggest the potential for relating ontogeny to demographic views of life histories.

PROSPECTUS

This chapter illustrates several key issues for the future of life history studies. First, we advocate a shift in life history studies toward ontogenetic data. Such data are absolutely essential to furthering our knowledge. We recommend a shift from the kinds of life history data analyzed by traditional studies to data that speak directly to ontogeny and development (see Shea 1990). A longitudinal perspective is likely to be especially valuable in this context (DeRousseau 1990a,b) because it provides an opportunity to measure the fitness consequences of events occurring prior to adulthood (see Altmann, S. A., 1998). This perspective readily accommodates ideas about genetic trade-offs. Second, traditional interspecific comparative studies are not likely to add to what we currently know about life history evolution in primates. However, it is clear that such analyses serve as a valuable foundation for large-scale life history theories (e.g., Charnov 1993). These kinds of analysis also provide insights into historical processes (Shea 1987). Third, theoreticians should follow Charnov's lead by investigating ontogenetic dimensions of maternal energetics, metabolic risk aversion, and differences in extrinsic mortality among phases of life history (including infanticide and predation) and the relation of these to population dynamics. This will require approaches that rely on quantitative genetics and demography. Fourth, we are sorely lacking in ontogenetic data from noncaptive populations. Recent analyses are beginning to solve this problem (Altmann, S. A., 1998; Johnson 2003, 2006; Stone 2004). Future researchers must be well versed in both ontogeny and population dynamics to productively extend these theories. We can note that ontogenetic studies of captive samples are also necessary, particularly given new imaging technologies that will revolutionize studies of ontogeny. Taken together, these advances present significant opportunities to develop life history theories that accommodate both ontogenetic and population phenomena.

CONCLUSIONS

Applications of life history theory to primates have generated significant insights at many levels. These theories prioritize understanding of how energetic resources are allocated either to growth or to reproduction and seek a general understanding of the mechanisms responsible for patterns of allocation and resulting life histories. Theories reviewed here, including Charnov's ideas, Janson and van Schaik's risk-aversion hypothesis, and Kozlowski and Weiner's optimization model, all contribute to our understanding of primate life history evolution. However, this theoretical plurality implies possibilities for a more robust life history theory.

Improvements in life history theory rely on a better understanding of trade-offs and development of models that establish stronger links between ontogenetic processes, quantitative genetics, and demography. Moreover, explicit recognition of the paradigm established by Williams can aid in advancing contemporary primate life history theory. Specifically, greater attention to selection during all phases of life history (e.g., Altmann, S. A., 1998; Johnson 2003) will improve our understanding of how changes during life evolve. Reconceptualizing primate life histories in terms of modes and phases offers new insight into patterns of variation across the primate order by fitting this priority more adequately than traditional approaches. Theories that address the modularity of development may be especially valuable in promoting theoretical advances. Of course, the goal of better theory also relies upon the generation and analysis of high-quality data from primate species and populations.

26

Primate Growth and Development
A Functional and Evolutionary Approach

Debra Bolter and Adrienne Zihlman

1. What are the advantages of a functional approach to growth and development, and how does natural selection factor into this approach?
2. Select two body systems discussed in the chapter. How is the development of each system related to behaviors observed in the wild?
3. How does considering changes in the body composition of an individual through its life provide a better understanding of a species' growth and development than relying on body mass alone?

INTRODUCTION

Growth and development are of concern for evolutionary studies in two major ways. From a phylogenetic perspective, different growth patterns produce adults that diverge morphologically, even when the early stages show considerable similarity. From a life history perspective, natural selection operates on immature individuals: Before any adult primate can reproduce, it must survive long infant and juvenile stages, when mortality is high. If a young animal does not reach maturity, reproductive success, longevity, and speciation are irrelevant.

We begin with a brief historical background of a few influential scientists of growth research, the questions they asked, and the methods they used. Early data collection on extant primates generally described one or two body systems without reference to function or evolution. More recent studies model growth of different body parts, for example, the face, neurocranium, or long bones. These studies speak of ontogenetic scaling, allometry, heterochrony, and neoteny; but meanings and usages vary among the researchers (see Shea 2000). Such studies do not seem relevant to field primatologists and behavioral ecologists because their conclusions do not consider function or adaptation. Furthermore, studies are cloaked in terminology for specialists only.

Our goal is to present a holistic view of individuals. We incorporate information from multiple systems in order to link the developing body to behaviors observed in the wild: the young primate that gains locomotor independence, the shift from mother's milk to adult diet, social integration, and the approach to reproductive maturity. We hope to convey the sense of flow between the immature and the adult and the ways that growth affects species adaptation. We draw on available data from several somatic systems that mature at different rates—for example, dental, neurological, skeletal, and muscular—and from a range of primate species, including lemurs, tamarins, langurs, vervet monkeys, baboons, macaques, gorillas, and chimpanzees.

HISTORICAL PERSPECTIVE: NINETEENTH AND EARLY TWENTIETH CENTURIES

An evolutionary framework is fundamental to the interpretation of immaturity as a developmental stage, with the adult as "end product." The individual—the whole individual—survives and reproduces, and the challenge is to integrate functionally the parts and the whole.

Charles Darwin's Contributions

In formulating the mechanism of natural selection, Charles Darwin discussed in Chapter 13 of *On the Origin of Species* the physical traits and adaptations of immature individuals. Darwin considered that his thoughts on embryology and development in *Origin* were among his most original (Darwin 1876:125). Very early stages of life were not under heavy selection pressure, and therefore, embryos of related groups of animals looked similar (Darwin 1859:449). Darwin argued that selection acted on an immature individual and subsequent features were modulated only when the young animal must provide for its own survival (Darwin 1859, 1860, 1876:125).

Adolph Schultz's Contributions

Adolph Schultz contributed some of the earliest systematic and comparative primate studies well before fieldwork on socioecology had begun (e.g., 1924, 1927, 1930). He gathered information on fetuses, neonates, and other immature individuals. Schultz, who was trained in human biology, drew on museum collections of wild-shot primates and on captive animals from zoos and breeding colonies.

Because dental eruption can be evaluated in both living primates and their skeletal remains, Schultz used dentition as the life stage marker: He wanted a dental "comparable age" across species as a baseline for growth and development. He determined an order of eruption of the deciduous (milk, temporary, or "baby") teeth and the timing of

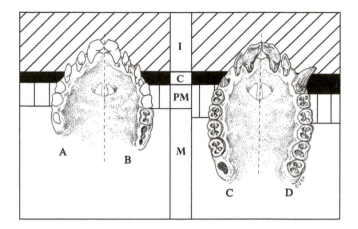

Figure 26.1 Dental stages of catarrhine primates. Drawn to scale. *A*: Infant, deciduous dentition. *B*: Juvenile 1, deciduous and permanent dentition. *C*: Juvenile 2, permanent teeth, not all erupted. *D*: Adult, all permanent teeth erupted. I, incisors; C, canines; PM, premolars, M, molars (drawing by Carol Underwood).

eruption of the first permanent molar (M1) from zoo records (e.g., for chimpanzees, *Pan troglodytes*, Schultz 1940; for orangutans, *Pongo pygmaeus*, Schultz 1941). Based on dental eruption, Schultz defined four life stages that are still used: infant, juvenile 1, juvenile 2, adult (e.g., 1940; cf. Krogman 1969). Infants have a partial or full set of deciduous teeth; juvenile 1 has a combination of deciduous and permanent teeth; juvenile 2 has only permanent teeth but not all of them erupted; and adults have all the permanent teeth erupted (Figure 26.1). Schultz's dental definition of the adult life stage with all permanent dentition erupted had no reference to skeletal growth or joint fusion. Based on these comparative data, he illustrated primate life stages, which remains one of the most widely recognized of his self-penned diagrams (e.g., Bogin 1999) (see Figure 26.2).

MID-TWENTIETH CENTURY: THE MODERN SYNTHESIS AND NEW APPROACHES TO GROWTH

New possibilities for studies of growth and development came with the "rediscovery" of Darwin's ideas about the individual. This came from biologists, the bridge builders, who integrated information from several fields: genetics and the importance of mutation as a source of variation, the concept of surviving and reproducing individuals as the focus for natural selection, and the population—its size, gene pool composition, and migration patterns. All these processes contribute to speciation and, therefore, to evolutionary change (see reviews in Mayr 1980, 1997, 2004).

Connecting the study of growth to the modern synthesis, G. G. Simpson, a mammalian paleontologist and one of the bridge builders, emphasized the individual as four-dimensional. Although usually conceptualized as a three-dimensional adult, Simpson maintained that an individual is not a "combination of tissues arrested at a moment in time, but a whole sequence of states and forms through which one animal passes from fertilized ovum to death" (1944:9). Thus, growth is the dynamic element which gives the individual its fourth dimension (see Figure 26.3).

In order to introduce evolutionary ideas into the study of human evolution, Sherwood Washburn and Theodosius Dobzhansky organized the "Origin of Man" conference at Cold Spring Harbor, New York, in 1950. Washburn (1951a) urged that anthropologists move away from mere description of bones, teeth, and traits and include information from living and fossil primates, in the same way that he emphasized the integration of bone, muscle, and joints into functional regions rather than describing isolated bones or taking nonbiological measurements (1946a, 1946b, 1947a, 1947b, 1951a; Detwiler and Washburn 1943). Washburn's importation of the modern synthesis into physical anthropology in the mid-twentieth century energized the entire field (Zihlman 2000).

Washburn recognized that growth had a role to play in an evolutionary approach. In his article "The New Physical Anthropology," Washburn (1951b) specifically noted growth and development as a main subarea for future research. This evolutionary framework has been slow in acceptance, in part because of the Darwinian ideas about selection pressures on the individual (Bowler 1988, Mayr 2004) and because developmental processes were not fully appreciated by those who supported the modern synthesis.

Kenneth Brizzee and William Dunlap (1986) of Tulane University reviewed the literature on primate growth studies through the mid-1980s, highlighting the absence of the modern synthesis and illustrating some of the pitfalls of growth studies: narrow description of one anatomical region and neglect of its functional assessment and effect in another anatomical region (see also Watts 1986). For example, they reviewed facial growth and dental eruption in two different sections but missed the correlation between the erupting teeth and the elongation of the face (refer to Figure 26.1).

THE SHIFT FROM DESCRIPTION TO FUNCTION AND EVOLUTION

Primate Immaturity in the Field

Primate field research began in the 1940s (e.g., Carpenter 1940). Longer-term primatological studies in the late 1950s and early 1960s depended on the identification of individuals

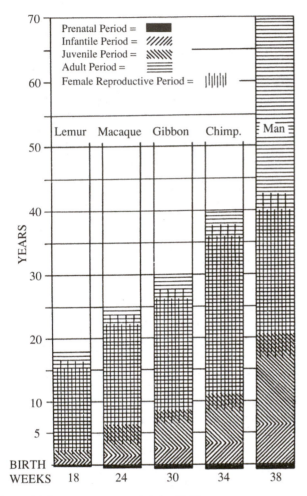

Figure 26.2 Life stages (after Schultz 1963).

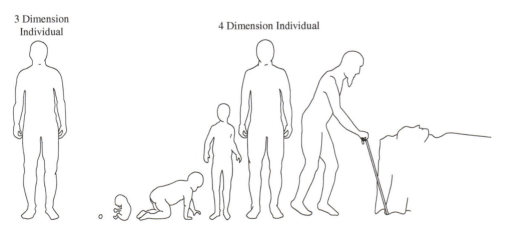

Figure 26.3 Third and fourth dimensions of the individual (after Simpson 1944).

3 Dimension Individual

4 Dimension Individual

and brought new possibilities for their study in Simpson's "fourth dimension" (through their lifetimes). As field studies extended beyond a few months, researchers could observe variation between individuals of different age/sex classes and diverse patterns of social organization (Washburn et al. 1965, Crook and Gartlan 1966). In the early studies, the high-profile males received the most attention; they could be readily identified by their size; infants and juveniles were less likely to be named and observed in much detail (Fedigan and Fedigan 1989, Fedigan 1992).

Two early projects focused on immature primates. A study on Indian langur monkeys (*Presbytis entellus*) described infant behavior and mother–infant interactions (Jay 1962, Dolhinow 1972a), which later led to studies on behavioral development in captive groups (e.g. Dolhinow and Murphy 1982). Early fieldwork coincided with emerging laboratory observations on infant monkey social development by psychologist Harry Harlow and his colleagues in the United States that revealed the importance of the mother–infant bond (Blum 2002).

In her studies on chimpanzees (*Pan troglodytes*) at Gombe National Park beginning in 1960, Jane Goodall named individuals and followed them through their lives. Early on, she charted the development of newborn infants (Goodall 1971, 1986, 1990). Focus on Gombe infant, juvenile, and adolescent chimpanzees continued in the research of Anne Pusey, Frans Plooij, and Hedwig van de Rijt-Plooij, which described physical and behavioral markers in these early immature stages (Pusey 1978, 1983, 1990; Plooij 1984; van de Rijt-Plooij and Plooij 1987). Other long-term studies have provided parallel life history data on wild chimpanzee populations of the Mahale Mountains at Bossou (e.g., Nishida et al. 1990, 2003; Sugiyama 1994) and at Taï (Boesch and Boesch-Achermann 2000).

Researchers of baboons, macaques, and langurs also catalogued age classes of wild populations—infant, juvenile, subadult, adult—based on physical appearance and relative size (*Papio hamadryas*, Kummer 1968a, Sigg et al. 1982; *Papio cynocephalus*, Altmann, J., et al. 1977; *Macaca fuscata*, A. Mori 1979; *Pr. entellus*, Dolhinow 1972a). This fieldwork on Old World monkeys showed the significance of life stages for understanding complex social behaviors in different age and sex classes. Jeanne Altmann (1980) linked infant development and survival to the mother's energetics in her research on the intense infant–mother relationship in yellow baboons (*Papio cynocephalus*). In this same population, Stuart Altmann (1991, 1998) summarized decades of research on wild infant baboon social and ecological strategies. Phyllis Lee (1984, 1987) investigated how differential maternal investment affected infant vervet social development (*Cercopithecus aethiops*). As research accumulated, a more detailed picture of immature primates began to emerge.

Infancy

Long-term research demonstrates that infancy is quite risky. Provisioned primates may experience only 10% infant mortality, whereas among wild primates, one-half to two-thirds of infants die (*Lemur catta*, Gould et al. 2003; *Papio cynocephalus*, Altmann, J., et al. 1977; *M. fuscata*, A. Mori 1979, Itoigawa et al. 1992; *Pan troglodytes*, Boesch and Boesch-Achermann 2000). Infants depend on the mother for food, protection, and companionship and are affected by maternal health and nutrition (A. Mori 1979). While still dependent, infants must learn foraging techniques, food choices, and social skills that will help them survive as adults (e.g., Mason 1979, Walters 1987b, Altmann, S., 1998). For example, in captive tarsiers (*Tarsius bancanus*), the infancy phase is extremely short compared to other haplorhines. During this accelerated infancy period, the babies perfect their hunting skills; and by 80 days, they are weaned and independent juveniles (Roberts 1994).

In field studies, infants may be easily distinguished from adults by hair and facial skin color. These physical markers signify behavioral transition to weaning, end of dependence, and differential social behaviors from mother and other group members. For example, infant hamadryas baboons (*Papio hamadryas*) differ from other group members by a black neonatal coat; as the infant matures, the color turns into the tan coat of a juvenile, which elicits different kinds of interaction from other group members. The tan coats "get away with less" and become potential food competitors with other juveniles and adults (Kummer 1968a, 1995). Crested langur infants (*Trachypithecus cristatus*) are born with neon orange hair but, at about 5 months of age, acquire the silvery coats of the adults (Wolf 1984). Proboscis monkey infants (*Nasalis larvatus*) have blackish coats and blue faces at birth; during the eighth month, faces become the fleshy color of adults and coat color transitions to silvery red (Pournelle 1967, Rüedi 1981).

Juvenility

The juvenile stage is one of marked physical and social change and has been little studied. To rectify this, Michael Pereira and Lynn Fairbanks (1993a) edited a volume to bring together a number of studies on juveniles. Juvenile primates travel and forage independently, while remaining socially connected to their mother and siblings. They are not adult in size; by definition, they are reproductively immature. They are vulnerable because they must compete with other group members for resources. In times of scarcity, only about half of the juveniles may survive (Pereira and Altmann 1985, Watanabe et al. 1992, Pereira and Fairbanks 1993b, Gould et al. 2003).

The risks of increasing the duration of immaturity are balanced by the skills and abilities that they acquire for later benefit. Through play and group activities, immature primates learn complex behaviors and social roles and develop flexibility in coping with stressful situations. They acquire knowledge of food location and territorial boundaries and gain experience in mothering and in motor coordination during play while their bodies gradually transform into adults (Mason 1979; Lancaster 1971; Pereira and Altmann

1985; Watts, E. S., 1985; Walters 1987b; Fairbanks 1993a, 2000a; Fagan 1993; Altmann, S., 1998).

Sexual Maturity

As immature individuals shift to sexual maturity, the duration of reproduction varies across species. Some primates have a distinct adolescent stage: marked physical and social behavioral divergence of females and males. This stage has not been well studied except for chimpanzees (*Pan troglodytes*). Pusey (1978, 1983, 1990) describes early and late adolescent stages and notes distinct physical and behavioral changes in individuals, as well as in the behavior directed to them by other group members. For example, in early adolescence, a female's first estrus is defined by the response of the adult males, though only in late adolescence does a female conceive (Pusey 1978, 1990). This phase of life, as in the other stages, is marked by changes in physiology, anatomy, physical appearance, and behavior among group members; no one component is independent of the others.

LEVELS OF LIFE HISTORY: WHOLE BODIES, MULTIPLE SYSTEMS

Whether one studies the maturing individual in captivity or in the wild or studies behavior, physiology, or anatomy, the process is time-dependent.

Temporal shifts over the course of life are called "life history" and can be studied on three "macro" levels (Fedigan 1997). First, life history may represent a personalized biography of an individual in terms of social rank, relations, and life experiences. The second level of analysis focuses on variation in the timing of growth events of individuals within a social group or local population, such as the range of ages at which infants are weaned or when females and males diverge in their development. Third, at the species level, the "average" age for major life events establishes a species pattern, such as age at weaning, locomotor independence, dispersal, sexual maturity, and first reproduction, as well as birth intervals and life span. These variables are among the life history features laid out by Harvey et al. (1987) and expanded in volumes edited by DeRousseau (1990) and Morbeck et al. (1997). The personalized biographies, population variation in maturation, and species patterns represent alternate survival and reproductive strategies employed across primate species. These approaches illustrate the different levels of this topic, with growth and development at the center.

At the "micro" level, body systems transform and interact to promote survival at the infant and juvenile stages. The individual must be able to function during day-to-day activities while its body adjusts, changes, and matures in its transformation to adulthood. Anatomical systems underpin behavior and provide physical markers for behavioral expression. Both anatomy and behavior shape survival. Therefore, a functional approach to growth and development integrates multiple lines of information, for example, timing of brain maturity and dental emergence, rate of skeletal growth and fusion, degree of change in body proportions and muscularity, and attainment of adult body mass.

The Part and the Whole

In courses and textbooks on human anatomy, the body is necessarily divided into systems—musculoskeletal, neural, reproductive, hormonal. These courses often begin with a discussion of cells, move into tissues, then rise up a level to individual systems. Dividing the body into finite pieces provides a way to study a complex machine. Growth and development studies have also followed this path, isolating the body into a series of systems, like dentition, brain size, and regions (e.g., pelvis, limbs, facial structure), as well as body mass, body proportions, and composition, and comparing the sexes in these areas. However, the body is integrated and these separations, although an aid in sorting through this dynamic process, oversimplify it.

Different systems in the body mature at different rates, a point illustrated in Scammon's (1930) classic graph of human growth. Physical anthropologists have demonstrated that the timing of physical growth events can also fingerprint nonhuman primate groups (e.g., Schultz 1927, 1937; Washburn 1943; Tappen and Severson 1971; Grand 1977b, 1992; Watts, E. S., 1985, 1990; Leigh 1994; Bolter and Zihlman 2002, 2003; Bolter 2003, 2004; King 2004, Dirks and Bowman 2007). We elaborate on this theme and offer ideas on relating physical changes to locomotion and other behaviors for future application to laboratory and field research. We begin with the dentition because hard tissues are consistent markers of immature individuals. We move on to the skeleton, body mass and sex differences, body proportions, and body composition. Although preliminary, our findings illustrate functional ways to look at growth.

Dentition

Life stage classifications of individuals, particularly skeletal specimens, are most often assessed based on the timing and order of tooth emergence. In the 1980s, Holly Smith highlighted the importance of dental eruptions in primate life histories, correlating the emergence of the first permanent molar (M1) with 90%–95% brain growth in primates (Smith 1989, 1991, 1994). Other studies of tooth enamel and the counting of incremental markings showed that crown formation of M1 begins about 2 weeks before birth. These findings make it possible to interpret growth and development by estimations of chronological age at death for immature specimens of extinct genera, extrapolation to the tempo of brain growth in extinct species (especially hominids), and evolution of a childhood stage in the life history pattern of *Homo sapiens* (e.g., Laird 1967, Bromage and Dean 1985, Dean et al. 1986, Bogin and Smith 1996, Leigh and Park 1998, Bogin 1999, Zollikofer et al. 1998, Dean and Reid 2001, Dean et al. 2001, Ponce de Leon and Zollikofer

2001, Kelley and Smith 2003, Ramirez Rozzi and De Castro 2004, Zihlman et al. 2004b, Zihlman and Bolter 2004).

An individual is labeled "infant" when no permanent teeth have yet erupted but there is a set of deciduous (milk) teeth: four incisors, two canines, and four (pre)molars in the lower and upper jaws. Typically, the incisors come in before the canines and premolars (Smith et al. 1994). The transition from infant to juvenile comes with the eruption of the first permanent (M1) tooth. The M1 is a "new" tooth, which does not replace any of the deciduous teeth. Permanent incisors replace deciduous incisors and permanent canines, deciduous canines. In the catarrhines, the permanent premolars replace the deciduous molars (Figure 26.1). The full eruption of M1 has been interpreted as signifying the timing of weaning and, therefore, the end of infancy. Weaning may be abrupt due to marked seasonality in reproduction in lemurs (Wright 1999) and in some platyrrhines where twinning is the norm (Garber and Leigh 1997). In catarrhines, weaning is gradual when young animals shift from maternal calories to independently acquired foods (e.g., Altmann, S., 1998).

Juveniles have a mixture of deciduous and permanent teeth. In catarrhines, the second molar (M2) emerges while the trunk and long bones continue to grow and body mass and musculature increase. During this stage, females and males diverge and juvenile males in some species leave their natal group.

The eruption and full occlusion of the last molar (M3s or M2s in callitrichids) serves as only one biological marker of adulthood. Skeletal mass and body mass are two examples of systems still undergoing considerable growth after M2/M3 occlusion (e.g., platyrrhines, Tappen and Severson 1971; colobines, Schultz 1942, Dirks 2003; cercopithecines, Turner et al. 1997, Bolter and Zihlman 2002; hominoids, Randall 1943, Zihlman et al. 2007).

Data on dental eruption are available for a few species (Smith et al. 1994) (Table 26.1). The species in Table 26.1 represent the major phylogenetic groups and demonstrate that a pattern of increased periods of development does exist, just as Schultz recognized. However, although timing in dental eruption is delayed from prosimians to anthropoids, from platyrrhines to catarrhines, from cercopithecoids to hominoids, body mass of the female does not follow this same trajectory. Timing of dental eruption and weight of adult females are disassociated. The table demonstrates the limited data across species on other aspects of growth. Dental information may be available for an individual, but facts about skeletal fusion stages, body mass, or behavioral markers like weaning age may not be known for the same individual (Watts 1986, Smith et al. 1994, Godfrey et al. 2001).

Table 26.1　Dental Emergence Time and Approximate Body Mass in Female Primates (Wild Weights Used from Postcranially Mature Individuals, Where Available)

SPECIES	M1 GINGIVAL EMERGENCE (YEARS)	M3 GINGIVAL EMERGENCE (YEARS)	FEMALE ADULT BODY MASS (KG)
Prosimian			
Lemur catta (ring-tailed lemur)	0.3	1.3	2.5
Anthropoid			
Platyrrhine			
Saguinus nigricollis	0.4	Absent in genera	0.5
Saimiri sciureus	0.4	1.7	0.6
Catarrhine			
Cercopithecoid			
Trachypithecus cristata	1	3.5	6.2
Semnopithecus entellus	1.4[1]	4.7	9.9
Cercopithecus aethiops	1.2	4.4	3.5
Macaca mulatta	1.3[1]	5.8[1]	7.1
Papio hamadryas	1.7	7.7	9.9
Hominoid			
Pan troglodytes	4	13	35.2
Homo sapiens	6	18.5	60

Sources: Long and Cooper 1968; Chase and Cooper 1969; Wolf 1984; Iwamoto et al. 1987; Uehara and Nishida 1987; Harvey et al. 1987; Phillips-Conroy and Jolly 1988; Smith et al. 1994; Zihlman 1997; Vancata et al. 2000; Bolter and Zihlman 2003; Liversidge 2003; Dirks 2003; Zihlman et al. 2004a,b; Dirks and Bowman 2007; Zihlman et al 2007.
[1] Emergence through the jawbone.

The Skeleton

The skeleton is the frame that provides a stable structure for body mass. It anchors the muscles and facilitates movement around the joints. Bones add length during development in the *growth plate*, or the area of cartilage that sits between the shaft (diaphysis) and the tip (epiphysis). The new bone pushes the epiphysis away and increases the length of body segments such as the arm, forearm, thigh, and leg. Fusion of the epiphysis to the bone shaft terminates growth and additional length (see Figure 26.4).

Washburn noted a consistent order of skeletal fusion. For example, in Old World monkeys—crested langurs (*Trachypithecus cristatus*) and crab-eating macaques (*Macaca fascicularis*)—the skeletal elements comprising the elbow, hip, and knee joints fuse first, then the ankle, wrist, and finally shoulder joints. He emphasized regional union of joints—for example, the elbow joint with several epiphyses in the distal humerus, proximal radius, and proximal ulna—as opposed to a single, isolated bone. Washburn (1942, 1943) implied that variation in fusion patterns between species was due to locomotor specialization.

The timing of skeletal fusion compared to the eruption of teeth can vary within sexually dimorphic species. In rhesus macaques (*Macaca mulatta*) and Japanese macaques (*M. fuscata*), females complete skeletal fusion of the long bones prior to male conspecifics of comparable dental development (van Wagenen and Asling 1958, Cheverud 1981, Kimura and Hamada 1990). In colobine monkeys—crested langurs (*T. cristatus*), maroon leaf monkeys (*Presbytis rubicunda*), and proboscis monkeys (*Nasalis larvatus*)—females tend to fuse pelvic and long bones earlier than males of the same dental age (Bolter 2004).

Body Systems as a Developmental Mosaic in Two Monkey Genera

The following case studies from the callitrichine New World monkeys (platyrrhines) and cercopithecine African monkeys (catarrhines) illustrate that primates do not synchronize maturity in the skeleton, dentition, reproductive anatomy, body mass, and behavioral events (see Watts, E., 1990). Furthermore, sexual maturity and reaching adult stage involve a whole pattern, one that differs for females and males (*bimaturation*). This combination of information integrates and highlights physical and behavioral elements of maturation and reminds fieldworkers of the dynamic anatomy that underpins the behavior of immature primates.

Tamarins. The saddle-back tamarins of South America (*Saguinus fuscicollis*) are among the smallest of anthropoid primates, with minimal sex difference in adult body mass and fast maturation. Tamarins are born with 12 deciduous teeth (eight incisors and four canines); at 3 months, they have all the deciduous molars and are weaned (Glassman 1983, Garber and Leigh 1997). At 5 months, permanent teeth appear; at 12 months, all of them have erupted (12 premolars and eight molars).

Skeletal fusion begins around the third month, at the time of weaning. Elbow and pelvic elements are the first bones to complete growth. Most elements of the hip, knee, and wrist joints (bones that contribute to the joints) fuse by 15 months, after all of the permanent teeth are erupted. The ankle and shoulder joints mature much later, at about

Figure 26.4 Immature chimpanzee humerus, showing proximal and distal ends; proximal end fusion indicates adult stage reached (drawing by Carol Underwood).

Table 26.2 Developmental Events of *Saguinus*[1]

AGE (MONTHS)	EVENT
At birth	12 deciduous teeth
2–3	All deciduous teeth
3	Skeletal fusion—elbow, pelvic elements—weaning complete
4–5	First permanent teeth, body mass 40% of adult
10–12	All permanent teeth (12 premolars/8 molars)
12	Body mass 85%, linear growth complete
15	Skeletal fusion—hip, knee, wrist joints
16	Sexual maturity of female, adult body mass
24	First birth, skeletal fusion—ankle, shoulder—social adults

Sources: Glassman 1983, Chase and Cooper 1969, Garber and Leigh 1997.

[1] Combines information from *Saguinus fuscicollis* and *S. nigricollis*.

24 months, around the time of social adulthood (Glassman 1983) (Table 26.2).

Sexual maturity of captive saddle-back tamarin females is reached as early as 16 months of age. Pregnancy may occur before the skeleton has finished growing but after the permanent teeth have erupted. The average age at first birth is 24 months, which follows a 5-month pregnancy, and they most commonly give birth to twins (Glassman 1983).

The closely related black-mantled tamarins (*S. nigricollis*) similarly have all of their deciduous teeth at 2 months and wean at age 3 months (Chase and Cooper 1969). In body mass, these young weanlings are only 40% of adult body mass. By 12 months, they have reached 85% of their adult weights and all of their adult linear proportions are achieved. Adult body mass is not reached until 16 months. Females, like their saddle-back relatives, have their first offspring at 24 months (Chase and Cooper 1969).

A "fast-track" timing of dental–skeletal maturation compared to other anthropoids may reflect callitrichine-unique reproductive and social strategies: twinning as the norm, with two smaller infants rather than one, heavier infant, as typical in most anthropoids; only one reproductively active female per social group; siblings and adult males that help carry the infants. These species features may influence their accelerated weight gains compared to other anthropoids (Garber and Leigh 1997) and quicker dental–skeletal maturation.

Vervet Monkeys (C. aethiops). These African monkeys exhibit a moderate sex difference in adult body mass and illustrate two aspects of growth that characterize many catarrhine primates: sex differences in pattern of development (skeletal fusion, body mass) and the order in which different body systems mature (Turner et al. 1997, Bolter and Zihlman 2003).

In contrast to tamarins, tooth eruption is later. Incisor, canine, and (pre)molar deciduous teeth erupt between 6 and 115 days, first permanent molars between 12 and 14 months, with all permanent teeth erupted by 67 months (L. Fairbanks personal communication). Females and males at the same dental age depart in timing of skeletal fusion; for example, female vervet monkeys fuse their pelvic bones (at the ischiopubic ramus and acetabulum) before males (Bolter and Zihlman 2003). Female vervets with erupted canines have completed pelvic growth, whereas males with erupted canines have not.

With all teeth erupted, the limb and trunk lengths and body mass of wild vervet females are within the adult female average for that population, although all elements of the post-crania (shoulder joint) have not fused. Males of comparable dental age have not completed their limb length and trunk growth and have achieved only 80% of adult body mass.

Other studies of wild vervet populations in Kenya document bimaturation patterns (Turner et al. 1997). After about 1.5 years, female and male vervet monkeys exhibit sex differences in growth rates for body mass, body length, chest girth, and several limb measurements. Males grow faster in juvenility (20–48 months old) compared to females and thereby achieve overall larger adult sizes.

A comparison of immature and adult vervet monkeys from one local Ugandan population illustrates the mosaic order in which body systems mature (Bolter and Zihlman 2003). This sample with data available on body mass and soft tissue (testicle weights and specific muscles) expands information normally available on growth patterns. As other studies have shown, brain volume matures early; linear growth of trunk and long bones follow, then dental eruption, skeletal fusion, and body mass; muscularity as measured by masticatory muscles is late (see Figure 26.5).

Body Mass

Body mass, commonly reported in kilograms, is seemingly a straightforward measure of growth. However, it can be unreliable as an assessment of stages of growth for many reasons, some of which are outlined here: (*1*) mass maturity is not synchronized with maturity in particular body systems; (*2*) increase in body mass during growth may have a different pattern in males and females; (*3*) body mass fluctuates over a lifetime, seasonally and reproductively and with illness; (*4*) average adult mass varies within a species by population; and (*5*) few weights are available for infants and juvenile individuals for populations where adult weights are available. Each point is considered in brief.

Nonsynchronicity. As illustrated with the black-mantled tamarins (*S. nigricollis*) and the vervet monkeys (*C. aethiops*), growth in body mass does not predict the level of anatomical maturity of dental eruption, linear proportions, and reproductive systems nor does body mass correlate with external attributes often used in behavioral studies like nipple size, mantle length, or coat coloration.

Figure 26.5 Percentage of growth completed between immature and mature vervet monkeys (*Cercopithecus aethiops*) (after Bolter and Zihlman 2003).

Percent growth completed

Cranial capacity (cc)	98
Tail, Foot, Hand (mm)	91
Femur, Trunk (mm)	86
Dentition	60
Body mass (g)	59
Skeleton	56
Testes (g)	43
Muscles (mastication) (g)	41

Bimaturity. In sexually dimorphic catarrhines, the timing to maturity of body mass raises another problem. In adult males but not females body mass is one of the last physical markers to mature. Female primates may stop growing earlier and shift energetics into nutritional transfer (lactation), carrying infants, and pregnancy (Altmann, J., 1980, 1986; Lee 1987). Larger-bodied males of the species continue growth in the skeleton, canine teeth, and body mass. For example, female olive baboons (*Papio anubis*) and gray-cheeked mangabeys (*Lophocebus albigena*) exhibit a period of declining body mass growth around menarche, while males have a spurt in growth between 4 and 6 years (Coelho 1985, Deputte 1992). In wild toque macaques (*Macaca sinica*), females finish muscle mass and weight growth at about 8 years of age and males, at age 12 years (Cheverud et al. 1992). In extremely dimorphic species, such as hamadryas baboons (*Papio hamadryas*), wild females achieve adult body mass around 5.5 years whereas males add body weight for another 5 years, by which time mass has doubled (Sigg et al. 1982).

Species Variability. There is a high degree of variability within species, a point that is often neglected (Smith and Jungers 1997; cf. Vancata et al. 2000). Variation in body mass has also been documented within a subspecies and across local populations. For example, *Pan troglodytes schweinfurthii* at Gombe Reserve average less than the Mahale Mountain populations—adult females at Gombe average 31.4 vs. Mahale 35.2 kg (Pusey et al. 2005, Uehara and Nishida 1987, Smith and Jungers 1997). Within the Gombe population, males that have sired offspring vary widely in body weight (Constable et al. 2001). Such differences may reflect ecological quality or a broader genetic variation.

Seasonal and Lifetime Variability. Information available from field studies demonstrates not only that body mass changes during growth but also that accurate measurements are further complicated because the individual's mass may fluctuate with reproductive status, social status, seasonality, and time of day (Bercovitch 1987a, Altmann, J., et al. 1993, Pusey et al. 2005). For example, a full stomach in a colobine leaf-eating monkey can weigh up to 1.4 kg (Washburn unpublished field notes 1937) or between 7% and 18% of the total body weight (Kay and Davies 1994).

Weights from the Wild. Body mass data collected in the wild are a more accurate reflection of age and sex and, therefore, are preferred (and specified in publications) over captive population weights. In most instances, field weights are recorded during expeditions and often associated with individual specimens in museums (e.g., Allen and Coolidge 1940). These specimens are not always complete and may be represented only by cranial remains. As a result of historical practices in categorizing masses in specimens according to dental eruption status, wild body masses for adults have been underestimated (e.g., Napier and Napier 1967, Harvey and Clutton-Brock 1985, Jungers 1985,

Brizzee and Dunlap 1986, Harvey et al. 1987, Smith and Jungers 1997).

For example, when calculating body mass estimates of adult *N. larvatus* from the Asiatic Primate Expedition collection of the Comparative Museum of Zoology at Harvard, Bolter (2004) noted that females that are dental adults but skeletally immature average only 9.1 kg, whereas dental plus skeletal adults (shoulder joint fused) average 10.4 kg; for males the difference is more extreme, 16.7 vs. 21.0 kg. Reliance on wild-collected, museum body weights raises questions about studies that depend on these "adult" masses as the size baseline for scaling of other somatic and life history variables through growth (limb stoutness, gestation length, etc.) (e.g., Jungers 1985).

A few field researchers have obtained accurate weights, although adults and not infants and juveniles were often the targets of the data collection. Animals may be captured and sedated to obtain weights and other anthropometrics (e.g., Turner et al. 1997), or feral individuals may be enticed to climb onto a scale to reach food or sit for a long enough period to be weighed (e.g., Altmann, J., et al. 1993; Johnson 2003; Pusey et al. 2005). One notable success is the collection of body weights on Gombe chimpanzees (*P. troglodytes*), measurements that span more than 30 years (Pusey et al. 2005). Another notable study, on wild toque macaques (*M. sinica*), has tracked body masses and other demographic and genetic variables over several generations of immatures and adults (Dittus 2004). Most often, field researchers do not have weights on their study populations and rarely on immature animals; captive animals provide the majority of known body mass data on immature animals (Brizzee and Dunlap 1986, Leigh 1994).

BEYOND BODY MASS: BODY COMPOSITION OF SOFT TISSUE

Increase in body mass during growth reflects an overall change but masks changes in the components, for example, muscle, bones, skin, brain, and fat. When the measure is only "body mass," the relative proportion of the components is smoothed into one all-encompassing number (cf., Grand 1977a, 1990). This number obscures the functional relationships between the systems at each stage. Differences in the relative proportions of these elements in age and sex classes reflect the functional aspects of survival: motor skills and locomotion, reproduction, perception and cognition.

Brain and Nervous System

An infant macaque (*M. mulatta*) has a relatively heavy head, 20% of body mass compared to the adult's 6% (Grand 1977b). Head mass consists primarily of the brain, which is 10 times heavier (proportional to body mass) than the adult's. A large brain at birth correlates with a young primate's early social interaction, though in physical skills it lags behind other mammals (Grand 1992). Well-developed brains during infancy function early as adaptive centers to process information from the visual, olfactory, auditory, and tactile senses (Altmann, J., 1980; Dominy and Lucas 2001). The expanded forebrain, especially the somatosensory, motor, and visual cortices, controls the functions of locomotion, manipulation, visual processing, memory, communication, social interaction, and emotional responses—critical to infant survival into the next life stage. Infant primates divert extensive somatic energy and resources into brain growth, which reaches adult mass around the time they transition to adult foods, as measured by the eruption of their permanent teeth as juveniles (Smith 1989).

Muscular System

Muscularity and proportions of muscle groups shift as newborn primates acquire motor skills. At birth, an infant macaque (*M. mulatta*) consists of 25% muscle compared to 43% in the adult (Grand 1977b). The hands and feet function for survival in the baby's ability to flex its digits (Hines 1942). This use is reflected in the infant's hand and foot proportions: Infant hands are 2.6% of body mass compared to the adult 1.2%; feet are 3.8% versus the adult 2.4% (Grand 1977b). An infant is therefore well equipped to grip its mother's hair tightly as she carries, feeds, and protects it.

As the infant achieves locomotor independence, its thigh increases in muscularity and it decreases emphasis on the grasping function of the foot; the thigh provides much of the propulsion in adult quadrupedal locomotion (Grand 1983, Bolter and Zihlman 2002). Thigh mass, particularly muscularity, is the best single indicator of locomotor independence in primates (Grand 1990).

Linear dimensions also reflect proportional changes in the forearm, leg, hands, and feet—the ability of the dependent infant to cling. Schultz (1937) noted that during fetal development the relative lengths of the hands and feet of near-term rhesus macaques (*M. mulatta*) comprise a very large overall proportion of the distal limb segments (90% and 130%, respectively) but in postnatal life both indexes drop markedly. Two important points are made with Schultz's observations: (*1*) linear skeletal proportions and the percentage of soft tissue mass per body mass give the same information but (*2*) linear dimensions without consideration of surrounding muscle and locomotor function give an incomplete picture of how the proportions equip the infant to survive.

Body Fat

Body fat, its metabolism and storage, is a mammalian attribute that plays a critical role in survival and reproduction (Pond 1977). In nonhuman primate adult females, levels of body fat affect ovulation, successful pregnancy, and ability to lactate (McFarland 1997). Body fat differences in two groups of a wild population of yellow baboons (*Papio cynocephalus*) were dramatic: 2% in wild-feeding females compared to 23% in garbage-feeding females, which traveled much less each day (Altmann, J., et al. 1993). The greater body fat stores in the garbage eaters correlated with a first conception 1 year earlier and accelerated weaning of infants than leaner females. Little is known about when during development body fat is acquired, though it probably parallels the timing in human females, who add body fat at puberty under hormonal influence (Short 1976, Frisch 2002).

Reproductive Organs

Rate and timing of maturation of reproductive anatomy diverge significantly in females and males of some species. For example, ovaries in yearling vervet monkey females (*C. aethiops*) are mature in structure but are not functional. In contrast, testicular development in vervet monkeys is relatively complete in yearlings and relatively rapid during juvenility in baboons (*Papio anubis* and *hamadryas*), chimpanzees (*Pan troglodytes*), and humans (Altmann, J., et al. 1981; Bolter and Zihlman 2003; Pusey 1990; Short 1976). The baboon testicular growth spurt during juvenility appears to vary with different forms of these species' social organizations (Jolly and Phillips-Conroy 2003). In colobine leaf monkeys (*T. cristatus*, *P. rubicunda*, and *N. larvatus*), the testicles descend into the scrotum before dental, skeletal, and body mass maturity (Bolter 2004). Male cercopithecine, chimpanzee, and human testicles are sexually mature (descend and produce sperm) before reaching adult body size and before males develop the social skills and strategies to mate successfully (Short 1976, Matsubayashi and Mochizuki 1982, Pusey 1990, Dixson 1998).

Case Studies: Body Composition in Tamarins, Gorillas, and Chimpanzees

Golden Lion Tamarins

Golden lion tamarins (*Leontopithecus rosalia*) of South America are small-bodied New World monkeys that produce twins as the norm. Gestation takes 140 days, and newborns weigh about 55 g, or 10% of adult body mass (Grand 1992). Brain mass at birth is about 50% of the adult size, but by 2 months of age the brain is fully grown and indicates the mature level of development of the sensory system. In contrast, muscle development is about half of the adult value at birth and continues to increase for 12 months to reach maturity. The mass and growth pattern indicate relatively delayed physical strength compared to their social responsiveness, which translates to more socially but less physically interactive juveniles (Grand 1992).

Distribution of musculature also differs between infant and adult tamarins. Newborns have heavier forelimb muscles and heavier flexors in the fingers and toes but less back extensor muscle mass—all related to an infant's primary anatomical function: clinging to its mother for transport. As adults, hindlimb and back extensors disproportionately increase, reflecting the shift to quadrupedalism (Grand 1992).

Gorillas and Chimpanzees

Gorilla (*Gorilla gorilla*) and chimpanzee (*Pan troglodytes*) infants are born at similar weights, but by 3 years of age captive gorillas are over three times the body mass of chimpanzees (Bolter and Zihlman 2002). Dental development in the gorilla is also on an accelerated schedule compared to that in the chimpanzee (Dean and Reid 2001). In a study of the skin, muscle, bone, and fat among infant, juvenile, and adult chimpanzees and gorillas, limb masses increased disproportionately in gorillas compared to chimpanzees. The infant gorilla, infant chimpanzee, and juvenile chimpanzee are about 30% muscle to total body mass. The juvenile gorilla muscle mass is more like an adult chimpanzee ratio at 38% of total body mass. In particular, the juvenile gorilla thigh musculature reaches adult proportions earlier than that of the chimpanzee.

Wild gorillas achieve locomotor independence between 2.5 and 3 years. In wild chimpanzees, independence is achieved much later, around 6 years (Doran 1997b). The juvenile gorilla's accelerated maturation of body mass and "adult" composition in thigh muscularity provide a possible functional explanation for the gorilla's accelerated acquisition of quadrupedal knuckle-walking compared to chimpanzee's.

ECOLOGICAL–SOCIAL ASPECTS: GROWTH RATES

Wild Versus Captive

Wild primates have extended growth periods compared to captive animals (baboons, e.g., Altmann, J., et al. 1981, Phillips-Conroy and Jolly 1988, Garcia et al. 2009; vervets, Bramblett 1980, Cheney et al. 1988; macaques, Cheverud et al. 1992; chimpanzees, Pusey 1978, Kimura and Hamada 1990, Watts 1993, Boesch and Boesch-Achermann 2000, Zihlman et al. 2004a, 2007). Compared to their wild counterparts, captive *Papio cynocephalus* baboons accelerate dental development (Altmann, J., et al. 1981; Phillips-Controy and Jolly 1988). Anubis baboon infants in captivity have growth rates nearly double that of wild infants (Garcia et al. 2009). Vervet females (*C. aethiops*) in captivity consistently reach sexual maturity earlier (Bramblett 1980, Cheney et al. 1988). Macaques (*M. sinica*) in Sri Lanka prolong growth by several years compared to laboratory populations (Cheverud et al. 1992).

Development is accelerated in captive versus wild chimpanzees (*Pan troglodytes*) (Zihlman et al. 2004a, 2007). Infants in captivity erupt M1s about 6 months ahead of their wild counterparts, and juveniles erupt their M2s about 2 years earlier. The skeleton follows a similar pattern, where growth differences become more marked during the juvenile stages; elements fuse up to 2 years earlier in captive chimpanzees compared to wild individuals (Watts 1993, Zihlman et al. 2007). The stages of dental and skeletal maturity parallel those of sexual maturity and age at first birth, which is up to 3 years earlier in captives versus wild relatives (Boesch and Achermann-Boesch 2000).

A number of variables might account for the observed accelerated development in captive immature primates: better nutrition during nursing and when weaned (e.g., Hinde et al. 2009), absence of predation, limited daily activity and travel, as well as less social uncertainty and stress. In wild versus captive chimpanzees the difference in dental emergence accelerates in juveniles over infants (2 years versus about 6 months). Stuart Altmann, writing about his long term study on immature baboons, noted that energy allocation for motor activity has been vastly underestimated for juveniles (1998). This prolongation in the juvenile stage may result from expending more energy in locomotor activity with less available for growth. The actual mechanisms regulating the growth patterns remain unknown, and point to an area of future research.

Seasonal Variation

Ring-tailed lemurs (*Lemur catta*), diurnal prosimians, reach behavioral adulthood at age 3, a younger age than most monkeys and apes (Gould et al. 2003). Three-year-old females begin reproducing seasonally. The schedules for conception, birth, and lactation correspond with optimal food resources (Sauther et al. 1999a,b). The 18-week gestation occurs during the dry season, when fruits (their main staple food) are scarce and, consequently, females have little opportunity to accumulate fat (Sauther et al. 1999a,b). As both mothers and infants rely directly on the resources in the environment and not on stored fat reserves, infant mortality is tied to the seasonal ecology (Sauther et al. 1999a,b; Gould et al. 2003). An infant nurses during the wet season, when increased rainfall promotes an abundant food supply for the mother. Mortality of infants can rise 50%–80% when food is scarce, for example, following a drought year; this is higher than the mortality encountered by anthropoids (Wright 1999, Gould et al. 2003).

Infant ring-tailed lemurs grow quickly and are weaned at 4 months during a second peak of food productivity, when fruit and young leaves are available to the weanling. This behavioral change coincides with the eruption of the first permanent molars (Smith et al. 1994). The permanent teeth complete eruption at about 15 months, when individuals are behavioral juveniles and their mothers are giving birth to the next offspring (Gould et al. 2003). The "subadult" phase is between 2 and 3 years, when reproductive systems mature and individuals learn adult social skills. Short life stages make the ring-tailed lemurs more vulnerable to rainfall and temperature yet allow them to begin reproductive life earlier than most of their anthropoid cousins.

Same Ecology, Variation in Strategy

The length of the juvenile stage in primates dramatically affects adult life and survival. Comparison of vervet monkeys (*C. aethiops*) and yellow baboons (*Papio cynocephalus*) illustrates the importance of this stage in learning and practicing skills. These closely related monkeys inhabit the same woodland areas of eastern Africa; they are also similar in dental and digestive anatomy; they differ in body size and in length of the juvenile stage. Vervet monkeys mature about 2 years earlier and, thus, reach adulthood before the larger baboons.

During the juvenile stage, both species learn to feed independently and to choose their foods. Stuart Altmann (1998) documents the adeptness of young baboons at finding and choosing quality foods; they exploit more than 250 species over their lifetimes, far more than vervets. Local vervet monkey populations have become extinct during periods of food scarcity, whereas baboons have survived (Lee and Hauser 1988). Narrow food choices and reliance on seasonally abundant foods limit vervet foraging, especially in times of reduced availability. Baboons, in contrast, are able to exploit less lush and less available plant foods. With a shorter juvenile stage, vervets essentially "give up" time during immaturity for learning, finish growth earlier, and begin reproducing. Attenuated immature stages with more opportunities for acquiring skills and life strategies make young animals more resistant to temperature, drought, and predation, therefore leading to higher survival rates.

Juvenile primates acquire not only ecological knowledge but also social skills that contribute to survival and, later in life, to reproductive outcome and form the basis for species-typical social organization. Although the acquisition of social skills may be more difficult to measure than the numbers of utilized plants, the time to learn and practice social behavior and communication is as critical to survival as mastering a complex ecology (Rowell 1988a,b).

SUMMARY AND FUTURE STUDIES

Growth and development began as a descriptive area within physical anthropology. Its emphasis on bones and teeth traces back to Schultz and to his training as a human osteologist. Although Washburn's bone growth and laboratory work demonstrated functional and dynamic aspects of immature animals, research on immature primates and the dynamics of growth processes has tended to remain atheoretical, nonfunctional, and divorced from field research.

New data, approaches, and syntheses offer possibilities for bridging informational and conceptual gaps based on laboratory, museum, captive, and field studies: dental and brain development (e.g., Smith 1989), dental development through incremental lines (e.g., Schwartz et al. 2006), field research tracking individuals through immaturity (Pusey 1978, Pusey et al. 2005), research on soft tissue (e.g., Grand 1977a, 1990; Bolter and Zihlman 2002, 2003), comparing captive and free-ranging populations (e.g., Phillips-Conroy and Jolly 1988; Kimura and Hamada, 1990; Zihlman et al., 2004a, 2007; Garcia et al. 2009), and long-term demographic field studies (e.g., Cheverud et al. 1992, Boesch and Boesch-Achermann 2000, Nishida et al. 2003, Dittus 2004).

Individual case studies that integrate multiple dimensions of growth and development begin to answer some of our questions about the timing of somatic growth events relative to a species' behavioral development (e.g., Zihlman et al. 1990; Fragaszy et al. 2004a,b). Furthermore these connections have implications for interpreting life stages in fossil species (e.g., Kelley and Smith 2003, Dean and Leakey 2004). A nonhuman primate growth and developmental framework is essential for interpreting nonadult life stages in immature fossils such as *Australopithecus africanus* (e.g., Taung), *Homo habilis* (e.g., OH7), *Homo erectus* (e.g., WT 15000 Nariokotome Boy, Mojokerto), *Homo neanderthalensis* (e.g., Devil's Tower). Understanding variation in growth patterns that are interconnected with an individual's survival and reproduction, especially foraging and traveling, enables a broader view of primate lives. This approach represents the beginning of a more holistic analysis of primate growth in behavioral as well as physical development. We are entering a new era in which Simpson's vision of primates, both extant and extinct, in the four dimensions of time and space is becoming a reality.

27

Social Beginnings

The Tapestry of Infant and Adult Interactions

Katherine C. MacKinnon

1. What is the most important social bond for an infant primate, and how may the quality of this bond influence the rest of its life?
2. What is allomothering, and how might it benefit all of the participants?
3. What are some of the variable ways in which adult males interact with infants?

INTRODUCTION

Among mammals, primates are characterized by having extended life history stages, particularly infancy. While there is much variation within the Primate order, from an infant stage of 2–3 months in some prosimians to 4 years in some great apes, all primates are born altricial and require an extended period of care. The relationships that form during this time are critical for a young primate's survival. Immature primates not only need help with transportation, feeding, and protection but also need social stimulation and physical contact (explicitly demonstrated by Harlow's cloth-mother experiments in the 1950s and 1960s: Harlow 1959, Harlow and Harlow 1962).

  For at least the first few months of life, the infant primate is the object of great interest and attention from others in the group. The earliest interactions, apart from those with its mother, involve those individuals with which the mother associates. In many species, all group members at some point approach, stare at, sniff, touch, and inspect the infant; but certain individuals are allowed longer and closer contact, depending on their relationship with the infant's mother and the species in question. Generally, young infants tend to reflect the social preferences of their mothers. Thus, in those species where females show marked preferences for matrilineal kin (e.g., Japanese macaques [*Macaca fuscata*], Kurland 1977), infants show such preferences as well (Berman 1982b, Thierry 1985, and see Berman 2004 for review; Chapters 14 and 38). This also holds true in species where females do not have strong leanings toward matrilineal kin (e.g., sooty mangabeys [*Cercocebus* spp.], Ehardt 1988, Gust and Gordon 1994).

  Once an infant is off of its mother, its social world changes. It can decide whom to approach and when to leave, in addition to being the recipient of invitations for play and other affiliative contacts (Dolhinow and Bishop 1970). Toward the end of infancy, individual foraging increases, play is common, strong bonds among peers develop, and preferential relationships with certain older animals are formed. Crucial to the way infants maneuver through this change from dependence to independence are their relationships with others in the group (their mothers, adult males, other females, and immatures). Older infants are able to seek out and form relationships with those individuals that do not reflect their mother's affiliative preferences. Infants are able to modify their relationships once they are more independent, and thus, complex variables are at work in determining the bonds younger animals may form with older members of the social group.

  In most primate taxa, animals live in social groups that are stable through time, individuals know one another well, and this familiarity is a major factor in facilitating and maintaining strong social bonds. Immature animals take an active role in learning about their physical and social environments and often are the determining force behind not only the nature of the social interactions themselves but also how others in the group perceive them (Dolhinow 1991, Suomi 2005).

  In this chapter, I review the interaction patterns of infants and their mothers, infants and nonmother females, and infants and adult males in order to highlight the extreme variability that exists within the Primate order. My aim is to give a broad survey of this variability; specific data sets, examples, and case studies are referenced throughout the text should the reader wish to explore these topics in more depth. Theoretical explanations are presented to better illuminate why primate infants have such a long period of dependence on their mothers and caregivers and why immatures and nonmother adults in the group interact socially. The majority of examples and discussion will come from infant and adult interactions across primate taxa in natural settings.

HISTORY OF INFANT STUDIES

No other dynamic interaction demonstrates the importance of social attachments to primate survival more than the complex bonds infants form with others in the group. Most notably, the mother–infant bond has long been studied in primates, in both captivity and field settings. Through early studies, such as Harlow's work on deprivation (e.g., Harlow 1959, Harlow and Harlow 1962; and see Blum 2002), we see that having the basics of food, water, and protection is simply not enough to ensure an infant's survival. Strong social bonds must be established and maintained in order for an infant to develop well physically, cognitively, and socially. Because of close attention, our knowledge of behavior in infant primates has grown considerably in the past 50 years.

From the time of Carpenter's descriptions of howler monkey infancy in 1934 and Ladygina-Kohts's descriptions of an infant chimpanzee first published in 1935 (reissued in 2002), the development of the infant primate has been detailed. While captive studies were the first sources of information, field studies on infants in the wild increased dramatically during the 1960s, 1970s, and 1980s. Table 27.1 includes a partial list of some of the notable scientists who have shed light on how young primates mature and interact with others in natural settings. Due to the legacy of such fieldworkers (and others not mentioned here), we have a richer understanding of infant sociality. The collective conclusion from these primatologists working on diverse taxa

Table 27.1 Some Earlier (Pre–1990) Field Studies that Include Infant Social Behavior and/or Development in Wild (Unless Otherwise Noted) Primate Populations

	SCIENTIFIC NAME	COMMON NAME	LOCATION OF STUDY	SOURCES
Apes	*Gorilla gorilla beringei*	Mountain gorilla	Virunga Volcanoes: Uganda, Rwanda, Democratic Republic of Congo	Schaller 1963; Fossey 1979; Watts 1985a,b
	Pan troglodytes schweinfurthii	Common chimpanzee	Gombe Stream Nature Reserve, Tanzania; Mahale, Tanzania	van Lawick-Goodall 1967; Goodall 1986; Nishida 1968, 1988
	Pongo pygmaeus	Orangutan	Tanjung Putang, Indonesia	Horr 1977; Galdikas 1981a,b
Old World monkeys	*Semnopithecus (formerly Presbytis) entellus*	Indian langur monkey	Several sites across north and central India	Jay 1963, 1965
	Cercopithecus aethiops	Vervet monkey	Zambia	Lancaster 1971
	Erythrocebus patas	Patas monkey	Uganda	Hall 1965a–c
	Papio cynocephalus	Yellow baboon	Amboseli Nature Reserve, Kenya	Altmann 1980
	Papio anubis	Olive baboon	Uganda	Ransom and Rowell 1972; Nash 1978
	Macaca mulatta	Rhesus macaque	Cayo Santiago, Puerto Rico (provisioned free-ranging groups)	Sade 1965, 1980; Berman, 1982a,b, 1984
	Macaca fuscata	Japanese macaque	Japan and United States (wild and provisioned free-ranging groups)	Itani 1959; Fedigan, 1976; Hiraiwa, 1981; Tanaka, 1989
New World monkeys	*Alouatta palliata*	Mantled howler monkey	Barro Colorado Island, Panama; Chiriqui, Panama; La Pacifica, Costa Rica	Carpenter 1934; Baldwin and Baldwin 1973; Clarke 1990
	Alouatta seniculus	Red howler monkey	Trinidad and Venezuela	Neville, 1972a,b
	Brachyteles arachnoides	Woolly spider monkey	Fazenda Montes Claros, Brazil	Strier 1986a,b
	Ateles geoffroyi	Spider monkey	Barro Colorado Island, Panama	Carpenter 1935
	Callicebus moloch	Titi monkey	Manu National Park, Peru	Wright 1984
	Aotus trivirgatus	Owl monkey	Barro Colorado Island, Panama; Manu National Park, Peru	Moynihan 1964; Wright 1984
	Saimiri sciureus	Squirrel monkey	Panama	Baldwin 1969; Baldwin and Baldwin 1972a,b
Prosimians	*Lemur catta*	Ring-tailed lemur	Berenty Reserve, Madagascar; Beza-Mahafaly Reserve, Madagascar	Jolly 1966; Sussman 1977; Gould 1990
	Eulemur fulvus	Brown lemur	Beza-Mahafaly Reserve, Madagascar	Sussman 1977
	Varecia variegata	Black and white ruffed lemur	Nosy Mangabe, Madagascar	Morland 1990

is that primate infant behavior and ontogeny are highly variable within and across species and that an extended period of infancy is a successful evolutionary strategy that allows for proper development and facilitates learning in a rich social context.

PRIMATE INFANT SOCIAL DEVELOPMENT

Primates are mammals and, as such, possess the traits associated with this taxonomic class. Among these is an enlarged brain given body size, especially in the cerebral cortex (Harvey and Clutton-Brock 1985). Another is having offspring born in a dependent state. Unlike the offspring of reptiles, amphibians, fish, and some birds, the young of virtually all mammals are born needing an extended period of constant care, sometimes lasting months or even years. This latter trait is especially striking in primates (see Chapter 26).

Long before an infant primate is independent of its mother, its social development has begun. Particularly, individuals the mother feels comfortable around are allowed to approach, sniff, and touch the young infant. The processes of infant socialization have been well studied in a wide variety of primates (e.g., Hall 1968, Hinde 1974b, Hinde et al. 1964, Poirier 1972, Ransom and Rowell 1972, Fragaszy and Mitchell 1974, Chevalier-Skolnikoff 1977, Sussman 1977, McKenna 1979b, Mason 1986, Clarke 1990, Suomi 2002, Parker 2004). An obvious component of mammalian development is play, and an infant's first play partners (after its mother) are usually other immatures in the group. Such early exposure to group members facilitates and enhances social learning. Dolhinow and Murphy (1982) describe the 3-month-old infant langur (*Semnopithecus* [formerly *Presbytis*] *entellus*) as an "active strategist" as it typically has numerous choices for social partners and behavioral actions and ventures away from its mother to explore its social world whenever possible. Many aspects of behavior that are crucial for survival are also learned in a social context: what to eat and where to find it, the group's home range, and what is safe and dangerous in the environment.

Primates display protracted life history stages relative to size compared to other mammals (Leigh and Blomquist, see also Chapter 25). These include extended periods of infant dependence with increased parental care, a delayed maturation to sexual maturity, greater gestation lengths, and longer life spans. Motor and cognitive abilities are not refined until long after birth, and social development continues into adolescence in many species (see also Chapter 28). The call for long-term studies in the wild still needs to be sounded. While we have excellent data on several species (e.g., *Pan troglodytes*, *Macaca* spp., *Papio* spp., *Lemur catta*), we still sorely need longitudinal data on the social development parameters for most primates in natural settings.

INTERACTIONS BETWEEN INFANTS AND MOTHERS

Perhaps no other area of primate behavior has been as closely examined as the interactions between infants and their mothers. With such a long period of dependence and development, the relationship between infants and mothers was an obvious subject for early fieldwork, especially projects that concentrated on easily seen, well-known individuals living in large (often terrestrial or semiterrestrial) groups (e.g., yellow baboons [*Papio cynocephalus*], Hanuman langurs [*Semnopithecus entellus*], and common chimpanzees [*Pan troglodytes*]). The mother/infant pair is a highly attractive duo in the troop (e.g., Richard 1976, Nakamichi and Koyama 2000) and one that is readily identifiable by group members (and researchers). For example, we see a typical scenario in wild moor macaques (*Macaca maurus*) and white-faced capuchin monkeys (*Cebus capucinus*): Females with infants are approached more often than those without infants, and mothers receive more grooming (and do less grooming themselves) when they are holding their infants (Matsumura 1997, Manson et al. 1999).

As Dolhinow (1984:67) notes, "The mother–infant relationship is an elegant and complex system of biologically motivated as well as socially learned interactions." A deep interconnectedness lasts throughout the period of infancy and well into the juvenile and subadult stages in some species. Indeed, in some primates such as common chimpanzees (*Pan troglodytes*), bonobos (*P. paniscus*), and macaques (e.g., *Macaca fuscata*), the mother–offspring connection is so strong that it is customary for adult individuals and their mothers to have dynamic, complex, and tightly bonded relationships across their entire lives. In bonobos, for example, mothers and adult sons support each other in conflicts and are intensely close (de Waal 1997a). On the other hand, Tolentino et al. (2008) found nonpreferential associations between muriqui (*Brachyteles hypoxanthus*) mothers and their older male offspring, which suggests that an extended maternal investment might not be of great value in this species' egalitarian society (compared with its value in more hierarchical primate societies).

Mother primates are able to provide situations in which socially mediated learning can occur (see Figure 27.1). For example, young chimpanzees first learn how to use sticks for anting, wands for termite fishing, and appropriate stones for nut cracking from their mothers; and it can take years of tutoring until the youngsters master the appropriate sequence of behaviors (Goodall 1986, Biro et al. 2003, Hirata and Celli 2003). Japanese macaque mothers have been found to exaggerate tool-using behavior when observed by their offspring (Masataka et al. 2009). Thus, one of the advantages of group living is that an individual animal, particularly an immature, does not have to "rediscover" certain behaviors for itself: It can learn from others in the group. Social learning, as Box (1984, 1994) and many others have noted, also allows for innovation by young and mature animals alike. That is, newly developed patterns of behavior might, over

Figure 27.1 Infant white-faced capuchins (*Cebus capucinus*), like all young primates, first learn about the environment in a social context with their mothers (photo by Katherine C. MacKinnon).

time, become established parts of a group's repertoire (see Perry et al. 2003 for cultural traditions in wild capuchins [*Cebus capucinus*] and Whiten et al. 2001 for cultural traditions in wild chimpanzees [*Pan troglodytes*]).

Great ape immatures show the longest period of attachment: Infants are dependent on their mothers for nourishment, travel, and support in social interactions for the first several years of life (e.g., Nowell and Fletcher 2007) (Figure 27.2). Old World monkey infants are usually dependent for the first year of life, with a few exceptions (e.g., the fast-maturing patas monkeys [*Erythrocebus patas*]). In many New World monkey species, infants rely on their mothers for locomotion, food, warmth, and social contact for generally the first 6 months and up to 1 year (e.g., capuchins [*Cebus* spp.], bearded sakis [*Chiropotes satanas*]); however, the small-bodied callitrichines mature faster, while some of the larger atelines mature more slowly. Prosimians have the shortest period of infant dependence (generally 1–4 months), but this is still an extended life history stage when compared to similar-sized nonprimate mammals (see Figure 27.3).

Mothering Styles and Infant Temperaments

Mothering styles are critical not only to whether or not an infant survives but also to the quality of its early life (Dolhinow 1991). As Dolhinow and DeMay (1982:392) note, "For the infant the mother is the filter to the world surrounding both of them, and her aptitudes or ineptitudes set a measure for the likelihood of infant success and even survival." Adult females can be permissive, restrictive, attentive, and detached in their mothering styles; and these need not be

consistent modes throughout an individual female's life (Dolhinow and Krusko 1984).

Whether or not an infant is a female's first offspring can have effects on the nature of the interactions (see Table 27.2). For example, in langurs (*Semnopithecus*, *Presbytis*) and baboons (*Papio*), younger and more inexperienced mothers may be more restrictive, while older females may be more relaxed (Altmann 1980, Dolhinow and Krusko 1984, Dolhinow 1991).

In blue monkeys (*Cercopithecus mitis*), infants spend more time in contact with primiparous versus multiparous mothers and multiparous females tend to reject infants at a later age than first-time mothers (Förster and Cords 2002). In observations on the Mysore slender loris (*Loris tardigradus*), primiparous females "parked" their infants 2 weeks earlier than multiparous mothers (at 2 versus 4 weeks of age). The difference suggests lack of experience for first-time mothers (Kappeler 1998a) as 2 weeks is quite young compared to the majority of observations. Maestripieri (1998) found that previous maternal experience has an important influence on mothering styles in pig-tailed macaques (*Macaca nemestrina*). He also investigated behavioral similarities in cross-fostered rhesus macaque females and their biological mothers and found similarities in rates of social contact and aggression between fostered infants and their biological mothers but no clear similarities between infants and their foster mothers. He suggests that there is heritable variation in female social behaviors that may be maintained by natural selection. This variability then leads to a range of adaptations to different sociological niches within macaque populations (Maestripieri 2003).

Figure 27.2 Among the apes, orangutans are more solitary as adults; but for at least the first 5 years of life, infants are in the constant presence of their mothers, siblings, and occasionally other females with offspring. Here, an orangutan (*Pongo pygmaeus pygmaeus*) female relaxes with her dependent infant and older juvenile in Gunung Palung National Park, Indonesia (photo by Tim Laman).

It is important to remember that the infants themselves are not passive "blank slates" but individuals that take an active role in their own social development. For example, in western lowland gorillas (*Gorilla gorilla gorilla*), infants encourage their mothers to share food, play, or follow them, not vice versa. Other than encouraging motor skills like walking (see Maestripieri 1995b and 1996 for discussion of this in captive rhesus and pig-tailed macaques), gorilla mothers do not show strong encouragement of their infants in other contexts. This suggests that gorilla infants are more active than their mothers in generating situations that increase knowledge acquisition (Maestripieri et al. 2002, and see Watts 1985b). Similarly, infant chimpanzees have demonstrated that they are often more active than their mothers in initiating food-sharing bouts, especially those involving higher-quality foods (Ueno and Matsuzawa 2004); however, orangutan infants do not solicit high-quality foods more than low-quality foods from their mothers (Jaeggi et al. 2008).

Through recent hormonal studies in captivity, we are now gaining a clearer understanding of the role certain physiological stressors play in infant–mother interactions and temperament formation in young primates. A study of pig-tailed macaques (*M. nemestrina*) demonstrates that maternal responsiveness develops during pregnancy and is probably hormonally regulated, instead of being solely determined by social and experiential factors (Maestripieri and Wallen 1995). In captive rhesus macaques (*M. mulatta*), Cleveland et al. (2003) found that low serotonin activity and high stress (measured by elevated plasma cortisol) are correlated with patterns of high maternal restrictedness in young adult females. Blood cortisol levels are a reliable indicator

Figure 27.3 Although prosimians generally have shorter life history stages than anthropoid primates, infants are still dependent on their mothers for the first few months of life. At 2 months of age, this infant ring-tailed lemur (*Lemur catta*) relies on its mother for nourishment, transportation, social interaction, and learning which foods to eat and which to avoid (photo by Renee Bauer).

Table 27.2 Four Variables and Their Possible Effects on Infant–Mother Relationships

VARIABLE	SCIENTIFIC NAME	COMMON NAME	COMMENTS	SOURCES
Primiparity	*Macaca mulatta*	Rhesus macaque	Primiparous mothers less restrictive than multiparous mothers	Berman 1984
	Papio anubis	Olive baboon	Primiparous mothers rejected infants earlier than multiparous mothers	Ransom and Rowell 1972
Multiparity	*Gorilla gorilla beringei*	Mountain gorilla	Multiparous mothers less restrictive than primiparous mothers	Fletcher 2001; Nowell and Fletcher 2007
	Cercopithecus mitis	Blue monkey	Multiparous females tend to reject infants at a later age than first-time mothers	Förster and Cords 2002
Dominance rank	*Papio cynocephalus*	Yellow baboon	Low-ranking females are much more restrictive of young infants than high-ranking females	Altmann 1980
Sex of infant	*Gorilla gorilla beringei*	Mountain gorilla	Females invest more in their male infants	Fletcher 2001

of psychosocial stress in primates, and infants do show a cortisol increase after being separated from their mothers (Nakamichi et al. 1994). However, the *social* context into which an infant is placed after maternal separation has an effect on how well it deals with that stress (Nakamichi et al. 1994, Bardi et al. 2003).

Detailed explorations of controlled variables are extremely difficult in wild populations of primates; captive work on physiological processes is thus necessary to further our understanding of the bidirectional pathways between hormones and behavior (e.g., see Abbott et al. 2003; Sapolsky 1993, 1999, 2000). However, we must be conscious of discerning the effects of captivity and aware that these results may not always be reliable indicators of similar physiological processes in wild individuals.

There is much variability in care patterns among female primates and their offspring, and we know that general flexible strategies of maternal care are one of the hallmarks of primate evolution. Not only do mothering styles vary among individual females in a group, but each female may show different mothering techniques over the course of her lifetime. Thus, females have the ability to change their mothering styles throughout their lives based on temperament, individual experiences, and the biosocial dynamics of the group in which they live (Dolhinow and Krusko 1984, Berman et al. 1997).

Infant Abuse and Infanticide by Mothers

The warm and tender image of a typical primate mother cuddling her vulnerable infant is in stark contrast to the maternal tendencies of some primate females. While relatively uncommon, some mothers ignore or abuse their offspring for reasons that are often unclear. The role of a hormonal influence has been examined by Maestripieri and Megna (2000a,b), who found the estradiol and progesterone profiles of abusive and nonabusive rhesus macaque

(*M. mulatta*) mothers to be similar, with minor differences in other behavioral patterns between the two groups. Maestripieri (1999a) also found that while abusive macaque mothers are highly attracted to infants in general, abuse is a phenomenon specific to their own offspring. He concluded that infant abuse is not an accidental by-product of infant handling or the result of maternal inexperience. Rather, it is likely related to reproductive competition among lactating females (Maestripieri 1999a). That is, reproductively experienced females may harass or kidnap other females' infants as a form of competition (Silk 1980; Maestripieri 1994b, 1999a). However, primate females may abuse their own infants due to inexperience or ecological factors or perhaps because something is wrong with the infant.

Infanticide by females toward their own offspring is uncommon in primates, but it occasionally occurs (see Casar et al. 2008). The phenomenon is better known in some rodents, lagomorphs, mongoose, and lions (e.g., mongoose, Clutton-Brock et al. 1998, 1999; rodents, Blumstein 2000); but when this behavior occurs in any taxa, it is usually in response to an environmental stressor such as drought, shortage of food, or disruption of the habitat (e.g., Andrews 1998 for *Eulemur macaco*) or to offspring being directly disturbed. For example, if captive loris (*Loris tardigradus nordicus*) mothers are bothered during the first few days of infants being born, they will sometimes kill their offspring (Schulze and Meier 1995b, Digby 2000). The majority of cases of female infanticide are by nonmother females and often for indeterminate reasons (see next section "Interactions Between Infants and Nonmother Females").

INTERACTIONS BETWEEN INFANTS AND NONMOTHER FEMALES

As most primates grow up in a dynamic social setting with multiple members, infants have the opportunity to interact with a variety of others from a young age. Subadult females in particular are usually intrigued by infants and often sit close to, stare at, touch, and attempt to hold (with varied results) the latest additions to their social worlds.

Allomothering

Allomothering, or taking care of an infant that is not one's own, is common in several primate species; but the phenomenon varies widely (McKenna 1979a, 1987) (see Table 27.3). Across taxa we find that nulliparous females are generally more attracted to infants and do more of the allomothering than multiparous females (Nicholson 1987; but see Manson 1999). Given that many species exhibit female philopatry, where females remain in their natal group for many years and often life, allomothering can function to provide younger females with experience taking care of a dependent infant. In species that are not female-bonded (e.g., spider monkeys [genus *Ateles*] and common chimpanzees [*Pan troglodytes*]), younger females can also benefit from allomothering before or after they transfer groups. In either scenario, allomothering provides the dependent infant with socialization opportunities other than those with its mother (see Figure 27.4).

In some colobine monkeys, such as langurs (e.g., *Semnopithecus entellus*), infants are passed around from the first day of life, with no objections from the mother (Dolhinow and Murphy 1982, Dolhinow and Krusko 1984). Females groom, inspect genitals, nuzzle, touch, and huddle with the infants (Jay 1965, Dolhinow and Murphy 1982). Across cercopithecine species, we generally see two patterns: Patas (*Erythrocebus patas*) and guenon (*Cercopithecus* spp.) mothers permit extensive contact with young infants, while most baboon (*Papio* spp., *Theropithecus* spp.) and some macaque (*Macaca* spp.) mothers restrict early contact (Chism 2000). The genus *Macaca* in particular offers a striking view of allomothering diversity: Japanese (*M. fuscata*), long-tailed (*M. fascicularis*), rhesus (*M. mulatta*), and pig-tailed (*M. nemestrina*) macaque mothers restrict the amount of allocare provided by other females, while Barbary (*M. sylvanus*), lion-tailed (*M. silenus*), Tonkean (*M. tonkeana*), bonnet (*M. radiata*), crested (*M. nigra*), and moor (*M. maura*) macaque mothers are permissive and allocare is common, with many females handling and carrying infants from a young age (Thierry 2000, 2004; see also Chapter 14). Generally, across cercopithecines, early allomothering is most widespread in species with distinct seasonal breeding and relaxed social relationships among adult females (Chism 2000).

In New World primates, allomothering is taken to the extreme in the callitrichines (e.g., *Saguinus* spp., *Callithrix* spp.), titi monkeys (*Callicebus* spp.), and owl monkeys

Table 27.3 Characteristics of Allomaternal Care by Females in Several Primate Species

VARIABLE	SPECIES	SOURCES
Kinship an important factor	*Macaca fuscata*	Schino et al. 2003, Hiraiwa 1981
	Alouatta seniculus	Crockett and Pope 1993
	Papio cynocephalus	Cheney 1978
	Cercopithecus aethiops	Lee 1983
Kinship not an important factor	*Semnopithecus entellus*	Hrdy 1977a,b, Dolhinow and Murphy 1982
Young infants preferred	*Alouatta palliata*	Clarke 1990
	Cercopithecus aethiops	Lee 1983
Infants of high-ranking mothers receive more allocare (abuse rare)	*Cebus capucinus*	Manson 1999
	Papio cynocephalus	Cheney 1978
	Cercopithecus aethiops	Lee 1983
Infants of low-ranking mothers receive more allocare (often with abuse)	*Macaca radiata*	Silk 1980
Dominance rank is not an important factor in terms of a female's success with infant handling	*Papio cynocephalus*	Bentley-Condit et al. 2001
	Cebus capucinus	Manson 1999
Adult (multiparous) females do most allocare	*Lemur catta*	Jolly 1966; Gould 1992
Subadult or juvenile (nulliparous) females do most allocare	*Tarsius spectrum*	Gursky 2000a,b
	Macaca fuscata	Schino et al. 2003; Hiraiwa 1981
	Cercopithecus aethiops	Lancaster 1971
No difference in allocare rates of nulliparous vs. multiparous females	*Cebus capucinus*	Manson 1999
	Erythrocebus patas	Nakagawa 1995

Figure 27.4 Young adolescent or subadult nulliparous females are the most common participants in allomothering behaviors across primate species. This adolescent female white-faced capuchin (*Cebus capucinus*) is looking after a younger female sibling (photo by Katherine C. MacKinnon).

(*Aotus* spp.), with nonbreeding females spending up to 60% of their time caring for other females' offspring (e.g., Chapters 7 and 10). By comparison, we see lower rates of allomothering by female howler monkeys (*Alouatta* spp.) and squirrel monkeys (*Saimiri* spp.), yet in capuchins (e.g., *Cebus capucinus*), infants are often carried and sometimes nursed by nonmother females in the group (Manson 1999, Perry 1996, MacKinnon 2002).

In prosimians, allomothering occurs in the diurnal (and thus better-studied) species (e.g., ring-tailed lemurs [*Lemur catta*], Jolly 1966, Gould 1992, Nakamichi and Koyama 2000; sifakas [*Propithecus verreauxi*], Richard 1976). In ring-tailed lemurs, other mothers alloparent each other's infants more than do subadult nulliparous females (Gould 1992 and personal communication), which is contrary to the typical scenario found in most primate species. Allomothering is also an aspect of nocturnal prosimian behavioral patterns. In slender lorises (*Loris tardigradus* and *L. lydekkerianus*), parked infants (see Figure 27.5) are approached and played with by others in the group (Nekaris 2003a), while in the spectral tarsier (*Tarsius spectrum*), female subadults share food, groom, play, and carry infants more than other age/sex classes (Gursky 2000a; see also Chapter 6).

As with mothering styles, the variability we see in individual females that allomother can reflect experience, temperament, dominance status, the relationship with the infant's mother, and age (McKenna 1987) (see Table 27.3). The proclivity for some females in a group to allomother, while others are uninterested, is sometimes difficult to explain. Relatedness and inclusive fitness may play a role in some species. For example, an older daughter may carry her mother's infant more than other females will (e.g., red

Figure 27.5 In nocturnal prosimians, infants are often "parked" in trees while adults forage. This young slender loris (*Loris lydekkerianus nordicus*) is still nursing, does not yet move efficiently on its own, and will be parked for hours until its mother returns (photo by K. A. I. Nekaris).

howlers [*Alouatta seniculus*], Crockett and Pope 1993), while a mother's female siblings may show more allomothering behavior than unrelated individuals (e.g., Japanese macaques [*M. fuscata*], Hiraiwa 1981; yellow baboons [*Papio cynocephalus*], Cheney 1978; vervets [*Cercopithecus aethiops*], Lee 1983). In other cases, unrelated females are given access to a female's infant with the same frequency as related females (e.g., Hanuman langurs [*Semnopithecus entellus*], Hrdy 1977,b, Dolhinow and Krusko 1984).

Infants of dominant females sometimes receive more allomothering attention than infants of lower-ranking females (Cheney 1978, Lee 1983), while the opposite is occasionally true (Silk 1980). In yellow baboons (*Papio cynocephalus*), Bentley-Condit et al. (2001) found that the dominance rank of females did not seem to offer any privileges in terms of access to other infants; however, Silk et al. (2003b) found that while female baboons appeared to be equally attracted to all infants, they had greater access to offspring of their relatives and subordinate females. In long-tailed macaques (*M. fascicularis*), Gumert (2007) found that grooming interactions between females were longer when infants were scarce in the social environment and suggested that grooming may be a sort of payment for infant handling. In langur monkeys (*S. entellus*), allomothering females show a wide variety of behavioral differences in their interactions with infants. Some prefer older or younger infants, and these preferences can change throughout an individual female's life in this species (Dolhinow and Krusko 1984). Infants also have their preferences for certain females as caregivers, and degree of relatedness is unreliable when trying to predict the level of interest in infants by females and vice versa (Dolhinow and Krusko 1984). In Japanese macaques (*M. fuscata*), due to a social structure with distinct matrilines, infants are handled mainly by related individuals, usually juveniles and nulliparous subadult females (Schino et al. 2003). Maternal responses when infants are being handled are typically quite similar toward kin and nonkin alike but show a great deal of interindividual variability based on age, rank, experience, and relations with others in the group.

If a dependent infant is rejected early on by its mother or loses her to predation, illness, or other factors, it may be adopted by another individual in the social group. In many species, adoptions are initiated and maintained by the infants themselves, but which female they choose as their new caregiver is not easily predicted or understood. For example, Dolhinow and DeMay (1982) found no preference for related females by infant langur monkeys, and neither was past social interaction an indicator of which female an infant would actively seek out if it lost its mother. In that study, infants chose adult females that were the least approached beforehand; a similar scenario is reported by Clarke et al. (1998) for mantled howler monkeys (*Alouatta palliata*).

What can account for the amazing variability we see in allocare patterns? One of the defining primate characteristics is an evolved flexibility in behavioral pathways. Such plasticity allows variable responses to a wide range of potential stimuli given unpredictable variation in group structure and composition, ecological factors, development patterns, and experiences in individual lives (McKenna 1979a,b, 1987). Yet, the question remains: Why invest time and energy in someone else's offspring (Trivers 1972)? Several reasons have been proposed to explain why allomothering behavior evolved: (*1*) benefits to the mother, in terms of reproductive competition among females; (*2*) benefits to the infant,

in terms of increased survival rates; and (*3*) benefits to the allomother, in terms of enhancing acquisition of maternal skills (Lancaster 1971, Quiatt 1979, Nicholson 1987, Mitani and Watts 1997). For example, Ross and MacLarnon (2000) found that among anthropoid primates, species with high frequencies of allocare grow rapidly and wean their infants at a younger age than species of the same body size with lower allocare levels. Thus, mothers can increase their reproductive output if they allow allocare of their infants (Mitani and Watts 1997).

However, there is still no consensus about the evolutionary significance of allomothering in primates. Silk (1999) reviewed the function of infant handling in bonnet macaques (*M. radiata*) and suggested that attraction to dependent infants may have evolved in response to the time period during which maternal care is most critical for infant survival. Females that are highly responsive to infants (usually) make good mothers, so the basic components of "strong attraction" may be favored by natural selection. Her data on this species of macaque did not support the "learning to mother" hypothesis (i.e., age was not a determining factor of female interest), nor did they support the "reproductive competition" hypothesis (i.e., infant handling did not correspond to patterns of harassment). Rather, her data fit the hypothesis that allomothering is a by-product of selection for the suite of behavioral characteristics that contribute to normal patterns of maternal care (Silk 1999).

In an examination of infant handling by nonmothers in three semi-free-ranging groups of Barbary macaques (*M. sylvanus*), Paul and Kuester (1996b) also found poor evidence for a functional basis for the behavior. First, experience with infant handling by nulliparous females did not affect the survival rate of their own firstborn offspring. Second, infant handling frequency did not improve infant survival or female reproductive success. Third, while "aunting to death" (Hrdy 1977b; and see following section "Infant Abuse and Infanticide by Nonmother Females") occurred, the hypothesis that infant handling serves to reduce the fitness of competitors was not supported (Paul and Kuester 1996b). They instead found more support for the hypothesis that infant handling evolved as a nonadaptive by-product of selection for mother–offspring bonding. They came to this conclusion because rates of infant handling were highest among females that experienced early infant loss; females that had their own infants handled other infants significantly less often than females without dependent offspring, and both "aunting to death" and a successful adoption occurred regardless of kinship (Paul and Kuester 1996b).

In white-faced capuchins (*Cebus capucinus*), Manson (1999) found that both nulliparous and parous females had similar rates of infant handling, which were unrelated to the dominance ranks of the mother and the handler. Rather, females tended to handle the infants of females with which they groomed and formed aggressive coalitions more often; he suggested that infant handling might serve to test bonds between such females and found no support for the "learning to mother" hypothesis. Nonetheless, there is evidence that young, nulliparous females do gain some experience through allomothering (e.g., Lancaster 1971; Clarke 1978; Caine and Mitchell 1980; Hiraiwa 1981; Nicholson 1987; Paul and Kuester 1996b). We also know from captive and wild populations that young females that do not have adequate contact with infants often display poor mothering abilities toward their own offspring (van Lawick-Goodall 1971, Altmann 1980, Suomi 1982). Generally, the fact that allomothering is performed far more commonly by younger females suggests that this behavior has some sort of developmental component (Lancaster 1971, Nicholson 1987).

A Note on Neonatal Coat Color

In most primates, infants are born with a coat color that is slightly different from the adult shade. Typically, infants are darker or more monochrome in pelage; but in the Old World subfamily Colobinae, we see some striking contrasts (i.e., "conspicuous natal coats," or CNCs). For example, in black and white colobus monkeys (*Colobus guereza*), infants are solid white, while in Francois' leaf monkeys (*Trachypithecus francoisi*), they are a bright pumpkin orange. Coat color typically changes to the adult pattern after a few months, but why is there such a conspicuous distinction at birth in these species? Hrdy (1976) suggests that CNCs evolved to facilitate handling of infants by attracting other females. Similarly, Gouzoules and Gouzoules (2002) suggest that CNCs might be adaptive in situations where predation pressure is just high enough to warrant increased protection by mothers and allomothers. Treves (1997) proposes a different scenario: paternity concealment, such that promiscuous mating systems with multiple unrelated males in a group should favor selection of CNCs. Bright coloring could thereby act as a "paternity cloak" and serve to warn nongroup males from approaching (Treves 1997). However, he notes that "we do not yet have the necessary direct evidence to test whether infants are obscuring paternity or deterring the approach of infanticidal males" (Treves 2000a:232–233, and see Brent et al. 2008). A lack of agreement on an adaptive reason for sharply distinct neonatal coat color remains (see also Ross and Regan 2000, Gouzoules and Gouzoules 2002).

Infant Abuse and Infanticide by Nonmother Females

Although abuse of infants by adult females is rare across primate taxa, it occurs more frequently than abuse by adult males (Digby 2000). This phenomenon takes a variety of forms: It can range from mild aggression to "aunting to death" to directed infanticide. Additionally, the behavioral hallmark of infants—insatiable curiosity—can get them into trouble. Even before infants are independent and off their mothers for extended periods of time, they can, and do, get underfoot of bigger, older animals in the group. Attention-grabbing activities such as juveniles playing, adults mating, and agonistic encounters are all worthy of investigation by infants. In many studies, infants have been observed to put themselves in harm's way and, in doing so, are at increased risk of being stepped on, swatted at, or bitten. Occasionally, they are the unfortunate recipients of misplaced or misdirected aggression when two or more individuals are fighting (Dolhinow 1999).

In some species, allomothering can take a turn for the worse, when dominant females take the infants of subordinates and keep them away from their lactating mothers for so long that they die from dehydration or starvation (Quiatt 1979, Kohda 1985, Thierry and Anderson 1986, Maestripieri 1994b, Paul and Kuester 1996b, Digby 2000). In rhesus macaques (*M. mulatta*), females are very attracted to young infants and often approach, hold, and groom immatures that are not theirs. In these scenarios, unskilled handling, abusive behavior, and long periods away from the mother that keep dependent youngsters from nursing are potentially significant risks for infants (Collins et al. 1984; Maestripieri and Carroll 1998, 2000; Digby 2000; Jovanovic and Gouzoules 2001). In one year of an intensive long-term field study of patas monkeys (*Erythrocebus patas*), three infants (of eight born) died from inept handling by females that "kidnapped" them from their mothers (Enstam et al. 2002).

Infanticide by nonmother female mammals is likely a more widespread phenomenon than male infanticide and has been documented in over 50 species, in five mammalian orders, with over 300 recorded cases (Digby 2000). In primates, one of the most bizarre cases of female infanticide and cannibalism was documented by Jane Goodall. A mother chimpanzee and her daughter killed and ate at least three (and up to seven inferred) infants of other females in the same community over a 2-year period (Goodall 1977, 1986). That it never happened again after the mother died and that something of this odd nature with a mother/daughter duo has never been witnessed at any of the other long-term chimpanzee study sites suggest a truly aberrant case. A different scenario occurs in the Callitrichinae, where female infanticide has also been reported (e.g., *Callithrix jacchus*, Digby 1995a, Bezerra et al. 2007). In these cases, it appears that females that are pregnant or lactating perform most of the infanticides and are typically closely related to the targets (see Chapter 7). The (as yet untested) hypothesis of intense reproductive competition is offered as a possible explanation.

INTERACTIONS BETWEEN INFANTS AND ADULT MALES

Across primate species, infants and adult males interact socially, and the majority of interactions are affiliative in nature. Most primate males are not involved in direct caretaking or protective behavior toward infants, but they are usually quite tolerant of the immature members of their social groups. In many species, certain adult males allow certain infants greater access and participation in social interactions. There are likely proximate mechanisms such as temperament factors and relationship histories that determine the attractiveness of a certain adult male to a younger animal and vice versa (Clarke and Boinski 1995; Byrne and Suomi 1995, 1998). Additional theoretical explanations for such interactions will be discussed.

The Range of Male Involvement

While primate infants are in continual contact with their mothers from the moment they are born, their relations with adult males are, with a few exceptions, much less intensive and consistent. However, infants and adult males do interact,

and if we look across the ~250 species in the Primate order (Groves 2001a), we see a great range in social patterns (see Whitten 1987). On the neutral and affiliative side of the spectrum, we see ignorance, tolerance, occasional interest, active interest, and direct caretaking of infants. On the agonistic side, we see mild and harsh aggression, abuse, and infanticide. By far the most common patterns are for adult males to tolerate or take occasional to active interest in young individuals (see Table 27.4). Unfortunately, we are still lacking such behavioral data for many wild species.

Among the apes, mountain gorilla males are extremely tolerant and patient with infants in their groups (Fossey 1983), and silverbacks will even nest with youngsters that have lost their mothers (Stewart 2001). Chimpanzee males will occasionally play with immatures and will also adopt

Table 27.4 Range of Infant and Adult Male Interactions in Several Primate Species (Data Primarily from Wild Studies)

	SCIENTIFIC NAME	COMMON NAME	TYPE[1]	COMMENTS	SOURCES
Apes	*Gorilla gorilla beringei*	Mountain gorilla	MA, T	Infants gain closer proximity to silverback as weaning occurs; silverback increases affiliative behavior toward orphaned young	Stewart 2001
	Pan troglodytes schweinfurthii	Common chimpanzee	MA, T	Males are generally tolerant of young	Goodall 1977, 1986
	Pongo pygmaeus	Orangutan	A/T	Males rarely interact with infants	Galdikas 1981b
	Hylobates lar	White-handed gibbon	MA, T	Males generally interact very little with young infants but are tolerant of them; males will play with older infants but do not carry them	Brockelman et al. (1998;; Reichard 2003a,b; Bartlett 2003
	Hylobates syndactylus	Siamang	MA	Males will sometimes carry young	Chivers 1974; Dielentheis et al. 1991
Old World monkeys	*Semnopithecus (formerly Presbytis) entellus*	Indian langur monkey	MA, T, U/A	Wide range of behavioral interactions reported across several sites (one-male vs. multimale groups)	Boggess 1982; Jay 1965; Hrdy 1977a,b; Borries et al. 1999a,b
	Presbytis potenziani	Mentawai leaf monkey	T, A/I	Males generally do not interact socially with infants	Fuentes 1995
	Erythrocebus patas	Patas monkey	MA, T	No protective behavior of infants but some play	Hall 1965a–c; Enstam et al. 2002
	Papio cynocephalus	Yellow baboon	SA, MA, T, A/I, UA	Much variation occurs across groups and sites, with males showing broad range of affiliative behaviors with infants; infants also used by males for agonistic buffering or in more mutualistic interactions	Altmann 1980; Stein 1984; Strum 1984; Stein and Stacey 1981
	Papio anubis	Olive baboon	SA, MA, T, A/I, UA	Variation occurs between one-male and multimale groups; males often show broad range of affiliative behaviors with infants; infants also used by males for agonistic buffering or in more mutualistic interactions	Strum 1984; Smuts 1985
	Theropithecus gelada	Gelada baboon	MA, U/A	Infants used by males for agonistic buffering or in more mutualistic interactions	Dunbar 1984a,b
	Macaca fuscata	Japanese macaque	SA, MA, T	Males may form special protective relations with particular infants	Itani 1959
	Macaca sylvanus	Barbary macaque	SA, U/A	Males often groom, handle, and carry infants; infants also used by males in mutualistic male-male interactions (typically without "abuse")	Taub 1980a,b, 1984
New World monkeys	*Alouatta palliata*	Mantled howler monkey	T, A/I	Some males show a passing interest in infants but are generally indifferent	Clarke et al. 1998

Table 27.4 (cont'd)

	SCIENTIFIC NAME	COMMON NAME	TYPE[1]	COMMENTS	SOURCES
	Cebus apella	Brown capuchin	T	Males are tolerant of young infants left near them	Valenzuela 1994
	Cebus capucinus	White-faced capuchin	MA, T	Males occasionally carry and play with infants	MacKinnon 2002
	Callicebus moloch	Titi monkey	IC	Strong attachment between infants and male in group (putative father)	Wright 1984; Mendoza and Mason 1986a
	Aotus trivirgatus	Owl monkey	IC	Male carries infant most of the time; much play, grooming, and food sharing	Wright 1984, 1994a,b; Fernandez-Duque 2004, (Chapter 10 this volume)
	Saimiri sciureus	Squirrel monkey	A/I	Infants and adult males rarely interact	Fragaszy et al. 1991; Boinski and Cropp 1999
	Saguinus fuscicollis	Saddle-back tamarin	IC	Males provide extensive allocare	Goldizen 1987a,b
	Saguinus oedipus	Cotton-top tamarin	IC	Males provide extensive allocare	Savage et al., 1996a,b
	Pithecia pithecia	White-faced saki	MA	Interest in infants by adult males increases as the infant matures; males play and share food with 4-month-old infants	(Chapter 9 this volume)
Prosimians	*Lemur catta*	Ring-tailed lemur	SA, MA	Most males show interest in infants and occasionally participate in allocare; dominance rank or ageclass of male does not affect rates of affiliative behavior with immatures	Gould 1997
	Eulemur mongoz	Mongoose lemur	IC	Adult males frequently carry the infant after it is 2 weeks old	Curtis and Zaramody 1997
	Eulemur rubriventer	Red-bellied lemur	IC	Males have been seen carrying infants up to 100 days	Overdorff, 1993a,b; Mittermeier et al. 1994
	Loris lydekkerianus	Slender loris	MA	Males occasionally approach, groom, and play with parked infants	Nekaris 2003a–c
	Tarsius spectrum	Spectral tarsier	MA	Adult and subadult males occasionally groom and play with infants	Gursky 2000a,b
	Lepilemur ruficaudatus	Red-tailed sportive lemur	A/I	Males only rarely in proximity to infants and no evidence for direct infant care	Hilgartner et al. 2008

[1] IC, intense caretaking (adult males are involved in the majority of caretaking behaviors (excluding nursing), maintain close proximity to infants, and spend much time with them) SA, strong affiliation (males are involved with some caretaking behaviors and often spend time with infants and maintain proximity to them) MA, mild affiliation (males are involved with very few or no direct caretaking behaviors, yet on occasion socially interact with infants) T, tolerance (males do not actively engage infants in social interactions but do not avoid them if approached, and agonistic behaviors occur) A/I, avoidance/indifference (males do not respond to infants, do not engage them in social interactions, and might actively avoid them—no agonistic behaviors occur) U/A, use/abuse (males engage infants in mutualistic exchanges with other males, use them for agonistic buffering, or harm them [excluding infanticide].)

orphaned infants (Goodall 1977, 1983). Gibbon males (*Hylobates*) do not perform the intense caretaking duties that characterize some primate males living in pair-bonded associations, but they are quite tolerant of infants, including those potentially sired via extrapair copulations or in polyandrous social structures (e.g., white-handed gibbons [*Hylobates lar*], Brockelman et al. 1998, Reichard 2003b and personal communication). Reichard (personal communication) notes that he has never observed an adult male display agonistic behavior toward an immature less than 5 years of age. Lappan (2008) found that siamang males in polyandrous groups provide care for infants but males in socially monogamous groups provided substantially more care than males in polyandrous groups. Siamang males will

carry infants, while males of the other gibbon species generally do not (Chivers 1974). However, all species display male parental care by way of playing, grooming, and protection (Brockelman 2004).

Old World monkey (e.g., *Papio*, *Macaca*) males show varying levels of interest in infants and often play with them and provide transportation. Barbary macaque (*M. sylvanus*) males in particular show strong interest in infants (kin and nonkin alike), and males are tolerated well by adult females in such contexts. Depending on the relationship with the infant's mother, a subadult or adult male may be allowed to take and hold or carry an infant for extended periods of time (Taub 1984) (see Figure 27.6). Male care of offspring in these macaques is not strongly tied to paternity, and the

Figure 27.6 In many species, male primates play an active role in the development of infants. This subadult male Barbary macaque (*Macaca sylvanus*) carries a young infant in a species-typical display of male allocare (photo by Katherine C. MacKinnon).

most active male caregivers tend to be subadults that mate with adult females less often (Taub 1984) (see subsection "Agonistic Buffering and Triadic Interactions" for continued discussion of Old World monkeys in this context).

In New World monkeys, brown capuchin (*Cebus apella*) infants older than 2 months are often left with the dominant male, who is tolerant as they play or huddle next to him (Valenzuela 1994). White-faced capuchin (*C. capucinus*) males are also highly tolerant of infants (see Figure 27.7) and occasionally carry, play with, and sit in contact with immatures (MacKinnon 2002). Interactions between infant and adult male howler monkeys have been documented (e.g., mantled howlers [*Alouatta palliata*], Clarke et al. 1998), and generally, howler males show a passing interest in infants and sometimes "kidnap" them for short periods of time until the infant cries for its mother and is retrieved. In the earliest report of wild howler monkey behavior, Carpenter (1934:128) described infants watching, following, crawling on, and playing with adult males and noted "In general males behave indifferently to young animals, but they may assist in retrieving fallen infants, protect them from predatory animals, and in unusual situations care for an infant in somewhat the same manner as a mother." Squirrel monkey (*Saimiri* spp.) infants rarely, if ever, interact socially with adult males in the group (Fragaszy et al. 1991). Infants travel independently early on and might not have as much exposure to adult males via their mothers. Squirrel monkeys also spend more time foraging alone than closely related species (e.g., *Cebus* spp.), and social interaction rates overall are quite low.

In sharp contrast, owl monkey (*Aotus*) and titi monkey (*Callicebus*) males are actively involved in parental

Figure 27.7 Even in species where adult males are not actively involved in direct caregiving, they still play a role in the social and physical development of immatures. Here, an adult male white-faced capuchin (*Cebus capucinus*) carries an unrelated dependent infant (photo by Katherine C. MacKinnon).

care (Fragaszy et al. 1982; Mendoza and Mason 1986a,b; Dixson 1994; Wright 1984; see Chapters 9 and 10). They also play with, groom, and share food with infants (e.g., Rotundo et al. 2000). In owl monkeys, independent infants return to the adult male if frightened or disturbed (Dixson 1994), and infants are much more attached to their putative fathers than to their mothers (Fernandez-Duque 2004, see also Chapter 10). Similarly, in titi monkeys (*Callicebus*), infants seem to be much more "tuned into" their male caregivers and show increased cortisol levels (indicating stress) upon separation from fathers but not mothers (Mason and Mendoza 1998, Hoffman 1998; see Chapter 9). Why has such intense male care evolved in these primates? The question remains largely unanswered, although Fernandez-Duque (e.g., Chapter 10) suggests that male care in these species may function as mating effort, especially as other hypotheses (e.g., creating increased foraging opportunities for females) are largely unsupported (see Fernandez-Duque et al. 2008b). Concomitantly, while the exact mechanisms are unknown, paternal care might have influenced the evolution of social monogamy at different stages and to varying degrees (Wittenberger and Tilson 1980, Fuentes 2000, Reichard 2003).

In captive Coquerel's sifaka (*Propithecus verreauxi coquereli*), Bastian and Brockman (2007) found that adult males groomed and held their infants but individual males varied in the time spent in these activities, which was likely due to relationship status with the infants' mothers. Finally, it bears mentioning that in some nocturnal prosimians, long believed (incorrectly) to be completely solitary in the wild, males also interact socially with immatures. For example, Nekaris (2003a) documented males approaching, grooming, and playing with parked infants throughout the night in three subspecies of slender loris (*Loris lydekkerianus lydekkerianus*, *L. l. nordicus*, and *L. tardigradus tardigradus*) in Sri Lanka and India. It is unknown whether or not these males were related to the infants. The results from this study are an excellent example of how additional fieldwork on the lesser-known members of the Primate order will help to expand our understanding of the social complexities in all gregarious species.

Callitrichinae and Cooperative Infant Care

The small-bodied marmosets and tamarins in the subfamily Callitrichinae (see Chapter 7) are perhaps best known for the extensive care males give to infants. About 80%–90% of Callitrichinae births are twin offspring, and young may be born twice a year (there is no lactational amenorrhea in most species). Infants are relatively large, with twins collectively weighing 14%–25% of the mother's weight at birth and up to 50% by the time they are weaned (Goldizen 1987a). Only one adult female in a group breeds, and there are varying suppression mechanisms at work on the reproductive endocrinology of the other females (Abbott and Hearn 1978, Abbott et al. 1993; see also Chapter 7). Generally speaking, the mother nurses the infants but then passes them on to other members of the group. Thus, multiple individuals take care of infants, and adult males play an active role in the physical and social development of young via carrying, playing, and sharing food (Goldizen and Terborgh 1986, Goldizen 1987a, Ziegler et al. 2009; see also Chapter 7). This caregiving system also helps with infant survival (see

Bales et al. 2000 for review) and is an adaptive response to the physiological demands of twins. For example, in the moustached tamarin monkey (*Saguinus mystax*), Garber et al. (1984) found a statistically significant relationship between the number of adult males in a group and surviving offspring.

Interestingly, there are high levels of the hormone prolactin in "helping" males in many bird and rodent species as well as in several primates (e.g., *Callithrix jacchus* and *Saguinus oedipus*) (Ziegler 2000, Ziegler et al. 2009). Prolactin, a protein hormone secreted from the anterior pituitary gland, is usually associated with stimulating breast development, milk production, and maternal behavior in females but can also be produced by males. In a study of paternal care in three species of New World monkeys—the titi monkey (*Callicebus cupreus*), common marmoset (*Callithrix jacchus*), and Goeldi's monkey (*Callimico goeldii*)—species differences in the patterns of prolactin levels were apparent and reflected varying paternal roles (Schradin et al. 2003). Mota and Sousa (2000) note that extra prolactin is produced in response to physical contact, may be associated with carrying behavior, and may be related to learning parental skills in some New World monkeys.

When examining the evolution of paternal care in primates, the reproductive effort of both males and females has to be taken into account. Key and Aiello (2000) note that when male reproductive costs are less than female reproductive costs, males cooperate with females even when females do not reciprocate. They suggest that such nonreciprocal altruism is akin to male investment in a female and her offspring.

Agonistic Buffering and Triadic Interactions

Baboons and some macaques are well known for their "use" of infants in a wide variety of contexts, which can involve a tense situation between a male and an infant (e.g., gelada baboons [*Theropithecus gelada*], Dunbar 1984a), agonistic situations between two or more adult males (e.g., yellow baboons [*Papio cynocephalus*], Strum 1984), or a more mutualistic interaction that strengthens bonds between males and/or between a male and an infant (e.g., Barbary macaques [*Macaca sylvanus*], Taub 1980b; yellow baboons, Strum 1984). In the literature this sort of interaction is labeled variously as "infant use" (Strum 1984), "triadic male–infant interactions" (Taub 1980a,b; Kümmerli and Martin, 2008), and "agonistic buffering" (Deag and Crook 1971). If a high-ranking baboon male threatens a lower-ranking male, that male might approach a female with an infant, pick up the infant, carry it toward the threatening male, and sit close to him while still holding the infant (Strum 1984). The threatening male then typically leaves; the infant is usually compliant during this exchange. In Barbary macaques, two males will often face each other with an infant in between them, clutch at each other while teeth-chattering, and then chatter to the infant. They will also nuzzle, groom, inspect, and touch the infant, which usually shows no distress at being intensely examined (Taub 1980b, 1984). In both baboons and macaques, if infants do show signs of stress, mothers (and others) will attack and chase off the male (Stein 1984, Strum 1984).

Male choice of an infant is specific and not random and does not reflect paternity. In fact, several studies have documented no male preference for infants that are likely their offspring (e.g., olive baboons [*Papio anubis*], Strum 1984; yellow baboons, Stein 1984). These males do not actively avoid their own offspring, nor do they actively seek them. In Barbary macaques, male handling of an infant does not result in increased mating opportunities with the infant's mother (Taub 1980b); the benefits males gain seem to be in the form of male/male negotiations, interactions, and alliances (Taub 1980b, 1984; Riechelmann et al. 1994). Kümmerli and Martin (2008) compared patterns of infant handling in a free-ranging group of 46 Barbary macaques on Gibraltar and found that males, females, and juveniles all participated extensively in triadic interactions using infants of more closely related females (above average). They also found that only males and juveniles were highly involved in dyadic interactions with infants of related females; females rarely handled infants other than their own. Across species there is a strong affiliation between the male and the infant before, during, and after such interactions (e.g., Dunbar 1984a); males that routinely care for infants increase the chance that those infants will cooperate in future triadic interactions (Strum 1984, Stein 1984).

In some species, males that interact with infants gain access to females for future consortships and mating opportunities or merely strengthen already existing relationships with certain females (e.g., olive baboons [*Papio anubis*], Smuts 1985). In ring-tailed lemurs (*Lemur catta*), males might gain benefits from interacting with immatures in the form of achieving more spatial centrality in the social group, which can then lead to increased chances to develop affiliative relationships with females (Gould 1997). Young lemurs can also benefit from these interactions in the form of enhanced predator detection, allocare, and increased opportunities to develop social skills (Gould 1997).

Adult males sometimes play a role in protecting infants; and in one group of Hanuman langurs (*Semnopithecus entellus*), Borries et al. (1999a) found that only genetic fathers or males that were in the group when an infant was conceived were seen to protect infants. However, in many species, there are reports of males adopting and/or caring for infants that are not their offspring (e.g., chimpanzees [*Pan troglodytes*], Goodall 1977, 1986; Japanese macaques [*M. fuscata*], Wolfe 1981; owl monkeys [*Aotus*], see Chapter 10).

Infant interactions with adult males occur across a wide range of contexts. Possible reasons for this type of social interaction are plural and not mutually exclusive. Variables such as adult male dominance rank and the infant's familiarity with certain males seem to be important proximate components in the formation of such relationships. Intense bonds develop between specific infants and specific males in many species, and kinship is not always a consistent predictor of the degree of affiliative interactions. The male's relationship with the infant's mother and his previous history with the infant are more reliable indicators of the degree of contact and interaction the two will share.

Infant Abuse and Infanticide by Males

Overall, the rates of abuse toward infants by adult males are extremely low. As mentioned, agonistic buffering may occur in some species in order to diffuse a tense situation or so that an individual can gain favor with higher-ranking individuals in the group. Such use of infants usually does not result in death, although infants may be indirectly injured. Generally, when fatal aggression by males toward infants is inferred, it is within the context of infanticide.

The majority of reported cases of male infanticide are concentrated in three mammalian orders: carnivores, rodents, and primates (see van Schaik 2000a for review). In many primate species, infanticide has been reported and described as an evolved reproductive strategy for adult males (e.g., chimpanzees [*Pan troglodytes*], Murray et al. 2007; Hanuman langurs [*Semnopithecus entellus*], Hrdy 1974, 1977a, 1979, Hrdy and Hausfater 1984a,b, Sommer 1987, 1994, Borries et al. 1999b, Borries and Koenig 2000, Koenig and Borries 2001; snub-nosed monkeys [*Rhinopithecus bieti*], Xiang and Grueter 2007; chacma baboons [*Papio hamadryas ursinus*], Palombit et al. 2000; howler monkeys [*Alouatta* spp.], Crockett 2003; white-faced capuchins [*Cebus capucinus*], Fedigan 2003; black capuchins [*Cebus apella nigritus*], Ramírez-Llorens et al. 2008; and for general review, van Schaik and Janson 2000). In certain cercopithecoid primates (e.g., *S. entellus*), infanticide is assumed to be so common (or was common in the recent past) that other aspects of sociality, including the social system itself, have evolved in response to the potential of infanticide occurring (van Schaik and Kappeler 1997, Kappeler and van Schaik 2002, Pradhan and van Schaik 2008).

The sexual selection hypothesis has become the dominant theoretical framework for interpreting infant deaths caused by males in primate groups. This hypothesis is based on the assumption that males practice infanticidal strategies to enhance their reproductive fitness. Other adaptive reasons for infanticide have been proposed (e.g., resource competition, exploitation, parental manipulation) (Hrdy 1979). Nonadaptive hypotheses (e.g., social pathology, generalized aggression) have also been put forth to explain infant deaths resulting from male behavior. These alternatives view the majority of infanticides as epiphenomena of heightened male aggression toward other group members (e.g., Curtin and Dolhinow 1978; and see Bartlett et al. 1993 for review). The importance of considering the context of such an event when inferring adaptive reasons (e.g., sexual selection hypothesis) for infanticide by adult males cannot be overstated as the number of infanticides actually witnessed (over several decades and hundreds of thousands of hours of field observations) is still quite low (<100). In some directly observed cases, it is clear that the aberrant behavior of the infant's low-ranking mother placed it unequivocally in harm's way (e.g., MacKinnon 2003b). In other cases, where cannibalism has followed infanticide, Melo et al. (2003) hypothesize that this could be a result of competition for scarce resources and the need for animal protein, brought on by forest degradation.

Infanticide by adult males remains a subject of debate in primatology (Bartlett et al. 1993, Sussman et al. 1995, Fuentes 1999, Dolhinow 1999, Sommer 2000, van Schaik 2000b). The main crux of the disagreement is not whether males occasionally kill infants—they do—but whether this behavior is an evolved reproductive strategy: a behavior (or set of behaviors) that has afforded certain males higher reproductive success over the course of their lifetimes and that has been selected for via the mechanism of natural selection (see Cheverud 2004 for a discussion of the natural selection of heritable variation and its application to social behaviors).

SUMMARY

From birth, a young primate's world is complex and constantly changing. Across the Primate order we see a remarkable variability in the characteristics of behavioral interactions between infants and the adult members of their social groups. We also see a degree of plasticity in development not found in any other group of mammals. A prolonged period of physical and emotional dependence by infants on others, set against a multifaceted and dynamic social landscape, results in an amazing array of infant–adult primate behaviors. With the ongoing development of molecular techniques more applicable in natural settings, future genetic work will elucidate some of the underlying kinship factors that may or may not be at play in the formation and maintenance of immature–adult relationships. We desperately need long-term fieldwork on young primates in their natural environments. Such data, when coupled with results from captive work where variables can be more easily manipulated, will give us a more comprehensive understanding of the complexities inherent in young primate lives.

28

Juvenile and Adolescent Primates
The Application of Life History Theory

Hogan M. Sherrow and Katherine C. MacKinnon

1. How does the primate pattern of slow, extended life histories affect juvenile and adolescent behavior?
2. How do juvenile and adolescent primates use adults for behavioral development?
3. What reasons have been proposed by researchers for the extended periods of immaturity, known as juvenility and adolescence, found in primates?

INTRODUCTION

The order Primates is characterized as having slow life histories that include prolonged maturation, heavy investment in altricial offspring, relatively long life spans, and the longest juvenile periods, relative to body size, of all mammals (Schultz 1960, Harvey and Clutton-Brock 1985, Kappeler 1996, Ross 1998; see Chapters 25 and 26). Because primates mature relatively slowly, they have extensive opportunities for brain development. For example, Joffe (1997) found a positive relationship between age at first reproduction and nonvisual neocortex ratios. These results led Joffe to suggest that the extended juvenile period in primates is related to increased brain size. One of the results of increased encephalization in primates is greater behavioral flexibility. While behavioral flexibility has allowed some species to occupy a wide range of habitats through modifications in behavior instead of morphology, it has also determined that primates are highly dependent on learned behaviors.

Dependence on learned behaviors puts selective pressure on primate populations because young primates need time to become skilled at the behaviors necessary to survive and successfully compete with conspecifics. As a result, some researchers have suggested that the need to develop social skills (Janson and van Schaik 1993, Joffe 1997, Pereira and Altmann 1985), foraging skills (Janson and van Schaik 1993, Johnson 2001), and sex-typical behaviors (Glick 1980, Pereira 1993) prior to adulthood has resulted in the extended juvenile and adolescent periods observed in primates. During extended periods of development, young primates face a series of trade-offs between current survival and future reproductive opportunities.

Young primates could maximize immediate gains by consuming as much food as possible and devoting energy to growth and development, while avoiding predation. However, competition with older conspecifics over food and other resources leads to agonistic interactions (Janson and van Schaik 1993). For young primates to maximize their food intake, they need to minimize competition with older conspecifics for food and, therefore, may need to spend little time in close proximity to them. If young animals can minimize direct competition with, and aggression from, older conspecifics while also associating with them, they can thereby benefit from the social information that adults provide (Janson and van Schaik 1993, Watts and Pusey 1993).

Maturation rates, offspring investment, relative brain size and dependence on learned behaviors vary considerably among primate species. Most lemurids reach sexual maturity relatively fast, have litters of up to three offspring, have relatively smaller brains, and rely less on learned behaviors than haplorhine primates (Charnov and Berrigan 1993, Jolly 1966, Kappeler 1996, Koyama et al. 2001, Byrne and Corp 2004). At the other end of the spectrum, the hominoids typically take up to 10 years or more to reach sexual maturity, have one offspring at a time, have relatively large brains, and rely heavily on learned behaviors (Charnov and Berrigan 1993, Goodall 1986, Watts and Pusey 1993, Byrne and Corp 2004, Leigh 2004; see Chapters 18–20).

Infancy is a critical life history stage, during which individuals learn certain behaviors and skills that impact success later in life, the most pressing of which is survival. Accordingly, there is ample literature on mother–infant interactions and the behavior of infant primates in captivity (e.g., Harlow 1959; Sauer 1967; Hinde 1977; Fairbanks 1988a,b; Maestripieri 1994a,b, 2001; Brown and Dixson 2000; Bales et al. 2000) and in free-ranging and wild populations (e.g., Itani 1959; Jay 1963; Jolly 1966; Nash 1978; Fossey 1979; Altmann 1980; Watts 1985b; Goodall 1986; Schino et al. 2003; Berman et al. 1997; see Chapter 27 for review). Despite its importance, infancy is a relatively short life history stage, often lasting months or a few years, while the other stages of immaturity can account for 25%–35% of an animal's life span. This is demonstrated in wild male chimpanzees, which can live 35–40 years. They spend the first 4 years of their lives as infants (10%) and the ensuing 12+ years as juveniles and adolescents (30%–34%). Despite the apparent importance of juvenility and adolescence as life history stages in primates, studies that focus on juveniles and adolescents to address questions about the ontogeny of social behavior in wild primates are particularly sparse.

Pereira and Fairbanks highlighted this gap in our knowledge in the edited volume *Juvenile Primates: Life History, Development and Behavior* (1993), when they called for additional studies on primates during the underrepresented life history stages of juvenility and adolescence. In the second edition of that same volume, Pereira and Fairbanks (2002) recognized the continued lack of studies focusing on juveniles and adolescents. They presented data showing that between 1990 and 1999 3% of articles published in the journal *Folia Primatologica* pertained to juvenile growth or sexual maturation, with another 3% focused on postweaning behavioral development (mainly on dispersal patterns). In the same decade, a mere six articles on social play appeared in the five major primate journals.

Juvenile and adolescent primates are typically independent of their mothers while foraging and often while traveling but are still physically and socially immature. As a result, they are particularly vulnerable to attacks from conspecifics and to predation. Consequently, juvenile and adolescent primates are extremely difficult subjects to study in the wild. Immature primates in some species can be difficult to habituate and to identify on an individual basis, let alone observe for the kind of time necessary for quality behavioral analyses. In addition, most studies on wild primates are limited in time and scope, making it difficult to collect the long-term data that are ideal when analyzing periods of maturation. The result has been a dearth in the number of successful studies on juvenile and adolescent primates in the wild. However, with patience, time, and proper habituation, behavioral data can be collected on individual animals, providing important information about these relatively little-known life history stages.

In this chapter we review and summarize some (though by no means all) of the studies that have successfully focused on juvenile or adolescent primates, especially those that have utilized a life history framework in their analyses, whether explicitly or implicitly. We then propose that a life history framework is one of the most promising approaches to understanding juvenility and adolescence in primates and that the social behaviors of juvenile and adolescent primates, in particular, are best understood through such a framework. Finally, we synthesize the results from previous studies, provide additional data on two species (*Pan troglodytes* and *Cebus capucinus*), and present a comprehensive picture of our current understanding of juvenile and adolescent primate behavior.

PREVIOUS STUDIES

While there have been few studies that have focused exclusively on free-ranging or wild juvenile or adolescent primates, a few research projects have been successful. Baboons have been well studied in the wild (e.g., Washburn and DeVore 1961; Kummer 1971; Altmann 1980; Strum 1987; Cheney et al. 2004; Jolly and Phillips-Conroy 2003, 2006), and studies that have focused on immature baboons are fairly common. Johnson (2001, 2003) studied life history interactions with the social environment in chacma baboons (*Papio hamadryas ursinus*) and showed that social and foraging efficiency are difficult for young baboons to acquire. Young baboons were subject to more aggression than other group members and they made more social errors than their older groupmates (Johnson 2001). They antagonized high-ranking conspecifics more than older groupmates and subsequently received more aggression from them. Alberts (1992) studied the variability of male life histories in yellow baboons (*P. h. cynocephalus*), focusing on age at reproduction and dispersal decisions. Results indicated that, on average, males dispersed from their natal groups while still growing (Alberts, 1992) and that their ability to mate in non-natal groups depended partly on what ranks they attained in their new groups (Alberts and Altmann 1995a).

Pereira's work on the behavioral development of juvenile yellow baboons (*P. h. cynocephalus*) shed light on a great many issues regarding the ontogeny of behavior and sex differences in this species (Pereira 1984, 1988a,b, 1989). He found that juveniles of both sexes spent less time close to other group members as they became more socially confident, that juvenile females were assisted by relatives in the acquisition of rank, and that juvenile males were less likely to receive aggression from adult males than were juvenile females, although males received much more aggression when they hit adolescence. Similarly, Charpentier et al. (2004) found a shift in social contacts and behavioral patterns in mandrill (*Mandrillus sphinx*) males upon reaching adolescence, with more aggression and less submission compared to younger individuals.

Pereira's work also showed that male and female juveniles differed in their levels of attraction to male and female social partners (Pereira 1988a, and see Förster and Cords 2005). Both male and female juveniles were significantly more likely to be in proximity to same-sex adults than opposite-sex adults. Further, they were significantly more likely to groom same-sex adults than opposite-sex adults. These results provided support for the "sex-typing" hypothesis, which states that selective attention toward same-sex adults and their interactions is an important component of the development of sex-typical behavior (Maccoby and Wilson 1957 in Pereira 1988a). The sex-typing hypothesis was originally designed for analyzing the behavior of children. Pereira's work showed that the general foundation of the hypothesis could be applied to the behavior of immature nonhuman primates (Pereira 1988a).

While the general form of the sex-typing hypothesis sufficiently explains why immature males and females differ in their general levels of attraction to social partners and the amount of attention they pay to adult members of the same sex (Pereira 1988a), it does not address variation among same-sex, immature primates. In particular, the sex-typing hypothesis does not deal with the potential for strategic choices made by individuals, the impacts of those choices, and which social partners are more valued than others.

In the late 1970s and 1980s two studies were conducted on wild chimpanzees (*Pan troglodytes schweinfurthii*) in Tanzania. Pusey (1978, 1983, 1990) studied the behavioral and morphological development of adolescents at Gombe, and Hayaki (1985, 1988) studied adolescent behavior at Mahale. Pusey's study provided critical baseline data on the growth and development of wild chimpanzees. Results showed that as males and females entered adolescence they began spending less time with their mothers, although for different reasons. Females usually emigrated to new communities or began establishing their own home ranges within a community's territory, while males began preferentially associating with adult and other adolescent males (Pusey 1990). Hayaki, who had the advantage of a larger sample size than Pusey, obtained results at Mahale similar to those from Gombe. Also, he noted that adolescent males were part of the adult male dominance hierarchy but ranked below all adult males and did not have dominance relationships with each other (Hayaki 1985, Hayaki et al. 1989).

Sherrow (2008) recently completed a detailed study on the variation in, and development of, social behavior in adolescent and other young male chimpanzees at Ngogo, Kibale National Park, Uganda. He found that as males matured and entered adolescence they redirected their social investments from their mothers to the adult males of the community (Figure 28.1), seeking them out as social partners, grooming them, traveling with them and joining them in coalitionary displays (Sherrow unpublished data). Further, as males matured, they began participating in male-typical behaviors, like forming dyadic dominance relationships, hunting mammalian prey, and patrolling territorial boundary areas (Sherrow unpublished data). Finally, he provided evidence of competition over mates, dominance relationships, and the impact of rank on mating success among adolescent male chimpanzees. Sherrow's work expanded on Pereira's application of the sex-typing hypothesis to immature primate behavior, using a life history theory framework to develop the behavioral role modeling hypothesis (BRMH) to explain the behaviors observed in immature male chimpanzees (Sherrow unpublished data).

In a study that examined whether developmental factors might explain age differences in foraging patterns in wild white-faced capuchins (*Cebus capucinus*) in northwestern Costa Rica, one of us (K. C. M.) found that younger individuals had an extremely high rate of the behavior "explore"

Figure 28.1 An adolescent male chimpanzee (*Pan troglodytes schweinfurthii*) grooming an adult male. Adolescent male chimpanzees invest heavily in social relationships with adult males (photo by Hogan M. Sherrow).

during the wet season and that in both wet and dry seasons this behavior had a tendency to decrease with age. In addition, infants made more mistakes while foraging compared to juveniles (e.g., higher rates of the behavior "grab and miss" when going after insects) during both seasons, as well as having the lowest rates of actually eating food items (MacKinnon 1995, 2006). Thus, experience may influence foraging success in that older individuals should be more efficient than younger (and smaller) individuals (Boinski and Fragaszy 1989, Fragaszy and Boinski 1995, and see Gunst et al. 2008). Other behaviors, however, were contradictory to this hypothesis. For example, complex behaviors were not always performed more often by older and more experienced individuals (e.g., the behaviors "break open with hands and/or mouth" and "probe," among others).

These data suggest that *C. capucinus* juveniles are accomplished foragers as younger immatures, even if certain behaviors are not yet refined, and that individuals do not develop complex foraging skills along a consistently progressive trajectory (MacKinnon 1995, 2006) (Figure 28.2). For brown capuchins (*C. apella*) in Suriname, Gunst et al. (2008) found that foraging competence relies on the gradual acquisition of skills influenced by both physical maturation and indirectt social input via foraging evidence left by conspecifics.

Long-term studies on the behavior of long-tailed macaques (*Macaca fascicularis*) in Sumatra have provided valuable information on the development of behavior in this species (e.g., van Noordwijk and van Schaik 1985, 2001). van Noordwijk and van Schaik's (1985) results provided insights into male strategies and indicated that males emigrated from their natal groups before reaching adult size and that age at emigration was not affected by maternal rank or time spent separate from the main party. Results

Figure 28.2 A juvenile white-faced capuchin (*Cebus capucinus*) investigates a potential food item in the hand of an adult female. Individuals in this species are accomplished foragers at a young age, and often learn about food sources in a social context (photo by Matthew A. Wyczalkowski).

also indicated that males were most successful at reproduction when they acquired high rank and that high rank was most often attained when a male was a long-term resident of a group (van Noordwijk and van Schaik 2001).

Setchell (2003) studied the sociosexual development of free-ranging adolescent male mandrills (*Mandrillus sphinx*) and showed that intrasexual competition and the social environment affected the development of secondary sexual characteristics in this flamboyant species. Males appeared to be suppressed in their development when in the presence of a high-ranking adult male and did not fully develop the elaborate secondary sexual characteristics so recognizable in mandrills until they became high-ranking. Further, the intensity of the coloration of male facial skin correlated with male rank.

The studies summarized here provided great insight into species-typical patterns and individual strategies across a number of primate taxa, but they also provided something else. These studies demonstrated how effectively a life history theory framework could be applied to the analysis of immature primate social behavior. Though most studies did not explicitly state their use of a life history framework, they employed one by assuming that immature primates were limited in their ability to acquire highly valued resources and that the behaviors they exhibited would be a reflection of the trade-offs they faced.

LIFE HISTORY THEORY

Although the concept of life histories in biology has been around since the early 1900s (Chapman 1905), life history theory as a conceptual tool was not formalized until the 1970s (Charnov 1979; Stearns 1977, 1980). Life history theory is based on the laws of thermodynamics and is concerned with the differential allocation of energy by organisms as they move through their life cycles (Stearns 1992). The application of life history theory allows researchers to investigate developmental and size-related (allometric) variation in key events in life histories (e.g. birth, age at maturity, life span length) that are shaped by evolutionary processes (Harvey et al. 1987; Godfrey et al. 2001, 2003, 2004; Harvey and Clutton-Brock 1985; Martin and MacLarnon 1990; Ross 1991, 2002, 2004; Shea 1987; Breuer et al. 2009; and see Chapter 25 for review).

Natural selection is expected to result in the continual refinement of traits in organisms, relative to a given ecology, resulting in the maximization of those traits. However, because there is a finite amount of energy available for any organism to extract from its environment and energy that is used for one purpose cannot be used for another, traits in organisms are often not maximized but, instead, the result of energetic trade-offs (Charnov 1991, Stearns 1992). Phenotypic correlations can confound situations involving energetic trade-offs. Phenotypic correlations occur when some traits in organisms are positively correlated with each other, such as when an organism in good condition also experiences early reproduction and high survival or fertility through time (Hill and Hurtado 1996).

Charnov and colleagues have demonstrated that choices made by males and females from taxa ranging from reptiles to mammals over reproductive effort are strongly impacted by life history variables (e.g., Charnov 1991, 1993, 1997; Charnov et al. 2007; Gross and Charnov 1980). In a recent study, Charnov and colleagues (2007) demonstrated that, as predicted by Williams (1966a), lifetime reproductive effort is a key component of fitness within populations and is critical for understanding life history evolution. Further, there

is a trade-off of the production of offspring, relative mass at independence, and survival in female vertebrates, which results in the lifetime production of offspring with a mass that is approximately 1.4 times that of their own body mass (Charnov et al. 2007).

Life history theory is concerned with the total quantitative assessment of the life of an organism (Bribiescas 2006, Stearns 1992). It allows researchers to understand the action of natural selection through the analysis of demography, quantitative genetics, and energetic trade-offs (Roff 2002, Bribiescas 2006) and is a powerful tool in the study of evolution (Stearns 1992, Leigh 2004). Life history studies have been successfully carried out, primarily, on organisms that have rapid maturation, short interbirth intervals, and short life spans (e.g., cotton rats, *Sigmodon hispidus*, Bergstrom and Rose 2004; bluegill sunfish, *Lepomis macrochirus*, Gross and Charnov, 1980; water bears, *Macrobiotus richtersi*, Guidetti et al. 2007; common frogs, *Rana temporaria*, Lardern and Loman 2003; wood tiger moths, *Parasemia plantaginis*, Ojala et al. 2007; painted turtles, *Chrysemys picta*, Rollinson and Brooks 2008; fruit flies, *Drosophila melanogaster*, Stearns et al. 2000; spiny rats, *Niviventer coxingi*, Yu and Lin, 1999). Those organisms provide opportunities for observing multiple generations over a relatively short period of time, allowing for complete tests of hypotheses derived from life history theory. For example, Stearns et al. (2000) studied the effects of extrinsic mortality on life history variables and schedules among fruit flies. Their results provided strong evidence in support of the hypothesis that extrinsic mortality has an important effect on growth, maturation, reproduction and aging in animal populations.

Life History Theory and Primates

While most of the research on life history variables has focused on animals with "fast" lifestyles (Sibley and Brown 2007), studies of life history variables can be conducted on slow-lived organisms, like primates, if sufficient data exist. Kappeler (1996) analyzed the causes and consequences of life history variation among 20 strepsirhine species, with particular attention to the metabolic rates and energy requirements of females. He found that "There was little evidence for an independent role of brain size and metabolic rate in determining life-history variation in traits reflecting maternal investment" (Kappeler 1996:875). Using life history data, Kappeler was able to comment on the long-standing debate regarding female dominance and maternal investment in some strepsirhine species. His conclusions did not support the hypothesis that high maternal investment is primarily responsible for female social dominance in lemurs.

Harvey and Clutton-Brock (1985) collected and extracted data from the literature on 17 life history variables of 135 primate species, primarily from wild populations. Their study was vital for later studies on primate life histories as it established numerous baselines for multiple species. For example, they found that subfamilies of primates shared basic life history traits including age at first birth, adult body size, and life span. Further, all of the variables analyzed except cycle length and number of offspring were highly correlated with each other, indicating that the relationship between the variables is complex and that many taxa probably experience phenotypic correlations. Finally, in regard to maturation, Harvey and Clutton-Brock's findings suggest "that postnatal brain development, learning skills, or social maturity may be intimately linked with physical maturity" (1985:577).

According to life history theory, organisms should minimize their prereproductive period in order to maximize their potential reproductive output (Hill and Hurtado 1996, Stearns 1992). Consequently, when organisms have extended prereproductive periods such as those found in primates, the phenomena observed should be the result of trade-offs or adaptations. *Trade-offs* are the differential allocation of limited resources between traits (Stearns 1992) and, in the case of maturation, involve the balance between current survival and future reproduction (Hill and Hurtado 1996). *Adaptations* are changes in phenotypes that occur in response to specific selective pressures and result in increased survival and reproduction (fitness) of organisms (Williams 1966a). Trade-offs and adaptations are not mutually exclusive, and adaptations can occur if the same trade-offs are experienced consistently over multiple generations.

By couching behavioral studies in a life history theory framework, researchers gain the ability to analyze whether or not social behaviors and maturation are the results of trade-offs, adaptations, or a combination of the two. For example, when Pereira (1988a) demonstrated that juvenile baboons were more attracted socially to same-sex adults, he also provided support for the hypothesis that juvenility in baboons is linked to the development of social skills and possibly part of a trade-off experienced by maturing baboons that balances current survival (via growth) with future reproductive opportunities (via social learning). It is most likely this trade-off that has resulted in the prolonged juvenile period observed in baboons and that baboon juvenility is part of an adaptive complex linked to social maturation.

The hypothesis that the extended period of juvenility and the presence of adolescence in primates is the result of selective pressures to learn social skills has received additional support from several recent studies, including those of the authors. Godfrey et al. (2003 and 2004) examined the correlates of ontogeny and diet in Malagasy lemurs and demonstrated how strepsirhine dental development is linked to the strong seasonality (and subsequent restricted food availability in some seasons) of Madagascar forests. They showed that among the folivorous indri (*Propithecus verreauxi*) dental and life history schedules were optimized to the mechanical challenges that weanlings faced. In other words, immature indris were prepared to process adult-type

foods long before they were physically or socially mature. Other studies have produced corroborating results.

In a study on immature baboons, Altmann (1991, 1998) found that by the time baboons were 70 weeks old, they consumed a diet that was identical to the adult diet (e.g., food types eaten as well as the proportion of different foods in the diet). Perhaps the most remarkable outcome of this study was the fact that most measures of female reproductive success (i.e., reproductive life span, age at first conception, number of infants, offspring survival, and individual survival) correlated with the quality of diet that the female immature consumed during the weaning period. In essence, female baboons established their foraging patterns at an early age.

Interestingly, data from captivity suggest that male chimpanzees that do not have to forage for resources, avoid predation, or avoid potentially dangerous older groupmates are capable of developing sex-typical behaviors much earlier (Anestis 2005, 2006). In this study, Anestis showed that as juveniles captive male chimpanzees are capable of competing for status and resources and forming dominance relationships (Anestis 2006). In her results, she recognized that captive peer-housed juveniles and adolescents are different from their wild counterparts. Indeed, wild chimpanzees face challenges associated with survival and development that are not faced by captive individuals, many of those affecting their ability to socialize and develop social skills. However, as Anestis (2006:543) stated, "The captive environment therefore affords a unique opportunity to explore what *could* happen if young wild chimpanzees were able to spend more time together."

Juvenile capuchins (*Cebus capucinus*) have been shown to be efficient foragers from 1–2 years of age, which is quite remarkable in an omnivorous species that has a complex extractive and manipulative foraging strategy (MacKinnon 2006). While overall the diets of the different capuchin age classes are very similar, there are differences that can be best explained by the reduced size and strength of small juveniles (e.g., small juveniles eat vertebrate prey, such as squirrels and birds, at a rate that is less than half that of either large juveniles or adults) (Figure 28.3). Thus, young capuchins' food choice may be constrained by the size and strength required to handle and process certain food items (Gibson 1986, Fragaszy and Boinski 1995).

Additional examples of foraging proficiency from a young age include mountain gorillas (*Gorilla gorilla beringei*), which consume an essentially adult-like diet when they are 3–4 years old (Watts 1985b); squirrel monkeys (*Saimiri sciureus*), which forage like adults at only 6–12 months of age (Boinski and Fragaszy 1989; Stone 2004, 2006); and brown lemurs (*Eulemur fulvus*), which consume an adult-like diet when they are 10–12 months old (Tarnaud 2004). While all of these studies indicate that immature primates are typically physically and socially immature when they become independent foragers, other studies indicate that foraging skills are important limiting factors in primate behavioral evolution.

Figure 28.3 A juvenile white-faced capuchin (*Cebus capucinus*) stays in close proximity, watching and waiting for scraps, while a subadult male processes a magpie jay (*Calocitta formosa*) (photo by Katherine C. MacKinnon).

Johnson and Bock (2004) found that juvenile chacma baboons (*Papio hamadryas ursinus*) were less adept at extracting food resources than were adults and that they took multiple years to become ecologically competent. Similarly, Krakauer (2006) found that immature aye-ayes (*Daubentonia madagascariensis*) only became ecologically competent after a long period of learning. This long period of dependence is inferred to be related to the aye-aye's unique percussive and extractive foraging techniques, which require fine motor coordination and skill and which take time and practice to develop proficiency (Krakauer 2006).

Chimpanzees (*Pan troglodytes*) and orangutans (*Pongo pygmaeus*) have some of the most complex foraging behaviors and techniques among the primates. Accordingly, immature chimpanzees and orangutans take years to attain adult foraging skills. For example, van Noordwijk and van Schaik (2005) found that immature orangutans usually took 8–9 years to be able to forage effectively and acquire basic tool use. They suggested that the extended period of juvenility and ecological dependence observed in orangutans may be a consequence of a solitary lifestyle and the relatively few opportunities available for social learning. Chimpanzees show a similar pattern, taking years to become proficient in tool use (Lonsdorf 2005; Lonsdorf et al 2004). Interestingly, while chimpanzees and orangutans develop at similar rates (van Noordwijk and van Schaik 2005; van Adrichem et al. 2006), immature chimpanzees become proficient tool users by 5.5 years of age (Lonsdorf 2005; Lonsdorf et al 2004), a full 2–3.5 years earlier than orangutans do. These results appear to support van Noordwijk and van Schaik's (2005) proposition regarding social learning and the acquisition of foraging knowledge.

Clearly, there is not a simple solution when it comes to analyzing what selective forces have driven the prolonged period of maturation in primates. All immature primates need to take in enough nutrition to survive and promote growth and development. However, nearly all of those same primates need to develop the social skills necessary to cooperate and compete with conspecifics. In some taxa, the need to acquire foraging knowledge and skill seems to have a strong impact on the rate of maturation (e.g. *Pongo pygmaeus*, van Noordwijk and van Schaik 2005). Still, in others, the acquisition of social skills and abilities seems to be extremely important (e.g. *Papio hamadryas cynocephalus*, Pereira 1988a, 1989). Collectively, the majority of the studies outlined here do not provide support for the hypothesis that prolonged maturation in primates is an adaptation to allow sufficient time to learn a foraging repertoire. However, it is important to keep in mind that some of our behavioral categories might not be explicit enough to parse out minute differences in foraging ability that may be operating (see MacKinnon 2006). In all likelihood the answer may be that the need to acquire skills in foraging and the social milieu are of varying importance depending upon the species considered, the ecological niche it occupies, and the local demographic and ecological contexts. All of those variables combine to establish and limit the life histories of species.

BEHAVIORAL ROLE MODELING AND JUVENILE AND ADOLESCENT PRIMATES

Regardless of whether or not immature primates are acquiring skills in foraging or social relationships, one thing is becoming increasingly clear: Many of them use older groupmates as role models. For example, Ottoni et al. (2005) found that in a group of semi-free-ranging capuchins (*Cebus apella*), younger, less proficient individuals actively observed older, more skilled group members when they used stones to crack open nuts. The choice of which group members the younger animals observed was not random: They preferentially watched the more successful tool users, resulting in greater scrounging payoffs and enhanced learning opportunities. In a study of white-faced capuchins (*Cebus capucinus*) in Costa Rica, Jack and Fedigan (2004a) found that rates of dispersal from natal groups for juvenile males (either with favored coresident males or into groups with known males) were very high (71%–82%). They suggest this pattern of dispersal may have evolved to increase inclusive fitness benefits in male-dispersed species and that it probably also increases survivorship in young males that are not yet physically mature.

In a study on wild young male chimpanzees (*Pan troglodytes*), one of us (H. M. S.) developed and tested the behavioral role modeling hypothesis (BRMH) (Sherrow 2008). The BRMH is an extension of the sex-typing hypothesis developed in psychology and employed by Pereira (1988a) and others in the study of immature nonhuman primates. The BRMH states that maturing primates use adults as role models for behavioral development and that they choose social partners strategically, favoring role models that can provide them with competitive advantages (Pereira 1988a). H. M. S. found support for the BRMH among young male chimpanzees of the Ngogo community in Kibale National Park, Uganda (Sherrow unpublished data). One of the many results from that study that provided support for the BRMH was that young male chimpanzees were strategic in their allocation of social effort, grooming high-ranking adult males more than middle- and low-ranking adult males (Figure 28.4).

Similarly, in a study of juvenile white-faced capuchins (*Cebus capucinus*) in Costa Rica, one of us (K. C. M.) found that dominance rank was a major predictor of which adult males in a multimale group the large juveniles (defined for this species and study as 3–5 years old) preferred to interact with (Figure 28.5); however, there were also individual differences with regard to preferences and other predictors such as familiarity and access to novel social or foraging situations; these categories were not mutually exclusive in many cases (MacKinnon 2002 and 2003a).

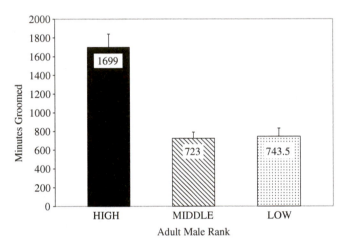

Figure 28.4 Comparison of grooming given by young male chimpanzees (*Pan troglodytes*) to adult males, relative to rank ($n = 3,165.5$ minutes, $F = 8.720$, $p = 008$).

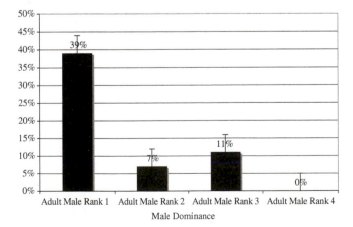

Figure 28.5 The percent of groom time when large juvenile white-faced capuchins (*Cebus capucinus*) are the actors, showing significant differences among the age/sex classes (K-W = 14.5, $p = 0.0128$).

KINSHIP AND JUVENILE AND ADOLESCENT PRIMATES

Kinship as a potentially influential factor in a wide variety of behavioral mechanisms and evolved strategies has generated much interest and debate (e.g., Trivers 1972, 1985; Wrangham 1980; Silk 1987; Gouzoules 1984; Gouzoules and Gouzoules 1987; Bernstein 1991a; Bernstein et al., 1993b; Clutton-Brock 2002; Kappeler and van Schaik 2002; Charpentier et al. 2007; Chapais and Berman 2004c; Strier 2004a,b; and see Chapters 14 and, 38 for review,). Chapais and Berman (2004), in particular, provided an excellent synthesis and assessment of current kinship research in nonhuman primates. They referred to kinship as a "black box" because there is still so much we do not understand about the relationship between the degree of genetic relatedness of individuals and social behavior. However, it is this very problem that encourages new ideas, new questions to answer, and new hypotheses to test. Macaques provide a perfect illustration of the complexity involved when assessing kinship and behavior. There is an incredibly wide variety of social interactions among related and nonrelated individuals in macaque societies. In some species, kinship has been shown to influence social behaviors to a great degree (e.g., Japanese macaques, *Macaca fuscata*, and rhesus macaques, *M. mulatta*, Sade 1980). However, in the Tonkean macaque, *Macaca tonkeana*, kinship did not affect social behavior patterns nearly as strongly (Thierry 2000).

A crucial point in this discussion, which has been stated and reiterated by many authors, is that while we attempt to theoretically address animal behavior in an evolutionary sense, we may ignore the actual issues the animals themselves are facing, that is, what is important in their view (e.g., see Altmann 1980:182, Dolhinow and DeMay 1982:415, Rendall 2004:298–299). Rendall (2004:299) sums up this problematic situation succinctly: "…although discriminating kinship may be central to *our accounts* of animal behavior, it may not be central to animal behavior" (emphasis added). Ehardt (1988) cautioned against using statements such as "kinship is the most cohesive force in social organization" and wondered if the effects of kinship on behavior have been overstated without adequate consideration of age factors and degree of relatedness. Rather, peer groups are assumed to be extremely important in the socialization of immatures, especially in social groups that have nonkin adults (Ehardt and Bernstein 1987, Ehardt 1988).

Although familiarity (via development or tenure in a group) is the most likely mechanism by which primates recognize kin (Chapais 2001, Rendall 2004), the process is dynamic and complex, socially controlled, and not strictly genetically determined. Indeed, many have argued that it is familiarity that is of prime importance in forming relationships and determining behavior, not degree of genetic relatedness (e.g., Bernstein and Ehardt 1986b,c; Bernstein, 1991b; Martin 1997; see Chapter 38). Also, strong positive correlations between degree of relatedness and social behaviors do not automatically imply cause and effect: that is, they may represent associations that are not necessarily *relationships* (Box 1984).

Despite this, there is solid evidence that kinship has an impact on the behavior of nonhuman primates. In many primate species, maternal rank is the main determinant of female rank (e.g., *Papio cynocephalus*, Hausfater et al. 1982; *Macaca arctoides*, Gouzoules 1975; *M. radiata*, Silk et al. 1981b; *M. nemestrina*, Bernstein 1969; *M. mulatta*, Ehardt and Bernstein 1987; *M. fuscata*, Koyama 1970, Mori et al. 1989, Takahata 1991, Nakamichi et al. 1995; but see also Kutsukake 2000 for interesting variation in this species). Typically, young females first attain ranks correlated with those of their mothers in peer interactions and then challenge and outrank older conspecifics subordinate to their mothers (e.g., Holekamp and Smale 1991). Some immatures seem to learn their ranks from observation, which occurs during direct aggression received from higher-ranking adults. Also, both kin and nonkin frequently form coalitions that may assist juveniles during rank reversals (Silk et al. 1981b).

In still other studies, researchers have shown that relatedness can have an impact on the behavior of adult male primates. For example, Buchan et al. (2003) provided evidence for true paternal care in yellow baboons (*Papio cynocephalus*). In their study they were able to show that adult male yellow baboons discriminated their own offspring from those of other males and provided care to them in the form of repeated support during agonistic encounters. More recently, Langergraber et al. (2007a) demonstrated that among wild chimpanzees maternal relatedness had a limited impact on adult male behavior. Overall, there have been far fewer clear reports on the impacts of kinship on juvenile or adolescent primate behavior (e.g. *Mandrillus sphinx*, Charpentier et al. 2007; *Macaca mulatta*, Ehardt and Bernstein 1987), and it remains an area that needs greater attention.

SUMMARY

In wild, unprovisioned primate populations at least 50% of infants die in their first year due to disease, predation, infection, and injury (e.g., *Lemur catta*, Gould et al. 2003; *Papio cynocephalus*, Altmann et al. 1977; *Pan troglodytes*, Boesch and Boesch-Achermann 2000; and see Debyser 1995a,b for review); just getting through that developmental stage is challenging enough, but then comes an even longer period of being a juvenile, followed by a (typically) shorter span of adolescence, and finally adulthood. All of these stages need to be considered in order to evaluate the behavioral strategies of individuals in a population over time (Rowell 1993b; Pereira and Fairbanks 1993a; Dolhinow 1991; see Chapter 26).

The extended periods of juvenility and adolescence have been given short shrift in field studies when compared to infancy. Mother–infant interactions are paramount in all taxa and have long held center stage in developmental theory. While the mother–infant relationship is fundamentally important to immature primates, development does not end abruptly with weaning. Later stages may be as important in the formation of adults capable of functioning and competing effectively. While we have solid data on several species in the wild (e.g., *Pan troglodytes*, Pusey 1978, Hayaki 1988, Sherrow 2008; *Cercopithecus* spp., Cords, 2000; *Macaca* spp., Hanya 2003, Perry and Manson 1995, Sprague 1998, Sprague et al. 1998, Tilford 1982; *Papio* spp., Altmann, J., et al. 1977; Altmann, S. A., 1998; *Saimiri sciureus*, Stone 2004, 2006; *Cebus capucinus*, Jack and Fedigan 2004a; MacKinnon 2006; *C. apella*, Agostini and Visalberghi 2005), we still sorely lack quantitative and qualitative longitudinal information on the social development parameters for most juvenile and adolescent primates in natural settings.

The studies outlined here are exceptions to this generality and have started to paint a clear picture regarding the variation and development of juvenile and adolescent primate behavior. Studies from the 1970s and 1980s demonstrated that juvenility and adolescence are important life history stages in primates and more than just holding patterns on the way to adulthood (e.g., van Noordwijk and van Schaik 1985; Pereira 1988a,b; Pusey 1978, 1983). Pereira and Fairbanks (1993a) called for more studies that focused on juveniles and adolescents, a call we echo here. While Pereira and Fairbank's suggestion received little attention overall, as demonstrated by the dearth of studies focused on juveniles and adolescents, some researchers took up the challenge and provided important insights into these little-known life history stages.

In the 1990s and early 2000s authors utilized a life history framework and showed that juvenile and adolescent primates are strategic in their behavioral choices (e.g., Alberts 1992, Fragaszy and Bard, 1997; Nishida 1990; Setchell et al. 2006; MacKinnon 2006 and unpublished data) and that they often use adult conspecifics as behavioral role models in development (e.g., Ottoni et al. 2005, Sherrow 2008 and unpublished data). Further, the social and ecological contexts in which primates develop can impact their rates of development (e.g., Maggioncalda et al. 2002, Setchell and Dixson 2001a). Finally, events that happen during juvenility or adolescence can have strong impacts on adulthood (e.g., Fairbanks et al. 2004, Setchell 2003).

It is clear that the extended period of juvenility and the presence of adolescence in primates are the results of an intricate system in which immature animals must invest in both current survival and future reproduction (Stearns 1992).

It is rarely the case that a juvenile or adolescent primate can focus on the attainment of either foraging or social knowledge, to the exclusion of the other. Instead, juveniles and adolescents must strike a delicate balance between investing heavily in current survival, via consumption of food and avoidance of dangerous older, larger groupmates (Janson and van Schaik 1993) and future reproduction via the attainment of ecological and social knowledge and the application of that knowledge. The end result is that juvenile and adolescent primates are under extremely strong, unique selective pressures that no other age groups face.

The emerging picture is one of a dynamic developmental system, in which individual juveniles and adolescents make strategic choices that impact their current survival and future reproduction in different ways depending on the social and ecological contexts in which they make

those choices. Despite our greater understanding of juvenility and adolescence, we have only scratched the surface of these critical periods in the life histories of primates. We need to better understand how relatedness impacts the behaviors of immature primates. We also need data on wild populations that focus on the proximate mechanisms (e.g., hormones) that control development and interact with and impact behavior. These elements need to be combined with more thorough behavioral data that focus on juveniles and adolescents in a life history framework to complete the picture of development and maturation in primates.

29

Primate Sexuality and Reproduction

Christina J. Campbell

1. It is unclear whether estrous cycles are characteristic of all primates. What is estrus, and why is its occurrence in some primates questionable?
2. Sexual skins are a prominent feature in some primate taxa. What are they, and what is their function(s)?
3. What is the priority of access model, and how does it relate to paternity in nonhuman primates?

INTRODUCTION

The title of this chapter suggests that the terms "sexuality" (sexual behavior) and "reproduction" (the production of offspring) refer to different avenues of primatological inquiry. Indeed, for many animal species, the terms refer to synonymous events; that is, sexual activity takes place only in the context of reproduction (although this by no means implies that every sexual encounter results in conception). Within the primate order, strepsirhine primates (lemurs and lorises) resemble nonprimate mammals in that sexual activity is largely restricted to times when conception is most likely (Shideler et al. 1983b, Dixson 1998, Koyama et al. 2001, although see Brockman and Whitten 1996 for an exception). The sexual behavior of tarsiers remains largely unstudied, but from the limited information available, they also appear to restrict sexual behavior to brief periods of time associated with midcycle ovulation (Wright et al. 1986). Many species of monkey and ape

(anthropoid primates), however, deviate from this pattern and engage in sexual activity at times when it is unlikely or even impossible for them to conceive (Loy 1987, Dixson 1998). For these primates, sexuality therefore is not necessarily synonymous with reproduction and may have been co-opted for functions other than the production of offspring. Mountings and, indeed, copulations have multiple functions, including dominant–subordinate interactions, appeasement, stress reduction, and the confirmation of social relationships. In reality, for many primate species, it may be extremely difficult for researchers to determine whether the function of a sexual encounter is reproductive or social in nature. Much of the primatological literature, however, treats sexual behavior as a reproductive event, often without strong evidence of it being so. My aim in this chapter is to discuss patterns and trends in primate sexuality. Separating out sociosexual and reproductive sexual behaviors in many cases, however, is beyond the scope of this chapter.

Biases in Research

In addition to the difficulty in consistently correlating sexual behavior with reproduction in many primate species, much of our knowledge of the sexual and reproductive biology of nonhuman primates comes from research on captive animals. Maintaining captive primates for medical and biological research requires extensive knowledge of their sexual and reproductive biology for successful breeding. Such knowledge is increasingly important as more and more species become endangered in the wild (see Chapter 45), forcing captive programs to breed all their animals "in house." In addition, captive situations lend themselves to studying many facets of sexual and reproductive biology far more than field settings. Studies investigating the interaction between female ovarian cycles and sexual behavior provide a good example. In the wild, such studies are extremely difficult due to the fact that females rarely cycle (i.e., they are usually acyclic as a result of pregnancy or lactation). In contrast, captive conditions allow researchers to prevent pregnancy, a condition that can result in prolonged periods of ovarian cyclicity. Detailed information on the sexual behavior and reproduction of species in more natural settings, however, is a fundamental requirement to our full understanding of the evolution and behavioral ecology of primate species. Given the precarious natural existence of many species, understanding their reproductive and sexual biology in the wild is also becoming more vital, to ensure the success of conservation endeavors.

Additional biases of research into primate sexuality and reproduction stem from the fact that the focus of early primatological inquiry was largely restricted to terrestrial species of Old World monkeys, for example, various species of macaques (*Macaca* spp.) and baboons (*Papio* spp.). These early studies provided extensive and highly useful information regarding the sexuality of these species (e.g., Carpenter 1942b, Conaway and Koford 1965, Kauffman 1965). Recently, it has become clear, however, that the data from these early studies and the theories regarding primate sexual behavior garnered from them were largely taxon-specific and did not reflect the great diversity in the reproductive and sexual morphology, physiology, and behavior of the primate order. For example, the taxonomic ubiquity of certain morphological features, such as sexual skins (see later section "Sexual Skins and Female Genitalia"), was overemphasized in the past due to a dearth of studies on New World monkeys and prosimians. In addition, differences in the sexual behavior of human and nonhuman primate females tended to, and often continue to, be overemphasized because of the continued use of incorrect terminology in the description of monkey and ape sexuality (see later section "The Question of Estrus").

The taxonomic imbalance in data concerning sexuality and reproduction stemmed from disparities in general research on certain primate groups but also from the fact that the ease of studying these topics specifically varies widely from species to species. For example, rhesus monkeys (*Macaca mulatta*) are largely terrestrial (making them easier to follow than an arboreal species), females have sexual skin coloration changes (allowing for visual determination of ovulation) (Zuckerman et al. 1938), and individuals copulate frequently during a defined breeding season and, in fact, within a given day (Manson 1992). As a result of these characteristics, researchers of rhesus monkey sexual behavior and reproduction can record large numbers of sexual encounters and visually estimate ovulation from sexual skin changes. Such forms of data allow investigations into a wide variety of topics, such as the relationship of sexual behavior to physiology (e.g., Herbert 1970, Michael and Zumpe 1970, Dierschke et al. 1974, Chambers and Phoenix 1987), mate choice (Manson 1992, see Chapter 30), and reproductive success (Bercovitch and Nürnberg 1996). In contrast, the primates that are the main focus of my research, spider monkeys (genus *Ateles*), spend nearly all of their time high in the forest canopy (Youlatos 2002) and often copulate in seclusion from other group members (Campbell 2006a, Campbell and Gibson 2008), making it highly unlikely that researchers will actually witness a copulatory event. In addition, female spider monkeys have no visual cue to ovulation, so this can be determined only from hormonal data (Hernández-López et al. 1998, Hodges et al. 1981, Campbell et al. 2001) and vaginal swab histology (McDaniel et al. 1993, Goodman and Wislocki 1935). Recent advances in noninvasive forms of collecting hormonal information (see Chapter 23) are slowly changing this disparity in research on certain taxonomic groups. However, the sexual and reproductive biology of many species remains largely unstudied in either the captive or the natural setting.

FEMALE MORPHOLOGY AND PHYSIOLOGY

Ovarian and Menstrual Cycles

After puberty, primate females experience cyclical changes in the excretion of certain hormones (e.g., estrogen and progesterone) as a result of changing reproductive states. Alternating periods of ovarian cycling (including ovulation), pregnancy, and lactation (in this order) each bring a unique hormonal and, in some cases, behavioral profile. When a female is lactating, most do not experience ovarian cycles and, consequently, do not ovulate, usually preventing females from conceiving when the nutritional requirements of their current offspring are still great (Valeggia and Ellison 2001). An exception to this general pattern is found among female callitrichids and some lorisiforms (e.g., *Arctocebus calabarensis*), where ovulation and conception may occur within a few days of giving birth, that is, when they are lactating (Dixson 1992, Bearder 1987). In all primates, ovarian cycling involves the cyclical buildup and regression of the uterine lining of blood capable of supporting a fertilized egg upon conception. In many primate species, uterine blood is reabsorbed during regression; and thus, no external bleeding is observed (Strassman 1996). Some New World monkey species (e.g., *Ateles geoffroyi*, Goodman and Wislocki 1935, Campbell et al. 2001) experience some level of external blood loss, but this is highly variable both within cycles and between females. In many species of Old World anthropoids, including *Homo sapiens*, copious externally visible bleeding occurs; and in such cases, these are referred to as "menstrual cycles."

The evolution of overt menstrual bleeding has received much attention in the primatological literature. Profet (1993) proposed that menstruation is a means of removing pathogens transported to the uterus via sperm. This hypothesis carries with it three predictions. First, pathogens should be more prevalent before, rather than after, menstrual bleeding. Second, menstruation should occur during life stages when pathogen loads are likely to be highest. Third, in species where promiscuous breeding is common, and hence the likelihood of exposure to multiple pathogen sources is increased, menstruation should be more prevalent. Strassman (1996) tested each of these predictions with data from the primatological literature and found that there is little to no evidence to support the pathogen-removal hypothesis. Alternatively, the regression and rebuilding of the lining is proposed to be energetically less expensive than maintaining it constantly (Strassman 1996). Strassman also hypothesizes that overt cyclical bleeding, as opposed to absorption of blood, may be a by-product of the different endometrial microvasculature in Old World anthropoids; that is, levels of endometrial blood in these primates may be too great to be reabsorbed effectively and, thus, are partially shed.

Sexual Skins and Female Genitalia

Female reproductive morphology is highly variable in the primate order. As already noted, early studies of many Old World monkeys emphasized sexual skins (swelling of the skin surrounding the anogenital region) in the signaling of ovulation to adult males. Where sex skins were absent, ovulation was considered "concealed," and various hypotheses were put forward to explain this concealment in both nonhuman primate (Burley 1979, Andelman 1987, Datta 1987) and human (Benshoff and Thornhill 1979, Lovejoy 1981, Strassman 1981, Daniels 1983, Turke 1984) females. Although a highly conspicuous event (Figure 29.1), the importance of these sexual skins in the reproductive and sexual biology of the primate order as a whole has been overemphasized as the trait is by no means widely distributed. Today, it is recognized that this characteristic has evolved at least three times, twice in the Cercopithecoidea and once in the Hominoidea. Only 25 species of Old World monkey and two species of ape (*Pan troglodytes* and *P. paniscus*) show exaggerated cyclical swelling or coloration changes of the perineal skin (Dixson 1983a). Given the limited distribution of this trait within the primate order, focus should, and in some cases has, changed to asking why such elaborate signals are present (e.g., Burt 1992, Nunn 1999), rather than why they are not, as a lack of strong visual signaling of female reproductive state appears to be the "norm" in the primate order.

Sexual skin swelling (*tumescence*) results from cyclical water retention in the skin surrounding the anogenital region in association with fluctuating hormone levels, particularly estrogen (Bullock et al. 1972, Wildt et al. 1977, Dixson

Figure 29.1 Chimpanzee female (*Pan troglodytes*) with a sexual swelling (photo by Ilka Herbinger).

1983a, Thomson et al. 1992). Swellings can be so exaggerated that they may result in significant changes in female body composition; for example, in some subspecies of red colobus monkeys (*Piliocolobus badius*), female body weight may increase by 25% when fully swollen (Struhsaker 1975), and the midcycle swelling of chimpanzee females (*Pan troglodytes*) can increase the depth of the female vaginal canal by 50% (Dixson and Anderson 2002). In addition to swelling, the skin often changes color, becoming bright pink or red as a result of increases in blood supply to the region (Collings 1926). In most species, maximal tumescence is reached close to or on the day of ovulation after being stimulated by increasing estrogen levels during the follicular phase of the cycle (the stage associated with development of the ovum prior to ovulation). This is by no means a perfect association, however; and females may actually ovulate within a number of days surrounding maximal tumescence (e.g., *Macaca tonkeana*, Thierry et al. 1996; *Cercocebus torquatus atys*, Whitten and Russell 1996; *Pan paniscus*, Reichert et al. 2002; *Papio hamadryas anubis*, Higham et al. 2008). After ovulation, the area detumesces, the color intensity reduces, and the cycle begins again (Wildt et al. 1977, Thomson et al. 1992).

Researchers have long suggested that sex skins act as strong visual stimulants to males, which find fully swollen and/or colored females highly attractive; and experimental work appears to lend strong support to this hypothesis (Bielert and Girolami 1986, Girolami and Bielert 1987). In a series of experiments involving captive chacma baboons (*Papio ursinus*), males showed a 20-fold increase in seminal ejaculations when shown an ovariectomized female with a false sexual swelling attached to her rump. A lack of ovarian function and invitational behavior in the experimental females strongly suggest that changes in male behavior resulted solely from the visual cue of the fake swelling (Girolami and Bielert 1987). Further experiments showed that the red coloration of a sex skin is also important in stimulating male sexual arousal (Bielert et al. 1989), although more recent research suggests that the informational content of swelling size and color may differ (Higham et al. 2008).

Several hypotheses have been proposed to explain the adaptive function of sex skins. Clutton-Brock and Harvey (1976) suggest that the visual stimulus of sex skins prompts adult males to compete for females, thus allowing females to choose "the best male" from among the competitors. Hrdy (1981b) proposes that, rather than instigating male/male competition, sex skins attract "multiple males," all of which may be potential mating partners (see also Hrdy and Whitten 1987, Wallis 1992b). Hamilton (1984) hypothesizes that sex skins are true indicators of the time of ovulation ("obvious ovulation"), thus allowing males to pinpoint the best time to copulate with a female (e.g., *Macaca sylvanus*, Brauch et al. 2007). van Noordwijk (1985) elaborates on the previous hypothesis by suggesting that the "obvious ovulation" indicator allows females to manipulate the behavior of males in a way that provides them protection from predators and conspecifics during consortships. Such consortships are more likely to occur around the time of maximal swelling. Two hypotheses gaining more attention are the "reliable indicator" hypothesis (Pagel 1994) and the "graded signal" hypothesis (Nunn 1999). Pagel (1994) suggests that sex skins are a "reliable indicator," providing information to males regarding the reproductive "quality" of the female, allowing males to predictably choose females that are likely to be successful mothers. A variation on the reliable indicator hypothesis suggests that swelling size indicates cycle to cycle variation in the likelihood of conception (Zinner et al. 2002, Emery and Whitten 2003). Thus, as females get closer to conception, their swellings should become larger (e.g., Deschner et al. 2003, 2004; Gesquiere et al. 2007; Higham et al. 2008). Nunn (1999) hypothesizes that sex skins act as a "graded signal" of the probability of ovulation. This hypothesis proposes that the imperfect association of maximal swelling with ovulation was selected for as it allows females to limit paternity to the more dominant males (which are more likely to mate-guard at maximum swelling) while at the same time attracting multiple males outside of peak swelling in order to confuse paternity (Nunn 1999, Deschner et al. 2004, Barelli et al. 2008). Zinner and Deschner (2000) suggest that sexual swellings occurring in female hamadryas baboons (*Papio hamadryas*) after a male take-over may act as a deceptive signal in order to prevent infanticide by the new breeding male. To date, most of the data appear to lend more support to the graded signal (Nunn 1999) than the reliable indicator (Pagel 1994) hypothesis (Deschner et al. 2003, 2004; Setchell and Wickings 2003; Setchell et al. 2006a; Gesquiere et al. 2007; Higham et al. 2008), although as indicated there are data supporting the notion that swelling size indicates likelihood of conception. It also appears unlikely, given the discontinuous taxonomic distribution of the trait, that one hypothesis will sufficiently explain its function across all species.

There is general lack of exaggerated physical change in the appearance of external female genital regions in the prosimians and New World monkeys. That said, females of many species without exaggerated sex skins show some swelling and color changes in their genitalia during the periovulatory phase of the cycle. Female ring-tailed lemurs (*Lemur catta*) show moderate swelling and reddening of the external genitalia during estrus (Jolly 1966), and spider monkey (*Ateles* spp.) females may show a small amount of swelling of the clitoris; however, this is highly variable within cycles and between individuals (Goodman and Wislocki 1935, C. J. Campbell personal observation). Strier (1996b) believes that in an arboreal habitat (as inhabited by all New World monkeys) visual cues of reproductive state would be of little use as visibility is often obscured. Thus, chemical or olfactory cues may be more important signals of female reproductive status for New World monkeys and prosimians than for the Old World anthropoids (Epple 1976).

While there is little evidence of exaggerated external changes in the genital morphology of female strepsirhines, some species show extreme internal changes in association with changing hormonal levels and ovulation. Vaginal closure by means of a "vaginal closure membrane" is associated with seasonal shortages in food (Dixson 1998) and has been documented in various species of galagos and lemurs (e.g., *Galago senegalensis*, Manley 1966, Butler 1974; *Microcebus murinus*, Peter-Rousseaux 1964; *Cheirogaleus major*, Peter- Rousseaux 1964; *Varecia variegata*, Shideler et al. 1983b). Such a feature means that copulation is physically impossible in these species except for around the time of ovulation when the vaginal closure membrane regresses.

The clitoris shows great variation in shape and size within the primate order. A hypertrophied clitoris is present in many species of New World monkey, and in some cases (e.g., members of the Atelini tribe) it is pendulous (Figure 29.2) and larger than the flaccid penis of the male (Wislocki 1936). The clitoris of capuchin monkeys (*Cebus* spp.) and some strepsirhine species actually contains a small bone, the os clitoris (Hill 1972, Carosi et al. 2001); however, the function of this bone, if any, remains unknown. The clitoris of some anthropoid species is capable of a certain level of erection (e.g., *Cercopithecus* spp., Hill 1966; *Macaca fuscata*, Wolfe 1984). However, these species are not reported to have an os clitoris, so it is unlikely that this bone functions in a similar way to the male baculum. Unlike in studies of the male penis, there has been little to no attempt to explain the variety in primate clitoral morphology in terms of ecological or social environments.

In many species of New World monkey, the labia are pseudoscrotal. This is extreme in some species of howler monkey (*Alouatta* spp.), where they are of similar color and location as the scrota of males. However, *Alouatta caraya* females do not show any genital mimicry or exaggeration (Thorington et al. 1984). Thus, even within a genus there can be considerable variation in reproductive morphology.

Female Orgasm

Long thought to be a unique feature of human female sexuality, there is now strong evidence for the presence of orgasms (as evidenced by the presence of uterine muscle contractions, increases in heartbeat rate, and climax faces) in female nonhuman primates (Goldfoot et al. 1980, Phoebus 1982, Slob et al. 1986, Triosi and Carosi 1998). The function of the female orgasm remains unclear however. Early work stressed its function in maintaining a pair-bond in humans by making sex with one partner enjoyable for the human female (Alexander 1979). The presence of an orgasm in nonhuman primates, especially in species with multimale/multifemale mating systems, appears to negate this possibility. It has been postulated that the female orgasm functions in one of two ways to ensure conception. Firstly, muscle contractions may enhance the movement of the seminal fluids and, hence, sperm up the vaginal canal (Morris 1967). Studies imaging the movement of seminal fluid injected with radio-active opaque materials, however, suggest that this is not the case (Masters and Johnson 1966). Secondly, female orgasm may aid in sperm retention (Morris 1967, Baker and Bellis 1993), and there is some evidence in humans that this may be the case (Baker and Bellis 1993). Female orgasm is not

Figure 29.2 Hypertrophied clitoris of a juvenile female spider monkey, *Ateles geoffroyi* (photo by author).

necessary for spermatic transport, however, and thus is unlikely to have evolved as an adaptation solely to ensure conception (Smith 1984, Baker and Bellis 1993). Work with Japanese macaques (*M. fuscata*) suggests that social stimuli may be important in the promotion of female orgasm in this species; that is, low-ranking females were more likely to experience orgasm than their high-ranking counterparts. In addition, females of all ranks were more likely to orgasm when copulating with a high-ranking male (Triosi and Carosi 1998). The authors of this study suggest their data provide evidence that the female orgasm may have evolved as a mate choice adaptation. Alternatively, the function of the female orgasm may be far more basic in that it may simply provide enjoyment for the female in order to ensure her participation in sexual encounters. Testing this possibility will prove difficult, however, as measuring "enjoyment" is beyond our current capabilities.

MALE MORPHOLOGY AND PHYSIOLOGY

Penile Morphology

Like the female clitoris, penis length and complexity are highly variable. Species that typically exhibit pair-bonded or polygynous breeding systems tend to have relatively shorter penises with relatively simple distal ends (Dixson 1987a). In contrast, where females typically mate with multiple males, penis length is on average greater and the distal end of the penis tends to be more complex in terms of general shape and the presence of specialized features such as penile spines (Dixson 1987a). Some descriptions of penile morphology in the literature may be misleading, however, as most come from inspections of the flaccid penis or by measuring different parts of the penis (e.g., Dahl 1994). For example, all references to the penis of male spider monkeys (*Ateles* spp.) come from one citation (Hill 1962) that describes the penis as being elongated with a relatively simple or undifferentiated distal end. However, from my observations of spider monkeys, it is clear that the distal end of the erect penis is actually mushroom-shaped (Figure 29.3), as was described by Hill (1962) for the closely related muriquis (*Brachyteles* spp.) and woolly monkeys (*Lagothrix* spp.).

The penis in primates typically contains a bone (os penis, os priapi, or baculum), as it does in several other orders of mammals. This bone is completely absent, however, from tarsiers, the Atelini primates, and humans (Hill 1972, Dixson 1987b). Bacula size, relative to body size, varies significantly (see Dixson 1987b, 1998, for a full list of bacula lengths); and this variation in length can be largely explained by taxonomy and copulatory pattern. In general, for their body size, there is great variation in baculum length in the strepsirhines, being greatest in the lorises. New World monkeys in general have smaller bacula than their Old World counterparts; indeed, in some species of New World monkey this tendency toward size reduction is so extreme

Figure 29.3 Erect penis of an adult male spider monkey (*Ateles geoffroyi*) soon after copulation (redrawn by Jody Biggers from photo by author). Note the distinctive shape of the distal end of the penis.

that the baculum is now absent. Among the Old World monkeys, the colobines have shorter bacula lengths in relation to body size than the cercopithecines. Hominoids have smaller relative bacula lengths than most of the Old World monkeys (Dixson 1987b).

Once differences in phylogeny have been controlled for, species that show prolonged intromission times tend to have greater relative bacula lengths (Dixson 1987a,b). It has been suggested that the bacula may assist in copulation when initial penetration is difficult either as a result of the male being much larger than the female or because of a complex penile morphology (Long and Frank 1968). Alternatively, Romer (1962) proposed that the role of the baculum during copulation might be as a supporting rod for the penis, allowing for the maintenance of erection during a prolonged intromission period. However, the relationship between bacula length and copulatory pattern is by no means concrete as the Atelini monkeys show extended copulation lengths but lack a baculum entirely.

The surface of the penis of many primate species is covered with androgen-dependent spines. Dixson (1998) classifies these spines into three different levels depending on their size, shape, and complexity. The most complex spines (type 3) are found in the prosimians. While the function of penile spines in nonhuman primates is not fully understood, they are suggested to provide tactile feedback to the male, aiding in intromission, prolonged copulations, and removal of sperm plugs (Dixson 1998), and possibly to the female, providing sensory feedback and possibly even increasing the chance of facultative reflex ovulation in females of some species (Bercovitch and Ziegler 2002), although there is no direct evidence of this.

The "lock-and-key" hypothesis has been proposed to explain the evolution of genital morphology in the primate order. This hypothesis suggests that penile morphology has evolved to complement female genitalia and eventually forms a species-isolating mechanism. In general, however, there is little support for this hypothesis. Penis morphology can often vary significantly within a genus and in some cases can be used to distinguish between species that are otherwise visually very similar, e.g., *Galagoides* spp. (Perkin 2007). The classic example of such intrageneric variation is found within the macaques (*Macaca* spp., Fooden 1976). The lock-and-key hypothesis predicts that different species within this genus would not be able to hybridize as the vaginal morphology of one species would not match the penile morphology of another. In fact, this is not the case, and many species have been demonstrated to hybridize in captivity (Bernstein and Gordon 1980). One species of macaque, *M. arctoides*, does appear to show complementary evolution of male and female genitalia (Fooden 1967), suggesting that, in isolated cases, the lock-and-key hypothesis may have some validity. The increased complexity of penile morphology (length, spines, shape, and baculum length) in species where females typically mate with multiple males supports Eberhard's (1985) hypothesis of sexual selection by female choice. The long filiform penis of chimpanzees (*Pan troglodytes*) also represents a likely case of cryptic female choice. Its length is necessary to successfully deliver sperm to the cervix when the female is maximally swollen, which can increase the depth of her vaginal canal by up to 50% (Dixson and Anderson 2002).

Testes Size and Sperm Competition

Schultz (1938b) first documented the wide variation of testes size relative to body size in the primate order. Much later, Parker (1970) hypothesized that male insects might compete by means of the quality or quantity of their sperm in species where females mate with multiple males. Primatological data lend support to this "sperm-competition" hypothesis in that relative testes size is generally greater in males of species where the females mate with multiple males and smallest in those species that typically exhibit pair-bonded or polygynous mating systems (Short 1979, Harcourt et al. 1981a, Dixson 1987a, Jolly and Phillips-Conroy 2003). Larger testes produce greater quantities of sperm (Møller 1988), a greater percentage of motile sperm (Møller 1988), a higher frequency of sperm that are "capacitated" or ready to interact with the ovum (Gomendio et al. 2006), and longer sperm (Dixson 1993) that tend to have higher motility rates (Gomendio and Roldan 1991). However, Gage and Freckleton (2003) have shown that when phylogeny is controlled across 83 mammalian species, increase in sperm length is not correlated with increased testicular size, suggesting that this might not be as important as once thought. Anderson and Dixson (2002) suggest that, rather than total length, the volume of the midpiece of the sperm is greater in species where females mate with multiple partners. This section contains the mitochondria that power sperm motility (Anderson and Dixson 2002), and thus, an increase in size may reflect an increase in mitochondrial density. Males

with larger testes are proposed to experience an increased likelihood of siring the offspring of the females with which they copulate. Investigations into the potential role of sperm competition and the evolution of testes morphology are typically comparative in nature, that is, comparing testes size and mating strategies over different species. Concrete evidence supporting the hypothesis, however, can come only from investigations within a species, that is, evidence showing that males with larger testes produce more sperm than males of the same species with smaller testicles. In addition, it must be shown that males with larger testes have higher reproductive success than males of the same species with smaller testes. Some evidence to support this comes from Pullen et al. (2000), who have shown that male lesser galagos (*Galago moholi*) with the largest testes during the mating season have the highest mating success; but no data are available to determine whether this translates to greater reproductive success. Intraspecific data from various nonprimates do provide support for the relationship between testes size and reproductive success however (see Gage and Freckleton 2003 for a review).

Copulatory Plugs

The semen of male primates coagulates after ejaculation to varying degrees. In the most extreme cases, the ejaculate forms a solid plug-like structure in the female vaginal canal. The adaptive function of copulatory plugs has traditionally been explained by one of several hypotheses, the two most common of which involve the prevention of sperm leakage out of the vaginal canal (Michener 1984) and a form of male/male competition by creating a physical barrier to subsequent matings (Parker 1970, Schwab 2000). Behavioral observations from various primate species, however, show that intromission by subsequent males may easily remove a copulatory plug from the female's vaginal canal (e.g., *Macaca nemestrina*, Busse and Estep 1984; *Brachyteles arachnoides*, Milton 1985a; *Lemur catta*, Parga 2003; *Microcebus murinus*, Eberle et al. 2007). Such observations suggest that copulatory plugs may be relatively ineffective at preventing subsequent matings. In ring-tailed lemurs (*L. catta*), removal of a copulatory plug appears to be a viable mating strategy for males that are not the first to mate with a female during her receptive period. Males that first mate with females during this time are more likely to practice postejaculatory mate-guarding (Sauther 1991, Parga 2003), thus reducing the possibility of subsequent males mating with the female. Once mate-guarding by the first mating male is terminated, the distinct penile morphology of this species (enlarged distal end with penile spines, Dixson 1987a) is thought to help non-first-mating males to remove copulatory plugs from a female's vaginal canal (Sauther 1991, Parga 2003). Evidence showing that copulatory plugs function to prevent leakage of semen (and consequently sperm) from the vaginal canal in primates is lacking.

SEXUAL BEHAVIOR

Patterns of Sexual Behavior Across the Female Ovarian Cycle

The Question of Estrus

In both seasonal and aseasonal breeders, the sexual behavior of female mammals is often limited to a short period of time coinciding with ovulation, making conception more likely to occur (Heape 1900). This period of female sexual activity is known widely as "heat" and/or estrus and was first defined by Heape (1900:6) as "the special period of sexual desire of the female; it is during oestrus and only at that time, the female is willing to receive the male and fruitful coition rendered possible in most, if not in all, mammals." Today, the term "estrus" is widely used by scientists and laypeople alike and is applicable to most mammalian species. Like many nonprimate mammals, sexual behavior in many lemur and loris species is strictly limited by hormonally controlled mechanisms, such as vaginal closure and the presence of lordotic postures (positional changes necessary for copulation). As such, sexual behavior in these species is seen only during limited periods of time in association with ovulation (*L. catta*, Evans and Goy 1968; *V. variegata*, Shideler et al. 1983b). In some lemur species (e.g., *Propithecus verreauxi*), the pattern is not so clear-cut and sexual behavior may take place under certain social conditions regardless of female hormonal status (Brockman and Whitten 1996). Indeed, Whitten and Brockman (2001) suggest that social factors may impact lemur and loris sexuality more than previously thought.

It is unclear whether or not the concept of estrus, given Heape's original definition, is applicable at all to the anthropoid primates. Female anthropoids of many species copulate at times when conception is not likely or possible, that is, throughout the ovarian cycle, when lactating, or when pregnant (e.g., *Ateles geoffroyi*, Campbell 2004; *Brachyteles arachnoides*, Strier and Ziegler 1997; *Cercocebus mitis*, *C. aethiops*, *C. neglectus*, Booth 1962; *Hylobates lar*, Barelli et al. 2008; *Miopithecus talapoin*, Rowell and Dixson 1975; *M. mulatta*, Michael and Zumpe 1970; *Papio ursinus*, Saayman 1970; *Pan troglodytes*, Tutin 1980; *P. paniscus*, Thompson-Handler et al. 1984). Such a major deviation from Heape's (1900) definition has resulted in a plethora of publications attempting to redefine the concept of "estrus" to fit the sexual behavior of the anthropoid primates. Heape (1900) himself noted that sexual behavior does occur during pregnancy in some species and termed such behavior "abnormal estrus." Other authors have coined terms such as "postconception estrus" (Conaway and Koford 1965), "perimenstrual estrus" (Loy 1970, Wolfe 1979), "pseudoestrus" (Hrdy 1977b), and "periabortive estrus" (Wolfe 1979) to describe the sexual behavior of female primates during times other than the time close to ovulation or the periovulatory period.

In addition, various authors differ in the factors they include in a definition of estrus. Loy (1970) states that the term "ovulation" must be removed from the definition. He coined the term "perimenstrual estrus" to describe increases in copulation frequency in rhesus macaques (*M. mulatta*) near menstruation. He noted that hormonally this sexual activity is similar to estrus in that it is associated with heightened levels of estrogen. Thus, he stresses the importance of endocrinology over histology. Other authors have suggested that researchers may be stressing the histological and endocrinological components of estrus at the expense of the behavioral component (e.g., Carpenter 1942b). Rowell (1972) shows that there is often little correlation between sexual behavior and stage of the menstrual cycle of Old World monkeys. She goes on to suggest that there are no histological components to estrus. Likewise, Hrdy and Whitten (1987) remove any relationship with ovulation from the concept of estrus.

Resulting from this confusion over how to define estrus in anthropoid primates, other authors have suggested that the term should not be used when discussing the sexual behavior of female monkeys and apes (Loy 1987, Dixson 1998, Campbell 2004). Removal of the term allows for a more detailed understanding of female sexuality. A clear example of how this term has clouded our vision of female sexuality relates to its ubiquitous use in reports on the female sexual behavior of monkeys and apes but not humans. By realizing that, in fact, monkey and ape females do not exhibit true estrous cycles, we can begin to understand that human female sexuality and its seeming lack of association with hormonal fluctuations may not be as different from that of other anthropoids as has long been postulated in the literature.

Alternatives to the Estrus Concept

Alternatives to the concept of estrus in female primates are available. Beach (1976) proposed a model that separates female sexual behaviors into three characteristics: proceptivity, receptivity, and attractivity. In the same year, Keverne (1976) proposed that female behaviors should be investigated in terms of hormonal influences on attractivity and receptivity. Behaviors and physiological conditions that are necessary and sufficient for copulation to take place indicate *receptivity*. For example, female black and white ruffed lemurs (*Varecia variegata*) are not receptive outside of the periovulatory period as their vaginal canal is closed, lordosis is not inducible, and therefore copulation is physically impossible (Shideler et al. 1983b). *Proceptive* behaviors are those that a female engages in to indicate her motivation to mate, such as presentation of the anogenital region, facial displays, and vocalizations. *Attractivity* refers to the stimulus effects of the female on the male, that is, how attractive she is to an interested male. Any changes in the attractiveness of a female primate can be observed by changes in the behavior of adult males toward the female in question, often in relation to olfactory (urinary scents) and visual (sexual swellings) cues. Changes in the frequency of sexual behaviors in anthropoids are usually attributed to changes in attractivity and proceptivity, rather than receptivity (Beach 1976, Keverne 1976); that is, monkey and ape females have no physical barrier to copulation at any time. These two models deal solely with the behavioral component of female sexuality, but they do provide a framework to investigate the association of these behavioral changes with underlying histological and endocrinological changes. While these terms are becoming more widely used in the primatological literature, they are not without their own problems. The most persistent of these is that it can often be difficult to determine whether certain behaviors are related to attractivity, proceptivity, or receptivity. For example, it is likely that a proceptive female is more attractive to a male because of the mere fact that she is, by definition, motivated to copulate. Thus, these behavioral categories are not necessarily independent.

Copulatory Patterns

Specific copulatory sequences and behaviors within the primate order vary widely and are beyond the scope of this chapter. All primates, however, can be classified as having one of three copulatory patterns (Dewsbury and Pierce 1989, Dixson 1998). These patterns are typically distinguished by whether each copulatory bout involves single or multiple *intromissions* (insertion of the penis into the vagina). In the case of multiple intromission patterns, such as that found in the rhesus monkey (*M. mulatta*), ejaculation occurs only after a series of brief intromissions, each merely a few seconds long (Carpenter 1942b). Single intromissions are classified as brief (e.g., mantled howler monkeys, *Alouatta palliata*, approximately 30 sec; C. J. Campbell personal observation) or prolonged (e.g., black-handed spider monkeys, *A. geoffroyi*, 12–24 min; Campbell 2000b, 2006a). Additional features, such as the presence or absence of a copulatory lock, are often used to distinguish among mammals in general; but it is unclear as to whether or not this phenomenon actually exists within the primate order (Dewsbury and Pierce 1989, Dixson 1998). Many species appear to maintain intromission past male ejaculation. This can be anywhere from a few minutes or less in species such as the black-handed spider monkey (*A. geoffroyi*, Campbell 2006a) to up to several hours in thick-tailed bushbabies (*Galago crassicaudatus*) (Eaton et al. 1973). Whether the male and female are actually locked together and unable to physically separate, as is the case in dogs, is not known however (Dewsbury and Pierce 1989, Dixson 1998).

Copulatory Postures

Primate copulatory postures are somewhat stereotyped, with a dorsoventral position being the "norm"; that is, the dorsal surface of the female meets with the ventral surface of the male (Figure 29.4). Variations on this dorsoventral posture

are related to different ways in which the animals are oriented in their environment and ways in which males and females hold on to each other. Members of the Lorisinae are usually suspended underneath a branch while copulating in an otherwise standard dorsoventral posture (Schulze and Meier 1995b). In many species of Old World monkey (e.g., baboons and macaques) and in some species of New World monkey (e.g., capuchins and squirrel monkeys) and strepsirhines (e.g., galagos), the dorsoventral position is modified slightly by the male "climbing" on the female and grasping her lower hindlegs with his feet, also referred to as a "double foot clasp" (Dixson 1998) (Figure 29.4). Members of the Atelini tribe (see Chapter 11) vary from the standard posture in that the male does not mount the female but, rather, sits upright on a branch and the female then backs onto him (Figure 29.5). Probably to stabilize the pair during extended copulation times, the male wraps his legs over those of the female in what has become known as a "leg-lock" (Klein 1971, Milton 1985a, Nishimura 1988). Copulatory postures in the apes appear to be more varied, especially in the bonobo, where ventroventral (face-to-face) copulations are often observed (26% of copulations in one study, Thompson-Handler et al. 1984). In the other apes, face-to-face copulations are observed less frequently and the more standard dorsoventral posture predominates (Harcourt et al. 1981b, Goodall 1986), although the arboreal orangutan is often seen copulating in a variety of suspensory positions (Nadler 1977).

Sociosexual Behavior

Homosexuality

While homosexual relationships are present in the human species, there has been little to no evidence to date of

Figure 29.4 Dorsoventral copulatory posture with male foot clasp in the Japanese macaque, *Macaca fuscata* (photo by Linda Wolfe).

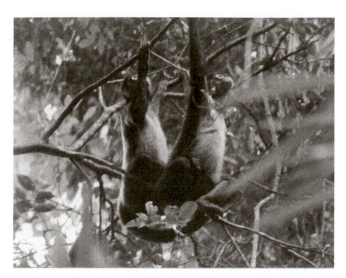

Figure 29.5 Dorsoventral copulatory posture of spider monkeys, *Ateles geoffroyi*. Note the "leg-lock" behavior of the male seated behind the female (photo by author).

a consistent choice for same-sex sexual partners among any nonhuman primate species (Dixson 1998, Kirkpatrick 2000). Indeed, there is some disagreement among authors as to how to define "homosexuality," and this probably adds to the confusion over whether homosexuality is a phenomenon specific to humans or one that applies to primates in general. Some define any behavioral patterns performed by members of the same sex that are common to heterosexual copulations as evidence for homosexuality, for example, mountings, presentations, and genital manipulation (Nadler 1990, Wolfe 1991a, Kano 1992, Vasey 1995, Fox 2001). Such behaviors are common throughout the primate order and often function in the formation and/or maintenance of social relationships, alliances, etc. Perhaps the most widely publicized example of such behavior among the nonhuman primates is the so-called genitogenital (g–g) rubbing of bonobos, where females rub their genitals together (Kano 1992). This behavior appears to act both as a tension reducer and as a social bonding mechanism among adult females (Kano 1992). In contrast, Wallen and Parsons (1997) and Dixson (1998) stipulate that homosexuality in humans is not defined by mere behavioral patterns but by the presence of a "preference" for members of the same sex as sexual partners. Thus, they define "homosexuality" as a preference for sexual encounters with members of the same sex. Using this definition, evidence for homosexuality in the nonhuman primates is generally lacking. However, female/female sexual behavior and consortships are common among certain populations of Japanese macaques (*Macaca fuscata*) (Vasey 2004) and some females do appear to prefer same-sex partners over males (Vasey and Gauthier 2000, Vasey 2004).

Masturbation

Among the Old World anthropoids, self-genital stimulation by males, which may or may not result in ejaculation,

appears to be relatively common (e.g., *Piliocolobus badius*, Struhsaker 1975; *Macaca mulatta*, Carpenter 1942b, Lindburg 1971; *Papio anubis*, Ransom 1981; *Pan troglodytes*, Goodall 1986). Reports of female masturbation are less common (e.g., *Papio anubis*, Ransom 1981; *Pan troglodytes*, Goodall 1986), and this may be partially due to the fact that male masturbation is likely to be more visually obvious to human observers. Among the New World monkeys, reports of masturbation are less common. Female spider monkeys (*A. geoffroyi*) are known to rub or hold their own clitoris, and both of these behaviors appear to occur more frequently in females that are reproductively cycling (Campbell 2004). There does not, however, appear to be a relationship of this behavior with any particular stage of the ovarian cycle (Campbell 2004). In addition, it is difficult to determine whether or not females are sexually stimulating themselves. A possible alternative explanation is that they are reacting to the presence of irritating insects such as mosquitoes that are likely to be attracted to the large hairless clitoris. Masturbation in prosimians has not been reported. Much remains unknown about nonhuman primate masturbation. For example, why the prosimians and New World monkeys differ from the Old World anthropoids in the occurrence of this behavior is unclear. In addition, whether or not masturbation has any adaptive function is unknown, and it is more likely that such behavior is a side effect of the increased tactile sensitivity of genitalia in response to their copulatory function.

REPRODUCTION

Conception

Young female primates typically go through a period of adolescent sterility or subfecundity (Knott 2001, Bercovitch and Ziegler 2002) where they do not ovulate even though they are experiencing gonadotropin pulsatility (Bercovitch and Ziegler 2002) or where seemingly normal hormonal cycling coincides with a complete lack of sexual behavior, a lack of interest by adult males, and a lack of conception (e.g., *Papio cynocephalus*, Altmann et al. 1977; *Gorilla gorilla*, Harcourt et al. 1980, Watts 1991b). Bercovitch and Ziegler (2002) suggest that this time might be adaptive by delaying ovulation to a time when socioecological factors that favor offspring survival are present. Although adult female primates in natural conditions rarely cycle (i.e., they are usually pregnant or lactating), data suggest that multiparous females (those who have previously reproduced) usually cycle for a few months before conception occurs (e.g., *A. geoffroyi*, three to six cycles, Campbell et al. 2001; *B. arachnoides*, three to six cycles, Strier and Ziegler 1997; *Papio cynocephalus*, 5 months, Altmann et al. 1977; *Papio hamadryas*, one to four cycles, Zinner and Deschner 2000). Females are sexually active during these nonconceptive cycles, and to date, there are no data available to explain why sexual encounters during these cycles do not result in conception. Given the huge number of variables related to a successful conception (effective sperm motility, ovulation, fertilization, implantation, etc.), it is not surprising that the majority of copulatory events are unsuccessful in this regard.

Paternity

When indeed a female does conceive, it is often of interest to primatologists to know which male sired the offspring. Historically, paternity in primates has been inferred by behavioral data; that is, males that were seen to copulate with a female around the presumed time of conception were considered to be likely fathers. With the increased availability of deoxyribonucleic acid (DNA) technology (see Chapter 24), reliance on observed copulations as a source of paternal information is no longer acceptable. Genetic data have illuminated many cases where behavioral observations do not provide correct information concerning paternity. For example, in some species thought to be monogamous, genetics have shown that extrapair males may in fact sire offspring. In the fat-tailed dwarf lemur (*Cheirogaleus medius*) the frequency of extrapair paternity was 44% (Fietz et al. 2000). Likewise, in one study of Japanese macaques, nontroop males sired 55% of offspring born during one study period (Hayakawa 2008). Additionally, in some species where females give birth to multiple offspring and mate with multiple partners, mixed paternity can occur; that is, offspring born in the same litter may have different fathers (e.g., gray mouse lemurs, *Microcebus murinus*, Eberle and Kappeler 2004a; moustached tamarins, *Saguinus mystax*, Huck et al. 2005).

Mating success and reproductive success may not be concordant for a number of reasons (Alberts et al. 2006). The first is a likely problem with many primate species that are difficult to observe; that is, what we observe may not be what is really happening. Certain males may be more likely to be observed copulating, while others may be more likely to participate in furtive copulations not seen by other males and, consequently, researchers. Data from my own research on spider monkeys (*Ateles geoffroyi*) likely suffer from this bias as most copulations are performed in seclusion from other group members and it appears as if dominant males are more likely to copulate in the presence of other group members (Campbell 2006a, Campbell and Gibson 2008). Alternatively, reproductive success may differ from mating success because some males may limit their mating activity to females that are more likely to be reproductively successful, disproportionately increasing their reproductive success in relation to their mating success (Alberts et al. 2006). Lastly, there may be postcopulation selection via sperm competition or sperm selection that skews the reproductive success of one male over another (Engelhardt et al. 2006, Brauch et al. 2008).

In social groups that have multiple males, the highest-ranking male is often assumed to sire most, if not all, of the offspring. However, genetic data have also shown that this may not always be the case. Dominant males may not monopolize all matings for various reasons. First, they may concede mating opportunities to subordinate males because of the benefit of having the continued presence of these males in the group in the defense of females or territories (Kokko and Johnstone 1999). Alternatively, there may be incomplete control by the dominant male and a tug-of-war between males may ensue, with variance in reproductive success reflecting variance in competitive ability (Bradley et al. 2005). The level of exclusive mating access experienced by top-ranking males may also be related to the number of additional males present in the group and the level of synchronicity in female fertile periods (known commonly as the "priority of access model") (Altmann 1962). Higher male-to-female ratios and synchronized female fertility potentially make it more difficult for dominant males to monopolize fertile females. In such situations, subordinate males are hypothesized to be more likely to mate with females and even to sire offspring (although see Charpentier et al. 2005). This model has been challenged by genetic information showing that subordinate and extragroup males indeed sire offspring and not always just because dominant males were not able to effectively mate-guard females (Fietz et al. 2000, Heistermann et al. 2001b, Vigilant et al. 2001, Bradley et al. 2005, Charpentier et al. 2005, Boesch et al. 2006, Alberts et al. 2006, Brauch et al. 2008, Hayakawa 2008). However, male reproductive success does usually correlate with dominance rank, and even though dominant males do not appear to have exclusive mating rights, they often do sire the majority of offspring (Pope 1990, Altmann et al. 1996, Gachot-Neveu et al. 1999, Vigilant et al. 2001, Huck et al. 2005, Alberts et al. 2006, Boesch et al. 2006, Engelhardt et al. 2006).

Seasonality

Not only are many species of nonhuman primate seasonal breeders but sexual behavior is also usually limited to this season. Individual females may only mate over a matter of hours (e.g., gray mouse lemurs, *Microcebus murinus*, Eberle and Kappeler 2004a), days (e.g., ring-tailed lemurs, *Lemur catta*, Jolly 1966), or a few months (e.g., *Macaca fuscata*, Kawai et al. 1967); and female mating periods may or may not be synchronized during a mating season. In such species, births are also strictly limited in their occurrence throughout the year. Some species may even exhibit multiple distinct birthing periods within a year, such as lesser galagos (*Galago moholi*, Pullen et al. 2000). Alternatively, in other species, there is no strict reproductive seasonality; thus, sexual behavior and conception can take place during most months of the year, although "birth peaks" may be recorded (e.g., *Ateles* spp., Milton 1981b, Symington 1987a, Chapman and Chapman 1990, Campbell 2000b; *Alouatta fusca*, Strier et al. 2001).

Most researchers agree that seasonal reproduction is a response to seasonal changes in food supply (triggered by changes in rainfall patterns and day length); however, species do differ in how they react to these changes. In general, it is thought that reproduction is timed so that the most energetically expensive stage of reproduction for females occurs when there is the most food available (Knott 2001). Sichuan snub-nosed monkeys (*Rhinopithecus roxellana*) time their births so that females begin to lactate when food becomes more abundant (Zhang et al. 2000, Bao-Ping et al. 2003). Squirrel monkeys (*Saimiri* spp.) are the most seasonal of the neotropical primates (Di Bitetti and Janson 2000), and their short birthing season appears to serve two functions: first, females give birth during the period of maximal food availability; second, the tightly synchronized birthing season may act to swamp the predator community (Boinski 1987b). For species with longer interbirth intervals, it is difficult to time any one reproductive stage to food availability (Bronson and Heideman 1988, Knott 2001). In such cases, seasonal variation in births is most likely related to female conditional changes in response to fluctuating food availability. A lack of conception during certain months is almost certainly due to poor female condition as a result of low food availability (Knott 1997b, Ziegler et al. 2000b, Strier et al. 2001). During such months, a reduction in food intake can negatively impact hormonal levels, leading to a cessation of ovulation and an inability to conceive (Bercovitch 1987a, Ziegler et al. 2000b). More folivorous primates, such as howler monkeys (*Alouatta* spp.) and mountain gorillas (*Gorilla gorilla beringei*), may be less likely to show seasonal peaks in births or conceptions as their ability to digest mature leaves prevents major seasonal changes in food availability (Watts 1998c, Di Bitetti and Janson 2000, Knott 2001, Strier et al. 2001). Thus, in species with shorter reproductive parameters (e.g., gestation, lactation), breeding can be timed to coincide with a certain level of food production and seasonal breeding is usually present. In species with longer reproductive parameters (usually larger species), any seasonal pattern is more likely to be an immediate interaction between food availability, available energy, and female reproductive condition ("ecological energetics hypothesis") (Knott 2001).

Males of some primate species undergo spermatogenesis during the breeding season only (Gordon et al. 1976, Wiebe et al. 1984, Rostal et al. 1986). For example, in the squirrel monkey (*Saimiri* spp.), a seasonal cycle of spermatogenesis is correlated with a seasonal fatted appearance of the male (Kinzey 1971, Baldwin and Baldwin 1981, Boinski 1987b; see also Chapter 8). During this time, the male is heavy, with subcutaneous fat in the upper torso, shoulders, and arms. This appearance is gradually lost at the end of the breeding season and is absent for much of the year (Kinzey 1971). Male mouse lemurs (*Microcebus murinus*) also gain weight prior to the mating season, probably in response to testicular tissue development and steroid action (Schmid and Kappeler 1998). In other species, spermatogenesis may continue throughout the year but sperm quality may differ. In male spider monkeys (*A. geoffroyi*), sperm quality appears (in terms of motility and percent of live sperm) to be better during months when females are more likely to conceive (Hernández-López et al. 2002), suggesting that declining female body condition may not be the only factor leading to reduced conception rates.

CONCLUSIONS

The information presented in this chapter briefly reviews the sexual and reproductive biology of the nonhuman primates. It should be obvious to the reader that the topics presented here are diverse and that the primate order also shows a great deal of diversity in many of the areas covered. Indeed, entire books have been written on the topics presented here, and I would refer anyone interested in further information to sources such as Dixson (1998) for a more extensive review. Future studies of primate sexuality and reproduction will benefit greatly from advancements in endocrinological and genetic techniques such as those discussed in Chapters 23 and 24 of this volume. In addition, an important avenue of future research is the increasing tendency within the primatological community to study species about which we currently know little. Data across all taxonomic categories will allow us to perform more complete comparative studies and help to broaden our knowledge in this fascinating field of study.

30

Mate Choice

Joseph H. Manson

1. What is sexual selection, and why does it predict that in most primate species females will be more choosy than males when selecting mates?
2. What are some of the criteria used by females in selecting male mating partners?
3. Although traditionally not considered to be particularly selective in choosing mating partners, what are the criteria that male primates may use to select female mates?

HISTORY AND THEORY

Background and Definitions

During the first few decades of behavioral primatology, mate choice among nonhuman primates was given almost no formal attention. As in biology generally, female mate choice was dismissed as unlikely or unimportant despite Darwin's (1871) and Fisher's (1930) theoretical arguments for its likely role in producing exaggerated male ornaments (Cronin 1991). Male mate choice was regarded as even less plausible than female mate choice. The rise of evolutionary social theory (Hamilton 1964; Williams 1966a; Trivers 1971, 1972; Maynard Smith 1974; Wilson 1975) and the growing prominence of female primatologists (Hrdy 1986, Fedigan 1982) put female primate mate choice on the map as a research topic by the 1980s. The past 15 years have witnessed the steady production of systematic studies of primate female mate choice, some of them accompanied by genetic paternity data that have greatly increased researchers' ability to test hypotheses. Few systematic studies of male mate choice have been carried out, but some inferences are possible based on available data. In this chapter, I will review the relevant theoretical concepts and use these as a framework within which to examine findings about primate mate choice. I will focus on studies conducted since 1990 because (*1*) few rigorous studies focusing on mate choice were conducted before then and (*2*) reviews by Fedigan (1982:282–285), Smuts (1987), Small (1989), and Keddy-Hector (1992) summarize the largely anecdotal information available at that time.

Mate choice may be defined as "any pattern of behavior, shown by members of one sex, that leads to their being more likely to mate with certain members of the opposite sex than others" (Halliday 1983:4). "Mating" here refers to the actual production of offspring, although many studies have used the occurrence or frequency of copulation as a proxy for reproduction. The strength of Halliday's definition is that it focuses attention on the evolutionary causes and consequences of repeated behavior patterns (i.e., their effects on the fitness of both the choosing and the chosen sex), rather than on the internal states (preferences, desires) of putative choosers, which are difficult to infer and in any case have evolutionary consequences only insofar as they are manifest in overt behavior. In contrast to mate *choice*, mate *preference* "denotes a disposition or propensity that an individual possesses whether or not it is exercised" (Heisler et al. 1987:99).

Sexual Selection Theory and Mate Choice

Intrasexual and Intersexual Selection

Darwin (1871) developed the concept of sexual selection to explain certain traits that apparently decrease their bearers' probability of survival, the prototypical example being the peacock's tail, which presumably increases vulnerability to predation. He proposed that such traits confer an advantage in acquiring mates that compensates for their negative effects on survival. Sexually selected traits may be weapons (e.g., antlers, large body size) used in contests over mates (*intrasexual selection*) or ornaments that increase their bearers' attractiveness to the opposite sex (*intersexual selection*). Darwin (1871) noted that, where the sexes differed in their endowments of weapons and/or ornaments (*sexual dimorphism*), it was usually the males that had more elaborate versions of sexually selected traits. Although intrasexual selection was accepted by biologists, female mate choice for male ornaments was, for the most part, decisively rejected (e.g., Huxley 1938), even after Fisher (1930) argued that an initially modest male ornament and a slight female preference for it could lead to the "runaway" evolution of a highly exaggerated ornament.

Parental Investment and Sexual Selection

Trivers (1972) played a major role in rehabilitating Darwin's theory of intersexual selection by proposing, as the source of the usual sex difference in weapons and ornaments, sex differences in parental investment. He defined *parental investment* as any investment in time, energy, or risk by a parent in an offspring that enhances the offspring's reproductive success (RS) and that detracts from the parent's ability to invest in other offspring. In this model, access to members of the more heavily investing sex (usually females, particularly in mammals) becomes the resource limiting the RS of the less heavily investing sex. In species with the typical sex difference in parental investment, males are then forced to compete against each other for access to females, either directly or indirectly through contests for dominance, producing selection for weaponry (in excess of female weaponry) and possibly for ornaments. Variance in male RS is hypothesized to exceed variance in female RS, and males try to acquire large numbers of mates, whereas females (whose RS cannot be increased by mating with additional males) focus on choosing mates that offer either genes that enhance offspring survivorship (possibly signaled by elaborate ornaments) or parental investment to supplement the mother's contribution (see Chapter 32 for additional discussion).

Although this model remains a cornerstone of sexual selection research, recent theoretical and empirical work has eroded the sharp distinction between "choosy" females and "promiscuous" males (reviewed by Drea 2005). Females may, in fact, increase their RS by mating with multiple males for a number of reasons: some males may be infertile, or a single male may not be able to produce enough sperm to impregnate all his mates (Small 1988, Smith 1984); a more genetically diverse set of offspring may experience higher survivorship on average than the offspring of a single male (Parker 1970, Keller 1994, Olsson et al. 1994, Williams 1975); and multiple males may be induced to invest in an offspring, or at least to refrain from harming it, by providing it with some probability of paternity (Hrdy 1981b; but see Soltis and McElreath 2001). On the other hand, males do not always benefit by indiscriminate mating, even in species where they provide little parental investment. Ejaculates may be costly to produce (Dewsbury 1982), male/male competition may be potentially damaging enough that it pays males to choose their contests carefully (Pagel 1994), and females may vary in their reproductive and parenting capacities and, hence, in the likely reproductive payoff from mating with them (Altmann 1997).

Mate Choice for Genes

Theorists generally agree that when males provide females with resources, parental care, or other phenotypic benefits, females should choose males on the basis of their ability and willingness to provide these resources (Andersson 1994). However, controversy continues regarding the adaptive value of choosing mates on the basis of their ornamental features such as bright plumage or elaborate displays. In Fisher's (1930, Lande 1981) formulation, sexually selected ornaments are arbitrary and the only benefit to females of preferring ornamented males is that their sons will inherit the ornament and therefore be attractive to females. Zahavi (1975) argued, in contrast, that by decreasing their bearers' survival prospects, inherited ornaments send the unfakeable message that their bearers' *other* genes increase viability: They tell prospective mates, in effect, "Having survived to breeding age with this ornament, I must be of superior genetic quality." This hypothesis (the "handicap principle") was overwhelmingly rejected by behavioral ecologists until Grafen (1990) produced a formal model that illuminated two aspects of the idea that Zahavi's verbal arguments had left unclear: first, an individual's level of ornamentation must be adjusted facultatively; second, low-quality individuals must suffer larger viability costs than high-quality individuals for producing the same level of ornamentation. Under these conditions, an evolutionarily stable state is possible in which males produce costly ornaments proportionately to their underlying quality and females use these ornaments as reliable indicators of male genetic quality. A large number of empirical studies in various animal taxa have supported predictions from the handicap principle (e.g., Petrie 1994, Kose et al. 1999), though support for Fisher's "runaway" process has also been found (e.g., Brooks 2000).

"Good genes" models, such as Zahavi's (1975), also faced the problem of explaining how heritable variation in fitness could be maintained over evolutionary time. Would selection not quickly eliminate this variation and thereby eliminate the basis of mate choice? Hamilton (1980, Hamilton and Zuk 1982) proposed that parasites, being selected to circumvent hosts' immune systems and having much shorter generation times than their hosts, create cyclically varying selection pressures, such that the most fit host genotypes change fairly rapidly. A large number of studies have found that sexually selected ornaments accurately display parasite resistance (e.g., Duffy and Ball 2003, Doucet and Montgomerie 2003). A related line of research has discovered evidence that heterozygotes, particularly at the loci of the major histocompatibility complex (MHC), which affects immune function, are better able to resist pathogens than homozygotes (Potts and Wakeland 1993, Brown 1997). Choice for genetically dissimilar or otherwise genetically compatible mates could potentially explain (*1*) apparently idiosyncratic mates choices as well as (*2*) systematic inbreeding avoidance, which has long been a hypothesized mate choice criterion because of the greatly increased probability that offspring of closely related mates will be homozygous for rare deleterious recessive alleles (e.g., Ralls and Ballou 1982). Because of females' greater investment in each offspring, they are predicted to be more averse to inbreeding than males (Clutton-Brock and Harvey 1976). Some evidence suggests that overly outbred offspring also suffer depressed fitness and that there is an optimal level of outbreeding (Bateson 1983).

Conflict Between the Sexes

Behavioral ecologists have recognized for some time that conflicts between the sexes may arise about mating decisions (Parker 1979). Smuts and Smuts (1993:2–3) proposed, as a third form of sexual selection, *sexual coercion*, which they defined as "the use by a male of force, or the threat of force, that functions to increase the chances that a female will mate with him at the time when she is likely to be fertile, and to decrease the chances that she will mate with other males, at some cost to the female." Sexual coercion includes sexually selected infanticide (Hrdy 1974, 1977a), in which males are hypothesized to kill unweaned infants that were sired by other males to hasten the victims' mothers' return to ovulatory cycling and thereby gain quicker sexual access to them. Although debate continues about the validity of this hypothesis (Sussman et al. 1995, van Schaik 2000b), many researchers use the concept of sexually selected infanticide as a platform from which to investigate various aspects of social and sexual behavior (van Schaik and Janson 2000). Most relevant to the focus of this chapter, Hrdy (1981b) proposed that females may seek out multiple mates, particularly newly arrived males, to confuse paternity and thereby reduce the likelihood of infanticide.

Sexual dialectics theory (Gowaty 1997) holds that conflict between the sexes is ubiquitous and that its resolution has profound effects on a species' evolution. Mating partners' genetic compatibility could affect the number and/or quality of the offspring they produce, and individuals may be able to assess prospective mates' genetic compatibility with themselves. If so, then social constraints imposed by males on free mate choice will decrease female fitness even as they benefit the competitively successful males. Some of a successful male's mates would have higher RS if they were allowed to mate with a preferred but competitively less successful male. In fruit flies (*Drosophila pseudoobscura*) (Gowaty et al. 2002) and house mice (*Mus domesticus*) (Drickamer et al. 2000, 2003), both quality and quantity of offspring are enhanced when individuals are mated with preferred partners.

Postcopulatory Sexual Selection

Neither male/male competition nor female mate choice necessarily ends with the initiation of copulation. Parker (1970) first drew attention to the possibility of *sperm competition* (see also Chapter 29), in which sperm from different males compete to fertilize a female's egg(s). *Cryptic female choice*, or *postcopulatory choice* (Thornhill and Alcock 1983, Eberhard 1996), is the process whereby a female's anatomy, physiology, or behavior selectively alters the probability of paternity among her copulatory partners. Reeder (2003) discusses the potential for cryptic female choice in primates. Schwensow et al. (2008a) found that

among wild gray mouse lemurs, MHC constitution was not associated with copulatory patterns but mating pairs that actually produced offspring were more genetically dissimilar than expected by chance, suggesting postcopulatory female choice for the sperm of MHC-dissimilar males (see later section "Traits that May Indicate 'Good Genes'"). Findings of discrepancies between numbers of copulations and genetically ascertained paternity in long-tailed macaques (Engelhardt et al. 2006) and Barbary macaques (Brauch et al. 2008) have been proposed to result from cryptic female choice. Copulation calls by female guinea baboons were (*1*) given selectively to particular males and (*2*) stimulated postcopulatory mate-guarding, suggesting to Maestripieri et al. (2005) that they constitute a form of postcopulatory female choice.

SOURCES OF EVIDENCE ABOUT MATE CHOICE

Observational Studies

The simplest but least precise kind of information about mate choice is a statistical description for a particular study group or population of which members have copulated or of which male/female pairs have associated closely while the female was sexually receptive ("consortships," see Manson 1997 for a critique of this concept). Because mate choice is defined as behavior that affects the relative probabilities of mating among different pairs, data describing these probabilities tell us something about mate choice, particularly when infants' paternities are determined using genetic methods (see Chapter 24). However, observed patterns of mating and parentage reflect the operation of several processes (see also Small 1989): female mate choice, male mate choice, male/male competition, female/female competition, and possibly male sexual coercion. Ascribing any observed pattern in the distribution of copulations or parentage among dyads (e.g., high-ranking females conceive more than expected by chance with high-ranking males) to just one or two of these processes is problematic. Furthermore, neither mating nor paternity data alone reveal anything about the mechanisms of mate choice—that is, the behaviors that alter dyads' likelihood of reproduction.

Some behaviors have strong *face validity*, or *apparent validity*, as forms of mate choice. If a sexually attractive/ proceptive (see Chapter 29 for definitions) female persistently approaches one adult male or a small subset of her group's adult males, we may infer that she is positively affecting her likelihood of mating with him or them. If she presents sexually to a male or engages in other species-typical sexual solicitation behaviors, we have even stronger grounds for regarding her behavior as a form of mate choice. Chimpanzee (*Pan troglodytes*) consortships, in which a cycling female and a male range together apart from the rest of their community for up to several weeks, are regarded as dependent on female mate choice because a female can vocalize to summon other males to break up an incipient consortship if she does not desire it (Tutin 1979). In female-transfer species (e.g., mountain gorillas, Sicotte 2001), females' transfer decisions are almost certainly a form of mate choice. However, many putative mate choice behaviors are subject to alternative interpretations. If a fertile female persistently avoids a particular male, is she engaging in negative mate choice or is she testing his ability to continue following her? Strong evidence for mate choice is produced when a putative mate choice behavior is linked statistically with an altered likelihood of parentage or, at least, an altered likelihood of copulation during the female's fertile period. On the other hand, the same behavior pattern may affect dyads' mating probabilities in some social or demographic contexts but not in others within the same species.

Experimental Studies

As in any area of behavioral research, experimental studies of mate choice permit stronger inferences about causality at some cost in confidence about ecological validity, particularly when experimental subjects are captive animals. Experimental subjects (usually fertile females) are typically given a choice between two or more prospective mating partners that differ along a theoretically relevant dimension. The differences themselves may be natural (e.g., Nikitopoulos et al., 2005) or manipulated (e.g., Waitt et al. 2003). Experiments involving pairs or trios of animals have particular value in the study of female mate choice because of the possibility that female behavior in normal group settings is constrained by male sexual coercion (Buck 1998, Schneider 1998, Nikitopoulos et al. 2005).

FEMALE MATE CHOICE

Female Choice Can Undermine the Effects of Male/Male Competition

Studies using genetic paternity assignment technologies have revealed that in multimale group-living primates dominant males usually produce more offspring than subordinates and almost never produce fewer offspring than subordinates (e.g., Pope 1990, Bercovitch and Nurnberg 1997, Gerloff et al. 1999, Constable et al. 2001, Alberts et al. 2006, Engelhardt et al. 2006, Muniz et al. 2006; but see Inoue and Takenaka 2008). However, dominant males often fail to monopolize reproduction to the extent predicted by Altmann's (1962) "priority-of-access" model (e.g., Alberts et al. 2003). Traditionally, it was assumed that females would choose the winners of male/male competition (see Fedigan 1982:273–277 for a critique). However, evidence from several primate taxa indicate that female choice for multiple mates or for male qualities uncorrelated with dominance rank may be partly responsible for observed deviations from the priority-of-access model.

Cercopithecines

Female aversion to mating with high-ranking males is particularly well documented in Japanese and rhesus macaques (*Macaca fuscata* and *M. mulatta*, respectively). Males frequently disrupt the consortships of lower-ranking males, usually by chasing or approaching the female of the pair; but females usually return to their previous consort partner within a short time and only rarely join the more dominant male (Huffman 1987, Manson 1992, Perloe 1992, Soltis et al. 1997b). Studies that quantified spatial proximity maintenance have found that greater female responsibility for proximity maintenance is sometimes associated with higher copulation rates during the female's fertile period (Manson 1992) or a greater likelihood of producing offspring (Soltis et al. 1997b). Sexually attractive/proceptive females have been observed to maintain proximity preferentially to lower-ranking males (Takahata 1982b, Manson 1992, Perloe 1992, Soltis et al. 2000b), though this result was not always statistically significant. As a result of these female choice behaviors, alpha and other high-ranking males are sometimes virtually excluded from mating (Huffman 1992, Perloe 1992, Manson 1992). Statistical comparisons of the effects of male dominance and attractiveness to females on male RS have sometimes shown that the two variables have independent significant effects (Manson 1992, Soltis et al. 1997a). Although these macaque results must be qualified by the consideration that they were obtained in provisioned or captive groups (in which, e.g., males may remain longer than they would under natural conditions), they at least show the potential for female choice to undermine the effects of male/male competition. Among captive sooty mangabeys (*Cercocebus torquatus*), female sexual solicitations and acceptance of male courtship positively affected male mating success independently of the effects of male dominance rank (Buck 1998). An experimental pair choice paradigm revealed no consistent choice for higher-ranking males among long-tailed macaques (Nikitopoulos et al. 2005).

Female choice also affects putative male RS, independently of the effects of male strategies, in a hybrid anubis/hamadryas baboon (*Papio anubis* × *P. hamadryas*) population studied by Bergman and Beehner (2003). Males varied in their employment of a hamadryas-like social strategy (i.e., spending time in close proximity to nonfecundable females; see Chapter 15). Hamadryas-like social behavior was positively correlated with mating success. However, after statistically controlling for male social strategy, males that received more grooming and hindquarter presents from females experienced greater mating success. In a classic experimental study of hamadryas baboons, Bachmann and Kummer (1980) showed that males are inhibited from challenging each other for the "possession" of a female if they have seen that the female prefers her current partner, even if he is subordinate to the would-be challenger.

Table 30.1 Studies of Female Mate Choice in Primates

SPECIES	TYPE OF EVIDENCE[1]	EFFICACY OF FEMALE CHOICE RELATIVE TO EFFECTS OF MALE/ MALE COMPETITION	CHOICE CRITERIA	SOURCES
Microcebus murinus (gray mouse lemur)	Premating behavior *Dist. copulations; Paternity	No reproductive advantage to dominant males	Inbreeding avoidance	Radespiel et al. 2002, Radespiel and Zimmermann 2003
	Paternity	Not measured	Heterozygosity, MHC dissimilarity (postcopulatory choice only)	Schwensow et al. 2008a
Cheirogaleus medius (fat-tailed dwarf lemurs)	Paternity	N.A. (social monogamy)	Heterozygosity, MHC dissimilarity	Schwensow et al. 2008b
Propithecus verreauxi (sifaka)	Premating behavior; Dist. copulations	Subordinates sometimes chosen	Inbreeding avoidance	Richard 1992, Brockman 1999
Eulemur fulvus rufus (red-fronted lemur)	Premating behavior	No male dominance relationships	Prior affiliative relationship (?)	Pereira and McGlynn 1997
Lemur catta (ring-tailed lemur)	Premating behavior *Dist. copulations *Paternity	Subordinates chosen if unfamiliar	Inbreeding avoidance	Pereira and Weiss 1991
	Premating behavior; Dist. copulations	Males of all ranks chosen	Inbreeding avoidance	Sauther 1991
Cebus apella (brown capuchin)	Premating behavior *Dist. copulations	Alpha male chosen, especially if food is monopolizable	Alpha male chosen, especially if food is monopolizable; inbreeding avoidance	Janson 1984, 1994, Di Bitetti and Janson 2001a,b Lynch Alfaro 2005
Saimiri oerstedii (Costa Rican squirrel monkey)	Premating behavior; Dist. copulations	No male dominance relationships	Largest (seasonally enlarged) male chosen	Boinski 1987a–c
Brachyteles arachnoides (muriqui)	Dist. copulations	No male dominance relationships	Inbreeding avoidance; sorting of mates by age	Strier 1997 Possamai et al. 2007
Cercocebus torquatus (sooty mangabey)	Premating behavior *Dist. copulations	Female choice affects male copulatory success independent of male rank	prior affiliative relationship(?)	Buck 1998
Erythrocebus patas (patas monkey)	Premating behavior †Dist. copulations	Neither female solicitations nor male fighting success affects male copulatory success	None found	Chism and Rogers 1997
Cercopithecus aethiops (vervet)	Experimental	Alpha male chosen	Alpha male chosen	Keddy 1986
	Premating behavior; Dist. copulations	Dominants have no mating advantage with fertile females	Older males chosen	Andelman 1987
Macaca cyclopis (Formosan macaque)	Premating behavior †Dist. copulations	Dominance rank, but not female proximity maintenance, enhances male mating success	Females maintain proximity preferentially to lower-ranking and extra group males	Birky 2002
Macaca sylvanus (Barbary macaque)	Premating behavior †Dist. copulations	Females incited male/male contests; males most successful at coalition formation had highest mating success	Inbreeding avoidance by both sexes; no other criteria	Paul and Kuester 1985, Kuester and Paul 1992, Kuester et al. 1994
	Premating behavior †Dist. copulations *Paternity	Reproductive advantage to dominant males	High-ranking males preferred	Brauch et al. 2008
Macaca fuscata (Japanese macaque)	Premating behavior *Dist. copulations	No reproductive advantage to dominant males	Particular middle-ranking young adult males preferred	Huffman 1987, 1992
	Premating behavior; Dist. copulations	Mating advantage to dominant males, though females maintained proximity most to low-rankers	Possible preference for males likely to retain or attain high rank in the future	Perloe 1992

Table 30.1 (cont'd)

SPECIES	TYPE OF EVIDENCE[1]	EFFICACY OF FEMALE CHOICE RELATIVE TO EFFECTS OF MALE/ MALE COMPETITION	CHOICE CRITERIA	SOURCES
Captive group	Premating behavior *Dist. copulations and premating behavior *Paternity	Female choice affects male copulatory success independent of male rank	Inbreeding avoidance; frequency of displays; idiosyncratic choices	Soltis et al. 1997a,b, 1999
Wild group	Premating behavior [†]Dist. copulations and premating behavior [†]Paternity	Reproductive advantage to dominant males, though female proximity maintenance was unrelated to male rank	Females seeking multiple mates	Soltis et al. 2001
Macaca nigra (Sulawesi crested black macaques)	Premating behavior; Dist. copulations	Dominant males preferred	Dominant males preferred	Reed et al. 1997
Macaca fascicularis (long-tailed macaque)	Premating behavior; Dist. copulations [†]Paternity	Reproductive advantage to dominant males	Females seeking multiple mates	Engelhardt et al. 2006
	Experimental	N.A.	Females seeking multiple mates	Nikitopoulos et al. 2005
Macaca mulatta (rhesus macaques)	Premating behavior *Dist. copulations	Female choice affects male copulatory success independent of male rank	Inbreeding avoidance; idiosyncratic choices	Manson 1992, 1994a,b, 1995, 1996, Manson and Perry 1993
	Dist. copulations	?	Choice for males that give copulation calls	Hauser 1993a,b
	Experimental	N.A.	Females preferred to look at red male faces	Waitt et al. 2003
Papio hamadryas anubis (anubis or olive baboon)	Premating behavior *Dist. Consortships	Dominant male preferred	Dominant male preferred	Seyfarth 1978a,b
	Dist. Consortships	No mating advantage to dominant males	Males with established affiliative relationships (Friends) preferred	Smuts 1985
	Premating behavior *Dist. consortships	Dominant males preferred	Young, high-ranking recent immigrants preferred; multiple mates sought	Bercovitch 1991, 1995
	Experimental	Mixed evidence for relative roles of males and females in consortship formation	Preference for frequent grooming partner	Schneider 1998
Papio hamadryas papio (Guinea baboon)	Dist. Copulations *Postmating behavior	Male rank not a predictor offemale copulation calls	Calling patterns suggest post-copulatory mate choice	Maestripieri et al. 2005
Papio hamadryas hamdryas (hamadryas baboons)	Experimental	Males inhibited from "taking" females from preferred subordinate males	?	Bachmann and Kummer 1980
Papio hamdryas hamadryas × *P. h. anubis* (hamadryas / anubis baboon hybrid)	Premating behavior *Dist. copulations	?	Preference for males of own phenotype	Bergman and Beehner 2003
Pongo pygmaeus (orangutan)	Premating behavior; Dist. fertile copulations	Locally dominant male preferred	Protection from harassment by subordinate males	Galdikas 1981a,b; Schürmann and van Hooff 1986; Fox 1998, 2002
Gorilla beringei beringei (mountain gorillas)	Premating behavior; Dist. copulations[2]	Dominant not always preferred	Inbreeding avoidance	Sicotte 2001
Pan troglodytes (chimpanzees) Gombe	Dist. consortships	Consortships not used by dominant males	Inbreeding avoidance; choice for males that groom and share meat with the choosing female	Tutin 1979; Pusey 1980

Table 30.1 (cont'd)

SPECIES	TYPE OF EVIDENCE[1]	EFFICACY OF FEMALE CHOICE RELATIVE TO EFFECTS OF MALE/ MALE COMPETITION	CHOICE CRITERIA	SOURCES
Kibale	Premating behavior	Not addressed	Unclear; different between peri-ovulatory and other periods	Pieta 2008
Mahale	Premating behavior *Dist. copulations	Dominant males preferred (?)	Dominant males preferred (?)	Matsumoto-Oda 1999
Taï	Premating behavior *Dist. fertile copulations	Alpha and omega males preferred in one group; alpha disfavored in a another group	Inbreeding avoidance; age sorting; idiosyncratic preferences	Stumpf and Boesch 2005
Captive group	Premating behavior †Dist. copulations † (?) Paternity	Dominant males preferred	Dominant males preferred	Klinkova et al. 2005

[1] Premating behavior, female choice inferred from premating behaviors (e.g., sexual solicitations, rejections of mount attempts) that have face validity as forms of mate choice; Dist. copulations, distribution of copulations among dyads; Dist. consortships, distribution of consortships among dyads; Paternity, paternity determined using genetic technology; * statistical association between the occurrence of a female behavior pattern and a measure of mating outcome (either distribution of copulations or genetically assigned paternity), † statistical tests failed to find an association between the occurrence of a female behavior pattern and a measure of mating outcome (When "Premating behavior" is separated from "Dist. copulations" and/or "Paternity" by a semicolon, these sources of evidence were used but no attempt was made to find a statistical association between female behavior and mating outcomes); Postmating behavior, female choice inferred from postmating behavior (copulatory vocalizations); Experimental, female given a choice of males in an experimental setting.

[2] In multimale groups.

Some studies have quantified putative female mate choice behaviors in cercopithecines and found that they had little effect on male RS. Soltis et al. (2001), using the same methods that revealed the effectiveness of female mate choice in a captive group of Japanese macaques, found that females of a wild group maintained proximity preferentially to males of various dominance ranks, yet high-ranking males sired most of the offspring that were sired by troop males. Free-ranging female Barbary macaques engaged in no more sexual presenting or solicitation toward the males that eventually sired their offspring compared to other males (Brauch et al. 2008). Proceptive female Formosan macaques (*M. cyclopis*) maintained proximity preferentially to lower-ranking and extragroup males, yet male copulatory success was positively correlated with dominance rank (Birky 2002). Among wild anubis baboons studied by Bercovitch (1991, 1995), rates of grooming and hindquarter presentations by sexually swollen females affected whether dyads formed consortships but these and other female behaviors had little effect on male sexual access to a female after they had formed a consortship. Persistent female refusal to allow a male to copulate did not decrease consortship duration, nor was mount rejection rate correlated with male ejaculatory success. Among wild patas monkeys (*Erythrocebus patas*), male copulatory success was correlated neither with rate of female sexual solicitations received nor with success in male/male fights (Chism and Rogers 1997). Female Barbary macaques (*M. sylvanus*) reject almost no copulation attempts (Small 1990), although they may incite male/male competition by leaving a mating partner and approaching another male (Kuester and Paul 1992). Because mating success was concentrated among those males that successfully

passed these female-initiated tests, female behavior in this species reinforces the effects of male/male competition (see also Brauch et al. 2008).

Lemuroids

Perhaps because female lemuroids are usually dominant to males (see Chapter 5), this taxon provides several examples of female choice partially or completely negating the predicted effects of male dominance on RS. In one study, female ring-tailed lemurs (*Lemur catta*) preferentially solicited newly immigrated males that were lower-ranking than most long-term residents and the newcomers sired a disproportionately large share of the offspring (Pereira and Weiss 1991). Among wild Verreaux's sifakas (*Propithecus verreauxi*), females generally prefer resident and older dominant males but sometimes vigorously resist dominants' mating attempts and successfully conceive with subordinates (Brockman 1999) or even incite male/male fights and then copulate with a third, uninvolved male (Richard 1992). In a captive population of gray mouse lemurs (*Microcebus murinus*), Radespiel and colleagues (Radespiel and Zimmermann 2003, Radespiel et al. 2002) found that (*1*) females terminated 40% of all copulations, always by biting, slapping, or fighting the male; (*2*) females sometimes failed to conceive at all during a mating season by refusing to mate; and (*3*) dominant males sired no more offspring than would be expected by chance.

Chimpanzees

In a study of proceptivity and sexual resistance by female wild chimpanzees in the Taï Forest, the females of one group showed the most resistance and the least proceptivity toward the highest-ranking among three males, whereas in the other group they preferred both the highest- and the lowest-ranking of four males (Stumpf and Boesch 2005). Among male/female dyads, female proceptivity rate was positively correlated with copulation frequency throughout female ovarian cycles, whereas female sexual rejection rate was negatively correlated with copulation frequency only during periovulatory periods (Stumpf and Boesch 2006). For the Kibale site, Pieta (2008) reports even greater differences between females' periovulatory and nonperiovulatory mate choices but no effect of male rank.

Female Choice May Contribute to Discrepancies Between Grouping Pattern and Mating System

Early accounts of primate social organization (e.g., Kummer 1971:31) equated species-typical grouping pattern with species-typical mating system. Females' potential mates were assumed to be limited to the male members of her social group. Long-term observation and paternity assignment studies have led primatologists to reject this assumption for many species, just as ornithologists earlier discovered that extrapair copulations (EPCs) are common in many socially monogamous avian species (e.g., Westneat 1987). EPCs and pair dissolutions are common in socially monogamous siamangs (*Hylobates syndactylus*) (Palombit 1994a,b), white-handed gibbons (*H. lar*) (Reichard 1995, Sommer and Reichard 2000), fat-tailed dwarf lemurs (*Cheirogaleus medius*) (Fietz et al. 2000), and possibly titi monkeys (*Callicebus molloch*) (Mason 1966). Seasonal influxes of males apparently prevent resident males from monopolizing reproduction in guenon (*Cercopithecus*) groups (Cords 1987b). In multimale groups, females sometimes mate with extragroup males (Agoramoorthy and Hsu 2000), and such males may sire offspring (e.g., Berard et al. 1993, Gerloff et al. 1999), sometimes via "sneaked" matings unseen by observers (Soltis et al. 2001). Although there are few systematic data quantifying the importance of female choice in producing these discrepancies between social system and mating system, some evidence suggests that female behavior plays a major role in these phenomena. Proceptive females of many species have been observed to solicit matings from extragroup males, sometimes moving considerable distances from their own group to do so (Hrdy 1977b, Henzi and Lucas 1980, Brereton 1981, Burton and Fukuda 1981, Richard 1985b, Wolfe 1986;, Sprague 1991, Manson 1995, Strier 1997c, Brockman 1999).

Mate Choice Can Be Costly to Females

If females incur costs by acting in ways that alter the probability of mating with a particular male, it is reasonable to infer that mate choice confers compensating benefits on females—if it did not, then selection would favor female indifference to a mate's identity. Furthermore, costs of female mate choice play a major role in models that attempt to determine the conditions under which Fisherian "runaway," as opposed to Zahavian "good genes," selection processes will prevail (e.g., Pomiankowski et al. 1991).

By seeking out matings with low-ranking, peripheral, and/or extragroup males, primate females may incur costs in the form of increased aggression from dominant males of their own group (Smuts and Smuts 1993). I found (Manson 1994a) that among free-ranging rhesus macaques sexually attractive/proceptive females experienced a higher rate of male escalated aggression (chases and bites) while in proximity to low-ranking males than while in proximity to high-ranking males. Aggressors in most cases were males dominant to the male that was accompanying the female. Among wild chimpanzees, male aggression, directed disproportionately often toward the most fecund females as a counterstrategy to female promiscuity, imposes physiological costs in the form of elevated stress responses (cortisol levels) (Muller et al. 2007).

Influences on the Efficacy of Female Mate Choice

Although the number of systematic studies of primate female mate choice is still too small to permit quantitative comparisons among species or populations, it is possible that female behavior has more effect (relative to male/male competition) on male reproductive success when more females are sexually receptive simultaneously (Soltis et al. 2001). This hypothesis is supported by comparisons between the two Japanese macaque populations studied by Soltis and colleagues (1997a, 2001) and by Paul's (1997) finding that breeding seasonality is associated, in cross-species comparisons, with a weakened association between male dominance and RS. Variation among species in degree of sexual dimorphism may also affect females' ability to influence the distribution of copulations among males (Barenstain and Wade 1983, Smuts and Smuts 1993, Kuester and Paul 1992; but see Bercovitch 1995).

Female Mate Choice Criteria

Choice for Dominant Males

Female choice for subordinate males is conspicuous to observers because of (*1*) the resultant clash with "conventional wisdom" and (*2*) the high drama created by events such as prolonged refusals to mate with alpha males and attacks by dominant males on courting pairs including subordinate males. However, several studies have clearly documented female choice for high-ranking males (Hausfater 1975a, Seyfarth 1978a, Janson 1984, Bercovitch 1991, Fox 2002; Lynch Alfaro 2005). Sexually swollen nonconsort female anubis baboons directed more grooming toward higher-ranking males, which also tended to be young and to be new immigrants (Bercovitch 1991). Interestingly, these males did not experience an overall mating advantage (Bercovitch 1986). In an experimental setting, female vervets (*Cercopithecus aethiops*) consistently chose to mate with their group's alpha male (Keddy 1986). Female Sulawesi crested black macaques (*Macaca nigra*) directed more sexual solicitations toward dominant than toward subordinate males (Reed et al. 1997), as did female Barbary macaques in one study (Brauch et al. 2008) and captive female chimpanzees at Arnhem Zoo (Klinkova et al. 2005). Among wild chimpanzees at Mahale, copulation rate was correlated with male dominance rank during females' periovulatory periods and female grooming rate was positively correlated with male mating rate in comparisons among dyads; however, it is unclear whether female grooming rate was correlated with male dominance rank (Matsumoto-Oda 1999).

Access to Resources

A possible benefit to females of choosing dominant males is to provide females and/or their offspring with access to reproductively valuable resources. Few empirical examples of this process have been discovered in primates, presumably because individuals rarely have exclusive control over food. In the clearest case, Janson (1984) found that female brown capuchins at Manu National Park in Peru showed more proceptive behavior toward the alpha male than other males, that females were largely responsible for maintaining proximity to the alpha male, and that the alpha male monopolized copulations during females' likely fertile periods. At this site, preferred food sources were easily monopolizable, so the alpha male could exclude juveniles that he was unlikely to have sired from access to valuable feeding sites. At another brown capuchin site, in contrast, preferred food was less monopolizable and females showed a weaker mating preference for the alpha male (Janson 1994). In one study of wild chimpanzees, the most preferred consort partners were males that shared meat most readily with females (Tutin 1979).

Inbreeding Avoidance

The most widely documented primate female mate choice criterion is avoidance of genetic relatives. Sex-biased dispersal usually separates adult opposite-sexed close kin into different social groups (Pusey and Packer 1987); however, when relatives are potential mates, they copulate much less than expected by chance, and females are usually primarily responsible for this pattern. Female chimpanzees resist their sons' and brothers' courtship attempts (Tutin 1979, Pusey 1980, Goodall 1986) and sometimes avoid high-ranking male relatives while sexually swollen by engaging in consortships with unrelated males. These tactics usually prevent incestuous conceptions (Constable et al. 2001). Among free-ranging rhesus macaques, males breeding in their natal group experienced copulatory success equal to that of immigrant males, but they received greater proximity maintenance and fewer sexual refusals from unrelated attractive/proceptive females than from related attractive/proceptive females (Manson and Perry 1993). These behavioral patterns are probably responsible for the observed

low frequency of incestuous conceptions in captive rhesus macaques (Smith 1995). Soltis et al.'s (1999) study of Japanese macaques also found that attractive/proceptive females avoided proximity to their male matrilineal kin and that related dyads copulated less than expected by chance and never during females' periovulatory periods. Female ring-tailed lemurs showed proceptivity toward unrelated males and aggression toward close relatives that courted them, thus avoiding being impregnated by kin (Pereira and Weiss 1991, Taylor and Sussman 1985, Sauther 1991). Similar results have been obtained from other lemuroids (Richard 1992, Brockman 1999, Radespiel et al. 2002, Radespiel and Zimmermann 2003). A sex difference in aversion to inbreeding is suggested by Chapais and Mignault's (1991) analysis of heterosexual and homosexual interactions in Japanese macaques. Female/female consortships were frequent yet never involved close kin, even though (*1*) female kin preferred each other for nonsexual interactions and (*2*) occasional incestuous heterosexual consortships were observed. In Barbary macaques, in contrast, inbreeding avoidance appears to show no obvious sex difference in intensity: Matings and conceptions occurred at much lower frequencies than expected in maternally related dyads, but females did not resist the few copulations between kin that were observed (Paul and Kuester 1985, Kuester et al. 1994). Matings and conceptions between paternally related kin did not occur at lower than expected frequencies. However, mating was inhibited in male/female dyads in which the male had had a close ("caretaking") relationship with the female when she was an infant. Taken together, these results support the hypothesis that individuals avoid inbreeding by using early social familiarity (which occurs among maternal, but not necessarily paternal, kin in this species) as a cue to genetic relatedness (Paul and Kuester 2004), a hypothesis originally proposed to account for human inbreeding avoidance (Westermarck 1891). However, yellow baboon (*Papio cynocephalus*) paternal half-siblings engage in less affiliative and sexual behavior while in consort than do nonsiblings (Alberts 1999), and dyads of similar age (which are statistically likely to be paternal half-sibs) engage in consortships at lower frequency than dyads of dissimilar age. These results suggest that in species in which dominant males temporarily monopolize paternity, selection can favor natal individuals' use of age proximity (in addition to early social familiarity) as a cue of relatedness.

Unfamiliarity

Females are predicted to choose unfamiliar males for five reasons. First, familiarity may be a reliable cue of relatedness. Second, altered sociodemographic conditions (e.g., provisioning leading to very large group size and extended male tenures) may cause unrelated or distantly related dyads to experience enough familiarity that they behave, maladaptively, as if they were kin (Takahata 1982a). Third, choice for unfamiliar males presumably leads to the production of a genetically diverse set of offspring, and this in turn may increase females' long-term RS in temporally or spatially variable environments (Parker 1970, Williams 1975). Fourth, choice for unfamiliar males increases the number of potential sires of the choosing female's subsequent offspring and, thus, reduces the number of potentially infanticidal males (Hrdy 1981b), assuming that males refrain from killing infants that they might have sired, as documented in a recent review (van Schaik 2000b). Fifth, choice for unfamiliar males could increase the number of potentially investing males (Hrdy 1981b; but see Soltis and McElreath 2001).

Female primates frequently approach and solicit matings from extragroup, peripheral, and newly immigrated males (Cords et al. 1986, Strier 1997c, Hrdy 1977b, Brockman 1999, Henzi and Lucas 1980, Richard 1985b, Brereton 1981, Burton and Fukuda 1981, Wolfe 1986, Manson 1995, Sprague 1991, Packer 1979a, Hayakawa 2007). Bercovitch (1991) found that newly immigrated male anubis baboons (that were also usually high-ranking, thus confounding the variables of rank and novelty) received more grooming from sexually swollen nonconsort females compared to long-term resident males. These behaviors almost certainly increase these males' chances of siring offspring (e.g., Berard et al. 1993, Gerloff et al. 1999, Soltis et al. 2001), and, thus, qualify as mate choice for novelty. However, interpreting these observations as evidence of female preference for novel males (Small 1993) is less straightforward (Manson 1995). Consider a case in which females prefer males independently of their novelty (i.e., using a criterion that is uncorrelated with novelty). Because achieving copulations with an extragroup or peripheral male probably requires more effort than mating with a central male of the female's own group, females will need to direct more choice behavior toward preferred novel males than toward preferred familiar males.

Protection of Self and/or Infants

Females might choose mates that provide protection from male aggression against themselves or their infants, either in courtship contexts only or at other times as well. This choice criterion is thought to be central to the phenomenon of male/female friendships in anubis baboons (Ransom and Ransom 1971, Strum 1983, Smuts 1985). These relationships consist of preferential proximity and grooming while the female is not sexually receptive, agonistic support of the female by the male, and preferential mating while the female has a sexual swelling. Smuts (1985) showed that some friend dyads form their relationships before mating and argued that females prefer to mate with males with which they already have an affiliative relationship. Other studies (e.g., Palombit et al. 1997, Bercovitch 1991, Busse and Hamilton 1981, Packer 1980, Noë and Sluijter 1990, Weingrill 2000) indicate that baboon friendships form only after consortships during females' fertile periods and function to enable the male to protect his likely offspring from the danger of infanticide. However, most baboon researchers have not examined, as Smuts (1985) did, whether males experience greater consorting success with friends than would be expected given their overall level of consorting activity. Macaque male/female friendships do not include preferential mating access (Fedigan and Gouzoules 1978, Enomoto 1978, Hill 1990, Manson 1994b). Among red-fronted lemurs (*Eulemur fulvus rufus*), one of the few non-female-dominant lemuroids, male/female friendships involve spatial association and mutual agonistic support (Pereira and McGlynn 1997). Although completed copulations were rarely observed in this study, all four female subjects accepted copulation attempts from male friends on likely days of ovulation, and two subjects aggressively rejected mating attempts by nonfriends on those same days. Among northern muriquis, limited evidence suggests that copulations occur more frequently in dyads that associate closely outside mating contexts (Possamai et al. 2007).

Orangutans (*Pongo pygmaeus*) are unique among nonhuman primates (but see Engelhardt et al. 2006) in that forced copulations are frequent (MacKinnon 1974; Galdikas 1981b). Subadult males and males using the arrested development strategy (see Chapter 18) are more likely than full-sized (flanged) males to encounter female sexual resistance and to force copulations on females (Galdikas 1985b, Mitani 1985a). Fox (1998, 2002) found that (*1*) female resistance to male mating attempts did not consistently decrease over time, suggesting that resistance is "honest" rather than a test of male health and vigor, and (*2*) females that maintained proximity to a full-sized male gained some protection from harassment by unflanged males in that a smaller proportion of unflanged males' copulation attempts were successful, although the overall rate of unflanged male copulations, including forced copulations, was not affected by the presence of a full-sized male.

Traits that May Indicate "Good Genes"

In contrast to researchers of other taxa (e.g., insects, Moore 1994), primatologists have gathered only a small number of data sets suggesting female choice for genes and have made almost no progress in determining whether any particular cases conform better to Fisherian or Zahavian models of female choice. The most compelling relevant results are the recent findings of Schwensow et al. (2008b) in a study of pair-living wild fat-tailed dwarf lemurs (*Cheirogaleus medius*). Only a minority of males formed social pairs with females, and almost 40% of offspring were sired by extrapair males. Both social fathers and genetic fathers had significantly more MHC alleles and MHC supertypes (i.e., greater heterozygosity) than other males.

"Choice for genes" includes choice for mates whose genotypes are compatible with (i.e., optimally dissimilar to) that of the chooser, in the sense that such matings would produce offspring with higher fitness, most likely by increasing heterozygosity (Brown 1997). In Schwensow et al.'s (2008b) study, female fat-tailed dwarf lemurs that mated with extra-pair males shared more MHC supertypes with their social partner compared to faithful females. As described, genetic analyses of parents and offspring among gray mouse lemurs produced a pattern suggestive of postcopulatory female choice for MHC-dissimilar males (Schwensow et al. 2008a). Furthermore, two studies of macaques have found that MHC genotypic dissimilarity increases offspring fitness. Among captive pig-tailed macaques (*Macaca nemestrina*), parental sharing of certain MHC antigens significantly predicted pregnancy wastage (Knapp et al. 1996); and among free-ranging rhesus macaques, males that were heterozygous at an MHC locus sired significantly more offspring than homozygotes, even though they did not live significantly longer (Sauermann et al. 2001).

Turning to indirect evidence for mate choice for genes, four studies have addressed whether male macaque copulation calls (loud vocalizations given exclusively while copulating) are costly, reliable (i.e., Zahavian) indicators of male quality. Specifically, they have tested the hypothesis that calling males incite aggression against themselves and thereby demonstrate their ability to tolerate harassment and wounds. Hauser (1993a) reported that, among male rhesus macaques, (*1*) the occurrence and frequency of calling were positively associated with the frequency of male aggression suffered and (*2*) calling males experienced higher mating success than noncallers. However, studying the same population, I found that (*1*) males were almost never attacked during or immediately after a copulation and their likelihood of being attacked was not affected by whether they called and (*2*) there was no evidence from analysis of female choice behavior (proximity maintenance, cooperation with mount attempts) to suggest that calling males were more attractive than silent males (Manson 1996). In one study of Formosan macaques, the occurrence of calling was variable among males yet males were never harassed or attacked during or shortly following copulations (Hsu et al. 2002). Another study of the same species found no evidence that calling males were more attractive to females than silent males (Birky 2002). Thus, the adaptive significance of macaque male copulation calls remains unclear.

In an experimental setting, estrous female gray mouse lemurs (*Microcebus murinus*) showed a strong trend toward preferring males that produced trill calls at higher rates (Craul et al. 2004). Because trill call rate is correlated with dominance rank (Zimmermann and Lerch 1993) and males provide no paternal care, this result suggests female choice for genes.

Male macaques engage in courtship displays consisting of shaking branches, leaping, and other conspicuous behaviors. Modahl and Eaton (1977) found that male Japanese macaque display rates did not correlate with dominance rank but did correlate with rate of completed copulations, suggesting that females choose males that display more

frequently. Consistent with this result, Soltis et al. (1999) report that periovulatory female Japanese macaques maintained proximity most actively to males that had displayed most frequently before the female's first periovulatory period of the mating season.

Several dozen primate species are sexually dichromatic, raising the possibility that females could use male color as a signal of genetic quality (Gerald 2003). Very little research has addressed this possibility, although some evidence indicates that male color variation plays a role in male/male competition (Wickings and Dixson 1992b, Gerald 2001, Setchell and Dixson 2001b). Using digitized images in an experimental setting, Waitt et al. (2003) found that female rhesus macaques spent more time looking at images of reddened male faces than at otherwise identical pale faces. Among mandrills (*Mandrillus sphinx*), the most strikingly sexually dichromatic primate, semi-free-ranging periovulatory females directed more proceptive and receptive behavior toward more brightly colored males, independent of the effects of male dominance rank (Setchell 2005). In the wild, male mandrills may lack the long-term social bonds of other male cercopithecines and may be forced to rely on displays of color to establish dominance over rival males and to induce females to choose them (Abernethy et al. 2002). Besides color, symmetry is a visual cue that could indicate genetic quality (Møller 1990). Suggestively, rhesus macaques looked longer and more frequently at manipulated symmetrical faces of opposite-sexed conspecifics than at asymmetrical versions of the same faces (Waitt and Little 2006). Genetic quality could be manifest in olfactory cues or signals. Charpentier et al. (2008) found that among male ring-tailed lemurs individual semiochemical diversity of scrotal secretions was correlated with individual genetic quality (heterozygosity) but only during the 3-month breeding season.

"Choice" for Multiple Mates

Hrdy (1986) and Small (1989) have cautioned against confusing female mate choice with generalized female sexual assertiveness, which could be an adaptation to prevent infanticide (Hrdy 1981b) or a nonadaptive by-product of sexual motivation (Small 1993). Several studies of multimale group-living primates show that as the duration of attractive/proceptive periods, or focal observation time on a female, increases, the number of her copulatory partners increases also (Small 1990, Manson 1992, Soltis et al. 2001), indicating that females seek out multiple mates. Nikitopoulos et al.'s (2005) experimental pair choice study found that, given complete control over mating decisions, female long-tailed macaques sought to distribute copulations evenly among males. However, with rare exceptions, such as Barbary macaques (Small 1990, Kuester and Paul 1992; but see Brauch et al. 2008), females of multimale primate groups reject the sexual advances of some males. Even in muriquis, which do not form dominance hierarchies and almost never engage in aggression, some males copulate more than others (Strier 1997c). Furthermore, some studies (Soltis et al. 1999, Stumpf and Boesch 2005) show that females become more selective when they are most likely to be ovulating. These results suggest that females of most group-living primate species pursue a mixed mating strategy, incorporating elements of both selectivity and promiscuity, a pattern sometimes present even in species traditionally thought to be monogamous (white-handed gibbons, Barelli et al. 2008). A similar view is supported by comparative analyses of sexual swellings (Nunn 1999).

MALE MATE CHOICE

Conditions likely to favor male discrimination among potential mates include (1) high costs of male/male competition (Pagel 1994, Domb and Pagel 2001), (2) high variation in female fertility and/or offspring survivorship (Altmann 1997), and (3) synchronized breeding that prevents even the most dominant or attractive males from mating with multiple females (Kuester and Paul 1996). Hypothesized male mate choice criteria include female dominance rank, age, and heritable quality.

Choice for Female Rank

Dominant females in multifemale primate groups often experience reproductive advantages such as shorter interbirth intervals and higher rates of infant survival (Silk 1993, Packer et al. 1995). Thus, males should be expected to choose high-ranking mates. Evidence that they do so (reviewed by Robinson 1982, Berenstain and Wade 1983, Silk and Boyd 1983) consists largely of patterns of assortative mating in which high-ranking females tend to mate with high-ranking or otherwise competitively successful males, leading to the inference that males compete most strongly for high-ranking females. Setchell and Wickings (2006) measured male mate choice more directly by examining mate-guarding by male mandrills and found that males guarded higher-ranking females more than low-ranking females. Parga (2006) found no relationship between male mating effort (as measured by male/male agonism) and the rank of estrous females in free-ranging ring-tailed lemurs.

Choice for Female Age

Adolescent female primates are typically (1) not as likely to conceive as adults and (2) less likely than adults to successfully raise an offspring to weaning age (see Chapter 29). Presumably for this reason, male primates find sexually receptive adolescents less attractive than adult females in a wide range of cercopithecines, colobines, and apes (Anderson 1986a, Anderson and Bielert 1994, Perry and Manson 1995, Setchell and Wickings 2006). Even in species

in which adolescents display exaggerated versions of sexual signals (e.g., sexual swellings), males disfavor them (Anderson and Bielert 1994). Muller et al. (2006) found that older female chimpanzees received more male copulatory approaches, attracted larger traveling parties of males, and mated with higher-ranking males and that the relationship between female age and attractiveness was linear, rather than merely a result of lower attractivity of nullipara. However, in ring-tailed lemurs, Parga (2006) found that the most attractive females were middle-aged (4–9 years), which corresponds to the period of highest fecundity and infant survival.

Sexual Swellings: A Signal of Heritable Quality?

The function(s) of primate female sexual swellings remains a puzzle (see Chapter 29). Most hypotheses proposed to account for them (Clutton-Brock and Harvey 1976; Hrdy 1981b, Hamilton 1984, Nunn 1999) assume that sexual swellings signal some probability that the female is ovulating. These hypotheses differ in the proposed benefit to females, but all agree that sexual swellings enable choosing males to make within-female, across-time comparisons only. Pagel (1994), in contrast, has hypothesized that female sexual swellings are Zahavian signals of quality and that males use them to assess interfemale differences in fertility. Among anubis baboons, (1) females with larger swellings reached sexual maturity sooner and produced more surviving offspring per year than females with smaller swellings and (2) males consorting with females with larger swellings received more aggression than males consorting with females with smaller swellings, suggesting that males are more attracted to the former (Domb and Pagel 2001). However, swelling size was not corrected for body size in this study (Zinner et al. 2002). Furthermore, (1) a comparative study by Nunn et al. (2001) found no evidence that exaggerated sexual swellings are associated with measures of intensity of female/female mate competition such as the adult sex ratio and estimated female mating synchrony and (2) Higham et al. (2008) found no evidence that baboon swelling size is an enduring female trait or that it varies between troops as a function of nutrition-mediated reproductive condition.

CONCLUSIONS

Over 20 years ago, Fedigan (1982:275–276) asked whether primatologists, like behavioral biologists generally, had inadvertently reduced female mate choice to Hobson's choice, that is, not a real choice but mere validation of the results of male/male competition. Theory and data generated since then have partially corrected the biases identified by Fedigan. Female choice sometimes undermines the effects of male/male competition, though we do not yet have the comparative data to determine the conditions under which this occurs. Inbreeding avoidance is a strong and wide-spread primate mate choice criterion, as in almost all mammals. Unsurprisingly, given the great diversity of primate ecological niches, social organizations, and cognitive abilities, other female choice criteria vary throughout the order. Smuts's (1985) proposal that nonsexual "friendships" affect primate mate choice has enjoyed only limited empirical validation, although it has not been tested properly as frequently as it deserves. Potential effects of genetic quality, including genetic compatibility, on male attractiveness have only recently begun to attract attention from primatologists. We may anticipate that in the coming years primatological data will begin to catch up with theory and with data from other taxa regarding these issues.

31

Menopause
Interspecific Comparisons of Reproductive Termination in Female Primates

Linda Marie Fedigan and Mary S. M. Pavelka

1. If living creatures are selected to reproduce maximally, why do human females cease to produce children well before they die?
2. What is unique about the human reproductive pattern, and what is shared with other primate species?
3. What good are Japanese macaque grandmothers to their descendants?

THE EVOLUTION OF MENOPAUSE

In the past 30 years or so, menopause has emerged from the private domain of aging women and into public view. As the postwar bulge in the populations of developed nations (the "baby boomers") reached middle age, science and the media alike began to take an increasing interest in understanding why human females universally cease to experience menstrual cycles and fertility around the age of 50. A recent check of a popular search engine with the keyword "menopause" garnered close to 17 million results on the World Wide Web, and databases for journal articles indicate more than 15,000 papers have been published on this subject in the biological literature alone. Clearly, this is no longer a topic that is discussed only rarely and guardedly among older women.

The vast majority of the scientific and popular literature on menopause concerns the treatment of its side effects. However, evolutionary biologists and biological anthropologists have also begun to address the issues of why menopause originated, what maintains it, and whether it is unique to humans or common in mammalian species. It was to be expected that the search for an evolutionary context and an animal model of menopause would turn to the accumulating data on life histories and reproductive senescence in the nonhuman primates, our closest relatives.

In this chapter, we first describe the competing explanations for the origin and maintenance (selection) of menopause in humans and then turn to the search for a larger understanding of reproductive senescence in female primates. As a life history characteristic of human females, menopause is universal, it happens halfway through the maximum life span of our species, and it consistently occurs at the average age of 50–52 years in different populations around the globe and throughout history. There is some debate about whether menopause is a discrete event or merely the end point of a long process of follicular depletion (Wood et al. 2001), whether the term might also apply to reproductive decline in men (Peccei 2001), whether individual variation in age at menopause is affected by genetic and/or environmental factors (Brambilla and McKinlay 1989), and whether increasing life expectancies in developed countries indicate that menopause is a recent artifact of modernization (Sherman 1998, Leidy 1999). We have described/addressed these debates in prior publications (e.g., Pavelka and Fedigan 1991). In this chapter we focus on competing explanations for the distinctive pattern called "menopause" in the human female—the pattern of permanent and universal cessation of menstrual cycles and reproduction in all women who live to the age of 60, an aging effect that occurs in middle age, long before the senescence of other somatic systems.

There are three primary evolutionary explanations for menopause, which, as noted by Peccei (2001), are not mutually exclusive: the adaptation explanation (the grandmother or maternal investment hypothesis) and the two epiphenomenon explanations, antagonistic pleiotropy (the trade-off hypothesis) and the physiological constraint model (the prolonged life hypothesis).

Taking these in reverse order, the physiological constraint model argues that during the course of human evolution there was strong selection for a longer life span, but the reproductive system of the human female could not keep pace (Weiss 1981). Judging by brain to body ratios in fossil hominids, the maximum human life span increased from about 50 years in early hominids to approximately 120 years in *Homo sapiens* (Bogin and Smith 1996, Hammer and Foley 1996). However, the age at which human females cease to experience ovarian cycles appears to have remained stationary at approximately 50 years, the same age at which great ape females in the wild cease to give birth (e.g., Nishida et al. 2003; Wich et al. 2004b; Robbins et al. 2006; Emery Thompson et al. 2007). This is also close to the maximum life span of great apes. Why does reproduction terminate in human females so far in advance of maximum life span? One answer is that human females are born with all the oocytes and primordial follicles they will ever possess (approximately 1 million), a mammalian pattern that is referred to as "semelgametogenesis." It is widely held that oocytes and follicles are depleted throughout a woman's life until they reach a minimum threshold below which hormonal signals fail, at which point cycling ceases (Armstrong 2001). However, there is some recent evidence from mice that it is actually the germline stem cells that deplete and that female mice may be able to produce new oocytes during their lifetimes (Johnson et al. 2004). Nonetheless, this explanation holds that due to the physiological constraints of what we might refer to as the "shelf life of eggs," it may simply not be possible to select for longer and longer reproductive life spans in mammalian females. Thus, menopause is viewed as an epiphenomenon or side effect of selection for a prolonged life span.

However, the epiphenomenon explanation that is more commonly endorsed by evolutionary biologists is known as antagonistic pleiotropy, or the trade-off hypothesis (e.g., Williams 1957, Rose 1991, Gosden and Faddy 1998). According to this model, patterns that have high adaptive value early in the life course (intense reproductive output) will be selected for even if they result in reduced fitness (follicular depletion) later in the life course. Certain genes, such as those governing semelgametogenesis may be selected if

they have pleiotropic fitness effects—beneficial effects at early ages and deleterious effects at later ages. In the wider research on somatic aging, it is commonly accepted that antagonistic pleiotropy is a good explanation for the evolution of senescence in general (Wood et al. 2001).

In a sense, pleiotropy *is* an adaptive explanation for menopause since it argues that there is an adaptive benefit to follicular depletion early in life that outweighs its deleterious effects later on. However, the grandmother hypothesis, developed by several theorists (e.g., Lancaster and King 1985; Hawkes et al., 1989, 1997; Alvarez 2000; Hawkes 2003), is the only one of the three explanations that argues for a direct benefit of midlife cessation of fertility in the human female and is widely regarded as "the adaptive explanation." This hypothesis proposes that selection has favored cessation of reproduction midway through the life span because as human females age there are greater benefits from investing care in existing children and grandchildren than from continuing to produce additional offspring. In other words, older women will be more reproductively successful if they put their energies toward ensuring the survival of their offspring and grandchildren than they would be by producing yet more children themselves. Within this directly adaptive school of thought, there are several hypotheses related through the underlying argument that a postreproductive life exists in order to facilitate enhanced parental investment. Peccei (2001) distinguished between a "mother hypothesis," focused on improving survivorship and fertility of the first generation of offspring, and a "grandmother hypothesis," directed toward increased survival rates in grandchildren. However, in terms of inclusive fitness, the benefits to females and their children and grandchildren are clearly interrelated. If a female protects or provisions her grandchildren, she may enhance their survival as well as the reproductive success of their parents (her children) and, thus, her own inclusive fitness. Hence, the "grandmother hypothesis" is often used as the general and more memorable term for postreproductive investment in any existing progeny, even though "enhanced maternal investment hypothesis" might be a more global and accurate name.

There is a growing body of literature that addresses the costs and benefits of grandmothers in human populations (e.g., see chapters in volumes edited by Voland et al. 2005, Hawkes and Paine 2006). A few studies have examined what are considered to be the costs of old-age reproduction in human females, such as age-related risks of dying in childbirth, risks to fetuses and neonates resulting from old-age pregnancies, risks to siblings of maternal death during childbirth, and fertility costs to offspring from sibling competition. According to Peccei (2001), the only clearly established cost of old-age reproduction is increased risk of fetal loss, stillbirths, and birth defects when mothers have pregnancies late in life. However, Mace (2000) modeled sibling competition and concluded that competition between siblings for food, status, territory, breeding opportunities, or kin support can greatly increase the cost of maintaining the mother's fertility beyond the age of 50 years.

Other studies have looked for fitness benefits of ceasing to reproduce, with mixed results. For example, Hill and Hurtado (1996) addressed the question of whether the presence of living, helpful grandmothers among the Ache of South America significantly enhanced the fertility of their children and the survival of grandchildren. These authors concluded that the presence of postreproductive mothers and grandmothers did not enable their children to raise more offspring.

Rogers (1993) developed a theoretical model that came to the same conclusion. On the other hand, Hawkes and collaborators (1989, 1997, 2003) provided evidence that postmenopausal Hadza grandmothers supply sufficient surplus calories and babysitting services to allow their daughters to successfully raise more offspring. Blurton-Jones et al. (1999, 2005) concluded that grandmothering facilitated the evolution of earlier weaning in hominids, and Mace (2000) showed that babies in The Gambia are more likely to survive if their maternal grandmother is alive. Lahdenpera et al. (2004) documented that premodern Finnish and Franco-Canadian women with a prolonged postreproductive life span have more grandchildren. There is scattered evidence from the ethnographic literature on a variety of societies that older women are a substantial help to their progeny, but whether or not this has a significant effect on inclusive fitness is still a matter of debate (e.g., compare Kaplan et al. 2000 to Hawkes 2003, 2004b).

There is further divergence of opinion between those who see the grandmother hypothesis as applicable only to the case of human females (Lancaster and King 1985, Hawkes et al. 1989) and those who would extend this adaptive explanation to other species of primates (Hrdy 1981a, Sommer et al. 1992, Paul et al. 1993a, Paul 2005). Although universal midlife termination of female reproduction is not known to occur in any primate species other than humans, there are reports of individual female monkeys and apes living part of their lives in a postreproductive state. This has generated interest in the possibility that some form of early reproductive termination may have been selected for in nonhuman primate females.

PRIMATE COMPARISONS: THE SEARCH FOR PARALLELS IN OLD-AGE REPRODUCTION

Endocrine Studies of Reproductive Senescence in Captive Female Primates

If menopause is not unique to humans but rather shared with the nonhuman primates as part of our phylogenetic heritage, then we do not need a special explanation for its existence in *Homo sapiens*. Furthermore, if we could find a nonhuman primate model suitable for laboratory research on aging and for testing various clinical aspects of ovarian decline, then we could learn more about this phenomenon

in our own species. Thus, it is not surprising that biomedical primatologists spent a good part of the 1970s and 1980s examining cases in which captive female primates lost ovarian function.

Macaques (genus *Macaca*) are the most common primate kept in biomedical labs, and several examinations of colony records have shown that particular aged female macaques ceased to experience ovarian cycles around the age of 22 years, if they lived to be that old (e.g., Hodgen et al. 1977, Graham et al. 1979, Graham 1986). Maximum life span in captive macaques is about 30–35 years, though it is very rare for individuals to live to this maximum, just as it is for individual humans to live to be 100–120. Chimpanzees (genus *Pan*) are the nonhuman primates most closely related to humans, and examination of reproductive decline in captive chimpanzees also turned up a few cases of very old females that ceased to cycle close to their deaths at 48–50 years of age, although the majority died while still cycling (Graham 1979, Gould et al. 1981).

The few studies available on the hormonal profiles of individual old female monkeys indicate that the decline of ovarian function in nonhuman primates parallels the hormonal events associated with menopause in women—prolonged follicular phases, failure to ovulate, breakthrough bleeding, high plasma luteinizing hormone concentrations, low estrogen levels and/or lack of patterned estrogen/pregnanediol-3-glucuronide dynamics, etc. (e.g., Tardif and Ziegler 1992, Nozaki et al. 1995, Gilardi et al. 1997, Bellino and Wise 2003). Studies of aged callitrichid females indicate there are some differences in ovarian aging between neotropical and Old World primates (Tardif and Ziegler 1992, Tardif et al. 2008), and we clearly have much more to learn on this topic. However, the overall pattern of ovarian hormonal senescence (physiological deteriorations leading to cycle termination) in human and nonhuman female primates appears similar across species.

Thus, from an endocrinological perspective, female reproductive decline may well follow a similar pattern in all primates, and we could therefore employ individual cases of postreproductive monkeys and apes as clinical models of the physiological basis for menopause in women. However, from an evolutionary and life history perspective, these studies fail to demonstrate similarity between reproductive senescence in nonhuman primates and menopause in the human female. Instead, they highlight two critical differences: (1) female macaques and chimpanzees that cease to cycle are very close to age at death, whereas human females cease to cycle less than halfway through the maximum life span, and (2) female macaques and chimpanzees cease to cycle on an idiosyncratic basis, whereas human females universally cease to cycle at the average age of approximately 50 years.

A recent book on aging in primates (Erwin and Hof 2002) argued that there is pressing need for endocrine monitoring of the oldest female primates in captivity to document patterns of reproductive senescence, and indeed there was surprisingly little published on this subject in the 1990s (for exceptions, see Tardif and Ziegler 1992, Nozaki et al. 1995, Walker 1995, Gilardi et al. 1997, Bellino and Wise 2003, Coleman and Kemnitz 1998). However, a recently edited volume focusing on captive primates (Atsalis et al. 2008) has begun to answer that call.

Demographic Studies of Reproductive Cessation in Socially Living Primates

Endocrine studies of reproductive senescence in captive female primates have the distinct advantage of being able to obtain direct hormonal documentation of the reproductive status of their subjects. They are disadvantaged, however, by small sample sizes (even most captive primates continue to reproduce until death) and by the fact that the animals are captive and may not be representative of their counterparts in natural settings. Thus, while valuable in the search for biomedical models, such studies have less to say about the

evolution of reproductive senescence. Evolutionary biologists and biological anthropologists are generally interested in such phenomena from a life history perspective in which the distribution of the trait in the population and in the life course is important. For this we must turn to demographic data in which the reproductive histories of free- or semi-free-ranging animals of known age at birth and death are available. Of course, here the problems include the relative lack of information on free-ranging and semi-free-ranging primates, from which it is difficult to obtain data on old age and reproduction, and, equally important, identifying reproductive cessation via demographic data in the absence of direct hormonal data.

In Table 31.1, we summarize the available data from demographic studies of reproductive cessation in socially living primates, including breeding colonies, free-ranging and semi-free-ranging provisioned populations, and wild study groups. There is relatively little available from the latter category; however, using very recent publications and

Table 31.1 Demographic and Behavioral Estimates of Life History Variables Related to Reproductive Termination in Primates

SPECIES	FREE RANGING/ CAPTIVE[1]	MAXIMUM LIFE SPAN (OLDEST KNOWN FEMALE) (YEARS)	MAXIMUM AGE AT LAST BIRTH (YEARS)	PERCENT LIFE SPAN COMPLETED AT TIME OF LAST BIRTH	SAMPLE SIZE	PERCENT SAMPLE REACHING REPRODUCTIVE TERMINATION[2]	METHOD FOR CALCULATING PRLS/ASSESSING REPRODUCTIVE TERMINATION[2]
Pan troglodytes[3]	F, FP	~63	~55	93.9	172 (88 dead/ analyzed)	21.6	6
Pongo abelii[4]	F	53	50	94	10	—	1
Gorilla gorilla beringei[5]	F	43.7	40.6	93	10	20	7
Semnopithecus entellus[6]	F	—	—	—	36	8.3	2
Cebus capucinus[7]	F	25	24	93	14	7.1	6
Propithecus verreauxi[8]	F	31	28	90	200	—	—
Papio cynocephalus[9]	F	27.7	23.6	93	96	1.0	7
Semnopithecus entellus[10]	FP	35.0	32	91.4	29	13.8	1
Macaca fuscata[11]	FP	32.7	25	67.3	32	25.0	4
Macaca mulatta[12]	SF	34.0	25	73.5	285	10	5
Macaca sylvanus[13]	SF	30.0	28	93.3	12	58.0	1
Pan troglodytes[14]	C	60.0	36	60.0	15	60.0	6
Pongo pygmaeus[14]	C	58.7	40	68.1	53	31.9	6
Gorilla gorilla gorilla[15]	C	55.0	41	74.5	379	40.0	6
Papio anubis × cynocephalus hybrids[14]	C	33.4	17	50.9	13	10.0	6
Macaca mulatta[16]	C	30.0	20	66.7	38	13.2	6
Macaca nemestrina[14]	C	28.9	20	69.2	209	25.6	6
Macaca radiata[14]	C	28.9	19	65.7	13	3.8	6
Cercopithecus aethiops[14]	C	25.4	17	66.9	12	0	6

Table 31.1 (cont'd)

SPECIES	FREE RANGING/ CAPTIVE[1]	MAXIMUM LIFE SPAN (OLDEST KNOWN FEMALE) (YEARS)	MAXIMUM AGE AT LAST BIRTH (YEARS)	MEAN % LIFE SPAN COMPLETED AT TIME OF LAST BIRTH	SAMPLE SIZE	% SAMPLE REACHING REPRODUCTIVE TERMINATION[2]	METHOD FOR CALCULATING PRLS/ASSESSING REPRODUCTIVE TERMINATION[2]
Saimiri scuireus[14]	C	21.0	19	90.5	28	32.1	6
Saguinus fuscicollis[14]	C	15.8	13	82.3	6	20.0	6
Leontopithecus rosalia[14]	C	24.7	12	48.6	21	47.4	6
Callithrix jacchus[14]	C	16.2	10	61.7	14	36.4	6
Lemur spp.[14]	C	28.7	22	76.7	30	44.8	6
Homo sapiens[17]	N/A	100–120	50	41.7–50.0	106	99.1	6

[1] F, free ranging, wild, not provisioned ; FP, free ranging but provisioned; SF, semi-free ranging and provisioned; C, breeding colonies (usually caged).
[2] 1, estimated age at death—age at last birth; 2, minimum time the female survived after last giving birth, female's age unknown; 3, cycling (or cessation thereof) inferred from perineal swellings; 4, LP - D-1.5years; 5, >25 years of age + no births for 2 years; 6, Caro's criterion: LP-D/(xIBI + 2 SD); 7, Modified Caro's Criterion, see Robbins et al. 2006; 8, ovarian cycling data from fecal hormones.

Sources (source for life span value listed first; source for other life history values in this species listed second):

[3] Thompson personal communication, Thompson et al. 2006.

[4] Wich et al. 2004a,b

[5] Robbins et al. 2006, A. Robbins personal communication

[6] Borries personal communication, life span unknown; Borries and Koenig 2008, Borries et al. 1991.

[7] L. M. Fedigan, unpublished data

[8] Beza Beza Mahafaly Sifaka Research Consortium personal communication.

[9] S. Alberts & J. Altmann (pers. comm.)

[10] Sommer et al. 1992.

[11] Takahata et al. 1995.

[12] Walker 1995, Johnson and Kapsalis 1998.

[13] Paul et al. 1993a,b, Paul 2005.

[14] Judge and Carey 2000, Caro et al. 1995.

[15] T. Stoinski & D. Wharton (pers. comm.) Caro et al. 1995.

[16] Walker 1995, Caro et al. 1995.

[17] Kaplan et al. 2000, Caro et al. 1995.

personal communications from their authors, we have put together all the extant field data we could find on female reproductive cessation in wild primates. Interpretation of these data is difficult due to inconsistencies as to how reproductive cessation was determined and how postreproductive life span (PRLS) was calculated.

Some researchers have simply regarded the time lag between the birth of the final infant (last parturition) and the death of the mother as "postreproductive," although this clearly does not pinpoint a postreproductive stage of life but only the time lag between last parturition and death. All females die in an interval between births regardless of their age, the length of time since they had the infant, or the normal interbirth interval for the species. Death within a normal interbirth interval is not evidence of the cessation of reproductive capabilities, and the time lag between last parturition and the death of the mother cannot be automatically regarded as "postreproductive." Using this method, a young adult female who died when her infant was 18 months old in a species with a 2-year interbirth interval would be classified as postreproductive for 1.5 years. A

very old mother of a healthy newborn could be regarded as reproductively terminated simply because she died. In fact, by this measure, all females would live to have a postreproductive life span. Chimpanzees were at one time reported to be postreproductive for 4.2 and 3.6 years (medians for two populations) despite a mean interbirth interval for all females in the population of 6 years (Nishida et al. 1990).

Other researchers have tried to find more precise ways of identifying reproductive termination from demographic data by requiring that females live longer than the average interbirth interval for the species without producing another infant before declaring them postreproductive. Others (e.g., Takahata et al. 1995, Nishida et al. 2003) subtract a value equal to the weaning age of an infant as part of the calculation so that age at death minus age at last parturition minus weaning age is considered the postreproductive period. Johnson and Kapsalis (1998) further require that female rhesus macaques be over 25 since that is the maximum age at which females in their population had given birth. Caro et al.'s (1995) method is increasingly recognized as the best for identifying reproductively terminated females using demographic data, when hormonal profiles, ovarian histology, and direct measures of menstrual activity are not available:

$$(LP - D) / (\bar{X}IBI + 2\ SD)$$

where LP is age at last parturition, D is age at death, and $\bar{X}$IBI is the mean length of a female's interbirth intervals across her lifetime. In this method a female is classified as postreproductive when the time lag between last parturition and the death of the mother exceeds two standard deviations of the female's own mean lifetime interbirth interval. Thus, it requires that an individual female live significantly longer than *her own* mean lifetime interbirth interval, and it reduces the likelihood of categorizing females as reproductively terminated simply because they died in a normal interbirth interval. Because it can apply to females at any age, it also allows researchers to explore questions about variation within the population in the age at which reproductive termination can occur. Recently, Robbins et al. (2006) modified the Caro method by excluding from analysis interbirth intervals that were shortened by the death of the former infant, thus avoiding variance that arises from infant mortality.

Despite considerable variation in the data presented in Table 31.1 (in part because of the differences in the way reproductive cessation and, thus, length of postreproductive life are measured), certain themes emerge, particularly in the data on wild and free-ranging populations. Column three summarizes the known maximum life span for females of the 21 species listed. Of particular interest here is the small difference between maximum life span for the species (column 3) and maximum age at last birth (column 4). The maximum life span of wild Sumatran orangutans is 53 years and maximum age at last birth is 50, leading Wich et al. (2004b) to conclude that orangutans do not terminate reproduction prior to death or experience anything like human menopause. Indeed, this 3-year gap is only one-third as long as the normal interbirth interval for this species (9.3 years, see Wich et al. 2004b). This pattern of a very short time period between age at last birth and death for the very longest-lived females is reported in diverse primate taxa: For mountain gorillas (*Gorilla gorilla beringei*), the difference is also only 3 years; for capuchin monkeys (*Cebus capucinus*), it is 1 year; for sifakas (*Propithecus verreauxi*), it is 3 years; and for baboons (*Papio cynocephalus*), it is 4 years. Column 5 expresses this time difference as the mean percent of the maximum life span completed before time of last birth, and these values are mostly above 90% for wild and free-ranging populations. Despite differences in the life span of these species, it is clear that females are able to reproduce until very close to their own species-specific maximum life span. Also, recall that maximum life span is the age of the oldest known individual and represents the extreme for the species' capacity to maintain life. Average life span or life expectancy values for all species will be much lower. Most animals are dead long before the ages

listed here, and conversely, most females will continue to reproduce until death. Thus, even where the mean percent of the maximum life span completed before time of last birth in this table is relatively low, such as for Japanese macaques, for which maximum life span is 32.7 and oldest age at last birth is 25, most females continue to reproduce until death, which comes before age 25. This is discussed further in the section "Case Study: Reproductive Termination in Japanese Monkeys."

The data on free-ranging primate populations in Table 31.1 thus show that most nonhuman primate females experience reproductive termination when a large proportion of their life span is already completed. In contrast, human females typically experience reproductive cessation when only 42% of their life span is completed (Table 31.1). Caro et al. (1995) report the example of a human population (eighteenth-century Germans) in which 99.1% of the adult females sampled reached reproductive termination and lived an average of more than 29 years in a postreproductive state.

The great range of variation in the data on captive colonies is harder to interpret. At a glance it appears that animals in captive colonies (often breeding colonies) cease to reproduce earlier relative to the maximum life span for their populations, and thus, the values for the mean percent of the maximum life span completed before maximum age at last birth are lower. Where these values are mostly above 90% for free-ranging populations, they are more typically in the 50%–60% range for captive colonies. At the extremes, the study of captive squirrel monkeys (*Saimiri sciureus*) reports values much like the free-ranging data for nonhuman primates, with females remaining reproductive until age 19 when the maximum life span is 21. The study of *Leontopithecus rosalia*, on the other hand, reports very human-like values, with a maximum age at last birth that is only half of the maximum life span for the species.

Despite this one exceptional report on *Leontopithecus*, our conclusion is that reproductive cessation in human females is quite distinct from the patterns of reproductive senescence and termination found in the nonhuman primates. It was the apparent difference between human menopause and other forms of primate reproductive termination and the lack of any substantial data on a large sample of nonhuman primate females that led us to begin a series of analyses on old-age reproduction in the Arashiyama West population of Japanese macaque females. In the rest of this chapter, we will outline our ongoing examination of reproductive cessation in these monkeys and how it compares to the human case.

CASE STUDY: REPRODUCTIVE TERMINATION IN JAPANESE MONKEYS

In our attempt to address problems of earlier cross-species comparisons, we used a large sample (*n* = 95) of completed lives, for which exact ages and reproductive histories are known, to explore a series of questions regarding reproductive termination and postreproductive life in female Japanese macaques. Data for our studies were collected on the Arashiyama West population of Japanese monkeys. These monkeys were studied in Japan from 1954 to 1972, then transplanted as an entire group to a large ranch in south Texas, where they were studied in semi-free-ranging conditions until 1996. (For more information on group history, management, demography, and environment, see Fedigan and Asquith 1991, Pavelka 1993.) Genealogical records were maintained between 1954 and 1996 on each individual born into the group, which include date of birth, reproductive history, and date of death or disappearance. Females first mate around 4.5 years of age and can produce their first infant at 5 years of age. Japanese macaques are seasonal breeders, with a fall mating season and a spring birth season (Fedigan and Griffin 1996). The youngest female in our sample died at the age of 5 years, and the oldest lived to 32.6 years.

Reproductively terminated adult females were identified by Caro et al.'s (1995) interbirth interval criterion, when $(LP - D)/(\bar{X}IBI + 2\ SD) > 1.0$; that is, when the time lag between last parturition and the death of the mother (LP-D) exceeds two standard deviations of the female's own mean lifetime interbirth interval. This criterion for identifying reproductive termination does not provide direct evidence for the cessation of reproductive capabilities because it relies on externally observable events to infer internal states. However, it is a good estimate of reproductive termination for populations in which hormonal profiles, ovarian histology, and direct measures of menstrual activity are not available. Out of the sample of 95 females, 70 individuals had given birth to at least three infants (the minimum required to calculate a mean interbirth interval and standard deviation).

Reproductive Termination and Age

From the sample of 70 females that had given birth to at least three infants, 20 (28.5%) were identified as having terminated reproduction using Caro's criterion. These 20 females ranged from 14.5 to 32.7 years of age, with a mean age at death of 24.6 years. The 50 females that continued to reproduce ranged in age at death from 8.8 to 25.7 years, with a mean age at death of 17.3 years. Females identified as reproductively terminated were significantly older than those that maintained reproduction until death (Pavelka and Fedigan 1999). No females under the age of 10 were identified as reproductively terminated. While possible, it is unlikely that a female under the age of 10 would meet the criterion for reproductive termination since she would need to produce three infants and then live significantly longer than her own interbirth interval and then die all within 5 years. Reproductive termination is uncommon in females between the ages of 10 and 20; only 12% of reproductively

terminated females were in this age group. Of particular interest are the females aged 20–25 years. Twenty years of age is uniformly regarded as aged for Japanese macaques (e.g., Pavelka 1991, Takahata et al. 1995) since it represents the beginning of the third trimester of the life span for this species. Yet 81% of these old females were still reproductive, with only 19% showing termination. Thus, old females between the ages of 20 and 25 that become reproductively terminated are not representative of their age class. Continued reproduction in these old females is the norm.

The maximum age at which a female gave birth in this population was 25. Given that there are no records of any female in this (or any Japanese macaque) population ever giving birth after the age of 25, we appear to be dealing with a biologically meaningful cut-off point in the reproductive lives of female Japanese macaques. This parallels the findings of Takahata et al. (1995) for another population of Japanese macaques and Walker (1995), Walker and Herndon (2008), and Johnson and Kapsalis (1998), who report reproductive termination after age 25 for the closely related rhesus macaque.

How Old Is a 25-Year-Old Japanese Macaque in Human Years?

Our finding that reproductive termination occurs at a low frequency in female Japanese macaques between the ages of 10 and 25, becoming a certainty after age 25, suggests some similarities between Japanese macaque and human females. Women too may stop producing infants and even occasionally experience full-blown menopause in the decades before age 50, with termination becoming universal and certain for women who live into their 50s. However, before we can conclude that Japanese macaques at age 25 experience something equivalent to the human female menopause, other factors must be considered. Menopause, the complete cessation of ovulation, menstruation, and reproductive capabilities, at 50 years of age is universal—not idiosyncratic—among women. Further, it occurs less than halfway through the species' life span of 120 years. This termination of reproductive capabilities in women does not occur in association with extreme old age or with advanced deterioration of the organism as it approaches the maximum life span of the species. Thus, we are led to the question: How old is a 25-year-old Japanese macaque, and how does she compare to a 50-year-old woman?

One way to draw such a comparison would be to use survivorship values to compare our monkey data with a human population whose survivorship values are unlikely to have been affected by the forces of modernization. For example, Howell (1979) reports a 2% survivorship to age 85 for the Dobe !Kung, a value which is comparable to the 2.3% survivorship to age 26 for the Japanese macaque females (Pavelka and Fedigan 1999). Based on this comparison, one might argue that in Japanese macaques reproductive termination is unlikely to occur before the equivalent human age of 85 years. However, the use of survivorship values to make comparisons between species is problematic, due to the fact that these values are environmentally dependent and vary widely among populations of the same species.

The question of how old a 25-year-old monkey is compared to a 50-year-old woman might be better approached using the following equation:

$$\frac{\text{Species A Age at Reproductive Termination}}{\text{Species A Maximum Life Span}} = \frac{\text{Species B Age at Reproductive Termination}}{\text{Species B Maximum Life Span}}$$

There are no reports of Japanese macaques living longer than the oldest female in our own sample; thus, age 32 is considered the maximum life span of Japanese macaques. The oldest woman of documented age at death lived to be 122 years of age (*Detroit News*, August 4, 1997). Using these maximum life span values in the equation, we could argue that a 25-year-old monkey is the equivalent human age of 95 years. Undoubtedly, the 122-year value represents an exceptionally rare outlier for human maximum life span. Nonetheless, this is the value for the oldest known individual, and there is no other agreed-upon value for maximum life span of humans. Estimates range from 90 (Weiss 1981) to 122 years. Using more modest values of 100 years for the maximum human life span and 30 years for the maximum Japanese macaque life span, we would argue that a 25-year-old monkey is the equivalent human age of 83 years.

Although it is difficult to obtain exact figures, it is clear that population-wide reproductive termination in Japanese macaques occurs very late in the life course. Reproductive termination at age 25 is much later in the life course of Japanese macaques than is menopause in women at age 50. Japanese macaques are widely regarded as old when they reach age 20 (only 7.9% of our population lived to this age), yet those aged 20–25 years are unlikely to experience reproductive termination: 81% of this age group is still reproductive.

Is Reproductive Termination Adaptive in Japanese Macaques?

Female macaques living in matrilineal societies are good candidates for an investigation of the grandmother hypothesis since they frequently engage in kin-directed affiliative behaviors. Pavelka (1991) found that old females in the Arashiyama West population continue to have active social lives and to interact affiliatively with their offspring, relatives, and friends. Grandmother macaques that have ceased to produce infants of their own would be in a position to offer a variety of supportive behaviors to their children and grandchildren, including carrying, retrieving when lost, predator protection, interventions in agonistic interactions, and alloparental supervision when the mother is absent. They could also lactate longer than usual for their final infant. In theory, these caregiving behaviors

might be sufficient to advantageously influence survivorship of descendents. We investigated the possible adaptive value of reproductive termination in Japanese macaques by comparing the survivorship of the descendents of our postreproductive females to those who continued reproducing until death. We compared reproductive ($n = 50$) and postreproductive ($n = 20$) females for three measures of offspring survivorship: (*1*) mean survival of all offspring, (*2*) survival of *final* offspring, and (*3*) survival of daughter's offspring (matrilineal "grandchildren"). Survival analyses did not reveal any significant differences between these two groups (Fedigan and Pavelka 2001). Infant survival rates to age 1 were remarkably similar for those offspring born to postreproductive and reproductive females (85% vs. 83%) and not significantly different for survival to age 5 (71% vs. 79%). Survivorship of the final infant was greater for those born to postreproductive females than to reproductive females (85% vs. 72% survival to age 1, 80% vs. 67% to age 5); however, this difference was not statistically significant. Finally, we found that the survival rates of a daughter's offspring did not differ between postreproductive and reproductive grandmothers (86% vs. 83% to age 1, 80% vs. 77% to age 5). In this sample of Japanese macaque females, the cessation of reproduction before death does not result in greater survivorship of immediate offspring or of daughters' offspring, and these tests failed to support the grandmother hypothesis. In spite of the fact that female Japanese macaques do direct differential caregiving behaviors to their descendents and other matrilineal relatives, these behaviors do not confer fitness benefits in the form of greater survivorship of descendents.

Other Differences Between Reproductive and Postreproductive Females

Next, we attempted to determine if there are other traits with possible adaptive value that distinguish females that cease to reproduce before death from those that do not and found that reproductive and postreproductive females were not significantly different in dominance, matriline affiliation, body weight, infant sex ratios, age at first birth, or lifetime reproductive success (number of infants surviving to reproductive age). They were, however, significantly different in age at death, the length of time between last parturition and death, cause of death, fecundity (number of infants produced), and reproductive life span (Fedigan and Pavelka 2001). Those 20 females classified as postreproductive lived on average 7 years longer (to age 24.6 compared to 17.4) and specifically lived five times longer after last parturition (60.0 vs. 12.1 months). These older postreproductive females were more likely to disappear, whereas reproductive (younger) females were more likely to succumb to infectious diseases, in line with a previous study with a much larger sample of mortality causes which found that death from infectious disease is more common in younger Japanese macaques (Fedigan and Zohar 1997).

Postreproductive females also had more infants (9.7 compared to 7.7) and a *longer* reproductive life span (age at final birth – age at first birth): 13.8 years compared to 10.1 years for those that died while still reproducing. Thus, postreproductive females experienced a significantly greater number of years during which they were producing infants than did reproductive females. This means that postreproductive females not only were longer-lived in total life spans but also had longer *reproductive* life spans than did those that died while still reproducing. Since age at first birth varied little, postreproductive females must have acquired these extra reproductive years at the end of the reproductive phase of their lives. A linear regression of fecundity against age at death showed that age at death is highly predictive of fecundity. We concluded that postreproductive females produced more offspring because, in spite of experiencing some postreproductive years at the end of their lives, they still lived through more years in which to give birth than did reproductive females, which died younger.

Why did postreproductive females have higher fecundity but not significantly greater lifetime reproductive success than reproductive females? Lifetime reproductive success was calculated as the number of offspring to reach breeding age at 5 years. We suggest that because living long enough to become postreproductive was *not* associated with greater survivorship of one's offspring, the greater fecundity of postreproductive females did not translate into significantly higher lifetime reproductive success.

Based on these analyses, we concluded that reproductive cessation in Japanese macaques is adaptive only in its indication of a set of females that have reached very old age and have lived long enough to produce many infants. However, the grandmother hypothesis predicts that postreproductive females are able to direct additional care to their grandchildren, and in the analysis we investigated all postreproductive females, without specifying which were grandmothers. Thus, we undertook an investigation of the *availability* and adaptive value of reproductive and postreproductive mothers *and* grandmothers (Pavelka et al. 2002). This allowed us to target not just postreproductive females in general (many of which may not have living adult daughters or grandchildren) but the theoretically important subcategories of postreproductive mothers and grandmothers in particular. Eight of the 70 females that could be categorized as reproductive or postreproductive did not produce any daughters (and hence grandchildren that were known to us), so this analysis is based on 62 grandmothers and their 175 daughters and 905 grandchildren. For the analyses in which we needed the daughters' death dates, we were able to use only 88 of the 175 daughters, and only 74 of these produced infants. Of the 905 grandchildren born, we have complete information on 886 individuals that were included in the survival analyses. Of these, 504 survived to age 5. The probability of survival to age 5 is calculated based on the survivorship of all of the 886 individuals in the sample.

Availability of Reproductive and Postreproductive Mothers and Grandmothers

Of the 175 Japanese macaque daughters in our sample, 64 (36.6%) had a deceased mother, 106 (60.5%) had a living reproductive mother, and only 5 (2.8%) had a living postreproductive mother at the time they first gave birth. Therefore, nearly two-thirds of daughters had their mother still alive at the time of their first birth—a mother that could potentially contribute to her daughter's production of infants or to the survivorship of those infants—but the vast majority of these grandmothers were themselves still producing infants. Very few daughters had a postreproductive mother available to them when they began to reproduce.

The mean length of reproductive life for the sample of 74 daughters was 9.71 years. Over half of the reproductive years of the daughters in our sample (5.26 years, 54.2%) were lived without a mother present at all. During the other half (4.45 years or 46% of the daughter's reproductive life), the grandmother was available but the vast majority of these years were spent with a grandmother that was herself still reproducing. Only 4.2% of the reproductive life span of the daughter was spent with a postreproductive grandmother available to help the daughter. This represents less than 5 months for the average female in the population.

This pattern is further reflected in the next generation in the time available for a postreproductive grandmother to have an impact on her grandchildren's survival—the essence of the grandmother hypothesis. Most of the grandchildren's first 5 years of life was spent without a living grandmother (3.6 years or 72%). Grandchildren that survived to age 5 had, on average, a living grandmother available to them for only 1.4 years, or 28% of that time, and during most of that time the grandmother had an infant of her own to care for. Only 0.2 years, or 4% of the first 5 years of life of the grandchildren in our sample—that is 2 months on average—was spent with a postreproductive grandmother available to offer the extra caregiving behavior predicted by the grandmother hypothesis.

Thus, grandmother Japanese macaques (both reproductive and especially postreproductive) proved to be far less common than we had expected. This study showed that while nearly two-thirds of newly adult females have a living mother when they first give birth, the mother will likely remain alive for less than half of her daughter's reproductive life. Postreproductive grandmothers are exceedingly rare. Only five out of 175 daughters had a postreproductive mother available to them when they reached reproductive age, and the average adult daughter had a postreproductive mother available to her for less than 6 months of her reproductive years. The average grandchild had a postreproductive grandmother available to it for less than 2 months. This surprisingly restricted availability of postreproductive grandmothers, even in a long-lived, multigenerational, provisioned population such as the Arashiyama West Japanese macaques, must limit opportunity for the classic maternal investment models to operate.

Adaptive Value of Reproductive and Postreproductive Mothers and Grandmothers

In the Arashiyama West population, most Japanese macaque females do not have their mothers around and free to help them with their infants because the living grandmothers are still reproducing themselves. Likewise, the vast majority of grandchildren do not have a postreproductive grandmother around to help take care of them. Still, it is interesting to explore the possible effects that a female (grandmother) can have on the reproduction of her daughter. Does having a live mother improve any aspect of her daughter's reproduction? For example, do females whose mother is alive start to give birth earlier, or do they have shorter interbirth intervals? Do the grandchildren survive better if grandmother is alive—especially in the rare case of a grandmother that is alive and postreproductive? In these Japanese macaque data, having a mother alive did appear to be associated with improvements in some of her daughter's reproductive parameters. First, females whose mother was alive were more likely to give birth at age 5 rather than age 6, although it did not matter if that mother was reproductive or postreproductive. This may be because the presence of the mother improves a young female's chance of having a "successful" first fertile cycle (one that results in a conception) at age 4.5. The first active estrous period for female Japanese macaques requires that these young females venture out of the tight female kinship units for the first time to establish contact with unrelated adult males—animals they would have had little need or opportunity to interact with previously. The inexperienced behavior of these young females increases the likelihood that adult males will target them for aggression (McDonald 1985). In vervet monkeys, adult daughters whose mothers were still living in the group received less aggression and were defended more often than young adult females whose mother had died (Fairbanks 1988b). Thus, the mother of the pubescent female, through the agonistic support she provides to her daughter, may help to increase confidence on the part of the daughter and/or reduce the frequency and intensity of serious aggression from adult males, thus increasing the likelihood of the daughter forming a successful consortship and producing an infant at age 5. Second, the presence of a mother also appears to be of some benefit to females in terms of shortening their interbirth intervals, from 19.2 months for females whose mother was dead to 18.1 months for females whose mother was alive. Moreover, if the mother was alive and postreproductive, the interbirth intervals of the daughters dropped to 16 months, although this trend is not quite statistically significant (Pavelka et al. 2002). Thus, despite being less common than expected, we found evidence that the presence of a living mother—most of which were still reproductive—is advantageous to the reproduction of their daughters.

The essence of the grandmother hypothesis is that grandmothers—specifically postreproductive grandmothers—are able to enhance the survivorship of their grandchildren. Do the grandchildren in this Japanese macaque population survive better if the grandmother is alive—and especially if she is alive and no longer producing infants of her own? Interestingly, given the initial survival analysis, we found that they do: There is a significant difference in survivorship to age 1 depending on the status of the grandmother. Specifically, and although there were very few in this category, grandchildren with a living postreproductive grandmother were significantly more likely to live to age 1 than were grandchildren with either a dead grandmother or a live one that was still herself reproducing. There are no differences in the survivorship of grandchildren to age 5 based on whether the grandmother was dead, alive and reproductive, or alive and postreproductive.

Our earlier test of all 20 reproductively terminated females compared with all 50 reproductive females (irrespective of whether these females were mothers or grandmothers) found no difference in the survivorship of first- or second-generation descendants. Reproductive termination was characteristic of only a small cohort of very old females, and postreproductive females were significantly older at time of death than were reproductive ones, supporting the conclusion that reproductive termination in this population is a by-product of selection favoring longevity. Yet in a subsequent analysis, in which we were able to target the theoretically important postreproductive mothers and grandmothers, we did see some tantalizing evidence for the adaptive value of postreproduction. Those few grandchildren that had a postreproductive grandmother present during their first year of life were significantly more likely to survive (95%) than were those that had a dead grandmother (85%) or a living and reproducing grandmother (89%). The first year of life is critical as most mortality of immatures occurs in this period (Fedigan and Zohar 1997) and most infants are weaned at 1 year. However, the presence of a postreproductive grandmother did not enhance the survival of grandchildren to age 5, so the effects appear to be restricted to the infant or preweaning stage of life. Recall, however, that by this stage almost 75% of the grandchildren no longer had a grandmother alive, postreproductive or otherwise. Grandmothers have a much better opportunity to affect survival to age 1 than to age 5 because they are more available during the earlier years of the grandchildren's lives.

SUMMARY

Human females are unique among the primates in the form of menopause that they experience—the universal permanent cessation of menstrual cycles and reproductive capability at only halfway through the maximum life span for the species. Biological anthropologists and evolutionary biologists interested in the evolution of this peculiar trait (generally speaking, natural selection should favor reproducing phenotypes) have developed two categories of explanation. Either human menopause is a by-product of increased life span or of traits with high selective value early in life or human menopause is directly adaptive because nonreproducing middle-aged women can better maximize their inclusive fitness by aiding their daughters and grandchildren in reproduction and survival than they can from continuing to reproduce themselves. A small number of individual monkeys and apes that have ceased to reproduce and that show endocrine changes similar to menopausal women have been identified in captivity; however, in addition to being rare individuals, they are very old and close to the known maximum life span for their species. Reproductive cessation has also been reported for a few free-ranging nonhuman primates, although the problem of distinguishing reproductive termination from death during an interbirth interval plagues these reports.

We investigated reproductive termination in a large sample of free-ranging Japanese macaques by using a female's own reproductive history to determine when she had lived significantly longer than her own average lifetime interbirth interval without reproducing. We found that reproductive termination was uncommon in individuals under 25 years of age but universal after that. A 25-year-old Japanese macaque, however, is a rare creature as only just over 2% of the population will live to this age. A 25-year-old Japanese macaque is estimated to be equivalent to an 85-year-old woman, at least. There was no difference in the survivorship of the grandchildren of these females that lived long enough to become postreproductive; however, this group did produce significantly more infants than those that died while reproducing because fecundity and longevity are positively correlated. For a Japanese macaque female in our sample, the key to having many infants was living a long life, even though this could mean living a few years beyond reproductive ability before death in very old age. In the very rare cases when a reproductively terminated female also had a live grandchild to which she could direct caregiving behaviors, those grandchildren did have an improved chance of surviving to age 1. We conclude that the reproductive termination that we have documented in female Japanese macaques of the Arashiyama West population probably occurs too late in life, with too few females reaching and remaining in this stage for any substantial proportion of their descendants' lives to have sufficient inclusive fitness effects to compensate for the loss of the grandmother's direct reproductive output, as theorized by the grandmother hypothesis. Our findings here suggest that increased survivorship of grandchildren during their first year of life and more rapid production of young by their daughters may be a secondary benefit enjoyed by those few postreproductive females that find themselves without an unweaned infant of their own but with a grandchild available to benefit from their free time and energy.

Thus far our investigation of Japanese macaques has focused on the benefits that might be associated with reproductive termination. However, the grandmother hypothesis is based on the assertion that the benefits of ceasing to reproduce will outweigh the costs of continuing to reproduce. Thus, we will continue this investigation by turning to the other side of the equation and considering the possible costs, for older females, of continuing to reproduce.

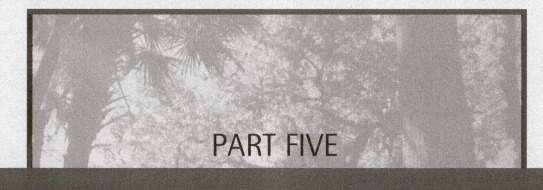

PART FIVE

Ecology

32

Social Systems and Socioecology
Understanding the Evolution of Primate Behavior

Agustín Fuentes

1. Why do we care about primate social systems as opposed to just studying individual groups of primates? What parameters make up a social system?
2. What are the basic assumptions of the socioecological model? How do the variables in the model interact to help us understand what a given social system will look like?
3. What are some of the complexities of actual primate behavior we need to keep in mind, and what are some of the new perspectives in evolutionary theory that we can include in order to enhance the basic socioecological model?

INTRODUCTION

Over 30 years ago, Phyllis Dolhinow proposed that we "focus on those aspects of life common to most primates and in so doing, see to what extent primate behavior is variable" (Dolhinow 1972b:352). Today, as primatology moves into its second century, the result of Dolhinow's call is a growing awareness of the complexity and variation in intra- and interspecific behavior across the primate order. This complexity and heightened awareness of variation presents itself as a major challenge to our abilities to model and envision primate patterns (Kappeler and van Schaik 2002; Strier 1994b, 2003b; Thierry 2008). In their influential review of the evolution of social systems, Kappeler and van Schaik (2002:708) state that "We therefore need a theoretical framework that relates fitness-relevant behavior of individuals, such as foraging, predator avoidance, mating and parental care, to the defining characters of a social system." In their review, Kappeler and van Schaik use the socioecological model (Crook 1970, Emlen and Oring 1977, Terborgh and Janson 1986) as the basal framework in their functional quest to link the individual (as unit of natural selection) to the higher-level facets of societies such as groups and populations. In this chapter we examine the basic precepts of socioecology and primate social systems, as well as a few relevant innovations in evolutionary theory, in order to develop a broad picture of the evolution of primate behavior.

Basal assumptions about primate social behavior and social systems were heavily influenced by Hamilton's (1964) proposal of kin selection and Wilson's (1975) notion of reciprocal alliances between related individuals as core facets of group-living organisms. In most primate species, one sex is *philopatric* (stays in the natal group) and, thus, has access to kin throughout their lives. Individuals that are around kin are able to enhance their inclusive fitness in a manner that dispersing individuals are not, so the philopatric sex has a benefit relative to the dispersing sex (see Isbell 2004). However, as Kappeler and van Schaik (2002) note, socioecological pressures also have great impact on shaping primate lives. The current socioecological model for primates is based on the concept that ecologies shape and structure primate social systems at the level of the individual and, thus, the social group. At the heart of this perspective is the distribution and defensibility of food, with evenly distributed foods being difficult to defend or monopolize and patchily distributed or clumped foods lending themselves to monopolization or at least some degree of defensibility. These parameters of ecological constraints on food then determine the type of groups that females form and the types of relationships between females. Males associate themselves with the females based on female distribution and grouping patterns, resulting in mixed-sex social groups (as summarized by Wrangham (1980), see Figure 32.1). van Schaik (1989) expanded this basic model to include the ecological variable of predation, acknowledging the core role that predation pressure has in influencing primates to live in groups. He posited that predation avoidance coupled with food distribution are cofactors in shaping of primate grouping patterns. He also described two types of competition regimes (scramble and contest) related to food distribution that could affect grouping and social structures. *Scramble competition* is when individuals are competing against time and space

Figure 32.1 Wrangham's basic model of female distribution related to food distribution.

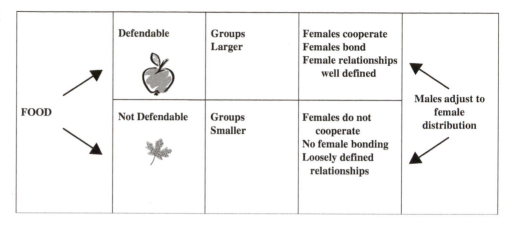

to acquire sufficient food given a nondefensible nature of the food source. *Contest competition* emerges when food sources can be monopolized/defended and, thus, individuals can compete for domination and/or priority of access to the food source. Further refinements of the model included Isbell's (2004) proposal that ecological constraints of population structure affect dispersal patterns and, thus, social systems and Sterck et al.'s (1997) focus on the character and relative asymmetry of intragroup relationships as a result of specific ecological pressures, including the possibility of infanticide by males.

Kappeler and van Schaik (2002:708) recognize that the "social organization and demographic conditions *created* by individual behaviors also impose constraints on the behavioral options of these same individuals leading to complex feedback loops" (italics mine). The evident complexity in primate social systems does call for the intensive focus on socioecology that Kappeler and van Schaik propose; however, their focus exclusively on functional aspects of behavior (as products of natural selection), while powerful, may overlook some salient factors in our ongoing quest for effective assessments of primate social evolution and behavior. The purpose of this chapter is threefold: (*1*) to review the basic concepts of primate socioecology; (*2*) to provide an overview of variation in primates' social behavior, including the impact of selection; and (*3*) to expand our horizon regarding the processes of evolution that can impact variation and patterns in primate social systems.

SOCIAL SYSTEMS AND SOCIOECOLOGY

Social systems are the focus of numerous books and major articles in primatology (Clutton-Brock 1974b, Crook and Gartlan 1966, DiFiore and Rendall 1994, Dolhinow 1993, Dunbar 1988, Hall 1965c, Kappeler and van Schaik 2002, Kummer 1971, Rowell 1993, van Schaik and van Hoof 1983; Wrangham 1980, 1987; Zuckerman 1932). Various definitions of "social system" have emerged over time, and it remains a concept with variable uses in primatology

(Dolhinow 1993, Dunbar 1988, Fuentes 1999a, Rowell 1993,) and the broader arena of animal behavior (Tinbergen 1953, Wilson 1975, McFarland 1987). Kappeler and van Schaik (2002) use the term "social system" to reflect the conflux of factors that make up primate societies, and that is how we will use it in this chapter.

What is the benefit of employing a concept such as "social system"? Why not simply discuss and measure the constituent factors of social systems, what primates are actually doing? Using this higher-level perspective of primates' social existence helps us remember that, as primatologists, we are in a dual undertaking: examining the behavior and ecology of individuals during their lifetimes (or some segment thereof) and hypothesizing as to the impacts of natural selection and other forces of evolution across time on groups and populations. That is, we examine individual behavior and ecologies but attempt to explain the observed behavioral patterns by invoking and/or constructing evolutionary histories (past action of selection, adaptations; Williams 1966a, Kappeler and van Schaik 2002). Thinking about primate behavior in the context of social systems allows us to maintain our focus on the individuals while at the same time emphasizing that they do not exist outside of the broader group and population. Behavioral patterns and interactions of individuals create the social environment in which primates exist, and social systems emerge from these interactions. Thus, the behavior exhibited by an individual primate reflects adaptive strategies, stochastic (chance) occurrences, and specific responses to the social context in which the individual finds itself. Social systems, then, change over time as their constituent factors change across generational time, at least in response to selection, genetic drift, gene flow, chance, and historical events. Individual primates do not exist outside of their social context, and understanding that social context and the patterns of social behavior in a primate group, population, or species allows us to better attempt broad-scale explanations for the occurrence of observed behavior.

Primate socioecology lays the basis for our understanding of the evolution of social systems. Since the 1980s there have been a number of major contributions that have built upon basic ecological assumptions and created the overall perspective that we refer to as "primate socioecology" today. Traditionally, biologists and ecologists present the following suite of pressures that face any organism: avoiding predation, finding and ingesting sufficient food and water, finding mates, and achieving effective thermoregulation. Obviously, each of these pressures will be different for organisms in distinct environments. For example, large mammals such as primates may, in some environments, suffer a lower risk of predation than other smaller mammals, or the same species of primate living in two different environments may suffer very different predation risks (see Chapter 35). For primates, living in a group adds to these pressures via the potential for competition within the group and its effects on changing the costs of obtaining food. At the same time, group living may also act to ameliorate the pressures of predation and intergroup feeding competition (generally contest competition) and to provide ready access to mates (van Schaik 1989, Sterck et al. 1997, Strier 2000a).

SUMMARY OF THE CURRENT SOCIOECOLOGICAL MODEL

Most researchers agree on a basic set of environmental and hypothesized social pressures that can result in differential reproductive success for individuals (Wilson 1975, Rubenstein and Wrangham 1988, Kappeler and van Schaik 2002). Additionally, in their overview of the evolution of social systems, Kappeler and van Schaik (2002) follow the socioecological paradigm and treat males and females differently in their explanations of selective pressures on individuals. They do this because under the general model male and female fitness (or relative potential reproductive success) is limited by different factors. That is, "the model assumes that the distribution of females is primarily determined by the distribution of risks and resources in the environment, whereas males distribute themselves primarily in response to the temporal and spatial distribution of receptive females" (Kappeler and van Schaik 2002:708). This orientation follows directly from the anisogamy argument of Bateman (1948) and its subsequent expansion and elaboration into models of parental and reproductive investment (Trivers 1972, 1974; Clutton-Brock 1991). This lays the baseline for the current socioecological model that most researchers utilize when describing the evolution of social systems and when predicting patterns of behavior in primates.

"The model assumes that variation in predation risk and the abundance and distribution of food resources shape the relationships of females by determining the competitive regimes to which they are submitted. When predation risk is high, animals form cohesive groups to reduce it" (Thierry 2008:94). Echoing Kappeler and van Schaik (2002), Sterck et al. (1997), and Isbell (2004), Thierry (2008) also notes that, given these basal assumptions, the relationships between variation in food distribution and population density can act to elicit the emergence of different types of social systems. Thierry draws together all of the socioecological model's main components and derives a set of predicted outcomes (which he calls "social syndromes" but which can be seen as equal to "social systems" as used here). Thierry (Following Sterck et al. 1997) divides selective pressures and outcomes relevant to the model into three main areas—ecological conditions, competitive regimes, and behavioral expression (see Table 32.1)—with four main social syndromes—dispersal-egalitarian, resident-egalitarian, resident-nepotistic, and resident-nepotistic-tolerant. Assuming that predation risk and abundance/distribution of food sources drive competitive regimes among females and their grouping patterns and that population densities affect within- and between-group relationships, the parameters of the socioecological model help us predict social systems by comparing across the possible combinations of various selective pressures.

For example, in a situation where predation pressure is high, food is dispersed, and there is low population density, we would expect a social system with high levels of within-group scramble competition and low overall contest competition because food is not defensible and there are few conspecifics with which to compete. This leads to a situation where dominance relationships are weak (not very asymmetrical or clearly defined), coalition formation is rare, and no nepotism occurs as there is no need for the formation of alliances and coalitions to improve one's chances in competitive contests. In such a social system we would also expect that females regularly disperse as there are not clear internal forces tying them to their natal groups. This social system is termed "dispersal-egalitarian" (Figure 32.2).

Table 32.1 Major Selective Conditions of the Socioecological Model[1]

A. Ecological conditions	Predation risk: high or low Food distribution: dispersed or clumped Population density: high or low
B. Competitive regime	Within-group scramble competition: high or low Within-group contest competition: high or low Between-group contest competition: high or low
C. Behavioral expression	Degree of dominance asymmetry: weak or variable or strong Coalition formation: rare or common Nepotism (favoring of close kin): rare or common Dispersal (females leave natal group): rare or common

[1] included in the conditions but not assigned to the three categories is the presence or absence of infanticide by males in the population.
Source: Derived from Thierry 2008 and Sterck et al. 1997.

Alternatively, if we have a scenario where predation risk is still high and foods are dispersed but there is high population density, we would expect similarity to the previous social system but with two major differences: reduced female dispersal and high between-group contest competition. The high population density would discourage female dispersal but within-group contest competition would stay low due to the dispersal of food sources, so we would not see clear dominance asymmetries or extensive coalition formation within groups. We would see higher between-group contest competition as the higher density of individuals leads to more or larger groups competing for space and for the exploitation of the dispersed food sources within that space. This is what is labeled "resident-egalitarian" (Figure 32.3).

A third system, resident-nepotistic (Figure 32.4), is predicted from ecological conditions wherein predation risk is still high and population density is low but food sources are clumped. The clumping of food sources makes them potentially defendable by fewer individuals, which elicits high levels of within-group contest competition for the resources, resulting in very asymmetrical dominance relationships, leading to frequent development of coalitions and alliances and, thus, a strong push for females to ally with their close relatives (nepotism). This bonding among females also reduces the occurrence of dispersal from natal groups as the benefits of staying put are high relative to leaving one's female kin.

A fourth system, resident-nepotistic-tolerant (Figure 32.5), results from clumped food sources with low predation pressure and high population density. These two variables add variation to the intensity of within-group contests and the relative strength of dominance asymmetry due to the lack of predation pressure on the individuals in the group but the

Figure 32.2 Dispersal-egalitarian system.

Figure 32.3 Resident-egalitarian system.

Figure 32.4 Resident-nepotistic system.

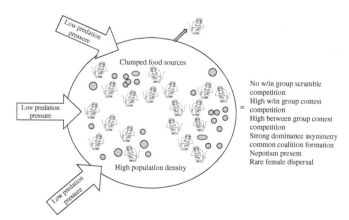

Figure 32.5 Resident-nepotistic-tolerant system.

maintenance of high levels of competition with other groups in the high-density population.

Included in Kappeler and van Schaik's (2002) overview of social systems is the hypothesis of sexually selected infanticide. Thierry also notes that infanticide can affect social systems in the socioecological model. Its occurrence can influence female coalitions (bonding) and, possibly, female dispersal patterns in ways similar to that of predation but more specifically related to male behavior and, thus, within-group sexual activity. The hypothesis proposes that the killing of infants by unrelated males is an adaptive strategy and a significant pressure on individual primates (especially females), equivalent to or greater than predation or food distribution pressures. While relatively popular among primatologists, this hypothesis remains contested in the scope of its impact and the factors involved in its recording and assessment. If assertions of the prevalence of sexually selected infanticide are correct (see Janson and van Schaik 2002), then the behavioral pattern of infanticide will be expressed differentially under differing social systems. That is, if infanticide is common due to past increased reproductive success by those males that carried heritable biological variants that facilitate the appearance of infanticide, then its expression will be dependent on certain demographic, ecological, and behavioral variables that make it a beneficial endeavor for current males with those variants (van Schaik 2002). In this case, the frequency and effectiveness of infanticidal events will impact individual primates; the overall pattern should then be apparent at the level of the population, and anti-infanticide behaviors by females should also become evident. However, the assertion that infanticide is truly widespread across primate taxa and represents an important pressure on primates remains contentious. The majority of infanticidal events have been reported in a few species and only at some of the study sites for those species (van Schaik 2002). Therefore, the actual impact of infant killing by males as a significant factor (evolutionary pressure) on the majority of primate populations remains unknown.

It is obvious from this overview of the predictions of the socioecological model that, given the variation in habitats, population patterns, and the mosaic nature of food sources and primate diets, there will be a degree of structural variation in each of the four main scenarios described (Figures 32.2–32.5). While the socioecological model continues to be valuable to develop basal predictions about primate behavior, there is a growing body of studies demonstrating that the expectations of the model do not always, or even frequently, do a particularly good job of describing what we actually see in primate behavior (although sometimes they do; Saj et al. 2007, Sterck et al. 1997, Thierry 2008). Variations in ecologies within species' ranges, diversity of historical influences and contingencies on the demography and behavior in primate groups (e.g., Sapolsky and Share 2004), and a higher than expected degree of plasticity in primate behavioral patterns all create divergences from the

expectations of the socioecological model in field studies of living primate groups. Because the basic assumptions of the model are constrained to relative measures ("high," "low," "common," "rare"; see Figures 32.2–32.5), it can be difficult to apply detailed measurements to (or to gain accurate measurements of) elements of the model, and most field studies yield mixed support for the various assumptions. Given this, it is worthwhile to revisit the structural basics of primate behavior and grouping patterns and ask, "What are the factors that make up primate social systems?" and "How might selection and other factors impact their evolution?" In this next section we will review the constituent factors of social systems so that we can create a more robust infrastructure to apply along with the socioecological model and to expand our efforts to include other evolutionary pressures and processes in addition to predation, food, population density, styles of competition, and infanticide.

THE CONSTITUENT FACTORS OF SOCIAL SYSTEMS

In order to fully assess individual primates' contexts for exhibiting behavior, we require a clear understanding of the constituent factors that make up the social system in which a primate exists. These factors are composed of a set of variable elements, and understanding the patterns of variation in these elements, in conjunction with the utilization and expansion on the socioecological model, will enhance our ability to describe and model the range, expression, and evolution of primate behavior.

Primate Grouping Patterns

A primate *group* is defined as that collection of individuals that engage in regular, patterned social contact over extended periods of time and use the same general spatial areas (Kummer 1971). Rather than seeing a group as a static entity, long-term research on a diverse array of species suggests that groups change over time along demographic lines as individuals die, are born, and migrate and as groups fission or merge. However, despite this variation in individual composition, most species do exhibit one or more basic general demographic patterns in their social groups (Crook 1970, Dolhinow 1972b). Table 32.2 describes the basic demographic compositions for primate groups. Some form of multiple adults of both sexes is the most common grouping pattern. There is substantial variation within this category, with differential representation of the variants across the families and subfamilies of Primates. One-adult male/multiple adult female and two-adult (one male and one female) groups are less common but present in both anthropoids and prosimians. One-adult female/multiple adult male groups are rather uncommon and found primarily among the callitrichid primates. There remains some debate as to whether "dispersed sociality" is truly a group-living situation; however, most, or all, individuals in dispersed social systems, be they apes (*Pongo pygmaeus*) or prosimians, interact with one another through visual, olfactory, vocal, or other contexts (see Chapters 4 and 18). That is, they engage in patterned social interactions in the same general spatial areas, even though they spend less time in spatial proximity. Therefore, we can safely say that the majority of primates live in social groups and that the composition of the social group (of various kinds) is the main unit of demography of social systems.

Given that there are general demographic patterns (age/sex compositions) that characterize primate groups but that these patterns vary, it is relatively important to recognize the patterns of intraspecific variation in grouping. Table 32.3 provides examples of intraspecific variation in grouping patterns. At the most basic level, this variation in grouping

patterns is a result of ecology and history. Different environments are going to have basic impacts on primate groups. As noted, the socioecological model predicts that the size and dispersal of food sources, the presence or absence of predators, and the impact of climatological factors will provide differential pressures impacting overall group size and issues of seasonality in reproduction, which in turn impact demographic factors of the groups (Kummer 1971, Dolhinow 1972b, Dunbar 1988, Sussman 1999a, Isbell 2004). Variation in group size can also be due to stochastic (chance) occurrences that can impact the sex ratio, age structure, and ranging patterns of groups (Kappeler and van Schaik 2002). In these cases, chance events and other historical factors (e.g., a predation event in a particular location

Table 32.2 Grouping Types (Demographic Labels) of Extant Primates

A. Multifemale/multimale	The group consists of greater than one adult female and greater than one adult male and immatures. Variants:
	1. *Cohesive group.* The majority of individuals within the group occupy the same general areas at the same time, generally within vocal and/or visual range for the majority of active time.
	2. *Fission–fusion.* The overall group breaks into smaller subgroups of variable composition but that use similar range. Subgroups can fuse into larger groups, or the entire group can occasionally occupy the same area at the same time. Fissioning or fusion may be seasonally and/or behaviorally driven.
	3. *Community.* Similar to a fission–fusion; however, subgroups more dispersed and less likely to coalesce into full group.
B. Unimale/mutlifemale	A cohesive group consisting of one adult male, greater than one adult female, and young.
C. Unifemale/multimale	A cohesive group consisting of one adult female, greater than one adult male, and young.
D. Two-adult group	A cohesive group consisting of one adult female, one adult male and immatures.
E. Dispersed sociality	Individual primates with individual ranges that overlap extensively. Interactions between individuals within the overlapping ranges can be frequent or infrequent, vocal and/or visual, and/or tactile. Occasionally multi-individual subgroups may form.

Table 32.3 Examples of intraspecific variation in grouping patterns

SPECIES	GROUP COMPOSITION	STUDY
Propithecus diadema edwardsi	Unimale/multifemale = 9 groups Unifemale/multimale = 11 groups Two-adult group = 12 groups	Pochron and Wright 2003
Semnopithecus entellus	Unimale/multifemale = 27 groups Multifemale/multimale = 229 groups All-male = 4 groups	Treves and Chapman 1996
Hylobates lar	Two-adult group = 3 groups Unifemale/multimale = 1 group	Brockelman et al. 1998
Gorilla gorilla beringei	7 groups observed Unimale/multifemale = 41.1% of observation time Multimale/multimale = 36.8% of observation time All-male = 22.1% of observation time	Robbins 1995
Saguinus mystax	Two-adult group= 27 group Unimale/multifemale = 30 groups Unifemale/multimale = 52 groups Multifemale/multimale = 163 groups	Sussman and Garber 1987

leading to members of the group avoiding that location) can have social impacts. That is to say, current social behavior that a primatologist is observing can be impacted by the demography (makeup) of a group and that demography can be impacted by both ecological and historical factors (Sapolsky and Share 2004, Wasser et al. 2004). For example, historical changes in the group sex ratio that increase or decrease the number of adult females will impact intrasexual competition, female bonding, mating opportunities, size of peer groups, and so on (see Isbell 2004). In other words, social strategies displayed by individual primates can vary depending on the demographic patterning in which they find themselves and that patterning can be affected not only by the basal factors of the socioecological model but also by historical and chance events.

Primate Mating and Rearing Patterns

Under the socioecological model, our measures of function in primate behavior revolve around assessments of fitness, or reproductive success. Thus, mating behavior is quite important. In primates, there is a wide array of variability surrounding individual behavior patterns related to mating, with more than one mating pattern frequently being found in a given species, population, or even group (Knott 1999a, Malone and Fuentes 2009, Alberts et al. 2003, Pochron and Wright 2003). The patterns of reproductive activity within a group, taken together, can be seen as a mating system for a species or population. There is abundant evidence that basic grouping patterns are not the sole determinant, or necessarily good indicators, of the mating patterns of individual primates (Palombit 1995, Digby 1999, Borries 2000, Cords 2000a, Sommer and Reichard 2000, Pochron and Wright 2003). Grouping patterns do not equal the mating pattern for most primates. Table 32.4 identifies and defines the general mating patterns found in primates. One or more of these patterns can be found in primate species, with polygamy (polygynandry) being the most common. True polyandry is relatively rare and found predominantly in species where an adult female can cause reproductive suppression in other females in the group or where there is only one adult female in the group (as in some callitrichids) (Rylands 1993a, Garber 1997; see also Chapter 7). Absolute monogamy is also quite rare, but monogamy as the primary mating

Table 32.4 Primate Mating Patterns

A. Monogamy: One female and one male mate exclusively during mating period

B. Polygamy (Polygynandry): Females and males mate with multiple partners during mating periods

C. Polyandry: One female mates with multiple males during mating period

D. Polygyny: One male mates with multiple females during mating period

pattern is frequently found in primates that exist mainly in two-adult groups (Fuentes 1999b, 2002b). Absolute polygyny is fairly rare as well, except in some highly successful mate-guarding species or species where there is a preponderance of one-male groups. The most important distinction between these mating patterns is the relative potential for variance in reproductive success between males and females. In polygynous and polygamous/polygynandrous patterns where male/male contest competition for access to females is high, males have a much higher variance due to this competition and the subsequent unequal distribution of fertilizations. That is, such a system potentially results in some males having many opportunities for reproduction and others having few. Polyandrous mating patterns also tend to result in high variance for both males and females as mating is usually limited by the alpha female and involves only a few males at most (Garber 1997). In monogamous patterns, variance will be relatively equal for both sexes depending on the amount of polygyny or polyandry that is also present. The relative distribution of the different mating patterns within a population can then lead to differential variances in reproductive success, impacting individuals' behavior patterns. Here, it is clear that demographic variables can directly impact the mating patterns via availability of potential mates, so population density and sex ratios of adults can be quite important.

Style of rearing and caring for young is also an important factor. In most primates, adult females carry and nurse their offspring; however, some also "park" infants for substantial portions of their active periods. Many species engage in variable levels of *alloparenting*, where infants/immatures are carried and protected for long periods by nonparents. In many cases alloparenting is related to nepotistic (kinship) relationships between females, with kin doing much of the alloparenting. However, there are also a number of species (such as the callitrichids) where females or other age/sex classes not necessarily the mothers also contribute allocare. There is also substantial variation in how much assistance in infant carrying, guarding, grooming, and other contact behaviors mothers receive from groupmates.

As mating and rearing patterns are a prime factor in social systems, and male–female differences in selective pressures form the basis of the socioecological model, it behooves us to briefly discuss current theoretical perspectives on these patterns and what the variance in reproductive access might mean (see Chapters 29 and 30). Bateman (1948) and subsequently Trivers (1972, 1974) and others have identified *anisogamy* (differential investment in gametes by males and females) combined with the greater investment by females in rearing offspring as the main driving force behind differences in male and female mating strategies. Recent reviews (Kokko and Jennions 2003, Borgerhoff-Mulder 2004) recognize the important, and often overlooked, concept that patterns and contexts of sexual selection drive sexual investment patterns, not merely baseline anisogamy (Tang-Martinez 2000). Bateman (1948) established a basic premise for current models: Females, with their large, costly eggs, have greater cost per offspring and should therefore invest more in each offspring, whereas males, with their (hypothetically) inexpensive sperm, should invest minimally in individual offspring and maximize their mating opportunities. Trivers (1972, 1974) and others translated this into models of parental investment and created behavioral predictions for many organisms including primates—namely, that females are constrained by the distribution of resources and that males are constrained by the distribution of females. This assumption suggests that male competition for mates will usually result in much higher variance in reproductive success for males than for females (since the limits on all females are relatively the same in a population: resource distribution). Kokko and Jennions (2003) suggest that the initial assumptions propagated by Trivers and many others overlook the fact that all offspring have a mother and a father, so if all males and females are mating with equal success, then their reproductive variance is equal and males should invest in the rearing of offspring. This suggests that anisogamy, in and of itself, does not create inherent differences in reproductive success. That is, maybe females are not simply limited by resources and males by females but, rather, female choice, male/male competition, and population densities also affect reproductive variance (above and beyond basic food distribution patterns). Thus, sexual selection, not inherent anisogamy, can be a main potential selective pressure for male and female investment differences. Behavioral patterns and demographic contexts resulting in sexual selection are malleable and vary across time; therefore, the relationship between the sexes is not necessarily one based on differential potential variance and investment but, rather, can be envisioned as one of negotiated interactions (and potentially differential investment patterns) resulting in the production and rearing of offspring.

The factors influencing the specifics of where, when, and how primate females and males achieve reproduction within a given demographic context and the manner in which infants are reared are going to have a direct and pronounced effect on social systems. Given the wide array of ecological and historical conditions that impact primate groups, we should expect to see a variety of mixed mating and rearing strategies across different local conditions, especially in long-term studies. Looking at the behavior of individuals, comparing them to the basic assumptions of the socioecological model, and seeing where they fit and do not fit, the predictions enhance our ability to assess the causes and impacts of primate behavior and, thus, understand social systems.

Intragroup Behavior

The majority of the primatological literature focuses on measuring degrees of asymmetry in *dyadic* interactions (interactions between two individuals) related to issues of dominance (priority of access to resources), aggression and conflict negotiation, affiliative (positive social relationships) patterns, etc. to provide descriptive and quantitative analyses of primate social lives (Hinde 1976). The within-group patterns of interaction form the daily nexus for the activity of any given primate. As individuals encounter different facets of their group's demographic and mating structures, they are forced to learn a series of patterned but malleable relationship characteristics that construct their day-to-day dyadic interactions (Morbeck 1997, Bernstein 1999b). These day-to-day patterns of behavior form the social structure, the interindividual framework, in which an individual behaves. Therefore, variations at the individual level impact the group. That is, different groups are going to be composed of distinct individuals that will vary in aspects of their behavior. Thus, within the same demographic, ecological, and mating parameters (as predicted by the socioecological model), two groups may exhibit different patterns of behavior due to the individual variation in their members (Dolhinow 1972b, Fuentes 1999a, Sapolsky and Share 2004). These differences in individual patterns can be attributable to different behavioral strategies (expressed adaptations), stochastic events, and/or historical factors. Obviously, the issue of *personality*, idiosyncratic behavior patterns that are relatively consistently expressed by individuals, may be relevant here. However, measuring personality per se is exceedingly difficult, and there is substantial debate as to its practicality as a concept in the study of nonhuman organisms; thus, for purposes of achieving measurable, quantitative assessments, one must refer to individual variation in measurable behavior units rather than to some aspect of primate psychology. By examining the behavior exhibited in a group (the general patterns) and as a collection of individuals (a variety of slightly variant patterns), we are provided with two distinct and complementary ways to envision and analyze the social structure of primate groups. This level of analysis happens under the level that is provided by the basics of socioecology. For example, two groups in the population described as resident-nepotistic (Figure 32.4) might differ in general behavior patterns due to group size (one large and one small), the presence of three major lineages of related females in one group and only one in the other group (changing within-group contest competition pattern), and the fact that one group lives in a more rocky area where the predator pressure is lower than in the flatlands where the other group lives (leading to different social activity and ranging patterns). These different contexts will result in different behavioral patterns by the individuals within the group, thus giving the groups different overall behavioral profiles: variants on a similar social system.

Intergroup Behavior

Group behavioral profiling is a strong tool in primatology. To achieve a temporally robust picture of a group's habitat use and relationships with other groups, one can characterize a given group using mean ranging, feeding, activity, etc. patterns in addition to its relationships with other groups within the population. In this sense, for the purposes of comparison across groups within a population, the individual is subsumed into the social structure in which it exists (e.g., Treves and Chapman 1996). These intergroup interactions are core to describing social systems (e.g., see differences in between-group contest competition from Figures 32.2–32.5). Demography, mating, rearing, and individual variation all impact population-level patterns. However the kinds and patterns of intergroup interactions (competitive, tolerant, affiliative) help to give us a string picture of population-wide aspects of social systems and how they fit or do not fit with our basic expectations.

SOCIAL SYSTEMS AND EVOLUTIONARY PROCESSES

The impact of variation in the constituent factors of demography, mating systems, and intra- and intergroup behavior is reflected in social systems. By comparing social systems of different species or populations and the patterns of their constituent factors, we can potentially identify the patterns of action for particular evolutionary processes across primate populations. All four of the evolutionary processes—natural selection, gene flow, genetic drift, and mutation—impact social systems by affecting changes in the patterns of the constituent factors. However, none of these factors is truly independent, thus making social systems an important tool for viewing broadly evolved patterns and changes in those patterns.

A substantial component of our attempts to understand the lives of primates lies in our ability to model the patterns and processes of evolutionary histories and ongoing evolutionary forces on primate populations. Taking the approach outlined here, examining the individual primate in the context of its demographic characteristics, mating and rearing patterns, inter- and intragroup behavior, and membership in a broader social system facilitates our attempt to elucidate evolutionary and other patterns of change.

Natural Selection: Caution with Adaptation and Social Behavior

> Functional labels should not substitute for a full analysis of the causes of behaviour. (Krebs and Davies 1997:11)

The current primatological approach is couched in the context of socioecological models and derives lines of inquiry from Tinbergen's four "why" questions about behavior (Table 32.5) (Tinbergen 1963; Krebs and Davies 1997; Bernstein 1999b). However, despite the recognized and affirmed importance of addressing all of the "why" elements of a behavior, all too frequently our focus in primatology is on the role of natural selection and our quest is solely for functional explanations. Many of our

Table 32.5 Four "Why" Questions Regarding a Behavior: A Behavior Has the Following Facets and Each Can Be Approached in Order to Elucidate the Particular Aspect of the Behavior in Question

A. Evolutionary cause underlying a behavior

B. Developmental or ontogenetic cause underlying behavior

C. Proximate stimulus eliciting behavior

D. Functional, or fitness-related, impact of behavior

arguments for the evolution of specific traits (be they behavioral or morphological) rely on demonstrating the potential strength of selection pressures on the trait in question. Underlying most estimates of the power of natural selection on a trait is the assumption that there is a causal link between fitness (or lifetime reproductive success) and the trait in question. Given the variable factors that influence any given primate's life, one must tread with caution when invoking a functional explanation for an observed behavior. It is at the level of the population and via examining changes in the constituent facets of social systems that we can most effectively assess actual evolutionary patterns. The focus on trait commonalities among individuals can provide us with insight. However, the inherent variation at the levels of demographics, mating and rearing patterns, and interindividual interactions makes it difficult to always effectively measure the actual impact on the reproductive success of any given heritable behavior or behavioral pattern. Our assumptions about fitness-enhancing values and the "adaptiveness" of certain primate behaviors may overlook the influence of environmental variation, historical factors, and possible stochastic factors. This can mislead us into assuming that a particular behavior of interest is an adaptation to the ecological or social pressures we see in the current system. However, despite the importance of caution against assumptions of the universality of natural selection as an explanation for traits, there is a broad and robust body of theory and multiple hypotheses that clearly outline how selection can impact the behavior of primates. Here, we summarize ideas of how selection impacts individuals, groups, and populations, all of which affect what we describe as "social systems."

While we cannot directly measure selection on individuals, we can model how particular behaviors may contribute to or detract from an individual's reproductive success and, thus, relative fitness. It is this relative fitness that is a core assumption underlying the predictions of the socioecological model. That is, primates will respond to pressures in particular ways to enhance their relative fitness. These responses are what the socioecological model predicts. Selection does not truly impact the individual, but factors that affect the individual's ability to reproduce affect the representation of that individual's unique genetic component in subsequent generations and are of interest in models of evolutionary patterns. Of course, this focus on the individual is relevant to selection models only if there is some heritable facet to the behavior. Alternatively, focus on the individual can also be very useful if the behaviors in question are learned/socially transmitted with no direct biologically heritable facet.

Williams's (1966a) criticism and rejection of group selection coupled with Hamilton's (1964) arguments for kin selection as an explanation of apparently altruistic behavior formed the basis for Trivers's (1972) and Wilson's (1975) emphasis on examining the individual as the focal point for understanding selection and subsequent adaptations. However, work by Wilson and Sober (1994) argues robustly for envisioning multiple levels in the action of selection (see also Wilson 2004, Wilson and Wilson 2007). As primates are primarily found in social groups for all, or a majority, of their lives, impacts on groups can have impacts on individuals. Interactions between groups, whether they are direct contests for resources or active partitioning of the environment (as predicted by facets of the socioecological model), can affect the reproductive success of individuals within the groups. Therefore, interactions between groups can have positive or negative fitness results for the individuals in those groups. For example, especially aggressive or heightened coalitionary behavior by a few individuals in a group may result in that group's increased access to contested food resources over another group in the same population, which in turn may enhance the reproductive opportunities or efficiency of all members in that group. This is not to say that the individuals acting did so "for" the group but, rather, that intergroup interactions can have costs and benefits even for those group members that do not engage directly in the contests. Obviously, the demographics of a population will have great impact on the ways in which groups interact (remember the predicted differences in Figures 32.2–32.5 based on population density), and the individual behavioral patterns of group members will also influence these relations. Living in a group also increases the potential for disease transmission, a major factor that has been generally left out of basic socioecological models. The current demographic composition of a group may dramatically alter the impact that a given pathogen may have on the individuals within the group, especially if it is highly contagious. In fact, disease itself can produce dramatic demographic shifts (even greater than predation) which alter the ways the remaining individuals can interact with their environment, other groups, and themselves. For example, Sapolsky and Share (2004) review a case where a tuberculosis outbreak killed the majority of males in a baboon group. The males that died happened to be highly aggressive, and prior to the disease event, all group males displayed a negative correlation between rank and physiological indicators of stress. The males that survived were atypically nonaggressive, and the overall pattern of behavior in the group was significantly different after the die-off. Over a decade later, the patterns of behavior in this group remain relatively nonaggressive, and there is no correlation between rank and physiological indicators of stress. A disease event changed the demography and the behavior and physiology of this group. Disease may be a major, if somewhat understudied, factor of selection on primates (Kappeler and van Schaik 2002, Nunn and Alitzer 2006).

Examining the impact of selection at the individual, group, and population levels provides a broad, complex picture of the facets of social organization. The three levels of the individual, the group, and the population are interconnected but also can be examined, in a quantitative format, independently to assess distinct patterns and trends (Figure 32.6). It is also worth noting that many of the behaviors and patterns we observe today are related not only to the results of selection but also to a wide array of historical and developmental factors that may or may not be reflected as adaptations. *Phylogenetic inertia*, or the presence of ancestral traits that no longer may be directly adaptive or responsive to current selection pressures, is a potentially common confounding factor in understanding extant behavior (DiFiore and Rendall 1994, Thierry 2008). This suggests that some of what we see at the individual, group, or population level may reflect past adaptations or past events that no longer act as pressures on the extant forms. For example, Dunbar (1992a) proposed that brain development may limit group size related to the ability of primates to effectively interact in complex social ways with a limited number of individuals. The ecological structure of the habitat and the climatological patterns in which a primate lives can also have an impact on locomotory patterns,

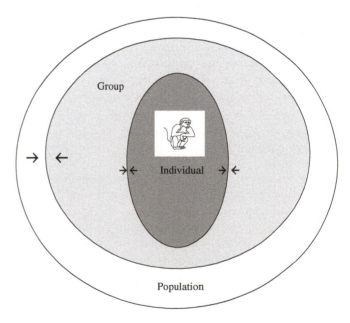

Figure 32.6 Interconnections between the individual, the group, and the population.

ranging patterns, and the type and frequency of particular behaviors (Dunbar 1988). Finally, gene flow and/or lack thereof (isolation) can affect the behavior of an individual primate via either the introduction of particular biologically based variants of traits or simply exposure (or lack of exposure) to other members of the species. Gene flow also has implications for the probabilities of disease pressure and for the patterns of demographic change or stability in a population.

The Social Evolution of Social Systems?

Numerous primatologists have shown great interest in the possibilities of nonhuman primate "culture" (McGrew 1998b, Whiten et al. 1999, Perry and Manson 2003, van Schaik et al. 2003a,; see Chapter 44). The existence of behaviors, functional and nonfunctional, that are acquired via social learning and transmitted across generations within some social groups and not others is becoming an increasingly common aspect of primatological natural histories. These behaviors vary from patterns of grooming to tool use to greeting behaviors and are reported for a number of primate species (McGrew 1998b; Perry and Manson 2003). The debate surrounding whether or not nonhuman primates have the same sort of culture as humans is well covered in other venues (McGrew 1998b; see Chapter 44), so I will not review it here. However, its relevance to this discussion lies in the fact that there is little contention that many species of primates do show variation in behavioral patterns, frequently apparently nonfunctional patterns, across sites, groups, and populations. This behavioral variation, or *social tradition*, can be seen as an indication that primates are fairly flexible in their behavioral potential and that living in social groups can facilitate the appearance and spread of social patterns that may have no direct impact on fitness. In other words, primate behavior can change over time because of social innovation and learning without a direct relationship to the impact on reproductive success of individuals (e.g., Sapolsky and Share 2004). Selection and other evolutionary factors mold the genetic variation in primate populations; however, day-to-day behavioral interactions among primates also actively shape the social environment. We must consider aspects of behavioral flexibility and plasticity and their potential for leading to social innovation and/or change.

The chapters in this volume effectively demonstrate that a major component of the environment for the majority of primates is the social environment. There is an emerging recognition that *phenotypic plasticity* (continuous and reversible transformations in behavior, physiology, and morphology in response to rapid environmental fluctuations) can be seen as an adaptation. Piersma and Drent (2003:228) suggest that "when environments change over shorter timescales than a lifetime, individuals that can show continuous, but reversible, transformations in behavior, physiology or morphology, might incur a selective

advantage." It is also the case that immediate environments not only impact organisms but also, in part, shape those environments and thus the selection pressures that they face. In a sense, we can see organisms engaged in a certain amount of niche construction (Laland et al. 2001, Stamps 2003). If niche construction is ongoing, then we can envision the social network (group and population) as a major component of the primate niche. This is a flexible and malleable environment, and thus, we might expect that individual primates are able to exploit a variety of behavioral patterns and strategies in order to effectively negotiate it (i.e., successfully interact on a day-to-day basis with their social partners). In fact, Flack et al. (2006) argue for a form of social niche construction in primates where social networks constitute the essential social resources in gregarious primate societies. They posit that "The structure of such networks plays a critical role in infant survivorship, emergence and spread of cooperative behavior, social learning and cultural traditions." Following Flack et al. (2006), we can view individuals engaging in social niche construction as individuals in primate groups negotiating these social networks and modifying their boundaries and internal landscapes in the context of changing demographic and, potentially, ecological variables.

Given the primacy of social interactions for primates, social complexity may itself be an adaptation, the result of existing in complex and variable social environments (or social niches). If this is the case, then we must include aspects of behavioral flexibility, including potentially nonfunctional behavioral patterns, in our assessments of primate social systems. It remains possible that local social variants can change a group's behavior in ways that have an effect on their demographics, mating, and patterns of interaction with one another or with other groups. Documenting social and behavioral variation across the life span is as important as measuring other life history variables (Kappeler and Pereira 2003).

MODERN EVOLUTIONARY BIOLOGY AND PRIMATE SOCIAL SYSTEMS

Kappeler and van Schaik (2002) express a concern that primatology and evolutionary biology are not as engaged with one another as they should be. They suggest that this may be because of differences in the mating systems of primates relative to many other animals, the inability to engage in much controlled experimental study with primates (for ethical and practical reasons), and the fact that primate life histories are different from those of many other animals and, thus, potentially not pertinent to some major questions in evolutionary biology (Kappeler and van Schaik 2002). These points emphasize the differences between primates and many of the avian or insect models that make up the bulk of the socioecological work in mainstream evolutionary biology. However, these differences are a strong reason to continue to expand and integrate evolutionary theory into the study of primates. Primates represent a mammalian taxon that is characterized by high levels of social and behavioral flexibility, with long periods of physical and social development. These patterns do not always easily fit into current socioecological models that have been developed with other organisms (Thierry 2008). In fact, it is the complexity inherent in primate social systems that requires a diffuse and multifaceted approach in our attempts to understand their evolution.

Emergent evolutionary biology theory and models allow for a broadening of our inquiry into evolutionary patterns and processes affecting primates. In addition to the basal socioecological model that allows us to assess and model the impacts of natural selection and ecological parameters on primates, we can integrate other prospects, such as phenotypic and developmental plasticity, niche construction, and social traditions, into our attempts to understand and model the complexity and variation in intra- and interspecific behavior across the primate order. These additional evolutionary perspectives can be encompassed under the rubric of developmental systems theory (DST) (Oyama et al. 2001). The main point of DST is that we need to view *evolution as construction*, that evolution is not a matter of organisms or populations simply being molded by their environments but of organism–environment systems changing over time. This perspective allows us to contextualize variability in primate social systems through an integration of the various processes (socioecological, niche construction, phenotypic plasticity, etc.), enabling multiple ways to answer questions about the function, origin, and maintenance of behavior. Social systems are constantly being constructed via their constituent elements of demography, mating and rearing systems, and intra- and intergroup behavior patterns and are affected by socioecological and other evolutionary processes. This constant construction is a major reason for the variation we see in primate societies at any given time. Integrating our current reliance on socioecological models with basic analyses of individual and group behavioral plasticity and social traditions, combined with niche construction and DST perspectives, may facilitate a greater and increasingly successful engagement between primatology and evolutionary biology.

THE CONTINUED IMPORTANCE OF NATURAL HISTORY DATA AND HYPOTHESIS TESTING IN INVESTIGATING SOCIAL SYSTEMS AND SOCIOECOLOGY

In combining the different paradigms of approach in primatology, we have a richer context from which to ask questions about the behavior and social systems of our subjects. Kappeler and van Schaik (2002) divide primatology into three phases: the early "descriptive inventory" phase, the second "correlational approach" phase, and the current, third phase dominated by "hypothetical–deductive" approaches. They counter Rodman's (1993) and Sussman's (1999a) call for caution in hypothesis formulation in the absence of natural history data and suggest that, in the absence of relevant data, socioecological hypotheses are created to provide focal points for field researchers to collect those data. The evident variability in individual behavior, and thus social systems, provides an arena for diverse approaches. Current information does facilitate the deductive creation of hypotheses that can result in specific a priori research projects. However, the variation, especially intraspecific variation, evident from primate studies also emphasizes the urgent need for continued and expanded natural history and baseline data collection and publication. Many of our current hypotheses are derived from, or supported/refuted by, data sets limited to a few social groups within a species and rarely drawn from different populations within that species. While Kappeler and van Schaik see the collection of natural history data as a beneficial by-product of hypothesis testing research, I suggest that descriptive research projects and protocols are required for a basal understanding of the patterns and processes in primate evolution. The complexity evident in primate social systems at the level of the individual, the group, and the population in mating, ranging, and foraging patterns is best tackled by a multifaceted program interweaving descriptive, correlational, and hypothetical–deductive approaches, all conducted with an ongoing integration into a dynamic evolutionary biology theory. Rather than being three distinct phases in primatology, I prefer to think that we learn from our past, look toward the future, and retain those features of our research program which act to benefit our "fitness" as a discipline.

33

Primate Nutritional Ecology
Feeding Biology and Diet
at Ecological and Evolutionary Scales

Joanna E. Lambert

The whole of nature is a conjugation of the verb to eat, in the active and the passive.
(Inge 1927:4)

1. How do primates meet their energetic and nutritional requirements?
2. What are the diversity of ecological, mechanical, and chemical challenges confronting primates as they acquire requisite energy, micro- and macronutrients? How do species behavioral, anatomical, and physiological adaptations influence how we define "dietary quality"?
3. How might information on a primate species' feeding biology and nutritional ecology be used in planning conservation tactics?

INTRODUCTION

Given the critical role of feeding in all aspects of an animal's biology, it is not surprising that research on foraging and diet has occupied a central position in the history of primate studies. In the landmark volume *Primate Societies*, Oates (1987) provided an excellent overview of the ecological aspects of primate feeding and paid particular attention to the patterns in which primate foraging is influenced by the spatial and temporal distribution of food resources. I offer a follow-up in this chapter with the benefit of over two decades during which we have seen remarkable progress in our understanding of foraging in wild primates and its implications for animal physiology (Ross 1992a; Messier and Stewart 1997, Knott 1998a, Conklin-Brittain et al. 2006, Rothman et al. 2009), morphology (e.g., McGraw 1998b, Vinyard et al. 2003, Lucas 2004, Wright 2005), ontogeny, growth and development (Leigh 1994, Dirks 2003, Bernstein et al. 2008), and ecology (Oates et al. 1990, Milton 1996, Shultz and Noe 2002) at individual (Altmann 1998; Curtis 2004; Garber 2004; Janson and Vogel 2006a,b), population (e.g., Isbell 1991; Chapman 1990b; White 1998; Mammides et al. 2009), and community (Fleagle et al. 1999, Brugiere et al. 2002, Reed and Bidner 2004) scales.

An understanding of the nutritional content of different food types is necessary for gaining insight into an animal's energetic requirements, which can then, in itself, provide a direct link to the complexities of feeding adaptations in primates. To wit: Food in its various forms is the only source

of energy for animals, and we can use an understanding of energy to define the oft-used term "dietary quality." Moreover, energetic requirements scale with body size; since body size correlates with a number of other ecological and behavioral variables, we can integrate our understanding of nutrients and diet to explain other components of an animal's behavior and biology.

However, nutrition and dietary quality do not exist in a void. Indeed, primates are confronted with an array of feeding challenges that influence the quality of food. These challenges can be *intrinsic*—representing some inherent chemical, nutritional, or mechanical feature of the food—or *extrinsic*—a function of the current ecological availability of that food and the costs (e.g., competition) associated with feeding on it as a consequence of that availability. Like all biological phenomena, questions regarding feeding challenges may be evaluated at two intersecting scales: proximate (ecological) and ultimate (evolutionary). *Proximate* mechanisms are short-term responses within the lifetime of an individual to immediate circumstances. For example, food availability influences which foods an individual eats, which can then influence such responses as increases in day range length or production of endogenous enzymes and gastrointestinal nutrient transporters depending on what food an animal finds and consumes. If an animal's response both has a heritable component and is under enough selective pressure (which presumably many components of diet and feeding would be, given their centrality in an animal's biology), then we can address feeding-related questions at an ultimate scale and evaluate animal responses in terms of evolved strategies and species' features. Obviously, these scales are synergistic and intrinsically related, and neither is the more correct scale at which to discuss and evaluate behavior and biology. They are, instead, complementary and facilitate answering both "how" questions regarding an animal's response to immediate conditions and "why" questions concerning the reasons that animals have evolved mechanisms to carry out these responses (Alcock 1989). The interplay between the proximate and ultimate is extraordinarily complex, as is the nexus of responses resulting in the diversity of feeding adaptations observed in primates. It is the job of primate nutritional ecologists to determine the nature of feeding challenges confronted by primates in acquiring their requisite nutrients and energy and to unravel the solutions exhibited by primate species.

I cannot give the spectrum of foraging, feeding, and diet full justice in this chapter. Instead, I focus on those aspects not largely discussed by Oates (1987), namely, nutritional biology. I start by evaluating feeding categories, primate nutritional requirements, and what wild primate diets offer with respect to micro- and macronutrients. After discussing metabolism, energy, and dietary quality, I outline the remainder of the discussion in terms of feeding challenges (ecological, chemical, and mechanical) to primates and provide several examples of solutions to these challenges. I conclude by arguing that questions regarding diet and nutrition are not just academic concerns but in fact have important implications for the conservation of wild species.

PRIMATE DIET: FEEDING CATEGORIES AND NUTRITIONAL REQUIREMENTS

Feeding Categories

The nutrients required by vertebrates are divided into six broad classes: carbohydrates, protein, lipids, vitamins, minerals, and water. The first three (carbohydrates, protein, lipids) are commonly referred to as *macronutrients* and are required in large quantities for the energy (typically measured in calories) used in growth and maintenance, while minerals and vitamins, or *micronutrients*, are not used for energy per se but are instead vitally important for innumerable physiological processes (Leonard 2000, McNab 2002, Karasov and Martinez del Rio 2007). The Council on Food and Nutrition has defined nutrition as "the science of food, the nutrients and substances therein, their action, interaction, and balance in relation to health and disease, and the process by which the organism ingests, absorbs, transports, utilizes, and excretes food substances" (cited in Wardlaw et al. 2004:3). *Nutritional ecology*, then, can be defined as the broader field of investigation into the means by which animals procure these nutrients from their habitat.

Evaluating how primate nutrition per se relates to a species' nutritional ecology can be vexing, largely because of the disjuncture between what we know from studies of primates in captivity and from those of primates in the wild.

On the one hand, we have a decent understanding of micro- and macronutritional requirements in a few commonly captive species as a consequence of work by the National Research Council (NRC). The nutritional requirements of nonhuman primates were first considered and outlined at the behest of the NRC's Committee on Animal Nutrition in 1972; this report was later expanded (1978). The current *Nutrient Requirements of Nonhuman Primates* represents a second revised edition (National Research Council 2003). The current NRC report is distinct from other animal nutrition reports in that, rather than focusing on the dietary requirements of a single domesticated species, it extrapolates requirements from a few taxa (i.e., *Saimiri*, *Cebus*, *Macaca*, and *Homo*) to an entire order exhibiting an extraordinary array of diet-related adaptations. Regardless of these caveats, we now have updated nutritional recommendations for primates for energy, fiber, protein, lipids, minerals, and vitamins (National Research Council 2003).

On the other hand, primate biologists today have both intensive knowledge on several groups of a few well-known wild species in well-studied communities and extensive but more superficial knowledge on many, if not all, wild species. For example, we have specific information regarding the diets of several groups of *Lophocebus albigena* (gray-cheeked mangabeys) and *Cercopithecus ascanius* (red-tail monkeys) living in the Kanyawara study area of Kibale National Park, Uganda, and can calculate dietary particulars for these groups for an almost 30-year period (Waser 1977; Struhsaker 1978a; Olupot et al. 1998; Conklin-Brittain et al. 1998; Wrangham et al. 1998; Lambert 2002a; Rode et al. 2003, 2006). Moreover, in a few cases of particularly well-studied primate populations, data are available for the nutritional components of these foods as well. For example, Conklin-Brittain et al. (1998) have described Kanyawara *C. ascanius* and *L. albigena* foods with regard to their macronutritional content, finding few differences with regard to levels of crude lipid, crude protein, soluble carbohydrates, and fiber. Such long-term comparisons of particular primate communities were not possible until fairly recently with continuing research in given populations.

The challenge to primate biologists is to link data on feeding and foraging in wild populations with knowledge regarding nutrition in captive primates. For example, despite our understanding of what a particular group (or even several groups) of wild primate species is eating at a given time, it is difficult to know whether these foods are or are not meeting the basic vitamin, mineral, and macronutrient needs of that wild population. This is in part a consequence of the difficulties of measuring the wild diet in the first place (Oftedal and Allen 1996). Primate diet is typically expressed as time spent foraging and feeding, although the quantity of food (e.g., g/min) actually ingested can vary greatly depending on food type, age/sex, food availability, etc. (Hladik 1977b, Milton 1984b, Oftedal 1991, Oftedal and Allen 1996). Moreover, details of ingestion can be extremely difficult to observe in arboreal primates living in tall, closed-canopy forest.

Integrating these bodies of information (i.e., data on natural diet and nutritional requirements) is where field-workers and nutritional biologists working together have the most to offer. For example, recent work by Rode and colleagues evaluated the mineral resources and consumption by colobus (Rode et al. 2003) and red-tail (Rode et al. 2006) monkeys in Kibale National Park, Uganda. They demonstrated that the sodium content of foods was extremely low and fell well below suggested NRC requirements. In addition, while intakes of copper and sodium correlated with red-tail population biomass, diets were deficient when compared with the requirements of *Macaca mulatta* (National Research Council 2003). The red-tails increased their intake of copper and sodium during the birthing season, and it was suggested that the effects of deficiency in these two minerals during the post–birthing season may have been most significant for red-tail population abundance.

Adding to the complexity of evaluating primate nutrition is the plethora of terms associated with feeding. Many of these terms are used in slightly different ways by different researchers. Animals that consume from only one trophic level are labeled either *herbivores* (consumers of plants only) or *carnivores* (consumers of animals only). Organisms that consume foods from more than one trophic level (e.g., both plants and animals) are technically *omnivores*, although in some cases authors use *omnivory* to suggest flexibility in some sense or another. With very few exceptions (e.g., *Loris* spp., *Tarsius* spp.; see Chapters 4 and 6), primates derive a majority of their energetic and nutritional requirements from plants; indeed, there are reports of virtually all components of plant anatomy consumed by different primate species around the world, including fruit (both ripe and unripe), seeds (immature and mature), leaves (all developmental phases from buds to mature leaves), petioles, corms (base of stem that gives rise to vegetative plant growth), rhizome (underground stems), bark, flowers (buds and petals), nectar, sap, and gum. Some primates are technically best defined as herbivores (e.g., colobines and indriids), although most primates consume both plants and animals. Harding (1981), for example, undertook a review of the diet of 131 species of wild primates and found that 90% of these species ate fruit, 79% ate soft foliar parts (i.e., buds, shoots), 69% consumed mature leaves, 41% consumed seeds, 65% ate invertebrates (primarily insects), and 37% ate other vertebrate prey (small vertebrates, eggs).

Omnivory is often taken to be synonymous with a *generalist* or *flexible* feeder. This is not necessarily always the case, and in fact, this term masks the extreme diversity of feeding strategies in primates; indeed, within this large category, a species may focus on some particular category of food, hence terms such as *frugivory* (fruit), *folivory* (leaves), *gummivory* (exudates), *gramnivory* (seeds), *insectivory* (insects), and *faunivory* (animals). However, these labels have not been rigorously defined and can mask variability. For example, it is common to encounter the idea that "frugivores" should behave in a particular way and "folivores" in

another. Such categorization can obfuscate the complexity of feeding, and many assumptions we make regarding what a frugivore or folivore should do are not so much a result of what these animals actually feed on but more a result of the labels that we apply. Such categorization also obfuscates the chemical complexity of the foods themselves. Indeed, in a recent analysis, Danish et al. (2006) provide a critical caution: Food category (e.g., "fruit" or "leaf") is not tantamount to dietary type or nutrient. In an analysis of red colobus and red-tail monkey diets in Kibale National Park, Uganda, these authors found that there is actually more sugar in the leaves consumed by the colobine than in the fruit consumed by the cercopithecine (Danish et al. 2006). Related to these complexities are issues such as the point at which an animal, population, or species should actually be considered, say, frugivorous and how this is determined (actual food intake, time spent feeding, anatomical adaptations of the teeth and/or gut, etc.). Different populations of the same species can differ markedly in their diet, as can different members of the same group and even the same individual at different times, depending on myriad factors, such as microhabitat, seasonal shifts in availability, community structure, age/rank. Indeed, it is rarely, if ever, possible to define what is "normal" for a species with regard to the foods it consumes or the nutrients it ingests (Oftedal and Allen 1996), suggesting that, at the very least, we must get a handle on intraspecific variation and how it relates to interspecific patterns (Chapman et al. 2003a, Strier 2003b).

Anyone perusing primary literature on primate feeding biology will inevitably encounter the food category of "other," which may include such foods as bark, fungus, sap, decaying wood, flowers, and petioles, among others. While this food category tends to receive the least attention by researchers, there is the potential that, of all food categories, it has the greatest impact on primate abundance and adaptation. For example, comments regarding minerals and primate population density are largely anecdotal despite the fact that in other literature phosphorus has been demonstrated to limit populations of large herbivores (e.g., *Equus asinus* in Australia; Freeland and Choquenot 1990). Janson and Chapman (1999) have thus suggested that this trace and little-studied primate mineral requirement may in fact limit folivorous primate populations. In the Bwindi Impenetrable National Park, Uganda, Rothman and colleagues (2006a) have demonstrated that gorillas will seek out decaying wood with a sodium content of 100–1,920 ppm (mean = 811 ppm) in a habitat where the average sodium content of all other foods was of 90 ppm (Rothman et al. 2006a). These examples highlight the fact that, with regard to nutritional factors that may influence population density and evolution of feeding traits, it is not so much what is required most of the time but what food resources are required during critical periods. For example, in an evaluation of the impact of the frequency and intensity of selection on adaptations, Lambert and colleagues (2004) argued that it is the difference in the mechanical properties of fallback foods during critical periods that may have served as the selective pressure for thick molar enamel in *Lophocebus albigena* relative to thinner-enameled cercopithecines (e.g., *Cercopithecus* spp.) (see also Lambert 2007). The fact that for the most part *L. albigena* consumes similar food to that fed on by sympatric cercopithecines (e.g., soft, fleshy fruit) but also has a diet comprising a small (but probably very important) percentage of mechanically tough foods, such as bark and seeds, suggests that the thick enamel of *L. albigena* serves a critical function (Rosenberger 1992, Kinzey 1978). That is, thick dental enamel was selected for because of its benefits during periods when preferred—and softer—foods were unavailable.

Nutritional Requirements

Regardless of the feeding and food categories we employ to describe primate diet, primates require the full suite of nutrients required by most mammals in general (45–47 in total of amino acids, fatty acids, vitamins, and minerals) (Oftedal and Allen 1996) from the principal classes of carbohydrates, protein, lipids, vitamins, and minerals. Here, I describe features of these basic nutritional classes and evaluate the major dietary sources of these nutrients for wild primate species.

Carbohydrates

Carbohydrates take the form of monosaccharide sugars, disaccharides, and polysaccharides. Monosaccharides (simple sugars), such as glucose or fructose, are readily absorbed by the body and utilized directly along ordinary metabolic pathways. Disaccharides consist of two monosaccharide units bonded together; sucrose, for example, is a common fruit sugar comprising the monosaccharides glucose and fructose. Along with monosaccharides, the disaccharides are often referred to as the "soluble sugars." Disaccharides must be hydrolyzed into simple sugars (in the small intestine) before they can be absorbed and utilized (Schmidt-Nielsen 1997). The most complex sugars are the polysaccharides, polymers of monosaccharides, which can be divided into the starch or starch-like polysaccharides (storage polysaccharides) and the nonstarch polysaccharides. The starch and starch-like polysaccharides are essentially plant energy reserves that can be broken down to varying degrees by the consumer. The enzyme amylase, for example, is secreted in the pancreas and saliva of Catarrhini. Its function is to reduce polysaccharide starch to disaccharide sugars, which can then be hydrolyzed in the small intestine; this process is initiated by amylase directly in the mouth. Cercopithecinae in particular have very high salivary amylase expression, even relative to modern populations of *Homo sapiens* that have been demonstrated to have undergone positive selection for multiple copies of the salivary amylase gene (*AMY1*) (Perry et al. 2007a).

The nonstarch polysaccharides are the fiber components of plants and can be further divided into the soluble nonstarch carbohydrates (soluble fiber) and the insoluble nonstarch polysaccharides (insoluble fiber). The insoluble, nonstarch polysaccharides comprise the structural components of the plant cell walls (hence, "structural polysaccharides") and include hemicellulose, cellulose, and lignin. Cellulose is the most abundantly distributed carbohydrate in the world (Sharon 1980) and represents a large proportion of the available energy content of plant foods (Blaxter 1962, Alexander 1993), although no vertebrate has the cellulose-digesting enzyme (cellulase) for breaking down this carbohydrate, an interesting evolutionary question given that primates have relied on plants heavily throughout their evolutionary history (Milton 1987b, 1993). The structural carbohydrates must instead be broken down with the assistance of protozoans or, more commonly, bacterial symbionts (Lambert 1998). Like cellulose, hemicellulose cannot be digested enzymatically but, instead, requires fermentation. There is however, some evidence that hemicellulose can be partially hydrolyzed in a low-pH stomach; *Pan troglodytes* (and *Homo sapiens*), for example, ferment hemicellulose more completely than cellulose (Milton and Demment 1987, National Research Council 2003). The nonstructural, nonstarch polysaccharides include some components that, like starch, represent sources and storage of energy for plant metabolism. Although they are not as digestible as starch, they are completely fermentable and include substances in plants known as "fructans." Pectin (a component of plant cell walls) is not a plant energy resource per se but is associated with the plant cell wall; it is closely associated with hemicellulose and is completely fermentable.

Because of its ubiquity in current analyses of wild animal nutrition, it is useful at this juncture to define the detergent analysis method for evaluating fiber fractions (Van Soest 1994). This method classifies fiber types in terms of their availability to fermenting microorganisms (primarily bacteria) found in mammal guts and yields two fractions containing either mainly cell contents (cytoplasm) or mainly cell wall constituents. In the laboratory, detergents are used first to emulsify and bind soluble cell components and yield "neutral detergent fiber" (NDF). NDF is then refluxed in an acid detergent to yield ADF (acid-insoluble remnant) and ADS (mass of original sample minus ADF). NDF and ADF are most commonly reported; NDF comprises cellulose, hemicellulose, and lignin; ADF is composed primarily of cellulose and lignin (and, in grasses, silica). A drawback of this method is its inability to yield an assessment of the pectin in plant foods (see Karasov and Martinez del Rio 2007 for an excellent interpretation of the method first described by Van Soest 1994).

In the diets of wild primates, fruit pulp is an excellent source of carbohydrate energy, providing about 50–100 kcal/g (Leonard 2000). In comparison to cultivated fruit, the sugar of wild fruits is hexose-dominated (i.e., some fructose and glucose) (Milton 1999). Fruits can also have more pectin than leaves, although gastrointestinal anaerobic bacteria very effectively ferment such pectin (National Research Council 2003).

The volatile short chain fatty acids thereby produced can provide considerable energetic benefits. A number of vegetative plant parts can also be fermented for their energy, including leaves, petioles, and bark. Exudates, too, are exploited by primates, although there are chemical and nutritional differences among them (Bearder and Martin 1980a, Nash 1986a). For example, gum is water-soluble and high in complex carbohydrates composed of nonstarch, multibranched polysaccharides (Power 1996, Caton et al. 1996). Saps, on the other hand, are typically high in relatively easy-to-digest, simple, and water-soluble carbohydrates (Nash 1986a).

Protein

Protein provides energy and is critical for growth and replacement of tissue in the body (Leonard 2000). Proteins are composed of amino acids; primates require 20 amino acids for body maintenance, nine of which are considered essential because they are not synthesized endogenously— i.e., they must be derived from dietary sources (Leonard 2000). Protein requirements are greatest during growth and reproduction and can increase by as much as 30% during such critical periods (Oftedal 1991). Oftedal (1991) has argued that protein deficiency is likely to be uncommon as protein requirements in the order generally are low and primates are generally characterized by slow life histories, particularly when compared to carnivorous mammals (Oftedal 1991, Oftedal and Allen 1996). This argument, however, is belied by the finding that an index of dietary quality protein:fiber is a useful predictor of leaf-eating primate abundance (Milton 1979, Chapman and Chapman 2002a; see later section "Chemical Challenges").

Approximately 75% of protein synthesis comes from protein breakdown in the body; the other 25% comes from dietary input. Isotope labeling techniques indicate a very high turnover rate of proteins in the body, and virtually all "marked" proteins are eventually incorporated throughout the body (Schmidt-Nielsen 1997). Not including water, protein comprises the highest percentage of lean primate body tissue (approximately 17% total body weight) and is critical for growth and replacement of tissue in the body (National Research Council 2003); however, animals require amino acids, not protein per se. Thus, the quality of a protein is defined by its ability to provide a balanced pattern of indispensable amino acids, and rapidly growing animals require more amino acids for tissue growth than do adults. The major sources of dietary protein in primate diets are leaves, insects, and other animal matter. With exceptions, immature leaves offer more protein than mature leaves since as they age structural components increase and protein is allocated to other parts of the plant (storage organs, seeds). Leaves in general offer approximately 10–20 kcal/g (Lueng 1968, Eaton and Konner 1985, Leonard 2000). Overall, fruit pulp is not high in protein, and few mammals and birds exist on an exclusively frugivorous diet even when this resource is abundant (Whiten et al. 1991, Leighton 1993). Indeed, while it has been argued that fruit availability generally limits

fitness and primate population density (Whiten et al. 1991, Plumptre and Reynolds 1994, Gupta and Chivers 1999), others argue that there are behavioral mechanisms for the selection of rare but protein- and mineral-rich resources during critical periods of growth and reproduction (Cords 1986, Sourd and Gautier-Hion 1986, Butynski 1988). For example, young primates or pregnant/lactating females often increase their intake of leaves and/or insects, and Cords (1986) has found that the size of juvenile *Cercopithecus ascanius* is proportional to the percentage of leaves and insects in the diet.

Several factors impact the availability of plant protein to primates. Tannins (polyphenolic compounds) are common in tropical plants and can render protein inaccessible to animals (Robbins et al. 1987; see later section "Chemical Challenges"). Condensed tannins readily bind to macromolecules such as proteins because of the large numbers of free hydroxyl groups that form hydrogen bonds (Mole and Waterman 1985). Hydrolyzable tannins tend to be more rare in distribution than condensed tannins; these too bind proteins but can be more toxic in that hydrolyzation can result in harmful end products that can cause serious liver damage. In addition, while endogenously circulating nitrogen is critical for protein synthesis in the body, nitrogen can be bound to the lignin and cellulose components of plant cell walls and, thus, may not be available to the primate consumer. This nitrogen may be part of the phenolic complex of lignin or of the proteins associated with plant cell walls (e.g., elastins; Van Soest 1994) and is substantial in many tropical plant parts (Milton et al. 1980, Milton and Dintzis 1981, Conklin-Brittain et al. 1999, J. M. Rothman personal communication) but typically is not included in the calculation to estimate crude protein in primate diets. Indeed, Rothman (2006) has demonstrated in mountain gorillas in Uganda that this unavailable portion of nitrogen was typically about 30% of the total protein in commonly eaten foods but, in some foods, represented more than 95% of the total nitrogen.

Lipids (or Fats)

Lipids (or fats) are the body's most highly concentrated source of energy, providing more than two times the energy per unit weight as either carbohydrates or proteins. Fats are divided into three main classes: simple, compound, and derived. The simple fats comprise triglycerides, which are made up of two types of molecule: glycerol and fatty acids. Fatty acids are divided into saturated and unsaturated types, based on differences in the bonding between carbon atoms (Leonard 2000). The major sources of fat for primates are insects and other animal matter, seeds, and the arils of some fruit species (e.g., *Virola* spp.). Insects, for example, provide upward of 100–200 kcal/g. We know very little about the consumption of lipids by primates, although wild plant foods tend to show a fairly equal balance between saturated and unsaturated fats (Chamberlain et al. 1993). Milton (1999) has analyzed the fatty acid composition of wild plant foods consumed by *Alouatta palliata* and found that the most common fatty acids are palmitic (30%), linoleic

(23%), /-linolenic (16%), and oleic (15%). Conklin-Brittain et al. (1998) report that lipid intake by anthropoids in Kibale National Park, Uganda, is marginal. As lipids influence neurotransmitter levels that regulate reproductive hormones, insufficient lipid intake can result in developmental and reproductive problems in mammals (Robbins 1993).

Minerals

Minerals perform functions essential to life and are generally divided into the macrominerals (calcium, phosphorus, magnesium, potassium, sodium, chlorine, sulfur; typically expressed as a percentage) and trace elements (iron, copper, manganese, zinc, iodine, selenium, chromium, cobalt; typically expressed as parts per million) (National Research Council 2003). Minerals are inorganic elements that serve as structural components in many biological molecules (e.g., iron in hemoglobin), as activators in hormonal and enzymatic processes, as regulators of cell activity, and as components of body fluids (National Research Council 2003). Collectively, minerals make up approximately 4% of body weight and are critical for animal health and for maintaining physiological function (Leonard 2000, Robbins 1993). Calcium and sodium, for example, are very important for lactating females as insufficient intake can result in slower growth and higher infant mortality (Buss and Cooper 1970, Power et al. 1999). We have few data on the distribution of important dietary minerals in natural habitats, although a number of studies have indicated that primates select specific plant foods for their mineral content and that the highest levels of minerals are in foliar material, followed by bark and fruit (Hladik 1978, Oates 1978, Nagy and Milton 1979a, Yeager et al. 1997, Waterman 1984, Milton 1999). Overall, there is high variability in the distribution of these minerals (Milton 1999). Rode et al. (2003) found that ripe and unripe fruits in Kibale National Park, Uganda, are lower in minerals than other primate plant foods and that rarely consumed foods, such as bark, petioles, and caterpillars, had the highest mineral levels. *Geophagy* (i.e., soil consumption) is quite common in primates, and the role of soil as a mineral supplement has been suggested by a number of authors (see review by Krishnamani and Mahaney 2000). Primates are likely to encounter the same mineral-related problems as other, better-studied primary consumers (Rode et al. 2003). Oates (1978), for example, has suggested that the presence of swampy areas containing sodium-rich plants partly explains the relatively high density of *Colobus guereza* in the Kanyawara area of Kibale National Park. Rode et al. (2003) have found that sodium is low in primate foods generally in Kibale and that the iron and copper content of primate foods is marginal. These are very similar findings to those of Nagy and Milton (1979a), who found that *Alouatta palliata* in Panama had a natural diet low in sodium and copper.

Vitamins

Vitamins, like minerals, facilitate the use of energy rather than providing energy per se (Leonard 2000). Vitamins are divided into those that are water-soluble (B, C) and those that are fat-soluble (A, D, E, K). In contrast to fat-soluble vitamins, primates should generally consume water-soluble vitamins daily because they cannot be stored in the body; any excess is excreted in urine. Very little is known about the vitamin requirements of nonhuman primates, and even less is known about the distribution of vitamins in primate diets, although it is thought that virtually all vitamins except for B_{12} can be met by a plant diet. B_{12} is provided by the activity of microbes in the gut as well as animal resources, including insect larvae found in fruit pulp (Milton 1999). E, K, and B complex (niacin, thiamine, riboflavin, pyridoxine) are widespread and found in seeds, nuts, and leaves; A is found mostly in leaves and C in fruit flesh, leaves, and seeds (Waterman 1984). Vitamin C (ascorbic acid) is of particular interest because, like other highly herbivorous mammals (e.g., some rodents, lagomorphs, and bats), primates cannot synthesize this important vitamin themselves (Oftedal and Allen 1996, Milton 1999). While there can be significant intra- and interspecific variation, the pulp of fleshy tropical fruit can be extremely high in vitamins. For example, Milton and Jenness (1987) have found that on average diets of *Alouatta palliata* and *Ateles geoffroyi* contain upward of 10 times the daily recommendation for *Homo sapiens*. Given the important health consequences of vitamins and minerals, we clearly have much more work to undertake on the variability in mineral and vitamin content of plants (Milton 1999, 2000).

An individual primate's access to energy and "quality" diet has important fitness consequences. Altmann (1991, 1998), for example, has demonstrated that the ability of yearling savanna baboons (*Papio anubis*) to obtain a diet rich in energy and protein is a robust predictor of lifetime fitness. Since the transition from mother's milk to independent feeding is a critical period, young primates may be more affected by energetic shortages than adults and may be particularly affected by the quality of their diet at this critical stage. Young animals need energy and nutritional input not only for maintenance but also for growth. For example, growing juvenile mountain gorillas consumed almost twice as much energy and protein and triple the amounts of some important minerals per 0.75 kg of body mass than adult males (Rothman et al. 2006b). Since growth rate is closely linked with age at first reproduction, fecundity, and survival (Altmann and Alberts 2005), young animals that are unable to meet these increased nutritional needs may face severe consequences. Energetic shortfalls can influence not only the growth and development of an individual but also that individual's reproductive maturation. For example, reduced energy and inadequate amino acid balance in captive primates can inhibit the production of luteinizing hormone and have a direct impact on reproductive maturation (Steiner 1987, Bercovitch 2000). Yet, despite the critical links among nutrition, energy, and reproduction, very few studies (cf. savanna baboons; Altmann 1991,

1998) have linked energy and specific nutrients to measures of lifetime fitness.

METABOLISM AND ENERGY, OR WHAT IS A "QUALITY" DIET?

With the exception of minerals and vitamins (i.e., micronutrients), all nutrients (macronutrients) are involved in the balance of energy in the body. Determinants of any primate's energy requirements are dietary *thermogenesis* (i.e., the energy required to process, absorb, and store nutrients), growth, and reproduction, as well as basal metabolism, or basic life processes such as respiration and regulation of body temperature (Krebs 1972, Begon et al. 1990, Schmidt-Nielsen 1997, Leonard 2000). Energy is required for all other components of a primate's behavioral repertoire as well, such as traveling, foraging, social interactions in all forms, and predator avoidance. The energetic costs associated with food handling (i.e., the thermic effect of food) make up a relatively small proportion of daily energy expenditure and are influenced by amount consumed and the composition of the diet (e.g., high-protein meals elevate dietary thermogenesis) (Leonard 2000).

To link nutrition with an animal's feeding adaptations, we must define "dietary quality"; to do so, we need an understanding of metabolism. *Metabolism* refers to all the chemical reactions of the body and represents a balancing act between two processes: anabolic, or synthetic, processes and catabolic, or degradative, processes (Schmidt-Nielsen 1997, Tortora and Anagnostakos 1987). *Anabolic* processes are chemical reactions that combine simple substances into more complex molecules, for example, building proteins out of amino acids. Such synthesis requires energy. *Catabolic* processes are those chemical reactions in the body that break down complex organic compounds into simple ones. Catabolic reactions release energy by breaking bonds between atoms of food molecules, for example, breaking down disaccharides to monosaccharides. These processes provide energy for the body. Thus, dietary quality is best defined in terms of its capacity to yield energy.

Among mammals, there are important and reasonably predictable relationships among metabolism (energetic needs), body mass, and dietary quality. The relationship between body mass and basal metabolism is a negatively allometric one. Smaller mammals thus require relatively more energy to maintain endothermic homeothermy and larger mammals, relatively less (Kleiber 1961; Bell 1971; Jarman 1974; Parra 1978; Gaulin 1979; Schmidt-Nielsen 1984, 1997; Harvey et al. 1987; Martin 1990). This negative relationship can have important implications for the quality of foods (defined by a food's nutrient density or capacity to yield energy) that primates consume (Bell 1971, Jarman 1974, Gaulin 1979). Most importantly, the relatively greater metabolic requirements of smaller mammals must be met by a diet with a capacity to yield energy readily, either as a function of how easily it is digested or by its energetic

density or both. However, since they do not have large absolute requirements for food, smaller primates can afford to consume foods that are spatially and temporally rare (Kleiber 1961, Parra 1978, Schmidt-Nielsen 1984, Van Soest 1982, Gaulin 1979, Kay 1985, Harvey et al. 1987, Martin 1990). This is in contrast to larger-bodied mammals that have lower energy requirements per unit body weight, and although they require absolutely more food and thus cannot rely on rare resources, they can afford to consume either more food with lower nutrient density or foods that require lengthy processing (via bacterial fermentation) (Gaulin 1979, Van Soest 1982). According to these generalizations, larger mammals theoretically have more options than smaller primates in that they experience fewer metabolic constraints; that is, they can be large and still maintain a high-quality diet (and just eat more of it) or rely on foods that require lengthy processing times (Milton 1984b). According to these predictions, the latter option is not open to smaller-bodied primates, which are more metabolically constrained and must instead rely on foods that yield more readily accessible energy (e.g., insects, nectar, ripe fruit).

HOW TO EVALUATE DIETARY QUALITY: FEEDING CHALLENGES AND WAYS AROUND THEM

Ecological Challenges

The initial challenge to all animal consumers is to find food in the first place, a not insignificant task given the extreme spatial and temporal variability in food availability (Oates 1987, White 1998). There are differences in availability among broad food types (Isbell 1991). Certainly, "a leaf is not a leaf is not a leaf" (Janson and Chapman 1999:246); however, on average within a given tropical habitat, leaves do tend to be a more abundant and predictably available resource relative to ripe fruit, exudates, and insects. Milton (1980), for example, quantified resource availability on Barro Colorado Island, Panama, and found that plant vegetative parts (e.g., leaves, petioles) are more evenly distributed than plant reproductive parts (e.g., fruit). In contrast, fruit as a primate food resource can be highly unpredictable. This is a function of the fact that most tropical forests are characterized by tree species with differences in seasonal and annual fruit production and by trees that produce fruit either in small quantities or in abundance over widely scattered individuals (Janzen 1967, Frankie et al. 1974, Whitmore 1990, van Schaik et al. 1993).

Primates exhibit a diversity of responses to the challenges of locating food. At an ecological or more proximate scale, one means by which to deal with the vagaries of food availability is to shift those features that are more labile than others, for example, the number of individuals in a foraging group. The relationship between foraging group size and food availability, particularly fruit, has been a focus of extensive research, particularly as it may shed light on patterns of intraspecific feeding competition (Gillespie and Chapman 2001). While there is disagreement over the degree to which group living and group size can be explained by avoidance of predators, conspecific threat, or foraging advantages (or some combination of these pressures), it is generally held that increased feeding competition with more group members can be a cost and that the size of a primate social group mitigates the costs of predation at the expense of increasing feeding competition (Struhsaker 1981a;, Terborgh and Janson 1986, Dunbar 1988, Garber 1988a, Janson and van Schaik 1988, van Schaik 1989, Isbell 1991, Janson 1992, van Schaik and Kappeler 1993, Chapman et al. 1994a, Janson and Goldsmith 1995, Treves and Chapman 1996, Sterck et al. 1997, Boinski et al. 2000b, Koenig 2002). The "ecological constraints model" posits that as group size increases, the food resources required by that group will also increase, a primary consequence being that day range length, and hence home range length, will also increase (Chapman 1990b; Gillespie and Chapman 2001; Chapman and Chapman 2002a). Different primate species, and different populations for that matter (because of differences in habitat), will experience these patterns to varying degrees. For example, it has been suggested that primates that rely heavily on fruit will exhibit a stronger relationship between group size and day range because of the temporally and spatially patchy nature of fruit distribution in forests. Also, primates that rely on fruit patches are seen to incur greater costs of contest competition because fruit is more monopolizable than other resources (e.g., leaves). Thus, fruit-eating primates are seen to have greater limitations on group size than primates that are not so reliant on fruit. Indeed, although there are exceptions, it has been demonstrated that more frugivorous primates in larger groups travel farther than those in smaller groups (van Schaik et al. 1983a, Chapman 1990b, Wrangham et al. 1993a, Chapman et al. 1995, O'Brien and Kinnaird 1997).

Less labile traits, such as morphological adaptations that have evolved over time (evolutionary, ultimate scale), may also represent adaptations to the variations in food availability. For example, cheek pouches are found ubiquitously within the subfamily Cercopithecinae (Hill 1966, Murray 1973, Fleagle 1999). Lambert (2005) has investigated the function of cheek pouches in closed-canopy cercopithecines and found that they were used more commonly by *Cercopithecus ascanius* and *Lophocebus albigena* when feeding on fruit and when the number of conspecifics increased in a feeding patch. For both species, cheek-pouch use increased in larger food patches, a result that corroborates those of Pruetz and Isbell (1999), who found that the size of the food patch has a direct effect on the potential for competition. In their work on *Erythrocebus patas* and *Cercopithecus aethiops* in Kenya, these authors found that small food sources may be used up too quickly to result in feeding competition since the finder of the food essentially uses up that food before other foragers begin exploiting the patch. Using cheek pouches may be more important in those settings where the patch size is large enough to engender

competition but not large enough to accommodate all group members.

Digestive strategies may also mitigate the costs of interspecific feeding competition over foods that vary in availability. Cercopithecines in particular are noted for their diverse and eclectic diet (Rudran 1978; Struhsaker 1978b, Gautier-Hion 1988a, Beeson 1989, Richard et al. 1989; Maisels 1993; Chapman et al. 2002c). *Papio cynocephalus* in Kenya, for example, ingest over 200 food types by the age of 1 year (Altmann 1998). Such dietary flexibility is almost certainly related to their digestive strategy: very long mean retention times relative to body size in conjunction with a simple, unspecialized stomach (Lambert 1998, 2002a,b). If a flexible digestive and dietary strategy means that more items can be used for food, then resource switching may provide a means to minimize intra- and interspecific contest, with switchers able to access a diversity of foods over a wider area (Lambert 2002a). An ability to consume many food types can facilitate larger feeding spheres, which may allow a species to avoid pushing forward (van Schaik et al. 1983a, Chapman and Chapman 2000a). That is, for species with digestive flexibility, the likelihood of encroaching on another individual's feeding space is decreased because a given area essentially holds greater food richness. Such a mechanism (digestive and dietary flexibility) may help to explain the coexistence of closely related cercopithecines in similar trophic space (Lambert 2002a).

Chemical Challenges

Many plants, as well as some insects, defend themselves from animal feeding by arming themselves with chemical compounds. Because they are the result of secondary processes involved in defense, rather than the primary metabolism important for basic plant processes such as reproduction and growth, these compounds are collectively known as "secondary metabolites" (Harborne 1988). Primates are thus confronted with the challenge of selecting a nutritionally balanced diet while minimizing their intake of plant secondary compounds which are toxic or inhibit digestion (chemical challenges). To date, over 100,000 secondary metabolites have been identified (Wink 1999), including nitrogen-free (terpenes, polyketides, saponins, and polyacetylenes) and nitrogen-containing (e.g., alkaloids, amines, cyanogenic glycosides, nonprotein amino acids, and glucosinolates) compounds (Levin 1971, 1976; Freeland and Janzen 1974; Rosenthal and Janzen 1979; Harborne 1988; McKey 1974; McKey et al. 1978; Gartlan et al. 1980; Wink 1999). These compounds are classically divided into two broad categories defined in terms of their impact on an animal consumer: (*1*) digestion inhibitors, which interfere with the efficiency with which nutrients are obtained by the animal, and (*2*) true toxins, which interfere with physiology (Feeny 1976, Waterman and Kool 1994).

As with ecological challenges, primates exhibit various ways around chemical defense. For example, one of the best-studied digestion inhibitors is the group of polyphe-nolic compounds known as "tannins" (both condensed and hydrolyzable forms). As discussed earlier (see section "Protein"), these compounds can interfere with protein uptake by the animals because they have the capacity to bind with plant proteins, forming insoluble complexes. Several studies have demonstrated that the presence of condensed tannins in plant material is negatively related to primate feeding behavior (e.g., Oates et al. 1977, Glander 1981, McKey et al. 1981, Wrangham and Waterman 1981). However, hydrolyzable tannins tend to be more rare than condensed tannins, and while they are generally more toxic than condensed tannins, overall we know much less about their impact on primate biology (Marks et al. 1988, *Macaca mulatta*). In fact, overall the degree to which the presence of tannins in plant food is problematic for animal feeding remains uncertain. *Surfactant* substances (i.e., agents that lower surface tension) found in the gastrointestinal tract of some insects and some mammals can interfere with the for-mation of tannin–protein complexes, and in some cases tan-nins have the potential to enhance the rate at which protein is hydrolyzed via the regeneration of denatured protein(s) (Mole and Waterman 1985, Waterman and Kool 1994). In addition, some primates, including humans, have "proline-rich" proteins in their saliva known as "tannin-binding salivary proteins" (TBSPs). These TBSPs have a higher than average affinity for binding with tannins, allowing for uptake of plant proteins in the presence of tannins (Foley and McArthur 1994, Waterman and Mole 1994).

Toxic compounds such as alkaloids and cyanogenic gly-cosides are typically absorbed through the gastrointestinal tract and have a specific toxic effect on the consumer (Foley and McArthur 1994). As first clarified by Freeland and Janzen (1974), there are two major ways in which mammals deal with toxic secondary metabolites: microbial activity and microsomal enzymes. In herbivorous mammals with specialized stomachs, such as the Colobinae, some toxic secondary metabolites may be broken down via a diverse bacterial and protozoan microflora supported in the anaero-bic, alkaline stomach environment. Essentially, secondary compounds are degraded during fermentation before they are absorbed by the animal (Freeland and Janzen 1974, Kay and Davies 1994). For primates without such stomach specializations, it is probable that a complex detoxification system relying largely on microsomal enzymes takes place after digesta (containing plant toxins) has left the stomach. Microsomal enzymes are produced in mammalian liver cells, and their major function is the degradation of for-eign chemicals, with most of their activity concentrated in the liver and kidneys (Freeland and Janzen 1974). While there is very little known about microsomal detoxification systems in primates, information stemming from research on domestic mammal species suggests that the intestinal mucosa may be an important site for detoxification of plant allelochemicals. For example, in sheep and cattle, the activity of one important allelochemical enzyme (uridine

diphosphate glucuronosyltransferase) was found to be three times as great in the intestinal wall as it was in the liver (Foley and McArthur 1994). After absorption in the gastro-intestinal tract, toxins can be chemically biotransformed. If the toxin is water-soluble (hydrophilic), it can be filtered by the kidneys and excreted in urine. If fat-soluble (lipophilic), toxins are biotransformed and rendered hydrophilic through one of two (or both) chemical reactions: phase I and phase II. Phase I reactions serve to increase the polarity and water solubility of toxins and to speed their excretion; phase II reaction pathways involve conjugating toxic metabolites to endogenous substrates (e.g., glucose), which are then excreted (Freeland and Janzen 1974, Foley and McArthur 1994, Karasov and Martinez del Rio 2007).

Rates of enzymatic activity scale negatively with mam-malian body size (Walker 1978, Freeland 1991). Freeland (1991) thus suggests that smaller mammals are at an advan-tage for detoxifying plant secondary metabolites and that the larger the mammal, the greater the preference for foods with low amounts of toxic plant metabolites. These asser-tions rest largely on results by Walker (1978), who has expressed rates of enzyme activity in rats as a function of liver mass relative to body mass. Thus, small body size in some primates may facilitate the consumption of plant foods high in secondary metabolites (Lambert 2002b), which may explain why cercopithecines are able to consume plant foods with greater secondary metabolite loads than plants consumed by sympatric chimpanzees (Waser 1977; Wrangham et al. 1998; Lambert 2000b). Indeed, Wrangham et al. (1998) reported that *Cercopithecus ascanius*, *C. mitis*, and *Lophocebus albigena* have absolutely more condensed tannins, monoterpenoids, and triterpenoids than chimpan-zees in their annual diet.

In addition to toxins and digestion inhibitors, primates must deal with other nonnutritive components of plants (e.g., fiber, especially lignin) in order to maximize nutri-tional intake. For example, the ratio of protein and fiber in leaves is known to be an important selection crite-rion for primates and may serve as a useful, quantifiable index for dietary quality (McKey 1978b, Milton 1979, Chapman and Chapman 2002a). Several studies have dem-onstrated a positive correlation between folivorous primate biomass and protein:fiber ratios, including colobines in Africa and Asia (Waterman et al. 1988; Oates et al. 1990; Davies 1994; Chapman and Chapman 2002a) and lemurs in Madagascar (Ganzhorn 1992). Chapman and Chapman (2002a) found that *Procolobus badius* (also known as *Piliocolobus badius*) in Kibale National Park, Uganda, selected young leaves over mature leaves and that young leaves had more protein, were more digestible, and had higher protein:fiber ratios than mature leaves. Selection of leaves was not related to levels of secondary metabolites as these authors found no difference in secondary metabolite load in young versus mature leaves. Indeed, one of the most preferred trees (*Prunus africana*) had the highest levels of toxins (cyanogenic glycosides and saponins) and yet was among the most commonly consumed plant species with high protein content in young leaves.

Mechanical Challenges

Feeding primates are presented with an array of mechanical challenges as well. In many cases, although not all, these are plant mechanisms for protecting against predation such as thick exo- or pericarps for protecting seeds (see Chapter 43 for a discussion of tool use to access foods). Primate anatomical adaptations for getting around such plant mechanisms tend to be less labile than responses to shifts in food availability (group size shift) or dietary selectivity to avoid secondary metabolites or fiber. For example, Kinzey and Norconk (1990, 1993) evaluated the responses of *Ateles paniscus* and *Chiropotes satanus* in Suriname to both chemical and mechanical challenges and found that these two species, while exhibiting a high degree of overlap in their fruit diet, had very different means to deal with these challenges. They found that the outer coverings of the fruit (pericarp) that *C. satanus* consume are significantly harder than those consumed by *A. paniscus*, while the seed covering (exocarp) consumed by *A. paniscus* was significantly harder. *C. satanus* consume the young seeds of hard fruits and, by doing so, avoid the secondary metabolites of mature seeds but are confronted with the mechanical challenge of difficult-to-penetrate hard pericarps—hence the evolution of small, procumbent lower incisors and larger, laterally splayed canines, which work together to crack open and gouge hard foods. *A. paniscus*, on the other hand, have no such dental adaptations. Also, while the pericarps of the fruits they consume are significantly softer, the seeds of consumed fruit are significantly harder. *A. paniscus* swallow fruits whole to remove adhesive pulp and defecate the nonmasticated seeds, thereby avoiding this mechanical challenge.

In another example of ways around mechanical challenges, Lambert (1999) studied the variable methods that *Pan troglodytes* and *Cercopithecus ascanius* employ to deal with the mechanical challenges of what to do with seeds when pulp flesh is the food and not seeds. The seeds of tropical fruits range in size from less than 1 mm in length to large palm nuts that can be greater than 30 cm in width and can account for more than half of the weight of fruits (van Roosmalen 1984, Garber 1986, Howe and Westley 1988, Whitmore 1990). Swallowed seeds can thus represent a significant cost to a fruit pulp consumer in that they not only increase an animal's body mass but may also displace more readily processed, nutritious digesta from the gut (Snow and Snow 1988, Fleming 1988, Corlett and Lucas 1990, Levey and Grajal 1991, Leighton 1993, Levey and Karasov 1994). The seeds of fruits are literally an unwanted mass, and the adhesive pulp must be removed and the seeds discarded in some way. Lambert (1999) found that *P. troglodytes* swallow most seeds while *C. ascanius* spit most seeds and argued that these seed-handling differences can be attributed to differences in digestive retention times, oral anatomy, and alternative

mechanisms by which to avoid the mechanical cost of seed ballast. Cercopithecine monkeys have among the longest (both absolutely and relatively) digestive retention times across the Primate order (Clemens and Phillips 1980; Clemens and Maloiy 1981; Maisels 1993; Lambert 1997, 1998, 2002b; see Warner 1981 for definitions of digestive passage times). Seed spitting is important for cercopithecines because if large-seeded fruit were swallowed whole to remove pulp, then the energetic cost of indigestible seed ballast would be incurred for an absolutely and relatively long time. Moreover, it could severely limit space for incoming food. On the other hand, *P. troglodytes* lack the oral adaptations of the Cercopithecinae (e.g., large postcanine teeth, high-crested bilophodont molars, and cheek pouches). In addition, they are large primates that exhibit the more typical emphasis of relatively rapid digestive processing common to frugivores (Milton 1984b; Milton and Demment 1987; Levey and Grajal 1991; Lambert 1998, 2002b). Because chimpanzees move seeds through the gastrointestinal tract relatively faster than cercopithecines, they are also less likely to incur such a cost in terms of seed ballast and displacement of nutritious digesta. Moreover, being a large animal, swallowed seeds will represent a smaller mass relative to chimpanzee body size. A potential result of this is that natural selection has been less intense in hominoids than in cercopithecines for evolving behavioral mechanisms to avoid the cost of seeds in the gut.

Fallback Food Strategies = Solutions to Ecological + Chemical + Mechanical Challenges?

Ecological, chemical, and mechanical challenges are central to a lively and current discussion regarding the ecological and evolutionary significance of fallback foods (Lambert et al. 2004; Laden and Wrangham 2005; Hemingway and Bynum 2005; Wright 2005; Ungar et al. 2006; Marshall and Wrangham 2007; Lambert 2007, 2009; see also recent volume on the significance of fallback foods: Constantino and Wright, 2009). Historically, authors have varied in their use of the term "fallback" and differentially focused on one or more variables important to optimal feeding and foraging more generally, including inherent energy yield of food (nutrient density), search time, and handling time (MacArthur and Pianka 1966). In theory, an animal should exhibit preference for the most profitable (in terms of energy yield) foods, that is, foods that are nutrient-dense, easy to find, and easy to access. Foods that are less nutrient-dense, harder to find, or more difficult to access are presumably less profitable and thus less preferred. As such, "fallback" has been used to indicate food preference and availability—absolute abundance or relative to preferred foods—or to simply refer to the absolute inherent quality and nutrient density of the food (Gautier-Hion and Michaloud 1989; Wrangham et al. 1996, 1999; Tutin et al. 1997; Yamakoshi 1998; Gursky 2000b; Furuichi et al. 2001; Fox et al. 2004; Lambert et al. 2004; Ungar 2004).

However, more recent discussions have focused on classification schemes proposed to characterize and provide operational definitions for the ecological and potential evolutionary significance of fallback foods. Lambert (2007), for example, suggested two broad categories of fallback foods that ultimately relate to different fallback *strategies,* including (*1*) fallback foods of lower nutritional density (lower inherent energy yield). These foods typically, although not always, comprise vegetative (nonreproductive) plant anatomy such as bark, leaves, petioles. Such foods often have the advantage of being more abundant (less search time); however, they inherently require more processing on the part of the animal (more handling time). It is in the use of these fallback foods that we tend to see anatomical adaptation (e.g., thicker dental enamel, longer molar shearing crests, longer digestive retention times); extant Cercopithecoidea and *Gorilla* spp. are good examples of this strategy. Alternatively, (2) fallback foods can be of higher nutritional density (greater inherent energy yield). These foods are commonly, although not exclusively, reproductive plant parts such as fruit and seeds. The advantage of these foods is that they can facilitate the maintenance of a high-energy-yield diet throughout the year. However, they often carry with them the disadvantages of being mechanically protected (longer handling time; e.g., palm nuts, termite mounds) and difficult to find (more search time). With these foods we may observe behavioral innovation and tool use for coping with seasonal shortages—*Pan troglodytes* and *Cebus* spp. are examples of such a strategy. It should be noted that neither strategy precludes the other so that, for example, species that use tools to access a high-energy but mechanically protected food may also have thick dental enamel and a robust masticatory apparatus (e.g., *Cebus* spp.).

The ecological and evolutionary implications of fallback foods for anatomical and behavioral adaptation have also been carefully evaluated by Marshall and Wrangham 2007. In their operational definitions and classification scheme, these authors suggest that fallback foods be classified into two subsets: staple and filler. *Staple* fallback foods are those food resources that are annually available and thus can be accessed throughout the year; during seasonal shortages of other foods, staple foods may comprise the majority of a primate species' diet. Because of this potential for meeting 100% of the diet during seasonal shortage, these authors suggested that staple fallback foods would result in digestive and socioecological adaptation. In contrast, *filler* fallback foods are either seasonally or annually available but never comprise a majority of the diet. As such, the authors suggest that the degree of physiological, behavioral, and ecological adaptation is less complete in response to filler fallback foods than to staple fallback foods.

CONCLUSIONS AND CONSERVATION IMPLICATIONS

As I have endeavored to illustrate in this chapter, diet and nutrition are tied to virtually every aspect of an animal's biology. Understanding the means by which primates solve feeding challenges can thereby lend invaluable insight into the ecology and evolution of species. However, such understanding is more than academic. Insight into nutrition and

its myriad consequences for animal biology is essential in at least three ways to an increasingly critical goal: conservation of wild primate populations. The sobering reality is that 96 of the world's extant 250 primate species are presently in the World Conservation Union's critically endangered or endangered category and could continue to decline to extinction within the next century (Struhsaker 1999, Wright and Jernvall 1999; see Chapter 45). Population size directly influences the intrinsic components of population decline and is hence critically related to species vulnerability (Cowlishaw and Dunbar 2000). Since food is commonly argued to be among the most important biotic variables limiting population size, an understanding of the direct and causal relationships among food resource availability, diet, nutrition, and population size should be part of any sound in situ conservation plan (Struhsaker 1976, Milton 1982, Gautier-Hion 1988a, Cowlishaw and Dunbar 2000; Chapman et al. 2003a). Yet, few studies have directly linked specific nutrients to changes in population size, despite the fact that such information is key for the implementation of conservation tactics (Chapman et al. 2003a). For example, as suggested by Chapman et al. (2003a), if it could be demonstrated that a wild primate population is limited by *x* nutrient, tree species rich in this nutrient could be left standing in selective logging operations. With careful planning, postfelling population decline could be lower and/ or recovery may be faster.

Information on diet and nutrition is critical for the successful ex situ management of endangered species in captivity as well. One response on the part of conservation managers to the decline of wild populations is to maintain endangered taxa in captive settings. *Procolobus badius waldroni* (also known as *Piliocolobus badius waldroni*) was declared extinct in 2000 (Oates et al. 2000a); yet, while some colobines (e.g., *Colobus guereza, Presbytis* spp.) do reasonably well in captivity, maintaining *P. badius* in captivity has proven extremely difficult. Why this is the case is not clear; however, *P. badius* has a quadripartite stomach and complex digestive ecology, and the biology of maintaining a healthy community of anaerobic gastric microbes in a delicately balanced stomach pH, along with unknown mineral/vitamin requirements, is almost certainly implicated.

Finally, nutritional information can also lend insight into dietary flexibility, which too can inform our decisions as conservation managers. For example, dietary flexibility and a capacity to fall back on low-quality foods may facilitate animal persistence in degraded landscapes and forest fragments. Are these then the same species whose populations are less impacted overall by anthropogenic change? Should they in turn receive greater conservation priority because they are likely to tolerate future human-induced change or less attention because they are presently tolerating anthropogenic stress? Such questions are central to conservation tactics, and we are yet to get a handle on what flexibility actually means in terms of nutritional requirements. These examples clearly highlight the fact that questions regarding diet and nutrition represent more than academic concerns.

34

Primate Seed Dispersal
Linking Behavioral Ecology with Forest Community Structure

Sabrina E. Russo and Colin A. Chapman

1. What is a seed shadow, and how do primates influence it?
2. How is seed dispersal by primates related to the conservation and/or regeneration of topical forests?
3. What are the basic models for understanding seed dispersal?

INTRODUCTION

The role of animals in seed dispersal is well recognized. As many as 75% of tropical tree species produce fruits presumably adapted for animal dispersal (Frankie et al. 1974, Howe and Smallwood 1982), and animals are estimated to move more than 95% of tropical seeds (Terborgh et al. 2002b). Some vertebrate groups may be particularly important seed dispersers. Primates, for example, comprise between 25% and 40% of the frugivore biomass in tropical forests (Chapman 1995), eat large quantities of fruit, and defecate or spit large numbers of viable seeds (Lambert 1999). Primate frugivory and seed dispersal have been quantified by studies in South America (Garber 1986, Julliot 1996b, Stevenson 2000, Dew 2001, Vulinec 2002), Central America (Estrada and Coates-Estrada 1984, 1986; Chapman 1989a), Africa (Gautier-Hion 1984; Gautier-Hion et al. 1985; Wrangham et al. 1994a; Chapman and Chapman 1996a; Kaplin and Moermond 1998; Lambert 1999; Voysey et al. 1999a,b; Gross-Camp et al. 2009), and Asia (Corlett and Lucas 1990, Davies 1991, Leighton 1993, Lucas and Corlett 1998, McConkey 2000, Stevenson 2007). This research has illustrated that primates disperse significant numbers of seeds. For example, on Borneo, a single gibbon group (*Hylobates muelleri × agilis*) dispersed a minimum of 16,400 seeds · km^{-2} · year^{-1} of 160 species; and since the survival rate of seeds to 1 year was 8%, a group of gibbons effectively dispersed 13 seedlings · ha^{-1} · year^{-1} (McConkey 2000).

Although it is clear that primates play an important role in dispersing many seeds throughout tropical forests, the ecological and evolutionary significance of these activities is not well understood. The objective of this chapter is to review the role that primate seed dispersal plays in shaping the ecology of tropical forests and to shed light on inconsistencies in the literature in order to point to new directions for research. We first evaluate how variation among primate species in traits such as digestive anatomy, body size, social structure, movement patterns, and diet influences the spatial distribution of dispersed seeds (the *seed shadow*) and thereby produces a diversity of seed shadows. We then consider how traits of fruiting species, including patch characteristics and fruit and seed traits, influence the seed shadow that different primates generate. Few studies have quantified primate-generated seed shadows. However, to understand their effect on plants, it is important to consider primate seed shadows in light of plant population and community ecological theory. To this end, we next review current theories and concepts and evaluate the role that primate seed dispersal plays in plant ecological processes. This analysis of variation in seed shadows points to two critical areas in which our understanding of primates' roles is currently lacking: the significance of primates dispersing seeds in clumped versus scattered patterns and the consequences of primate seed dispersal to different distances away from the parent tree. We conclude by discussing potential conservation implications related to changes in the nature of primate–plant interactions.

VARIATION IN SEED SHADOWS GENERATED BY PRIMATES

Dispersal of seeds by primates results in a seed shadow. The shape of the seed shadow is defined by the dispersal processes that produce it, namely, what proportion of seeds are dispersed from parent tree crowns, which depends on the visitation rate and the number of seeds dispersed per visit of a dispersal agent, how far those seeds are moved from the parent (*seed dispersal distance*), and the locations and densities at which seeds are deposited (*dispersal sites* and *seed density*). The interaction between traits of primates and of the fruit-producing plant species that they feed upon influences these dispersal processes, thereby shaping the seed shadow. The shape of the seed shadow is important because it influences the survival of seeds to the seedling stage (Janzen 1970, McCanny 1985), which is a critical demographic process in the life cycle of a plant and may ultimately influence the spatial distribution of plants in later life stages (Nathan and Muller-Landau 2000, Wang and Smith 2002).

Traits of Primates

As seed-dispersal agents, frugivorous primates have high functional diversity in traits influencing dispersal processes, such as digestive anatomy, body size, social structure, movement patterns, and diet, all of which generate heterogeneity in seed shadows. Few studies have quantified how variation in these factors affects seed-deposition patterns. However, data are becoming available that identify trends in these relationships, which should be further evaluated in future studies.

Primates largely disperse seeds by endozoochory, but once an animal has located and acquired fruit, there remains the challenge of what to do with the seeds. Protective seed coats are typically difficult to digest, and seeds themselves can also account for more than half of the weight of fruits consumed by primates (van Roosmalen 1984, Waterman and Kool 1994). Swallowed seeds can thus represent a significant cost in that they not only increase an animal's body mass but may also displace more readily processed, nutritious digesta from the gut. Given these constraints, it is somewhat surprising that seed swallowing is by far the most common means of primate seed dispersal in the neotropics (Estrada and Coates-Estrada 1984, Chapman 1989a, Andresen 1999). In the paleotropics, many primates also swallow seeds (Lieberman et al. 1979, Corlett and Lucas 1990, Wrangham et al. 1994a, Lambert 1999, Kaplin and Lambert 2002); however, seed spitting is common in African and Asian cheek-pouched monkeys (Cercopithecinae, Gautier-Hion 1980; Rowell and Mitchell 1991; Kaplin and Moermond 1998; Lambert 1999, 2000a). The cheek pouches of cercopithecines have nearly the same capacity as their stomachs (Fleagle 1999) and allow these monkeys to store multiple fruits and extract the pulp without incurring the costs of ingested seeds (Lambert 1999). They may process the fruits and spit seeds at or away from parent trees.

Other primates consistently prey upon seeds as ingested seeds do not appear in feces or are fragmented. These include black and white colobus (*Colobus guereza*, Poulsen et al. 2001b), saki (*Pithecia pithecia* and *Chiropotes satanas*,

Kinzey and Norconk 1993), and red leaf (*Presbytis rubicunda*, Davies 1991) monkeys. The extent to which primates act as seed predators is likely underreported since some authors lump seed predation with the ingestion of fruit. Primates' seed-handling strategies depend on the interactions between their digestive anatomy and the traits of the fruiting species (Norconk et al. 1998).

Variation in seed-handling strategies, social structure, and movement patterns shapes the seed shadows primates generate (Figure 34.1). Some of these characters are predictably related to body size. Primates range in body mass from the 30 g pygmy mouse lemur (*Microcebus myoxinus*) to the 200 kg gorilla (*Gorilla gorilla*, Fleagle 1999). Larger primates typically have larger gut capacities, longer digesta passage times, and longer day ranges than smaller primates, although substantial variation in these relationships exists (Kay and Davies 1994, Lambert 1998, Nunn and Barton 2000). In particular, similarly sized species may have different ranging patterns due to different social systems. These characters, in turn, influence the quantity of seeds dispersed as well as their sizes, passage times, dispersal distances, and spatial distribution (Table 34.1). However, based on the limited available data (Table 34.1), we found no statistically significant relationship between either the mean or the maximum seed-dispersal distance of primate species and body size (Spearman rank correlation $r = 0.009$, $p = 0.979$, $n = 11$, mean distance; $r = 0.108$, $p = 0.818$, $n = 7$, maximum distance).

Social structure can also affect primate seed shadows. Primates with more fluid social structures, such as the fission–fusion societies of spider monkeys (*Ateles* spp.) and chimpanzees (*Pan troglodytes*) and the multispecies foraging groups of capuchins (*Cebus* spp.), cover large areas in their daily ranges; and individuals in groups are often widely dispersed (Terborgh 1983, McFarland 1986). Thus, one might expect primates with fluid social structures to have a more scattered pattern of seed dispersal than species with more cohesive ones. For example, Wehncke et al. (2003) demonstrated that white-faced capuchins (*Cebus capucinus*) on Barro Colorado Island, Panama, spend only 10 min feeding in individual trees, travel between 1 and 3 km a day, and defecate seeds in small clumps throughout the day. Repeated use of the same sleeping site may alter this relationship (Figure 34.1). Seeds defecated diurnally by spider monkeys (*Ateles paniscus*) in Peru were more widely scattered and had longer dispersal distances than did those defecated at sleeping sites, which were closer to parent trees and more clumped (Russo and Augspurger 2004, Russo et al. 2006). No data are available to compare the extent of clumping in seed shadows generated by primates with fluid versus cohesive social structure. However, based on the data in Table 34.1, there is no difference in the mean dispersal distances of species with fluid (e.g., spider monkeys) or dispersed (e.g., red-tail monkeys) groups in comparison to primates whose social structures are more cohesive (e.g., howler monkeys; $t = 1.024$, $p = 0.333$, $n = 12$). Furthermore, in species that do

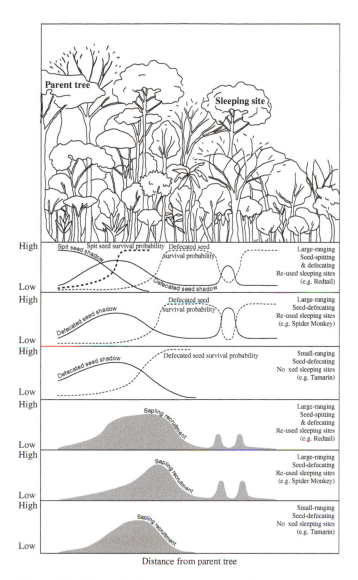

Figure 34.1 The predicted patterns of seed deposition, seed survival, and sapling recruitment for a hypothetical tree species dispersed by primates with different seed-handling strategies and ranging patterns. For many forest tree species, including the hypothetical one depicted here, some habitats may be more suitable for recruitment than others. Hence, the shape of the survival curve will depend on the plant species, but this source of variation is not depicted in the figure. Actual recruitment probabilities are entirely dependent on the actual seed densities relative to realized per capita survival probabilities, but in this schematic we show possible qualitative relationships. The relative difference in the height of the seed shadow curve and the survival curve will determine whether clumped seed-deposition patterns remain clumped as individuals age. This will partly be determined by the strength of distance- and density-dependent mortality at the seed and seedling stages, as well as other sources of mortality that are nonrandom in space or time, relative to the spatial scale and strength of clumping of seed deposition.

not have cohesive group structures, sex and age may influence seed shadows. In spider monkeys and woolly monkeys (*Lagothrix lagotricha*), males have greater day ranges and tend to disperse seeds longer distances than do females (Symington 1987b, Stevenson 2000, Dew 2001). Furthermore, route-based travel and shared routes can spatially restrict seed

dispersal, even for wide-ranging primates (Di Fiore and Suarez 2007).

The interaction between diurnal variation in movement patterns and passage time affects both dispersal distance and the spatial distribution of seeds. For example, seeds ingested by spider monkeys (*Ateles* spp.) and woolly monkeys (*Lagothrix lagotricha*) in the morning or early afternoon tend to be dispersed the same day. In contrast, seeds ingested later in the afternoon tend to be dispersed at sleeping sites at night or the next morning (Stevenson 2000, Dew 2001, Russo et al. 2006). In addition, foraging movements in the morning and early afternoon tend to be faster than those in the late afternoon (Chapman and Chapman 1991). As a result, seeds ingested early in the day are likely dispersed longer distances than those ingested later (Stevenson 2000, Dew 2001, Russo et al. 2006).

Variation in diet can also influence seed shadows, not only by determining which fruiting species are dispersed but also because the quantity of nonfruit foods in the diet typically increases seed-passage time (Milton 1981c, Kay and Davies 1994, Lambert 1998). Primates that consume substantial quantities of leaves, such as howler monkeys (*Alouatta* spp.), have greater colon volume than do similarly sized close relatives (e.g., spider monkeys [*Ateles* spp.]; Milton 1981c, 1986). As a result, howler monkeys have longer passage times than do spider monkeys (Milton 1981c, 1986). However, this difference does not necessarily translate into longer seed-dispersal distances by howler monkeys as they tend to move slower and have smaller day ranges than do spider monkeys (Milton 1981c).

Primate Responses to Traits of Fruiting Species

Primate species respond differently to traits of fruiting species, and this response generates variation in seed shadows. These traits range from those at the community level, such as the abundance and distribution of fruit resources, to those at the individual plant or patch level, such as fruit production, to those at the level of a single fruit or seed, such as seed size. Primates' responses to fruiting plant traits are mediated by many factors, including their foraging decisions, digestive anatomy, seed-handling strategies, social structure, and ranging behavior. Taken together, these plant–primate interactions determine the variation in the seed shadows that primates generate, which in turn influences plant demography and spatial distribution. Here, we consider how variations in (*1*) the distribution and abundance of fruit resources, (*2*) the characteristics of fruit patches, and (*3*) fruit and seed traits influence primates' seed shadows.

Distribution and Abundance of Fruit Resources

Primates' foraging decisions, and hence visitation and seed dispersal of fruiting species, are sensitive to the community-wide abundance and distribution of fruit resources (Agetsuma and Noma 1995, Garber and Paciulli 1997,

Table 34.1 Estimates of the Straight-Line Distance (m) that Primate Species Disperse Seeds Away from the Parent Plant

SPECIES	WEIGHT[1] (KG)	SOCIAL STRUCTURE[2]	PASSAGE TIME[3] (HR)	DISTANCE (M)	MODE (M)	RANGE (M)	REFERENCE
Seeds swallowed							
Saguinus spp.	0.5	1	3.3	–	100–200	34–513	Garber 1986
Cebus apella	2.5	3	3.5	390 (215)	–	–	Zhang and Wang 1995
C. capucinus[4]	3.3	3	3.5	–	–	200–1,000	Rowell and Mitchell 1991
C. capucinus[4]	3.3	3	3.5	208 (113)	–	20–844	Wehncke et al. 2003
Alouatta palliata[4]	6.2	1	20.4	–	–	10–811	Estrada and Coates-Estrada 1984
A. seniculus	7.3	1	35.0	225 (129)	–	0–550	Julliot 1996a,b
Ateles. paniscus	6.2	2	5.3	151 (241)	–	11–1,119	Russo 2003a
A. paniscus	6.2	2	5.3	254 (145)	–	–	Zhang and Wang 1995
A. belzebuth	6.0	2	–	245	–	50–500	Dew 2001
Lagothrix lagotricha	6.3	2	6.8	300	–	0–390	Stevenson 2000
L. lagotricha	6.3	2	6.8	354 (213)	300–400	0–989	Stevenson 2000
L. lagotricha	6.3	2	6.8	245	–	50–500	Dew 2001
Cercopithecus ascanius	3.6	3	20.2	1,178			Lambert 1997
Hylobates muelleri muit agilis	5.9	1	–	220 (215)		0–1,250	McConkey 2000
Pan troglodytes	36.4	2	23.2	3,000			Lambert 1997
Seeds spit out							
Cercopithecus ascanius	3.6	3	20.2	2	10	0–100	Lambert 1999
C. mitis	6.0	3	21.3	–	–	30–50	Rowell and Mitchell 1991
Pan troglodytes	36.4	2	23.2	4	1	0–20	Lambert 1997, 1999

[1] Average of male and female weight (from Harvey et al. 1987).

[2] Social structures are 1, cohesive; 2, fission–fusion; 3, dispersed foraging groups.

[3] Average passage time through the digestive tract (from Lambert 1998).

[4] Estimates from passage time and path length.

Janson 1998b). Fruit is a spatially and temporally patchy resource (Frankie et al. 1974, van Schaik et al. 1993). As such, any particular fruiting species may not be reliable, especially given the rarity of tropical plant species in general and their tendency to have supra-annual fruiting schedules (Chapman et al. 1999, 2009). Such variability in fruit resources likely contributes to the often-observed pattern of dietary plasticity and helps explain the great diversity of plant species that primates disperse (Chapman 1987, 1995; Chapman et al. 2002a; Russo et al. 2005). A comparative study of diets in spider monkeys (*Ateles* spp.) indicated that, despite their catholic diets, spider monkeys appear to prefer relatively abundant fruiting species that consistently produce annual crops (Russo et al. 2005).

The community-wide abundance of fruit may also affect dispersal distances and the frequency of seed predation. During periods of community-wide fruit scarcity, daily path lengths of spider and woolly monkeys decreased and the proportion of time spent resting increased relative to periods of fruit abundance (Symington 1987b, Stevenson 2000). As a result, dispersal distances may decrease during periods of fruit scarcity. In addition, the consumption of unripe fruit by some primates, which likely represents seed predation, often increases during periods of fruit scarcity (Kaplin et al. 1998, Stevenson 2000). Cercopithecine monkeys alternate between acting predominantly as seed spitters, seed predators, and seed defecators depending on fruit resource availability (Kaplin et al. 1998).

Since the vast majority of primate field studies are relatively short-term (i.e., 1 year or less), it is difficult to assess the importance of periods of fruit scarcity. Yet, these times may be important periods of selection for both the primate dispersers and the trees. Fruit availability has been quantified in Kibale National Park, Uganda, for over 12 years (Chapman et al. 2009). Over this period, temporal variation in fruit availability was high; the proportion of trees per month with ripe fruit varied from 0.1% to 15.9%.

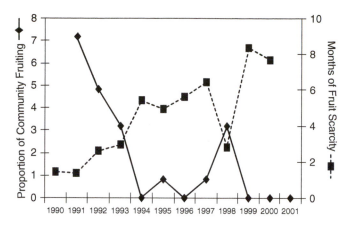

Figure 34.2 The average monthly proportion of the tree community that fruited each year in Kibale National Park, Uganda, and the number of months of fruit scarcity, defined as months with less than 1% of the entire population bearing fruit.

In addition, there was dramatic interannual variation in fruit availability: In 1990, on average only 1.1% of trees bore ripe fruit each month, while in 1999 an average of 6.7% of trees bore fruit each month (Figure 34.2). If a month of fruit scarcity is defined as one with less than 1% of monitored trees bearing ripe fruit, there is considerable interannual variation in how often frugivores experienced food shortages (Figure 34.2). For example, 9 of the 12 months in 1990 had less than 1% of the trees with fruit, while in 2000 no month had less than 1% of trees fruiting. This level of variability means that for a tree that fruits during a period of general community-wide fruit scarcity, the probability of its seeds being dispersed is likely very different from that in a subsequent year when a greater number of species are attempting to attract the services of dispersers.

Characteristics of Fruit Patches

Characteristics of fruit resource patches, such as size and density of fruits, affect visitation by primates (Leighton 1993). Greater numbers of ripe fruits in the wild nutmeg tree (*Virola calophylla*, Myristicaceae) increased both visitation and the number of seeds ingested by spider monkeys (*Ateles paniscus*, Russo 2003b). Similarly, orangutans (*Pongo pygmaeus*) preferred to forage in trees with large crop size (Leighton 1993). Furthermore, as the size of primate foraging parties increases, they tend to deplete resource patches completely, thereby increasing travel between patches (Symington 1987a, Chapman and Chapman 2000b, Gillespie and Chapman 2001). Seed-dispersal distances may therefore increase, all else being equal, as foraging party size increases.

Fruit and Seed Traits

Despite dietary plasticity, primates forage selectively based on traits of fruits and seeds, and these traits have been the basis of describing a primate seed-dispersal "syndrome" (van der Pijl 1982, Gautier-Hion et al. 1985). Suites of traits (e.g., color, size, protection) have been interpreted as coadapted features of plants that govern the choice of fruits by dispersers. While these syndromes may be broadly meaningful, responses of different primate species to these traits are nonetheless highly variable, likely due in part to variability in the floristic composition of their habitats and interspecific interactions (Gautier-Hion et al. 1993; Chapman and Chapman 2002b, Russo et al. 2005; Gross-Camp et al. 2009). Fruits considered, based on their traits, to be primarily bird-dispersed can be extensively dispersed by primates (Julliot 1996b, Russo 2003b). In fact, it may be more difficult for a plant to exclude primates as dispersers, for example, compared to birds since primates are better able to manipulate even fruits protected by hard outer coverings. Gautier-Hion et al. (1985) went so far as to say that these suites of traits exist despite an outstanding lack of specificity between fruit and consumer species. This lack of specificity has led some to question the generality of such syndromes, and they have pointed out large differences in how syndromes have been defined (Fischer and Chapman 1993). For example, based on observations in Peru, Janson (1983) described a primate syndrome as fruits that are large and yellow, orange, brown, or green with a woody outer covering, whereas based on observations in Gabon, Gautier-Hion et al. (1985) described the primate syndrome as fruits intermediate in size; bright red, orange, or yellow; and either dehiscent with arillate seeds or succulent and fleshy. Research on fruit syndromes incorporating models of primate vision has found that fruit color can be a reliable signal to different frugivore taxa, but that the colors of primate-dispersed fruits are less contrasting to primates than those of bird-dispersed fruits (Lomascolo + Schaefer 2010).

Despite this controversy over the generality of syndromes, primates are obviously selective in their choice of fruits. Choice of fruiting species can be based on morphology (i.e., color, protection, size) as well as nutrient, caloric, and secondary chemical content. Bornean orangutans preferred high pulp mass per diaspore (Leighton 1993). In contrast, spider monkeys (*Ateles paniscus*) did not respond to the aril:seed ratio of the fruits of *Virola calophylla* in terms of either visitation or seed removal (Russo 2003b). Similarly, long-tailed macaques (*Macaca fascicularis*) did not appear to select fruit on the basis of seed or fruit size (Corlett and Lucas 1990), although howler monkeys (*Alouatta seniculus*, Julliot 1996b) and four species of lemurs in Madagascar appeared to select fruits in part based on size (Dew and Wright 1998). Fruit selection may also be based upon macronutrient balancing and protein, rather than caloric, content (Altmann 2009; Felton et al. 2009; Ganzhorn et al. 2009).

The hardness of the fruit's outer covering and its seed plays an important role in the foraging decisions of primates, thereby governing which primate species disperse or prey upon seeds of a particular species. When sympatric primates compete for fruit resources, these axes of trait variation may provide means of niche separation. In Suriname and Venezuela, the bearded saki monkey (*Chiropotes satanas*) consumes significantly harder fruits than does the spider monkey (*Ateles belzebuth*, Kinzey and Norconk 1990). Furthermore, seeds masticated by saki monkeys are softer than those dispersed by spider monkeys (Kinzey and Norconk 1990). The fruit consumed by red-tail monkeys (*Cercopithecus ascanius*) and mangabeys (*Lophocebus albigena*) did not differ in dietary hardness during periods of fruit abundance, but mangabeys exploited a diet more resistant to puncture than red-tail monkeys during periods of fruit scarcity (Lambert et al. 2004). This suggests that it is the differences in the mechanical properties of fallback foods during periods of resource scarcity that may facilitate niche separation.

Many primates pass ingested seeds, resulting in few, if any, negative impacts on germination (Estrada and Coates-Estrada 1984; Dew and Wright 1998; McConkey 2000; Lambert 2001; Poulsen et al. 2001a,b; Stevenson et al. 2002). However, even among primate species that rarely prey upon seeds, seed predation may nonetheless occur, albeit at low frequency, and properties of fruits and seeds influence the likelihood of seed predation. Seeds having relatively soft coats or that provide no pulp or aril appear more likely to be subject to predation (Corlett and Lucas 1990, Russo et al. 2005). The shape and size of the seed may also influence seed handling (Chapman 1995). Primates tend to ingest and disperse a higher number of smaller than larger seeds (Corlet and Lucas 1990, Lucas and Corlet 1998, Lambert 1999, Oliveira and Ferrari 2000) and more ovoid than round seeds (Garber 1986, McConkey 2000). In olive baboons (Papio anubis), there is evidence of fruit handling that depends on seed size: spot seeds were larger than seeds from randomly collected fruits, compared to swallowed seeds (Kunz + Linsenmair 2008). When pulp or aril is easily removed from the seed, gibbons (*Hylobates muelleri × agilis*) and long-tailed macaques (*Macaca fascicularis*) tend to spit seeds under the parent tree (McConkey 2000), although some observations indicate this effect may be overridden by the effect of seed size (Corlett and Lucas 1990).

The effect of fruit and seed traits on the probability of being ingested has consequences for seed shadows because seeds that are ingested by primates usually are dispersed longer distances than seeds that are spit-dispersed (Lambert 1999, 2001) (Table 34.1). Furthermore, spit or dropped seeds are likely to be deposited on the forest floor singly as fruits are processed one by one (Lambert 1999). However, swallowed seeds can be deposited either in a high-density clump or as just a few seeds, depending on animal size and position in the canopy, defecation size, and the intensity of the feeding bout during which the seeds were swallowed (Wrangham et al. 1994a, Kaplin and Moermond 1998, Andresen 1999, Lambert 1999). In Cameroon, for example, Poulsen et al. (2001b) found that gorillas and chimpanzees averaged 18 and 41 large (>2 cm) seeds per fecal sample, respectively, while average numbers of large seeds per fecal sample for four frugivorous monkeys were less, ranging 1.0–2.1 seeds. Some primates that largely ingest seeds, however, do not always disperse them in clumps (e.g., *Cebus capucinus*, Wehncke et al. 2003). In addition, seeds ingested by spider monkeys

(*Ateles paniscus*) that are defecated diurnally tend to be dispersed at lower densities than those that are defecated at sleeping sites (Russo and Augspurger 2004). A similar pattern has been observed for woolly monkeys (Stevenson 2000) and gorillas (Rogers et al. 1998).

Although size can influence the passage time of ingested seeds, this effect may not necessarily translate into dispersal distances, which also vary with seed size (Garber 1986, Stevenson 2000, but see McConkey 2000). In fact, in some primates, passage time correlates poorly with dispersal distance, primarily because movements are rarely unidirectional and because seeds dispersed at sleeping sites are retained in the gut overnight (Stevenson 2000), although in primates the relationship is positive (Garber 1986).

Because fruit pulp attracts animals that may disperse a plant's seed, it generally presents less of a challenge to ingest and digest than other primate foods (Janson and Chapman 1999). In general, most fruits are high in sugars but low in fats and protein. As a result, protein content seems to have little to do with fruit choice (Leighton 1993) since the animals often obtain protein from other sources. Immature fruits are often chemically defended against insect or mammalian herbivory, and some species continue to be defended even when the pulp is ripe (Fenner 1992, Cipollini and Levey 1997). Some fruits have evolved chemical traits that appear to restrict dispersal of seeds to a fraction of the frugivore community. For example, the chemicals that make red peppers spicy to humans and other mammals are not perceived by birds (Janson and Chapman 1999). An interesting avenue for future research would involve investigations of whether some of these apparently defensive chemicals can alter gut-passage times in primates and, thus, dispersal distances, as in birds (Murray et al. 1994).

SEED-DISPERSAL MODELS

Quantifying seed-deposition patterns created by primates and the relationship between these patterns and the behavioral processes that produce them is a very challenging research agenda. The challenge partly arises due to the difficulty of quantifying the complex movement patterns of primates, which often occur over large spatial scales involving convoluted routes, in relation to their effects on plant population and community dynamics. However, the link between animal behavior and its consequences for seed-dispersal curves and seed shadows generated by primates is successfully being approached through several modeling methods.

Two modeling approaches, inverse and mechanistic models, have made fundamental contributions to understanding seed dispersal (Nathan and Muller-Landau 2000). The inverse-modeling approach estimates parameter values for dispersal functions that result in the best fit to seed-dispersal data from seed traps, given a model of dispersal, or dispersal kernel, describing the seed shadow of individual, mapped parent trees (Ribbens et al. 1994, Clark et al. 1999). A limitation of inverse modeling is that it does not directly incorporate information about the behavior of dispersal agents and therefore does not provide an explicit link between seed-deposition patterns and disperser behavior. Furthermore, seed dispersal by more terrestrial primates cannot be assessed using seed traps. In addition, there are logistic difficulties in applying this approach to primates that range over large spatial scales (e.g., chimpanzees and baboons). Thus, inferences based on inverse modeling of the role of primate dispersers in plant population and community dynamics are limited.

Mechanistic models predict seed-deposition patterns directly from the traits of plants and their dispersal agents (Nathan and Muller-Landau 2000). They can be developed from the primate or plant perspective. Models from the primate perspective focus on describing features of the environment and the distribution of resources that determine the movement patterns of primates. However, dispersal agents are not monotonic in their generation of dispersal patterns, and there is evidence of considerable variation among dispersal kernels generated by the same disperser, depending on the tree species (Westcott et al. 2005). In modeling seed shadows from the plant's perspective, the multiple dispersal agents that contribute to the seed shadow must be accounted for in a total dispersal kernel (Nathan 2007), which is more complex than considering a single disperser and requires substantial amounts of field data for parameterization.

From the primate perspective, basic mechanistic models predict seed-dispersal distances based on passage times of seeds and movement rates (Wehncke et al. 2003). However, such models do not incorporate ecological data on the timing or directionality of movements and behaviors of primates that often cause them to disperse seeds repeatedly in the same locations, such as sleeping sites. Hence, models based on seed-passage or -handling times likely underestimate clumping of primate-dispersed seeds. Advances in this modeling approach from the plant's perspective, however, have recently been made. Dennis and Wescott (2007) have modified this approach to estimate the total dispersal kernel of the Australian tree *Elaeocarpus grandis*. Although their approach still requires considerable observational data on passage times and movement rates for each dispersal agent, they achieve a reduction in complexity by using a functional classification of dispersal agents and fruits. Based on a similar approach using functional classifications, a generalized mathematical formulation of the total dispersal kernel has also recently been developed (Nathan 2007, Nathan et al. 2008). Given that the seeds of most plant species are dispersed by multiple agents including primates, these approaches to modeling complex dispersal systems will likely prove informative for quantifying the influence that primates have on seed shadows, relative to other dispersers.

Field observations and field experiments manipulating the distribution of food resources suggest that primates can forage based on memory of a spatiotemporal map (Garber 1989, Janson 1998b) and routinely use particular routes, often defined by topographic features of their habitat (Di Fiore and Suarez 2007), or resting sites (Chapman 1989b, Julliot 1997, Schupp et al. 2002, Russo et al. 2006). It is important for models of primate-generated seed shadows to incorporate such behaviors. For example, spider monkeys (*Ateles geoffroyi*) in Yucatan, Mexico, were observed to orient and travel directly to fruiting trees from distances of up to 1,500 m, which are presumably out of the range of sight in closed-canopy forest (Valero and Byrne 2007). From the primate perspective, a foraging model based on the simple strategy of minimizing the traveled distance while maximizing food intake has been fitted to movement data from that spider monkey population (Boyer et al. 2006, Ramos-Fernandez et al. 2006). The model makes the fundamental assumptions that the monkeys know the location and size of all potential food patches in their home range, that the sizes of food patches are distributed as an inverse power law, and that the patches themselves are randomly distributed in space as a Poisson process. They found that the distribution of sojourns by spider monkeys could be described by a Levy random walk, which has a higher probability of reaching new, unvisited sites in shorter times and a lower probability of reaching previously visited sites, relative to a Brownian random walk (Boyer et al. 2006). Variation in the patch-size frequency distribution can produce a rich variety of movement behaviors, including ones that fit the real movements of spider monkeys. Their model is elegant in its simplicity, but it is based on far-reaching assumptions that require further validation. Furthermore, simply because a model is capable of generating patterns that correspond to those in nature does not necessarily mean that it has correctly identified the underlying processes responsible for generating those patterns (Bell 2001).

To understand the link between primate activity and seed-deposition patterns, an alternative is to develop more complex mechanistic models. These models should be structured based on spatially explicit field observations of the behavior and movement patterns of primates combined with data on the fecundity of mapped parent trees. One such model has been developed for spider monkeys (*Ateles paniscus*) dispersing *Virola calophylla* in Peru (Russo et al. 2006). It demonstrated that the seed-deposition pattern of *V. calophylla* displayed considerable spatial heterogeneity as a direct result of spider monkeys' behavior of dispersing some seeds in a scattered pattern and others in a highly clumped pattern. This heterogeneity was reflected in the multimodality of the seed-dispersal curve, which was long-tailed and did not display a monotonic decrease with distance from the source tree. Thus, the shape of the seed-dispersal curve was a direct consequence of primate dispersal processes, highlighting the need to incorporate primate behavior directly in seed-dispersal models. The inter- and intraspecific variation in primate traits combined with their interactions with traits of fruiting species suggests that primates produce a diversity of seed shadows. Although developing more complex mechanistic models is highly informative, few models have quantified primate-generated seed shadows in relation to these characters, perhaps because detailed field data are required to parameterize them, which can limit generality. Nevertheless, it is this linkage and understanding the subsequent fate of the primate-dispersed seed, which can be highly variable over space and time (Balcomb and Chapman 2003, Russo and Augspurger 2004), that will ultimately illuminate the effect of primate seed dispersal on plant populations and communities.

SEED DISPERSAL, PLANT POPULATION, AND COMMUNITY ECOLOGY: A REVIEW OF THEORY AND CONCEPTS

Seed dispersal influences plant ecological processes in many ways, primarily because it results in colonization of potential recruitment sites and establishes the initial template of offspring spatial distribution. First, the spatial extent of seed dispersal restricts the suite of potential sites for recruitment (Howe and Smallwood 1982). Second, recruitment of seedlings and saplings often depends on the initial spatial pattern of seed dispersal, such as when survival depends on the distance from a conspecific adult or density of seeds, seedlings, and/or saplings (Janzen 1970, Connell 1971). Such nonrandom survival produces spatial patterns of juvenile and adult plants that differ from what would result simply from random or uniform thinning of the initial template of dispersed seed (Augspurger 1983). As a result, the spatial pattern of dispersed seeds and the consequences of that pattern for seed and seedling survival affect the density and spatial distribution of recruits (Hubbell 1980, McCanny 1985). Furthermore, theory suggests that, when survival is negatively density-dependent, spatial variation in density among subpopulations can modulate population fluctuations at larger scales (Chesson 1996, 1998). In the case of seed dispersal, as seeds become more clumped, the number of seeds surviving to the seedling stage decreases since more seeds experience the higher densities that decrease survival rate (Lloyd 1967), with negative consequences for seedling population growth rate (P. Chesson and S. E. Russo unpublished data). Large-scale spatial variation in seed densities resulting from seed dispersal by primates can thereby influence the demographic stability of plant populations. Thus, studies of primate seed dispersal should be designed to estimate variance in seed density as well as the dispersal processes that produce it. Third, the availability and spatial distribution of sites suitable for establishment combined with the interaction between the environment and a plant's regeneration niche (Grubb 1977) can determine the spatial pattern of seedling and sapling recruitment. This is likely particularly critical for small-seeded and light-demanding species, which recruit best in gaps (Dalling and Hubbell 2002, Dalling et al. 2002), but also for larger-seeded,

shade-tolerant species that have, for example, elevation- or soil-related habitat requirements (Russo et al. 2007).

At the community level, the spatial distribution of individual plants mediates intra- and interspecific interactions, such as competition and density-dependent mortality; and the balance of these interactions affects species coexistence (Pacala and Tilman 1994, Chesson 2000). Limitations on seed dispersal, for example, may promote species coexistence if the inability of a better competitor to disperse to a particular site permits poorer competitors to establish there instead (Levin 1974, Tilman 1994, Hurtt and Pacala 1995). If such dispersal limitation occurs on a community-wide basis, then competitive exclusion and loss of species from a community may be slowed, thereby contributing to the maintenance of plant species diversity (Hurtt and Pacala 1995). In addition, theory has suggested and simulations have demonstrated that the spatial scale of seed dispersal affects properties of plant communities, in particular species richness and the shapes of species-accumulation curves and rank-abundance distributions at community and regional scales (Hubbell 2001, Chave et al. 2002). Thus, understanding the role of seed dispersal in the development of spatial pattern in plant populations is critical to explanations of the structure of plant communities (Levin 1974, Tilman 1994, Hurtt and Pacala 1995, Chesson 2000).

At landscape and larger scales, long-distance seed dispersal mediates the colonization of new habitats (Howe and Smallwood 1982; Cain et al. 2000, 2003). Although long-distance dispersal is notoriously difficult to document (Nathan et al. 2003, 2008), it nonetheless has important implications for migration of plant species under global climate change (Cain et al. 2000, Clark et al. 2003), rates of population spread of invasive species (Clark et al. 2003), and recolonization of fragmented or degraded habitats (Duncan and Chapman 1999, Kaplin and Lambert 2002). Long-distance seed dispersal by primates could occur with wide-ranging species like chimpanzees (*Pan troglodytes*) and mandrills (*Mandrillus sphinx*), when young animals disperse to new groups and carry seeds in their guts from their natal home range, and during seasonal movements.

Seed dispersal at all scales has consequences for population genetics of plants. Ecological mechanisms affecting gene movement play an important role in the development of genetic structure; and in seed plants, gene movement is accomplished by pollen and seed dispersal (Loveless and Hamrick 1984, Sork et al. 1999). Although gene flow from pollen movement can be extensive (Loveless and Hamrick 1984), genetic structure nonetheless may develop within populations as a result of nonrandom patterns of seed dispersal (Epperson and Alvarez-Buylla 1997, Schnabel et al. 1998, Hu and Ennos 1999). Genetic structure has important evolutionary consequences for plants, including effects on viability selection through herbivore- or pathogen-mediated effects on survival (Augspurger and Kelly 1984, Schmitt and Antonovics 1986, Sork et al. 1993), levels of inbreeding depression (Ellstrand and Elam 1993), and effective population size, which influences the dynamics of genetic change (Hedrick 2000).

Population genetic theory predicts that the more spatially restricted seed dispersal is, the more likely plant populations are to develop local genetic differentiation (Wright 1943, Loveless and Hamrick 1984). Many primates disperse seeds long distances, and those that are large-bodied can move large quantities of seeds (McConkey 2000, Stevenson 2000, Dew 2001, Poulsen et al. 2001b, Russo 2003a). Such gene flow would homogenize plant genetic variation at large scales and reduce the probability of local differentiation of plant populations, thus decreasing the probability of the formation of tight coevolutionary relationships (Herrera 2002).

Primates have high functional diversity in terms of the seed-dispersal services they provide. The diversity and heterogeneity of seed shadows they generate means that primate–plant interactions have the potential to influence nearly every aspect of plant population and community ecology.

CONSEQUENCES OF VARIATION IN SEED–DEPOSITION PATTERNS

As demonstrated thus far, many factors influence the initial seed shadows that primates generate. The spatial distribution of dispersed seeds has important consequences for plants in terms of demography and the spatial pattern of recruitment. Two aspects of seed dispersal that warrant special consideration are the scatter versus clump dispersal dichotomy (Howe 1989) and the distances that seeds are dispersed from parent trees.

Scatter Versus Clump Dispersal

Because some primate species spit or defecate seeds in low-density piles and others defecate seeds into high-density clumps, primates lend themselves to an evaluation of the scatter versus clump dispersal hypothesis suggested by Howe (1989) for all fruit-eating animals. Howe (1989) proposed that many tree species are scatter-dispersed by small frugivores that regurgitate, spit, or defecate seeds singly. These offspring recruit as isolated individuals and are unlikely to experience selection for resistance to herbivores, pathogens, or other sources of density-dependent seed or seedling mortality. In contrast, other species are dispersed by large frugivores that deposit large numbers of seeds in a single location. Howe (1989) proposed that these clump-dispersed seeds germinate in close proximity to one another and thus evolve chemical or morphological defenses against seedling predators, pathogens, and herbivores that act in a density-dependent fashion. These processes should therefore be reflected in the spatial distribution of adults, with scatter-dispersed species being randomly or widely dispersed and clump-dispersed species being highly aggregated.

Howe's (1989) hypothesis rests on one critical assumption: that the initial seed-deposition pattern persists after the seed stage so that density-dependent factors can play a role in determining the spatial pattern of recruitment. Given that processes acting after dispersal may alter patterns generated by the dispersers (Herrera et al. 1994, Jordano and Herrera 1995, Schupp and Fuentes 1995, Kollman et al. 1998, Rey and Alcantara 2000, Balcomb and Chapman 2003), this assumption should be evaluated.

Postdispersal seed predation and secondary dispersal can dramatically affect seed survival and, ultimately, seedling recruitment and the dynamics of plant demography (Feer and Forget 2002). In the paleotropics, Lambert (1997, 2001, 2002c) experimentally determined that most seeds do not remain at the site of deposition and that postdispersal processes obscure the initial seed-deposition pattern. Numerous studies cite the importance of rodent seed predation on the fate of dispersed seeds and tree regeneration. For example, DeSteven and Putz (1984) documented the influence of mammalian seed predation on the recruitment of a tropical canopy tree (*Dipteryx panamensis*, Leguminosae) on Barro Colorado Island, Panama. They found that predation of unprotected *D. panamensis* seeds and seedlings exceeded 90% and suggested that predation by seed-eating mammals can be so extensive that even dispersed seeds have little chance of escape (DeSteven and Putz 1984). At Santa Rosa National Park, Costa Rica, 98% of the seeds placed at experimental stations were removed or killed within 70 days (Chapman 1989a). In Peru, 99% of *Virola calophylla* seeds that had been naturally dispersed by spider monkeys (*Ateles paniscus*) or had fallen below the parent tree were preyed upon within 15 months (Russo and Augspurger 2004).

Initial seed-deposition patterns by primates can also be altered by secondary seed dispersal by dung beetles and rodents that cache seeds. Andresen (1999) investigated the fate of monkey-dispersed seeds in Peru and found that 27 species of dung beetle visited the dung of spider monkeys (*Ateles paniscus*) and howler monkeys (*Alouatta seniculus*) and buried 41% of the seeds they encountered. In Uganda, Shepherd and Chapman (1998) documented that dung beetles buried 31% of the seeds placed in experimental primate dung. Clumps of seeds found in chimpanzee defecations rarely remain at the site of deposition after a single day (C. A. Chapman personal observation).

Secondary seed dispersal by rodents, which cache seeds for future consumption, may also alter the initial seed-deposition pattern. However, an important consideration is that the relatively short distances that seeds are secondarily dispersed may not result in substantial modification of the primary seed shadow (Wenny 1999, Brewer and Webb 2001, Russo 2005), although high-density clumps of seeds may be thinned by secondary dispersal. Larger rodents, such as acouchis (*Myoprocta acouchy*), may secondarily disperse seeds up to 124 m away from their original location, although most distances were 20 m or less (Jansen et al. 2002). The dominant effect of secondary dispersal by dung beetles and rodents may be mediated by the burial, which can increase seed survival (Andresen 2001, Brewer 2001), rather than the actual movement of the seeds. Andresen and Levey (2004) documented that burial of seeds by dung beetles decreased the probability of seed predation by rodents by threefold and increased the probability of establishment by twofold. Further quantification of how secondary dispersal modifies primate seed shadows in terms of seed distribution and survival is clearly needed; particularly, more emphasis needs to be placed on distinguishing rodent seed predation from secondary dispersal.

Some experimental studies suggest that whether seeds are scatter- or clump-dispersed may not always predictably influence seed fate. Forget and Milleron (1991) investigated the fate of experimentally dispersed *Virola surinamensis* seeds on Barro Colorado Island, Panama. Using thread-marking methods, these authors observed that agoutis (*Dasyprocta punctata*, a large rodent) scatter-hoarded *V. surinamensis* seeds that they found both singly and in clumps. Seed-removal and seed-burial rates were strongly affected by features of forest habitats, such as *V. surinamensis* tree abundance and/or forest age, but not by seed-dispersal treatment (scattered vs. clumped). Predation of unburied seeds by weevils was independent of habitat and dispersal treatment.

Although some experimental studies suggest that clump-dispersed seeds may not remain in a clumped distribution, a different picture emerges when the fate of seeds that were naturally dispersed by primates is linked to the spatial distributions of plants in later life stages. In a study of *Virola calophylla* in Peru, a clumped pattern of seed deposition was generated by the key dispersal agent, the spider monkey (*Ateles paniscus*) (Russo and Augspurger 2004), which dispersed an estimated maximum of 92% of all dispersed seeds of this tree species (Russo 2003b). The initially clumped seed-deposition pattern was largely maintained through recruitment to the sapling stage. The density of seedlings and saplings was highest where seed fall was greatest, that is, underneath *V. calophylla* parents and at the sleeping sites of spider monkeys. This clumped pattern was maintained even though per capita seed survival to 15 months at these two dispersal site types was the lowest, relative to seeds that spider monkeys dispersed diurnally at low density. In addition, per capita seed survival was negatively dependent on the density of *V. calophylla* seeds and positively dependent on their distance from the nearest female *V. calophylla* tree. Therefore, the clumped recruitment pattern of seedlings and saplings arose despite considerable distance- and density-dependent seed mortality. Thus, spatially aggregated seed dispersal by spider monkeys appeared to play a dominant role in the spatial structuring of this tropical tree population. The interesting challenge that remains is to determine under what conditions the initial seed-deposition pattern generated by primates will persist and what conditions will cause

its disruption. The balance likely lies in the strength of distance- and density-dependent mortality at the seed and seedling stages relative to the spatial scale and strength of seed clumping and in the propensity of primates to disperse seeds to habitats consistent with a plant species' regeneration niche, in addition to interactions with a plant species' competitors and mutualists.

Given that many primates use one or a small number of sleeping sites repeatedly over a number of years (Anderson 1984, Chapman 1989b, Chapman et al. 1989b, Julliot 1997, Russo and Augspurger 2004), these sites are clearly areas of high seed deposition. A similar pattern of repeated defecation may also exist for feeding trees or any other frequently used location (McConkey 2000, Schupp et al. 2002). Such patterns have led some researchers to speculate that defecation at such sites may alter the distribution of adult trees, depending on how many other dispersers contribute to the seed shadow of a tree species. For example, Milton (1980) suggested that the clumped pattern of seeds dispersed by howler monkeys (*Alouatta palliata*) could account for the patchy distribution of food trees in the animals' home ranges. Lieberman et al. (1979) suggested that by defecating near sleeping sites, baboons (*Papio anubis*) increase the concentration of food plants within their core area. However, these studies have not examined the long-term fate of seeds dispersed to sleeping sites, particularly after the seedling and sapling stages. Thus, speculating on the impact of these activities with respect to adult distributions is premature. A study done on hornbills (large, fruit-eating African birds) examined the long-term fate of seeds deposited near nests and found that directed seed dispersal to nests over a decade did not substantially alter sapling community composition at these sites (Paul 2001).

Seed-Dispersal Distance

Janzen (1970) and Connell (1971) suggested that dispersal away from the parent plant enhances survival by removing offspring from mortality factors (e.g., competition, predation, pathogens) acting in a distance- or density-dependent fashion. In one community, negative density-dependent effects on seedling recruitment have been found for every one of the 53 species examined (Harms et al. 2000), but evidence supporting distance- and density-dependent effects is mixed (Clark and Clark 1984, Condit et al. 1992, Hille Ris Lambers et al. 2002, Hyatt et al. 2003). These ideas have led to investigations of the consequences of primates dispersing seeds different distances away from the parent plant (Stevenson 2000, Dew 2001).

Primates disperse seeds to a range of distances, depending in part on seed handling. Seeds that are dispersed through spitting are on average moved only a few meters from the parent tree, whereas those defecated are often moved hundreds of meters from the parent (Figure 34.1, Table 34.1). However, the effect of seed-dispersal distance on the probability of a seed germinating and surviving will likely be variable and species-specific. For example, several species-specific studies examining seedling survival under parent trees have found little or no recruitment under parent trees (Augspurger 1983, 1984), whereas other studies reveal relatively small differences in the probability of survival between seeds under parent trees and those dispersed farther away (Condit et al. 1992; DeSteven and Putz 1984; Chapman and Chapman 1995, 1996a). Howe et al. (1985) found nearly complete mortality (99.96%) within only 12 weeks for *Virola surinamensis* fruit that dropped under the parent. Similarly, Schupp (1988) documented only 7% survival of *Faramea occidentalis* seeds under the crown over 30 weeks in comparison to 24% survival 5 m away from the parent tree. In contrast, Chapman and Chapman (1996a) investigated primate-dispersed trees in Kibale National Park, Uganda, and found that *Uvariopsis congensis* experienced 56% more seed predation when dispersed away from parent trees versus directly under a parent tree.

Hence, dispersal by primates away from parent trees may not always guarantee a higher probability of seed survival. Per capita survival of *Virola calophylla* seeds at sleeping sites where spider monkeys disperse large numbers of seeds is just as low as it is underneath parents (Russo and Augspurger 2004). Although sleeping sites were distant from the nearest female *V. calophylla* tree, the high seed densities that occur there appear to attract seed predators, indicating that distance- and density-dependent mortality can have independent effects on seed survival. In fact, the effect of seed density on seed survival was stronger than that of distance. Thus, for seeds being dispersed to sleeping sites, the value of being dispersed from a parent is reduced in this system.

Understanding whether the strength of distance- or density-dependent survival varies with scale is critical for evaluating the significance of variation among primate species in seed-dispersal distance, particularly that associated with different seed-handling strategies. At least one study has found that the strength of distance- and density-dependent survival drops off dramatically once seeds are beyond the parent crown (Schupp 1988). Furthermore, seed predators, the agents of distance- and density-dependent survival, are likely to respond to different scales of variation of their food resources, depending on their own natural history (Schupp 1992).

Evidence points to substantial interspecific variation in the degree to which distance and density dependence influence seed fate, the scale at which they do so, and the degree to which these factors act independently. Studies should not only determine whether density and distance have separate, significant effects but also quantify and compare the strength of each. Possibly a more significant implication of variation in dispersal distances is a seed's increasing ability to colonize new habitats with increasing dispersal distance (Howe and Smallwood 1982). Newly colonized habitats provide either favorable or unfavorable conditions for establishment, so the underlying spatial arrangement of habitats in a landscape and the degree to which primates cross ecotone boundaries may influence whether longer dispersal distances result in increased seed survival and seedling establishment.

CONSERVATION IMPLICATIONS

Within the last 5–6 years the number of reports that have dealt with implications of forest fragmentation and/or hunting on the seed-dispersal process has increased by at least an order of magnitude. This is particularly well illustrated by a special section of the journal *Biotropica* in 2007 that dealt with hunting of large mammals in tropical forests and its consequences on plant recruitment patterns (Wright et al. 2007a,b). Many of the chapters in this special section used primate examples since primates are one of the dominant large seed dispersers in the tropics (Chapman 1995, Nunes-Iturri and Howe 2007), and many dealt with primates in forest fragments because hunting and fragmentation often go hand in hand.

We view this increased interest as a result of two realizations. First, there has been a realization of the extent of loss of tropical forests. It is estimated that, during the 1990s, 16.1 million ha of forest was lost globally each year to deforestation, of which 15.2 million ha occurred in the tropics (FAO 2005). This will have had significant impacts on primate populations. Other populations were affected by forest degradation (logging and fire), habitat fragmentation, and hunting (Chapman et al. 2003a,b; Peres 1990b). Ironically, while evidence has been available for a long time (Martin, G. H. G., 1983), the magnitude of the impact of hunting has been realized only recently (Fa et al. 2002). For example, Chapman and Peres (2001) estimate that 3.8 million primates are consumed annually in the Brazilian Amazon. The concern about hunting of seed dispersers has been reinforced by a renewed interest in human demographic patterns in the tropics (Jacob et al. 2008, Wright and Muller-Landau 2006). In the past 50 years the human population has grown by 3.7 billion (Potts 2007), and in the next 50 years it is estimated to reach between 8 and 10 billion (United Nations 2005). Between now and 2050, 90% of this growth will occur in developing countries, many of which are tropical and sustain the greatest proportion of the world's biodiversity (Potts 2007). Second, there has been a realization that the effects of fragmentation and hunting on plant recruitment can be significant and that this is a complex issue, one in which academic studies can help provide information useful for conservation (Wright 2003). Nunes-Iturri and Howe (2007) illustrated the potential significance of this process by studying the composition of the juvenile tree community in hunted and nonhunted areas in Peru. In this region hunting exterminates large primates and reduces populations of intermediate-sized species by 80%. This corresponds with a 55% reduction in tree species richness and a 60% reduction in density of species dispersed by large- and medium-bodied primates. Similar patterns of reduced tree species diversity and effects on large-seeded tree species have been found in forests with anthropogenically reduced ateline primate population densities in Colombia (Stevenson and Aldana 2008).

Illustrating the complexity of the issues, Wright (2003) suggests that there are three interrelated issues that determine how severely the hunting of seed dispersers will impact plant communities. First, the types of species hunted and the intensity of hunting will be critical in determining impact. If only large seed-dispersing mammals, such as large-bodied primates, are hunted or disappear from a fragment, then evidence suggests that there will be a decline in large seeded species (Chapman and Onderdonk 1998, Pacheco and Simonetti 2000, Wright et al. 2007b). In contrast, in Mexico and Panama, seedling densities were higher in areas with more completely depleted mammalian communities (Dirzo and Miranda 1991, Wright et al. 2000) because terrestrial seed predators were hunted out and only small rodents remained, which do not prefer large seeds (Dirzo and Mendoza 2007). Second, the impact of fragmentation and/or hunting will depend on the presence of other seed dispersers that can expand their niche to fill the role of the heavily exploited species (Wright 2003). Although there can be overlap in the assemblages of fruiting species that different agents disperse, this overlap may not be enough to promote redundancy when a group of dispersers, such as primates, is lost as a result of hunting (Poulsen et al. 2002). Furthermore, density compensation by smaller primates when densities of larger primates fall (Peres and Dolman 2000) is unlikely to make up for the decline in functional diversity of dispersal services. Evidence of seedling recruitment from hunted and nonhunted sites in Panama suggests that smaller, nonhunted animals are not able to expand their niche to disperse large seeded species. As a result, the seedling communities in hunted areas would become dominated by species dispersed by birds, bats, and wind (Wright, et al. 2007b). Thus, with the loss of primate dispersal agents, rates of seed dispersal may inevitably decrease, seed shadows may become more stereotyped, and tree species diversity may decline to the extent that declines in primate populations also result in decreases in the functional diversity of the dispersal services that they provide

(Stoner et al. 2007). Whether this results in changes in plant populations and communities depends on the dynamics of postdispersal survival underneath parent trees versus at sites where primates would have dispersed seeds. For example, in Uganda and Bolivia, reduced numbers of large-bodied primates were correlated with lower seedling densities of large-seeded forest trees species (Chapman and Onderdonk 1998, Pacheco and Simonetti 2000) and greater seedling aggregation around parent trees (Pacheco and Simonetti 2000). In contrast, in Mexico and Panama, seedling densities were higher in areas with depleted mammalian communities (Dirzo and Miranda 1991, Wright et al. 2000). Third, hunting can alter the future plant community if the harvest changes mutualisms or pest/disease dynamics with particular species. For example, Wright and Duber (2001) presented evidence that the influence of rodent and beetle seed predation changed as a function of distance between hunted and nonhunted sites so that there was increased recruitment near parent trees in hunted areas. They suggested that this increased recruitment near conspecifics could in time reduce tree species diversity through lack of colonization of sites with low mortality. In addition to the demographic consequences, plants may also experience changes in spatial population genetic structure as a result of extinction—caused reductions in the dispersal services of primates (Pachelo + Simonetti 2000, Voigt et al. 2009).

Several authors have argued that the role of primates as seed dispersers is probably particularly important for large-seeded or hard-husked fruit species, which may be inaccessible to smaller, arboreal taxa (Andresen 2000, Kaplin and Lambert 2002). The conservation of primates is therefore argued to be critical to maintaining effective seed dispersal of such species (Andresen 2000). However, in most situations, variation in postdispersal seed fate makes it very difficult to predict accurately how a particular tree species will respond to the removal of its primate dispersers (Balcomb and Chapman 2003). Thus, if models are developed to predict changes in plant populations and communities with the loss of primate dispersal services, they must account not only for changes in the seed shadow but also for the resulting alterations in the transition probabilities between seed, seedling, sapling, and adult stages; the spatial component of these transitions; and the consequences for interspecific interactions among plants.

35

Predation on Primates
Past Studies, Current Challenges, and Directions for the Future

Lynne E. Miller and Adrian Treves

1. Why do we consider predation to be an important evolutionary force in primates? Is predator pressure more important than actual predation itself?
2. Why do we have trouble compiling large data sets of predation occurrences? What kinds of organisms are the main predators on primates?
3. What kind of antipredator behaviors do we see in primates?

INTRODUCTION

Predation has long been thought to be a major selective pressure shaping the morphology and behavior of primates (Alexander 1974, Cheney and Wrangham 1987, Clutton-Brock and Harvey 1977, Crook and Gartlan 1966, Eisenberg et al. 1972, Goss-Custard et al. 1972, Rodman 1973, van Schaik 1983, Struhsaker 1969, Terborgh and Janson 1986). Specific morphological traits, such as cheek pouches in cercopithecines and leaping musculature in galagines, have been attributed to predator avoidance (e.g., Crompton and Sellers 2007, Lambert 2005, Nekaris et al. 2007, Smith, L. W., et al. 2008), along with entire referential communication systems (e.g., Arnold et al. 2008, Blumstein 2001, Kirchhof and Hammerschmidt 2006, Zuberbuhler 2000). Isbell (2006, 2009) has argued that the selective pressures imposed by snake predation have shaped the primates' unique visual and neurological functioning. However, the many challenges of studying primate predator–prey interactions have left us with a paucity of reliable data on actual patterns of predation (cf., Boinski and Chapman 1995, Boinski et al. 2000b, Isbell 1994, Janson 1998c, Treves 1999) and have limited our ability to make rigorous tests of hypothetical relationships. Certainly, more data become available every year, including a recent rise in creatively designed experimental work meant to tease apart the impact of interacting variables, so that the literature available on predators and primate prey seems to be growing exponentially compared with just a decade ago. Such contributions enhance our capacity to understand the subtle nuances of how and when predators exert directional selection and when predation is minor relative to other selective pressures. However, there are many questions left to answer, and therefore, this chapter will again be a call for further attention to this important area of research.

This chapter begins with a brief review of the reports of predation on primates, along with some caveats about using such data for quantitative analyses. We then move on to methodological issues, including the challenges of studying predator–primate interactions, and experimental protocols that might allow researchers to overcome some of these challenges. From there, we will discuss the growing body of data regarding the behaviors associated with predation risk, including the formation of social groups; manipulation of foraging, sleeping, and ranging patterns; and vigilance and alarm calling. This discussion will include reference to the variables that may mediate an individual's vulnerability to predation, which may in turn influence its willingness to take risks. By this approach, we hope to encourage multifactorial analyses of predator–primate interactions and, thus, begin to develop a more realistic model of this complex relationship. The chapter will close with a brief discussion of the role of humans as predators of nonhuman primates, addressing the extent to which humans represent a special category of predators. Although hard data on predator–primate interactions are few, there is an enormous body of literature on the myriad factors related to predation. In a chapter of this length, it is impossible to review all of these studies; thus, our aim is to provide an overview of current issues, along with a review of classic and recent publications in each area.

ACCOUNTS OF PREDATOR–PRIMATE INTERACTIONS

There are numerous reports of predatory attacks on primates. Recent publications have summarized such accounts

(Boinski et al. 2000b; see also Enstam 2007 on cerco-pithecines, Urbani 2005 on platyrrhines), and Hart (2007, and Hart and Sussman 2005) has added to the available publications with data from surveys of field primatologists. Table 35.1 presents an updated list of prey and predator taxa

from the published literature. We have included accounts of successful and unsuccessful attacks, along with a few data based upon scat and nest debris (though we recognize that these remains may represent scavenging behavior rather than predation). We have not included accounts of attacks

Table 35.1 Reports of Primates as Prey

PREY GENUS	PREDATOR SPECIES	REFERENCES
Lemur	Red-fronted lemur (*Eulemur fulvus rufus*)	Jolly et al. 2000
	Fossa (*Cryptoprocta ferox*)	Goodman et al. 1993
	Civet (*Civettictis civetta*)	Goodman et al. 1993
	Feral cat (unknown)	Goodman et al. 1993
Eulemur	Henst's goshawk (*Accipiter henstii*)	Goodman et al. 1998 Karpanty 2003 Overdorff and Strait 1995 Schwab 1999
	Fossa (*Cryptoprocta ferox*)	Wright et al. 1998
Hapalemur	Madagascar harrier-hawk (*Polyboroides radiatus*)	Karpanty 2003
	Henst's goshawk (*Accipiter henstii*)	Karpanty 2003
	boa constrictor (*Boa manditra*)	Rakotondravony et al. 1998
	fossa (*Cryptoprocta ferox*)	Wright et al. 1998
Avahi	Henst's goshawk (*Accipiter henstii*)	Goodman et al. 1998 Karpanty 2003
	Madagascar buzzard (*Buteo* spp.)	Wright et al. 1998
Propithecus	Madagascar harrier-hawk (*Polyboroides radiatus*)	Brockman 2003
	Boa constrictor (*Acrantophis madagascariensis*)	Burney 2002
	Wild cat (*Felis silvestris*)	Brockman et al. 2008
	Fossa (*Cryptoprocta ferox*)	Wright et al. 1997 Wright et al. 1998
Lepilemur	Madagascar harrier-hawk (*Polyboroides radiatus*)	Schulke and Ostner 2001
Cheirogaleus	Madagascar harrier-hawk (*Polyboroides radiatus*)	Gilbert and Tingay 2001
	Henst's goshawk (*Accipiter henstii*)	Karpanty 2003
	Madagascar buzzard (*Buteo* sp.)	Wright et al. 1998
	Madagascar boa (*Sanzinia madagascariensis*)	Wright et al. 1998
	Ring-tailed mongoose (*Galidia elegans*)	Wright et al. 1998
Microcebus	Madagascar harrier-hawk (*Polybroides radiatus*)	Karpanty 2003 Wright et al. 1998
	Henst's goshawk (*Accipiter henstii*)	Karpanty 2003
	Ring-tailed mongoose (*Galidia elegans*)	Wright et al. 1998 Deppe et al. 2008

Table 35.1 (cont'd)

PREY GENUS	PREDATOR SPECIES	REFERENCES
Galago	Genet (*Genetta* sp.)	Andrews and Evans 1979
	Chimpanzee (*Pan troglodytes*)	Gaspersic and Pruetz 2004
	Bonobo (*Pan paniscus*)	Hohmann and Fruth 2008
Nycticebus	Reticulated python (*Python reticulatus*)	Wiens and Zitzmann 1999
	Orangutan (*Pongo pygmaeus*)	Utami and van Hooff 1997
Tarsius	Reticulated python (*Python reticulatus*)	Gursky 2002a
Callithrix	Pit viper (*Bothrops jararaca*)	Correa and Coutinho 1997
	Tayra (*Eira barbara*)	Bezerra et al. 2008
Saguinus	Ornate hawk eagle (*Spizaeus ornatus*)	Terborgh 1983
	Bicolored hawk (*Accipiter bicolor*)	Terborgh 1983
	Crested eagle (*Morphnus guianensis*)	Oversluijs Vasquesz and Heymann 2001
	Anaconda (*Eunectes* spp.)	Heymann 1987
	Boa constrictor (*Boa constrictor*)	Tello et al. 2002
	Ocelot (*Felis pardalis*)	Heymann 1990
	Tayra (*Eira barbara*)	Buchanan-Smith 1990 Galef et al. 1976
Saimiri	Harpy eagle (*Harpia harpyja*)	Terborgh 1983
	Guianan crested eagle (*Morphnus guianensis*)	Terborgh 1983
	Ornate hawk eagle (*Spizaeus ornatus*)	Terborgh 1983
	Black-and-white hawk eagle (*Spizastur melanoleucos*)	Terborgh 1983
	Slate-colored hawk (*Leucopternis shistacea*)	Terborgh 1983
	Bicolored hawk (*Accipiter bicolor*)	Terborgh 1983
	Tayra (*Eira barbara*)	Galef et al. 1976
Callicebus	Guianan crested eagle (*Morphnus guianensis*)	Terborgh 1983
	Ornate hawk eagle (*Spizaeus ornatus*)	Terborgh 1983
	Tufted capuchin (*Cebus apella*)	Sampaio and Ferrari 2005
	Boa constrictor (*Boa constrictor*)	Cisneros-Herediam et al. 2005
Cebus	Harpy eagle (*Harpia harpyja*)	Ford and Boinski 2007 Fowler and Cope 1964 Izor 1985 Terborgh 1983
	Guianan crested eagle (*Morphnus guianensis*)	Terborgh 1983
	Boa constrictor	Chapman 1986

Table 35.1 (cont'd)

PREY GENUS	PREDATOR SPECIES	REFERENCES
	(*Boa constrictor*)	Bianchi and Mendes 2007
	Ocelot	
	(*Leopardus pardalis*)	Defler 1980
	Tayra	Phillips 1985
	(*Eira barbara*)	Oliveira et al. 2008
	Domestic dog	
	(*Canis lupus familiaris*)	
Chiropotes	Harpy eagle	Martins et al. 2005
	(*Harpia harpyja*)	
Pithecia	Harpy eagle	Ford and Boinski 2007
	(*Harpia harpyja*)	Rettig 1978
	Crested eagle	Gilbert 2000
	(*Morphnus guianensis*)	
Ateles	Crested eagle	Julliot 1994
	(*Morphnus guianensis*)	
	Jaguar	Emmons 1987
	(*Panthera onca*)	Matsuda and Izawa 2008
	Puma	Chinchilla 1997
	(*Felis concolor*)	Matsuda and Izawa 2008
Brachyteles	Jaguar	Olmos 1994
	(*Panthera onca*)	
	Ocelot	Bianchi and Mendes 2007
	(*Leopardus pardalis*)	
Alouatta	Harpy eagle	Eason 1989
	(*Harpia harpyja*)	Fowler and Cope 1964
		Izor 1985
		Peres 1990a
		Rettig 1978
		Sherman 1991
		Terborgh 1983
	Jaguar	Chinchilla 1997
	(*Panthera onca*)	Cuaron 1997
		Peetz et al. 1992
	Puma	Chinchilla 1997
	(*Felis concolor*)	Ludwig et al. 2007
	Ocelot	Bianchi and Mendes 2007
	(*Leopardus pardalis*)	
	Tayra	Phillips 1985
	(*Eira barbara*)	
Cercopithecus	African crowned eagle	Brown 1962 and 1971
(*Chlorocebus*)	(*Stephanoaetus coronatus*)	Maisels et al. 1993
		Sanders and Mitani 2000
		Sanders et al. 2003
		Schultz et al. 2004
		Struhsaker and Leakey 1990
	Leopard	Cheney et al. 1981
	(*Panthera pardus*)	Isbell 1990
		Zuberbuhler and Jenny 2002
	Chimpanzee	Gaspersic and Pruetz 2004
	(*Pan troglodytes*)	Goodall 1986
		Mitani and Watts 1999
		Newton-Fisher et al. 2002
		Nishida et al. 1992
		Pobiner et al. 2007
	Bonobo	Surbeck and Hohmann 2008
	(*Pan paniscus*)	
Cercocebus	African crowned eagle	Sanders and Mitani 2000
(*Lophocebus*)	(*Stephanoaetus coronatus*)	Schultz et al. 2004
	Yellow baboon	Struhsaker and Leakey 1990
	(*Papio cynocephalus*)	

Table 35.1 (cont'd)

PREY GENUS	PREDATOR SPECIES	REFERENCES
	Leopard (*Panthera pardus*)	Hausfater 1976
	Chimpanzee (*Pan troglodytes*)	Zuberbuhler and Jenny 2002
		Pobiner et al. 2007
Papio	Black eagle (*Ictinaetus malayensis*)	Boshoff et al. 1991
	Spotted hyena (*Crocuta crocuta*)	Stelzner and Strier 1981
	Leopard (*Panthera pardus*)	Busse 1980
	Lion (*Panthera leo*)	Busse 1980
	Chimpanzee (*Pan troglodytes*)	Condit and Smith 1994 Pobiner et al. 2007
Macaca	Crocodile (*Crocodylus porosus*)	Galdikas and Yeager 1984
	Leopard (*Panthera pardus*)	Ramakrishnan et al. 1999 Seidensticker 1983.
Colobus	Crowned eagle (*Stephanoaetus coronatus*)	Sanders and Mitani 2000 Struhsaker and Leakey 1990
	Leopard (*Panthera pardus*)	Zuberbuhler and Jenny 2002
	Chimpanzees (*Pan troglodytes*)	Mitani and Watts 1999 Newton-Fisher et al. 2002 Pobiner et al. 2007
Procolobus	Crowned eagle (*Stephanoaetus coronatus*)	Sanders and Mitani 2000
	Leopard (*Panthera pardus*)	Zuberbuhler and Jenny 2002
	Chimpanzee (*Pan troglodytes*)	Boesch and Boesch 1989 Boesch 1994a–c Goodall 1986 Mitani and Watts 1999 Nishida et al. 1992 Pobiner et al. 2007 Stanford et al. 1994 Wrangham and van Zinnicq BergmannRiss 1990
Presbytis (*Semnopithecus*) (*Trachypithecus*)	Python (*Python reticulatus*)	Shine et al. 1998
	Jackal (*Canis aureus*)	Stanford 1989
	Leopard (*Panthera pardus*)	Ramakrishnan et al. 1999 Seidensticker 1983
Nasalis	Clouded leopard (*Neofelis nebulosa*)	Jackson and Nowell 1996 Matsuda et al. 2008
Hylobates	Leopard (*Panthera pardus*)	Rabinowitz 1989
Symphalangus	Python (*Python* spp.)	Schneider 1906, cited in Uhde and Sommer 2002
Pan	Leopard (*Panthera pardus*)	Boesch 1991 D'Amour et al. 2006 Furuichi 2000 Zuberbuhler and Jenny 2002
	Lion (*Panthera leo*)	Tsukahara 1993
Gorilla	Leopard (*Panthera pardus*)	Fay et al. 1995 Henschel et al. 2005

among conspecifics (e.g., accounts of intraspecific infanticide) for these fall outside of our proposed discussion; however, we have included those cases where primates have preyed upon members of other primate species, for example, chimpanzees (*Pan troglodytes*) preying upon red colobus monkeys (*Procolobus badius*, a.k.a. *Piliocolobus badius*). We have chosen not to include human predators in this summary, postponing this issue for the end of the chapter. Besides the accounts of attacks summarized here, many additional publications have described other types of interaction between primates and potential predators, interactions that might include flight, mobbing, alarm calling, etc. Those are not included in the table but are discussed below.

We hope that this summary table is useful to our readers. It certainly demonstrates that diverse primate taxa (from *Microcebus* to *Gorilla*) are attacked by a variety of predators (especially large raptors, felids, and reptiles) and that it is reasonable to predict an enormous array of antipredator adaptations. However, while such a summary may serve to generate hypotheses, there are major problems in using it to test such hypotheses. First, the data set is surely incomplete because we must have missed some relevant publications. Second, while it summarizes the genera of primates and their predators, it does not indicate the number of predatory incidents for any given genus because such data are often lacking. Third, it only weakly represents the actual distribution of predators across primate taxa because of the rarity of observation of predation (particularly for rare, arboreal, or nocturnal primates and those facing predators wary of humans) and the fact that more predation events are likely to have been observed for those primate taxa that have been the subject of long-term observational research. For example, a glance at the table would suggest that *Alouatta* is much more vulnerable than *Callithrix* or *Tarsius*, as indicated by the relative numbers of predation accounts; however, this disparity more probably represents a difference in observer hours in the field. Thus, we must be very cautious about drawing conclusions from such anecdotal data.

A fourth major difficulty with analyses based upon anecdotal accounts is the lack of contextual data. Such accounts rarely include details about the predator's attack strategy or relevant information about the age, sex, social status, microhabitat, or behavior of the victims and/or the survivors; therefore, these accounts shed little light on the relative vulnerability of the victims or survivors. Furthermore, even a perfectly observed and fully documented event might reveal nothing about the selective pressure imposed by predation because the event might be random with regard to the genotype or phenotype of the prey. Only with large samples of predator encounters can we hope to understand how predation imposes directional selection on a population and therefore might shape the evolution of behavior. Even studies that have generated adequate sample sizes (such as Stanford's work with red

colobus predation by chimpanzees: Stanford 1995, 1998b, 2002) often provide no more than a glimpse of whether a particular behavior or morphological feature predisposes certain individuals to predation or loss of offspring, prerequisites for selection to be acting. Overall, collections of anecdotal data are unlikely to contribute significantly to our understanding of the selective role of predation.

Despite these problems, several authors have used published data on rates of predation and/or predator encounters in selected populations to produce cross-taxonomic analyses, particularly to isolate and elucidate the role of variables like body size and group size (Anderson 1986a,b; Hill and Dunbar 1998; Hill and Lee 1998; Isbell 1994; Lycett et al. 1998; see also Boinski and Chapman 1995 for a critique of these approaches). These attempts have produced some conflicting results. One factor that may explain conflicting results is the relative selective impact of predation rate versus predation risk (Cowlishaw 1997b, Dunbar 1988, Hill and Dunbar 1998, Janson 1998c, Young 1994). Predation *rate* is the frequency of successful predatory attacks for a given group, population, species, etc. Predation *risk* represents the number of encounters, including unsuccessful attacks. Predation rates are hugely variable, across taxa and over time. Cheney and Wrangham (1987) estimated that predation rates vary from 0% to 15%, with an average of 3%, though Isbell (1990) witnessed a period in which 45% of vervets were preyed upon by leopards and Stanford (1995) documented 25% mortality caused by predation in a population of red colobus (see also Fourrier et al. 2008, Teelen 2008). However, even where rates of successful predation are low, frequent encounters will increase an individual's perception of risk, and this alone might stimulate significant behavioral response. In addition to predation rate and perceived risk, individual vulnerability must be considered. *Vulnerability* reflects the individual's experience, wariness, and a variety of biological and social variables not necessarily shared by associates (Miller 2002a). There is an element of chance in predator encounters such that any individual can potentially fall prey but vulnerable ones are more likely to be taken. It is likely that the significant impact predation can have on the inclusive fitness of slowly reproducing animals serves to make primates highly responsive to predators (Abrams 1994, Altmann 1974, Hill and Dunbar 1998, Janson 1992, Lima 1998, Lima and Dill 1990).

THE STUDY OF PREDATOR–PRIMATE INTERACTIONS

As we have noted, predation is notoriously difficult to observe. Events happen quickly, and many predators are active at night, when observation conditions are poor. In addition, human observers may have a significant influence on predator behavior, deterring approach and attack (Isbell and Young 1993a). Another challenge to studying predation and its impact on primate evolution is the modern, altered environments in which we are making our observations; these often differ significantly from the environments in which primates evolved. Human transformation of primate habitats and active elimination of many predators (Berger et al. 2001, McDougal 1987, Treves and Naughton-Treves 1999) have surely altered patterns of predation. Human modification of habitats or depletion of predator populations can raise, lower, or leave unchanged the rate and risk of predation. For example, in south Asia, leopards preyed on primates less frequently after human activities eliminated tigers from the ecosystem (Seidensticker 1983). Humans sometimes remove virtually all the predators of primates, leading to widespread reduction in risk (Southwick et al. 1965), while at other times humans replace wild predators and impose unprecedented rates of predation on primate populations (Naughton-Treves et al. 2003). Given such variation, generalizations are not possible. We must achieve a more thorough and nuanced understanding of predator–prey ecology if we are to understand the evolutionary role of predation. Moreover, we cannot assume that morphological or behavioral characteristics that exist today, even those that clearly serve antipredator functions, were originally selected to serve this purpose (Byers 1997, Rodman 1988).

Although several projects have produced important data on predation of certain primate taxa (Cheney et al. 1988 on vervets; Wright et al. 1998 on prosimians; Newton-Fisher et al. 2002, Pobiner et al. 2007, Stanford et al. 1994 on predation by chimpanzees on other primate species), a full understanding of predation as a selective pressure in the evolution of primate behavior and morphology continues to be impeded by a lack of comprehensive data and detailed understanding of ecosystem complexity. Accurate accounts of predator–primate encounters, even those that result in no mortality, are very informative. Primatologists should be encouraged to make detailed observations of predator attack patterns and the microhabitats in which they occur. Important factors include the age, sex, condition, rank, social context, and spatial location of as many victims and survivors as possible. Long-lasting changes in the behavior of survivors, such as avoidance of that portion of the range, increased group cohesion, vigilance, or alarm calling, and whether such postattack behaviors diminish over time are also highly relevant.

Insights gained by careful observation of primates as prey can be enhanced by observing the predators that threaten them; thus, we encourage field researchers to follow and observe the predators themselves, rather than limiting study to the prey. These data can be augmented by studies of raptor nest remains and carnivore scat (though such remains could represent scavenging rather than predation). Recent efforts include those by Karpanty (2003, 2006), Karpanty and Goodman (1999), McGraw et al. (2006), Mitani et al. (2001), Sanders et al. (2003), Shultz and Noe (2002), Shultz et al. (2004), and Zuberbuehler (2007). Observers have been able to estimate the percentage of a predator's diet represented by primates (Rettig 1978,

Skorupa 1989, Struhsaker and Leakey 1990), the proportion of a primate population annually consumed by predators (Goodman, S. M., et al. 1993, 1998; Struhsaker and Leakey 1990), and even the correlation between predator rates and prey morphology and behavior (Shultz et al. 2004, Pobiner et al. 2007, Zuberbuehler 2007). Such quantitative approaches provide valuable data for population-specific tests of theory and the formulation of new hypotheses; they have the potential to revolutionize our understanding of predation on primates.

Quasi-experimental studies are those in which the research fortuitously coincides with predator introduction or removal or a significant change in predation pressure. Several such cases have been used to understand prey responses among both primate and nonprimate taxa (e.g., Anderson 1989, 1992; Berger et al. 2001; Gil-da-Costa 2007; Gouzoules et al. 1975; Hunter and Skinner 1998; Isbell 1990; Isbell and Young 1993b; Seidensticker 1983; Southwick et al. 1965; Terborgh et al. 2002a). Another approach involves a comparison across sites with different levels of predation pressure (cf., Marsh 1979a,b; Pruetz et al. 2008; van Schaik and Horstermann 1994; van Schaik and van Noordwijk 1985; Stephan and Zuberbuehler 2008; Struhsaker 2000). Of course, care must be taken to assure that the prey are responding to predation and not some other variable in the environment. Thus, extensive ecological data are essential to understand the impacts of predators in such quasi-experimental situations.

True experimental manipulation of predation risk offers a unique opportunity to isolate the role of predators in prey behavior. Simply put, the observer controls the timing and perhaps the location of encounters. When encounters are simulated, confounding variables must be monitored carefully. These studies need controls that present nonpredatory stimuli in a similar modality or they may fail to discriminate antipredatory behavior from xenophobic or neophobic behavior. There has been ample research on this topic, including the landmark work of Cheney and Seyfarth (1990a). More recent studies include those by Arnold et al. (2008), Bshary and Noë (1997a,b), Chapman and Chapman (1996b), Coss et al. (2005, 2007), Gebo et al. (1994), Gil-da-Costa et al. (2003), Gursky (2005b, 2007b), Miller (2008), Papworth et al. (2008), Ramakrishnan et al. (2005), Stone (2007), Treves (1999), Wheeler (2008), Zuberbuhler (2000), and Zuberbuehler et al. (1997, 1999). Predator simulations can be time-consuming and expensive, and they must be used cautiously to avoid the confounding effects of social learning by the subjects; however, simulations have the greatest potential for explicit discrimination of hypotheses about current selective pressures or the adaptive function of behavior.

BEHAVIORS RELATED TO PREDATOR PRESSURE

Predator–prey interaction may involve a series of moves and countermoves as the predator seeks to capture and consume prey and the prey seeks to evade consumption (Bednekoff and Lima 1998, Kerfoot and Sih 1987, Sih 1987). However, there is no reason to presuppose that a given predator will employ the same attack strategy for all prey species in its range (Lima 2002). Consequently, it is possible that different prey species will respond differently to the same predator. Effective responses will depend on a suite of biological characteristics (such as body size or visual acuity), ecological factors (such as distribution of refuges), and social factors (such as group size and collective response), all of which can vary significantly across species, populations, groups, and even individuals. Moreover, it is possible that an individual primate will respond differently to different predators (Cheney and Seyfarth 1990a; Goodman et al. 1993; Gursky 2003a, 2007b; Macedonia and Evans 1993; van Schaik and Kappeler 1996; Treves 1999; Zuberbuehler 2007). Prey responses will depend in part on characteristics of the predator's attack pattern; obviously, an animal will respond differently to avoid an aerial raptor versus a large, terrestrial felid. Prey species with many potential predators may have developed either an extensive repertoire of specific antipredator behaviors or a few generalized responses that are effective against a variety of predators. Some of these will be more labile than others (e.g., alarm calling can be used when effective and stifled when ineffective, whereas a rapid change of group size is generally less readily accomplished). The multiplicity of variables that are relevant to predator–prey interactions place heavy burdens on researchers to collect sufficient data to conduct the multifactorial analyses that are appropriate to this complex problem.

Predation and Social Structure

Predation has long been thought to play a major role in the formation of social groups (Anderson 1986b, Busse 1977, Jarman 1974, Pulliam and Caraco 1984, van Schaik 1983, van Schaik and van Hooff 1983). Individuals in larger social groups are expected to be safer than those in smaller groups because of dilution effects (Hamilton 1971); the added vigilance of many eyes, leading to improved predator detection (Elgar 1989, Pulliam 1973, Rodman 1973, Triesman 1975); or the improved success of mobbing (Altmann 1956, Curio 1978). While many have asserted a strong relationship, others have questioned the role of predation in the formation of primate groups (van Schaik and Kappeler 1996, Smuts 1985, Treves 2000b, Treves and Chapman 1996, Wrangham 1979b, Wright 1998).

Scientists working with nonprimate species have produced elegant tests of the hypothesis that predation results in aggregation (Caraco et al. 1980, Cresswell 1994, Lindstrom 1989, Wrona and Dixon 1991). On the other hand, some studies, including a few on primates, have reported disaggregation following an increase in predation risk (Anderson 1992, Fitzgibbon 1990, Lima 1993, Treves 1999, Wolf 1985). In any case, the data for primates are largely circumstantial;

that is, many studies have explored the extent to which group size is related to antipredator behaviors, such as vigilance, but it has been difficult to isolate the relationship between group size and predation itself. Classic studies include those comparing primate populations in areas of high versus low predator density (Marsh 1979a, van Schaik and Horstermann 1994, van Schaik and van Noordwijk 1985, Struhsaker 2000) or the response to inferred changes in predator pressure (Isbell 1990, Isbell and Young 1993b). Further attempts to demonstrate a link between predation and group size include cross-taxonomic analyses of published data, some of which have found the predicted positive correlation (Dunbar 1988, Hill and Lee 1998, Janson and Goldsmith 1995, van Schaik 1983) while others have not (Hill and Dunbar 1998, Isbell 1994). For all such meta-analyses, confounding variables (such as the specific array of predators or the distribution of food resources) often cloud these patterns. The differences often depend on methods of assessing predation rate and predation risk and the types of taxa included in the analysis. In certain socioecological settings, larger groups may actually impose greater predation risk and rate (Wright 1998, Zuberbuehler 2007). In one of the most clearly documented investigations of predator–prey interaction, Stanford (1998b, 2002) has shown that larger groups of red colobus monkeys are attacked more frequently by chimpanzees and that high rates of successful predation upon immatures actually depress group size for this prey species (Stanford 1995, Krüger et al. 1998). On the other hand, Shultz et al. (2004) demonstrated that, for one African primate community, the relationship between group size and predation rate depended heavily on the type of predator and the specific prey habitat. Thus, we must avoid overgeneralizations. It is inappropriate to conclude that predator pressure is always positively correlated with group size or that large groups are solely attributable to predator pressure. We must develop a more sophisticated model of predator–prey relationships as they pertain to the adaptive value of social aggregations.

A clearer (though not undisputed) relationship exists between predation risk and the formation of mixed-species groups. Mixed-species groups have been observed among various primate taxa, most notably the callitrichids and the cercopithecids. This phenomenon has commonly been attributed to foraging efficiency and/or to predation risk (Caine 1993; Chapman and Chapman 1996b; Cords 1990a,b; Garber 1988a; Gautier-Hion et al. 1983; Hardie 1998; Hardie and Buchanan-Smith 1997; Heymann 1990; Heymann and Buchanan-Smith 2000; Norconk 1990; Peres 1993a; Porter and Garber 2007; Struhsaker 1981a, 2000; Teelen 2007; Terborgh 1990; Waser 1987). Garber and Bicca-Marques (2002) have argued that the literature on tamarins reveals no consistent differences in predation risk for members of single- vs. mixed-species troops (see also Chapter 7). However, Bshary and Noë (1997b) have demonstrated experimentally that red colobus monkeys are more likely to initiate and maintain associations with

Diana monkeys (*Cercopithecus diana*) when under conditions of heightened predation risk (see also Noë and Bshary 1997, McGraw and Bshary 2002). In addition, Sauther (2002) found that smaller groups of ring-tailed lemurs (*Lemur catta*) were more likely to join groups of Verreaux's sifaka (*Propithecus verreauxi*), particularly when vulnerable infants were present (see also Gould and Sauther 2007). Thus, some primates may achieve safety in numbers by joining with members of other species. However, the variability described here may reflect variation in the predators' methods of prey selection, the vulnerability of prey in different habitats, and the behavioral tactics used for attack and defense (Fitzgibbon 1990).

Similarly, early works proposed that predation pressure played a role in the formation of multimale groups (Anderson 1986b, Crook and Gartlan 1966, Devore and Washburn 1963, Hamilton et al. 1978). Males may serve an important role in vigilance and defense (Anderson 1986b, Baldellou and Henzi 1992, Chapman and Chapman 1996b, Gautier-Hion and Tutin 1988, Gursky 2005b, Rose 1994a, Rose and Fedigan 1995, van Schaik and van Noordwijk 1989, Stanford 1995). However, there is still debate about the relationship between predation risk and the proportion of males in a group. Cheney and Wrangham (1987) found no evidence that predation rate influenced the number of males in groups of vervets (*Cercopithecus aethiops*) or red colobus; Struhsaker (2000) showed the same for certain populations of African forest monkeys. van Schaik and Horstermann (1994) attributed multimale groups of colobus and howlers to eagle predation and the unimale groups of langurs to relaxed predation; however, many other ecological variables could provide alternative explanations for this disparity. Hill and Lee (1998) did a multivariate analysis of 121 populations of 39 primate species and found that, for multimale groups, the proportion of males was positively correlated with predation risk. Stanford (1998b) found a similar pattern in a handful of red colobus populations. Mitani et al. (1996) concluded, from a cross-species analysis, that the number of males in a group was most closely related to the number of females but that this, in turn, may be influenced by multiple ecological factors including, but not limited to, predator risk. Group composition is influenced by many ecological and social variables (Andelman 1986, Dunbar 1988, Hill and Lee 1998, Mitani et al. 1996, Treves and Chapman 1996, Van Belle and Estrada 2008). Isolating and evaluating the role of just one of these variables is unlikely without experimental controls.

The view that predation has influenced primate social evolution is persistent despite vexing difficulties in producing convincing scientific tests. We share the view that primates associate for safety, but we encourage appropriate tests of several basic assumptions. First, do larger primate groups detect predators earlier? This is by no means clear for nocturnal primates or for the large groups of arboreal primates found across the world, where vigilant animals are continually distracted by associates and where visual obstructions from the trees often occlude clear lines of sight (Treves 2002). An effective test of this basic relationship will require inventive field experiments because we cannot equate detection with flight (Ydenberg and Dill 1986). Other measures of threat detection, such as individual heart rate or pupil dilation, will be more convincing. Second, are alarm calls produced more often or more quickly by larger groups, and do they improve the safety of listeners? The first proposition should be easy to test, but the trick will always be to distinguish antipredator alarms from anticonspecific alarms or false alarms. Furthermore, determining if alarm signals improve the rate of escape will require sophisticated field experiments, comparing situations where an alarm is given with situations where danger is equal but no alarm is given, perhaps requiring the videotaped responses of many individuals. Third, do larger groups actually attract predators? Because many primatologists can find large primate groups more easily than small ones from their noise or other signs, it has long been assumed that predators do the same. Collaboration with predator researchers will be needed to assess this relationship for any given population. Fourth, does per capita predation rate decline with group size, and does it do so for all age/sex classes similarly? This question may be testable only in a population suffering high rates of predation until we devise methods of experimentally simulating the predatory targeting of wild primates.

Risk, Vulnerability, and Predator-Sensitive Behaviors

Predator-sensitive behaviors are, in effect, antipredator behaviors in the absence of a predator. Individuals may alter their patterns of foraging, sleeping, and ranging to reduce the probability of encountering a predator (reducing risk). The extent to which they do so depends on multiple biological and social variables that relate to individual vulnerability. The interplay between risk and vulnerability produces a complex set of adaptive behaviors.

Certain environmental variables are clearly linked to risk, including abundance of refuges, height and density of grass, the presence of protective versus obstructive cover, and the intensity of moonlight (Cheney and Wrangham 1987; Cowlishaw 1997a; Dunbar 1988, 1996; Enstam 2007; Gursky 2003a; Hill and Weingrill 2007; Isbell 1994; Nash 2007; Treves 2002). As individuals move through the microhabitats of their ranges, their risk may change throughout the day.

Many variables have been hypothetically linked to vulnerability, but few have been unequivocally linked to predation rate. For example, it has long been suggested that smaller-bodied species are taken by predators more often than larger-bodied species, and this may be true in the context of smaller-bodied and/or aerial predators (e.g., Karpanty and Wright 2007); however, several recent studies suggest that larger-bodied and/or terrestrial predators may actually target larger-bodied prey species (Dollar et al. 2007, Pobiner et al. 2007, Zuberbuehler 2007, Zuberbühler and Jenny 2002, Urbani 2005). Similarly, immatures are expected be more vulnerable than adults; this pattern has been supported by data on red colobus monkeys (Pobiner et al. 2007, Stanford 2002, Struhsaker and Leakey 1990, Tappen and Wrangham 2000), but studies show no such bias for other primate species (Cheney et al. 1981, Mitani et al. 2001, Struhsaker and Leakey 1990). Females may be more vulnerable than males, in part because of smaller body size. Cowlishaw (1997a) showed greater risk to female baboons (genus *Papio*), and Pobiner et al. (2007) showed that female red colobus were taken by chimpanzees more frequently than expected, given their representation in the population; however, Mitani et al. (2001) showed that males more than females were represented in crowned eagle nest remains (see also Olupot and Waser 2001a). Females may also be vulnerable because of their proximity to and encumbrance by immatures (Busse 1984, Collins 1984), but so far only one study supports this hypothesis (Boinski 1987b, 1992). Social variables such as rank may influence vulnerability. Cheney et al. (1988) showed that high-ranking male vervets were more likely to be eaten at water holes than were their lower-ranking conspecifics. Group size and composition may also affect vulnerability, but convincing quantitative data are lacking. An individual's vulnerability may change over the course of its lifetime, and predator-sensitive decisions are expected to shift accordingly.

Foraging is one behavior that is predicted to be predator-sensitive; that is, animals may sacrifice foraging success in order to minimize predation risk, and more vulnerable animals might show greater predator sensitivity (Abrams 1994; Altmann 1974; Lima 1987; Lima and Dill 1990; Schoener 1971; Sih 1987, 1992). Many studies of nonprimate taxa have shown clear correlations between foraging patterns and specific variables of vulnerability (see Miller 2002a for review). The data for primates are less conclusive but still compelling. Cowlishaw (1997a) demonstrated that baboons left refuges principally when traveling and feeding; additional work (Cowlishaw 1997b) showed that baboons opted for low-risk but low-quality feeding sites, rather than exploiting high-quality but high-risk patches. Similarly, Stone (2007) found that juvenile squirrel monkeys eschewed high-risk feeding platforms (those with lower cover), even when a high-quality reward was available. In contrast, Prescott and Buchanan-Smith (2002) showed that red-bellied tamarins (*Saguinus labiatus*) in a captive setting would forage close to the ground (which was assumed to represent higher risk) when a greater food reward was offered. In experimental work, Miller (2008) showed that, for wedge-capped capuchin monkeys, latency to feed increased significantly with greater risk (exposure to a model predator) and that, in the highest-risk conditions, members of larger groups were quicker to feed than were those in smaller groups. Cheney et al. (1988) showed that water holes were visited out of necessity despite the high risk of predation. Such trade-offs are the central issue in studies of predator-sensitive foraging. Numerous anecdotal reports support the existence of

such a trade-off. For example, Uhde and Sommer (2002) reported that, in the wake of a tiger encounter, gibbons (*Hylobates lar*) clustered together and delayed foraging for several hours. While many studies have inferred predator impact upon foraging behavior and morphology (Cords 2002a, Gould and Sauther 2007, Hill and Cowlishaw 2002, Lambert 2005, Miller 2002b, Overdorff et al. 2002, Sauther 2002), few have examined the trade-off experimentally to control the multiple potentially confounding factors that affect daily behavior. As primatologists increase their attention to these complex patterns, more fine-tuned testing of the specific hypotheses may be achieved.

Primates may also seek to reduce encounters with predators by altering their sleeping habits. Predator pressure has been said to affect sleeping positions (Anderson, J. R., 1984, 1998; Smith et al. 2007); the timing of sleep (Caine 1987, Dawson 1979, Gibbons and Menzel 1980, Hankerson and Caine 2004); the amount of vigilance and noise before retirement (Caine 1987, Heymann 1995, Peres 1993a, Terborgh 1983); the choice of sleeping site, with regard to height, cover, and/or access (Caine et al. 1992; Cowlishaw 1994; Day and Elwood 1999; Fruth and Hohmann 1993; Pruetz et al. 2008; Ramakrishnan and Coss 2001a,b; van Schaik et al. 1996b; Setz 1999; Tenaza and Tilson 1985; Wright 1998); the speed of travel and/or choice of path to the sleeping site (Ferrari and Lopes-Ferrari 1990b, Franklin et al. 2007); and the use of sentinels before sleeping (Zullo and Caine 1988). Observations of predatory attacks during primate sleep are rare (Busse 1980, Wright 1998, Wright and Martin 1995). Experimental design, as seen in several of the studies cited here, helps to elucidate the nuances of predator-sensitive sleeping.

Predation is also expected to influence primate ranging (reviewed in Boinski et al. 2000b). Surely, primates will avoid moving through areas known to be dangerous, unless there is considerable incentive to do otherwise. This topic has been reviewed by Treves (2002) for arboreal primates and by Cowlishaw (1997a,b) for terrestrial baboons. On a smaller scale, primates are expected to modify their group cohesion in response to perceived risk (Altmann and Altmann 1970, Hamilton 1971, Vine 1971); cohesion might, in turn, impact other behaviors, such as vigilance and foraging (Cords 2002a, Cowlishaw 1998, Robinson 1981, Treves 1998). Predator-related cohesion has been shown in various primate species (Boinski et al. 2000b; Byrne et al. 1989; Gursky 2003a; Sigg 1980; Stanford 1995, 2002; Tutin et al. 1983; Uhde and Sommer 2002); however, this is not a universal response (Isbell 1990, Treves 1999), nor can cohesion be related solely to predation pressure (Cowlishaw 1999a, Isbell and Enstam 2002). There is no reason to assume a priori that clumping is the best strategy for avoiding a predator. In some cases, it might be better to split up and thereby confuse the predator or hide oneself at the expense of associates. Finally, individual primates may change their location within the group, which might affect not only risk but also foraging success. Hypothetically, the periphery of a group presents greater predation risk than do central locations (Devore 1965a, Hamilton 1971, Vine 1971). The command of "safer" locations may depend on sex and/or size (Wright 1998), reproductive state (Busse 1984, Collins 1984 on lactating females), and/or social variables such as rank (Gould et al. 1997; Hall and Fedigan 1997; Janson 1990a,b; Robinson 1981; Ron et al. 1996). Individuals, especially those that perceive themselves to be more vulnerable, may opt for safer locations despite a probable drop in foraging success (Chapman and Chapman 2000a–c, Janson 1990a). Again, these trade-offs are the central focus of predator-sensitive behaviors.

In sum, primates have at their disposal a wide variety of options when it comes to predator-sensitive behaviors, and univariate models do little to convey the complexity of these patterns. Moreover, without direct observations of predator encounters to confirm elevated risk, simple observations of behavior consistent with predator avoidance generate equivocal conclusions about antipredator behavior. For example, individuals must respond not only to the threat of predation but also to the distribution of mates, food resources, and shelter from climatic extremes. The relative power of these sometimes competing selective pressures is hard to assess by simple observation of the prey species (Boinski et al. 2000b, Chapman and Chapman 1999, Treves et al. 2003, Ydenberg and Dill 1986). Experimental or quasi-experimental manipulation of predation risk is required. We surely need more observations of verified encounters and experimental simulations.

Antipredator Behaviors

The most fundamental antipredator behavior is to avoid the initial encounter. Predator avoidance may be improved by cryptic habits, including nocturnality, forming smaller and quieter groups, and foraging alone; but few studies have examined these as antipredator strategies because controls are so difficult to establish. Such strategies may be easier for individuals with smaller bodies, populations that can hide effectively by day, and those that have traditions or evolved patterns of cooperation in small group settings (Horrocks and Hunte 1986, Pages-Feuillade 1988, Savage et al. 1996b, Wahome et al. 1993, Watanabe 1981, Wright 1998, Wright and Martin 1995). Large-bodied, diurnal species may also behave cryptically when risk is perceived to be high (Fay et al. 1995 on gorillas, *Gorilla gorilla*; Slatkin and Hausfater 1976 on an adult male baboon).

Predator avoidance is useful, but it does not eliminate chance encounters with predators or prevent ambush at sites used repeatedly. From the perspective of prey, there is a great deal of uncertainty about the location, number, ability, and hunger of predators. This uncertainty can be resolved in part by gathering information using all the senses. Virtually nothing is known about olfactory or auditory vigilance in primates (but see Jaenicke and Ehrlich 1982). However, we know a great deal about visual information gathering or

vigilance (for review, see Treves 2000b). First, long-term observational studies indicate that vigilance is intermittent and frequent throughout the foraging period of primates (Cowlishaw et al. 2004, Jack 2001, Treves et al. 2001). Studies of sleeping birds indicate that vigilance may also recur during sleep (Lendrem 1983, 1984). This is reasonable as information gathering must be repeated to compensate for unpredictable movements of predators. Correspondingly, one expects that prey vigilance will increase when entering unfamiliar areas. This prediction has never been tested explicitly with time-sensitive and spatially explicit information on the movements of primates into areas that vary in familiarity. We encourage tests of this prediction with real or simulated predator encounters. The key variable will be the length of time elapsing between revisits to the same site as longer intervals imply greater uncertainty and, hence, greater risk of predator encounter.

Measures of vigilance without evidence for the target or the severity of risk are of little use for inferring antipredator function and utility of the vigilance. Indeed, vigilance serves a very important social purpose, which can conflict with antipredator vigilance (Biben et al. 1989; Brick 1998; Treves 1999, 2000b; Treves and Pizzagalli 2002). Our own human experience reveals the multifunctional and flexible nature of vigilance such that it need not be intended to detect a predator to nevertheless detect one (Treves 2000b, Treves and Pizzagalli 2002). With this in mind, we have found a few vigilance studies that experimentally simulated and varied the risk of predation or related vigilance to real predator encounters (reviewed in Bshary and Noë 1997a; Cheney and Seyfarth 1990a,b; Treves 1999; see also recent work by Barros et al. 2008, Dacier et al. 2006). The studies are concordant: Vigilance rises when predation risk is elevated, presumably as primates attempt to gather visual information that will reduce individual vulnerability.

Primates and many other animals use the early detection of predators to escape or signal this detection. Several brilliant studies have documented the use of signals in the predator context. The best-documented research concerns vocal alarms (Arnold et al. 2008; Blumstein 2001; Bshary and Noë 1997a,b; Cheney and Seyfarth 1990a; Coss et al. 2007; Fichtel and Kappeler 2002; Kirchhof and Hammerschmidt 2006; Macedonia and Evans 1993; Struhsaker 1967a. Wheeler 2008; Zuberbuehler et al. 1999; Zuberbuehler 2000), but visual signals have also been noted (Horrocks and Hunte 1986, Savage et al. 1996b). These studies have shown that, in some primate populations, individuals engage in elaborate signaling following the detection of predators and associates often use the semantic information encoded in the signals to respond appropriately. In addition, some vocal signals warn predators that they have been detected, thereby potentially deterring ambush predators from further efforts (Zuberbuehler et al. 1997, Zuberbuehler 2000).

Antipredator signals could make groups advantageous—beyond the benefit of dilution—via collective detection

(Bednekoff and Lima 1998) if signals improve the efficiency of escape by nondetectors. On the other hand, if signals such as vocal alarms predominantly function as deterrents—warnings to predators that they have been detected and attack will fail (e.g., Bearder 2007, Wheeler 2008, Zuberbuehler et al. 1997, Zuberbuehler 2000)—then the role of the group and the benefit of alarm signals to associates is less clear. For example, Chapman (1986), Ross (1993), Srivastava (1991), and many other authors have reported protracted vocalizations following encounters or even following successful attacks in which the predator no longer poses a danger. Persistent vocalizations seem ill-suited to collective detection—that is, improving associates' chances of escape—because they could attract additional predators or permit the initial attacker to monitor the prey at a distance. Persistent calls are better suited to warning the predator or distant animals, neither of which can be easily tied to the evolution of group living. We advocate that primatologists devote attention to whether and when alarm signals improve escape, to distinguish those alarm signaling systems whose primary role and evolutionary history reflect predator warnings from those signaling systems whose function and evolution reflect collective detection for gregarious prey. Primatologists wishing to investigate these questions might learn a great deal from studies done on birds (Lima 1994, 1995a,b; Lima and Zollner 1996).

Regarding the use of alarms to help associates escape more quickly, Bednekoff and Lima (1998) emphasized that inefficient transmission and reception of predator-related information can nullify the benefits of collective detection for group-living animals. An alarm signal produced too late or responded to slowly offers little benefit to associates, but the efficiency of escape following alarm signals has never, to our knowledge, been measured in primates. This may be because primatologists have often assumed that production of an alarm signal is instantaneously followed by appropriate responses. This assumption may be false. Collective detection also lessens in value when predators are highly selective of prey (Bednekoff and Lima 1998). In effect, selective predators reduce the number of effective alarm signalers to that subset that are vulnerable themselves or are sensitive to the vulnerability of others. The latter condition may hold in kin-based primate groups and those involving high levels of reciprocity, but this conjecture remains to be tested.

Not all predators rely on surprise. Coursing predators, such as wild dogs, hyenas, and chimpanzees, may pursue despite detection; and the predators themselves may gain valuable information by observing the flight of their quarry (Fanshawe and Fitzgibbon 1993, Fitzgibbon and Lazarus 1995, Holekamp et al. 1997b, Treves 2000b). Some raptors may behave in this way as well, but the data are not as clear (Lima 1993). When confronted by coursing predators, prey require rapid escape or refuges that are inaccessible to their pursuers; vigilance and alarm calling are of limited use against such predators. This emphasizes our point that the use of vocal signals in the predator context requires further

study before we can conclude that the evolution of antipredator signals is tied to the evolution of group living.

To escape from predators, all primates seek refuge (with the possible exception of resident adult male patas monkeys; Chism and Rowell 1988, Hall 1965a). Refuges for terrestrial taxa include trees and cliffs, while arboreal forms rapidly change level (reviewed in Treves 2002). Prey species might also perform evasive behaviors such as coordinated attempts to confuse the predator; this has been observed in some fish, birds, and open-country ungulates (Edmunds 1974, Lima 1993) but not in primates.

More rarely, primates stand their ground to counterattack or mob predators. Of the two forms, mobbing appears to be the less dangerous for the prey and is more common among smaller primates. It involves two or more prey animals making repeated advances on a predator, usually while vocalizing and displaying in a conspicuous fashion. The predator is often distracted or repelled by persistent approaches (Baeninger et al. 1977, Eason 1989, Gursky 2005b, Lloyd et al. 2006). Adult males, acting alone or in small parties, are more likely to attack predators than are other classes of individuals (Chapman and Chapman 1996b, Cowlishaw 1994). The likelihood of counterattack by primates appears to depend on the size difference between predator and prey (Bshary and Noë 1997a, Stanford 1995).

If all the antipredator strategies fail (avoidance, vigilance-assisted early detection and collective detection, alarm signals, escape, counterattack), the prey is usually doomed. However, close spatial cohesion of prey may allow one individual to interpose an associate between itself and a predator at the last moment . Another last-ditch struggle is the distress call of a captured prey animal. Distress calls in birds are proposed as a tactic to startle the predator into dropping the captured prey, rather than to alert associates (Conover 1994). Similar startle responses may exist in lorisids, along with toxic secretions that may deter the final attack (see Chapter 4). Primates have evidently evolved a wide repertoire of morphologies and behaviors to avoid, escape, and otherwise elude their predators.

HUMANS AS PREDATORS, PAST AND PRESENT

Paleontological evidence suggests that humans have been predators of nonhuman primates for hundreds of thousands of years. For example, fossil accumulations from Europe include cercopithecoid remains that may be attributed to hominid activity (Freeman 1994, Klein 1987). Human hunting has long been implicated in the early Holocene extinction of North and South American megafauna (Kurten and Anderson 1980, Martin 1984), along with similar patterns in Africa (Klein 1984) and Eurasia (Vereshchagin and Baryshnikov 1984, see also Janzen 1983). Perhaps the most extreme example of human impact on primate populations comes from Madagascar. Fossil data indicate that approximately 2,000 years ago humans arrived on this island and rapidly drove some 15 primate species into extinction (Richard and

Dewar 1991). Prehistoric hunting may also account for much of today's distribution of primates (Peres 1990b, Struhsaker 1999). In addition, ethnographic research has shown that primates represent a small but significant part of the diet for human cultures around the world (Alvard 1993, Bailey and Peacock 1988, de Garine 1993, Fitzgibbon et al. 1995, Hart 1978, Hill and Padwe 2000). From this point of view, primates have long had to adapt to humans as predators.

On the other hand, it could be argued that humans represent a fundamentally unique type of predator. Humans armed with projectile weapons differ from other predators because they can kill at a distance, approximating or even exceeding the detection distance of their prey. Thus, a human armed with a projectile is like an ambush predator with an enormous striking range, especially if the human uses bait or traps. However, unlike other ambush predators, humans also resemble coursing predators in that they are rarely deterred by alarm calls and are hard to outrun (at least for primates). Like coursing predators, humans with projectiles target not only the infirm and the unwary but also prime adults (Brain 1981, Faraizl and Stiver 1996, Hill and Hurtado 1996, Urbani 2005). Hence, humans with projectiles share characteristics of both ambush and coursing predators, making them especially deadly. Moreover, human intelligence and social transmission of cultural knowledge may give these predators greater capacity to study the habits of their prey over time and circumvent prey defenses. In combination, these predatory traits are evolutionary novelties, imposing unique selective pressures on prey species. It is possible that primates have developed specialized morphologies and behaviors to avoid human predators, but we know of no studies testing this idea; it is probable that no such specialized adaptations exist. While humans have long represented a part of primate ecological communities, in more recent millennia they may have become a selective pressure superseding the prey species' capacity to adapt; thus, today's primate morphology and behavior are likely only weakly related to humans as predators.

Clearly, this has become the case in recent decades. The impact of human hunting on primate populations has reached record proportions assisted by logging roads; four-wheel-drive vehicles, which give access to previously unexploited tracts of forest; and high-powered rifles, which can easily kill even the largest of primates. The bushmeat trade has been noted for many years (Anadu et al. 1988, Chardonnet et al. 1995, Martin 1983, Njiforti 1996); however, it is a growing problem, reaching unsustainable levels in many parts of the globe, particularly in major regions of Africa

(Walsh et al. 2003, Wilkie and Carpenter 1999). Primates are not the only taxa in peril, but they are especially vulnerable due to their large size (relative to many other forest species) and their tendency to form conspicuous social groups (Bodmer et al. 1988, Struhsaker 1999). Although tribal custom prevents some human cultures from hunting primates, the economics of the commercial bushmeat trade is causing rapid change in long-standing traditions (Struhsaker 1999). There is a great deal of recent debate about the possibility and even value of sustainable hunting (Bodmer et al. 1988, 1994; Freese 1997; Hart 2000; Hill and Padwe 2000; Hill et al. 2003; Robinson 2000; Robinson and Bennett 2000; Robinson and Redford 1994; Slade et al. 1998; Waite 2007), but few can argue with the fact that current levels of bushmeat hunting are leading many primate species toward extinction (Naughton-Treves et al. 2003, Walsh et al. 2003, Wilkie and Carpenter 1999).

CONCLUSION

In closing, primates show remarkable diversity in their morphological and behavioral adaptations, and surely some of these relate to predator avoidance. Even where successful predation is rare, a single predatory event may limit an individual's lifetime reproductive success in a profound way; thus, there is likely to be considerable selective pressure to avoid becoming prey. Although we recognize the compelling logic behind predation as an influence on primate evolution, we also see the challenge of demonstrating this impact unequivocally and exploring alternatives. The next generation of primatologists can learn from the previous one by conducting rigorous tests of mutually exclusive hypotheses and by wariness toward hasty conclusions based on a few variables. This will demand primatologists who can study both primates and predators with the sophistication of specialists so that we may understand predator behavior in relation to prey vulnerability. It will also take primatologists with facility in experimental design, who can simulate predator attacks in innovative ways and measure antipredator behavior unambiguously. These are not easy tasks, but we are encouraged by the energy, determination, and creativity of the new generation of primatologists. Finally, our chapter calls attention to the plight of primates subject to human predation. In some areas, the first battle must be conservation, lest we lose populations or entire species; basic research, though fascinating, is a luxury we must earn.

36

Primate Locomotor Positional
Behavior and Ecology

Paul A. Garber

No group of vertebrates shows a wider range of locomotor adaptations to arboreality than do living primates. (Fleagle 1978:243)

1. What are the distinctions between positional behavior and locomotor behavior, and why are these distinctions important in evaluating primate functional morphology?
2. Although body mass is an important variable in understanding species differences in positional behavior, what other factors contribute to the fact that body mass is not as strong a predictor of substrate use, leaping, and climbing as first thought?
3. What do species differences in the onset of adult-like patterns of positional behavior tell us about primate ontogeny?

INTRODUCTION

Primates represent a successful radiation of primarily arboreal, tropical forest mammals that vary considerably in body mass, limb proportions, postcranial anatomy, and positional adaptations. These differences reflect important evolutionary distinctions in the manner in which species exploit their environment, obtain access to food resources, and avoid predators, as well as in the energetic costs of travel. The study of primate locomotion is central to an understanding of primate adaptive diversity because major changes in the ability of primate lineages to exploit their environment are associated with evolutionary changes in positional behavior and positional morphology. These include the evolution of claw-like nails and vertical clinging in tamarins and marmosets (subfamily Callitrichinae); elongated forelimbs and suspensory behaviors in subfossil sloth lemurs (Paleopropithecidae), lesser apes (*Hylobatidae*), and great apes; prehensile tails in atelines and capuchins; and hindlimb elongation and vertical clinging and leaping in some prosimians (Niemitz 1984b; Cant 1986, 1990; Oxnard et al. 1990, Hunt 1991; Garber 1992; Anemone 1993; Crompton et al. 1993; Davis 1996; Terranova 1996; Larson 1998a; Godfrey and Jungers 2003). In the case of human evolution, for example, adaptations associated with bipedal locomotion have been associated with fundamental changes in hominid social organization, food sharing, tool use, hunting, and patterns of habitat utilization (Lovejoy 1981, Ruff 1987, Foley and Elton 1998, Hunt 1998).

Early in their evolution, the ancestors of modern primates (euprimates) evolved a unique suite of locomotor traits associated with exploiting an arboreal environment. These include

grasping feet and hands, an opposable big toe (hallux) bearing a flattened nail (and eventually nails replacing claws on most or all digits), and changes in the visual system (orbital convergence, color vision, visual images from the retinae crossing at the optic chiasm, enhanced depth perception; see Martin 1990). Several theories have been proposed to account for the evolution of these traits, including the possibility that the ancestors of modern primates were visual predators of insects in the low shrub layer of the forest canopy (Cartmill 1972, 1974, 1992), exploited resources found on terminal branches in the periphery of the tree crown (Charles-Dominique and Martin 1970, Rasmussen 1990), or played an important role in the evolution of angiosperms and tropical forests by serving as agents of pollination and seed dispersal (Sussman and Raven 1977; Sussman 1991b). Each of these theories highlights the fact that the evolutionary history of modern primates is closely linked to their ability to travel on, grasp, and exploit small, flexible arboreal supports. In fact, recent fossil evidence (*Carpolestes simpsoni*) dated at 56 million years ago (Sargis 2002) indicates that changes in the grasping capabilities of the feet and hands of primate ancestors (nail replacing a claw on the big toe) preceded changes in visual acuity. These initial primate positional adaptations provide a framework for examining subsequent changes in primate locomotor anatomy and behavior (Schmitt and Lemelin 2002). Derived trait complexes such as prehensile tails, divergent and opposable thumbs, mobile shoulder girdle, elongated forelimbs, the evolution of an elongated tarsus, and tibia–fibula fusion, as well as forms of hindlimb-dominated locomotion (i.e., leaping, hindlimb grasping, bipedalism), appear to represent functional and adaptive solutions enabling modern primates to obtain resources and exploit environments in new and novel ways. As expressed by Jolly and Plog (1979:114), "almost every major adaptive radiation within the primate order has involved changes in locomotor patterns that opened up a new array of econiches."

In the remainder of this chapter, I examine diversity in patterns of posture and locomotion among living primates and the challenges that primates of different body mass, limb proportions, and diet face in exploiting resources in arboreal and terrestrial environments. Specifically, I focus on distinctions between posture and locomotion, definitions of positional categories, the role of the primate tail in balance and weight support, the effects of body mass on patterns of substrate preference and positional behavior, and evidence for seasonal, site-specific, and sex-based differences in locomotion and posture. In general, within a given species, patterns of positional behavior are more conservative during travel than during feeding. This may reflect seasonal changes in diet and foraging strategies that characterize most primate taxa. Although smaller- and larger-bodied primates face different challenges in maintaining balance and weight support and in the cost of travel, body mass is not a strong predictor of positional behavior. This may reflect the fact that both smaller- and larger-bodied primates have evolved particular morphological adaptations that enable them to exploit similar sets of resources in different ways.

POSTURE AND LOCOMOTION

The study of primate locomotion is more accurately described as the study of primate positional behavior (Prost 1965). Positional behavior consists of two components, posture and locomotion. *Posture* is defined as a set of behaviors in which an individual's center of mass remains relatively stable and the limb segments proper, rather than the entire skeleton, are being manipulated for some end (Prost 1965). Postural activity tends to dominate the positional repertoire during feeding, resting, and various socially directed behaviors (Rose 1974, Garber 1984b, Hunt 1992, Walker 1996). Sitting, hanging, and standing represent common postural behaviors. In most studies of primates in the wild, postural behavior accounts for a greater proportion of the positional repertoire than does locomotor behavior (Rose 1974, McGraw 1998a).

Locomotor behaviors differ from postural behaviors in that "body mass (as opposed to limb mass) is grossly displaced relative to its physical surroundings" (Prost 1965:1200). Locomotor behaviors are associated with such common activities as traveling, foraging, and predator avoidance and represent a primary component of a species' adaptive pattern. Locomotion is a means to an end—for example, traveling to a food patch or resting site, fleeing a predator, or stalking insect prey—and, as such, feeds back onto other social/maintenance activities (Ripley 1967). Running, climbing, and leaping are common locomotor behaviors.

In mechanical terms, postural and locomotor modes may represent integrated points on a behavioral continuum, for example, when quadrupedal standing transitions into quadrupedal walking or quadrupedal running (Prost 1965,

Kinzey 1976). During postural behavior, many of the forces acting directly on the musculoskeletal system may be analogous to those involved in locomotion. Conversely, the skeleton may be exposed to forces that are infrequent but extreme during locomotion and not commonly encountered during postural behavior. In the case of many vertical clinging and leaping prosimians, the high take-off and landing forces generated when leaping to and from vertical trunks (Peters and Preuschoft 1984; Preuschoft 1985; Crompton et al. 1993; Demes et al. 1995; Warren and Crompton 1998a,b) have no biomechanical counterpart during postural behavior. Thus, distinctions between posture and locomotion are important because each of these sets of positional behaviors may select for alternative forms of musculoskeletal design.

Positional behavior is activity- and context-specific. That is, certain positional behaviors occur more frequently during particular activities (feeding vs. traveling vs. resting) and on supports of certain diameter, orientation, and weight-bearing capacity (Ripley 1967; Kinzey 1967; Garber 1980, 1998; Bergeson 1998). The selection of particular substrates may be a response to biomechanical factors (i.e., due to forces of torque and gravity, it is more stable to walk on top of a relatively large horizontal branch than it is to walk on top of a relatively small horizontal branch) (Grand 1972, Cartmill and Milton 1977), as well as the physical structure and geometry of the forest canopy (Dunbar and Badam 2000). Many fruit, flower, leaf, and insect resources exploited by primates are located on thin, flexible branches in the periphery of the tree crown (Sussman 1991b). Access to these resources commonly is associated with prehensile, grasping, and below-branch suspensory postures (Grand 1972, Meldrum et al. 1997). For example, gibbons (*Hylobates lar*) take advantage of the downward displacement of a branch during forelimb suspensory behavior to expand their feeding sphere and gain access to below-branch resources (Grand 1972). Other foods exploited by primates may be associated with a large branch niche. It has been argued that the evolution of claw-like nails in trunk-foraging primates in Africa (*Euoticus*, needle-clawed galago), Madagascar (*Phaner*, forked-tooth lemur), and Central and South America (*Cebuella*, *Callithrix*, *Mico*, *Saguinus*, *Leontopithecus*, *Callimico*; tamarins, marmosets, and Goeldi's monkey) represents parallel postural adaptations, enabling these small-bodied taxa to obtain resources such as exudates, insects, and fungi by clinging to large vertical supports that would otherwise be too large to be spanned by their diminutive hands and feet (Charles-Dominique 1977a; Garber 1980, 1992). If one assumes that individuals preferentially adopt particular positional modes and select particular substrates that offer them increased efficiency and stability in an arboreal environment, then factors such as body mass and ontogenetic changes in motor control and limb and body proportions should play an important role in a species' positional repertoire (Cartmill and Milton 1977, Fleagle 1978, Fleagle and Mittermeier 1980). A major goal of the study of positional behavior is to understand the relationship between form and function in a species' natural environment (Bock and von Wahlert 1965).

CHALLENGES IMPOSED BY THE ARBOREAL AND TERRESTRIAL ENVIRONMENTS

Primates exploiting an arboreal environment face different challenges from primates exploiting a terrestrial environment. These include exposure to different types of predators, different strategies for escape, alternative conditions of temperature and sunlight, and access to different types and distributions of food resources (McGraw and Bshary 2002). In terms of positional behavior, terrestrial environments present foragers with a more uniform and stable set of substrates as well as greater opportunity for continuous pathways of travel than do arboreal environments. Several primate lineages (African apes, savanna baboons, mandrills, geladas, patas monkeys, macaques, Asian colobines, ring-tailed lemurs, humans) commonly travel and feed on the ground (there is no evidence of terrestrial adaptations in any extant New World monkey). Patterns of terrestrial travel in Asian great apes (orangutan, fist-walking), African great apes (knuckle-walking in gorillas, chimpanzees, and bonobos), and humans (bipedalism) are quite distinctive from those of Old World monkeys and represent alternative solutions to problems of balance and weight-bearing. For example, African apes exhibit several derived features of the distal radius that extend and stabilize the wrist during knuckle-walking as well as permit climbing and suspensory behavior in the trees (Richmond and Strait 2000). In the case of human ancestors, evidence of bipedal adaptations appears in fossil hominids by at least 4 million years ago (Leakey et al. 1995). However, despite an expanding fossil record, there is no consensus on whether human ancestors shared a knuckle-walking ancestry with African apes or whether human bipedalism evolved from an arboreal ancestor that emphasized climbing and other suspensory positional behaviors (Richmond and Strait 2000).

In contrast to terrestrial environments, the arboreal canopy is characterized by substrates that are discontinuous, mobile, contoured in a cylindrical shape, orientated at various angles to the ground, and variable in size and weight-bearing capacity (Cartmill 1974, 1985; Grand 1984; Cant 1992). Moreover, tree branches become smaller and more unstable as one approaches the outer or food-bearing zones of the canopy; therefore, animals traveling in an arboreal environment face significant challenges of maintaining balance and support on branches that bend, sway, and possibly break under their weight (Cartmill and Milton 1977, Cant 1992). Arboreal primates attempt to solve these problems by distributing their body mass over a range

of different supports, lowering their center of gravity nearer to the support, adopting flexed and abducted limb postures, and maintaining stability by engaging in suspensory or below-branch postures (Figure 36.1). Given that primates from the size of orangutans to mouse lemurs travel and feed principally in the trees, species differences in body mass are expected to be an important factor in primate positional behavior and musculoskeletal design (Fleagle 1976a,b; Cartmill and Milton 1977; Fleagle and Mittermeier 1980; Jungers 1985; Fleagle and Meldrum 1988; Demes and Gunther 1989; Jungers and Burr 1994; Preuschoft et al. 1996; Polk 2002).

PRIMATE QUADRUPEDALISM

Virtually all primates walk quadrupedally during slow horizontal progression on the ground or in the trees, and for many species quadrupedal walking is the most common mode of travel (Tables 36.1–36.6). Kay et al. (1997) argue that the

Figure 36.1 Patterns of locomotion and posture among a community of primates in Suriname. From Fleagle (1988).

Table 36.1 Intraspecific Variation in Positional Behavior[1] in Common Chimpanzees, *Pan troglodytes schweinfurthi* and *Pan troglodytes verus*

FIELD SITE	QUAD	CLIMB/SCRAMBLE	SUSPENSORY	BIPEDAL	LEAP	SIT/LIE	STAND	ARM-HANG	N
Gombe	16.9	0.9	0	0.4	0	73.2	3.6	4.8	2,700
Mahale	16.1	0.9	0.3	0.3	0	76.8	1.4	4.4	11,393
Taï Forest	12.1	1.5	0.2	0.2	0.1	79.5	5.0	1.3	10,079

Sources: Adapted from Doran 1993, Hunt 1992.
[1] Data represent percentage of time engaged in each positional behavior. Hunt 1992 collected data using a 2 min instantaneous point-sampling technique. Doran 1993 collected data using a 1 min instantaneous point sampling technique.

Table 36.2 Positional Behavior in Seven Sympatric Primate Species in Suriname[1]

SPECIES	BODY MASS[2]	CLIMB	LEAP	QUAD	TAIL–ARM SUSPEND	N
Travel						
Ateles paniscus	7.7	31.1	4.2	25.4	38.6	743
Alouatta seniculus	6.0	16.0	4.0	80.0	0	818
Cebus apella	3.0	5.0	10.0	84.0	0	363
Chiropotes satanas	3.0	Rare	18.0	80.0	0	452
Pithecia pithecia	1.8	Unknown	75.0	Unknown	0	40
Saimiri sciureus	0.63	4.0	42.0	55.0	0	224
Saguinus midas	0.45	0	24.0	76.0	0	625
Feed/forage						
Ateles paniscus	7.7	44.9	2.6	24.8	27.7	274
Alouatta seniculus	6.0	41.0	0	59.0	0	194
Cebus apella	3.0	8.0	4.0	88.0	0	438
Chiropotes satanas	3.0	2.0	9.0	88.0	0	282
Pithecia pithecia	1.8	Unknown	Unknown	Unknown	0	0
Saimiri sciureus	0.63					
Feed		2.0	11.0	87.0	0	158
Forage		3.0	22.0	75.0	0	463
Saguinus midas	0.45	2.0	12.0	87.0	0	128

Sources: From Fleagle and Mittermeier 1980.

[1] Data represent percentage of time engaged in each positional behavior.

[2] Adult female body mass in kilograms.

Table 36.3 Patterns of Positional Behavior and Substrate Use[1] in Taï Forest Monkeys During all Activities

		LOCOMOTOR BEHAVIOR					
SPECIES	BODY MASS (KG)	ARM SWING	CLIMB	LEAP	WALK	RUN	N
Colobus polykomos	8.3	0	14.3	14.5	41.8	29.4	918
Colobus badius	8.2	3.9	17.0	17.8	53.1	8.2	1,466
Cercocebus atys	6.2	0	12.5	1.0	80.7	5.7	466
Colobus verus	4.2	0	12.0	20.4	45.2	22.4	508
Cercopithecus diana	3.9	<1	19.4	10.4	59.3	10.8	1,553
Cercopithecus petaurista	2.9	0	18.8	10.1	61.3	9.8	512
Cercopithecus campbelli	2.7	0	14.5	5.2	72.6	7.7	596

		POSTURAL BEHAVIOR					
SPECIES	BODY MASS (KG)	SIT	QUAD STAND	SPRAWL	LIE	STAND/FORAGE	N
Colobus polykomos	8.3	89.5	<1	2.9	6.3	<1	2,495
Colobus badius	8.2	87.0	1.4	4.2	6.1	<1	2,690
Cercocebus atys	6.2	80.9	15.7	0	0	1.9	854
Colobus verus	4.2	90.7	1.3	3.8	3.7	<1	1,081
Cercopithecus diana	3.9	59.6	22.1	<1	<1	15.2	1,908
Cercopithecus petaurista	2.9	81.9	12.9	0	<1	3.1	1,530
Cercopithecus campbelli	2.7	70.9	21.9	<1	<1	4.8	838

Sources: McGraw 1998a,b, 2000.

[1] Data represent percentage of time engaged in each positional behavior.

Table 36.4 Positional Behavior[1] in Five Species of Primates in Kibale Forest, Uganda

SPECIES	BODY MASS (KG)	CLIMB	LEAP	QUAD	QUAD STAND	BIMANUAL	VERTICAL BOUND	N
Travel								
Colobus badius	8.2	29	30	34	<1	3	3	3,204
Colobus guereza	8.0	11	44	39	<1	<1	5	2,056
Cercocebus albigena	6.4	31	21	46	0	1	<1	2,328
Cercopithecus mitis	3.5	29	18	51	<1	<1	1	1,367
Cercopithecus ascanius	3.0	27	25	41	0	<1	7	1,124
Feed								
Colobus badius	8.2	37	16	4	<1	2	1	1,745
Colobus guereza	8.0	18	33	43	0	1	4	2,087
Cercocebus albigena	6.4	40	11	48	<1	1	<1	1,583
Cercopithecus mitis	3.5	38	6	55	0	<1	<1	2,046
Cercopithecus ascanius	3.0	50	11	38	0	<1	71	2,529

Sources: Gebo and Chapman 1995a.

[1] Data represent percentage of time engaged in each positional behavior.

Table 36.5 Positional Behavior[1] in Malagasy Prosimians in Ranomafana National Park

SPECIES	BODY MASS (KG)	QUAD	LEAP	CLIMB	OTHER	N	SIT	STAND	SUSPEND	VERTICAL CLING	LIE
Propithecus diadema (site Talatakely)	6.5										
Dry season		0.8	88.6	8.9	1.7	1,795	72.3	0.5	4.1	17.8	0.7
Wet season		0.5	89.1	8.6	1.7	1,220	76.4	0.4	1.8	21.0	0.4
Varecia variegata (site Vatoharanana)	3.8										
Dry season		29.0	55.5	12.5	0.0	1,001	83.5	6.6	4.9	0.2	4.8
Wet season		44.0	48.0	6.5	0.0	666	80.1	8.6	2.7	0.7	8.0
Eulemur fulvus (site Talatakely)	2.5										
Dry season		18	73.3	7.1	0.7	597	74.5	3.3	0.1	1.4	20.6
Wet season		25.8	62.7	8.5	1.9	978	80.4	10.5	0.2	1.6	7.3
Eulemur rubriventer (site Talatakely)	2.3										
Dry season		18.0	69.3	12.1	0.7	583	91.9	4.2	0.1	3.8	0.2
Wet season		27.3	59.7	11.3	2.0	1,505	79.6	15.1	0.1	2.6	1.7
Eulemur rubriventer (site Vatoharanana)	2.3										
Dry season		27.5	53.5	15.5	4.0	583	62.0	0.3	0.3	1.1	36.4
Wet season		20.0	64.0	15.5	1.0	1,505	78.9	9.4	0.1	7.5	4.2

Sources: Adapted from Dagosto 1995.

[1] Data represent percentage of time engaged in each positional behavior.

Table 36.6 Positional Behavior[1] (Travel, Feed, and Forage) in Nine Species of Primates in Ecuador

SPECIES	BODY MASS (KG)	QWALK	QBOUND	CLIMB/CLAMBER	BRIDGE	SUSPEND	LEAP	SCANSORIAL	N
Lagothrix lagotricha[2]	5.7–8.7	28.7		44.5	10.2	11.4[3]	3.9	0.0	3,926
Ateles belzebuth[2]	7.7	20.8		40.7	10.7	23.1[3]	2.3	0.0	3,760
Ateles paniscus	7.7	20.1		28.1	8.8	34.2[3]	2.6	0.0	1,089
Cebus albifrons	2.8	41.5	7.0	17.7	5.7	0.8	27.3	0.0	243
Pithecia monachus	2.4	38.4	8.4	22.9	1.5	0.4	28.4	0.0	449
Callicebus moloch	1.0	45.2	8.9	26.7	1.2	0.3	17.7	0.0	325
Saimiri sciureus	0.7	37.9	7.2	25.2	4.4	0.9	25.4	0.0	569
Saguinus tripartitus	0.4	26.1	14.2	9.7	0.3	0.4	33.7	15.6	626
Cebuella pygmaea	0.1	14.8	12.3	5.2	0.3	0.3	24.4	42.7	709

Sources: Youlatos 1999, 2002, and Cant et al. 2001.

QWalk, quadrupedal walk; QBound, quadrupedal bound; scansorial, vertical quadrupedal locomotion on tree trunks in which claw-like nails are embedded in the bark.

[1] Data represent percentage of time engaged in each positional behavior.

[2] Data from Cant et al. 2001.

[3] Primarily arm–tail suspension.

earliest anthropoids emphasized quadrupedal walking over leaping or other forms of progression. Primates are distinctive among mammals in their use of a *diagonal sequence–diagonal couplet gait* (i.e., a gait sequence in which movement of the left hindfoot is followed by movement of the right forelimb, followed by right hindfoot and left forelimb) when walking quadrupedally (Hildebrand 1967, Larson 1998b). In addition, during quadrupedal walking, the primate forelimb adopts a more protracted orientation at touchdown (primate forelimb is angled at >90° relative to the axis of the body), and peak vertical substrate reaction forces acting on the forelimb are reduced relative to those acting on the hindlimb (Demes et al. 1994, Schmitt and Lemelin 2002). According to Larson (1998b:166), "the reduction in the use of forelimb propulsive muscles is a component of the basic adaptation of reaching out with the clawless grasping hand in an arboreal setting that arose early in the evolution of primates." Thus, there appears to have been a major shift in primates from a primarily weight-bearing role of the forelimb to a greater role in grasping, exploration, and object manipulation. Greater forelimb mobility and enhanced cortical control of forelimb movement in primates are consistent with the mechanical demands of exploiting resources in a fine-branch setting (Larson 1998b; Schmitt 1998, 2003; Schmitt and Lemelin 2002).

Relative to most mammals, primates are characterized by elongated forelimbs. During quadrupedal walking, arboreal primates (Figure 36.2, upper) adopt a crouching or flexed-elbow gait that places their center of mass nearer to the substrate. This serves to improve balance and stability, especially on small or flexible branches (Schmitt 2003). A crouched position also may facilitate rapid changes in travel direction and substrate use and facilitate opportunistic leaping (Schmitt 1998). In contrast, primates that walk quadrupedally on the ground (Figure 36.2, lower) tend to have their limbs more adducted (nearer to the body) and elbows extended and to experience vertical reaction forces more equally distributed between forelimbs and hindlimbs (Larson 1998b, Polk 2002, Schmitt and Lemelin 2002). According to Fleagle (1988:247), compared to their more arboreal relatives, terrestrial Old World monkeys are characterized by "a narrow, deep trunk and relatively long limbs, designed for long strides and speed, and their tails are often short or absent." These changes in anatomical design facilitate rapid travel in a terrestrial environment (Fleagle 1988).

DEFINITION OF POSITIONAL CATEGORIES

One of the major challenges facing the study of primate positional behavior is constructing precise and functionally meaningful categories of posture and locomotion. There has been a tendency for researchers to group what may be a set of biomechanically diverse behaviors into a small number of discrete categories (Walker 1998). For example, many studies (Tables 36.2 and 36.3) collapse a species' entire locomotor repertoire into four or five activities, such as climb, leap, quadrupedalism, tail–arm suspend, or some other form of bimanual or suspensory locomotor behavior. Clearly, a species' positional repertoire is considerably more diverse, and the inclusion of behaviors that are biomechanically distinct and exert different forces on the skeleton into a single category is problematic (Walker 1998). For example, the category "quadrupedalism" may include three distinct behaviors: quadrupedal walking (which in most primates is associated with a slow diagonal sequence–diagonal couplet gait with footfall placements at regular intervals), quadrupedal running (a more rapid form of pronograde travel which is mechanically similar to quadrupedal walking), and bounding (a hindlimb-dominated, asymmetrical form of progression in which the hindlimbs provide powerful and rapid extension and the forelimbs may act as struts, providing stability and orientation in redirecting the line of progression). Bounding also is distinct from running in that it includes a brief in-air phase of stride. Lumping these modes into a single category obscures potentially important species, sex, ontogenetic, seasonal, and biomechanical differences in positional behavior. Similarly, the category "leaping" is often inadequately defined. In general, long hindlimbs are advantageous to leapers because

Arboreal quadruped

Long Tail

Narrow Thorax
Laterally placed Scapula

Long Olecranon Process

Deep Ulna

Grasping Foot

Short, Similar–length Forelimb and Hindlimb

Terrestrial quadruped

Reduced Tail

Narrow Thorax

Restricted Shoulder Joint

Posteriorly Extended Olecranon Process

Robust Radius

Short Digits

Short Digits

Long, Similar–length Forelimb and Hindlimb

Figure 36.2 Positional behavior and anatomical features in arboreal (*upper*) and terrestrial (*lower*) primates (see text for a discussion of these features). From Fleagle (1988).

they extend the duration of time over which the body can be accelerated during take-off (Niemitz 1984b, Emerson 1985, Anemone 1993, Crompton et al. 1993). However, the same primate species may leap to and from large, noncompliant (rigid) vertical trunks in the undercanopy, which requires powerful hindlimb extension and mechanisms for absorbing significant force on impact (vertical cling and leap); leap from compliant supports located in the periphery of the canopy onto a mass of terminal branches; leap by dropping down into the lower levels of the canopy, which requires no hindlimb propulsion; or leap as a transition from high-speed quadrupedal progression requiring powerful and rapid extension of the hindlimbs (Garber 1991, Garber and Pruetz 1995, Terranova 1996). Leaping should be divided into a range of distinct positional categories based on support compliancy (e.g., when taking off from compliant supports, a leaper may be able to take advantage of elastic rebound to generate additional power) (Demes et al. 1995), branch orientation and size, distance traversed, forces acting on the forelimbs and hindlimbs during take-off and landing, and adjustments of the body, tail, and center of mass (Dunbar 1988; Demes et al. 1995; Garber and Leigh 2001; Garber et al. 2009). Finally, the locomotor category "climbing" often has been used as a "catch-all" for a range of prehensile locomotor behaviors involving various combinations of limbs and tail (McGraw 2000). For example, Fleagle and Mittermeier (1980:302) define climbing as "progression along continuous supports using various combinations of three or more limbs, frequently in irregular, opportunistic sequences." The description of a lateral-sequence walking gait (Dunbar and Badam 2000) used by juvenile bonnet macaques when feeding and foraging on small-diameter branches also may fit this definition. Others have restricted climbing to various modes of vertical progression only (Hunt et al. 1996) or use the term "clamber" to describe cautious movements in which "the body is orthograde with the head superior, and various combinations of all the four appendages attach to substrates in different ways, including suspension by the forelimbs from above" (Cant 1992:279).

Hunt and colleagues (1996) called for a standardized method for describing primate positional behavior. These authors highlighted the fact that we have been limited in our ability to identify relationships between positional behavior and positional anatomy because "important behavioral information is lost in a categorical system wherein a critical positional mode may be conflated with one set of behaviors in one study and a different set in another study" (Hunt et al. 1996:365). Clearly, it is difficult to develop a standard set of positional behaviors that is consistent with the behavior of primate species that vary greatly in body mass (from 35 g in the mouse lemur to 180,000 g in an adult male gorilla); are characterized by significant variation in limb and body proportions; have prehensile tails, unspecialized grasping tails, or no fuctional tail; and have an opposable thumb, a nonopposable thumb, or a vestigial thumb. Moreover, two primate species of equal body size, one whose lineage has undergone

an evolutionary increase in size and one whose lineage has undergone an evolutionary decrease in size, may be characterized by very different allometric growth trajectories, different body proportions, and different ways to locomote (Godfrey et al. 1998). Nevertheless, Hunt et al. (1996) have defined 126 positional behaviors (52 postural and 74 locomotor) that characterize living primates. Although this inventory will need to be amended as new patterns of positional behavior are defined, the goal of standardization is valid and represents an important step forward in comparative studies linking positional behavior with positional anatomy.

THE PRIMATE TAIL

In several species of nonhuman primate, the tail has come to serve an important role in balance, weight support, and arboreal locomotion (Meldrum 1998). Prehensile tails have evolved independently in at least 15 families of mammals, including once in the common ancestor of spider monkeys (*Ateles*), howlers (*Alouatta*), woolly spider monkeys (*Brachyteles*), and woolly monkeys (*Lagothrix*) (subfamily Atelinae) and a second time in capuchins (*Cebus*, subfamily Cebinae, Emmons and Gentry 1983, Rosenberger and Strier 1989, Lemlin 1995, Garber and Rehg 1999). Given significant differences in diet, body mass, and ranging patterns that characterize prehensile-tailed platyrrhines, the prehensile tail likely solves a range of locomotor and postural problems encountered in an arboreal setting (Figure 36.1).

The ateline prehensile tail is distinct from the capuchin prehensile tail in several ways (see Garber and Rehg 1999: Table 1). Data from Grand (1977a) indicate that in *Alouatta* and *Ateles* the tail accounts for 5.5%–7.8% of total body mass. In capuchins, the prehensile tail accounts for 5.4% of body mass. In addition, the ateline tail is elongated, possesses a hairless sensitive pad on the ventrodistal surface containing specialized tactile receptors (possibly Meissner corpuscles), and has an expanded caudal sensorimotor area compared to capuchins (Falk 1980, Niemitz 1990, Bergeson 1996, Garber and Rehg 1999). Despite differences in tail musculoskeletal anatomy, both atelines and capuchins employ their prehensile tails in a variety of postural and locomotor behaviors and can support their full body weight hanging by the tail alone (albeit briefly in capuchins). In *Ateles*, tail–arm suspensory locomotion (tail-assisted brachiation) represents a form of high-speed locomotor behavior enabling an individual to rapidly travel long distances following relatively straight-line pathways across gaps in the forest canopy (Figure 36.1). Youlatos (2002) reports that black spider monkeys (*Ateles paniscus*) use their prehensile tails during 61.5% of travel bouts and 54.2% of locomotor feeding bouts.

In contrast, tail–arm suspensory locomotion is absent from the *Alouatta* positional repertoire (Fleagle and Mittermeier 1980, Gebo 1992, Bergeson 1996). Despite having an ateline prehensile tail, howler monkeys locomote principally using slow, cautious, above-branch quadrupedal walking augmented by climbing and bridging (Fleagle and Mittermeier

1980, Gebo 1992, Bergeson 1996) (Figure 36.1, Table 36.2). During quadrupedal walking, the howler prehensile tail serves an important role in balance and in providing support by grasping onto branches located above the plane of travel.

Cebus is distinct from the atelines in that it engages in more rapid forms of quadrupedal travel: leaping and climbing. Garber and Rehg (1999) report that, during travel, leaping accounted for 24.7% and climbing 9.4% of the white-faced capuchin (*Cebus capucinus*) positional repertoire. During travel, the capuchin prehensile tail served a weight-bearing role 13.3% of the time. However, during feeding and foraging, tail-assisted suspensory postures accounted for 40.6% of capuchin behavior (Garber and Rehg 1999). White-faced capuchins also have been observed to adopt a crouched bipedal posture with the tail anchored around a moderate-sized support when exploiting palm fruits and during destructive insect foraging. Garber and Rehg (1999:335) state "the prehensile tail in *C. capucinus* serves a broad adaptive role in both above and below-branch foraging providing access to resources located on small and medium sized supports in the perimeter and central regions of the tree crown."

In a discussion of the primate tail, Meldrum (1998:147) highlights the fact that several species of Old World monkeys, New World monkeys, and prosimians have been observed to use their tails for "seizing and grasping" arboreal supports. In the case of the bearded saki monkey (*Chiropotes satanas*) and the red-ruffed lemur (*Varecia variegata*), Meldrum (1998) argues that the tail serves an important function during tail-assisted hindlimb suspensory behavior. In this context, their relatively long and fluffy tail is draped over a branch and acts as a brace while the forager hangs vertically downward. Meldrum (1998:154) hypothesizes that "tail bracing and twining, in conjunction with hind limb suspension selected for greater caudal robusticity and prehensile capabilities in some primate lineages, especially ancestral atelines and *Cebus*." However, many primate taxa engage in hindlimb suspension without tail assistance (Meldrum et al. 1997). Moreover, given major differences in *Cebus* and ateline behavior and ecology, it is likely that the evolution of a prehensile tail in these lineages is the result of alternative selective pressures.

Finally, many primates use their tail as a counterbalance during quadrupedal walking and leaping. For example, Dunbar and Badam (2000) report that in juvenile bonnet macaques (*Macaca radiata*) the tail may be hooked around a branch to control descent when traveling on small-diameter branches. Moreover, in several species of prosimians, trunk-to-trunk leaping requires mechanical adjustments of limbs, tail, upper torso, and center of gravity during the take-off, in-air, and landing phases of travel. In particular, movement of the tail plays an important role, along with aerial rotation, enabling the hindlimbs to be brought forward and to strike the landing platform well in advance of the forelimbs (Dunbar 1988). Why hominoids lost an external tail, why some primate lineages have a reduced tail, or why no lineage of Old World primate (prosimian or higher primate) evolved

a prehensile tail remains unclear. Abitbol (1987) has identified differences between hominoids and Old World monkeys in the development and weight-bearing role of the sacrum that may provide some insights into why ape ancestors lost their tails. He argues that "the tendency of the hominoid sacrum to incorporate the last lumbar vertebrae and to widen markedly... offers a firm base of support for the trunk... and stabilization of the body during bipedal posture and locomotion" (Abitbol 1987:65). Moreover, Pilbeam (2004:254) notes that whereas cercopithecoids are relatively conservative in the number of vertebral elements and appear to be subject to strong stabilizing selection to maintain a fixed pattern, hominoids are characterized by considerably greater axial skeleton variability and "tolerate a broader range of phenotypes." The habitual use of vertically oriented (orthograde) postures associated with increased weight-bearing via forelimb suspension and/or hindlimb compression may have resulted in a reorientation and revised function of lumbar, sacral, and coccygeal vertebrae in early hominoids. This argument, however, remains speculative.

EFFECTS OF BODY MASS ON PRIMATE POSITIONAL BEHAVIOR AND HABITAT UTILIZATION

Based on biomechanical principles, larger- and smaller-bodied primates are expected to encounter different problems of weight support and progression in the forest canopy (Cartmill and Milton 1977, Fleagle and Mittermeier 1980, Cant 1992). Body mass and limb and body proportions are critical aspects of a species' adaptive pattern, affect among other things basal metabolic rate and locomotor costs, and have significant biomechanical implications for terrestrial and arboreal progression (Larson 1998b, Schmitt 1998, Polk 2002). Firstly, in terms of the arboreal environment, it has been hypothesized that smaller-bodied animals encounter more gaps or discontinuities in the forest "that could only be crossed by leaping," whereas "a larger animal, by virtue of its longer dimensions, would see a relatively greater number that could be crossed by bridging or climbing" (Fleagle and Mittermeier 1980:309). Secondly, it is expected that "a larger animal must find it easier to hang below a relatively small branch than balance atop it" (Cartmill and Milton 1977:269). Thirdly, larger animals are expected to be more restricted in their use of smaller supports than are smaller-bodied animals and, in order to maintain balance, distribute their weight over a greater number of supports (Fleagle and Mittermeier 1980). Finally, given their relatively shorter forelimbs (Jungers 1985), smaller-bodied primates are expected to face greater difficulty than larger-bodied primates when grasping and climbing on large vertical or sharply inclined supports (Cant 1992).

In this regard, Fleagle and Mittermeier (1980) conducted a study of positional behavior in seven species of sympatric primates in Suriname (Table 36.2). These primates varied in adult body mass from 500 to 8,000 g. Studying a set of closely related taxa in the same forest represents an important methodological tool in controlling for phylogeny and ecology in order to isolate the potential effects that species differences in body size have on positional behavior. Although these authors concluded that the empirical evidence on positional behavior in neotropical primates was consistent with theoretical predictions, there were many exceptions. For example, *Pithecia pithecia* (white-faced saki) leaped more than expected for its body mass, *Saguinus midas* (golden-handed tamarin) leaped less than expected for its body mass, and *Chiropotes satanas* (bearded saki) climbed less than expected for its body mass (Table 36.2). With respect to support size, *S. midas* was found to use smaller supports less frequently than expected and *Ateles paniscus* (black spider monkey) was found to use smaller supports more frequently than predicted for its body mass. In black spider monkeys, elongated forelimbs and a prehensile tail facilitate arm–tail suspensory behavior on relatively small supports. In the golden-handed tamarin, the evolution of claw-like nails enables exploitation of larger vertical supports than expected based on body size alone. Thus, although body mass is an important factor in studies of positional behavior and substrate preference, many lineages have evolved behavioral and anatomical adaptations that enable them to override the predicted effects of body size in exploiting food resources and locomotor pathways in an arboreal environment.

In Tables 36.3–36.7, additional data are presented on body mass and patterns of positional behavior in several primate communities. McGraw (1998a,b, 2000) conducted extensive research on positional behavior in seven species of Old World monkeys inhabiting the Taï Forest, Ivory Coast. As indicated in Table 36.3, larger-bodied species did not consistently leap less, climb more, or exploit larger arboreal supports than did smaller-bodied forms. Similar results have been reported in studies of positional behavior in primate communities in Uganda (Gebo and Chapman 1995a, Table 36.4), Madagascar (Dagosto 1994, Table 36.5), and South America (Youlatos 1999, 2002; Cant et al. 2001; Tables 36.6 and 36.7). Overall, patterns of positional behavior and substrate utilization in prosimians, monkeys, and apes appear to be more closely related to species-specific differences in diet (and associations between the physical structure of the canopy and access

to particular food resources), foraging strategies, and morphological adaptations than solely to body mass (Ripley 1967, Fleagle 1978, Rose 1978, Garber 1984b, Meldrum et al. 1997, Dunbar and Badam 2000, Schmitt 2003).

HOW VARIABLE IS POSITIONAL BEHAVIOR?

Studies of the relationship between positional behavior and postcranial anatomy in living primates provide important models for identifying evolutionary and ecological relationships in extinct lineages. If a goal of primatology is to reconstruct the lifestyle of fossil primates, then major challenges to the study of locomotion and posture include establishing functional relationships between behavior and anatomy, identifying elements of the positional repertoire that are stable from those that are more variable, and determining more precisely how positional behavior is influenced by habitat, physiology, and the mechanical costs of movement (Garber 1998). It is generally assumed that locomotor patterns are strongly constrained by anatomy. If, however, patterns of positional behavior are found to be highly variable within a species, then functional anatomists will need to incorporate concepts of behavioral plasticity into models of adaptation and anatomy (Dagosto 1995, Gebo and Chapman 1995a, Garber 1998). Figure 36.3 outlines a set of physiological, nutritional, ecological, behavioral, and social factors that are likely to influence the variability in positional behavior, habitat utilization, and substrate preference within a species.

EVIDENCE OF SEX-BASED DIFFERENCES IN POSITIONAL BEHAVIOR

Given male/female differences in body mass, the nutritional costs of reproduction, social dominance, access to resources, and the role of adult males and females in infant care and protection, significant sex-based differences in positional behavior are expected. However, among many primate taxa, male/female differences in positional behavior are extremely subtle. For example, in the case of common chimpanzees (*Pan troglodytes*, males weigh approximately

Table 36.7 SUBSTRATE UTILIZATION[1] (TRAVEL, FEED, AND FORAGE) IN SEVEN SPECIES OF SYMPATRIC PRIMATES IN ECUADOR[2]

SPECIES	BODY MASS (KG)	SMALL	MEDIUM	LARGE	VERY LARGE	N
Ateles belzebuth	7.7	41.4	40.6	16.0	1.8	1,355
Cebus albifrons	2.8	45.7	44.9	2.9	1.6	243
Pithecia monachus	2.4	24.4	54.7	16.3	2.6	449
Callicebus moloch	1.0	32.0	63.7	2.7	1.2	325
Saimiri sciureus	0.7	52.2	27.9	17.2	0.7	569
Saguinus tripartitus	0.4	31.1	46.6	10.5	11.2	626
Cebuella pygmaea	0.1	14.1	54.5	17.2	13.1	709

Sources: Youlatos 1999, 2002.

[1] Substrate size in diameter: small, ≤2 cm; medium, >2 but ≤10 cm; large, >10 but ≤20 cm; very large, >20 cm.

[2] Data represent percentage of time supports of particular size were utilized.

Figure 36.3 Factors influencing patterns of within-species variability in substrate preference and positional behavior.

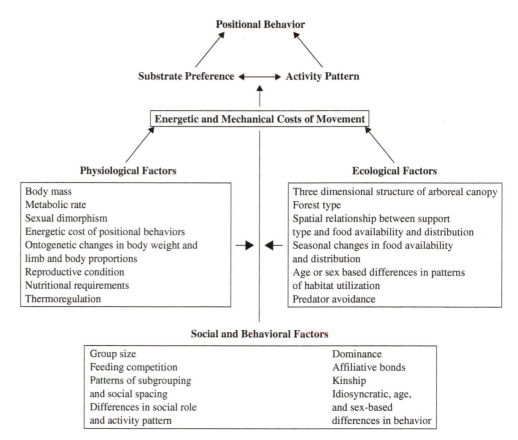

Table 36.8 Evidence of Sex-Based Differences in Positional Behavior in Taï Forest Chimpanzees[1] (Pan troglodytes verus)

	BODY MASS (KG)	ALL LOCOMOTION					
		QUAD	CLIMB/SCRAMBLE	SUSPENSORY	BIPEDAL	LEAP	N
Males	46	86.6	11.1	1.1	1.2	0	732
Females	34	85.6	10.9	1.4	1.2	0.6	685

	ARBOREAL LOCOMOTION				
	QUAD	CLIMB/SCRAMBLE	SUSPENSORY	BIPEDAL	LEAP	N
Males	11.7	76.7	5.8	5.8	0.0	103
Females	30.3	59.8	7.4	0.8	1.6	122

	ALL POSTURES			
	SIT	STAND	LIE	ARM-HANG	N
Males	70.8	6.1	22.1	1.0	4,293
Females	80.6	5.6	11.6	2.1	4,367

	ARBOREAL POSTURES			
	SIT	STAND	LIE	ARM-HANG	N
Males	82.0	1.6	13.5	2.8	1,524
Females	83.3	1.0	12.4	3.6	2,383

Source: Adapted from Doran 1993.

[1] Data represent percentage of time engaged in each positional behavior.

Table 36.9 Evidence of Sex-Based Differences in Patterns of Arboreal Positional Behavior in Western Lowland[1] Gorillas (Gorilla gorilla)[2]

		ARBOREAL LOCOMOTION						
SEX	BODY MASS (KG)	QUAD	CLIMB	SCRAMBLE/CLAMBER	SUSPEND	BIPED	OTHER	N
Females	71	23.2	40.0	24.0	4.2	4.2	4.2	95
Males	169	21.5	42.1	23.7	7.9	2.6	7.9	38

		ARBOREAL FEEDING POSTURES							
SEX	BODY MASS (KG)	SIT	SQUAT	SUSPEND	BIPED	TRIPED	QUAD STAND	LIE	N
Females	71	62.0	27.0	3.0	3.4	1.5	3.8	0.4	269
Males	169	54.6	36.1	1.3	0.8	1.1	1.9	0.3	366

Source: Adapted from Remis 1998.

[1] Lowland gorillas are from Bai Hokou, Central African Republic Remis 1998.

[2] Data represent percentage of time engaged in each positional behavior.

30% more than females), Doran (1993:102) found "no significant sex differences in the frequency of locomotor activities performed during either feeding or travel" (Table 36.8). This occurred in part because the dominant locomotor activity of both male and female chimpanzees was quadrupedal travel (knuckle-walking) on the ground. Doran (1993) does report, however, that males and females differ in arboreal feeding locomotion, with males characterized by an increase in climbing and females characterized by an increase in quadrupedal locomotion. However, arboreal feeding locomotion accounted for only 2.2% of the chimpanzee positional repertoire; thus, in general, sex-based differences in chimpanzee positional behavior were minimal.

In Table 36.9, data are presented on arboreal locomotion in western lowland gorillas (*Gorilla gorilla gorilla*). Despite the fact that adult males weigh over twice as much as adult females and females spend several years nursing and caring for young, patterns of positional behavior were virtually identical (Remis 1998). Climbing was the most common locomotor behavior for both males and females, followed by scramble/clamber, quadrupedal travel, and suspension. Male and female lowland gorillas also had a similar positional profile during arboreal feeding postures. The primary difference was that, although sitting was the most common feeding posture for both sexes (Table 36.9), males were found to squat more frequently than females. Overall, females did exploit smaller supports and spend more time in the periphery of the tree crown than did adult males. However, this difference appeared to relate more to social interactions and social role than to mechanical problems associated with body mass and weight support (Remis 1998).

Orangutans (*Pongo pygmaeus*) are the largest habitually arboreal mammal, with males being approximately 2.2 times heavier than females (84 vs. 38 kg). Cant (1987b) conducted a study of positional behavior in male and female Sumatran orangutans exploiting two large fig trees over the

course of 11 days. He found that males tended to exploit larger supports than females but that, contrary to theoretical expectations, smaller-bodied females engaged in suspensory postures more frequently than did larger-bodied males (35% vs. 23%). Suspension most commonly occurred on small supports, but even when controlling for support size, females suspended more than males. Males were found to sit and stand more during feeding.

Tables 36.10–36.12 provide a comparison of male and female patterns of positional behavior in five species of Old World monkeys (Gebo and Chapman 1995b) and two species of New World monkeys (Gebo 1992). In the case of arboreal cercopithecids, Gebo and Chapman (1995b) found virtually no differences in postural behavior, locomotor behavior, or substrate utilization (size and orientation) between males and females, even in highly dimorphic species (Table 36.10). These authors conclude that, either within or between species, "changes in body size do not correspond closely to changes in locomotor frequency" (Gebo and Chapman 1995a:71). A generally similar pattern was found in African guenons (McGraw 2002) and New World primates (Gebo 1992). Taken together, data on apes, Old World monkeys, and New World monkeys provide only limited evidence for significant sex-based differences in locomotion, posture, and substrate utilization. This supports the contention that for many primate species patterns of positional behavior are highly conservative, at least among adult group members.

EVIDENCE OF ONTOGENETIC OR AGE-BASED DIFFERENCES IN POSITIONAL BEHAVIOR

Data on ontogenetic changes in primate positional behavior are limited but offer critical insight into how factors such as body mass, motor skills, and developmental trajectories affect ecological and dietary distinctions between adult and immature individuals. In Table 36.13, I summarize data published by Bezanson (2006) comparing adult

Table 36.10 Evidence of Sex–Based Differences in Positional Behavior[1] in Five Species of Primates in Kibale Forest, Uganda

SPECIES	BODY MASS (KG)	CLIMB	LEAP	QUAD	QUAD STAND	BIMANUAL	VERTICAL BOUND	N
					TRAVEL			
Colobus guereza								
Males	10.1	11	44	39	<1	<1	6	751
Females	8.0	10	45	39	0	<1	5	1,252
Cercocebus albigena								
Males	8.9	33	21	46	0	<1	<1	821
Females	6.4	31	22	46	0	<1	<1	1,491
Colobus badius								
Males	8.2	29	31	35	0	3	2	1,194
Females	8.2	29	30	33	<1	3	4	1,995
Cercopithecus mitis								
Males	6.0	29	18	50	<1	1	2	576
Females	3.5	29	18	51	0	<1	1	789
Cercopithecus ascanius								
Male	4.1	27	26	42	0	<1	5	313
Females	3.0	27	24	41	0	<1	7	808

Source: Gebo and Chapman 1995b.

[1] Data represent percentage of time engaged in each positional behavior.

Table 36.11 Evidence of Sex–Based Differences in Positional Behavior[1] in *Alouatta palliata* in Costa Rica

	BODY MASS (KG)	QUAD	CLIMB	LEAP	QUAD STAND	BRIDGE	BIMANUALISM	N
				LOCOMOTOR BEHAVIOR				
Males	6–7	48	35	5	2	9	1	2,400
Females	4–5	45	38	4	2	10	1	2,577

	BODY MASS (KG)	SIT	RECLINE	QUAD SUSPEND	HINDLIMB SUSPEND+TAIL	QUAD STAND	BIPED STAND	N
				POSTURAL BEHAVIOR				
Males	6–7	53	12	2	9	21	1	1,338
Females	4–5	52	12	3	11	19	2	1,274

Source: Adapted from Gebo 1992.

[1] Data represent percentage of time engaged in each positional behavior.

Table 36.12 Evidence of Sex–Based Differences in Positional Behavior[1] in *Cebus capucinus* in Costa Rica

	BODY MASS (KG)	QUAD	CLIMB	LEAP	QUAD STAND	BRIDGE	BIMANUALISM	N
				LOCOMOTOR BEHAVIOR				
Males	3.7	58	22	17	<1	2	<1	454
Females	2.5	49	29	15	<1	4	2	546

	BODY MASS (KG)	SIT	RECLINE	QUAD SUSPEND	HINDLIMB SUSPEND+TAIL	QUAD STAND	BIPED STAND	N
				POSTURAL BEHAVIOR				
Males	3.7	38	19	1	2	37	4	189
Females	2.5	50	8	4	5	26	6	219

Source: Gebo 1992.

[1] Data represent percentage of time engaged in each positional behavior.

Table 36.13 Evidence of Age-Based Differences in Locomotor Behavior[1] in Two Sympatric Primates

	LOCOMOTOR BEHAVIOR					
SPECIES[2]	QW	QR	BRIDGE	VERTICAL CLIMB	LEAP	BIMANUAL PULL-UP
Alouatta palliata						
Juvenile 1	64.7	7.9	3.1	10.8	6.1	2.8
Juvenile 2	62.4	5.7	12.2	9.2	7.0	2.1
Adult	77.3	1.8	10.2	5.9	2.4	1.1
Cebus capucinus						
Juvenile 1	55.9	17.2	<1	9.2	15.5	1.1
Juvenile 2	66.7	11.7	<1	6.4	10.8	1.1
Adult	76.6	4.3	<1	6.8	10.3	<1

Source: Adapted from Bezanson 2006.

[1] Data represent percentage of time engaged in each locomotor behavior.

[2] J1 howlers are estimated to be 6–12 months of age, J2 howlers are estimated to be 12–24 months of age, and adult howlers are estimated to be greater than 48 months of age.

J1 capuchins are estimated to be 5–12 months of age, J2 howlers are estimated to be 12–36 months of age, and adult howlers are estimated to be greater than 72 months of age.

QW, quadupedal walk, QR, quadrupedal run.

and juvenile locomotor behavior in white-faced capuchin monkeys (*Cebus capucinus*) and mantled howler monkeys (*Alouatta palliata*). Both species inhabited a wet tropical rain forest, La Suerte Biological Research Station, in northeastern Costa Rica. Relative to howler monkeys, capuchins are reported to exhibit a wide range of anatomical and behavioral traits associated with a slow life history and an extended period of juvenile development (Fragaszy et al. 2004b). For example, adult female white-faced capuchins weigh approximately 2.6 kg but do not have their first offspring until age 7, whereas adult mantled howler monkey females weigh 4.7–6.4 kg and give birth at age 4 (Glander 2006). In terms of motor development, Bezanson (2006:341) reports that both capuchins and howlers exhibit adult-like patterns of locomotion early in development, generally at 5–24 months of age. However, "juvenile white-faced capuchins exhibited adult-like locomotion earlier in development" (5 months of age) than did juvenile mantled howler monkeys (6–24 months of age). This is consistent with data on feeding and foraging behavior collected by Mackinnon (2006). She found that juvenile white-faced capuchins 1–2 years old were adult-like in their ability to locate, process, and consume plant and animal prey. An early onset of adult-like locomotor and feeding behavior similarly has been reported in Costa Rican squirrel monkeys (*Saimiri oerstedii*) (Boinski 1989a). These data suggest that in some primate taxa there exists a disassociation between body mass, social maturation, and the development of locomotor and foraging skills.

In a series of publications Thorpe and Crompton (2005, 2006, 2009) examined evidence for age- and sex-based differences in positional behavior in two adult male, three adult female, two subadult male, and three immature wild Sumatran orangutans (*Pongo pygmaeus abelii*). Similar to Cant's (1992) orangutan study, Thorpe and Crompton found no significant differences in the frequency of suspensory behaviors between adult males, adult females, and immatures (adult male weight is approximately three times that of immatures and twice that of adult females) (Thorpe and Crompton, 2005) (Table 36.14). Moreover, Thorpe and Crompton (2005, 2006, 2009) report that adults and immatures have very similar patterns of locomotor behavior, postural behavior, and support use and that neither body mass nor age (adult male vs. adult female, immature vs. adult) is a strong predictor of orangutan positional behavior (Table 36.14). Thorpe and Crompton (2009) did note, however, certain differences in patterns of positional behavior and habitat use among adults and immatures. They found that immature orangutans more frequently engaged in "faster" locomotor modes than did adults and that, although flanged males acquired fruits on small supports, they more often moved to a more stable support to feed than did adult females and immatures.

Thus, taken together, data on apes, Old World monkeys, and New World monkeys provide only limited evidence for significant age- and sex-based differences in positional behavior and substrate utilization. This supports the contention that, for many primates, intraspecific patterns of positional behavior are highly conservative and constrained more by anatomy and neural development than body mass. For example, in some primate species it is more efficient for males, females, and immatures of the same species to solve problems of balance and weight support on small arboreal branches either by lowering their center of gravity close to the support, shifting their center rearward and maintaining a strong grasp with their hindfeet, or grasping the support with a combination of hands, feet, and tail to stabilize the body's center of mass over the support (Cartmill 1985). For other

Table 36.14 Evidence of Age-Based Differences in Positional Behavior[1] in Sumatran Orangutans (*Pongo pygmaeus abelii*)

AGESEX	POSTURAL BEHAVIOR						LOCOMOTOR BEHAVIOR					
	SIT	LIE	FHS	OS	PSU	PST	TOS	Q	VC	VD	B	SWY
Adolescent	53	17	7	6	2	4	35	22	16	7	6	9
Adult female	58	8	10	7	4	4	38	16	15	13	8	7
Adult male/ subadult male	53	17	12	4	5	2	34	17	17	9	7	5

Source: Adapted from Thorpe and Crompton 2006.

Values are estimated from Figures 1 and 2 of Thorpe and Crompton 2006 and therefore do not represent actual values.

[1] Study is based on two immature males estimated by be 7–8 years of age, one immature female estimated to be 12 years of age, three adult females 24–34 years of age, two subadult males 21–31 years of age, and two adult phalanged males of unknown age.

FHS, forelimb/hindlimb suspend; OS, orthograde stand; PSU, pronograde suspend; PST, pronograde stand; TOS, torso-orthograde suspensory locomotion; Q, quadrupedal walk; VC, vertical climb; VD, vertical descent; B, bipedal walk; SWY, sway.

primates, problems of balance and weight support are more effectively addressed by suspensory postures and locomotor modes using some combination of forelimbs, hindlimbs, and tail. Finally, based on current evidence, it appears that in many primate species, including taxa that are characterized by a relatively short juvenile period and those characterized by a relatively long juvenile period, immatures exhibit adult-like patterns of positional behavior at a relatively early age. A similar pattern is reported in humans. Human children generally begin to walk by 12 months of age and over the course of the next 6–12 months begin to resemble an adult-like pattern of gait stability and relative stride length (Polk et al. 2008). In a study of 309 normal human children (Sutherland et al. 1988), limb stance, walking velocity, step length, and the ratio of pelvic span to ankle spread were scored as adult-like by age 3 (2 years of walking experience). This early age of locomotor development in human children contrasts markedly with other aspects of human development, including a late adolescent growth spurt and late age at first reproduction (Leigh 2001, Walker et al. 2006).

EVIDENCE OF SEASONAL AND SITE-SPECIFIC DIFFERENCES IN POSITIONAL BEHAVIOR

Gebo and Chapman (1995a) examined intraspecific variability in positional behavior of red colobus monkeys (*Piliocolobus badius*, also known as *Procolobus badius*) in three forests in Uganda. These authors found that the red colobus positional repertoire was consistent from site to site and that common postural and locomotor behaviors at one site also were common at other sites. However, the frequency of positional behaviors was more variable during feeding than during traveling. This also has been reported in several other primate species (Crompton 1984, Dagosto 1995, Garber 1998, McGraw 1998c). Marked consistency in patterns of positional behavior during travel is likely to reflect the fact that when moving between feeding sites individuals may reuse

particular arboreal pathways and/or select substrates that offer them safety and efficiency. A higher degree of variability in the positional repertoire during feeding may be associated with seasonal changes in diet, foraging strategies, and the availability and distribution of feeding sites (Grand 1972; Charles-Dominique 1977b; Garber 1980, 1984b; Crompton 1984; Cant 1987a; Boinski 1989a; McGraw 1998a).

Table 36.15 presents data on site-specific differences in positional behavior in two species of spider monkey (Youlatos 2002). *Ateles paniscus* was studied by Youlatos (2002) in French Guiana and by Mittermeier (1978) in Suriname. Despite differences in methods of data collection (instantaneous vs. bout sampling) and definitions of behavioral categories, there was remarkable consistency between studies. At both sites, tail–arm suspension was the most common positional behavior. At both sites, leaping and bridging were the least common positional behaviors. The greatest difference found between sites was in the frequency of clambering/climbing behavior (28.1% in French Guiana and 17.0 in Suriname).

In the case of *Ateles geoffroyi* (black-handed spider monkey), however, site-specific differences in positional behavior were more marked. In four studies, the frequency of quadrupedal progression ranged from 22.0% to 52.0% and climbing varied from 6% to 24%. Although some of this variation is likely related to differences in methodology, a study by Bergeson (1996) on *A. geoffroyi* using the same methodology at two different sites (a dry tropical forest and a rain forest) found considerable within-species variability during gap crossings. In the wetter and taller forest of La Selva, Costa Rica, *A. geoffroyi* was found to use bridging significantly more and tail–arm suspension significantly less than in the drier forest of Santa Rosa, Costa Rica. All other forms of positional behavior were used at similar frequencies. Although the orientations and sizes of substrates were generally similar between sites, at La Selva spider monkeys descended oblique supports at greater frequency, ascended oblique supports at lower frequency, and tended to use smaller supports more frequently than at Santa Rosa (Bergeson 1996).

Table 36.15 Evidence of Within- and Between-Species Variation in Positional Behavior[1] During Travel in Spider Monkeys

SPECIES	LOCOMOTOR BEHAVIOR					SUBSTRATE SIZE			
	QUAD	CLAMBER/CLIMB	BRIDGE	TAIL–ARM SUSPEND	LEAP	SMALL	MEDIUM	LARGE	N
Ateles paniscus (Youlatos 2002)	20.1	28.1	8.8	34.2	2.6	41.9	40.3	17.8	1,089
Ateles paniscus (Mittermeier 1978)	25.4	17.0	8.6	38.6	4.2	38.2	49.5	12.3	743
Ateles geoffroyi (Mittermeier 1978)	22.0	24.2	4.6	25.7	10.9	28.0	51.6	20.4	1,000
Ateles geoffroyi (Fontaine 1990)	51.0	6.0	1.6	21.9	10.9	16.5	60.7	22.8	1,962
Ateles geoffroyi (Cant 1986)	52.0	19.0	4.0	25.0	1.0	50.0	42.0	8.0	1,685
Ateles geoffroyi (Bergeson 1996)	36.1	18.2	3.0	36.3	2.5	42.2	35.7	22.1	647

Source: Adapted from Youlatos 2002.

[1] Data represent percentage of time engaged in each positional behavior and exploitation of supports of particular size.

Finally, Dagosto (1995) provides data on seasonal and site-specific patterns of positional behavior in Malagasy prosimians (Table 36.5). Based on her study of *Eulemur fulvus* (brown lemur), *Eulemur rubriventer* (red-bellied lemur), *Propithecus diadema* (diademed sifaka), and *Varecia variegata* (ruffed lemur), Dagosto (1995:811) concluded that "locomotion during travel appears to be fairly conservative while differences during feeding contribute substantially to the overall differences observed."

CONCLUSION

Throughout its development the study of primate locomotion has been characterized by its interdisciplinary focus combining ecology and behavioral observations with physiological and morphometric studies to understand both how and why primates use different patterns of locomotion and posture. (Fleagle 1998:109)

Evidence presented in this chapter on primate positional behavior points to several important conclusions. First, closely related primates living in the same ecological community are often characterized by significant interspecific differences in patterns of positional behavior, body mass, and positional anatomy. Second, although smaller- and larger-bodied arboreal primates face different challenges in maintaining balance and weight support, body mass appears not to be as strong a predictor of positional behavior as is often assumed. Individual species of different body mass and body size are characterized by particular behavioral and morphological adaptations that enable them to exploit similar sets of resources in different ways. Third, patterns of positional behavior within a species are generally conservative, especially during travel, and vary little among adult males and adult females. In contrast, within-species patterns of positional behavior during feeding are more variable and likely to reflect seasonal changes in diet, day range, foraging strategies, activity pattern, and social interactions. In this regard, Dagosto (1995:832) cautions that "differences of as much as 15–20% between species in any particular positional behavioral category may not be meaningful for interpreting morphology since differences of such magnitude may be observed within a single species and a result of temporal or geographic variation."

A major challenge for the study of primate positional behavior is to continue to develop new methodologies in both field and laboratory settings for identifying functional relationships between patterns of movement, habitat utilization, and anatomical structure. The expanded use of digital video and other imaging technologies, standardized and biomechanically meaningful definitions of positional categories, and more detailed information on the ecological and social referents of positional behavior are critical to this effort. Morphological design sets limits or boundaries on the range and mechanical efficiency of particular locomotor and postural patterns. For a particular species, these limits may be broad or narrow depending on phylogenetic, ontogenetic, ecological, and behavioral factors. We know very little about the ontogeny of positional behavior in primates and the degree to which age-related changes in motor skills, limb and body proportions, and body mass affect an individual's ability to exploit its environment (Doran 1997b, Bezanson 2003). In addition, there is evidence that in some primate species neuromuscular development associated with locomotor skills may have become dissociated from neuromuscular development required for fine motor control and object manipulation. For example, many primate species attain locomotor independence within a few months after birth but still lack the fine motor skills required for extractive foraging and prey manipulation. In the case of tamarin monkeys, individuals reach locomotor independence by 3 months of age but are still provisioned with insects by adult caretakers at 9 months of age (Garber and Leigh 1997). Similarly, chimpanzees attain an adult-like locomotor profile at approximately 5 years of age (Doran 1997b). However, according to Boesch and Boesch-Achermann (2000), it was not until age 7 that juveniles reached a level of efficiency in cracking *Panda* nuts such that energy intake for cracking exceeded energy intake by sharing nuts opened by their mothers. Moreover, adult levels of *Panda* nut cracking were not reached until age 11. In other primate species, locomotor skills and fine manipulative skills appear to develop at approximately the same time (e.g., *Saimiri oerstedii*, Boinski and Fragaszy 1989). In this regard, studies of primate positional behavior should be placed within the context of primate life history strategies and evaluated in terms of patterns of somatic and neural growth and development, age-related survivorship, and the

requirements of efficiently exploiting particular resources and forest habitats.

Finally, greater attention must be paid to identifying the energetic costs of locomotion and posture. Many models of primate socioecology and behavioral ecology assume high energetic costs of increased travel (Janson and Goldsmith 1995, Chapman and Chapman 2000). However, recent physiological studies suggest that for both arboreal and terrestrial primates the distance traveled each day represents only 3.5%–10% of daily energy expenditure (Steudel 2000).

37

Primate Self-Medication

Michael A. Huffman

1. What drives the individual to self-medicate?
2. What ways can the individual avoid, suppress, and/or treat the causes of illness?
3. How widespread is self-medication in the primates? Are there similarities in how non-human and human primates self-medicate?

INTRODUCTION

Whether prosimian or great ape, a primate's existence is closely intertwined with that of parasites and pathogens (Huffman and Chapman 2009). There is a great diversity of host–parasite relationships across the animal kingdom (see Clayton and Moore 1997, Moore 2002). Parasites and pathogens have a strong influence on their hosts, as is attested by the variety of effects they have on the overall behavior and reproductive fitness of individuals infected by them (e.g., Beckage 1997, Hart 1990, Holmes and Zohar 1990). While parasites are considered to be of little problem to the host (e.g., Ashford et al. 2000), when not kept in check the consequences can be great (cf. Allison 1982, Brack 1987, Huffman et al. 1996b, Toft et al. 1991). Hence, when the well-being of an individual is compromised to the point that daily activities no longer become possible or when reproduction and infant rearing are affected, it is unquestionably in that individual's interest to respond.

Growing awareness and interest in the way that hosts are able to suppress or prevent the deleterious effects of parasitism and other causes of illness via behavioral means has led to increased attention on the evolution and operation

of such a system (e.g., Hart 1988; Huffman 1997, 2001; Lozano 1998; Engel 2002). This field of study is referred to as "zoopharmacognosy" (Rodriguez and Wrangham 1993, Berry et al. 1995) or "self-medication" (e.g., Huffman 1997, 2001; Lozano 1998). The original definition given by Rodriguez and Wrangham (1993) included only the use of medicinal plants and their pharmacological properties, hence the root word "pharmacognosy." Advancement of this field in recent years has shed new light on known and new behaviors incorporating nonplant materials and indeed nonpharmacological means of parasite control using plants, making this term's general applicability obsolete. In this chapter, as in my previous works, *self-medication* is used, defined as those behavioral strategies by which animals avoid or suppress disease transmission and treat or control disease and/or its symptoms, thereby directly or indirectly enhancing their health and reproductive fitness. The study of self-medication is devoted to understanding how animals respond to potential threats to their health and how these behaviors are maintained within a population.

Behavioral strategies for health maintenance typically form what Hart (1990) calls "the front-line defense" against invasion by agents of disease such as nematodes, microparasites (protozoa, bacteria, viruses, fungi), ectoparasites, and biting insects. I have argued that such behavioral strategies come into play when physiological adaptations are insufficient and behavioral avoidance, limited contact with, or a direct response to illness is warranted (Huffman and Caton 2001). From either perspective, the maintenance of behavioral strategies that insure basic survival and, thus, indirectly or directly enhance reproductive fitness is universally adaptive to all animals.

Self-medication may be operated by innate mechanisms, socially transmitted from generation to generation, or indeed a combination of the two (Huffman and Hirata 2003). At the ultimate level, behavioral propensities to perform such actions or basic dietary preferences may be selected for their direct benefits to health maintenance and increased reproductive fitness. Lozano (1998) divided self-medication into prophylactic and therapeutic modes and predicted that the level of occurrence and modes of acquisition should be different. The use of the term "prophylactic" implies intentionality and an understanding of the cause of illness and how to prevent it. Such capabilities have not yet been demonstrated in nonhuman animals, and as Lozano himself admits, the use of this classification is premature as future research may well show that some presumptive prophylactic behaviors are in fact examples of therapeutic self-medication.

At our current level of understanding, health-maintenance and self-medicative behaviors can be classified into five levels: (*1*) *sick behaviors*, such as lethargy, depression, anorexia, reduction in grooming, behavioral fever (self-induced rise in body temperature, e.g., sun basking), basking behavior (Kluger et al. 1975, Hart 1988); (*2*) *optimal avoidance or reduction of the possibility for disease transmission*, such as avoidance of feces-contaminated food, water, substrates; (*3*) *dietary selection of items with a preventative or health-maintenance effect*, such as items eaten routinely in small amounts or on a limited basis; (*4*) *ingestion of a substance for the curative treatment of a disease or the symptoms thereof*, such as use of toxic or otherwise biologically active items at low frequency or in small amounts, having little or no nutritional value; and (*5*) *external application of a substance to the body for the treatment or control of a disease or condition*.

Some of the most extensively documented examples of animal self-medication come from primates. This chapter reviews the current evidence for self-medication and other possible forms of health maintenance in primates. Behavioral categories to be discussed in this chapter fall within levels 2–5.

BEHAVIORAL AVOIDANCE OF DISEASE TRANSMISSION

Perhaps the most fundamental behavioral strategy for maintaining health is to avoid or decrease the likelihood of disease transmission. This can occur in various ways. Waterborne diseases pose a threat to primates directly dependent on streams and ponds as their main water source. *Papio hamadryas* living near the city of Taif, Saudi Arabia, are known to dig drinking holes in the sand directly adjacent to the murky, algae-tainted watering sites of livestock. They patiently wait for the filtered water to seep through the sand (A. Bough and A. Mori personal communications). This appears to be a common behavior of this species at many locations (e.g., Nelson 1960, Kummer 1995). *Theropithecus gelada* living in the Simien highlands of Ethiopia often drink from the cliff springs where they spend their nights, instead of down in the riverbeds. In addition to the obvious benefits of filtering out unwanted slime, dirt, and debris, Mori and Bough believe this is likely an efficient way for baboons to avoid fecal contamination and water-transmitted parasites such as the blood fluke *Schistosoma mansoni*, found in stagnant ponds and slow-running streams in these areas.

Hausfater and Meade (1982) describe the routine changing of sleeping grove sites by *Papio cynocephalus* in Kenya's Amboseli National Park as a strategy for avoiding parasite infection via restricted contact with infective ova and larvae of nematodes (*Oesophagostomum*, *Strongyloides*, and *Trichostrongylus*). *Cercocebus albigena* defecate throughout the day and appear not to be able to avoid, or even try to avoid, contaminating vegetation that they later use for feeding, traveling, or sleeping (Freeland 1980). Freeland (1980) proposed an alternative way that these monkeys may avoid infection from intestinal protozoa. He showed that *C. albigena* ranged more widely during dry periods than rainy periods, when parasite-contaminated fecal material was likely to stay longer on the leaves at Ngogo in the Kibale Forest, Uganda. In this study, neither food density nor an inhibition to travel in the rain could explain this difference (Freeland 1980). Interestingly, 10 km away in Kanyawara, Olupot et al. (1997) found that this species traveled significantly longer distances during rainy than dry periods and that fruit density did influence travel patterns (Olupot et al. 1997). Differences in food distribution and availability between the two sites are cited as a likely reason for this difference, suggesting that either food takes priority over risk of parasite infection or risk of infection may vary between even neighboring habitats. A study by Gillespie (2004) in the same region showed that risk of parasite infection differs significantly for some species between neighboring logged and unlogged habitats.

DIETARY SELECTION AND HEALTH MAINTENANCE

Medicinal Foods

Selecting a proper diet incorporating nonnutritional bioactive elements, so-called medicinal foods (Huffman 1997, Huffman et al. 1998), should be an important part of disease control and prevention. Johns (1990) argues that the herbal medicines and modern pharmaceuticals used by humans today replaced the nonnutritive secondary compounds typically present in the diets of animals in the wild.

In spite of the potential ill effects of secondary plant compounds (Howe and Westley 1988, Wink et al. 1993), primates commonly ingest them (e.g., Carrai et al. 2003; Chapman and Chapman 2002a; Glander 1975, 1982; Janzen 1978; McKey 1978b; Milton 1979; Oates et al. 1977, 1980; Wrangham et al. 1998; Wrangham and Waterman 1983). Janzen (1978) first suggested the possibility that their incidental ingestion by nonhuman primates might help to combat parasite infection. Lozano (1991) proposed that parasites have a significant impact on the foraging decisions of animals, comparable to that of basic nutritional requirements.

A preliminary look at the diet of several primate species suggests that there may be a continuum from mildly bioactive and highly nutritious to minimally nutritious and highly bioactive items. Glander (1994) suggested that *Alouatta palliata* might benefit from the antiparasitic properties of various fig species in their diet. Freeland (1983) also noted figs as a possible mediating dietary factor in primate host–parasite interactions. Clinical trials show that *Ficus* spp. are effective against helminths in both humans (Hansson et al. 1986) and nonhuman animals (de Amorin et al. 1999). The susceptible parasites include *Ascaris* and *Trichuris* (Hansson et al. 1986), both of which have been noted in wild *A. palliata* (Stuart et al. 1998). Ficin is a proteolytic enzyme present in all figs (*Ficus* spp.) (Harborne et al. 1999), potentially active against intestinal parasites. Hansson et al. (1986) reported that concentrations as low as 0.05% of *F. glabrata* latex destroyed the cuticle of *Ascaris* and caused other lethal alterations in the parasites' morphology.

Pan and *Gorilla* across Africa frequently eat the pith and fruit of *Aframomum* (Zingiberaceae). In an extensive literature survey of the pharmacological properties of *Aframomum* eaten by *Gorilla*, a wide range of significant pharmacological properties, including potent bactericidal activities, were found (Cousins and Huffman 2002). Broad fungicidal and pathogen growth-inhibitory activities have also been reported (Adegoke and Skura 1994, Oloke et al. 1988).

Parasites were recognized to have a significant seasonal impact on the health and behavior of *Pan troglodytes schweinfurthii* at Mahale (Huffman et al. 1993, 1996b, 1997). This prompted myself and colleagues to form a multidisciplinary research consortium, The C.H.I.M.P.P. Group (Chemoethology of Hominoid Interactions with Medicinal Plants and Parasites; see Huffman 1994, Jacobsen and Hamel 1996), based at Kyoto University with affiliates worldwide, to conduct research on the pharmacologically based antiparasitic activity of plant foods in the diet of chimpanzees (e.g., Huffman et al. 1998; Koshimizu et al. 1993, Ohigashi et al. 1991b, 1994; Ohigashi 1995). Out of 172 native plant species eaten by chimpanzees at Mahale (Nishida and Uehara 1983), we found that 43 are used to treat parasitic or gastrointestinal illnesses traditionally by humans in Africa (Huffman et al. 1998). In 16 of these species, the same plant part(s) was utilized by both humans and chimpanzees, and 33% (20/63 plant parts) of the items ingested from them by chimpanzees corresponded to the parts utilized in

ethnomedicine for the treatment of intestinal nematodes, dysentery, malaria, colic, and diarrhea and/or as an antiseptic (Huffman et al. 1998). Chimpanzees of M group eat these items only occasionally and typically in small amounts but significantly more frequently during rainy-season months. This coincides with the identified period of reinfection by *Oesophagostomum stephanostomum*, the nodular worm associated with two therapeutic forms of self-medicative behavior at Mahale. Of interest, in vitro and in vivo investigations of the activity of such "medicinal food" items detected significant antimalarial and antischistosomal activity from extracts of *Combretum molle* (Combretaceae), *Erythrina abyssinica* (Papilionaceae), and *Stephania abyssinica* (Menispermaceae) (Huffman et al. 1998).

Krief and colleagues conducted a similar study on the health and pharmacological properties of the diet of Kanyawara group chimpanzees in the Kibale Forest, Uganda (Krief 2003, Krief et al. 2005). Significantly high activity was noted in compounds isolated from plant parts of three species only rarely eaten at Kanyawara, *Diospyros abyssinica* (Ebenaceae), *Uvariopsis congensis* (Annonaceae), and *Trichilia rubescens* (Meliaceae). From *T. rubescens* leaves, two novel compounds possessing significant antimalarial activities were discovered (Krief et al. 2004).

Phillips-Conroy (1996) described the ingestion of *Balanites aegyptiaca* (Balanitaceae) fruit by *Papio hamadryas* and *P. anubis* × *P. hamadryas* hybrid troops living in areas of acute susceptibility to infection by the blood fluke *Schistosoma mansoni* along the Awash River, Ethiopia. While the trees were equally abundant all along the course of the river in their study area, the fruits were eaten only by those troops in areas with the most likely risk of contact with the infective-stage cercaria that persist in warmer, slow-moving water sources where the secondary host snail is found. This led Phillips-Conroy to hypothesize a medicinal function for this food item. Cautious of not overinterpreting results, she left open both possibilities that this is either a dietary strategy with prophylactic effect or a curative treatment.

Subsequent laboratory experiments were conducted by Phillips-Conroy and Knopf (1986), feeding schistosome-infected mice with the active principle of the fruit, diosgenin (a plant hormone). Attempts were made to simulate both prophylactic ingestion (access to diosgenin in the diet before the infection continuing up to the end of the study) and curative ingestion (access only subsequent to infection up to the end of the study). The experiment was inconclusive in demonstrating an antischistosomal effect in either mode of ingestion. In fact, it was shown that liver weight (a pathology of the disease) and female schistosome egg output were greater for diosgenin-treated mice than for control mice (Phillips-Conroy and Knopf 1986). The reader is cautioned, however, that these results do not disprove the original hypothesis that *Papio* benefit from the consumption of these fruits. As the investigators note, the interactions between parasite, hormone, and host are likely to be unique for each species. One important contribution this study has provided is a useful model for future testing.

The tendency for some, but not all, food items in the diet of primates to contain active secondary compounds of potential health-maintenance value suggests that a category of "medicinal foods" exists. It is possible that some examples described here have remained obscure for lack of detailed behavioral and health-related data.

Geophagy

A detailed review of geophagy in primates has been presented by Krishnamani and Mahaney (2000); thus, the subject need only be briefly reviewed here. Clay contains none of the essential components of nutrition: protein, carbohydrates, lipids, and vitamins. Clay soils are fine-grained mineral deposits consisting mainly of the hydrous silicates of aluminum magnesium and/or iron. The chief groups of clay minerals are kaolinite, halloysite, illite, montmorillonite, and vermiculite, whose outstanding property is the capacity for holding water.

Geophagy is practiced by many nonhuman animal species and humans worldwide (e.g., Aufreiter et al. 1997, Bateson and LeBray 1978, Halstead 1968, Johns 1990, Klaus et al. 1998, Kreulen and Jager 1984, Wiley and Katz 1998). From the ethnographic literature, clay is consumed for its effectiveness as an antidiarrheal and is useful for the elimination of bitter taste and the prevention of stomachaches or vomiting after the consumption of large quantities of "famine foods" high in secondary compounds (Johns 1990). Currently, there are at least 38 primate species for which geophagy has been reported (Krishnamani and Mahaney 2000). Similar to humans, in nonhuman primates it has been described as functioning to absorb tannins and other toxins acquired from a diet rich in secondary compounds, to relieve indigestion from a diet of restricted variation, and possibly to mediate endoparasitic infection (Davies and Baillie 1988, Knezevich 1998, Müller et al. 1997, Oates 1978, Wakibara et al. 2001). Overconsumption of clay, however, can lead to serious nutritional deficiencies and illness (e.g., Halstead 1968, Kreulen 1985). Ingestion of termite mound soil in particular has been proposed to aid in the acquisition, enrichment, or expansion of the pool of cellulose-digesting flora in the gut of specialized leaf eaters (Norscia et al. 2005). This may also apply to more generalist primates, like chimpanzees, that also consume termite mound soil.

William Mahaney and colleagues formed the Geophagy Research Project at York University (Toronto, Ontario, Canada) and have begun extensive studies in Africa (Rwanda, Uganda, Tanzania, Guinea), Asia (Sumatra, Borneo, Japan), and the Americas (Puerto Rico) to analyze soils ingested by primates. Thus far, studies of the mountain gorilla (*Gorilla gorilla beringei*) and rhesus monkey (*Macaca mulatta*) suggest that they ingest soil for rare minerals of possible nutritional value or to relieve diarrhea caused by dietary changes

or parasitic infection (Mahaney et al. 1990, 1995, 1996, 1997). All clay ingested by humans and nonhuman primates analyzed thus far has been shown to have components closely resembling those of Kaopectate® (Pharmacia & Upjohn, Kalamazoo, MI), a pharmaceutical commercially sold to treat gastrointestinal upset (Mahaney et al. 1990, 1995, 1996, 1997). Analogous in function to geophagy is charcoal eating, a behavior described for *Piliocolobus kirkii* on the island of Zanzibar, Tanzania, by Cooney and Struhsaker (1997) but also known to occur in other primates and, indeed, other mammals (Engel 2002).

THERAPEUTIC SELF-MEDICATIVE BEHAVIOR

At present, the African great apes offer the most detailed work on possible therapeutic self-medication in nonhuman primates. The hypothesis being developed from these investigations is that certain behaviors aid in the control of intestinal parasites and/or provide relief from related gastrointestinal upset. The two most clearly described behaviors are bitter pith chewing and leaf swallowing. Among the African great apes, one or both of these proposed therapeutic behaviors have been documented from 15 sites and 24 communities at locations spanning their entire geographical distribution (Huffman 2001).

The majority of details discussed here for these two behaviors come from three study areas, Mahale (M group, now extinct K group), Gombe (Kasakela, now extinct Kahama), and Kibale (Kanyawara, Ngogo), where research into chimpanzee self-medication began.

Parasites, Infection, and General Health at the Time of Self-Medication

A first step toward demonstrating therapeutic self-medication is identifying the disease being treated. In a longitudinal investigation of the intestinal parasite fauna of *Pan troglodytes schweinfurthii* conducted on the Mahale M group (Huffman et al. 1997), we showed that they are infected by three parasite species from three genera of nematode, *Strongyloides fuelleborni* (threadworm), *Trichuris trichiura* (whipworm), and *Oesophagostomum stephanostomum* (nodular worm); one genus of trematode, *Dicrocoelium lanceatum* (lancet fluke); and four genera of protozoa, *Entamoeba coli* (amoeba), *Endolimax nana* (amoeba), *Iodamoeba buetschlii* (amoeba), and *Giardia lamblia* (amoeba which causes giardiasis). All are parasites common to *Pan paniscus* and the other *Pan* subspecies in the wild (e.g., Ashford et al. 2000, File et al. 1976, Landsoud-Soukate et al. 1995, Murray et al. 2000).

Individuals of M group were monitored over time to detect weekly, monthly, and yearly changes in infection levels. Among all group members monitored in 1991–1992 and 1993–1994, a statistically significant seasonal difference was recognized only for the number of individuals infected by nodular worms (Fisher's exact test $p = 0.0001$

and $p = 0.01$, respectively) (Huffman et al. 1997). Among all three nematode species detected, nodular worm infections were associated significantly more frequently with bitter pith chewing and leaf swallowing during the same two study periods than either whipworm or threadworm infections (Fisher's exact test for nodular and whipworms $p = 0.0142$ and for nodular and threadworms $p = 0.005$) (Huffman et al. 1997).

Oesophagostomum stephanostomum is the most hazardous species of nodular worm found in the great apes (Brack 1987). Repeated infection occurs in the wild and causes significant complications including secondary bacterial infection, diarrhea, severe abdominal pain, weight loss, and weakness, resulting in high mortality (Brack 1987). Whipworm and threadworm infections usually are not serious and may go unnoticed in mild to moderate cases (Brack 1987). The serious effects of nodular worms on the host, however, suggest that it may be an important stimulus for bitter pith chewing and other antiparasite behaviors.

Bitter Pith Chewing

Behavioral Ecology

Another important step toward demonstrating therapeutic self-medication is distinguishing its use from that of everyday food items and providing evidence for recovery from symptoms identified with the illness associated with its use. Bitter pith chewing is proposed to aid in the control of intestinal nematode infection via pharmacological action and relief from gastrointestinal upset (see Huffman 1997 for detailed review). The hypothesis that bitter pith chewing has medicinal value for chimpanzees was first proposed from detailed behavioral observations and parasitological and phytochemical analyses of patently ill individuals ingesting *Vernonia amygdalina* (Compositae) at Mahale (Huffman and Seifu 1989, Huffman et al. 1993).

Within the home range of M group, *V. amygdalina* is neither abundant nor evenly distributed and usually occurs singly along or near streams. When ingesting the pith from young shoots, the outer bark and leaves are meticulously removed to chew on the exposed pith, from which the bitter juice is extracted. The amount of pith ingested in a single bout is relatively small, ranging from portions of 5 to 120 × 1 cm. The entire process, depending on the amount ingested, takes anywhere from less than 1 to 8 min (Huffman 1997). Often, mature conspecifics in proximity to an individual chewing *Vernonia* bitter pith show no interest in ingesting the pith (Huffman and Seifu 1989, Huffman et al. 1997). Infants, however, have been observed on occasion to taste small amounts of the pith discarded by their ill mothers. The ingestion of leaves or bark, the most abundant parts of the plant available for consumption by chimpanzees, is extremely rare.

At Mahale, use of *V. amygdalina* has been recorded in all months except June and October (late dry season), demonstrating its year-round availability (Nishida and Uehara 1983). Despite this, plant use is highly seasonal and rare (Huffman 1997). General observations on the state of health at the time bitter pith was chewed were reported for four cases recorded between 1987 and 1996 in M group. Ill health was evidenced in all cases by the presence of diarrhea, malaise, and nematode infection (Huffman et al. 1997). Recovery within 20–24 hr from a lack of appetite, malaise, and constipation or diarrhea has also been reported in detail twice (Huffman and Seifu 1989, Huffman et al. 1993).

In one case, the intensity of the infection could be measured and was found to have dropped from an egg per gram (EPG) count of 130 to 15 within 20 hr. This was quite unusual when compared to seven other individuals with nodular worm infections monitored over the same period. Their nodular worm EPG levels conversely increased over time. The average increase was 69.9 EPG (standard deviation = 84, range 5–236) and reflects the rainy-season trend for increased reinfection by nodular worms of chimpanzees at Mahale (Huffman et al. 1997).

Ethnomedicinal Evidence

Along with direct measurable evidence for a change in health conditions following a proposed self-medicative behavior, ethnomedicinal reports of plant activity and direct pharmacological analysis of compounds extracted from the specific plants used by the sick animals are essential steps toward demonstrating therapeutic self-medication. For numerous African ethnic groups, a concoction made from *V. amygdalina* is a prescribed treatment for malarial fever, schistosomiasis, amoebic dysentery, several other intestinal parasites, and stomachaches (Burkill 1985, Dalziel 1937, Huffman et al. 1996a, Watt and Breyer-Brandwinjk 1962).

The noted recovery time of 20–24 hr after bitter pith chewing in two individuals of M group is comparable to that of local human inhabitants, the Tongwe, who use cold concoctions of this plant as a treatment for malaria, intestinal parasites, diarrhea, and stomach upset. These observations encouraged us to further investigate the possible contribution of plant secondary compounds in *V. amygdalina* against parasitic infection. Phytochemical analysis of *V. amygdalina* samples collected at Mahale in 1989 and 1991 from plants used by sick chimpanzees revealed the presence of two major classes of bioactive compounds. From this work, a total of four known sesquiterpene lactones (vernodalin, vernolide, hydroxyvernolide, vernodalol), seven new stigmastane-type steroid glucosides (vernonioside A1–A4, B1–B3), and two freely occurring aglycones of these glucosides (vernoniol A1, B1) were isolated (Ohigashi et al. 1991a; Jisaka et al. 1992a,b, 1993a,b).

Phytochemical Evidence

The sesquiterpene lactones present in *V. amygdalina* are well known in many species of *Vernonia* for their antihelminthic, antiamoebic, antitumor, and antibiotic properties (Asaka et al. 1977; Gasquet et al. 1985; Jisaka et al. 1992a, 1993b; Kupchan et al. 1969; Toubiana and Gaudemer 1967).

Koshimizu et al. (1993) also found inhibition of tumor promotion and immunosuppressive activities.

In vitro tests on the antischistosomal activity of the pith's most abundant steroid glucoside (vernonioside B1) and sesquiterpene lactone (vernodalin) showed significant inhibition of egg-laying capacity consistent with the observed decline in nodular worm EPG level 20 hr after a sick adult female in M group ingested *V. amygdalina* pith (Jisaka et al. 1992b, Huffman et al. 1993).

In vitro tests on the antiparasitic activity of both the sesquiterpene lactones and the steroid glucosides using *Leishmania infantum* (produces visceral leishmaniasis), *Entamoeba histolytica* (produces amoebic dysentery), and a K1 multidrug-resistant strain of *Plasmodium falciparum* (produces falciparum malaria) have also been conducted and showed significant plasmodicidal activity (Ohigashi et al. 1994). In total, the evidence from parasitological, pharmacological, and ethnomedicinal observations is substantial and lends support to the hypothesis that bitter pith chewing is a therapeutic form of self-medication stimulated by, and controlling, nodule worm infection.

Leaf Swallowing

Behavioral Ecology

Leaf swallowing is currently proposed to control nodular worm infection (Huffman et al. 1996b) and relieve potential pain caused by tapeworm infection (Wrangham 1995). The substantiated mode of parasite control is a physical mechanism for the expulsion of parasites via the self-induced increase in gut motility that acts as a purge (Huffman and Caton 2001).

Leaf-swallowing behavior in the African great apes was first recorded for chimpanzees at Gombe and Mahale (Wrangham 1977, Wrangham and Nishida 1983). It came to the attention of Wrangham and Nishida at both sites that leaf swallowing was unlikely to provide any nutritional value after they found the folded, undigested leaves of *Aspilia mossambicensis* (Compositae), *A. pluriseta*, and *A. rudis* in the dung. The interest generated in this behavior by the claim that *Aspilia* spp. leaves contained the powerful antibiotic compound thiarubrine A (Rodriguez et al. 1985; but see Page et al. 1997) stimulated many field researchers to look for leaf swallowing at their ape study sites (e.g., Boesch 1995, Dupain et al. 2002, Matsuzawa and Yamakoshi 1996, Pruetz and Johnson-Fulton 2003, Takasaki and Hunt 1987).

Leaf-swallowing behavior has since been observed in numerous populations of *Pan troglodytes schweinfurthii*, *P. t. troglodytes*, *P. t. verus*, *P. paniscus*, and *Gorilla gorilla graueri* across Africa (Huffman 2001). The number continues to increase yearly as fieldworkers venture into new study sites or look for the presence of folded, undigested leaves in the dung. This behavior appears not to be limited to the African great apes. Recent observations of white-handed gibbons (*Hylobates lar*) in the Khao Yai National Park, Thailand, suggest that they also expel parasites with leaf-swallowing behavior (C. Barelli, W. Brockelman, and M. Huffman unpublished data).

The behavior of leaf swallowing is quite distinct (see Huffman and Hirata 2004, online http://dx.doi.org/10.1007/s10329–003-0065–5 for an mpg clip of this behavior). Leaves are taken from the distal end one at a time and carefully folded by the tongue and palate as they are slowly pulled into the mouth and then swallowed whole. The leaves' roughness makes them difficult to swallow, so folding them with the tongue and palate before swallowing is a necessary part of ingestion. An individual may swallow anywhere from one to 100 leaves in one bout, at a median rate of five leaves per minute (range 2.4–15.4 min, $n = 7$ bouts; data from Wrangham and Goodall 1989). An individual may do so more than once in a day and over several days (Huffman et al. 1996b, Wrangham and Nishida 1983). Leaves are typically swallowed within the first few hours after leaving the sleeping nest before the first meal (Boesch 1995, Wrangham and Goodall 1989, Wrangham and Nishida 1983, Huffman et al. 1996b) and, thus, on a relatively empty stomach. At Mahale, Gombe, Kibale, Lomako (Congo), and Fongoli (Senegal), where longitudinal data exist, leaf swallowing is an extremely rare behavior (Dupain et al. 2002, Pruetz and Johnson-Fulton 2003, Wrangham and Nishida 1983, Huffman 1997). Average rates of once every 69.0 hr ($n = 18$ bouts during 1,242 hr) and once every 81.9 hr ($n = 11$ bouts during 901.37 hr) for *Aspilia mossambicensis* at Mahale and once every 102.6 hr ($n = 10$ bouts during 1,026 hr) for *Aspilia* spp. at Gombe have been reported (Wrangham and Nishida 1983, Huffman 1997).

All 12 species whose leaves are swallowed whole at both Gombe and Mahale are available year-round. Based on botanical descriptions, this is expected to be the case for the more than 40 different species that were confirmed to be swallowed at sites across Africa. Nonetheless, interannual variation in the frequency of leaf swallowing at Mahale and Gombe is substantial. At Mahale, use is most common after the beginning of the rainy season (November–May), with peak frequencies in January and February being 10–12 times higher than those in other months (Wrangham and Nishida 1983).

At Kibale, direct observations were made only three times, but systematic dung collection from December 1987 to May 1994 provided evidence ($n = 44$) for leaf swallowing by *Pan troglodytes schweinfurthii* in both Kanyawara and Ngogo (Wrangham 1995). Unlike the distinct dry-season (June–October) and rainy-season (November–May) patterns of Gombe and Mahale, rain falls more evenly throughout the year at Kanyawara, with two "wet" periods (approximately March–May and August–November) and two drier "transitional" periods (approximately December–February and June–July) but no extended dry season (Struhsaker 1997). Based on these criteria, no significant difference was detected in the frequency of leaf swallowing between the wet and transitional periods at Kanyawara (estimated by the author from Fig. 1 of Wrangham 1995).

Parasite Infection and General Health

Information on parasite infection and health at the time of leaf swallowing is so far available only from observations at Mahale and Kibale. This information varies in a number of important aspects, so these two sites will be discussed separately.

Mahale. Behavioral profiles were collected concurrently with all cases of leaf swallowing observed during a 4-month study from December 1993 to February 1994 at Mahale (Huffman et al. 1996b). As part of this procedure, dung samples were systematically inspected macroscopically during focal-animal and *ad libitum* observations to check for the presence of whole leaves and parasites in the dung. General health was also monitored by stool type (diarrheal or normal) and collection of dung samples for microscopic parasitological examination. Nematode infection was demonstrated in 83% of all cases of leaf swallowing documented ($n = 10/12$ cases, 11 individuals, five from dung samples and seven from direct observation; Huffman et al. 1996b). Multiple-species infections were common, but nodular worm infections (78%) were most commonly associated with leaf-swallowing behavior, followed by those of threadworms (56%) and whipworms (33%). Infection by other intestinal parasite species was rare and inconsequential during this study period. As noted, during the rainy season there is a significant trend for an increase in the number of individuals infected with nodular worms at Mahale. Threadworm infections remained low year-round, and whipworm infections were inconsistent from year to year in their rainy season–dry season prevalence (Huffman et al. 1997). The presence of these two nematode species may have intensified symptoms of illness, but alone their impact on chimpanzees was not found to be substantial.

Symptoms which may be caused by infections from these nematodes (diarrhea, malaise, abdominal pain; Brack 1987, Anderson 1992) were verified from direct observation in seven of the eight chimpanzees at the time that leaves were swallowed (Huffman et al. 1996b, 1997). Nodular worms (body ca. 2.5 cm) were found in 3.7% ($n = 9/254$) of the dung collected from focal and nonfocal individuals. The occurrence of worms in the dung was rare and limited to individuals that displayed symptoms of malaise and diarrhea around the time leaf swallowing was observed.

Kibale. The most frequently occurring parasites in Kibale, Gombe, Mahale, and Budongo chimpanzees have been reported to be a nodular worm species, followed by threadworms and whipworms (Barrows 1996, File et al. 1976, Huffman et al. 1997, Rodriguez and Wrangham 1993). During a 78-month monitoring period of chimpanzee dung between 1987 and 1994 at Kanyawara, tapeworm proglottid fragments (*Bertiella studeri*) were found in 1% (16/1,696) of the total chimpanzee dungs inspected (Wrangham 1995). This occurrence of tapeworm fragments in the dung was restricted to a 6-month period (transitional–wet–transitional), referred to by Wrangham as the "period of tapeworm

infection." Note, however, that the absence of tapeworm fragments in the dung does not rule out the possibility of tapeworm infection. Infection can be preceded by the presence of tapeworms for as long as 2 years before proglottids appear in the feces of humans (Beaver et al. 1984).

Wrangham (1995) directly observed leaf swallowing at Kibale only three times. Parasite infection and a general evaluation of health around the time of leaf swallowing could only be inferred from macroscopic dung analysis. Of 12 dungs containing tapeworm fragments collected during this 6-month period, three were diarrheal and another three were scored as intermediate between diarrheal and normal. From this it was suggested that some individuals egesting tapeworm fragments may experience digestive disturbances, but it was not mentioned if these dungs contained leaves or not (Wrangham 1995).

Parasites and the Possible Functions and Mechanisms of Leaf Swallowing

The Mahale nodular worm hypothesis. The nematode-control hypothesis derived from the Mahale leaf-swallowing data complements that of bitter pith chewing in function and target but differs in its proposed mechanism. At Mahale, both behaviors are sometimes displayed by the same individual on the same day and together may have a synergetic effect on nodular worm infection. In 1993–1994, six of the nine dungs ($n = 254$) found to contain worms also contained whole undigested leaves of *Aspilia mossambicensis*, *Trema orientalis*, or *Aneilema aequinoctiale*. The remaining three dungs containing worms but no leaves were collected from three of the individuals whose dung contained both leaves and worms earlier that day (Huffman et al. 1996b). Worms may continue to be expelled for a short while after leaves have been egested. In only one case were worms not found in the dung when leaves were present during the same study period. The relationship between the presence of both leaves and nodular worms in the dung was highly significant (Fisher's exact test, two-sided, $p = 0.0001$; Huffman et al. 1996b). From this it was concluded that a strong relationship between leaf swallowing and the expulsion of nodular worms exists.

The proposed mechanism for nodular worm expulsion is based solely on data collected from Mahale (Huffman and Caton 2001). Leaf swallowing appears to affect only nodular worms. Nodular worms inhabit the large intestine, where they attach by the buccal capsule. This attachment is not permanent, and worms move around the intestine in search of food and mates (Anderson 1992). Adult threadworms (2 mm) and whipworms (30–40 mm), on the other hand, are smaller and burrow into the mucosa of the small intestine and cecum, respectively, where they embed themselves firmly (Anderson 1992) and, thus, are not considered to be susceptible to mechanical removal by the leaves. The "control of nematode infection hypothesis" predicts that since nodular worm infections are typically self-limiting (Beaver et al. 1984), the total infection may be controllable if a chimpanzee periodically swallows leaves during the most likely period of reinfection. In general, for all nodular worm species, the peak period of reinfection is the rainy season (Krepel et al. 1995). At Mahale, the peak period of reinfection by nodular worms was found to occur 1–2 months after the onset of the rainy season around December or January (Huffman et al. 1997). This is also when chimpanzees tend to swallow leaves (and chew bitter pith) most frequently at Mahale and Gombe (Huffman et al. 1996b, 1997; Wrangham and Goodall 1989; Wrangham and Nishida 1983). The continued expulsion of adult worms over time from successive bouts of leaf swallowing may be partially responsible for the observed decline in total nodular worm egg burdens (directly correlated to the number of egg-laying adults) later in the rainy season (Huffman et al. 1997).

Nodular worms produce abdominal pain along with bowel irritation and diarrhea (Brack 1987), symptoms that could provide ample stimulus for leaf-swallowing behavior in chimpanzees infected with this nematode (Huffman and Seifu 1989; Huffman et al. 1993, 1996b).

The Kibale tapeworm hypothesis. In 1993, significantly more leaves (8.5% of 271 dungs) were found in dungs during the 6-month period that tapeworm fragments were also found than in the dungs outside of this period (1.3% of 1,425 dungs collected in 1987–1992) (Mann-Whitney $z = 4.32$, $p < 0.001$; Wrangham 1995). Also, the proportion of dungs with both tapeworm fragments and whole leaves (21.7%, $n = 23$) was significantly higher than the proportion of dungs with whole leaves and no tapeworm fragments (3.6%, $n = 248$, binomial test $z = 4.46$, $p < 0.001$; Wrangham 1995). From this, Wrangham concluded that whole leaves in the gut increase the probability of tapeworm fragments being shed.

Whole leaves were found in dung during 21 months of the survey, while leaves and tapeworms were found together in only 33% of these months. It is possible that some tapeworm fragments went undetected. The greater overall occurrence of leaves in dung without tapeworms, however, suggests that the expulsion of proglottid fragments may not be the only function of leaf swallowing. Wrangham (1995) proposes that the stimuli for leaf swallowing as observed at Kibale could be abdominal pain caused by tapeworm infections and, thus, may function to alleviate pain.

Therapeutic self-medication by definition assumes the relief of discomfort as a driving motive behind its practice. In this light, further observational data from Kibale will help to expand upon our understanding of the antiparasitic function of leaf-swallowing behavior with regard to tapeworms.

Phytochemical Evidence Reevaluated

Thiarubrine A, a powerful antibiotic, antifungal, and antihelminthic secondary plant compound (Towers et al. 1985), was reported by Rodriguez et al. (1985) to be in the leaves of *Aspilia mossambicensis* and *A. pluriseta*. Rodriguez and Wrangham (1993) later proposed that thiarubrine A could

act as a nematocide. Two alternative hypotheses were proposed to explain the function of swallowing *Aspilia* leaves based on the role of thiarubrine A: (*1*) buccal absorption (Newton and Nishida 1990) and (*2*) unchewed leaves as thiarubrine A–protective capsules (Rodriguez and Wrangham 1993). Both were proposed ad hoc to resolve the dilemma of how thiarubrine A would make it to the site of treatment without being destroyed by gastric acids.

However, after numerous attempts to replicate these phytochemical findings, Page and colleagues (1992, 1997) could detect no trace of thiarubrine A, B, D, E, or their corresponding thiophenes in the leaves of 27 samples of three relevant *Aspilia* species collected from Mahale and Gombe; Nanyuki, Kenya; the Kampala District in Uganda; and greenhouse-grown specimens from Vancouver, Canada. Furthermore, analysis revealed no detectable antifungal activity against *Candida albicans* (Page et al. 1997), a human pathogenic yeast sensitive to thiarubrines (Towers et al. 1985). Messner and Wrangham (1996) found no nematocidal activity against infective L3-stage *Strongyloides* nematode larvae in the leaf extracts of *Rubia cordifolia* (Compositae). Neither of these phytochemical hypotheses provided an adequate universal explanation for leaf swallowing, and they have been abandoned (Huffman et al. 1996b, Page et al. 1997).

EXTERNAL APPLICATION OF SUBSTANCES TO THE BODY

Fur Rubbing

The health-maintenance and therapeutic self-medicative behaviors discussed to this point all have in common the ingestion of substances for internal action. Fur rubbing differs from these behaviors in that substance use is limited to external application. Both medicative and social functions of the behavior have been proposed in primates. Considerations of possible nutritional gain are irrelevant as the items used in these behaviors are applied to the body and not eaten. The wide taxonomic and geographic distribution of primates that are reported to display fur rubbing suggests that it could be even more widespread than currently reported among New World primates and prosimians and one great ape species (Table 37.1).

Behavioral Ecology

Fur rubbing has been observed both in the wild and under captive conditions (Table 37.1). To date, it has been reported for *Cebus capucinus*, *C. olivaceus*, *C. apella* (e.g., Baker 1996; Buckley 1983; Hill 1960; Ludes and Anderson 1995; Oppenheimer 1968; Valderrama et al. 2000b), *Ateles geoffroyi*, *At. belzebuth* (Campbell 2000a, Richard 1970, Klein 1972), *Aotus boliviensis*, *A. lemurinus griseimembra*, *A. nancymaae* (Zito et al. 2003), and *Eulemur macaco* (Birkinshaw 1999), as well as one great ape species, *Pongo pygmaeus* (Morrogh-Bernard 2008).

Table 37.1 Fur-rubbing and Its Proposed Functions in Primates

SPECIES	SUBSTANCE USED	PARTS/CHEMICALS USED	LOCATION (SETTINGS)	PROPOSED FUNCTION
White-faced capuchin, *Cebus capucinus*	*Citrus* spp. *Soloanea teniflora* *Clematis dioica* *Piper marginatum*	Fruit Leaves, stem Leaves, stem Leaves, stem	Nicoya Peninsula Costa Rica (wild)[1]	Insect repellent
	Camponatus sericeiventris	Ant	Corcovado National Park Costa Rica (wild)[2]	Skin stimulation
	Gardening peat		Strasbourg, France (captive)[3]	Skin/pelage maintenance
Wedge-capped capuchin, *Cebus olivaceus* 2-Methoxy-3-Methyl-1-benzoquinone	Millipede *Orthoporus dorsovittatus*	Whole insect 2 Methyl-1,4-benzoquinone	Ilanos, Venezuela (wild)[4]	Insect repellent
Black-handed spider Monkey *Ateles geoffroyi*	*Citrus aurantifolia* *Zanthozylum procerum* *Zanthozylum belizense*	Leaves Leaves Leaves	Barro Colorado Is and, Panama (wild)[5,6]	Scent marking Insect repellent
	Brongniartia alamosana *Cecropia obtusifolia* *Apium graveolens*	Leaves Leaves Leaves	Mexico[7]	Male scent marking
Owl monkeys	*Piper marginatum*	Plant extract	DuMond Conservancy, Florida (captive)[8]	Insect repellent
Aotus boliviensis	Milliped	Whole insect		
A. lemurinus griseimembra	*Anadenobolus monilicornis*			
A. nancymaae	Millipede extract	Benzoquinone, toluquinone, 2-methyl-1, 4–benzoquinine, hydroquinone extracts placed on filter paper		
	Onions, garlic and chives			
Black lemur, *Eulemur macaco*	Millipede *Charactopygus* spp.	Whole insect Aldehydes, quinones Phenols, chlorine	Lokobe Forest– Nosy Be, Madagascar (wild)[9]	Control of ectoparasites
Bornean orangutan, *Pongo pygmaeus*	*Commelina* spp.	Leaves	Sabangau Peat swamp forest central Kalimantan, Borneo[10]	Anti-inflammation, antibacterial

Sources:
[1] Baker 1996.
[2] Longino 1984.
[3] Ludes and Anderson 1995.
[4] Valderrama et al 2000b.
[5] Campbell 2000a.
[6] Richard 1970.
[7] Laska et al. 2007.
[8] Zito et al. 2003.
[9] Birkinshaw 1999.
[10] Morrogh-Bernard 2008.

Plant material or insects can be applied directly or chewed and mixed with saliva, then rubbed frenziedly into the fur all over the body in a group and/or solitarily for some species. Baker (1996) studied fur rubbing in capuchins on the southeastern edge of the Nicoya Peninsula of Costa Rica. Fifty-three bouts were observed during a total of 1,140 hr (i.e., 1 bout/19.36 hr). Duration of time spent fur rubbing ranged from 10 sec to 41 min (mean 6.3 min). Solitary and group rubbing were observed. Capuchins used at least five different plant genera including the fruits of four introduced *Citrus* spp. (Rutaceae) and the leaves and stems of *Sloanea terniflora* (Elaeocarpaceae), *Clematis dioica* L. (Ranunculaceae), and *Piper marginatum* (Piperaceae). *Citrus* (lemon, lime, orange) was the most frequently used genus, but use was often combined with other plant types. Citrus fruits were used exclusively during the dry season; however, fur rubbing occurred significantly more often in the rainy season than the dry season (chi squared = 15.8, $p < 0.05$; Baker 1996). The trend for an increase in fur rubbing during the rainy season was suggested to reflect the increase in risk of bacterial and fungal infection due to a rise in humidity. This suggests, but alone does not demonstrate, that the behavior is performed for the relief of symptoms caused by these ailments.

Aotus (Zito et al. 2003), *Cebus* (Longino 1984; Valderrama et al. 2000b), and *Eulemur* (Birkinshaw 1999) are unique among primates for their use of millipede species (*Anadenobolus monilicornis*, *Charactopygus* spp., *Orthoporus dorsovittatus*) and ants as material for fur rubbing. Millipedes secrete caustic benzoquinones when threatened or attacked by predators. Valderrama et al. (2000b) investigated fur rubbing in *C. olivaceus* at Fundo Pecuario Masaguaral in the llanos of central Venezuela and analyzed the chemistry and biological activities of the benzoquinone secretions in the laboratory. As many as four individuals will share one millipede, rolling on it, putting it gently in the mouth, and drooling with eyes apparently glazed over. Body rubbing with the apparent intent of picking up millipede secretions from conspecifics is also observed. The behavior can last from 5 sec to over 2 min. Individuals may anoint several times a day. There is not an age/sex class bias in individuals performing this behavior. The behavior occurred most consistently during periods of high rainfall. The two benzoquinones chemically isolated in the laboratory from the millipede secretions are known to be effective at repelling insects and to have antimicrobial properties.

Any benefits gained from applying these substances to the skin are presumed to be immediate. As was found for Baker's study site, the wet, humid conditions of the rainy season may present stressful environmental conditions for *Cebus*. Valderrama et al. (2000b) proposed that the capuchins were anointing themselves as protection possibly from insects. Positive social effects of this behavior in *Cebus apella* have also been described (Gilbert et al. 1998).

Ateles geoffroyi in the BCI group on Barro Colorado Island and Parque de Flora y Fauna Silvestre Tropical, located near Catemaco, Veracruz, Mexico, perform fur rubbing solitarily, applying plant material only to the sternal and axillary regions of the body (Campbell 2000a, Laska et al. 2007). Differences in the behavior between *Cebus* and this species led Campbell (2000a) and Laska et al. (2007) to reject a medicinal hypothesis. Citing as evidence against the medicinal hypothesis were three main factors: (*1*) males tend to perform the behavior more frequently than females; (*2*) only a limited part of the body is anointed, and therefore, it is unlikely to be effective against insects; and (*3*) there is a lack of seasonal variation in the occurrence of the behavior in spite of strong seasonality in the prevalence of mosquitoes and ticks (Campbell 2000a). Laska et al. (2007) suggest that the behavior is involved in social communication.

Pongo pygmaeus in the Sabangau peat swamp forest of central Kalimantan, Borneo, have been observed to chew the leaves of *Commelina* species and deliberately rub themselves with the resultant greenish white lather mixed with saliva on specific parts of the body. Unlike capuchins, the individuals appear calm and take considerable time to rub the substance onto arms or legs (Morrogh-Bernard 2008).

Phytochemical and Ethnobotanical Evidence

A number of the plant species identified by Baker (1996) as being used by capuchins for fur rubbing are known to contain secondary compounds with insecticidal, antiseptic, fungistatic, anti-inflammatory, anesthetic, and general dermatological activities. Richard (1970) suggested that the use of citrus fruits by *Ateles* might repel insects or act as an astringent. All plants share the quality of being pungent. Fur rubbing by capuchins and lemurs includes the use of a variety of pungent items, such as tobacco, onions, garlic (in captivity), millipedes, and formic acid (in the wild) (cf. Baker 1996, Valderrama et al. 2000b). More investigation is needed for the case of orangutan fur rubbing, but the plant reported to be used is used by humans in the area as a treatment for joint and muscle pains. Morrogh-Bernard (2008) speculates that it may function as an anti-inflammatory or antibacterial.

It seems likely that this behavior has multiple functions based on differing environmental pressures. Because the behavior can be induced readily under captive conditions in New World monkeys, it seems that these species have a behavioral predisposition to rub scented objects into the fur. Some reports suggest an almost hedonic pleasure derived from its practice. This ability to study the behavior easily under controlled conditions could provide important opportunities for future investigations into aspects of social learning and ontogenetic development as well as clinical evaluation of the effects of the behavior on health conditions that are not possible in the wild. Further direct evidence for a medicinal or insect-repellent effect of this behavior for more species at different sites is needed in order to clarify its role in health maintenance. Further research into the taxonomic distribution of this behavior in primates is also needed.

SUMMARY AND FUTURE DIRECTIONS

The evidence described in this chapter suggests that a variety of behavioral adaptations exist for the control of internal and possibly external parasites. Much of the detailed evidence in support of the developing hypotheses on self-medication in the African great apes comes from their response to changing levels of parasite infection. The applicability of this model to other primate species needs to be confirmed. Three of the greatest constraints on field investigations of self-medicative behavior are (*1*) the unpredictability of the behaviors' occurrence, (*2*) the unreliability of being able to consistently follow and observe sick individuals over time, and (*3*) restraints on subject manipulation. These constraints must somehow be overcome to test relevant hypotheses.

Systematic monitoring of individuals throughout the year for fluctuating levels of parasite infection is one effective means of identifying the predominant parasites and their likely period of increased impact on the host (e.g., Huffman et al. 1997). Relevant pathogens other than parasites also need to be investigated, but traditionally this has required capturing animals for the collection of blood and other tissues. The development of noninvasive techniques with wide applicability will allow for a better assessment of the range and impact of self-medicative behaviors and provide a wide range of useful tools for conservation biology.

Close long-term monitoring of the overall health of group members using detailed analysis of activity budgets and systematic inspection of general health is necessary to identify illness when it occurs and to understand the proximate effects of that illness on the individual as well as the function and effectiveness of any proposed self-medicative behavior (e.g., Huffman et al. 1997, Kaur and Huffman 2004, Krief et al. 2005). These procedures are straightforward and compatible with most primate field study designs. Experimental behavioral studies on captive primates can be useful to answer some questions that are not possible or are extremely difficult to answer in the wild, for example, individual acquisition and transmission of behaviors within the group (e.g., Huffman and Hirata 2004, Sgaravatti et al. 2004).

This chapter has shown than self-medication can take many forms and occurs widely among primates. An important task that lies ahead is to identify and evaluate not only the immediate outcome of such behaviors but also the long-term impact on individual fitness and reproductive success. While the current direct evidence for primate self-medication comes largely from work on the great apes, we should avoid the temptation to conclude that advanced cognitive ability is a prerequisite for exhibiting such behavior. Indeed, it would be unwise not to expect it across the entire animal kingdom (Engel 2002). Further efforts to better understand the diversity, function, and mechanisms of self-medicative behavior in all primates will be rewarded with a richer appreciation for our primate ancestors and no doubt provide important lessons for the future of our own species.

PART SIX

Social Behavior and Cognition

38

Kinship
Family Ties and Social Behavior
Carol M. Berman

1. Why do some primates show strong preferences for their genetic relatives while others show only mild preferences or none at all?
2. How well does kin-selection theory explain kin preferences in primates?
3. How do primates avoid mating with close kin?

INTRODUCTION

Several basic features of primate life history set the stage for social organization to be based around kin relationships. Primates have long life spans and overlapping generations, most species live in permanent groups, and dispersal is typically sex-biased such that one sex tends to remain in its natal group for life while the other sex disperses during puberty or early adulthood. This leads to the accumulation of either related females or related males within groups and makes it possible for extended networks of biological kin (matrilineal or patrilineal) to interact on a regular basis (Chapais and Berman 2004a, Strier 2008). These features, combined with enduring bonds between parents and offspring and the potential selective advantages of cooperating with genetic relatives (Hamilton 1964, Dugatkin 1997), increase the likelihood that kin will favor one another within groups (Silk 2006).

The importance of kin relatedness in the social interactions of group members was first recognized in the late 1950s and 1960s by M. Kawai (1958) and S. Kawamura (1958) in wild but provisioned Japanese macaques (*Macaca fuscata*) and by D. S. Sade (1965, 1967) in free-ranging, provisioned rhesus macaques (*M. mulatta*). Their studies established kinship as a central organizing principle of primate social structure, along with sex, age, and dominance, and it remains so to the present day. In this chapter, I discuss some basic factors that shape the potential for individuals to display preferential behavior toward genetic relatives (i.e., kin bias). I then survey what we know so far about patterns of matrilineal and patrilineal kin bias and incest avoidance in various species, emphasizing their variation. Finally, I describe some current ecological/evolutionary models about the origins and variability of kin bias and discuss evidence for and against kin selection as an explanation

for the evolution and maintenance of kin bias in nonhuman primate groups.

KINSHIP STRUCTURE, KIN RECOGNITION, KIN SELECTION, AND THE POTENTIAL FOR KIN BIAS

Dispersal Patterns

Species-typical dispersal patterns largely determine the classes of kin that are present and absent within groups, generally limiting the potential for kin bias to those that are present. Since female matrilineal kin tend to accumulate in female-philopatric/male-dispersal groups, female members often reside with a wide range of female maternal kin throughout their lives, and they may also reside with fathers, paternal siblings (sisters and undispersed brothers), and other patrilineal kin (females and undispersed males). Males reside with the same categories of maternal and paternal kin before they disperse, but after dispersal male contact with kin is usually limited to a small number of males (and their descendants) that disperse into the same social groups. Recent genetic studies of relatedness within female-philopatric/male-dispersal groups reveal kin structures that are consistent with observed dispersal patterns (reviews in Morin and Goldberg 2004, Kappeler 2008). For example, in baboon and long-tailed macaque groups, adult females are more closely related to one another than are males, and females are more closely related to members of their own matrilines than to females in different matrilines (Altmann et al. 1996, de Ruiter and Geffen 1998).

In male-philopatric/female-dispersal groups, adult males, but not adult females, are expected to reside with a wide range of male paternal kin (fathers, sons, grandfathers, grandsons, brothers, nephews, uncles, cousins, etc). If females disperse before reproducing, both sexes should reside with only a limited number of categories of matrilineal kin—males with mothers, brothers, and undispersed sisters, and females with sons and undispersed daughters. Both sexes may also live with the male and undispersed female descendants of male maternal kin and other matrilineal kin that may have dispersed into the same group (Chapais and Berman 2004a). As predicted, genetic studies of woolly monkeys (*Lagothrix lagotricha*), a female-dispersal species, suggest that most females within a group are unrelated (Di Fiore 2003a). In contrast, genetic analyses do not find high levels of male relatedness within male-philopatric bonobo (*Pan paniscus*) groups (Gerloff et al. 1999). Nor are male chimpanzees more closely related to one another than are females (Vigilant et al. 2001, Lukas et al. 2005). In an attempt to resolve this apparent discrepancy between observed dispersal patterns and relatedness, Lukas et al. (2005) constructed a mathematical model of relatedness that takes not only dispersal patterns but also reproductive skew and group size into consideration. The model predicts that male relatedness should actually be high only in relatively small groups, a prediction that appears to be consistent with published data for a variety of primate and nonprimate taxa.

When both sexes disperse, kin structures can be variable. Recent genetic studies suggest that, in many cases, one sex is more likely to disperse or tends to disperse farther than the other (review in Kappeler 2008). For example, genetic data indicate that males tend to disperse farther than females in western lowland gorillas (*Gorilla gorilla gorilla*) (Douadi et al. 2007, but see Bradley et al. 2004), orangutans (*Pongo pygmaeus*) (van Schaik and van Hooff 1996, Singleton and van Schaik 2002, Krutzen et al. 2008), and Coquerel's dwarf lemurs (*Mirza coquereli*) (Kappeler et al. 2002), producing a dispersed matrilineal population structure. Conversely, related male siamangs (*Symphalangus syndactylus*) are more often found in adjacent groups than related females, presumably because females disperse farther than males (Lappan 2007).

Female Reproductive Rates and Mortality Patterns

Another factor affecting the potential for kin bias through its effect on kinship structure is female reproductive rate. Female reproductive rates are determined by female ages at first reproduction, intervals between infants, and female life spans. These factors, along with mortality patterns, influence the number of generations found within groups, the categories of kin residing at the same time, as well as the size and number of matrilines. Although little empirical work has been done on the relationship between reproductive rates, mortality, and kinship structure in primates, simulation studies of populations with high and low reproductive rates suggest that matriline size and other characteristics can vary widely (Dunbar 1988), profoundly affecting the availability of kin of various kinds. For example, when reproductive rates are slow, fathers and older siblings may disperse before younger siblings are born. Similarly, mothers and other

close kin may die before they are able to provide youngsters with the support necessary to attain high rank (Datta and Beauchamp 1991). Even when kin co-reside, their availability to one another may influence their propensities for kin bias (Berman et al. 2008, Perry et al. 2008). For example, female baboons (*Papio cynocephalus*) with no maternal sisters form stronger bonds with paternal sisters than they would otherwise (Silk et al. 2006). Thus, the extent to which females favor paternal sisters may depend on the availability of maternal kin (Widdig et al. 2001; Smith, K., et al. 2003; Silk et al. 2006). Availability is also hypothesized to affect the costs and benefits of favoring various categories of kin, particularly when time for social interaction is limited (Altmann, S., 1979; Chapais and Belisle 2004). For example, when individuals have many kin, they should receive the greatest payoff in terms of inclusive fitness by investing all available time and energy into their closest kin, whereas when they have few kin, their optimal investment is likely to include a wider range of kin categories.

Kin Recognition and Reproductive Skew

The potential for kin bias is also limited by the ability of individuals to recognize their relatives. Individuals living in groups with extended kin networks, either paternal or maternal, are expected to associate preferentially with kin only if they can distinguish them from unrelated counterparts. Most researchers agree that maternal kin recognize each other primarily through familiarity. Mothers and infants have been selected to form intense emotional bonds that promote close association and mutual attraction (Bowlby 1969, Hinde 1974b). In primates these bonds typically endure throughout the juvenile period and, in many cases, throughout adulthood. In strongly kin-structured societies, close association with mothers brings offspring into affiliative interaction with maternal siblings and other maternal kin in direct proportion to their degrees of relatedness through maternal lines. Thus, familiarity, maternal kinship, and affiliation all become linked (Chapais 2001, Rendall 2004, Berman 2004).

The precise mechanisms involved in the development of kin bias via familiarity are not well understood, but a combination of maternal transmission and independent learning is likely to be involved (review in Berman 2004). Primate infants typically begin by showing affiliative patterns toward matrilineal kin and nonkin that reflect the relationships of their mothers. Hence, infants in species with strong matrilineal bonds also show strong tendencies to associate with close matrilineal kin, whereas those in species with weaker matrilineal bonds display weaker or no kin preferences. This initial mirroring of maternal kinship networks is undoubtedly due to the tendencies for mothers and infants to remain in close proximity to each other and, hence, to the mother's close associates. In addition, in some species mothers exert a large degree of control over infants' interactions with others, typically displaying more tolerance for their close associates than for others. Hence, initial forms of kin bias are most likely the result of passive behavior on the part of the infant (de Waal 1996, Berman 2004). As infants become progressively more independent from their mothers, they generally continue to pursue relationships with many of the same individuals with whom they interacted previously, often taking the initiative. In this sense, the development of an active form of kin bias can be seen as a process of social transmission through the mother. Transmission may be accomplished by differential exposure and interaction with kin and nonkin (MacKenzie et al. 1985, Welker et al. 1987, Martin 1997), observational learning using the mother as a model (Altmann 1980, Hikami et al. 1990), and/or active and selective intervention in the infant's interactions with others by the mother (Maestripieri 1995a). By late infancy and juvenescence, maternal transmission appears to gradually give way to other processes based on independent learning through direct interaction with kin and nonkin. Infants' networks no longer exactly duplicate those of their mothers (Berman and Kapsalis 1999) and may increasingly reflect attractions to similarly aged peers (possible paternal siblings) or differences in the quality of interaction between older kin and nonkin. In strongly kin-biased groups, older kin are more likely than nonkin to initiate friendly and supportive interaction and less likely to threaten or harass them (review in Berman 2004).

With one exception that could not be replicated (Wu et al. 1980), experimental studies of captive primates that separated familiarity from maternal and/or paternal relatedness support the primacy of familiarity over genetic relatedness per se as a mechanism underlying kin-biased behavior (stump-tailed macaques [*Macaca arctoides*], MacKenzie et al. 1985; pig-tailed macaques [*Macaca nemestrina*], Fredrickson and Sackett 1984, Sackett and Frederickson 1987, Martin 1997; baboons, Erhart et al. 1997; crab-eating macaques [*Macaca fascicularis*], Welker et al. 1987, review in Rendall 2004). Based on these studies, the potential for patrilineal kin bias was thought to be limited to a minority of pair-bonded and single-male/multifemale species in which most infants in a group are sired by a single male, making familiarity as well as paternity certainty high between fathers and offspring. In other species, fathers were thought to be unable to distinguish their own offspring from others, rendering kin-biased favoritism impossible. An inability for fathers and offspring to recognize one another would in turn preclude other kinds of patrilineal kin recognition (e.g., paternal siblings, cousins, uncles, and nephews) through association with the father. However, with the recent application of genetic methods to determine paternity, researchers are beginning to find evidence of patrilineal kin recognition in some wild and free-ranging multimale primate groups. For example, adult female rhesus monkeys and baboons prefer to associate with paternal sisters over unrelated females (Widdig et al. 2001, 2002; Smith

et al 2003; Silk et al. 2006b), and adult males favor their immature offspring over unrelated youngsters in a variety of interactions (mandrills [*Mandrillus sphinx*], Charpentier et al. 2007; chimpanzees, Lehmann et al. 2006; baboons, Buchan et al. 2003).

How primates recognize paternal kin is not well understood (review in Widdig 2007). Some hypotheses are based on findings that male reproductive skew can be high even in groups that contain more than one adult male (e.g., Pope 1990, Cowlishaw and Dunbar 1991, Altmann et al. 1996). If one or a few high-ranking males sire most infants, the probability is high that the care they provide is directed toward their kin. Reproductive skew also leads to the production of similarly aged paternal siblings (Altmann, J., 1979), a finding that has been confirmed genetically in several species (baboons, Altmann et al. 1996, Silk et al. 2006b; macaques, Widdig et al. 2002; Hanuman langurs, Launhardt et al. 2001; mandrills, Charpentier et al. 2005; capuchin monkeys [*Cebus capucinus*], Muniz et al. 2006, Perry et al. 2008; great apes, Gerloff et al. 1999, Vigilant et al. 2001, Bradley et al. 2005, Robbins and Robbins 2005, review in Widdig 2007). Thus, offspring have the potential to use age similarity as a cue for paternal relatedness with an accuracy approximating the degree of reproductive skew. When the tenures of high-ranking males are short, (e.g., 1 year), cohorts of paternal siblings should be confined to offspring born in the same year; but when they are longer, paternal siblings should span wider age ranges and full siblings should be common (Altmann et al. 1996).

Several studies document preferential relationships between peers (Pereira 1988b; Janus 1992; Watts and Pusey 1993; Widdig et al. 2001, 2002; Silk et al. 2006b; Nakamichi and Yamada 2007, but see Charpentier et al. 2007, Schino et al. 2007b), and researchers suggest that they may be mediated by a general attraction to similarly aged partners (de Waal and Luttrell 1986). However, other studies reveal preferences for paternal siblings over similarly aged unrelated group members (Widdig et al. 2001, Lehman et al. 2006), suggesting that more specific mechanisms are involved in some cases. Several hypotheses are based on familiarity, recognition of physical similarities such as odor or appearance (i.e., phenotypic matching) (Lacy and Sherman 1983), or combinations of the two (review in Widdig 2007): Fathers in multimale groups may recognize their own offspring through phenotypic matching or through association with mothers with whom they mated frequently or at peak periods of fertility (Buchan et al. 2003). Paternal siblings may become familiar with one another when mothers that mated with the same male bring their infants into contact. Fathers that recognize their offspring by whatever means may jointly associate with them. Infants may be attracted to males or other infants with whom they share inherited physical similarities. Finally, paternal kin may be particularly compatible because they have similar inherited temperaments (Widdig et al. 2001).

Kin Selection or Inclusive Fitness

A final factor affecting the potential for kin bias concerns its effects on fitness. Assuming kin are available and recognizable, Hamilton's (1964) inclusive fitness or kin selection theory provides a powerful evolutionary framework for predicting when kin preferences are likely to be expressed. Hamilton's theory posits that individuals that behave in ways that promote the reproductive success not only of their own offspring but also of offspring of other genetic relatives will be evolutionarily selected. This is because individuals that are related through a common ancestor share a proportion of their genes in common. Indeed, one may promote the reproduction of one's own genes by promoting the reproduction of one's kin, even in cases in which helping one's kin reproduce decreases one's own production of offspring (i.e., when the act is altruistic). Hamilton identified the conditions in which altruistic acts to kin should be selected in a formal mathematical model, since called "Hamilton's rule": $br > c$, where b represents the benefit of an act to the recipient (the amount that the act increases the recipient's reproductive success), c represents the cost of the act to the performer (the amount the act decreases the performers own reproductive success), and r represents the average degree of relatedness between the performer and the recipient due to common descent. Hamilton's rule posits that altruistic acts should be selected only when the cost to the performer is smaller than the benefit to the recipient discounted by their degree of relatedness. Although Hamilton's rule has provided a major impetus to the study of kinship in primates, testing its predictions and ruling out alternative explanations have been challenging (see later section "The Question of Kin Selection").

A SURVEY OF KIN PREFERENCES

Solitary or Dispersed Social Organizations

Many small, nocturnal prosimians generally forage alone. Nevertheless, behavioral and genetic evidence is accumulating in some species that maternal kinship structures their societies within larger dispersed communities (Nash 2004). For example, gray mouse lemur (*Microcebus murinus*) females disperse only a short distance from their natal areas, and as a result, matrilineally related females occupy overlapping home ranges (e.g., Radespiel 2006). They forage solitarily during the night and sleep in hollow trees by day, sharing nests with two to five other matrilineally related females. Females raise their young cooperatively in the nests, favoring their own but grooming and nursing all infants, particularly when the infants' mothers are out foraging. Interestingly, mothers carry only their own young, suggesting that they do not provide care to others merely because they fail to recognize their own (Eberle and Kappeler 2006). Males, in contrast, disperse farther away from their natal areas, usually sleep alone, and apparently

interact little with maternal kin (e.g., Eberle and Kappeler 2004a,b). Coquerel's dwarf lemurs display similar matrilineal ranging patterns and social structure but do not share nests or interact cooperatively with maternal kin (e.g., Kappeler et al. 2002, Wimmer et al. 2002). Similar matrilineal social structures have also been attributed to several galago species (Galaginae) (Clark 1978b, Bearder 1987) and to orangutans (Galdikas 1988, van Schaik and van Hooff 1996, Singleton and van Schaik 2002, Krutzen et al. 2008), but data are incomplete in most cases (reviewed in Nash 2004, Morin and Goldberg 2004, Kappeler 2008).

Female Philopatry/Male Dispersal

Females

Many Old World monkeys, including rhesus macaques, Japanese macaques, crab-eating macaques, savanna baboons, and vervet monkeys, display high degrees of matrilineal kin bias and socially cohesive matrilines. Generally, close maternal kin spend disproportionate amounts of time together and preferentially groom, feed, huddle, sleep, support, and reconcile with one another (e.g., Hunte and Horrocks 1987, Judge 1991, Aureli et al. 1997, Nakamichi and Shazawa 2003, Majolo et al. 2005, Silk et al. 2006b, Wada et al. 2007). Largely as a result of support from mothers and other maternal kin, they also display "classic" matrilineally based dominance hierarchies in which daughters acquire ranks immediately below their mothers and in reverse order of age (youngest ascendancy rule) (see recent reviews in Chapais 2001; Silk 2001; 2002a,2006; Berman 2004; Kapsalis 2004; Strier 2008). Although maternal kin may show relatively high rates of aggression, it is generally relatively mild in intensity (Bernstein 1991b). Similarly, social play tends to be less maternally kin-biased than other interactions because immatures tend to prefer same-sex partners of similar size when available. However, close kin may be frequent play partners when more suitable partners are not available or when well-matched kin are available (review in Walters 1987a,b). Close maternal kin also engage in disproportionate amounts of allocare (review in Berman 2004) and are likely to adopt related infants if they are orphaned (e.g., Berman 1982a, Hasegawa and Hiraiwa 1980, Nicholson 1987). Maternal kin are preferred models for social learning (review in Chauvin and Berman 2004). The social transmission of novel foods (Leca et al. 2007b), techniques of food processing (Kawai et al. 1992, Hirata et al. 2001) and grooming (Tanaka 1995, 1998), affiliative patterns (de Waal 1996, Berman and Kapsalis 1999), and maternal styles (Berman 1990, 1997) often follow lines of kinship and affiliation. Finally, when social groups fission, they typically split along matrilineal lines, with different lineages settling in different daughter groups and close female kin remaining together (Koyama 1970, Chepko-Sade and Sade 1979, Oi 1988, Dittus 1988, Prud'homme 1991, Li et al. 1996, Kuester and Paul 1997, Widdig et al. 2006a, van Horn et al. 2007, review in Okamoto 2004).

Matrilineal kin in these species appear to discriminate gradations of relatedness, favoring mothers and offspring over siblings and grandmothers over aunts, nieces, and others (review in Kapsalis 2004). Several studies point to a limit for kin bias, that is, a degree of relatedness (r) below which individuals no longer favor matrilineal kin over unrelated group members (rhesus, Kapsalis and Berman 1996a, Rendall et al. 1996; Japanese macaques, Bélisle and Chapais 2001, Chapais et al. 2001). Chapais and colleagues found similar limits ($r = 0.25-0.125$) for support, tolerated co-feeding, and homosexual behavior (Chapais et al. 1997, Bélisle and Chapais 2001), leading initially to speculation that macaques may be cognitively limited in their abilities to recognize certain categories of kin. However, additional research indicated that limits indeed vary across different behaviors. For example, Rendall et al. (1996) identified a limit of $r = 0.25-0.125$ for proximity kin bias among female rhesus monkeys on Cayo Santiago (see also Kapsalis and Berman 1996a) but found a broader limit ($r = 0.063-0.016$) for responses to recorded vocalizations of kin. Limits may instead reflect variations in the balance between the costs and benefits of different behaviors such that only some behaviors are profitable with more distant categories of kin (Chapais 2006).

Recent studies of adult rhesus monkeys (Widdig et al. 2001, 2002), savanna baboons (Smith, K., et al. 2003; Silk et al. 2006b), and juvenile mandrills (Charpentier et al. 2007) suggest that adult paternal sisters also affiliate more with one another than with unrelated adult females. However, there is no evidence that paternal siblings engage in less aggressive behavior with one another than with unrelated females (rhesus, Widdig et al. 2002; mandrills, M. Charpentier personal communication cited in Widdig 2007), nor do they support them more in agonistic encounters, although there is some evidence that they refrain from supporting their opponents (Widdig et al. 2006b). Paternal kin remain together when groups fission in some cases (savanna baboons, Smith, K., 2000) but not in others (Barbary macaques, Kuester and Paul 1997). Results for rhesus monkeys are inconsistent (Widdig et al. 2006a).

We know less about females of other Old World monkeys with female philopatry/male dispersal, and what we know suggests more moderate and less consistent roles for kin bias in their social behavior. For example, females of several macaques, including Barbary (*Macaca sylvanus*), stump-tailed, bonnet (*Macaca radiata*), and the various Sulawesi macaque species, show discernible but weaker tendencies to affiliate, co-feed, reconcile with, and/or support matrilineal kin (Rosenblum et al. 1975, Defler 1978, Ehardt 1988, de Waal and Luttrell 1989, Butovskaya 1993, Aureli et al. 1997, Matsumura and Okamoto 1997, reviews in Thierry 2000, Thierry and Aureli 2006, Cords and Simpson 2006, Thierry et al. 2008). Similarly, tendencies to split along matrilineal lines may also be weaker or absent (Okamoto and Matsumura 2001, Menard et al. 2006). Maternal kin bias may also vary over time as a function of group size (Tibetan

macaques, Berman et al. 2008; rhesus macaques, Berman and Kapsalis 2009; capuchins, Perry et al. 2008) or other factors (Japanese macaques, Nakamichi 2004a). Other species may show strong maternal kin bias in some behaviors but not in others. For example, Barbary macaque females in one study showed substantial kin bias in affiliation but relatively little in reconciliation (Aureli et al. 1997). When support is rare or follows less strict matrilineal lines, support from unrelated individuals may be key in determining ranks within families. As a result, dominance structures may deviate from the "classic" matrilineal model to varying degrees. For example, older Barbary macaque mothers lose rank to their daughters, and it is common for older sisters to outrank younger sisters (Paul and Kuester 1987, Prud'homme and Chapais 1993a, Thierry and Aureli 2006, Paul 2006). The latter is not attributed to a lack of kin support in favor of younger sisters but rather to low rates of support from either kin or nonkin (Chapais 2004, see also Hill and Okayasu 1996). Mother/daughter rank reversal may take place before old age in Tonkean macaques (*Macaca tonkeana*) (Thierry 2000), some savanna baboons (Combes and Altmann 2001), and bonnet macaques (Silk et al. 1981b). When this occurs, mothers cannot help younger daughters gain rank over older daughters, and as a result, sisters may not rank in any consistent order (Chapais 2004).

Among New World monkeys, we know most about white-faced capuchin monkeys (*Cebus capucinus*), which, like many Old World monkeys, are highly social and female-bonded. These females preferentially groom and support matrilineal kin, but mother/daughter pairs and sisters affiliate at similar rates. Their dominance hierarchies do not follow the youngest sister ascendancy rule, nor do paternal sisters associate with one another any more than with unrelated females (Perry et al. 2008). Among diurnal prosimians, we know most about ring-tailed lemurs (*Lemur catta*), which are also highly social and female-bonded. Mother/daughter pairs and maternal sisters groom each other more than expected by chance (Nakamichi and Koyama 1997, see also Oda 1996), and groups tend to split along matrilineal lines (review in Gould and Sauther 2007). Participation in territorial conflicts is also related to kinship; in a study of two semi-free-ranging groups, females with fewer living daughters in the group participated less than those with more (Nunn and Deaner 2004). On the other hand, mothers rarely support daughters and, perhaps as a result, daughters' ranks are not similar to those of their mothers (Nakamichi and Koyama 1997).

Males

Male relationships in female-philopatric/male-dispersal societies are also structured in part by matrilineal kinship, particularly before males disperse from their natal groups. Females of many species favor related natal adult and juvenile males, affiliating, supporting, reconciling, and co-feeding preferentially with close kin (Pereira and Fairbanks 1993a, Fairbanks 2000b, Schino et al. 2007b).

Similarly, juvenile male macaques preferentially affiliate with and/or support maternal brothers and other close kin before they disperse (rhesus, Colvin 1983; bonnets, Silk 1992b; Barbarys, Widdig et al. 2000; see also Barbaranelli 1994 cited in Paul 2006), although kin bias may be weaker for support given to others than support received (Schino et al. 2007b). There is also evidence that matrilineal kin facilitate rank acquisition in natal males (e.g., Kuester and Paul 1988, Paul 2006).

Typically, a male's focus on kin weakens as he matures. Nevertheless, many males disperse into groups in which maternal or paternal male kin already reside (Cheney and Seyfarth 1983, but see Alberts and Altmann 1995a). There is some evidence that doing so may ease their entry into new groups and increase their chances of achieving high rank or at least of receiving support (Meikle and Vessey 1981; van Noordwijk and van Schaik 1985; Pusey and Packer 1987; but see Melnick et al. 1984b, Colvin 1986). In other cases, small groups of similarly aged males may disperse together and cooperate with one another in aggressive encounters with other males (e.g., Peruvian squirrel monkeys [*Saimiri boliviensis*], Mitchell, C. L., 1994; ring-tailed lemurs, Sussman 1992, Sauther et al. 1999a; vervet monkeys, Cheney and Seyfarth 1983; long-tailed macaques [*Macaca fascicularis*], van Noordwijk and van Schaik 1985). In species with high reproductive skew, dispersal partners are likely to be paternal brothers, but it is not yet clear whether they preferentially choose paternal kin in these cases or whether alliances with more closely related males are more successful or more enduring than those with fewer close kin.

Although males in many species form special relationships with particular infants or juveniles, in many cases the infants are not their genetic offspring and male care can be better explained as mating effort (Smuts 1985, van Schaik and Paul 1996) or agonistic buffering (Strum 1984, Stein 1984, review in MacKinnon 2007). However, Buchan et al. (2003) reported that male savanna baboons preferentially intervene on behalf of their own juvenile offspring when they get into fights with other youngsters. Although rhesus males do not similarly support their offspring (Widdig 2007), juvenile mandrills appear to recognize their own fathers and affiliate with them more than with unrelated males (Charpentier et al. 2007). There is also evidence that putative fathers attempt to protect their own offspring from infanticide by other males (Hanuman langurs, Borries et al. 1999a; long-tailed macaques, de Ruiter et al. 1994; Japanese macaques, Soltis et al. 2000a; chacma baboons (*Papio cynocephalus ursinus*), Weingrill 2000, review in Widdig 2007).

Male Philopatry/Female Dispersal

Females

Females that disperse often have limited access to matrilineal kin. Yet some associate preferentially with those that are available. Although most chimpanzee females disperse

when they reach sexual maturity, some remain in their natal groups or return after a period of time away (Goodall 1986, Nishida 1989, Stumpf 2007). At Gombe, long-term records indicate that mothers that reside in the same community with adolescent or adult daughters spend much more time together than do unrelated females (Williams et al. 2002a). Mothers also maintain ties with adult sons and may influence their dominance ranks (Goodall 1986; Boesch and Boesch-Achermann 2000; but see Williams et al. 2002a). Juveniles enthusiastically care for and play with younger maternal siblings before adolescence and are likely to adopt younger ones if their mother dies (Goodall 1986). On the other hand, maternal sisters living in the same group in Mahale fail to maintain strong bonds as adults (Nishida 1989).

Among bonobos, mother/son and female/female relationships are among the most prominent in the group (Furuichi 1989; White 1996b; but see Hohmann et al. 1999). Mothers' support for sons increases the sons' chances of attaining high ranks (Furuichi 1989, Ihobe 1992). Female/female support, affiliation, and sociosexual behavior are common (White 1996b); but these relationships are not based on kinship (Hashimoto et al. 1996, Gerloff et al. 1999).

Mother/daughter muriqui pairs (*Brachyteles hypoxanthus*) occasionally find themselves together in the same social group, and although they tend to spend more time with one another than expected by chance, they seldom form strong affiliative bonds with either them or any other female (Strier 1993). Nor do mothers maintain long-term preferential bonds with their sons, with whom they typically reside for long periods of time (Tolentino et al. 2008). Female kin do not generally care for each other's infants, although one exceptional grandmother that did not have an infant of her own carried her daughter's infant about as much as did the daughter (Assuncao et al. 2007).

Males

As noted, males in some, but not all, male-philopatric/female-dispersal species maintain long-term ties to their mothers. Although male chimpanzees generally show only low levels of care of youngsters, Lehmann et al. (2006) found that genetically identified fathers played longer with their offspring and directed less aggression toward them than toward unrelated juveniles. Close relationships between maternal brothers are prominent (Goodall 1986, Nishida 1989), and recent genetic analyses confirm that maternal brothers preferentially affiliate and cooperate with one another (Langergraber et al. 2007a). However, most cooperation and affiliation takes place between unrelated or distantly related males (Langergraber et al. 2007a). Earlier genetic studies using mitochondrial DNA analysis indicate that kin bias does not extend to matrilineal kin in general (Goldberg and Wrangham 1997; Mitani et al. 2000, 2002a), nor do paternal brothers show any preference for one another (Langergraber et al. 2007a; but see Lehmann et al. 2006). Male/male bonds are generally weak among bonobos (White 1996b; but see Furuichi and Ihobe 1994), and at this point, there is no firm evidence that bonds between either maternal or paternal siblings are stronger than those between other males (Furuichi and Ihobe 1994, but see Hohmann et al. 1999). Although male muriquis associate closely, cooperate in group defense, and show high levels of mating tolerance, there is no evidence that maternal brothers prefer one another over other males (Strier et al. 2002). Similarly, male spider monkeys have strong affiliative and cooperative relationships, but it is not known whether they favor close kin over others (see Chapter 11).

Both Sexes Disperse

When both sexes disperse, access to kin may vary considerably. Nevertheless, both males and females commonly favor matrilineal kin when available. Although most mountain gorilla females (*Gorilla beringei beringei*) disperse from their natal groups, about 80% spend part of their adult lives with other maternally or paternally related adult females, either because they remain in their natal groups or join groups containing female kin (Watts 1996, 2001). When that occurs, related adult females tend to display preferential affiliative and supportive bonds with one another but are more likely to do so when they are related maternally than paternally (Stewart and Harcourt 1987, Watts 1994c). In contrast, maternal sibling relationships among both male and female immature gorillas are not consistently stronger than those with other immatures (many of which are likely to be paternal siblings) (Watts and Pusey 1993). Maternally related brothers may co-reside in multimale/ multifemale groups or in all-male groups (Robbins 2007), but there is little information on how their relationships differ from those of other males. Silverback males sire most of the infants born into their groups and are remarkably tolerant of infants and juveniles. Most sons emigrate as young blackbacks, but some continue to reside with their fathers into adulthood. Thus, some males in multimale groups are father/son pairs or brothers (paternal and/or maternal) (Robbins 1995, Bradley et al. 2005). Males in multimale groups cooperate during intergroup encounters but generally coexist by tolerating or avoiding one another rather than by forming highly affiliative or hostile relationships (review in Robbins 2007). Nevertheless, it has been suggested that related males may form more affiliative relationships than unrelated males (Harcourt 1979b, Harcourt and Stewart 1981).

Less is known about social relationships among western lowland gorillas. Genetic analyses suggest that, like mountain gorillas, females have opportunities to interact with other female kin after they disperse: Females disperse shorter distances than males (Douadi et al. 2007), and about 40% reside in groups with other female relatives (Bradley et al. 2007). Similarly, immature offspring often stay with aging silverback fathers when their breeding groups break up, or they may join nonbreeding groups that contain related silverbacks (Levrero et al. 2006). Breuer et al. (2009) suggest that breeding groups based on father/ son relationships are less common in western lowland gorillas than in mountain gorillas because western lowland males take longer to mature. As a result, they may be less likely to reach maturity in a breeding group headed by their fathers.

Hamadryas baboons (*Papio cynocephalus hamadryas*) live in one-male groups within multilevel societies composed of associated one-male units (clans), associated clans (bands), and associated bands (troops). Females disperse from their natal units and live in a series of units during their lives with other adult females and their young. In general, adult females within each unit are not closely related (although they may be members of the same clan), and females focus their interactions on the unit male and on their own offspring (Kummer 1968a, 1990; Sigg et al. 1982; Abegglen 1984). However, there is evidence that mother/daughter pairs and paternal sisters, but not maternal sisters, are found together and groom and support each other more often than expected by chance (Colmenares 1992, 2004). There is also evidence that related males co-reside in the same clans more than expected by chance. Males cooperate in decisions over clan movement and show high propensities to reconcile after disputes, but it is not known whether they favor their close kin over other clan males. Some units also contain a second, younger related follower male that may eventually take over the unit (review in Colmenares 2004). In captivity, clans are more cohesive when they contain father/son pairs, full or maternal brothers, or other maternally related males (Colmenares 1992).

Callitrichid females generally give birth to two sets of twins per year. Their ability to do this depends on receiving help with infant care and home range defense from adult males and nonbreeding group members. Nonbreeding helpers are frequently, but not exclusively, close relatives of the breeding pair, often older brothers and sisters of the infants. Those that are old enough to breed on their own are typically suppressed reproductively via physiological and/or behavioral mechanisms (see Chapter 7). In most cases, there is only one breeding female per group, but in one population of golden lion tamarins (*Leontopithecus rosalia*) a minority of females share reproduction with subordinate females. Females are more likely to share reproduction with daughters than with sisters and least likely to do so with unrelated females (Dietz 2004). In groups that contain both related and unrelated helpers, the effect of kinship on helping varies by species and sex. In golden lion tamarins, males, but not females, help care for related infants more than unrelated infants (Baker 1991 cited in Dietz 2004). In contrast, related and unrelated cotton-top tamarins (*Saguinus oedipus*) carry infants similar amounts both in the wild and in captivity (see also Wamboldt et al. 1988 for pygmy marmosets [*Cebuella pygmaea*], review in Dietz 2004).

INCEST AVOIDANCE

Data from a wide range of taxa indicate that inbreeding tends to lower fitness by allowing the expression of deleterious recessive homozygous alleles and by decreasing genetic variation. Decreased genetic variation diminishes a population's ability to effectively adapt to changed circumstances and to engage in coevolutionary races with parasites and disease organisms (Hamilton et al. 1990, Hamilton 1993). Thus, it is not surprising that many taxa, including primates, have evolved one or more mechanisms to avoid inbreeding. The potential for inbreeding in primates is greatly diminished by dispersal (Pusey and Packer 1987). Generally, individuals disperse initially around puberty, and males often continue to disperse at intervals that correspond roughly to the ages of maturity of potential daughters. However, some opportunities for inbreeding remain when some adults of the dispersing sex remain in their natal groups, when individuals begin breeding before dispersing, when dispersed adults remain in groups after their opposite-sex offspring mature, when close kin of the opposite sex disperse into the same groups (but see Bradley et al. 2007), or when individuals join groups from which their own parents dispersed. In spite of opportunities for incest, both quantitative studies of mating and genetic studies of reproductive success indicate that close kin in a wide range of primate species rarely reproduce together even when they interact regularly (reviews in Pusey 1990, Wolf 1995, Paul and Kuester 2004, Manson 2007). Results are particularly consistent for matrilineal kin and father/daughter pairs (e.g., gorillas, Stewart and Harcourt 1987, Watts 1990b; capuchins, Di Bitetti and Janson 2001a, Muniz et al. 2006) but vary by species for other patrilineal kin. For example, wild yellow baboon paternal siblings form consortships but are less likely to complete matings than unrelated pairs (Alberts 1999), whereas free-ranging or captive paternal siblings in other species show no evidence of incest avoidance (Barbary macaques, Kuester et al. 1994; Japanese macaques, Inoue et al. 1990; rhesus macaques, Smith 1995; ring-tailed lemurs, Pereira and Weiss 1991). Several studies indicate that females are less inclined to mate with close kin than males, a finding that is consistent with the idea that inbreeding is potentially more costly to female reproductive success (Paul and Kuester 2004, Setchell and Wickings 2006, Manson 2007). Females frequently avoid or resist attempts at courtship by sons, brothers, and other matrilineal kin, especially during peak periods of fertility; and they initiate few matings with close kin (e.g., rhesus macaques, Manson and Perry 1993; Japanese macaques, Soltis et al. 1999; chimpanzees, Goodall 1986, Pusey 1990; ring-tailed lemurs, Pereira and Weiss 1991, Sauther 1991).

The data on human and nonhuman primate incest and incest avoidance largely support the idea that individuals avoid mating specifically with others with which they have had close associations as infants and juveniles. Edward Westermark (1891) first proposed this hypothesis, suggesting that humans have been selected to develop sexual aversions to close kin in this manner. In many species, both degree of inbreeding avoidance and early familiarity correlate with maternal kinship (review in Paul and Kuester 2004). In addition, several studies suggest that early familiarity leads to sexual avoidance regardless of actual kinship (Coe et al. 1979, Smith 1995, Wolf 1995, Erickson 2006, review in Paul and Kuester 2004) and, conversely, that kin that associate closely only later in life experience no sexual inhibition (Kuester et al. 1994, Erickson 2006). Mechanisms based on phenotypic matching are also possible, particularly for paternal siblings; but further investigation is needed (e.g. Alberts 1999). Finally, age proximity may provide cues to paternal relatedness, although, as Paul and Kuester (2004) point out, age proximity often correlates with early familiarity.

EVOLUTIONARY/ECOLOGICAL MODELS OF KIN BIAS

Socioecological Explanations

Socioecological models of primate sociality hypothesize that patterns of affiliative kin bias among diurnal females are linked to patterns of competition for monopolizable food resources (e.g., Sterck et al. 1997). The model posits that females form groups with one another when doing so effectively reduces risks of predation and infanticide. However, group living inevitably leads to competition for food resources that limit female reproductive success. When females compete directly for monopolizable resources, they are expected to form alliances to increase their access to food. Females that experience high within-group contest competition (WGC) and low between-group competition (BGC) for food resources (i.e., resident-nepotistic groups) are expected to form within-group coalitions with kin and with high-ranking individuals to increase their access to food. Alliances with kin are expected given both long-term familiarity and potential inclusive fitness gains. Females are expected to display strong tendencies to both support and affiliate with kin, and given the value of alliances, they should remain philopatric.

In contrast, females that experience only high BGC (resident-egalitarian groups) are expected to participate in group-level coalitions only and to display much weaker tendencies to favor kin. The same is true for females that experience both high WGC and BGC (resident-nepotistic-tolerant) because they are expected to display more tolerance toward nonkin than resident-nepotistic females in order to maintain their cooperation in group-level alliances. Females that do not experience either high WGC or BGC (dispersal-egalitarian groups) are not expected to form alliances, display high degrees of kin bias, or be philopatric. While some studies have demonstrated links between WGC and tendencies for females to affiliate and form coalitions with one another (e.g., Mitchell et al. 1991; Barton et al.

1996; Hill 1999, 2004), others have not (e.g., Boinski 1999, Koenig 2002, see also Ménard 2004). No direct links between levels of WGC and intensities of affiliative and supportive kin bias have been found (Berman et al. 2008).

Dominance Style

More success has been found linking variations in kin bias to species-typical, sex-specific dominance styles (e.g., Thierry 2000). The dominance style concept posits a continuum of species ranging from despotic to relaxed or egalitarian, based on the intensity and unidirectionality of aggression and on conciliatory tendencies following aggression. Several other aspects of social structure, including kin bias, appear to co-vary with dominance style. Notably, species described as resident-nepotistic by Sterck et al. (1997) are generally classified as despotic and display intense kin bias, whereas those classified as resident-nepotistic-tolerant or resident-egalitarian are generally classified as relaxed or egalitarian and display weak kin bias (e.g., Thierry 2000). Explanations for the co-variation of dominance style and kin bias also focus on tendencies for kin to form within-group agonistic alliances (Thierry 1990a,b). Chapais (2004), like Sterck et al. (1997), hypothesizes direct links between affiliative kin bias, the profitability of alliances, and the strength of WGC. Others view dominance styles as emergent properties of self-organizing social systems (Hemelrijk 1999; Thierry 1990a,b) driven by species-specific variations in kin bias or intensities of aggression. Rather than reflecting current conditions, dominance styles are hypothesized to reflect past ecological conditions and competitive regimes and to represent evolutionarily stable strategies that render them somewhat independent of current ecological conditions (Matsumura and Kobayashi 1998). Thus, current dominance styles are thought to be outcomes of past biogeographical factors (Thierry 2004) and phylogenetic inertia (Matsumura and Kobayashi 1998; Matsumura 1999; Thierry et al. 2000b). A recent analysis of captive macaques shows significant linkage and phylogenetic signaling for some core indicators of dominance style but only weak evidence of a phylogenetic signal for kin bias (Thierry et al. 2008).

Time Constraints

A related body of theory hypothesizes indirect links between WGC and affiliative networks based on time constraints (e.g., Dunbar 1988, 1992b). When groups increase in size, WGC is expected to increase (Sterck et al. 1997), along with the need for large within-group alliances, maintained through affiliation. However, when social time is limited, as it is likely to be in groups with substantial competition for food, females may not have enough time to maintain increasing numbers of affiliative relationships (Dunbar 1984b). Under these circumstances, affiliative behavior should become increasingly focused on a small proportion of group members, most likely one's closest kin (Altmann, S., 1979;

Chapais 2001). So far, studies of rhesus macaques (Berman et al. 1997, Berman and Kapsalis 2009), Tibetan macaques (*Macaca thibetana*) (Berman et al. 2008), and capuchin monkeys (Perry et al. 2008) are consistent with this prediction: Kin bias is more intense when groups are large than when they are small.

Dispersal Costs/Foraging Efficiency

Isbell and colleagues (reviewed in Isbell 2004) have proposed a model in which groups of female kin form by default when there is an advantage to living in groups. Females remain on their mothers' home ranges when dispersal costs are high enough to make reproduction unlikely in a new area and when their mothers can maintain their own reproductive output while sharing her range with another female. Mothers vary in the extent to which they are energetically constrained both physiologically and ecologically. This leads in turn to variations in the extent to which they can expand their ranges to accommodate daughters. In some cases, daughters can contribute to home range defense or expansion, facilitating the formation of kin groups and opening the door to inclusive fitness benefits through kin bias. In this sense, female kin bias is seen as a secondary outcome of group living rather than a driving force for group formation, and the model predicts that it should vary systematically with dispersal costs and foraging behavior related to energetic constraints.

REPRODUCTIVE BENEFITS OF KIN BIAS

It is tempting to assume that the many forms of favoritism and cooperation shown by kin to one another are adaptive, but researchers are only beginning to document clear links between kin bias and components of reproductive success. This is partly because it is difficult to quantify individual reproductive success in long-lived animals and even more difficult to quantify the costs and benefits of particular behaviors in the short term, let alone over a lifetime. Some of the best evidence that kin bias leads to increased fitness comes from red howler monkeys (*Alouatta seniculus*), where juvenile and adolescent females attempt to stay within their natal groups and reproduce, but often face eviction when the number of adult females reaches a limit (Pope 2000a). While still in their natal groups, young females vigorously pursue supportive relationships with their mothers and older sisters in an apparent effort to avoid eviction and mothers intervene on behalf of their daughters. Those without support from a high-ranking mother must invariably disperse, sometimes over long distances to find unrelated females with whom they can jointly establish a territory. This leads to delays in reproduction of about 2 years, higher infant mortality compared with females in well-established groups, and an overall correlation between female reproductive success and relatedness within groups (Crockett and Pope 1993, Pope 2000a).

Juvenile male red howler monkeys also pursue alliances, choosing kin (maternal and paternal) when available. As adolescents, they disperse together and cooperate to invade or defend another bisexual group. Once they become established in a group, the dominant male monopolizes reproduction but still requires the cooperation of the male coalition to defend the group from take-overs by other males. Coalitions that are composed of related males last on average 8.2 years, whereas those made up of unrelated males last only 2.3 years. In this way, kin coalitions provide greater reproductive opportunities for the dominant male and appear to provide indirect benefits to his male kin (Pope 1990, 2000b).

Other researchers have found lower infant mortality when individuals are raised in the presence of grandmothers (Fairbanks 1988b, Pavelka et al. 2002) and earlier sexual maturity in females that mature in the presence of mothers and maternal sisters (Charpentier et al. 2008). However, it is not clear whether these advantages are due to the presence of kin or whether the presence of kin is the result of higher fitness that runs in families for other reasons. It has also been argued that when support from kin is important in attaining high rank, kin relationships lead to increased fitness (e.g., Silk 2006).

THE QUESTION OF KIN SELECTION

Evidence of a link between kin bias and increased fitness is necessary to infer kin selection, but it is by no means sufficient. Kin-biased behavior could be the result of a number of processes. Most simply, given that close kin tend to occupy similar ranks in many species, kin bias could be an artifact of processes that tend to bring individuals of similar rank into close association and cooperation. Seyfarth (1983) suggested that an interaction between attraction to high rank and competition for access to high-ranking individuals may contribute to high rates of grooming between adjacently ranked female baboons. Later, de Waal and Luttrell (1986) suggested that females may be attracted directly to individuals that are similar to themselves in rank, age, and kinship. Subsequent studies indicate, however, that rank and age effects are not sufficient to explain kin bias. Female rhesus and bonnet macaques show strong kinship effects even when rank or age differences are controlled (Silk 1982; de Waal 1991; Kapsalis and Berman 1996a,b). Indeed, kinship effects are often stronger than effects of rank distance or age (bonnet macaques, Silk 1982; rhesus macaques, Kapsalis and Berman 1996a,b; baboons, Silk et al. 1999). Moreover, kin-biased behavior is found in species in which kinship and dominance ranks are uncorrelated (Hanuman langurs, Borries et al. 1994; gorillas, Watts and Pusey 1993).

It has also been suggested that kin bias may merely be a by-product of mother/infant association patterns (Chapais 2001). If relationships between infants and other maternal kin develop out of their common association with the mother, it is reasonable to ask whether their interaction is merely an outcome of common proximity or whether they represent independent relationships. While this question has not been addressed in a wide range of species, a few studies suggest that females show strong kin bias even when amount of time in proximity is controlled (e.g., Bernstein et al. 1993b, Berman et al. 2004) and that by several months of age infant rhesus macaques pursue relationships with others without regard to maternal proximity (Berman 1982b). In baboons, maternal sisters (but not maternal aunts) maintain strong relationships after their mother dies; indeed, their bonds tend to become stronger (Silk et al. 2006b), further supporting the idea that their relationships are not dependent on the mother. Finally, the fact that paternal kin engage in preferential relationships suggests that kin bias cannot be explained entirely as a by-product of extended maternal ties.

Another possibility is that bonds between kin are "real" but that, rather than being products of kin selection, they are based on mutualism or reciprocity (Chapais 2001, 2006; Chapais and Belisle 2004). In mutualism and reciprocity, cooperating individuals receive direct benefits from their interaction, although they are delayed in the case of reciprocity. Kin selection, in contrast, operates through indirect benefits derived from augmenting the reproduction of genetic relatives. In many cases, kin interaction yields both direct and indirect benefits, and it is difficult to determine which are primary. For example, when young tamarins remain within their natal groups to help their parents rear their younger siblings, they may receive indirect benefits from helping kin but they may also receive direct benefits from being able to remain on their parents' territory, particularly when their chances of establishing a territory and reproducing on their own are small (Dietz 2004). The facts that some helpers are unrelated to the parents and offspring and that help is often distributed without regard to kinship suggest that direct benefits may be sufficient to explain helping behavior in some cases. In other cases, however, both the amount of help given (Baker 1991 cited in Dietz 2004) and the probability that a dominant female shares reproduction with a subordinate (Baker et al. 2002) are correlated with degree of relatedness, suggesting a role for indirect benefits as well.

While direct benefits may be sufficient to explain some forms of cooperation, individuals that form mutualistic or reciprocal relationships preferentially with kin should theoretically be at an advantage over those that cooperate with nonkin because they are able to reap both direct and indirect benefits (Wrangham 1982). However, other factors may offset some or all of the benefits of cooperating with kin. Chapais (2006) suggests that kin may not be optimal choices for interaction that requires partners with specific attributes. For example, it may be more beneficial to form an alliance with an unrelated high-ranking individual than with a low-ranking relative because an alliance with a powerful ally is more likely to be effective. Similarly, it may be beneficial for juveniles to choose play partners that are well matched by size and sex over poorly matched related partners. In other cases, competition with kin may offset the benefits of cooperating with them. For example, Japanese macaque sisters generally support one another in agonistic interactions against nonkin, except when younger sisters attempt to rise in rank over older sisters. In these challenges, younger sisters typically solicit support from higher-ranking kin or nonkin, and this support is generally pivotal in determining whether a rank reversal takes place. As such, when a targeted older sister is challenged by an unrelated dominant female, a younger sister is just as likely to side with the unrelated female as with her sister (Chapais et al. 1994). In these cases, competition for rank with the older sister appears to conflict with the younger sister's usual tendency to support a close relative.

Because it is difficult to measure and compare the reproductive benefits received from kin versus the benefits received directly from cooperation versus the costs of competing with kin, it is difficult to come to firm conclusions about the role of kin selection versus mutualism or reciprocity in most cooperative behavior. This is less of a problem, however, with unilaterally altruistic behavior, that is, behavior that increases the fitness of another individual at a cost to the actor and is directed in only one direction— from the altruist to the recipient (Chapais and Belisle 2004). Here, the actor receives no immediate benefit from the act and incurs a clear cost for which it is not reciprocated at another time. Allowing a subordinate individual access to food can be considered a form of unilateral altruism, and observational studies support the idea that both passive co-feeding (Yamada 1963, Furuichi 1983, de Waal 1986a, Ihobe 1989, Bélisle and Chapais 2001, Berman et al. 2004, but see Pastor-Nieto 2001) and active food sharing (McGrew 1975, Silk 1979, Feistner and Price 1990, Perry and Rose 1994) are highly kin-biased. However, these studies do not definitively demonstrate kin selection because they do not control for variations in motivation (e.g., hunger) and costs of defending food (e.g., injury, loss of efficiency). One experimental study that provided these controls found that co-feeding in female Japanese macaques was indeed correlated with maternal degree of relatedness (Bélisle and Chapais 2001). Mothers and daughters ($r = 0.5$) co-fed more than grandmother/granddaughter or sister/sister pairs ($r = 0.25$), which in turn co-fed more than aunt/niece pairs ($r = 0.125$) and unrelated females ($r = 0$). In a second experiment, Bélisle (2002) reduced the costs of defending food and found that females no longer favored their sisters or grandmothers, extending feeding tolerance only to daughters. These results support the notion of kin selection in unilateral altruism but suggest that when the costs of defending food are minimal, favoritism may extend only to one's closest kin (Chapais and Belisle 2004).

Some forms of support are also considered examples of unilateral altruism (Chapais and Belisle 2004). These include instances in which supporters intervene on behalf of immatures that are unable to effectively reciprocate, and in which support is risky because it is directed against individuals that rank higher than the supporter. Several studies indicate that support against dominant opponents is highly kin-biased (Japanese macaques, Kurland 1977, Watanabe 1979, Chapais et al. 1997; bonnet macaques, Silk 1982; rhesus macaques, Kaplan 1978, Chapais 1983; baboons, Walters 1980, Pereira 1989; vervet monkeys, Hunte and Horrocks 1987; long-tailed macaques, Netto and van Hooff 1986; gorillas, Harcourt and Stewart 1989) and that, in these cases, supporters are likely to support close kin over distant kin (Datta 1983, Chapais et al. 1997). Chapais and colleagues experimentally confirmed unilateral support kin bias on behalf of juvenile Japanese macaques, while controlling for several possibly confounding factors, including the availability of kin that require help, the availability of potential helpers, and the ability to gain access to needy kin (Chapais 1988a,b; Chapais et al. 1997, 2001; Chapais and Belisle 2004).

A final approach to testing predictions for kin selection comes from studies of "consensual" rank reversals between baboon mothers and their mature daughters. Reversing ranks may benefit aging mothers indirectly when their reproductive value drops below that of their daughters as they lose fertility and have a shorter span of time to continue reproducing. Combes and Altmann (2001) tested the idea that such reversals are products of kin selection by using long-term demographic data and Hamilton's rule to predict precise ages at which they should take place. With some exceptions, the actual ages at rank reversal were consistent with the predictions.

CONCLUSIONS

Patterns of kin bias vary widely both between and within species and are constrained by factors that determine the kinship structure of groups, the availability and/or distribution of limiting resources, and the mechanisms available for recognizing kin. Within groups, kinship also interacts with rank, age, and sex. Nevertheless, kin bias and incest avoidance appear to be reoccurring themes throughout the primate order, seemingly surfacing repeatedly when circumstances allow.

While some researchers see variation in the expression of kin bias as indicative of a shrinking role for kinship in the study of primate social behavior, it is also possible, and perhaps more accurate, to see an expanding role in the challenge of understanding how kinship interacts with a multiplicity of biological, demographic, ecological, and social factors to bring about its variable expression (Chapais and Berman 2004a, Strier 2008). Certainly, the discovery of paternal kin recognition represents an area of enormous potential and expansion, aided by the increasing use and resolution of genetic analyses of kinship relationships. Indeed, we are only beginning to learn about the extent to which various species recognize patrilineal kin and favor them and/or avoid reproducing with them. At this point, the available data suggest that patrilineal kinship is generally weaker in shaping relationships than matrilineal kinship, even in cases in which patrilineal kin are available and able to recognize one another. The reasons for this pose intriguing questions for future research.

Kin selection continues to pose intriguing questions and challenges. Primate researchers have benefited enormously from skeptics who have criticized the acceptance of evidence of kin bias alone as evidence for kin selection (Chapais 2001, Clutton-Brock 2002). Researchers who have been stimulated to examine alternative hypotheses for kin

bias have uncovered several examples of kin-biased behavior that can be better explained as kin-based mutualism or reciprocity, for example, support against lower-ranking opponents. At the same time, new experimental (Chapais and Belisle 2004) and long-term field-based approaches (Combes and Altmann 2001) are being developed to probe more precise predictions of Hamilton's rule and to rule out alternative explanations. These studies have provided both stronger support for the operation of kin selection than we have seen previously as well as indications that the limits of kin selection may be narrow, rarely going beyond parent/offspring and sibling relationships.

39

Cooperation, Collective Action, and Competition in Primate Social Interactions

Robert W. Sussman and Paul A. Garber

1. Why is it important to study the context of aggressive and cooperative interactions?
2. Do you think that intense aggression and mild aggression occur in the same contexts and among the same individuals?
3. Are patterns of aggression and cooperation the same in different species?

INTRODUCTION

Virtually all species of diurnal primates live in bisexual social groups composed of individuals of all age classes. Although changes in group size and composition are affected by a variety of events, including immigration, emigration, births, deaths, and fissioning, it is common for the same set of related and unrelated individuals to remain in a group for periods of months, years, and in some cases decades. Group living requires that individuals form predictable social relationships and develop affiliative social bonds. As an order, primates are characterized by an extremely broad range of social grouping patterns, including nuclear families, extended matrilines, extended patrilines, multifemale/multimale units, single-male/multifemale groups, and fission–fusion groups. In many primate species, group size, composition, and cohesiveness are highly flexible and vary in response to local demographic, ecological, and historical factors.

WHY BE SOCIAL?

Several theories exist concerning the costs and benefits to individuals of interacting in cooperative ways: traveling together, exploiting a common set of resources, defending a common range, and forming alliances. These include sociobiological explanations of kinship and inclusive fitness as well as models of individual selection, mutualism, and multilevel selection (van Schaik and van Hooff 1983, Dugatkin 1997, Wilson 2000, Clutton-Brock 2002, Cheverud 2004; see Chapter 40). In general, it has been argued that the primary benefits of group living include enhanced predator detection, increased access to sexual partners, opportunities for information sharing, avoidance of male infanticide, enhanced food sharing, and cooperative behavior (Terborgh and Janson 1986, van Schaik and Horstermann 1994, Chapman and Chapman 2000, Wilson 2000). The primary costs of sociality are assumed to be increased levels of feeding competition and aggressive interactions, especially with increasing group size (van Schaik 1983, 1989; Janson and Goldsmith 1995; Sterck et al. 1997; Chapman and Chapman 2000; Janson 2000a; Wittig and Boesch 2003a; see Chapter 40). In this regard, the idea that competition and aggression over access to food and sexual partners are central to an understanding of the origins of group living and sociality in human and nonhuman primates remains a dominant theory in primatology and behavioral ecology. In part, this may relate to a commonly held idea that most social interactions represent a zero-sum game in which all benefits accrue to a single winner. Using this paradigm, competitive and aggressive interactions are expected to be widespread among members of neighboring social groups as well as among members of the same group (Wrangham 1980, van Schaik 1989, Sterck et al. 1997, Wittig and Boesch 2003a).

In most primate species, however, cooperative, coordinated, and affiliative interactions are considerably more common than agonistic interactions and may play a more formative role in structuring behavioral interactions (Sussman and Garber 2004, Erhart and Overdorff 2008). For example, after 15 years of studying baboons (*Papio cynocephalus*), Strum (2001:158) observed that "aggression was not as pervasive or important an influence in evolution as had been thought, and that social strategies and social reciprocity were extremely important." Similarly, Sussman et al. (2003a) report that adult female ring-tailed lemurs (*Lemur catta*) spent on average less than 1 min per day in agonistic interactions. This pattern of extremely low frequencies of aggressive behaviors and high frequencies of affiliative behaviors characterizes individuals in most primate species (Sussman and Garber 2004). Furthermore, there can be considerable advantages to both kin and nonkin group members in developing dyadic, polyadic, and group-level affiliative and cooperative behaviors in which several participants receive collective or positive benefits (Dugatkin 1997; Clutton-Brock 2002; Johnson et al. 2002; Korstjens et al. 2002; Stephens et al. 2002; Silk et al. 2003a,b; Cheverud 2004; Snowdon 2004). Thus, one goal of this chapter is to more clearly examine the relative roles and social functions of agonistic behavior and affiliative behavior in primate sociality. Agonistic, competitive, and affiliative behaviors are factors of social life. That they exist requires no particular explanation. How these behaviors function in promoting individual fitness, group cohesion, and social strategies does require explanation. We argue that primate sociality and affiliative, cooperative, and agonistic behaviors are best understood in terms of the individual and collective advantages of living in a functional social unit.

QUESTIONS CONCERNING THE CONTEXT OF PRIMATE SOCIAL INTERACTIONS

Data on the contexts, functions, and effectiveness of affiliative and agonistic behaviors in wild primates are needed to test theories of primate sociality. For example, how much time do different primate species spend in overt social interactions? How much of this behavior is friendly, and how much is agonistic? How variable are rates of affiliation and agonism among different groups of the same species? How do group size and composition affect rates of social interactions? Are highly sexually dimorphic species or species that live in larger social groups more agonistic or do they experience higher rates of feeding competition than other species? What are the differences in the frequency and quality of social interactions among kin, friends, and non-kin? Although at the present time it is not possible to answer these questions fully, we begin by addressing a few more basic questions: How much time do diurnal, social-living prosimians, New World monkeys, Old World monkeys, and apes spend in social behavior? How much of this time is affiliative? And how much of this time is agonistic?

We present data from field studies on the frequency, rate, and nature of social behavior in almost 60 species of primates (Table 39.1). We acknowledge several limitations of the data. For example, researchers have used a variety of alternative definitions for similar behavioral categories and have collected data using different sampling procedures (i.e., focal animal sampling, scan sampling, ad libitum or all-occurrence data; see Chapter 21). Moreover, different species and different individuals of the same species are likely to vary considerably in the expression and conspicuousness of social interactions. Thus, conspicuous behaviors (loud, visual) may be overrepresented in the data and more subtle forms of social interaction (displacement or avoidance) may be underrepresented. Similarly, studies may focus on only certain age or sex classes and, therefore, can bias the results (i.e., in a species in which females groom more than males, a concentration on female behavior may overrepresent the overall frequency and nature of social interactions). Finally, most individual primates spend the vast majority of their day in peaceful and close proximity to conspecifics. Although in some species spatial proximity may represent the strongest measure of social affiliation (Crockett and Eisenberg 1987), it is often not included in studies of primate sociality.

With this in mind, we view published percentages and rates of social interaction as general values that are likely to have considerable variance. However, we have confidence that the data represent reasonable approximations of the true activity patterns for these primates. Furthermore, since these data were collected using widely different methods and trait definitions, we do not attempt to apply complex meta-analysis statistics to them but, rather, use them simply for comparative and illustrative purposes. Using these data, our main purpose is to show that group-living primates generally spend 5%–10% of their day in active social interactions and that extremely few of these social interactions involve agonistic behavior. We hope this chapter will stimulate further research on these questions and the development of more comparable methodologies. However, given the patterns we observe in our large data set, we predict that our general conclusions will stand.

Table 39.1 Activity Budget and Rates of Agonism in Diurnal Primates

SPECIES	% TIME SOCIAL	% AFFILIATIVE	AGONISTIC	REFERENCE
Diurnal prosimians				
Varecia variegata	<1.0	Rare	Rare	Vasey 1997
Eulemur fulvus	<1.0	Rare	Rare	Vasey 1997
Eulemur fulvus	1.0	1.0	Rare	Overdorff 1991
Eulemur rubriventer	2.0	2.0	Rare	Overdorff 1991
Lemur catta	2.6	2.1	0.5%	Gould 1994
Lemur catta	2.6	1.9	0.7%	Sussman unpublished
Eulemur fulvus	2.8	2.5	0.3%	Sussman unpublished
Varecia variegata	3.1	3.1	0.02/hr male	Morland 1991
			0.17/hr female	
Eulemur fulvus	3.5	3.5	Rare	Tattersall 1977a–c
Propithecus verreauxi	3.8	n.d.	0.35/hr	Richard 1978
Eulemur mongoz	4.0	4.0	ND	Curtis 1997
Eulemur coronatus	4.3	4.0	0.3%	Freed 1996
Propithecus diadema	5.0	4.5	0.12/hr	Hemingway 1995
Eulemur fulvus	5.8	5.4	0.4%	Freed 1996
Eulemur fulvus	8.5	7.7	0.8%	Sussman unpublished
Lemur catta	8.6	7.1	1.4%	Sussman unpublished
Overall mean	3.7 (± 2.3, mean weighted by species = 3.68)			
New World monkeys				
Alouatta palliata	0.8			Estrada et al. 1999
Alouatta palliata	0.9			Smith 1977
Brachyteles arachnoides	0.9			Milton 1984a,b
Callicebus torquatus	0.9			Kinzey 1981
Ateles paniscus	0.9			Symington 1988a
Alouatta palliata	1.0			Stoner 1996a,b
Cebus olivaceus	1.3			Miller 1992, 1996
Alouatta seniculus	1.7			Gaulin and Gaulin 1982
Saguinus mystax	1.8	1.4	0.41%	Castro 1991
Alouatta palliata	1.9		0.003/(ind · hr)	Larose 1996
Callicebus torquatus	1.9		0.0006/(ind · hr)	Easley 1982
Alouatta palliata	2.0			Stoner 1996a,b
Cebus olivaceus	2.1			Miller 1992, 1996
Saguinus fuscicollis	2.2	1.9	0.35%	Castro 1991
Brachyteles arachnoides	2.7		0.0006/(ind · hr)	Strier 1986a,b, 1987a–c
Saguinus fuscicollis	2.8			Peres 1991a,b
Callithrix geoffroyi	2.8			Passamani 1998
Saimiri sciureus	2.9		0.0047/(ind · hr)	Mitchell 1990
Leontopithecus rosalia	3.1			Dietz et al. 1997
Saguinus mystax	3.1			Peres 1991a,b
Saguinus mystax	3.5		0.20%	Garber 1997
			0.0066/(ind · hr)	
Alouatta pigra	3.9		0.10%	Silver et al. 1998
Lagothrix lagotricha	4.0	3.1	0.60%	Stevenson 1998
Lagothrix lagotricha	4.8			Defler 1995
Alouatta caraya	4.9		0.38%	Bicca-Marques 1993
			0.019/(ind · hr)	
Callithrix humeralifer	5.0	4.2		Rylands 1982

Table 39.1 (cont'd)

SPECIES	% TIME SOCIAL	% AFFILIATIVE	AGONISTIC	REFERENCE
Leontopithecus rosalia	6.1			Peres 1986
Leontopithecus rosalia	7.0			Dietz et al. 1997
Leontopithecus chrysomelas	9.1	8.5		Rylands (1982)
Cebus capucinus	9.9		0.016/(ind · hr)	Mitchell 1989
Cebus apella	9.9			Zhang 1995
Cebus capucinus	12.5	7.7	0.034/(ind · hr)	Fedigan 1993
Saimiri oerstedii	13.9		0.000009/(ind · hr)	Boinski 1986
Callithrix jacchus	14.0	11.1	0.051/(ind · hr)	Digby 1994
Ateles geoffroyi	17.0		0.18%	Fedigan and Baxter 1984
Ateles geoffroyi	22.0		0.92%	Fedigan and Baxter 1984
Alouatta palliata			0.007/(ind · hr)	Jones 1980a,b
Ateles geoffroyi			0.0043/(ind · hr)	Klein and Klein 1977
Cebus apella			0.042/(ind · hr)	Janson 1984, 1988a,b
Cebus apella			0.007/(ind · hr)	Janson 1984, 1988a,b
Leontopithecus rosalia			0.0012/(ind · hr)	Baker 1991
Saguinus fuscicollis			0.01/(ind · hr)	Goldizen et al. 1996
Saguinus nigricollis			0.20%	de la Torre et al. 1995
Overall mean	5.1% (± 5.1, mean weighted by species = 5.76)			
Old World monkeys				
Cercopithecus diana	1.2			McGraw 1998a–c
Colobus badius	1.9	1.8		Decker 1994
Presbytis pontenziani	1.9			Fuentes 1996
Colobus polykomos	2.0			Teichroeb et al. 2003
Colobus vellerosus	2.0			Teichroeb et al. 2003
Colobus vellerosus	2.0			Teichroeb et al. 2003
Macaca silenus	2.4			Kurup and Kumar 1993
Colobus badius	2.6	2.5		Decker 1994
Cercopithecus campbelli	2.8			McGraw 1998a–c
Macaca silenus	3.4			Menon and Poirer 1996
Colobus vellerosus	4.0			Teichroeb et al. 2003
Cercopithecus petaurista	4.3			McGraw 2000
Colobus vellerosus	5.0			Teichroeb et al. 2003
Colobus polykomos	5.3			McGraw 1998
Colobus guereza	5.9	5.7	0 in 7,793 scans	Fashing 2001
Colobus badius	6.3			McGraw 1998a–c
Colobus verus	6.7			McGraw 1998a–c
Presbytis entellus	7.4	6.7		Fashing 2001a–c
Cercocebus atys	7.9			McGraw 1998a–c
Colobus badius	8.0	7.6		Decker 1994
Colobus badius	8.2	7.9		Decker 1994
Colobus guereza	8.3	8.3	0 in 8,917 scans	Newton 1992
Macaca silenus	8.4			Menon and Poirer 1996
Colobus badius	8.5	8.5		Decker 1994
Macaca sylvanus	10.0			Menard and Vallet 1997
Papio anubis	10.4	5.2		Harding 1980
Cercopithecus mitis	10.4			Kaplin and Moermond 2000
Cercopithecus l'hoesti	11.4			Kaplin and Moermond 2000

Table 39.1 (cont'd)

SPECIES	% TIME SOCIAL	% AFFILIATIVE	AGONISTIC	REFERENCE
Macaca sylvanus	11.5			Menard and Vallet 1997
Colobus satanas	13.0			Teichroeb et al. 2003
Rhinopithecus bieti	13.1	9.8		Kirkpatrick et al. 1998
Macaca nigra	18.7			O'Brien and Kinnaird 1997
Macaca fuscata	21.7	18.9		Agetsuma 1995a,b
Macaca nigra	23.1			O'Brien and Kinnaird 1997
Macaca nigra	23.5			O'Brien and Kinnaird 1997
Presbytis francoisi	27.9	27.5		Burton et al. 1995
Presbytis entellus			0.084/(ind · hr) females	Borries et al. 1991
Presbytis entellus			0.01/(ind · hr) males	Laws and Laws 1984
Papio cynocephalus			0.14/(ind · hr) males	Strum 1982
Papio cynocephalus			0.11/(ind · hr) males	Noë and Sluijter 1995
Papio cynocephalus			0.079/(ind · hr) males	Noë and Sluijter 1995
Papio cynocephalus			0.14/(ind · hr) males	Noë and Sluijter 1995
Cercopithecus aethiops			0.0007/(ind · hr)	Pruetz 1999
Erythrocebus patas			0.0007/(ind · hr)	Pruetz 1999
Papio anubis			0.084/(ind · hr) males	Harding 1980
Papio cynocephalus			0.037/(ind · hr) males	Smuts 1985
Overall mean	8.6 % (± 6.8, mean weighted by species = 9.38)			
Apes				
Pongo pygmaeus	1.6[1]			Rodman 1973
Hylobates lar	3.0			Gittens and Raemaekers 1980
Gorilla gorilla	3.6			Watts 1988
Hylobates muelleri	4.0			Leighton 1987
Pongo pygmaeus	5.4[1]			Galdikas 1985a–c
Gorilla gorilla	7.0	6.7	0.3%	Olejniczak personal communication
Hylobates lar	11.0		0.009/hr	Bartlett 1999a,b
Hylobates syndactylus	15.0		0.15/hr	Chivers 1974
Pan troglodytes	22.0	9.0% (groom) (includes resting)	0.067/(ind · hr) males	Boesch and Boesch-Achermann 2000
Pan troglodytes	24.9	16.8 (groom)		Teleki 1981a,b
Gorilla gorilla		0.39/hr	0.20/hr	Schaller 1963
Pan troglodytes			0.03/hr	Bygott 1974
Pan troglodytes			0.05/hr	Ghigliari 1984
Pan troglodytes			0.016/hr males	Goodall 1986
			0.009/hr females	Goodall 1986
Overall mean apes	9.7% (± 8.2, mean weighted by species = 9.7)			

%, percent of total activity budget; ND, no data are provided.

[1] Orangutan data include any proximity between adults (76) or all individuals (77), not all of which would include active social interaction.

TIME SPENT IN AGONISTIC BEHAVIOR

Overall, we found that diurnal prosimians (3.7% ± 2.3%), New World monkeys (5.1% ± 5.1%), Old World monkeys (8.6% ± 6.8%), and apes (9.7% ± 8.2%) devoted less than 10% of their daily activity budget to direct social interactions. The overwhelming majority of these interactions were affiliative and cooperative behaviors, such as grooming, food sharing, huddling, and alliance formation (Tables 39.1 and 39.2). In contrast, agonism, which included mild spats, displacements, threats, stares, and fighting, was rare and episodic, typically accounting for less than 1% of all social interactions. For example, even among sexually dimorphic species of Old World monkeys, rates of agonism and aggression were extremely low. In 10 studies of social behavior in Hanuman langurs (*Semnopithecus entellus*, formerly *Presbytis entellus*), baboons (*Papio* spp.), vervet monkeys (*Cercopithecus aethiops*), and patas monkeys (*Erythrocebus patas*), hourly rates of agonism varied from 0.14 episodes per individual to 0.0007 per individual (Table 39.1). Although Old World monkeys spent more time than prosimians and New World monkeys engaged in social interactions, this was fully accounted for by an increased level of affiliative behaviors such as grooming.

As in prosimians and monkeys, data on ape social interactions are extremely variable. This is likely to reflect significant differences in social structure among all species of living apes. Gibbons and siamangs (lesser apes, *Hylobates* spp.) principally live in small pair-bonded groups. In contrast, mountain gorillas (*Gorilla gorilla beringei*) live in cohesive multifemale/one-male and multifemale/multimale social groups, common chimpanzees (*Pan troglodytes*) and bonobos (*Pan paniscus*) live in fission–fusion communities, and orangutans (*Pongo pygmaeus*) tend to be more solitary (except for mother/offspring pairs) and only rarely meet in the wild. Time spent in social interactions for apes ranged from 3.6% of the activity budget of the mountain gorilla to 25% in some studies of chimpanzees. Similar to prosimians and monkeys, rates of agonism in apes are low, ranging from 0.009–0.016 events per hour in chimpanzees to 0.20 per hour in gorillas (Table 39.1).

Based on the expectation that within-group feeding and reproductive competition increase with increasing group size, where possible, we corrected the rate of agonism by the number of potential interactions in the group. Although it is important to acknowledge that not all group members are equally involved in aggressive interactions, there is reason to assume that the number of aggressive events occurring per hour of observation is influenced by the number of potential interactants. One can use the analogy of billiard balls on a table: If there are 15 billiard balls on a table and they are set in random motion, the likelihood of some number of them coming into contact or occupying the same space is much greater than if only two balls were set in motion on the table. We feel this analogy also is informative because it highlights the fact that some agonistic interactions that occur among group members are likely to be random—that is, not the result of a particular social strategy but simply of

the fact that two individuals cannot jointly occupy the same space at the same time.

Our review of the literature indicates that overall rates of agonism per adult group member averaged less than one per day in Old World monkeys to approximately one every 2 days in New World monkeys and that such events were extremely rare in apes and prosimians (on average less than one per month, Table 39.2). The highest frequency of agonism (mild spats, displacements, and threats, as well as more severe forms of violent physical aggression) per individual group member per week was 10–11 times in adult male *Papio cynocephalus*.

Quantitative data on the costs and benefits of social cooperation, mutualism, and affiliative behaviors other than grooming, playing, or huddling are not commonly reported in the literature, although qualitative accounts of these behaviors are common. In part, this may reflect difficulties in quantifying both the short-term and long-term benefits and costs of cooperative behaviors. It also may reflect the fact that researchers have tended to underestimate the importance of various forms of mutual and cooperative behavior in primate sociality. Notable examples of affiliative and cooperative behaviors in primates include communal infant care (Garber 1997), food sharing (Boesch and Boesch-Achermann 2000), shared vigilance and other behaviors associated with predator protection and defense (Treves and Chapman 1996, Treves 1998, Miller 2002b), alliance and friendship formation (Smuts 1985, de Waal 2000b, Silk et al. 2003a), coordinated hunting (Boesch 1994c, Boesch and Boesch-Achermann 2000, Mitani and Watts 2001), and group-level range and resource defense (Wrangham 1980, Peres 2000b).

We found that in prosimians affiliative interactions were eight times more common than agonistic interactions. In New World and Old World monkeys, affiliative behaviors commonly accounted for over 80%–90% of direct social interactions. In a study of lowland gorillas (*Gorilla gorilla gorilla*), C. Olejniczak (personal communication) reports that over 95% of all social interactions were affiliative. Clearly, affiliative interactions represent the overwhelming majority of primate social interactions and form the basis of individual social bonds. For example, data presented by Smuts (1985) on male and female anubis baboon social

Table 39.2 Summary of Agonistic and Affiliative Social Interactions Across All Activities in a Sample of 41 Species of Diurnal Primates

TAXA	% MEAN TIME SOCIAL	% MEAN TIME AFFILIATIVE	AGONISM (EVENTS/HR)	AGONISM (INDIVIDUAL EVENTS/WEEK)
Diurnal prosimians (10 species)	3.7 (± 2.3)	91.7	0.16	0.002
New World monkeys (21 species)	5.1 (± 5.1)	80.0	0.60	3.6
Old World monkeys (14 species)	10.7 (± 7.1)	89.9	0.58	6.3
Lesser and great apes (6 species)	9.7 (± 8.2)	81.6	0.09	0.0001

Source: Adapted from Sussman and Garber (2004).

interactions and spatial proximity strongly support the idea that affiliative social interactions by newly resident males play a critical role in whether they are able to integrate themselves into a troop or must try their luck elsewhere.

Even in species in which social interactions typically account for only 2%–4% of the activity budget and adult group members are not related (i.e., mantled howler monkeys, *Alouatta palliata*), individuals are found to exhibit consistent partner preferences from year to year (Bezanson et al. 2002). These preferences are based on patterns of spatial proximity and affiliation, enabling individuals to feed together in the same food patch and to develop social and mating bonds.

NEW PERSPECTIVES ON COSTS AND BENEFITS OF GROUP LIVING

Traditional models of primate sociality have focused on issues of feeding competition and the costs to individuals of living in social groups (Wrangham 1980; van Schaik 1983, 1989). In general, it has been suggested that individuals should live in groups that are small enough to avoid the costs of aggression at feeding sites and large enough to benefit from cooperative predator detection and avoidance or infanticide (van Schaik 1983, 1989; Janson 1988b; Isbell 1991; Crockett and Janson 2000). These models assume that groups in the population maintain maximal or optimal size and, therefore, that the addition of one or a small number of new group members results in significant costs in feeding competition, travel, increased group spread, and aggression (van Schaik 1983, Janson and Goldsmith 1995, Chapman and Chapman 2000). Fewer studies have been designed to identify nonpredation benefits of group living, such as allomaternal care, cooperative range and resource defense, social learning, friendships, locating feeding sites, and other forms of cooperative behavior.

SOCIOECOLOGICAL MODEL: FEEDING COMPETITION

Several researchers have argued that feeding competition is a pervasive cost of social group living and has played a determining role in structuring the size, composition, dominance hierarchy, and social system of group-living primates (van Schaik 1989, Janson and Goldsmith 1995, Sterck et al. 1997). Under conditions in which food is limited, the frequency of agonistic interactions, especially among females, is expected to be high, with low-ranking females experiencing costs in reduced fitness. This may result from a decrease in nutrition, fertility, or infant survivorship; an increase in energy costs associated with traveling to more distant feeding sites; or increased vulnerability to predators by foraging apart from other group members (Boinski et al. 2000b). This perspective is referred to as the "socioecological model."

Two forms of feeding competition have been identified, contest competition and scramble competition. *Contest competition* is a direct form of social interaction in which a single or small number of individuals displace or aggressively exclude other individuals from a feeding site. This

may take the form of high-cost behaviors such as attacking, chasing, and intimidating displays or more subtle behaviors such as staring, displacement, and avoidance. Contest competition is expected to occur at high-quality, small, or defensible food patches, especially among females, and to result in significant fitness gains for the winner and significant fitness costs for the loser. Given the highly visual and vocal nature of agonistic interactions, they are unlikely to go unnoticed in primate field studies. *Scramble competition* is a more subtle form of social interaction in which an individual that locates a feeding site first is likely to gain a finder's advantage and, in this way, outcompete individuals that might otherwise exclude the finder from a food patch. The finder's advantage is expected to be high when resource patches are depletable, high-quality, ephemeral, and scattered (Ranta et al. 1996). The extent of the finder's advantage is expected to be low when resources are found in intermediate or large clumps or are of low quality (Ranta et al. 1996). However, scramble competition is extremely difficult to document (Chapman and Chapman 2000).

> there is no empirical basis for predicting if a later arriving animal would have obtained full, partial, or no access to the resource, or in fact whether it would have visited the feeding site at all. Given that, in most cases, scramble competition can only be inferred, its role in the evolution of individual foraging and social strategies remains unclear. (Sussman and Garber 2004:162–163)

Finally, Crockett and Janson (2000:76) acknowledge that some primate populations actually experiencing little food competition may be vulnerable to increased mortality of infants in larger groups due to factors unrelated to food, a pattern erroneously suggesting relations between group size and female fecundity. They argue that in the case of red howlers (*Alouatta seniculus*) infanticide and male mating competition are likely to be greater in groups with four adult females than in groups with fewer than four adult females.

It has been hypothesized that, due to their increased reproductive requirements, adult females engage in food-related contests more than adult males (van Schaik 1989, Sterck et al. 1997). In particular, van Schaik (1989) has suggested that in species with linear, despotic female dominance hierarchies, aggression and displacements at feeding sites are frequent. However, even in species exhibiting a linear dominance hierarchy, evidence of resource-based aggression among females is extremely low (Table 39.3). For example, in a group of 53–75 Peruvian squirrel monkeys (*Saimiri sciureus*), Mitchell (1990:116) reported that only 9% of "dyadic agonistic interactions involved food resources." Despite the presence of 23 adult female group members in the troop, over the course of 12 months only 28 cases of female/female food-related agonistic interactions were observed. Similarly, in a study of feeding competition among a community of 14 female chimpanzees (*Pan troglodytes verus*), Wittig and Boesch (2003a:860) reported only 103 food conflicts over 1,028 hr of observation. Although these authors argue that "Taï female chimpanzees used

Table 39.3 Rate of Female/Female Food Conflicts During Feeding in Seven Species of Diurnal Primates

SPECIES	CONFLICT RATE	FEMALES IN GROUP (N)	CONTESTS (N)	DOMINANCE STYLE
Sooty mangabey, *Cercocebus atys* (Range and Noe 2002)	1 per female every 5.7 days	24	ND	Linear
Patas, *Erythrocebus patas* (Pruetz and Isbell 2000)	1 per female every 83.3 days	13–14	75	Egalitarian
Vervet, *Cercopithecus aethiops* (Pruetz and Isbell 2000)	1 per female every 47.6 days	8–9	55	Linear
Squirrel monkey, *Saimiri sciureus* (Mitchell 1990)	1 per female every 25.6 days	23	28	Linear
Common chimp, *Pan troglodytes* (Wittig and Boesch 2003 a,b)	1 per female every 47.6 days	14	103	Linear
Anubis baboon, *Papio anubis* (Barton et al. 1996)	1 every 17.5 days	ND	(Group size 103)	Linear
Mountain baboon, *Papio ursinus* (Barton et al. 1996)	1 every 150 days	ND	(Group size 22)	Linear

ND, no data.

aggression to keep resources or to gain them from other females" and were characterized by despotic dominance relationships and a linear hierarchy, the individual rate of food-related agonism was 0.007/hr. Assuming a 12 hr pattern of daily activity and an activity budget in which 25% of the day is devoted to feeding and foraging, this translates into one food-related agonistic interaction per female every 2–3 months. Moreover, in 35% of these conflicts, the dominant individual did not obtain access to the food reward.

In Table 39.3, we present published data on food conflicts in five primate species. The data represent food conflicts per female per hour of feeding (not over all activities) and indicate extremely low rates of contests even among species forming linear dominance hierarchies (see Erhart and Overdorff 2008 for similar results). Taken together these data indicate that the frequency of food-related agonism among females occurred at a rate that ranged on average from once per 17 hr of feeding (once every 3–5 days) in mangabeys to once per 143 hr of feeding (once every 1–2 months) in chimpanzees and vervet monkeys. Thus, food-related agonism among females may not be as important a factor in shaping daily social interactions and individual relationships as commonly assumed; therefore, the socioecological model of primate sociality needs to be reevaluated. We agree with Thierry (2008:93): "Despite enduring inconsistencies, this model drives most research on primate ecology.... Though (this model) has been truly heuristic for awhile,... it now impedes the development of new lines of thought and the time has come to ask where the Emperor's clothes are."

ECOLOGICAL CONSTRAINTS MODEL: TRAVEL COSTS

Chapman et al. (1995) and Chapman and Chapman (2000) have proposed an ecological constraints model to evaluate the costs of increased group size on within-group feeding competition in primates. This model builds on an analysis by Janson and Goldsmith (1995) which suggested that a decrease in resource availability or quality or an increase in group size will result in greater travel distance in order to maintain a constant rate of food intake. Individuals in larger groups, therefore, are expected to experience increased travel costs in searching over greater distances to locate a

sufficient number of feeding sites to satisfy their energetic needs. Thus, even in the absence of contests at feeding sites, individuals may suffer high costs of scramble competition by increased travel. In this regard, Janson and Goldsmith (1995) reported a positive relationship between group size and daily path length in 16 of 21 primate species studied. The effect was greatest in frugivorous primate species and least or absent in more folivorous species. However, data provided by Chapman and Chapman (2000) and Dias and Strier (2003) indicate that, across a range of different primate species, individuals in larger groups did not travel greater distances than individuals in smaller groups and that even in those cases when travel distance did increase, the additional distance traveled was generally less than 200 m. These data are highlighted in Table 39.4. Although factors other than group size (i.e., seasonal or yearly changes in food availability, range defense) can affect day range, the data indicate that in 50% (8/16) of cases individuals in larger groups of a given species had a day range that was less than that of individuals of the same species living in smaller social groups. In only 18% (3/16) of cases did individuals in larger groups have a mean day range that exceeded by 200 m or more that of individuals of the same species living in a smaller group. In one of these cases, a group of chacma baboons (*Papio hamadryas ursinus*) consisting of 45 individuals traveled an average distance of 4,670 m/day, whereas a group of 47 chacma baboons traveled a distance of 10,460 m/day. However, it is unlikely that a doubling of day range in this species was the direct consequence of adding two additional group members (Table 39.4).

Costs of Additional Travel

Travel costs can be in the form of increased time and energy moving, reduced time feeding and resting, increased conspicuousness to predators, or injury. If these costs are high and individuals in larger groups are required to travel greater distances to obtain resources, then we may expect a decrease in individual fitness. Recent physiological studies of the cost of locomotion in primates indicate that the cost of additional travel represents only a small fraction of an individual's daily energy budget (Warren and Crompton 1998a, Steudel

Table 39.4 Relationship Between Group Size and Distance Traveled per Day Within a Primate Species

SPECIES	GROUP SIZE	DISTANCE TRAVELED (M)	DIFFERENCE (M)
Saguinus fuscicollis	4.7	1,370	
Saguinus fuscicollis	6.5	1,220	−150
Callicebus moloch	3.2	570	
Callicebus moloch	4.2	670	+100
Alouatta palliata	9.1	120	
Alouatta palliata	12.2	600	+480
Alouatta palliata	15.5	440	−160
Alouatta seniculus	7.1	540	
Alouatta seniculus	9.0	710	+170
Alouatta seniculus	9.5	390	−320
Brachyteles arachnoides	24.5	1,280	
Brachyteles arachnoides (Dias and Strier 2003)	57–63	1,313	+33
Procolobus badius	20	600	
Procolobus badius	34	560	−40
Presbytis melalophos	9.3	1,150	
Presbytis melalophos	14	610	−540
Cercopithecus ascanius	26.2	1,540	
Cercopithecus ascanius	32.5	1,450	−90
Cercopithecus mitis	18.7	1,300	
Cercopithecus mitis	32.6	1,140	−160
Erythrocebus patas	20.6	2,250	
Erythrocebus patas	35.5	4,330	+2,080
Papio hamadryas ursinus	45	4,670	
Papio hamadryas ursinus	47.2	10,460	+5,790
Hylobates syndactylus	3.0	740	
Hylobates syndactylus	3.8	930	+190
Hylobates syndactylus	4.0	790	−140
Hylobates syndactylus	5.0	970	+160

Source: Adapted from Chapman and Chapman (2000).

2000). For example, Steudel (2000) calculated that in traveling 6.1 km/day, a female yellow baboon expends 6% of her daily energy budget. Similarly, hamadryas baboons travel approximately 9.5 km/day, and this accounts for 9.1%–9.5% of their daily energy budget. Thus, increases in day range of 100 m, 200 m, or more add relatively little to the energetic costs of travel in these primates. Similarly, Warren and Crompton (1998a) calculated the relative cost of locomotion in *Lepilemur edwardsi* (sportive lemur), *Avahi occidentalis* (woolly lemur), *Tarsius bancanus* (Bornean tarsier), *Galago moholi* (South African lesser bushbaby), and *Otolemur crassicaudatus* (large-eared greater bushbaby) based on an estimate of field metabolic rate over a 24 hr period. These prosimians expended 0.9%–3% of their daily metabolic

energy budget in locomotion (based on night travel distances, locomotor frequencies, and travel height changes in different locomotor modes). Similarly, in mantled howler monkeys, Nagy and Milton (1979b) calculated that the relative cost of locomotion was 2% of daily energy, and Steudel (2000) reported that adult Costa Rican squirrel monkeys (*Saimiri oerstedii*) traveling 3,500 m/day expend only 4.4% of their total daily energy on locomotion. Based on these data, added travel distance is likely to exert only a minor constraint on the energetic costs of primate foraging.

In summary, the empirical data on the frequency and nature of primate social interactions; relationships between group size, day range, and the energetic costs of travel; and female feeding competition in species with linear dominance hierarchies are best interpreted as follows:

1. Agonism occurs at very low frequencies among diurnal primates living in the same social group. This does not fit the predictions of the socioecological model.
2. Agonism at feeding sites does not occur more frequently among female primates that form linear dominance hierarchies than among females that form egalitarian social relationships. This does not fit the feeding competition model.
3. Individuals in larger groups of primates do not necessarily travel greater distances than individuals of the same species living in smaller groups. This does not fit the ecological constraints model.
4. In those cases when individuals in larger groups experience increased travel, the energetic cost of travel for arboreal or terrestrial primates is expected to be relatively low. This does not fit the ecological constraints model.

ALTERNATIVE MODEL

We present an alternative model of primate sociality and group size which we feel is better supported by the available data (Figure 39.1). This model is based on three principles.

First, resources in tropical forests are found in dispersed, heterogeneous patches. This is supported by recent information on the distribution of plant species in Amazonian (Tuomisto et al. 2003) and Central American (Wehncke et al. 2003) forests, as well as theoretical models of variance in spatiotemporal changes of food availability and resource distributions (Kotliar and Wiens 1990, Johnson et al. 2002). According to Johnson et al. (2002:564), the existence of both nearby and distant food patches of various size and quality scattered across the landscape limits opportunity for feeding competition and increases the probability that several individuals have access to feeding sites, which "shifts the cost–benefit balance towards group living."

Second, compared to many groups of animals, primates are characterized by tremendous dietary breadth in the types of food exploited (insects, fruits, flowers, leaves, seeds, small vertebrates) and the number of plant, insect, and vertebrate species consumed (Garber 1987). It is not uncommon for primates to exploit resources from 50 to over 100 plant species

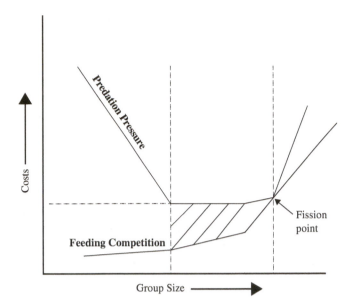

Figure 39.1 Predation pressure, feeding competition, and group size in primates.

over the course of a single year. Finally, in many primate taxa, grouping patterns are highly flexible and individuals may form subgroups as a facultative response to local ecological and social conditions (Symington 1988a,b, 1990; Chapman 1990a,b; Kinzey and Cunningham 1994; Chapman et al. 1995; Jamieson 1998; Garber et al. 1999; Dias and Strier 2003). The ability to temporarily subgroup in response to changes in the availability, density, and distribution of food resources minimizes opportunity for contest competition at feeding sites. We feel this type of social flexibility is common among a range of primate species and not simply restricted to those taxa exhibiting a fission–fusion system.

Predation is a critical component of all models of sociality. Predation risk in our model is depicted as a U-shaped curve (Figure 39.1). Predation risk is assumed to be high at low group size, low and relatively constant at intermediate group sizes, and high at larger group size. Although data on the effects of predation on primate sociality are improving (Miller 2002b, see Chapter 35), given the limited information on predation rates and predation success (Hill and Dunbar 1998, Zuberbuhler 2004, but see Hart 2000, Hart and Sussman 2005), there is little empirical basis at present to critically evaluate this assumption. However, once groups attain large size, it is likely that (1) a large group may attract a greater number of predators and (2) if large groups are less cohesive due to increased levels of feeding competition and social stress, then individuals that forage apart from the main body of a large group may become more vulnerable to predator attack (Janson 1988b). Our model predicts that, across a relatively wide range of group sizes, feeding competition (Figure 39.1) and the associated aggression among conspecifics at feeding sites are expected to be low and to have minimal fitness consequences and levels of predator detection are expected to be high (Figure 39.1, crosshatched area). This occurs in response to the heterogeneous distribution of food patches in tropical forests, the behavioral

flexibility primates show in short-term changes in diet and subgrouping patterns, and the relatively low energetic cost of increased travel. We expect that within-group feeding competition principally becomes a significant problem to individual fitness only at larger group size or among group members living in marginal or degraded habitats.

The third part of the model involves the question of group demography and the process by which groups increase and decrease in size. A basic assumption of the competition models is that most groups in a population are at or approaching their maximal size. Therefore, within-group feeding competition is expected to be relatively high, and the inclusion of additional individuals results in significant increases in feeding competition, aggression, and travel costs. However, there are reasons to suggest that groups in a population are commonly below their maximal size and, therefore, that the effects of feeding competition are low across a range of group sizes. If new groups typically form by a process of fissioning, in which a set of individuals bud off from an established group, then we expect relatively few large groups (they are unstable and fission over a relatively limited period of time), many small groups (newly fissioned elements), and a majority of intermediate-sized groups (those in which a small set of animals have recently left). There is evidence of group fissioning among a wide range of primate taxa, including diurnal lemurs (Sussman 1992, Gould et al. 2003, *Lemur catta*), New World monkeys (Garber 1988a, *Saguinus mystax*; Fedigan et al. 1998, *Alouatta palliata*), Old World monkeys (Altmann, S. A., 1974, *Papio anubis*; Hrdy 1977b, *Presbytis entellus*; Duggleby 1977, Cheverud et al. 1978, *Macaca mulatta*; Cords and Rowell 1986, *Cercopithecus mitis*; Dunbar 1988, *Theropithecus gelada*; Bennett and Davies 1994, *Trachypithecus vetulus*; Oates 1994, *Colobus vellerosus*; Windfelder and Lwanga 2002, *Cercopithecus ascanius*), and apes (Watts 1998a, *Gorilla gorilla*). For example, Fedigan et al. (1998) report that over a 6-year period four of nine new mantled howler (*A. palliata*) groups formed by fissioning. Similarly, Sussman (1992:408–409) argues that in ring-tailed lemurs (*L. catta*) "group fission may be more important than male transfer in adjusting population size over larger areas." It is likely that the dynamics of social living, patterns of group fission, opportunities for subgrouping, and flexibility in the size, composition, and structure of many primate social groups limit the set of conditions under which feeding competition results in significant fitness costs to individuals residing in the same social group. Moreover, individuals that reside together in the same social groups may receive critical fitness benefits by interacting in ways that enhance social cohesion, predator protection, social learning and information exchange, and access to and defense of resources desired by individuals in other social groups.

It is important to point out that we are not suggesting that group size is unrelated to the quality of resources within a home range or that feeding competition is not a factor in primate behavioral ecology and individual fitness. For example, Chapman et al. (2004) found evidence of a significant relationship between the protein-to-fiber ratio of mature leaves and the biomass of African colobines. What we are suggesting is that, in general, rain forest habitats are characterized by food patches that vary considerably in size, richness, and spatial–temporal distribution (Bacon et al. 1991). As species exploit more broad-based diets (increase in both the number of plant species and food types exploited), the potential number of food patches available to group members increases dramatically and is sufficient to support individuals living in groups that vary across a range of sizes (Johnson et al. 2002). Thus, at small and intermediate group sizes, feeding competition is expected to play a minimal role in primate social interactions, whereas in large groups it is likely to result in an increase in short-term rates of agonism and eventual fissioning (group size categories being different for each species and habitat).

COLLECTIVE ACTION

Animals living in social groups commonly confront what has been described as a "collection action problem." The collective action problem represents an economic model of social interactions designed to "investigate how the distribution of benefits within groups motivates individual action" (Nunn 2000: 192). The model examines and describes the costs and benefits to individuals of joint or co-ordinated action (alliance formation, cooperation, resource, mate, or predator defense, collective foraging and hunting, food calls, cooperative infant caregiving), as well as the costs and benefits to those group members who act as 'free-riders' or receive the immediate benefits of successful group action, but by not joining the action incur none of the costs.   However, asymmetrical benefits of social group living are common in many primate species (Jack and Fedigan 2006, van Noordwijk and van Schaik 1999, van Schaik 1989, Noe et al. 1991, Cooper et al. 2004, Kitchen and Beehner 2007), and even in the absence of direct coercion or aggressive retribution targeting free-riders, it is unlikely that individuals who fail to engage in cooperative behavior gain greater long-term benefits than individuals who actively participate, especially in cases in which the number of collective actors increases the success of the action (Kitchen and Beeher 2007) or under conditions in which females exercise mate choice based on male cooperative behavior (Garber 1997). Collective action resulting in enhanced benefits to individuals residing in an effective, cooperative, and coordinated social unit is an important component of theoretical models of primate sociality (Goodnight et al. 1992, Nunn 2000, Altmann 1990, Isbell, 1991 Garber and Kowalewski, in press). In the case of many species of atelines (howlers, spider monkeys, muriquis, and woolly monkey), for example, within-group male social interactions are affiliative and cooperative, and resident males act collectively via patrolling or intergroup encounters to exclude neighboring males and extragroup males from the group (Nunn 2000, Kowalewski et al. 2006, Strier, 2000; Di Fiore and Campbell 2007, Di Fiore 2009). In other primate taxa, the collective and coordinated actions of several group members, such as cooperative hunting in chimpanzees may provide advantages (food sharing) to both participants and nonparticipants (Mitani et al. 2002).

BENEFITS OF GROUP LIVING

If the costs of feeding competition and group living are less than previously thought, then researchers need to focus on the benefits of affiliation, cooperation, and mutualism in understanding the evolution of primate sociality. Several recent studies of primate socioecology have highlighted the role of cooperation and affiliation in determining the benefits to individuals of forming groups or subgroups of a particular size and composition (Clutton-Brock 2002, Silk 2002a, Silk et al. 2003a, Sussman and Garber 2004).

Cooperation and affiliation represent behavioral tactics that can be used by individual group members to obtain resources, maintain or enhance social position, or increase reproductive opportunities (Garber 1997). Many affiliative or cooperative behaviors can be explained by individual actions that also may benefit several group members. In acts of cooperation, both participants may receive immediate benefits. Coordinated behaviors such as joint resource defense, range defense, cooperative hunting, alliance formation, cooperative food harvesting, and predator vigilance can be explained in terms of immediate benefits to participating individuals. Acts that appear to benefit recipients may also benefit actors. These benefits need not be equal for each individual. If the cost to the actors of affiliative behavior is low, even if the rewards are low and/or variable, we should expect affiliation and cooperation to be common. Thus, many types of social interaction may be best understood in terms of a non-zero-sum game with multiple winners. According to Watts (2002b:366), "Formation of low-risk coalitions in which all participants stand to make immediate net gains is widespread in primates and may even incorporate much presumed 'altruism' among kin." Taken together, these data may help to explain observations that nonhuman primates live in relatively stable, cohesive social groups and solve the problems of everyday life in a generally cooperative fashion.

NEUROENDOCRINE MECHANISMS FOR COOPERATION

Animals of many kinds are social.... It has often been assumed that animals were in the first place rendered social, and that they feel as a consequence uncomfortable when separated from each other, and comfortable whilst together, but it is a more probable view that these sensations were first developed, in order that those animals which would profit by living in society, should be induced to live together, in the same manner as the sense of hunger and the pleasure of eating were, no doubt, first acquired in order to induce animals to eat. The feeling of pleasure from society is probably an extension of the parental or filial affections, since the social instinct seems to be developed by the young remaining for a long time with their parents; and this extension may be attributed in part to habit, but chiefly to natural selection. (Darwin 1874:97, 102)

As we have stressed, most nonhuman primates are highly social and are members of social groups throughout all or nearly all of their long lives. Investigations into the evolution of primate sociality have tended to focus on aggression and processes by which primates reconcile their conflicts with one another to retain their allies' support, emphasizing competition instead of cooperation. However, many results from behavioral, hormonal, and brain imaging studies offer a new perspective about primates and their proclivities for cooperation, sociality, and peace.

For example, in a report based on 16 years of data on the behavior and ecology of wild savanna baboons, Silk et al. (2003a:1231) conclude as follows:

> Our data indicate that social integration has positive effects on the reproductive performance of female baboons. Females who had more social contact with other adult group members and were more fully socially integrated into their groups were more likely than other females to rear infants successfully. These effects were independent of female dominance rank and variation in ecological conditions....These data provide direct empirical evidence that sociality enhances the fitness of nonhuman primate females.

Recently, researchers have identified a set of mechanisms that might lead to cooperative behavior among related and nonrelated individuals that do not necessitate selfish genes, complex calculations of kin recognition or relationships, or accurate cost–benefit predictions of future reciprocity. In experiments using magnetic resonance imaging (MRI, a technique similar to X-ray imaging that uses radio waves, rather than radiation, to elucidate deep structure), mutual cooperation has been associated with consistent activation in two areas of the brain that have been linked with reward processing (the anteroventral striatum and the orbitofrontal cortex). It has been proposed that activation of this neural network positively reinforces cooperative social interactions (Rilling et al. 2002). Even more compelling, the strength of the neural response increases with the persistence of mutual cooperation over successive trials. This activation of the brain's reward center may account for why we tend to feel good when we cooperate and provides a neural mechanism for understanding the positive reinforcement that sustains cooperative social relationships.

Both the anteroventral striatum and the orbitofrontal cortex are rich in neurons that respond to dopamine, the neurotransmitter known for its role in addictive behaviors. The dopamine system evaluates rewards—both those that flow from the environment and those conjured up in the brain. When the stimulus is positive, dopamine is released. "The dopamine system works unconsciously and globally, providing guidance for making decisions when there is not time to think things through" (Blakeslee 1999:347). In experiments with rats, for example, in which electrodes are placed in the striatum, the animals continue to press a bar to stimulate the electrodes, apparently receiving such pleasurable feedback that they will starve to death rather than stop pressing the bar. With these systems, investigators believe that they have identified a pattern of neural activation "that may be involved in sustaining cooperative social relationships, perhaps by labeling cooperative social interactions as rewarding" (Rilling et al. 2002:403).

Another physiological mechanism related to affiliation and nurturing is the neuroendocrine circuitry associated with maternal responses in mammals. Orchestrating the broad suite of these biobehavioral responses is the hormone oxytocin. Oxytocin has been related to every type of animal bonding: parental, fraternal, sexual, and even the capacity to soothe one's self (Angier 1999, Carter 1999, Taylor et al. 2000, Carter and Cushing 2004). Research conducted primarily on monogamous mammals, capable of pair-bond formation and biparental care, has revealed that, along with other chemicals, oxytocin influences parental behavior, general sociality, and the capacity to form social bonds. These neurochemicals and their receptors are regulated in early life by epigenetic factors, such as early social experiences, producing lifelong behavioral responses, which may help to account for individual and species-typical variations in social interactions (Bales et al. 2004). *Epigenesis* is the study of heritable changes in gene function that occur without changes in the sequence of nuclear deoxyribonucleic acid (DNA), including how environmental factors affecting a parent can result in changes in gene expression in the offspring (Waterland and Jirtle 2003). In the case of oxytocin, it has been suggested that although its primary role may have been in forging the mother/infant bond, its ability to influence brain circuitry may have been co-opted to serve other affiliative purposes that allowed the formation of alliances and partnerships, thus facilitating the evolution of cooperative behaviors (Carter 1999, Taylor et al. 2000).

Results from both behavioral endocrinology and functional MRI studies on cotton-top tamarin monkeys (*Saguinus oedipus*) reveal other hormonal mechanisms critical to cooperation and affiliative behavior (Snowdon 2004). In these primates, males and other nonmaternal helpers provide essential infant care (see Chapter 7). Elevated levels of the hormone prolactin, usually associated with lactation, have been implicated in predisposing individuals to exhibiting nonmaternal infant care. New findings also indicate that prolactin can be considered a "relation hormone" because its levels fluctuate in response to the quality of a male's relationship with his mate. Furthermore, the same regions of the brain associated with positive emotional states (such as mutual cooperation) in humans are also activated in male monkeys when exposed to cues from familiar mates and unfamiliar but ovulating females (Snowdon 2004).

If cooperation and spatial proximity among group-living animals are rewarding in a variety of environmental and social circumstances and if physiological and neurological feedback systems reinforce social tolerance and cooperative behavior, then social group living can persist in the absence of any conscious recognition that material gains might also flow from mutual cooperation. Social animals appear to be wired to cooperate and to reduce stress by seeking each other's company (Carter 1999, Taylor et al. 2000, Rilling et al. 2002, Carter and Cushing 2004, Snowdon 2004). Social affiliation and cooperative behaviors provide psychological, physiological, and ecological benefits to social primates; are reinforced by hormonal and neurological systems; and provide internal positive reinforcement to the participants.

Sociality has evolved independently in many diverse groups of animals. Among primates, sociality may have its origin in the general benefits of mutual cooperation, strong mother/infant bonds, and the evolution of an extended juvenile period in which developing individuals are dependent on other group members. Specifically, neurological, hormonal (oxytocin), and endogenous opioid mechanisms may be at the core of innate cooperative social responses (Carter 1999, Taylor et al. 2000). This could explain the evolution not only of cooperation among nonrelatives but also of "nonselfish" altruistic behavior. Again, we acknowledge the important role of aggression and competition in understanding primate social interactions. Our perspective, however, is that affiliation, cooperation, and social tolerance associated with the long-term benefits of mutualism form the core of social group living. In most instances, aggression and competition are better understood as social tactics and individual adjustments to the immediate and ephemeral conditions of particular social situations.

Finally, we highlight the importance of collecting systematic data on the frequency and context of social behavior to better understand the mechanisms that govern everyday interactions within social groups. We must better understand who does what to whom, how often, and when. As stressed by Silk (2002c:440), "we need to pay more attention to methodological details, such as how we should interpret information about the content, frequency, quality and patterns of social interactions." Furthermore, since active social behavior generally takes up less than 10% of an individual's time, social interactions must be understood within a wider context, such as general activity pattern, life history of the individual, group and population demography, and recent perturbations to the ecosystem that may affect the individual, group, or population. Until we have a better understanding of these mechanisms, hypotheses concerning evolutionary explanations of cooperation, agonism, and sociality may be misleading. In any case, we agree with Clutton-Brock (2002:72):

> if mutualism proves to be important in maintaining cooperative animal societies, the benefits of cooperation in animals may be more similar to those of cooperation in humans than has been previously supposed. In humans, unrelated individuals commonly assist each other...[and] generalized reciprocity appears to be important in maintaining many social institutions...[these] trends appear to have close parallels in other cooperative animals.

40

Social Mechanisms in the Control of Primate Aggression

Irwin S. Bernstein

---

1. What information is obtained during an agonistic encounter that influences the progress or termination of the encounter?
2. How may agonistic episodes lead to dominance relationships, and how do such relationships influence future agonistic episodes?
3. How do each of the four alternative mechanisms work to produce animal densities roughly proportional to the density of resources?

INTRODUCTION

Adolph Schultz (1969) said that the special adaptation of the primates was to be understood by recognizing three attributes, none of which was exclusive to the order. First, primates are "smart"; that is, they have the ability to modify their behavior based on past experience. Second, they are social. Third,

they have extremely long periods of biological dependence for their body size. (The period of biological dependence is considered to be the time that an individual cannot survive on its own and depends on others for necessities such as nourishment, thermoregulation, etc. Although the end of this period is not clearly marked, infants and young juveniles are thought to be biologically dependent on others for survival.)

This combination means that (*1*) primates will learn as a function of experience, (*2*) much of this learning will be social learning, and (*3*) due to their long periods of biological dependence, young primates will have the opportunity to learn from parents and caregivers, thus setting up the possibility of nongenetic means of transmission of information from generation to generation. Information need not be discovered anew each generation and can accumulate over time. This permits the development of traditions and even culture (see Chapter 44).

Sociality in primates means that they not only establish and maintain social bonds but use social techniques in response to ecological problems, that is, joint action on the environment (both physical and social). This is what Kummer (1971) considered to be their only special ecological adaptation. By "joint action on the environment," he meant that, rather than respond to an environmental challenge individually, primates usually enlist or join others to face a challenge. If a predator threatens, they vocalize, look to see what others are doing, and then may jointly face the predator. If the challenge is temperature, they will huddle with others. If it requires searching for resources, they will join others rather than forage individually. If they are threatened by a conspecific (the social environment), they enlist others to aid them to meet the threat.

Not all primates show the same degree of sociality, but as mammals, all show at least the obligate level of sociality required for mating and care of the young. Long developmental periods mean that even the least sociable primates typically spend many years in association with at least one other, be it parent, siblings, or offspring. The more sociable primates spend all of their lives as members of a social group, although members of one or both sexes may spend brief periods in transit between groups. Even when moving between two heterosexual groups, many will form unisexual bands or even travel in company with another species (Bernstein 1967).

DEFINING THE SOCIAL UNIT

What does it mean to be a member of a social unit as opposed to simply being in the same place as another, as in an aggregation or some other such grouping? Although different investigators may emphasize different points, I believe that most at least discuss the following five criteria:

1. *Spatial.* If animals were never together, we would have little reason to ask if they were members of a social unit. What constitutes "proximity," however, is a matter of judgment, and some may require physical contact or distances within 1 m or 5 or 10. Others stress being within arm's reach or one body length or one pace (biological rather than physical units).

2. *Temporal.* If proximity is fleeting and not regular, we might think that the individuals were like "ships that

pass in the night," with no special attachment or attraction. Again, however, there is no consensus as to how much time they must spend together, nor how often. Some are together nearly constantly, whereas others may part and reunite regularly but within a specified community of others.

3. *Recognition of group members.* If animals treat all conspecifics in the same way, it would perhaps suggest general sociality but not specific social units. Social units require boundaries such that group members are treated differently from nongroup members. In many cases, behavior directed toward nongroup members is described as "xenophobic" (Scott 1955). This label was suggested when it was noted that the difference between the way group members and nonmembers are treated is largely a matter of the frequency of expressions of agonistic behavior in encounters. Agonistic responses are much more likely in encounters with nongroup members.

4. *Communication and coordination.* Social units will require coordination of activities such that individuals cannot do whatever they want whenever they want. To have each animal travel, eat, drink, rest, etc. without regard to others would quickly result in the spatial dispersal of the group. Coordination requires communication, including social facilitation where the activity of one stimulates the same activity in others. (It is the similarity of the behavior of initiator and others that marks "social facilitation" compared to other interactions in which an observer may respond in a variety of ways to the behavior of another.) Active communication, where animals exchange information and each modifies the behavior of the other, is also required. Indeed, Altmann (1965) suggested that an operational definition of social units could be based on the measurement of frequencies of communication: The area of high density of communication is the social unit and is bounded by areas of much lower density of communication.

5. *Division of labor.* Not every activity in the group will be displayed by every individual. Of course, we might expect individual, age, and sex differences; but when individuals of the same class show consistent differences in behavioral patterns in specified situations, we suspect a division of labor (or social roles). Sarbin (1954) defined *society* as a network of interrelated social roles. Wilson (1975) suggested that social complexity might be measured by the principle of minimum specification—that is, the number of individuals that have to be studied to see all of the major patterns of behavior in a social group. The term "division of labor" may be construed to be broader than "social role" in that "social role" usually implies a direct set of social interactions and "division of labor" simply means that some actions seen in a group are seen only for some individuals (although it is implied that such actions serve beneficial functions) (see Chapter 32 for a further discussion of what it means to be social).

SELECTIVE PRESSURES THAT ACT ON SOCIAL LIVING

Overdorff and Parga (Chapter 2) review various theories concerning the effects of the ecology on social organization. Although different theories stress the importance of different aspects of ecology, all acknowledge that individuals are influenced by the ecological conditions that influence genetic fitness. Particular ecological factors may favor social living or solitary living. If social living is favored, it can be more beneficial to live in smaller or larger groups with various social structures, depending on the combination and strength of interacting ecological factors.

Most primates are social. Clumped and locally abundant food can favor social living since there is enough in one place to feed multiple individuals and the group can jointly fend off other conspecifics trying to displace them at the food source. Moreover, if food clumps are widely scattered, a social unit has many individuals to search for the next food clump. Sharing information upon finding clumped food would not cost an individual if the clump were large enough to feed multiple group members; all would benefit. On the other hand, if food were sparse or the clumps small, then sociality would not be favored because of intragroup competition. If food were evenly scattered (or locally superabundant) but slowly renewing, social feeding might be favored in that groups would not return to a feeding area too soon after exploiting it, whereas individuals feeding alone would likely crisscross each other's paths, wasting time visiting already depleted patches.

Social living permits larger groups to displace smaller groups at resources, but larger groups experience more intragroup competition, theoretically requiring greater travel distances to obtain sufficient resources for all. Attempts to empirically demonstrate what seems a logical and obvious conclusion have met with only mixed success (Chapman and Chapman 2000), but it may well be that group size influences travel distances in a more complex fashion. For example, one might assume that baboons (*Papio cynocephalus*) might travel farther to find food when food is scarce, but Altmann and Altmann (1970) found that they used more of their home range in the wet season, when food was more abundant, than in the dry season. Travel in search of food in the dry season was constrained by the necessity to stay within the area of the home range that contained permanent water sources.

Social living provides many sensory systems to detect predators and the possibility of joint defense against predators (mobbing), but it also makes it difficult to be cryptic in that all are betrayed by the most noticeable individual. Predators may also find it profitable to continually track a large group so that, even if an individual benefits by being unlikely to be the one attacked, predator attacks may become regular and frequent.

Finding a potential mate is facilitated by living in heterosexual social groups. In contrast, solitaries must search and advertise for mates, thereby exposing themselves to predators as well as potential mates. The same activities that attract mates can also attract predators, or predators can use mate-advertisement signals to attract prey. On the other hand, social living means that rivals as well as potential mates may already be present.

Thermoregulation may be possible in social groups huddling together, but close social proximity can also favor the transmission of disease and parasites. Sociality will permit ready transfer of information such that the location of resources known to one individual is available to all. In this way, long-term storage of information is based on the memories of the oldest individuals in the group. The sharing of such knowledge insures both the simultaneous use of resources and the presence of potential rivals for those resources. On the other hand, group living means sacrificing personal agendas concerning when to go where to exploit what. Individuals will have to delay satisfaction of some needs or travel to and consume resources at times other than they would have based on personal desires in order to maintain continuous contact with the group. Efforts must be made to coordinate the satisfaction of one's own needs with the satisfaction of those same needs in others. Some social units do temporarily divide in the pursuit of resources, but social living requires the development of means to reassemble or continuously monitor the location of others not in immediate proximity. The information provided on the location of others could also be used by predators searching for prey.

Social living may foster alloparental care and facilitate adoption of orphaned young, but it may also facilitate infanticidal breeding strategies. Social living may facilitate the synchrony of breeding using social communication when there is an advantage to coordinated production of young (as in "swamping" predators by producing all of the infants at one time so that a predator will not have a year-round supply of vulnerable infants to exploit). Breeding synchrony, however, may mean that a female cannot wait until she is in optimal condition to breed. Communication requirements may be increased to facilitate interactions in large groups, and within-group competition may result in agonistic contests that would disrupt the group if not countered by other mechanisms.

Although not exhaustive, this list indicates that social living will be influenced by multiple selective pressures and that specific ecological conditions will not only select for or against social living but also influence the specific organization of social units when sociality is favored. Various investigators have focused on the different selective pressures that operate on social units and the problems generated by social living but acknowledge that sociality is influenced by all (see Chapter 2 for a historical analysis of theories of socioecology).

CONFLICT, COMPETITION, AND AGGRESSION

My research has focused on the expression and control of *agonistic behavior* (defined as aggression and the usual responses to aggression; Scott 1955) in social groups. Although the expression or control of aggression is important in maintaining social groups, it need not be the central principle behind the organization of social units. Sociality cannot be understood based on the study of any one factor or selective pressure (see Altmann, S. A., 1974 for a discussion of the multiple pressures that shape baboon activity profiles and energy budgets). Social living can indeed promote conflicts, competition, and aggression; and these can be understood as three separate problems (Mason 1993). *Conflicts* (two mutually incompatible goals or demands within the same individual or between two or more individuals) are inevitable when multiple individuals want to do different things at the same time and yet all want to remain together. A conflict exists even in situations where one partner wants to groom and the other wants to play, and yet we do not expect such conflicts to be resolved by aggression. *Competition* (defined to exist whenever a necessary resource is available in insufficient supply for all who require it) is inevitable when a necessary resource is limited and an individual is in proximity to others with the same requirement. Whereas animals may fight over a limited resource, there is little point in fighting over access to each kernel of food when it is widely scattered in small packets, like grains of rice on the ground, and scramble competition is likely to prevail. Although neither conflicts nor competitions need to be resolved using aggressive contests, some conflicts will end in agonistic encounters and some competitions will be resolved by aggressive contests. Aggression can also be generated in social groups, despite the absence of conflict or competition, when individuals under attack by one redirect to another or come to the defense of others by attacking an animal with which they neither had an immediate conflict nor were in competition at the moment. Close proximity can also generate the appearance of conflict or competition when there is none. Hall (1964) stated that the most common cause of aggression in primate groups was a perceived violation of the social code. For example, if an individual does not tolerate others in immediate proximity while feeding and another walks too close in passing, it might be perceived as approaching the food of the feeding individual. This perceived violation of the social code (the expectation that others would not come into proximity during feeding) could result in aggression to repel the violator and bring it into conformity with expectations.

HOW AGGRESSION IS EXPRESSED AND CONTROLLED IN SOCIAL GROUPS

Theoreticians have argued at great length concerning the causes of aggression and even as to how to define "aggression" (Bernstein 1991a). Pain- or fear-induced or irritable aggression is clearly due to a different cause from instrumental aggression used to achieve a particular goal. Aggression is also often identified by function, such as predatory and antipredatory, defensive, sexual, territorial, and weaning aggression. Parental punishment and even moralistic aggression have been listed, and some believe that aggression may be motivated by striving for dominance (but see Mason 1993). Although aggression has many causes, merely living in a social unit will generate conflicts and competitions, some of which will be resolved by aggressive contests. In these cases, aggression is being used as an instrumental act to achieve a goal.

Since aggression is costly (in terms of not only energy expenditure and social relationships but also the injuries that can be sustained by resistance or retaliation), natural selection should favor aggression only in those cases where the value of the incentive outweighs the costs. Moreover, since there can be, at most, one victor in an aggressive contest (either one individual in a dyadic contest or one side when multiple individuals fight with multiple others), individuals will be selected for who can accurately assess not only the relative value of the incentive to the cost but also the likelihood of victory in a particular agonistic encounter. Since agonistic encounters impose costs on both the aggressor and victim, natural selection will favor both victims and aggressors that minimize the frequency and severity of agonistic contests but simultaneously favor individuals that "correctly" use aggressive contests to gain access to resources or to resolve conflicts in their favor.

The decision to initiate aggression then will involve assessing the value of the incentive to be obtained, the costs of the aggression used to obtain the incentive, and the likelihood that the incentive will be obtained. The cost of aggression is not an absolute value but dependent on the relative abilities of the opponent (as is the likelihood of successfully obtaining the incentive); agonistic encounters should, therefore, include a period for assessment of the opponent. It is in the interest of both parties to provide the other maximal information about their aggressive abilities and their willingness to use those abilities in the contest situation. Natural selection will favor individuals that do not respond to false signals and respond appropriately to honest signals. Similarly, individuals that can deter resistance by advertising their abilities and their determination to use those abilities will be favored over individuals that attack immediately and always incur the costs of combat. Each will, therefore, benefit if it can repel a rival without resorting to combat or yield to a rival that it cannot defeat in combat. Each will be subject to selection that favors appropriate responding to honest signals. Some signals, like piloerection to increase apparent body size, may be basically deceptive but do, in fact, convey accurate information about body size. If all individuals use the same device, relative sizes are the same as they would be if no individual did so. Piloerection will be favored only because of the miscommunication of size that would result in animals that failed to do so. A miscommunication may mean that individuals that would have retreated from an individual of a particular size will not do so, requiring escalation to more costly forms of contest. Even if one wins such a contest after escalation, there has been an additional cost, and the loser has also needlessly paid the costs of combat in a losing fight.

The onset of an aggressive encounter will, therefore, include a period of signaling and assessment. If initial information on size, strength, and determination (as well as number of allies and their abilities and willingness to participate) is insufficient to convince either that they are likely to lose, then additional information will be required, which can be obtained by further escalation of signals. Each escalation will be more energy-intensive and closer to actual physical combat. Even the onset of physical combat is an opportunity to gain further information on the aggressive powers of the opponent, and at any time that a contestant is convinced that the cost of further combat will exceed the value of the incentive or is unlikely to result in obtaining the incentive, it is in the interest of both parties to terminate the contest. The loser does not gain the incentive but incurs no additional costs, and the winner does obtain the incentive but would incur additional costs with no further benefit if the fight were prolonged beyond the point at which the rival yielded and indicated that it would no longer contest the outcome. Continuing such a fight due to "anger" or "rage" or a failure to respond to a signal indicating the intention of the rival to break off the contest and retreat may prolong a fight with additional costs and no gain to either contestant. Clear signals of such a decision and appropriate cessation of further hostilities will be favored for both winners and losers, and we may expect that vocalizations, facial expressions, bodily postures, and as many other devices as possible will be used to communicate submission in addition to the simple act of physical withdrawal. Individuals capable of detecting early signs that they cannot win a contest will suffer fewer injuries and defeats, and those willing to allow opponents to withdraw will suffer fewer injuries inflicted by opponents defending themselves against attack. Aggression is thus likely to be ritualized in that agonistic responses will be favored that convey maximal information with minimal costs (including injuries). We thus see the marvelously ritualized combats between animals in the mating season, where few injuries are actually incurred despite the lethal potential of the weaponry used. Primates do not have antlers, horns, and other specializations of many animals but, males especially, may have canine teeth that rival those of the largest carnivores. Nonhuman primate combat, however, rarely results in deaths, and the most vulnerable anatomical sites (throat and ventrum) are injured far less frequently than would be expected based on their relative sizes (Bernstein and Gordon 1974, Owens 1975, Ruehlmann et al. 1987).

DOMINANCE

Inasmuch as primates are smart, repeated fights between the same opponents will be influenced by learning and memories of previous fight outcomes. Recognition of individuals allows primates to learn which individuals have defeated them in the past and which individuals they have defeated. Knowledge of previous fight outcomes can be used to predict the outcome of the next fight, but appropriate action will still depend on evaluation of the incentive and assessment of the determination of the opponent to contest that incentive using aggressive means. Only the fighting ability of the opponent can be stored in memory; the value of an incentive and the determination of the rival must be assessed each time. I use the word "dominance" to describe relationships between individuals where decisions to fight or retreat are largely based on having learned about the outcomes of previous encounters with the same opponent and where the subordinate immediately displays those submissive signals that it displayed at the end of the last agonistic encounter (Bernstein 1981). Note that the loser of a previous fight does not automatically yield every incentive to the past winner but may still decide that the incentive is so valuable that the costs of fighting are overcome (such as a mother defending her baby against an opponent that has defeated her in past fights) or that the opponent is not signaling a high level of motivation to contest the incentive (as when the incentive is food and the rival is already sated) and that it will not be met by aggressive challenge if it takes the incentive. In such cases, a subordinate may successfully obtain the incentive despite the presence of another that has been victorious in the past. If a dominant individual gives no sign of interest in an incentive or objection to an activity, then a subordinate can obtain the incentive or engage in the activity. Priority of access to incentives thus correlates with dominance but is not synonymous with the learned relationship that predicts which of two equally motivated opponents will pursue aggressive contests and which will submit.

If at the very start of an agonistic encounter one individual withdraws, we can invoke the concept of dominance to explain why agonistic signals were unidirectional instead of symmetrical. Such an explanation infers the existence of learning based on a past history of encounters and an assessment of the current situation that the subordinate uses to predict the likelihood of an agonistic encounter and its outcome. At times, however, it appears as if individuals that have never previously met already have a dominance relationship in that one submits to the other virtually immediately. The assessment that results in submission in such a case does not involve knowledge of a past history of agonistic encounters with that opponent, but the rapidity of the decision to submit must mean that sufficient information is available at the start concerning the agonistic potential of the opponent. This need not involve a self-concept, such that an individual compares its own size and strength to an opponent, but may be based on generalization. Whereas each animal may be recognized as a distinct individual, some classification and generalizations are to be expected. If a new opponent seems to share those characteristics that are common to the individuals that defeated one in the past, then one can generalize and assume that this individual will also be capable of winning. Socially sophisticated individuals with sufficient experience with others may develop concepts regarding the characteristics of opponents that they can and cannot defeat. Such generalizations may make it appear as if dominance was an attribute of an individual that others could recognize immediately or that individuals had self-concepts that they could use to compare against any new opponent, but both explanations seem unlikely. Since all, except the most dominant and most subordinate, are dominant to some and subordinate to others, being dominant (or subordinate) cannot be an absolute attribute of an individual. Moreover, since it is possible for A to dominate B and B to dominate C and still have C dominant to A (as in the child's game of "paper–scissors–rock"), dominance is not even a property that can always be plotted on a linear scale. The possibility of comparing an opponent to oneself seems unlikely when we realize that monkeys do not seem to have well-developed self-concepts in that they fail to recognize that their mirror image is a representation of self (Gallup 1982, 1987).

It has been suggested that primates may infer their dominance relationship to another without direct experience by observing the outcomes of fights between other individuals and then reasoning that if one defeats another that has defeated them, then the first can also defeat them (Cheney and Seyfarth 1990a). Surely, such a comparison would yield valuable data and would reduce the number of direct contests required, but evidence for such abilities is lacking. Chapais (1992) has demonstrated that when a younger member of a high-ranking matriline is isolated with all of the members of a lower-ranking matriline, the individual will lose to the intact family and assume subordinate status. The introduction of an adult member of the isolated individual's matriline does not immediately restore dominance, suggesting that merely witnessing submission of the resident matriline to one's relative is not enough. Status is restored to the subject only when its relative comes to its aid against the residents and it personally experiences successful aggression against the other matriline. This suggests that direct experience, rather than passive observation, is required to establish new learned relationships in this domain.

Primate sociality often results in complex agonistic interactions with multiple parties. Complex interactions, however, can be seen as resulting from individual interactions. They add and multiply, but the same principles apply. Fights may be context-specific, and winning and losing may be as well. Dominance, however, is an intervening variable and should not be context-specific. Winning one time is not a permanent condition, but the past influences the future when animals can learn and remember. Even dominance can change as learning is ongoing and new information can

change a relationship. If it changes so readily that it allows no prediction, then the intervening variable does not allow prediction and it is pointless to invoke it.

FIGHT INTERFERENCE

Agonistic contests in nonhuman primates often involve more than two individuals. Animals with strong social bonds will use social techniques in facing any challenge, be it from the physical or social environment. At the onset of an agonistic encounter, we should expect animals to seek help from allies and to assess the likelihood that others will aid their opponent. Alliances and coalitions may be in the mutual interest of allies that repel the same opponent (as when groups repel other groups at resources), or individuals with strong social bonds may habitually aid one another regardless of whether they have a conflict or are in competition with the opponent (as when mothers aid their offspring). Such joint action may be balanced or not (a mother may aid her infant more often than her offspring aids her [Kaplan 1978]), but it is the social bond that dictates aid rather than any exact reciprocation of services. Whereas reciprocal altruism can be invoked to explain some instances of one animal aiding another, a market analogy (see Noë 2001) may be misleading (e.g., if one thinks that mothers are aiding their young in expectation of payback). Although fight interference on behalf of kin can be explained in terms of classical and inclusive fitness benefits, closely bonded individuals need not be kin. Unrelated animals may aid, groom, and socially interact based on social bonding rather than on elaborate mental bookkeeping of the value of services, favors, and ideas of equality (Matheson and Bernstein 2000). Inasmuch as there is very little evidence that primates have a distinct concept of kinship (e.g., Alberts 1999, Silk 2002a) but rather end up preferentially associating with kin based on mechanisms such as familiarity that correlate with kinship (Bernstein 1988, 1991b, 1999a; Bernstein et al. 1993b; Fredrickson and Sackett 1984; Sackett and Fredrickson 1987), it might be expected that behavior toward kin that could be explained based on inclusive fitness benefits might also be directed toward unrelated individuals that are socially bonded for the same reasons (e.g., long-term association in the same social unit). Martin (1997) has demonstrated that the degree of association among members of a captive pig-tailed monkey (*Macaca nemestrina*) group was correlated nearly perfectly with years of association and the intensity of that association. Even though all sisters were genetically equally related to one another on average, those that were of more similar age spent more years dependent on the same mother and were more closely bonded. Sisters with a living mother were also more closely bonded than orphaned sisters, and old matriarchs spent as much time with each other, although not genetically related, as they spent with younger relatives that had been in the group for only a few years.

CONTROL ROLE BEHAVIOR

Aggression has the potential to inflict injuries on group members and thereby degrade their abilities to contribute to group defense against external threats such as predators and rival groups. It is therefore advantageous to every animal in the group to limit the damage done in intragroup fighting. Whereas ritualization of agonistic encounters will reduce actual fighting, there will be times when equally matched individuals must escalate encounters to the most intense levels to decide which will be the winner or loser. (The "fairest" fights are often the most dangerous.) Fights may also occur due to miscommunications and to redirection and displacement. Regardless of the cause, fights between group members represent a threat to all members of the group. Fight interference may involve family alliances (subgroups within the group), but such interference has as much potential to escalate as to terminate the encounter. Fight interference by a clearly superior group member will likely terminate the encounter as all opponents yield to the interfering individual. I have (Bernstein 1964a, 1966a) previously referred to this as "control role behavior" and indicated that the interfering individual will usually be dominant to the contestants but need not be the most dominant animal in the group and that several individuals may show the same role behavior. Moreover, the control animal may attack either the victim or the aggressor with equal effect in that its aggression will disrupt the initial fight and both combatants will leave the scene on the approach of an aggressor dominant to them both. In fact, the control animal can attack an uninvolved third party in the immediate vicinity of the combatants with the same result; the two combatants (as well as the innocent third party) will flee the vicinity of the aggressive dominant individual. Attacks on uninvolved third parties may be so common and directed to specific classes of individuals that they may selectively socialize individuals. Bernstein and Ehardt (1986a) have suggested that adult male rhesus monkeys (*Macaca mulatta*) interfere in fights between adult females or juveniles by attacking the nearest subordinate male or adolescent male. Such consistent behavior results in subordinate males avoiding the vicinity of noisy agonistic signals produced by females and immature animals and thus curbs their own aggression toward females and immature animals that respond to aggression with noisy agonistic signals. Fight interference, however, is not only seen by animals dominant to the combatants; subordinates that interfere can also be effective at limiting aggression between group members (Boehm 1994).

Whereas the control role is a social technique to control aggression in the group, control role behavior is also seen when individuals protect group members against external sources of disturbance (Bernstein 1966b). Although the same individual may do both, these may be two independent roles (Wilson 1975). Other mechanisms also exist to control the social damage done by intragroup agonistic encounters, and a rich literature on "peacemaking" (de Waal 1989b) has developed.

Contests with other social units may be expected whenever social units come into conflict or are in direct competition with one another. Whereas a control animal may take a position between an external threat and potential sources of disturbance, conflicts and competitions are resolved by joint action of as many group members as possible. Again, whereas size and strength may be important in deciding such contests, the sheer number of individuals on each side may be decisive in that few primates can successfully fight multiple opponents simultaneously. In addition, the coordination of individual support and the determination of group members may influence the outcome of an intergroup agonistic encounter. Each individual in a group will then have to assess the opposing group, the relative strength and determination of its own group, and the value of the incentive. This is a more demanding task than assessment of individuals, but primates do learn to recognize other groups and to contest other groups, or not, based on a history of past encounters as well as current relative determination and incentive values. Recognition is unlikely to include recognition of every member of a rival group but may be based on recognition of a few significant group members. Assessment of relative numbers and determination may be based on the behavior of those individuals in one's immediate vicinity and may be, therefore, somewhat imperfect; but the nature of intergroup fights suggests that these assessments are continual during fights and that the changing course of combat reflects new information provided by which individuals join or flee the fight and the successes and failures of key encounters. Once again, it is in the interest of both groups to engage in fighting only when the incentive is worth the cost and when the incentive is likely to be obtained by contest. Nevertheless, it is still in the interest of both groups to minimize the costs of combat. As long as the opponent still has the potential to inflict damage in self-defense, it is best to let an opponent yield an incentive without further combat when a clear signal is provided.

ANIMAL DENSITIES APPROXIMATE RESOURCE DENSITIES

In this section, I consider intergroup encounters that seem to disperse groups in space roughly in proportion to the density of resources. These encounters usually represent the results of competitions for resources, but several different patterns may be identified, although all serve to prevent all groups from congregating at the richest resources. The four patterns that I have identified are territoriality, intergroup aggression, intergroup dominance, and intergroup avoidance.

Territoriality

Territoriality is defined as the defense of a specific geographic area against a specified class of others (modified from Burt 1943 to permit joint defense of a territory by mixed-species groups as described by Garber 1988a). This may result in the defense of resources (but note that males defending an area or a territory in a lek are not defending any identifiable "resource") but will not result in exclusive use of the territory inasmuch as without invasion there would be no evidence of defense. Territorial defense requires that intruders be detected before they can be repelled, so some intruders may succeed in temporary incursions if undetected. The defense of a specific geographic area by a group (or subclass of members in the group) will usually mean that the class of others being defended against will not have regular access to the resources in the territory. Sequential use of the same resources is thus precluded in animals that maintain territories.

Territorial defense is feasible only when the area being defended is small enough to make it possible to detect intruders and when the number of intruders is not so great that the time and energy required in repelling them exceed the value of the territory to the defenders. If territorial behavior evolves in a species, then there will be an advantage to all that hold and defend territories and individuals or groups not occupying a territory will try to establish one. They will likely go to the richest and most suitable areas first, but if such areas are already occupied, they will be repelled by the occupants. They will then proceed to the next most suitable areas until they find an attractive location without the negative of an opposing occupying group. When all suitable territories are claimed, there will be no recourse but to try to invade an already occupied territory. The choice of area to attempt invasion will be based on the value of the area and the ability (strength and determination) of the occupants to resist. If there is relatively little difference in defensive abilities among groups but there are large differences in the quality of areas, then we would expect the richest areas to be the most attractive. Given that there are no alternative areas available, the invading group can lose and possibly die without a suitable territory; win and displace the residents, which will then have to become invaders elsewhere; or partition the territory with the defenders. Partitioning will be possible only if the resources contained are at least sufficient to support both groups. Thus, only the richest territories can be partitioned as partitioning poor territories, barely able to support one group, will result in two nonviable groups. With partitioning, large and rich territories become smaller and more easily defended but contain lesser absolute quantities of resources. The end result of successive invasions of nonestablished groups should be a series of territories each of the size necessary to contain sufficient total resources to sustain the residents. This will mean that areas with the densest resources will have the smallest territories and that the density of animals will be roughly proportional to the density of resources. A "rich" territory will be rich not because of the total quantity of resources available but because the necessary resources are found in a small area. Richer or poorer areas all end up with animals having access to quantities of resources that closely match their needs. Territories will be maintained by active defense, but once groups have learned the boundaries that other groups defend, little fighting need actually occur. Intruders that know the boundaries may still intrude but will withdraw beyond the boundaries whenever challenged by the territory holders. This differs from dominance in that, although it is a learned relationship that influences the directionality of agonistic encounters, it is specific to a location. Neighboring groups can, in fact, take turns chasing each other in a zigzag pattern at a territorial boundary (Bernstein 1968).

Aggressive Contest

Not every habitat is suited to the establishment of a system of territories. If resources must be obtained from scattered areas such that a group cannot defend all of the sectors required to sustain it, then the group will not be able to prevent others from using resources in areas that they do not currently occupy; that is, they will not be able to prevent others from using resources in a sector that is far away and where they cannot detect intrusion. Sequential use of resources will now be possible, but simultaneous use of a resource may still be precluded. Groups may fight one another when they meet at a resource so that the loser must leave; dominant groups may repel subordinates that have learned, based on past fights, that resistance is futile against this particular rival; or groups may avoid the immediate proximity of any group, thereby avoiding the costs of fighting to obtain resources.

When groups fight over resources, one would expect that each group would first be attracted to the most desirable area. If such an area is already occupied, then the two groups will contest the area and the loser will be forced to withdraw and seek an alternative area. It will be expected to go to the next most desirable area and to either use the resource there, if the area is unoccupied, or contest the resource if the area if occupied. After each defeat, the losing group will move to the next most desirable resource until it either finds an available resource or there are no alternatives left. If defeated at all available sites, a group has no choice but to return to a site where it had been defeated and wait for the occupying group to leave. They should thus choose a site that is most attractive, with regard to resources, and least aversive due to the group already there. Since any group that defeats another is aversive, the decision will be largely based on the richness of the resource. If there are more groups than resource sites, then more than one group will be in waiting. The decision as to where to wait will be based on the value of the resource and the number of groups in the area that successfully defend it. Groups will therefore congregate at the richest resources with the fewest groups present that can defeat them. This will result in the distribution of groups in space roughly in proportion to the density of resources; that is, there will be more groups waiting at the areas of richest resources and fewer willing to wait at less attractive areas occupied by groups feeding or waiting that can defeat them. Such a system works only when groups must abandon resources in one area to obtain different resources in another area (e.g., leave a feeding area to travel to a source of water some distance away).

Intergroup Dominance

If groups are capable of recognizing other groups and remembering the outcomes of past agonistic encounters, then rather than fight a group at a resource to learn if it can displace that group or not, a group can decide to contest or yield based on its established relative dominance relationships (i.e., the members will not contest groups that they can identify and remember having lost to in previous contests). Groups will still be attracted to the most attractive resources first and repelled by groups present that are dominant. When all sites are occupied, groups will be most attracted to the richest sites with the fewest groups feeding or waiting to which they are subordinate. Once again, the density of animals in the vicinity of a resource will be roughly proportional to the density of the resource (i.e., rich areas will have more groups waiting). Sequential use is possible as long as the users cannot permanently occupy the site. If all resources are concentrated in a small area, then animals can remain at the same site permanently and territoriality may result. When the various needs of a group must be met by traveling distances that preclude the defense of all sites that the group utilizes, then sequential use of resources is possible, as is contest at those sites for use of the resources. Since animals have multiple needs, a group cannot permanently remain at a food site when it needs water or shelter or protection from the elements more than it needs food. If it must travel elsewhere to satisfy these other needs, then groups whose need for food is greater than their needs for the other resources will have an opportunity to feed when the most dominant group departs. The order of feeding will be based primarily on relative dominance, and there will be less fighting than if the order was determined by a fight on each encounter. Dominance relationships, however, require repeated contacts with the same groups and sufficient learning ability to identify groups and store memories of past encounters with those groups.

Avoidance

Finally, animals can entirely escape the costs of aggressive competition if they always avoid other groups. In this case, the group travels to the richest resource first but withdraws if the site is occupied and moves to the next richest resource. It will keep moving until it finds an unoccupied site or that all sites are occupied. It will now be equally repelled by a group at every site but most attracted to the richest site and may be expected to return to that location and wait for the occupying group to vacate the area. If other groups are already waiting, then their presence is also aversive; thus, a group finding the richest site already occupied with one

or more other groups also waiting in the area may move to a site that is less attractive but has fewer waiting groups. There will be a balance between the attraction of the site and the aversion to the number of waiting groups such that animals will once again be distributed in space roughly in proportion to the density of resources. Note that if groups can detect the location of other groups without actually traveling to the site, then considerable time and energy may be saved. Loud vocalizations or prominent visual displays that advertise location may thus benefit all groups as they will not congregate at the same locations and can move directly to sites that no group currently occupies.

All four systems will result in animals being distributed in space roughly in proportion to the density of resources, but only in the case of territoriality is sequential use of resources prohibited. Obviously, there will be ecological conditions that will favor one over another. I have already commented on the conditions that might favor territoriality. Aggressive contests may occur if groups do not regularly meet the same opponents and thus cannot establish dominance relationships. Although aggressive encounters each time are more costly, dominance relationships require an opportunity to learn or make relative assessments without resorting to combat. This will require learning ability, the ability to identify other groups, memory of past encounters, and the ability to assess groups in each situation. Avoiding all others is the least costly in terms of aggressive consequences but is likely to evolve only when resources are so widely distributed that a group can always move to an alternative site and when the quality of each site is roughly similar to all other sites such that no location is sufficiently superior so that the extra cost of aggressive competition is compensated for by the increase in quality. This results in scramble, rather than contest, competition. It is my impression (Bernstein 1964b) that the howlers (*Alouatta palliata*) of Barro Colorado live in such an ecological situation and demonstrate scramble competition based on avoidance, but others have suggested that their spacing is based on territoriality or aggressive competitions and even dominance. It should be possible to empirically test these alternatives.

CONCLUDING COMMENTS

In this chapter, I have focused on the expression and control of aggression that might be expected to be provoked by social living. Such aggression may be triggered between individuals in a social unit or between social units, but the means of expression and control are similar in both cases. Whereas aggression can function to maintain the integrity of groups (defense against external and internal sources of disturbance) and integrate individuals into a group (as in socialization processes where punishment of unacceptable behavior results in individuals learning group standards), uncontrolled aggression has the potential to destroy individuals and groups (Bernstein and Gordon 1974). Parental punishment, defense, and the socialization of newcomers may be essential to the maintenance of social units. Nonhuman primates, lacking the language skills to explain what behavior (or absence of what behavior) resulted in a reward, may rely heavily on punishment of unacceptable behavior in socialization. The behavior of infant rhesus monkeys (*Macaca mulatta*) is modified by frequent punishment of unacceptable behavior on the part of mothers and other kin and close associates (Bernstein and Ehardt 1986b). The behavior of new group members is also modified by selective aggressive responses to their behavior.

Aggression may be essential to social life for primates, but it must also be controlled. I have taken the position that natural selection will favor the control of aggressive contests in socially living primates and that such control is essential to sociality but does not define social systems. Dominance need not be a universal attribute of social units, nor does the expression of dominance necessarily characterize every aspect of the social unit. On the other hand, every social unit may be expected to experience aggression and to have evolved mechanisms to control aggression and its consequences.

41

Postconflict Reconciliation

Kate Arnold, Orlaith N. Fraser, and Filippo Aureli

1. Why do primates reconcile after aggressive conflicts?
2. How do reconciliation levels vary across species, and what factors might account for this variation?
3. How do characteristics of the conflict and the opponents' relationship influence the occurrence of reconciliation?

INTRODUCTION

Conflicts of interest between group members occur in many contexts. The most common contexts are competition for access to limited resources, such as mates or food (Darwin 1871, Trivers 1972, Clutton-Brock 1989, van Schaik 1989), and disagreement about decisions, such as the allocation of time to different activities or the direction of travel (van Schaik and van Noordwijk 1986, Menzel 1993, Boinski 2000). Conflicts of interest may escalate into aggression and may compromise the benefits of group living if they induce the loser to leave the group (Janson 1992). Even without such a consequence, aggression may jeopardize future cooperation between the former opponents (de Waal 1986c). Thus, behavioral mechanisms that mitigate conflict, prevent aggressive escalation, and resolve disputes should have been strongly selected in group-living animals (Aureli and de Waal 2000).

Although behavioral mechanisms for conflict management before aggressive escalation are of paramount importance, little theoretical and empirical research on nonhuman primates has focused on this aspect (Aureli and Smucny 1998, Cords and Killen 1998, Aureli and de Waal 2000). Most systematic research has dealt with the mechanisms of conflict resolution that occur after aggressive escalation, especially friendly postconflict reunions between former opponents (Kappeler and van Schaik 1992, Silk 1997, de Waal 2000a, Aureli et al. 2002, Judge 2003). Descriptive accounts of such reunions have been reported for various primate species (see de Waal 2000b for a review), but their systematic study started when de Waal and van Roosmalen (1979) showed that, contrary to the common view, chimpanzee (*Pan troglodytes*) opponents actually spent more time in close proximity and exchanging friendly behavior in the aftermath of an aggressive conflict than beforehand. They called these reunions "reconciliations," a heuristic term that clearly implies a specific function. Their pioneering work was instrumental in the shift from simple descriptions of friendly postconflict reunions to the systematic study of conflict resolution in primates.

In this chapter, we review the knowledge accumulated on postconflict friendly reunions in primates in the last 25 years. Most of the studies used the term "reconciliation," so we have also adopted it and discuss its functional implications. We start with reviewing methodological issues regarding collection and analysis of postconflict data. We then review the evidence for reconciliation across the Primate order, the benefits of reconciliation in reducing the costs of aggressive conflict, the variation and ontogeny of reconciliation patterns, and the various hypotheses put forward to explain its function and distribution across individuals, groups, and species.

MEASURING RECONCILIATION

Following the first study of reconciliation in chimpanzees, which was based only on postconflict observations (de Waal and van Roosmalen 1979), a second study introduced control observations that allowed direct comparison of postconflict behavior with behavior exhibited under "baseline" conditions (i.e., not affected by a recent conflict) (de Waal and Yoshihara 1983). This procedure, known as the "PC–MC method," has become the standard methodology and has been used in the majority of postconflict studies (for reviews see Kappeler and van Schaik 1992, Cords 1993, Veenema et al. 1994, Veenema 2000).

Postconflict observations (PCs) begin immediately after the end of the conflict and usually consist of 10 min continuous focal observations on either the aggressor or victim, during which particular attention is paid to the nature and timing of social interactions. For each PC, a corresponding matched-control observation (MC) of the behavior of the same individual is recorded. This is usually carried out on the next possible day and at the same time as the PC in order to control for seasonal and diurnal activity patterns and to reflect the current social situation, which may change over time. To ensure that MCs are not influenced by a recent conflict, they are postponed until the next day if aggression involving the subject of the MC closely precedes the planned start time. This procedure results in an equal number of PCs and MCs, which are then compared with respect to the timing of the first friendly contact between former opponents.

Further controls have been used in some studies to ensure that PCs and MCs are comparable in other respects. The distance between former opponents at the beginning of corresponding observations has been taken into account in a number of captive studies (de Waal and Ren 1988, York and Rowell 1988, Kappeler 1993, Swedell 1997, Call 1999, Rolland and Roeder 2000). Some studies of wild populations have employed additional controls, selecting MCs which are comparable to PCs in terms of former opponent distance, group activity, and subgroup composition (Aureli 1992; Watts 1995a,b; Matsumura 1996; Arnold and Whiten 2001; Cooper and Bernstein 2002).

If former opponents make friendly contact earlier in the PC than in the corresponding MC, they are considered to be attracted to one another as a result of the conflict. In the reverse situation, where former opponents contact each other earlier (or only) in the MC, the conflict is considered to have resulted in the dispersal of the former opponents. Thus, PC–MC pairs are labeled "attracted" or "dispersed." PC–MC pairs in which no contact between former opponents occurs during either observation or contact occurs at the same time are labeled "neutral" (de Waal and Ren 1988). If postconflict contacts between two individuals are randomly distributed throughout PCs (as would be expected in MCs), then the ratio of attracted to dispersed pairs should be 1:1 (de Waal and Yoshihara 1983). If this ratio is significantly higher in favor of attracted pairs, then it is concluded that a high proportion of affiliative contacts between former opponents occurred sooner after a conflict than at other times and that reconciliation is demonstrated.

Variations on this method have been used in some studies, where MCs were taken from the period immediately preceding the conflict (Cheney and Seyfarth 1989) or latency to first affiliative contact was calculated as the mean interval between consecutive affiliative interactions from a number of baseline focal observations (Wittig and Boesch 2003b). Other studies have compared the rate, instead of the first occurrence, of affiliative behaviors between former opponents during PCs with that during MCs (the *rate method*) (Judge 1991), with MC rates calculated using standard next-day observations (Judge 1991), baseline data collected throughout the study period (de Waal 1987), or the period immediately preceding the conflict (Silk et al. 1996).

A second and complementary approach to demonstrating the occurrence of reconciliation, known as the *time rule*, was developed by Aureli et al. (1989). This method compares the cumulative distribution of first affiliative contacts during all PCs and MCs over time (Figure 41.1) and identifies the time period during which the frequencies of contact are significantly elevated during PCs compared to MCs. Within this time window, affiliative contacts can be operationally defined as reconciliation and used to study functional aspects by comparing PCs with and those without reconciliation. The relative advantages and disadvantages of the PC–MC method, the rate method, and the time rule are discussed in detail elsewhere (Veenema et al. 1994, Veenema 2000).

de Waal and Yoshihara (1983) also developed a measure of conciliatory tendency (CT) for use in comparative analyses (comparisons between species, groups, or subsets of individuals—e.g., kin vs. nonkin). They defined the CT as the proportion of the total number of PC–MC pairs that were attracted. This measure was later refined to fully correct for baseline levels of affiliation by removing a proportion of attracted pairs, which might have occurred by chance (Veenema et al. 1994). The revised measure, known as the corrected conciliatory tendency (CCT), is defined

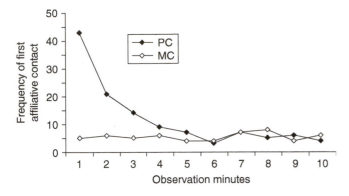

Figure 41.1 The frequency of the first affiliative contact between former opponents in the 10 min of postconflict (PC) and matched control (MC) observations in long-tailed macaques. Modified from Aureli et al. (1989).

as (number of attracted pairs – number of dispersed pairs)/ total number of PC–MC pairs.

EVIDENCE FOR RECONCILIATION AS A WIDESPREAD PHENOMENON IN PRIMATES

Almost all studies of primates have demonstrated that friendly interactions between former opponents occur sooner after aggression than during control periods (Table 41.1). Exceptions are red-bellied tamarins (*Saguinus labiatus*), which are highly cooperative and have very secure social bonds (Schaffner and Caine 2000), and black lemurs (*Eulemur macaco*), whose relationships are either very hostile or very friendly, to the point that they are unlikely to be disrupted by conflict (Roeder et al. 2002).

Table 41.1 Studies in Which the Occurrence of Reconciliation Was Investigated in Primates and the Associated Corrected Conciliatory Tendencies (CCTs) Were Given in the Study or Where Data Were Available to Calculate Them

SPECIES	LOCATION	RECONCILIATION	SELECTIVE ATTRACTION[1]	CCT/MEAN CCT %[2]	SOURCE
Prosimians					
Lemur catta	Captive	No	✓	—	Kappeler 1993
	Captive	Yes[3]		—	Rolland and Roeder 2000
	Captive	Yes/no[4]		41.5	Palagi et al. 2005
Eulemur fulvus rufus	Captive	Yes		6.4	Kappeler 1993
	Captive	Yes		26.5	Roeder et al. 2002
Eulemur macaco	Captive	No		—	Roeder et al. 2002
Propithecus verreauxi	Wild	Yes		48.8, 38.0 (two groups)	Palagi et al. 2008
New World monkeys					
Callithrix jacchus	Captive	Yes	✓	—	Westlund et al. 2000
Sanguinus labiatus	Captive	No		—	Schaffner and Caine 2000
Cebus apella	Captive	Yes/no	✓	14.3 (non-food-related conflicts only)	Verbeek and de Waal 1997
	Captive	Yes		25.0–50.0 (immature–unrelated adults)	Weaver and de Waal 2003
Cebus capucinus	Captive	Yes	✓	18.5	Leca et al. 2002
	Wild	Yes/no[4]	✗[5]	27.0	Manson et al. 2005
Saimiri sciureus	Captive	Yes		77.0	Pereira et al. 2000
Old World monkeys					
Cercopithecus aethiops	Wild	Yes		—	Cheney and Seyfarth 1989
Erythrocebus patas	Captive	Yes[3]	✓	—	York and Rowell 1988
Cercocebus torquatus atys	Captive	Yes		—	Gust and Gordon 1993
Rhinopithecus bieti	Captive	Yes	✓	54.5	Grüter 2004
Macaca assamensis	Free-ranging[6]	Yes	✓		

Table 41.1 (cont'd)

SPECIES	LOCATION	RECONCILIATION	SELECTIVE ATTRACTION[1]	CCT/MEAN CCT %[2]	SOURCE
Macaca arctoides	Captive	Yes	✓	—	de Waal and Ren 1988
	Captive	Yes		—	Perez-Ruiz and
	Captive	Yes		34.9	Mondragon-Ceballos 1994
					Call et al. 1999
Macaca fascicularis	Captive	Yes		—	Cords 1988a,b
	Captive	Yes	✓	13.9	Aureli et al. 1989
	Wild	Yes	✓	—	Aureli 1992
Macaca fuscata	Captive	Yes	✓	—	Aureli et al. 1993
	Captive	Yes[3]		22.4	Chaffin et al. 1995
	Captive	Yes		7.9–30.3	Petit et al. 1997
	Captive	Yes		8.4	Schino et al. 1998
	Free-ranging[6]	Yes		26.8	Koyama 2001
	Free-ranging[6]	Yes	✓	14.0	Kutsukake and Castles 2001
	Captive	Yes		—	Abegg et al. 2003
	Wild	Yes		17.5	Majolo et al. 2005
Macaca maurus	Wild	Yes		40.0	Matsumura 1996
Macaca mulatta	Captive	Yes	✓	—	de Waal and Yoshihara 1983
	Captive	Yes		8.1 (low density)	Call et al. 1996
	Captive	Yes		7.1 (high density)	Demaria and Thierry 2001
				—	
Macaca nemestrina	Captive	Yes		—	Judge 1991
	Captive	Yes		41.9, 20.4 (two groups)	Castles et al. 1996
Macaca nigra	Captive	Yes	✓	40.4	Petit and Thierry 1994a,b
Macaca radiata	Wild	Yes	✓	29.3	Cooper et al. 2007
Macaca silensus	Captive	Yes	✓	—	Abegg et al. 1996
Macaca sylvanus	Captive	Yes	✓	—	Aureli et al. 1994
Macaca tonkeana	Captive	Yes		—	Demaria and Thierry 2001
Macaca thibetana	Free-ranging	Yes		7.9	Berman et al. 2006
Theropithecus gelada	Captive	Yes		29.8	Swedell 1997
Papio anubis	Wild	Yes	✓	15.6	Castles and Whiten 1998a,b
	Captive	Yes		27.3	Meishvili et al. 2005
Papio papio	Captive	Yes		27.0	Petit and Thierry 1994a,b
Papio hamadryas	Captive	Yes	✓	19.1	Romero et al. 2009
Papio ursinus	Wild	Yes		—	Silk et al. 1996
Rhinopithecus roxellanae	Captive	Yes		43.4	Ren et al. 1991
Semnopithecus entellus	Wild	Yes/no[7]	✓	—	Sommer et al. 2002
Colobus guereza	Captive	Yes	✓	45.1	Bjornsdotter et al. 2000
Great apes					
Gorilla gorilla berengei	Wild	Yes/No[8]	✓	—	Watts 1995a,b
Gorilla gorilla gorilla	Captive	Yes		14.4	Cordoni et al. 2006
	Captive	Yes[3]		29.0	Mallavarapu et al. 2006
Pan paniscus	Captive	Yes		—	de Waal 1987
	Wild	Yes		—	Hohmann and Fruth 2000
	Captive	Yes		35.6	Palagi et al. 2004

Table 41.1 *(cont'd)*

SPECIES	LOCATION	RECONCILIATION	SELECTIVE ATTRACTION[1]	CCT/MEAN CCT %[2]	SOURCE
Pan troglodytes	Captive	Yes			de Waal and van Roosmalen 1979
	Wild	Yes	✓	*12.3*	Arnold and Whiten 2001
	Captive	Yes	✓	*41.2*	Preuschoft et al. 2002
	Captive	Yes		*17.3*	Fuentes et al. 2002
	Wild	Yes		*15.9*	Wittig and Boesch 2005
	Wild	Yes	✓	*14.4*	Kutsukake and Castles 2004
	Captive	Yes		*29.8*	Palagi et al. 2004
	Captive	Yes	✓	*21.6*	Koski et al. 2007a
	Captive	Yes		*47.5*	Fraser et al. 2008a,b

[1] ✓ denotes selective attraction demonstrated.

[2] Numbers represent the revised measure of conciliatory tendency at the group level (Veenema et al. 1994). Numbers in italics are the mean CCTs across individuals in a group. Other measures of the tendency to reconcile are not included in order to avoid confusion in making comparisons.

[3] Proximity alone is included as conciliatory behavior.

[4] Reconciliation was demonstrated in only one of the two groups included in the study.

[5] Selective attraction was tested for but not demonstrated.

[6] "Free-ranging" is used for wild populations that are food-provisioned.

[7] Only female/female conflicts were reconciled, although statistical demonstration was not carried out.

[8] Only male/female conflicts were reconciled.

Many studies have also shown that former opponents are selectively attracted to one another as opposed to simply seeking contact in an indiscriminate fashion, thus strengthening the case for such contacts performing a specific function that is directly related to the recent conflict (see Table 41.1).

The vast majority of studies (74%, Table 41.1) have been carried out in captivity, where space is often limited. It could be argued that reconciliation might be an artifact of an artificially intense social environment with no possibility of escape. This has not proven to be the case since a number of studies have demonstrated reconciliation in large enclosures where prolonged avoidance was possible and in the wild where the option to leave the group, at least temporarily, was available (reviewed in Colmenares 2006).

In some species, particularly those with high conciliatory tendencies, reconciliation can involve conspicuous behavior patterns (Figure 41.2) that are rarely used in other contexts (i.e., *explicit reconciliation*), while others employ affiliative behaviors that are regularly observed in other contexts, such as grooming or making body contact (i.e., *implicit reconciliation*) (de Waal 1993) (Table 41.2). However, there is some variation in this respect where species reported as demonstrating explicit behaviors in their reconciliatory repertoire have not been shown to do so in all groups (pig-tailed macaques [*Macaca nemestrina*], Castles et al. 1996; chimpanzees, Arnold and

Whiten 2001, Fuentes et al. 2002, Kutsukake and Castles 2004). Noncontact behaviors might also be sufficient to restore relationships in species that do not have an extensive repertoire of friendly behavior. Sooty mangabeys (*Cercocebus torquatus atys*) reconcile by merely presenting their hindquarters (Gust and Gordon 1993), and York and Rowell (1988) found reconciliation in patas monkeys (*Erythrocebus patas*) only when they included close proximity as a friendly postconflict interaction between former opponents. Proximity was also found to have a conciliatory function in experiments with pairs of long-tailed macaques, a species that usually uses affiliative behaviors involving body contact (Cords 1993, also see Call 2000). Vocal signals also appear to facilitate friendly interactions following conflicts in chacma baboons (*Papio ursinus*, Cheney et al. 1995, Silk et al. 1996).

Comparisons of species characteristics such as the intensity of aggression, the degree of symmetry in contests, affiliation rates, and conciliatory tendencies have allowed classification of species according to their "dominance styles" (de Waal 1989a, Thierry 2000). For example, long-tailed and rhesus macaques (*Macaca mulatta*), which are relatively aggressive, have steep power gradients and reconcile few conflicts, leading to their classification as "despotic." By contrast, "tolerant" Tonkean (*Macaca tonkeana*) and stump-tailed (*Macaca arctoides*) macaques exhibit low-intensity aggression with frequent bidirectional conflicts and high levels of affiliation and reconciliation (de Waal

Figure 41.2 Hold-bottom ritual in stump-tailed macaques, which is often used for reconciliation (photograph by Josep Call).

Table 41.2 Species that Have Been Shown to Reconcile Conflicts "Explicitly" or "Implicitly" and the Behaviors Observed in This Context

SPECIES	BEHAVIORS USED	SOURCE
Explicit		
Cebus capucinus	Hold-bottom	Leca et al. 2002
Macaca arctoides	Hold-bottom	de Waal and Ren 1988
Macaca nemestrina	Standing grasp	Castles et al. 1996
Macaca tonkeana	Clasp, mount	Demaria and Thierry 2001
Rhinopithecus bieti	Hold lumbar, mount	Grüter 2004
Rhinopithecus roxellanae	Hold lumbar	Ren et al. 1991
Trachypithecus obscura	Ventro-ventro embrace	Arnold and Barton 2001a
Colobus guereza	Low-intensity mouth opening	Björnsdotter et al. 2000
Gorilla gorilla beringei, male/female dyads	Embrace, grumble, reach-toward	Watts 1995a,b
Pan paniscus	Genital massage	de Waal 1987
Pan troglodytes	Kiss, embrace	de Waal and van Roosmalen 1979, Fraser and Aureli 2008
Implicit		
Gorilla gorilla gorilla		Cordoni et al. 2006, Mallavarapu et al. 2006
Macaca fascicularis		Cords 1988a,b; Aureli et al. 1989
Macaca fuscata		Aureli et al. 1993
Macaca maurus		Matsumura 1996
Macaca mulatta		de Waal and Yoshihara 1983
Macaca nigra	e.g., grooming, gentle touch, body contact	Petit and Thierry 1994a
Macaca radiata		Cooper et al. 2007
Theropithecus gelada		Swedell 1997
Papio anubis		Castles and Whiten 1998a
Papio papio		Petit and Thierry 1994b
Pan troglodytes		Arnold and Whiten 2001, Kutsukake and Castles 2004

and Luttrell 1989, Thierry 1990a, Petit et al. 1997). Tolerant species have been shown to use explicit gestures during reconciliation and to have higher conciliatory tendencies than despotic species, which use implicit gestures, even after controlling for phylogeny (Thierry et al. 2008). Covariation between reconciliation frequencies and type of gestures occurs even between groups of the same species: Studies of pig-tailed macaques (Castles et al. 1996) and chimpanzees (de Waal and van Roosmalen 1979, Arnold and Whiten 2001, Fuentes et al. 2002, Kutsukake and Castles 2004, Fraser and Aureli 2008) reported that while some groups reconcile often and with explicit gestures, other groups reconcile less frequently and without the use of specific gestures.

THE COSTS OF AGGRESSIVE CONFLICT AND THE BENEFITS OF RECONCILIATION

Although conflicts of interest may be managed without escalation (Aureli and de Waal 2000), they often result in aggression and the loss of access to a contested resource by one of the participants. Similarly, aggression may serve to establish or reinforce dyadic dominance relationships between individuals. In both cases, the cost to the loser of the conflict is evident. In addition, conflicts themselves are energetically costly and can result in injury. Once aggression is instigated, one or more of these costs are almost impossible to avoid. There are also less obvious costs that can be incurred in the aftermath of conflict. Recipients of aggression are more likely to be the subject of renewed attacks either by the original aggressor (York and Rowell 1988; Aureli and van Schaik 1991; Aureli 1992; Cords 1992; Watts 1995a,b; Silk et al. 1996; Castles and Whiten 1998b; Das et al. 1998; Schino et al. 1998, Kutsukake and Castles 2001) or by third parties (Aureli et al. 1989, Aureli and van Schaik 1991, Kutsukake and Castles 2001). In wild long-tailed macaques, recipients of aggression have also been shown to forage less in postconflict situations than during control periods, perhaps because they then spend more time monitoring the social environment in anticipation of further aggression (Aureli 1992).

Both recipients of aggression and, perhaps more surprisingly, aggressors appear to experience increased levels of anxiety following conflicts. A growing body of pharmacological evidence shows that elevated levels of anxiety result in increased rates of self-directed behaviors (e.g., scratching, self-grooming) (Ninan et al. 1982; Schino et al. 1991, 1996; Barros et al. 2000). This together with circumstantial evidence suggests that self-directed behavior can be used as a noninvasive tool for quantifying levels of anxiety in primates (Maestripieri et al. 1992, Troisi 2002). A number of studies have demonstrated increased rates of self-directed behaviors following conflicts in one or both opponents (Aureli et al. 1989, 1992; Aureli and van Schaik 1991; Aureli 1997; Castles and Whiten 1998b; Das et al.

1998; Schino et al. 1998, 2007a; Kutsukake and Castles 2001; Cooper et al. 2007; Koski et al. 2007a; Fraser et al. 2008b). Although a prolonged stress response in recipients of aggression is likely to be adaptive in that they are primed for action in the event of further attacks or more predisposed to behave submissively (Leshner 1983), it can have negative physiological and developmental consequences over the long term (Henry 1982, Von Holst 1985, Kaplan 1986). However, anxiety following aggression appears to affect both former opponents, and it has been proposed that this is due to uncertainty about relationships that may have been damaged by a conflict (Aureli and van Schaik 1991).

For both parties, a reduction in the potential for an ongoing, mutually beneficial relationship would be a highly negative mid- and long-term consequence of aggressive conflict, so the ability to approach or to be approached by a former opponent for future affiliation is crucial. In a study of free-ranging Japanese macaques (*Macaca fuscata*), in which former opponents were monitored for 10 days after conflicts, grooming, time spent in proximity, and approaches by the victim were significantly reduced relative to baseline levels after nonreconciled conflicts (Koyama 2001).

Reconciliation serves to reduce the costs of conflict, both directly and indirectly. For the victim of aggression, the risk of further attacks by the former aggressor or other group members is significantly decreased after reconciliation (Aureli and van Schaik 1991; Cords 1992; de Waal 1993; Watts 1995a,b; Silk et al. 1996; Castles and Whiten 1998b; Koyama 2001; Kutsukake and Castles 2001; Wittig and Boesch 2005; Palagi et al. 2008). In Koyama's (2001) study of the long-term effects of reconciliation, the rate of aggression between former opponents was lower during the 10 days following reconciled conflicts than during baseline or control observations. The anxiety experienced by former opponents following conflicts, expressed by increased rates of self-directed behaviors, returns to baseline levels once reconciliation has taken place (Aureli and van Schaik 1991, Castles and Whiten 1998b, Das et al. 1998, Kutsukake and Castles 2001, Fraser et al. 2010). A similar effect was found in schoolchildren, measuring postconflict stress using cortisol levels (Butovskaya 2008). Heart rates of former opponents, which increase after a conflict, are also reduced following reconciliation (Smucny et al. 1997, Aureli and Smucny 2000). Interestingly, reconciliation may be more effective than affiliative contact with other group members in reducing self-directed behaviors and heart rate (Smucny et al. 1997, Das et al. 1998, Aureli and Smucny 2000), although postconflict affiliation from a bystander who shares a valuable relationship with the victim has been shown to have a similar stress-reducing effect to reconciliation in chimpanzees (Fraser et al. 2008b).

Cords (1992) provided strong evidence that the function of reconciliation is relationship repair. In a series of experiments, she showed that after conflicts subordinate long-tailed macaques were deterred from approaching dominants in order to drink simultaneously from closely adjacent containers and received more aggression from them than in control periods. Thus, conflicts appeared to have had negative consequences for their relationships. However, when the pair made postconflict affiliative contact before being given access to the containers, the effect of the conflict was nullified in that the pair then accessed the containers with no further avoidance or aggressive behaviors. The results of this study demonstrated that tolerance was restored by reconciliation and implied that the former opponents were no longer uncertain about their relationships. Further support that reconciliation reduces uncertainty between former opponents is provided by playback experiments involving friendly grunts in wild baboons (Cheney and Seyfarth 1997). After playbacks of grunts from the former opponent, recipients of aggression approached former aggressors and tolerated their approaches more often than during postconflict periods without such playbacks.

THE VARIATION IN PATTERNS OF RECONCILIATION

Although reconciliation is a widespread phenomenon in primates, only a proportion of all conflicts are reconciled

(Aureli et al. 2002). A number of factors have been considered that might affect the probability that reconciliation will occur, and most studies have investigated one or more of these (see Table 41.3).

Conflict Characteristics

Conflict Intensity

Conflicts vary in intensity and can range from threats to full contact aggression, sometimes resulting in wounding. It seems reasonable to assume that the more intense the conflict, the more disturbing to the relationship it will be. If so, conflicts of high intensity should be reconciled more often than those of low intensity. However, this appears not to be the case. Of the 21 studies in which the effect of conflict intensity has been investigated, only five clearly showed higher conciliatory tendencies following high-intensity conflicts. The remainder reconciled low-intensity conflicts more often or exhibited no difference in their tendency to reconcile conflicts of different intensities (Table 41.3). These results are difficult to explain, although it is likely that a number of factors are involved. The effect of conflict

Table 41.3 Factors Influencing the Probability of Reconciliation

SPECIES	RELATIONSHIP CHARACTERISTICS						CONFLICT CHARACTERISTICS			SOURCE
	KINSHIP	FRIENDLINESS	SEX	AGE	RANK	SUPPORT	INTENSITY	COUNTER AGGRESSION	INITIATION	
Cebus capucinus	Kin		m–f only				nd	nd	agg/d	Leca et al., 2002
								nd		Manson et al. 2005
Saimiri sciureus	Kin	✓	f–f most				High most			Pereira et al. 2000
Cercopithecus aethiops	Kin									Cheney and Seyfarth 1989
Erythrocebus patas	nd								nd	York and Rowell 1988
Cercocebus torquatus atys									vict	Gust and Gordon 1993
Macaca assamensis			m–m most				nd		agg/s	Cooper and Bernstein 2002
		✓ (f–f only)				✓ (f–f only)				Cooper et al. 2005
Macaca arctoides	Kin								agg/s	de Waal and Ren 1988
	Kin			Complex	Complex					Perez-Ruiz and Mondragon-Ceballos 1994
	Kin	✓		nd	nd		nd	nd	nd	Call et al. 1999

Table 41.3 (cont'd)

SPECIES	KINSHIP	FRIENDLINESS	SEX	AGE	RANK	SUPPORT	INTENSITY	COUNTER AGGRESSION	INITIATION	SOURCE
		RELATIONSHIP CHARACTERISTICS					CONFLICT CHARACTERISTICS			
Macaca fascicularis	Non-kin								agg/s	Cords 1988a,b
	Kin	✓						nd	vict	Aureli et al. 1989
									vict	Aureli 1992
									nd	Cords 1993
	nd	✓	jf–af > jm–af + jm–jm > jf–jf	nd		nd	nd	nd	vict/s	Cords and Aureli 1993
	Kin									Aureli et al. 1997
Macaca fuscata									nd	Aureli et al. 1993
	Kin									Aureli et al. 1997
			m–m most							Petit et al. 1997
	Kin	✓	m–f least	j most			Low most			Schino et al. 1998
	Kin									Kutsukake and Castles 2002
						✓				Koyama 2002
	nd	✓	mf least	nd	nd		nd	nd		Majolo et al. 2009
Macaca maurus			m > f						vict	Matsumura 1996
Macaca mulatta	Kin									de Waal and Yoshihara 1983
	Kin									Call et al. 1996
	Kin								agg	Demaria and Thierry 2001
Macaca nemestrina	Kin				✓				vict	Judge 1991
	Kin	✓					nd	nd	nd, agg/s	Castles et al. 1996
Macaca nigra							nd	✓		Petit and Thierry 1994a,b
			nd							Petit et al. 1997
Macaca radiata			f–f > f–m				High most			Cooper et al. 2007
Macaca sylvanus	nd									Aureli et al. 1997
Macaca thibetana	nd	✓	m–m	nd	nd	nd		nd		Berman et al. 2006
Macaca tonkeana	nd							nd	nd	Demaria and Thierry 2001
Theropithecus gelada									nd	Swedell 1997
Papio anubis							nd	nd	nd	Castles and Whiten, 1998a
Papio hamadryas		✓				✓				Romero et al. 2009
Papio papio							nd	nd		Petit and Thierry 1994a,b

Table 41.3 (*cont'd*)

SPECIES	KINSHIP	FRIENDLINESS	SEX	AGE	RANK	SUPPORT	INTENSITY	COUNTER AGGRESSION	INITIATION	SOURCE
			m–f most							Petit et al. 1997
Papio ursinus	Kin				Closely ranked + high vict rank				agg	Silk et al. 1996
Rhinopithecus bieti									nd	Grüter 2004
Trachypithecus obscura										
Trachypithecus obscura	nd	✓		a–j least				nd	vict/s, agg/s	Arnold and Barton 2001a
Colobus guereza									vict	Bjornsdotter et al. 2000
Lemur catta		nd					High most, low most		Agg,	Palagi et al. 2005
Propithecus verreauxi		✓	ns		High rank difference		Low most		Vict	Palagi et al. 2008
Gorilla gorilla beringei	Kin	✓	m–f only							Watts 1995a,b
Gorilla gorilla gorilla			m–f only				High most			Cordoni et al. 2006
									nd	Mallavarapu et al. 2006
Pan paniscus							High most		agg	de Waal 1997a,b
		✓	f–f most						nd	Palagi et al. 2004
Pan troglodytes			m–m > m–f > f–f	nd					nd	de Waal and van Roosmalen 1979
		✓	m–m, m–f > f–f	a–j less	nd		Low most (trend)	nd	f/d + vict/d	Arnold and Whiten 2001
			f–f most, m–m least	nd						Fuentes et al. 2002
		✓	m–f most, f–f least			nd	Low most		Social, nd Food, vict	Wittig and Boesch 2003b
		nd	nd	nd			High most			Kutsukake and Castles 2004
		✓				✓				Watts 2006
		✓	m–m > f–f				nd	nd		Koski et al. 2007b
	nd	nd	f–f > m–m				nd			Fraser et al. 2010

✓, significantly increases the probability of reconciliation; f, female; m, male; a, adult; j, juvenile; nd, has no effect on the probability of reconciliation; agg, aggressor; vict, victim; s/d, same/opposite pattern observed in nonconflict contexts. A comma separates findings concerning different groups of the same species reported in the same study.

intensity on resultant interopponent distances was investigated in wild chimpanzees; as might be expected, low-intensity conflicts had a less dispersive effect on opponents than high-intensity conflicts, and these conflicts were also reconciled more often (Arnold and Whiten 2001; Wittig and Boesch 2003b; but see Kutsukake and Castles 2004). Similarly, conflicts among rhesus macaques often involve chases over 2 m and are rarely reconciled, whereas stump-tailed macaques tend to remain stationary while slapping or threatening their opponents and have a high conciliatory tendency (Call 1999, Call et al. 1999); however, the cause–effect relation between postconflict distance and likelihood of reconciliation has not been directly examined. Victims of more severe conflicts might be deterred from contacting former opponents (Petit et al. 1997), especially if the likelihood of renewed attacks after a high-intensity conflict is greater, although this has not yet been established. However, if their relationship is valuable to one or both former opponents, then they should still be motivated to communicate their interest in maintaining the relationship by attempting to reconcile the conflict despite the potential risks (Aureli et al. 1989). Based on this view and on the contradictory findings, conflict intensity may not be a major factor affecting the likelihood of reconciliation.

Counteraggression and Conflict Outcome

In general, the proportion of conflicts that involve direct retaliation (or counteraggression) is higher in more tolerant species. Weak power asymmetries allow a wide distribution of affiliative behaviors across group members (Thierry et al. 1990, 1994; Petit et al. 1992). Familiarity with the former opponent and the ability to approach dominants without negative consequences in other contexts (de Waal and Luttrell 1989) combined with the low cost of conflict are likely to increase the probability of reconciliation, especially if tolerant relationships within the group are relatively common. Counteraggression either has not been shown to have any effect on the probability of reconciliation (stump-tailed macaques, Call et al. 1999; Tonkean macaques, Demaria and Thierry 2001; spectacled leaf monkeys, *Trachypithecus obscura*, Arnold and Whiten 2001; white-faced capuchins, *Cebus capucinus*, Leca et al. 2002, Japanese macaques, Majolo et al. 2009) or has resulted in higher conciliatory tendencies (black macaques, *Macaca nigra*, Petit and Thierry 1994a) in more tolerant species.

The outcome of a conflict can also have an effect on the likelihood of following reconciliation. Aureli et al. (1989) found that undecided conflicts (i.e., those involving counteraggression and no clear signs of submission) in long-tailed macaques were reconciled more often than decided conflicts and that the initiative to reconcile was taken equally often by the aggressor and victim. They suggested that the uncertainty associated with undecided conflicts resulted in a greater motivation to reconcile in both opponents, whereas decided conflicts rarely resulted in aggressor-initiated reconciliation. As expected, counteraggression was

rare in this despotic species but accounted for 25% of conflicts in a study of wild olive baboons (*Papio anubis*, Castles and Whiten 1998a), which have a similar dominance style. In this case, unidirectional conflicts and those involving counteraggression (also called "bidirectional conflicts") were reconciled equally often. A very similar pattern was reported for Guinea baboons (*Papio papio*, Petit and Thierry 1994b). Castles and Whiten (1998a) suggested that bidirectional conflicts might not result in greater uncertainty if counteraggression is common. By contrast, Wittig and Boesch (2003b) found that wild chimpanzees rarely reconciled undecided conflicts and tended to renew aggression instead despite the fact that undecided conflicts were uncommon.

Relationship Characteristics

Kinship

The importance of kinship in determining patterns of social interactions has been intensively studied and shown to be central in many primate species (Gouzoules and Gouzoules 1987, Silk 2002a). Accordingly, the influence of kinship on patterns of reconciliation has been widely investigated. Species in which cooperative and affiliative relationships occur most often within matrilines reconcile more conflicts with their kin than with unrelated opponents (e.g., *Macaca* spp., see Table 41.3). Kinship was not found to influence reconciliation in species with more open relationship networks (patas monkeys, York and Rowell 1988; spectacled leaf monkeys, Arnold and Barton 2001a), in immature long-tailed macaques (Cords 1988b, Cords and Aureli 1993), and in macaque species whose social systems are less kin-biased (Tonkean macaques, Demaria and Thierry 2001; Barbary macaques, *Macaca sylvanus*, Aureli et al. 1997).

Friendliness

Dyads that engage in high levels of affiliation are said to be "friendly" (Cords 1997). Studies that have compared conciliatory tendencies among dyads which are more friendly and less friendly than average have found, without exception, that the most friendly dyads are more conciliatory (see Table 41.3). Primates living in kin-biased societies tend to direct most of their affiliation toward close relatives. In such cases, the influence of friendliness can be confounded by kinship, and a number of studies have overcome this problem by analyzing data on kin and nonkin dyads separately (de Waal and Yoshihara 1983, Aureli et al. 1989) or by using multivariate analyses, which identify variables that independently contribute to differences in reconciliation in dyads of differing types (Schino et al. 1998, Call et al. 1999, Majolo et al. 2009, Fraser et al. 2010).

Sex

Sex differences in the occurrence of reconciliation should depend on the nature of the society of the species under study and the role of the sexes within that society. For example, chimpanzees live in *male-philopatric* societies, in which males remain in their natal community and females emigrate (Goodall 1986). Consequently, most males are related to some degree and have lifelong social relationships with other males in the community. Males also cooperate with one another in community defense and in coalition formation. Females have few kin relations within the community and are relatively solitary. These broad sex differences in bondedness have been reflected in higher levels of reconciliation between male/male dyads in some studies of both wild (Arnold and Whiten 2001) and captive (de Waal 1986c, Koski et al. 2007a) chimpanzee populations, whereas other studies have found that mixed-sex dyads (Wittig and Boesch 2003b) or female/female dyads (Fuentes et al. 2002, Fraser and Aureli 2008) are more likely to reconcile or that sex combination has no effect on reconciliation rates (Kutsukake and Castles 2004). This suggests that sex combination may not be the most important factor in determining reconciliation rates in chimpanzees. In mountain gorillas (*Gorilla gorilla beringei*), bonds between females and between males are weak and reconciliation within these classes of dyads could not be demonstrated (Watts 1995b). However, female mountain gorillas rely on males for protection and agonistic support (Watts 1991c, 1994b). As a consequence, reconciliation between males and females was common. Among female-bonded species, conflicts between females were reconciled more frequently than male/female conflicts in some cases (rhesus macaques, Thierry 1986; Japanese macaques, Schino et al. 1998; squirrel monkeys, *Saimiri sciureus*, Pereira et al. 2000; Assamese macaques, *Macaca assamensis*, Cooper and Bernstein 2002; bonnet macaques, *Macaca radiata*, Cooper et al. 2007), but in other cases male/male conflicts were reconciled at least as often as female/female conflicts (Tonkean macaques, Thierry 1986; juvenile long-tailed macaques, Cords and Aureli 1993; Japanese macaques, Petit et al. 1997, Schino et al. 1998; Assamese macaques, Cooper and Bernstein 2002).

Age

The effect of age on the likelihood of reconciliation has been investigated in some studies. Conflicts between adults and juveniles appear to be reconciled less often than conflicts within age classes (Schino et al. 1998, Arnold and Barton 2001a, Arnold and Whiten 2001). In their study of juvenile long-tailed macaques, Cords and Aureli (1993) found that juvenile females reconciled more conflicts with unrelated adult females than juvenile males did and attributed this difference to the life histories of females, which remain in their natal troops in this species and have long-term relationships with other females. Males, on the other hand, emigrate to other groups and should have less interest in maintaining relationships with unrelated adult females. Perez-Ruiz and Mondragon-Ceballos (1994) found that in stump-tailed macaques juvenile females that attacked infants often initiated reconciliation with them. They suggested that this might serve to appease adult females and

reduce the probability of being attacked themselves, particularly by mothers.

Dominance Rank

The effect of dominance rank on reconciliation has received little attention to date. When the roles of aggressor and victim were not distinguished, the influence of rank and rank distance on conciliatory tendencies has not been consistent. In chacma baboons, individuals that were close in rank reconciled conflicts more often than individuals that were more distantly ranked (Silk et al. 1996). Close kin often occupy adjacent ranks in this species, and when the confounding effect of kinship was removed, this effect disappeared. However, the effect did remain in pig-tailed macaques when kin were removed from the analysis (Judge 1991). Similar analyses did not reproduce these findings in other species (Call et al. 1999, Arnold and Barton 2001a, Arnold and Whiten 2001). In a study in which the roles of aggressor and victim were distinguished, reconciliation was more frequent in cases where lower-ranking pig-tailed macaques attacked higher-ranking individuals than the reverse (Judge 1991). This finding supports Aureli et al.'s (1989) suggestion that greater uncertainty associated with the outcome of conflict promotes reconciliation between former opponents.

Support

Intuitively, one of the most valuable relationships a socially living primate can have is a coalition partnership (Harcourt and de Waal 1992). A reliable supporter would be a great asset in competitions for any resource, so individuals should be highly motivated to restore such relationships after potentially damaging conflicts. Remarkably, the early studies that investigated this aspect found that individuals did not reconcile more often with individuals that supported them (captive chimpanzees, Preuschoft et al. 2002; juvenile long-tailed macaques, Cords and Aureli 1993). Cords and Aureli (1993) pointed out that supporters of juveniles may not necessarily have good relationships with them or benefit from the relationship in the short term. Their primary motivation might, in some cases, be to direct aggression at the opponent rather than to aid the victim. They suggested that reconciliation should be most common where both former opponents perceive the relationship to be beneficial and less likely where asymmetries exist. Preuschoft et al. (2002) also suggested that their negative result might be due to the fact that there was little overlap between affiliation and support networks and that support relationships were likely to be highly asymmetrical. Wittig and Boesch (2003b) found that dyads that supported one another and shared food reconciled more often than other types of dyad, although analyses of the effect of either of these factors alone did not yield significant results. Fraser et al. (2010) confirmed that partners that shared food and supported one another were more likely to reconcile, but this effect disappeared when the analysis included more species-specific variables. Cooper et al. (2005) did find higher rates of reconciliation among regular supporters than among dyads that never supported one another in wild female Assamese macaques (*Macaca assamensis*). A similar result was found in Japanese macaques even when the effect of kinship was controlled (Koyama unpublished data). Contrary to previous studies, there was a significant overlap between dyadic relationships that were highly affiliative and those in which agonistic support was given. Watts (2006) also found that likelihood to reconcile was positively correlated with both the level of agonistic support and the amount of time spent grooming among adult male wild chimpanzees, suggesting that valuable allies are more likely to reconcile in order to repair their relationship and restore the benefits that their relationship affords.

Other Factors

Subgroup Association

In societies characterized by a high degree of fission–fusion dynamics, the formation of temporary subgroups means that group members are rarely all together (Goodall 1986, Nishida and Hiraiwa-Hasegawa 1987, Chapman et al. 1993). Thus, subgroup association preferences could be reflected in conciliatory tendencies. If chimpanzees choose to associate with certain individuals over others, it might be expected that reconciliation following conflicts with these individuals should be relatively common. The current evidence is mixed. Whereas Wittig and Boesch (2003b) found that chimpanzees that often associate with one another are more likely to reconcile than those that associate rarely, Arnold and Whiten (2001) did not find such a pattern.

Mating Seasonality

Certain species have a defined mating season (Dixson 1998). Aureli and colleagues (1993, 1994) confirmed the occurrence of reconciliation in both mating and nonmating seasons in Japanese and Barbary macaques. The mating season is often a time of increased levels of aggression (Schino et al. 1998, Abegg et al. 2003). Japanese macaques have been found to reconcile fewer conflicts during this period, suggesting a general deterioration of social relationships as a result of increased competition and tension (Schino et al. 1998, Majolo and Koyama 2006). Conversely, Palagi et al. (2005) was able to demonstrate reconciliation in a group of captive ring-tailed lemurs (*Lemur catta*) that was studied only during the mating season, a period characterized by higher tolerance levels than during the birth season. Other studies have failed to find seasonal differences in conciliatory tendencies (Kutsukake and Castles 2001, Abegg et al. 2003). Arnold and Whiten (2001) also detected no difference in conciliatory tendencies after male/female conflicts involving females that were sexually receptive as opposed to conflicts involving females that were not.

Foraging

The majority of studies that have compared the occurrence of reconciliation in feeding and nonfeeding contexts have found that monkeys do not reconcile conflicts over food as often as conflicts in other contexts (de Waal 1984, Aureli 1992, Matsumura 1996, Verbeek and de Waal 1997, Castles and Whiten 1998a, Westlund et al. 2000, but see Cooper and Bernstein 2002). It is likely that many non-food-related conflicts have a more directly social dimension, and it has been suggested that social relationships might not be disturbed by food contest competition (Aureli 1992). Verbeek and de Waal (1997) predicted that chimpanzees might not show the same pattern of reduced reconciliation in feeding contexts as they employ a wide range of reassurance and appeasement behaviors at clumped food sources (Goodall 1986, de Waal 1989a), which suggests that social relationships are at stake. Data on wild populations of chimpanzees support this prediction as reconciliation was at least as common during foraging as it was in other contexts (Arnold and Whiten 2001, Wittig and Boesch 2003b).

INITIATIVE TO RECONCILE

Whether it is the aggressor or the recipient of aggression that is responsible for the initiation of reconciliation should depend on a number of factors. Most relationships are asymmetrical to some extent, and one party will have more to gain from restoring the relationship than the other (Cords and Aureli 2000). Therefore, the individual that has most to gain should be the one that initiates conciliatory contact. However, approaching a former opponent can be risky, especially for the victim. In despotic species, approaching a dominant individual can have negative consequences, even outside of agonistic contexts, and is relatively rare. For this reason, victims of aggression may be deterred from attempting reconciliation with a higher-ranking former opponent even when it is in their interest to do so. Aggressors may have less to gain from reconciliation than the victim, but they must still maintain their social networks and, thus, should also be motivated to resolve at least some of their conflicts. Both parties should also benefit from the stress-reducing effect of reconciliation and the elimination of uncertainty regarding their future relationship, but again the relative benefit is likely to be different for each opponent.

In addition to methodological difficulties in detecting genuine attempts to reconcile (Aureli and Smucny 2000, Cords and Aureli 2000), the combination of variation in the motivation to reconcile and the risks involved in approaching a former opponent is reflected in inconclusive findings regarding the initiation of reconciliation (Table 41.3). For example, aggressors were primarily responsible for reconciliation in one group of rhesus macaques (Demaria and Thierry 2001), a species in which it might be expected that victims would be reluctant to approach former opponents, and in more tolerant species such as stump-tailed macaques (de Waal and Ren 1988) and bonobos (*Pan paniscus*, de Waal 1987). Variations in patterns of initiation have also been reported for different groups of the same species (e.g., long-tailed macaques, Aureli et al. 1989, Aureli 1992, Cords 1993). A number of other studies have reported no difference in patterns of initiation where biases in favor of one type of opponent might be expected according to their dominance styles (Table 41.3).

The majority of studies that have considered the identities of individuals that take the initiative to reconcile within any particular dyad have found that the individuals that initiate reconciliation are also primarily responsible for friendly contacts in general (Cords 1988b, de Waal and Ren 1988, Aureli et al. 1993, Cords and Aureli 1993, Castles et al. 1996, Arnold and Barton 2001a, Cooper and Bernstein 2002). In a few cases, the postconflict pattern was different. In white-faced capuchins, aggressors initiated reconciliation more than affiliation in MCs (Leca et al. 2002), whereas in long-tailed macaques and a newly established group of pig-tailed macaques, victims initiated reconciliation more often than affiliation in MCs (Aureli et al. 1989, Castles et al. 1996). Wild chimpanzee victims were shown to initiate a little more than half of their affiliative contacts with former opponents in normal social contexts, but this figure was significantly elevated in postconflict conditions (Arnold and Whiten 2001). Such changes in patterns of behavior in postconflict situations can be viewed as an indication of the relative importance of reestablishing relationships and the differential motivation to do so between opponents.

Almost all studies have defined the initiator of reconciliation as the individual that approaches the former opponent. Silk and colleagues have usefully pointed out that in some cases chacma baboons grunt before approaching former opponents and suggest that this might serve as reassurance that subsequent interactions will not be antagonistic (Silk 1996, 1997; Silk et al. 1996). A number of species produce vocalizations during reconciliation (stump-tailed macaques, de Waal and Ren 1988; mountain gorillas, Watts 1995b; brown capuchins, *Cebus apella*, Verbeek and de Waal 1997; spectacled leaf monkeys, Arnold and Barton 2001a; white-faced capuchins, Leca et al. 2002), although details of whether or not these vocalizations begin prior to the act of reconciliation and which individuals produce such signals have not been provided except in the case of chacma baboons. Other forms of more subtle signaling, such as changes in body posture or facial expressions, may also have been overlooked by researchers (Cords and Aureli 2000). It is possible that the responsibility for initiating reconciliation could have been ascribed wrongly in some cases if such signals are involved in soliciting contact or communicating nonaggressive intentions and are therefore instrumental in bringing about reconciliation (Arnold and Barton 2001b).

THE ONTOGENY OF RECONCILIATION

The interspecific and interindividual variation in conciliatory tendencies and conciliatory styles can be quite marked, as illustrated. While some of this variation is likely to result from differences in genetically based dispositions, there is good evidence that the social environment has an important role in influencing reconciliation behavior. The absence of social experience can result in an inability to reconcile at all. Rhesus macaques which had been removed from their mothers before weaning, hand-reared, and housed singly for extended periods did not show any form of reconciliation, in contrast to individuals that after separation from the mother were housed in a peer group at an early age (Ljungberg and Westlund 2000). That early socialization seems sufficient to develop conciliatory behavior is supported by a study on chimpanzees (Spijkerman et al. 1997) where adolescents reared in peer groups did not differ in reconciliation rates from those reared with mothers in a zoo group.

Variation in reconciliation was found between two groups of long-tailed macaques, which were formed by dividing a group according to the female dominance hierarchy: One group contained the higher-ranking females and the other, the lower-ranking females (Butovskaya et al. 1996). Reconciliation was more common in the low-ranking group, suggesting that the difference could have been due to exposure to a more tolerant social environment during development or to genetic predisposition given the relatedness of the females. Similarly, when juvenile rhesus macaques were cohoused with juvenile stump-tailed macaques over a period of 6 weeks, the conciliatory tendency of the rhesus macaques rose gradually until it matched the much higher conciliatory tendency typical of the more tolerant stump-tailed macaques (de Waal and Johanowicz 1993). Interestingly, the rhesus macaques did not extend their social learning to the specific patterns of behavior, such as hold-bottom, typically used by stump-tailed macaques during reconciliation (Table 41.2, Figure 41.2), which suggests that some aspects of reconciliation might not be modifiable.

A second line of research has explored the influence of the mother/offspring relationship in the development of reconciliatory behavior. Infant, weanling, and juvenile brown capuchin monkeys were classified as having either secure or insecure relationships with their mothers depending on the relative rate of affiliation and agonism observed between them (Weaver and de Waal 2000, 2003). Reconciliation began to emerge in infancy irrespective of the security of the mother/offspring relationship. The quality of the mother/offspring relationship did, however, affect the way in which reconciliation was accomplished. Secure offspring were not intimidated by their mother's aggression and were similarly relaxed after conflicts with unrelated adults. They performed calm appeasement behaviors toward unrelated adult aggressors and were often approached for reconciliation by former opponents. Insecure offspring, by contrast, were often noisy and agitated after conflicts.

They had a higher conciliatory tendency with unrelated adults than secure offspring, but this was a result of a high proportion of offspring-initiated reconciliation. Insecure offspring were more inclined to seek reconciliation that involved directly comforting behaviors involving body contact and rarely had contact with their mothers in post-conflict periods. It appears that the style of individual reconciliation behavior develops from the mother/infant attachment behavior system in this species (Weaver and de Waal 2000, 2003).

HYPOTHESES CONCERNING THE FUNCTION AND DISTRIBUTION OF RECONCILIATION

Since 1979, a large amount of comparative data have been accumulated which have enabled the generation of a number of hypotheses to explain why reconciliation is such a wide-spread feature of primate social life and to account for the form and distribution of reconciliation within and between species. Early hypotheses (reviewed by Kappeler and van Schaik 1992) include the *minimum cognitive capacity hypothesis* (Gallup 1982), which states that the ability to reconcile requires self-awareness and the capacity to empathize and attribute mental states to others. In fact, there is no reason to assume that such high-order cognitive capacities are necessary since individual recognition, memory, and an affiliative disposition in postconflict situations are sufficient (de Waal and Yoshihara 1983, Castles 2000). Thus, reconciliation is expected to be a widespread phenomenon among social animals. Operational reconciliation has been demonstrated in captive bottlenose dolphins (*Tursiops truncatus*, Weaver 2003), domestic goats (*Capra hircus*, Schino 1998), spotted hyenas (*Crocuta crocuta*, Hofer and East 2000, Wahaj et al. 2001), dogs (*Canis familiaris*, Cools et al. 2008), wolves (*Canis lupus*, Cordoni and Palagi 2008), and even between cleaner fish and their clients (*Labroides dimidiatus*, Bshary and Würth 2001).

The *social evolution hypothesis* (de Waal 1989b) states that reconciliation is necessary to maintain group cohesion in the face of the dispersive effects of aggression and should be found in all species which live in social groups. Kappeler's (1993) early failure to demonstrate reconciliation in ring-tailed lemurs appeared not to support this hypothesis, but later studies suggest that reconciliation may occur in this species (Rolland and Roeder 2000, Palagi et al. 2005). Since then, reconciliation has been reported to be absent in black lemurs and red-bellied tamarins (Roeder et al. 2002, Schaffner and Caine 2000), but explanations have been suggested which do not undermine this hypothesis (Aureli et al. 2002, Schaffner et al. 2005).

de Waal (1986c) proposed the *reconciled hierarchy hypothesis*, which is based on the notion that dominance relationships promote peaceful coexistence and that reconciliation is dependent on the subordinate signaling its status to the satisfaction of the dominant following a conflict. Kappeler and van Schaik (1992) interpreted this hypothesis as predicting that reconciliation should be more easily achieved in species with formalized dominance hierarchies where ritualized submission on the part of one opponent can terminate conflict and promote friendly interactions. This has proven not to be the case as such species have among the lowest reported conciliatory tendencies (Kappeler and van Schaik 1992), and in an analysis of reconciliation patterns in 15 groups from nine *Macaca* species, Thierry et al. (2008) found that species exhibiting a high degree of dominance asymmetry showed lower conciliatory tendencies than species with low dominance asymmetry. Furthermore, Arnold and Whiten (2001) found that only about 50% of conciliatory contacts were preceded by submissive "pant-grunts" in wild chimpanzees, which suggests that status signaling is not necessary in more tolerant species at least.

One of the most robust hypotheses proposed to date is the *valuable relationships hypothesis* (de Waal and Aureli 1997). This has been developed from earlier hypotheses which stressed the importance of relationship quality in determining which kinds of dyad are most likely to reconcile (de Waal and Yoshihara 1983, Aureli et al. 1989, Kappeler and van Schaik 1992, Cords and Aureli 1993). This hypothesis assumes the relationship repair function of reconciliation and predicts that conflicts within dyads that have a relationship which has important positive fitness benefits for both parties are more likely to be reconciled than conflicts within dyads that have less valuable relationships (for review see Cords and Aureli 2000, Watts 2006). Support for this hypothesis is abundant and can be found in the earlier sections of this chapter.

Cords and Aureli (1993, 2000) emphasized three dimensions of relationship quality that are likely to influence the probability that reconciliation will occur. The first is *value*, that is, what an individual can potentially offer, how disposed the individual is to offer it, and how accessible that individual is as a social partner. Value is not an easily defined or measurable quality. It is made easier if it is based on behavioral exchanges that result in clear fitness advantages. A coalition partner fits this description, and although direct evidence that reconciliation is more frequent among coalition partners is currently mixed (Cords and Aureli 1993, Preuschoft et al. 2002, Wittig and Boesch 2003b, Cooper et al. 2005, Watts 2006; Fraser et al. 2010), indirect evidence comes from studies that show high rates of reconciliation within dyads that are known to be the most likely to offer coalitionary support. For example, in species in which matrilineal kin are most often supporters, reconciliation is kin-biased. In chimpanzees, males typically form coalitions with one another for within- and between-group conflicts and male/male conflicts are reconciled more often than female/female conflicts (de Waal 1986c, Goodall 1986, Arnold and Whiten 2001, Koski et al. 2007a). Direct evidence that value influences reconciliation comes from an experimental study in which pairs of long-tailed macaques were trained to cooperate with one another to gain access to food (Cords and Thurnheer 1993). The conciliatory tendencies of these dyads rose after their value was increased during the training phase.

The second dimension of relationship quality is *compatibility*, defined as the general tenor of social interactions between individuals which derives from both the temperaments of the partners and their shared history of social interactions (Cords and Aureli 1993, 2000). This is usually measured as the relative frequencies of friendly interactions such as grooming, contact sitting, or approaches leading to affiliative behavior compared to aggressive interactions. Reconciliation has usually been found to be more common within dyads that have friendly relationships than within those that do not, even after controlling for their higher baseline levels of affiliation (Table 41.3). Additionally, groups with a wide distribution of compatible or friendly partners have higher overall conciliatory tendencies than groups which are more cliquish and restrict affiliation to a small subset of group members (Call et al. 1996, Castles et al. 1996). Similarly, species that show high rates of baseline affiliation are particularly conciliatory (de Waal and Luttrell 1989, Thierry 2000).

The third dimension of relationship quality that should influence patterns of reconciliation is *security*, defined as the perceived probability that the relationship will not change for the worse and measured by the consistency of the partners' behavioral response (Cords and Aureli 1993, 2000). The more secure a relationship is, the less likely it is to need restoration via reconciliation. Attempts to directly measure relationship security have been made using measures based on the outcome of approaches (Cooper et al. 2005) and the variation, consistency, and reciprocity of affiliation (Fraser et al. 2008a, Majolo et al. 2009); but no significant effects have yet been found. There is, however, some indirect evidence that relationship security affects levels of reconciliation. Peer groups of juvenile long-tailed macaques reconciled with nonkin more often than with kin, and this difference was interpreted as reflecting greater uncertainty about these relationships in a situation where both kin and nonkin were similar in their value as social partners (Cords and Aureli 1993). Immature brown capuchin monkeys that generalized their insecure relationships with their mothers to other unrelated adults were more conciliatory than their more secure counterparts (Weaver and de Waal 2000, 2003). Furthermore, the absence of reconciliation in cooperatively breeding callitrichids has been attributed to the secure nature of their relationships and the associated resistance to postconflict disturbance (Schaffner et al. 2005).

While there is at least some evidence to support the predicted relationships between each of the dimensions of relationship quality and reconciliation, it is likely that the inconsistency in results obtained is at least in part due to the variation in the measures used to assess the dimensions and the degree to which those measures are capable of accurately representing the appropriate quality of the relationship. While dyads can be classified into broad categories that are likely to represent different aspects of relationship quality, such as sex combination or kinship, such measures do not allow the variation within those categories to be detected. Furthermore, evidence for the attribution of relationship characteristics to such categories may be limited. Male philopatry and male/male alliances in wild chimpanzees (Goodall 1986, Mitani et al. 2000, Watts 2006), for example, are often used as a basis for classifying male/male dyads as having highly valuable relationships (de Waal 1986c, Watts 2006, Koski et al. 2007a). Evidence from a number of studies, however, suggests that male/male dyads may not always have more valuable relationships than other sex combinations (Baker and Smuts 1994, Fraser et al. 2008a, Lehmann and Boesch 2009) and, thus, may not always be expected to have the highest conciliatory tendencies (Fraser and Aureli 2008). The other common method for assessing relationship quality is to use levels of particular behaviors deemed to represent particular aspects of relationship quality. This method allows greater variability in the quality of relationships to be detected, but basing assessments of relationship quality on the level of a single behavioral variable may limit their accuracy. Furthermore, the assignment of behaviors to a particular relationship's quality dimensions is not always straightforward. Grooming, for example, may be used to represent relationship value (Majolo et al. 2005, Watts 2006) or relationship compatibility (Arnold and Whiten 2001, Koski et al. 2007b). These problems may be overcome by using principal component analysis to extract composite behavioral components from a number of behavioral variables, thus eliminating the need to base measures on relative frequencies of single behaviors or to make subjective judgments about which dimension of relationship quality each behavior or demographic category represents (Fraser et al. 2008a).

Silk (1996) offers an alternative view of the function of reconciliation. In chacma baboons, dominant former aggressors often approach their victims and grunt to them (Silk et al. 1996). Grunting facilitates friendly contact and the ability of former aggressors to handle the infants of their victims in particular. Silk (1996) views grunts as signals of "benign intent" that communicate a friendly attitude in the former aggressor and serve to reduce anxiety in the victim. She views reconciliation not as a mechanism for relationship repair but as facilitation of friendly interactions in the short term, leading to the fulfillment of immediate goals (Silk 1996, 1997). This distinction is a subtle one but rides on the lack of evidence that relationships are improved following reconciliation compared to preconflict periods, which Silk (1996) believes is the claim of the relationship repair hypothesis. However, as Cords and Aureli (1996) point out, the relationship repair hypothesis predicts a change in the relationship but only in so far as it restores the relationship to its preconflict state, which will have consequences for the maintenance of the relationship into the future but might not necessarily improve it. There is clearly evidence for such restoration.

A second line of research has focused on the short-term benefits of reconciliation in reducing stress levels in former opponents. As mentioned, the rate of self-directed behaviors increases in both aggressors and victims after conflicts and is reduced to baseline levels when reconciliation is achieved. Following the first studies of this phenomenon, Aureli and collaborators (Aureli et al. 1989, Aureli and van Schaik 1991) proposed the *uncertainty-reduction hypothesis*, which predicts that self-directed behaviors should decrease after reconciliation and interprets this reduction as a decrease in uncertainty in the former opponents about the status of the relationship. This hypothesis has been supported by all subsequent studies of cercopithecines that have investigated the effects of reconciliation on postconflict anxiety of either opponent (Aureli et al. 1989, Aureli and van Schaik 1991, Castles and Whiten 1998b, Das et al. 1998, Kutsukake and Castles 2001, Cooper et al. 2007), although studies of apes have reported mixed results (Arnold and Whiten 2001, Koski et al. 2007a, Fraser et al. 2010). A recent multivariate analysis on the determinants of postconflict anxiety in Japanese macaque aggressors and victims indicated that for both opponents the main predictor of postconflict scratching was the opponent's postconflict scratching, suggesting that victims and aggressors share the same uncertainty about the state of their relationship (Schino et al. 2007a). Conversely, Koski et al. (2007b) found that victims, but not aggressors, showed increased postconflict scratching in chimpanzees; and postconflict scratching in bonnet macaques was higher for victims than for aggressors (Cooper et al. 2007), suggesting that victims may experience greater postconflict anxiety, possibly as a result of increased risk of renewed aggression.

The most recent development integrates the valuable relationships hypothesis and the uncertainty-reduction hypothesis in order to take into account the variation and flexibility of reconciliation. The *integrated hypothesis* (Aureli 1997) suggests that postconflict anxiety is due to uncertainty about the future of the relationship between the former opponents and predicts that levels of postconflict anxiety mediate the occurrence of reconciliation. Higher levels of postconflict anxiety are expected after conflicts between valuable partners because of the potential loss of benefits. Thus, conflicts that disturb valuable relationships are expected to result in higher anxiety and then reconciliation. In support of this hypothesis, Aureli (1997) found that long-tailed macaque victims scratched more after conflicts with individuals with which they

shared strong affiliative bonds than with other group members; they also reconciled more with such individuals (Aureli et al. 1989). Moreover, Kutsukake and Castles (2001) found that conflicts among Japanese macaque kin resulted in higher rates of self-directed behavior and reconciliation than conflicts among nonkin. In both studies, the risk of renewed attacks was not associated with such relationship-dependent variation; thus, the higher anxiety levels were attributed to the disturbance of more valuable relationships. Further support for the integrated hypothesis is found in chimpanzees as victims of male/male conflicts were found to show both increased postconflict anxiety and a higher conciliatory tendency compared to victims of female/female conflicts, which are likely to have less valuable relationships (Koski et al. 2007b). Variation in postconflict anxiety may even mediate differential rates of reconciliation, as female/female opponents in bonnet macaques (*Macaca radiata*) show both increased levels of postconflict anxiety and higher conciliatory tendencies than other sex combinations (Cooper et al. 2007). Hamadryas baboon (*Papio hamadryas*) aggressors also exhibited higher levels of self-directed behavior and reconciliation after conflicts with high-quality partners (Romero et al. 2009). Such emotional mediation could serve as a proximate mechanism at the basis of the variation and flexibility of reconciliation depending on the quality of the relationship between former opponents (Aureli and Smucny 2000, Aureli et al. 2002). In support of this, female/female bonnet macaque opponents showed increased anxiety prior to reconciliation but not after unreconciled conflicts (Cooper et al. 2007).

CONCLUSIONS

To date, over 60 empirical studies have investigated postconflict reconciliation in nonhuman primates (Table 41.1). After demonstrating the phenomenon, many of them reported on various factors affecting reconciliation patterns (Table 41.3). Given the overall similarity in the methodology across studies, it has been possible to make meaningful inferences from comparisons of the available evidence (de Waal 1989b, 2000a; Thierry 1990a, 2000; Kappeler and van Schaik 1992; Cords and Killen 1998; Silk 1997; Aureli and de Waal 2000; Aureli et al. 2002; Judge 2003). In contrast, only a minority of studies have examined the function of reconciliation. This is certainly a shortcoming as reconciliation may function differently in different species and under different conditions. The next generation of studies on reconciliation should focus on functional aspects, checking first whether the aggressive conflict actually disturbed the social relationship between former opponents rather than simply assuming the need for reconciliation under any circumstance (Aureli et al. 2002).

Another aspect that should receive more attention is the patterning of behavioral exchanges that lead to reconciliation. It is well established that certain species, but not others, use explicit forms of reconciliation characterized by conspicuous affiliative patterns rarely displayed in other contexts. However, variation in the use of these patterns within the same species has been reported (Castles et al. 1996, Arnold and Whiten 2001, Fraser and Aureli 2008), and new studies need to focus on the reasons behind the variation. Furthermore, the role each opponent plays in achieving reconciliation is still poorly understood. Empirical data are needed to test the various predictions based on asymmetries between opponents in partner value and in postconflict negative consequences. Better empirical evidence for such testing requires more attention to subtle signals and vocalizations (Silk et al. 1996), but measuring initiative in reconciliation will be challenging because of the difficulty in recognizing subtle cues (Cords and Aureli 2000). Further research may use indicators of emotional responses, such as self-directed behavior, or attention to former opponents to examine how the opponents themselves perceive the postconflict situation and to infer the opponents' relative interest, motivation, and expected initiative in reconciliation (Aureli and Smucny 2000).

A critical aspect of the study of postconflict behavioral exchanges between former opponents is to establish whether reconciliation attempts are successful or not. Reconciliation is a dyadic phenomenon in which both partners need to cooperate (Aureli and Schaffner 2006). Only attempts that are accepted by the receiver are expected to achieve the beneficial effects implied in the term "reconciliation." No study on nonhuman primates has explicitly addressed this issue, probably due to the difficulty in determining the actual acceptance of a friendly overture (but see Ljungberg et al. 1999 for a child study). More effort should be put toward examining the responses of the receiver of the conciliatory attempt. For example, attempts that are reciprocated by the partner (e.g., mutual grooming) or that require direct participation from the partner (e.g., hold-bottom in stumptailed macaques, Figure 41.2) might be more effective in terms of restoring the relationship than attempts that are characterized only by unidirectional friendly contacts. It is also likely that the variation in effectiveness of behaviors used in reconciliation attempts is linked to the quality of the opponents' relationship. Indeed, relationship quality is likely to play a central role in mediating much of the variation we have observed in the patterns of postconflict behavior, and it is important therefore that reliable measures of each dimension of relationship quality are incorporated into future studies.

Our review of the various hypotheses put forward to explain the function and distribution of reconciliation shows that there is a general consensus in favor of the valuable relationships hypothesis. The uncertainty-reduction hypothesis has received even more support, but more research on the emotional responses of former aggressors is needed. These data would help in producing tests for the integrated hypothesis, which combines the valuable relationship hypothesis and the uncertainty-reduction hypothesis to explain the variation and flexibility of reconciliation.

Although more studies are needed to gather the detailed data to fully understand the phenomenon of reconciliation in nonhuman primates, much has been accomplished since the pioneering work of de Waal and van Roosmalen (1979). The wealth of knowledge accumulated so far can be used to predict patterns of postconflict behavior in other primate and nonprimate species, to facilitate more systematic investigation, and to develop a framework for the evolution of conflict resolution in animal societies (Aureli et al. 2002).

42

The Conundrum of Communication

Harold Gouzoules and Sarah Gouzoules

1. Does the study of nonhuman primate vocal communication provide insights into the evolution of human language? How do researchers look for precursors to human language in the communication systems of monkeys and apes?
2. Why are the issues of deception and honesty of special theoretical interest to researchers studying nonhuman primate communication?
3. A conundrum is a difficult, intricate problem with an elusive solution. Why might primate communication be considered a conundrum?

INTRODUCTION

Communication underpins social behavior: Without it, only the most basic forms of sociality are possible. Species that show complex social interactions are likely to have relatively rich communication repertoires (reviewed in Hauser 1996:644, Blumstein and Armitage 1997), and many primate taxa appear to follow this pattern. Important as it is, however, communication is not easily defined. Bradbury and Vehrencamp (1998:5), in an authoritative 882-page overview, dauntingly construe the book's entirety as a thorough definition of animal communication. Minimally, communication has taken place when a sender emits a signal that is detected by a receiver whose behavior might consequently change. Signals, which can take many and varied forms, have two sources: They can evolve over time (i.e., be the direct target of natural selection), or they can be acquired through learning and convention. Most researchers hold that signals convey information to receivers (Bradbury and Vehrencamp 1998:2), although definitions of what constitutes information vary (Dusenbery 1992). An ongoing issue to which we will return is the question of what goes on following signal detection by receivers: Do they decode or cognitively process information, or is this level of analysis anthropomorphic and unnecessary for nonhuman primate communication?

Primates rely on four major modalities (visual, auditory, olfactory, and tactile) used by most mammals for communication. Species vary in terms of relative emphasis of each form (Hauser 1996). To date, vocal communication has received the most attention from researchers; nonetheless, it is noteworthy that not even a single species' vocal repertoire has been adequately and fully documented, so much basic research still needs to be done. Early studies of nonhuman primate communication focused on issues such as cataloging the repertoires of various species (Altmann 1962, Chevalier-Skolnikoff 1974, Marler 1965, Rowell and Hinde 1962, Struhsaker 1967a), establishing the neurological bases of communication (Jürgens 1979, 1988; Müller-Preuss 1988; Ploog 1967), comparisons with human language using perceptual and psycholinguistic approaches (Snowdon 1982), the socioecological functions of communication systems (Gautier and Gautier 1977, Green 1975, Marler 1968, Struhsaker 1967a), the impact of habitat on signal structure (Waser and Waser 1977), and the capacity of apes to acquire artificial language (reviewed in Wallman 1992). While many of the studies were guided by evolutionary theory in a broad comparative sense (i.e., the view that nonhuman primates were especially appropriate animals to provide insights into human behavior and evolution), very few examined communication systems with respect to narrower, more specific evolutionary hypotheses. Exceptions include studies that examined the ecological and social influences on signal structure (Waser 1982b, Waser and Waser 1977, Waser and Brown 1984) and a few that searched for homologies among species (e.g., van Hooff 1967).

Space constraints prevent a comprehensive review of the large literature on primate communication published over the last 40 years. Such an inventory-taking summary is perhaps not the most instructive approach to take in order to assess the state of and future directions for research. Instead, we examine the literature selectively, with an emphasis on material relevant to current evolutionary issues, and evaluate the available data as they relate to these ideas. We emphasize vocal communication in discussing empirical research, but in many cases, the evolutionary theories to which these studies relate have relevance to other modalities as well. Certain areas that have been recently reviewed and thoroughly debated, for example, laboratory studies of the linguistic abilities of great apes (Wallman 1992, Savage-Rumbaugh et al. 1998, Savage-Rumbaugh and Fields 2000), are not considered here. Other areas neglected here, such as the ontogeny of primate communication (e.g., Seyfarth and Cheney 1997, Snowdon and Hausberger 1997, Snowdon 1999) and facial and gestural communication (e.g., Preuschoft and van Hooff 1995, 1997; van Hooff and Preuschoft 2003; Maestripieri 1999b), are also treated elsewhere.

THE ADAPTIVE NATURE OF SIGNAL STRUCTURE

Signal Structure and the Ecological Environment

Identification of the factors involved in the evolution of signal design and form is a critical area in animal communication research, including that on primates. Ecological and biological conditions that impact the physical nature of signals include the distance over which communication must be effective, the energetic costs of communicating, body size and structure, physical properties of the habitat, the timing of communication, the number and phylogenetic relationships of sympatric species, the presence and behavior of both conspecific competitors and those of other species, and costs imposed by predation (reviewed in Gerhardt 1983, Wiley 1983, Brown and Waser 1988, Brown et al. 1995, Hauser 1996, Bradbury and Vehrencamp 1998).

For some conditions, the impact on design structure has been more significant for signals used over relatively large distances. The diverse habitats of primates both shape and constrain the design of communication systems (Waser and Brown 1984, 1986; Brown and Gomez 1992). Habitats may vary in the level of background noise and in how signals produced in them are modified or distorted.

Sound transmission, for example, is frequency-dependent, and the nature of the habitat will affect the distance over which calls will be effective (Marten and Marler 1977; Marten et al. 1977; Piercy et al. 1977; Waser and Waser 1977; Michelsen 1978; Wiley and Richards 1978; Waser and Brown 1984, 1986). Different habitats can also distort important components of vocal signals, including the frequency, time, and amplitude domains (Brown and Waser 1988, Brown and Gomez 1992). Brown et al. (1995) noted that behavioral motor patterns that govern signal features such as duration, rise and fall time, bandwidth, rate and slope of frequency modulation, amplitude variation, or pulse repetition rate can be adjusted to minimize distortion associated with specific habitat types. They conducted field experiments to test Gish and Morton's (1981) *local adaptation hypothesis*, which predicts that the characteristics of the environment in which a signal is normally broadcast have imposed selection pressures on signals, resulting in designs that reduce distortion. The hypothesis was tested for calls broadcast in "appropriate" and "inappropriate" habitats of blue monkeys (*Cercopithecus mitis*) and gray-cheeked mangabeys (*Cercocebus albigena*), both of which are rain forest species, and two savanna species, vervet monkeys

(*Cercopithecus aethiops*) and yellow baboons (*Papio cynocephalus*). Exemplars of common call categories in each species' repertoire were broadcast and rerecorded at both short and long distances for comparisons between point of origin and transmitted signals that involved acoustical analyses of digitized calls. Overall, the results supported the idea that the general form of the entire repertoire of the vocalizations of forest monkeys had been selected to reduce the significance of distortion. In contrast, the intensity of selection for resistance to distortion posed by habitat acoustics appears to have been relaxed for savanna monkey calls. In the savanna, signalers and receivers can communicate visually, and it is possible that the availability of an unobstructed visual channel has meant that habitat acoustics have imposed fewer constraints on call structure.

Brown and colleagues (1995:958) point out that it is not known whether or not the physical distortion of calls that is measured in their analyses would map faithfully to perceived distortion by the monkeys. As is the case for many studies of primate vocal communication that rely on acoustical analyses, whether or not statistically significant acoustical differences (or the lack of differences) are biologically relevant to the animals themselves is often not known.

Signal Structure and "Motivational–Structural Rules"

In theory, signal structure for shorter-distance communication will be less influenced by selection imposed by the problems of production, effective transmission, and detection in natural habitats. Nonetheless, shorter-range signals are also thought to take forms that may be understood from an evolutionary perspective, with some ideas dating back to Darwin (1872). Following Darwin's notions, Morton (1977, 1982) outlined a set of "motivational–structural rules" for vocalizations after observing that harsh (broadband), lower-frequency vocalizations are widely used in hostile and aggressive contexts, while tonal, higher-frequency calls tend to be employed in appeasing or friendly situations. Morton (1977) noted that the ability to produce harsh sounds of lower frequency is linked to body size and that, in many species, size tends to determine the outcome of aggressive encounters. Owings and Morton (1998:111) have further suggested that more complicated use of vocalizations evolved when the direct and largely fixed sound symbolism of amphibian vocalizations was modified by motivational factors in warm-blooded vertebrates. Vocalizations came to reflect differences in motivation, with harsh, low-frequency sounds linked to aggressive states. Appeasing, submissive, or friendly vocalizations, either through selection based on the principle of antithesis or, perhaps, selection to mirror the higher-pitched sounds produced by infants, tend to be tonal and of higher frequency. Primate studies have provided general support for Morton's rules (Hauser 1993b, 1996). Hauser (1993b) looked at several hundred vocalizations from 43 primate species and found patterns consistent with

Morton's predictions. However, exceptions to the predictions of motivational–structural rules have been noted (Gouzoules and Gouzoules 2000). That some primate vocalizations in certain contexts deviate from Morton's motivational–structural rules is noteworthy but perhaps not unexpected if the rules are viewed as evolutionary "starting points" from which species might depart and deviate depending on new selective pressures. For example, species in which alliances and coalitions have a greater influence on dominance status than does body size (see Harcourt and de Waal 1992) might rely less on "expressive sound symbolism."

Signal Structure and "Honest Communication"

Zahavi (1975, 1977, 1980, 1987, 1993; Zahavi and Zahavi 1997) presented arguments for yet a different perspective on signal evolution, one that has received much attention. Zahavi contends that there should be selection for receivers to ignore signals that are not "honest" (i.e., accurate and reliable); receivers are therefore the "ultimate" determiners of how evolution shapes communication. He reasoned that honest signals could evolve only if they took forms that required considerable cost to produce, a condition that would result in ineffective communication if the sender could not bear that cost. These costly signals were referred to as "handicaps." More extreme expression of these expensive signals results in higher cost, and this expenditure is relatively greater for less fit or poorer-condition individuals. Costliness, according to Zahavi's theory, is essential to the evolution of honesty.

Gouzoules and Gouzoules (2002) reviewed the nonhuman primate literature with respect to Zahavi's ideas on costly signaling. Female sexual swellings, copulation calls, rhesus monkey grunts and girneys, and a number of other possible examples of costly signaling were considered. Game theory models and the empirical studies reviewed suggest that costly signals are not often likely to be the basis for honest communication in nonhuman primates. Costly signals are not evolutionarily obligatory in many contexts where communication occurs among related animals (Bergstrom and Lachmann 1997, 1998; Lachmann and Bergstrom 1998), a prominent feature of many nonhuman primate species. Another condition where inexpensive signals are possible is when sender and receiver both benefit from coordinated interactions (Farrell and Rabin 1996). Also, if individuals interact repeatedly and can use past interactions to assess the honesty of signals and modify future response to signals, then low-cost signals can evolve (Silk et al. 2000). Clearly, much of the communication nonhuman primates perform occurs under these circumstances.

There are, however, contexts in which costly signaling might evolve in primates. These are likely to include overtly competitive interactions where unrelated individuals have limited access to information about one another. Such situations would center around animals with little or no history of interaction and would involve a restricted set of signals.

An example might be the long vocal exchanges that male howler monkeys (*Alouatta* spp.) perform when they encounter other troops and solitary males. Sekulic (1982b) and Sekulic and Chivers (1986) suggested that the primary function of male howler roaring is to deter outside males from attempting to gain entrance into the caller's group. They proposed that a high roaring rate might be costly and indicate fighting ability to rivals. While some of the evidence on male howler roaring is consistent with this hypothesis, there has been no direct test examining whether male qualities affect roaring and whether the calls are effective in contests with challengers. Furthermore, da Cunha and Jalles-Filho (2007) suggest multiple and variable functions for howler roars and caution against narrow interpretations.

Another possible example with more empirical support is the *wahoo* call that male baboons produce in contexts such as dawn choruses, intergroup interactions, harassment of females, and fights with other males. Fischer and colleagues (2004) found that acoustic features of these calls potentially provide information about a male's health and physical state. *Wahoos* are often produced while males are exerting themselves by running or moving rapidly through trees and therefore, variation among callers in some acoustic features might reveal the signaler's stamina and competitive ability.

The "red-white-and-blue" display of male vervet monkeys might also represent a case of costly signaling. During dominance interactions, adult males perform a genital display that involves approaching a subordinate with tail erect, then exposing the red perianal region, the bright blue scrotum, and the white hair between the scrotum and perineum. Gerald (2001) conducted an experiment on the relationship between scrotal color and social status in adult males. She analyzed the social interactions of paired unfamiliar males, matched for size but differing in scrotal color. Difference in scrotal color predicted eventual dominance status: Males with darker scrota came to dominate those with paler ones. Experimental modification of scrotal color revealed that this signal alone did not signal status; males with scrota of similar color were more antagonistic than were those with dissimilar color. Scrotal color pales under stressful conditions (Gartlan and Brain 1968, Isbell 1995) and therefore could reflect male condition. Furthermore, the display, which involves exposure of the genitals to a potential aggressor, might qualify as risky and thus costly. Males transferring into new groups might rely more heavily on scrotal color as a way to evaluate rivals.

In another possible example, coloration might represent a costly signal in the striking natal coats that infants of some nonhuman primate species bear during the first months of life. Ross and Regan (2000) investigated possible relationships between conspicuous natal coats and factors that have been thought to be linked to its evolution—namely, allocare, predation risk, and type of mating system. They point out that a conspicuous natal coat (CNC) very likely enhances visibility to predators and may advertise an infant's vulnerability, lack of experience, and age, which would also be detrimental in species where infanticide occurs. Five hypotheses were tested with data from 82 species (28 of which have prominent natal coats). CNC was examined with respect to (*1*) the degree of allocare, (*2*) the level of competition among infants for allocare, (*3*) the risk of predation on infants, (*4*) terrestrial habitats, and (*5*) the type of social structure as measured by relative size of the testes (presumed to correlate with the level of competition among males for females). Ross and Regan (2000) concluded none of the hypotheses clearly accounted for the distribution of CNCs in anthropoid primates, so the evolution of this feature remains unexplained. While acknowledging the likely costs associated with CNCs, Ross and Regan did not consider the implications of costly signal theory in their hypotheses. Gouzoules and Gouzoules (2002) suggested that CNCs may be adaptive for the infants that bear them under conditions where the threat of predation is significant enough to require increased protection and attention from mothers and other group members, but not so great as to represent too extravagant a cost. Zahavi and Zahavi (1997:121) consider distinctive and prominent colors or markings of the infants of various nonprimate species as costly handicaps that can be interpreted as a form of "extortion" of parents by infants—in effect, an infant's gambit in a specific example of Trivers's (1974) theory of parent/offspring conflict. Thinking about primate CNCs in this light might result in several of the hypotheses examined by Ross and Regan becoming complementary rather than competing.

Primate olfactory communication might also be an area where costly honest signals are used (Kappeler 1998b). Gosling and Roberts (2001) note that while there has been little formal investigation of the costs associated with any olfactory signal, some forms, such as scent marking, are likely to be energetically expensive and, in some nonprimate mammals, have been shown to attract predators and parasites. Scent-marking behavior is very common in some primate species, such as *Lemur catta*, where its use varies by age, sex, and social status (Kappeler 1990). It is used at the boundaries of group home ranges and among group members. Costly signaling theory would predict that in order for scents deposited at the boundaries of territories to effectively deter intruders, they would have to be reliable indicators of a signaler's quality, a likely relationship due to the fact that such signals are linked to physiological conditions (Kappeler 1998b).

HOW PRIMATES DEAL WITH DECEPTION

Several major overviews of the empirical data on primate deception are available, as are various theoretical interpretations of deceptive behavior (de Waal 1986b; Byrne and Whiten 1988; Whiten and Byrne 1988; Cheney and Seyfarth 1990a,b, 1991; Hauser 1996, 1997b; Tomasello and Call 1997). Our concern here is with deception as a problem for honest communication: If nonhuman primates

only rarely rely on costly signaling to ensure honesty, what evidence is there that they instead use past interactions to assess the honesty of signals and modify future responses to signals, as predicted by the models discussed here (e.g., Silk et al. 2000)? We briefly summarize the kinds of deception with which nonhuman primates are confronted and then discuss two possible mechanisms that have been suggested as solutions to the problem of dishonesty: punishment (Clutton-Brock and Parker 1995) and skeptical responding (Moynihan 1982, Smith 1986).

Whiten and Byrne (1988) proposed the term "tactical deception" for short-term changes between honest and deceptive behaviors exhibited by an animal in the context of its social group. Key features of tactical deception are that the behavior itself must be part of the normal "honest" repertoire and the deceptive behavior must be rare, subtle, and used in different contexts from those in which it would ordinarily appear. Accounts of functionally deceptive behavior range from the withholding of information (alarm calls, Cheney and Seyfarth 1990a,b; food calls, Hauser 1992, 1997b; excitement calls, Goodall 1986, Boesch and Boesch 1989; body postures and facial signals, de Waal 1986b) to the use of signals in apparently inappropriate circumstances (alarm calls, Byrne and Whiten 1991, Cheney and Seyfarth 1990a,b; screams, Byrne and Whiten 1988).

Evidence from nonhuman primates for punishment in response to deception is scarce, despite theoretical arguments concerning its importance in animal societies. Experiments conducted on Cayo Santiago rhesus monkeys suggested the occurrence of punishment when information was withheld (Hauser 1992; Hauser and Marler 1993a,b). Individuals discovering food (coconut or monkey chow) provided by the researchers gave "food calls" in fewer than half of the trials, and the calling behavior of discoverers was the most significant predictor of aggression they received. Those that produced at least one food call received fewer aggressive acts, which were also of shorter duration. Hauser (1996:582) argued that this evidence suggests that rhesus monkeys withholding information about food are punished. While these data constitute the most formal test of punishment as a response to deception to date, stronger support would come from results showing that individuals that do call are more likely than silent monkeys not to receive any aggression at all. If callers receive any aggression, as they apparently do, other factors must also be involved. Additional evidence that punishment for withholding information was occurring would come from data revealing the consequences of having received aggression: Are animals that failed to call and were therefore targets of aggression more likely to call subsequently when they encounter food?

One context in which punishment in nonhuman primates has been described is parent/offspring conflict, especially during the weaning period (Altmann 1980). Begging signals that young of various species use to elicit feeding from parents have received considerable attention within the framework of honest communication (Hauser

1986, Kilner and Johnstone 1997). This is a particularly interesting context in which to examine nonhuman primate communication: The individuals involved are related and kinship, as previously noted, is one factor that should promote honest signaling (Bergstrom and Lachmann 1997, 1998; Lachmann and Bergstrom 1998). However, a potential conflict of interest between the infant and parent exists as well. Trivers (1974) suggested that weaning behavior in mammals resulted from parent/offspring conflict over the amount of investment (usually milk) that the offspring should receive. In nonhuman primates, however, it appears that conflicts are related to the timing of investment in the offspring. Conflicts occur as a female restricts her infant's suckling to those times that do not interfere with her ability to engage in other activities, including mating (e.g., baboons, Altmann 1980; Japanese macaques, Worlein et al. 1988; rhesus macaques, Gomendio 1991; gelada baboons, Barrett et al. 1995). At these times, negative behavior directed toward the infants tends to increase, as do maternal rejections and punishment. Concurrently, the infants tend to display regressive behavior and signs of distress, including vigorous tantrums. No studies to date have examined the exact coordination of these interactions with respect to the signals used by infants.

Skeptical responding is another mechanism for dealing with deception. Any communication system open to deception will be inherently unstable due to selection pressure on both the deceptive signaler and the receiver to become more proficient (Smith 1986, 1991). In the case of deceptive use of some signals, for example, alarm calls by competent adults, the false alarm is likely to be identical, or nearly so, in terms of its acoustic features to honest alarms. Under these conditions, selection on receivers might promote enhanced capacity to assess the reliability of the signaler, rather than the signal itself. The solution, Moynihan (1982:9) suggested, is that animals "must assess the information proffered by companions and rivals of their own species with some considerable degree of skepticism," an ability labeled "skeptical responding" by Smith (1986). Signalers that misinform sufficiently often may become devalued as sources of information.

Moynihan (1982) noted that skepticism, and any comparisons involved in assessing reliability, cost the receiver in terms of delays and energy expenditures. However, skepticism may be less costly if animals are not equally skeptical of all conspecific signals. An ability to recognize individuals (and their signals) and recall past encounters with them allows skepticism to be restricted to those animals most likely to be deceptive. Playback experiments with vocalizations have been used to assess how monkeys respond to inaccurate or false information (Cheney and Seyfarth 1990a,b; Gouzoules et al. 1996). Not surprisingly, monkeys habituate to these calls: They become less responsive over time. Interestingly, with respect to the skeptical responding hypothesis, their decreased response is specific to particular contexts and individuals. For example, a study of vervet monkeys (Cheney and Seyfarth 1988) used playback of false calls (intergroup *wrrs* and *chutters*, eagle and leopard alarm calls) in habituation experiments (calls repeated eight times at 20 min intervals). Receivers treated calls from a particular individual as unreliable in that referential context but not in others. In other words, skepticism regarding the caller was restricted to the context relevant for the call used in the experiment.

In another study of skeptical responding, Gouzoules et al. (1996) played tape-recorded alarm calls of high- and low-ranking female rhesus monkeys to their groups in a feeding context. We had previously observed what appeared to be false alarm calls from low-ranking monkeys: In each instance, while other monkeys ran and climbed into trees, the caller remained put and continued to eat. In our study, initial response to alarm playback was strong for the calls of both high- and low-ranking animals. By the eighth day, response had diminished greatly; however, habituation was faster and more extensive to the calls of low-ranking monkeys, those presumably with the most to gain through deception in this context.

PRIMATE COMMUNICATION, LINGUISTIC MODELS, AND THE EVOLUTION OF LANGUAGE

Exactly what studies of natural nonhuman primate communication and the ape language projects reveal about the evolution of human language has been debated extensively (e.g., Cheney and Seyfarth 1990a, Bickerton 1990, Wallman 1992, Hauser 1996, Deacon 1997, King 1999, Snowdon 2001) and will certainly not be resolved by us here. The continuing debates center on what qualifies as language (in ape linguistic capacity studies, e.g., Savage-Rumbaugh and Fields 2000) and what should be considered as evolutionary precursors to language (in the case of natural communication). There is no doubt, however, that over the last two decades many studies have pointed to the vital role vocal communication plays in primates' lives (reviewed in Cheney and Seyfarth 1990a, 2007; Hauser 1996; Seyfarth and Cheney 2003).

Some researchers, however, have expressed concerns that certain of these results have been overinterpreted, in part due to the use of what they consider to be anthropomorphic and anthropocentric linguistic models that have come to dominate animal communication research, in particular that on nonhuman primates (Owren and Rendall 1997, 2001, Rendall et al. 2009). The view of communication as a process of conveying information to other individuals has been questioned and alternative models and concepts have been suggested; in effect, the critics have advocated a major paradigm shift in the study of animal communication (Owren and Rendall 1997, 2001; Owings and Morton 1997, 1998; Thompson 1997). A rejection of linguistic models in nonhuman primate communication research effectively eliminates the search for precursors to language. In this section, we assess concerns about the "informational perspective" in animal communication (Owren and Rendall 1997, 2001; Owings and Morton 1998) and some proposed alternatives as they apply to examples of referential vocalizations in primate vocal communication. We begin by reviewing some of the conclusions and remaining uncertainties concerning the best-known example of calls with external referents, the alarm calls of vervet monkeys. We then briefly review another example of putative external reference, macaque agonistic screams. Screams are among the vocalizations that opponents of the informational perspective have targeted for alternative interpretations. Finally, we will come back to the issue of precursors to language in the communication systems of nonhuman primates and argue that recent pessimism about evidence is unwarranted.

Vervet Alarm Calls and External Reference: Evaluating the Evidence

What data have been used to document external reference in primate vocalizations and how do they meet the criteria proposed as necessary to demonstrate this property in animal vocal communication? Only after Seyfarth et al. (1980) focused attention on the alarm calls of vervet monkeys did studies of additional call classes in this and other primate species begin to explore the possibility of more widespread occurrence of externally referential signals. Early on, researchers overemphasized differences between the majority of calls, thought to reflect an individual's internal state or emotions, and a smaller subset of calls that were considered to have referents external to the signaler (e.g., Marler 1984, Gouzoules et al. 1985). This emphasis was not meant to suggest that internal and external referents were in any sense antithetical but, instead, to promote awareness and thinking about the existence of the latter. Recently, more care is taken to be clear that animal vocalizations are more likely to fall along a continuum, with signals at the extremes mostly reflecting the internal state of the sender at one end and referential signals, with relatively little emotional content, at the other (e.g., Marler et al. 1992). Over a similar period, the question of what evidence is sufficient for calls to be considered referential has also been approached in more complex and sophisticated ways (Marler et al. 1992, Macedonia and Evans 1993). Criteria center on both the calling behavior of the signaler and the effects of the vocalizations on conspecifics.

Evans (1997), for example, proposed that, with respect to evidence on calling behavior, referential signals should both be structurally discrete (i.e., belong to an acoustically distinct class) and have a degree of stimulus specificity, meaning that eliciting stimuli should be members of a distinct category. He reasoned that one would not expect the same class of referential signal to be generated in response to stimuli that are members of qualitatively different categories. Importantly, we should keep in mind that it is from the animal's perspective, and not the investigator's, that these

criteria should hold: What constitutes a structurally discrete class of calls for a researcher, who has examined selected acoustical variables and determined uniformity on the basis of statistical patterns, might not correspond to the animal's perception of the same vocalizations; relevant data for the animal's perspective are rare (e.g., Le Prell et al. 2002). The prerequisite that eliciting stimuli should be members of a distinct category requires investigators to have essentially complete knowledge about how the animals categorize their world; here, too, the data are unfortunately lacking and elusive, something to which we will return for further comment. The next step in demonstrating external reference for a vocalization has been to establish that calls are capable of eliciting an appropriate response in a receiver deprived of any additional real-time information about the putative referent that might be available from the external environment (Evans et al. 1993, Macedonia and Evans 1993, Evans 1997). (This leaves open the likely possibility that past experience with contextually specific information is available through memory and can influence response.) Evidence of this kind is most convincing for the alarm calls of vervet monkeys (summarized in Cheney and Seyfarth 1990a). Through field experiments in which tape recordings of the three types of alarm call were broadcast from hidden speakers, Seyfarth et al. (1980) determined that the vocalizations were sufficient to elicit adaptive, predator-specific escape responses. Evidence of this sort demonstrates what Marler et al. (1992) and Macedonia and Evans (1993) have labeled "functional reference." Nothing is revealed about the nature of the mental representation that the vervets are using. It is possible that upon hearing a leopard alarm a vervet conjures up a mental image of a large spotted cat armed with sharp teeth and claws and runs for the nearest tree. On the other hand, these playback data alone would also be consistent with the idea that the representation of the alarm call in an animal's auditory system simply might directly trigger motor neurons that control escape behavior (Lloyd 1989, Bickerton 1990). However, the playback data, combined with observations that the monkeys vary their responses to a specific alarm call depending on the setting in which it is heard, suggest that the simpler explanation is unlikely.

Typically, researchers decide what constitutes an appropriate response to an experimentally broadcast vocalization on the basis of contextual analyses or interpretations of the situations as they occur in nature. The alarm calls of vervet monkeys are excellent examples of calls that have demonstrably adaptive and functional responses associated with them. The nature of any stored representation linked to such vocalizations is not easily established through such an approach, however. The exact knowledge or information that either the signaler possesses or the receiver acquires by hearing the call is difficult to assess because this, too, requires the investigator to have a thorough understanding of how the species classifies its world (Cheney and Seyfarth 1990a). Seyfarth et al. (1980) were able to establish, through alarm call playback experiments, that vervet monkeys react differently to calls given in response to leopards and those in response to martial eagles; but such data reveal nothing about what features are critical for the monkeys to distinguish a leopard from an eagle or a leopard from a lion; in fact, such information would be next to impossible to obtain from animals under natural conditions. Thus, Cheney and Seyfarth (1990a:171) noted that, from such evidence, "the external referents of primate vocalizations are best defined in relation to the external referents of other calls within the animals' repertoire."

In additional experiments designed to test whether vervets are able to assess calls according to the things they denote, Cheney and Seyfarth (1988) obtained results supporting the existence of semanticity in what they argue is a stronger sense. Two sets of acoustically different vocalizations were used in a habituation/dishabituation procedure. One pair of calls included the leopard and eagle alarms, while the other pair was composed of two calls the monkeys use during encounters with neighboring groups of vervets. *Wrrs* and *chutters* are thought to be acoustically different calls. (Note that Evans' [1997:120] concern about whether the monkeys "partition the acoustic space" identically to the investigators is an important issue regarding the interpretation of the results of these experiments.) *Wrrs* are typically given when another group has been spotted in the distance, and *chutters* are used during aggressive contact with a different group. Baseline responses of 10 subjects to recordings of both the *chutters* and eagle alarms of one of the group's females were measured. The subjects were then presented with two separate series of eight repetitions of other recorded calls of that same female. In one series, the female's *wrrs* were played and in the other, her leopard alarms. Following the eighth presentation, the other member of the pair of calls was tested: The repeated *wrrs* were followed by a *chutter*, while after the leopard sequence, an eagle alarm was broadcast. Subjects rapidly habituated to both of the repeated calls. Habituation was transferred in the case of the subsequently played *chutter* but not to the eagle alarm that followed the leopard alarms. Cheney and Seyfarth concluded that these results were consistent with the view that the monkeys were assessing calls on the basis of their referents rather than their acoustical structure. They (Cheney and Seyfarth 1990a:154–155) suggest that transfer of habituation occurred in the case of the *wrr–chutter* pair because these calls had "roughly the same referent," while eagle and leopard alarms denoted significantly different things. Interestingly, when the *chutters* of one animal were tested in a similar fashion with a sequence of *wrrs* from a different monkey, habituation was not transferred to the *chutter*. Thus, call meaning and the identity of the signaler were important determinants of the subjects' responses (see also Owren 1990).

Wrrs and *chutters* seem to be members of a superordinate categorical class that links events concerning other vervet groups, though the precise nature of the similarities is not clear. That habituation does not transfer from an individual monkey's leopard alarm to its eagle alarm suggests that these two basic categories are not members of a superordinate class, even though leopard alarms are occasionally given to stooping eagles about to grasp a victim in its talons (Cheney and Seyfarth 1990a:107). Leopard alarms are sometimes also given to attacking snakes (Macedonia and Evans 1993). Macedonia and Evans (1993:183) have interpreted the relative lack of production specificity for leopard alarms as an indication that the calls are "less specific and hence might be placed further from the referential pole of a 'motivational-to-referential continuum' (Marler et al. 1992) than eagle and snake alarms." Another possible explanation for the more varied use of leopard calls might be that the call's referent is, in fact, primarily external but what is denoted is something shared by, but not restricted to, the spotted cat that is the most frequent elicitor of the vocalization.

In another series of experiments, Seyfarth and Cheney (1990) conducted habituation/dishabituation tests using vervet eagle and leopard alarms paired with the terrestrial predator and raptor alarms of the sympatric superb starling (*Spero superbus*) to test whether the monkeys assessed starling alarm calls on the basis of their meaning. If this were the case, vervet habituation to the raptor (or terrestrial predator) alarm call of one species should transfer to the corresponding call of the other. The starling's alarms, especially its terrestrial predator call, are given to a much wider range of species than are vervet alarm calls, although included among the eliciting species are those of concern to the vervets. Thus, Seyfarth and Cheney suspected that transfer of habituation might occur more readily with the raptor/eagle pairing.

Results of the starling/vervet alarm call playback experiments suggested that the monkeys respond to starling raptor calls as relatively precise signals that are similar in meaning to their own eagle alarms. Subjects transferred habituation from one species' alarm call to the other's when the calls broadcast had the same referent. Transfer of habituation also occurred for tests pairing starling terrestrial predator calls and vervet leopard alarms, though the response disparity between control and test calls was not as great as it was for the avian predator pairing. Seyfarth and Cheney (1990) interpret this difference as an indication that, while vervets treat starling terrestrial predator calls and their own leopard alarms as similar, the calls are judged to be less synonymous than starling raptor calls and their own eagle alarms.

Finally, one other interesting outcome of these experiments deserves attention because it reveals something about how the information from calls may vary in different situations. Recall that for the experiment involving habituation to *wrrs* transfer of habituation extended to *chutters* when the identity of the caller was the same for both vocalizations but this did not occur when one monkey's *wrr* was tested with another individual's *chutter*. Cheney and Seyfarth (1990a:201) note that subjects "did not, however, transfer their skepticism to a different individual's calls, suggesting that doubts about one individual's reliability did not generalize to include other group members." (Here, Cheney and Seyfarth use the habituation phase as a "cry wolf" experiment, where a monkey was made to appear unreliable to group members by repeating, through playbacks, unfounded warnings about danger.) In the case of the starling/vervet alarm call tests, however, habituation was transferred, even though members of two different species were involved. The monkeys appear not to attend to information about the identity of the caller when generalizing across species. Why the monkeys should show skepticism to the alarm call of a group member that has not shown itself to be unreliable following exposure to false starling alarms is not clear.

Zuberbühler (2003) employed similar playback techniques to extend the general conclusions of the vervet monkey research to other species in the genus *Cercopithecus*. Cheney and Seyfarth (2007) recently reviewed this work along with many other studies that have explored primates' assessment of meaning through the use of playback experiments, including their own elegant and extensive research on baboon vocalizations, and concluded that the evidence points to a "rich conceptual structure in which calls are linked both to objects and relations in the world and to other calls in the species' repertoire. When responding to calls, monkeys act as if they recognize individuals and have concepts like *leopard, eagle, close associate*, and so on" (p. 262). Nonetheless, the exact nature of the external referents for any primate call remains an elusive but important issue, to which we shall return after illustrating the challenges further with another set of primate vocalizations.

External Reference and Macaque Agonistic Screams: Navigating a Tangled Web

In many species of primates, screams are a prominent feature of aggressive interactions. Studies of agonistic aiding in free-ranging rhesus (Kaplan 1977, 1978) and numerous other species (reviewed in Walters and Seyfarth 1987, Cheney and Seyfarth 1990a:25–44) revealed a complex behavioral system in which alliance formation and recruitment is critical to both the development and maintenance of dominance status in social groups. Vocalizations were frequently noted to play a significant role in eliciting support from allies within the group. Successful recruitment might occur because of vocal recognition of the caller and information provided by the scream about the caller's emotional state or arousal level. Another hypothesis is that screams have external referents and make available more specific information about the agonistic event, information that might be of use to spatially distant allies in making decisions about intervention tactics. While these two hypotheses are not mutually exclusive, our research has focused primarily on this latter possibility.

Gouzoules et al. (1984) found that in rhesus monkeys on Cayo Santiago, victims of attack usually made use of one of five acoustically distinct screams. The major scream types were each associated with a particular class of opponent,

defined in terms of relative dominance rank and matrilineal relatedness to the signaler, as well as the severity of the aggression, differentiated as to whether or not physical contact had occurred. Noteworthy here are "noisy" and "arched" screams. Acoustically atonal noisy screams were most often given when the opponent was a higher-ranking group member and the incident involved contact aggression such as biting. Arched screams were typically given to lower-ranking opponents during noncontact aggression. They were highly unlikely to occur during serious aggression with higher-ranking opponents or during aggression from matrilineal kin. The different types of scream did not provide a good prediction of the caller's future behavior. Screams can be heard over considerable distances, and this led us to reason that selection had promoted the evolution of call features that enable messages to be conveyed to remote group members; given the energy costs and potential danger in attracting predators, such features would not be predicted for these calls if the primary receivers were opponents at close range (but see later section "Alternative Perspectives").

Playback experiments (Gouzoules et al. 1984) revealed that mothers responded most strongly to noisy screams, which are associated with contact aggression from higher-ranking opponents. Responses were measured in terms of latency to look to, and the duration of looking in the direction of, a concealed speaker. Such measures reflect, most directly, the ability of the calls to attract the attention of the subject. Naturally occurring screams have been observed to elicit approach and intervention by an ally, and for playback experiments, this might lead to habituation. Thus, short exemplars of calls were used to minimize the likelihood that a subject would approach the concealed speaker. Competing hypotheses based on the probable information made available by the vocalizations were used to predict subjects' interest in the broadcast calls.

Arched screams, associated with noncontact aggression from lower-ranking opponents, elicited the next strongest response from mothers. This finding was pivotal because serious aggression with a risk of injury (i.e., fights involving contact with the opponent) was extremely rare when these screams were produced, and it did not seem likely that caller fear levels should be high. Thus, the physical consequences for a victim of attack from a lower-ranking opponent were minimal, and the high level of response elicited from callers' mothers could not be explained by concern over protecting their offspring from injury. Instead, preservation of the matrilineal dominance rank seems a more probable basis for the allies' reactions to the arched screams.

Additional playback experiments were used to examine whether rhesus monkeys responded differently to calls of dependent offspring as opposed to those from comparably aged collateral kin (e.g., sisters, nieces, and cousins) (Gouzoules et al. 1986). The results supported our earlier conclusions that, for allies, the key information provided by the screams and influencing the likelihood of intervention concerns the nature of the aggression and opponent. While subjects responded more strongly to the noisy screams of their offspring compared to those of more distantly related juveniles, no such difference emerged from a comparison of playbacks involving arched screams from the same callers. When responses to only collateral kin were examined, the same patterns were found: Relatedness influenced response to noisy screams but not to arched screams. It may be that the intervention on behalf of a more distant relative whose opponent is higher-ranking (noisy screams) is more costly. We also found that the sex of the caller influenced the intensity of response during playback experiments with the noisy and arched screams of collaterally related juveniles. While there were no differences in response strength to noisy screams of male and female kin, arched screams from young female kin elicited significantly stronger responses than did those from male relatives. This sex difference may be related to male and female life history patterns: Pubescent males generally emigrate from the natal group, whereas females remain for life and are dependent on members of their matriline for rank preservation.

Interestingly, rhesus macaque screams would fail a strict application of the production criterion for external reference suggested by Macedonia and Evans (1993). Screams are often used in agonistic contexts other than those with which they are primarily associated. However, there are several sources of variation in the use of calls that can account for an apparent lack of specificity and yet allow the possibility of external reference. For example, our analyses of pig-tailed macaque screams (Gouzoules and Gouzoules 1989) suggested that juvenile monkeys under 3 years of age are far more likely than older individuals to make mistakes in the contextual use of calls but that improvement comes with experience and age (Gouzoules and Gouzoules 1995). Production specificity, then, is significantly greater for adult pig-tailed macaques.

Another factor that might also account for an apparent lack of production specificity is a failure by researchers to accurately describe or recognize the objects or events that are indexed by calls (Gouzoules and Gouzoules 1995, Gouzoules et al. 1998). What might initially appear to be different eliciting stimuli could well have common and relevant features to the monkeys. A further complicating issue is that many studies have approached the problem as though the animals classify items with a logic system having only two truth values, 1 or 0 (true or false). Investigators typically have followed the "essential first step" (Marler et al. 1992) for determining the meaning of a signal: Assess correspondence between vocalizations and putative referents. As we have seen, a subsequent stage in demonstrating external reference has been to show that a call is capable of eliciting, in a receiver denied additional information about the referent, a response appropriate to the referent. Both of these steps require a decision by the researcher about the nature of the category to which a particular referent might belong; often, as a starting point,

these presumptive categories have been very simple ones. For example, in our studies of macaque agonistic screams (e.g., Gouzoules and Gouzoules 1989), we classified aggressors in agonistic encounters as either of higher rank or of lower rank than the victim and the intensity of the conflict as involving contact aggression or noncontact aggression. Similarly, Cheney and Seyfarth (1982) differentiated vervet grunts as to whether the recipient was dominant or subordinate to the caller. Studies of primate food calls, which have also been proposed as referential, have categorized food as high-quality or low-quality, divisible or not, preferred or not (e.g., Dittus 1984; Benz 1993; Hauser and Marler 1993a,b; Hauser et al. 1993). Although many of these studies have established associations between different vocalizations and referents by employing this dichotomous approach, the mapping between call and referent has been rudimentary. This bisecting of the animal's world was done for logistical and analytical reasons, of course; and we are not suggesting that any investigator has viewed this as reflecting reality from the animal's perspective. Such bivalent logic is probably not the actual basis for the categorization of many things relevant to, and communicated about, by monkeys. For some aspects of the nonhuman primate world, a multivalued or fuzzy logic system, with more than two truth values, might provide a more appropriate and realistic way to conceptualize this problem (Gouzoules et al. 1995, 1998). A fuzzy set approach to categorizing the referents of primate calls—dominance of opponents, severity of aggression, aspects of discovered food, or even a predator—would hold that members of the set of objects or events are members to some degree, in contrast to classical set theory where membership is all or none. Thus, for example, an attack from the alpha male or alpha female of the group for a female rhesus monkey usually would have a truth value of close to 1 for the primate proposition "attack from a higher-ranking opponent." The truth value for such a proposition might be considerably less for an incident involving a different monkey that ranks closer to the victim on the ordinal form of dominance hierarchy typically constructed by observers to assess dominance relationships.

Predictions can be made as to how vocal behavior might be reflected by this more complex classification. If, for example, particular calls have as referents the relative dominance status of another group member, as has been argued for vervet monkey grunts (Cheney and Seyfarth 1982) and macaque screams (Gouzoules et al. 1984; Gouzoules and Gouzoules 1989, 1995), those calls should be more reliably used during interactions with individuals that are good members of that set, a prediction that stems from the human psychological literature on fuzzy categories (Mervis and Rosch 1981, Lakoff 1987) and has been supported empirically (Gouzoules et al. 1998). Thus, a strict application of the production criterion for external reference suggested by Macedonia and Evans (1993) would perhaps not be appropriate for call systems with fuzzy categories.

Alternative Perspectives: Criticisms of the Informational Perspective and a Reexamination of Macaque Screams

For some researchers, the issue of reference itself has become problematic (Owren 2000; Rendall et al. 2009). Owren and Rendall (1997, 2001) were the first to rebel against the view that animal calls provide information of any sort and have proposed an affect-conditioning (or "affect-induction") model of nonhuman primate vocal communication in which they argue that a framework involving roles for both Pavlovian and operant conditioning could account for important aspects of how calls function to influence the behavior of conspecific receivers. They, with Owings and Morton (1998), have moved away from the informational perspective common in communication studies, that signals make available information to which a receiver may respond. The arguments marshaled against the idea that animal displays and signals provide information to receivers take a number of forms. At a general evolutionary level, linguistically based notions about information are discounted as anthropocentric and anthropomorphic; they are viewed as being at odds with, or at least ignoring, two "conceptual pillars of causality in evolutionary analysis," form and function (Owren and Rendall 2001:60). An example of a "form and function" account of animal communication would be Morton's motivational–structural rules. The stance that animals provide information through communication is also suggested to be group-selectionist and not consistent with modern "selfish gene" (Darwinian) perspectives.

These arguments against reference and information, purportedly based in modern evolutionary theory, are not entirely sound, however. Evolutionary theory readily accommodates cooperation, reciprocity, altruism, and game theoretic approaches, all of which can help explain conditions under which selection would favor providing others with information (Bradbury and Vehrencamp 1998). The anti-information position also claims that the concept of reference implies that senders produce signals specifically to provide information to receivers. This depends on psychological processing in receivers that is conceptual in nature, paralleling mental activity that occurs with meaning extraction in humans. Again, the critics believe this to be anthropomorphic and unnecessary for adequate accounts of animal communication. However, with respect to such concerns about the cognitive bases for primate referential communication, it remains unclear what the minimum cognitive requirements for reference are. Some recent brain imaging studies have suggested that macaques have the neural circuitry to support conceptual knowledge (Gil-da-Costa et al. 2004, 2006). The affect-conditioning model, in contrast, holds that individuals vocalize to influence the behavior of others (not to inform them); conditioning determines how sound might have an impact.

Owren and Rendall's model attempts to account for the basic design and function of primate vocalizations by viewing the sounds as stimuli that senders use to elicit simple affective responses in receivers. They envisage roles for both unconditioned and conditioned responses in the manner by which behaviors are elicited by vocalizations, but with respect to monkey screams, which are singled out for attention, the former class of learning is suggested to be strongest. They reason that, due to the general properties of the mammalian auditory system, a call can elicit negative or positive reactions; negative reactions are thought to be associated with high amplitude and overall noisiness in calls. They suggest primate screams as an example of calls that produce unconditioned, negative responses in receivers, something they argue would discourage impending or ongoing aggression from a higher-ranking attacker. A subordinate monkey's most effective vocal solution to the problem of attack from a higher-ranking opponent would therefore be to "use sheer magnitude and raw features of acoustic signals for inducing aversive unconditioned responses in opponents" (Owren and Rendall 1997:330).

While a comprehensive evaluation of Owren and Rendall's model for understanding animal vocalizations is not possible within the constraints of this chapter, a few pertinent observations can be made (and others have been offered by Seyfarth and Cheney 2003, Zuberbühler 2003, Gouzoules 2005). One issue relevant to evaluating Owren and Rendall's hypothesis with respect to macaque screams pertains to the general acoustic structure of the calls. Although not a direct test of their ideas, some of our comparative research on macaque screams has yielded data we think are inconsistent with their predictions. For example, Gouzoules and Gouzoules (2000) compared screams of four species of macaques (rhesus monkey, *Macaca mulatta*; pig-tailed monkey, *M. nemestrina*; Sulawesi crested black macaque, *M. nigra*; stump-tailed macaque, *M. arctoides*) by looking at calls produced by victims of attack that involved contact aggression (pulling, pushing, slapping, grappling, and biting) from a higher-ranking opponent. For each macaque species, we digitized 100 screams from adult females and measured acoustic features for each call. Discriminant function analysis was used to determine whether the 400 vocalizations could be assigned to the correct caller species on the basis of their acoustic structure. Calls were assigned to the correct species at a significantly higher rate (93.5%) than expected by chance. These analyses revealed that each of the four macaque species used acoustically distinct screams in a shared context, receiving contact aggression from a higher-ranking opponent (Figure 42.1). Additional analyses as well as previous work on rhesus and pig-tailed macaques (Gouzoules et al. 1984, 1998; Gouzoules and Gouzoules 1989) suggested that each of the four species is physically capable of producing screams that are similar to those found in the repertoires of the other species, which is not surprising given the similarity among members of the genus in the mechanisms underlying phonation (Hauser 1996:479).

That the four macaque species we examined use distinctly different screams is a finding at odds with Owings and Morton's (1998:115) observation that animal screams

Figure 42.1 Screams of four species of macaques (genus *Macaca*): *A* pig-tailed macaque, *B* black crested macaque, *C* rhesus macaque, *D* stump-tailed macaque.

often lack species-typical attributes. Their argument was based on the fact that screams in many species are produced in response to attacks from predators that are usually of a different species. It might be adaptive for such calls to be of a similar acoustic structure because they serve to startle the predator, attract others (conspecifics or not) to mob the predator, or lure an even larger predator that might chase the first one away.

A significant difference, then, for macaque screams is that they are most prominent and common during intraspecific aggression, and suggestions for understanding their acoustic structure must relate to the nature of this largely within-group conflict. Although fear and pain are likely consequences for the victims of severe aggression in each macaque species, there appears to be no simple correspondence among the four species in the acoustic structure of their screams. While all four macaque species' screams are loud, the variation in call structure among species does not appear to be explained by Owren and Rendall's affect-conditioning model. Based on the starting assumption of their model, that general properties of the mammalian auditory system predispose the monkeys to react negatively to certain call properties, it is not clear why the four closely related macaque species we studied should have evolved such acoustically distinct screams if they serve as unconditioned stimuli that are naturally repugnant to aggressors.

Concluding Remarks on Language Precursors

Hauser et al. (2002:1576) reviewed the topic of language evolution and concluded the following:

Without pursuing the matter here, it appears that many of the elementary properties of words, including those that enter into referentiality, have only weak analogs or homologs in natural animal communication systems, with only slightly better evidence from the training studies with apes and dolphins. Future research must therefore provide either stronger support for the precursor position, or abandon this hypothesis.

These conclusions are very reminiscent of those expressed in the 1960s (e.g., Lancaster 1968). Yet, in their review, Hauser and colleagues acknowledge that the empirical research in this area over the past 35 years, some of which we have discussed in the present chapter, supports the following conclusions:

1. Monkeys produce certain acoustically distinctive calls in response to functionally significant contexts.
2. The acoustic morphology of the signal, which can be arbitrary in terms of its association with a particular context, is sufficient to enable listeners to respond appropriately without requiring any other contextual information.
3. The number of such signals in the repertoire is small; referents are restricted to objects and events experienced in the present, with no evidence of creative production of new sounds for new situations.
4. The acoustic morphology of the calls is more or less fixed, appearing early in development, with experience playing a larger role in refining the range of objects or events that elicit the calls.
5. While calling is under voluntary control, there is no evidence that it is intentional in the sense of taking into account what other individuals believe, want, or need to know.

In contrast to the conclusions Hauser and colleagues drew, these findings seem to be strong evidence in support of limited forms of referentiality in primate vocal communication that might indeed represent broad precursors to human symbolic capacities. The symbolic capacities of our own species are sometimes hard to define, recognize, and demonstrate, especially in young children (Tomasello 1999, 2003; Namy 2005), something that those working with nonhuman primates perhaps should keep in mind.

GENERAL CONCLUSIONS

Knowledge concerning nonhuman primate communication has increased dramatically since 1980. Repertoires used in communication are clearly much larger than previously imagined, and studies have revealed a radically different view of the role communication plays in nonhuman primates' lives. In the 1960s, Jane Lancaster (1968:442) summarized the prevalent view for vocal communication as follows:

field and laboratory workers have emphasized that vocalizations do not carry the major burden of meaning in most social interactions, but function instead either to call visual attention to the signaler or to emphasize or enhance the effect of visual or tactile signals. In other words, a blind monkey would be greatly handicapped in his social interactions whereas a deaf one would probably be able to function almost normally.

Our suspicion is that the complexity and richness of nonhuman primate communication is still underestimated. The degree to which multimodal communication regulates and modifies social interactions is, without doubt, an area that needs more exploration (Partan and Marler 1999, Partan 2002, Parr 2004). Brain-imaging technologies, combined with socially and evolutionarily relevant hypotheses, are providing valuable insights regarding enduring controversies such as those that surround the issue of language

precursors and the continuity of conceptual systems across species (e.g., Poremba et al. 2004; Gil-da-Costa et al. 2004, 2006). Linguistic models continue to be useful and should not be rejected outright as anthropomorphic, but their use should not dominate to the point of excluding other evolutionary approaches and hypotheses, especially those that, for nonhuman primates, have received relatively little attention, for example, signal evolution through receiver sensory bias or exploitation (Burley and Symanski 1998). Consistent with Bradbury and Vehrencamp's (1998:5) extended definition of animal communication, continued progress in our understanding of nonhuman primate communication will require a very broad range of empirical approaches and theoretical perspectives.

43

Tool Use and Cognition in Primates

Tatyana Humle and Dorothy M. Fragaszy

1. What is our current understanding of primates' perception of themselves and of the behavioral tendencies and mental states (e.g., knowledge and desires) of others?
2. What is known about primates' understanding of space and objects and their various spatial, causal, and featural interrelations?
3. What kinds of tools do nonhuman primates use, and how can we understand variations in tool use across individuals and species?

INTRODUCTION

Cognition, the act or process of knowing, is a broad concept employed by many different disciplines, ranging from psychology, ethology, linguistics, neuroscience, and anthropology to philosophy, computer science, and artificial intelligence. From a biological perspective, cognition supports behavioral adaptation to challenges arising in a variable social and physical environment. Studies of cognition concern attention, perception, memory, judgments, decision making, problem solving, planning, and reasoning. Although many different paradigms are used to study the broad domain of cognition in humans, few paradigms in fact have been applied to the study of cognition in nonhuman animals. Thus, we know far less about cognition in nonhuman species than in humans.

Studies of cognition in nonhuman animals have focused on virtually all orders but most prominently rodents, birds, cetaceans, and primates. Due to our close evolutionary proximity with nonhuman primates, this group has received during the course of the last century remarkable attention from researchers interested in cognition. Although studying the minds of nonhuman primates has yielded fascinating insights, it has also proved quite a challenging affair. Since we cannot ask nonverbal subjects why they do certain things, empirically oriented scientists have had to devise indirect means to explore cognition in nonhuman animals, as we have had to do for human infants.

It is not our intention to review the full range of findings on cognition in nonhuman primates as this would deserve a book in its own right; we refer the reader to existing reviews for this purpose (Tomasello and Call 1997, Washburn 2007).

Our aim in this chapter is thus threefold: (*1*) summarize some of the most recent topics, approaches, findings, and debates in primate cognition; (*2*) discuss how these topics relate to our understanding of tool use in primates and compare theoretical frameworks to the study of cognition and tool use in primates and other animals; and (*3*) briefly introduce the relatively new empirical approach to metacognition in primates.

COGNITION IN THE SOCIAL WORLD

Chapters 38–44 in this volume address in detail different aspects of social cognition in nonhuman primates related to their social knowledge and strategies. We thus focus here on three additional topics central to social cognition not addressed elsewhere: the concept of self, social attention, and theory of mind.

The Concept of Self

There has been much interest in the recognition of self in animals. Whereas in human children, observations of self-conscious emotions (such as embarrassment), self-referencing, and self-evaluation through language have been used to evaluate the concept of self, the mirror-test was the first method used to investigate the concept of self in nonhuman subjects (Gallup 1970). In the standard mirror-test, typically the subject's forehead or another visually imperceptible body part is marked in some way while asleep, distracted, or sedated. The subject is then allowed to see its reflection in a mirror. If the individual spontaneously acts on the mark, this self-directed behavior is taken as an indication that it recognizes the image in the mirror as its own reflection. Although studies in primates have yielded conflicting results, some individuals of all great ape species—chimpanzees (*Pan troglodytes*), bonobos (*Pan paniscus*), orangutans (*Pongo* spp.), and gorillas (*Gorilla* spp.)—show mirror self-recognition, while monkeys, prosimians, and lesser apes typically fail the standard mark-test (see Tomasello and Call 1997).

However, the story is not as straightforward as it seems. Some authors have argued that the methodological particularities of the traditional mark-test prevent reliable distinction between true-negative and false-negative results (e.g., Mitchell 1993, Heschl and Burkart 2006). Further, it remains puzzling that many species of monkeys are capable of effectively using mirrors instrumentally (e.g., to locate hidden food) while failing at the standard mark-test (Heschl and Burkart 2006). Some authors contend that the instrumental use of mirrors reflects self-recognition in some sense because it suggests the animal has an understanding of the details of the image in the mirror as reflecting its own position in space in relation to external features of the environment (e.g., Mitchell 1993). Some authors have also argued that success at mirror self-recognition tasks parallels capacities for facial imitation and facial pretense, to date observed among primates only in humans and apes (e.g., Lewis and

Brooks-Gunn 1979, Meltzoff 1990, Parker et al. 2005). Therefore, it has been suggested that the discrepancy in the results in monkeys between success on instrumental mirror tasks and failure in the mark-test may be due to the absence of kinesthetic–visual matching elicited by the spatiotemporal contingency between facial movements and their reflection in these species. The concept of self in nonhuman primates and other animals remains a field of study open for enquiry. Progress in this area requires the use of newer paradigms, such as response to real-time vs. delayed video images of one's own body movements (e.g., Anderson et al. 2009).

Social Attention: Joint Attention and Gaze Following

Social attention is a key feature of social cognition in humans. Studies of social attention in nonhuman primates may help us understand the evolution and development of theory of mind (the ability to attribute mental states, i.e., intents, desires, knowledge, beliefs, etc., to others) (Tomasello 1999). Since the mid-1990s, experimental psychologists have spent much time investigating the abilities of nonhuman primates for gaze following and joint attention, whereby individual A follows individual B's direction of attention toward a particular object (for review see Emery 2000). In naturalistic settings with conspecifics, Tomasello et al. (1998) found evidence for gaze following in Old World monkeys and chimpanzees. In addition, by employing a sophisticated telemetric gaze-tracking device, Shepherd and Platt (2008) recently reported evidence for gaze following in a prosimian primate (*Lemur catta*) in its natural social environment. Therefore, the ability to orient toward and follow the attention of other conspecifics appears to be quite widespread in primates.

Although studies in chimpanzees suggest that they follow the gaze direction of conspecifics or humans and that they respond to the attention shift of another by changing their angle of view to investigate the other's target of attention (e.g., Bräuer et al. 2005; Call et al. 1998; Krachun and Call 2008; Tomasello et al. 1999, 2001), it remains unclear whether nonhuman primates are capable of taking the perspective of others when shifting their attention by using gaze cues. Indeed, the ability to follow the gaze of another does not necessarily imply an understanding of what the other sees, knows, or intends to do. While some authors argue that nonhuman primates do not engage in shared attention (i.e., forming a referential triangle, which requires that both individual A and individual B have an understanding of the direction of the other's attention, thus checking that what the other is attending to is the same as what they are looking at [e.g., Tomonaga et al. 2004]), others contend otherwise (e.g., Krachun and Call 2008, Tomasello and Call 2009).

The Question of Theory of Mind

Premack and Woodruff (1978)'s seminal paper entitled "Does the Chimpanzee Have a Theory of Mind?" sparked a series of studies seeking evidence for "theory of mind" in nonhuman primates, especially in chimpanzees. This line of research has raised much controversy, and several authors argue that behavioral and contextual cues are sufficient to predict the future behavior of others without mentalistic understanding of others (Povinelli et al. 1998).

One paradigm used to test for theory of mind, developed by human developmental psychologists, is based upon recognizing false belief (Wimmer and Perner 1983). A sample experiment with children goes like this: *A child witnesses a scene in which a character, Maxi, puts chocolate in a drawer and leaves the room. While Maxi is away, his mother takes a bit of chocolate for cooking and then replaces the rest somewhere else and leaves. Then Maxi comes back, and the experimenter asks the child, "Where will Maxi look for the chocolate?"* In such a scenario, children with false-belief understanding will answer that Maxi will look in the drawer where he placed the chocolate as the child has an understanding that Maxi lacks knowledge of the chocolate's new location. This task has been modified to study false-belief understanding in nonhuman species, principally chimpanzees. Numerous studies have shown that chimpanzees fail on false-belief tasks where the active participants in the scenario are humans (Call and Tomasello 1999, Povinelli and Eddy 1996).

Researchers working with nonhuman primates therefore have designed new paradigms requiring subjects to compete over food when paired with a conspecific or a human experimenter, rather than requiring them to cooperate and/ or communicate with a human experimenter (e.g., Kaminski et al. 2004; Liebal et al. 2004a,b; Hare et al. 2006; Melis et al. 2006). Subordinate individuals were more likely to retrieve food when it was safer for them do so, as when they knew that the dominant individual they were paired with had no prior knowledge of its location. Additionally, dominant individuals, when presented with two food sources (one visible to and one hidden from view of the subordinate), most often first retrieved the visible reward before the one that was occluded from the subordinate's view (Hare et al. 2000, Bräuer et al. 2008) (Figure 43.1). These experiments show that chimpanzees can determine what others can see or not see, know or not know, and that they can potentially deceive others by concealing their actions when it is advantageous for them to do so.

However, an understanding of what others perceive or know does not necessarily imply an understanding of what others intend to do (Tomasello and Call in press). A series of studies requiring chimpanzees to discriminate between an intentional versus an accidental action (Call and Tomasello 1998), between a willing and an unwilling actor (Call et al. 2004), or to help another achieve his or her goal (Warneken and Tomasello 2006) suggests that chimpanzees do understand others' goal-directed intentions. These experiments could not, however, confidently rule out the simpler interpretation, that through the observation of others' actions, subjects could derive a set of "behavioral rules"

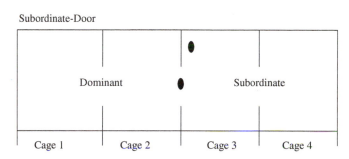

Figure 43.1 Example of experimental design presenting chimpanzees (*Pan troglodytes*) with a competitive food-retrieval situation aimed at testing the ability to understand what others know and do not know. Reproduced from Hare et al. (2000) with permission of the publisher.

not necessitating any understanding of others goals or intentions (Povinelli and Vonk 2006). In a series of experiments designed to refute the possibility of deriving sets of rules in response to behavior, the experimenter performed the same actions in both the control and test conditions but in differing contexts leading to different interpretations of what the experimenter was trying to do (Buttelmann et al. 2008a). This study provides further evidence for chimpanzees' abilities to understand others' goals and intentions.

The current consensus is that chimpanzees understand what others see, hear (Melis et al. 2006), and know (Vonk and Subiaul 2009), as well as others' intentions and goals, but not their beliefs (Kaminski et al. 2008, Call and Tomasello 2008, Tomasello and Call in press). Ongoing studies suggest a similar pattern in other great ape species (Tomasello and Call in press) and that rhesus monkeys (*Macaca mulatta*) understand something of the visual and auditory perception of others (see Flombaum and Santos 2005, Santos et al. 2006a). In contrast, at 3–4 years old children understand false beliefs (Kaminski et al. 2008). Thus, humans appear to differ from nonhuman primates in that from about 3 years of age they comprehend that others' actions are driven by their beliefs, while nonhuman primates are incapable of understanding unobserved causal forces, although they seem to understand something about the interrelationship between perception, goals, and actions.

COGNITION IN THE PHYSICAL WORLD

It is generally assumed that cognitive skills evolved in the context of finding and processing food, locating other resources (including mates), caring for young, and avoiding predators. All these challenges involve coping with varying circumstances in the physical world. Accordingly, many studies have focused on nonhuman primates' understanding of space and objects and their various spatial, causal, and featural interrelations (reviewed in Tomasello and Call 1997). Other challenges include identifying and quantifying foods and other entities, potentially via concept formation and categorization as well as numerical judgments. Tool use is a pertinent behavior embodying both physical and social aspects of cognition. The relevance of the study of tool use and recent advances in this domain in nonhuman primates, as well as other species, are addressed in the section on Tool use which follows this section. Here, we summarize recent findings in other areas of physical cognition.

Spatial Cognition

Experimental studies indicate that primates can move objects or icons through alleys or around detours (e.g., Menzel 2005, Fragaszy et al. 2003), infer the location of hidden objects based on their past association with specific landmarks, and follow the successive displacement of objects behind a series of barriers (e.g., Beran et al. 2005, Okamoto-Barth and Call 2008). Nonhuman primates may not use spatial configurations as adult humans do to solve these kinds of problems in the laboratory but, rather, may solve them using egocentric (body-based) spatial frames of reference (Gouteux et al. 2001, Poti et al. 2005, Sutton et al. 2000). Recent studies indicate that primates navigate large-scale spaces (e.g., traveling to different locations in their home ranges) efficiently and strategically (e.g., Janson 2007; Noser and Byrne 2007a,b; Golledge 1999), but precisely how they do so is not yet clear.

Concept Formation and Categorization

The ability to categorize elements or objects, including novel ones, is not unique to primates. The main research question has therefore focused on whether object categorization in nonhuman primates is based on abstract conceptual representation (e.g., recognizing that an object belongs to a category) or stimulus generalization based on certain perceptual features or cues (Chater and Heyes 1994, Fagot

and Berbet 2006). Fagot and Berbet (2006) describe how baboons (*Papio papio*) recognize objects primarily on the basis of local features (e.g., small squares displayed in such a way as to form a large circle), whereas adult humans recognize objects primarily on the basis of global features (e.g., the large circle as opposed to the small squares), a perceptual difference that these authors suggest has important consequences for how each species manages a variety of other tasks, such as perceptually grouping objects into categories. Leighty et al. (2008) show that adult chimpanzees can use information about object form and compositional structure from a two-dimensional image to guide their search in three dimensions, as can 4-year-old children (but not 3-year-old children), suggesting that there may be a quantitative, if not a qualitative, difference across primate taxa in the bias to use local features to classify visual stimuli.

However, experimental research on natural concepts has often been plagued by difficulties in teasing conceptual representation and stimulus generalization apart. Identifying the cues to which subjects attend when performing discrimination tasks, modifying salient features of stimuli, and investigating the types of errors performed during test trials have proved somewhat useful in circumventing this problem (D'Amato and Van Sant 1988, Jitsumori and Delius 2001). These methodological and analytical procedures do not, however, always help (Marsh and MacDonald 2008). The artificiality of the stimuli used, whether pictures of human faces, flowers, animals, conspecifics, objects, etc., renders conclusions about nonhuman primates' natural concepts rather problematic. Nevertheless, the consensus so far is that great apes and rhesus monkeys can engage in abstract conceptual mental representation, behaving in ways consistent with conceptual mediation (e.g., orangutans, Marsh and MacDonald 2008; chimpanzees, Tomonaga 1995, Murai et al. 2005; rhesus macaques, Washburn and Rumbaugh 1991). On a methodological note, the use of noninvasive eye-tracking devices is likely to improve our understanding of how nonhuman primates and other animals perceive familiar and novel items in their environment (e.g., Hall-Haro et al. 2008, Kano and Tomonaga 2009).

Relational concepts, such as identity, oddity, and sameness, have been widely investigated using a matching-to-sample testing paradigm (Figure 43.2). In this paradigm, the subject is first presented with a target sample stimulus and then required to select from two or more test stimuli. The subject first learns which items to select with a smallish set of stimuli, seen repeatedly. For example, during training trials, the subject may learn to attend to the color of a sample display and then to select an icon of the matching color from two displays. Once a high level of performance on the training set is reached, transfer trials follow, presenting the subject this time with a completely novel set of stimuli. For example, the subject may first learn to select an item matching the color of the sample (say, red). The question tested in the transfer trials is whether the subject will continue to match red with red when the stimuli are novel or match a sample of a novel color (say, indigo) with an indigo icon when both icons are novel. Maintenance of a high proportion of correct choices on the first appearance of novel sets of stimuli is assumed to reflect the use of a relational concept. Alternatively, if performance on the first trial drops to that seen during initial training, then it is inferred that the subject has not learned the relational concept. Many studies suggest that there are no major qualitative differences among primates in their abilities to form relational concepts whether using social or physical stimuli (see Tomasello and Call 1997), although there are findings suggesting that rhesus monkeys are challenged by some problems where the sample embodies a relation. More complicated designs are used to evaluate relational concepts. Flemming et al. (2007) showed that rhesus monkeys readily learned to label arrays of two, four, six, or eight stimulus pairs in a

computerized display as "same" or "different." However, the monkeys failed to solve a more complicated form of the problem, called "relational matching to sample," where they had to match a sample that was either two identical images ("same") or two differing images ("different") with one of two sets of images with the same property ("same" or "different") (Figure 43.3). On the other hand, Spinozzi et al. (2004) showed that capuchin monkeys (*Cebus apella*) could match the spatial relations "above" and "below." For example, if the sample displayed a square above a horizontal line, the monkey chose the display of an icon positioned above a line from two choices displayed. *Analogical reasoning* (recognizing similarities among object relations when the objects themselves are dissimilar) is a powerful kind of relational reasoning that supports effective problem solving. To date, chimpanzees (Thompson et al. 1997) and capuchin monkeys (Hoy Kennedy and Fragaszy 2008) have mastered analogical problems. However, the details of how relational

Figure 43.2 Orangutan subject, named Azy, performing a matching-to-sample task using a touch-screen computer. Symbols used can match any number of items (e.g., familiar foods or objects). Azy's performance on such matching-to-sample tasks is nearly 100% on each trial session. The configuration of the screen changes with every trial, in terms of both the number of symbols presented (ranging from 2 to 28) as well as the arrangement of the symbols in order to avoid the potential for memorizing and responding to specific locations (photo courtesy of R. Shumaker).

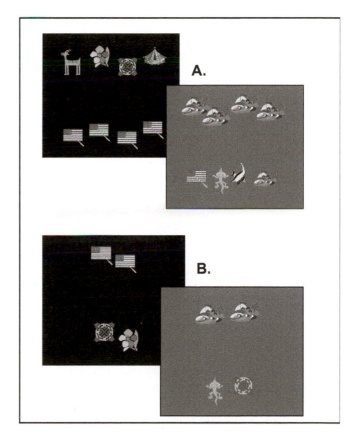

Figure 43.3 Representative displays used by Flemming et al. (2007) to study rhesus monkeys' (*Macaca mulatta*) relational reasoning. *Top*: The monkeys learned readily to indicate if the depicted images were "same" or "different" as a function of the background. For example, if the background was gray, the monkey should select the set of "same" objects; if the background was black, the monkey should select the "different" set. They did this for displays with different numbers of images (illustrated as *A* and *B*). *Bottom*: The monkeys did not master matching pairs of images depicting a particular relation (same or different) with other pairs depicting the equivalent relation. Reproduced from Flemming et al. (2007) with permission of the publisher.

(A)   (B)

Figure 43.4 A ring-tailed lemur (*Lemur catta*) works on a transitive inference problem presented on a touch screen (MacLean et al. 2008). A: The lemur starts the trial by touching the bottom portion of the screen. B: The lemur makes a choice by touching one of two images. (Photos courtesy of E. MacLean.)

problems are presented make a substantial difference in nonhuman primates' success at mastering them.

Many species of primates show evidence for transitive inference (if A > B and B > C, then A > C) (Vasconcelos 2008). Transitive inference underlies logical deductive reasoning, and it may be of high adaptive value in primate species with social organizations based on hierarchical relationships between individuals (e.g., Bovet and Washburn 2003, Cheney and Seyfarth 2007). Some authors have argued that transitive inference in animals has evolved in response to selective pressures related to feeding ecology and spatial representation (e.g., Davis 1992, Lazareva et al. 2000, Roberts and Phelps 1994, Terrace and McGonigle 1994). Others have suggested that social complexity has acted as an important selective force in the evolution of cognitive abilities that support transitive reasoning (e.g., prosimians, MacLean et al. 2008) (Figure 43.4).

Numerical Abilities and Quantity Judgments

The ability to understand and estimate quantities enables individuals to make effective choices and decisions in diverse circumstances, for example, during foraging (choosing between different sources of food with different quantities located at different travel distances) and during agonistic or predatory encounters (recognizing the number of allies, opponents, prey, or predators). There are different ways of making such estimates, and such abilities do not necessarily require a human-like concept of number or complex arithmetic skills.

Animals in diverse orders, including birds, rodents, and primates, demonstrate some understanding of numerosity (Brannon 2006). Evidence gathered so far in nonhuman primates—capuchin monkeys (*Cebus apella*), rhesus macaques, and chimpanzees—indicates that they share

a similar degree of competence in ordering small and large numerosities as represented by items in an array, Arabic numerals, or tokens (e.g., capuchins, Addessi et al. 2007, Beran et al. 2008, Judge et al. 2005; chimpanzees, Tomonaga and Matsuzawa 2000; rhesus monkeys, Brannon and Terrace 2000, Cantlon and Brannon 2006). Trained chimpanzees can exhibit impressive skills in ordering numerals in any combination from 1 through 9 in ascending order, an ability thus far not yet demonstrated in other nonhuman primate species (Biro and Matsuzawa 1999). In an experiment known as the "masking task," after touching the first numeral on the touch-sensitive monitor, all the other numerals are replaced by white squares and the chimpanzee is then required to remember which numerals appeared in which location and to touch them in ascending order (Kawai and Matsuzawa 2000) (Figure 43.5). By varying the time interval between first touch and numeral replacement with white squares, Inoue and Matsuzawa (2007) revealed that expert chimpanzees have a remarkable short-term photographic memory in this task, surpassing that of untrained humans.

Several studies indicate that nonhuman primates are sensitive to quantitative differences between sets of visible stimuli (Addessi et al. 2008; Beran 2007a, 2008; Muncer 1983; Suda and Call 2005; Tomonaga 2008). Individuals can maintain their performance even if the quantities are presented sequentially rather than simultaneously, thus indicating that they are able to remember each quantity (e.g., Beran and Beran 2004, Evans et al. 2008, Hauser et al. 2000). Research also suggests that monkeys, apes, and humans share a system for adding and subtracting quantities (e.g., chimpanzees, Beran 2001; rhesus macaques, Cantlon and Brannon 2007, Sulkowski and Hauser 2001) and that quantity judgments can also be modality-independent (e.g., rhesus macaques, Jordan et al. 2005; chimpanzees, Wilson

(A)   (B)

Figure 43.5 Ordering numerals from memory. *A*: A chimpanzee (*Pan troglodytes*) looks at a monitor displaying Arabic numerals 1–9. The chimpanzee is about to touch the first numeral. *B*: Upon touching the first numeral, the locations of the other numerals become squares (the numerals disappear and are replaced by squares). The chimpanzee then touches the square where the number 2 was previously displayed, etc., until all numbers are ordered based on memory of their initial location. (Photos courtesy of T. Matsuzawa.)

(A)   (B)

Figure 43.6 A female brown lemur (*Eulemur fulvus*) indicates her choice between two possible food items. In this experiment, when the lemur chooses the smaller quantity of raisins (*A*), she is then given the larger quantity for reward (*B*). When she chooses the larger quantity, she is given the smaller one. This is called a "reverse reward contingency" and is thought to tap an individual's ability to regulate a prepotent behavior (reach for the preferred item or greater quantity), which is considered an aspect of self-control. (Photos courtesy of E. Genty.)

et al. 2002). Overall, research on numerical cognition has revealed similar performance in monkeys and apes.

Inhibition, Delayed Gratification, and Anticipation

Problem solving, such as obtaining an item located beyond the individual's reach, may require inhibiting the prepotent response (e.g., to reach for the object immediately and directly) in favor of generating an alternative behavior, such as making a detour or using an object to extend the arm's reach. Several studies have focused on the process of inhibition using a paradigm developed by Boysen and Berntson (1995) known as the "reverse contingency task." In this task, two different quantities of food are offered to an individual and the individual receives the quantity that it did *not* indicate as its choice. Therefore, to receive the array

with the greatest quantity of food, this task demands that the subject reach toward the smaller quantity. So far, great apes and rhesus macaques have succeeded in this task in its original structure (e.g., Murray et al. 2005, Shumaker et al. 2001). Japanese monkeys (*Macaca fuscata*) initially failed (Silberberg and Fujita 1996) but succeeded when presented with a large-to-none contingency and when the position of the food array remained unchanged until there was a successful trial (Anderson et al. 2000). With similar modifications, similar results have been reported in squirrel monkeys (*Saimiri sciureus*) (Anderson et al. 2000), cotton-top tamarins (*Saguinus oedipus*) (Kralik 2005), white-crowned mangabeys (*Cercocebus torquatus lunulatus*) (Albiach-Serrano et al. 2007), and two species of lemur (*Eulemur fulvus* and *E. macaco*) (Genty et al. 2004) (Figure 43.6). In general, substituting food items with symbols or colored dishes, providing a symbolic representation of quantity, improved performance in all species tested (e.g., Boysen and Berntson 1995, Vlamings et al. 2006). Some studies have revealed long-term retention of the reverse contingency over several months, even up to a year, thus providing evidence for long-term memory for a specific solution to a choice problem (Albiach-Serrano et al. 2007, Anderson et al. 2004, Genty et al. 2004, Kralik 2005).

Using a series of five tasks assessing motor and temporal (delayed gratification) aspects of inhibition, Amici et al. (2008) showed that a sample of primate taxa with a fission–fusion social system (*Pan troglodytes*, *P. paniscus*, *Pongo pygmaeus*, *Ateles* spp.) exhibit greater inhibitory skills than a sample of taxa with more cohesive social dynamics (*Gorilla gorilla*, *Cebus* spp., and *Macaca fascicularis*). This finding indicates that inhibitory skills may have evolved from selection for monitoring conspecifics and adopting flexible strategies in changing social environments, rather than selection pressures imposed by feeding ecology (e.g., Rosati et al. 2007; Stevens et al. 2005a,b).

Inhibiting an action in favor of a future outcome suggests anticipation of the future, sometimes referred to as "mental time travel." Naqshbandi and Roberts (2006) showed that squirrel monkeys can alter their choice to a smaller quantity of dry food when that quantity is linked to earlier access to water (vs. a larger quantity of food and delayed access to water). This finding indicates that monkeys can anticipate a future need state (i.e., thirst), evidence for mental time travel.

TOOL USE AND COGNITION IN ACTION

Tool Use Illustrates Cognition in Action

Tool use in nonhumans holds an abiding interest for behavioral scientists because it is regarded as flexible, adaptive behavior that expands a species' ecological niche, it is relevant for understanding the exuberant elaboration of technology in ancestral humans, and it embodies reasoning and problem solving in a manner that is familiar to humans in their daily lives. It is this last aspect of tool use that we address here. Tool-using activities incorporate memory, attention, planning, spatial reasoning, fine motor coordination, and other cognitive processes in the service of problem solving.

Observers noted long ago that many species of nonhuman animals use "tools," in a folk notion of that term. To retain human uniqueness in this domain, manufacture, transport, re-use, and other characteristics were suggested to reflect "intelligent" (i.e., human-unique and, later, primate-unique) tool use (reviewed in Beck 1980). This view, in its simple form, became untenable with respect to humans versus nonhuman primates when Goodall (1964) reported that wild chimpanzees use modified wands of vegetation to "fish" for termites. The split between primates and other animals is also increasingly untenable as

we learn more about tool use in other orders, including birds (e.g., Hunt 1996; Hunt and Gray 2003, 2004a,b) and cetaceans (Mann et al. 2008). Tool use remains valuable to study because it affords an ecologically valid expression of cognition in action, not because it defines an exclusive club of "intelligent" creatures (Panger 2007).

Immediately, we encounter the problem of definition because "tool use" is not a natural category of behavior, like "foraging," and "tool" is not a natural category of object, like "rock" (Preston 1998). Thus, even deciding what counts as a tool and as tool use is debated. The commonly used definition, in its short form, is that a *tool* is an object used as a functional extension of the body to act on another object or surface to attain an immediate goal (Beck 1980). Thus, *tool use* is a form of problem solving characterized by a particular way of producing and managing spatial relations between objects. Some add the proscription that the tool must be a detached object (Parker and Gibson 1977) or that the actor must produce the relation between tool and another object or surface (Fragaszy and Cummins-Sebree 2005). However, we cannot treat "tool use" as a single category for purposes of understanding its relation to cognition. We need more conceptually precise formulations to explain why (and predict how) tool use varies in form and frequency across individuals and populations and why some forms of tool use are more common than others.

Here, we provide selected examples of tool use in nonhuman primates that illustrate typical situations. We direct the reader to recent reviews for fuller treatments (Fragaszy and Cummins-Sebree 2005, McGrew 2004, Tomasello and Call 1997, Visalberghi and Fragaszy 2006, Sanz and Morgan in press, Humle in press a).

Examples of Nonhuman Primates Using Tools

Captive individuals from various primate taxa have been reported to use objects as tools. Folivorous species (e.g., colobine monkeys and howler monkeys [*Alouatta* spp.]) are least represented. The most common examples concern using a long implement (stick or rake) to pull in food items otherwise out of reach (e.g., Beck 1980; long-tailed macaques, Westergaard 1988, Zuberbuehler et al. 1996; lemurs, Santos et al. 2005). Other common examples include using objects as percussors (hammers), boards or poles as climbing aids, bowls or cups as containers, paper or leaves as absorbents (to collect liquids), and probes to extract viscous foods (such as peanut butter or syrup) (e.g. macaques [*Macaca* spp.] and baboons [*Papio* spp.], Westergaard and Lindquist 1987; long-tailed macaques, Westergaard 1988; baboons, Westergaard 1992; Japanese macaques, Machida 1990; capuchins, reviewed in Fragaszy et al. 2004a). Captive individuals rarely have an opportunity to manufacture tools, but manufacture by subtraction (removing sticks from a branch and leaves from a stick, breaking large sticks) and by accumulation (combining paper and leaves into a spongy mass) has been observed in captive apes and monkeys (e.g., Westergaard and Fragaszy 1987, Pouydebat et al. 2005).

Fewer species of primates have been seen to use tools in natural environments. This may reflect differential opportunities to see uncommon behaviors or true differences in the rate of performance. The most frequent and most diverse observations relate to feeding and come from four species—the bearded capuchin monkey (*Cebus libidinosus*), the Sumatran orangutan (*Pongo pygmaeus*), the chimpanzee (*Pan troglodytes*), and the long-tailed macaque (*Macaca fascicularis*). Sumatran orangutans are known to manufacture and/or use twigs, sticks, or branches to extract insects, their products (e.g., honey), or water from tree holes or seeds from the protected *Neesia* fruit (Fox et al. 1999; van Schaik et al. 2003b). Long-tailed macaques have been observed using stones as tools on Piak Nam Yai Island, Thailand (Malaivijitnond et al. 2007, Gumert et al. 2009). Gumert et al. report that *M. f. aurea* use stone hammers to crack open five different types of shelled marine animals (oysters, crabs, barnacles, clam-like bivalves, and sea snails) and at least one species of nut (Figure 43.7). Upon examination of the stones at the sites of use, Gumert et al. established that the monkeys use stones of different sizes and shapes for different purposes. They chip oysters with the sharp point of relatively smaller stones and crack hard-shelled objects with larger, rounder stones. Visalberghi et al. (2007) previously used a similar method to evaluate the hammer stones and anvils used by wild bearded capuchin monkeys.

We now know that tool use is geographically widespread in bearded capuchins (Ottoni and Izar 2008), as it is in chimpanzees (Whiten et al. 2001). The most common form of tool use described for wild capuchins is using stones to percuss foods, usually nuts or seeds, on an anvil surface (usually a stone) (Figure 43.8). Bearded capuchin monkeys in some regions of Brazil crack very tough palm nuts with large stones (Fragraszy et al. 2004a). Field experiments have shown that these monkeys select stones by weight and composition to transport for use as hammer stones (Visalberghi et al. 2009). At another site, wild bearded capuchins use sets of sharp and rounded stones to scrape earth and then to extract invertebrates or underground plant storage organs (such as tubers). At the same site, the monkeys use sticks to probe crevices in tree limbs for invertebrate prey or possibly honey and to flush lizards from rock crevices, which they then capture and eat (Mannu and Ottoni 2009) (see Figure 43.9).

Tool use by wild chimpanzees has been observed for many decades at several long-term research sites. Overviews of observations from the long-term research projects in Gombe, Tanzania (Goodall 1986, McGrew 1992); Bossou, Guinea, (Matsuzawa et al. 2006); and Taï, Côte d'Ivoire, (Boesch and Boesch-Achermann 2000) are available. Nevertheless, our knowledge of tool use in chimpanzees continues to grow as these primates are observed in other regions. For example, Morgan and Abwe (2006) reported nut cracking in chimpanzees on the eastern side of the Sassandra River, thus refuting the idea that only chimpanzees west of that river engaged in this behavior (Boesch et al. 1994, McGrew et al. 1997). A common theme in several recent reports about tool use in wild chimpanzees concerns the structure of their activities, motivated by an interest in understanding the social processes involved in tool-use acquisition (see Chapter 44) and the complexity of the skill (see in particular Sanz and Morgan in press).

Figure 43.7 Long-tailed macaques (*Macaca fascicularis aurea*) on Piak Nam Yai Island in Thailand use stone hammers to crack different types of shelled marine animals (oysters, crabs, barnacles, clam-like bivalves, and sea snails) and at least one species of nut as well as axe stones to chip oysters off rock surfaces (Gumert et al. 2009). In this photo, the adult female monkey in the foreground uses an axe hammer to chip oysters, which cover the entire rock. A subadult male is behind her, using the point of the stone to chip the oysters. Patterns of usage leave clear scars on the stones so that researchers can later determine what area of the stone was being used to strike the food item. (Photo courtesy of M. D. Gumert.)

(A)

(B)

(C)

Figure 43.8 A wild adult female capuchin monkey (*Cebus libidinosus*) in the Boa Vista research site of EthoCebus, Piauí, Brazil, cracking a palm nut using a hammer stone and a log as an anvil. *A*: She transports the stone and nut to the anvil. *B*: She places the nut into a pit on the anvil and strikes it with the hammer stone. *C*: She eats the kernel. If she has time, she will crack the second nut, which she has also placed on the anvil. (© EthoCebus Research Group, photos courtesy of Barth Wright.)

Cognitive Models of Tool Use

Field studies have presented evidence of complexity in several dimensions of tool use by nonhuman primates, and these provide a window onto the cognitive processes that might support these actions. One aspect of complexity is the use by the same individuals of diverse techniques to acquire the same kind of prey. In Bossou, individual chimpanzees use sticks of different lengths and different techniques to capture army ants (*Dorylus* spp.) (Humle and Matsuzawa 2002; Humle 2006, in press b). Another aspect of complexity is the sequential use by an individual of more than one object directed at a single goal (i.e., a tool set) (Sugiyama 1997). At the Goualougo site located in the Nouabalé-Ndoki National Park, Republic of Congo, chimpanzees use sturdy sticks to breach subterranean termite nests, then use thinner probes, one end of which they "fringe" in advance of use, to gather termites (Sanz and Morgan 2007, in press). They also use two sizes of sticks to collect honey (Figure 43.10). The use of two different shapes of stones to excavate and then remove objects from the soil by capuchin monkeys (Mannu and Ottoni 2009) is another example of such tool sets. Yet another dimension of complexity is the number of steps involved in preparing and using the tool. Pruetz and Bertolani (2007) report that chimpanzees in Senegal probe with sticks into holes to capture trapped vertebrate prey. The chimpanzees prepare the sticks used for this activity by trimming off leaves and projecting branchlets and trimming the end, producing sequences of up to four actions prior to using the tool (see Figure 43.11). These recent findings expand our knowledge of the range and diversity of tool use by wild nonhuman primates in important ways. They indicate, for example, preparation of a tool in advance of use, the use two or more tools in sequence, and planning in selection and transport.

Researchers explore cognitive aspects of tool use from different perspectives, in keeping with the diversity of these behaviors. Some experimental work has focused on the skilled actions with the tool object. Bril et al. (2009) describe adjustments in striking actions made by a captive chimpanzee using objects of different weights to crack nuts. Liu et al. (2009) describe the impressively athletic actions of wild capuchins standing bipedally to lift proportionally very heavy stones to shoulder height and to strike them with precision on nuts on an anvil surface. This line of research promises insights into the dimensions of skill in actions with objects, a critical aspect of technology (see Roux and Bril 2005 for an introduction).

A second line of experimental work concerns whether nonhuman primates recognize objects that can be effective at retrieving food. It seems that many species, including many that have not been seen to use sticks as tools in natural settings, choose to pull food in with an object that is long and rigid enough, rather than objects that are less appropriate for the task at hand (e.g., Hauser 1997a, Santos et al. 2006b; see Hauser and Santos 2007 for review). This characteristic appears to reflect primate-general perceptual

Figure 43.9 A wild subadult male capuchin monkey (*Cebus libidinosus*) in Serra de Capivara National Park, Brazil, uses a stick to retrieve honey. (Photos by Tiago Falótico. Reprinted from Ottoni and Mannu [2009] with permission of the publisher.)

biases. Some of these tasks do not involve tool use (as the actor merely pulls a preplaced stick); but in some cases, the subjects place the stick themselves to use it (e.g., lemurs, Santos et al. 2005; capuchins, Cummins-Sebree and Fragaszy 2005), and these cases qualify as tool use.

Another body of work concerns the extent to which tool use by nonhuman species demonstrates *causal reasoning*, defined as "knowledge about interactions between objects" (Call in press). Some studies adopt a choice paradigm, presenting either a choice of objects, choice of a substrate to pull a food over (one side with a gap), or choice of a side of an apparatus to insert a stick in a "trick" apparatus (e.g., Buttelman et al. 2008c, Povinelli 2000, Visalberghi and Limongelli 1994) (Figure 43.12). Interpreting findings from these studies is problematic, however, because procedural differences not linked to the experimental question appear to affect performance substantively; that is, processes other than, or in addition to, causal reasoning drive choice behavior in these tasks. This

appears to be the case in humans as well as in nonhumans (Silva et al. 2005, 2008). Call (in press), after reviewing a large number of studies meant to assess causal knowledge as evident in tool selection in apes from several points of view, eloquently summarizes the current impasse. He suggests a "composite hypothesis," that a variety of cognitive processes contribute to behavior in tasks that experimenters design to test just one (causal reasoning). We agree with Call's view that causal reasoning can be studied more effectively with other kinds of problems (such as inference) rather than tool-choice studies.

Another approach to understanding cognitive aspects of tool use concerns the organization of action sequences and the incorporation of objects into these actions (e.g., Fragaszy and Cummins-Sebree 2005, Greenfield 1991, Matsuzawa 1996, Sanz and Morgan (in press). Greenfield (1991) and Matsuzawa (1996) suggested similar hierarchical models that describe a tool-using problem in terms of either the number of different actions or the number of object

(A)

(B)

Figure 43.10 A chimpanzee (*Pan troglodytes*) uses a tool set to extract honey from a hive of stingless bees (Meliponini) on the underside of a large branch in the forest canopy in the Nouabalé-Ndoki National Park, Republic of Congo. The chimpanzee first uses a pounding club to open the entrance of the beehive (*A*). She then repeatedly pounds the hive entrance with this large branch before manufacturing a second pounding tool. Between pounding bouts, she modifies the length and shape of the second pounding tool by pulling on either end of the tool with her hands while using her foot to break it in the middle section. After opening the hive entrance, she uses a dipping stick to extract honey from the hive (*B*). (Photos courtesy of M. Nichols and the National Geographic Society [all rights reserved]. Reprinted from Sanz and Morgan [2007], with permission from the publisher and the National Geographic Society.)

combinations involved in its solution and the fixed order in which these actions must be performed or these objects must be combined. Matsuzawa's "tree structure analysis" model, which focuses on the number of objects used and their combinatorial relationships, is shown in Figure 43.13. In this model, a direct action on an object is listed as level 0; in level 1, an object is used in some way as an intermediary between the body and the goal object, such as when using a wand to dip for army ants. In level 2, using the example of nut cracking, the nut and the anvil stone are connected at one node and the hammer stone is linked to these (joined) elements. A level 2 tool-use behavior therefore involves the appropriate combination of three objects. Combinations occurring later in the sequence are shown as higher nodes.

The sequential hierarchical models of Greenfield (1991) and Matsuzawa (1996) delineate two important features of the actions in tool use: the number of spatial relations produced by the actor and the order in which they occur. Thus, they provide a principled basis to evaluate different forms of tool use (such as cracking nuts, fishing for termites, or using a stick to lever open a fruit), and in this way they support comparison across ages and species in a way that representational theories do not. Analyses of tool-using actions by nonhuman primates using Matsuzawa's

tree analysis structure suggest that most forms of tool use are level 1, involving one object acting directly on the goal (as in using a probe to gather termites). Such a hierarchical model of reasoning applies equally well to language and to social interactions as to tool use, making it an integrative explanation for a variety of features of behavior (Greenfield 1991, Matsuzawa 2001b).

Fragaszy and Cummins-Sebree (2005; see also Visalberghi and Fragaszy 2006; Fragaszy et al. 2004b) proposed a spatiotemporal (Sp-Tm) model of spatial reasoning to analyze goal-directed actions with objects and applied the model to tool use. This model addresses the relations among actions, objects, and surfaces embodied in using a tool that, from the perspective of ecological theory (Gibson and Pick 2000, Gibson 1979, Lockman 2000), have a substantive impact on the problem. From an ecological perspective, the process of using a tool unfolds in time and in space, through actions performed by the body and in accord with the physical context of action (e.g., the nature of the objects used and the supporting surfaces). The Sp-Tm model, like the hierarchical models, takes into account the number of spatial relationships that must be managed in a given problem. The Sp-Tm model additionally takes into account the temporal order of relationships with each other (sequential or concurrent) and properties of each relationship: (*1*) specificity (with respect to position, orientation, or location), (*2*) relation to the body, and (*3*) duration over which it must be maintained through action (see Table 43.1). These relationships are assessed from the actor's point of view, in which the tool is treated as an extension of the body. This

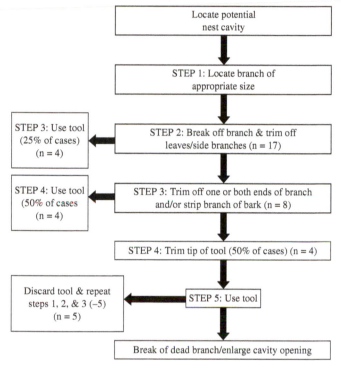

Figure 43.11 The number of steps involved in preparing and using the tool is one dimension of complexity in the use of tools. This flow diagram illustrates the different steps Fongoli chimpanzees in Senegal use while manufacturing a hunting tool. (Reprinted from Pruetz and Bertolani [2007] with permission of the publisher.)

model has to date been used retrospectively to organize a scattered literature about tool use (Fragaszy and Cummins-Sebree 2005), but key hypotheses are readily testable—for example, that managing concurrent relations is more challenging than managing sequential relations. Kuroshima and colleagues (2008) have shown that chimpanzees and capuchin monkeys are more challenged than young children 2 years old or older when they must manage two concurrent spatial relations to place an object into a matching groove, as predicted by the Sp-Tm theory.

A third approach examines the sequence of actions, comparing the frequency of particular transitions to a random sequence of the same elements and identifying recursive elements (i.e., repeated sequences nested within a longer sequence; e.g., Tonooka 2001, Byrne et al. 2001). Using this approach, Sanz and Morgan (in press) determined that tool-using behaviors performed by Goualougo chimpanzees ranged from more standardized (sponging for water, fishing for termites at elevated nests) to more variable (gathering honey) (Figure 43.10). Using leaves to sponge for water has a simple structure, as evaluated by several methods; and it is thus not surprising that it is commonly observed in chimpanzees across their range and appears earlier in ontogeny than other forms of tool use (Sanz and Morgan in press). Using two objects and performing different actions with each one (e.g., sticks to perforate a termite nest and then to collect termites), a more variable action set, is much rarer.

In sum, researchers approach cognitive aspects of tool use from diverse perspectives. In general, tool use by nonhuman primates, compared to tool use by humans older than 2 years, is structurally and kinematically simple. For example, we have not observed nonhuman primates spontaneously using two objects concurrently, as humans use a knife and fork. Nevertheless, descriptions of behavior by wild individuals indicate that monkeys and apes use

Figure 43.12 A tufted capuchin monkey (*Cebus apella*) using a stick to push a food item out of a clear tube. If she pushes the food into the center of the tube, it will fall into a "trap." This task was developed by E. Visalberghi and colleagues in the early 1990s, and since then variations of it have been presented to several species of nonhuman primates and to children in an effort to study causal reasoning. (Photo courtesy of E. Visalberghi.)

Level 1 Tool Use   Level 2 Tool Use   Level 3 Tool Use

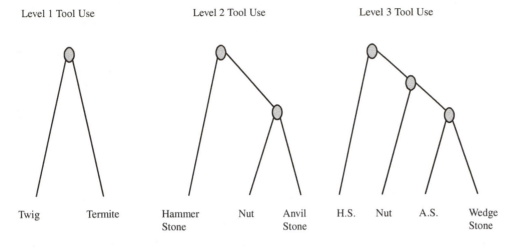

Figure 43.13 Tree structure analysis model developed by T. Matsuzawa to describe tool use as a nested hierarchy of actions with objects. H.S., hammer stone; A.S., anvil stone. (Reproduced from Matsuzawa [2001] with permission of the publisher.)

Twig   Termite   Hammer   Nut   Anvil   H.S.   Nut   A.S.   Wedge
                    Stone   Stone   Stone

Table 43.1 The Spatiotemporal Model: Properties of Spatial Relations Produced, Used, or Embodied in Action

PROPERTY	VARIANTS
Number of Relations	Variable (1, 2, 3, etc.)
Specificity	Permissive vs Specific
Temporal order of production	Sequential vs Concurrent
Relation to body	Direct vs Indirect
Temporal nature of control	Static[1] vs Dynamic[2]

Note: For each property, the alternative ends of the spectrum of possible variants are given.

[1] Static = unchanging. A static relation may vary along a temporal scale from Momentary (as in placing an object and releasing it) or b) Continuing in a constant manner (as in continued even pressure to hold an object in place against gravity). Momentary is easier to manage than Continuing.

[2] Dynamic = varying over time, as in tracking a moving object with a pointer.

Source: As proposed by Fragaszy and Cummins-Sebree 2005.

different natural objects (stones, plant material, shells, etc.) as tools for diverse purposes, that they anticipate their use, and that, when well-practiced, they act skillfully with these objects. Empirical studies have shown that the physical context of the problem makes a difference in how monkeys and apes approach a problem in ways that a purely representational model (e.g., causal reasoning) cannot predict. Understanding why this is so and how the individual manages the spatial relations and sequences of actions with objects while using an object as a tool, and in other kinds of problems involving objects, remain challenges for behavioral science.

METACOGNITIVE ABILITIES IN NONHUMAN PRIMATES

Recent research has investigated whether nonhuman animals, especially nonhuman primates, are able to monitor

and control their own cognitive activities, an ability known as *metacognition* (Nelson and Narens 1990, 1994; see Smith, J. D., et al. 2003, Kornell 2009 for reviews). One testing paradigm, the uncertainty-monitoring paradigm, used in these studies involves manipulating the degree of certainty with which the subject is able to generate the correct choice. The subject is typically given three kinds of response choices: (*1*) correct response, yielding an immediate reward; (*2*) wrong response, yielding no food reward and often a time-out; (*3*) uncertainty response ("escape" response), providing an option for the subject to quit the current trial and eventually receive a reward after a time delay or to obtain a lesser reward (e.g., smaller quantity of food than if the correct answer is given). Studies using this kind of paradigm have found that rats, dolphins, rhesus monkeys, and orangutans choose the uncertain response when their uncertainty is high (e.g., rats, Foote and Crystal 2007; dolphins, Browne 2004; rhesus macaques, Hampton et al. 2004, Kornell et al. 2007, Washburn et al. 2006; orangutans, Suda-King 2008). Other studies have used a search paradigm, where the subject chooses to search (or not) several potential hiding places when it cannot see where a food has been hidden (e.g., capuchins, Basile et al. 2008; rhesus macaques, Hampton et al. 2004; chimpanzees, Krachun and Call 2008). It remains unclear whether nonhuman animals truly evaluate their knowledge (certainty and uncertainty judgments) and adopt strategies to supplement what they do not know. Some authors who have critically examined these studies suggest that simpler mechanisms may underlie the response strategy of the subjects (e.g., Carruthers 2008). In order to convince the skeptics, experimental psychologists must develop more robust paradigms to test for metacognitive abilities in nonhuman animals (e.g., Smith et al. 2006).

CONCLUSION AND PERSPECTIVES

Primate taxa largely share fundamental aspects of cognition, in accord with shared phylogeny. Nevertheless, diversity of cognitive processes across primate taxa should be expected, in accord with evolutionary radiation into diverse ecological niches. Close examination of the relationships among social systems, ecology, and cognition across the primate order is likely to provide invaluable insights into the evolution of cognition. We have not yet examined cognition across taxa with appropriate resolution to identify many such differences, and this is one of the important tasks facing primatologists studying cognition. Some recent studies provide examples of where we might look. For example, as reviewed here, a few studies have presented subjects of different species with identical tasks in both the social and physical domains, with the goal of evaluating the association of ecological and social variables with disparities in performance (e.g., Amici et al. 2008, Herrmann et al. 2007). This kind of experimental work is challenging because the researchers must ensure that individuals of each species face comparable tasks, which means that the tasks must be adjusted for species-specific characteristics of motivation, action biases, etc. Moreover, we have limited opportunities to conduct studies with more than one species at a time. An alternative approach to comparing cognition across species is to carry out a meta-analysis of published data, selecting studies providing (presumably) comparable paradigms and data sets (e.g., Deaner and van Schaik 2001). This approach provides a broad comparison across many species but is inevitably constrained by the limited number and range of comparable data sets available for analysis and by the distance of the researchers from the primary data (i.e., methodological details important for comparative purposes are often unclear in the original publications). Convergent evidence from diverse approaches is our best hope for identifying differences across taxa and for understanding these differences in ecological and evolutionary contexts.

A second important task facing primatologists is to understand within-species variation in cognitive skills. Here, a developmental perspective seems promising (for a general introduction, see Munakata and Johnson 2006). How does experience during development shape cognition? How does experience affect acquisition of species-normal social and technological skills? Recent studies suggest that individual variation in skills can be large and that social circumstances can play a large role in the development of skills (e.g., Humle and Snowdon 2008; Humle et al. in press; van Schaik et al. 2003b). The developmental approach is powerful, and developmental studies provide important linkages between primatology and other life sciences. Experimental studies that examine cognitive development in relation to neural development (e.g., Weed et al. 2008) promise important insights and link primate studies with developmental neuroscience. On the other hand, longitudinal designs are difficult to implement, particularly in field settings. Once again, convergent approaches may be our best solution.

Finally, development of coherent theoretical approaches to the study of tool use that link to other areas of inquiry within primatology, and more widely with other life sciences, promise to invigorate this area.

One big question primatologists want to answer is why humans have distinctive cognitive skills not displayed by other primates, for example, language, complex physical reasoning, and advanced abilities to infer "the mental states of others as causes of behavior" (Herrmann et al. 2007:1365). Much remains to be learned through a fusion of approaches and disciplines, only a sample of which have been described here. Advances in the fields of cognitive neuroscience and artificial intelligence, for example, are likely to provide important insights and to stimulate novel hypotheses, approaches, and paradigms.

44

Social Learning in Monkeys and Apes
Cultural Animals?

Christine A. Caldwell and Andrew Whiten

1. Do monkeys and apes show evidence of imitative abilities?
2. What is the evidence for cultural variation between natural populations of monkeys and apes?
3. What are the best methods for testing for imitation and culture in nonhuman primates?

INTRODUCTION

Claims of cultural behaviors in nonhuman primates are becoming ever more common, based on both observational field data and experimental approaches. However, skepticism is still expressed about whether the behavioral variation observed in wild primates is maintained through social learning (Galef 1992; Heyes 1993; Laland and Galef 2009; Laland and Hoppitt 2003; Laland and Janik 2006; Tomasello 1990, 1994). Why is such skepticism still widespread in the face of an ever-increasing number of empirical studies on the topic? We address this fundamental question of whether or not there is evidence for what we might label "culture" in apes and monkeys. To do this, we shall review and evaluate several of the more prevalent approaches which have been applied to the investigation of culture in nonhuman primates.

Prior to examining the evidence, however, it is worth emphasizing the significance of the question of whether there is evidence of culture in nonhuman species. To begin with, culture is one of the key traits (alongside others such as language and tool use) that have been proposed, at one time or another, to be unique to humans. To describe an animal as possessing culture challenges traditional preconceptions about humankind's position in relation to the rest of the animal kingdom.

Culture has also fascinated scientists in its role as a nongenetic evolutionary process (Blackmore 1999, Bonner 1980, Mesoudi et al. 2004). Cultural transmission offers another, faster-operating method by which traits can spread, which has to date been much less intensely studied than genetic transmission.

These considerations make the study of culture in any animal important, but there are additional reasons that primates have been the focus of an unusually high proportion of such research. One is simply that some of the earliest, and certainly the most widely cited, examples of apparently cultural behaviors in nonhumans came from primates. Sweet potato washing by Japanese macaques, *Macaca fuscata* (Itani and Nishimura 1973, Kawai 1965, Kawamura 1959), is the classic example. The macaques on Japan's Koshima Islet were regularly provisioned with sweet potatoes, and one particular individual, Imo, developed a tendency to wash her potatoes prior to consumption. Over several years, this behavior spread until most of the rest of the group were also regularly washing their potatoes. Shortly afterward, Jane Goodall (1973) suggested that many of the behaviors exhibited by chimpanzees, *Pan troglodytes*, at Gombe in Tanzania were cultural variants, clearly distinct from behaviors exhibited at other field sites. Further, and more subtle, variants have been documented as observations have continued to the present time. Whiten et al. (1999) noted that some of the differences between communities represented alternative versions of otherwise similar patterns. Chimpanzees from the Taï Forest in the Ivory Coast use a short stick to fish ants from their nests, whereas those from Gombe use a much longer stick and a more efficient bimanual technique to achieve the same goal. Through such observations, primatologists have understandably developed a long-standing tradition of their own: a fascination with social learning.

Primates may also be particularly likely, in comparison with other animals, to exhibit culture. We humans are undeniably cultural, and it seems plausible that our closest evolutionary relatives would be the most likely to show some semblance of culture—the question being, to what extent? This close relationship to humans also makes primates particularly good research subjects in terms of providing insight into the evolutionary history of human culture. Monkeys and apes also share with humans several traits which may be associated with a propensity for cultural transmission. A long life span and lengthy period of infancy with high levels of parental investment mean that an ability to learn socially could be particularly valuable in these animals (Roper 1986). Indeed, these features are sometimes proposed to have evolved alongside cultural tendencies because, once a capacity for social learning exists, the fitness of offspring can be increased by maximizing the opportunities for learning from elder relatives (Barrett-Lennard et al. 2001).

Understanding how wild primates develop their full behavioral repertoire may also prove crucial to their protection in an increasingly hostile world. Around half of our 250 primate species are considered to be of conservation concern by the World Conservation Union (IUCN), and recent assessments suggest that over 30% of all primates should be considered seriously threatened with extinction. Although a large number of successful captive breeding programs exist for many endangered species, reintroductions are notoriously unsuccessful (Beck 1994). Captive-bred individuals appear to lack the survival skills possessed by their wild-born counterparts (Beck et al. 2002, Stoinski et al. 2003). If, under natural conditions, a large proportion of learning is dependent on social interaction with skilled conspecifics, then this needs to be taken into account when planning reintroduction programs (Beck 1995, Box 1991, Custance et al. 2002).

DEFINING CULTURE

"Culture" has been defined in many different ways, ranging from the very broad and inclusive to the extremely narrow and restrictive (Kroeber and Kluckhorn 1963, McGrew 1998b, Rendell and Whitehead 2001, Whiten et al. 1999). Analyzing the possible points of reference for culture is beyond the scope of the present chapter (see Whiten et al. 2003 for a fuller discussion), but those interested in culture from a nonhuman perspective typically agree on two central criteria (Boesch and Tomasello 1998, Fragaszy and Perry 2003, Laland and Hoppitt 2003, McGrew 1998b, Russell and Russell 1990, Whiten et al. 1999): Cultural behaviors are those which are (1) specific to members of a group ("group" is used intentionally loosely here as cultural variation could occur on any social scale) and (2) transmitted via some form of social learning. We have adopted this as our definition for this chapter.

The first criterion, group-specific behavior, can be relatively easy to identify. The examples of the potato washing of the Koshima macaques and the chimpanzee tool-use behaviors at Taï and Gombe illustrate admirably that primate behavior can show both intriguing uniformity within groups as well as striking variability between them. However, the second criterion is more elusive: How do we know for sure that behaviors have been socially learned? Galef (1990) has pointed out that geographic variation could arise from a number of possible sources: (1) heritable differences that influence behavior, (2) environmental differences

between the habitats of the populations which result in systematic differences in the reinforcement individuals receive for engaging in various behaviors, and (3) differences in the behavior of population members that influence the behavioral development of new recruits (i.e., some form of social transmission). Teasing these alternative explanations apart can prove quite problematic.

We shall describe and evaluate three principal approaches that have been taken to the question of whether apparent local traditions really represent evidence for culture in monkeys and apes. These are depicted in Figure 44.1. As shown in this illustration, the first two approaches, systematic field site surveys and analysis of diffusion patterns, look directly at the naturally expressed behaviors that have been proposed to be socially learned in the wild. The final approach, experimental research, involves direct manipulation of an individual's experiences. Such research allows for definite conclusions to be made about whether or not social learning is involved. However, this type of research may not necessarily reveal much about natural behaviors. Experimental research can focus on natural behaviors, involving transfer of individuals between social groups or alternative environmental conditions; but these manipulations are extremely difficult. Experiments with primates tend only to *simulate* the type of behaviors that might be socially learned in the wild.

SYSTEMATIC SURVEYS OF MULTIPLE FIELD SITES

Ape Field-Site Surveys

One of the most influential approaches to date for addressing the issue of primate culture is the systematic field-site survey. Such approaches, which compile data from multiple social groups of the same species, have existed in limited form for many years. For example, Goodall's (1973) classic chapter on chimpanzee cultural behavior includes tables detailing the presence or absence of certain behaviors identified at Gombe (and proposed by Goodall to be socially learned) at several other locations in Africa. McGrew (1992) and Boesch (1996b) include ever-larger tables, comparing behaviors observed at different chimpanzee field sites. This method of looking for behavioral differences between groups has been referred to in the literature by a variety of different labels, including "group contrast" (Fragaszy and Perry 2003), "regional contrast" (Dewar 2003), and the "method of elimination" (van Schaik 2003). Such between-group variation provides perhaps the most intuitively compelling evidence of culture in any animal, although, as mentioned, variation could arise from factors other than social learning. The validity of this method therefore relies on the accumulation of data in order to, as far as possible, assess the likelihood of either genetic or environmental influence.

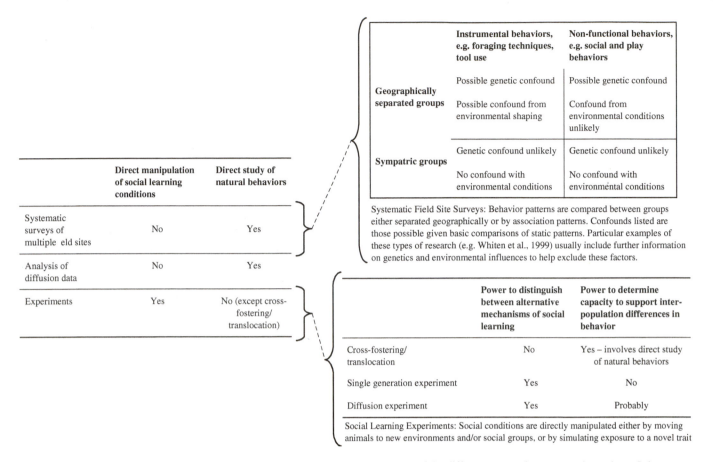

	Direct manipulation of social learning conditions	Direct study of natural behaviors
Systematic surveys of multiple eld sites	No	Yes
Analysis of diffusion data	No	Yes
Experiments	Yes	No (except cross-fostering/ translocation)

	Instrumental behaviors, e.g. foraging techniques, tool use	Non-functional behaviors, e.g. social and play behaviors
Geographically separated groups	Possible genetic confound	Possible genetic confound
	Possible confound from environmental shaping	Confound from environmental conditions unlikely
Sympatric groups	Genetic confound unlikely	Genetic confound unlikely
	No confound with environmental conditions	No confound with environmental conditions

Systematic Field Site Surveys: Behavior patterns are compared between groups either separated geographically or by association patterns. Confounds listed are those possible given basic comparisons of static patterns. Particular examples of these types of research (e.g. Whiten et al., 1999) usually include further information on genetics and environmental influences to help exclude these factors.

	Power to distinguish between alternative mechanisms of social learning	Power to determine capacity to support inter-population differences in behavior
Cross-fostering/ translocation	No	Yes – involves direct study of natural behaviors
Single generation experiment	Yes	No
Diffusion experiment	Yes	Probably

Social Learning Experiments: Social conditions are directly manipulated either by moving animals to new environments and/or social groups, or by simulating exposure to a novel trait

Figure 44.1 Where is the evidence for culture in primates? A diagrammatic summary of the different approaches to assessing culture. It is difficult to simultaneously study primate social learning and natural behaviors in a direct fashion as the types of experiments that permit these sorts of manipulation are difficult to carry out in these species. Evidence for culture in primates therefore remains, up to a point, circumstantial, based on the existence of interpopulation differences in behavior that cannot be readily attributable to environmental or genetic factors and on the results of experiments that show primates can learn from others.

Capitalizing on the potential power of this method, therefore, more recent attempts have taken a more systematic approach. Whiten and collaborators (1999, 2001) pooled data from the seven longest-running chimpanzee field sites. The first stage of this process was to collate a list of proposed cultural variants, that is, behaviors the researchers had observed at their field site that they suspected might be transmitted socially. The leaders of each of the field sites involved were then asked to summarize their knowledge about their chimpanzees' performance of each of these behaviors. For each of the behaviors, there were six possible options: *customary* (observed in all or most able-bodied members of at least one age/sex class), *habitual* (not customary but observed repeatedly in several individuals, consistent with some degree of social transmission), *present* (neither customary nor habitual but clearly identified), *absent* (never recorded), *ecological explanation* (never recorded but absence was explicable because of a local ecological feature), and *unknown* (never recorded but possibly due to an inadequacy of relevant observational opportunities).

Of the 65 original candidate behaviors, only those that were found to be either customary or habitual at one site at least and absent from at least one other (without an

apparent ecological explanation) joined the list of behaviors proposed to be cultural variants. A total of 39 different behaviors remained following this process (see Figure 44.2 for a graphical summary of behaviors found at different field sites). These spanned a range of contexts, some social, such as grooming and courtship behaviors, and some instrumental, such as foraging behaviors and food-processing skills. It remains possible that in these geographically separated populations inconspicuous (possibly extremely subtle) ecological differences could be responsible for the observed variation in some of the food-related behaviors. However, for social behaviors, like hand-clasp grooming, this is much less of a concern. To assess the hypothesis that the regional differences could instead have a genetic basis, Lycett et al. (2007, 2009) subjected the data of Whiten et al. (2001) to what are referred to as 'phylogenetic' or 'cladistic' numerical analyses, originally developed to identify branching evolutionary patterns among biological taxa but more recently applied to human cultural evolution. Lycett et al. showed that the cultural evolutionary tree structure that best fits the chimpanzee data is not consistent with a genetic explanation, and they concluded that it supports a cultural hypothesis.

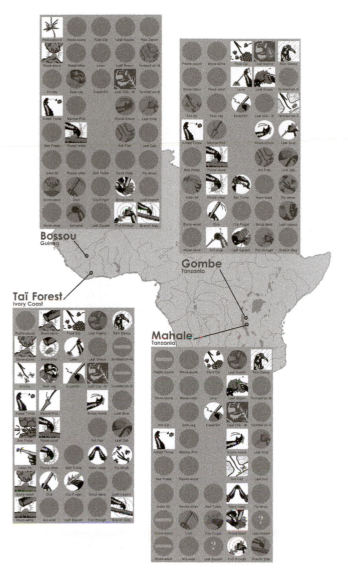

Figure 44.2 Distribution of behavior patterns across four African study sites. Square icons, customary; circular icons on white background, habitual; square and circular icons on dark backgrounds, present; no icon or blank circular icon, absent; horizontal bar, absent with ecological explanation; question mark, answer uncertain (see Whiten et al. 1999).

This same survey methodology was later employed by researchers at six different orangutan (*Pongo pygmaeus*) field sites in Sumatra and Borneo (van Schaik et al. 2003a), identifying a total of 24 putative cultural variants. Again, the behaviors were wide-ranging, including social communicatory signals, foraging techniques, and tool-use behaviors. van Schaik et al. (2003a) also investigated the relationship between behavioral similarity and geographical proximity. Sites with similar patterns of cultural variation, they hypothesized, would also be ones that were geographically close, supporting a model of innovation followed by social diffusion of the behavior patterns. The cultural difference between any pair of sites was calculated as a percentage of all customary and habitual local

variants, and this was found to be significantly correlated with the distance between the sites, suggesting that there was diffusion between localities. This particular finding does not contribute a great deal toward eliminating the possible alternatives of environmental or genetic influence as we would expect ecological and genetic divergence also to correlate with distance. However, the researchers also found that the variety of cultural behaviors was related to the opportunities available for social learning. Repertoire size was measured as the number of customary and habitual variants expressed at each site. This was found to be significantly related to the percentage of time that nondependent animals spent in association. This result seems to be much less easily dismissed as a potential confound of either environment or genetics. The correlation with association data allowed the researchers to look more directly at social influence on behavior, rather than merely indirectly, through having ruled out the other plausible alternatives.

Monkey Field-Site Surveys

Similar methods have now also been applied to the behavior of monkeys as well as apes. Panger et al. (2002b) pooled data from three long-term white-faced capuchin (*Cebus capucinus*) field sites in Costa Rica. Their approach (based on the logic that the majority of chimpanzee and orangutan variants involved foraging techniques) was to compile a list of the foods from the diet of this species and compare how the same food was processed by the capuchins from the different sites. Out of the 61 foods that formed part of the diet at more than one site, 20 were found to be processed in a different manner between sites. *Pithecellobium saman* fruits, for example, a species that is broken open by the monkeys for the beetle larvae within, is processed by tapping (rhythmic tapping with the fingertips) at Lomas Barbudal, whereas the capuchins at Palo Verde use a *fulcrum* behavior (the application of force on an object working against a substrate). Individuals from the Santa Rosa site use a variety of techniques with this particular food species, including rubbing and pounding, as well as the fulcrum behavior observed at Palo Verde.

Even more persuasively, data from the Palo Verde field site allowed for an analysis of the relative levels of association of those dyads that shared particular processing techniques. Pairs of animals that had behaviors in common had higher proximity scores than nonmatched pairs. As illustrated in Figure 44.1, when the groups being compared are sympatric, rather than geographically separated, it makes it considerably less likely that the contrast is produced by environmental shaping alone. Given that the foods being processed are liable to be precisely the same species (possibly even fruits or leaves picked from the very same plant), differences in, for example, hardness or texture are much less likely to be responsible for the differences in behavior within sites than they would be across sites.

It is possible, however, that those individuals that associate closely are typically more closely related compared with pairs that rarely associate. These association measures could therefore be confounded by genetic factors. When dealing with sympatric groups, it should nonetheless be possible, in principle, to assess genetic influence in relation to social factors. If relatedness is known (or can be determined through, e.g., deoxyribonucleic acid [DNA] testing), this can be directly compared with the association data to determine which is more closely correlated with measures of behavioral similarity. If the association data explain a greater amount of the variation in behavior compared with the relatedness data, then a social learning interpretation is supported over a genetic one.

As well as foraging behaviors, a separate analysis (Perry et al. 2003) of possible social traditions in capuchin monkeys was carried out using data from the same three field sites plus one additional site, Curú. As with the previous surveys of this kind, the researchers were able to identify behaviors that were present in certain social groups and absent in others. As discussed in relation to the chimpanzee field-site data, social behaviors are less susceptible to the criticism that they could arise as an artifact of environmental conditions. There is nonetheless a slim possibility that, when comparing groups at different field sites, local conditions in some way affect these behaviors (e.g., display behaviors might need to be louder and more conspicuous at more densely forested sites).

However, in addition to having this information about static group differences, the researchers were able to document the patterns of emergence, expansion, and decline of these behaviors within the groups. The three social behaviors under investigation were hand sniffing (bouts of sniffing of another individual's hand), body part sucking (sucking of the fingers, toes, ears, or tail of another individual), and "games" (which involved the retrieval of an object from another individual's mouth). The dyadic nature of these activities meant that researchers could also chart the probable transfer of the behavior between individuals in the group. When individuals were first observed playing one of the "games," for example, it was assumed that they had acquired the behavior from the particular monkey with which they were interacting. In this way, it was possible to document multiple links in the transmission chain, where former partners of the original innovator went on to engage in game playing with other, previously naive individuals. It is difficult to see how one could explain such diffusion patterns as anything other than some kind of social transmission.

The field-site survey method has also recently been applied to Japanese macaque monkeys (*Macaca fuscata*). Apparent traditions of "stone handling" have been documented in populations of macaques for some years (Huffman 1984). Stone handling is a form of solitary play which involves the manipulation of stones in various ways, including clacking two together or pounding one onto

another surface. Leca et al. (2007a) have compared the specific behaviors used while engaging in this activity across a total of 10 troops of Japanese macaques. A total of 45 different behavior patterns were distinguished, and these were classified according to the same terms used by Whiten et al. (1999), that is, customary, habitual, present, or absent. Leca et al. (2007a) found that many of the behaviors were patchily distributed across the 10 populations compared. For example, a behavior classified as "flinting" stones was identified as customary in one of the populations and habitual in two others, while being completely absent from three troops.

ANALYSIS OF DIFFUSION DATA

When comparing static patterns of group behavior, we are restricted to making post-hoc inferences about the learning processes that must have produced them. However, in some cases, it is possible to record the diffusion process as it is taking place. Although it is generally not possible to determine from exactly which individual an animal has acquired a trait (cf. Perry et al. 2003), some researchers have used data on the rate of spread of a novel behavior through a population in an attempt to determine whether or not social transmission was likely involved.

The analysis of diffusion data is an approach that has been used by a number of researchers to assess social learning in animals (Galef 1990; Lefebvre 1995a,b; Reader and Laland 2000; Rendell and Whitehead 2001). The typical assumption is that if a behavior is socially learned, the number of individuals performing the behavior will increase at an accelerating rate. This is based on the logic that the more potential demonstrators (skilled individuals) there are in a group, the greater the chance that a naive individual will learn the new behavior.

Galef (1990) initially used this reasoning to argue that the potato-washing behavior of the Koshima macaques was unlikely to have been socially learned. He plotted the number of individuals reported to wash potatoes at a given time against the year of data collection. The resulting graph showed a relatively steady increase in the number of potato washers over time (see Figure 44.3). This led Galef (1990) to suggest that the potato-washing behavior was in fact being individually learned by each monkey through trial and error.

Since Galef's (1990) proposal, more comprehensive mathematical approaches have been used to support similar arguments. Lefebvre (1995a) analyzed the diffusion patterns of a number of behavioral innovations that had been documented in primates. A number of theoretical models of trait diffusion, under conditions of social transmission, exist (e.g., Boyd and Richerson 1985, Cavalli-Sforza and Feldman 1981). Lefebvre (1995a) compared such models against the observed cumulative distributions of the reported behaviors. A total of five theoretical models were utilized, three of which (logistic, hyperbolic sine, and positive exponential

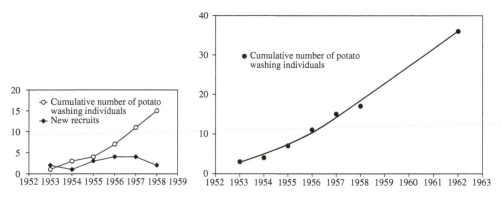

Figure 44.3 Data on the number of Koshima macaques engaging in sweet potato–washing behavior, as plotted by Galef (1990, *left*) and Lefebvre (1995a, *right*). Galef (1990) suggests that the relatively consistent number of new recruits per year to the potato-washing behavior is indicative of asocial learning processes. Lefebvre (1995a) finds that an accelerating function best fits the cumulative data, suggesting social learning processes. The contrasting interpretations appear to be primarily attributable to minor differences in the methods used to plot the data reported by Kawai (1965), including the inclusion of an extra data point by Lefebvre (1995a).

equations) featured the accelerating functions assumed to be consistent with social transmission. The other two functions were linear (representing a constant rate of increase of skilled individuals) and logarithmic (featuring a decelerating rate of learning).

Lefebvre (1995a) analyzed data on Japanese macaque innovations. These included the potato-washing behavior as well as seven others, such as the adoption of novel foods into the diet. For four out of these eight behaviors, the best-fit model was one of the accelerating functions. The remaining four were better accounted for by the linear and decelerating equations. Surprisingly, potato washing, the very behavior that had been identified by Galef (1990) as having a distribution inconsistent with social learning, was found by Lefebvre (1995a) to correspond most closely to the (accelerating) logistic function. Lefebvre (1995a) attributes this primarily to his inclusion of a key extra data point, omitted by Galef (1990). The contrasting data sets can be seen in Figure 44.3.

Such data should, however, be treated with extreme caution. Quite apart from the fact that analyses may be extremely fragile (dependent even on the inclusion or exclusion of a single data point, as illustrated by the contrasting interpretations of Galef [1990] and Lefebvre [1995a]), there are good reasons why counterintuitive diffusion patterns might arise. Social learning may not necessarily result in accelerating learning curves and, equally, asocial learning may under some circumstances produce accelerating cumulative distributions.

As noted by Lefebvre (1995a) (and more recently by Day et al. 2001 and Laland and Kendal 2003), an S-shaped curve similar to those produced by some acceleratory functions is exactly what would be expected under circumstances of individual learning, given a normal distribution of learning latencies within the population. An accelerating cumulative distribution of learners is therefore not sufficient to conclude that social learning is involved in transmission.

Likewise, the absence of an accelerating curve may not be indicative of an absence of social transmission processes. Laland et al. (1996) have modeled a number of social learning scenarios and note that, under certain circumstances, apparently linear diffusion curves may result. The higher the proportion of skilled individuals in the group, the greater the opportunities for *scrounging* (exploiting of food resources found by another individual). Hence, naive individuals may adopt a strategy of scrounging rather than performing the behavior themselves.

Opportunities for scrounging behavior would therefore be expected to have a dampening effect on transmission curves. Such effects are certainly liable to operate within primate groups. Fragaszy and Visalberghi (1990) reviewed several of their studies of the learning of novel extractive foraging behaviors within groups of capuchin monkeys (*Cebus apella*) and noted that individuals that scrounged from others typically did not go on to perform the skill themselves.

These points clearly illustrate the need for caution in interpreting particular cumulative distributions as indicative of either social or asocial learning processes. However, that is not to say that this approach should necessarily be abandoned altogether. In certain studies there may be other methods by which the presence or absence of social learning can be established, and these can be used to determine which cumulative distributions are genuinely characteristic of social learning. Kendal et al. (2007) have taken just such an approach, and their results suggest that this could be a promising avenue for future research. They analyzed data from a series of naturalistic observational studies in which novel extractive foraging tasks were introduced to groups of callitrichid monkeys. Since each individual's behavior was sampled at prespecified time intervals, the presence of processes such as stimulus enhancement could be verified through testing whether (in the case of stimulus enhancement) success on the task by one individual at time t increased the likelihood of approach by other individuals at time $t + 1$. It was then possible to determine whether social processes like this were related to success on the task. Cumulative distributions were then plotted for trials on which four or more individuals had solved the task, and the best-fit function was determined. The majority of the diffusion curves were found to be deceleratory. Coupled with Kendal et al.'s (2007) conclusion that the tasks were apparently solved through predominantly asocial learning processes, this finding alone perhaps has little power to elucidate which particular patterns may be indicative of social learning. However, further studies elaborating on this approach may be able to identify whether there are indeed particular distributions which can be regarded as indicative of social or asocial learning.

EXPERIMENTAL STUDIES OF LEARNING MECHANISM

Experimental Studies of Natural Behaviors

For some group-specific behaviors, it is possible to carry out experiments which can demonstrate conclusively the role of social interaction in acquisition. For example, in Terkel's classic research on the pinecone-stripping behavior of black rats (Aisner and Terkel 1992; Terkel 1995, 1996), it was possible to demonstrate that this skill was maintained in the population through social interaction, by carrying out a cross-fostering experiment.

However, as pointed out by Laland and Hoppitt (2003), under circumstances in which behavioral variation has been found between two geographically separated populations, even cross-fostering experiments tell only half the story. Genetic factors can be conclusively excluded, but the role of environmental shaping is still uncertain. Translocation experiments are required: Given two behaviorally distinct populations, each group in its entirety is transferred from its own environment to the former habitat of the other. Truly cultural behaviors would persist within the populations, whereas those that were artifacts of environmental shaping would be expected to die out or, indeed, mutate into the form previously observed in the former residents.

On this basis, Laland and Hoppitt (2003) suggested that the best evidence of culture to date, in any animal, comes from fish as these are the only species with which such experiments have been conducted. Helfman and Schultz (1984) found that French grunts, *Haemulon flavolineatum*, transferred into an existing population adopted the same schooling sites and migration routes as the residents, whereas a control group, introduced after the resident population was removed, did not show these same preferences.

These kinds of experimental design, although extremely informative, are unfortunately of restricted application within the field of primate behavior. Although such studies could certainly be illuminating on a long-lived, large-bodied, and in many cases endangered animal, cross-fostering and translocation not only would be costly and time-consuming but also would involve taking serious risks with an already threatened population. To date, therefore, experimental research on social learning in primates has typically involved simulations in captivity rather than direct study of natural behaviors. These experiments allow researchers to establish that social learning mechanisms exist in these animals, and from this we can infer that some of the group-specific behaviors observed in the wild may be supported by such processes.

The Significance of Imitation

Experimental research on social learning in monkeys and apes dates back to the beginning of the century (e.g., Haggerty 1909; and see reviews in Hall 1963a, Tomasello and Call 1997, Whiten and Ham 1992), and many primate species have shown clear evidence of social learning effects under experimental conditions. Studies such as those carried out by Tomasello et al. (1987) and Warden and Jackson (1935) with chimpanzees and rhesus macaques (*Macaca mulatta*), respectively, clearly established that subjects that observed a skilled demonstrator performing a simple action (e.g., a lever press for food or raking in of a food reward) acquired the behavior faster than those that had seen no demonstration.

However, it has been proposed by some researchers that not all social learning mechanisms have the capacity to support stable behavioral traditions such as those proposed to exist in the wild. Whiten and Ham (1992) pointed out that many of the early experiments with primates did not have the power to distinguish the mechanism of imitation from other forms of social learning. In particular, results could readily be attributed to stimulus enhancement (Galef 1988, Spence 1937, Whiten and Ham 1992). Stimulus enhancement is said to have occurred when the presence of an individual merely draws an observer's attention to a particular object or part of an object, thus enhancing the observer's opportunity to learn about the object. The behavior itself is therefore not being learned through observation, although similar behavior patterns may often be the result.

Imitation, by contrast, has been defined as "learning to do an act from seeing it done" (Thorndike 1898, Whiten and Ham 1992). Observation of the behavior is therefore both necessary and sufficient for a naive individual to reproduce it through imitation. For this reason, imitation has been singled out by a number of researchers as being particularly suited, among mechanisms of social learning, to support culture (Boyd and Richerson 1985, Dawkins 1976, Galef 1992, Whiten and Ham 1992). As other social learning processes require some interaction between the naive individual and the environment for learning to take place, it is assumed that these processes would be much more susceptible to interference from environmental shaping. Hence, it is proposed that these mechanisms could not support stable, socially learned, behavioral variation between groups as learners would typically settle on techniques no different from those they would have adopted under conditions of asocial learning (albeit their arrival at these solutions may be made more likely or happen more rapidly).

Furthermore, Galef (1992) has argued that culture should be defined in such a way that it is homologous, rather than simply analogous, to human culture. In Galef's (1992) view this entails a reliance on the same learning mechanisms, specifically imitation and teaching. It should be noted that a reliance on imitation is not an aspect of culture which features in every definition (indeed, this would considerably narrow the definition given at the beginning of this chapter), but the issue of homology is noteworthy nonetheless. Whether or not the behavioral traditions observed in non-human animals can be described as homologous to human culture may be of little importance to some researchers, but to those interested in the question of the evolution of human culture (and primatologists may form a particularly high proportion of this group), this is clearly a significant issue (Whiten et al. 2003).

A number of reasons have therefore been proposed regarding the significance of imitation to cultural transmission. Fortunately, the kinds of experiment which researchers have been able to carry out with primates are ideally suited to determining social learning mechanisms (a goal that is extremely difficult to achieve in other types of research, which focus on natural behaviors).

Imitation and the Two-Action Method

There have been a large number of studies of imitation in primates, so we are unable to provide an exhaustive review here (see Whiten et al. 2004 for a more thorough review of the literature on imitation in apes). A variety of methodologies have been employed, the most successful being "do-as-I-do" ("Simon says") tests and what have become known as "two-action" experiments. Here, we review the two-action tests as these are most relevant to the question of whether the group-specific behaviors observed in wild primates are socially learned. While do-as-I-do experiments involve training a subject to respond to a cue to imitate (therefore posing the question of whether they can imitate), two-action experiments rely on spontaneous behaviors (thus revealing whether they actually do imitate) (Horowitz 2003, Whiten and Custance 1996). A further reason for choosing to focus on two-action studies is that, within imitation research, the two-action design has become something of a methodological gold standard. The majority of scholars in the field, while often disagreeing on other aspects of research design and interpretation, generally agree on this as a preferred methodology (see review in Caldwell and Whiten 2002, but see Byrne 2002 for an alternative view).

In two-action experiments, subjects observe one of two possible actions on a particular object. If they subsequently produce behaviors more like the version that they saw demonstrated, it is concluded that imitation must be implicated. Two-action designs can control for other social learning effects, such as stimulus enhancement, as these would not be expected to generate differences between the groups in terms of the particular actions displayed.

Experimental Studies of Imitation in Apes

Whiten et al. (1996) used the two-action design to test for imitative effects in chimpanzees. They designed an "artificial fruit," a box containing a food reward, protected by several defenses (see Figure 44.4). For each of the defensive locks, there were two alternative methods of release. A pair of bolts could be either twisted or poked out, and a handle could be either lifted out or turned around. Chimpanzee subjects were each shown one of the methods of opening by a human demonstrator. Following the demonstrations, the chimpanzees used actions similar to the versions that they had been shown. This effect showed clear evidence that the chimpanzees were reproducing the details of behaviors rather than just showing a more generalized stimulus enhancement effect.

The artificial fruit methodology has also been employed to test for imitation in other ape species. Stoinski et al. (2001) found that, as with the chimpanzees, gorilla subjects (*Gorilla gorilla gorilla*) showed a significant tendency to copy the technique they had observed from the human demonstrator. However, in orangutans, results were less clear. Both Custance et al. (2001) and Stoinski and Whiten (2003) used the artificial fruit methodology with orangutan subjects, finding no significant tendency to match the observed behaviors. However, these results contrast strongly with those described by Russon and Galdikas

Figure 44.4 An "artificial fruit" (*above*) with bolts, handle, and pin defenses, which was used to test for imitative tendencies in chimpanzees.

(1993, 1995), who documented many cases of spontaneous imitation in a rehabilitant population of orangutans. Perhaps consistent with this, Buttelmann et al. (2008a) have shown orangutans copying whichever body part a human used to operate a device. They did this as long as the model's hands were free and not if the hands were occupied and in this way demonstrated what appeared to be rationally selective imitation.

Research on imitation in apes has also addressed the question of whether the sequence of demonstrated actions is preserved as well as the behaviors themselves. Whiten (1998) was able to test this in chimpanzees using a variation on the artificial fruit methodology. Subjects were again shown one of two different methods of accessing the food reward, although in this version the component elements were ordered differently in each. The chimpanzees showed a significant tendency to unfasten the locks in the order they had seen demonstrated (Figure 44.4). The imitation of complex sequences of actions may be particularly relevant to the issue of culture in primates as many of the behaviors proposed to be socially learned in the wild are relatively complex, multistage procedures (Byrne and Russon 1998). Experiments that incorporate multiple sequential elements may therefore be extremely revealing in terms of understanding how the natural behaviors are learned.

Experimental Studies of Imitation in Monkeys

Although earlier reviews suggested the possibility of a monkey–ape divide in terms of imitative ability (Visalberghi and Fragaszy 1990, Whiten and Ham 1992), evidence for imitation in monkeys is mounting. Human demonstrators have typically been used in much of the ape imitation research, but this methodology is less suitable for experiments with monkeys, due to the greater discrepancies in body size and morphology. However, recent research has addressed this problem by using trained conspecific demonstrators.

Common marmoset subjects have been tested using the two-action method (Voelkl and Huber 2000). This particular task involved opening photographic film canisters, and two trained marmoset demonstrators were used, one of which removed the lid with its hand while the other used its mouth. Voelkl and Huber (2000) found a significant difference between the two observer groups in the method used to open the canisters: All subjects that saw the hand method used their hands every time, and of the subjects that saw the mouth method, most used their mouth for at least some of the test trials. Similarly, in our research using the artificial fruit methodology (Caldwell and Whiten 2004), marmoset subjects matched some aspects of the actions they saw demonstrated by conspecifics.

There is also now good evidence for imitative ability in capuchin monkeys (*Cebus apella*) from a number of studies. The artificial fruit method has been used with capuchins (Custance et al. 1999). In this experiment,

human demonstrators were used as the capuchins were "enculturated" and thus liable to respond well to a human demonstrator. The capuchins produced actions similar to the ones they had seen demonstrated, albeit not perfectly matched. Fredman and Whiten (2007) tested the same population of human-reared capuchins with another two-action task, this time involving a container with a lid which could be either levered off or poked through using a tool. Once again a human demonstrator was used, and the monkeys showed a tendency to match the alternative that they had seen demonstrated. More recent work (Dindo et al. 2008) has shown that nonenculturated capuchins also show good evidence of faithful transmission using two-action tests. Dindo et al. (2008) began by training two capuchin demonstrators to access food from a box by either sliding or lifting a door. These trained individuals were each used as demonstrators for one subject. The test subjects, having learned the action, then went on to perform as demonstrator for a further subject. This process was continued until the "lift" behavior had been successfully transmitted along a chain of five (untrained) subjects and the "slide" behavior had been transmitted along a chain of four. The fidelity with which these two options were maintained illustrates that capuchins are clearly capable of reproducing demonstrated actions.

Although the majority of evidence for imitation in monkeys to date comes from capuchins and marmosets (both common lab species), this does not necessarily mean that other species are less able in this regard. Presented with appropriate tests, other monkey species may well show positive evidence for imitation too. Price and Caldwell (2007) tested for imitation in zoo-housed colobus monkeys (*Colobus guereza*) using a two-action design. Two separately housed groups were presented with videotaped demonstrations of a neighboring group interacting spontaneously with the apparatus, which had a door that could be either pushed or pulled. One of the test groups was shown clips in which the door was pushed, and the other test group was shown clips in which the door was pulled. There was a significant tendency for the observers to match the action they saw demonstrated.

It is perhaps worth noting at this point that the vast majority of experimental studies of social learning in primates involve foraging or otherwise food-related behaviors, while many of the proposed behavioral traditions (particularly in monkeys) are not food-related, such as social customs (Perry et al. 2003), or apparently functionless behaviors like stone handling (Huffman 1984). Watson and Caldwell (2009) have drawn attention to this discrepancy, noting that experimental studies of non-food-related behaviors may be particularly illuminating in order to understand the mechanisms that may underlie non-food-related traditions. Clearly, although there is now good evidence of imitation in a variety of primate species, there are still significant gaps in our knowledge concerning the contexts in which this learning mechanism may be used.

CULTURE AND CORRUPTION

Social Learning and Corruption from Individual Learning

It is clear then that if it is imitation we are particularly interested in, there is increasing evidence from primates. Additionally, the field data clearly show divergent patterns of behavior between different localities and social groups. However, skeptics remain unconvinced that there is evidence for culture in primates (Galef 2003, Laland and Janik 2006). The problem is, essentially, that it is very difficult to show that it is these same mechanisms, which have been identified in experimental research, that support the behavioral variation observed in the field. Field data typically identify social learning only indirectly, through eliminating alternatives, and experimental data (while directly manipulating learning conditions) allow us to conclude only that certain social learning mechanisms are available to a given species. Whether or not these are exploited in the development of natural behaviors is a different matter.

To put forward such an argument might appear to some to be nitpicking to an excessive degree. In the face of such a huge amount of circumstantial evidence, why should one be inclined to draw any other conclusion? In fact, a number of authors have offered persuasive arguments for why inter-population differences in behavior, such as those observed in chimpanzees, would be unlikely to be the result of the sorts of social learning mechanism identified in experiments.

Galef (1995) makes the point that socially learned behaviors will be either maintained within or extinguished from the repertoire of an individual as a consequence of the reinforcement contingencies that the actor experiences. In other words, asocial learning will interact with the socially learned information, and behaviors will persist only if reinforced. On this basis, therefore, Galef (1995) argues that those behaviors that are maintained in populations (such as those commonly labeled as culture in primates) must be locally adaptive, in the sense that they must result in greater reinforcement relative to alternatives. Such an argument obviously poses a fairly major problem for the interpretation of the contrasting behaviors at different primate field sites. If it is the individually learned reinforcement patterns that determine whether a behavior is maintained in an individual's repertoire (irrespective of whether any form of social learning was involved in the initial acquisition), then the divergent patterns of behavior observed at different field sites almost certainly have their origins in environmental shaping.

Heyes (1993) puts forward a similar argument, proposing that the particular transmission mechanism involved is in fact irrelevant to the question of whether or not a behavior will extinguish. Behaviors learned via imitation, like any other, will be subject to modification from trial-and-error learning. Heyes (1993) cites Galef et al.'s (1986) data from a two-action study on budgerigars (*Melopsittacus undulatus*): Subjects that observed conspecifics accessing hidden food

using either the beak or the foot tended to match the demonstrated technique for only the first two trials postdemonstration. In later trials, the difference between the groups disappeared.

What is more pertinent to the issue of culture, proposes Heyes (1993), is a separate insulating mechanism which protects the socially learned information against the influence of trial-and-error learning. Heyes (1993) takes birdsong as an example and notes that socially learned, interpopulation differences in song patterns exist not because they are learned via imitation but because other mechanisms exist which either prevent exposure to or resist the influence of other song types.

Diffusion Experiments

In order to support culture, it is not crucial that a particular behavioral variant remains in an individual's repertoire indefinitely; the real issue is whether or not it persists long enough to be passed on to another individual or individuals. Experiments which simulate social transmission over multiple learner generations or show how these can be maintained within a group are therefore likely to prove particularly illuminating with regard to understanding whether socially learned behaviors can endure within a population (see Figure 44.1).

Experiments such as this have been carried out in both humans (Mesoudi and Whiten 2008) and a range of nonhuman species (Whiten and Mesoudi 2008). Here, we focus specifically on those experiments which have been carried out with nonhuman primates.

Traditionally, experiments on social learning typically involved exposing individual subjects to a demonstrated behavior and recording their responses performed in a relatively short time period following this demonstration (e.g., our "artificial fruit" experiments). It is easy to see how such designs are vulnerable to the criticisms of Heyes (1993) and Galef (1995). Perhaps the learning effects documented in such experiments are too transitory to sustain lasting, stable differences in behavior between groups. However, recently experiments on social learning in primates have shifted focus somewhat. Whiten and Mesoudi (2008) describe several different types of "diffusion" experiments, each of which has its own strengths in terms of its power to reveal a little more about the longevity of socially transmitted behaviors compared with a traditional dyadic experiment.

One such design is described as the "open group diffusion" method. In these studies a novel behavior is introduced to a group (ideally experimentally, through the intentional training of a particular demonstrator individual or some other method of exposure to the target behavior), and this group's behavior can be compared either to that of a control group with no such experience or (even more convincingly) to that of a group exposed to an alternative behavior. Following the initial experimental manipulation, the groups' behaviors can then be recorded over a relatively long time period in order to determine whether the differences are maintained over time or, indeed, if they become corrupted by the effects of individual learning. Whiten et al. (2005) carried out just such an experiment with chimpanzee subjects, comparing the behavior of two groups, each of which had been exposed to a different method of manipulating a foraging apparatus. Whiten et al. (2005) found that the variation between the two groups was maintained over several days of testing and was still apparent even when the chimpanzees were tested again 2 months later. A similar design was used by Price and Caldwell (2007) in the study of imitation in colobus monkeys, described previously. The difference in behavior between the two groups was maintained even after the demonstration video footage was removed and was still apparent when testing ended several days later. Both of these studies strongly suggest that social learning in primates does have the capacity to support natural behavioral variation as the researchers showed that these alternatives were maintained over time within the social groups being compared.

Whiten and Mesoudi (2008) also describe the "linear chain" method. These experiments start off very much as a traditional dyadic social learning experiment would (i.e., with an experimentally controlled demonstration from either a human or a trained conspecific), but following this initial test, the individual that took the role of observer on one trial then goes on to perform the role of demonstrator on the subsequent trial for a new naive subject. In this manner researchers can document exactly how "corrupted" behaviors can become through the process of repeated transmission between multiple individuals. Such an experiment has already been described in the earlier section on imitation in monkeys as Dindo et al. (2008) found that two alternative behaviors could be transmitted along chains of four and five capuchin monkeys. This design has also been used with chimpanzees (Horner et al. 2006) with very similar results. The two alternative behaviors were transmitted with very high fidelity, right up to the final individuals tested, over chains consisting of five and six naive subjects.

CONCLUSIONS

We find the case for culture in primates to be a convincing one. Although some skeptics will maintain that the evidence from primate research is less rigorous than that from other species, the literature is full of evidence which suggests that certain nonhuman primates may be more "cultured" than any animal other than ourselves.

From the field research data, the sheer number of apparent cultural variants identified in primates is noteworthy. Although group-specific behaviors have been documented in many different species, from birds to rodents (Lefebvre and Palameta 1988), the variety of such behaviors identified in chimpanzees, orangutans, and capuchin monkeys is striking. Isolated cases may be dependent on relatively inflexible, hard-wired mechanisms; but the multiple cultural variants identified in chimpanzees, and now other primates too, may point to a more general underlying social learning ability or tendency toward adopting the behavior of conspecifics.

All the same, it must be noted that comparable diversity may yet be identified in animals other than monkeys and apes. Whiten et al.'s (1999) survey of chimpanzee culture (once apparently unique in the diversity it revealed) is now echoed to varying extents in Panger et al.'s (2002b), Perry et al.'s (2003), and van Schaik et al.'s (2003a) studies of capuchin monkeys and orangutans, as well as Rendell and Whitehead's (2001) review of the literature from cetacean field observations and laboratory experiments. Given appropriate research effort, it may be only a matter of time before other phyletic groups are shown to have similarly rich patterns of behavioral variation.

There are clearly still plenty of opportunities for research on this topic. Approaches such as Panger et al.'s (2002b) and van Schaik et al.'s (2003a), which make use of association data, will help to determine the role of social contact in the development of particular behaviors. Approaches such as Kendal et al.'s (2007), which aim to determine whether certain rates of diffusion are characteristic of social learning, will also help to validate conclusions based on observational data. In terms of experimental approaches, diffusion experiments are already providing important insights into how behavioral variation can be maintained between groups, but further data are still required. Studies could be carried out over even longer time periods, and a wider range of primate species should be tested in this manner. Also, there is little experimental research which is specifically targeted toward understanding the mechanisms underlying social traditions as to date the majority of experimental research has focused on food-related behaviors (Watson and Caldwell 2009). Such studies are only just beginning in primates (Bonnie et al. 2006) but are likely to prove particularly illuminating.

The case for culture in primates is a strong one. However, there are clearly profitable directions for future research that will enable conclusions about the role of social learning in supporting behavioral variation to be made much more firmly.

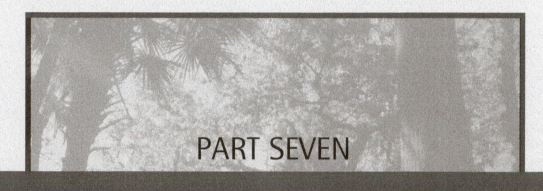

PART SEVEN

The Future

45

Conservation

Karen B. Strier

1. What are the major threats to primates today, and how does the behavioral ecology of different primates affect their vulnerability to these threats?
2. What factors are involved in establishing conservation priorities for primates and their habitats?
3. What are the greatest challenges that primate conservationists must overcome to succeed in their efforts to protect endangered populations and species?

INTRODUCTION

Of more than 600 species and subspecies of primates recognized today, nearly half are in grave danger of going extinct (Mittermeier et al. 2009). This is an alarming increase over estimates from less than a decade ago (Mittermeier et al. 2000b, Konstant et al. 2002), and although some of the increase can be attributed to the rarity of taxa that have recently been discovered or reclassified (Groves 2001a; Mittermeier et al. 2004, 2008b), it also exemplifies the rapidity with which extinction risks can rise (Konstant et al. 2002). Some of the primates that are threatened with extinction can still be saved if efforts to protect them and their habitats succeed. Others, however, may already be reduced to such small populations and confined to such limited or altered habitats that the chances for their long-term survival are slim (Cowlishaw 1999b). One West African primate, known as Miss Waldron's red colobus monkey (*Procolobus badius waldronae*), may already be extinct (Oates et al. 2008k). Primate conservation is about preserving the diversity of species and subspecies whose futures are endangered and preventing those taxa that are still relatively secure from becoming endangered themselves.

This chapter is structured around an initial review of the steps that are involved in identifying conservation priorities and developing and implementing informed conservation programs. Informed conservation requires an understanding of the major threats to primates and how the dynamics of small populations affect different taxa in different ways depending on their degree of ecological and behavioral specialization and intrinsic reproductive rates. All conservation efforts call for establishing and enforcing prohibitions on unsustainable hunting, protecting remaining habitats, and developing targeted research and education programs, but the relative importance of each of these tactics and the

feasibility of implementing them depend on the taxa and the local conditions that affect them (Brook et al. 2006). In the final section, I consider some of the many challenging choices that need to be made about the role of zoos and other captive breeding facilities in conserving primates and whether wild populations and disturbed habitats can, or should, be actively managed and restored. These choices raise important questions about the trade-offs involved in meeting conservation priorities and will ultimately determine the long-term prognosis for many of the world's most endangered primates.

IDENTIFYING PRIORITIES

The first step in conserving primates is to identify priorities, which requires an assessment of each taxon's current status in the wild. The Primate Specialist Group of the World Conservation Union (IUCN) regularly updates these assessments using the Red List criteria, which are based on quantitative estimates of the total number of individuals, the size and distribution of their populations, the rate at which their occurrence has declined, and the causes of these declines (Baillie et al. 2004). These kinds of data generally come from surveys and censuses conducted by experienced fieldworkers, but the quality of the data can still vary greatly depending on the accessibility of the sites, the sampling methods employed, the history of human contact, and the habits of the primates themselves. Nonetheless, they represent the best, and in many cases the only, available sources of information about the abundance and distribution of the world's extant primates, and there are still a considerable number of taxa for which data on status in the wild are deficient (Hilton-Taylor and Rylands 2002; Mittermeier et al. 2008b).

Recent evaluations of primate extinction risks have contributed to revisions in primate taxonomy, with the result that many previously recognized species or subspecies have been split into two or more distinct taxa. These revisions include newly recognized genera, species, and subspecies (Mittermeier et al. 2000b, 2008b; Groves 2001a; Brandon-Jones et al. 2004; Grubb et al. 2003). Although some of these revisions are still subject to debate, there are merits to recognizing primate diversity at the subspecific level for the purpose of conservation because calling attention to variant forms insures that none is overlooked in assessments of conservation priorities (Rylands et al. 1995, Purvis and Hector 2000, Groves 2001a, Agapow et al. 2004, Mace 2004).

The 2008 IUCN Red List of Threatened Species includes 303 primate taxa that have suffered especially severe population and habitat losses in recent years (Mittermeier et al. 2009). The degree and rate of these losses determine whether a taxon is classified as critically endangered, endangered, or vulnerable (Table 45.1). All primates that meet the criteria for any of these three categories are considered threatened, but the 206 endangered and critically endangered taxa face the highest extinction risks of all. Asian primates account for about 43% of endangered and critically endangered taxa, followed by neotropical (24%), African (20%), and Malagasy (13%) taxa (Mittermeier et al. 2008b). More than 80% of the world's 56 critically endangered primates occur in just 12 countries (Figure 45.1). More than 70% of the endangered and critically endangered primates occur in just six of the world's biodiversity "hotspots," which have been identified on the basis of their high levels of plant endemism and the severity of habitat losses and degradation (Myers et al. 2000, Konstant et al. 2002).

Taxa with the highest extinction risks are justifiably top priorities for conservation efforts at local, national, and regional scales. Nonetheless, assessments of these extinction risks apply the same population and habitat criteria equally to all taxa regardless of their phylogenetic distinctiveness or evolutionary uniqueness. Indices of evolutionary uniqueness take the number of closely related species or subspecies and their phylogenetic distance from one another into account to determine the amount of evolutionary history that would be lost if the taxon were to go extinct (Nee and May 1997, Hacker et al. 1998, Owens and Bennett 2000, Purvis and Hector 2000, Faith 2008).

Assuming similar rates of evolution (Cowlishaw and Dunbar 2000), the extinction of a species belonging to a monotypic genus (or higher taxonomic unit) would result in a greater loss of evolutionary history than the extinction of a species belonging to a polytypic genus (Purvis et al. 2000). Consider, for example, the aye-aye (*Daubentonia madagascariensis*), the sole surviving species and genus of the family Daubentoniidae, which is classified in its own infraorder, Chiromyiformes (Groves 2001a). Extinction of the aye-aye would represent the loss of the distinctive evolutionary history of an entire infraorder. Although the extinction of a taxon such as Miss Waldron's red colobus monkey would obviously result in the loss of biodiversity, it is also the case that less evolutionary history would be lost

Table 45.1 International Union for Conservation of Nature (IUCN) Criteria for Threatened Taxa

Critically Endangered (CR): Facing an extremely high risk of extinction in the wild due to any of the following:

  A. Reduction in population size of 80%–90% over the last 10 years or three generations depending on the causes and reversibility of the reductions

  B. Extent of occurrence <100 km² or area of occupancy <10 km²

  C. Population size estimated to number fewer than 250 mature individuals and to be declining or unevenly distributed

  D. Population size estimated to number fewer than 50 mature individuals

  E. Probability of extinction within 10 years or three generations at least 50%

Endangered (EN): Facing a very high risk of extinction in the wild due to any of the following:

  A. Reduction in population size of 50%–70% over the last 10 years or three generations depending on the causes and reversibility of the reductions

  B. Extent of occurrence <5,000 km² or area of occupancy <500 km²

  C. Population size estimated to number fewer than 2,500 mature individuals and to be declining or unevenly distributed

  D. Population size estimated to number fewer than 250 mature individuals

  E. Probability of extinction at least 20% within 20 years or five generations

Vulnerable (VU): Facing a high risk of extinction in the wild due to any of the following:

  A. Reduction in population size of 30%–50% over the last 10 years or three generations depending on the causes and reversibility of the reductions

  B. Extent of occurrence <20,000 km² or area of occupancy <2,000 km²

  C. Population size estimated to number fewer than 10,000 mature individuals and to be declining or unevenly distributed

  D. Population size estimated to number fewer than 1,000 mature individuals or restricted area of occupancy or number of locations

  E. Probability of extinction at least 10% within 100 years or three generations

Source: Simplified and condensed from IUCN Standards and Petitions Working Group, 2008.

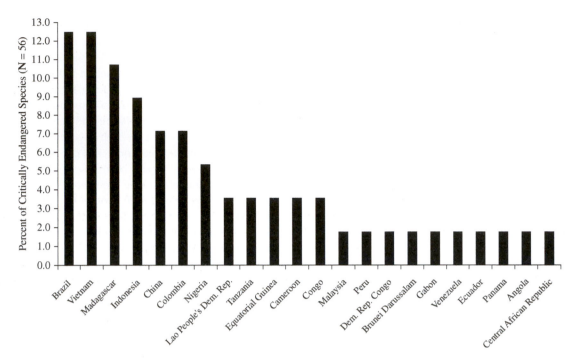

Figure 45.1 Distribution of critically endangered species by country. Drawn from data in IUCN/SSC Primate Specialist Group, IUCN Red List 2008: Threatened Primates by Habitat Country (www.primate-sg.org/RL08.countries.htm). All countries with ≥1 species classified as critically endangered are shown here.

because the phylogenetic distance between this taxon and other species and subspecies of red colobus is shorter than that between the aye-aye and its closest living relatives in other infraorders.

Primate evolutionary history is also more heavily concentrated in biodiversity hotspots than would be expected from the numbers of endemic primate taxa, or the *taxon richness*, that these areas support (Sechrest et al. 2002). Prioritizing these hotspots in conservation efforts may therefore be justified based on both the number of endemic taxa and the amount of evolutionary history they support.

MAJOR THREATS TO PRIMATES

Independent of whether conservation priorities emphasize preserving the greatest number of taxa or the most evolutionary history, developing informed conservation plans also requires an understanding of why primates become endangered in the first place and why these threats impact different taxa in different ways. The major threats to primates come from human activities that have reduced the size and viability of local populations. Hunting and other forms of harvesting are serious problems for many primates, but it is the destruction, degradation, and fragmentation of their habitats that pose the most ubiquitous threats of all. Overhunting and harvesting directly deplete local prey populations, while the destruction and deterioration of habitats affect the distribution of primates and further exacerbate

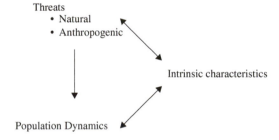

Figure 45.2 Interactions among threats to primates, population dynamics, and intrinsic characteristics of primates.

local population declines. The impact of these human activities varies widely in different regions of the world, and primates vary in their susceptibility to different kinds of threats. It is the interactions among the major threats, dynamics of small populations, and intrinsic characteristics of primates that affect their extinction risks (Figure 45.2).

Population and Habitat Dynamics

The IUCN's criteria for assessing extinction risks emphasize the status of populations and their habitats because extinction risks for small or rapidly declining populations often reflect declines in the total area and quality of habitat. Expanding human populations have increased the scope and intensity of human activities that negatively impact primates, particularly in the tropics, where most of the world's endangered primates are found.

Hunting represents one of the most serious threats to wild primate populations in many regions of the world, but other forms of overharvesting due to poaching or live captures can also severely deplete local populations. For example, rhesus macaques (*Macaca mulatta*) were nearly exterminated in some parts of India as a result of overharvesting to meet international demands for experimental subjects in biomedical research. It was not until the 1970s, when India introduced strict prohibitions on these live exports, that the depleted populations began to show signs of recovering (Southwick and Siddiqi 1988). Live captures for zoos, circuses, and other public exhibitions were also routine until strong international restrictions on wildlife trade were established. The Convention on International Trade in Endangered Species of Wild Fauna and Flora, widely known as CITES, went into effect in 1975; and as of 2008 there were some 174 participating countries. Yet, even with strict laws and enforcement efforts, mortality rates during the illegal capture and transport of live primates continue to be unconscionably high (Cowlishaw and Dunbar 2000). By one estimate, from Sierra Leone, five chimpanzees (*Pan troglodytes*) may be killed for every infant captured and only one of five captives is likely to survive to reach its destination alive (Teleki 1989).

Habitat loss restricts the total geographic distribution of primates and often leads to the fragmentation of whatever habitat remains. Fragmentation is an insidious problem because it isolates populations from one another and thereby disrupts natural dispersal processes and the corresponding gene flow that contributes to the maintenance of genetic variation within populations. Small, isolated populations may not be able to recover without human intervention, in the form of reintroductions from captive stock or translocations from other wild populations. Even then, habitat fragments may be too small to sustain viable populations of primates for very long. Forest fragments also suffer from additional ecological damage due to wind-induced tree falls, colonization from invasive species around their edges, and the elimination of other important members of their ecosystems, such as large predators and pollinators or seed dispersers (Marsh 2003, Ferraz et al. 2003).

Causes and Consequences of Population Declines

The dynamics of small, isolated populations fuel what has been described as an "extinction vortex" because of the negative feedback mechanisms that inevitably ensue (Lacy 1993). Small populations are equivalent to founder populations, which are likely to possess only a fraction of the genetic variation represented by an entire taxon. Small populations are also more likely to lose variation through the stochastic, or random, process of genetic drift. Limited dispersal not only interferes with gene flow but also reduces access to unrelated mates and thereby increases the risks

of deleterious close inbreeding and compounds the loss of genetic variation.

There are many examples of dramatic population decline due to natural phenomena, such as hurricanes, droughts, and epidemics. Recently, for example, researchers studying black howler monkeys (*Alouatta pigra*) along the coast of Belize have documented the decimation of the population following Hurricane Iris (Pavelka et al. 2003). Before the hurricane struck the site in October 2001, the 53 black howler monkeys in this study area were distributed among eight stable social groups. After the hurricane, which completely defoliated the canopy and uprooted many trees, the howler monkey population was reduced by 49% and the established social groups had disintegrated.

Other striking population declines due to prolonged droughts, storms, or epidemics have been documented in a wide diversity of primates, ranging from mantled howler monkeys (*Alouatta palliata*) on Barro Colorado Island, Panama (Collias and Southwick 1952, Milton 1996), to red howler monkeys (*A. seniculus*) in Venezuela (Rudran and Fernandez-Duque 2003) to ring-tailed lemurs (*Lemur catta*) in Madagascar (Gould et al. 1999) to brown lemurs (*Eulemur fulvus fulvus*) on the Indian Ocean island of Mayotte (Tarnaud and Simmen 2002) to eastern chimpanzees at Gombe National Park, Tanzania (Goodall 1986), to both western and central chimpanzees as well as lowland gorillas (*Gorilla gorilla*) (Leroy et al. 2004). Although the ecological and demographic damage caused by natural catastrophes can be severe, it is also acute and tends to be geographically localized. Human impact, by contrast, tends to be much greater in both its temporal persistence and spatial scale.

The impact of hunting and harvesting on primate populations also differs from that of natural catastrophes in at least three important respects. First, in contrast to droughts and disease, which generally take their greatest tolls on the youngest and weakest members of a population, human hunters either fail to discriminate among their targets or preferentially aim for the largest, most robust adults. They also often target adult females carrying infants, which are then kept or sold as pets if they survive the mother's death (Mittermeier and Cheney 1987). Entire mountain gorilla (*Gorilla gorilla beringei*) families have been killed for the illegal capture of a single infant (Fossey 1983), and the persistent availability of infant orangutans (*Pongo pygmaeus*) and other primates through illegal channels is an inevitable sign that their mothers were killed to procure them (Galdikas 1995a). The number of reproductive females in a population directly determines the potential and rate of population growth in a way that the number of adult males does not. Consequently, the loss of adult females from hunting and poaching reduces the potential for population recovery. In one computer simulation for northern muriquis (*Brachyteles hypoxanthus*), the harvesting of a single adult female per year was shown to significantly reduce the population's growth rate and, therefore, its ability to recover to a viable size (Rylands et al. 1998). A second factor that makes hunting such a serious threat is that hunters typically live on or near their hunting grounds and therefore exert chronic pressures on local populations of their prey. By one calculation, a single family of hunters in the Brazilian Amazon was estimated to have killed up to 500 large primates, including woolly monkeys (*Lagothrix*), spider monkeys (*Ateles*), and howler monkeys (*Alouatta*), within a 2-year period (Peres 1990b). In the Peruvian Amazon, local populations of large primates such as these are especially vulnerable when hunters use shotguns instead of bows (Alvard et al. 1997). The effects of subsistence hunting are even more severe in small, isolated forest fragments than in larger, more contiguous Amazonian forests because there are no opportunities for immigrants to "rescue" an isolated population that has been depleted by local hunters (Peres 2001). A third distinction that makes hunting such a serious threat is that it often extends beyond a local family's or community's subsistence needs to satisfy commercial markets (Naughton-Treves et al. 2003). Demands for bushmeat are increasing wherever local human populations are expanding, especially where international logging and development operations have opened new roads into remote areas that were previously inaccessible to hunters (Rose 2002). Bushmeat hunting may be responsible for the recent extinction or near extinction of Miss Waldron's red colobus monkey (Oates et al. 2000a), and unsustainable levels of bushmeat hunting of other taxa have been documented in many parts of western and central Africa (Fa et al. 1995, Milner-Gulland and Akçakaya 2001, Rose 2002).

Although population size is a critical predictor of long-term viability, in most taxa only a fraction of individuals in a population reproduce. The number of reproductive members of a population, or its effective population size (N_e), is a more accurate predictor of the population's viability than its total size because N_e describes the number of individuals contributing genes to the next generation (Franklin 1980). Estimates of minimum viable population sizes have ranged from 2,000 to 5,500 individuals necessary to achieve a minimum viable N_e of 200–500 breeding individuals, and more recent analyses suggest that as many as 7,000 breeding-age adults may be necessary for a 99% probability of persistence for 40 generations (Reed et al. 2003). Critically endangered taxa have much smaller total numbers, and even smaller populations, than this. For example, fewer than 1,000 northern muriquis are estimated to survive in a dozen forest fragments (Mendes et al. 2005), and the single largest population includes fewer than 300 individuals (Strier et al. 2006). Similarly, only about 700 mountain gorillas have been counted in recent censuses, distributed between two populations in the Bwindi Impenetrable National Park, Uganda, and the Virunga Volcanoes of Rwanda, Uganda, and the Democratic Republic of Congo (Robbins and Williamson 2008). Although such small total populations and correspondingly lower N_e sizes do not inevitably lead to extinction, the probability of their persistence declines steeply over time.

Both northern muriquis and mountain gorillas ranked among the world's top 25 most endangered primate taxa until 2006, when other even more critically endangered taxa replaced them (Table 45.2). The largest populations of each of these taxa have been the subjects of intensive long-term field studies. They have also been the beneficiaries of intensive conservation efforts, which have succeeded in reducing hunting pressures and protecting their remaining habitats. Increases in their populations have been documented during the decades they have been monitored, but it is still not clear whether these protected populations are sufficiently large to rebound from population declines caused by stochastic demographic processes or unpredictable catastrophic environmental events.

Habitat loss and fragmentation compound the problems of small populations because the areas that remain may not be large enough to support viable populations. In the Atlantic Forest of Brazil, for example, the area required to sustain viable populations of endangered species, such as northern muriquis, has been estimated to be at least 20,000 ha in size (Chiarello 2003). This is more than 20 times the size of the forest at the federally protected, private reserve in Minas Gerais, where one of the largest known populations of northern muriquis resides. There are other tracts of Atlantic Forest within the northern muriquis' geographic distribution that do meet these minimum size estimates; but illegal hunting is difficult to eliminate in these large national and state parks, and the population densities of northern muriquis and other sympatric primates found in them are correspondingly low (Pinto et al. 1993, Strier and Fonseca 1996/1997, Mendes et al. 2005).

Causes and Consequences of Habitat Loss and Fragmentation

Human demands for timber and land for agricultural and development purposes are the major causes of habitat loss and fragmentation. By 1985, only about a third of eastern Madagascar's rain forests remained, and recent assessments of southern and southwestern forests there show similar, if not faster, rates of loss today than in the recent past (Green and Sussman 1990; Sussman et al. 2003b). During the twentieth century alone, more than 90% of the Brazilian Atlantic Forest was destroyed (Câmara 2003). Although the Amazon Basin still has over 50% of its rain forest, it suffers from the highest absolute rate of deforestation in the world (Laurance et al. 2002). The vast Indonesian forests of Borneo and Sumatra, like those in western and central Africa, are now seriously threatened by logging concessions. Even protected areas, such as Borneo's vast Gunung Palung National Park, are suffering irreversible losses and fragmentation due to logging for timber and clearing for oil palm plantations, which devastated the 10 km buffer zone surrounding the park and then began encroaching on the so-called protected forest within the park itself. Since 1999, nearly 10% of the forest within the park's boundaries has been destroyed each year (Curran et al. 2004).

Table 45.2 Top 25 Most Endangered Primates, 2008–2010[1]

COMMON NAME	LATIN NAME	COUNTRY
Greater bamboo lemur	*Prolemur simus*	Madagascar
Grey-headed lemur	*Eulemur cinereiceps*	Madagascar
Silky sifaka	*Propithecus candidus*	Madagascar
Sclater's lemur	*Eulemur flavifrons*	Madagascar
Northern sportive lemur	*Lepilemur septentrionalis*	Madagascar
Rondo dwarf galago	*Galagoides rondoensis*	Tanzania
Roloway monkey	*Cercopithecus diana roloway*	Côte d'Ivoire, Ghana
Tana River red colobus	*Procolobus rufomitratus*	Kenya
Kipunji	*Rungwecebus kipunji*	Tanzania
Niger Delta red colobus	*Procolobus pennantii epieni*	Nigeria
Cross River gorilla	*Gorilla gorilla diehli*	Cameroon, Nigeria
Siau Island tarsier	*Tarsius tumpara*	Indonesia
Javan slow loris	*Nycticebus javanicus*	Indonesia
Pig-tailed langur	*Simias concolor*	Indonesia
Delacour's langur	*Trachypithecus delacouri*	Vietnam
Golden-headed or Cat Ba langur	*Trachypithecus p. poliocephalus*	Vietnam
Western purple-faced langur	*Semnopithecus vetulus nestor*	Sri Lanka
Gray-shanked douc	*Pygathrix cinerea*	Vietnam
Tonkin snub-nosed monkey	*Rhinopithecus avunculus*	Vietnam
Western Hoolock gibbon	*Hoolock hoolock*	Bangladesh, India, Myanmar
Cao-Vit or eastern black-crested gibbon	*Nomascus nasutus*	China, Vietnam
Sumatran orangutan	*Pongo abelli*	Indonesia
Variegated spider monkey	*Ateles hydribus*	Colombia, Venezuela
Peruvian yellow-tailed wooly monkey	*Oreonax flavicauda*	Peru
Cotton-top tamarin	*Saguinus oedipus*	Colombia

[1] From Mittermeier et al. (2009); see also Mittermeier et al. (2004, 2007) for earlier listings.

Selective logging involves the harvesting of valuable species of trees instead of clear-cutting entire forest tracts. While the disruptive effects of selective logging may not be as immediately apparent as clear-cutting, they nonetheless cause long-term ecological damage. Selectively logged forests in Indonesia, for example, are more susceptible to damage from fires caused by droughts associated with El Niño Southern Oscillation events (Siegert et al. 2001). There and elsewhere many of the trees being logged are often the same mature, fruit-producing species that serve as important food sources for frugivorous primates, including orangutans. Because many primates, such as chimpanzees, are also responsible for dispersing a considerable number of seeds from the species of fruits they ingest, reductions in the densities of frugivorous primates can have deleterious effects on forest recovery and regeneration (Chapman and Chapman 1996a, Chapman and Onderdonk 1998; see Chapter 34).

More than 20 years after selective logging ceased in Kibale National Park, Uganda, the densities of frugivorous primates are still significantly lower in the previously logged areas than in areas that had not been disturbed (Chapman et al. 2000). Male gray-cheeked mangabeys (*Lophocebus albigena*) were also found to be significantly heavier, and presumably healthier, in unlogged forests than in logged areas (Olupot 2000). At the same time, however, the densities of folivorous taxa, such as black and white colobus monkeys (*Colobus guereza*), in these selectively logged forests do not appear to have declined and may even be elevated relative to undisturbed areas. The apparent ability of folivorous primates to thrive in disturbed forests here and elsewhere may be attributed to the higher protein-to-fiber ratios of foliage found in the vegetation that colonizes open gaps within the forests and forest edges (Ganzhorn 1995, Ganzhorn et al. 1999, Oates 1996).

The noise and disturbances caused by logging operations can also cause primates to shift their established ranges into neighboring areas, where encounters with hostile conspecifics can have lethal consequences. The decline of one chimpanzee population in Gabon has been directly attributed to the lethal territorial clashes that occurred when chimpanzees fleeing from logging disturbances in their own ranges were attacked by the residents of the surrounding communities (White and Tutin 2001). By the time the loggers had left the area, the density of chimpanzees, as estimated from nest counts, had declined.

Like the folivorous taxa in the logged forests in Kibale National Park, local populations of western lowland gorillas did not fare as badly as those of sympatric chimpanzees during the logging operations in Gabon. One reason for this difference may be that gorillas are not territorial and are therefore more tolerant toward conspecific refugees that move into their ranges. The ability of gorillas to rely heavily on foliage may also contribute to their tolerance of conspecifics at higher densities (Yamagiwa 1999).

Like human harvesting of primates, anthropogenic impacts on primate habitats tend to have longer-term and farther-reaching consequences than natural ecological processes, such as the annual or super-annual fluctuations in rainfall that affect the availability of food and water. Nonetheless, the short-term impact of natural disasters that affect primate habitats cannot be ignored. For example, extended droughts can reduce standing sources of water, forcing primates, such as vervet monkeys or grivets (*Chlorocebus aethiops*) at Amboseli National Park, Kenya, during the 1970s, to make risky incursions into one another's ranges and exacting high mortality on low-ranking and less competitive individuals (Wrangham 1981). Droughts also reduce food availability, which negatively impacts survival and fertility. One population of ring-tailed lemurs that suffered high mortality due to a 2-year drought did not show signs of recovering until 4 years after the drought had ended (Gould et al. 1999). Smaller, more isolated populations may be unable to recover so quickly from these natural ecological fluctuations (Gould et al. 2003).

Many tropical plant species reproduce resulting in super-annual cycles, or masting events, such as the 4-year fruiting cycles of the Dipterocarpaceae trees in Indonesia, which have, at least until recently, coincided with El Niño Southern Oscillation events (Curran et al. 2004). Orangutan energy balance and fertility are strongly influenced by the timing of these mast fruiting events, with both energy intake and fertility declining in the intervening years when the lipid-rich fruits that many dipterocarps produce are scarce (Knott 1998a).

Historical cycles of mast fruiting events or of the unpredictable, harsh climatic conditions on islands such as Madagascar are ones to which primates inhabiting these areas have had generations to adapt (Wright 1999). Human impacts on primate habitats, however, represent new selection pressures; and the complete destruction of habitats is the most extreme and irreversible type. Forests can never be restored to their original conditions after they have been repeatedly cleared for planting crops or grazing livestock, flooded during the construction of hydroelectric dams, or asphalted for urban areas. The viability of primates in forest fragments is related to the size of the fragments and their distance from contiguous habitat, and small fragments can

lose significant components of their biodiversity within a few decades of being isolated (Marsh 2003). Acute disturbances, such as those caused by selective logging or other forest product extractions, can also have long-term deleterious impacts on habitat quality because these kinds of activity are so targeted in their scopes (Cowlishaw et al. 2009).

Variation in Primate Vulnerabilities to Different Threats

Pressures from human hunting and habitat destruction are not evenly distributed throughout the world, nor do they impact all primates to similar degrees. Chimpanzees, for example, are protected in some areas by local taboos against hunting animals so similar to humans, whereas sympatric gorillas and other primates are considered edible meat (Sabater-Pi and Groves 1972).

Habitat disturbances also affect different primates in different ways, depending on their degree of ecological specialization. Habitat alterations almost always have negative ecological impacts on specialized niches, with devastating consequences on the primates that occupy them.

Densities and Ecology

Sympatric taxa live at very different population densities, even in relatively undisturbed habitats. These differences reflect a combination of extrinsic factors, such as levels of mortality from predators and infectious disease and competition for food and other essential resources from other species, including nonprimates, and intrinsic factors, such as their respective energetic and nutritional requirements. Food and other critical resources are thought to determine the carrying capacities of different habitats for different taxa, but we still know very little about the carrying capacities at which most primates can live (Cowlishaw and Dunbar 2000).

Many primates have historically restricted geographic distributions and have evolved specializations to survive in narrow, highly specialized ecological niches. Others occur much more widely and can survive in a diversity of habitat types by adjusting their diets, ranging, and grouping patterns in response to the distribution and availability of their food resources. Geographic barriers that isolate populations are responsible for historical speciation events, such as the divergence between bonobos (*Pan paniscus*) and chimpanzees that occurred approximately 2.2 million years ago, when the Congo River became a barrier defining the boundaries of the bonobos' geographic distribution. Chimpanzees occupy a much greater diversity of habitat types throughout sub-Saharan Africa, from rain forest to semideciduous forest to woodland to arid savanna, than either gorillas, which occur sympatrically with chimpanzees (but not bonobos) in some parts of their ranges, or Indonesia's orangutans, which are restricted to the rain forests of Borneo and Sumatra.

Baboons (*Papio* spp.) and macaques are sympatric with these great apes in many African and Asian forests, respectively. However, unlike these apes, baboons and macaques have much more generalized adaptations, and the wider diversity of habitats they occupy is consistent with their more flexible ecological niches. Forest-dwelling baboons in Africa and macaques in Asia are much more frugivorous than conspecific or congeneric populations found in more open or seasonal habitats (Rowell 1966, Richard et al. 1989), just as central lowland gorillas are more arboreal and frugivorous than mountain gorillas, whose present high-altitude habitats necessitate the inclusion of more foliage in their diets (Remis 1997, Doran and McNeilage 1998).

Primates that can subsist on regenerating vegetation, which tends to occur in disturbed forests and along forest edges, may also be more adept at raiding human crops. Rhesus macaques have been described as just such a "weed species" because they appear to thrive in the altered habitats associated with human agricultural expansion (Richard et al. 1989). Brown capuchin monkeys (*Cebus apella*) may be the New World ecological equivalent of a weed species because of their extractive foraging techniques and skills at crop raiding (Fragaszy et al. 2004b).

The greater sensitivity of ecological specialists, such as frugivorous spider monkeys or frugivorous–insectivorous blue monkeys (*Cercopithecus mitis*) and red-tail monkeys (*C. ascanius*), to habitat disturbances cannot be decoupled from the roles these primates typically play within their ecological communities. Because specialists tend to be so sensitive to disturbances, they are more likely to suffer local extinctions when their populations are isolated in disturbed fragments of habitat (Brashares 2003). The loss of such primates can also negatively impact the plant species they eat, as appears to be the case with spider monkeys and the mimosoids (*Inga ingoides*) that rely on these monkeys for dispersing their seeds (Pacheco and Simonetti 2000).

At a more global scale, recent analyses have indicated that the diverse Malagasy lemurs play a much larger and more unique set of roles in their respective ecosystems than primates in much of South America or continental Africa, where other groups of animals such as birds, bats, and small or medium-sized mammals contribute to their ecological communities. The extinction of endangered lemurs from Madagascar would leave their ecosystems more severely compromised than the extinction of endangered primates elsewhere (Jernvall and Wright 1998).

BEHAVIOR AND THE COMPOSITION OF GROUPS AND POPULATIONS

The age/sex structure and the degree of relatedness among members of social groups and populations also significantly affect the viability of populations and the ability of small populations to recover to viable sizes. Most primates spend most, if not all, of their lives as members of social groups. The composition of these groups is affected by sex-biased or bisexual dispersal patterns, which also affect the demographic composition of populations. Although models in primate behavioral ecology have tended to focus on the size and composition of social groups, genetic considerations and conservation priorities require focusing on the size and composition of populations as well (Strier 1997a,2003c,d).

Populations with unbalanced age structures grow faster or slower than those with stable age distributions depending on the relative proportion of adults and immatures. Similarly, those that contain more males than females will have slower rates of growth than those with more females. The growth of the northern muriqui population at the Estação Biológica de Caratinga (now known as the RPPN Feliciano Miguel Abdala) has been attributed to the 2-to-1 female bias in infant births that occurred in one large group over more than a decade (Strier 1993/1994, 2000b). An increase in the number of male births and survivorship relative to females in recent years has led to projections for slower, if not negative, population expansion in the future (Strier et al. 2006).

The composition of large, relatively stable populations can fluctuate due to stochastic demographic processes, such as sex-biased infant sex ratios or mortality rates within groups. The impacts of these fluctuations on population growth rates, genetic structure, and even behavior may not become evident until members of one or more birth cohorts reach reproductive age. Slow maturation rates can result in lags of more than a decade before the demographic consequences of these events are observed. Dispersal can buffer the effects of fluctuations on group demography by the redistribution of individuals across populations, which is another reason that small, isolated populations with limited or no dispersal opportunities are so vulnerable to stochastic demographic events.

Sex-biased or bisexual dispersal patterns also directly affect the degrees of relatedness among and between groups, with potentially important consequences on behavior. For example, the unusually high levels of male/male competition and mate-guarding observed in the Ngogo chimpanzees at Kibale National Park, Uganda, have been associated with the disproportionately high number of males in this community (Watts 1998b, Mitani and Watts 1999). No one knows why this chimpanzee community is so large or has so many males, but the behavior patterns of its members cannot be compared to those of other chimpanzees without controlling for the effects of its unusual demography, which must be at least partially due to historically male-biased sex ratios or survivorship in their male-philopatric society (Strier 2000c).

Levels of male/male competition vary across different taxa as well as among populations of the same taxa, and in both cases they affect the degree to which fertilizations can be monopolized by individual males or are more equitably distributed among different males (Strier 2003d). While dispersal permits males of many species, such as ring-tailed lemurs (Sussman 1992) and vervet monkeys (Cheney and Seyfarth 1983), to adjust their reproductive opportunities by transferring between groups, male philopatry reduces the ability of males to escape from unfavorable demographic conditions in their natal groups (Strier 2000c). Although the transfers of two adult male bonobos from their natal community into an adjacent community with fewer males may be an indication that even philopatric males respond to population-wide demographic conditions when opportunities to disperse exist (Hohmann 2001), it is also the case that philopatric tendencies affect the genetic composition of local groups.

Reproductive Rates

Compared to most other mammals, primates have slow reproductive rates relative to body size. Taxa with rapid reproductive rates can sustain greater harvesting pressures than those with slower reproductive rates, which also tend to be larger in body size within each of the major clades and are therefore preferentially hunted for their meat (Oates 1996). Sustainability implies that population losses due to hunting or harvesting can be offset by replacements from births or immigrations, but when net losses exceed gains, populations decline and extinction risks rise (Bennett and Robinson 2000). Comparing the prevalence of different taxa in bushmeat markets provides a way of estimating their minimum harvest rates, which can then be weighed against their respective reproductive rates to evaluate whether hunting pressures are occurring at sustainable or unsustainable rates (Fa et al. 1995). This method works well only for taxa that are still relatively plentiful, however, because hunting success also declines as populations are depleted (Cowlishaw and Dunbar 2000).

Reproductive rates determine the potential for small populations to recover once the pressures that contributed to their declines are relaxed. Consider the difference in the reproductive rates of most marmosets (*Callithrix* spp.), which begin reproducing at about 2 years of age and give birth to twins at roughly 6-month intervals, versus orangutans, which begin reproducing at 8–10 years of age and give birth to single infants at 9-year intervals. The two decades it takes a female orangutan to produce two surviving offspring is longer than the average life span of a female marmoset, which can produce two infants in a single pregnancy.

Comparing marmoset and orangutan reproductive rates is an extreme and distorted contrast because it ignores the strong relationship between reproductive rates, on the one hand, and body size and phylogeny, and to a lesser degree ecology, on the other hand (Ross and Jones 1999).

Among closely related taxa, intrinsic reproductive rates tend to decline and generation lengths increase as body size increases and, relative to body size, reproductive rates also tend to decline with increasing brain size. Forest-dwelling primates also tend to reproduce more slowly than closely related taxa that occur in more seasonal habitats or are subject to less predictable ecological conditions (Ross 1998, Ross and Jones 1999, Wright 1999).

Slow reproductive rates and long generations represent life history trade-offs for increased infant survivorship (Charnov and Berrigan 1993). The relatively fast maturation rates and 1- to 2-year birth intervals of many lemurs, for example, have been attributed to selection pressures favoring fast life histories under the unpredictable and harsh seasonal environmental conditions that characterize much of Madagascar (Wright 1999, Gould et al. 2003, but see Dewar and Richard 2007). Taxa that have evolved slower life histories under more stable and predictable environmental conditions in their evolutionary past may be initially more resistant to environmental perturbations because they are more likely to forfeit reproduction than to die but ultimately more vulnerable to being drawn into the extinction vortex if their populations decline below minimal sizes because of their slower intrinsic reproductive rates.

APPROACHES TO CONSERVATION

Conservationists have long realized that their efforts must proceed at multiple levels depending on whether the major threats to primates occur at local, regional, national, or international scales. There is also increasing sensitivity to the necessity of working within the political, economic, and social systems in place at each level. Solutions vary depending on whether the primates are at risk from hunters, habitat destruction, or combinations of the two; but even then, protective policies are almost always much easier to establish than they are to enforce.

Primate conservation is a global concern, and most conservation efforts involve international collaborations between nongovernmental organizations (NGOs), policy makers, law enforcers, researchers, and local people whose lives and livelihoods intersect with those of primates. Most of the conservation NGOs in the United States have international offices or cooperative arrangements with NGOs based in the countries that primates inhabit. The various NGOs that are involved in primate conservation efforts also fill different niches based on their respective commitments to lobbying governmental agencies in support of conservation legislation, securing protected lands, or supporting research and educational activities. However, whatever their particular emphases and target groups, all conservation tactics for primates include protecting remaining populations and their habitats and, when appropriate, active management efforts on their behalf.

The hunting and poaching of primates represent quite different kinds of problems because the markets for meat versus live primates or primate products are so varied. International conservation initiatives, such as those regulated by CITES, and local enforcement and education can help to curtail poaching and illegal trade; but prohibitions on the hunting of primates for their meat require that alternative sources of protein are available to the people who need them (Rose 2002).

Even with controls on hunting, primates generally suffer when humans live at high densities nearby. As of 1995, nearly 20% of all humans lived within the world's biodiversity hotspots, and human population growth rates in these areas are higher than elsewhere (Cincotta et al. 2000). The rapid emergence and spread of infectious diseases, such as Ebola and human immunodeficiency virus disease, have been attributed to the encroachment of humans into primate habitats (Daszak et al. 2000, Walsh et al. 2003). In the Republic of Congo, Ebola outbreaks have caused multiple mass mortality among both species, with the recent toll on gorillas estimated to be 3,500–5,500 fatalities (Bermejo et al. 2006). Also, because a greater proportion of females died than males, the impact of these losses on the population is severe (Caillaud et al. 2006).

The setting aside of protected areas as sanctuaries for remaining populations of endangered primates has been a major achievement in many parts of the world. There is evidence that these protected areas can fulfill their functions of protecting primates and their habitats when there are sufficient financial resources for local enforcement and economic alternatives to hunting and habitat destruction are available for the local people (Bruner et al. 2001). Yet, there are still many officially protected areas throughout the tropics that are little more than "parks on paper." Like the lowland forests of Indonesia, which may be gone within decades despite their protected status (Jepson et al. 2001), human encroachment on national parks represents a serious and urgent problem in many other parts of the world. For example, the 1,000 km² buffer zone surrounding a 2,300 km² national park in Madagascar has already been penetrated by subsistence rice farmers, who are projected to reach the park's boundaries within a few years if their rate of expansion persists (Kremen et al. 2000).

Other protected areas in Madagascar, such as the Ranomafana National Park, have more promising futures. When Ranomafana National Park was inaugurated in 1991, it was only the fourth national park in the country (Wright 1992). International funds and cooperative programs for sustaining ecotourism and research have contributed to the success of Ranomafana National Park, which is now a World Heritage Site recognized by the United Nations.

Ecotourism can provide economic incentives to increase the participation of local people in conservation. The long-term economic value of ecotourism to visit the mountain gorillas in Rwanda is believed to have contributed to the gorillas' survival during a recent civil war that forced foreign researchers to evacuate the area (Weber and Vedder 2001). Profit sharing with local communities has also been critical to the success of ecotourism in Nepal (Bookbinder et al. 1998). Similarly, increasing the local market value of forest products would provide incentives for local communities in Honduras to conserve their surrounding forests because of the higher returns they would gain by conserving instead of exploiting these products (Godoy et al. 2000).

Research can also play a key role in conservation efforts, both through the critical information that researchers obtain about the primates and ecological communities they study and through the local and international attention that their scientific discoveries attract (Strier 2007). Long-term studies provide invaluable insights into ecological, demographic, and reproductive processes; and it is often these data, derived from the small number of taxa that have been studied over multiple decades, that provide the parameters for modeling extinction risks for other, less well-studied taxa (Strier and Mendes 2009). Computer programs known as "population viability analyses," or PVAs, are now widely used to simulate extinction risks based on known or estimated demographic and reproductive variables that affect population dynamics (Brook et al. 2000). However, surveys and more rapid assessments are still essential for evaluating the status of the majority of wild populations that have not been the subjects of long-term studies and about which little else is known.

CHALLENGES AND CHOICES

In addition to protecting primate populations and their habitats from irreversible losses, there are a number of more active, hands-on initiatives that conservationists can adopt. Examples include captive breeding programs and various ways of managing wild populations to increase and maintain their viability and of restoring altered habitats through the planting or maintenance of corridors. Debates about how much time and how many resources should be dedicated to these kinds of activity raise as many questions about the assumptions underlying conservation priorities as they do about which conservation plans should be implemented and how well they are likely to fare.

Captive Breeding and Reintroduction Programs

The stated priority of the Primate Specialist Group is the in situ preservation of species and communities (Cowlishaw and Dunbar 2000). In situ preservation requires protected habitats and sufficient population sizes, which may already be threatened or no longer exist. In such cases, the maintenance of viable captive breeding populations to preserve the gene pools of endangered taxa can serve as insurance against their extinction in the wild. Nonetheless, genetic and phenotypic changes occur in the necessarily small populations that can be maintained in captivity, and the most successful captive breeding programs are those that can reinforce in situ conservation efforts to maintain and restore wild populations (Snyder et al. 1996, Earnhardt et al. 2001).

The most successful example of a captive breeding program for an endangered primate involves the golden lion tamarin (*Leontopithecus rosalia*), many of which have been reintroduced into protected areas of their native Brazilian Atlantic Forest. By 1975, fewer than 200 golden lion tamarins were estimated to survive in just a few small patches of Atlantic Forest in the state of Rio de Janeiro; but by 2000, their numbers in the wild were estimated at about 1,000, with another 500 or so housed in zoos (Kleiman and Rylands 2002). Approximately one-third of all golden lion tamarins living in the wild today either were born in captivity and then reintroduced or are the descendents of captive-born tamarins that subsequently reproduced after being released in the wild (Kierulff et al. 2008). The first captive-born tamarins were reintroduced to the Poço das Antas Biological Reserve, a 5,200 ha forest established to protect the largest remaining wild population of this species. Reintroductions began there in 1984, and management efforts since 1994 have expanded to include the translocation of more than 120 individuals from 16 groups from small, isolated forests to larger, protected areas (Kierulff et al. 2002a).

The translocated individuals had higher survival rates and adapted more rapidly than the reintroduced monkeys, although the success rates for reintroduced tamarins increased as prerelease training and reintroduction procedures were refined (Kleiman 1996). Translocations were also much more economical, costing only about half as much as the reintroductions from the time of release (Kierulff et al. 2002a).

Space limitations and high housing and maintenance costs impose limits on the number of animals that captive facilities can maintain (Snyder et al. 1996). Small captive populations also require careful management because they are susceptible to the same genetic risks that threaten small populations in the wild. The Species Survival Commission (SSC), Captive Breeding Specialist Group (CBSG), and societies such as the American Association of Zoological Parks and Aquariums (AAZPA) maintain records of all registered individuals for management purposes. Transferring individuals from one facility to another can reduce the genetic risks of inbreeding in the captive "stock," and often separate housing facilities or birth-control measures are necessary to keep captive populations at a manageable size.

The success of the Golden Lion Tamarin Project can be attributed to a number of factors, including the self-sustaining population that could be maintained in captivity due to the small size and high reproductive rates of these monkeys and the natural tendencies of both sexes to disperse from their natal groups. Translocating entire family groups together and pairing captive-born individuals with wild-born mates proved to be highly effective procedures. There were also still suitable available habitats into which golden lion tamarins could be released. Protection of these habitats and accompanying research and education efforts have been essential to the ongoing conservation efforts on behalf of this species.

Managing Wild Populations and Habitats

Although less expensive than reintroductions, translocations of primates from unprotected to protected habitats, or among small and isolated populations to increase gene flow, are not entirely risk-free. As with reintroductions, care must be taken with translocations to prevent the spread of infectious diseases and pathogens and to minimize social and ecological disruptions in the communities into which the individuals or groups are being moved (Yeager and Silver 1999).

An increasingly attractive alternative to translocating individuals from isolated populations is to establish habitat corridors between fragments so that the animals can move safely through them on their own. The creation of corridors between isolated populations assumes that natural dispersal processes will increase gene flow, thereby decreasing the deleterious effects of small population size (Lens et al. 2002). Genetic studies indicate that one small population of Scandinavian gray wolves (*Canis lupus*) was "rescued" by the immigration of a single, reproductively successful individual (Vila et al. 2002), suggesting that it may not require many such dispersal events for small populations to recover.

Corridors that connect isolated populations also pose potential risks, such as the spread of infectious diseases (Stoner 1996b) or the synchronization of population dynamics that can lead to simultaneous extinctions (Earn et al. 2000). Nonetheless, establishing corridors to increase connectivity among isolated populations can at least partially offset the effects of habitat loss. Many African countries are predicted to lose up to 30% of their primate fauna due to the "extinction debts" caused by the long time lags between habitat losses and population extinctions, and restoring connectivity among fragmented habitats is recommended before the populations show signs of collapsing (Cowlishaw 1999b).

Prioritizing Charismatic Taxa or Habitats

Most conservation education programs focus on highly charismatic taxa, which then serve as flagships for attracting attention and resources to conservation efforts that benefit other fauna and flora in the process. Primates, along with other mega-fauna, such as the World Wildlife Fund's famous panda emblem, hold a special appeal. Among primates, the great apes have the unique distinction of being most closely related to humans, and this has contributed to the establishment of special initiatives, such as the Great Ape Conservation Act of 2000. In 2007 alone, the U.S. Congress appropriated $1.4 million to the Great Ape Conservation Fund.

Nonetheless, the challenges of establishing conservation priorities extend beyond particular charismatic taxa or their endangered status and evolutionary uniqueness because the largest populations of the most endangered primates do not always persist in the most pristine habitats. For example, some of the highest densities of endangered primates endemic to Brazil's Atlantic Forest are found in small, degraded forest fragments, instead of in the few tracts of large protected forests that remain. Should the protection of these large populations in disturbed habitats take priority over that of smaller populations in undisturbed habitats? Alternatively, should the most pristine ecosystems be given priority, even if they support smaller populations of endangered taxa living at lower densities?

Comparative analyses of the biogeography of endangered species suggest that although populations are often most fragmented at the extremes of their geographic ranges, these may also be the populations with the best prospects for survival because they are more removed from the disturbances responsible for population losses in the first place (Channell and Lomolino 2000). Higher rates of evolution may also occur at the edges of a species' distribution because of the more extreme ecological conditions to which they are subjected (Thomas et al. 2001), and these changes are expected to become more pronounced from the alterations of habitats due to global warming (Pounds and Puschendorf

2004, Thomas et al. 2004). Indeed, some conservationists have already posed the uncomfortable question of whether the parks and reserves established to protect endangered primates today will continue to provide suitable habitat as global climatic conditions change (Cowlishaw and Dunbar 2000).

THE FUTURE OF PRIMATE BIODIVERSITY

Between the inevitable effects of global warming on the world's endangered ecosystems and the ongoing expansion of human populations in and around the world's biodiversity hotspots, it is difficult to foresee how the future of primates—and other animals—that are threatened with extinction can be protected. Global losses of biodiversity and ecosystem changes are predicted to occur by 2050, and major primate extinctions may occur even sooner than this because rates of deforestation in countries such as Indonesia and Madagascar are so high (Jenkins 2003).

There is no question that human pressures are accelerating the extinction risks for many primate taxa. Whether through direct actions, such as unsustainable hunting and habitat destruction, or indirect activities, such as the far-reaching effects of atmospheric pollution on global climate, the impact of humans on other primates today is much greater than it has been in the past (Fuentes and Wolfe 2002). Yet, despite the depressing forecast for primates, increased awareness about the status of the world's endangered primates has fueled intensified international conservation efforts. It is too soon to tell whether these efforts will ultimately succeed in securing the futures of all endangered taxa, but there is no doubt that they are helping gain essential time in what for many primates is now an urgent race against extinction.

46

Ethnoprimatology
Contextualizing Human and Nonhuman Primate Interactions

Erin P. Riley, Linda D. Wolfe, and Agustín Fuentes

Every group of non-human primates is increasingly dependent on its nearest human neighbours. What the humans think of these animals, the manner in which they are willing to live with them, whether or not they exploit them, ignore them, or provide for them determines the ultimate well-being and survival of those non-human primates. Until recently, ecological questions regarding non-human primates have been treated as if the ecosystem in which they lived was self-perpetuating, and separate from the human ecosystem. Clearly this is not case. Humans and non-human primates are engaged in a complex set of interactions. (Burton and Carroll 2005:25)

1. What is ethnoprimatology, and why is it becoming an important area of primatological studies?
2. What are some of the basic patterns in human–nonhuman primate interactions?
3. How have nonhuman primates been included in human cultures and mythologies? Why are these historical and sociocultural elements relevant to the study of primates?

INTRODUCTION

The subject of this chapter is *ethnoprimatology*, the multifaceted interactions of human and nonhuman primates (Sponsel 1997, Sponsel et al. 2002). Ethnoprimatology should not be confused with *cultural primatology*, the study of primate cultural traditions (McGrew 1992, Perry 2006). Instead, ethnoprimatology draws from both biological and cultural anthropology to explore the interface of primate and human ecology. Drawing from *folkbiology* (i.e., the way people understand and categorize plants and animals; Medin and Atran 1999) and *ethnoecology* (i.e., people's knowledge, beliefs, and values of their environment; Sponsel 1997), ethnoprimatology is concerned with how people conceptualize environmental domains such as nonhuman primates but also with how these conceptualizations shape people's behavior toward primates and nature in general. As such, ethnoprimatology provides a holistic view of human and nonhuman primates and, hopefully, will facilitate the survival of nonhuman primates, many of which are facing extinction (see Chapter 45).

This holistic approach to the human–primate interface includes examinations of human hunting of nonhuman primates for food, keeping nonhuman primates as pets, bidirectional pathogen exchange, the impacts of habitat alteration/destruction, overlapping resource use (including crop raiding), indigenous knowledge of nonhuman primate behavior and biology, and the incorporation of nonhuman primates into myths, folklore, and other narratives. Space does not permit an examination of all of these subjects, but there are three recent contributions with extensive bibliographies and chapters/articles devoted to various topics concerning the interaction between human and nonhuman primates (e.g., Fuentes and Wolfe 2002, Paterson and Wallis 2005, Fuentes 2006a,b). Hunting, habitat destruction, crop raiding, and conservation are among the older concerns of primatologists and conservationists (see Chapter 45). The bidirectional spread of infectious diseases and the consequences of ecotourism and primate pet ownership are more recent concerns. Janette Wallis and Lisa Jones-Engel have been instrumental in bringing these important issues to our attention (Wallis and Lee 1999, Jones-Engel et al. 2005).

In this chapter, we first identify the overall ecological and temporal context of the human and nonhuman primate interface. We then provide a broad overview of the basic patterns of interconnections between human and nonhuman primates, including temple/urban monkeys, primates as prey, primates as pets, primates as resource competitors, and primates in myth and folklore. The latter includes a number of in-depth examples of the ways in which nonhuman primates appear in the myth, folklore, and behavioral attitudes of peoples in different regions of the world.

HUMAN AND NONHUMAN PRIMATE INTERCONNECTIONS IN CONTEXT

While human and nonhuman primates share a number of interconnections, at the most basic level there is a distinction introduced by geography. When envisioning the globe, there are relatively clear demarcations between zones of *sympatry* (geographic overlap) and zones of *allopatry* (lack of geographic overlap) (Figure 46.1). The zones of sympatry fall primarily in Africa; southern, eastern, and southeastern Asia; and South/Central America and reflect the distribution of nonhuman primates for at least the last 10 millennia. The zones of allopatry are primarily above ~30° north of the equator (above the Tropic of Cancer) except in East Asia, where two species of macaque monkeys (*Macaca mulatta* and *M. fuscata*) range relatively far north in China and Japan, and in North America, where nonhuman primate populations reach the Tropic of Cancer (23.5° north).

Understanding the relationships between human and nonhuman primates requires a temporal context as well as a geographic one. Long-term sympatry, especially sympatry that involves common usage of habitat, can result in a form of coecology. That is, ecological pressures impact mammals in particular ways, and mammals that share many morphological and physiological facets, such as the anthropoid primates (monkeys, apes, and humans), may also share similar adaptations at a variety of levels. This is important in understanding the interconnections between human and nonhuman primates because long-term overlap and similarity in behavior/other modes of adaptation (even if slight) can impact human conceptualizations of "nature" and act to facilitate distinct patterns of integration/engagement between humans and other primates. Alternatively, in areas with less time overlap, especially in zones of allopatry where overlap is mitigated by captivity or other forms of limited/selective exposure, specific patterns of strong association and incorporation are expected to be uncommon. However, as with all things human, this dichotomy is far from complete. With rapid urbanization in areas of sympatry and cultural shifts in the context and conceptualizations of nonhuman animals (especially apes) in zones of allopatry, these distinctions can become blurred or even reversed.

PATTERNS OF HUMAN–MONKEY INTERCONNECTIONS

Throughout much of southern and southeastern Asia, at least three species of macaques appear to coexist well alongside human populations. Two of these species, the rhesus

Figure 46.1 Zones of sympatry (■) and allopatry (□).

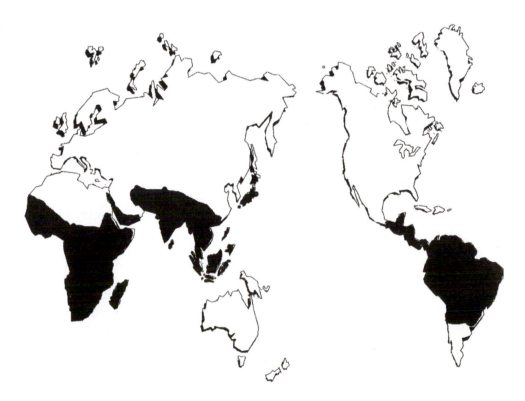

macaque (*Macaca mulatta*) and the long-tailed macaque (*M. fascicularis*), may in fact be increasing their numbers in areas of direct habitat overlap with humans relative to forested or other areas considered "natural" habitat for them. These increases are tied to both human-directed changes to the local ecologies which provide favorable habitat/ecological features for the macaques and specific aspects of the macaques' "fit" within human cultural practice such that humans may also derive specific cultural benefits from the macaques (Fuentes et al. 2005). In some areas of Amazonia, local peoples differentially treat specific species of monkeys such that they are brought into human cultural practice and offered relative protection from predation (Cormier 2002, 2003; Lizarralde 2002; Shepard 2002). These behavioral and physical overlaps and interactions also create an environment where more than space is shared; there may also be commingling of pathogens. While it is generally assumed that human–nonhuman primate pathogen "sharing" has deleterious repercussions, our understanding of the patterns and contexts of these copathogen environments and their evolution remains rather incomplete (Engel et al. 2002, Jones-Engel et al. 2005, Chapman et al. 2006). Regardless, if human and nonhuman primates overlap, gain "benefits" from (physiological, ecological, behavioral, cultural, or otherwise), exert "costs" on (physiological, ecological, behavioral, cultural, or otherwise), and impact one another over time via their sympatry, then one can envision each of the species having effects on the other, especially under conditions where the temporal variable is deep. There are four specific examples of interconnections that deserve attention in this context: temple/urban monkeys, primates as prey, primates as pets, and primates as resource competitors.

Temple/Urban Monkeys

Temple/urban monkeys exhibit extensive sympatry with human populations and tend to reside around specific temple structures either in or around areas of human habitation. These situations can be in human cities, small local shrines and temples in villages and towns, or specific temple sites that double as sacred places and tourist locations. This pattern is primarily found in southern and southeastern Asia and is strongly associated with cultures that practice Hinduism and/or Buddhism and their associated rituals and temple construction patterns (Aggimarangee 1992, Fuentes et al. 2005, Wolfe 2002). However, in the case of purely, or primarily, tourist locations, these sites can also include nontemple sites where monkeys are residents. Such tourist sites can be found in Gibraltar in Europe (macaques); many places in Kenya, South Africa, and Tanzania (baboons and vervet monkeys); in Japan (macaques); in Singapore (macaques); and even in the U.S. state of Florida (macaques). The primary ecological concern that arises from the interactions at tourist sites is disease transmission (Fuentes et al. 2008, Engel et al. 2008).

While construction of Hindu and Buddhist temples reflects a time depth of no more than one to four millennia (depending on location), human and macaque overlap in these areas has a much greater temporal depth. Evidence for habitat overlap between orangutans, macaques, and humans (and possible predation by humans) in southeastern Asia is clear from remains at Niah Cave in Borneo over 25,000 years ago (see Chapter 18; Sponsel et al. 2002). Given the prominent role of monkeys in Hindu and Buddhist myth (see later section "Primates in Human Myth and Folklore") and the high degree of security and nutritional assistance that humans provide to extant temple macaque populations, one can argue that there are human cultural benefits to this arrangement (Fuentes et al. 2005). This situation is strongly exemplified by the pattern of interactions and coexistence between *Macaca fascicularis* and humans on the island of Bali, Indonesia.

The macaques on Bali are found throughout the island, except in the extreme urban areas in the regency of Badung (such as the capital city, Denpasar). There appear to be at least 63 sites where macaques reside on the island (Fuentes et al. 2005). Each of these sites has one to three groups of macaques that either range fully within the site or use the site as part of their total home range. Each site has between 15 and over 300 monkeys. Macaque densities at these sites range from one to over 20 individuals/km^2, while human densities average over 500 individuals/km^2 across the island. Most importantly, over 68% of these sites are associated with a temple or shrine. These religious complexes can be as small as a simple shrine consisting of a few stones and an altar to elaborate temple complexes that are heavily used by Balinese and, in some cases, foreigners (Fuentes et al. 2005). Many of these macaque groups receive some provisioning at sites that contain a shrine or temple complex and have temple ceremonies several times per year. A component of Balinese Hinduism is the regular placement of offerings at shrines and temples. In most cases, these offerings consist of at least 30% edible elements. This suggests that a large percentage of the macaque groups on Bali are food-enhanced (see Fa and Southwick 1988, Wheatley 1999), especially if one includes crop raiding as food enhancement. Presumably tied to this food availability, temple-site populations tend to have larger group sizes than non-temple-site populations. Importantly, it appears that specific land-use patterns and wet-rice agriculture combined with the complex temple and irrigation systems of the Balinese (Lansing 1991) have resulted in a mosaic of riparian forest corridors and small forest islands throughout much of Bali (Fuentes et al. 2005). This type of landscape fits remarkably well with the macaques' patterns of using riparian habitats and small forest clusters for residence, foraging, and dispersal. This landscape has been formed over at least the last few millennia, and the pattern of distribution of macaque populations across the island suggests that the macaques are exploiting it (Fuentes et al. 2005). In this scenario, the human niche construction, or creation of Balinese *place* (socially impacted space, such as the temple complexes), creates an ecology and habitat that is beneficial to the macaques via protection, nutrition, and ecology. Simultaneously, the humans gain specific cultural and economic benefits from the presence of the macaques in and around their temple sites (Fuentes et al. 2005, Wheatley 1999). This sympatry impacts the behavior and possibly other physiological facets of both primate species. Finally, there is also growing evidence of cross-species pathogen exchange and possibly coevolution to similar pathogen environments between humans and macaques facilitated via extensive sympatry resulting in mutual habitat use and behavioral interactions (Engel et al. 2002).

Primates as Prey, Humans as Predators

Humans are typically excluded from analyses of predation ecology primarily because, traditionally, they are not considered a natural component of the faunal community of forest ecosystems (Sponsel 1997). Anthropological primatologists, however, are increasingly pointing to the potential significance of human predation on nonhuman primates across the world (Tutin and Oslisly 1995, Sponsel et al. 2002, Godfrey and Jungers 2003, Reed and Bidner 2004). Second, research has demonstrated that primates are not only key subjects of prey in contemporary areas of human–nonhuman primate sympatry but also captured from allopatric zones for various human uses ranging from ingredients in traditional medicines to subjects of biomedical research (Mittermeier 1987, Alvard et al. 1997, Wallis and Lee 1999, Shepard 2002, Cormier 2003, Urbani 2005).

As relatively large mammals, especially in the neotropics, primates are a common choice for hunters (Alvard et al. 1997, Amman et al. 2000, Lizarralde 2002, Shepard 2002). In fact, in a recent review on predation on New World primates by humans, Urbani (2005) argues that **Homo** constitutes the main predator of nonhuman primate communities in the Neotropics. There is evidence that humans do not take nonhuman primates randomly or solely in respect to optimal prey return models (Shepard 2002). In fact, it is common for human hunters to selectively hunt specific primates over others for very cultural reasons (Cormier 2002, 2003; Fuentes 2002a; Lizarralde 2002; Shepard 2002). For example, the Matsigenka, who live in the Manu region of Peru, preferentially hunt larger monkeys, so much so that in areas where shotgun use is practiced, the larger species (spider [genus *Ateles*] and woolly [genus *Lagothrix*] monkeys) have become locally extinct (Alvard and Kaplan 1991). However, the howler monkeys (*Alouatta seniculus*) in Matsigenka territory, similar in size to the spider and woolly monkeys, remain at viable population levels and are taken about half as often as the capuchin monkeys, which are one-half their size (Shepard 2002). The Matsigenka have two reasons for this differential predation: The howlers are less "tasty" and are considered slow and lazy, characteristics that it is believed are passed on to children who eat the meat (Shepard 2002). However, this aversion to howlers is not common to all Amazonian

peoples; in fact, some preferentially consume them (also for practical and cultural reasons) (Cormier 2002, 2003). These examples and others suggest that humans selectively prey on primates, such that some species see a greater or lesser impact than others. Given these cases, one would assume various models proposing that selective predation by humans impacts the prey populations of nonhuman primates. In fact, recent overviews have demonstrated substantial behavioral and possibly morphological changes in primates due to extant or even past predation risks caused by nonhuman predators (Miller 2002a). However, aside from general impacts on prey densities and local extinction events, there are few studies that have attempted to model the impact of human predation on nonhuman primate behavior, physiology, and morphology in a manner similar to the modeling of other animals' predation on primates (Alvard and Kaplan 1991). Unfortunately, this suggests that long-term, sustained predation by humans at a level below extinction-causing may impact the behavior and morphology of nonhuman primates, but we currently do not have sufficient data to reach a conclusion. It is also possible that selective hunting for body parts or other features that are used in traditional medicines may impact primate populations.

Primates as Pets

Ethnoprimatological research has documented tremendous variability in conceptualizations and manifestations of the concept of "pet." We therefore broadly define this category to include animals kept for companionship, enjoyment, or status; for sociocultural inclusion; and for use as economic tools. Today, it is largely in areas of sympatry that we see primates as pets. This is due to the laws regarding the import/export of animals across international boundaries and the impact of the World Conservation Union (IUCN, see Chapter 45) classifications regarding the endangered/threatened status of a variety of primate species. However, the present situation is largely recent as nonhuman primates, especially small monkeys, were widely available in Europe during the early centuries following colonial expansion and both monkeys and apes were kept as pets in private menageries and homes during that time. Due to the costs of transport and the high levels of mortality during transport, it is unlikely that the zones of allopatry ever had the levels of primate pet ownership that we see in zones of sympatry. Today, pet ownership of select nonhuman primate species is widespread in southeastern Asia and Amazonia and occurs in southern Asia and other parts of South America and Africa as well. In Asia, the majority of primates owned as pets belong to the genera *Macaca* and *Nycticebus* (the slow loris) (Malone et al. 2002); in South America, nearly all of the species found in Amazonia are kept as pets by at least some indigenous groups; and in sub-Saharan Africa, most pets are the medium–small monkeys such as vervets (genus *Chlorocebus*), guenons (genus *Cercopithecus*), and galagos (genera *Galago* and *Galagoides*). In North Africa, the sole African macaque, *Macaca sylvanus*, is kept as a pet;

and this species has also had a small resurgence as a pet in western Europe as well (although it remains illegal). Apes are held as status pets in Asia and Africa but have fairly low rates of survivorship in captivity as pets and are quite expensive and energy-intensive to keep relative to other nonhuman primates.

It is not clear that the keeping of pets impacts the "wild" populations of monkeys/apes in a way distinct from predation as pets are frequently by-products of hunting and their capture functions similarly to extraction via predation. Because the pet tradition is quite old in southeastern Asia (and probably South America), it is also quite likely that, as with the temple monkey overlap or even more so, there is substantial bidirectional pathogen transmission between humans and their primate pets (Jones-Engel et al. 2005). It is possible that there are physiological changes, even adaptations, in populations of human and nonhuman primates that overlap extensively, such as pets, humans, and the local populations from which the pet primates come, as a result of these intensive interactions.

Although not analogous to pet keeping, some nonhuman primates are currently raised in captivity and selectively bred for human biomedical experimentation. These colonies (primarily of macaque monkeys and a few other species) are selectively bred to create specific pathogen-free breeding colonies to provide test subjects for a variety of human disorders ranging from obesity to human immunodeficiency virus/acquired immunodeficiency syndrome. Many of these colonies are in zones of allopatry (primarily the United States and Europe), but increasing numbers are being established and run in zones of sympatry (e.g., Indonesia, Nepal, and Japan).

Sociocultural inclusion refers to the integral inclusion of the nonhuman primate as a participant in cultural facets of human society. While there are a number of human societies where this pattern takes place to a greater or lesser extent, the Guaja of Brazil are among the strongest examples (Cormier 2002, 2003). The Guaja, a Tupi-Guarani-speaking group in Amazonia, display a remarkable and intense inclusion of nonhuman primates, of many species, into their social fabric. Young monkeys are kept as pets, trained by children, and cared for by women and represent a significant culturally valued image of fertility (Cormier 2002, 2003). Young monkeys whose parents are killed for food will be brought to the village and "adopted." Women will bathe, breast-feed, and "wear" the monkeys; and throughout much of their development, the monkeys are carried and cared for by members of the Guaja village (Cormier 2002, 2003). The Guaja have a complex cosmology that involves ritual cannibalism (of howler monkeys), an emphasis on fertility and lactation, and the "wearing" of monkeys as signs of physical and social prowess for women. While patterns of carrying and raising monkeys have a considerable cost in terms of energy, movement, and foodstuffs, the returns go beyond the purely social. Young girls (and to a lesser extent boys) frequently practice mothering not only with

their siblings but also with infant monkeys, with some girls as young as 5 years being the primary caretakers of infant monkeys. Boys gain important hunting experience through exposure to monkey movement and behavioral patterns including vocalizations (Cormier 2002, 2003). However, the mortality rate for the captive monkeys is fairly high at the infant level; and if they live into adulthood, they are no longer as tightly wound into the fabric of the Guaja's social lives and generally begin to display an array of aberrant and aggressive behaviors (although some do reproduce in the Guaja villages with others of their own species) (Cormier 2002, 2003).

Nonhuman primates are used as economic tools in southern, southeastern, and northeastern Asia. While there is reference to crop-picking macaques across much of southern Asia, the best current studies come from Thailand, where males of one species of macaque (*Macaca nemestrina*) are kept, raised, and trained to pick coconuts (Sponsel et al. 2002). In this case, humans in southern Thailand capture macaques as young individuals and then invest substantial time and energy to train them to be effective coconut pickers. This system can be highly efficient for the humans as a well-trained macaque can harvest 500–1,000 coconuts per day and do so with a lower cost and higher return than a human could (the other harvesters are young human boys). Sponsel et al. (2002) present this situation as an example of a possible adaptive shift from a predator–prey relationship to a cooperative human–monkey economic relationship. They argue that cultural and agricultural systems have favored a move from conflict between humans and macaques over crops to a synergistic relationship wherein the humans capture, train, and maintain the macaques as they serve their economic role (job). Obviously, the impact on the macaque population may be negative in the sense of individuals taken out (similar to predation); but the overlap between the humans and the trained macaques, in this case, can be seen almost as a direct functional domestication, or at least the use of a "wild" animal in a domestic capacity.

Monkey performance is a blend of the economic and social inclusion facets of the monkey–human relationship. Monkey performances include a variety of theatrical performances by trained macaques, where the monkeys mimic human cultural behavior via a series of staged interactions with their trainer, the audience, and in some cases other monkeys. Generally, the audience provides monetary contributions at the conclusion of the performance. These funds are the income of the trainer and are used to maintain both the monkey(s) and the trainer's family. While these performances are widespread across Asia, they also occur in northern Africa and historically in Europe (Janson 1952). The most in-depth description of these practices comes from Emiko Ohnuki-Tierney's work on Japanese monkey performances. Ohnuki-Tierney (1987, 1995) presents the notion that in Japan the monkey performances, and the macaque monkeys themselves, act as a mirror for humanity, "playing a powerful role in their [humans'] deliberations of who they are as humans vis-à-vis animals and as a people vis-à-vis other peoples" (Ohnuki-Tierney 1995:297). As monkeys are the only animal addressed as *san*, the adult human address, and referred to as "humans minus three pieces of hair," there is an aspect of social inclusion, at least in the context of a special role for the monkeys, in Japanese culture that is distinct from the roles of other animals (Ohnuki-Tierney 1987, 1995). The integration of economic and cultural roles for macaques in Japan results in a distinct and complex cultural relationship between humans and monkeys that exists at the same time that increasing conflict over land and crops emerges as a predominant pattern of interaction between humans and "wild" macaques (Sprague 2002).

Synecologies: Primates as Resource Competitors

A key feature of the ethnoprimatological approach is envisioning human and nonhuman primates as members of a dynamic ecosystem (Fuentes and Wolfe 2002). Currently, in many areas of the world, these dynamic ecosystems are characterized by an increased likelihood of conflict due to overlapping resource use. The transformation of forest into agroforested and agricultural lands, for example, can lead to overlapping resource use, given that such areas often result in the creation of new potential sources of food for nonhuman primates. Crop raiding by primates is particularly problematic in Asia and Africa, with the most common (and successful) "pest" species including members from the genera *Macaca*, *Papio*, and *Cercopithecus* (Else 1991, Strum 1994, Hill 1997, Paterson and Wallis 2005, Riley 2007). The extent to which crop raiding is a problem typically depends on the location of crops in relation to wildlife habitats and the distance of the crops from such habitats, the types of crops grown, and people's ability to protect their crops from invading animals (Hill 1997, Riley 2007). Because crop raiding poses a serious challenge to the future conservation of many primate species (Strum 1994) and conservation inherently involves the values and actions of humans, research in this area is increasingly addressing both sides of the conflict; that is, examining crop raiding as part of a primate's ecological strategy *and* how this strategy affects the livelihoods of local farmers (e.g., Naughton-Treves 1996, Hill 2005, Lee and Priston 2005, Riley 2007a).

Human–nonhuman primate conflict and competition can also emerge as a result of overlapping use of forest resources. The conservation significance of overlapping resource use is contingent upon the importance of the forest product for the diets of the primate species and for the livelihoods of the human users. At the Tana River Reserve in Kenya, Kinnaird (1992) found that the palm, *Phoenix reclinata*, both is heavily exploited by people of the Tana River district and provides an important food source for the endangered Tana River crested mangabey. Thus, in this case, from a conservation perspective, it is critical that human use be monitored so that the palm can persist for future use by both the mangabeys

and human residents. In Lore Lindu National Park, Sulawesi, there is overlapping use of both forest and cultivated (e.g., cacao) resources between Tonkean macaques and villagers (Riley 2007a). One primary forest resource for which there exists overlap is the *Arenga pinnata* palm. While humans benefit from multiple parts of the tree, including the sap as a fermented wine, the hair fibers for household items, and the palm fronds for roofing, the macaques rely on the fruits of the palm as a key resource, particularly in human-modified habitats (e.g., the palm's fruits comprised 52% of the diet of one of the macaque study groups; Riley 2007b). Riley (2007) points out, however, that many of these culturally salient items from the palm can be removed without felling the tree. In fact, the manner in which villagers exploit the palm ensures its persistence and, thus, may *positively* impact the Tonkean macaque's ability to persist in human-modified landscapes. In terms of cultivated resources, there was considerable discrepancy between farmers' perceptions regarding the severity of cacao crop raiding by macaques and the actual amount of cacao fruits consumed by the macaques (Riley 2007a). Although the macaques were responsible for significantly more raiding than other forest mammals, Riley (2007a) suggests that crop loss caused by macaques is likely to increase in the future as a result of further habitat modification by humans. In this case, as well as in many others, the ultimate question becomes, is future coexistence of human and nonhuman primates possible?

Wright et al. (2005) also examined overlapping resource use between Tanala village residents and the lemurs of Ranomafana National Park, Madagascar. The authors found that of the 235 taxa of plants used by the Tanala, only 56 of them were consumed by lemurs. However, human use of 60% of those 56 taxa involved felling and destruction of the plant, thereby inhibiting future use by lemurs. That being said, the authors acknowledge that if plant extraction for household use is maintained at traditional, low levels, then it will be unlikely to pose a threat to lemur survival. Instead, destruction and conversion of forest ecosystems to *large-scale* agricultural lands is seen as the primary source of human–nonhuman primate conflict and, hence, the primary threat to lemur conservation.

PRIMATES IN HUMAN MYTH AND FOLKLORE

Shepard (2002:114) suggests the following:

> Myths concerning the origins of biodiversity are a source of both entertainment and philosophical speculation. Animal species are anthropomorphized in folktales, spiritual beliefs, and personal anecdotes. Metaphors drawn from ecological processes are used to understand illness, death, and other aspects of the human condition, while human intentions and emotions are projected into the ecological sphere.

The symbolic worldviews of people who are sympatric with nonhuman primates make up an important aspect of the study of the primates themselves. Myths, epic poems, folklore, and so forth are important aspects of culture and impact the lives of the primates that exist in and around the humans. It is possible that aspects of human cultures might be used to create projects and programs to ensure the future survival of nonhuman primates. However, it is not always clear how, and if, primates' presence in these cultural narratives influences human attitudes and treatment of nonhuman primates or how such views might be used to promote conservation. Part of the problem is that these facets of human culture (myths and the like) are often analyzed independently of observations of the behavior and attitudes of people toward nonhuman primates. Because of the way academia has constructed its disciplines, moreover, conservationists and primatologists can be educated separately from folklorists and cultural anthropologists (and vice versa); and therefore, one discipline possesses very little of the knowledge held by another discipline. The hope of ethnoprimatology is that a more holistic view of human and nonhuman primates will suggest new opportunities to develop better programs to avert the extinction of nonhuman primates.

The remainder of this chapter presents a very brief view, by region, of some myths, epic poems, folklore, and human perspectives concerning nonhuman primates. Each region is differentially represented; and given space limitations, rather than attempt a broad survey in each zone, we selected specific points that we hope are fodder for possible incorporation into, or stimulation of, research, management, or conservation projects and perspectives. A more holistic view of the interconnections between human and nonhuman primates must include the human folkloric perspective and potentially stimulate the reader to attempt integration of these disparate components into analyses of human and nonhuman primate relations.

African Folktales and Some Current Perspectives

Africa is a large continent with many different cultures, and ethnographers of the nineteenth and twentieth centuries recorded some African folktales and myths related to nonhuman primates. Like those across the globe, the folktales of Africa often explain the world as it is and may have a moral or set of instructions for appropriate human behavior. Unfortunately, the many ethnographic accounts of African folklore do not tell us much about how folktales may have influenced the attitudes toward or treatment of nonhuman primates.

For example, the cosmology of ancient Egypt featured several *therianthropes* (animal–human hybrids), including human forms with heads of birds, jackals, baboons, and so forth. Hapy, a son of Horus, is the baboon-headed therianthrope present on the canopic jars that held the lungs of a mummy. Animals, including baboon pets, were also mummified when the owner died (Hamilton-Paterson and Andrews 1978); but it is unclear how these historical contexts might impact any current behavior by humans in the same region.

Many folktales invoke the biobehavioral proximity between other primates and humans. Herskovits and Herskovits (1958) recorded many folktales of the Dahomean (a West African people). One tale explains why monkeys are similar to humans but still animals. Mawu, the creator god, after creating all the animals, told Monkey to work some clay and promised that if he did he would become a man. Monkey became excited and forgot to work the clay. Consequently, Monkey had to stay an animal. Another story is about a monkey that tried to fool a diviner and was killed by a leopard as a consequence. There are also tales of why monkeys (perhaps baboons) have red buttocks, such as the folktale that states that a monkey received red hindquarters by sitting on a hot hoe to keep a woman from coming out after he swallowed her. The Nuer of Sudan have a story that starts with the premise that God made the monkey to be like a person. The monkeys are described as acting like humans by cooking their potatoes, watching their offspring play, and talking among themselves. When a crocodile killed a monkey, the other monkeys grieved (Huffman 1970).

Honey (1969) records a variety of South African tales of monkeys doing very human-like things. One tells of a monkey that received a magic bow and arrow and a fiddle. A wolf accused the monkey of stealing his bow and arrow, and a jury agreed. The wolf took the bow and arrow from the monkey. The monkey started to play his magic fiddle, and everyone started to dance. Exhausted, everyone agreed to give the monkey back his bow and arrow if he would just stop playing his fiddle. In another folktale, a jackal was caught in a trap set by a farmer who wanted to stop the killing of his lamb. The jackal convinced the monkey to let him down and try hanging in the trap. The jackal ran away, and the farmer shot the monkey. In yet another tale, a tailor is upset that Mouse had torn his clothes. Baboon took care of the problem and the tailor thanked him. Subsequently, Baboon changed his name from Jan to Baboon and engaged in foolish behavior. The tale ends "Since that time Baboon walks on all fours, having probably lost the privilege of walking erect through his foolish judgment" (Honey 1969:120).

Other folktales reflect interconnections related to foods. For example, a Kikuyu (Kenya) folktale, as recorded by Njururi (1966), begins with baboons crop raiding and the men of the village hunting them down. Peace was eventually made between the people and the baboons. The baboons, however, broke the peace agreement, which caused the men to start hunting them once again. The men chased the baboons away from the crops and up into the trees, where baboons and monkeys remained forever. In this tale, the baboons could talk and enter into an agreement with the people.

None of the above examples can truly inform us of how humans might behave toward nonhuman primates based on the folktales. However, Courlander (1975) reports on a Yoruba (Nigeria) legend relating monkeys and twins. The belief is that twins are sent by monkeys. It is important that the twins are well treated; otherwise, misfortune could come to the parents. On the other hand, "the arrival of twins is always welcomed because of their power to better the lot of the parents" (Courlander 1975:233). Because of the connection between monkeys and twins, twins and their parents are prohibited from eating the flesh of monkeys. Also, Morris (1998) reports that in Malawi baboons and monkeys are in an ambiguous category but recognized as being like humans. They are eaten but perhaps not as frequently as other animals. Women, however, "are less inclined [than men] to eat baboon meat" (Morris 1998:200). Some recent studies of the interaction between human and nonhuman primates include discussions of how African worldviews influence the treatment of sympatric nonhuman primates. For example, Saj et al. (2005b) published evidence that local folklore can affect the treatment of nonhuman primates. They report that prior to the 1970s ursine colobus (*Colobus vellerosus*) were protected from hunting by the beliefs of villagers in Ghana of an association between their gods and the local monkeys. During this time, the monkeys were plentiful. In the early 1970s, a Christian sect known as the Savior Church of Ghana encouraged converts to hunt monkeys, and the number of monkeys drastically declined. In 1974, village elders asked the Wildlife Division of Ghana for help in protecting the monkeys. Wildlife officers stepped in, and the monkeys received protection.

Like parts of Asia, Africa also has diverse ways in which humans view other primates. In Madagascar, Jolly (1980) reports visiting Lavalohalika Island to see the black lemurs (*Lemur macaco*) that guard an old temple. The people of Lavalohalika Island feed the lemurs bananas as part of their religious ceremonies, and killing the lemurs is forbidden. However, the aye-ayes (genus *Daubentonia*) of Madagascar are a dramatically distinct-looking primate with their "bat ears, beaver teeth, two skeleton's fingers, and a black and silver tail like an overgrown ostrich feather" (Jolly 1980:60). Perhaps because of their strange appearance, the Malagasy fear the aye-aye and believe that they cause bad luck or death (Colquhoun 2004). If an aye-aye comes to a village, it must be killed and the village burned and relocated. In this case, cultural perceptions result in two very different treatments of two nonhuman primate species.

In the chapter "Reflections on the Concept of Nature and Gorillas in Rwanda: Implications for Conservation," Sicotte and Uwengeli (2002) report on interviews with people who live near the Parc National des Volcans, the home of Rwanda's gorillas. The authors were interested in the "cultural construction of nature" in Rwanda, including the animals that inhabit "wild spaces." Sicotte and Uwengeli (2002:174) note that gorillas are absent from the folklore of the Rwandans despite "the fact that the Virungas and the spirits that inhabit them are the objects of stories." Rwandans do, however, recognize the similarity of gorillas and humans and are aware that many foreigners are very interested in gorillas. The people of the Virungas compare their care of cows with that given to gorillas by white people and their Rwandan employees. As a consequence, the gorillas may now have a "symbolic presence" they lacked in the past (Sicotte and Uwengeli 2002:177). It is this pattern of approach and inclusion of human perspective, especially in the context of folklore, which may result in a more engaged understanding of the human and nonhuman primate interconnections in any given locale.

Views from the New World: Past, Present, and Future

As in parts of Africa, many indigenous people of Central and South America lacked a system of writing; therefore, there are few written records of their folklore and myths. Archeologists, however, have documented the prevalence of monkeys, particularly *Ateles*, within the iconography of Mesoamerican and some South American cultures (Bruner and Cucina 2005). Moreover, ethnographers have provided through their studies a record of some of the folklore concerning nonhuman primates. As we covered a few examples of Amazonian human and nonhuman primate interconnections earlier, we will give a brief review of those same people's myths regarding the sympatric monkeys.

According to Lizarralde (2002:85), until recently the Barí of the Parque Nacional Sierra de Perijá, Venezuela, "have eaten monkeys, kept them as pets, and used their teeth for decoration." Lizarralde (2002) reports that they believe that their creator god turned two Barí into monkeys as punishment for their misdeeds. Because the Barí want to acquire the characteristics of spider monkeys (genus *Ateles*) and capuchins (genus *Cebus*), they prefer to eat these monkeys and use their teeth for decoration. Howler monkeys (genus *Alouatta*), because they are considered slow and of low intelligence, are eaten only when no other monkeys can be found, and their teeth are not used for decoration. Infant spider monkeys are kept as pets and given family names.

As reviewed earlier, Loretta Cormier (2002) presents the relationship between the Guaja and the monkeys they hunt and eat as complex and related to both the enhanced image of the fertility of women and symbolic endocannibalism. Cormier (2002:80) suggests the following:

> Guaja identification with the physical and behavioral similarities between themselves and the monkeys is prerequisite for infant monkeys to fill their social role as substitute human beings and in their symbolic role as a version of the self to be cannibalized. Howler monkeys are the preferred game because they are considered to be most like the Guaja.

The Matsigenka of the Manu National Park (Peru), while slash-and-burn agriculturalists, also hunt and eat monkeys (Shepard 2002). The Matsigenka prefer the woolly (genus *Lagothrix*) and spider monkeys over the smaller capuchins and howler monkeys. Matsigenka mythology posits that all animals were originally humans and were "devolved" by the first shaman. Monkeys, however, are believed to be less devolved and more like humans than other animals.

In a different approach, Burton and Carroll (2005) provide an example of what they refer to as "by-product mutualism" in Honduras, where they describe a complex relationship between figs, howler monkeys (*Alouatta palliata*), and humans and their domesticants (pigs, bovids). The monkeys drop figs as they forage in the fig trees (*Ficus*), and the domesticants eat the figs that have fallen to the ground. Thus, the domesticants have access to a nutritious food source that would otherwise be beyond their reach. Everyone benefits in that the seeds of the fig trees are dispersed, the monkeys and the domesticants have fruit to eat, and the humans have less need to provide food to their domesticants. Burton and Carroll do not explicitly discuss the local people's perspectives on the foraging habits of their domesticants. They suggest that this complex interaction between monkeys, figs, and domesticants might be used to create an awareness, a "modern folktale" possibly, that might convince local people of the importance of conserving monkeys and fig trees.

Asian Folklore in Perspective

Asia has long had a system of writing, and there are many old documents on folklore, epic poems, and myths of various peoples' origins and history. Many of these documents have been translated and are available to speakers of modern languages. At the same time, many folktales remain largely within the realm of oral tradition. For example, Riley (in press) documented a folktale regarding the relationship between humans and Tonkean macaques in the Lindu highland enclave in Lore Lindu National Park, central Sulawesi, Indonesia. The story describes how a father left behind his teenage daughter to guard their house and gardens from raiding monkeys while he went to the lake to fish. As soon as the father was gone, the monkeys approached their land and the young girl began yelling and screaming at them. The monkeys ignored her and began raiding their crops, stealing all of their chicken eggs, and ultimately kidnapping the young girl. The father heard all of the commotion from the lake and grabbed his *parang* (machete) and went after them, crossing six mountains before he was able to catch up to them. With the *parang*, he fought the monkeys to rescue his daughter, leaving only a male and pregnant female macaque alive. He then brought his daughter safely home and held a large party for which he slaughtered a water buffalo in celebration. The moral of the story, as understood by the Lindu, is that they may not speak harshly or behave negatively toward the monkeys or they will cause havoc in their lives. Riley (in press) found, however, that this folktale could only be recounted in its entirety by elders of the indigenous To Lindu ethnic group from one of the villages in the enclave. This finding illustrates an important feature of people's attitudes and beliefs: They are not cultural fixed points, eternal and unchanging (Knight 1999, Hill 2002). The use of cultural arguments for primate conservation is seen as problematic for this very reason. For example, among the

Iban in Kalimantan, Indonesia, Wadley et al. (1997) found that conversion to Christianity has resulted in many people abandoning previous taboos against the killing and eating of orangutans. Riley (2005) documented a similar pattern in the Lindu highlands. In the past, the felling of fig trees was taboo because evil spirits were believed to reside within the trees. This taboo has had a conservation outcome because it resulted in the protection of strangling forms of the genus *Ficus*, species that are important Tonkean macaque foods (Riley 2007b) and that are considered the single most important food source for Sulawesi's fruit-eating birds and mammals (Kinnaird et al. 1999). Today, many villagers indicate that the predominance of religion (i.e., Christianity) now negates the belief in evil spirits. A changing sociocultural environment has therefore resulted in the relaxing of a cultural taboo that may ultimately negatively affect the survivability of wildlife that depend on fig fruits. Riley (in press) notes, however, that human–macaque folklore in Lindu appears to not necessarily have been affected by the conversion of the Lindu to Christianity. Instead, the inability of younger residents to tell the story may reflect limited transmission of knowledge across generations, as well as changing ecological and socioeconomic conditions in Lindu (i.e., shift from traditional *ladang* agriculture to wet-rice agriculture, cash crops, and fishing as the major sources of livelihood). Nonetheless, what does remain salient among younger Lindu residents is not so much the details of the story but rather its essence (i.e., "If you talk bad to them [monkeys], you will become their enemy" or "If monkeys come to our gardens we can only shoo them away…you can't be harsh because otherwise they will get angry"); a belief that continues to guide people's behavior as they relate to monkeys, even if they are unaware of the origin of the belief. This story likely explains why the Lindu tolerated crop raiding by macaques in the past and continue to tolerate it, and although only remnants of the story remain, these remnants of human–macaque folklore may help to ensure the persistence of macaques.

Asian Buddhist and Hindu Influences

The island of Bali (Indonesia) is a Hindu society with substantial differences in beliefs, artistic styles, and ceremonies from Hindu societies in southern Asia (such as India). However, as in India, Hanuman, the demigod monkey character from the epic poem *Ramayana*, is revered as a hero and loyal devotee of Lord Rama. As mentioned previously, the macaques of Bali are generally tolerated and provided with space and food at such tourist sites as the Wanara Wana Monkey Forest and Hindu temple complex located in the Padangtegal village of Ubud (Fuentes et al. 2005, Stephenson et al. 2002). Wheatley (1999) and Fuentes et al. (2005) suggest that in the temple areas human religious beliefs and experiential perspectives result in a high tolerance and interaction with the macaques. This tolerance by Balinese Hindu, however, does not appear to extend beyond those monkeys associated with temple sites, explaining why translocated Balinese Hindu consume macaque monkey meat in Sulawesi (Jones-Engel et al. 2005).

At many Buddhist temples in China, macaques are traditional inhabitants and are often fed by the monks. Buddhists believe that Buddha had at least one incarnation as a monkey. There is, however, more than just the occurrence of an incarnation of Buddha as a monkey; there are the moral lessons that are connected to the accounts of the incarnations. The folktales of the incarnations of Buddha as different animals, including as a monkey, are known as *jatakas* (Khoroche 1989:xiv):

> adaptations of popular tales that are used to teach Buddhist morals. The ethical and moral superiority of animals over men is a typical theme. There are over five hundred *jatakas*. . . . And some if not all of these were already in existence in the third/second centuries BC.

A *jataka* entitled "The King of the Monkeys" is typical of the folktales involving the rebirth of the Buddha. In this folktale, the king of the monkeys (i.e., the Buddha) sacrifices himself so that the arrows of the men of the local king would not kill the monkeys. The king asked the king of the monkeys why he did not run away and save himself instead of injuring himself beyond healing to save the monkeys. The king of the monkeys responded that he was paying his debt for the blessing he had received from his subjects. Upon telling this to the king, the Buddha died and went to heaven. The folktale ends "So, then those who make a practice of good behavior can win over the hearts even of their enemies. With this in mind, he who wants to win people over completely should follow the ways of the good" (Khoroche 1989:192).

As reported by Chinese primatologist Zhao (1991), the macaques at the Buddhist temples can be dangerous to tourists who do not know how to act around them. Rather than eliminating the monkeys, however, it is important to find ways to protect the tourists and use the presence of the monkeys to educate people as to the value of biodiversity (Wolfe 1991b, Zhao 2005). Moreover, because the *jatakas* are used to teach Buddhist morality, it is important that care is taken not to disparage the folktales. Understanding their contexts may enable researchers to work with the local people, governments, religious leaders, etc. to protect the monkeys and other wildlife that occupy areas around Buddhist temples while at the same time providing for the safety of temple visitors.

Burton (2002) recently explored the Chinese epic *Journey to the West* in which the Monkey King, or Sun WuKong (in Mandarin), is a central character. *Journey to the West* is based in part on the travels of Hsuan-tsang, a Buddhist priest who traveled from China to India in the seventh century to bring back documents of the teachings of Buddha (Dudbridge 1970). The round trip to and from India took 16 years (629–645 CE). The priest traveled alone and published a "geographical and ethnical account

of the lands to the west" (Dudbridge 1970:11). Drawing on the travels of Hsuan-tsang and folklore presented in song and on stage, *Journey to the West*, or the *Hsiyu Chi*, was written by Wu Ch'Eng-En (c. 1505–1580) in the sixteenth century (Wu Ch'Eng-En 1943). The first seven chapters tell of the early life and mischievousness of Monkey. The subsequent chapters concern the trials and tribulations of the trip from China to India and the heroism of Monkey. The *Hsiyu Chi* combines elements of the travels of Hsuan-tsang, Buddhism, the *jatakas*, Daoism, Confucianism, and the monkey character Sun WuKong. Dudbridge (1970) examines the origin of the idea of Sun WuKong and suggests that Wu Ch'Eng-En based Sun WuKong on the demigod Hanuman of the Hindu epic poem the *Ramayana*. The similarities between Sun WuKong and Hanuman include immortality and the ability to fly and change shape. Because the story combines many aspects of the religions found in China, it is widely known and used to teach values to children. The image of Sun WuKong, moreover, is used in advertising in China; and Burton (2002) suggests that the image of the monkey king could also be used to promote conservation and assist in the survival of the monkeys of China.

In India, the epic poem the *Ramayana* is central in the Hindu belief system. In the *Ramayana*, the monkeys are among the main heroes and Hanuman, the loyal servant of Rama and possessor of supernatural abilities, is the most important of the monkey heroes (for more details see Wolfe 2002). Generally, monkeys in India are not killed outright; rather, their presence is tolerated, and people bring food to the monkeys that inhabit areas in and around temples (Wolfe 2002 personal observation). The monkeys are not, however, otherwise provisioned. While people do bring the monkeys food, it is not consistent throughout the week. On Tuesdays and Saturdays (the two days associated with the good deeds of Hanuman) and on religious occasions, the monkeys of the temple are brought more than they can eat, and on the other days little food is carried to them. For the most part, the monkeys are given food so that Hanuman and Lord Rama might look favorably upon their human providers.

In the 1980s, one of us (Linda Wolfe) studied the rhesus monkeys that live in and around the city of Jaipur, which is located in the state of Rajasthan in north-central India south of New Delhi (Wolfe 1992). Early in the twentieth century, as Jaipur began to grow outside of the walls of the city, the forest that surrounded the walled city was cut down. Unfortunately, the forest is where the monkeys had lived. By cutting down the forest, the people forced the monkeys to become more dependent on them for food. Nothing was done to preserve any of the forest or its wildlife. Moreover, there was no systematic attempt to monitor the health or demographics of the monkeys of Jaipur.

There was a dominance hierarchy among troops, moreover; and not all troops had access to the temple where the most food was deposited. As a consequence, there were monkeys that appeared to be in poor health and troops in which no offspring were born during the 1985 birth season. Our point is that here is a belief system in which the monkeys are valued because of their association with a popular legend but that association does not necessarily mean that they will be well treated. The legend of Hanuman is helpful to use as a rationale to start the process of the care of the monkeys of India, but that care needs to be given more systemically as they need both veterinary attention and a consistent food source.

While monkey performance has at times been an important facet of Japanese culture, Japan seems to lack major epic narratives involving monkeys. While devoted Buddhists may know of the *jatakas* and the legend of the "journey to the west," the *jatakas* of Buddhism do not seem to be part of the cultural consciousness of the people of Japan. Ohnuki-Tierney (1987) argues that the meaning of the entertainment provided by the monkey dances and their trainers in modern Japan is culturally complex and can vary from individual to individual and city to city.

Wolfe and Gray (1982) report on an observation that occurred during a field study of Japanese macaques (*Macaca fuscata*) at the tourism site of Arashiyama, Japan. Despite articles in the popular literature that described the important role of females in the social system of Japanese macaques, the Japanese public was inordinately interested in the *zaru bosu* (or monkey boss). The authors argue that the interest in the monkey boss was twofold: First, Japan is a patriarchal society, and the interest in the monkey boss reflected that reality; second, Japanese society functions by consensus, and the perception of the individual decision-making power of the monkey boss contradicts that of society and, therefore, separates the natural animal world from the cultural life of humans.

In the literature on the creation myths of the Tibetan people, there are several variations on the account of their creation. Snellgrove (1987) has suggested that an old creation myth was later modified when Buddhism came to the Tibetan highlands sometime around 600 CE. The old creation myth begins with the meeting of a wise and compassionate monkey and a rock demoness. At first, the monkey, which wanted to maintain his chastity, rejected the advances of the rock demoness. Eventually, however, the monkey succumbed to the requests of the rock demoness, and the results were the six original ancestors of the people we know today as the Tibetans (Boord 1993). With the arrival of Buddhists, the saintly monkey became associated with Avalokiteshvara (a bodhisattva) and the rock demoness with the Buddhist goddess Tara. Avalokiteshvara is today manifested as the Dalai Lama. The folktale of the monkey and the rock demoness persists today, although in the form altered by the more recent conversion of Tibetans to Buddhism.

It is abundantly evident that human cultural narratives play a very important role in establishing the context for human and nonhuman primate interconnections in Asia.

However, these patterns are neither uniform nor ubiquitous across the broad expanse of the region.

A Modern North American Case

A final view of humans' cultural perceptions can be found in the United States of America. In 1938, the late Colonel Tooey released several rhesus monkeys (*Macaca mulatta*) along the Silver River near Ocala, Florida. Silver Springs, the headwater of the Silver River, is a popular recreation park. Tooey ran a boat ride and released the monkeys to enhance the excitement of the boat ride and increase his revenues (Wolfe 2002). The size of the monkey population increased, and between about 1982 and 1995 the Florida Game and Fresh Water Fish Commission (FGFWFC) began to oversee the trapping and disposal of the monkeys. Most of the local people opposed the trapping of the monkeys. The people of Silver Springs liked observing the monkeys and particularly enjoyed watching the mothers care for their infants. Although the FGFWFC did remove perhaps as many as 500 monkeys during the 1980s, there are still rhesus monkeys living in the area of the Silver River. One of us (Wolfe 2002) attributed the survival of the rhesus macaques of the Silver River in large part to the objections of the local people to the trapping of the monkeys.

SOME CONCLUDING THOUGHTS

Human and nonhuman primates living in sympatry affect each other's lifeways. People who share a habitat with nonhuman primates tend to include those primates in their myths and folklore. It is the premise of ethnoprimatology that folklore, myth, and a broader understanding of the complex array of interconnections between human and nonhuman primates can be used to encourage diverse groups of people to cooperate with government officials and conservation and management groups to ensure the survival of nonhuman primates. At the very least, understanding and contextualizing myths and folklore combined with detailed assessments of the actual patterns of interactions can give people from places of sympatry and the researchers involved in primatology access to possible reasons why they might want to see that the animals around them survive into the future. These reasons themselves may also be key components in effective management and conservation strategies.

The ethnoprimatological approach might help to overcome the cultural and perceptive isolation of conservationists and primatologists from those who work with local peoples in areas of sympatry. Holistic projects based on knowledge of the interaction between human and nonhuman primates, plans that include both the enhancement of the lives of local peoples and the needs of wildlife, are required if nonhuman primates are to survive in the face of the continuing growth of the human population.

47

Where We Have Been, Where We Are, and Where We Are Going
The Future of Primatological Research

Christina J. Campbell, Agustín Fuentes, Katherine C. MacKinnon, Melissa Panger, Rebecca M. Stumpf, and Simon K. Bearder

THE PLACE OF PRIMATOLOGY: EMBRACING INTERACTIONS WITH OTHER DISCIPLINES

Primatology is, by its very nature, interdisciplinary. Field studies and captive work are undertaken from intellectual foundations as varied as anthropology, psychology, behavioral ecology, zoology, anatomy, ethnology, genetics, epidemiology, evolutionary biology, conservation biology, and veterinary medicine. Indeed, it is the object of study—the primates themselves—that best defines *primatology*, as opposed to a restrictive academic philosophy. Thus, questions addressing topics as diverse as the biomechanics of locomotion, behavioral variation in maternal care patterns, bidirectional pathogen transmission between primate species, and the selective pressures acting on primate social systems, to name a few, can all be effectively addressed from under the encompassing umbrella of our field.

Historically, this interdisciplinary nature has been of primary importance to the field; that is, the relevance of nonhuman primates to inquiries about human evolution and behavior is one of the foundations of primate studies. Pioneers Robert Yerkes (a psychologist), Clarence Ray Carpenter (an ethologist), Solly Zuckerman (an anatomist), and Kinji Imanishi (an entomologist, ecologist, and anthropologist) implemented the first studies of nonhuman primate behavior, thereby helping to shape primatology into a multidisciplinary field. Importantly, Sherwood Washburn's call for a comparative primatology (which supplemented studies of anatomy and human behavioral evolution with comparative data from primates living in the wild) stimulated several generations of field primatologists. Additionally, the paleoanthropologist Louis Leakey promoted the long-term ape research projects of Birute Galdikas, Diane Fossey, and Jane Goodall, which provided the basis for the growth in ape research. The last half of the twentieth century was indeed a fruitful time in our field, and as the number of primate species studied grew, the data increasingly revealed the massive variability within the primate order. Would this productive trajectory of investigations into the lives of nonhuman primates have occurred if the field did not have such an interdisciplinary background and character? Probably not, for the simple reason that many of the research questions asked could not have sprung forth from a more insular origin.

Modern evolutionary theory, specifically current behavioral ecological approaches, and more than a half-century of primatological data have placed us, at the beginning of the twenty-first century, in a position to make greater strides into the complexities of primate biology than ever before. Primatology continues to be wonderfully positioned to examine a wide range of topics and, in doing so, can draw from a wealth of information from varied data sets. This philosophical position provides not only insight but also a complex prism through which we can view the interwoven facets of primate behavior, ecology, and physiology.

We have seen in a number of chapters in this volume how a multidisciplinary focus is the modern standard in primate work. Whether researchers' home bases lie in anthropology, conservation biology, epidemiology, behavioral ecology, or genetics, a "cross-pollination" approach to research can only continue to strengthen primatology. Exciting results have come about from multidisciplinary collaborations, and we urge primatologists-to-be to keep this in mind as they begin their studies. In addition, we implore more seasoned veterans to incorporate varied lines of evidence when tackling their research questions. Being interdisciplinary is primatology's legacy, and further communication between and among the various disciplines involved is our continued charge.

NATURAL HISTORY AND THE SCIENTIFIC METHOD

Natural history, the study and description of organisms and natural objects (especially their origins, evolution, and interrelationships), has played a pivotal role in the biological sciences including primatology. Many of the earliest primatologists were natural historians who set out to broadly describe the behavior and ecology of primates living under natural conditions. Such studies are vital for any scientific field of study since the collection of general descriptive data represents the first step in the scientific method, which involves observing phenomena, detecting patterns from those observations, and developing testable hypotheses to try to explain the patterns observed.

Natural history data were the focus of primatology until the 1970s and 1980s, when there was a major shift toward hypothesis-driven research (the collection of a limited set of data used to test specific hypotheses). During this time, the two forms of research, natural history and hypothesis-driven research, were no longer considered simply steps in the scientific process but instead dichotomized, with greater value placed on hypothesis-driven research. This switch in focus occurred in many fields of study and was partly the result of what has been called "physics envy." As a result of this shift, natural history data are currently undervalued, and it can be challenging in primatology (and other biological fields) to obtain funding specifically for the collection of broad behavioral and ecological data and to get natural history information published in leading primatological journals.

Natural history data, however, are important for a variety of reasons. For example, as stated, collecting general descriptive data is an important first step in the scientific method and the information retrieved from broad-based studies drives hypothesis-driven research. The chapters in this book clearly illustrate that while we currently know a great deal about many primate species, we have only really scratched the surface of what there is to know about many of our primate cousins. Related to this, broad behavioral and ecological research is also important because as we learn more about prosimians, monkeys, and apes, exciting and novel hypotheses will be developed that we cannot currently fathom. Natural history data could provide important information for later studies as new "hot topics" and research goals emerge in the future. For example, recent studies of culture in a variety of primates have relied heavily on natural history data (see Chapter 44). The long-term data used in these studies were not collected to test specific hypotheses related to culture but as part of larger, broad-based studies, yet the data have been successfully used to support an argument for culture in nonhuman primates. Finally, we feel that natural history data are important because many primate populations are rapidly disappearing (see chapter 45) and, thus, there is an urgent need to collect as much information as possible from the declining populations before they go extinct. This is akin to the philosophy of "salvage ethnography" that played an important role in the collection of ethnographic information from a variety of cultures in the mid-part of the last century before those cultures were lost due to assimilation into an increasingly modernized world.

We do not wish for the primate research pendulum to swing completely away from hypothesis-driven research, but we believe the value of both types of research should be recognized. Ideally, funding opportunities to conduct natural history research on understudied taxa should be increased. However, at the very least, primatologists and primatologists-to-be would gain much by recognizing the importance of broad-based data and regularly collecting them while conducting hypothesis-driven research. In short, natural history has played an important role in the history of primatology, and we think it has an important role to play in the future of the field.

INTERACTIONS BETWEEN RESEARCHERS

Although researchers from diverse fields of study are carrying out ongoing primatological inquiry, their interactions with each other are often not as productive as we would hope. An interesting and informative parallel between the behavior of the animals that we study and our own reaction to some of our colleagues is competition. How can we arrange our research programs so as to minimize conflict, aggression, and territoriality and maximize collaboration? The answer is linked to the theory of competitive exclusion: If we occupy different niches, then we are less likely to compete. Indeed, it may be possible to arrange research themes that are complementary and, thus, benefit in much the same way as species that live together symbiotically. An effective way to achieve this comes from our ability to communicate over great distances and to keep each other informed about what we are doing and why. Now that the global network makes communication between researchers easier, such cooperation should be even more prominent. There are many present-day collegial situations among researchers. A fine example involves primatologists from a number of fields who work on capuchin monkeys (*Cebus* spp.). The "capuchin people" are famously cohesive, often overlap in terms of species and study topics, and share data (anecdotal and otherwise) back and forth. Multiauthored papers have been written as a result of fruitful collaborations, and a special satellite conference of the 2004 International Primatological Society meetings was organized, entitled "Capuchins: The State of the Art"; this conference brought together capuchin researchers from all over the world, contacts were made and reestablished between new and old friends, and as a result, future research ideas were stimulated.

The primatological field will surely be strengthened if we appreciate the myriad research niches that are available and the range of strengths and abilities of the researchers who may queue to fill them. Even people working on the same species and topic can work in different study areas or on different captive populations, and they are likely to produce synergistic results. Those interested in a particular taxonomic group can diversify by concentrating on different species or subspecies. Ironically, this becomes ever easier as the number of primate taxa increases year by year as a result of our more detailed understanding of cryptic species and more thorough exploration of wild places. An exciting shift in taxonomic concentration is currently under way as more and more people enter our exciting field. Gone are the days of everyone wanting to study the "famous" primate species (e.g., gorillas, chimpanzees, baboons). We are now witnessing an influx of those who are taking on the difficult task of studying species that have received less attention in the past, for example, many nocturnal species. A less species-centric approach is enormously rewarding and opens up previously unexplored research niches, enabling the discovery of new species and, for example, a dramatic new appreciation of phylogenetic relationships through new techniques such as molecular analysis. For those who are setting out in this fascinating field of study, we suggest keeping an open mind about where to go, what to do, and what to study. Try to avoid areas of work that are already heavily populated, but think instead of how to add extra pieces to the fascinating puzzle. Work on topics that complement the work of others and thereby benefit from the give and take of true collaboration. Also, of course, publish or publicize as much as possible to ensure that the results reach the widest public.

TECHNOLOGY

In addition to the increasing ability of researchers to collaborate due to faster and more efficient means of communication, one of the most exciting advances in recent primatological inquiry is the employment of new technologies. No longer is it sufficient to infer paternity from behavioral observations of copulations or to infer female reproductive state from behavior alone. Today, field researchers commonly employ techniques providing information on paternity and genetic relatedness through the analysis of genetic data (see Chapter 24). In addition, hormonal information indicating female reproductive state, ovarian cycle stage, male testosterone levels, and stress levels as indicated by cortisol levels is obtainable by those carrying out field studies (see Chapter 23). Both sources of information have been available for a while, but not until they could be obtained from sources collected noninvasively (e.g., from fecal samples) were these methods to be widely employed by field researchers. With increasing prevalence in primatological studies, such technologies will undoubtedly provide data that will challenge current theoretical ideologies. Already, genetic data are challenging the ubiquity of the "priority of access" model of male reproductive success. In addition, hormonal data are allowing us to reassess the uniqueness of human female sexuality and often suggest that female monkeys and apes do not show estrous cycles as they are repeatedly reported to in the literature (see Chapter 29). The expense of such technologies is the main reason they are not employed in all studies, but in the future

it is reasonable to expect that the costs of performing such studies will decrease as the technologies become more common and more refined.

Technology is also playing a part in easing one of the more difficult components of primatological fieldwork: maintaining contact with the animals on a day-to-day basis. Those who study species such as howler monkeys (*Alouatta* spp.) benefit from being able to rely on the animals making a great deal of noise every morning, thus making them easy to locate on a regular basis. In most cases, however, there is no such daily announcement of the study group's location. In the past, this has meant that, in order to maintain contact, field primatologists have had to perform what has become the ritual dawn-to-dusk follow: waking up and being in the field prior to dawn to locate the animals where they were left the night before and following them all day until they settle down to sleep, only to repeat the process the next day. Clearly, continuing this for any great length of time leads to extreme fatigue, both mental and physical. Today and in the future, primatologists can benefit greatly from the increased use and availability of radiotelemetry and global positioning system (GPS) technology (see Chapter 21). Radio and GPS-based collars for primates have become smaller and less obtrusive, batteries have a longer life span, and drugs used when anesthetizing wild primates have become safer. Many still question the danger associated with darting wild animals, especially those that are high in the forest canopy; and we are by no means implying that researchers should go out darting their animals without careful consideration of the many risks involved. The benefits of radiotracking and GPS-based data extend beyond merely locating the study group however. Contact can be maintained with multiple groups by one researcher, allowing for studies on intergroup contacts and interactions. Following cryptic nocturnal species in many cases would be near to impossible if the animals were not fitted with radiotracking devices. Future advances in the field of radiotracking are potentially very exciting to primatologists. Further decreases in collar and battery size and the potential for subcutaneous devices will hopefully decrease the immediate impact that wearing a collar has on the daily life of a primate. Advances in the technologies used to measure ecological variables (see Chapter 22) have also been made. Data on ecological parameters are precise and provide information that is vital to understanding behavioral variations and environmental changes.

We are not suggesting that future primatological research will be conducted purely through technological means and that behavioral observations will not be important. Nor are we implying that long hours in the field should be dropped from the primatologist's repertoire. It is our collective experience that one only begins to see the intricate details of social interactions after hundreds, if not thousands, of hours in the field. We believe that technologies such as those described here can greatly benefit us as field primatologists, allowing us to test hypotheses that in the past we could only dream of testing without making assumptions unsupported by any empirical data. It is indeed an exciting time to be entering the primatological field!

NEW DIRECTIONS IN PRIMATOLOGICAL RESEARCH

Now that primatology has entered its second century of sustained research, we can take a brief look back at what has been accomplished and preview some of the new perspectives and directions for primate research. Initially, field research targeted a few primate populations of baboons and apes, initially chosen for their potential to provide comparative models of human evolution (Chapter 1). Primates remain essential referents for understanding the biology of our fossil ancestors, but the scope of field primatology has expanded dramatically as data have accrued from multiple decades of research on a broad sample of species in wild and captive settings. Current behavioral and ecological research is pursued at the individual, group, and community levels, with primary goals being the development and testing of models to explain behavioral, life history, demographic, and ecological differences within and among groups.

Comparative studies of primates remain fundamentally important for understanding human nature and the extent to which humans are unique. Many factors that were previously used to distinguish humans from other primates (e.g., tool use, meat eating, cooperative hunting, and cooperative childrearing) have fallen one after another as more observational data have been obtained. Even the notion that "culture" is a uniquely human construct has been thrown into debate through recent analyses of behavioral variation within species (Chapter 44). While the study of primatology has yielded important comparative socioecological information, we hope that the future brings a greater understanding of human socioecology.

Looking to the future, we can also anticipate some of the major new conceptual directions for primatology, where there are many opportunities for theoretical and empirical research, at both the macro- and microevolutionary scales. An integrated primatology that synthesizes organism, ecology, and evolution is essential as primates are defined by these influences and cannot be understood apart from them. Although we know how ecology, phylogeny, and demography may affect the behavior of individual populations, we have little insight into how groups interact directly (or indirectly) with conspecifics on larger scales. We have only preliminary information about intergroup interactions, including intergroup aggression and conflict, decisions regarding dispersal, and genetic relationships between groups. Interspecies interactions at all ecological levels are also valuable as they can provide more information on the interdependence of species in particular habitats. Of critical importance, however, are human/primate interactions as humans continue to encroach on primate populations and habitats and issues of conservation and disease transmission become imperative.

These are but a small sample of new topics and directions. The opportunities for pursuing primate behavioral and ecological research are growing, and we look forward to continued efforts to protect primate populations and to the contributions of the next generation of primatologists.

CONSERVATION AND THE FUTURE OF THE PRIMATES THEMSELVES

Human and nonhuman primates share intertwined destinies. Nonhuman primates are our closest evolutionary relatives and are integrated into our mythologies, diets, and economies. There is no longer the possibility of studying a group or population of nonhuman primates without some element of human interaction. Nearly 90% of the world's primates are found in tropical forests, and these forests are being altered by human use faster and more dramatically than any other habitats on earth. Fifty percent of all primate species are a conservation concern to the Species Survival Commission and the World Conservation Union, and 20% are considered endangered or critically endangered. Human and nonhuman primates have been interacting for millions of years, and it is now up to humans to decide how much longer the other primates will be around to continue this pattern.

Accordingly, practically every practicing primatologist will agree that conservation is among the most important aspects of the field. Karen Strier eloquently lays out the current case for primate conservation, its practice and patterns, and the urgent need to act regarding endangered and threatened primate populations in Chapter 45. However, we stress two additional aspects related to primate conservation in this concluding chapter: the role of disease and the role of source-country primatologists.

Changing human social, economic, and political conditions and associated deforestation result in nonhuman primates coming into increasingly regular contact with humans. Hunting, logging, farming, the pet trade, ecotourism, and research all provide contact opportunities in which diseases can jump the species barrier. While research has demonstrated the devastating effect that human diseases (such as tuberculosis, polio, and respiratory viruses) can have on nonhuman primates in captivity, only recently has research begun to focus on the effects of these, and other, diseases on populations of wild nonhuman primates. It is most likely that one result of increased interactions with humans is that many wild nonhuman primate populations will be negatively impacted by human disease. This can be especially dangerous for those populations or species that are currently endangered or threatened with extinction. In the other direction, it has become apparent that many human diseases may have origins in nonhuman primates. For example, recent research suggests that the chimpanzee simian immunodeficiency virus (SIVcpz) made the jump to humans and mutated into the human immunodeficiency virus (HIV-1) at some point in the last century.

The incorporation of tools from pathobiology, geographic information systems (GIS), and virulence monitoring through global Internet surveillance can improve our understanding of disease transmission and ecosystem health (e.g., Wolfe and Switzer 2009, Bermejo et al. 2006, Pedersen et al. 2007, Köndgen et al. 2008). From a micro-ecological and microevolutionary perspective, studies of primate host/microbe and host/parasite interactions can provide new insight into coevolutionary adaptation and primate evolution .

Source countries are those where nonhuman primate populations currently exist. The last few decades have seen some increase in the amount of trained and practicing primatologists from countries where nonhuman primates are endemic. It is our contention that this trend must continue and, in fact, needs to intensify if conservation actions are to become truly sustainable at the local level. Given the diverse and complex cultural, social, and political boundaries and hierarchies that characterize the modern world, effective and long-lived conservational activities are most likely to succeed if local populations and source-country citizens are, in general, deeply invested and involved in their creation and implementation. It is our hope that the future of primatology continues to enhance the infrastructure and professional opportunities for primatologists from source countries as a major component of our conservation paradigm.

CONCLUSION

We hope that readers of this volume have become excited about the potential for future primatological study. It was our intent when putting these chapters together to provide a volume broad in theoretical and topical issues, and we are confident that we have achieved this goal. We thank all those who have contributed to this volume (both the first edition and this updated second edition) and, indeed, all those who have contributed to the myriad primatological studies discussed in each chapter. To those contemplating entering the primatological community, we wish you well in your studies and hope to see results from your future research incorporated into updated volumes of this kind.

GLOSSARY

Abiotic: Not associated with or derived from living organisms. Abiotic factors in an environment include sunlight, temperature, wind patterns, and precipitation.

Adolescent: The developmental period between the onset of puberty and the attainment of full adult body size, appearance, and behavior.

Adult: The developmental period often defined by the attainment of full body size.

Affiliative: Any behavior that promotes group cohesion (e.g., prosocial behaviors such as grooming, touching, and hugging).

Agonistic: Any behavior associated with aggressive interactions, including the actions and responses of the actor and recipient of the aggression.

Alarm call: Any vocalization given in response to the presence of danger (e.g., a predator).

Allogrooming: The grooming of an individual by another individual.

Allomothering: Any mothering-type behavior (carrying, provisioning, nursing) shown toward an infant by a female individual other than the infant's own biological mother. Also referred to as "aunting" behavior.

Alloparent: An individual other than the mother or father that helps care for young.

Allopatric: Species or populations of organisms that exist in distinct (noncontiguous) geographic areas.

Altricial: Young animals that are born helpless and remain so for a substantial period following birth.

Altruism: A behavior that benefits the receiver in terms of reproductive success at a cost to the reproductive success of the individual performing it. The term "altruism" implies nothing about the proximate causes of the behavior or its motivation but refers only to the effects of the behavior on the fitness of the actor and receiver.

Antagonistic pleiotropy: Patterns that have high adaptive value early in the life course will be selected for even if they result in reduced fitness later in the life course.

Anthropogenic: Changes caused by human activity.

Anthropoid: Monkeys, apes, and humans, both living and extinct.

Anthropomorphism: The description of animal behavior using human terms (e.g., emotions).

Ape: A common name given to members of the superfamily Hominoidea. Modern apes are united by a number of morphological characteristics including, but not limited to, absence of a tail, an increase in brain size, Y-5 molars, and a variety of features that enable brachiation.

Arboreal: Refers to trees. Arboreal primates are those that live in trees.

Behavioral role modeling hypothesis (BRMH): A hypothesis suggesting that maturing primates use adults as role models for behavioral development and that they choose social partners strategically, favoring role models that can provide them with competitive advantages.

Bilophodont molars: A dental characteristic of modern Old World monkeys in which two sets of transverse ridges are present between four cusps on the molars, creating an effective shearing surface.

Biologically active (or bioactive): Describes a substance, usually a chemical, which acts upon or influences the bodily functions of an organism.

Biome (also ecosystem): A community of organisms defined by geographic and climatic variables.

Biotic: Of or having to do with life or living organisms; produced or caused by living organisms.

Bipedalism: Walking on two legs.

Bitter pith chewing: A form of self-medication practiced by chimpanzees, where an ill animal removes the outer bark and leaves of a plant, *Vernonia amygdalina*, to chew on the exposed, bitter pith, which has medicinal properties.

Brachiation: A form of arboreal locomotion that involves swinging by the arms between tree limbs. This mode of locomotion is limited to members of the Hominoidea, although members of the Hylobatidae are the only habitual brachiators (see also semi-brachiation).

Carnivory: The consumption of meat (usually excluding insects).

Caro method: A method for identifying reproductively terminated or postreproductive female primates using reproductive records. The formula $LP - D/(\bar{x}IBI + 2 \text{ SD})$ is used where LP = age at last parturition, D = age at death, and $\bar{x}IBI$ = mean length of a female's interbirth intervals across her lifetime. Using this method, a female is classified as postreproductive when the time lag between final parturition and the death of the mother exceeds 2 standard deviations of the female's own mean lifetime interbirth interval.

Catarrhine: Within anthropoids, the catarrhines are the monkeys of Africa and Asia (the Old World monkeys) and the apes (including humans).

Cathemeral: An activity pattern in which an animal can be active both at day and at night.

Cecum: A pouch connecting at the beginning of the large intestine that acts as a fermentation chamber.

Cellulose: The structural component of the cell wall of plants. Cellulose can be broken down only by the enzyme cellulase, which is not present in the digestive system of mammals; thus, cellulose can be broken down only in the presence of bacteria, whose cellulase aids in the fermentation process.

Cheek pouch: A pouch inside the cheeks used to carry food, found in cercopithecine monkeys.

Coalition: An alliance involving cooperation among individuals or groups.

Cognition: Process(es) by which organisms gain, store, and retrieve knowledge or become aware of conditions, events, or objects and use that knowledge to make decisions and to guide action.

Commensalism: A relationship between two species or populations where one benefits and the other is not impacted (either positively or negatively).

Concept: An unobservable entity, such as a category, about which one can reason.

Consortship: A temporary and exclusive grouping of a male and female (and her dependent young) involving sexual activity.

Contest (or interference) competition: When animals compete for a resource that is monopolizable.

Copulatory plug: A vaginal plug resulting from the coagulation of seminal fluid.

Core area: Preferred area of range use; the boundaries of a range where an individual spends ≥75% of its time in a year.

Corrected conciliatory tendency (CCT): A relative measure of the occurrence of reconciliation after controlling for baseline levels of affiliation.

Crepuscular: An activity pattern in which an animal is primarily active at dawn and at dusk.

Cryptic female choice: The process whereby a female's anatomy, physiology, or behavior selectively alters the probability of paternity among her copulatory partners.

Cusp: A pointed or rounded projection on the chewing surface of a tooth.

Day range: The distance an individual travels in the course of a 24 hr period.

Delayed gratification: Forgoing an immediate reward (physical or emotional) in order to gain a more substantial one in the future.

Dental comb: A strepsirhine feature made up of horizontally protruding lower incisors and canines, used for grooming and scraping.

Diagonal sequence–diagonal couplet gait: A patterned gait sequence in which movement of the left hindfoot is followed by movement of the right forelimb, followed by right hindfoot and left forelimb, during quadrupedal progression.

Diameter at breast height (DBH): Diameter of a tree at breast height, or 4.5 feet above ground level, the accepted point of diameter measurement for most trees.

Dichromatic color vision: The ability to perceive differences in two color fields, black–white, and blue–yellow. Dichromatic individuals are "red–green" colorblind.

Dispersal: Departure from an existing population or natal group. See also immigrate, emigrate.

Dispersal kernel: A probability density function describing the probability of a seed being deposited per unit area at locations measured with respect to the parent plant. See also Nathan and Muller-Landau (2000).

Diurnal: Active during the daytime.

Dominance: A social relationship defined by the ability of one individual (the dominant) to solicit submissive behaviors from another individual (the subordinate).

Emigrate: Dispersal from a group in which an animal has been an established member (often, but not always, the natal group) (see also immigrate).

Endocrinology: The study of the secretion, storage, chemistry, and function of hormones.

Endozoochory: Seed dispersal in which seeds are transported inside an animal. For primates, this can occur when seeds are ingested into the gut and passed (*defecation dispersal*) or when seeds are processed in cheek pouches and later spat (*spit dispersal*).

Epic: According to Pentikainen (1997:223), "In folkloristics, epic is a narrative couched in poetic language and subject to special rules of texture, structure, style and form. Usually, epics contain hundreds or thousands of verses and present a complex narrative full of wonders and heroism, often centered around the exploits of a main personage."

Estrus: A term first defined by Walter Heape in 1900. It is the period in the female ovarian cycle where females are willing (and able) to participate in sexual intercourse—which in most mammals coincides with ovulation.

Estrous cycle: The female physiological and behavioral cycle associated with hormonal and behavioral changes related to sexual activity (see also estrus). While prosimian females typically experience estrous cycles, the relationship between sexual behavior and physiological state in female anthropoids is less clear.

Ethnomedicine: The study of the beliefs and practices concerning illness in different, traditional human populations.

Ethnoprimatology: The study of the multifarious interaction of human and nonhuman primates including, but not exclusive to, the image of primates in folklore, legends, and myths; the influence of cultural beliefs on the hunting of primates; conservation ecology; and the management of primate populations in their natural habitats.

Explicit reconciliation: Behaviors used for reconciliation are conspicuous and rarely seen outside of a postconflict context.

Extant: Still in existence (i.e., not extinct).

Extinct: No longer in existence.

Extractive foraging: Locating and processing, sometimes with the use of objects, embedded food items. Such items are not readily visible or accessible, for example, insects embedded in dead wood, underground roots, or hard-husked fruits.

Fecundity: The potential reproductive output or ability to produce live offspring of an organism or population (influenced by gamete production, fertilization, and ability to carry pregnancy to term). Unlike fertility, fecundity cannot be measured directly.

Fertility: The actual number of live offspring produced by an individual or population over a lifetime.

Folivory: The consumption of leaves.

Folkbiology: A cross-cultural and interdisciplinary study of how ordinary people understand, perceive, and categorize the natural world (see Medin and Atran, 1999). Folkbiologists are also concerned with the relationship between local worldviews, globalization, and conservation.

Folklore: A word coined in 1846 by English scholar William J. Thoms. There are several common features that comprise folklore: (*1*) usually the tales are transmitted orally, (*2*) the elements are traditional in form, (*3*) different versions or variants exist, and (*4*) the tales are anonymous as to origin (Brunvand 1996).

Folktale: "A fictional narrative varied in length and rich in symbolic and metaphorical meaning, oral in origin but now found more often in printed collections. . . . Folktales are filled with fantastic creatures, events and objects. . . . Folktales arise in an oral tradition; thus, one cannot identify original or authoritative texts. There are as many texts of any particular folktale as there are people who have told it over many generations" (Stone 1996:294).

Foraging proficiency: The ability to forage at an adult level, consuming the same foods, with the same ecological knowledge.

Frugivory: The consumption of fruit.

Gestation: The period between conception and birth.

GIS (geographic information system): A series of tools that allow the user to integrate and arrange spatial information, edit maps, and analyze and display geographic information.

GPS (global positioning system): A tool used in navigation, land survey, map making, etc. that uses signals from satellites to provide an exact location.

Grandmother hypothesis: The hypothesis that menopause in humans evolved because nonreproducing older females gain fitness advantages by helping their offspring and grand-offspring survive and reproduce.

Great apes: A term used to refer to the largest of the modern apes (i.e., chimpanzees, bonobos, orangutans, and gorillas).

Gregarious: Living in a social group, interacting socially with others.

Grooming claw: A specialized claw present on one or more digits of the foot in prosimians. Also referred to as a "toilet claw."

Guiana Shield: An ancient geological formation (Precambrian) of the northern neotropics where basement rock is close to the surface and supports specialized forests and black-water rivers or creates tabletop mountains. Includes portions of Colombia, Venezuela, Brazil, French Guiana, Guiana, and Suriname. Spanish: *Guayana*.

Gular: Pertaining to the neck region and the location of scent glands for some platyrrhine monkeys.

Gummivory: The consumption of tree exudates (gum and sap).

Habitat: from the Latin for "dwelling," habitat refers to the immediate as well as general physical, environmental setting in which a population is found. Populations interact with each other, forage for food, and prey or are preyed upon in their habitat. Habitat describes where an organism's life history occurs but not how it lives (see niche).

Habitat generalist: Species that occur in a wide variety of habitats and can make do with the resources found there.

Habitat specialist: Species that occur in a narrow subset of habitats. Habitat specialists are typically detrimentally impacted by habitat destruction.

Handicap principle: A term proposed in 1975 by biologist Amotz Zahavi to explain the evolution of honest signaling. Sexually selected "ornamental" traits are attractive because they are costly to individuals that have them; only a high-quality individual could afford to maintain these costly ornaments.

Haplorhini: The "dry-nosed" primates (monkeys, apes, and tarsiers).

Hidden/cryptic ovulation: Ovulation with no cues, either morphological or behavioral, that are discernible to human observers (note: they may, however, be discernible to other animals).

Hominoid: Apes (including humans) and their close ancestors.

Igapó: Seasonally flooded forests along black-water (nutrient-poor) rivers, particularly the Rio Negro in Brazil. Also see várzea.

Immigrate: The act of joining and becoming integrated into a new group (see also emigrate).

Implicit reconciliation: Behaviors used for reconciliation are regularly seen in other contexts.

Inclusive fitness: A theoretical model that measures fitness in terms of an individual's fitness plus the impact that individual has on the fitness of kin, devalued by the proportion of shared genes (or the likelihood of sharing a given gene/allele).

Infant: The developmental period between birth and the end of the weaning process. In primates the infancy period can last months or years, depending on the taxa.

Infanticide: The killing of an infant.

Insectivory: The consumption of insects.

Interbirth interval (IBI): The time period between the births of two successive infants. Although inter- and intraindividual variations exists, species have typical interbirth intervals.

Inhibition: Stopping or delaying performance of a previously routine behavior.

Intermembral index (IMI): [(Length of humerus + radius)/(Length of femur + tibia)] × 100. Serves as an indicator of the relative emphasis of the forelimb vs. hindlimb during locomotion.

In vitro: Literally "in glass," meaning a reaction, process, or experiment in a metaphorical test tube rather than in a living organism.

In vivo: Literally "in life," meaning a reaction, process, or experiment in a living organism.

Ischial callosities: Thickened areas of hairless skin on the anal region of Old World monkeys and members of the Hylobatidae family (gibbons and siamangs) that function as sitting pads.

Jatakamala: Folktales "that are used to teach Buddhist morals" (Khoroche 1989:xiv). Literally it means "stories of the rebirth of Buddha."

Juvenile: The developmental period between the end of weaning and the onset of puberty.

Kin bias: The tendency for individuals to favor genetically related individuals within groups or populations that contain individuals varying in degrees of relatedness to one another. Thus, kin bias, also called "nepotism," refers to observable behavior favoring closer kin over more distantly related and unrelated individuals. It is more specific than kin discrimination, a term that includes avoidance of kin.

Kin recognition: The ability to identify genetically related individuals and distinguish them from unrelated individuals.

Kin selection: W. D. Hamilton's (1964) theory for the evolution of altruism among kin. This theory posits that behaviors that promote the reproductive success not only of an individual's own offspring but also of the offspring of other genetic relatives will be evolutionarily selected, provided the behavior benefits the relative in terms of reproductive success more than it costs the actor, discounted by their degree of relatedness.

Knuckle walking: A form of quadrupedal locomotion used by chimpanzees, bonobos, and gorillas where weight borne by the front limbs is concentrated on the intermediate phalanges of the hands, thus placing pressure on the knuckles.

Leaf area index (LAI): The ratio of standing vegetation divided by the surface area of the earth beneath.

Leaf-swallowing: The slow and deliberate swallowing, one at a time without chewing, of whole leaves that are folded between tongue and palate and pass through the grastrointestinal tract visibly unchanged. The behavior has been recorded in chimpanzees.

Legend: A localized and historicized story or narrative usually related orally about a "specific place and time with named characters (Tangherlini 1996:437).

Lesser apes: A term used to refer to the smallest of the modern apes (i.e., gibbons and siamangs).

Life history theory: A special class of evolutionary theory used to explain the evolution of changes during the life course by analyzing demography, genetics, behavior, and morphology in a developmental and, typically, quantitative context.

Life span: The species-specific maximum age at death; genetic capacity for individuals of a species to maintain life. Human life span is estimated to be 100 years, although exceptionally rare individuals have lived beyond 100.

Life expectancy: Average life span of individuals of a given population at a given point in time. Life expectancy values are heavily influenced by infant mortality and vary considerably in different environments and different time periods.

Locomotor behavior: Differs from postures in that "body mass (as opposed to limb mass) is grossly displaced relative to its physical surroundings" (Prost 1965:1200). Locomotor behaviors are associated with such common activities as traveling, foraging, and predator avoidance and represent a primary component of a species' adaptive pattern.

Lordosis: A sexual position assumed by females of some mammalian species that involves a dorsally concave curvature of the distal portion of the vertebral column and is sensitive to hormonal fluctuations.

Lunar philia: Attraction to bright moonlight.

Lunar phobia: Avoidance of bright moonlight.

Major histocompatibility complex: A highly variable portion of the genome that plays an important role in immune function and possibly in mate choice.

Male fattening: Present in *Saimiri* spp., where males experience an accumulation of water stores throughout the body (particularly in the muscles around the shoulders, back, and arms) just prior to the mating season, causing a "fattened" appearance. This process is linked to increasing levels of testosterone.

Mammary glands: Found only in members of the Mammalia, mammary glands are exocrine glands that in females produce milk.

Mandibular symphysis: A joint present in the middle of the mandible of prosimians.

Mate choice: Any pattern of behavior, shown by members of one sex, that leads to their being more likely to mate with certain members of the opposite sex than others.

Matrilines: Individuals that are related to one another through maternal lines (i.e., that can trace their ancestry through females back to a common female relative) (see

also patrilines). Unlike social anthropologists, primatologists do not use these terms to infer anything about residence patterns, exogamy, or inheritance rules.

Metopic suture: A line of union between the two halves of the front bone of the skull, found in prosimians.

Menopause: The universal midlife cessation of menstrual cycling and reproductive ability that typically occurs in women at around 50 years of age.

Menstrual cycle: The cyclical buildup and release of the blood-based lining of the uterus.

Metacognition: Monitoring and willfully managing one's own cognitive activities.

Mixed-species association: When groups of two different species combine to travel and forage together. The association may not always be mutual. Also known as "polyspecific association."

Modularity: Varying levels of integration during development among parts of the body. Highly integrated parts form modules, while others are dissociated.

Monochromatic color vision: The ability to distinguish between black and white. Monochromatic individuals are considered "colorblind."

Monogamy: A mating system in which one male mates with one female.

Monophyletic: A taxonomic grouping descended from a single common ancestor and that contains all descendents of that ancestor.

Monospecific: Describes a genus containing a single species.

Mutualism: An evolutionary explanation for cooperative behavior that provides net benefits in terms of reproductive success for both (all) interacting individuals.

Myth: According to Farrer (1997:577–578), a myth is different from other forms of narrative in that it is a "sacred narrative that is most often believed to be true. Because myth is of such a sacred nature, many people object to the very word myth when it is attached to their own sacred narratives. For those in whose culture the myth is embedded, the truth of the myth may be literal or figurative. Literalists believe fully in the absolute truth of myth, whereas figurativists believe in myth as metaphor."

Natal coat color: The presence of a distinct color pattern (from the typical adult pattern) in infants. Often seen in members

of the Colobinae subfamily of Old World monkeys.

Natal dispersal: The primary dispersal event for an individual, which involves emigration from the birth group.

Natal group: The group into which an individual was born.

Nematodes (or roundworms): Worms with an unsegmented body, which may be parasitic or free-living.

Neogene: A chronostratigraphic unit of time referring to the Miocene (23—5 million years ago) and Pliocene (5—1.8 million years ago) epochs. Current uses of the term also include the Pleistocene epoch (1.8 million—~20,000 years ago).

Neotropics: The area of the New World (i.e., the Americas) between the tropics of Capricorn and Cancer.

New World monkeys: Monkeys belonging to the infraorder Platyrrhini and the superfamily Ceboidea that are native to the New World (southern Mexico, Central and South America).

Niche: The combination of habitat and behavior that make a species' way of life distinct from that of all other species. A niche cannot be defined by a singular feature or parameter but rather (and only) by including all functional aspects of an organism's existence: environmental setting, foraging habits, intra- and interspecific social systems, and life history.

Nocturnal: Active during the nighttime.

Old World monkeys: Monkeys belonging to the infraorder Catarrhini and the superfamily Cercopithecoidea that are native to the Old World (Africa, Asia, and Europe).

Omnivory: The consumption of a variety of different food types.

Ontogeny: Growth (size change) and development (shape change) with age including physical and behavioral aspects.

Ovulation: The release of a mature egg (ovum) from the ovary.

Paleogene: A chronostratigraphic unit of time referring to the beginning of the Tertiary era (65—23 million years ago). The Paleogene encompasses the Paleocene (65—55 million years ago), Eocene (55—35 million years ago), and Oligocene (35—23 million years ago) epochs.

Paleotropics: The regions of the Old World (i.e., Asia, Africa) between the tropics of Capricorn and Cancer.

Parallel dispersal: Dispersal with co-resident males and/or females into groups containing familiar males.

Parapatric: Species or populations of organisms that exist next to one another geographically but that are nonoverlapping.

Paraphyly: A taxonomic group descended from a single common ancestor but that does not include all descendents of that ancestor (see also monophyletic).

Parasite: Something that lives in, with, or on another organism (the host) and obtains benefits from that organism. A parasite is detrimental to the host in varying degrees.

Parental investment: Any investment in time, energy, or risk by a parent in an offspring that enhances the offspring's reproductive success and detracts from the parent's ability to invest in other offspring.

Parent–offspring conflict: A conflict that arises as a result of the fact that a parent's inclusive fitness is maximized by ceasing to invest in current offspring (for future offspring) sooner than when an offspring's inclusive fitness would be maximized by gaining siblings.

Pathogenic: Any agent causing disease, especially a microorganism.

Patriline: A grouping of primates that includes individuals that can trace their ancestry through males back to a common male relative.

Periovulatory period: A period of time prior to, including, and following ovulation when conception is most likely. The duration varies by species.

Phenotypic matching: A mechanism by which genetically related individuals are able to recognize one another. In phenotypic matching, individuals use physical similarities such as odor or appearance.

Philopatry: Remaining in the social groups in which one was born for life. Philopatric individuals generally do not disperse to other groups or live solitarily.

Phylogeny: The true evolutionary history of a lineage, which remains unknown except through the interpretations of indirect data such as the fossil record, the genome of living species, and comparative anatomy of living species.

Platyrrhine: An infraorder of primates; monkeys living in the New World (southern Mexico, Central and South America).

Polyandry: A mating system in which one female mates with multiple males.

Polygyny: A mating system in which one male mates with multiple females. Can refer to single-male groups or multimale groups where each male mates with multiple females.

Polygynandry: A mating system in which both males and females mate with multiple members of the opposite sex.

Polyspecific association: A mixed-species association.

Positional behavior: A combination of postural and locomotor behaviors.

Postorbital bar: A complete ring of bone around the eye socket, characteristic of all primates.

Postorbital closure: A full bony enclosure of the back of the eye socket, found in anthropoids and tarsiers.

Postreproductive life span: The number of years between reproductive cessation and death. In human females, the postreproductive life span is up to 50 years. Not to be confused with the time period between last parturition (birth) and death as this time is not necessarily postreproductive.

Postural behavior: A set of actions in which an individual's center of mass remains relatively stable and the limb segments proper, rather than the entire skeleton, are being manipulated for some end (Prost 1965). Postural activity tends to dominate the positional repertoire during feeding, resting, and various socially directed behaviors.

Precocial: Young animals that are born fairly independent (i.e., able to move about and forage at a very early age).

Prehensile tail: An elongated, highly muscularized, and vascularized tail that functionally serves as a fifth limb due to its ability to grasp and/or hold objects. Prehensile tails have evolved twice during primate evolution (both times in the New World monkeys), once in the common ancestor of capuchins (*Cebus* spp.) and once in the common ancestor of atelines (*Alouatta* spp., *Ateles* spp., *Brachyteles* spp., and *Lagothrix* spp.). The capuchin and ateline tails differ significantly in morphology.

"Priority-of-access" model: An idealized model of mating patterns in which the ability of the dominant male to monopolize mating with females is inversely proportional to the number of females that are simultaneously sexually receptive.

Procumbent incisors: Tendency of incisors in some primates to be orientated in a horizontal plane; functionally related to grooming or procuring food.

Prosimian: A group of living primates and their ancestors that includes lemurs, lorises, galagos, and tarsiers, as distinct from the monkeys, apes, and humans (the anthropoids). Prosimians are restricted to Africa and Asia and are relatively small, less encephalized, and more likely to be nocturnal than are anthropoids.

Pseudo-opposable thumbs: Thumbs that flex in parallel with other digits but can be used to form a power grip.

Puberty: Process of physical and physiological changes that change the nonreproductive body of a juvenile to one that is capable of reproduction.

Quadrumanuous clambering: A form of locomotion used often by orangutans that involves the use of the four limbs in various combinations by grasping supports in different directions, both above and below the animal.

Quadrupedal: Various forms of locomotion that involve the use of all four limbs.

r-K selection: Historically influential idea in life history theory that species can be arrayed along a continuum between short-lived, fast reproducers typically found in more volatile habitats (r-selected) and long-lived, slower reproducers expected to be found in stable habitats (K-selected). More recently referred to as a "fast–slow continuum."

Radiation: Rapid diversification of a lineage into multiple species that adapt to different ecological or behavioral niches.

Reciprocity or reciprocal altruism: Robert Trivers' (1971) explanation for the evolution and maintenance of altruistic behavior regardless of the genetic relationship between the interactors. It posits that altruistic behavior will be selected when individuals that behave altruistically to others receive similar benefits from their partners at some time in the future, provided the net benefits exceed the costs for each individual and that cheaters (nonreciprocators) can be detected and avoided.

Reconciliation: A postconflict affiliative reunion between former opponents; it is more accurate to use this term only when the function of such reunions in repairing the social relationship between former opponents is demonstrated.

Regeneration niche: A term coined by Grubb (1977) referring to the biotic and abiotic requirements that ensure a high probability of success in the replacement of one mature individual of a plant species by another mature individual of that species. When seeds of a species are dispersed to a habitat outside of its regeneration niche, these seeds have a low probability of germinating, establishing seedlings, or reaching adulthood.

Relationship value: A measure of the direct benefits afforded by a relationship to the two partners.

Relationship compatibility: A measure of the degree of tolerance between the two partners.

Relationship security: The predictability of interactions between the two partners or the consistency of their interactions over time.

Relative humidity: The ratio of the partial pressure of water vapor in the mixture to the saturated vapor pressure of water at a prescribed temperature.

Reproductive life span: The total number of years that an individual is reproductive. In human females it is the number of years between puberty and menopause; in other primates it is usually the number of years between age at first reproduction and age at death since most nonhuman primates are reproductive until death.

Reproductive cessation: The termination of reproductive ability.

Reproductive fitness: A measure of reproductive success; the relative contribution of an individual to the gene pool of the next generation or measured by how many offspring are produced which live to reproduce themselves.

Reproductive senescence: The deterioration and decline of reproductive ability with age.

Reproductive skew: Differential (uneven) reproductive success among individuals within a group.

Reproductive success: The relative production of fertile offspring.

Reverse contingency task: A task requiring selection of the nonpreferred icon or object to obtain the preferred object; aimed at tapping into an individual's ability to regulate a pre-potent behavior (e.g., reaching for a preferred food item or greater quantity).

Rhinarium: The wet, naked surface around the nostrils of the nose in most mammals including the strepsirhine primates.

"Runaway" sexual selection: A model proposed by R. A. Fisher (1915) to explain evolutionary process in which genes for a sexually selected ornament and genes for a preference for the ornament become correlated, producing a rapid increase in the ornament's magnitude, even though it signals nothing about an individual's underlying "quality."

Sacculated stomach: A stomach with "saccules," or sack-like compartments, in which bacteria slowly break down the cellulose found in leaves. Found in monkeys of the Colobinae subfamily.

Scent gland: Glands found in various parts of the body (e.g., anal region, chest, wrists, arm pits) that release odiferous fluids.

Schistosomiasis (or bilharzias): A disease caused by a blood fluke of the genus *Schistosoma*, a type of flatworm parasite. The intermediate host is a snail, in which cercariae (larvae) develop and migrate out into water; the cercariae penetrate the skin of hosts that make contact with the water. Symptoms depend on species causing infection but can include rash, fever, aching, cough, diarrhea, and liver and spleen enlargement.

Scramble competition: When animals compete for a resource that cannot be monopolized but is exploited by the first to utilize the resource.

Secondary compound: A plant chemical that does not seem to directly function in the processes of plant growth and development.

Seed density: The number of seeds in a unit area. Seed density is often used as an indicator of the extent to which dispersed seeds are clumped (higher seed density in a small area) or scattered (lower seed density in that area). See Howe (1989).

Seed dispersal: The movement of seeds away from a parent plant. *Primary seed dispersal* for animal-dispersed seeds refers to the movement of the seed away from the parent by the first dispersal agent, aside from gravity. *Secondary dispersal* refers to subsequent movement of seeds after they have been dispersed by primary dispersal. Secondary dispersal is often by dung beetles (Scarabeidae). When primary dispersal agents defecate seeds, dung beetles accidentally roll seeds up in the dung balls in which they lay their eggs, moving the seeds short distances and burying them in the process. Secondary dispersal also occurs by rodents that cache seeds under the leaf litter or soil (scatter-hoarding and larder-hoarding) for later consumption (see *seed predation*). However, some seeds cached by rodents escape predation because they are not relocated and are, thus, dispersed.

Seed-dispersal curve: The frequency distribution of distances that seeds are moved from the parent plant. Seed-dispersal curves may also be expressed in terms of the density of seeds in annuli (rings) of a particular width surrounding a parent plant, which accounts for the area over which seeds are deposited at each distance from a parent.

Seed-dispersal distance: The straight-line distance a seed travels away from its parent.

Seed passage time: The amount of time that a seed takes to pass through the gut of its animal dispersal agent.

Seed predation: The consumption of seeds, in particular the endosperm, cotyledons, or embryo, rendering the seed, by definition, nonviable for germination. Seed predation is often by either arboreal or terrestrial birds and mammals (particularly rodents) but also by insects that lay eggs inside seeds, where the larvae develop and consume the seed, or by pathogens, such as fungi.

Seed shadow: The spatial distribution of seeds dispersed from a single plant. Seed shadows from multiple individuals may be summed over an area to represent the spatial distribution of seeds for a population of plants.

Self-medication: (pertaining to nonhuman animals) The use of plants, soils, insects, fungi, or other substances by animals in ways that treat illness and/or relieve symptoms.

Semi-brachiation: A form of locomotion that is similar to brachiation in that it involves swinging beneath the branches; however, unlike brachiation, in semi-brachiation the prehensile tail is also used. Found in the ateline primates.

Semiprehensile tail: The tail of capuchin (*Cebus* spp.) monkeys adapted for seizing and grasping and commonly used as an anchor or for balance. Unlike the fully prehensile tail of the atelines, the semiprehensile tail is completely covered in hair, unable to support the weight of an adult, and not equivalent to another limb.

Senescence: Decline in bodily function with age leading to increased frailty and risk of death.

Sex-typing hypothesis: A hypothesis suggesting that selective attention toward same-sex adults by immatures is critical for the development of sex-typical behaviors.

Sexual coercion: The use of force (or the threat of force) by males that functions to increase the likelihood that a female will mate with a male at a time when she is likely to be fertile and to decrease the chances that she will mate with other males, at some cost to the female.

Sexual conflict: A conflict between the evolutionary interests of individuals of the two sexes. Adaptations that increase fitness in one sex may reduce or constrain fitness in the opposite sex.

Sexual dichromatism: Differences in coat color in adult males and females.

Sexual dimorphism: Physical differences in adult males and females (e.g., in size, color).

Sexual selection: A theory proposed by Charles Darwin (1871) to explain the evolution of certain traits that grant greater success in acquiring mates, even if these traits diminish an individual's survival chances. *Intersexual selection* is selection for traits that enhance the ability to manipulate members of the opposite sex for increased mating opportunities (e.g., ornaments, coercion, female choice); *intrasexual selection* is selection for traits that enhance the ability to compete against members of the same sex for mates (i.e., weapons).

Sexual swelling: The cyclical tumescence of the perineum in females. Sexual swellings are often accompanied by color changes and usually occur around the time of ovulation.

Socialization: The behavioral modification of an individual resulting from its interactions with other members of its society.

Social proficiency: The ability to operate in a social world, competing for resources and mates and interacting, on an adult level. Socioecology: The study of how environmental variables influence group size, composition, and social dynamics.

Sociosexual: Sexual behaviors used in behavioral/social contexts outside of copulation.

Sonogram: A visual display of the frequency content of a sound.

Sperm competition: The competition between individual sperm of one or more males in the female reproductive tract to fertilize an ovum.

Standard mark test: The subject's forehead or another visually imperceptible body part is marked in some way while it is asleep, distracted, or sedated; and the subject is then allowed to see its reflection in a mirror. If the individual spontaneously acts on the mark, this self-directed behavior is taken as an indication that it recognizes the image in the mirror as its own reflection.

Strepsirhini: The "wet-nosed" primates (lemuriforms and lorisiforms).

Suspensory behaviors: Associated with a range of below-branch postures and locomotor modes (grasping, forelimb suspension, hindlimb suspension, tail hang, arm–tail suspension) that aid in maintaining balance and weight support while taking advantage of the downward displacement of flexible supports to obtain access to an increased sphere of food resources and areas of the tree crown.

Sympatric: Species or populations of organisms that exist in geographically overlapping areas.

Tapetum lucidum: A reflective layer in the eye of most nocturnal mammals, including the nocturnal strepsirhines.

Taxonomic classification: The hierarchical classification of organisms based on genetics, similar physical structures, or likely phylogeny (see phylogeny).

Torpor: A state of decreased physical and physiological activity in an animal that is usually associated with a decrease in core body temperature and metabolic rate.

Terminal branches: Food-bearing, thin, flexible supports located in the periphery of the tree crown.

Terrestrial: Referring to the ground. Terrestrial primates are those that live primarily on the ground.

Theory of mind: Attributing mental states (i.e., intents, desires, knowledge, beliefs, etc.) to others.

Tool use: Producing a spatial relation between one object and another so that the first object serves as a functional extension of the body to act on the other object or surface to attain an immediate goal.

Tool set: Two or more types of tools used sequentially to attain an objective.

Trade-off: A relationship (physiological and/or genetic) between variables that prevents selection from acting on each trait independently such that they arrive at values that would yield highest fitness (e.g., current vs. future reproduction, offspring quality vs. quantity).

Transitive inference: Inferring a relation between two elements from preliminary information establishing the relations of each of those two elements to a third.

Trichromatic color vision: The ability to distinguish three color fields: black–white, blue–yellow, and red–green.

Tumescence: The state of being swollen, usually as a result of blood or water buildup.

Várzea: Seasonally flooded forests along the Amazon River and its tributaries (i.e., white-water, nutrient-rich rivers). Also see igapó.

Vertical clinging and leaping: A form of locomotion common among the prosimians in which animals cling to a vertical substrate and then move by leaping between these substrates.

Vigilance: Being alert and/or watchful for danger.

Xerophytic forest: A dry forested habitat adapted to long periods of drought.

Y-5 molars: A Y-shaped pattern present in the five cusps of the lower molars of members of the Hominoidea.

Zoopharmacognosy: The study of how animals use medicinal substances. Interchangeably used by some with the term animal "self-medication."

REFERENCES

Abbott, D. H., and Hearn, J. P. (1978). Physical, hormonal and behavioural aspects of sexual development in the marmoset monkey (*Callithrix jacchus*). *J. Reprod. Fertil.* 53:155–166.

Abbott, D. H., Hodges, J. K., and George, L. M. (1988). Social status controls LH secretion and ovulation in female marmoset monkeys (*Callithrix jacchus*). *J. Endocrinol.* 117:329–339.

Abbott, D. H., Saltzman, W., Schultz-Darken, N. J., and Smith, T. E. (1997). Specific neuroendocrine mechanisms not involving generalized stress mediate social regulation of female reproduction in cooperatively breeding marmoset monkeys. *Ann. N. Y. Acad. Sci.* 807:219–238.

Abbott, D. H., Saltzman, W., Schultz-Darken, N. J., and Tannenbaum, P. L. (1998). Adaptations to subordinate status in female marmoset monkeys. *Comp. Biochem. Physiol.* 119:261–274.

Abbott, D. H., Barrett, J., and George, L. M. (1993). Comparative aspects of the social suppression of reproduction in female marmosets and tamarins. In: Rylands, A. B. (ed.), *Marmosets and Tamarins: Systematics, Behaviour, and Ecology.* Oxford University Press, Oxford. pp. 152–163.

Abbott, D. H., Keverne, E. B., Bercovitch, F. B., Shively, C. A., Mendoza, S. P., Saltzman, W., Snowdon, C. T., Ziegler, T. E., Banjevic, M., Garland, T., Jr., and Sapolsky, R. M. (2003). Are subordinates always stressed? A comparative analysis of rank differences in cortisol levels among primates. *Horm. Behav.* 43:67–82.

Abegg, C., and Thierry, B. (2002a). Macaque evolution and dispersal in insular Southeast Asia. *Biol. J. Linn. Soc.* 75:555–576.

Abegg, C., and Thierry, B. (2002b). The phylogenetic status of Siberut macaques: hints from the bared-teeth display. *Primate Rep.* 63:63–72.

Abegg, C., Petit, O., and Thierry, B. (2003). Variability in behavior frequencies and consistency in transactions across seasons in captive Japanese macaques (*Macaca fuscata*). *Aggress. Behav.* 29:81–93.

Abegglen, J. J. (1984). *On Socialization in Hamadryas Baboons: A Field Study.* Bucknell University Press, Lewisburg, PA.

Abernethy, K. A., and White, L. J. T (in press). Mandrill (Mandrillus sphinx). In: Butynski, T. M., Kingdon, J. S., and Kalina, J. (eds.) The Mammals of Africa. Vol. 2. Primates. Academic Press, Amsterdam.

Abernethy, K. A., White, L. J. T., and Wickings, E. J. (2002). Hordes of mandrills (*Mandrillus sphinx*): extreme group size and seasonal male presence. *J. Zool.* 258:131–137.

Abitbol, M. M. (1987). Evolution of the sacrum in hominoids. *Am. J. Phys. Anthropol.* 74:65–82.

Abrams, P. A. (1994). Should prey overestimate the risk of predation? *Am. Nat.* 144:317–328.

Adams, E. J., and Parham, P. (2001). Species-specific evolution of MHC class I genes in the higher primates. *Immunol. Rev.* 183:41–64.

Adams-Curtis, L. E. (1990). Conceptual learning in capuchin monkeys. *Folia Primatol.* 54:129–137.

Addessi, E., Crescimbene, L., and Visalberghi, E. (2007). Do capuchin monkeys (*Cebus apella*) use tokens as symbols? *Proc. R. Soc. Lond. B* 274:2579–2585.

Addessi, E., Crescimbene, L., and Visalberghi, E. (2008). Food and token quantity discrimination in capuchin monkeys (*Cebus apella*). *Anim. Cogn.* 11:275–282.

Adegoke, G. O., and Skura, B. J. (1994). Nutritional profile and antimicrobial spectrum of the spice *Aframomum danielli*. *Plant Foods Hum. Nutr.* 45:175–182.

Agapow, P. M., Bininda-Emonds, O. R. P., Crandall, K. A., Gittleman, J. L., Mace, G. M., Marshall, J. C., and Purvis, A. (2004). The impact of species concept on biodiversity studies. *Q. Rev. Biol.* 79:161–179.

Agetsuma, N. (1995a). Foraging strategies of Yakushima macaques (*Macaca fuscata yakui*). *Int. J. Primatol.* 16:595–610.

Agetsuma, N. (1995b). Foraging synchrony in a group of Yakushima macaques (*Macaca fuscata yakui*). *Folia Primatol.* 64:167–179.

Agetsuma, N., and Nakagawa, N. (1998). Effects of habitat differences on feeding behaviors of Japanese monkeys: comparison between Yakushima and Kinkazan. *Primates* 39:275–289.

Agetsuma, N., and Noma, N. (1995). Rapid shifting of foraging pattern by Yakushima macaques (*Macaca fuscata yakui*) in response to heavy fruiting of *Myrica rubra*. *Int. J. Primatol.* 16:247–260.

Aggimarangee, N. (1992). Survey for semi-tame colonies of macaques in Thailand. *Nat. Hist. Bull. Siam Soc.* 40:103–166.

Agoramoorthy, G., and Hsu, M. J. (2000). Extragroup copulation among wild red howler monkeys in Venezuela. *Folia Primatol.* 71:147–151.

Agoramoorthy, G., and Hsu, M. J. (2005). Occurrence of infanticide among wild proboscis monkeys (*Nasalis larvatus*) in Sabah, North Borneo. *Folia Primatol.* 76:177–179.

Agoramoorthy, G., and Lohmann, R. (1999). Population and conservation status of the black-and-gold howler monkeys, *Alouatta caraya*, along the Rio Riachuelo, Argentina. *Neotrop. Primates* 7:43–44.

Agoramoorthy, G., and Rudran, R. (1993). Male dispersal among free-ranging howler monkeys (*Alouatta seniculus*) in Venezuela. *Folia Primatol.* 61:92–96.

Agoramoorthy, G., and Rudran, R. (1995). Infanticide by adult and subadult males in free-ranging red howler monkeys, *Alouatta seniculus*, in Venezuela. *Ethology* 99:75–88.

Agostini, I., and Visalberghi, E. (2005). Social influences on the acquisition of sex-typical foraging patterns by juveniles in a group of wild tufted capuchin monkeys (*Cebus nigritus*). *Am. J. Primatol.* 65:335–351.

Aguilar, J. M., and Lacher, T. E., Jr. (2003). On the morphological distinctiveness of *Callithrix humilis* van Roosmalen et al., 1998. *Neotrop. Primates* 11:11–17.

Aguirre, A. C. (1971). *O Mono* Brachyteles arachnoides. Academia Brasiliera de Ciencias, Rio de Janeiro.

Agusti, J., de Siria, A. S., and Garces, M. (2003). Explaining the end of the hominoid experiment in Europe. *J. Hum. Evol.* 45:145–153.

Ah-King, M., and Tullberg, B. S. (2000). Phylogenetic analysis of twinning in Callitrichinae. *Am. J. Primatol.* 51:135–146.

Ahsan, M. F. (1994). Feeding ecology of the primates of Bangladesh. In: Thierry, B., Anderson, J. R., Roeder, J. J., and Herrenschmidt, N. (eds.), *Current Primatology. Ecology and Evolution,* vol. I. Université Louis Pasteur, Strasbourg. pp. 79–86.

Ahsan, M. F. (2001). Socio-ecology of the hoolock gibbon (*Hylobates hoolock*) in two forests of Bangladesh. In: *The Apes: Challenges for the 21st Century. Conference Proceedings.* Chicago Zoological Society, Chicago. pp. 286–299.

Ahumada, J. A. (1989). Behavior and social structure of free ranging spider monkeys (*Ateles belzebuth*) in La Macarena. *Field Studies New World Monkeys La Macarena Colombia* 2:7–31.

Aidara, D., Badawi, M., Tahiri-Zagret, C., and Robyn, C. (1981). Changes in concentrations of serum prolactin, FSH, oestradiol and progesterone and of the sex skin during the menstrual cycle in the mangabey monkey (*Cercocebus atys lunulatus*). *J. Reprod. Fertil.* 62:475–481.

Aiello, L. (1986). The relationship of the Tarsiiformes: a review of the case for the Haplorhini. In: Wood, B. A., Martin, L. B., and Andrews, P. J. (eds.), *Major Topics in Primate and Human Evolution,* Cambridge University Press, Cambridge. pp 47–65.

Aiello, L. C., and Wheeler, P. (1995). The expensive tissue hypothesis. *Curr. Anthropol.* 36:199–222.

Aisner, R., and Terkel, J. (1992). Ontogeny of pine cone opening behaviour in the black rat, *Rattus rattus*. *Anim. Behav.* 44:327–336.

Aitken, N., Smith, S., Schwarz, C., and Morin, P. A. (2004). Single nucleotide polymorphism (SNP) discovery and genotyping in mammals: a targeted-gene approach. *Mol. Ecol.* 13:1423–1431.

Albernaz, A. L., and Magnusson, W. E. (1999). Home-range size of the bare-ear marmoset (*Callithrix argenta*) at Alter do Chão, Central Amazonia, Brazil. *Int. J. Primatol.* 20:665–677.

Alberts, S. (1987). Parental care in captive siamangs (*Hylobates syndactylus*). *Zoo Biol.* 6:401–406.

Alberts, S. C. (1999). Paternal kin discrimination in wild baboons. *Proc. R. Soc. Lond. B* 266:1501–1506.

Alberts, S. C., and Altmann, J. (1995a). Balancing costs and opportunities: dispersal in male baboons. *Am. Nat.* 145:279–306.

Alberts, S. C., and Altmann, J. (1995b). Preparation and activation: determinants of age at reproductive maturity in male baboons. *Behav. Ecol. Sociobiol.* 36:397–406.

Alberts, S. C., and Altmann, J. (2001). Immigration and hybridization patterns of yellow and anubis baboons in and around Amboseli, Kenya. *Am. J. Primatol.* 53:139–154.

Alberts, S. C., and Altmann, J. (2006). The evolutionary past and the research future: environmental variation and life history flexibility in a primate lineage. In: Swedell, L., and Leigh, S. R. (eds.) *Reproduction and Fitness in Baboons: Behavioral, Ecological, and Life History Perspectives*. Springer, New York. pp. 277–303.

Alberts, S. C., Sapolsky, R. M., and Altmann, J. (1992). Behavioral, endocrine, and immunological correlates of immigration by an aggressive-male into a natural primate group. *Horm. Behav.* 26(2):167–178.

Alberts, S. C., Altmann, J., Wilson, M. L. (1996). Mate guarding constrains foraging activity of male baboons. *Anim. Behav.* 51:1269–1277.

Alberts, S., Watts, H. E., and Altmann, J. (2003). Queuing and queue-jumping: long-term patterns of reproductive skew in male savannah baboons, *Papio cynocephalus*. *Anim. Behav.* 65:821–840.

Alberts, S. C., Buchan, J. E., and Altmann, J. (2006). Sexual selection in wild baboons: from mating opportunities to paternity success. *Anim. Behav.* 72:1177–1196.

Albiach-Serrano, A., Guillen-Salazar, F., and Call, J. (2007). Mangabeys (*Cercocebus torquatus lunulatus*) solve the reverse contingency task without a modified procedure. *Anim. Cogn.* 10:387–396.

Albiach-Serrano, A., Guillen-Salazar, F., and Call, J. (2008). Inhibition of a prepotent response in mangabeys (*Cercocebus torquatus lunulatus*). *Folia Primatol.* 79:306.

Albignac, R. (1981a). Variabilité dans l'organisation territoriale et l'écologie de Avahi laniger (Lémurien nocturne de Madagascar). *C. R. Acad Sci. III* 292:331–334.

Albignac, R. (1981b). Lemurine social and territorial organization in a northwestern Malagasy forest (restricted area of Ampijoroa). In: Chiarelli, A. B., and Corruccini, R. S. (eds.), *Primate Behavior and Sociobiology*. Springer-Verlag, Berlin. pp. 25–29.

Alcock, J. (1989). *Animal Behavior: An Evolutionary Approach*. Sinauer Associates, Sunderland, MA.

Aldrich, B. C., Milleson, L., and Nekaris, K. A. I. (2008). Vocalizations as a conservation tool: an auditory survey of the Andean titi monkey *Callicebus oenanthe* Thomas, 1924 (Mammalia: Primates: Pitheciidae) at Tarangue, northern Peru. *Contrib. Zool.* 77:1–6.

Aldrich-Blake, F. P. G. (1968). A fertile hybrid between two *Cercopithecus* spp. in the Budongo Forest, Uganda. *Folia Primatol.* 9:15–21.

Aldrich-Blake, F. P. G., and Chivers, D. J. (1973). On the genesis of a group of siamang. *Am. J. Phys. Anthropol.* 38:631–636.

Aldrich-Blake, F. P. G., Bunn, T. K., Dunbar, R. I. M., and Headley, P. M. (1971). Observations on baboons, *Papio anubis*, in an arid region in Ethiopia. *Folia Primatol.* 15:1–35.

Alencar, A., Sousa, M., Abbott, D., and Yamamoto, M. (2006). Contested dominance modifies the anovulatory consequences of social subordination in female marmosets. *Braz. J. Med. Biol. Res.* 39:647–58.

Alexander, R. D. (1974). The evolution of social behavior. *Annu. Rev. Ecol. Syst.* 5:324–383.

Alexander, R. D. (1979). Sexuality and sociality in humans and other primates. In: Katchadourian, A. (ed.), *Human Sexuality: A Comparative and Developmental Perspective*. University of Californian Press, Berkeley. pp. 81–97.

Alexander, R. M. (1993). The relative merits of foregut and hindgut fermentation. *J. Zool. Lond.* 231:391–401.

Allen, G. M., and Coolidge, H. J. (1940). Mammal and bird collections of the Asiatic primate expedition: mammals. *Bull. Mus. Comp. Zool. Harv. Univ.* 87:131–166.

Allison, A. C. (1982). Co-evolution between hosts and infectious disease agents and its effect on virulence. In: Anderson, R. M., and May, R. M. (eds.), *Population Biology of Infectious Diseases*. Springer-Verlag, Berlin. pp. 245–267.

Allman, J., and Hasenstaub, A. (1999). Brain maturation times, and parenting. *Neurobiol. Aging* 20:447–454.

Allman, J. M., McLaughlin, T., and Hakeem, A. (1993). Brain structures and life-span in primate species. *Proc. Natl. Acad. Sci. USA* 90:3559–3563.

Almeida Noronha, M., Silva, J. D. S., Spironello, W. R., and Ferreira, D. C. (2007). New occurrence records of *Mico acariensis* (Primates, Callitrichidae). *Neotrop. Primates* 14:140–141.

Almeida Noronha, M., Silva, J. D. S., Spironello, W. R., and Ferreira, D. C. (2008a). New occurence records for Maues marmoset, *Mico mauesi* (Primates, Callitrichidae). *Neotrop. Primates* 15:24–26.

Almeida Noronha, M., Spironello, W. R., and Ferreira, D. C. (2008b). New occurrence records for *Mico melanurus* (Primates, Callitrichidae). *Neotrop. Primates* 15:26–28.

Alonso, C., and Langguth, A. (1989). Ecologia e comportamento de *Callithrix jacchus* (Primates: Callitrichidae) numa ilha de foresta Atlântica. *Rev. Nordestina Biol.* 6:107–137.

Al-Safadi, M. M. (1994). The hamadryas baboon, *Papio hamadryas* (Linnaeus, 1758) in Yemen (Mammalia: Primates: Cercopithecidae). *Zool. Midd. East* 10:5–16.

Alterman, L. (1995). Toxins and toothcombs: potential allospecific chemical defenses in *Nycticebus* and *Perodicticus*. In: Alterman, L., Doyle, G. A., and Izard, M. K. (eds.), *Creatures of the Dark: The Nocturnal Prosimians*. Plenum Press, New York. pp. 413–424.

Altmann, J. (1974). Observational study of behaviour: sampling methods. *Behaviour* 49:227–265.

Altmann, J. (1979). Age cohorts as paternal sibships. *Behav. Ecol. Sociobiol.* 6:161–164.

Altmann, J. (1980). *Baboon Mothers and Infants*. Harvard University Press, Cambridge, MA.

Altmann, J. (1986). Adolescent pregnancies in non-human primates: an ecological and developmental perspective. In: Lancaster, J., and Hamburg, B. (eds.), *School-Age Pregnancy and Parenthood: Biosocial Dimensions*. Aldine Press, New York. pp. 247–262.

Altmann, J. (1990). Primate males go where the females are. *Anim. Behav.* 39:193–195.

Altmann, J. (1997). Mate choice and intrasexual reproductive competition: contributions to reproduction that go beyond acquiring more mates. In: Gowaty, P. A. (ed.), *Feminism and Evolutionary Biology: Boundaries, Intersections, and Frontiers*. Chapman and Hall, New York. pp. 320–333.

Altmann, J. (2000). Models of outcome and process: predicting the number of males in primate groups. In: Kappeler, P. M. (ed.), *Primate Males: Causes and Consequences of Variation in Group Composition*. Cambridge University Press, Cambridge. pp 236–247.

Altmann, J., and Alberts, S. C. (2003a). Intraspecific variability in fertility and offspring survival in a nonhuman primate: behavioral control of ecological and social sources. In: Wachter, K. W., and Bulatao, R. A. (eds.), *Offspring: The Biodemography of Fertility and Family Behavior* National Academy Press, Washington DC. pp. 140–169.

Altmann, J., and Alberts, S. C. (2003b). Variability in reproductive success viewed from a life-history perspective in baboons. *Hum. Biol.* 15:401–409.

Altmann, J., and Alberts, S. C. (2005). Growth rates in a wild primate population: ecological influences and maternal effects. *Behav. Ecol. Sociobiol.* 57:490–501.

Altmann, J., and Muruthi, P. (1988). Differences in daily life between semiprovisioned and wild-feeding baboons. *Am. J. Primatol.* 15:213–221.

Altmann, J., Altmann, S. A., Hausfater, G., and McCuskey, S. A. (1977). Life history of yellow baboons: physical development, reproductive parameters, and infant mortality. *Primates* 18:315–330.

Altmann, J., Altmann, S., and Hausfater, G. (1981). Physical maturation and age estimates of yellow baboons, *Papio cynocephalus*, in Amboseli National Park, Kenya. *Am. J. Primatol.* 1:389–399.

Altmann, J., Hausfater, G., and Altmann, S. A. (1985). Demography of Amboseli baboons, 1963–1983. *Am. J. Primatol.* 8:113–125.

Altmann, J., Hausfater, G., and Altmann, S.A. (1988). Determinants of reproductive success in savannah baboons, *Papio cynocephalus*. In: Clutton-Brock, T. H. (ed.), *Reproductive Success: Studies of Individual Variation in Contrasting Breeding Systems*. University of Chicago Press, Chicago. pp 403–418.

Altmann, J., Schoeller, D., Altmann, S., Muruthi, P., and Sapolsky, R. (1993). Body size and fatness of free-living baboons reflect food availability and activity levels. *Am. J. Primatol.* 30:149–161.

Altmann, J., Alberts, S. C., Haines, S. A., Dubach, J., Muruthi, P., Coote, T., Geffen, E., Cheesman, D. J., Mututua, R. S., Saiyalele, S. N., Wayne, R. K., Lacy, R. C., and Bruford, M. W. (1996). Behavior predicts genetic structure in a wild primate group. *Proc. Natl. Acad. Sci. USA* 93:5797–5801.

Altmann, S. A. (1956). Avian mobbing behavior and predator recognition. *Condor* 58:241–253.

Altmann, S. A. (1959). Field observations on the howler monkey society. *J. Mammal.* 40:317–330.

Altmann, S. A. (1962). A field study on the sociobiology of rhesus monkeys (*Macaca mulatta*). *Ann. N. Y. Acad. Sci.* 102:338–435.

Altmann, S. A. (1965). Sociobiology of rhesus monkeys. II. Stochastics of social communication. *J. Theor. Biol.* 8:490–522.

Altmann, S. A. (ed.) (1967). *Social Communication Among Primates*. University of Chicago, Chicago.

Altmann, S. A. (1970). The pregnancy sign in savannah baboons. *Lab. Anim. Digest* 6:7–10.

Altmann, S. A. (1974). Baboons: space, time, and energy. *Am. Zool.* 14:221–248.

Altmann, S. (1979). Altruistic behaviour: the fallacy of kin deployment. *Anim. Behav.* 27:958–962.

Altmann, S. A. (1991). Diets of yearling female primates (*Papio cynocephalus*) predict lifetime fitness. *Proc. Natl. Acad. Sci. USA* 88:420–423.

Altmann, S. A. (1998). *Foraging for Survival: Yearling Baboons in Africa.* University of Chicago Press, Chicago.

Altmann, S. A. (2009). Fallback foods, eclectic omnivores, and the packaging problem. *Am. J. Phys. Anthropol.* 140:615–629.

Altmann, S. A., and Altmann, J. (1970). *Baboon Ecology.* University of Chicago Press, Chicago.

Altmann, S., and Altmann, J. (1979). Demographic constraints on behavior and social organization. In: Bernstein, I. and Smith, E. O. (eds.), *Primate Ecology and Human Origins.* Garland STPM Press, New York. pp. 47–63.

Altmann, S. A., Post, D. G., Klein, D. F. (1987). Nutrients and toxins of plants in Amboseli, Kenya. *Afr. J. Ecol.* 25:279–293.

Altrichter, H. (2001). Playback zur Erkennung und Diskrimination von Informationskategorien in den Sozialrufen von Edward-Wieselmakis (*Lepilemur edwardsi*) in NW-Madagaskar [diplomarbeit]. Universität Osnabrück, Osnabrück.

Alvard, J., and Kaplan, H. (1991). Procurement technology and prey mortality among indigenous neotropical hunters. In: Stines, M. C. (ed.), *Human Predators and Prey Mortality.* Westview Press, Boulder, CO. pp. 335–387.

Alvard, M. S. (1993). Testing the "ecologically noble savage" hypothesis: interspecific prey choice by Piro hunters of Amazonian Peru. *Hum. Ecol.* 21:355–387.

Alvard, M. S., Robinson, J. G., Redford, K. H., and Kaplan, H. (1997). The sustainability of subsistence hunting in the neotropics. *Conserv. Biol.* 11:977–982.

Alvarez, H. P. (2000). Grandmother hypothesis and primate life histories. *Am. J. Phys. Anthropol.* 113:435–450.

Ambrose, L. (1999). Species diversity in West and central African galagos (Primates, Galagonidae): the use of acoustic analysis [PhD diss.]. Oxford Brookes University, Oxford.

Ambrose, L. (2003). Three acoustic forms of Allen's galagos (Primates; Galagonidae) in the central African region. *Primates* 44:25–39.

Ambrose, L. (2006). A survey of prosimians in the national parks and forest reserves of Uganda. In: Newton-Fisher, N., Notman, H., Paterson, J. D., and Reynolds, V. (eds.), *Primates of Western Uganda.* Springer, New York. pp. 329–343.

Ambrose, L., and Perkin, A. W. (2000). A survey of nocturnal prosimians at Moca on Bioko Island, Equatorial Guinea. *Afr. Primates* 4:4–10.

Amici, F., Aureli, F., and Call, J. (2008). Fission–fusion dynamics, behavioral flexibility, and inhibitory control in primates. *Curr. Biol.* 18:1415–1419.

Amman, K., Pearce, J., and Williams, J. (2000). *Bushmeat: Africa's Conservation Crisis.* World Society for the Protection of Animals, London.

Amrine-Madsen, H., Koepfli, K. P., Wayne, R. K., and Soringer, M. S. (2003). A new phylogenetic marker APOB provides compelling evidence for eutherian relationships. *Mol. Phylogenet. Evol.* 28:225–240.

Anadu, P. A., Elamah, P. O., and Oates, J. F. (1988). The bushmeat trade in southwestern Nigeria: a case study. *Hum. Ecol.* 16:199–208.

Ancrenaz, M., Calaque, R., and Lackman-Ancrenaz, I. (2004a). Orangutan nesting behavior in disturbed forest of Sabah, Malaysia: implications for nest census. *Int. J. Primatol.* 25:983–1000.

Ancrenaz, M., Goossens, B., Gimenez, O., Sawang, A., and Lackman-Ancrenaz, I. (2004b). Determination of ape distribution and population size using ground and aerial surveys: a case study with orang-utans in lower Kinabatangan, Sabah, Malaysia. *Anim. Conserv.* 7:375–385.

Ancrenaz, M., Gimenez, O., Ambu, L., Ancrenaz, K., Andau, P., Goossens, B., Payne, J., Sawang, A., Tuuga, A., and Lackman-Ancrenaz, I. (2005). Aerial surveys give new estimates for orangutans in Sabah, Malaysia. *PLoS Biol.* 3(1):e3.

Ancrenaz, M., Marshall, A., Goossens, B., van Schaik, C., Sugardjito, J., Gumal, M., and Wich, S. (2008). *Pongo pygmaeus.* In: IUCN 2008. 2008 IUCN Red List of Threatened Species, www.iucnredlist.org (accessed April 21, 2009).

Andelman, S. J. (1986). Ecological and social determinants of cercopithecine mating patterns. In: Rubenstin, D. I., and Wrangham, R. W. (eds.), *Ecological Aspects of Social Evolution: Birds and Mammals.* Princeton University Press, Princeton, NJ. pp. 201–216.

Andelman, S. J. (1987). Evolution of concealed ovulation in vervet monkeys (*Cercopithecus aethiops*). *Am. Nat.* 129:785–799.

Andelman, S. J., Else, J. G., Hearn, J. P., and Hodges, J. K. (1985). The non-invasive monitoring of reproductive events in wild vervet monkeys (*Cercopithecus aethiops*) using urinary pregnanediol-3-alpha-glucuronide and its correlation with behavioural observations. *J. Zool.* A205:467–477.

Anderson, C. M. (1981a). Intertroop relations of chacma baboons. *Int. J. Primatol.* 2:285–310.

Anderson, C. M. (1981b). Subtrooping in a chacma baboon (*Papio ursinus*) population. *Primates* 22:445–458.

Anderson, C. M. (1982). Baboons below the Tropic of Capricorn. *J. Hum. Evol.* 11:205–217.

Anderson, C. M. (1986a). Female age–male-preference and reproductive success in primates. *Int. J. Primatol.* 7:305–326.

Anderson, C. M. (1986b). Predation and primate evolution. *Primates* 27:15–39.

Anderson, C. M. (1989). The spread of exclusive mating in a chacma baboon population. *Am. J. Phys. Anthropol.* 78:355–360.

Anderson, C. M. (1990). Desert, mountain and savanna baboons: a comparison with special reference to the Suikerbosrand population. In: Thiago de Mello, M., Whiten, A., and Byrne, R. W. (eds.), *Baboons: Behaviour and Ecology. Use and Care. Selected Proceedings of the XIIth Congress of the International Primatological Society.* University of Brasilia, Brasilia, Brazil. pp. 89–103.

Anderson, C. M. (1992). Male investment under changing conditions among chacma baboons at Suikerbosrand. *Am. J. Phys. Anthropol.* 87:479–495.

Anderson, C. M., and Bielert, C. F. (1994). Adolescent exaggeration in female catarrhine primates. *Primates* 35:283–300.

Anderson, D., Nordheim, E. V., Boesch, C., and Moermond, T. C. (2002). Factors influencing fission–fusion grouping in chimpanzees in the Taï National Park, Cote d'Ivoire. In: Boesch, C., Hohmann, G., and Marchant, L. F. (eds.), *Behavioral Diversity in Chimpanzees and Bonobos.* Cambridge University Press, Cambridge. pp. 90–101.

Anderson, D. P., Nordheim, E. V., and Boesch, C. (2006). Environmental factors influencing the seasonality of estrus in chimpanzees. *Primates* 47:43–50.

Anderson, H. B., Emery Thompson, M., Knott, C., and Perkins, L. (2008). Fertility and mortality patterns of captive Bornean and Sumatran orangutans: is there a species difference in life history? *J. Hum. Evol.* 54:34–42.

Anderson, J. R. (1984). Ethology and ecology of sleep in monkeys and apes. *Adv. Study Behav.* 14:166–229.

Anderson, J. R. (1985). Development of tool-use to obtain food in a captive group of *Macaca tonkeana*. *J. Hum. Evol.* 14:637–645.

Anderson, J. R. (1990). Use of objects as hammers to open nuts by capuchin monkeys (*Cebus apella*). *Folia Primatol.* 54:138–145.

Anderson, J. R. (1998). Sleep, sleeping sites, and sleep-related activities: awakening to their significance. *Am. J. Primatol.* 46:63–75.

Anderson, J. R., and McGrew, W. C. (1984). Guinea baboons (*Papio papio*) at a sleeping site. *Am. J. Primatol.* 6:1–14.

Anderson, J. R., Awazu, S., and Fijita, K. (2000). Can squirrel monkeys (*Saimiri sciureus*) learn self-control? A study using food array selection tests and reversed reward contingency. *J. Exp. Psychol. Anim. Behav. Proc.* 26:87–97.

Anderson, J. R., Kuroshima, H., Kuwahata, H., Fujita, K., and Vick, S. J. (2001). Training squirrel monkeys (*Saimiri sciureus*) to deceive: acquisition and analysis of behavior toward cooperative and competitive trainers. *J. Comp. Psychol.* 115:282–293.

Anderson, J. R., Awazu, S., and Fujita, K. (2004). Squirrel monkeys (*Saimiri sciureus*) choose smaller food arrays: long-term retention, choice with non-preferred food, and transposition. *J. Comp. Psychol.* 118:58–64.

Anderson, J. R., Kuroshima, H., Paukner, A., and Fujita, K. (2009). Capuchin monkeys (*Cebus apella*) respond to video images of themselves. *Anim. Cogn.* 12:55–62.

Anderson, M. (1998). Comparative morphology and speciation in galagos. *Folia Primatol.* 69(suppl. 1):325–331.

Anderson, M. J. (2000). Penile morphology and classification of bush babies (subfamily Galagoninae). *Int. J. Primatol.* 21:815–836.

Anderson, M. J. (2001). The use of hair morphology in the classification of galagos (Primates, subfamily Galagonidae). *Primates* 42:113–121.

Anderson, M. J., and Dixson, A. F. (2002). Motility and the midpiece in primates. *Nature* 416:496.

Anderson, M. J., Ambrose, L., Bearder, S. K., Dixson, A. F., and Pullen, S. (2000). Intraspecific variation in the vocalizations and hand pad morphology of southern lesser bush babies (*Galago moholi*): a comparison with *G. senegalensis*. *Int. J. Primatol.* 21:537–555.

Anderson, M., Chapman, S., Videan, E., Fritz, J., Stoinski, T., Dixson, A. F., and Gagneux, P. (2007). Sperm bioenergetics in humans and chimpanzees—functional evidence for sperm competition. *Am. J. Phys. Anthropol.* 134:274–280.

Anderson, R. C. (1992). *Nematode Parasites of Vertebrates. Their Development and Transmission*. C.A.B. International, Wallingford.

Andersson, M. B. (1994). *Sexual Selection*. Princeton University Press, Princeton, NJ.

Ando, C., Y. Iwata., and Yamagiwa, J. (2008). Progress of habituation of western lowland gorillas and their reaction to observers in Moukalaba-Doudou National Park, Gabon. *African Study Monographs* (suppl. 39):55–69.

Andres, M., Solignac, M., and Perret, M. (2003). Mating system in mouse lemurs: theories and facts, using analysis of paternity. *Folia Primatol.* 74:355–366.

Andresen, E. (1999). Seed dispersal by monkeys and the fate of dispersed seeds in a Peruvian rain forest. *Biotropica* 31:145–158.

Andresen, E. (2000). Ecological roles of mammals: the case of seed dispersal. In: Entwistle, A., and Dunstone, N. (eds.), *Priorities for the Conservation of Mammalian Diversity*. Cambridge University Press, Cambridge. pp. 2–26.

Andresen, E. (2001). Effects of dung presence, dung amount and secondary dispersal by dung beetles on the fate of *Micropholis guyanensis* (Sapotaceae) seeds in central Amazonia. *J. Trop. Ecol.* 17:61–78.

Andresen, E., and Levey, D. J. (2004). Effects of dung and seed size on secondary dispersal, seed predation, and seedling establishment of rain forest trees. *Oecologia* 139:45–54.

Andrews, J. (1998). Infanticide by a female black lemur, *Eulemur macaco,* in disturbed habitat on Nosy Be, north-western Madagascar. *Folia Primatol.* 69(suppl. 1):14–17.

Andrews, J. R., and Birkinshaw, C. R. (1998). A comparison between the day-time and night-time diet, activity and feeding height of the black lemur *Eulemur macaco* (Primates: Lemuridae), in Lokobe Forest, Madagascar. *Folia Primatol.* 69(suppl. 1):175–182.

Andrews, P., and Evans, E. M. (1979). Small mammal bone accumulations produced by mammalian carnivores. *Palaeobiology* 9:289–307.

Andrews, T. D., Jermiin, L. S., and Easteal, S. (1998). Accelerated evolution of cytochrome *b* in simian primates: adaptive evolution in concert with other mitochondrial proteins? *J. Mol. Evol.* 47:249–257.

Andriaholinirina, N., Fausser, J.-L., Roos, C., Ravaoarimanana, I., Zinner, D., Thalmann, U., Rabarivola, C., Ravoarimanana, I., Ganzhorn, J. U., Meier, B., Hilgartner, R., Walter, L., Zaramody, A., Langer, C., Hahn, T., Zimmermann, E., Radespiel, U., Craul, M., Tomiuk, J., Tattersall, I., and Rumpler, Y. (2006). Molecular phylogeny and taxonomic revision of the sportive lemurs (*Lepilemur,* Primates). *BMC Evol. Biol.* 6:17.

Andrianarivo, A. J. (1981). Étude comparée de l'organisation sociale chez *Microcebus coquereli* [diss.]. University of Madagascar, Antananarivo.

Andriantompohavana, R., Zaonarivelo, J. R., Engberg, S. E., Randriamampionona, R., McGuire, S. M., Shore, G. D., Rakotonomenjanahary, R., Brenneman, R. A., and Louis, E. E. (2006). Mouse lemurs of northwestern Madagascar with a description of a new species at Lokobe Special Reserve. *Occas. Pap. Museum Tex. Tech. Univ.* 259:1–24.

Andriantompohavana, R., Lei, R., Zaonarivelo, J. R., Engberg, S. E., Nalanirina, G., McGuire, S. M., Shore, G. D., Andrianasolo, J., Herrington, K., Brenneman, R. A., and Louis, E. E. (2007). Molecular phylogeny and taxonomic revision of the woolly lemurs, genus *Avahi* (Primates: Lemuriformes). *Museum Tex. Tech Univ. Spec. Publ.* 51:1–59.

Anemone, R. L. (1993). The functional anatomy of the hip and thigh in primates. In: Gebo, D. L. (ed.), *Postcranial Adaptation in Nonhuman Primates*. Northern Illinois University Press, Dekalb. pp. 150–174.

Anestis, S. F. (2004). Female genito-genital rubbing in a group of captive chimpanzees. *Int. J. Primatol.* 25:477–488.

Anestis, S. F. (2006). Testosterone in juvenile and adolescent male chimpanzees (*Pan troglodytes*): effects of dominance rank, aggression, and behavioral style. *Am. J. Phys. Anthropol.* 130:536–545.

Angier, N. (1999). Illuminating how bodies are built for sociality. In: Sussman, R. W. (ed.), *The Biological Basis of Human Behavior: A Critical Review*. Prentice Hall, Upper Saddle River, NJ. pp. 350–352.

Angst, W. (1975). Basic data and concepts on the social organization of *Macaca fascicularis*. In: Rosenblum, L. A. (ed.), *Primate Behavior: Developments in Field and Laboratory Research,* vol. 4. Academic Press, New York, pp. 325–388.

Ankel-Simons, F., and Rasmussen, D. T. (2008). Diurnality, nocturnality, and the evolution of primate visual systems. *Am. J. Phys. Anthropol.* 51:100–117.

Antinucci, F. (1989). *Cognitive Structure and Development in Nonhuman Primates*. Lawrence Earlbaum, Hillsdale, NJ.

Anzenberger, G. (1993). Social conflict in two monogamous New World primates: pairs and rivals. In: Mason, W. A., and Mendoza, S. P. (eds.),

Primate Social Conflict. State University of New York Press, New York. pp. 291–329.

Anzenberger, G., and Gossweiler, H. (1993). How to obtain individual urine samples from undisturbed marmoset families. *Am. J. Primatol.* 31:223–230.

Apetrei, C., Robertson, D. L., and Marx, P. A. (2004). The history of SIVS and AIDS: epidemiology, phylogeny and biology of isolates from naturally SIV infected non-human primates (NHP) in Africa. *Front. Biosci.* 9:225–254.

Appanah, S. (1985). General flowering in the climax rain forests of South-east Asia. *J. Trop. Ecol.* 1:225–250.

Aquino, R. (1998). Some observations on the ecology of *Cacajao calvus ucayalii* in the Peruvian Amazon. *Primate Conserv.* 18:21–24.

Aquino, R., and Encarnación, F. (1986a). Characteristics and use of sleeping sites in *Aotus* (Cebidae: Primates) in the Amazon lowlands of Perú. *Am. J. Primatol.* 11:319–331.

Aquino, R., and Encarnación F. (1986b). Population structure of *Aotus nancymai* (Cebidae: Primates) in Peruvian Amazon lowland forest. *Am. J. Primatol.* 11:1–7.

Aquino, R., and Encarnación, F. (1988). Population densities and geographic distribution of night monkeys (*Aotus nancymai* and *Aotus vociferans*) (Cebidae: Primates) in northeastern Perú. *Am. J. Primatol.* 14:375–381.

Aquino, R., and Encarnación, F. (eds.) (1989). *Aspectos de la dinámica poblacional de* Aotus nancymai *(Cebidae: Primates)*. IUCN/CSE-World Wildlife Fund, Washington DC.

Aquino, R., and Encarnación, F. (1994). Owl monkey populations in Latin America: field work and conservation. In: Baer, J. F., Weller, R. E., and Kakoma, I. (eds.), Aotus: *The Owl Monkey*. Academic Press, San Diego. pp. 59–95.

Aquino, R., and Encarnación, C. F. (1999). Observaciones preliminares sobre la dieta de *Cacajao calvus ucayalii* en el Nor-Oriente Peruano. *Neotrop. Primates* 7:1–5.

Aquino, R., Puertas, P. E., and Encarnación, F. (1990). Supplemental notes on population parameters of northeastern Peruvian night monkeys, genus *Aotus* (Cebidae). *Am. J. Primatol.* 21:215–221.

Aquino, R., Puertas, P. E., and Encarnación, F. (1992a). Evaluacion post-captura de *Aotus vociferans* y *Aotus nancymae* en bosques de la amazonia peruana. *Folia Amazon.* 4:141–151.

Aquino, R., Puertas, P. E., and Encarnación, F. (1992b). Las poblaciones de *Aotus vociferans* y *Aotus nancymae* en la Amazonía Peruana. *Alma Mater* 3:77–82.

Aquino, R., Puertas, P. E., and Encarnación, F. (1993). Effects of cropping on the *Aotus nancymae* population in a forest of Peruvian Amazonia. *Primate Rep.* 37:31–40.

Aquino, R., Terrones, W., Cornejo, F., and Heymann, E.W. (2008). Geographic distribution and possible taxonomic distinction of *Callicebus torquatus* populations (Pitheciidae: Primates) in Peruvian Amazonia. *Am. J. Primatol.* 70:1181–1186.

Araripe, J., Tagliaro, C., Rego, P., Sampaio, I., Ferrari, S., and Schneider, H. (2008). Molecular phylogenetics of large-bodied tamarins, *Saguinus* spp. (Primates, Platyrrhini). *Zoologica Scripta* 37:461–467.

Araújo, A., and de Sousa, M. B. C. (2008). Testicular volume and reproductive status of wild *Callithrix jacchus*. *Int. J. Primatol.* 29:1355–1364.

Arcadi, A. C., and Wrangham, R. W. (1999). Infanticide in chimpanzees: review of cases and a new within-group observation from the Kanyawara study group in Kibale National Park. *Primates* 40:337–351.

Arditi, S. I. (1992). Variaciones estacionales en la actividad y dieta de *Aotus azarae* y *Alouatta caraya* en Formosa, Argentina. *Bol. Primatol. Latinoam.* 3:11–30.

Arditi, S. I., and Placci, G. L. (1990). Hábitat y densidad de *Aotus azarae* y *Alouatta caraya* en Riacho Pilagá, Formosa. *Bol. Primatol. Latinoam.* 2:29–47.

Arlet, M. E., Molleman, F., and Chapman, C. (2007). Indications for female mate choice in grey-cheeked mangabeys *Lophocebus albigena johnstoni* in Kibale National Park, Uganda. *Acta Etholog.* 10:89–95.

Arlet, M. E., Molleman, F., and Chapman, C. (2008). Mating tactics in male grey-cheeked mangabeys (*Lophocebus albigena*). *Ethology* 114:851–862.

Arlet, M. E., Grote, M. N., Molleman, F., Isbell, L. A., and Carey, J. R. (2009). Reproductive tactics influence cortisol levels in individual male gray-cheeked mangabeys (*Lophocebus albigena*). *Horm. Behav.* 55:210–216.

Armstrong, D. T. (2001). Effects of maternal age on oocyte developmental competence. *Theriogeneology* 55:1303–1322.

Arnason, U., Gullberg, A., Janke, A., and Xu, X. (1996). Pattern and timing of evolutionary divergences among hominoids based on analyses of complete mtDNAs. *J. Mol. Evol.* 43:650–661.

Arnason, U., Gullberg, A., and Janke, A. (1998). Molecular timing of primate divergences as estimated by two nonprimate calibration points. *J. Mol. Evol.* 47:718–727.

Arnold, K., and Barton, R. A. (2001a). Postconflict behavior of spectacled leaf monkeys (*Trachypithecus obscurus*). I. Reconciliation. *Int. J. Primatol.* 22:243–266.

Arnold, K., and Barton, R. A. (2001b). Postconflict behavior of spectacled leaf monkeys (*Trachypithecus obscurus*). II. Contact with third parties. *Int. J. Primatol.* 22:267–286.

Arnold, K., and Whiten, A. (2001). Post-conflict behaviour of wild chimpanzees (*Pan troglodytes schweinfurthii*) in the Budongo Forest, Uganda. *Behaviour* 138:649–690.

Arnold, K., and Whiten, A. (2003). Grooming interactions among the chimpanzees of the Budongo Forest, Uganda: tests of five explanatory models. *Behaviour* 140:519–552.

Arnold, K., Pohlner, Y., and Zuberbuehler, K. (2008). A forest monkey's alarm call series to predator models. *Behav. Ecol. Sociobiol.* 62(4): 549–559.

Arnold, N., Stanyon, R., Jauch, A., O'Brien, P., and Wienberg, J. (1996). Identification of complex chromosome rearrangements in the gibbon by fluorescent in situ hybridization (FISH) of a human chromosome 2q specific microlibrary, yeast artificial chromosomes, and reciprocal chromosome painting. *Cytogenet. Cell Genet.* 74:80–85.

Arruda, M. F., Araújo, A., Sousa, M. B. C., Albuquerque, F. S., Albuquerque, A. and Yamamoto, M. E. (2005). Two breeding females within free-living groups may not always indicate polygyny: alternative subordinate female strategies in common marmosets (*Callithrix jacchus*). *Folia Primatol.* 76:10–20.

Asanov, S. S. (1972). Comparative features of the reproductive biology of hamadryas baboons (Papio hamadryas), grivet monkeys (Cercopithecus aethiops) and rhesus monkeys (Macaca mulatta). Acta Endocrin. Suppl. 71 (Suppl. 166): 458–472.

Asaka, Y., Kubota, T., and Kulkarni, A. B. (1977). Studies on a bitter principle from *Vernonia anthelmintica*. *Phytochemistry* 16:1838–1839.

Asensio, N., Korstjens, A. H, Schaffner, C. M., and Aureli, F. (2008). Intragroup aggression, fission–fusion dynamics and feeding competition in spider monkeys. *Behaviour* 145:983–1001.

Asensio, N., Korstjens, A. H., and Aureli, F. (2009). Fissioning minimizes ranging costs in spider monkeys: a multiple-level approach. *Behav. Ecol. Sociobiol.* 6:649–659.

Ashford, R. W., Reid, G. D. F., and Wrangham, R. W. (2000). Intestinal parasites of the chimpanzee *Pan troglodytes* in Kibale Forest, Uganda. *Ann. Trop. Med. Parasitol.* 84:337–340.

Ashton, P. S. (1988). Dipterocarp biology as a window to the understanding of tropical forest structure. *Annu. Rev. Ecol. Syst.* 19:347–370.

Ashton, P. S., Givnish, T. J., and Appanah, S. (1988). Staggered flowering in the *Dipterocarpaceae:* new insights into floral induction and the evolution of mast fruiting in the aseasonal tropics. *Am. Nat.* 132:44–66.

Asquith, P. J. (1989). Provisioning and the study of free-ranging primates: history, effects, and prospects. *Ybk. Phys. Anthropol.* 32:129–158.

Asquith, P. (2000). Negotiating science: internationalization and Japanese primatology. In: Strum, S. C., and Fedigan, L. M. (eds.), *Primate Encounters: Models of Science, Gender, and Society.* University of Chicago Press, Chicago. pp. 296–309.

Assuncao, M. L., Mendes S. L., and Strier, K. B. (2007). Grandmaternal infant carrying in wild northern muriquis (*Brachyteles hypoxanthus*). *Neotrop. Primates* 14:120–122.

Astaras, C., Mühlenberg, M., and Waltert, M. (2008). Note on drill (*Mandrillus leucophaeus*) ecology and conservation status in Korup National Park, southwest Cameroon. *Am. J. Primatol.* 70:306–310.

Atsalis, A. (2000). Spatial distribution and population composition of the brown mouse lemur (*Microcebus rufus*) in Ranomafana National Park, Madagascar, and its implications for social organization. *Am. J. Primatol.* 51:61–78.

Atsalis, S. (1998). Seasonal fattening and changes in activity levels in the brown mouse lemur (*Microcebus rufus*) in Ranomafana National Park, Madagascar. *Am. J. Primatol.* 45:165.

Aufreiter, S., Hancock, R., Mahaney, W. C., Stambolic-Robb, A., and Sanmugadas, K. (1997). Geochemistry and mineralogy of soils eaten by humans. *Int. J. Food Sci. Nutr.* 48:292–305.

Augspurger, C. K. (1983). Offspring recruitment around tropical trees: changes in cohort distance with time. *Oikos* 40:189–196.

Augspurger, C. K. (1984). Seedling survival of tropical tree species: interactions of dispersal distance, light-gaps, and pathogens. *Ecology* 65:1705–1712.

Augspurger, C. K., and Kelly, C. K. (1984). Pathogen mortality of tropical tree seedlings: experimental studies of the effects of dispersal distance, seedling density, and light conditions. *Oecologia* 61:211–217.

Aujard, F., Heistermann, M., Thierry, B., and Hodges, J. K. (1998). Functional significance of behavioral, morphological, and endocrine correlates across the ovarian cycle in semifree ranging female Tonkean macaques. *Am. J. Primatol.* 46:285–309.

Aureli, F. (1992). Post-conflict behaviour among wild long-tailed macaques (*Macaca fascicularis*). *Behav. Ecol. Sociobiol.* 31:329–337.

Aureli, F. (1997). Post-conflict anxiety in nonhuman primates: the mediating role of emotion in conflict resolution. *Aggress. Behav.* 23:315–328.

Aureli, F., and de Waal, F. B. M. (eds.) (2000). *Natural Conflict Resolution.* University of California Press, Berkeley.

Aureli, F., and Schaffner, C. M. (2006). Causes, consequences and mechanisms of reconciliation: the role of cooperation. In: Kappeler, P. M., and van Schaik, C. P. (eds.), *Cooperation in Primates and Humans.* Springer, Berlin. pp. 121–136.

Aureli, F. and Schaffner, C. M. (2008). Social interactions, social relationships and the social system of spider monkeys. In: Campbell, C. J. (ed.), *Spider Monkeys: Behavior, Ecology and Evolution of the Genus Ateles.* Cambridge University Press, Cambridge. pp. 236–265.

Aureli, F., and Smucny, D. A. (1998). New directions in conflict resolution research. *Evol. Anthropol.* 6:115–119.

Aureli, F., and Smucny, D. A. (2000). The role of emotion in conflict and conflict resolution. In: Aureli, F., and de Waal, F. B. M. (eds.), *Natural Conflict Resolution.* University of California Press, Berkeley. pp. 199–224.

Aureli, F., and van Schaik, C. P. (1991). Post-conflict behaviour in long-tailed macaques (*Macaca fascicularis*): I. The social events. *Ethology* 89:89–100.

Aureli, F., van Schaik, C. P., and van Hooff, J. A. R. A. M. (1989). Functional aspects of reconciliation among captive long-tailed macaques (*Macaca fascicularis*). *Am. J. Primatol.* 19:39–51.

Aureli, F., Cozzolino, R., Cordischi, C., and Scucchi, S. (1992). Kin oriented redirection among Japanese macaques: an expression of a revenge system. *Anim. Behav.* 44:283–291.

Aureli, F., Veenema, H. C., van Panthaleon van Eck, C. J., and van Hooff, J. A. R. A. M. (1993). Reconciliation, consolation, and redirection in Japanese macaques (*Macaca fuscata*). *Behaviour* 124:1–21.

Aureli, F., Das, M., Verleur, D., and van Hooff, J. A. R. A. M. (1994). Postconflict social interactions among Barbary macaques (*Macaca sylvanus*). *Int. J. Primatol.* 15:471–485.

Aureli, F., Das, M., and Veenema, H. C. (1997). Differential kinship effect on reconciliation in three species of macaques (*Macaca fascicularis, M. fuscata, M. sylvanus*). *J. Comp. Psychol.* 111:91–99.

Aureli, F., Cords, M., and van Schaik, C. P. (2002). Conflict resolution following aggression in gregarious animals: a predictive framework. *Anim. Behav.* 64:325–343.

Aureli, F., Schaffner, C. M., Verpooten, J., Slater, K., and Ramos-Fernandez, G. (2006). Raiding parties of male spider monkeys: insights into human warfare?. *Am. J. Phys. Anthropol.* 131:486–497.

Aureli, F., Schaffner, C.M., Boesch, C., Bearder, S. K., Call, J., Chapman, C. A., Connor, R. C., Di Fiore, A., Dunbar, R. I. M., Henzi, S. P., Holekamp, K., Korstjens, A. H., Layton, R., Lee, P., Lehmann, J., Manson, J. H., Ramos-Fernandez, G., Strier, K. B, and van Schaik, C. P. (2008). Fission–fusion dynamics: new research frameworks. *Curr. Anthropol.* 49:627–654.

Auricchio, P. (1995). *Primatas do Brasil.* Terra Brasilis Comercio de Material Didático e Editora Letda, São Paulo.

Austad, S. N., and Fischer, K. E. (1992). Primate longevity: its place in the mammalian scheme. *Am. J. Primatol.* 28:251–262.

Avise, J. C. (2004). *Molecular Markers, Natural History, and Evolution,* 2nd ed. Sinauer Associates, Sunderland, MA.

Ayres, J. M. (1981). Observações sobre a ecologia e o comportamento dos cuxiús (*Chiropotes albinasus* e *Chiropotes satanas,* Cebidae: Primates). Fundação Universidade do Amazonas, Manaus, Brazil.

Ayres, J. M. (1985). On a new species of squirrel monkey, genus *Saimiri,* from Brazilian Amazonia (Primates, Cebidae). *Pap. Avul. Zool.* 36:147–164.

Ayres, J. M. (1986). Uakaris and Amazonian flooded forest [PhD diss.]. Cambridge University, Cambridge.

Ayres, J. M. (1989). Comparative feeding ecology of the uakari and bearded saki, *Cacajao* and *Chiropotes. J. Hum. Evol.* 18:697–716.

Ayres, J. M., and Nessimian, J. L. (1982). Evidence for insectivory in *Chiropotes satanas. Primates* 22:458–459.

Ayres, J. M., Alves, A. R., de Queiroz, H. L., Marmontel, M., Moura, E., de Magalhães Lima, D., Azevedo, A., Reis, M., Santos, P., da Silveira, R., and Masterson, D. (1999). Mamirauá: the conservation of biodiversity in an Amazonian flooded forest. In: Padoch, C., Ayres, J. M., Pinedo-Vasquez, M., and Henderson, A. (eds.), *Várzea: Diversity, Development, and Conservation of Amazonia's Whitewater Floodplains*. New York Botanical Garden Press, Bronx. pp. 203–216.

Baba, M. L., Darga, L. L., and Goodman, M. (1979). Immunodiffusion systematics of the primates. Part V. The Platyrrhini. *Folia Primatol.* 32:207–238.

Babb, P. L., Gagneux, P., Fernandez-Duque, E., and Schurr, T. G. (2008). Genetic variation and population structure in the owl monkey, *Aotus azarai. Am. J. Phys. Anthropol* 135:62.

Bach, A., Raboy, B., and Dietz, J. (2001). Birth seasonality in wild golden-headed lion tamarins (*Leontopithecus chrysomelas*) in Una Reserve, Bahia state, Brazil. *Am. J. Primatol.* 54:69.

Bachmann, C., and Kummer, H. (1980). Male assessment of female choice in hamadryas baboons. *Behav. Ecol. Sociobiol.* 6:315–321.

Bacon, A.-M., and Long, V. T. (2001). The first discovery of a complete skeleton of a fossil orang-utan in a cave of the Hoa Binh Province, Vietnam. *J. Hum. Evol.* 41:227–241.

Bacon, P. J., Ball, F., and Blackwell, P. (1991). A model for territory and group formation in a heterogeneous habitat. *J. Theor. Biol.* 148:445–468.

Badrian, A., and Badrian, N. (1984). Social organization of *Pan paniscus* in the Lomako Forest, Zaire. In: Susman, R. L. (ed.), *The Pygmy Chimpanzee: Evolutionary Biology and Ecology*. Plenum Press, New York. pp. 325–346.

Badrian, N., and Malenky, R. K. (1984). Feeding ecology of *Pan paniscus* in the Lomako Forest, Zaire. In: Susman, R. L. (ed.), *The Pygmy Chimpanzee*. Plenum, New York. pp. 275–299.

Badrian, N., Badrian, A., and Susman, R. (1981). Preliminary observations of the feeding behavior of *Pan paniscus* in the Lomako Forest of Central Zaire. *Primates* 22:173–181.

Baeninger, R., Estes, R., and Baldwin, S. (1977). Anti-predator behaviour of baboons and impalas toward a cheetah. *East Afr. Wildl. J.* 15:327–329.

Bahr, N. I., Martin, R. D., and Pryce, C. R. (2001). Peripartum sex steroid profiles and endocrine correlates of postpartum maternal behavior in captive gorillas (*Gorilla gorilla gorilla*). *Horm. Behav.* 40:533–541.

Bailey, J. A., and Eichler, E. E. (2006). Primate segmental duplications: crucibles of evolution, diversity and disease. *Nat. Rev. Genet.* 7:552–564.

Bailey, J. A., Yavor, A. M., Massa, H. F., Trask, B. J., and Eichler, E. E. (2001). Segmental duplications: organization and impact within the current human genome project assembly. *Genome Res.* 11:1005–1017.

Bailey, J. A., Gu, Z., Clark, R. A., Reinert, K., Samonte, R. V., Schwartz, S., Adams, M. D., Myers, E. W., Li, P. W., and Eichler, E. E. (2002). Recent segmental duplications in the human genome. *Science* 297:1003–1007.

Bailey, R. C., and Peacock, N. R. (1988). Efe pygmies of northeast Zaire: subsistence strategies in the Ituri Forest. In: de Garine, I., and Harrison, G. A. (eds.), *Uncertainty in the Food Supply*. Cambridge University Press, Cambridge. pp. 88–117.

Baillie, J. E. M., Hilton-Taylor, C., and Stuart, S. N. E. (2004). *2004 IUCN Red List of Threatened Species. A Global Species Assessment*. IUCN, Gland, Switzerland, and Cambridge, UK.

Bajpai S., Kay, R.F., Williams, B. A., Das, D. P., Kapur, V. V., and Tiwari, B. N. (2008). The oldest Asian record of Anthropoidea. *Proc. Natl. Acad. Sci. USA* 105:11093–11098.

Bakeman, R., and Gottman, J. (1986). *Observing Interaction: An Introduction to Sequential Analysis*. Cambridge University Press, Cambridge.

Baker, A. J. (1991). Evolution of the social system of the golden lion tamarin (*Leontopithecus rosalia*): mating system, group dynamics and cooperative breeding [PhD diss.]. University of Maryland, College Park.

Baker, A. J. (ed.) (2000). *Molecular Methods in Ecology*. Blackwell Scientific, Oxford.

Baker, A. J., and Dietz, J. M. (1996). Immigration in wild groups of golden lion tamarins (*Leontopithecus rosalia*). *Am. J. Primatol.* 38:47–56.

Baker, A. J., and Woods, F. (1992). Reproduction of the emperor tamarin (*Saguinus imperator*) in captivity, with comparisons to cotton-top and golden lion tamarins. *Am. J. Primatol.* 26:1–10.

Baker, A. J., Dietz, J. M., and Kleiman, D. G. (1993). Behavioural evidence for monopolization of paternity in multi-male groups of golden lion tamarins. *Anim. Behav.* 46:1091–1103.

Baker, A. J., Bales, K., and Dietz, J. M. (2002). Mating system and group dynamics in lion tamarins. In: Kleiman, D. G., and Rylands, A. B. (eds.), *Lion Tamarins: Biology and Conservation*. Smithsonian Institution Press, Washington DC. pp. 188–212.

Baker, J. V., Abbott, D. H., and Saltzman, W. (1999). Social determinants of reproductive failure in male common marmosets housed with their natal family. *Anim. Behav.* 58:501–513.

Baker, M. (1996). Fur rubbing: use of medicinal plants by capuchin monkeys (*Cebus capucinus*). *Am. J. Primatol.* 38:263–270.

Baker, K. C., and Smuts, B. B. (1994). Social relationships of female chimpanzees: diversity between captive social groups. In: Wrangham, R. W., McGrew, W. C., de Waal, F. B. M., and Heltne, P. G. (eds.), *Chimpanzee Cultures*. Harvard University Press, Cambridge, MA. pp. 227–242.

Baker, R. R., and Bellis, M. A. (1993). Human sperm competition: ejaculate manipulation by females and a function for the female orgasm. *Anim. Behav.* 46:887–909.

Balcomb, S. R., and Chapman, C. A. (2003). Bridging the seed dispersal gap: consequences of seed deposition for seedling recruitment in primate-tree interactions. *Ecol. Monogr.* 73:625–642.

Baldellou, M., and Henzi, P. (1992). Vigilance, predator detection and the presence of supernumerary males in vervet monkey troops. *Anim. Behav.* 43:451–461.

Balding, D. J., Bishop, M., and Cannings, C. (eds.) (2007). *Handbook of Statistical Genetics*, 3rd ed. Wiley, New York.

Baldwin, J. D. (1969). The ontogeny of social behavior of squirrel monkeys (*Saimiri sciureus*) in a seminatural environment. *Folia Primatol.* 11:35–79.

Baldwin, J. D., and Baldwin, J. (1972a). The ecology and behavior of squirrel monkeys (*Saimiri oerstedi*) in a natural forest in western Panama. *Folia Primatol.* 18:161–184.

Baldwin, J. D., and Baldwin, J. I. (1972b). Population density and use of space by howling monkeys (*Alouatta villosa*) in southwestern Panama. *Primates* 13:371–379.

Baldwin, J. D., and Baldwin, J. (1973). Interactions between adult female and infant howling monkeys (*Alouatta palliata*). *Folia Primatol.* 20:27–71.

Baldwin, J. D., and Baldwin, J. I. (1976). Primate populations in Chiriqui, Panama. In: Thorington, R. W., Jr., and Heltne, P. G. (eds.), *Neotropical Primates: Field Studies and Conservation*. National Academy of Sciences, Washington DC. pp. 20–31.

Baldwin, J. D., and Baldwin, J. I. (1978). Exploration and play in howler monkeys (*Alouatta palliata*). *Primates* 19:411–422.

Baldwin, J. D., and Baldwin, J. I. (1981). The squirrel monkeys, genus *Saimiri*. In: Coimbra-Filho, A. F., and Mittermeier, R. A. (eds.), *Ecology and Behavior of Neotropical Primates*, vol. 1. Academia Brasileira de Ciências, Rio de Janeiro. pp. 277–330.

Baldwin, P. J., McGrew, W. C., and Tutin, C. E. G. (1982). Wide-ranging chimpanzees at Mt. Assirik, Senegal. *Int. J. Primatol.* 3:367–385.

Bales, K. L., Dietz, J. M., Baker, A. J., Miller, K., and Tardif, S. D. (2000). Effects of allocare-givers on fitness of infants and parents in callitrichid primates. *Folia Primatol.* 71:27–38.

Bales, K., French, J. A., and Dietz, J. M. (2002). Explaining variation in maternal care in a cooperatively breeding mammal. *Anim. Behav.* 63:453–461.

Bales, K. L., Pfeifer, L. A., and Carter, S. C. (2004). Sex differences and developmental effects of manipulations of oxytocin on alloparenting and anxiety in prairie voles. *Dev. Psychobiol.* 44:123–131.

Bales, K. L., Mason, W. A., Catana, C., Cherry, S. R., and Mendoza, S. P. (2007). Neural correlates of pair-bonding in a monogamous primate. *Brain Res.* 1184:245–253.

Balko, E. A. (1998). A behaviorally plastic response to forest composition and logging disturbance by *Varecia variegata variegata* in Ranomafana National Park, Madagascar [PhD diss.]. State University of New York, Syracuse.

Balldelou, M., and Henzi, S. P. (1992). Vigilance, predator detection and the presence of supernumerary males in vervet monkey troops. *Anim. Behav.* 117:220–241.

Bao-Ping, R., Shu-Yi, Z., Shu-Zhong, X., Qing-Fen, L., Bing, L., and Min-Qiang, L. (2003). Annual reproductive behavior of *Rhinopithecus roxellana*. *Int. J. Primatol.* 24:575–589.

Barbaranelli, G. (1994). Sozialbeziehungen Zwishen Adulten Berberaffenmannchen (*Macaca sylvanus L. 1758*) [thesis]. University of Gottingen, Gottingen.

Bardi, M., Shimizu, K., Barrett, G. M., Borgognini-Tarli, S. M., and Huffman, M. A. (2003). Peripartum cortisol levels and mother–infant interactions in Japanese macaques. *Am. J. Phys. Anthropol.* 120:298–304.

Barelli, C., Heistermann, M., Boesch, C., and Reichard, U. H. (2007). Sexual swellings in wild whitehanded gibbon females (*Hylobates lar*) indicate the probability of ovulation. *Horm. Behav.* 51:221–230.

Barelli, C., Boesch, C., Heistermann, M., and Reichard, U. H. (2008). Female white-handed gibbons (*Hylobates lar*) lead group movements and have priority of access to food resources. *Behaviour* 145:965–981.

Barelli, E., Heistermann, M., Boesch, C., and Reichard, D. H. (2008). Mating patterns and sexual swellings in pair-living and multimale groups of white-handedgibbons, *Hylobates lar. Anim. Behav.* 75:991–1001.

Barenstain, L., and Wade, T. D. (1983). Intrasexual selection and male mating strategies in baboons and macaques. *Int. J. Primatol.* 4:201–235.

Barnes, M. R. (ed.) (2007). *Bioinformatics for Geneticists: A Bioinformatics Primer for the Analysis of Genetic Data,* 2nd ed. Wiley, New York.

Barnett, A.A. (2005). *Cacajao melanocephalus. Mamm. Species* 776:1–6.

Barnett, A. A., and Brandon-Jones, D. (1997). The ecology, biogeography and conservation of the uakaris, *Cacajao* (Pitheciinae). *Folia Primatol.* 68:223–235.

Barnett, A. A., Borges, S. H., Castilho, C. V. D., Neri, F. M., and Shapley, R. L. (2002). Primates of the Jaú National Park, Amazonas, Brazil. *Neotrop. Primates* 10:65–70.

Barnett, A. A., Volkmar de Castilho, C., Shapley, R. L., and Anicácio, A. (2005). Diet, habitat selection and natural history of *Cacajao melanocephalus ouakary* in Jaú National Park, Brazil. *Int. J. Primatol.* 26:949–970.

Barnett, D. T., and Stohlgren, T. J. (2003). A nested-intensity design for surveying plant diversity. *Biodivers. Cons.* 12:255–278.

Barnicot, N. A., and Hewett-Emmett, D. (1972). Red cell and serum proteins of *Cercocebus, Presbytis, Colobus* and certain other species. *Folia Primatol.* 17:442–457.

Barre, V., Lebac, A., Petter, J. J., and Albignac, R. (1988). Étude du Microcèbe par radiotracking dans la forêt de l'Ankarafantsika. In: Rakotovao, L., Barre, V., and Sayer, J. (eds.), *L'Equilibre des Ecosystèmes Forestiers a Madagascar: Actes d'un Séminaire International.* IUCN, Gland, Switzerland. pp. 61–71.

Barreto, C. E. (1996). Comportamento de fêmeas reproductivas em grupos poligínicos de *Callithrix jacchus* (Primates: Callitrichidae) no ambiente natural: perfil das interaços afiliativas, agonísticas e da marcação de cheiro [masters thesis]. Universidade Federal do Rio Grande do Norte, Natal, Brazil.

Barrett, E. (1981). The present distribution and status of the slow loris in peninsular Malaysia. *Malays. Appl. Biol.* 10:205–211.

Barrett, E. (1984). The ecology of some nocturnal, arboreal mammals in the rainforests of peninsular Malaysia [PhD diss.]. Cambridge University, Cambridge.

Barrett, J., Abbott, D. H., and George, L. M. (1990). Extension of reproductive suppression by pheromonal cues in subordinate female marmoset monkeys, *Callithrix jacchus. J. Reprod. Fertil.* 90:411–418.

Barrett, L., and Henzi, S. P. (1997). An inter-population comparison of body Weight in chacma baboons. South Afr. J. Sci. 93: 436–438.

Barrett, L., and Henzi, P. (2002). Constraints on relationship formation among female primates. *Behaviour* 139:263–289.

Barrett, L., Dunbar, R. I. M., and Dunbar, P. (1995). Mother–infant contact as contingent behaviour in gelada baboons. *Anim. Behav.* 49:805–810.

Barrett, L., Henzi, S. P., Weingrill, T., Lycett, J. E., and Hill, R. A. (1999). Market forces predict grooming reciprocity in female baboons. *Proc. R. Soc. Lond. B* 266:665–670.

Barrett, L., Henzi, S. P., Weingrill, T., Lycett, J. E., and Hill, R. A. (2000). Female baboons do not raise the stakes but they give as good as they get. *Anim. Behav.* 59:763–770.

Barrett, L., Gaynor, D., and Henzi, S. P. (2002). A dynamic interaction between aggression and grooming reciprocity among female chacma baboons. *Anim. Behav.* 63:1047–1053.

Barrett, L., Gaynor, D., Rendall, D., Mitchell, D., and Henzi, S. P. (2004). Habitual cave use and thermoregulation in chacma baboons. *J. Hum. Evol.* 46:215–222.

Barrett, L., Henzi, S. P., and Lycett, J. E. (2006). Whose life is it anyway? Maternal investment, developmental trajectories, and life history strategies in baboons. In: Swedell, L., and Leigh, S. R. (eds.), *Reproduction and Fitness in Baboons: Behavioral, Ecological, and Life History Perspectives.* Springer, New York. pp. 199–224.

Barrett-Lennard, L. G., Deecke, V. B., Yurk, H., and Ford, J. K. B. (2001). A sound approach to the study of culture. *Behav. Brain Sci.* 24:325–326.

Barros, M., Boere, V., Huston, J. P., and Toaz, C. (2000). Measuring fear and anxiety in the marmoset (*Callithrix penicillata*) with a novel predator confrontation model: effects of diazepam. *Behav. Brain Res.* 108:205–211.

Barros, M., Alencar, C., de Souza Silva, M. A., and Tomaz, C. (2008). Changes in experimental conditions alter anti-predator vigilance and sequence predictability in captive marmosets. *Behav. Proc.* 77(3):351–356.

Barrows, M. (1996). A survey of the intestinal parasites of the primate in Budongo Forest, Uganda. B. Vet. Med. project, Glasgow University.

Barthold, J., Fitchel, C., and Kappeler, P. (2009). What is it going to be? Pattern and potential function of natal coat change in sexually dichromatic redfronted lemurs (*Eulemur fulvus rufus*). *Am. J. Phys Anthropol.* 138:1–10.

Bartholomew, G. A., Jr., and Birdsell, J. B. (1953). Ecology and the protohominids. *Am. Anthropol.* 55:481–498.

Bartlett, T. Q. (1999a). Feeding and ranging behavior of the white-handed gibbon (*Hylobates lar*) in Khao Yai National Park, Thailand [PhD diss.]. Washington University, St. Louis.

Bartlett, T. Q. (1999b). The gibbons. In: Dolhinow, P., and Fuentes, A. (eds.), *The Nonhuman Primates.* Mayfield, Mountain View, CA. pp. 44–49.

Bartlett, T. Q. (2003). Intragroup and intergroup social interactions in white-handed gibbons. *Int. J. Primatol.* 24:239–259.

Bartlett, T. Q. (2009a). Seasonal home range use and defendability in white-handed gibbons (*Hylobates lar*) in Khao Yai National Park, Thailand. In: The Lappan, S. M., and Whittaker, D. (eds.), *Gibbons: New Perspectives on Small Ape Socioecology and Population Biology.* Springer, Berlin. pp. 265–278.

Bartlett, T. Q. (2009b). *The Gibbons of Khao Yai: Seasonal Variation in Behavior and Ecology.* Pearson Prentice Hall, Upper Saddle River, NJ.

Bartlett, T. Q., Sussman, R. W., and Cheverud, J. M. (1993). Infant killing in primates: a review of observed cases with specific reference to the sexual selection hypothesis. *Am. Anthropol.* 95:958–990.

Barton, R. (1990). Feeding, reproduction and social organisation in female olive baboons (*Papio anubis*). In: Thiago de Mello, M., Whiten, A., and Byrne, R. W. (eds.), *Baboons: Behaviour and Ecology. Use and Care. Selected Proceedings of the XIIth Congress of the International Primatological Society.* University of Brasilia, Brasilia. pp. 29–37.

Barton, R. A. (1993). Sociospatial mechanisms of feeding competition in female olive baboons, *Papio anubis. Anim. Behav.* 46:791–802.

Barton, R. (2000). Socioecology of baboons: the interaction of male and female strategies. In: Kappeler, P. M. (ed.), *Primate Males: Causes and Consequences of Variation in Group Composition.* Cambridge University Press, Cambridge. pp 97–107.

Barton, R. A., and Whiten, A. (1993). Feeding competition among female olive baboons, *Papio anubis. Anim. Behav.* 46:777–789.

Barton, R. A., and Whiten, A. (1994). Reducing complex diets to simple rules: food selection by olive baboons. *Behav. Ecol. Sociobiol.* 35:283–293.

Barton, R. A., Whiten, A., Strum, S. C., Byrne, R. W., and Simpson, A. J. (1992). Habitat use and resource availability in baboons. *Anim. Behav.* 43:831–844.

Barton, R. A., Whiten, A., Byrne, R. W., and English, M. (1993). Chemical composition of baboon plant foods: implications for the interpretation of intra- and interspecific differences in diet. *Folia Primatol.* 61:1–20.

Barton, R. A., Byrne, R. W., and Whiten, A. (1996). Ecology, feeding competition and social structure in baboons. *Behav. Ecol. Sociobiol.* 38:321–329.

Basabose, A. K. (2002). Diet composition of chimpanzees inhabiting the montane forest of Kahuzi, Democratic Republic of Congo. *Am. J. Primatol.* 58:1–21.

Basabose, A. K., and Yamagiwa, J. (2002). Factors affecting nesting site choice in chimpanzees at Tshibati, Kahuzi-Biega National Park: influence of sympatric gorillas. *Int. J. Primatol.* 23:263–282.

Basile, B. M., Hampton, R. R., Suomi, S. J., and Murray, E. A. (2008). An assessment of memory awareness in tufted capuchin monkeys (*Cebus apella*). *Anim. Cogn.* 12(1):169–180.

Bastian, M. L., and Brockman, D. K. (2007). Paternal care in *Propithecus verreauxi coquereli. Int. J. Primatol.* 28:305–313.

Bateman, A. J. (1948). Intra-sexual selection in *Drosophila. Heredity* 2:349–368.

Bateson, E. M., and LeBray, J. (1978). Clay eating by aboriginals of the Northern Territory. *Med. J. Aust.* 1(suppl.):60–61.

Bateson, P. (1983). Optimal outbreeding. In: Bateson, P. (ed.), *Mate Choice.* Cambridge University Press, Cambridge. pp. 257–278.

Bauchot, R. (1982). Brain organization and taxonomic relationships in Insectivora and Primates. In: Armstrong, E. (ed.), *Primate Brain Evolution.* Plenum Press, New York. pp. 163–174.

Bauers, K. A., and Hearn, J. P. (1994). Patterns of paternity in relation to male social rank in the stumptailed macaque, *Macaca arctoides. Behaviour* 129:149–176.

Bavister, B. D., and Boatman, D. E. (1993). IVF in nonhuman primates: current status and future directions. In: Wolf, D. P., Stouffer, R. L., and Brenner, R. M. (eds.), *In Vitro Fertilization and Embryo Transfer in Primates.* Spring-Verlag, New York. pp. 30–45.

Bayes, M. K. (1998). A molecular phylogenetic study of the galagos, strepsirhine primates and archontan mammals [PhD diss.]. Oxford Brookes University, Oxford.

Beach, F. A. (1976). Sexual attractivity, proceptivity and receptivity in female mammals. *Horm. Behav.* 7:105–138.

Beadle, L. C. (1981). *The Inland Waters of Tropical Africa.* Longman, London.

Beard, K. (1998). A new genus of Tarsiidae (Mammalia; Primates) from the middle Eocene of Shanxi Province, China with notes on the historical biogeography of tarsiers. *Bull. Carnegie Mus. Nat. Hist.* 34:260–277.

Beard, K. C. (2002). Basal anthropoids. In: Hartwig, W. C. (ed.), *The Primate Fossil Record.* Cambridge University Press, Cambridge. pp. 133–150.

Beard, K., and Wang, B. (1990). Phylogenetic and biogeographic significance of tarsiiform primate *Asiomomys changbaicus* from the Eocene of Jilin Province, People's Republic of China. *Am. J. Phys. Anthropol.* 85:159–166.

Beard, K. C., and Wang, J. (2004). The eosimiid primates (Anthropoidea) of the Heti Formation, Yuanqu Basin, Shanxi and Henan Provinces, People's Republic of China. *J. Hum. Evol.* 46:401–432.

Bearder, S. K. (1969). Territorial and intergroup behaviour of the lesser bushbaby, *Galago senegalensis moholi* (A Smith), in semi-natural conditions and in the field [MSc diss.]. University of Witwatersrand, Johannesburg.

Bearder, S. K. (1974). Aspects of the ecology and behaviour of the thick-tailed bushbaby *Galago crassicaudatus* [PhD diss.]. University of Witwatersrand, Johannesburg.

Bearder, S. K. (1987). Lorises, bushbabies and tarsiers: diverse societies in solitary foragers. In: Smuts, B. B., Cheney, D. L., Seyfarth, R. M., Wrangham, R. W., and Struhsaker, T. T. (eds.), *Primate Societies.* University of Chicago Press, Chicago. pp. 11–24.

Bearder, S. K. (1999). Physical and social diversity among nocturnal primates: a new view based on long term research. *Primates* 40:267–282.

Bearder, S. K. (2007). A comparison of calling patterns in two nocturnal primates, *Otolemur crassicaudatus* and *Galago moholi* as a guide to predation risk. In: Gursky, S. L., and Nekaris, K. A. I. (eds.), *Primate Anti-Predator Strategies.* New York: Springer. pp. 206–221.

Bearder, S. K., and Doyle, G. A. (1974a). Field and laboratory studies of social organisation in bushbabies (*Galago senegalensis*). *J. Hum. Evol.* 3:37–50.

Bearder, S. K., and Doyle, G. A. (1974b). Ecology of bushbabies *Galago senegalensis* and *Galago crassicaudatus,* with some notes on their behaviour in the field. In: Martin, R. D., Doyle, G. A., and Walker, A. C. (eds.), *Prosimian Biology.* Duckworth Press, London. pp. 109–130.

Bearder, S. K., and Martin, R. D. (1980a). Acacia gum and its use by bushbabies, *Galago senegalensis* (Primates: Lorisidae). *Int. J. Primatol.* 1:103–128.

Bearder, S. K., and Martin, R. D. (1980b). The social organisation of a nocturnal primate revealed by radio tracking. In: Amlaner, C. J., and MacDonald, D. W. (eds.), *A Handbook on Biotelemetry and Radio Tracking.* Pergamon Press, London. pp. 633–648.

Bearder, S. K., Honess, P. E., and Ambrose, L. (1995). Species diversity among galagos with special reference to mate recognition. In: Alterman, L., Doyle, G., and Izard, M. K. (eds.), *Creatures of the Dark: The Nocturnal Prosimians.* Plenum Press, New York. pp. 331–352.

Bearder, S. K., Nekaris, K. A. I., and Buzzell, C. A. (2002). Dangers of the night: are some primates afraid of the dark? In: Miller, L. E. (ed.), *Eat or Be Eaten: Predator Sensitive Foraging in Primates.* Cambridge University Press, Cambridge. pp. 21–43.

Bearder, S. K., Ambrose, L., Harcourt, C., Honess, P., Perkin, A., Pimley, E., Pullen, S., and Svoboda, N. (2003). Species-typical patterns of infant contact, sleeping site use and social cohesion among nocturnal primates in Africa. *Folia Primatol.* 74:337–354.

Bearder, S. K., Nekaris, K. A. I., and Curtis, D. J. (2006). A re-evaluation of the role of vision in the activity and communication of nocturnal primates. *Folia Primatol.* 77:50–71.

Beaudrot, L. H., Kahlenberg, S. M., and Marshall, A. J. (2009). Why male orangutans don't kill infants. *Behav. Ecol. Scoiobiol.* 63: 1549–1562.

Beaumont, M. A. (1999). Detecting population expansion and decline using microsatellites. *Genetics* 153:2013–2029.

Beaver, P. C., Jung, R. C., and Cupp, E. W. (1984). *Clinical Parasitology,* 9th ed. Lea & Febiger, Philadelphia.

Beck, B. (1980). *Animal Tool Behavior: The Use and Manufacture of Tools by Animals.* Garland Press, New York.

Beck, B. B. (1982). Fertility in North American lowland gorillas. *Am. J. Primatol.* 1(suppl.):7–11.

Beck, B. B. (1994). Reintroduction of captive-born animals. In: Olney, P. J. S., Mace, G. M., and Feistner, A. T. C. (eds.), *Creative Conservation.* Chapman & Hall, London. pp. 265–286.

Beck, B. B. (1995). Reintroduction, zoos, conservation and animal welfare. In: Norton, B. G., Hutchins M., Stevens, E. F., and Maple, T. L. (eds.), *Ethics on the Ark: Zoos, Animal Welfare, and Wildlife Conservation.* Smithsonian Institution Press, Washington DC. pp. 155–163.

Beck, B. B., Castro, M. I., Stoinski, T. S., and Ballou, J. (2002). The effects of pre-release environments on survivorship in reintroduced golden lion tamarins. In: Kleiman, D. G., and Rylands, A. B. (eds.), *The Lion Tamarins of Brazil: Twenty-Five Years of Research and Conservation.* Smithsonian Institution Press, Washington DC. pp. 283–300.

Beckage, N. (1997). *Parasites and Pathogens—Effects on Host Hormones and Behavior.* International Thomson Publishing, New York.

Bednekoff, P. A., and Lima, S. L. (1998). Re-examining safety in numbers: interactions between risk dilution and collective detection depend upon predator targeting behaviour. *Proc. R. Soc. Lond.* B 265:2021–2026.

Beehner, J. C., and Bergman, T. J. (2008). Infant mortality following male takeovers in wild geladas. *Am. J. Primatol.* 70:1152–1159.

Beehner, J. C., Bergman, T. J., Cheney, D. L., Seyfarth, R. M., and Whitten, P. L. (2005). The effect of new alpha males on female stress in free-ranging baboons. *Anim. Behav.* 69:1211–1221.

Beehner, J. C., Onderdonk, D. A., Alberts, S. C., and Altmann, J. (2006). The ecology of conception and pregnancy failure in wild baboons. *Behav. Ecol.* 17:741–750.

Beeson, M. (1989). Seasonal dietary stress in a forest monkey (*Cercopithecus mitis*). *Oecologia* 78:565–570.

Beeson, M., Tame, S., Keeming, E., and Lea, S. E. G. (1996). Food habits of guenons (*Cercopithecus* spp.) in Afro-montane forest. *Afr. J. Ecol.* 34:202–210.

Begon, M., Harper, J. L., and Townsend, C. R. (1990). *Ecology: Individuals, Populations, and Communities,* 2nd ed. Blackwell Scientific, New York.

Begun, D. R. (2002a). Pliopithecoidea. In: Hartwig, W. C. (ed.), *The Primate Fossil Record.* Cambridge University Press, Cambridge. pp. 221–240.

Begun, D. R. (2002b). European hominoids. In: Hartwig, W. C. (ed.), *The Primate Fossil Record.* Cambridge University Press, Cambridge. pp. 339–368.

Begun, D. R. (2004). Knuckle-walking and the origin of human bipedalism. In: Meldrum, D. J., and Hilton, C. E. (eds.), *From Biped to Strider: The Emergence of Modern Human Walking, Running and Resource Transport.* Kluwer, New York. pp. 9–33.

Bekoff, M. (2002). Animal reflections. *Nature* 419:255.

Bélisle, P. (2002). Apparentement et co-alimentation chez le macaque Japonais (*Macaca fuscata*) [PhD diss.]. Université de Montréal, Montreal.

Bélisle, P., and Chapais, B. (2001). Tolerated co-feeding in relation to degree of kinship in Japanese macaques. *Behaviour* 138:487–509.

Belkhir, K., Borsa, P., Chikhi, L., Raufaste, N., and Bonhomme, F. (2004). GENETIX 4.05, Logiciel sous Windows TM pour la génétique des populations. Laboratoire Génome, Populations, Interactions, CNRS UMR 5171, Université de Montpellier II, Montpellier, France.

Bell, G. (2001). Neutral macroecology. *Science* 293:2413–2418.

Bell, R. H. V. (1971). A grazing ecosystem in the Serengeti. *Sci. Am.* 225:86–93.

Bellem, A. C., Monfort, S. L., and Goodrowe, K. L. (1995). Monitoring reproductive development, menstrual cyclicity, and pregnancy in the lowland gorilla (*Gorilla gorilla*) by enzyme immunoassay. *J. Zoo Wildl. Med.* 26:24–31.

Bellino, F. L., and Wise, P. M. (2003). Nonhuman primate models of menopause workshop. *Biol. Reprod.* 68:10–18.

Benefit, B. R., and McCrossin, M. L. (2002). The Victoriapithecidae, Cercopithecoidea. In: Hartwig, W. C. (ed.), *The Primate Fossil Record.* Cambridge University Press, Cambridge. pp. 241–253.

Benneton, C., and Noe, R. (2004). Reproductive tactics of adult male sooty mangabeys in Taï National Park, Ivory Coast [abstract]. *Folia Primatol.* 75:169.

Bennett, C. L., Leonard, S., and Carter, S. (2001). Abundance, diversity, and patterns of distribution of primates on the Tapiche River in Amazonian Peru. *Am. J. Primatol.* 54:119–126.

Bennett, E. L. (1983). The banded langur: ecology of a colobine in west Malaysian rain-forest [PhD diss.]. Cambridge University, Cambridge.

Bennett, E. L. (1986). Environmental correlates of ranging behaviour in the banded langur, *Presbytis melalophus.* *Folia Primatol.* 47:26–38.

Bennett, E. L., and Davies, A. G. (1994). The ecology of Asian colobines. In: Davies, A. G., and Oates, J. F. (eds.), *Colobine Monkeys: Their Ecology, Behaviour and Evolution.* Cambridge University Press, Cambridge. pp. 129–172.

Bennett, E. L., and Robinson, J. G. (2000). Carrying capacity limits to sustainable hunting in tropical forests. In: Robinson, J. G., and Bennett, E. L. (eds.), *Hunting for Sustainability in Tropical Forests.* Columbia University Press, New York. pp. 13–30.

Bennett, E. L., and Sebastian, A. C. (1988). Social organization and ecology of proboscis monkeys (*Nasalis larvatus*) in mixed coastal forest in Sarawak. *Int. J. Primatol.* 9:233–255.

Benshoff, L., and Thornhill, R. (1979). The evolution of monogamy and concealed ovulation in humans. *J. Soc. Biol.* 2:95–106.

Bentley, D. R., Balasubramanian, S., Swerdlow, H. P., et al. (2008). Accurate whole human genome sequencing using reversible terminator chemistry. *Nature* 456:53–59.

Bentley, G. R. (1999). Aping our ancestors: comparative aspects of reproductive ecology. *Evol. Anthropol.* 7:175–185.

Bentley-Condit, V. K. (2009). Food choices and habitat use by the Tana River yellow baboons (*Papio cynocephalus*): a preliminary report on five years of data. *Am. J. Primatol.* 71:432–436.

Bentley-Condit, V. K., and Smith, E. O. (1997). Female reproductive parameters of Tana River yellow baboons. *Int. J. Primatol.* 18:581–596.

Bentley-Condit, V. K., and Smith, E. O. (1999). Female dominance and female social relationships among yellow baboons (*Papio hamadryas cynocephalus*). *Am. J. Primatol.* 47:321–334.

Bentley-Condit, V. K., Moore, T., and Smith, E. O. (2001). Analysis of infant handling and the effects of female rank among Tana River adult female yellow baboons (*Papio cynocephalus cynocephalus*) using permutation/randomization tests. *Am. J. Primatol.* 55:117–130.

Benz, J. J. (1993). Food-elicited vocalizations in golden lion tamarins: design features for representational communication. *Anim. Behav.* 45:443–455.

Beran, M. J. (2001). Summation and numerousness judgments of sequentially presented sets of items by chimpanzees (*Pan troglodytes*). *J. Comp. Psychol.* 115:181–91.

Beran, M. J. (2007a). Rhesus monkeys (*Macaca mulatta*) succeed on a computerized test designed to assess conservation of discrete quantity. *Anim. Cogn.* 10:37–45.

Beran, M. J. (2007b). Rhesus monkeys (*Macaca mulatta*) enumerate large and small sequentially presented sets of items using analog numerical representations. *J. Exp. Psychol. Anim. Behav. Proc.* 33:42–54.

Beran, M. J., and Beran, M. M. (2004). Chimpanzees remember the results of one-by-one addition of food items to sets over extended time periods. *Psychol. Sci.* 15:94–99.

Beran, M., Gibson, K., and Rumbaugh, D. (1999). Predicting hominid intelligence from brain size. In: Corballis, M. C., and Lea, S. E. G. (eds.), *The Descent of Mind: Psychological Perspectives on Hominid Evolution.* Oxford University Press, New York. pp. 88–97.

Beran, M. J., Beran, M. M., and Menzel, C. R. (2005). Spatial memory and monitoring of hidden items through spatial displacements by chimpanzees (*Pan troglodytes*). *J. Comp. Psychol.* 119:14–22.

Beran, M. J., Harris, E. H., Evans, T. A., Klein, E. D., Chan, B., Flemming, T. M., and Washburn, D. A. (2008). Ordinal judgments of symbolic stimuli by capuchin monkeys (*Cebus apella*) and rhesus monkeys (*Macaca mulatta*): the effects of differential and nondifferential reward. *J. Comp. Psychol.* 122:52–61.

Berard, J. D. (1999). A four-year study of the association between male dominance rank, residency status, and reproductive activity in rhesus macaques (*Macaca mulatta*). *Primates* 40:159–175.

Berard, J. D., Nürnberg, P., Epplen, J. T., and Schmidtke, J. (1993). Male rank, reproductive behavior, and reproductive success in free-ranging rhesus macaques. *Primates* 34:481–489.

Berard, J. D., Nürnberg, P., Epplen, J. T., and Schmidtke, J. (1994). Alternative reproductive tactics and reproductive success in male rhesus macaques. *Behaviour* 129:177–201.

Bercovitch, F. B. (1983). Time budgets and consortships in olive baboons (*Papio anubis*). *Folia primatol.*, 41:180–190.

Bercovitch, F. B. (1985). Reproductive tactics in adult female and adult male olive baboons [PhD diss.]. University of California, Los Angeles.

Bercovitch, F. B. (1986). Male rank and reproductive activity in savanna baboons. *Int. J. Primatol.* 7:533–550.

Bercovitch, F. B. (1987a). Female weight and reproductive condition in a population of olive baboons (*Papio anubis*). *Am. J. Primatol.* 12:189–195.

Bercovitch, F. B. (1987b). Reproductive success in male savanna baboons. *Behav. Ecol. Sociobiol.* 21:163–172.

Bercovitch, F. B. (1988). Coalitions, cooperation and reproductive tactics among adult male baboons. *Anim. Behav.* 36:1198–1209.

Bercovitch, F. B. (1989). Body size, sperm competition, and determinants of reproductive success in male savanna baboons. *Evolution* 43:1507–1521.

Bercovitch, F. B. (1991). Mate selection, consortship formation, and reproductive tactics in adult female savanna baboons. *Primates* 32:437–452.

Bercovitch, F. B. (1992). Sperm competition, reproductive tactics, and paternity in savanna baboons and rhesus macaques. In: Martin, R. D., Dixson, A. F., and Wickings, E. J. (eds.), *Paternity in Primates: Genetic Tests and Theories.* Karger, Basel. pp. 225–237.

Bercovitch, F. B. (1993). Dominance rank and reproductive maturation in male rhesus macaques (*Macaca mulatta*). *J. Reprod. Fertil.* 99:113–120.

Bercovitch, F. B. (1995). Female cooperation, consortship maintenance, and male mating success in savanna baboons. *Anim. Behav.* 50:137–149.

Bercovitch, F. B. (1997). Reproductive strategies of rhesus macaques. *Primates* 38:247–263.

Bercovitch, F. B. (2000). Behavioral ecology and socioendocrinology of reproductive maturation in cercopithecine monkeys. In: Whitehead, P. F., and Jolly, C. J. (eds.), *Old World Monkeys.* Cambridge University Press, Cambridge, pp. 298–320.

Bercovitch, F. B., and Berard, J. D. (1993). Life history costs and consequences of rapid reproductive maturation in female rhesus macaques. *Behav. Ecol. Sociobiol.* 32:103–109.

Bercovitch, F. B., and Harding, R. S. O. (1993). Annual birth patterns of savanna baboons (*Papio cynocephalus anubis*) over a ten-year period at Gilgil, Kenya. *Folia Primatol.* 61:115–122.

Bercovitch, F. B., and Harvey, N. (2004). Reproductive life history. In: Thierry, B., Singh, M., and Kaumanns, W. (eds.), *Macaque Societies: A Model for the Study of Social Organization.* Cambridge University Press, Cambridge, pp. 61–80.

Bercovitch, F. B., and Nürnberg, P. (1996). Socioendocrine and morphological correlates of paternity in rhesus macaques. *J. Reprod. Fertil.* 107:59–68.

Bercovitch, F. B., and Nurnberg, P. (1997). Genetic determination of paternity and variation in male reproductive success in two populations of rhesus macaques. *Electrophoresis* 18:1701–1705.

Bercovitch, F. B., and Strum, S. C. (1993). Dominance rank, resource availability, and reproductive maturation in female savanna baboons. *Behav. Ecol. Sociobiol.* 33:313–318.

Bercovitch, F. B., and Ziegler, T. E. (2002). Recent topics in primate socioendocrinology. *Annu. Rev. Anthropol.* 31:45 67.

Berenstain, L., and Wade, T. D. (1983). Intrasexual selection and male mating strategies in baboons and macaques. *Int. J. Primatol.* 4:201–235.

Berger, J., Stacey-Peter, B., Bellis, L., and Johnson, M. P. (2001). A mammalian predator–prey imbalance: grizzly bear and wolf extinction affect avian neotropical migrants. *Ecol. Appl.* 11:947–960.

Berger, M. E. (1972). Live-weights and body measurements of olive baboons (*Papio anubis*) in Laikipia District of Kenya. *J. Mammal.* 53:404–406.

Bergeson, D. J. (1996). The positional behavior and prehensile tail use of *Alouatta palliata, Ateles geoffroyi,* and *Cebus capucinus* [PhD diss.]. Washington University, St. Louis.

Bergeson, D. J. (1998). Patterns of suspensory feeding in *Alouatta palliata, Ateles geoffroyi,* and *Cebus capucinus.* In: Strasser, E., Fleagle, J. G., Rosenberger, A., and McHenry, H. (eds.), *Primate Locomotion. Recent Advances.* Plenum Press, New York. pp. 45–60.

Bergl, R. A., and Vigilant, L. (2007). Genetic analysis reveals population structure and recent migration within the highly fragmented range of the Cross River gorilla (*Gorilla gorilla diehli*). *Mol. Ecol.* 16:501–516.

Bergl, R., Bradley, B. J., Nsubuga, A., and Vigilant, L. (2008). Effects of habitat fragmentation, population size and demographic history on genetic diversity: the Cross River gorilla in a comparative context. *Am. J. Primatol.* 70:848–859.

Bergman, T. J., and Beehner, J. C. (2003). Hybrid zones and sexual selection: insights from the Awash baboon hybrid zone (*Papio hamdryas anubis* × *P. h. hamadryas*). In: Jones, C. B. (ed.), *Sexual Selection and Reproductive Competition in Primates: New Perspectives and Directions.* American Society of Primatologists, Norman, OK. pp. 503–537.

Bergman, T. J., Beehner, J. C., Cheney, D. L., Seyfarth, R. M., and Whitten, P. L. (2005). Correlates of stress in free-ranging male chacma baboons, *Papio hamadryas ursinus. Anim. Behav.* 70:703–713.

Bergstrom, C. T., and Lachmann, M. (1997). Signaling among relatives. I. When is signaling too costly? *Philos. Trans. R. Soc. Lond. B* 352:609–617.

Bergstrom, C. T., and Lachmann, M. (1998). Signaling among relatives: III. Talk is cheap. *Proc. Natl. Acad. Sci. USA* 95:5100–5105.

Bering, J. M., Bjorklund, D. F., and Ragan, P. (2000). Deferred imitation of object-related actions in human-reared juvenile chimpanzees and orangutans. *Dev. Psychobiol.* 36:218–232.

Berkovitch, F. R. (1992). Estradiol concentrations, fat deposits, and reproductive strategies in male rhesus macaques. *Horm. Behav.* 26:272–282.

Berman, C. M. (1982a). The social development of an orphaned rhesus infant on Cayo Santiago: male care, foster mother–orphan interaction and peer interaction. *Am. J. Primatol.* 3:131–141.

Berman, C. M. (1982b). The ontogeny of social relationships with group companions among free-ranging infant rhesus monkeys. I. Social networks and differentiation. *Anim. Behav.* 30:149–162.

Berman, C. M. (1984). Variation in mother–infant relationships: traditional and nontraditional factors. In: Small, M. F. (ed.), *Female Primates: Studies by Women Primatologists.* Alan R. Liss, New York. pp. 17–36.

Berman, C. M. (1986). Maternal lineages as tools for understanding infant social development and social structure. In: Rawlins, R. G. and Kessler, M. J. (eds.), *The Cayo Santiago Macaques: History, Behavior and Biology.* SUNY Press, Albany. pp. 73–92.

Berman, C. M. (1997). An animal model for the intergenerational transmission of maternal style? *Family Systems* 3:125–140.

Berman, C. M. (2004). Developmental aspects of kin bias in behavior. In: Chapais, B., and Berman, C. M. (eds.), *Kinship and Behavior in Primates.* Oxford University Press, New York. pp. 317–346.

Berman, C. M., and Kapsalis, E. (1999). Development of kin bias among rhesus monkeys: maternal transmission or individual learning? *Anim. Behav.* 58:883–894.

Berman, C. M., and Kapsalis, E. (2009). Variation over time in grooming kin bias among female rhesus macaques on Cayo Santiago supports the time constraints hypothesis. *Am. J. Phys. Anthropol.* 138(suppl.). In press.

Berman, C. M., Rasmussen, K. L. R., and Suomi, S. J. (1997). Group size, infant development and social networks in free-ranging rhesus monkeys. *Anim. Behav.* 53:405–421.

Berman, C. M., Ionica, C. S., and Li, J. H. (2004). Dominance style among *Macaca thibetana* on Mt. Huangshan, China. *Int. J. Primatol.* 25:1283–1312.

Berman, C. M., Ionica, C. S., Dorner, M., and Li, J. H. (2006). Postconflict affiliation between former opponents in *Macaca thibetana* on Mt. Huangshan, China. *Int. J. Primatol.* 27:827–854.

Berman, C. M., Ionica, C., and Li, J. H. (2007). Supportive and tolerant relationships among male Tibetan macaques at Huangshan, China. *Behaviour* 144:631–661.

Berman, C. M., Ogawa, H., Ionica, C. S., Yin, H., and Li, J. (2008). Variation in kin bias over time in a group of Tibetan macaques at Huangshan, China: contest competition, time constraints or stress response? *Behaviour* 145:863–896.

Bermejo, M. (1999). Status and conservation of primates in Odzala National Park, Republic of the Congo. *Oryx* 33:323–331.

Bermejo, M. (2004). Home range use and inter-group encounters in western gorillas (*Gorilla g. gorilla*) at Lossi Forest, north Congo. *Am. J. Primatol.* 64:223–232.

Bermejo, M., Rodriguez-Teijeiro, J. D., Illera, G., Barroso, A., Vila, C., and Walsh, P. D. (2006). Ebola outbreak killed 5000 gorillas. *Science* 314:1564.

Bernede, L. (2009). Social organisation, ecology and conservation of *Loris tardigradus tardigradus* (Lorisiformes, Primates) [PhD diss.]. Oxford Brookes University, Oxford.

Bernede, L., Beresford, A. S., Nekaris, K. A. I., and Gunawardene, A. (2008). Home range use of the red slender loris (*Loris tardigradus tardigradus*) in Masmullah Proposed Forest Reserve, Sri Lanka. *Folia Primatol.* 79:311–311.

Bernstein, I. (1964a). Role of the dominant male rhesus in response to external challenges to the group. *J. Comp. Physiol. Psychol.* 57:404–406.

Bernstein, I. (1964b). A field study of the activities of howler monkeys. *Anim. Behav.* 12:84–91.

Bernstein, I. (1966a). Analysis of a key role in a capuchin (*Cebus albifrons*) group. *Tulane Stud. Zool.* 13:49–54.

Bernstein, I. (1966b). An investigation of the organization of pigtail monkey groups through the use of challenges. *Primates* 7:471–480.

Bernstein, I. (1967). Intertaxa interactions in a primate community. *Folia Primatol.* 7:198–207.

Bernstein, I. (1968). The lutong of Kuala Selangor. *Behaviour* 32:1–16.

Bernstein, I. (1969). Stability of the status hierarchy in a pigtail monkey group (*Macaca nemestrina*). *Anim. Behav.* 17:452–458.

Bernstein, I. (1976). Dominance, aggression and reproduction in primate societies. *J. Theor. Biol.* 60:459–472.

Bernstein, I. (1981). Dominance: the baby and the bathwater. *Behav. Brain Sci.* 4:419–457.

Bernstein, I. (1988). Kinship and behavior in nonhuman primates. *Behav. Genet.* 18:511–524.

Bernstein, I. (1991a). Aggression. In: Dubecco, R. (ed.), *Encyclopedia of Human Biology.* Academic Press, San Diego. pp. 113–118.

Bernstein, I. (1991b). The correlation between kinship and behavior in non-human primates. In: Hepper, P. (ed.), *Kin Recognition.* Cambridge University Press, Cambridge. pp. 6–29.

Bernstein, I. (1999a). Kinship and the behavior of nonhuman primates. In: Dolhinow, P. (ed.), *The Nonhuman Primates.* Manfield, Mountain View, CA. pp. 202–205.

Bernstein, I. (1999b). The study of behavior. In: Dolhinow, P., and Fuentes, A. (eds.), *The Nonhuman Primates.* Mayfield, Mountain View, CA. pp. 176–180.

Bernstein, I., and Ehardt, C. (1986a). Selective interference in rhesus monkey (*Macaca mulatta*) intragroup agonistic episodes by age–sex class. *J. Comp. Psychol.* 100:380–384.

Bernstein, I., and Ehardt, C. (1986b). The influence of kinship and socialization on aggressive behaviour in rhesus monkeys (*Macaca mulatta*). *Anim. Behav.* 34:739–747.

Bernstein, I. S., and Ehardt, C. L. (1986c). Modification of aggression through socialization and the special case of adult and adolescent male rhesus monkeys (*Macaca mulatta*). *Am. J. Primatol.* 10:213–227.

Bernstein, I., and Gordon, T. (1974). The function of aggression in primate societies. *Am. Sci.* 62:304–311.

Bernstein, I. S., and Gordon, T. P. (1980). Mixed taxa introductions, hybrids and macaque systematics. In: Lindburg, D. G. (ed.), *The Macaques: Studies in Ecology, Behavior and Evolution.* Van Nostrand Reinhold, New York, pp. 125–147.

Bernstein, I. S., Judge, P. G., and Ruehlmann, T. E. (1993a). Sex differences in adolescent rhesus monkey (*Macaca mulatta*) behavior. *Am. J. Primatol.* 31(2):197–210

Bernstein, I., Judge, P., and Ruehlmann, T. (1993b). Kinship, association, and social relationships in rhesus monkeys (*Macaca mulatta*). *Am. J. Primatol.* 31:41–45.

Bernstein, R. M., Leigh S. R., Donovan S. M., and Monaco, M. H. (2008). Hormonal correlates of ontogeny in baboons (*Papio hamadryas anubis*) and mangabeys (*Cercocebus atys*) *Am. J. Phys. Anthropol.* 136(2):156–168.

Berrigan, D., Charnov, E. L., Purvis, A., and Harvey, P. H. (1993). Phylogenetic contrasts and the evolution of mammalian life histories. *Evol. Ecol.* 7:270–278.

Berry, J. P., McFarren, M. A., and Rodriguez, E. (1995). Zoopharmacognosy: a "biorational" strategy for phytochemical prospecting. In: Gustine, D. L., and Flores, H. E. (eds.), *Phytochemicals and Health.* American Society of Plant Physiologists, Rockville, MD. pp. 165–178.

Berson, S. A., Yalow, R. S., Bauman, A., Rothchild, M. A., and Newerly, K. (1956). Insulin-I131 metabolism in human subjects: demonstration of insulin binding globulin in the circulation on insulin-treated subjects. *J. Clin. Invest.* 35:170–190.

Bert, J., Ayats, H., Martino, A., and Collomb, H. (1967). Le sommeil nocturne chez le babouin *Papio papio. Folia Primatol.* 6:28–43.

Bertolani, P., and Boesch, C. (2008). Habituation of wild chimpanzees (*Pan troglodytes*) of the South Group at Taï Forest, Côte d'Ivoire: empirical measure of progress. *Folia Primatol.* 79:162–171.

Bertrand, M. (1969). *The Behavioral Repertoire of the Stumptail Macaque. Bibliotheca Primatologica*, vol. 11. S. Karger, Basel.

Bettinger, T., Kuhar, C., Sironen, A., and Laudenslager, M. (1998). Behavior and salivary cortisol in gorillas housed in an all male group. In: *Proceedings of the American Zoological Association Annual Conference.* Tulsa OK. pp. 242–246.

Bezanson, M. F. (2003). Patterns of positional behavior in juvenile and adult white-faced capuchins (*Cebus capucinus*). *Am. J. Phys. Anthropol.* 120:66.

Bezanson, M. F. (2006). Leap, bridge, or ride? Ontogenetic influences on positional behavior in *Cebus* and *Alouatta*. In: Estrada, A., Garber, P. A., Pavelka, M., and Luecke, L. (eds.), *New Perspectives in the Study of Mesoamerican Primates: Distribution, Ecology, Behavior, and Conservation.* Springer Science+Business Media, New York. pp. 333–348.

Bezanson, M., Garber, P. A., Rutherford, J. R., and Cleveland, A. (2002). Patterns of subgrouping, social affiliation and social networks in Nicaragua mantled howler monkeys (*Alouatta palliata*). *Am. J. Phys. Anthropol.* 32(suppl.):44.

Bezerra, B. M., Da Silva Souto, A., and Schiel, N. (2007). Infanticide and cannibalism in a free-ranging plurally breeding group of common marmosets (*Callithrix Jacchus*). *Am. J. Primatol.* 69:945–952.

Bezerra, B. M., Barnett, A. A., Souto, A., and Jones, G. (2008). Predation by the tayra on the common marmoset and the pale-throated three-toed sloth. *J. Ethol.* 27:91–96.

Bianchi, R. C., and Mendes, S. L. (2007). Ocelot (*Leopardus pardalis*) predation on primates in Caratinga Biological Station, Southeast Brazil. *Am. J. Primatol.* 69:1173–1178.

Biben, M., Symmes, D., and Bernhards, D. (1989). Vigilance during play in squirrel monkeys. *Am. J. Primatol.* 17:41–49.

Bicca-Marques, J. C. (1993). Padrão de atividades diárias do bugio-preto *Alouatta caraya* (Primates: Cebidae): uma análise temporal e bioenergética. *Primatol. Brasil* 4:35–49.

Bicca-Marques, J. C. (1994). Padrão de utilização de uma ilha de mata por *Alouatta caraya* (Primates: Cebidae). *Rev. Bras. Biol.* 54:161–171.

Bicca-Marques, J. C., and Calegaro-Marques, C. (1994). Feeding behavior of the black howler monkey (*Alouatta caraya*) in a seminatural forest. *Acta Biol. Leopold.* 16:69–84.

Bicca-Marques, J. C., and Garber P. A. (2001). Cognitive ecology and within-patch foraging decisions in tamarins. *Am. J. Primatol.* 32(suppl.):40.

Bicca-Marques, J. C., and Garber, P. A. (2003). Experimental field study of the relative costs and benefits to wild tamarins (*Saguinus imperator* and *Saguinus fuscicollis*) of exploiting contestable food patches as single- and mixed-species troops. *Am. J. Primatol.* 60:139–153.

Bickerton, D. (1990). *Language and Species.* University of Chicago Press, Chicago.

Bielert, C., and Anderson, C. M. (1985). Baboon sexual swellings and male response: a possible operational mammalian supernormal stimulus and response interaction. *Int. J. Primatol.* 6:377–393.

Bielert, C., and Girolami, C. (1986). Experimental assessments of behavioral and anatomical components of female chacma beboon (*Papio ursinus*) sexual attractiveness. *Psychoneuroendocrinology* 11:75–90.

Bielert, C., Girolami, L., and Anderson, C. (1986). Male chacma baboon (*Papio ursinus*) sexual arousal—studies with adolescent and adult females as visual stimuli. *Dev. Psychobiol.* 19:368–383.

Bielert, C., Girolami, L., and Jowell, S. (1989). An experimental examination of the colour component in visually mediated sexual arousal of the male chacma baboon (*Papio ursinus*). *J. Zool.* 219:569–579.

Bigoni, F., Stanyon, R., Wimmer, R., and Schempp, W. (2003). Chromosome painting shows that the proboscis monkey (*Nasalis larvatus*) has a derived karyotype and is phylogenetically nested within Asian colobines. *Am. J. Primatol.* 60:85–93.

Bigoni, F., Houck, M. L., Ryder, O. A., Wienberg, J., and Stanyon, R. (2004). Chromosome painting shows that *Pygathrix nemaeus* has the most basal karyotype among Asian Colobinae. *Int. J. Primatol.* 4:679–688.

Bingham, H. C. (1932). *Gorillas in a Native Habitat*, vol. 426. Carnegie Institute, Washington DC. pp. 1–66.

Biquand, S., Biquand-Guyot, V., Boug, A., and Gautier, J.-P. (1992a). The distribution of *Papio hamadryas* in Saudi Arabia: ecological correlates and human influence. *Int. J. Primatol.* 13:223–243.

Biquand, S., Biquand-Guyot, V., Boug, A., and Gautier, J.-P. (1992b). Group composition in wild and commensal hamadryas baboons: a comparative study in Saudi Arabia. *Int. J. Primatol.* 13:533–543.

Biquand, S., Biquand-Guyot, V., Boug, A., and Gautier, J.-P. (1994). Management of commensal baboons in Saudi Arabia. *Rev. Ecol. (Terre Vie)* 49:213–222.

Birkinshaw, C. R. (1999). Use of millipedes by black lemurs to anoint their bodies. *Folia Primatol.* 70:170–171.

Birky, W. A. (2002). Mating patterns and social structure in a wild group of Formosan macaques [PhD diss.]. Rutgers University, Piscataway, NJ.

Biro, D., and Matsuzawa, T. (1999). Numerical ordering in a chimpanzee (*Pan troglodytes*): planning, executing and monitoring. *J. Comp. Psychol.* 113:178–85.

Biro, D., Inoue-Nakamura, N., Tonooka, R., Yamakoshi, G., Sousa, C., and Matsuzawa, T. (2003). Cultural innovation and transmission of tool use in wild chimpanzees: evidence from field experiments. *Anim. Cogn.* 6:213–223.

Bishop, N. H. (1979). Himalayan langurs: temperate colobines. *J. Hum. Evol.* 8:251–281.

Bishop, N., Hrdy, S. B., Teas, J., and Moore, J. (1981). Measures of human influence in habitats of South Asian monkeys. *Int. J. Primatol.* 2:153–167.

Bjorklund, D. F., Bering, J. M., and Ragan, P. (2000). A two-year longitudinal study of deferred imitation of object manipulation in an encultured juvenile chimpanzee (*Pan troglodytes*) and orangutan (*Pongo pygmaeus*). *Dev. Psychobiol.* 37:229–237.

Bjornsdotter, M., Larsson, L., and Ljungberg, T. (2000). Post-conflict affiliation in two captive groups of black-and-white guereza (*Colobus guereza*). *Ethology* 106:289–300.

Blackmore, S. (1999). *The Meme Machine.* Oxford University Press, Oxford.

Blakeslee, S. (1999). How brain uses a simple dopamine system. In: Sussman, R. W. (ed.), *The Biological Basis of Human Behavior: A Critical Review.* Prentice Hall, Upper Saddle River, NJ. pp. 347–349.

Blaxter, K. L. (1962). *The Energy Metabolism of Ruminants.* Hutchison, London.

Bleier, R. (1986). *Feminist Approaches to Science.* Pergamon Press, New York.

Bleisch, W. V., and Chen, N. (1991). Ecology and behavior of wild black-crested gibbons (*Hylobates concolor*) in China with a reconsideration of evidence for polygyny. *Primates* 32:539–548.

Bleisch, W. V., and Xie, J. H. (1998). Ecology and behavior of the Guizhou snub-nosed langur (*Rhinopithecus [Rhinopithecus] brelichi*), with a discussion of socioecology in the genus. In: Jablonski, N. G. (ed.), *The Natural History of the Doucs and Snub-Nosed Monkeys.* World Scientific Press, Singapore. pp. 217–240.

Bleisch, W. V., Cheng, A. S., Ren, X. D., and Xie, J. H. (1993). Preliminary results from a field study of wild Guizhou snub-nosed monkeys (*Rhinopithecus brelichi*). *Folia Primatol.* 60:72–92.

Bloch, J. I., and Boyer, D. M. (2002). Grasping primate origins. *Science* 298:1606–1610.

Blom, A., Cipolletta, C., and Brunsting, A. M. H. (2004). Behavioral responses of gorillas to habituation in the Dzanga-Ndoki National Park, Central African Republic. *Int. J. Primatol.* 25:179–196.

Blomquist, G. E. (2009). Population regulation and the life history studies of LaMont Cole. *Hist. Philos. Life Sci.* 29:495–516.

Blum, D. (2002). *Love at Goon Park: Harry Harlow and the Science of Affection.* Perseus, Cambridge.

Blumstein, D. T. (2000). The evolution of infanticide in rodents: a comparative analysis. In: van Schaik, C. P., and Janson, C. H. (eds.), *Infanticide by Males and Its Implications.* Cambridge University Press, Cambridge. pp. 178–197.

Blumstein, D. T. (2001). The evolution of functionally referential alarm communication: multiple adaptations, multiple constraints. *Evol. Comm.* 3(2):135–147.

Blumstein, D. T., and Armitage, K. B. (1997). Does sociality drive the evolution of communicative complexity? A comparative test with ground-dwelling sciurid alarm calls. *Am. Nat.* 150:179–200.

Blurton-Jones, N., Hawkes, K., and O'Connell, J. F. (1999). Some current ideas about evolution of the human life history. In: Lee, P. C. (ed.), *Comparative Primate Socioecology.* Cambridge University Press, Cambridge. pp. 140–166.

Bocian, C. M. (1997). Niche separation of black-and-white colobus monkeys (*Colobus angolensis* and *C. guereza*) in the Ituri Forest [PhD diss.]. City University of New York, New York.

Bock, W., and von Wahlert, G. (1965). Adaptation and the form–function complex. *Evolution* 19:269–299.

Bodmer, R. E., Fang, T. G., and Moya Ibanez, L. (1988). A comparison of susceptibility to hunting. *Primate Conserv.* 9:79–83.

Bodmer, R. E., Fang, T. G., Moya Ibanez, L., and Gill, R. (1994). Managing wildlife to conserve Amazonian forests: population biology and economic considerations of game hunting. *Biol. Conserv.* 67:29–35.

Bodry-Sanders, P. (1998). *African Obsession: The Life and Legacy of Carl Akeley.* Batax Museum Publishing, Jacksonville.

Boehm, C. (1994). Pacifying interventions at Arnhem Zoo and Gombe. In: Wrangham, R., McGrew, W., and de Waal, F. (ed.), *Chimpanzee Cultures.* Harvard University Press, Cambridge, MA. pp. 211–226.

Boenisch, T. (ed.) (2001). *Immunochemical Staining Methods*, 3rd ed. Dako, Carpinteria, CA.

Boes, A., Probst, B., and Erkert, H. G. (1993). Urinary estradiol-17-beta excretion in common marmosets, *Callithrix jacchus*: diurnal pattern and relationship between creatinine-related values and excreted amount. *Comp. Biochem. Physiol.* A105:287–292.

Boesch, C. (1991). The effects of leopard predation on grouping patterns in forest chimpanzees. *Behaviour* 117:220–242.

Boesch, C. (1994a). Chimpanzees–red colobus monkeys: a predator–prey system. *Anim. Behav.* 47:1135–1148.

Boesch, C. (1994b). Cooperative hunting in wild chimpanzees. *Anim. Behav.* 48:653–667.

Boesch, C. (1994c). Hunting strategies of Gombe and Taï chimpanzees. In: Wrangham, R. W., McGrew, W. C., de Waal, F. B. M., and Heltne. P. G. (eds.), *Chimpanzee Cultures.* Harvard University Press, Cambridge, MA. pp. 77–91.

Boesch, C. (1995). Innovation in wild chimpanzees (*Pan troglodytes*). *Int. J. Primatol.* 16:1–16.

Boesch, C. (1996a). Social grouping in Taï chimpanzees. In: McGrew, W., Marchant, L., and Nishida, T. (eds.), *Great Ape Societies.* Cambridge University Press, Cambridge. pp. 101–113.

Boesch, C. (1996b). Three approaches for assessing chimpanzee culture. In: Russon, A. E., Bard, K. A., and Parker, S. (eds.), *Reaching into Thought: The Minds of the Great Apes.* Cambridge University Press, Cambridge. pp. 404–429.

Boesch, C. (1997). Evidence for dominant wild female chimpanzees investing more in sons. *Anim. Behav.* 54:811–815.

Boesch, C. (2002). Behavioural diversity in *Pan*. In: Boesch, C., Hohmann, G., and Marchant, L. F. (eds.), *Behavioural Diversity in Chimpanzees and Bonobos.* Cambridge University Press, Cambridge. pp. 1–8.

Boesch, C., and Boesch, H. (1981). Sex differences in the use of natural hammers by wild chimpanzees: a preliminary report. *J. Hum. Evol.* 10:585–593.

Boesch, C., and Boesch, H. (1989). Hunting behavior of wild chimpanzees in the Taï National Park. *Am. J. Phys. Anthropol.* 78:547–573.

Boesch, C., and Boesch, H. (1990). Tool use and tool making in wild chimpanzees. *Folia Primatol.* 54:86–99.

Boesch, C., and Boesch-Achermann, H. (2000). *The Chimpanzees of the Taï Forest: Behavioral Ecology and Evolution.* Oxford University Press, Oxford.

Boesch, C., and Tomasello, M. (1998). Chimpanzee and human cultures. *Curr. Anthropol.* 39:591–614.

Boesch, C., Marchesi, P., Marchesi, N., Fruth, B., and Joulian, F. (1994). Is nut cracking in wild chimpanzees a cultural behaviour? *J. Hum. Evol.* 26:325–338.

Boesch C., Uehara, S., and Ihobe, H. (2002). Variations in chimpanzee–red colobus interactions. In: Boesch, C., Hohmann, G., and Marchant, L. (eds.), *Behavioural Diversity in Chimpanzees and Bonobos.* Cambridge University Press, Cambridge. pp. 221–230.

Boesch C., Kohou, G., Néné, H., and Vigilant, L. (2006). Male competition and paternity in wild chimpanzees of the Taï forest. *Am. J. Phys. Anthropol.* 130:103–115.

Boesch, C., Head, J., Tagg, N., Arandjelovic, M., Vigilant, L., and Robbins, M. M. (2007). Fatal chimpanzee attack in Loango National Park, Gabon. *Int. J. Primatol.* 28:1025–1034.

Boesch, C., Crockford, C., Herbinger, I., Wittig, R., Moebius, Y., and Normand E. (2008). Intergroup conflicts among chimpanzees in Taï National Park: lethal violence and the female perspective. *Am. J. Primatol.* 70: 519–532.

Boese, G. K. (1973). Behavior and social organization of the guinea baboon (*Papio papio*) [PhD diss.]. Johns Hopkins University, Baltimore.

Boese, G. K. (1975). Social behavior and ecological considerations of West African baboons (*Papio papio*). In: Tuttle, R. H. (ed.), *Socioecology and Psychology of Primates.* Mouton, The Hague, pp. 205–230.

Boggess, J. (1982). Immature male and male interactions in bisexual langur (*Presbytis entellus*) troops. *Folia Primatol.* 38:19–38.

Bogin, B. B. (1999). *Patterns of Human Growth.* Cambridge University Press, Cambridge.

Bogin, B., and Smith, B. H. (1996). Evolution of the human life cycle. *Am. J. Hum. Biol.* 8:703–716.

Bohlman, S. A., Adams, J. B., Smith, R. O., and Peterson, D. L. (1998). Seasonal foliage changes in the eastern Amazon basin detected from Landsat Thematic Mapper satellite images. *Biotropica* 30:376–391.

Boinski, S. (1987a). Mating patterns in squirrel monkeys (*Saimiri oerstedi*): implications for seasonal sexual dimorphism. *Behav. Ecol. Sociobiol.* 21:13–21.

Boinski, S. (1987b). Birth synchrony in squirrel monkeys (*Saimiri oerstedii*): a strategy to reduce neonatal predation. *Behav. Ecol. Sociobiol.* 21:393–400.

Boinski, S. (1987c). Habitat use by squirrel monkeys (*Saimiri oerstedii*) in Costa Rica. *Folia Primatol.* 49:151–167.

Boinski, S. (1988a). Sex differences in foraging behavior of squirrel monkeys: ecological implications. *Behav. Ecol. Sociobiol.* 21:177–186.

Boinski, S. (1988b). Use of a club by a wild white-faced capuchin (*Cebus capucinus*) to attack a venemous snake (*Bothrops asper*). *Am. J. Primatol.* 14:177–179.

Boinski, S. (1988c). Sex differences in the foraging behavior of squirrel monkeys in a seasonal habitat. *Behav. Ecol. Sociobiol.* 23:177–186.

Boinski, S. (1989a). The positional behavior and substrate use of squirrel monkeys: ecological implications. *J. Hum. Evol.* 18:659–677.

Boinski, S. (1989b). Why don't *Saimiri oerstedii* and *Cebus capucinus* form mixed-species groups? *Int. J. Primatol.* 10:103–114.

Boinski, S. (1991). The coordination of spatial position: a field study of the vocal behavior of adult female squirrel monkeys. *Anim. Behav.* 41:89–120.

Boinski, S. (1992). Monkeys with inflated sex appeal. *Nat. Hist.* 101:42–49.

Boinski, S. (1994). Affiliation patterns among male Costa Rican squirrel monkeys. *Behaviour* 130:191–209.

Boinski, S. (1998). Monkeys with inflated sex appeal. In: Ciochon, R. L., and Nisbett, R. A. (eds.), *The Primate Anthology: Essays on Primate Behavior, Ecology, and Conservation from Natural History.* Prentice Hall, Englewood Cliffs, NJ. pp. 174–179.

Boinski, S. (1999). The social organization of squirrel monkeys: implications for ecological models of social evolution. *Evol. Anthropol.* 8:101–112.

Boinski, S. (2000). Social manipulation within and between troops mediates primate group movement. In: Boinski, S., and Garber, P. A. (eds.), *On the Move—How and Why Animals Travel in Groups.* University of Chicago Press, Chicago. pp. 421–446.

Boinski, S. (2005). Dispersal patterns among three species of squirrel monkeys (*Saimiri oerstedii, S. boliviensis and S. sciureus*): III. Cognition. *Behaviour* 142(5):679–699.

Boinski, S., and Chapman, C. A. (1995). Predation on primates: where are we and what's next? *Evol. Anthropol.* 4:1–3.

Boinski, S., and Cropp, S. J. (1999). Disparate data sets resolve squirrel monkey (*Saimiri*) taxonomy: implications for behavioural ecology and biomedical usage. *Int. J. Primatol.* 20:237–256.

Boinski, S., and Fragaszy, D. M. (1989). The ontogeny of foraging in squirrel monkeys, *Saimiri oerstedii. Anim. Behav.* 37:415–428.

Boinski, S., and Garber, P. A. (2000). *On the Move: How and Why Animals Travel in Groups.* University of Chicago Press, Chicago.

Boinski, S., and Mitchell, C. L. (1994). Male resident and association patterns in Costa Rican squirrel monkeys (*Saimiri oerstedii*). *Am. J. Primatol.* 34:157–169.

Boinski, S., Quatrone, R., and Swats, H. (2000a). Substrate and tool use by brown capuchins in Suriname: ecological contexts and cognitive bases. *Am. Anthropol.* 102:741–761.

Boinski, S., Treves, A., and Chapman, C. A. (2000b). A critical evaluation of the influence of predators on primates: effects on group travel. In: Boinski, S., and Garber, P. A. (eds.), *On the Move: How and Why Animals Travel in Groups.* University of Chicago Press, Chicago. pp. 43–72.

Boinski, S., Sughrue, K., Selvaggi, L., Quatrone, R., Henry, M., and Cropp, S. (2002). An expanded test of the ecological model of primate social evolution: competitive regimes and female bonding in three species of squirrel monkeys (*Saimiri oerstedii, S. boliviensus, and S. sciureus*). *Behaviour* 139:227–261.

Boinski, S., Kauffman, L., Westoll, A., Stickler, C. M., Cropp, S., and Ehmke, E. (2003). Are vigilance, risk from avian predators and group size consequences of habitat structure? A comparison of three species of squirrel monkey (*Saimiri oerstedii, S. boliviensis, and S. sciureus*). *Behaviour* 140:1421–1467.

Boinski, S., Kauffman, L., Ehmke, E., Schet, S., and Vreedzaam, A. (2005a). Dispersal patterns among three species of squirrel monkeys (*Saimiri oerstedii, S. boliviensis and S. sciureus*): I. Divergent costs and benefits. *Behaviour* 142(5):525–632.

Boinski, S., Ehmke, E., Kauffman, L., Schet, S., and Vreedzaam, A. (2005b). Dispersal patterns among three species of squirrel monkeys (*Saimiri oerstedii, S. boliviensis and S. sciureus*): II. Within-species and local variation. *Behaviour* 142(5):633–677.

Bolin, I. (1981). Male parental behavior in black howler monkeys (*Alouatta palliata pigra*) in Belize and Guatemala. *Primates* 22:349–360.

Bolter, D. R. (2003). Anatomical growth and development in hamadryas baboons (*Papio hamadryas*) compared with the closely related vervet monkeys (*Cercopithecus aethiops*). *Am. J. Phys. Anthropol.* 36(suppl.):70.

Bolter, D. R. (2004). Anatomical growth patterns in colobine monkeys and implications for primate evolution [PhD diss.]. University of California, Santa Cruz.

Bolter, D. R., and Zihlman, A. L. (2002). Growth and development in body tissues and proportions in African apes (*Gorilla gorilla* and *Pan troglodytes*): a preliminary report. *Am. J. Phys. Anthropol.* 34(suppl.):46.

Bolter, D. R., and Zihlman, A. L. (2003). Morphometric analysis of growth and development in wild-collected vervet monkeys (*Cercopithecus aethiops*) with implications for growth patterns in Old World monkeys, apes and humans. *J. Zool. Lond.* 260:99–110.

Bolwig, N. (1959a). A study of the behaviour of the chacma baboon. *Behaviour* 14:136–163.

Bolwig, N. (1959b). A study of the nests built by mountain gorilla and chimpanzee. *S. Afr. J. Sci.* 55:286–291.

Bond, A. B., Kamil, A. C., and Balda, R. P. (2003). Social complexity and transitive inference in corvids. *Anim Behav.* 65:479–487.

Bond, A. B., Kamil, A. C., and Balda, R. P. (2007). Serial reserval learning and the evolution of behavioral flexibility in three species of North American corvids (*Gymnorhinus cyanocephalus, Nucifraga columbiana, Aphelocoma californica*). *J. Comp. Psychol.* 121:372–379.

Bongers, F., Charles-Dominique, P., Forget, P.-M., and Théry, M. (2001). *Nouragues: Dynamics and Plant–Animal Interactions in a Neotropical Rainforest.* Kluwer Academic Publishers, Dordrecht.

Bonner, J. T. (1980). *The Evolution of Culture in Animals.* Princeton University Press, Princeton, NJ.

Bonney, R. C., and Kingsley, S. (1982). Endocrinology of pregnancy in the orang-utan (*Pongo pygmaeus*) and its evolutionary significance. *Int. J. Primatol.* 34:431–444.

Bonney, R. C., Dixson, A. F., and Fleming, D. (1979). Cyclic changes in the circulating and urinary levels of ovarian steroids in adult owl monkeys. *J. Reprod. Fertil.* 56:271–280.

Bonney, R. C., Dixson, A. F., and Fleming, D. (1980). Plasma concentrations of oestradiol-17-beta, estrone, progesterone and testosterone during the ovarian cycle of the owl monkey (*Aotus trivirgatus*). *J. Reprod. Fertil.* 60:101–107.

Bonnie, K. E., Horner, V., Whiten, A., and de Waal, F. B. M. (2006). Spread of arbitrary customs among chimpanzees: a controlled experiment. *Proc. R. Soc. B* 274:367–372.

Bonvicino, C. R. (1989). Ecologia e comportamento de *Alouatta belzebul* (Primates: Cebidae) na mata Atlântica. *Rev. Nordest. Biol.* 6:149–179.

Bonvicino, C. R., Boubli, J. P., Otazú, I. B., Almeida, F. C., Nascimento, F. F., Coura, J. R., and Seuánez, H. N. (2003). Morphologic, karyotypic, and molecular evidence of a new form of *Chiropotes* (Primates, Pitheciinae). *Am. J. Primatol.* 61:123–133.

Bookbinder, M. P., Dinerstein, E., Rijal, A., Cauley, H., and Rajouria, A. (1998). Ecotourism's support of biodiversity conservation. *Conserv. Biol.* 12:1399–1404.

Boonratana, R. (1994). The ecology and behaviour of the proboscis monkey (*Nasalis larvatus*) in the Lower Kinabatangan, Sabah [PhD diss.]. Mahidol University, Bangkok.

Boonratana, R., and Le, X. C. (1998). Preliminary observations of the ecology and behavior of Tonkin snub-nosed monkey (*Rhinopithecus [Presbytiscus] avunculus*) in northern Vietnam. In: Jablonski, N. G. (ed.), *The Natural History of the Doucs and Snub-Nosed Monkeys*. World Scientific Press, Singapore. pp. 207–216.

Boord, M. (1993). Myths of origin. In: Willis, R. (ed.), *World Mythology*. Henry Holt and Company, New York. pp. 102–109.

Booth, A. H. (1956). The distribution of primates in the Gold Coast. *J. W. Afr. Sci. Assoc.* 2:122–133.

Booth, A. H. (1957). Observations on the natural history of the olive colobus monkey, *Procolobus verus* (van Beneden). *Proc. Zool. Soc. Lond.* 129:421–430.

Booth, C. (1962). Some observations on behaviour of *Cercopithecus* monkeys. *Ann. N. Y. Acad. Sci.* 102:477–487.

Borgerhoff-Mulder, M. (2004). Are men and women really so different? *Trends Ecol. Evol.* 19:3–6.

Borries, C. B. (1997). Infanticide in seasonally breeding multimale groups of Hanuman langurs (*Presbytis entellus*) in Ramnagar (south Nepal). *Behav. Ecol. Sociobiol.* 41:139–150.

Borries, C. (2000). Male dispersal and mating influxes in Hanuman langurs living in multimale groups. In: Kappeler, P. M. (ed.), *Primate Males*. Cambridge University Press, Cambridge. pp. 146–158.

Borries, C., and Koenig, A. (2000). Infanticide in Hanuman langurs: social organization, male migration, and weaning age. In: van Schaik, C. P., and Janson, C. H. (eds.), *Infanticide by Males and Its Implications*. Cambridge University Press, Cambridge. pp. 99–122.

Borries, C., Sommer, V., and Srivastave, A. (1991). Dominance, age, and reproductive success in free-ranging female Hanuman langurs (*Presbytis entellus*). *Int. J. Primatol.* 12:231–258.

Borries, C. B., Sommer, V., and Srivastava, A. (1994). Weaving a tight social net: allogrooming in free-ranging female langurs (*Presbytis entellus*). *Int. J. Primatol.* 15:421–443.

Borries, C. B., Laumhardt, K., Epplen, C., Epplen, J. T., and Winkler, P. (1999a). Males as infant protectors in Hanuman langurs (*Presbytis entellus*) living in multimale groups— defense pattern, paternity and sexual behavior. *Behav. Ecol. Sociobiol.* 46:350–356.

Borries, C. B., Laumhardt, K., Epplen, C., Epplen, J. T., and Winkler, P. (1999b). DNA analyses support the hypothesis that infanticide is adaptive in langur monkeys. *Proc. R. Soc. Lond. B* 266:901–904.

Borries, C. B., Koenig, A., and Winkler, P. (2001). Variation of life history traits and mating patterns in female langur monkeys (*Semnopithecus entellus*). *Behav. Ecol. Sociobiol.* 50:391–402.

Boshoff, A. F., Palmer, N. G., Avery, G., Davies, R. A. G., and Jarvis, M. J. F. (1991). Biogeographical and topographical variation in the prey of the black eagle in the Cape Province, South Africa. *Ostrich* 62:59–72.

Bossuyt, F. (2002). Natal dispersal of titi monkeys (*Callicebus moloch*) at Cocha Cashu, Manu National Park, Peru. *Am. J. Phys. Anthropol.* 34(suppl):47.

Boubli, J. P. (1999). Feeding ecology of black-headed uacaris (*Cacajao melanocephalus melanocephalus*) in Pico de Neblina National Park, Brazil. *Int. J. Primatol.* 20:719–749.

Boubli, J. P. (2002). Western extension of the range of bearded sakis: a possible new taxon of *Chiropotes* sympatric with *Cacajao* in the Pico da Neblina National Park, Brazil. *Neotrop. Primates* 10:1–4.

Boubli, J. P. (2005). Floristics, primary productivity and primate diversity in Amazonia: contrasting a eutrophic várzea forest and an oligotrophic caatinga forest in Brazil. In: Dew, J. L., and Boubli, J. P. (eds.), *Tropical Fruits and Frugivores: The Search for Strong Interactors*. Springer, Dordrecht, the Netherlands, pp. 59–71.

Boubli, J. P., Silva, M. N. F., Amado M. V., Hrbek, T., and Pontual, F. (2008). A taxonomic reassessment of *Cacajao melanocephalus* Humboldt (1811), with the description of two new species. *Int. J. Primatol.* 29:723–741.

Boug, A., Biquand, S., Biquand-Guyot, V., and Kamal, K. (1994a). Home range and daily march of commensal *Papio hamadryas* in the Alhada Mountain of Saudi Arabia [abstract]. 15th Congress of the International Primatological Society, Bali, Indonesia Abstracts:148.

Boug, A., Biquand, S., Biquand-Guyot, V., and Kamal, K. (1994b). The response of commensal hamadryas baboons to seasonal rteduction in food provisioning. *Rev. Ecol. (Terre Vie)* 49:307–319.

Bourliere, F. (1964). *The Natural History of Mammals*, 3rd ed. Knopf, New York.

Bourlière, F., Hunkeler, C., and Bertrand, M. (1970). Ecology and behaviour of Lowe's guenon (*Cercopithecus cambelli lowei*) in the Ivory Coast. In: Napier, J. R., and Napier, P. H. (eds.), *Old World Monkeys: Evolution, Systematics, and Behaviour*. Academic Press, New York. pp. 397–350.

Bovet, D., and Washburn, D. A. (2003). Rhesus macaques (*Macaca mulatta*) categorize unknown conspecifics according to their dominance relations. *J. Comp. Psychol.* 117:400–405.

Bovine Genome Sequencing and Analysis Consortium (2009). The genome sequence of taurine cattle: a window to ruminant biology and evolution. *Science* 324:522–528.

Bowen-Jones, E. (1998). A review of the commercial bushmeat trade with emphasis on central/West Africa and the great apes. In: *Report for the Ape Alliance*. Fauna & Flora International, Cambridge.

Bowlby, J. (1969). *Attachment and Loss*, vol. 1. *Attachment*. Basic Books, New York.

Bowler, M., Barton, C., McCann-Wood, S., Puertas, P., and Bodmer, R. (2009). Annual variation in breeding success and changes in population density of *Cacajao calvus ucayalii* in the Lago Preto Conservation Concession, Peru. In: Barnett, A. A., Veiga, L. M., Ferrari, S. F., and Norconk, M. A. (eds.), *Evolutionary Biology and Conservation of Titis, Sakis, and Uacaris*. Cambridge University Press, Cambridge. In press.

Bowler, P. (1988). *The Non-Darwinian Revolution: Reinterpreting a Historical Myth*. Johns Hopkins University Press, Baltimore.

Box, H. O. (1984). *Primate Behaviour and Social Ecology*. Chapman and Hall, London.

Box, H. (1991). Training for life after release: simian primates as examples. In: Gipps, J. H. W. (ed.), *Beyond Captive Breeding: Reintroducing Endangered Mammals to the Wild. Symposium of the Zoological Society of London*. Clarendon Press, Oxford. pp. 111–123.

Box, H. O. (1994). Comparative perspectives in primate social learning: new lessons for old traditions. In: Roeder, J. J., Thierry, B., Anderson, J. B., and Herrenschmidt, N. (eds.), *Current Primatology. Social Development, Learning and Behaviour*, vol. 2. Université Louis Pasteur, Strasbourg. pp. 321–327.

Boyd, R., and Richerson, P. J. (1985). *Culture and the Evolutionary Process*. University of Chicago Press, Chicago.

Boyd, R., and Silk, J. B. (2003). *How Humans Evolved*. W. W. Norton, New York.

Boyer, D., Ramos-Fernandez, G., Miramontes, O., Mateos, J. L., Cocho, G., Larralde, H., Ramos, H., and Rojas, F. (2006). Scale-free foraging by primates emerges from their interaction with a complex environment. *Proc. R. Soc. B. Biol. Sci.* 273:1743–1750.

Boyle, S. (2008a). The effects of forest fragmentation on primates in the Brazilian Amazon [PhD diss.]. Arizona State University, Tempe.

Boyle, S. (2008b). Human impacts on primate conservation in central Amazonia. *Trop. Conserv. Sci.* 1:6–17.

Boyle, S. A., Lourenco, W. C., da Silva, L. R., and Smith, A. T. (2009a). Home range estimates vary with sample size and methods. *Folia Primatol.* 80:33–42.

Boyle, S. A., Lourenco, W. C., da Silva, L. R., and Smith, A. T. (2009b). Travel and spatial patterns changes when bearded sakis (*Chiropotes satanas chiropotes*) inhabit forest fragments. *Int. J. Primatol.* 30:515–531.

Boysen, S. T., and Bernston, G. G. (1995). Response to quantity: perceptual versus cognitive mechanisms in chimpanzees (*Pan troglodytes*). *J. Exp. Psychol. Anim. Behav. Proc.* 21:82–86.

Brack, M. (1987). *Agents Transmissible from Simians to Man*. Springer-Verlag, Berlin.

Bradbury, J. W., and Vehrencamp, S. L. (1998). *Principles of Animal Communication*. Sinauer Associates, Sunderland, MA.

Bradley, B. J. (1999). Levels of selection, altruism, and primate behavior. *Q. Rev. Biol.* 74:171–194.

Bradley, B. J. (2003). Molecular ecology of wild gorillas [PhD diss.]. State University of New York, Stony Brook.

Bradley, B. J., and Lawler, R. R. (in press). Primatology in the age of genomics. *Evol. Anthropol.* In press.

Bradley, B. J., Chambers, K. E., and Vigilant, L. (2001a). Accurate DNA-based sex identification of apes using non-invasive samples. *Conserv. Genet.* 2:179–181.

Bradley, B. J., Robbins, M. M., Williamson, E. A., Boesch., C., and Vigilant, L. (2001b). Assessing male reproductive success in wild mountain gorillas using DNA analysis. *Primate Rep.* 60:16–17.

Bradley, B. J., Doran-Sheehy, D. M., Lukas, D., Boesch, C., and Vigilant, L. (2004). Dispersed male networks in western gorillas. *Curr. Biol.* 14:510–514.

Bradley, B. J., Robbins, M. M., Williamson, E. A., Steklis, H. D., Gerald-Steklis, N., Eckhardt, N., Boesch, C., and Vigilant, L. (2005). Mountain gorilla tug-of-war: silverbacks have limited control over reproduction in multi-male groups. *Proc. Natl. Acad. Sci. USA* 102:9418–9423.

Bradley, B. J., Doran-Sheehy, D. M., and Vigilant, L. (2007). Potential for female kin associations in wild western gorillas despite female dispersal. *Proc. R. Soc. Lond. B* 274:2179–2185.

Bradley, M. H. (1922). *On the Gorilla Trail*. D. Appleton, London.

Brain, C. K. (1981). *The Hunters or the Hunted? An Introduction to African Cave Taphonomy*. University of Chicago Press, Chicago.

Brain, C. (1990). Spatial usage of a desert environment by baboons (*Papio ursinus*). *J. Arid Environ.* 18:67–73.

Brain, C. (1991). Activity of water-deprived baboons (*Papio ursinus*) during inter-troop encounters. *S. Afr. J. Sci.* 87:221–222.

Brain, C., and Mitchell, D. (1999). Body temperature changes in free-ranging baboons (*Papio hamadryas ursinus*) in the Namib Desert, Namibia. *Int. J. Primatol.* 20:585–598.

Brambilla, D. J., and McKinlay, S. M. (1989). A prospective study of factors affecting age at menopause. *J. Clin. Epidemiol.* 42:1031–1039.

Bramblett, C. A. (1980). Model for development of social behavior in vervet monkeys. *Dev. Psychol.* 13:205–223.

Bramblett, C. A., Pejaver, L. D., and Drickman, D. J. (1975). Reproduction in captive vervet and Sykes' monkeys. *J. Mammal.* 56:940–946.

Brandon-Jones, D. (1996). The Asian colobinae (Mammalia: Cercopithecidae) as indicators of Quaternary climatic change. *Biol. J. Linn. Soc.* 59:327–350.

Brandon-Jones, D., Eudey, A. A., Geissmann, T., Groves, C. P., Melnick, D. J., Morales, J. C., Shekelle, M., and Stewart, C. B. (2004). Asian primate classification. *Int. J. Primatol.* 25:97–164.

Brannon, E. M. (2006). The representation of numerical magnitude. *Curr Opin Neurobiol.* 16:222–229.

Brannon, E. M., and Terrace, H. (2000). Representation of the numerosities 1–9 by rhesus macaques (*Macaca mulatta*). *J. Exp. Psychol. Anim. Behav. Proc.* 26:31–49.

Brashares, J. S. (2003). Ecological, behavioral, and life-history correlates of mammalian extinctinons in West Africa. *Conserv. Biol.* 17:733–743.

Brauch, K., Pfefferle, D., Hodges, K., Möhle, U., Fischer, J., and Heistermann, M. (2007). Female sexual behavior and sexual swelling size as potential cues for males to discern the female fertile phase in free-ranging Barbary macaques (*Macaca sylvanus*) of Gibraltar. *Horm. Behav.* 52:375–383.

Brauch, K., Hodges, K., Engelhardt, A., Fuhrmann, K., Shaw, E., and Heistermann, M. (2008). Sex-specific reproductive behaviours and paternity in free-ranging Barbary macaques (*Macaca sylvanus*). *Behav. Ecol. Sociobiol.* 62:1453–1466.

Bräuer, J., Call, J., and Tomasello, M. (2005). All four great ape species follow gaze around barriers. *J. Comp. Psychol.* 119:145–154.

Bräuer, J., Call, J., and Tomasello, M. (2008). Chimpanzees do not take into account what others can hear in a competitive situation. *Anim. Cogn.* 11:175–178.

Braune, P., Schmidt, S., and Zimmermann, E. (2005). Spacing and group coordination in a nocturnal primate, the golden brown mouse lemur (*Microcebus ravelobensis*): the role of olfactory and acoustic signals. *Behav. Ecol. Sociobiol.* 58:587–596.

Bravo, S. P., and Sallenave, A. (2003). Foraging behavior and activity patterns of *Alouatta caraya* in the northeastern Argentinean flooded forest. *Int. J. Primatol.* 24:825–846.

Bray, M. S., Boerwinkle, E., and Doris, P. A. (2001). High-throughput multiplex SNP genotyping with MALDI-TOF mass spectrometry: practice, problems and promise. *Hum. Mutat.* 17:296–304.

Braza, F., Alvarez, F., and Azcarate, T. (1981). Behaviour of the red howler monkey (*Alouatta seniculus*) in the llanos of Venezuela. *Primates* 22:459–473.

Braza, F., Alvarez, F., and Azcarate, T. (1983). Feeding habits of the red howler monkey (*Alouattta seniculus*) in the llanos of Venezuela. *Mammalia* 42:205–215.

Brennan, E. J., Else, J. G., and Altmann, J. (1985). Ecology and behaviour of a pest primate: vervet monkeys in a tourist-lodge habitat. *Afr. J. Ecol.* 23:35–44.

Brent, L. J. N., Teichroeb, J. A., and Sicotte, P. (2008). Preliminary assessment of natal attraction and infant handling in wild *Colobus vellerosus*. *Am. J. Primatol.* 70:101–105.

Brereton, A. (1981). Intertroop consorting by a free-ranging female rhesus monkey. *Primates* 22:417–423.

Brett, F. L., Jolly, C. J., Socha, W., and Wiener, A. S. (1976). Human-like ABO blood groups in wild Ethiopian baboons. *Ybk. Phys. Anthropol.* 20:276–289.

Breuer, T., Hockemba, M., Olejniczak, C., Parnell, R. J., and Stokes, E. J. (2009). Physical maturation, life-history classes and age estimates of free-ranging western gorillas—insights from Mbeli Bai, Republic of Congo. *Am. J. Primatol.* 71:106–119.

Brewer, S. M., and McGrew, W. C. (1990). Chimpanzee use of a tool-set to get set honey. *Folia Primatol.* 54:100–104.

Brewer, S. W. (2001). Predation and dispersal of large and small seeds of a tropical palm. *Oikos* 92:245–255.

Brewer, S. W., and Webb, M. A. H. (2001). Ignorant seed predators and factors affecting the seed survival of a tropical palm. *Oikos* 93:32–41.

Bribiescas, R. G. (2006). *Men: Evolutionary and Life History*. Harvard University Press, Cambridge, MA.

Brick, O. (1998). Fighting behaviour, vigilance and predation risk in the cichlid fish, *Namacara anomala*. *Anim. Behav.* 56:309–317.

Bricknell, S. J. (1999). Hybridisation and behavioral variation: a socio-ecological study of hybrid gibbons (*Hylobates agilis albibarbis* × *H. muelleri*) in central Kalimantan, Indonesia [PhD diss.]. Australian National University, Canberra.

Brigham, C. C. (1923). *A Study of American Intelligence*. Princeton University Press, Princeton, NJ.

Bril, B., Dietrich, G., Foucart, J., Fuwa, K., and Hirata, S. (2009). Tool use as a way to assess cognition: how do captive chimpanzees handle the weight of the hammer when cracking a nut? *Anim. Cogn.* 11:175–178.

Brinkman-van der Linden, E. C., Sjoberg, E. R., Juneja, L. R., Crocker, P. R., Varki, N., and Varki, A. (2000). Loss of *N*-glycolylneuraminic acid in human evolution. Implications for sialic acid recognition by siglecs. *J. Biol. Chem.* 275:8633–8640.

Britt, A. (2000). Diet and feeding behaviour of the black-and-white ruffed lemur (*Varecia variegata variegata*) in the Betampona Reserve, eastern Madagascar. *Folia Primatol.* 71:133–141.

Britt, A., Randriamandatonirina, N. J., Glasscock, K. D., and Iambana, B. R. (2001a). Diet and feeding behaviour of *Indri indri* in a low-altitude rain forest. *Folia Primatol.* 73:225–239.

Britt, A., Welch, C., and Katz, A. (2001b). The impact of *Cryptoprocta ferox* on the *Varecia variegata* reinforcement project at Betampona. *Lemur News* 6:35–37.

Brizzee, K., and Dunlap, W. (1986). Growth. In: Dukelow, W., and Erwin, J. (eds.), *Comparative Primate Biology: Reproduction and Development*. Alan R. Liss, New York. pp. 363–413.

Broadfield, D. C., Delson, E., and Atsalis, S. (1994). Cercopithecid fossils from the later Pleistocene of Taung, South Africa [abstract]. *Am. J. Phys. Anthropol.* (suppl. 18):59–60.

Brockelman, W.Y. (2009). Ecology and the social system of gibbons. In: Lappan, S., and Whitttaker, D.J. (eds.), The Gibbons: New Perspectibes on Small Ape Socioecology and Population Biology. Springer, New York. Pp 211–240.

Brockelman, W. Y. (2004). Inheritance and selective effects of color phase in white-handed gibbons (*Hylobates lar*) in central Thailand. *Mamm. Biol.* 69:73–80.

Brockelman, W. Y., and Gittins, S. P. (1984). Natural hybridization in the *Hylobates lar* species group: implications for speciation in gibbons. In: Preuschoft, H., Chivers, D. J., Brockelman, W. Y., and Creel, N. (eds.), *The Lesser Apes: Evolutionary and Behavioural Biology*. Edinburgh University Press, Edinburgh. pp. 498–532.

Brockelman, W. Y., and Schilling, D. (1984). Inheritance of stereotyped gibbon calls. *Nature* 312:634–636.

Brockelman, W. Y., and Srikosamatara, S. (1984). Maintenance and evolution of social structure in gibbons. In: Preuschoft, H., Chivers, D. J., Brockelman, W. Y., and Creel, N. (eds.), *The Lesser Apes: Evolutionary and Behavioural Biology.* Edinburgh University Press, Edinburgh. pp. 298–323.

Brockelman, W., and Srikosamatara, S. (1993). Estimation of density of gibbon groups by use of loud songs. *Am. J. Primatol.* 29:93–108.

Brockelman, W. Y., Reichard, U., Treesucon, U., and Raemaekers, J. J. (1998). Dispersal, pair formation and social structure in gibbons (*Hylobates lar*). *Behav. Ecol. Sociobiol.* 42:329–339.

Brockett, R. C., Horwich, R. H., and Jones, C. B. (1999). Disappearance of infants following male takeovers in the Belizean black howler monkey (*Alouatta pigra*). *Neotrop. Primates* 7:86–88.

Brockett, R. C., Horwich, R. H., and Jones, C. B. (2000a). Female dispersal in the Belizean black howling monkey (*Alouatta pigra*). *Neotrop. Primates* 8:32–34.

Brockett, R. C., Horwich, R. H., and Jones, C. B. (2000b). Reproductive seasonality in the Belizean black howling monkey (*Alouatta pigra*). *Neotrop. Primates* 8:136–138.

Brockman, D. K. (1999). Reproductive behavior of female *Propithecus verreauxi* at Beza Mahafaly, Madagascar. *Int. J. Primatol.* 20:375–398.

Brockman, D. K. (2003). *Polyboroides radiatus* predation attempts on *Propithecus verreauxi. Folia Primatol.* 74:71–74.

Brockman, D. K., and Whitten, P. L. (1996). Reproduction in free-ranging *Propithecus verreauxi:* estrus and the relationship between multiple partner matings and fertilization. *Am. J. Phys. Anthropol.* 100:57–69.

Brockman, D. K., Whitten, P. L., Richard, A. F., and Schneider, A. (1998). Reproduction in free-ranging male *Propithecus verreauxi:* the hormonal correlates of mating and aggression. *Am. J. Phys. Anthropol.* 105:137–151.

Brockman, D. K., Godfrey, L. R., Dollar, L. J., and Ratsirarson, J. (2008). Evidence of invasive *Felis silvestris* predation on *Propithecus verreauxi* at Beza Mahafaly Special Reserve, Madagascar. *Int. J. Primatol.* 29:135–152.

Bromage, T., and Dean, M. C. (1985). Re-evaluation of the age at death of immature fossil hominids. *Nature* 317:525–527.

Bronikowski, A. M., and Altmann, J. (1996). Foraging in a variable environment: weather patterns and the behavioral ecology of baboons. *Behav. Ecol. Sociobiol.* 39:11–25.

Bronson, F. H., and Heideman, P. D. (1988). Seasonal regulation of reproduction in mammals. In: Knobil, E., and Neill, J. D. (eds.), *The Physiology of Reproduction.* Raven Press, New York. pp. 541–582.

Brook, B. W., O'Grady, J. J., Chapman, A. P., Burgman, M. A., Akçakayas, H. R., and Frankham, R. (2000). Predictive accuracy of population viability analysis in conservation biology. *Nature* 404:385–387.

Brook, B. W., Traill, L. W., and Bradshaw, C. J. (2006). Minimum viable population sizes and global extinction risk are unrelated. *Ecol. Lett.* 9:375–82.

Brooks, D. M. (1996). Some observations of primates in Paraguay. *Neotrop. Primates* 4:15–19.

Brooks, R. (2000). Negative genetic correlation between male sexual attractiveness and survival. *Nature* 406:67–70.

Brotherton, P. N. M., and Komers, P. E. (2003). Mate guarding and the evolution of social monogamy in mammals. In: Reichard, U. H., and Boesch, C. (eds.), *Monogamy: Mating Strategies and Partnerships in Birds, Humans, and Other Mammals.* Cambridge University Press, Cambridge. pp. 42–58.

Brower, J., Zar, J., and von Ende, C. (1990). *Field and Laboratory Methods for General Ecology.* Wm. C. Brown Publishers, Dubuque.

Brown, A. D., and Zunino, G. E. (1994). Hábitat, densidad y problemas de conservación de los primates en Argentina. *Vida Silv. Neotrop.* 3:30–40.

Brown, A. K., and Brown, P. (2005). Ultra low-power GPS recorder (TrackTag®). In: Proceedings of the Institute of Navigation 61st Annual Meeting, Cambridge, MA (June 2005). pp. 726–732.

Brown, C. H., and Gomez, R. (1992). Functional design features in primate vocal signals: the acoustic habitat and sound distortion. In: Nishida, T., McGrew, W. C., Marler, P., Pickford, M., and de Waal, F. B. M. (eds.), *Topics in Primatology. Human Origins,* vol. 1. University of Tokyo Press, Tokyo. pp. 177–198.

Brown, C. H., and Waser, P. M. (1988). Environmental influences on the structure of primate vocalizations. In: Todt, D., Goedeking, P., and Symmes, D. (eds.), *Primate Vocal Communication.* Springer-Verlag, Berlin. pp. 51–68.

Brown, C. H., Gomez, R., and Waser, P. M. (1995). Old World monkey vocalizations: adaptation to the local habitat? *Anim. Behav.* 50:945–961.

Brown, J. L. (1997). A theory of mate choice based on heterozygosity. *Behav. Ecol.* 8:60–65.

Brown, L. H. (1962). The prey of the crowned hawk eagle *Stephanoaetus coronatus* in central Kenya. *Scopus* 6:91–94.

Brown, L. H. (1971). The relations of the crowned eagle, *Stephanoaetus coronatus* and some of its prey animals. *Ibis* 113:240–243.

Browne, D. (2004). Do dolphins know their own minds? *Biol. Philos.* 19:633–53.

Bruce, E., and Ayala, F. (1979). Phylogenetic relationships between man and the apes: electrophoretic evidence. *Evol. Hum. Behav.* 33:1040–1056.

Bruford, M. W., Hanotte, O., and Burke, T. (1998). Multi- and single-locus DNA fingerprinting. In: Hoelzel, A. R. (ed.), *Molecular Genetic Analysis of Populations—A Practical Approach,* 2nd ed. ILR Press, Oxford. pp. 225–269.

Brugière, D., Gautier, J.-P., Moungazi, A., and Gautier-Hion, A. (2002). Primate diet and biomass in relation to vegetation composition and fruiting phenology in a rain forest in Gabon. *Int. J. Primatol.* 23:999–1024.

Brumback, R. A. (1973). Two distinctive types of owl monkeys (*Aotus*). *J. Med. Primatol.* 2:284–289.

Brumback, R. A. (1974). A third species of the owl monkey (*Aotus*). *J. Hered.* 65:321–323.

Brumback, R. A., Staton, R. D., Benjamin, S.A., and Lang, C. M. (1971). The chromosomes of *Aotus trivirgatus* Humboldt 1812. *Folia Primatol.* 15:264–273.

Brumloop, A., Homburg, I., Peetz, A., and Riehl, R. (1994). Gular scent glands in adult female white-faced saki, *Pithecia pithecia pithecia,* and field observations on scent-marking behaviour. *Folia Primatol.* 63:212–215.

Bruner, A. G., Gullison, R. E., Rice, R. E., and Fonseca, G. A. B. (2001). Effectiveness of parks in protecting tropical biodiversity. *Science* 291:125–128.

Bruner, E., and Cucina, A. (2005). *Alouatta, Ateles,* and the ancient mesoamerican cultures. *J. Anthropol. Sci.* 83:111–117.

Brunvand, J. H. (1996). Folklore. In: Brunvand, J. H. (ed.), *American Folklore: An Encylopedia.* Garland Publishing, New York. pp. 285–287.

Brush, J. A., and Norconk, M. A. (1999). Early behavioral development in a wild white-faced saki monkey (*Pithecia pithecia*). *Am. J. Phys. Anthropol.* 28(suppl.):99.

Bshary, R., and Noë, R. (1997a). Anti-predation behaviour of red colobus monkeys in the presence of chimpanzees. *Behav. Ecol. Sociobiol.* 41:321–333.

Bshary, R., and Noë, R. (1997b). Red colobus and diana monkeys provide mutual protection against predators. *Anim. Behav.* 54:1461–1474.

Bshary, R., and Würth, M. (2001). Cleaner fish (*Labroides dimidiatus*) manipulate client reef fish providing tactile stimulation. *Proc. R. Soc. Lond. B* 286:1495–1501.

Buchan, J., Alberts, S. C., Silk, J. B., and Altmann, J. (2003). True paternal care in a multi-male primate society. *Nature* 425:179–181.

Buchanan-Smith, H. M. (1990). Polyspecific association of two tamarin species, *Saguinus labiatus* and *Saguinus fuscicollis,* in Bolivia. *Am. J. Primatol.* 22:205–214.

Buchanan-Smith, H. M., Hardie, S. M., Caceres, C., and Prescott, M. J. (2000). Distribution and forest utilization of *Saguinus* and other primates of the Pando department, northern Bolivia. *Int. J. Primatol.* 21:353–379.

Buck, M. R. (1998). Female mate choice in sooty mangabeys: social constraints on mating behavior [PhD diss.]. Emory University, Atlanta, GA.

Buckley, J. S. (1983). The feeding behavior, social behavior, and ecology of the white-faced monkey *Cebus capucinus* at Trujillo, northern Honduras, Central America [PhD diss.]. University of Illinois, Urbana.

Budnitz, N. (1978). Feeding behavior of *Lemur catta* in different habitats. In: Bateson, P. P. G., and Klopfer, P. H. (eds.), *Perspectives in Ethology. Social Behavior,* vol. 3. Plenum Press, New York. pp. 85–108.

Budnitz, N., and Dainis, K. (1975). *Lemur catta:* ecology and behavior. In: Tattersall, I., and Sussman, R. W. (eds.), *Lemur Biology,* Plenum Press, New York. pp. 219–236.

Buettner-Janusch, J. (ed.) (1962). The relatives of man: modern studies on the relation of the evolution of nonhuman primates to human evolution. *Ann. N. Y. Acad. Sci.* 102:108–514.

Buffon, G. (1765). *Histoire Naturelle, Generale et Particuliere,* vol. 13. L'Imprimerie du Roi, Paris. pp. 87–91.

Buij, R., Singleton, I., Krakauer, E., and van Schaik, C. P. (2003). Rapid assessment of orangutan density. *Biol. Conserv.* 114:103–113.

Bulger, J. G. (1993). Dominance rank and access to estrous females in male savanna baboons. *Behaviour* 127:67–103.

Bulger, J., and Hamilton, W. J. I. (1987). Rank and density correlates of inclusive fitness measures in a natural chacma baboon (*Papio ursinus*) troop. *Int. J. Primatol.* 8:635–650.

Bullock, D. W., Paris, C. A., and Goy, R. W. (1972). Sexual behavior, swelling of the sex skin and plasma progesterone in the pigtail macaque. *J. Reprod. Fertil.* 31:225–236.

Bunn, H. T., and Kroll, E. M. (1986). Systematic butchery by Plio-Pleistocene hominids at Olduvai Gorge, Tanzania. *Curr. Anthropol.* 27:431–452.

Burity, C., da Cruz, L., Rocha, V., da Conceicao, N., da Luz, D., da Silva Santos, D., Campos, D., and Pissinatti, A. (2007). Golden lion tamarins, *Leontopithecus rosalia* (Linnaeus, 1766) in the Taquara Municipal Natural Park (Duque de Caxias, RJ): a southern extension of the known range. *Neotrop. Primates* 14:30–31.

Burke, T., Dolf, G., Jeffreys, A. J., and Wolff, R. (eds.) (1991). *DNA Fingerprinting: Approaches and Applications.* Birkhäuser Verlag, Basel.

Burkill, H. M. (1985). *The Useful Plants of West Tropical Africa,* 2nd ed., vol. 1. Royal Botanical Gardens, Kew.

Burley, N. (1979). The evolution of concealed ovulation. *Am. Nat.* 114:835–858.

Burley, N. T., and Symanski, R. (1998). "A taste for the beautiful": latent aesthetic mate preferences for white crests in two species of Australian grassfinches. *Am. Nat.* 152:792–802.

Burney, D. A. (1997). Theories and facts regarding Holocene environmental change before and after human colonization. In: Goodman, S. M., and Patterson, B. D. (eds.), *Natural Change and Human Impact in Madagascar.* Smithsonian Institution Press, Washington DC. pp. 75–89.

Burney, D. A. (2002). Sifaka predation by a large boa. *Folia Primatol.* 73:144–145.

Burrell, A. S. (2009). Phylogenetics and population genetics of central African baboons [PhD diss.]. New York University, New York.

Burrell, A. S., Jolly, C. J., Tosi, A. J., and Disotell, T. R. (2009). Mitochondrial evidence for the hybrid origin of the kipunji, *Rungwecebus kipunji* (Primates: Papionini). *Mol. Phylogen. Evol.* doi:10.1016/j.ympev.2009.02.004.

Burrows, W., and Ryder, O. A. (1997). Y-chromosome variation in great apes. *Nature* 385:125–126.

Burt, A. (1992). "Concealed ovulation" and sexual signals in primates. *Folia Primatol.* 58:1–6.

Burt, W. (1943). Territoriality and home range concepts as applied to mammals. *J. Mammal.* 24:346–352.

Burton, F. D. (2002). Monkey king in China: basis for a conservation policy. In: Fuentes, A., and Wolfe, L. D. (eds.), *Primates Face to Face.* Cambridge University Press, Cambridge.

Burton, F., and Carroll, A. (2005). By-product mutualism: conservation implications amongst monkeys, figs, humans, and their domesticants in Honduras. In: Paterson, J. (ed.), *Commensalism and Conflict: The Human–Primate Interface.* American Society of Primatology, Norman, OK. pp. 24–39.

Burton, F. D., and Fukuda, F. (1981). On female mobility: the case of the Yugawara-T group of *Macaca fuscata. J. Hum. Evol.* 10:381–386.

Burton, F. D., Snarr, K. A., and Harrison, S. E. (1995). Preliminary report on *Presbytis francoisi leucocephalus. Int. J. Primatol.* 16:311–327.

Bush, E. C., and Allman, J. M. (2004). The scaling of frontal cortex in primates and carnivores. *Proc. Natl. Acad. Sci. USA* 101:3962–3966.

Buskirk, W. H., Buskirk, R. E., and Hamilton, W. J. III (1974). Troop-mobilizing behavior of adult male chacma baboons. *Folia Primatol.* 22:9–18.

Buss, D. H., and Cooper, R. W. (1970). Composition of milk from talapoin monkeys. *Folia Primatol.* 13:196–206.

Busse, C. D. (1977). Chimpanzee predation as a possible factor in the evolution of red colobus monkey social organization. *Evolution* 31:907–911.

Busse, C. D. (1980). Leopard and lion predation upon chacma baboons living in the Moremi Wildlife Reserve. *Botswana Notes Rec.* 12:15–20.

Busse, C. D. (1984). Spatial sructure of chacma baboon groups. *Int. J. Primatol.* 5:247–261.

Busse, C. D., and Estep, D. Q. (1984). Sexual arousal in male pigtailed monkeys (*Macaca nemestrina*); effects of serial matings by two males. *J. Comp. Psychol.* 98:227–231.

Busse, C., and Hamilton, W. J. (1981). Infant carrying by male chacma baboons. *Science* 212:1281–1283.

Butler, H. (1974). Evolutionary trends in primate sex cycles. In: *Contributions to Primatology,* vol. 3. Karger, Basel. pp. 2–35.

Butovskaya, M. (1993). Kinship and different dominance styles in groups of three species of the genus *Macaca* (*M. arctoides, M. mulatta, M. fascicularis*). *Folia Primatol.* 60:210–224.

Butovskaya, M. (2008). Reconciliation, dominance and cortisol levels in children and adolescents (7–15-year-old boys). *Behaviour* 145:1557–1576.

Butovskaya, M., Kozintsev, A., and Welker, C. (1996). Conflict and reconciliation in two groups of crab-eating monkeys differing in social status by birth. *Primates* 37:261–270.

Buttelmann, D., Call, J., and Tomasello, M. (2008a). Behavioral cues that great apes use to forage for hidden food. *Anim. Cogn.* 11:117–128.

Buttelmann, D., Carpenter, M., Call, J., and Tomasello, M. (2008b). To go or not to go: chimpanzees' understanding of a human's goals. *Int. J. Psychol.* 43:87–87.

Buttelmann, D., Carptenter, M., Call, J., and Tomasello, M. (2008c). Rational tool use and tool choice in human infants and great apes. *Child Dev.* 79:609–626.

Butynski, T. M. (1982). Vertebrate predation by primates: a review of hunting patterns and prey. *J. Hum. Evol.* 11:421–430.

Butynski, T. M. (1988). Guenon birth seasons and correlates with rainfall and food. In: Gautier-Hion, A., Bourlière, F., Gautier, J. P., and Kingdon, J. (eds.), *A Primate Radiation: Evolutionary Biology of the African Guenons.* Cambridge University Press, New York. pp. 284–322.

Butynski, T. M. (1990). Comparative ecology of blue monkeys (*Cercopithecus mitis*) in high- and low-density subpopulations. *Ecol. Monogr.* 60:1–26.

Butynski, T. M. (1996/1997). African primate conservation—the species and the IUCN/SSC primate specialist group network. *Primate Cons.* 17:87–100.

Butynski, T. M. (2001). Africa's great apes. In: Beck, B. B., Stoinski, T. S., Hutchins, M., Maple, T. L., Norton, B., Rowan, A., Stevesn, E. F., and Arluke, A. (eds.), *Great Apes & Humans: The Ethics of Coexistence.* Smithsonian Institute Press, Washington DC. pp. 3–56.

Butynski, T. M. (2002). The guenons: an overview of diversity and taxonomy. In: Glenn, M. E., and Cords, M. (eds.), *The Guenons: Diversity and Adaptation in African Monkeys.* Kluwer Academic/Plenum Publishers, New York. pp. 3–13.

Butynski, T. (2003). The robust chimpanzee, *Pan troglodytes:* taxonomy, distribution, abundance and conservation status. In: Kormos, R., Boesch, C., Bakarr, M. I., and Butynski, T. (eds.), *West African Chimpanzees. Status Survey and Conservation Action Plan.* IUCN/SSC Primate Specialist Group, Gland, Switzerland. pp. 31–39.

Butynski, T. H. (In press). The white-naped mangabey (*Cercocebus lunulatus*). In: Kingdon, J., Happold, D., and Butynski, T. (eds.), *The Mammals of Africa.* University of California Press, Berkeley, CA.

Butynski, T. M., and de Jong, Y. A. (2004). Natural history of the Somali lesser galago (*Galago gallarum*). *J. East Afr. Nat. Hist.* 93:23–38.

Butynski, T. M., and de Jong, Y. A. (2007). Distribution of the potto *Perodicticus potto* (Primates: Lorisidae) in Eastern Africa, with a description of a new subspecies from Mount Kenya. *J. East Afr. Nat. Hist.* 96:113–147.

Butynski, T. M., and Kalina, J. (1998). Gorilla tourism: a critical look. In: Milner-Gulland, E. J., and Mace, R. (eds.), *Conservation of Biological Resources.* Blackwell Science, Oxford. pp. 280–300.

Butynski, T. M., and Sanderson, J. G. (2007). A rapid survey of the primates and large mammals of Lokuto. *RAP Bull. Biol. Assess.* 46:47–51.

Butynski, T. M., Ehardt, C. L., and Struhsaker, T. T. (1998). Notes on two dwarf galagos (*Galagoides udzungwensis* and *Galagoides orinus*) in the Udzungwa Mountains, Tanzania. *Primate Cons.* 18:69–75.

Butynski, T. M., de Jong, Y. A., Perkin, A. W., Bearder, S. K., and Honess, P. E. (2006). Taxonomy, distribution, and conservaton status of three species of dwarf galagos (Galagoides) in Eastern Africa. *Primate Cons.* 21:63–79.

Butynski, T. M., Gippoliti, S., Kingdon, J., and de Jong, Y. (2008). *Chlorocebus djamdjamensis.* In: IUCN 2009. IUCN Red List of Threatened Species, www.iucnredlist.org (accessed June 23, 2009).

Buzzard, P. J. (2006). Ecological partitioning of *Cercopithecus campbelli, C. petaurista,* and *C. diana* in the Tai Forest. *Int. J. Primatol.* 27:529–558.

Buzzard, P., and Eckhardt, W. (2007) The social systems of the guenons. In: McGraw, W. S., Zuberbuhler, K., and Noe, R. (eds.), *Monkeys of the Tai Forest.* Cambridge University Press, New York. pp. 51–71.

Buzzell, C. A., and Brush, J. A. (2000). Ontogeny of independence in wild and captive white-faced saki monkeys (*Pithecia pithecia*). *Am. J. Primatol.* 51(suppl. 1):49–50.

Byers, J. A. (1997). *American Pronghorn: Social Adaptations and the Ghosts of Predators Past.* University of Chicago Press, Chicago.

Bygott, J. D. (1974). Agonistic behaviour and dominance in wild chimpanzees [PhD diss.]. Cambridge University, Cambridge.

Bygott, J. D. (1979). Agonistic behavior, dominance, and social structure in wild chimpanzees of the Gombe National Park. In: Hamburg, D. A., and

McCown, E. R. (eds.), *The Great Apes*. Benjamin/Cummings, Menlo Park, CA. pp. 405–427.

Bynum, E. L., Bynum, D. Z., and Supriatna, J. (1997). Confirmation and location of the hybrid zone between wild populations of *Macaca tonkeana* and *Macaca hecki* in central Sulawesi, Indonesia. *Am. J. Primatol.* 43:181–209.

Byrne, G., and Suomi, S. J. (1995). Development of activity patterns, social interactions, and exploratory behavior in infant tufted capuchins (*Cebus apella*). *Am. J. Primatol.* 35:255–270.

Byrne, G., and Suomi, S. J. (1998). Relationship of early infant state measures to behavior over the first year of life in the tufted capuchin monkey (*Cebus apella*). *Am. J. Primatol.* 44:43–56.

Byrne, J. M. (1993). Evolving a theory of mind: the nature of non-verbal mentalism in other primates. In: Baron-Cohen, S., Tager Flusberg, H., and Cohen, D. J. (eds.), *Understanding Other Minds*. Oxford University Press, Oxford. pp. 367–396.

Byrne, J. M. (1994). The evolution of intelligence. In: Slater, P. J. B., and Halliday, T. R. (eds.), *Behaviour and Evolution*. Cambridge University Press, Cambridge. pp. 223–265.

Byrne, J. M. (1996). The misunderstood ape: cognitive skills of the gorilla. In: Russon, A. E., Bard, K. A., and Parker, S. T. (eds.), *Reaching into Thought: The Minds of the Great Apes*. Cambridge University Press, Cambridge. pp. 111–130.

Byrne, J. M., and Whiten, A. (1992). Cognitive evolution in primates: evidence from tactical deception. *Man* 27:609–627.

Byrne, R. W. (2002). Imitation of novel complex actions: what does the evidence from animals mean? *Adv. Study Behav.* 31:77–105.

Byrne, R. W., and Byrne, J. M. E. (1993). Complex leaf-gathering skills of mountain gorillas (*Gorilla g. beringei*): variability and standardization. *Am. J. Primatol.* 31:241–261.

Byrne, R. W., and Russon, A. E. (1998). Learning by imitation: a hierarchical approach. *Behav. Brain Sci.* 21:667–721.

Byrne, R. W., and Whiten, A. (eds.) (1988a). *Machiavellian Intelligence: Social Expertise and the Evolution of Intellect in Monkeys, Apes and Humans*. Clarendon Press, Oxford.

Byrne, R. W., and Whiten, A. (1988b). Tactical deception of familiar individuals in baboons. In: Byrne, R. W., and Whiten, A. (eds.), *Machiavellian Intelligence: Social Expertise and the Evolution of Intellect in Monkeys, Apes and Humans*. Clarendon Press, Oxford. pp. 201–210.

Byrne, R. W. and Whiten, A. (1990). Tactical deception in primates: the 1990 database. *Primate Rep.* 27:1–101.

Byrne, R. W., and Whiten, A. (1991). Computation and mindreading in primate tactical deception. In: Whiten, A. (ed.), *Natural Theories of Mind*. Basil Blackwell, Oxford. pp. 127–142.

Byrne, R. W., Whiten, A., and Henzi, S. P. (1987). One-male groups and intergroup interactions of mountain baboons. *Int. J. Primatol.* 8:615–633.

Byrne, R. W., White, A., and Henzi, S. P. (1989). Social relationships of mountain baboons: leadership and affiliation in a non-female-bonded monkey. *Am. J. Primatol.* 18:191–207.

Byrne, R. W., Whiten, A., Henzi, S. P., and McCulloch, F. M. (1993). Nutritional constraints on mountain baboons (*Papio ursinus*): implications for baboon socioecology. *Behav. Ecol. Sociobiol.* 33:233–246.

Byrne, R. W., Corp, N., and Byrne, J. M. E. (2001). Estimating the complexity of animal behaviour: how mountain gorillas eat thistles. *Behaviour* 138:525–557.

Cabrera, A. (1958). Catálogo de los mamíferos de America del Sur. *Rev. Museo Arg. Cien. Nat. Bern. Riv.* 26:1–307.

Caccone, A., and Powell, J. R. (1989). DNA divergence among hominoids. *Evolution* 43:925–942.

Cáceres, M., Lachuer, J., Zapala, M. A., Redmond, J. C., Kudo, L., Geschwind, D. H., Lockhart, D. J., Preuss, T. M., and Barlow, C. (2003). Elevated gene expression levels distinguish human from non-human primate brains. *Proc. Natl. Acad. Sci. USA* 100:13030–13035.

Caillaud, D., Levréro, F., Cristescu, R., Gatti, S., Dewas, M., Douadi, M., Gautier-Hion, A., Raymond, M., and Ménard, N. (2006). Gorilla susceptibility to Ebola virus: the cost of sociality. *Curr. Biol.* 16:489–491.

Caillaud, D., Levréro, F., Gatti, S., Ménard, N., and Raymond, M. (2008). Influence of male morphology on male mating status and behavior during interunit encounters in western lowland gorillas. *Am. J. Phys. Anthropol.* 135:379–388.

Cain, M. L., Milligan, B. G., and Strand, A. E. (2000). Long-distance seed dispersal in plant populations. *Am. J. Bot.* 87:1217–1227.

Cain, M. L., Nathan, R., and Levin, S. A. (2003). Long-distance dispersal. *Ecology* 84:1943–1944.

Caine, N. G. (1987). Vigilance, vocalizations, and cryptic behavior at retirement in captive groups of red-bellied tamarins (*Saguinus labiatus*). *Am. J. Primatol.* 12:241–250.

Caine, N. G. (1993). Flexibility and co-operation as unifying themes in *Saguinus* social organization and behaviour: the role of predation pressures. In: Rylands, A. B. (ed.), *Marmosets and Tamarins: Systematics, Behaviour, and Ecology*. Oxford University Press, Oxford. pp. 200–219.

Caine, N. G., and Mitchell, G. (1980). Species differences in the interest shown in infants by juvenile female macaques (*Macaca radiata* and *M. mulatta*). *Int. J. Primatol.* 1:323–332.

Caine, N. G., Potter, M. P., and Mayer, K. E. (1992). Sleeping site selection by captive tamarins (*Saguinus labiatus*). *Ethology* 90:63–71.

Caldecott, J. O. (1986a). *An Ecological and Behavioral Study of the Pig-tailed Macaque*. Karger, Basel.

Caldecott, J. O. (1986b). Mating patterns, societies and the ecogeography of macaques. *Anim. Behav.* 34:208–220.

Caldecott, J., and Miles, L. (2005). *The World Atlas of Great Apes and their Conservation*. University of California Press, Berkeley.

Caldwell, C. A., and Whiten, A. (2002). Evolutionary perspectives on imitation: is a comparative psychology of social learning possible? *Anim. Cogn.* 5:193–208.

Caldwell, C. A., and Whiten, A. (2004). Testing for social learning and imitation in common marmosets, *Callithrix jacchus,* using an "artificial fruit." *Anim. Cogn.* 7:77–85.

Calegaro-Marques, C., and Bicca-Marques, J. C. (1993). Reprodução de *Alouatta caraya* Humboldt, 1812 (Primates, Cebidae). *Primatol. Brasil* 4:51–66.

Calegaro-Marques, C., and Bicca-Marques, J. C. (1996). Emigration in a black howling monkey group. *Int. J. Primatol.* 17:229–237.

Caljan, V. G., Meisvili, N. V., and Vancatova, M. A. (1987). Sexual behaviour of hamadryas baboons. *Anthropologie* 25:183–187.

Call, J. (1999). The effect of inter-opponent distance on the occurrence of reconciliation in stumptail (*Macaca arctoides*) and rhesus macaques (*Macaca mulatta*). *Primates* 40:515–523.

Call, J. (2000). Distance regulation in macaques: a form of implicit reconciliation? In: Aureli, F., and de Waal, F. B. M. (eds.), *Natural Conflict Resolution*. University of California Press, Berkeley. pp. 191–193.

Call, J. (2006). Inferences by exclusion in the great apes: the effect of age and species. *Anim. Cogn.* 9:393–403.

Call, J. (In Press). Trapping the minds of apes: causal knowledge and inferential reasoning about object–object interactions. In: Lonsdorf, E. V., Ross, S., and Matsuzawa, T. (eds.), *The Mind of the Chimpanzee: Ecological and Experimental Perspectives*. Chicago University Press, Chicago.

Call, J., and Rochat, P. (1996). Liquid conservation in orangutans (*Pongo pymaeus*) and humans (*Homo sapiens*): individual differences and perceptual strategies. *J. Comp. Psychol.* 110(3):219–232.

Call, J., and Tomasello, M. (1994a). Production and comprehension of referential pointing by orangutans (*Pongo pygmaeus*). *J. Comp. Psychol.* 108(4):307–317.

Call, J., and Tomasello, M. (1994b). The social learning of tool use by orangutans (*Pongo pygmaeus*). *Hum. Evol.* 9:297–313.

Call, J., and Tomasello, M. (1995). Use of social information in the problem solving of orangutans (*Pongo pygmaeus*) and human children (*Homo sapiens*). *J. Comp. Psychol.* 109:308–320.

Call, J., and Tomasello, M. (1999). A nonverbal false belief task: the performance of children and great apes. *Child Dev.* 70:381–395.

Call, J., and Tomasello, M. (2008). Does the chimpanzee have a theory of mind? 30 years later. *Trends Cogn. Sci.* 12:187–192.

Call, J., Judge, P. G., and de Waal, F. B. M. (1996). Influence of kinship and spatial density on reconciliation and grooming in rhesus monkeys. *Am. J. Primatol.* 39:35–45.

Call, J., Hare, B., and Tomasello, M. (1998). Chimpanzee gaze following in an object choice task. *Anim Cogn.* 1:89–100.

Call, J., Aureli, F., and de Waal, F. B. M. (1999). Reconciliation patterns among stumptail macaques: a multivariate approach. *Anim. Behav.* 58:165–172.

Call, J., Hare, B., Carpenter, M., and Tomasello, M. (2004). "Unwilling" versus "unable": chimpanzees' understanding of human intentional action. *Dev. Sci.* 7:488–498.

Calle, P. P., Chaudhuri, M., and Bowen, R. (1990). Menstrual cycle characterization and artificial insemination in the black mangabey (*Cercocebus aterrimus*). *Zoo Biol.* 9:11–24.

Calvert, J. J. (1985). Food selection by western gorillas (*G. g. gorilla*) in relation to food chemistry. *Oecologia* 65:236–246.

Câmara, I. G. (2003). Brief history of conservation in the Atlantic Forest. In: Galindo-Leal, C., and Câmara, I. G. (eds.), *The Atlantic Forest of South America*. Island Press, Washington DC. pp. 31–42.

Camelleri, C., and Shine, R. (1990). Sexual dimorphism and dietary divergences: differences in trophic morphology between male and female snakes. *Copeia* 1990:649–658.

Campbell, A. F., and Sussman, R. W. (1994). The value of radio tracking in the study of neotropical rain forest monkeys. *Am. J. Primatol.* 32:291–301.

Campbell, C. J. (2000a). Fur rubbing behavior in free-ranging black-handed spider monkeys (*Ateles geoffroyi*) in Panama. *Am. J. Primatol.* 51:205–208.

Campbell, C. J. (2000b). The reproductive biology of the black-handed spider monkey: integrating behavior and endocrinology [PhD diss.]. University of California, Berkeley.

Campbell, C. J. (2002). The influence of a large home range on the social structure of free ranging spider monkeys (*Ateles geoffroyi*) on Barro Colorado Island, Panama. *Am. J. Phys. Anthropol.* S34:51–52.

Campbell, C. J. (2003). Female-directed aggression in free-ranging *Ateles geoffroyi*. *Int. J. Primatol.* 24:223–237.

Campbell, C. J. (2004). Patterns of behavior across reproductive states of free-ranging female black-handed spider monkeys (*Ateles geoffroyi*). *Am. J. Phys. Anthropol.* 124:166–176.

Campbell, C. J. (2006a). Copulatory behavior in free-ranging black-handed spider monkeys (*Ateles geoffroyi*). *Am. J. Primatol.* 68:507–511.

Campbell, C. J. (2006b). Lethal intragroup aggression by adult male spider monkeys (*Ateles geoffroyi*). *Am. J. Primatol.* 68:1197–1201.

Campbell, C. J., and Gibson, K. N. (2008). Sexuality and reproduction. In: Campbell, C. J. (ed.), *Spider Monkeys: Behavior, Ecology and Evolution of the Genus* Ateles. Cambridge University Press, Cambridge. pp. 266–287.

Campbell, C. J., Shideler, S. E., Todd, H. E., and Lasley, B. L. (2001). Fecal analysis of ovarian cycles in female black-handed spider monkeys (*Ateles geoffroyi*). *Am. J. Primatol.* 54:79–89.

Campbell, C. J., Aureli, F., Chapman, C. A., Ramos-Fernández, G., Matthews, K., Russo, S. E., Suarez, S. A., and Vick, L. (2005). Terrestrial behavior of *Ateles* spp. *Int. J. Primatol.* 26:1039–1051.

Camperio Ciani, A., Stanyon, R., Scheffrahn, W., and Sampurno, B. (1989). Evidence of gene flow between Sulawesi macaques. *Am. J. Primatol.* 17:257–270.

Campos, F. A., and Fedigan, L. M. (2009). Behavioral adaptations to heat stress and water scarcity in white-faced capuchins (*Cebus capucinus*) in Santa Rosa National Park, Costa Rica. *Am. J. Phys. Anthropol.* 138:101–111.

Campos, J. M., Benítez, I., and Meritt, D. A. J. (2004). On the occurrence of the owl monkey (*Aotus azarai*) in Cerro León, Chaco, Paraguay. *Neotrop. Prim.* 12(2):55–56.

Canavez, F. C., Moreira, M. A. M., Ladasky, J. J., Pissinatti, A., Parham, P., and Seuánez, H. N. (1999a). Molecular phylogeny of New World primates (Platyrrhini) based on beta2–microglobulin DNA sequences. *Mol. Phylogenet. Evol.* 12:74–82.

Canavez, F. C., Moreira, M. A. M., Simon, F., Parham, P., and Seuánez, H. N. (1999b). Phylogenetic relationships of the Callitrichinae (Platyrrhini, Primates) based on beta2-microglobin DNA sequences. *Am. J. Primatol.* 48:225–236.

Cann, R. L., Stoneking, M., and Wilson, A. C. (1987). Mitochondrial DNA and human evolution. *Nature* 325:31–36.

Cannon, C. H., and Leighton, M. (1994). Comparative locomotor ecology of gibbons and macaques: selection of canopy elements for crossing gaps. *Am. J. Phys. Anthropol.* 93:505–524.

Cannon, C. H., Curran, L. M., Marshall, A. J., and Leighton, M. (2007). Long-term reproductive behaviour of woody plants across seven Bornean forest types in the Gunung Palung National Park (Indonesia): suprannual synchrony, temporal productivity and fruiting diversity. *Ecol. Lett.* 10:956–969.

Cant, J. G. H. (1977). Ecology, locomotion, and social organization of spider monkeys (*Ateles geoffroyi*) [PhD diss.]. University of California, Davis.

Cant, J. G. H. (1986). Locomotion and feeding postures of spider and howling monkeys: field study and evolutionary interpretation. *Folia Primatol.* 46:1–14.

Cant, J. G. H. (1987a). Positional behavior of female Bornean orangutans. *Am. J. Primatol.* 12:71–90.

Cant, J. G. H. (1987b). Effects of sexual dimorphism in body size on feeding postural behavior of Sumatran orangutans (*Pongo pygmaeus*). *Am. J. Phys. Anthropol.* 74:143–148.

Cant, J. G. H. (1990). Feeding ecology of spider monkeys (*Ateles geoffroyi*) at Tikal, Guatemala. *Hum. Evol.* 5:269–281.

Cant, J. G. H. (1992). Positional behavior and body size of arboreal primates: a theoretical framework for field studies and an illustration of its application. *Am. J. Phys. Anthropol.* 88:273–283.

Cant, J. G. H., Youlatos, D., and Rose, M. D. (2001). Locomotor behavior of *Lagothrix lagothricha* and *Ateles belzebuth* in Yasuní National Park, Ecuador: general patterns and non- suspensory modes. *J. Hum. Evol.* 41:141–166.

Cant, J. G. H., Youlatos, D., and Rose, M. D. (2003). Suspensory locomotion of *Lagothrix lagothricha* and *Ateles belzebuth* in Yasuní National Park, Ecuador. *J. Hum. Evol.* 44:685–699.

Cantlon, J. F., and Brannon, E. M. (2006). Shared system for oredring small and large numbers in monkeys and humans. *Psychol. Sci.* 17:401–406.

Cantlon, J. F., and Brannon, E. M. (2007). How much does number matter to a monkey (*Macaca mulatta*)? *J. Exp. Psychol. Anim. Behav. Proc.* 33:32–41.

Caraco, T., Martindale, S., and Pulliam, H. R. (1980). Avian flocking in the presence of a predator. *Nature* 285:400–401.

Carlson, A. A., and Isbell, L. A. (2001). Causes and consequences of single-male and multi-male mating in free-ranging patas monkeys, *Erythrocebus patas*. *Anim. Behav.* 62:1047–1058.

Carlson, A. A., Ziegler, T. E., and Snowdon, C. T. (1997). Ovarian function of pygmy marmoset daughters (*Cebuella pygmaea*) in intact and motherless families. *Am. J. Primatol.* 43:347–355.

Carnegie, S. D., Fedigan, L. M., and Ziegler, T. E. (2005). Behavioral indivators of ovarian phase in white-faced capuchins (*Cebus capucinus*). *Am. J. Primatol.* 67(1):51–68.

Carnegie, S. D., Fedigan, L. M., and Ziegler, T. E. (2006). Post-conceptive mating in white-faced capuchins, *Cebus capucinus*: hormonal and sociosexual patterns of cycling, non-cycling and pregnant females. In: Estrada, A., Garber, P., Luecke, L., and Pavelka, M. S. M. (eds.), *New Perspectives in the Study of Mesoamerican Primates: Distribution, Ecology, Behavior, and Conservation*. Kluwer Academic/Plenum, New York. pp. 387–409.

Caro, T. M., Sellen, D. W., Parish, A., Frank, R., Brown, D. M., Voland, E., and Borgerhoff Mulder, M. (1995). Termination of reproduction in nonhuman and human female primates. *Int. J. Primatol.* 16:205–220.

Carosi, M., and Visalberghi, E. (2002). Analysis of tufted capuchin (*Cebus apella*) courtship and sexual behavior repertoire: changes throughout the female cycle and female interindividual differences. *Am. J. Phys. Anthropol.* 118:11–24.

Carosi, M., Ulland, A. E., Gerald, M. S., and Suomi, S. J. (2001). Male-like external genitalia in female tufted capuchins (*Cebus apella*) and the presence of a clitoral bone (baubellum): a cross sectional study. *Folia Primatol.* 72:149.

Carpenter, C. R. (1934). A field study of the behavior and social relations of howling monkeys. *Comp. Psychol. Monogr.* 10:1–168.

Carpenter, C. R. (1935). Behavior of red spider monkeys in Panama. *J. Mammal.* 16:171–180.

Carpenter, C. R. (1940). A field study in Siam of the behavior and social relations of the gibbon (*Hylobates lar*). *Comp. Psychol. Monogr.* 16:1–212.

Carpenter, C. R. (1942a). Characteristics of social behavior in non-human primates. *Trans. N. Y. Acad. Sci.* 2:248–258.

Carpenter, C. R. (1942b). Sexual behavior of free ranging rhesus monkeys (*Macaca mulatta*). I. Specimens, procedures and behavioral characteristics of estrus. *J. Comp. Psychol.* 33:118–142.

Carpenter, C. R. (1954). Tentative generalizations on the grouping behavior of non-human primates. *Hum. Biol.* 26:269–276.

Carpenter, C. R. (1965). The howlers of Barro Colorado Island. In: DeVore, I. (ed.), *Primate Behavior: Field Studies of Monkeys and Apes*. Holt, Rinehart, and Winston, New York. pp. 250–291.

Carrai, V., Silvana, M., Borgognini-Tarli, S., Huffman, M. A., and Massimo, B. (2003). Increase in tannin consumption by sifaka (*Propithecus verreauxi verreauxi*) females during the birth season: a case for self medication in prosimians? *Primates* 44:61–66.

Carrillo-Bilbao, G., Di Fiore, A., and Fernandez-Duque, E. (2005). Dieta, forrajeo y presupresto de tiempo en contoncillos (*Callicebus discolor*) del Parque Nacional Yasuni en la Amazonia Ecuatoriana. *Neotrop. Primates* 13: 7–11.

Carruthers, P. (2008). Meta-cognition in animals: a skeptical look. *Mind Lang.* 23:58–89.

Cartegni, L., Chew, S. L., and Krainer, A. R. (2002). Listening to silence and understanding nonsense: exonic mutations that affect splicing. *Nat. Rev. Genet.* 3:285–298.

Cartelle, C., and Hartwig, W. C. (1996). A new extinct primate among the Pleistocene megafauna of Bahia, Brazil. *Proc. Natl. Acad. Sci. USA* 93:6405–6409.

Carter, S. (ed.) (1999). *Hormones, Brain and Behavior: Integrative Neuroendocrinology of Affiliation.* MIT Press, Boston.

Carter, S., and Cushing, B. S. (2004). Proximate mechanisms regulating sociality and social monogamy, in the context of evolution. In: Sussman, R. W., and Chapman, A. R. (eds.), *Origins and Nature of Sociality.* Aldine de Gruyter, New York. pp. 99–121.

Cartmill, E., and Byrne, R. (2007). Orangutans modify their gestural signaling according to their audience's comprehension. *Curr. Biol.* 17:1345–1348.

Cartmill, M. (1972). Arboreal adaptations and the origin of the order Primates. In: Tuttle, R. (ed.), *The Functional and Evolutionary Biology of Primates.* Aldine, Chicago. pp. 97–122.

Cartmill, M. (1974). Rethinking primate origins. *Science* 184:436–443.

Cartmill, M. (1985). Climbing. In: Hildebrand, M., Bramble, D. M., Liem, K. F., and Wake, D. B. (eds.), *Functional Vertebrate Morphology.* Belknap Press, Cambridge, MA. pp. 73–88.

Cartmill, M. (1992). New views on primate origins. *Evol. Anthropol.* 1:105–111.

Cartmill, M., and Kay, R. (1978). Cranio-dental morphology, tarsier affinities, and primate suborders. In: Chivers, D., and Joysey, K. (eds.), *Recent Advances in Primatology. Evolution,* vol. 3. Academic Press, London. pp. 205–214.

Cartmill, M., and Milton, K. (1977). The lorisiform wrist joint and the evolution of "brachiating" adaptations in the Hominoidea. *Am. J. Phys. Anthropol.* 47:249–272.

Casado, F., Bonvicino, C.R., and Seuánez, H.N. (2007). Phylogeographic analyses of *Callicebus lugens* (Platyrrhini, Primates). *J. Hered.* 98(1):88–92.

Casar, C., Franco, E. S., de Castro Nogueira Soares, G., and Young, R. J. (2008). Observed case of maternal infanticide in a wild group of black-fronted titi monkeys (*Callicebus nigrifrons*). *Primates* 49:143–145.

Caspersson, T., Farber, S., Foley, G. E., Kudynowski, J., Modest, E. J., Simonsson, E., Wagh, U., and Zeck, L. (1968). Chemical differentiation along metaphase chromosomes. *Exp. Cell Res.* 49:219–222.

Castaño, J. H., and Cardona, D. M. (2005). Presencia del mono nocturno andino (*Aotus lemurinus* I. Geoffroy-St. Hilaire, 1843) en fragmentos de bosque de la cuenca media del Río Cauca. *Mus. Hist. Nat.* 9:111–117.

Castellanos, H. G. (1995). Feeding behaviour of *Ateles belzebuth* E. Geoffroy, 1806 (Cebidac: Atelinae) in Tawadu Forest southern Venezuela [PhD diss.]. University of Exeter, Exeter, UK.

Castellanos, H. G., and Chanin, P. (1996). Seasonal differences in food choice and patch preference of long-haired spider monkeys (*Ateles belzebuth*). In: Norconk, M. A., Rosenberger, A. L., and Garber, P. A. (eds.), *Adaptive Radiations of Neotropical Primates.* Plenum Press, New York. pp. 451–466.

Castles, D. L. (2000). Triadic versus dyadic resolutions. In: Aureli, F., and Waal, F. B. M. (eds.), *Natural Conflict Resolution.* University of California Press, Berkeley. pp. 289–291.

Castles, D. L., and Whiten, A. (1998a). Post-conflict behaviour of wild olive baboons. I. Reconciliation, redirection and consolation. *Ethology* 104:126–147.

Castles, D. L., and Whiten, A. (1998b). Post-conflict behaviour of wild olive baboons. II. Stress and self-directed behaviour. *Ethology* 104:148–160.

Castles, D. L., Aureli, F., and de Waal, F. B. M. (1996). Variation in conciliatory tendency and relationship quality across groups of pigtail macaques. *Anim. Behav.* 52:389–403.

Castro, C. (2000). Ecology and behavior of the common marmoset, *Callithrix jacchus. Neotrop. Primates* 8:50–51.

Castro, N. R. (1991). Behavioral ecology of two coexistent tamarin species (*Saguinus fuscicollis nigrifrons* and *Saguinus mystax mystax,* Callitrichidae, Primates) in Amazonian Peru [PhD diss.]. Washington University, St. Louis, MO.

Catchpole, H., and Fulton, J. (1943). The oestrus cycle in *Tarsius. J. Mammal.* 24:90–93.

Caton, J. M. (1999). Digestive strategy of the Asian colobine genus *Trachypithecus. Primates* 40:311–325.

Caton, J. M., Hill, D. M., Hume, I. D., and Crook, G. A. (1996). The digestive strategy of the common marmoset, *Callithrix jacchus. Comp. Biochem. Physiol.* A114:1–8.

Cavalli-Sforza, L. L., and Feldman, M. W. (1981). *Cultural Transmission and Evolution: A Quantitative Approach.* Princeton University Press, Princeton, NJ.

Cave, A. J. (1973). The primate nasal fossa. *Biol. J. Linn. Soc.* 5:377–387.

Cavigelli, S. A., and Pereira, M. E. (2000). Mating season aggression and fecal testosterone levels in male ring-tailed lemurs (*Lemur catta*). *Horm. Behav.* 37:246–255.

Cekan, S. Z., Goncharov, N., Aso, T., and Diczfalusy, E. (1977). Circulating steroid levels in male and female baboons (*Papio hamadryas*). In: Prasad, M. R. N., and Anand Kumar, T. C. (eds.), *Use of Non-Human Primates in Biomedical Research.* Indian National Science Academy, New Delhi. pp. 234–241.

Chalmers, N. R. (1968a). Group composition, ecology and daily activities of free living mangabeys in Uganda. *Folia Primatol.* 8:247–262.

Chalmers, N. R. (1968b). The social behaviour of free living mangabeys in Uganda. *Folia Primatol.* 8: 263–281.

Chalmers, N. R., and Rowell, T. E. (1971). Behaviour and female reproductive cycles in a captive group of mangabeys. *Folia Primatol.* 14:1–14.

Chalyan, V. G., Lapin, B. A., and Meishvili, N. V. (1994). Kinship and troop structure formation in baboons in Gumista Reserve. *Congress of the International Primatological Society* (Bali, Indonesia) 15:238.

Chambers, K. C., and Phoenix, C. H. (1987). Differences among ovariectomized female rhesus macaques in the display of sexual behavior without and with estradiol treatment. *Behav. Neurosci.* 101:303–308.

Chance, M. R. A. (1967). Attention structure as the basis of primate rank orders. *Man* 2:503–518.

Chancellor, R. L., and Isbell, L. A. (2009). Female grooming markets in a population of gray-cheeked mangabeys (*Lophocebus albigena*). *Behav. Ecol.* 20:79–86.

Channell, R., and Lomolino, M. V. (2000). Dynamic biogeography and conservation of endangered species. *Nature* 403:84–86.

Chaoui, N. J., and Hasler-Gallusser, S. (1997). Incomplete sexual suppression in *Leontopithecus chrysomelas*: a behavioural and hormonal study in a semi-natural environment. *Folia Primatol.* 70:47–54.

Chapais, B. (1983). Dominance, relatedness and the structure of female relationships in rhesus monkeys. In: Hinde, R. A (ed.), *Primate Social Relationships: An Integrated Approach.* Blackwell, Oxford. pp. 208–217.

Chapais, B. (1986). Why do adult male and female rhesus monkeys affiliate during the birth season? In: Rawlins, R. G., and Kessler, M. J. (eds.), *The Cayo Santiago Macaques: History, Behavior and Biology.* State University of New York Press, Albany. pp. 173–200.

Chapais, B. (1988a). Experimental matrilineal inheritance of rank in female Japanese macaques. *Anim. Behav.* 36:1025–1037.

Chapais, B. (1988b). Rank maintenance in female Japanese macaques: experimental evidence for social dependency. *Behaviour* 104:41–59.

Chapais, B. (1992). The role of alliances in social inheritance of rank among female primates. In: de Waal, F., and Harcourt, A. (ed.), *Coalitions and Alliances in Humans and Other Animals.* Oxford University Press, Oxford. pp. 29–59.

Chapais, B. (1995). Alliances as a means of competition in primates: evolutionary, developmental and cognitive aspects. *Ybk. Phys. Anthropol.* 38:115–136.

Chapais, B. (2001). Primate nepotism: what is the explanatory value of kin selection? *Int. J. Primatol.* 22:203–229.

Chapais, B. (2004). How kinship generates dominance structures: a comparative perspective. In: Thierry, B., Singh, M., and Kaumanns, W. (eds.), *Macaque Societies: A Model for the Study of Social Organization.* Cambridge University Press, Cambridge. pp. 186–204.

Chapais, B. (2006). Kinship, competence and cooperation in primates. In: Kappeler, P. M., and van Schaik, C.P. (eds.), *Cooperation in Primates and Humans.* Springer, New York. pp. 47–64.

Chapais, B. (2008). *Primeval Kinship.* Harvard University Press, Cambridge, MA.

Chapais, B., and Belisle, P. (2004). Constraints on kin selection in primate groups. In: Chapais, B., and Berman, C. M. (eds.), *Kinship and Behavior in Primates.* Oxford University Press, New York. pp. 365–386.

Chapais, B., and Berman, C. M. (2004a). Variation in nepotistic regimes and kin recognition: a major area for future research. In: Chapais, B., and Berman, C. M. (eds.), *Kinship and Behavior in Primates.* Oxford University Press, New York. pp. 477–489.

Chapais, B., and Berman, C. M. (2004b). Introduction: the kinship black box. In: Chapais, B., and Berman, C. M. (eds.), *Kinship and Behavior in Primates.* Oxford University Press, New York. pp. 3–11.

Chapais, B., and Berman, C. M. (2004c). *Kinship and Behavior in Primates.* Oxford University Press, New York.

Chapais, B., and Lecomte, M. (1995). Induction of matrilineal rank instability by the alpha male in a group of Japanese macaques. *Am. J. Primatol.* 36:299–312.

Chapais, B., and Mignault, C. (1991). Homosexual incest avoidance among females in captive Japanese macaques. *Am. J. Primatol.* 23:171–183.

Chapais, B., Girard, M., and Primi, G. (1991). Nonkin alliances, and the stability of matrilineal dominance relations in Japanese macaques. *Anim. Behav.* 41:481–491.

Chapais, B., Prud'homme, J., and Teijeiro, S. (1994). Dominance competition among siblings in Japanese macaques: constraints on nepotism. *Anim. Behav.* 48:1335–1347.

Chapais, B., Gauthier, C., and Prud'homme, J. (1995). Dominance competition through affiliation and support in Japanese macaques: an experimental study. *Int. J. Primatol.* 16:521–536.

Chapais, B., Gauthier, C., Prud'homme, J., and Vasey, P. (1997). Relatedness threshold for nepotism in Japanese macaques. *Anim. Behav.* 53:1089–1101.

Chapais B., Savard, L., and Gauthier, C. (2001). Kin selection and the distribution of altruism in relation to degree of kinship in Japanese macaques. *Behav. Ecol. Sociobiol.* 49:493–502.

Chapman, C. A. (1986). *Boa constrictor* predation and group response in white faced *Cebus* monkeys. *Biotropica* 18:171–172.

Chapman, C. (1987). Flexibility in diets of three species of Costa Rican primates. *Folia Primatol.* 49:90–105.

Chapman, C. (1988a). Patch use and patch depletion by the spider and howling monkeys of Santa Rosa National Park, Costa Rica. *Behaviour* 105:99–116.

Chapman, C. (1988b). Patterns of foraging and range use by three species of neotropical primates. *Primates* 29:177–194.

Chapman, C. A. (1989a). Primate seed dispersal: the fate of dispersed seeds. *Biotropica* 21:148–154.

Chapman, C. A. (1989b). Spider monkey sleeping sites: use and availability. *Am. J. Primatol.* 18:53–60.

Chapman, C. A. (1990a). Association patterns of spider monkeys: the influence of ecology and sex on social organization. *Behav. Ecol. Sociobiol.* 26:409–414.

Chapman, C. A. (1990b). Ecological constraints on group size in three species of neotropical primates. *Folia Primatol.* 55:1–9.

Chapman, C. A. (1995). Primate seed dispersal: coevolution and conservation implications. *Evol. Anthropol.* 4:74–82.

Chapman, C. A., and Balcomb, S. R. (1998). Population characteristics of howlers: ecological conditions or group history. *Int. J. Primatol.* 19:385–403.

Chapman, C. A., and Chapman, L. J. (1990). Reproductive biology of captive and free-ranging spider monkeys. *Zoo Biol.* 9:1–9.

Chapman, C. A., and Chapman, L. J. (1991). The foraging itinerary of spider monkeys: when to eat leaves? *Folia Primatol.* 56:162–166.

Chapman, C. A., and Chapman, L. J. (1995). Survival without dispersers: seedling recruitment under parents. *Conserv. Biol.* 9:675–678.

Chapman, C. A., and Chapman, L. J. (1996a). Frugivory and the fate of dispersed and non-dispersed seeds of six African tree species. *J. Trop. Ecol.* 12:491–504.

Chapman, C. A., and Chapman, L. J. (1996b). Mixed-species primate groups in the Kibale Forest: ecological constraints on association. *Int. J. Primatol.* 17:31–50.

Chapman, C. A., and Chapman, L. J. (1999). Implications of small scale variation in ecological conditions for the diet and density of red colobus monkeys. *Primates* 40:215–231.

Chapman, C. A., and Chapman, L. J. (2000a). Constraints on group size in red colobus and red-tailed guenons: examining the generality of the ecological constraints model. *Int. J. Primatol.* 21:565–585.

Chapman, C. A., and Chapman, L. J. (2000b). Determinants of group size in primates: the importance of travel costs. In: Boinski, S., and Garber, P. A. (eds.), *On the Move: How and Why Animals Travel in Groups.* University of Chicago Press, Chicago. pp. 24–42.

Chapman, C. A., and Chapman, L. J. (2000c). Interdemic variation in mixed-species association patterns: common diurnal primates of Kibale National Park, Uganda. *Behav. Ecol. Sociobiol.* 47:129–139.

Chapman, C. A., and Chapman, L. J. (2002a). Foraging challenges of red colobus monkeys: influence of nutrients and secondary compounds. *Comp. Biochem. Physiol.* 133:861–875.

Chapman, C. A., and Chapman, L. J. (2002b). Plant–animal coevolution: is it thwarted by spatial and temporal variation in animal foraging? In: Levey, D. J., Silva, W. R., and Galetti, M. (eds.), *Seed Dispersal and Frugivory: Ecology, Evolution, and Conservation.* CABI Publishing, New York. pp. 275–290.

Chapman, C. A., and Onderdonk, D. A. (1998). Forests without primates: primate/plant codependency. *Am. J. Primatol.* 45:127–141.

Chapman, C. A., and Peres, C. A. (2001). Primate conservation in the new millennium: the role of scientists. *Evol. Anthropol.* 10:16–33.

Chapman, C. A., and Pavelka, C. A. (2005). Group size in folivorous primates: ecological constraints and the possible influence of social factors. *Primates* 46:1–9.

Chapman, C., and Wrangham, R. W. (1993). Range use of the forest chimpanzees of Kibale: implications for the understanding of chimpanzee social organization. *Am. J. Primatol.* 31:263–273.

Chapman, C. A., Chapman, L. J., and Glander, K. E. (1989a). Primate population in northwestern Costa Rica: potential for recovery. *Primates* 10:37–44.

Chapman, C. A., Chapman, L. J., and McLaughlin, R. L. (1989b). Multiple central place foraging by spider monkeys: travel consequences of using many sleeping sites. *Oecologia* 79:506–511.

Chapman, C. A., Fedigan, L. M., Fedigan, L., and Chapman, L. J. (1989c). Post-weaning resource competition and sex ratios in spider monkeys. *Oikos* 54:315–319.

Chapman, C. A., Chapman, L. J., Wrangham, R. W., Hunt, K., Gebo, D., and Gardner, L. (1992). Estimators of fruit abundance of tropical trees. *Biotropica* 24:527–531.

Chapman, C. A., White, F. J., and Wrangham, R. W. (1993). Defining subgroup size in fission–fusion societies. *Folia Primatol.* 61:31–34.

Chapman, C. A., White, F. J., and Wrangham, R. W. (1994a). Party size in chimpanzees and bonobos. In: Wrangham, R. W., McGrew, W. C., de Waal, F. B. M., Heltne, P. G., and Marquardt, L. A. (eds.), *Chimpanzee Cultures.* Harvard University Press, Cambridge, MA. pp. 41–57.

Chapman, C. A., Wrangham, R., and Chapman, L. J. (1994b). Indexes of habitat-wide fruit abundance in tropical forests. *Biotropica* 26:160–171.

Chapman, C. A., Wrangham, R. W., and Chapman, L. J. (1995). Ecological constraints on group sizes: an analysis of spider monkey and chimpanzee subgroups. *Behav. Ecol. Sociobiol.* 36:59–70.

Chapman, C. A., Wrangham, R. W., Chapman, L. J., Kennard, D. K., and Zanne, A. E. (1999). Fruit and flower phenology at two sites in Kibale National Park, Uganda. *J. Trop. Ecol.* 15:189–211.

Chapman, C. A., Balcomb, S. R., Gillespie, T. R., Skorupa, J. P., and Struhsaker, T. T. (2000). Long-term effects of logging on African primate communities: a 28-year comparison from Kibale National Park, Uganda. *Conserv. Biol.* 14:207–217.

Chapman, C. A., Chapman, L. J., and Gillespie, T. R. (2002a). Scale issues in the study of primate foraging: red colobus of Kibale National Park. *Am. J. Phys. Anthropol.* 117:349–363.

Chapman, C. A., Chapman, L. J., Bjorndal, K. A., and Onderdonk, D. A. (2002b). Application of protein-to-fiber ratios to predict colobine abundance on different spatial scales. *Int. J. Primatol.* 23:283–310.

Chapman, C. A., Chapman, L. J., Gautier-Hion, A., Lambert, J. E., Rode, K., Tutin, C. E. G., and White, L. J. T. (2002c). Variation in the diet of *Cercopithecus* monkeys: differences within forests, among forests, and across species. In: Glenn, M., and Cords, M. (eds.), *The Guenons: Diversity and Adaptation in African Monkeys.* Kluwer Academic, New York. pp. 319–344.

Chapman, C. A., Chapman, L. J., Rode, K. D., Hauck, E. M., and McDowell, L. R. (2003a). Variation in the nutritional value of primate foods: among trees, time periods, and areas. *Int. J. Primatol.* 24:317–333.

Chapman, C. A., Chapman, L. J., Vulinec, K., Zanne, A. E., and Lawes, M. J. (2003b). Fragmentation and alteration to seed dispersal processes: an initial evaluation of dung beetles, seed fate, and seedling diversity. *Biotropica* 35:382–393.

Chapman, C. A., Lawes, M. J., Naughton-Treves, L., and Gillespie, T. R. (2003c). Primate survival in community-owned forest fragments: are metapopulation models useful amidst intensive use? In: Marsh, L. K. (ed.), *Primates in Fragments: Ecology and Conservation.* Kluwer Academic/Plenum, New York. pp. 63–78.

Chapman, C. A., Chapman, L. J., Naugthon-Treves, L., Lawes, M. J., and McDowell, L. R. (2004). Predicting folivorous primate abundance: validation of a nutritional model. *Am. J. Primatol.* 62:55–69.

Chapman, C. A., Chapman, L. J., Struhsaker, T. T., Zanne, A. E., Clark, C. J., and Poulsen, J. R. (2005). A long-term evaluation of fruiting phenology: importance of climate change. *J. Trop. Ecol.* 21:31–45.

Chapman, C. A., Speirs, M. L., Gillespie, T. R., Holland, T., and Austad, K. M. (2006). Life on the edge: gastrointestinal parasites from the forest edge and interior primate groups. *Am. J. Primatol.* 68:397–409.

Chapman, C. A., Chapman, L. J., Struhsaker, T. T., Zanne, A. E., Clark, C. J., and Poulsen, J. R. (2005). A thirty-year evaluation of fruit phenology: importance of climate change. *J. Trop. Ecol.* 21:1–14.

Chapman, F. (1905). *A Contribution to the Life History of the American Flamingo,* 1st ed. American Museum of Natural History, New York.

Chardonnet, P., Fritz, H., Zorzi, N., and Feron, E. (1995). Current importance of traditional hunting and major contrasts in wild meat consumption in sub-Saharan Africa. In: Bissonette, J. A., and Krausman, P. R. (eds.),

Integrating People and Wildlife for a Sustainable Future. Wildlife Society, Bethesda, MD. pp. 304–307.

Charles-Dominique, P. (1972). Ecologie et vie sociale de *Galago demidovii* (Fisher 1808, Prosimii). *Z. Tierpsychol. Suppl.* 9:7–41.

Charles-Dominique, P. (1974a). Vie sociale de *Perodicticus potto* (Primates Lorisides). Étude de terrain en forêt equatorial de l'ouest africain au Gabon. *Mammalia* 38:355–379.

Charles-Dominique, P. (1974b). Ecology and feeding behaviour of five sympatric lorisids in Gabon. In: Martin, R. D., Doyle, G. A., and Walker, A. C. (eds.), *Prosimian Biology.* Duckworth Press, London. pp. 131–150.

Charles-Dominique, P. (1977a). *Ecology and Behaviour of Nocturnal Primates.* Duckworth Press, London.

Charles-Dominique, P. (1977b). Urine marking and territoriality in *Galago alleni* (Waterhouse, 1837-Lorisoidea, Primates)—a field study by radiotelemetry. *Z. Tierpsychol.* 43:113–138.

Charles-Dominique, P. (1978). Solitary and gregarious prosimians: evolution of social structures in primates. In: Chivers, D. J., and Joysey, K. A. (eds.), *Recent Advances in Primatology,* vol. 3. Academic Press, New York. pp. 139–149.

Charles-Dominique, P., and Bearder, S. K. (1979). Field studies of lorisid behavior: methodological aspects. In: Doyle, G. A., and Martin, R. D. (eds.), *The Study of Prosimian Behavior.* Academic Press, London. pp. 567–629.

Charles-Dominique, P., and Hladik, C. M. (1971). Le *Lepilemur* du sud de Madagascar: écologie, alimentation et vie sociale. *Terre Vie* 25:3–66.

Charles-Dominique, P., and Martin, R. D. (1970). Evolution of lorises and lemurs. *Nature* 227:257–260.

Charles-Dominique, P., and Petter, J.-J. (1980). Ecology and social life of *Phaner furcifer.* In: Charles-Dominique, P., Cooper, H. M., Hladik, A., Hladik, C. M., Pages, E., Pariente, G. F., Petter-Rousseaux, A., Petter, J.-J., and Schilling, A. (eds.), *Nocturnal Malagasy Primates: Ecology, Physiology and Behavior.* Academic Press, New York. pp. 75–95.

Charnov, E. L. (1979). The genetical evolution of patterns of sexuality: Darwinian fitness. *Am. Nat.* 113(4):465.

Charnov, E. L. (1991). Evolution of life history variation among female mammals. *Proc. Natl. Acad. Sci. USA* 88:1134–1137.

Charnov, E. L. (1993). *Life History Invariants.* Oxford University Press, Oxford.

Charnov, E. L. (2001). Evolution of mammalian life histories. *Evol. Ecol. Res.* 3:521–535.

Charnov, E. L., and Berrigan, D. (1993). Why do female primates have such long lifespans and so few babies? or life in the slow lane. *Evol. Anthropol.* 1:191–194.

Charpentier, M., Peignot, P., Hossaert-McKey, M., and Wickings, E. J. (2004). Changes in social interactions during adolescence in male mandrills (*Mandrillus sphinx*). *Am. J. Primatol.* 63:63–73.

Charpentier, M., Peignot, P., Hossaert-McKey, M., Gimenez, O., Setchell, J. M., and Wickings, E. J. (2005). Constraints on control: factors influencing reproductive success in male mandrills (*Mandrillus sphinx*). *Behav. Ecol.* 16:614–623.

Charpentier, M., Peignott, P., Hossaert-McKey, M., and Wickings, E. J. (2007). Kin discrimination in juvenile mandrills, *Mandrillus sphinx.* *Anim. Behav.* 73:37–45.

Charpentier, M. J. E., Boulet, M., and Drea, C. M. (2008). Smelling right: the scent of male lemurs advertises genetic quality and relatedness. *Mol. Ecol.* 17:3225–3233.

Chase, J. E., and Cooper, R. W. (1969). *Saguinus nigricollis*—physical growth and dental eruption in a small population of captive-born individuals. *Am. J. Phys. Anthropol.* 30:111–116.

Chater, N., and Heyes, C. (1994). Animal concepts: content and discontents. *Mind Lang.* 9:209–246.

Chauvin, C., and Berman, C. M. (2004). Intergenerational transmission of behavior. In: Thierry, B., Singh, M., and Kaumanns, W. (eds.), *Macaque Societies: A Model for the Study of Social Organization.* Cambridge University Press, Cambridge. pp. 209–230.

Chave, J., Muller-Landau, H. C., and Levin, S. A. (2002). Comparing classical community models: theoretical consequences for patterns of diversity. *Am. Nat.* 159:1–23.

Chen, F. C., and Li, W. H. (2001). Genomic divergences between humans and other hominoids and the effective population size of the common ancestor of humans and chimpanzees. *Am. J. Hum. Gen.* 68:444–456.

Chen, F.-C., Chen, C. J., Li, W.-H., and Chuang, T.-J. (2006). Human-specific insertions and deletions inferred from mammalian genome sequences. *Genome Res.* 17:16–22.

Chen, J. J., Smith, E. R., Gray, G. D., and Davidson, J. M. (1981). Seasonal changes in plasma testosterone and ejaculatory capacity in squirrel monkeys (*Saimiri sciureus*). *Primates* 22:253–260.

Chen, N. (1995). Ecology of the black-crested gibbon (*Hylobates concolor*) in the Ailao Mt. Reserve, Yunnan, China [MA thesis]. Mahidol University, Bangkok, Thailand.

Cheney, D. L. (1977). The acquisition of rank and the development of reciprocal alliances among free-ranging immature baboons. *Behav. Ecol. Sociobiol.* 2:303–318.

Cheney, D. L. (1978). Interactions of immature male and female baboons with adult females. *Anim. Behav.* 26:389–408.

Cheney, D. L. (1981). Intergroup encounters among free-ranging vervet monkeys. *Folia Primatol.* 35:124–146.

Cheney, D. L. (1987). Interactions and relationships between groups. In: Smuts, B. B., Cheney, D. L., Seyfarth, R. M., Wrangham, R. W., and Struhsaker, T. T. (eds.), *Primate Societies.* University of Chicago Press, Chicago. pp. 267–281.

Cheney, D. L. (1992). Within-group cohesion and inter-group hostility: the relation between grooming distributions and inter-group competition among female primates. *Behav. Ecol.* 3:334–345.

Cheney, D. L., and Seyfarth, R. M. (1982). How vervet monkeys perceive their grunts: field playback experiments. *Anim. Behav.* 30:739–751.

Cheney, D. L., and Seyfarth, R. M. (1983). Non-random dispersal in free-ranging vervet monkeys: social and genetic consequences. *Am. Nat.* 122:392–412.

Cheney, D. L., and Seyfarth, R. M. (1988). Assessment of meaning and the detection of unreliable signals by vervet monkeys. *Anim. Behav.* 36:477–486.

Cheney, D. L., and Seyfarth, R. M. (1989). Redirected aggression and reconciliation among vervet monkeys, *Cercopithecus aethiops.* *Behaviour* 110:258–275.

Cheney, D. L., and Seyfarth, R. M. (1990a). *How Monkeys See the World.* University of Chicago Press, Chicago.

Cheney, D. L., and Seyfarth, R. M. (1990b). The representation of social relations by monkeys. *Cognition* 37:167–196.

Cheney, D. L., and Seyfarth, R. M. (1991). Reading minds or reading behaviour? Tests for a theory of mind in monkeys. In: Whiten, A. (ed.), *Natural Theories of Mind.* Basil Blackwell, Oxford. pp. 175–194.

Cheney, D. L., and Seyfarth, R. M. (1992). Meaning, reference, and intentionality in the natural vocalizations of monkeys. In: Nishida, T., McGrew, W. C., Marler, P., Pickford, M., and de Waal, F. B. M. (eds.), *Topics in Primatology. Human Origins,* vol. 1. University of Tokyo Press, Tokyo.

Cheney, D. L., and Seyfarth, R. M. (1997). Reconciliatory grunts by dominant female baboons influence victim's behaviour. *Anim. Behav.* 54:409–418.

Cheney, D. L., and Seyfarth, R. M. (2007). *Baboon Metaphysics:* Chicago University Press, Chicago.

Cheney, D. L., and Wrangham, R. W. (1987). Predation. In: Smuts, B. B., Cheney, D. L., Seyfarth, R. M., Wrangham, R. W., and Struhsaker, T. T. (eds.), *Primate Societies.* University of Chicago Press, Chicago. pp. 227–239.

Cheney, D. L., Lee, P. C., and Seyfarth, R. M. (1981). Behavioral correlates of non-random mortality among free-ranging female vervet monkeys. *Behav. Ecol. Sociobiol.* 9:153–161.

Cheney, D. L., Seyfarth, R. M., Andelman, S. J., and Lee, P. C. (1988). Reproductive success in vervet monkeys. In: Clutton-Brock, T. H. (ed.), *Reproductive Success: Studies of Individual Variation in Contrasting Breeding Systems.* University of Chicago Press, Chicago. pp. 384–402.

Cheney, D. L., Seyfarth, R. M., and Silk, J. B. (1995). The role of grunts in reconciling opponents and facilitating interactions among adult female baboons. *Anim. Behav.* 50:249–257.

Cheney, D. L., Seyfarth, R. M., Fischer, J., Beehner, J. C., Bergman, T. J., Johnson, S. E., Kitchen, D. M., Palombit, R. A., Rendall, D., and Silk, J. B. (2004). Factors affecting reproduction and mortality among baboons in the Okavango Delta, Botswana. *Int. J. Primatol.* 25:401–428.

Cheney, D. L., Seyfarth, R. M., Fischer, J., Beehner, J. C., Bergman, T. J., Johnson, S. E., Kitchen, D. M., Palombit, R. A., Rendall, D., and Silk, J. B. (2006). Reproduction, mortality, and female reproductive success in chacma baboons in the Okavango Delta, Botswana. In: Swedell, L., and Leigh, S. R. (eds.), *Reproduction and Fitness in Baboons: Behavioral, Ecological, and Life History Perspectives.* Springer, New York. pp. 147–176.

Cheng, Z., Ventura, M., She, W., Khaitovich, P., Graves, T., Osoegawa, K., Church, D., DeJong, P., Wilson, R. K., Pääbo, S., Rocchi, M., and Eichler, E. E. (2005). A genome-wide comparison of recent chimpanzee and human segmental duplications. *Nature* 437:88–93.

Chepko-Sade, B. D., and Sade, D. S. (1979). Patterns of group spitting within matrilineal kinship groups: a study of social group structure in *Macaca mulatta* (Cercoptithecidae: Primates). *Behav. Ecol. Sociobiol.* 5:67–87.

Chesson, P. (1996). Matters of scale in the dynamics of populations and communities. In: Floyd, R. B., Sheppard, A. W., and De Barro, P. J. (eds.), *Frontiers in Population Ecology.* CSIRO Publishing, Melbourne. pp. 353–368.

Chesson, P. (1998). Spatial scales in the study of reef fishes: a theoretical perspective. *Aust. J. Ecol.* 23:209–215.

Chesson, P. (2000). General theory of competitive coexistence in spatially-varying environments. *Theor. Popul. Biol.* 58:211–237.

Chevalier-Skolnikoff, S. (1974). The ontogeny of communication in the stumptail macaque (*Macaca arctoides*). In: *Contributions to Primatology.* Karger Press, Basel.

Chevalier-Skolnikoff, S. (1977). A piagetian model for describing and comparing socialization in monkey, ape and human infants. In: Chevalier-Skolnikoff, S., and Poirier, F. E. (eds.), *Primate Bio-Social Development: Biological, Social, and Ecological Determinants.* Garland Publishing, New York. pp. 159–187.

Chevalier-Skolnikoff, S. (1983). Sensorimotor development in orang-utans and other primates. *J. Hum. Evol.* 12:545–561.

Chevalier-Skolnikoff, S. (1990). Tool use by wild *Cebus* monkeys at Santa Rosa National Park, Costa Rica. *Primates* 31:375–383.

Cheverud, J. M. (1981). Epiphyseal union and dental eruption in *Macaca mulatta. Am. J. Phys. Anthropol.* 56:157–167.

Cheverud, J. M. (2004). Darwinian evolution by natural selection of heretible variation: definitions of parameters and application to social behavior. In: Sussman, R. W., and Chapman, A. R. (eds.), *Origins and Nature of Sociality.* Aldine de Gruyter, New York. pp. 140–157.

Cheverud, J. M., Buettner-Janusch, J., and Sade, D. (1978). Social group fission and the origin of intergroup genetic differentiation among the rhesus monkeys of Cayo Santiago. *Am. J. Phys. Anthropol.* 49:449–456.

Cheverud, J. M., Wilson, P., and Dittus, W. (1992). Primate population studies at Polonnaruwa. III. Somatometric growth in a natural population of toque macaques (*Macaca sinica*). *J. Hum. Evol.* 23:51–77.

Cheyne, S. M. (2007). Effects of meteorology, astronomical variables, location and human disturbance on the singing apes: *Hylobates albibarbis. Am. J. Primatol.* 70:1–7.

Cheyne, S. M., Thompson, C. J. H., Phillips, A. C., Hill, R. M. C., and Limin, S. H. (2008). Density and population estimate of gibbons (*Hylobates albibarbis*) in the Sabangau catchment, Central Kalimantan, Indonesia. *Primates* 49(1):50–56.

Chiarello, A. G. (1993a). Activity pattern of the brown howler monkey, *Alouatta fusca,* Geoffroy 1812, in a forest fragment of southeastern Brazil. *Primates* 34:289–293.

Chiarello, A. G. (1993b). Home range of the brown howler monkey, *Alouatta fusca,* in a forest fragment of southeastern Brazil. *Folia Primatol.* 60:173–175.

Chiarello, A. G. (1994). Diet of the brown howler monkey *Alouatta fusca* in semi-deciduous forest fragment of southeastern Brazil. *Primates* 35:25–34.

Chiarello, A. G. (1995a). Grooming in brown howler monkeys, *Alouatta fusca. Am. J. Primatol.* 35:73–81.

Chiarello, A. G. (1995b). Role of loud calls in brown howlers, *Alouatta fusca. Am. J. Primatol.* 36:213–222.

Chiarello, A. G. (2003). Primates of the Brazilian Atlantic Forest: the influence of forest fragmentation on survival. In: Marsh, L. K. (ed.), *Primates in Fragments.* Kluwer Academic/Plenum, New York. pp. 99–121.

Chiarello, A. G., and de Melo, F. R. (2001). Primate population densities and sizes in Atlantic forest remnants of northern Espirito Santo, Brazil. *Int. J. Primatol.* 22:379–396.

Chimpanzee Sequencing and Analysis Consortium (2005). Initial sequence of the chimpanzee genome and comparison with the human genome. *Nature* 437:69–87.

Chinchilla, F. A. (1997). La dieta del Jaguar (*Panthera onca*), el Puma (*Felis concolor*) yel manigordo (*Felis pardalis*) (Carnivoria: Felidae) en el Parque Nacional Corcovado, Costa Rica. *Rev. Biol. Trop.* 45:1223–1229.

Chism, J. (1978). Behavior of group members other than the mother toward captive patas infants. In: Chivers, D., and Herbert, J. (eds.), *Recent Advances in Primatology,* vol. 1. Academic Press, New York. pp. 173–176.

Chism, J. (1999). Intergroup encounters in wild patas monkeys (*Erythrocebus patas*) in Kenya. *Am. J. Primatol.* 49:43.

Chism, J. (2000). Allocare patterns among cercopithecines. *Folia Primatol.* 71:55–66.

Chism, J. (2005). Round up the usual suspects—conflict between monkeys and farmers in East and West Africa. In: Paterson, J. D., and Wallis, J. (eds.), *Commensalism and Conflict: The Human–Primate Interface.* American Society of Primatologists, Norman, OK. pp. 338–349.

Chism, J., and Rogers, W. (1997). Male competition, mating success and female choice in a seasonally breeding primate (*Erythrocebus patas*). *Ethology* 103:109–126.

Chism, J., and Rogers, W. (2002). Grooming and social cohesion in patas monkeys and other guenons. In: Glenn, M. E., and Cords, M. (eds.), *The Guenons: Diversity and Adaptation in African Monkeys.* Kluwer Academic/Plenum, New York. pp. 233–244.

Chism, J., and Rowell, T. E. (1986). Mating and residence patterns of male patas monkeys. *Ethology* 72:31–39.

Chism, J., and Rowell, T. E. (1988). The natural history of patas monkeys. In: Gautier-Hion, A., Bourlière, F., Gautier, J. P., and Kingdon, J. (eds.), *Primate Radiation: Evolutionary Biology of the African Guenons.* Cambridge University Press, Cambridge. pp. 412–438.

Chism, J., Olson, D. K., and Rowell, T. E. (1983). Diurnal births and perinatal behavior among wild patas monkeys: evidence of an adaptive pattern. *Int. J. Primatol.* 4:167–184.

Chism, J., Rowell, T., and Olson, D. (1984). Life history patterns of female patas monkeys. In: Small, M. F. (ed.), *Female Primates: Studies by Women Primatologists.* Alan R. Liss, New York. pp. 175–190.

Chivers, D. J. (1969). On the daily behavior and spacing of howling monkey groups. *Folia Primatol.* 10:48–102.

Chivers, D. J. (1974). The siamang in Malaya: a field study of a primate in tropical rain forest. *Contrib. Primatol.* 4:1–335.

Chivers, D. J. (1977). The lesser apes. In: Prince Rainier of Monaco and Bourne, G. H. (eds.), *Primate Conservation.* Academic Press, New York. pp. 539–598.

Chivers, D. J. (1980). Introduction. In: Chivers, D. J. (ed.), *Malayan Forest Primates. Ten Years' Study in Tropical Rain Forest.* Plenum Press, New York. pp. 1–27.

Chivers, D. J. (1984). Feeding and ranging in gibbons: a summary. In: Preuschoft, H., Chivers, D. J., Brockelman, W. Y., and Creel, N. (eds.), *The Lesser Apes: Evolutionary and Behavioural Biology.* Edinburgh University Press, Edinburgh. pp. 267–281.

Chivers, D. J. (1994). Functional anatomy of the gastrointestinal tract. In: Davies, A. G., and Oates, J. F. (eds.), *Colobine Monkeys: Their Ecology, Behavior and Evolution.* Cambridge University Press, Cambridge. pp. 205–227.

Chivers, D. J. (2001). The swinging singing apes: fighting for food and family in the far-east forests. In: *The Apes: Challenges for the 21st Century. Conference Proceedings.* Chicago Zoological Society, Chicago. pp. 1–28.

Chivers, D. J., and Hladik, C. M. (1980). Morphology of the gastrointestinal tract in primates: comparisons with other mammals in relation to diet. *J. Morphol.* 166:337–386.

Chivers, D. J., and Raemaekers, J. J. (1986). Natural and synthetic diets of Malayan gibbons. In: Else, J. G., and Lee, P. C. (eds.), *Primate Ecology and Conservation.* Cambridge University Press, New York. pp. 39–56.

Christen, A. (1999). Survey of Goeldi's monkeys (*Callimico goeldii*) in northern Bolivia. *Folia Primatol.* 70:107–111.

Cincotta, R. P., Wisnewski, J., and Engelman, R. (2000). Human population in the biodiversity hotspots. *Nature* 404:990–992.

Ciochon, R. L., and Gunnell, G. F. (2002). Chronology of primate discoveries in Myanmar: influences on the anthropoid origins debate. *Ybk. Phys. Anthropol.* 45:2–35.

Cipolletta, C. (2003). Ranging patterns of a western gorilla group during habituation to humans in the Dzanga-Ndoki National Park, Central African Republic. *Int. J. Primatol.* 24:1207–1226.

Cipolletta, C. (2004a). Consistency and variability in the ranging patterns of a western gorilla group (*Gorilla gorilla gorilla*) at Bai Hokou, Central African Republic. *Am. J. Primatol.* 64:193–205.

Cipolletta, C. (2004b). Effects of group dynamics and diet on the ranging patterns of a western gorilla group (*Gorilla gorilla gorilla*) at Bai Hokou, Central African Republic. *Am. J. Primatol.* 64:193–205.

Cipollini, M. L., and Levey, D. J. (1997). Secondary metabolites of fleshy vertebrate-dispersed fruits: adaptive hypotheses and implications for seed dispersal. *Am. Nat.* 150:346–372.

Cisneros-Herediam, D. F., Leon-Reyes, A., and Seger, S. (2005). *Boa constrictor* predation on a titi monkey, *Callicebus discolor. Neotrop. Primates* 13(3):11–12.

Clark, A. B. (1978a). Olfactory communication, *Galago crassicaudatus,* and the social life of prosimians. In: Chivers, D. J., and Joysey, K. A. (eds.),

Recent Advances in Primatology. Evolution, vol. 3. Academic Press, New York. pp. 109–117.

Clark, A. B. (1978b). Sex ratio and local resource competition in a prosimian primate. *Science* 201:163–165.

Clark, A. B. (1982a). Scent marks as social signals in *Galago crassicaudatus* I. Sex and reproductive status as factors in signals and responses. *J. Chem. Ecol.* 8:1133–1151.

Clark, A. B. (1982b). Scent marks as social signals in *Galago crassicaudatus*. II. Discrimination between individuals by scent. *J. Chem. Ecol.* 8:1153–1165.

Clark, A. B. (1985). Sociality in a nocturnal "solitary" prosimian: *Galago crassicaudatus. Int. J. Primatol.* 6:581–600.

Clark, D. A., and Clark, D. B. (1984). Spacing dynamics of a tropical rain forest tree: evaluation of the Janzen-Connell model. *Am. Nat.* 124:769–788.

Clark, J. S., Silman, M., Kern, R., Macklin, E., and HilleRisLambers, J. (1999). Seed dispersal near and far: patterns across temperate and tropical forests. *Ecology* 80:1475–1494.

Clark, J. S., Lewis, M., McLachlan, J. S., and HilleRisLambers, J. (2003). Estimating population spread: what can we forecast and how well? *Ecology* 84:1979–1988.

Clarke, A. S., and Boinski, S. (1995). Temperament in nonhuman primates. *Am. J. Primatol.* 37:103–125.

Clarke, E. A., Reichard, U. H., and Zuberbühler, K. (2006). The syntax and meaning of wild gibbon songs. PLoS ONE 1:e73. doi:10.1371/journal. pone.0000073.

Clarke, M. R. (1978). Social interactions of juvenile female bonnet monkeys, *Macaca radiata. Primates* 19:517–524.

Clarke, M. R. (1983). Infant-killing and infant disappearance following male takeovers in a group of free-ranging howling monkeys (*Alouatta palliata*) in Costa Rica. *Am. J. Primatol.* 5:241–247.

Clarke, M. R. (1990). Behavioral development and socialization of infants in a free-ranging group of howling monkeys (*Alouatta palliata*). *Folia Primatol.* 54:1–15.

Clarke, M. R., and Glander, K. E. (1984). Female reproductive success in a group of free-ranging howling monkeys (*Alouatta palliata*) in Costa Rica. In: Small, M. F. (ed.), *Female Primates: Studies by Women Primatologists.* Alan R. Liss, New York. pp. 111–126.

Clarke, M. R., and Zucker, E. L. (1994). Survey of the howling monkey population at La Pacifica: a seven-year follow-up. *Int. J. Primatol.* 15:61–73.

Clarke, M. R., Zucker, E. L., and Scott, N. J., Jr. (1986). Population trends of the mantled howler groups of La Pacifica, Guanacaste, Costa Rica. *Am. J. Primatol.* 11:79–88.

Clarke, M. R., Zucker, E. L., and Glander, K. E. (1994). Group takeover by a natal male howling monkey (*Alouatta palliata*) and associated disappearance and injuries of immatures. *Primates* 35:435–442.

Clarke, M. R., Glander, K. E., and Zucker, E. L. (1998). Infant–nonmother interactions of free-ranging mantled howlers (*Alouatta palliata*) in Costa Rica. *Int. J. Primatol.* 19:451–472.

Clarke, M. R., Crockett, C. M., Zucker, E. L., and Zaldivar, M. (2002). Mantled howler population of Hacienda la Pacifica, Costa Rica, between 1991 and 1998: effects of deforestation. *Am. J. Primatol.* 56:155–163.

Clayton, D. H., and Moore, J. (1997). *Host–Parasite Evolution. General Principles and Avian Models.* Oxford University Press, Oxford.

Clemens, E. T., and Moloiy, G. M. O. (1981). Organic acid concentrations and digesta movement in the gastrointestinal tract of the bushbaby (*Galago crassicaudatus*) and vervet monkey (*Cercopithecus aethiops*). *J. Zool. Lond.* 193:487–497.

Clemens, E. T., and Phillips, B. (1980). Organic acid production and digesta movement in the gastrointestinal tract of the baboon and sykes monkey. *Comp. Biochem. Physiol.* 66:529–532.

Cleveland, A., Westergaard, G. C., Trenkle, M. K., and Higley, J. D. (2003). Physiological predictors of reproductive outcome and mother–infant behaviors in captive rhesus macaque females (*Macaca mulatta*). *Neuropsychopharmacology* 29:891–910.

Clifford, S. L., Anthony, N. M., Bawe-Johnson, M., Abernethy, K. A., Tutin, C. E. G., White, L. J. T, Bermejo, M., Goldsmith, M. L., McFarland, K., Jeffery, K. J., Bruford, M. W., and Wickings, E. J. (2004). Identification of polymorphic microsatellite loci in the gorilla (*Gorilla gorilla gorilla*) using human primers: application to noninvasively collected hair samples. *Mol. Ecol.* 8:1556–1558.

Clobert, J., Danchin, E., Dohndt, A. A., and Nichols, J. D. (2001). *Dispersal.* Oxford University Press, Oxford.

Clutton-Brock, T. H. (1974a). Activity patterns of red colobus (*Colobus badius tephrosceles*). *Folia Primatol.* 21:161–187.

Clutton-Brock, T. H. (1974b). Primate social organization and ecology. *Nature* 250:539–542.

Clutton-Brock, T. H. (1975a). Feeding behaviour of red colobus and black and white colobus in East Africa. *Folia Primatol.* 23:165–207.

Clutton-Brock, T. H. (1975b). Ranging behavior of red colobus (*Colobus badius tephrosceles*) in the Gombe National Park. *Anim. Behav.* 23:706–722.

Clutton-Brock, T. H. (1977). Some aspects of intraspecific variation in feeding and ranging behavior in primates. In: Clutton-Brock, T. H. (ed.), *Primate Ecology.* Academic Press, New York. pp. 539–556.

Clutton-Brock, T. H. (1989). Mammalian mating systems. *Proc. R. Soc. Lond. B* 236:339–372.

Clutton-Brock, T. H. (1991). *The Evolution of Parental Care.* Princeton University Press, Princeton, NJ.

Clutton-Brock, T. (2002). Breeding together: kin selection and mutualism in cooperative vertebrates. *Science* 296:69–72.

Clutton-Brock, T. H., and Harvey, P. H. (1976). Evolutionary rules and primate societies. In: Bateson, P. P. G., and Hinde, R. A. (eds.), *Growing Points in Ethology.* Cambridge University Press, Cambridge. pp. 195–237.

Clutton-Brock, T. H., and Harvey, P. H. (1977a). Primate ecology and social organization. *J. Zool. Lond.* 183:1–39.

Clutton-Brock, T. H., and Harvey, P. H. (1977b). Species differences in feeding and ranging behaviour in primates. In: Clutton-Brock, T. H. (ed.), *Primate Ecology: Studies of Feeding and Ranging Behavior in Lemurs, Monkeys and Apes.* Academic Press, New York. pp. 557–584.

Clutton-Brock, T. H., and Harvey, P. H. (1980). Primates, brains and ecology. *J. Zool. Lond.* 190:309–323.

Clutton-Brock, T. H., and Parker, G. A. (1995). Punishment in animal societies. *Nature* 373:209–216.

Clutton-Brock, T. H., Brotherton, P. N. M., Smith, R., McIlrath, G. M., Kansky, R., Gaynor, D., O'Riain, M. J., and Skinner, J. D. (1998). Infanticide and expulsion of females in a cooperative mammal. *Proc. R. Soc. Lond. B* 265:2291–2295.

Clutton-Brock, T. H., Gaynor, D., McIlrath, G. M., Maccoll, A. D. C., Kansky, R., Chadwick, P., Manser, M., Skinner, J. D., and Brotherton, P. N. M. (1999). Predation, group size and mortality in a cooperative mongoose, *Suricata suricatta. J. Anim. Ecol.* 68:672–683.

Coan, J. A., and Allen, J. J. B. (eds.) (2007). *Handbook of Emotion Elicitation and Assessment:* Oxford University Press, New York.

Cochran, W. W., and Lord, R. D. (1963). A radio tracking system for wild animals. *J. Wildlife Manage.* 27:9–24.

Cochran, W. W., Warner, D. W., Tester, J. R., and Kuechle, V. B. (1965). Automatic radio-tracking system for monitoring animal movements. *Bioscience* 15:98–100.

Cocks, L. (2007). Factors affecting mortality, fertility, and well-being in relation to species differences in captive orangutans. *Int. J. Primatol.* 28:421–428.

Codron, D., Lee-Thorp, J. A., Sponheimer, M., de Ruiter, D., and Codron, J. (2008). What insights can baboon feeding ecology provide for early hominin niche differentiation? *Int. J. Primatol.* 29:757–772.

Coe, C. L., Conolly, A. C., Kraemer, H. C., and Levine, S. (1979). Reproductive development and behavior of captive female chimpanzees. *Primates* 20:571–582.

Coelho, A. M. (1985). Baboon dimorphism: growth in weight, length and adiposity from birth to 8 years of age. In: Watts, E. S. (ed.), *Nonhuman Primate Models for Human Growth and Development.* Alan R. Liss, New York. pp. 125–159.

Coehlo, A. M., Jr., Bramblett, C. A., Quick, L. B., and Bramblett, S. S. (1976a). Resource availability and population density in primates: a socio-bioenergetic analysis of energy budgets of Guatemalan howler and spider monkeys. *Primates* 17:63–80.

Coehlo, A. M., Jr., Coehlo, L. S., Bramblett, C. A., Bramblett, S. S., and Quick, L. B. (1976b). Ecology, population characteristics, and sympatric association in primates: a socio-bioenergetic analysis of howler and spider monkeys in Tikal, Guatemala. *Ybk. Phys. Anthropol.* 20:96–135.

Cole, L. C. (1954). The population consequences of life history phenomena. *Q. Rev. Biol.* 29:103–137.

Coleman, R. J., and Kemnitz, J. W. (1998). Aging experiments using nonhuman primates. In: Yu, B. P. (ed.), *Methods in Aging Research.* CRC Press, Boca Raton, FL. pp. 249–267.

Coles, R., and Talebi, M. G. (2004). Spatial relations in wild southern muriquis (*Brachyteles arachnoides*): choices of nearest neighbour and proximity. *Folia Primatol.* 75(suppl. 1):365–366.

Collard, M., and O'Higgins, P. (2001). Ontogeny and homoplasy in the papionin monkey face. *Evol. Dev.* 3:322–331.

Collias, N., and Southwick, C. (1952). A field study of population density and social organization in howling monkeys. *Proc. Am. Phil. Soc.* 96:143–156.

Collings, M. R. (1926). A study of the cutaneous reddening and swelling about the genitalia of the monkey *Macacus rhesus. Anat. Rec.* 33:271–287.

Collins, A. C. (2001). The importance of sampling for reliable assessment of phylogenetics and conservation among neotropical primates: a case study in spider monkeys (*Ateles*). *Primate Rep.* 61:9–30.

Collins, A. C. (2003). Atelinae phylogenetic relationships: the trichotomy revisited? *Am. J. Phys. Anthropol.* S36:78.

Collins, A. C., and Dubach, J. M. (2000a). Biogeographic and ecological forces responsible for speciation in *Ateles. Int. J. Primatol.* 21:421–444.

Collins, A. C., and Dubach, J. M. (2000b). Phylogenetic relationships of spider monkeys (*Ateles*) based on mitochondrial DNA variation. *Int. J. Primatol.* 21:381–420.

Collins, D. A. (1981). Social behaviour and patterns of mating among adult yellow baboons (*Papio c. cynocephalus.* L. 1766) [PhD diss.]. University of Edinburgh, Edinburgh.

Collins, D. A. (1984). Spatial pattern in a troop of yellow baboons (*Papio cynocephalus*) in Tanzania. *Anim. Behav.* 32:536–553.

Collins, D. A., Busse, C. D., and Goodall, J. (1984). Infanticide in two populations of savanna baboons. In: Hrdy, S. B., and Hausfater, G. (eds.), *Infanticide: Comparative and Evolutionary Perpectives.* Aldine de Gruyter, New York. pp. 193–215.

Collins, R., and Nekaris, K. A. I. (2008). Release of greater slow lorises (*Nycticebus coucang*), confiscated from the pet trade, to Batutegi Protected Forest, Sumatra, Indonesia. In: Sorrae, P. S. (ed.), *IUCN Global Re-introduction Perspectives: Re-introduction Case Studies from Around the Globe.* IUCN/SSC Reintroduction Specialist Group, Abu Dhabi, UAE. pp. 192–195.

Colmenares, F. (1990). Greeting behaviour in male baboons, I: communication, reciprocity and symmetry. *Behaviour* 113:81–116.

Colmenares, F. (1991). Greeting behaviour between male baboons: oestrus females, rivalry and negotiation. *Anim. Behav.* 41:49–60.

Colmenares, F. (1992). Clans and harems in a colony of hamadryas and hybrid baboons: male kinship, familiarity and the formation of brother-teams. *Behaviour* 121:61–94.

Colmenares, F. (2004). Kinship structure and its impact on behavior in multilevel societies. In: Chapais, B. and Berman, C.M. (eds.), *Kinship and Behavior in Primates.* Oxford University Press, New York. pp. 242–270.

Colmenares, F. (2006). Is postconflict affiliation in captive nonhuman primates an artifact of captivity? *Int. J. Primatol.* 27:1311–1335.

Colmenares, F., and Gomendio, M. (1988). Changes in female reproductive condition following male take-overs in a colony of hamadryas and hybrid baboons. *Folia Primatol.* 50:157–174.

Colquhoun, I. C. (1993). The socioecology of *Eulemur macaco.* In: Kappeler, P. M., and Ganzhorn, J. U. (eds.), *Lemur Social Systems and Their Ecological Basis.* Plenum Press, New York. pp. 11–23.

Colquhoun, I. C. (1998). Cathemeral behavior of *Eulemur macaco macaco* at Ambato Massif, Madagascar. *Folia Primatol.* 69(suppl. 1):22–34.

Colquhoun, I. C. (2004). Primates in the forest: Sakalava ethnoprimatology and synecological relations with black lemurs at Ambato Massif, Madagascar. In: Paterson, J. (ed.), *Commensalism and Conflict: The Human–Primate Interface.* American Society of Primatologists, Norman, OK.

Colquhoun, I. C. (2006). Predation and cathemerality. *Folia Primatol.* 77:143–165.

Colvin, J. D. (1983). Familiarity, rank and the structure of rhesus male peer networks. In: Hinde, R.A (ed.), *Primate Social Relationships: An Integrated Approach.* Blackwell, Oxford. pp. 190–200.

Colvin, J. D. (1986). Proximate causes of male emigration at puberty in rhesus monkeys. In: Rawlins, R. G., and Kessler, M. J. (eds.), *The Cayo Santiago Macaques: History, Behavior and Biology.* SUNY Press, Albany. pp. 131–157.

Colyn, M. (1994). Données pondérales sur les primates Cercopithecidae d-Afrique Central (Bassin du Zaire/Congo). *Mammalia* 58:483–487.

Colyn, M., Gautier-Hion, A., and Thys van den Audenaerde, D. (1991). *Cercopithecus dryas* Schwarz 1932 and *C. solango* Thys van den Audenaerde 1977 are the same species with an age-related coat pattern. *Folia Primatol.* 56:167–170.

Combes, S. L., and Altmann, J. (2001). Status change during adulthood: life-history by-product or kin selection based on reproductive value? *Proc. R. Soc. Lond.* 268:1367–1373.

Conaway, C. H., and Koford, C. B. (1965). Estrous cycles and mating behavior in a free-ranging band of rhesus monkeys. *J. Mammal.* 45:577–588.

Condit, R., Hubbell, S. P., and Foster, R. B. (1992). Recruitment near conspecific adults and the maintenance of tree and shrub diversity in a neotropical forest. *Am. Nat.* 140:261–286.

Condit, V. K., and Smith, E. O. (1994). Predation on a yellow baboon (*Papio cynocephalus cynocephalus*) by a lioness in the Tana River National Primate Reserve, Kenya. *Am. J. Primatol.* 33:57–64.

Conklin-Brittain, N. L., Wrangham, R. W., and Hunt, K. D. (1998). Dietary response of chimpanzees and cercopithecines to seasonal variation in fruit abundance: II. Nutrients. *Int. J. Primatol.* 19:971–987.

Conklin-Brittain, N. L., Dierenfeld, E. S., Wrangham, R. W., Norconk, M., and Silver, S. C. (1999). Chemical protein analysis: a comparison of Kjeldahl crude protein and total ninhydrin protein from wild, tropical vegetation. *J. Chem. Ecol.* 25:2601–2622.

Conklin-Brittain, N., Knott, C. D., and Wrangham, R. W. (2001). The feeding ecology of apes. In: *Apes: Challenges for the 21st Century.* Chicago Zoological Society, Brookfield, IL. pp. 167–174.

Conklin-Brittain, N. L., Knott, C. D., and Wrangham, R. W. (2006). Energy intake by wild chimpanzees and orangutans: methodological considerations and a preliminary comparison. In: Hohmann G., Robbins, M. M., and Boesche, C. (eds.), *Feeding Ecology in Apes and Other Primates: Ecological, Physiological, and Behavioural Aspects.* Cambridge University Press, Cambridge. pp. 445–472.

Connell, J. H. (1971). On the role of natural enemies in preventing competitive exclusion in some marine animals and rain forest trees. In: dem Boer, P. J., and Gradwell, G. R. (eds.), *Dynamics of Populations.* Centre for Agricultural Publishing and Documentation, Wageningen. pp. 298–312.

Conover, M. R. (1994). Stimuli eliciting distress calls in adult passerines and response of predators and birds to their broadcast. *Behaviour* 131:19–37.

Constable, J. L., Packer, C., Collins, D. A., and Pusey, A. E. (1995). Nuclear DNA from primate dung. *Nature* 373:393.

Constable, J. L., Ashley, M. V., Goodall, J., and Pusey, A. E. (2001). Noninvasive paternity assignment in Gombe chimpanzees. *Mol. Ecol.* 10:1279–1300.

Constantino, P. J., and Wright, B. W. (2009). Introduction to the symposium issue: the importance of fallback foods in primate ecology and evolution. *Am. J. Phys. Anthropol.* 140(4): 599–602.

Converse, L. J., Carlson, A. A., Ziegler, T. E., and Snowdon, C. T. (1995). Communication of ovulatory state to mates by female pygmy marmosets, *Cebuella pygmaea. Anim. Behav.* 49:615–621.

Coolidge, H. J. (1933). *Pan paniscus,* pygmy chimpanzee from south of the Congo River. *Am. J. Phys. Anthropol.* 18:1–59.

Cools, A. K. A., van Hout, A. J. M., and Nelissen, M. H. J. (2008). Canine reconciliation and third-party-initiated postconflict affiliation: do peacemaking social mechanisms in dogs rival those of higher primates? *Ethology* 114:53–63.

Coombs, J. A., Letcher, B. H., and Nislow, K. H. (2008). CREATE: a software to create input files from diploid genotypic data for 52 genetic software programs. *Mol. Ecol. Resources* 8:578–580.

Cooney, D. O., and Struhsaker, T. T. (1997). Adsorptive capacity of charcoals eaten by Zanzibar red colobus monkeys: implications for reducing dietary toxins. *Int. J. Primatol.* 18:235–246.

Cooper, L. R., and Harlow, H. F. (1961). Note on a cebus monkey's use of a stick as a weapon. *Psychol. Rep.* 8:418.

Cooper, M. A. (2004). Inter-group relationships. In: Thierry, B., Singh, M., and Kaumanns, W. (eds.), *Macaque Societies: A Model for the Study of Social Organization.* Cambridge University Press, Cambridge. pp. 204–208.

Cooper, M. A., and Bernstein, I. S. (2002). Counter aggression and reconciliation in Assamese macaques (*Macaca assamensis*). *Am. J. Primatol.* 56:215–230.

Cooper, M. A., and Bernstein, I. S. (2008). Evaluating dominance styles in Assamese macaques and rhesus macaques. *Int. J. Primatol.* 29:225–243.

Cooper, M. A., Aureli, F., and Singh, M. (2004a). Between-group encounters among bonnet macaques (*Macaca radiata*). *Behav. Ecol. Sociobiol.* 56:217–227.

Cooper, M. A., Chaitra, M. S., and Singh, M. (2004b). Effect of dominance, reproductive state, and group size on body mass in *Macaca radiata. Int. J. Primatol.* 25:165–178.

Cooper, M. A., Bernstein, I. S., and Hemelrijk, C. K. (2005). Reconciliation and relationship quality in Assamese macaques (*Macaca assamensis*). *Am. J. Primatol.* 65:269–282.

Cooper, M., Aureli, F., and Singh, M. (2007). Sex differences in reconciliation and post-conflict anxiety in bonnet macaques. *Ethology* 113:26–38.

Corbin, G. D., and Schmid, J. (1995). Insect secretions determine habitat use patterns by a female lesser mouse lemur (*Microcebus murinus*). *Am. J. Primatol.* 37:317–324.

Cordeiro, N. J., and Howe, H. F. (2003). Low recruitment of trees dispersed by animals in African forest fragments. *Conserv. Biol.* 15:1733–1741.

Cordoni, G., and Palagi, E. (2007). Response of captive lowland gorillas (*Gorilla gorilla gorilla*) to different housing conditions: testing the aggression-density and coping models. *J. Comp. Psych.* 121:171–80.

Cordoni, G., and Palagi, E. (2008). Reconciliation in wolves (*Canis lupus*): new evidence for a comparative perspective. *Ethology* 114:298–308.

Cordoni, G., Palagi, E., and Tarli, S. (2006). Reconciliation and consolation in captive western gorillas. *Int. J. Primatol.* 27:1365–1382.

Cords, M. (1986). Interspecific and intraspecific variation in diet of two forest guenons, *Cercopithecus ascanius* and *C. mitis*. *J. Anim. Ecol.* 55:811–827.

Cords, M. (1987a). *Mixed-Species Association of* Cercopithecus *Monkeys in the Kakamega Forest, Kenya*. University of California Publications in Zoology, vol. 17. University of California Press, Berkeley.

Cords, M. (1987b). Forest guenons and patas monkeys: male–male competition in one-male groups. In: Smuts, B. B., Cheney, D. L., Seyfarth, R. M., Wrangham, R. W., and Struhsaker, T. T. (eds.), *Primate Societies*. University of Chicago Press, Chicago. pp. 98–111.

Cords, M. (1988a). Mating systems of forest guenons: a preliminary review. In: Gautier-Hion, A., Bourlière, F., Gautier, J. P., and Kingdon, J. (eds.), *A Primate Radiation: Evolutionary Biology of the African Guenons*. Cambridge University Press, New York. pp. 323–339.

Cords, M. (1988b). Resolution of aggressive conflicts by immature long-tailed macaques, *Macaca fascicularis*. *Anim. Behav.* 36:1124–1135.

Cords, M. (1990a). Mixed-species association of East African guenons: general patterns or specific examples? *Am. J. Primatol.* 21:101–114.

Cords, M. (1990b). Vigilance and mixed-species associations of some East African forest monkeys. *Behav. Ecol. Sociobiol.* 26:297–300.

Cords, M. (1992). Postconflict reunions and reconciliation in long-tailed macaques. *Anim. Behav.* 44:57–61.

Cords, M. (1993). On operationally defining reconciliation. *Am. J. Primatol.* 29:255–267.

Cords, M. (1997). Friendships, alliances, reciprocity and repair. In: Whiten, A., and Byrne, R. W. (eds.), *Machiavellian Intelligence II: Evaluations and Extensions*. Cambridge University Press, Cambridge. pp. 24–49.

Cords, M. (2000a). The number of males in guenon groups. In: Kappeler, P. M. (ed.), *Primate Males*. Cambridge University Press, Cambridge. pp.84–96.

Cords, M. (2000b). The agonistic and affiliative relationships of adult females in a blue monkey group. In: Whitehead, P., and Jolly, C. (eds.), *Old World Monkeys*. Cambridge University Press, Cambridge. pp. 453–479.

Cords, M. (2000c). Grooming partners of immature blue monkeys (*Cercopithecus mitis*) in the Kakamega Forest, Kenya. *Int. J. Primatol.* 21(2):239–254.

Cords, M. (2002a). Foraging and safety in adult female blue monkeys in the Kakamega Forest, Kenya. In: Miller, L. E. (ed.), *Eat or Be Eaten: Predator Sensitive Foraging Among Primates*. Cambridge University Press, Cambridge. pp. 205–221.

Cords, M. (2002b). When are there influxes in blue monkey groups? In: Glenn, M. E., and Cords, M. (eds.), *The Guenons: Diversity and Adaptation in African Monkeys*. Kluwer Academic/Plenum Press, New York. pp. 189–201.

Cords, M. (2002c) Friendship among adult female blue monkeys (*Cercopithecus mitis*). *Behaviour* 139:291–314.

Cords, M., and Aureli, F. (1993). Patterns of reconciliation among juvenile long-tailed macaques. In: Pereira, M. E., and Fairbanks, L. A. (eds.), *Juvenile Primates: Life History, Development and Behavior*. Oxford University Press, New York. pp. 271–284.

Cords, M., and Aureli, F. (1996). Reasons for reconciling. *Evol. Anthropol.* 5:42–45.

Cords, M., and Aureli, F. (2000). Reconciliation and relationship qualities. In: Aureli, F., and de Waal, F. B. M. (eds.), *Natural Conflict Resolution*. University of California Press, Berkeley. pp. 177–198.

Cords, M., and Killen, M. (1998). Conflict resolution in human and nonhuman primates. In: Langer, J., and Killen, M. (eds.), *Piaget, Evolution, and Development*. Lawrence Erlbaum, Mahwah, NJ. pp. 193–219.

Cords, M., and Rowell, T. E. (1986). Group fission in blue monkeys of the Kakamega Forest, Kenya. *Folia Primatol.* 46:70–82.

Cords, M., and Rowell, T. E. (1987). Birth intervals of *Cercopithecus* monkeys of the Kakamega Forest, Kenya. *Primates* 28:277–281.

Cords, M., and Simpson, S. (2006). Limits to kin bias in wild blue monkey society. *Int. J. Primatol.* 27(suppl. 1):266.

Cords, M., and Thurnheer, S. (1993). Reconciling with valuable partners by long-tailed macaques. *Ethology* 93:315–325.

Cords, M., Mitchell, B. J., Tsingalia, H. M., and Rowell, T. E. (1986). Promiscuous mating among blue monkeys in the Kakamega Forest, Kenya. *Ethology* 72:214–226.

Corlett, R. T. (1998). Frugivory and seed dispersal by vertebrates in the Oriental (Indomalayan) region. *Biol. Rev.* 73:413–448.

Corlett, R. T., and Lucas, P. W. (1990). Alternative seed-handling strategies in primates: seed-spitting by long-tailed macaques (*Macaca fascicularis*). *Oecologia* 82:166–171.

Cormier, L. A. (2002). Monkey as food, monkey as child: Guaja symbolic cannibalism. In: Fuentes, A., and Wolfe, L. D. (eds.), *Primates Face to Face*. Cambridge University Press, Cambridge. pp. 63–84.

Cormier, L. A. (2003). *Kinship with Monkeys: The Guaja Foragers of Eastern Amazonia*. Columbia University Press, New York.

Cornejo, F. M., Aquino, R., and Jiménez, C. (2008). Notes on the natural history, distribution and conservation status of the Andean night monkey, *Aotus miconax* Thomas, 1927. *Prim. Conserv.* 23: 1–4.

Corrêa, H. K. M., and Coutinho, P. E. G. (1997). Fatal attack of a pit viper, *Bothrops jararaca*, on an infant buffy-tufted ear marmoset (*Callithrix aurita*). *Primates* 38:215–217.

Corrêa, H. K. M., Coutinho, P. E. G., and Ferrari, S. F. (2000). Between-year differences in the feeding ecology of highland marmosets (*Callithrix aurita* and *Callithrix flaviceps*) in southeastern Brazil. *J. Zool.* 252:421–427.

Corrêa, H. K. M., Coutinho, P. E. G., and Ferrari, S. F. (2002). Dieta de grupos de *Mico argentatus* em fragmentos naturais de Alter do Chão, Santarém, Pará. Livro de Resumos do X Congresso Brasileiro de Primatologia, 10–15 November, 2002. Belém, Brazil p. 46.

Cortés-Ortiz, L., Bermingham, E., Rico, C., Rodríguez-Luna, E., Sampaio, I., and Ruiz-García, M. (2003). Molecular systematics and biogeography of the neotropical monkey genus, *Alouatta*. *Mol. Phylogenet. Evol.* 26:64–81.

Coss, R. G., Ramakrishnan, U., and Schank, J. (2005). Recognition of partially concealed leopards by wild bonnet macaques (*Macaca radiata*) the role of the spotted coat. *Behav. Proc.* 68(2):145–163.

Coss, R. G., McCowan, B., and Ramakrishnan, U. (2007). Threat-related acoustical differences in alarm calls by wild bonnet macaques (*Macaca radiata*) elicited by python and leopard models. *Ethology* 113(4):352–367.

Costello, R. K., Dickinson, C., Rosenberger, A. L., Boinski, S., and Szalay, F. S. (1993). Squirrel monkey (genus *Saimiri*) taxonomy: a multidisciplinary study of the biology of species. In: Kimbell, W. H., and Martin, L. B. (eds.), *Species, Species Concepts, and Primate Evolution*. Plenum Press, New York. pp. 177–210.

Cottam, G., and Curtis, J. T. (1956). The use of distance measures in phytosociological sampling. *Ecology* 37:451–460.

Coultas, D. S. (2002). Bioacoustic analysis of the loud call of two species of slender loris (*Loris tardigradus* and *L. lydekkerianus nordicus*) from Sri Lanka [MSc thesis]. Oxford Brookes University, Oxford.

Courlander, H. (1975). *A Treasury of African Folklore*. Crown Publishers, New York.

Courtenay, D. O., and Bearder, S. K. (1989). The taxonomic status and distribution of bushbabies in Malawi with emphasis on the significance of vocalisations. *Int. J. Primatol.* 10:17–34.

Courtenay, J., Groves, C., and Andrews, P. (1988). Inter- or intra-island variation? An assessment of the differences between Bornean and Sumatran orang-utans. In: Schwartz, J. H. (ed.), *Orang-utan Biology*. Oxford University Press, Oxford. pp. 19–29.

Cousins, D., and Huffman, M. A. (2002). Medicinal properties in the diet of gorillas—an ethnopharmacological evaluation. *Afr. Stud. Monogr.* 23:65–89.

Coussi-Korbel, S., and Fragaszy, D. M. (1995). On the social relation between social dynamics and social learning. *Anim. Behav.* 50:1441–1453.

Coutinho, P. E. G., and Corrêa, H. K. M. (1995). Polygyny in a free-raning group of buffy-tufted-ear marmosets, *Callithrix aurita*. *Folia Primatol.* 65:25–29.

Covert, H. H. (2002). The earliest fossil primates and the evolution of prosimians: introduction. In: Hartwig, W. C. (ed.), *The Primate Fossil Record*. Cambridge University Press, Cambridge. pp. 13–20.

Cowlishaw, G. (1994). Vulnerability to predation in baboon populations. *Behaviour* 131:293–304.

Cowlishaw, G. (1997a). Refuge use and predation risk in a desert baboon population. *Anim. Behav.* 54:241–253.

Cowlishaw, G. (1997b). Trade-offs between foraging and predation risk determine habitat use in a desert baboon population. *Anim. Behav.* 53:667–686.

Cowlishaw, G. (1998). The role of vigilance in the survival and reproductive strategies of desert baboons. *Behaviour* 135:431–452.

Cowlishaw, G. (1999a). Ecological and social determinants of spacing behaviour in desert baboon groups. *Behav. Ecol. Sociobiol.* 45:67–77.

Cowlishaw, G. (1999b). Predicting the decline of African primate diversity. *Conserv. Biol.* 13:1183–1193.

Cowlishaw, G., and Dunbar, R. I. M. (1991). Dominance rank and mating success in male primates. *Anim. Behav.* 41:1045–1056.

Cowlishaw, G., and Dunbar, R. I. M. (2000). *Primate Conservation Biology*. University of Chicago Press, Chicago.

Cowlishaw, G. O., and O'Connell, S. M. (1996). Male–male competition, paternity certainty and copulation calls in female baboons. *Anim. Behav.* 51:235–238.

Cowlishaw, G., Lawes, M. J., Lightbody, M., Martin, A., Pettifor, R., and Rowcliffe, J. M. (2004). A simple rule for the costs of vigilance: empirical evidence from a social forager. *Proc. R. Soc. Lond. B* 271:27–33.

Cowlishaw, G., Pettifor, R. A., and Isaac, N. J. (2009). High variability in patterns of population decline: the importance of local processes in species extinctions. *Proc. Biol. Sci.* 276:63–69.

Cox, C. R., and Le Boeuf, B. J. (1977). Female incitation of male competition: a mechanism in sexual selection. *Am. Nat.* 111:317–335.

Cramer, D. L. (1977). Craniofacial morphology of *Pan paniscus*: a morphological and evolutionary appraisal. *Contrib. Primatol.* 10:1–64.

Cranfield, M. R. (2008). Mountain gorilla research: the risk of disease transmission relative to the benefit form the perspective of ecosystem health. *Am. J. Primatol.* 70:751–754.

Cranfield, M. R., Bavister, B. D., Boatman, D. E., Berger, N. G., Schaffer, N., Kempski, S. E., Ialeggio, D. M., and Smart, J. (1993). Assisted reproduction in the propagation management of the endangered lion-tailed macaque (*Macaca silenus*). In: Wolf, D. P., Stouffer, R. L., and Brenner, R. M. (eds.), *In Vitro Fertilization and Embryo Transfer in Primates*. Spring-Verlag, New York. pp. 331–348.

Craul, M., Zimmermann, E., and Radespiel, U. (2003). Experimental evidence for female mate choice in the grey mouse lemur (*Microcebus murinus*). *Folia Primatol.* 74:187.

Craul, M., Zimmermann, E., and Radespiel, J. (2004). First experimental evidence for female mate choice in a nocturnal primate. *Primates* 45:271–274.

Craul, M., Zimmermann, E., Rasoloharijaona, S., Randrianambinina, B., and Radespiel, U. (2007). Unexpected species diversity of Malagasy primates (Lepilemur spp.) in the same biogeographical zone: a morphological and molecular approach with the description of two new species. *BMC Evol. Biol.* 7:83.

Cresswell, W. (1994). Flocking is an effective anti-predation strategy in redshanks, *Tringa totanus*. *Anim. Behav.* 47:433–442.

Crile, G., and Quiring, D. P. (1940). A record of the body weight and certain organ and gland weights of 3690 animals. *Ohio. J. Sci.* 40:219–241.

Crissey, S. D., Oftedal, O. T., Currier, J. A., and Rudran, R. (1990). Gastrointestinal tract capacity, food passage rates and the possible role of fiber in diets fed to captive red howler monkeys (*Alouatta seniculus*) in Venezuela. *Am. Assoc. Zoo Vet. Annu. Proc.* 1990:81–86.

Crockett, C. M. (1984). Emigration by female red howler monkeys and the case for female competition. In: Small, M. F. (ed.), *Female Primates: Studies by Women Primatologists*. Alan R. Liss, New York. pp. 159–173.

Crockett, C. M. (1996). The relation between red howler monkey (*Alouatta seniculus*) troop size and population growth in two habitats. In: Norconk, M. A., Rosenberger, A. L., and Garber, P. A. (ed.), *Adaptive Radiations of Neotropical Primates*. Plenum, New York. pp. 489–510.

Crockett, C. M. (2003). Re-evaluating the sexual selection hypothesis for infanticide by *Alouatta* males. In: Jones, C. B. (ed.), *Selection and Reproductive Competition in Primates: New Perspectives and Directions*. American Society of Primatologists, Norman, OK. pp. 327–365.

Crockett, C. M., and Eisenberg, J. F. (1987). Howlers: variation in group size and demography. In: Smuts, B. B., Cheney, D. L., Seyfarth, R. M., Wrangham, R. W., and Struhsaker, T. T. (eds.), *Primate Societies*. University of Chicago Press, Chicago. pp. 54–68.

Crockett, C. M., and Janson, C. H. (2000). Infanticide in red howlers: female group size, male membership and a possible link to folivory. In: van Schaik, C. P., and Janson, C. H. (eds.), *Infanticide by Males and Its Implications*. Cambridge University Press, Cambridge. pp. 75–98.

Crockett, C. M., and Pope, T. R. (1988). Inferring patterns of aggression from red howler monkey injuries. *Am. J. Primatol.* 14:1–21.

Crockett, C. M., and Pope, T. R. (1993). Consequences of sex differences in dispersal for juvenile red howler monkeys. In: Pereira, M. E., and

Fairbanks, L. A. (eds.), *Juvenile Primates: Life History, Development, and Behavior*. Oxford University Press, New York. pp. 104–118.

Crockett, C. M., and Rudran, R. (1987a). Red howler monkey birth data. I: Seasonal variation. *Am. J. Primatol.* 13:347–368.

Crockett, C. M., and Rudran, R. (1987b). Red howler monkey birth data. II: Interannual, habitat, and sex comparisons. *Am. J. Primatol.* 13:369–384.

Crockett, C. M., and Sekulic, R. (1982). Gestation length in red howler monkeys. *Am. J. Primatol.* 3:291–294.

Crockett, C. M., and Sekulic, R. (1984). Infanticide in red howler monkeys (*Alouatta seniculus*). In: Hausfater, G., and Hrdy, S. B. (eds.), *Infanticide: Comparative and Evolutionary Perspectives*. Aldine, New York. pp. 173–191.

Crockett, C. M., and Wilson, W. L. (1980). The ecological separation of *Macaca nemestrina* and *M. fasicularis* in Sumatra. In: Lindburg, D. (ed.), *The Macaques. Studies in Ecology, Behavior, and Evolution*. van Nostrand Rheinhold, New York. pp. 148–181.

Crofoot, M. C., and Knott, C. D. (in press). What we do and don't know about orangutan male dimorphism. In: Galdikas, B. M. F., Briggs, N., Sheeran, L. K., and Shapiro, G. L. (eds.), *Great and Small Apes of the World. Orangutans and Gibbons, vol. 2*. Kluwer, New York.

Crofoot, M. C., Gilby, I. C., Wikelski, M. C., and Kays, R. W. (2008). Interaction location outweighs the competitive advantage of numerical superiority in *Cebus capucinus* intergroup contests. *Proc. Natl. Acad. Sci. USA* 105:577–581.

Crompton, R. H. (1980). A leap in the dark: locomotor behaviour and ecology in *Galago senegalensis* and *Galago crassicaudatus* [PhD diss.]. Harvard University, Cambridge, MA.

Crompton, R. H. (1983). Age differences in locomotion in two subtropical Galaginae. *Primates* 24:241–259.

Crompton, R. H. (1984). Foraging, habitat structure, and locomotion in two species of *Galago*. In: Rodman, P. S., and Cant, J. G. H. (eds.), *Adaptations for Foraging in Non-Human Primates*. Columbia University Press, New York. pp. 73–111.

Crompton, R. (1989). Mechanisms for speciation in *Galago* and *Tarsius*. *Hum. Evol.* 4:105–116.

Crompton, R., and Andau, P. (1986). Locomotion and habitat utilization in free ranging *Tarsius bancanus*: a preliminary report. *Primates* 27:337–355.

Crompton, R., and Andau, P. (1987). Ranging, activity rhythms, and sociality in free-ranging *Tarsius bancanus*: preliminary report. *Int. J. Primatol.* 8:43–71.

Crompton, R. H., and Sellers, W. I. (2007). A consideration of leaping locomotion as a means of predator avoidance in prosimian primates. In: Gursky, S. L., and Nekaris, K. A. I. (eds.), *Primate Anti-Predator Strategies*. Springer, New York. pp. 127–145.

Crompton, R. H., Sellers, W. I., and Gunther, M. M. (1993). Energetic efficiency and ecology as selective factors in saltatory adaptation of prosimian primates. *Proc. R. Soc. Lond. B* 254:41–45.

Cronin, H. (1991). *The Ant and the Peacock: Altruism and Sexual Selection from Darwin to Today*. Cambridge University Press, Cambridge.

Crook, H. H. (1970). The socio-ecology of primates. In: Crook, J. H. (ed.), *Social Behaviour in Birds and Mammals*. Academic Press, Cambridge. pp. 84–96.

Crook, J. H. (1972). Sexual selection, dimorphism, and social organization in the primates. In: Campbell, B. (ed.), *Sexual Selection and the Descent of Man*. Aldine, Chicago. pp. 231–281.

Crook, J. H., and Aldrich-Blake, P. (1968). Ecological and behavioural contrasts between sympatric ground dwelling primates in Ethiopia. *Folia Primatol.* 8:192–227.

Crook, J. H., and Gartlan, J. C. (1966). Evolution of primate societies. *Nature* 210:1200–1203.

Crook, J. H., Ellis, J. E., and Goss-Custard, J. D. (1976). Mammalian social systems. *Anim. Behav.* 24:261–274.

Cropp, S., and Boinski, S. (2000). The Central American squirrel monkey (*Saimiri oerstedii*): introduced hybrid or endemic species? *Mol. Phylogenet. Evol.* 16:350–365.

Crovella, S. J. C., Masters, J. C., and Rumpler, Y. (1994). Highly repeated DNA sequences as phylogenetic markers among Galaginae. *Am. J. Primatol.* 32:177–185.

Csermely, D. (1996). Antipredator behavior in lemurs: evidence of an extinct eagle on Madagascar or something else? *Int. J. Primatol.* 17:349–354.

Cuaron, A. D. (1997). Conspecific aggression and predation: costs for a solitary mantled howler monkey. *Folia Primatol.* 68:100–105.

Cui, L. W. (2003). A note on an interaction between *Rhinopithecus bieti* and a buzzard at Baima Snow Mountain. *Folia Primatol.* 74:51–53.

Cummins-Sebree, S. E., and Fragaszy, D. M. (2005). Capuchins as stone-knappers? An evaluation of the evidence. In: Roux, V., and Bril, B. (eds.), *Stone Knapping: The Necessary Conditions for a Uniquely Hominid Behavior*. McDonald Institute for Archaeological Research, Cambridge. pp. 171–182.

Cunningham, E., and Janson, C. (2007). A socioecological perspective on primate cognition, past and present. *Anim. Cogn.* 10:273–281.

Curio, E. (1978). The adaptive significance of avian mobbing. *Z. Tierpsychol.* 48:175–183.

Curnow, E. C., Kuleshova, L., Shaw, J. M., and Hayes, E. S. (2002). Comparison of slow- and rapid-cooling protocols for early-cleavage-stage *Macaca fascicularis* embryos. *Am. J. Primatol.* 58:169–174.

Curran, L. M., and Leighton, M. (2000). Vertebrate responses to spatio-temporal variation in seed production by mast-fruiting Bornean Dipterocarpaceae. *Ecol. Monogr.* 70:101–128.

Curran, L. M., Caniago, I., Paoli, G. D., Astiani, D., Kusneti, M., Leighton, M., Nirarita, C. E., and Haeruman, H. (1999). Impact of El Niño and logging on canopy tree recruitment in Borneo. *Science* 286:2184–2188.

Curran, L. M., Trigg, S. N., McDonald, A. K., Astiani, D., Hardiono, Y. M., Siregar, P., Caniago, I., and Kasischke, E. (2004). Lowland forest loss in protected areas of Indonesian Borneo. *Science* 303:1000–1003.

Curtin, R. A. (1975). The socioecology of the common langur, *Presbytis entellus*, in the Nepal Himalayas [PhD diss.]. University of California, Berkeley.

Curtin, R., and Dolhinow, P. (1978). Primate social behavior in a changing world. *Am. Sci.* 66:468–475.

Curtin, S. H. (1976). Niche separation in sympatric Malaysian leaf-monkeys (*Presbytis obscura* and *Presbytis melalophos*). *Ybk. Phys. Anthropol.* 20:421–439.

Curtin, S. H. (1980). Dusky and banded leaf monkeys. In: Chivers, D. J. (ed.), *Malayan Forest Primates*. Plenum Press, New York. pp. 107–145.

Curtin, S. H. (2002). Diet of the Roloway monkey, *Cercopithecus diana roloway* in Bia National Park, Ghana. In: Glenn, M. E., and Cords, M. (eds.), *The Guenons: Diversity and Adaptation in African Monkeys*. Kluwer Academic/Plenum, New York. pp. 351–371.

Curtis, D. J. (1997). The mongoose lemur (*Eulemur mongoz*): a study in behavior and ecology [PhD diss.]. University of Zurich, Zurich.

Curtis, D. J. (2004). Diet and nutrition in wild mongoose lemurs (*Eulemur mongoz*) and their implications for the evolution of female dominance and small group size in lemurs. *Am. J. Phys. Anthropol.* 124:234–247.

Curtis, D. J. (2006). Cathemerality in lemurs. In: Gould, L., and Sauther, M. L. (eds.), *Lemurs: Ecology and Adaptation*. Springer, New York. pp. 133–158.

Curtis, D. J., and Rasmussen, M. A. (2002). Cathemerality in lemurs. *Evol. Anthropol.* 1(suppl.):83–86.

Curtis, D. J., and Rasmussen, M. A. (2006). The evolution of cathemerality in primates and other mammals: a comparative and chronoecological approach. *Folia Primatol.* 77:178–93.

Curtis, D. J., and Zaramody, A. (1997). Monogamy and mate monopolization by females in *Eulemur mongoz*. *Primate Rep.* 48:16–17.

Curtis, D. J., and Zaramody, A. (1998). Group size, home range use, and seasonal variation in the ecology of *Eulemur mongoz*. *Int. J. Primatol.* 19:811–835.

Curtis, D. J., and Zaramody, A. (1999). Social structure and seasonal variation in the behaviour of *Eulemur mongoz*. *Folia Primatol.* 70:79–96.

Curtis, D. J., Velo, A., Raheliarisoa, E. O., Zaramody, A., and Mueller, P. (1998). Surveys on *Propithecus verreauxi deckeni*, a melanistic variant, and *P.v. coronatus* in north-west Madagascar. *Oryx* 32:157–164.

Curtis, D. J., Zaramody, A., and Martin, R. D. (1999). Cathemerality in the mongoose lemur, *Eulemur mongoz*. *Am. J. Primatol.* 47:279–298.

Custance, D., Whiten, A., and Fredman, T. (1999). Social learning of an artificial fruit task in capuchin monkeys (*Cebus apella*). *J. Comp. Psychol.* 113:13–23.

Custance, D. M., Whiten, A., Sambrook, T., and Galdikas, B. (2001). Testing for social learning in the "artificial fruit" processing of wildborn orangutans (*Pongo pygmaeus*), Tanjung Puting, Indonesia. *Anim. Cogn.* 4:305–313.

Custance, D. M., Whiten, A., and Fredman, T. (2002). Social learning and primate reintroduction. *Int. J. Primatol.* 23:479–499.

Czekala, N. M., and Robbins, M. M. (2001). Assessment of reproduction and stress through hormone analysis in gorillas. In: Robbins, M. M., Sicotte, P., and Stewart, K. J. (eds.), *Mountain Gorillas: Three Decades of Research at Karisoke*. Cambridge University Press, Cambridge. pp. 317–339.

Czekala, N. M., and Sicotte, P. (2000). Reproductive monitoring of free-ranging female mountain gorillas by urinary hormone analysis. *Am. J. Primatol.* 51:209–215.

Czekala, N. M., Benirschke, K., McClure, H., and Lasley, B. L. (1983). Urinary estrogen excretion during the pregnancy in the gorilla (*Gorilla gorilla*), orangutan (*Pongo pygmaeus*) and the human (*Homo sapiens*). *Biol. Reprod.* 28:289–294.

Czekala, N. M., Gallusser, S., Meier, J. E., and Lasley, B. L. (1986). The development and application of an enzyme immunoassay for urinary estrone conjugates. *Zoo Biol.* 5:1–6.

Czekala, N. M., Mitchell, W. R., and Lasley, B. L. (1987). Direct measurements of urinary estrone conjugates during the normal menstrual cycle of the gorilla (*Gorilla gorilla*). *Am. J. Primatol.* 12:223–229.

Czekala, N. M., Roser, J. F., Mortensen, R. B., Reichard, T., and Lasley, B. L. (1988). Urinary hormone analysis as a diagnostic tool to evaluate the ovarian function of female gorillas (*Gorilla gorilla gorilla*). *J. Reprod. Fertil.* 82:255–261.

Dacier, A., Maia, R., Agustinho, D. P., and Barros, M. (2006). Rapid habituation of scan behavior in captive marmosets following brief predator encounters. *Behav. Proc.* 71:66–69.

da Cunha, R. G. T., and Jalles-Filho, E. (2007). The roaring of southern brown howler monkeys (*Alouatta guariba clamitans*) as a mechanism of active defence of borders. *Folia Primatol.* 78:259–271.

Dagosto, M. (1994). Testing positional behavior of Malagasy lemurs: a randomization approach. *Am. J. Phys. Anthropol.* 94:189–202.

Dagosto, M. (1995). Seasonal variation in positional behavior of Malagasy lemurs. *Int. J. Primatol.* 16:807–834.

Dagosto, M. (2002). The origin and diversification of anthropoid primates. In: Hartwig, W. C. (ed.), *The Primate Fossil Record*. Cambridge University Press, Cambridge. pp. 125–132.

Dagosto, M., and Gebo, D. (1998). A preliminary study of the Philippine tarsier (*Tarsius syrichta*) in Leyte. *Am. J. Phys. Anthropol.* 26(suppl.):73.

Dagosto, M., Gebo, D., and Dolino, C. (2001). Positional behavior and social organization of the Philippine tarsier (*Tarsius syrichta*). *Primates* 42:233–243.

Dahl, J. (1994). Size and form of the penis in orangutans. *J. Mammal.* 75:1–9.

Dahl, K. D., and Hseuh, A. J. (1987). Use of the granulosa cell aromatase bioassay for measurement of bioactive follicle- stimulating hormone in urine and serum samples of diverse species. *Steroids* 50:375–392.

Dalling, J. W., and Hubbell, S. P. (2002). Seed size, growth rate and gap microsite conditions as determinants of recruitment success for pioneer species. *J. Ecol.* 90:557–568.

Dalling, J. W., Muller-Landau, H. C., Wright, S. J., and Hubbell, S. P. (2002). Role of dispersal in the recruitment limitation of neotropical pioneer species. *J. Ecol.* 90:714–727.

Dallmann, R., and Geissmann, T. (2001). Different levels of variability in the female song of wild silvery gibbons (*Hylobates moloch*). *Behaviour* 138:629–648.

Daly, M., Behrends, P., Wilson, M., and Jacobs, L. (1992). Behavioral modulation of predation risk: moonlight avoidance and crepuscular compensation in a nocturnal desert rodent, *Dipodomys merriami*. *Anim. Behav.* 44:1–9.

Dalziel, J. M. (1937). The useful plants of west tropical Africa. In: Hutchinson, J., and Dalziel, J. M. (eds.), *Appendix to Flora of West Tropical Africa*. Whitefriars Press, London.

D'Amato, M. R., and Van Sant, P. (1988). The person concept in monkeys (*Cebus apella*). *J. Exp. Psychol. Anim. Behav. Proc.* 14:43–55.

D'Amour, D. E., Hohmann, G., and Fruth, B. (2006). Evidence of leopard predation on bonobos (*Pan paniscus*). *Folia Primatol.* 77(3):212–217.

Daniels, D. (1983). The evolution of concealed ovulation and self-deception. *Ethol. Sociobiol.* 4:69–87.

Danish, L., Chapman, C. A., Hall, M. B., Rode, K. D., and O'Driscoll Worman, C. (2006). The role of sugar in diet selection in redtail and red colobus monkeys. In: Hohmann G., Robbins, M. M., and Boesche, C. (eds.), *Feeding Ecology in Apes and Other Primates: Ecological, Physiological, and Behavioural Aspects*. Cambridge University Press, Cambridge. pp. 473–488.

Darwin, C. (1859). *On the Origin of Species by Means of Natural Selection*. John Murray, London. Reprinted 1964, Harvard University Press, Cambridge, MA.

Darwin, C. (1860). Letter written to Asa Gray from Down on September 10. Darwin Correspondance Project, Letter # 2930. http://www.darwinproject.ac.uk/entry-2930.

Darwin, C. (1871). *The Descent of Man and Selection in Relation to Sex*. John Murray, London.

Darwin, C. (1872). *The Expression of the Emotions in Man and Animals*. J. Murray, London. Reprinted 1965, University of Chicago Press, Chicago.

Darwin, C. (1874). *The Descent of Man and Selection in Relation to Sex,* 2nd ed. Henneberry, Chicago.

Darwin, C. (1876). *The Autobiography of Charles Darwin 1809–1882.* Reprinted 1969, W. W. Norton, New York.

Das, M., Penke, Z., and van Hooff, J. A. R. A. M. (1998). Postconflict affiliation and stress-related behavior of long-tailed macaque aggressors. *Int. J. Primatol.* 19:53–71.

Das, N. (2009). Conservation status and survey of Bengal slow loris *Nycticebus bengalensis* in Assam, north-eastern India. Proceedings of the 10th Student Conference on Conservation Science. 24 -26 March, 2009, University of Cambridge, Cambridge. p. 4.

da Silva, E. C., Jr. (1981). A preliminary survey of brown howler monkeys (*Alouatta fusca*) at the Catareira Reserve, São Paulo, Brazil. *Rev. Brasil. Biol.* 41:897–909.

Dasilva, G. L. (1989). The ecology of the western black and white colobus (*Colobus polykomos polykomos* Zimmerman 1780) on a riverine island in south-eastern Sierra Leone [PhD diss.]. University of Oxford, Oxford.

Dasilva, G. L. (1992). The western black-and-white colobus as a low-energy strategist: activity budgets, energy expenditure and energy intake. *J. Anim. Ecol.* 61:79–91.

Dasilva, G. L. (1993). Postural changes and behavioural thermoregulation in *Colobus polykomos:* the effect of climate and diet. *Afr. J. Ecol.* 31:226–241.

Dasilva, G. L. (1994). Diet of *Colobus polykomos* on Tiwai Island: selection of food in relation to its seasonal abundance and nutritional quality. *Int. J. Primatol.* 15:655–680.

Daszak, P., Cunningham, A. A., and Hyatt, A. D. (2000). Emerging infectious diseases of wildlife—threats to biodiversity and human health. *Science* 287:443–449.

Datta, S. B. (1983). Relative power and the acquisition of rank. In: Hinde, R. A. (ed.), *Primate Social Relationships: An Integrated Approach.* Blackwell, Oxford. pp. 93–103.

Datta, S. B. (1987). The evolution of concealed ovulation in vervet monkeys. *Trends Ecol. Evol.* 2:323–324.

Datta, S. B. (1988). The acquisition of dominance among free-ranging rhesus monkey siblings. *Anim. Behav.* 36:754–772.

Datta, S. B. (1992). Effects of availability of allies on female dominance structure. In: Harcourt, A. H., and de Waal, F. B. M. (eds.), *Coalitions and Alliances in Humans and Other Animals.* Oxford University Press, Oxford, pp. 61–82.

Datta, S. B., and Beauchamp, G. (1991). Effects of group demography on dominance relationships among female primates. I. Mother–daughter and sister–sister relations. *Am. Nat.* 138:201–226.

Dausmann, K. H. (2005). Measuring body temperature in the field—evaluation of external vs. implanted transmitters in a small mammal. *J. Therm. Biol.* 30(3):195–202.

Davenport, T. R. B., and Butynski, T. H. (in press). The Kipunji (*Rungwecebus kipunji*). In: Kingdon, J., Happold, D., Butynski, T. (eds.), *The Mammals of Africa.* University of California Press, Berkeley, CA.

Davenport, T. R. B., Stanley, W. T., Sargis, E. J., De Luca, D. W., Mpunga, N. E., Machaga, S. J., and Olson, L. E. (2006). A new genus of African monkey, *Rungwecebus:* morphology, ecology, and molecular phylogenetics. *Science* 312:1378–1381.

Davenport, T. R. B., De Luca, D. W., Jones, T., Mpunga, N. E., Machaga, S. J., Kitegile, A., and Phillipps, G. P. (2008). The critically endangered kipunji *Rungwecebus kipunji* of southern Tanzania: first census and conservation status assessment. *Oryx* 42:352–359.

Davidge, C. (1978a). Activity patterns of chacma baboons (*Papio ursinus*) at Cape Point. *Zool. Afr.* 13:143–155.

Davidge, C. (1978b). Ecology of baboons (*Papio ursinus*) at Cape Point. *Zool. Afr.* 13:329–350.

Davies, A. G. (1984). An ecological study of the red leaf monkey (*Presbytis rubicunda*) in the dipterocarp forest of northern Borneo [PhD diss.]. Cambridge University, Cambridge.

Davies, A. G. (1991). Seed-eating by red leaf monkeys (*Presbytis rubicunda*) in dipterocarp forest of northern Borneo. *Int. J. Primatol.* 12:119–144.

Davies, A. G. (1994). Colobine populations. In: Davies, A. G., and Oates, J. F. (eds.), *Colobine Monkeys: Their Ecology, Behaviour and Evolution.* Cambridge University Press, Cambridge. pp. 285–310.

Davies, A. G., Bennett, E. L., and Waterman, P. G. (1988). Food selection by two South-east Asian colobine monkeys (*Presbytis rubicunda* and *Presbytis melalophos*) in relation to plant chemistry. *Biol. J. Linn. Soc.* 34:33–56.

Davies, A. G., Oates, J. F., and Dasilva, G. L. (1999). Patterns of frugivory in three West African colobine monkeys. *Int. J. Primatol.* 20:327–357.

Davies, G. (1991). Seed-eating by red leaf monkeys (*Presbytis rubicunda*) in dipterocarp forest of northern Borneo. *Int. J. Primatol.* 12:119–144.

Davies, N. D., and Baillie, I. C. (1988). Soil-eating in red leaf monkeys (*Presbytis rubicunda*) in Saba, northern Borneo. *Biotropica* 20:252–258.

Davis, H. (1992). Transitive inference in rats (*Rattus norvegicus*). *J. Comp. Psychol.* 106:342–349.

Davis, L. C. (1996). Functional and phylogenetic implications of ankle morphology in Goeldi's monkey (*Callimico goeldii*). In: Norconk, M., Rosenberger, A. L., and Garber, P. A. (eds.), *Adaptive Radiations of Neotropical Primates.* Plenum Press, New York. pp. 133–156.

Davis, L. C., and Walker-Pacheco, S. E. (in press). Functional morphology and positional behavior in the Pitheciini. In Barnett, A. A., Veiga, L.M., Ferrari, S.F., and Norconk, M.A. (eds.), *Evolutionary Biology and Conservation of Titis, Sakis, and Uacaris.* Cambridge: Cambridge University Press.

Dawkins, M., and Gosling, M. (1991). *Ethics in Research on Animal Behaviour.* Academic Press, London.

Dawkins, R. (1976). *The Selfish Gene.* Oxford University Press, Oxford.

Dawson, G. A. (1978). Composition and stability of social groups of the tamarin, *Saguinus oedipus geoffroyi,* in Panama: ecological and behavioral implications. In: Kleiman, D. G. (ed.), *The Biology and Conservation of the Callitrichidae.* Smithsonian Institution Press, Washington DC. pp. 23–37.

Dawson, G. A. (1979). The use of time and space by the Panamanian tamarin, *Saguinus oedipus. Folia Primatol.* 31:253–284.

Day, R. T., and Elwood, R. W. (1999). Sleeping site selection by the golden-handed tamarin *Saguinus midas midas:* the role of predation risk, proximity to feeding sites, and territorial defense. *Ethology* 105:1035–1051.

Day, R. L., Kendal, J., and Laland, K. N. (2001). Validating cultural transmission in cetaceans. *Behav. Brain Sci.* 24:330–331.

Day, R. L., Coe, R. L., Kendal, J. R., and Laland, K. N. (2003). Neophilia, innovation and social learning: a study of intergeneric differences in callitrichid monkeys. *Anim. Behav.* 65:559–571.

Deacon, T. W. (1997). *The Symbolic Species: The Co-evolution of Language and the Brain.* W. W. Norton, New York.

Deag, J. M. (1980). Interactions between males and unweaned Barbary macaques: testing the agonistic buffering hypothesis. *Behaviour* 75:54–81.

Deag, J. M., and Crook, J. H. (1971). Social behaviour and "agonistic buffering" in the wild Barbary macaque *Macaca sylvana* L. *Folia Primatol.* 15:183–200.

de Amorin, A., Borba, H. R., Carauta, J. P. P., Lopes, D., and Kaplan, M. A. (1999). Antihelmintic activity of the latex of *Ficus* species. *J. Ethnopharm.* 64:255–258.

Dean, C., Stringer, C., and Bromage, T. (1986). Age at death of the Neanderthal child from Devil's Tower, Gibraltar and the implications for studies of general growth and development in Neanderthals. *Am. J. Phys. Anthropol.* 70:301–309.

Dean, F. B., Hosono, S., Fang, L., Wu, X., Faruqi, A. F., Bray-Ward, P., Sun, Z., Zong, Q., Du, Y., Du, J., Driscoll, M., Song, W., Kingsmore, S. F., Egholm, M., and Lasken, R. S. (2002). Comprehensive human genome amplification using multiple displacement amplification. *Proc. Natl. Acad. Sci. USA* 99:5261–5266.

Dean, M. C., and Leakey, M. G. (2004). Enamel and dentine development and the life history profile of *Victoriapithecus macinnesim* from Maboko Island, Kenya. *Ann. Anat.* 186:405–412.

Dean, M. C., and Reid, D. J. (2001). Perikymata spacing and distribution on hominid anterior teeth. *Am. J. Phys. Anthropol.* 116:209–215.

Dean, M. C., Leakey, M. G., Reid, D., Schrenk, F., Schwartz, G. T., Stringer, C., and Walker, A. (2001). Growth processes in teeth distinguish modern humans from *Homo erectus* and earlier hominins. *Nature* 414:627–631.

Deaner, R. O., and Platt, M. L. (2003). Reflexive social attention in monkeys and humans. *Curr Biol.* 13:1609–1613.

Deaner RO, van Schaik CP, Johnson V. 2006. Do some taxa have better domain cognition than others? A meta-analysis of nonhuman primate studies. Evol Psych 4: 149–196 .

Deaner, R. O., Nunn, C. L., and van Schaik, C. P. (2000). Comparative tests of primate cognition: different scaling methods produce different results. *Brain Behav. Evol.* 55:44–52.

Deaner, R. O., Barton, R. A., and van Schaik, C. P. (2002). Primate brains and life histories: renewing the connection. In: Kappeler, P. M., and Pereira, M. E. (eds.), *Primate Life Histories and Socioecology.* University of Chicago Press, Chicago. pp. 233–265.

Deblauwe, I., and Janssens, G. (2008). New insights in insect prey choice by chimpanzees and gorillas in southeast Cameroon: the role of nutritional value. *Am. J. Phys. Anthropol.* 135:42–55.

Deblauwe, I., Dupain, J., Nguenang, G. M., Werdenich, D., and Van Elsacker, L. (2003). Insectivory by *Gorilla gorilla gorilla* in southeast Cameroon. *Int. J. Primatol.* 24:493–501.

Debyser, I. W. J. (1995a). Platyrrhine juvenile mortality in captivity and in the wild. *Int. J. Primatol.* 16(6):909–933.

Debyser, I. W. J. (1995b). Catarrhine juvenile mortality in captivity, under seminatural conditions, and in the wild. *Int. J. Primatol.* 16(6):935–969.

Decanini, D. P., and Macedo, R. H. (2008a). Sociality in *Callithrix penicillata*: I. Intragroup male profile. *Int. J. Primatol.* 29:433–447.

Decanini, D. P., and Macedo, R. H. (2008b). Sociality in *Callithrix penicillata*: II. Individual strategies during intergroup encounters. *Int. J. Primatol.* 29:627–639.

de Carvalho, S., Ferrari, S. F., and Strier, K. B. (2004). Diet of a muriqui group (*Brachyteles arachnoides*) in continuous primary forest. *Primates* 45:201–204.

de Castro, C. S. S., de Menezes, A. L., and Moreira, L. F. D. (2003). Locomotor activity rhythm in free-ranging common marmosets (*Callithrix jacchus*). *Biol. Rhythm Res.* 34:23–30.

Decker, B. S. (1994). Effects of habitat disturbance on the behavioral ecology and demographics of the Tana River red colobus (*Colobus badius rufomitratus*). *Int. J. Primatol.* 15:703–737.

de Faria, D. S. (1986). Tamanho, composição de um grupo social e área de vivência (home-range) do sagüi *Callithrix jacchus penicillata* na mata ciliar do córrego capetinga, Brasília, DF. In: Thiago de Mello, M. (ed.), *A Primotologia no Brasil*, vol. 2. Vitória, ES Brazil pp. 87–105.

Defler, T. R. (1978). Allogrooming in two species of macaque (*Macaca nemestrina* and *Macaca radiata*). *Primates* 19:153–167.

Defler, T. R. (1979a). On the ecology and behavior of *Cebus albifrons* in eastern Colombia: I. Ecology. *Primates* 20:475–490.

Defler, T. R. (1979b). On the ecology and behavior of *Cebus albifrons* in eastern Colombia: II. Behavior. *Primates* 20:491–502.

Defler, T. R. (1980). Notes on interactions between the tayra (*Eira barbara*) and the white fronted capuchin (*Cebus albifrons*). *J. Mammal.* 61:156.

Defler, T. R. (1981). The density of *Alouatta seniculus* in the eastern llanos of Colombia. *Primates* 22:564–569.

Defler, T. R. (1982). A comparison of intergroup behavior in *Cebus albifrons* and *C. apella*. *Primates* 23:385–392.

Defler, T. R. (1983). Some population characteristics of *Callicebus torquatus lugens* (Humboldt, 1812) (Primates: Cebidae) in eastern Colombia. *Lozania (Acta Zool. Colomb.)* 38:1–9.

Defler, T. R. (1989). Recorrido y uso del espacio en un grupo de *Lagothrix lagothricha* (Primates: Cebidae) mono lanudo churuco en la Amazonia Colombiana. *Trianea* 3:183–205.

Defler, T. R. (1994). *Callicebus torquatus* is not a white-sand specialist. *Am. J. Primatol.* 33:149–154.

Defler, T. R. (1995). The time budget of a group of wild woolly monkeys (*Lagothrix lagotricha*). *Int. J. Primatol.* 16:107–120.

Defler, T. R. (1996). Aspects of the ranging pattern in a group of wild woolly monkeys (*Lagothrix lagothricha*). *Am. J. Primatol.* 38:289–302.

Defler, T. R. (1999a). Fission–fusion in the black-headed uacari (*Cacajao melanocephalus*) in eastern Colombia. *Neotrop. Primates* 7:5–8.

Defler, T. R. (1999b). Locomotion and posture in *Lagothrix lagotricha*. *Folia Primatol.* 70:313–327.

Defler, T. R. (2001). *Cacajao melanocephalus ouakary* densities on the lower Apaporis River, Colombian Amazon. *Primate Rep.* 61:31–36.

Defler, T. R. (2003a). *Primates de Colombia*. Conservation International, Bogotá.

Defler, T. R. (2003b). Densidad de especies y organización espacial de una comunidad de primates: Estación Biológica Caparú, Deparamento de Vaupés, Colombia. In: Pereira-Bengoa, V., Nassar-Montoya, F., and Savage, A. (eds.), *Primatología del Nuevo Mundo: Biología, Medicina, Manejo y Conservación*. Centro de Primatología Araguatos, Bogotá. pp. 23–39.

Defler, T. R. (2004). *Primates of Colombia*. Conservación Internacional, Bogotá.

Defler, T. R., and Bueno, M. L. (2003). Karyological guidelines for *Aotus* taxonomy. *Am. J. Primatol.* 60:134–135.

Defler, T. R., and Bueno, M. L. (2007). *Aotus* diversity and the species problem. *Primate Conserv.* 22:55–70.

Defler, T. R., and Defler, S. B. (1996). Diet of a group of *Lagothrix lagothricha lagothricha* in southeastern Colombia. *Int. J. Primatol.* 17:161–190.

Defler, T. R., Bueno, M. L., and Hernández-Camacho, J. I. (2001). Taxonomic status of *Aotus hershkovitzi*: its relationship to *Aotus lemurinus lemurinus*. *Neotrop. Primates* 9:37–52.

de Garine, I. (1993). Food resources and preferences in the Cameroonian forest. In: Hladik, C. M., Hladik, A., Linares, O. F., Pagezy, H., Semple, A., and Hadley, M. (eds.), *Tropical Forests, People and Food: Biocultural Interactions and Applications to Development*. UNESCO, Paris. pp. 561–574.

de Granville, J.-J. (1982). Rain forest and xeric flora refuges in French Guiana. In: Prance, G. T. (ed.), *Biological Diversification in the Tropics*. Columbia University Press, New York. pp. 159–181.

Deinard, A., and Smith, D. G. (2001). Phylogenetic relationships among the macaques: evidence from the nuclear locus *NRAMP1*. *J. Hum. Evol.* 41:45–59.

de Jong, G. (1993). Covariances between traits deriving from successive allocations of a resource. *Funct. Ecol.* 7:75–83.

de Jong, G., and van Noordwijk, A. J. (1992). Acquisition and allocation of resources: genetic (co)variances, selection, and life histories. *Am. Nat.* 139:749–770.

de Laguerie, P., Olivieri, I., Atlan, A., and Gouyon, (1993). Analytic and simulation models predicting positive genetic correlations between traits linked by tradeoffs. *Evol. Ecol.* 5:361–369.

de la Torre, S., Campos, F., and de Vries, T. (1995). Home range and birth seasonality of *Saguinus nigricollis* in Ecuadorian Amazonia. *Am. J. Primatol.* 37:39–56.

de la Torre, S., Snowdon, C. T., and Bejarana, M. (2000). Effects of human activities on wild pygmy marmosets in Ecuadorian Amazonia. *Biol. Conserv.* 94:153–163.

Delgado, R. A. (2006). Sexual selection in the loud calls of male primates: signal content and function. *Int. J. Primatol.* 27:5–25.

Delgado, R. A. (2007). Geographic variation in the long calls of male orangutans (*Pongo* spp). *Ethology* 113:487–98.

Delgado, R. A., and van Schaik, C. P. (2000). The behavioral ecology and conservation of the orangutan (*Pongo pygmaeus*): a tale of two islands. *Evol. Anthropol.* 9:201–218.

de Lima, M. M. C., Sampaio, I., Vieira, R. D. S., and Schneider, H. (2007). Spider monkey, muriqui and woolly monkey relationships revisited. *Primates* 48:55–63.

DelPero, M., Masters, J. C., Zuccon, D., Cervella, P., Crovella, S., and Ardito, G. (2000). Mitochondrial sequences as indicators of generic classification in bush babies. *Int. J. Primatol.* 21:889–904.

Delson, E. (1980). Fossil macaques, phyletic relationships and a scenario of deployment. In: Lindburg, D. G. (ed.), *The Macaques: Studies in Ecology, Behavior and Evolution*. Van Nostrand Reinhold, New York, pp. 10–30.

Delson, E., Terranova, C. J., Jungers, W. L., Sargis, E. J., Jablonski, N. G., and Dechow, P. C. (2000). Body mass in Cercopithecidae (Primates, Mammalia): estimation and scaling in extinct and extant taxa. *Anthropol. Pap. Am. Mus. Nat. Hist.* 83:1–159.

De Luca, D. W., Picton Phillipps, G., Machaga, S. J., Davenport, T. R. B. (2009). Home range, core areas and movement in the 'critically endangered' kipunji (Rungwecebus kipunji) in southwest Tanzania. Afr. J. Ecol. doi:10.1111/j.1365-2028.2009.01189.x.

DeLuycker, A. M. (2006). Preliminary report and conservation status of the Rio Mayo titi monkey, *Callicebus oenanthe* Thomas, 1924, in the Alto Mayo Valley, northeastern Peru. *Primate Conserv.* 21:33–39.

DeLuycker, A. M. (2007). Notes on the yellow-tailed woolly monkey (*Oreonax flavicauda*) and its status in the protected forest of Alto Mayo, northern Peru. *Primate Conserv.* 22:41–47.

Demaria, C., and Thierry, B. (2001). A comparative study of reconciliation in rhesus and Tonkean macaques. *Behaviour* 138:397–410.

de Marques, A. A. B. (1995). O Bugio-Ruivo *Alouatta fusca clamitans* (Cabrera, 1940) (Primates: Cebidae) na Estação Ecológica de Aracuri, RS; Variações Sazonias de Forregeamento [MSc thesis]. Pontifícia Universidade Católica do Rio Grande do Sul, Porto Alegre.

de Marques, A. A. B. (1996). Ecology and behavior of brown howlers in Araucaria Pine Forest, southern Brazil. *Neotrop. Primates* 4:90–91.

de Melo Junior, T., and Zara, F. (2007). Black-tufted-ear marmoset *Callithrix penicillata* (Primates: Callitrichidae) following the army ant *Labidus praedator* (Formicidae: Ecitoninae) in the Cerrado and the Atlantic Forest, Brazil. *Neotrop. Primates* 14:32–33.

Demes, B., and Gunther, M. M. (1989). Biomechanics and allometric scaling in primate locomotion and morphology. *Folia Primatol.* 53:125–141.

Demes, B., Larson, S. G., Stern, J. T., Jr., Jungers, W. L., Biknevicius, A. R., and Schmitt, D. (1994). The kinetics of "hind limb drive" reconsidered. *J. Hum. Evol.* 26:353–374.

Demes, B., Jungers, W. L., Gross, T. S., and Fleagle, J. G. (1995). Kinetics of leaping primates: influence of substrate orientation and compliance. *Am. J. Phys. Anthropol.* 96:419–430.

Demes, B., Fleagle, J. G., and Lemelin, P. (1998). Myological correlates of prosimian leaping. *J. Hum. Evol.* 34:385–399.

de Moraes, P. L. R., Oswaldo de Carvalho, J., and Strier, K. B. (1998). Population variation in patch and party size in muriquis (*Brachyteles arachnoides*). *Int. J. Primatol.* 19:325–337.

Dene, H. T., Goodman, M., and Prychodko, W. (1976a). Immunodiffusion evidence on the phylogeny of the primates. In: Goodman, M., Tashian, R. E., and Tashian, J. H. (eds.), *Molecular Anthropology: Genes and Proteins in the Evolutionary Ascent of the Primates*. Plenum Press, New York. pp. 171–195.

Dene, H., Goodman, M., Prychodko, W., and Moore, G. W. (1976b). Immunodiffusion systematics of the primates. III. The Strepsirhini. *Folia Primatol.* 25:35–61.

Dene, H., Goodman, M., and Prychodko, W. (1980). Immunodiffusion systematics of the primates. IV. Lemuriformes. *Mammalia* 44:211–223.

Denham, W. W. (1971). Energy relations and some basic properties of primate social organization. *Am. Anthropol.* 73:77–95.

Dennis, A. J., and Wescott, D. A. (2007). Estimating dispersal kernels produced by a diverse community of vertebrates. In: Dennis, A. J., Schupp, E. W., Green, R., and Westcott, D. A. (eds.), *Seed Dispersal: Theory and Its Application in a Changing World*. C.A.B. International, Wallington, UK. pp. 201–228.

de Oliveira, M. S., Lopes, F. A., Alonso, C., and Yamamoto, M. E. (1999). The mother's participation in infant carrying in captive groups of *Leontopithecus chrysomelas* and *Callithrix jacchus*. *Folia Primatol.* 70:146–153.

Depeine, C. C., Rotundo, M., Juárez, C. P., and Fernandez-Duque, E. (2008). Hoot calling in owl monkeys (*Aotus azarai*) of Argentina: sex differences and function. *Am. J. Primatol.* 70:69.

Deppe, A. M., Randriamiarisoa, M., Kasprak, A. H., and Wright, P. C. (2008). Predation on the brown mouse lemur (*Microcebus rufus*) by a diurnal carnivore, the ring-tailed mongoose (*Galidia elegans*). *Lemur News* 13:17–18.

Deputte, B. L. (1991). Reproductive parameters of captive grey-cheeked mangabeys. *Folia Primatol.* 57:57–69.

Deputte, B. L. (1992). Life history of captive gray-cheeked mangabeys: physical and sexual development. *Int. J. Primatol.* 13:509–531.

DeRosseau, C. J. (1990). Life-history thinking in perspective. In: DeRosseau, C. J. (ed.), *Primate Life History and Evolution*. Wiley-Liss, New York. pp. 1–13.

deRousseau, C. J. (ed.) (1990). *Primate Life History and Evolution*. Wiley-Liss, New York.

de Ruiter, J. R. (1992). Male social rank and reproductive success in wild long-tailed macaques: paternity exclusions by blood protein analysis and DNA-fingerprinting. In: Martin, R. D., Dixson, A. F., and Wickings, E. J. (eds.), *Paternity in Primates: Genetic Tests and Theories*. Karger, Basel. pp. 175–191.

de Ruiter, J. R. (2004). Genetic markers in primate studies: elucidating behavior and its function. *Int. J. Primatol.* 25:1173–1189.

de Ruiter, J. R., and Geffen, E. (1998). Relatedness of matrilines, dispersing males and social groups in long-tailed macaques (*Macaca fascicularis*). *Proc. R. Soc. Lond. B* 265:79–87.

de Ruiter, J. R., Scheffrhan, W., Trommelen, G. J. J. M., Uitterlinden, A. G., and Martin, R. D. (1992). Male social rank and reproductive success in wild long-tailed macaques. In: Martin, R. D., Dixson, A. F., and Wickings, E. J. (eds.), *Paternity in Primates: Genetic Tests and Theories*. Karger, Basel. pp. 175–191.

de Ruiter, J., van Hooff, J. A. R. A. M., and Scheffrahn, W. (1994). Social and genetic aspects of paternity in wild long-tailed macaques (*Macaca fascicularis*). *Behaviour* 129:203–224.

Deschner, T., Heistermann, M., Hodges, K., and Boesch, C. (2003). Timing and probability of ovulation in relation to sex skin swelling in wild West African chimpanzees, *Pan troglodytes verus*. *Anim. Behav.* 66:551–560.

Deschner, T., Heistermann, M., Hidges, K., and Boesch, C. (2004). Female sexual swelling size, timing of ovulation, and male behavior in wild West African chimpanzees. *Horm. Behav.* 46:204–215.

de Sousa, M. B. C., Naciemento Lopes, M. C., and Albuquerque, A. C. S. R. (2002). Fecal collection in free-ranging common marmosets, *Callithrix jacchus*. *Neotrop. Primates* 10:21–24.

de Sousa e Silva, J., Jr., and Nunes, A. (1995). Geographic distribution of night monkeys, *Aotus*, in northern Brazil: new data and a correction. *Neotrop. Primates* 3:72–74.

DeSteven, D., and Putz, F. E. (1984). Impact of mammals on early recruitment of a tropical canopy tree, *Dipteryx panamensis*, in Panama. *Oikos* 43:207–216.

Dettling, A. C. (2002). Reproduction and development in Goeldi's monkey (*Callimico goeldii*). *Evol. Anthropol.* 1(suppl.):207–210.

Dettling, A., and Pryce, C. R. (1999). Hormonal monitoring of age at sexual maturation in female Goeldi's monkeys (*Callimico goeldii*) in their family groups. *Am. J. Primatol.* 48:77–83.

Detwiler, S. R., and Washburn, S. L. (1943). An experiment bearing on the problems of physical anthropology. *Am. J. Phys. Anthropol.* 1:171–190.

de Veer, M. W., and van den Bos, R. (1999). A critical review of methodology and interpretation of mirror self-recognition research in non-human primates. *Anim. Behav.* 58:459–468.

De Vleeschouwer, K., Heistermann, M., Van Elsacker, L., and Verheyen, R. F. (2000a). Signaling of reproductive status in captive female golden-headed lion tamarins (*Leontopithecus chrysomelas*). *Int. J. Primatol.* 21:445–465.

De Vleeschouwer, K., Van Elsacker, L., Heistermann, M., and Leus, K. (2000b). An evaluation of the suitability of contraceptive methods in golden-headed lion tamarins (*Leontopithecus chrysomelas*), with emphasis on melengestrol acetate (MGA) implants: (II) endocrinological and behavioural effects. *Anim. Welf.* 9:385–401.

De Vleeschouwer, K., Van Elsacker, L., and Leus, K. (2001). Multiple breeding females in captive groups of golden-headed lion tarmarins (*Leontopithecus chrysomelas*): causes and consequences. *Folia Primatol.* 72:1–10.

De Vleeschouwer, K., Leus, K. K., and Van Elsacker, L. (2003). Characteristics of reproductive biology and proximate factors regulating seasonal breeding in captive golden-headed lion tamarins (*Leontopithecus chrysomelas*). *Am. J. Primatol.* 60:123–137.

DeVore, I. (1965a). Male dominance and mating behavior in baboons. In: Beach, F. A. (ed.), *Sex and Behavior*. John Wiley and Sons, New York. pp. 266–289.

DeVore, I. (ed.) (1965b). *Primate Behavior: Field Studies of Monkeys and Apes*. Holt, Rinehart and Winston, New York.

DeVore, I., and Hall, K. R. L. (1965). Baboon ecology. In: DeVore, I. (ed.), *Primate Behavior: Field Studies of Monkeys and Apes*. Holt, Rinehart and Winston, New York. pp. 20–52.

DeVore, I., and Washburn, S. L. (1963). Baboon ecology and human evolution. In: Howell, F. C., and Bourliere, F. (eds.), *African Ecology and Human Evolution*. Aldine, Chicago. pp. 335–367.

Devos, C., Sanz, C., Morgan, D., Onononga, J. R., Laporte, N., and Huynen, M. C. (2008). Comparing ape densities and habitats in northern Congo: surveys of sympatric gorillas and chimpanzees in the Odzala and Ndoki regions. *Am. J. Primatol.* 70:439–451.

Dew, J. L. (2001). Synecology and seed dispersal in woolly monkeys (*Lagothrix lagotricha poeppigii*) and spider monkeys (*Ateles belzebuth belzebuth*) in Parque Nacional Uasuni, Ecuador [PhD diss.]. University of California, Davis.

Dew, J. L., and Wright, P. (1998). Frugivory and seed dispersal by four species of primates in Madagascar's eastern rain forest. *Biotropica* 30:425–437.

de Waal, F. (1982). *Chimpanzee Politics: Power and Sex Among Apes*. Johns Hopkins University Press, Baltimore, MD.

de Waal, F. B. M. (1984). Coping with social tension: sex differences in the effect of food provision to small rhesus monkey groups. *Anim. Behav.* 32:765–773.

de Waal, F. B. M. (1986a). Class structure in a rhesus monkey group: the interplay between dominance and tolerance. *Anim. Behav.* 34:1033–1040.

de Waal, F. B. M. (1986b). Deception in the natural communication of chimpanzees. In: Mitchell, R. W., and Thompson, N. S. (eds.), *Deception: Perspectives on Human and Nonhuman Deceit*. State University of New York Press, Albany. pp. 221–244.

de Waal, F. B. M. (1986c). The integration of dominance and social bonding in primates. *Q. Rev. Biol.* 61:459–479.

de Waal, F. B. M. (1987). Tension regulation and nonreproductive functions of sex in captive bonobos (*Pan paniscus*). *Natl. Geogr. Res.* 3:318–335.

de Waal, F. B. M. (1988). The communicative repertoire of captive bonobos (*Pan paniscus*), compared to that of chimpanzees. *Behaviour* 106:183–251.

de Waal, F. B. M. (1989a). Behavioral contrasts between bonobos and chimpanzees. In: Helte, P. G., and Marquardt, L. A. (eds.), *Understanding Chimpanzees*. Harvard University Press, Cambridge, MA. pp. 154–175.

de Waal, F. B. M. (1989b). *Peacemaking Among Primates*. Harvard University Press, Cambridge, MA.

de Waal, F. B. M. (1989c). Food sharing and reciprocal obligations among chimpanzees. *J. Hum. Evol.* 18:433–459.

de Waal, F. B. M. (1991). Rank distance as a central feature of rhesus monkey social organisation: a sociometric analysis. *Anim. Behav.* 41:383–395.

de Waal, F. B. M. (1993). Reconciliation among primates: a review of empirical evidence and unresolved issues. In: Mason, W. A., and Mendoza, S. P. (eds.), *Primate Social Conflict*. State University of New York Press, Albany. pp. 111–144.

de Waal, F. B. M. (1996). Macaque social culture: development and perpetuation of affiliative networks. *J. comp. psychol.* 110:147–154.

de Waal, F. B. M. (1997a). *Bonobo: The Forgotten Ape*. University of California Press, Berkeley.

de Waal, F. B. M. (1997b). Foreword. In: Mitchell, R. W., Thompson, N. S., and Miles, H. L. (eds.), *Anthropomorphism, Anecdotes, and Animals.* State University of New York Press, New York.

de Waal, F. B. M. (1998). Reply to Stanford (1998) The social behavior of chimpanzees and bonobos—empirical evidence and shifting assumptions. *Curr. Anthropol.* 39:407–408.

de Waal, F. B. M. (2000a). Primates—a natural heritage of conflict resolution. *Science.* 289:586–590.

de Waal, F. B. M. (2000b). The first kiss: foundations of conflict resolution research in animals. In: Aureli, F., and de Waal, F. B. M. (eds.), *Natural Conflict Resolution.* University of California, Berkeley. pp. 15–33.

de Waal, F. B. M., and Aureli, F. (1997). Conflict resolution and distress alleviation in monkeys and apes. In: Carter, C. S., Kirkpatrick, B., and Lenderhendler, I. (eds.), *The Integrative Neurobiology of Affiliation.* Annals of the New York Academy of Sciences, New York. pp. 317–328.

de Waal, F. B. M., and Johanowicz, D. L. (1993). Modification of reconciliation behavior through social experience—an experiment with 2 macaque species. *Child Dev.* 64:897–908.

de Waal, F. B. M., and Lanting, F. (1997). *Bonobos: The Forgotten Ape.* University of California Press, Berkeley.

de Waal, F. B. M., and Luttrell, L. M. (1985). The formal hierarchy of rhesus monkeys: an investigation of the bared-teeth display. *Am. J. Primatol.* 9:73–85.

de Waal, F. B. M., and Luttrell, L. M. (1986). The similarity principle underlying social bonding among female rhesus monkeys. *Folia Primatol.* 46:215–234.

de Waal, F. B. M., and Luttrell, L. M. (1989). Toward a comparative socioecology of the genus *Macaca*: different dominance styles in rhesus and stumptail monkeys. *Am. J. Primatol.* 19:83–109.

de Waal, F. B. M., and Ren, R. M. (1988). Comparison of the reconciliation behavior of stumptail and rhesus macaques. *Ethology* 78:129–142.

de Waal, F. B. M., and van Roosmalen, A. (1979). Reconciliation and consolation among chimpanzees. *Behav. Ecol. Sociobiol.* 5:55–66.

de Waal, F. B. M., and Yoshihara, D. (1983). Reconciliation and redirected affection in rhesus monkeys. *Behaviour* 85:224–241.

Dewar, G. (2003). The cue reliability approach to social transmission: designing tests for adaptive traditions. In: Fragaszy, D. M., and Perry, S. (eds.), *The Biology of Traditions: Models and Evidence.* Cambridge University Press, Cambridge. pp. 127–158.

Dewar, R. E. (1997). Were people responsible for the extinction of Madagacar's subfossils, and how will we ever know? In: Goodman, S. M., and Patterson, B. D. (eds.), *Natural Change and Human Impact in Madagascar.* Smithsonian Institution Press, Washington DC. pp. 364–377.

Dewar, R. E., and Richard, A. F. (2007). Evolution in the hypervariable environment of Madagacar. *Proc. Natl. Acad. Sci. USA* 104:13723–13727.

Dewsbury, D. A. (1982). Ejaculate cost and male choice. *Am. Nat.* 119:601–610.

Dewsbury, D. A., and Pierce, J. D. (1989). Copulatory patterns of primates as viewed in a broad mammalian perspective. *Am. J. Primatol.* 15:51–72.

Diamond, E. J., Aksel, S., Hazelton, J. M., Jennings, R. A., and Abee, C. R. (1987). Serum oestradiol, progesterone, chorionic gonadotropin and prolactin concentrations during pregnancy in the Bolivian squirrel monkey (*Saimiri sciureus*). *J. Reprod. Fertil.* 80:373–381.

Dias, L. G., and Strier, K. B. (2003). Effects of group size on ranging patterns in *Brachyteles arachnoides hypoxanthus*. *Int. J. Primatol.* 24:209–221.

Di Bitetti, M. S., and Janson, C. H. (2000). When will the stork arrive? Patterns of birth seasonality in neotropical primates. *Am. J. Primatol.* 50:109–130.

Di Bitetti, M. S., and Janson, C. H. (2001a). Reproductive socioecology of tufted capuchins (*Cebus apella nigritus*) in northeastern Argentina. *Int. J. Primatol.* 22:127–142.

Di Bitetti, M. S., and Janson, C. H. (2001b). Social foraging and the finder's share in capuchin monkeys, *Cebus apella. Anim. Behav.* 62:47–56.

Dielentheis, T. F., Zaiss, E., and Geissmann, T. (1991). Infant care in a family of siamangs (*Hylobates syndactylus*) with twin offspring at Berlin Zoo. *Zoo Biol.* 10:309–317.

Dierschke, D. S., Weiss, G., and Knobil, E. (1974). Sexual maturation in the female rhesus monkeys and the development of estrogen-induced gonadotropic hormone release. *Endocrinology* 94:198–206.

Dietz, J. M. (2004). Kinship structure and reproductive skew in cooperatively breeding primates. In: Chapais, B., and Berman, C. M. (eds.), *Kinship and Behavior in Primates.* Oxford University Press, New York. pp. 223–241.

Dietz, J. M., and Baker, A. J. (1993). Polygyny and female reproductive success in golden lion tamarins, *Leontopithecus rosalia. Anim. Behav.* 46:1067–1078.

Dietz, J. M., Baker, A. J., and Miglioretti, D. (1994a). Seasonal variation in reproduction, juvenile growth, and adult body mass in golden lion tamarins (*Leontopithecus rosalia*). *Am. J. Primatol.* 34:115–132.

Dietz, J. M., de Sousa, N. F., and da Silva, J. R. (1994b). Population structure and territory size in golden-headed lion tamarins, *Leontopithecus chrysomelas. Neotrop. Primates* 2(suppl.):21–23.

Dietz, J. M., Pere, C. A., and Pinder, L. (1997). Foraging ecology and use of space in wild golden lion tamarins (*Leontopithecus rosalia*). *Am. J. Primatol.* 41:289–305.

Di Fiore, A. (1997). Ecology and behavior of lowland woolly monkeys (*Lagothrix lagotricha poeppigii*, Atelinae) in eastern Ecuador [PhD diss.]. University of California, Davis.

Di Fiore, A. (2002). Molecular perspectives on dispersal in lowland woolly monkeys (*Lagothrix lagotricha poeppigii*). *Am. J. Phys. Anthropol.* S34:63.

Di Fiore, A. (2003a). Molecular genetic approaches to the study of primate behavior, social organization, and reproduction. *Ybk. Phys. Anthropol.* 46:62–99.

Di Fiore, A. (2003b). Ranging behavior and foraging ecology of lowland woolly monkeys (*Lagothrix lagotricha poeppigii*) in Yasuní National Park, Ecuador. *Am. J. Primatol.* 59:47–66.

Di Fiore, A. (2004). Diet and feeding ecology of woolly monkeys in a western Amazonian rain forest. *Int. J. Primatol.* 24:767–801.

Di Fiore, A. (2009a). Genetic approaches to the study of dispersal and kinship in New World primates. In: Garber, P. A., Estrada, A., Bicca-Marques, J. C., Heymann, E. W., and Strier, K. B. (eds.), *South American Primates: Comparative Perspectives in the Study of Behavior, Ecology, and Conservation.* Springer, New York. pp. 211–250.

Di Fiore, A. (2009b). A rapid genetic method for sex assignment in nonhuman primates. *Conserv. Genet.* 6: 1053–1058.

Di Fiore, A., and Campbell, C. J. (2007). The Atelines: variations in ecology, behavior, and social organization. In: Campbell C. J., Fuentes, A., MacKinnon, K. C., Panger, M., and Bearder, S. K. (eds.), *Primates in Perspective,* 1st ed. Oxford University Press, New York. pp. 155–185.

Di Fiore, A., and Fleischer, R. C. (2005). Social behavior, reproductive strategies, and population genetic structure of *Lagothrix lagotricha poeppigii. Int. J. Primatol.* 26:1137–1173.

Di Fiore, A., and Gagneaux, P. (2007). Molecular primatology. In: Campbell, C. J., Fuentes, A., MacKinnon, K. C., Panger, M., and Bearder, S. K. (eds.), *Primates in Perspective,* 1st ed. Oxford University Press, New York. pp. 369–393.

Di Fiore, A., and Rendall, D. (1994). Evolution of social organization: a reappraisal for primates by using phylogenetic methods. *Proc. Natl. Acad. Sci. USA* 91:9941–9945.

Di Fiore, A., and Rodman, P. S. (2001). Time allocation patterns of lowland woolly monkeys (*Lagothrix lagotricha poeppigii*) in a neotropical terra firma forest. *Int. J. Primatol.* 22:449–480.

Di Fiore, A., and Strier, K. B. (2004). Flexibility in social organisation in ateline primates. *Folia Primatol.* 75(suppl. 1):140–141.

Di Fiore, A., and Suarez, S. (2007). Route-based travel and shared routes in sympatric spider and woolly monkeys: cognitive and evolutionary implications. *Anim. Cogn.* 10:317–329.

Di Fiore, A., Fernandez-Duque, E., and Hurst, D. (2007). Adult replacement in socially monogamous equatorial saki monkeys (*Pithecia aequatorialis*). *Folia Primatol.* 78:88–98.

Di Fiore, A., Link, A., and Dew, J. L. (2008). Diets of wild spider monkeys. In: Campbell, C. J. (ed.), *Spider Monkeys: Behavior, Ecology and Evolution of the Genus* Ateles. Cambridge University Press, Cambridge. pp. 81–137.

Di Fiore, A., Link, A. L., Schmitt, C. A., and Spehar, S. N. (2009). Dispersal patterns in sympatric woolly and spider monkeys: integrating molecular and observational data. *Behaviour* 146:437–470.

Digby, L. (1992). Intruders in the wild: intergroup encounters in a natural population of *Callithrix jacchus*. Abstracts of the XIVth Congress of the International Primatological Society, Strasbourg, France, p. 293.

Digby, L. J. (1995a). Infant care, infanticide, and female reproductive strategies in polygynous groups of common marmosets (*Callithrix jacchus*). *Behav. Ecol. Sociobiol.* 37:51–61.

Digby, L. J. (1995b). Social organization in a wild population of *Callithrix jacchus*. Part II: Intragroup social behavior. *Primates* 36:361–375.

Digby, L. J. (1999). Sexual behavior and extra-group copulations in a wild population of common marmosets (*Callithrix jacchus*). *Folia Primatol.* 70:136–145.

Digby, L. J. (2000). Infanticide by female mammals: implications for the evolution of social systems. In: van Schaik, C. P., and Janson, C. H. (eds.),

Infanticide by Males and Its Implications. Cambridge University Press, Cambridge. pp. 423–446.

Digby, L. J., and Barreto, C. E. (1993). Social organization in a wild population of *Callithrix jacchus.* Part I: Group composition and dynamics. *Folia Primatol.* 61:123–134.

Digby, L. J., and Barreto, C. E. (1996). Activity and ranging patterns in common marmosets (*Callithrix jacchus*): implications for reproductive strategies. In: Norconk, M., Rosenberger, A., and Garber, P. A. (eds.), *Adaptive Radiations of Neotropical Primates,* Plenum Press, New York, pp. 173–185.

Digby, L. J., and Barreto, C. E. (1998). Vertebrate predation in common marmosets. *Neotrop. Primates* 6:124–126.

Digby, L. J., and Ferrari, S. F. (1994). Multiple breeding females in free-ranging groups of *Callithrix jacchus. Int. J. Primatol.* 15:389–397.

Digby, L., and Saltzman, W. (2009). Balancing cooperation and competition in callitrichine primates: examining the relative risk of infanticide across species. In: Ford, S., Porter, L., and Davis, L. (eds.), *Smallest Anthropoids: The Marmoset/Callimico Radiation.* Springer, New York. pp.172–201.

Digby, L. J., Ferrari, S. F., and Saltzman, W. (2007). Callitrichines: the role of competition in cooperatively breeding species. In: Campbell, C. J., Fuentes, A., MacKinnon, K. C., Panger, M., and Bearder, S. K. (eds.), *Primates in Perspective,* 1st ed. Oxford University Press, New York. pp. 85–106.

DiGiano, L., Nagle, C. A., Quiroga, S., Paul, N., Farinati, Z., Torres, M., and Mendizabal, A. F. (1992). Salivary progesterone for the assessment of the ovarian function in the capuchin monkey (*Cebus apella*). *Int. J. Primatol.* 13:113–123.

Dindo, M., Thierry, B., and Whiten, A. (2008). Social diffusion of novel foraging methods in brown capuchin monkeys (*Cebus apella*). *Proc. R. Soc. Lond. B* 275:187–193.

Dirks, W. A. (2003). Effect of diet on dental development in four species of catarrhine primates. *Am. J. Primatol.* 61:29–40.

Dirks, W., and Bowman, J. (2007) Life history theory and dental development in four species of catarrhine primates. *J. Hum. Evol.* 53:309–320.

Dirzo, R., and Mendoza, E. (2007). Size-related differential seed predation in a heavily defaunated neotropical rain forest. *Biotropica* 39:355–362.

Dirzo, R., and Miranda, A. (1991). Altered patterns of herbivory and diversity in the forest understory: a case study of the possible consequences of contemporary defaunation. In: Price, P., Lewinsohn, T. M., Fernandes, G. W., and Benson, W. W. (eds.), *Plant–Animal Interactions: Evolutionary Ecology in Tropical and Temperate Regions.* John Wiley and Sons, New York. pp. 273–287.

Disotell, T. R. (1994). Generic level relationships of the Papionini (Cercopithecoidea). *Am. J. Phys. Anthropol.* 94:47–57.

Disotell, T. R. (2000). Molecular systematics of the Cercopithecidae. In: Whitehead, P. F., and Jolly, C. J. (eds.), *Old World Monkeys.* Cambridge University Press, Cambridge. pp. 29–56.

Disotell, T. R. (2008). Primate phylogenetics. In: *Encyclopedia of Life Sciences.* John Wiley & Sons, New York. pp. 1–5.

Dit Durell, S. E. A., Goss-Custard, J. D., and Caldow, R. W. G. (1993). Sex-related differences in diet and feeding method in the oystercatcher *Haematopus ostralegus. J. Anim. Ecol.* 62:205–215.

Dittus, W. P. J. (1975). Population dynamics of the toque monkey, *Macaca sinica.* In: Tuttle, R. H. (ed.), *Socioecology and Psychology of Primates.* Mouton Publishers, The Hague, pp. 125–151.

Dittus, W. P. J. (1984). Toque macaque food calls: semantic communication concerning food distribution in the environment. *Anim. Behav.* 32:470–477.

Dittus, W. P. J. (1988). Group fission among wild toque macaques as a consequence of female resource competition and environmental stress. *Anim. Behav.* 36:1626–1645.

Dittus, W. P. J. (2004). Demography: a window to social evolution. In: Thierry, B., Singh, M., and Kaumanns, W. (eds.), *Macaque Societies: A Model for the Study of Social Organization.* Cambridge University Press, Cambridge, pp. 87–112.

Dixson, A. F. (1982). Some observations on the reproductive physiology and behaviour of the owl monkey. *Int. Zoo Ybk.* 22:115–119.

Dixson, A. F. (1983a). Observations on the evolution and behavioral significance of "sexual skin" in female primates. *Adv. Study Behav.* 13:63–106.

Dixson, A. F. (1983b). The owl monkey (*Aotus trivirgatus*). In: Hearn, J. P. (ed.), *Reproduction in New World Primates. New Models in Medical Sciences.* International Medical Publishers, Lancaster. pp. 69–113.

Dixson, A. F. (1986). Plasma testosterone concentrations during postnatal development in the male common marmoset. *Folia Primatol.* 47:166–170.

Dixson, A. F. (1987a). Observations on the evolution of the genitalia and copulatory behaviour in male primates. *J. Zool. Lond.* 213:423–443.

Dixson, A. F. (1987b). Baculum length and copulatory behavior in primates. *Am. J. Primatol.* 13:51–60.

Dixson, A. F. (1989). Effects of sexual selection upon the genitalia and copulatory behaviour in male primates. *Int. J. Primatol.* 10:47–55.

Dixson, A. F. (1992). Observations on postpartum changes in hormones and sexual behavior in callitrichid primates: do females exhibit post partum "estrus"? In: Itoigawa, N., Sugiyama, Y., Sackett, G. P., and Thompson, R. K. R. (eds.), *Topics in Primatology,* vol. 2. University of Tokyo Press, Tokyo. pp. 141–149.

Dixson, A. F. (1993). Sexual selection, sperm competition and the evolution of sperm length. *Folia Primatol.* 61:221–227.

Dixson, A. F. (1994). Reproductive biology of the owl monkey. In: Baer, J. F., Weller, R. E., and Kakoma, I. (eds.), *Aotus: The Owl Monkey.* Academic Press, San Diego. pp. 113–132.

Dixson, A. F. (1995). Sexual selection and the evolution of copulatory behavior in nocturnal prosimians. In: Alterman, L., Doyle, G. A., and Izard, M. K. (eds.), *Creatures of the Dark: The Nocturnal Prosimians.* Plenum Press, New York. pp. 93–118.

Dixson, A. F. (1998). *Primate Sexuality: Comparative Studies of the Prosimians, Monkeys, Apes, and Humans.* Oxford University Press, Oxford.

Dixson, A. F., and Anderson, M. J. (2002). Effects of sexual selection upon sperm morphology and sexual skin morphology in primates. *Int. J. Primatol.* 25:1159–1171.

Dixson, A. F., and Fleming, D. (1981). Parental behaviour and infant development in owl monkeys (*Aotus trivirgatus griseimembra*). *J. Zool. Lond.* 194:25–39.

Dixson, A. F., and Gardner, J. S. (1981). Diurnal variations in plasma testosterone in a male nocturnal primate, the owl monkey (*Aotus trivirgatus*). *J. Reprod. Fertil.* 62:83–86.

Dixson, A. F., and George, L. (1982). Prolactin and parental behaviour in a male New World primate. *Nature* 299:551–553.

Dixson, A. F., and Herbert, J. (1977). Gonadal hormones and sexual behavior in groups of adult talapoin monkeys (*Miopithecus talapoin*). *Horm. Behav.* 8:141–154.

Dixson, A. F., Martin, R. D., Bonney, R. C., and Fleming, D. (1979). Reproductive biology of the owl monkey, *Aotus trivirgatus griseimembra.* In: *7th Congress of the International Primatological Society.* Karger, Bangalore. pp. 61–68.

Dixson, A. F., Gardner, J. S., and Bonney, R. C. (1980). Puberty in the male owl monkeys (*Aotus trivirgatus griseimembra*): a study of physical and hormonal development. *Int. J. Primatol.* 1:129–139.

Dixson, A. F., Bossi, T., and Wickings, E. J. (1993). Male dominance and genetically determined reproductive success in the mandrill (*Mandrillus sphinx*). *Primates* 34:525–532.

Djojosudharmo, S., and van Schaik, C. P. (1992). Why are orangutans so rare in the highlands?: altitudinal changes in a Sumatran forest. *Trop. Biodivers.* 1:11–22.

Dolhinow, P. J. (1972a). The north Indian langur. In: Dolhinow, P. J. (ed.), *Primate Patterns.* Holt, Rinehart and Winston, New York. pp. 181–238.

Dolhinow, P. (1972b). Primate patterns. In: Dolhinow, P. (ed.), *Primate Patterns.* Holt, Rinehart, and Winston, New York. pp. 352–392.

Dolhinow, P. (1978). A behavior repertoire for the Indian langur monkey (*Presbytis entellus*). *Primates* 19:449–472.

Dolhinow, P. (1984). The primates: age, behavior, and evolution. In: Kertzer, D. I., and Keith, J. (eds.), *Age and Anthropological Theory.* Cornell University Press, Ithaca, NY. pp. 45–81.

Dolhinow, P. (1991). Tactics of primate immaturity. In: Robinson, M. H., and Tiger, L. (eds.), *Man and Beast Revisited.* Smithsonian Institution Press, Washington DC. pp. 139–157.

Dolhinow, P. (1993). Social systems and the individual. *Evol. Anthropol.* 3:73–74.

Dolhinow, P. (1999). A mystery: explaining behavior. In: Strum, S. C., Lindburg, D. G., and Hamburg, D. (eds.), *The New Physical. Anthropology: Science, Humanism, and Critical Reflection.* Prentice Hall, Upper Saddle River, NJ. pp. 119–131.

Dolhinow, P. J., and Bishop, N. (1970). The development of motor skills and social relationships among primates through play. *Minn. Symp. Child Psychol.* 4:141–198.

Dolhinow, P., and DeMay, M. G. (1982). Adoption: the importance of infant choice. *J. Hum. Evol.* 11:391–420.

Dolhinow, P. D., and Fuentes, A. (1999). *The Nonhuman Primates.* Mayfield-McGraw-Hill, New York.

Dolhinow, P., and Krusko, N. (1984). Langur monkey females and infants: the female's point of view. In: *Female Primates: Studies by Women Primatologists.* Alan R. Liss, New York. pp. 37–57.

Dolhinow, P., and Murphy, G. (1982). Langur monkey (*Presbytis entellus*) development. The first 3 months of life. *Folia Primatol.* 39:305–331.

Dollar, L. (1999). Preliminary report on the status, activity cycle, and ranging of *Cryptoprocta ferox* in the Malagasy rain forest, with implications for conservation. *Sm. Carniv. Conserv.* 20:7–10.

Dollar, L., Ganzhorn, J. U., and Goodman, S. M. (2007). Primates and other prey in the seasonally variable diet of *Cryptoprocta ferox* in the dry deciduous forest of western Madagascar. In: Gursky, S. L., and Nekaris, K. A. I. (eds.), *Primate Anti-Predator Strategies.* Springer, New York. pp. 63–76.

Domb, L. G., and Pagel, M. (2001). Sexual swellings advertise female quality in wild baboons. *Nature* 410:204–206.

Domingo-Roura, X., Marmi, J., Andrés, O., Yamagiwa, J., and Terradas, J. (2004). Genotyping from semen of wild japanese macaques (*Macaca fuscata*). *Am. J. Primatol.* 62:31–42.

Dominy, N. J., and Duncan, B. (2001). GPS and GIS methods in an African rain forest: applications to tropical ecology and conservation. *Conserv. Ecol.* 5:6, http://www.consecol.org/vol5/iss2/art6.

Dominy, N., and Lucas, P. W. (2001). Ecological importance of trichromatic vision to primates. *Nature* 410:363–366.

Dominy, N., Garber, P. A., Bicca-Marques, J. C., and Azevedo-Lopes, M. A. O. (2003). Do female tamarins use visual cues to detect fruit rewards more successfully than do males? *Anim. Behav.* 66:829–837.

Donati, G., and Borgognini Tarli, S. M. (2002a). Feeding ecology of the collared brown lemur, *Eulemur fulvus collaris,* in the Sainte Luce Littoral Forest. *Folia Primatol.* 73:315.

Donati, G., and Borgognini Tarli, S. M. (2002b). The role of abiotic factors in influencing cathemeral activity of collared brown lemurs (*E. fulvus collaris*) in the Sainte Luce Littoral Forest. *Folia Primatol.* 73:305–306.

Donati, G., Lunardini, A., and Kappeler, P. M. (1999). Cathemeral activity of red-fronted brown lemurs (*Eulemur fulvus rufus*) in the Kirindy Forest. In: Rakotosamimanana, B., Rasamimanana, H., Ganzhorn, J. U., and Goodman, S. M. (eds.), *New Directions in Lemur Studies.* Kluwer/Plenum, New York. pp. 119–137.

Donati, G., Lunardini, A., Kappeler, P. M., and Borgognini Tarli, S. M. (2001). Nocturnal activity in the cathemeral red-fronted lemur (*Eulemur fulvus rufus*) with observations during a lunar eclipse. *Am. J. Primatol.* 53:69–78.

Doran, D. (1993). Sex differences in adult chimpanzee positional behavior: the influence of body size on locomotion and posture. *Am. J. Phys. Anthropol.* 91:99–116.

Doran, D. (1997a). Influence of seasonality on activity patterns, feeding behavior, ranging and grouping patterns in Taï chimpanzees. *Int. J. Primatol.* 18:183–206.

Doran, D. (1997b). Ontogeny of locomotion in mountain gorilla and chimpanzees. *J. Hum. Evol.* 32:323–344.

Doran, D. M., and McNeilage, A. (1998). Gorilla ecology and behavior. *Evol. Anthropol.* 6:120–131.

Doran, D. M., and McNeilage, A. (2001). Subspecific variation in gorilla behavior: the influence of ecological and social factors. In: Robbins, M. M., Sicotte, P., and Stewart, K. J. (eds.), *Mountain Gorillas: Three Decades of Research at Karisoke.* Cambridge University Press, Cambridge. pp. 123–149.

Doran, D., Jungers, W. L., Sugiyama, Y., Fleagle, J. G., and Heesy, C. P. (2002a). Multivariate and phylogenetic approaches to understanding chimpanzee and bonobo behavioral diversity. In: Boesch, C., Hohmann, G., and Marchant, L. (eds.), *Behavioural Diversity in Chimpanzees and Bonobos.* Cambridge University Press, Cambridge. pp. 14–34.

Doran, D. M., McNeilage, A., Greer, D., Bocian, C., Mehlman, P. T., and Shah, N. (2002b). Western lowland gorilla diet and resource availability: new evidence, cross-site comparisons, and reflections on indirect sampling methods. *Am. J. Primatol.* 58:91–116.

Doran-Sheehy, D. M., Greer, D., Mongo, P., and Schwindt, D. (2004). Impact of ecological and social factors on ranging in western gorillas. *Am. J. Primatol.* 64:207–222.

Doran-Sheehy, D. M., Derby, A. M., Greer, D., and Mongo, P. (2007). Habituation of western gorillas: the process and factors that influence it. *Am. J. Primatol.* 69:1354–1369.

Douadi, M. I., Gatti, S., Leverero, F., Duhamel, G., Bermejo, M., Vallet, D., Menard, N., and Petit, E. (2007). Sex-biased dispersal in western lowland gorillas (*Gorilla, gorilla gorilla*). *Mol. Ecol.* 16:2247–2259.

Doucet, S. M., and Montgomerie, R. (2003). Multiple sexual ornaments in satin bowerbirds: ultraviolet plumage and bowers signal different aspects of male quality. *Behav. Ecol.* 14:503–509.

Doyle, G. A. (1979). Development of behaviour in prosimians with special reference to the lesser bushbaby, *Galago senegalensis moholi.* In: Doyle, G. A., and Martin, R. D. (eds.), *The Study of Prosimian Behaviour.* Academic Press, London. pp. 157–189.

Drea, C. M. (2005). Bateman revisited: the reproductive tactics of female primates. *Integr. Comp. Biol.* 45:915–923.

Drickamer, L. C. (1974). Social rank, observability and sexual behavior of rhesus monkeys (*Macaca mulatta*). *J. Reprod. Fertil.* 37:117–120.

Drickamer, L. C., and Vessey, S. H. (1973). Group changing in free ranging male rhesus monkeys. *Primates* 14:359–368.

Drickamer, L. C., Gowaty, P. A., and Holmes, C. M. (2000). Free female mate choice in house mice affects reproductive success and offspring viability and performance. *Anim. Behav.* 59:371–378.

Drickamer, L. C., Gowaty, P. A., and Wagner, D. M. (2003). Free mutual mate preferences in house mice affect reproductive success and offspring performance. *Anim. Behav.* 65:105–114.

Duckworth, J. W. (1994). Field sighting of the pygmy loris (*Nycticebus pygmaeus*) in Laos. *Folia Primatol.* 63:99–101.

Ducoing, A. M., and Thierry, B. (1998). An experimental study of deception in semi-free-ranging Tonkean macaques. *Am. J. Primatol.* 45:177–178.

Dudbridge, G. (1970). *A Study of Antecedents to the Sixteenth-Century Chinese Novel.* Cambridge University Press, Cambridge.

Duffy, D. L., and Ball, G. F. (2003). Song predicts immunocompetence in male European starlings (*Sturnus vulgaris*). *Proc. R. Soc. Biol. Sci. B* 269:847–852.

Dugatkin, L. A. (1997). *Cooperation Among Animals: An Evolutionary Perspective.* Oxford University Press, New York.

Duggleby, C. R. (1977). Blood group antigens and the population genetics of *Macaca mulatta* on Cayo Santiago. II Effects of social group division. *Ybk. Phys. Anthropol.* 20:263–271.

Dukelow, W. R. (1993). Assisted reproduction in New World primates. In: Wolf, D. P., Stouffer, R. L., and Brenner, R. M. (eds.), *In Vitro Fertilization and Embryo Transfer in Primates.* Springer-Verlag, New York. pp. 73–84.

DuMond, F. V., and Hutchinson, T. C. (1967). Squirrel monkey reproduction: the "fatted" male phenomenon and seasonal spermatogenesis. *Science* 158:1067–1070.

Dunbar, D. C. (1988). Aerial maneuvers of leaping lemurs: the physics of whole body rotations while airborne. *Am. J. Primatol.* 16:291–303.

Dunbar, R. I. M. (1977). Feeding ecology of gelada baboons: a preliminary report. In: Clutton-Brock, T. H. (ed.), *Primate Ecology: Studies in Feeding and Ranging Behaviour of Lemurs, Monkeys, and Apes.* Academic Press, London. pp. 251–273.

Dunbar, R. I. M. (1978). Sexual behaviour and social relationships among gelada baboons. *Anim. Behav.* 26:167–178.

Dunbar, R. I. M. (1980a). Demographic and life history variables of a population of gelada baboons (*Theropithecus gelada*). *J. Anim. Ecol.* 49:485–506.

Dunbar, R. I. M. (1980b). Determinants and evolutionary consequences of dominance among female gelada baboons. *Behav. Ecol. Sociobiol.* 7:253–265.

Dunbar, R. I. M. (1984a). Infant-use by male gelada in agonistic contexts: agonistic buffering, progeny protection, or soliciting support? *Primates* 25:28–35.

Dunbar, R. I. M. (1984b). *Reproductive Decisions: An Economic Analysis of Gelada Baboon Social Strategies.* Princeton University Press, Princeton, NJ.

Dunbar, R. I. M. (1987). Habitat quality, population dynamics, and group composition in a colobus monkey (*Colobus guereza*). *Int. J. Primatol.* 8:299–329.

Dunbar, R. I. M. (1988). *Primate Social Systems.* Cornell University Press, Ithaca, NY.

Dunbar, R. I. M. (1989). Reproductive strategies of female gelada baboons. In: Rasa, A. E., Vogel, C., and Voland, E. (eds.), *The Sociobiology of Sexual and Reproductive Strategies.* Chapman and Hall, New York. pp 74–92.

Dunbar, R. I. M. (1992a). Neocortex size as a constraint on group size in primates. *J. Hum. Evol.* 28:287–296.

Dunbar, R. I. M. (1992b). Time: a hidden constraint on the behavioural ecology of baboons. *Behav. Ecol. Sociobiol.* 33:35–49.

Dunbar, R. I. M. (1993a). Ecological constraints on group size in baboons. *Phys. Ecol. Jpn. B.* 29:221–236.

Dunbar, R. I. M. (1993b). Social organization of the gelada. In: Jablonski, N. G. (ed.), *Theropithecus: The Rise and Fall of a Primate Genus.* Cambridge University Press, Cambridge. pp 425–439.

Dunbar, R. M. (1995). Neocortex size and group size in primates: a test of the hypothesis. *J. Hum. Evol.* 28:287–296.

Dunbar, R. I. M. (1996). Determinants of group size in primates: a general model. *Proc. Br. Acad.* 88:33–57.

Dunbar, R. I. M. (1998). Impact of global warming on the distribution and survival of the gelada baboon: a modelling approach. *Glob. Change Biol.* 4:293–304.

Dunbar, R. I. M. (2000). Male mating strategies: a modeling approach. In: Kappeler, P. M. (ed.), *Primate Males*. Cambridge University Press, Cambridge. pp. 259–268.

Dunbar, R. I. M. (2002). Modeling primate behavioral ecology. *Int. J. Primatol.* 23:785–819.

Dunbar, D. C., and Badam, G. L. (2000). Locomotion and posture during terminal branch feeding. *Int. J. Primatol.* 21:649–669.

Dunbar, R., and Barrett, L. (2000). *Cousins: Our Primate Relatives*. Bookman-Huntingdon, London.

Dunbar, R. I. M., and Dunbar, E. P. (1974a). Ecological relations and niche separation between sympatric terrestrial primates in Ethiopia. *Folia Primatol.* 21:36–60.

Dunbar, R. I. M., and Dunbar, E. P. (1974b). The reproductive cycle of the gelada baboon. *Anim. Behav.* 22:203–210.

Dunbar, R. I. M., and Dunbar, E. P. (1974c). Ecology and population dynamics of *Colobus guereza* in Ethiopia. *Folia Primatol.* 21:188–208.

Dunbar, R., and Dunbar, P. (1975). *Social Dynamics of Gelada Baboons*. Basel, S. Karger.

Dunbar, R. I. M., and Dunbar, E. P. (1976). Contrasts in social structure among black-and-white colobus monkey groups. *Anim. Behav.* 24:84–92.

Dunbar, R. I. M., and Dunbar, E. P. (1977). Dominance and reproductive success among female gelada baboons. *Nature* 266:351–352.

Dunbar, R. I. M., and Dunbar, E. P. (1980). The pairbond in klipspringer. *Anim. Behav.* 28:219–229.

Dunbar, R. I. M., and Nathan, M. F. (1972). Social organization of the Guinea baboon, *Papio papio. Folia Primatol.* 17:321–334.

Dunbar, R. I. M., and Sharman, M. (1983). Female competition for access to males affects birth rate in baboons. *Behav. Ecol. Sociobiol.* 13:157–159.

Duncan, R. S., and Chapman, C. A. (1999). Seed dispersal and potential forest succession in abandoned agriculture in tropical Africa. *Ecol. App.* 9:998–1008.

Dupain, J., Van Elsacker, L., Nell, C., Garcia, P., Ponce, F., and Huffman, M. A. (2002). New evidence for leaf swallowing and *Oesophagostomum* infection in bonobos (*Pan paniscus*). *Int. J. Primatol.* 23:1053–1062.

Dusenbery, D. B. (1992). *Sensory Ecology*. Freeman and Company, New York.

Dutrillaux, B., and Rumpler, Y. (1988). Absence of chromosomal similarities between tarsiers (*Tarsius syrichta*) and other primates. *Folia Primatol.* 50:130–133.

Dykyj, D. (1980). Locomotion of the slow loris in a designed substrate context. *Am. J. Phys. Anthropol.* 52:577–586.

Earn, D. J. D., Levin, S. A., and Rohani, P. (2000). Coherence and conservation. *Science* 290:1360–1364.

Earnhardt, J. M., Thomson, S. D., and Marhevsky, E. A. (2001). Interactions of target population size, population parameters, and program management on viability of captive populations. *Zoo Biol.* 20:169–183.

Easley, S. P. (1982). Ecology and behavior of *Callicebus torquatus*: Cebidae, Primates [PhD diss.]. Washington University, St. Louis.

Easley, S. P., and Kinzey, W. G. (1986). Territorial shift in the yellow-handed titi monkey (*Callicebus torquatus*). *Am. J. Primatol.* 11:307–318.

Eason, P. (1989). Harpy eagle attempts predation on adult howler monkey. *Condor* 91:469–470.

Eaton, G. G., and Resko, J. A. (1974). Ovarian hormones and sexual behavior in *Macaca nemestrina. J. Comp. Physiol. Psychol.* 86:919–925.

Eaton, G. G., Slob, A., and Resko, J. A. (1973). Cycles of mating behaviour, oestrogen and progesterone in the thick-tailed bushbaby (*Galago crassicaudatus*) under laboratory conditions. *Anim. Behav.* 21:309–315.

Eaton, S. B., and Konner, M. (1985). Paleolithic nutrition: a consideration of its nature and current implications. *N. Engl. J. Med.* 312:283–289.

Eberhard, W. G. (1985). *Sexual Selection and Animal Genitalia*. Harvard University Press, Cambridge, MA.

Eberhard, W. G. (1996). *Female Control: Sexual Selection by Cryptic Female Choice*. Princeton University Press, Princeton, NJ.

Eberle, M., and Kappeler, P. M. (2002). Mouse lemurs in space and time: a test of the socioecological model. *Behav. Ecol. Sociobiol.* 51:131–139.

Eberle, M., and Kappeler, P. M. (2004a). Sex in the dark: determinants and consequences of mixed male mating tactics in *Microcebus murinus,* a small solitary nocturnal primate. *Behav. Ecol. Sociobiol.* 57:77–90.

Eberle, M., and Kappeler, P. M. (2004b). Selected polyandry: female choice and inter-sexual conflict in a small nocturnal solitary primate (*Microcebus murinus*). *Behav. Ecol. Sociobiol.* 57:91–100.

Eberle, M., and Kappeler, P. M. (2006). Family insurance: kin selection and cooperative breeding in a solitary primate (*Microcebus murinus*). *Behav. Ecol. Sociobiol.* 57:60:582–588.

Eberle, M., Perret, M., and Kappeler, P. K. (2007). Sperm competition and optimal timing of matings in *Microcebus murinus. Int. J. Primatol.* 28:1267–1278.

Eckardt, W., and Zuberbuhler, K. (2004). Cooperation and competition in two forest monkeys. *Behav. Ecol.* 15:400–411.

Edmunds, M. (1974). *Defence in Animals: A Survey of Anti-Predator Defences*. Longman, London.

Edwards, J. E. (1990). Validity of the doubly labeled water method of measuring energy expenditure with rest and exercise in nonhuman primates (*Erythrocebus patas*). [PhD diss.]. Indiana University, Bloomington.

Ehardt, C. L. (1988). Absence of strongly kin-preferential behavior by adult female sooty mangabeys (*Cercocebus atys*). *Am. J. Phys. Anthropol.* 76:233–243.

Ehardt, C. L. (in press). The red-capped mangabey (*Cercocebus torquatus*). In: Kingdon, J., Happold, D., and Butynski, T. (eds.), *The Mammals of Africa*. University of California Press, Berkeley, CA.

Ehardt, C. L., and Butynski, T. M. (2006). The recently described highland mangabey, *Lophocebus kipunji* (Cercopithecoidea, Cercopithecinae): current knowledge and conservation assessment. *Primate Conserv.* 21:81–87.

Ehardt, C. L., and Butynski, T. M. (in press a). The Golden-bellied mangabey (*Cercocebus chrysogaster*). In: Kingdon, J., Happold, D., and Butynski, T. (eds.), *The Mammals of Africa*. University of California Press, Berkeley, CA.

Ehardt, C. L., and Butynski, T. M. (in press b). The Sanje mangabey (*Cercocebus sanjei*). In: Kingdon, J., Happold, D., and Butynski, T. (eds.), *The Mammals of Africa*. University of California Press, Berkeley, CA.

Ehardt, C. L., Jones, T. P., and Butynski, T. M. (2005). Protective status, ecology and strategies for improving conservation of Cercocebus sanjei in the Udzungwa Mountains, Tanzania. *Int. J. Primatol.* 26: 557–583.

Ehrlich, A., and Macbride, L. (1989). Mother–infant interactions in captive slow lorises (*Nycticebus coucang*). *Am. J. Primatol.* 19:217–228.

Ehrlich, A., and Musicant, A. (1977). Social and individual behaviors in captive slow lorises (*Nycticebus coucang*). *Behaviour* 60:195–220.

Eisenberg, J. (1973). Reproduction in two species of spider monkeys, *Ateles fusciceps* and *Ateles geoffroyi. J. Mammal.* 54:955–957.

Eisenberg, J. F. (1979). Habitat, economy, and society: some correlations and hypotheses for the neotropical primates. In: Bernstein, I. S., and Smith, E. O. (eds.), *Primate Ecology and Human Origins: Ecological Influences on Social Organization*. Garland STPM Press, New York. pp. 215–262.

Eisenberg, J. F. (1989). *Mammals of the Neotropics. The Northern Neotropics Panamá, Colombia, Venezuela, Guyana, Suriname, French Guiana,* vol. 1. University of Chicago Press, Chicago.

Eisenberg, J. F., and Kuehn, R. E. (1966). The behavior of *Ateles geofffroyi* and related species. *Smithsonian Misc. Coll.* 151:1–63.

Eisenberg, J. F., and Redford, K. H. (1999). *Mammals of the Neotropics: The Central Neotropics. Ecuador, Peru, Bolivia, Brazil,* vol. 3. University of Chicago Press, Chicago.

Eisenberg, J. F., Muckenhirn, N. A., and Rudran, R. (1972). The relation between ecology and social structure in primates. *Science* 176:863–874.

Eizirik, E., Murphy, W. J., and O'Brien, S. J. (2001). Molecular dating and biogeography of the early placental mammal radiation. *J. Hered.* 92:212–219.

Ejidike, B. N., and Okosodo, F. E. (2007). Food and feeding habits of the thick-tailed galago (*Otolemur crassicaudatus*) in Okumu National Park, Edo State. *J. Fisheries Int.* 2:231–233.

Eley, R. M., Strum, S. C., Muchemi, G., and Reid, G. D. F. (1989a). Nutrition, body condition, activity patterns, and parasitism of free-ranging troops of olive baboons (*Papio anubis*) in Kenya. *Am. J. Primatol.* 18:209–220.

Eley, R. M., Tarara, R. P., Worthman, C. M., and Else, J. G. (1989b). Reproduction in the vervet monkey (*Cercopithecus aethiops*): III. The menstrual cycle. *Am. J. Primatol.* 17:1–10.

Elgar, M. A. (1986). House sparrows establish foraging flocks by giving chirrup calls if the resources are divisible. *Anim. Behav.* 34:169–174.

Elgar, M. (1989). Predator vigilance and group size in mammals and birds: a critical review of the empirical evidence. *Biol. Rev.* 64:13–33.

Ellefson, J. (1974). A natural history of white-handed gibbons in the Malayan Peninsula. In: Rumbaugh, D. M. (ed.), *Gibbon and Siamang,* vol. 3. Karger, Basel. pp. 1–136.

Elliot, D. G. (1913). *A Review of the Primates. Monograph Series,* vol. II. American Museum of Natural History, New York.

Elliot, O., and Elliot, M. (1967). Field notes on the slow loris in Malaya. *J. Mammal.* 48:497–498.

Ellstrand, N. C., and Elam, D. R. (1993). Population genetics of small population size: implications for plant conservation. *Annu. Rev. Ecol. Syst.* 24:217–242.

Else, J. G. (1991). Nonhuman primates as pests. In: Box, H. (ed.), *Primate Responses to Environmental Change.* Chapman and Hall, London. pp. 155–165.

Else, J. G., and Eley, D. (1985) Don't feed the monkeys. *Swara* 8: 31–32.

Elton, S. (2006). Forty years on and still going strong: the use of hominin-cercopithecid comparisons in palaeoanthropology. *J. R. Anthropol. Inst. (N.S.)* 12:19–38.

Emerson, S. B. (1985). Jumping and leaping. In: Hildebrand, M. H., Bramble, D. M., Liem, K. R., and Wake, D. B. (eds.), *Functional Vertebrate Morphology.* Belknap Press of Harvard University, Cambridge, MA. pp. 59–72.

Emery, M. A., and Whitten, P. L. (2003). Size of sexual swellings reflects ovarian function in wild chimpanzees (*Pan troglodytes*). *Behav. Ecol. Sociobiol.* 54:340–351.

Emery, N. J. (2000). The eyes have it: the neuroethology, function and evolution of social gaze. *Neurosci. Biobehav. R* 24:581–604.

Emery Thompson, M., and Knott, C. D. (2008). Urinary C-peptide of insulin as a non-invasive measure of energy balance in wild orangutans. *Horm. Behav.* 53:526–534.

Emlen, S. T., and Oring, L. W. (1977). Ecology, sexual selection and the evolution of mating systems. *Science* 197:215–223.

Emmons, L. H. (1987). Comaparative feeding ecology of felids in a neotropical rainforest. *Behav. Ecol. Sociobiol.* 20:271–283.

Emmons, L. H., and Gentry, A. (1983). Tropical forest structure and the distribution of gliding and prehensile-tailed vertebrates. *Am. Nat.* 12:513–524.

Enard, W., Khaitovich, P., Klose, J., Zöllner, S., Heissig, F., Giavalisco, P., Nieselt-Struwe, K., Muchmore, E., Varki, A., Ravid, R., Doxiadis, G. M., Bontrop, R. E., and Pääbo, S. (2002). Intra- and interspecific variation in primate gene expression patterns. *Science* 296:340–343.

Engel, C. (2002). *Wild Health.* Houghton Mifflin, Boston.

Engel, G. A., Jones-Engel, L., Suaryana, K. G., Arta Putra, I. G. A., Schilliaci, M. A., Fuentes, A., and Henkel, R. (2002). Human exposures to herpes B seropositive macaques in Bali, Indonesia. *Emerg. Infect. Dis.* 8:789–795.

Engel G. A., Pizarro, M., Shaw, E., Cortes, J., Fuentes, A., Barry, P., Lerche, N., Grant, R., Cohn, D., Jones-Engel, L. (2008). Unique pattern of enzootic primate viruses in Gibraltar macaques. *Emerg. Infect. Dis.* 14:1112–1115.

Engelhardt, A., Pfeifer, J. B., Heistermann, M., Niemitz, C., van Hooff, J. A. R. A. M., and Hodges, J. K. (2004). Assessment of female reproductive status by male long-tailed macaques, *Macaca fascicularis,* under natural conditions. *Anim. Behav.* 67:915–924.

Engelhardt, A., Hodges, J. K., Niemitz, C., and Heistermann, M. (2005). Female sexual behavior, but not sex skin swelling, reliably indicates the timing of the fertile phase in wild long-tailed macaques (*Macaca fascicularis*). *Horm. Behav.* 47:195–204.

Engelhardt, A., Heistermann, M., Hodges, J. K., Nürnberg, P., and Niemitz, C. (2006). Determinants of male reproductive success in wild long-tailed macaques (*Macaca fascicularis*)—male monopolization, female mate choice or post-copulatory mechanisms? *Behav. Ecol. Sociobiol.* 59:740–752.

Engh, A. E., Beehner, J. C., Bergman, T. J., Whitten, P. L., Hoffmeier, R. R., Seyfarth, R. M., and Cheney, D. L. (2006). Female hierarchy instability, male immigration, and infanticide increase glucocorticoid levels in female chacma baboons. *Anim. Behav.* 71:1227–1237.

Engqvist, A., and Richard, A. (1991). Diet as a possible determinant of cathemeral activity patters in primates. *Folia Primatol.* 57:169–172.

Enoksson, B. (1988). Age- and sex-related differences in dominance and foraging behaviour of nuthatches *Sitta europaea. Anim. Behav.* 36:231–238.

Enomoto, T. (1978). On social preference in sexual behavior of Japanese monkeys (*Macaca fuscata*). *J. Hum. Evol.* 7:283–293.

Ensminger, A. L., and Hoffman, S. M. G. (2002). Sex identification assay useful in great apes is not diagnostic in a range of other primate species. *Am. J. Primatol.* 56:129–134.

Enstam, K. L. (2007). Effects of habitat structure on perceived risk of predation and anti-predator behavior of vervet (*Cercopithecus aethiops*) and patas (*Erythrocebus patas*) monkeys. In: Gursky, S. L., and Nekaris, K. A. I. (eds.), *Primate Anti-Predator Strategies.* Springer, New York. pp. 308–338.

Enstam, K. L., and Isbell, L. A. (2002). Comparison of responses to alarm calls by patas (*Erythrocebus patas*) and vervet (*Cercopithecus aethiops*) monkeys in relation to habitat structure. *Am. J. Phys. Anthropol.* 119:3–14.

Enstam, K. L., Isbell, L. A., and de Maar, T. W. (2002). Male demography, female mating behavior, and infanticide in wild patas monkeys (*Erythrocebus patas*). *Int. J. Primatol.* 23:85– 104.

Epperson, B. K., and Alvarez-Buylla, E. R. (1997). Limited seed dispersal and genetic structure in life stages of *Cecropia obtusifolia. Evolution* 51:275–282.

Epple, G. (1976). Chemical communication and reproductive processes in nonhuman primates. In: Doty, R. L. (ed.), *Mammalian Olfaction, Reproductive Processes, and Behavior.* Academic Press, New York. pp. 257–282.

Epple, G., and Katz, Y. (1980). Social influences on first reproductive success and related behaviors in the saddle-back tamarin (*Saguinus fuscicollis,* Callitrichidae). *Int. J. Primatol.* 1:171–183.

Epple, G., and Katz, Y. (1983). The saddle back tamarin and other tamarins. In: Hearn, J. (ed.), *Reproduction in New World Primates: New World Models in Medical Science.* MTP Press, Lancaster. pp. 115–148.

Epple, G., and Katz, Y. (1984). Social influences on estrogen excretion and ovarian cyclicity in saddle back tamarins (*Saguinus fuscicollis*). *Am. J. Primatol.* 6:215–227.

Eppley, T. M., and Donati, G. (2009). Dietary flexibility: subsistence of the southern gentle lemur *Hapalemur meridionalis* on a low quality diet in the Mandena littoral forest, SE Madagascar [abstract]. *Am. J. Phys. Anthropol.* 48(suppl. 1):125.

Erdelen, W. (1988). Forest ecosystems and nature conservation in Sri Lanka. *Biol. Conserv.* 43:115–135.

Erhart, E. M. (1996). Growing up: defining adolescence for female savanna baboons (*Papio cynocephalus*). [PhD diss.]. University of Texas, Austin.

Erhart, E. M., and Overdorff, D. J. (1998). Infanticide in *Propithecus diadema edwardsi:* an evaluation of the sexual selection hypothesis. *Int. J. Primatol.* 19:73–81.

Erhart, E. M., and Overdorff, D. J. (2008). Rates of agonism by diurnal lemuroids: implications for female social relationships. *Int. J. Primatol.* 29:1227–1247.

Erhart, E. M., Coelho, A. M., and Bramblett, C. A. (1997). Kin recognition by paternal half-siblings in captive *Papio cynocephalus. Am. J. Primatol.* 43:147–157.

Erickson, C. J. (1991). Percussive foraging in the aye-aye, *Daubentonia madagascariensis. Anim. Behav.* 41:793–801.

Erickson, C. J. (1994). Tap-scanning and extractive foraging in aye-ayes, *Daubentonia madagascariensis. Folia Primatol.* 62:125–135.

Erickson, G. E. (1963). Brachiation in New World monkeys and in anthropoid apes. *Symp. Zool. Soc. Lond.* 10:135–164.

Erickson, M. (2006). Nature disrupted: evolution, kinship and child sexual abuse. *Clin. Neuropsychiatr.* 3:110–120.

Erkert, H. G. (1974). Der einflub des mondlichtes auf die aktivitatsperiodik nachtaktiver saugetiere. *Oecologia* 14:269–287.

Erkert, H. G. (1976). Lunarperiodic variation of the phase-angle difference in nocturnal animals under natural zeitgeber-conditions near the equator. *Int. J. Chronobiol.* 4:125–138.

Erkert, H. G. (1991). Influence of ambient temperature on circadian rhythms in Colombian owl monkeys, *Aotus lemurinus griseimembra.* In: Ehard, A., Kimura, T., Takenaka, O., and Iwamoto, M. (eds.), *Primatology Today.* Elsevier Science, Amsterdam. pp. 435–438.

Erkert, H. G. (1999). Owl monkeys. In: Poole, T. (ed.), *The UFAW Handbook on the Care and Management of Laboratory Animals.* Blackwell Science, London. pp. 574–590.

Erkert, H. G. (2004). Chronobiological background to cathemerality. *Int. J. Primatol.* 75(suppl. 1):65.

Erkert, H. G. (2008). Diurnality and nocturnality in nonhuman primates: comparative chronobiological studies in laboratory and nature. *Biol. Rhythm Res.* 39:229–267.

Erkert, H. G., and Grober, J. (1986). Direct modulation of activity and body temperature of owl monkeys (*Aotus lemurinus griseimembra*) by low light intensities. *Folia Primatol.* 47:171–188.

Erkert, H. G., and Kappeler, P. M. (2004). Arrived in the light: diel and seasonal activity patterns in wild Verreaux's sifakas (*Propithecus v. verreauxi;* Primates: Indriidae). *Behav. Ecol. Sociobiol.* 57:174–186.

Erkert, H. G., and Thiemann-Jager, A. (1983). Dark switch in the entrainment of circadian activity rhythms in night monkeys, *Aotus trivirgatus* Humboldt. *Comp. Biochem. Physiol.* 74A:307–310.

Erwin, J. M., and Hof, P. R. (eds.) (2002). *Aging in Nonhuman Primates. Interdisciplinary Topics in Gerontology,* vol. 31. S. Karger, Basel.

Erxleben, J. (1777). *Systema Regni Animalis.* Impensis Weygandianis, Liepzig.

Escobar-Páramo, P. (2000). Inbreeding avoidance and the evolution of male mating strategies [PhD diss.]. State University of New York, Stonybrook.

Estrada, A. (1982). Survey and census of howler monkeys (*Alouatta palliata*) in the rain forest of "Los Tuxtlas," Veracruz, Mexico. *Am. J. Primatol.* 2:363–372.

Estrada, A. (1984). Resource use by howler monkeys (*Alouatta palliata*) in the rain forest of Los Tuxtlas, Veracruz, Mexico. *Int. J. Primatol.* 5:105–131.

Estrada, A., and Coates-Estrada, R. (1984). Fruit eating and seed dispersal by howling monkeys (*Alouatta palliata*) in the tropical rain forest of Los Tuxtlas, Mexico. *Am. J. Primatol.* 6:77–91.

Estrada, A., and Coates-Estrada, R. (1986). Frugivory in howling monkeys (*Alouatta palliata*) at Los Tuxtlas, Mexico: dispersal and fate of seeds. In: Estrada, A., and Fleming, T. H. (eds.), *Frugivores and Seed Dispersers*. Dr. W. Junk Publishers, Dordrecht. pp. 94–104.

Estrada, A., Juan-Solano, S., Ortíz Martínez, T., and Coates-Estrada, R. (1999). Feeding and general activity patterns of a howler monkey (*Alouatta palliata*) troop living in a forest fragment at Los Tuxtlas, Mexico. *Am. J. Primatol.* 48:167–183.

Estrada, A., Castellanos, L., Garcia, Y., Franco, B., Muñoz, D., Ibarra, A., Rivera, A., Fuentes, E., and Jimenez, C. (2002a). Survey of the black howler monkey, *Alouatta pigra*, population at the Mayan site of Palenque, Chiapas, Mexico. *Primates* 43:51–58.

Estrada, A., Luecke, L., van Belle, S., French, K., Muñoz, D., García, Y., Castellanos, L., and Mendoza, A. (2002b). The black howler monkey (*Alouatta pigra*) and spider monkey (*Ateles geoffroyi*) in the Mayan site of Yaxchilán, Chiapas, Mexico: a preliminary survey. *Neotrop. Primates* 10:89–95.

Estrada, A., Luecke, L., van Belle, S., Barrueta, E., and Meda, M. R. (2004). Survey of black howler (*Alouatta pigra*) and spider (*Ateles geoffroyi*) monkeys in the Mayan sites of Calakmul and Yaxchilán, Mexico and Tikal, Guatemala. *Primates* 45:33–39.

Eudey, A. A. (1980). Pleistocene glacial phenomena and the evolution of Asian macaques. In: Lindburg, D. G. (ed.), *The Macaques: Studies in Ecology, Behavior and Evolution*. Van Nostrand Reinhold, New York, pp. 52–83.

Evans, B. J., Morales, J. C., Supriatna, J., and Melnick, D. J. (1999). Origin of the Sulawesi macaques (Cercopithecidae: *Macaca*) as suggested by mitochondrial DNA phylogeny. *Biol. J. Linn. Soc.* 66:539–560.

Evans, B. J., Supriatna, J., and Melnick, D. J. (2001). Hybridization and population genetics of two macaque species in Sulawesi, Indonesia. *Evolution* 55:1986–1702.

Evans, C. S. (1997). Referential signals. In: Owings, D. H., Beecher, M., and Thompson, N. (eds.), *Perspectives in Ethology. Communication*, vol. 12. Plenum Press, New York. pp. 99–143.

Evans, C. S., and Goy, R. W. (1968). Social behaviour and reproductive cycles in captive ring-tailed lemurs (*Lemur catta*). *J. Zool. Lond.* 156:181–197.

Evans, C. S., Macedonia, J. M., and Marler, P. (1993). Effects of apparent size and speed on the response of chickens, *Gallus gallus*, to computer-generated simulations of aerial predators. *Anim. Behav.* 46:1–11.

Evans, T. A., Beran, M. J., Harris, E. H., and Rice, D. F. (2008). Quantity judgments of sequentially presented food items by capuchin monkeys (*Cebus apella*). *Anim Cogn.* doi:10.1007/s10071–008–0174-z.

Fa, J. E. (1991). Provisioning of Barbary macaques on the Rock of Gibraltar. In: Box, H. O. (ed.), *Primate Responses to Environmental Change*. Chapman and Hall, New York. pp. 137–154.

Fa, J. E., and Lindburg, D. G. (eds.) (1996). *Evolution and Ecology of Macaque Societies*. Cambridge University Press, Cambridge.

Fa, J. E., and Southwick, C. H. (ed.) (1988). *Ecology and Behavior of Food-Enhanced Groups*. Alan R. Liss, New York.

Fa, J. E., Juste, J., Perez del Val, J., and Castroviejo, J. (1995). Impact of market hunting on mammal species in Equatorial Guinea. *Conserv. Biol.* 9:1107–1115.

Fa, J. E., Peres, C. A., and Meeuwig, J. (2002). Bushmeat exploitation in tropical forests: an intercontinental comparison. *Conserv. Bio.* 16:232–237.

Fagan, R. (1993). Primate juveniles and primate play. In: Pereira, M., and Fairbanks, L. (eds.), *Juvenile Primates*. Oxford University Press, New York. pp. 182–196.

Fagot, J., and Berbet, I. (2006). Grouping and segmentation of visual objects by baboons (*Papio papio*) and humans (*Homo sapiens*). In: Wasserman, E. A., and Zentall, T. R. (eds.), *Comparative Cognition: Experimental Explorations of Animal Intelligence*. Oxford University Press, New York. pp. 15–28

Fagot, J., and Deruelle, C. (2002). Perception of pictorial eye gaze by baboons (*Papio papio*). *J. Exp. Psychol. Anim. Behav. Proc.* 28:298–308.

Fairbanks, L. (1988a). Mother–infant behavior in vervet monkeys: response to failure of last pregnancy. *Behav. Ecol. Sociobiol.* 23(3):157–165.

Fairbanks, L. A. (1988b). Vervet monkey grandmothers: effects on mother–infant relationships. *Behaviour* 104:176–188.

Fairbanks, L. (1993a). Juvenile vervet monkeys: establishing relationships and practicing skills for the future. In: Pereira, M., and Fairbanks, L. (eds.), *Juvenile Primates*. Oxford University Press, Oxford. pp. 211–227.

Fairbanks, L. (1993b). Risk-taking by juvenile vervet monkeys. *Behaviour* 124:57–72.

Fairbanks, L. A. (2000a). The developmental timing of primate play: a neural selection model. In: Parker, S. T., Langer, J., and McKinney, M. L. (eds.), *Biology, Brains and Behavior: The Evolution of Human Development*. SAR Press, Santa Fe. pp. 131–158.

Fairbanks, L. A. (2000b). Maternal investment throughout the life span in Old World monkeys. In: Whitehead, P. F., and Jolly, C. F. (eds.), *Old World Monkeys*. Cambridge University Press, Cambridge. pp. 341–367.

Fairbanks, L. A., Jorgensen, M. J., Huff, A., Blau, K., Hung, Y.-Y., and Mann, J. J. (2004). Adolescent impulsivity predicts adult dominance attainment in male vervet monkeys. *Am. J. Primatol.* 64:1–17.

Faircloth, B. C. (2008). MSATCOMMANDER: detection of microsatellite repeat arrays and automated, locus-specific primer design. *Mol. Ecol. Resources* 8:92–94.

Faith, D. P. (2008). Threatened species and the potential loss of phylogenetic diversity: conservation scenarios based on estimated extinction probabilities and phylogenetic risk analysis. *Conserv. Biol.* 22:1461–1470.

Falk, D. (1980). Comparative study of the endocranial casts of New and Old World monkeys. In: Ciochon, R., and Chiarelli, A. (eds.), *Evolutionary Biology of the New World Monkeys and Continental Drift*. Plenum Press, New York. pp. 275–292.

Falk, D. (2000). *Primate Diversity*. Norton, New York.

Fan, P. F., and Jiang, X. L. (2008). Effects of food and topography on ranging behavior of black crested gibbon (*Nomascus concolor jingdongensis*) in Wuliang Mountain, Yunnan, China. *Am. J. Primatol.* 70:871–878.

Fan, P. F., and Jiang, X. L. (2009). Predation on giant flying squirrels (*Petaurista philippensis*) by black crested gibbons (*Nomascus concolor jingdongensis*) at Mt. Wuliang, Yunnan, China. *Primates* 50:45–49.

Fan, P.F., Jiang, X.L., Liu, C.M., and Luo, W.S. (2006). Polygynous mating system and behavioural reason of black crested gibbon (*Nomascus concolor jingdongensis*) at Dazhaizi, Mt Wuliang, Yunnan, China. *Zool Res* 27(2):216–220

Fan, P. F., Ni, Q. Y., Sun, G. Z., Huang, B., and Jiang, X. L. (2008). Seasonal variations in the activity budget of black-crested gibbons (*Nomascus concolor jingdongensis*) at Mt. Wuliang, Central Yunnan, China: effects of diet and temperature. *Int. J. Primatol.* 29(4):1047–1057.

Fan, P. F., Ni, Q. Y., Sun, G. Z., Huang, B., and Jiang, X. L. (2009). Gibbons under seasonal stress: the diet of the black crested gibbons (*Nomascus concolor*) on Mt. Wuliang, Central Yunnan. *Primates* 50:37–44.

Fanshawe, J. H., and Fitzgibbon, C. D. (1993). Factors influencing the hunting success of an African wild dog pack. *Anim. Behav.* 45:479–490.

FAO (2005). Global Forest Resources Assessment 2005: progress towards sustainable forest management. FAO Forestry Paper 147, Rome.

Faraizl, S. D., and Stiver, S. J. (1996). A profile of depredating mountain lions. *Proc. Vert. Pest Conf.* 17:88–90.

Farrell, J., and Rabin, M. (1996). Cheap talk. *J. Econ. Perspect.* 10:110–118.

Farrer, C. R. (1997). Myth. In: Green, T. A. (ed.), *Folklore: An Encyclopedia of Beliefs, Customs, Tales, Music and Art*, vol. 2. ACB-Clio, Santa Barbara.

Fashing, P. J. (1999). The behavioral ecology of an African colobine monkey: diet, range use, and patterns of intergroup aggression in eastern black and white colobus monkeys (*Colobus guereza*) [PhD diss.]. Columbia University, New York.

Fashing, P. J. (2001a). Activity and ranging patterns of guerezas in the Kakamega Forest: intergroup variation and implications for intragroup feeding competition. *Int. J. Primatol.* 22:549–577.

Fashing, P. J. (2001b). Feeding ecology of guerezas in the Kakamega Forest, Kenya: the importance of Moraceae fruit in their diet. *Int. J. Primatol.* 22:579–609.

Fashing, P. J. (2001c). Male and female strategies during intergroup encounters in guerezas (*Colobus guereza*): evidence for resource defense mediated through males and a comparison with other primates. *Behav. Ecol. Sociobiol.* 50:219–230.

Fashing, P. J. (2002). Population status of black and white colobus monkeys (*Colobus guereza*) in the Kakamega Forest, Kenya: are they really on the decline? *Afr. Zool.* 37:119–126.

Fashing, P. J. (2004). Mortality trends in the African cherry (*Prunus africana*) and the implications for colobus monkeys (*Colobus guereza*) in Kakamega Forest, Kenya. *Biol. Conserv.* 120:449–459.

Fashing, P. J., and Cords, M. (2000). Diurnal primate densities and biomass in the Kakamega Forest: an evaluation of census methods and a comparison with other forests. *Am. J. Primatol.* 50:139–152.

Fashing, P. J., and Oates, J. F. (in press). *Colobus guereza.* In: Kingdon, J., Happold, D., and Butynski, T. (eds.), *The Mammals of Africa.* University of California Press, Berkeley, CA.

Fashing, P. J., Mulindahabi, F., Gakima, J.-B., Masozera, M., Mununura, I., Plumptre, A. J., and Nguyen, N. (2007). Activity and ranging patterns of Angolan black-and-whitecolobus (*Colobus angolensis ruwenzorii*) in Nyungwe Forest, Rwanda: possible costs of large group size. *Int. J. Primatol.* 28:529–550.

Faulkes, C. G., Arruda, M. F., and Monteiro da Cruz, A. O. M. (2003). Matrilineal genetic structure within and among populations of the cooperatively breeding common marmoset, *Callithrix jacchus. Mol. Ecol.* 12:1101–1108.

Fawcett, K., and Muhumuza, G. (2000). Death of a wild chimpanzee community member: possible outcome if intense sexual competition. *Am. J. Primatol.* 51:243–247.

Fay, J. C., and Wu, C. I. (2000). Hitch-hiking under positive Darwinian selection. *Genetics* 155:1405–1413.

Fay, M., Carroll, R. W., Kerbis Peterhans, J. C., and Harris, D. (1995). Leopard attack on and consumption of gorillas in the Central African Republic. *J. Hum. Evol.* 29:93–99.

Fazleabas, A. T., Hild-Petito, S., Donnelly, K. M., Mavrogianis, P., and Verhage, H. G. (1993). Interactions between the embryo and uterine endometrium during implantation and early pregnancy in the baboon (*Papio anubis*). In: Wolf, D. P., Stouffer, R. L., and Brenner, R. M. (eds.), *In Vitro Fertilization and Embryo Transfer in Primates.* Springer-Verlag, New York. pp. 169–181.

Fedigan, L. M. (1976). A study of roles in the Arashiyama West troop of Japanese monkeys (*Macaca fuscata*). *Contrib. Primatol.* 9:1–95.

Fedigan, L. M. (1982). *Primate Paradigms: Sex Roles and Social Bonds.* Eden Press, Montreal.

Fedigan, L. M. (1983). Dominance and reproductive success in primates. *Ybk. Phys. Anthropol.* 26:91–129.

Fedigan, L. M. (1986). Demographic trends in the *Alouatta palliata* and *Cebus capucinus* populations of Santa Rosa National Park, Costa Rica. In: Else, J. G., and Lee, P. C. (eds.), *Primate Ecology and Conservation.* Cambridge University Press, Cambridge. pp. 285–293.

Fedigan, L. M. (1990). Vertebrate predation in *Cebus capucinus:* meat eating in a neotropical monkey. *Folia Primatol.* 54:196–205.

Fedigan, L. M. (1992). *Primate Paradigms: Sex Roles and Social Bonds.* Eden Press, Montreal.

Fedigan, L. M. (1993). Sex differences and intersexual relations in adult white-faced capuchins (*Cebus capucinus*). *Int. J. Primatol.* 14:853–877.

Fedigan, L. (1997). Changing views of female life histories. In: Morbeck, M. E., Galloway, A., and Zihlman, A. L. (eds.), *The Evolving Female: A Life History Perspective.* Princeton University Press, Princeton, NJ. pp. 15–26.

Fedigan, L. M. (2000). Gender encounters. In: Strum, S. C., and Fedigan, L. M. (eds.), *Primate Encounters: Models of Science, Gender and Society.* University of Chicago Press, Chicago. pp. 498–520.

Fedigan, L. M. (2003). Impact of male takeovers on infant deaths, births, and conceptions in *Cebus capucinus* at Santa Rosa, Costa Rica. *Int. J. Primatol.* 24:723–741.

Fedigan, L. M., and Asquith, P. J. (eds.) (1991). *The Monkeys of Arashiyama: Thirty-Five Years of Research in Japan and the West.* State University of New York, Albany. pp. 54–73.

Fedigan, L. M., and Baxter, M. J. (1984). Sex differences and social organization in free-ranging spider monkeys (*Ateles geoffroyi*). *Primates* 25:279–294.

Fedigan, L. M., and Fedigan, L. (1989). Gender and the study of primates. In: Morgan, S. (ed.), *Gender and Anthropology. Critical Reviews for Teaching and Research.* American Anthropological Association, Washington DC. pp. 41–64.

Fedigan, L. M., and Gouzoules, H. (1978). The consort relationship in a troop of Japanese monkeys. In: Chivers, D. J., and Herbert, J. (eds.), *Recent Advances in Primatology.* Academic Press, New York. pp. 493–495.

Fedigan, L. M., and Griffin, L. (1996). Determinants of reproductive seasonality in the Arashiyama West Japanese macaques. In: Fa, J. E., and Lindburg, D. G. (eds.), *Evolution and Ecology in Macaque Societies.* Cambridge University Press, Cambridge. pp. 369–388.

Fedigan, L. M., and Jack, K. (2001). Neotropical primates in a regenerating Costa Rican dry forest: a comparison of howler and capuchin population patterns. *Int. J. Primatol.* 22:689–713.

Fedigan, L. M., and Pavelka, M. S. M. (2001). Is there adaptive value to reproductive termination in Japanese macaques? A test of maternal investment hypotheses. *Int. J. Primatol.* 22:109–125.

Fedigan, L. M., and Rose, L. M. (1995). Interbirth interval variation in three sympatric species of neotropical monkey. *Am. J. Primatol.* 37:9–24.

Fedigan, L. M., and Zohar, S. (1997). Sex differences in mortality of Japanese macaques: twenty-one years of data from the Arashiyama West population. *Am. J. Phys. Anthropol.* 102:161–175.

Fedigan, L. M., Fedigan, L., and Chapman, C. (1985). A census of *Alouatta palliata* and *Cebus capucinus* in Santa Rosa National Park, Costa Rica. *Brenesia* 23:309–322.

Fedigan, L. M., Fedigan, L., Gouzoules, S., Gouzoules, H., and Koyama, N. (1986). Lifetime reproductive success in female Japanese macaques. *Folia Primatol.* 47:143–157.

Fedigan, L. M., Fedigan, L., Chapman, C., and Glander, K. E. (1988). Spider monkey home ranges: a comparison of radio telemetry and direct observation. *Am. J. Primatol.* 16:19–29.

Fedigan, L. M., Rosenberger, A. L., Boinski, S., Norconk, M. A., and Garber, P. A. (1996). Critical issues in cebine evolution and behavior. In: Norconk, M. A., Rosenberger, A. L., and Garber, P. A. (eds.), *Adaptive Radiations of Neotropical Primates.* Plenum Press, New York. pp. 219–228.

Fedigan, L. M., Rose, L. M., and Avila, R. M. (1998). Growth of mantled howler groups in a regenerating Costa Rican dry forest. *Int. J. Primatol.* 19:405–432.

Feeny, P. (1976). Plant apparency and chemical defense. *Recent Adv. Phytochem.* 10:1–40.

Feer, F., and Forget, P. M. (2002). Spatio-temporal variation in post-dispersal seed fate. *Biotropica* 34:555–566.

Feeroz, M. M., Islam, M. A., and Kabir, M. M. (1994). Food and feeding behaviour of hoolock gibbon (*Hylobates hoolock*) capped langur (*Presbytis pileata*) and pigtailed macaque (*Macaca nemestrina*) of Lawachara. *Bangladesh J. Zool.* 22:123–132.

Feged, A., Wolowich, C., and Evans, S. (2002). Food sharing in owl monkeys. *Am. J. Primatol.* 57(suppl. 1):26.

Feistner, A. T., and Ashbourne, C. J. (1994). Infant development in a captive-bred aye-aye (*Daubentonia madagascariensis*) over the first year of life. *Folia Primatol.* 62(1–3):74–92.

Feistner, A., and Price, E. C. (1990). Food-sharing in cotton-top tamarins (*Saguinus oedipus*). *Folia Primatol.* 54:34–45.

Feistner, A. T. C., and Schmid, J. (1999). Lemurs of the Reserve Naturelle Integrale d'Andohahela, Madagascar. *Field. Zool.* 94:269–284.

Felsenstein, J. (1985). Phylogenies and the comparative method. *Am. Nat.* 125:1–15.

Felsenstein, J. (2004a). PHYLIP (Phylogeny Inference Package), version 3.6. Department of Genome Sciences, University of Washington, Seattle (http://evolution.genetics.washington.edu/phylip.html).

Felsenstein, J. (2004b). *Inferring Phylogenies.* Sinauer Associates, Sunderland, MA.

Felton, A. M., Engstrom, L. M., Felton, A., and Knott, C. D. (2003). Orangutan population density, forest structure and fruit availability in hand-logged and unlogged peat swamp forests in West Kalimantan, Indonesia. *Biol. Conserv.* 114:91–101.

Felton, A., Felton, A. M., Wallace, R. B., and Gómez, H. (2006). Identification, behavioral observations, and notes on the distribution of the titi monkeys *Callicebus modestus* Lönnberg, 1939 and *Callicebus olallae,* Lönnberg 1939. *Primate Conserv.* 20:41–46.

Felton, A. M., Felton, A., Wood, J. T., and Lindenmayer, D. B. (2008). Diet and feeding ecology of *Ateles chamek* in a Bolivian semihumid forest: the importance of *Ficus* as a staple food resource. *Int. J. Primatol.* 29:379–403.

Felton, A. M., Felton, A., Wood, J. T., Foley, W. J., Raubenheimer, D., Wallis, I. R., and Lindenmayer, D. B. (2009). Nutritional ecology of Ateles chamek in lowland Bolivia: How macronutrient balancing influences food choices. *Int. J. Primatol.* 30:675–696.

Fenner, M. (1992). *Seeds: The Ecology of Regeneration in Plant Communities.* C.A.B. International, Wallingford, UK.

Fernandes, M. E. B. (1991). Tool use and predation of oysters (*Crassostrea rhizophorea*) by the tufted capuchin, *Cebus apella apella,* in brackish water mangrove swamp. *Primates* 32:529–531.

Fernandes, M. E. B. (1993a). New field records of night monkeys, genus *Aotus,* in northern Brazil. *Neotrop. Primates* 1:6–7.

Fernandes, M. E. B. (1993b). Tail-wagging as a tension relief mechanism in pithecines. *Folia Primatol.* 61:52–56.

Fernandez-Duque, E. (2003). Influences of moonlight, ambient temperature and food availability on the diurnal and nocturnal activity of owl monkeys (*Aotus azarai*). *Behav. Ecol. Sociobiol.* 54:431–440.

Fernandez-Duque, E. (2004). High levels of intrasexual competition in sexually monomorphic owl monkeys. *Int. J. Primatol.* 75(suppl. 1):260.

Fernandez-Duque, E. (2009). Natal dispersal in monogamous owl monkeys (*Aotus azarai*) of the Argentinean Chaco. *Behaviour* 146, 583–606.

Fernandez-Duque, E., and Bravo, S. (1997). Population genetics and conservation of owl monkeys (*Aotus azarai*) in Argentina: a promising field site. *Neotrop. Primates* 5:48–50.

Fernandez-Duque, E., and Erkert, H. G. (2006). Cathemerality and lunar periodicity of the activity rhytms in owl monkeys of the Argentinian Chaco. *Folia Primatol.* 77:123–138.

Fernandez-Duque, E., and Huntington, C. (2002). Disappearances of individuals from social groups have implications for understanding natal dispersal in monogamous owl monkeys (*Aotus azarai*). *Am. J. Primatol.* 57:219–225.

Fernandez-Duque, E., and Rotundo, M. (2003). Are female owl monkeys imposing social monogamy on males? *Rev. Etol.* 5(suppl.):92.

Fernandez-Duque, E., Rotundo, M., and Sloan, C. (2001). Density and population structure of owl monkeys (*Aotus azarai*) in the Argentinean Chaco. *Am. J. Primatol.* 53:99–108.

Fernandez-Duque, E., Rotundo, M., and Ramírez-Llorens, P. (2002). Environmental determinants of birth seasonality in owl monkeys (*Aotus azarai*) of the Argentinean Chaco. *Int. J. Primatol.* 23:639–656.

Fernandez-Duque, E., Di Fiore A., and Carrillo-Bilbao, G. (2008a). Behavior, ecology and demography of *Aotus vociferans* in Yasuní National Park, Ecuador. *Int. J. Primatol.* 29(2):421–431.

Fernandez-Duque, E., Juárez, C. P., and Di Fiore, A. (2008b). Adult male replacement and subsequent infant care by male and siblings in socially monogamous owl monkeys (*Aotus azarai*). *Primates* 49:81–84.

Fernandez-Duque, E., Valeggia, C. R., and Mendoza, S. P. (2009). The biology of paternal care. *Ann. Rev. Anthropol.* Volume 38:115–130..

Ferrari, S. F. (1988). The behaviour and ecology of the buffy-headed marmoset, *Callithrix flaviceps* (O. Thomas, 1903) [PhD diss.]. University College, London.

Ferrari, S. F. (1993). Ecological differentiation in the Callitrichidae. In: Rylands, A. B. (ed.), *Marmosets and Tamarins: Systematics, Ecology and Behaviour.* Oxford University Press, Oxford. pp. 314–328.

Ferrari, S. F. (1995). Observations on *Chiropotes albinasus* from the Rio dos Marmelos, Amazonas, Brazil. *Primates* 36:289–293.

Ferrari, S. F., and Diego, V. H. (1992). Long-term changes in a wild marmoset group. *Folia Primatol.* 58:215–218.

Ferrari, S. F., and Digby, L. J. (1996). Wild *Callithrix* groups: stable extended families? *Am. J. Primatol.* 38:19–27.

Ferrari, S. F., and Lopes, M. A. (1996). Primate populations in eastern Amazonia. In: Norconk, M., Rosenberger, A. L., and Garber, P. A. (eds.), *Adaptive Radiations of Neotropical Primates.* Plenum Press, New York. pp. 53–67.

Ferrari, S. F., and Lopes Ferrari, M. A. (1989). A re-evaluation of the social organization of the Callitrichidae, with references to the ecological differences between genera. *Folia Primatol.* 52:132–147.

Ferrari, S. F., and Lopes Ferrari, M. A. (1990a). A survey of primates in central Pará. *Bol. Mus. Para. Emílio Goeldi Série Zool.* 6:169–179.

Ferrari, S. F., and Lopes-Ferrari, M. A. (1990b). Predator avoidance behaviour in the buffy-headed marmoset, *Callithrix flaviceps*. *Primates* 31:323–338.

Ferrari, S. F., and Martins, E. S. (1992). Gummivory and gut morphology in two sympatric callitrichids (*Callithrix emiliae* and *Saguinus fuscicollis weddelli*) from western Brazilian Amazonia. *Am. J. Phys. Anthropol.* 88:97–103.

Ferrari, S. F., Lopes, M. A., and Krause, E. A. K. (1993). Gut morphology of *Callithrix nigriceps* and *Saguinus labiatus* from western Brazilian Amazonia. *Am. J. Phys. Anthropol.* 90:487–493.

Ferrari, S. F., Corrêa, H. K. M., and Coutinho, P. E. G. (1996). Ecology of the "southern" marmosets (*Callithrix aurita* and *Callithrix flaviceps*). In: Norconk, M. A., Rosenberger, A. L., and Garber, P. A. (eds.), *Adaptive Radiations of Neotropical Primates.* Plenum Press, New York. pp. 157–171.

Ferrari, S. F., Iwanaga, S., Coutinho, P. E. G., Messias, M. R., Cruz Neto, E. H. D., Ramos, E. M., and Ramos, P. C. S. (1999a). Zoogeography of *Chiropotes albinasus* (Platyrrhini, Atelidae) in southwestern Amazonia. *Int. J. Primatol.* 20:995–1004.

Ferrari, S. F., Emidio-Silva, C., Lopes, M. A., and Bobadilla, U. L. (1999b). Bearded sakis in south-eastern Amazonia—back from the brink? *Oryx* 33:346–351.

Ferrari, S. F., Iwanaga, S., Ramos, E. M., Messias, M. R., Ramos, P. C. S., and Cruz Neto, E. H. (1999c). Expansion of the known distribution of Goeldi's monkey (*Callimico goeldii*) in southwestern Brazilian Amazonia. *Folia Primatol.* 70:112–116.

Ferrari, S. F., Iwanaga, S., Messias, M. R., Ramos, E. M., Ramos, P. C. S., Cruz Neto, E. H. D., and Coutinho, P. E. G. (2000). Titi monkeys (*Callicebus* spp., Atelidae: Platyrrhini) in the Brazilian state of Rondônia. *Primates* 41:229–234.

Ferrari, S. F., Pereira, W. L. A., Santos, R. R. , and Veiga, L. M. (2004). Fatal attack of a *Boa constrictor* on a bearded saki (*Chiropotes satanas utahicki*). *Folia Primatol.* 75:111–113.

Ferrari, S. F., Veiga, L. M., and Urbani, B. (2008). Geophagy in New World monkeys (Platyrrhini): ecological and geographic patterns. *Folia Primatol.* 79:402–415.

Ferraris, J. D., and Palumbi, S. R. (eds.) (1996). *Molecular Zoology: Advances, Strategies, and Protocols.* Wiley-Liss, New York.

Ferraz, G., Russell, G. J., Stouffer, P. C., Bierregaard, R. O. J., Pimm, S. L., and Lovejoy, T. E. (2003). Rates of species loss from Amazonian forest fragments. *Proc. Natl. Acad. Sci. USA* 100:14069–14073.

Ferreira, L. V., and Prance, G. T. (1998). Structure and species richness of low-diversity floodplain forest on the Rio Tapajós, eastern Amazonia, Brazil. *Biodivers. Conserv.* 7:585–596.

Ferris, S. D., Brown, W. M., Davidson, W. S., and Wilson, A. C. (1981). Extensive polymorphism in the mitochondrial DNA of apes. *Proc. Natl. Acad. Sci. USA* 78:6319–6323.

Fichtel, C., and Kappeler, P. M. (2002). Anti-predator behavior of group-living Malagasy primates: mixed evidence for a referential alarm call system. *Behav. Ecol. Sociobiol.* 51:262–275.

Field, M.Y. (2008). So hot you could fry a monkey: behavioral thermoregulation in human-commensal white-fronted capuchins (*Cebus albifrons*). *Am. J. Phys. Anthropol. Suppl* 46: 95–96.

Fietz, J. (1999a). Mating system of *Microcebus murinus*. *Am. J. Primatol.* 48:127–133.

Fietz, J. (1999b). Monogamy as a rule rather than exception in nocturnal lemurs: the case of the fat-tailed dwarf lemur, *Cheirogaleus medius*. *Ethology* 105:259–272.

Fietz, J., and Dausmann, K. H. (2006). Big is beautiful: fat storage and hibernation as a strategy to cope with marked seasonality in the fat-tailed dwarf lemur (*Cheirogaleus medius*). In: Gould, L., and Sauther, M. L. (eds.), *Lemurs: Ecology and Adaptation.* Springer, New York. pp. 97–110.

Fietz, J., Zischler, H., Schwieg, C., Tomiuk, J., Dausmann, K. H., and Ganzhorn, J. U. (2000). High rates of extra-pair young in the pair-living fat-tailed dwarf lemur, *Cheirogaleus medius*. *Behav. Ecol. Sociobiol.* 49:8–17.

Figueiredo, W. (2006). Estimativas de tempos de divergencia em platirrinos e filogenia molecular e filogeografia dos uacaris, parauacus e cuxius [diss.]. University of Para, Brazil.

File, S. K., McGrew, W. C., and Tutin, C. E. G. (1976). The intestinal parasites of a community of feral chimpanzees, *Pan troglodytes schweinfurthii*. *J. Parasitol.* 62:259–261.

Fimbel, C., Vedder, A., Dierenfeld, E., and Mulindahabi, F. (2001). An ecological basis for large group size in *Colobus angolensis* in the Nyungwe Forest, Rwanda. *Afr. J. Ecol.* 39:83–92.

Firos, S. (2001). Absence of intragroup coalitions in adult male red colobus (*Colobus badius tephrosceles*) in the Kibale National Park, Uganda. *Folia Primatol.* 72:54–56.

Fischer, F., Gross, M., and Kunz, B. K. (1999–2000). Primates of the Comoé National Park, Ivory Coast. *Afr. Primates* 4:10–15.

Fischer, A., Wiebe, V., Paabo, S., and Przeworski, M. (2004). Evidence for a complex demographic history of chimpanzees. *Mol. Biol. Evol.* 21(5):799–808.

Fischer, A., Pollack, J., Thalmann, O., Nickel, B., and Pääbo, S. (2006). Demographic history and genetic differentiation in apes. *Curr. Biol.* 16: 1133–1138.

Fischer, J. O., and Geissmann, T. (1990). Group harmony in gibbons: comparison between white-handed gibbon (*Hylobates lar*) and siamang (*H. syndactylus*). *Primates* 31:481–494.

Fischer, J., Kitchen, D. M., Seyfarth, R. M., and Cheney, D. L. (2004). Baboon loud calls advertise male quality: acoustic features and their relation to rank, age and exhaustion. *Behav. Ecol. Sociobiol.* 56:140–148.

Fischer, K., and Chapman, C. A. (1993). Frugivores and fruit syndromes: differences in patterns at the genus and species levels. *Oikos* 66:472–482.

Fisher, H. S., Swaisgood, R. R., and Fitch-Snyder, H. (2003a). Odor familiarity and female preferences for males in a threatened primate, the pygmy loris *Nycticebus pygmaeus*: applications for genetic management of small populations. *Naturwissenschaften* 90:509–512.

Fisher, H. S., Swaisgood, R. R., and Fitch-Snyder, H. (2003b). Countermarking by male pygmy lorises (*Nycticebus pygmaeus*): do females use odor cues to select mates with high competitive ability? *Behav. Ecol. Sociobiol.* 53:123–130.

Fisher, P. J., Gardner, R. C., and Richardson, T. E. (1996). Single locus microsatellites isolated using 5′-anchored PCR. *Nucleic Acids Res.* 24:4369–4371.

Fisher, R. A. (1930). *The Genetical Theory of Natural Selection.* Clarendon Press, Oxford.

Fitch-Snyder, H., and Carter, J. (1993). Tool use to acquire drinking water by free-ranging lion-tailed macaques (*Macaca silenus*). *Lab. Primate Newslett.* 32:1–2.

Fitch-Snyder, H., and Ehrlich, A. (2003). Mother–infant interactions in slow lorises (*Nycticebus bengalensis*) and pygmy lorises (*Nycticebus pygmaeus*). *Folia Primatol.* 74:259–271.

Fitch-Snyder, H., and Jurke, M. (2003). Reproductive patterns in pygmy lorises (*Nycticebus pygmaeus*): behavioral and physiological correlates of gonadal activity. *Zoo Biol.* 22:15– 32.

Fitch-Snyder, H., and Vu, N. T. (2002). A preliminary survey of lorises (*Nycticebus* spp.) in northern Vietnam. *Asian Primates* 8:1–3.

Fitzgibbon, C. D. (1990). Why do hunting cheetahs prefer male gazelles? *Anim. Behav.* 40:837–845.

Fitzgibbon, C. D., and Lazarus, J. (1995). Antipredator behavior of Serengeti ungulates: individual differences and population consequences. In: Sinclair, A. R. E. (ed.), *Serengeti II: Dynamics, Management and Conservation of an Ecosystem.* University of Chicago Press, Chicago. pp. 274–296.

Fitzgibbon, C. D., Mogaka, H., and Fanshawe, J. H. (1995). Subsistence hunting in Arabuko-Sokoke Forest, Kenya and its effects on mammal populations. *Conserv. Biol.* 9:1116–1126.

Fitzpatrick, M. J., Ben-Shahar, Y., Smid, H. M., Vet, L. E. M., Robinson, G. E., and Sokolowski, M. B. (2005). Candidate genes for behavioral ecology. *Trends Ecol. Evol.* 20:96–104.

Fleagle, J. F., Janson, C. H., and Reed, K. E. (1999). *Primate Communities.* Cambridge University Press, Cambridge.

Fleagle, J. G. (1976a). Locomotion and posture of the Malayan siamang and implications for hominoid evolution. *Folia Primatol.* 26:245–269.

Fleagle, J. G. (1976b). Locomotor behavior and skeletal anatomy of sympatric Malaysian leaf-monkeys (*Presbytis obscura* and *Presbytis melalophos*). *Ybk. Phys. Anthropol.* 20:440–453.

Fleagle, J. G. (1978). Locomotion, posture, and habitat utilization in two sympatric, Malaysian leaf-monkeys (*Presbytis obscura* and *Presbytis melalophos*). In: Montgomery, G. G. (ed.), *The Ecology of Arboreal Folivores.* Smithsonian Institution Press, Washington DC. pp. 243–252.

Fleagle, J. G. (1988). *Primate Adaptations and Evolution.* Academic Press, New York.

Fleagle, J. G. (1998). Introduction to part II. In: Strasser, E., Fleagle, J. G., Rosenberger, A., and McHenry, H. (eds.), *Primate Locomotion: Recent Advances.* Plenum Press, New York. pp. 109–110.

Fleagle, J. G. (1999). *Primate Adaptation and Evolution.* Academic Press, New York.

Fleagle, J. G., and Kay, R. F. (eds.) (1994). *Anthropoid Origins.* Plenum Press, New York.

Fleagle, J. G., and McGraw, W. S. (1999). Skeletal and dental morphology supports diphyletic origin of baboons and mandrills. *Proc. Natl. Acad. Sci. USA* 96:1157–1161.

Fleagle, J. G., and McGraw, W. S. (2002). Skeletal and dental morphology of African papionins: unmasking a cryptic clade. *J. Hum. Evol.* 42:267–292.

Fleagle, J. G., and Meldrum, D. J. (1988). Locomotor behavior and skeletal morphology of two sympatric pitheciine monkeys, *Pithecia pithecia* and *Chiropotes satanas. Am. J. Primatol.* 16:227–249.

Fleagle, J. G., and Mittermeier, R. A. (1980). Locomotor behavior, body size, and comparative ecology of seven Surinam monkeys. *Am. J. Phys. Anthropol.* 52:301–314.

Fleagle, J. G., and Reed, K. E. (2004). The evolution of primate ecology: patterns of geography and phylogeny. In: Anapol, F., German, R. Z., and Jablonski, N. (eds.), *Shaping Primate Evolution: Form, Function and Behavior.* Cambridge University Press, New York. pp. 353–367.

Fleagle, J. G., and Tejedor, M. F. (2002). Early platyrrhines of southern South America. In: Hartwig, W. C. (ed.), *The Primate Fossil Record.* Cambridge University Press, Cambridge. pp. 161–174.

Fleming, T. H. (1988). *The Short-Tailed Fruit Bat: A Study in Plant–Animal Interactions.* University of Chicago Press, Chicago.

Fleming, T. H., Breitwisch, R., and Whitesides, G. H. (1987). Patterns of tropical vertebrate frugivore diversity. *Annu. Rev. Ecol. Syst.* 18:91–109.

Flemming, T. M., Beran, M. J., and Washburn, D. A. (2007). Disconnect in concept learning by rhesus monkeys (*Macaca mulatta*): judgment of relations and relations-between-relations. *J. Exp. Psychol. Anim. Behav. Proc.* 33:55–63.

Fletcher, A. W. (1994). The social development of immature mountain gorillas (*Gorilla gorilla beringei*) [PhD diss.]. Bristol University, Bristol.

Fletcher, A. W. (2001). Development of infant independence from the mother in wild mountain gorillas. In: Robbins, M. M., Sicotte, P., and Stewart, K. J. (eds.), *Mountain Gorillas: Three Decades of Research at Karisoke.* Cambridge University Press, Cambridge. pp. 153–182.

Fleury, M. C., and Gautier-Hion, A. (1997). Better to live with allogenerics than to live alone? The case of single male *Cercopithecus pogonias* in troops of *Colobus santanas. Int. J. Primatol.* 18:967–974.

Fleury, M. C., and Gautier-Hion, A. (1999). Seminomadic ranging in a population of black colobus (*Colobus satanas*) in Gabon and its ecological correlates. *Int. J. Primatol.* 20:491–509.

Flombaum, J. I., and Santos, L. R. (2005). Rhesus monkeys attribute perceptions to others. *Curr. Biol.* 15: 447–52.

Florez, J. (2007). Choice of nest species in Bornean orangutans: a role for parasite repellent? [masters thesis]. University of Zurich, Zurich.

Flower, S. S. (1931). Contributions to our knowledge of the duration of life in vertebrate animals. *Proc. Zool. Soc. Lond.* 1931:145–233.

Fobes, J. L., and King, J. E. (1982a). Complex learning by primates. In: Fobes, J. L., and King, J. E. (eds.), *Primate Behavior.* Academic Press, New York. pp. 327–360.

Fobes, J. L., and King, J. E. (1982b). Measuring primate learning abilities. In: Fobes, J. L., and King, J. E. (eds.), *Primate Behavior.* Academic Press, New York. pp. 289–326.

Foerg, R. (1982). Reproduction in *Cheirogaleus medius. Folia Primatol.* 39:49–62.

Fogden, M. L. (1974). A preliminary field study of the western tarsier, *Tarsius bancanus* Horsfield. In: Martin, R. D., Doyle, G. A., and Walker, A. C. (eds.), *Prosimian Biology.* Duckworth Press, London.

Foley, R. A., and Elton, S. (1998). Time and energy: the ecological context for the evolution of bipedalism. In: Strasser, E., Fleagle, J. G., Rosenberger, A., and McHenry, H. (eds.), *Primate Locomotion: Recent Advances.* Plenum Press, New York. pp. 419–433.

Foley, W. J., and McArthur, C. (1994). The effects and costs of allelochemicals for mammalian herbivores: an ecological perspective. In: Chivers, D. J., and Langer, P. (eds.), *The Digestive System of Mammals.* Cambridge University Press, Cambridge. pp. 370–391.

Fontaine, B., Moisson, P. Y., and Wickings, E. J. (1995). Observations of spontaneous tool making and tool use in a captive group of western lowland gorillas (*Gorilla gorilla gorilla*). *Folia Primatol.* 65:219–223.

Fontaine, R. (1981). The uakaris, genus *Cacajao*. In: Coimbra-Filho, A. F., and Mittermeier, R. A. (eds.), *Ecology and Behavior of Neotropical Primates,* vol. 1. Academia Brasileira de Ciências, Rio de Janeiro. pp. 443–493.

Fooden J. (1963). A revision of the woolly monkeys (genus *Lagothrix*). *J. Mammal.* 44:213–247.

Fooden, J. (1967). Complementary specialization of male and females reproductive structures in the bear macaque, *Macaca arctoides. Nature* 214:939–941.

Fooden, J. (1969). *Taxonomy and Evolution of the Monkeys of Celebes.* Karger, Basel.

Fooden, J. (1971). Color and sex in gibbons. *Bull. Field Mus. Nat. Hist.* 42:2–7.

Fooden, J. (1976). Provisional classification and key to the living species of macaques (Primates: *Macaca*). *Folia Primatol.* 25:225–236.

Fooden, J. (1982). Ecogeographic segregation of macaque species. *Primates* 23:574–579.

Fooden, J. (1986). Taxonomy and evolution of the sinica group of macaques: 5. Overview of natural history. *Field. Zool.* 29:1–22.

Fooden, J. (1990). The bear macaque, *Macaca arctoides:* a systematic review. *J. Hum. Evol.* 19:607–686.

Fooden, J. (1995). Systematic review of Southeast Asian longtailed macaques, *Macaca fascicularis* (Raffles, [1821]). *Field. Zool.* 81:1–206.

Fooden, J. (1997). Tail length variation in *Macaca fascicularis* and *M. mulatta. Primates* 38:221–231.

Fooden, J. (2000). Systematic review of the rhesus macaque *Macaca mulatta* (Zimmermann, 1780). *Field. Zool.* 96:1–179.

Fooden, J. (2007). Systematic review of the Barbary macaque *Macaca sylvanus* (Linnaeus, 1758). *Field. Zool.* 113:1–57.

Foote, A. L., and Crystal, J. D. (2007). Metalcognition in the rat. *Curr. Biol.* 17:551–555.

Foote, M., Hunter, J. P., Janis, C. M., and Sepkoski, J. J., Jr. (1999). Evolutionary and preservational constraints on origins of biologic groups: divergence times of eutherian mammals. *Science* 283:1310–1314.

Ford, S. M. (1986). Systematics of the New World monkeys. In: Swindler, D. R., and Erwin, J. (eds.), *Comparative Primate Biology. Systematics, Evolution, and Anatomy*, vol. 1. Alan R. Liss, New York. pp. 73–135.

Ford, S. M. (1994a). Evolution of sexual dimorphism in body weight in platyrrhines. *Am. J. Primatol.* 34:221–244.

Ford, S. M. (1994b). Taxonomy and distribution of the owl monkey. In: Baer, J. F., Weller, R. E., and Kakoma, I. (eds.), *Aotus: The Owl Monkey*. Academic Press, San Diego. pp. 1–57.

Ford, S. M., and Boinski, S. (2007). Primate predation by harpy eagles in the Central Suriname Nature Reserve [abstract]. *Am. J. Phys. Anthropol.* (suppl. 44):109.

Ford, S. M., and Davis, L. C. (1992). Systematics and body size: implications for feeding adaptations in New World monkeys. *Am. J. Phys. Anthropol.* 88:415–468.

Ford, S. M., and Hobbs, D. G. (1996). Species definition and differentiation as seen in the postcranial skeleton of *Cebus*. In: Norconk, M. A., Rosenberger, A. L., and Garber, P. A. (eds.), *Adaptive Radiations of Neotropical Primates*. Plenum Press, New York. pp. 229–249, 540–541.

Forget, P. M., and Milleron, T. (1991). Evidence for secondary seed dispersal by rodents in Panama. *Oecologia* 87:596–599.

Forman, L., Kleiman, D. G., Bush, R. M., Dietz, J. M., Ballou, J. D., Phillips, L. G., Coimbra-Filho, A. F., and O'Brien, S. J. (1987). Genetic variation within and among lion tamarins. *Am. J. Phys. Anthropol.* 71:1–11.

Förster, S., and Cords, M. (2002). Development of mother–infant relationships and infant behavior in wild blue monkeys (*Cercopithecus mitis stuhlmanni*). In: Glenn, M., and Cords, M. (eds.), *The Guenons: Diversity and Adaptation in African Monkeys*. Kluwer Academic/Plenum, New York. pp. 245–272.

Forster, T., Roy, D., and Ghazal, P. (2003). Experiments using microarray technology: limitations and standard operating procedures. *J. Endocrinol.* 178:195–204.

Forthman, D. L., Strum, S. C., and Muchemi, G. M. (2005). Applied conditioned taste aversion and the management and conservation of crop-raiding primates. In: Paterson, J. D., and Wallis, J. (eds.), *Commensalism and Conflict: The Human–Primate Interface*. American Society of Primatologists, Norman, OK. pp. 420–443.

Forthman Quick, D. L. (1986). Activity budgets and the consumption of human food in two troops of baboons, *Papio anubis*, at Gilgil, Kenya. In: Else, J. G., and Lee, P. C. (eds.), *Primate Ecology and Conservation*. Cambridge University Press, Cambridge. pp. 221–228.

Forthman Quick, D. L., and Demment, M. W. (1988). Dynamics of exploitation: differential energetic adaptations of two troops of baboons to recent human contact. In: Fa, J. E., and Southwick, C. H. (eds.), *Ecology and Behavior of Food-Enhanced Primate Groups*. Alan R. Liss, New York. pp. 25–51.

Fortman, J. D., Herring, J. M., Miller, J. B., Hess, D. L., Verhage, H. G., and Fazleabas, A. T. (1993). Chorionic gonadotropin, estradiol, and progesterone levels in baboons (*Papio anubis*) during early pregnancy and spontaneous abortion. *Biol. Reprod.* 49:737–742.

Fossey, D. (1979). Development of the mountain gorilla (*Gorilla gorilla beringei*): the first thirty-six months. In: Hamburg, D. A., and McCown, E. R. (eds.), *Perspectives on Human Evolution. The Great Apes*, vol. 5. Benjamin/Cummings, Menlo Park, CA. pp. 139–184.

Fossey, D. (1983). *Gorillas in the Mist*. Houghton Mifflin, Boston.

Fossey, D. (1984). Infanticide in mountain gorillas (*Gorilla gorilla berengei*) with comparative notes on chimpanzees. In: Hausfater, G., and Hrdy, S. B. (eds.), *Infanticide: Comparative and Evolutionary Perspectives*. Aldine, Hawthorne, NY. pp. 217–235.

Fossey, D., and Harcourt, A. H. (1977). Feeding ecology of free ranging mountain gorillas (*Gorilla gorilla beringei*). In: Clutton-Brock, T. H. (ed.), *Primate Ecology: Studies of Feeding and Ranging Behavior in Lemurs, Monkeys and Apes*. Academic Press, London. pp. 539–556.

Foster, J. B. (1982). The seasonal rhythm of fruitfall on Barro Colorado Island. In: Leigh, E. G., Rand, A. S., and Windsor, D. M. (eds.), *The Ecology of a Neotropical Forest: Seasonal Rhythms and Long-Term Changes*. Smithsonian Institution, Washington DC. pp. 151–172.

Fourrier, M., Sussman, R. W., Kippen, R., and Childs, G. (2008). Demographic modeling of a predator–prey system and its implication for the Gombe population of *Procolobus rufomitratus* tephrosceles. *Int. J. Primatol.* 29:497–508.

Fowler, A. M., and Sommer, V. (2007). Subsistence technology of Nigerian chimpanzees. *Int. J. Primatol.* 28:997–1023.

Fowler, J. M., and Cope, J. B. (1964). Notes on the harpy eagle in British Guiana. *Auk* 81:257–273.

Fox, E. A. (1998). The function of female mate choice in the Sumatran orangutan (*Pongo pygmaeus abelii*) [PhD diss.]. Duke University, Durham, NC.

Fox, E. A. (2001). Homosexual behavior in wild Sumatran orangutans (*Pongo pygmaeus abelii*). *Am. J. Primatol.* 55:177–181.

Fox, E. A. (2002). Female tactics to reduce sexual harassment in the Sumatran orangutan (*Pongo pygmaeus abelii*). *Behav. Ecol. Sociobiol.* 52:93–101.

Fox, E. A., Sitompul, A. F., and van Schaik, C. P. (1999). Intelligent tool use in wild Sumatran orangutans. In: Parker, S. T., Mitchell, R. W., and Miles, H. L. (eds.), *The Mentalities of Gorillas and Orangutans*. Cambridge University Press, Cambridge. pp. 99–116.

Fox, E. A., van Schaik, C. P., Sitompul, A. F., and Doniele, N. W. (2004). Intra- and inter-populational differences in orangutan (*Pongo pygmaeus*) activity and diet: implications for the invention of tool use. *Am. J. Phys. Anthropol.* 125:162–174.

Fox, R. (2004). Primate kin and human kinship. In: Parker, R., and Stone, L. (eds.), *Kinship and the Family: An Anthropological Reader*. Blackwell, Malden, MA. pp. 424–437.

Fragaszy, D. (1990). Sex and age differences in the organization of behavior in wedge-capped capuchins, *Cebus olivaceus*. *Behav. Ecol.* 1:81–94.

Fragaszy, D. M., and Bard, K. (1997). Comparison of development and life history in *Pan* and *Cebus*. *Int. J. Primatol.* 18(5):683–701.

Fragaszy, D. M., and Boinski, S. (1995). Patterns of individual diet choice and efficiency of foraging in wedge-capped capuchin monkeys (*Cebus olivaceus*). *J. Comp. Psychol.* 109:339–348.

Fragaszy, D. M., and Cummins-Sebree, S. E. (2005). Relational spatial reasoning by a nonhuman: the example of capuchin monkeys. *Behav. Cogn. Neurosci. Rev.* 4:282–306.

Fragaszy, D. M., and Mitchell, G. (1974). Infant socialization in primates. *J. Hum. Evol.* 3:563–574.

Fragaszy, D. M., and Perry, S. (2003). Towards a biology of traditions. In: Fragaszy, D. M., and Perry, S. (eds.), *The Biology of Traditions: Models and Evidence*. Cambridge University Press, Cambridge. pp. 1–32.

Fragaszy, D. M., and Visalberghi, E. (1990). Social processes affecting the appearance of innovative behaviours in capuchin monkeys. *Folia Primatol.* 54:155–165.

Fragaszy, D., and Visalberghi, E. (1996). Social learning in monkeys: primate "primacy" reconsidered. In: Heyes, C. M., and Galef, B. G. (eds.), *Social Learning in Animals: The Roots of Culture*. Academic Press, San Diego. pp. 65–84.

Fragaszy, D. M., Schwarz, S., and Shimosaka, D. (1982). Longitudinal observations of care and development of infant titi monkeys (*Callicebus moloch*). *Am. J. Primatol.* 2:191–200.

Fragaszy, D. M., Visalberghi, E., and Robinson, J. G. (1990). Variability and adaptability in the genus *Cebus*. *Folia Primatol.* 54:114–118.

Fragaszy, D. M., Baer, J., and Adams-Curtis, L. (1991). Behavioral development and maternal care in tufted capuchins (*Cebus capucinus*) and squirrel monkeys (*Saimiri sciureus*) from birth through seven months. *Dev. Psychobiol.* 24:375–393.

Fragaszy, D., Johnson-Pynn, J., Hirsh, E., and Brakke, K. (2003). Strategic navigation of two-dimensional alley mazes: comparing capuchin monkeys and chimpanzees. *Anim. Cogn.* 6:149–160.

Fragaszy, D., Izar, P., Visalberghi, E., Ottoni, E. B., and De Oliveira, M. G. (2004a). Wild capuchin monkeys (*Cebus libidinosus*) use anvils and stone pounding tools. *Am. J. Primatol.* 64:359–366.

Fragaszy, D. M., Visalberghi, E., and Fedigan, L. M. (2004b). *The Complete Capuchin. The Biology of the Genus Cebus*. Cambridge University Press, Cambridge.

Frank, L. G. (1986). Social organization of the spotted hyena (*Crocuta crocuta*). II. Dominance and reproduction. *Anim. Behav.* 34:1510–1527.

Frankie, G. W., Baker, H. G., and Opler, P. A. (1974). Comparative phenological studies of trees in tropical wet and dry forests in the lowlands of Costa Rica. *J. Ecol.* 62:881–919.

Franklin, I. R. (1980). Evolutionary change in small populations. In: Soulé, M. E., and Wilcox, B. A. (eds.), *Conservation Biology: An Evolutionary–Ecological Perspective*. Sinauer Associates, Sunderland, MA. pp. 135–150.

Franklin, S. P., Hankerson, S. J., Baker, A. J., and Dietz, J. M. (2007). Golden lion tamarin sleeping-site use and pre-retirement behavior during intense predation. *Am. J. Primatol.* 69:325–335.

Frantzen, M. A. J., Silk, J. B., Ferguson, J. W. H., Wayne, R. K., and Kohn, M. H. (1998). Empirical evaluation of preservation methods for faecal DNA. *Mol. Ecol.* 7:1423–1428.

Fraser, O. N., and Aureli, F. (2008). Reconciliation, consolation and postconflict behavioral specificity in chimpanzees. *Am. J. Primatol.* 70:1114–1123.

Fraser, O. N., Schino, G., and Aureli, F. (2008a). Components of relationship quality in chimpanzees. *Ethology* 114:834–843.

Fraser, O. N., Stahl, D., and Aureli, F. (2008b). Stress reduction through consolation in chimpanzees. *Proc. Natl. Acad. Sci. USA* 105:8557–8562.

Frazão, E. (1991). Insectivory in free-ranging bearded saki (*Chiropotes satanas chiropotes*). *Primates* 32:243–245.

Fredman, T., and Whiten, A. (2007). Observational learning from tool using models by human-reared and mother-reared capuchin monkeys (*Cebus apella*). *Anim. Cogn.* 11:295–309.

Fredrickson, W., and Sackett, G. (1984). Kin preferences in primates (*Macaca nemestrina*): relatedness or familiarity? *J. Comp. Psychol.* 98:29–34.

Fredsted, T., and Villesen, P. (2004). Fast and reliable sexing of prosimian and human DNA. *Am. J. Primatol.* 64:345–350.

Freed, B. Z. (1996). Co-occurrence among crowned lemurs (*Eulemur coronatus*) and Sanford's lemurs (*Eulemur fulvus sanfordi*) of Madagascar [PhD diss.]. Washington University, St. Louis.

Freed, B. Z. (1999). An introduction to the ecology of daylight-active lemurs. In: Dolhinow, P., and Fuentes, P. A. (eds.), *The Non-Human Primates*. Mayfield Press, Mountain View, CA. pp. 123–132.

Freed, B.Z. (2006). Polyspecific associations of crowned lemurs and Sanford's lemurs in Madagascar. In: Gould, L., and Sauther, M.L. (eds.), *Lemurs: Ecology and Adaptation*. Springer, New York, pp.111–131.

Freeland, J.R. (2005). *Molecular Ecology*. J. Wiley & Sons, Chichester.

Freeland, W. J. (1976). Pathogens and the evolution of primate sociality. *Biotropica* 8:12–24.

Freeland, W. J. (1979). Mangabey (*Cercocebus albigena*) social organization and population density in relation to food use and availability. *Folia Primatol.* 32:108–124.

Freeland, W. J. (1980). Mangabey (*Cercocebus albigena*) movement patterns in relation to food availability and fecal contamination. *Ecology* 61:1297–1303.

Freeland, W. J. (1983). Parasites and the coexistence of animal host species. *Am. Nat.* 121:223–236.

Freeland, W. J. (1991). Plant secondary metabolites. Biochemical coevolution with herbivores. In: Palo, R. T., and Robbins, C. T. (eds.), *Plant Defenses Against Mammalian Herbivory*. CRC Press, Boca Raton, FL. pp. 61–82.

Freeland, W. J., and Choquenot, D. (1990). Determinants of herbivore carrying capacity: plants, nutrients and *Equus asinus* in northern Australia. *Ecology* 701:589–597.

Freeland, W. J., and Janzen, D. H. (1974). Strategies in herbivory by mammals: the role of plant secondary compounds. *Am. Nat.* 108:269–289.

Freeman, L. G. (1994). Torralba and Ambrona: a review of discoveries. In: Corruccini, R. S., and Chiochon, R. L. (eds.), *Integrative Paths to the Past: Paleoanthropological Advances in Honor of F. Clark Howell*. Prentice Hall, Englewood Cliffs, NJ. pp. 597–637.

Freese, C. (1976). Censusing *Alouatta palliata, Ateles geoffroyi*, and *Cebus capucinus* in the Costa Rican dry forest. In: Thorington, R. W., Jr., and Heltne, P. G. (eds.), *Neotropical Primates: Field Studies and Conservation*. National Academy of Sciences, Washington DC. pp. 4–9.

Freese, C. H. (1983). *Cebus capucinus,* mono cara blanca, white-faced capuchin. In: Janzen, D. H. (ed.), *Costa Rica Natural History*. University of Chicago Press, Chicago. pp. 458–460.

Freese, C. H. (1997). *Harvesting Wild Species: Implications for Biodiversity Conservation*. Johns Hopkins University Press, Baltimore, MD.

Freese, C. H., and Oppenheimer, J. R. (1981). The capuchin monkey, genus *Cebus*. In: Coimbra-Filho, A. F., and Mittermeier, R. H. (eds.), *Ecology and Behaviour of Neotropical Primates,* vol. 1. Academia Brasilia, Rio de Janeiro. pp. 331–390.

French, J. A. (1997). Proximate regulation of singular breeding in Callitrichid primates. In: Solomon, N. G., and French, J. A. (eds.), *Cooperative Breeding in Mammals*. Cambridge University Press, Cambridge. pp. 34–75.

French, J. A., and Stribley, J. A. (1985). Patterns of urinary oestrogen excretion in female golden lion tamarins (*Leontopithecus rosalia*). *J. Reprod. Fertil.* 75:537–546.

French, J. A., Abbott, D. H., Scheffler, G., Robinson, J. A., and Goy, R. W. (1983). Cyclic excretion of urinary oestrogens in female tamarins (*Saguinus oedipus*). *J. Reprod. Fertil.* 68:177–184.

French, J. A., deGraw, W. A., Hendricks, S. E., Wegner, F., and Bridson, W. E. (1992). Urinary and plasma gonadotropin concentrations in golden lion tamarins (*Leontopithecus r. rosalia*). *Am. J. Primatol.* 26:53–59.

French, J. A., Brewer, K. J., Schaffner, C. M., Schalley, J., Hightower-Merritt, D. L., Smith, T. E., and Bell, S. M. (1996). Urinary steroid and gonadotropin excretion across the reproductive cycle in female Wied's black tufted-ear marmosets (*Callithrix kuhli*). *Am. J. Primatol.* 40:231–245.

French, J. A., De Vleeschouwer, K., Bales, K., and Heistermann, M. (2002). Lion tamarin reproductive biology. In: Kleiman, D. G., and Rylands, A. B. (eds.), *Lion Tamarins: Biology and Conservation*. Smithsonian Institution Press, Washington DC. pp. 133–156

French, J. A., Bales, K. L., Baker, A. J., and Dietz, J. M. (2003). Endocrine monitoring of wild dominant and subordinate female *Leontopithecus rosalia*. *Int. J. Primatol.* 24:1281–1300.

Frisch, R. E. (2002). *Female Fertility and the Body Fat Connection*. Chicago University Press, Chicago.

Froelich, J. W., Thorington, R. W., and Otis, J. S. (1981). The demography of howler monkeys (*Alouatta palliata*) on Barro Colorado Island, Panamá. *Int. J. Primatol.* 2:207–236.

Frost, S. R., Marcus, L. F., Bookstein, F. L., Reddy, D. P., and Delson, E. (2003). Cranial allometry, phylogeography, and systematics of large-bodied Papionins (Primates: Cercopithecinae) inferred from geometric morphometric analysis of landmark data. *Anat. Rec. A* 275:1048–1072.

Fruth, B. (1995). *Nests and Nest Groups in Wild Bonobos* (Pan paniscus): *Ecological and Behavioural Correlates*. Ludwig-Maximilians-Universität, Munich.

Fruth, B. (1998). Reply to Stanford (1998) "The social behavior of chimpanzees and bonobos—empirical evidence and shifting assumptions." *Curr. Anthropol.* 39:408–409.

Fruth, B., and Hohmann, G. (1993). Ecological and behavioral aspects of nest-building in wild bonobos (*Pan paniscus*). *Ethology* 94:113–126.

Fruth, B., and Hohmann, G. (1996). Nest building behavior in the great apes: the great leap forward? In: McGrew, W. C., Marchant, L. F., and Nishida, T. (eds.), *Great Ape Societies*. Cambridge University Press, Cambridge. pp. 225–240.

Fruth, B., and Hohmann, G. (2002). How bonobos handle hunts and harvests: why share food? In: Boesch, C., Hohmann, G., and Marchant, L. (eds.), *Behavioural Diversity in Chimpanzees and Bonobos*. Cambridge University Press, Cambridge. pp. 231–243.

Fry, B. G., and Fry, A. (2003). The loris: a venomous primate. *Fauna* 4:8–11.

Fu, Y.-X., and Li, W.-H. (1993). Statistical tests of neutrality of mutations. *Genetics* 133:693–709.

Fuentes, A. (1992). Object rubbing in Balinese macaques (*Macaca fascicularis*). *Lab. Primate Newslett.* 31:14–15.

Fuentes, A. (1995). The socioecology of the Mentawai Island langur (*Presbytis potenziani*) [PhD diss.]. University of California, Berkeley.

Fuentes, A. (1996). Feeding and ranging in the Mentawai Island langur (*Presbytis potenziani*). *Int. J. Primatol.* 17:525–548.

Fuentes, A. (1999a). Variable social organization in primates: what can looking at primate groups tell us about the evolution of plasticity in primate societies? In: Dolhinow, P., and Fuentes, A. (eds.), *The Nonhuman Primates*. Mayfield Publishing, Mountain View, CA. pp. 183–189.

Fuentes, A. (1999b). Re-evaluating primate monogamy. *Am. Anthropol.* 100:890–907.

Fuentes, A. (2000). Hylobatid communities: changing views on pair bonding and social organization in hominoids. *Ybk. Phys. Anthropol.* 43:33–60.

Fuentes, A. (2002a). Monkeys, humans and politics in the Mentawai Islands: no simple solutions in a complex world. In: Fuentes, A., and Wolfe, L. D. (eds.), *Primates Face to Face: The Conservation Implications of Human–Non-Human Primate Interconnections*. Cambridge University Press, Cambridge. pp. 187–207.

Fuentes, A. (2002b). Patterns and trends in primate pair bonds. *Int. J. Primatol.* 23:953–978.

Fuentes, A. (2006a). Human–nonhuman primate interconnections and their relevance to anthropology. *Ecol. Environ. Anthropol.* 2:1–11.

Fuentes, A. (2006b). Human culture and monkey behavior: assessing the contexts of potential pathogen transmission between macaques and humans. *Am. J. Primatol.* 68(9):880–896.

Fuentes, A., and Wolfe, L. D. (2002). *Primates Face to Face: The Conservation Implications of Human–Non-Human Primate Interconnections*. Cambridge University Press, Cambridge.

Fuentes, A., Malone, N., Sanz, C., Matheson, M., and Vaughan, L. (2002). Conflict and post-conflict behavior in a small group of chimpanzees. *Primates* 43:223–235.

Fuentes, A., Southern, M., and Suaryana, K. G. (2005). Monkey forests and human landscapes: is extensive sympatry sustainable for *Homo sapiens* and *Macaca fascicularis* in Bali? In: Paterson, J., and Wallis, J. (eds.), *Commensalism and Conflict: The Human–Primate Interface*. American Society of Primatology, Norman, OK. pp. 168–195.

Fuentes, A., Kalchik S., Gettler, L., Kwiatt, A., Konecki, M., and Jones-Engel, L. (2008). Characterizing human–macaque interactions in Singapore. *Am. J. Primatol.* 70(9):879–883.

Fujita, S., Mitsunaga, F., Sugiura, H., and Shimuzu, K. (2001). Measurement of urinary and fecal steroid metabolites during the ovarian cycle in captive and wild Japanese macaques, *Macaca fuscata*. *Am. J. Primatol.* 53:167–176.

Fukuda, F. (1989). Habitual fission–fusion and social organization of the Hakone troop T of Japanese macaques in Kanagawa Prefecture, Japan. *Int. J. Primatol.* 10:419–439.

Furuichi, T. (1983). Interindividual distance and influence of dominance on feeding in a natural Japanese macaque troop. *Primates* 24:445–455.

Furuichi, T. (1987). Sexual swelling, receptivity, and grouping of wild pygmy chimpanzee females at Wamba Zaire. *Primates* 28:309–318.

Furuichi, T. (1989). Social interactions and the life history of female *Pan paniscus* in Wamba, Zaire. *Int. J. Primatol.* 10:173–197.

Furuichi, T. (1992). The prolonged oestrus of females and factors influencing mating in a wild group of bonobos (*Pan paniscus*) in Wamba, Zaire. In: Itoigawa, N., Sugiyama, Y., Sackett, G. P., and Thompson, R. K. R. (eds.), *Topics in Primatology,* vol. 2. University of Tokyo Press, Tokyo. pp. 179–190.

Furuichi, T. (1997). Agonistic interactions and matrifocal dominance rank of wild bonobos (*Pan paniscus*) at Wamba. *Int. J. Primatol.* 18:855–875.

Furuichi, T. (2000). Possible case of predation on a chimpanzee by a leopard in the Petit Loango Reserve, Gabon. *PanAfr. News* 7:21–23.

Furuichi, T., and Hashimoto, C. (2002). Social relationships between cycling females and adult males in Mahale chimpanzees. In: Boesch, C., Hohmann, G., and Marchant, L. (eds.), *Behavioural Diversity in Chimpanzees and Bonobos.* Cambridge University Press, Cambridge. pp. 156–167.

Furuichi, T., and Ihobe, H. (1994). Variation in male relationships in bonobos and chimpanzees. *Behaviour* 130:211–228.

Furuichi, T., Idani, G., Ihobe, H., Kuroda, S., Kitamura, K., Mori, A., Enomoto, T., Okayasu, N., Hashimoto, C., and Kano, T. (1998). Population dynamics of wild bonobos (*Pan paniscus*) at Wamba. *Int. J. Primatol.* 19:1029–1043.

Furuichi, T., Hashimoto, C., and Tashiro, Y. (2001). Fruit availability and habitat use by chimpanzees in the Kalinzu Forest, Uganda: examination of fallback foods. *Int. J. Primatol.* 22:929–945.

Furuya, Y. (1961). The social life of silvered leaf monkeys. *Primates* 3:41–60.

Gachot-Neveu, H., and Ménard, N. (2004). Gene flow, dispersal patterns and social organization. In: Thierry, B., Singh, M., and Kaumanns, W. (eds.), *Macaque Societies: A Model for the Study of Social Organization.* Cambridge University Press, Cambridge, pp. 117–131.

Gachot-Neveu, H., Petit, M., and Roeder, J. J. (1999). Paternity determination in two groups of *Eulemur fulvus mayottensis*: implications for understanding mating strategies. *Int. J. Primatol.* 20:107–119.

Gachot-Neveu, H., Fazio, G., and Thierry, B. (2004). Evolution of the genetic structure in a group of Tonkean macaques: consequences of the mating system [abstract]. *Folia Primatol.* 75:170.

Gadsby, E. L., and Jenkins, P. D. (1997–1998). The drill—integrated in situ and ex situ conservation. *Afr. Primates* 3:12–18.

Gage, M. J. G., and Freckleton, R. P. (2003). Relative testis size and sperm morphometry across mammals: no evidence for an association between sperm competition and sperm length. *Proc. R. Soc. Lond. B* 270:625–632.

Gagneux, P., and Varki, A. (2001). Genetic differences between humans and great apes. *Mol. Phylogenet. Evol.* 18:2–13.

Gagneux, P., Boesch, C., and Woodruff, D. S. (1997a). Microsatellite scoring errors associated with noninvasive genotyping based on nuclear DNA amplified from shed hair. *Mol. Ecol.* 6:861–868.

Gagneux, P., Woodruff, D. S., and Boesch, C. (1997b). Furtive mating by female chimpanzees. *Nature* 387:327–328.

Gagneux, P., Boesch, C., and Woodruff, D. S. (1999a). Female reproductive strategies, paternity and community structure in wild West African chimpanzees. *Anim. Behav.* 57:19–32.

Gagneux, P., Wills, C., Gerloff, U., Tautz, D., Morin, P. A., Boesch, C., Fruth, B., Hohmann, G., Ryder, O. A., and Woodruff, D. S. (1999b). Mitochondrial sequences show diverse evolutionary histories of African hominoids. *Proc. Natl. Acad. Sci. USA* 96:5077–5082.

Gagneux, P., Amess, B., Diaz, S., Moore, S., Patel, T., Dillmann, W., Parekh, R., and Varki, A. (2001a). Proteomic comparison of human and great ape blood plasma reveals conserved glycosylation and differences in thyroid hormone metabolism. *Am. J. Phys. Anthropol.* 115:99–109.

Gagneux, P., Gonder, M. K., Goldberg, T. L., and Morin, P. A. (2001b). Gene flow in wild chimpanzee populations: what genetic data tell us about chimpanzee movement over space and time. *Phil. Trans. R. Soc. Lond. B* 356:889–897.

Gagneux, P., Cheriyan, M., Hurtado-Ziola, N., van der Linden, E. C., Anderson, D., McClure, H., Varki, A., and Varki, N. M. (2003). Human-specific regulation of alpha 2–6-linked sialic acids. *J. Biol. Chem.* 278:48245–48250.

Galat-Luong, A., and Galat, G. (1979). Quelques observations sur l'écologie de *Colobus pennanti oustaleti* en Empire Centrafricain. *Mammalia* 43:309–312.

Galat-Luong, A., and Galat, G. (2005). Conservation and survival adaptations of Temmick's red colobus (*Procolobus badius temmincki*), in Senegal. *Int. J. Primatol.* 26:585–603.

Galat-Luong, A., and Galat, G. (in press). The Guinea baboon (*Papio papio*). In: Kingdon, J., Happold, D., and Butynski, T. (eds.), *The Mammals of Africa.* University of California Press, Berkeley, CA.

Galat-Luong, A., Galat, G., and Hagell, S. (2006). The social and ecological flexibility of Guinea baboons: implications for Guinea baboon social organization and male strategies. In: Swedell, L. and Leigh, S. (eds.), *Reproduction and Fitness in Baboons: Behavioral, Ecological, and Life History Perspectives.* Kluwer, New York. pp. 105–121.

Galbreath, G. J. (1983). Karyotypic evolution in *Aotus*. *Am. J. Primatol.* 4:245–251.

Galdikas, B. M. F. (1979). Orangutan adaptation at Tanjung Puting Reserve: mating and ecology. In: Hamburg, D. A., and McCown, E. R. (eds.), *The Great Apes.* Benjamin/Cummings, Menlo Park, CA. pp. 194–233.

Galdikas, B. M. F. (1981a). Wild orangutan studies at Tanjung Puting Reserve, central Indonesian Borneo, 1971–1977. *Nat. Geogr. Soc. Res. Rep.* 13:1–10.

Galdikas, B. M. F. (1981b). Orangutan reproduction in the wild. In: Graham, C. E. (ed.), *Reproductive Biology of the Great Apes: Comparative and Biomedical Perspectives.* Academic Press, New York. pp. 281–300.

Galdikas, B. M. F. (1982a). An unusual instance of tool-use among wild orang-utans in Tanjung Puting Reserve, Indonesian Borneo. *Primates* 23:138–139.

Galdikas, B. M. F. (1982b). Orang-utan tool-use at Tanjung Puting Reserve, central Indonesian Borneo. *J. Hum. Evol.* 10:19–33.

Galdikas, B. M. F. (1983). The orangutan long call and snag crashing at Tanjung Puting Reserve. *Primates* 24:371–384.

Galdikas, B. M. F. (1985a). Orangutan sociality in Tanjung Puting. *Am. J. Primatol.* 9:101–119.

Galdikas, B. M. F. (1985b). Adult male sociality and reproductive tactics among orangutans at Tanjung Puting. *Folia Primatol.* 45:9–24.

Galdikas, B. M. F. (1985c). Subadult male orangutan sociality and reproductive behavior at Tanjung Puting. *Am. J. Primatol.* 8:87–99.

Galdikas, B. M. F. (1988). Orangutan diet, range, and activity at Tanjung Puting, central Borneo. *Int. J. Primatol.* 9:1–35.

Galdikas, B. M. F. (1995a). *Reflections of Eden: My Year with the Orangutans of Borneo.* Little, Brown, Boston.

Galdikas, B. M. F. (1995b). Social and reproductive behavior of wild adolescent female orangutans. In: Nadler, R. D., Galdikas, B. M. F., Sheeran, L. K., and Rosen, N. (eds.), *The Neglected Ape.* Plenum Press, New York. pp. 163–182.

Galdikas, B. M. F., and Vasey, P. (1992). Why are orangutans so smart? Ecological and social hypothesis. In: Burton, F. D. (ed.), *Social processes and mental abilities in nonhuman primates.* Edwin Mellen Press, Lewiston, NY. pp. 183–224.

Galdikas, B. M. F., and Wood, J. W. (1990). Birth spacing patterns in humans and apes. *Am. J. Phys. Anthropol.* 83:185–191.

Galdikas, B. M. F., and Yeager, C. P. (1984). Crocodile predation on a crab-eating macaque in Borneo. *Am. J. Primatol.* 6:49–51.

Galef, B. G., Jr. (1988). Imitation in animals: history, definition, and interpretation of data from the psychological laboratory. In: Zentall, T. R., and Galef, B. G., Jr. (eds.), *Social Learning: Psychological and Biological Perspectives.* Lawrence Erlbaum, Hillsdale, NJ. pp. 3–28.

Galef, B. G., Jr. (1990). Tradition in animals: field observations and laboratory analyses. In: Beckoff, M., and Jamieson, D. (eds.), *Interpretations and Explanations in the Study of Behaviour: Comparative Perspectives.* Westview Press, Boulder, CO. pp. 74–95.

Galef, B. G., Jr. (1992). The question of animal culture. *Hum. Nat.* 3:157–178.

Galef, B. G., Jr. (1995). Why behavior patterns that animals learn socially are locally adaptive. *Anim. Behav.* 49:1325–1334.

Galef, B. G., Jr. (2003). "Traditional" foraging behaviours of brown and black rats (*Rattus norvegicus* and *Rattus rattus*). In: Fragaszy, D. M., and Perry, S. (eds.), *The Biology of Traditions: Models and Evidence.* Cambridge University Press, Cambridge. pp. 159–186.

Galef, B. G., Mittermeir, J. R. A., and Bailey, R. C. (1976). Predation by the tayra (*Eira barbara*). *J. Mammal.* 57:760–761.

Galef, B. G., Jr., Manzig, L. A., and Field, R. M. (1986). Imitation learning in budgerigars: Dawson and Foss 1965 revisited. *Behav. Proc.* 13:191–202.

Galetti, M., Pedroni, F., and Paschoal, M. (1994). Infanticide in the brown howler monkey, *Alouatta fusca*. *Neotrop. Primates* 2:6–7.

Gallup, G. G. (1970). Chimpanzee: self-recognition. *Science* 167:86–87.

Gallup, G. G. J. (1982). Self-awareness and the emergence of mind in primates. *Am. J. Primatol.* 2:237–248.

Gallup, G. (1987). Self-awareness. In: Mitchell, G., and Erwin, J. (eds.), *Comparative Primate Biology. Behavior, Cognition, and Motivation*, vol. 2, part B. Alan R. Liss, New York. pp. 3–16.

Gallup, G. G. (1998). Can animals empathize? *Sci. Am.* 9:66–71.

Ganas, J., and Robbins, M. M. (2004). Intrapopulation differences in ant eating in the mountain gorillas of Bwindi Impenetrable National Park, Uganda. *Primates* 45:275–279.

Ganas, J., and Robbins, M. M. (2005). Ranging behavior of the mountain gorillas (*Gorilla beringei beringei*) in Bwindi Impenetrable National Park, Uganda: a test of the ecological constraints model. *Behav. Ecol. Sociobiol.* 58:277–288.

Ganas, J., Robbins, M. M., Nkurunungi, J. B., Kaplin, B. A., and McNeilage, A. (2004). Dietary variability of mountain gorillas in Bwindi Impenetrable National Park, Uganda. *Int J. Primatol.* 25:1043–1072.

Ganas, J., Ortmann, S., and Robbins, M. M. (2008). Food preferences of wild mountain gorillas. *Am. J. Primatol.* 70:1–12.

Ganas, J., Nkurunungi, J.B., and Robbins, M.M. (2009a). Temporal and spatial variability of herbaceous vegetation consumed by mountain gorillas in an Afromontane rainforest. *Biotropica* 41:37–46.

Ganas, J., Ortmann, S., and Robbins, M. M. (2009b). Food choices of the mountain gorilla in Bwindi Impenetrable National Park, Uganda: the influence of nutrients, phenolics, and availability. *J. Trop. Biol.* 25:123–134.

Ganzhorn, J. U. (1988). Food partitioning among Malagasy primates. *Oecologica* 75:436–450.

Ganzhorn, J. U. (1992). Leaf chemistry and the biomass of folivorous primates in tropical forests: tests of a hypothesis. *Oecologia* 91:540–547.

Ganzhorn, J. U. (1993). Flexibility and constraints of *Lepilemur* ecology. In: Kappeler, P. M., and Ganzhorn, J. U. (eds.), *Lemur Social Systems and Their Ecological Basis*. Plenum Press, New York. pp. 153–166.

Ganzhorn, J. U. (1995). Low-level forest disturbance effects on primary production, leaf chemistry, and lemur populations. *Ecology* 76:2084–2096.

Ganzhorn, J. (2003). Habitat description and phenology. In: Curtis, D., and Setchell, J. (eds.), *Field and Laboratory Methods in Primatology*. Cambridge University Press, Cambridge. pp. 40–56.

Ganzhorn, J. U., and Wright, P. C. (1994). Temporal patterns in primate leaf eating: the possible role of leaf chemistry. *Folia Primatol.* 63:203–208.

Ganzhorn, J. U., Abraham, J. P., and Razanahoera-Rakotomalala, M. (1985). Some aspects of the natural history and food selection of *Avahi laniger*. *Primates* 26:452–463.

Ganzhorn, J. U., Wright, P. C., and Ratsimbazafy, J. (1999). Primate communities: Madagascar. In: Fleagle, J. G., Janson, C. H., and Reed, K. E. (eds.), *Primate Communities*. Cambridge University Press, Cambridge. pp. 75–89.

Ganzhorn, J. U., Arrigo-Nelson, S., Boinski, S., Bollen, A., Carrai, V., Derby, A., Donati, G., Koenig, A., Kowalewski, M., Lahann, P., Norscia, I., Polowinsky, S. Y., Schwitzer, C., Stevenson, P. R., Talebi, M. G., Tan, C., Vogel, E. R., and Wright, P. C. (2009). Possible fruit protein effects on primate communities in Madagascar and the Neotropics. PLoS ONE 4(12): e8253.

Garber, P. A. (1980). Locomotor behavior and feeding ecology of the Panamanian tamarin (*Saguinus oedipus geoffroyi*, Callitrichidae, Primates). *Int. J. Primatol.* 1:185–201.

Garber, P. A. (1984a). Proposed nutritional importance of plant exudates in the diet of the Panamanian tamarin, *Saguinus Oedipus geoffroyi*. *Int. J. Primatol.* 5:1–15.

Garber, P. A. (1984b). Use of habitat and positional behavior in a neotropical primate, *Saguinus oedipus*. In: Rodman, P. S., and Cant, J. G. H. (eds.), *Adaptations for Foraging in Nonhuman Primates: Contributions to an Organismal Biology of Prosimians, Monkeys, and Apes*. Columbia University Press, New York. pp. 112–133.

Garber, P. A. (1986). The ecology of seed dispersal in two species of callitrichid primates (*Saguinus mystax* and *Saguinus fuscicollis*). *Am. J. Primatol.* 10:155–170.

Garber, P. A. (1987). Foraging strategies among living primates. *Annu. Rev. Anthropol.* 16:339–364.

Garber, P. A. (1988a). Diet, foraging patterns, and resource defense in a mixed species troop of *Saguinus mystax* and *Saguinus fuscicollis* in Amazonian Peru. *Behaviour* 105:18–34.

Garber, P. A. (1988b). Foraging decisions during nectar feeding by tamarin monkeys (*Saguinus mystax* and *Saguinus fuscicollis*, Callitrichidae, Primates) in Amazonian Peru. *Biotropica* 20:100–106.

Garber, P. A. (1989). The role of spatial memory in primate foraging patterns: *Saguinus mystax* and *Saguinus fuscicollis*. *Am. J. Primatol.* 19:203–216.

Garber, P. A. (1991). A comparative study of positional behavior in three species of tamarin monkeys. *Primates* 32:219–230.

Garber, P. A. (1992). Vertical clinging, small body size, and the evolution of feeding adaptations in the Callitrichinae. *Am. J. Phys. Anthropol.* 88:469–482.

Garber, P. A. (1993a). Feeding ecology and behaviour of the genus *Saguinus*. In: Rylands, A. B. (ed.), *Marmosets and Tamarins: Systematics, Behaviour, and Ecology*. Oxford University Press, Oxford. pp. 273–295.

Garber, P. A. (1993b). Seasonal patterns of diet and ranging in two species of tamarin monkeys: stability versus variability. *Int. J. Primatol.* 14:145–166.

Garber, P. A. (1997). One for all and breeding for one: cooperation and competition as a tamarin reproductive strategy. *Evol. Anthropol.* 5:187–199.

Garber, P. A. (1998). Within- and between-site variability in moustached tamarin (*Saguinus mystax*) positional behavior during food procurement. In: Strasser, E., Fleagle, J. G., Rosenberger, A., and McHenry, H. (eds.), *Primate Locomotion: Recent Advances*. Plenum Press, New York. pp. 61–78.

Garber, P. A. (2004). New perspectives in primate cognitive ecology. *Am. J. Primatol.* 62:133–137.

Garber, P. A., and Bicca-Marques, J. C. (2002). Evidence of predator sensitive foraging in traveling in single- and mixed-species tamarin troops. In: Miller, L. E. (ed.), *Eat or Be Eaten: Predator Sensitive Foraging Among Primates*. Cambridge University Press, Cambridge. pp. 138–153.

Garber, P. A., and Brown, E. (2002). Experimental field study of tool use in wild capuchins (*Cebus capucinus*): learning by association or learning by insight? *Am. J. Phys. Anthropol.* (suppl. 34):74–75.

Garber, P. A., and Kowalewski, M. M. (2009). Male cooperation in Pitheciines: the reproductive costs and benefits to individuals of forming large mulitmale and multifmale groups. In: Veiga, L., Barnett, A., and Norconk, M. A. (eds.), *Evolutionary Biology and Conservation of Titis, Sakis and Uakaris*. Cambridge, Cambridge University Press. In press.

Garber, P. A., and Leigh, S. R. (1997). Ontogenetic variation in small-bodied New World primates: implications for patterns of reproduction and infant care. *Folia Primatol.* 68:1–22.

Garber, P. A., and Leigh, S. R. (2001). Patterns of positional behavior in mixed species groups of *Callimico goeldii*, *Saguinus labiatus*, and *Saguinus fuscicollis* in northwestern Brazil. *Am. J. Primatol.* 55:17–31.

Garber, P. A., and Paciulli, L. M. (1997). Experimental field study of spatial memory and learning in wild capuchin monkeys (*Cebus capucinus*). *Folia Primatol.* 68:236–253.

Garber, P. A., and Pruetz, J. D. (1995). Positional behavior in moustached tamarin monkeys: effects of habitat on locomotor variability and locomotor stability. *J. Hum. Evol.* 28:411–426.

Garber, P. A., and Rehg, J. A. (1999). The ecological role of the prehensile tail in white-faced capuchins (*Cebus capucinus*). *Am. J. Phys. Anthropol.* 110:325–339.

Garber, P. A., Moya, L., and Malaga, C. (1984). A preliminary field study of the moustached tamarin monkey (*Saguinus mystax*) in northeastern Peru: questions concerned with the evolution of a communal breeding system. *Folia Primatol.* 42:17–32.

Garber, P. A., Encarnacion, F., Moya, L., and Pruetz, J. D. (1993). Demographic and reproductive patterns in moustached tamarin monkeys (*Saguinus mystax*): implications for reconstructing platyrrhine mating systems. *Am. J. Primatol.* 29:235–254.

Garber, P. A., Pruetz, J. D., Lavallee, A. C., and Lavallee, S. G. (1999). A preliminary study of mantled howling monkey (*Alouatta palliata*) ecology and conservation on Isla de Ometepe, Nicaragua. *Neotrop. Primates* 7:113–117.

Garber, P. A., Sallenave, A., Blomquist, G., and Anzenberger, A. (2009). A comparative study of the kinematics of leaping *in Callimico goeldii*, *Callithrix jacchus*, and *Cebuella pygmaea*. In: Ford, S., Porter, L., and Davis, L. (eds.), *The Smallest Anthropoids: The Marmoset/Callimico Radiation*. Springer Science+Business Media. New York.pp. 259–277.

Garcia, C., Lee, P., and Rosetta, L. (2009). Growth in colony living anubis baboon infants and its relationship with maternal activity budgets and reproductive status. *Am. J. Phys. Anthropol.* 138:123–135.

García, J. E., and Braza, F. (1987). Activity rhythms and use of space of a group of *Aotus azarae* in Bolivia during the rainy season. *Primates* 28:337–342.

García, J. E., and Braza, F. (1989). Density comparisons using different analytic-methods in *Aotus azarae*. *Primate Rep.* 25:45–52.

Garcia, J. E., and Braza, F. (1993). Sleeping sites and lodge trees of the night monkey (*Aotus azarae*) in Bolivia. *Int. J. Primatol.* 14:467–477.

García, J. E., and Tarifa, T. (1988). Primate survey of the Estación Biológica Beni, Bolivia. *Primate Conserv.* 9:97–100.

Garcia Yuste, J. E. (1989). Patrones etológicos y ecológicos del mono nocturno, *Aotus azarae boliviensis. Diss. Abstr. Int.* 50:473-C.

Garland, T., and Adolph, S. C. (1994). Why not to do two-species comparative studies: limitations on inferring adaptation. *Physiol. Zool.* 67:797–828.

Garner, R. L. (1892). *The Speech of Monkeys.* Charles L. Webster, New York.

Garner, R. L. (1896). *Gorillas and Chimpanzees.* Osgood, McIlvaine, London.

Gartlan, J. S. (1970). Preliminary notes on the ecology and behavior of the drill, *Mandrillus leucophaeus* Ritgen, 1824. In Napier, J. R., and Napier, P. H. (eds.), *Old World Monkeys: Evolution, Systematics, and Behavior.* Academic Press, New York. pp. 445–480.

Gartlan, J. S., and Brain, C. K. (1968). Ecology and social variability in *Cercopithecus aethiops* and *Cercopithecus mitis.* In: Jay, P. C. (ed.), *Primates, Studies in Adaptation and Variability.* University of California Press, Berkeley. pp. 253–292.

Gartlan, J. S., McKey, D. B., Waterman, P. G., Mbi, C. N., and Struhsaker, T. T. (1980). A comparative study of the chemistry of two African rain forests. *Biochem. Syst. Ecol.* 8:401–422.

Gaspersic, M., and Pruetz, J. D. (2004). Predation on a monkey by savanna chimpanzees at Fongoli, Senegal. *PanAfr. News* 11:8–10.

Gasquet, M., Bamba, D., Babadjamian, A., Balansard, G., Timon-David, P., and Metzger, J. (1985). Action amoebicide et anthelminthique du vernolide et de l'hydroxyvernolide isoles des feuilles de *Vernonia colorata* (Willd.) Drake. *Eur. J. Med. Chem. Theory* 2:111–115.

Gatinot, B. L. (1976). Les milieux fréquentés par le colobe bai d'Afrique de l'ouest (*Colobus badius temmincki*, Kuhn 1820) en Sénégambie. *Mammalia* 40:1–12.

Gatinot, B. L. (1977). Le régime alimentaire du colobe bai au Sénégal. *Mammalia* 41:373–402.

Gatinot, B. L. (1978). Characteristics of the diet of the West African red colobus. In: Chivers, D. J., and Herbert, J. (eds.), *Recent Advances in Primatology,* vol. 1. Academic Press, London. pp. 253–255.

Gatti, S., Levréro, F., Ménard, N., Petit, E., and Gautier-Hion, A. (2003). Bachelor groups of western lowland gorillas (*Gorilla gorilla gorilla*) at Lokoué Clearing, Odzala National Park, Republic of Congo. *Folia Primatol.* 74:195–196.

Gatti, S., Levréro, F., Ménard, N., and Gautier-Hion, A. (2004). Population and group structure of western lowland gorillas (*Gorilla gorilla gorilla*) at Lokoué, Republic of Congo. *Am. J. Primatol.* 63:111–123.

Gaulin, S. J. C. (1979). A Jarman/Bell model of primate feeding niches. *Hum. Ecol.* 7:1–20.

Gaulin, S. J. C., and Gaulin, C. K. (1982). Behavioral ecology of *Alouatta seniculus* in Andean Cloud Forest. *Int. J. Primatol.* 3:1–32.

Gauthier, C. A. (1999). Reproductive parameters and paracallosal skin color changes in captive female Guinea baboons, *Papio papio. Am. J. Primatol.* 47:67–74.

Gautier, J.-P. (1985). Quelques caracteristique ecologiques du singe des marais: *Allenopithecus nigroviridis* Lang 1923. *Terre Vie* 40:331–342.

Gautier, J.-P., and Gautier, A. (1977). Communication in Old World monkeys. In: Sebeok, T. E. (ed.), *How Animals Communicate.* Indiana University Press, Bloomington. pp. 890–964.

Gautier-Hion, A. (1971). L'ecologie du talapoin du Gabon. *Terre Vie* 25:427–490.

Gautier-Hion, A. (1973). Social and ecological features of talapoin monkeys—comparisons with sympatric cercopithecines. In: Michael, R. P., and Crook, J. H. (eds.), *Comparative Ecology and Behaviour of Primates.* Academic Press, London, pp. 147–170.

Gautier-Hion, A. (1978). Food niches and coexistence in sympatric primates in Gabon. In: Chivers, D. J., and Herbert, J. (eds.), *Recent Advances in Primatology.* Academic Press, London. pp. 269–286.

Gautier-Hion, A. (1980). Seasonal variation of diet related to species and sex in a community of *Cercopithecus* monkeys. *J. Anim. Ecol.* 49:237–269.

Gautier-Hion, A. (1984). La dissemination des graines par les cercopithecides forestiers Africains. *Terre Vie* 39:159–165.

Gautier-Hion, A. (1988a). Diet and dietary habits of forest guenons. In: Gautier-Hion, A., Bourlière, F., Gautier, J. P., and Kingdon, J. (eds.), *A Primate Radiation: Evolutionary Biology of the African Guenons.* Cambridge University Press, New York. pp. 257–283.

Gautier-Hion, A. (1988b). Polyspecific associations among forest guenons: ecological, behavioural and evolutionary aspects. In: Gautier-Hion, A., Bourlière, F., Gautier, J. P., and Kingdon, J. (eds.), *A Primate Radiation:*

Evolutionary Biology of the African Guenons. Cambridge University Press, New York. pp. 452–576.

Gautier-Hion, A. F. M. (1994). Mutualism between a leguminous tree and large African monkeys as pollinators. *Behav. Ecol. Sociobiol.* 34: 203–210.

Gautier-Hion, A., and Gautier, J. P. (1974). Les associations polyspécifiques du plateau de M'passa, Gabon. *Folia Primatol.* 22:134–177.

Gautier-Hion, A., and Gautier, J. P. (1976). Croissance, maturité sociale et sexuelle, reproduction chez les cercopithecinés forestiers arboricoles. *Folia Primatol.* 4:103–118.

Gautier-Hion, A., and Gautier, J. P. (1978). Le singe de Brazza: une stratégie originale. *Z. Tierpsychol.* 46:84–104.

Gautier-Hion, A., and Maisels, F. (1994). Mutualism between a leguminous tree and large African monkeys as pollinators. *Behav. Ecol. Sociobiol.* 34: 203–210.

Gautier-Hion, A., and Michaloud, G. (1989) Are figs always keystone resources for tropical frugivorous vertebrates? A test in Gabon. *Ecology* 70:1826–1833.

Gautier-Hion, A., Gautier, J. P., and Quris, R. (1981). Forest structure and fruit availability as complementary factors influencing habitat use by a troop of monkeys (*Cercopithecus cephus*). *Rev. Ecol.* 35:511–536.

Gautier-Hion, A., Quris, R., and Gautier, J. P. (1983). Monospecific vs. polyspecific life: a comparative study of foraging and anti-predatory tactics in a community of *Cercopithecus* monkeys. *Behav. Ecol. Sociobiol.* 12:325–335.

Gautier-Hion, A., Duplantier, J. M., Quris, R., Feer, F., Sourd, C., Decous, J. P., Doubost, G., Emmons, L., Erard, C., Hecketsweiler, P., Moungazi, A., Roussilhon, C., and Thiollay, J. M. (1985). Fruit characters as a basis of fruit choice and seed dispersal in a tropical forest vertebrate community. *Oecologia* 65:324–337.

Gautier-Hion, A., and Tutin, C. E. G. (1988). Simultaneous attack by adult males of a polyspecific troop of monkeys against a crowned hawk eagle. *Folia Primatol.* 51:149–151.

Gautier-Hion, A., Gautier, J.-P., and Maisels, F. (1993). Seed dispersal versus seed predation: an intersite comparison of two related African monkeys. *Vegetation* 107/108:237–244.

Gautier-Hion, A., Gautier, J.-P., and Moungazi, A. (1997). Do black colobus in mixed-species groups benefit from increased foraging. *C. R. Acad. Sci. Paris Sci.* 320:67–71.

Gavan, J. A. (ed.) (1955). *The Nonhuman Primates and Human Evolution.* Wayne State University, Detroit.

Gavan, J. A., and Swindler, D. R. (1966). Growth rates and phylogeny in primates. *Am. J. Phys. Anthropol.* 24:181–190.

Gaynor, D. (1994). Foraging and feeding behaviour of chacma baboons in a woodland habitat [PhD diss.]. University of Natal, Natal, South Africa.

Geary, D. C., Byrd-Craven, J., Hoard, M. K., Vigil, J., and Numtee, C. (2003). Evolution and development of boys' social behavior. *Dev. Rev.* 23(4):444–470.

Gebo, D. L. (1986). Miocene lorisids: the foot evidence. *Folia Primatol.* 47:217–225.

Gebo, D. L. (1992). Locomotor and postural behavior in *Alouatta palliata* and *Cebus capucinus. Am. J. Primatol.* 26:277–290.

Gebo, D. L. (2002). Adapiformes: phylogeny and adaptation. In: Hartwig, W. C. (ed.), *The Primate Fossil Record.* Cambridge University Press, Cambridge. pp. 21–44.

Gebo, D. L., and Chapman, C. A. (1995a). Positional behavior in five sympatric Old World monkeys. *Am. J. Phys. Anthropol.* 97:49–76.

Gebo, D. L., and Chapman, C. A. (1995b). Habitat, annual and seasonal effects on positional behavior in red colobus monkeys. *Am. J. Phys. Anthropol.* 96:73–82.

Gebo, D. L., and Chapman, C. A. (2000). Locomotor behavior in Ugandan monkeys. In: Whitehead, P. F., and Jolly, C. J. (eds.), *Old World Monkeys.* Cambridge University Press, Cambridge. pp. 480–495.

Gebo, D. L., Chapman, C. A., Chapman, L. J., and Lambert, J. (1994). Locomotor reesponse to predator threat in red colobus monkeys. *Primates* 35:219–223.

Geissmann, T. (1984). Inheritance of song parameters in the gibbon song, analysed in 2 hybrid gibbons (*Hylobates pileatus* × *H. lar*). *Folia Primatol.* 42:216–235.

Geissmann, T. (1991). Reassessment of age of sexual maturity in gibbons. *Am. J. Primatol.* 23:11–22.

Geissmann, T. (1993). Evolution of communication in gibbons (Hylobatidae) [PhD diss.]. Universitaet Zurich, Zurich.

Geissmann, T. (1995). Gibbon systematics and species identification. *Int. Zoo News* 42:467–501.

Geissmann, T. (2002a). Duet-splitting and the evolution of gibbon songs. *Biol. Rev.* 77:57–76.

Geissmann, T. (2002b). Taxonomy and evolution of gibbons. *Evol. Anthropol. Suppl.* 1:28–31.

Geissmann, T. (2004). The gibbons (Hylobatidae): an introduction, http://www.tiho-hannover.de/gibbons/main/index.html (accessed September 14, 2004).

Geissmann, T., and Orgeldinger, M. (2000). The relationship between duet songs and pair bonds in siamangs, *Hylobates syndactylus*. *Anim. Behav.* 60:805–809.

Genoud, M., Martin, R. D., and Glaser, D. (1997). Rate of metabolism in the smallest simian primate, the pygmy marmoset (*Cebuella pygmaea*). *Am. J. Primatol.* 41:229–245.

Genty, E., and Roeder, J. J. (2007). Transfer of self-control in lemurs: choice of a less-preferred food item under a reverse-reward contingency. *J. Comp. Psychol.* 121:354–362.

Genty, E., Palmier, C., and Roeder, J. J. (2004). Learning to suppress responses to the larger of two rewards in two species of lemurs, *Eulemur fulvus* and *E. macaco*. *Anim Behav.* 67:925–932.

Gerald, C. N. (1995). Demography of the Virunga mountain gorilla (*Gorilla gorilla beringei*) [MSc thesis]. Princeton University, Princeton, NJ.

Gerald, M. S. (2001). Primate colour predicts social status and aggressive outcome. *Anim. Behav.* 61:559–566.

Gerald, M. S. (2003). How color may guide the primate world: possible relationships between sexual selection and sexual dichromatism. In: Jones, C. B. (ed.), *Sexual Selection and Reproductive Competition in Primates: New Perspectives and Directions,* vol. 3. American Society of Primatologists, Norman, OK. pp. 141–171.

Gerber, S., Chabrier, P., and Kremer, A. (2003). FaMoz: a software for parentage analysis using dominant, codominant and uniparentally inherited markers. *Mol. Ecol. Notes* 3:479–481.

Gerhardt, H. C. (1983). Communication and environment. In: Halliday, T. R., and Slater, P. J. B. (eds.), *Animal Behaviour: Communication.* W. H. Freeman, New York. pp. 82–113.

Gerloff, U., Hartung, B., Fruth, B., Hohmann, G., and Tautz, D. (1999). Intracommunity relationships, dispersal pattern and paternity success in a wild living community of bonobos (*Pan paniscus*) determined from DNA analysis of faecal samples. *Proc. R. Soc. Lond. B* 266:1189–1195.

Gerson, J. S. (2001). Social relationships in wild red-fronted brown lemurs (*Eulemur fulvus rufus*) [PhD diss.]. Duke University, Durham, NC.

Gesquiere, L. R., Wango, E. O., Alberts, S. C., and Altmann, J. (2007). Mechanisms of sexual selection: sexual swellings and estrogen concentrations as fertility indicators and cues for male consort decisions in wild baboons. *Horm. Behav.* 51:114–125.

Gesquiere, L. R., Khan, M., Shek, L., Wango, T. L., Wango, E. O., Alberts, S. C., and Altmann, J. (2008). Coping with a challenging environment: effects of seasonal variability and reproductive status on glucocorticoid concentrations of female baboons (*Papio cynocephalus*). *Horm. Behav.* 54:410–416.

Gevaerts, H. (1992). Birth seasons of *Cercopithecus, Cercocebus* and *Colobus* in Zaire. *Folia Primatol.* 59:105–113.

Ghigliari, M. P. (1984). *The Chimpanzees of Kibale Forest.* Columbia University Press, New York.

Gibbons, E. F., Jr., and Menzel, E. W. (1980). Rank orders or arising, eating, and retiring in a family group of tamarins (*Saguinus fuscicollis*). *Primates* 21:44–52.

Gibbs, R. A., Weinstock, G. M., Metzker, M. L., et al. (2004). Genome sequence of the brown Norway rat yields insights into mammalian evolution. *Nature* 428:493–521.

Gibson, E. J., and Pick, A. D. (2000). *An Ecological Approach to Perceptual Learning and Development.* Oxford University Press, Oxford.

Gibson, G., and Muse, S. V. (2009). *A Primer of Genome Science,* 3rd ed. Sinauer Associates, Sunderland, MA.

Gibson, J. J. (1979). *The Ecological Approach to Visual Perception.* Houghton Mifflin, Boston.

Gibson, K. R. (1986). Cognition, brain size, and the extraction of embedded food resources. In: Else, J. G., and Lee, P. C. (eds.), *Primate Ontogeny, Cognition, and Social Behavior.* Cambridge University Press, Cambridge. pp. 93–103.

Gibson, K., Rumbaugh, D., and Beran, M. (2001). Bigger is better: primate brain size in relationship to cognition. In: Falk, D., and Gibson, K. (eds.), *Evolutionary Anatomy of the Primate Cerebral Cortex.* Cambridge University Press, Cambridge. pp. 79–93.

Gibson, K. N., Vick, L. G., Palma, A. C., Carrasco, F. M., Taub, D., and Ramos-Fernández, G. (2008). Intra-community infanticide and forced copulation in spider monkeys: a multi-site comparison between Cocha Cashu, Peru and Punta Laguna, Mexico. *Am. J. Primatol.* 70:485–489.

Gibson, S. V., Tustin, G. W., Williams, L. E., Schuler, A. M., Brady, A. G., and Abee, C. R. (2008). Reproduction in an owl monkey breeding colony (*Aotus sp.*). *Am. J. Primatol.* 70:29.

Gilardi, K. G., Shideler, S. E., Valverde, C., Roberts, J. A., and Lasley, B. L. (1997). Characterization of the onset of menopause in the rhesus macaque. *Biol. Reprod.* 57:335–340.

Gilbert, A. (1986). Mammary number and litter size in Rodentia: the one-half rule. *Proc. Natl. Acad. Sci. USA* 83:4828–4830.

Gilbert, K. A. (2000). Attempted predation on a white-faced saki in the central Amazon. *Neotrop. Primates* 8:103–104.

Gilbert, K. A., and Setz, E. Z. (2001). Primates in a fragmented landscape. Six species in central Amazonia. In: Bierregaard, R. O., Jr., Gascon, C., Lovejoy, T. E., and Mesquita, R. (eds.), *Lessons from Amazonia: The Ecology and Conservation of a Fragmented Forest.* Yale University Press, New Haven. pp. 262–270.

Gilbert, K. A., and Tingay, R. E. (2001). Predation of a fat-tailed dwarf lemur (*Cheirogaleus medius*) by a Madagascar Harrier-hawk (*Polyboroides radiatus*): an incidental observation. *Lemur News* 6:6.

Gilbert, T. M., Brown, D. A., and Boysen, S. T. (1998). Social effects on behavior in capuchins (*Cebus apella*) [abstract]. *Am. J. Primatol.* 45:182.

Gilby, I. C., Eberly, L. E., Pintea, L., and Pusey, A. E. (2006). Ecological and social influences on the hunting behaviour of wild chimpanzees, *Pan troglodytes schweinfurthii*. *Anim. Behav.* 72:169–180.

Gil-da-Costa, R. (2007). Howler monkeys and harpy eagles: a communication arms race. In: Gursky, S. L., and Nekaris, K. A. I. (eds.), *Primate Anti-Predator Strategies.* Springer, New York. pp. 289–307.

Gil-da-Costa, R., Palleroni, A., Hauser, M. D., Touchton, J., and Kelley, J. P. (2003). Rapid acquisition of an alarm response by a neotropical primate to a newly introduced avian predator. *Proc. R. Soc. Lond. B* 270:605–610.

Gil-da Costa, R., Braun, A., Lopes, M., Carson, R., Herscovitch, P., and Martin, A. (2004). Toward an evolutionary perspective on conceptual representation: species-specific calls activate visual and affective processing systems in the macaque. *Proc. Natl. Acad. Sci. USA* 101:17516–17521.

Gil-da-Costa, R., Martin, A., Lopes, A., Muñoz, M., Fritz, J., and Braun, A. (2006). Species-specific calls activate homologs of Broca's and Wernicke's areas in the macaque. *Nat. Neurosci.* 8:1064–1070.

Gillespie, T. (2004). Effects of human disturbance on primate–parasite dynamics [PhD diss.]. University of Florida, Gainsville.

Gillespie, T. R., and Chapman, C. A. (2001). Determinants of group size in the red colobus monkey (*Procolobus badius*): an evaluation of the generality of the ecological-constraints model. *Behav. Ecol. Sociobiol.* 50:329–338.

Gilmore, H. A. (1981). From Radcliffe-Brown to sociobiology: some aspects of the rise of primatology within physical anthropology. *Am. J. Phys. Anthropol.* 56:387–392.

Giménez, M. C. (2004). Dieta y comportamiento de forrajeo en verano e invierno del mono mirikiná (*Aotus azarai azarai*) en bosques secos y húmedos del Chaco Argentino [thesis]. University of Buenos Aires, Buenos Aires.

Giménez, M., and Fernandez-Duque, E. (2003). Summer and winter diet of night monkeys in the gallery and thorn forests of the Argentinean Chaco. *Rev. Etol.* 5(suppl.):164.

Gingerich, P. D., and Martin, R. D. (1981). Cranial morphology and adaptations in Eocene Adapidae II: the Cambridge skull of *Adapis parisiensis*. *Am. J. Phys. Anthropol.* 56:235–257.

Ginsburg, L., and Mein, P. (1986). *Tarsius thailandica nov. sp.,* Tarsiidae (Primates, Mammalia) fossile d'Asie. *C. R. Acad. Sci.* 304:1213–1215.

Ginther, A. J., Ziegler, T. E., and Snowdon, C. T. (2001). Reproductive biology of captive male cottontop tamarin monkeys as a function of social environment. *Anim. Behav.* 61:65–78.

Ginther, A. J., Carlson, A. A., Ziegler, T. E., and Snowdon, C. T. (2002). Neonatal and pubertal development in males of a cooperatively breeding primate, the cotton-top tamarin (*Saguinus oedipus oedipus*). *Biol. Reprod.* 66:282–290.

Giraldo, A., Bueno, M. L., Silva, E., Ramírez, J., Umaña, J., and Espinal, C. (1986). Estudio citogenético de 288 *Aotus* Colombianos. *Biomédica* 6:5–13.

Girolami, C., and Bielert, C. (1987). Female perineal swelling and its effects on male sexual arousal: on apparent sexual releaser in the chacma baboon (*Papio ursinus*). *Int. J. Primatol.* 8:651–661.

Gish, S. L., and Morton, E. S. (1981). Structural adaptations to local habitat acoustics in Carolina wren songs. *Z. Tierpsychol.* 56:74–84.

Gittens, G. P., and Raemaekers, J. J. (1980). Siamang, lar and agile gibbons. In: Chivers, D. J. (ed.), *Malayan Forest Primates*. Plenum Press, New York. pp. 63–105.

Gittins, S. P. (1978). The species range of the gibbon *Hylobates agilis*. In: Chivers, D. J., and Joysey, K. A. (eds.), *Recent Advances in Primatology. Evolution*, vol. 3. Academic Press, New York. pp. 319–321.

Gittins, S. P. (1979). The behaviour and ecology of the agile gibbon (*Hylobates agilis*) [PhD diss.]. Cambridge University, Cambridge.

Gittins, S. P. (1980). Territorial behavior in the agile gibbon. *Int. J. Primatol.* 1:381–399.

Gittins, S. P. (1982). Feeding and ranging in the agile gibbon. *Folia Primatol.* 38:39–71.

Gittins, S. P. (1984). Territorial advertisement and defense in gibbons. In: Preuschoft, H., Chivers, D. J., Brockelman, W. Y., and Creel, N. (eds.), *The Lesser Apes: Evolutionary and Behavioural Biology*. Edinburgh University Press, Edinburgh. pp. 420–424.

Gittins, S. P., and Raemaekers, J. J. (1980). Siamang, lar and agile gibbons. In: Chivers, D. J. (ed.), *Malayan Forest Primates: Ten Years' Study in Tropical Rain Forest*. Plenum Press, New York. pp. 63–105.

Glander, K. E. (1975). Habitat description and resource utilization: a preliminary report on mantled howler monkey ecology. In: Tuttle, R. H. (ed.), *Socioecology and Psychology of Primates*. Mouton Press, The Hague. pp. 37–57.

Glander, K. E. (1978). Howling monkey feeding behavior and plant secondary compounds: a study of strategies. In: Montgomery, G. G. (ed.), *The Ecology of Arboreal Folivores*. Smithsonian Institution Press, Washington DC. pp. 561–573.

Glander, K. E. (1980). Reproduction and population growth in free-ranging mantled howling monkeys. *Am. J. Phys. Anthropol.* 53:25–36.

Glander, K. E. (1981). Feeding patterns in mantled howling monkeys. In: Kamil, A. C., and Sargent, T. D. (eds.), *Foraging Behavior: Ecological, Ethological, and Psychological Approaches*. Garland STPM Press, New York. pp. 231–257.

Glander, K. E. (1982). The impact of plant secondary compounds on primate feeding behavior. *Ybk. Phys. Anthropol.* 25:1–18.

Glander, K. E. (1992). Dispersal patterns in Costa Rican mantled howling monkeys. *Int. J. Primatol.* 13:415–436.

Glander, K. E. (1994). Nonhuman primate self-medication with wild plant foods. In: Etkin, N. L. (ed.), *Eating on the Wild Side: The Pharmacologic, Ecologic, and Social Implications of Using Noncultigens*. University of Arizona Press, Tuscon. pp. 239–256.

Glander, K. E. (2006). Average body weight for mantled howling monkeys (*Alouatta palliata*): as assessment of average values. In: Estrada, A., Garber, P. A., Pavelka, M., and Luecke, L. (eds.), *New Perspectives in the Study of Mesoamerican Primates: Distribution, Ecology, Behavior, and Conservation*. Springer Science+Business Media, New York. pp. 247–264.

Glander, K. E., and Powzyk, J. A. (1998). Morphometrics of wild *Indri indri* and *Propithecus diadema diadema*. *Folia Primatol.* 69(suppl. 1):399.

Glander, K. E., Wright, P. C., Seigler, D. S., and Randriansolo, B. (1989). Consumption of cyanogenic bamboo by a newly discovered species of bamboo lemur. *Am. J. Primatol.* 19:199–124.

Glander, K. E., Wright, P. C., Daniels, P. S., and Merenlender, A. M. (1992). Morphometrics and testicle size of rain forest lemur species from southeastern Madagascar. *J. Hum. Evol.* 22:1–17.

Glassman, D. M. (1983). Growth and development in the saddle-back tamarin: the sequence and timing of dental eruption and epiphyseal union. *Am. J. Phys. Anthropol.* 5:51–60.

Glassman, D. M., and Wells, J. P. (1984). Positional and activity behavior in a captive slow loris: a quantitative assessment. *Am. J. Primatol.* 7:121–132.

Glassman, D. M., Coelho, A. M., Jr., Carey, K. D., and Bramblett, C. A. (1984). Weight growth in savannah baboons: a longitudinal study from birth to adulthood. *Growth* 48:425–433.

Gleason, T. M. (1998). The ecology of olfactory communication in Venezuelan white-faced sakis. *Am. J. Primatol.* 45:183.

Glick, B. B. (1980). Ontogenetic and psychobiological aspects of the mating activities of male *Macaca radiata*. In: Lindburg, D. G. (ed.), *The Macaques: Studies in Ecology, Behavior and Evolution*. Van Nostrand Reinhold, New York, pp. 345–369.

Godfrey, L. R., and Irwin, M. T. (2007). The evolution of extinction risk: past and present anthropogenic impacts on the primate communities of Madagascar. *Folia Primatol.* 78:405–419.

Godfrey, L. R., and Jungers, W. L. (2002). Quaternary fossil lemurs. In: Hartwig, W. C. (ed.), *The Primate Fossil Record*. Cambridge University Press, Cambridge. pp. 97–121.

Godfrey, L. R., and Jungers, W. L. (2003). The extinct sloth lemurs of Madagascar. *Evol. Anthropol.* 12:252–263.

Godfrey, L. R., Jungers, W. L., Reed, K. E., Simons, E. L., and Chatrath, P. S. (1997). Subfossil lemurs: inferences about past and present primate communities. In: Goodman, S. M., and Patterson, B. D. (eds.), *Natural Change and Human Impact in Madagascar*. Smithsonian Institution Press, Washington DC. pp. 218–256.

Godfrey, L. R., King, S. J., and Sutherland, M. R. (1998). Heterochronic approaches to the study of locomotion. In: Strasser, E., Fleagle, J. G., Rosenberger, A., and McHenry, H. (eds.), *Primate Locomotion: Recent Advances*. Plenum Press, New York. pp. 277–307.

Godfrey, L.R., Lyon, S.K., and Sutherland, M.R. (1993). Sexual dimorphism in large-bodied primates: the case of the subfossil lemurs. *Am. J. Phys. Anthropol.* 90:315–334

Godfrey, L. R., Samonds, K. E., Jungers, W. L., and Sutherland, M. R. (2001). Teeth, brains, and primate life histories. *Am. J. Phys. Anthropol.* 114:192–214.

Godfrey, L. R., Samonds, K. E., and Jungers, W. L. (2002). Dental development and primate life histories. In: Kappeler, P. M., and Pereira, M. E. (eds.), *Primate Life Histories and Socioecology*. University of Chicago Press, Chicago. pp. 177–203.

Godfrey, L. R., Samonds, K. E., Jungers, W. L., Sutherland, M. R., and Irwin, M. T. (2004a). Ontogenetic correlates of diet in Malagasy lemurs. *Am. J. Phys. Anthropol.* 123:250–276.

Godfrey, L. R., Semprebon, G. M., Jungers, W. L., Sutherland, M. R., Simons, E. L., and Solounias, N. (2004b). Dental use wear in extinct lemurs: evidence of diet and niche differentiation. *J. Hum. Evol.* 47:145–169.

Godfrey, L. R., Jungers, W. L., and Schwartz, G. T. (2006). Ecology and extinction of Madagascar's subfossil lemurs. In: Gould, L., and Sauther, M. L. (eds.), *Lemurs: Ecology and Adaptation*. Springer, New York. pp. 41–64.

Godoy, R., Wildke, D., Overman, H., Cubas, A., Cubas, G., Demmer, J., McSweeney, K., and Brokaw, N. (2000). Valuation of consumption and sale of forest goods from a Central American rain forest. *Nature* 406:62–63.

Goerlitz, H. R., and Siemers, B. M. (2007). Sensory ecology of prey rustling sounds: acoustical features and their classification by wild grey mouse lemurs. *Funct. Ecol.* 21:143–153.

Goldberg, T. L., and Ruvolo, M. (1997a). Molecular phylogenetics and historical biogeography of East African chimpanzees. *Biol. J. Linn. Soc.* 61:301–324.

Goldberg, T. L., and Ruvolo, M. (1997b). The geographic apportionment of mitochondrial genetic diversity in East African chimpanzees, *Pan troglodytes schweinfurthii*. *Mol. Biol. Evol.* 14:976–984.

Goldberg, T. L., and Wrangham, R. W. (1997). Genetic correlates of social behaviour in wild chimpanzees: evidence from mitochondrial DNA. *Anim. Behav.* 54:559–570.

Goldfoot, D. A., Westerborg-Van loon, H., Groeneveld, W., and Slob, A. K. (1980). Behavioral and physiological evidence of sexual climax in the female stump-tailed macaque (*Macaca arctoides*). *Science* 208:1477–1479.

Goldizen, A. W. (1987a). Tamarins and marmosets: communal care of offspring. In: Smuts, B. B., Cheney, D. L., Seyfarth, R. M., Wrangham, R. W., and Struhsaker, T. T. (eds.), *Primate Societies*. University of Chicago Press, Chicago. pp. 34–43.

Goldizen, A. W. (1987b). Facultative polyandry and the role of infant-carrying in wild saddle-back tamarins (*Saguinus fuscicollis*). *Behav. Ecol. Sociobiol.* 20:99–109.

Goldizen, A. W. (1989). Social relationships in a cooperatively polyandrous group of tamarins (*Saguinus fuscicollis*). *Behav. Ecol. Sociobiol.* 24:79–89.

Goldizen, A. W. (2003). Social monogamy and its variations in callitrichids: do these relate to the costs of infant care? In: Reichard, U. H., and Boesch, C. (eds.), *Monogamy: Mating Strategies and Partnerships in Birds, Humans, and Other Mammals*. Cambridge University Press, Cambridge. pp. 232–247.

Goldizen, A. W., and Terborgh, J. (1986). Cooperative polyandry and helping behavior in saddle-backed tamarins (*Saguinus fuscicollis*). In: Else, J. G., and Lee, P. C. (eds.), *Primate Ecology and Conservation*. Cambridge University Press, New York. pp. 191–198.

Goldizen, A. W., Terborgh, J., Cornejo, F., Porras, D. T., and Evans, R. (1988). Seasonal food shortage, weight-loss and the timing of births in saddle-back tamarins (*Saguinus fuscicollis*). *J. Anim. Ecol.* 57:893–901.

Goldizen, A. W., Mendelson, J., van Vlaardingen, M., and Terborgh, J. (1996). Saddle-back tamarin (*Saguinus fuscicollis*) reproductive strategies: evidence from a thirteen-year study of a marked population. *Am. J. Primatol.* 38:57–84.

Goldman, D., Giri, P. R., and O'Brien, S. J. (1987). A molecular phylogeny of the hominoid primates as indicated by two-dimensional protein electrophoresis. *Proc. Natl. Acad. Sci. USA* 84:3307–3311.

Goldsmith, M. L. (1999). Ecological constraints on the foraging effort of western gorillas (*Gorilla gorilla gorilla*) at Bai Hokou, Central African Republic. *Int. J. Primatol.* 20:1–23.

Goldsmith, M. L. (2003). Comparative behavioral ecology of a lowland and highland gorilla population: where do Bwindi gorillas fit? In: Taylor, A. B., and Goldsmith, M. L. (eds.), *Gorilla Biology: A Multidisciplinary Perspective*. Cambridge University Press, Cambridge. pp. 358–384.

Goldstein, S. J., and Richard, A. F. (1989). Ecology of rhesus macaques (*Macaca mulatta*) in northwest Pakistan. *Int. J. Primatol.* 10:531–567.

Golledge, R. G. (ed.) (1999). *Wayfinding behavior: cognitive mapping and other spatial processes*. Johns Hopkins University Press, Baltimore, MD.

Gomendio, M. (1991). Parent–offspring conflict and maternal investment in rhesus macaques. *Anim. Behav.* 42:993–1005.

Gomendio, M., and Roldan, E. R. S. (1991). Sperm competition influences sperm size in mammals. *Proc. R. Soc. Lond. B* 243:181–185.

Gomendio, M., and Roldan, E. R. S. (1993). Mechanisms of sperm competition: linking physiology and behavioral ecology. *Trends Ecol. Evol.* 8:95–100.

Gomendio, M., Martin-Coello, J., Crespo, C., Magaña, C., and Roldan, E. R. S. (2006). Sperm competition enhances functional capacity of mammalian spermatozoa. *Proc. Natl. Acad. Sci. USA* 103:15113–15117.

Gomes, C. M., and Boesch, C. (2009). Wild chimpanzees exchange meat for sex on a long-term basis. *PLoS ONE* 4(4):e5116. doi:10.1371/journal.pone.0005116.

Gonçalves, E. C., Ferrari, S. F., Silva, A., Coutinho, P. E. G., Menezes, E. V., and Schneider, M. P. C. (2003). Effects of habitat fragmentation on the genetic variability of silvery marmosets, *Mico argentatus*. In: Marsh, L. K. (ed.), *Primates in Fragments*. Kluwer Academic, New York. pp. 17–28.

Gonder, K. M. (2000). Evolutionary genetics of chimpanzees (*Pan troglodytes*) in Nigeria and Cameroon [PhD diss.]. City University of New York, New York.

Gonder, K. M., Oates, J. F., Disotell, T. R., Forstner, M. R., Morales, J. C., and Melnick, D. J. (1997). A new West African chimpanzee subspecies? *Nature* 388:337.

Gonder, K. M., Disotell, T. R., and Oates, J. F. (2006). New genetic evidence on the evolution of chimpanzee populations, and implications for taxonomy. *Int. J. Primatol.* 27(4):1103–1127.

Gonzalez-Kirchner, J. P. (1997). Behavioural ecology of two sympatric colobines on Bioko Island, Equatorial Guinea. *Folia Zool.* 46:97–104.

Gonzalez-Kirchner, J. P. (1998). Group size and population density of the black howler monkey (*Alouatta pigra*) in Muchukux Forest, Quintana Roo, Mexico. *Folia Primatol.* 9:260–265.

Gonzalez-Kirchner, J. P., and Sainz de la Maza, M. (1996). Preliminary notes on the ecology of the drill (*Mandrillus leucophaeus*) on Bioko Island, Rep. Equatorial Guinea. *Garcia Orta Ser. Zool. Lisboa* 21:1–5.

González-Solís, J., Guix, J. C., Mateos, E., and Lorrens, L. (2001). Population density of primates in a large fragment of the Brazilian Atlantic rainforest. *Biodiv. Conserv.* 10:1267–1282.

Goodall, J. (1964). Tool-using and aimed throwing in a community of free-living chimpanzees. *Nature* 201:1264–1266.

Goodall, J. (1971). *In the Shadow of Man*. Delta Books, New York.

Goodall, J. (1973). Cultural elements in a chimpanzee community. In: Menzel, E. W. J. (ed.), *Precultural Primate Behaviour*. Karger, Basel. pp. 144–184.

Goodall, J. (1977). Infant killing and cannibalism in free-living chimpanzees. *Folia Primatol.* 28:259–282.

Goodall, J. (1983). Population dynamics during a 15 year period in one community of free-living chimpanzees in the Gombe National Park, Tanzania. *Z. Tierpsychol.* 61:1–60.

Goodall, J. (1986). *The Chimpanzees of Gombe: Patterns of Behavior*. Harvard University Press, Cambridge, MA.

Goodall, J. (1990). *Through a Window: My Thirty Years with the Chimpanzees of Gombe*. Houghton Mifflin, Boston.

Goodman, L. M., and Wislocki, G. B. (1935). Cyclical uterine bleeding in a New World monkey (*Ateles geoffroyi*). *Anat. Rec.* 61:379–387.

Goodman, M. (1976). Toward a geneological description of the primates. In: Goodman, M., Tashian, R. E., and Tashian, J. H. (eds.), *Molecular Anthropology: Genes and Proteins in the Evolutionary Ascent of the Primates*. Plenum Press, New York. pp. 321–353.

Goodman, M., and Moore, G. W. (1971). Immunodiffusion systematics of the primates. I. The catarrhini. *Syst. Zool.* 20:19–62.

Goodman, M., Farris, W., Jr., Moore, W., Prychodko, W., Poulik, E., and Sorenson, M. (1974). Immunodiffusion systematics of the primates: II. Findings on *Tarsius*, Lorisidae and Tupaiidae. In: Martin, R. D., Doyle, G. A., and Walker, A. C. (eds.), *Prosimian Biology*. Duckworth, London. pp. 881–890.

Goodman, M., Porter, C. A., Czelusniak, J., Page, S. L., Schneider, H., Shoshani, J., Gunnell, G., and Groves, C. P. (1998). Toward a phylogenetic classification of primates based on DNA evidence complemented by fossil evidence. *Mol. Phylogenet. Evol.* 9:585–598.

Goodman, M., McConkey, E. H., and Page, S. L. (2002). Reconstructing human evolution in the age of genomic exploration. In: Harcourt, C. S., and Sherwood, B. R. (eds.), *New Perspectives in Primate Evolution and Behavior*. Westbury, Otley, UK. pp. 47–70.

Goodman, S. J. (1997). RST CALC: a collection of computer programs for calculating unbiased estimates of genetic differentiation and gene flow from microsatellite data and determining their significance. *Mol. Ecol.* 6:881–886.

Goodman, S. M. (1994). The enigma of antipredator behavior in lemurs: evidence of a large extinct eagle on Madagascar. *Int. J. Primatol.* 15:129–134.

Goodman, S. M., and Benstead, J. P. (eds.) (2003). *The Natural History of Madagascar*. University of Chicago Press, Chicago.

Goodman, S. M., and Langrand, O. (1996). A high mountain population of the ringtailed lemur *Lemur catta* on the Andringitra Massif, Madagascar. *Oryx* 30:259–268.

Goodman, S. M., O'Connor, S., and Langrand, O. (1993). A review of predation on lemurs: implications for the evolution of social behavior in small, nocturnal primates. In: Kappeler, P. M., and Ganzhorn, J. U. (eds.), *Lemur Social Systems and Their Ecological Basis*. Plenum Press, New York. pp. 51–66.

Goodman, S. M., Rene de Roland, L. A., and Thorstrom, R. (1998). Predation on the eastern woolly lemur (*Avahi laniger*) and other vertebrates by Henst's goshawk (*Accipiter henstii*). *Lemur News* 3:14–15.

Goodnight, K. F., and Queller, D. C. (1999). Computer software for performing likelihood tests of pedigree relationship using genetic markers. *Mol. Ecol.* 8:1231–1234.

Goodnight, C., Schwartz, J., and Stevens, L. (1992). Contextual analysis of models of group selection, soft selection, hard selection, and the evolution of altruism. *Am. Nat.* 140:743–761.

Goodwin, W., Linacre, A., and Hadi, Sibte. (2007). *An Introduction to Forensic Genetics*. Wiley, New York.

Goossens, B., Anthony, N., Jeffrey, K., Johnson-Bawe, M., and Bruford, M. W. (2003). Collection, storage, and analysis of non-invasive genetic material in primate biology. In: Setchell, J. M., and Curtis, D. M. (eds.), *Field and Laboratory Methods in Primatology: A Practical Guide*. Cambridge University Press, Cambridge. pp. 295–308.

Goossens, B., Chikhi, L., Ancrenaz, M., Lackman-Ancrenaz, I., Andau, P., and Bruford, M. W. (2006a). Genetic signature of anthropogenic population collapse in orang-utans. *PLoS Biol* 4(2):e25.

Goossens, B., Setchell, J. M., James, S. S., Funk, S. M., Chikhi, L., Abulani, A., Acrenaz, M., Lackman, Acrenaz, I., and Bruford, M. W. (2006b). Philopatry and reproductive success in Bornean orang-utans (*Pongo pygmaeus*). *Mol. Ecol.* 15:2577–2588.

Goossens, B., Chikhi, L., Jalil, M. F., James, S., Ancrenaz, M., Lackman-Ancrenaz, I., and Bruford, M. W. (2009). Taxonomy, geographic variation and population genetics of Bornean and Sumatran orangutans. In: Wich, S. A., Utami, S. S., Mitra Setia, T., and van Schaik, C. (eds.), *Orangutans: Geographic Variation in Behavioral Ecology and Conservation*. Oxford University Press, Oxford. pp. 1–14.

Gordon, T. P., Rose, R., and Bernstein, I. S. (1976). Seasonal rhythm in plasma testosterone levels in the rhesus monkey (*Macaca mulatta*): a three-year study. *Horm. Behav.* 7:229–243.

Gosden, R. G., and Faddy, M. J. (1998). Biological bases of premature ovarian failure. *Reprod. Fertil. Dev.* 10:644–661.

Goss-Custard, J. D., Dunbar, R. I. M., and Aldrich-Blake, F. P. G. (1972). Survival, mating and rearing strategies and the evolution of primate social structure. *Folia Primatol.* 17:1–19.

Goudet, J. (1995). FSTAT (version 1.2): a computer program to calculate F-statistics. *J. Hered.* 86:485–486.

Goudet, J. (2001). FSTAT, a program to estimate gene diversity and fixation indices (version 2.9.3). Institute of Ecology, Laboratory for Zoology, University of Lausanne, Lausanne.

Gould, K. G., and Dahl, J. F. (1993). Assisted reproduction in the great apes. In: Wolf, D. P., Stouffer, R. L., and Brenner, R. M. (eds.), *In Vitro Fertilization and Embryo Transfer in Primates*. Spring-Verlag, New York. pp. 46–72.

Gould, K. G., Flint, M., and Graham, C. E. (1981). Chimpanzee reproductive senescence: a possible model for evolution of the menopause. *Maturitas* 3:157–166.

Gould, L. (1990). The social development of free-ranging infant *Lemur catta* at Berenty Reserve, Madagascar. *Int. J. Primatol.* 11:297–318.

Gould, L. (1992). Alloparental care in free-ranging *Lemur catta* at Berenty Reserve, Madagascar. *Folia Primatol.* 58:72–83.

Gould, L. (1994). Patterns of affiliative behavior in adult male ringtailed lemurs (*Lemur catta*) at the Beza-Mahafaly Reserve, Madagascar [PhD diss.]. Washington University, St. Louis.

Gould, L. (1997). Affiliative relationships between adult males and immature group members in naturally occurring ringtailed lemurs (*Lemur catta*). *Am. J. Phys. Anthropol.* 103:163–171.

Gould, L. (2006). *Lemur catta* ecology: what we know and what we need to know. In: Gould, L., and Sauther, M. L. (eds.), *Lemurs: Ecology and Adaptation,* Springer, New York. pp. 255–274.

Gould, L., and Sauther, M. L. (2007). Anti-predator strategies in a diurnal prosimian, the ring-tailed lemur (Lemur catta), at the Beza Mahafaly Special Reserve, Madagascar. In: Gursky, S. L., Nekaris, K. A. I. (eds.), *Primate Anti-Predator Strategies.* Springer, New York. pp. 275–288.

Gould, L., Fedigan, L. M., and Rose, L. M. (1997). Why be vigilant? The case of the alpha animal. *Int. J. Primatol.* 18:401–414.

Gould, L., Sussman, R. W., and Sauther, M. L. (1999). Natural disasters and primate populations: the effects of a two-year drought on a naturally occurring population of ringtailed lemurs in southwestern Madagascar. *Int. J. Primatol.* 20:69–84.

Gould, L., Sussman, R. W., and Sauther, M. L. (2003). Demographic and life-history patterns in a population of ring-tailed lemurs (*Lemur catta*) at Beza Mahafaly Reserve, Madagascar: a 15-year perspective. *Am. J. Phys. Anthropol.* 120:182–194.

Gould, S. J. (1977). *Ontogeny and Phylogeny.* Belknap Press, Cambridge, MA.

Gouteux, S., Thinus-Blanc, C., and Vauclair, J. (2001). Rhesus monkeys use geometric and nongeometric information during a reorientation task. *J. Exp. Psychol. Gen.* 130:505–519.

Gouzoules, H. (1975). Maternal rank and early social interactions of infant stumptail macaques, *Macaca arctoides. Primates* 16:405–418.

Gouzoules, H. (2005). Monkeying around with symbolism: are vocalizations simple symbols...or more like cymbals? In: Namy, L. (ed.), *Symbolic Use and Symbolic Representation.* Erlbaum, Mahwah, NJ. pp. 245–263.

Gouzoules, H., and Gouzoules, S. (1989). Design features and developmental modification of pigtail macaque, *Macaca nemestrina,* agonistic screams. *Anim. Behav.* 37:383–401.

Gouzoules, H., and Gouzoules, S. (1995). Recruitment screams of pigtail monkeys (*Macaca nemestrina*): ontogenetic perspectives. *Behaviour* 132:431–450.

Gouzoules, H., and Gouzoules, S. (2000). Agonistic screams differ among four species of macaques: the significance of motivation–structural rules. *Anim. Behav.* 59:501–512.

Gouzoules, H., and Gouzoules, S. (2002). Primate communication: by nature honest, or by experience wise? *Int. J. Primatol.* 23:821–848.

Gouzoules, H., Fedigan, L. M., and Fedigan, L. (1975). Responses of a transplanted troop of Japanese macaques (*Macaca fuscata*) to bobcat (*Lynx rufus*) predation. *Primates* 16:335–349.

Gouzoules, H., Gouzoules, S., and Marler, P. (1985). External reference and affective signaling in mammalian vocal communication. In: Zivin, G. (ed.), *The Development of Expressive Behavior (Biology–Environment Interactions).* Academic Press, New York. pp. 77–101.

Gouzoules, H., Gouzoules, S., and Miller, K. (1996). Skeptical responding in rhesus monkeys (*Macaca mulatta*). *Int. J. Primatol.* 17:549–568.

Gouzoules, H., Gouzoules, S., and Tomaszycki, M. (1998). Agonistic screams and the classification of dominance relationships: are monkeys fuzzy logicians? *Anim. Behav.* 55:51–60.

Gouzoules, S. (1984). Primate mating systems, kin associations, and cooperative behavior: evidence for kin recognition? *Ybk. Phys. Anthropol.* 27:99–134.

Gouzoules, S., and Gouzoules, H. (1987). Kinship. In: Smuts, B. B., Cheney, D. L., Seyfarth, R. M., Wrangham, R. W., and Struhsaker, T. T. (eds.), *Primate Societies.* University of Chicago Press, Chicago. pp. 299–305.

Gouzoules, S., Gouzoules, H., and Marler, P. (1984). Rhesus monkey (*Macaca mulatta*) screams: representational signalling in the recruitment of agonistic aid. *Anim. Behav.* 32:182–193.

Gouzoules, S., Gouzoules, H., and Marler, P. (1986). Vocal communication: a vehicle for the study of social relationships. In: Rawlins, R. G., and Kessler,

M. J. (eds.), *The Cayo Santiago Macaques.* State University of New York Press, Albany. pp. 111–129.

Gowaty, P. A. (1996). Battles of the sexes and origins of monogamy. In: Black, J. M. (ed.), *Partnerships in Birds: The Study of Monogamy.* Oxford University Press, Oxford. pp. 21–52.

Gowaty, P. A. (1997). Sexual dialectics, sexual selection, and variation in reproductive behavior. In: Gowaty, P. A. (ed.), *Feminism and Evolutionary Biology: Boundaries, Intersections, and Frontiers.* Chapman and Hall, New York. pp. 351–384.

Gowaty, P. A., Steinichen, R., and Anderson, W. W. (2002). Mutual interest between the sexes and reproductive success in *Drosophila pseudoobscura. Evolution* 56:2537–2540.

Gozalo, A., and Montoya, E. (1990). Reproduction in the owl monkey (*Aotus nancymai*) (Primates: Cebidae) in captivity. *Am. J. Primatol.* 21:61–68.

Grafen, A. (1990). Sexual selection unhandicapped by the Fisher process. *J. Theor. Biol.* 144:473–516.

Graham, C. E. (1970). Reproductive physiology of the chimpanzee. In: Bourne, G. H. (ed.), *The chimpanzee,* vol. 3. Karger, Basel. pp. 183–220.

Graham, C. E. (1979). Reproductive function in aged female chimpanzees. *Am. J. Phys. Anthropol.* 50:291–300.

Graham, C. E. (1981). Menstrual cycle of the great apes. In: Graham, C. E. (ed.), *Reproductive Biology of the Great Apes: Comparative and Biomedical Perspectives.* Academic Press, New York. pp. 1–43.

Graham, C. E. (1986). Endocrinology of reproductive senescence. In: Dukelow, W. R., and Erwin, J. (eds.), *Comparative Primate Biology. Reproduction and Development,* vol. 3. Alan R. Liss, New York. pp. 93–99.

Graham, C. E., and Nadler, R. D. (1990). Socioendocrine interactions in great ape reproduction. In: Ziegler, T. E., and Bercovitch, F. B. (eds.), *Socioendocrinology of Primate Reproduction.* Wiley-Liss, New York. pp. 33–58.

Graham, C. E., Kling, O. R., and Steiner, R. A. (1979). Reproductive senescence in female nonhuman primates. In: Bowden, D. J. (ed.), *Aging in Nonhuman Primates.* Van Nostrand Reinhold, New York. pp. 183–209.

Graham-Jones, O., and Hill, W. C. O. (1962). Pregnancy and parturition in a Bornean orangutan. *Proc. Zool. Soc. Lond.* 139:503–510.

Grand, T. I. (1972). A mechanical interpretation of terminal feeding. *J. Mammal.* 53:198–201.

Grand, T. I. (1977a). Body weight: its relationships to tissue composition, segment distribution, and motor function. I. Interspecific comparisons. *Am. J. Phys. Anthropol.* 47:211–240.

Grand, T. I. (1977b). Body weight: its relation to tissue composition, segment distribution, and motor function II. Development of *Macaca mulatta. Am. J. Phys. Anthropol.* 47:241–247.

Grand, T. I. (1983). Body weight: its relationship to tissue composition, segmental distribution of mass, and motor function III. The Didelphidae of French Guiana. *Aust. J. Zool.* 31:299–312.

Grand, T. I. (1984). Motion economy within the canopy: four strategies for mobility. In: Rodman, P., and Cant, J. G. H. (eds.), *Adaptations for Foraging in Nonhuman Primates.* Columbia University Press, New York. pp. 54–72.

Grand, T. I. (1990). The functional anatomy of body mass. In: Damuth, J., and MacFadden, B. J. (eds.), *Body Size in Mammalian Paleobiology: Estimation and Biological Implications.* Cambridge University Press, Cambridge. pp. 39–47.

Grand, T. I. (1992). Altricial and precocial mammals: a model of neural and muscular development. *Zoo Biol.* 11:3–15.

Grassi, C. (2001). The behavioral ecology of *Hapalemur griseus griseus:* the influence of microhabitat and population density on this small-bodied prosimian folivore [PhD diss.]. University of Texas, Austin.

Grassi, C. (2006). Variability in habitat, diet, and social structure of *Hapalemur griseus* in Ranomafana National Park, Madagascar. *Am. J. Phys. Anthropol.* 131:50–63.

Graur, D., and Li, W.-H. (2000). *Fundamentals of Molecular Evolution,* 2nd ed. Sinauer Associates, Sunderland, MA.

Graur, D., and Martin, W. (2004). Reading the entrails of chickens: molecular timescales of evolution and the illusion of precision. *Trends Genet.* 20:80–86.

Graves, G. R., and O'Neill, J. P. (1980). Notes on the yellow-tailed woolly monkey (*Lagothrix flavicauda*) of Peru. *J. Mammal.* 61:345–347.

Gray, J. E. (1863). Revision of the species of lemuroid animals, with the description of some new species. *Proc. Zool. Soc. Lond.* 1863:129–152.

Gray, M., McNeilage, A, Fawcett, K, Robbins, MM, Ssebide, B, Mbula, D, & Uwingeli, P.(2010). Censusing the mountain gorillas in the Virunga Volcanoes: complete sweep method vs. monitoring. *Afr. J. Ecol.* In press.

Green, G., and Sussman, R. W. (1990). Deforestation history of the eastern rain forests of Madagascar from satellite images. *Science* 248:212–215.

Green, K. M. (1978). Primate censusing in northern Colombia: a comparison of two techniques. *Primates* 19:537–550.

Green, S. (1975). Communication by a graded vocal system in Japanese monkeys. In: Rosenblum, L. A. (ed.), *Primate Behavior,* vol. 4. Academic Press, New York. pp. 1–102.

Greenfield, P. M. (1991). Language, tools and the brain: the ontogeny and phylogeny of hierarchically organized sequential behavior. *Behav. Brain Sci.* 14:531–95.

Greenwood, P. J. (1980). Mating systems, philopatry, and dispersal in birds and mammals. *Anim. Behav.* 28:1140–1162.

Gresl, T. A., Baum, S. T., and Kemnitz, J. W. (2000). Glucose regulation in captive *Pongo pygmaeus abelii, P. p. pygmaeus,* and *P. p. abelii* × *P. p. pygmaeus* orangutans. *Zoo Biol.* 19:193–208.

Grimes, K. H. (2000). Guereza dietary and behavioural patterns at the Entebbe Botanical Gardens [PhD diss.]. University of Calgary, Calgary.

Gros-Louis, J. J., Perry, S. E., Fichtel, C., Wikberg, E., Gilkenson, H., Wofsy, S., and Fuentes, A. (2008). Vocal repertoire of *Cebus capucinus:* acoustic structure, context, and usage. *Int. J. Primatol.* 29:641–670.

Gross, J. J. (ed.), (2007), *Handbook of Emotion Regulation:* Guilford Press, New York.

Gross-Camp, N. D., Masozera, M., and Kaplin, B. A. (2009). Chimpanzee seed dispersal quantity in a tropical montane forest of Rwanda. *Am. J. Primatol.* 71:901–911.

Groves, C. P. (1970). The forgotten leaf-eaters, and the phylogeny of the Colobinae. In: Napier, J. R., and Napier, P. H. (eds.), *Old World Monkeys: Evolution, Systematics, and Behavior.* Academic Press, New York. pp. 555–587.

Groves, C. P. (1972). Systematics and phylogeny of gibbons. *Gibbon Siamang* 1:1–89.

Groves, C. P. (1973). Notes on the ecology and behaviour of the Angola colobus (*Colobus angolensis* P. L. Sclater 1860) in N. E. Tanzania. *Folia Primatol.* 20:12–26.

Groves, C. P. (1978). Phylogenetic and population systematics of the mangabeys (Primates: Cercopithecoidea). *Primates* 19: 1–34.

Groves, C. P. (1984). A new look at the taxonomy and phylogeny of the gibbons. In: Preuschoft, H., Chivers, D. J., Brockelman, W. Y., and Creel, N. (eds.), *The Lesser Apes: Evolutionary and Behavioural Biology.* Edinburgh University Press, Edinburgh. pp. 542–561.

Groves, C. P. (1986). Systematics of the great apes. In: Swindler, D. R. (ed.), *Comparative Primate Biology.* Alan R. Liss, New York. pp. 187–217.

Groves, C. P. (1989). *A Theory of Human and Primate Evolution.* Clarendon Press, Oxford.

Groves, C. P. (1993). Order Primates. In: Wilson, D. E., and Reeder, D. M. (eds.), *Mammal Species of the World: A Taxonomic and Geographic Reference.* Smithsonian Institution Press, Washington DC. pp. 243–277.

Groves, C. P. (1998). Systematics of tarsiers and lorises. *Primates* 39:13–27.

Groves, C. P. (2000). The genus *Cheirogaleus:* unrecognized biodiversity in dwarf lemurs. *Int. J. Primatol.* 21:943–962.

Groves, C. P. (2001a). *Primate Taxonomy.* Smithsonian Institution Press, Washington DC.

Groves, C. (2001b). Getting to know the tarsiers: yesterday, today, and tomorrow. Paper presented at the International Society of Primatologists, 14th Congress, Adelaide, Australia, January 7–12, 2001.

Groves, C. (2003). A history of gorilla taxonomy. In: Taylor, A. B., and Goldsmith, M. L. (eds.), *Gorilla Biology: A Multidisciplinary Perspective.* Cambridge University Press, Cambridge. pp. 15–34.

Groves, C. P. (2005) Order Primates. In: Wilson, D. E., and Reeder, D. M. (eds.), *Mammal Species of the World: A Taxonomic and Geographic Reference,* vol. 1. Johns Hopkins University Press, Baltimore, MD. pp. 111–184.

Groves, C. (2008). *Extended Family: Long Lost Cousins. A Personal Look at the History of Primatology.* Conservation International, Arlington.

Groves, C. P., and Tattersall, I. (1991). Geographical variation in the fork-marked lemur, *Phaner furcifer* (Mammalia, Primates). *Folia Primatol.* 56:39–49.

Groves, C. P., Westwood, C., and Shea, B. T. (1992). Unfinished business: mahalanois and a clockwork orang. *J. Hum. Evol.* 22:327–340.

Groves, C. P., Butynski, T., and Kingdon J. (in press). Genus *Lophocebus.* In: Kingdon, J., Happold, D., and Butynski, T. (eds.), *The Mammals of Africa.* University of California Press, Berkeley, CA.

Grow, N., and Gursky, S. (2009a). Rediscovery of the pygmy tarsier. *Am. J. Phys. Anthropol.* 138(S48):140.

Grubb, P. (1973). Distribution, divergence and speciation of the drill and mandrill. *Folia Primatol.* 20:161–177.

Grubb, P. J. (1977). The maintenance of species-richness in plant communities: the importance of the regeneration niche. *Biol. Rev.* 52:107–145.

Grubb, P. J., Butynski, T. M., Oates, J. F., Bearder, S. K., Disotell, T. R., Groves, C. P., and Struhsaker, T. T. (2003). Assessment of the diversity of African primates. *Int. J. Primatol.* 24:1301–1357.

Guimarães, A. (1998). Ecology and social behavior of buffy-headed marmosets, *Callithrix flaviceps. Neotrop. Primates* 6:51–52.

Gumert, M. D. (2007). Grooming and infant handling interchange in *Macaca fascicularis:* the relationship between infant supply and grooming payment. *Int. J. Primatol.* 28:1059–1074.

Gumert, M. D., Kluck, M., and Malaivijitnond, S. (2009). The physical characteristics and usage patterns of stone axe and pounding hammers used by long-tailed macaques in the Andaman Sea region of Thailand. *Am. J. Primatol.* doi:10.1002/ajp.20694.

Gunnell, G. F., and Rose, K. D. (2002). Tarsiiformes: evolutionary history and adaptation. In: Hartwig, W. C. (ed.), *The Primate Fossil Record.* Cambridge University Press, Cambridge. pp. 45–82.

Gunst, N., Boinski, S., and Fragaszy, D. M. (2008). Acquisition of foraging competence in wild brown capuchins (*Cebus apella*), with special reference to conspecifics' foraging artefacts as an indirect social influence. *Behaviour* 145:195–229.

Gupta, A. K., and Chivers, C. J. (1999). Biomass and use of resources in south and southeast Asian primate communities. In: Fleagle, J. G., Janson, C., and Reed, K. E. (eds.), *Primate Communties.* Cambridge University Press, Cambridge. pp. 38–54.

Gurmaya, K. J. (1986). Ecology and behavior of *Presbytis thomasi* in northern Sumatra. *Primates* 27:151–172.

Gursky, S. (1995). Group size and composition. *Trop. Biodivers.* 3:57–62.

Gursky, S. (1997). Modeling maternal time budgets: the impact of lactation and gestation on the behavior of the spectral tarsier, *Tarsius spectrum* [PhD diss.]. State University of New York, Stony Brook.

Gursky, S. (1998a). Conservation status of the spectral tarsier *Tarsius spectrum:* population density and home range size. *Folia Primatol.* 69:191–203.

Gursky, S. (1998b). The conservation status of two Sulawesian tarsier species: *Tarsius spectrum* and *Tarsius dianae. Primate Conserv.* 18:88–91.

Gursky, S. (1999). The effect of moonlight on the behavior of spectral tarsiers. *Am. J. Phys. Anthropol.* 18(suppl.):142.

Gursky, S. (2000a). Allocare in a nocturnal primate: data on the spectral tarsier, *Tarsius spectrum. Folia Primatol.* 71:39–54.

Gursky, S. (2000b). The effect of seasonality on the behavior of an insectivorous primate, *Tarsius spectrum. Int. J. Primatol.* 21:477–495.

Gursky, S. (2002a). Predation on a wild spectral tarsier (*Tarsius spectrum*) by a snake. *Folia Primatol.* 73:60–62.

Gursky, S. (2002b). The behavioral ecology of the spectral tarsier, *Tarsius spectrum. Evol. Anthropol.* 11:226–234.

Gursky, S. (2002c). Determinants of gregariousness in the spectral tarsier (Prosimian: *Tarsius spectrum*). *J. Zool.* 256:1–10.

Gursky, S. (2003a). Lunar philia in a nocturnal primate. *Int. J. Primatol.* 24:351–367.

Gursky, S. (2003b). Predation experiments on infant spectral tarsiers (*Tarsius spectrum*). *Folia Primatol.* 74:272–284.

Gursky, S. (2003c). The effect of moonlight on a nocturnal prosimian primate. *Int. J. Primatol.* 24:351–367.

Gursky, S. (2005a). Associations between adult spectral tarsier. *Am. J. Phys. Anthropol.* 128:74–83.

Gursky, S. (2005b). Predator mobbing in *Tarsius spectrum. Int. J. Primatol.* 26:207–221.

Gursky, S. (2006). Function of snake mobbing in spectral tarsiers. *Am. J. Phys. Anthropol.* 129:601–608.

Gursky, S. (2007a). *The Spectral Tarsiers.* In: Vasey, N., and Sussman, R. (eds.), Primate Field Studies Series. New York: Prentice Hall.

Gursky, S. L. (2007b). The response of spectral tarsiers toward avian and terrestrial predators. In: Gursky, S. L., and Nekaris, K. A. I. (eds.), *Primate Anti-Predator Strategies.* Springer, New York. pp. 241–252.

Gursky, S., and Nekaris, K. A. I. (2003). An introduction to mating, birthing and rearing systems of nocturnal prosimians. *Folia Primatol.* 74:272–284.

Gursky, S. L., and Nekaris, K. A. I. (2007). *Primate Anti-Predator Strategies.* Springer Science+Business Media, New York.

Guschanski, K., Caillaud, D., Robbins, M. M., and Vigilant, L. (2008). Females shape the genetic structure of a gorilla population. *Curr. Biol.* 18:1809–1814.

Guschanski, K., Vigilant, L., McNeilage, A., Gray, M., Kagoda, E., and Robbins, M. M. (2009). Counting elusive animals: comparing field and genetic census of the entire mountain gorilla population of Bwindi Impenetrable National Park, Uganda. *Biol. Conserv.* 142:290–300.

Gust, D. A. (1994). A brief report on the social behavior of the crested mangabey (*Cercocebus galeritus galeritus*) with a comparison to the sooty mangabey (*C. torquatus atys*). *Primates* 35:375–383.

Gust, D. A. (1995). Moving up the dominance hierarchy in young sooty mangabeys. *Anim. Behav.* 50:15–21.

Gust, D. A., and Gordon, T. P. (1993). Conflict resolution in sooty mangabeys. *Anim. Behav.* 46:685–694.

Gust, D. A., and Gordon, T. P. (1994). The absence of a matrilineally based dominance system in sooty mangabeys (*Cercocebus torquatus atys*). *Anim. Behav.* 47:589–594.

Gust, D. A., Busse, C. D., and Gordon, T. P. (1990). Reproductive parameters in the sooty mangabey (*Cercocebus torquatus atys*). *Am. J. Primatol.* 22:241–250.

Gut, I. G. (2001). Automation in genotyping of single nucleotide polymorphisms. *Hum. Mutat.* 17:475–492.

Guy, F., Brunet, M., Schmittbuhl, M., Viriot, L. (2003). New approaches in hominoid taxonomy: morphometrics. *Am. J. Phys. Anthropol.* 121(3):198–218.

Gwenzi, D., Katsvanga, C. A. T., Ngorima, G. T., Mupangwa, J. F., and Valintine, S. (2007). Baboon *Papio ursinus* ranging patterns and troop size relative to bark stripping in the Chimanimani Pine Plantations of Zimbabwe. *Acta Zool. Sin.* 53:777–782.

Hacia, J. G., Fan, J.-B., Ryder, O., Jin, L., Edgemon, K., Ghandour, G., Mayer, R. A., Sun, B., Hsie, L., Robbins, C. M., Brody, L. C., Wang, D., Lander, E. S., Lipshutz, R., Fodor, S. P. A., and Collins, F. S. (1999). Determination of ancestral alleles for human single nucleotide polymorphisms using high-density oligonucleotide arrays. *Nat. Genet.* 22:164–167.

Hacker, J. E., Cowlishaw, G., and Williams, P. H. (1998). Patterns of African primate diversity and their evaluation for the selection of conservation areas. *Biol. Conserv.* 84:251–262.

Haddow, A. J. (1952). Field and laboratory studies on an African monkey, *Cercopithecus ascanius schmidti* Matschie. *Proc. Zool. Soc. Lond.* 122:297–394.

Hadidian, J., and Bernstein, I. S. (1979). Female reproductive cycles and birth data from an Old World monkey colony. *Primates* 20:429–442.

Haff, L., and Smirnov, I. (1997). Single nucleotide polymorphism identification assays using a thermostable DNA polymerase and MALDI-TOF MS. *Genome Res.* 7:378–388.

Hager, R., and Johnstone, R. A. (2004). Infanticide and control of reproduction in cooperative and communal breeders. *Anim. Behav.* 67:941–949.

Haggerty, M. E. (1909). Imitation in monkeys. *J. Comp. Neurol. Psychol.* 19:337–455.

Hahn, B. H., Shaw, G. M., De Cock, K. M., and Sharp, P. M. (2000). AIDS as a zoonosis: scientific and public health implications. *Science* 287:607–614.

Haimoff, E. H., Yang, X.-J., He, S.-J., and Chen, N. (1987). Preliminary observations of wild black-crested gibbons (*Hylobates concolor concolor*) in Yunnan Province, People's Republic of China. *Primates* 28:319–335.

Hakeem, A., Sandoval, R. G., Jones, M., and Allman, J. (1996). Brain and life span in primates. In: Birren, J. E., and Schaie, K. W. (eds.), *Handbook of the Psychology of Aging*, 4th ed. Academic Press, San Diego. pp. 78–104.

Hall, B. (2007). *Phylogenetic Trees Made Easy: A How-To Manual*, 3rd ed. Sinauer Associates, Sunderland, MA.

Hall, C. L., and Fedigan, L. M. (1997). Spatial benefits afforded by high rank in white-faced capuchins. *Anim. Behav.* 53:1069–1082.

Hall, J. S., White, L. J. T., Inogwabini, B. I., Omari, I., Morland, H. S., Williamson, E. A., Saltonstall, K., Walsh, P., Sikubwabo, C., Bonny, D., Kiswele, K. P., Vedder, A., and Freeman, K. (1998). Survey of Grauer's gorillas (*Gorilla gorilla graueri*) and eastern chimpanzees (*Pan troglodytes schweinfurthii*) in the Kahuzi-Biega National Park lowland sector and adjacent forest in eastern Democratic Republic of Congo. *Int. J. Primatol.* 19:207–235.

Hall, K. R. L. (1962a). Numerical data, maintenance activities, and locomotion of the wild chacma baboon, *Papio ursinus*. *Proc. Zool. Soc. Lond.* 139:181–220.

Hall, K. R. L. (1962b). The sexual, agonistic and derived social behaviour patterns of the wild chacma baboon, *Papio ursinus*. *Proc. Zool. Soc. Lond.* 139: 283–327.

Hall, K. R. L. (1963a). Observational learning in monkeys and apes. *Br. J. Psychol.* 54:201–226.

Hall, K. R. L. (1963b). Tool-using performances as indicators of behavioral adaptability. *Curr. Anthropol.* 4:479–487.

Hall, K. R. L. (1963c). Variations in the ecology of the chacma baboon, *Papio ursinus*. *Symp. Zool. Soc. Lond.* 10:1–28.

Hall, K. (1964). Aggression in monkey and ape societies. In: Carthy, J., and Ebling, F. (eds.), *The Natural History of Aggression*. Academic Press, London. pp. 51–64.

Hall, K. R. L. (1965a). Behaviour and ecology of the wild patas monkey, *Erythrocebus patas*, in Uganda. *J. Zool.* 148:15–87.

Hall, K. R. L. (1965b). Experiment and quantification in the study of baboon behavior in its natural habitat. In: Vagtborg, H. (ed.), *The Baboon in Medical Research*. University of Texas, San Antonio. pp. 43–61.

Hall, K. R. L. (1965c). Social organization of the Old World monkeys and apes. *Symp. Zool. Soc. Lond.* 14:265–289.

Hall, K. R. L. (1968). Social learning in monkeys. In: Jay, P. C. (ed.), *Primates. Studies in Adaptation and Variability*. Holt, Rinehart and Winston, New York. pp. 383–397.

Hall, K. R. L., and DeVore, I. (1965). Baboon social behavior. In: DeVore, I. (ed.), *Primate Behavior: Field Studies of Monkeys and Apes*. Holt, Rinehart and Winston, New York. pp. 53–110.

Hall, L. M., Jones, D. S., and Wood, B. A. (1998). Evolution of gibbon subgenera inferred from cytochrome *b* DNA sequence data. *Mol. Phylogenet. Evol.* 10:281–286.

Hall, R. D., and Hodgen, G. D. (1979). Pregnancy diagnosis in owl monkeys (*Aotus trivirgatus*): evaluation of the hemaglutination inhibition test for urinary chorionic gonadotropin. *Lab. Anim. Sci.* 29:345–348.

Hall-Haro, C., Johnson, S. P., Price, T. A., Vance, J. A., and Kiorpes, L. (2008). Development of object concepts in macaque monkeys. *Dev. Psychobiol.* 50:278–87.

Halliday, T. R. (1983). The study of mate choice. In: Bateson, P. (ed.), *Mate Choice*. Cambridge University Press, Cambridge. pp. 3–22.

Halstead, J. A. (1968). Geophagia in man: its nature and nutritional effects. *Am. J. Clin. Nutr.* 21:1384–1393.

Haltenorth, T., and Diller, H. (1988). *The Collins Field Guide to the Mammals of Africa, Including Madagascar*. Stephen Greene Press, Lexington, MA.

Ham, R. (1994). Behaviour and ecology of grey-cheeked mangabeys (*Cercocebus albigena*) in the Lopé Reserve, Gabon. [Ph.D. desis.]. University of Stirling, Stirling.

Hamada, Y., and Udono, T. (2002). Longitudinal analysis of length growth in the chimpanzee (*Pan troglodytes*). *Am. J. Phys. Anthropol.* 118(3):268–284.

Hamai, M., Nishida, T., Takasaki, H., and Turner, L. (1992). New records of within-group infanticide and cannibalism in wild chimpanzees. *Primates* 33:151–162.

Hamilton, M. B., Pincus, E. L., Di Fiore, A., and Fleischer, R. C. (1999). Universal linker and ligation procedures for construction of genomic DNA libraries enriched for microsatellites. *Biotechniques* 27:500–507.

Hamilton, W. D. (1964). The genetical evolution of social behaviour. I, II. *J. Theor. Biol.* 7:1–52.

Hamilton, W. D. (1971). Geometry for the selfish herd. *J. Theor. Biol.* 31:295–311.

Hamilton, W. D. (1980). Sex versus non-sex versus parasite. *Oikos* 35:282–290.

Hamilton, W. D. (1993). Inbreeding in Egypt and in this book: a childish perspective. In: Thornhill, N. (ed.), *The Natural History of Inbreeding and Outbreeding. Theoretical and Empirical Perspectives*. University of Chicago Press, Chicago. pp. 429–450.

Hamilton, W. D., and Zuk, M. (1982). Heritable true fitness and bright birds: a role for parasites? *Science* 218:384–387.

Hamilton, W. D., Axelrod, R., and Tanese, R. (1990). Sexual reproduction as an adaptation to resist parasites. *Proc. Natl. Acad. Sci. USA* 87:3566–3573.

Hamilton, W. J. III (1982). Baboon sleeping site preferences and relationships to primate grouping patterns. *Am. J. Primatol.* 3:41–53.

Hamilton, W. J. III (1984). Significance of paternal investment by primates to the evolution of male–female associations. In: Taub, D. M. (ed.), *Primate Paternalism*. Van Nostrand Reinhold, New York. pp. 309–355.

Hamilton, W. J. III (1985). Demographic consequences of a food and water shortage to desert chacma baboons, *Papio ursinus*. *Int. J. Primatol.* 6:451–462.

Hamilton, W. J. III, and Arrowood, P. C. (1978). Copulatory vocalizations of chacma baboons (*Papio ursinus*), gibbons (*Hylobates hoolock*), and humans. *Science* 200:1405–1409.

Hamilton, W. J. III, and Bulger, J. (1987). Interpretations of one-male savanna baboon troops. *Int. J. Primatol.* 8:466.

Hamilton, W. J. III, and Bulger, J. (1992). Facultative expression of behavioral differences between one-male and multimale savanna baboon groups. *Am. J. Primatol.* 28:61–71.

Hamilton, W. J. III, and Bulger, J. (1993). Origin and fate of a one-male savanna baboon group formed by fissioning. *Int. J. Primatol.* 14:131–143.

Hamilton, W. J. III, and Busse, C. D. (1982). Social dominance and predatory behavior of chacma baboons. *J. Hum. Evol.* 11:567–573.

Hamilton, W. J. III, Buskirk, R. E., and Buskirk, W. H. (1976). Defense of space and resources by chacma (*Papio ursinus*) baboon troops in an African desert and swamp. *Ecology* 57:1264–1272.

Hamilton, W. J. III, Buskirk, R. E., and Buskirk, W. H. (1978). Omnivory and utilization of food resources by chacma baboons, *Papio ursinus. Am. Nat.* 112:911–924.

Hamilton-Paterson, J., and Andrews, C. (1978). *Mummies.* Penguin Books, New York.

Hammer, M. F., and Zegura, S. L. (2002). The human Y chromosome haplogroup tree: nomenclature and phylogeography of its major divisions. *Annu. Rev. Anthropol.* 31:303–321.

Hammer, M. L. A., and Foley, R. A. (1996). Longevity and life history in hominid evolution. *Hum. Evol.* 11:61–66.

Hammond, R. L., Lawson-Handley, L. J., Winney, B. J., Bruford, M. W., and Perrin, N. (2006). Genetic evidence for female-biased dispersal and gene flow in a polygynous primate. *Proc. R. Soc. Lond. B* 273:479–484.

Hampton, R. R., Zivin, A., and Murray, E. A. (2004). Rhesus monkeys (*Macaca mulatta*) discriminate between knowing and not knowing and collect information as needed before acting. *Anim Cogn.* 7:239–254.

Hamrick, M. A. (2001). Primate origins: evolutionary change in digital ray patterning and segmentation. *J. Hum. Evol.* 40:339–351.

Hancock, A. M., and Di Rienzo, A. (2008). Detecting the genetic signature of natural selection in human populations: models, methods, and data. *Annu. Rev. Anthropol.* 37:197–217.

Hand, J. L. (1986). Resolution of social conflicts: dominance, egalitarianism, spheres of dominance and game theory. *Q. Rev. Biol.* 61:201–220.

Hankerson, S. E. J., and Caine, N. G. (2004). Pre-retirement predator encounters alter the morning behavior of captive marmosets, *Callithrix geoffroyi. Am. J. Primatol.* 63:75–85.

Hansson, A., Veliz, G., Naquira, C., Amren, M., Arroyo, M., and Arevalo, G. (1986). Preclinical and clinical studies with latex from *Ficus glabrata* Hbk, a traditional intestinal antehelminthic in the Amazon area. *J. Ethnopharm.* 17:105–138.

Hapke, A., Zinner, D., and Zischler, H. (2001). Mitochondrial DNA variation in Eritrean hamadryas baboons (*Papio hamadryas hamadryas*): life history influences population genetic structure. *Behav. Ecol. Sociobiol.* 50:483–492.

Happold, D., and Happold, M. (1992). Termites as food for the thick-tailed bushbaby (*Otolemur crassicaudatus*) in Malawi. *Folia Primatol.* 58:118–120.

Haraway, D. (1986). Primatology is politics by other means. In: Bleier, R. (ed.), *Feminist Approaches to Science.* Pergamon Press, New York. pp. 77–117.

Haraway, D. J. (1989). *Primate Visions.* Routledge, Chapman, and Hall, New York.

Harborne, J. B. (1988). *Introduction to Ecological Biochemistry.* Academic Press, New York.

Harborne, J. B., Moss, G. P., and Baxter, H. (1999). *Phytochemical Dictionary.* Taylor & Francis, London.

Harcourt, A. H. (1978). Strategies of emigration and transfer by female primates, with special reference to mountain gorillas. *Z. Tierpsychol.* 48:401–420.

Harcourt, A. H. (1979a). Social relationships among adult female mountain gorillas. *Anim. Behav.* 27:251–264.

Harcourt, A. H. (1979b). Social relationships between adult male and female mountain gorillas. *Anim. Behav.* 27:325–342.

Harcourt, A. H. (1979c). Contrasts between male relationships in wild gorilla groups. *Behav. Ecol. Sociobiol.* 5:39–49.

Harcourt, A. H. (1998). Does primate socioecology need non-primate socioecology? *Evol. Anthropol.* 7:3–7.

Harcourt, A. H., and de Waal, F. B. M. (eds.) (1992). *Coalitions and Alliances in Humans and Other Animals.* Oxford University Press, New York.

Harcourt, A. H., and Greenberg, J. (2001). Do gorilla females join males to avoid infanticide? A quantitative model. *Anim. Behav.* 62:905–915.

Harcourt, A. H., and Harcourt, S. A. (1984). Insectivory by gorillas. *Folia Primatol.* 43:229–233.

Harcourt, A. H., and Stewart, K. J. (1981). Gorilla male relationships: can differences during immaturity lead to contrasting reproductive tactics in adulthood? *Anim. Behav.* 29:206–210.

Harcourt, A. H., and Stewart, K. J. (1987). The influence of help in contests on dominance rank in primates: hints from gorillas. *Anim. Behav.* 35:182–190.

Harcourt, A. H., and Stewart, K. J. (1989). Functions of alliances in contests within wild gorilla groups. *Behaviour* 109:176–190.

Harcourt, A.H., and Stewart, K. J. (2007a). Gorilla society: conflict, compromise, and cooperation between the sexes Chicago and London: University of Chicago Press.

Harcourt, A. H., and Stewart, K. J. (2007b). Gorilla society: what we know and don't know. *Evol. Anthropol.* 16:147–158.

Harcourt, A. H., Stewart, K. J., and Fossey, D. (1976). Male emigration and female transfer in wild mountain gorilla. *Nature* 263:226–227.

Harcourt, A. H., Fossey, D., Stewart, K. J., and Watts, D. P. (1980). Reproduction in wild gorillas and some comparisons with chimpanzees. *J. Reprod. Fertil.* 28(suppl.):59–70.

Harcourt, A. H., Harvey, P. H., Larson, S. G., and Short, R. V. (1981a). Testis weight, body weight and breeding system in primates. *Nature* 293:55–57.

Harcourt, A. H., Stewart, K. J., and Fossey, D. (1981b). Gorilla reproduction in the wild. In: Graham, C. E. (ed.), *Reproductive Biology of the Great Apes: Comparative and Biomedical Perspectives.* Academic Press, New York. pp. 265–318.

Harcourt, C. S. (1980). Behavioural adaptations of South African galagos [MSc thesis]. University of Witwatersrand, Johannesburg.

Harcourt, C. S. (1981). An examination of the function of urine washing in *Galago senegalensis. Z. Tierpsychol.* 55:119–128.

Harcourt, C. S. (1984). The behaviour and ecology of galagos in Kenyan coastal forest [PhD diss.]. University of Cambridge, Cambridge.

Harcourt, C. S. (1986a). Seasonal variation in the diet of South African galagos. *Int. J. Primatol.* 7:491–506.

Harcourt, C. (1986b). *Galago zanzibaricus:* birth seasonality, litter size and perinatal behaviour of females. *J. Zool.* 210:451–457.

Harcourt, C. S. (1987). Brief trap/retrap study of the brown mouse lemur (*Microcebus rufus*). *Folia Primatol.* 49:209–211.

Harcourt, C. (1988). *Avahi laniger:* a study in inactivity. *Primate Eye* 35:9.

Harcourt, C. S., and Bearder, S. K. (1989). A comparison of *Galago moholi* in South Africa with *Galago zanzibaricus* in Kenya. *Int. J. Primatol.* 10:35–45.

Harcourt, C. S., and Nash, L. T. (1986a). Species differences in substrate use and diet between sympatric galagos in two Kenyan coastal forests. *Primates* 27:41–52.

Harcourt, C. S., and Nash, L. T. (1986b). Social organization of galagos in Kenyan coastal forests, I. *Galago zanzibaricus. Am. J. Primatol.* 10:339–356.

Hardie, S. M. (1998). Mixed-species tamarin groups (*Saguinus fuscicollis* and *Saguinus labiatus*) in northern Bolivia. *Primate Rep.* 50:39–62.

Hardie, S. M., and Buchanan-Smith, H. M. (1997). Vigilance in single- and mixed-species groups of tamarins (*Saguinus labiatus* and *Saguinus fuscicollis*). *Int. J. Primatol.* 18:217–234.

Harding, R. D., Hulme, M. J., Lunn, S. F., Henderson, C., and Aitken, R. J. (1982). Plasma progesterone levels throughout the ovarian cycle of the common marmoset (*Callithrix jacchus*). *J. Med. Primatol.* 11:43–51.

Harding, R. S. O. (1975). Meat-eating and hunting in baboons. In: Tuttle, R. H. (ed.), *Socioecology and Psychology of Primates.* Mouton, The Hague. pp. 245–257.

Harding, R. S. O. (1976). Ranging patterns of a troop of baboons (*Papio anubis*) in Kenya. *Folia Primatol.* 25:143–185.

Harding, R. S. O. (1980). Agonism, ranking, and the social behavior of adult male baboons. *Am. J. Phys. Anthropol.* 53:203–216.

Harding, R. S. O. (1981). An order of omnivores: nonhuman primate diets in the wild. In: Harding, R. S. O., and Teleki, G. (eds.), *Omnivorous Primates: Gathering and Hunting in Human Evolution.* Columbia University Press, New York. pp. 191–214.

Harding, R. S. O. (1984). Primates of the Killini area, northwest Sierra Leone. *Folia Primatol.* 42:96–114.

Harding, R. S. O., and Olson, D. K. (1986). Patterns of mating among male patas monkeys (*Eyrthrocebus patas*) in Kenya. *Am. J. Primatol.* 11:343–358.

Hardus, M. E., Lameira, A. R., Singleton, I., Morrogh-Bernard, H., Knott, C. D., Ancrenaz, M., Utami-Atmoko, S. S. and Wich, S. A. (2009). A description of the orangutan's vocal and sound repertoire, with a focus on geographical variation. In: Wich, S. A., Utami, S. S., Mitra Setia, T., and van Schaik, C. (eds.), *Orangutans: Geographic Variation in Behavioral Ecology and Conservation.* Oxford University Press, Oxford. pp. 49–64.

Hare, B., Call, J., Agnetta, B., and Tomasello, M. (2000). Chimpanzees know what conspecifics do and do not see. *Anim. Behav.* 59:771–785.

Hare, B., Call, J., and Tomasello, M. (2006). Chimpanzees deceive a human by hiding. *Cognition.* 101:495–514.

Haring, D., and Wright, P. (1989). Hand raising a Philippine tarsier, *Tarsius syrichta*. *Zoo Biol.* 8:265–274.

Harlow, C. R., Gems, S., Hodges, J. K., and Hearn, J. P. (1983). The relationship between plasma progesterone and the timing of ovulation and early embryonic development in the marmoset monkey (Callitrichidae). *J. Zool.* 201:273–282.

Harlow, C. R., Hearn, J. P., and Hodges, J. K. (1984). Ovulation in the marmoset monkeys: endocrinology, prediction, and detection. *J. Endocrinol.* 103:17–24.

Harlow, H. F. (1959). Love in infant monkeys. *Sci. Am.* 200:68–74.

Harlow, H. F., and Harlow, M. K. (1962). Social deprivation in monkeys. *Sci. Am.* 207:136–146.

Harms, K. E., Wright, S. J., Calderon, O., Hernandez, A., and Herre, E. A. (2000). Pervasive density-dependent recruitment enhances seedling diversity in a tropical forest. *Nature* 404:493–495.

Harrington, J. E. (1975). Field observations of social behavior of *Lemur fulvus fulvus* (E. Geoffroy 1812). In: Tattersall, I., and Sussman, R. W. (eds.), *Lemur Biology*. Plenum Press, New York. pp. 259–279.

Harrington, J. E. (1978). Diurnal behavior of *Lemur mongoz* at Ampijoroa, Madagascar. *Folia Primatol.* 29:291–302.

Harris, J. W. K. (1983). Cultural beginnings: Plio-Pleistocene archaeological occurrences from the Afar, Ethiopia. *Afr. Archaeol. Rev.* 1:3–31.

Harris, T. R., and Monfort, S. L. (2003). Behavioral and endocrine dynamics associated with infanticide in a black and white colobus monkey (*Colobus guereza*). *Am. J. Primatol.* 61:135–142.

Harris, T. R., Caillaud, D., Chapman, C. A., and Vigilant, L. (2009). Neither genetic nor observational data alone are sufficient for understanding sex-biased dispersal in a social-group-living species. *Mol. Ecol.* 18:1777–1790.

Harrison, M. J. (1983a). Age and sex differences in the diet and feeding strategies of the green monkey, *Cercopithecus sabaeus*. *Anim. Behav.* 31:969–977.

Harrison, M. J. S. (1983b). Territorial behavior in the green monkey, *Cercopithecus sabaeus*: seasonal defense of local food supplies. *Behav. Ecol. Sociobiol.* 12:85–94.

Harrison, M. J. S. (1986). Feeding ecology of black colobus, *Colobus satanas*, in central Gabon. In: Else, J. G., and Lee, P. C. (eds.), *Primate Ecology and Conservation*. Cambridge University Press, Cambridge. pp. 31–37.

Harrison, M. J. S. (1988). The mandrill in Gabon's rain forest—ecology, distribution and status. *Oryx* 22:218–228.

Harrison, M. J. S., and Hladik, C. M. (1986). Un primate granivore: le colobe noir dans la forêt du Gabon; potentialité d'évolution du comportement alimentaire. *Rev. Ecol. (Terre Vie)* 41:281–298.

Harrison, T. (1999). The palaeoecological context at Niah Cave, Sarawak: evidence from the primate fauna. *Bull. Indo-Pac. Prehist. Assoc.* 1964:521–530.

Harrison, T. (2002). Late Oligocene to middle Miocene catarrhines from Afro-Arabia. In: Hartwig, W. C. (ed.), *The Primate Fossil Record*. Cambridge University Press, Cambridge. pp. 311–338.

Hart, B. L. (1988). Biological basis of the behavior of sick animals. *Neurosci. Biobehav. Rev.* 12:123–137.

Hart, B. L. (1990). Behavioral adaptations to pathogens and parasites: five strategies. *Neurosci. Biobehav. Rev.* 14:273–294.

Hart, D. L. (2000). Primates as prey: ecological, morphological and behavioral relationships between primate species and their predators [PhD diss.]. Washington University, St. Louis.

Hart, D. (2007). Predation on primates: a biogeographical analysis. In: Gursky, S. L., and Nekaris, K. A. I. (eds.). *Primate Anti-Predator Strategies*. Springer, New York. pp. 27–59.

Hart, D. L., and Sussman, R. W. (2005). *Man the Hunted: Primates, Predators, and Human Evolution*. Westview Press, Boulder, CO.

Hart, D., and Sussman, R. W. (2009). *Man the Hunted: Primates, Predators, and Human Evolution*, exp. ed. Westview Press, Boulder, CO.

Hart, J. A. (1978). From subsistence to market: a case study of the Mbuti net hunters. *Hum. Ecol.* 6:32–53.

Hart, J. (2000). Impact and sustainability of indigenous hunting in the Ituri Forest, Congo-Zaire: a comparison of unhunted and hunted duiker populations. In: Robinson, J., and Bennett, E. (eds.), *Hunting for Sustainability in Tropical Forests*. Columbia University Press, New York. pp. 106–153.

Hart, J., and Butynski, T.M. (2008). *Cercopithecus hamlyni*. In: IUCN 2009. IUCN Red List of Threatened Species, www.iucnredlist.org (accessed June 23, 2009).

Hart, J. A., Katembo, M., and Punga, K. (1996). Diet, prey selection and ecological relations of leopard and golden cat in the Ituri Forest, Zaire. *Afr. J. Ecol.* 34:364–379.

Hart, J., Butynski, T. M., and Hall, J. (2008a). *Cercopithecus lhoesti*. In: IUCN 2009. IUCN Red List of Threatened Species, www.iucnredlist.org (accessed June 23, 2009).

Hart, J., Butynski, T. M., and Hurley, M. (2008b). *Cercopithecus dryas*. In: IUCN 2009. IUCN Red List of Threatened Species, www.iucnredlist.org (accessed June 23, 2009).

Hartwig, W. C. (1994). Patterns, puzzles and perspectives on platyrrhine origins. In: Corruccini, R. S., and Ciochon, R. L. (eds.), *Integrative Paths to the Past: Paleoanthropological Essays in Honor of F. Clark Howell*. Prentice-Hall, Englewood Cliffs, NJ. pp. 69–94.

Hartwig, W. C. (1995). Effect of life history on the squirrel monkey (Platyrrhini, *Saimiri*) cranium. *Am. J. Phys. Anthropol.* 97:435–449.

Hartwig, W. C. (1996). Perinatal life history traits in New World monkeys. *Am. J. Primatol.* 40:99–130.

Hartwig, W. C. (ed.) (2002). *The Primate Fossil Record*. Cambridge University Press, Cambridge.

Hartwig, W. C., and Cartelle, C. (1996). A complete skeleton of the giant South American primate *Protopithecus*. *Nature* 381:307–311.

Hartwig, W. C., and Meldrum, D. J. (2002). Miocene platyrrhines of the northern neotropics. In: Hartwig, W. C. (ed.), *The Primate Fossil Record*. Cambridge University Press, Cambridge. pp. 175–188.

Harvey, P. H., and Clutton-Brock, T. H. (1985). Life history variation in primates. *Evolution* 39:559–581.

Harvey, P. H., and Krebs, J. R. (1990). Comparing brains. *Science* 249:140–146.

Harvey, P. H., Martin, R. D., and Clutton-Brock, T. H. (1987). Life histories in comparative perspective. In: Smuts, B. B., Cheney, D. L., Seyfarth, R. M., Wrangham, R. W., and Struhsaker, T. T. (eds.), *Primate Societies*. University of Chicago Press, Chicago. pp. 181–196.

Hasegawa, T., and Hiraiwa, M. (1980). Social interactions of orphans observed in a free-ranging troop of Japanese monkeys. *Folia Primatol.* 33:129–158.

Hasegawa, T., and Hiraiwa-Hasegawa, M. (1983). Opportunistic and restrictive matings among wild chimpanzees in the Mahale Mountains, Tanzania. *J. Ethol.* 1:75–85.

Hasegawa, T., and Hiraiwa-Hasegawa, M. (1990). Sperm competition and mating behavior. In: Nishida, T. (ed.), *The Chimpanzees of the Mahale Mountains: Sexual and Life History Strategies*. University of Tokyo Press, Tokyo. pp. 115–132.

Hashimoto, C., Furuichi, T., and Takenaka, O. (1996). Matrilineal kin relationships and social behavior of wild bonobos (*Pan paniscus*): sequencing the D-loop region of mitochondrial DNA. *Primates* 37:305–318.

Hashimoto, C., Tashiro, Y., Kimura, D., Enomoto, T., Ingmanson, E. J., Idani, G., and Furuichi, T. (1998). Habitat use and ranging of wild bonobos (*Pan paniscus*) at Wamba. *Int. J. Primatol.* 19:1045–1060.

Hastings, M. L., and Krainer, A. R. (2001). Pre-mRNA splicing in the new millennium. *Curr. Opin. Cell Biol.* 13:302–309.

Hauser, M. D. (1986). Parent–offspring conflict: care elicitation behaviour and the "cry-wolf" syndrome. In: Else, J. G., and Lee, P. C. (eds.), *Primate Ontogeny, Cognition and Social Behaviour*. Cambridge University Press, Cambridge. pp. 193–203.

Hauser, M. D. (1992). Costs of deception: cheaters are punished in rhesus monkeys (*Macaca mulatta*). *Proc. Natl. Acad. Sci. USA* 89:12137–12139.

Hauser, M. D. (1993a). Rhesus monkey copulation calls: honest signals for female choce? *Proc. R. Soc. Lond. B* 254:93–96.

Hauser, M. D. (1993b). The evolution of non-human primate vocalizations: effects of phylogeny, body-weight, and social context. *Am. Nat.* 142:528–542.

Hauser, M. D. (1996). *The Evolution of Communication*. MIT Press, Cambridge, MA.

Hauser, M.D. (1997a). Artificial kinds and functional design features: what a primate understands without language. *Cognition* 64:285–308.

Hauser, M. D. (1997b). Minding the behaviour of deception. In: Whiten, A., and Byrne, R. W. (eds.), *Machiavellian Intelligence II*. Cambridge University Press, Cambridge. pp. 112–143.

Hauser, M. D., and Marler, P. (1993a). Food-associated calls in rhesus macaques (*Macaca mulatta*): I. Socioecological factors. *Behav. Ecol.* 4:194–205.

Hauser, M. D., and Marler, P. (1993b). Food-associated calls in rhesus macaques (*Macaca mulatta*): II. Costs and benefits of call production and suppression. *Behav. Ecol.* 4:206–212.

Hauser, M. D., and Santos, L. R. (2007). The evolutionary ancestry of our knowledge of tools: from precepts to concepts. In: Margolis, E., and Laurence, S. (eds.), *Creations of the Mind: Theories of Artifacts and Their Representation*. Oxford University Press, New York. pp. 267–288.

Hauser, M. D., Cheney, D. L., and Seyfarth, R. M. (1986). Group extinction and fusion in free-ranging vervet monkeys. *Am. J. Primatol.* 11:63–77.

Hauser, M. D., Teixidor, P., Field, L., and Flaherty, R. (1993). Food-elicited calls in chimpanzees: effects of food quantity and divisibility. *Anim. Behav.* 45:817–819.

Hauser, M. D., Kralik, J., Botto-Mahan, C., Garrett, M., and Oser, J. (1995). Self-recognition in primates: phylogeny and the salience of species-typical features. *Proc. Natl. Acad. Sci. USA* 92:10811–10814.

Hauser, M. D., Carey, S., and Hauser, L. B. (2000). Spontaneous number representation in semi-free-ranging rhesus monkeys. *Proc. R. Soc. Lond. B.* 267:829–833.

Hauser, M. D., Miller, C. T., Liu, K., and Gupta, R. (2001). Cotton-top tamarins (*Saguinus oedipus*) fail to show mirror-guided self-exploration. *Am. J. Primatol.* 53:131–137.

Hauser, M. D., Chomsky, N., and Tecumseh Fitch, W. T. (2002). The faculty of language: what is it, who has it, and how did it evolve? *Science* 298:1569–1579.

Hausfater, G. (1975a). *Dominance and Reproduction in Baboons* (Papio cynocaphalus). Karger, Basel.

Hausfater, G. (1975b). Estrous females: their effects on the social organization of the baboon group. *Proceedings from the Symposia of the Fifth Congress of the International Primatological Society (Nagoya, Japan, August 1974).* pp. 117–127.

Hausfater, G. (1976). Predatory behavior of yellow baboons. *Behaviour* 56:44–68.

Hausfater, G., and Meade, B. J. (1982). Alternation of sleeping groves by yellow baboons (*Papio cynocephalus*) as a strategy for parasite avoidance. *Primates* 23:287–297.

Hausfater, G., Altmann, J., and Altmann, S. (1982). Long-term consistency of dominance relations among female baboons (*Papio cynocephalus*). *Science* 217(4561):752–755.

Hawkes, K. (2003). Grandmothers and the evolution of human longevity. *Am. J. Hum. Biol.* 15:380–400.

Hawkes, K. (2004a). Mating, parenting and the evolution of human pair bonds. In: Chapais, B., and Berman, C. M. (eds.), *Kinship and Behavior in Primates.* Oxford University Press, New York. pp. 443–473.

Hawkes, K. (2004b). The grandmother effect. *Nature* 428:128–129.

Hawkes, K., O'Connell, J. F., and Blurton Jones, N. G. (1989). Hardworking Hadza grandmothers. In: Standon, V., and Foley, R. (eds.), *Comparative Socioecology of Mammals and Man.* Blackwell, London. pp. 341–366.

Hawkes, K., O'Connell, J. F., and Blurton-Jones, N. G. (1997). Hadza women's time allocation, offspring provisioning, and the evolution of long postmenopausal life Spans. *Curr. Anthropol.* 18:551–577.

Hawkes, K., O'Connell, J. F., and Blurton-Jones, N. G. (2002). Human life histories: primate trade-offs, grandmothering socioecology, and the fossil record. In: Kappeler, P. M., and Pereira, M. E. (eds.), *Primate Life Histories and Socioecology.* University of Chicago Press, Chicago. pp. 204–232.

Hawkins, A. F. A., Chapman, P., Ganzhorn, J. U., Bloxam, Q. M., Barlow, S. C., and Tonge, S. J. (1990). Vertebrate conservation in Ankarana Special Reserve, northern Madagascar. *Biol. Conserv.* 54:83–110.

Hayakawa, S. (2007). Female defensibility in small troops of Japanese macaques vis-à-vis nontroop males and copulation on the periphery of the troop. *Int. J. Primatol.* 28:73–96.

Hayakawa, S. (2008). Male–female mating tactics and paternity of wild Japanese macaques (*Macaca fuscata yakui*). *Am. J. Primatol.* 70:986–989.

Hayakawa, S., and Takenaka, O. (1999). Urine as another potential source for template DNA in polymerase chain reaction (PCR). *Am. J. Primatol.* 48:299–304.

Hayaki, H. (1985). Copulation of adolescent male chimpanzees, with special reference to the influence of adult males, in the Mahale National Park, Tanzania. *Folia Primatol.* 44(3–4):148–160.

Hayaki, H. (1988). Association partners of young chimpanzees in the Mahale Mountains National Park, Tanzania. *Primates* 29(2):147–161.

Hayashi, S., Hayasaka, K., Takenaka, O., and Horai, S. (1995). Molecular phylogeny of gibbons inferred from mitochondrial DNA sequences: preliminary report. *Mol. Evol.* 41:359–365.

He, Y. M., Pei, Y. J., Zou, R. J., and Ji, W. Z. (2001). Changes of urinary steroid conjugates and gonadotropin excretion in the menstrual cycle and pregnancy in the Yunnan snub-nosed monkey (*Rhinopithecus bieti*). *Am. J. Primatol.* 55:223–232.

Heape, W. (1900). The "sexual season" of mammals and the relation of the "pro-oestrum" to menstruation. *Q. J. Microsc. Sci.* 44:1–70.

Hedrick, P. W. (2000). *Genetics of Populations.* Jones and Bartlett, Sudbury.

Hein, J. (1994). TreeAlign. In: Griffin, A. M., and Griffin, H. G. (eds.), *Methods in Molecular Biology. Computer Analysis of Sequence Data,* Part II, vol. 25. Humana Press, Totowa, NJ. pp. 349–364.

Heisler, L., Andersson, M. B., Arnold, S. J., Boake, C. R., Borgia, G., Hausfater, G., Kirkpatrick, M., Lande, R., Maynard Smith, J., O'Donald, P., Thornhill, R., and Weissing, F. J. (1987). The evolution of mating preferences and sexually selected traits. In: Bradbury, J. W., and Andersson, M. B. (eds.), *Sexual Selection: Testing the Alternatives.* John Wiley and Sons, Chichester. pp. 97–118.

Heistermann, M., and Hodges, J. K. (1995). Endocrine monitoring of the ovarian cycle and pregnancy in the saddle-back tamarin (*Saguinus fuscicollis*) by measurement of steroid conjugates in urine. *Am. J. Primatol.* 35:117–127.

Heistermann, M., Proeve, E., Wolter, H. J., and Mika, G. (1987). Urinary oestrogen and progesterone excretion before and during pregnancy in a pied bare-faced tamarin (*Saguinus bicolor bicolor*). *J. Reprod. Fertil.* 80:635–640.

Heistermann, M., Tari, S., and Hodges, J. K. (1993). Measurement of faecal steroids for monitoring ovarian function in New World primates, Callitrichidae. *J. Reprod. Fertil.* 99:243–251.

Heistermann, M., Finke, M., and Hodges, J. K. (1995a). Assessment of female reproductive status in captive-housed Hanuman langurs (*Presbytis entellus*) by measurement of urinary and fecal steroid excretion patterns. *Am. J. Primatol.* 37:275–284.

Heistermann, M., Moestl, E., and Hodges, J. K. (1995b). Non-invasive endocrine monitoring of female reproductive status: methods and applications to captive breeding and conservation of exotic species. In: Ganslosser, U., and Hodges, J. K. (eds.), *Research and Captive Propagation.* Filander Verlag, Fuerth. pp. 36–48.

Heistermann, M., Möhle, U., Vervaecke, H., van Elsacker, L., and Hodges, J. K. (1996). Application of urinary and fecal steroid measurements for monitoring ovarian function and pregnancy in the bonobo (*Pan paniscus*) and evaluation of perineal swelling patterns in relation to endocrine events. *Biol. Reprod.* 55:844–853.

Heistermann, M., Uhrigshardt, J., Husung, A., Kaumanns, W., and Hodges, J. K. (2001a). Measurement of faecal steroid metabolites in the lion-tailed macaque (*Macaca silenus*): a non-invasive tool for assessing female ovarian function. *Primate Rep.* 59:27–42.

Heistermann, M., Ziegler, T., van Schaik, C. P., Launhardt, K., Winkler, P., and Hodges, J. K. (2001b). Loss of oestrus, concealed ovulation and paternity confusion in free-ranging Hanuman langurs. *Proc. R. Soc. Lond. B* 268:2445–2451.

Heistermann, M., Brauch, K., Möhle, U., Pfefferle, D., Dittami, and Hodges, K. (2008). Female ovarian cycle phase affects the timing of male sexual activity in free-ranging Barbary macaques (*Macaca sylvanus*) of Gibraltar. *Am. J. Primatol.* 70:44–53.

Helfman, G. S., and Schultz, E. T. (1984). Social traditions in a coral reef fish. *Anim. Behav.* 32:379–384.

Heltne, P. G. (1977). *Census of* Aotus *in Northern Colombia.* Panamerican Health Organization, Washington DC. pp. 1–11.

Heltne, P. G., Turner, D. C., and Scott, N. J., Jr. (1976). Comparison of census data on *Alouatta palliata* from Costa Rica and Panama. In: Thorington, R. W., Jr., and Heltne, P. G. (eds.), *Neotropical Primates: Field Studies and Conservation.* National Academy of Sciences, Washington DC. pp. 10–19.

Hemelrijk, C. K. (1999). An individual-orientated model of the emergence of despotic and egalitarian societies. *Proc. R. Soc. Lond. B* 266(1417):361–369.

Hemelrijk, C. K. (2002a). Self-organization and natural selection in the evolution of complex despotic societies. *Biol. Bull.* 202:283–288.

Hemelrijk, C. K. (2002b). Self-organizing properties of primate social behavior: a hypothesis for intersexual rank overlap in chimpanzees and bonobos. *Evol. Anthropol.* 11:91–94.

Hemingway, C. A. (1995). Feeding and reproductive strategies of the Milne-Edwards sifaka, *Propithecus diadema edwardsi* [PhD diss.]. Duke University, Durham, NC.

Hemingway, C. A. (1996). Morphology and phenology of seeds and whole fruit eaten by Milre-Edwards sifaka. *Propithecus diadema edwardsi* in Ranomafana National Park, Madagascar. *Int. J. Primatol.* 17:637–659.

Hemingway, C. A., and Bynum, N. (2005) The influence of seasonality on primate diet and ranging. In: Brockman, D. K., and Van Schaik, C. P. (eds.), *Seasonality in Primates: Studies of Living and Extinct Human and Non-human Primates.* Cambridge University Press, Cambridge. pp. 57–104.

Henderson, P. (2003). *Practical Methods in Ecology*. Blackwell Publishing, Malden.

Hendrickx, A. G., and Kraemer, D. C. (1969). Observations on the menstrual cycle, optimal mating time and pre-implantation embryos of the baboon, *Papio anubis* and *Papio cynocephalus*. *J. Reprod. Fertil.* 6:119–128.

Hennemann, W. W. (1983). Relationship among body mass, metabolic rate, and the intrinsic rate of natural increase in mammals. *Oecologia* 56:419–424.

Hennemann, W. W. (1984). Basal metabolic rate and the intrinsic rate of natural increase: an empirical and theoretical reexamination. *Oecologia* 64:421–423.

Hennessy, M. D., Mendoza, S. P., Mason, W. A., and Moberg, G. P. (1995). Endocrine sensitivity to novelty in squirrel monkeys and titi monkeys: species differences in characteristic modes of responding to the environment. *Physiol. Behav.* 57:331–338.

Henschel, P., Abernethy, K., and White, L. J. T. (2005). Leopard food habits in the Lope National Park, Gabon, Central Africa. *Af. J. Ecol.* 43(1):21–28.

Henry, J. P. (1982). The relation of social to biological processes in disease. *Soc. Sci. Med.* 16:369–380.

Henzi, S. P. (1996). Copulation calls and paternity in chacma baboons. *Anim. Behav.* 51:233–234.

Henzi, S. P., and Barrett, L. (1999). The value of grooming to female primates. *Primates* 40:47–59.

Henzi, S. P., and Barrett, L. (2003). Evolutionary ecology, sexual conflict, and behavioral differentiation among baboon populations. *Evol. Anthropol.* 12:217–230.

Henzi, S. P., and Barrett, L. (2005). The historical socioecology of savanna baboons (*Papio hamadryas*). *J. Zool. Lond.* 265:215–226.

Henzi, S. P., and Lawes, M. J. (1987). Breeding influxes and the behaviour of adult male samango monkeys (*Cercopithecus mitis albogularis*). *Folia Primatol.* 48:125–136.

Henzi, S. P., and Lawes, M. J. (1988). Strategic responses of male samango monkeys (*Cercopithecus mitis*) to a decline in the number of receptive females. *Int. J. Primatol.* 9:479–495.

Henzi, S. P., and Lucas, J. W. (1980). Observations on the intertroop movement of adult vervet monkeys (*Cercopithecus aethiops*). *Folia Primatol.* 33:220–235.

Henzi, S. P., and Lycett, J. E. (1995). Population structure, demography and dynamics of mountain baboons: an interim report. *Am. J. Primatol.* 35:155–163.

Henzi, S. P., Dyson, M. L., and Deenik, A. (1990). The relationship between altitude and group size in mountain baboons (*Papio cynocephalus ursinus*). *Int. J. Primatol.* 11:319–325.

Henzi, S. P., Byrne, R. W., and Whiten, A. (1992). Patterns of movement by baboons in the Drakensberg Mountains: primary responses to the environment. *Int. J. Primatol.* 13:601–629.

Henzi, S. P., Lycett, J. E., and Piper, S. E. (1997a). Fission and troop size in a mountain baboon population. *Anim. Behav.* 53:525–535.

Henzi, S. P., Lycett, J. E., and Weingrill, T. (1997b). Cohort size and the allocation of social effort by female mountain baboons. *Anim. Behav.* 54:1235–1243.

Henzi, S. P., Lycett, J. E., Weingrill, T., Byrne, R., and Whiten, A. (1997c). The effect of troop size on travel and foraging in mountain baboons. *S. Afr. J. Sci.* 93:333–335.

Henzi, S. P., Lycett, J. E., and Weingrill, T. (1998). Mate guarding and risk assessment by male mountain baboons during inter-troop encounters. *Anim. Behav.* 55:1421–1428.

Henzi, S. P., Weingrill, T., and Barrett, L. (1999). Male behaviour and the evolutionary ecology of chacma baboons. *S. Afr. J. Sci.* 95:240–242.

Henzi, S. P., Barrett, L., Weingrill, T., Dixon, P., and Hill, R. A. (2000). Ruths amid the alien corn: males and the translocation of female baboons. *S. Afr. J. Sci.* 96: 61–62.

Herbert, J. (1970). Hormones and reproductive behavior in rhesus and talapoin monkeys. *J. Reprod. Fertil.* 11:119–140.

Herbinger, I. (2004). Inter-group interactions among three communities of wild West African chimpanzees (*Pan troglodytes verus*) [PhD diss.]. University of Leipzig, Leipzig.

Herbinger, I., Boesch, C., and Rothe, H. (2001). Territory characteristics among three neighboring chimpanzee communities. *Int. J. Primatol.* 22:143–167.

Hernández-Camacho, J., and Cooper, R. W. (1976). The non-human primates of Colombia. *Neotropical Primates: Field Studies and Conservation*. In: Thorington, R. W., and Heltne, J. P. G. (eds.). National Academy of Sciences, Washington DC. 35–69.

Hernández-Camacho, J., and Defler, T. R. (1985). Some aspects of the conservation of non-human primates in Colombia. *Primate Conserv.* 6:42–50.

Hernández-López, L., Mayagoita, L., Esquivel-Lacroix, C., Rojas-Maya, S., and Mondragon-Ceballos, R. (1998). The menstrual cycle of the spider monkey (*Ateles geoffroyi*). *Am. J. Primatol.* 44:183–195.

Hernández-López, L., Parra, G. C., Cerda-Molina, A. L., Pérez-Bolaños, S. C., Sánchez, V. D., and Mondragón-Ceballos, R. (2002). Sperm quality differences in captive black-handed spider monkeys (*Ateles geoffroyi*). *Am. J. Primatol.* 57:35–41.

Herrera, C. M. (2002). Seed dispersal by vertebrates. In: Herrera, C. M., and Pellmyr, O. (eds.), *Plant–Animal Interactions: An Evolutionary Approach*. Blackwell Science, Oxford. pp. 185–210.

Herrera, C. M., Jordano, P., Lopez-Soria, L., and Amat, J. A. (1994). Recruitment of a mast-fruiting, bird-dispersed tree: bridging frugivore activity and seedling establishment. *Ecol. Monogr.* 64:315–344.

Herrick, J. R., Agoramoorthy, G., Rudran, R., and Harder, J. D. (2000). Urinary progesterone in free-ranging red howler monkeys (*Alouatta seniculus*): preliminary observations of the estrous cycle and gestation. *Am. J. Primatol.* 51:257–263.

Herrmann, E., Call, J., Hernandez-Lloreda, M. V., Hare, B., and Tomasello, M. (2007). Humans have evolved specialized skills of social cognition: the cultural intelligence hypothesis. *Science* 317:1360–1366.

Herrnstein, R. J., Loveland, D. H., and Cable, C. (1976). Natural concepts in pigeons. *J. Exp. Psychol. Anim. Behav. Proc.* 2:285–302.

Hershkovitz, P. (1949). Mammals of northern Colombia. Preliminary report 4. Monkeys (Primates) with taxonomic revisions of some forms. *Proc. U.S. Natl. Mus.* 98:323–427.

Hershkovitz, P. (1963). A systematic and zoogeographic account of the monkeys of the genus *Callicebus* (Cebidae) of the Amazonas and Orinoco River basins. *Mammalia* 27:1–80.

Hershkovitz, P. (1969). The recent mammals of the neotropical region: a zoogeographic and ecological review. *Q. Rev. Biol.* 44:1–70.

Hershkovitz, P. (1977). *Living New World Monkeys (Platyrrhini), with an Introduction to Primates*. vol. 1. Chicago University Press, Chicago.

Hershkovitz, P. (1983). Two new species of night monkeys, genus *Aotus* (Cebidae: Platyrrhini): a preliminary report on *Aotus* taxonomy. *Am. J. Primatol.* 4:209–243.

Hershkovitz, P. (1985). A preliminary taxonomic review of the South American bearded saki monkeys genus *Chiropotes* (Cebidae, Platyrrhini), with the description of a new subspecies. *Fieldiana* 27(1363):1–46.

Hershkovitz, P. (1987a). Uacaries, New World monkeys of the genus *Cacajao* (Cebidae, Platyrrhini): a preliminary taxonomic review with the description of a new subspecies. *Am. J. Primatol.* 12:1–53.

Hershkovitz, P. (1987b). The taxonomy of South American sakis, genus *Pithecia* (Cebidae, Platyrrhini): a preliminary report and critical review with the description of a new species and a new subspecies. *Am. J. Primatol.* 12:387–468.

Hershkovitz, P. (1990). Titis, New World monkeys of the genus *Callicebus* (Cebidae, Platyrrhini): a preliminary taxonomic review. *Fieldiana* 55:1–109.

Hershkovitz, P. (1993). Male external genitalia of non-prehensile tailed South American monkeys. Part 1: Subfamily Pitheciinae, family Cebidae. *Field. Zool. (New Series)* 73:1–17.

Herskovits, M. J., and Herskovits, S. (1958). *Dahomean Narrative: A Cross-Cultural Analysis*. Northwestern University Press, Evanston, IL.

Heschl, A., and Burkart, J. (2006). A new mark test for mirror self-recognition in non-human primates. *Primates* 47:187–198.

Hewett-Emmett, D., Cook, C. N., and Barnicot, N. A. (1976). Old World monkey hemoglobins: deciphering phylogeny from complex patterns of molecular evolution. In: Goodman, M., Tashian, R. E., and Tashian, J. H. (eds.), *Molecular Anthropology: Genes and Proteins in the Evolutionary Ascent of the Primates*. Plenum Press, New York. pp. 257–275.

Heyes, C. M. (1993). Imitation, culture and cognition. *Anim. Behav.* 46:999–1010.

Heyes, C. M. (1998). Theory of mind in nonhuman primates. *Behav. Brain Sci.* 21:101–148.

Heymann, E. W. (1987). Field observation of predation on a moustached tamarin (*Saguinus mystax*) by an anaconda. *Int. J. Primatol.* 8:193–195.

Heymann, E. W. (1990). Reactions of wild tamarins *Saguinus mystax* and *Saguinus fuscicollis* to avian predators. *Int. J. Primatol.* 11:327–337.

Heymann, E. W. (1995). Sleeping habits of tamarins, *Saguinus mystax* and *Saguinus fuscicollis* (Mammalia; Primates; Callitrichidae), in north-eastern Peru. *J. Zool.* 237:211–226.

Heymann, E. W. (1996). Social behavior of wild moustached tamarins, *Saguinus mystax,* at the Estación Biológica Quebrada Blacno, Peruvian Amazonia. *Am. J. Primatol.* 38:101–113.

Heymann, E. W. (2000). The number of adult males in callitrichine groups and its implications for callitrichine social evolution. In: Kappeler, P. M. (ed.), *Primate Males: Causes and Consequences of Variation in Group Composition.* Cambridge University Press, Cambridge. pp. 64–71.

Heymann, E. W. (2001). Interspecific variation of scent-marking behaviour in wild tamarins, *Saguinus mystax* and *Saguinus fuscicollis. Folia Primatol.* 72:253–267.

Heymann, E. W., and Buchanan-Smith, H. (2000). The behavioral ecology of mixed-species troops of callitrichine primates. *Biol. Rev.* 75:169–190.

Heymann, E. W., and Smith, A. C. (1999). When to feed on gums: temporal patterns of gummivory in wild tamarins, *Saguinus mystax* and *Saguinus fuscicollis* (Callitrichinae). *Zoo Biol.* 18:459–471.

Heymann, E. W., and Soini, P. (1999). Offspring number in pygmy marmosets, *Cebuella pygmaea,* in relation to group size and number of adult males. *Behav. Ecol. Sociobiol.* 46:400–404.

Heymann, E. W., Encarnación, C. F., and Canaquin, Y. J. E. (2002). Primates of the Río Curaray, northern Peruvian Amazon. *Int. J. Primatol.* 23:191–201.

Hick, U. (1968). Erstmalig gelungene Zucht eines Bartsakis (Vater: Rotrückensaki) *Chiropotes chiropotes* (Humboldt 1811). Mutter: Weissnasensaki, *Chiropotes albinasus* (Geoffroy et Deville 1848) im Kölner Zoo. *Freunde Kölner Zoo* 11(2):35–41.

Hicks, T. C., Fouts, R. S., and Fouts, D. H. (2005). Chimpanzee (*Pan troglodytes troglodytes*) tool use in the Ngotto Forest, Central African Republic. *Am. J. Primatol.* 65:221–237.

Higgins, D. G., Thompson, J. D., and Gibson, T. J. (1996). Using CLUSTAL for multiple sequence alignments. *Methods Enzymol.* 266:383–402.

Higham, J. P., Heistermann, M., Ross, C., Semple, S., and MacLarnon, A. (2008). The timing of ovulation with respect to sexual swelling detumescence in wild olive baboons. *Primates* 49:295–299.

Higham, J. P., Semple, S., MacLarnon, A., Heistermann, M., and Ross, C. (2009a). Female reproductive signaling, and male mating behavior, in the olive baboon. *Horm. Behav.* 55:60–67.

Higham, J. P., Warren, Y., Adanu, J., Umaru, B. N., MacLarnon, A. M., Sommer, V., and Ross, C. (2009b). Living on the edge: life-history of olive baboons at Gashaka-Gumti National Park, Nigeria. *Am. J. Primatol.* 71:293–304.

Higuchi, R., von Berholdingen, C. H., Sensabaugh, G. F., and Erlich, H. A. (1988). DNA typing from single hairs. *Nature* 332:543–546.

Hikami, K., Hasegawa, Y., and Matsuzawa, T. (1990). Social transmission of food preferences in Japanese monkeys (*Macaca fuscata*) after mere exposure or aversion training. *J. Comp. Psychol.* 104:233–237.

Hildebrand, M. (1967). Symmetrical gaits in primates. *Am. J. Phys. Anthropol.* 26:119–130.

Hilgartner, R., Zinner, D., and Kappeler, P. M. (2008). Life history traits and parental care in *Lepilemur ruficaudatus. Am. J. Primatol.* 70:2–11.

Hill, C. M. (1994). The role of female diana monkeys, *Cercopithecus diana,* in territorial defence. *Anim. Behav.* 47:425–431.

Hill, C. M. (2000). Conflict of interest between people and baboons: crop raiding in Uganda. *Int. J. Primatol.* 21:299–315.

Hill, C. M. (2005). People, crops, and primates: a conflict of interests. In: Paterson, J. D., and Wallis, J. (eds.), *Commensalism and Conflict: The Human–Primate Interface.* American Society of Primatologists, Norman, OK. pp. 40–59.

Hill, D. A. (1990). Social relationships between adult male and female rhesus macaques: II. Non-sexual affiliative behavior. *Primates* 31:33–50.

Hill, D. A. (1997). Seasonal variation in the feeding behavior and diet of Japanese macaques (*Macaca fuscata yakui*) in lowland forest of Yakushima. *Am. J. Primatol.* 43:305–322.

Hill, D. A. (1999). Effects of provisioning on the social behavior of Japanese and rhesus macaques: implications for socioecology. *Primates* 40:187–198.

Hill, D. A. (2004). Intraspecific variation: implications for interspecific comparisons. In: Thierry, B., Singh, M., and Kaumanns, W. (eds.), *How Societies Arise: The Macaque Model,* Cambridge University Press, Cambridge, pp. 262–266.

Hill, D. A., and Okayasu, N. (1995). Absence of "youngest ascendancy" in the dominance relations of sisters in wild Japanese macaques (*Macaca fuscata yakui*). *Behaviour* 132:367–379.

Hill, D. A., and Okayasu, N. (1996). Determinants of dominance among female macaques: nepotism, demography and danger. In: Fa, J. E., and Lindburg, D. G. (eds.), *Evolution and Ecology of Macaque Societies.* Cambridge University Press, Cambridge. pp. 459–472.

Hill, K., and Hurtado, A. M. (1996). *Ache Life History. The Ecology and Demography of a Foraging People.* Aldine de Gruyter, New York.

Hill, K., and Padwe, J. (2000). Sustainability of Ache hunting in the Mbaracayu Reserve, Paraguay. In: Robinson, J. G., and Bennett, E. L. (eds.), *Hunting for Sustainability in Tropical Forests.* Columbia University Press, New York. pp. 79–105.

Hill, K., Boesch, C., Goodall, J., Pusey, A., Williams, J., and Wrangham, R. (2001). Mortality rates among wild chimpanzees. *J. Hum. Evol.* 40:437–450.

Hill, K., McMillan, G., and Farina, R. (2003). Hunting-related changes in game encounter rates from 1994 to 2001 in the Mbaracayu Reserve, Paraguay. *Conserv. Biol.* 17:1312–1323.

Hill, R. A. (2006). Thermal constraints on activity scheduling and habitat choice in baboons. *Am. J. Phys. Anthropol.* 129:242–249.

Hill, R. A., and Cowlishaw, G. (2002). Foraging female baboons exhibit similar patterns of antipredator vigilance across two populations. In: Miller, L. E. (ed.), *Eat or Be Eaten: Predator Sensitive Foraging Among Primates.* Cambridge University Press, Cambridge. pp. 187–204.

Hill, R. A., and Dunbar, R. I. M. (1998). An evaluation of the roles of predation rate and predation risk as selective pressures on primate grouping behaviour. *Behaviour* 135:411–430.

Hill, R. A., and Dunbar, R. I. M. (2002). Climatic determinants of diet and foraging behaviour in baboons. *Evol. Ecol.* 16:579–593.

Hill, R. A., and Lee, P. C. (1998). Predation risk as an influence on group size in cercopithecoid primates: implicatoins for social structure. *J. Zool. Lond.* 245:447–456.

Hill, R. A., and Weingrill, T. (2007). *Predation risk and habitat use in Chacma baboons (Papio hamadryas ursinus).* In: Gursky, S. L., and Nekaris, K. A. I. (eds.), *Primate Anti-Predator Strategies.* Springer, New York. pp. 339–354.

Hill, R. A., Lycett, J. E., and Dunbar, R. I. M. (2000). Ecological and social determinants of birth intervals in baboons. *Behav. Ecol.* 11:560–564.

Hill, R. A., Barrett, L., Gaynor, D., Weingrill, T., Dixon, P., Payne, H., and Henzi, S. P. (2003). Day length, latitude and behavioural (in)flexibility in baboons (*Papio cynocephalus ursinus*). *Behav. Ecol. Sociobiol.* 53:278–286.

Hill, W. C. (1955). *Primates Haplorhini, Tarsoidea.* Interscience, New York.

Hill, W. C. O. (1957). *Primates, Comparative Anatomy and Taxonomy. Hapalidae,* vol. 3. University of Edinburgh Press, Edinburgh.

Hill, W. C. O. (1960). *Primates: Comparative Anatomy and Taxonomy, Cebidae,* vol. 4, pt. A. Edinburgh University Press, Edinburgh.

Hill, W. C. O. (1962). *Primates: Comparative Taxonomy and Anatomy, Cebidae,* vol. 5, pt. B. Edinburgh University Press, Edinburgh.

Hill, W. C. O. (1966). *Primates: Comparative Anatomy and Taxonomy. Catarrhini, Cercopithecoidea, Cercopithecinae,* vol. 6. Edinburgh University Press, Edinburgh.

Hill, W. C. O. (1969). The vascular supply of the face in long-snouted primates (*Papio* and *Mandrillus*). Preliminary notes. In: Hofer, H. O. (ed.), *Proceedings of the Second International Congress of Primatology. Recent Advances in Primatology,* vol. 2. S. Karger, Basel. pp. 155–159.

Hill, W. C. O. (1972). *Evolutionary Biology of the Primates.* Academic Press, London.

Hill, W. C. O., Appleby, H. M., and Auber, L. (1959). The specialised area of skin glands in *Aotus* Humboldt (Simiae Platyrrhini). *Trans. R. Soc. Edinburgh* 63:535–551.

Hille Ris Lambers, J., Clark, J. S., and Beckage, B. (2002). Density-dependent mortality and the latitudinal gradient in species diversity. *Nature* 417:732–735.

Hillis, D. M., Moritz, C., and Mable, B. K. (eds.) (1996). *Molecular Systematics,* 2nd ed. Sinauer Associates, Sunderland, MA.

Hilton-Taylor, C. (2002). *2002 IUCN Red List of Threatened Species.* World Conservation Union, Species Survival Commission, Gland, Switzerland.

Hilton-Taylor, C., and Rylands, A. B. (2002). The 2002 IUCN Red List of threatened species. *Neotrop. Primates* 10:149–153.

Hinde, R. A. (1974a). *Biological Bases of Human Social Behaviour.* New York: McGraw-Hill.

Hinde, R. A. (1974b). Mother/infant relations in rhesus monkeys. In: White, N. F. (ed.), *Ethology and Psychiatry.* University of Toronto Press, Toronto. pp. 29–46.

Hinde, R. A. (1976). Interactions, relationships and social structure. *Man* 11:1–17.

Hinde, R. A. (1977). Mother–infant separation and the nature of inter-individual relationships: experiments with rhesus monkeys. *Proc. R. Soc. Lond. B* 196(1122):29–50.

Hinde, K., Power, M., and Oftedal, O. (2009). Rhesus macaque milk: magnitude, sources, and consequences of individual variation over lactation. *Am. J. Phys. Anthropol.* 138:148–157.

Hinde, R. A., Rowell, T. E., and Spencer-Booth, Y. (1964). Behaviour of socially living rhesus monkeys in their first six months. *Proc. Zool. Soc. Lond.* 143:609–649.

Hines, M. (1942). The development and regression of reflexes, postures and progression in the young macaque. *Contrib. Embryol.* 30:153–209.

Hiraiwa, M. (1981). Maternal and alloparental care in a troop of free-ranging Japanese monkeys. *Primates* 22:309–329.

Hiraiwa-Hasegawa, M. (1987). Infanticide in primates and a possible case of male-biased infanticide in chimpanzees. In: Ito, Y., Brown, J. L., and Kikkawa, J. (eds.), *Animal Societies: Theories and Facts.* Japan Scientific Societies Press, Tokyo. pp. 125–139.

Hiraiwa-Hasegawa, M. (1988). Adaptive significance of infanticide in primates. *Trends. Ecol.* 3:102–105.

Hiraiwa-Hasegawa, M. (1989). Sex differences in the behavioral development of chimpanzees at Mahale. In: Heltne, P. G., and Marquardt, L. A. (eds.), *Understanding Chimpanzees.* Harvard University Press, Cambridge, MA. pp. 104–115.

Hiraiwa-Hasegawa, M., Hasegawa, T., and Nishida, T. (1984). Demographic study of a large-sized unit-group of chimpanzees in the Mahale Mountains, Tanzania: a preliminary report. *Primates* 25:401–413.

Hirata, S., and Celli, M. L. (2003). Role of mothers in the acquisition of tool-use behaviours by captive infant chimpanzees. *Anim. Cogn.* 6:235–244.

Hirata, S., and Matsuzawa, T. (2001). Tactics to obtain a hidden food item in chimpanzee pairs (*Pan troglodytes*). *Anim. Cogn.* 4:285–295.

Hirata, S., Watanabe, K., and Kawai, M. (2001). "Sweet-potato washing" revisited. In: Matsuzawa, T. (ed.) *Primate Origins of Human Cognition and Behavior.* Springer-Verlag, Tokyo. pp. 487–508.

Hladik, A. (1980). *The Dry Forest of the West Coast of Madagascar: Climate, Phenology, and Food Available for Prosimians.* Academic Press, New York.

Hladik, C. (1973). Alimentation et activite d'un groupe de chimpanzes reintroduits en foret Gabonaise. *Terre Vie* 27:343–413.

Hladik, C. M. (1977a). A comparative study of the feeding strategies of two sympatric species of leaf monkeys: *Presbytis senex* and *Presbytis entellus*. In: Clutton-Brock, T. H. (ed.), *Primate Ecology: Studies of Feeding and Ranging Behaviour in Lemurs, Monkeys and Apes.* Academic Press, New York. pp. 323–353.

Hladik, C. M. (1977b). Chimpanzees of Gabon and chimpanzees of Gombe: some comparative data on the diet. In: Clutton-Brock, T. H. (ed.), *Primate Ecology: Studies of Feeding and Ranging Behavior in Lemurs, Monkeys and Apes.* Academic Press, London. pp. 481–503.

Hladik, C. M. (1978). Adaptive strategies of primates in relation to leaf-eating. In: Montgomery, G. G. (ed.), *The Ecology of Arboreal Folivores.* Smithsonian Institution Press, Washington DC. pp. 373–395.

Hladik, C. M. (1979). Diet and ecology of prosimians. In: Doyle, G. A., and Martin, R. D. (eds.), *The Study of Prosimian Behavior.* Academic Press, New York. pp. 307–357.

Hladik, C. M., and Charles-Dominique, P. (1974). The behavior and ecology of the sportive lemur (*Lepilemur mustelinus*) in relation to its dietary peculiarities. In: Martin, R. D., Doyle, G. A., and Walker, A. C. (eds.), *Prosimian Biology.* Duckworth Press, London. pp. 23–37.

Hladik, C. M., Hladik, A., Bousset, J., Valdegbouze, P., Viroben, G., and Delort'Laval, J. (1971). Le régime alimentaire del primates de l'oele de Barro Colorado (Panamá): résultats des analyses quantitatives. *Folia Primatol.* 16:85–122.

Hladik, C. M., Charles-Dominique, P., and Petter, J. J. (1980). Feeding strategies of five nocturnal prosimians in the dry forest of the west coast of Madagascar. In: Charles-Dominique, P., Cooper, H. M., Hladik, A., Hladik, C. M., Pages, E., Pariente, G. F., Petter-Rousseaux, A., Petter, J. J., and Schilling, A. (eds.), *Nocturnal Malagasy Primates: Ecology, Physiology and Behavior.* Academic Press, New York. pp. 41–73.

Hockings, K. J., Humle, T., Anderson, J. R., et al. (2007). Chimpanzees Share Forbidden Fruit. *PLoS ONE* 2(9):e886. doi:10.1371/journal.pone.0000886.

Hodgen, G. D., Goodman, A. L., O'Conner, A., and Johnson, D. K. (1977). Menopause in rhesus monkeys: model for study of disorders in the human climacteric. *Am. J. Obstet. Gynecol.* 127:581–584.

Hodges, J. K. (1985). The endocrine control of reproduction. *Sym. Zool. Soc. Lond.* 54:149–168.

Hodges, J. K. (1996). Determining and manipulating female reproductive parameters. In: Kleiman, D. G., Allen, M. E., Thompson, K. V., and Lumpkin, S. (eds.), *Wild Mammals in Captivity.* University of Chicago Press, Chicago. pp. 418–428.

Hodges, J. K., and Eastman, S. A. K. (1984). Monitoring ovarian function in marmosets and tamarins by the measurement of urinary oestrogen metabolites. *Am. J. Primatol.* 6:187–197.

Hodges, J. K., and Green, D. G. (1989). A simplified enzyme immunoassay for urinary pregnanediol-3-glucuronide: applications to reproductive assessment of exotic species. *J. Zool.* 219:89–99.

Hodges, J. K., and Heistermann, M. (2003). Field endocrinology: monitoring hormonal changes in free-ranging primates. In: Setchell, J. M., and Curtis, D. M. (eds.), *Field and Laboratory Methods in Primatology: A Practical Guide.* Cambridge University Press, Cambridge. pp. 282–294.

Hodges, J. K., Czekala, N. M., and Lasley, B. L. (1979). Estrogen and luteinizing hormone secretion in diverse primate species from simplified urinary analysis. *J. Med. Primatol.* 3:349–364.

Hodges, J. K., Gulick, B. A., Czekala, N. M., and Lasley, B. L. (1981). Comparison of urinary oestrogen excretion in South American primates. *J. Reprod. Fertil.* 61:83–90.

Hodges, J. K., Tarara, R., Hearn, J. P., and Else, J. G. (1986). The detection of ovulation and early pregnancy in the baboon by direct measurement of conjugated steroids in urine. *Am. J. Primatol.* 10:329–338.

Hodgson, J. A., Sterner, K. N., Matthews, L. J., Burrell, A. S., Jani, R. A., Raaum, R. L., Stewart, C.-B., and Disotell, T. R. (2009). Successive radiations, not stasis, in the South American primate fauna. *Proc. Natl. Acad. Sci. USA* 106:5534–5539.

Hoelzel, A. R. (ed.) (1998). *Molecular Genetic Analysis of Populations—A Practical Approach,* 2nd ed. ILR Press, Oxford.

Hoelzer, G. A., and Melnick, D. J. (1996). Evolutionary relationships of the macaques. In: Fa, J. E., and Lindburg, D. G. (eds.), *Evolution and Ecology of Macaque Societies.* Cambridge University Press, Cambridge. pp. 3–19.

Hoelzer, G. A., Morales, J. C., and Melnick, D. J. (2004). Dispersal and population genetics of primate species. In: Chapais, B., and Berman, C. M. (eds.), *Kinship and Behavior in Primates.* Oxford University Press, Oxford. pp. 109–131.

Hofer, H., and East, M. L. (2000). Conflict management in female-dominated spotted hyenas. In: Aureli, F., and de Waal, F. B. M. (eds.), *Natural Conflict Resolution.* University of California Press, Berkeley. pp. 232–234.

Hoffman, K. A. (1998). Transition from juvenile to adult stages of development in titi monkeys (*Callicebus moloch*) [PhD diss.]. University of California, Davis.

Hoffman, K. A., Mendoza, S. P., Hennessy, M. B., and Mason, W. A. (1995a). Responses of infant titi monkeys, *Callicebus moloch,* to removal of one or both parents: evidence for paternal attachment. *Dev. Psychobiol.* 28:399–407.

Hoffman, K. A., Valeggia, C. R., Mensoza, S. P., and Mason, W. A. (1995b). Responses to mature titi monkey daughters following a one month separation from their family groups. *Am. J. Primatol.* 36:128.

Hoffman, M., Grubb, P., Groves, C.P., Hutterer, R., van der Straeten, E., Simmons, N., and Bergmans, W. (2009). A synthesis of African and western Indian Ocean Island mammal taxa (Class: Mammalia) described between 1988 and 2008: an update to Allen (1939) and Ansell (1989). *Zootaxa* 2205: 1–36.

Hohmann, G. (1989). Group fission in Nilgiri langurs (*Presbytis johnii*). *Int. J. Primatol.* 10:441–454.

Hohmann, G. (2001). Association and social interactions between strangers and residents in bonobos (*Pan paniscus*). *Primates* 42:91–99.

Hohmann, G., and Fruth, B. (1993). Field observations on meat sharing among bonobos (*Pan paniscus*). *Folia Primatol.* 60:225–229.

Hohmann, G., and Fruth, B. (1996). Food sharing and status in unprovisioned bonobos. In: Wiessner, P., and Schiefenhoevel, W. (eds.), *Food and the Status Quest.* Berghahn, Oxford. pp. 47–67.

Hohmann, G., and Fruth, B. (2000). Use and function of genital contacts among female bonobos. *Anim. Behav.* 60:107–120.

Hohmann, G., and Fruth, B. (2002). Dynamics in social organization of bonobos (*Pan paniscus*). In: Boesch, C., Hohmann, G., and Marchant, L. (eds.), *Behavioural Diversity in Chimpanzees and Bonobos.* Cambridge University Press, Cambridge. pp. 138–155.

Hohmann, G., and Fruth, B. (2003a). Culture in bonobos? Between-species and within-species variation in behavior. *Curr. Anthropol.* 44:563–571.

Hohmann, G., and Fruth, B. (2003b). Intra- and inter-sexual aggression by bonobos in the context of mating *Behaviour* 140:1389–1413.

Hohmann, G., and Fruth, B. (2008). New records on prey capture and meat eating by bonobos at Lui Kotale, Salonga National Park, Democratic Republic of Congo. *Folia Primatol.* 79(2):103–110.

Hohmann, G., Gerloff, U., Tautz, D., and Fruth, B. (1999). Social bonds and genetic ties: kinship association and affiliation in a community of bonobos (*Pan paniscus*). *Behaviour* 136:1219– 1235.

Holbrook, G. D., Chambers, C. M., and Evans, S. (2004). Seasonal breeding in captive owl monkeys. *Folia Primatol.* 75:276.

Holekamp, K. E., and Smale, L. (1991). Dominance acquisition during mammalian social development: the "inheritance" of maternal rank. *Am. Zool.* 31(2):306–317.

Holekamp, K. E., and Smale, L. (1998). Behavioral development in the spotted hyena. *BioScience* 48(12):997–1005.

Holekamp, K. E., Cooper, S. M., Katona, C. I., Berry, N. A. Frank, L. G., and Smale, L. (1997a). Patterns of association among female spotted hyenas (*Crocuta crocuta*). *J. Mammal.* 78(1):55–64.

Holekamp, K. E., Smale, L., Berg, R., and Cooper, S. M. (1997b). Hunting rates and hunting success in the spotted hyena (*Crocuta crocuta*). *J. Zool. Lond.* 242:1–15.

Holenweg, A.-K., Noë, R., and Schabel, M. (1996). Waser's gas model applied to associations between red colobus and diana monkeys in the Taï National Park, Ivory Coast. *Folia Primatol.* 67:125–136.

Holmes, J. C., and Zohar, S. (1990). Pathology and host behavior. In: Barnard, C. J., and Behnke, J. M. (eds.), *Parasitism and Host Behavior.* Taylor and Frances, London. pp. 34–63.

Homburg, I. (1997). Ökologie and sozialverhalten einer gruppe von weißgesicht-sakis (*Pithecia pithecia pithecia* Linnaeus 1766) im Estado Bolívar, Venezuela [PhD diss.]. Universität Bielefeld, Bielefeld.

Homewood, K. M. (1976). Ecology and behaviour of the Tana mangabey [PhD diss.], University of London, London.

Homewood, K. M. (1978). Feeding strategy of the Tana mangabey (*Cercocebus galeritus galeritus*) (Mammalia: Primates). *J. Zool. Lond.* 186:375–391.

Homewood, K. M., and Rodgers, A. (1985). Tanzania's newest primate. *Anim. King.* 88:14–17.

Höner, O. P., Leumann, L., and Noë, R. (1997). Dyadic associations of red colobus and diana monkey groups in the Taï National Park, Ivory Coast. *Primates* 38:281–291.

Honess, P. E. (1996). Speciation among galagos (Primates, Galagidae) in Tanzanian forests [PhD diss.]. Oxford Brookes University, Oxford.

Honess, P. E., and Bearder, S. K. (1996). Descriptions of the dwarf galago species of Tanzania. *Afr. Primates* 2:75–79.

Honess, P. E., and MacDonald, D. W. (2003). Marking and radio-tracking primates. In: Setchell, J. M., and Curtis, D. J. (eds.), *Field and Laboratory Methods in Primatology: A Practical Guide.* Cambridge University Press, Cambridge. pp. 158–173.

Honess, P. E., Perkin A. W., and Bearder, S.K. (2007). Primates in peril: the world's top 25 most endangered primates 2006–2008—Rondo dwarf galago, Rungwe mountain galago. *Primate Conserv.* 22:12–13.

Honey, J. A. (1969). *South-African Folk-Tales.* Negro Universities Press, New York.

Hood, L. C., and Jolly, A. (1995). Troop fission in female *Lemur catta* at Berenty Reserve, Madagascar. *Int. J. Primatol.* 16:997–1016.

Hoogland, J. L. (1985). Infanticide in prairie-dogs: lactating females kill offspring of close kin. *Science* 230:1037–1040.

Hooijer, D. A. (1960). The orang-utan in Niah cave pre-history. *Sarawak Mus. J.* 9:408–421.

Hooton, E.A. (1942). *Man's Poor Relations.* Doubleday, New York.

Hooton, E.A. (1954). The importance of primate studies in anthropology. *Hum. Biol.* 26:179–88.

Hoppe-Dominik, B. (1984). Étude du spectre des proies de la panthère, *Panthera pardus,* dans le Parc National de Taï en Côte d'Ivoire. *Mammalia* 48:477–487.

Horai, S., Hayasaka, K., Kondo, R., Tsugane, K., and Takahata, N. (1995). Recent African origin of modern humans revealed by complete sequences of hominoid mitochondrial DNAs. *Proc. Natl. Acad. Sci. USA* 92:532–536.

Horn, A. D. (1987a). The socioecology of the black mangabey (*Cercocebus aterrimus*) near Lake Tumba, Zaire. *Am. J. Primatol.* 12:165–180.

Horn, A. D. (1987b). Taxonomic assessment of the allopatric gray-cheeked mangabey (*Cercocebus albigena*) and black mangabey (*C. aterrimus*): comparative socioecological data and the species concept. *Am. J. Primatol.* 12:181–187.

Horner, V., and Whiten, A. (2005). Causal knowledge and imitation/emulation switching in chimpanzees (*Pan troglodytes*) and humans (*Homo sapiens*). *Anim. Cogn.* 8:164–181.

Horner, V., Whiten, A., Flynn, E., and de Waal, F. B. M. (2006). Faithful replication of foraging techniques along cultural transmission chains by chimpanzees and children. *Proc. Natl. Acad. Sci. USA* 103:13878–13883.

Horovitz, I., and MacPhee, R. D. E. (1999). The Quaternary Cuban platyrrhine *Paralouatta varonai* and the origin of Antillean monkeys. *J. Hum. Evol.* 36:33–68.

Horowitz, A. (2003). Do humans ape? Or do apes human? Imitation and intention in humans (*Homo sapiens*) and other animals. *J. Comp. Psychol.* 117:325–336.

Horr, D. A. (1977). Orang-utan maturation: growing up in a female world. In: Chevalier-Skolnikoff, S., and Poirier, F. E. (eds.), *Primate Bio-Social Development: Biological, Social, and Ecological Determinants.* Garland Publishing, New York. pp. 289–321.

Horrocks, J. A., and Hunte, W. (1986). Sentinel behaviour in vervet monkeys: who sees whom first? *Anim. Behav.* 34:1566–1567.

Horsfield, J. (1821). *Zoological Research in Java and in the Neighboring Islands.* Kingsbury, Parbury and Allen, London.

Horvath, J. E., Weisrock, D. W., Embry, S. L., Fiorentino, I., Balhoff, J. P., Kappeler, P., Wray, G. A., Willard, H. F., and Yoder, A. D. (2008). Development and application of a phylogenomic toolkit: resolving the evolutionary history of Madagascar's lemurs. *Genome Res.* 18:489–499.

Horwich, R., and Gebhard, K. (1983). Roaring rhythms in black howler monkeys (*Alouatta pigra*) of Belize. *Primates* 24:290–296.

Horwich, R. H., and Manski, D. (1975). Maternal care and infant transfer in two species of *Colobus* monkeys. *Primates* 16:49–73.

Horwich, R. H., Brockett, R. C., and Jones, C. B. (2000). Alternative male reproductive behaviors in the Belizean black howler monkey (*Alouatta pigra*). *Neotrop. Primates* 8:95–98.

Hoshino, J. (1985). Feeding ecology of mandrills (*Mandrillus sphinx*) in Campo Animal Reserve, Cameroon. *Primates* 26:248–273.

Hoshino, J., Mori, A., Kudo, H., and Kawai, M. (1984). Preliminary report on the grouping of mandrills (*Mandrillus sphinx*) in Cameroon. *Primates* 25:295–307.

Houck, M. L., Ryder, O. A., Kumamoto, A. T., and Benirschke, K. (1995). Cytogenetics of the Rhinocerotidae. *Verhandlungsbericht des 37. Internationalen Symposiums über die Erkrankungen der Zootiere.* Dresden. pp. 25–32.

Houle, D. (1991). Genetic covariance of fitness correlates: what genetic correlations are made of and why it matters. *Evolution* 45:630–645.

Hourani, P. E., Vinyard, C. J., and Lemelin, P. (2003). Forelimb forces during gouging and other behaviors on vertical substrates in common marmosets. *Am. J. Phys. Anthropol.* 36(suppl.):118.

Howe, H. F. (1989). Scatter- and clump-dispersal and seedling demography: hypothesis and implications. *Oecologia* 79:417–426.

Howe, H. F., and Smallwood, J. (1982). Ecology of seed dispersal. *Annu. Rev. Ecol. Syst.* 13:201–228.

Howe, H. F., and Westley, L. C. (1988). *Ecological Relationships of Plants and Animals.* Oxford University Press, New York.

Howe, H. F., Schupp, E. W., and Westley, L. C. (1985). Early consequences of seed dispersal for a neotropical tree (*Virola surinamensis*). *Ecology* 66:781–791.

Howell, F. C., and Bourliere, F. (eds.) (1963). *African Ecology and Human Evolution.* Aldine, Chicago.

Howell, N. (1979). *The Demography of the Dobe !Kung.* Academic Press, New York.

Howells, W. W. (ed.) (1962). *Ideas on Human Evolution: Selected Essays 1949–1961.* Harvard University, Cambridge, MA.

Hrdy, S. B. (1974). Male–male competition and infanticide among the langurs (*Presbytis entellus*) of Abu, Rajasthan. *Folia Primatol.* 22:19–58.

Hrdy, S. B. (1976). The care and exploitation of non-human primate infants by conspecifics other than the mother. In: Rosenblatt, J. S., Hinde, R. A., Shaw, E., and Beer, C. (eds.), *Advances in the Study of Behaviour,* vol. 6. Academic Press, New York. pp. 101–158.

Hrdy, S. B. (1977a). Infanticide as a primate reproductive strategy. *Am. Sci.* 65:40–49.

Hrdy, S. B. (1977b). *The Langurs of Abu: Female and Male Strategies of Reproduction.* Harvard University Press, Cambridge, MA.

Hrdy, S. B. (1979). Infanticide among animals: a review, classification, and examination of the implications for the reproductive strategies of females. *Ethol. Sociobiol.* 1:13–40.

Hrdy, S. B. (1981a). Nepotists and altruists: the behavior of old females among macaques and langur monkeys. In: Amoss, P. T., and Harrell, S. (eds.), *Other Ways of Growing Old: Anthropological Perspectives.* Stanford University Press, Stanford, CA. pp. 59–96.

Hrdy, S. B. (1981b). *The Woman that Never Evolved.* Harvard University Press, Cambridge, MA.

Hrdy, S. B. (1986). Empathy, polyandry, and the myth of the coy female. In: Bleier, R. (ed.), *Feminist Approaches to Science.* Pergamon Press, New York. pp. 119–146.

Hrdy, S. B., and Hausfater, G. (1984a). Comparative and evolutionary perspectives on infanticide: introduction and overview. In: Hausfater, G. and Hrdy, S. B. (eds.), *Infanticide: Comparative and Evolutionary Perspectives.* Aldine de Gruyter, New York. pp. xiii–xxxv.

Hausfater, G. and Hrdy, S. B. (eds.) (1984b). *Infanticide: Comparative and Evolutionary Perpectives.* Aldine de Gruyter, New York.

Hrdy, S. B., and Hrdy, D. B. (1976). Hierarchical relations among female Hanuman langurs (Primates: *Colobinae, Presbytis entellus*). *Science* 193:913–915.

Hrdy, S. B., and Whitten, P. L. (1987). Patterning of sexual activity. In: Smuts, B. B., Cheney, D. L., Seyfarth, R. M., Wrangham, R. W., and Struhsaker, T. T. (eds.), *Primate Societies.* University of Chicago Press, Chicago. pp. 370–384.

Hrdy, S. B., Janson, C., and van Schaik, C. (1995). Infanticide: let's not throw out the baby with the bath water. *Evol. Anthropol.* 3:151–154.

Hsu, M. J., and Lin, J. F. (2001). Troop size and structure in free-ranging Formosan macaques (*Macaca cyclopis*) at Mt Longevity, Taiwan. *Zool. Stud.* 40:49–60.

Hsu, M. J., Lin, J. F., Chen, L. M., and Agoramoorthy, G. (2002). Copulation calls in male Formosan macaques: honest signals of male quality? *Folia Primatol.* 73:220–223.

Hu, X. S., and Ennos, R. A. (1999). Impacts of seed and pollen flow on population genetic structure for plant genomes with three contrasting modes of inheritance. *Genetics* 152:441–450.

Hu, Y., Xu, H., and Yang, D. (1989). The studies on ecology of *Hylobates leucogenys* [in Chinese]. *Zool. Res.* 10(suppl.):61–66.

Hu, Y., Xu, H., and Yang, D. (1990). Feeding ecology of the white-cheek gibbon (*Hylobates concolor leucogenys*) [in Chinese]. *Acta Ecol. Sin.* 10:155–159.

Huang, C. M., Wei, F. W., Li, M., Li, Y. B., and Sun, R. Y. (2003). Sleeping cave selection, activity pattern and time budget of white-headed langurs. *Int. J. Primatol.* 24:813–824.

Hubbell, S. P. (1980). Seed predation and the coexistence of tree species in tropical forests. *Oikos* 35:214–229.

Hubbell, S. P. (2001). *The Unified Neutral Theory of Biodiversity and Biogeography.* Princeton University Press, Princeton, NJ.

Huck, M., Löttker, P., Böhle, U., and Heymann, E. W. (2005). Paternity and kinship patterns in polyandrous moustached tamarins (*Saguinus mystax*). *Am. J. Phys. Anthropol.* 127:449–464.

Huck, M., Roos, C., and Heymann, E. W. (2007). Spatio-genetic population structure in mustached tarnarins, *Saguinus mystax. Am. J. Phys. Anthropol.* 132:576–583.

Hudson, R. R., Kreitman, R., and Aguade, M. (1987). A test of neutral molecular evolution based on nucleotide data. *Genetics* 116:153–159.

Huffman, M. A. (1984). Stone-play of *Macaca fuscata* in Arashiyama B troop: transmission of a non-adaptive behaviour. *J. Hum. Evol.* 13:725–735.

Huffman, M. A. (1987). Consort intrusion and female mate choice in Japanese macaques (*Macaca fuscata*). *Ethology* 75:221–234.

Huffman, M. A. (1991). Mate selection and partner preferences in female Japanese macaques. In: Fedigan, L. M. and Asquith, P. J. (eds.), *The Monkeys of Arashiyama: Thirty-five Years of Research in Japan and the West.* State University of New York Press, Albany. pp. 101–122.

Huffman, M. A. (1992). Influences of female partner preference on potential reproductive outcome in Japanese macaques. *Folia Primatol.* 59:77–88.

Huffman, M. A. (1994). The C.H.I.M.P.P. Group: a mutli-disciplinary investigation into the use of medicinal plants by chimpanzees. *PanAfr. News* 1:3–5.

Huffman, M. A. (1997). Current evidence for self-medication in primates: a multidisciplinary perspective. *Ybk. Phys. Anthropol.* 40:171–200.

Huffman, M. A. (2001). Self-medicative behavior in the African great apes—an evolutionary perspective into the origins of human traditional medicine. *Bioscience* 51:651–661.

Huffman, M. A., and Caton, J. M. (2001). Self-induced increase of gut motility and the control of parasitic infections in wild chimpanzees. *Int. J. Primatol.* 22:329–346.

Huffman, M. A., and Hirata, S. (2003). Biological and ecological foundations of primate behavioral traditions. In: Fragaszy, D. M., and Perry, S. (eds.), *The Biology of Traditions.* Cambridge University Press, Cambridge. pp. 267–296.

Huffman, M. A., and Hirata, S. (2004). An experimental study of leaf swallowing in captive chimpanzees—insights into the origin of a self-medicative behavior and the role of social learning. *Primates* 45:113–118.

Huffman, M. A., and Seifu, M. (1989). Observations on the illness and consumption of a possibly medicinal plant *Vernonia amygdalina* (Del.), by a wild chimpanzee in the Mahale Mountains National Park, Tanzania. *Primates* 30:51–63.

Huffman, M. A., Gotoh, S., Izutsu, D., Koshimizu, K., and Kalunde, M. S. (1993). Further observations on the use of *Vernonia amygdalina* by a wild chimpanzee, its possible effect on parasite load, and its phytochemistry. *Afr. Stud. Monogr.* 14:227–240.

Huffman, M. A., Koshimizu, K., and Ohigashi, H. (1996a). Ethnobotany and zoopharmacognosy of *Vernonia amygdalina,* a medicinal plant used by humans and chimpanzees. In: Caligari, P. D. S., and Hind, D. J. N. (eds.), *Compositae: Biology and Utilization,* vol. 2. Royal Botanical Gardens Press, Kew. pp. 351–360.

Huffman, M. A., Page, J. E., Sukhdeo, M. V. K., Gotoh, S., Kalunde, M. S., Chandrasiri, T., and Towers, G. H. N. (1996b). Leaf-swallowing by chimpanzees, a behavioral adaptation for the control of strongyle nematode infections. *Int. J. Primatol.* 72:475–503.

Huffman, M. A., Gotoh, S., Turner, L. A., Hamai, M., and Yoshida, K. (1997). Seasonal trends in intestinal nematode infection and medicinal plant use among chimpanzees in the Mahale Mountains, Tanzania. *Primates* 38:111–125.

Huffman, M. A., Ohigashi, H., Kawanaka, M., Page, J. E., Kirby, G. C., Gasquet, M., Murakami, A., and Koshimizu, K. (1998). African great ape self-medication: a new paradigm for treating parasite disease with natural medicines? In: Ebizuka, Y. (ed.), *Towards Natural Medicine Research in the 21st Century.* Elsevier Science, Amsterdam. pp. 113–123.

Huffman, R. (1970). *Nuer Customs & Folk-Lore.* Frank Cass, London.

Hughes, K. (2003). The global positioning system, geographical information systems and remote sensing. In: Curtis, D., and Setchell, J. (eds.), *Field and Laboratory Methods in Primatology.* Cambridge University Press, Cambridge. pp. 57–73.

Huijbergts, B., De Wachter, P., Ndong Obiang, L. S., and Ella Akou, M. (2003). Ebola and the decline of gorilla (*Gorilla gorilla*) and chimpanzee (*Pan troglodytes*) populations in Minkebe Forest, north-eastern Gabon. *Oryx* 37:437–443.

Humle, T. (2006). Ant-dipping in chimpanzees: an example of how microecological variables, tool-use and culture reflect the cognitive abilities of chimpanzees. In: Matsuzawa, T., Tomonaga, M., and Tanaka, M. (eds.), *Cognitive Development in Chimpanzees.* Springer-Verlag, Tokyo. pp. 452–475.

Humle, T. (2010 in press). Primate material culture. In: Hicks, D., and Beaudry, M. C. (eds.), *Oxford Handbook of Material Culture Studies.* Oxford University Press, Oxford. In press.

Humle, T. (2010 in press). How are army ants shedding new light on culture in chimpanzees? In: Lonsdorf, E. V., Ross, S., and Matsuzawa, T. (eds.), The Mind of the Chimpanzee: Ecological and Experimental Perspectives. Chicago University Press, Chicago. In press.

Humle, T., and Matsuzawa, T. (2002). Ant dipping among the chimpanzees of Bossou, Guinea, and comparisons with other sites. *Am. J. Primatol.* 58:133–148.

Humle, T., and Snowdon, C. T. (2008). Socially biased learning in the acquisition of a complex foraging task in juvenile cottontop tamarins (*Saguinus oedipus*). *Anim. Behav.* 75:267–77.

Humle, T., Snowdon, C. T., and Matsuzawa, T. (2009). Social influences on the acquisition of ant-dipping among the wild chimpanzees (*Pan troglodytes verus*) of Bossou, Guinea, West Africa. *Anim Cogn.* Suppl 1:S37–48.

Hunkeler, C., Bourlière, F., and Bertrand, M. (1972). Le comportement de la Mone de Lowe (*Cercopithecus campbelli lowei*). *Folia Primatol.* 17:218–236.

Hunt, G. R. (1996). Manufacture and use of hook-tools by New Caledonian crows. *Nature* 379:249–251.

Hunt, G. R., and Gray, R. D. (2003). Diversification and cumulative evolution in New Caledonian crow tool manufacture. *Proc. R. Soc. Lond. B.* 270:867–874.

Hunt, G. R., and Gray, R. D. (2004a). The crafting of tool hooks by wild New Caledonian crows. *Proc. R. Soc. Lond. B. (Suppl).* 270:S88–S90.

Hunt, G. R., and Gray, R. D. (2004b). Direct observations of pandanus-tool manufacture and use by a New Caledonian crow (*Corvus moneduloides*). *Anim. Cogn.* 7:114–120.

Hunt, K. D. (1991). Positional behavior in the Hominoidea. *Int. J. Primatol.* 12L:95–118.

Hunt, K. D. (1992). Positional behavior of *Pan troglodytes* in the Mahale Mountains and Gombe Stream National Parks, Tanzania. *Am. J. Phys. Anthropol.* 87:83–105.

Hunt, K. D. (1998). Ecological morphology of *Australopithecus afarensis*: traveling terrestrially, eating arboreally. In: Strasser, E., Fleagle, J. G., Rosenberger, A., and McHenry, H. (eds.), *Primate Locomotion: Recent Advances.* Plenum Press, New York. pp. 387–418.

Hunt, K. D. (2004). Chimpanzee socioecology in a hot, dry habitat: implications for early hominin evolution. Conference on African Great Apes: Evolution, Diversity, and Conservation. Kyoto University, Kyoto, March 4, 2004.

Hunt, K. D., and McGrew, W. (2002). Chimpanzees in the dry habitats of Assirik, Senegal and Semliki Wildlife Reserve, Uganda. In: Boesch, C., Hohmann, G., and Marchant, L. (eds.), *Behavioural Diversity in Chimpanzees and Bonobos.* Cambridge University Press, Cambridge. pp. 35–51.

Hunt, K. D., Cant, J. G. H., Gebo, D. L., Rose, M. D., Walker, S. E., and Youlatos, D. (1996). Standardized description of primate locomotor and postural modes. *Primates* 37:363–387.

Hunt, K. D., Cleminson, A. J. M., Latham, J., Weiss, R. I., and Grimmond, S. (1999). A partly habituated community of dry-habitat chimpanzees in the Semliki Valley Wildlife Reserve, Uganda. *Am. J. Phys. Anthropol. Suppl.* 28:157.

Hunte, W., and Horrocks, J. A. (1987). Kin and non-kin interventions in the aggressive disputes of vervet monkeys. *Behav. Ecol. Sociobiol.* 20:257–263.

Hunter, A. J. (1981). *Chemical Communication, Aggression and Sexual Behavior in the Owl Monkey* (Aotus trivirgatus griseimembra) [PhD diss.]. University of London, London.

Hunter, A. J., and Dixson, A. F. (1983). Anosmia and aggression in male owl monkeys (*Aotus trivirgatus*). *Physiol. Behav.* 30:875–879.

Hunter, A. J., Fleming, D., and Dixson, A. F. (1984). The structure of the vomeronasal organ and nasopalatine ducts in *Aotus trivirgatus* and some other primate species. *J. Anat.* 138:217–225.

Hunter, C. (2001). Ecological determinants of gelada ranging patterns (*Theropithecus gelada*) [PhD diss.]. University of Liverpool, Liverpool.

Hunter, J., Martin, R. D., Dixson, A. F., and Rudder, B. C. C. (1979). Gestation and interbirth intervals in the owl monkey (*Aotus trivirgatus griseimembra*). *Folia Primatol.* 31:165–175.

Hunter, T. B., and Skinner, J. D. (1998). Vigilance behaviour in African ungulates: the role of predation pressure. *Behaviour* 135:195–211.

Huntington, C., and Fernandez-Duque, E. (2001). Natal dispersal in the monogamous owl monkey (*Aotus azarai*) of Formosa, Argentina. *Am. J. Phys. Anthropol.* 32(suppl.):83–84.

Hurtt, G. C., and Pacala, S. W. (1995). The consequences of recruitment limitation: reconciling chance, history, and competitive differences between plants. *J. Theor. Biol.* 176:1–12.

Husson, S. J., Wich, S. A., Marshall, A. J., Dennis, R. D., Ancrenaz, M., Brassey, R., Gumal, M., Hearn, A. J., Meijaard, E., Simorangkir, T., and Singleton, I. (2009). Orangutan distribution, density, abundance and impacts of disturbance. In: Wich, S. A., Utami, S. S., Setia, T. M., and van Schaik, C. (eds.), *Orangutans: Geographic Variation in Behavioral Ecology and Conservation.* Oxford University Press, Oxford. pp. 77–96.

Hutchinson, G. E. (1957). Concluding remarks. *Cold Spring Harb. Symp. Quant. Biol.* 22:415–427.

Huxley, J. (1938). Darwin's theory of sexual selection and the data subsumed by it, in light of recent research. *Am. Nat.* 72:416–433.

Huxley, T. H. (1863). *Man's Place in Nature.* Williams & Norwood, London. Reprinted 1959, University of Michigan, Ann Arbor.

Hyatt, C. W., and Hopkins, W. H. (1994). Self-awareness in bonobos and chimpanzees: a comparative perspective. In: Parker, S. T., Mitchell, R. W., and Boccia, M. L. (eds.), *Self-Awareness in Animals and Humans: Developmental Perspectives.* Cambridge University Press, Cambridge. pp. 248–253.

Hyatt, L. A., Rosenberg, M. S., Howard, T. G., Bole, G., Fang, W., Anastasia, J., Brown, K., Grella, R., Hunman, K., Kurdziel, J. P., and Gurevitch, J. (2003). The distance dependence prediction of the Janzen-Connell hypothesis: a meta-analysis. *Oikos* 103:590–602.

Idani, G. (1990). Relations between unit-groups of bonobos at Wamba, Zaire: encounters and temporary fusions. *Afr. Study Monogr.* 11:153–186.

Idani, G. (1991). Social relationships between immigrant and resident bonobo (*Pan paniscus*) females at Wamba. *Folia Primatol.* 57:83–95.

Ihobe, H. (1989). How social relationships influence a monkey's choice of feeding sites in the troop of Japanese macaques (*Macaca fuscata fuscata*) on Koshima Islet. *Primates* 30:17–25.

Ihobe, H. (1990). Interspecific interactions between wild pygmy chimpanzees (*Pan paniscus*) and red colobus (*Colobus badius*). *Primates* 31:109–112.

Ihobe, H. (1992). Male–male relationships among wild bonobos (*Pan paniscus*) at Wamba, Republic of Zaire. *Primates* 33:163–179.

Imanishi, K. (1941). *A Japanese View of Nature: The World of Living Things.* Asquith, P., Kawakatsu, H., Yagi, S., and Takasaki, H. (trans.). Reprinted 2002, Routledge Curzon, New York.

Imanishi, K. (1953). Social behavior in Japanese monkeys, *Macaca fuscata. Psychologia* 1:47–54.

InfoNatura (2004). *Birds, Mammals, and Amphibians of Latin America,* version 3.2. Arlington, VA, NatureServe, http://www.natureserve.org/infonatura (accessed July 1, 2004).

Inge, W. R. (1927). *Confessio fidei: Outspoken Essays,* 2nd ser. Longmans, Green, New York.

Ingmanson, E. J. (1996). Tool-using behavior in wild *Pan paniscus:* social and ecological considerations. In: Russon, A. E., Bard, K. A., and Parker, S. T. (eds.), *Reaching into Thought: The Minds of the Great Apes.* Cambridge University Press, Cambridge. pp. 190–210.

Ingmanson, E. J. (1998). Reply to Stanford (1998) "The social behavior of chimpanzees and bonobos—empirical evidence and shifting assumptions." *Curr. Anthropol.* 39:409–410.

Inoue, E., and Takenaka, O. (2008). The effect of male tenure and female mate choice on paternity in free-ranging Japanese macaques. *Am. J. Primatol.* 70:62–68.

Inoue, M., Takenata, A., Tanaka, S., Kominami, R., and Tanaka, O. (1990). Paternity discrimination in a Japanese macaque group by DNA fingerprinting. *Primates* 31:563–570.

Inoue, S., and Matsuzawa, T. (2007). Working memory of numerals in chimpanzees. *Curr. Biol.* 17:R1004–R1005.

Inoue-Nakamura, N. (1997). Mirror self-recognition in nonhuman primates: a phylogenetic approach. *Jpn. Psychol. Res.* 39:266–275.

Inoue-Nakamura, N., and Matsuzawa, T. (1997). Development of stone tool use by wild chimpanzees (*Pan troglodytes*). *J. Comp. Psychol.* 111:159–173.

International Human Genome Sequencing Consortium (2001). Initial sequencing and analysis of the human genome. *Nature* 409:860–921.

International Human Genome Sequencing Consortium (2004). Finishing the euchromatic sequence of the human genome. *Nature* 431:931–945.

International SNP Map Working Group (2001). A map of human genome sequence variation containing 1.42 million single nucleotide polymorphisms. *Nature* 409:928–933.

Irwin, M. T. (2006a). Ecologically enigmatic lemurs: the sifakas of the eastern forests (*Propithecus candidus, P. diadema, P. edwardsi, P. perrieri,* and *P. tattersalli*). In: Gould, L. and Sauther, M. L. (eds.), *Lemurs: Ecology and Adaptation.* Springer, New York. pp. 305–326.

Irwin, M. T. (2006b). Leaner diet, less energy: diademed sifakas (*Propithecus diadema*) in forest fragments show reduced ranging and altered activity patterns [abstract]. *Am. J. Phys. Anthropol.* 42(suppl.):108.

Irwin, M. T. (2007). Living in forest fragments reduces group cohesion in diademed sifakas (*Propithecus diadema*) in eastern Madagascar, by reducing food patch size. *Am. J. Primatol.* 69:434–447.

Isabirye-Basuta, G. (1988). Food competition among individuals in a free-ranging chimpanzee community in Kibale Forest, Uganda. *Behaviour* 105:135–147.

Isabirye-Basuta, G. (1989). Feeding ecology of chimpanzees in the Kibale Forest, Uganda. In: Heltne, P., and Marquardt, L. (eds.), *Understanding Chimpanzees.* Harvard University Press, Cambridge, MA. pp. 116–127.

Isbell, L. A. (1990). Sudden short-term increase in mortality of vervet monkeys (*Cercopithecus aethiops*) due to leopard predation in Amboseli National Park, Kenya. *Am. J. Primatol.* 21:41–52.

Isbell, L. A. (1991). Contest and scramble competition: patterns of female aggression and ranging behaviour among primates. *Behav. Ecol.* 2:143–155.

Isbell, L. A. (1994). Predation on primates: ecological patterns and evolutionary consequences. *Evol. Anthropol.* 3:61–71.

Isbell, L. A. (1995). Seasonal and social correlates of changes in hair, skin, and scrotal condition in vervet monkeys (*Cercop-ithecus aethiops*) of Amboseli National Park, Kenya. *Am. J. Primatol.* 36:61–70.

Isbell, L. A. (1998). Diet for a small primate: insectivory and gummivory in the (large) patas monkey (*Erythrocebus patas pyrrhonotus*). *Am. J. Primatol.* 45:381–389.

Isbell, L. A. (2004). Is there no place like home? Ecological bases of female dispersal and philopatry and their consequences for the formation of kin groups. In: Chapais, B., and Berman, C. (eds.), *Kinship and Behavior in Primates.* Oxford University Press, New York. pp. 71–108.

Isbell, L. A. (2006). Snakes as agents of evolutionary change in primate brains. *J. Hum. Evol.* 51(1):1–35.

Isbell, L. A. (2009a). *The Fruit, the Tree, and the Serpent: Why We See So Well.* Harvard University Press, Cambridge, MA.

Isbell, L. A. (In press). Patas monkeys (*Erythrocebus patas*). In: Kingdon, J., Happold, D., and Butynski, T. (eds.), *The Mammals of Africa.* University of California Press, Berkeley, CA.

Isbell, L. A., and Enstam, K. L. (2002). Predator (in)sensitive foraging in sympatric female vervets (*Cercopithecus aethiops*) and patas monkeys (*Erythrocebus patas*): a test of ecological models of group dispersion. In: Miller, L. E. (ed.), *Eat or Be Eaten: Predator Sensitive Foraging Among Primates.* Cambridge University Press, Cambridge. pp. 154–168.

Isbell, L. A., and Pruetz, J. D. (1998). Differences between vervets (*Cercopithecus aethiops*) and patas monkeys (*Erythrocebus patas*) in agonistic interactions between adult females. *Int. J. Primatol.* 19:837–855.

Isbell, L. A., and Young, T. P. (1993a). Human presence reduces predation in a free-ranging vervet monkey population in Kenya. *Anim. Behav.* 45:1233–1235.

Isbell, L. A., and Young, T. P. (1993b). Social and ecological influences on activity budgets of vervet monkeys and their implications for group living. *Behav. Ecol. Sociobiol.* 32:377–385.

Isbell, L. A., and Young, T. P. (2002). Ecological models of female social relationships in primates: similarities, disparities, and some directions for future clarity. *Behaviour* 139:177–202.

Isbell, L. A., and Young, T. P. (2007). Interspecific and temporal variation of ant species within *Acacia drepanolobium* ant domatia, a staple food of patas monkeys (*Erythrocebus patas*) in Laikipia, Kenya. *Am. J. Primatol.* 69:1387–1398.

Isbell, L. A., Cheney, D. L., and Seyfarth, R. M. (1991). Group fusions and minimum group sizes in vervet monkeys (*Cercopithecus aethiops*). *Am. J. Primatol.* 25:57–65.

Isbell, L. A., Pruetz, J. D., and Young, T. P. (1998). Movements of vervets (*Cercopithecus aethiops*) and patas monkeys (*Eyrthrocebus patas*) as estimators of food resource size, density, and distribution. *Behav. Ecol. Sociobiol.* 42:123–133.

Isbell, L. A., Pruetz, J. D., Nzuma, B. M., and Young, T. P. (1999). Comparing measures of travel distances in primates: methodological considerations and socioecological implications. *Am. J. Primatol.* 48:87–98.

Isbell, L. A., Cheney, D. L., and Seyfarth, R. M. (2002). Why vervet monkeys (*Cercopithecus aethiops*) live in multimale groups. In: Glenn, M. E., and Cords, M. (eds.), *The Guenons: Diversity and Adaptation in African Monkeys.* Kluwer Academic/Plenum, New York. pp. 173–187.

Isbell, L. A., Young, T. P., Jaffe, K. E., Carlson, A. A., and Chancellor, R. L. (2009). Demography and life histories of sympatric patas monkeys (*Erythrocebus patas*) and vervets (*Cercopithecus aethiops*) in Laikipia, Kenya. *Int. J. Primatol.* 30:103–124.

Ishida, H., Hirasaki, E., and Matano, S. (1992). Locomotion of the slow loris between discontinuous substrates. In: Matano, S., Tuttle, R. H., and Ishida, H. (eds.), *Topics in Primatology. Evolutionary Biology, Reproductive Endocrinology and Virology,* vol. 3. University of Tokyo Press, Tokyo. pp. 139–152.

Ishida, H., Kunimatsu, Y., Takano, T., Nakano, Y., and Nakatsukasa, M. (2004). *Nacholapithecus* skeleton from the middle Miocene of Kenya. *J. Hum. Evol.* 46:69–103.

Islam, M. A., and Feeroz, M. M. (1992). Ecology of hoolock gibbon of Bangladesh. *Primates* 33:451–464.

Itani, J. (1958). On the acquisition and propagation of new food habits in the troop of Japanese monkeys at Takasakiyama. *Primates* 1:84–98.

Itani, J. (1959). Paternal care in the wild Japanese monkey, *Macaca fuscata fuscata. Primates* 2:61–93.

Itani, J., and Nishimura, A. (1973). The study of infrahuman culture in Japan. In: Menzel, E. W. J. (ed.), *Precultural Primate Behaviour.* Karger, Basel. pp. 26–50.

Itoigawa, N., Tanaka, T., Ukai, N., Fujii, H., Kurokawa, T., Koyama, T., Ando, A., Watanabe, Y., and Imakawa, S. (1992). Demography and reproductive parameters of a free-ranging group of Japanese macaques (*Macaca fuscata*) at Katsuyama. *Primates* 33:49–68.

IUCN (2004). 2004 IUCN Red List of Threatened Species, www. redlist.org.

IUCN (2009). IUCN Red List of Threatened Species, version 2009.1, www. iucnredlist.org (accessed June 23, 2009).

IUCN Species Survival Commission (2000). *Red List of Threatened Species.* IUCN–World Conservation Union, Gland, Switzerland.

IUCN Species Survial Commission (2008). IUCN Red List, http://www. redlist.org/.

IUCN Standards and Petitions Working Group. (2008). Guidelines for Using the IUCN Red List Categories and Criteria. Version 7.0. Prepared by the Standards and Petitions Working Group of the IUCN SSC Biodiversity Assessments Sub-Committee in August 2008. Downloadable from http:// intranet.iucn.org/webfiles/doc/SSC/RedList/RedListGuidelines.pdf.

Iwamoto, M., Watanabe, T., and Hamada, Y. (1987). Eruption of permanent teeth in Japanese monkeys (*Macaca fuscata*). *Primate Res.* 3:18–28.

Iwamoto, T., (1979). Feeding ecology. In: Kawai, M (ed.), *Ecological and Sociological Studies of Gelada Baboons.* Karger, Basel, pp. 280–330.

Iwamoto, T. (1993). The ecology of *Theropithecus gelada.* In: Jablonski, N. G. (ed.), *Theropithecus: The Rise and Fall of a Primate Genus.* Cambridge University Press, Cambridge. pp. 441–452.

Iwamoto, T., and Dunbar, R. I. M. (1983). Thermoregulation, habitat quality and the behavioural ecology of gelada baboons. *J. Anim. Ecol.* 52:357–366.

Iwano, T., and Iwakawa, C. (1988). Feeding behaviour of the aye-aye (*Daubentonia madagascariensis*) on nuts of Ramy (*Canarium madagascariensis*). *Folia Primatol.* 50:136–142.

Izard, M. K., and Rasmussen, D. T. (1985). Reproduction in the slender loris (*Loris tardigradus malabaricus*). *Am. J. Primatol.* 8:153–165.

Izard, M., Wright, P., and Simons, E. (1985). Gestation length in *Tarsius bancanus. Am. J. Primatol.* 9:327–331.

Izawa, K. (1976). Group sizes and compositions of monkeys in the upper Amazon Basin. *Primates* 17:367–399.

Izawa, K. (1980). Social behavior of the wild black-capped capuchin (*Cebus apella*). *Primates* 20:57–76.

Izawa, K. (1993). Soil-eating by *Alouatta* and *Ateles. Int. J. Primatol.* 14:229–242.

Izawa, K. (1994). Group division of wild black-capped capuchins. *Field Studies of the New World Monkeys La Macarena, Colombia* 9:5–14.

Izawa, K., and Nishimura, A. (1988). Primate fauna at the study site, La Macarena, Colombia. *Field Studies of New World Monkeys La Macarena, Colombia* 1:5–11.

Izor, R. J. (1985). Sloths and other mammalian prey of the harpy eagle. In: Montgomery, G. G. (ed.), *The Evolution and Ecology of Armadillos, Sloths, and Vermilinguas.* Smithsonian Institution Press, Washington DC. pp. 343–346.

Jablonski, N. G. (1998a). The evolution of the doucs and snub-nosed monkeys and the question of the phyletic unity of the odd-nosed colobines. In: Jablonski, N. G. (ed.), *The Natural History of the Doucs and Snub-Nosed Monkeys.* World Scientific Press, Singapore. pp. 13–52.

Jablonski, N. G. (1998b). The response of catarrhine primates to Pleistocene environmental fluctuations in east Asia. *Primates* 39:29–37.

Jablonski, N. G. (2002). Fossil Old Worl d monkeys: the late Neogene radiation. In: Hartwig, W. C. (ed.), *The Primate Fossil Record.* Cambridge University Press, Cambridge. pp. 255–299.

Jablonski, N., Whitfort, M., Roberts-Smith, J. N., and Qinqi, X. (2000). The influence of life history and diet on the distribution of catarrhine primates during the Pleistocene in eastern Asia. *J. Hum. Evol.* 39:131–157.

Jack, K. M. (2001). Effect of male emigration on the vigilance behavior of coresident males in white-faced capuchins (*Cebus capucinus*). *Int. J. Primatol.* 22:715–732.

Jack, K. (2003a). Affiliative relationships among male white-faced capuchins (*Cebus capucinus*): evidence of male bonding in a female bonded species. *Folia Primatol.* 74:1–16.

Jack, K. (2003b). Males on the move: evolutionary significance of secondary dispersal in male nonhuman primates. *Primate Rep.* 67:61–84.

Jack, K.M. (2005). Conservation status of *Cebus albifrons aequatorialis* in forest fragments of coastal Manabi, Ecuador. *Am. J. Primatol.* 66 (suppl 1): 132

Jack, K., and Fedigan, L. M. (2003). Male dominance and reproductive success in white-faced capuchins (*Cebus capucinus*). *Am. J. Phys. Anthropol.* 36(suppl.):121–122.

Jack, K. M., and Fedigan, L. M. (2004a). Male dispersal patterns in white-faced capuchins (*Cebus capucinus*). Part 1: Patterns and causes of natal emigration. *Anim. Behav.* 67:761–769.

Jack, K. M., and Fedigan, L. M. (2004b). Male dispersal patterns in white-faced capuchins (*Cebus capucinus*). Part 2: Patterns and causes of secondary dispersal. *Anim. Behav.* 67(4):771–782.

Jack, K. M., and Fedigan, L. M. (2006). Why be alpha male? Dominance and reproductive success in wild white-faced capuchins (*Cebus capucinus*). In: Estrada, A., Garber, P., Luecke, L., and Pavelka, M. S. M. (eds.), *New Perspectives in the Study of Mesoamerican Primates: Distribution, Ecology, Behavior, and Conservation.* Kluwer Academic/Plenum, New York.

Jackson, P., and Nowell, K. (1996). *Wild Cats: Status Survey and Conservation Action Plan.* IUCN, Gland, Switzerland.

Jacob, A. L., Vaccaro, I., Sengupta, R., Hartter, J., and Chapman, C. A. (2008). How can conservation biology best prepare for declining rural population and ecological homogenization? *Trop. Conserv. Sci.* 4:307–320.

Jacobs, G. H. (2007). New world monkeys and color. *Int. J. Primatol.* 28:729–759.

Jacobsen, L., and Hamel, R. (1996). *International Directory of Primatology.* Wisconsin Regional Primate Research Center, Madison.

Jacobshagen, B. (1979). Morphometrics studies in the taxonomy of the orang-utan (*Pongo pygmaeus*, L. 1760). *Folia Primatol.* 32:29–34.

Jaeggi, A. V., van Noordwijk, M. A., and van Schaik, C. P. (2008). Begging for information: mother–offspring food sharing among wild bornean orang-utans. *Am. J. Primatol.* 70:533–541.

Jaenicke, C., and Ehrlich, A. (1982). Effects of animate vs. inanimate stim-uli on curiosity behavior in greater galago and slow loris. *Primates* 23:95–104.

Jaffe, K. E., and Isbell, L. A. (2009). After the fire: benefits of reduced ground cover for vervet monkeys (*Cercopithecus aethiops*). *Am. J. Primatol.* 71:252–260.

Jaffe, K. E. and Isbell, L. A. (in press). Changes in ranging and agonistic behavior of vervet monkeys (Cercopithecus aethiops) after predator-induced group fusion. *Am. J. Primatol.*

Jagannadha Rao, A., Kotagi, S. G., and Moudgal, N. R. (1984). Serum con-centrations of chorionic gonadotropin, oestradiol-17beta and progesterone during early pregnancy in the south Indian bonnet monkey (*Macaca radiata*). *J. Reprod. Fertil.* 70:449–455.

Jalles-Filho, E., Grassetto, R., de Cunha, T., and Salm, R. A. (2001). Transport of tools and mental representation: is capuchin monkey tool behaviour a useful model of Plio-Pleistocene hominid technology? *J. Hum. Evol.* 40:365–377.

Jamieson, R. W. (1998). The effects of seasonal variation in fruit availability on social and foraging behavior in *Macaca fascicularis* in Mauritius [PhD diss.]. Washington University, St. Louis.

Janczewski, D. N., Goldman, D., and O'Brien, S. J. (1990). Molecular genetic divergence of orangutan (*Pongo pygmaeus*) subspecies based on isozyme and two-dimensional gel electrophoresis. *J. Hered.* 81:375–387.

Janecka, J. E., Miller, W., Pringle, T. H., Wiens, F., Zitzmann, A., Helgen, K. M., Springer, M. S., and Murphy, W. J. (2007). Molecular and genomic data identify the closest living relative of primates. *Science* 318:792–794.

Janmaat, K. R. L., Byrne, R. W., and Zuberbuhler, K. (2006). Primates take weather into account when searching for fruits. *Curr. Biol.* 16:1232–1237.

Janmaat, K. R. L., Olupot, W., Chancellor, R. L., Arlet, M. E., and Waser, P. M. (2009). Long-term site fidelity and individual home range shifts in Lophocebus albigena. *Int. J. Primatol.* 30: 443–466.

Jansen, C. H., and van Schaik, C. P. (2002). The behavioral ecology of infan-ticide by males. In: van Schaik, C. P., and Jansen, C. H. (eds.), *Infanticide by Males and Its Implications.* Cambridge University Press, Cambridge. pp. 469–494.

Jansen, P. A., Bartholomeus, M., Bongers, F., Elzinga, J. A., den Ouden, J., and Van Wieren, S. E. (2002). The role of seed size in dispersal by a scatter-hoarding rodent. In: Levey, D. J., Silva, W. R., and Galetti, M. (eds.), *Seed Dispersal and Frugivory: Ecology, Evolution, and Conservation.* CABI Publishing, New York. pp. 209–232.

Janson, C. H. (1983). Adaptation of fruit morphology to dispersal agents in a neotropical forest. *Science* 219:187–189.

Janson, C. H. (1984). Female choice and mating system of the brown capuchin monkey (*Cebus apella*) (Primates: Cebidae). *Z. Tierpyschol.* 65:177–200.

Janson, C. H. (1985a). Ecological and social consequences of food competi-tion in brown capuchin monkeys [PhD diss.]. University of Washington, Seattle.

Janson, C. H. (1985b). Aggressive competition and individual food consump-tion in wild brown capuchin monkeys (*Cebus apella*). *Behav. Ecol. Sociobiol.* 18:125–138.

Janson, C. H. (1986). The mating system as a determinant of social evolution in capuchin monkeys (*Cebus*). In: Else, J. G., and Lee, P. C. (eds.), *Primate Ecology and Conservation.* Cambridge University Press, Cambridge. pp. 169–179.

Janson, C.H. (1987). Ecological correlates of aggression in brown capuchin monkeys. *Int. J. Primatol.* 8:431–431.

Janson, C. H. (1988a). Intra-specific food competition and primate social struc-ture: a synthesis. *Behaviour* 105:1–17.

Janson, C. H. (1988b). Food competition in brown capuchin monkeys (*Cebus apella*): quantitative effects of group size and tree productivity. *Behaviour* 105:53–76.

Janson, C. H. (1990a). Ecological consequences of individual spatial choice in foraging groups of brown capuchin monkeys, *Cebus apella. Anim. Behav.* 40:922–934.

Janson, C. H. (1990b). Social correlates of individual spatial choice in forag-ing groups of brown capuchin monkeys, *Cebus apella. Anim. Behav.* 40:910–921.

Janson, C. H. (1992). Evolutionary ecology of primate social structure. In: Smith, E. A., and Winterhalder, B. (eds.), *Evolutionary Ecology and Human Behavior.* Aldine de Gruyter, New York. pp. 95–130.

Janson, C. H. (1994). Comparison of mating system across two populations of brown capuchin monkeys. *Am. J. Primatol.* 33:217.

Janson, C. H. (1998a). Capuchin counterpoint. In: Ciochon, R. L., and Nisbett, R. A. (eds.), *The Primate Anthology. Essays on Primate Behavior, Ecology, and Conservation from Natural History.* Prentice Hall, Englewood Cliffs, NJ. pp. 153–159.

Janson, C. H. (1998b). Experimental evidence for spatial memory in foraging wild capuchin monkeys, *Cebus apella. Anim. Behav.* 55:1229–1243.

Janson, C. H. (1998c). Testing the predation hypothesis for vertebrate sociality: prospects and pitfalls. *Behaviour* 135:389–410.

Janson, C. H. (2000a). Primate socio-ecology: the end of a golden age. *Evol. Anthropol.* 9:73–86.

Janson, C. H. (2000b). Spatial movement strategies: theory, evidence and challenges. In: Boinski, S., and Garber, P. A. (eds.), *On The Move: How and Why Animals Travel in Groups.* Chicago University Press, Chicago. pp. 165–203.

Janson, C. H. (2001). Field experiments in primate ecology: the monkeys are always right. *Am. J. Primatol.* 54:107.

Janson, C. H. (2007). Experimental evidence for route intgration and strategic planning in wild capuchin monkeys. *Anim. Cogn.* 10:341–356.

Janson, C. H., and Boinski, S. (1992). Morphological and behavioral adapta-tions for foraging in generalist primates: the case of the cebines. *Am. J. Phys. Anthropol.* 88:483–498.

Janson, C. H., and Chapman, C. A. (1999). Resources and the determination of primate community structure. In: Fleagle, J. G., Janson, C. H., and Reed, K. (eds.), *Primate Communities.* Cambridge University Press, Cambridge. pp. 237–267.

Janson, C. H., and Di Bitetti, M. S. (1997). Experimental analysis of food detection in capuchin monkeys: effects of distance, travel speed, and resource size. *Behav. Ecol. Sociobiol.* 41:17–24.

Janson, C. H., and Goldsmith, M. L. (1995). Predicting group size in primates: foraging costs and predation risks. *Behav. Ecol.* 6:326–336.

Janson, C. H., and van Schaik, C. P. (1988). Recognizing the many faces of primate food competition: methods. *Behaviour* 105:165–186.

Janson, C. H., and van Schaik, C. P. (1993). Ecological risk aversion in juve-nile primates: slow and steady wins the race. In: Pereira, M. E., and Fairbanks, L. A. (eds.), *Juvenile Primates.* Oxford University Press, New York. pp. 57–74.

Janson, C. H., and Verdolin, J. (2005). Seasonality of primate births in relation to climate. In: Brockman, D. K., and van Schaik, C. P. (eds.), *Seasonality in Primates: Studies of Living and Extinct Human and Non-human Primates.* Cambridge University Press, New York. pp. 307–350.

Janson, C. H., and Vogel, E. R. (2006a). Estimating the effects of hunger on primate social ecology. In: Hohmann, G., Robbins, M., and Boesch, C. (eds.), *Feeding Ecology in Apes and Other Primates.* Cambridge University Press, Cambridge. pp. 285–312.

Janson, C.H., and Vogel, E. (2006b). Hunger and aggression in capuchin monkeys. In: Hohmann, G., Robbins, M. M., and Boesche, C. (eds.), *Feeding Ecology in Apes and Other Primates: Ecological, Physiological, and Behavioural Aspects.* Cambridge University Press, Cambridge. pp. 285–312.

Janson, H. W. (1952). *Apes and Ape Lore in the Middle Ages and the Renaissance.* Studies of the Warburg Institute, vol. 20. University of London, London.

Jantschke, B., Welker, C., and Klaiber-Schuh, A. (1996). Rearing without paternal help in the Bolivian owl monkey *Aotus azarae boliviensis*: a case study. *Folia Primatol.* 69:115–120.

Janus, M. (1992). Interplay between various aspects in social relationships of young rhesus monkeys: dominance, agonistic help, and affiliation. *Am. J. Primatol.* 26:291–308.

Janzen, D. H. (1967). Synchronization of sexual reproduction of trees within the dry season in Central America. *Evolution* 21:620–637.

Janzen, D. H. (1970). Herbivores and the number of tree species in tropical forests. *Am. Nat.* 104:501–528.

Janzen, D. H. (1972). *Jacquinia pungens*, a heliophile from the understory of a tropical deciduous forest. *Biotropica* 2:112–119.

Janzen, D. H. (1978). Complications in interpreting the chemical defenses of trees against tropical arboreal plant-eating vertebrates. In: Montgomery, G. G. (ed.), *The Ecology of Arboreal Folivores.* Smithsonian Institution Press, Washington DC. pp. 73–84.

Janzen, D. H. (1983). The Pleistocene hunters had help. *Am. Nat.* 121:598–599.

Jaquish, C. E., Toal, R. L., Tardif, S. D., and Carson, R. L. (1995). Use of ultrasound to monitor prenatal growth and development in the common marmoset (*Callithrix jacchus*). *Am. J. Primatol.* 36:259–275.

Jarman, P. J. (1974). The social organisation of antelope in relation to their ecology. *Behaviour* 48:215–267.

Jay, P. C. (1962). Aspects of maternal behavior among langurs. *Ann. N. Y. Acad. Sci.* 102:468–476.

Jay, P. (1963). Mother–infant relations in langurs. In: Rheingold, H. R. (ed.), *Maternal Behavior in Mammals.* John Wiley & Sons, New York. pp. 282–304.

Jay, P. (1965). The common langur of north India. In: DeVore, I. (ed.), *Primate Behavior: Field Studies of Monkeys and Apes.* Holt, Reinhart and Winston, New York. pp. 197–249.

Jay, P. C. (1968a). Primate field studies and human evolution. In: Jay, P. C. (ed.), *Primates: Studies in Adaptation and Variability.* Holt, Rinehart and Winston, New York. pp. 487–519.

Jay, P. C. (ed.) (1968b). *Primates: Studies in Adaptation and Variability.* Holt, Rinehart and Winston, New York.

Jenkins, M. (2003). Prospects for biodiversity. *Science* 302:1175–1188.

Jenkins, P. D. (1987). Catalogue of *Primates in the British Museum (Natural History) and Elsewhere in the British Isles Part IV: Suborder Strepsirrhini, Including the Subfossil Madagascar Lemurs and Family Tarsiidae.* British Museum (Natural History), London. p. 189.

Jensen-Seaman, M. I., and Li, W. H. (2003). Evolution of the hominoid semenogelin genes, the major proteins of ejaculated semen. *J. Mol. Evol.* 57:261–270.

Jepson, P., Jarvie, J. K., MacKinnon, K., and Monk, K. A. (2001). The end for Indonesia's lowland forest? *Science* 292:859–861.

Jerison, H. J. (1973). *Evolution of the Brain and Intelligence.* Academic Press, San Diego.

Jerison, H. J. (2000). The evolution of intelligence. In: Sternberg, R. J. (ed.), *Handbook of Intelligence.* Cambridge University Press, Cambridge. pp. 216–244.

Jernvall, J., and Wright, P. C. (1998). Diversity components of impending primate extinctions. *Proc. Natl. Acad. Sci. USA* 95:11279–11283.

Jerusalinsky, L., Oliveira, M. M., Pereira, R. F., Santana, V., Bastos, P. C. R., and Ferrari, S. F. (2006). Preliminary evaluation of the conservation status of *Callicebus coimbrai* Kobayashi & Langguth, 1999 in the Brazilian state of Sergipe. *Primate Conserv.* 21:25–32.

Jewell, P. A., and Oates, J. F. (1969). Ecological observations of the lorisoid primates of African lowland forest. *Zool. Afr.* 4:231–248.

Jiang, X., Ma, S., Wang, Y., Sheeran, L. K., and Poirier, F. E. (1994). Human encounter and predator avoidance behavior in black-crested gibbon, *Hylobates concolor. Acta Anthropol. Sin.* 13(2):182–188.

Jiang, X,L,, Wang, Y.X., and Wang, Q. (1999). Coexistence of monogamy and polygyny in black-crested gibbon (*Hylobates concolor*). *Primates* 40(4):607–611

Jiang, Z., Sugita, M., Kitahara, M., Takatsuki, S., Goto, T., and Yoshida, Y. (2008). Effects of habitat feature, antenna position, movement, and fix interval on GPS radio collar performance in Mount Fuji, central Japan. *Ecol. Res.* 23:581–588.

Jisaka, M., Kawanaka, M., Sugiyama, H., Takegawa, K., Huffman, M. A., Ohigashi, H., and Koshimizu, K. (1992a). Antischistosomal activities of sesquiterpene lactones and steroid glucosides from *Vernonia amygdalina,* possibly used by wild chimpanzees against parasite-related diseses. *Biosci. Biotech. Biochem.* 56:845–846.

Jisaka, M., Ohigashi, H., Takagaki, T., Nozaki, H., Tada, T., Hirota, M., Irie, R., Huffman, M. A., Nishida, T., Kaji, M., and Koshimizu, K. (1992b). Bitter steroid glucosides, vernonisoides A1, A2, and A3 and related B1 from a possible medicinal plant *Vernonia amygdalina,* used by wild chimpanzees. *Tetrahedron* 48:625–632.

Jisaka, M., Ohigashi, H., Takegawa, K., Hirota, M., Irie, R., Huffman, M. A., and Koshimizu, K. (1993a). Steroid glucosides from *Vernonia amygdalina,* a possible chimpanzee medicinal plant. *Phytochemistry* 34:409–413.

Jisaka, M., Ohigashi, H., Takegawa, K., Huffman, M. A., and Koshimizu, K. (1993b). Antitumor and antimicrobial activities of bitter sesquiterpene lactones of *Vernonia amygdalina,* a possible medicinal plant used by wild chimpanzees. *Biosci. Biotech. Biochem.* 57:833–834.

Jitsumori, M., and Delius, J. D. (2001). Object recognition and object categorization in animals. In: Matsuzawa, T. (ed.), *Primate Origin of Human Cognition and Behavior.* Springer, Tokyo. pp. 269–293.

Joffe, T. H. (1997). Social pressures have selected for an extended juvenile period in primates. *J. Hum. Evol.* 32(6):593–605.

Johns, T. (1990). *With Bitter Herbs They Shall Eat It.* University of Arizona Press, Tucson.

Johnson, A., Knott, C., Pamungkas, B., Pasaribu, M., and Marshall, A. (2005). A survey of the orangutan *Pongo pygmaeus pygmaeus* population in and around Gunung Palung National Park, West Kalimantan, Indonesia based on nest counts. *Biol. Conserv.* 121:495–507.

Johnson, B., Kern, J., Wisdom, M., Findholt, S., and Kie, J. (2000). Resource selection and spatial separation of mule deer and elk during spring. *J. Wild. Manage.* 64:685–697.

Johnson, D. P., Kays, R., Blackwell, P. G., and Macdonald, D. W. (2002). Does the resource dispersion hypothesis explain group living? *Trends Ecol. Evol.* 17:563–570.

Johnson, J., Canning, J., Kaneko, T., Pru, J. K., and Tilly, J. L. (2004). Germline stem cells and follicular renewal in the postnatal mammalian ovary. *Nature* 428:145–150.

Johnson, K. G., Wang, W., Reid, D. G., and Hu, J. C. (1993). Food habits of Asiatic leopards (*Panthera pardus fusea*) in Wolong Reserve, Sichuan, China. *J. Mamm.* 74:646–650.

Johnson, M. E., Viggiano, L., Bailey, J. A., Abdul-Rauf, M., Goodwin, G., Rocchi, M., and Eichler, E. E. (2001). Positive selection of a gene family during the emergence of humans and African apes. *Nature* 413:514–519.

Johnson, R. B. (1990). The effects of plant chemistry on food selection by adult male yellow baboons (*Papio cynocephalus*). In: Thiago de Mello, M., Whiten, A., and Byrne, R. W. (eds.), *Baboons: Behaviour and Ecology. Use and Care.* Selected Proceedings of the 12th Congress of the International Primatological Society Brasilia, Brasil. pp. 23–28.

Johnson, R. L., and Kapsalis, E. (1998). Menopause in free-ranging rhesus macaques: estimated incidence, relation to body condition, and adaptive significance. *Int. J. Primatol.* 19:751–765.

Johnson, S. E. (2001). Life history and the social environment rank, risk-taking, and resource acquisition among chacma baboons (*Papio hamadryas ursinus*) [PhD diss.]. University of New Mexico, Alberquerque.

Johnson, S. (2002). Ecology and speciation in brown lemurs: white-collared lemurs (*Eulemur albocollaris*) and hybrids (*Eulemur albocollaris* × *Eulemur fulvus rufus*) in southeastern Madagascar [PhD diss.]. University of Texas at Austin, Austin.

Johnson, S. E. (2003). Life history and the competitive environment: trajectories of growth, maturation, and reproductive output among chacma baboons. *Am. J. Phys. Anthropol.* 120:83–98.

Johnson, S. E. (2006). Maternal characteristics and offspring growth in chacma baboons: a life-history perspective. In: Swedell, L., and Leigh, S. R. (eds.), *Reproduction and Fitness in Baboons: Behavioral, Ecological, and Life History Perspectives.* Springer, New York. pp. 177–199.

Johnson, S. E., and Bock, J. (2004). Trade-offs in skill acquisition and time allocation among juvenile chacma baboons. *Hum. Nature* 15(1):45–62.

Johnson, S., and Overdorff, D. J. (2002). Scarce season diet and keystone resources in three brown lemur (*Eulemur fulvus* spp.) populations. *Am. J. Primatol.* 56:65–66.

Johnson, S. E., Lei, R. H., Martin, S. K., Irwin, M. T., and Louis, E. E. (2008). Does *Eulemur cinereiceps* exist? Preliminary evidence from genetics and ground surveys in southeastern Madagascar. *Am. J. Primatol.* 70:372–385.

Johnston, C. (1998). *Geographic Information Systems in Ecology.* Blackwell Science, Malden.

Jolly, A. (1966). *Lemur Behavior: A Madagascar Field Study.* University of Chicago Press, Chicago.

Jolly, A. (1980). *A World Like Our Own.* Yale University Press, New Haven.

Jolly, A. (1984). The puzzle of female feeding priority. In: Small, M. (ed.), *Female Primates: Studies by Women Primatologists.* Alan R. Liss, New York. pp. 197–215.

Jolly, A. (2000). The bad old days of primatology? In: Strum, S. C., and Fedigan, L. M. (eds.), *Primate Encounters: Models of Science, Gender, and Society.* University of Chicago Press, Chicago. pp. 71–84.

Jolly, A., and Sussman, R. W. (2006). Notes on the history of ecological studies of Malagasy lemurs. In: Gould, L., and Sauther, M. L. (eds.), *Lemurs: Ecology and Adaptation.* Springer, New York. pp. 19–39.

Jolly, A., Rasamimanana, H. R., Kinnaird, M. F., O'Brien, T. G., Crowley, H. M., Harcourt, C. S., Gardner, S., and Davidson, J. (1993). Territoriality in *Lemur catta* groups during the birth season at Berenty, Madagascar. In: Kappeler, P. M., and Ganzhorn, J. U. (eds.), *Lemur Social Systems and Their Ecological Basis.* Plenum Press, New York. pp. 85–109.

Jolly, A., Caless, S., Cavigelli, S., Gould, L., Pereira, M. E., Pitts, A., Pride, R. E., Rabenandrasana, H. D., Walker, J. D., and Zafison, T. (2000). Infant killing, wounding, and predation in *Eulemur* and *Lemur*. *Int. J. Primatol.* 21:21–40.

Jolly, A., Dobson, A., Rasaminmanana, H., Walker, J., O'Connor, S., Solberg, M., and Perel, V. (2002). Demography of *Lemur catta* at Berenty Reserve, Madagascar: effects of troop size, habitat and rainfall. *Int. J. Primatol.* 23:327–353.

Jolly, A., Sussman, R.W., Koyama, N., and Rasamimanana, H. (2006). *Ringtailed Lemur Biology*: Lemur catta *in Madagascar*. Springer Science+Business Media, New York.

Jolly, C. J. (1963). A suggested case of evolution by sexual selection in primates. *Man* 63:177–178.

Jolly, C. J. (1970a). *Hadropithecus*: a lemuroid small-object feeder. *Man* 5:619–626.

Jolly, C. J. (1970b). The large African monkeys as an adaptive array. In: Napier, J. R., and Napier, P. H. (eds.), *Old World Monkeys*. Academic Press, New York. pp. 139–174.

Jolly, C. J. (1970c). The seed-eaters: a new model of hominid differentiation based on a baboon analogy. *Man* 5:1–26.

Jolly, C. J. (1993). Species, subspecies, and baboon systematics. In: Kimbel, W. H., and Martin, L. B. (eds.), *Species, Species Concepts, and Primate Evolution*. Plenum Press, New York. pp. 67–107.

Jolly, C. J. (2001). A proper study for mankind: analogies from the papionin monkeys and their implications for human evolution. *Ybk. Phys. Anthropol.* 44:177–204.

Jolly, C. J. (2003). Cranial anatomy and baboon diversity. *Anat. Rec. A Discov. Mol. Cell. Evol. Biol.* 275:1043–1047.

Jolly, C. J. (2007). Baboons, mandrills, and mangabeys: Afro-papionin socioecology in a phylogenetic perspective. In: Campbell, C. J., Fuentes, A., MacKinnon, K. C., Panger, M., and Bearder, S. K. (eds.), *Primates in Perspective*. Oxford University Press, New York. pp. 240–251.

Jolly, C. J. (In press). Genus *Papio*. In: Kingdon, J., Happold, D., and Butynski, T. (eds.), *The Mammals of Africa*. University of California Press, Berkeley, CA.

Jolly, C. J., and Brett, F. L. (1973). Genetic markers and baboon biology. *J. Med. Primatol.* 2:85–99.

Jolly, C. J., and Phillips-Conroy, J. (2003). Testicular size, mating system, and maturation schedules in wild anubis and hamadryas baboons. *Int. J. Primatol.* 24:125–142.

Jolly, C. J., and Phillips-Conroy, J. E. (2006). Testicular size, developmental trajectories, and male life history strategies in four baboon taxa. In: Swedwell, L., and Leigh, S. R. (eds.), *Reproduction and Fitness in Baboons: Behavioral, Ecological, and Life History Perspectives*. Springer, New York. pp. 257–275.

Jolly, C. J., and Plog, F. (1979). *Physical Anthropology and Archaeology*, 2nd ed. Alfred A. Knopf, New York.

Jolly, C. J., Woolley-Barker, T., Beyene, S., Disotell, T. R., and Phillips-Conroy, J. E. (1997). Intergeneric hybrid baboons. *Int. J. Primatol.* 18:597–627.

Jones, C. B. (1980a). The functions of status in the mantled howler monkey, *Alouatta palliata* Gray: intraspecific competition for group membership in a folivorous neotropical primate. *Primates* 21:389–405.

Jones, C. B. (1980b). Seasonal parturition, mortality, and dispersal in the mantled howler monkey, *Alouatta palliata* Gray. *Brenesia* 17:1–10.

Jones, C. B. (1981). The evolution and socioecology of dominance in primate groups: a theoretical formulation, classification, and assessment. *Primates* 22:70–83.

Jones, C. B. (1985). Reproductive patterns in mantled howler monkeys: estrus, mate choice and copulation. *Primates* 26:130–142.

Jones, C. B. (1996). Relative reproductive success in the mantled howler monkey: implications for conservation. *Neotrop. Primates* 4:21–23.

Jones, C., and Sabater Pi, J. (1968). Comparative ecology of *Cercocebus albigena* (Gray) and *Cercocebus torquatus* (Kerr) in Rio Muni, West Africa. *Folia Primatol.* 9:99–113.

Jones, K. E., and MacLarnon, A. (2001). Bat life histories: testing models of mammalian life-history evolution. *Evol. Ecol. Res.* 3:465–476.

Jones, M. L. (1969). The geographical races of orangutan. *Proc. 2nd Int. Congr. Primatol.* 2:217–223.

Jones, T., Ehardt, C. L., Butynski, T. M., Davenport, T. R. B., Mpunga, N. E., Machaga, S. J., and De Luca, D. W. (2005). The highland mangabey *Lophocebus kipunji*: a new species of African monkey. *Science* 308:1161–1164.

Jones, W. T., and Bush, B. B. (1988) Movement and reproductive behavior of solitary male redtail guenons (*Cercopithecus ascanias*). *Am. J. Primatol.* 14:203–222.

Jones-Engel, L., Schillaci, M., Engel, G., Paputungan, U., and Froehlich, J. (2005). Characterizing primate pet ownership in Sulawesi: implications for disease transmission. In: Paterson, J. (ed.), *Commensalism and Conflict: The Human–Primate Interface*. American Society of Primatologists, Norman, OK.

Jones-Engel, L., May, C. C., Engel, G. A., Steinkraus, K. A., Schillaci, M. A., Fuentes, A., Rompis, A., Chalise, M. K., Aggimarangsee, N., Feeroz, M. M., Grant, R., Allan, J. S., Putra, A., Wandia, I. N., Watanabe, R., Kuller, L., Thongsawat, S., Chaiwarith, R., Kyes, R. C., and Linial, M. L. (2008). Diverse contexts of zoonotic transmission of simian foamy viruses in Asia. *Emerg. Infect. Dis.* 14(8):1200–1208.

Jordan, C. (1982). Object manipulation and tool-use in captive pygmy chimpanzees (*Pan paniscus*). *J. Hum. Evol.* 11:35–39.

Jordan, K. E., MacLean, E. L., and Brannon, E. M. (2008). Monkeys match and tally quantities across senses. *Cognition* 108:617–625.

Jordano, P., and Herrera, C. M. (1995). Shuffling the offspring: uncoupling and spatial discordance of multiple stages in vertebrate seed dispersal. *Ecoscience* 2:230–237.

Jorgensbye, H. (2007). Mountain galago *Galagoides orinus* in the evergreen fragments on the Dabaga Plateau, Tanzania. *Nature E. Afr. EANHS Bull.* 35:7–12.

Joshi, P. K., Singh, S., Agarwal, S., and Roy, P. S. (2001). Forest cover assessment in western Himalayas, Himachal Pradesh using IRS 1C/1D WiFS data. *Curr. Sci.* 80:941–947.

Joslin, J. O., Weissman, W. D., Johnson, K., Forster, M., Wasser, S., and Collins, D. (1995). In vitro fertilization of Bornean orangutan (*Pongo pygmaeus pygmaeus*) gamete followed by embryo transfer into a surrogate hybrid orangutan (*Pongo pygmaeus*). *J. Zoo Wildl. Med.* 26:32–42.

Jouventin, P. (1975a). Les rôles des colorations du mandrill (*Mandrillus sphinx*). *Z. Tierpsychol.* 39:455–462.

Jouventin, P. (1975b). Observations sur la socioecologie du mandrill. *Terre Vie* 29:439–532.

Jovanovic, T., and Gouzoules, H. (2001). Effects of nonmaternal restraint on the vocalizations of infant rhesus monkeys (*Macaca mulatta*). *Am. J. Primatol.* 53:33–45.

Juárez, C., Rotundo, M., and Fernandez-Duque, E. (2003). Behavioral sex differences in the socially monogamous night monkeys of the Argentinean Chaco. *Rev. Etol.* 5(suppl.):174.

Judge, D. S., and Carey, J. R. (2000). Postreproductive life predicted by primate patterns. *J. Gerontol.* 55A:B201–B209.

Judge, P. G. (1991). Dyadic and triadic reconciliation in pigtail macaques (*Macaca nemestrina*). *Am. J. Primatol.* 23:225–237.

Judge, P. G. (2003). Conflict resolution. In: Maestripieri, D. (ed.), *Primate Psychology*. Harvard University Press, Cambridge, MA. pp. 41–68.

Judge, P. G., Evans, T. A., and Was, D. K. (2005). Ordinal representation of numeric quantities by brown Capuchin monkeys (*Cebus apella*). *J. Exp. Psychol. Anim. Behav. Proc.* 31:79–94.

Julliot, C. (1994). Predation of a young spider monkey (*Ateles paniscus*) by a crested eagle (*Morphnus guianensis*). *Folia Primatol.* 63:75–77.

Julliot, C. (1996a). Fruit choice by red howler monkeys (*Alouatta seniculis*) in a tropical forest. *Am. J. Primatol.* 40:261–282.

Julliot, C. (1996b). Seed dispersal by red howling monkeys (*Alouatta seniculus*) in the tropical rain forest of French Guiana. *Int. J. Primatol.* 17:239–258.

Julliot, C. (1997). Impact of seed dispersal by red howler monkeys *Alouatta seniculus* on the seedling population in the understory of tropical rain forest. *J. Ecol.* 85:431–440.

Julliot, C., and Sabatier, D. (1993). Diet of the red howler monkeys (*Alouatta seniculus*) in French Guiana. *Int. J. Primatol.* 14:527–550.

Jungers, W. L. (1979). Locomotion, limb proportions and skeletal allometry in lemurs and lorises. *Folia Primatol.* 32:8–28.

Jungers, W. L. (1985). Body size and scaling in limb proportions in primates. In: Jungers, W. L. (ed.), *Size and Scaling in Primate Biology*. Plenum Press, New York. pp. 345–381.

Jungers, W. L., and Burr, D. B. (1994). Body size, long bone geometry, and locomotion in quadrupedal monkeys. *Z. Morphol. Anthropol.* 80:89–97.

Jungers, W. L., and Susman, R. L. (1984). Body size and skeletal allometry in African apes. In: Susman, R. L. (ed.), *The Pygmy Chimpanzee: Evolutionary Biology and Behavior*. Plenum, New York. pp. 131–177.

Jungers, W. L., Godfrey, L. R., Simons, E. L., Wunderlich, R. E., Richmond, B. G., and Chatrath, P. S. (2002). Ecomorphology and behavior of giant extinct lemurs from Madagascar. In: Plavcan, J. M., Kay, R. F., Jungers, W. L., and van Schaik, C. P. (eds.), *Reconstructing Behavior in the Primate Fossil Record*. Kluwer, New York. pp. 371–411.

Jungers, W. L., Demes, B., and Godfrey, L. R. (2007). How big were the "giant" extinct lemurs of Madagascar? In: Fleagle, J. G., and Gilbert, C. C. (eds.), *Elwyn Simons: A Search for Origins*. Springer, New York. pp. 343–360.

Junk, W. J. (1989). Flood tolerance and tree distribution in central Amazonian floodplains. In: Holm-Nielsen, L. B., Nielsen, I. C., and Balslev, H. (eds.), *Tropical Forests: Botanical Dynamics, Speciation and Diversity*. Acadmic Press, London. pp. 47–73.

Jürgens, U. (1979). Neural control of vocalization in non-human primates. In: Steklis, H. D., and Raleigh, M. J. (eds.), *Neurobiology of Social Communication in Primates*. Academic Press, New York. pp. 11–44.

Jürgens, U. (1988). Central control of monkey calls. In: Todt, D., Goedeking, P., and Symmes, D. (eds.), *Primate Vocal Communication*. Springer-Verlag, Berlin. pp. 162–167.

Jurke, M. H., Pryce, C. R., Dobeli, M., and Martin, R. D. (1994). Non-invasive detection and monitoring of pregnancy and the postpartum period in Goeldi's monkey (*Callimico goeldii*) using urinary pregnanediol-3-alpha-glucuronide. *Am. J. Primatol.* 34:319–331.

Jurke, M. H., Hagey, L. R., Jurke, S., and Czekala, N. M. (2000). Monitoring hormones in urine and feces of captive bonobos (*Pan paniscus*). *Primates* 41:311–319.

Jurke, M. H., Hagey, L. R., Czekala, N. M., and Harvey, N. C. (2001). Metabolites of ovarian hormones and behavioral correlates in captive female bonobos (*Pan paniscus*). In: Galdikas, B. M. G., Briggs, N. E., Sheeran, L. K., Shapiro, F. L., and Goodall, J. (eds.), *All Apes Great and Small. African Apes*, vol. 1. Kluwer Academic/Plenum, New York. pp. 217–229.

Kaas, J. H., and Preuss, T. M. (eds.) (2007). *Evolution of the Nervous Systems: A Comprehensive Reference. Primates*, vol. 4. Elsevier Academic Press, Boston.

Kaessmann, H., Wiebe, V., and Paabo, S. (1999). Extensive nuclear DNA sequence diversity among chimpanzees. *Science* 286:1159–1162.

Kahlenberg, S. M., Emery Thompson, M., and Wrangham, R. W. (2008a). Evidence for female competition over core areas in *Pan troglodytes*, Kibale National Park, Uganda. *Int. J. Primatol.* 29:931–947.

Kahlenberg, S. M., Emery Thompson, M., Muller, M. N., and Wrangham, R. W. (2008b). Immigration costs for female chimpanzees and male protection as an immigrant counterstrategy to intrasexual aggression. *Anim. Behav.* 76:1497–1509.

Kalinowski, S. T., Wagner, A. P., and Taper, M. L. (2006). ML-Relate: a computer program for maximum likelihood estimation of relatedness and relationship. *Mol. Ecol. Notes* 6:576–579.

Kalinowski, S. T., Taper, M. L., and Marshall, T. C. (2007). Revising how the computer program CERVUS accommodates genotyping error increases success in paternity assignment. *Mol. Ecol.* 16:1099–1006.

Kalpers, J., Williamson, E. A., Robbins, M. M., McNeilage, A., Nzamurambaho, A., Ndakasi, L., and Mugiri, G. (2003). Gorillas in the crossfire: assessment of population dynamics of the Virunga mountain gorillas over the past three decades. *Oryx* 37:326–337.

Kamal, K. B., Boug, A., Brain, P. F. (1997). Effects of food provisioning on the behaviour of commensal hamadryas baboons, *Papio hamadryas*, at Al Hada Mountain in Western Saudi Arabia. *Zool. Mid. East* 14:11–22.

Kaminski, J., Call, J., and Tomasello, M. (2004). Body orientation and face orientation: two factors controlling apes' begging behavior from humans. *Anim. Cogn.* 7:216–223.

Kaminski, J., Call, J., and Tomasello, M. (2008). Chimpanzees know what others know, but not what they believe. *Cognition* 109:224–234.

Kanga, E. M. (2001). Survey of black and white colobus monkeys (*Colobus angolensis palliatus*) in Shimba Hills National Reserve and Maluganji Sanctuary, Kenya. *Am. Soc. Primatol. Bull.* 25:8–9.

Kanga, E. M., and Heidi, C. M. (1999/2000). Survey of the Angolan black-and-white colobus monkey *Colobus angolensis palliatus* in the Diani forests, Kenya. *Afr. Primates* 4:50–54.

Kano, T. (1971). The chimpanzees of Filabanga, western Tanzania. *Primates* 12:229–246.

Kano, T. (1980). Social behavior of wild pygmy chimpanzees *Pan paniscus* of Wamba: a preliminary report. *J. Hum. Evol.* 9:243–260.

Kano, T. (1982). The social group of pygmy chimpanzees (*Pan paniscus*) of Wamba. *Primates* 23:171–188.

Kano, T. (1989). The sexual behavior of pygmy chimpanzees. In: Heltne, P. G., and Marquardt, L. (eds.), *Understanding Chimpanzees*. Harvard University Press, Cambridge, MA. pp. 176–183.

Kano, T. (1992). *The Last Ape: Pygmy Chimpanzee Behavior and Ecology*. Stanford University Press, Stanford, CA.

Kano, T. (1996). Male rank order and copulation rate in a unit-group of bonobos at Wamba, Zaire. In: McGrew, W. C., Marchant, L. F., and Nishida, L. (eds.), *Great Ape Societies*. Cambridge University Press, Cambridge. pp. 135–145.

Kano, T. (1998). Reply to Stanford (1998) "The social behavior of chimpanzees and bonobos—empirical evidence and shifting assumptions." *Curr. Anthropol.* 39:410–411.

Kano, T., and Mulavwa, M. (1984). Feeding ecology of *Pan paniscus* at Wamba. In: Susman, R. L. (ed.), *The Pygmy Chimpanzee: Evolutionary Biology and Behavior*. Plenum, New York. pp. 233–274.

Kano, F., and Tomonaga, M. (2009) How chimpanzees look at pictures: a comparative eye-tracking study. *Proc R. Soc. Lond. B.* doi:10.1098/rspb.2008.1811.

Kansky, R., and Gaynor, D. (2000). Baboon management strategy for the Cape Peninsula. Final report ZA 568. Cape Town, South Africa, WWF South Africa. p. 149.

Kaplan, H., Hill, K., Lancaster, J., and Hurtado, A. M. (2000). A theory of human life history evolution: diet, intelligence and longevity. *Evol. Anthropol.* 9:156–185.

Kaplan, J. (1977). Patterns of fight interference in free-ranging rhesus monkeys. *Am. J. Phys. Anthropol.* 47:279–288.

Kaplan, J. R. (1978). Fight interference and altruism in rhesus monkeys. *Am. J. Phys. Anthropol.* 49:241–250.

Kaplan, J. R. (1986). Psychological stress and behavior in nonhuman primates. In: Mitchell, G., and Erwin, J. (eds.), *Comparative Primate Biology*. Alan R. Liss, New York. pp. 455–492.

Kaplin, B. A. (2001). Ranging behavior of two species of guenons (*Cercopithecus lhoesti* and *C. mitis doggetti*) in the Nyungwe Forest Reserve, Rwanda. *Int. J. Primatol.* 22:521–548.

Kaplin, B. A., and Lambert, J. E. (2002). Effectiveness of seed dispersal by *Cercopithecus* monkeys: implications for seed input into degraded areas. In: Levey, D. J., Silva, W. R., and Galetti, M. (eds.), *Seed Dispersal and Frugivory: Ecology, Evolution, and Conservation*. CABI Publishing, New York. pp. 351–364.

Kaplin, B. A., and Moermond, T. C. (1998). Variation in seed handling by two species of forest monkeys in Rwanda. *Am. J. Primatol.* 45:83–101.

Kaplin, B. A., and Moermond, T. C. (2000). Foraging ecology of the mountain monkey (*Cercopithecus lhoesti*): implications for its evolutionary history and use of disturbed forest. *Am. J. Primatol.* 50:227–246.

Kaplin, B. A., Munyaligoga, V., and Moermond, T. C. (1998). The influence of temporal changes in fruit availability on diet composition and seed handling in blue monkeys (*Cercopithecus mitis doggetti*). *Biotropica* 30:56–71.

Kappeler, M. (1984). Diet and feeding behavior of the moloch gibbon. In: Preuschoft, H., Chivers, D. J., Brockelman, W. Y., and Creel, N. (eds.), *The Lesser Apes: Evolutionary and Behavioural Biology*. Edinburgh University Press, Edinburgh. pp. 228–241.

Kappeler, P. M. (1990). Social status and scent-marking behaviour in *Lemur catta. Anim. Behav.* 40:774–776.

Kappeler, P. M. (1993). Reconciliation and post-conflict behaviour in ring-tailed lemurs, *Lemur catta*, and redfronted lemurs, *Eulemur fulvus rufus. Anim. Behav.* 45:901–915.

Kappeler, P. M. (1996). Causes and consequences of life-history variation among strepsirhine primates. *Am. Nat.* 148(5):868–891.

Kappeler, P. M. (1997). Intrasexual selection in *Mirza coquereli*: evidence for scramble competition polygyny in a solitary primate. *Behav. Ecol. Sociobiol.* 45:115–127.

Kappeler, P. M. (1998a). Nests, tree holes, and the evolution of primate life histories. *Am. J. Primatol.* 46:7–33.

Kappeler, P. M. (1998b). To whom it may concern: the transmission and function of chemical signals in *Lemur catta. Behav. Ecol. Sociobiol.* 42:411–421.

Kappeler, P. M. (2000a). Lemur origins: rafting by groups of hibernators? *Folia Primatol.* 71:422–425.

Kappeler, P. M. (2000b). *Primate Males*. Cambridge University Press, Cambridge.

Kappeler, P. M. (2008). Genetic and ecological determinants of primate social systems. In: Korb, J., and Heinze, J. (eds.), *Ecology of Social Evolution*. Springer, London. pp. 225–243.

Kappeler, P. M., and Erkert, H. G. (2003). On the move around the clock: correlates and determinants of cathemeral activity in wild redfronted lemurs (*Eulemur fulvus rufus*). *Behav. Ecol. Sociobiol.* 54:359–369.

Kappeler, P., and Ganzhorn, J. (1993). *Lemur Social Systems and Their Ecological Basis*. Plenum Press, New York.

Kappeler, P. M., and Pereira, M. E. (eds.) (2003). *Primate Socioecology and Life History.* University of Chicago Press, Chicago.

Kappeler, P. M., and van Schaik, C. P. (1992). Methodological and evolutionary aspects of reconciliation among primates. *Ethology* 92:51–69.

Kappeler, P. M., and van Schaik, C. P. (2002). Evolution of primate social systems. *Int. J. Primatol.* 23:707–740.

Kappeler, P. M., Wimmer, B., Zinner, D., and Tautz, D. (2002). The hidden matrilineal structure of a solitary lemur: implications for primate social evolution. *Proc. R. Soc. Lond. B* 269:1755–1763.

Kappeler, P. M., Rasoloarison, R. M., Razafimanantsoa, L., Walter, L., and Roos, C. (2005). Morphology, behaviour, and molecular evolution of giant mouse lemurs (*Mirza* spp.) Gray, 1870, with description of a new species. *Primate Rep.* 71:3–26.

Kapsalis, E. (2004). Matrilineal kinship and primate behavior. In: Chapais, B., and Berman, C. M. (eds.), *Kinship and Behavior in Primates.* Oxford University Press, New York. pp. 153–176.

Kapsalis, E., and Berman, C. M. (1996a). Models of affiliative relationships among free-ranging rhesus monkeys, *Macaca mulatta.* I. Criteria for kinship. *Behaviour* 133:1209–1234.

Kapsalis, E., and Berman, C. M. (1996b). Models of affiliative relationships among free-ranging rhesus monkeys, *Macaca mulatta.* II. Testing predictions for three hypothesized organizing principles. *Behaviour* 133:1235–1263.

Karaman, M. W., Houck, M. L., Chemnick, L. G., Nagpal, S., Chawannakul, D., Sudano, D., Pike, B. L., Ho, V. V., Ryder, O. A., and Hacia, J. G. (2003). Comparative analysis of gene-expression patterns in human and African great ape cultured fibroblasts. *Genome Res.* 13:1619–1630.

Karanth, P. K., Delefosse, T., Rakotosamimanana, B., Parsons, T. J., and Yoder, A. D. (2005). Ancient DNA from giant extinct lemurs confirms single origin of Malagasy primates. *Proc. Natl. Acad. Sci. USA* 102:5090–5095.

Karasov, W. H., and Martinez del Rio, C. (2007) *Physiological Ecology: How Animals Profess Energy, Nutrients, and Toxins.* Princeton University Press, Princeton, NJ.

Karere, G. M., Oguge, N. O., Kirathe, J., Muoria, P. K., Moinde, N. N., and Suleman, M. A. (2004). Population sizes and distribution of primates in the lower Tana River forests, Kenya. *Int. J. Primatol.* 25:351–365.

Karesh, W. B., Wallace, R. B., Painter, R. L. E., Rumiz, D., Braselton, W. E., Dierenfeld, E. S., and Puche, H. (1998). Immobilization and health assessment of free-ranging black spider monkeys (*Ateles paniscus chamek*). *Am. J. Primatol.* 44:107–123.

Kar Gupta, K. (2007). Socioecology and conservation of the slender loris in Southern India [PhD diss.]. Arizona State University, Tempe.

Karpanty, S. M. (2003). Rates of predation by diurnal raptors on the lemur community of Ranomafana National Park, Madagascar. *Am. J. Phys. Anthropol.* 36(suppl.):126–127.

Karpanty, S. M. (2006). Direct and indirect impacts of raptor predation on lemurs in southeastern Madagascar. *Int. J. Primatol.* 27:239–261.

Karpanty, S. M., and Goodman, S. M. (1999). Diet of the Madagascar harrier-hawk *Polyboroides radiatus,* in southeastern Madagascar. *J. Rapt. Res.* 33:313–316.

Karpanty, S. M., and Grella, R. (2001). Lemur responses to diurnal raptor calls in Ranomafana National Park, Madagascar. *Folia Primatol.* 72:100–103.

Karpanty, S. M., and Wright, P. C. (2007). Predation on lemurs in the rainforest of Madagascar by multiple predator species: observations and experiments. In: Gursky, S. L., and Nekaris, K. A. I. (eds.), *Primate Anti-Predator Strategies.* Springer, New York. pp. 77–99.

Kauffman, J. H. (1965). A three-year study of mating behavior in a freeranging band of rhesus monkeys. *Ecology* 46:500–512.

Kauffman, L., Ehmke, E., and Boinski, S. (2004). Increased male–male cooperation among brown capuchin monkeys (*Cebus apella*) in Suriname. *Am. J. Phys. Anthropol.* 23:123.

Kaumanns, W., Rohrhuber, B., and Zinner, D. (1989). Reproductive parameters in a newly established colony of hamadryas baboons (*Papio hamadryas*). *Primate Rep.* 24:25–33.

Kaur, T., and Huffman, M. A. (2004). Descriptive urological record of chimpanzees (*Pan troglodytes schweinfurthii*) in the wild and limitations associated with using multi-reagent dipstick test strips. *J. Med. Primatol.* 33:187–196.

Kavanaugh, M., and Dresdale, L. (1975). Observations on the woolly monkey (*Lagothrix lagothricha*) in northern Colombia. *Primates* 16:285–294.

Kawai, M. (1958). On the system of social ranks in a natural troop of Japanese monkeys: I, II [in Japanese]. *Primates* 1/2:111–148. (English translation in Imanishi, K., and Altmann, S. A. [eds.] [1965]. *Japanese Monkeys: A Collection of Translations.* Emory University, Atlanta. pp. 66–104.)

Kawai, M. (1965). Newly-acquired pre-cultural behavior of the natural troop of Japanese monkeys on Koshima Islet. *Primates* 6:1–30.

Kawai, M., and Iwamoto, T. (1979). Nomadism and activities. In: Kawai, M. (ed.) *Ecological and Sociological Studies of Gelada Baboons,* vol. 16. Karger, Basel. pp. 251–278.

Kawai, M., Azuma, S., and Yoshiba, K. (1967). Ecological studies of reproduction in Japanese monkeys (*Macaca fuscata*). 1. Problems of the birth season. *Primates* 8:35–74.

Kawai, M., Dunbar, R., Ohsawa, H., and Mori, U. (1983). Social organization of gelada baboons: social units and definitions. *Primates* 24:13–24.

Kawai, M., Watanabe, K., and Mori, A. (1992). Pre-cultural behaviors observed in free-ranging Japanese monkeys on Koshima Islet over the past 25 years. *Primate Rep.* 32:143–153.

Kawai, N., and Matsuzawa, T. (2000). Cognition—numerical memory span in a chimpanzee. *Nature* 403:39–40.

Kawamura, S. (1958). Matriarchal social ranks in the Minoo-B troop: a study of the rank system of Japanese macaques [in Japanese]. *Primates* 1:149–156. (English translation in Imanishi, K., and Altmann, S. A. [eds.] [1965]. *Japanese Monkeys: A Collection of Translations.* Emory University, Atlanta, pp. 105–112.)

Kawamura, S. (1959). The process of sub-culture propagation among Japanese macaques. *Primates* 2:43–60.

Kawanaka, K. (1981). Infanticide and cannibalism in chimpanzees, with special reference to the newly observed case in the Mahale Mountains. *Afr. Study Monogr.* 1:69–99.

Kawanaka, K. (1984). Association, ranging, and the social unit in chimpanzees of the Mahale Mountains, Tanzania. *Int. J. Primatol.* 5:411–434.

Kay, R. F. (1981). The nut-crackers—a new theory of the adaptations of the Ramapithecinae. *Am. J. Phys. Anthropol.* 55:141–151.

Kay, R. F. (1990). The phyletic relationships of extant and fossil Pitheciinae (Platyrrhini, Anthropoidea). *J. Hum. Evol.* 19:175–208.

Kay, R. F. (2003). Review of the primate fossil record. *Am. J. Hum. Biol.* 15:839–840.

Kay, R. F., and Cozzuol, M. A. (2006). New platyrrhine monkeys from the Solimões Formation (late Miocene, Acre State, Brazil). *J. Hum. Evol.* 50:673–686.

Kay, R. F., and Kirk, E. C. (2000). Osteological evidence for the evolution of activity pattern and visual acuity in primates. *Am. J. Phys. Anthropol.* 113:235–262.

Kay, R. F., Plavcan, J. M., Glander, K. E., and Wright, P. C. (1988). Sexual selection and canine dimorphism in New World monkeys. *Am. J. Phys. Anthropol.* 77:385–397.

Kay, R. F., Ross, C., and Williams, B. A. (1997a). Anthropoid origins. *Science* 275:797–804.

Kay, R. F., Madden, R. H., Cifelli, R. L., and Flynn, J. J. (eds.) (1997b). *Vertebrate Paleontology in the Neotropics.* Smithsonian Institution Press, Washington DC.

Kay, R. F., Fleagle, J. G., Mitchell, T. R. T., Colbert, M., Bown, T., and Powers, D. W. (2008). The anatomy of *Dolichocebus gaimanensis,* a stem platyrrhine monkey from Argentina. *J. Hum. Evol.* 54:323–382.

Kay, R. N. B. (1985). Comparative studies of food propulsion in ruminants. In: Ooms, L. A. A., Degryse, A. D., and van Miert, A. S. J. A. M. (eds.), *Physiological and Pharmacological Aspects of the Reticulo-Rumen.* Martinus Nijhoff, Dordrecht. pp. 155–170.

Kay, R. N. B., and Davies, A. G. (1994). Digestive physiology. In: Davies, A. G., and Oates, J. F. (eds.), *Colobine Monkeys: Their Ecology, Behavior and Evolution.* Cambridge University Press, Cambridge. pp. 229–259.

Keane, B., Dittus, W. P. J., and Melnick, D. J. (1997). Paternity assessment in wild groups of toque macaques *Macaca sinica* at Polonnaruwa, Sri Lanka using molecular markers. *Mol. Ecol.* 6:267–282.

Keddy, A. C. (1986). Female mate choice in vervet monkeys (*Cercopithecus aethiops sabaeus*). *Am. J. Primatol.* 10:125–134.

Keddy-Hector, A. C. (1992). Mate choice in non-human primates. *Am. Zool.* 32:62–70.

Keller, L. (1994). Rewards of promiscuity. *Nature* 372:229–230.

Kelley, E. A., and Sussman, R. W. (2007). An academic genealogy on the history of American field primatology. *Am. J. Phys. Anthropol.* 132:406–425.

Kelley, J. (2002). The hominoid radiation in Asia. In: Hartwig, W. C. (ed.), *The Primate Fossil Record.* Cambridge University Press, Cambridge. pp. 369–384.

Kelley, J. (2004). Life history and cognitive evolution in the apes. In: Russon, A. E., and Begun, D. (eds.), *The Evolution of Thought.* Cambridge University Press, New York. pp. 280–297.

Kelley, J., and Smith, T. (2003). Age at first molar emergence in early Miocene *Afropithecus turkanensis* and life-history evolution in the Hominoidea. *J. Hum. Evol.* 44:307–329.

Kelley, J., Andrews, P., and Alpagut, B. (2008). A new hominoid species from the middle Miocene site of Paşalar, Turkey. *J. Hum. Evol.* 54:455–479.

Kemnitz, J. W., Sapolsky, R. M., Altmann, J., Muruthi, P., Mott, G. E., and Stefanick, M. L. (2002). Effects of food availability on serum insulin and lipid concentrations in free-ranging baboons. *Am. J. Primatol.* 57:13–19.

Kendal, J. R., Kendal, R. L., and Laland, K. N. (2007). Quantifying and modelling social learning in monkey populations. *Int. J. Psychol. Psychol. Therapy* 7:123–138.

Kendal, R. L., Coe, R. L., and Laland, K. N. (2005). Age differences in neophilia, exploration, and innovation in family groups of Callitrichid monkeys. *Am. J. Primatol.* 66:167–188.

Kendrick, K. M., and Dixson, A. F. (1983). The effect of the ovarian cycle on the sexual behaviour of the common marmoset (*Callithrix jacchus*). *Physiol. Behav.* 30:735–742.

Kennedy, E. H., and Fragaszy, D. M. (2008). Analogical reasoning in a capuchin monkey (*Cebus apella*). *J. Comp. Psychol.* 122(2):167–175.

Kenward, R. E. (1978). Hawks and doves: factors affecting success and selection in goshawk attacks on wild pigeons. *J. Anim. Ecol.* 47:449–460.

Kenwood, R. (2001). *A Manual for Wildlife Radio Tracking.* London: Academic Press.

Kerfoot, W. C., and Sih, A. (1987). Introduction. In: Kerfoot, W. C., and Sih, A. (eds.), *Predation.* University Press of New England, Hanover. pp. vii–viii.

Kessler, P. (1998). Primate densities in the natural reserve of Nourague, French Guiana. *Neotrop. Primates* 6:45–46.

Keverne, E. B. (1976). Sexual receptivity and attractiveness in the female rhesus monkey. *Adv. Study Behav.* 7:155–196.

Key, C., and Aiello, L. C. (2000). A prisoner's dilemma model of the evolution of paternal care. *Folia Primatol.* 71:77–92.

Khaitovich, P., Muetzel, B., She, X., Lachmann, M., Hellmann, I., Dietzsch, J., Steigele, S., Do, H.-H., Weiss, G., Enard, W., Heissig, F., Arendt, T., Nieselt-Struwe, K., Eichler, E. E., and Pääbo, S. (2004). Regional patterns of gene expression in human and chimpanzee brains. *Genome Res.* 14:1462–1473.

Khaitovich, P., Hellman, I., Enard, W., Nowick, K., Leinweber, M., Franz, H., Weiss, G., Lachmann, M., and Pääbo, S. (2005). Parallel patterns of evolution in the genomes and transcriptomes of humans and chimpanzees. *Science* 309:1850–1854.

Khan, S. A., Katzija, G., Lindberg, M., and Dicsfalusy, E. (1985). Radioimmunoassay of luteinizing hormone in the baboon (*Papio hamadryas*). *J. Med. Primatol.* 14:143–157.

Kholkute, S. D., Joseph, R., Joshi, U. M., and Munshi, S. R. (1981). Diurnal variations of serum testosterone levels in the male bonnet monkey (*Macaca radiata*). *Primates* 22:427–430.

Khoroche, P. (1989). *Once the Buddha Was a Monkey: Arya Sura's Jatakamala.* University of Chicago Press, Chicago.

Kibunjia, M., Roche, H., Brown, F., and Leakey, R. E. F. (1992). Pliocene and Pleistocene archaeological sites west of Lake Turkana. *J. Hum. Evol.* 23:431–438.

Kierulff, M. C. M. (2000). Ecology and behaviour of translocated groups of golden lion tamarins (*Leontopithecus rosalia*) [PhD diss.]. University of Cambridge, Cambridge.

Kierulff, M. C. M., Oliveira, P. P., Beck, B. B., and Martins, A. (2002a). Reintroduction and translocation as conservation tools for golden lion tamarins. In: Kleiman, D. G., and Rylands, A. B. (eds.), *Lion Tamarins.* Smithsonian Institution Press, Washington DC. pp. 271–282.

Kierulff, M. C. M., Raboy, B. E., Procópio de Oliveira, P., Miller, K., Passos, F. C., and Prado, F. (2002b). Behavioral ecology of lion tamarins. In: Kleiman, D. G., and Rylands, A. B. (eds.), *Lion Tamarins: Biology and Conservation.* Smithsonian Institution Press, Washington DC. pp. 157–187.

Kierulff, M. C. M., Rylands, A. B., Mendes, S. L., and de Oliveira, M. M. (2008). *Leontopithecus chrysomelas.* In: 2008 IUCN Red List of Threatened Species, www.iucnredlist.org (accessed January 25, 2009).

Kilner, R., and Johnstone, R. A. (1997). Begging the question: are offspring solicitation behaviours signals of need? *Trends Ecol. Evol.* 12:11–15.

Kimura, T., and Hamada, Y. (1990). Development of epiphyseal union in Japanese macaques of known chronological age. *Primates* 31:79–93.

Kindleman, P., Dixon, A. F. G., and Dostalkova, I. (1999). Does body size optimization result in skewed body size distribution on a logarithmic scale? *Am. Nat.* 153:445–447.

King, B. J. (ed.) (1999). *The Origins of Language.* School of American Research Press, Santa Fe, NM.

King, S. J. (2004). Relative timing of ontogenetic events in primates. *J. Zool. Lond.* 264:267–280.

Kingdon, J. S. (1971). *East African Mammals: An Atlas of Evolution in Africa,* vol. 1. Academic Press, London.

Kingdon, J. (1974). *East African Mammals: An Atlas of Evolution in Africa.* University of Chicago Press, Chicago.

Kingdon, J. (1987). What are face patterns and do they contribute to reproductive isolation in guenons? In: Gautier-Hion, A., Bourlière, F., Gautier, J. P., and Kingdon, J. (eds.), *A Primate Radiation: Evolutionary Biology of the African Guenons.* Cambridge University Press, New York. pp. 227–245.

Kingdon, J. (1997). *The Kingdon Field Guide to African Mammals.* Academic Press, London.

Kingdon, J., and Butynski, T. M. (2008). *Chlorocebus aethiops.* In: IUCN 2009. IUCN Red List of Threatened Species, version 2009.1, www.iucnredlist.org (accessed June 23, 2009).

Kingdon, J., Gippoliti, S., Butynski, T. M., Lawes, M. J., Eeley, H., Lehn, C., and de Jong, Y. (2008a). *Cercopithecus mitis.* In: IUCN 2009. IUCN Red List of Threatened Species, version 2009.1, www.iucnredlist.org (accessed June 23, 2009).

Kingdon, J., Butynski, T. M., and de Jong, Y. (2008b). *Erythrocebus patas.* In: IUCN 2009. IUCN Red List of Threatened Species, version 2009.1, www.iucnredlist.org (accessed June 23, 2009).

Kingdon, J., Happold, D., and Butynski, T. (in press). *The Mammals of Africa.* University of California Press, Berkeley, CA.

Kingsley, S. R. (1982). Causes of non-breeding and the development of secondary sexual characteristics in the male orangutan: a hormonal study. In: de Boer, L. E. M. (ed.), *The Orangutan: Its Biology and Conservation.* Dr. W. Junk Publishers, The Hague. pp. 215–229.

Kingsley, S. R. (1988). Physiological development of male orang-utans and gorillas. In: Schwartz, J. H. (ed.), *Orang-utan Biology.* Oxford University Press, New York. pp. 123–131.

Kinnaird, M. F. (1990a). Behavioral and demographic responses to habitat change by the Tana River crested mangabey (*Cercocebus galeritus galeritus*) [PhD diss.]. University of Florida, Gainesville.

Kinnaird, M. F. (1990b). Pregnancy, gestation and parturition in free-ranging Tana River crested mangabeys. *Am. J. Primatol.* 22:285–289.

Kinnaird, M. F. (1992). Variable resource defense by the Tana River crested mangabey. *Behav. Ecol. Sociobiol.* 31:115–122.

Kinnaird, M. F., and O'Brien, T. G. (2000).Comparative movement patterns of two semiterrestrial cercopithecine primates: the Tana River crested mangabey and the Sulawesi crested black macaque. In: Boinski, S., and Garber, P. A. (eds.), *On the Move: How and Why Animals Travel in Groups.* University of Chicago Press, Chicago. pp. 327–350.

Kinnaird, M. F., O'Brien, T. G., and Suryadi, S. (1999). The importance of figs to Sulawesi's imperiled wildlife. *Trop. Biodivers.* 6:5–18.

Kinzey, W. G. (1967). Preface. *Am. J. Phys. Anthropol.* 26:115–118.

Kinzey, W. G. (1971). Male reproductive system and spermato-genesis. In: Hafez, E. S. E. (ed.), *Comparative Reproduction of Nonhuman Primates.* Charles C. Thomas, Springfield, IL. pp. 85–114.

Kinzey, W. G. (1976). Positional behavior and ecology in *Callicebus torquatus. Ybk. Phys. Anthropol.* 20:468–480.

Kinzey, W. G. (1978). Feeding behaviour and molar features in two species of titi monkey. In: Chivers, D. J., and Herbert, J. (eds.), *Recent Advances in Primatology,* vol. 1. Academic Press, London. pp. 373–385.

Kinzey, W. G. (1981). The titi monkeys, genus *Callicebus.* In: Coimbra-Filho, A. F., and Mittermeier, R. A. (eds.), *Ecology and Behavior of Neotropical Primates,* vol. 1. Academia Brasileira de Ciências, Rio de Janeiro. pp. 241–276.

Kinzey, W. G. (1986). New World primate field studies: what's in it for anthropology? *Annu. Rev. Anthropol.* 15:121–148.

Kinzey, W. G. (1992). Dietary and dental adaptations in the Pitheciinae. *Am. J. Phys. Anthropol.* 88:499–514.

Kinzey, W. G. (1997a). *Callicebus.* In: Kinzey, W. G. (ed.), *New World Primates: Ecology, Evolution and Behavior.* Aldine de Gruyter, New York. pp. 213–221.

Kinzey, W. G. (1997b). New World primate studies. In: Spencer, F. (ed.), *History of Physical Anthropology: An Encyclopedia.* Garland, New York. pp. 743–748.

Kinzey, W. G. (1997c). *New World Primates. Ecology, Evolution and Behavior.* Aldine Press, New York.

Kinzey, W. G. (1997d). *Aotus.* In: Kinzey, W. G. (ed.), *New World Primates: Ecology, Evolution and Behavior.* Aldine de Gruyter, New York. pp. 186–191.

Kinzey, W. G., and Becker, M. (1983). Activity pattern of the masked titi monkey, *Callicebus personatus. Primates* 24:337–343.

Kinzey, W. G., and Cunningham, E. P. (1994). Variability in platyrrhine social organization. *Am. J. Primatol.* 34:185–198.

Kinzey, W. G., and Gentry, A. H. (1979). Habitat utilization in two species of *Callicebus*. In: Sussman, R. W. (ed.), *Primate Ecology: Problem Oriented Field Studies.* John Wiley and Sons, New York. pp. 89–100.

Kinzey, W. G., and Norconk, M. A. (1990). Hardness as a basis of fruit choice in two sympatric primates. *Am. J. Phys. Anthropol.* 81:5–15.

Kinzey, W. G., and Norconk, M. A. (1993). Physical and chemical properties of fruit and seeds eaten by *Pithecia* and *Chiropotes* in Suriname and Venezuela. *Int. J. Primatol.* 14:207–227.

Kinzey, W. G., and Robinson, J. G. (1983). Intergroup loud calls, range size, and spacing in *Callicebus torquatus*. *Am. J. Phys. Anthropol.* 60:539–544.

Kirchhof, J., and Hammerschmidt, K. (2006). Functionally referential alarm calls in tamarins (*Saguinus fuscicollis* and *Saguinus mystax*)—evidence from playback experiments. *Ethology* 112(4):346–354.

Kirk, E. C. (2006). Eye morphology in cathemeral lemurids and other mammals. *Folia Primatol.* 77:27–49.

Kirkness, E. F., Bafna, V., Halpern, A. L., Levy, S., Remington, K., Rusch, D. B., Delcher, A. L., Pop, M., Wang, W., Fraser, C. M., and Venter, J. C. (2003). The dog genome: survey sequencing and comparative analysis. *Science* 301:1898–1903.

Kirkpatrick, R. C. (1996). Ecology and behavior of the Yunnan snub-nosed langur (*Rhinopithecus bieti*, Colobinae) [PhD diss.]. University of California, Davis.

Kirkpatrick, R. C. (1998). Ecology and behavior of the snub-nosed and douc langurs. In: Jablonski, N. G. (ed.), *The Natural History of the Doucs and Snub-Nosed Monkeys.* World Scientific Press, Singapore. pp. 155–190.

Kirkpatrick, R. C. (1999). Colobine diet and social organization. In: Dolhinow, P., and Fuentes, A. (eds.), *The Nonhuman Primates.* Mayfield, Mountain View, CA. pp. 93–105.

Kirkpatrick, R. C. (2000). The evolution of human homosexual behavior. *Curr. Anthropol.* 41:385–413.

Kirkpatrick, R. C., Long, Y. C., Zhong, T., and Xiao, L. (1998). Social organization and range use in the Yunnan snub-nosed. *Int. J. Primatol.* 19:13–52.

Kirkpatrick, R. C., Gu, H. J., and Zhou, X. P. (1999). A preliminary report on the ecology of *Rhinopithecus roxellana* at Baihe Nature Reserve. *Folia Primatol.* 70:117–120.

Kitahara-Frisch, J., and Norikoshi, K. (1982). Spontaneous sponge-making in captive chimpanzees. *J. Hum. Evol.* 11:41–47.

Kitchen, D. M., and Beehner, J. C. (2007). Factors affecting individual participation in group-level aggression among non-human primates. *Behaviour* 144:1551–1581.

Kitchen, D. M., Seyfarth, R. M., Fischer, J., and Cheney, D. L. (2003). Loud calls as indicators of dominance in male baboons (*Papio cynocephalus ursinus*). *Behav. Ecol. Sociobiol.* 53:374–384.

Kitchen, D. M., Cheney, D. L., and Seyfarth, R. M. (2004). Factors mediating inter-group encounters in savannah baboons (*Papio cynocephalus ursinus*). *Behaviour* 141:197–218.

Kitchen, D. M., Beehner, J. C., Bergman, T. J., Cheney, D. L., Crockford, C., Engh, A. L., Fischer, J., Seyfarth, R. M., and Wittig, R. M. (2009). The causes and consequences of male aggression directed at female chacma baboons. In: Muller, M. N., and Wrangham, R. W. (eds.), *Sexual Coercion in Primates and Humans: An Evolutionary Perspective on Male Aggression Against Females.* Harvard University Press, Cambridge, MA. pp. 128–156.

Kitopeni, R., Kasagula, M., and Turner, L. A. (1995). Ntologi falls? *PanAfr. News* 2:9–11.

Klaus, G., Klaus-Hugi, C., and Schmid, B. (1998). Geophagy by large mammals at natural licks in the rain forest of the Dzanga National Park, Central African Republic. *J. Trop. Ecol.* 14:207–227.

Klecha, F., and McCann, C. (2001). A profile of progesterone excretion during pregnancy in gelada baboons (*Theropithecus gelada*). In: Chan, S. D. (ed.), American Association of Zoo Keepers 27th National Conference. Columbus Zoo and Aquarium, Powell. pp. 62–71.

Kleiber, M. (1961). *The Fire of Life: An Introduction to Animal Energetics.* John Wiley & Sons, New York.

Kleiman, D. (1977). Monogamy in mammals. *Q. Rev. Biol.* 52:39–69.

Kleiman, D. G. (1996). Reintroduction programs. In: Kleiman, D. G., Allen, M. E., Thompson, K. V., and Lumpkin, S. (eds.), *Wild Mammals in Captivity.* University of Chicago Press, Chicago. pp. 297–305.

Kleiman, D. G., and Rylands, A. B. (2002). *Lion Tamarins: Biology and Conservation.* Smithsonian Institution Press, Washington DC.

Klein, L. L. (1971). Observations on copulation and seasonal reproduction of two species of spider monkey, *Ateles belzebuth* and *A. geoffroyi. Folia Primatol.* 15:233–248.

Klein, L. L. (1972). The ecology and social behavior of the spider monkey, *Ateles belzebuth* [PhD diss.]. University of California, Berkeley.

Klein, L., and Klein, D. (1971). Aspects of social behavior in a colony of spider monkeys *Ateles geoffroyi* at the San Francisco Zoo. *Int. Zoo Ybk.* 11:175–181.

Klein, L. L., and Klein, D. J. (1976). Neotropical primates: aspects of habitat usage, population density, and regional distribution in La Macarena, Colombia. In: Thorington, R. W., Jr., and Heltne, P. G. (eds.), *Neotropical Primates: Field Studies and Conservation.* National Academy of Sciences, Washington DC. pp. 70–78.

Klein, L. L., and Klein, D. J. (1977). Feeding behavior of the Colombian spider monkey, *Ateles belzebuth.* In: Clutton-Brock, T. H. (ed.), *Primate Ecology: Studies of Feeding and Ranging Behaviour in Lemurs, Monkeys, and Apes.* Academic Press, London. pp. 153–181.

Klein, R. G. (1984). Mammalian extinctions and Stone Age people in Africa. In: Martin, P. S., and Klein, R. G. (eds.), *Quaternary Extinctions: A Prehistoric Revolution.* University of Arizona Press, Tucson. pp. 553–573.

Klein, R. G. (1987). Problems and prospects in understanding how early people exploited animals. In: Nitecki, M. H., and Nitecki, D. V. (eds.), *The Evolution of Human Hunting.* Plenum Press, New York. pp. 11–45.

Klinger, H. (1963). The somatic chromosomes of some primates: *Tupaia glis, Nycticebus coucang, Tarsius bancanus, Cercocebus aterrimus, Symphalangus syndactylus. Cytogenetics* 2:140–151.

Klinkova, E., Hodges, J. K., Fuhrmann, K., de Jong, T., and Heistermann, M. (2005). Male dominance rank, female mate choice and male mating and reproductive success in captive chimpanzees. *Int. J. Primatol.* 26:357–384.

Kloss, C. B. (1908). The white-handed gibbon. *J. Straits Branch R. Asia. Soc.* 50:79–80.

Kluger, M. J., Ringler, D. H., and Anver, M. R. (1975). Fever and survival. *Science* 188:166–168.

Klüver, H. (1933). *Behavior Mechanisms in Monkeys.* University of Chicago Press, Chicago.

Klüver, H. (1937). Re-examination of implement-using behavior in a cebus monkey after an interval of three years. *Acta Psychol.* 2:347–395.

Knapp, L. A., Ha, J. C., and Sackett, G. P. (1996). Parental MHC antigen sharing and pregnancy wastage in captive pigtailed macaques. *J. Reprod. Immunol.* 32:73–88.

Knapp, R. (1984). Sampling methods and taxon analysis in vegetation science. In: *Handbook of Vegetation Science*, vol. 1, pt. 4. Dr. W. Junk Publishers. The Hague, The Netherlands.

Knezevich, M. (1998). Geophagy as therapeutic mediator of endoparasitism in a free-ranging group of rhesus macaques (*Macaca mulatta*). *Am. J. Primatol.* 44:71–82.

Knight, J. (1999). Monkeys on the move: the natural symbolism of people-macaque conflict in Japan. *J. Asian Stu.* 58:622–647.

Knogge, C., and Heymann, E. W. (2003). Seed dispersal by sympatric tamarins, *Saguinus mystax* and *Saguinus fuscicollis*: diversity and characteristics of plant species. *Folia Primatol.* 74:33–47.

Knop, E., Ward, P. I., and Wich, S. A. (2004). A comparison of orangutan density in a logged and unlogged forest in Sumatra. *Biol. Conserv.* 120:197–192.

Knopff, K. H., Knopff, A. R. A., and Pavelka, M. S. M. (2004). Observed case of infanticide committed by a resident male Central American black howler monkey (*Alouatta pigra*). *Am. J. Primatol.* 63:239–244.

Knott, C. D. (1997a). Field collection and preservation of urine in orangutans and chimpanzees. *Trop. Biodivers.* 4:95–102.

Knott, C. D. (1997b). Interactions between energy balance, hormonal patterns, and mating behavior in wild Borean orangutans (*Pongo pygmaeus*). *Am. J. Primatol.* 42:124.

Knott, C. D. (1998a). Changes in orangutan diet, caloric intake and ketones in response to fluctuating fruit availability. *Int. J. Primatol.* 19:1061–1079.

Knott, C. D. (1998b). Orangutans in the wild. *Natl. Geogr.* 194:30–57.

Knott, C. D. (1998c). Social system dynamics, ranging patterns and male and female strategies in wild Bornean orangutans (*Pongo pgymaeus*). *Am. J. Phys. Anthropol.* (suppl. 26):140.

Knott, C. D. (1999a). Orangutan behavior and ecology. In: Dolhinow, P., and Fuentes, A. (eds.), *The Nonhuman Primates.* Mayfield Press, Mountain View, CA. pp. 50–57.

Knott, C. D. (1999b). Reproductive, physiological and behavioral responses of orangutans in Borneo to fluctuations in food availability [PhD diss.]. Harvard University, Cambridge, MA.

Knott, C. D. (2001). Female reproductive ecology of the apes: implications for human evolution. In: Ellison, P. (ed.), *Reproductive Ecology and Human Evolution.* Aldine de Gruyter, New York. pp. 429–463.

Knott, C. D. (2004). Orangutans: the largest canopy dwellers. In: Lowman, M. D., and Rinker, H. B. (eds.), *Forest Canopies*. Academic Press, San Diego. pp. 322–324.

Knott, C. D. (2005a). Energetic responses to food availability in the great apes: implications for hominin evolution. In: Brockman, D. K., and van Schaik, C. P. (eds.), *Primate Seasonality: Implications for Human Evolution*. Cambridge University Press, Cambridge. pp. 351–378.

Knott, C. D. (2005b). Radioimmunoassay of estrone conjugates from urine dried on filter paper. *Am. J. Primatol.* 67(1):121–35.

Knott, C.D. (2009). Orangutans:sexual coercion without sexual violence. In: Muller M. N., and Wrangham, R. W. (eds.), *Sexual Coercion in Primates: An Evolutionary Perspective on Male Aggression against Females*. Cambridge, MA: Harvard University Press.Pp 81–111.

Knott, C. D., and Kahlenberg, S. (2007). Orangutans in perspective: forced copulations and female mating resistance. In: Campbell C. J., Fuentes, A., MacKinnon, K. C., Panger, M., and Bearder, S. (eds.) *Primates in Perspective*. Oxford University Press, Oxford. pp. 290–305.

Knott, C. D., Beaudrot, L., Snaith, T., White, S., Tschauner, H., and Planansky, G. (2008). Female–female competition in Bornean orangutans. *Int. J. Primatol.* 29:975–997.

Knott, C. D., Emery Thompson, M., and Wich, S.A. (2009b). The ecology of reproduction in wild orangutans. In: Wich, S. A., Utami, S. S., Mitra Setia, T., and van Schaik, C. (eds.), *Orangutans: Geographic Variation in Behavioral Ecology and Conservation*. Oxford University Press, Oxford. pp. 171–188.

Koehler, U., Arnold, N., Wienberg, J., Tofanelli, S., and Stanyon, R. (1995a). Genomic reorganization and disrupted chromosomal synteny in the siamang (*Hylobates syndactylus*) revealed by fluorescence in situ hybridization. *Am. J. Phys. Anthropol.* 97:37–47.

Koehler, U., Bigoni, F., Wienberg, J., and Stanyon, R. (1995b). Genomic reorganization in the concolor gibbon (*Hylobates concolor*) revealed by chromosome painting. *Genomics* 30:287–292.

Koenders, L., Rumpler, Y., and Ratisirarson, J. (1985). *Lemur macaco flavifrons*: a rediscovered subspecies of primate. *Folia Primatol.* 44:210–215.

Koenig, A. (1995). Group size, composition, and reproductive success in wild common marmosets (*Callithrix jacchus*). *Am. J. Primatol.* 35:311–317.

Koenig, A. (2002). Competition for resources and its behavioral consequences among female primates. *Int. J. Primatol.* 23:759–783.

Koenig, A., and Borries, C. (2001). Socioecology of Hanuman langurs: the story of their success. *Evol. Anthropol.* 10:122–137.

Koenig, A., Beise, J., Chalise, M. K., and Ganzhorn, J. U. (1998). When females should contest for food-testing: hypotheses about resource density, distribution, size, and quality with Hanuman langurs (*Presbytis entellus*). *Behav. Ecol. Sociobiol.* 42:225–237.

Kofler, R., Schlotterer, C., and Lelley, T. (2007). SciRoKo: a new tool for whole genome microsatellite search and investigation. *Bioinfor. Appl. Note* 23:1683–1685.

Kohda, M. (1985). Allomothering behaviour of New and Old World monkeys. *Primates* 26:28–44.

Kohler, W. (1925). *The Mentality of Apes*. Kegan Paul, Trench, Trubner, London.

Kokko, H., and Jennions, M. (2003). It takes two to tango. *Trends Ecol. Evol.* 18:103–104.

Kokko, H., and Johnstone, R.A. (1999). Social queuing in animal societies: a dynamic model of reproductive skew. *Proc. R. Soc. Lond. B.* 266:571–578.

Kollmann, J., Coomes, D. A., and White, S. M. (1998). Consistencies in post-dispersal seed predation of temperate fleshy-fruited species among seasons, years, and sites. *Funct. Ecol.* 12:683–690.

Köndgen, S., Kühl, H., N'Goran, P. K., Walsh, P. D., Schenk, S., Ernst, N., Biek, R., Formenty, P., Mätz-Rensing, K., Schweiger, B., Junglen, S., Ellerbrok, H., Nitsche, A., Briese, T., Lipkin, W. I., Pauli, G., Boesch, C., and Leendertz, F. H. (2008). Pandemic human viruses cause decline of endangered great apes. *Curr. Biol.* 18:260–264.

Konrad, R., and Geissmann, T. (2006). Vocal diversity and taxonomy of *Nomascus* in Cambodia. *Int. J. Primatol.* 27: 713–745.

Konstant, W. R., Mittermeier, R. A., Rylands, A. B., Butynski, T. M., Eudey, A. A., Ganzhorn, J., and Kormos, R. (2002). The world's 25 most endangered primates–2002. *Neotrop. Primates* 10:128–131.

Kool, K. M. (1992). Food selection by the silver leaf monkey, *Trachypithecus auratus sondaicus*, in relation to plant chemistry. *Oecologia* 90:527–533.

Kool, K. M. (1993). The diet and feeding behavior of the silver leaf monkey (*Trachypithecus auratus sondaicus*) in Indonesia. *Int. J. Primatol.* 14:667–700.

Koops, K., Humle, T., Sterck, E. H. M., and Matsuzawa, T. (2007). Ground-nesting by the chimpanzees of the Nimba Mountains, Guinea: environmentally or socially determined? *Am. J. Primatol.* 69:407–419.

Kordos, L., and Begun, D. R. (2002). Rudabanya: a late Miocene subtropical swamp deposit with evidence of the origin of the African apes and humans. *Evol. Anthropol.* 11:45–57.

Kornell, N. (2009). Metacognition in humans and animals. *Curr. Dir. Psychol. Sci.* 18:11–15.

Kornell, N., Son, L. K., and Terrace, H. S. (2007). Transfer of metacognitive skills and hint seeking in monkeys. *Psychol Sci.* 18:64–71.

Korstjens, A. H. (2001). The mob, the secret sorority, and the phantoms: an analysis of the socio-ecological strategies of the three colobines at Taï [PhD diss.]. University of Utrecht, Utrecht.

Korstjens, A. H., and Noë, R. (2004). Mating system of an exceptional primate, the olive colobus (*Procolobus verus*). *Am. J. Primatol.* 62:261–273.

Korstjens, A. H., and Schippers, E. Ph. (2003). Dispersal patterns among olive colobus in Taï National Park. *Int. J. Primatol.* 24:515–539.

Korstjens, A. H., Sterck, E. H. M., and Noe, R. (2002). How adaptive or phylogenetically inert is primate social behaviour? A test with two sympatric colobines. *Behaviour* 139:203–225.

Korstjens, A. H., Bergman, K., Deffernez, C., Krebs, M., Nijssen, E. C., van Oirschot, B. M. A., Paukert, C., and Schippers, E. P. (2007). How small-scale differences in food competition lead to different social systems in three closely related sympatric colobines. In: McGraw, S., Zuberbühler, K., and Noë, R. (eds.), *Monkeys of the Taï Forest: An African Primate Community*. Cambridge University Press, Cambridge. pp. 72–108.

Kose, M., Maend, R., and Møller, A. P. (1999). Sexual selection for white tail spots in the barn swallow in relation to habitat choice by feather lice. *Anim. Behav.* 58:1201–1205.

Koski, S. E., Koops, K., and Sterck, E. H. M. (2007a). Reconciliation, relationship quality, and postconflict anxiety: testing the integrated hypothesis in captive chimpanzees. *Am. J. Primatol.* 69:158–172.

Koski, S. E., de Vries, H., van den Tweel, S. W., and Sterck, E. H. M. (2007b). What to do after a fight? The determinants and inter-dependency of post-conflict interactions in chimpanzees. *Behaviour* 144:529–555.

Koshimizu, K., Ohigashi, H., Huffman, M. A., Nisihda, T., and Takasaki, H. (1993). Physiological activities and the active constituents of potentially medicinal plants used by wild chimpanzees of the Mahale Mountains, Tanzania. *Int. J. Primatol.* 14:345–356.

Kostrub, C. E. (1997). Preliminary field observations of golden-mantled tamarins, *Saguinus tripartitus*, in eastern Ecuador. *Neotrop. Primates* 5:102–103.

Kostrub, C. E. (2003). The social organization and behavior of golden-mantled tamarins, *Saguinus tripartitus*, in eastern Ecuador [PhD diss.]. University of California, Davis.

Kotler, B., Brown, J., and Hasson, O. (1993). Factors affecting gerbil foraging, behavior and rates of owl predation. *Ecology* 72:2249–2260.

Kotliar, N. B., and Wiens, J. A. (1990). Multiple scales of patchiness and patch structure: a hierarchical framework for the study of heterogeneity. *Oikos* 59:253–260.

Kowaleski, M., and Zunino, G. E. (2004). Birth seasonality in *Alouatta caraya* in northern Argentina. *Int. J. Primatol.* 25:283–400.

Kowalewski, M. M., Garber, P. A., Zunino, G. E. (2006). Alternative mating strategies in black and gold howler monkeys. *Am. J. Primatol.* 68(suppl. 1): 95.

Koyama, N. (1970). Changes in dominance rank and division of a wild Japanese monkey troop in Arashiyama. *Primates* 11:335–390.

Koyama, N. (1991). Troop division and inter-troop relationships of ring-tailed lemurs (*Lemur catta*) at Berenty, Madagascar. In: Ehara, A., Kimura, T., and Iwamoto, M. (eds.), *Primatology Today*. Elsevier, Amsterdam. pp. 173–176.

Koyama, N. F. (2001). The long-term effects of reconciliation in Japanese macaques (*Macaca fuscata*). *Ethology* 107:975–987.

Koyama, N. F. (2002). Reconciliation and coalition formation in Japanese macaques (*Macaca fuscata*). Abstract presented at the First European Conference on Behavioural Biology. p. 61.Muenster, Germany, July 31-August 4, 2002

Koyama, N., Takahata, Y., Huffman, M. A., Norikoshi, K., and Suzuki, H. (1992). Reproductive parameters of female Japanese macaques: thirty years data from the Arashiyama troops, Japan. *Primates* 33:33–47.

Koyama, N., Nakamichi, M., Oda, R., Miyamoto, N., Ichino, S., and Takahata, Y. (2001). A ten-year summary of reproductive parameters for ring-tailed lemurs at Berenty, Madagascar. *Primates* 42:1–14.

Koyama, N., Nakamichi, M., Ichino, S., and Takahata, Y. (2002). Population and social dynamics changes in ring-tailed lemur troops at Berenty, Madagascar, between 1989–1999. *Primates* 43:291–314.

Kozlowski, J., and Gawelczyk, A. T. (2002). Why are species' body size distributions usually skewed to the right? *Funct. Ecol.* 16:419–432.

Kozlowski, J., and Wiener, J. (1997). Interspecific allometries are by-products of body size optimization. *Am. Nat.* 149:352–380.

Krachun, C., and Call, J. (2008). Chimpanzees (*Pan troglodytes*) know what can be seen from where. *Anim Cogn.* 12:317–331.

Krachun, C., Carpenter, M., Call, J., and Tomasello, M. (2008). Children read minds but apes read reaches in a competitive nonverbal false belief task. *Anim. Cogn.* 109:224–234.

Krakauer, E. B. (2006). Social influence on the development of foraging in aye-ayes (*Daubentonia madagascariensis*) and ruffed lemurs (*Varecia variegata rubra*) [abstract]. *Am. J. Phys. Anthropol.* (suppl. 42):116.

Kralik, J. D. (2005). Inhibitory control and response selection in problem solving: how cotton-top tamarins (*Saguinus oedipus*) overcome a bias for selecting the larger quantity of food. *J. Comp. Psychol.* 119:78–89.

Krebs, C. J. (1972). *Ecology: The Experimental Analysis of Distribution and Abundance.* Harper and Row, New York.

Krebs, C. J. (1999). *Ecological Methodology,* 2nd ed. Benjamin/Cummings, Menlo Park, CA.

Krebs, C. J. (2000). *Ecology: The Experimental Analysis of Distribution and Abundance,* 5th ed. Pearson Addison Wesley, Boston.

Krebs, J., and Davies, N. (1984). *Behavioral Ecology.* Blackwell Scientific, Oxford.

Krebs, J. R., and Davies, N. B. (eds.) (1997). *Behavioural Ecology: An Evolutionary Approach.* Blackwell Science, Oxford.

Kremen, C., Niles, J. O., Dalton, M. G., Daily, G. C., Ehrlich, P. R., Fay, J. P., Grewal, D., and Guillery, R. P. (2000). Economic incentives for rain forest conservation across scales. *Science* 288:1828–1832.

Krepel, H. P., Baeta, S., Kootstra, C. J., and Polderman, A. M. (1995). Reinfection patterns of *Oesophagostomum bifurcum* and hookworm after anthelmintic treatment. *Trop. Geogr. Med.* 47:160–163.

Kreulen, D. A. (1985). Lick use by large herbivores: a review of benefits and banes of soil consumption. *Mammal Rev.* 15:107–123.

Kreulen, D. A., and Jager, T. (1984). The significance of soil ingestion in the utilization of arid rangelands by large herbivores with special reference to natural licks on the Kalahari pan. In: Gilchrist, F. M. C., and Mackie, R. I. (eds.), *Herbivore Nutrition in the Subtropics and Tropics.* Science Press, Johannesburg. pp. 204–221.

Krief, S. (2003). Métabolites secondaires des plantes et comportement animal: surveillance sanitaire et observations de l'alimentation de chimpanzés (*Pan troglodytes schweinfurthii*) en Ouganda. Activités biologiques et étude chimique de plantes consommées [diss.]. Museum National d'Histoire Naturelle, Paris.

Krief, S., Martin, M.-T., Grellier, P., Kasenene, J., and Sévenet, T. (2004). Novel antimalarial compounds isolated in a survey of self-medicative behavior of wild chimpanzees in Uganda. *Antimicrob. Agents Chemother.* 48:3196–3199.

Krief, S., Huffman, M. A., Sévenet, T., Guillot, J., Bories, C., Hladik, C. M., and Wrangham, R. W. (2005). Non-invasive monitoring of the health of *Pan troglodytes schweinfurthii* in the Kibale National Park, Uganda. *Int. J. Primatol.* 26:467–490.

Krishnamani, R., and Mahaney, W. C. (2000). Geophagy among primates: adaptive significance and ecological consequences. *Anim. Behav.* 59:899–915.

Kroeber, A. L., and Kluckhorn, C. (1963). *Culture: A Critical Review of Concepts and Definitions.* Random House, New York.

Krogman, W. M. (1969). Growth changes in skull, face, jaws, and teeth of the chimpanzee. In: Bourne, G. H. (ed.), *The Chimpanzee.* University Park Press, Baltimore, MD. pp. 104–164.

Krüger, O., Affeldt, E., Brackmann, M., and Milhahn, K. (1998). Group size and composition of *Colobus guereza* in Kyambura Gorge, southwest Uganda, in relation to chimpanzee activity. *Int. J. Primatol.* 19:287–297.

Kruglyak, S., Durrett, R. T., Schug, M. D., and Aquadro, C. F. (1998). Equilibrium distribution of microsatellite repeat length resulting from a balance between slippage events and point mutations. *Proc. Natl. Acad. Sci. USA* 95:10774–10778.

Krutzen, M., Arora, N., Nater, A., and van Schaik, C. P. (2008). Genetic estimates of male and female dispersal in multiple populations of Bornean and Sumatran orangutans (*Pongo spp.*). *Primate Eye* 96: Abstract 371.

Kruuk, L. E. B., Merila, J., and Sheldon, B. C. (2003). When environmental variation short-circuits natural selection. *Trends Ecol. Evol.* 18:207–209.

Kudo, H. (1987). The study of vocal communication of wild mandrills in Cameroon in relation to their social structure. *Primates* 28:289–308.

Kuederling, I., and Heistermann, M. (1997). Ultrasonographic and hormonal monitoring of pregnancy in the saddle back tamarin, *Saguinus fuscicollis.* *J. Med. Primatol.* 26:299–306.

Kuederling, I., Evans, C. S., Abbott, D. H., Pryce, C. R., and Epple, G. (1995). Differential excretion of urinary oestrogen by breeding females and daughters in the red-bellied tamarin (*Saguinus labiatus*). *Folia Primatol.* 64:140–145.

Kuehl, H., Todd, A., Boesch, C., and Walsh, P. D. (2007). Manipulating decay time for efficient large mammal density estimation: gorillas and dung height. *Ecol. Appl.* 17:2403–2414.

Kuehl, H., Maisels, F., Ancrenaz, M., and Williamson, E.A. (2008). *Best Practice Guidelines to Surveys and Monitoring of Great Ape Populations.* IUCN SSC Primate Specialist Group, Gland, Switzerland.

Kuester, J., and Paul, A. (1988). Rank relations of juvenile and subadult natal males of Barbary macaques (*Macaca sylvanus*) at Affenberg Salem. *Folia Primatol.* 51:33–44.

Kuester, J., and Paul, A. (1992). Influence of male competition and female mate choice on male mating success in Barbary macaques (*Macaca sylvanus*). *Behaviour* 120:192–217.

Kuester, J., and Paul, A. (1996). Female–female competition and male mate choice in Barbary macaques (*Macaca sylvanus*). *Behaviour* 133:763–790.

Kuester, J., and Paul, A. (1997). Group fission in Barbary macaques (*Macaca sylvanus*) at Affenberg Salem. *Int. J. Primatol.* 18:941–966.

Kuester, J., Paul, A., and Arnemann, J. (1994). Kinship, familiarity and mating avoidance in Barbary macaques, *Macaca sylvanus. Anim. Behav.* 48:1183–1194.

Kuhn, T. S. (1962). *The Structure of Scientific Revolutions.* University of Chicago Press, Chicago.

Kumar, A., Singh, M., Kumara, H. N., Sharma, A. K., and Bertsch, C. (2001). Male migration in lion-tailed macaques. *Primate Rep.* 59:5–17.

Kumar, S., Dudley, J., Nei, M., and Tamura, K. (2008). MEGA: a biologist-centric software for evolutionary analysis of DNA and protein sequences. *Brief. Bioinform.* 9:299–306.

Kumara, H. N., Singh, M., and Kumar, S. (2006). Distribution, habitat correlates, and conservation of *Loris lydekkerianus* in Karnataka, India. *Int. J. Primatol.* 27:941–969.

Kummer, H. (1968a). *Social Organization of Hamadryas Baboons.* University of Chicago Press, Chicago.

Kummer, H. (1968b). Two variations in the social organization of baboons. In: Jay, P. C. (ed.), *Primates: Studies in Adaptation and Variability.* Holt, Rinehart and Winston, New York. pp 293–312.

Kummer, H. (1971). *Primate Societies: Group Techniques of Ecological Adaptation.* Aldine, Chicago.

Kummer, H. (1973). Dominance versus possession: an experiment on hamadryas baboons. In: Menzel, E. W. J. (ed.), *Symposia of the IVth International Congress of Primatology,* vol. 1. *Precultural Primate Behavior.* Karger, Basel. pp. 226–231.

Kummer, H. (1984). From laboratory to desert and back: a social system of hamadryas baboons. *Anim. Behav.* 32:965–971.

Kummer, H. (1990). The social system of hamadryas baboons and its presumable evolution. In: de Mello, T., Whiten, A., and Byrne, R. W. (eds.), Baboons: behaviour and ecology, use and care. *Selected Proceedings of the XIIth Congress of the International Primatological Society.* Brasilia, Brazil pp. 43–60.

Kummer, H. (1995). *In Quest of the Sacred Baboon—A Scientist's Journey.* Princeton University Press, Princeton, NJ.

Kummer, H., Götz, W., and Angst, W. (1974). Triadic differentiation: an inhibitory process protecting pair bonds in baboons. *Behaviour* 49:62–87.

Kummer, H., Banaja, A. A., Abo-Khatwa, A. N., and Ghandour, A. M. (1981). Mammals of Saudi Arabia: primates: a survey of hamadryas baboons in Saudi Arabia. *Fauna Saudi Arabia* 3:441–471.

Kümmerli, R., and Martin, R. D. (2008). Patterns of infant handling and relatedness in Barbary macaques (*Macaca sylvanus*) on Gibraltar. *Primates* 49:271–282.

Kunz, B. K., and Linsenmair, K. E. (2008). The disregarded west: diet and behavioural ecology of olive baboons in the Ivory Coast. *Folia Primatol.* 79:31–51.

Kupchan, S. M., Hemingway, R. J., Karim, A., and Werner, D. (1969). Tumor inhibitors XLVII. Vernodalin and vernomygdin, two new cytotoxic sesquiterpene lactones from *Vernonia amygdalina* Del. *J. Organ. Chem.* 34:3908–3911.

Kurland, J. A. (1977). *Kin Selection in the Japanese Monkey.* Karger, Basel.

Kuroda, S. (1979). Grouping of the pygmy chimpanzees. *Primates* 20:161–183.

Kuroda, S. (1980). Social behavior of the pygmy chimpanzees. *Primates* 21:181–197.

Kuroda, S. (1992). Ecological interspecies relationships between gorillas and chimpanzees in the Ndoki-Nouabale Reserve, northern Congo. In:

Itoigawa, N., Sugiyama, Y., Sackett, G. P., and Thompson, R. K. R. (eds.), *Topics in Primatology. Behavior, Ecology, and Conservation,* vol. 2. Tokyo University Press, Tokyo. pp. 385–394.

Kuroda, S., Nishihara, T., Suzuki, S., and Oko, R. A. (1996). Sympatric chimpanzees and gorillas in the Ndoki Forest, Congo. In: McGrew, M. C., Marchant, L. F., and Nishida, T. (eds.), *Great Ape Societies.* Cambridge University Press, Cambridge. pp. 71–81.

Kuroshima, H., Scott, N., Stone, B., Menzel, C., and Fragaszy, D. (2008). How young humans, apes and monkeys align objects to surfaces. *Primate Eye.* 96(Sp CD-ROM iss—IPS 2008):Abstract 429.

Kurten, B., and Anderson, E. (1980). *Pleistocene Mammals of North America.* Columbia University Press, New York.

Kurup, G. U., and Kumar, A. (1993). Time budget and activity patterns of lion-tailed macaque (*Macaca silenus*). *Int. J. Primatol.* 14:27–39.

Kutsukake, N. (2000). Matrilineal rank inheritance varies with absolute rank in Japanese macaques. *Primates* 41:321–335.

Kutsukake, N., and Castles, D. L. (2001). Reconciliation and variation in post-conflict stress in Japanese macaques (*Macaca fuscata fuscata*): testing the integrated hypothesis. *Anim. Cogn.* 4:259–268.

Kutsukake, N., and Castles, D. L. (2004). Reconciliation and post-conflict third-party affiliation among wild chimpanzees in the Mahale Mountains, Tanzania. *Primates* 45:157–165.

Kutsukake, N., and Matsusaka, T. (2002). Incident of intense aggression by chimpanzees against an infant from another group in Mahale Mountains National Park, Tanzania. *Am. J. Primatol.* 58:175–180.

Kuze, N., Malim, T. P., and Kohshima, S. (2005). Developmental changes in the facial morphology of the Borneo orangutan (*Pongo pygmaeus*): possible signals in visual communication. *Am. J. Primatol.* 65:353–376.

Kuze, N., Sipangkui, S., Malim, T., Bernard, H., Ambu, L., and Kohshima, S. (2008). Reproductive parameters over a 37-year period of free-ranging female Borneo orangutans at Sepilok Orangutan Rehabilitation Centre. *Primates* 49:126–134.

Lachmann, M., and Bergstrom, C. T. (1998). Signaling among relatives. II. Beyond the Tower of Babel. *Theor. Popul. Biol.* 54:146–160.

Lack, D. (1954). *The Natural Regulation of Animal Numbers.* Oxford University Press, Oxford.

Lacy, R. C. (1993). VORTEX: a computer simulation model for population viability analysis. *Wildl. Res.* 20:45–65.

Lacy, R. C., and Sherman, P. W. (1983). Kin recognition by phenotypic matching. *Am. Nat.* 121:489–512.

Laden, G., and Wrangham, R. W. (2005). The rise of the hominids as an adaptive shift in fallback foods: plant underground storage organs (USOs) and australopith origins. *J. Hum. Evol.* 49:482–498.

Ladygina-Kohts, N. N. (2002). *Infant Chimpanzee and Human Child: A Classic 1935 Comparative Study of Ape Emotions and Intelligence.* Oxford University Press, New York.

Lagercrantz, E. (2008). eBioX version 1.5.1.

Lagreca, L., Marotta, E., and Vales, L. (1992). Relationship between sow's teat number and the litter size at birth and at weaning. *An. Vet. Murica* 1992:863–867.

Lahdenpera, M., Lummaa, V., Helle, S., Tremblay, M., and Russell, A. F. (2004). Fitness benefits of prolonged post-reproductive lifespan in women. *Nature* 428:178–181.

Lahm, S. A. (1985). Mandrill ecology and the status of Gabon's rainforests. *Primate Conserv.* 6:32–33.

Lahm, S. A. (1986). Diet and habitat preference of *Mandrillus sphinx* in Gabon: implications of foraging strategy. *Am. J. Primatol.* 11:9–26.

Laird, A. K. (1967). Evolution of the human growth curve. *Growth* 31:345–355.

Lakoff, G. (1987). *Women, Fire, and Dangerous Things: What Categories Reveal About the Mind.* University of Chicago Press, Chicago.

Laland, K. N., and Galef, B. G. (2009). *The Question of Animal Culture.* Harvard University Press, Cambridge, MA.

Laland, K. N., and Hoppitt, W. (2003). Do animals have culture? *Evol. Anthropol.* 12:150–159.

Laland, K. N., and Janik, V. M. (2006). The animal cultures debate. *Trends Ecol. Evol.* 21:542–547.

Laland, K. N., and Kendal, J. (2003). What the models say about social learning. In: Fragaszy, D. M., and Perry, S. (eds.), *The Biology of Traditions: Models and Evidence.* Cambridge University Press, Cambridge. pp. 33–55.

Laland, K. N., and Williams, K. (1997). Shoaling generates social learning of foraging information in guppies. *Anim. Behav.* 53:1161–1169.

Laland, K. N., and Williams, K. (1998). Social transmission of maladaptive information in the guppy. *Behav. Ecol.* 9:493–499.

Laland, K. N., Richerson, P. J., and Boyd, R. (1996). Developing a theory of animal social learning. In: Heyes, C. M., and Galef, B. G., Jr. (eds.), *Social Learning in Animals: The Roots of Culture.* Academic Press, London. pp. 129–154.

Laland, K. N., Odling-Smee, F. J., and Feldman, M. W. (2001). Niche construction, ecological inheritance, and cycles of contingency in evolution. In: Oyama, S., Griffiths, P. E., and Gray, R. D. (eds.), *Cycles of Contingency: Developmental Systems and Evolution,* MIT Press, Cambridge, MA. pp. 117–126.

Lambert, J. E. (1997). Digestive strategies, fruit processing, and seed dispersal in the chimpanzees (*Pan troglodytes*) and redtail monkeys (*Cercopithecus ascanius*) of Kibale National Park, Uganda [PhD diss.]. University of Illinois, Urbana-Champaign.

Lambert, J. E. (1998). Primate digestion: interactions among anatomy, physiology, and feeding ecology. *Evol. Anthropol.* 7:8–20.

Lambert, J. E. (1999). Seed handling in chimpanzees (*Pan troglodytes*) and redtail monkeys (*Cercopithecus ascanius*): implications for understanding hominoid and cercopithecine fruit processing strategies and seed dispersal. *Am. J. Phys. Anthropol.* 109:365–386.

Lambert, J. E. (2000a). The fate of seeds dispersed by African apes and cercopithecines. *Am. J. Phys. Anthropol.* (suppl. 30):204.

Lambert, J. E. (2000b). Urine drinking in wild *Cercopithecus ascanius,* evidence of nitrogen balancing? *Afr. J. Ecol.* 389:360–363.

Lambert, J. E. (2001). Red-tailed guenons (*Cercopithecus ascanius*) and *Strychnos mitis*: evidence for plant benefits beyond seed dispersal. *Int. J. Primatol.* 22:189–201.

Lambert, J. E. (2002a). Resource switching in guenons: a community analysis of dietary flexibility. In: Glenn, M., and Cords, M. (eds.), *The Guenons: Diversity and Adaptation in African Monkeys.* Kluwer Academic, New York. pp. 303–317.

Lambert, J. E. (2002b). Digestive retention times in forest guenons with reference to chimpanzees. *Int. J. Primatol.* 26:1169–1185.

Lambert, J. E. (2002c). Exploring the link between animal frugivory and plant strategies: the case of primate fruit-processing and post-dispersal seed fate. In: Levey, D. J., Silva, W. R., and Galetti, M. (eds.), *Seed Dispersal and Frugivory: Ecology, Evolution and Conservation.* CABI Publishing, New York. pp. 365–379.

Lambert, J. E. (2005). Competition, predation, and the evolutionary significance of the cercopithecine cheek pouch: the case of *Cercopithecus* and *Lophocebus. Am. J. Phys. Anthropol.* 126:183–192.

Lambert, J. E. (2007). Seasonality, fallback strategies, and natural selection: a chimpanzee versus cercopithecoid model for interpreting the evolution of hominin diet. In: Ungar, P. (ed): *Evolution of Human Diet: The Known, the Unknown, and the Unknowable.* Oxford University Press, Oxford. pp. 324–343.

Lambert, J. E. (2009). Fallback food consumption as adaptive plasticity. *Am. J. Phys. Anthropol..* 140:759–766.

Lambert, J. E., Chapman, C. A., Wrangham, R. W., and Conklin-Brittain, N. L. (2004). Hardness of cercopithecine foods: implications for the critical function of enamel thickness in exploiting fallback foods. *Am. J. Phys. Anthropol.* 125:363–368.

Lambert, T. D., Kays, R. W., Jansen, P. A., Aliaga-Rossel, E., and Wikelski, M. (2009). Nocturnal activity by the primarily diurnal Central American agouti (*Dasyprocta punctata*) in relation to environmental conditions, resource abundance and predation risk. *J. Trop. Ecol.* 25:211–215.

Lameira, A., and Wich, S. (2008). Orangutan long call degradation and individuality over distance: a playback approach. *Int. J. Primatol.* 29:615–625.

Lan, D. Y. (1989). Preliminary study on the group composition, behavior and ecology of the black gibbons (*Hylobates concolor*) in southwest Yunnan. *Zool. Res.* 10(suppl.):119–126.

Lan, D. Y. (1993). Feeding and vocal behaviours of black gibbons (*Hylobates concolor*) in Yunnan: a preliminary study. *Folia Primatol.* 60:94–105.

Lan, D. Y., and Sheeran, L. K. (1995). The status of black gibbons (*Hylobates concolor jingdongensis*) at Xiaobahe Wuliang Mountains, Yunnan Province, China. *Asian Primates* 5:2–4.

Lancaster, J. (1968). Primate communications and the emergence of human language. In: Jay, P. C. (ed.), *Primates, Studies in Adaptation and Variability.* Holt, Rinehart and Winston, New York. pp. 439–457.

Lancaster, J. (1971). Play-mothering: the relations between juvenile females and young infants among free-ranging vervet monkeys (*Cercopithecus aethiops*). *Folia Primatol.* 15:161–182.

Lancaster, J. B., and King, B. J. (1985). An evolutionary perspective on menopause. In: Brown, J. K., and Kerns, V. (eds.), *In Her Prime: A New View of Middle-Aged Women.* Bergin and Garvey Press, S. Hadley, MA. pp. 13–20.

Lande, R. (1981). Models of speciation by sexual selection on polygenic traits. *Proc. Natl. Acad. Sci. USA* 78:3721–3725.

Lande, R. (1982). A quantitative genetic theory of life history evolution. *Ecology* 63:607–615.

Lander, E. S., Linton, L. M., Birren, B., et al. (2001). Initial sequencing and analysis of the human genome. *Nature* 409:860–921.

Landsoud-Soukate, J., Tutin, C. E. G., and Fernandez, M. (1995). Intestinal parasites of sympatric gorillas and chimpanzees in the Lope Reserve, Gabon. *Ann. Trop. Med. Parasitol.* 89:73–79.

Landsteiner, K. (1900). Zur Kenntnis der antifermentativen, lytischen und agglutinierenden Wirkungen des Blutserums und der Lymphe. *Zentralbl. Bakteriol. Parasit. Infekt.* 27:357–362.

Langergraber, K. E., Mitani, J. C., and Vigilant, L. (2007a). The limited impact of kinship on cooperation in wild chimpanzees. *Proc. Natl. Acad. Sci. USA* 104:7786–7790.

Langergraber, K. E., Siedel, H., Mitani, J. C., Wranghman, R. W., Reynolds, V., Hunt, K., and Vigilant, L. (2007b). The genetic signature of sex-biased migration in patrilocal chimpanzees and humans. *PLoS ONE* 2:e973.

Lansing, S. J. (1991). *Priests and Programmers: Technologies and Power in the Engineered Landscape of Bali.* Princeton University Press, Princeton, NJ.

Lapenta, M. J., de Oliveira, P. P., and Nogueira-Neto, P. (2007). Daily activity period, home range and sleeping sites of golden lion tamarins (*Leontopithecus rosalia*) translocated to the Uniao Biological Reserve, RJ-Brazil. *Mammalia* 71:131–137.

Lappan, S. (2007). Patterns of dispersal in Sumatran siamangs (*Symphalangus syndactylus*): preliminary mtDNA evidence suggests more frequent male than female dispersal to adjacent groups. *Am. J. Primatol.* 69:692–698.

Lappan, S. (2008). Male care of infants in a siamang (*Symphalangus syndactylus*) population including socially monogamous and polyandrous groups. *Behav. Ecol. Sociobiol.* 62:1307–1317.

Lappan, S., and Whittaker, D. J. (2004). Trends in the study of gibbon populations, 1940–2004 [abstract]. Presented at the International Primatological Society Congress, Torino, Italy, August 24, 2004.

Larose, F. (1996). Foraging strategies, group size, and food competition in the mantled howler monkey, *Alouatta palliata* [PhD diss.]. University of Alberta, Edmonton.

Larson, S. G. (1998a). Parallel evolution in the hominoid trunk and forelimb. *Evol. Anthropol.* 6:87–99.

Larson, S. G. (1998b). Unique aspects of quadrupedal locomotion in non-human primates. In: Strasser, E., Fleagle, J. G., Rosenberger, A., and McHenry, H. (eds.), *Primate Locomotion: Recent Advances.* Plenum Press, New York. pp. 157–173.

Lasley, B. L. (1985). Methods for evaluating reproductive function in exotic species. *Adv. Vet. Sci. Comp. Med.* 30:209–228.

Lasley, B. L., and Kirkpatrick, J. F. (1991). Monitoring ovarian function in captive and free-ranging wildlife by means of urinary and fecal steroids. *J. Zoo Wildl. Med.* 22:23–31.

Lasley, B. L., and Shideler, S. E. (1993). Methods for assessing reproduction in nondomestic species. In: Fowler, M. E. (ed.), *Zoo and Wild Animal Medicine: Current Therapy,* 3rd ed. W. B. Saunders, Philadelphia. pp. 79–86.

Lasley, B. L., Bogart, M. H., and Shideler, S. E. (1978). A comparison of lemur ovarian cycles. In: Alexander, N. J. (ed.), *Animal Models for Research on Contraception and Fertility.* Harper and Row, Hagerstown. pp. 417–424.

Lasley, B. L., Hodges, J. K., and Czekala, N. M. (1980a). Monitoring the female reproductive cycle of great apes and other species by estrogen and bioactive luteinizing hormone determinations in small volumes of urine. *J. Reprod. Fertil.* 28:121–129.

Lasley, B., Presley, S., and Czekala, N. (1980b). Monitoring ovulation in great apes; urinary immunoreactive estrogen and bioactive luteinizing hormone. *Annu. Proc. Am. Soc. Zoo Vet.* 37–40.

Lasley, B. L., Lindburg, D. G., Robinson, P. T., and Benirschke, K. (1981a). Captive breeding of exotic species. *J. Zoo Anim. Med.* 12:67–73.

Lasley, B. L., Monfort, S. L., Hodges, J. K., and Czekala, N. M. (1981b). Comparison of urinary estrogens during pregnancy in diverse species. In: Novy, M. J., and Resko, J. A. (eds.), *Fetal Endocrinology.* Academic Press, New York. pp. 111–126.

Lau, J. (2002). Characterization of microsatellite loci in three owl monkey species (*Aotus azarai, A. lemurinus, A. nancymaae*) [MSc thesis]. University of California, San Diego.

Lau, J., Fernandez-Duque, E., Evans, S., Dixson, A. F., and Ryder, O. A. (2004). Heterologous amplification and diversity of microsatellite loci in three owl monkey species (*Aotus azarai, A. lemurinus, A. nancymaae*). *Conserv. Genet.* 5(5):727–731.

Launhardt, K., Borries, C., Hardt, C., Epplen, J. T., and Winkler, P. (2001). Paternity analysis of alternative male reproductive routes among the langurs (*Semnopithecus entellus*) of Ramnagar. *Anim. Behav.* 61:53–64.

Laurance, W. F., Lovejoy, T. E., Vanconcelos, H. L., Bruna, E. M., Didham, R. K., Souffer, P. C., Gascon, C., Bierregaard, R. O., Laurance, S. G., and Sampaio, E. (2002). Ecosystem decay of Amazonian forest fragments: a 22-year investigation. *Conserv. Biol.* 16:605–618.

Lawes, M. J. (1991). Diet of samango monkeys (*Cercopithecus mitis erthrarchus*) in the Cape Vidal dune forest, South Africa. *J. Zool. Lond.* 224:149–173.

Lawes, M. J., Henzi, S. P., and Perrin, M. R. (1990). Diet and feeding behaviour of samango monkeys (*Cercopithecus mitis labiatus*) in Ngoye Forest, South Africa. *Folia Primatol.* 54:57–69.

Lawler, R. R., Richard, A. F., and Riley, M.A. (2003). Genetic population structure of the white sifaka (*Propithecus verreauxi verreauxi*) at Beza Mahafaly Special Reserve, southwest Madagascar (1992–2001). *Mol. Ecol.* 12:2307–2317.

Lawler, R. R., Richard, A. F., and Riley, M.A. (2005). Intrasexual selection in Verreaux's sifaka (*Propithecus verreauxi verreauxi*). *J. Hum. Evol.* 48:259–277.

Lawrence, D. C., and Leighton, M. (1996). Ecological determinants of feeding bout length in long-tailed macaques (*Macaca fascicularis*). *Trop. Biodivers.* 3:227–242.

Lawrence, J. M. (2003). Preliminary report on the natural history of brown titi monkeys (*Callicebus brunneus*) at the Los Amigos Research Station, Madre de Dios, Peru. *Am. J. Phys. Anthropol.* (suppl. 36):136.

Laws, J. W., and Laws, J. V. H. (1984). Social interactions among adult male langurs (*Presbytis entellus*) at Rajaji Wildlife Sanctuary. *Int. J. Primatol.* 5:31–50.

Lazareva, O. F., Smirnova, A. A., Raevskii, V. V., and Zorina, Z. A. (2000). Transitive inference formation in hooded crows: preliminary data. *Dokl. Akad. Nauk.* 370:698–700.

Lazaro-Perea, C. (2001). Intergroup interactions in wild common marmosets, *Callithrix jacchus*: territorial defense and assessment of neighbors. *Anim. Behav.* 62:11–21.

Lazaro-Perea, C., Castro, C. S. S., Harrison, R., Araujo, A., Arruda, M. F., and Snowdon, C. T. (2000). Behavioral and demographic changes following the loss of the breeding female in cooperatively breeding marmosets. *Behav. Ecol. Sociobiol.* 48:137–146.

Lazaro-Perea, C., Arruda, M. F., and Snowdon, C. T. (2004). Grooming as a reward? Social function of grooming between females in cooperatively breeding marmosets. *Anim. Behav.* 67:627–636.

Leakey, L. S. B. (1935). *The Stone Age Races of Kenya.* Oxford University Press, London.

Leakey, M. C., Fiebel, C. S., McDougall, I., and Walker, A. (1995). New four million year old hominid species from Kanapoi and Allia Bay, Kenya. *Nature* 376:565–571.

Le Boeuf, B. J., Crocker, D. E., Costa, D. P., Blackwell, S. B., Webb, P. M., and Houser, D. S. (2000). Foraging ecology of northern elephant seals. *Ecol. Monogr.* 70:353–382.

Leca, J. B., Fornasieri, I., and Petit, O. (2002). Aggression and reconciliation in *Cebus capucinus*. *Int. J. Primatol.* 23:979–998.

Leca, J. B., Gunst, N., and Huffman, M. A. (2007a). Japanese macaque cultures: inter- and intra-troop behavioural variability of stone handling patterns across 10 troops. *Behaviour* 144:251–281.

Leca, J. B., Gunst, N., Watanabe, K., and Huffman, M. (2007b). A new case of fish-eating in Japanese macaques: implications for social constraints on the diffusion of feeding innovation. *Am. J. Primatol.* 69:821–828.

Lee, P. C. (1983). Caretaking of infants and mother–infant relationships. In: Hinde, R. A. (ed.), *Primate Social Relationships: An Integrated Approach.* Sinauer Associates, Sunderland, MA. pp. 145–151.

Lee, P. (1984). Early infant development and maternal care in free-ranging vervet monkeys. *Primates* 25:36–47.

Lee, P. C. (1987). Nutrition, fertility and maternal investment in primates. *J. Zool. Lond.* 213:409–422.

Lee, P. C. (1996). The meanings of weaning: growth, lactation, and life history. *Evol. Anthropol.* 5:87–96.

Lee, P. C., and Hauser, M. D. (1988). Long-term consequences of changes in territory quality on feeding and reproductive strategies of vervet monkeys. *J. Anim. Ecol.* 67:347–358.

Lee, P. C., Brennan, E. J., Else, J. G., and Altmann, J. (1986). Ecology and behaviour of vervet monkeys in a tourist lodge habitat. In: Else, J. G., and Lee, P. C. (eds.), *Primate Ecology and Conservation.* Cambridge University Press, Cambridge. pp 229–235.

Lee, P. C., Majluf, P., and Gordon, I. (1991). Growth, weaning and maternal investment from a comparative perspective. *J. Zool. Lond.* 225:99–114.

Lefebvre, L. (1995a). Culturally-transmitted feeding behavior in primates: evidence for accelerating learning rates. *Primates* 36:227–239.

Lefebvre, L. (1995b). The opening of milk bottles by birds: evidence for accelerating learning rates, but against the wave-of-advance model of cultural transmission. *Behav. Proc.* 34:43–53.

Lefebvre, L., and Palameta, B. (1988). Mechanisms, ecology, and population diffusion of socially learned, food-finding behaviour in feral pigeons. In: Zentall, T. R., and Galef, B. G., Jr. (eds.), *Social Learning: Psychological and Biological Perspectives.* Lawrence Erlbaum, Hillsdale, NJ. pp. 141–163.

Le Gros Clark, W. (1956). *A Miocene Lemuroid Skull from East Africa.* Fossil Mammals of Africa, 9. British Museum of Natural History, London. p. 6.

Le Gros Clark, W. E. (1959). *The Antecedents of Man.* Harper and Row, New York.

Lehman, S. M., Prince, W., and Mayor, M. (2001). Variations in group size in white-faced sakis (*Pithecia pithecia*): evidence for monogamy or seasonal cogregations? *Neotrop. Primates* 9:96–101.

Lehmann, J., and Boesch, C. (2003). Social influences on ranging patterns among chimpanzees (*Pan troglodytes verus*) in the Taï National Park, Cote d'Ivoire. *Behav. Ecol.* 14:642–649.

Lehmann, J., and Boesch, C. (2004). To fission or to fusion: effects of community size on wild chimpanzee (*Pan troglodytes verus*) social organization. *Behav. Ecol. Sociobiol.* 56:207–216.

Lehmann, J., and Boesch, C. (2005). Bisexually bonded ranging in chimpanzees (*Pan troglodytes verus*). *Behav. Ecol. Sociobiol.* 57:525–535.

Lehmann, J., and Boesch, C. (2008). Sexual differences in chimpanzee sociality. *Int. J. Primatol.* 29:65–81.

Lehmann, J., and Boesch, C. (2009). Sociality of the dispersing sex: the nature of social bonds in West African female chimpanzees, *Pan troglodytes. Anim. Behav.* 77:377–387.

Lehmann, J., Fickenscher, G., and Boesch, C. (2006). Kin biased investment in wild chimpanzees. *Behaviour* 143:931–955.

Lehner, P. N. (1996). *Handbook of Ethological Methods.* Cambridge University Press, Cambridge.

Lei, R., Engberg, S. E., Andriantompohavana, R., McGuire, S. M., Mittermeier, R. A., Zaonarivelo, J. R., Brenneman, R. A., and Louis, E. E. (2008). Nocturnal lemur diversity at Masoala National Park. *Spec. Publ. Mus. Tex. Tech Univ.* 53:1–41.

Leidy, L. E. (1999). Menopause in evolutionary perspective. In: Trevathan, W. R., Smith, E. O., and McKenna, J. T. (eds.), *Evolutionary Medicine.* Oxford University Press, New York. pp. 407–427.

Leigh, S. R. (1994). Ontogenetic correlates of diet in anthropoid primates. *Am. J. Phys. Anthropol.* 94:499–522.

Leigh, S. R. (1996). Evolution of human growth spurts. *Am. J. Phys. Anthropol.* 101:455–474.

Leigh, S. R. (2001). Evolution of human growth. *Evol. Anthropol.* 10:223–236.

Leigh, S. R. (2004). Brain growth, life history, and cognition in primate and human evolution. *Am. J. Primatol.* 62:139–164.

Leigh, S. R. (2006). Brain ontogeny and life history in *Homo erectus. J. Hum. Evol.* 50:104–108

Leigh, S. R., and Bernstein, R. M. (2006). Ontogeny, life history, and maternal investment in baboons. In: Swedell, L., and Leigh, S. R. (eds.), *Reproduction and Fitness in Baboons: Behavioral, Ecological, and Life History Perspectives.* Springer, New York. pp. 225–256.

Leigh, S. R., and Bernstein, R. M. (2006). Ontogeny, life history, and maternal reproductive strategies in baboons. In: Swedell, L., and Leigh, S. R. (eds.), *Life History, Reproductive Strategies, and Fitness in Baboons.* Kluwer/Academic Press, New York.Pp.225–256.

Leigh, S. R., and Park, P. B. (1998). Evolution of human growth prolongation. *Am. J. Phys. Anthropol.* 107:331–350.

Leigh, S. R., and Shea, B. T. (1995). Ontogeny and the evolution of adult body size dimorphism in apes. *Am. J. Primatol.* 36:37–60.

Leigh, S. R., and Shea, B. T. (1996). The ontogeny of body size variation in African apes. *Am. J. Phys. Anthropol.* 99:43–65.

Leighton, D. R. (1987). Gibbons: territoriality and monogamy. In: Smuts, B., Cheney, D., Seyfarth, R., Wrangham, R., and Struhsaker, T. (eds.), *Primate Societies.* University of Chicago Press, Chicago. pp. 135–145.

Leighton, M. (1993). Modeling dietary selectivity by Bornean orangutans: evidence for integration of multiple criteria in fruit selection. *Int. J. Primatol.* 14:257–313.

Leighton, M., and Leighton, D. (1983). Vertebrate responses to fruiting seasonality within a Bornean rain forest. In: Sutton, S. L., Whitmore, T. C., and Chadwick, A. C. (eds.), *Tropical Rain Forest: Ecology and Management.* Blackwell Scientific, Boston. pp. 181–196.

Leighty, K., Menzel, C., and Fragaszy, D. (2008). How young children and chimpanzees (*Pan troglodytes*) perceive objects in a 2D display: putting an assumption to the test. *Dev. Sci.* 11(5):778–792

Leinfelder, I., de Vries, H., Deleu, R., and Nelissen, M. (2001). Rank and grooming reciprocity among females in a mixed-sex group of captive hamadryas baboons. *Am. J. Primatol.* 55:25–42.

Le Maho, Y., Goffart, M., Rochas, A., Felbabel, H., and Chatonnet, J. (1981). Thermoregulation in the only nocturnal simian: the night monkey *Aotus trivirgatus. Am. J. Physiol.* 240:R156–R165.

Lemelin, P. (1999). Morphological correlates of substrate use in didelphid marsupials: implications for primate origins. *J. Zool.* 247:165–175.

Lemlin, P. (1995). Comparative and functional myology of the prehensile tail in New World monkeys. *J. Morphol.* 224:1–18.

Lemos de Sá, R. M. (1988). Situação de uma população de Mono Carvoeiro, *Brachyteles arachnoides,* em um fragmento de Mata Atlântica (M.G.) e implicações para sua conservação [MSc thesis]. Universidade de Brasília, Brasília.

Lemos de Sá, R. M., and Glander, K. E. (1993). Capture techniques and morphometrics for the woolly spider monkey, or muriqui (*Brachyteles arachnoides* E. Geoffroy 1806). *Am. J. Primatol.* 29:145–153.

Lemos de Sá, R. M., Pope, T. R., Glander, K. E., Struhsaker, T. T., and da Fonseca, G. A. B. (1990). A pilot study of genetic and morphological variation in the muriqui (*Brachyteles arachnoides*). *Primate Conserv.* 11:26–30.

Lendrem, D. W. (1983). Sleeping and vigilance in birds I. Field observations of the mallard (*Anas platyrhynchos*). *Anim. Behav.* 31:532–538.

Lendrem, D. W. (1984). Sleeping and vigilance in birds II. An experimental study of the barbary dove (*Streptopelia risoria*). *Anim. Behav.* 32:243–248.

Lens, L., Van Dongen, S., Norris, K., Githiru, M., and Matthysen, E. (2002). Avian persistence in fragmented rainforest. *Science* 298:1236–1238.

Leo Luna, M. (1980). First field study of the yellow-tailed woolly monkey. *Oryx* 15:386–389.

Leo Luna, M. (1982). Conservation of the yellow-tailed woolly monkey. *Int. Zoo Ybk.* 22:47–52.

Leo Luna, M. (1984). The effects of hunting, selective logging and clear-cutting on the conservation of the yellow-tailed woolly monkey (*Lagothrix flavicauda*) [MA thesis]. University of Florida, Gainesville.

Leonard, W. R. (2000). Human nutritional evolution. In: Stinson, S., Bogin, B., Huss-Ashmore, R., and O'Rourke, D. (eds.), *Human Biology: An Evolutionary and Biocultural Approach.* Wiley-Liss, New York. pp. 295–343.

Le Prell, C. G., Hauser, M. D., and Moody, D. B. (2002). Discrete or graded variation within rhesus monkey screams? Psychophysical experiments on classification. *Anim. Behav.* 63:47– 62.

Lerchl, A., Kuederling, I., and Epple, G. (1991). Diurnal variations of urinary testosterone excretion in the male saddle-back tamarin (*Saguinus fuscicollis*). *Primate Rep.* 30:17–23.

Leroy, E. M., Rouquet, P., Formenty, P., Souquière, S., Kilbourne, A., Froment, J. M., Bermejo, M., Smit, S., Karesh, W., Swanepoel, R., Zaki, S. R., and Rollin, P. E. (2004). Multiple ebola virus transmission events and rapid decline of central African wildlife. *Science* 303:387–390.

Leshner, A. I. (1983). Pituitary–adrenocortical effects on intermale agonistic behavior. In: Svare, S. B. (ed.), *Hormones and Aggressive Behavior.* Plenum Press, New York. pp. 27–38.

Leskes, A., and Acheson, N. H. (1971). Social organization of a free-ranging troop of black and white colobus monkeys (*Colobus abyssinicus*). In: Kummer, H. (ed.), *Proceedings of the Third International Congress of Primatology, Zurich 1970. Behavior,* vol. 3. S. Karger, Basel. pp. 22–31.

Lethmate, J. (1982). Tool-using skills of orangutans. *J. Hum. Evol.* 11:49–64.

Leutenegger, W., and Lubach, G. (1987) Sexual dimorphism, mating system, and effect of phylogeny in De Brazza's monkey (*Cercopithecus neglectus*). *Am. J. Primatol.* 13:171–179.

Levey, D. J., and Grajal, A. (1991). Evolutionary implications of fruit-processing limitations in cedar waxwings. *Am. Nat.* 138:171–189.

Levey, D. J., and Karasov, W. H. (1994). Gut passage of insects by European starlings and comparison with other species. *Auk* 111:478–481.

Levin, D. A. (1971). Plant phenolics: an ecological perspective. *Am. Nat.* 105:157–181.

Levin, D. A. (1976). Alkaloid-bearing plants: an ecogeographic perspective. *Am. Nat.* 110:261–284.

Levin, S. A. (1974). Dispersion and population interactions. *Am. Nat.* 108:207–228.

Levinson, G., and Gutman, G. A. (1987). Slipped-strand mispairing: a major mechanism for DNA sequence evolution. *Mol. Biol. Evol.* 4:203–221.

Levrero, F., Gatti, S., Menard, N., Petit, E., Caillaud, D., and Gautier-Hion, A. (2006). Living in nonbreeding groups: an alternative strategy for maturing gorillas. *Am. J. Primatol.* 68:275–291.

Levy, S., Sutton, G., Ng, P. C., et al. (2007). The diploid genome sequence of an individual human. *PLoS Biol.* 5:e254.

Lewis, M., and Brooks-Gunn, J. (1979). *Social Cognition and the Acquisition of Self.* Plenum Press, New York.

Lewis, R. J. (2002). Beyond dominance: the importance of leverage. *Q. Rev. Biol.* 77:149–164.

Ley, R. E., Hamady, M., Lozupone, C., Turnbaugh, P. J., Ramey, R. R., Bircher, J. S., Schlegel, M. L., Tucker, T. A., Schrenzel, M. D., Knight, R., and Gordon, J. I. (2008). Evolution of mammals and their gut microbes. *Science* 320:1647–1651.

Li, B., Chen, C., Ji, W., and Ren, B. (1999). Seasonal home range changes of the Sichuan snub-nosed monkey (*Rhinopithecus roxellana*) in the Qinling Mountains of China. *Folia Primatol.* 71:375–386.

Li, D. Y., Grueter, C. C., Ren, B. P., Long, Y. C., Li, M., Peng, Z. S., and Wei, F. W. (2008). Ranging of *Rhinopithecus bieti* in the Samage Forest, China. II. Use of land cover types and altitudes. *Int. J. Primatol.* 29(5):1147–1173.

Li, J. H., Wang, Q. S., and Han, D. M. (1996). Fission in a free-ranging Tibetan macaque troop at Huangshan Mountain, China. *Kexue Tongbao Chin. Sci. Bull.* 41:1377–1381.

Liebal, K., Call, J., and Tomasello, M. (2004a). The use of gesture sequences by chimpanzees. *Am. J. Primatol.* 64:377–396.

Liebal, K., Pika, S., Call, J., and Tomasello, M. (2004b). To move or not to move: how apes adjust to the attentional state of others. *Interact. Stud.* 5:199–219.

Liebal, K., Pika, S., and Tomasello, M. (2006). Gestural communication of orangutans (*Pongo pygmaeus*). *Gesture* 6:1–38.

Lieberman, D., Hall, J. B., Swaine, M. D., and Lieberman, M. (1979). Seed dispersal by baboons in the Shai Hills, Ghana. *Ecology* 60:65–75.

Lima, S. L. (1987). Vigilance while feeding and its relation to the risk of predation. *J. Theor. Biol.* 125:303–316.

Lima, S. L. (1993). Ecological and evolutionary perspectives on escape from predatory attack: a survey of North American birds. *Wilson Bull.* 105:1–47.

Lima, S. L. (1994). On the personal benefits of anti-predatory vigilance. *Anim. Behav.* 48:734–736.

Lima, S. L. (1995a). Back to the basics of anti-predatory vigilance: the group-size effect. *Anim. Behav.* 49:11–20.

Lima, S. L. (1995b). Collective detection of predatory attack by social foragers: fraught with ambiguity? *Anim. Behav.* 50:1097–1108.

Lima, S. L. (1998). Stress and decision making under the risk of predation: recent developments from behavioral, reproductive, and ecological perspectives. *Adv. Study Behav.* 27:215–290.

Lima, S. L. (2002). Putting predators back into behavioral predator–prey interactions. *Trends Ecol. Evol.* 17:70–75.

Lima, S. L., and Dill, L. M. (1990). Behavioral decisions made under the risk of predation: a review and prospectus. *Can. J. Zool.* 68:619–640.

Lima, S. L., and Zollner, P. A. (1996). Anti-predatory vigilance and the limits to collective detection: visual and spatial separation between foragers. *Behav. Ecol. Sociobiol.* 38:355–363.

Lindblad-Toh, K., Wade, C. M., Mikkelsen, T. S., et al. (2005) Genome sequence, comparative analysis and haplotype structure of the domestic dog. *Nature* 438:803–819.

Lindburg, D. G. (1971). *The Rhesus Monkey in Northern India: An Ecological and Behavioral Study.* Academic Press, New York.

Lindburg, D. G. (1983). Mating behavior and estrus in the Indian rhesus monkey. In: Seth, P. K. (ed.), *Perspectives in Primate Biology.* Today & Tomorrow, New Delhi, pp. 45–61.

Lindstrom, A. (1989). Finch flock size and risk of hawk predation at a migratory stopover site. *Auk* 106:225–232.

Link, A., Di Fiore, A., and Spehar, S. (2006). Predation risk affects subgroup size in spider monkeys (*Ateles belzebuth*) at Yasuni National Park, Ecuador. *Folia Primatol.* 77:318.

Link, A., Di Fiore, A., and Spehar, S.N. (2009). Female-directed aggression and social control in spider monkeys. In: Muller, M. N., and Wrangham, R. W. (eds.), *Sexual Coercion in Primates: An Evolutionary Perspective on Male Aggression Against Females.* Harvard University Press, Cambridge, MA. pp 157–183.

Linneaus, C. (1758). *Systema Naturae per Regna Tria Naturae, Secundum Classes, Ordines, Genera, Species, cum Characteribus et Differentiis,* 10th ed. L. Salvii, Holmiae.

Lippold, L. K. (1977). The douc langur: a time for conservation. In: Rainier, H. S. H. P., and Bourne, G. H. (eds.), *Primate Conservation.* Academic Press, New York. pp. 513–538.

Lippold, L. K. (1988). A census of primates in Cabo Blanco Absolute Nature Reserve, Costa Rica. *Brenesia* 29:101–105.

Lippold, L. K. (1989). Primates in Cabo Blanco Absolute Nature Reserve, Costa Rica. *Primate Conserv.* 10:23–25.

Lipschitz, D. L. (1996). A preliminary investigation of the relationship between ovarian steroids, LH, reproductive behaviour and vaginal changes in lesser bushbabies (*Galago moholi*). *J. Reprod. Fertil.* 107:164–174.

Liu, Q., Simpson, K., Izar, P., Ottoni, E., Visalberghi, E., and Fragaszy, D. (2009). Kinematics and energetics of nut-cracking in wild capuchin monkeys (*Cebus libidinosus*). *Am. J. Phys. Anthropol.* 138(2):210–220.

Liversidge, H. (2003). Variation in modern human dental development. In: Thompson, J. L., Krovitz, G. E., and Nelson, A. J. (eds.), *Patterns of Growth and Development in the Genus* Homo. Cambridge University Press, Cambridge. pp. 73–113.

Lizarralde, M. (2002). Ethnoecology of monkeys among the Bari of Venezuela: perception, use and conservation. In: Fuentes, A., and Wolfe, L. D. (eds.), *Primates Face to Face.* Cambridge University Press, Cambridge. pp. 85–100.

Ljungberg, T., and Westlund, K. (2000). Impaired reconciliation in rhesus macaques with a history of early weaning and disturbed socialization. *Primates* 41:79–88.

Ljungberg, T., Westlund, K., and Lindqvist Forsberg, A. J. (1999). Conflict resolution in 5-year-old boys: does postconflict affiliative behaviour have a reconciliatory role? *Anim. Behav.* 58:1007–1016.

Lloyd, E., Kreetiyutanont, K., Prabnasuk, J., Grassman, L. I., Jr., and Borries, C. (2006). Observation of Phayre's leaf monkeys mobbing a clouded leopard at Phu Khieo Wildlife Sanctuary (Thailand). *Mammalia* 70(1–2):158–159.

Lloyd, L. B. (1989). *Simple Minds.* MIT Press, Cambridge, MA.

Lloyd, M. (1967). Mean crowding. *J. Anim. Ecol.* 36:1–30.

Lockman, J. J. (2000). A perception–action perspective on tool use development. *Child Dev.* 71:137–144.

Lohiya, N. K., Sharma, R. S., Puri, C. P., David, G. F. X., and Anand Kumar, T. C. (1988). Reproductive exocrine and endocrine profile of female langur monkeys, *Presbytis entellus.* *J. Reprod. Fertil.* 82:485–492.

Lohiya, N. K., Sharma, R. S., Manivannan, B., and Anand Kumar, T. C. (1998). Reproductive exocrine and endocrine profiles and their seasonality in male langur monkeys (*Presbytis entellus entellus*). *J. Med. Primatol.* 27:15–20.

Lomascolo, S. B. and Schaefer, H. M. (2010). Signal convergence in fruits: a result of selection by frugivores? *J. Evol. Biol.* 23:614–624.

Long, C. A., and Frank, T. (1968). Morphometric function and variation in the baculum with comments on the correlation of parts. *J. Mammal.* 49:32–43.

Long, J. O., and Cooper, R. W. (1968). Physical growth and dental eruption in captive-bred squirrel monkeys, *Saimiri sciureus* (letica, colombia). In: Rosenblum, L. A., and Cooper, R. W. (eds.), *The Squirrel Monkey.* Academic Press, New York. pp. 193–205.

Longino, J. T. (1984). True anting by the capuchin, *Cebus capucinus.* *Primates* 25:243–245.

Lonsdorf, E. V., Eberly, L. E., and Pusey, A. E. (2004). Sex differences in learning in chimpanzees. *Nature* 428(6984):715–716.

Lopes, M. A., and Ferrari, S. F. (1994). Foraging behaviour of a tamarin group (*Saguinus fuscicollis weddelli*), and interactions with marmosets (*Callithrix emiliae*). *Int. J. Primatol.* 15:373–387.

Lorini, V. G., and Persson, M. L. (1990). Uma nova espécie de *Leontopihecus* Lesson, 1840, do sul do Brasil (Primates, Callitrichidae). *Bol. Mus. Nac. R. J. Zool.* 338:1–14.

Löttker, P., Huck, M., Zinner, D. P., and Heymann, E. W. (2007). Grooming relationships between breeding females and adult group members in cooperatively breeding moustached tamarins (*Saguinus mystax*). *Am. J. Primatol.* 69:1159–1172.

Louis, E. E., Coles, M. S., Andriantompohavana, R., Sommer, J. A., Engberg, S. E., Zaonarivelo, J. R., Mayor, M. I., and Brenneman, R. A. (2006a). Revision of the mouse lemurs (*Microcebus*) of eastern Madagascar. *Int. J. Primatol.* 27:347–389.

Louis, E. E., Engberg, S. E., Lei, R., Geng, H., Sommer, J. A., Randriamampionona, R., Randriamanana, J. C., Zaonarivelo, J. R., Andriantomphohavana, R., Randria, G., Prosper, Ramaromilanto, B., Rakotoarisoa, G., Rooney, A., and Brenneman, R. A. (2006b). Molecular and morphological analyses of the sportive lemurs (family Megaladapidae: genus *Lepilemur*) reveals 11 previously unrecognized species. *Spec. Publ. Mus. Tex. Tech Univ.* 49:1–47.

Louis, E. E., Engberg, S. E, McGuire, S. M., McCormick, M. J., Randriamampionona, R., Ranaivoarisoa, J. F., Bailey, C. A., Mittermeier, R. A., and Lei, R. (2008). Revision of the mouse lemurs, *Microcebus* (Primates, Lemuriformes), of northern and northwestern Madagascar with descriptions of two new species at Montagne d'Ambre National Park and Antafondro Classified Forest. *Primate Conserv.* 23:19–38.

Lovejoy, C. O. (1981). The origin of man. *Science* 211:341–350.

Loveless, M. D., and Hamrick, J. L. (1984). Ecological determinants of genetic structure in plant populations. *Annu. Rev. Ecol. Syst.* 15:65–95.

Lovern, M. B. (2000). Behavioral ontogeny in free-ranging juvenile male and female green anoles, *Anolis carolinensis,* in relation to sexual selection. *J. Herpetol.* 34:274–281.

Low, M. (2005). Female resistance and male force: context and patterns of copulation in the New Zealand stitchbird *Notiomystis cincta*. *J. Avian Biol.* 36:436–448.

Lowen, J. C. L., Bartrina, C. R. P., and Tobias, J. A. (1996). *Biological surveys and conservation priorities in eastern Paraguay.* CSB Conservation Publications, Cambridge.

Loy, J. (1970). Peri-menstrual sexual behavior among rhesus monkeys. *Folia Primatol.* 13:286–297.

Loy, J. (1987). The sexual behavior of African monkeys and the question of estrus. In: Zucker, E. L. (ed.), *Comparative Behavior of African Monkeys.* Alan R. Liss, New York. pp. 175–195.

Lozano, G. A. (1991). Optimal foraging theory: a possible role for parasites. *Oikos* 60:391–395.

Lozano, G. A. (1998). Parasitic stress and self-medication in wild animals. *Adv. Study Behav.* 27:291–317.

Lu, F. E. (1999). Changes in subsistence patterns and resource use of the Huaorani Indians in the Ecuadorian Amazon [PhD diss.]. University of North Carolina, Chapel Hill.

Lucas, P. W. (2004). *Dental Functional Morphology: How Teeth Work.* Cambridge University Press, Cambridge, UK.

Lucas, P. W., and Corlett, R. T. (1998). Seed dispersal by long-tailed macaques. *Am. J. Primatol.* 45:29–44.

Luckett, W. (1976). Comparative development and evolution of the placenta in primates. *Contrib. Primatol.* 3:142–234.

Ludes, E., and Anderson, J. R. (1995). "Peat-bathing" by captive white-faced capuchin monkeys (*Cebus capucinus*). *Folia Primatol.* 65:38–42.

Ludwig, G., Aguiar, L. M., Miranda, J. M. D., Teixeira, G. M., Svoboda, W. K., Malanski, L. S., Shiozawa, M. M., Hilst, C. L. S., Navarro, I. T., and Passos, F. C. (2007). Cougar predation on black-and-gold howlers on Mutum Island, Southern Brazil. *Int. J. Primatol.* 28(1):39–46.

Lukas, D., Reynolds, V., Boesch, C., and Vigilant, L. (2005). To what extent does living in a group mean living with kin? *Mol. Ecol.* 14:2181–2196.

Lunt, D. H., Hutchinson, W. F., and Carvalho, G. R. (1999). An efficient method for PCR-based identification of microsatellite arrays (PIMA). *Mol. Ecol.* 8:893–894.

Lycett, J. E., Henzi, S. P., and Barrett, L. (1998). Maternal investment in mountain baboons and the hypothesis of reduced care. *Behav. Ecol. Sociobiol.* 42:49–56.

Lycett, J. E., Weingrill, T., Henzi, S. P. (1999). Birth patterns in the Drakensberg Mountain Baboons (*Papio cynocephalus ursinus*). *S. Afr. J. Sci.* 95:354–356.

Lycett, S. J., Collard, M., and McGrew, W. C. (2007). Phylogenetic analyses of behavior support existence of culture among wild chimpanzees. *Proc. Natl. Acad. Sci. USA* 104:17588–17592.

Lydekker, R. (1905). Colour evolution in guereza monkeys. *Proc. Zool. Soc. Lond.* 2:325–329.

Lyles, A. M., and Dobson, A. P. (1988). Dynamics of provisioned and unprovisioned primate populations. In: Fa, J. E., and Southwick, C. H. (eds.), *Ecology and Behavior of Food-Enhanced Primate Groups.* Alan R. Liss, New York, pp. 167–198.

Lynch, J. W., Ziegler, T. E., and Strier, K. B. (2002). Individual and seasonal variation in fecal testosterone and cortisol levels in wild male tufted capuchin monkeys, *Cebus apella nigritus*. *Horm. Behav.* 41:275–287.

Lynch Alfaro, J. W. (2005). Male mating strategies and reproductive constraints in a group of wild tufted capuchin monkeys (*Cebus apella nigritus*). *Am. J. Primatol.* 67:313–328.

Ma, N. S. F. (1981). Chromosome evolution in the owl monkey, *Aotus*. *Am. J. Phys. Anthropol.* 54:293–303.

Ma, N. S. F., Elliot, M. W., Morgan, L. M., Miller, A. C., and Jones, T. C. (1976a). Translocation of Y chromosome to an autosome in the Bolivian owl monkey, *Aotus*. *Am. J. Phys. Anthropol.* 45:191–202.

Ma, N. S. F., Jones, T. C., Miller, A., Morgan, L., and Adams, E. (1976b). Chromosome polymorphism and banding patterns in the owl monkey (*Aotus*). *Lab. Anim. Sci.* 26:1022–1036.

Ma, N. S. F., Jones, T. C., Bedard, M. T., Miller, A. C., Morgan, L. M., and Adams, E. A. (1977). The chromosome complement of an *Aotus* hybrid. *J. Hered.* 68:409–412.

Ma, N. S. F., Rossan, R. N., Kelley, S. T., Harper, J. S., Bedard, M. T., and Jones, T. C. (1978). Banding patterns of the chromosomes of two new karyotypes of the owl monkey, *Aotus,* captured in Panamá. *J. Med. Primatol.* 7:146–155.

Ma, N. S. F., Aquino, R., and Collins, W. E. (1985). Two new karyotypes in the Peruvian owl monkey (*Aotus trivirgatus*). *Am. J. Primatol.* 9:333–341.

MacArthur, R. H., and Pianka, E. R. (1966). An optimal use of a patchy environment. *Am. Nat.* 100:603–609.

MacArthur, R. H., and Wilson, E. O. (1967). *The Theory of Island Biogeography.* Princeton University Press, Princeton, NJ.

MacDonald, D. (2001). *The New Encyclopedia of Mammals.* Oxford University Press, Oxford.

MacDonald, E. A., Fernandez-Duque, E., Evans, S., and Hagey, L.R. (2008). Sex, age, and family differences in the chemical composition of owl monkey (*Aotus nancymaae*) subcaudal scent secretions. *Am. J. Primatol.* 70:12–18.

Mace, G. M. (2004). The role of taxonomy in species conservation. *Philos. Trans. R. Soc. Lond. B* 359:711–719.

Mace, R. (2000). Evolutionary ecology of human life history. *Anim. Behav.* 59:1–10.

Macedonia, J. M., and Evans, C. S. (1993). Variation among mammalian alarm call systems and the problem of meaning in animal signals. *Ethology* 93:177–197.

Machida, S. (1990). Standing and climbing a pole by members of a captive group of Japanese monkeys. *Primates* 31:291–298.

MacKenzie, M. M., McGrew, W. C., and Chamove, A. S. (1985). Social preferences in stumptailed macaques (*Macaca actoides*): effects of companionship, kinship and rearing. *Dev. Psychobiol.* 18:115–123.

MacKinnon, J. R. (1974). The behaviour and ecology of wild orang-utans (*Pongo pygmaeus*). *Anim. Behav.* 22:3–74.

MacKinnon, J. R. (1977). A comparative ecology of Asian apes. *Primates* 18:747–772.

MacKinnon, J. (1979). Reproductive behavior in wild orangutan populations. In: Hamburg, D. A., and McCown, E. R. (eds.), *The Great Apes.* Benjamin/Cummings, Menlo Park, CA. pp. 257–273.

MacKinnon, J. R., and MacKinnon, K. S. (1980a). Niche differentiation in a primate community. In: Chivers, D. J. (ed.), *Malayan Forest Primates: Ten Years Study in Tropical Rain Forest.* Plenum Press, New York. pp. 167–191.

MacKinnon, J., and MacKinnon, K. (1980b). The behavior of wild spectral tarsiers. *Int. J. Primatol.* 1:361–379.

MacKinnon, K. C. (1995). Age differences in foraging patterns and spatial associations of the white-faced capuchin monkey (*Cebus capucinus*) in Costa Rica [MA thesis]. University of Alberta, Edmonton.

MacKinnon, K. C. (2002). Social development of wild white-faced capuchin monkeys (*Cebus capucinus*) in Costa Rica: an examination of social interactions between immatures and adult males [PhD diss.]. University of California, Berkeley.

MacKinnon, K. C. (2003a). Behavioral interactions between small juvenile and adult male white-faced capuchin monkeys (*Cebus capucinus*) in Costa Rica. *Am. J. Phys. Anthropol.* (suppl 36):143.

MacKinnon, K. C. (2003b). The context of an observed infant-killing event in *Cebus capucinus* at Santa Rosa National Park, Costa Rica. *Am. J. Primatol.* 60(suppl. 1):60–61.

MacKinnon, K. C. (2006). Food choice by juvenile capuchn monkeys (*Cebus capucinus*) in a tropical dry forest. In: Estrada, A., Garber, P. A., Pavelka, M., and Luecke, L. (eds.), *New Perspectives in the Study of Mesoamerican Primates: Distribution, Ecology, Behavior, and Conservation.* Springer Science+Business Media, New York. pp. 349–365.

MacKinnon, K. C. (2007). Social beginnings: the tapestry of infant and adult interactions. In: Campbell, C. J., Fuentes, A., MacKinnon, K. C., Panger, M., and Bearder, S. K. (eds.), *Primates in Perspective,* 1st ed. Oxford University Press, New York. pp. 571–591.

MacLatchy, L. M. (1996). Another look at the australopithecine hip. *J. Hum. Evol.* 31:455–476.

MacLatchy, L. (2004). The oldest ape. *Evol. Anthropol.* 13:90–103.

Maclean, E. L., Merritt, D. J., and Brannon, E. M. (2008). Social complexity predicts transitive reasoning in prosimian primates. *Anim. Behav.* 76:479–486.

MacPhee, R., and Cartmill, M. (1986). Basicranial structures and primate systematics. In: Swindler, D., and Erwin, J. (eds.), *Comparative Primate*

Biology, Systematics, Evolution and Anatomy, vol. 1. Alan R. Liss, New York. pp. 219–275.

MacPhee, R. D. E., and Horovitz, I. (2002). Extinct Quaternary platyrrhines of the Greater Antilles and Brazil. In: Hartwig, W. C. (ed.), *The Primate Fossil Record.* Cambridge University Press, Cambridge. pp. 189–199.

MacPhee, R. D. E., and Horovitz, I. (2004). New craniodental remains of the Quaternary Jamaican monkey *Xenothrix mcgregori* (Xenotrichini, Callicebinae, Pitheciidae), with a reconsideration of the *Aotus* hypothesis. *Am. Mus. Novit.* 3434:1–51.

Maddison, W. P., and Maddison, D. R. (2004). Mesquite: a modular system for evolutionary analysis, version 1.01, http://mesquiteproject.org.

Madsen, O., Scally, M., Douady, C. J., Kao, D. J., DeBry, R. W., Adkins, R., Amrine, H. M., Stanhope, M. J., de Jong, W. W., and Springer, M. S. (2001). Parallel adaptive radiations in two major clades of placental mammals. *Nature* 409:610–614.

Maestripieri, D. (1994a). Mother–infant relationships in 3 species of macaques (*Macaca mulatta, M. nemestrina, M. arctoides*). 2: The social environment. *Behaviour* 131:97–113.

Maestripieri, D. (1994b). Social structure, infant handling, and mothering styles in group-living Old World monkeys. *Int. J. Primatol.* 15:531–553.

Maestripieri, D. (1995a). Assessment of danger to themselves and their infants by rhesus macaque (*Macaca mulatta*) mothers. *J. Comp. Psychol.* 109:416–420.

Maestripieri, D. (1995b). First steps in the macaque world: do rhesus mothers encourage their infants' independent locomotion? *Anim. Behav.* 49:1541–1549.

Maestripieri, D. (1996). Maternal encouragement of infant locomotion in pigtail macaques, *Macaca nemestrina. Anim. Behav.* 51:603–610.

Maestripieri, D. (1998). Social and demographic influences on mothering style in pigtail macaques. *Ethology* 104:379–385.

Maestripieri, D. (1999a). Fatal attraction: interest in infants and infant abuse in rhesus macaques. *Am. J. Phys. Anthropol.* 110:17–25.

Maestripieri, D. (1999b). Primate social organization, gestural repertoire size, and communication dynamics: a comparative study of macaques. In: King, B. J. (ed.), *The Origins of Language: What Nonhuman Primates Can Tell Us.* School of American Research Press, Santa Fe, NM. pp. 55–78.

Maestripieri, D. (2001). Is there mother–infant bonding in primates? *Dev. Rev.* 21(1):93–120.

Maestripieri, D. (2003). Similarities in affiliation and aggression between cross-fostered rhesus macaque females and their biological mothers. *Dev. Psychobiol.* 43:321–327.

Maestripieri, D., and Carroll, K. A. (1998). Risk factors for infant abuse and neglect in group-living rhesus monkeys. *Psychol. Sci.* 9:143–145.

Maestripieri, D., and Carroll, K. A. (2000). Causes and consequences of infant abuse and neglect in monkeys. *Aggr. Viol. Behav.* 5:245–254.

Maestripieri, D., and Megna, N. L. (2000a). Hormones and behavior in rhesus macaque abusive and nonabusive mothers. 1. Social interactions during late pregnancy and early lactation. *Physiol. Behav.* 71:35–42.

Maestripieri, D., and Megna, N. L. (2000b). Hormones and behavior in rhesus macaque abusive and nonabusive mothers. 2. Mother–infant interactions. *Physiol. Behav.* 71:43–49.

Maestripieri, D., and Wallen, K. (1995). Interest in infants varies with reproductive condition in group-living female pigtail macaques (*Macaca nemestrina*). *Physiol. Behav.* 57:353–358.

Maestripieri, D., Schino, G., Aureli, F., and Troisi, A. (1992). A modest proposal: displacement activities as an indicator of emotions in primates. *Anim. Behav.* 44:967–979.

Maestripieri, D., Ross, S. K., and Megna, N. L. (2002). Mother–infant interactions in western lowland gorillas (*Gorilla gorilla gorilla*): spatial relationships, communication, and opportunities for social learning. *J. Comp. Psychol.* 116:219–227.

Maestripieri, D., Leoni, M., Raza, S. S., Hirsch, E. J., and Whitham, J. C. (2005). Female copulation calls in guinea baboons: evidence for postcopulatory female choice? *Int. J. Primatol.* 26:737–758.

Maestripieri, D., Mayhew, J., Carlson, C. L., Hoffman, C., and Radtke, J. M. (2007). One-male harems and female social dynamics in Guinea baboons. *Folia Primatol.* 78:56–68.

Maggioncalda, A. N., Sapolsky, R. M., and Czekala, N. M. (1999). Reproductive hormone profiles in captive male orangutans: implications for understanding developmental arrest. *Am. J. Phys. Anthropol.* 109:19–32.

Maggioncalda, A. N., Czekala, N. M., and Sapolsky, R. M. (2000). Growth hormone and thyroid stimulating hormone concentrations in captive male orangutans: implications for understanding developmental arrest. *Am. J. Primatol.* 50:67–76.

Maggioncalda, A. N., Sapolsky, R. M., and Czekala, N. M. (2002). Male orangutan subadulthood: a new twist on the relationship between chronic stress and developmental arrest. *Am. J. Phys. Anthropol.* 118:25–32.

Magliocca, F., and Gautier-Hion, A. (2002). Mineral content as a basis for food selection by western lowland gorillas in a forest clearing. *Am. J. Primatol.* 57:67–77.

Magliocca, F., Querouil, S., and Gautier-Hion, A. (1999). Population structure and group composition of western lowland gorillas in north-western Republic of Congo. *Am. J. Primatol.* 48:1–14.

Magurran, A. E., and Macias Garcia, C. (2000). Sex differences in behaviour as an indirect consequence of mating system. *J. Fish Biol.* 57:839–857.

Mahaney, W. C., Watts, D. P., and Hancock, R. G. V. (1990). Geophagia by mountain gorillas (*Gorilla gorilla beringei*) in the Virunga Mountains, Rwanda. *Primates* 31:113–120.

Mahaney, W. C., Hancock, R. G. V., and Aufreiter, S. (1995). Mountain gorilla geophagy: a possible strategy for dealing with intestinal problems. *Int. J. Primatol.* 16:475–487.

Mahaney, W. C., Hancock, R. G. V., Aufreiter, S., and Huffman, M. A. (1996). Geochemistry and clay minerology of termite mound soil and the role of geophagy in chimpanzees of the Mahale Mountains, Tanzania. *Primates* 37:121–134.

Mahaney, W. C., Milner, M. W., Sanmugadas, K., Hancock, R. G. V., Aufreiter, S., Wrangham, R. W., and Pier, H. W. (1997). Analysis of geophagy soils in Kibale Forest, Uganda. *Primates* 38:159–176.

Maher, C. R., and Lott, D. F. (1995). Definitions of territoriality used in the study of variation in vertebrate spacing systems. *Anim. Behav.* 49:1581–1597.

Maier, W., Alonso, C., and Langguth, A. (1982). Field observations on *Callithrix jacchus jacchus. Z. Saugetierk.* 47:334–346.

Maisels, F. (1993). Gut passage rate in guenons and mangabeys: another indicator of a flexible dietary niche? *Folia Primatol.* 61:35–37.

Maisels, F. G., Gautier, J. P., Cruickshank, A., and Bosefe, J. P. (1993). Attacks by crowned hawk eagles (*Stephanoaetus coronatus*) on monkeys in Zaire. *Folia Primatol.* 61:157–159.

Maisels, F., Gautier-Hion, A., and Gautier, J.-P. (1994). Diets of two sympatric colobines in Zaire: more evidence on seed-eating in forests on poor soils. *Int. J. Primatol.* 15:681–701.

Majolo, B., and Koyama, N. (2006). Seasonal effects on reconciliation in *Macaca fuscata yakui. Int. J. Primatol.* 27:1383–1397.

Majolo, B., Schino, G., and Troisi, A. (2005). Towards thirty years of ethological research on the Japanese macaque (*Macaca fuscata*) colony of the Rome Zoo: a review. *J. Anthropol. Sci.* 83:43–60.

Majolo, B., Ventura, R., and Koyama, N. (2009). A statistical modelling approach to the occurrence and timing of reconciliation in wild Japanese macaques. *Ethology* 115:152–166.

Málaga, C. A., Weller, R. E., and Buschbom, R. L. (1991). Twinning in the karyotype I night monkey (*Aotus nancymai*). *J. Med. Primatol.* 20:370–372.

Málaga, C. A., Weller, R. E., Buschbom, R. L., Baer, J. F., and Kimsey, B. B. (1997). Reproduction of the owl monkey (*Aotus* spp.) in captivity. *J. Med. Primatol.* 26:147–152.

Malaivijitnond, S., Lekprayoon, C., Tandavanittj, N., Panha, S., Cheewatham, C., and Hamada, Y. (2007). Stone-tool usage by Thai long-tailed macaques (*Macaca fascicularis*). *Am. J. Primatol.* 69:227–233.

Malenky, R. (1990). Ecological factors affecting food choice and social organization in *Pan paniscus* [PhD diss.]. State University of New York, Stony Brook.

Malenky, R. K., and Stiles, E. W. (1991). Distribution of terrestrial herbaceous vegetation and its consumption by *Pan paniscus* in the Lomako Forest, Zaire. *Am. J. Primatol.* 23:153–169.

Malenky, R. K., and Wrangham, R. W. (1994). A quantitative comparison of terrestrial herbaceous food-consumption by *Pan paniscus* in the Lomako Forest, Zaire, and *Pan troglodytes* in the Kibale Forest, Uganda. *Am. J. Primatol.* 32:1–12.

Mallinson, J. (1984). Breeding of great apes at the Jersey Wildlife Preservation Trust and a look into the future. *Zoo Biol.* 3:1–11.

Malone, N., Purnama, A. R., Wedana, M., and Fuentes, A. (2002). Assessment of the sale of primates at Indonesian bird markets. *Asian Primates* 8:7–11.

Mammides, C., Cords, M., Peters, M. K. (2009). Effects of habitat disturbance and food supply on population densities of three primate species in the Kakamega Forest, Kenya. *Afr. J. Ecol.* 47(1):87–96.

Mangel, M. (1990). Resource divisibility, predation and group formation. *Anim. Behav.* 39:1163–1172.

Manley, G. H. (1966). Reproduction in lorisoid primates. In: Rowlands, I. W. (ed.), *Comparative Biology of Reproduction in Mammals. Symp. Zool. Soc. Lond.* 15: 493–509.

Manley, G. H. (1967). Gestation periods in the Lorisidae. *Int. Zool. Ybk.* 7:80–81.

Mann, A., and Weiss, M. (1996). Hominoid phylogeny and taxonomy: a consideration of the molecular and fossil evidence in an historical perspective. *Mol. Phylogen. Evol.* 5(1):169–181.

Mann, D. R., Blank, M. S., Gould, K. G., and Collins, D. C. (1982). Validation of radioimmunoassays for LH and FSH in the sooty mangabey (*Cercocebus atys*): characterization of LH and the response to GnRH. *Am. J. Primatol.* 2:275–283.

Mann, G. F. (1956). Efecto del frío en mamíferos Amazónicos. *Invest. Zool. Chil.* 3:155.

Mann, J., Sargeant, B., Watson-Capps, J., Gibson, Q., Heithaus, M., Connor, R., and Patterson, E. (2008). Why do dolphins carry sponges? *PLoS ONE* 3(12). doi:10.1371/journal.pone.0003868.

Mannu, M., and Ottoni, E. B. (2009). The enhanced tool-kit of two groups of wild bearded capuchin monkeys in the Caatinga: tool making, associative use, and secondary tools. *Am J Primatol.* 71:242–251.

Manoukis, N. C. (2007). FORMATOMATIC: a program for converting diploid allelic data between common formats for population genetic analysis. *Mol. Ecol. Notes* 7:592–593.

Mansfield, K. (2003). Marmoset models commonly used in biomedical research. *Comp. Med.* 53:383–392.

Manson, J. H. (1992). Measuring female mate choice in Cayo Santiago rhesus macaques. *Anim. Behav.* 44:405–416.

Manson, J. H. (1994a). Male aggression: a cost of female mate choice in Cayo Santiago rhesus macaques. *Anim. Behav.* 48:473–475.

Manson, J. H. (1994b). Mating patterns, mate choice, and birth season heterosexual relationships in free-ranging rhesus macaques. *Primates* 35:417–433.

Manson, J. H. (1995). Do female rhesus macaques choose novel males? *Am. J. Primatol.* 37:285–296.

Manson, J. H. (1996). Rhesus macaque copulation calls: reevaluating the "honest signal" hypothesis. *Primates* 37:145–154.

Manson, J. H. (1997). Primate consortships: a critical review. *Curr. Anthropol.* 38:353–374.

Manson, J. (1999). Infant handling in wild *Cebus capucinus*: testing bonds between females? *Anim. Behav.* 57:911–921.

Manson, J. H. (2007). Mate choice. In: Campbell, C. J., Fuentes, A., MacKinnon, K. C., Panger, M., and Bearder, S. K. (eds.), *Primates in Perspective*, 1st ed. Oxford University Press, New York. pp. 447–463.

Manson, J. H., and Perry, S. E. (1993). Inbreeding avoidance in rhesus macaques: whose choice? *Am. J. Phys. Anthropol.* 90:335–344.

Manson, J. H., and Wrangham, R. W. (1991). Intergroup aggression in chimpanzees and humans. *Curr. Anthropol.* 32:369–390.

Manson, J. H., Perry, S., and Parish, A. R. (1997). Nonconceptive sexual behavior in bonobos and capuchins. *Int. J. Primatol.* 18:767–786.

Manson, J. H., Rose, L. M., Perry, S., and Gros-Louis, J. (1999). Dynamics of female–female relationships in wild *Cebus capucinus*: data from two Costa Rican sites. *Int. J. Primatol.* 20:679–706.

Manson, J. H., Perry, S., and Stahl, D. (2005). Reconciliation in wild white-faced capuchins (*Cebus capucinus*). *Am. J. Primatol.* 65:205–219.

Manzolillo, D. L. (1986). Factors affecting intertroop transfer by adult male *Papio anubis*. In: Else, J. G., and Lee, P. C. (eds.), *Primate Ontogeny, Cognition and Social Behaviour*. Cambridge University Press, Cambridge. pp. 371–380.

Maple, T. (1980). *Orangutan Behavior*. van Nostrand Reinhold, New York.

Maples, W. R., Maples, M. K., Greenhood, W. R., and Walek, M. L. (1976). Adaptations of crop-raiding baboons in Kenya. *Am. J. Phys. Anthropol.* 45:309–316.

Marais, E. (1968). *The Soul of the Ape*. Atheneum, New York.

Mardis, E. R. (2008a). The impact of next-generation sequencing technology on genetics. *Trends Genet.* 24:133–141.

Mardis, E. R. (2008b). Next-generation DNA sequencing methods. *Annu. Rev. Genomics Hum. Genet.* 9:487–402.

Marín-Gómez, O. H. (2008). Consumo de néctar por *Aotus lemurinus* y su rol como posible polinizador de las flores de *Inga edulis* (Fabales: Mimosoideae). *Neotrop. Primates* 15:30–32.

Marivaux, L., Chaimanee, Y., Ducrocq, S., Marandat, B., Sudre, J., Soe, A. N., Tun S. T., Htoon, W., and Jaeger, J. J. (2003). The anthropoid status of a primate from the late middle Eocene Pondaung Formation (central Myanmar): tarsal evidence. *Proc. Natl. Acad. Sci. USA* 100:13173–13178.

Markham, C. A., and Altmann, J. (2008). Remote monitoring of primates using automated GPS technology in open habitats. *Am. J. Primatol.* 70:1–5.

Markham, R. J. (1990). Breeding orangutans at Perth zoo: twenty years of appropriate husbandry. *Zoo Biol.* 9:171–182.

Markham, R., and Groves, C. P. (1990). Weights of wild orangutans. *Am. J. Phys. Anthropol.* 81:1–3.

Markolf, M., Kappeler, P. M., and Rasoloarison, R. (2008). Distribution and conservation status of *Mirza zaza*. *Lemur News* 13:37–40.

Marks, D. L., Swain, T., Goldstein, S., Richard, A., and Leighton, M. (1988). Chemical correlates of rhesus monkey food choice: the influence of hydrolyzable tannins. *J. Chem. Ecol.* 14:213–235.

Marler, P. (1965). Communication in monkeys and apes. In: Devore, I. (ed.), *Primate Behavior*. Holt, Rinehart, and Winston, New York. pp. 544–584.

Marler, P. (1968). Aggregation and dispersal: two functions in primate communication. In: Jay, P. C. (ed.), *Primates, Studies in Adaptation and Variability*. University of California Press, Berkeley. pp. 420–438.

Marler, P. (1969). *Colobus guereza*: territoriality and group composition. *Science* 163:93–95.

Marler, P. (1972). Vocalizations of East African monkeys II: black and white colobus. *Behaviour* 42:175–197.

Marler, P. (1984). Animal communication: affect or cognition? In: Scherer, K. R., and Ekman, P. (eds.), *Approaches to Emotion*. Erlbaum, Hillsdale, NJ. pp. 345–365.

Marler, P., Evans, C. S., and Hauser, M. D. (1992). Animal signals: motivational, referential, or both? In: Papousek, H., and Jürgens, U. (eds.), *Nonverbal Communication: Comparative and Developmental Approaches*. Cambridge University Press, Cambridge.

Marquès-Bonet, T., Cáceres, M., Bertranpetit, J., Preuss, T. M., Thomas, J. W., and Navarro, A. (2004). Chromosomal rearrangements and the genomic distribution of gene-expression divergence in humans and chimpanzees. *Trends Genet.* 20:524–529.

Marquès-Bonet, T., Kidd, J. M., Ventura, M., Graves, T. A., Cheng, Z., Hillier, L. W., Jiang, Z., Baker, C., Malfavon-Borja, R., Fulton, L. A., Alkan, C., Aksay, G., Girirajan, S., Siswara, P., Chen, L., Cardone, M. F., Navarro, A., Mardis, E. R., Wilson, R. K., and Eichler, E. E. (2009). A burst of segmental duplications in the genome of the African great ape ancestor. *Nature* 457:877–881.

Marroig, G., and Cheverud, J. M. (2001). A comparison of phenotypic variation and covariation patterns and the role of phylogeny, ecology, and ontogeny during cranial evolution of New World monkeys. *Evolution* 55:2576–2600.

Marroig, G., and Cheverud, J. M. (2004). Cranial evolution in sakis (*Pithecia*, Platyrrhini) I: interspecific differentiation and allometric patterns. *Am. J. Phys. Anthropol.* 125:266–278.

Marsh, C. (1978a). Comparative activity budgets of red colobus. In: Chivers, D. J., and Herbert, J. (eds.), *Recent Advances in Primatology*, vol. 1. Academic Press, London. pp. 249–251.

Marsh, C. (1978b). Ecology and social organization of the Tana River red colobus (*Colobus badius rufomitratus*) [PhD diss.]. University of Bristol, Bristol.

Marsh, C. W. (1979a). Comparative aspects of social organization in the Tana River red colobus, *Colobus badius rufomitratus*. *Z. Tierpsychol.* 51:337–362.

Marsh, C. W. (1979b). Female transference and mate choice among Tana River red colobus. *Nature* 281:568–569.

Marsh, C. W. (1981a). Diet choice among red colobus (*Colobus badius rufomitratus*) on the Tana River, Kenya. *Folia Primatol.* 35:147–178.

Marsh, C. W. (1981b). Ranging behaviour and its relation to diet selection in Tana River red colobus (*Colobus badius rufomitratus*). *J. Zool. Lond.* 195:473–492.

Marsh, C. W. (1981c). Time budget of Tana River red colobus. *Folia Primatol.* 35:30–50.

Marsh, H. L., and MacDonald, S. E. (2008). The use of perceptual features in categorization by orangutans (*Pongo abelli*). *Anim. Cogn.* 11:569–585.

Marsh, L. K. (2003). *Primates in Fragments*. Kluwer Academic/Plenum, New York.

Marshall, A. J., and Hohmann, G. (2005). Urinary testosterone levels of wild male bonobos (*Pan paniscus*) in the Lomako Forest, Democratic Republic of Congo. *Am. J. Primatol.* 65:87–92.

Marshall, A. J., and Leighton, M. (2006). How does food availability limit the population density of agile gibbons? In: Hohmann, G., Robbins, M., and Boesch, C. (eds.), *Feeding Ecology of the Apes*. Cambridge University Press, Cambridge. pp. 313–335.

Marshall, A. J., and Wrangham, R. W. (2007). Evolutionary consequences of fallback foods. *Int. J. Primatol.* 28:1219–1226.

Marshall, A. J., Nardiyono, Engstrom, L. M., Pamungkas, B., Palapa, J., Meijaard, E., and Stanley, S. A. (2006). The blowgun is mightier than the chainsaw in determining population density of Bornean orang-utans

(*Pongo pygmaeus morio*) in the forests of East Kalimantan. *Biol. Conserv.* 129:566–578.

Marshall, A. J., Ancrenaz, M., Brearley, F. Q., Fredriksson, G. M., Ghaffar, N., Heydon, M., Husson, S. J., Leighton, M., McConkey, K. R., Morrough-Bernard, H. C., Proctor, J., van Schaik, C. P., Yeager, C. P., and Wich, S. A. (2009a). The effects of forest phenology and floristics on populations of Bornean and Sumatran orangutans: are Sumatran forests better orangutan habitats than Bornean forests? In: Wich, S. A., Utami Atmoko, S. S., Mitra Setia, T., and van Schaik, C. P. (eds.), *Orangutans: Geographic Variation in Behavioral Ecology and Conservation*. Oxford University Press, New York. pp. 97–117.

Marshall, A. J., Lacy, R., Ancrenaz, M., Byers, O., Husson, S. J., Leighton, M., Meijaard, E., Rosen, N., Singleton, I., Stephens, S., Traylor-Holzer, K., Utami Atmoko, S. S., van Schaik, C. P., and Wich, S. A. (2009b). Orangutan population biology, life history, and conservation. In: Wich, S. A., Utami Atmoko, S. S., Mitra Setia, T., and van Schaik, C. (eds.), *Orangutans: Geographic Variation in Behavioral Ecology and Conservation*. Oxford University Press, New York. pp. 311–326.

Marshall, J. T., and Marshall, E. R. (1976). Gibbons and their territorial songs. *Science* 193:235–238.

Marshall, J., and Sugardjito, J. (1986). Gibbon systematics. In: Swindler, D. R., and Erwin, J. (eds.), *Comparative Primate Biology. Systematics, Evolution, and Anatomy,* vol. 1. Alan R. Liss, New York. pp. 137–185.

Marshall, T. C., Slate, J., Kruuk, L. E. B., and Pemberton, J. M. (1998). Statistical confidence for likelihood-based paternity inference in natural populations. *Mol. Ecol.* 7:639–655.

Marshall, V. S., Tannenbaum, P., Browne, M. A., Knowles, L., Kalishman, J. K., and Thomson, J. A. (1999). Assisted reproductive technologies in a New World primate, the common marmoset monkey (*Callithrix jacchus*). *American Association of Zoo Veterinarians Annual Conference Proceedings.* pp. 93–94.

Marten, K., and Marler, P. (1977). Sound transmission and its significance for animal vocalization. I. Temperate habitats. *Behav. Ecol. Sociobiol.* 2:271–290.

Marten, K., Quine, D., and Marler, P. (1977). Sound transmission and its significance for animal vocalization. II. Tropical forest habitats. *Behav. Ecol. Sociobiol.* 2:291–302.

Martin, D. (1997). Kinship bias: a function of familiarity in pigtailed macaques (*Macaca nemestrina*) [PhD diss.]. University of Georgia, Athens.

Martin, G. H. G. (1983). Bushmeat in Nigeria as a natural resource with environmental implications. *Environ. Conserv.* 10:125–134.

Martin, L. B., Olejniczak, A. J., and Maas, M. C. (2003). Enamel thickness and microstructure in pitheciin primates, with comments on dietary adaptations of the middle Miocene hominoid *Kenyapithecus*. *J. Hum. Evol.* 45:351–367.

Martin, P. S. (1984). Prehistory overkill: the global model. In: Martin, P. S., and Klein, R. G. (eds.), *Quaternary Extinctions: A Prehistoric Revolution.* University of Arizona Press, Tucson. pp. 354–403.

Martin, P., and Bateson, P. (1986). *Measuring Behaviour: An Introductory Guide.* Cambridge University Press, Cambridge.

Martin, P., and Bateson, P. (1993). *Measuring Behavior.* Cambridge University Press, Cambridge.

Martin, P. and Bateson, P. (2009). *Measuring Behaviour: An Introductory Guide,* 3rd ed. Cambridge University Press, Cambridge.

Martin, R. D. (1972a). A preliminary field study of the lesser mouse lemur (*Microcebus murinus* J. F. Miller 1777). *Z. Tierpsychol.* 9(suppl.):43–89.

Martin, R. D. (1972b). Adaptive radiation and behaviour of the Malagasy lemurs. *Phil. Trans. R. Soc. Lond. B* 264:295–352.

Martin, R. D. (1973). A review of the behaviour and ecology of the lesser mouse lemur (*Microcebus murinus* J. F. Miller 1777). In: Michael, R. P., and Crook, J. H. (eds.), *Comparative Ecology and Behaviour of Primates.* Academic Press, London. pp. 1–68.

Martin, R. D. (1982). Allometric approaches to the evolution of the primate nervous system. In: Armstrong, E., and Falk, D. (eds.), *Primate Brain Evolution: Methods and Concepts.* Plenum Press, New York. pp. 39–56.

Martin, R. D. (1983). *Human Brain Evolution in an Ecological Context. 52nd James Arthur Lecture on the Evolution of the Human Brain.* American Museum of Natural History, New York.

Martin, R. D. (1984). Dwarf and mouse lemurs. In: MacDonald, D. (ed.), *The Encyclopaedia of Mammals,* vol. 1. George Allen and Unwin, London. p. 331.

Martin, R. D. (1990). *Primate Origins and Evolution: A Phylogenetic Reconstruction.* Princeton University Press, Princeton, NJ.

Martin, R. D. (1993). Primate origins: plugging the gaps. *Nature* 363:223–234.

Martin, R. D. (1996). Scaling of the mammalian brain: the maternal energy hypothesis. *News Physiol. Sci.* 11:149–156.

Martin, R. D. (2000). Origins, diversity, and relationships of lemurs. *Int. J. Primatol.* 21:1021–1049.

Martin, R. D. (2003). Combing the primate record. *Nature* 422:388–391.

Martin, R. D., and MacLarnon A. M. (1990). Reproductive patterns in primates and other mammals: the dichotomy between altricial and precocial offspring. In: DeRousseau, C. J. (ed.), *Primate Life History and Evolution.* Wiley-Liss, New York. pp. 47–80.

Martin, R. D., Soligo, C., and Tavaré, S. (2007) Primate origins: implications of a Cretaceous ancestry. *Folia Primatol.* 78:277–296.

Martinez, J., and Wallace, R. B. (2007). Further notes on the distribution of endemic Bolivian titi monkeys, *Callicebus modestus* and *Callicebus olallae*. *Neotrop. Primates* 14:47–54.

Martins, M. M. (1998). Feeding ecology of *Callithrix aurita* in a forest fragment of Minas Gerais. *Neotrop. Primates* 6:126–127.

Martins, M. M. (2008). Fruit diet of *Alouatta guariba* and *Brachyteles arachnoides* in southeastern Brazil: comparison of fruit type, color, and seed size. *Primates* 49:1–8.

Martins, M. M., and Setz, E. Z. F. (2000). Diet of buffy tufted-eared marmosets (*Callithrix aurita*) in a forest fragment in southeastern Brazil. *Int. J. Primatol.* 21:467–476.

Martins, S. S., Lima, E. M., and Silva, J. S., Jr. (2005). Predation of bearded saki (*Chiropotes utahicki*) by a harpy eagle (*Harpia harpja*). *Neotrop. Primates* 13(1):7–10.

Martins, W. P., and Strier, K. B. (2004). Age at first reproduction in philopatric female muriquis (*Brachyteles arachnoides hypoxanthus*). *Primates* 45:63–67.

Maryanto, I., and Yani, M. (2008). The third record of pygmy tarsier (*Tarsius pumilus*) from Lore Lindu National Park, central Sulawesi, Indonesia. *Trop. Biodivers.* 8(2):79–85.

Masataka, N., Koda, H., Urasopon, N., and Watanabe, K. (2009). Free-ranging macaque mothers exaggerate tool-using behavior when observed by offspring. *PLoS ONE* 4(3):e4768.

Masi, S. (2008). Social influence on foraging strategies, activity and energy budgets of western lowland gorillas (*Gorilla gorilla gorilla*) in Bai-Hokou, Central African Republic [PhD diss.]. University of Rome La Sapienza, Italy.

Masi, S., Cipolletta, C., and Robbins, M. M. (2009). Western lowland gorillas (*Gorilla gorilla gorilla*) change their activity patterns in response to frugivory. *Am. J. Primatol.* 70:1–10.

Mason, W. A. (1966). Social organization of the South American monkey, *Callicebus moloch*: a preliminary report. *Tulane Stud. Zool.* 13:23–28.

Mason, W. A. (1968). Use of space by *Callicebus* groups. In: Jay, P. C. (ed.), *Primates: Studies in Adaptation and Variability.* Holt, Rinehart and Winston, New York. pp. 200–216.

Mason, W. A. (1971). Field and laboratory studies of social organization in *Saimiri* and *Callicebus*. In: Rosenblum, L. A. (ed.), *Primate Behavior,* vol. 2. Academic Press, New York. pp. 107–137.

Mason, W. A. (1974). Comparative studies of social behavior in *Callicebus* and *Saimiri*: behavior of male–female pairs. *Folia Primatol.* 22:1–8.

Mason, W. (1979). Ontogeny of social behavior. In: Marler, P., and Vandenbergh, J. G. (eds.), *Handbook of Behavioral Neurobiology.* Plemium Press, New York. pp. 1–28.

Mason, W. A. (1986). Early socialization. In: Benirschke, K. (ed.), *Primates: The Road to Self-Sustaining Populations.* Springer-Verlag, New York. pp. 321–329.

Mason, W. (1993). The nature of social conflict: a psychoethological perspective. In: Mason, W., and Mendoza, S. (ed.), *Primate Social Conflict.* State University of New York Press, Albany. pp. 13–48.

Mason, W. A., and Mendoza, S. P. (1998). Generic aspects of primate attachments: parents, offspring and mates. *Psychoneuroendocrinology* 23:765–778.

Masters, A., and Markham, R. J. (1991). Assessing reproductive status in orangutans by using urinary estrone. *Zoo Biol.* 10:197–208.

Masters, J. C. (1988). Speciation in the greater galagos (Prosimii: Galaginae): a review and synthesis. *Biol. J. Linn. Soc.* 34:149–174.

Masters, J. C. (1991). Loud calls of *Galago crassicaudatus* and *G. garnettii* and their relation to habitat structure. *Primates* 32:153–167.

Masters, J. C. (1998). Speciation in the lesser galagos. *Folia Primatol.* 69(suppl. 1):357–370.

Masters, J. C., and Brothers, D. J. (2002). Lack of congruence between morphological and molecular data in reconstructing the phylogeny of the Galagonidae. *Am. J. Phys. Anthropol.* 117:79–93.

Masters, J. C., Rayner, H., Ludewig, H., Zimmermann, E., Molez-Verriere, F., Vincent, F., and Nash, L. T. (1994). Phylogenetic relationships among the Galaginae as indicated by erythrocytic allozymes. *Primates* 35:177–190.

Masters, J. C., Anthony, N. M., de Wit, M. J., and Mitchell, A. (2005). Reconstructing the evolutionary history of the Lorisidae using morphological, molecular and geological data. *Am. J. Phys. Anthropol.* 127:465–480.

Masters, W. H., and Johnson, V. A. (1966). *Human Sexual Response*. J. and A. Churchill, London.

Masterson, T. J. (1995). Morphological relationships between the Ka'apor capuchin (*Cebus kaapori* Queiroz 1992) and other male *Cebus* crania: a preliminary report. *Neotrop. Primates* 3:165–169.

Mather, R. J. (1992). A field study of hybrid gibbons in central Kalimantan, Indonesia [PhD diss.]. University of Cambridge, Cambridge.

Matheson, M., and Bernstein, I. (2000). Grooming, social bonding, and agonistic aiding in rhesus monkeys. *Am. J. Primatol.* 51:177–186.

Matsubayashi, K., and Enomoto, T. (1983). Longitudinal studies on annual changes in plasma testosterone, body weight and spermatogenesis in adult Japanese monkeys (*Macaca fuscata fuscata*) under laboratory conditions. *Primates* 24:521–529.

Matsubayashi, K., and Mochizuki, K. (1982). Growth of male reproductive organs with observation of their seasonal morphological changes in Japanese monkeys (*Macaca fuscata*). *Jpn. J. Vet.* 44:891–902.

Matsuda, I., and Izawa, K. (2008). Predation of wild spider monkeys at La Macarena, Colombia. *Primates* 49(1):65–68.

Matsuda, I., Tuuga, A., and Higashi, S. (2008). Clouded leopard (*Neofelis diardi*) predation on proboscis monkeys (*Nasalis larvatus*) in Sabah, Malaysia. *Primates* 49(3):227–231.

Matsumoto-Oda, A. (1999). Female choice in the opportunistic mating of wild chimpanzees (*Pan troglodytes schweinfurthii*) at Mahale. *Behav. Ecol. Sociobiol.* 46:258–266.

Matsumoto-Oda, A. (2002). Social relationships between cycling females and adult males in Mahale chimpanzees. In: Boesch, C., Hohmann, G., and Marchant, L. (eds.), *Behavioural Diversity in Chimpanzees and Bonobos*. Cambridge University Press, Cambridge. pp. 168–180.

Matsumoto-Oda, A., and Hayashi, Y. (1999). Nutritional aspects of fruit choice by chimpanzees. *Folia Primatol.* 70:154–162.

Matsumoto-Oda, A., and Oda, R. (1998). Changes in the activity budget of cycling female chimpanzees. *Am. J. Primatol.* 46:157–166.

Matsumoto-Oda, A., Hosaka, K., Huffman, M. A., and Kawanaka, K. (1998). Factors affecting party size in chimpanzees of the Mahale Mountains. *Int. J. Primatol.* 19:999–1011.

Matsumura, S. (1996). Postconflict affiliative contacts between former opponents among wild moor macaques (*Macaca maurus*). *Am. J. Primatol.* 38:211–219.

Matsumura, S. (1997). Mothers in a wild group of moor macaques (*Macaca maurus*) are more attractive to other group members when holding their infants. *Folia Primatol.* 68:77–85.

Matsumura, S. (1999). The evolution of "egalitarian" and "despotic" social systems among macaques. *Primates* 40:23–31.

Matsumura, S. (2001). The myth of despotism and nepotism: dominance and kinship in matrilineal societies of macaques. In: Matsuzawa, T. (ed.), *Primate Origins of Human Cognition and Behavior*. Springer, Tokyo, pp. 441–462.

Matsumura, S., and Kobayashi, T. (1998). A game model for dominance relations among group-living animals. *Behav. Ecol. Sociobiol.* 42:77–84.

Matsumura, S., and Okamoto, K. (1997). Factors affecting proximity among member of a wild group of moor macaques during feeding, moving, and resting. *Int. J. Primatol.* 18:929–940.

Matsuzawa, T. (1996). Chimpanzee intelligence in nature and in captivity: isomorphism of symbol use and tool-use. In: McGrew, W. C., Marchant, L. F., and Nishida, T. (eds.), *Great Ape Societies*. Cambridge University Press, Cambridge. pp. 197–209.

Matsuzawa, T. (2001a). Primate foundations of human intelligence: a view of tool use in nonhuman primates and fossil hominids. In: Matsuzawa, T. (ed.), *Primate Origins of Human Cognition and Behavior*. Springer-Verlag, Hong Kong. pp. 3–25.

Matsuzawa, T. (ed.) (2001b). *Primate Origins of Human Cognition and Behavior*. Springer-Verlag, Tokyo.

Matsuzawa, T., and Yamakoshi, G. (1996). Comparison of chimpanzee material culture between Bossou and Nimba, West Africa. In: Russon, A. E., Bard, K. A., and Taylor Parker, S. (eds.), *Reaching into Thought. The Minds of the Great Apes*. University of Cambridge Press, Cambridge. pp. 211–232.

Matsuzawa, T., Tomonaga, M., and Tanaka, M. (eds.) (2006). *Cognitive Development in Chimpanzees*. Springer-Verlag, Tokyo.

Matteri, R. L., Roser, J. F., Baldwin, D. M., Lipovetsky, V., and Papkoff, H. (1987). Characterization of a monoclonal antibody which detects luteinizing hormone from diverse mammalian species. *Dom. Anim. Endocrinol.* 4:157–165.

Matthews, L. H. (1956). The sexual skin of the gelada baboon (*Theropithecus gelada*). *Trans. Zool. Soc. Lond.* 28:543–552.

Matthews, L.J. (2008). Activity patterns, home range size, and intergroup encounters in *Cebus albifrons* support existing models of capuchin socioecology. *Int. J. Primatol.* 30:709–728.

Matthews, L. J., and Rosenberger, A. L. (2008). Taxon combinations, parsimony analysis (PAUP*), and the taxonomy of the yellow-tailed woolly monkey, *Lagothrix flavicauda*. *Am. J. Phys. Anthropol.* 137:245–255.

Mattle, F. M. E., Pryce, C. R., and Anzenberger, G. (2008). Different ovarian responses to potential mates underlie species-specific breeding strategies in common marmoset and Goeldi's monkey. *Horm. Behav.* 54:302–311.

Mau, M., Südekum, K-H., Johann, A., Sliwa, A., and Kaiser, T. M. (2009). Saliva of the graminivorous Theropithecus gelada lacks proline-rich proteins and tannin-binding capacity. *Am. J. Primatol.* 71: 663–669.

Maxim, P. E., and Buettner-Janusch, J. (1963). A field study of the Kenya Baboon. *Am. J. Phys. Anthropol.* 21:165–180.

Maynard Smith, J. (1974). The theory of games and the evolution of animal conflicts. *J. Theor. Biol.* 47:209–221.

Mayr, E. (1980). Prologue. Some thoughts on the history of the evolutionary synthesis. In: Mayr, E., and Provine, W. B. (eds.), *The Evolutionary Synthesis. Perspectives on the Unification of Biology*. Harvard University Press, Cambridge, MA. pp. 1–48.

Mayr, E. (1997). *This Is Biology: The Science of the Living World*. Harvard University Press, Cambridge, MA.

Mayr, E. (2004). *What Makes Biology Unique? Considerations on the Autonomy of a Scientific Discipline*. Cambridge University Press, Cambridge.

Mbora, D. N. M., and McGrew, W. C. (2002). Extra-group sexual consortship in the Tana River red colobus (*Procolobus rufomitratus*)? *Folia Primatol.* 73:210–213.

Mbora, D. N. M., and Meikle, D. B. (2004a). Forest fragmentation and the distribution, abundance and conservation of the Tana River red colobus (*Procolobus rufomitratus*). *Biol. Conserv.* 118:67–77.

Mbora, D. N. M., and Meikle, D. B. (2004b). The value of unprotected habitat in conserving the critical endangered Tana River red colobus (*Procolobus rufomitratus*). *Biol. Conserv.* 120:91–99.

McCann, C. M. (1995). Social factors affecting reproductive success in female gelada baboons (*Theropithecus gelada*) [PhD diss.]. City University of New York, New York.

McCanny, S. J. (1985). Alternatives in parent–offspring relationships in plants. *Oikos* 45:148–149.

McConkey, K. R. (1999). Gibbons as seed dispersers in the rain forests of central Borneo [PhD diss.]. Cambridge University, Cambridge.

McConkey, K. R. (2000). Primary seed shadow generated by gibbons in the rain forests of Barito Ulu, central Borneo. *Am. J. Primatol.* 52:13–29.

McConkey, K. R., and Chivers, D. J. (2007). Influence of gibbon ranging patterns on seed dispersal distance and deposition in a Bornean forest. *J. Trop. Ecol.* 23:269–275.

McConkey, K. R., Aldy, F., Ario, A., and Chivers, D. J. (2002). Selection of fruit by gibbons (*Hylobates muelleri* × *agilis*) in the rain forests of central Borneo. *Int. J. Primatol.* 23:123–145.

McConkey, K. R., Ario, A., Aldy, F., and Chivers, D. J. (2003). Influence of forest seasonality on gibbon food choice in the rain forests of Barito Ulu, central Kalimantan. *Int. J. Primatol.* 24:19–32.

McConnell, I. W., Atwell, R., Renquist, D., and Gunnels, R. (1990). Serum concentrations of progesterone and estrogens during the reproductive cycle of owl monkeys (*Aotus trivirgatus*). *Georgia J. Sci.* 48:96–101.

McCrossin, M. L. (1992). New species of bushbaby from the middle Miocene of Maboko Island, Kenya. *Am. J. Phys. Anthropol.* 89:215–233.

McDaniel, P. S., Janzow, F. T., Porton, I., and Asa, C. S. (1993). The reproductive and social dynamics of captive *Ateles geoffroyi* (black-handed spider monkey). *Am. Zool.* 33:173–179.

McDermott, W. C. (1938). *The Apes in Antiquity*. Johns Hopkins Press, Baltimore, MD.

McDonald, J. H., and Kreitman, M. (1991). Adaptive protein evolution at the ADH locus in Drosophila. *Nature* 351:652–654.

McDonald, M. (1985). Courtship behavior of female Japanese monkeys. *Can. Rev. Phys. Anthropol.* 4:67–75.

McDougal, C. (1987). The man-eating tiger in geographical and historical perspective. In: Tilson, A. U. S. S. R. L. (ed.), *Tigers of the World*. Park City, Noyes. pp. 435–448.

McFarland, D. (ed.) (1987). *The Oxford Companion to Animal Behavior*. Oxford University Press, Oxford.

McFarland, M. J. (1986). Ecological determinants of fission–fusion sociality in *Ateles* and *Pan*. In: Else, J. G., and Lee, P. C. (eds.), *Primate Ecology and Conservation*. Cambridge University Press, Cambridge. pp. 181–190.

McFarland, R. (1997). Primate females: fit or fat. In: Morbeck, M. E., Galloway, A., and Zihlman, A. L. (eds.), *The Evolving Female: A Life-History Perspective*. Princeton University Press, Princeton, NJ. pp. 163–175.

McGraw, S. (1994). Census, habitat preference, and polyspecific associations of six monkeys in the Lomako Forest, Zaire. *Am. J. Primatol.* 34:295–307.

McGraw, S. W., Plavacan, J. M., and Adachi-Kanazawa, K. (2002). Adult female *Cercopithecus diana* employ canine teeth to kill another adult female *C. diana*. *Int. J. Primatol.* 23:1301–1330.

McGraw, S. W., Cooke, C., and Shultz, S. (2006). Primate remains from African crowned eagle (*Stephanoaetus*) nests in Ivory Coast's Tai Forest: implications for primate predation and early hominid taphonomy in South Africa. *Am. J. Phys. Anthropol.* 131:151–165.

McGraw, W. S. (1998a). Posture and support use of Old World monkeys (Cercopithecideae): the influence of foraging strategies, activity patterns, and the spatial distribution of preferred food items. *Am. J. Primatol.* 46:229–250.

McGraw, W. S. (1998b). Comparative locomotion and habitat use of six monkeys in the Taï Forest, Ivory Coast. *Am J. Phys. Anthropol.* 105:493–510.

McGraw, W. S. (1998c). Locomotion, support use, maintenance activities, and habitat structure: the case of the Taï Forest cercopithecids. In: Strasser, E., Fleagle, J. G., Rosenberger, A., and McHenry, H. (eds.), *Primate Locomotion: Recent Advances*. Plenum Press, New York. pp. 79–94.

McGraw, W. S. (2000). Positional behavior of *Cercopithecus petaurista*. *Int. J. Primatol.* 21:151–182.

McGraw, W. S. (2002). Diversity of guenon positional behavior. In: Glenn, M., and Cords, M. (eds.), *The Guenons: Diversity and Adaptation in African Monkeys*. Kluwer, New York, pp. 113–131.

McGraw, W. S. (2005). Update on the search for Miss Waldron's red colobus monkey (*Procolobus badius waldroni*). *Int. J. Primatol.* 26:605–620.

McGraw, W. S. (in press). The sooty mangabey (*Cercocebus atys*). In: Kingdon, J., Happold, D., and Butynski, T. (eds.), *The Mammals of Africa*. University of California Press, Berkeley, CA.

McGraw, W. S., and Bshary, R. (2002). Association of terrestrial mangabeys (*Cercocebus atys*) with arboreal monkeys: experimental evidence for the effects of reduced ground predator pressure on habitat use. *Int. J. Primatol.* 23:311–326.

McGraw, W. S., and Oates, J. F. (2002). Evidence for a surviving population of Miss Waldron's red colobus. *Oryx* 36:223.

McGraw, W. S., and Zuberbuhler, K. (2007). The monkeys of the Tai forest: an introduction. In: McGraw, W. S., Zuberbuhler, K., and Noe, R. (eds.). *Monkeys of the Tai Forest*. Cambridge University Press, New York. pp. 1–50.

McGraw, W. S., Cooke, C., and Shultz, S. (2006). Primate remains from African crowned eagle (*Stephanoaetus coronatus*) nests in Ivory Coast's Tai Forest: implications for primate predation and early hominid taphonomy in South Africa. *Am. J. Phys. Anthropol.* 131(2):151–165.

McGrew, W. C. (1974). Tool use by wild chimpanzees in feeding upon driver ants. *J. Hum. Evol.* 3:501–508.

McGrew, W. C. (1975). Patterns of plant food sharing by wild chimpanzees. In: Kondo, S., Kawai, M., and Ehara, A. (eds.), *Contemporary Primatology, Proceedings of the Fifth International Congress of Primatology*. Karger, Basel. pp. 304–309.

McGrew, W. C. (1979). Evolutionary implications of sex differences in chimpanzee predation and tool use. In: Hamburg, D. A., and McCown, E. R. (eds.), *The Great Apes*. Benjamin/Cummings, Menlo Park, CA. pp. 440–463.

McGrew, W. C. (1992). *Chimpanzee Material Culture: Implications for Human Evolution*. Cambridge University Press, Cambridge.

McGrew, W. C. (1993). The intelligent use of tools: twenty propositions. In: Gibson, K., and Ingold, T. (eds.), *Tools, Language and Cognition in Human Evolution*. Cambridge University Press, Cambridge. pp. 151–170.

McGrew, W. (1998a). Reply to Stanford (1998) "The social behavior of chimpanzees and bonobos—empirical evidence and shifting assumptions." *Curr. Anthropol.* 39:411.

McGrew, W. C. (1998b). Culture in nonhuman primates? *Annu. Rev. Anthropol.* 27:301–328.

McGrew, W. C. (2004). *The Cultured Chimpanzee: Reflections on Cultural Primatology*. Cambridge University Press, New York.

McGrew, W. C., and Barnett, J. (2008). Lifetime reproductive output in a female common marmoset (*Callithrix jacchus*): a case study. *Am. J. Phys. Anthropol.* (suppl. 43):152.

McGrew, W. C., and Marchant, L. F. (1992). Chimpanzees, tools, and termites: hand preference or handedness? *Curr. Anthropol.* 33:114–119.

McGrew, W. C., and Marchant, L. F. (1997). Using the tools at hand: manual laterality and elementary technology in *Cebus* spp. and *Pan* spp. *Int. J. Primatol.* 18:787–810.

McGrew, W., Baldwin, P. J., and Tutin, C. E. (1979). Chimpanzees, tools and termites: cross-cultural comparisons of Senegal, Tanzania and Rio Muni. *Man* 14:185–214.

McGrew, W., Baldwin, P., and Tutin, C. E. G. (1981). Chimpanzees in a hot, dry and open habitat, Mt. Assirik, Senegal, West Africa. *J. Hum. Evol.* 10:227–244.

McGrew, W., Baldwin, P. J., and Tutin, C. E. (1988). Diet of wild chimpanzees (*Pan troglodytes verus*) at Mt. Assirik, Senegal: I. Composition. *Am. J. Primatol.* 16:213–226.

McGrew, W. C., Marchant, L., and Nishida, T. (1996). *Great Ape Societies*. Cambridge University Press, Cambridge.

McGrew, W. C., Ham, R. M., White, L. J. T., Tutin, C. E. G., and Fernandez, M. (1997). Why don't chimpanzees in Gabon crack nuts? *Int. J. Primatol.* 18:353–374.

McGrew, W. C., Ensminger, A. L., Marchant, L., Pruetz, J., and Vigilant, L. (2004). Genotyping aids field study of unhabituated wild chimpanzees. *Am. J. Primatol.* 63:87–93.

McGuire, A. L., Cho, M. K., McGuire, S. E., and Caulfield, T. (2007). The future of personal genomics. *Science* 317:1687.

McKee, J. K. (1993). Taxonomic and evolutionary affinities of *Papio izodi* fossils from Taung and Sterkfontein. *Palaeontol. Afr.* 30:43–49.

McKenna, J. J. (1979a). Aspects of infant socialization, attachment, and maternal caregiving patterns among primates: a cross-disciplinary review. *Ybk. Phys. Anthropol.* 22:250–286.

McKenna, J. J. (1979b). The evolution of allomothering behavior among colobine monkeys: function and opportunism in evolution. *Am. Anthropol.* 81:818–840.

McKenna, J. J. (1981). Primate infant caregiving behavior: origins, consequences, and variability with emphasis on the common Indian langur monkey. In: Gubernick, D. J., and Klopfer, P. H. (eds.), *Parental Care in Mammals*. Plenum Press, New York. pp. 389–416.

McKenna, J. J. (1987). Parental supplements and surrogates among primates: cross-species and cross-cultural comparisons. In: Lancaster, J., Altmann, J., and Rossi, A. (eds.), *Parenting Across the Life Span: Biosocial Dimensions*. Aldine de Gruyter, New York. pp. 143–184.

McKenna, P., Hoffmann, C., Minkah, N., Aye P. P., Lackner, A., Liu, Z., Lozupone, C. A., Hamady, M., Knight, R., and Bushman, F. D. (2008). The macaque gut microbiome in health, lentiviral infection, and chronic enterocolitis. *PLoS Pathog.* 4:e20.

McKey, D. B. (1974). Adaptive patterns in alkaloid physiology. *Am. Nat.* 108:305–320.

McKey, D. B. (1978a). Plant chemical defenses and the feeding and ranging behavior of colobus monkeys in African rainforests [PhD diss.]. University of Michigan, Ann Arbor.

McKey, D. B. (1978b). Soils, vegetation, and seed-eating by black colobus monkeys. In: Montgomery, G. G. (ed.), *The Ecology of Arboreal Folivores*. Smithsonian Institution Press, Washington DC. pp. 423–437.

McKey, D. B., and Waterman, P. G. (1982). Ranging behavior of a group of black colobus (*Colobus satanas*) in the Douala-Edea Reserve, Cameroon. *Folia Primatol.* 39:264–304.

McKey, D. B., Waterman, P. G., Gartlan, J. S., and Struhsaker, T. T. (1978). Phenolic content of vegetation in two African rain forests: ecological implications. *Science* 202:61–64.

McKey, D. B., Gartlan, J. S., Waterman, P. G., and Choo, G. M. (1981). Food selection by black colobus monkeys (*Colobus satanas*) in relation to food chemistry. *Biol. J. Linn. Soc.* 16:115–146.

McNab, B. (1978). The energetics of arboreal folivores: the problem of feeding on a ubiquitous food source. In: Montgomery, G. G. (ed.), *The Ecology of Arboreal Folivores*. Smithsonian Institution Press, Washington DC. pp. 153–162.

McNab, B. K. (2002). *The Physiological Ecology of Vertebrates: A View from Energetics*. Cornell University Press, Ithaca, NY.

McNeilage, A. (1995). Mountain gorillas in the Virunga Volcanoes: ecology and carrying capacity [PhD diss.]. University of Bristol, Bristol.

McNeilage, A. (2001). Diet and habitat use of two mountain gorilla groups in contrasting habitats in the Virungas. In: Robbins, M. M., Sicotte, P., and Stewart, K. J. (eds.), *Mountain Gorillas: Three Decades of Research at Karisoke*. Cambridge University Press, Cambridge. pp. 265–292.

McNeilage, A., Plumptre, A. J., Brock-Doyle, A., and Vedder, A. (2001). Bwindi Impenetrable National Park, Uganda: gorilla census 1997. *Oryx* 35:39–47.

McNeilage, A., Robbins, M. M., Gray, M., Olupot, W., Babaasa, D., Bitariho, R., Kasangaki, A., Rainer, H., Asuma, S., Mugiri, G., and Baker, J. (2006). Census of the mountain gorilla population in Bwindi Impenetrable National Park, Uganda. *Oryx* 40:419–427.

McWhorter, T. J., and Karasov, W. H. (2007). Paracellular nutrient absorption in a gum-feeding new world primate, the common marmoset *Callithrix jacchus*. *Am. J. Primatol.* 69:1399–1411.

Meador, D. M., Rumbaugh, D., Pate, J. L., and Bard, K. A. (1987). Learning, problem solving, cognition, and intelligence. In: Mitchell, G., and Erwin, J. (eds.), *Comparative Primate Biology*. Alan R. Liss, New York. pp. 17–83.

Medawar, P. B. (1946). Old age and natural death. *Med. Q.* 1:30–56.

Medawar, P. B. (1952). *An Unsolved Problem in Biology*. H. K. Lewis, London.

Medin, D. L., and Atran, S. (eds.) (1999). *Folkbiology*. MIT Press, Cambridge, MA.

Medway, L. (1972). Phenology of a tropical rain forest in Malaya. *Biol. J. Linn. Soc.* 4:117–146.

Medway, L. (1977). The Niah excavations and an assessment of the impact of early man on mammals in Borneo. *Asian Persp.* 20:51–69.

Megantara, E. N. (1989). Ecology, behavior and sociality of *Presbytis femoralis* in east central Sumatra. *Comp. Primatol. Monogr.* 2:171–301.

Mehlman, P. (1986). Male intergroup mobility in a wild population of the Barbary macaque (*Macaca sylvanus*), Ghomaran Rif Mountains, Morocco. *Am. J. Primatol.* 10:67–81.

Mehlman, P. T. (2008). Current status of wild gorilla populations and strategies for their conservation. In: Stoinski, T. S., Steklis, H. D., and Mehlman, P. T. (eds.), *Conservation in the 21st Century: Gorillas as a Case Study*. Springer, New York. pp. 3–53.

Meier, B., and Albignac, R. (1991). Rediscovery of *Allocebus trichotis* Günther 1875 (Primates) in north east Madagascar. *Folia Primatol.* 56:57–63.

Meier, B., and Rumpler, Y. (1987). Preliminary survey of *Hapalemur simus* and of a new species of *Hapalemur* in eastern Betsileo, Madagascar. *Primate Conserv.* 8:40–43.

Meier, B., Albignac, R., Peyrieras, A., Rumpler, Y., and Wright, P. C. (1987). A new species of *Hapalemur* (Primates) from southeast Madagascar. *Folia Primatol.* 48:211–215.

Meijaard, E., and Wich, S. A. (2007). Putting orang-utan population trends into perspective. *Curr. Biol.* 17:R540.

Meikle, D. B., and Vessey, S. H. (1981). Nepotism among rhesus monkey brothers. *Nature* 294:160–161.

Meireles, C. M., Czelusniak, J., Schneider, M. P. C., Muniz, J. A. P. C., Brigido, M. C., Ferreira, H. S., and Goodman, M. (1999). Molecular phylogeny of ateline New World monkeys (Platyrrhini, Atelinae) based on g-globin gene sequences: evidence that *Brachyteles* is the sister group of *Lagothrix*. *Mol. Phylogenet. Evol.* 12:10–30.

Meirmans, P. G., and Van Tienderen, P. H. (2004). GENOTYPE and GENODIVE: two programs for the analysis of genetic diversity of asexual organisms. *Mol. Ecol. Notes* 4:792–794.

Meishvili, N. V., Chalyan, V. G., and Butovskaya, M. L. (2005). Studies of reconciliation in anubis baboons. *Neurosc. Behav. Physiol.* 35:913–916.

Meldrum, D. J. (1998). Tail-assisted hind limb suspension as a transitional behavior in the evolution of the platyrrhine prehensile tail. In: Strasser, E., Fleagle, J. G., Rosenberger, A., and McHenry, H. (eds.), *Primate Locomotion: Recent Advances*. Plenum Press, New York. pp. 145–156.

Meldrum, D. J., Dagosto, M., and White, J. (1997). Hindlimb suspension and hind foot reversal in *Varecia variegata* and other arboreal mammals. *Am J. Phys. Anthropol.* 103:85–102.

Melis, A. P., Hare, B., and Tomasello, M. (2006). Engineering cooperation in chimpanzees: tolerance constraints on cooperation. *Anim. Behav.* 72:275–286.

Meller, A. P. (1990). Fluctuating assymetry in male sexual ornaments may reliably reveal male quality. *Anim. Behav.* 40:1185–1187.

Melnick, D. J., and Hoelzer, G. A. (1996). The population genetic consequences of macaque social organization and behaviour. In: Fa, J. E., and Lindburg, D. G. (eds.), *Evolution and Ecology of Macaque Societies*. Cambridge University Press, Cambridge. pp. 413–443.

Melnick, D. J., Jolly, C. J., and Kidd, K. K. (1984a). The genetics of a wild population of rhesus monkeys (*Macaca mulatta*): I. Genetic variability within and between social groups. *Am. J. Phys. Anthropol.* 63:341–360.

Melnick, D. J., Pearl, M. C., and Richard, A. F. (1984b). Male migration and inbreeding avoidance in wild rhesus monkeys. *Am. J. Primatol.* 7:229–243.

Melo, L., Mendes Pontes, A. R., and Monteiro da Cruz, M. A. (2003). Infanticide and cannibalism in wild common marmosets. *Folia Primatol.* 74:48–50.

Meltzoff, A. N. (1990). Towards a developmental cognitive science—the implications of cross-modal matching and imitation for the development of representation and memory in infancy. *Ann. N. Y. Acad. Sci.* 608:1–37.

Mena, P., Stallings, J. R., Regalado, J. B., and Cueva, L. R. (2000). The sustainability of hunting practices by the Huaorani. In: Robinson, J. G., and Bennett, E. L. (eds.), *Hunting for Sustainability in Tropical Forests*. Columbia University Press, New York. pp. 57–78.

Ménard, N. (2004). Do ecological factors explain variation in social organization? In: Thierry, B., Singh, M., and Kaumanns, W. (eds.), *Macaque Societies: A Model for the Study of Social Organization*. Cambridge University Press, Cambridge. pp. 237–262.

Ménard, N., and Vallet, D. (1996). Demography and ecology of Barbary macaques (*Macaca sylvanus*) in two different habitats. In: Fa, J. E., and Lindburg, D. G. (eds.), *Evolution and Ecology of Macaque Societies*. Cambridge University Press, Cambridge, pp. 106–131.

Menard, N., and Vallet, D. (1997). Behavioral responses of Barbary macaques (*Macaca sylvanus*) to variations in environmental conditions in Algeria. *Am. J. Primatol.* 43:285–304.

Ménard, N., Hecham, R., Vallet, D., Chikhi, H., and Gautier-Hion, A. (1990). Grouping patterns of a mountain population of *Macaca sylvanus* in Algeria—a fission–fusion system? *Folia Primatol.* 55:166–175.

Ménard, N., von Segesser, F., Scheffrahn, W., Pastorini, J., Vallet, D., Gaci, B., Martin, R. D., and Gautier-Hion, A. (2001). Is male-infant caretaking related to paternity and/or mating activities in wild Barbary macaques (*Macaca sylvanus*)? *C. R. Acad. Sci. Paris Life Sci.* 324:601–610.

Menard, N., Lathuilliere, M., Petit, E., Vallet, D., and Crouau-Roy, B. (2006). Philopatry of female macaques in relation to group dynamics and the distribution of genes—the case of the Barbary macaque (*Macaca sylvanus*). In: Hodges, J. K., and Cortes, J. (eds.), *The Barbary Macaque: Biology, Management and Conservation*. Nottingham University Press, Nottingham. pp. 129–147.

Mendes, N., Hanus, D., and Call, J. (2007). Raising the level: orangutans use water as a tool. *Biol. Lett.* 3:453–455.

Mendes, S. L. (1989). Estudo ecológico de *Alouatta fusca* (Primates: Cebidae) na Estaçao Biológica de Caratinga, MG. *Rev. Nordest. Biol.* 6:71–104.

Mendes, S. L., Melo, F. R., Boubli, J. P., Dias, L. G., Strier, K. B., Pinto, L. P. S., Fagundes, V., Cosenza, B., and De Marco, P. J. (2005). Directives for the conservation of the northern muriqui, *Brachyteles hypoxanthus* (Primates, Atelidae). *Neotrop. Primates* 13(suppl.):7–18.

Mendes Pontes, A. R., and Monteiro da Cruz, M. A. O. (1995). Home range, intergroup transfers, and reproductive status of common marmosets, *Callithrix jacchus,* in a forest fragment in north-eastern Brazil. *Primates* 36:335–347.

Mendoza, S. P., and Mason, W. A. (1986a). Parental division of labour and differentiation of attachments in a monogamous primate (*Callicebus moloch*). *Anim. Behav.* 34:1336–1347.

Mendoza, S. P., and Mason, W. A. (1986b). Parenting within a monogamous society. In: Else, J. G. and Lee, P. C. (eds.), *Primate Ontogeny, Cognition and Social Behaviour*. Cambridge University Press, Cambridge. pp. 255–266.

Mendoza, S. P., and Mason, W. A. (1994). Constitution and context: the social modulation of temperament. In: Thierry, B., Anderson, J. R., Roeder, J. J., and Herrenschmidt, N. (eds.), *Current Primatology. Social Development, Learning and Behaviour*, vol 2. University Louis Pasteur, Strasbourg, Germany. pp. 251–256.

Mendoza, S. P., Lowe, E. L., Davidson, J. M., and Levine, S. (1978). Annual cyclicity in the squirrel monkey (*Saimiri sciureus*): the relationship between testosterone, fatting, and sexual behavior. *Horm. Behav.* 11:295–303.

Menon, S., and Poirier, F. E. (1996). Lion-tailed macaques (*Macaca silenus*) in a disturbed forest fragment: activity patterns and time budget. *Int. J. Primatol.* 17:969–985.

Menzel, C. R. (1986). An experimental study of territory maintenance in captive titi monkeys (*Callicebus moloch*). In: Else, J. G., and Lee, P. C. (eds.), *Primate Ecology and Conservation*. Cambridge University Press, Cambridge. pp. 133–143.

Menzel, C. R. (1993). Coordination and conflict in *Callicebus* social groups. In: Mason, W. A., and Mendoza, S. P. (eds.), *Primate Social Conflict*. State University of New York Press, New York. pp. 253–290.

Menzel, C. (2005). Progress in the study of chimpanzee recall and episodic memory. In: Terrace, H. S., and Metcalfe, J. (eds.), *The Missing Link in Cognition: Origins of Self-Reflective Consciousness*. Oxford University Press, New York. pp. 188–224.

Mercader, J., Panger, M., and Boesch, C. (2002). Excavation of a chimpanzee stone tool site in the African rainforest. *Science* 296:1452–1455.

Meritt, D. A. J. (1980). Captive reproduction and husbandry of the Douroucouli and the titi monkey. *Int. Zoo Ybk.* 20:52–59.

Merker, S. (2006). Habitat-specific ranging patterns of Dian's tarsiers (*Tarsius dianae*) as revealed by radiotracking. *Am. J. Primatol.* 68(2):111–125.

Merker, S., and Groves, C. (2006). *Tarsius lariang*: a new primate species from western central Sulawesi. *Int. J. Primatol.* 27(2):465–485.

Merker, S., and Muhlenberg, M.. 2000. Traditional land use and tarsiers–human influences on population densities of *Tarsius dianae*. *Folia Primato.a* 71 (6): 426–428.

Merker, S., and Yustian, I. (2008). Habitat use analysis of Dian's tarsier (*Tarsius dianae*) in a mixed-species plantation in Sulawesi, Indonesia. *Primates* 49(2):161–164.

Merker, S., Yustian, I., and Muhlenberg, M. (2005). Responding to forest degradation: altered habitat use by Dian's tarsier *Tarsius dianae* in Sulawesi, Indonesia. *Oryx* 39(2):189–195.

Merker, S., Driller, C., Perwitasari-Farajallah, D., Zahnar, R., and Zischler, H. (2006). Isolation and characterization of 12 microsatellite loci for population studies of Sualwesi tarsiers (*Tarsius spp.*). *Mol. Ecol. Notes* 7(6):1216–1218.

Mervis, C., and Rosch, E. (1981). Categorization of natural objects. *Annu. Rev. Psychol.* 32:89–115.

Mesoudi, A., and Whiten, A. (2008). The multiple roles of cultural transmission experiments in understanding human cultural evolution. *Phil. Trans. R. Soc. B* 363:3489–3501.

Mesoudi, A., Whiten, A., and Laland, K. N. (2004). Is human cultural evolution Darwinian? Evidence reviewed from the perspective of *The Origin of Species*. *Evolution* 58:1–11.

Messier, W., and Stewart, C. B. (1997). Episodic adaptive evolution of primate lysozymes. *Nature* 385:151–154.

Messner, E. J., and Wrangham R. W. (1996). In vitro testing of the biological activity of *Rubia cordifolia* leaves on primate *Strongyloide* species. *Primates* 37:105–108.

Meyers, D. M. (1993). The effects of resource seasonality on the behavior and reproduction of the golden-crowned sifaka (*Propithecus tattersalli*, Simons, 1988) in three Malagasy forests [PhD diss.]. Duke University, Durham, NC.

Meyers, D. M., and Ratsirarson, J. (1989). Distribution and conservation of two endangered sifakas in northern Madagascar. *Primate Conserv.* 10:82–87.

Meyers, D. M., and Wright, P.C. (1993). Resource tracking: food availability and *Propithecus* seasonal reproduction. In: Kappeler, P. M., and Ganzhorn, J. U. (eds.), *Lemur Social Systems and Their Ecological Basis*. Plenum Press, New York. pp. 179–192.

Michael, R. P., and Zumpe, D. (1970). Rhythmic changes in the copulatory frequencies of rhesus monkeys (*Macaca mulatta*) in relation to the menstrual cycle, and a comparison with the human cycle. *J. Reprod. Fertil.* 21:199–201.

Michael, R. P., Bonsall, R. W., and Zumpe, D. (1987). Testosterone and its metabolites in male cynomolgus monkeys (*Macaca fascicularis*): behavior and biochemistry. *Physiol. Behav.* 40:527–537.

Michelsen, A. (1978). Sound reception in different environments. In: Ali, B. A. (ed.), *Perspectives in Sensory Ecology*. Plenum Press, New York. pp. 345–373.

Michener, G. R. (1984). Copulatory plugs in Richardsons' ground squirrels. *Can. J. Zool.* 62:267–270.

Mikkelsen, T. S., Hillier, L. W., Eichler, E. E., et al. (2005). Initial sequence of the chimpanzee genome and comparison with the human genome. *Nature* 437:69–87.

Miles, H. L. (1994). ME CHANTEK: the development of self-awareness in a signing orangutan. In: Parker, S. T., Mitchell, R. W., and Boccia, M. L. (eds.), *Self-Awareness in Animals and Humans: Developmental Perspectives*. Cambridge University Press, Cambridge. pp. 254–272.

Mill, R. R. (1995). Regional overview: Indian subcontinent. In: *Centres of Plant Diversity: A Guide to Strategy for Their Conservation. Asia, Australia and the Pacific,* vol. 2. World Wildlife Fund for Nature and IUCN Press, Cambridge. pp. 62–135.

Millar, R. (1972). *The Piltdown Men*. St. Martin's Press, New York.

Miller, G., and Hollister, F. (1921). A review of *Tarsius spectrum* from Celebes. *Proc. Biol. Soc. Wash.* 34:103–228.

Miller, L. E. (1992). Socioecology of the wedge-capped capuchin monkey (*Cebus olivaceus*) [PhD diss.]. University of California, Davis.

Miller, L. E. (1996). The behavioral ecology of wedge-capped capuchin monkeys (*Cebus olivaceus*). In: Norconk, M. A., Rosenberger, A. L., and Garber, P. A. (eds.), *Adaptative Radiations of Neotropical Primates*. Plenum Press, New York. pp. 271–288.

Miller, L. E. (2002a). *Eat or Be Eaten: Predator Sensitive Foraging Among Primates*. Cambridge University Press, Cambridge.

Miller, L. E. (2002b). An introduction to predator sensitive foraging. In: Miller, L. E. (ed.), *Eat or Be Eaten: Predator Sensitive Foraging Among Primates*. Cambridge University Press, Cambridge. pp. 1–17.

Miller, L. E. (2002c). The role of group size in predator sensitive foraging decisions for wedge-capped capuchin monkeys (*Cebus olivaceus*). In: Miller, L. E. (ed.), *Eat or Be Eaten: Predator Sensitive Foraging Among Primates*. Cambridge University Press, Cambridge. pp. 95–106.

Miller, L. E. (2008). Predator-sensitive foraging in wedge-capped capuchins (*Cebus olivaceus*): experimentation to investigate vulnerability and risk. *Primate Eye* 96:Abstract 344.

Miller, L. E., and Treves, A. (2007). Predation on primates: past studies, current challenges, and directions for the future. In: Campbell, C. J., Fuentes, A., Mackinnon, K. C., Panger, M., and Bearder, S. K. (eds.), *Primates in Perspective*. Oxford University Press, New York. pp. 525–543.

Milner-Gulland, E. J., and Akçakaya, H. R. (2001). Sustainability indices for exploited populations. *Trends Ecol. Evol.* 16:686–692.

Miller-Rushing, A. J., and Primack, R. B. (2008). Global warming and flowering times in Thoreau's Concord: a community perspective. *Ecology* 89:332–341.

Milton, K. (1979). Factors influencing leaf choice by howler monkeys: a test for some hypotheses of food selection by generalist herbivores. *Am. Nat.* 114:362–378.

Milton, K. (1980). *The Foraging Strategy of Howler Monkeys: A Study in Primate Economics*. Columbia University Press, New York.

Milton, K. (1981a). Distribution patterns of tropical plant foods as an evolutionary stimulus to primate mental development. *Am. Anthropol.* 83:534–548.

Milton, K. (1981b). Estimates of reproductive parameters for free-ranging *Ateles geoffroyi*. *Primates* 22:574–579.

Milton, K. (1981c). Food choice and digestive strategies by two sympatric primate species. *Am. Nat.* 117:496–505.

Milton, K. (1982). Dietary quality and demographic regulation in a howler monkey population. In: Leigh, E. G., Jr., Rand, A. S., and Windsor, D. M. (eds.), *The Ecology of a Tropical Forest: Seasonal Rhythms and Long-Term Changes*. Smithsonian Institution Press, Washington DC. pp. 273–289.

Milton, K. (1984a). Habitat, diet, and activity patterns of free-ranging woolly spider monkeys (*Brachyteles arachnoides* E. Geoffroy 1806). *Int. J. Primatol.* 5:491–514.

Milton, K. (1984b). The role of food processing factors in primate food choice. In: Rodman, P. S., and Cant, J. G. H. (eds.), *Adaptations for Foraging in Nonhuman Primates: Contributions to an Organismal Biology of Prosimians, Monkeys and Apes*. Columbia University Press, New York. pp. 249–279.

Milton, K. (1985a). Mating patterns of woolly spider monkeys, *Brachyteles arachnoides*: implications for female choice. *Behav. Ecol. Sociobiol.* 17:53–59.

Milton, K. (1985b). Multimale mating and absence of canine tooth dimorphism in woolly spider monkeys (*Brachyteles arachnoides*). *Am. J. Phys. Anthropol.* 68:519–523.

Milton, K. (1986). Digestive physiology in primates. *News Phys. Sci.* 1:76–79.

Milton, K. (1987a). Mating behaviors in woolly spider monkeys (*Brachyteles arachnoides*). *Int. J. Primatol.* 8:460.

Milton, K. (1987b). Primate diets and gut morphology: implications for hominid evolution. In: Harris, M., and Ross, E. B. (eds.), *Food and Evolution: Toward a Theory of Human Food Habits*. Temple University Press, Philadelphia. pp. 93–115.

Milton, K. (1993). Diet and primate evolution. *Sci. Am.* 269:86–93.

Milton, K. (1996). Effects of bot fly (*Alouattamyia baeri*) parasitism on a free-ranging howler monkey (*Alouatta palliata*) population in Panama. *J. Zool. Lond.* 239:39–63.

Milton, K. (1998). Physiological ecology of howlers (*Alouatta*): energetic and digestive considerations and comparison with the Colobinae. *Int. J. Primatol.* 19:513–548.

Milton, K. (1999). Nutritional characteristics of wild primate foods: do the diets of our closest living relatives have lessons for us? *Nutrition* 15:488–498.

Milton, K. (2000). Back to basics: why foods of wild primates have relevance for modern human health. *Nutrition* 16:480–483.

Milton, K., and Demment, M. W. (1987). Digestion and passage kinetics of chimpanzees fed high and low fiber diets and comparison with human data. *J. Nutr.* 118:1082–1088.

Milton, K., and Dintzis, F. (1981). Nitrogen-to-protein conversion factors for tropical plant samples. *Biotropica* 12:177–181.

Milton, K., and Jenness, R. (1987). Ascorbic acid content of neotropical plant parts available to wild monkeys and bats. *Experientia* 43:339–342.

Milton, K., and May, M. L. (1976). Body weight, diet, and home range area in primates. *Nature* 259:459–462.

Milton, K., Van Soest, P. J., and Robertson, J. B. (1980). Digestive efficiencies of wild howler monkeys. *Physiol. Zool.* 53:402–409.

Mishra, C., and Sinha, A. (2008). A voucher specimen for *Macaca munzala*: interspecific affinities, evolution, and conservation of a newly discovered primate. *Int. J. Primatol.* 29:743–756.

Missakian, E. A. (1972). Genealogical and cross-genealogical dominance relations in a group of free-ranging rhesus monkeys (*Macaca mulatta*) on Cayo Santiago. *Primates* 13:169–180.

Mitani, J. C. (1984). The behavioral regulation of monogamy in gibbons (*Hylobates muelleri*). *Behav. Ecol. Sociobiol.* 15:225–229.

Mitani, J. C. (1985a). Location-specific responses of gibbons (*Hylobates muelleri*) to male songs. *Z. Tierpsychol.* 70:219–224.

Mitani, J. C. (1985b). Mating behaviour of male orangutans in the Kutai Game Reserve, Indonesia. *Anim. Behav.* 33:392–402.

Mitani, J. C. (1985c). Sexual selection and adult male orangutan long calls. *Anim. Behav.* 33:272–283.

Mitani, J. C. (1987). Species discrimination of male song in gibbons. *Am. J. Primatol.* 13:413–423.

Mitani, J. C. (1989). Orangutan activity budgets: monthly variation and the effects of body size, parturition, and sociality. *Am. J. Primatol.* 18:87–100.

Mitani, J. C. (1990). Demography of agile gibbons (*Hylobates agilis*). *Int. J. Primatol.* 11:411–424.

Mitani, J. C. (1991). Niche overlap and polyspecific association among sympatric *Cercopithecus* in the Campo Animal Reserve, southwest Cameroon. *Primates* 32:137–151.

Mitani, J. C. (2006). Demographic influences on the behavior of chimpanzees. *Primates* 47:6–13.

Mitani, J. C. (2009). Male chimpanzees form enduring and equitable social bonds. *Anim. Behav.* 77:633–640.

Mitani, J. C., and Rodman, P. S. (1979). Territoriality: the relation of ranging pattern and home range size to defendability, with an analysis of territoriality among primate species. *Behav. Ecol. Sociobiol.* 5:241–251.

Mitani, J. C., and Watts, D. (1997). The evolution of non-maternal caretaking among anthropoid primates: do helpers help? *Behav. Ecol. Sociobiol.* 40:213–220.

Mitani, J. C., and Watts, D. P. (1999). Demographic influences on the hunting behavior of chimpanzees. *Am. J. Phys. Anthropol.* 109:439–454.

Mitani, J., and Watts, D. (2001). Why do chimpanzees hunt and share meat. *Anim. Behav.* 61:915–924.

Mitani, J. C., and Watts, D. P. (2005). Correlates of territorial boundary patrol behaviour in wild chimpanzees. *Anim. Behav.* 70:1079–1086.

Mitani, J. C., Gros-Louis, J., and Manson, J. H. (1996). Number of males in primate groups: comparative tests of competing hypotheses. *Am. J. Primatol.* 38:315–332.

Mitani, J. C., Merriwether, D. A., and Zhang, C. (2000). Male affiliation, cooperation and kinship in wild chimpanzees. *Anim. Behav.* 59:885–893.

Mitani, J. C., Sanders, W. J., Lwanga, J. S., and Windfelder, T. L. (2001). Predatory behavior of craned hawk-eagles (*Stephanoaetus coronatus*) in Kibale National Park, Uganda. *Behav. Ecol. Sociobiol.* 49:187–195.

Mitani, J. C., Watts, D. P., Pepper, J. W., and Merriwether, D. A. (2002a). Demographic and social constraints on male chimpanzee behaviour. *Anim. Behav.* 64:727–737.

Mitani, J., Watts, D. P., and Muller, M. N. (2002b). Recent developments in the study of wild chimpanzee behavior. *Evol. Anthropol.* 11:9–25.

Mitani, J., Watts, D., and Lwanga, J. S. (2002c). Ecological and social correlates of chimpanzee party size and composition. In: Boesch, C., Hohmann, G., and Marchant, L. F. (eds.), *Behavioural Diversity in Chimpanzees and Bonobos.* Cambridge University Press, Cambridge. pp. 102–111.

Mitani, M. (1989). *Cercocebus torquatus*: adaptive feeding and ranging behaviors related to seasonal fluctuations of food resources in the tropical rain forest of south-western Cameroon. *Primates* 30:307–323.

Mitchell, A. H. (1994). Ecology of Hose's langur, *Presbytis hosei*, in mixed logged and unlogged dipterocarp forest of northeast Borneo [PhD diss.]. Yale University, New Haven.

Mitchell, B. J. (1989). Resources, group behavior, and infant development in white-faced capuchin monkeys, *Cebus capucinus* [PhD diss.]. University of California, Berkeley.

Mitchell, C. L. (1990). The ecological basis for female social dominance: a behavioral study of the squirrel monkeys (*Saimiri sciureus*) [PhD diss.]. Princeton University, Princeton, NJ.

Mitchell, C. L. (1994). Migration alliances and coalitions among adult male South American squirrel monkeys (*Saimiri sciureus*). *Behavior* 130:169–190.

Mitchell, C. L., Boinski, S., and van Schaik, C. P. (1991). Competitive regimes and female bonding in two species of squirrel monkeys (*Saimiri oerstedii* and *S. sciureus*). *Behav. Ecol. Sociobiol.* 28:55–60.

Mitchell, R. W. (1993). Mental models of mirror self-recognition": two theories. *New Ideas Psychol.* 11:295–325.

Mitchell, R. W., and Anderson, J. R. (1997). Pointing, withholding information, and deception in capuchin monkeys (*Cebus apella*). *J. Comp. Psychol.* 111:351–361.

Mitchell, R. W., Thompson, N. S., and Miles, H. L. (eds.) (1997). *Anthropomorphism, Anecdotes, and Animals.* State University of New York Press, New York.

Mitchell, W. R., Presely, S., Czekala, N. M., and Lasley, B. L. (1982). Urinary immunoreactive estrogen and pregnanediol-3-glucuronide during the normal menstrual cycle of the female lowland gorilla (*Gorilla gorilla*). *Am. J. Primatol.* 2:167–175.

Mitra Setia, T., and van Schaik, C. P. (2007). The response of adult orang-utans to flanged male long calls: inferences about their function. *Folia Primatol.* 78:215–226.

Mitra Setia, T., Delgado, R. A., Utami Atmoko, S. S., Singleton, I., and van Schaik, C. P. (2009). Social organization and male–female relationships. In: Wich, S. A., Utami, S. S., Mitra Setia, T., and van Schaik, C. (eds.), *Orangutans: Geographic Variation in Behavioral Ecology and Conservation.* Oxford University Press, Oxford. pp. 245–254.

Mittermeier, R. A. (1973). Group activity and population dynamics of the howler monkey on Barro Colorado Island. *Primates* 14:1–19.

Mittermeier, R. A. (1977). Distribution, synecology and conservation of Surinam monkeys [PhD diss.]. Harvard University, Cambridge, MA.

Mittermeier, R. A. (1978). Locomotion and posture in *Ateles geoffroyi* and *Ateles paniscus. Folia Primatol.* 30:161–193.

Mittermeier, R. A. (1987). Effects of hunting on rain forest primates. In: Marsh, C. W., and Mittermeier, R. A. (eds.). *Primate Conservation in the Tropical Rain Forest.* Alan R. Liss. New York. pp. 109–146.

Mittermeier, R. A., and Cheney, D. L. (1987). Conservation of primates and their habitats. In: Smuts, B. B., Cheney, D. L., Seyfarth, R. M., Wrangham, R. W., and Struhsaker, T. T. (eds.), *Primate Societies.* University of Chicago Press, Chicago. pp. 477–490.

Mittermeier, R. A., and Roosmalen, M. G. M. (1981). Preliminary observations on habitat utilization and diet in eight Surinam monkeys. *Folia Primatol.* 36:1–39.

Mittermeier, R. A., Konstant, W. R., Ginsberg, H., van Roosmalen, M. G. M., and Cordeiro da Silva, E., Jr. (1983). Further evidence of insect consumption in the bearded saki monkeys, *Chiropotes satanas chiropotes. Primates* 24:602–605.

Mittermeier, R. A., Tattersall, I., Konstant, W. R., Meyers, D. M., and Mast, R. B. (1994). *Lemurs of Madagascar. Conservation International Tropical Field Guide Series.* Conservation International, Washington DC.

Mittermeier, R. A., Rylands, A. B., and Konstant, W. R. (1999). Introduction. In: Nowak, R. M. (ed.), *Walker's Primates of the World.* Johns Hopkins University Press, Baltimore, MD. pp. 1–52.

Mittermeier, R. A., Konstant, W. R., and Rylands, A. B. (2000a). The world's top 25 most endangered primates. *Neotrop. Primates* 8:49.

Mittermeier, R. A., Rylands, A. B., Konstant, W. R., Eudey, A., Butynski, T., Ganzhorn, J. U., and Rodríguez-Luna, E. (2000b). Primate specialist group. *Species* 34:82–88.

Mittermeier, R. A., Konstant, W. R., Rylands, A. B., Butynski, T. M., Eudey, A. A., Ganzhorn, J. U., and Kormos, R. (2002). The world's top 25 most endangered primates—2002. *Neotrop. Primates* 10:128–131.

Mittermeier, R. A., Rylands, A. B., Konstant, W. R., Kormos, R., Eudey, A. A., Walker, S., and Aguiar, J. M. (2004). *IUCN/SSC Primate Specialist Group Report 2001–2004.* Conservation International, Washington DC.

Mittermeier, R. A., Valladares-Pádua, C., Rylands, A. B., Eudey, A. A., Butynski, T. M., Ganzhorn, J. U., Kormos, R., Aguiar, J. M., and Walker,

S. (2006). Primates in peril: the world's 25 most endangered primates, 2004—2006. *Primate Conserv.* 20:1–28.

Mittermeier, R. A., Ratsimbazafy, J., Rylands, A. B., Williamson, L., Oates, J. F., Mbora, D., Ganzhorn, J. U., Rodríguez-Luna, E., Palacios, E., Heymann, E. W., Kierulff, M. C. M., Yongcheng, L., Supriatna, J., Roos, C., Walker, S., and Aguiar, J. M. (2007). Primates in peril: the world's 25 most endangered primates 2006–2008. *Primate Conserv.* 22:1–40.

Mittermeier, R., Ganzhorn, J., Konstant, W., Glander, K., Tattersall, I., Groves, C., Rylands, A., Hapke, A., Ratsimbazafy, J., Mayor, M., Louis, E., Rumpler, Y., Schwitzer, C., and Rasoloarison, R. (2008a). Lemur diversity in Madagascar. *Int. J. Primatol.* 29:1607–1656.

Mittermeier, R. A., Oates, J. F., Williamson, E. A., Ganzhorn, J. U., Waslker, S. R., Rodrigues-Luna, E., Palacios, E., Heymann, E. W., Kierulff, M. C. M., Rylands, A. B., Mbora, D., Supriatna, J., Roos, C., Yongcheng, L., and Aguiar, J. M. (2008b). *IUCN/SSC Primate Specialist Group: Report 2005–2008. Report to the International Union for Conservation of Nature and Natural Resources (IUCN).* IUCN Species Survival Commission, Gland, Switzerland.

Mittermeier, R. A., Wallis, J., Rylands, A. B., Ganzhorn, J. U., Oates, J. F., Williamson, E. A., Palacios, E., Heymann, E. W., Kierulff, M. C. M., Long Yongcheng, Supriatna, J., Roos, C., Walker, S., Cortés-Ortiz, L., and Schwitzer, C. (2009). Primates in Peril: the world's 25 most endangered primates 2008-2010. *Primate Cons.* (24): 1–57.

Miyamoto, S., Chen, Y., Kurotori, H., Sankai, T., Yoshida, T., and Machida, T. (2001). Monitoring the reproductive status of female gorillas (*Gorilla gorilla gorilla*) by measuring the steroid hormones in fecal samples. *Primates* 42:291–299.

M' Kirera, F., and Unger, P. (2003). Occlusal relief changes with molar wear in *Pan t. troglodytes* and *Gorilla g. gorilla. Am. J. Primatol.* 60:31–41.

Modahl, K. B., and Eaton, G. G. (1977). Display behavior in a confined troop of Japanese macaques. *Anim. Behav.* 25:525–535.

Modrek, B., and Lee, C. (2002). A genomic view of alternative splicing. *Nat. Genet.* 30:13–19.

Möhle, U., Heistermann, M., Palme, R., and Hodges, J. K. (2002). Characterization or urinary and fecal metabolites of testosterone and their measurement for assessing gonadal endocrine function in male non-human primates. *Gen. Comp. Endocrinol.* 129:135–145.

Mole, S., and Waterman, P. G. (1985). Stimulatory effects of tannins and cholic acid on tryptic hydrolysis of proteins: ecological implications. *J. Chem. Ecol.* 11:1323–1332.

Møller, A. P. (1988). Ejaculate quality, testes size and sperm competition in primates. *J. Hum. Evol.* 17:479–488.

Monfort, S. L., Hess, D. L., Shideler, E. S., Samuels, S. J., Hendrickx, A. G., and Lasley, B. L. (1987). Comparison of serum estradiol to urinary estrone conjugates in rhesus macaque (*Macaca mulatta*). *Biol. Reprod.* 37:832–837.

Monfort, S. L., Bush, M., and Wildt, D. E. (1996). Natural and induced ovarian synchrony in golden lion tamarins (*Leontopithecus rosalia*). *Biol. Reprod.* 55:875–882.

Montagu, A. (1959). Introduction to the Ann Arbor paperback edition. In: *Man's Place in Nature.* University of Michigan Press, Ann Arbor.

Montoya, E. G., Moro, J., Gozalo, A., and Samame, H. (1995). Reproducción de *Aotus vociferans* (Primates: Cebidae) en cautiverio. *Rev. Invest. Pecuarias IVITA* 7(2):122–126.

Mooney, J. C., and Lee, P. C. (1999). Reproductive parameters in captive woolly monkeys (*Lagothrix lagotricha*). *Zoo Biol.* 18:421–427.

Moore, A. J. (1994). Genetic evidence for the "good genes" process of sexual selection. *Behav. Ecol. Sociobiol.* 35:235–241.

Moore, J. (1996). Savanna chimpanzees, referential models and the last common ancestor. In: McGrew, W. C., Marchant, L., and Nishida, T. (eds.), *Great Ape Societies.* Cambridge University Press, Cambridge. pp. 275–292.

Moore, J. (2002). *Parasites and the Behavior of Animals.* Oxford University Press, Oxford.

Moor-Jankowski, J., Wiener, A. S., Socha, W. W., Gordon, E. B., and Mortelmans, J. (1972). Blood groups of the dwarf chimpanzee (*Pan paniscus*). *J. Med. Primatol.* 1:90–101.

Moor-Jankowski, J., Wiener, A. S., Socha, W. W., Gordon, E. B., Mortelmans, J., and Sedgwick, C. J. (1975). Blood groups of pygmy chimpanzees (*Pan paniscus*): human-type and simian-type. *J. Med. Primatol.* 4:262–267.

Moorman, E. A., Mendoza, S. P., Shideler, S. E., and Lasley, B. L. (2002). Excretion and measurement of estradiol and progesterone metabolites in the feces and urine of female squirrel monkeys (*Saimiri sciureus*). *Am. J. Primatol.* 57:79–90.

Mootnick, A. R. (2006). Gibbon (Hylobatidae) species identification recommended for rescue or breeding centers. *Primate Conserv.* 21:103–138.

Mootnick, A., and Groves, C. (2005). A new generic name for the hoolock gibbon (Hylobatidae). *Int. J. Primatol.* 26:971–976.

Morales, J. C., and Melnick, D. J. (1998). Phylogenetic relationships of the macaques (*Cercopithecidae: Macaca*), as revealed by high resolution restriction site mapping of mitochondrial ribosomal genes. *J. Hum. Evol.* 34:1–23.

Morbeck, M. E. (1997). Life history, the individual and evolution. In: Morbeck, M. E., Galloway, A., and Zihlman, A. L. (eds.), *The Evolving Female.* Princeton University Press, Princeton, NJ. pp. 3–14.

Morbeck, M. E., Galloway, A., and Zihlman, A. L. (eds.) (1997). *The Evolving Female: A Life-History Perspective.* Princeton University Press, Princeton, NJ.

Moreno-Black, G. S., and Bent, E. F. (1982). Secondary compounds in the diet of *Colobus angolensis. Afr. J. Ecol.* 20:29–36.

Morgan, B. J., and Abwe, E. E. (2006). Chimpanzees use stone hammers in Cameroon. *Curr. Biol.* 16:R632-R633.

Morgan, D., and Sanz, C. (2003). Naive encounters with chimpanzees of the Gouaougo Triangle, Republic of Congo. *Int. J. Primatol.* 24:369–381.

Morgan, D., Sanz, C., Onononga, J. R., and Strindberg, S. (2006). Ape abundance and habitat use in the Goualougo Triangle, Republic of Congo. *Int. J. Primatol.* 27:147–179.

Mori, A. (1979). Analysis of population changes by measurement of body weight in the Koshima troop of Japanese monkeys. *Primates* 20:371–397.

Mori, A., and Belay, G. (1990). The distribution of baboon species and a new population of gelada baboons along the Wabi-Shebeli River, Ethiopia. *Primates* 31:495–508.

Mori, A., Watanabe, K., and Yamaguchi, N. (1989). Longitudinal changes of dominance rank among the females of the Koshima group of Japanese monkeys. *Primates* 30(2):147–173.

Mori, A., Iwamoto, T., and Bekele, A. (1997). A case of infanticide in a recently found gelada population in Arsi, Ethiopia. *Primates* 38:79–88.

Mori, S. A. (1989). Diversity of Lecythidaceae in the Guianas. In: Holm-Nielsen, L. B., Nielsen, I. C., and Balslev, H. (eds.), *Tropical Forests: Botanical Dynamics, Speciation and Diversity.* Academic Press, London. pp. 319–322.

Mori, U. (1979a). Development of sociability and social status. In: Kawai, M. (ed.), *Ecological and Sociological Studies of Gelada Baboons*, vol. 16. Karger, Basel. pp. 125–155.

Mori, U. (1979b). Individual relationships within a unit. In: Kawai, M. (ed.), *Ecological and Sociological Studies of Gelada Baboons*, vol. 16. Karger, Basel. pp. 93–124.

Mori, U. (1979c). Inter-unit relationships. In: Kawai, M. (ed.), *Ecological and Sociological Studies of Gelada Baboons*, vol. 16. Karger, Basel. pp. 83–93.

Mori, U. (1979d). Reproductive behaviour. In: Kawai, M. (ed.), *Ecological and Sociological Studies of Gelada Baboons*, vol. 16. Karger, Basel. pp. 183–197.

Mori, U. (1979e). Social structure of gelada baboons. In: Kawai, M. (ed.), *Ecological and Sociological Studies of Gelada Baboons*, vol. 16. Karger, Basel. pp. 243–250.

Mori, U. (1979f). Unit formation and the emergence of a new leader. In: Kawai, M. (ed.), *Ecological and Sociological Studies of Gelada Baboons*, vol. 16. Karger, Basel. pp. 156–182.

Morin, P. A. (1994). Decoding chimp genes and lives. *Science* 265:1172–1173.

Morin, P. A., and Goldberg, T. L. (2004). Determination of genealogical relationships from genetic data: a review of methods and applications. In: Chapais, B., and Berman, C. M. (eds.), *Kinship and Behavior in Primates.* Oxford University Press, New York. pp. 15–45.

Morin, P. A., and Woodruff, D. S. (1992). Paternity exclusion using multiple hypervariable microsatellite loci amplified from nuclear DNA of hair cells. In: Martin, R. D., Dixson, A. F., and Wickings, E. J. (eds.), *Paternity in Primates: Genetic Tests and Theories.* Karger, Basel. pp. 63–81.

Morin, P. A., Moore, J. J., Chakraborty, R., Jin, L., Goodall, J., and Woodruff, D. S. (1994). Kin selection, social structure, gene flow, and the evolution of chimpanzees. *Science* 265:1193–1201.

Morin, P. A., Chambers, K. E., Boesch, C., and Vigilant, L. (2001). Quantitative polymerase chain reaction analysis of DNA from noninvasive samples for accurate microsatellite genotyping of wild chimpanzees (*Pan troglodytes*). *Mol. Ecol.* 10:1835–1844.

Morland, H. S. (1990). Parental behavior and infant development in ruffed lemurs (*Varecia variegata*) in a northeast Madagascar rain forest. *Am. J. Primatol.* 20:253–265.

Morland, H. S. (1991). Preliminary report on the social organization of ruffed lemurs (*Varecia variegata variegata*) in a northeast Madagascar rain forest. *Folia Primatol.* 56:157–161.

Morland, H. S. (1993). Seasonal behavioral variation and its relationship to thermoregulation in ruffed lemurs (*Varecia variegata variegata*). In: Kappeler, P. M., and Ganzhorn, J. U. (eds.), *Lemur Social Systems and Their Ecological Basis.* Plenum Press, New York. pp. 193–203.

Morrell, J. M., Nubbemeyer, R., Heistermann, M., Rosenbusch, J., Kuederling, I., Holt, W., and Hodges, J. K. (1998). Artificial insemination in *Callithrix jacchus* using fresh or cryopreserved sperm. *Anim. Reprod. Sci.* 52:165–174.

Morris, B. (1998). *The Power of Animals.* Berg, Oxford.

Morris, D. (1967). *The Naked Ape: A Zoologist's Study of the Human Animal.* Johnathan Cape, London.

Morrison, D. (1978). Lunar phobia in a neotropical fruit bat, *Artibeus jamaicensis* (Chiroptera, Phyllostomidae). *Anim. Behav.* 26:852–855.

Morrison, P., and Simoes, J. (1962). Body temperatures in two Brazilian primate. *Zoologia* 24:167–178.

Morrogh-Bernard, H. C., Husson, S., Page, S. E., and Rieley, J. O. (2003). Population status of the Bornean orang-utan (*Pongo pygmaeus*) in the Sebangau peat swamp forest, Central Kalimantan, Indonesia. *Biol Conserv.* 110:141–152.

Morrogh-Bernard, H. C., Husson, S. J., Knott, C. D., Wich, S. A., van Schaik, C. P., van Noordwijk, M., Lackman-Ancrenaz, I., Marshall, A. J., Kanamori, T., Kuze, N., and bin Sakong, R. (2009). Orangutan activity budgets and diet. In: Wich, S. A., Utami, S. S., Mitra Setia, T., and van Schaik, C. (eds.), *Orangutans: Geographic Variation in Behavioral Ecology and Conservation.* Oxford University Press, Oxford. pp. 119–134.

Morton, E. S. (1977). On the occurrence and significance of motivation–structural rules in some bird and mammal sounds. *Am. Nat.* 111:855–869.

Morton, E. S. (1982). Grading, discreteness, redundancy, and motivation–structural rules. In: Kroodsma, D. E., and Miller, E. H. (eds.), *Acoustic Communication in Birds,* vol. 1. Academic Press, New York. pp. 183–212.

Mota, M. T., and Sousa, M. B. C. (2000). Prolactin levels of fathers and helpers related to alloparental care in common marmosets, *Callithrix jacchus.* *Folia Primatol.* 71:22–26.

Mota, M. T. D., Franci, C. R., and de Sousa, M. B. C. (2006). Hormonal changes related to paternal and alloparental care in common marmosets (*Callithrix jacchus*). *Horm. Behav.* 49:293–302.

Mouse Genome Sequencing Consortium (2002). Initial sequencing and comparative analysis of the mouse genome. *Nature* 420:520–561.

Mowry, C. B., Decker, B. S., and Shure, D. J. (1996). The role of phytochemistry in dietary choices of Tana River red colobus monkeys (*Procolobus badius rufomitratus*). *Int. J. Primatol.* 17:63–84.

Moya, L., Encarnación, F., Aquino, R., Tapia, J., Ique, C., and Puertas, P. (1990). El estado de las poblaciones naturales de primates y los beneficios de las cosechas sostenidas. In: Arámbulo, P. III, Encarnación, F., Estupinan, J., Samame, H., Watson, C. R., and Weller, R. E. (eds.), *Primates de las Américas.* Battelle Press, Columbus, OH. pp. 173–192.

Moynihan, M. (1964). Some behavior patterns of playtyrrhine monkeys. I. The night monkey (*Aotus trivirgatus*). *Smith. Misc. Coll.* 146:1–84.

Moynihan, M. (1976). *The New World Primates. Adaptive Radiation and the Evolution of Social Behavior, Languages, and Intelligence.* Princeton University Press, Princeton, NJ.

Moynihan, M. (1982). Why is lying about intentions rare during some kinds of contests? *J. Theor. Biol.* 97:7–12.

Mturi, F. A. (1991). The feeding ecology and behavior of the red colobus monkey (*Colobus badius kirkii*) [PhD diss.]. University of Dares Salaam, Dares Salaam.

Mturi, F. A. (1993). Ecology of the Zanzibar red colobus monkey, *Colobus badius kirkii* (Gray, 1968), in comparison with other red colobines. In: Lovett, J. C., and Wasser, S. K. (eds.), *Biogeography and Ecology of the Rain Forests of Eastern Africa.* Cambridge University Press, Cambridge. pp. 243–266.

Muckenhirn, N. A., Mortensen, B. K., Vessey, S., Fraser, C. E. O., and Singh, B. (1975). *Report on a Primate Survey in Guyana, 1975.* Pan American Health Organization, Washington DC. pp. 1–49.

Mudakikwa, A. B., Cranfield, M. R., Sleeman, J. M., and Eilenberger, U. (2001). Clinical medicine, preventive health care and research on mountain gorillas in the Virunga Volcanoes region. In: Robbins, M. M., Sicotte, P., and Stewart, K. J. (eds.), *Mountain Gorillas: Three Decades of Research at Karisoke.* Cambridge University Press, Cambridge. pp. 341–360.

Mudry, M. D., Slavutsky, I., and Labal de Vinuesa, M. (1990). Chromosome comparison among five species of platyrrhini (*Alouatta caraya, Aotus azarae, Callithrix jacchus, Cebus apella, Saimiri sciureus*). *Primates* 31:415–420.

Mudry de Pargament, M., Colillas, O. J., and de Salum, S. B. (1984). The *Aotus* from northern Argentina. *Primates* 25:530–537.

Mueller, A. E. (1999). Social organization of the fat-tailed dwarf lemur (*Cheirogaleus medius*) in northwestern Madagascar. In: Rakotosamimanana, B., Rasaminanana, H., and Ganzhorn, J. U. (eds.), *New Directions in Lemur Studies.* Plenum Press, New York. pp. 139–157.

Mugambi, K. G., Butynski, T. M., Suleman, M. A., and Ottichilo, W. (1997). The vanishing De Brazza's monkey (*Cercopithecus neglectus*) in Kenya. *Int. J. Primatol.* 18:995–1004.

Muir, C., Galdikas, B. M. F., and Beckenbach, A. T. (1995). Genetic variability in orangutans. In: Nadler, R. D., Galdikas, B. M. F., Sheeran, L. K., and Rosen, N. (eds.), *The Neglected Ape.* Plenum Press, New York. pp. 267–272.

Muir, C., Galdikas, B. M. F., and Beckenbach, A. T. (1998). Is there sufficient evidence to elevate the orangutan of Borneo and Sumatra to separate species? *J. Mol. Evol.* 46:378–379.

Muir, C. C., Galdikas, B. M. F., and Beckenbach, A. T. (2000). mtDNA sequence diversity of orangutans from the island of Borneo and Sumatra. *J. Mol. Evol.* 51:471–480.

Mukherjee, R. P. (1978). Further observations on the golden langur (*Presbytis geei* Khajuria, 1956), with a note to capped langur (*Presbytis pileatus* Blyth, 1843) of Assam. *Primates* 19:737–747.

Mukherjee, R. P., and Saha, S. S. (1974). The golden langurs (*Presbytis geei* Khajuria, 1956) of Assam. *Primates* 15:327–340.

Mulcahy, N. J., Call, J., and Dunbar, R. I. M. (2005). Gorillas (*Gorilla gorilla*) and orangutans (*Pongo pygmaeus*) encode relevant problem features in a tool-using task. *J. Comp. Psychol.* 119:23–32.

Müller, A. E. (1998). A preliminary report on the social organisation of *Cheirogaleus medius* (Cheirogaleidae; Primates) in north-west Madagascar. *Folia Primatol.* 69:160–166.

Müller, A. E. (1999). Aspects of social life in the fat-tailed dwarf lemur (*Cheirogaleus medius*): inferences from body weights and trapping data. *Am. J. Primatol.* 49:265–280.

Müller, A. E., and Thalmann, U. (2000). Origin and evolution of primate social organisation: a reconstruction. *Biol. Rev. Camb. Philos. Soc.* 75:405–435.

Müller, E. F., Kamau, J. M. Z., and Maloiy, G. M. O. (1983). A comparative study of basal metabolism and thermoregulation in a folivorous (*Colobus guereza*) and an omnivorous (*Cercopithecus mitis*) primate species. *Comp. Biochem. Physiol.* 74:319–322.

Müller, H. J., and Rabbitt, P. M. A. (1989). Reflexive and voluntary orienting of visual attention: time course of activation and resistance to interruption. *J. Exp. Psychol. Human.* 15:315–330.

Müller, K.-H. (1996). Diet and feeding ecology of masked titis (*Callicebus personatus*). In: Norconk, M. A., Rosenberger, A. L., and Garber, P. A. (eds.), *Adaptive Radiations of Neotropical Primates.* Plenum Press, New York. pp. 383–401.

Müller, K.-H., Ahl, C., and Hartman, G. (1997). Geophagy in masked titi monkeys (*Callicebus personatus melanochir*). *Primates* 38:69–77.

Muller, M. N. (2002). Agonistic relations among Kanyawara chimpanzees. In: Boesch, C., Hohmann, G., and Marchant, L. (eds.), *Behavioural Diversity in Chimpanzees and Bonobos.* Cambridge University Press, Cambridge. pp. 112–123.

Muller, M., and Mitani, J. C. (2005). Conflict and cooperation in wild chimpanzees. In: Slater, P. J. B., Rosenblatt, J., Snowdon, C., Roper, T., and Naguib, M. (eds.), *Advances in the Study of Behavior.* Academic Press, New York. pp. 275–331.

Muller, M. N., and Wrangham, R. (2001). The reproductive ecology of male hominoids. In: Ellison, P. (ed.), *Reproductive Ecology and Human Evolution.* Aldine de Gruyter, New York. pp. 397–427.

Muller, M. N., and Wrangham, R. W. (2004a). Dominance, aggression and testosterone in wild chimpanzees: a test of the "challenge hypothesis." *Anim. Behav.* 67:113–123.

Muller, M. N., and Wrangham, R. W. (2004b). Dominance, cortisol and stress in wild chimpanzees (*Pan troglodytes schweinfurthii*). *Behav. Ecol. Sociobiol.* 55:332–340.

Muller, M. N., and Wrangham, R. W. (eds.) (2009). *Sexual Coercion in Primates and Humans: An Evolutionary Perspective on Male Aggression Against Females.* Harvard University Press, Cambridge, MA.

Muller, M. N., Emery Thompson, M., and Wrangham, R. W. (2006). Male chimpanzees prefer mating with older males. *Curr. Biol.* 16:2234–2238.

Muller, M. N., Kahlenberg, S. M., Emery Thompson, M., and Wrangham, R. W. (2007). Male coercion and the costs of promiscuous mating for female chimanzees. *Proc. R. Soc. B* 274:1009–1014.

Müller, P., Velo, A., Raheliarisoa, E., Zaramody, A., and Curtis, D. J. (2002). Surveys of sympatric lemurs at Anjamena, north-west Madagascar. *Afr. J. Ecol.* 38:248–257.

Müller, S., O'Brien, P. C. M., Ferguson-Smith, M. A., and Wienberg, J. (1997). Reciprocal chromosome painting between human and prosimians (*Eulemur macaco macaco* and *E. fulvus mayottensis*). *Cytogenet. Cell Genet.* 78:260–271.

Müller, S., Hollatz, M., and Wienberg, J. (2003). Chromosomal phylogeny and evolution of gibbons (Hylobatidae). *Hum. Genet.* 113:493–501.

Müller-Preuss, P. (1988). Neural basis of signal detection. In: Todt, D., Goedeking, P., and Symmes, D. (eds.), *Primate Vocal Communication*. Springer-Verlag, Berlin. pp. 154–161.

Mullis, K., and Faloona, F. (1987). Specific synthesis of DNA in vitro via a polymerase catalyzed chain reaction. *Methods Enzymol.* 155:335–350.

Munakata, Y., and Johnson, M. (eds.) (2006). *Processes of Change in Brain and Cognitive Development. Attention and Performance XXI.* Oxford University Press, New York.

Munds, R. A., Collins, R., Nijman, V., and Nekaris, K. A. I. (2008). Abundance estimates of three slow loris taxa in Sumatra (*N. coucang*), Java (*N. javanicus*), and Borneo (*N. menagensis*). *Primate Eye* 96:Abstract 902.

Muniz, L., Perry, S., Manson, J. H., Gilkenson, H., Gros-Louis, J., and Vigilant, L. (2006). Father–daughter inbreeding avoidance in a wild primate population. *Curr. Biol.* 16:R156-R157.

Murai, C., Kosugi, D., Tomonaga, M., Tanaka, M., Matsuzawa, T., and Itakura, S. (2005). Can chimpanzee infants (*Pan troglodytes*) form categorical representations in the same manner as human infants (*Homo sapiens*)? *Dev. Sci.* 8:240–254.

Murphy, W. J., Eizirik, E., Johnson, W. E., Zhang, Y. P., Ryder, O. A., and O'Brien, S. J. (2001). Molecular phylogenetics and the origins of placental mammals. *Nature* 409:614–618.

Murray, C. M., Eberly, L. E., and Pusey, A. E. (2006). Foraging strategies as a function of season and rank among wild female chimpanzees (*Pan troglodytes*). *Behav. Ecol.* 17:1020–1028.

Murray, C. M., Wroblewski, E., and Pusey, A. E. (2007). New case of intragroup infanticide in the chimpanzees of Gombe National Park. *Int. J. Primatol.* 28:23–37.

Murray, E. A., Kralik, J. D., and Wise, S. P. (2005). Learning to inhibit prepotent responses: successful performance by rhesus macaques, *Macaca mulatta,* on the reversed-contingency task. *Anim. Behav.* 69:991–998.

Murray, K. G., Russell, S., Picone, C. M., Winnett Murray, K., Sherwood, W., and Kuhlmann, M. L. (1994). Fruit laxatives and seed passage rates in frugivores: consequences for plant reproductive success. *Ecology* 75:989–994.

Murray, P. (1973). The anatomy and adaptive significance of cheek pouches (bursae buccales) in the Cercopithecinae, Cercopithecoidea [PhD diss.]. University of Chicago, Chicago.

Murray, S., Stem, C., Boudreau, B., and Goodall, J. (2000). Intestinal parasites of baboons (*Papio cynocephalus anubis*) and chimpanzees (*Pan troglodytes*) in Gombe National Park. *J. Zoo Wildl. Med.* 31:176–178.

Murton, R. K., Isaacson, A. J., and Westwood, N. J. (1971). Significance of gregarious feeding behaviour and adrenal stress in a population of woodpigeons *Columba palumbus*. *J. Zool.* 165:53.

Muruthi, P., Altmann, J., and Altmann, S. A. (1991). Resource base, parity, and reproductive condition affect females' feeding time and nutrient intake within and between groups of a baboon population. *Oecologia* 87:467–472.

Musser, G., and Dagosto, M. (1987). The identity of *Tarsius pumilus*, a pygmy species endemic to montane mossy forests of central Sulawesi. *Am. Mus. Novit.* 2867:1–53.

Mutschler, T. (2002). Alaotran gentle lemur: some aspects of its behavioral ecology. *Evol. Anth. Suppl.* 1:101–104.

Mutschler, T., Nievergelt, C. M., and Feistner, A. T. C. (2000). Social organization of the Alaotran gentle lemur (*Hapalemur griseus alaotrensis*). *Am J. Primatol.* 50:9–24.

Myers, N., Mittermeier, R. A., Mittermeier, C. G., Fonseca, G. A. B., and Kent, J. (2000). Biodiversity hotspots for conservation priorities. *Nature* 403:853–858.

Myers Thompson, J. (2002). Bonobos of the Lukuru Wildlife Research Project. In: Boesch, C., Hohmann, G., and Marchant, L. (eds.), *Behavioural Diversity in Chimpanzees and Bonobos*. Cambridge University Press, Cambridge. pp. 61–70.

Myers-Thompson, J. A. (2003). A model of the biogeographical journey from Proto-*Pan* to *Pan paniscus. Primates* 44(2):191–197.

Mylius, S. D., and Diekmann, O. (1995). On evolutionary stable life histories, optimisation, and the need to be specific about density dependence. *Oikos* 74:218–224.

Mzilikazi, N., Masters, J. C., and Lovegrove, B. G. (2006). Lack of torpor in free-ranging southern lesser galago, *Galago moholi*: ecological and physiological considerations. *Folia Primatol.* 77:465–476.

Nadjafzadeh, M. N., and Heymann, E. W. (2008). Prey foraging of red titi monkeys, *Callicebus cupreus,* in comparison to sympatric tamarins, *Saguinus mystax* and Saguinus *fuscicollis*. *Am. J. Phys. Anthropol.* 135:56–63.

Nadler, R. D. (1977). Sexual behavior of captive orang-utans. *Arch. Sex. Behav.* 6:457–475.

Nadler, R. D. (1982). Reproductive behavior and endocrinology of orangutans. In: de Boer, L. E. M. (ed.), *The Orangutan: Its Biology and Conservation*. Dr. W. Junk Publishers, The Hague. pp. 231–248.

Nadler, R. D. (1988). Sexual and reproductive behavior. In: Schwartz, J. H. (ed.), *Orang-utan Biology*. Oxford University Press, New York. pp. 105–116.

Nadler, R. D. (1990). Homosexual behavior in nonhuman primates. In: McWhirter, D., Sanders, P., and Reinisch, J. (eds.), *Homosexuality/ Heterosexuality: Concepts of Sexual Orientation*. Oxford University Press, New York. pp. 138–170.

Nadler, R. D., Graham, C. E., Gosselin, R. E., and Collins, D. C. (1985). Serum levels of gonadotropin and gonadal steroids, including testosterone, during the menstrual cycle of the chimpanzee (*Pan troglodytes*). *Am. J. Primatol.* 9:273–284.

Nagel, U. (1973). A comparison of anubis baboons, hamadryas baboons and their hybrids at a species border in Ethiopia. *Folia Primatol.* 19:104–165.

Nagell, K., Olguin, R. S., and Tomasello, M. (1993). Processes of social-learning in the tool use of chimpanzees (*Pan troglodytes*) and human children (*Homo sapiens*). *J. Comp. Psychol.* 107(2):174–186.

Nagle, C. A., and Denari, J. H. (1982). The reproductive biology of capuchin monkeys. *Int. Zoo Ybk.* 22:143–150.

Nagle, C. A., Paul, N., Mazzoni, I., Quiroga, S., Torres, M., Mendizabal, A. F., and Farinati, Z. (1989). Interovarian relationship in the secretion of progesterone during the luteal phase of the capuchin monkey (*Cebus apella*). *J. Reprod. Fertil.* 85:389–396.

Nagy, K. A., and Milton, K. (1979a). Aspects of dietary quality, nutrient assimilation and water balance in wild howler monkeys (*Alouatta palliata*). *Ecology* 60:475–480.

Nagy, K. A., and Milton, K. (1979b). Energy metabolism and food consumption by wild howler monkeys (*Alouatta palliata*). *Ecology* 60:475–480.

Nakagawa, N. (1989). Activity budget and diet of patas monkeys in Kala Maloue National Park, Cameroon: a preliminary report. *Primates* 30:27–34.

Nakagawa, N. (1995). A case of infant kidnapping and allo-mothering by members of a neighbouring group in patas monkeys. *Folia Primatol.* 64:62–68.

Nakagawa, N. (1999). Differential habitat utilization by patas monkeys (*Erythrocebus patas*) and tantalus monkeys (*Cercopithecus aethiops tantalus*) living sympatrically in northern Cameroon. *Am. J. Primatol.* 49:243–264.

Nakagawa, N. (2008). Despotic wild patas monkeys (*Erythrocebus patas*) in Kala Maloue, Cameroon. *Am. J. Primatol.* 70:238–246.

Nakagawa, N., and Okamoto, K. (2003). Van Schaik's socioecological model: developments and problems. *Reichorui Kenkyu Primate Res.* 19:243–264.

Nakagawa, N., Ohsawa, H., and Muroyama, Y. (2003). Life-history parameters of a wild group of West African patas monkeys (*Erythrocebus patas patas*). *Primates* 44:281–290.

Nakamichi, M. (2004a). Grooming interactions of adult females in a free-ranging group of Japanese monkeys: comparison with those recorded 10 years ago. *Folia Primatol.* 75:311–312.

Nakamichi, M. (2004b). Tool-use and tool-making by captive, group-living orangutans (*Pongo pygmaeus abelii*) at an artificial termite mound. *Behav. Proc.* 65:87–93.

Nakamichi, M., and Koyama, N. (1997). Social relationships among ring-tailed lemurs (*Lemur catta*) in two free-ranging troops at Berenty Reserve, Madagascar. *Int. J. Primatol.* 18:73–93.

Nakamichi, M., and Koyama, N. (2000). Intra-troop affiliative relationships of females with newborn infants in wild ring-tailed lemurs (*Lemur catta*). *Am. J. Primatol.* 50:187–203.

Nakamichi, M., and Shazawa, Y. (2003). Distribution of grooming among adult females in a large, free-ranging group of Japanese macaques. *Int. J. Primatol.* 24:607–625.

Nakamichi, M., and Yamada, K. (2007). Long-term grooming partnerships between unrelated adult females in a free-ranging group of Japanese monkeys (*Macaca fuscata*). *Am. J. Primatol.* 69:652–663.

Nakamichi, M., Kanazawa, T., and Terao, K. (1994). Behavior and serum cortisol in infant Japanese monkeys following maternal separation and grouping of infants. In: Roeder, J. J., Thierry, B., Anderson, J. B., and Herrenschmidt, N. (eds.), *Current Primatology. Social Development, Learning and Behaviour,* vol. 2. Université Louis Pasteur, Strasbourg. pp. 257–263.

Nakamichi, M., Itoigawa, N., Imakawa, S., and Machida, S. (1995). Dominance relations among adult females in a free-ranging group of Japanese monkeys at Katsuyama. *Am. J. Primatol.* 37(3):241–251.

Nakasukasa, M. (2004). Acquisition of bipedalism: the Miocene hominoid record and modern analogues for bipedal protohominids. *J. Anal.* 204:385–402.

Namy, L. (ed.) (2005). *Symbolic Use and Symbolic Representation.* Erlbaum, Mahwah, NJ.

Nanda Kumar, N. V., Rajasekhar, M., and Ameer Basha, S. (2004). Slender loris habitat analysis in Sri Venkateswara National Park area, Andrah Pradesh by remote sensing. *Ecol. Environ. Conserv.* 10:501–503.

Napier, J. R., and Barnicot, N. A. (eds.) (1963). The primates. *Symp. Lond. Zool. Soc.* 10.

Napier, J. R., and Napier, P. H. (1967). *A Handbook of Living Primates.* Academic Press, New York.

Napier, J. R., and Napier, P. H. (1985). *The Natural History of the Primates.* Cambridge, MA: MIT Press.

Napier, P. H. (1981). *Catalogue of Primates in the British Museum (Natural History). Part 2: Family Cercopithecidae, Subfamily Cercopithecinae.* British Museum of Natural History, London.

Naqshbandi, M., and Roberts, W. A. (2006). Anticipation of future events in squirrel monkeys (*Saimiri sciureus*) and rats (*Rattus norvegicus*): tests of the Bischof-Kohler hypothesis. *J. Comp. Psychol.* 120(4):345–357.

Nash, L. T. (1976). Troop fission in free-ranging baboons in the Gombe Stream National Park, Tanzania. *Am. J. Phys. Anthropol.* 44:63–77.

Nash, L. T. (1978). The development of the mother–infant relationship in wild baboons (*Papio anubis*). *Anim. Behav.* 26:746–759.

Nash, L. T. (1986a). Dietary, behavioral, and morphological aspects of gummivory in primates. *Ybk. Phys. Anthropol.* 29:113–137.

Nash, L. T. (1986b). Influence of moonlight level on traveling and calling patterns in two sympatric species of *Galago* in Kenya. In: Taub, D. M., and King, F. A. (eds.), *Current Perspectives in Primate Social Dynamics.* Van Nostrand Reinhold, New York. pp. 357–367.

Nash, L. T. (1989). Galagos and gummivory. *Hum. Evol.* 4:199–206.

Nash, L. T. (1993). Juveniles in nongregarious primates. In: Pereira, M. E., and Fairbanks, L. A. (eds.), *Juvenile Primates: Life History, Development, and Behavior.* Oxford University Press, Oxford. pp. 119–137.

Nash, L. T. (2004). Kinship and behavior among nongregarious noctournal prosimians. In: Chapais, B., and Berman, C. M. (eds.), *Kinship and Behavior in Primates.* Oxford University Press, New York. pp. 200–222.

Nash, L. T. (2007). Moonlight and behavior in nocturnal and cathemeral primates, especially *Lepilemur leucopus*: illuminating possible anti-predator efforts. In: Gursky, S. L., and Nekaris, K. A. I. (eds.), *Primate Anti-Predator Strategies.* Springer, New York. pp. 173–205.

Nash, L. T., and Harcourt, C. H. (1986). Social organization of galagos in Kenyan coastal forest II: *Galago garnettii. Am. J. Primatol.* 10:357–369.

Nash, L. T., and Whitten, P. L. (1998). Preliminary observations on the role of *Acacia* gum chemistry in *Acacia* utilization by *Galago senegalensis* in Kenya. *Am. J. Primatol.* 17:27–39.

Nash, L. T., Bearder, S. K., and Olson, T. (1989). Synopsis of galago species characteristics. *Int. J. Primatol.* 10:57–80.

Nash, V. J. (1982). Tool use by captive chimpanzees at an artificial termite mound. *Zoo Biol.* 1:211–221.

Nathan, R. (2007). Total dispersal kernels and the evoluation of diversity an dsimilarity in complex dispersal systems. In: Dennis, A., Schupp, E. W., Green, R., and Westcott, D. (eds.), *Seed Dispersal: Theory and Its Application in a Changing World.* CAB International, Wallingford, UK. pp. 252–276.

Nathan, R., and Muller-Landau, H. C. (2000). Spatial patterns of seed dispersal, their determinants and consequences for recruitment. *Trends Ecol. Evol.* 15:278–285.

Nathan, R., Perry, G., Cronin, J. T., Strand, A. E., and Cain, M. L. (2003). Methods for estimating long-distance dispersal. *Oikos* 103:261–273.

Nathan, R., Schurr, F. M., Spiegel, O., Steinitz, O., Trakhtenbrot, A., and Tsoar, A. (2008). Mechanisms of long-distance seed dispersal. *Trends Ecol. Evol.* 23:638–647.

National Research Council (2003). *Nutrient Requirements of Nonhuman Primates.* National Academies Press, Washington DC.

Naughton-Treves, L. (1996). Uneasy neighbors: wildlife and farmers around Kibale National Park, Uganda [PhD diss.]. University of Florida, Gainsville.

Naughton-Treves, L. (1998). Predicting patterns of crop damage by wildlife around Kibale National Park, Uganda. *Conserv. Biol.* 12:156–168.

Naughton-Treves, L. M., Mena, J. L., Treves, A., Alvarez, N., and Radeloff, V. C. (2003). Wildlife survival beyond park boundaries: the impact of slash-and-burn agriculture and hunting on mammals in Tambopata, Peru. *Conserv. Biol.* 17:1106–1117.

Navarro, A., and Barton, N. H. (2003). Chromosomal speciation and molecular divergence—accelerated evolution in rearranged chromosomes. *Science* 300:321–324.

Nee, S., and May, R. M. (1997). Extinction and the loss of evolutionary history. *Science* 278:692–694.

Nekaris, K. A. I. (1997). A preliminary survey of the slender loris (*Loris tardigradus*) in south India. *Am. J. Phys. Anthropol.* (suppl. 24):176–177.

Nekaris, K. A. I. (2000). The socioecology of the Mysore slender loris (*Loris tardigradus lydekkerianus*) in Dindigul, Tamil Nadu, south India [PhD diss.]. Washington University, St. Louis, MO.

Nekaris, K. A. I. (2001). Activity budget and positional behavior of the Mysore slender loris (*Loris tardigradus lydekkarianus*): implications for "slow climbing" locomotion. *Folia Primatol.* 72:228–241.

Nekaris, K. A. I. (2002). Slender in the night. *Nat. Hist.* 2:54–59.

Nekaris, K. A. I. (2003a). Observations on mating, birthing and parental care in three taxa of slender loris in India and Sri Lanka (*Loris tardigradus* and *Loris lydekkerianus*). *Folia Primatol.* 74(suppl.):312–336.

Nekaris, K. A. I. (2003b). Spacing system of the Mysore slender loris (*Loris lydekkerianus lydekkerianus*). *Am. J. Phys. Anthropol.* 121:86–96.

Nekaris, K. A. I. (2003c). Rediscovery of the slender loris in Horton Plains National Park, Sri Lanka. *Asian Primates* 8:1–7.

Nekaris, K. A. I. (2005). Foraging behaviour of the slender loris: implications for theories of primate origins. *J. Hum. Evol.* 49:289–300.

Nekaris, K. A. I. (2006). The social lives of Mysore slender lorises. *Am. J. Primatol.* 68:1–12.

Nekaris, K. A. I. (2009). Visual predation in the slender loris. *J. Hum. Evol.* In press.

Nekaris, K. A. I., and Jaffe, S. (2007). Unexpected diversity of slow lorises (*Nycticebus* spp.) within the Javan pet trade: implications for slow loris taxonomy. *Contrib. Zool.* 76:187–196.

Nekaris, K. A. I., and Jayewardene, J. (2003). Pilot study and conservation status of the slender loris (*Loris tardigradus* and *Loris lydekkerianus*) in Sri Lanka. *Primate Conserv.* 19:83–90.

Nekaris, K. A. I., and Jayewardene, J. (2004). Distribution of slender lorises in four ecological zones in Sri Lanka. *J. Zool.* 262:1–12.

Nekaris, K. A. I., and Nijman, V. (2007). CITES proposal highlights rarity of Asian nocturnal primates (Lorisidae: *Nycticebus*). *Folia Primatol.* 78:211–214.

Nekaris, K. A. I., and Rasmussen, D. T. (2003). Diet of the slender loris. *Int. J. Primatol.* 24:33–46.

Nekaris, K. A. I., and Schulze, H. (2004). Historical and recent developments of human–loris interactions in south and southeast Asia. Invited lecture for Primate Society of Great Britain winter meeting, London. *Primate Eye* 84:17–18.

Nekaris, K. A. I., and Stephens, N. J. (2007). All lorises are not slow: rapid arboreal locomotion in the newly recognised red slender loris (*Loris tardigradus tardigradus*) of southwestern Sri Lanka. *Am. J. Primatol.* 69:112–120.

Nekaris, K. A. I., Liyanage, W. K. D. D., and Gamage, S. (2005). Relationship between forest structure and floristic composition and population density of the southwestern Ceylon slender loris (*Loris tardigradus tardigradus*) in Masmullah Forest, Sri Lanka. *Mammalia.* 69:1–10.

Nekaris, K. A. I., Pimley, E. R., and Ablard, K. (2007a). Anti-predator behaviour of lorises and pottos. In: Gursky, S. G., and Nekaris, K. A. I. (eds.), *Primates and Their Predators.* Springer, New York. pp. 220–238.

Nekaris, K. A. I., Pimley, E. R., and Ablard, K. M. (2007b). Predator defense by slender lorises and pottos. In: Gursky, S. L., and Nekaris, K. A. I. (eds.). *Primate Anti-Predator Strategies.* Springer, New York. pp. 222–240.

Nekaris, K. A. I., Blackham, G., and Nijman, V. (2008). Conservation implications of low encounter rates of five nocturnal primate species (*Nycticebus* spp.) in Asia. *Biodivers. Conserv.* 17:733–747.

Nekaris, K. A. I., Munds, R., and Ford, S. (2010). Unmasking diversity: the slow lorises (Primates: Lorisidae: *Nycticebus*) of Indonesia. In: Gursky, S. and Supriatna, J. (eds.),Indonesian Primates . Springer, New York..

Nekaris, K. A. I., Starr, C., Collins, R. L., and Navarro-Montes, A. (in press). Comparative ecology of exudate feeding by lorises (*Nycticebus, Loris, Perodicticus*). In: Burrows, A. and Nash, L. (eds.), *The Evolution of Exudativory in Primates*. Springer, New York. In press.

Nellemann, C., Miles, L., Kaltenborn, B. P., Virtue, M., and Ahlenius, H. (eds.) (2007). *The Last Stand of the Orangutan—State of Emergency: Illegal Logging, Fire and Palm Oil in Indonesia's National Parks*. UN Environment Programme, GRID-Arendal, Arendal, Norway.

Nelson, G. S. (1960). Schistosome infections as zoonoses in Africa. *Transcrip. R. Soc. Trop. Med. Hyg.* 54:301–314.

Nelson, T. O., and Narens, L. (1990). Metamemory: a theoretical framework and new findings. In: Bower, G. H. (ed.), *The Psychology of Learning and Motivation*, vol. 26. Academic Press, New York. pp. 125–141.

Nelson, T. O., and Narens, L. (1994). Why investigate metacognition? In: Metcalfe, J., and Shimamura, A. P. (eds.), *Metacognition: Knowing About Knowing*. MIT Press, Cambridge, MA. pp. 1–25.

Neri-Arboleda, I. (2001). Ecology and behavior of *Tarsius syrichta* in Bohol, Philippines: implications for conservation [MSc thesis]. University of Adelaide, Adelaide.

Neris, N., Colmán, F., Ovelar, E., Sukigara, N., and Ishii, N. (2002). *Guía de Mamíferos Medianos y Grandes del Paraguay. Distribución, Tendencia Poblacional y Utilización*. Secretaría del Ambiente, Asunción, Paraguay.

Netto, J. W., and van Hooff, J. A. R. A. M. (1986). Conflict interference and the development of dominance relationships in immature *Macaca fascicularis*. In: Else, J. G., and Lee, P. C. (eds.), *Primate Ontogeny, Cognition and Social Behaviour*. Cambridge University Press, Cambridge. pp. 291–300.

Neusser, M., Stanyon, R., Bigoni, F., Wienberg, J., and Müller, S. (2001). Molecular cytotaxonomy of New World monkeys (Platyrrhini)—comparative analysis of five species by multi-color chromosome painting gives evidence for a classification of *Callimico goeldii* within the family of Callitrichidae. *Cytogenet. Cell Genet.* 94:206–215.

Neville, M. K. (1972a). The population structure of red howler monkeys (*Alouatta seniculus*) in Trinidad and Venezuela. *Folia Primatol.* 17:56–86.

Neville, M. K. (1972b). Social relations within troops of red howler monkeys (*Alouatta seniculus*). *Folia Primatol.* 18:47–77.

Neville, M. (1976). The population and conservation of howler monkeys in Venezuela and Trinidad. In: Thorington, R. W., Jr., and Heltne, P. G. (eds.), *Neotropical Primates: Field Studies and Conservation*. National Academy of Sciences, Washington DC. pp. 101–109.

Neville, M. K., Glander, K. E., Braza, F., and Rylands, A. B. (1988). The howling monkeys, genus *Alouatta*. In: Mittermeier, R. A., Rylands, A. B., Coimbra-Filho, A. F., and da Fonseca, G. A. B. (eds.), *Ecology and Behavior of Neotropical Primates*, vol. 2. World Wildlife Fund, Washington DC. pp. 349–453.

Newman, T. K., Jolly, C. J., and Rogers, J. (2004). Mitochondrial phylogeny and systematics of baboons (*Papio*). *Am. J. Phys. Anthropol.* 124:17–27.

Newman, T. L., Tuzun, E., Morrison, V. A., Hayden, K. E., Ventura, M., McGrath, S. D., Rocchi, M., and Eichler, E. E. (2005). A genome-wide survey of structural variation between human and chimpanzee. *Genome Res.* 15:1344–1356.

Newton, P. N. (1987). The social organization of forest Hanuman langurs (*Presbytis entellus*). *Int. J. Primatol.* 8:199–232.

Newton, P. N. (1992). Feeding and ranging patterns of forest Hanuman langurs (*Presbytis entellus*). *Int. J. Primatol.* 13:245–285.

Newton, P. N. (1994). Social stability and change among forest Hanuman langurs (*Presbytis entellus*). *Primates* 35:489–498.

Newton, P. N., and Dunbar, R. I. M. (1994). Colobine monkey society. In: Davies, A. G., and Oates, J. F. (eds.), *Colobine Monkeys: Their Ecology, Behavior and Evolution*. Cambridge University Press, Cambridge. pp. 311–346.

Newton, P. N., and Nishida, T. (1990). Possible buccal administration of herbal drugs by wild chimpanzees, *Pan troglodytes*. *Anim. Behav.* 39:798–801.

Newton-Fisher, N. E. (1999a). The diet of chimpanzees in the Budongo Forest Reserve, Uganda. *Afr. J. Ecol.* 37:344–354.

Newton-Fisher, N. E. (1999b). Infant killers of Budongo. *Folia Primatol.* 70:167–169.

Newton-Fisher, N. E. (2002a). Relationships of male chimpanzees in the Budongo Forest, Uganda. In: Boesch, C., Hohmann, G., and Marchant, L. F. (eds.), *Behavioural Diversity in Chimpanzees and Bonobos*. Cambridge University Press, Cambridge. pp. 125–137.

Newton-Fisher, N. E. (2002b). Ranging patterns of male chimpanzees in the Budongo Forest, Uganda: range structure and individual differences. In: Harcourt, C. S., and Sherwood, B. (eds.), *New Perspectives in Primate Evolution and Behaviour*. Westbury Academic & Scientific Publishing, London. pp. 287–308.

Newton-Fisher, N. E. (2003). The home range of the Sonso community of chimpanzees from the Budongo Forest, Uganda. *Afr. J. Ecol.* 41:150–156.

Newton-Fisher, N. E. (2006). Female coalitions against male aggression in wild chimpanzees of the Budongo Forest. *Int. J. Primatol.* 27:1589–1599.

Newton-Fisher, N. E., Reynolds, V., and Plumptre, A. (2000). Food supply and chimpanzee (*Pan troglodytes schweinfurthii*) in the Budongo Forest Reserve, Uganda. *Int. J. Primatol.* 21:615–628.

Newton-Fisher, N. E., Notman, H., and Reynolds, V. (2002). Hunting of mammalian prey by Budongo Forest chimpanzees. *Folia Primatol.* 73:281–283.

Neyman, P. F. (1978). Aspects of the ecology and social organization of free-ranging cotton-top tamarins (*Saguinus oedipus*) and the conservation status of the species. In: Kleiman, D. G. (ed.), *The Biology and Conservation of the Callitrichidae*. Smithsonian Institution Press, Washington DC. pp. 39–71.

Nguyen, N., Van Horn, R. C., Alberts, S. C., and Altmann, J. (2009). "Friendships" between new mothers and adult males: adaptive benefits and determinants in wild baboons (Papio cynocephalus). *Behav. Ecol. Sociobiol.* 63:1331–1344.

Ni, X., Wang, Y., Hu, Y., and Li, C. (2004). A euprimate skull from the early Eocene of China. *Nature* 427:65–68.

Nicholson, A. J. (1954). An outline of the dynamics of animal populations. *Aust. J. Zool.* 2:9–65.

Nicholson, N. A. (1987). Infants, mothers and other females. In: Smuts, B. B., Cheney, D. L., Seyfarth, R. M., Wrangham, R. W., and Struhsaker, T. T. (eds.), *Primate Societies*. University of Chicago Press, Chicago. pp. 330–342.

Nielsen, R. (2005). Molecular signatures of natural selection. *Annu. Rev. Genet.* 39:197–218.

Nielsen, R., and Beaumont, R. A. (2009). Statistical inferences in phylogeography. *Mol. Ecol.* 18:1034–1047.

Niemitz, C. (1972). Puzzle about *Tarsius*. *Sarawak Mus. J.* 20:329–337.

Niemitz, C. (1974). A contribution to the postnatal behavioral development of *Tarsius bancanus*, studied in two cases. *Folia Primatol.* 21:250–276.

Niemitz, C. (1979). Results of a field study on the western tarsier in Sarawak. *Sarawak Mus. J.* 27:171–228.

Niemitz, C. (1984a). *Biology of Tarsiers*. Gustav Fischer, Stuttgart.

Niemitz, C. (1984b). Locomotion and posture of *Tarsius bancanus*. In: Niemitz, C. (ed.), *Biology of Tarsiers*. Gustav Fischer Verlag, New York. pp. 191–226.

Niemitz, C. (1990). The evolution of primate skin structures in relation to gravity and locomotor patterns. In: Jouffroy, F., Stack, M., and Niemitz, C. (eds.), *Gravity, Posture, and Locomotion in Primates*. Editrice Il Sedicesimo, Florence. pp. 129–156.

Niemitz, C., Nietsch, A., Warter, S., and Rumpler, Y. (1991). *Tarsius dianae*: a new primate species from central Sulawesi (Indonesia). *Folia Primatol.* 56:105–116.

Nietsch, A. (1999). Duet vocalizations among different populations of Sulawesi tarsiers. *Int. J. Primatol.* 20:567–583.

Nietsch, A., and Kopp, M. (1998). The role of vocalization in species differentiation of Sulawesi tarsiers. *Folia Primatol.* 69:371–378.

Nietsch, A., and Niemitz, C. (1992). Indication for facultative polygamy in free-ranging *Tarsius spectrum*, supported by morphometric data. In: *International Primatological Society Abstracts*. International Primatological Society, Strasbourg. p. 318.

Nieuwenhuijsen, K., de Neef, K. J., and Slob, A. K. (1986). Sexual behaviour during ovarian cycles, pregnancy and lactation in group-living stumptail macaques (*Macaca arctoides*). *Hum. Reprod.* 1:159–169.

Nieuwenhuijsen, K., Bonke-Jansen, M., Broekhuijzen, E., de Neef, K. J., van Hooff, J. A. R. A. M., van der Werff ten Bosch J. J., and Slob, A. K. (1988). Behavioral aspects of puberty in group-living stumptail monkeys (*Macaca arctoides*). *Physiol. Behav.* 42:255–264.

Nievergelt, C. M., and Martin, R. D. (1999). Energy intake during reproduction in captive common marmosets (*Callithrix jacchus*). *Physiol. Behav.* 65:849–854.

Nievergelt, C. M., Mutschler, T., and Feistner, A. T. C. (1998). Group encounters and territoriality in wild Alaotran gentle lemurs (*Hapalemur griseus alaotrensis*). *Am. J. Primatol.* 50:9–24.

Nievergelt, C. M., Digby, L. J., Ramakrishnan, U., and Woodruff, D. S. (2000). Genetic analysis of group composition and breeding system in a wild common marmoset (*Callithrix jacchus*) population. *Int. J. Primatol.* 21:1–20.

Nigi, H., and Torii, R. (1983). Periovulatory time courses of plasma estradiol and progesterone in the Japanese monkey (*Macaca fuscata*). *Primates* 24:410–418.

Nigi, H., Hayama, S. I., and Torii, R. (1989). Rise in age of sexual maturation in male Japanese monkeys in Takasakiyama in relation to nutritional conditions. *Primates* 30:571–575.

Nijman, V. (2005). *Hanging in the Balance: An Assessment of Trade in Orangutans and Gibbons in Kalimantan, Indonesia*. TRAFFIC, Southeast Asia, Petaling Jaya, Malaysia.

Nikitopoulos, E., Heistermann, M., de Vries, H., van Hooff, J. A. R. A. M., and Sterck, E. H. M. (2005). A pair choice test to identify female mating pattern relative to ovulation in longtailed macaques, Macaca fascicularis. *Anim. Behav.* 70:1283–1296.

Ninan, P. T., Insel, T. M., Cohen, R. M., Cook, J. M., Skolnick, P., and Paul, S. M. (1982). Benzodiazepine receptor–mediated experimental "anxiety" in primates. *Science* 218:1332–1334.

Nishida, T. (1968). The social group of wild chimpanzees in the Mahali Mountains. *Primates* 9:167–224.

Nishida, T. (1979). The social structure of chimpanzees of the Mahale Mountains. In: Hamburg, D. A., and McCown, E. R. (eds.), *The Great Apes*. Benjamin/Cummings, Menlo Park, CA. pp. 73–121.

Nishida, T. (1983). Alpha status and agonistic alliance in wild chimpanzees (*Pan troglodytes schweinfurthii*). *Primates* 24:318–336.

Nishida, T. (1988). Development of social grooming between mother and offspring in wild chimpanzees. *Folia Primatol.* 50:109–123.

Nishida, T. (1989). Social interactions between resident and immigrant female chimpanzees. In: Heltne, P. G., and Marquardt, L. A. (eds.), *Understanding Chimpanzees*. Harvard University Press, Cambridge, MA. pp. 68–89.

Nishida, T. (1990a). A quarter century of research in the Mahale Mountains: an overview. In: Nishida, T. (ed.), *The Chimpanzees of the Mahale Mountains: Sexual and Life History Strategies*. University of Tokyo Press, Tokyo. pp. 3–36.

Nishida, T. (1990b). *The Chimpanzees of the Mahale Mountains: Sexual and Life History Strategies*. University of Tokyo Press, Tokyo.

Nishida, T. (1997). Sexual behavior of adult male chimpanzees of the Mahale Mountains National Park, Tanzania. *Primates* 38:379–398.

Nishida, T., Hasegawa, T., Hayaki, H., Takahata, Y., and Uehara, S. (1992). Meat-sharing as a coalition strategy by an alpha male chimpanzee? In: Nishida, T., McGrew, W., Marler, P., Pickgord, M., and deWall, F. (eds.), *Topics in Primatology. Human Origins,* vol. 1. Tokyo University Press, Tokyo. pp. 159–174.

Nishida, T., and Hiraiwa-Hasegawa, M. (1987). Chimpanzees and bonobos: cooperative relationships among males. In: Smuts, B. B., Cheney, D. L., Seyfarth, R. M., Wrangham, R. W., and Struhsaker, T. T. (eds.), *Primate Societies*. University of Chicago Press, Chicago. pp. 165–178.

Nishida, T., and Hosaka, K. (1996). Coalition strategies among adult male chimpanzees of the Mahale Mountains, Tanzania. In: McGrew, W. C., Marchant, L. F., and Nishida, T. (eds.), *Great Ape Societies*. Cambridge University Press, Cambridge. pp. 114–134.

Nishida, T., and Kawanaka, K. (1972). Inter-unit-group relationships among wild chimpanzees of the Mahale Mountains. *Kyoto Univ. Afr. Stud.* 7:131–169.

Nishida, T., and Uehara, S. (1983). Natural diet of chimpanzees (*Pan troglodytes schweinfurthii*): long-term record from the Mahale Mountains, Tanzania. *Afr. Study Monogr.* 3:109–130.

Nishida, T., and Van Hooff, J. (2002). UNESCO World Heritage status for the great apes. Caring for Primates. Abstracts of the XIXth Congress, International Primatological Society, Mammalogical Society of China, Beijing, August 2002. p. 194.

Nishida, T., Uehara, S., and Nyundo, R. (1979). Predatory behavior among wild chimpanzees of the Mahale Mountains. *Primates* 20:1–20.

Nishida, T., Hiraiwa-Hasegawa, M., Hasegawa, T., and Takahata, Y. (1985). Group extinction and female transfer in wild chimpanzees in the Mahale Mountains. *Z. Tierpsychol.* 67:284–301.

Nishida, T., Takasaki, H., and Takahata, Y. (1990). Demography and reproductive profiles. In: Nishida, T. (ed.), *The Chimpanzees of the Mahale Mountains: Sexual and Life History Strategies*. University of Tokyo Press, Tokyo. pp. 63–97.

Nishida, T., Corp, N., Hamai, M., Hasegawa, T., Hiraiwa-Hasegawa, M., Hosaka, K., Hunt, K. D., Itoh, N., Kawanaka, K., Matsumoto-Oda, A., Mitani, J. C., Nakamura, M., Norikoshi, K., Sakamaki, T., Turner, L., Uehara, S., and Zamma, K. (2003). Demography, female life history, and reproductive profiles among the chimpanzees of Mahale. *Am. J. Primatol.* 59:99–121.

Nishida, T., Matsusaka, T., and McGrew, W. C. (2009). Emergence, propagation or disappearance of novel behavioral patterns in the habituated chimpanzees of Mahale: a review. *Primates* 50:23–36.

Nishihara, T. (1995). Feeding ecology of western lowland gorillas in the Nouable-Ndoki National Park, Congo. *Primates* 36:151–168.

Nishimura, A. (1988). Mating behavior of woolly monkeys, *Lagothrix lagotricha*, at La Macarena, Colombia. *Field Stud. New World Monk. La Macarena Col.* 1:19–27.

Nishimura, A. (1990a). A sociological and behavioral study of woolly monkeys, *Lagothrix lagotricha*, in the Upper Amazon. *Sci. Eng. Rev. Doshisha Univ.* 31:87–121.

Nishimura, A. (1990b). Mating behavior of woolly monkeys (*Lagothrix lagotricha*) at La Macarena, Colombia (II): mating relationships. *Field Stud. New World Monk. La Macarena Col.* 3:7–12.

Nishimura, A. (1994). Social interaction patterns of woolly monkeys (*Lagothrix lagotricha*): a comparison among the atelines. *Sci. Eng. Rev. Doshisha Univ.* 35:236–254.

Nishimura, A. (2003). Reproductive parameters of wild female *Lagothrix lagotricha*. *Int. J. Primatol.* 24:707–722.

Nishimura, A., Wilches, A. V., and Estrada, C. (1992). Mating behaviors of woolly monkeys, *Lagothrix lagotricha*, at La Macarena, Colombia (III): reproductive parameters viewed from a longterm study. *Field Stud. New World Monk. La Macarena Col.* 7:1–7.

Nissen, H. W. (1931). A field study of the chimpanzee. *Comp. Psychol. Monogr.* 8:1–122.

Njiforti, H. L. (1996). Preference and present demand for bushmeat in northern Cameroon: some implications for wildlife conservation. *Environ. Conserv.* 23:149–155.

Njururi, N. (1966). *Agikuyu*. Oxford University Press, London.

Nkurunungi, J. B. (2004). The availability and distribution of fruit and non-fruit resources in Bwindi: their influence on gorilla habitat and food choice [PhD diss.]. Makerere University, Makerere, Uganda.

Nkurunungi, J. B., Ganas, J., Robbins, M. M., and Stanford, C. B. (2004). A comparison of two mountain gorilla habitats in Bwindi Impenetrable National Park, Uganda. *Afr. J. Ecol.* 42:289–297.

Noë, R. (1986). Lasting alliances among adult male savannah baboons. In: Else, J. G., and Lee, P. C. (eds.), *Primate Ontogeny, Cognition and Social Behaviour*. Cambridge University Press, Cambridge. pp. 381–392.

Noë, R. (1992). Alliance formation among male baboons: shopping for profitable partners. In: Harcourt, A. H., and de Waal, F. B. M. (eds.), *Coalitions and Alliances in Humans and Other Animals*. Oxford University Press, Oxford. pp. 285–321.

Noë, R. (2001). Biological markets: partner choice as the driving force behind the evolution of mutualisms. In: Noe, R., van Hooff, J., and Hammerstein, P. (eds.), *Economics in Nature: Social Dilemmas, Mate Choice and Biological Markets*. Cambridge University Press, Cambridge. pp. 93–118.

Noë, R., and Bshary, R. (1997). The formation of red colobus–diana monkey associations under predation pressure from chimpanzees. *Proc. R. Soc. Lond.* 264:253–259.

Noë, R., and Sluijter, A. A. (1990). Reproductive tactics of male savanna baboons. *Behaviour* 113:117–170.

Noë, R., and Sluijter, A. A. (1995). Which adult male savanna baboons form coalitions. *Int. J. Primatol.* 16:77–106.

Noe, R., van Schaik, C. P., van Hooff, J. A. R. A. M. (1991). The market effect: an explanation for pay-off asymmetries among collaborating animals. *Ethology* 87:97–118.

Norconk, M. A. (1990). Introductory remarks: ecological and behavioural correlates of polyspecific primate troops. *Am. J. Primatol.* 21:81–85.

Norconk, M. A. (1996). Seasonal variation in the diets of white-faced and bearded sakis (*Pithecia pithecia* and *Chiropotes satanas*) in Guri Lake, Venezuela. In: Norconk, M. A., Rosenberger, A. L., and Garber, P. A. (eds.), *Adaptive Radiations of Neotropical Primates*. Plenum Press, New York. pp. 403–423.

Norconk, M. A. (2006). A long-term study of group dynamics and female reproduction in Venezuelan *Pithecia pithecia*. *Int. J. Primatol.* 27:653–674.

Norconk, M. A., and Conklin-Brittain, N. L. (2004). Variation on frugivory: the diet of Venezuelan white-faced sakis. *Int. J. Primatol.* 25:1–26.

Norconk, M. A., and Kinzey, W. G. (1994). Challenge of neotropical frugivory: travel patterns of spider monkeys and bearded sakis. *Am. J. Primatol.* 34:171–183.

Norconk, M. A., Rosenberger, A. L., and Garber, P. A. (1996a). *Adaptive Radiations of Neotropical Primates*. Plenum Press, New York.

Norconk, M. A., Sussman, R. W., and Phillips-Conroy, J. E. (1996b). Primates of Guyana Shield forests. In: Norconk, M., Rosenberger, A. L., and Garber, P. A. (eds.), *Adaptive Radiations of Neotropical Primates*. Plenum Press, New York. pp. 69–83.

Norconk, M. A., Grafton, B. W., and Conklin-Brittain, N. L. (1998). Seed dispersal by neotropical seed predators. *Am. J. Primatol.* 45:103–126.

Norconk, M. A., Oftedal, O. T., Power, M. L., Jakubasz, M., and Savage, A. (2002). Digesta passage and fiber digestibility in captive white-faced sakis (*Pithecia pithecia*). *Am. J. Primatol.* 58:23–34.

Norconk, M. A., Raghanti, M. A., Martin, S. K., Grafton, B. W., Gregory, L. T., and DiDijn, B. P. E. (2003). Primates of Brownsberg Natuurpark, Suriname, with particular attention to the pitheciins. *Neotrop. Primates* 11:94–100.

Norconk, M. A., Grafton, B. W., and McGraw, W. S. (In Press). Morphological and ecological adaptations to seed predation- a primate-wide perspective. In: Barnett, A. A., Veiga, L. M., Ferrari, S. F., and Norconk, M. A. (eds.), *Evolutionary Biology and Conservation of Titis, Sakis, and Uacaris.* Cambridge University Press, Cambridge.

Norconk, M. A., Wright, B. W., Conklin-Brittain, N. L., and Vinyard, C. J. (2009). Mechanical and nutritional properties of food as factors in platyrrhine dietary adaptations. In: Garber, P. A., Estrada, A., Bicca-Marques, J.C., Heymann, E.W., and Strier, K.B. (Eds.) *South American primates: Comparartive Perspectives in the Study of Behavior, Ecology, and Conservation..* Springer Science+Business Media, New York. pp. 279–319.

Norman, R. L., Brandt, H., and Van Horn, R. N. (1978). Radioimmunoassay for luteinizing hormone (LH) in the ring-tailed lemur (*Lemur catta*) with antiovine LH and ovine 125I-LH1. *Biol. Reprod.* 19:1119–1124.

Norris, J. (1988). Diet and feeding behavior of semi-free ranging mandrills in an enclosed Gabonais forest. *Primates* 29: 449–463.

Norscia, I. (2008). Pilot survey of avahi population (woolly lemurs) in littoral forest fragments of southeast Madagascar. *Primates* 49:85–88.

Norscia, I., and Borgognini-Tarli, S. B. (2008). Ranging behavior and possible correlates of pair-living in southeastern Avahis (Madagascar). *Int. J. Primatol.* 29:153–171.

Northrop, F. S. C. (1965). *The Logic of the Sciences and Humanities.* Meridian, Cleveland, OH.

Norton, G. W., Rhine, R. J., Wynn, G. W., and Wynn, R. D. (1987). Baboon diet: a five-year study of stability and variability in the plant feeding and habitat of the yellow baboons (*Papio cynocephalus*) of Mikumi National Park, Tanzania. *Folia Primatol.* 48:78–120.

Noser, R., and Byrne, R. W. (2007a). Travel routes and planning of visits to out-of-sight resources in wild chacma baboons, *Papio ursinus. Anim. Behav.* 73:257–266.

Noser, R., and Byrne, R. W. (2007b). Mental maps in chacma baboons, *Papio ursinus*: using inter-group encounters as a natural experiment. *Anim. Cogn.* 10:331–340.

Nowell, A., and Fletcher, A. (2007). Development of independence from the mother in *Gorilla gorilla gorilla. Int. J. Primatol.* 28:441–455.

Nozaki, M. (2009). Grandmothers care for orphans in a provisioned troop of Japanese macaques (*Macaca fuscata*). *Primates* 50:85–88.

Nozaki, M., Mitsunaga, F., and Shimizu, K. (1995). Reproductive senescence in female Japanese monkeys (*Macaca fuscata*): age and season-related changes in hypothalamic–pituitary–ovarian functions and fecundity rates. *Biol. Reprod.* 52:1250–1257.

Nsubuga, A. M. (2005). Genetic analysis of the social structure in wild mountain gorillas (*Gorilla beringei beringei*) of Bwindi Impenetrable National Park, Uganda [PhD diss.]. University of Leipzig, Leipzig.

Nsubuga, A. M., Robbins, M. M., Vigilant, L., and Boesch, C. (2001). A non-invasive genetic assessment of the social structure and male reproductive success of mountain gorillas in Bwindi Impenetrable National Park, Uganda. *Primate Eye* 75:21–22.

Nsubuga, A. M., Robbins, M. M., Roeder, A. D., Morin, P. A., Boesch, C., and Vigilant, L. (2004). Factors affecting the amount of genomic DNA extracted from ape faeces and the identification of an improved sample storage method. *Mol. Ecol.* 13:2089–2094.

Nsubuga, A. M., Robbins, M. M., Boesch, C., and Vigilant, L. (2008). Patterns of paternity and group fission in wild multimale mountain gorilla groups. *Am. J. Phys. Anthropol.* 135:265–274.

Nunes, A. (1995). Foraging and ranging patterns in white-bellied spider monkeys. *Folia Primatol.* 65:85–99.

Nunes, A. (1998). Diet and feeding ecology of *Ateles belzebuth belzebuth* at Maracá Ecological Station, Roraima, Brazil. *Folia Primatol.* 69:61–76.

Nunes, S., Brown, C., and French, J. A. (2002). Variation in circulating and excreted estradiol associated with testicular activity in male marmosets. *Am. J. Primatol.* 546:27–42.

Nunes-Iturri, G., and Howe, H. F. (2007). Bushmeat and the fate of trees with seeds dispersed by large primates in a lowland rain forest in western Amazonia. *Biotropica* 39:348–354.

Nunn, C. L. (1999). The evolution of exaggerated sexual swellings in primates and the graded-signal hypothesis. *Anim. Behav.* 58:229–246.

Nunn, C. L. (2000). Collective benefits, free-riders and male extra-group confiict. In: Kappeler, P. M. (ed.), *Primate Males: Causes and Consequences of Variation in Group Composition.* Cambridge University Press, Cambridge. pp. 192–204.

Nunn, C. L. (2002). Spleen size, disease risk, and sexual selection: a comparative study in primates. *Evol. Ecol. Res.* 4:91–107.

Nunn, C. L. (2003a). Behavioural defenses against sexually transmitted diseases in primates. *Anim. Behav.* 66:37–48.

Nunn, C. L. (2003b). Sociality and disease risk: a comparative study of leukocyte counts in primates. In: de Waal, F. B. M., and Tyack, P. L. (eds.), *Animal Social Complexity: Intelligence Culture, and Individualized Societies.* Harvard University Press, Cambridge, MA. pp. 26–31.

Nunn, C. L., and Altizer, S. M. (2005). The global mammal parasite database: an online resource for infectious disease records in wild primates. *Evol. Anthropol.* 14:1–2.

Nunn, C. L., and Altizer, S. (2006). *Infectious Diseases in Primates: Behavior, Ecology and Evolution.* Oxford University Press, Oxford.

Nunn, C. L., and Barton, R. A. (2000). Allometric slopes and independent contrasts: a comparative test of Kleiber's law in primate ranging patterns. *Am. Nat.* 156:519–533.

Nunn, C. L., and Barton, R. A. (2001). Comparative methods for studying primate adaptation and allometry. *Evol. Anthropol.* 10:81–98.

Nunn, C. L., and Deaner, R. O. (2004). Patterns of participation and free riding in territorial conflicts among ringtailed lemurs (*Lemur catta*). *Behav. Ecol. Sociobiol.* 57:50–61.

Nunn, C. L., and Dokey, A. T.-W. (2006). Ranging patterns and parasitism in primates. *Biol. Lett.* 2:351–354.

Nunn, C. L., and Heymann, E. W. (2005). Malaria infection and host behavior: a comparative study of neotropical primates. *Behav. Ecol. Sociobiol.* 59:30–37.

Nunn, C. L., and Lewis, R. J. (2001). Cooperation and collective action in animal behaviour. In: Noe, R., van Hooff, J. A. R. A. M., and Hammerstein, P. (eds.), *Economics in Nature: Social Dilemmas, Mate Choice, and Biological Markers.* Cambridge University Press, Cambridge. pp. 42–66.

Nunn, C. L., and van Schaik, C. P. (2000). Social evolution in primates: the relative roles of ecology and intersexual conflict. In: van Schaik, C. P., and Janson, C. H. (eds.), *Infanticide by Males and Its Implications.* Cambridge University Press, Cambridge. pp. 388–419.

Nunn, C. L., Gittleman, J. L., and Antonovics, J. (2000). Promiscuity and the primate immune system. *Science* 290:1168–1170.

Nunn, C. L., van Schaik, C. P., and Zinner, D. (2001). Do exaggerated sexual swellings function in female mating competition in primates? A comparative test of the reliable indicator hypothesis. *Behav. Ecol.* 12:646–654.

Nuttal, G. H. F. (1904). *Blood Immunity and Blood Relationship.* Cambridge University Press, Cambridge.

Oates, J. F. (1974). The ecology and behaviour of the black-and-white colobus monkey (*Colobus guereza* Ruppell) in East Africa [PhD diss.]. University of London, London.

Oates, J. F. (1977a). The guereza and its food. In: Clutton-Brock, T. H. (ed.), *Primate Ecology: Studies of Feeding and Ranging Behavior in Lemurs, Monkeys and Apes.* Academic Press, New York. pp. 275–321.

Oates, J. F. (1977b). The guereza and man. In: Prince Ranier III and Bourne, G. H. (eds.), *Primate Conservation.* Academic Press, London. pp. 419–467.

Oates, J. F. (1977c). The social life of a black and white colobus monkey, *Colobus guereza. Z. Tierpsychol.* 45:1–60.

Oates, J. F. (1978). Water–plant and soil consumption by guereza monkeys (*Colobus guereza*): a relationship between minerals and toxins in the diet? *Biotropica* 10:241–253.

Oates, J. F. (1984). The niche of the potto, *Perodicticus potto. Int. J. Primatol.* 5:51–61.

Oates, J. F. (1985). The Nigerian guenon *Cercopithecus erythrogaster*: ecological, behavioural, systematic, and historical observations. *Folia Primatol.* 45:25–43.

Oates, J. F. (1987). Food distribution and foraging behavior. In: Smuts, B. B., Cheyney, D. L., Seyfarth, R. M., Wrangham, R. W., and Struhsaker, T. T. (eds.), *Primate Societies.* University of Chicago Press, Chicago. pp. 197–209.

Oates, J. F. (1988a). The diet of the olive colobus monkey, *Procolobus verus*, in Sierra Leone. *Int. J. Primatol.* 9:457–478.

Oates, J. F. (1988b). The distribution of *Cercopithecus* monkeys in West African forests. In: Gautier-Hion, A., Bourlière, F., Gautier, J. P., and

Kingdon, J., (eds.), A *Primate Radiation: Evolutionary Biology of the African Guenons.* Cambridge University Press, Cambridge. pp. 79–103.

Oates, J. F. (1994). The natural history of African colobines. In: Davies, A. G., and Oates, J. F. (eds.), *Colobine Monkeys: Their Ecology, Behavior and Evolution.* Cambridge University Press, Cambridge. pp. 75–128.

Oates, J. F. (1996). Habitat alteration, hunting and the conservation of folivorous primates in African forests. *Aust. J. Ecol.* 21:1–9.

Oates, J. F. (1999). *Myth and Reality in the Rain Forest: How Conservation Strategies Are Failing in West Africa.* University of California Press, Berkeley.

Oates, J. F. (2006). Is the chimpanzee, *Pan troglodytes,* an endangered species? It depends on what "endangered" means. *Primates* 47(1): 102–112.

Oates, J. F. (2008). *Cercopithecus preussi.* In: IUCN 2009. IUCN Red List of Threatened Species, version 2009.1, www.iucnredlist.org (accessed June 23, 2009).

Oates, J. F., and Bearder, S. (2008). *Cercopithecus solatus.* In: IUCN 2009. IUCN Red List of Threatened Species, version 2009.1, www.iucnredlist.org (accessed June 23, 2009).

Oates, J. F., and Davies, A. G. (1994). What are the colobines? In: Davies, A. G., and Oates, J. F. (eds.), *Colobine Monkeys: Their Ecology, Behavior and Evolution.* Cambridge University Press, Cambridge. pp. 1–9.

Oates, J. F., and Groves, C. P. (2008a). *Cercopithecus nictitans.* In: IUCN 2009. IUCN Red List of Threatened Species, version 2009.1, www.iucnredlist.org (accessed June 23, 2009).

Oates, J. F., and Groves, C. P. (2008b). *Allenopithecus nigroviridis.* In: IUCN 2009. IUCN Red List of Threatened Species, version 2009.1, www.iucnredlist.org (accessed June 23, 2009).

Oates, J. F., and Groves, C. P. (2008c). *Miopithecus ogouensis.* In: IUCN 2009. IUCN Red List of Threatened Species, version 2009.1, www.iucnredlist.org (accessed June 23, 2009).

Oates, J. F., and Groves, C. P. (2008d). *Miopithecus talapoin.* In: IUCN 2009. IUCN Red List of Threatened Species, version 2009.1, www.iucnredlist.org (accessed June 23, 2009).

Oates, J., and Jewell, P. A. (1967). Westerly extent of the range of three African lorisoid primates. *Nature* 215:778–779.

Oates, J. F., and Trocco, T. F. (1983). Taxonomy and phylogeny of black-and-white colobus monkeys: inferences from an analysis of loud call variation. *Folia Primatol.* 40:83–113.

Oates, J. F., and Whitesides, G. H. (1990). Association between olive colobus (*Procolobus verus*), diana guenons (*Cercopithecus diana*), and other forest monkeys in Sierra Leone. *Am. J. Primatol.* 21:129–146.

Oates, J. F., Swain, T., and Zantovska, J. (1977). Secondary compounds and food selection by colobus monkeys. *Biochem. Syst. Ecol.* 5:317–321.

Oates, J. F., Waterman, P. G., and Choo, G. M. (1980). Food selection by the south Indian leaf-monkey, *Presbytis johnii,* in relation to leaf chemistry. *Oecologia* 45:45–56.

Oates, J. F., Whitesides, G. H., Davies, A. G., Waterman, P. G., Green, S. M., Dasilva, G. L., and Mole, S. (1990). Determinants of variation in tropical forest primate biomass: new evidence from West Africa. *Ecology* 71:328–343.

Oates, J. F., Davies, A. G., and Delson, E. (1994). The diversity of living colobines. In: Davies, A. G., and Oates, J. F. (eds.), *Colobine Monkeys: Their Ecology, Behavior and Evolution.* Cambridge University Press, Cambridge. pp. 45–73.

Oates, J. F., Abedi-Lartey, M., McGraw, W. S., Struhsaker, T. T., and Whitesides, G. H. (2000a). Extinction of a West African red colobus monkey. *Conserv. Biol.* 14:1526–1532.

Oates, J. F., Bocian, C. M., and Terranova, C. J. (2000b). The loud calls of black-and-white colobus monkeys: new information and a reappraisal of their phylogenetic and functional significance. In: Whitehead, P. F., and Jolly, C. J. (eds.), *Old World Monkeys.* Cambridge University Press, Cambridge. pp. 431–452.

Oates, J. F., McFarland, K. L., Groves, J. L., Bergl, R. A., Linder, J. M., and Disotell, T. R. (2003). The Cross River gorilla: natural history and status of a neglected and critically endangered subspecies. In: Taylor, A. B., and Goldsmith, M. L. (eds.), *Gorilla Biology: A Multidisciplinary Perspective.* Cambridge University Press, Cambridge. pp. 472–497.

Oates, J. F., Hart, J., Groves, C. P., and Butynski, T. M. (2008a). *Cercopithecus ascanius.* In: IUCN 2009. IUCN Red List of Threatened Species, version 2009.1, www.iucnredlist.org (accessed June 23, 2009).

Oates, J. F., Gippoliti, S., and Groves, C. P. (2008b). *Cercopithecus campbelli.* In: IUCN 2009. IUCN Red List of Threatened Species, version 2009.1, www.iucnredlist.org (accessed June 23, 2009).

Oates, J. F., Gippoliti, S., and Bearder, S. (2008c). *Cercopithecus cephus.* In: IUCN 2009. IUCN Red List of Threatened Species, version 2009.1, www.iucnredlist.org (accessed June 23, 2009).

Oates, J. F., Gippoliti, S., and Groves, C. P. (2008d). *Cercopithecus diana.* In: IUCN 2009. IUCN Red List of Threatened Species, version 2009.1, www.iucnredlist.org (accessed June 23, 2009).

Oates, J. F., Gippoliti, S., and Bearder, S. (2008e). *Cercopithecus erythrogaster.* In: IUCN 2009. IUCN Red List of Threatened Species, version 2009.1, www.iucnredlist.org (accessed June 23, 2009).

Oates, J. F., Gippoliti, S., and Groves, C. P. (2008f). *Cercopithecus erythrotis.* In: IUCN 2009. IUCN Red List of Threatened Species, version 2009.1, www.iucnredlist.org (accessed June 23, 2009).

Oates, J. F., Gippoliti, S., and Groves, C. P. (2008g). *Cercopithecus mona.* In: IUCN 2009. IUCN Red List of Threatened Species, version 2009.1, www.iucnredlist.org (accessed June 23, 2009).

Oates, J. F., Gippoliti, S., and Groves, C. P. (2008h). *Cercopithecus petaurista.* In: IUCN 2009. IUCN Red List of Threatened Species, version 2009.1, www.iucnredlist.org (accessed June 23, 2009).

Oates, J. F., Hart, J., Butynksi, T., and Groves, C. P. (2008i). *Cercopithecus pogonias.* In: IUCN 2009. IUCN Red List of Threatened Species, version 2009.1, www.iucnredlist.org (accessed June 23, 2009).

Oates, J. F., Baker, L. R., and Tooze, Z. J. (2008j). *Cercopithecus sclateri.* In: IUCN 2009. IUCN Red List of Threatened Species, version 2009.1, www.iucnredlist.org (accessed June 23, 2009).

Oates, J. F., Struhsaker, T., and McGraw, S. (2008k). *Procolobus badius* ssp. waldroni. In: 2008 IUCN Red List of Threatened Species, http://www.iucnredlist.org/ (accessed January 25, 2009).

Oates, J. F., Groves, C. P., and Jenkins, P. D. (2009). The type locality of *Pan troglodytes* vellerosus (Gray, 1862), and implications for the nomenclature of West African chimpanzees. *Primates* 50:78–80.

O'Brien, T. C. (1991). Female–male social interactions in wedge-capped capuchin monkeys: benefits and costs of group living. *Anim. Behav.* 41:555–567.

O'Brien, T. C., and Robinson, J. G. (1991). Allomaternal care by female wedge-capped capuchin monkeys: effects of age, rank and relatedness. *Behaviour* 119:30–50.

O'Brien, T. G., and Kinnaird, M. G. (1997). Behavior, diet, and movements of the Sulawesi crested black macaque (*Macaca nigra*). *Int. J. Primatol.* 18:321–351.

O'Brien, T. G., Kinnaird, M. F., Nurcahyo, A., Prasetyaningrum, M., and Iqbal, M. (2003). Fire, demography and the persistence of siamang (*Symphalangus syndactylus*: Hylobatidae) in a Sumatran rainforest. *Anim. Conserv.* 6:115–121.

O'Brien, T. G., Kinnaird, M. F., Nurcahyo, A., Iqbal, M., and Rusmanto, M. (2004). Abundance and distribution of sympatric gibbons in a threatened Sumatran rain forest. *Int. J. Primatol.* 25:267–284.

O'Connell, S. M., and Cowlishaw, G. (1994). Infanticide avoidance, sperm competition and mate choice: the function of copulation calls in female baboons. *Anim. Behav.* 48:687–694.

O'Connell, S. M., and Cowlishaw, G. (1995). The post-copulation withdrawal response in female baboons: a functional analysis. *Primates* 36:441–446.

Oda, R. (1996). Effects of contextual and social variables on contact call production in free-ranging ringtailed lemurs (*Lemur catta*). *Int. J. Primatol.* 17:191–205.

Oda, R., and Masataka, N. (1992). Functional significance of female Japanese macaque copulatory calls. *Folia Primatol.* 58:146–149.

Oerke, A. K., Heistermann, M., Kuderling, I., Martin, R. D., and Hodges, J. K. (2002). Monitoring reproduction in Callitrichidae by means of ultrasonography. *Evol. Anthropol.* 11(suppl. 1):183.

Off, E. C., and Gebo, D. L. (2005). Galago locomotion in Kibale National Park, Uganda. *Am J Primatol.* 66:189–195.

Off, E. C., Isbell, L. A., and Young, T. P. (2008). Population density and habitat preferences of the Kenya lesser galago (*Galago senegalensis braccatus*) along the Ewaso Nyiro River, Lakipia, Kenya. *J. East Afr. Nat. Hist.* 97:109–116.

Oftedal, O. T. (1991). The nutritional consequences of foraging in primates: the relationship of nutrient intakes to nutrient requirements. *Philos. Trans. R. Soc. Lond. B* 334:161–170.

Oftedal, O. T., and Allen, A. E. (1996). The feeding and nutrition of omnivores with emphasis on primates. In: Kleiman, D. G., Allen, M. E., Thompson, K. V., Lumpkin, S., and Harris, H. (eds.), *Wild Mammals in Captivity: Principles and Techniques.* University of Chicago Press, Chicago. pp. 148–157.

Ogawa, H. (1995). Bridging behavior and other affiliative interactions among male Tibetan macaques (*Macaca thibetana*). *Int. J. Primatol.* 16:707–727.

Ogonuki, N., Tsuchiya, H., Hirose, Y., Okada, H., Ogura, A., and Sankai, T. (2003). Pregnancy by the tubal transfer of embryos developed after injection of round spermatids into oocyte cytoplasm of the cynomolgus monkey (*Macaca fascicularis*). *Hum. Reprod.* 18:1273–1280.

Ohigashi, H. (1995). Plants used medicinally by primates in the wild and their physiologically active constituents. Report to the Ministry of Science, Education and Culture for 1994 Grant- in-Aid for Scientific Research (06303012). JapanOhigashi, H., Jisaka, M., Takagaki, T., Nozaki, H., Tada, T., Huffman, M. A., Nishida, T., Kaji, M., and Koshimizu, K. (1991a). Bitter principle and a related steroid glucoside from *Vernonia amygdalina*, a possible medicinal plant for wild chimpanzees. *Agricult. Biol. Chem.* 55:1201–1203.

Ohigashi, H., Takagaki, T., Koshimizu, K., Nishida, T., Huffman, M. A., Takasaki, H., Jato, J., and Muanza, D. N. (1991b). Biological activities of plant extracts from tropical Africa. *Afr. Stud. Monogr.* 12:201–210.

Ohigashi, H., Huffman, M. A., Izutsu, D., Koshimizu, K., Kawanaka, M., Sugiyama, H., Kirby, G. C., Warhurst, D. C., Allen, D., Wright, C. W., Phillipson, J. D., Timmon-David, P., Delnas, F., Elias, R., and Balansard, G. (1994). Toward the chemical ecology of medicinal plant-use in chimpanzees: the case of *Vernonia amygdalina* Del. A plant used by wild chimpanzees possibly for parasite-related diseases. *J. Chem. Ecol.* 20:541–553.

Ohnuki-Tierney, E. (1987). *The Monkeys as Mirror*. Princeton University Press, Princeton, NJ.

Ohnuki-Tierney, E. (1995). Representations of the monkey (Saru) in Japanese culture. In: Corby, R., and Theunissen, B. (eds.), *Evaluative Proceedings of the Symposium Ape, Man, Apeman: Changing Views Since 1600, Leiden, the Netherlands, 28 June–1 July, 1993*. Department of Prehistory, Leiden University, Leiden. pp. 297–308.

Ohsawa, H. (1979a). Herd dynamics. In: Kawai, M. (ed.), *Ecological and Sociological Studies of Gelada Baboons*, vol. 16. Karger, Basel. pp. 47–80.

Ohsawa, H. (1979b). The local gelada population and environment of the Gich Area. In: Kawai, M. (ed.), *Ecological and Sociological Studies of Gelada Baboons*, vol. 16. Karger, Basel. pp. 3–45.

Ohsawa, H., Inoue, M., and Takenaka, O. (1993). Mating strategy and reproductive success of male patas monkeys (*Erythrocebus patas*). *Primates* 34:533–544.

Oi, T. (1988). Sociological study on the troop fission of wild Japanese monkeys (*Macaca fuscata yakui*) on Yakushima Island. *Primates* 29:1–19.

Oi, T. (1996). Sexual behaviour and mating system of the wild pig-tailed macaque in West Sumatra. In: Fa, J. E., and Lindburg, D. G. (eds.), *Evolution and Ecology of Macaque Societies*. Cambridge University Press, Cambridge, pp. 342–368.

Okamoto, K. (2004). Patterns of group fission. In: Thierry, B., Singh, M., and Kaumanns, W. (eds.), *Macaque Societies: A Model for the Study of Social Organization*. Cambridge University Press, Cambridge. pp. 112–116.

Okamoto, K., and Matsumura, S. (2001). Group fission in moor macaques (*Macaca maurus*). *Int. J. Primatol.* 22:481–493.

Okamoto, K., Matsumura, S., and Watanabe, K. (2000). Life history and demography of wild moor macaques (*Macaca maurus*): summary of ten years of observations. *Am. J. Primatol.* 52:1–11.

Okamoto-Barth, S., and Call, J. (2008). Tracking and inferring spatial rotation by children and great apes. *Dev. Psychol.* 44:1396–1408.

Okecha, A., and Newton-Fisher, N. (2006). The diet of olive baboons (*Papio anubis*) in the Budongo Forest Reserve, Uganda. In: Newton-Fisher, N., Notman, H., Paterson, J. D., and Reynolds, V. (eds.), *Primates of Western Uganda*. Springer, New York. pp. 61–73.

Okra, T., and Takenaka, O. (2001). Wild gibbons' parentage tested by non-invasive DNA sampling and PCR-amplified polymorphic microsatellites. *Primates* 42:67–73.

Oldham, M. C., Horvath, S., and Geschwind, D. H. (2006). Conservation and evolution of gene coexpression networks in human and chimpanzee brains. *Proc. Natl. Acad. Sci. USA* 103:17973–17978.

Oliveira, A. C. M., and Ferrari, S. F. (2000). Seed dispersal by black-handed tamarins, *Saguinus midas niger* (Callitrichinae, Primates): implications for the regeneration of degraded forest habitats in eastern Amazonia. *J. Trop. Ecol.* 16:709–716.

Oliveira, A., and Ferrari, S. (2008). Habitat exploitation by free-ranging *Saguinus niger* in eastern Amazonia. *Int. J. Primatol.* 29:1499–1510.

Oliveira, J. M. S., Lima, M. C., Bonvincino, C., Ayres, J. M., and Fleagle, J. G. (1985). Preliminary notes on the ecology and behavior of the Guianan saki (*Pithecia pithecia*, Linnaeus 1766; Cebidae, Primate). *Acta Amazon.* 15:249–263.

Oliveira, V. B., Linares, A. M., Correa, G. L. C., and Chiarello, A. G. (2008). Predation on the black capuchin monkey *Cebus nigritus* (Primates: Cebidae) by domestic dogs *Canis lupus familiaris* (Carnivora: Canidae), in the Parque Estadual Serra do Brigadeiro, Minas Gerais, Brazil. *Rev. Bras. Zool.* 25(2):376–378.

Olivieri, G., Zimmermann, E., Randrianambinina, B., Rassoloharijaona, S., Rakotondravony, D., Guschanski, K., and Radespiel, U. (2007). The ever-increasing diversity in mouse lemurs: three new species in north and northwestern Madagascar. *Mol. Phylogenet. Evol.* 43:309–327.

Olmos, F. (1994). Jaguar preation on muriqui *Brachyteles arachnoides*. *Neotrop. Primates* 2:16.

Oloke, J. K., Kolawole, D. O., and Erhun, W. O. (1988). The antibacterial and antifungal activities of certain components of *Afromomum melegueta* fruits. *Fitoterapia* LIX:384–388.

Olson, D. K. (1980). Male interactions and troop split among black-and-white colobus monkeys (*Colobus polykomosvellerosus*). Paper presented at the Eighth Congress of the International Primatological Society, Florence, Italy. July 4–6 1980.

Olson, D. K. (1986). Determining range size for arboreal monkeys: methods, assumptions, and accuracy. In: Taub, D. M., and King, F. A. (eds.), *Current Perspectives in Primate Social Dynamics*. Van Nostrand Reinhold, New York. pp. 212–227.

Olson, L. E., Sargis, E. J., Stanley, W. T., Hildebrandt, K. B. P., and Davenport, T. R. B. (2008). Additional molecular evidence strongly supports the distinction between the recently described African primate *Rungwecebus kipunji* (Cercopithecidae, Paionini) and *Lophocebus*. *Mol. Phylogenet. Evol.* 48:789–794.

Olson, T. R. (1979). Studies on aspects of the morphology and systematics of the genus *Otolemur* (Coquerel, 1859) (Primates: Galagidae) [PhD diss.]. University of London, London.

Olsson, M., Madsen, T., Shine, R., Gullberg, A., and Tegelström, H. (1994). Can female adders multiply? *Nature* 369:528.

Olupot, W. (2000). Mass differences among male mangabey monkeys inhabiting logged and unlogged forest compartments. *Conserv. Biol.* 14:833–843.

Olupot, W., and Waser, P. M. (2001a). Activity patterns, habitat use and mortality risks of mangabey males living outside social groups. *Anim. Behav.* 61:1227–1235.

Olupot, W., and Waser, P. M. (2001b). Correlates of intergroup transfer in male grey-cheeked mangabeys. *Int. J. Primatol.* 22:169–181.

Olupot, W., and Waser, P. M. (2005). Patterns of male residency and intergroup transfer in gray-cheeked mangabeys (*Lophocebus albigena*). *Am. J. Primatol.* 66:331–349.

Olupot, W., and Waser, P. (in press). The grey-cheeked mangabey (*Lophocebus albigena*). In: Kingdon, J., Happold, D., and Butynski, T. (eds.), *The Mammals of Africa*. University of California Press, Berkeley, CA.

Olupot, W., Chapman, C. A., Brown, C. H., and Waser, P. M. (1994). Mangabey (*Cercocebus albigena*) population density, group size, and ranging: a twenty-year comparison. *Am. J. Primatol.* 32:197–205.

Olupot, W., Chapman, C. A., Waser, P. M., and Isabirye-Basuta, G. (1997). Mangabey (*Cercocebus albigena*) ranging patterns in relation to fruit availability and the risk of parasite infection in Kibale National Park, Uganda. *Am. J. Primatol.* 43:65–78.

Olupot, W., Waser, P. M., and Chapman, C. A. (1998). Fruit finding by mangabeys (*Lophocebus albigena*): are monitoring of fig trees and use of sympatric frugivore calls possible strategies? *Int. J. Primatol.* 19:339–353.

Onderdonk, D. A. (2000). Infanticide of a newborn black-and-white colobus monkey (*Colobus guereza*) in Kibale National Park, Uganda. *Primates* 41:209–212.

Onderdonk, D. A., and Chapman, C. A. (2000). Coping with forest fragmentation: the primates of Kibale National Park, Uganda. *Int. J. Primatol.* 21:587–611.

O'Neill, A. C., Fedigan, L. M., and Ziegler, T. E. (2004). Relationship between ovarian cycle phase and sexual behavior in female Japanese macaques (*Macaca fuscata*). *Am. J. Phys. Anthropol.* 125:352–362.

Oppenheimer, J. R. (1968). Behavior and ecology of the white-faced monkey *Cebus capucinus*, on Barro Colorado Island, CZ [PhD diss.]. University of Illinois, Urbana.

Orians, G. H. (1969). On the evolution of mating systems in birds and mammals. *Am. Nat.* 103:589–603.

Osman Hill, W. C. (1953). *Primates. Comparative Anatomy and Taxonomy. I. Strepsirhini*. Edinburgh University Press, Edinburgh.

Ostner, J., Nunn, C. L., and Schülke, O. (2008). Female reproductive synchrony predicts skewed paternity across primates. *Behav. Ecol.* 19:1150–1158.

Ostro, L. E. T., Silver, S. C., Koontz, F. W., Horwich, R. H., and Brockett, R. (2001). Shifts in social structure of black howler (*Alouatta pigra*) groups associated with natural and experimental variation in population density. *Int. J. Primatol.* 22:733–748.

Osvath, M., and Osvath, H. (2008). Chimpanzee (*Pan troglodytes*) and orangutan (*Pongo abelii*) forethought: self-control and pre-experience in the face of future tool use. *Anim. Cogn.* 11:661–674.

Ottoni, E., and Izar, P. (2008). Capuchin monkey tool use: overview and implications. *Evol. Anthropol.* 17:171–178.

Ottoni, E. B., de Resende, B. D., and Izar, P. (2005). Watching the best nut-crackers: what capuchin monkeys (*Cebus apella*) know about others' tool-using skills. *Anim. Cogn.* 8(4):215–219.

Overdorff, D. J. (1988). Preliminary report on the activity cycle and diet of the red-bellied lemur (*Eulemur rubriventer*) in Madagascar. *Am. J. Primatol.* 16:143–153.

Overdorff, D. J. (1991). Ecological correlates to social structure in two prosimian primates: *Eulemur fulvus rufus* and *Eulemur rubriventer* in Madagascar [PhD diss.]. Duke University, Durham, NC.

Overdorff, D. J. (1992). Differential patterns of flower feeding by *Eulemur fulvus rufus* and *Eulemur rubriventer* in Madagascar. *Am. J. Primatol.* 28:191–203.

Overdorff, D. J. (1993a). Ecological and reproductive correlates to range use in red-bellied lemurs (*Eulemur rubriventer*) and rufous lemurs (*Eulemur fulvus rufus*). In: Kappeler, P. M., and Ganzhorn, J. U. (eds.), *Lemur Social Systems and Their Ecological Basis.* Plenum Press, New York. pp. 167–178.

Overdorff, D. J. (1993b). Similarities, differences, and seasonal patterns in the diets of *Eulemur rubriventer* and *Eulemur fulvus rufus* in the Ranomafana National Park, Madagascar. *Int. J. Primatol.* 14:721–753.

Overdorff, D. J. (1996a). Ecological correlates to activity and habitat use of two prosimian primates: *Eulemur rubriventer* and *Eulemur fulvus rufus* in Madagascar. *Am. J. Primatol.* 40:327–342.

Overdorff, D. J. (1996b). Ecological correlates to social structure in two lemur species in Madagascar. *Am. J. Phys. Anthropol.* 100:487–506.

Overdorff, D. J. (1998). Are *Eulemur* species pair-bonded? Social organization and mating strategies in *Eulemur fulvus rufus* from 1988–1995 in southeast Madagascar. *Am. J. Phys. Anthropol.* 105:153–166.

Overdorff, D. J., and Erhart, E. M. (2001). Social and ecological influences on female dominance in day-active prosimian primates. *Am. J. Phys. Anthropol.* 32(suppl.):116.

Overdorff, D. J., and Johnson, S. E. (2003). *Eulemur*, true lemurs. In: Goodman, S. M., and Benstead, J. P. (eds.), *The Natural History of Madagascar.* University of Chicago Press, Chicago. pp. 1320–1324.

Overdorff, D., and Parga, J. (2007). The new era of primate socioecology: ecology and intersexual conflict. In: Campbell, C. J., Fuentes, A., Mackinnon, K. C., Panger, M., and Bearder, S. K. (eds.), *Primates in Perspective.* Oxford University Press, New York, pp. 466–481.

Overdorff, D. J., and Rasmussen, M. A. (1995). Determinants of nighttime activity in "diurnal" lemurid primates. In: Alterman, L., Doyle, G. A., and Izard, M. K. (eds.), *Creatures of the Dark: The Nocturnal Prosimians.* Plenum Press, New York. pp. 61–74.

Overdorff, D. J., and Strait, S. C. (1995). Life history and predation in *Eulemur rubriventer* in Madagascar. *Am. J. Phys. Anthropol.* 100:487–506.

Overdorff, D. J., Merenlender, A. M., Talata, P., Telo, A., and Forward, Z. A. (1999). Life history of *Eulemur fulvus rufus* from 1988–1998 in southeastern Madagascar. *Am. J. Phys. Anthropol.* 108:295–310.

Overdorff, D. J., Strait, S. G., and Seltzer, R. G. (2002). Species differences in feeding in Milne Edwards sifakas (*Propithecus diadema edwardsi*), rufus lemurs (*Eulemur fulvus rufus*), and red-bellied lemurs (*Eulemur rubriventer*) in southern Madagascar: implications for predator avoidance. In: Miller, L. E. (ed.), *Eat or Be Eaten: Predator Sensitive Foraging Among Primates.* Cambridge University Press, Cambridge. pp. 126–137.

Overdorff, D. J., Erhart, E. M., and Mutschler, T. (2003). Fission–fusion in *Eulemur fulvus rufus* in southeastern Madagascar. *Am. J. Primatol.* 60(suppl. 1):42–43.

Oversluijs Vasques, M. R., and Heymann, E. W. (2001). Crested eagle (*Morphnus guianensis*) predation on infant tamarins (*Saguinus mystax* and *Saguinus fuscicollis*, Callitrichinae). *Folia Primatol.* 72:301–303.

Owens, I. P. F., and Bennett, P. M. (2000). Quantifying biodiversity: a phenotypic perspective. *Conserv. Biol.* 14:1014–1022.

Owens, N. W. (1975). Comparison of aggressive play and aggression in free living baboons (*Papio anubis*). *Anim. Behav.* 23:757–765.

Owings, D. H., and Morton, E. S. (1997). The role of information in communication: an assessment/management approach. In: Owings, D. H., Beecher, M., and Thompson, N. (eds.), *Perspectives in Ethology. Communication,* vol. 12. Plenum Press, New York.

Owings, D. H., and Morton, E. S. (1998). *Animal Vocal Communication: A New Approach.* Cambridge University Press, Cambridge.

Owren, M. J. (1990). Acoustic classification of alarm calls by vervet monkeys (*Cercopithecus aethiops*) and humans (*Homo sapiens*): I. Natural calls. *J. Comp. Psychol.* 104:20–28.

Owren, M. J. (2000). Standing evolution on its head: the uneasy role of evolutionary theory in comparative cognition and communication. *Rev. Anthropol.* 29:55–69.

Owren, M. J., and Rendall, D. (1997). An affect-conditioning model of non-human primate vocal signaling. In: Owings, D. H., Beecher, M., and Thompson, N. (eds.), *Perspectives in Ethology. Communication,* vol. 12. Plenum Press, New York. pp. 299–346.

Owren, M. J., and Rendall, D. (2001). Sound on the rebound: bringing form and function back to the forefront in understanding nonhuman primate vocal signaling. *Evol. Anthropol.* 10:58–71.

Oxnard, C. E., Crompton, R. H., and Liebermann, S. S. (1990). *Animal Lifestyles and Anatomies: The Case of the Prosimian Primates.* University of Washington Press, Seattle.

Oyama, S., Griffiths, P. E., and Gray, R. D. (2001). Introduction. What is developmental systems theory? In: Oyama, S., Griffiths, P. E., and Gray, R. D. (eds.), *Cycles of Contingency: Developmental Systems and Evolution.* MIT Press, Cambridge, MA. pp. 1–12.

Oyen, O. J. (1979). Tool-use in free-ranging baboons of Nairobi National Park. *Primates* 20:595–597.

Pacala, S. W., and Tilman, D. (1994). Limiting similarity in mechanistic and spatial models of plant competition in heterogeneous environments. *Am. Nat.* 143:222–257.

Paccagnella, S. G. (1991). Censo da população de monos (*Brachyteles arachnoides*) do Parque Estadual Carlos Botelho, Estado de São Paulo. *Primatol. Brasil* 3:225–233.

Pacheco, L. F., and Simonetti, J. A. (2000). Genetic structure of a mimosoid tree deprived of its seed disperser, the spider monkey. *Conserv. Biol.* 14:1766–1775.

Packer, C. (1977). Reciprocal altruism in *Papio anubis. Nature* 265:441–443.

Packer, C. (1979a). Inter-troop transfer and inbreeding avoidance in *Papio anubis. Anim. Behav.* 27:1–36.

Packer, C. (1979b). Male dominance and reproductive activity in *Papio anubis. Anim. Behav.* 27:37–45.

Packer, C. (1980). Male care and exploitation of infants in *Papio anubis. Anim. Behav.* 28:512–520.

Packer, C., Collins, D. A., Sindimwo, A., and Goodall, J. (1995). Reproductive constraints on aggressive competition in female baboons. *Nature* 373:60–63.

Packer, C., Tatar, M., and Collins, A. (1998). Reproductive cessation in female mammals. *Nature* 392:807–811.

Paetkau, D. (1999). Microsatellites obtained using strand extension: an enrichment protocol. *Biotechniques* 26:690–697.

Page, J. E., Balza, F. F., Nishida, T., and Towers, G. H. N. (1992). Biologically active diterpenes from *Aspilia mossambicensis,* a chimpanzee medicinal plant. *Phytochemistry* 31:3437–3439.

Page, J. E., Huffman, M. A., Smith, V., and Towers, G. H. N. (1997). Chemical basis for medicinal consumption of *Aspilia* leaves by chimpanzees: a re-analysis. *J. Chem. Ecol.* 23:2211–2225.

Pagel, M. (1994). The evolution of conspicuous oestrous advertisement in Old World monkeys. *Anim. Behav.* 47:1333–1341.

Pagel, M. D., and Harvey, P. H. (1993). Evolution of the juvenile period in mammals. In: Pereira, M. E., and Fairbanks, L. A. (eds.), *Juvenile Primates: Life History, Development, and Behavior.* University of Chicago Press, Chicago. pp. 28–37.

Pagès, E. (1980). Ethoecology of *Microcebus coquereli* during the dry season. In: Charles-Dominique, P., Cooper, H. M., Hladik, A., Hladik, C. M., Pagès, E., Pariente, C. F., Petter-Rousseaux, A., Peter, J. J., and Schilling, A. (eds.), *Nocturnal Malagasy Primates: Ecology, Physiology and Behavior.* Academic Press, New York. pp. 17–116.

Pages-Feuillade, E. (1988). Spatial distribution and inter-individual relationships in a nocturnal Malagasy lemur, *Microcebus* murinus. *Folia Primatol.* 50:204–220.

Pagès-Feuillade, E. (1989). Modalités de l'occupation de l'espace et interindividuelles chez un prosimien nocturne malagache (*Microcebus murinus*). *Folia Primatol.* 50:204–220.

Palacios, E., and Rodriguez, A. (2001). Ranging pattern and use of space in a group of red howler monkeys (*Alouatta seniculus*) in a southeastern Colombian rainforest. *Am. J. Primatol.* 55(4):233–51.

Palagi, E., Paoli, T. and Borgognini Tarli, S. (2004). Reconciliation and consolation in captive bonobos (*Pan paniscus*). *Am. J. Primatol.* 62:15–30.

Palagi, E., Paoli, T., and Borgognini Tarli, S. M. (2005). Aggression and reconciliation in two captive groups of *Lemur catta*. *Int. J. Primatol.* 26:279–294.

Palagi, E., Cordoni, G., and Borgognini Tarli, S. (2006). Possible roles of consolation in captive chimpanzees (*Pan troglodytes*). *Am. J. Phys. Anthropol.* 129:105–111.

Palagi, E., Antonacci, D., and Norscia, I. (2008). Peacemaking on treetops: first evidence of reconciliation from a wild prosimian (*Propithecus verreauxi*). *Anim. Behav.* 76:737–747.

Pallas, P. S. (1778). *Novae Species quad e Glirium Ordinae cum Illustrationibus Variis Complurium ex hoc Ordinae Animalium*. W. Walther, Erlangen.

Palombit, R. A. (1992). Pair bonds and monogamy in wild siamang (*Hylobates syndactylus*) and white-handed gibbon (*Hylobates lar*) in northern Sumatra [PhD diss.]. University of California, Davis.

Palombit, R. A. (1993). Lethal territorial aggression in a white-handed gibbon. *Am. J. Primatol.* 31:311–318.

Palombit, R. A. (1994a). Dynamic pair bonds in hylobatids: implications regarding monogamous social systems. *Behaviour* 128:65–101.

Palombit, R. A. (1994b). Extra-pair copulations in a monogamous ape. *Anim. Behav.* 47:721–723.

Palombit, R. A. (1995). Longitudinal patterns of reproduction in wild female siamang (*Hylobates syndactylus*) and white-handed gibbons (*Hylobates lar*). *Int. J. Primatol.* 16:739–760.

Palombit, R. A. (1996). Pair bonds in monogamous apes: a comparison of the siamang *Hylobates syndactylus* and the white-handed gibbon *Hylobates lar*. *Behaviour* 133:321–356.

Palombit, R. A. (1997). Inter- and intraspecific variation in the diets of sympatric siamang (*Hylobates syndactylus*) and lar gibbons (*Hylobates lar*). *Folia Primatol.* 68:321–337.

Palombit, R. A. (1999). Infanticide and the evolution of pair bonds in nonhuman primates. *Evol. Anthropol.* 7:117–129.

Palombit, R. A. (2000). Infanticide and the evolution of male–female bonds in animals. In: van Schaik, C. P., and Janson, C. H. (eds.), *Infanticide by Males and Its Implications*. Cambridge University Press, Cambridge. pp. 239–268.

Palombit, R. A. (2003). Male infanticide in wild savanna baboons: adaptive significance and intraspecific variation. In: Jones, C. B. (ed.), *Sexual Selection and Reproductive Competition in Primates: New Perspectives and Directions*. American Society of Primatologists, Norman, OK. pp. 3–47.

Palombit, R. A. (in press). The olive baboon (*Papio anubis*). In: Kingdon, J., Happold, D., and Butynski, T. (eds.), *The Mammals of Africa*. University of California Press, Berkeley, CA.

Palombit, R. A., Seyfarth, R. M., and Cheney, D. L. (1997). The adaptive value of "friendships" to female baboons: experimental and observational evidence. *Anim. Behav.* 54:599–614.

Palombit, R. A., Cheney, D. L., Fischer, J., Johnson, S., Rendall, D., Seyfarth, R. W., and Silk, J. B. (2000). Male infanticide and defense of infants in chacma baboons. In: van Schaik, C. P., and Janson, C. H. (eds.), *Infanticide by Males and Its Implications*, Cambridge University Press, Cambridge. pp. 123–152.

Panger, M. A. (1998). Object-use in free-ranging white-faced capuchins (*Cebus capucinus*) in Costa Rica. *Am. J. Phys. Anthropol.* 106:311–321.

Panger, M. (2007). Tool use and cognition in primates. In: Campbell, C. J., Fuentes, A., MacKinnon, K. C., Panger, M. A., and Beader S. K. (eds.), *Primates in Perspective*, 1st ed. Oxford University Press, New York. pp. 665–677.

Panger, M. A., Brooks, A., Richmond, B. G., and Wood, B. (2002a). Older than the Oldowan? Rethinking the emergence of hominin tool use. *Evol. Anthropol.* 11:235–245.

Panger, M. A., Perry, S., Rose, L., Gros-Louis, J., Vogel, E., MacKinnon, K. L., and Baker, M. (2002b). Cross-site differences in foraging behavior of white-faced capuchins (*Cebus capucinus*). *Am. J. Phys. Anthropol.* 119:52–66.

Paoli, G. D., and Curran, L. M. (2007). Soil nutrients limit fine litter production and tree growth in mature lowland forest of southwestern Borneo. *Ecosystems* 10(3):503–518.

Papworth, S., Boese, A. S., Barker, J., Schel, A. M., and Zuberbuehler, K. (2008). Male blue monkeys alarm call in response to danger experienced by others. *Biol. Lett.* 4(5):472–475.

Parga, J. A. (2003). Copulatory plug displacement evidences sperm competition in *Lemur catta*. *Int. J. Primatol.* 24:889–899.

Parga, J. A. (2006). Male mate choice in *Lemur catta*. *Int. J. Primatol.* 27:107–131.

Pariente, G. (1979). The role of vision in prosimian behaviour. In: Doyle, G. A., and Martin, R. D. (eds.), *The Study of Prosimian Behaviour*. Academic Press, London. pp. 411–459.

Parish, A. R. (1994). Sex and food control in the uncommon chimpanzee: how bonobo females overcome a phylogenetic legacy of male-dominance. *Ethol. Sociobiol.* 15:157–179.

Parish, A. R. (1996). Female relationships in bonobos (*Pan paniscus*)—evidence for bonding, cooperation, and female dominance in a male-philopatric species. *Hum. Nat.* 7:61–96.

Parish, A. R. (1998). Reply to Stanford (1998) "The social behavior of chimpanzees and bonobos—empirical evidence and shifting assumptions." *Curr. Anthropol.* 39:413–414.

Parish, T. J., and Nekaris, K. A. I. (2009). Identifying the CITES Appendix I–listed slow loris: a training programme for enforcement officials and rescue centres in Southeast Asia. *Endang. Sp. Res.* In press.

Parker, G. A. (1970). Sperm competition and its evolutionary consequences in insects. *Biol. Rev. Cambr. Philos. Soc.* 45:525–567.

Parker, G. A. (1979). Sexual selection and sexual conflict. In: Blum, M. S., and Blum, N. A. (eds.), *Sexual Selection and Reproductive Competition in Insects*. Academic Press, New York. pp. 123–166.

Parker, S. T. (2004). The cognitive complexity of social organization and socialization in wild baboons and chimpanzees: guided participation, socializing interactions, and event representation. In: Russon, A. E., and Begun, D. R. (eds.), *The Evolution of Thought: Evolutionary Origins of Great Ape Intelligence*. Cambridge University Press, Cambridge. pp. 45–60.

Parker, S. T., and Gibson, K. (1977). Object manipulation, tool use and sensorimotor intelligence as feeding adaptations in *Cebus* monkeys and great apes. *J. Hum. Evol.* 6:623–641.

Parker, S. T., and McKinney, M. L. (1999). *Origins of Intelligence: The Evolution of Cognitive Development in Monkeys, Apes, and Humans*. Johns Hopkins University Press, Baltimore, MD.

Parker, S. T., and Mitchell, R. W. (1994). Evolving self-awareness. In: Parker, S. T., Mitchell, R. W., and Boccia, M. L. (eds.), *Self-Awareness in Animals and Humans: Developmental Perspectives*. Cambridge University Press, Cambridge. pp. 413–428.

Parker, S. T., Kerr, M., Markowitz, H., and Gould, J. (1999). A survey of tool use in zoo gorillas. In: Parker, S. T., Mitchell, R. W., and Miles, H. L. (eds.), *The Mentalities of Gorillas and Orangutans*. Cambridge University Press, Cambridge. pp. 188–193.

Parker, S. T., Langer, J., and McKinney, M. C. (eds.) (2000). *Biology, Brains and Behavior: The Evolution of Human Development*. SAR Press, Santa Fe.

Parker, S. T., Langer, J., and Milbrath, C. (eds.) (2005). *Biology and Knowledge Revisited: From Neurogenesis to Psychogenesis*. Lawrence Erlbaum, Mahwah, NJ.

Parnell, R. J. (2002a). Group size and structure in western lowland gorillas (*Gorilla gorilla gorilla*) at Mbeli Bai, Republic of Congo. *Am. J. Primatol.* 56:193–206.

Parnell, R. J. (2002b). The social structure and behaviour of western lowland gorillas (*Gorilla gorilla gorilla*) at Mbeli Bai, Republic of Congo [PhD diss.]. University of Stirling, Stirling.

Parolin, P. (2000). Seed mass in Amazonian floodplain forests with contrasting nutrient supplies. *J. Trop. Ecol.* 16:417–428.

Parolin, P. (2001). Seed germination and early establishment of 12 tree species from nutrient-rich and nutrient-poor central Amazonian floodplains. *Aquat. Bot.* 70:89–103.

Parr, L. A. (2004). Perceptual biases for multimodal cues in chimpanzee (*Pan troglodytes*) affect recognition. *Anim. Cogn.* 7:171–178.

Parr, L. A., and Waller, B. (2007). The evolution of human emotion. In: Kaas, J. H., and Preuss, T. M. (eds.), *Evolution of Nervous Systems: A Comprehensive Reference. Primates*, vol. 4. Elsevier Academic Press, Boston. pp. 447–472.

Parra, R. (1978). Comparison of foregut and hindgut fermentation in herbivores. In: Montgomery, G. G. (ed.), *The Ecology of Arboreal Folivores*. Smithsonian Institution Press, Washington DC. pp. 205–229.

Partan, S. R. (2002). Single and multichannel signal composition: facial expressions and vocalizations of rhesus macques (*Macaca mulatta*). *Behaviour* 139:993–1027.

Partan, S., and Marler, P. (1999). Communication goes multimodal. *Science* 283:1272–1273.

Passamani, M. (1998). Activity budget of Geoffroy's marmoset (*Callithrix geoffroyi*) in an Atlantic Forest in southeastern Brazil. *Am. J. Primatol.* 46:333–340.

Passamani, M., and Rylands A. B. (2000). Feeding behavior of Geoffroy's marmoset (*Callithrix geoffroyi*) in an Atlantic Forest fragment of southeastern Brazil. *Primates* 41:27–38.

Passingham, R. E. (1975). The brain and intelligence. *Brain Behav. Evol.* 11:1–15.

Passos, F. C. (1994). Behavior of the black lion tamarin, *Leontopithecus chrysopygus*, in different forest levels in the Caetetus Ecological Station, São Paulo, Brazil. *Neotrop. Primates* 2(suppl.):40–41.

Passos, F. C. (1997). Padrão de atividades, dieta e uso do espaço em um grupo de mico-leão-preto (*Leontopithecus chrysopygus*) na Estação Ecológica do Caetetus, SP [PhD diss.]. Univerdade Federal de São Carlos, São Carlos.

Passos, F. C. (1998). Ecology of the black lion tamarin. *Neotrop. Primates* 6:128–129.

Passos, F. C. (1999). Dieta de um grupo de mico-leão-preto, *Leontopithecus chrsyopygus* (Mikan) (Mammalia, Callitrichidae), na Estação Ecológica do Caetetus, São Paolo. *Rev. Brasil. Zool.* 16(suppl. 2):269–278.

Pastinen, T., Raitio, M., Lindroos, K., Tainola, P., Peltonen, L., and Syvänen, A.-C. (2000). A system for specific, high-throughput genotyping by allele-specific primer extension on microarrays. *Genome Res.* 10:1031–1042.

Pastorini, J., Martin, R. D., Ehresmann, P., Zimmermann, E., and Forstner, M. R. J. (2001). Molecular phylogeny of the lemur family Cheirogaleidae (primates) based on mitochondrial DNA sequences. *Mol. Phylogenet. Evol.* 19:45–56.

Pastorini, J., Thalmann, U., and Martin, R. D. (2003). A molecular approach to comparative phylogeography of extant Malagasy lemurs. *Proc. Natl. Acad. Sci. USA* 100:5879–5884.

Pastor-Nieto, R. (2001). Grooming, kinship and co-feeding in captive spider monkeys (*Ateles geoffroyi*). *Zoo Biol.* 20:293–303.

Patel, E. R. (2006). Activity budget, ranging, and group size in silky sifakas (*Propithecus candidus*). *Lemur News* 11:42–45.

Paterson, H. E. H. (1985). The recognition concept of species. In: Vbra, E. S. (ed.), *Species and Speciation.* Transvaal Museum, Pretoria. pp. 21–29.

Paterson, J. (2001). *Primate Behavior: An Exercise Workbook,* 2nd ed. Waveland Press, Prospect Heights, IL.

Paterson, J. (2006). Aspects of diet, foraging, and seed predation in Ugandan forest baboons. In: *Primates of Western Uganda.* Springer, New York. p. 75.

Paterson, J. D., and Wallis, J. (eds.) (2005). *Commensalism and Conflict: The Human–Primate Interface.* American Society of Primatologists, Norman, OK.

Patino, E., Borda, J. T., and Ruiz, J. C. (1996). Sexual maturity and seasonal reproduction in captive *Cebus apella. Lab. Primate Newslett.* 35:8–10.

Patterson, F. G., and Cohn, R. H. (1994). Self-recognition and self-awareness in lowland gorillas. In: Parker, S. T., Mitchell, R. W., and Boccia, M. L. (eds.), *Self-Awareness in Animals and Humans: Developmental Perspectives.* Cambridge University Press, Cambridge. pp. 273–290.

Paul, A. (1997). Breeding seasonality affects the association between dominance and reproductive success in non-human male primates. *Folia Primatol.* 68:344–349.

Paul, A. (2002). Sexual selection and mate choice. *Int. J. Primatol.* 23:877–904.

Paul, A. (2004). Dominance and paternity. In: Thierry, B., Singh, M., and Kaumanns, W. (eds.), *Macaque Societies: A Model for the Study of Social Organization.* Cambridge University Press, Cambridge. pp. 131–134.

Paul, A. (2006). Kinship and behavior in Barbary macaques. In: Hodges, J. K., and Cortes, J. (eds.), *The Barbary Macaque: Biology, Management and Conservation.* Nottingham University Press, Nottingham. pp. 47–61.

Paul, A., and Kuester, J. (1985). Intergroup transfer and incest avoidance in semifree-ranging barbary macaques (*Macaca sylvanus*) at Salem (FRG). *Am. J. Primatol.* 8:317–322.

Paul, A., and Kuester, J. (1987). Dominance, kinship and reproductive value in female Barbary macaques (*Macaca sylvanus*) at Affenberg Salem (FRG). *Behav. Ecol. Sociobiol.* 21:323–331.

Paul, A., and Kuester, J. (1996a). Differential reproduction in male and female Barbary macaques. In: Fa, J. E., and Lindburg, D. G. (eds.), *Evolution and Ecology of Macaque Societies.* Cambridge University Press, Cambridge, pp. 293–317.

Paul, A., and Kuester, J. (1996b). Infant handling by female Barbary macaques (*Macaca sylvanus*) at Affenberg Salem: testing functional and evolutionary hypotheses. *Behav. Ecol. Sociobiol.* 39:133–145.

Paul, A., and Kuester, J. (2004). The impact of kinship on mating and reproduction. In: Chapais, B., and Berman, C. M. (eds.), *Kinship and Behavior in Primates.* Oxford University Press, Oxford. pp. 271–291.

Paul, A., Kuester, J., and Podzuweit, D. (1993a). Reproductive senescence and terminal investment in female Barbary macaques (*Macaca sylvanus*) at Salem. *Int. J. Primatol.* 14:105–124.

Paul, A., Kuester, J., Timme, A., and Arnemann, J. (1993b). The association between rank, mating effort, and reproductive success in male Barbary macaques (*Macaca sylvanus*). *Primates* 34:491–502.

Paul, A., Kuester, J., and Arnemann, J. (1996). The sociobiology of male–infant interactions in Barbary macaques, *Macaca sylvanus. Anim. Behav.* 51:155–170.

Paul, A., Preuschoft, S., and van Schaik, C. P. (2000). The other side of the coin: infanticide and the evolution of affiliative male–infant interactions in Old World primates. In: van Schaik, C. P., and Janson, C. H. (eds.), *Infanticide by Males and Its Implications.* Cambridge University Press, Cambridge. pp. 269–292.

Paul, J. R. (2001). Patterns of seed dispersal by animals: influence on sapling composition in a tropical forest [PhD diss.]. University of Florida, Gainesville.

Pavelka, M. S. M. (1991). Sociability in old female Japanese monkeys, human versus nonhuman primate aging. *Am. Anthropol.* 93:588–598.

Pavelka, M. S. M. (1993). *Monkeys of the Mesquite: The Social Life of the South Texas Snow Monkey.* Kendall/Hunt Publishing, Dubuque.

Pavelka, M. S. M., and Fedigan, L. M. (1991). Menopause: a comparative life history perspective. *Ybk. Phys. Anthropol.* 34:13–38.

Pavelka, M. S. M., and Fedigan, L. M. (1999). Reproductive termination in female Japanese monkeys: a comparative life history perspective. *Am. J. Phys. Anthropol.* 109:455–464.

Pavelka, M. S. M., Fedigan, L. M., and Zohar, S. (2002). Availability and adaptive value of reproductive and postreproductive Japanese macaque mothers and grandmothers. *Anim. Behav.* 64:407–414.

Pavelka, M. S. M., Brusselers, O. T., Nowak, D., and Behie, A. M. (2003). Population reduction and social disorganization in *Alouatta pigra* following a hurricane. *Int. J. Primatol.* 24:1037– 1055.

Payne, H. F. P., Lawes, M. J., and Henzi, S. P. (2003). Fatal attack on an adult female *Cercopithecus mitis erythrarchus*: implications for female dispersal in female-bonded societies. *Int. J. Primatol.* 24:1245–1250.

Pazol, K. (2003). Mating in the Kakamega Forest blue monkeys (*Cercopithecus mitis*): does female sexual behavior function to manipulate paternity assessment? *Behaviour* 140:473–499.

Peakall, R., and Smouse, P. E. (2006). GENALEX 6: genetic analysis in Excel. Population genetic software for teaching and research. *Mol. Ecol. Notes* 6:288–295.

Pearse, D. E., Eckerman, C. M., Janzen, F. J., and Avise, J. C. (2001). A genetic analogue of "mark–recapture" methods for estimating population size: an approach based on molecular parentage assessments. *Mol. Ecol.* 10:2711–2718.

Peccei, J. S. (2001). Menopause: adaptation or epiphenomenon? *Evol. Anthropol.* 10:43–57.

Peedicayil, J. (2006). Chimpanzee epigenome project? *Med. Hypotheses* 66:1250–1251.

Peeters, M., Courgnaud, V., Abela, B., Auzel, P., Pourrut, X., Bibollet-Ruche, F., Loul, S., Liegeois, F., Butel, C., Koulagna, D., Mpoudi-Ngole, E., Shaw, G. M., Hahn, B. H., and Delaporte, E. (2002). Risk to human health from a plethora of simian immunodeficiency viruses in primate bushmeat. *Emerg. Infect. Dis.* 8:451–457.

Peetz, A. (2001). Ecology and social organization of the bearded saki *Chiropotes satanas chiropotes* (Primates: Pitheciinae) in Venezuela. *Ecotrop. Monogr.* 1:1–170.

Peetz, A., Norconk, M. A., and Kinzey, W. G. (1992). Predation by jaguar on howler monkyes (*Alouatta seniculus*) in Venezuela. *Am. J. Primatol.* 28:223–228.

Pei, W. C. (1935). Fossil mammals from the Kwangsi caves. *Bull. Geol. Soc. China* 14:413–425.

Peláez, F. G., Gil-Burmann, C., and Sánchez, S. (2000). Feeding tactics in juvenile rhesus monkeys (*Macaca mulatta*). *Am. J. Primatol.* 51:147–152.

Pentikainen, J. (1997). Epic. In: Green, T. A. (ed.), *Folklore: An Encyclopedia of Beliefs, Customs, Tales, Music and Art,* vol. 1. ACB-Clio, Santa Barbara. pp. 222–225.

Pepper, J., Mitani, J., and Watts, D. (1999). General gregariousness and specific social preferences among wild chimpanzees. *Int. J. Primatol.* 20:613–632.

Pereira, M. E. (1983). Abortion following the immigration of an adult male baboon (*Papio cynocephalus*). *Am. J. Primatol.* 4:93–98.

Pereira, M. E. (1984). Age changes and sex differences in the social behavior of juvenile yellow baboons (*Papio cynocephalus*) [PhD diss.]. University of Chicago, Chicago.

Pereira, M. E. (1986). Support during agonistic interactions given and received by free-ranging juvenile male and female baboons. *Am. J. Primatol.* 10:421–422.

Pereira, M. E. (1988a). Agonistic interactions of juvenile savannah baboons I. fundamental features. *Ethology* 79:195–217.

Pereira, M. E. (1988b). Effects of age and sex on intra-group spacing behaviour in juvenile savannah baboons, *Papio cynocephalus cynocephalus*. *Anim. Behav.* 36:184–204.

Pereira, M. E. (1989). Agonistic interactions of juvenile savanna baboons. II. agonistic support and rank acquisition. *Ethol.* 80:152–171.

Pereira, M. E. (1993a). Agonistic interactions, dominance relations and ontogenetic trajectories in ringtailed lemurs. In: Pereira, M. E., and Fairbanks, L. A. (eds.), *Juvenile Primates: Life History, Development and Behavior.* Oxford University Press, New York. pp. 285–305.

Pereira, M. E. (1993b). Seasonal adjustment of growth rate and adult body weight in ringtailed lemurs. In: Kappeler, P. M., and Ganzhorn, J. U. (eds.), *Lemur Social Systems and Their Ecological Basis.* Plenum Press, New York. pp. 205–221.

Pereira, M. E., and Altmann, J. (1985). Development of social behavior of free-living nonhuman primates. In: Watts, E. S. (ed.), *Nonhuman Primate Models for Human Growth and Development.* Alan R. Liss, New York. pp. 217–309.

Pereira, M., and Fairbanks, L. (eds.) (1993a). *Juvenile Primates.* Oxford University Press, Oxford.

Pereira, M., and Fairbanks, L. (1993b). What are juvenile primates all about? In: Pereira, M., and Fairbanks, L. (eds.), *Juvenile Primates.* Oxford University Press, New York. pp. 3–12.

Pereira, M. E., and Fairbanks, L. A. (2002). Forward. In: Pereira, M. E., and Fairbanks, L. A. (eds.), *Juvenile Primates: Life History, Development, and Behavior,* 2nd ed. New York: Oxford University Press, pp. vii–x.

Pereira, M. E., and Leigh, S. R. (2003). Modes of primate development. In: Kappeler, P. M., and Pereira, M. E. (eds.), *Primate Life Histories and Socioecology.* University of Chicago Press, Chicago. pp. 149–176.

Pereira, M. E., and McGlynn, C. A. (1997). Special relationships instead of female dominance for redfronted lemurs, *Eulemur fulvus rufus*. *Am. J. Primatol.* 43:239–258.

Pereira, M. E., and Weiss, M. L. (1991). Female mate choice, male migration, and the threat of infanticide in ringtailed lemurs. *Behav. Ecol. Sociobiol.* 28:141–152.

Pereira, M. E., Klepper, A., and Simons, E. L. (1987). Tactics of care for young infants by forest-living ruffed lemurs (*Varecia variegata variegata*): ground nests, parking, and biparental guarding. *Am. J. Primatol.* 13:129–144.

Pereira, M. E., Kaufman, R., Kappeler, P. M., and Overdorff, D. J. (1990). Female dominance does not characterize all of the Lemuridae. *Folia Primatol.* 55:96–103.

Pereira, M. E., Schill, J. L., and Charles, E. P. (2000). Reconciliation in captive Guyanese squirrel monkeys (*Saimiri sciureus*). *Am. J. Primatol.* 50:159–167.

Peres, C. A. (1986). Costs and benefits of territorial defense in golden lion tamairns (*Leontopithecus rosalia*) [MA thesis]. University of Florida, Gainesville.

Peres, C. A. (1989). Costs and benefits of territorial defense in wild golden lion tamarins, *Leontopithecus rosalia*. *Behav. Ecol. Sociobiol.* 25:227–233.

Peres, C. A. (1990a). A harpy eagle successfully captures an adult male red howler monkey. *Wilson Bull.* 102:560–561.

Peres, C. A. (1990b). Effects of hunting on western Amazonian primate communities. *Biol. Conserv.* 54:47–59.

Peres, C. A. (1991a). Ecology of mixed-species groups of tamarins in Amazonian terra firme forests [PhD diss.]. Cambridge University, Cambridge.

Peres, C. A. (1991b). Humboldt's woolly monkeys decimated by hunting in Amazonia. *Oryx* 25:89–95.

Peres, C. A. (1992). Prey-capture benefits in a mixed-species group of Amazonian tamarins, *Saguinus fuscicollis* and *S. mystax*. *Behav. Ecol. Sociobiol.* 31:339–347.

Peres, C. A. (1993a). Anti-predation benefits in a mixed-species group of Amazonian tamarins. *Folia Primatol.* 61:61–76.

Peres, C. A. (1993b). Notes on the ecology of buffy saki monkeys (*Pithecia albicans*, Gray 1860): a canopy seed-predator. *Am. J. Primatol.* 31:129–140.

Peres, C. (1993c). Notes on the primates of the Juruá River, western Brazilian Amazonia. *Folia Primatol.* 61:97–103.

Peres, C. A. (1994a). Diet and feeding ecology of gray woolly monkeys (*Lagothrix lagotricha cana*) in central Amazonia: comparisons with other atelines. *Int. J. Primatol.* 15:333–372.

Peres, C. A. (1994b). Which are the largest New World monkeys? *J. Hum. Evol.* 26:245–249.

Peres, C. A. (1996). Use of space, spatial group structure, and foraging group size of gray woolly monkeys (*Lagothrix lagotricha cana*) at Urucu, Brazil. In: Norconk, M. A., Rosenberger, A. L., and Garber, P. A. (eds.), *Adaptive Radiations of Neotropical Primates.* Plenum Press, New York. pp. 467–488.

Peres, C. A. (1999). Effects of subsistence hunting and forest types on the structure of Amazonian primate communities. In: Fleagle, J. G., Janson, C. H., and Reed, K. E. (eds.), *Primate Communities.* Cambridge University Press, Cambridge. pp. 268–283.

Peres, C. A. (2000a). Effects of subsistence hunting on vertebrate community structure in Amazonian forests. *Conserv. Biol.* 14:240–253.

Peres, C. A. (2000b). Territorial defense and the ecology of group movements in small-bodied neotropical primates. In: Boinski, S., and Garber, P. A. (eds.), *On the Move: How and Why Animals Travel in Groups.* University of Chicago Press, Chicago. pp. 100–123.

Peres, C. A. (2001). Synergistic effects of subsistence hunting and habitat fragmentation on Amazonian forest vertebrates. *Conserv. Biol.* 15:1490–1505.

Peres, C. A., and Dolman, P. M. (2000). Density compensation in neotropical primate communities: evidence from 56 hunted and nonhunted Amazonian forests of varying productivity. *Oecologia* 122:175–189.

Peres, C. A., and Lake, I. R. (2003). Extent of nontimber resource extraction in tropical forests: accessibility to game vertebrates by hunters in the Amazon Basin. *Conserv. Biol.* 17:521–535.

Peres, C. A., and Palacios, E. (2007) Basin-wide effects of game harvest on vertebrate population densities in Amazonian forests: implications for animal-mediated seed dispersal. *Biotropica* 39:304–315.

Perez, L., Czekala, N. M., Weisenseel, A., and Lasley, B. L. (1988). Excretion of radiolabeled estradiol metabolites in the slow loris (*Nycticebus coucang*). *Am. J. Primatol.* 16:321–330.

Perez-Ruiz, A. L., and Mondragon-Ceballos, R. (1994). Rates of reconciliatory behaviors in stumptail macaques: effects of age, sex, rank and kinship. In: Roeder, J. J., Thierry, B., Anderson, J. R., and Herrenschmidt, N. (eds.), *Current Primatology. Social Development, Learning and Behavior,* vol. 2. Presses de l'Université Louis Pasteur, Strasbourg. pp. 147–155.

Perkin, A. W. (2000). Bushbabies of Tanzania: an update. *Miombo* 20:4.

Perkin, A. W. (2001a). A field study on the conservation status and species diversity of galagos in the West Kilombero Scarp Forest Reserve. In: Doody, K. Z., Howell, K. M., and Fanning, E. (eds.), *West Kilombero Scarp Forest Reserve—Zoological Report.* Matumizi Endeleyu Mazingira, Iringa, Tanzania. pp. 149–159.

Perkin, A. W. (2001b). The taxonomic status and distribution of bushbabies (galagos) in the Uluguru Mountains, Tanzania. *Miombo* 23:5–7.

Perkin, A. W. (2002). The Rondo galago *Galagoides rondoensis* (Honess & Bearder, 1996): a primate conservation priority. *Primate Eye* 77:14–15.

Perkin, A. W. (2007). Comparative penile morphology of East African galagos of the genus *Galagoides* (family Galagidae): implications for taxonomy. *Am. J. Primatol.* 69:16–26.

Perkin, A. W., and Butynski, T. (2002–2003). A note on the Somali galago *Galago gallarum*. *Afr. Primates* 6:48–49.

Perkin, A. W., Bearder, S. K., Butynski, T., Bytebier, B., and Agwanda, B. (2002). The Taita Mountain dwarf galago Galagoides sp.: a new primate for Kenya. *J. East Afr. Nat. Hist.* 91:1–13.

Perloe, S. I. (1992). Male mating competition, female choice and dominance in a free ranging group of Japanese macaques. *Primates* 33:289–304.

Perret, M. (1986). Social influences on oestrous cycle length and plasma progesterone concentrations in the female lesser mouse lemur (*Microcebus murinus*). *J. Reprod. Fertil.* 77:303–311.

Perry, G. H., Tchinda, J., McGrath, S. D., Zhang, J. Z., Picker, S. R., Cáceres, A. M., Iafrate, A. J., Tyler-Smith, C., Scherer, S. W., and Eichler, E. E. (2006). Hotspots for copy number variation in chimpanzees and humans. *Proc. Natl. Acad. Sci. USA* 103:8006–8011.

Perry, G. H., Dominy, N. J., Claw, K. G., Lee, A. S., Fiegler, H., Redon, R., Werner, J., Villanea, F. A., Mountain, J. L., Misra, R., Carter, N. P., Lee, C., and Stone, A. C. (2007a). Diet and the evolution of human amylase gene copy number variation. *Nat. Genet.* 39(10):1255–1260.

Perry, G. H., Martin, R. D., and Verrelli, B. C. (2007b). Signatures of functional constraint at aye-aye opsin genes: the potential of adaptive color vision in a nocturnal primate. *Mol. Biol. Evol.* 24:1963–1970.

Perry, G. H, Yang, F., Marquès-Bonet, T., Murphy, C., Fitzgerald, T., Lee, A. S., Hyland, C., Stone, A. C., Hurles, M. E., Tyler-Smith, C., Eichler, E. E., Carter, N. P., Lee, C., and Redon, R. (2008). Copy number variation and evolution in humans and chimpanzees. *Genome Res.* 18:1698–1710.

Perry, S. (1996). Female–female social relationships in wild white-faced capuchin monkeys, *Cebus capucinus. Am. J. Primatol.* 40:167–182.

Perry, S. (1997). Male–female social relationships in wild white-faced capuchins (*Cebus capucinus*). *Behaviour* 134:477–510.

Perry, S. (1998a). A case report of a male rank reversal in a group of wild white-faced capuchins (*Cebus capucinus*). *Primates* 39:51–70.

Perry, S. (1998b). Male–male social relationships in wild white-faced capuchins, *Cebus capucinus. Behaviour* 135:139–172.

Perry, S. (2006). What cultural primatology can tell anthropologists about the evolution of culture. *Annu. Rev. Anthropol.* 35:171–190.

Perry, S., and Manson, J. H. (1995). A comparison of the mating behavior of adolescent and adult female rhesus macaques (*Macaca mulatta*). *Primates* 36:27–39.

Perry, S., and Manson, J. (2003). Traditions in monkeys. *Evol. Anthropol.* 12:71–81.

Perry, S., and Rose, L. M. (1994). Begging and transfer of coati meat by white-faced capuchin monkeys, *Cebus capucinus. Primates* 35:409–415.

Perry, S., Baker, M., Fedigan, L. M., Gros-Louis, J., Jack, K., MacKinnon, K. C., Manson, J. H., Panger, M., Pyle, K., and Rose, L. M. (2003). Social conventions in wild white-faced capuchins: evidence for traditions in a neotropical primate. *Curr. Anthropol.* 44:241–268.

Perry, S., Manson, J. H., Muniz, L., Gros-Louis, J., and Vigilant, L. (2008). Kin-biased social behaviour in wild adult female white-faced capuchins, *Cebus capucinus. Anim. Behav.* 76:187–199.

Persons, M. H., and Uetz, G. W. (1999). Age and sex-based differences in the use of prey sensory cues in wolf spiders (Araneae: Lycosidae). *J. Insect Behav.* 12:723–736.

Peters, A., and Preuschoft, H. (1984). External biomechanics of leaping in *Tarsius* and its morphological and kinematic consequences. In: Niemitz, C. (ed.), *Biology of Tarsiers.* Gustav Fischer Verlag, New York. pp. 227–255.

Peters, C., and O'Brien, E. (1981). The early hominid plant-food niche: insights from an analysis of plant exploitation by *Homo, Pan,* and *Papio* in eastern and southern Africa. *Curr. Anthropol.* 22:127–140.

Petit, O., and Thierry, B. (1992). Affiliative function of the silent bared-teeth display in moor macaques (*Macaca maurus*): further evidence for the particular status of Sulawesi macaques. *Int. J. Primatol.* 13:97–105.

Petit, O., and Thierry, B. (1994a). Reconciliation in a group of black macaques. *Dodo J. Wildl. Preserv. Trusts* 30:89–95.

Petit, O., and Thierry, B. (1994b). Reconciliation in a group of Guinea baboons. In: Roeder, J. J., Thierry, B., Anderson, J. R., and Herrenschmidt, N. (eds.), *Current Primatology. Social Development, Learning and Behavior,* vol. 2. Presses de l'Université Louis Pasteur, Strasbourg. pp. 137–145.

Petit, O., Desportes, C., and Thierry, B. (1992). Differential probability of "coproduction" in two species of macaque (*Macaca tonkeana, M. mulatta*). *Ethology* 90:107–120.

Petit, O., Abegg, C., and Thierry, B. (1997). A comparative study of aggression and conciliation in three cercopithecine monkeys (*Macaca fuscata, Macaca nigra, Papio papio*). *Behaviour* 134:415–432.

Petit, O., Bertrand, F., and Thierry, B. (2008). Social play in crested and Japanese macaques: testing the covariation hypothesis. *Dev. Psychobiol.* 50:399–408.

Petiver, J. (1705). *Gazophylacii Naturae et Artis.* (1702–1706)London. Smith & Bateman.

Petrie, M. (1994). Improved growth and survival of offspring of peacocks with more elaborate trains. *Nature* 371:598–599.

Petter, J. J. (1962). Recherches dur l'ecologie et l'ethologie des lemuriens malagashes. *Mém. Mus. Natl. Hist. Nat. Paris* 27:1–146.

Petter, J. J. (1978). Ecological and physiological adaptations of five sympatric nocturnal lemurs to seasonal variation in food production. In: Chivers, D. J., and Herbert, J. (eds.), *Recent Advances in Primatology. Behaviour,* vol. 1. Academic Press, London. pp. 211–223.

Petter, J. J., and Charles-Dominique, P. (1979). Vocal communication in prosimians. In: Doyle, G. A., and Martin, R. D. (eds.), *The Study of Prosimian Behavior.* Academic Press, New York. pp. 247–305.

Petter, J. J., and Hladik, C. M. (1970). Observations sur le domaine vital et la densité de population de *Loris tardigradus* dans les forêts de Ceylan. *Mammalia* 34:394–409.

Petter, J. J., and Petter-Rousseaux, A. (1959). Contribution to the study of the aye-aye. *Nat. Malagache* 11:153–164.

Petter, J. J., and Petter-Rousseaux, A. (1979). Classification of the prosimians. In: Doyle, G. A., and Martin, R. D. (eds.), *The Study of Prosimian Behavior.* Academic Press, London. pp. 1–44.

Petter, J. J., and Peyriéras, A. (1970a). Nouvelle contribution a l'étude d'un lémurien malagache, le aye-aye (*Daubentonia madagascariensis* E. Geoffroy). *Mammalia* 34:167–193.

Petter, J. J., and Peyriéras, A. (1970b). Observation écoétholgiques sur les lemuriens malagaches du genre *Hapalemur. Terre Vie* 24:356–382.

Petter, J. J., Schilling, A., and Pariente, G. (1971). Observations écoéthologiques sur deux lémuriens malagaches nocturnes: *Phaner furcifer* et *Microcebus coquereli. Terre Vie* 25:287–327.

Petter, J. J., Schilling, A., and Pariente, G. (1975). Observations on behavior and ecology of *Phaner furcifer.* In: Tattersall, I., and Sussman, R. W. (eds.), *Lemur Biology.* Plenum Press, New York. pp. 209–218.

Petter, J. J., Albignac, R., and Rumpler, Y. (1977). Lemurine mammals (Primates, Prosimians). *Faune Madagascar* 44:1–513.

Petter-Rousseaux, A. (1964). Reproductive physiology and behavior of the Lemuroidea. In: Buettner-Janusch, J. (ed.), *Evolutionary and Genetic Biology of Primates,* vol. 2. Academic Press, New York. pp. 91–132.

Petter-Rousseaux, A. (1980). Seasonal activity rhythms, reproduction, and body weight variations in five sympatric nocturnal prosimians, in simulated light and climatic conditions. In: Charles-Dominique, P., Cooper, H. M., Hladik, A., Hladik, C. M., Pages, E., Pariente, G. F., Petter-Rousseaux, A., Petter, J. J., and Schilling, A. (eds.), *Nocturnal Malagasy Primates: Ecology, Physiology and Behavior.* Academic Press, New York. pp. 137–152.

Phillips, K. (1985). Differing responses to a predator (*Eira barbara*) by *Alouatta* and *Cebus. Neotrop. Primates* 3:45–46.

Phillips, K. A. (1995). Resource patch size and flexible foraging in white-faced capuchins (*Cebus capucinus*). *Int. J. Primatol.* 16:509–519.

Phillips, K. (1998). Tool use in wild capuchin monkeys (*Cebus albifrons trinitatis*). *Am. J. Primatol.* 46:259–261.

Phillips, K. A., and Abercrombie, C. L. (2003). Distribution and conservation status of the primates of Trinidad. *Primate Conserv.* 19:19–22.

Phillips, K. A., and Newlon, K. (2000). Female–female social relationships in white-fronted capuchins (*Cebus albifrons*): testing hypotheses about resource size and quality. *Am. J. Primatol.* 51(suppl.):81.

Phillips, K. A., Elvey, C. R., and Abercrombie, C. L. (1998). Applying GPS to the study of primate ecology: a useful tool? *Am. J. Primatol.* 46:167–172.

Phillips-Conroy, J. E. (1996). Baboons, diet, and disease: food plant selection and schistosomiasis. In: Taub, D. M., and King, F. A. (eds.), *Current Perspectives in Primate Social Dynamics.* Van Nostrand Reinhold, New York. pp. 287–304.

Phillips-Conroy, J. E., and Jolly, C. J. (1981). Sexual dimorphism in two subspecies of Ethiopian baboons (*Papio hamadryas*) and their hybrids. *Am. J. Phys. Anthropol.* 56:115–129.

Phillips-Conroy, J. E., and Jolly, C. J. (1986). Changes in the structure of the baboon hybrid zone in the Awash National Park, Ethiopia. *Am. J. Phys. Anthropol.* 71:337–350.

Phillips-Conroy, J., and Jolly, C. J. (1988). Dental eruption schedules of wild and captive baboons. *Am. J. Primatol.* 15:17–29.

Phillips-Conroy, J. E., and Jolly, C. J. (2004). Male dispersal and philopatry in the Awash baboon hybrid zone. *Primate Rep.* 68:27–52.

Phillips-Conroy, J. E., and Knopf, P. M. (1986). The effects of ingesting plant hormones on schistosomiasis in mice: an experimental study. *Biochem. Syst. Ecol.* 14:637–645.

Phillips-Conroy, J. E., Jolly, C. J., and Brett, F. L. (1991). Characteristics of hamadryas-like male baboons living in anubis troops in the Awash Hybrid Zone, Ethiopia. *Am. J. Phys. Anthropol.* 86:353–368.

Phillips-Conroy, J. E., Jolly, C.J., Nystrom, P., and Hemmalin, H. A. (1992). Migration of male hamadryas baboons into anubis groups in the Awash National Park, Ethiopia. *Int. J. Primatol.* 13:455–476.

Phoebus, E. C. (1982). Primate female orgasm. *Am. J. Primatol.* 2:223–224.

Piaget, J. (1952). *The Origins of Intelligence in Children.* W. W. Norton, New York.

Piaget, J. (1954). *The Construction of Reality in the Child.* W. W. Norton, New York.

Pianka, E. R. (1970). On "r" and "K" selection. *Am. Nat.* 102:592–597.

Pieczarka, J. C., de Souza Barros, R. M., de Faria, F. M., and Nagamachi, C. Y. (1993). *Aotus* from the southwestern Amazon region is geographically and chromosomally intermediate between *A. azarae boliviensis* and *A. infulatus. Primates* 34:197–204.

Pielou, E. C. (1969). *An Introduction to Mathematical Ecology*. Wiley Interscience, New York.

Piercy, J. E., Embleton, T. F. W., and Sutherland, L. C. (1977). Review of noise propagation in the atmosphere. *J. Acoust. Soc. Am.* 61:143–148.

Piersma, T., and Drent, J. (2003). Phenotypic flexibility and the evolution of organismal design. *Trends Ecol. Evol.* 18:228–233.

Pieta, K. (2008). Female mate preferences among *Pan troglodytes schweinfurthii* of Kanyawara, Kibale National Park, Uganda. *Int. J. Primatol.* 29:845–864.

Piggott, M. P., Bellemain, E., Taberlet, P., and Taylor, A. C. (2004). A multiplex pre-amplification method that significantly improves microsatellite amplification and error rates for faecal DNA in limiting conditions. *Conserv. Genet.* 5:417–420.

Pilbeam, D. (1982). New hominoid skull material from the Miocene of Pakistan. *Nature* 295:232–234.

Pilbrow, V. (2006). Population systematics of chimpanzees using molar morphometrics. *J. Hum. Evol.* 51(6):646–662.

Pimley, E. R. (2002). The behavioural ecology and genetics of the potto (*Perodicticus potto edwardsi*) and Allen's bushbaby (*Galago alleni cameronensis*) [PhD diss.]. University of Cambridge, Cambridge.

Pimley, E. R., Bearder, S. K., and Dixson, A. F. (2005a). Home range analysis of *Perodicticus potto edwardsi* and *Sciurocheirus cameronensis*. *Int. J. Primatol.* 26:191–206.

Pimley, E. R., Bearder, S. K., and Dixson, A. F. (2005b). Social organization of the Milne-Edward's potto *Perodicticus potto edwardsi*. *Am. J. Primatol.* 66:317–330.

Pimley, E. R., Bearder, S. K., and Dixson, A. F. (2005). Examining the social organization of the Milne-Edward's potto *Perodicticus potto edwardsi*. *Am. J. Primatol.* Volume 66(4): 317–330.

Pinto, L. P. (2002a). Diet, activity, and home range of *Alouatta belzebul discolor* (Primates: Atelidae) in Paranaíta, northern Mato Grosso. *Neotrop. Primates* 10:98–99.

Pinto, L. P. (2002b). Dieta, padrão de atividades e área de vida de *Alouatta belzebul discolor* (Primates: Atelidae) em Paranaíta, norte do Mato Grosso [MSc thesis]. Universidade Estadual de Campinas (UNICAMP), São Paulo.

Pinto, L., and E. Setz. (2004). Diet of *Alouatta belzebul discolor* in an Amazonian rain forest of northern Mato Grosso State, Brazil. *Int. J. Primatol.*, 25: 1197–1211.

Pinto, L. P. S., Costa, C. M. R., Strier, K. B., and da Fonseca, G. A. B. (1993). Habitats, density, and group size of primates in the Reserva Biologica Augusto Ruschi (Nova Lombardia), Santa Teresa, Brazil. *Folia Primatol.* 61:135–143.

Piry, S., Alapetite, A., Cornuet, J.-M., Paetkau, D., Baudouin, L., and Estoup, A. (2004). GeneClass2: a software for genetic assignment and first-generation migrant detection. *J. Hered.* 95:536–539.

Plant, T. M. (1981). Time courses of concentrations of circulating gonadotropin, prolactin, testosterone, and cortisol in adult male rhesus monkeys (*Macaca mulatta*) throughout the 24h light–dark cycle. *Biol. Reprod.* 25:244–252.

Plavcan, J. (2001). Sexual dimorphism in primate evolution. *Ybk. Phys. Anthropol.* 44:25–53.

Pliwsungnoen, M. (2007). Spatial use, social structure, and food habits of Bengal slow loris *Nycticebus bengalensis* in different logged habitats. *Nat. Hist. Bull. Siam Soc.* 55:46.

Ploog, D. (1967). The behavior of squirrel monkeys (*Saimiri sciureus*) as revealed by sociometry, bioacoustics, and brain stimulation. In: Altmann, S. A. (ed.), *Social Communication Among Primates*. University of Chicago Press, Chicago. pp. 149–184.

Plooij, F. (1984). *The Behavioral Development of Free-Living Chimpanzee Babies and Infants*. Ablex Publishing, Norwood.

Plotnik, J. M., de Waal, F. M. B., and Reiss, D. (2006). Self recognition in an Asian elephant. *Proc. Natl. Acad. Sci. USA* 103:17053–17057.

Plummer, T. (2004). Flaked stones and old bones: biological and cultural evolution at the dawn of technology. *Ybk. Phys. Anthropol.* 47:118–164.

Plumptre, A. J., and Reynolds, V. (1994). The effect of selective logging on the primate populations in the Budongo Forest Reserve, Uganda. *J. Appl. Ecol.* 31:631–641.

Plumptre, A. J., and Williamson, E. A. (2001). Conservation-oriented research in the Virunga region. In: Robbins, M. M., Sicotte, P., and Stewart, K. J. (eds.), *Mountain Gorilla: Three Decades of Research at Karisoke*. Cambridge University Press, Cambridge. pp. 361–389.

Plumptre, A. J., McNeilage, A., Hall, J. S., and Williamson, E. A. (2003). The current status of gorillas and threats to their existence at the beginning of a new millennium. In: Taylor, A. B., and Goldsmith, M. L. (eds.). *Gorilla Biology: A Multidisciplinary Perspective*. Cambridge University Press, Cambridge. pp. 414–431.

Pobiner, B. L., DeSilva, J., Sanders, W. J., and Mitani, J. C. (2007). Taphonomic analysis of skeletal remains from chimpanzee hunts at Ngogo, Kibale National Park, Uganda. *J. Hum. Evol.* 52(6):614–636.

Pochron, S. T., and Wright, P. C. (2003). Variability in adult group compositions of a prosimian primate. *Behav. Ecol. Sociobiol.* 54:285–293.

Pochron, S. T., Tucker, W. T., and Wright, P. C. (2004). Demography, life history and social structure in *Propithecus diadema edwardsi* from 1986 to 2000 in Ranomafana National Park, Madagascar. *Am. J. Phys. Anthropol.* 125:61–72.

Pocock, R. I. (1918). On the external characteristics of the lemurs and of *Tarsius*. *Proc. Zool. Soc. Lond.* 1918:19–53.

Poirier, F. E. (1970). The Nilgiri langur (*Presbytis johnii*) of south India. In: Rosenblum, R. A. (ed.), *Primate Behavior: Developments in Field and Laboratory Research*, vol. 1. Academic Press, New York. pp. 251–383.

Poirier, F. E. (ed.) (1972). *Primate Socialization*. Random House, New York.

Polk, J. D. (2002). Adaptive and phylogenetic influences on musculoskeletal design in cercopithecine primates. *J. Exp. Biol.* 205:3399–3412.

Polk, J. D., Spencer-Smith, J., DiBerardino, L., Ellis, D., Downen, M., and Rosengren, K. S. (2008). Quantifying variability in phase portraits: application to gait ontogeny. *Infant Behav. Dev.* 31:382–306.

Pollard, K. S., Salama, S. R., Lambert, N., Lambot, M. A., Coppens, S., Pedersen, J. S., Katzman, S., King, B., Onodera, C., Siepel, A., Kern, A. D., Dehay, C., Igel, H., Ares, M., Jr., Vanderhaeghen, P., and Haussler, D. (2006). An RNA gene expressed during cortical development evolved rapidly in humans. *Nature* 443:167–172.

Pollock, J. I. (1975). Field observations on *Indri indri*: a preliminary report. In: Tattersall, I., and Sussman, R. W. (eds.), *Lemur Biology*. Plenum Press, New York. pp. 287–311.

Pollock, J. I. (1977). The ecology and socioecology of feeding in *Indri indri*. In: Clutton-Brock, T. (ed.), *Primate Ecology: Studies of Feeding and Ranging Behavior in Lemurs, Monkeys and Apes*. Academic Press, London. pp. 37–69.

Pombo, A. R., Waltert, M., Mansjoer, S. S., Mardiasti, A., and Mühlenberg, M. (2004). Home range, diet and behaviour of the Tonkean macaque (*Macaca tonkeana*) in Lore Lindu National Park, Sulawesi. In: Gerold, G., Fremerey, M., and Guhardja, E. (eds.), *Land Use, Nature Conservation and the Stability of Rainforest Margins in Southeast Asia*. Springer, Berlin, pp. 313–325.

Pomiankowski, A., Iwasa, Y., and Nee, S. (1991). The evolution of costly mate preferences: 1. Fisher and biased mutation. *Evolution* 45:1422–1430.

Ponce de Leon, M. S., and Zollikofer, C. P. E. (2001). Neanderthal cranial ontogeny and its implications for late hominid diversity. *Nature* 412:534–538.

Pond, C. M. (1977). The significance of lactation in the evolution of mammals. *Evolution* 31:177–199.

Pontius, J. U., Mullikin, J. C., Smith, D. R., et al. (2007). Initial sequence and comparative analysis of the cat genome. *Genome Res.* 17:1675–1689.

Pook, A. G., and Pook, G. (1981). A field study of the status and socioecology of Goeldi's monkey (*Callimico goeldii*) in northern Bolivia. *Folia Primatol.* 35:288–312.

Pook, A. G., and Pook, G. (1982). Polyspecific association between *Saguinus fuscicollis*, *Saguinus labiatus*, *Callimico goeldii* and other primates in north-western Bolivia. *Folia Primatol.* 38:196–216.

Poorman, P., Cartmill, M., and MacPhee, R. (1985). The G banded karotype of *Tarsius bancanus* and its implications for primate phylogeny. *Am. J. Phys. Anthropol.* 66:215.

Pop, M., and Salzberg, S. L. (2008). Bioinformatics challenges of new sequencing technology. *Trends Genet.* 24:142–149.

Pope, C. E., Dresser, B. L., Chin, N. W., Liu, J. H., Loskutoff, N. M., Behnke, E. J., Brown, C., McRae, M. M., Sinovway, C. E., Campbell, M. K., Cameron, K. N., Owens, O. M., Johnson, C. A., Evans, R. R., and Cedars, M. I. (1997). Birth of a western lowland gorilla (*Gorilla gorilla gorilla*) following in vitro fertilization and embryo transfer. *Am. J. Primatol.* 41:247–260.

Pope, T. R. (1989). The influence of mating systems and dispersal patterns on the genetic structure of red howler monkey populations [PhD diss.]. University of Florida, Gainesville.

Pope, T. R. (1990). The reproductive consequences of male cooperation in the red howler monkey: paternity exclusion in multi-male and single-male troops using genetic markers. *Behav. Ecol. Sociobiol.* 27:439–446.

Pope, T. R. (1992). The influence of dispersal patterns and mating systems on genetic differentiation within and between populations of the red howler monkey (*Alouatta seniculis*). *Evolution* 46:1112–1128.

Pope, T. R. (1998a). Effects of demographic change on group kin structure and gene dynamics of populations of red howling monkeys. *J. Mammal.* 79:692–712.

Pope, T. R. (1998b). Genetic variation in remnant populations of the woolly spider monkey (*Brachyteles arachnoides*). *Int. J. Primatol.* 19:95–109.

Pope, T. R. (2000a). The evolution of male philopatry in neotropical monkeys. In: Kappeler, P. M. (ed.), *Primate Males: Causes and Consequences of Variation in Group Composition*. Cambridge University Press, Cambridge. pp. 219–235.

Pope, T. R. (2000b). Reproductive success increases with degree of kinship in cooperative coalitions of female red howler monkeys (*Alouatta seniculus*). *Behav. Ecol. Sociobiol.* 27:439–446.

Popp, J. L. (1983). Ecological determinism in the life histories of baboons. *Primates* 24:198–210.

Poremba, A., Malloy, M., Saunders, R. C., Carson, R. E., Herscovitch, P., and Mishkin, M. (2004). Species-specific calls evoke asymmetric activity in the monkey's temporal poles. *Nature* 427:448–451.

Port-Carvalho, M., and Ferrari, S. F. (2004). Occurrence and diet of the black bearded saki (*Chiropotes satanas satanas*) in the fragmented landscape of western Maranhao, Brazil. *Neotrop. Primates* 12:17–21.

Porter, C. A., Page, S. L., Czelusniak, J., Schneider, H., Schneider, M. P. C., Sampaio, I., and Goodman, M. (1997). Phylogeny and evolution of selected primates as determined by sequences of the e-globin locus and 5′ flanking regions. *Int. J. Primatol.* 18:261–295.

Porter, J., Arzberger, P., Braun, H.-W., Bryant, P., Gage, S., Hansen, T., Hanson, P., Lin, C.-C., Lin, F.-P., Kratz, T., Michener, W., Shapiro, S., and Williams, T. (2005). Wireless sensor networks for ecology. *BioScience* 55:7561–7572.

Porter, L. M. (2001a). Social organization, reproduction and rearing strategies of *Callimico goeldii*: new clues from the wild. *Folia Primatol.* 72:69–79.

Porter, L. M. (2001b). Dietary differences among sympatric Callitrichinae in northern Bolivia: *Callimico goeldii, Saguinus fuscicollis* and *S. labiatus*. *Int. J. Primatol.* 22:961–992.

Porter, L. M. (2001c). Benefits of polyspecific associations for the Goeldi's monkey (*Callimico goeldii*). *Am. J. Primatol.* 54:143–158.

Porter, L. M. (2004). Forest use and activity patterns of *Callimico goeldii* in comparison to two sympatric tamarins, *Saguinus fuscicollis* and *Saguinus labiatus*. *Am. J. Phys. Anthropol.* 124:139–153.

Porter, L. M. (2007). *The Behavioral Ecology of Callimicos and Tamarins in Northwestern Bolivia*. Pearson/Prentice Hall, Upper Saddle River, NJ.

Porter, L. M., and Garber, P. A. (2004). Goeldi's monkey: a primate paradox. *Evol Anthropol.* 13:104–115.

Porter, L. M., and Garber, P. A. (2007). Niche expansion of a cryptic primate, *Callimico goeldii*, while in mixed species troops. *Am. J. Primatol.* 69:1340–1353.

Porter, L. M., Hanson, A. M., and Becerra, E. N. (2001). Group demographies and dispersal in a wild group of Goeldi's monkeys (*Callimico goedii*). *Folia Primatol.* 72:108–110.

Porter, L. M., Sterr, S. M., and Garber, P. A. (2007). Habitat use and ranging behavior of *Callimico goeldii*. *Int. J. Primatol.* 28:1035–1058.

Porter, L. M., Garber, P. A., and Nacimento, E. (2009). Exudates as a fallback food for *Callimico goeldii*. *Am. J. Primatol.* 71:120–129.

Poss, S. R., Kuhar, C., Stoinski, T. S., and Hopkins, W. D. (2006). Differential use of attentional and visual communicative signaling by orangutans (*Pongo pygmaeus*) and gorillas (*Gorilla gorilla*) in response to the attentional status of a human. *Am. J. Primatol.* 68:978–992.

Possamai, C. B., Young, R. J., Mendes, S. L., and Strier, K. B. (2007). Socio-sexual behavior of female northern muriquis (*Brachyteles hypoxanthus*). *Am. J. Primatol.* 69:766–776.

Post, D. G. (1981). Activity patterns of yellow baboons (*Papio cynocephalus*) in the Amboseli National Park, Kenya. *Anim. Behav.* 29:357–374.

Post, D. G. (1982). Feeding behavior of yellow baboons (*Papio cynocephalus*) in the Amboseli National Park, Kenya. *Int. J. Primatol.* 3:403–430.

Post, D. G., Hausfater, G., and McCuskey, S. A. (1980). Feeding behavior of yellow baboons (*Papio cynocephalus*): relationship to age, gender and dominance rank. *Folia Primatol.* 34:170–195.

Post, E. S., Pedersen, C., Wilmers, C. C., and Forchhammer, M. C. (2008). Phenological sequences reveal aggregate life history response to climatic warming. *Ecology* 89:363–370.

Poti, P., Bartolommei, P., and Saporiti, M. (2005). Landmark use by *Cebus apella*. *Int. J. Primatol.* 26:921–948.

Potts, M. (2007). Population and environment in the twenty-first century. *Pop. Environ.* 28:204–211.

Potts, W. K., and Wakeland, E. K. (1993). Evolution of MHC genetic diversity: a tale of incest, pestilence and sexual preference. *Trends Genet.* 9:408–412.

Potts, W. K., Manning, C. J., and Wakeland, E. K. (1991). Mating patterns in seminatural populations of mice influenced by MHC genotype. *Nature* 352:619–621.

Poulsen, J. R., Clark, C. J., and Smith, T. B. (2001a). Seasonal variation in the feeding ecology of the grey-cheeked mangabey (*Lophocebus albigena*) in Cameroon. *Am. J. Primatol.* 54:91– 105.

Poulsen, J. R., Clark, C. J., and Smith, T. B. (2001b). Seed dispersal by a diurnal primate community in Dja Reserve, Cameroon. *J. Trop. Ecol.* 17:787–808.

Poulsen, J. R., Clark, C. J., Connor, E. F., and Smith, T. B. (2002). Differential resource use by primates and hornbills: implications for seed dispersal. *Ecology* 83:228–240.

Pounds, J. A., and Puschendorf, R. (2004). Clouded futures. *Nature* 427:107–108.

Pournelle, G. H. (1967). Observations on reproductive behaviour and early postnatal development of the proboscis monkey (*Nasalis larvatus orientalis*). *Int. Zoo Ybk.* 7:90–92.

Pouydebat, E., Berge, C., Gorce, P., and Coppens, Y. (2005). Use and manufacture of tools to extract food by captive *Gorilla gorilla gorilla*: experimental approach. *Folia Primatol.* 76:180–183.

Pouydebat, E., Gorce, P., Bels, V., and Coppens, Y. (2006). Substrate optimization in nut cracking by capuchin monkeys (*Cebus apella*). *Am. J. Primatol.* 68:1017–1024.

Povinelli, D. (2000). *Folk Physics for Apes: A Chimpanzee's Theory of How the World Works*. Oxford University Press, Oxford.

Povinelli, D. J., and Cant, J. G. H. (1995). Arboreal clambering and the evolution of self-conception. *Q. Rev. Biol.* 70:393–421.

Povinelli, D., and Eddy, T. J. (1996). What young chimpanzees know about seeing. *Monogr. Soc. Res. Child.* 61:1–191.

Povinelli, D. J., and Vonk, J. (2006). We don't need a microscope to explore the chimpanzee's mind. In: Hurley, S. (ed.), *Rational Animals*. Oxford University Press, Oxford. pp. 385–412.

Povinelli, D. J., Rulf, A. B., Landau, K. R., and Bierschwale, D. T. (1993). Self-recognition in chimpanzees (*Pan troglodytes*): distribution, ontogeny, and patterns of emergence. *J. Comp. Psychol.* 107:347–372.

Povinelli, D., Perilloux, H., Reaux, J., and Bierschwale, D. (1998). Young chimpanzees' reactions to intentional versus accidental and inadvertent actions. *Behav. Proc.* 42:205–218.

Powell, G. V. N. (1974). Experimental analysis of the social value of flocking by starlings (*Sturnus vulgaris*) in relation to predation and foraging. *Anim. Behav.* 22:501–505.

Power, M. L. (1996). The other side of callitrichine gummivory: digestibility and nutritional value. In: Norconk, M. A., Rosenberger, A. L., and Garber, P. A. (eds.), *Adaptive Radiations of Neotropical Primates*. Plenum Press, New York. pp. 97–110.

Power, M. L., and Oftedal, O. T. (1996). Differences among captive callitrichids in the digestive responses to dietary gum. *Am. J. Primatol.* 40:131–144.

Power, M. L., Oftedal, O. T., Savage, A., Blumer, E. S., Soto, L. H., Chen, T. C., and Holick, M. F. (1997). Assessing vitamin D status of callitrichids: baseline data from wild cotton-top tamarins (*Saguinus oedipus*) in Colombia. *Zoo Biol.* 16:39–46.

Power, M. L., Tardif, S. D., Layne, D. G., and Schulkin, J. (1999). Ingestion of calcium solutions by common marmosets (*Callithrix jacchus*). *Am. J. Primatol.* 47:255–261.

Power, M. L., Tardif, S. D., Power, R. A., and Layne, D. G. (2003). Resting energy metabolism of Goeldi's monkey (*Callimico goeldi*) is similar to that of other callitrichids. *Am. J. Primatol.* 60:57–67.

Powzyk, J. A. (1997). The socioecology of two sympatric indriids: *Propithecus diadema diadema* and *Indri indri*, a comparison of feeding strategies and their possible repercussions on species-specific behavior [PhD diss.]. Duke University, Durham, NC.

Powzyk, J. A., and Mowry, C. B. (2003). Dietary and feeding differences between sympatric *Propithecus diadema diadema* and *Indri indri*. *Int. J. Primatol.* 24:1143–1162.

Pradhan, G. R., and van Schaik, C. (2008). Infanticide-driven intersexual conflict over matings in primates and its effects on social organization. *Behaviour* 145:251–275.

Prado, F. (1999a). Ecology and behavior of black-faced lion tamarins. *Neotrop. Primates* 7:137.

Prado, F. (1999b). Ecologia, comportamento e conservação do mico-leão-de-cara-preta (*Leontopithecus caissara*) no Parque Nacional de Superagüi,

Guaraqueçaba Paraná [MSc thesis]. Universidade Estadual Paulista, Botucatu.

Prasetyo, D., Ancrenaz, M., Morrogh-Bernard, H. C., Utami Atmoko, S. S., Wich, S. A., and van Schaik, C. P. (2009). Nest building in orangutans. In: Wich, S. A., Utami, S. S., Mitra Setia, T., and van Schaik, C. (eds.), *Orangutans: Geographic Variation in Behavioral Ecology and Conservation*. Oxford University Press, Oxford. pp. 269–278.

Premack, D., and Woodruff, G. (1978). Does the chimpanzee have a theory of mind. *Behav. Brain Sci.* 1:515–526.

Prescott, M. J., and Buchanan-Smith, H. M. (2002). Predation sensitive foraging in captive tamarins. In: Miller, L. E. (ed.), *Eat or Be Eaten: Predator Sensitive Foraging Among Primates*. Cambridge University Press, Cambridge. pp. 44–57.

Preston, B. (1998). Cognition and tool use. *Mind Lang.* 13:513–547.

Preuschoft, H. (1985). On the quality and magnitude of mechanical stresses in the locomotor system during rapid movement. *Z. Morphol. Anthropol.* 75:245–262.

Preuschoft, H., Chivers, D. J., Brockelman, W. Y., and Creel, N. (eds.) (1984). *The Lesser Apes: Evolutionary and Behavioural Biology*. Edinburgh University Press, Edinburgh.

Preuschoft, H., Witte, H., Christian, A., and Fischer, M. (1996). Size influences primate locomotion and body shape, with special emphasis on the locomotion of "small mammals." *Folia Primatol.* 66:93–112.

Preuschoft, S. (1995). "Laughter" and "smiling" in macaques: an evolutionary perspective [PhD diss.]. University of Utrecht, Utrecht.

Preuschoft, S., and van Hooff, J. A. R. A. M. (1995). Homologizing primate facial displays: a critical review of methods. *Folia Primatol.* 65:121–137.

Preuschoft, S., and van Hooff, J. A. R. A. M. (1997). The social function of "smile" and "laughter": variations across primate species and societies. In: Segerstrale, U., and Molnar, P. (eds.), *Nonverbal Communication: Where Nature Meets Culture*. Erlbaum, Hove, UK. pp. 171–190.

Preuschoft, S., Wang, X., Aureli, F., and de Waal, F. B. M. (2002). Reconciliation in captive chimpanzees: a reevaluation with controlled methods. *Int. J. Primatol.* 23:29–50.

Preuss, T. M., Cáceres, M., Oldham, M. C., and Geschwind, D. H. (2004). Human brain evolution: insights from microarrays. *Nat. Rev. Genet.* 5:850–860.

Price, E., and Caldwell, C. A. (2007). Artificially generated cultural variation between two groups of captive monkeys, *Colobus guereza kikuyuensis*. *Behav. Proc.* 74:13–20.

Price, E. C., and Evans, S. (1991). Terminology in the study of callitrichid reproductive strategies. *Anim. Behav.* 34:31–41.

Price, E. C., and Piedade, H. M. (2001). Ranging behavior and intraspecific relationships of masked titi monkeys (*Callicebus personatus personatus*). *Am. J. Primatol.* 53:87–92.

Price, E. C., Piedade, H. M., and Wormell, D. (2002). Population densities of primates in a Brazilian Atlantic forest. *Folia Primatol.* 73:54–56.

Prior, H., Schwarz, A., and Güntürkün, O. (2008). Mirror-induced behavior in the magpie (*Pica pica*): evidence of self-recognition. *PLoS Biol.* 6:e202. doi:10.1371/journal. pbi0.0060202.

Pritchard, J. K., Stephens, M., and Donnelly, P. (2000). Inference of population structure using multilocus genotype data. *Genetics* 155:945–959.

Profet, M. (1993). Menstruation as a defense against pathogens transported by sperm. *Q. Rev. Biol.* 68:335–386.

Promislow, D. E. L., and Harvey, P. H. (1990). Living fast and dying young: a comparative analysis of life-history variation among mammals. *J. Zool.* 220:417–437.

Prost, J. H. (1965). A definitional system for the classification of primate locomotion. *Am. Anthropol.* 67:1198–1214.

Prothero, J., and Jürgens, K. D. (1987). Scaling of maximal lifespan in mammals: a review. In: Woodhead, A. D., and Thompson, K. H. (eds.), *Evolution of Longevity in Animals*. Plenum Press, New York. pp. 49–74.

Prouty, L. A., Buchanan, P. D., Pollitzer, W. S., and Mootnick, A. R. (1983a). A presumptive new hylobatid subgenus with 38 chromosomes. *Cytogenet. Cell Genet.* 35:141–142.

Prouty, L. A., Buchanan, P. D., Pollitzer, W. S., and Mootnick, A. R. (1983b). *Bunopithecus*: a genus-level taxon for the hoolock gibbon (*Hylobates hoolock*). *Am. J. Primatol.* 5:83–87.

Prud'homme, J. (1991). Group fission in a semifree-ranging population of Barbary macaques (*Macaca sylvanus*). *Primates* 32:9–22.

Prud'homme, J., and Chapais, B. (1993a). Aggressive interventions and matrilineal dominance relations in semifree-ranging Barbary macaques (*Macaca sylvanus*). *Primates* 34:271–283.

Prud'homme, J., and Chapais, B. (1993b). Rank relations among sisters in semi-free ranging Barbary macaques (*Macaca sylvanus*). *Int. J. Primatol.* 14:405–420.

Pruetz, J. D. (1999). Socioecology of adult female vervet (*Cercopithecus aethiops*), and patas monkeys (*Erythrocebus patas*) in Kenya: food availability, feeding competition, and dominance relationships [PhD diss.]. University of Illinois, Urbana.

Pruetz, J. D. (2004). Feeding and ranging ecology of chimpanzees in an open environment at Fongoli, Senegal. Feeding Ecology Conference, Max Planck Institute for Evolutionary Anthropology, Leipzig, Germany, October 2004.

Pruetz, J. D., and Bertolani, P. (2007). Savanna chimpanzees, *Pan troglodytes verus*, hunt with tools. *Curr. Biol.* 17:412–417.

Pruetz, J. D., and Isbell, L. A. (1999). What makes a food contestable? Food properties and contest competition in vervets and patas monkeys in Laikipia, Kenya. *Am. J. Phys. Anthropol.* 28(suppl.):225–226.

Pruetz, J. D., and Isbell, L. A. (2000). Correlations of food distribution and patch size with agonistic interactions in female vervets (*Chlorocebus aethiops*) and patas monkeys (*Erythrocebus patas*) living in simple habitats. *Behav. Ecol. Sociobiol.* 49:38–47.

Pruetz, J. D., and Johnson-Fulton, S. (2003). Evidence for leaf-swallowing behavior by savanna chimpanzees in Senegal—a new site record. *PanAfr. News* 10:14–16.

Pruetz, J. D., Marchant, L. F., Arno, J., and McGrew, W. C. (2002). Survey of savanna chimpanzees (*Pan troglodytes verus*) in southeastern Senegal. *Am. J. Primatol.* 58:35–43.

Pruetz, J. D., Fulton, S. J., Marchant, L. F., McGrew, W. C. Schiel, M., and Waller, M. (2008). Arboreal nesting as anti-predator adaptation by savanna chimpanzees (*Pan troglodytes verus*) in southeastern Senegal. *Am. J. Primatol.* 70(4):393–401.

Pryce, C. R., Jurke, M., Shaw, H. J., Sandmeier, I. G., and Doebeli, M. (1993). Determination of ovarian cycle in Goeldi's monkey (*Callimico goeldii*) via the measurement of steroids and peptides in plasma and urine. *J. Reprod. Fertil.* 99:427–435.

Pryce, C. R., Schwarzenberger, F., and Doebeli, M. (1994). Monitoring fecal samples for estrogen excretion across the ovarian cycle in Goeldi's monkey (*Callimico goeldii*). *Zoo Biol.* 13:219–230.

Pryce, C. R., Schwarzenberger, F., Doebeli, M., and Etter, K. (1995). Comparative study of oestrogen excretion in female New World monkeys: an overview of non-invasive ovarian monitoring and a new application in evolutionary biology. *Folia Primatol.* 64:107–123.

Puertas, P. E., Aquino, R., and Encarnación, F. (1992). Uso de alimentos y competición entre el mono nocturno *Aotus vociferans* y otros mamíferos, Loreto, Perú. *Folia Amazon.* 4:135–144.

Pullen, S. L. (2000). Behavioural and genetic studies of the mating system in a nocturnal primate: the lesser galago (*Galago moholi*) [PhD diss.]. University of Cambridge, Cambridge.

Pullen, S. L., Bearder, S. K., and Dixson, A. F. (2000). Preliminary observations on sexual behavior and the mating system in free-ranging lesser galagos (*Galago moholi*). *Am. J. Primatol.* 51:79–88.

Pulliam, H. R. (1973). On the advantages of flocking. *J. Theor. Biol.* 38:419–422.

Pulliam, H. R., and Caraco, T. (1984). Living in groups: is there an optimal group size? In: Krebs, K. R., and Davies, N. (eds.), *Behavioural Ecology*. Blackwell Scientific, Oxford. pp. 122–147.

Purvis, A. (1995). A composite estimate of primate phylogeny. *Phil. Trans. R. Soc. Lond.* B348:405–421.

Purvis, A., and Harvey, P. H. (1995). Mammal life-history evolution: a comparative test of Charnov's model. *J. Zool.* 237:259–283.

Purvis, A., and Hector, A. (2000). Getting the measure of bio-diversity. *Nature* 405:212–219.

Purvis, A., Agapow, P., Gittleman, J. L., and Mace, G. M. (2000). Nonrandom extinction and the loss of evolutionary history. *Science* 288:328–330.

Purvis, A., Webster, A. J., Agapow, P. M., Jones, K. E., and Isaac, N. J. B. (2003). Primate life histories and phylogeny. In: Kappeler, P. M., and Pereira, M. E. (eds.), *Primate Life Histories and Socioecology*. University of Chicago Press, Chicago. pp. 24–40.

Pusey, A. (1978). The physical and social development of wild adolescent chimpanzees (*Pan trolgodytes schweinfurthii*) [PhD diss.]. Stanford University, Palo Alto, CA.

Pusey, A. E. (1979). Inter-community transfer of chimpanzees in Gombe National Park. In: Hamburg, D. A., and McCown, E. R. (eds.), *The Great Apes*. Benjamin/Cummings, Menlo Park, CA. pp. 465–479.

Pusey, A. E. (1980). Inbreeding avoidance in chimpanzees. *Anim. Behav.* 28:543–552.

Pusey, A. (1983). Mother–offspring relationships in chimpanzees after weaning. *Anim. Behav.* 31:363–377.

Pusey, A. (1990). Behavioural changes at adolescence in chimpanzees. *Behaviour* 115:203–246.

Pusey, A. E. (2001). Of genes and apes: chimpanzee social organization and reproduction. In: de Waal, F. (ed.), *The Tree of Origin.* Harvard University Press, Cambridge, MA. pp. 9–37.

Pusey, A. E., and Packer, C. (1987). Dispersal and philopatry. In: Smuts, B. B., Cheney, D. L., Seyfarth, R. M., Wrangham, R. W., and Struhsaker, T. T. (eds.), *Primate Societies.* University of Chicago Press, Chicago. pp. 250–266.

Pusey, A. E., Williams, J., and Goodall, J. (1997). The influence of dominance rank on the reproductive success of female chimpanzees. *Science* 277:828–831.

Pusey, A. E., Oehlert, G., Williams, J., and Goodall, J. (2005). Influence of ecological and social factors on body mass of wild chimpanzees. *Int. J. Primatol.* 26:3–31.

Pusey, A., Murray, C., Wallauer, W., Wilson, M., Wroblewski, E., and Goodall, J. (2008). Severe aggression among female *Pan troglodytes schwienfurthii* at Gombe National Park, Tanzania. *Int. J. Primatol.* 29:949–973.

Pyke, G. H., Pulliam, H. R., and Charnov, E. L. (1977). Optimal foraging: a selective review of theory and tests. *Q. Rev. Biol.* 52:137–154.

Qi, J. F. (1986). The Chinese golden monkey: husbandry and reproduction. In: Benirschke, K. (ed.), *Primates: The Road to Self-Sustaining Populations.* Springer-Verlag, New York. pp. 837–843.

Queiroz, H. L. (1992). A new species of capuchin monkey, genus *Cebus* Erxleben 1977 (Cebidae, Primates), from eastern Brazil Amazonia. *Goeld. Zool.* 15:1–3.

Queller, D. C., and Goodnight, K. F. (1989). Estimating relatedness using genetic markers. *Evolution* 43:258–275.

Quiatt, D. (1979). Aunts and mothers: adaptive implications of allomaternal behaviour of non-human primates. *Am. Anthropol.* 81:311–319.

Quiatt, D. (1985). The "household" in non-human primate evolution: a basic linking concept. *Anthropol. Contemp.* 3:187–193.

Quiatt, D. (1987). *A-group: one example of gibbon social organization.* Videotape available from the Primate Center Library, Wisconsin Regional Primate Research Center, Madison.

Quiatt, D., and Kelso, J. (1987). The concept of the household: linking behavior and genetic analyses. *Hum. Evol.* 2:429–435.

Quinn, A., and Wilson, D. E. (2004). *Daubentonia madagascariensis. Mammal. Species* 740:1–6.

Quris, R. (1975). Ecologie et organisation sociale de *Cercocebus galeritus agilis* dans le nord-est due Gabon. *Terre Vie* 29:337–398.

Quris, R. (1976). Données comparatis sure la socio-ecologie de huit especes de cercopithecidae vivant dans une meme zone de fôret primitive periodiquement inondee (nord-est due Gabon). *Terre Vie* 30:193–209.

Raaum, R. L., Sterner, K. N., Noviello, C. M., Stewart, C. B., and Disotell, T. R. (2005). Catarrhine primate divergence dates estimated from complete mitochondrial genomes: concordance with fossil and nuclear DNA evidence. *J. Hum. Evol.* 48:237–257.

Rabarivola, C., Meyers, D., and Rumpler, Y. (1991). Distribution and morphological characteristics of intermediate forms between the black lemur (*Eulemur macaco macaco*) and the Sclater's lemur (*E. m. flavifrons*). *Primates* 32:269–273.

Rabarivola, C., Prosper, P., Zaramody, A., Andriaholinirina, N., & Hauwy, M. (2007). Cytogenetics and taxonomy of the genus *Hapalemur. Lemur News, 12,* 46–49.

Rabinowitz, A. (1989). The density and behavior of large cats in a dry tropical forest mosaic in Huai Kha Khaeng Wildlife Sanctuary, Thailand. *Nat. Hist. Bull. Siam Soc.* 37:235–251.

Raboy, B. E., and Dietz, J. M. (2004). Diet, foraging, and use of space in wild golden-headed lion tamarins. *Am. J. Primatol.* 63:1–15.

Raboy, B. E., Canale, G. R., and Dietz, J. M. (2008). Ecology of *Callithrix kuhlii* and a review of eastern Brazilian marmosets. *Int. J. Primatol.* 29:449–467.

Radespiel, U. (2000). Sociality in the gray mouse lemur (*Microcebus murinus*) in northwestern Madagascar. *Am. J. Primatol.* 51:21–40.

Radespiel, U. (2006). Ecological diversity and seasonal adaptations of mouse lemurs (*Microcebus spp.*). In: Gould, L., and Sauther, M. L. (eds.), *Ecology and Adaptation.* Springer, New York. pp. 211–234.

Radespiel, U., and Zimmermann, E. (2003). The influence of familiarity, age, experience, and female mate choice on pregnancies in captive grey mouse lemurs. *Behaviour* 140:301–318.

Radespiel, U., Cepok, S., Zimmermann, E., and Zietemann, V. (1998). Sex-specific usage patterns of sleeping-sites in grey mouse lemurs (*Microcebus murinus*) in northwestern Madagascar. *Am. J. Primatol.* 46:77–84.

Radespiel, U., Funk, S. M., Zimmermann, E., and Bruford, M. W. (2001). Isolation and characterization of microsatellite loci in the grey mouse lemur (*Microcebus murinus*) and their amplification in the family Cheirogaleidae. *Mol. Ecol. Notes* 1:16–18.

Radespiel, U., Dal Secco, V., Drogemuller, C., Braune, P., Labes, E., and Zimmermann, E. (2002). Sexual selection, multiple mating and paternity in grey mouse lemurs, *Microcebus murinus. Anim. Behav.* 63:259–268.

Radespiel, U., Ehresmann, P., and Zimmerman, E. (2003a). Species-specific usage of sleeping sites in two sympatric mouse lemur species (*Microcebus murinus* and *M. ravelobensis*) in northwestern Madagascar. *Am. J. Primatol.* 59:139–151.

Radespiel, U., Lutermann, H., Schmelting, B., Bruford, M. W., and Zimmerman, E. (2003b). Patterns and dynamics of sex-biased dispersal in a nocturnal primate, the grey mouse lemur, *Microcebus murinus. Anim. Behav.* 65:709–719.

Radespiel, U., Olivieri, G., Rasolofoson, D. W., Rakotontratsimba, G., Rakotonirainy, O., Rasoloharijanona, S., Randrianambinena, B., Ratsimbazafy, J. H., Ratelolahy, F., Randriamboavonjy, T., Rasolofoharivelo, T., Craul, M., Rakotozafy, L., and Randriananrison, R. M. (2008). Exceptional diversity of mouse lemurs (*Microcebus* spp.) in the Makira region with the description of one new species. *Am. J. Primatol.* 70:1033–1046.

Radinsky, L. (1982). Some cautionary notes on making inferences about relative brain size. In: Armstrong, E., and Falk, D. (eds.), *Primate Brain Evolution: Methods and Concepts.* Plenum Press, New York. pp. 29–37.

Raemaekers, J. J. (1979). Ecology of sympatric gibbons. *Folia Primatol.* 31:227–245.

Raemaekers, J. J. (1980). Causes of variation between months in the distance traveled daily by gibbons. *Folia Primatol.* 34:46–60.

Raemaekers, J. J., and Chivers, D. J. (1980). Socio-ecology of Malayan forest primates. In: Chivers, D. J. (ed.), *Malayan Forest Primates: Ten Years' Study in Tropical Rain Forest.* Plenum Press, New York. pp. 279–316.

Raemaekers, J. J., and Raemaekers, P. M. (1984). Vocal interactions between two male gibbons, *Hylobates lar. Nat. Hist. Bull. Siam Soc.* 32:95–106.

Raemaekers, J. J., and Raemaekers, P. M. (1985). Field playback of loud calls to gibbons (*Hylobates lar*): territorial, sex-specifc, and species-specific responses. *Anim. Behav.* 33:481–493.

Raemaekers, J. J., Raemaekers, P. M., and Haimoff, E. H. (1984). Loud calls of the gibbon (*Hylobates lar*): repertoire, organisation and context. *Behaviour* 91:147–189.

Raff, E. C., and Raff, R. A. (2000). Dissociability, modularity, evolvability. *Evol. Dev.* 2:235–237.

Raff, R. A. (1996). *The Shape of Life.* University of Chicago Press, Chicago.

Rafferty, K. L., Teaford, M. F., and Jungers, W. L. (2002). Molar microwear of subfossil lemurs: improving the resolution of dietary inferences. *J. Hum. Evol.* 43:645–657.

Rahm, U. H. (1970). Ecology, zoogeography and systematics of some African forest monkeys. In: Napier, J. R., and Napier, P. M. (eds.), *Old World Monkeys: Evolution, Systematics and Behavior.* Academic Press, New York. pp. 589–626.

Raich, J. W., Rastetter, E. B., Melillo, J. M., Kicklighter, D. W., Steudler, P. A., Peterson, B. J., Grace, A. L., Moore, B., Vorosmarty, C. J. (1991). Potential net primary productivity in South America—application of a global-model. *Ecol. Appl.* 1(4):399–429.

Rajalingam, R., Hong, M., Adams, E. J., Shum, B. P., Guethlein, L. A., and Parham, P. (2001). Short *KIR* haplotypes in pygmy chimpanzee (bonobo) resemble the conserved framework of diverse human *KIR* haplotypes. *J. Exp. Med.* 193:135–146.

Rajanathan, R., and Bennett, E. L. (1990). Notes on the social behavior of wild proboscis monkeys (*Nasalis larvatus*). *Malay. Nat. J.* 44:35–44.

Rakotondravony, D., Goodman, S. M., and Soarimalala, V. (1998). Predation on *Hapalemur griseus griseus* by *Boa manditra* (Boidae) in the littoral forest of eastern Madagascar. *Folia Primatol.* 69:405–408.

Ralls, K., and Ballou, J. (1982). Effects of inbreeding on infant mortality in captive primates. *Int. J. Primatol.* 3:491–505.

Ramakrishnan, U., and Coss, R. G. (2001a). A comparison of the sleeping behavior of three sympatric primates. *Folia Primatol.* 72:51–53.

Ramakrishnan, U., and Coss, R. G. (2001b). Strategies used by bonnet macaques (*Macaca radiata*) to reduce predation risk while sleeping. *Primates* 42:193–206.

Ramakrishnan, U., Coss, R. G., and Pelkey, N. W. (1999). Tiger decline caused by the reintroduction of large ungulate prey: evidence from a study of leopard diets in southern India. *Biol. Conserv.* 89:113–120.

Ramakrishnan, U., Coss, R. G., Schank, J., Dharawat, A., and Kim S. (2005). Snake species discrimination by wild bonnet macaques (*Macaca radiata*). *Ethology* 111(4):337–356.

Ramanamanjato, J. B., and Ganzhorn, J. U. (2001). Effects of forest fragmentation, introduced *Rattus rattus* and the role of exotic tree plantations and secondary vegetation for the conservation of an endemic rodent and a small lemur in littoral forests of southeastern Madagascar. *Anim. Conserv.* 4:175–183.

Ramaromilanto, B., Lei, R., Engberg, S. E., Johnson, S. E., Sitzmann, B. D., and Louis, E. E. (2009). Sportive lemur diversity at Mananara-Nord Biosphere Reserve, Madagascar. *Occas. Pap. Mus. Tex. Tech Univ.* 286:1–22.

Ramirez, M. (1980). Grouping patterns of the woolly monkey, *Lagothrix lagothricha,* at the Manu National Park, Peru. *Am. J. Phys. Anthropol.* 52:269.

Ramirez, M. (1988). The woolly monkeys, genus *Lagothrix.* In: Mittermeier, R. A., Rylands, A. B., Coimbra-Filho, A. F., and da Fonseca, G. A. B. (eds.), *Ecology and Behavior of Neotropical Primates,* vol. 2. World Wildlife Fund, Washington DC. pp. 539–575.

Ramirez, M. F., Freese, C. H., and Revilla, J. C. (1977). Feeding ecology of the pygmy marmoset, *Cebuella pygmaea,* in northeastern Peru. In: Kleiman, D. G. (ed.), *The Biology and Conservation of the Callitrichidae.* Smithsonian Institution Press, Washington DC. pp. 91–104.

Ramírez-Llorens, P., Di Bitetti, M. S., Baldovino, M. C., and Janson, C. H. (2008). Infanticide in black capuchin monkeys (*Cebus apella nigritus*) in Iguazú National Park, Argentina. *Am. J. Primatol.* 70:473–484.

Ramirez Rozzi, F. V., and De Castro, J. M. B. (2004). Surprisingly rapid growth in Neanderthals. *Nature* 428:936–939.

Ramos-Fernández, G., and Ayala-Orozco, B. (2003). Population size and habitat use of spider monkeys at Punta Laguna, Mexico. In: March, L. K. (ed.), *Primates in Fragments: Ecology and Conservation.* Kluwer Academic/Plenum, New York. pp. 191–209.

Ramos-Fernandez, G., Boyer, D. and Gomez, V. P. (2006). A complex social structure with fission–fusion properties can emerge from a simple foraging model. *Behav. Ecol. Soc.* 60:536–549.

Ramsey, G., Bastian, M., and van Schaik, C. (2007). Animal innovation defined and operationalized. *Behav. Brain Sci.* 30:393–437.

Randall, F. E. (1943). The skeletal and dental development and variability of the gorilla. *Hum. Biol.* 15:236–254.

Randrianambinina, B., Rakotondravony, P., Radespiel, U., and Zimmermann, E. (2003). Seasonality in general activity, body mass and reproduction of the golden brown mouse lemur (*Microcebus ravelobensis*) in northwestern Madagascar and the brown mouse lemur (*M. rufus*) in eastern Madagascar. *Primates* 44:321–331.

Range, F. (2006). Social behavior of free-ranging juvenile sooty mangabeys (*Cercocebus torquatus atys*). *Behav. Ecol. Sociobiol.* 59: 511–520.

Range, F., and Noe, R. (2002). Familiarity and dominance relations among female sooty mangabeys in the Taï National Park. *Am. J. Primatol.* 56:137–153.

Ransom, T. W. (1981). *Beach Troop of the Gombe.* Bucknell University Presses, Lewisburg, PA.

Ransom, T. W., and Ransom, B. S. (1971). Adult male–infant relations among baboons (*Papio anubis*). *Folia Primatol.* 16:179–195.

Ransom, T. W., and Rowell, T. E. (1972). Early social development of feral baboons. In: Poirier, F. (ed.), *Primate Socialization.* Random House, New York. pp. 105–144.

Ranta, E., Peuhkuri, N., Laurila, A., Rita, H., and Metcalfe, N. B. (1996). Producers, scroungers and foraging group structure. *Anim. Behav.* 51:171–175.

Rao, M., and van Schaik, C. P. (1997). The behavioral ecology of Sumatran orangutans in logged and unlogged forest. *Trop. Biodivers.* 4:173–185.

Rappold, I., and Erkert, H. G. (1994). Re-entrainment, phase-response and range of entrainment of circadian rhythms in owl monkeys (*Aotus lemurinus*) of different age. *Biol. Rhythm Res.* 25:133–152.

Rasmussen, D. R. (1978). Environmental and behavioral correlates of changes in range use in a troop of yellow (*Papio eynocephalus*) and a troop of, olive (*P. anubis*) baboons. [PhD diss.]. University of California, Riverside.

Rasmussen, D. R. (1981). Communities of baboon troops (*Papio cynocephalus*) in Mikumi National Park, Tanzania: a preliminary report. *Folia Primatol.* 36:232–242.

Rasmussen, D. T. (1986). Life history and behavior of slow lorises and slender lorises [PhD diss.]. Duke University, Durham, NC.

Rasmussen, D. T. (1990). Primate origins: lessons from a neotropical marsupial. *Am. J. Primatol.* 22:263–278.

Rasmussen, D. T. (2002a). The origin of primates. In: Hartwig, W. C. (ed.), *The Primate Fossil Record.* Cambridge University Press, Cambridge. pp. 5–10.

Rasmussen, D. T. (2002b). Early catarrhines of the African Eocene and Oligocene. In: Hartwig, W. C. (ed.), *The Primate Fossil Record.* Cambridge University Press, Cambridge. pp. 203–220.

Rasmussen, D. T., and Izard, M. K. (1988). Scaling of growth and life-history traits relative to body size, brain size and metabolic rate in lorises and galagos (Lorisidae, Primates). *Am. J. Phys. Anthropol.* 75:357–367.

Rasmussen, D. T., and Nekaris, K. A. I. (1998). Evolutionary history of the lorisiform primates. *Folia Primatol.* 69(suppl. 1):250–285.

Rasmussen, K. L. R. (1983). Influence of affiliative preferences upon the behaviour of male and female baboons during sexual consortships. In: Hinde, R. A. (ed.), *Primate Social Relationships.* Sinauer Associates, Sunderland, MA. pp. 116–120.

Rasmussen, K. L. R. (1985). Changes in the activity budgets of yellow baboons (*Papio cynocephalus*) during sexual consortships. *Behav. Ecol. Sociobiol.* 17:161–170.

Rasmussen, K. L. R. (1986). Spatial patterns and peripheralisation of yellow baboons (*Papio cynocephalus*) during sexual consortship. *Behaviour* 97:161–180.

Rasmussen, M. A. (1999). Ecological influences on activity cycle in two cathemeral primates, the mongoose lemur (*Eulemur mongoz*) and the common brown lemur (*Eulemur fulvus fulvus*) [PhD diss.]. Duke University, Durham, NC.

Rasoazanabary, E. A. (2004). Preliminary study of mouse lemurs in the Bezamahafaly Special Reserve, southwest Madagascar. *Lemur News* 9:4–7.

Rasoloarison, R. M., Goodman, S. M., and Ganzhorn, J. U. (2000). Taxonomic revision of mouse lemurs (*Microcebus*) in the western portions of Madagascar. *Int. J. Primatol.* 21:963–1019.

Rasoloharijaona, S. (1998). Vocal communication in a nocturnal lemur in northwestern Madagascar. XVIIth Congress of the International Primatological Society, Antananarivo, Madagascar August 10–14. Abstract 170.

Rasoloharijaona, S. B., Rakotosamimanana, B., Randrianambinina, B., and Zimmermann, E. (2003). Pair-specific usage of sleeping sites and their implications for social organization in a nocturnal Malagasy primate, the Milne Edwards sportive lemur (*Lepilemur edwardsi*). *Am. J. Phys. Anthropol.* 122:251–258.

Rasoloharijaona, S., Randrianambinina, B., Braune, P., and Zimmermann, E. (2006). Loud calling, spacing, and cohesiveness in a nocturnal primate, the Milne Edwards' sportive lemur (*Lepilemur edwardsi*). *Am. J. Phys. Anthropol.* 129:591–600.

Rasoloharijaona, S., Randrianambinina, B., and Zimmermann, E. (2008). Sleeping site ecology in a rain-forest dwelling nocturnal lemur (*Lepilemur mustelinus*): implications for sociality and conservation. *Am. J. Primatol.* 70:247–253.

Ratajszczak, R. (1998). Taxonomy, distribution and status of the lesser slow loris *Nycticebus pygmaeus* and their implications for captive management. *Folia Primatol.* 69(suppl. 1):171–174.

Rathbun, G. B., and Gache, M. (1980). Ecological survey of the night monkey, *Aotus trivirgatus,* in Formosa Province, Argentina. *Primates* 21:211–219.

Rathcke, R. B., and Lacey, E. P. (1985). Phenological patterns of terrestrial plants. *Annu. Rev. Ecol. Syst.* 16:179–214.

Ratsimbazafy, J. H. (2001). On the brink of extinction and the process of recovery: responses of black-and-white ruffed lemurs (*Varecia variegata variegata*) to disturbance in Manombo Forest, Madagascar [PhD diss.]. State University of New York, Stony Brook.

Ratsimbazafy, J. H. (2002). How do black and white ruffed lemurs still survive in a highly disturbed habitat? *Lemur News* 7:7–10.

Ratsimbazafy, J. H. (2006). Diet composition, foraging, and feeding behavior in relation to habitat disturbance: implications for the adaptability of ruffed lemurs (*Varecia v. editorium*) in Manombo forest, Madagascar. In: Gould, L., and Sauther, M. L. (eds.), *Lemurs: Ecology and Adaptation.* Springer, New York. pp. 403–422.

Ratsirarson, J., and Rumpler, Y. (1988). Contribution à l'étude comparée de l'écoéthologie de deux espèces de lémuriens, *Lepilemur mustelinus* (I. Geoffroy 1850), *Lepilemur septentrionalis* (Rumpler and Albignac 1975). In: Rakotovao, L., Barre, V., and Sayer, J. (eds.), *L'Equilibre des Ecosystèmes forestiers à Madagascar. Actes d'un Séminaire International.* IUCN, Gland, Switzerland. pp. 100–102.

Ratsirarson, J., Anderson, J., Warter, S., and Rumpler, Y. (1987). Notes on the distribution of *Lepilemur septentrionalis* and *L. mustelinus* in northern Madagascar. *Primates* 28:119–122.

Rauth-Widmann, B., Thiemann-Jager, A., and Erkert, H. G. (1991). Significance of nonparametric light effects in entrainment of circadian rhythms in owl monkeys (*Aotus lemurinus griseimembra*) by light–dark cycles. *Chronobiol. Int.* 8:251–266.

Rauth-Widmann, B., Fuchs, E., and Erkert, H. G. (1996). Infradian alteration of circadian rhythms in owl monkeys (*Aotus lemurinus griseimembra*): an effect of estrus? *Physiol. Behav.* 59:11–18.

Ravosa, M. J., and Dagosto, M. (eds.) (2006). *Primate Origins*. New York: Springer.

Ravosa, M. J., and Savakova, D. G. (2004). Euprimate origins: the eyes have it. *J. Hum. Evol.* 46:357–364.

Raxworthy, C. J., and Rakotondraparany, F. (1988). Mammals report. In: Quansah, N. (ed.), *Manongarivo Special Reserve (Madagascar), 1987/88 Expedition Report*. Madagascar Environmental Research Group, London.

Ray, D. A., Xing, J. C., Hedges, D. J., Hall, M. A., Laborde, M. E., Anders, B. A., White, B. R., Stoilova, N., Fowlkes, J. D., Landry, K. E., Chemnick, L. G., Ryder, O. A., and Batzer, M. A. (2005). *Alu* insertion loci and platyrrhine primate phylogeny. *Mol. Phylogenet. Evol.* 35:117–126.

Raymond, M., and Roussett, M. (1995). GENEPOP (version 1.2), a population genetics software for exact tests and ecumenicism. *J. Hered.* 86:248–249.

Razanahoera-Rakotomalala, M. (1981). Les adaptations alimentaires comparées de deux lémuriens folivores sympatriques: *Avahi Jourdan*, 1834—*Lepilemur* I. Geoffroy 1851 [PhD diss.]. University of Madagascar, Antananarivo.

Reader, S. M., and Laland, K. N. (2000). Diffusion of foraging innovations in the guppy. *Anim. Behav.* 60:175–180.

Reader, S. M., and Laland, K. N. (2002). Social intelligence, innovation, and enhanced brain size in primates. *Proc. Natl. Acad. Sci. USA* 99:4436–4441.

Reddy, S. M., Ray, A. K., and Gupta, P. (1992). The assisted reproductive technologies (ART) in conserving endangered wildlife: a review. *Proc. Zool. Soc. Calcutta* 45(suppl. A):149–159.

Redford, K. H. (1992). The empty forest. *Bioscience* 42:412–422.

Reed, B. C., Brown, J. F., Vanderzee, D., Loveland, T. R., Merchant, J. W., and Ohlen, D. O. (1994). Measuring phenological variability from satellite imagery. *J. Veg. Sci.* 5(5):703–714.

Reed, B. C., White, M., and Brown, J. F. (2003). Remote sensing phenology. In: Schwartz, M. D. (ed.), *Phenology: An Integrative Environmental Science*. Kluwer Academic, Dordrecht, The Netherlands. pp. 365–379

Reed, C., O'Brien, T. G., and Kinnaird, M. F. (1997). Male social behavior and dominance hierarchy in the Sulawesi crested black macaque (*Macaca nigra*). *Int. J. Primatol.* 18:247–260.

Reed, D. H., O'Grady, J. J., Brook, B. W., Ballou, J. D., and Frankham, R. (2003). Estimates of minimum viable population sizes for vertebrates and factors influencing those estimates. *Biol. Conserv.* 113:23–34.

Reed, K. E., and Bidner, L. R. (2004). Primate communities: past present and possible future. *Ybk. Phys. Anthropol.* 47:2–39.

Reeder, D. M. (2003). The potential for cryptic female choice in primates: behavioral, anatomical, and physiological considerations. In: Jones, C. B. (ed.), *Sexual Selection and Reproductive Competition in Primates: New Perspectives and Directions*. American Society of Primatologists, Norman, OK. pp. 255–303.

Rehg, J.A. (2006). Seasonal variation in polyspecific associations among *Callimico goeldii*, *Saguinus labiatus*, and *S. fuscicollis* in Acre, Brazil. *Int. J. Primatol.* 27(5):1399–428.

Reich, P. B., and Borchert, R. (1984). Water stress and tree phenology in a tropical dry forest in the lowlands of Costa Rica. *J. Ecol.* 72:61–74.

Reichard, U. (1995). Extra-pair copulation in a monogamous gibbon (*Hylobates lar*). *Ethology* 100:99–112.

Reichard, U. (2003a). Monogamy: past and present. In: Reichard, U. H., and Boesch, C. (eds.), *Monogamy: Mating Strategies and Partnerships in Birds, Humans, and Other Mammals*. Cambridge University Press, Cambridge. pp. 3–25.

Reichard, U. (2003b). Social monogamy in gibbons: the male perspective. In: Reichard, U. H., and Boesch, C. (eds.), *Monogamy: Mating Strategies and Partnerships in Birds, Humans, and Other Mammals*. Cambridge University Press, Cambridge. pp. 190–213.

Reichard, U. H., and Barelli, C. (2008). Life history and reproductive strategies of Khao Yai white-handed gibbon females (*Hylobates lar*). *Int. J. Primatol.* 29:823–844.

Reichard, U., and Sommer, V. (1997). Group encounters in wild gibbons (*Hylobates lar*): agonism, affiliation, and the concept of infanticide. *Behaviour* 134:1135–1174.

Reichert, K. E., Heistermann, M., Hodges, J. K., Boesch, C., and Hohmann, G. (2002). What females tell males about their reproductive status: are morphological and behavioural cues reliable signals of ovulation in bonobos (*Pan paniscus*)? *Ethology* 108:583–600.

Reimann, E. W. (2002). Coexistence and feeding ecology in female grey and golden-brown mouse lemurs (*Microcebus murinus* and *M. ravelobensis*) in north-western Madagascar [diss.]. Tierärztliche Hochschule, Hannover.

Reinartz, G. E., Karron, J. D., Phillips, R. B., and Weber, J. L. (2000). Patterns of microsatellite polymorphism in the range-restricted bonobo (*Pan paniscus*): considerations for interspecific comparison with chimpanzees (*P. troglodytes*). *Mol. Ecol.* 9:315–328.

Reiss, D., and Marino, L. (2001). Mirror self-recognition in the bottlenose dolphin: a case of cognitive convergence. *Proc. Natl. Acad. Sci. USA* 98:5937–5942.

Remis, M. J. (1997). Western lowland gorillas (*Gorilla gorilla gorilla*) as seasonal frugivores: use of variable resources. *Am. J. Primatol.* 43:87–109.

Remis, M. (1998). The gorilla paradox: the effects of body size and habitat on the positional behavior of lowland and mountain gorillas. In: Strasser, E., Fleagle, J. G., Rosenberger, A., and McHenry, H. (eds.), *Primate Locomotion: Recent Advances*. Plenum Press, New York. pp. 95–106.

Remis, M. J. (2003). Are gorillas vacuum cleaners of the forest floor? The roles of body size, habitat, and food preferences on dietary flexibility and nutrition. In: Taylor, A. B., and Goldsmith, M. L. (eds.), *Gorilla Biology: A Multidisciplinary Perspective*. Cambridge University Press, Cambridge. pp. 385–404.

Remis, M. J., and Kerr, M. E. (2002). Taste response to fructose and tannic acid among gorillas (*Gorilla gorilla gorilla*). *Int. J. Primatol.* 23:251–261.

Remis, M. J., Dierenfeld, E. S., Mowry, C. B., and Carroll, R. W. (2001). Nutritional aspects of western lowland gorilla (*Gorilla gorilla gorilla*) diet during seasons of fruit scarcity at Bai Hokou, Central African Republic. *Int. J. Primatol.* 22:807–836.

Rempel, R. S., Rodgers, A. R., and Abraham, K. F. (1995). Performance of a GPS animal location system under boreal forest canopy. *J. Wildl. Manage.* 59:543–551.

Ren, R. M., Yan, K. H., Su, Y. J., Qi, H. J., Liang, B., Bao, W. Y., and de Waal, F. B. M. (1991). The reconciliation behavior of golden monkeys (*Rhinopithecus roxellanae roxellanae*) in small breeding groups. *Primates* 32:321–327.

Ren, R. M., Yan, K. H., Su, Y. J., Qi, H. J., Liang, B., Bao, W. Y., and de Waal, F. B. M. (1995). The reproductive behavior of golden monkeys in captivity (*Rhinopithecus roxellana roxellana*). *Primates* 36:135–143.

Ren, R. M., Su, Y. J., Yan, K. H., Li, J. J., Yin, Z., Zhu, Z. Q., Hu, Z. L., and Hu, Y. F. (1998). Preliminary survey of the social organization of *Rhinopithecus roxellana* in Shennongjia National Natural Reserve, Hubei, China. In: Jablonski, N. G. (ed.), *The Natural History of the Doucs and Snub-Nosed Monkeys*. World Scientific Press, Singapore. pp. 269–278.

Rendall, D. (2004). "Recognizing kin": mechanisms, media, minds, modules and muddles. In: Chapais, B., and Berman, C. M. (eds.), *Kinship and Behavior in Primates*. Oxford University Press, New York. pp. 295–316.

Rendall, D., and Di Fiore, A. D. (1994). Evolution of social organization: a reappraisal for primates by using phylogenetic methods. *Proc. Natl. Acad. Sci. USA* 91:9941–9945.

Rendall, D., Rodman, P. S., and Emond, R. (1996). Vocal recognition of individuals and kin in free-ranging rhesus monkeys. *Anim. Behav.* 51:1007–1015.

Rendell, L., and Whitehead, H. (2001). Culture in whales and dolphins. *Behav. Brain Sci.* 24:309–324.

Rettig, N. L. (1978). Breeding behavior of the harpy eagle (*Harpia harpyja*). *Auk* 95:629–643.

Rey, P. J., and Alcantara, J. M. (2000). Recruitment dynamics of a fleshy-fruited plant (*Olea europaea*): connecting patterns of seed dispersal to seedling establishment. *J. Ecol.* 88:622–633.

Reynolds, V. (1992). Chimpanzees in the Budongo Forest, 1962–1992. *J. Zool.* 228:695–699.

Rhadakrishna, S. (2001). The social behavior of the Mysore slender loris (*Loris tardigradus lydekkerianus*) [PhD diss.]. University of Mysore, Manasagangotri.

Rhadakrishna, S., and Singh, M. (2002). Social behaviour of the slender loris (*Loris tardigradus lydekkerianus*). *Folia Primatol.* 73:181–196.

Rhadakrishna, S., and Singh, M. (2004). Reproductive biology of the slender loris (*Loris lydekkerianus lydekkerianus*). *Folia Primatol.* 75:1–13.

Rhadakrishna, S., Goswami, A. B., and Sinha, A. (2006). Distribution and conservation of *Nycticebus bengalensis* in northeastern India. *Int. J. Primatol.* 27:971–982.

Rhesus Macaque Genome Sequencing and Analysis Consortium (2007). Evolutionary and biomedical insights from the rhesus macaque genome. *Science* 316:222–234.

Rhine, R. J., Wasser, S. K., and Norton, G. W. (1988). Eight-year study of social and ecological correlates of mortality among immature baboons of Mikumi National Park, Tanzania. *Am. J. Primatol.* 16:199–212.

Rhine, R. J., Norton, G. W., and Wasser, S. K. (2000). Lifetime reproductive success, longevity, and reproductive life history of female yellow baboons (*Papio cynocephalus*) of Mikumi National Park, Tanzania. *Am. J. Primatol.* 51:229–241.

Ribbens, E., Silander, J. A., and Pacala, S. W. (1994). Seedling recruitment in forests: calibrating models to predict patterns of tree seedling dispersion. *Ecology* 75:1794–1806.

Rice, P., Longden, I., and Bleasby, A. (2000). EMBOSS: the European Molecular Biology Open Software Suite. *Trends Genet.* 16:276–277.

Richard, A. (1970). A comparative study of the activity patterns and behavior of *Alouatta villosa* and *Ateles geoffroyi*. *Folia Primatol.* 12:241–263.

Richard, A. F. (1973). Social organisation and ecology of *Propithecus verreauxi*, Grandidier 1867 [PhD diss.]. University of London, London.

Richard, A. F. (1974). Intra-specific variation in the social organization and ecology of *Propithecus verreauxi*. *Folia Primatol.* 22:178–207.

Richard, A. F. (1976). Preliminary observations on the birth and development of *Propithecus verreauxi* to the age of six months. *Primates* 17:357–366.

Richard, A. F. (1978). *Behavioral Variation: Case Study of a Malagasy Lemur.* Bucknell University Press, Lewisburg.

Richard, A. F. (1984). Lemurs. In: MacDonald, D. (ed.), *The Encyclopedia of Mammals*, vol. 1. George Allen and Unwin, London. pp. 330–331.

Richard, A. F. (1985a). *Primates in Nature*. W. H. Freeman, New York.

Richard, A. F. (1985b). Social boundaries in a Malagasy prosimian, the sifaka (*Propithecus verreauxi*). *Int. J. Primatol.* 6:553–568.

Richard, A. F. (1987). Malagasy prosimians: female dominance. In: Smuts, B. B., Cheney, D. L., Seyfarth, R. M., Wrangham, R. W., and Struhsaker, T. T. (eds.), *Primate Societies*. University of Chicago Press, Chicago. pp. 227–239.

Richard, A. F. (1992). Aggressive competition between males, female-controlled polygyny and sexual monomorphism in a Malagasy primate, *Propithecus verreauxi*. *J. Hum. Evol.* 22:395–406.

Richard, A. (2003). *Propithecus*, sifakas. In: Goodman, S. M., and Benstead, J. P. (eds.), *The Natural History of Madagascar*. University of Chicago Press, Chicago. pp. 1345–1348.

Richard, A. F., and Dewar, R. E. (1991). Lemur ecology. *Annu. Rev. Ecol. Syst.* 22:145–175.

Richard, A., and Schulman, S. R. (1982). Sociobiology: primate field studies. *Annu. Rev. Anthropol.* 11:231–255.

Richard, A. F., Goldstein, S. J., and Dewar, R. E. (1989). Weed macaques: the evolutionary implications of macaque feeding ecology. *Int. J. Primatol.* 10:569–594.

Richard, A. F., Rakotomanga, P., and Schwartz, M. (1993). Dispersal by *Propithecus verreauxi* at Beza Mahafaly, Madagascar: 1984–1991. *Am. J. Primatol.* 30:1–20.

Richard, A. F., Dewar, R. E., Schwartz, M., and Ratsirarson, J. (2002). Life in the slow lane: demography and life histories of male and female sifaka (*Propithecus verreauxi verreauxi*). *J. Zool.* 256:421–436.

Richards, P. W. (1996). *The Tropical Rain Forest,* 2nd ed. Cambridge: Cambridge University Press.

Richmond, B. G., and Strait, D. S. (2000). Evidence that humans evolved from a knuckle-walking ancestor. *Nature* 404:382–385.

Ricklefs, R. E. (2003). Is rate of ontogenetic growth constrained by resource supply or tissue growth potential? A comment on West et al.'s model. *Funct. Ecol.* 17:384–393.

Riechelmann, C., Hultsch, H., and Todt, D. (1994). Early development of social relationships in Barbary macaques (*Macaca sylvanus*): trajectories of alloparental behaviour during an infant's first three months of life. In: Roeder, J. J., Thierry, B., Anderson, J. B., and Herrenschmidt, N. (eds.), *Current Primatology. Social Development, Learning and Behaviour,* vol. II. Université Louis Pasteur, Strasbourg. pp. 279–286.

Ried, T., Arnold, N., Ward, D. C., and Wienberg, J. (1993). Comparative high-resolution mapping of human and primate chromosomes by fluorescence in situ hybridization. *Genomics* 18:381–386.

Rigamonti, M. M. (1993). Home range and diet in red ruffed lemurs (*Varecia variegata rubra*) on the Masoala Peninsula, Madagascar. In: Kappeler, P. M., and Ganzhorn, J. U. (eds.), *Lemur Social Systems and Their Ecological Basis*. Plenum Press, New York. pp. 25–39.

Rijksen, H. D. (1978). *A Field Study on Sumatran Orang-utans (*Pongo pygmaeus abelii*, Lesson 1827): Ecology, Behavior, and Conservation.* Veenaman and Zonen, Wageningen, The Netherlands. pp. 157–323.

Rijksen, H. D., and Meijaard, E. (1999). *Our Vanishing Relative: The Status of Wild Orang-utans at the Close of the Twentieth Century.* Kluwer Academic Publishers, Boston.

Riley, E. P. (2007). Flexibility in diet and activity patterns of *Macaca tonkeana* in response to anthropogenic habitat alteration. *Int. J. Primatol.* 28:107–133.

Rilling, J. K., Gutman, D. A., Zeh, T. R., Pagnoni, G., Berns, G. S., and Kilts, D. (2002). A neural basis for social cooperation. *Neuron* 35:395–405.

Ripley, S. (1967). The leaping of langurs: a problem in the study of locomotor adaptation. *Am. J. Phys. Anthropol.* 26:149–170.

Risler, L., Wasser, S. K., and Sackett, G. P. (1987). Measurement of excreted steroids in *Macaca nemestrina*. *Am. J. Primatol.* 12:91–100.

Robbins, A. M., and Robbins, M. M. (2005). Fitness consequences of dispersal decisions for male mountain gorillas (*Gorilla beingei beringei*). *Behav. Ecol. Sociobiol.* 58:295–309.

Robbins, A. M., Robbins M. M., Gerald-Steklis, N., and Steklis, H. D. (2006). Age-related patterns of reproductive success among female mountain gorillas. *Am. J. Phys. Anthropol.* 131:511–521.

Robbins, A. M., Stoinski, T. S., Fawcett, K. A., and Robbins, M. M. (2009a). Leave or conceive: natal dispersal and philopatry of female mountain gorillas in the Virunga volcano region. *Anim. Behav.* 77:831–838.

Robbins, A. M., Stoinski, T. S., Fawcett, K. A., and Robbins, M. M. (2009b). Socioecological influences on the dispersal of female mountain gorillas—evidence of a second folivore paradox. *Behav. Ecol. Sociobiol.* 63:477–489.

Robbins, C. T. (1993). *Wildlife Feeding and Nutrition*. Academic Press, New York.

Robbins, C. T., Mole, S., Hagerman, A. E., and Hanley, T. A. (1987). Role of tannins in defending plants against ruminants: reduction in protein availability. *Ecology* 68:1606–1615.

Robbins, A. M., Stoinski, T. S., Fawcett, K. A., and Robbins, M. M. (2009a). Leave or conceive: natal dispersal and philopatry of female mountain gorillas in the Virunga volcano region. *Anim. Behav.* 77: 831–838.

Robbins, A. M., Stoinski, T. S., Fawcett, K. A., and Robbins, M. M. (2009b). Socioecological influences on the dispersal of female mountain gorillas- evidence of a second folivore paradox. *Behav. Ecol. Sociobiol.* 63:477–489.

Robbins, M. M. (1995). A demographic analysis of male life history and social structure of mountain gorillas. *Behaviour* 132:21–47.

Robbins, M. M. (1996). Male–male interactions in heterosexual and all-male wild mountain gorilla groups. *Ethology* 102:942–965.

Robbins, M. M. (1999). Male mating patterns in wild multimale mountain gorilla groups. *Anim. Behav.* 57:1013–1020.

Robbins, M. M. (2001). Variation in the social system of mountain gorillas: the male perspective. In: Robbins, M. M., Sicotte, P., and Stewart, K. J. (eds.), *Mountain Gorillas: Three Decades of Research at Karisoke*. Cambridge University Press, Cambridge. pp. 29–58.

Robbins, M. M. (2003). Behavioral aspects of sexual selection in mountain gorillas. In: Jones, C. L. (ed.), *Sexual Selection & Reproductive Competition in Primates: New Perspectives and Directions*. American Society of Primatologists, Norman, OK. pp. 477–501.

Robbins, M. M. (2007). Gorillas: diversity in ecology and behavior. In: Campbell, C. J., Fuentes, A., MacKinnon, K. C., Panger, M., and Bearder, S. K. (eds.), *Primates in Perspective*, 1st ed. Oxford University Press, New York. pp. 305–321.

Robbins, M. M. (2008). Feeding competition and female social relationships in mountain gorillas of Bwindi Impenetrable National Park, Uganda. *Int. J. Primatol.* 29:999–1018.

Robbins, M. M. (2009). Male aggression towards females in mountain gorillas: courtship or coercion? In: Muller, M. N., and Wrangham, R. (eds.), *Sexual Coercion in Primates: An Evolutionary Perspective on Male Aggression Against Females*. Harvard University Press, Cambridge, MA.

Robbins, M. M., and Czekala, N. M. (1997). A preliminary investigation of urinary testosterone and cortisol levels in wild male mountain gorillas. *Am. J. Primatol.* 43:51–64.

Robbins, M. M., and McNeilage, A. (2003). Home range and frugivory patterns of mountain gorillas in Bwindi Impenetrable National Park, Uganda. *Int. J. Primatol.* 24:467–491.

Robbins, M. M., and Sawyer, S. C. (2007). Intergroup encounters in mountain gorillas of Bwindi Impenetrable National Park, Uganda. *Behaviour* 144:1497–1519.

Robbins, M., and Williamson, L. (2008). *Gorilla beringei*. In: 2008 IUCN Red List of Threatened Species, www.iucnredlist.org (accessed January 25, 2009).

Robbins, M. M., Bermejo, M., Cipolletta, C., Magliocca, F., Parnell, R. J., and Stokes, E. (2004). Social structure and life history patterns in western gorillas (*Gorilla gorilla gorilla*). *Am. J. Primatol.* 64:145–159.

Robbins, M. M., Robbins, A. M., Gerald-Steklis, N., and Steklis, H. D. (2005). Long-term dominance relationships in female mountain gorillas: strength, stability and determinants of rank. *Behaviour* 142:779–809.

Robbins, M. M., Robbins, A. M., Gerald-Steklis, N., and Steklis, H. D. (2007). Female reproductive success of the Virunga mountain gorillas: comparisons with the socioecological model. *Behav. Ecol. Sociobiol.* 61:919–931.

Robbins, M. M., Gray, M., Kagoda, E., and Robbins, A. M. (2009). Population dynamics of the Bwindi mountain gorillas. *Biol. Cons.* 142:2886–2895.

Roberts, M. (1994). Growth, development and parental care in the western tarsier (*Tarsius bancanus*) in captivity: evidence for a "slow" life history and nonmonogamous mating system. *Int. J. Primatol.* 15:1–28.

Roberts, M., and Kohn, F. (1993). Habitat use, foraging behavior and activity patterns in reproducing western tarsiers, *Tarsius bancanus,* in captivity: a management synthesis. *Zoo Biol.* 12:217–232.

Roberts, T. E., Davenport, T. R. B., Hildebrandt, K. B. P., Jones, T., Stanley, W. T., Sargis, E. J., and Olson, L. E. (2009). The biogeography of introgression in the critically endangered African monkey Rungwecebus kipunji. Biol. Letters doi:10.1098/rsbl.2009.0741.

Roberts, W. A., and Phelps, M. T. (1994). Transitive inference in rats—a test of the spatial coding hypothesis. *Psychol. Sci.* 5:368–374.

Robertson Yarrow, J. M., and van Schaik, C. P. (2001). Causal factors underlying the dramatic decline of the Sumatran orang-utan. *Oryx* 35:26–38.

Robinson, J. G. (1979). Vocal regulation of use of space by groups of titi monkeys, *Callicebus moloch. Behav. Ecol. Sociobiol.* 5:1–15.

Robinson, J. G. (1981). Spatial structure in foraging groups of wedge-capped capuchin monkeys, *Cebus nigrivittatus. Anim. Behav.* 29:1036–1056.

Robinson, J. G. (1982). Intrasexual competition and mate choice in primates. *Am. J. Primatol.* Supplement 1:131–144.

Robinson, J. G. (1986). Seasonal variation in use of time and space by the wedge-capped capuchin monkey, *Cebus olivaceus*: implications for foraging theory. *Smithson. Contrib. Zool.* 431:1–60.

Robinson, J. G. (1988a). Demography and group structure in wedge-capped capuchin monkeys, *Cebus olivaceus. Behaviour* 104:202–232.

Robinson, J. G. (1988b). Group size in wedge-capped capuchin monkeys *Cebus olivaceus* and the reproductive success of males and females. *Behav. Ecol. Sociobiol.* 23:187–197.

Robinson, J. G. (2000). Calculating maximum sustainable harvests and percentage offtakes. In: Robinson, J. G., and Bennett, E. L. (eds.), *Hunting for Sustainability in Tropical Forests.* Columbia University Press, New York. pp. 521–524.

Robinson, J. G., and Bennett, E. L. (eds.) (2000). *Hunting for Sustainability in Tropical Forests.* Columbia University Press, New York.

Robinson, J. G., and Janson, C. H. (1987). Capuchins, squirrel monkeys, and atelines: socioecological convergence with Old World primates. In: Smuts, B. B., Cheney, D. L., Seyfarth, R. M., Wrangham, R. W., and Struhsaker, T. T. (eds.), *Primate Societies.* University of Chicago Press, Chicago. pp. 69–82.

Robinson, J. G., and Redford, K. H. (1994). Community-based approaches to wildlife conservation in neotropical forests. In: Western, D., Wright, R. M., and Strum, S. C. (eds.), *Natural Connections: Perspectives in Community-Based Conservation.* Island Press, Washington DC. pp. 300–319.

Robinson, J. G., Wright, P. C., and Kinzey, W. G. (1987). Monogamous cebids and their relatives: intergroup calls and spacing. In: Smuts, B. B., Cheney, D. L., Seyfarth, R. M., Wrangham, R. W., and Struhsaker, T. (eds.), *Primate Societies.* University of Chicago Press, Chicago. pp. 44–53.

Roche, H., Delagnes, A., Brugal, J. P., Feibel, C., Kibunjia, M., Mourrel, V., and Texier, P. J. (1999). Early hominid stone tool production and technical skill 2.34 myr ago in West Turkana, Kenya. *Nature* 399:57–60.

Rocheta, M., Dionisio, F. M., Fonseca, L., and Pires, A. M. (2007). Paternity analysis in Excel. *Comp. Meth. Prog. Biomed.* 88:234–238.

Roda, S. A., and Mendes Pontes, A. R. (1998). Polygyny and infanticide in common marmosets in a fragment of Atlantic Forest of Brazil. *Folia Primatol.* 69:372–376.

Rode, K. D., Chapman, C. A., Chapman, L. J., and McDowell, L. R. (2003). Mineral resource availability and consumption by colobus in Kibale National Park, Uganda. *Int. J. Primatol.* 24:541–573.

Rode, K. D., Chapman, C. A., McDowell, L. R., and Stickler, C. (2006). Nutritional correlates of population density across habitats and log-

ging intensities in redtail monkeys (*Cercopithecus ascanius*). *Biotropica* 38:625–634.

Rodgers, A. R. (2001). Recent telemetry technology. In: Millspaugh, J. J., and Marzluff, J. M. (eds.), *Radio Tracking and Animal Populations.* Academic Press, San Diego. pp. 82–121.

Rodman, P. S. (1973). Population composition and adaptative organization among orang-utans of the Kutai Reserve. In: Michael, R. P., and Crook, J. H. (eds.), *Comparative Ecology and Behavior of Primates.* Academic Press, New York. pp. 171–209.

Rodman, P. S. (1977). Feeding behavior of orangutans in the Kutai Reserve, East Kalimantan. In: Clutton-Brock, T. H. (ed.), *Primate Ecology.* Academic Press, London. pp. 383–413.

Rodman, P. S. (1978). Diets, densities, and distributions of Bornean primates. In: Montgomery, G. G. (ed.), *The Ecology of Arboreal Folivores.* Smithsonian Institution Press, Washington DC. pp. 465–478.

Rodman, P. S. (1979a). Individual activity profiles and the solitary nature of orangutans. In: Hamburg, D. A., and McCown, E. R. (eds.), *The Great Apes.* Benjamin/Cummings, Menlo Park, CA. pp. 234–255.

Rodman, P. S. (1979b). Skeletal differentiation of *Macaca fascicularis* and *Macaca nemestrina* in relation to arboreal and terrestrial quadrupedalism. *Am. J. Phys. Anthropol.* 51:51–62.

Rodman, P. (1984). Foraging and social systems of orangutans and chimpanzees. In: Rodman, P., and Cant, J. (eds.), *Adaptations for Foraging in Nonhuman Primates.* Columbia University Press, New York. pp. 135–156.

Rodman, P. S. (1988). The ecology of social behavior. In: Slobodchikoff, C. N. (ed.), *Resources and Group Sizes of Primates.* Academic Press, San Diego. pp. 83–108.

Rodman, P. (1993). The human origins program and evolutionary ecology in anthropology today. *Evol. Anthropol.* 2:215–224.

Rodman, P. S., and Mitani, J. C. (1987). Orangutans: sexual dimorphism in a solitary species. In: Smuts, B. B., Cheney, D. L., Seyfarth, R. M., Wrangham, R. W., and Struhsaker, T. T. (eds.), *Primate Societies.* University of Chicago Press, Chicago. pp. 146–152.

Rodrigues de Moraes, P. L., de Carvalho, O., and Strier, K. B. (1998). Population variation in patch and party size in muriquis (*Brachyteles arachnoides*). *Int. J. Primatol.* 19:325–337.

Rodriguez, E., and Wrangham, R. W. (1993). Zoopharmacognosy: the use of medicinal plants by animals. In: Downum, K. R., Romeo, J. T., and Stafford, H. (eds.), *Recent Advances in Phytochemistry. Phytochemical Potential of Tropic Plants,* vol. 27. Plenum Press, New York. pp. 89–105.

Rodriguez, E., Aregullin, M., Nishida, T., Uehara, S., Wrangham, R. W., Abramowski, Z., Finlayson, A., and Towers, G. H. N. (1985). Thiarubrin A, a bioactive constituent of *Aspilia* (Asteraceae) consumed by wild chimpanzees. *Experientia* 41:419–420.

Rodríguez, G. A. C., and Boher, S. (1988). Notes on the biology of *Cebus nigrivittatus* and *Alouatta seniculus* in northern Venezuela. *Primate Conserv.* 9:61–66.

Rodseth, L., Wrangham, R. W., Harrigan, A. M., and Smuts, B. B. (1991). The human community as a primate society. *Curr. Anthropol.* 32:221–254.

Roeder, A. D., Archer, F. I., Poinar, H. N., and Morin, P. A. (2004). A novel method for collection and preservation of faeces for genetic studies. *Mol. Ecol. Notes* 4:761–764.

Roeder, J. J., Fornasieri, I., and Gosset, D. (2002). Conflict and postconflict behaviour in two lemur species with different social organizations (*Eulemur fulvus* and *Eulemur macaco*): a study on captive groups. *Aggress. Behav.* 28:62–74.

Roff, D. A. (1992). *The Evolution of Life Histories.* Chapman and Hall, New York.

Roff, D. A. (2002). *Life History Evolution,* 2nd ed. Sinauer Associates, Sunderland, MA.

Rogers, A. R. (1993). Why menopause? *Evol. Ecol.* 7:406–420.

Rogers, M. E., Maisels, F., Williamson, E. A., Tutin, C. E. G., and Fernandez, M. (1990). Gorilla diet in the Lope Reserve, Gabon: a nutritional analysis. *Oecologia* 84:326–339.

Rogers, M. E., Abernethy, K. A., Fontaine, B., Wickings, E. J., White, L. J. T., and Tutin, C. E. G. (1996). Ten days in the life of a mandrill horde in the Lopé Reserve, Gabon. *Am. J. Primatol.* 40: 297–313.

Rogers, M. E., Voysey, B. C., McDonald, K. E., Parnell, R. J., and Tutin, C. E. G. (1998). Lowland gorillas and seed dispersal: the importance of nest sites. *Am. J. Primatol.* 45:45–68.

Rogers, M. E., Abernethy, K., Bermejo, M., Cipolletta, C., Doran, D., McFarland, K., Nishihara, T., Remis, M., and Tutin, C. E. G. (2004). Western gorilla diet: a synthesis from six sites. *Am. J. Primatol.* 64:173–192.

Rolland, N., and Roeder, J. J. (2000). Do ringtailed lemurs (*Lemur catta*) reconcile in the hour post-conflict? A pilot study. *Primates* 41:223–227.

Romero, T., Colmenares, F., and Aureli, F. (2009). Testing the function of reconciliation and third-party affiliation for aggressors in hamadryas baboons (*Papio hamadryas hamadryas*). *Am. J. Primatol.* 71:60–69.

Romero-Herrera, A. E., Lehmann, H., Joysey, K. A., and Friday, A. E. (1976). Evolution of myoglobin amino acid sequences in primates and other vertebrates. In: Goodman, M., Tashian, R. E., and Tashian, J. H. (eds.), *Molecular Anthropology: Genes and Proteins in the Evolutionary Ascent of the Primates*. Plenum Press, New York. pp. 289–300.

Ron, T., Henzi, S. P., and Motro, U. (1996). Do female chacma baboons compete for a safe spatial position in a southern woodland habitat? *Behaviour* 133:475–490.

Rönn, A. C., Andrés, O., Bruford, M. W., Crouau-Roy, B., Doxiadis, G., Domingo-Roura, X., Roeder, A. D., Verschoor, E., Zischler, H., and Syvänen, A.-C. (2006). Multiple displacement amplification for generating an unlimited source of DNA for genotyping in nonhuman primate species. *Int. J. Primatol.* 27:1145–1169.

Ronquist, F., and Huelsenbeck, J. P. (2003). MrBayes 3: bayesian phylogenetic inference under mixed models. *Bioinformatics* 19:1572–1574.

Roonwal, M. L., and Mohnot, S. M. (1977). *Primates of South Asia: Ecology, Sociobiology and Behavior*. Harvard University Press, Cambridge, MA.

Roos, C. (2003). Molekulare Phylogenie der Halbaffen, Schlankaffen, und Gibbons [diss.]. University of Munich, Munich.

Roos, C., and Geissmann, T. (2001). Molecular phylogeny of the major hylobatid divisions. *Mol. Phylogenet. Evol.* 19:486–494.

Roos, C., Schmitz, J., and Zischler, H. (2004). Primate jumping genes elucidate strepsirrhine phylogeny. *Proc. Natl. Acad. Sci. USA* 101:10650–10654.

Roper, T. J. (1986). Cultural evolution of feeding behaviour in animals. *Sci. Prog.* 70:751–783.

Rorhrer-Ertl, O. (1988). Research history, nomenclature, and taxonomy of the orang-utan. In: Schwartz, J. H. (ed.), *Orang-utan Biology*. Oxford University Press, New York. pp. 7–18.

Rosati, A. G., Stevens, J. R., Hare, B., and Hauser, M. D. (2007). The evolutionary origins of human patience: temporal preferences in chimpanzees, bonobos, and human adults. *Curr. Biol.* 17:1663–1668.

Rose, A. L. (2002). Conservation must pursue human–nature biosynergy in the era of social chaos and bushmeat commerce. In: Fuentes, A., and Wolfe, L. D. (eds.), *Primates Face to Face*. Cambridge University Press, Cambridge. pp. 208–239.

Rose, L. M. (1994a). Benefits and costs of resident males to females in white-faced capuchins, *Cebus capucinus*. *Am. J. Primatol.* 32:235–248.

Rose, L. M. (1994b). Sex differences in diet and foraging behavior in white-faced capuchins, *Cebus capucinus*. *Int. J. Primatol.* 15:63–82.

Rose, L. M. (1998). Behavioral ecology of white-faced capuchins (*Cebus capucinus*) in Costa Rica [PhD diss.]. Washington University, St. Louis, MO.

Rose, L. M., and Fedigan, L. M. (1995). Vigilance in white-faced capuchins (*Cebus capucinus*) in Costa Rica. *Anim. Behav.* 49:63–70.

Rose, M. D. (1974). Postural adaptations in New and Old World monkeys. In: Jenkins, F. A., Jr. (ed.), *Primate Locomotion*. Academic Press, New York. pp. 201–222.

Rose, M. D. (1976). Bipedal behavior of olive baboons (*Papio anubis*) and its relevance to an understanding of the evolution of human bipedalism. *Am. J. Phys. Anthropol.* 44:247–262.

Rose, M. D. (1978). Feeding and associated positional behavior of black and white colobus monkeys (*Colobus guereza*). In: Montgomery, C. G. (ed.), *The Ecology of Arboreal Folivores*. Smithsonian Institution Press, Washington DC. pp. 253–262.

Rose, M. R. (1991). *Evolutionary Biology of Aging*. Oxford University Press, Oxford.

Rosenberger, A. L. (1977). *Xenothrix* and ceboid phylogeny. *J. Hum. Evol.* 6:461–481.

Rosenberger, A. L. (1981). Systematics of the higher taxa. In: Coimbra-Filho, A. F., and Mittermeier, R. A. (eds.), *Ecology and Behavior of Neotropical Primates*, vol. 1. Academia Brasileira de Ciências, Rio de Janeiro. pp. 9–27.

Rosenberger, A. L. (1992). Evolution of feeding niches in New World monkeys. *Am. J. Phys. Anthropol.* 88:525–562.

Rosenberger, A. L. (2002). Platyrrhine paleontology and systematics: the paradigm shifts. In: Hartwig, W. C. (ed.), *The Primate Fossil Record*. Cambridge University Press, Cambridge. pp. 151–160.

Rosenberger, A. L., and Matthews, L. J. (2008). *Oreonax*—not a genus. *Neotrop. Primates* 15:8–12.

Rosenberger, A. L., and Strier, K. B. (1989). Adaptive radiation of the ateline primates. *J. Hum. Evol.* 18:717–750.

Rosenberger, A., and Szalay, F. (1980). On the tarsiiform origins of the Anthropoidea. In: Ciochon, R., and Chiarelli, A. (eds.), *Evolutionary Biology of the New World Monkeys and Continental Drift*. Plenum Press, New York. pp. 139–157.

Rosenblum, L. A., Kaufman, I. C., and Stynes, A. J. (1964). Individual distance in two species of macaque. *Anim. Behav.* 12:338–342.

Rosenblum, L. A., Coe, C. L., and Bromley, L. J. (1975). Peer relations in monkeys: the influence of social structure, gender, and familiarity. *Orig. Behav.* 4:67–98.

Rosenblum, L. L., Supriatna, J., and Melnick, D. J. (1997a). Phylogeographic analysis of pigtail macaque populations (*Macaca nemestrina*) inferred from mitochondrial DNA. *Am. J. Phys. Anthropol.* 104:35–45.

Rosenblum, L. L., Supriatna, J., Hasan, M. N., and Melnick, D. J. (1997b). High mitochondrial DNA diversity with little structure within and among leaf monkey populations (*Trachypithecus cristatus* and *Trachypithecus auratus*). *Int. J. Primatol.* 18:1005–1028.

Rosenthal, G. A., and Janzen, D. H. (1979). *Herbivores: Their Interaction with Secondary Plant Metabolites*. Academic Press, New York.

Ross, C. (1988). The intrinsic rate of natural increase and reproductive effort in primates. *J. Zool. Lond.* 214:199–219.

Ross, C. (1991). Life history patterns in New World monkeys. *Int. J. Primatol.* 12:481–502.

Ross, C. (1992a). Basal metabolic rate, body weight and diet in primates: an evaluation of the evidence. *Folia Primatol.* 58:7–23.

Ross, C. (1992b). Environmental correlates of the intrinsic rate of natural increase in primates. *Oecologia* 90:383–390.

Ross, C. (1993). Predator mobbing by an all-male band of Hanuman langurs (*Presbytis entellus*). *Primates* 34:105–107.

Ross, C. (1994). The craniofacial evidence for anthropoid and tarsier relationships. In: Fleagle, J., and Kay, R. (eds.), *Anthropoid Origins*. Plenum Press, New York. pp. 469–547.

Ross, C. (1996). Adaptive explanation for the origins of the Anthropoidea (Primates). *Am. J. Primatol.* 40:205–230.

Ross, C. (1998). Primate life histories. *Evol. Anthropol.* 6:54–63.

Ross, C. (2002). Life history, infant care strategies, and brain size in primates. In: Kappeler, P. M., and Pereira, M. E. (eds.), *Primate Life Histories and Socioecology*. University of Chicago Press, Chicago. pp. 266–284.

Ross, C. (2004). Life history and the evolution of large brain size in great apes. In: Russon, A. E., and Begun, D. (eds.), *The Evolution of Thought*. Cambridge University Press, New York. pp. 122–139.

Ross, C., and Jones, K. E. (1999). Socioecology and the evolution of primate reproductive rates. In: Lee, P. C. (ed.), *Comparative Primate Socioecology*. Cambridge University Press, Cambridge. pp. 73–110.

Ross, C. F., and Kay, R. F. (eds.) (2004). *Anthropoid Origins: New Visions*. New York: Springer.

Ross, C., and MacLarnon, A. (2000). The evolution of non-maternal care in anthropoid primates: a test of the hypotheses. *Folia Primatol.* 71:93–113.

Ross, C., and Regan, G. (2000). Allocare, predation risk, social structure and natal coat colour in anthropoid primates. *Folia Primatol.* 71:67–76.

Ross, M., Menzler, S., and Zimmermann, E. (2008). Rapid facial mimicry in orangutan play. *Biol. Lett.* 4:27–30.

Rostal, D. C., Glick, B. B., Eaton, G. G., and Resko, J. A. (1986). Seasonality of adult male Japanese macaques (*Macaca fuscata*): androgens and behavior in a confined troop. *Horm. Behav.* 20:452–462.

Rothman, J. M. (2006). Nutritional ecology and parasite dynamics of mountain gorillas [PhD diss.]. Cornell University, Ithaca, NY.

Rothman, J. M., Van Soest, P. J., and Pell, A. N. (2006a). Decaying wood is a sodium source for mountain gorillas. *Biol. Lett.* 2:321–324.

Rothman, J. M. Dierenfeld, E. S., Molina, D. O., Shaw, A. V., Hintz, H. F., Pell, A. N. (2006b). Nutritional chemistry of foods eaten by gorillas in Bwindi Impenetrable National Park, Uganda. *Am. J. Primatol.* 68:675–691.

Rothman, J. M., Chapman, C. A., Twinomgisha, D., Wasserman, M. D., Lambert, J. E., and Goldberg, T. L. (2008). Measuring physical traits of primates remotely: the use of parallel lasers. *Am. J. Primatol.* 12:1191–1195.

Rothman, J. M., Dusinberre, K., and Pell, A. N. (2009). Condensed tannins in the diets of primates: a matter of methods? *Am. J. Primatol.* 71:70–76.

Rotundo, M., Sloan, C., and Fernandez-Duque, E. (2000). Cambios estacionales en el ritmo de actividad del mono mirikiná (*Aotus azarai*) en Formosa Argentina. In: Cabrera, E., Mércolli, C., and Resquin, R. (eds.), *Manejo de Fauna Silvestre en Amazonía y Latinoamérica.* Fundación Moise's Bertoni, Asunción, Paraguay. pp. 413–417.

Rotundo, M., Fernandez-Duque, E., and Giménez, M. (2002). Cuidado biparental en el mono de noche (*Aotus azarai azarai*) de Formosa, Argentina. *Neotrop. Primates* 10:70–72.

Rotundo, M., Fernandez-Duque, E., and Dixson, A. F. (2005). Infant development and parental care in free-ranging groups of owl monkeys (*Aotus azarai azarai*) in Argentina. *Int. J. Primatol.* 26(6):1459–1473.

Roux, V., and Bril, B. (eds.) (2005). *Stone Knapping: The Necessary Conditions for a Uniquely Hominin Behavior.* McDonald Institute Monograph Series. MCDonald Institute Monographs series. Cambridge: McDonald Institute for Archaeological Research,. pp. 171–182.

Rowe, N. (1996). *The Pictorial Guide to the Living Primates.* Pogonias Press, East Hampton, NY.

Rowell, T. E. (1966). Forest living baboons in Uganda. *J. Zool. Soc. Lond.* 147:344–364.

Rowell, T.E. (1967). A quantitative comparison of the behaviour of a wild and a caged baboon group *Anim. Behav.,* 15, 1967, 499–509

Rowell, T. E. (1970a). Baboon menstrual cycles affected by social environment. *J. Reprod. Fertil.* 21:133–141.

Rowell, T. E. (1970b). Reproductive cycles of two *Cercopithecus* monkeys. *J. Reprod. Fertil.* 22:321–328.

Rowell, T. E. (1971). Social organization of wild talapoin monkeys. *Am. J. Phys. Anthropol.* 38:593–598.

Rowell, T. E. (1972). Female reproduction cycles and social behavior in primates. In: Lehrmen, D. S., Hinde, R. A., and Shaw, E. (eds.), *Advances in the Study of Behavior,* vol. 4. Academic Press, New York. pp. 69–105.

Rowell, T. (1988a). The social system of guenons, compared with baboons, macaques and mangabeys. In: Gautier-Hion, A., Bourliere, F., and Gautier, J. (eds.), *A Primate Radiation: Evolutionary Biology of the African Guenons.* Cambridge University Press, Cambridge. pp. 439–451.

Rowell, T. (1988b). Beyond the one-male group. *Behaviour* 104:189–201.

Rowell, T. E. (1993a). Reification of social systems. *Evol. Anthropol.* 2:215–224.

Rowell, T. E. (1993b). Forward. In: Pereira, M. E., and Fairbanks, L. A. (eds.), *Juvenile Primates: Life History, Development, and Behavior.* Oxford University Press, New York. pp. vii–ix.

Rowell, T. E. (2000). A few peculiar primates. In: Strum, S. C., and Fedigan, L. M. (eds.), *Primate Encounters: Models of Science, Gender, and Society.* University of Chicago Press, Chicago. pp. 57–70.

Rowell, T. E., and Dixson, A. F. (1975). Changes in the social organization during the breeding seasons of wild talapoin monkeys. *J. Reprod. Fertil.* 43:419–434.

Rowell, T. E., and Hinde, R. A. (1962). Vocal communication by the rhesus monkey (*Macaca mulatta*). *Proc. Zool. Soc. Lond.* 138:279–294.

Rowell, T. E., and Mitchell, B. J. (1991). Comparison of seed dispersal of guenons in Kenya and capuchins in Panama. *J. Trop. Ecol.* 7:269–274.

Rowell, T. E., and Richards, S. M. (1979). Reproductive strategies of some African monkeys. *J. Mammal.* 60:58–69.

Rozas, J., Sánchez-DelBarrio, J. C., Messeguer, X., and Rozas, R. (2004). DnaSP, DNA polymorphism analyses by the coalescent and other methods. *Bioinformatics* 19:2496–2497.

Rubenstein, D. I., and Wrangham, R. W. (1988). *Ecological Aspects of Social Evolution: Birds and Mammals.* Princeton University Press, Princeton, NJ.

Rudran, R. (1973). The reproductive cycles of two subspecies of purple-faced langurs (*Presbytis senex*) with relation to environmental factors. *Folia Primatol.* 19:41–60.

Rudran, R. (1978). *Socioecology of the Blue Monkeys of the Kibale Forest, Uganda.* Smithsonian Contributions to Zoology, 249. Smithsonian Institution Press, Washington DC.

Rudran, R. (1979). The demography and social mobility of a red howler (*Alouatta seniculus*) population in Venezuela. In: Eisenberg, J. F. (ed.), *Vertebrate Ecology of the Northern Neotropics.* Smithsonian Institution Press, Washington DC. pp. 107–126.

Rudran, R., and Fernandez-Duque, E. (2003). Demographic changes over thirty years in a red howler population in Venezuela. *Int. J. Primatol.* 24:925–947.

Rüedi, D. (1981). Hand-rearing and reintegration of a caesarian-born proboscis monkey *Nasalis larvatus* at Basle Zoo. *Int. Zoo Ybk.* 21:225–229.

Ruehlmann, T., Bernstein, I., Gordon, T., and Balcaen, P. (1987). Wounding patterns in three species of captive macaques. *Am. J. Primatol.* 14:125–134.

Ruff, C. B. (1987). Sexual dimorphism in human lower limb bone structure: relationship to subsistence strategy and sexual division of labor. *J. Hum. Evol.* 16:391–416.

Ruhiyat, Y. (1983). Socio-ecological study of *Presbytis aygula* in west Java. *Primates* 24:344–359.

Rumiz, D. I. (1990). *Alouatta caraya*: population density and demography in northern Argentina. *Am. J. Primatol.* 21:279–294.

Rumpler, Y., and Albignac, R. (1975). Intraspecific chromosome variability in a lemur from the north of Madagascar: *Lepilemur septentrionalis,* species nova. *Am. J. Phys. Anthropol.* 42:425–429.

Ruschi, A. (1964). Macacos do estado do Espírito Santo. *Bol. Mus. Mello-Leitão* 23:1–18.

Russell, C., and Russell, W. M. S. (1990). Cultural evolution of behaviour. *Neth. J. Zool.* 40:745–762.

Russell, R. J. (1977). The behavior, ecology, and environmental physiology of a nocturnal primate, *Lepilemur mustelinus* (Strepsirhini, Lemuriformes, Lepilemuridae) [PhD diss.]. Duke University, Durham, NC.

Russell, R. J. (1980). The environmental physiology and ecology of *Lepilemur ruficaudatus* (*L. leucopus*) in arid southern Madagascar. *Am. J. Phys. Anthropol.* 52:273–274.

Russo, S. E. (2003a). Linking spatial patterns of seed dispersal and plant recruitment in a neotropical tree, *Virola calophylla* (Myristicaceae) [PhD diss.]. University of Illinois, Urbana.

Russo, S. E. (2003b). Responses of dispersal agents to tree and fruit traits in *Virola calophylla* (Myristicaceae): implications for selection. *Oecologia* 136:80–87.

Russo, S. E. (2005). Linking seed fate to natural dispersal patterns: factors affecting predation and scatter-hoarding of *Birola calophylla* seeds in Peru. *J. Trop. Ecol.* 21:243–253.

Russo, S. E., and Augspurger, C. K. (2004). Aggregated seed dispersal by spider monkeys limits recruitment to clumped patterns in *Virola calophylla*. *Ecol. Lett.* 7:1058–1067.

Russo, S. E., Campbell, C. J., Dew, J. L., Stevenson, P. R., and Suarez, S. (2005). A multi-site comparison of dietary preferences and seed dispersal by spider monkeys (*Ateles* spp.). *Int. J. Primatol.* 26:1017–1037.

Russo, S. E., Portnoy, S., and Augspurger, C. K. (2006). Incorporating animal behavior into seed dispersal models: implications for seed shadows. *Ecology* 87:3160–3174.

Russo, S. E., Potts, M. D., Davies, S. J., and Tan, S. (2007). Determinants of tree species distributions: comparing the roles of dispersal, seed size, and soil specialization in a Bornean rain forest. In: Dennis, A., Schupp, E. W., Green, R., and Westcott, D. (eds.), *Seed Dispersal: Theory and Its Application in a Changing World.* CAB International, Wallingford, UK. pp. 499–518.

Russon, A. E. (2002). Return of the native: cognition and site-specific expertise in orangutan rehabilitation. *Int. J. Primatol.* 23:461–78.

Russon, A. E., and Galdikas, B. M. F. (1993). Imitation in ex-captive orangutans (*Pongo pygmaeus*). *J. Comp. Psychol.* 107:147–161.

Russon, A. E., and Galdikas, B. M. F. (1995). Constraints on great apes' imitation: model and action selectivity in rehabilitant orangutan (*Pongo pygmaeus*) imitation. *J. Comp. Psychol.* 109:5–17.

Russon, A. E., Erman, A., and Sinaga, P. (2001). The population and distribution of orangutans (*Pongo pygmaeus pygmaeus*) in and around the Danau Sentarum wetlands reserve, West Kalimantan, Indonesia. *Biol. Conserv.* 97:21–28.

Russon, A. E., Handayani, D. P., Kuncoro, P., and Ferisa, A. (2007). Orangutan leaf-carrying for nest-building: toward unraveling cultural processes. *Anim. Cogn.* 10:189–202.

Russon, A. E., van Schaik, C. P., Kuncoro, P., Ferisa, A., Hadayani, D. P., and van Noordwijk, M. A. (2009a). Innovation and intelligence in orangutans. In: Wich, S. A., Utami, S. S., Mitra Setia, T., and van Schaik, C. (eds.), *Orangutans: Geographic Variation in Behavioral Ecology and Conservation.* Oxford University Press, Oxford. pp. 279–298.

Russon, A. E., Wich, S. A., Ancrenaz, M., Kanamori, T., Knott, C. D., Kuze, N., Morrogh-Bernard, H. C., Pratje, P., Ramlee, H., Rodman, P., Sawang, A., Sidiyasa, K., Singleton, I., and van Schaik, C. P. (2009b). Geographic variation in orangutan diets. In: Wich, S. A., Utami, S. S., Mitra Setia, T., and van Schaik, C. (eds.), *Orangutans: Geographic Variation in Behavioral Ecology and Conservation.* Oxford University Press, Oxford. pp. 135–156.

Rusterholz, H., and Erhardt, A. (2000). Can nectar properties explain sex-specific preferences in the adonis blue butterfly *Lysandra bellargus?* *Ecol. Entomol.* 25:81–90.

Rutberg, A. T. (1983). The evolution of monogamy in primates. *J. Theor. Biol.* 104:93–112.

Ruvolo, M. (1997a). Genetic diversity in hominoid primates. *Annu. Rev. Anthropol.* 26:515–540.

Ruvolo, M. (1997b). Molecular phylogeny of the hominoids: inferences from multiple independent DNA sequence data sets. *Mol. Biol. Evol.* 14:248–265.

Ruvolo, M., Pan, D., Zehr, S., Goldberg, T., Disotell, T. R., and von Dornum, M. (1994). Gene trees and hominoid phylogeny. *Proc. Natl. Acad. Sci. USA* 91:8900–8904.

Ryan, K. M. (1995). Preliminary report on social structure and alloparental care on an island in Guri Reservoir. *Am. J. Phys. Anthropol.* 20(suppl.):187.

Ryan, M. J., and Rand, A. S. (1993). Species recognition and sexual selection as a unitary problem in animal communication. *Evolution* 47:647–657.

Ryder, O. A., and Chemnick, L. G. (1993). Chromosomal and mitochondrial DNA variation in orangutan. *J. Hered.* 84:405–409.

Rylands, A. B. (1982). The behaviour and ecology of three species of marmosets and tamarins (Callitrichidae, Primates) in Brazil [PhD diss.]. University of Cambridge, Cambridge.

Rylands, A. B. (1986a). Ranging behaviour and habitat preference of a wild marmoset group, *Callithrix humeralifer* (Callitrichidae, Primates). *J. Zool. Lond.* 210:489–514.

Rylands, A. B. (1986b). Infant-carrying in a wild marmoset group, *Callithrix humeralifer*: evidence for a polyandrous mating system. In: Thiago de Mello, M. (ed.), *A Primatologia No Brasil*, vol. 2. Sociedade Brasileira de Primatologia, Brasilia. pp. 131–144.

Rylands, A. B. (1989). Sympatric Brazilian callitrichids—the black-tufted-ear marmoset, *Callithrix kuhli*, and the golden-headed lion tamarin, *Leontopithecus chrysomelas. J. Hum. Evol.* 18:679–695.

Rylands, A. B. (1993a). *Marmosets and Tamarins: Systematics, Behaviour, and Ecology*. Oxford University Press, Oxford.

Rylands, A. B. (1993b). The ecology of the lion tamarins, *Leontopithecus*: some intrageneric differences and comparisons with other callitrichids. In: Rylands, A. B. (ed.), *Marmosets and Tamarins: Systematics, Behaviour, and Ecology*. Oxford University Press, Oxford. pp. 296–313.

Rylands, A. B., and Chiarello, A. G. (2003). Official list of Brazilian fauna threatened with extinction—2003. *Neotrop. Primates* 11:43–49, http://www.mma.gov.br/port/sbf/fauna/index.cfm.

Rylands, A. B., and de Faria, D. S. (1993). Habitats, feeding ecology, and home range size in the genus *Callithrix*. In: Rylands, A. B. (ed.), *Marmosets and Tamarins: Systematics, Behaviour, and Ecology*. Oxford University Press, Oxford. pp. 262–272.

Rylands, A., and Mittermeier, R. (2008). The diversity of the New World primates (Platyrrhini): an annotated taxonomy. In: Garber, P., Estrada, A., Bicca-Marques, J. C., Heymann, E. W., and Strier, K. B. (eds.), *South American Primates: Comparative Perspectives in the Study of Behavior, Ecology and Conservation*. Springer, New York. pp. 23–54.

Rylands, A. B., Mittermeier, R. A., and Luna, E. R. (1995). A species list for the New World primates (Platyrrhini): distribution by country, endemism, and conservation status according to the Mace-Land system. *Neotrop. Primates* 3(suppl.):113–160.

Rylands, A., Strier, K., Mittermier, R., Borovansky, J., and Seal, U. S. (1998). *Population and Habitat Viability Assessment Workshop for the Muriqui (*Brachyteles arachnoides*)*. Conser- vation Breeding Specialist Group, Apple Valley, MN.

Rylands, A. B., Schneider, H., Langguth, A., Mittermeier, R. A., Groves, C. P., and Rodríguez-Luna, E. (2000). An assessment of the diversity of New World primates. *Neotrop. Primates* 8:61–93.

Saayman, G. S. (1970). The menstrual cycle and sexual behaviour in a troop of free-ranging chacma baboons (*Papio ursinus*). *Folia Primatol.* 12:81–110.

Saayman, G. S. (1971). Behaviour of the adult males in a troop of free-ranging chacma baboons (*Papio ursinus*). *Folia Primatol.* 15:36–57.

Saayman, G. S. (1972). Effects of ovarian hormones upon the sexual skin and mounting behaviour in the free-ranging chacma baboon (*Papio ursinus*). *Folia Primatol.* 17:297–303.

Saayman, G. S. (1973). Effects of ovarian hormones on the sexual skin and behavior of ovariectomized baboons (*Papio ursinus*) under free-ranging conditions. In: Phoenix, C. H. (ed.), *Symposia of the Fourth International Congress of Primatology. Primate Reproductive Behavior*, vol. 2. Karger, Basel. pp. 64–98.

Saayman, G. (1975). The influence of hormonal and ecological factors upon sexual behavior and social organization in Old World Primates. In: Tuttle, R. H. (ed.), *Socioecology and Psychology of Primates*. Mouton, The Hague. pp. 181–204.

Sabater Pi, J. (1972). Contribution to the ccology of *Mandrillus sphinx* Linnaeus 1758 of Rio Muni (Republic of Equatorial Guinea). *Folia Primatol.* 17:304–319.

Sabater-Pi, J. (1979). Feeding behaviour and diet of chimpanzees (*Pan troglodytes troglodytes*) in the Okorobik Mountains of Rio Muni (West Africa). *Z. Tierpsychol.* 50:265–281.

Sabater-Pi, J., and Groves, C. (1972). The importance of higher primates in the diet of the Fang of Rio Muni. *Man* 7:239–243.

Sabeti, P. C., Reich, D. E., Higgins, J. M., Levine, H. Z. P., Richter, D. J., Schaffner, S. F., Gabriel, S. B., Platko, J. V., Patterson, N. J., and McDonald, G. J. (2002). Detecting recent positive selection in the human genome from haplotype structure. *Nature* 419:832–837.

Sacher, G. A. (1959). The relation of life span to brain weight and body weight in mammals. In: Wolstenholme, G. E. W., and O'Connor, M. (eds.), *CIBA Foundation Colloquia on Aging. The Lifespan of Animals*, vol. 5. Churchill, London. pp. 115–133.

Sacher, G. A., and Staffeldt, E. F. (1974). Relation of gestation time to brain weight for placental mammals. *Am. Nat.* 108:593–615.

Sackett, G., and Fredrickson, W. (1987). Social preferences by pigtailed macaques: familiarity versus degree and type of kinship. *Anim. Behav.* 35:603–606.

Sade, D. S. (1965). Some aspects of parent–offspring and sibling relations in a group of rhesus monkeys, with a discussion of grooming. *Am. J. Phys. Anthropol.* 23:1–17.

Sade, D. S. (1967). Determinants of dominance in a group of free-ranging rhesus monkeys. In: Altmann, S. (ed.), *Social Communication Among Primates*. University of Chicago Press, Chicago. pp. 99–114.

Sade, D. S. (1968). Inhibition of son-mother mating among free-ranging rhesus monkeys. *Sci. Psychoanal.* 12:18–38.

Sade, D. S. (1972). Sociometrics of *Macaca mulatta*. I. Linkages and cliques in grooming matrices. *Folia Primatol.* 18:196–223.

Sade, D. S. (1980). Population biology of free-ranging rhesus monkeys on Cayo Santiago, Puerto Rico. In: Cohen, M. N., Malpass, R. S., and Klein, H. G. (eds.), *Mechanisms of Population Regulation*. Yale University Press, New Haven. pp. 171–187.

Sade, D. S. (1990). Intrapopulation variation in life history parameters. In: DeRousseau, C. J. (ed.), *Primate Life History and Evolution*. Wiley-Liss, New York. pp. 181–194.

Sade, D. S., Chepko-Sade, B. D., Schneider, J. M., Roberts, S. S., and Richtsmeier, J. T. (1985). *Basic Demographic Observations on Free-Ranging Rhesus Monkeys*. Human Relations Area Files, New Haven.

Saiki, R. K., Scharf, S. J., Faloona, F., Mullis, K., Horn, G. T., Erlich, H. A., and Arnheim, N. (1985). Enzymatic amplification of beta-globin genomic sequences and restriction site analysis for diagnosis of sickle-cell anemia. *Science* 230:1350–1354.

Saiki, R. K., Gelfand, D. H., Stoffel, S., Scharf, S. J., Higuchi, R., Horn, G. T., Mullis, K., and Erlich, H. A. (1988). Primer-directed enzymatic amplification of DNA with a thermostable DNA polymerase. *Science* 239:487–491.

Saj, T. L., and Sicotte, P. (2005). Male takeover in *Colobus vellerosus* at Boabeng-Fiema Monkey Sanctuary, central Ghana. *Primates* 46:211–214.

Saj, T. L., Teichroeb, J. A., and Sicotte, P. (2005a). The population status and habitat quality of the Geoffroy's pied colobus (*Colobus vellerosus*) at Boabeng-Fiema sacred grove, Ghana. In: Paterson, J. D. (ed.), *Commensalism and Conflict: Human–Primate Interface*. American Society of Primatology, Norman, OK. pp. 264–287.

Saj, T. L., Teichroeb, J. A., and Sicotte, P. (2005b). The population status of the ursine colobus (*Colobus vellerosus*) at Boabeng-Fiema, Ghana. In: Paterson, J. (ed.), *Commensalism and Conflict: The Human–Primate Interface*. American Society of Primatologists, Norman, OK. pp. 350–374.

Sakura, O., and Matsuzawa, T. (1991). Flexibility of wild chimpanzee nut-cracking behavior using stone hammers an anvils: an experimental analysis. *Ethology* 87:237–248.

Saltzman, W. (2003). Reproductive competition among female common marmosets (*Callithrix jacchus*): proximate and ultimate causes. In: Jones, C. (ed.), *Sexual Selection and Reproductive Competition in Primates: New Perspectives and Directions*. American Society of Primatologists, Norman, OK. pp. 197–229.

Saltzman, W., Schultz-Darken, N. J., Scheffler, G., Wegner, F. H., and Abbott, D. H. (1994). Social and reproductive influences on plasma cortisol in female marmoset monkeys. *Physiol. Behav.* 56:801–810.

Saltzman, W., Schultz-Darken, N. J., and Abbott, D. H. (1997). Familial influences on ovulatory function in common marmosets (*Callithrix jacchus*). *Am. J. Primatol.* 41:159–177.

Saltzman, W., Schultz-Darken, N. J., Wegner, F. H., Wittwer, D. J., and Abbott, D. H. (1998). Suppression of cortisol levels in subordinate female marmosets: reproductive and social contributions. *Horm. Behav.* 33:58–74.

Saltzman, W., Pick, R. R., Salper, O. J., Liedl, K. J., and Abbott, D. H. (2004). Onset of plural cooperative breeding in common marmoset families following replacement of the breeding male. *Anim. Behav.* 68:59–73.

Saltzman, W., Liedl, K. J., Salper, O. J., Pick, R. R., and Abbott, D. H. (2008). Post-conception reproductive competition in cooperatively breeding common marmosets. *Horm. Behav.* 53:274–286.

Saltzman, W., Digby, L. J., and Abbott, D. H. (2009). Reproductive skew in female common marmosets: what can proximate mechanisms tell us about ultimate causes? *Proc. R. Soc. Biol. Sci.* 276:389–399.

Sambrook, J., and Russell, D. (2001). *Molecular Cloning: A Laboratory Manual,* 3rd ed. Cold Spring Harbor Laboratory Press, Cold Spring Harbor, NY.

Samonte, R. V., and Eichler, E. E. (2002). Segmental duplications and the evolution of the primate genome. *Nat. Rev. Genet.* 3:65–72.

Sampaio, D. T., and Ferrari, S. F. (2005). Predation of an infant titi monkey (*Callicebus moloch*) by a tufted capuchin (*Cebus apella*). *Folia Primatol.* 76:113–115.

Samuels, A., and Altmann, J. (1986). Immigration of a *Papio anubis* male into a group of *Papio cynocephalus* baboons and evidence for an *anubis- cynocephalus* hybrid zone in Amboseli, Kenya. *Int. J. Primatol.* 7(2):131–8.

Samuels, A., Silk, J. B., and Altmann, J. (1987). Continuity and change in dominance relationships among female baboons. *Anim. Behav.* 35:785–792.

Sanders, W. J., and Mitani, J. C. (2000). Taphonomic aspects of eagle predation of primates from Kibale Forest, Uganda. *Am. J. Phys. Anthropol.* (suppl. 30):267–268.

Sanders, W. J., Trapani, J., and Mitanti, J. C. (2003). Taphonomic aspects of crowned hawk-eagle predation on monkeys. *J. Hum. Evol.* 44:87–105.

Sanfey, P. (2003). Study of re-released semi-free ranging slow and pygmy lorises in Thailand [MSc thesis]. Oxford Brookes University, Oxford.

Sanger, F., Nicklen, S., and Coulson, A. R. (1977). DNA sequencing with chain-terminating inhibitors. *Proc. Natl. Acad. Sci. USA* 74:5463–5467.

Sano, M. (1993). Foraging activities and diets of males and females in a hare-mic sandperch (Pisces: Pinguipedidae). *Mar. Ecol. Prog. Ser.* 98:55–59.

Santiago, M. L., Lukasik, M., Kamenya, S., Li, Y., Bibollet-Ruche, F., Bailes, E., Muller, M. N., Emery, M., Goldenberg, D. A., Lwanga, J. S., Ayouba, A., Nerrienet, E., McClure, H. M., Heeney, J. L., Watts, D. P., Pusey, A. E., Collins, D. A., Wrangham, R. W., Goodall, J., Brookfield, J. F., Sharp, P. M., Shaw, G. M., and Hahn, B. H. (2003). Foci of endemic simian immunodeficiency virus infection in wild-living eastern chimpanzees (*Pan troglodytes schweinfurthii*). *J. Virol.* 77:7545–7562.

Santos, L. R., Mahajan, N., and Barnes, J. L. (2005). How prosimian primates represent tools: experiments with two lemur species (*Eulemur fulvus* and *Lemur catta*). *J. Comp. Psychol.* 119:394–403.

Santos, L. R., Nissen, A. G., and Ferrugia, J. A. (2006a). Rhesus monkeys, *Macaca mulatta,* know what others can and cannot hear. *Anim. Behav.* 71(5):1175–1181.

Santos, L. R., Pearson, H. M., Spaepen, G. M., Tsao, F., and Hauser, M. D. (2006b). Probing the limits of tool competence: experiments with two non-tool using species (*Cercopithecus aethiop* and *Saguinus oedipus*). *Anim. Cogn.* 9:94–109.

Santos-Mello, R., and Thiago de Mello, M. (1986). Cariótipo de *Aotus trivirgatus* (Macaco-da-Noite) das proximidades de Manaus, Amazonas. Nota preliminar. In: Thiago de Mello, M. (ed.), *A Primatologia no Brasil,* vol. 2. Sociedade Brasileira de Primatologia, Brasilia. p. 388.

Sanz, C. M., and Morgan, D. B. (2007). Chimpanzee tool technology in the Goualougo Triangle, republic of Congo. *J Hum Evol.* 52:420–433.

Sanz, C. M., and Morgan, D. B. (2010). Complexity of chimpanzee tool using behaviors. In: Lonsdorf, E. V., Ross, S., and Matsuzawa, T. (eds.), *The Mind of the Chimpanzee: Ecological and Experimental Perspectives.* Chicago University Press, Chicago. In press.

Sanz, C., Morgan, D., and Gulick, S. (2004). New insights into chimpanzees, tools, and termites from the Congo Basin. *Am. Nat.* 164:567–581.

Sanz, C., Call, J., and Morgan, D. (2009). Design complexity in termite-fishing tools of chimpanzees (*Pan troglodytes*). *Biol. Letters* 5:293–296.

Sapolsky, R. M. (1982). The endocrine stress-response and social status in the wild baboon. *Horm. Behav.* 16:279–292.

Sapolsky, R. M. (1992). Cortisol concentrations and the social significance of rank instability among wild baboons. *Psychoneuroendocrinology* 17:701–709.

Sapolsky, R. M. (1993). The physiology of dominance in stable versus unstable social hierarchies. In: Mason, W. A., and Mendoza, S. P. (eds.), *Primate Social Conflict.* State University of New York Press, Albany. pp. 171–204.

Sapolsky, R. M. (1996). Why should an aged male baboon ever transfer troops? *Am. J. Primatol.* 30:149–157.

Sapolsky, R. M. (1999). Hormonal correlates of personality and social contexts: from non-human to human primates. In: Panter-Brick, C., and Worthman, C. M. (eds.), *Hormones, Health, and Behavior: A Socio-Ecological and Lifespan Perspective.* Cambridge University Press, New York. pp. 18–46.

Sapolsky, R. M. (2000). Physiological correlates of individual dominance style. In: Aureli, F., and de Waal, F. B. M. (eds.), *Natural Conflict Resolution.* University of California Press, Berkeley. pp. 114–116.

Sapolsky, R. M., and Share, L. J. (2004). A pacific culture among wild baboons: its emergence and transmission. *PLoS Biol.* 2:534–541.

Sarbin, T. (1954). Role theory. In: Gardner, L. (ed.), *Handbook of Social Psychology,* vol. 1. Addison-Wesley, Cambridge, MA.

Sargis, E. J. (2002). Primate origins nailed. *Science* 298:1564–1565.

Sargis, E. J. (2004). New views on tree shrews: the role of tupaiids in primate supraordinal relationships. *Evol. Anthropol.* 13:56–66.

Sarich, V. M., and Cronin, J. E. (1976). Molecular systematics of the primates. In: Goodman, M., Tashian, R. E., and Tashian, J. H. (eds.), *Molecular Anthropology: Genes and Proteins in the Evolutionary Ascent of Primates.* Plenum Press, New York. pp. 141–170.

Sarich, V. M., and Wilson, A. C. (1967). Immunological time scale for hominid evolution. *Science* 179:1144–1147.

Sarich, V. M., Schmid, C. W., and Marks, J. (1989). DNA hybridization as a guide to phylogenies: a critical analysis. *Cladistics* 5:3–32.

Sarmiento, E. (1998). The validity of *Pseudopotto martini. Afr. Primates* 3:44–45.

Sarmiento, E. E., Butynski, T. M., and Kalina, J. (1996). Gorillas of Bwindi-Impenetrable Forest and the Virunga Volcanoes: taxonomic implications of morphological and ecological differences. *Am. J. Primatol.* 40:1–21.

Satta, Y., Klein, J., and Takahata, N. (2000). DNA archives and our nearest relative: the trichotomy problem revisited. *Mol. Phylogenet. Evol.* 14:259–275.

Sauer, E. G. (1967). Mother–infant relationships in galagos and the oral child-transport among primates. *Folia Primatol.* 7(2):127–149.

Sauermann, U., Nurnberg, P., Bercovitch, F. B., Berard, J. D., Trefilov, A., Widdig, A., Kessler, M., Schmidtke, J., and Krawczak, M. (2001). Increased reproductive success of MHC class II heterozygous males among free-ranging rhesus macaques. *Hum. Genet.* 108:249–254.

Sauther, M. L. (1991). Reproductive behavior of free-ranging *Lemur catta* at Beza Mahafaly Special Reserve, Madagascar. *Am. J. Phys. Anthropol.* 84:463–477.

Sauther, M. L. (1992). The effect of reproductive state, social rank and group size on resource use among free-ranging ringtailed lemurs (*Lemur catta*) of Madagascar [PhD diss.]. Washington University, St. Louis, MO.

Sauther, M. L. (1993). Resource competition in wild populations of ringtailed lemurs (*Lemur catta*): implications for female dominance. In: Kappeler, P. M., and Ganzhorn, J. U. (eds.), *Lemur Social Systems and Their Ecological Basis.* Plenum Press, New York. pp. 135–152.

Sauther, M. L. (1994). Wild plant use by pregnant and lactating ringtailed lemurs, with implications for early hominid foraging. In Etkin, N. L. (ed.), *Eating on the Wild Side: The Pharmacologic, Ecologic, and Social Implications of Using Non-Cultigens,* University of Arizona Press, Tucson, pp. 240–256.

Sauther, M. L. (1998). The interplay of phenology and reproduction in ringtailed lemurs: implications for ringtailed lemur conservation. *Folia Primatol.* 69(suppl. 1):309–320.

Sauther, M. L. (2002). Group size effects on predation sensitive foraging in wild ring-tailed lemurs (*Lemur catta*). In: Miller, L. E. (ed.), *Eat or Be Eaten: Predator Sensitive Foraging Among Primates.* Cambridge University Press, Cambridge. pp. 107–125.

Sauther, M. L., and Sussman, R. W. (1993). A new interpretation of the social organization and mating system of the ringtailed lemur (*Lemur catta*). In: Kappeler, P. M., and Ganhorn, J. U. (eds.), *Lemur Social Systems and Their Ecological Basis.* Plenum Press, New York. pp. 111–122.

Sauther, M. L., Sussman, R. W., and Gould, L. (1999a). The socioecology of the ringtailed lemur: thirty-five years of research. *Evol. Anthropol.* 8:120–132.

Sauther, M. L., Sussman, R. W., and Gould, L. (1999b). The myth of the typical *Lemur catta*: 30 years of research on the ringtailed lemur. *Evol. Anthropol.* 8:120–132.

Sauther, M. L., Fish, K. D., Cuozzo, F. P., Miller, D. S., Hunter-Ishikawa, M., and Culbertson, H. (2006). Patterns of health, disease, and behavior among wild ringtailed lemurs, *Lemur catta*: effects of habitat and sex. In: Jolly,

A., Sussman, R. W., Koyama, N., and Rasamimanana, H. (eds.), *Ringtailed Lemur Biology: Lemur catta in Madagascar.* Springer Science+Business Media, New York. pp. 313–331.

Savage, A., Ziegler, T. E., and Snowdon, C. T. (1988). Sociosexual development, pair bond formation, and mechanisms of fertility suppression in female cotton-top tamarins (*Saguinus oedipus oedipus*). *Am. J. Primatol.* 14:345–359.

Savage, A., Shideler, S. E., Moorman, E. A., Ortuño, A. M., Whittier, C. A., Casey, K. K., and McKinney, J. (1992). The reproductive biology of the white-faced saki (*Pithecia pithecia*) in captivity. Abstracts of the XIVth Congress of the International Primatological Society, Strasbourg, France. pp. 59–60. August 16–21.

Savage, A., Lasley, B. L., Vecchio, A. J., Miller, A. E., and Shideler, S. E. (1995). Selected aspects of female white-faced saki (*Pithecia pithecia*) reproductive biology in captivity. *Zoo Biol.* 14:441–452.

Savage, A., Giraldo, L. H., Soto, L. H., and Snowdon, C. T. (1996a). Demography, group composition, and dispersal in wild cotton-top tamarins (*Saguinus oedipus*) groups. *Am. J. Primatol.* 38:85–100.

Savage, A., Snowdon, C. T., Giraldo, L. H., and Soto, L. H. (1996b). Parental care patterns and vigilance in wild cotton-top tamarins (*Saguinus oedipus*). In: Norconk, M. A., Rosenberger, A. L., and Garber, P. A. (eds.), *Adaptive Radiations of Neotropical Primates.* Plenum Press, New York, pp. 187–199.

Savage, A., Shideler, S. E., Soto, L. H., Causado, J., Giraldo, L. H., Lasley, B. L., and Snowdon, C. T. (1997). Reproductive events of wild cotton-top tamarins (*Saguinus oedipus*) in Colombia. *Am. J. Primatol.* 43:329–337.

Savage-Rumbaugh, E. S. (1986). *Ape Language: From Conditioned Response to Symbol.* Columbia University Press, New York.

Savage-Rumbaugh, E. S., and Fields, W. M. (2000). Linguistic, cultural and cognitive capacities of bonobos (*Pan paniscus*). *Cult. Psychol.* 6:131–153.

Savage-Rumbaugh, S., Shanker, S. G., and Taylor, T. J. (1998). *Apes, Language, and the Human Mind.* Oxford University Press, New York.

Savini, T., Boesch, C., and Reichard, U. H. (2008). Home-range characteristics and the influence of seasonality on female reproduction in white-handed gibbons (*Hylobates lar*) at Khao Yai National Park, Thailand. *Am. J. Phys. Anthropol.* 135:1–12.

Scammon, R. E. (1930). The measurement of the body in childhood. In: Harris, J. A., Jackson, C. M., Paterson, D. G., and Scammon, R. E. (eds.), *The Measurement of Man.* University of Minnesota Press, Minneapolis. pp. 174–215.

Scanlon, C. E., Chalmers, N. R., and Monteiro da Cruz, M. A. O. (1988). Changes in the size, composition and reproductive condition of wild marmoset groups (*Callithrix jacchus jacchus*) in north east Brazil. *Primates* 29:295–305.

Schaffner, C. M., and Caine, N. G. (2000). The peacefulness of cooperatively breeding primates. In: Aureli, F., and de Waal, F. B. M. (eds.), *Natural Conflict Resolution.* University of California Press, Berkeley. pp. 155–169.

Schaffner, C. M., and French, J. A. (1997). Group size and aggression: "recruitment incentives" in a cooperatively breeding primate. *Anim. Behav.* 54:171–180.

Schaffner, C. M., and French, J. A. (2004). Behavioral and endocrine responses in male marmosets to the establishment of multimale breeding groups: evidence for non-monopolizing facultative polyandry. *Int. J. Primatol.* 25:709–732.

Schaffner, C. M., Aureli, F., and Caine, N. G. (2005). Following the rules: why small groups of tamarins do not reconcile conflicts. *Folia Primatol.* 76:67–76.

Schaller, G. (1963). *The Mountain Gorilla: Ecology and Behavior.* University of Chicago Press, Chicago.

Schaller, G. B. (1964). *The Year of the Gorilla.* University of Chicago, Chicago.

Schenkel, R., and Schenkel-Hulliger, L. (1967). On the sociology of free-ranging colobus (*Colobus guereza caudatus* Thomas 1885). In: Starck, D., Schneider, R., and Kuhn, H. J. (eds.), *Progress in Primatology.* Karger, Basel. pp. 185–194.

Scheumann, M., and Call, J. (2006). Sumatran orangutans and a yellow-cheeked crested gibbon know what is where. *Int. J. Primatol.* 27:575–602.

Scheumann, M., Rabesandratana, A., and Zimmermann, E. (2007). Predation, communication, and cognition in lemurs. In: Gursky, S. L., and Nekaris, K. A. I. (eds.), *Primate Anti-Predator Strategies.* Springer, New York. pp. 100–126.

Schick, K. D., and Toth, N. (1993). *Making Silent Stones Speak.* Simon and Schuster, New York.

Schick, K. D., Toth, N., Garufi, G., Savage-Rumbaugh, E. S., Rumbaugh, D., and Sevcik, R. (1999). Continuing investigations into the stone tool-making and tool-using capabilities of a bonobo (*Pan paniscus*). *J. Archaeol. Sci.* 26:821–832.

Schilling, A. (1979). Olfactory communication in prosimians. In: Doyle, G. A., and Martin, R. D. (eds.), *The Study of Prosimian Behaviour.* Academic Press, London. pp. 461–542.

Schiml, P. A., Mendoza, S. P., Saltzman, W., Lyons, D. M., and Madon, W. A. (1996). Seasonality in squirrel monkeys (*Saimiri sciureus*): social facilitation by females. *Physiol. Behav.* 60:1105–1113.

Schiml, P. A., Mendoza, S. P., Saltzman, W., Lyons, D. M., and Madon, W. A. (1999). Annual physiological changes in individually housed squirrel monkeys (*Saimiri sciureus*). *Am. J. Primatol.* 47:93–103.

Schino, G. (1998). Reconciliation in domestic goats. *Behaviour* 135:343–356.

Schino, G. (2007). Grooming and agonistic support: a meta-analysis of primate reciprocal altruism. *Behav. Ecol.* 18:115–120.

Schino, G., Troisi, A., Perretta, G., and Monaco, V. (1991). Measuring anxiety in nonhuman primates: effects of lorazepam on macaque scratching. *Pharmacol. Biochem. Behav.* 38:889–891.

Schino, G., Perretta, G., Taglioni, A. M., Monaco, V., and Troisi, A. (1996). Primate displacement activities as an ethopharmacological model of anxiety. *Anxiety* 22:186–191.

Schino, G., Rosati, L., and Aureli, F. (1998). Intragroup variation in conciliatory tendencies in captive Japanese macaques. *Behaviour* 135:897–912.

Schino, G., Speranza, L., Ventura, R., and Troisi, A. (2003). Infant handling and maternal response in Japanese macaques. *Int. J. Primatol.* 24:627–639.

Schino, G., Aureli, F., Ventura, R., and Troisi, A. (2004). A test of the cross-generational transmission of grooming preferences in macaques. *Ethology* 110:137–146.

Schino, G., Rosati, L., Geminiani, S., and Aureli, F. (2007a). Post-conflict anxiety in Japanese macaques (*Macaca fuscata*): aggressor's and victim's perspectives. *Ethology* 113:1081–1088.

Schino, G., Tiddi, B., and Polizzi di Sorrentino, E. (2007b). Agonistic support in juvenile Japanese macaques: cognitive and functional implications. *Ethology* 113:1151–1157.

Schlotterhausen, L. (2000). Town monkeys, country monkeys: a socioecological comparison of a human commensal and wild group of bonnet macaques (*Macaca radiata*) [PhD diss.]. New York University, New York.

Schmelting, B., Ehresmann, P., Lutermann, H., Randrianambinina, B., and Zimmermann, E. (2000). Reproduction of two sympatric mouse lemur species (*Microcebus murinus* and *Microcebus ravelobensis*) in northwest Madagascar: first results of a long term study. In: Lourenco, W. R., and Goodmann, S. M. (eds.), *Diversity and Endemism in Madagascar.* Mémoires de la Société de Biogéographie, Paris. pp. 165–175.

Schmid, J. (1998). Tree holes used for resting by gray mouse lemurs (*Microcebus murinus*) in Madagascar: insulation capacities and energetic consequences. *Int. J. Primatol.* 19:797–809.

Schmid, J. (2000). Daily torpor in the gray mouse lemur (*Microcebus murinus*) in Madagascar: energetic consequences and biological significance. *Oecologia* 123(2):175–183.

Schmid, J. (2001). Daily torpor in free-ranging gray mouse lemurs (*Microcebus murinus*) in Madagascar. *Int. J. Primatol.* 22(6):1021–1031.

Schmid, J., and Kappeler, P. M. (1994). Sympatric mouse lemurs (*Microcebus* spp.) in western Madagascar. *Folia Primatol.* 63:162–170.

Schmid, J., and Kappeler, P. M. (1998). Fluctuating sexual dimorphism and differential hibernation by sex in a primate, the grey mouse lemur (*Microcebus murinus*). *Behav. Ecol. Sociobiol.* 43:125–132.

Schmid, J., and Speakman, J. R. (2000). Daily energy expenditure of the grey mouse lemur (*Microcebus murinus*): a small primate that uses torpor. *J. Comp. Physiol. [B]* 170(8):633–641.

Schmidt, H. A., Strimmer, K., Vingron, M., and von Haeseler, A. (2002). TREE-PUZZLE: maximum likelihood phylogenetic analysis using quartets and parallel computing. *Bioinformatics* 18:502–504.

Schmidt-Nielsen, K. (1984). *Scaling: Why Is Animal Size so Important?* Cambridge University Press, Cambridge.

Schmidt-Nielsen, K. (1997). *Animal Physiology: Adaptation and Environment.* Cambridge University Press, Cambridge.

Schmitt, D. (1998). Forelimb mechanics during arboreal and terrestrial quadrupedalism in Old World monkeys. In: Strasser, E., Rosenberger, A., McHenry, H., and Fleagle, J. G. (eds.), *Primate Locomotion: Recent Advances.* Plenum, New York. pp. 175–200.

Schmitt, D. (2003). Substrate size and primate forelimb mechanics: implications for understanding the evolution of primate locomotion. *Int. J. Primatol.* 24:1023–1036.

Schmitt, D., and Lemelin, P. (2002). The origins of primate locomotion: gait mechanics of the woolly opossum. *Am. J. Phys. Anthropol.* 118:231–238.

Schmitt, J., and Antonovics, J. (1986). Experimental studies of the evolutionary significance of sexual reproduction: III. Maternal and paternal effects during seedling establishment. *Evolution* 40:817–829.

Schnabel, A., Nason, J. D., and Hamrick, J. L. (1998). Understanding the population genetic structure of *Gleditsia triacanthos* L.: seed dispersal and variation in female reproductive success. *Mol. Ecol.* 7:819–832.

Schneider, G. (1906). Ergebnisse zoologischer Forschungsreisen in Sumatra. *Zool. Jahr. Abt. Syst. Geogr. Biol. Tiere* 23:1–172.

Schneider, H. (2000). The current status of the New World monkey phylogeny. *Ann. Acad. Brasil. Ciênc.* 72:165–172.

Schneider, H., and Rosenberger, A. L. (1996). Molecules, morphology and platyrrhine systematics. In: Norconk, M. A., Rosenberger, A. L., and Garber, P. A. (eds.), *Adaptive Radiations of Neotropical Primates*. Plenum Press, New York. pp. 3–19.

Schneider, H., Martins, E. S., and Leão, V. (1987). Syntopy and troops association between *Callithrix* and *Saguinus* from Rondônia. *Int. J. Primatol.* 8:527.

Schneider, H., Schneider, M. P. C., Sampaio, I., Harada, M. L., Barroso Stanhopes, M., Czelusniak, J., and Goodman, M. (1993). Molecular phylogeny of the New World monkeys (Platyrrhini, Primates). *Mol. Phylogenet. Evol.* 2:225–242.

Schneider, H., Sampaio, I., Harada, M. L., Barroso, C. M. L., Schneider, M. P. C., Czelusniak, J., and Goodman, M. (1996). Molecular phylogeny of the New World monkeys (Platyrrhini, Primates) based on two unlinked nuclear genes: IRBP intron 1 and epsilon-globin sequences. *Am. J. Phys. Anthropol.* 100:153–179.

Schneider, H., Canavez, F. C., Sampaio, I., Moreira, F. A. M., Tagliaro, C. H., and Seuanez, H. N. (2001). Can molecular data place each neotropical monkey in its own branch? *Chromosoma* 109:515–523.

Schneider, R. D. (1998). Female social preferences and mating behavior in captive group-living baboons (*Papio cynocephalus anubis*): an experimental study [PhD diss.]. University of Michigan, Ann Arbor.

Schneider, S., Roessli, D., and Excoffier, L. (2000). *Arlequin: a software for population genetics data analysis* (version 2.000). Genetics and Biometry Lab, Department of Anthropology, University of Geneva, Geneva.

Schoener, T. W. (1971). Theory of feeding strategies. *Annu. Rev. Ecol. Syst.* 2:369–404.

Schoof, V. M., Jack, K., and Isbell, L. (2009). What traits promote male parallel dispersal in primates? *Behaviour* 146:701–726.

Schradin, C., and Anzenberger, G. (2001). Infant carrying in family groups of Goeldi's monkeys (*Callimico goeldii*). *Am. J. Primatol.* 53:57–67.

Schradin, C., Reeder, D. M., Mendoza, S. P., and Anzenberger, G. (2003). Prolactin and paternal care: comparison of three species of monogamous new world monkeys (*Callicebus cupreus, Callithrix jacchus,* and *Callimico goeldii*). *J. Comp. Psychol.* 117:166–175.

Schrago, C. G. (2007). On the time scale of New World primate diversification. *Am. J. Phys. Anthropol.* 132:344–354.

Schreier, A. (2009). The influence of resource distribution on the social structure and travel patterns of wild hamadryas baboons (*Papio hamadryas*) at Filoha, Awash National Park, Ethiopia [PhD diss.]. City University of New York, New York.

Schreier, A., and Swedell, L. (2009). The fourth level of social structure in hamadryas baboons: clans, their possible functions, and maintenance by both males and females. *Am. J. Primatol.* 71:948–955.

Schulke, O., and Kappeler, P. M. (2003). So near and yet so far: territorial pairs but low cohesion between pair partners in a nocturnal lemur, *Phaner furcifer. Anim. Behav.* 65:331–343.

Schulke, O., and Ostner, J. (2001). Predation on *Lepilemur* by a harrier hawk and implications for sleeping site quality. *Lemur News* 6:5.

Schulke, O., and Ostner, J. (2007). Physiological ecology of cheirogaleid primates: variation in hibernation and torpor. *Acta Etholog.* 10:13–21.

Schultz, A. H. (1924). Observations on colobus fetuses. *Bull. Am. Mus. Nat. Hist.* 49:443–457.

Schultz, A. H. (1927). Studies on the growth of gorilla and of other higher primates with special reference to a fetus of gorilla, preserved in the Carnegie Museum. *Mem. Carnegie Mus.* 11:1–87.

Schultz, A. H. (1930). The skeleton of the trunk and limbs of higher primates. *Hum. Biol.* 2:303–438.

Schultz, A. H. (1937). Growth and development. In: Hartman, C. G., and Straus, W. L. (eds.), *The Anatomy of the Rhesus Macaque*. Hafner, New York. pp. 10–27.

Schultz, A. H. (1938a). Genital swelling in the female orang-utan. *J. Mammal.* 19:363–366.

Schultz, A. H. (1938b). The relative weights of the testes in primates. *Anat. Rec.* 72:387–394.

Schultz, A. H. (1940). Growth and development of the chimpanzee. *Contrib. Embryol.* 28:3–63.

Schultz, A. H. (1941). Growth and development of the orang-utan. *Contrib. Embryol.* 29:57–110.

Schultz, A. H. (1942). Growth and development of the proboscis monkey. *Bull. Mus. Comp. Zool.* 89:279–323.

Schultz, A. (1948). The number of young at a birth and the number of nipples in primates. *Am. J. Phys. Anthropol.* 6:1–23.

Schultz, A. H. (1963). Age changes, sex differences, and variability as factors in the classification of primates. *Viking Fund Publ. Anthropol.* 37:85–115.

Schultz, A. H. (1969a). *The Life of Primates*. Universe Press, New York.

Schultz, A. (1969b). *The Life of Primates*. The Weidenfeld & Nicolson Natural History. Weidenfeld & Nicolson, London. pp. 281.

Schultz, S., Noë, R., McGraw, W. S., and Dunbar, R. I. M. (2004). A community-level evaluation of the impact of prey behavioural and ecological characteristics on predator diet composition. *Proc. R. Soc. Lond. B* 271:725–732.

Schulze, H., and Groves, C. P. (2004). Asian lorises: taxonomic problems caused by illegal trade. In: Nadler, T., Streicher, U., and Thang Long, H. (eds.), *Conservation of Primates in Vietnam*. Frankfurt Zoological Society, Frankfurt. pp. 33–36.

Schulze, H., and Meier, B. (1995a). The subspecies of *Loris tardigradus* and their conservation status: a review. In: Alterman, L., Doyle, G. A., and Izard, M. K. (eds.), *Creatures of the Dark: The Nocturnal Prosimians*. Plenum Press, New York. pp. 193–209.

Schulze, H., and Meier, B. (1995b). Behaviour of captive *Loris tardigradus nordicus*: a qualitative description including some information about morphological bases of behavior. In: Alterman, L., Doyle, G. A., and Izard, M. K. (eds.), *Creatures of the Dark: The Nocturnal Prosimians*. Plenum Press, New York. pp. 221–250.

Schupp, E. W. (1988). Seed and early seedling predation in the forest understory and in treefall gaps. *Oikos* 51:71–78.

Schupp, E. W. (1992). The Janzen-Connell model for tropical tree diversity: population implications and the importance of spatial scale. *Am. Nat.* 140:526–530.

Schupp, E. W., and Fuentes, M. (1995). Spatial patterns of seed dispersal and the unification of plant-population ecology. *Ecoscience* 2:267–275.

Schupp, E. W., Milleron, T., and Russo, S. E. (2002). Dissemination limitation and the origin and maintenance of species-rich tropical forests. In: Levey, D. J., Silva, W. R., and Galetti, M. (eds.), *Seed Dispersal and Frugivory: Ecology, Evolution, and Conservation*. CABI Publishing, New York. pp. 19–34.

Schürmann, C. L. (1981). Courtship and mating behavior of wild orangutans in Sumatra. In: Chiarelli, A. B., and Corruccini, R. S. (eds.), *Primate Behavior and Sociobiology*. Springer-Verlag, Berlin. pp. 130–135.

Schürmann, C. (1982). Mating behaviour of wild orangutans. In: de Boer, L. E. M. (ed.), *The Orangutan: Its Biology and Conservation*. Dr. W. Junk Publishers, The Hague. pp. 269–284.

Schürmann, C. L., and van Hooff, J. A. R. A. M. (1986). Reproductive strategies of the orang-utan: new data and a reconsideration of existing sociosexual models. *Int. J. Primatol.* 7:265–287.

Schwab, D. (1999). Predation on *Eulemur fulvus* by *Accipiter henstii* (Henst's goshawk). *Lemur News* 4:34.

Schwab, D. (2000). A preliminary study of spatial distribution and mating system of pygmy mouse lemurs (*Microcebus myoxinus*). *Am. J. Primatol.* 51:41–60.

Schwartz, E. (1929). Das vorkommen des Schimpansen auf den linken Kongo-Ufer. *Rev. Zool. Bot. Afr.* 16:425–433.

Schwartz, G. T., Reid, D. J., Dean, M. C., and Zihlman, A. L. (2006). A faithful record of stress life events in the dental developmental record of a juvenile gorilla. *Int. J. Primatol.* 27(4):1201–1219.

Schwartz, J. H. (1996). *Pseudopotto martini*: a new genus and species of extant lorisiform primate. *Anthropol. Pap. Am. Mus. Nat. Hist.* 78:1–14.

Schwartz, J. H., Long, V. T., Cuong, N. L., Kha, L. T., and Tattersall, I. (1994). A diverse hominoid fauna from the late Middle Pleistocene brecia cave of

Tham Khuyen, Socialist Republic of Vietnam. *Anthropol. Pap. Am. Mus. Nat. Hist.* 74:1–11.

Schwartz, J. H., Long, V. T., Cuong, N. L., Kha, L. T., and Tattersall, I. (1995). A review of the Pleistocene hominoid fauna of the socialist republic of Vietnam (excluding Hylobatidae). *Anthropol. Pap. Am. Mus. Nat. Hist.* 76:1–24.

Schwarz, E. (1929). On the local races and distribution of black and white colobus monkeys. *Proc. Zool. Soc. Lond.* 1929:585–598.

Schwensow, N., Eberle, M., and Sommer, S. (2008a). Compatibility counts: MHC-associated mate choice in a wild promiscuous primate. *Proc. R. Soc. B* 275:555–564.

Schwensow, N., Fietz, L., Dausmann, K., and Sommer, S. (2008b). MHC-associated mating strategies and the importance of overall genetic diversity in an obligate pair-living primate. *Evol. Ecol.* 22:617–636.

Schwindt, D. M., Carrillo, G. A., Bravo, J. J., Di Fiore, A., and Fernandez-Duque, E. (2004). Comparative socioecology of monogamous primates in the Amazon and Gran Chaco. *Int. J. Primatol.* 75(suppl. 1):412.

Scott, J. (1955). *Animal Behavior.* University of Chicago Press, Chicago.

Scott, L. M. (1984). Reproductive behavior of adolescent female baboons (*Papio anubis*) in Kenya. In: Small, M. F. (ed.), *Female Primates: Studies by Women Primatologists.* Alan R. Liss. New York. pp. 77–100.

Sechrest, W., Brooks, T. M., Fonseca, G. A. B., Konstant, W. R., Mittermeier, R. A., Purvis, A., Rylands, A. B., and Gittleman, J. L. (2002). Hotspots and the conservation of evolutionary history. *Proc. Natl. Acad. Sci. USA* 99:2067–2071.

Seidensticker, J. (1983). Predation by *Panthera* cats and measures of human influence in habitats of south Asian monkeys. *Int. J. Primatol.* 4:323–326.

Seiffert, E. R. (2007). Early evolution and biogeography of lorisiform strepsirrhines. *Am. J. Primatol.* 69:27–35.

Seiffert, E. R., Simon, E. L., and Attia, Y. (2003). Fossil evidence for an ancient divergence of lorises and galagos. *Nature* 422:421–424.

Sekulic, R. (1982a). Daily and seasonal patterns of roaring and spacing in four red howler (*Alouatta seniculus*) troops. *Folia Primatol.* 39:22–48.

Sekulic, R. (1982b). The function of howling in red howler monkeys (*Alouatta seniculus*). *Behaviour* 81:38–54.

Sekulic, R., and Chivers, D. J. (1986). The significance of call duration in howler monkeys. *Int. J. Primatol.* 7:183–190.

Sellers, W. (1996). A biomechanical investigation into the absence of leaping in the locomotor repertoire of the slender loris (*Loris tardigradus*). *Folia Primatol.* 67:1–14.

Sellers, W. I., and Crompton, R. H. (2004). Automatic monitoring of primate locomotor behaviour using accelerometers. *Folia Primatol.* 75:279–293.

Semaw, S., Renne, P., Harris, J. W. K., Feibel, C. S., Bernor, R. L., Fesseha, N., and Mowbray, K. (1997). 2.5-million-year-old stone tools from Gona, Ethiopia. *Nature* 385:333–336.

Semple, S. (2001). Individuality and male discrimination of female copulation calls in the yellow baboon. *Anim. Behav.* 61:1023–1028.

Semple, S., McComb, K., Alberts, S. C., and Altmann, J. (2002). Information content of female copulation calls in yellow baboons. *Am. J. Primatol.* 56:43–56.

Sept, J. (1998). Shadows on a changing landscape: comparing nesting patterns of hominids and chimpanzees since their last common ancestor. *Am. J. Primatol.* 46:85–101.

Setchell, D., and Curtis, J. (eds.) (2003). *Field and Laboratory Methods in Primatology.* Cambridge University Press, Cambridge.

Setchell, J. M. (1999). Socio-sexual development in the male mandrill (*Mandrillus sphinx*). [PhD diss.]. University of Cambridge, Cambridge.

Setchell, J. M. (2003). Behavioural development in male mandrills (*Mandrillus sphinx*): puberty to adulthood. *Behaviour* 140:1053–1089.

Setchell, J. M. (2005). Do female mandrills (*Mandrillus sphinx*) prefer brightly colored males? *Int. J. Primatol.* 26:715–735.

Setchell, J. M., and Dixson, A. F. (2001a). Arrested development of secondary sexual adornments in subordinate adult male mandrills (*Mandrillus sphinx*). *Am. J. Phys. Anthropol.* 115:245–252.

Setchell, J. M., and Dixson, A. F. (2001b). Changes in the secondary sexual adornments of male mandrills (*Mandrillus sphinx*) are associated with gain and loss of alpha status. *Horm. Behav.* 39:177–184.

Setchell, J. M., and Dixson, A. F. (2002). Developmental variables and dominance rank in adolescent male mandrills (*Mandrillus sphinx*). *Am. J. Primatol.* 56:9–25.

Setchell, J. M., and Wickings, E. J. (2003). Sexual swelling in mandrills (*Mandrillus sphinx*): a test of the reliable indicator hypothesis. *Behav. Ecol.* 15:438–445.

Setchell, J. M., and Wickings, E. J. (2004). Social and seasonal influences on the reproductive cycle in female mandrills (*Mandrillus sphinx*). *Am. J. Phys. Anthropol.* 125:73–84.

Setchell, J. M., and Wickings, E. J. (2005). Dominance, status signals and coloration in male mandrills (*Mandrillus sphinx*). *Ethology* 111:25–50.

Setchell, J. M., and Wickings, E. J. (2006). Mate choice in male mandrills (*Mandrillus sphinx*). *Ethology* 112:91–99.

Setchell, J. M., Lee, P. C., Wickings, E. J., and Dixson, A. F. (2001). Growth and ontogeny of sexual size dimorphism in the mandrill (*Mandrillus sphinx*). *Am. J. Phys. Anthropol.* 115:349–360.

Setchell, J. M., Lee, P. C., Wickings, E. J., and Dixson, A. F. (2002). Reproductive parameters and maternal investment in mandrills (*Mandrillus sphinx*). *Int. J. Primatol.* 23:51–68.

Setchell, J. M., Charpentier, M., and Wickings, E. J. (2005a). Mate guarding and paternity in mandrills: factors influencing alpha male monopoly. *Anim. Behav.* 70:1105–1120.

Setchell, J. M., Charpentier, M., and Wickings, E. J. (2005b). Sexual selection and reproductive careers in mandrills. *Behav. Ecol. Sociobiol.* 58:474–485.

Setchell, J. M., Charpentier, M. J. E., Bedjabaga, I. B., Reed, P., Wickings, E. J., and Knapp, L. A. (2006a). Secondary sexual characters and female quality in primates. *Behav. Ecol. Sociobiol.* 61:305–315.

Setchell, J. M., Wickings, E. J., and Knapp, L. A. (2006b). Life history in male mandrills (*Mandrillus sphinx*): physical development, dominance rank, and group association. *Am. J. Phys. Anthropol.* 131:498–510.

Setchell, J. M., Smith, T., Wickings, E. J., and Knapp, L. A. (2008). Social correlates of testosterone and ornamentation in male mandrills. *Horm. Behav.* 54:365–372.

Setchell, K. D. R., and Bonney, R. C. (1981). The excretion of urinary steroids by the owl monkey (*Aotus trivirgatus*) studied using open tubular capillary column gas chromatography and mass spectrometry. *J. Steroid Biochem.* 14:37–43.

Setz, E. Z. F. (1993). Ecologia alimentar de um grupo de parauacus (*Pithecia pithecia chrysocephala*) em um fragmento florestal na Amazónia central. Ciências Biológicas (Ecologia) [PhD diss.]. Instituto Biologica, UNICAMP, Campinas, Brazil.

Setz, E. Z. F. (1999). Sleeping habits of the golden-faced sakis in a forest fragment in the central Amazon. *Am. J. Primatol.* 49(suppl.):100.

Setz, E. Z. F., and Gaspar, D. A. (1997). Scent-marking behaviour in free-ranging golden-faced saki monkeys, *Pithecia pithecia chrysocephala*: sex differences and context. *J. Zool. Lond.* 241:603–611.

Setz, E. Z. F., Enzweiler, J., Soiferini, V. N., Amêndola, M. P., and Berton, R. S. (1999). Geophagy in the golden-faced saki monkey (*Pithecia pithecia chrysocephala*) in the central Amazon. *J. Zool.* 247:91–103.

Seuánez, H. N., Di Fiore, A., Moreira, Â. M., da Silva Almeida, C. A., and Canavez, F. C. (2002). Genetics and evolution of lion tamarins. In: Kleiman, D. G., and Rylands, A. B. (eds.), *Lion Tamarins: Biology and Conservation.* Smithsonian Institution Press, Washington DC. pp. 117–132.

Seutin, G., White, B. N., and Boag, P. T. (1991). Preservation of avian blood and tissue samples for DNA analysis. *Can. J. Zool.* 69:82–90.

Seyfarth, R. M. (1976). Social relationships among adult female baboons. *Anim. Behav.* 24:917–938.

Seyfarth, R. M. (1977). A model of social grooming among adult female monkeys. *J. Theor. Biol.* 65: 671–698.

Seyfarth, R. M. (1978a). Social relationships among adult male and female baboons. I. Behaviour during sexual consortship. *Behaviour* 64:204–226.

Seyfarth, R. M. (1978b). Social relationships among adult male and female baboons. II. behaviour throughout the female reproductive cycle. *Behaviour* 64:227–247.

Seyfarth, R. M. (1983). Grooming and social competition in primates. In: Hinde, R. A. (ed.), *Primate Social Relationships: An Integrated Approach.* Blackwell, Oxford. pp. 182–190.

Seyfarth, R. M., and Cheney, D. L. (1990). The assessment by vervet monkeys of their own and another species' alarm calls. *Anim. Behav.* 40:754–764.

Seyfarth, R. M., and Cheney, D. L. (2003). Signalers and receivers in animal communication. *Annu. Rev. Psychol.* 54:145–173.

Seyfarth, R. M., Cheney, D. L., and Marler, P. (1980). Vervet monkey alarm calls: semantic communication in a free-ranging primate. *Anim. Behav.* 28:1070–1094.

Sgaravatti, A., Spiezio, C., Grassi, D., and Huffman, M. A. (2004). How do chimpanzees learn leaf-swallowing behavior? *Folia Primatol.* 75(suppl. 1):333.

Shah, N. F. (2003). Foraging strategies of two sympatric mangabey species [PhD diss.]. State University of New York, Stony Brook.

Shah, N. F. (in press). The agile mangabey (*Cercocebus agilis*). In: Kingdon, J., Happold, D., and Butynski, T. (eds.), *The Mammals of Africa*. University of California Press, Berkeley, CA.

Shaikh, A. A., Celaya, C. L., Gomez, I., and Shaikh, S. A. (1982). Temporal relationship of hormonal peaks to ovulation and sex skin deturgescence in the baboon. *Primates* 23:444–452.

Shanee, N., Shanee, S., and Maldonado, A. M. (2007). Conservation assessment and planning for the yellow tailed woolly monkey (*Oreonax flavicauda*) in Peru. *Wildl. Biol. Pract.* 3:1–10.

Shankar, S., Mohapatra, B., Verma, S., Selvi, R., Jagadish, N., and Suri, A. (2004). Isolation and characterization of a haploid germ cell specific sperm associated antigen 9 (SPAG9) from the baboon. *Mol. Reprod. Dev.* 69:186–193.

Sharma, N. E., Fernandez, E., and Lawrance, S. K. (2003). The major histocompatibility complex of the owl monkey, *Aotus azarai*: a model system for the study of mate choice. *Ohio J. Sci.* 1031:A–9.

Sharman, M. J. (1981). Feeding, ranging, and social behavior of the Guinea baboon [PhD diss.]. University of St. Andrews, Fife.

Sharon, M. (1980). Carbohydrates. *Sci. Am.* 243:90–116.

Shea, B. T. (1984). An allometric perspective on the morphological and evolutionary relationships between pygmy (*Pan paniscus*) and common (*Pan troglodytes*) chimpanzees. In: Susman, R. L. (ed.), *The Pygmy Chimpanzee. Evolutionary Biology and Behavior*. Plenum, New York. pp. 89–130.

Shea, B. T. (1987). Reproductive strategies, body size, and encephalization in primate evolution. *Int. J. Primatol.* 8:139–156.

Shea, B. T. (1990). Dynamic morphology: growth, life history, and ecology in primate evolution. In: DeRousseau, C. J. (ed.), *Primate Life History and Evolution*. Wiley-Liss, New York. pp. 325–352.

Shea, B. T. (2000). Current issues in the investigation of evolution by heterochrony, with emphasis on the debate over human neoteny. In: Taylor Parker, S., Langer, J., and McKinney, M. L. (eds.), *Biology, Brains and Behavior: The Evolution of Human Development*. School of American Research Press, Santa Fe, NM. pp. 181–213.

Shea, B. T., and Coolidge, H. J. (1988). Craniometric differentiation and systematics in the genus *Pan. J. Hum. Evol.* 17:671–685.

Shea, B. T., Leigh, S. R., and Groves, C. P. (1993). Multivariate craniometric variation in chimpanzees: implications for species identification in paleoanthropology. In: Kimbel, W. H., and Martin, L. B. (eds.), *Species, Species Concepts, and Primate Evolution*. Plenum, New York. pp. 265–296.

Sheeran, L. K. (1997). Asian apes. In: Spencer, F. (ed.), *History of Physical Anthropology: An Encyclopedia*. Garland, New York. pp. 112–121.

Sheeran, L. K., Yongzu, Z., Poirier, F. E., and Dehua, Y. (1998). Preliminary report on the behavior of the Jingdong black gibbon (*Hylobates concolor jingdongensis*). *Trop. Biodivers.* 5:113–125.

Shekelle, M. (2003). Taxonomy and biogeography of eastern tarsiers [PhD diss.]. Washington University, St. Louis, MO.

Shekelle, M., Leksono, S., Ischwan, L., and Masala, Y. (1997). The natural history of the tarsiers of north and central Sulawesi. *Sulawesi Primate News* 4:4–11.

Shepard, G. H. (2002). Primates in Matsigenka: subsistence and world view. In: Fuentes, A., and Wolfe, L. D. (eds.), *Primates Face to Face*. Cambridge University Press, Cambridge. pp. 101–136.

Shepherd, S. V., and Platt, M. L. (2008). Spontaneous social orienting and gaze following in ringtailed lemurs (*Lemur catta*). *Anim. Cogn.* 11:13–20.

Shepherd, V. E., and Chapman, C. A. (1998). Dung beetles as secondary seed dispersers: impact on seed predation and germination. *J. Trop. Ecol.* 14:199–215.

Sheppard, D. J. (2000). Ecology of the Budongo Forest redtail: patterns of habitat use and population density in primary and regenerating forest sites [MSc thesis]. University of Calgary, Alberta.

Sherman, P. T. (1991). Harpy eagle predation on a red howler monkey. *Folia Primatol.* 56:53–56.

Sherman, P. (1998). The evolution of menopause. *Nature* 392:759–761.

Sherman, P., Braude, S., and Jarvis, J. (1999). Litter sizes and mammary numbers of naked mole-rats: breaking the one-half rule. *J. Mammal.* 80:720–733.

Sherrow, H. M., and Amsler, S. J. (2007). New intercommunity infanticides by the chimpanzees of Ngogo, Kibale National Park, Uganda. *Int. J. Primatol.* 28:9–22.

Shideler, S. E., and Lasley, B. L. (1982). A comparison of primate ovarian cycles. *Am. J. Primatol.* (suppl. 1):171–180.

Shideler, S. E., Czekala, N. M., Kasman, L. H., Lindburg, D. G., and Lasley, B. L. (1983a). Monitoring ovulation and implantation in the lion-tailed macaque (*Macaca silenus*) through urinary estrone conjugate evaluations. *Biol. Reprod.* 29:905–911.

Shideler, S. E., Lindburg, D. G., and Lasley, B. L. (1983b). Estrogen–behavior correlates in the reproductive physiology and behavior of the ruffed lemur (*Lemur variegatus*). *Horm. Behav.* 17:249–263.

Shideler, S. E., Mitchell, W. R., Lindburg, D. G., and Lasley, B. L. (1985). Monitoring luteal function in the lion-tailed macaque (*Macaca silenus*) through urinary progesterone metabolite measurements. *Zoo Biol.* 4:65–73.

Shideler, S. E., Munro, C. J., Tell, L., Owiti, G., Laughlin, L., Chatterton, R., Jr., and Lasley, B. L. (1990). The relationship of serum estradiol and progesterone concentrations to the enzyme immunoassay measurements of urinary estrone conjugates and immunoreactive pregnanediol-3-glucuronide in *Macaca mulatta*. *Am. J. Primatol.* 22:113–122.

Shideler, S. E., Ortuno, A. M., Moran, F. M., Moorman, E. A., and Lasley, B. L. (1993a). Simple extraction and enzyme immunoassays for estrogen and progesterone metabolites in the feces of *Macaca fascicularis* during nonconceptive and conceptive ovarian cycles. *Biol. Reprod.* 48:1290–1298.

Shideler, S. E., Shackleton, C. H. L., Moran, F. M., Stauffer, P., Lohstroh, P. N., and Lasley, B. L. (1993b). Enzyme immunoassays for ovarian steroid metabolites in the urine of *Macaca fascicularis*. *J. Med. Primatol.* 22:301–312.

Shideler, S. E., Savage, A., Ortuño, A. M., Moorman, E. A., and Lasley, B. L. (1994). Monitoring female reproductive function by measurement of fecal estrogen and progesterone metabolites in the white-faced saki (*Pithecia pithecia*). *Am. J. Primatol.* 32:95–108.

Shideler, S. E., Munro, C. J., Johl, H. K., Taylor, H. W., and Lasley, B. L. (1995). Urine and fecal sample collection on filter paper for ovarian hormone evaluations. *Am. J. Primatol.* 37:305–315.

Shigehara, N., Takai, M., Kay, R. F., Aung, A. K., Soe, A. N., Tun, S. T., Tsubamoto, T., and Thein, T. (2002). The upper dentition and face of *Pondaungia cotteri* from central Myanmar. *J. Hum. Evol.* 43:143–166.

Shimizu, K., Lohstroh, P. N., Laughlin, L. S., Gee, N. A., Todd, H., Shideler, S. E., and Lasley, B. L. (2001). Noninstrumented enzyme-linked immunosorbant assay for detection of early pregnancy in macaques. *Am. J. Primatol.* 54:57–62.

Shimooka, Y. (2003). Seasonal variation in association patterns of wild spider monkeys (*Ateles belzebuth belzebuth*) at La Macarena, Colombia. *Primates* 44:83–90.

Shimooka, Y. (2005). Sexual differences in ranging of *Ateles belzebuth belzebuth* at La Macarena, Colombia. *Int. J. Primatol.* 26:385–406.

Shimooka, Y., Campbell, C. J., Di Fiore, A., Felton, A. M., Izawa, K., Link, A., Nishimura, A., Ramos-Fernandez, G., and Wallace R. B. (2008). Demography and group composition in *Ateles*. In: Campbell, C. J. (ed.), *Spider Monkeys: Behavior, Ecology and Evolution of the Genus* Ateles. Cambridge University Press, Cambridge. pp. 329–348.

Shine, R., Harlow, P. S., Keogh, J. S., and Boeadi (1998). The influence of sex and body size on food habits of a giant tropical snake, *Python reticulatus*. *Funct. Ecol.* 12:248–258.

Shoemaker, A. H. (1982). Notes on the reproductive biology of the white-faced saki in captivity. *Int. Zoo Ybk.* 22:124–127.

Shoener T. W. (1971). Theory of feeding strategies. *Annu. Rev. Ecol. Syst.* 2:369–404.

Sholley, C. (1991). Conserving gorillas in the midst of guerrillas. American Association of Zoological Parks and Aquariums Annual Conference Proceedings, Wheeling, WV. pp. 30–37.

Shopland, J. M. (1987). Food quality, spatial deployment, and the intensity of feeding interference in yellow baboons (*Papio cyanocephalus*). *Behav. Ecol. Sociobiol.* 21:149–156.

Short, R. V. (1976). The evolution of human reproduction. *Proc. R. Soc. Lond.* 195:3–24.

Short, R. V. (1979). Sexual election and its component parts, somatic and genital selection, as illustrated by man and the great apes. *Adv. Study Behav.* 9:131–158.

Shultz, S. (2001). Notes on interactions between monkeys and African crowned eagles in Taï National Park, Ivory Coast. *Folia Primatol.* 72:248–250.

Shultz, S., and Noe, R. (2002). The consequences of crowned eagle central-place foraging on predation risk in monkeys. *Proc. R. Soc. Lond. B* 269:1797–1802.

Shultz, S., Noe, R., McGraw, W. S., and Dunbar, R. I. M. (2004). A community-level evaluation of the impact of prey behavioural and ecological characteristics on predator diet composition. *Proc. R. Soc. Lond. B* 271:725–732.

Shumaker, R. W., Palkovitch, A. M., Beck, B. B., Guagnano, G. A., and Morowitz, H. (2001). Spontaneous use of magnitude discrimination and ordination by the orangutans (*Pongo pygmaeus*). *J. Comp. Psychol.* 15:385–391.

Sibley, C. G., and Ahlquist, J. E. (1984). The phylogeny of the hominoid primates as indicated by DNA–DNA hybridization. *J. Mol. Evol.* 20:2–15.

Sibley, C. G., and Ahlquist, J. E. (1987). DNA hybridization evidence of hominoid phylogeny: results from an expanded data set. *J. Mol. Evol.* 26:99–121.

Sicotte, P. (1993). Inter-group encounters and female transfer in mountain gorillas: influence of group composition on male behavior. *Am. J. Primatol.* 30:21–36.

Sicotte, P. (1994). Effects of male competition on male–female relationships in bi-male groups of mountain gorillas. *Ethology* 97:47–64.

Sicotte, P. (2000). A case of mother–son transfer in mountain gorillas. *Primates* 41:95–103.

Sicotte, P. (2001). Female mate choice in mountain gorillas. In: Robbins, M. M., Sicotte, P., and Stewart, K. J. (eds.). *Mountain Gorillas: Three Decades of Research at Karisoke.* Cambridge University Press, Cambridge. pp. 59–87.

Sicotte, P., and MacIntosh, A. J. (2004). Inter-group encounters and male incursions in *Colobus vellerosus* in central Ghana. *Behaviour* 141:533–553.

Sicotte, P., and Uwengeli, P. (2002). Reflections on the concept of nature and gorillas in Rwanda: implications for conservation. In: Fuentes, A., and Wolfe, L. D. (eds.), *Primates Face to Face.* Cambridge University Press, Cambridge. pp. 163–182.

Sicotte, P., Teichroeb, J. A., and Saj, T. L. (2007). Aspects of male competition in *Colobus vellerosus*: preliminary data on male and female loud calling, and infant deaths after a takeover. *Int. J. Primatol.* 28:627–636.

Siegert, F., Ruecker, G., Hinrichs, A., and Hoffmann, A. (2001). Increased damage from fires in logged forests during droughts caused by El Niño. *Nature* 414:437–440.

Siex, K. S. (2003). Effects of population compression on the demography, ecology, and behavior of the Zanzibar red colobus monkey (*Procolobus kirkii*) [PhD diss.]. Duke University, Durham, NC.

Siex, K. S., and Struhsaker, T. T. (1999a). Colobus monkeys and coconuts: a study of perceived human–wildlife conflicts. *J. Appl. Ecol.* 36:1009–1020.

Siex, K. S., and Struhsaker, T. T. (1999b). Ecology of the Zanzibar red colobus monkey: demographic variability and habitat stability. *Int. J. Primatol.* 20:163–192.

Sigg, H. (1980). Differentiation of female positions in hamadryas one-male-units. *Z. Tierpsychol.* 53:265–302.

Sigg, H., and Falett, J. (1985). Experiments on respect of possession and property in hamadryas baboons (*Papio hamadryas*). *Anim. Behav.* 33:978–984.

Sigg, H., and Stolba, A. (1981). Home range and daily march in a hamadryas baboon troop. *Folia Primatol.* 36:40.

Sigg, H., Stolba, A., Abegglen, J.-J., and Dasser, V. (1982). Life history of hamadryas baboons: physical development, infant mortality, reproductive parameters and family relationships. *Primates* 23:473–487.

Sih, A. (1987). Predators and prey lifestyles: an evolutionary and ecological overview. In: Kerfoot, W. C., and Sih, A. (eds.), *Predation.* University Press of New England, Hanover. pp. 203–224.

Sih, A. (1992). Prey uncertainty and the balancing of antipredator and feeding needs. *Am. Nat.* 139:1052–1069.

Silberberg, A., and Fujita, K. (1996). Pointing at smaller food amounts in an analogue of Boysen and Berntson's (1995) procedure. *J. Exp. Anal. Behav.* 66:143–47.

Silk, J. B. (1978). Patterns of food sharing among mother and infant chimpanzee at Gombe National Park, Tanzania. *Folia Primatol.* 29:129–141.

Silk, J. B. (1979). Feeding, foraging, and food sharing behavior of immature chimpanzees. *Folia Primatol.* 31:123–142.

Silk, J. B. (1980). Kidnapping and female competition among captive bonnet macaques. *Primates* 21:100–110.

Silk, J. B. (1982). Altruism among adult female bonnet macaques: explanation and analysis of patterns of grooming and coalition formation. *Anim. Behav.* 79:162–187.

Silk, J. B. (1987). Social behavior in evolutionary perspective. In: Smuts, B. B., Cheney, D. L., Seyfarth, R. M., Wrangham, R. W., and Struhsaker, T. T. (eds.), *Primate Societies.* University of Chicago Press, Chicago. pp. 318–329.

Silk, J. B. (1992a). The patterning of intervention among male bonnet macaques: reciprocity, revenge, and loyalty. *Curr. Anthropol.* 33:318–325.

Silk, J. B. (1992b). Patterns of intervention in agonistic contests among male bonnet macaques. In: Harcourt, A. H., and de Waal, F. B. M. (eds.), *Coalitions and Alliances in Humans and Other Animals.* Oxford University Press, Oxford. pp. 215–232.

Silk, J. B. (1993). The evolution of social conflict among female primates. In: William, A., and Mason, S. P. M. (eds.), *Primate Social Conflict.* State University of New York Press, Albany. pp. 49–83.

Silk, J. B. (1996). Why do primates reconcile? *Evol. Anthropol.* 5:39–42.

Silk, J. B. (1997). The function of peaceful post-conflict contacts among primates. *Primates* 38:265–279.

Silk, J. B. (1999). Why are infants so attractive to others? The form and function of infant handling in bonnet macaques. *Anim. Behav.* 57:1021–1032.

Silk, J. B. (2001). Ties that bond: the role of kinship in primate societies. In: Stone, L. (ed.), *New Directions in Anthropological Kinship.* Rowman & Littlefield, New York. pp. 71–92.

Silk, J. (2002a). Kin selection in primate groups. *Int. J. Primatol.* 23:849–875.

Silk, J. (2002b). Introduction. What are friends for? The adaptive value of social bonds in primate groups. *Behaviour* 139:173–176.

Silk, J. (2002c). Using the "F"-word in primatology. *Behaviour* 139:421–446.

Silk, J. B. (2006). Practicing Hamilton's rule: kin selection in primate groups. In: Kappeler, P. M., and van Schaik, C. P. (eds.), *Cooperation in Primates and Humans.* Springer, New York. pp. 25–46.

Silk, J. B., and Boyd, R. (1983). Cooperation, competition, and mate choice in matrilineal macaque groups. In: Wasser, S. K. (ed.), *Social Behavior of Female Vertebrates.* Academic Press, New York. pp. 315–347.

Silk, J. B., Samuels, A., and Rodman, P. (1981a). Hierarchical organization of female *Macaca radiata* in captivity. *Primates* 22:84–95.

Silk, J. B., Samuels, A., and Rodman, P. S. (1981b). The influence of kinship, rank, and sex on affiliation and aggression between adult female and immature bonnet macaques (*Macaca radiata*). *Behaviour* 78:111–137.

Silk, J. B., Cheney, D. L., and Seyfarth, R. M. (1996). The form and function of post-conflict interactions between female baboons. *Anim. Behav.* 52:259–268.

Silk, J. B., Seyfarth, R. M., and Cheney, D. L. (1999). The structure of social relationships among female savanna baboons in Moremi Reserve, Botswana. *Behaviour* 136:679–703.

Silk, J. B., Kaldor, E., and Boyd, R. (2000). Cheap talk when interests conflict. *Anim. Behav.* 59:423–432.

Silk, J. B., Alberts, S. C., and Altmann, J. (2003a). Social bonds of female baboons enhance infant survival. *Science* 302:1231–1234.

Silk, J. B., Rendall, D., Cheney, D. L., and Seyfarth, R. M. (2003b). Natal attraction in adult female baboons (*Papio cynocephalus ursinus*) in the Moremi Reserve, Botswana. *Ethology* 109:627–644.

Silk, J. B., Alberts, S. C., and Altmann, J. (2004). Patterns of coalition formation by adult female baboons in Amboseli, Kenya. *Anim. Behav.* 67:573–582.

Silk, J. B., Alberts, S. C., and Altmann, J. (2006a). Social relationships among adult female baboons (*Papio cynocephalus*) II. variation in the quality and stability of social bonds. *Behav. Ecol. Sociobiol.* 61:197–204.

Silk, J. B., Altmann, J., and Alberts, S. C. (2006b). Social relationships among adult female baboons (*Papio cynocephalus*) I. Variation in the strength of social bonds. *Behav. Ecol. Sociobiol.* 61:183–195.

Silva, F. J., Page, D. M., and Silva, K. M. (2005). Methodological–conceptual problems in the study of chimpanzees' folk physics: how studies with adult humans can help. *Learn. Behav.* 33:47–58.

Silva, F. J., Silva, K. M., Cover, K. R., Leslie, A. L., and Rubalcaba, M. A. (2008). Humans' folk physics is sensitive to physical connection and contact between a tool and reward. *Behav. Proc.* 77:327–333.

Silva, J. S., Jr., and Figueiredo, W. M. B. (2002). Revisão sistemática dos cuxiús, genero *Chiropotes* Lesson 1840 (Primates, Pithecidae). In: *XIX Congresso Brasileiro de Zoologia*, Belem, Brazil. p. 173.

Silva, J. S., Jr., Figueiredo, W. M. B., and Ferrari, S. F. (2009). Taxonomy and geographic distribution of the Pitheciidae. In: Barnett, A. A., Veiga, L. M., Ferrari, S. F., and Norconk, M. A. (eds.), *Evolutionary Biology and Conservation of Titis, Sakis, and Uacaris.* Cambridge University Press, Cambridge. In press.

Silva, S. S. B., and Ferrari, S. F. (2009). Behavior patterns of southern bearded sakis (*Chiropotes satanas*) in the fragmented landscape of eastern Brazilian Amazonia. *Am J. Primatol.* 71:1–7.

Silver, S. C., Ostro, L. E. T., Yeager, C. P., and Horwich, R. (1998). Feeding ecology of the black howler monkey (*Alouatta pigra*) in northern Belize. *Am. J. Primatol.* 45:263–279.

Simonds, P. E. (1992). Primate behavioral dynamics. In: Seth, P. K., and Seth, S. (eds.), *Perspectives in Primate Biology*. Today & Tomorrow, New Dehli, pp. 95–109.

Simons, E. L. (1972). *Primate Evolution: An Introduction to Man's Place in Nature*. MacMillan, New York.

Simons, E., and Bown, T. (1985). *Afrotarsius chatrathi*, the first tarsiiform primate (Tarsiidae) from Africa. *Nature* 313:475–477.

Simons, E. L., and Rasmussen, D. T. (1994). A whole new world of ancestors: Eocene anthropoideans from Africa. *Evol. Anthropol.* 3:128–139.

Simons, E. L., Burney, D. A., Chatrath, P. S., Godfrey, L. R., Jungers, W. L., and Rakotosamimanana, B. (1995). AMS 14C dates for extinct lemurs from caves in the Ankarana Massif, northern Madagascar. *Quat. Res.* 43:249–254.

Simpson, G. G. (1944). Paleontology. The role of the individual in evolution. *J. Wash. Acad. Sci.* 31:1–20.

Simpson, G. G. (1945). The principles of classification and a classification of mammals. *Bull. Am. Mus. Nat. Hist.* 85:1–350.

Singh, M., and Sinha, A. (2004). Life-history traits: ecological adaptations or phylogenetic relics? In: Thierry, B., Singh, M., and Kaumanns, W. (eds.), *Macaque Societies: A Model for the Study of Social Organization*. Cambridge University Press, Cambridge. pp. 80–83.

Singh, M., Lindburg, D. G., Udhayan, A., Kumar, M. A., and Kumara, H. N. (1999). Status survey of the slender loris in Dindigul, Tamil Nadu, India. *Oryx* 33:31–37.

Singh, M., Kumar, M. A., Kumara, H. N., and Mohnot, S. M. (2000). Distribution and conservation of slender lorises in southern Andhra Pradesh, south India. *Int. J. Primatol.* 21:721–730.

Singh, M., Kumara, H. N., Kumar, M. A., Singh, M., and Cooper, M. (2006). Male influx, infanticide, and female transfer in *Macaca radiata radiata*. *Int. J. Primatol.* 27:515–528.

Singleton, I. S., and van Schaik, C. P. (2001). Orangutan home range size and its determinants in a Sumatran swamp forest. *Int. J. Primatol.* 22:877–911.

Singleton, I. S., and van Schaik, C. P. (2002). The social organisation of a population of Sumatran orang-utans. *Folia Primatol.* 73:1–20.

Singleton, I., Wich, S. A., Husson, S., Stephens, S., Utami Atmoko, S. S., Leighton, M., Rosen, N., Traylor-Holzer, K., Lacy, R., and Byers, O. (2004). *Orangutan Population and Habitat Viability Assessment*. IUCN/SSC Conservation Breeding Specialist Group, Apple Valley, MN.

Singleton, I., Knott, C. D., Morrogh-Bernard, H. C., Wich, S. A., and van Schaik, C. P. (2009). Ranging behavior of orangutan females and social organization In: Wich, S. A., Utami, S. S., Mitra Setia, T., and van Schaik, C. (eds.), *Orangutans: Geographic Variation in Behavioral Ecology and Conservation*. Oxford University Press, Oxford. pp. 205–214.

Sinha, A. (1997). Complex tool manufacture by a wild bonnet macaque, *Macaca radiata*. *Folia Primatol.* 68:23–25.

Sinha, A., Mukhopadhyay, K., Datta-Roy, A., and Ram, S. (2005). Ecology proposes, behaviour disposes: ecological variability in social organisation and male behavioural strategies among wild bonnet macaques. *Curr. Sci.* 89:11–66.

Skorupa, J. P. (1986). Responses of rainforest primates to selective logging in Kibale Forest, Uganda: a summary report. In: Benirschke, K. (ed.), *Primates: The Road to Self-Sustaining Populations*. Springer-Verlag, New York. pp. 57–70.

Skorupa, J. P. (1987). Crowned eagles *Stephanoaetus coronatus* in rainforest: observations on breeding chronology and diet at a nest in Uganda. *Ibis* 131:294–298.

Slade, N. A., Gomulkiewicz, R., and Alexander, H. M. (1998). Alternatives to Robinson and Redford's method for assessing overharvest from incomplete demographic data. *Conserv. Biol.* 12:148–155.

Slate, J. (2005). Quantitative trait locus mapping in natural populations: progress, caveats and future directions. *Mol. Ecol.* 14:363–379.

Slater, K. (2002). Dominance style, grooming and biological markets in the buton macaque (*Macaca ochreata brunnescens*) [MSc thesis]. University of Liverpool, Liverpool.

Slater, K. Y., Schaffner, C. M., and Aureli, F. (2008). Female-directed male aggression in wild *Ateles geoffroyi yucatanensis*. *Int. J. Primatol.* 29:1657–1669

Slater, K. Y., Schaffner, C. M., and Aureli, F. (2009). Sex differences in the social behavior of wild spider monkeys (*Ateles geoffroyi yucatanensis*). *Am. J. Primatol.* 71:21–29.

Slatkin, M., and Hausfater, G. (1976). A note on the activities of a solitary male baboon. *Primates* 17:311–322.

Sly, D. L., Harbaugh, S. W., London, W. T., and Rice, J. M. (1983). Reproductive performance of a laboratory breeding colony of patas monkeys (*Erythrocebus patas*). *Am. J. Primatol.* 4:23–32.

Small, M. F. (1988). Female primate sexual behavior and conception: are there really sperm to spare? *Curr. Anthropol.* 29:81–100.

Small, M. F. (1989). Female choice in nonhuman primates. *Ybk. Phys. Anthropol.* 32:103–127.

Small, M. F. (1990). Promiscuity in Barbary macaques (*Macaca sylvanus*). *Am. J. Primatol.* 20:267–282.

Small, M. F. (1993). *Female Choices: Sexual Behavior of Female Primates*. Cornell University Press, Ithaca.

Smeenk, C., Godfrey, L. R., and Williams, F. L. (2006). The early specimens of the potto *Perodicticus potto* (Statius Mueller, 1776) in the National Museum of Natural History, Leiden, with the selection of a neotype. *Zool. Mededlongen* 80:139–164.

Smith, A. C. (2000). Composition and proposed nutritional importance of exudates eaten by saddleback (*Saguinus fuscicollis*) and mustached (*Saguinus mystax*) tamarins. *Int. J. Primatol.* 21:69–83.

Smith, A. C., Tirado Herrara, E. R., and Buchanan-Smith, H. M. (2001). Multiple breeding females and allo-nursing in a wild group of moustached tamarins (*Saguinus mystax*). *Neotrop. Primates* 9:67–69.

Smith, A. C., Buchanan-Smith, H. M., Surridge, A. K., and Mundy, N. I. (2005). Factors affecting group spread within wild mixed-species troops of saddleback and mustached tamarins. *Int. J. Primatol.* 26:337–355.

Smith, A. C., Knogge, C., Huck, M., Lottker, P., Buchanan-Smith, H. M., and Heymann, E. W. (2007). Long-term patterns of sleeping site use in wild saddleback (*Saguinus fuscicollis*) and mustached tamarins (*S. mystax*): effects of foraging, thermoregulation, predation, and resource defense constraints. *Am. J. Phys. Anthropol.* 134(3):340–353.

Smith, B. H. (1989). Dental development as a measure of life history in primates. *Evolution* 43:683–688.

Smith, B. H. (1991). Dental development and the evolution of life history in Hominidae. *Am. J. Phys. Anthropol.* 86:157–174.

Smith, B. H. (1994). Patterns of dental development in *Homo*, *Australopithecus*, *Pan* and *Gorilla*. *Am. J. Phys. Anthropol.* 94:307–325.

Smith, B. H., Crummett, T. L., and Brandt, K. L. (1994). Age of eruption of primate teeth: a compendium for aging individuals and comparing life histories. *Ybk. Phys. Anthropol.* 37:177–231.

Smith, C. C. (1977). Feeding behavior and social organization in howling monkeys. In: Clutton-Brock, T. H. (ed.), *Primate Ecology*. Academic Press, San Francisco. pp. 97–129.

Smith, D. G. (1995). Avoidance of close consanguineous inbreeding in captive groups of rhesus macaques. *Am. J. Primatol.* 35:31–40.

Smith, J. D., Shields, W. E., and Washburn, D. A. (2003). The comparative psychology of uncertainty monitoring and metacognition. *Behav. Brain Sci.* 26:317–373.

Smith, J. D., Redford, J. S., Beran, M. J., and Washbum, D. A. (2006). Dissociating uncertainty responses and reinforcement signals in the comparative study of uncertainty monitoring. *J. Exp. Psychol. Gen.* 135:282–297.

Smith, J. D., Beran, M. J., Couchman, J. J., and Coutinho, M. V. C. (2008). The comparative study of metacognition: sharper paradigms, safer inferences. *Psychon. Bull. Rev.* 15:679–691.

Smith, K. (2000). Paternal kin matter: the distribution of social behavior among wild adult female baboons [PhD diss.]. University of Chicago, Chicago.

Smith, K., Alberts, S. C., and Altmann, J. (2003). Wild female baboons bias their social behaviour towards paternal half-sisters. *Proc. R. Soc. Lond. B.* 270:503–510.

Smith, L. W., Link, A., and Cords, M. (2008). Cheek pouch use, predation risk, and feeding competition in blue monkeys (*Cercopithecus mitis stuhlmanni*). *Am. J. Phys. Anthropol.* 137(3):334–341.

Smith, R. (1984). Human sperm competition. In: Smith, R. (ed.), *Sperm Competition and the Evolution of Animal Mating Systems*. Academic Press, New York. pp. 601–659.

Smith, R. J., and Cheverud, J. M. (2002). Scaling of sexual dimorphism in body mass: a phylogenetic analysis of Rensch's rule in primates. *Int. J. Primatol.* 23:1095–1135.

Smith, R. J., and Jungers, W. L. (1997). Body mass in comparative primatology. *J. Hum. Evol.* 32:523–559.

Smith, R. J., and Leigh, S. R. (1998). Sexual dimorphism in primate neonatal body mass. *J. Hum. Evol.* 34:173–201.

Smith, R. L. (1984). Human sperm competition. In: Smith, R. L. (ed.), *Sperm Competition and the Evolution of Animal Mating Systems*. Wildwood House, London. pp. 601–660.

Smith, T. E., and French, J. A. (1997a). Social and reproductive conditions modulate urinary cortisol excretion in black tufted-ear marmosets (*Callithrix kuhli*). *Am. J. Primatol.* 42:253–267.

Smith, T. E., and French, J. A. (1997b). Psychosocial stress and urinary cortisol excretion in marmoset monkeys (*Callithrix kuhli*). *Physiol. Behav.* 62:225–232.

Smith, T. E., Schaffner, C. M., and French, J. A. (1997). Social and developmental influences on reproductive function in female Wied's black tufted-ear marmosets (*Callithrix kuhli*). *Horm. Behav.* 31:159–168.

Smith, W. J. (1986). An "informational" perspective on manipulation. In: Mitchell, R. W., and Thompson, N. S. (eds.), *Deception: Perspectives on Human and Nonhuman Deceit.* State University of New York Press, Albany. pp. 71–86.

Smith, W. J. (1991). Animal communication and the study of cognition. In: Ristau, C. A. (ed.), *Cognitive Ethology: The Minds of Other Animals (Essays in Honor of Donald R. Griffin).* Erlbaum, Hillsdale, NJ. pp. 209–230.

Smucny, D. A., Price, C. S., and Byrne, E. A. (1997). Post-conflict affiliation and stress reduction in captive rhesus macaques. *Adv. Ethol.* 32:157.

Smuts, B. B. (1983a). Dynamics of "special relationships" between adult male and female olive baboons. In: Hinde, R. A. (ed.), *Primate Social Relationships.* Sinauer Associates, Sunderland, MA. pp 112–116.

Smuts, B. B. (1983b). Special relationships between adult male and female olive baboons: selective advantages. In Hinde, R. A. (ed.), *Primate Social Relationships.* Sinauer Associates, Sunderland, MA. pp. 262–266.

Smuts, B. B. (1985). *Sex and Friendship in Baboons.* Aldine de Gruyter, New York.

Smuts, B. B. (1987). Sexual competition and mate choice. In: Smuts, B. B., Cheney, D. L., Seyfarth, R. M., Wrangham, R. W., and Struhsaker, T. T. (eds.), *Primate Societies.* University of Chicago Press, Chicago. pp. 385–399.

Smuts, B., and Nicolson, N. (1989). Reproduction in wild female olive baboons. *Am. J. Primatol.* 19:229–246.

Smuts, B. B., and Smuts, R. W. (1993). Male aggression and sexual coercion of females in nonhuman primates and other mammals: evidence and theoretical implications. *Adv. Study Behav.* 22:1–63.

Smuts, B. B., Cheney, D. L., Seyfarth, R. M., Wrangham, R. W., and Struhsaker, T. T. (eds.) (1987). *Primate Societies.* University of Chicago Press, Chicago.

Snaith, T. V., Chapman, C. A., Rothman, J. M., and Wasserman, M. D. (2008). Bigger groups have fewer parasites and similar cortisol levels: a multi-group analysis in red colobus monkeys. *Am. J. Primatol.* 70:1072–1080.

Snellgrove, D. (1987). *Indo-Tibetan Buddhism.* Shambhala, Boston.

Snow, B. K., and Snow, D. W. (1988). *Birds and Berries.* T&AD Poyser, Calton, UK.

Snowdon, C. T. (1982). Linguistic and psycholinguistic approaches to primate communication. In: Snowdon, C. T., Brown, C. H., and Petersen, M. R. (eds.), *Primate Communication.* Cambridge University Press, Cambridge. pp. 212–238.

Snowdon, C. T. (1996). Infant care in cooperatively breeding species. *Adv. Study Behav.* 25:648–689.

Snowdon, C. T. (1999). An empiricist view of language evolution and development. In: King, B. J. (ed.), *The Origins of Language: What Nonhuman Primates Can Tell Us.* School of American Research Press, Santa Fe, NM. pp. 79–114.

Snowdon, C. T. (2001). From primate communication to human language. In: de Waal, F. (ed.), *Tree of Origin: What Primate Behavior Can Tell Us About Human Social Evolution.* Harvard University Press, Cambridge, MA. pp. 193–227.

Snowdon, C. (2004). Affiliative processes and male primate social behavior. Paper presented at the Annual Meeting of the American Association for the Advancement of Science, Seattle, WA, February 2004.

Snowdon, C. T., and Soini, P. (1988). The tamarins, genus *Saguinus.* In: Mittermeier, R. A., Rylands, A. B., Coimbra-Filho, A. F., and da Fonseca, G. A. B. (eds.), *Ecology and Behavior of Neotropical Primates,* vol. 2. World Wildlife Fund, Washington DC. pp. 223–298.

Snyder, N. F. R., Derrickson, S. F., Beissinger, S. R., Wiley, J. W., Smith, T. B., Toone, W. D., and Miller, B. (1996). Limitation of captive breeding in endangered species recovery. *Conserv. Biol.* 10:338–348.

Soini, P. (1982). Ecology and population dynamics of the pygmy marmoset, *Cebuella pygmaea. Folia Primatol.* 39:1–21.

Soini, P. (1986). A synecological study of a primate community in the Pacaya-Samiria National Reserve, Peru. *Primate Conserv.* 7:63–71.

Soini, P. (1987). Sociosexual behavior of a free-ranging *Cebuella pygmaea* Callitrichidae, Platyrrhini troop during post-partum estrus of its reproductive female. *Am. J. Primatol.* 13:223–230.

Soini, P. (1988). The pygmy marmoset, genus *Cebuella.* In: Mittermeier, R. A., Rylands, A. B., Coimbra-Filho, A. F., and da Fonseca, G. A. B. (eds.), *Ecology and Behavior of Neotropical Primates,* vol. 2. World Wildlife Fund, Washington DC. pp. 79–129.

Soini, P. (1990). Ecologia y dinámica poblacional del "choro" (*Lagothrix lagotricha,* Primates) en el Rio Pacaya, Peru. In: *La Primatologia en el Perú.* Proyecto Peruano de Primatologia, Lima. pp. 382–396.

Soini, P. (1993). The ecology of the pygmy marmoset, *Cebuella pygmaea*: some comparisons with two sympatric tamarins. In: Rylands, A. B. (ed.), *Marmosets and Tamarins: Systematics, Behaviour, and Ecology.* Oxford University Press, Oxford. pp. 257–261.

Sokal, R., and Rohlf, F. J. (1994). *Biometry: The Principles and Practice of Statistics in Biological Research,* 5th ed. W. H. Freeman, New York.

Solano, P. (1995). Patrón de actividad y área de acción del mico nocturno *Aotus brumbacki hershkovitz,* 1983 (Primates: Cebidae) Parque Nacional Natural Tinigua, Meta, Colombia [PhD diss.]. Pontificia Universidad Javeriana, Bogotá.

Soligo, C., and Martin, R. D. (2006). Adaptive origins of primates revisited. *J. Hum. Evol.* 50:414–430.

Solomon, N. G., and French, J. A. (1997). *Cooperative Breeding in Mammals.* Cambridge University Press, Cambridge.

Soltis, J. (2002). Do female primates gain nonprocreative benefits by mating with multiple males? Theoretical and empirical considerations. *Evol. Anthropol.* 11:187–197.

Soltis, J. (2004). Mating systems. In: Thierry, B., Singh, M., and Kaumanns, W. (eds.), *Macaque Societies: A Model for the Study of Social Organization.* Cambridge University Press, Cambridge. pp. 135–151.

Soltis, J., and McElreath, R. (2001). Can females gain extra paternal investment by mating with multiple males? A game theoretic approach. *Am. Nat.* 158:519–529.

Soltis, J., Mitsunaga, F., Shimizu, K., Nozaki, M., Yanagihara, Y., Domingo Roura, X., and Takenaka, O. (1997a). Sexual selection in Japanese macaques: 2. Female mate choice and male–male competition. *Anim. Behav.* 54:737–746.

Soltis, J., Mitsunaga, F., Shimizu, K., Yanagihara, Y., and Nozaki, M. (1997b). Sexual selection in Japanese macaques: 1. Female mate choice or male sexual coercion? *Anim. Behav.* 54:725– 736.

Soltis, J., Mitsunaga, F., Shimizu, K., Yanagihara, Y., and Nozaki, M. (1999). Female mating strategy in an enclosed group of Japanese macaques. *Am. J. Primatol.* 47:263–278.

Soltis, J., Thomsen, R., Matsubayashi, K., and Takenaka, O. (2000a). Infanticide by resident males and female counter-strategies in wild Japanese macaques (*Macaca fuscata*). *Behav. Ecol. Sociobiol.* 48:195–202.

Soltis, J., Thomsen, R., and Takenaka, O. (2000b). Reproductive strategies and paternity in wild Japanese macaques on Yakushima Island, Japan. *Am. J. Phys. Anthropol.* (suppl. 30):287.

Soltis, J., Thomsen, R., and Takenaka, O. (2001). The interaction of male and female reproductive strategies and paternity in wild Japanese macaques, *Macaca fuscata. Anim. Behav.* 62:485–494.

Sommer, V. (1987). Infanticide among free-ranging langurs (*Presbytis entellus*) at Jodhpur (Rajasthan/India): recent observations and a reconsideration of hypotheses. *Primates* 28:163–197.

Sommer, V. (1994). Infanticide among langurs of Jodhpur: testing the sexual selection hypothesis with a long-term record. In: Parmigiani, S., and vom Saal, F. S. (eds.), *Infanticide and Parental Care.* Harwood, New York. pp. 155–198.

Sommer, V. (2000). The holy wars about infanticide. Which side are you on? And why? In: van Schaik, C. P., and Janson, C. H. (eds.), *Infanticide by Males and Its Implications.* Cambridge University Press, Cambridge. pp. 9–26.

Sommer, V., and Rajpurohit, L. S. (1989). Male reproductive success in harem troops of Hanuman langurs (*Presbytis entellus*). *Int. J. Primatol.* 10:293–317.

Sommer, V., and Reichard, U. (2000). Rethinking monogamy: the gibbon case. In: Kappeler, P. M. (ed.), *Primate Males: Causes and Consequences of Variation in Group Composition.* Cambridge University Press, Cambridge. pp. 159–168.

Sommer, V., Srivastava, A., and Borries, C. (1992). Cycles, sexuality and conception in free-ranging langurs (*Presbytis entellus*). *Am. J. Primatol.* 28:1–27.

Sommer, V., Adanu, J., Faucher, I., and Fowlera, A. (2004). Nigerian chimpanzees (*Pan troglodytes vellerosus*) at Gashaka: two years of habituation efforts. *Folia Primatol.* 75:295–316.

Sonnenburg, J. L., Altheide, T. K., and Varki, A. (2004). A uniquely human consequence of domain-specific functional adaptation in a sialic acid–binding receptor. *Glycobiology* 14:339–346.

Sork, V. L., Stowe, K. A., and Hochwender, C. (1993). Evidence for local adaptation in closely adjacent subpopulations of northern red oak

(*Quercus rubra* L.) expressed as resistance to leaf herbivores. *Am. Nat.* 142:928–936.

Sork, V. L., Nason, J., Campbell, D. R., and Fernandez, J. F. (1999). Landscape approaches to historical and contemporary gene flow in plants. *Trends Ecol. Evol.* 14:219–224.

Sourd, C., and Gautier-Hion, A. (1986). Fruit selection by a forest guenon. *J. Anim. Ecol.* 55:235–244.

Sousa, M. B. C., Albuquerque, A., Albuquerque, F. D., Araujo, A., Yamamoto, M. E., and Arruda M. D. (2005). Behavioral strategies and hormonal profiles of dominant and subordinate common marmoset (*Callithrix jacchus*) females in wild monogamous groups. *Am. J. Primatol.* 67:37–50.

Southwick, C. H. (ed.) (1963). *Primate Social Behavior*. Van Nostrand Reinhold, New York.

Southwick, C. H., and Siddiqi, M. F. (1988). Partial recovery and a new population estimate of rhesus monkey populations in India. *Am. J. Primatol.* 16:187–197.

Southwick, C. H., Beg, M. A., and Siddiqi, M. R. (1965). Rhesus monkeys in north India. In: DeVore, I. (ed.), *Primate Behavior: Field Studies of Monkeys and Apes*. Holt, Rinehart and Winston, New York. pp. 111–162.

Souto, A., Bezerra, B. M., Schiel, N., and Huber, L. (2007). Saltatory search in free-living *Callithrix jacchus*: environmental and age influences. *Int. J. Primatol.* 28:881–893.

Spence, K. W. (1937). Experimental studies of learning and higher mental processes in infrahuman primates. *Psychol. Bull.* 34:806–850.

Sperling, S. (1991). Baboons with briefcases vs. langurs in lipstick: feminism and functionalism in primate studies. In: Leonardo, M. (ed.), *Gender at the Crossroads of Knowledge: Feminist Anthropology in the Postmodern Era*. University of California Press, Berkeley. pp. 204–234.

Spijkerman, R. P., van Hooff, J. A. R. A. M., Dienske, H., and Jens, W. (1997). Differences in subadult behaviors of chimpanzees living in peer groups and in a family group. *Int. J. Primatol.* 18:439–454.

Spinozzi, G., Lubrano, G., and Truppa, V. (2004). Categorization of above and below spatial relations by tufted capuchin monkeys (*Cebus apella*). *J. Comp. Psychol.* 118:403–412.

Sponsel, L. E. (1997). The human niche: exploration in ethnoprimatology. In: Kinzey, W. G. (ed.), *New World Primates*. Walter de Gruyter, New York. pp. 143–165.

Sponsel, L. E., Ruttanadakul, N., and Natadecha-Sponsel, P. (2002). Monkey business? The conservation implications of macaque ethnoprimatology in southern Thailand. In: Fuentes, A., and Wolfe, L. D. (eds.), *Primates Face to Face*. Cambridge University Press, Cambridge. pp. 288–309.

Sprague, D. S. (1991). Mating by nontroop males among the Japanese macaques of Yakushima Island. *Folia Primatol.* 57:156–158.

Sprague, D. S. (1992). Life history and male intertroop mobility among Japanese macaques (*Macaca fuscata*). *Int. J. Primatol.* 13:437–454.

Sprague, D. S. (1998). Age, dominance rank, natal status, and tenure among male macaques. *Am. J. Phys. Anthropol.* 105:511–521.

Sprague, D. S. (2002). Monkey in the backyard: encroaching wildlife in rural communities in Japan. In: Fuentes, A., and Wolfe, L. D. (eds.), *Primates Face to Face: The Conservation Implications of Human–Non-Human Primate Interconnections*. Cambridge University Press, Cambridge. pp. 254–272.

Sprague, D. (2004). GPS collars for monkeys: the state of the technology. *Am. J. Phys. Anthropol. Suppl.* 186.

Sprague, D. S., Suzuki, S., Takahashi, H., and Sato, S. (1998). Male life history in natural populations of Japanese macaques: migration, dominance rank, and troop participation of males in two habitats. *Primates* 39(3):351–363.

Sprague, D. S., Kabaya, H., and Hagihara, K. (2004). Field testing a global positioning system (GPS) collar on a Japanese monkey: reliability of automatic GPS positioning in a Japanese forest. *Primates* 45:151–154.

Springer, M. S., Murphy, W. J., Eizirik, E., and O'Brien, S. J. (2003). Placental mammal diversification and the Cretaceous–Tertiary boundary. *Proc. Natl. Acad. Sci. USA* 100:1056–1061.

Spurlock, L. B. (2002). Reproductive suppression in the pygmy marmoset *Cebuella pygmaea*. *Diss. Abst. Int.* B62:3872.

Srikosamatara, S. (1984). Ecology of pileated gibbons in south-east Thailand. In: Preuschoft, H., Chivers, D. J., Brockelman, W. Y., and Creel, N. (eds.), *The Lesser Apes: Evolutionary and Behavioural Biology*. Edinburgh University Press, Edinburgh. pp. 242–257.

Srivastava, A. (1991). Cultural transmission of snake-mobbing in free-ranging Hanuman langurs. *Folia Primatol.* 56:117–120.

Srivastava, A., Biswas, J., Das, J., and Bujarbarua, P. (2001). Status and distribution of golden langurs (*Trachypithecus geei*) in Assam, India. *Am. J. Primatol.* 55:15–23.

Stabenfeldt, G. H., and Hendrickx, A. G. (1973). Progesterone levels in the sooty mangabey (*Cercocebus atys*) during the menstrual cycle, pregnancy, and parturition. *J. Med. Primatol.* 2:1–10.

Stacey, P. B. (1986). Group size and foraging efficiency in yellow baboons. *Behav. Ecol. Sociobiol.* 18:175–187.

Stallings, J. R., West, L., Hahn, W., and Gamarra, I. (1989). Primates and their relation to habitat in the Paraguayan Chaco. In: Redford, K. H., and Eisenberg, J. F. (eds.), *Advances in Neotropical Mammalogy*. Sandhill Crane Press, Gainesville, FL. pp. 425–442.

Stammbach, E. (1987). Desert, forest and montane baboons: multilevel societies. In: Smuts, B. B., Cheney, D. L., Seyfarth, R. M., Wrangham, R. W., and Struhsaker, T. T. (eds.), *Primate Societies*. University of Chicago Press, Chicago. pp. 112–120.

Stamps, J. (2003). Behavioural processes affecting development: Tinbergen's fourth question comes of age. *Anim. Behav.* 66:1–13.

Stanford, C. B. (1989). Predation on capped langurs (*Presbytis pileata*) by cooperatively hunting jackals (*Canis aureus*). *Am. J. Primatol.* 19:53–56.

Stanford, C. B. (1991a). The capped langur in Bangladesh: behavioral ecology and reproductive tactics. *Contrib. Primatol.* 26:1–179.

Stanford, C. B. (1991b). The diet of the capped langur (*Presbytis pileata*) in a moist deciduous forest in Bangladesh. *Int. J. Primatol.* 12:199–216.

Stanford, C. B. (1991c). Social dynamics of intergroup encounters in the capped langur (*Presbytis pileata*). *Am. J. Primatol.* 25:35–47.

Stanford, C. B. (1995). The influence of chimpanzee predation on group size and antipredator behavior in red colobus monkeys. *Anim. Behav.* 49:577–587.

Stanford, C. B. (1998a). *Chimpanzee and Red Colobus: The Ecology of Predator and Prey*. Harvard University Press, Cambridge, MA.

Stanford, C. B. (1998b). Predation and male bonds in primate societies. *Behaviour* 135:513–533.

Stanford, C. B. (1998c). The social behavior of chimpanzees and bonobos—empirical evidence and shifting assumptions. *Curr. Anthropol.* 39:399–420.

Stanford, C. B. (2001). The subspecies concept in primatology: the case of mountain gorillas. *Primates* 42:309–318.

Stanford, C. B. (2002). Avoiding predators: expectations and evidence in primate antipredator behavior. *Int. J. Primatol.* 23:741–757.

Stanford, C. B., and Nkurunungi, J. B. (2003). Behavioral ecology of sympatric chimpanzees and gorillas in Bwindi Impenetrable National Park, Uganda: diet. *Int. J. Primatol.* 24:901–918.

Stanford, C. B., Wallis, J., Matama, H., and Goodall, J. (1994). Patterns of predation by chimpanzees on red colobus monkeys in Gombe National Park, Tanzania, 1982–1991. *Am. J. Phys. Anthropol.* 94:213–228.

Stanyon, R., Consigiliere, S., Müller, S., Moreschalchi, A., Neusser, M., and Wienberg, J. (2000). Fluorescence in situ hybridization (FISH) maps chromosomal homologies between the dusky titi and squirrel monkey. *Am. J. Primatol.* 50:95–107.

Stanyon, R., Consigiliere, S., Bigoni, F., Ferguson-Smith, M., O'Brien, P. C. M., and Wienberg, J. (2001). Reciprocal chromosome painting between a New World primate, the woolly monkey, and humans. *Chromosome Res.* 9:97–106.

Stanyon, R., Koehler, U., and Consigiliere, S. (2002). Chromosome painting reveals that galagos have highly derived karyotypes. *Am. J. Phys. Anthropol.* 117:319–326.

Starin, E. D. (1978). A preliminary investigation of home range use in the Gir Forest langur. *Primates* 19:551–568.

Starin, E. D. (1988). Gestation and birth-related behaviors in Temminck's red colobus. *Folia Primatol.* 51:161–164.

Starin, E. D. (1989). Threats to the monkeys of the Gambia. *Oryx* 23:208–214.

Starin, E. D. (1991). Socioecology of the red colobus monkey in the Gambia with particular reference to female–male differences and transfer patterns [PhD diss.]. City University of New York, New York.

Starin, E. D. (1993). The kindness of strangers. *Nat. Hist.* 102:44–50.

Starin, E. D. (1994). Philopatry and affiliation among red colobus. *Behaviour* 130:253–270.

Starr, C. R., Streicher, U., Nekaris, K. A. I., and Leung, L. (2008). The distribution and conservation of the pygmy loris (*Nycticbeus pygmaeus*) in Mondulkiri Province, Cambodia. *Primate Eye* 96. Abstract 216.

Starr, C. R., Streicher, U., Nekaris, K. A. I., and Leung, L.. The use of lorises (*Nycticebus bengalensis* and *N. pygmaeus*) in traditional Cambodian medicines: implications for conservation. *Endang. Sp. Res.* In press.

Stavisky, R., Russell, E., Stallings, J., Smith, E. O., Worthman, C., and Whitten, P. L. (1995). Fecal steroid analysis of ovarian cycles in free-ranging baboons. *Am. J. Primatol.* 36:285–297.

Stearns, S. C. (1992). *Life History Evolution*. Yale University Press, New Haven.

Steenbeek, R. (2000). Infanticide by males and female choice in wild Thomas's langurs. In: van Schaik, C. P., and Janson, C. H. (eds.), *Infanticide by Males and Its Implications*. Cambridge University Press, Cambridge. pp. 153–177.

Steenbeek, R., Sterck, E. H. M., de Vries, H., and van Hooff, J. A. R. A. M. (2000). Costs and benefits of the one-male, age-graded, and all-male phases in wild Thomas's langur groups. In: Kappeler, P. M. (ed.), *Primate Males: Causes and Consequences of Variation in Group Composition*. Cambridge University Press, Cambridge. pp. 130–145.

Stein, D. M. (1984). *The Sociobiology of Infant and Adult Male Baboons*. Albex, Norwood.

Stein, D. M., and Stacey, P. B. (1981). A comparison of infant–adult male relations in a one-male group with those in a multi-male group for yellow baboons (*Papio cynocephalus*). *Folia Primatol.* 36:264–276.

Steiner, R. A. (1987). Nutritional and metabolic factors in the regulation of reproductive hormone secretion in the primate. *Proc. Nutr. Soc.* 46:159–175.

Steiper, M. E. (2006). Population history, biogeography, and taxonomy of orangutans (genus: *Pongo*) based on a population genetic meta-analysis of multiple loci. *J. Hum. Evol.* 50:509–522.

Steiper, M. E., and Ruvolo, M. (2003). Genetic sex identification of orang-utans. *Anthropol. Anz.* 61:1–5.

Steiper, M. E., and Young, N. M. (2008). Timing primate evolution: lessons from the discordance between molecular and paleontological evidence. *Evol. Anthropol.* 17:179–188.

Stelzner, J. K. (1988). Thermal effects on movement patterns of yellow baboons. *Primates* 29:91–105.

Stelzner, J. K., and Hausfater, G. (1986). Posture, microclimate, and thermoregulation in yellow baboons. *Primates* 27:449–463.

Stelzner, J., and Strier, K. (1981). Hyena predation on an adult male baboon. *Mammalia* 45:269–260.

Stephan, C., and Zuberbuehler, K. (2008). Predation increases acoustic complexity in primate alarm calls. *Biol. Lett.* 4(6):641–644.

Stephen, H., Baron, G., and Frahm, H. D. (1988). Comparative size of brains and brain components. In: Steklis, H. D., and Erwin, J. (eds.), *Comparative Primate Biology. Neurosciences,* vol. 4. Wiley-Liss, New York. pp. 1–38.

Stephens, D. W., McLinn, C. M., and Stevens, J. R. (2002). Discounting and reciprocity in an iterated prisoner's dilemma. *Science* 298:2216–2218.

Stephenson, R. A., Kurashina, H., Iverson, T. J., and Chiang, L. N. (2002). Visitor's perception of cultural improprieties in Bali, Indonesia. *J. Natl. Park* 12:156–169.

Sterck, E. H. M. (1995). Females, foods and fights: a socioecological comparison of the sympatric Thomas langur and long-tailed macaque [PhD diss.]. University of Utrecht, Utrecht.

Sterck, E. H. M. (1999). Variation in langur social organization in relation to the socioecological model, human habitation alteration, and phylogenetic constraints. *Primates* 40:199–213.

Sterck, E. H. M. (2002). Predator sensitive foraging in Thomas langurs. In: Miller, L. E. (ed.), *Eat or Be Eaten: Predator Sensitive Foraging Among Primates*. Cambridge University Press, Cambridge. pp. 74–91.

Sterck, E. H. M., and Korstjens, A. H. (2000). Female dispersal and infanticide avoidance in primates. In: van Schaik, C. P., and Janson, C. H. (eds.), *Infanticide by Males and Its Implications*. Cambridge University Press, Cambridge. pp. 293–321.

Sterck, E. H. M., and Steenbeek, R. (1997). Female dominance relationships and food competition in the sympatric Thomas langur and long-tailed macaque. *Behaviour* 134:749–774.

Sterck, E. H. M., and van Hooff, J. A. R. A. M. (2000). The number of males in langur groups: monopolizability of females or demographic processes? In: Kappeler, P. M. (ed.), *Primate Males: Causes and Consequences of Variation in Group Composition*. Cambridge University Press, Cambridge. pp. 120–129.

Sterck, E. H. M., Watts, D. P., and van Schaik, C. P. (1997). The evolution of female social relationships in nonhuman primates. *Behav. Ecol. Sociobiol.* 41:291–309.

Sterling, E. J. (1992). Timing the reproduction of aye-ayes (*Daubentonia madagascariensis*) in Madagascar. *Am. J. Primatol.* 27:59–60.

Sterling, E. J. (1993). Patterns of range use and social organization in aye-ayes (*Daubentonia madagascariensis*) on Nosy Mangabe. In: Kappeler, P. M., and Ganzhorn, J. U. (eds.), *Lemur Social Systems and Their Ecological Basis*. Plenum Press, New York. pp. 1–10.

Sterling, E. J., and Richard, A. F. (1995). Social organization in the aye-aye (*Daubentonia madagascariensis*) and the perceived distinctiveness of nocturnal primates. In: Alterman, L., Doyle, G. A., and Izard, M. K. (eds.), *Creatures of the Dark: The Nocturnal Prosimians*. Plenum Press, New York. pp. 439–451.

Sterling, E. J., Dierenfeld, E. S., Ashbourne, C. G., and Feistner, A. T. C. (1994). Dietary intake, food composition and nutrient intake in wild and captive populations of *Daubentonia madagascariensis*. *Folia Primatol.* 62:115–124.

Sterling, E. J., Nguyen, N., and Fashing, P. (2000). Spatial patterning in nocturnal prosimians: a review of methods and relevance to studies of sociality. *Am. J. Primatol.* 51:3–19.

Sterner, K. N., Raaum, R. L., Zhang, Y. P., Stewart, C.-B., and Disotell, T. R. (2006). Mitochondrial data support an odd-nosed colobine clade. *Mol. Phylogenet. Evol.* 40:1–7.

Steudel, K. (2000). The physiology and energetics of movement: effects on individuals and groups. In: Boinski, S., and Garber, P. A. (eds.), *On the Move: How and Why Animals Travel in Groups*. University of Chicago Press, Chicago. pp. 9–23.

Stevens, J. R., Hallinan, E.V., and Hauser, M.D. (2005a). The ecology and evolution of patience in two New World monkeys. *Biol Lett.* 1pp. 223–226.

Stevens, J. R., Rosati, A. G., Ross, K. R., and Hauser, M. D. (2005b). Will travel for food: spatial discounting in two New World monkeys. *Curr. Biol.* 15:1855–1860.

Stevenson, M. F., and Rylands, A. B. (1988). The marmosets, genus *Callithrix*. In: Mittermeier, R. A., Rylands, A. B., Coimbra-Filho, A. F., and da Fonseca, G. A. B. (eds.), *Ecology and Behavior of Neotropical Primates,* vol. 2. World Wildlife Fund, Washington DC. pp. 131–222.

Stevenson, P. R. (1992). Diet of woolly monkeys (*Lagothrix lagotricha*) at La Macarena, Colombia. *Field Stud. New World Monk. La Macarena Col.* 6:3–14.

Stevenson, P. (1997). Notes on the mating behavior of woolly monkeys (*Lagothrix lagotricha*) at Tinigua National Park, Colombia. *Field Stud. Fauna Flora La Macarena Col.* 10:13–15.

Stevenson, P. R. (1998). Proximal spacing between individuals in a group of woolly monkeys (*Lagothrix lagotricha*) in Tinigua National Park, Colombia. *Int. J. Primatol.* 19:299–312.

Stevenson, P. R. (2000). Seed dispersal by woolly monkeys (*Lagothrix lagothricha*) at Tinigua National Park, Colombia: dispersal distance, germination rates, and dispersal quantity. *Am. J. Primatol.* 50:275–289.

Stevenson, P. R. (2001). The relationship between fruit production and primate abundance in neotropical communities. *Biol. J. Linn. Soc.* 72:161–178.

Stevenson, P. R. (2002). Frugivory and seed dispersal by woolly monkeys at Tinigua National Park, Colombia [PhD diss.]. State University of New York, Stony Brook.

Stevenson, P. R. (2006). Activity and ranging patterns of Colombian woolly monkeys in north-western Amazonia. *Primates* 47:239–247.

Stevenson, P. R. (2007). Estimates of the number of seeds dispersed by a population of primates in a lowland forest in western Amazonia. In: Dennis, A., Schupp, E. W., Green, R., and Westcott, D. (eds.), *Seed Dispersal: Theory and Its Application in a Changing World*. CAB International, Wallingford, UK. pp. 340–362.

Stevenson, P. R. and Aldana. A. M. (2008). Potential effects of ateline extinction and forest fragmentation on plant diversity and composition in the western Orinoco basin, Colombia. *Int. J. Primatol.* 29:365–377.

Stevenson, P. R., and Castellanos, M. C. (2000). Feeding rates and daily path range of the Colombian woolly monkeys as evidence for between- and within-group competition. *Folia Primatol.* 71:399–408.

Stevenson, P. R., Quiñones, M. J., and Ahumada, J. A. (1994). Ecological strategies of woolly monkeys (*Lagothrix lagotricha*) at Tinigua National Park, Colombia. *Am. J. Primatol.* 32:123–140.

Stevenson, P. R., Quiñones, M. J., and Ahumada, J. A. (1998). Annual variation of fruiting pattern using two different methods in a lowland tropical forest at Tinigua National Park, Colombia. *Biotropica* 30:129–134.

Stevenson, P. R., Castellanos, M. C., Pizarro, J. C., and Garavito, M. (2002). Effects of seed dispersal by three ateline monkey species on seed germination at Tinigua National Park, Colombia. *Int. J. Primatol.* 23:1187–1204.

Stewart, C.-B., and Disotell, T. R. (1998). Primate evolution—in and out of Africa. *Curr. Biol.* 8:R582–R588.

Stewart, C.-B., Schilling, J. W., and Wilson, A. C. (1987). Adaptive evolution in the stomach lysozymes of foregut fermenters. *Nature* 330:401–404.

Stewart, K. J. (1988). Suckling and lactational anoestrus in wild gorillas (*Gorilla gorilla*). *J. Reprod. Fertil.* 83:627–634.

Stewart, K. J. (2001). Social relationships of immature gorillas and silverbacks. In: Robbins, M. M., Sicotte, P., and Stewart, K. J. (eds.), *Mountain Gorillas: Three Decades of Research at Karisoke*. Cambridge University Press, Cambridge. pp. 183–213.

Stewart, K. J., and Harcourt, A. H. (1987). Gorillas: variation in female relationships. In: Smuts, B. B., Cheney, D. L., Seyfarth, R. M., Wrangham, R. W., and Struhsaker, T. T. (eds.), *Primate Societies*. University of Chicago Press, Chicago. pp. 155–164.

Stohlgren,T. J., Bull, K. A., and Otsuki, Y. (1998). Comparison of rangeland vegetation sampling techniques in the Central Grasslands. *J. Range Manag.* 51:164–172.

Stoinski, T. S., and Whiten, A. (2003). Social learning by orangutans (*Pongo abelii* and *Pongo pygmaeus*) in a simulated food-processing task. *J. Comp. Psychol.* 117:272–282.

Stoinski, T. S., Wrate, J. L., Ure, N., and Whiten, A. (2001). Imitative learning by captive western lowland gorillas (*Gorilla gorilla gorilla*) in a simulated food-processing task. *J. Comp. Psychol.* 115:272–281.

Stoinski, T. S., Beck, B., Bloomsmith, M. A., and Maple, T. L. (2003). A behavioural comparison of captive-born, reintroduced golden lion tamarins and their wild-born offspring. *Behaviour* 140:137–160.

Stokes, E. J. (2004). Within-group social relationships amongst females and adult males in wild western lowland gorillas (*Gorilla gorilla gorilla*). *Am. J. Primatol.* 64:233–246.

Stokes, E. J., Parnell, R. J., and Olejniczak, C. (2003). Female dispersal and reproductive success in wild western lowland gorillas (*Gorilla gorilla gorilla*). *Behav. Ecol. Sociobiol.* 54:329–339.

Stoltz, L. P., and Saayman, G. S. (1970). Ecology and behaviour of baboons in the northern Transvaal. *Ann. Transv. Mus.* 26:99–143.

Stone, A. (2004). Juvenile feeding ecology and life history in a neotropical primate, the squirrel monkey (*Samiri sciureus*) [PhD diss.]. University of Illinois, Urbana.

Stone, A. C., Griffiths, R. C., Zegura, S. L., and Hammer, M. F. (2002). High levels of Y-chromosome nucleotide diversity in the genus *Pan*. *Proc. Natl. Acad. Sci. U.S.A.* 99:43–48.

Stone, A. I. (2006). Foraging ontogeny is not linked to delayed maturation in squirrel monkeys (*Saimiri sciureus*). *Ethology* 112(2):105–115.

Stone, A. I. (2007). Age and seasonal effects on predator-sensitive foraging in squirrel monkeys (*Saimiri sciureus*): a field experiment. *Am. J. Primatol.* 69(2):127–141.

Stone, K. F. (1996). Folktales. In: Brunvand, J. H. (ed.), *American Folklore: An Encyclopedia*. Garland, New York. pp. 294–295.

Stoner, K. E. (1994). Population density of the mantled howler monkey (*Alouatta palliata*) at La Selva Biological Reserve, Costa Rica: a new technique to analyze census data. *Biotropica* 26:332–340.

Stoner, K. E. (1996a). Habitat selection and seasonal patterns of activity and foraging in mantled howling monkeys (*Alouatta palliata*) in northeastern Costa Rica. *Int. J. Primatol.* 17:1–30.

Stoner, K. E. (1996b). Prevalence and intensity of intestinal parasites in mantled howling monkeys (*Alouatta palliata*) in northeastern Costa Rica: implications for conservation biology. *Conserv. Biol.* 10:539–546.

Stoner, K. E., Riba-Hernandez, P., Vulinec, K., and Lambert, J. E. (2007). The role of mammals in creating a modifying seed shadows in tropical forests and some possible consequences of their elimination. *Biotropica* 39:316–327.

Storr, G. (1780). *Prodromus methodi mammalium*. Litteris Reissianis, Tubingae, Germany

Storz, J. F., Beaumont, M. A., and Alberts, S. C. (2002a). Genetic evidence for long-term population decline in a savannah-dwelling primate: inferences from a hierarchical bayesian model. *Mol. Biol. Evol.* 19:1981–1990.

Storz, J. F., Ramakrishnan, U., and Alberts, S. C. (2002b). Genetic effective size of a wild primate population: influences of current and historical demography. *Evolution* 56:817–829.

Strangeways, I. (2000). *Measuring the Natural Environment*. Cambridge University Press, Cambridge.

Strassmann, B. T. (1981). Sexual selection, paternal care, and concealed ovulation in humans. *Ethol. Sociobiol.* 2:31–40.

Strassmann, B. T. (1996). The evolution of endometrial cycles and menstruation. *Q. Rev. Biol.* 71:181–220.

Strassmann, J. E., Solís, C. R., Peters, J. M., and Queller, D. C. (1996). Strategies for finding and using highly polymorphic DNA microsatellite loci for studies of genetic relatedness and pedigrees. In: Ferraris, J. D., and Palumbi, S. R. (eds.), *Molecular Zoology: Advances, Strategies, and Protocols*. Wiley-Liss, New York. pp. 163–180.

Streicher, U. (2003). Saisonale Veränderungen in Fellzeichnung und Fellfärbung beim Zwergplumplori *Nycticebus pygmaeus* und irhe taxonomische Bedeutung. *Zool. Garten N. F.* 73:368–373.

Streicher, U. (2004). Aspects of the ecology and conservation of the pygmy loris *Nycticebus pygmaeus* in Vietnam [inaugural diss.]. Ludwig-Maximilians University, Munich.

Streicher, U., Collins, R., Navarro-Montes, A., and Nekaris, K. A. I. (2009). Observations on the feeding preferences of slow lorises (*N. pygmaeus, N. javanicus, N. coucang*) confiscated from the trade. In: Masters, J., Genin, F., and Crompton, R. (eds.), *Leaping Ahead*. Springer, New York. In press.

Strier, K. B. (1986a). The behavior and ecology of the woolly spider monkey, or muriqui (*Brachyteles arachnoides*, E. Geoffroy 1806) [PhD diss.]. Harvard University, Cambridge, MA.

Strier, K. B. (1986b). Reprodução de *Brachyteles arachnoides* (Primates: Cebidae). *Primatol. Brasil* 2:163–175.

Strier, K. B. (1987a). Activity budgets of woolly spider monkeys, or muriquis (*Brachyteles arachnoides*). *Am. J. Primatol.* 13:385–395.

Strier, K. B. (1987b). Demographic patterns in one group of muriquis. *Primate Conserv.* 8:73–74.

Strier, K. B. (1987c). Ranging behavior of woolly spider monkeys, or muriquis, *Brachyteles arachnoides*. *Int. J. Primatol.* 8:575–591.

Strier, K. B. (1988). Behavioral correlates of food patch size in muriquis (*Brachyteles arachnoides*) at Fazenda Montes Claros, Minas-Gerais, Brazil. *Am. J. Phys. Anthropol.* 75:276.

Strier, K. B. (1989). Effects of patch size on feeding associations in muriquis (*Brachyteles arachnoides*). *Folia Primatol.* 52:70–77.

Strier, K. B. (1990). New World primates, new frontiers: insights from the woolly spider monkey, or muriqui (*Brachyteles arachnoides*). *Int. J. Primatol.* 11:7–19.

Strier, K. B. (1991a). Demography and conservation of an endangered primate, *Brachyteles arachnoides*. *Conserv. Biol.* 5:214–218.

Strier, K. B. (1991b). Diet in one group of woolly spider monkeys, or muriquis (*Brachyteles arachnoides*). *Am. J. Primatol.* 23:113–126.

Strier, K. B. (1992a). Atelinae adaptations: behavioral strategies and ecological constraints. *Am. J. Phys. Anthropol.* 88:515–524.

Strier, K. B. (1992b). Causes and consequences of nonaggression in the woolly spider monkey, or muriqui (*Brachyteles arachnoides*). In: Silverberg, J., and Gray, J. P. (eds.), *Aggression and Peacefulness in Humans and Other Primates*. Oxford University Press, New York. pp. 100–116.

Strier, K. B. (1992c). *Faces in the Forest: The Endangered Muriqui Monkeys of Brazil*. Oxford University Press, New York.

Strier, K. B. (1993). Growing up in a patrifocal society: sex differences in the spatial relations of immature muriquis. In: Pereira, M. E., and Fairbanks, L. A. (eds.), *Juvenile Primates: Life History, Development, and Behavior*. Oxford University Press, New York. pp. 138–147.

Strier, K. B. (1993/1994). Viability analyses of an isolated population of muriqui monkeys (*Brachyteles arachnoides*): implications for primate conservation and demography. *Primate Conserv.* 14/15:43–52.

Strier, K. B. (1994a). Brotherhoods among atelins: kinship, affiliation, and competition. *Behaviour* 130:151–167.

Strier, K. B. (1994b). The myth of the typical primate. *Ybk. Phys. Anthropol.* 37:233–271.

Strier, K. B. (1996a). Male reproductive strategies in New World primates. *Hum. Nat.* 7:105–123.

Strier, K. B. (1996b). Reproductive ecology of female muriquis (*Brachyteles arachnoides*). In: Norconk, M. A., Rosenberger, A. L., and Garber, P. A. (eds.), *Adaptive Radiations of Neotropical Primates*. Plenum Press, New York. pp. 511–352.

Strier, K. B. (1997a). Behavioral ecology and conservation biology of primates and other animals. *Adv. Study Behav.* 26:101–158.

Strier, K. B. (1997b). Subtle cues of social relations in male muriqui monkeys (*Brachyteles arachnoides*). In: Kinzey, W. G. (ed.), *New World Primates: Ecology, Evolution, and Behavior*. Aldine de Gruyter, New York. pp. 109–118.

Strier, K. B. (1997c). Mate preferences of wild muriqui monkeys (*Brachyteles arachnoides*): reproductive and social correlates. *Folia Primatol.* 68:120–133.

Strier, K. B. (1999). Why is female kin bonding so rare? Comparative sociality of neotropical primates. In: Lee, P. C. (ed.), *Comparative Primate Socioecology*. Cambridge University Press, Cambridge. pp. 300–319.

Strier, K. B. (2000a). *Primate Behavioral Ecology*. Allyn and Bacon, Boston.

Strier, K. B. (2000b). Population viabilities and conservation implications for muriquis (*Brachyteles arachnoides*) in Brazil's Atlantic Forest. *Biotropica* 32:903–913.

Strier, K. B. (2000c). From binding brotherhoods to short-term sovereignty: the dilemma of male Cebidae. In: Kappeler, P. M. (ed.), *Primate Males: Causes and Consequences of Variation in Group Composition*. Cambridge University Press, Cambridge. pp. 72–83.

Strier, K. B. (2001). Reproductive ecology of New World monkeys. In: Ellison, P. T. (ed.), *Reproductive Ecology and Human Evolution*. Aldine de Gruyter, New York. pp. 351–367.

Strier, K. B. (2003a). *Primate Behavioral Ecology,* 2nd ed. Allyn and Bacon, Boston.

Strier, K. B. (2003b). Primatology comes of age: 2002 AAPA luncheon address. *Ybk. Phys. Anthropol.* 46:2–13.

Strier, K. B. (2003c). Primate behavioral ecology: from ethnography to ethology and back. *Am. Anthropol.* 105:16–27.

Strier, K. B. (2003d). Demography and the temporal scale of sexual selection. In: Jones, C. B. (ed.), *Sexual Selection and Reproductive Competition in Primates.* American Society of Primatologists, Norman, OK. pp. 45–63.

Strier, K. B. (2004a). Patrilineal kinship and primate behavior. In: Chapais, B., and Berman, C. M. (eds.), *Kinship and Behavior in Primates.* Oxford University Press, New York pp. 177–199.

Strier, K. B. (2004b). Sociality among kin and nonkin in nonhuman primate groups. In: Sussman, R. W., and Chapman, A. R. (eds.), *The Origins and Nature of Sociality.* Hawthorne, NY: Aldine de Gruyter, pp. 191–214.

Strier, K. B. (2007). *Primate Behavioral Ecology,* 3rd ed. Allyn & Bacon, Boston.

Strier, K. B. (2008). The effects of kin on primate life histories. *Annu. Rev. Anthropol.* 37:21–36.

Strier, K. B., and Fonseca, G. A. B. (1996/1997). The endangered muriquis in Brazil's Atlantic Forest. *Primate Conserv.* 17:131–137.

Strier, K. B. and Mendes, S. L. (2009). Long-term field studies of South American primates. In: Garber, P. A., Estrada, A., Bicca-Marques, J. C., Heymann, E., and Strier, K. B. (eds.), *South American Primates: Comparative Perspectives in the Study of Behavior, Ecology, and Conservation.* Springer, New York, pp. 139–155.

Strier, K. B., and Ziegler, T. E. (1997). Behavioral and endocrine characteristics of the reproductive cycle in wild muriqui monkeys, *Brachyteles arachnoides. Am. J. Primatol.* 42:299–310.

Strier, K. B., and Ziegler, T. E. (2000). Lack of pubertal influences on female dispersal in muriqui monkeys, *Brachyteles arachnoides. Anim. Behav.* 59:849–860.

Strier, K. B., Mendes, F. D. C., Rimoli, J., and Rimoli, A. O. (1993). Demography and social structure of one group of muriquis (*Brachyteles arachnoides*). *Int. J. Primatol.* 14:513–526.

Strier, K. B., Ziegler, T. E., and Wittwer, D. J. (1999). Seasonal and social correlates of fecal testosterone and cortisol levels in wild male muriquis (*Brachyteles arachnoides*). *Horm. Behav.* 35:125–134.

Strier, K. B., Carvalho, D. S., and Bejar, N. O. (2000). Prescription for peacefulness. In: Aureli, F., and de Waal, F. B. M. (eds.), *Natural Conflict Resolution.* University of California Press, Los Angeles. pp. 315–317.

Strier, K. B., Mendes, S. L., and Santos, R. R. (2001). Timing of births in sympatric brown howler monkeys (*Alouatta fusca clamitans*) and northern muriquis (*Brachyteles arachnoides*). *Am. J. Primatol.* 55:87–100.

Strier, K. B., Dib, L. T., and Figueria, J. E. C. (2002). Social dynamics of male muriquis (*Brachyteles arachnoides hypoxanthus*). *Behaviour* 139:315–342.

Strier, K. B., Lynch, J. W., and Ziegler, T. E. (2003). Hormonal changes during the mating and conception seasons of wild northern muriquis (*Brachyteles arachnoides hypoxanthus*). *Am. J. Primatol.* 61:85–99.

Strier, K. B., Boubli, J. P., Possami, C. B., and Mendes, S. L. (2006). Population demography of northern muriquis (*Brachyteles hypoxanthus*) at the Estação Biológica de Caratinga/Reserva Patrimônio Particular Natural-Feliciano Miguel Abdula, Miras Gerais, Brazil. *Am. J. Phys. Anthropol.* 130:227–237.

Struhsaker, S., Oates, J. F., Hart, J., and Butynski, T. M. (2008). *Cercopithecus neglectus.* In: IUCN 2009. IUCN Red List of Threatened Species, version 2009.1, www.iucnredlist.org (accessed June 23, 2009).

Struhsaker, T. T. (1967a). Auditory communication among vervet monkeys (*Cercopithecus aethiops*). In: Altmann, S. A. (ed.), *Social Communication Among Primates.* University of Chicago Press, Chicago. pp. 238–324.

Struhsaker, T. T. (1967b). Ecology of vervet monkeys (*Cercopithecus aethiops*) in the Masai-Amboseli Game Reserve, Kenya. *Ecology* 48:891–904.

Struhsaker, T. T. (1967c). Social structure among vervet monkeys (*Cercopithecus aethiops*). *Behaviour* 29:83–121.

Struhsaker, T. T. (1969). Correlates of ecology and social organization among African cercopithecines. *Folia Primatol.* 11:80–118.

Struhsaker, T. T. (1970). Phylogenetic implications of some vocalizations of *Cercopithecus* monkeys. In: Napier, J. R., and Napier, P. H. (eds.), *Old World Monkeys: Evolution, Systematics, and Behaviour.* Academic Press, New York. pp. 367–344.

Struhsaker, T. T. (1974). Correlates of ranging behavior in a group of red colobus monkeys (*Colobus badius tephrosceles*). *Am. Zool.* 14:177–184.

Struhsaker, T. T. (1975). *The Red Colobus Monkey.* University of Chicago Press, Chicago.

Struhsaker, T. T. (1976). A further decline in numbers of Amboseli vervet monkeys. *Biotropica* 8:211–214.

Struhsaker, T. T. (1978a). Food habits of five monkey species in the Kibale Forest, Uganda. In: Chivers, D. J., and Herbert, J. (eds.), *Recent Advances in Primatology. Behaviour,* vol. 1. Academic Press, London. pp. 225–248.

Struhsaker, T. T. (1978b). Food habits of five monkey species in the Kibale Forest, Uganda. In: Chivers, D. J., and Herbert, J. (eds.), *Recent Advances in Primatology. Conservation,* vol. 2. Academic Press, London. pp. 87–94.

Struhsaker, T. T. (1981a). Polyspecific associations among tropical rain forest primates. *Z. Tierpsychol.* 57:268–304.

Struhsaker, T. T. (1981b). Vocalizations, phylogeny and palaeogeography of red colobus monkeys (*Colobus badius*). *Afr. J. Ecol.* 19:265–283.

Struhsaker, T. T. (1988). Male tenure, multi-male influxes, and reproductive success in redtail monkeys (*Cercopithecus ascanius*). In: Gautier-Hion, A., Bourlière, F., Gautier, J. P., and Kingdon, J., (eds.), *A Primate Radiation: Evolutionary Biology of the African Guenons.* Cambridge University Press, Cambridge. pp. 340–363.

Struhsaker, T. T. (1997). *Ecology of an African Rain Forest: Logging in Kibale and the Conflict Between Conservation and Exploitation.* University Press of Florida, Gainesville.

Struhsaker, T. T. (1999). Primate communities in Africa: the consequence of long-term evolution or the artifact of recent hunting? In: Fleagle, J. G., Janson, C., and Reed, K. E. (eds.), *Primate Communities.* Cambridge University Press, Cambridge. pp. 289–294.

Struhsaker, T. T. (2000). The effects of predation and habitat quality on the socioecology of African monkeys: lessons from the islands of Bioko and Zanzibar. In: Whitehead, P. F., and Jolly, C. J. (eds.), *Old World Monkeys.* Cambridge University Press, Cambridge. pp. 393–430.

Struhsaker, T. T., and Gartlan, J. S. (1970). Observations on the behaviour and ecology of the patas monkey (*Erythrocebus patas*) in the Waza Reserve, Cameroon. *J. Zool. Lond.* 161:49–63.

Struhsaker, T. T., and Leakey, M. (1990). Prey selectivity by crowned-hawk eagles on monkeys in the Kibale Forest, Uganda. *Behav. Ecol. Sociobiol.* 26:435–443.

Struhsaker, T. T., and Leland, L. (1979). Socioecology of five sympatric monkey species in the Kibale Forest, Uganda. In: Rosenblatt, J. S., Hinde, R. A., Beer, C., and Busnel, M. C. (eds.), *Advances in the Study of Behavior.* Academic Press, New York. pp. 159–228.

Struhsaker, T. T., and Leland, L. (1985). Infanticide in a patrilineal society of red colobus monkeys. *Z. Tierpsychol.* 69:89–132.

Struhsaker, T. T., and Leland, L. (1987). Colobines: infanticide by adult males. In: Smuts, B. B., Cheney, D. L., Seyfarth, R. M., Wrangham, R. W., and Struhsaker, T. T. (eds.), *Primate Societies.* University of Chicago Press, Chicago. pp. 83–97.

Struhsaker, T. T., and Leland, L. (1988). Group fission in redtail monkeys (*Cercopithecus ascanius*) in the Kibale Forest, Uganda. In: Gautier-Hion, A., Bourlière, F., Gautier, J. P., and Kingdon, J. (eds.), *A Primate Radiation: Evolutionary Biology of the African Guenons.* Cambridge University Press, Cambridge. pp. 364–388.

Struhsaker, T. T., and Oates, J. F. (1975). Comparison of the behavior and ecology of red colobus and black-and-white colobus monkeys in Uganda: a summary. In: Tuttle, R. H. (ed.), *Socioecology and Psychology of Primates.* Mouton, The Hague. pp. 103–123.

Struhsaker, T. T., Butynski, T. M., and Lwanga, J. S. (1988). Hybridization between redtail (*Cercopithecus ascanius schmidti*) and blue (*C. mitis stuhlmanni*) monkeys in the Kibale Forest, Uganda. In: Gautier-Hion, A., Bourlière, F., Gautier, J. P., and Kingdon, J. (eds.), *A Primate Radiation: Evolutionary Biology of the African Guenons.* Cambridge University Press, Cambridge. pp. 477–497.

Struhsaker, T. T., Cooney, D. O., and Siex, K. S. (1997). Charcoal consumption by Zanzibar red colobus monkeys: its function and its ecological and demographic consequences. *Int. J. Primatol.* 18:61–72.

Strum, S. C. (1982). Agonistic dominance in male baboons: an alternative view. *Int. J. Primatol.* 3:175–202.

Strum, S. C. (1983). Use of females by male olive baboons (*Papio anubis*). *Am. J. Primatol.* 5:93–109.

Strum, S. C. (1984). Why males use infants. In: Taub, D. M. (ed.), *Primate Paternalism.* van Nostrand Reinhold, New York. pp. 146–185.

Strum, S. C. (1986). A role for long-term primate field research in source countries. In: Else, J. G., and Lee, P. C. (eds.), *Primate Ecology and Conservation.* Cambridge University Press, Cambridge. pp. 215–220.

Strum, S. C. (1987). *Almost Human.* W. W. Norton, New York.

Strum, S. C. (1994). Prospects for management of primate pests. *Rev. Ecol. (Terre Vie)* 49:295–306.

Strum, S. C. (2000). Science encounters. In: Strum, S. C., and Fedigan, L. M. (eds.), *Primate Encounters: Models of Science, Gender, and Society*. University of Chicago Press, Chicago. pp. 475–497.

Strum, S. C. (2001). *Almost Human: A Journey into the World of Baboons*. University of Chicago Press, Chicago.

Strum, S. C. (2005). Measuring success in primate translocation: a baboon case study. *Am. J. Primatol.* 65:117–140.

Strum, S. C., and Fedigan, L. M. (1999). Theory, method, gender, and culture: what changes our views of primate sociality? In: Strum, S. C., Lindburg, D. G., and Hamburg, D. (eds.), *The New Physical Anthropology: Science, Humanism, and Critical Reflection*. Prentice Hall, Upper Saddle River, NJ. pp. 67–105.

Strum, S. C., and Fedigan, L. M. (2000). Changing views of primate society: a situated North American view. In: Strum, S. C., and Fedigan, L. M. (eds.), *Primate Encounters: Models of Science, Gender, and Society*. University of Chicago Press, Chicago. pp. 1–49.

Strum, S. C., and Mitchell, W. (1987). Baboon models and muddles. In: Kinzey, W. G. (ed.), *The Evolution of Human Behavior: Primate Models*. State University of New York Press, Albany. pp 87–104.

Strum, S. C., and Western, J. D. (1982). Variations in fecundity with age and environment in olive baboons (*Papio anubis*). *Am. J. Primatol.* 3:61–76.

Stuart, M. D., and Strier, K. B. (1995). Primates and parasites: a case for a multidisciplinary approach. *Int. J. Primatol.* 16:577–593.

Stuart, M., Pendergast, V., Rumfelt, S., Peirberg, S., Greenspan, L., Glander, K., and Clarke, M. (1998). Parasites of wild howlers (*Alouatta* spp.). *Int. J. Primatol.* 19:493–512.

Stucki, B. R., Dow, M. M., and Sade, D. S. (1991). Variance in intrinsic rates of growth among free-ranging rhesus monkey groups. *Am. J. Phys. Anthropol.* 84:181–192.

Stumpf, R. M. (2004). Female reproductive strategies of chimpanzees of the Taï National Park, Cote d'Ivoire [PhD diss.]. State University of New York, Stony Brook.

Stumpf, R. (2007). Chimpanzees and bonobos: diversity within and between species. In: Campbell, C. J., Fuentes, A., MacKinnon, K. C., Panger, M., and Bearder, S. K. (eds.), *Primates in Perspective*, 1st ed. Oxford University Press, New York. pp. 321–344.

Stumpf, R. M., and Boesch, C. (2005). Does promiscuous mating preclude female choice? Female sexual strategies and mate preferences in chimpanzees of the Taï Forest, Côte d'Ivoire. *Behav. Ecol. Sociobiol.* 57:511–524.

Stumpf, R. M., and Boesch, C. (2006). The efficacy of female choice in chimpanzees of the Taï Forest, Cote d'Ivoire. *Behav. Ecol. Sociobiol.* 60:749–765.

Stumpf, R. M., Emery Thompson, M., and Knott, C. (2008). A comparison of female mating strategies in *Pan troglodytes* and *Pongo* spp. *Int. J. Primatol.* 29:865–884.

Stumpf, R. M., Emery Thompson, M., Muller, M. N., and Wrangham, R. W. (2009). The context of dispersal in Kanyawara chimpanzees. *Behaviour* 146:629–656.

Su, H. H., and Lee, L. L. (2001). Food habits of Formosan rock macaques (*Macaca cyclopis*) in Jentse, northeastern Taiwan, assessed by fecal analysis and behavioral observation. *Int. J. Primatol.* 22:359–377.

Su, Y. J., Ren, R. M., Yan, K. H., Li, J. J., Yin, Z., Zhu, Z. Q., Hu, Z. L., and Hu, Y. F. (1998). Preliminary survey of the home range and ranging behavior of golden monkeys (*Rhinopithecus [Rhinopithecus] roxellana*) in Shennongjia National Natural Reserve, Hubei, China. In: Jablonski, N. G. (ed.), *The Natural History of the Doucs and Snub-Nosed Monkeys*. World Scientific Press, Singapore. pp. 255–268.

Suarez, S. A. (2003). Spatio-temporal foraging skills of white-bellied spider monkeys (*Ateles belzebuth belzebuth*) in the Yasuní National Park, Ecuador [PhD diss.]. State University of New York, Stony Brook.

Suarez, S. A. (2006). Diet and travel costs for spider monkeys in a nonseasonal, hyperdiverse environment. *Int. J. Primatol.* 27:411–436.

Suarez, S. D., and Gallup, G. G. (1981). Self-recognition in chimpanzees and orangutans, but not gorillas. *J. Hum. Evol.* 10:175–188.

Suda, C., and Call, J. (2005). Piagetian conservation of discrete quantities in bonobos (*Pan paniscus*), chimpanzees (*Pan troglodytes*), and orangutans (*Pongo pygmaeus*). *Anim. Cogn.* 8:220–235.

Suda-King, C. (2008). Do orangutans (*Pongo pygmaeus*) know when they do not remember? *Anim Cogn.* 11:21–42.

Sugardjito, J., and Nurhada, N. (1981). Meat-eating behavior in wild orangutan *Pongo pygmaeus*. *Primates* 22:414–416.

Sugardjito, J., te Boekhorst, I. J. A., and van Hooff, J. A. R. A. M. (1987). Ecological constraints on the grouping of wild orang-utans (*Pongo pygmaeus*) in the Gunung Leuser National Park, Sumatra, Indonesia. *Int. J. Primatol.* 8:17–41.

Sugiyama, Y. (1964). Group composition, population density and some sociological observations of Hanuman langurs (*Presbytis entellus*). *Primates* 5:7–37.

Sugiyama, Y. (1976). Life history of male Japanese macaques. *Adv. Stud. Behav.* 7:255–284.

Sugiyama, Y. (1988). Grooming interactions among adult chimpanzees at Bossou, Guinea, with special reference to social structure. *Int. J. Primatol.* 9:393–407.

Sugiyama, Y. (1989). Population dynamics of chimpanzees at Bossou, Guinea. In: Heltne, P. G., and Marquardt, L. G. (eds.), *Understanding Chimpanzees*. Harvard University Press, Cambridge, MA. pp. 134–145.

Sugiyama, Y. (1994). Age-specific birth rate and lifetime reproductive success of chimpanzees at Bossou, Guinea. *Am. J. Primatol.* 32:311–318.

Sugiyama, Y. (1997). Social tradition and the use of tool-composites by wild chimpanzees. *Evol. Anthropol.* 6:23–27.

Sugiyama, Y. (1999). Socio-ecological factors of male chimpanzee migration at Bossou, Guinea. *Primates* 40:61–68.

Sugiyama, Y. (2004). Demographic parameters and life history of chimpanzees at Bossou, Guinea. *Am. J. Phys. Anthropol.* 124:154–165.

Sugiyama, Y., and Koman, J. (1979). Social structure and dynamics of wild chimpanzees at Bossou, Guinea. *Primates* 20:323–339.

Sugiyama, Y., and Koman, J. (1992). The flora of Bossou: its utilization by chimpanzees and humans. *Afr. Study Monogr.* 13:127–169.

Sugiyama, Y., and Ohsawa, Y. (1982). Population dynamics of Japanese monkeys with special reference to the effect of artificial feeding. *Folia Primatol.* 39:238–263.

Sugiyama, Y., Kawamoto, S., Takenaka, O., Kumazaki, K., and Miwa, N. (1993). Paternity discrimination and inter-group relationships of chimpanzees at Bossou. *Primates* 34:545–552.

Sulkowski, G. M., and Hauser, M. D. (2001). Can rhesus monkeys spontaneously subtract? *Cognition* 79:239–262.

Sullivan, K., Walton, A., Kimpton, C., Tully, G., and Gill, P. (1993). A rapid and quantitative DNA sex test: fluorescence-based PCR analysis of X–Y homologous gene amelogenin. *Biotechniques* 15:637–641.

Suomi, S. J. (1982). The development of social competence by rhesus monkeys. *Ann. Ist. Sup. Sanita* 18:193–202.

Suomi, S. J. (2002). Parents, peers, and the process of socialization in primates. In: Borkowski, J. G., Ramey, S. L., and Bristol-Power, M. (eds.), *Parenting and the Child's World: Influences on Academic, Intellectual, and Social–Emotional Development*. Erlbaum, Mahwah, NJ. pp. 265–279.

Suomi, S. J. (2005). Mother–infant attachment, peer relationships, and the development of social networks in rhesus monkeys. *Hum. Dev.* 48:67–79.

Supriatna, J., Manullang, B. O., and Soekara, E. (1986). Group composition, home range, and diet of the maroon leaf monkey (*Presbytis rubicunda*) at Tanjung Puting Reserve, central Kalimantan, Indonesia. *Primates* 27:185–190.

Supriatna, J., Froehlich, J. W., Erwin, J. M., and Southwick, C. H. (1992). Population, habitat and conservation status of *Macaca maurus, Macaca tonkeana* and their putative hybrids. *Trop. Biodivers.* 1:31–48.

Surbeck, M., and Hohmann, G. (2008). Primate hunting by bonobos at LuiKotale, Salonga National Park. *Curr. Biol.* 18(19):R906-R907.

Surbeck, M., Fowler, A., Deimel, C., and Hohmann, G. (2009). Evidence for the consumption of arboreal, diurnal primates by bonobos (*Pan paniscus*). *Am. J. Primatol.* 71:171–174.

Surridge, A. K., Suarez, S. S., Buchanan-Smith, H. M., Smith, A. C., and Mundy, N. I. (2005). Color vision pigment frequencies in wild tamarins (*Saguinus* spp.). *Am. J. Primatol.* 67:463–470.

Susman, R. L. (ed.) (1984). *The Pygmy Chimpanzee: Evolutionary Biology and Behavior*. Plenum, New York.

Sussman, R. W. (1972). An ecological study of two Madagascan primates: *Lemur fulvus rufus* Audebert and *Lemur catta* Linnaeus [PhD diss.]. Duke University, Durham, NC.

Sussman, R. W. (1974). Ecological distinctions in sympatric species of *Lemur*. In: Martin, R. D., Doyle, G. A., and Walker, A. C. (eds.), *Prosimian Biology*. Duckworth, London. pp. 75– 108.

Sussman, R. W. (1977). Socialization, social structure and ecology of two sympatric species of lemur. In: Chevalier-Skolnikoff, S., and Poirier, F. E. (eds.), *Primate Bio-Social Development: Biological, Social, and Ecological Determinants*. Garland, New York. pp. 515–528.

Sussman, R. W. (ed.) (1979). *Primate Ecology: Problem-Oriented Field Studies*. Wiley, New York.

Sussman, R. W. (1991a). Demography and social organization of free-ranging *Lemur catta* in the Beza Mahafaly Reserve, Madagascar. *Am. J. Phys. Anthropol.* 84:43–58.

Sussman, R. W. (1991b). Primate origins and the evolution of angiosperms. *Am. J. Primatol.* 23:209–223.

Sussman, R. W. (1992). Male life history and intergroup mobility among ring-tailed lemurs (*Lemur catta*). *Int. J. Primatol.* 13:395–413.

Sussman, R. W. (1999a). *Primate Ecology and Social Structure. Lorises, Lemurs and Tarsiers,* vol. 1. Pearson, Needham Heights, MA.

Sussman, R. W. (1999b). Species-specific dietary patterns in primates and human dietary adaptations. In Sussman, R. W. (ed.), *The Biological Basis of Human Behavior: a critical review.* Prentice Hall, Saddle River, NJ. pp. 143–155.

Sussman, R. W. (2000). Piltdown Man: the father of American field primatology. In: Strum, S., and Fedigan, L. (eds.), *Primate Encounters: Models of Science, Gender, and Society.* University of Chicago, Chicago. pp. 85–103.

Sussman, R. W. (2003a). *Primate Ecology and Social Structure. Lorises, Lemurs, and Tarsiers,* vol. 1. Pearson Custom, Boston.

Sussman, R. W. (2003b). *Primate Ecology and Social Structure. New World Monkeys,* vol. 2. Pearson Custom, Boston.

Sussman, R. W., and Garber, P. (1987). A new interpretation of the social organization and mating system of the Callitrichidae. *Int. J. Primatol.* 8:73–92.

Sussman, R. W., and Garber, P. A. (2004). Rethinking sociality: cooperation and aggression among primates. In: Sussman, R. W., and Chapman, A. (eds.), *The Origin and Nature of Sociality.* Aldine de Gruyter, New York. pp. 161–190.

Sussman, R. W., and Raven, P. H. (1977). Pollination by lemurs and marsupials: an archaic coevolutionary system. *Science* 200:731–736.

Sussman, R. W., and Richard, A. F. (1986). Lemur conservation in Madagascar: the status of lemurs in the south. *Primate Conserv.* 7:85–92.

Sussman, R. W., and Tattersall, I. (1976). Cycles of activity, group composition and diet of *Lemur mongoz* Linnaeus 1766 in Madagascar. *Folia Primatol.* 26:270–283.

Sussman, R. W., and Tattersall, I. (1981). Behavior and ecology of *Macaca fascicularis in* Mauritius: a preliminary study. *Primates* 22:192–205.

Sussman, R. W., Cheverud, J. M., and Barlett, T. Q. (1995). Infant killing as an evolutionary strategy: reality or myth. *Evol. Anthropol.* 3:149–151.

Sussman, R. W., Andrianasolondraibe, O., Soma, T., and Ichino, S. (2003a). Social behavior and aggression among ringtailed lemurs. *Folia Primatol.* 74:168–172.

Sussman, R. W., Green, G. M., Porton, I., Andrianasolondraibe, O. L., and Ratsirarson, J. (2003b). A survey of the habitat of *Lemur catta* in southwestern and southern Madagascar. *Primate Conserv.* 19:32–57.

Sussman, R. W., Garber, P. A., and Cheverud, J. M. (2005). Importance of cooperation and affiliation in the evolution of primate sociality. *Am. J. Phys. Anthropol.* 1281:84–97.

Sutherland, D. H., Olshen, R. A., Biden, E. N., and Wyatt, M. P. (1988). *The Development of Mature Walking.* Mac Keith Press, London.

Sutton, J. E., Olthof, A., and Roberts, W. A. (2000). Landmark use by squirrel monkeys (*Saimiri sciureus*). *Anim. Learn. Behav.* 28:28–42.

Suwanvecho, U. (2003). Ecology and interspecific relations of two sympatric *Hylobates* species (*H. lar* and *H. pileatus*) in Khao Yai National Park, Thailand [PhD diss.]. Mahidol University, Bangkok, Thailand.

Suzuki, A. (1969). An ecological study of chimpanzees in a savanna woodland. *Primates* 10:103–148.

Suzuki, A. (1979). The variation and adaptation of social groups of chimpanzees and black and white colobus monkeys. In: Bernstein, I. S., and Smith, E. O. (eds.), *Primate Ecology and Human Origins.* Garland STPM Press, New York. pp. 153–173.

Swapna, N. (2008). Assessing the feeding ecology of the Bengal slow loris (*Nycticebus bengalensis*) in Trishna Wildlife Sanctuary, Tripura [MSc thesis]. Manipal University, Manipal, India.

Swartz, K., Shumaker, R., and Himmamen, S. (2007). Response strategies in list learning by orangutans (*Pongo pygmaeus × P-abelii*). *J. Comp. Psychol.* 121:260–269.

Swedell, L. (1997). Patterns of reconciliation among captive gelada baboons (*Theropithecus gelada*): a brief report. *Primates* 38:325–330.

Swedell, L. (2000). Two takeovers in wild hamadryas baboons. *Folia Primatol.* 71:169–172.

Swedell, L. (2002a). Affiliation among females in wild hamadryas baboons (*Papio hamadryas hamadryas*). *Int. J. Primatol.* 23:1205–1226.

Swedell, L. (2002b). Ranging behavior, group size and behavioral flexibility in Ethiopian hamadryas baboons (*Papio hamadryas hamadryas*). *Folia Primatol.* 73:95–103.

Swedell, L. (2005). Dispersal by force: residence patterns of wild female hamadryas baboons [abstract]. *Am. J. Phys. Anthropol.* (suppl. 40):202.

Swedell, L. (2006). *Strategies of Sex and Survival in Hamadryas Baboons: Through a Female Lens.* Pearson Prentice Hall, Upper Saddle River, NJ.

Swedell, L., and Saunders, J. (2006). Infant mortality, paternity certainty, and female reproductive strategies in hamadryas baboons. In: Swedell, L., and Leigh, S. R. (eds.), *Reproduction and Fitness in Baboons: Behavioral, Ecological, and Life History Perspectives.* Springer, New York. pp. 19–51.

Swedell, L., and Schreier, A. (2009). Male aggression towards females in hamadryas baboons: conditioning, coercion, and control. In: Muller, M. N., and Wrangham, R. W. (eds.), *Sexual Coercion in Primates and Humans: An Evolutionary Perspective on Male Aggression Against Females.* Harvard University Press, Cambridge, MA. pp. 244–268.

Swedell, L., and Tesfaye, T. (2003). Infant mortality after takeovers in wild Ethiopian hamadryas baboons. *Am. J. Primatol.* 60:113–118.

Swedell, L., Hailemeskel, G., and Schreier, A. (2008). Composition and seasonality of diet in wild hamadryas baboons: preliminary findings from Filoha. *Folia Primatol.* 79:476–490.

Swofford, D. L. (2002). *PAUP*: Phylogenetic Analysis Using Parsimony (and Other Methods),* 4.0 Beta. Sinauer Associates, Sunderland, MA.

Symington, M. M. (1987a). Ecological and social correlates of party size in the black spider monkey, *Ateles paniscus chamek* [PhD diss.]. Princeton University, Princeton, NJ.

Symington, M. M. (1987b). Sex ratio and maternal rank in wild spider monkeys: when daughters disperse. *Behav. Ecol. Sociobiol.* 20:421–425.

Symington, M. M. (1988a). Demography, ranging patterns and activity budgets of black spider monkeys (*Ateles paniscus chamek*) in the Manu National Park, Peru. *Am. J. Primatol.* 15:45–67.

Symington, M. M. (1988b). Environmental determinants of population densities in *Ateles. Primate Conserv.* 9:74–79.

Symington, M. M. (1988c). Food competition and foraging party size in the black spider monkey (*Ateles paniscus chamek*). *Behaviour* 105:117–134.

Symington, M. M. (1990). Fission–fusion social organization in *Ateles* and *Pan. Int. J. Primatol.* 11:47–61.

Syvänen, A.-C. (2001). Accessing genetic variation: genotyping single nucleotide polymorphisms. *Nat. Rev. Genet.* 2:930–942.

Szalay, F. S., and Delson, E. (1979). *Evolutionary History of the Primates.* Academic Press, New York.

Taberlet, P., Griffin, S., Goossens, B., Questiau, S., Manceau, V., Escaravage, N., Waits, L., and Bouvet, J. (1996). Reliable genotype of samples with very low DNA quantities using PCR. *Nucleic Acids Res.* 24:3189–3194.

Taberlet, P., Walts, L. P., and Luikart, G. (1999). Noninvasive genetic sampling: look before you leap. *Trends Ecol. Evol.* 14:321–325.

Tagliaro, C. H., Schneider, M. P. C., Schneider H., Sampaio, I. C., and Stanhope, M. J. (1997). Marmoset phylogenetics, conservation perspectives, and evolution of the mtDNA control region. *Mol. Biol. Evol.* 14:674–684.

Tagliaro, C. H., Schneider, M. P. C., Schneider, H., Sampaio, I., and Stanhope, M. J. (2000). Molecular studies of *Callithrix pygmaea* (Primates, Platyrrhini) based on transferrin intronic and ND1 regions: implications for taxonomy and conservation. *Genet. Mol. Biol.* 23:729–737.

Tajima, F. (1989). Statistical method for testing the neutral mutation hypothesis by DNA polymorphism. *Genetics* 123:585–595.

Takacs, Z., Morales, J. C., Geissmann, T., and Melnick, D. J. (2005). A complete species-level phylogeny of the Hylobatidae based on mitochondrial ND3-ND4 gene sequences. *Mol. Phylogenet. Evol.* 36:456–467.

Takahata, Y. (1982a). Social relations between adult males and females of Japanese monkeys in the Arashiyama B troop. *Primates* 23:1–23.

Takahata, Y. (1982b). The socio-sexual behavior of Japanese monkeys. *Z. Tierpsychol.* 59:89–108.

Takahata, Y. (1985). Adult male chimpanzees kill and eat a male newborn infant: newly observed intragroup infanticide and cannibalism in Mahale National Park Tanzania. *Folia Primatol.* 44:161–170.

Takahata, Y. (1990). Social relationships among adult males. In: Nishida, T. (ed.), *The Chimpanzees of the Mahale Mountains: Sexual and Life History Strategies.* University of Tokyo Press, Tokyo. pp. 149–170.

Takahata, Y. (1991). Diachronic changes in the dominance relations of adult female Japanese monkeys of the Arashiyama B group. In: Fedigan, L. M., and Asquith, P. J. (eds.), *The Monkeys of Arashiyama.* SUNY Press, Albany. pp. 123–139.

Takahata, Y. (1995). A genetic perspective on the origin and history of humans. *Annu. Rev. Ecol. Syst.* 26:343–372.

Takahata, Y., Koyama, N., and Suzuki, S. (1995). Do the old aged females experience a long post-reproductive life span? The cases of Japanese macaques and chimpanzees. *Primates* 36:169–180.

Takahata, Y., Ihobe, H., and Idani, G. (1996). Comparing copulations of chimpanzees and bonobos: do females exhibit proceptivity or receptivity?

In: McGrew, W. C., Marchant, L. F., and Nishida, T. (eds.), *Great Ape Societies.* Cambridge University Press, Cambridge. pp. 146–155.

Takai, M., and Anaya, F. (1996). New specimens of the oldest fossil platyrrhine, *Branisella boliviana,* from Salla, Bolivia. *Am. J. Phys. Anthropol.* 99:301–317.

Takai, M., Anaya, F., Shigehara, N., and Setoguchi, T. (2000). New fossil materials of the earliest New World monkey, *Branisella boliviana,* and the problem of platyrrhine origins. *Am. J. Phys. Anthropol.* 111:263–282.

Takai, M., Shigehara, N., Aung, A. K., Tun, S. T., Soe, A. N., Tsubamoto, T., and Thein T. (2001). A new anthropoid from the latest middle Eocene of Pondaung, central Myanmar. *J. Hum. Evol.* 40:393–409.

Takai, M., Setoguchi, T., and Shigehara, N. (2003). New aotine fossils from the middle Miocene of La Venta, Colombia. *Am. J. Phys. Anthropol.* 35(suppl.):205.

Takasaki, H. (2000). Traditions of the Kyoto school of field primatology in Japan. In: Strum, S. C., and Fedigan, L. M. (eds.), *Primate Encounters: Models of Science, Gender, and Society.* University of Chicago Press, Chicago. pp. 151–164.

Takasaki, H., and Hunt, K. (1987). Further medicinal plant consumption in wild chimpanzees? *Afr. Stud. Monogr.* 8:125–128.

Takasaki, H., and Takenaka, O. (1991). Paternity testing in chimpanzees with DNA amplification from hairs and buccal cells in wadges: a preliminary note. In: Ehara, A., Kimura, T., Takanaka, O., and Iwamoto, M. (eds.), *Primatology Today.* Elsevier Scientific, Amsterdam. pp. 613–616.

Talebi, M. (2005). Factors Affecting Food Choice of the Endangered Southern Muriquis (*Brachyteles arachnoides,* Primates, E. Geoffroy, 1806) in the Brazilian Atlantic Forest. [PhD diss.]. University of Cambridge.

Talebi, M., Bastos, A., and Lee, P. C. (2005). Diet of southern muriquis in continuous Brazilian Atlantic Forest. *Int. J. Primatol.* 26(5): 1175–1187.

Tamura, K., Dudley, J., Nei, M., and Kumar, S. (2007). MEGA4: Molecular Evolutionary Genetics Analysis (MEGA) software version 4.0. *Mol. Biol. Evol.* 24:1596–1599.

Tan, C. L. (1999). Group composition, home range size, and diet of three sympatric bamboo lemur species (genus *Hapalemur*) in Ranomafana National Park, Madagascar. *Int. J. Primatol.* 20:547–566.

Tan, C. L. (2001). Behavior and ecology of three sympatric bamboo lemur species (genus *Hapalemur*) in Ranomafana National Park, Madagascar [PhD diss.]. State University of New York, Stony Brook.

Tan, C. L., and Drake, J. H. (2001). Evidence of tree gouging and exudate eating in pygmy slow lorises (*Nycticebus pygmaeus*). *Folia Primatol.* 72:37–39.

Tanaka, I. (1989). Variability in the development of mother–infant relationships among free-ranging Japanese macaques. *Primates* 30:477–491.

Tanaka, I. (1995). Matrilineal distribution of louse egg-handling techniques during grooming in free-ranging Japanese macaques. *Am. J. Phys. Anthropol.* 98:197–201.

Tanaka, I. (1998). Social diffusion of modified louse egg-handling techniques during grooming in free-ranging Japanese macaques. *Anim. Behav.* 56:1229–1236.

Tang-Martinez, Z. (2000). Paradigms and primates: Bateman's principle, passive females, and perspectives from other taxa. In: Strum, S. C., and Fedigan, L. M. (eds.), *Primate Encounters: Models of Science, Gender, and Society.* University of Chicago Press, Chicago. pp. 261–274.

Tanner, J. M. (1962). *Growth at Adolescence, with a General Consideration of the Effects of Hereditary and Environmental Factors upon Growth and Maturation from Birth to Maturity.* Blackwell Scientific, Oxford.

Tanner, J. M. (1978). *Foetus into Man.* Harvard University Press, Cambridge, MA.

Tappen, M., and Wrangham, R. (2000). Recognizing hominoid-modified bones: the taphonomy of colobus bones partially digested by free-ranging chimpanzees in Kibale Forest, Uganda. *Am. J. Phys. Anthropol.* 113(2):217–234.

Tappen, N. C., and Severson, A. (1971). Sequence of eruption of permanent teeth and epiphyseal union in New World monkeys. *Folia Primatol.* 15:293–312.

Tardif, S. D. (1994). Relative energetic cost of infant care in small-bodied neotropical primates and its relation to infant-care patterns. *Am. J. Primatol.* 34:133–143.

Tardif, S. D. (1997). The bioenergetics of parental behavior and the evolution of alloparental care in marmosets. In: Solomon, N. G., and French, J. A. (eds.), *Cooperative Breeding in Mammals.* Cambridge University Press, Cambridge. pp. 11–33.

Tardif, S. D., and Ziegler, T. E. (1992). Features of female reproductive senescence in tamarins (*Saguinus* spp.), a New World primate. *J. Reprod. Fertil.* 94:411–421.

Tardif, S. D., Richter, C. B., and Carson, C. L. (1984). Reproductive performance of three species of Callitrichidae. *Lab. Anim. Sci.* 34:272–275.

Tardif, S. D., Harrison, M. L., and Simek, M. A. (1993). Communal infant care in marmosets and tamarins: relation to energetics, ecology, and social organization. In: Rylands, A. B. (ed.), *Marmosets and Tamarins: Systematics, Behaviour, and Ecology.* Oxford University Press, Oxford. pp. 220–234.

Tardif, S. D., Santos, C. V., Baker, A. J., van Elsacker, L., Ruiz-Miranda, C. R., Moura, A. C. A., Passos, F. C., Price, E. C., Rapaport, L. G., and De Vleeschouwer, K. (2002). Infant care in lion tamarins. In: Kleiman, D. G., and Rylands, A. B. (eds.), *Lion Tamarins: Biology and Conservation.* Smithsonian Institution Press, Washington DC. pp. 213–232.

Tardif, S. D., Smucny, D. A., Abbott, D. H., Mansfield, K., Schultz-Darken, N., and Yamamoto, M. E. (2003). Reproduction in captive common marmosets (*Callithrix jacchus*). *Comp. Med.* 53:364–368.

Tardif, S., Araujo, A., Arruda, M., French, J., Sousa, M., and Yamamoto, M. (2008). Reproduction and aging in marmosets and tamarins. In: Atsalis, S., Margulis, S., and Hof, P. (eds.), *Primate Reproductive Aging: Cross-Taxon Perspectives on Reproduction.* Karger, Basel. pp. 29–48.

Tarnaud, L. (2004). Ontogeny of feeding behavior of *Eulemur fulvus* in the dry forest of Mayotte. *Int. J. Primatol.* 25(4):803–824.

Tarnaud, L., and Simmen, B. (2002). A major increase in the population of brown lemurs on Mayotte since the decline reported in 1987. *Oryx* 36:297–300.

Tashian, R. E., Goodman, M., Ferrell, R. E., and Tanis, R. J. (1976). Evolution of carbonic anhydrase in primates and other mammals. In: Goodman, M., Tashian, R. E., and Tashian, J. H. (eds.), *Molecular Anthropology: Genes and Proteins in the Evolutionary Ascent of the Primates.* Plenum Press, New York. pp. 301–319.

Tattersall, I. (1976). Group structure and activity rhythm in *Lemur mongoz* (Primates, Lemuriformes) on Anjouan and Mohéli Islands, Comoro Archipelago. *Anthropol. Pap. Am. Mus. Nat. Hist.* 53:257–261.

Tattersall, I. (1977a). Distribution of the Malagasy lemurs, part 1. The lemurs of northern Madagascar. *Ann. N. Y. Acad. Sci.* 293:160–169.

Tattersall, I. (1977b). The lemurs of the Comoro Islands. *Oryx* 13:445–448.

Tattersall, I. (1977c). Ecology and behavior of *Lemur fulvus mayottensis* (Primates, Lemuriformes). *Anthropol. Pap. Am. Mus. Nat. Hist.* 52:195–216.

Tattersall, I. (1982). *The Primates of Madagascar.* Columbia University Press, New York.

Tattersall, I. (1987). Cathemeral activity in primates: a definition. *Folia Primatol.* 49:200–202.

Tattersall, I. (2004). Comments and conclusions. The historical biogeography of the strepsirhini: understanding the colonization of Madagascar. 2004 IPS Symposium. *Folia Primatol.* 75(suppl. 1):120.

Tattersall, I. (2006a). The concept of cathemerality: history and definition. *Folia Primatol.* 77:7–14.

Tattersall, I. (2006b). Origin of the Malagasy strepsirhine primates. In: Gould, L., and Sauther, M. L. (eds.), *Lemurs: Ecology and Adaptation.* Springer, New York. pp. 3–18.

Taub, D. M. (1980a). Female choice and mating strategies among wild Barbary macaques (*Macaca sylvanus* L.). In: Lindburg, D. G. (ed.), *The Macaques: Studies in Ecology, Behavior and Evolution.* Van Nostrand Reinhold, New York, pp. 287–344.

Taub, D. M. (1980b). Testing the "agonistic buffering" hypothesis. I. The dynamics of participation in the triadic interaction. *Behav. Ecol. Sociobiol.* 6:187–197.

Taub, D. M. (1984). Male caretaking behaviour among wild Barbary macaques (*Macaca sylvanus*). In: Taub, D. M. (ed.), *Primate Paternalism.* Van Nostrand Reinhold, New York. pp. 20–55.

Taussky, H. H. (1954). A microcolorimetric determination of creatinine in urine by the Jaffe reaction. *J. Biol. Chem.* 208:853–861.

Tavaré, S., Marshall, C. R., Will, O., Soligo, C., and Martin, R. D. (2002). Using the fossil record to estimate the age of the last common ancestor of extant primates. *Nature* 416:726–729.

Tavares, L. I., and Ferrari, S. F. (2002). Diet of the silvery marmoset (*Callithrix argentata*) at CFPn: seasonal and longitudinal variation. In: Lisboa, P. L. B. (ed.), *Caxiuanã, Populações Tradicionais, Meio Físico e Diversidade Biológica.* CNPq/MCT, Belém. pp. 707–719.

Tax, S. (ed.) (1960). *The Evolution of Man.* University of Chicago, Chicago.

Taylor, A. B. (2006). Feeding behavior, diet, and the functional consequences of jaw form in orangutans, with implications for the evolution of *Pongo. J. Hum. Evol.* 50:377–393.

Taylor, A. B. (2009). The functional significance of variation in jaw form in orangutans. In: Wich, S. A., Utami, S. S., Mitra Setia, T., and van Schaik,

C. (eds.), *Orangutans: Geographic Variation in Behavioral Ecology and Conservation.* Oxford University Press, Oxford. pp. 15–32.

Taylor, A. B., and Goldsmith, M. A. (2003). *Gorilla Biology: A Multidisciplinary Perspective.* Cambridge University Press, New York.

Taylor, A. B., and Groves, C. P. (2003). Patterns of mandibular variation in *Pan* and *Gorilla* and implications for African ape taxonomy. *J. Hum. Evol.* 44:529–561.

Taylor, A., and Vinyard, C. (2008). The relationship between jaw-muscle architecture and feeding behavior in primates: tree-gouging and nongouging gummivorous callitrichids as a natural experiment. In Vinyard, C., Ravosa, M., and Wall, C. (eds.), *Primate Craniofacial Function and Biology.* Springer, New York. pp. 241–262.

Taylor, L. (1986). Kinship, dominance, and social organization in a semi-free ranging group of ringtailed lemurs (*Lemur catta*) [PhD diss.]. Washington University, St. Louis, MO.

Taylor, L., and Sussman, R. W. (1985). A preliminary study of kinship and social organization in a semi-free-ranging group of *Lemur catta. Int. J. Primatol.* 6:601–614.

Taylor, S. E., Cousino, L., Klein, B., Gruenewals, T. L., Gurung, R. A. R., and Updegraff, J. A. (2000). Biobehavioral responses to stress in females: tend-and-befriend, nor fight-or-flight. *Psychol. Rev.* 107:411–429.

Teaford, M. F., and Glander, K. E. (1996). Dental microwear and diet in a wild population of mantled howling monkeys (*Alouatta palliata*). In: Norconk, M. A., Rosenberger, A. L., and Garber, P. A. (eds.), *Adaptive Radiations of Neotropical Primates.* Plenum Press, New York. pp. 433–449.

te Boekhorst, I. J. A., Schürmann, C. L., and Sugardjito, J. (1990). Residential status and seasonal movements of wild orang-utans in the Gunung Leuser Reserve (Sumatra, Indonesia). *Anim. Behav.* 39:1098–1109.

Teelen, S. (2007). Influence of chimpanzee predation on associations between red colobus and red-tailed monkeys at Ngogo, Kibale National Park, Uganda. *Int. J. Primatol.* 28(3):593–606.

Teelen, S. (2008). Influence of chimpanzee predation on the red colobus population at Ngogo, Kibale National Park, Uganda. *Primates.* 49(1): 41–49.

Teichroeb, J. A., Saj, T. L., Paterson, J. D., and Sicotte, P. (2003). Effect of group size on activity budgets of *Colobus vellerosus* in Ghana. *Int. J. Primatol.* 24:743–758.

Teichroeb, J. A., and Sicotte, P. (2009). Test of the ecological-constraints model on ursine colobus monkeys (*Colobus vellerosus*) in Ghana. *Am. J. Primatol.* 71:49–59.

Tejedor, M. (1998). La posición de *Aotus* y *Callicebus* en la filogenia de los primates platirrinos. *Bol. Primatol. Latinoam.* 7:13–29.

Tejedor, M. (2001). *Aotus* y los atelinae: nuevas evidencias en la sistemática de los primates platirrinos. *Mastozool. Neotrop.* 8:41–57.

Tejedor, M. F. (2002). Primate canines from the early Miocene Pinturas formation, southern Argentina. *J. Hum. Evol.* 43:127–141.

Tejedor, M. F. (2003). New fossil primate from Chile. *J. Hum. Evol.* 44:515–520.

Teleki, G. (1981a). C. Raymond Carpenter, 1905–1975. *Am. J. Phys. Anthropol.* 56:383–386.

Teleki, G. (1981b). The omnivorous diet and eclectic feeding habits of chimpanzees in Gombe National Park, Tanzania. In: Harding, R. S. O., and Teleki, G. (eds.), *Omnivorous Primates: Gathering and Hunting in Human Evolution.* Columbia University Press, New York. pp. 303–343.

Teleki, G. (1989). Population status of wild chimpanzees (*Pan troglodytes*) and threats to survival. In: Heltne, P. G., and Marquardt, L. A. (eds.), *Understanding Chimpanzees.* Harvard University Press, Cambridge, MA. pp. 312–353.

Tello, N. S., Huck, M., and Heymann, E. W. (2002). *Boa constrictor* attack and successful group defense in moustached tamarins, *Saguinus mystax. Folia Primatol.* 73:146–148.

Tenaza, R. R. (1975). Territory and monogamy among Kloss' gibbons (*Hylobates klossii*) in Siberut Island, Indonesia. *Folia Primatol.* 24:60–80.

Tenaza, R. R. (1985). Songs of hybrid gibbons (*Hylobates lar* × *H. muelleri*). *Am. J. Primatol.* 8:249–253.

Tenaza, R. R. (1989). Female sexual swellings in the Asian colobine *Simias concolor. Am. J. Primatol.* 17:81–86.

Tenaza, R. R., and Fuentes, A. (1995). Monandrous social organization of pigtailed langurs (*Simias concolor*) in the Pagai Islands, Indonesia. *Int. J. Primatol.* 16:295–310.

Tenaza, R., and Tilson, R. L. (1985). Human predation and Kloss's gibbon (*Hylobates klossii*) sleeping trees in Siberut Island, Indonesia. *Am. J. Primatol.* 8:299–308.

Terborgh, J. (1983). *Five New World Primates: A Study of Comparative Ecology.* Princeton University Press, Princeton, NJ.

Terborgh, J. (1990). Mixed flocks and polyspecific associations: costs and benefits of mixed groups to birds and monkeys. *Am. J. Primatol.* 21:87–100.

Terborgh, J., and Goldizen, A. W. (1985). On the mating system of the cooperatively breeding saddle-backed tamarin (*Saguinus fuscicollis*). *Behav. Ecol. Sociobiol.* 16:293–299.

Terborgh, J., and Janson, C. H. (1986). The socioecology of primate groups. *Annu. Rev. Ecol. Syst.* 17:111–135.

Terborgh, J., Lopez, L., Nu-ez, P., Rao, M., Shahabudin, G., Orihuela, G., Riveros, M., Ascanio, R., Adler, G. H., Lambert, T. D., and Balbas, L. (2002a). Ecological meltdown in predator-free forest fragments. *Science* 294:1923.

Terborgh, J., Pitman, N., Silman, M., Schichter, H., and Nunez, P. V. (2002b). Maintenance of tree diversity in tropical forests. In: Levey, D. J., Silva, W. R., and Galetti, M. (eds.), *Seed Dispersal and Frugivory: Ecology, Evolution and Conservation.* CABI Publishing, New York. pp. 351–364.

Terkel, J. (1995). Cultural transmission in the black rat: pine cone feeding. *Adv. Study Behav.* 24:119–154.

Terkel, J. (1996). Cultural transmission of feeding behaviour in the black rat (*Rattus rattus*). In: Heyes, C. M., and Galef, B. G., Jr. (eds.), *Social Learning in Animals: The Roots of Culture.* Academic Press, San Diego. pp. 17–47.

Terrace, H. S., and McGonigle, B. (1994). Memory and representation of serial order by children, monkeys, and pigeons. *Curr. Dir. Psychol. Sci.* 3:180–185.

Terrace, H. S., and Metcalfe, J. (eds.) (2005). *The Missing Link in Cognition: Origins of Self-Reflective Consciousness:* Oxford University Press, New York.

Terranova, C. J. (1996). Variation in the leaping of lemurs. *Am. J. Primatol.* 40:145–165.

Terranova, C. J., and Coffman, B. S. (1997). Body weights of wild and captive lemurs. *Zoo Biol.* 16:17–30.

Thalmann, O., Fischer, A., Lankester, F., Paabo, S., and Vigilant, L. (2007). The complex evolutionary history of gorillas: insights from genomic data. *Mol. Biol. Evol.* 24:146–158.

Thalmann, U. (2001). Food resource characteristics in two nocturnal lemurs with different social behaviour: *Avahi occidentali*s and *Lepilemur edwardsi. Int. J. Primatol.* 22:287–324.

Thalmann, U. (2006). Behavioral and ecological adaptations in two small folivorous lemurs with different social organization: *Avahi* and *Lepilemur.* In: Gould, L., and Sauther, M. L. (eds.), *Lemurs: Ecology and Adaptation.* Springer, New York. pp. 327–352.

Thalmann, U., and Geissmann, T. (2000). Distributions and geographic variation in the western woolly lemur (*Avahi occidentalis*) with description of a new species (*A. unicolor*). *Int. J. Primatol.* 21:915–941.

Thalmann, U., and Geissmann, T. (2005). New species of woolly lemur *Avahi* (Primates: Lemuriformes) in Bemaraha (central western Madagascar). *Am. J. Primatol.* 67:371–376.

Thierry, B. (1985). Social development in three species of macaque (*Macaca mulatta, M. fascicularis, M. tonkeana*): a preliminary report on the first ten weeks of life. *Behav. Proc.* 11:89–95.

Thierry, B. (1986). A comparative study of aggression and response to aggression in three species of macaque. In: Else, J. G., and Lee, P. C. (eds.), *Primate Ontogeny, Cognition and Social Behaviour.* Cambridge University Press, Cambridge. pp. 307–313.

Thierry, B. (1990a). Feedback loop between kinship and dominance: the macaque model. *J. Theor. Biol.* 145: 511–522.

Thierry, B. (1990b). The state of equilibrium among agonistic behavior patterns in a group of Japanese macaques (*Macaca fuscata*). *C. R. Acad. Sci. Paris* 310:35–40.

Thierry, B. (2000). Covariation of conflict management patterns across macaque species. In: Aureli, F., and de Waal, F. B. M. (eds.), *Natural Conflict Resolution.* University of California Press, Berkeley. pp. 106–128.

Thierry, B. (2004). Social epigenesis. In: Thierry, B., Singh, M., and Kaumanns, W. (eds.), *Macaque Societies: A Model for the Study of Social Organization.* Cambridge University Press, New York. pp. 267–290.

Thierry, B. (2007). Unity in diversity: lessons from macaque societies. *Evol. Anthropol.* 16:224–238.

Thierry, B. (2008). Primate socioecology, the lost dream of ecological determinism. *Evol. Anthropol.* 17:93–96.

Thierry, B., and Anderson, J. R. (1986). Adoption in anthropoid primates. *Int. J. Primatol.* 7:191–216.

Thierry, B., and Aureli, F. (2006). Barbary but not barbarian: social relations in a tolerant macaque. In: Hodges, J. K., and Cortes, J. (eds.), *The*

Barbary Macaque: Biology, Management and Conservation. Nottingham University Press, Nottingham. pp. 29–45.

Thierry, B., and Gachot-Neveu, H. (2004). The balance of power between males influences mating tactics in Tonkean macaques [abstract]. *Folia Primatol.* 75(suppl. 1):345.

Thierry, B., Demaria, C., Preuschoft, S., and Desportes, C. (1989). Structural convergence between silent bared-teeth display and relaxed open-mouth display in the Tonkean macaque (*Macaca tonkeana*). *Folia Primatol.* 52:178–184.

Thierry, B., Gauthier, C., and Peignot, P. (1990). Social grooming in Tonkean macaques (*Macaca tonkeana*). *Int. J. Primatol.* 11:357–375.

Thierry, B., Anderson, J. R., Demaria, C., Desportes, C., and Petit, O. (1994). Tonkean macaque behaviour from the perspective of the evolution of Sulawesi macaques. In: Roeder, J. J., Thierry, B., Anderson, J. R., and Herrenschmidt, N. (eds.), *Current Primatology. Social Development, Learning and Behavior,* vol. 2. Presses de l'Université Louis Pasteur, Strasbourg. pp. 103–117.

Thierry, B., Heistermann, M., Aujard, F., et al. (1996). Long-term data on basic reproductive parameters and evaluation of endocrine, morphological, and behavioral measures for monitoring reproductive status in a group of semifree-ranging Tonkean macaques (*Macaca tonkeana*). *Am. J. Primatol.* 39:47–62.

Thierry, B., Bynum, E. L., Baker, S., Kinnaird, M. F., Matsumura, S., Muroyama, Y., O'Brien, T. G., Petit, O., and Watanabe, K. (2000a). The social repertoire of Sulawesi macaques. *Primate Res.* 16:203–226.

Thierry, B., Iwaniuk, A. N., and Pellis, S. M. (2000b). The influence of phylogeny on the social behaviour of macaques (Primates: Cercopithecidae, genus *Macaca*). *Ethology* 106:713–728.

Thierry, B., Singh, M., and Kaumanns, W. (eds.) (2004). *Macaque Societies: A Model for the Study of Social Organization.* Cambridge University Press, Cambridge.

Thierry, B., Aureli, F., Nunn, C., Petit, O., Abegg, C., and de Waal, F. B. M. (2008). A comparative study of conflict resolution in macaques: insights into the nature of trait covariation. *Anim. Behav.* 75:847–860.

Thomas, C. D., Bodsworth, E. J., Wilson, J. R., Simmons, A. D., Davies, Z. G., Musche, M., and Conradt, L. (2001). Ecological and evolutionary processes at expanding range margins. *Nature* 411:577–581.

Thomas, C. D., Cameron, A., Green, R. E., Bakkenes, M., Beaumont, L. J., Collingham, Y. C., Erasmus, B. F. N., Ferreira de Siqueira, M., Grainger, A., Hannah, L., Hughes, L., Huntley, B., van Jaarsveld, A. S., Midgley, G. F., Miles, L., Ortega-Huerta, M. A., Townsend Peterson, A., Phillips, O. L., and Williams, S. E. (2004). Extinction risk from climate change. *Nature* 427:145–148.

Thomas, J. A., Hess, D. L., Dahl, K. D., Iliff-Sizemore, S. A., Stouffer, R. L., and Wolf, D. P. (1992). The Sulawesi crested black macaque (*Macaca nigra*) menstrual cycle: changes in perineal tumescence and serum estradiol, progesterone, follicle-stimulating hormone, and luteinizing hormone levels. *Biol. Reprod.* 46:879–884.

Thompson, C. L., and Norconk, M. A. (in press). Testing models of social behavior with regard to inter- and intra-troop interactions in free-ranging white-faced sakis (*Pithecia pithecia*). In: Barnett, A. A., Veiga, L. M., Ferrari, S. F., and Norconk, M. A. (eds.), *Evolutionary Biology and Conservation of Titis, Sakis, and Uacaris.* Cambridge: Cambridge University Press. In press.

Thompson, J. (2002). Bonobos of the Lukuru Wildlife Research Project. In: Boesch, C., Hohmann, G., and Marchant, L. (eds.), *Behavioural Diversity in Chimpanzees and Bonobos.* Cambridge University Press, Cambridge. pp. 61–70.

Thompson, M. E., and Wrangham, R. W. (2008). Diet and reproductive function in wild female chimpanzees (*Pan troglodytes schweinfurthii*) at Kibale National Park, Uganda. *Am. J. Phys. Anthropol.* 135:171–181.

Thompson, N. S. (1997). Communication and natural design. In: Owings, D. H., Beecher, M., and Thompson, N. (eds.), *Perspectives in Ethology,* vol. 12. Plenum Press, New York. pp. 391–415.

Thompson, R. K. R., Oden, D. L., and Boysen, S. T. (1997). Language-naïve chimpanzees (*Pan troglodytes*) judge relations between relations in a conceptual matching-to-sample task. *J. Exp. Psychol. Anim. Behav. Proc.* 23:31–43.

Thompson-Handler, N. E. (1990). The pygmy chimpanzee: sociosexual behaviour, reproductive biology and life history [PhD diss.]. Yale University, New Haven.

Thompson-Handler, N., Malenky, R. K., and Badrian, N. (1984). Sexual behavior of *Pan paniscus* under natural conditions in the Lomako Forest, Zaire. In: Susman, R. L. (ed.), *The Pygmy Chimpanzee: Evolutionary Biology and Behavior.* Plenum, New York. pp. 347–368.

Thomson, J. A., Hess, D. L., Dahl, K. D., Illiff-Sizemore, S. A., Stouffer, R. L., and Wolfe, D. P. (1992). The Sulawesi crested black macaque menstrual cycle; changes in perineal tumescence and serum estradiol, progesterone, follicle-stimulating hormone, and lutenizing hormone levels. *Biol. Reprod.* 46:879–884.

Thorington, R. W., Jr., Ruiz, J. C., and Eisenberg, J. F. (1984). A study of a black howling monkey (*Alouatta caraya*) population in northern Argentina. *Am. J. Primatol.* 6:357–366.

Thorn, J. S., Nijman, V., Smith, D., and Nekaris, K. A. I. (2009). Ecological niche modelling as a technique for assessing threats and setting conservation priorities for Asian slow lorises (Primates: *Nycticebus*). *Divers. Distrib.* 15:289–298.

Thorndike, E. L. (1898). Animal intelligence: an experimental study of the associative process in animals. *Psychol. Rev. Monogr.* 2.

Thornhill, R., and Alcock, J. (1983). *The Evolution of Insect Mating Systems.* Harvard University Press, Cambridge, MA.

Thorpe, S. K. S., and Crompton, R. H. (2005). Locomotor ecology of wild orangutans (*Pongo pygmaeus abelii*) in the Gunung Lesser Ecosystem, Sumatra, Indonesia: a multivariate analysis using log-linear modeling. *Am. J. Phys. Anthropol.* 127:58–78.

Thorpe, S. K. S., and Crompton, R. H. (2006). Orangutan positional behavior and the nature of arboreal locomotion in Hominoidea. *Am. J. Phys. Anthropol.* 131:384–401.

Thorpe, S. K. S., and Crompton, R. H. (2009). Orangutan positional behavior. In: Wich, S. A., Atomoko, S. S. U., Setia, T. M., and van Schaik, C. P. (eds.), *Orangutans: Geographic Variation in Behavioral Ecology and Conservation.* Oxford University Press, New York. pp. 33–47.

Thorpe, W. H. (1963). *Learning and Instinct in Animals.* Methuen, London.

Tiefenbacher, S., Lee, B., Meyer, J. S., and Spealman, R. D. (2003). Noninvasive technique for the repeated sampling of salivary free cortisol in awake, unrestrained squirrel monkeys. *Am. J. Primatol.* 60:69–75.

Tilden, C. (1993). Reproductive energetics of prosimian primates [PhD diss.]. Duke University, Durham, NC.

Tilden, C. D., and Oftedal, O. T. (1995). The bioenergetics of reproduction in prosimian primates: is it related to female dominance? In: Izard, M. K., Alterman, L., and Doyle, G. A. (eds.), *Creatures of the Dark: The Nocturnal Prosimians.* Plenum Press, New York. pp. 119–131.

Tilford, B. L. (1982). Seasonal rank changes for adolescent and sub-adult natal males in a free-ranging group of rhesus monkeys. *Int. J. Primatol.* 3:484–490.

Tilman, D. (1994). Competition and biodiversity in spatially structured habitats. *Ecology* 75:2–16.

Tilson, R. L. (1981). Family formation strategies of Kloss's gibbons. *Folia Primatol.* 35:259–287.

Tilson, R., Seal, U. S., Soemarna, K., Ramono, W., Sumardja, E., Poniran, S., van Schaik, C., Leighton, M., Rijksen, H., and Eudey, A. (1993). Orangutan population and habitat viability analysis report. Presented at Orangutan Population and Habitat Viability Analysis Workshop, Medan, North Sumatra, Indonesia, October 20, 1993.

Tinbergen, N. (1953). *Social Behaviour in Animals.* Chapman and Hall, New York.

Tinbergen, N. (1963). On aims and methods of ethology. *Z. Tierpsychol.* 20:410–433.

Ting, N., Tosi, A. J., Li, Y., Zhang, Y.-P., and Disotell, T. R. (2008). Phylogenetic incongruence between nuclear and mitochondrial markers in the Asian colobines and the evolution of the langurs and leaf monkeys. *Mol. Phylogenet. Evol.* 46:466–474.

Ting, S., Hartley, S., and Burns, K. C. (2008). Global patterns in fruiting seasons. *Global Ecol. Biogeogr.* 17:648–657.

Tjio, H. J., and Levan, A. (1956). The chromosome numbers of man. *Hereditas* 42:1–6.

Tobler, M. W. (2009). New GPS technology improves fix success for large mammal collars in dense tropical forests. *J. Trop. Ecol.* 25:217–221.

Toft, C. A., Aeschlimann, A., and Bolis, L. (1991). *Parasite–Host Associations: Coexistence or Conflict?* Oxford University Press, Oxford.

Tokida, E., Tanaka, I., Takefushi, H., and Hagiwara, T. (1994). Tool-using Japanese macaques: use of stones to obtain fruit from a pipe. *Anim. Behav.* 47:1023–1030.

Tolentino, K., Roper, J. J., Passos, F. C., and Strier, K. B. (2008). Mother–offspring associations in northern muriquis, *Brachyteles hypoxanthus.* *Am. J. Primatol.* 70:301–305.

Tomasello, M. (1990). Cultural transmission in tool use and communicatory signalling of chimpanzees. In: Parker, S., and Gibson, K. (eds.), *"Language" and Intelligence in Monkeys and Apes: Comparative Developmental Perspectives.* Cambridge University Press, Cambridge. pp. 274–311.

Tomasello, M. (1994). The question of chimpanzee culture. In: Wrangham, R. W., McGrew, W. C., de Waal, F. M. B., and Heltne, P. G. (eds.), *Chimpanzee Cultures.* Harvard University Press, Cambridge, MA. pp. 301–317.

Tomasello, M. (1999). *The Cultural Origins of Human Cognition.* Harvard University Press, Cambridge, MA.

Tomasello, M. (2003). *Constructing a Language: A Usage-Based Theory of Language Acquisition.* Harvard University Press, Cambridge, MA.

Tomasello, M., and Call, J. (eds.) (1997). *Primate Cognition.* Oxford University Press, New York.

Tomasello, M., and Call, J. (2009). Chimpanzees social cognition. In: Lonsdorf, E. V., Ross, S., and Matsuzawa, T. (eds.), *The Mind of the Chimpanzee: Ecological and Experimental Perspectives.* Chicago University Press, Chicago. In press.

Tomasello, M., Davis-Dasilva, M., Camak, L., and Bard, K. A. (1987). Observational learning of tool use by young chimpanzees. *Hum. Evol.* 2:175–183.

Tomasello, M., Call, J., and Hare, B. (1998). Five primate species follow the visual gaze of conspecifics. *Anim. Behav.* 55:1063–1069.

Tomasello, M., Hare, B., and Agnetta, B. (1999). Chimpanzees, *Pan troglodytes,* follow gaze direction geometrically. *Anim. Behav.* 58:769–77.

Tomasello, M., Hare, B., and Fogleman, T. (2001). The ontogeny of gaze following in chimpanzees, *Pan troglodytes,* and rhesus macaques, *Macaca mulatta. Anim Behav.* 61:335–343.

Tomonaga, M. (1995). Transfer of odd-item search performance in a chimpanzee (*Pan troglodytes*). *Percept. Mot. Skills* 80:35–42.

Tomonaga, M. (2006). Development of chimpanzee social cognition in the first 2 years of life. In: Matsuzawa, T., Tomonaga, M., and Tanaka, M. (eds.), *Cognitive Development in Chimpanzees.* Springer, New York. pp. 182–197.

Tomonaga, M. (2007). Is chimpanzee (*Pan troglodytes*) spatial attention reflexively triggered by the gaze cue? *J. Comp. Psychol.* 121:156–170.

Tomonaga, M. (2008). Relative numerosity discrimination by chimpanzees (*Pan troglodytes*): evidence for approximate numerical representations. *Anim. Cogn.* 11:43–57.

Tomonaga, M. (2009). Do chimpanzees eyes have it? In: Lonsdorf, E. V., Ross, S., and Matsuzawa, T. (eds.), *Understanding Chimpanzees: The Mind of the Chimpanzee.* Chicago University Press, Chicago. In press.

Tomonaga, M., and Matsuzawa, T. (2000). Sequential responding to Arabic numerals with wild cards by the chimpanzee (*Pan troglodytes*). *Anim. Cogn.* 3:1–11.

Tomonaga, M., Myowa-Yamakoshi, M., Mizuno, Y., Okamoto, S., Yamaguchi, M. K., Kosugi, D., Bard, K. A., Tanaka, M., and Matsuzawa, T. (2004). Development of social cognition in infant chimpanzees (*Pan troglodytes*): face recognition, smiling, gaze, and the lack of triadic interactions. *Jpn. Psychol. Res.* 46:227–235.

Tonooka, R. (2001). Leaf-folding behavior for drinking water by wild chimpanzees (*Pan troglodytes verus*) at Bossou, Guinea. *Anim. Cogn.* 4:325–334.

Torigoe, T. (1985). Comparison of object manipulation among 74 species of nonhuman primates. *Primates* 26:182–194.

Torii, R., Koizumi, H., Tanioka, Y., Inaba, T., and Mori, J. (1987). Serum LH, progesterone and estradiol-17beta levels throughout the ovarian cycle, during the early stage of pregnancy and after the parturition and the abortion in the common marmoset, *Callithrix jacchus. Primates* 28:229–238.

Torii, R., Hosoi, Y., Masuda, Y., Iritani, A., and Nigi, H. (2000). Birth of the Japanese monkey (*Macaca fuscata*) infant following in-vitro fertilization and embryo transfer. *Primates* 41:39–47.

Torres de Assumpção, C. (1983). Ecological and behavioural information on *Brachyteles arachnoides. Primates* 24:584–593.

Torres, O. M., Enciso, S., Ruiz, F., Silva, E., and Yunis, I. (1998). Chromosome diversity of the genus *Aotus* from Colombia. *Am. J. Primatol.* 44:255–275.

Tortora, G. J., and Anagnostakos, N. P. (1987). *Principles of Physiology.* Harper and Row, Cambridge.

Tosi, A. J., Morales, J. C., and Melnick, D. J. (2000). Comparison of Y chromosome and mtDNA phylogenies leads to unique inferences of macaque evolutionary history. *Mol. Phylogenet. Evol.* 17:133–144.

Tosi, A. J., Morales, J. M., and Melnick, D. J. (2002). Y-chromosome and mitochondrial markers in *Macaca fascicularis* indicate introgression with Indochinese *M. mulatta* and a biogeographic barrier in the Isthmus of Kra. *Int. J. Primatol.* 23:161–178.

Tosi, A. J., Morales, J. C., and Melnick, D. J. (2003). Paternal, maternal, and biparental molecular markers provide unique windows onto the evolutionary history of macaque monkeys. *Evolution* 57:1419–1435.

Toth, G., Gaspari, Z., and Jurka, J. (2000). Microsatellites in different eukaryotic genomes: survey and analysis. *Genome Res.* 10:967–981.

Toth, N. (1997). The artefact assemblages in the light of experimental studies. In: Isaac, G. L. (ed.), *Koobi Fora Research Project. Plio-Pleistocene Archaeology,* vol. 5. Clarendon Press, Oxford. pp. 361–388.

Toth, N., Schick, K. D., Savage-Rumbaugh, E. S., Sevcik, R., and Rumbaugh, D. (1993). *Pan* the tool-maker: investigations into the stone tool-making and tool-using capabilities of a bonobo (*Pan paniscus*). *J. Archaeol. Sci.* 20:81–91.

Toubiana, R., and Gaudemer, A. (1967). Structure du vernolide, nouvel ester sesquiterpique isole de *Vernonia colorata. Tetrahedron Lett.* 14:1333–1336.

Towers, G. H. N., Abramowski, Z., Finlayson, A. J., and Zucconi, A. (1985). Antibiotic properties of thiarubrine-A, a naturally occurring dithiacyclohedadiene polyine. *Planta Med.* 3:225–229.

Townsend, S. W., Slocombe, K. E., Thompson, M. E., and Zuberbühler, K. (2007). Female-led infanticide in wild chimpanzees. *Curr. Biol.* 17:355–356.

Trask, B. J. (2002). Human cytogenetics: 46 chromosomes, 46 years and counting. *Nat. Rev. Genet.* 3:769–778.

Tremble, M., Muskita, Y., and Supriatna, J. (1993). Field observations of *Tarsius dianae* at Lore Lindu National Park, central Sulawesi, Indonesia. *Trop. Biodivers.* 1:67–76.

Trent, B., Tucker, M., and Lockard, J. (1977). Activity changes with illumination in slow loris *Nycticebus coucang. Appl. Anim. Ethnol.* 3:281–286.

Treves, A. (1997). Primate natal coats: a preliminary analysis of distribution and function. *Am. J. Phys. Anthropol.* 104:47–70.

Treves, A. (1998). The influence of group size and near neighbor on vigilance in two species of arboreal primates. *Behaviour* 135:453–482.

Treves, A. (1999). Has predation shaped the social systems of arboreal primates? *Int. J. Primatol.* 20:35–67.

Treves, A. (2000a). Prevention of infanticide: the perspective of infant primates. In: van Schaik, C. P., and Janson, C. H. (eds.), *Infanticide by Males and Its Implications.* Cambridge University Press, Cambridge. pp. 223–238.

Treves, A. (2000b). Theory and method in studies of vigilance and aggregation. *Anim. Behav.* 60:711–722.

Treves, A. (2002). Predicting predation risk for foraging, arboreal monkeys. In: Miller, L. E. (ed.), *Eat or Be Eaten: Predator Sensitive Foraging Among Primates.* Cambridge University Press, Cambridge. pp. 222–241.

Treves, A., and Chapman, C. A. (1996). Conspecific threat, predation avoidance, and resource defense: implications for grouping in langurs. *Behav. Ecol. Sociobiol.* 39:43–53.

Treves, A., and Naughton-Treves, L. (1999). Risk and opportunity for humans coexisting with large carnivores. *J. Hum. Evol.* 36:275–282.

Treves, A., and Palmqvist, P. (2007). Reconstructing hominin interactions with mammalian carnivores (6.0–1.8 Ma). In: Gursky, S. L., and Nekaris, K. A. I. (eds.), *Primate Anti-Predator Strategies.* Springer, New York. pp. 355–381.

Treves, A., and Pizzagalli, D. (2002). Vigilance and perception of social stimuli: views from ethology and social neuroscience. In: Bekoff, C. A. M., and Burghardt, G. (eds.), *The Cognitive Animal: Empirical and Theoretical Perspectives on Animal Cognition.* MIT Press, Cambridge, MA. pp. 463–469.

Treves, A., Drescher, A., and Ingrisano, N. (2001). Vigilance and aggregation in black howler monkeys (*Alouatta pigra*). *Behav. Ecol. Sociobiol.* 50:90–95.

Treves, A., Drescher, A., and Snowdon, C. T. (2003). Maternal watchfulness in black howler monkeys (*Alouatta pigra*). *Ethology* 109:135–146.

Triesman, M. (1975). Predation and the evolution of gregariousness. I. Models for concealment and evasion. *Anim. Behav.* 23:799–900.

Triosi, A., and Carosi, M. (1998). Female orgasm rate increases with male dominance in Japanese macaques. *Anim. Behav.* 56:1261–1266.

Trivers, R. L. (1971). The evolution of reciprocal altruism. *Q. Rev. Biol.* 46:35–57.

Trivers, R. L. (1972). Parental investment and sexual selection. In: Campbell, B. (ed.), *Sexual Selection and the Descent of Man.* Aldine, Chicago. pp. 136–179.

Trivers, R. L. (1974). Parent–offspring conflict. *Am. Zool.* 14:249–264.

Trivers, R. L. (1985). *Social Evolution.* Benjamin/Cummings, Menlo Park, CA.

Troisi, A. (2002). Displacement activities as a behavioural measure of stress in nonhuman primates and human subjects. *Stress* 5:47–54.

Trouson, A. (1993). State of the art and future directions in human IVF. In: Wolf, D. P., Stouffer, R. L., and Brenner, R. M. (eds.), *In Vitro Fertilization and Embryo Transfer in Primates.* Springer-Verlag, New York. pp. 3–29.

Tsingalia, H. M., and Rowell, T. E. (1984). The behavior of adult male blue monkeys. *Z. Tierpsychol.* 64:253–268.

Tsukahara, T. (1993). Lions eat chimpanzees: the first evidence of predation by lions on wild chimpanzees. *Am. J. Primatol.* 29:1–11.

Tuomisto, H., Ruokolainen, K., and Yli-Halla, M. (2003). Dispersal, environment, and floristic variation of western Amazonian forests. *Science* 299:241–244.

Turbaugh, P. J., Ley, R. E., Hamady, M., Fraser-Ligget, C. M., Knight, R., and Gordon, J. I. (2007). The human microbiome project. *Nature* 449:804–810.

Turke, P. W. (1984). Effects of ovulatory concealment and synchrony on proto-hominid mating systems and parental roles. *Ethol. Sociobiol.* 5:33–44.

Turner, T. R., Anapol, F., and Jolly, C. J. (1997). Growth, development, and sexual dimorphism in vervet monkeys (*Cercopithecus aethiops*) at four sites in Kenya. *Am. J. Phys. Anthropol.* 103:19–35.

Tutin, C. E. G. (1979). Mating patterns and reproductive strategies in a community of wild chimpanzees (*Pan troglodytes schweinfurthii*). *Behav. Ecol. Sociobiol.* 6:29–38.

Tutin, C. E. G. (1980). Reproductive behaviour of wild chimpanzees in the Gombe National Park, Tanzania. *J. Reprod. Fertil.* 28(suppl.):43–57.

Tutin, C. E. G. (1996). Ranging and social structure of lowland gorillas in the Lopé Reserve, Gabon. In: McGrew, W. C., Marchant, L. F., and Nishida, T. (eds.), *Great Ape Societies*. Cambridge University Press, Cambridge. pp. 58–70.

Tutin, C. E. G., and Fernandez, M. (1991). Responses of wild chimpanzees and gorillas to arrival of primatologists: behaviour observed during habituation. In Box, H. O. (ed.), *Primates responses to Environmental Change*. Chapman & Hall, London. pp. 187–196.

Tutin, C. E. G., and Fernandez, M. (1993). Composition of the diet of chimpanzees and comparisons with that of sympatric lowland gorillas in the Lope Reserve, Gabon. *Am. J. Primatol.* 30:195–211.

Tutin, C. E. G., and McGinnis, P. R. (1981). Chimpanzee reproduction in the wild. In: Graham, C. E. (ed.), *Reproductive Biology of the Great Apes*. Academic Press, New York. pp. 239–264.

Tutin, C. E. G., and Oslisly, R. (1995). *Homo, Pan and Gorilla*: co-existence over 60,000 years at Lope in central Gabon. *J. Hum. Evol.* 28:597–602.

Tutin, C. E. G., McGrew, W. C., and Baldwin, P. J. (1983). Social organization of savanna-dwelling chimpanzees, *Pan troglodytes verus*, at Mt. Assirik, Senegal. *Primates* 24:154–173.

Tutin, C. E., Fernandez, M., Pierce, A. H., and Williamson, E. A. (1984). Foods consumed by sympatric populations of *Gorilla g. gorilla* and *Pan t. troglodytes* in Gabon. *Int. J. Primatol.* 6:27–43.

Tutin, C. E. G., Fernandez, M., Rogers, M. E., Williamson, E. A., and McGrew, W. C. (1991). Foraging profiles of sympatric lowland gorillas and chimpanzees in the Lope Reserve, Gabon. *Phil. Trans. R. Soc. Lond. B* 334:179–186.

Tutin, C., Parnell, R., White, L., and Fernandez, M. (1995). Nest building by lowland gorillas in the Lope Reserve, Gabon: environmental influences and implications for censusing. *Int. J. Primatol.* 16:53–76.

Tutin, C. E. G., Ham, R. M., White, L. J. T., and Harrison, M. J. S. (1997). The primate community of the Lopé Reserve, Gabon: diets, responses to fruit scarcity, and effects on biomass. *Am. J. Primatol.* 42:1–24.

Twinomugisha, D., and Chapman, C. A. (2007). Golden monkey ranging in relation to spatial and temporal variation in food availability. *Afr. J. Ecol.* 46(4):585–593.

Tyler, D. E. (1991). The evolutionary relationships of *Aotus*. *Folia Primatol.* 56:50–52.

Uchida, A. (1998). Variation in tooth morphology of *Pongo pygmaeus*. *J. Hum. Evol.* 34:71–79.

Uddin, M., Wildman, D. E., Liu, G., Xu, W., Johnson, R. M., Hof, P. R., Kapatos, G., Grossman, L. I., and Goodman, M. (2004). Sister grouping of chimpanzees and humans as revealed by genome-wide phylogenetic analysis of brain gene expression profiles. *Proc. Natl. Acad. Sci. USA* 101:2957–2962.

Uehara, S. (1990). Utilization patterns of a marsh grassland within the tropical rain-forest by the bonobos (*Pan-paniscus*) of Yalosidi, Republic of Zaire. *Primates* 31:311–322.

Uehara, S. (1997). Predation on mammals by chimpanzees (*Pan troglodytes*). *Primates* 38:193–214.

Uehara, S., and Nishida, T. (1987). Body weights of wild chimpanzees in the Mahale Mountains National Park, Tanzania. *Am. J. Phys. Anthropol.* 72:315–321.

Ueno, A., and Matsuzawa, T. (2004). Food transfer between chimpanzee mothers and their infants. *Primates* 45:231–239.

Uhde, N. L., and Sommer, V. (2002). Antipredatory behavior in gibbons (*Hylobates lar*, Khao Yai/Thailand). In: Miller, L. E. (ed.), *Eat or Be Eaten: Predator Sensitive Foraging Among Primates*. Cambridge University Press, New York. pp. 268–291.

Ullrey, D. E., Bernard, J. B., Peter, G. K., Lu, Z., Chen, T. C., Sikarski, J. G., and Holick, M. F. (2000). Vitamin D intakes by cotton-top tamarins (*Saguinus oedipus*) and associated serum 25-hydroxyvitamin D concentrations. *Zoo Biol.* 18:473–480.

Ulmer, F. A. (1963). Observations on the tarsier in captivity. *Zool. Garten* 27:106–121.

Underhill, P. A., Shen, P., Lin, A. A., Jin, L., Passarino, G., Yang, W. H., Kauffman, E., Bonné-Tamir, B., Bertranpetit, J., Francalacci, P., Ibrahim, M., Jenkins, T., Kidd, J. R., Mehdi, S. Q., Seielstad, M. T., Wells, R. S., Piazza, A., Davis, R. W., Feldman, M. W., Cavalli-Sforza, L. L., and Oefner, P. J. (2000). Y chromosome sequence variation and the history of human populations. *Nat. Genet.* 26:358–361.

Underhill, P. A., Passarino, G., Lin, A. A., Shen, P., Lahr, M. M., Foley, R. A., Oefner, P. J., and Cavalli-Sforza, L. L. (2001). The phylogeography of Y chromosome binary haplotypes and the origins of modern human populations. *Ann. Hum. Genet.* 65:43–62.

Ungar, P. S. (1995). Fruit preferences of four sympatric primate species at Ketambe, northern Sumatra, Indonesia. *Int. J. Primatol.* 16:221–245.

Ungar, P. S. (1996). Feeding height and niche separation in sympatric Sumatran monkeys and apes. *Folia Primatol.* 67:163–168.

Ungar, P. S. (2004). Dental topography and diets of *Australopithecus afarensis* and early *Homo*. *J. Hum. Evol.* 46:605–622.

Ungar, P. S., Grine, F. E., Teaford, M. F., and el Zaatari, S. (2006). Dental microwear and diets of African early *Homo*. *J. Hum. Evol.* 50(1):78–95.

United Nations (2005). *World Population Prospects: The 2004 Revision*. United Nations, New York.

Urbani, B. (2005). The targeted monkey: a re-evaluation of predation on New World primates. *J. Anthropol. Sci.* 83:89–109.

Usongo, L. I., and Amubode, F. O. (2001). Nutritional ecology of Preuss's red colobus monkey (*Colobus badius preussi* Rahm 1970) in Korup National Park, Cameroon. *Afr. J. Ecol.* 39:121–125.

Utami, S. S., and Mitra Setia, T. (1995). Behavioral changes in wild male and female orangutans (*Pongo pygmaeus abelii*) during and following a resident male take-over. In: Nadler, R. D., Galdikas, B. M. F., Sheeran, L. K., and Rosen, N. (eds.), *The Neglected Ape*. Plenum Press, New York. pp. 183–190.

Utami, S. S., and van Hooff, J. A. R. A. M. (1997). Meat-eating by adult female Sumatran orangutans (*Pongo pygmaeus abelii*). *Am. J. Primatol.* 43:159–165.

Utami, S. S., Wich, S. A., Sterck, E. H. M., and van Hooff, J. A. R. A. M. (1997). Food competition between wild orangutans in large fig trees. *Int. J. Primatol.* 18:909–927.

Utami, S. S., Goossens, B., Bruford, M. W., de Ruiter, J. R., and van Hooff, J. A. R. A. M. (2002). Male bimaturism and reproductive success in orangutans. *Behav. Ecol.* 13:643–652.

Utami Atmoko, S. S. (2000). Bimaturism in Orang-utan Males: Reproductive and Ecological Strategies [PhD diss.]. University of Utrecht, Utrecht.

Utami Atmoko, S. S., and van Hooff, J. A. R. A. M. (2004). Alternative male reproductive strategies: male bimaturism in orangutans. In: Kappeler, P. M., and van Schaik, C. P. (eds.), *Sexual Selection in Primates: New and Comparative Perspectives*. Cambridge University Press, Cambridge. pp. 196–207.

Utami Atmoko, S. S., Goossens, B., Bruford, M. W., de Ruiter, J. R., and van Hooff, J. A. R. A. M. (2002). Male bimaturism and reproductive success in Sumatran orangutans. *Behav. Ecol.* 13(5):643–652.

Utami Atmoko, S. S., Mitra Setia, T., Goossens, B., James, S. S., Knott, C. D., Morrogh-Bernard, H. C., van Schaik, C. P., and van Noordwijk, M. A. (2009a). Orangutan mating behavior and strategies. In: Wich, S. A., Utami, S. S., Mitra Setia, T., and van Schaik, C. (eds.), *Orangutans: Geographic Variation in Behavioral Ecology and Conservation*. Oxford University Press, Oxford. pp. 235–244.

Utami Atmoko, S. S., Singleton, I., van Noordwijk, M. A., van Schaik, C. P., and Mitra Setia, T. (2009b). Male–male relationships in orangutans. In: Wich, S. A., Utami, S. S., Mitra Setia, T., and van Schaik, C. (eds.), *Orangutans: Geographic Variation in Behavioral Ecology and Conservation*. Oxford University Press, Oxford. pp. 225–234.

Valderrama, X., Karesh, W. B., Wildman, D. E., and Melnick, D. J. (1999). Noninvasive methods for collecting fresh hair tissue. *Mol. Ecol.* 8:1749–1752.

Valderrama, X., Robinson, J. G., and Melnick, D. J. (2000a). Females control male reproductive success in wedge-capped capuchins, based on genetic and behavioral data. *Am. J. Phys. Anthropol.* (suppl. 30):308.

Valderrama, X., Robinson, J. G., Attygalle, A. B., and Eisner, T. (2000b). Seasonal anointment with millipedes in a wild primate: a chemical defense against insects. *J. Chem. Ecol.* 26:2781–2790.

Valeggia, C. R., and Ellison, P. T. (2001). Lactation, energetics, and postpartum fecundity. In: Ellison, P. T. (ed.), *Reproductive Ecology and Human Evolution*. Aldine de Gruyter, New York. pp. 85–106.

Valeggia, C. R., Mendoza, S. P., Fernandez-Duque, E., Mason, W. A., and Lasley, B. (1999). Reproductive biology of female titi monkeys (*Callicebus moloch*) in captivity. *Am. J. Primatol.* 47:183–195.

Valenzuela, N. (1994). Early behavioral development in three wild infant *Cebus apella* in Colombia. In: Roeder, J. J., Thierry, B., Anderson, J. B., and Herrenschmidt, N. (eds.), *Current Primatology. Social Development, Learning and Behaviour,* vol. 2. Université Louis Pasteur, Strasbourg. pp. 297–302.

Valero, A., and Byrne, R. W. (2007). Spider monkey ranging patterns in Mexican subtropical forest: do travel routes reflect planning? *Anim. Cogn.* 10:305–315.

Valero, A., Schaffner, C. M., Vick, L. G., Aureli, F., and Ramos-Fernandez, G. (2006). Intragroup lethal aggression in wild spider monkeys. *Am. J. Primatol.* 68:732–737.

Valière, N. (2002). GIMLET: a computer program for analysing genetic individual identification data. *Mol. Ecol. Notes* 2:377–379.

Valière, N., and Taberlet, P. (2000). Urine collected in the field as a source of DNA for species and individual identification. *Mol. Ecol.* 9:2150–2152.

Valladares-Padua, C. (1993). The ecology, behavior and conservation of the black lion tamarin (*Leontopithecus chrysopygus,* Mikan, 1823) [PhD diss.]. University of Florida, Gainesville.

Valladares-Padua, C., and Cullen, L., Jr. (1994). Distribution, abundance and minimum viable metapopulation of the black lion tamarin (*Leontopithecus chrysopygus*). *Dodo J. Wildl. Preserv. Trusts* 30:80–88.

van Adrichem, G. G. J., Utami, S. S., Wich, S. A., can Hooff, J. A. R. A. M., and Sterck, E. H. M. (2006). The development of wild immature Sumatran orangutans (*Pongo abelii*) at Ketambe. *Primates* 47:300–309.

Van Belle, S., and Estrada, A. (2008). Group size and composition influence male and female reproductive success in black howler monkeys (*Alouatta pigra*). *Am. J. Primatol.* 70(6):613–619.

van Bemmel, A. C. V. (1968). Contributions to the knowledge of the geographical races of *Pongo pygmaeus* (Hoppius). *Bijdr. Dierk.* 38:13–15.

Vancata, V., Vancatova, M., Chalyan, V., and Meishvili, N. (2000). Longitudinal study of growth and body mass changes in ontogeny in captive rhesus macaques (*Macaca mulatta*) from the Institute of Medical Primatology. *Sochi. Var. Evol.* 8:51–81.

van de Rijt-Plooij, H. H. C., and Plooij, F. X. (1987). Growing independence, conflict and learning in mother–infant relations in free-ranging chimpanzees. *Behaviour* 101:1–86.

van der Pijl, L. (1982). *Principles of Dispersal in Higher Plants.* Springer-Verlag, Berlin.

van der Zee, D., and Skinner, J. D. (1977). Preliminary observations on samango and vervet monkeys near Lake Sibayi. *S. Afr. J. Sci.* 73:381–382.

van Doorn, A. (2009). The interface of socioecology and management in chacma baboons [PhD diss.]. University of Cape Town, Cape Town.

van Doorn, A., O'Riain, M.J., and Swedell, L. (2009). The effects of seasonality and day length on the activity budget and diet of semi-commensal chacma baboons (*Papio ursinus*) in the Cape Peninsula of South Africa. *Am. J. Primatol.* On line first DOI: 10.1002/ajp.20759

van Geel, M., Eichler, E. E., Beck, A. F., Shan, Z. H., Haaf, T., van der Maarel, S. M., Frants, R. R., and de Jong, P. J. (2002). A cascade of complex subtelomeric duplications during the evolution of the hominoid and Old World monkey genomes. *Am. J. Hum. Genet.* 70:269–278.

van Hooff, J. A. R. A. M. (1967). The facial displays of the catarrhine monkeys and apes. In: Morris, D. (ed.), *Primate Ethology.* Anchor Books, New York. pp. 9–88.

van Hooff, J. A. R. A. M. (2000). Relationships among non-human primate males: a deductive framework. In: Kappeler, P. M. (ed.), *Primate Males: Causes and Consequences of Variation in Group Composition.* Cambridge University Press, Cambridge. pp. 183–191.

van Hooff, J. A. R. A. M., and Preuschoft, S. (2003). Laughter and smiling: the intertwining of nature and culture. In: de Waal, F., and Tyack, P. (eds.), *Animal Social Complexity: Intelligence, Culture, and Individualized Societies.* Harvard University Press, Cambridge, MA. pp. 260–287.

van Hooff, J. A. R. A. M., and van Schaik, C. P. (1994). Male bonds: affiliative relationships among nonhuman primate males. *Behaviour* 130:309–337.

van Horn, R. C., Buchan, J. C., Altmann, J., and Alberts, S. C. (2007). Divided destinies: group choice by female savannah baboons during social group fission. *Behav. Ecol. Sociobiol.* 61:1823–1837.

Van Horn, R. N., and Eaton, G. G. (1979). Reproductive physiology and behavior in prosimians. In: Doyle, G. A., and Martin, R. D. (eds.), *The Study of Prosimian Behavior.* Academic Press. New York. pp. 79–122.

van Lawick-Goodall, J. (1967). Mother–offspring relationships in free-ranging chimpanzees. In: Morris, D. (ed.), *Primate Ethology.* Weidenfeld and Nicolson, London. pp. 287–346.

van Lawick-Goodall, J. (1968). The behaviour of free-living chimpanzees in the Gombe Stream Reserve. *Anim. Behav. Monogr.* 1:165–311.

van Lawick-Goodall, J. (1971). Some aspects of mother–infant relationships in a group of wild chimpanzees. In: Schaffer, H. R. (ed.), *The Origins of Human Social Relations.* Academic Press, London. pp. 115–128.

van Lawick-Goodall, J., van Lawick, H., and Packer, H. (1973). Tool-use in free living babbons in the Gombe National Park. *Nature* 241:212–213.

van Noordwijk, A. J., and de Jong, G. (1986). Acquisition and allocation of resources: their influence on variation in life history tactics. *Am. Nat.* 128:137–142.

van Noordwijk, M. A. (1985). Sexual behaviour of Sumatran long-tailed macaques (*Macaca fascicularis*). *Z. Tierpsychol.* 70:277–296.

van Noordwijk, M. A., and van Schaik, C. P. (1985). Male migration and rank acquisition in wild long-tailed macaques (*Macaca fascicularis*). *Anim. Behav.* 33:849–861.

van Noordwijk, M. A., and van Schaik, C. P. (1987). Competition among female long-tailed macaques, *Macaca fascicularis. Anim. Behav.* 35:577–589.

van Noordwijk, M. A., and van Schaik, C. P. (1999). The effects of dominance rank and group size on female lifetime reproductive success in wild long-tailed macaques, *Macaca fascicularis. Primates* 40:105–130.

van Noordwijk, M. A., and van Schaik, C. P. (2001). Career moves: transfer and rank challenges decisions by male long-tailed macaques. *Behaviour* 138:359–395.

van Noordwijk, M. A., and van Schaik C. P. (2004). Sexual selection and the careers of primate males: paternity concentration, dominance-acquisition and transfer decisions. In: Kappeler, P. M., and van Schaik, C. P. (eds.), *Sexual Selection in Primates.* Cambridge University Press, Cambridge, pp. 208–229.

van Noordwijk, M. A., and van Schaik, C. P. (2005). Development of ecological competence in Sumatran orangutans. *Am. J. Phys. Anthropol.* 127:79–94.

van Noordwijk, M. A., and van Schaik, C. (2009). Intersexual food transfer among orangutans: do females test males for coercive tendency? *Behav. Ecol. Sociobiol.* 63:883–890.

van Noordwijk, M. A., Sauren, S. E. B., Nuzuar, Abulani, A., Morrogh-Bernard, H. C., Utami Atmoko, S. S., and van Schaik, C. P. (2009). Development of independence. In: Wich, S. A., Utami, S. S., Mitra Setia, T., and van Schaik, C. (eds.), *Orangutans: Geographic Variation in Behavioral Ecology and Conservation.* Oxford University Press, Oxford. pp. 189–204.

van Roosmalen, M. G. M. (1984). Subcategorizing foods in primates. In: Chivers, D. J., Wood, B. A., and Bilsborough, A. (eds.), *Food Acquisition and Processing in Primates.* Plenum Press, New York. pp. 167–176.

van Roosmalen, M. G. M. (1985). Habitat preferences, diet, feeding strategy and social organization of the black spider monkey (*Ateles paniscus paniscus* Linnaeus 1758) in Surinam. *Acta Amazon.* 15:1–238.

van Roosmalen, M. G. M., and Klein, L. L. (1988). The spider monkeys, genus *Ateles.* In: Mittermeier, R. A., Rylands, A. B., Coimbra-Filho, A. F., and da Fonseca, G. A. B. (eds.), *Ecology and Behavior of Neotropical Primates.* World Wildlife Fund, Washington DC. pp. 455–537.

van Roosmalen, M. G. M., and van Roosmalen, T. (1997). An eastern extension of the geographical range of the pygmy marmoset, *Cebuella pygmaea. Neotrop. Primates* 5:3–6.

van Roosmalen, M. G. M., and van Roosmalen, T. (2003). The description of a new marmoset genus, *Callibella* (Callitrichinae, Primates), including its molecular phylogentic status. *Neotrop. Primates* 11:1–10.

van Roosmalen, M. G. M., Mittermeier, R. A., and Milton, K. (1981). The bearded sakis, genus *Chiropotes.* In: Coimbra-Filho, A. F., and Mittermeier, R. A. (eds.), *Ecology and Behavior of Neotropical Primates,* vol. 1. Academia Brasileira de Ciências, Rio de Janeiro. pp. 419–441.

van Roosmalen, M. G. M., Mittermeier, R. A., and Fleagle, J. G. (1988). Diet of the northern bearded saki (*Chiropotes satanas chiropotes*): a neotropical seed predator. *Am. J. Primatol.* 14:11–36.

van Roosmalen, M. G. M., van Roosmalen, T., Mittermeier, R. A., and Rylands, A. B. (1998). A new and distinctive species of marmoset (Callitrichidae, Primates) from the lower Rio Aripuanã, state of Amazonas, central Brazilian Amazonia. *Goeld. Zool.* 22:1–27.

van Roosmalen, M. G. M., van Roosmalen, T., and Mittermeier, R. A. (2002). A taxonomic review of the titi monkeys, genus *Callicebus* Thomas, 1903, with the description of two new species, *Callicebus bernhardi* and *Callicebus stephennashi* from Brazilian Amazonia. *Neotrop. Primates* 10(suppl.):1–52.

van Schaik, C. P. (1983). Why are diurnal primates living in groups? *Behaviour* 87:120–144.

van Schaik, C. P. (1986). Phenological changes in a Sumatran rainforest. *J. Trop. Ecol.* 2:327–347.

van Schaik, C. P. (1989). The ecology of social relationships amongst female primates. In: Standen, V., and Foley, R. A. (eds.), *Comparative Socioecology: The Behavioral Ecology of Humans and Other Mammals.* Blackwell, Oxford. pp. 195–218.

van Schaik, C. P. (1996). Social evolution in primates: the role of ecological factors on male behavior. In: Runciman, W. G., Smith, J. M., and Dunbar,

R. I. M. (eds.), *Evolution of Social Behaviour Patterns in Primates and Man*. Oxford University Press, Oxford. pp. 9–31.

van Schaik, C. P. (1999). The socioecology of fission–fusion sociality in orangutans. *Primates* 40:73–90.

van Schaik, C. P. (2000a). Vulnerability to infanticide by males: patterns among mammals. In: van Schaik, C., and Janson, C. (eds.), *Infanticide by Males & Its Implications*. Cambridge University Press, Cambridge. pp. 61–71.

van Schaik, C. P. (2000b). Infanticide by male primates: the sexual selection hypothesis revisited. In: van Schaik, C. P., and Janson, C. H. (eds.), *Infanticide by Males and Its Implications*. Cambridge University Press, Cambridge. pp. 27–60.

van Schaik, C. P. (2002). Fragility of traditions: the disturbance hypothesis for the loss of local traditions in orangutans. *Int. J. Primatol.* 23:527–538.

van Schaik, C. P. (2003). Local traditions in orangutans and chimpanzees: social learning and social tolerance. In: Fragaszy, D. M., and Perry, S. (eds.), *The Biology of Traditions: Models and Evidence*. Cambridge University Press, Cambridge. pp. 297–328.

van Schaik, C. P. (2004). *Among Orangutans: Red Apes and the Rise of Human Culture*. Belknap Press of Harvard University, Cambridge, MA.

van Schaik, C. P., and Dunbar, R. I. M. (1990). The evolution of monogamy in large primates: a new hypothesis and some crucial tests. *Behaviour* 115:30–62.

van Schaik, C. P., and Horstermann, M. (1994). Predation risk and the number of adult males in a primate group: a comparative test. *Behav. Ecol. Sociobiol.* 35:261–272.

van Schaik, C. P., and Janson, C. H. (2000). *Infanticide by Males and Its Implications*. Cambridge University Press, Cambridge.

van Schaik, C. P., and Kappeler, P. M. (1993). Life history, activity period and lemur social systems In: Kappeler, P. M., and Ganzhorn, J. U. (eds.), *Lemur Social Systems and Their Ecological Basis*. Plenum, New York. pp. 241–260.

van Schaik, C. P., and Kappeler, P. M. (1996). The social systems of gregarious lemurs: lack of convergence with anthropoids due to evolutionary disequilibrium? *Ethology* 102:915–941.

van Schaik, C. P., and Kappeler, P. M. (1997). Infanticide risk and the evolution of male–female association in primates. *Proc. R. Soc. Lond. B* 264:1687–1694.

van Schaik, C. P., and Kappeler, P. M. (2003). The evolution of social monogamy in primates. In: Reichard, U. H., and Boesch, C. (eds.), *Monogamy: Mating Strategies and Partnerships in Birds, Humans, and Other Mammals*. Cambridge University Press, Cambridge. pp. 59–80.

van Schaik, C. P., and Knott, C. D. (2001). Geographic variation in tool use on *Neesia* fruits in orangutans. *Am. J. Phys. Anthropol.* 114:331–342.

van Schaik, C. P., and Paul, A. (1996). Male care in primates: does it ever reflect paternity? *Evol. Anthropol.* 5:152–156.

van Schaik, C. P., and Pfannes K. R. (2005). Tropical climates and phenology: a primate perspective. In: Brockman, D. K., and van Schaik, C. P. (eds.), *Seasonality in Primates: Studies of Living and Extinct Human and Non-Human Primates*. Cambridge University Press, New York. pp. 23–54.

van Schaik, C. P., and van Hoof, J. A. R. A. M. (1983). On the ultimate causes of primate social systems. *Behaviour* 85:91–117.

van Schaik, C. P., and van Hooff, J. A. R. A. M. (1996). Towards an understanding of the orangutan's social system. In: McGrew, W. C., Marchant, L. F., and Nishida, T. (eds.), *Great Ape Societies*. Cambridge University Press, Cambridge. pp. 3–15.

van Schaik, C. P., and van Noordwijk, M. A. (1985). Evolutionary effect of the absence of felids on the social organization of the macaques on the island of Simeulue (*Macaca fascicularis*, Miller 1903). *Folia Primatol.* 44:138–147.

van Schaik, C. P., and van Noordwijk, M. A. (1986). The hidden costs of sociality: intra-group variation in feeding strategies in Sumatran long-tailed macaques (*Macaca fascicularis*). *Behaviour* 99:296–315.

van Schaik, C. P., and van Noordwijk, M. A. (1988). Scramble and contest in feeding competition among female long-tailed macaques (*Macaca fascicularis*). *Behaviour* 105:77–98.

van Schaik, C. P., and van Noordwijk, M. A. (1989). The special role of male *Cebus* monkeys in predation avoidance and its effects on group composition. *Behav. Ecol. Sociobiol.* 24:265–272.

van Schaik, C. P., van Noordwijk, M. A., de Boer R. J., and den Tonkelaar, I. (1983a). The effect of group size on time budgets and social behaviour in wild long-tailed macaques (*Macaca fascicularis*). *Behav. Ecol. Sociobiol.* 13:173–181.

van Schaik, C. P., van Noordwijk, M. A., Warsono, B., and Sutriono, E. (1983b). Party size and early detection of predators in Sumatran forest primates. *Primates* 24:211–221.

van Schaik, C. P., van Noordwijk, M. A., and van Bragt, T. (1991). A pilot study of the social correlates of levels of urinary cortisol, prolactin, and testosterone in wild long-tailed macaques (*Macaca fascicularis*). *Primates* 32:345–356.

van Schaik, C. P., Assink, P. R., and Salafsky, N. (1992). Territorial behavior in Southeast Asian langurs: resource defense or mate defense? *Am. J. Primatol.* 26:233–242.

van Schaik, C. P., Terborgh, J. W., and Wright, S. J. (1993). The phenology of tropical forests: adaptive significance and consequences for primary consumers. *Annu. Rev. Ecol. Syst.* 24:353–377.

van Schaik, C. P., Priatna, A., and Priatna, D. (1995). Population estimates and habitat preferences of orangutans based on line transects of nests. In: Nadler, R. D., Galdikas, B. M. F., Sheeran, L. K., and Rosen, N. (eds.), *The Neglected Ape*. Plenum Press, New York. pp. 129–147.

van Schaik, C. P., Fox, E. A., and Sitompul, A. F. (1996a). Manufacture and use of tools in wild Sumatran orangutans. *Naturwissen* 83:186–188.

van Schaik, C. P., van Amerongen, A., and van Noordwijk, M. A. (1996b). Riverine refuging by wild Sumatran long-tailed macaques (*Macaca fascicularis*). In: Fa, J. E., and Lindburg, D. G. (eds.), *Evolution and Ecology of Macaque Societies*. Cambridge University Press, Cambridge. pp. 160–181.

van Schaik, C. P., Deaner, R. O., and Merrill, M. Y. (1999a). The conditions for tool use in primates: implications for the evolution of material culture. *J. Hum. Evol.* 36:719–741.

van Schaik, C. P., van Noordwijk, M. A., and Nunn, C. L. (1999b). Sex and social evolution in primates. In: Lee, P. C. (ed.), *Comparative Primate Socioecology*. Cambridge University Press, Cambridge. pp. 204–231.

van Schaik, C. P., Hodges, J. K., and Nunn, C. L. (2000). Paternity confusion and the ovarian cycles of female primates. In: van Schaik, C. P., and Janson, C. H. (eds.), *Infanticide by Males and Its Implications*. Cambridge University Press, Cambridge. pp. 361–387.

van Schaik, C. P., Monk, K., and Yarrow Robertson, J. M. (2001). Dramatic decline in orang-utan numbers in the Leuser Ecosystem, northern Sumatra. *Oryx* 35:14–25.

van Schaik, C. P., Ancrenaz, M., Borgen, G., Galdikas, B., Knott, C., Singleton, I., Suzuki, A., Utami, S., and Merrill, M. (2003a). Orangutan cultures and the evolution of material culture. *Science* 299:102–105.

van Schaik, C. P., Fox, E. A., and Fechtman, L. T. (2003b). Individual variation in the rate of use of tree-hole tools among wild orang-utans: implications for hominin evolution. *J. Hum. Evol.* 44:11–23.

van Schaik, C. P., Pradhan, G. R., and van Noordwijk, M. A. (2004). Mating conflict in primates: infanticide, sexual harassment, and female sexuality. In: Kappeler, P. M., and van Schaik, C. P. (eds.), *Sexual Selection in Primates: New and Comparative Perspectives*. Cambridge University Press, Cambridge. pp. 131–150.

van Schaik, C. P., van Noordwijk, M. A., and Wich, S. A. (2006). Innovation in wild Bornean orangutans (*Pongo pygmaeus wurmbii*). *Behaviour* 143:839–876.

van Schaik, C. P., Ancrenaz, M., Djojoasmoro, R., Knott, C. D., Morrogh-Bernard, H. C., Nuzuar Odom, K., Utami Atmoko, S. S., and van Noordwijk, M. A. (2009a). Orangutan cultures revisited. In: Wich, S. A., Utami, S. S., Mitra Setia, T., and van Schaik, C. (eds.), *Orangutans: Geographic Variation in Behavioral Ecology and Conservation*. Oxford University Press, Oxford. pp. 299–310.

van Schaik, C. P., Marshall, A. J., and Wich, S. A. (2009b). Geographic variation in orangutan behavior and biology. In: Wich, S. A., Utami, S. S., Mitra Setia, T., and van Schaik, C. (eds.), *Orangutans: Geographic Variation in Behavioral Ecology and Conservation*. Oxford University Press, Oxford. pp. 351–362.

van Schaik, C. P., van Noordwijk, M. A., and Vogel, E. R. (2009c). Ecological sex differences in wild orangutans. In: Wich, S. A., Utami, S. S., Mitra Setia, T., and van Schaik, C. (eds.), *Orangutans: Geographic Variation in Behavioral Ecology and Conservation*. Oxford University Press, Oxford. pp. 255–268.

Van Soest, P. J. (1982). *Nutritional Ecology of the Ruminant*. O & B Books, Corvallis, OR.

van Wagenen, G., and Asling, C. W. (1958). Roentgenographic estimation of bone age in the rhesus monkey (*Macaca mulatta*). *Am. J. Anat.* 103:163–185.

van Wagenen, G., and Catchpole, H. R. (1956). Physical growth of the rhesus monkey (*Macaca mulatta*). *Am. J. Phys. Anthropol.* 14:245–273.

Vargas, A., Jimenez, I., Palomares, F., and Palacios, M. J. (2002). Distribution, status, and conservation needs of the golden-crowned sifaka (*Propithecus tattersalli*). *Biol. Conserv.* 108:325–334.

Varki, A., Geschwind, D. H., and Eichler, E. E. (2008). Human uniqueness: genome interactions with environment, behaviour and culture. *Nat. Rev. Genet.* 9:749–763.

Varón, A., Vinh, L. S., Bomash, I., and Wheeler. W. C. (2009). POY 4.1.1. American Museum of Natural History, http://research.amnh.org/scicomp/projects/poy.php.

Vasconcelos, M. (2008). Transitive inference in non-human animals: an empirical and theoretical analysis. *Behav. Proc.* 78:313–334.

Vasey, N. (1997). Community ecology and behavior of *Varecia variegata* and *Lemur fulvus albifrons* on the Masoala Peninsula, Madagascar [PhD diss.]. Washington University, St. Louis.

Vasey, N. (2000). Niche separation in *Varecia variegata rubra* and *Eulemur fulvus albifrons*: I. Interspecific patterns. *Am. J. Phys. Anthropol.* 112:411–431.

Vasey, N. (2002). Niche separation in *Varecia variegata rubra* and *Eulemur fulvus albifrons*: II. Intraspecific patterns. *Am. J. Phys. Anthropol.* 118:169–183.

Vasey, N. (2006). Impact of seasonality and reproduction on social structure, ranging patterns, and fission–fusion social organization in red ruffed lemurs. In: Gould, L., and Sauther, M. L. (eds.), *Lemurs: Ecology and Adaptation.* Springer, New York. pp. 275–304.

Vasey, P. L. (1995). Homosexual behavior in primates: a review of evidence and theory. *Int. J. Primatol.* 61:173–204.

Vasey, P. L. (2004). Sex differences in sexual partner acquisition, retention and harassment during female homosexual consortships in Japanese macaques. *Am. J. Primatol.* 64:397–409.

Vasey, P. L., and Gauthier, C. (2000). Skewed sex ratios and female homosexual activity in Japanese macaques: an experimental analysis. *Primates* 41:17–25.

Vedder, A. L. (1984). Movement patterns of a free-ranging group of mountain gorillas (*Gorilla gorilla beringei*) and their relation to food availability. *Am. J. Primatol.* 7:73–88.

Vedder, A., and Fashing, P. J. (2002). Diet of a 300-member Angolan colobus monkey (*Colobus angolensis*) supergroup in the Nyungwe Forest, Rwanda. *Am. J. Phys. Anthropol.* 117(suppl. 34):159–160.

Veenema, H. C. (2000). Methodological progress in post-conflict research. In: Aureli, F., and de Waal, F. B. M. (eds.), *Natural Conflict Resolution.* University of California Press, Berkeley. pp. 21–23.

Veenema, H. C., Das, M., and Aureli, F. (1994). Methodological improvements for the study of reconciliation. *Behav. Proc.* 31:29–37.

Vehrencamp, S. L. (1983). A model for the evolution of despotic versus egalitarian societies. *Anim. Behav.* 31:667–682.

Veiga, L. (2006). A ecologia e o comportamento do cuxiú-preto (*Chiropotes satanas*) na paisagem fragmentada da Amazônia Oriental [PhD diss.]. Universidade Federal do Pará, Belém, Brazil.

Veiga, L. M., and Ferrari, S. F. (2006). Predation of arthropods by southern bearded sakis (*Chiropotes satanas*) in eastern Brazilian Amazonia. *Am. J. Primatol.* 68:209–215.

Veiga, L. M., and Silva, S. S. B. (2005). Relatives or just good friends? Affiliative relationships among male bearded sakis (*Chiropotes satanas*) in Tucuruí, Brazil. *Livro de Resumos, XI Congresso Brasileiro de Primatologia,* Porto Alegre, February 13–18, 2005. p. 174.

Venter, J. C., Adams, M. D., Myers, E. W., et al. (2001). The sequence of the human genome. *Science* 291:1304–1351.

Veracini, C. (1997). O comportamento alimentar de *Callithrix argentata* (Linnaeus 1771) (Primata Callitrichinae). In: Lisboa, P. L. B. (ed.), *Caxiuana.* MCT/CNPq, Belem, Brazil. pp. 437–446.

Veracini, C. (2000). Dado preliminaries sobre a ecologia de *Saguinus niger* na Estação Científica Ferreira Penna, Caxiuaná, Brasil. *Neotrop. Primates* 8:108–113.

Verbeek, P., and de Waal, F. B. M. (1997). Post-conflict behavior of captive brown capuchins in the presence and absence of attractive food. *Int. J. Primatol.* 18:703–725.

Vereshchagin, N. K., and Baryshnikov, G. G. (1984). Quaternary mammalian extinctions in northern Eurasia. In: Martin, P. S., and Klein, R. G. (eds.), *Quaternary Extinctions: A Prehistoric Revolution.* University of Arizona Press, Tucson. pp. 483–516.

Verstraete, M. M., Gobron, N., Aussedat, O., Robustelli, M., Pinty B., Widlowski, J., and Taberner, M. (2008). An automatic procedure to identify key vegetation phenology events using the JRC-FAPAR products. *Adv. Space Res.* 41(11):1773–1783.

Vervaecke, H., Dunbar, R., van Elsacker, L., and Verheyen, R. (1992). Interactions with and spatial proximity to the males in relation to rank of captive female adult hamadryas baboons (*Papio hamadryas*). *Acta Zool. Pathol. Antverp.* 82:61–77.

Vervaecke, H., De Vries, H., and Van Elsacker, L. (1999). An experimental evaluation of the consistency of competitive ability and agonistic dominance in different social contexts in captive bonobos. *Behaviour* 136:423–442.

Vervaecke, H., de Vries, H., and van Elsacker, L. (2000). Dominance and its behavioral measures in a captive group of bonobos (*Pan paniscus*). *Int. J. Primatol.* 21:47–68.

Vervaecke, H., Stevens, J., and Van Elsacker, L. (2003). Interfering with others: female–female reproductive competition in *Pan paniscus*. In: Jones, C. P. (ed.), *Sexual Selection and Reproductive Competition in Primates: New Perspectives and Directions.* American Society of Primatologists, Norman, OK. pp. 231–253.

Vick, S. J., Toxopeus, I., and Anderson, J. R. (2006). Pictorial gaze cues do not enhance long-tailed macaques' performance on a computerised object-location task. *Behav. Proc.* 73:308–314.

Vie, J.-C., Richard-Hansen, C., and Fournier-Chambrillon, C. (2001). Abundance, use of space, and activity patterns of white-faced sakis (*Pithecia pithecia*) in French Guiana. *Am. J. Primatol.* 55:203–221.

Vigilant, L., Stoneking, M., Harpending, H., Hawkes, K., and Wilson, A. C. (1991). African populations and the evolution of human mitochondrial DNA. *Science* 253:1503–1507.

Vigilant, L., Hofreiter, M., Seidel, H., and Boesch, C. (2001). Paternity and relatedness in wild chimpanzee communities. *Proc. Natl. Acad. Sci. USA* 98:12890–12895.

Vila, C., Sundqvist, A., Flagstad, O., Sedden, J., Bjornerfeldt, S., Kojola, I., Casulli, A., Sand, H., Wabakken, P., and Ellegren, H. (2002). Rescue of a severely bottlenecked wolf (*Canis lupus*) population by a single immigrant. *Proc. R. Soc. Lond. B* 270:91–97.

Villavicencio Galindo, J. M. (2003). Distribución geográfica de los primates del género *Aotus* en el Departamento Norte de Santander, Colombia. In: Pereira-Bengoa, V., Nassar-Montoya, F., and Savage, A. (eds.), *Primatología del Nuevo Mundo.* Centro de Primatología Araguatos, Bogotá.

Vine, I. (1970). Communication by facial visual signals. In: Crook, J. H. (ed.), *Social Behaviour in Birds and Mammals.* Academic Press, New York. pp. 279–354.

Vine, I. (1971). Risk of visual detection and pursuit by a predator and the selective advantage of flocking behaviour. *J. Theor. Biol.* 30:405–422.

Vinyard, C. J., Wall, C. E., Williams, S. H., Schmitt, D., and Hylander, W. L. (2001). A preliminary report on the jaw mechanics during tree gouging in common marmosets (*Callithrix jacchus*). In: Brook, A. (ed.), *Dental Morphology: 12th International Symposium on Dental Morphology.* Sheffield Academic Press, Sheffield. pp. 283–297.

Vinyard, C. J., Wall, C. E., Williams, S. H., and Hylander, W. L. (2003). Comparative functional analysis of skull morphology of tree-gouging primates. *Am. J. Phys. Anthropol.* 120:153–170.

Visalberghi, E. (1987). Acquisition of nut-cracking behaviour by 2 capuchin monkeys (*Cebus apella*). *Folia Primatol.* 49:168–181.

Visalberghi, E. (1990). Tool use in *Cebus. Folia Primatol.* 54:146–154.

Visalberghi, E., and Fragaszy, D. (1990). Do monkeys ape? In: Parker, S. T., and Gibson, K. (eds.), *"Language" and Intelligence in Monkeys and Apes.* Cambridge University Press, Cambridge. pp. 247–273.

Visalberghi, E., and Fragaszy, D. (2006). What is challenging about tool use? The capuchin's perspective. In: Wasserman, E. A., and Zentall, T. R. (eds.), *Comparative Cognition: Experimental Explorations of Animal Intelligence.* Oxford University Press, Oxford. pp. 529–552.

Visalberghi, E., and Limongelli, L. (1994). Lack of comprehension of cause–effect relations in tool-using capuchin monkeys (*Cebus apella*). *J. Comp. Psychol.* 108:15–22.

Visalberghi, E., and Tomasello, M. (1998). Primate causal understanding in the physical and psychological domains. *Behav. Proc.* 42:189–203.

Visalberghi, E., and Trinca, L. (1989). Tool use in capuchin monkeys: distinguishing between performing and understanding. *Primates* 30:511–521.

Visalberghi, E., Fragaszy, D. M., and Savage-Rumbaugh, S. (1995). Performance in a tool-using task by common chimpanzees (*Pan troglodytes*), bonobos (*Pan paniscus*), an orangutan (*Pongo pygmaeus*), and capuchin monkeys (*Cebus apella*). *J. Comp. Psychol.* 109:52–60.

Visalberghi, E., Fragaszy, D., Ottoni, E., Izar, P., de Oliveira, M. G., and Andrade, F. R. D. (2007). Characteristics of hammer stones and anvils used by wild bearded capuchin monkeys (*Cebus libidinosus*) to crack open palm nuts. *Am. J. Phys. Anthropol.* 132:426–444.

Visalberghi, E., Addessi, E., Truppa, V., Spagnletti, N., Ottoni, E., Izar, P., and Fragaszy, D. (2009). Selection of effective stone tools by wild bearded capuchin monkeys. *Curr. Biol.* 19:213–217.

Vlamings, P., Uher, J., and Call, J. (2006). How the great apes (*Pan troglodytes, Pongo pygmaeus, Pan paniscus,* and *Gorilla gorilla*) perform on the reversed contingency task: the effects of food quantity and food visibility. *J. Exp. Psychol. Anim. Behav. Proc.* 32:60–70.

Voelkl, B., and Huber, L. (2000). True imitation in marmosets. *Anim. Behav.* 60:195–202.

Vogel, E. R. (2004). The ecological basis of aggression in white-faced capuchin monkeys, *Cebus capucinus,* in a Costa Rican dry forest [PhD diss.]. State University of New York, Stony Brook, NY.

Vogel, E. R. (2005). Rank differences in energy intake rates in white-faced capuchin monkeys (*Cebus capucinus*): the effects of contest competition. *Behav. Ecol. Sociobiol.* 58:333–344.

Vogel, E. R., and Janson, C. H. (2003). The role of food patches in primate socioecology: a monkey's eye view. *Am. J. Phys. Anthropol.* 60(suppl. 1):43–44.

Vogel, E. R., and Janson, C. H. (2007). Predicting the frequency of food-related agonism in white-faced capuchin monkeys (*Cebus capucinus*), using a novel focal-tree method. *Am. J. Primatol.* 69(5):533–550.

Vogel, E. R., Munch, S. B., and Janson, C. H. (2007). Understanding escalated aggression over food resources in white-faced capuchin monkeys. *Anim. Behav.* 74:71–80.

Vogel, E. R., van Woerden, J. T., Lucas, P. W., Utami Atmoko, S. S., van Schaik, C. P., and Dominy, N. J. (2008). Functional ecology and evolution of hominoid molar enamel: *Pan troglodytes schweinfurthii* and *Pongo pygmeus wurmbii. J. Hum. Evol.* 55(1):60–74.

Voigt, F. A., Arafeh, R., Farwig, N., Griebeler, E. M., and Böhning-Gaese, K. (2009). Linking seed dispersal and genetic structure of trees: A biogeographical approach. J. Biogeo. 36:242–254.

von Dornum, M., and Ruvolo, M. (1999). Phylogenetic relationships of the New World monkeys (Primates, Platyrrhini) based on nuclear G6PD DNA sequences. *Mol. Phylogenet. Evol.* 11:459–476.

Von Holst, D. (1985). Coping behavior and stress physiology in male tree shrews (*Tupaia belangeri*). *Fortschr. Zool.* 31:461–470.

Vonk, J., and Subiaul, F. (2009). Do chimpanzees know what others can and cannot do? Reasoning about "capability." *Anim. Cogn.* 12:267–286.

Voysey, B. C., McDonald, K. E., Rogers, M. E., Tutin, C. E. G., and Parnell, R. J. (1999a). Gorillas and seed dispersal in the Lope Reserve, Gabon. I: Gorilla acquisition by trees. *J. Trop. Ecol.* 15:23–38.

Voysey, B. C., McDonald, K. E., Rogers, M. E., Tutin, C. E. G., and Parnell, R. J. (1999b). Gorillas and seed dispersal in the Lope Reserve, Gabon. II: Survival and growth of seedlings. *J. Trop. Ecol.* 15:39–60.

Vrba, E. (1995). The fossil record of African antelopes (Mammalia, Bovidae) in relation to human evolution and paleoclimate. In: Vrba, E., Denton, G., Burckle, L., and Partridge, T. (eds.), *Paleoclimate and Evolution with Emphasis on Human Origins.* Yale University Press, New Haven. pp. 385–424.

Vulinec, K. (2002). Dung beetle communities and seed dispersal in primary forest and disturbed land in Amazonia. *Biotropica* 34:297–309.

Wachter, B., Schabel, M., and Noë, R. (1997). Diet overlap and polyspecific associations of red colobus and diana monkeys in the Taï National Park, Ivory Coast. *Ethology* 103:514–526.

Wada, K., Tokida, E., and Ogawa, H. (2007). The influence of snowfall, temperature and social relationships on sleeping clusters of Japanese monkeys during winter in Shiga Heights. *Primates* 48:130–139.

Wadley, R. L., Colfer, C. J. P., and Hood, I. G. (1997). Hunting primates and managing forests: the case of Iban forest farmers in Indonesian Borneo. *Hum. Ecol.* 25:243–271.

Wahaj, S. A., Guse, K. R., and Holekamp, K. E. (2001). Reconciliation in the spotted hyena (*Crocuta crocuta*). *Ethology* 107:1057–1074.

Wahome, J. M., Rowell, T. E., and Tsingalia, H. M. (1993). The natural history of de Brazza's monkey in Kenya. *Int. J. Primatol.* 14:445–466.

Wahungu, G. M. (2001). Common use of sleeping sites by two primate species in Tana River, Kenya. *Afr. J. Ecol.* 39:18–23.

Waite, T. A. (2007). Revisiting evidence for sustainability of bushmeat hunting in West Africa. *Environ. Manage.* 40(3):476–480.

Waits, L. P., and Paetkau, D. (2005). Noninvasive genetic sampling tools for wildlife biologists: a review of applications and recommendations for accurate data collection. *J. Wildl. Manage.* 69:1419–1433.

Waitt, C., and Little, A. C. (2006). Preferences for symmetry in conspecific facial shape among *Macaca mulatta. Int. J. Primatol.* 27:133–145.

Waitt, C., Little, A. C., Wolfensohn, S., Honess, P., Brown, A. P., Buchanan-Smith, H. M., and Perrett, D. I. (2003). Evidence from rhesus macaques suggests that male coloration plays a role in female primate mate choice. *Proc. R. Soc. Biol. Sci. B* 270(suppl.):S144–S146.

Wakibara, J. V., Huffman, M. A., Wink, M., Reich, S., Aufreiter, S., Hancock, R. G. V., Sodhi, R., Mahaney, W. C., and Russell, S. (2001). Adaptive significance of geophagy for Japanese macaques (*Macaca fuscata*) at Arashiyama, Japan. *Int. J. Primatol.* 22:495–520.

Walker, A. (1967). Patterns of extinction among the subfossil Madagascan lemuroids. In: Martin, P. S., and Wright, H. E., Jr. (eds.), *Pleistocene Extinctions: The Search for a Cause.* Yale University Press, New Haven. pp. 425–432.

Walker, A. C. (1969). The locomotion of the lorises, with special reference to the potto. *E. Afr. Wildl. J.* 7:1–5.

Walker, C. H. (1978). Species differences in microsomal monooxygenase activity and their relationship to biological half-lives. *Drug Metab. Rev.* 7:295–310.

Walker, M. L. (1995). Menopause in female rhesus monkeys. *Am. J. Primatol.* 35:59–71.

Walker, R., Hill, K., Burger, O., and Hurtado, A. M. (2006). Life in the slow lane revisited: ontogenetic separation between chimpanzees and humans. *Am. J. Phys. Anthropol.* 129:577–583.

Walker, S. E. (1996). Evolution of positional behavior in the saki/uakaris (*Pithecia, Chiropotes, Cacajao*). In: Norconk, M. A., Rosenberger, A. L., and Garber, P. A. (eds.), *Adaptive Radiations of Neotropical Primates.* Plenum Press, New York. pp. 335–367.

Walker, S. E. (1998). Fine-grained differences within positional categories: a case study of *Pithecia* and *Chiropotes.* In: Strasser, E., Fleagle, J. G., Rosenberger, A., and McHenry, H. (eds.), *Primate Locomotion: Recent Advances.* Plenum Press, New York. pp. 31–44.

Walker, S. E., and Ayres, J. M. (1996). Positional behavior of the white uakari (*Cacajao calvus calvus*). *Am. J. Phys. Anthropol.* 101:161–172.

Walker, S. E., Strasser, M. E., and Field, L. P. (2004). Reproductive parameters and life-history variables in captive golden-bellied mangabeys (*Cercocebus agilis chrysogaster*). *Am. J. Primatol.* 64:123–131.

Wallace, G. (2006). Identification, abundance, and behaviour of galagos in Mulanje and Thyolo districts, Malawi [MSc thesis]. Oxford Brookes University, Oxford.

Wallace, R. B. (2001). Diurnal activity budgets of black spider monkeys, *Ateles chamek,* in a southern Amazonian tropical forest. *Neotrop. Primates* 9:101–107.

Wallace, R. B. (2005). Seasonal variations in diet and foraging behavior of *Ateles chamek* in a southern Amazonian tropical forest. *Int. J. Primatol.* 26:1053–1075.

Wallace, R. B. (2006). Seasonal variations in black-faced black spider monkey (*Ateles chamek*) habitat use and ranging behavior in a southern Amazonian tropical forest. *Am. J. Primatol.* 68:313–332.

Wallace, R. B. (2007). Towing the party line: territoriality, risky boundaries and male group size in spider monkey fission–fusion societies. *Am. J. Primatol.* 69:1–16.

Wallace, R. B., Painter, L. E., Rumiz, D. I., and Taber, A. B. (2000). Primate diversity, distribution and relative abundance in the Rios Blanco y Negro Wildlife Reserve, Santa Cruz Department, Bolivia. *Neotrop. Primates* 8:24–28.

Wallace, R. B., Gomez, H., Felton, A., and Felton, A. M. (2006). On a new species of titi monkey, genus *Callicebus* Thomas (Primates, Pitheciidae), from western Bolivia with preliminary notes on distribution and abundance. *Primate Conserv.* 20:29–39.

Wallen, K., and Parsons, W. A. (1997). Sexual behavior in same-sexed nonhuman primates: is it relevant to understanding human homosexuality. *Annu. Rev. Sex. Res.* 8:195–224.

Wallis, J. (1992a). Chimpanzee genital swelling and its role in the pattern of sociosexual behavior. *Am. J. Primatol.* 28:101–113.

Wallis, J. (1992b). Socioenvironmental effects on timing of first postpartum cycles in chimpanzees. In: Nishida, T., McGrew, W. C., Marler, P., Pickford, M., and de Waal, F. M. (eds.), *Topics in Primatology.* University of Tokyo Press, Tokyo. pp. 119–130.

Wallis, J. (1995). Seasonal influence on reproduction in chimpanzees of Gombe National Park. *Int. J. Primatol.* 16:435–451.

Wallis, J. (1997). A survey of reproductive parameters in the free-ranging chimpanzees of Gombe National Park. *J. Reprod. Fertil.* 109:297–307.

Wallis, J. (2002). Seasonal aspects of reproduction and sexual behavior in two chimpanzee populations: a comparison of Gombe (Tanzania) and Budongo (Uganda). In: Boesch, C., Hohmann, G., and Marchant, L. (eds.), *Behavioural Diversity in Chimpanzees and Bonobos.* Cambridge University Press, Cambridge. pp. 181–191.

Wallis, J., and Lee, D. R. (1999). Primate conservation: the prevention of disease transmission. *Int. J. Primatol.* 20:803–826.

Wallis, S. J. (1983). Sexual behavior and reproduction of *Cercocebus albigena johnstonii* in Kibale Forest, western Uganda. *Int. J. Primatol.* 4:153–166.

Wallman, J. (1992). *Aping Language.* Cambridge University Press, New York.

Walsh, P. D., Abernethy, K. A., Bermejo, M., Beyers, R., Wachter, P. D., Akou, M. E., Huijbregts, B., Mambounga, D. I., Toham, A. K., Kilbourn, A. M., Lahm, S. A., Latour, S., Maisels, F., Mbina, C., Mihindou, Y., Oblang, S. N., Effa, E. N., Starkey, M. P., Telfer, P., Thibault, M., Tutin, C. E. G., White, L. J. T., and Wildie, D. S. (2003). Catastrophic ape decline in western equatorial Africa. *Nature* 422:611–614.

Walters, J. (1980). Interventions and the development of dominance relationships in female baboons. *Folia Primatol.* 34:61–89.

Walters, J. R. (1987a). Kin recognition in nonhuman primates. In: Fletcher, D. F., and Michener, C. D. (eds.), *Kin Recognition in Animals.* John Wiley, New York. pp. 359–393.

Walters, J. R. (1987b). Transition to adulthood. In: Smuts, B., Cheney, D., Seyfarth, R., Wrangham, R., and Struhsaker, T. T. (eds.), *Primate Societies.* University of Chicago Press, Chicago. pp. 358–369.

Walters, J. R., and Seyfarth, R. M. (1987). Conflict and cooperation. In: Smuts, B. B., Cheney, D. L., Seyfarth, R. M., Wrangham, R. W., and Struhsaker, T. T. (eds.), *Primate Societies.* University of Chicago Press, Chicago. pp. 306–317.

Wamboldt, M. Z., Gelhard, R. E., and Insel, T. R. (1988). Gender differences in caring for infant *Cebuella pygmaea*: the role of infant age and relatedness. *Dev. Psychobiol.* 21:187–202.

Wang, B. C., and Smith, T. B. (2002). Closing the seed dispersal loop. *Trends Ecol. Evol.* 17:379–385.

Wang, E., and Milton, K. (2003). Intragroup social relationships of male *Alouatta palliata* on Barro Colorado Island, Republica of Panama. *Int. J. Primatol.* 24:1227–1243.

Ward, S. (1997). The taxonomy and phylogenetic relationships of *Sivapithecus* revisited. In: Begun, D. R., Ward, C. V., and Rose, M. D. (eds.), *Function, Phylogeny, and Fossils: Miocene Hominoid Evolution and Adaptations.* Plenum Press, New York. pp. 269–290.

Ward, S. C., and Duren, D. L. (2002). Middle and late Miocene African hominoids. In: Hartwig, W. C. (ed.), *The Primate Fossil Record.* Cambridge University Press, Cambridge. pp. 385–397.

Warden, C., and Jackson, T. (1935). Imitative behaviour in the rhesus monkey. *J. Genet. Psychol.* 46:103–125.

Wardlaw, G. M., Hampl, J. S., and DiSilvestro, R. A. (2004). *Perspectives in Nutrition.* McGraw-Hill, Boston.

Warneken, F., and Tomasello, M. (2006). Altruistic helping in human infants and young chimpanzees. *Science.* 311:1301–1303.

Warner, A. C. I. (1981). Rate of passage of digesta through the gut of mammals and birds. *Nutr. Abstr. Rev. B* 51:789–820.

Warren, K. S., Nijman, I. J., Lenstra, J. A., Swan, R. A., Heriyanto, and Den Boer, M. H. (2000). Microsatellite DNA variation in Bornean orang-utans, *Pongo pygmaeus. J. Med. Primatol.* 29:57–62.

Warren, K. S., Nijman, I. J., Lenstra, J. A., Swan, R. A., Heriyanto, and Den Boer, M. H. (2001). Speciation and intrasubspecific variation of Bornean orangutans, *Pongo pygmaeus pygmaeus. Mol. Biol. Evol.* 18:472–480.

Warren, R. (1994). Lazy leapers: a study of the locomotor ecology of two species of saltatory nocturnal lemur in sympatry at Ampijoroa, Madagascar [PhD diss.]. University of Liverpool, Liverpool.

Warren, R. D., and Crompton, R. H. (1997). A comparative study of the ranging behaviour, activity rhythms and sociality of *Lepilemur edwardsi* (Primates, Lepilemuridae) and *Avahi occidentalis* (Primates, Indriidae) at Ampijoroa, Madagascar. *J. Zool. Lond.* 243:397–415.

Warren, R. D., and Crompton, R. H. (1998a). Diet, body size, and energy costs of locomotion in saltatory primates. *Folia Primatol.* 69:86–100.

Warren, R. D., and Crompton, R. H. (1998b). Locomotor ecology of *Lepilemur edwardsi* and *Avahi occidentalis. Am. J. Phys. Anthropol.* 104:471–486.

Warren, W. C., Hillier, L. W., Graves, J. A. M., et al. (2008). Genome analysis of the platypus reveals unique signatures of evolution. *Nature* 453:175–183.

Warren, Y. (2003). Olive baboons (*Papio cynocephalus anubis*): behaviour, ecology and human conflict in Gashaka-Gumti National Park, Nigeria [PhD diss.]. University of Surrey, Roehampton.

Warter, S., Randrianasolo, G., Dutrillaux, B., and Rumpler, Y. (1987). Cytogenetic study of a new subspecies of *Hapalemur griseus. Folia Primatol.* 48:50–55.

Warter, S., Hauwy, M., Dutrillaux, B., and Rumpler, Y. (2005). Application of molecular cytogenetics for chromsomal evolution of the Lemuriformes (Prosimians). *Cytogenet. Genome Res.* 108:197–203.

Waser, P. (1975). Monthly variations in feeding and activity patterns of the mangabey, *Cercocebus albigena* (Lydekker). *E. Afr. Wildl. J.* 13:249–263.

Waser, P. (1976). *Cercocebus albigena*: site attachment, avoidance and intergroup spacing. *Am. Nat.* 110:911–917.

Waser, P. (1977). Feeding, ranging and group size in the mangabey *C. albigena.* In: Clutton-Brock, T. H. (ed.), *Primate Ecology: Studies of Feeding and Ranging Behavior in Lemurs, Monkeys and Apes.* Academic Press, New York. pp. 183–222.

Waser, P. M. (1978). Postreproductive survival and behavior in a free-ranging female mangabey. *Folia Primatol.* 29:142–160.

Waser, P. M. (1980). Polyspecific associations of *Cercocebus albigena*: geographic variation and ecological correlates. *Folia Primatol.* 33:57–76.

Waser, P. M. (1982a). Primate polyspecific associations: do they occur by chance? *Anim. Behav.* 30:1–8.

Waser, P. M. (1982b). The evolution of male loud calls among mangabeys and baboons. In: Snowdon, C. T., Brown, C. H., and Petersen, M. R. (eds.), *Primate Communication.* Cambridge University Press, Cambridge. pp. 117–143.

Waser, P. (1984). Ecological differences and behavioral contrasts between two mangabey species. In: Rodman, P. S., and Cant, J. G. H. (eds.), *Adaptations for Foraging in Non-human Primates.* Columbia University Press, New York. pp. 195–216.

Waser, P. M. (1987). Interactions among primate species. In: Smuts, B. B., Cheney, D. L., Seyfarth, R. M., Wrangham, R. W., and Struhsaker, T. T. (eds.), *Primate Societies.* University of Chicago Press, Chicago. pp. 210–226.

Waser, P. M., and Brown, C. H. (1984). Is there a sound window for primate communication? *Behav. Ecol. Sociobiol.* 15:73–76.

Waser, P. M., and Brown, C. H. (1986). Habitat acoustics and primate communication. *Am. J. Primatol.* 10:135–156.

Waser, P. M., and Waser, M. S. (1977). Experimental studies of primate vocalization: specializations for long-distance propagation. *Z. Tierpsychol.* 43:239–263.

Washburn, D. A. (ed.) (2007). *Primate Perspectives on Behavior and Cognition.* American Psychological Association, Washington DC.

Washburn, D. A., and Rumbaugh, D. M. (1991). Rhesus monkeys' (*Macaca mulatta*) complex learning skills reassessed. *Int. J. Primatol.* 12:377–388.

Washburn, D. A., Smith, J. D., and Shields, W. E. (2006). Rhesus monkeys (*Macaca mulatta*) immediately generalize the uncertain response. *J. Exp. Psychol. Anim. Behav. Proc.* 32:185–189.

Washburn, S. L. (1942). Skeletal proportions of adult langurs and macaques [abridged version of doctoral diss., Harvard University]. *Hum. Biol.* 14:444–472.

Washburn, S. L. (1943). The sequence of epiphyseal union in Old World monkeys. *Am. J. Anat.* 72:339–360.

Washburn, S. L. (1946a). The effect of facial paralysis on the growth of the skull of rat and rabbit. *Anat. Rec.* 94:163–168.

Washburn, S. L. (1946b). The effect of removal of the zygomatic arch in the rat. *J. Mammal.* 27:121–124.

Washburn, S. L. (1947a). The relation of the temporal muscle to the form of the skull. *Anat. Rec.* 99:239–248.

Washburn, S. L. (1947b). The biological basis of measurement. *Am. J. Phys. Anthropol.* 5:237.

Washburn, S. L. (1951a). The analysis of primate evolution with particular reference to the origin of man. In: Washburn, S. L., and Dobshansky, T. (eds.), *Cold Spring Harbor Symposia on Quantitative Biology, Origin and Evolution of Man,* vol. 15. Cold Spring Harbor Laboratory Press, Cold Spring Harbor, NY. pp. 67–78.

Washburn, S. L. (1951b). The new physical anthropology. *Trans. N. Y. Acad. Sci.* 2:298–304.

Washburn, S. L. (ed.) (1961). *Social Life of Early Man.* Aldine, Chicago.

Washburn, S. L. (ed.) (1963). *Classification and Human Evolution.* Aldine, Chicago.

Washburn, S. L. (1973). The promise of primatology. *Am. J. Phys. Anthropol.* 38:177–182.

Washburn, S. L., and DeVore, I. (1961). The social life of baboons. *Sci. Am.* 204:62–71.

Washburn, S. L., Jay, P. C., and Lancaster, J. (1965). Field studies of Old World monkeys and apes. *Science* 150:1541–1547.

Wasser, S. K. (1983). Reproductive competition and cooperation among female yellow baboons. In: Wasser, S. K. (ed.), *Social Behavior of Female Vertebrates.* Academic Press, New York. pp. 349–390.

Wasser, S. K. (1996). Reproductive control in wild baboons measured by fecal steroids. *Biol. Reprod.* 55:393–399.

Wasser, S. K., and Barash, D. P. (1983). Reproductive suppression among female mammals: implications for biomedicine and sexual selection theory. *Q. Rev. Biol.* 58:513–538.

Wasser, S. K., and Starling, A. K. (1986). Reproductive competition among female yellow baboons. In: Else, J. G., and Lee, P. C. (eds.), *Primate Ontogeny, Cognition and Social Behaviour,* vol. 3. Cambridge University Press, Cambridge. pp. 344–354.

Wasser, S. K., and Starling, A. K. (1988). Proximate and ultimate causes of reproductive suppression among female yellow baboons at Mikumi National Park, Tanzania. *Am. J. Primatol.* 16:97–121.

Wasser, S. K., Risler, L., and Steiner, R. A. (1988). Excreted steroids in primate feces over the menstrual cycle and pregnancy. *Biol. Reprod.* 39:862–872.

Wasser, S. K., Monfort, S. L., and Wildt, D. E. (1991). Rapid extraction of faecal steroids for measuring reproductive cyclicity and early pregnancy in free-ranging yellow baboons (*Papio cynocephalus cynocephalus*). *J. Reprod. Fertil.* 92:415–423.

Wasser, S. K., Norton, G. W., Kleindorfer, S., and Rhine, R. J. (2004). Population trend alters the effects of maternal dominance rank of lifetime reproductive success in yellow baboons (*Papio cynocephalus*). *Behav. Ecol. Sociobiol.* 56:338–345.

Wasserman, M. D., and Chapman, C. A. (2003). Determinants of colobine monkey abundance: the importance of food energy, protein and fibre content. *J. Anim. Ecol.* 72:650–659.

Watanabe, K. (1979). Alliance formation in a free-ranging troop of Japanese macaques. *Primates* 20:459–474.

Watanabe, K. (1981). Variations in group composition and population density of two sympatric Mentawaian leaf-monkeys. *Primates* 22:145–160.

Watanabe, K., Mori, A., and Kawai, M. (1992). Characteristic features of the reproduction of Koshima monkeys, *Macaca fuscata fuscata*: a summary of thirty-four years of observation. *Primates* 33:1–32.

Watanabe, S., Izawa, M., Kato, A., Ropert-Coudert, Y., and Naito, Y. (2005). A new technique for monitoring the detailed behaviour of terrestrial animals: a case study with the domestic cat *App. Anim. Behav. Sci.* 94:117–131.

Waterland, R. A., and Jirtle, R. L. (2003). Transposable elements: targets for early nutritional effects on epigenetic gene regulation. *Mol. Cell. Biol.* 23:5293–5300.

Waterman, P. G. (1984). Food acquisition and processing as a function of plant chemistry. In: Chivers, D. J., Wood, B. A., and Bilsborough, A. (eds.), *Food Acquisition and Processing in Primates*. Plenum, New York. pp. 177–212.

Waterman, P. G., and Kool, K. (1994). Colobine food selection and plant chemistry. In: Davies, A. G., and Oates, J. K. (eds.), *Colobine Monkeys: Their Ecology, Behavior and Evolution*. Cambridge University Press, Cambridge. pp. 251–284.

Waterman, P. G., Choo, G. M., Vedder, A. L., and Watts, D. (1983). Digestibility, digestion-inhibitors and nutrients of herbaceous foliage and green stems from an African montane flora and comparison with other tropical flora. *Oecologia* 60:244–249.

Waterman, P. G., Ross, J. A. M., Bennett, E. L., and Davies, A. G. (1988). A comparison of the floristics and leaf chemistry of the tree flora in two Malaysian rain forests and the influence of leaf chemistry on populations of colobine monkeys in the Old World. *Biol. J. Linn. Soc.* 34:1–32.

Watkinson, A. R., and White, J. (1985). Some life-history consequences of modular construction in plants. *Phil. Trans. R. Soc. Lond. B* 313:31–51.

Watson, C. F. I., and Caldwell, C. A. (2009). Understanding behavioral traditions in primates: are current experimental approaches too focused on food? *Int. J. Primatol.* 30:143–167.

Watt, J. M., and Breyer-Brandwinjk, M. G. (1962). *The Medicinal and Poisonous Plants of Southern and East Africa*. E. and S. Livingstone, Edinburgh.

Watts, D. P. (1984). Composition and variability of mountain gorilla diets in the central Virungas. *Am. J. Primatol.* 7:323–356.

Watts, D. P. (1985a). Relations between group size and composition and feeding competition in mountain gorilla groups. *Anim. Behav.* 33:72–85.

Watts, D. P. (1985b). Observations on the ontogeny of feeding behavior in mountain gorillas (*Gorilla gorilla beringei*). *Am. J. Primatol.* 8(1):1–10.

Watts, D. P. (1988). Environmental influences on mountain gorilla time budgets. *Am. J. Primatol.* 15:195–211.

Watts, D. P. (1989a). Ant eating behavior of mountain gorillas. *Primates* 30:121–125.

Watts, D. P. (1989b). Infanticide in mountain gorillas: new cases and a reconsideration of the evidence. *Ethology* 81:1–18.

Watts, D. P. (1990a). Mountain gorilla life histories, reproductive competition, and sociosexual behavior and some implications for captive husbandry. *Zoo Biol.* 9:185–200.

Watts, D. P. (1990b). Ecology of gorillas and its relation to female transfer in mountain gorillas. *Int. J. Primatol.* 11:21–44.

Watts, D. P. (1991a). Strategies of habitat use by mountain gorillas. *Folia Primatol.* 56:1–16.

Watts, D. P. (1991b). Mountain gorilla reproduction and sexual behavior. *Am. J. Primatol.* 24:211–225.

Watts, D. P. (1991c). Harassment of immigrant female mountain gorillas by resident females. *Ethology* 89:135–153.

Watts, D. P. (1992). Social relationships of immigrant and resident female mountain gorillas, I. Male–female relationships. *Am. J. Primatol.* 28:159–181.

Watts, D. P. (1994a). The influence of male mating tactics on habitat use in mountain gorillas (*Gorilla gorilla beringei*). *Primates* 35:35–47.

Watts, D. P. (1994b). Agonistic relationships of female mountain gorillas. *Behav. Ecol. Sociobiol.* 34:347–358.

Watts, D. P. (1994c). Social relationships of immigrant and resident female mountain gorillas, II. Relatedness, residence, and relationships between females. *Am. J. Primatol.* 32:13–30.

Watts, D. P. (1995a). Post-conflict social events in wild mountain gorillas, 1. Social interactions between opponents. *Ethology* 100:158–174.

Watts, D. P. (1995b). Post-conflict social events in wild mountain gorillas. II. Redirection, side direction, and consolation. *Ethology* 100:158–174.

Watts, D. P. (1996). Comparative socioecology of gorillas. In: McGrew, W. C., Marchant, L. F., and Nishida, T. (eds.), *Great Ape Societies*. Cambridge University Press, Cambridge. pp. 16–28.

Watts, D. P. (1997). Agonistic interventions in wild mountain gorilla groups. *Behaviour* 134:23–57.

Watts, D. P. (1998a). Chimpanzee male aggression and sexual coercion at Ngogo, Kibale National Park, Uganda. *Am. J. Phys. Anthropol.* (suppl. 26):227.

Watts, D. P. (1998b). Coalitionary mate guarding by male chimpanzees at Ngogo, Kibale National Park, Uganda. *Behav. Ecol. Sociobiol.* 44:43–55.

Watts, D. P. (1998c). Seasonality in the ecology and life histories of mountain gorillas (*Gorilla gorilla beringei*). *Int. J. Primatol.* 19:929–948.

Watts, D. P. (1998d). Long term habitat use by mountain gorillas (*Gorilla gorilla beringei*) 1. Consistency, variation, and home range size and stability. *Int. J. Primatol.* 19:651–680.

Watts, D. P. (2000a). Causes and consequences of variation in male mountain gorilla life histories and group membership. In: Kappeler, P. M. (ed.), *Primate Males*. Cambridge University Press, Cambridge. pp. 169–179.

Watts, D. P. (2000b). Grooming between male chimpanzees at Ngogo, Kibale National Park. I. Partner number and diversity and grooming reciprocity. *Int. J. Primatol.* 21:189–210.

Watts, D. P. (2000c). Grooming between male chimpanzees at Ngogo, Kibale National Park. II. Influence of male rank and possible competition for partners. *Int. J. Primatol.* 21:211–238.

Watts, D. P. (2001). Social relationships of female mountain gorillas. In: Robbins, M. M., Sicotte, P., and Stewart, K. J. (eds.), *Mountain Gorillas: Three Decades of Research at Karisoke*. Cambridge University Press, Cambridge. pp. 215–240.

Watts, D. P. (2002a). Interchange of grooming and coalitionary support by wild male chimpanzees. *Am. J. Phys. Anthropol.* (suppl. 34):162.

Watts, D. P. (2002b). Reciprocity and interchange in the social relationships of wild male chimpanzees. *Behaviour* 139:343–370.

Watts, D. P. (2003). Gorilla social relationships: a comparative overview. In: Taylor, A. B., and Goldsmith, M. L. (eds.), *Gorilla Biology: A Multidisciplinary Perspective*. Cambridge University Press, Cambridge. pp. 302–327.

Watts, D. P. (2004). Intracommunity coalitionary killing of an adult male chimpanzee at Ngogo, Kibale National Park, Uganda. *Int. J. Primatol.* 25:507–521.

Watts, D. P. (2006). Conflict resolution in chimpanzees and the valuable-relationships hypothesis. *Int. J. Primatol.* 27:1337–1364.

Watts, D. P. (2007). Effects of male group size, parity, and cycle stage on female chimpanzee copulation rates at Ngogo, Kibale National Park, Uganda. *Primates* 48:222–231.

Watts, D. P., and Mitani, J. C. (2000). Infanticide and cannibalism by male chimpanzees at Ngogo, Kibale National Park, Uganda. *Primates* 41:357–365.

Watts, D. P., and Mitani, J. C. (2001). Boundary patrols and inter-group encounters in wild chimpanzees. *Behaviour* 138:299–327.

Watts, D. P., and Mitani, J. C. (2002). Hunting behavior of chimpanzees at Ngogo, Kibale National Park, Uganda. *Int. J. Primatol.* 23:1–28.

Watts, D. P., and Pusey, A. E. (1993). Behavior of juvenile and adolescent great apes. In: Pereira, M. E., and Fairbanks, L. A. (eds.), *Juvenile Primates*. Oxford University Press, New York. pp. 148–167.

Watts, D. P., Muller, M., Amsler, S. J., Mbabazi, G., and Mitani, J. C. (2006). Lethal intergroup aggression by chimpanzees in Kibale National Park, Uganda. *Am. J. Primatol.* 68:161–180.

Watts, E. S. (1985). Adolescent growth and development of monkeys, apes and humans. In: Watts, E. S. (ed.), *Nonhuman Primate Models for Human Growth and Development*. Alan R. Liss, New York. pp. 41–65.

Watts, E. S. (1986). Skeletal development. In: Dukelow, W., and Erwin, J. (eds.), *Comparative Primate Biology: Reproduction and Development*. Alan R. Liss, New York. pp. 415–439.

Watts, E. (1990). Evolutionary trends in primate growth and development. In: DeRousseau, C. J. (ed.), *Primate Life History and Evolution*. Wiley-Liss, New York. pp. 89–104.

Watts, E. S. (1993). Epiphyseal union in captive chimpanzees. *Am. J. Phys. Anthropol.* (suppl. 16):206.

Watts, E. S., and Gavan, J. A. (1982). Postnatal growth of non-human primates: the problem of the adolescent spurt. *Hum. Biol.* 54:53–70.

Weaver, A. (2003). Conflict and reconciliation in captive bottlenose dolphins, *Tursiops truncatus*. *Mar. Mammal Sci.* 19:836–846.

Weaver, A. C., and de Waal, F. B. M. (2000). The development of reconciliation in brown capuchins. In: Aureli, F., and de Waal, F. B. M. (eds.),

Natural Conflict Resolution. University of California Press, Berkeley. pp. 216–218.

Weaver, A., and de Waal, F. B. M. (2003). The mother–offspring relationship as a template in social development: reconciliation in captive brown capuchins (*Cebus apella*). *J. Comp. Psychol.* 117:101–110.

Webb, C. O., and Peart, D. R. (2000). Habitat associations of trees and seedlings in a Bornean rain forest. *J. Ecol.* 88:464–478.

Webber, A. D., Hill, C. M., and Reynolds, V. (2007). Assessing the failure of a community-based human–wildlife conflict mitigation project in Budongo Forest Reserve, Uganda. *Oryx* 41:177–184.

Weber, A. W., and Vedder, A. (1983). Population dynamics of the Virunga gorillas: 1959–1978. *Biol. Conserv.* 26:341–366.

Weber, B., and Vedder, A. (2001). Afterword. Mountain gorillas at the turn of the century. In: Robbins, M. M., Sicotte, P., and Stewart, K. J. (eds.), *Mountain Gorillas*. Cambridge University Press, Cambridge. pp. 413–423.

Weed, M. R., Bryant, R., and Perry, S. (2008). Cognitive development in macaques: attentional set-shifting in juvenile and adult rhesus monkeys. *Neuroscience* 157:22–28.

Wehncke, E. V., Hubbell, S. P., Forester, R. B., and Dalling, J. W. (2003). Seed dispersal patterns produced by white-faced monkeys: implications for the dispersal limitation of neotropical tree species. *J. Ecol.* 91:677–685.

Wehrenberg, W. B., Dyrenfurth, I., and Ferin, M. (1980). Endocrine characteristics of the menstrual cycle in the Assamese monkey (*Macaca assamensis*). *Biol. Reprod.* 23:522–525.

Weidt, A., Hagenah, N., Randrianambinina, B., Radespiel, U., and Zimmermann, E. (2004). Social organization of the golden brown mouse lemur (*Microcebus ravelobensis*). *Am. J. Phys. Anthropol.* 123:40–51.

Weingrill, T. (2000). Infanticide and the value of male–female relationships in mountain chacma baboons. *Behaviour* 138:337–359.

Weingrill, T., Lycett, J. E., and Henzi, S. P. (2000). Consortship and mating success in chacma baboons (*Papio cynocephalus ursinus*). *Ethology* 106:1033–1044.

Weingrill, T., Lycett, J. E., Barrett, L., Hill, R. A., and Henzi, S. P. (2003). Male consortship behaviour in chacma baboons: the role of demographic factors and female conceptive probabilities. *Behaviour* 140:405–427.

Weinstein, T. A. R., and Capitanio, J. P. (2008). Individual differences in infant temperament predict social relationships of yearling rhesus monkeys, *Macaca mulatta*. *Anim. Behav.* 76:455–465.

Weir, A. A. S., Chappell, J., and Kacelnik, A. (2002). Shaping of hooks in New Caledonian crows. *Nature* 297:981.

Weir, B. S. (1996). *Genetic Data Analysis II*. Sinauer Associates, Sunderland, MA.

Weisenseel, K., Chapman, C. A., and Chapman, L. J. (1993). Nocturnal primates of Kibale Forest: effects of selective logging on prosimian densities. *Primates* 34:445–450.

Weisenseel, K. A., Izard, M. K., Nash, L. T., Ange, R. L., and Poorman-Allen, P. (1998). A comparison of reproduction in two species of *Nycticebus*. *Folia Primatol.* 69(suppl. 1):321–324.

Weiss, K. M. (1981). Evolutionary perspectives on human aging. In: Amoss, P. T., and Harrell, S. (eds.), *Other Ways of Growing Old: Anthropological Perspectives*. Stanford University Press, Stanford, CA. pp. 25–58.

Weiss, M. K. (2000). A view on the science: physical anthropology at the millennium. *Am. J. Phys. Anthropol.* 111:295–299.

Welker, C. (1973). Ethologic significance of the urine washing by *Galago crassicaudatus* E. Geoffroy, 1812 (Lorisiformes: Galagidae). *Folia Primatol.* 20:429–452.

Welker, C., Schwibbe, M. H., Schaefer-Witt, C., and Visalberghi, E. (1987). Failure of kin recognition in *Macaca fascicularis*. *Folia Primatol.* 49:216–221.

Wenny, D. G. (1999). Two-stage dispersal of *Guarea glabra* and *G. kunthiana* (Meliaceae) in Monteverde, Costa Rica. *J. Trop. Ecol.* 15:481–496.

Werre, J. L. R. (2000). Ecology and behavior of the Niger Delta red colobus (*Procolobus badius epieni*) [PhD diss.]. City University of New York, New York.

West, G. B., Brown, J. H., and Enquist, B. J. (2001). A general model for ontogenetic growth. *Nature* 413:628–631.

West, K. (1981). The behavior and ecology of the siamang in Sumatra [MA thesis]. University of California, Davis.

Westcott, D., Bentrupperbäumer, J., Bradford, M., and McKeown, A. (2005). Incorporating patterns of disperser behaviour into models of seed dispersal and its effects on estimated dispersal curves. *Oecologia* 146:57–67.

Westergaard, G. C. (1988). Lion-tailed macaques (*Macaca silenus*) manufacture and use tools. *J. Comp. Psychol.* 102:152–159.

Westergaard, G. C. (1992). Object manipulation and the use of tools by infant baboons (*Papio cynocephalus anubis*). *J. Comp. Psychol.* 106:398–403.

Westergaard, G. C., and Fragaszy, D. M. (1987). The manufacture and use of tools by capuchin monkeys (*Cebus apella*). *J. Comp. Psychol.* 101:159–168.

Westergaard, G. C., and Linquist, T. (1987). Manipulation of objects in a captive group of lion-tailed macaques (*Macaca silenus*). *Am. J. Primatol.* 12:231–234.

Westergaard, G. C., and Suomi, S. J. (1996). Hand preference for stone artefact production and tool-use by monkeys: possible implications for the evolution of right-handedness in hominids. *J. Hum. Evol.* 30:291–298.

Westergaard, G. C., Greene, J. A., Babitz, M. A., and Suomi, S. J. (1995). Pestle use and modification by tufted capuchins (*Cebus apella*). *Int. J. Primatol.* 16:643–651.

Westergaard, G. C., Lundquist, A. L., Kuhn, H. E., and Suomi, S. J. (1997). Ant-gathering with tools by captive tufted capuchins (*Cebus apella*). *Int. J. Primatol.* 18:95–103.

Westergaard, G. C., Lundquist, A. L., Haynie, M. K., Kuhn, H. E., and Suomi, S. J. (1998). Why some capuchin monkeys (*Cebus apella*) use probing tools (and others do not). *J. Comp. Psychol.* 112:207–211.

Westergaard, G. C., Suomi, S. J., Higley, J. D., and Mehlman, P. T. (1999). CSF 5-HIAA and aggression in female macaque monkeys: species and interindividual differences. *Psychopharmacology* 146:440–446.

Westermarck, E. (1891). *The History of Human Marriage*. Macmillan, London.

Westlund, K., Ljungberg, T., Borefelt, U., and Abrahamsson, C. (2000). Post-conflict affiliation in common marmosets (*Callithrix jacchus jacchus*). *Am. J. Primatol.* 52:31–46.

Westneat, D. F. (1987). Extra-pair copulations in a predominantly monogamous bird: observations of behaviour. *Anim. Behav.* 35:865–876.

Wheatley, B. P. (1980). Feeding and ranging of east Bornean *Macaca fascicularis*. In: Lindburg, D. G. (ed.), *The Macaques: Studies in Ecology, Behavior and Evolution*. Van Nostrand Reinhold, New York. pp. 215–246.

Wheatley, B. P. (1999). *The Sacred Monkeys of Bali*. Waveland Press, Prospect Heights, IL.

Wheeler, B. C. (2008). Selfish or altruistic? An analysis of alarm call function in wild capuchin monkeys, *Cebus apella nigritus*. *Anim. Behav.* 76(5):1465–1475.

Wheeler, D. A., Srinivasan, M., Egholm, M., et al. (2008). The complete genome of an individual by massively parallel DNA sequencing. *Nature* 452:872–876.

Wheeler, W. C., and Gladstein, D. G. (1994). MALIGN: a multiple sequence alignment program. *J. Hered.* 85:417–418.

White, E. C. (2007). Ecology of *Mandrillus sphinx*: ranging, diet and social structure of a mandrill horde in Lopé National Park, Gabon [PhD diss.]. University of Exeter, Exeter.

White, F. J. (1986). Behavioral ecology of the pygmy chimpanzee [PhD diss.]. State University of New York, Stony Brook.

White, F. J. (1988). Party composition and dynamics in *Pan paniscus*. *Int. J. Primatol.* 9:179–193.

White, F. J. (1989a). Diet, ranging behavior and social organization of the black and white ruffed lemur, *Varecia variegata variegata*, in southeastern Madagascar [abstract]. *Am. J. Phys. Anthropol.* 78:323.

White, F. J. (1989b). Ecological correlates of pygmy chimpanzee social structure. In: Standen, V., and Foley, R. A. (eds.), *Comparative Socioecology: The Behavioral Ecology of Humans and Other Mammals*. Blackwell, Oxford. pp. 151–164.

White, F. J. (1992). Pygmy chimpanzee social organization: variation with party size and between study sites. *Am. J. Primatol.* 26:203–214.

White, F. J. (1996a). Comparative socio-ecology of *Pan paniscus*. In: McGrew, W. C., Marchant, L. F., and Nishida, T. (eds.), *Great Ape Societies*. Cambridge University Press, Cambridge. pp. 29–41.

White, F. J. (1996b). *Pan paniscus* 1973 to 1996: twenty-three years of field research. *Evol. Anthropol.* 5:11–17.

White, F. J. (1998). Seasonality and socioecology: the importance of variation in fruit abundance to bonobo sociality. *Int. J. Primatol.* 19:1013–1027.

White, F. J., and Lanjouw, A. (1992). Feeding competition in Lomako bonobos: variation in social cohesion. *Topics Primatol.* 1:67–79.

White, F. J., and Wrangham, R. W. (1988). Feeding competition and patch size in the chimpanzee species, *Pan paniscus* and *Pan troglodytes*. *Behaviour* 105:148–163.

White, L. J. T. (1994). Biomass of rain forest mammals in the Lopé Reserve, Gabon. *J. Anim. Ecol.* 63:499–512.

White, L. J. T., and Tutin, C. E. G. (2001). Why chimpanzees and gorillas respond differently to logging: a cautionary tale from Gabon. In: Weber, B., White, L. J. T., Vedder, A., and Simons-Morland, H. (eds.), *African Rain Forest Ecology and Conservation*. Yale University Press, New Haven. pp. 449–462.

Whitehead, P. F., and Jolly, C. J. (eds.) (2000). *Old World Monkeys*. Cambridge University Press, Cambridge.

Whitehead, P. J., and Tschirner, K. (1992). Sex and age related variation in foraging strategies of magpie geese *Anseranas semipalmata*. *Emu* 92:28–32.

Whiten, A. (1998). Imitation of the sequential structure of actions by chimpanzees (*Pan troglodytes*). *J. Comp. Psychol.* 112:270–281.

Whiten, A., and Byrne, R. W. (1988). Tactical deception in primates. *Behav. Brain Sci.* 11:233–273.

Whiten, A., and Custance, D. M. (1996). Studies of imitation in chimpanzees and children. In: Heyes, C. M., and Galef, B. G., Jr. (eds.), *Social Learning in Animals: The Roots of Culture*. Academic Press, New York. pp. 291–318.

Whiten, A., and Ham, R. (1992). On the nature and evolution of imitation in the animal kingdom: reappraisal of a century of research. *Adv. Study Behav.* 21:239–283.

Whiten, A., and Mesoudi, A. (2008). Establishing an experimental science of culture: animal social diffusion experiments. *Phil. Trans. R. Soc. B* 363:3477–3488.

Whiten, A., and van Schaik, C. P. (2007). The evolution of animal "cultures" and social intelligence. *Philos. Trans. R. Soc. B* 362:603–620.

Whiten, A., Byrne, R. W., and Henzi, S. P. (1987). The behavioral ecology of mountain baboons. *Int. J. Primatol.* 8:367–388.

Whiten, A., Byrne, R. W., Waterman, P. G., Henzi, S. P., and McCullough, F. M. (1990). Specifying the rules underlying selective foraging in wild mountain baboons, *P. ursinus*. In: Thiago de Mello, M., Whiten, A., and Byrne, R. W. (eds.), *Baboons: Behaviour and Ecology. Use and Care. Selected Proceedings of the XIIth Congress of the International Primatological Society*. Brasilia, Brasil. pp. 5–22.

Whiten, A., Byrne, R. W., Barton, R. A., Waterman, P. G., and Henzi, S. P. (1991). Dietary and foraging strategies of baboons. *Philos. Trans. R. Soc. Lond. B* 344:187–197.

Whiten, A., Custance, D. M., Gomez, J. C., Texidor, P., and Bard, K. A. (1996). Imitative learning of artificial fruit processing in children (*Homo sapiens*) and chimpanzees (*Pan troglodytes*). *J. Comp. Psychol.* 110:3–14.

Whiten, A., Goodall, J., McGrew, W. C., Nishida, T., Reynolds, V., Sugiyama, Y., Tutin, C. E. G., Wrangham, R. W., and Boesch, C. (1999). Cultures in chimpanzees. *Nature* 399:682–685.

Whiten, A., Goodall, J., McGrew, W. C., Nishida, T., Reynolds, V., Sugiyama, Y., Tutin, C. E. G., Wrangham, R. W., and Boesch, C. (2001). Charting cultural variation in chimpanzees. *Behaviour* 138:1481–1516.

Whiten, A., Horner, V., and Marshall-Pescini, S. (2003). Cultural panthropology. *Evol. Anthropol.* 12:92–105.

Whiten, A., Horner, V., Litchfield, C., and Marshall-Pescini, S. (2004). How do apes ape? *Learn. Behav.* 32:36–52.

Whiten, A., Horner, V., and de Waal, F. B. M. (2005). Conformity to cultural norms of tool use in chimpanzees. *Nature* 437:737–740.

Whitesides, G. H. (1989). Interspecific associations of diana monkeys, *Cercopithecus diana*, in Sierra Leone, West Africa: biological significance or chance? *Anim. Behav.* 37:760–776.

Whitesides, G. H., Oates, J. F., Green, S. M., and Kluberdanz, R. P. (1988). Estimating primate densities from transects in a West African rain forest: a comparison of techniques. *J. Anim. Ecol.* 57:345–367.

Whitham, J. C., and Maestripieri, D. (2003). Primate rituals: the function of greetings between male Guinea baboons. *Ethology* 109:847–859.

Whitington, C., and Treesucon, U. (1991). Selection and treatment of food plants by white-handed gibbons (*Hylobates lar*) in Khao Yai National Park, Thailand. *Nat. Hist. Bull. Siam Soc.* 39:111–122.

Whitmore, T. C. (1990). *An Introduction to Tropical Rain Forests*. Clarendon Press, Oxford.

Whitten, A. J. (1982a). Diet and feeding behaviour of Kloss gibbons on Siberut Island, Indonesia. *Folia Primatol.* 37:177–208.

Whitten, A. J. (1982b). Home range use by Kloss gibbons (*Hylobates klossii*) on Siberut Island, Indonesia. *Anim. Behav.* 30:182–198.

Whitten, A. J. (1984). Ecological comparisons between Kloss gibbons and other small gibbons. In: Preuschoft, H., Chivers, D. J., Brockelman, W. Y., and Creel, N. (eds.), *The Lesser Apes: Evolutionary and Behavioural Biology*. Edinburgh University Press, Edinburgh. pp. 219–227.

Whitten, P. L. (1983). Diet and dominance among female vervet monkeys (*Cercopithecus aethiops*). *Am. J. Primatol.* 5:139–159.

Whitten, P. L. (1984). Competition among female vervet monkeys. In: Small, M. F. (ed.), *Female Primates: Studies by Women Primatologists*. Alan R. Liss, New York. pp. 127–140.

Whitten, P. L. (1987). Infant and adult males. In: Smuts, B. B., Cheney, D. L., Seyfarth, R. M., Wrangham, R. W., and Struhsaker, T. T. (eds.), *Primate Societies*. University of Chicago Press, Chicago. pp. 343–359.

Whitten, P. L. (1988). Effects of patch quality and feeding subgroup size on feeding success in vervet monkeys (*Cercopithecus aethiops*). *Behaviour* 105:5–52.

Whitten, P. L., and Brockman, D. K. (2001). Strepsirhine reproductive ecology. In: Ellison, P. T. (ed.), *Reproductive Ecology and Human Evolution*. Aldine de Gruyter, New York. pp. 321–350.

Whitten, P. L., and Russell, E. (1996). Information content of sexual swellings and fecal steroids in sooty mangabeys (*Cercocebus torquatus atys*). *Am. J. Primatol.* 40:67–82.

Wich, S. A., and Nunn, C. L. (2002). Do male "long-distance calls" function in mate defense? A comparative study of long-distance calls in primates. *Behav. Ecol. Sociobiol.* 52:474–484.

Wich, S. A., and van Schaik, C. P. (2000). The impact of El Niño on mast fruiting in Sumatra and elsewhere in Malesia. *J. Trop. Ecol.* 16:563–577.

Wich, S. A., Fredriksson, G., and Sterck, E. H. M. (2002). Measuring fruit patch size for three sympatric Indonesian primate species. *Primates* 43:19–27.

Wich, S. A., Singleton, I., Utami-Atmoko, S. S., Geurts, M. L., Rijksen, H. D., and van Schaik, C. P. (2003). The status of the Sumatran orang-utan, *Pongo abelii*: an update. *Oryx* 37(1):49–54.

Wich, S. A., Buij, R., and van Schaik, C. P. (2004a). Determinants of orangutan density in the dryland forests of the Leuser Ecosystem. *Primates* 45:177–182.

Wich, S. A., Utami-Atmoko, S. S., Setia, T. M., Rijksen, H. R., Schurmann, C., van Hooff, J. A. R. A. M., and van Schaik, C. P. (2004b). Life history of wild Sumatran orangutans (*Pongo abelii*). *J. Hum. Evol.* 47:385–398.

Wich, S. A., Geurts, M. L., Mitra Setia, T., and Utami-Atmoko, S. S. (2006). Influence of fruit availability on Sumatran orangutan sociality and reproduction. In: Hohmann, G., Robbins, M., and Boesch, C. (ed.), *Feeding Ecology in Apes and Other Primates*. Cambridge University Press, Cambridge. pp. 337–358.

Wich, S. A., Meijaard, E., Marshall, A. J., Husson, S., Ancrenaz, M., Lacy, R. C., van Schaik, C. P., Sugardjito, J., Simorangkir, T., Traylor-Holzer, K., Doughty, M., Supriatna, J., Dennis, R., Gumal, M., Knott, C. D., and Singleton, I. (2008). Distribution and conservation status of the orang-utan (*Pongo* spp.) on Borneo and Sumatra: how many remain? *Oryx* 42:329–339.

Wich, S. A., de Vries, H., Ancrenaz, M., Perkins, L., Shumaker, R. W., Suzuki, A., and van Schaik, C. P. (2009). Orangutan life history variation. In: Wich, S. A., Utami, S. S., Mitra Setia, T., and van Schaik, C. P. (eds.), *Orangutans: Geographic Variation in Behavioral Ecology and Conservation*. Oxford University Press, Oxford. pp. 65–76.

Wickings, E. J., and Dixson, A. F. (1992a). Development from birth to sexual maturity in a semi-free-ranging colony of mandrills (*Mandrillus sphinx*) in Gabon. *J. Reprod. Fertil.* 95:129–138.

Wickings, E. J., and Dixson, A. F. (1992b). Testicular function, secondary sexual development and social status in male mandrills (*Mandrillus sphinx*). *Physiol. Behav.* 52:909–916.

Wickings, E. J., Bossi, T., and Dixson, A. F. (1993). Reproductive success in the mandrill, *Mandrillus sphinx*: correlations of male dominance and mating success with paternity, as determined by DNA fingerprinting. *J. Zool. Lond.* 231:563–574.

Wickings, E. J., Ambrose, L., and Bearder, S. K. (1998). Sympatric populations of *Galagoides demidoff* and *G. thomasi* in the Haut Ogooué region of Gabon. *Folia Primatol.* 69(suppl. 1):389–393.

Wickler, W. (1980). Vocal duetting and the pair bond. *Z. Tierpsychol.* 52:201–209.

Wickler, W., and Seibt, U. (1983). Monogamy: an ambiguous concept. In: Bateson, P. (ed.), *Mate Choice*. Cambridge University Press, Cambridge.

Widdig, A. (2007). Paternal kin discrimination: the evidence and likely mechanisms. *Biol. Rev.* 82:319–334.

Widdig, A., Streich, W. J., and Rembrock, G. (2000). Coalition formation among male Barbary macaques (*Macaca sylvanus*). *Am. J. Primatol.* 50:37–51.

Widdig, A., Nürnberg, P., Krawczak, M., Streich, W. J., and Bercovitch, F. B. (2001). Paternal relatedness and age proximity regulate social relationships among adult female rhesus macaques. *Proc. Natl. Acad. Sci. USA* 98:13769–13773.

Widdig, A., Nurnberg, P., Krawczak, M., Streich, W. J., and Bercovitch, F. B. (2002). Affiliation and aggression among adult female rhesus macaques: a genetic analysis of paternal cohorts. *Behaviour* 139:371–391.

Widdig, A., Nurnberg, P. Bercovitch, F. B., Trefilov, A., Berard, J. B., Kessler, M. J., Schmidtke, J. Streich, W. J., and Krawczak, M. B. (2006a). Consequences of group fission for the patterns of relatedness among rhesus macaques. *Mol. Ecol.* 15:3825–3832.

Widdig, A., Streich, W. J., Nurnberg, P., Croucher, P. J. P., Bercovitch, F. B., and Krawczak, M. (2006b). Paternal kin bias in agonistic interventions of adult female rhesus macaques (*Macaca mulatta*). *Behav. Ecol. Sociobiol.* 61:205–214.

Widowski, T. M., Ziegler, T. E., Elowson, A. M., and Snowdon, C. T. (1990). The role of males in the stimulation of reproductive function in female cotton-top tamarins, *Saguinus o. oedipus. Anim. Behav.* 40:731–741.

Widowski, T. M., Porter, T. A., Ziegler, T. E., and Snowdon, C. T. (1992). The stimulatory effect of males on the initiation but not the maintenance of ovarian cycling in cotton-top tamarins (*Saguinus oedipus*). *Am. J. Primatol.* 26:97–108.

Wiebe, R. H., Diamond, E., Askel, S., Liu, M. D., and Abee, C. R. (1984). Diurnal variation in androgens in sexually mature Bolivian squirrel monkeys (*Saimiri sciureus*) studied longitudinally. *Acta Endocrinol.* 87:424–433.

Wieczkowski, J. (2004). Ecological correlates of abundance in the Tana mangabey (*Cercocebus galeritus*). *Am. J. Primatol.* 63:125–138.

Wieczkowski, J. (2005). Examination of increased annual range of a Tana River mangabey (*Cercocebus galeritus*) group. *Am. J. Phys. Anthropol.* 128:381–388.

Wieczkowski, J., and Butynski, T. M. (in press). The Tana River mangabey (*Cercocebus galeritus*). In: Kingdon, J., Happold, D., and Butynski, T. (eds.), *The Mammals of Africa*. University of California Press, Berkeley, CA.

Wieczkowski, J., and Mbora, D. N. M. (1999/2000). Increasing threats to the conservation of endemic endangered primates and forests of the lower Tana River, Kenya. *Afr. Primates* 4:32–40.

Wienberg, J., and Stanyon, R. (1995). Chromosome painting in mammals as an approach to comparative genomics. *Curr. Opin. Genet. Dev.* 5:792–797.

Wienberg, J., and Stanyon, R. (1997). Comparative painting of mammalian chromosomes. *Curr. Opin. Genet. Dev.* 7:784–791.

Wienberg, J., and Stanyon, R. (1998). Comparative chromosome painting of primate genomes. *ILAR J.* 39:77–91.

Wiens, F. (1995). Verhaltensbeobachtungen am plumplori, *Nycticebus coucang* (Primates: Lorisidae) im Freiland [diplomarbeit]. Universität Frankfurt am Main, Frankfurt am Main.

Wiens, F. (2002). Behavior and ecology of wild slow lorises (*Nycticebus coucang*): social organisation, infant care system and diet [PhD diss.]. Bayreuth University, Bayreuth.

Wiens, F., and Zitzmann, A. (1999). Predation on a wild slow loris (*Nycticebus coucang*) by a reticulated python (*Python reticulatus*). *Folia Primatol.* 70:362–364.

Wiens, F., and Zitzmann, A. (2003a). Social dependence of infant slow lorises to learn diet. *Int. J. Primatol.* 24:1007–1021.

Wiens, F., and Zitzmann, A. (2003b). Social structure of the solitary slow loris *Nycticebus coucang* (Lorisidae). *J. Zool.* 261:35–46.

Wikelski, M., Kays, R. W., Kasdin, N. J., Thorup, K., Smith, J. A., and Swenson, G. W. (2007). Going wild: what a global small-animal tracking system could do for experimental biologists. *J. Exper. Biol.* 210:181–186.

Wild, C., Morgan, B. J., and Dixson, A. F. (2005). Conservation of drill populations in Bakossiland, Cameroon: historial trends and current status. *Int. J. Primatol.* 26:759–773.

Wildman, D. E., Bergman, T. J., al-Aghbari, A., Sterner, K. N., Newman, T. K., Phillips-Conroy, J. E., Jolly, C. J., and Disotell, T. R. (2004). Mitochondrial evidence for the origin of hamadryas baboons. *Mol. Phylogenet. Evol.* 32:287–296.

Wildt, D. E., Doyle, L. L., Stone, S. C., and Harrison, R. M. (1977). Correlation of perineal swelling with serum ovarian hormone levels, vaginal cytology, and ovarian follicular development during the baboon reproductive cycle. *Primates* 18:261–270.

Wiley, A. S., and Katz, S. H. (1998). Geophagy in pregnancy: a test of a hypothesis. *Curr. Anthropol.* 39:532–544.

Wiley, R. H. (1983). The evolution of communication: information and manipulation. In: Halliday, T. R., and Slater, P. J. B. (eds.), *Animal Behaviour: Communication*. W. H. Freeman, New York. pp. 156–189.

Wiley, R. H., and Richards, D. G. (1978). Physical constraints on acoustical communication in the atmosphere: implications for the evolution of animal vocalizations. *Behav. Ecol. Sociobiol.* 3:69–94.

Wilkie, D. S., and Carpenter, J. F. (1999). Bushmeat hunting in the Congo Basin: an assessment of impacts and options for mitigation. *Biodivers. Conserv.* 8:927–955.

Wilks, J. W. (1977). Endocrine characterization of the menstrual cycle of the stumptailed macaque (*Macaca arctoides*). *Biol. Reprod.* 16:474–478.

Williams, G. C. (1957). Pleiotropy, natural selection, and the evolution of senescence. *Evolution* 11:398–411.

Williams, G. C. (1966a). *Adaptation and Natural Selection: A Critique of Some Current Evolutionary Thought*. Princeton University Press, Princeton, NJ.

Williams, G. C. (1966b). Natural selection, the costs of reproduction and a refinement of Lack's principle. *Am. Nat.* 100:687–690.

Williams, G. C. (1975). *Sex and Evolution*. Princeton University Press, Princeton, NJ.

Williams, J. (1999). Female strategies and the reasons for territoriality: lessons from three decades of research at Gombe [PhD diss.]. University of Minnesota, Minneapolis.

Williams, J. B. (1978). *Tool Use in Nonhuman Primates*. Primate Information Center, Seattle.

Williams, J. B. (1984). *Tool Use by Nonhuman Primates: A Bibliography, 1975–1983*. Primate Information Center, Seattle.

Williams, J. B. (1992). *Tool Use by Nonhuman Primates: A Bibliography, 1983–1992*. Primate Information Center, Seattle.

Williams, J., Liu, H.-Y., and Pusey, A. (2002a). Costs and benefits of grouping for female chimpanzees at Gombe. In: Boesch, C., Hohmann, G., and Marchant, L. (eds.), *Behavioural Diversity in Chimpanzees and Bonobos*. Cambridge University Press, Cambridge. pp. 192–203.

Williams, J. M., Pusey, A. E., Carlis, J. V., Farms, B. P., and Goodall, J. (2002b). Female competition and male territorial behaviour influence female chimpanzees' ranging patterns. *Anim. Behav.* 63:347–360.

Williams, J. M., Oehlert, G. W., Carlis, J. V., and Pusey, A. E. (2004). Why do male chimpanzees defend a group range? *Anim. Behav.* 68:523–532.

Williams, L. (1967). Breeding Humboldt's woolly monkey *Lagothrix lagotricha* at Murrayton Monkey Sanctuary. *Int. Zoo Ybk.* 7:86–89.

Williams, L. (1974). *Monkeys and the Social Instinct*. Monkey Sanctuary Publications, Looe, UK.

Williams-Blangero, S., Vandeberg, J. L., Blangero, J., Konigsberg, L., and Dyke, B. (1990). Genetic differentiation between baboon subspecies: relevance for biomedical research. *Am. J. Primatol.* 20:67–81.

Williams-Guillén, K. (2003). The behavioral ecology of mantled howling monkeys (*Alouatta palliata*) living in a Nicaraguan shade coffee plantation [PhD diss.]. New York University, New York.

Williamson, E., and Feistner, A. (2003). Habituating primates: processes, techniques, variables and ethics. In: Curtis, D., and Setchell, J. (eds.), *Field and Laboratory Methods in Primatology*. Cambridge University Press, Cambridge. pp. 25–39.

Williamson, E. A., Tutin, C. E. G., Rogers, M. E., and Fernandez, M. (1990). Composition of the diet of lowland gorillas at Lope in Gabon. *Am. J. Primatol.* 21:265–277.

Wilson, D. S. (2000). Animal movement as a group-level adaptation. In: Boinski, S., and Garber, P. A. (eds.), *On the Move: How and Why Animals Travel in Groups*. University of Chicago Press, Chicago. pp. 238–258.

Wilson, D. S. (2004). What is wrong with individual fitness? *Trends Ecol. Evol.* 19:245–248.

Wilson, D. S., and Sober, E. (1994). Reintroducing group selection to the human behavioral sciences. *Behav. Brain Sci.* 17:585–608.

Wilson, E. O. (1975). *Sociobiology: The New Synthesis*. Harvard University Press, Cambridge, MA.

Wilson, J. F., and Erlandsson, R. (1998). Sexing of human and other primate DNA. *Biol. Chem.* 379:1287–1288.

Wilson, J. M., Stewart, P. D., Ramangason, G. S., Denning, A. M., and Hutchings, M. S. (1989). Ecology and conservation of the crowned lemur, *Lemur coronatus*, at Ankarana, N. Madagascar. *Folia Primatol.* 53:1–26.

Wilson, M. L. (2001). Imbalances of power: how chimpanzees respond to the threat of intergroup aggression (*Pan troglodytes*) [PhD diss.]. Harvard University, Cambridge, MA.

Wilson, M. L., and Wrangham, R. W. (2003). Intergroup relations in chimpanzees. *Annu. Rev. Anthropol.* 32:363–392.

Wilson, M. L., Hauser, M. D., and Wrangham, R. W. (2001). Does participation in inter-group conflict depend on numerical assessment, range location, or rank for wild chimpanzees? *Anim. Behav.* 61:1203–1216.

Wilson, M. L., Britton, N. F., and Franks, N. R. (2002). Chimpanzees and the mathematics of battle. *Proc. R. Soc. Lond. B.* 269:1107–1112.

Wilson, M. L., Wallauer, W. R., and Pusey, A. E. (2004). New cases of intergroup violence among chimpanzees in Gombe National Park, Tanzania. *Int. J. Primatol.* 25:523–549.

Wimmer, H., and Perner, J. (1983). Beliefs about beliefs—representation and constraining function of wrong beliefs in young children's understanding of deception. *Cognition* 13:103–128.

Wimmer, B., Tautz, D., and Kappeler, P. M. (2002). The genetic population structure of the gray mouse lemur (*Microcebus murinus*), a basal primate from Madagascar. *Behav. Ecol. Sociobiol.* 52:166–175.

Winarti, I. (2008). Field research on Javan slow loris' population in Sukakerta Ciamis and Kawungsari Tasikmalaya,West Java, Indonesia. Report to IAR Indonesia, Ciapus, Bogor, Indonesia. p. 7.

Windfelder, T. L., and Lwanga, J. S. (2002). Group fission in red-tail monkeys (*Cercopithecus ascanius*) in Kibale National Park, Uganda. In: Glenn,

M. E., and Cords, M. (eds.), *The Guenons: Diversity and Adaptation in African Monkeys.* Kluwer Academic/Plenum, New York. pp. 147–160.

Wink, M. (1999). *Functions of Plant Secondary Metabolites and Their Exploitation in Biotechnology, Annual Plant Reviews,* vol. 3. CRC Press, Boca Raton, FL.

Wink, M., Hofer, A., Bilfinger, M., Englert, E., Martin, M., and Schneider, D. (1993). Geese and dietary allelochemicals—food palatability and geophagy. *Chemoecology* 4:93–107.

Winn, R. M. (1989). The aye-ayes, *Daubentonia madagascariensis,* at the Paris Zoological Gardens: maintenance and preliminary behavioural observations. *Folia Primatol.* 52:109– 123.

Winterer, J., Palmer, A. E., Cicmanec, J., Davis, E., Harbaugh, S., and Loriaux, D. L. (1985). Endocrine profile of pregnancy in the patas monkey (*Erythrocebus patas*). *Endocrinology* 116:1090–1093.

Wislocki, G. B. (1936). The external genitalia of the simian primates. *Hum. Biol.* 8:309–347.

Wittenberger, J. F. (1980). Group size and polygamy in social mammals. *Am. Nat.* 115:197–222.

Wittenberger, J. F., and Tilson, R. L. (1980). The evolution of monogamy: hypothesis and evidence. *Annu. Rev. Ecol. Syst.* 11:197–232.

Wittig, R. M., and Boesch, C. (2003a). Food competition and linear dominance hierarchy among female chimpanzees of the Taï National Park. *Int. J. Primatol.* 24:847–867.

Wittig, R. M., and Boesch, C. (2003b). The choice of post-conflict interactions in wild chimpanzees (*Pan troglodytes*). *Behaviour* 140:1527–1559.

Wittig, R. M., and Boesch, C. (2005). How to repair relationships—reconciliation in wild chimpanzees (*Pan troglodytes*). *Ethology* 111:736–763.

Wolf, A. P. (1995). *Sexual Attraction and Childhood Association: A Chinese Brief for Edward Westermarck.* Stanford University Press, Stanford.

Wolf, K. E. (1984). *Reproductive Competition Among Co-resident Male Silvered Leaf Monkeys* (Presbytis cristata). Yale University Press, New Haven.

Wolf, K., and Fleagle, J. G. (1977). Adult male replacement in a group of silvered leaf monkeys (*Presbytis cristata*) at Kuala Selangor, Malaysia. *Primates* 18:949–955.

Wolf, N. G. (1985). Odd fish abandon mixed-species groups when threatened. *Behav. Ecol. Sociobiol.* 17:47–52.

Wolfe, J., and Summerlin, C. (1989). The influence of lunar light on nocturnal activity of the old field mouse. *Anim. Behav.* 37:410–414.

Wolfe, L. (1979). Behavioral patterns of estrous females of the Arashiyama West troop of Japanese macaques (*Macaca fuscata*). *Primates* 20:525–534.

Wolfe, L. D. (1981). A case of male adoption in a troop of Japanese monkeys (*Macaca fuscata fuscata*). In: Chiarelli, A. B., and Corruccini, R. S. (eds.), *Primate Behavior and Sociobiology.* Springer-Verlag, New York. pp. 156–160.

Wolfe, L. D. (1984). Japanese macaque female sexual behavior: a comparison of Arashiyama East and West. In: Small, M. F. (ed.), *Female Primates: Studies by Women Primatologists.* Alan Liss, New York. pp. 141–157.

Wolfe, L. D. (1986). Sexual strategies of female Japanese macaques (*Macaca fuscata*). *Hum. Evol.* 1:267–275.

Wolfe, L. D. (1991a). Human evolution and the sexual behavior of female primates. In: Loy, J., and Peters, C. B. (eds.), *Understanding Behavior: What Primate Studies Tell Us About Human Behavior.* Oxford University Press, New York. pp. 121–151.

Wolfe, L. D. (1991b). Macaques, pilgrims, and tourists revisited. *Nat. Geogr. Res. Expl.* 7:241.

Wolfe, L. D. (1992). Feeding habits of the rhesus monkeys (*Macaca mulatta*) of Jaipur and Galta, India. *Hum. Evol.* 7:43–54.

Wolfe, L. D. (1995). Current research in field primatology. In: Boaz, N. T., and Wolfe, L. D. (eds.), *Biological Anthropology: The State of the Science.* International Institute for Human Evolutionary Research, Bend, OR. pp. 149–157.

Wolfe, L. D. (2002). Rhesus macaques: a comparative study of two sites, Jaipur, India and Silver Springs, Florida. In: Fuentes, A., and Wolfe, L. D. (eds.), *Primates Face to Face.* Cambridge University Press, Cambridge. pp. 310–330.

Wolfe, L. D., and Fuentes, A. (2007). Ethnoprimatology: contextualizing human and nonhuman primate interactions. In: Campbell, C. J., Fuentes, A., MacKinnon, K. C., Panger, M., and Bearder, S. K. (eds.), *Primates in Perspective,* 1st ed. Oxford University Press, New York. pp 691–702.

Wolfe, L. D., and Gray, J. P. (1982). Japanese monkeys and popular culture. *J. Pop. Cult.* 16:97–105.

Wolfe, N. D., Escalante, A. A., Karesh, W. B., Kilbourn, A. M., Spielman, A., and Lal, A. A. (1998). Wild primate populations in emerging infectious disease research: the missing link? *Emerg. Infect. Dis.* 4:149–158.

Wolfheim, J. H. (1983). *Primates of the World: Distribution, Abundance, and Conservation.* University of Washington Press, Seattle.

Wolin, L., and Massoupust, L. (1970). Morphology of the primate retina. In: Noback, C. R., and Montagna, W. (eds.), *The Primate Brain.* Appleton-Century-Crofts, New York. pp. 1–27.

Wolovich, C. K., and Evans, S. (2007). Sociosexual behavior and chemical communication of *Aotus nancymaae. Int. J. Primatol.* 28:1299–1313.

Wolovich, C. K., Feged, A., Evans, S., and Green, S. M. (2006). Social patterns of food sharing in monogamous owl monkeys. *Am. J. Primatol.* 68:663–674.

Wolovich, C. K., Evans, S., and French, J. A. (2008a). Dads do not pay for sex but do buy the milk: food sharing and reproduction in owl monkeys (*Aotus* spp.). *Anim. Behav.* 75:1155–1163.

Wolovich, C. K., Perea-Rodriguez, J. P., and Fernandez-Duque, E. (2008b). Food transfers to young and mates in wild owl monkeys (*Aotus axarai*). *Am. J. Primatol.* 70:211–221.

Wolters, S., and Zuberbühler, K. (2003). Mixed-species associations of diana and Campbell's monkeys: the costs and benefits of a forest phenomenon. *Behaviour* 140:371–385.

Wong, S. N. P., and Sicotte, P. (2006). Population size and density of Colobus vellerosus at the Boabeng-Fiema Monkey Sanctuary and surrounding forest fragments in Ghana. *Am. J. Primatol.* 68:465–476.

Wood, B., and Richmond, B. G. (2000). Human evolution: taxonomy and paleobiology. *J. Anat.* 197:19–60.

Wood, J. W., O'Conner, K. A., Holman, D. J., Brindle, E., Barson, S. H., and Grimes, M. A. (2001). The evolution of menopause by antagonistic pleiotropy. Working paper 01–04. Center for Studies in Demography and Ecology, University of Washington, Seattle, http://csde.washington.edu/pubs/wps/01–04.pdf.

Woodruff, D. S. (2004). Noninvasive genotyping and field studies of free-ranging nonhuman primates. In: Chapais, B., and Berman, C. M. (eds.), *Kinship and Behavior in Primates.* Oxford University Press, Oxford. pp. 46–68.

Woolley-Barker, T. (1999). Social organization and genetic structure in a baboon hybrid zone [PhD diss.]. New York University, New York.

Worlein, J. M., Eaton, G. G., Johnson, D. F., and Glick, B. B. (1988). Mating season effects on mother–infant conflict in Japanese macaques, *Macaca fuscata. Anim. Behav.* 36:1472–1481.

Worley, A. C., Houle, D., and Barrett, S. C. H. (2003). Consequences of hierarchical allocation for the evolution of life-history traits. *Am. Nat.* 161:153–167.

Wormell, D., and Price, E. (2001). Reproduction and management of black lion tamarin, *Leontopithecus chrysopygus* at Jersey Zoo. *Dodo J. Wildl. Preserv. Trusts* 37:34–40.

Wrangham, R. W. (1976). Aspects of feeding and social behaviour in gelada baboons. In: *A Report to the Science Research Council of the U.K.* pp. 1–60.

Wrangham, R. W. (1977). Feeding behavior of chimpanzees in Gombe National Park, Tanzania. In: Clutton-Brock, T. H. (ed.), *Primate Ecology.* London: Academic Press. pp. 504–538.

Wrangham, R. W. (1979a). Sex differences in chimpanzee dispersion. In: Hamburg, D. A., and McCown, E. R. (eds.), *The Great Apes.* Benjamin/Cummings, Menlo Park, CA. pp. 481–489.

Wrangham, R. W. (1979b). On the evolution of ape social systems. *Soc. Sci. Inf.* 18:335–368.

Wrangham, R. W. (1980). An ecological model of female-bonded primate groups. *Behaviour* 75:262–300.

Wrangham, R. W. (1981). Drinking competition in vervet monkeys. *Anim. Behav.* 29:904–910.

Wrangham, R. W. (1982). Mutualism, kinship, and social evolution. In: *Current Problems in Sociobiology.* Cambridge University Press, Cambridge. pp. 270–289.

Wrangham, R. W. (1986). Ecology and social relationships in two species of chimpanzee. In: Rubenstein, D. I., and Wrangham, R. W. (eds.), *Ecological Aspects of Social Evolution: Birds and Mammals.* Princeton University Press, Princeton, NJ. pp. 352–378.

Wrangham, R. W. (1987). Evolution of social structure. In: Smuts, B. B., Cheney, D. L., Seyfarth, R. M., Wrangham, R. W., and Struhsaker, T. T. (eds.), *Primate Societies.* University of Chicago Press, Chicago. pp. 282–296.

Wrangham, R. W. (1993). The evolution of sexuality in chimpanzees and bonobos. *Hum. Nat.* 4:47–79.

Wrangham, R. W. (1995). Relationship of chimpanzee leaf-swallowing to a tapeworm infection. *Am. J. Primatol.* 37:297–303.

Wrangham, R. W. (1999). Evolution of coalitionary killing. *Ybk. Phys. Anthropol.* 42:1–30.

Wrangham, R. (2000). Why are male chimpanzees more gregarious than mothers? A scramble competition hypothesis. In: Kappeler, P. M. (ed.), *Primate Males.* Cambridge University Press, Cambridge. pp. 248–258.

Wrangham, R. (2002). The cost of sexual attraction: is there a trade-off in female *Pan* between sex appeal and received coercion? In: Boesch, C., Hohmann, G., and Marchant, L. (eds.), *Behavioural Diversity in Chimpanzees and Bonobos*. Cambridge University Press, Cambridge. pp. 204–215.

Wrangham, R. W., and Goodall, J. (1989). Chimpanzee use of medicinal leaves. In: Heltne, P. G., and Marquardt, L. A. (eds.), *Understanding Chimpanzees*. Harvard University Press, Cambridge, MA. pp. 22–37.

Wrangham, R. W., and Nishida, T. (1983). *Aspilia* spp. leaves: a puzzle in the feeding behavior of wild chimpanzees. *Primates* 24:276–282.

Wrangham, R. W., and Peterson, D. (1996). *Demonic Males*. Houghton Mifflin, New York.

Wrangham, R. W., and Riss, E. (1990). Rates of predation on mammals by Gombe chimpanzees, 1972–1975. *Primates* 31:157–170.

Wrangham, R. W., and Smuts, B. B. (1980). Sex difference in the behavioral ecology of chimpanzees in the Gombe National Park, Tanzania. *J. Reprod. Fertil. Suppl.* 28:13–31.

Wrangham, R. W., and van Zinnicq Bergmann Riss, E. (1990). Rates of predation on mammals by Gombe chimpanzees, 1972–1975. *Primates* 31:157–170.

Wrangham, R. W., and Waterman, P. G. (1981). Feeding behavior of vervet monkeys on *Acacia tortilis* and *Acacia xanthphloea*: with special reference to reproductive strategies and tannin productions. *J. Anim. Ecol.* 50:715–731.

Wrangham, R. W., and Waterman, P. G. (1983). Condensed tannins in fruits eaten by chimpanzees. *Biotropica* 15:217–222.

Wrangham, R. W., Clark, A. P., and Isabirye-Basuta, G. (1992). Female social relationships and social organization of Kibale Forest chimpanzees. In: Nishida, T., McGrew, W. C., Marler, P., Pickford, M., and de Waal, F. B. M. (eds.), *Topics in Primatology, Human Origins,* vol. 1. University of Tokyo Press, Tokyo. pp. 81–98.

Wrangham, R. W., Conklin, N. L., Etot, G., Obua, J., Hunt, K. D., Hauser, M. D., and Clark, A. P. (1993a). The value of figs to chimpanzees. *Int. J. Primatol.* 14:243–256.

Wrangham, R. W., Gittleman, J. L., and Chapman, C. A. (1993b). Constraints on group size in primates and carnivores: population density and day range as assays of exploitation competition. *Behav. Ecol. Sociobiol.* 32:199–209.

Wrangham, R. W., Chapman, C. A., and Chapman, L. J. (1994a). Seed dispersal by forest chimpanzees in Uganda. *J. Trop. Ecol.* 10:355–368.

Wrangham, R. W., McGrew, W., de Waal, F., and Heltne, P. G. (1994b). *Chimpanzee Cultures*. Harvard University Press, Cambridge, MA.

Wrangham, R. W., Chapman, C. A., Clark-Arcadi, A. P., and Isabirye-Basuta, G. (1996). Social ecology of Kanyawara chimpanzees: implications for understanding the costs of great ape groups. In: McGrew, W. C., Marchant, L., and Nishida, T. (eds.), *Great Ape Societies*. Cambridge University Press, Cambridge. pp. 45–57.

Wrangham, R. W., Conklin-Brittain, N. L., and Hunt, K. D. (1998). Dietary response of chimpanzees and cercopithecines to seasonal variation in fruit abundance: I. Antifeedants. *Int. J. Primatol.* 19:949–970.

Wrangham, R. W., Jones, J. H., Laden, G., Pilbeam, D., and Conklin-Brittain, N. L. (1999). The raw and the stolen: cooking and the ecology of human origins. *Curr. Anthropol.* 40:567–594.

Wrangham, R., Crofoot, M., Lundy, R., and Gilby, I. (2007). Use of overlap zones among group-living primates: a test of the risk hypothesis. *Behaviour* 144:1599–1619.

Wright, B. W. (2005). Craniodental biomechanics and dietary toughness in the genus *Cebus*. *J. Human Evol.* 48(5):473–492.

Wright, J. S., and Duber, H. C. (2001). Poachers and forest fragmentation alter seed dispersal, seed survival, and seedling recruitment in the palm *Attalea butyraceae,* with implications for tropical tree diversity. *Biotropica* 33:583–595.

Wright, J. S., Hernandez, A., and Condit, R. (2007). The bushmeat harvest alters seedling banks by favoring lianas, large seeds and seeds dispersed by bats, birds, and wind. *Biotropica* 39:363–371.

Wright, P. C. (1978). Home range, activity patterns, and agonistic encounters of a group of night monkeys (*Aotus trivirgatus*) in Perú. *Folia Primatol.* 29:43–55.

Wright, P. C. (1981). The night monkeys, genus *Aotus*. In: Coimbra-Filho, A., and Mittermeier, R. A. (eds.), *Ecology and Behavior of Neotropical Primates*. Academia Brasileira de Ciencias, Rio de Janeiro. pp. 211–240.

Wright, P. C. (1984). Biparental care in *Aotus trivirgatus* and *Callicebus moloch*. In: Small, M. (ed.), *Female Primates: Studies by Women Primatologists*. Alan R. Liss, New York. pp. 59–75.

Wright, P. C. (1985). The costs and benefits of nocturnality for *Aotus trivirgatus* (the night monkey) [PhD diss.]. City University of New York, New York.

Wright, P. C. (1986). Ecological correlates of monogamy in *Aotus* and *Callicebus*. In: Else, J. G., and Lee, P. C. (eds.), *Primate Ecology and Conservation*. Cambridge University Press. New York. pp. 159–167.

Wright, P. C. (1989). The nocturnal primate niche in the New World. *J. Hum. Evol.* 18:635–658.

Wright, P. C. (1990). Patterns of paternal care in primates. *Int. J. Primatol.* 11:89–102.

Wright, P. C. (1992). Primate ecology, rainforest conservation, and economic development: building a national park in Madagascar. *Evol. Anthropol.* 1:25–33.

Wright, P. C. (1993). The evolution of female dominance and biparental care among non-human primates. In: Miller, B. (ed.), *Sex and Gender Hierarchies*. Cambridge University Press, Cambridge. pp. 127–147.

Wright, P. C. (1994a). The behavior and ecology of the owl monkey. In: Baer, J. F., Weller, R. E., and Kakoma, I. (eds.), *Aotus: The Owl Monkey*. Academic Press, San Diego. pp. 97–112.

Wright, P. C. (1994b). Night watch on the Amazon. *Nat. Hist.* 103:45–51.

Wright, P. C. (1995). Demography and life history of free-ranging *Propithecus diadema edwardsi* in Ranomafana National Park, Madagascar. *Int. J. Primatol.* 16:835–854.

Wright, P. C. (1996). The neotropical primate adaptation to nocturnality. In: Norconk, M. A., Rosenberger, A. L., and Garber, P. A. (eds.), *Adaptive Radiations of Neotropical Primates*. Plenum Press, New York. pp. 369–382.

Wright, P. C. (1998). Impact of predation risk on the behaviour of *Propithecus diadema edwardsi* in the rain forest of Madagascar. *Behaviour* 135:483–512.

Wright, P. C. (1999). Lemur traits and Madagascar ecology: coping with an island environment. *Ybk. Phys. Anthropol.* 42:31–72.

Wright, P. C. (2009). *Callicebus* in Manu National Park: territory, resources, scent-marking and vocalizations. In: Barnett, A. A., Veiga, L. M., Ferrari, S. F., and Norconk, M. A. (eds.), *Evolutionary Biology and Conservation of Titis, Sakis, and Uacaris*. Cambridge University Press, Cambridge. In press.

Wright, P. C., and Jernvall, J. (1999). The future of primate communities: a reflection of the present? In: Fleagle, J. G., Janson, C., and Reed, K. E. (eds.), *Primate Communities*. Cambridge University Press, Cambridge. pp. 38–54.

Wright, P. C., and Martin, L. B. (1995). Predation, pollination and torpor in two nocturnal prosimians: *Cheirogaleus major* and *Microcebus rufus* in the rain forest of Madagascar. In: Alterman, L., Doyle, G. A., and Izard, M. K. (eds.), *Creatures of the Dark: The Nocturnal Prosimians*. Plenum Press, New York. pp. 45–60.

Wright, P. C., Toyama, T. M., and Simons, E. I. (1986). Courtship and copulation in *Tarsius bancanus*. *Folia Primatol.* 46:142–148.

Wright, P. C., Daniels, P. S., Meyers, D. M., Overdorff, D. J., and Rabesoa, J. (1987). A census and study of *Hapalemur* and *Propithecus* in southeastern Madagascar. *Primate Conserv.* 8:84–88.

Wright, P. C., Heckscher, S. K., and Dunham, A. E. (1997). Predation on Milne-Edwards' sifaka (*Propithecus diadema edwardsi*) by the fossa (*Cryptoprocta ferox*) in the rain forest of southeastern Madagascar. *Folia Primatol.* 68:34–43.

Wright, P. C., Heckscher, K., and Dunham, A. (1998). Predation on rain forest prosimians in Ranomafana National Park, Madagascar. *Folia Primatol.* 69(suppl. 1):401.

Wright, P. C., Simons, E. L., and Gursky, S. (eds.) (2003). *Tarsiers: Past, Present and Future*. Rutgers University Press, New Brunswick, NJ.

Wright, P. C., Andriamihaja, B. R., and Raharimiandra, S. A. (2005). Tanala syn-ecological relations with lemurs in southeastern Madagascar. In: Paterson, J., and Wallis, J. (eds.), *Commensalism and Conflict: The Human–Primate Interface*. American Society of Primatologists, Norman, OK. pp. 119–145.

Wright, R. V. S. (1972). Imitative learning of a flaked stone technology—the case of an orangutan. *Mankind* 8:296–306.

Wright, S. (1943). Isolation by distance. *Genetics* 28:114–138.

Wright, S. J. (2003). The myriad consequences of hunting for vertebrates and plants in tropical forests. *Persp. Plant Ecol. Evol. Syst.* 6:73–86.

Wright, S. J., and Muller-Landau, H. C. (2006). The future of tropical forest species. *Biotropica* 38:287–301.

Wright, S. J., and van Schaik, C. P. (1994). Light and the phenology of tropical trees. *Am. Nat.* 143:192–199.

Wright, S. J., Zeballos, H., Dominguez, I., Gallardo, M. M., Moreno, M. C., and Ibanez, R. (2000). Poachers alter mammal abundance, seed dispersal, and seed predation in a neotropical forest. *Conserv. Biol.* 14:227–239.

Wright, S. J., Stoner, K. E., Beckman, N. Corlet, R. T., Dirzo, R., Muller-Landau, H. C., Nunes-Iturri, G., Peres, C. A., and Wang, B. C. (2007). The plight of large animals in tropical forest and the consequences for plant regeneration. *Biotropica* 39:289–291.

Wroegemann, D., and Zimmermann, E. (2001). Aspects of reproduction in the eastern rufous mouse lemur (*Microcebus rufus*) and their implications for captive management. *Zoo Biol.* 20:157–167.

Wrona, F. J., and Dixon, R. W. J. (1991). Group size and predation risk: a field analysis of encounter and dilution effects. *Am. Nat.* 137:186–201.

Wu, C. C., MacCoss, M. J., Howell, K. E., and Yates, J. R., 3rd (2003). A method for the comprehensive proteomic analysis of membrane proteins. *Nat. Biotechnol.* 21:532–538.

Wu Ch'Eng-En (1943). *Monkey.* Waley, A. (trans.). Grove Press, New York.

Wu, H. M. H., Holmes, W. G., Medina, S. R., and Sackett, G. P. (1980). Kin preference in infant *Macaca nemestrina*. *Nature* 285:225–227.

Xiang, Z. F., and Grueter, C. C. (2007). First direct evidence of infanticide and cannibalism in wild snub-nosed monkeys (*Rhinopithecus bieti*). *Am. J. Primatol.* 69:249–254.

Xiao, W., and Oefner, P. J. (2001). Denaturing high-performance liquid chromatography: a review. *Hum. Mutat.* 17:439–474.

Xu, X., and Arnason, U. (1996). The mitochondrial DNA molecule of Sumatran orangutan and a molecular proposal for two (Bornean and Sumatran) species of orangutan. *J. Mol. Evol.* 43:431–437.

Yalden, D. W., Largen, M. J., and Kock, D. (1977). Catalogue of the mammals of Ethiopia. 3. Primates. *Mon. Zool. Ital. (N. S.) Suppl.* 9:1–52.

Yamada, M. (1963). A study of blood-relationship in the natural society of the Japanese macaque—an analysis of co-feeding, grooming, and playmate relationships in Minoo-B-troop. *Primates* 4:43–65.

Yamagiwa, J. (1983). Diachronic changes in two eastern lowland gorilla groups (*Gorilla gorilla graueri*) in the Mt. Kahuzi region, Zaire. *Primates* 24:174–183.

Yamagiwa, J. (1987a). Male life history and the social structure of wild mountain gorillas (*Gorilla gorilla beringei*). In: Kawano, S., Connell, J. H., and Hidaka, T. (eds.), *Evolution and Coadaptation in Biotic Communities.* University of Tokyo Press, Tokyo. pp. 31–51.

Yamagiwa, J. (1987b). Intra- and inter-group interactions of an all-male group of Virunga mountain gorillas (*Gorilla gorilla berengei*). *Primates* 28:1–30.

Yamagiwa, J. (1999). Socioecological factors influencing population structure of gorillas and chimpanzees. *Primates* 40:87–104.

Yamagiwa, J., and Hill, D. A. (1998). Intraspecific variation in the social organization of Japanese macaques: past and present scope of field studies in natural habitats. *Primates* 39:257–273.

Yamagiwa, J., and Kahekwa, J. (2001). Dispersal patterns, group structure, and reproductive parameters of eastern lowland gorillas at Kahuzi in the absence of infanticide. In: Robbins, M. M., Sicotte, P., and Stewart, K. J. (eds.), *Mountain Gorillas: Three Decades of Research at Karisoke.* Cambridge University Press, Cambridge. pp. 89–122.

Yamagiwa, J., and Kahekwa, J. (2004). First observations of infanticide by a silverback in Kahuzi-Biega. *Gorilla J.* 29:6–9.

Yamagiwa, J., Mwanza, N., Yumoto, T., and Maruhashi, T. (1991). Ant eating by eastern lowland gorillas. *Primates* 32:247–253.

Yamagiwa, J., Mwanza, N., Yumoto, T., and Maruhashi, T. (1992). Travel distances and food habits of eastern lowland gorillas: a comparative analysis. In: Itoigawa, N., Sugiyama, Y., and Sackett, G. P. (eds.), *Topics in Primatology. Behavior, Ecology and Conservation,* vol. 2. University of Tokyo Press, Tokyo. pp. 267–281.

Yamagiwa, J., Mwanza, N., Spangenberg, A., Maruhashi, T., Yumoto, T., Fischer, A., and Steinhauer-Bukart, B. (1993). A census of the eastern lowland gorilla (*Gorilla gorilla graueri*) in the Kahuzi-Biegu National Park with reference to mountain gorillas (*G. g. beringei*) in the Virunga region, Zaire. *Biol. Conserv.* 64:83–89.

Yamagiwa, J., Mwanza, N., Yumoto, T., and Maruhashi, T. (1994). Seasonal change in the composition of the diet of eastern lowland gorillas. *Primates* 35:1–14.

Yamagiwa, J., Maruhashi, T., Yumoto, T., and Mwanza, N. (1996). Dietary and ranging overlap in sympatric gorillas and chimpanzees in Kahuzi-Biega National Park, Zaire. In: McGrew, W. C., Marchant, L. F., and Nishida, T. (eds.), *Great Ape Societies.* Cambridge University Press, Cambridge. pp. 82–98.

Yamagiwa, J., Basabose, K., Kaleme, K., and Yumoto, T. (2003a). Within-group feeding competition and socioecological organization of gorillas in the Kahuzi-Biega National Park, Democratic Republic of Congo. In: Taylor, A. B., and Goldsmith, M. L. (eds.), *Gorilla Biology: A Multidisciplinary Perspective.* Cambridge University Press, Cambridge. pp. 328–357.

Yamagiwa, J., Kahekwa, J., and Basabose, A. K. (2003b). Intra-specific variation in social organization of gorillas: implications for their social evolution. *Primates* 44:359–369.

Yamakoshi, G. (1998). Dietary responses to fruit scarcity of wild chimpanzees at Bossou, Guinea: possible implications for ecological importance of tool use. *Am. J. Phys. Anthropol.* 106:283–295.

Yamamoto, M. (2005). Infant care in callitrichids: cooperation and competition. *Annu. Rev. Biomed. Sci.* 7:149–160.

Yamamoto, M. E., and Box, H. O. (1997). The role of non-reproductive helpers in infant carrying in captive *Callithrix jacchus*. *Ethology* 103:760–771.

Yamamoto, M. E., Box, H. O., Albuquerque, F. S., and Arruda, M. F. (1996). Carrying behavior in captive and wild marmosets (*Callithrix jacchus*): a comparison between two colonies and a field site. *Primates* 37:297–304.

Yamamoto, M. E., Albuquerque, F. S., Lopes, N. A., and Ferreira, E. S. (2008). Differential infant carrying in captive and wild common marmosets (*Callithrix jacchus*). *Acta Ethol.* 11(2):95–99.

Yamanaka, S. (2009). A fresh look at iPS cells. *Cell* 137:13–17.

Yamane, A., Shotake, T., Mori, A., Boug, A. I., and Iwamoto, T. (2003). Extra-unit paternity of hamadryas baboons (*Papio hamadryas*) in Saudi Arabia. *Ethol. Ecol. Evol.* 15:379–387.

Yamashita, N. (2002). Diets of two lemur species in different micro-habitats in Beza Mahafaly Special Reserve, Madagascar. *Int. J. Primatol.* 23:1025–1051.

Yamigawa, J. (1999). Socioecologic factors influence population structure of gorillas and chimpanzees. *Primates* 40:87–101.

Yang, Z. (1997). PAML: a program package for phylogenetic analysis by maximum likelihood. *CABIOS* 13:555–556.

Ydenberg, R. C., and Dill, L. M. (1986). The economics of fleeing from predators. *Adv. Study Behav.* 16:229–251.

Ye, F., Li, M.-S., Taylor, J. D., Nguyen, Q., Colton, H. M., Casey, W. M., Wagner, M., Weiner, M. P., and Chen, J. (2001). Fluorescent microsphere-based readout technology for multiplexed human single nucleotide polymorphism analysis and bacterial identification. *Hum. Mutat.* 17:305–316.

Yeager, C. P. (1989). Feeding ecology of the proboscis monkey (*Nasalis larvatus*). *Int. J. Primatol.* 10:497–530.

Yeager, C. P. (1990). Proboscis monkey (*Nasalis larvatus*) social organization: group structure. *Am. J. Primatol.* 20:95–106.

Yeager, C. P. (1991a). Possible antipredator behavior associated with river crossings by proboscis monkeys (*Nasalis larvatus*). *Am. J. Primatol.* 24:61–66.

Yeager, C. P. (1991b). Proboscis monkey (*Nasalis larvatus*) social organization: intergroup patterns of association. *Am. J. Primatol.* 23:73–86.

Yeager, C. P. (1996). Feeding ecology of the long-tailed macaque (*Macaca fascicularis*) in Kalimantan Tengah, Indonesia. *Int. J. Primatol.* 17:51–62.

Yeager, C. P., and Kirkpatrick, R. C. (1998). Asian colobine social structure: ecological and evolutionary constraints. *Primates* 39:147–155.

Yeager, C. P., and Kool, K. (2000). The behavioral ecology of Asian colobines. In: Whitehead, P. F., and Jolly, C. J. (eds.), *Old World Monkeys.* Cambridge University Press, Cambridge. pp. 496–521.

Yeager, C. P., and Silver, S. C. (1999). Translocation and rehabilitation as primate conservation tools: are they worth the cost? In: Dolhinow, P., and Fuentes, A. (eds.), *The Nonhuman Primates.* Mayfield, Mountain View, CA. pp. 164–169.

Yeager, C. P., Silver, S. C., and Dierenfeld, E. S. (1997). Mineral and phytochemical influences on foliage selection by the proboscis monkey (*Nasalis larvatus*). *Am. J. Primatol.* 41:117–128.

Yeo, G. W., Coufal, N. G., Liang, T. Y., Peng, G. E., Fu, X. D., and Gage, F. H. (2009). An RNA code for the FOX2 splicing regulator revealed by mapping RNA–protein interactions in stem cells. *Nat. Struct. Mol. Biol.* 16:130–137.

Yeoman, R. R., Wegner, F. H., Gibson, S. V., Williams L. E., Abbot, D. H., and Abee, C. R. (2000). Midcycle and luteal elevations of follicle stimulating hormone in squirrel monkeys (*Saimiri boliviensis*) during the estrous cycle. *Am. J. Primatol.* 52:207–211.

Yépez, P., De la Torre, S., and Snowdon, C. T. (2005). Interpopulation differences in exudate feeding of pygmy marmosets in Ecuadorian Amazonia. *Am. J. Primatol.* 66:145–158.

Yerkes, R. M. (1943). *Chimpanzees: A Laboratory Colony.* Yale University Press, New Haven.

Yerkes, R. M., and Yerkes, A. (1929). *The Great Apes: A Study in Anthropoid Life.* Yale University Press, New Haven.

Yoder, A. D. (1997). Back to the future: a synthesis of strepsirhine systematics. *Evol. Anthropol.* 6:11–22.

Yoder, A. D., and Yang, Z. H. (2004). Divergence dates for Malagasy lemurs estimated from multiple gene loci: geological and evolutionary context. *Mol. Ecol.* 13:757–773.

Yoder, A. D., Cartmill, M., Ruvolo, M., Smith, K., and Vilgalys, R. (1996). Ancient single origin of Malagasy primates. *Proc. Natl. Acad. Sci. USA* 93:5122–5126.

Yoder, A. D., Rasoloarison, R. M., Goodman, S. M., Irwin, J. A., Atsalis, S., Ravosa, M. J., and Ganzhorn, J. U. (2000). Remarkable species diversity in Malagasy mouse lemurs (Primates, *Microcebus*). *Proc. Natl. Acad. Sci. USA* 97:11325–11330.

Yoder, A. D., Irwin, J. D., and Payseur, B. A. (2001a). Failure of the ILD to determine data combinability for slow loris phylogeny. *Syst. Biol.* 50:408–424.

Yoder, A. D., Rasoloarison, R. M., Goodman, S. M., Irwin, J. A., Atsalis, S., Ravosa, M., and Ganzhorn, J. U. (2001b). Remarkable species diversity in Malagasy mouse lemurs (Primates, *Microcebus*). *Proc. Natl. Acad. Sci. USA* 97:11325–11330.

Yoder, A. D., Burns, M. M., Zehr, S., Delefosse, T., Veron, G., Goodman, S. M., and Flynn, J. J. (2003). Single origin of Malagasy carnivora from an African ancestor. *Nature* 421:734–737.

York, A. D., and Rowell, T. E. (1988). Reconciliation following aggression in patas monkeys, *Erythrocebus patas*. *Anim. Behav.* 36:502–509.

Yost, J. A., and Kelley, P. M. (1983). Shotguns, blowguns, and spears: the analysis of technological efficiency. In: Hames, R. B., and Vickers, W. T. (eds.), *Adaptive Responses of Native Amazonians*. Academic Press, New York. pp. 189–224.

Youlatos, D. (1999). Comparative locomotion of six sympatric primates in Ecuador. *Ann. Sci. Nat.* 20:161–168.

Youlatos, D. (2002). Positional behavior of black spider monkeys (*Ateles paniscus*) in French Guiana. *Int. J. Primatol.* 23:1071–1094.

Young, A. L., Richard, A. F., and Aiello, L. C. (1990). Female dominance and maternal investment in strepsirhine primates. *Am. Nat.* 135:473–488.

Young, T. P. (1994). Predation risk, predation rate, and the effectiveness of anti-predator traits. *Evol. Anthropol.* 3:67.

Young, T. P., Stubblefield, C. H., and Isbell, L. A. (1997). Ants on swollen-thorn acacias: species coexistence in a simple system. *Oecologia* 109:98–107.

Yu, N., Jensen-Seaman, M. I., Chemnick, L., Kidd, J. R., Deinard, A. S., Ryder, O., Kidd, K. K., and Li, W. H. (2003). Low nucleotide diversity in chimpanzees and bonobos. *Genetics* 164:1511–1518.

Yunger, J. L., and Bjorklund, D. F. (2004). An assessment of generalization of imitation in two enculturated orangutans (*Pongo pygmaeus*). *J. Comp. Psychol.* 118:242–246.

Yunis, J. J., and Prakash, O. (1982). The origin of man: a chromosomal pictorial legacy. *Science* 215:1525–1530.

Zabel, C. J., Glickman, S. E., Frank, L. G., Woodmansee, K. B., and Keppel, G. (1992). Coalition formation in a colony of prepubertal spotted hyenas. In: Harcourt, A. H., and deWaal, F. B. M. (eds.), *Coalitions in Humans and Other Animals*. Oxford University Press, New York. pp. 113–135.

Zahavi, A. (1975). Mate selection: a selection for a handicap. *J. Theor. Biol.* 53:205–214.

Zahavi, A. (1977). The cost of honesty (further remarks on the handicap principle). *J. Theor. Biol.* 67:603–605.

Zahavi, A. (1980). Ritualization and the evolution of movement signals. *Behaviour* 72:77–81.

Zahavi, A. (1987). The theory of signal selection and some of its implications. In: Delfino, V. P. (ed.), *Proceedings of the International Symposium on Biological Evolution, Bari, Italy*. Adriatica Editrica, Rome. pp. 305–327.

Zahavi, A. (1993). The fallacy of conventional signaling. *Philos. Trans. R. Soc. Lond. B* 338:227–230.

Zahavi, A., and Zahavi, A. (1997). *The Handicap Principle*. Oxford University Press, New York.

Zane, L., Bargelloni, L., and Patarnello, T. (2002). Strategies for microsatellite isolation: a review. *Mol. Ecol.* 11:1–16.

Zaramody, A., Fausser, J.-L., Roos, C., Zinner, D., Andriaholinirina, N., Rabarivola, C., Norscia, I., Tattersall, I., and Rumpler, Y. (2006). Molecular phylogeny and taxonomic revision of the eastern woolly lemurs (*Avahi laniger*). *Primate Rep.* 74:9–22.

Zeeve, S. R. (1991). Behavior and ecology of primates in the Lomako Forest, Zaire [PhD diss.]. State University of New York, Stony Brook.

Zeltner, E., Klein, T., and Huber-Eicher, B. (2000). Is there social transmission of feather pecking in groups of laying hen chicks? *Anim. Behav.* 60:211–216.

Zentall, T. R. (2000). Animal intelligence. In: Sternberg, R. J. (ed.), *Handbook of Intelligence*. Cambridge University Press, Cambridge. pp. 197–215.

Zhang, C., Bailey, D. K., Awad, T., Liu, G., Xing, G., Cao, M., Valmeekam, V., Retief, J., Matsuzaki, H., and Taub, M. (2006). A whole genome long-range haplotype (WGLRH) test for detecting imprints of positive selection in human populations. *Bioinformatics* 22:2122–2128.

Zhang, J., Zhang, Y. P., and Rosenberg, H. F. (2002). Adaptive evolution of a duplicated pancreatic ribonuclease gene in a leaf-eating monkey. *Nat. Genet.* 30:411–415.

Zhang, J., Wang, X., and Podlaha, O. (2004). Testing the chromosomal speciation hypothesis for humans and chimpanzees. *Genome Res.* 14:845–851.

Zhang, S. Y. (1995). Activity and ranging patterns in relation to fruit utilization by brown capuchins (*Cebus apella*) in French Guiana. *Int. J. Primatol.* 16:489–508.

Zhang, S. Y., and Wang L. X. (1995). Fruit consumption and seed dispersal of *Ziziphus cinamomum* (Rhamnaceae) by two species of primates (*Cebus apella* and *Ateles paniscus*) in French Guiana. *Biotropica* 27:397–401.

Zhang, S., Ren, B., and Li, B. (1999). A juvenile Sichuan golden monkey (*Rhinopithecus roxellana*) predated by a goshawk (*Accipiter gentiles*) in the Qinling Mountains. *Folia Primatol.* 70:175–176.

Zhang, S., Liang, B., and Wang, L. (2000). Seasonality of matings and births in captive Sichuan golden monkeys (*Rhinopithecus roxellana*). *Am. J. Primatol.* 51:265–269.

Zhao, Q. K. (1991). Macaques and tourists at Mt. Emei, China. *Nat. Geogr. Res. Expl.* 7:115–116.

Zhao, Q. K. (1996). Etho-ecology of Tibetan macaques at Mount Emei, China. In: Fa, J. E., and Lindburg, D. G. (eds.), *Evolution and Ecology of Macaque Societies*. Cambridge University Press, Cambridge, pp. 263–289.

Zhao, Q. K. (2005). Management strategies and effects. In: Paterson, J., and Wallis, J. (eds.), *Commensalism and Conflict: The Human–Primate Interface*. American Society of Primatologists, Norman, OK. pp. 376–399.

Zhi, L., Karesh, W. B., Janczewski, D. N., Frazier-Taylor, H., Sajuthi, D., Gombek, F., Andau, M., Martenson, S., and O'Brien, S. J. (1996). Genomic differentiation among natural populations of orangutan (*Pongo pygmaeus*). *Curr. Biol.* 6:326–1336.

Zhou, J., Wei, F., Li, M., Lok, C. B. P., and Wang, D. (2008). Reproductive characters and mating behaviour of wild *Nomascus hainanus*. *Int. J. Primatol.* 29:1037–1046.

Ziegler, T. (2000). Hormones associated with non-maternal infant care: a review of mammalian and avian studies. *Folia Primatol.* 71:6–21.

Ziegler, T. E., and Snowdon, C. T. (1997). Role of prolactin in paternal care in a monogamous New World primate, *Saguinus oedipus*. *Ann. N. Y. Acad. Sci.* 807:599–601.

Ziegler, T. E., Bridson, W. E., Snowdon, C. T., and Eman, S. (1987a). Urinary gonadotropin and estrogen excretion during the postpartum estrus, conception, and pregnancy in the cotton-top tamarin (*Saguinus oedipus oedipus*). *Am. J. Primatol.* 12:127–140.

Ziegler, T. E., Savage, A., Scheffler, G., and Snowdon, C. T. (1987b). The endocrinology of puberty and reproductive functioning in female cotton-top tamarins (*Saguinus oedipus*) under varying social conditions. *Biol. Reprod.* 37:618–627.

Ziegler, T. E., Snowdon, C. T., and Bridson, W. E. (1990a). Reproductive performance and excretion of urinary estrogens and gonadotropins in the female pygmy marmoset (*Cebuella pygmaea*). *Am. J. Primatol.* 22:191–203.

Ziegler, T. E., Snowdon, C. T., Warneke, M., and Bridson, W. E. (1990b). Urinary excretion of oestrone conjugates and gonadotrophins during pregnancy in the Goeldi's monkey (*Callimico goeldii*). *J. Reprod. Fertil.* 89:163–168.

Ziegler, T. E., Widowski, T. M., Larson, M. L., and Snowdon, C. T. (1990c). Nursing does affect the duration of the post-partum to ovulation interval in cotton-top tamarins (*Saguinus oedipus*). *J. Reprod. Fertil.* 90:563–570.

Ziegler, T. E., Matteri, R. L., and Wegner, F. H. (1993a). Detection of urinary gonadotropins in callitrichid monkeys with a sensitive immunoassay based upon a unique monoclonal antibody. *Am. J. Primatol.* 31:181–188.

Ziegler, T. E., Wittwer, D. J., and Snowdon, C. T. (1993b). Circulating and excreted hormones during the ovarian cycle in the cotton-top tamarin, *Saguinus oedipus*. *Am. J. Primatol.* 31:55–65.

Ziegler, T. E., Scheffler, G., and Snowdon, C. T. (1995). The relationship of cortisol levels to social environment and reproductive functioning in female cotton-top tamarins, *Saguinus oedipus*. *Horm. Behav.* 29:407–424.

Ziegler, T. E., Scheffler, G., Wittwer, E. J., Schultz-Darken, N., Snowdon, C. T., and Abbott, D. H. (1996). Metabolism of reproductive steroids during the ovarian cycle in two species of callitrichids, *Saguinus oedipus* and *Callithrix jacchus*, and estimate of the ovulatory period from fecal steroids. *Biol. Reprod.* 54:91–99.

Ziegler, T. E., Santos, C. V., Pissinatti, A., and Strier, K. B. (1997). Steroid excretion during the ovarian cycle in captive and wild muriquis, *Brachyteles arachnoids*. *Am. J. Primatol.* 42:311–321.

Ziegler, T. E., Carlson, A. A., Ginther, A. J., and Snowdon, C. T. (2000a). Gonadal source of testosterone metabolites in urine of male cotton-top tamarin monkeys (*Saguinus oedipus*). *Gen. Comp. Endocrinol.* 118:332–343.

Ziegler, T., Hodges, K., Winker, P., and Heistermann, M. (2000b). Hormonal correlates of reproductive seasonality in wild female Hanuman langurs (*Presbytis entellus*). *Am. J. Primatol.* 51:119–134.

Ziegler, T. E., Wegner, F. H., Carlson, A. A., Lazaro-Perea, C., and Snowdon, C. T. (2000c). Prolactin levels during the periparturitional period in the biparental cotton-top tamarin (*Saguinus oedipus*): interactions with gender, androgen levels, and parenting. *Horm. Behav.* 38:111–122.

Ziegler, T., Abegg, C., Meijaard, E., Perwitasari-Farajallah, D., Walter, L., Hodges, J. K., and Roos, C. (2007). Molecular phylogeny and evolutionary history of Southeast Asian macaques forming the *silenus* group. *Mol. Phylogenet. Evol.* 42:807–816.

Ziegler, T. E., Prudom, S. L., and Zahed, S. R. (2009). Variations in male parenting behavior and physiology in the common marmoset. *Am. J. Hum. Biol.* 21:739–744.

Zihlman, A. (1984). Body build and tissue composition in *Pan paniscus* and *Pan troglodytes* with comparison to other hominoids. In: Susman, R. (ed.), *The Pygmy Chimpanzee*. Plenum Press, New York. pp. 179–200.

Zihlman, A. (1996). Reconstructions reconsidered: chimpanzee models and human evolution. In: McGrew, W. C., Marchant, L. F., and Nishida, T. (eds.), *Great Ape Societies*. Cambridge University Press, Cambridge. pp. 293–304.

Zihlman, A. L. (1997). Natural history of apes: life history features in females and males. In: Morbeck, M. E., Galloway, A., and Zihlman, A. L. (eds.), *The Evolving Female*. Princeton University Press, Princeton, NJ. pp. 86–104.

Zihlman, A. L. (2000). A skeletal survey of physical anthropology in the U.S. *Riv. Anthropol. (Roma)* 78:57–66.

Zihlman, A. L., and Bolter, D. R. (2004). Mammalian and primate roots of human sociality. In: Sussman, R. W., and Chapman, A. R. (eds.), *The Origins and Nature of Sociality*. Aldine de Gruyter, New York. pp. 23–52.

Zihlman, A. L., and Cramer, D. L. (1978). A skeletal comparison between pygmy (*Pan paniscus*) and common chimpanzees (*Pan troglodytes*). *Folia Primatol.* 29:86–94.

Zihlman, A. L., Morbeck, M. E., and Goodall, J. (1990). Skeletal biology and individual life history of Gombe chimpanzees. *J. Zool. Lond.* 221:37–61.

Zihlman, A. L., Bolter, D. R., and Boesch, C. (2004a). Skeletal and dental development in wild chimpanzees from the Taï National Forest, Ivory Coast, and Gombe Reserve, Tanzania. *Am. J. Phys. Anthropol.* (suppl. 38):215.

Zihlman, A. L., Bolter, D. R., and Boesch, C. (2004b). Wild chimpanzee dentition and its implications for assessing life history in immature hominin fossils. *Proc. Natl. Acad. Sci. USA* 101:10541–10543.

Zihlman A., Bolter D., and Boesch, C. (2007) Skeletal and dental growth in chimpanzees (Pan troglodytes) of Taï National Park, Côte d'Ivoire. *J. Zool. Lond.* 273(1):63–73.

Zimmermann, E. (1985). Vocalisations and associated behaviours in adult slow loris (*Nycticebus coucang*). *Folia Primatol.* 44:52–64.

Zimmermann, E. (1990). Differentiation of vocalisations in bushbabies (Galaginae, Prosimii, Primates) and the significance for assessing phylogenetic relationships. *Z. Zool. Syst. Evol.* 28:217–239.

Zimmermann, E. (1995a). Loud calls in nocturnal prosimians: structure, evolution and ontogeny. In: Zimmermann, E., Newman, J. D., and Jürgens, U. (eds.), *Current Topics in Primate Vocal Communication*. Plenum Press, New York. pp. 47–72.

Zimmermann, E. (1995b). Acoustic communication in nocturnal prosimians. In: Alterman, L., Doyle, G. A., and Izard, M. K. (eds.), *Creatures of the Dark: The Nocturnal Prosimians*. Plenum Press, New York. pp. 311–330.

Zimmermann, E. (1998). Nachtgeister im Tropenwald: die nachtaktiven Lemuren Madagaskars. *Biol. Zeit.* 28:294–303.

Zimmermann, E., and Lerch, C. (1993). The complex acoustic design of an advertisement call in male mouse lemurs (*Microcebus murinus*, Prosimii, Primates) and sources of its variation. *Ethology* 93:211–224.

Zimmermann, E., Bearder, S. K., Doyle, G. A., and Anderson, A. B. (1988). Variations in vocal patterns of Senegal and South African lesser bushbabies and their implications for taxonomic relationships. *Folia Primatol.* 51:87–105.

Zimmermann, E., Cepok, S., Rakotoarison, N., Zietemann, V., and Radespiel, U. (1998). Sympatric mouse lemurs in north-west Madagascar: a new rufous mouse lemur species. *Folia Primatol.* 69:106–114.

Zingeser, M. R. (1973). Dentition of *Brachyteles arachnoides* with reference to alouattine and ateline affinities. *Folia Primatol.* 20:351–390.

Zinner, D., and Deschner, T. (2000). Sexual swellings in female hamadryas baboons after male take-overs: "deceptive" swellings as a possible female counter-strategy against infanticide. *Am. J. Primatol.* 52:157–168.

Zinner, D., Schwibbe, M. H., and Kaumanns, W. (1994). Cycle synchrony and probability of conception in female hamadryas baboons *Papio hamadryas*. *Behav. Ecol. Sociobiol.* 35:175–183.

Zinner, D., Hindahl, J., and Kaumanns, W. (2001a). Experimental intergroup encounters in lion-tailed macaques (*Macaca silenus*). *Primate Rep.* 59:77–92.

Zinner, D., Pelaez, F., and Torkler, F. (2001b). Distribution and habitat associations of baboons (*Papio hamadryas*) in central Eritrea. *Int. J. Primatol.* 22:397–413.

Zinner, D., Peláez, F., and Torkler, F. (2001c). Group composition and adult sex-ratio of hamadryas baboons (*Papio hamadryas hamadryas*) in central Eritrea. *Int. J. Primatol.* 22:415–430.

Zinner, D., Alberts, S., Nunn, C. L., and Altmann, J. (2002). Significance of primate sexual swellings. *Nature* 420:142–143.

Zinner, D., Hilgartner, R. D., Kappeler, P. M., Pietsch, T., and Ganzhorn, J. U. (2003). Social organization of *Lepilemur ruficaudatus*. *Int. J. Primatol.* 24:869–888.

Zinner, D., Krebs, E., Schrod, A., and Kaumanns, W. (2006). Early sexual maturity in male hamadryas baboons (*Papio hamadryas hamadryas*) and its reproductive implications. *Am. J. Phys. Anthropol.* 129:584–590.

Zinner, D., Groeneveld, L. F., Keller, C., and Roos, C. (2009). Mitochondrial phylogeography of baboons (*Papio* spp.)—indication for introgressive hybridization? *BMC Evol. Biol.* 9:83. doi:10.1186/1471-2148-1189-1183.

Zito, M., Evans, S., and Weldon, P. J. (2003). Owl monkeys (*Aotus* spp.) self-anoint with plants and millipedes. *Folia Primatol.* 74:159–161.

Zollikofer, C. P. E., Ponce de Leon, M. S., and Martin, R. (1998). Computer-assisted paleoanthropology. *Evol. Anthropol.* 6:41–54.

Zuberbuhler, K. (2000). Interspecies semantic communication in two forest primates. *Proc. R. Soc. Lond. B* 267:713–718.

Zuberbühler, K. (2001). Predator-specific alarm calls in Campbell's monkeys, *Cercopithecus campbelli*. *Behav. Ecol. Sociobiol.* 50:414–422.

Zuberbühler, K. (2003). Referential signaling in non-human primates: cognitive precursors and limitations for the evolution of language. *Adv. Study Behav.* 33:265–307.

Zuberbühler, K. (2004). In search of the evolutionary impact on predation in primate behavior. *Am. J. Primatol.* 63:37–38.

Zuberbühler, K. (2007). Predation and primate cognitive evolution. In: Gursky, S. L., and Nekaris, K. A. I. (eds.), *Primate Anti-Predator Strategies*. Springer, New York. pp. 3–26.

Zuberbühler, K., and Jenny, D. (2002). Leopard predation and primate evolution. *J. Hum. Evol.* 43:873–886.

Zuberbuehler, K., Gygax, L., Harley, N., and Kummer, H. (1996). Stimulus enhancement and spread of a spontaneous tool use in a colony of long-tailed macaques. *Primates.* 37:1–12.

Zuberbuhler, K., Noe, R., and Seyfarth, R. M. (1997). Diana monkey long-distance calls: messages for conspecifics and predators. *Anim. Behav.* 53:589–604.

Zuberbühler, K., Jenny, D., and Bshary, R. (1999). The predator deterrence function of primate alarm calls. *Ethology* 105:477–490.

Zucker, E. L., and Clarke, M. R. (1998). Agonistic and affiliative relationships of adult female howlers (*Alouatta palliata*) in Costa Rica over a 4-year period. *Int. J. Primatol.* 19:433–449.

Zucker, E. L., and Kaplan, J. R. (1981). Allomaternal behavior in a group of free-ranging patas monkeys. *Am. J. Primatol.* 1:57–64.

Zucker, E. L., Clarke, M. R., and Glander, K. E. (1997). Latencies to first births by immigrating adult female howling monkeys (*Alouatta palliata*) in Costa Rica. *Am. J. Primatol.* 42:158.

Zuckerman, S. (1932). *The Social Life of Monkeys and Apes*. Routledge, London.

Zuckerman, S. (1998). *The Ape in Myth and Art*. Zoological Society of London, London.

Zuckerman, S., and Parkes, A. S. (1939). Observations on secondary sexual characters in monkeys. *J. Endocrinol.* 1:430–439.

Zuckerman, S., Van Wagenen, G., and Gardner, R. H. (1938). The sexual skin of the rhesus monkey. *Proc. Zool. Soc. Lond.* 108:385–401.

Zullo, J., and Caine, N. (1988). The use of sentinels by captive red-bellied tamarins. *Am. J. Primatol.* 14:455.

Zunino, G. E., and Rumiz, D. I. (1986). Observaciones sobre el comportamiento territorial del mono aullador negro (*Alouatta caraya*). *Bol. Primatol. Arg.* 4:36–52.

Zunino, G. E., Galliari, C. A., and Colillas, O. J. (1985). Distribución y conservación del mirikiná (*Aotus azarae*) en Argentina: resultados preliminares. A. Primatologia No. Brasil. S. B. d. Primatologia. *Campinas* 2:305–316.

Zunino, G. E., Chalukian, S. C., and Rumiz, D. I. (1986). Infanticide and infant disappearance related to male takeover in groups of *Alouatta caraya*. *Primatol. Brasil* 2:185–190.